NASA

NASA

PICTURES FROM SPACE taken by American astronauts. Major Edward White took a twenty-minute "walk in space" on June 3, 1965 (see also frontispiece). The photograph of the Florida peninsula was taken from the Gemini V spacecraft. Cape Kennedy, in the foreground, can be seen projecting into the Atlantic Ocean. Astronauts Edward White and James McDivitt (*below*) exit from their spacecraft during training exercises held in the Gulf of Mexico.

NASA

VAST DISTANCES separate our earth and its neighboring planets revolving in the solar system. The chart above indicates relative sizes of the nine planets and their respective distances from the sun in millions of miles, ranging from the closest, Mercury, to far-off Pluto. The asteroid belt between Mars and Jupiter is cluttered with cosmic debris, from pebbles to rocks as large as mountains.

ORBITAL PHASE

1—Unmanned space observatory

2—Multi-passenger spaceship in orbit

3—Weather satellite system operational

4—Radio-TV relay satellite system operational

5—Meeting of two satellites in space

INITIAL STAGES OF UNITED STATES ORBITAL AND INTERPLANETARY SPACE PROGRAM

INTERPLANETARY PHASE

1—We crash-land instruments on the moon.

2—Soft landing of lunar instruments achieved.

3—Multi-passenger spaceship circles the moon.

4—First men land on moon's surface and return.

5—Manned spaceship circles Mars and returns.

6—Manned spaceship circles Venus and returns.

7—First men land on Mars and begin exploring.

MOON

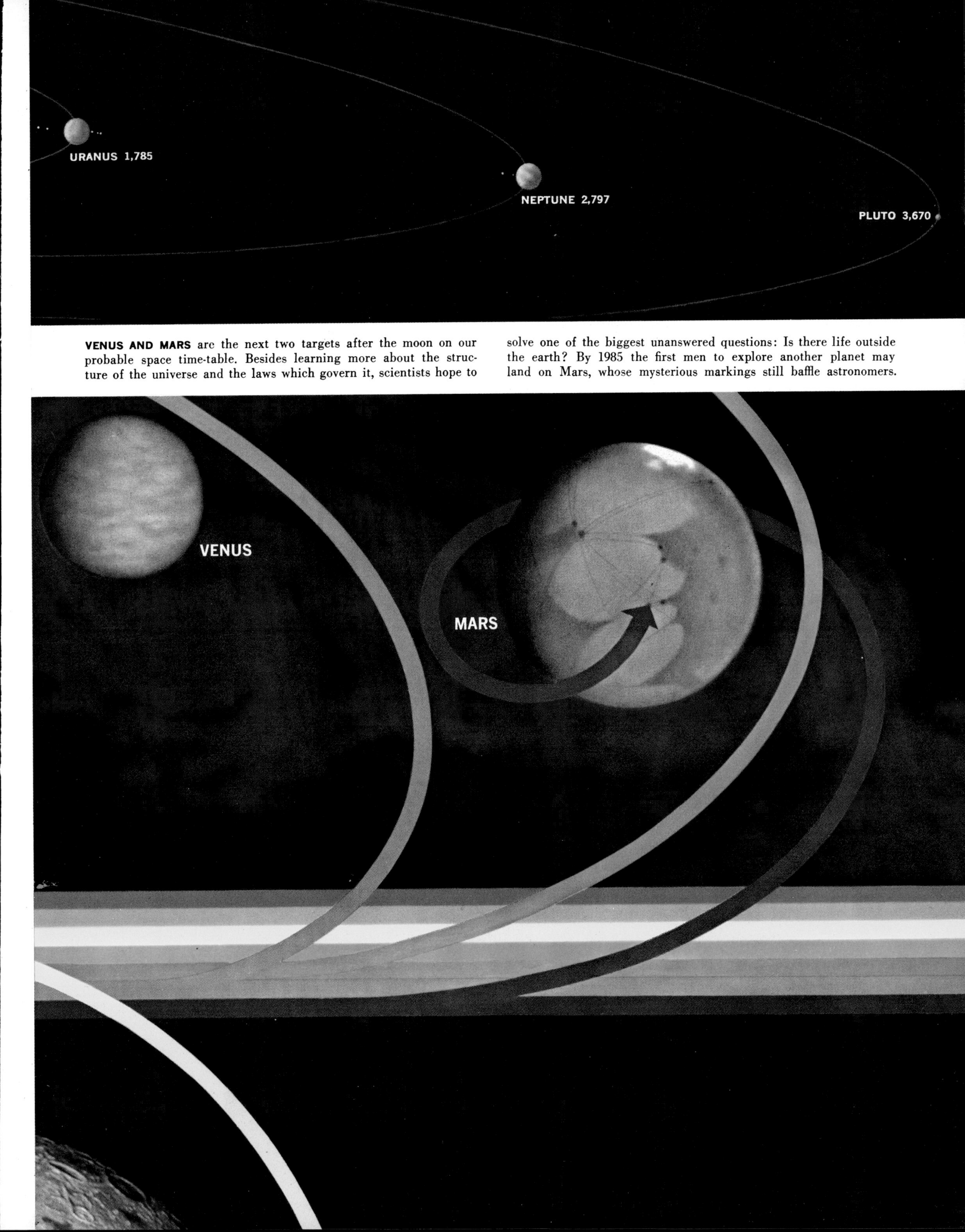

VENUS AND MARS are the next two targets after the moon on our probable space time-table. Besides learning more about the structure of the universe and the laws which govern it, scientists hope to solve one of the biggest unanswered questions: Is there life outside the earth? By 1985 the first men to explore another planet may land on Mars, whose mysterious markings still baffle astronomers.

MT. WILSON AND PALOMAR OBSERVATORIES

LICK OBSERVATORY

LICK OBSERVATORY

PLANETS, GALAXYS, AND NEBULA as seen through the telescope. The markings on Saturn (*above left*) and Jupiter (*above right*) can be clearly seen. The Great Nebula in Orion (*below*) and the North American Nebula (*center* and *left*) have been observed for many centuries.

ABOVE & BELOW. MT. WILSON AND PALOMAR OBSERVATORIES

COWLES ENCYCLOPEDIA OF SCIENCE, INDUSTRY AND TECHNOLOGY

COWLES ENCYCLOPEDIA OF SCIENCE, INDUSTRY AND TECHNOLOGY

NEW ENLARGED EDITION

from
Cowles Volume Library

COWLES BOOK COMPANY, INC.
NEW YORK

COWLES ENCYCLOPEDIA OF SCIENCE,
INDUSTRY AND TECHNOLOGY

NEW ENLARGED EDITION

S.B.N. 402-01081-7

Library of Congress Catalog Card Number 69-17310

Published simultaneously in Canada by
General Publishing Company, Ltd.,
30 Lesmill Road, Don Mills, Toronto,
Ontario

Printed in the United States of America
69A

PREFACE

Our world is literally exploding with innovations in science and technology. Until the 1940's, the store of scientific knowledge had doubled about every fifty years; from the '40's to the '60's, scientific knowledge had doubled every ten years; presently, it is doubling every seven years. To profit most from this age in which we live, we need to know as much as possible about the scientific past, science today, and the scientific and technological developments that will influence our world of tomorrow.

In the pages that follow you will discover how increasing knowledge has strengthened the relationship between the sciences, occasionally creating a new discipline. For example, man's greater knowledge of physics and astronomy—subjects as old as recorded history—led to the science of astronautics, the science of human activities in space. Although speculation about man's activity in space goes back to before the birth of Christ, astronautics did not become a true science until the 1950s, when the Space Age burst upon us.

Industry and technology apply scientific concepts to human welfare, thus freeing man's time from the pursuit of his basic needs. Improved farming methods, achieved through agronomy, botany, and chemistry, enable a small fraction of America's population to feed the entire nation and more. Mechanization and computerization—made possible through new developments in electronics, chemistry, and physics—have simplified and enhanced our daily life.

One hundred and sixty expert authors have contributed articles to *Cowles Encyclopedia of Science, Industry and Technology.* These include Dr. Lloyd V. Berkner, former Director, Southwest Center for Advanced Study; Howard I. Chapelle, Curator of Transportation, Smithsonian Institution; Dr. James Kip Finch, former Dean of Engineering, Columbia University; Dr. Julius H. Hlavaty, President, National Council of Teachers of Mathematics; Willy Ley, space authority; and Dr. Lorus J. Milne, Professor of Zoology, University of New Hampshire.

To aid your exploration of Science, Industry and Technology and increase the usefulness of this book as a reference, we have included an index section of more than 10,000 entries with page and column numbers to help you locate information quickly and easily; specialized glossaries accompanying key articles to explain and discuss technical terms and historical developments; a bibliography at the end of each topic to suggest further reading on the wide range of subjects covered under that topic.

Whether you are making a systematic study of a subject or just browsing, you will find the pages that follow packed with interest and information. Your *Cowles Encyclopedia of Science, Industry and Technology* is as current, accurate, and thorough as modern research can make it.

CONTENTS

PART 1

PART 2

PART 3

COWLES ENCYCLOPEDIA OF SCIENCE, INDUSTRY AND TECHNOLOGY

CONTRIBUTORS

ABBOTT, ROBERT E., B.S.M.E. Managing editor, *Product Engineering*. SCIENCE: Physics – Mechanics.

ALEXANDER, GEORGE W., B.S.E.E. Manager, High Voltage Laboratory, General Electric Company. INDUSTRY AND TECHNOLOGY: Energy and Power Sources – Transformer.

ALLEGRI, LINDA, Ph.D. Associate Professor of Education, Hunter College; Instructor, Department of Mathematics, Teachers College, Columbia University; Assistant editor, *Mathematics Education in the Americas*. SCIENCE: Mathematics – History of Mathematics.

ASHBURN, ANDERSON, B.S.E. Editor, *American Machinist*. INDUSTRY AND TECHNOLOGY: Machines and Processes – Industrial Control, Machine Tools, Quality Control.

BAINER, ROY, M.S. Dean of Engineering, University of California at Davis. INDUSTRY AND TECHNOLOGY: Food and Agriculture – Agricultural Engineering.

BAKER, LAURENCE H., Ph.D. Director, Computing Department, Pioneer Hi-Bred Corn Company. INDUSTRY AND TECHNOLOGY: Food and Agriculture – Animal Breeding, Plant Breeding, Corn.

BALAMUTH, LEWIS, Ph.D. Vice-President, Research and Development, Cavitron Corporation. INDUSTRY AND TECHNOLOGY: Energy and Power Sources – Ultrasonic Motor.

BARTHOLD, L. O., B.S. Manager, AC Transmission Engineering, General Electric Company. INDUSTRY AND TECHNOLOGY: Energy and Power Sources – Electric Power Transmission.

BATTISON, EDWIN A. Associate curator, Division of Mechanical and Civil Engineering, Museum of History and Technology, Smithsonian Institution. INDUSTRY AND TECHNOLOGY: Communications and Transportation – Clocks and Watches.

BERKEBILE, DON H. Museum specialist, Division of Transportation, Smithsonian Institution. INDUSTRY AND TECHNOLOGY: Communications and Transportation – Motor Vehicle, Roads and Highways.

BERKNER, LLOYD V. Late director, Southwest Center for Advanced Studies. INDUSTRY AND TECHNOLOGY: Introduction.

BERRY, E. WILLARD, Ph.D. Late Professor of Geology, Duke University. SCIENCE: Earth Sciences – Geology – Physical Geology.

BRAYNARD, FRANK O., M.A. Editor, *Tow Line Magazine*, Moran Towing & Transportation Company. INDUSTRY AND TECHNOLOGY: Communications and Transportation – Lighthouse, Marine Engineering, Marine Signaling, Ship.

BROOKS, MARVIN C., Ph.D. Commercial development manager, United States Rubber Tire Company. INDUSTRY AND TECHNOLOGY: Machines and Processes – Rubber Manufacture.

BROWN, ROBERT U., B.A. Publisher and editor, *Editor & Publisher*. INDUSTRY AND TECHNOLOGY: Communications and Transportation – Newspaper.

BUCHTA, J. W., Ph.D. Late president, Minnesota Academy of Science. SCIENCE: Physics – Introduction, Properties of Matter.

BURNLEY, ROSE MARIE, M.S. Home economics consultant. INDUSTRY AND TECHNOLOGY: Machines and Processes – Household Appliances, Sewing Machine.

BURR, HOWARD O. Assistant to the President, Cambridge Instrument Company, Inc. INDUSTRY AND TECHNOLOGY: Machines and Processes – Voting Machine.

BUTTFIELD, HELEN, A.M. Nature writer and photographer. SCIENCE: Life Science – Plant Life.

BYERLY, THEODORE C., Ph.D. Administrator, Cooperative State Research Service, United States Department of Agriculture. INDUSTRY AND TECHNOLOGY: Food and Agriculture – Agricultural Science.

CAMPBELL, JEROME, LL.B. Textile marketing consultant and trade publications editor; Assistant Corporation Counsel of the City of New York. INDUSTRY AND TECHNOLOGY: Materials and Structures – Synthetic Fibers.

CARMICHAEL, LEONARD, Ph.D. Vice-President of Research and Exploration, National Geographic Society. SCIENCE: Life Science – Psychology and Psychiatry.

CHAMBERLIN, ROBERT H., B.S. Account executive, Beaumont, Heller & Sperling, Inc. INDUSTRY AND TECHNOLOGY: Materials and Structures – Aluminum.

CHAPELLE, HOWARD I. Senior Historian (technical), Museum of History and Technology, Smithsonian Institution. INDUSTRY AND TECHNOLOGY: Machines and Processes – Shipbuilding.

COHEN, MARSHALL H., M.A. Economist, U.S. Department of Agriculture. INDUSTRY AND TECHNOLOGY: Food and Agriculture – Cacao, Coffee, Tea.

COLEMAN, JOHN W., Ph.D. Engineering leader, Scientific Instruments Engineering, Radio Corporation of America. INDUSTRY AND TECHNOLOGY: Machines and Processes – Electron Microscope.

COON, CARLETON S., Ph.D. Professor Emeritus of Anthropology, Harvard University and University of Pennsylvania; Research Curator, University Museum, University of Pennsylvania. SCIENCE: Life Science – Anthropology.

CUNNINGHAM, DONALD E., Ph.D., Dean for Research, Miami University, Ohio. SCIENCE: Physics – Nuclear Physics.

DABOLL, H. DAVIS, B.S. Former New York branch manager, Columbian Rope Company. INDUSTRY AND TECHNOLOGY: Machines and Processes – Rope.

DALRYMPLE, DANA G., Ph.D. Economist, International Agricultural Development Service, U.S. Department of Agriculture. INDUSTRY AND TECHNOLOGY: Food and Agriculture – Fruit, Grape, Vegetable.

DASMANN, RAYMOND F., Ph.D. Director, Environmental Studies, The Conservation Foundation. SCIENCE: Life Science – Biological Relationships.

DAVIS, FRANCIS K., Jr., Ph.D. Head, Physics Department, Drexel Institute of Technology. SCIENCE: Earth Sciences – Meteorology.

DENNSTEDT, F. D., B.S. Manager, Operations Coordination, Logistics Coordinating Department, Standard Oil Company. INDUSTRY AND TECHNOLOGY: Machines and Processes – Petroleum Refining.

DETWILER, SAMUEL B., Jr., M.A. Assistant to the Deputy Administrator for Nutrition, Consumer, and Industrial Use Research, Agricultural Research Service, U.S. Department of Agriculture. INDUSTRY AND TECHNOLOGY: Food and Agriculture – Agricultural Chemistry.

DUNBAR, ROBERT G., Ph.D. Professor of History, Montana State University. INDUSTRY AND TECHNOLOGY: Food and Agriculture – Alfalfa, Soils, Wheat.

EDGAR, ROBERT FERGUSON, M.S. Electrical engineer, General Electric Company. SCIENCE: Physics – Electricity and Magnetism.

EMILIANI, CESARE, Ph.D. Professor, Institute of Marine Sciences, University of Miami (Florida). SCIENCE: Earth Sciences – Oceanography.

ENGLEBARDT, STANLEY L., B.S. Free-lance science writer. INDUSTRY AND TECHNOLOGY: Communications and Transportation – Data Processing Systems, Duplication Machines; Machines and Processes – Automation, Calculating Machines, Computer.

FEINBERG, SAMUEL. Columnist, *Women's Wear Daily*. INDUSTRY AND TECHNOLOGY: Machines and Processes – Clothing Industry.

FINCH, JAMES K., D.Sc. Late Dean of Engineering, Columbia University. INDUSTRY AND TECHNOLOGY: Materials and Structures; Machines and Processes – Construction Equipment, Industrial Engineering, Surveying.

FISHER, DOUGLAS ALAN. Former public relations writer, United States Steel Corporation. INDUSTRY AND TECHNOLOGY: Materials and Structures – Iron and Steel; Machines and Processes – Iron and Steel Production.

FITTS, JAMES WALTER, Ph.D. Director, International Soil Testing Project and Professor of Soil Science, North Carolina State University; President, Soil Science Society of America. INDUSTRY AND TECHNOLOGY: Food and Agriculture – Agronomy.

FOREMAN, W. L. Director of public relations, National Cotton Council of America. INDUSTRY AND TECHNOLOGY: Food and Agriculture – Cotton.

FOSBURG, PHILIP L., B.S. Regional engineering services manager, Elevator Division, Westinghouse Electric Corporation. INDUSTRY AND TECHNOLOGY: Communications and Transportation – Elevator, Escalator, Moving Walk.

FOX, JOHN CAMERON, B.Sc. Secretary, Society of Mining Engineers of AIME. INDUSTRY AND TECHNOLOGY: Machines and Processes – Assaying, Mining, Ore Treatment.

FREEMAN, JIM, B.J. Director of Press Relations, American Telephone and Telegraph Company. INDUSTRY AND TECHNOLOGY: Communications and Transportation – Telephone, Teletypewriter.

FUSSELL, G. E. Formerly in the British Ministry of Agriculture. INDUSTRY AND TECHNOLOGY: Food and Agriculture – Alcoholic Beverages, Barley, Clover, Rye.

GARVIN, CLIFTON C., Jr., M.S. Director and Vice-president, Standard Oil Company. INDUSTRY AND TECHNOLOGY: Materials and Structures – Petrochemicals.

GIFFORD, RICHARD P., A.B. General Manager, Communication Products Department, General Electric Company. INDUSTRY AND TECHNOLOGY: Communications and Transportation – Future Communications.

GILLESPIE, PHILIP R.,B.S. Manager, Transportation Systems Sales, Transportation Division, Westinghouse Electric Corporation. INDUSTRY AND TECHNOLOGY: Communications and Transportation – Electric Transit.

GILLETT, CHARLES A., M.F. Consultant, American Forest Institute. INDUSTRY AND TECHNOLOGY: Machines and Processes – Lumber Industry, Paper.

GIORDANO, FELIX M., Ph.D. Editor, *The Tool and Manufacturing Engineer*. INDUSTRY AND TECHNOLOGY: Machines and Processes – Electroplating, Engraving, Etching, Metal Coating, Welding.

GLICKSMAN, ABRAHAM M., M.A. Teacher of mathematics, Bronx High School of Science and Polytechnic Institute of Brooklyn. SCIENCE: Mathematics – Trignometry.

GORDON, MATTHEW, B.S. Director of Information, Communications Satellite Corporation. INDUSTRY AND TECHNOLOGY: Communications and Transportation – Communications Satellite.

GRAHAM, GORDON F., B.A. Secretary, National Association of Wool Manufacturers. INDUSTRY AND TECHNOLOGY: Food and Agriculture – Wool.

GUNNELS, L. O., B.A. Senior Information Specialist, Battelle Memorial Institute. INDUSTRY AND TECHNOLOGY: Energy and Power Sources – Nuclear Reactor.

HANSEN, VIGGO P., Ph.D. Associate professor of Mathematics, San Fernando Valley State College. SCIENCE: Mathematics – Arithmetic.

HELMERS, RAYMOND A., Sr. Editor, *Furniture Design and Manufacturing*. INDUSTRY AND TECHNOLOGY: Machines and Processes – Furniture Manufacturing, Wood Finishing, Woodworking Tools.

HERON, S. DUNCAN, Jr., Ph.D. Associate Professor of Geology and Department Chairman, Duke University. Editor in Chief, *Southeastern Geology*. SCIENCE: Earth Sciences – Geology – Physical Geology.

HESS, CARL W., Ph.D. Chief, Poultry Research Branch, U.S. Department of Agriculture. INDUSTRY AND TECHNOLOGY: Food and Agriculture – Chicken, Duck, Egg Production, Goose, Guinea Fowl, Pheasant, Pigeon, Turkey.

HESTER, ALBERT S., B.S. Market research analyst, American Cyanamid Company. INDUSTRY AND TECHNOLOGY: Food and Agriculture – Fertilizer.

HILL, CHARLES G., Jr., Sc.D. Assistant Professor of Chemical Engineering, University of Wisconsin. SCIENCE: Physics – Heat.

HIRSCH, CHARLES J., E.E. Former engineering consultant, Radio Corporation of America. INDUSTRY AND TECHNOLOGY: Communications and Transportation – Television.

HLAVATY, JULIUS H., Ph.D. President, National Council of Teachers of Mathematics. SCIENCE: Mathematics – Analytic Geometry, Calculus.

HUGHES, HUNTER, B.S.M.E. Editorial Director, *Consulting Engineer Magazine*. INDUSTRY AND TECHNOLOGY: Energy and Power Sources; Machines and Processes – Bearing, Cable, Furnace.

HUNTER, L. N., B.A.Sc. Managing director, Air-Conditioning and Refrigeration Institute. INDUSTRY AND TECHNOLOGY: Machines and Processes – Refrigeration.

HYNEK, J. ALLEN, Ph.D. Professor of Astronomy and Department Chairman, Northwestern University; Director, Dearborn Observatory and Lindheimer Astronomical Research Center. SCIENCE: Space Sciences – Astronomy.

JACOBUS, WILLIAM W., Jr. Senior editor, Water Resources Section, *Engineering News-Record*. INDUSTRY AND TECHNOLOGY: Communications and Transportation; Machines and Processes; Materials and Structures.

JAHN, EDGAR A., B.M.E. Assistant director, Utilization Bureau, American Gas Association. INDUSTRY AND TECHNOLOGY: Materials and Structures – Gas.

JENKINS, FRANCES B., B.L.S., Ph.D. Professor of Library Science, Graduate School of Library Science, University of Illinois. INDUSTRY AND TECHNOLOGY: Bibliography.

JOHNSTON, S. PAUL, D.Sc. Director, National Air and Space Museum, Smithsonian Institution. INDUSTRY AND TECHNOLOGY: Communications and Transportation – Airship, Aviation.

KAREL, MARCUS, Ph.D. Associate professor of food engineering, Massachusetts Institute of Technology. INDUSTRY AND TECHNOLOGY: Food and Agriculture – Fish and Seafood, Food Additives, Food Engineering, Food Manufacturing, Food Preservation.

KELLEHER, JOSEPH J., B.S. Mechanical engineer, Associate editor, *Product Engineering*. INDUSTRY AND TECHNOLOGY: Communications and Transportation; Energy and Power Sources; Machines and Processes; Materials and Structures; SCIENCE: Physics – Mechanics.

CONTRIBUTORS

KILGOUR, FREDERICK G., A.B. Director, The Ohio College Library Center. INDUSTRY AND TECHNOLOGY: Machines and Processes – Introduction.

KNIGHT, ARTHUR, B.A. Professor, Cinema Department, University of Southern California; Contributing editor, *Saturday Review*. INDUSTRY AND TECHNOLOGY: Communications and Transportation: Motion Picture Industry.

KOCZY, FRIEDRICH F., Ph.D. Late Professor and Chairman, Physical Science Division, Institute of Marine Science, University of Miami. SCIENCE: Earth Sciences – Oceanography.

KOFF, RICHARD M., M.M.E. Administrative editor, *Playboy*. SCIENCE: Physics – Sound; Mathematics – Advanced Mathematics.

KOMINUS, NICHOLAS, B.S. Director of Information, United States Cane Sugar Refiners' Association. INDUSTRY AND TECHNOLOGY: Machines and Processes – Sugar Processing; Food and Agriculture – Artificial Sweeteners, Sugar.

KRANZBERG, MELVIN, Ph.D. Professor of History, Case Western Reserve University; Editor in Chief, *Technology and Culture*. INDUSTRY AND TECHNOLOGY: Introduction; Communications and Transportation – Introduction.

LACY, DAN, Litt.D. Senior Vice-President, McGraw-Hill Book Company. INDUSTRY AND TECHNOLOGY: Communications and Transportation – Book Publishing.

LAPORT, EDMUND A. Former director, Communication Engineering, Radio Corporation of America. INDUSTRY AND TECHNOLOGY: Communications and Transportation – Radio.

LEARY, JOHN S., Jr. Chief Staff Officer, Pharmacology, United States Department of Agriculture. INDUSTRY AND TECHNOLOGY: Food and Agriculture – Pesticides.

LEITNER, GORDON F., B.S. Vice-President, Aqua-Chem, Inc. INDUSTRY AND TECHNOLOGY: Machines and Processes – Desalinization.

LEVY, ALAN D., B.S. Senior Associate Programming Writer/Analyst, Systems Development Division, IBM Corporation. SCIENCE: Chemistry – Introduction, Table of Elements.

LEY, WILLY, L.H.D. Professor, Long Island University. INDUSTRY AND TECHNOLOGY: Machines and Processes – Guided Missile; SCIENCE: Space Sciences – Astronautics, Space Biology.

LYNCH, REV. JOSEPH J., Ph.D. Director, Seismic Observatory, Professor Emeritus, Physics, Fordham University. INDUSTRY AND TECHNOLOGY: Machines and Processes – Seismograph.

MACKAY-SMITH, ALEXANDER, LL.B. Editor, *The Chronicle of the Horse*. INDUSTRY AND TECHNOLOGY: Food and Agriculture – Horse.

MACKEY, EDWARD F., M.S. U.S. Government Engineer, Defense Contract Administration Services. SCIENCE: Physics – Light.

MacKENZIE, VERNON G., B.S. Assistant Surgeon General, Deputy Director, Bureau of Disease Prevention and Environmental Control, U.S. Public Health Service. INDUSTRY AND TECHNOLOGY: Machines and Processes – Air Pollution.

MAURO, J. A., Op.D. Consulting optics engineer, General Electric Company. SCIENCE: Physics – Light.

MAYHEW, ZEB, B.A. President, Esso Exploration, Inc. INDUSTRY AND TECHNOLOGY: Materials and Structures – Petroleum.

McBEE, RICHARD H., Ph.D. Dean, College of Letters and Science, Montana State University. SCIENCE: Life Science – Microbiology.

McCANN, HIRAM, B.A. Late editorial consultant, Society of Plastics Engineers. INDUSTRY AND TECHNOLOGY: Materials and Structures – Plastics.

McGANNON, HAROLD E. Technical editor, Research and Technology, United States Steel Corporation. INDUSTRY AND TECHNOLOGY: Materials and Structures – Alloys; Machines and Processes – Electric Furnace.

McGREGOR, SAMUEL E., M.S. Chief, Apiculture Research Branch, Entomology Research Division, U.S. Department of Agriculture. INDUSTRY AND TECHNOLOGY: Food and Agriculture – Beekeeping.

McGUIRE, ROBERT L., B.S. Promotion Coordinator, R. R. Donnelley & Sons Company. INDUSTRY AND TECHNOLOGY: Communications and Transportation—Printing; Machines and Processes – Bookbinding.

McLELLAN, GEORGE W., B.A. Director, Technical Information Service, Corning Glass Works. INDUSTRY AND TECHNOLOGY: Materials and Structures – Glass.

McNEIRNEY, FRANCIS A., B.A. Editor, *American Dyestuff Reporter Magazine*. INDUSTRY AND TECHNOLOGY: Machines and Processes – Dyeing.

MILLER, HERBERT F., Jr., M.S. Product Planning Engineer, Deere and Company. INDUSTRY AND TECHNOLOGY: Machines and Processes – Farm Machinery.

MILLER, STANLEY L., Ph.D. Associate Professor of Chemistry, University of California at San Diego. SCIENCE: Life Science – Origin of Life.

MILNE, LORUS J., Ph.D. Professor of Zoology, University of New Hampshire. SCIENCE: Life Science – Animal Life.

MILNE, MARGERY, Ph.D. Lecturer in Nature Recreation and Zoology, University of New Hampshire. SCIENCE: Life Science – Animal Life.

MITCHELL, JOHN W., Ph.D., D.Sc. Leader, Plant Hormone and Regulator Pioneering Research Laboratory, U.S. Department of Agriculture, Crops Research Division. INDUSTRY AND TECHNOLOGY: Food and Agriculture – Gibberellic Acid.

MONROE, DANIEL, M.D. Physician. SCIENCE: Life Science – Physiology, Glossary of Diseases.

MUREN, JAMES F., Ph.D. Medicinal Research Chemist, Charles Pfizer & Company, Inc. SCIENCE: Chemistry – Organic Chemistry.

MUSHRUSH, R. S., B.S. Manager, Direct Energy Conversion Operation, General Electric Company. INDUSTRY AND TECHNOLOGY: Energy and Power Sources – Fuel Cell.

NICHOLSON, THOMAS D., Ph.D. Assistant Director, The American Museum of Natural History. SCIENCE: Space Science – Astronomical Instruments.

O'BRINE, JOHN. Public Relations Consultant. INDUSTRY AND TECHNOLOGY: Communications and Transportation – Phonograph.

OSLIN, GEORGE P., B.A. Former public relations director, Western Union Telegraph Company. INDUSTRY AND TECHNOLOGY: Communications and Transportation – Stock Ticker, Telegraph.

PAVELIS, GEORGE A., Ph.D. Chief, Water Resources Branch, Natural Resource Economics Division, U.S. Department of Agriculture. INDUSTRY AND TECHNOLOGY: Food and Agriculture – Irrigation.

PETERSON, HAROLD L., M.A. Chief Curator, National Park Service, U.S. Department of the Interior. INDUSTRY AND TECHNOLOGY: Machines and Processes – Firearms.

PHILLIPS, C. J., M.A. Professor of Ceramics, Rutgers University. INDUSTRY AND TECHNOLOGY: Materials and Structures – Ceramics, Pottery; Machines and Processes – Kiln.

PIZZUTO, ANTHONY E., B.A. Advertising Manager, Water Treatment Products, Calgon Corporation. INDUSTRY AND TECHNOLOGY: Machines and Processes – Water Treatment.

PRESSLEY, RICHARD B. Senior editor, *Textile World*. INDUSTRY AND TECHNOLOGY: Machines and Processes – Tufting, Weaving.

PRICE, JOHN, B.A. Features editor, *Chemical Week*. SCIENCE: Chemistry – Analytical Chemistry, Physical Chemistry, Nuclear Chemistry; INDUSTRY AND TECHNOLOGY: Machines and Processes – Pharmaceutical Industry; Materials and Structures.

RAGLAND, DOUGLAS, B.S. Chief Engineer, Humble Oil & Refining Co. INDUSTRY AND TECHNOLOGY: Materials and Structures – Petroleum Engineering.

RASMUSSEN, WAYNE D., Ph.D. Chief, Agricultural History Branch, U.S. Department of Agriculture. INDUSTRY AND TECHNOLOGY: Machines and Processes – Canning and Preserving; Food and Agriculture – Introduction.

REGENSBURG, ALICE. Director, National Shoe Institute. INDUSTRY AND TECHNOLOGY: Machines and Processes – Shoe Manufacture.

REID, ROBERT C., M.S., Sc.D. Professor of Chemical Engineering, Massachusetts Institute of Technology. SCIENCE: Physics – Heat.

RICHARDSON, DEUEL, B.A. Public Relations Manager, National Fire Protection Association. INDUSTRY AND TECHNOLOGY: Machines and Processes – Fire Detection, Fire Prevention.

ROBERTS, KEN. Advertising Manager, Mosler Safe Company. INDUSTRY AND TECHNOLOGY: Machines and Processes – Safes.

ROGERS, CHARLES E., Ph.D. Former Information Officer, U.S. Department of Agriculture. INDUSTRY AND TECHNOLOGY: Food and Agriculture – Chicory, Coconut Palm, Cover Crops, Fiber Crops, Oats, Peanuts, Rice, Soybeans, Spices, Tobacco, Yeast.

RUEBENSAAL, CLAYTON F., B.Sc. Director of Corporate Planning, Uniroyal, Inc. INDUSTRY AND TECHNOLOGY: Materials and Structures – Rubber.

SCHLEBECKER, JOHN T., Ph.D. Curator in charge, Division of Agriculture and Forest Products, Smithsonian Institution. INDUSTRY AND TECHNOLOGY: Food and Agriculture – Camel, Casein, Cattle, Dairy Products, Fur Farming, Goat, Sheep, Swine; Materials and Structures – Fur and Leather.

SCHLUMPF, LESTER W., M.S. Principal, John Adams High School, New York City. SCIENCE: Mathematics – Algebra.

SHANER, DORIS D., B.A. Vice-president, Shaner-Grandelis Associates. SCIENCE: Earth Sciences – Geology – Mineralogy, Historical Geology, Economic Geology.

SHERIDAN, EUGENE T., B.S. Mineral Specialist, Bureau of Mines, U.S. Department of the Interior. INDUSTRY AND TECHNOLOGY: Materials and Structures – Coal, Coal Tar, Coke.

SIGFORD, JOHN V., B.E.E. President, Sigford and Associates. INDUSTRY AND TECHNOLOGY: Machines and Processes – Automatic Control Systems.

SILAGI, SELMA, Ph.D. Assistant Professor of Genetics, Cornell University Medical College. SCIENCE: Life Science – Genetics.

SITOMER, HARRY, M.A. Adjunct Associate Professor, C. W. Post College. SCIENCE: Mathematics – Geometry.

SMITH, WILLIAM V., Ph. D. Manager, Physics, IBM Corporation. SCIENCE: Physics – Masers and Lasers.

SNIVELY, HOWARD D., B.S. Manager, Advance Engineering, Medium AC Motor Department, General Electric Company. INDUSTRY AND TECHNOLOGY: Energy and Power Sources – Electric Generator, Electric Motor.

SPARLING, DOROTHY K., B.S. Merchandise Coordinator, Belgian Linen Association. INDUSTRY AND TECHNOLOGY: Materials and Structures – Linen.

SPENCER, MARTIN E., M.A. Associate Editor, *Encyclopedia International*. SCIENCE: Life Science – Physiology, Glossary of Diseases.

SPORN, PHILIP, E.E. Director and Consultant, American Electric Power Company. INDUSTRY AND TECHNOLOGY: Energy and Power Sources – Introduction.

STILES, WILLIAM W., M.D., M.P.H. Professor of Public Health, University of California at Berkeley. SCIENCE: Life Science – Public Health.

STUART, NEIL W., Ph.D. Research Plant Physiologist, U.S. Department of Agriculture. INDUSTRY AND TECHNOLOGY: Food and Agriculture – Hydroponics.

SWARBECK, GEORGE. Editor and Publisher, *The Northwestern Miller*. INDUSTRY AND TECHNOLOGY: Machines and Processes – Flour Milling.

SWARTZ, CLIFFORD E., Ph.D. Professor of Physics, State University of New York (Stony Brook); editor, *The Physics Teacher*. SCIENCE: Physics.

THOMAS, H. LAVERNE, M.A. Associate Professor of Mathematics, State University of New York (Oneonta). SCIENCE: Mathematics – Arithmetic.

TROJAN, MICHAEL A. Manager, Public and Industrial Relations, New York Trap Rock Corporation. INDUSTRY AND TECHNOLOGY: Machines and Processes – Quarrying.

TUGMAN, JAMES L., Ph.B. Former publicist, General Electric Company. INDUSTRY AND TECHNOLOGY: Energy and Power Sources – Lighting.

TURNER, WILLIAM J., Ph.D. Manager, Research Staff Operations, IBM Corporation. SCIENCE: Physics – Semiconductors.

VACZEK, LOUIS, B.Sc. Senior editor, Science, *Encyclopedia Britannica*. SCIENCE: Chemistry – Inorganic Chemistry.

VALENTINE, MARGOT, A.B. Former editor, technical papers, The Babcock & Wilson Company. INDUSTRY AND TECHNOLOGY: Machines and Processes – Boiler.

VATERLAUS, HANS. Executive Vice-president, International Silk Association, Inc. INDUSTRY AND TECHNOLOGY: Materials and Structures – Sericulture; Machines and Processes – Silk Manufacturing.

VERGARA, WILLIAM C., B.E.E. Director, Physical Electronics Department, Bendix Communication Division, Bendix Corporation. INDUSTRY AND TECHNOLOGY: Communications and Transportation; Energy and Power Sources; Machines and Processes; Materials and Structures – Acoustical Engineering.

VINCENT, EMIL P. Product Manager, Audio Systems, Visual Electronics Corporation. INDUSTRY AND TECHNOLOGY: Communications and Transportation – Sound Recording and Reproduction.

VOLK, OLIVER, R., B.S. Staff chemist, E. I. duPont de Nemours & Company, Inc. INDUSTRY AND TECHNOLOGY: Materials and Structures – Paint.

WAHBA, ISAAC J., Ph. D. Senior Research Chemist, General Mills, Inc. INDUSTRY AND TECHNOLOGY: Food and Agriculture – Gelatin.

WALLINGTON, G. GRANTLY. Free-lance writer and consultant. INDUSTRY AND TECHNOLOGY: Communications and Transportation – Photography; Machines and Processes – Camera, Camera Accessories.

WATSON, HOWARD C., B.S. Director of Public Relations, Magazine Publishers Association. INDUSTRY AND TECHNOLOGY: Communications and Transportation – Magazines.

WESTBROOK, WILMER C. Technical writer, Saco-Lowell Shops. INDUSTRY AND TECHNOLOGY: Machines and Processes – Spinning.

WHATMORE, MARVIN C., L.H.D. President, Cowles Communications, Inc. INDUSTRY AND TECHNOLOGY: Communications and Transportation – Xograph.

WHEELER, DONALD H., Ph.D. Principal scientist, General Mills Central Research Laboratories. INDUSTRY AND TECHNOLOGY: Food and Agriculture – Fats and Oils.

WHITE, JOHN HOXLAND, Jr., B.A. Curator of Transportation, Smithsonian Institution. INDUSTRY AND TECHNOLOGY: Communications and Transportation – Railroad, Railroad Engineering, Railroad Signaling.

WHITE, ROBERT M., Ph.D. Chief, U.S. Weather Bureau; Chief, Meteorological Development Laboratory; Research associate, Massachusetts Institute of Technology. SCIENCE: Earth Sciences – Meteorology – The Weather Bureau.

WILLIAMS, L. PEARCE, Ph.D. Professor of the History of Science, Cornell University. SCIENCE: History of Science.

WOOTON, ROGER O., M.S.E. Nuclear Engineer, Battelle Memorial Institute. INDUSTRY AND TECHNOLOGY: Energy and Power Sources – Nuclear Engineering, Nuclear Power.

ZOBLER, LEONARD, Ph.D. Professor and Chairman, Department of Geology and Geography, Barnard College, Columbia University. SCIENCE: Earth Sciences.

PART 1
Science

HISTORY

Definition.—The basic motive behind all scientific investigation—the compelling need to reduce the observable world to comprehensible terms—is as old as mankind. Science, however, is a relatively recent result of this desire to understand, and it would be well to come to some understanding of what science is before attempting to view its history.

Science is surprisingly difficult to define. One definition states that it is an organized body of knowledge. This explanation seems satisfactory until it is realized that by this definition a telephone directory becomes a respectable scientific publication and the *Handbook of Chemistry and Physics* appears worthy of the Nobel Prize. Yet it does contain an important element of truth. It serves to emphasize that science deals with facts—facts arranged in an orderly way, permitting easy access to them.

Another concept of science singles out its ability to predict the future. There is something appealing in the austerity of this definition but there is also something lacking. Bookmakers, after all, stake their livelihoods on the accuracy of their predictions, but few would insist that "bookies" are scientists. And accuracy of prediction, alone, is never sufficient proof of the validity of a theory; an explanation of how and why it works is necessary. It should be noted, however, that in the history of science there have been systems of considerable sophistication (such as the Babylonian lunar theory and Ptolemaic astronomy) which eschewed explanation and relied for their intellectual authority upon their ability to predict certain specific phenomena with a high degree of accuracy.

The greatest part of the history of science is concerned with the explanation of observed facts—explanations of how or why a particular effect or group of effects is produced. For more than 90 per cent of the period during which science has existed, this type of explanation has made up the body of scientific theory. There are, however, various kinds of explanations. There is a world of difference between an explanation of the motion of the moon that relies upon the desires of a moon-goddess and one that appeals to a law of universal attraction. For what follows, it is of the utmost importance to realize precisely what this difference is. It is definitely not one that involves either the description of the moon's position or the prediction of its future points in space. In both theories, description and prediction may be exactly the same. Where they do differ is in vulnerability to criticism. The use of a moon-goddess protects the theorist from any and all lunar aberrations. If the moon suddenly went shooting off at a tangent the priest of the moon-goddess would immediately call attention to the fact that his patroness had an urgent appointment elsewhere and decided to leave. The theory remains intact. The poor Newtonian astronomer is in much worse shape. Should the moon suddenly depart from the neighborhood of the earth,

DETAILED AND ACCURATE OBSERVATIONS are the basis upon which the scientist builds hypotheses about the nature of the universe. Tycho Brahe, a sixteenth-century Danish astronomer, using a great steel quadrant with a radius of over six feet, made the extremely precise measurements from which Kepler later devised his theories of planetary motion.

NEW YORK PUBLIC LIBRARY

PIX, INC.

ECLIPSES were regarded with awe and terror in ancient times; the Chinese, who first recorded a solar eclipse in 2137 B.C., believed it was caused by a dragon swallowing the sun (*top left*). Yet the successful prediction of eclipses was one of the earliest triumphs of scientific investigation. The drawing (*left*) shows the moon moving across the face of the sun during a partial eclipse.

the theory of universal attraction would have to be abandoned unless some other body could be detected whose gravitational force would account for the moon's strange behavior. Thus, the faith placed in scientific conclusions is not the result of the ability of scientific theories to answer all questions, but their ability to withstand the most severe criticism, both theoretical and experimental. If a theory is inherently irrefutable, it may be true but it does not have to be scientific. God's will may always be done, but it is rarely open to human comprehension. The history of science is, in good part, the gradual evolution of theories open to criticism.

One further point in connection with scientific theories is worth mentioning. If we require of them only that they be capable of empirical refutation, then the way is left open for the almost infinite multiplication of *sources* of theory. As long as an idea can be tested by actual experiment, it makes no difference whether the idea came from long years of laboratory work or from a daydream. Experiment, it should be noted, is used here not as an instrument of discovery but the most powerful critical tool the scientist possesses.

What, then, is science? It certainly contains accurate factual description. Many sciences, such as botany or mineralogy, were nothing but factual description until the nineteenth century. It also is predictive, for if a theory is to be taken seriously, it must give some insight into what we may expect in the future. Above all, it is explanatory, telling us not only what happens, but why. In the essay that follows all three of these elements will be traced, although the most attention will be focused upon the changing status of scientific theories. Two phases will be distinguished. In the first, extending in time from antiquity until the scientific revolution of the seventeenth century, the basic theory was essentially theological and, therefore, not entirely open to empirical refutation. The scientific revolution consisted essentially in the introduction of the principle of refutability, and from that time on, scientific theories have fallen because of the criticisms directed at them. Grasping these threads, the tour through the centuries of the history of science can now begin.

■BEGINNINGS OF SCIENCE.—From the remotest antiquity, man has wrestled with his environment in an effort simply to stay alive. Facts and their classification were of obvious importance. The hunting of wild game and the gathering of vegetable foods required this minimal intellectual activity. Certain regularities in nature, too, permitted the prediction of the more obvious aspects of natural phenomena. Explanations were magical and mystical, although it is probable that even this stage was reached only after millennia of intellectual effort.

The discovery of agriculture and the evolution of the social organism of the agrarian village created entirely new conditions of both practical and intellectual life. A new complexity developed in human affairs, the handling of which required specialists. Both in Egypt and Mesopotamia, a priestly caste, engaged in the regulation of social activity, emerged. In order for taxes to be assessed, land to be worked, and the whole complex social machinery kept in smooth working order, various new skills had to be developed. Among these were a method of keeping records and some kind of basic mathematics.

EGYPTIAN MATHEMATICS.—Egyptian mathematics was relatively simple and straightforward and was used primarily for practical computations. Through its use, the Egyptians were able to compute complicated areas and volumes. There is little doubt that they could calculate the volume of a cylinder and the area of a sphere. Their greatest achievement in practical mathematics was the calculation of the volume of the frustum of a pyramid. The Egyptians, however, went beyond such purely practical matters and were able to solve linear and quadratic equations containing one unknown.

MESOPOTAMIAN MATHEMATICS.—Mathematics in Mesopotamia was quite another matter. Earliest mathematical texts date from c. 1600 B.C. Of these, one group, called Table Texts, involved the tabulation of the results of various computations for handy reference. The other, the Problem Texts, were clearly intended for teaching mathematics to Babylonian scribes and presented the learner with exercises with which he could sharpen his skills.

The Table Texts are computations made on the basis of a sexagesimal system in which the basic unit is 60 rather than 10. The Table Texts comprise ordinary multiplication tables, tables of reciprocals, squares, square roots, cubes, and cube roots. With these tables handy, it was possible for the Babylonians to attack and solve quite advanced mathematical problems. A millennium before Pythagoras, they knew the Pythagorean theorem and worked with Pythagorean numbers with facility. They enjoyed numbers for numbers' sake and developed mathematical relationships of great complexity more than a thousand years before these numerical relationships were applicable to any physical process.

The Mesopotamians also created a mathematical astronomy of a very high order. By using approximations, a creditable lunar theory was devised by which the lunar calendar could be adjusted to the solar. The invention of the zodiac (an imaginary belt in the heavens) provided an essential astronomical reference band and, together with the use of mathematics, permitted the creation of a respectable planetary theory. When amalgamated with the Greek tradition, it was to produce an astronomical system not overthrown until the sixteenth century.

Classical Greek Science.—It was in the city of Miletus in Ionia that the first school of Greek scientific speculation emerged. Thales (c. 600 B.C.) is traditionally considered the first natural philosopher of Greece but it should be noted that absolutely nothing is known about him directly. No writings of his have survived, and the account which credits him with the prediction of a solar eclipse in 585 B.C. is suspect. What does seem certain is that the tradition Thales did create with Anaximander (611–547 B.C.) and Anaximenes (c. 570 B.C.) was a radical new departure in the history of thought.

To the Ionians, the god-ridden world of Homer was no fit place for a rational man. The harmony and regularity of the universe belied the capriciousness of the Homeric deities and the Ionians searched for the principle of unity and order behind the seeming diversity of observable reality. They thus initiated a quest that still endures. There must, they felt, be some *thing* that remained essentially itself through all time and whose modifications produced the world of constant flux. To Thales, this thing was water, for water could freeze and become like stone, or vaporize and become like air, and thus take up all the guises of common materials. Anaximenes chose air as his essential substratum. When air is compressed it becomes warm; when it is rarefied it becomes cold. Hence physical process gave rise to physical qualities. For Anaximander, an observable material thing seemed to be a result rather than a cause. The true essence underlying all reality must itself be without qualities but capable of assuming them all. The necessity for the invention of this metaphysical entity—what Anaximander called the "boundless"—was obvious. Water was *essentially* wet. No matter what you did to it, it retained the quality of wetness. Water, therefore, could not enter into substances that were essentially dry. The "boundless," however, suffered from no such difficulties for it could adopt the quality of dryness as easily as it could that of wetness.

All these schemes were but tentative beginnings on the road to science. They could give vague qualitative explanations of events, but they could not stand up under intense critical scrutiny. The most important of the critics of the Ionians was Parmenides of Elea (c. 540–450 B.C.). His point was a simple, but devastating, one. If water, for example, was constantly appearing in different forms, how could a rational man speak of it as water? For something to be and to be known, it must not change; it must exist in and of itself through eternity. Change, Parmenides therefore concluded, is mere illusion. Reality consists in what is unchanging and eternal; the universe, therefore, is really an unchanging sphere. While the logic of Parmenides' position was impeccable, by rejecting the observable world it also rejected any attempt at devising a meaningful explanation of that world.

In the period before Socrates altered the main course of Greek philosophy, two major attempts were made to depict the essential reality behind the observable world. Leucippus (fl. c. 475 B.C.) and Democritus (c. 470–400 B.C.) proposed an atomic theory that could satisfy both Parmenides' conditions of changelessness and the requirements of a hypothesis intended to explain the world of experience. The metaphysical world consisted of atoms, which were described as eternal, unchanging spheres moving through a void. The world of experience resulted from the collision and clumping of atoms and was constantly in flux and changing.

PYTHAGOREANISM.—The name of Pythagoras (c. 550 B.C.) is associated with the other departure from Ionian materialism. Like Thales, Pythagoras the man is shrouded in the mists of tradition and little, if anything, is known about him as a person. Upon his entry into history, we find a man drunk with numbers and filled with the mysterious essences of Eastern religion. He founded an ascetic sect quite foreign to the Greek way of life and placed number mysticism at its very center. The universe, said Pythagoras, was made up essentially of numbers. Relations between things were really relations between numbers. The flux of the world was no more than the flux of number, changing through infinite combinations, but reducible in the last analysis to the unchanging reality of the integer. Algebra, that is, Babylonian algebra, one might insist, was the key which all Greece was seeking.

The Pythagorean theory of numbers was, as such, merely an interesting competitor with the other theories of reality then spreading throughout Greece. It contained one element, however, that set it apart from these and was to give it an extraordinary influence in the history of science. Pythagoras (or his followers, for it is difficult to distinguish between them) did more than assume that relations between physical objects were essentially mathematical: they also proved it. It was the triumph of the Pythagoreans to show that the musical note produced by a stretched string was dependent (all other things being equal), in a simple mathematical way, upon the length of the string. It is by no means obvious that a musical tone and the length of a lyre string should be related by a simple mathematical ratio. Having discovered that this was so, the Pythagoreans were then in a position to create a mathematical musical theory in which they could calculate musical tones from algebraic functions. The full import of this should be clearly seen. The Pythagoreans had been successful in translating physical things into mathematical quantities. They could then operate upon these new functions by mathematical means to derive new physical relations. A tool of hitherto undreamed of power was hereby made available to the natural philosopher. More important than this, however, was the vision of the universe that the Pythagorean system made prevalent. If the cosmos were essentially mathematical, then the main task of the natural philosopher was to seek out the mathematical beauty that was to inspire men for centuries after the Pythagoreans as a sect had vanished.

The Pythagoreans were also responsible for the direction taken by Greek mathematics. Their original inspiration had been algebraic, stemming from Babylonia. By insisting that number was, in some mysterious sense, physical, the Pythagoreans committed themselves to a universe in which all magnitudes could be represented by the ratio of integers. Then, to the horror of the members of the sect, it was found that the diagonal of a unit square was incommensurable. It is impossible to represent $\sqrt{2}$ by the ratio of two integers but it is, of course, possible to represent it by the diagonal of a unit square, that is, geometrically. Hence, the algebraic way was abandoned in favor of the geometric, and geometry was to become uniquely Greek.

■PLATONIC SCIENCE.—The course of physical speculation in Greece was abruptly altered by the advent of Socrates. It was not that Socrates was opposed to the search for a principle of physical reality; he simply felt that if this search did not include man and the moral universe, it was irrelevant to the proper concern of men for their own moral and political condition. Socrates' great disciple, Plato (427–347 B.C.), set himself the task of bringing physics, morality, and politics together into a single system. The glue that was to hold these disparate elements together was a theology from which Plato deduced the basic elements of his philosophy.

Like his predecessors, Plato had to come to grips with the dualistic tension between *being* and *becoming* created by Parmenides. For Parmenides' unchanging sphere, Plato substituted the *Ideas,* those ideal, unchanging forms from which the observable world drew its principles of organization. The principle of change was uniquely the property of brute matter which strove constantly to return to chaos. Form and harmony in the observable world were the result of the influence (or the participation) of the Ideas in matter. To apprehend the Ideas required that the mind, drugged, as it were, by its association with brute matter, be revived by the contemplation of the purest of earthly forms. These, Plato insisted, were the geometrical forms, and mathematics, therefore, was considered to be the most efficacious of all intellectual purges.

Mathematics also provided a key to the system of the world, for Plato took over the Pythagorean belief in the essential mathematical harmony of the world. He also adopted the elements of their cosmology which, with numerous modifications, was to reign supreme until Copernicus.

Plato considered the heavens to be literally divine, and divinity, of course, implied perfection. Ordinarily, one would not expect a perfect thing to change in any way since change would suggest that the body was *not* perfect and was trying to improve itself. The stars did undoubtedly move, hence they were not *quite* perfect. The next best thing to not moving at all was to move in perfect motion. The Pythagoreans had already remarked that circular motion was the only motion which could go on for an eternity without changing. Hence the stars moved in circles. The planets, however, presented a problem. They, too, were almost perfect but they manifestly did not move in circles. Plato, therefore, set his disciples the great astronomical problem of antiquity: How could the motion of the planets be reduced to circular motion in order "to save the phenomena," that is, in order to force the

NEW YORK PUBLIC LIBRARY

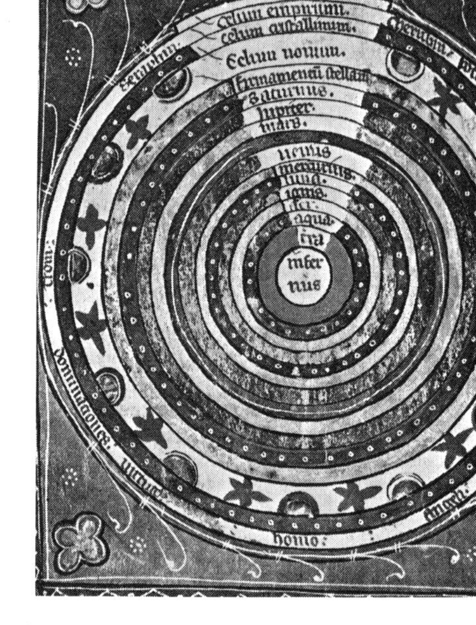

THROUGH MYTH, MAGIC, AND SCIENCE, men have tried to comprehend the universe and man's relation to it. The Egyptians imagined the world to be a flat disc, above which arched the sky-goddess Nut, her body spangled with stars. In Ptolemy's universe, myth gives way to mathematics, and the earth is enclosed in a finite series of geometric circles.

planetary orbits into those paths dictated for them by philosophy. The solutions were many and ingenious. Eudoxus (409–356 B.C.) suggested a scheme of spheres, one within the other, having a common center but revolving at different speeds around different axes. The resultant of all these motions was that of the planet. Still another solution, to become the mathematical model of Claudius Ptolemy (fl. 140 A.D.), involved the motion of a point upon a circle whose center lay upon another circle. The speed of rotation of the deferent (upon which the center of the first circle lay) and of the epicycle (upon which the point was fixed) could be adjusted arbitrarily so that the resultant motion closely resembled that of any one of the planets.

■ARISTOTELIAN SCIENCE. — Astronomy was one of the few scientific subjects Plato considered worthy of study. Those other subjects which brought man into too close proximity with matter could serve only to degrade the intellect. This viewpoint was vehemently rejected by Aristotle (384–322 B.C.), Plato's student and one of the greatest minds of all time. To Aristotle, there was no sense in Plato's separation of the Ideas from the matter they influenced. Not only did this introduce the clumsy intermediary of the Demiurgos, a kind of cosmic mechanic, who molded matter along the lines of the Ideas, but it seems to require the multiplication of the number of Ideas beyond the realm of probability. What Aristotle did was to unite form and matter within the object. Change was no longer a degradation of form, but a working out of the form inherent in the body undergoing change. Aristotle illustrated his thought by an appeal to four causes. The *material cause* was the matter of which a body was composed. The *efficient cause* was that cause which accounted for the shaping of the matter. It might be the chisel of a sculptor or an inherent principle, such as that which determined the evolution of an oak tree from an acorn. The *formal cause* was the principle of form which determined the action of the efficient cause. In the case of the sculptor, it would be his idea of what the sculpture should look like; in the case of the acorn, it would be the form of the oak tree inherent in the acorn. The fourth cause, the *final cause,* was the ultimate purpose for which the object was intended. To the artist, it was his concept of beauty which determined the form of the statue, guided his hand, and dictated his choice of material. The final cause was, therefore, the single most important factor in explaining any natural phenomenon. Once one found out the cosmic purpose of some event, the rest followed rather simply.

Aristotle generally accepted the scheme of the universe as depicted by Eudoxus. The Aristotelian universe was divided into two essentially different parts at the sphere of the moon. Above the moon, all was perfect and unchanging as befitted the divine nature of the heavenly bodies. Below the moon, generation and corruption, rectilinear motion and general chaotic change took place. The analysis of sublunar motion was of particular importance for the future of science for it was here that the Aristotelian system was to be mortally wounded. Aristotle distinguished two kinds of local motion: natural and violent. Natural motion was that type of motion brought about by a natural, inherent cause. Thus a stone fell to the ground because its natural place was at the earth, or center of the universe, and, when displaced, it strove to regain its natural place. The other elements, water, air, and fire, would in similar fashion move to their natural places. When a stone was hurled through the air, however, it did not, at once, fall to the ground. Its flight through the air was contrary to its natural tendencies and this was why it was called violent motion. Why, Aristotle asked, did the stone continue to move contrary to its natural tendency when it left the hand and there was nothing there to push it? Obviously, a body could not keep moving when the cause of its motion (the hand) was removed. Something, therefore, must continue pushing it. This "something," Aristotle suggested, was the air which was pushed out in front of the stone and rushed back behind it in order to prevent a vacuum from being formed. The air then continued to push the stone along.

PROGRESS IN MEDICINE.—The Egyptian and Babylonian civilizations had made little progress in the study of sickness and health as natural occurrences. It was the Greeks, particularly those associated with the school of Hippocrates of Cos (c. 400 B.C.), who first insisted that medicine should be divorced from the supernatural. Disease was not a demonic visitation, but the result of natural causes. What these causes were, whether a disharmony of the four humors, or some hitherto undiscovered disruption of physiological function, was not so easily answered. Nevertheless, a giant step had been made, for by reducing medicine to the circumference of man's reason, the precondition was created for development of a rational medical science.

HELLENISTIC SCIENCE.—The period of Greek civilization before the conquests of Alexander the Great (356–323 B.C.) produced the overall philosophical framework within which science was to exist for the next millennium and a half. During the Hellenistic period (roughly 330 B.C.–150 A.D.) many details of this picture were filled in. The Museum of Alexandria, founded by the new Greek rulers of Egypt, became the intellectual center of the ancient world. The astronomical works of the later Babylonians, through the extraordinary efforts of Hipparchus (c. 190–120 B.C.) and Claudius Ptolemy, were combined with the Greek astronomical tradition to create a mathematical astronomy of hitherto unattainable precision. The application of Babylonian mathematics to the Greek system of epicycles permitted the calculation of astronomical positions to a degree unsurpassed until the late sixteenth century. The addition of new and precise observational data also contributed to the exactness of Ptolemaic astronomy.

The Museum of Alexandria was also the center of physiological activity. Human dissections were performed and a number of discoveries made. The greatest physiologist of antiquity, Galen of Pergamum (131–201 A.D.), was not, however, a member of the Museum. He was a distinguished physician who utilized the physiological and anatomical discoveries of his predecessors, together with the results of his own dissections of apes and pigs, to formulate a system of physiology of enormous influence. Through the hypothesis of three "spirits," the *natural* (which provided nutrition), the *vital* (which was the principle of vitality or movement), and the *animal* (the basis for sensation and thought), Galen was able to tie the various organs of the body together into a single, harmonious whole. Furthermore, Galen was able to utilize the Aristotelian four causes to advantage and give his physiology the proper philosophical foundation.

Early in the Hellenistic period, physics reached its highest point in antiquity. The works of Archimedes of Syracuse (287–212 B.C.) were the culmination of the mathematical tradition of the Pythagoreans, without the number mysticism of this sect. His work on the lever and on hydrostatics showed that the translation of physical entities into mathematical quantities (that is, length, weight, and density) was even more fruitful than the Pythagoreans had dreamed. Just as important, Archimedes also showed that it was not necessary to be a numerologist to discover the mathematical laws of nature. In 212 B.C. Archimedes was slain by a Roman soldier in the sack of Syracuse. There is something symbolic in this murder, for as Rome cast its pall over the Mediterranean world, the sciences stagnated and died. In spite of their considerable administrative skill and great engineering ability (or perhaps because of them), the Romans took little interest in the theoretical constructions of their Greek subjects. By the end of the Empire, the great Greek achievement was known to but a few and the manuscripts in which almost a thousand years of this sustained, brilliant intellectual effort was recorded, lay moldering, unused, in a few scattered libraries. The march of the barbarians and the sack of Rome provided the proper funeral for the death of ancient science.

Survival of Ancient Science.—The collapse of Rome left behind many handsome ruins which were to inspire the generations that followed. Part of the legacy were the thousands of manuscripts that recorded the heights to which Greek science had ascended and which the Romans, by and large, had ignored. That such precious and delicate remnants of antiquity did not perish in the upheavals that accompanied the dissolution of the Roman Empire is to the eternal credit of the Christian Church, the Byzantines, and the Moslem conquerors.

The early Christians were not profound philosophers. Their religion had enormous appeal to the poor and the oppressed but it could make little headway against the sophistication of the ancient philosophers. In order to demonstrate its superiority to the philosophical systems of antiquity, the philosophers themselves had to be studied. A text concerning Epicurus might literally reek with heresy but if it were to be refuted it had first to be studied. And, in order for it to be studied properly by those competent to refute it, it had to be copied. So it happened that many of the profane writings of antiquity were passed on to future generations by monks writing in their scriptoria.

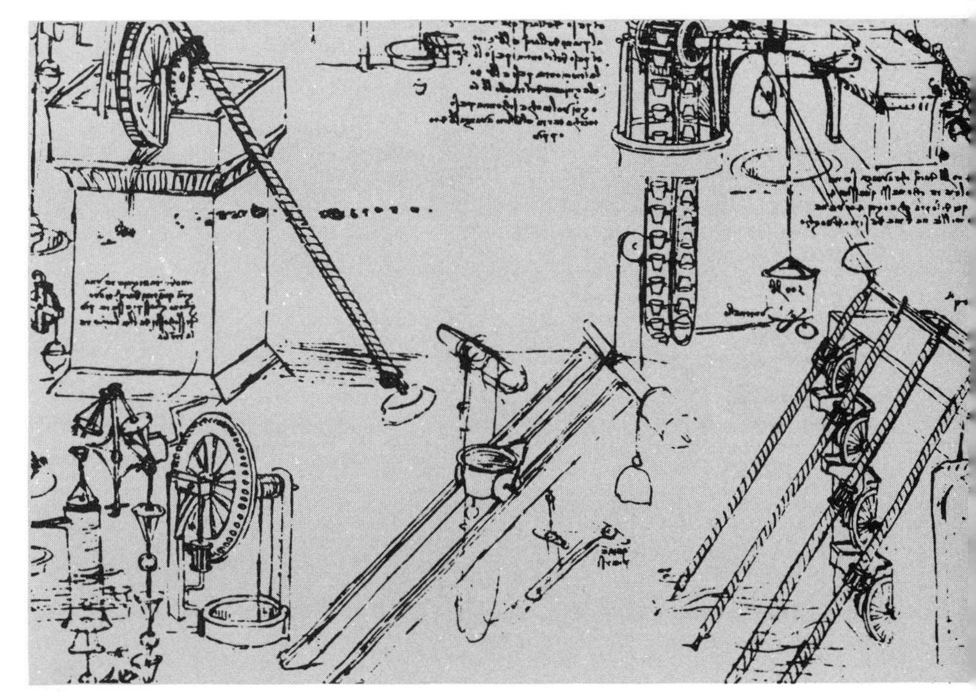

THE GENIUS OF ARCHIMEDES in mathematics and the invention of mechanical contrivances was not equaled until the Renaissance. Leonardo da Vinci used many of Archimedes' ideas, such as the water screw and wheel, in his numerous scientific and engineering drawings.

NEW YORK PUBLIC LIBRARY

THE ALCHEMISTS' DREAM of transmuting lead into gold was never realized, but their mystic treatises amassed much chemical information and their endless combinations of substances did result in chance discoveries such as gunpowder (*right*), attributed to a German monk.

The most important channel for the transmission of ancient science was the hordes of Islam. Erupting from the Arabian Peninsula in the seventh century, Islam rapidly conquered most of the ancient world. An aura still surrounded the name of Rome and the invaders preserved much of that civilization which they set out to conquer. No doubt the stubborn refusal of Byzantium to bow before them intensified Arabic respect for the higher culture of antiquity. As so often happens, the conquerors were conquered by their own subject peoples. Ancient art, literature, and science far surpassed anything Islam possessed, and these were eagerly studied by her rulers and scholars. A regular translating and copying industry was created to make ancient learning available in the Arabic tongue. Plato and Aristotle were, of course, put into Arabic almost immediately. The great tradition of Islamic medicine resulted from the translation of the ancient medical treatises and the commentaries thereon by Arabic medical philosophers. The *Canon of Medicine* by Ali ibn-Sina (980–1037), whose name was Latinized by Western Schoolmen into Avicenna, was the main channel through which this tradition reached the West. Ancient astronomy was kept alive in Islam, particularly in Baghdad, where observatories were supported by the Caliph and mathematicians and astronomers were honored. Mathematics flourished under Islam. From their far-flung conquests, the Arabs brought back the Hindu system of numerals. The combination of these simple numbers with the positional numeration of the Babylonian tradition made for a most powerful and simple computational system. It also restored an interest in numbers rather than lines, and Islamic algebra did much to restore the mathematical balance between geometry and algebra upset by the Pythagoreans.

The only area in which Islam introduced anything really new into science was in optics. Here the work of Ibn al-Haytham, or Alhazen (965–1038), as he was known to the West, was fundamental. He opposed Euclid and Claudius Ptolemy on the origin of vision, considering sight as the result of something impinging upon the eye, rather than as an effect of something passing from the eye to the object. He investigated problems of reflection, and even dealt with the paths of light rays reflected from convex mirrors which required the solution of an equation of the fourth degree. He also studied refraction and clearly comprehended, in qualitative terms, the effect of different media upon light. His study of lenses was unsurpassed for centuries and went far beyond that of the Greeks. It was for very good reason that Western scholars held Alhazen in the highest esteem.

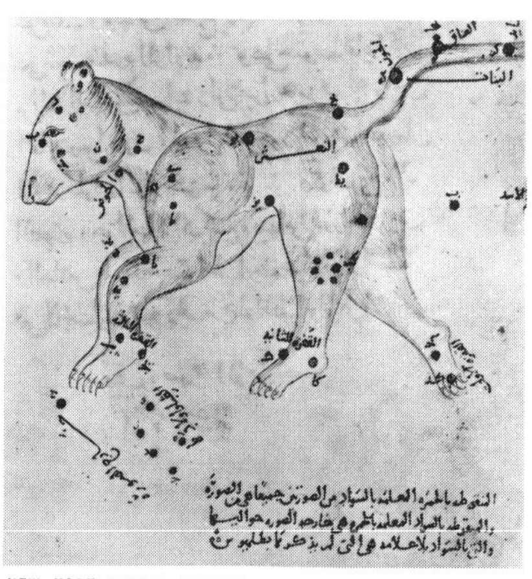

NEW YORK PUBLIC LIBRARY

GREAT BEAR CONSTELLATION, in Arabic.

Beginnings of the New Science.—By the twelfth century the political instability and economic parochialism that had marked European civilization since the fall of Rome were disappearing. Powerful political entities had emerged, and trade, stimulated by the Crusades, was increasing rapidly. The Church, in particular, found itself facing a world in which its traditions seemed out of date. A new confidence filled the air; what yesterday had seemed a rock of faith today revealed itself as a heap of sand. "I believe that I may know" had formed more than one saint's credo in those centuries when knowledge seemed futile in the face of God's displeasure and the uncertainty of events. "I know that I may believe" was Peter Abelard's (1079–1142) proud, almost arrogant, challenge to the past and summons to his own century. It was no coincidence that the twelfth century witnessed the creation of the University of Paris and the first flood of translations of ancient learning.

One of the fascinations of the medieval period lies in the order and dimensions of the dreams medieval man dreamed. Conscious of his fall he built cathedrals that sought divinity in every stone of their architure; while barons revolted time and again against their kings, he had visions of a unified empire that might recapture some of the grandeur that was Rome; conscious of his intellectual inferiority, he zealously collected every scrap of ancient knowledge. Only in his faith did he feel secure and the heir to something antiquity had not possessed.

Medieval science must be viewed in this double context. Men like Roger Bacon aspired for ultimate knowledge and power while, at the same time, they saw knowledge of this world as an aid in the real goal of all knowledge, that of God.

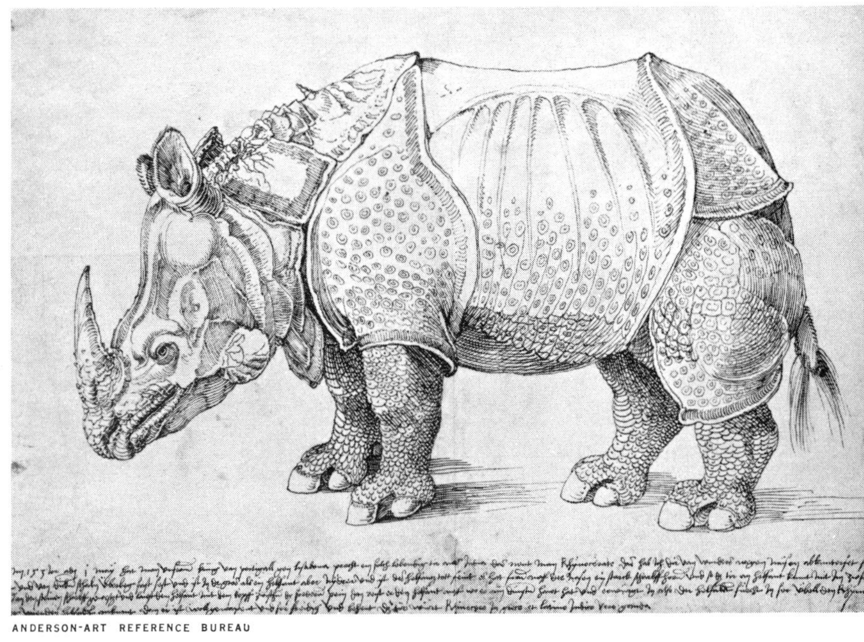

ANDERSON-ART REFERENCE BUREAU

NATURAL HISTORY developed from the fantasy of the Middle Ages into a new realism during the Renaissance. Fabulous beasts, such as the griffin (*left*), disappeared and artists examined the natural world with the scientific curiosity reflected in Dürer's *Rhinoceros,* 1515.

ALCHEMY.—Alchemy began as a more or less rational investigation of the properties of matter and the principles of their interaction. The Aristotelian basis of all existing bodies was a formless *materia prima* whose only property was the ability to take on qualities. If one defined gold as a heavy, yellow metal all that was necessary was to implant the qualities of heaviness and yellowness in a baser metal, say lead or mercury. The qualities themselves were considered to be separate entities which could be abstracted from or added to the *materia prima* if the alchemist knew the correct procedures. The correct procedures, however, were singularly difficult to find and the search for them was what lead alchemy into demonology and all its accompanying nonsense. The philosopher's stone was the magic wand that would accomplish the miracle of transmutation. Its value was enhanced when the idea was introduced from China that the stone or elixir would also provide its discoverer with eternal life.

The theological impetus to science formed two distinct traditions in the twelfth and thirteenth centuries. Neoplatonism, in a somewhat vague form, had survived the Dark Ages and was refreshed and revivified by the twelfth-century Renaissance. Neoplatonic philosophy viewed the material world as the result of a series of divine emanations, gradually becoming degraded in a scale reaching from the Godhead to brute matter. In this system light played a particularly important part. It was itself immaterial, but it existed in the material world, thus bridging the gap between the divine and the mundane.

THE FRANCISCANS.—The Franciscans, whose order was founded in 1210, were fascinated by light and saw in it a nearness to God, hence worthy of study. The work of Alhazen pointed the way. Robert Grosseteste (c. 1175–1253), although not himself a Franciscan, taught them and was the leading figure in the creation of medieval optics. He did some work on mirrors and lenses but, most importantly, dimly glimpsed the role of experiment in science. It was this aspect that Roger Bacon forcefully called to his contemporary's attention. It is, therefore, a bit surprising to discover that, while there was a great deal of very clever talk and discussion of the role of experiments, very few actual experiments were performed.

THE DOMINICANS.—The Dominican order was founded by Saint Dominic in 1205 specifically to combat heresy. It was largely because of their orthodoxy that they were given the task of examining the flood of translations, especially of Aristotle, that were produced in the twelfth century. The first contact with Aristotle came as a shock to Western Christendom. The greatest philosopher of antiquity was found to contradict Scripture: The world, said Aristotle in flat opposition to *Genesis,* was eternal. Moreover, and this was even more disquieting, he was able to prove it by a logic far superior to anything the West could offer in rebuttal. The Church moved swiftly and forbade the reading of Aristotle until such time as his "errors" could be reconciled with Revelation. It was the Dominicans who performed this invaluable service. Albertus Magnus (1206–1280) was the first great medieval student of Aristotle. He initiated the studies by which Aristotle was made not only acceptable, but fundamental, to the Catholic faith. In his treatment of Aristotle there was none of that uncritical awe which was later to become the fashion. Aristotle, St. Albertus knew, was only human and could err.

THOMISM.—The assimilation of Aristotelian philosophy was completed by Albertus Magnus' great pupil, St. Thomas Aquinas (1225–1274). With extraordinary skill, St. Thomas was able to work most of Aristotle's system into the fabric of Catholic theology. The result was the monumental *Summa Theologica,* which still stands today as the Catholic Church's most comprehensive and closely reasoned statement of its theology.

The effect of St. Thomas' achievement on science was twofold. In the Thomist system, the duality between matter and spirit which had been a marked feature of earlier theology was banished. The world of nature led by degrees to the realm of God; there were no abysses separating the two. The study of nature, therefore, was by no means a worthless, or even theologically dangerous, occupation as many of the Church fathers had insisted. Instead, it was one of the paths by which it was possible to rise to an apprehension of divinity. Thomism, then, was and is not hostile to science so long as the study of science leads one closer to God.

Thomism was not, however, a scientific system. It was, specifically, a theology, and natural science was clearly subordinate to theological issues. In theory, points of natural science were always open to question, but, in fact, once the *Summa* was accepted as official doctrine it became increasingly difficult to challenge any part of it successfully. The pyramid stood or fell as a whole. To reject one stone, no matter where it came from, was to weaken the entire edifice. Thomism, therefore, was both unfavorable and favorable to the development of science. So long as there was but one church, a rather wide latitude of opinion could be tolerated. When the Church's monopoly of salvation was successfully challenged by the Protestant revolt of the sixteenth century, attitudes hardened, perspectives narrowed, and science suffered.

YERKES OBSERVATORY

NO PICTURE OF THE UNIVERSE could explain observed planetary motion as long as the earth remained unmoving at the center. Copernicus dared to assert that the earth and other planets revolve around the sun.

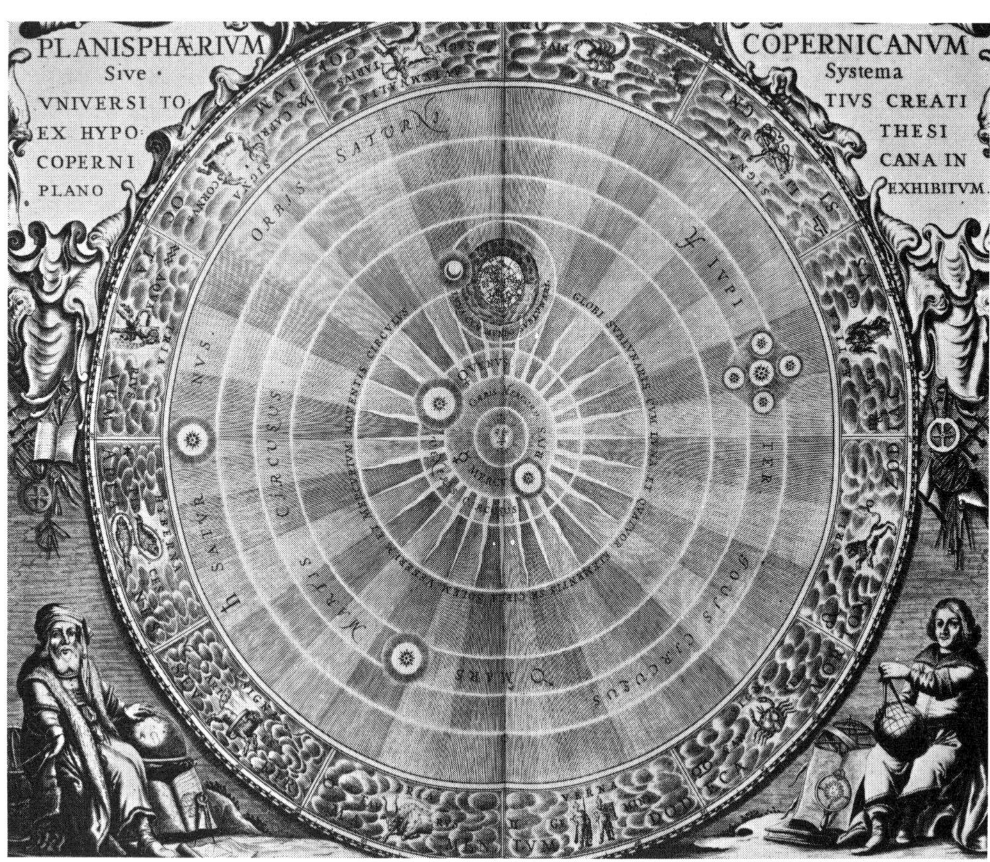

BETTMANN ARCHIVE

Scientific Revolution.—Origins of the seventeenth-century scientific revolution have been disputed by historians for the past two centuries. Some have seen it as the result of the liberation of the European mind by the Protestant Reformation; others have attributed it to the rise of the middle class, with its characteristic curiosity and quantitative penchant. Still a third school prefers to view modern science as a philosophical revolution in which the elements of science were entirely redefined. Finally, there is the view that modern science arose from a technological revolution which, by focusing attention upon the exploitation of nature, stimulated the search for natural laws. There are elements of truth in all these interpretations. While the Protestants were no more friendly or hostile to science than the Catholics, the very proliferation of Protestant sects made it possible for the holders of heterodox views to gain a hearing. The middle class, too, made its contribution. From its ranks came much of the driving force behind the attack on nature, which was provided by the vision of economic reward attendant upon eventual victory.

More important than these factors, however, was the reorientation in philosophy. The Copernican system owed far more to philosophical presuppositions than it did to anything else and classical dynamics was founded by challenging a fundamental tenet of Aristotelian philosophy. The technological innovations of the Renaissance were also to make an important contribution to seventeenth-century science. By the seventeenth century fairly complicated machinery was in use in a number of industries. They provided the analogy upon which science was to call for two centuries. The universe and its component processes could be likened to a machine and, like a machine, the universe could be understood in terms of the separate operations of its parts. The advantage of this approach over the organic analogy used by Aristotle is obvious. An organism is more than the sum of all its parts and there is always the temptation to call upon occult powers to account for this excess. Machines, on the other hand, are rigidly determined; understand the parts and one understands the machine. There is no room for mystery and the very lack of mystery is an indicator of success. The equations must always balance; mass and energy must be conserved. Such is the mechanical view and it was within this framework that science developed until the end of the nineteenth century.

■**COPERNICAN ASTRONOMY.**—Nikolaus Copernicus' (1473–1543) work, *On the Revolutions of the Heavenly Orbs,* was a strange one, combining the wildest speculations with page after page of tedious calculations. The purpose behind it was a laudable one in the eyes of his contemporaries, but few believed it could be attained. There had been, for over a century, a gradually increasing tension within astronomy. The Aristotelian universe had long been accepted as giving a true *physical* picture, for it could explain *why* the planets and stars moved. Although Claudius Ptolemy's epic work, the *Almagest,* had been translated in the twelfth century, there were few who could really work with its mathematics until the fifteenth century. The Ptolemaic system, with its deferents and epicycles, provided an excellent *mathematical* model of the universe, for from it *how* the planets moved could be calculated with considerable accuracy. The two systems, however, were incompatible with one another. One or both must be wrong and it was his search for a single model that led Copernicus to suggest moving the earth from its central position. His primary reasons for this revolutionary step were philosophical and aesthetic rather than logical or empirical. As a student in Italy, he had become imbued with the Neoplatonism of the Italian Renaissance and its metaphysics of light. There, too, he had become acquainted with Pythagorean thought and its emphasis upon mathematical harmonies. The emphasis upon light focused his attention on the sun as the source of this pure emanation; the apparent gain in mathematical simplicity which resulted from placing the sun at the center of the planetary system convinced him that this was where the sun really belonged.

Furthermore, a heliocentric approach removed some of the most difficult problems with which astronomy had been wrestling since antiquity. The observed change of brightness in the planets, for example, followed directly from a heliocentric system. So, too, did the deviation of the planets from circular motion. When one realized that the earth also moved around the sun, then the observed paths of the planets were the resultants of the motions of the earth and the planets.

The Copernican system was mathematically equivalent to the Ptolemaic. Indeed, one of the reasons for its slow acceptance was that it was only equivalent, not superior, to it. Physically, the Copernican system was attractive in that it accounted for a number of mysterious phenomena in a most satisfactory way. On one point, however, it remained inferior to the Aristotelian model, at least in the eyes of many of Copernicus' contemporaries: Copernicus could offer no sound reason for planetary motion. It is no exaggeration to say that the scientific revolution consisted in the search for an explanation that would preserve the Copernican arrangement of the heavenly bodies and, at the same time, explain why the planets moved about the sun.

■CONTRIBUTIONS OF KEPLER.—Johannes Kepler (1571–1630) clearly recognized the necessity of accounting for the earth's annual revolution around the sun. Kepler was a passionate adherent of Copernicus' doctrine because he, too, was impressed by the beauty of a heliocentric universe. Kepler was also a Pythagorean, convinced that mathematical harmonies gave the real clue to physical reality. His entire life was spent in searching for them. He was certain that his greatest discovery was the fact that the five regular Platonic solids—the cube, tetrahedron, octahedron, dodecahedron, and icosahedron—could be inscribed within planetary orbs.

Fortunately for Kepler, he was associated for some time with the greatest observational astronomer of the sixteenth century, Tycho Brahe (1546–1601). Kepler's mathematical flights of fancy were, therefore, strictly controlled by Brahe's accurate data. It was from these data that Kepler computed the orbit of Mars and found, somewhat to his horror, that it was *not* a circle but an ellipse. Although difficult to fit into the Copernican system, it was fatal to the Aristotelian, for even the cleverest philosopher found it impossible to suggest a reason why planetary souls moved in ellipses. Kepler, however, could at least fumble toward an explanation based upon the power of the sun to hold his numerous family together. Kepler's second law of planetary motion stated that the line drawn from the sun to a planet (the radius vector) sweeps out equal areas in equal times. To put it another way, a planet moves faster when it is nearer to the sun than when it is farther out. His third law stated that the square of the period of revolution of a planet around the sun is proportional to the cube of its mean distance from the sun. For example, if two planetary periods are compared, it is obvious that the one farther out moves more slowly. From these laws, Kepler postulated the existence of some entity, perhaps a force of some kind, emanating from the sun like light, that pushed the planets in their orbits.

■GALILEO'S WORK.—The same problem of accounting for planetary motion in a heliocentric system also occupied the man who may rightly be called the first of the modern mathematical physicists, Galileo Galilei (1564–1642). Paradoxically, his solution led him to reject Kepler's first law that planets move in ellipses. Like Kepler, Galileo was a Copernican for largely aesthetic reasons. A heliocentric universe was simply neater than a geocentric one. There was not, with Galileo, the number mysticism that marked Kepler's approach. Galileo's hero was Archimedes, not Pythagoras.

The problem of motion had intrigued Galileo since his youth. He had early become suspicious of Aristotle's whole account of both the *how* and the *why* of motion. While investigating falling bodies he had been led by experiments on inclined planes to the discovery of the law of free fall which he first enunciated in 1604. *Why* a body fell according to the law, however, escaped him for 30 years. His work with inclined planes permitted Galileo to formulate a mental experiment of fundamental importance for the progress of Copernicanism. From it, he deduced that, if there were not resistance to its motion, once put in motion a sphere would move in a circle till eternity. To the anti-Copernican who scornfully asked why the earth should move around the sun, Galileo answered, why shouldn't it? Circular motion had no cause; it simply continued forever. The consequence of this simple mental experiment was revolutionary. To Aristotle, only rest needed no causal explanation; all other motion or change had to be accounted for. Now, a certain kind of motion had achieved the same status as rest. It, too, needed no explanation; it simply was. Motion was as natural as rest—what needed explanation henceforth was not motion but *change* of motion. The static cosmos of the Greeks had been transformed into the dynamic universe of classical mechanics.

The final blow to Aristotelian physics came with Galileo's analysis of uniformly accelerated motion. This problem had been extensively studied in the Middle Ages when it was recognized that Aristotle's explanation was inadequate. Granted that a heavy body wished to return to its natural place, why did it go

NEW YORK PUBLIC LIBRARY

INTERNATIONAL BUSINESS MACHINES CORPORATION

MECHANICS AND MATHEMATICS became the dominant image of the seventeenth-century scientific revolution and the means by which the universe could be understood. The wheels of Pascal's algebraic calculator (*left*) reduced the mysterious to precise and measurable mathematical units, while the machines of industry (*right*) turned each new scientific discovery into a technological tool.

MEDIEVAL PHYSICISTS believed that a missile was carried upward in a straight line until gravity overcame impetus and the missile dropped straight to the ground.

AXIOMATA
SIVE
LEGES MOTUS

Lex. I.

Corpus omne perſeverare in ſtatu ſuo quieſcendi vel movendi uniformiter in directum, niſi quatenus a viribus impreſſis cogitur ſtatum illum mutare.

PRojectilia perſeverant in motibus ſuis niſi quatenus a reſiſtentia aeris retardantur & vi gravitatis impelluntur deorſum. Trochus, cujus partes cohærendo perpetuo retrahunt ſeſe a motibus rectilineis, non ceſſat rotari niſi quatenus ab aere retardatur. Majora autem Planetarum & Cometarum corpora motus ſuos & progreſſivos & circulares in ſpatiis minus reſiſtentibus factos conſervant diutius.

Lex. II.

Mutationem motus proportionalem eſſe vi motrici impreſſæ, & fieri ſecundum lineam rectam qua vis illa imprimitur.

Si vis aliqua motum quemvis generet, dupla duplum, tripla triplum generabit, ſive ſimul & ſemel, ſive gradatim & ſucceſſive impreſſa fuerit. Et hic motus quoniam in eandem ſemper plagam cum vi generatrice determinatur, ſi corpus antea movebatur, motui ejus vel conſpiranti additur, vel contrario ſubducitur, vel oblique oblique adjicitur, & cum eo ſecundum utriuſq; determinationem componitur.

Lex. III.

Actioni contrariam ſemper & æqualem eſſe reactionem: ſive corporum duorum actiones in ſe mutuo ſemper eſſe æquales & in partes contrarias dirigi.

Quicquid premit vel trahit alterum, tantundem ab eo premitur vel trahitur. Siquis lapidem digito premit, premitur & hujus digitus a lapide. Si equus lapidem funi allegatum trahit, retrahetur etiam & equus æqualiter in lapidem: nam funis utrinq; diſtentus eodem relaxandi ſe conatu urgebit Equum verſus lapidem, ac lapidem verſus equum, tantumq; impediet progreſſum unius quantum promovet progreſſum alterius. Si corpus aliquod in corpus aliud impingens, motum ejus vi ſua quomodocunq; mutaverit, idem quoque viciſſim in motu proprio eandem mutationem in partem contrariam vi alterius (ob æqualitatem preſſionis mutuæ) ſubibit. His actionibus æquales fiunt mutationes non velocitatum ſed motuum, (ſcilicet in corporibus non aliunde impeditis:) Mutationes enim velocitatum, in contrarias itidem partes factæ, quia motus æqualiter mutantur, ſunt corporibus reciproce proportionales.

Corol. I.

Corpus viribus conjunctis diagonalem parallelogrammi eodem tempore deſcribere, quo latera ſeparatis.

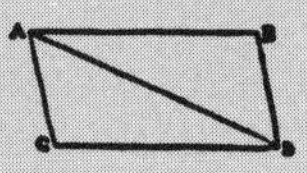

Si corpus dato tempore, vi ſola *M*, ferretur ab *A* ad *B*, & vi ſola *N*, ab *A* ad *C*, compleatur parallelogrammum *ABDC*, & vi utraq; feretur id eodem tempore ab *A* ad *D*. Nam quoniam vis *N* agit ſecundum lineam *AC* ipſi *BD* parallelam, hæc vis nihil mutabit velocitatem accedendi ad lineam illam *BD* a vi altera genitam. Accedet igitur corpus eodem tempore ad lineam *BD* ſive vis *N* imprimatur, ſive non, atq; adeo in fine illius temporis reperietur alicubi in linea illa

THE THEORY OF IMPETUS dominated medieval concepts of physics until the time of Galileo. Newton defined new laws of motion in his *Principia Mathematica;* the pages above show the first three laws, which define inertia, force, and the equality of action and reaction.

faster the nearer to earth it came? Increase in speed required, according to Aristotelian ideas, an increase in the cause of the speed. Or, decrease in speed, as when a body was thrown upward, required that the cause somehow exhaust itself. In order to handle this problem medieval philosophers had devised a theory based on the concept of impetus. Impetus was a substantial form added to a thrown body by the action of the thrower; as the body traveled upward, the impetus gradually wore itself out in fighting against the natural tendencies of the body to fall. When the impetus just balanced the gravity, the body came to a brief halt and then began to fall. As it fell, its motion generated impetus (in some mysterious fashion) which increased its impetus, which increased its speed, and so on. The *cause* of motion remained internal as the body moved in a polarized space, that is, toward or away from that point in the universe where it belonged.

Galileo was perfectly aware of this medieval work and, in his youth, was imbued with the ideas of the impetus theory. Its difficulties were obvious: It required that an effect (acceleration) generate its own cause (impetus). Furthermore, the theory implied that impetus, and therefore acceleration, depended upon either the speed of the body at every instant or the distance the body had traveled. Neither alternative could stand the test of experiment. For years, Galileo sought an explanation of his law of free fall and finally, in 1638, in his *Discourses on Two New Sciences,* Galileo announced the key. Acceleration depended upon time, not distance or speed. By thus removing the cause of change of motion from a spatial to a temporal dimension, Galileo radically altered the definition of space itself. Instead of having qualities (such as a center toward which the earth tended), space was turned into the space of Euclidean geometry. No direction had any priority over any other direction. Space was homogeneous; bodies moved through this space according to simple mathematical laws.

■**WORK OF NEWTON.**—A new tack was taken by Isaac Newton (1642–1727). From Galileo he learned the principles of the new dynamics. From Kepler he received the idea of force but purified it to rid it of its mystical connotations. With these as a foundation, Newton deduced that a body attracted by another body by a force acting along the line connecting their centers and varying inversely as the distance would travel in an ellipse. However, it was not until his friend Edmund Halley (1656–1742) discovered what he had deduced and nagged him ceaselessly that Newton put his ideas in order. The result was perhaps the most important scientific book ever written, *Principia mathematica philosophiae naturalis (The Mathematical Principles of Natural Philosophy).* Its appearance in 1687 announced the culmination of the scientific revolution.

In a severely mathematical style Newton cautiously laid the groundwork for his three laws of motion in the *Principia.* The first law is the principle of inertia, which defines the steady state of the dynamic universe. The second law defines force in terms of mass and acceleration, thus turning an intangible and vague concept into an experimentally determinable quantity. The third law states the equality of action and reaction, which is the very basis of statics and plays an essential role in dynamics. From these three laws, all ordinary problems of statics and dynamics could successfully be attacked. Even more important was the application of these laws to one specific problem: planetary motion. Here Newton had to postulate a gravitational force varying inversely as the square of the distance and directly as the product of the masses involved. The result was stunning. The system of the world fell neatly into place. For two centuries the *Principia* was both the boldest manifesto of the scientific revolution and the cornerstone of the new science. It did far more than lay down laws; it provided a vision of physical reality.

If the scientific revolution is properly to be understood, the world of the *Principia* must be compared with that of Aristotle. The difference can most clearly be seen by comparing what was considered to be an adequate explanation in each system. For an Aristotelian, the four causes provided a complete knowledge of an object or an event. The final cause—the purpose—was the key cause but this was often impossible to ascertain. In physics, particularly, processes had no obvious purpose, they must be sought in the mind of God. Physics, therefore, had almost inevitably to become theology and thereby leave the arena of experimental criticism. To the Newtonian an explanation consisted in giving the mathematical laws of action and in providing a physical picture of the process in terms of matter, motion, and force. It has often been said that the scientific revolution banished the question "why?" from science and substituted "how?" for it.

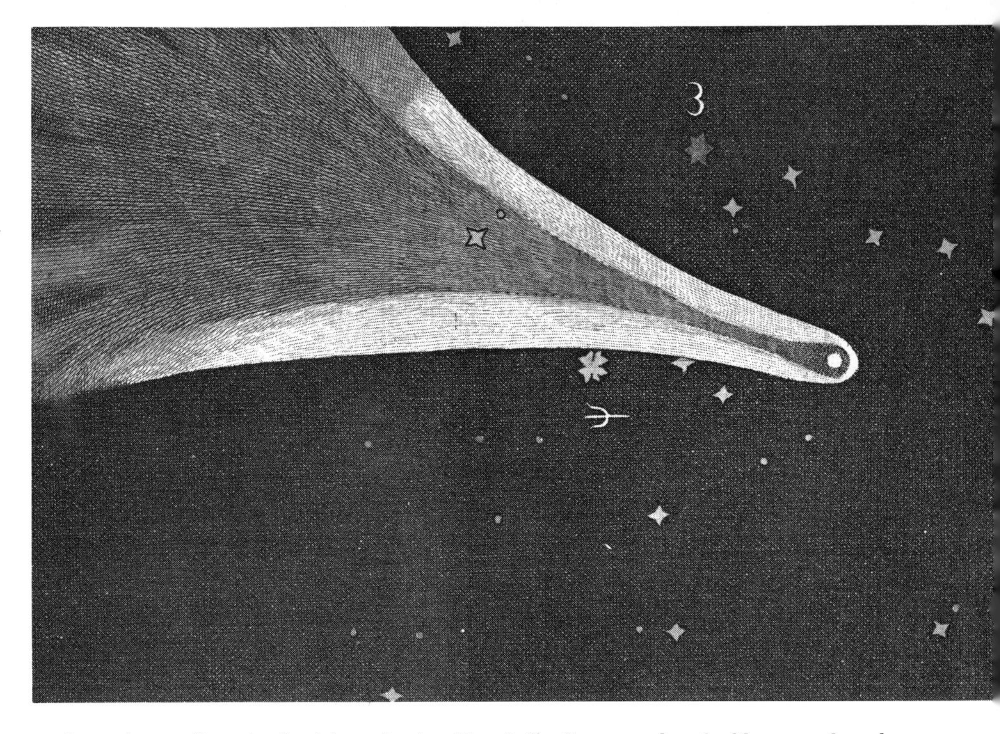

NEW YORK PUBLIC LIBRARY

NEWTON'S LAWS OF MOTION became the key to the understanding of celestial mechanics. Herschel's discovery that double stars obey the law of gravitation and the accurate prediction of the cycles of Halley's comet and the comet of 1811 (*right*) substantiated Newton's accuracy.

In the case of an ultimate "why?" this is true, but we should be careful not to eliminate the "why?" on a lesser level. "How?" is answered by a mathematical equation; "why?", in a physical sense, is answered by appealing to the mechanical philosophy. Why certain effects are produced can be answered by applying the Newtonian laws of motion to matter under given, clearly defined conditions. Thus, Newtonian science also answers "why?" but without having recourse to the Deity.

■MEDICAL PROGRESS.—There is one last aspect of the scientific revolution that requires mention. The authority of antiquity was finally broken by men of the seventeenth century. The realization that the ancients were often simply factually mistaken liberated science from the dead hand of tradition. This aspect is seen most clearly in the biological sciences where the issues did not involve a radical new picture of the universe.

In 1543, Andreas Vesalius (1514–1564) published his *De corporis humani fabrica (On the Structure of the Human Body)*, intended as an atlas of human anatomy. Vesalius' primary concern was an accurate description of the human body, not a new theory of physiology. He was unable to free himself completely from Galen's physiology, but he did point out a number of anatomical errors made by his great predecessor. As anatomical investigation became more and more intensive, further faults were found until, by the early seventeenth century, anatomical knowledge forced a revision of Galenic physiology. This was William Harvey's (1578–1657) great contribution. The anatomical discovery of valves in the veins led Harvey to question the ebb and flow of blood through the veins demanded by Galen's theory. Harvey showed that the blood could only flow in one direction through the veins to the heart. From this, it followed that the blood *must* circulate, even though Harvey could not find the connection he was certain must exist between the veins and the arteries. To Harvey, the court of last resort was not what the ancients had said but what Nature herself showed. This aspect of the scientific revolution took science out of the library and into the laboratory. Ideas need not arise from experiments—they rarely do—but they must somewhere, sometime, meet the acid test of experimental attack.

Newtonian Science.

—The scientific revolution led to an ever-increasing pace of scientific discovery and development which would take volumes to detail. All that can be done here is to trace out certain threads to indicate some of the triumphs and failures of the new science.

■ASTRONOMY.—Having begun in astronomy, the new science soon revolutionized this study. Newton's laws of motion and principle of gravitation provided the necessary foundation for an understanding of celestial mechanics. The prediction of the return of the great comet now named after Edmund Halley (1656–1742) was the first dramatic triumph of Newtonian theory. The discovery of double stars by Sir William Herschel (1738–1822) and the fact that they, too, obeyed the Newtonian law of gravitation showed that this principle and the other laws expressed in the *Principia* were truly universal.

The apex of Newtonian astronomy was reached with the discovery of Neptune by Urbain Leverrier (1811–1877) and John Couch Adams (1819–1892) in 1846. From small perturbations in the orbit of Uranus, both men, independently of one another, mathematically deduced the existence and location of a new planet. If there were any who doubted the Newtonian synthesis, this discovery converted them. There was, nevertheless, one small astronomical problem that served to perturb a few people. The orbit of Mercury stubbornly refused to obey Newtonian laws. One small planet was not enough, however, to shake confidence in classical dynamics and it was not until Albert Einstein (1879–1955) seriously modified Newtonian mechanics by introducing the theory of relativity that Mercury rejoined the family of planets, moving in accordance with scientific law.

■PHYSICS.—The Newtonian synthesis bore its most abundant fruit in areas quite distinct from astronomy. Before Newton, the phenomena of heat, light, electricity, and magnetism were almost total mysteries. The laws of action of light had only been barely glimpsed. The classical treatise on the magnet by William Gilbert (1540–1603) had only lifted a corner of the veil covering this subject. Heat and electricity were almost entirely outside the pale of science.

Newton's work on optics laid a firm foundation for the study of this subject. Considering light to be composed of minute imponderable particles of different sizes (size determined color), Newton treated light as the result of the passage of these particles through space, following the ordinary laws of motion. His *Opticks*, published in 1704, was to be the textbook on light for a century.

BURNDY LIBRARY

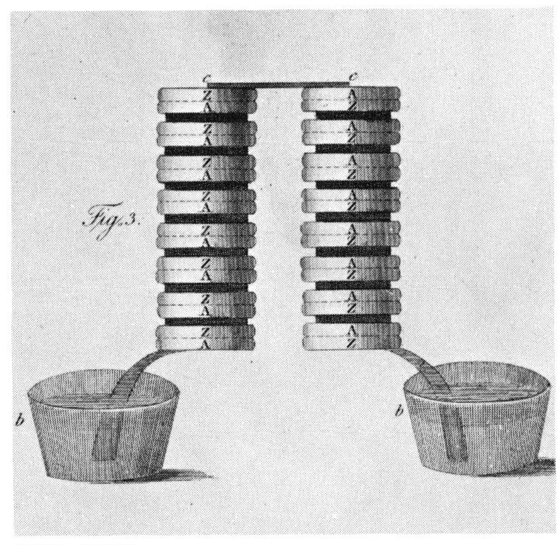

BURNDY LIBRARY

THE FIRST ELECTRIC PILE or battery (*right*), made by Alessandro Volta, had zinc and silver discs separated by brine-soaked paper; it was the result of Galvani's experiments on "animal electricity" in the frog.

Even Newton was puzzled by electricity, magnetism, and heat. While the latter might be considered the result of the "intestine motion of the particles of a body," this did not really help much in understanding heat flow and other thermal phenomena. Similarly, the supposition of electrical and magnetic effluvia permitted one to form a vague mental picture of electrical and magnetic phenomena but did not permit these two areas to be treated in detail. In the eighteenth century, these subjects were to become an integral part of Newtonian science.

The theory of heat profited greatly from the invention of fairly accurate thermometers. The mercury thermometer was introduced about 1715 by Gabriel Daniel Fahrenheit (1686–1736) and the new science of thermometry permitted an objective investigation of events often before beclouded by subjective factors.

The study of heat was greatly advanced by Joseph Black (1728–1799) of Scotland. Black's observations of the slowness with which the winter's snow melted led him to the discovery of latent and specific heats in the 1750's. This discovery, in turn, forced the formulation of a new theory of heat since latent heat, in particular, seemed irreconcilable with a hypothesis of heat as molecular motion. So was introduced what Black called the "matter of heat" and Antoine Laurent Lavoisier (1743–1794) later christened "caloric." Caloric was an imponderable fluid whose particles repelled one another (thus causing the expansion of bodies to which caloric was added) and which could exist in two different ways in matter. When free, the addition of caloric raised the temperature of a body; when combined with matter, the caloric became latent and addition of more caloric caused a change of state but not a change of temperature. It was this aspect that gave the caloric theory long tenacious life. The famous experiments of Count Rumford (Benjamin Thompson) in the 1790's "proving" that heat *had* to be a mode of motion were ignored for 50 years because latent heat was seemingly inexplicable in this theory.

Electricity and magnetism were also reduced to Newtonian terms of attraction and repulsion in the same century. In the 1780's Charles Augustin de Coulomb (1736–1806) suggested that electricity consisted of two imponderable fluids, positive and negative. Positive particles repelled one another and repulsions adhered to the Newtonian inverse-square law. Similarly, the two magnetic fluids also obeyed this law. The only difference between the electrical and magnetic fluids (and it was an important one) was that the electrical fluids could move between the particles of ponderable matter whereas the magnetic fluids were confined within these particles. When, in 1803, the English chemist John Dalton showed that the particles of ponderable matter were differentiated from one another by specific weights, the system seemed complete. From the largest star to the smallest atom, including even matter so subtle that it had no gravitational attraction for gross matter, the Newtonian laws held. It was this seeming simplicity that led Pierre Simon de Laplace (1749–1827) to suggest that, if one knew the position and momentum of every particle in the universe and could make the necessary computations, the entire future course of the cosmos could be predicted exactly.

■**CONVERSION OF FORCES.**—At the very time that Laplace was triumphantly declaring the finality of Newtonian science, there were rumblings of an earthquake that would ultimately bring this edifice tumbling down. The German philosopher Immanuel Kant (1724–1804) had already attacked the very foundations of Newtonian science in the 1780's by denying the existence of matter as something hard and atomic. All we can know about matter, Kant declared, are the forces associated with it. To assume that a force must arise from a material substratum is a metaphysical step that has little justification in experience. Furthermore, it may blind one to the interconnectedness of phenomena and thus really prevent scientific progress. For example, if the imponderable fluids are separate entities, there is no reason to suspect that they can interact or be transmuted into one another. If, on the other hand, they are merely different manifestations of the basic forces of attraction and repulsion, then the conversion of one force into another should cause no surprise.

Kant's suggestions created the school of *Naturphilosophie* in Germany and led to fantastic speculations as well as to some solid scientific achievements. Great impetus was given to the idea of the convertibility of forces with discovery by Alessandro Volta (1745–1827) of the voltaic pile in 1800. In the same year, it was found that an electric current could decompose water. This tied the force of chemical affinity and that of electricity closely together. In 1832, Michael Faraday (1791–1867) showed that the conversion of electrical force into chemical affinity, and vice versa, followed precise laws.

Within fifty years of Volta's discovery, the cycle of the conversion of forces was almost complete. Hans Christian Oersted (1777–1851) showed that electricity could be converted into magnetism in 1820; Michael Faraday revealed the opposite effect in 1831 with his discovery of electromagnetic induction. The conversion of electricity into heat was recognized from the very beginning of electrodynamics; thermoelectricity was discovered by Thomas Johann Seeback (1770–1831) in 1821. The effect of magnetism on light was discovered by Michael Faraday in 1845. He sought in vain for a similar effect due to electricity but this was not discovered until 1875 by John Kerr. The conversion of light into electricity (the photoelectric effect) was discovered in 1879.

■ENERGY.—To a large extent, these conversions were sought by those persons who believed in the unity of the forces of nature. To them, nothing was lost in the conversion of one force into another. What was lacking, however, was a common measure of force. This was supplied in the 1840's in the writings of Hermann Ludwig von Helmholtz (1821–1894), Julius Robert Mayer (1814–1878), and James Prescott Joule (1818–1889). The result was the concept of energy and by 1850 the principle of conservation of energy had been accepted as a fundamental law of nature.

The focus upon force, rather than matter, led to the development of new theories of the ways in which energy could be transmitted. Newtonian physics depended upon two forces: attraction or repulsion acting at a distance and the impact of one body on another. In the early 1800's, Thomas Young (1773–1829) in England and Augustin Fresnel (1788–1827) in France suggested another way in which energy, at least the energy of light, could be transmitted. Light, they argued, could be viewed as wave motion in an ether. Here the energy lay in the medium, not in any particle or mass. This concept was the leading string in the researches of Michael Faraday. He extended it to electricity and magnetism in a series of classical researches. These researches, in turn, were the basis for the mathematical field theory of James Clerk Maxwell (1831–1879). Problems raised by Maxwell's treatment of the field, in turn, led to Albert Einstein's theory of relativity and general field theory in the twentieth century. Pierre de Laplace would have found Einstein's universe quite different from the one he had envisioned in the early 1800's.

■CHEMISTRY. — The development of chemistry in the post-Newtonian period was less dramatic than that of physics. Except for John Dalton's atomic theory, attempts to apply Newtonian principles were not successful. The chemical revolution, associated with the name of Antoine Laurent Lavoisier, was largely a recognition of this fact. Chemical theories of the eighteenth century had all appealed to essentially metaphysical entities, whether these were Epicurean atoms or the even more elusive phlogiston of the German chemists. Lavoisier's revolution consisted in making the elements of chemistry those substances, and only those substances, which were palpable and capable of being fitted, by weight, into a chemical equation. The discovery that combustion was the result of the combination of a ponderable gas, oxygen, with a combustible and not the release of the imponderable, almost undetectable, phlogiston provided the dramatic proof of the values of this method. Lavoisier's approach, when added to John Dalton's atomic theory (which provided substances with the basic atomic parameter of weight), created the foundations of analytical chemistry. Throughout the nineteenth century, chemists could be relatively certain that they knew what their retorts were filled with.

The mechanics of chemical reactions were almost totally veiled in mystery throughout most of the nineteenth century. Electrochemistry indicated that electricity played a vital role, but precisely what this role was remained unknown until the twentieth century. Edward Frankland's suggestion in 1852 that elements had a specific combining ratio or valence accorded with experimental fact but could not be explained until the present century. Even the periodic law, enunciated by Dmitri Mendeleev (1834–1907) in 1869, was more an ordering of facts than a mature theory. Explanation of the periodic properties of the elements had to await the development of the modern theory of atomic structure.

The most important contribution to nineteenth-century chemical theory (besides Dalton's) was the concept of the tetravalent carbon atom whose bonds were directed toward the corners of a regular tetrahedron. With this idea, August Kekulé (1829–1896) was able to bring some order into the chaos of organic chemistry. In the hands of Joseph Achille Le Bel (1847–1930) and Jacobus Henricus van't Hoff (1852–1911) this hypothesis grew into stereochemistry and could relate chemical qualities (in part) to molecular structure.

■BIOLOGY.—The last of the subjects with which the Greeks dealt seriously to develop into a mature scientific discipline was biology. By and large, biological sciences prior to the nineteenth century had dealt with classification and the problems of health and disease. The enormous growth in purely factual knowledge of living forms in the seventeenth century demanded the invention of some classificatory system that would simply permit the student to find descriptions of specimens and recognize what he saw. The most successful taxonomic attempt was that of Carolus Linnaeus (or Karl von Linne, 1707–1778). Although his system of classification of plants was based on the now rejected primacy of the sexual organs, his binomial nomenclature made it possible to classify huge numbers of specimens conveniently. When the doctrine of evolution revealed the genetic basis of classification the Linnaean system was easily adapted.

The problems of health and disease were still posed in terms of classical physiology. Attempts to introduce seventeenth-century atomism into medicine, such as that of Hermann Boerhaave (1668–1738), were ingenious but notably unsuccessful

NEW YORK PUBLIC LIBRARY

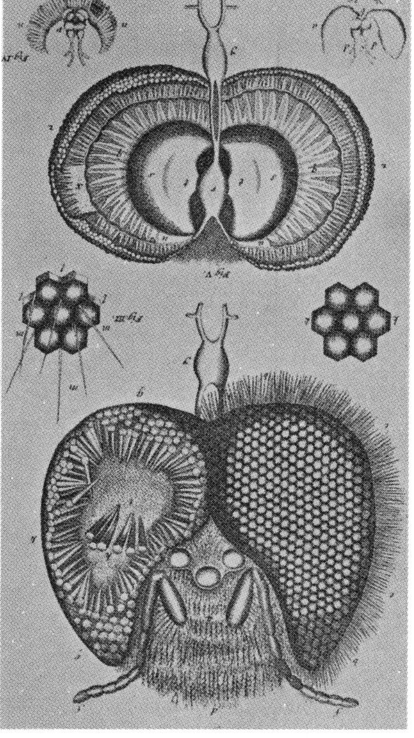

BURNDY LIBRARY

EXTENDING THE SCOPE of man's vision toward the infinitely far, the telescope revealed new facts that changed the picture of the world. The same principle, applied to the infinitely small, produced the microscope, invented by Galileo and greatly improved by Robert Hooke. With Hooke's compound microscope (*left*), observers such as Jan Swammerdam found new wonders in the previously invisible complexities of the eye of the bee (*right*).

NEW YORK PUBLIC LIBRARY

AMERICAN MUSEUM OF NATURAL HISTORY

CHARLES DARWIN'S theory of organic evolution represents one of the greatest examples of the scientific method; accurate observation and a massive accumulation of data led to the creation of a hypothesis relating the fossilized trilobite (*right*) to man himself.

from the patient's point of view. It was not until Louis Pasteur's (1822–1895) classical work established the germ theory of disease that a truly scientific basis of medicine was created.

Most notable biological achievements of the eighteenth and nineteenth centuries were the theory of evolution through natural selection and the creation of biology as an experimental science. In the eighteenth century, geographical distribution of living creatures and the paleontological record inspired ideas of the gradual evolution of species instead of their fixity. Jean Baptiste de Lamarck (1744–1829) vividly described the constant flux in nature, but his psychological, almost mystical, causes for the transformation of species were unacceptable to the majority of scientists. It was Charles Robert Darwin's (1809–1882) use of the mechanism of natural selection that led to his theory being taken seriously. Unfavorable variations were weeded out by the mechanical operation of the environment. No vital spirits or will power were needed to explain the change of forms over millennia. Given the expanded time scale made possible by nineteenth-century geology, it required no great intellectual effort to visualize evolution by means of natural selection. Armed with this principle, the interconnections of the animate world could begin to be seen. The theory of evolution was the periodic table of biology.

During the nineteenth century, biology became a full-fledged experimental science. The major obstacle to the achievement of this aim was the strength of the feeling that living matter somehow did not obey the ordinary laws of inorganic matter. The discovery of the circulation of the blood did not remove this feeling. The fact that the blood obeyed the laws of hydrodynamics did not unduly impress the vitalists. The arteries and veins, after all, were pipes and therefore the blood acted like any other fluid going through pipes. What were forever hidden from view, it was felt, were the processes of life, such as digestion or nervous function. This was where the "vital spirits" came in and this principle of life was, by definition, not susceptible to scientific or experimental study.

Assaults on this position were made in the eighteenth century, but it was in the nineteenth century that the full battle began. In the laboratories of such men as Johannes Purkinje (1787–1869), Johannes Müller (1801–(1858), and especially Claude Bernard (1813–1878) the mysteries of living tissue were gradually revealed. There it was showr that living matter, while complicated, was no less subject to scientific analysis than any other. It was possible to formulate hypotheses and then subject them to the same type of experimental test as those applied in physics and chemistry. By the end of the nineteenth century, an overall picture of life was possible in physicochemical terms. The development of finer techniques of physical and chemical analysis made it increasingly possible to probe down to the very molecular basis of life, and what was discovered was in accordance with the same laws that were found operative in other sciences.

This survey ends with the nineteenth century, for the modern scientific revolution that began with relativity and the quantum theory demands separate treatment. The beginning of the twentieth century also witnessed something quite new in the history of science. Where before there had been a number of different sciences with but few common elements, there now existed a single unifying foundation. The great principles of the conservation of matter and energy apply as well to the living cell as to the exploding atomic nucleus or an ordinary chemical reaction. Living forms are not the only ones considered to evolve; cosmologists today look to the evolution of the universe and its parts. The culmination of thousands of years of intellectual effort which we have traced so briefly in these pages has been the creation of one science with many branches forming a series of separate disciplines. It is this essential unity that is perhaps the greatest achievement of the human mind.

—L. Pearce Williams

BIBLIOGRAPHY

ASIMOV, ISAAC. *The Intelligent Man's Guide to Science.* Basic Books, Inc., 1963.

CONANT, JAMES BRYANT. *On Understanding Science: An Historical Approach.* Yale University Press, 1947.

DAMPIER, WILLIAM C. *History of Science.* Cambridge University Press, 1963.

FEUER, LEWIS SAMUEL. *The Scientific Intellectual.* Basic Books, Inc., 1963.

FORBES, ROBERT J. and DIJKSTERHUIS, E. J. *A History of Science and Technology.* 2 vols. Penguin Books, Inc., 1963.

GARDNER, ELDON JOHN. *History of Life Science.* Burgess Publishing Co., 1963.

GILLISPIE, CHARLES COULSTON. *The Edge of Objectivity.* Princeton University Press, 1960.

GOOD, IRVING JOHN and others (eds.). *The Scientist Speculates: An Anthology of Partly Baked Ideas.* Basic Books, Inc., 1963.

HALL, ALFRED RUPERT. *From Galileo to Newton: 1630–1720.* Harper & Row, Publishers, 1963.

NEWMAN, JAMES ROY. *Science and Sensibility.* Simon and Schuster, Inc., 1960.

SARTON, GEORGE. *History of Science and the New Humanism.* George Braziller, Inc., 1956.

SARTON, GEORGE. *The Life of Science: Essays in the History of Civilization.* Abelard-Schuman Ltd., 1948.

TAYLOR, FRANK SHERWOOD. *An Illustrated History of Science.* Frederick A. Praeger, Inc., 1955.

CHEMISTRY

History.—The first use of chemistry predates recorded history. When man began tanning hides for clothing, mixing clay to make pottery, and preparing medicines and dyes from plants, roots, and herbs, he employed elementary chemical processes. The development of chemistry as a science, however, came much later. By no stretch of the imagination could the prehistoric tanners, potters, and medicine men be considered scientists. Although they knew how to invoke chemical reactions, their knowledge resulted from the chance experience of trial and error. They did not experiment in the true sense of the word, and made no effort to discover *why* something would happen.

The Greek philosophers were the first to ask, Why? About 2,500 years ago, in the fifth century B.C., Democritus (c. 460–c. 370 B.C.) and Leucippus proposed the idea that everything was made of atoms and that these atoms were indivisible. At about the same time, Empedocles (c. 495–c. 435 B.C.) said that all matter was made of one fundamental substance called *ylem,* or *hyle,* which was the same in all objects. He went on to say that the differences between various bodies was caused by the presence of varying amounts of four elements—earth, air, fire, and water—in each body. Aristotle (384–322 B.C.), who was the most famous scientist of his day, accepted Empedocles' theory and added to it. Because of Aristotle's influence, the theory of Democritus and Leucippus remained obscured for over 2,000 years.

Unfortunately, the Greek philosophers engaged in no experimentation, depending instead upon speculation. They paid little attention to chemical processes, thus limiting the scope of their inquiries. The Egyptian and Phoenician scientists of the same period, on the other hand, developed their chemical technology by experimenting with chemical processes; however, they exhibited little desire to find the underlying principles.

Alchemy.—However, when Aristotle's pupil, Alexander the Great, conquered Egypt about 332 B.C., a merging of the Greek and Egyptian sciences resulted. From these emerged a prescientific chemistry called *alchemy,* whose primary goal was the transmutation of one metal into another. Many of the processes of early metallurgy seemed to support the idea that a *base metal,* such as iron or lead, when treated with an unknown material, would be changed (or transmuted) into silver or gold. The alchemists strove for eighteen centuries to find this unknown material, which they called the *philosopher's stone.* In addition, the alchemists searched for the *elixir of life,* a drink that would supposedly rejuvenate the aged; and a *universal solvent,* which would enable them to dissolve anything on the earth. As far-fetched as the ideas of a philosopher's stone, an elixir of life, and a universal solvent may seem today, many of the leading men of early science were sure that these were practicable.

Men such as Albertus Magnus (c. 1193–1280) and Roger Bacon (c. 1214–1294) were brilliant alchemists. Through the fourteenth and fifteenth centuries, the alchemists were the sole practitioners of chemistry. By the early sixteenth century, however, alchemy began its death throes, for it was in this period that the school of *iatrochemistry* (medical chemistry) developed. Through the work of such men as Philippus Aureolus Paracelsus (1493–1541), Jan Baptista van Helmont (1577–1644), and Sylvius (1614–1672), the cornerstone of chemical physiology was set in place. During this period, too, the work of such men as Agricola (1494–1555) and Johann Rudolf Glauber (1604–1668) expanded the knowledge of chemical materials and processes.

Founding of Modern Chemistry.—The seventeenth century might well be considered as the begining of scientific chemistry. By this time the work of the iatrochemists and their contemporaries had completely destroyed alchemy. The work of Francis Bacon (1561–1626) in chemistry, Galileo Galilei (1564–1642) in astronomy and physics, William Harvey (1578–1657) in anatomy, and William Gilbert (1540–1603) in physics during the latter part of the sixteenth century had established the superiority of the *scientific method* (inductive reasoning) over the *Aristotelian method* (deductive reasoning). In the Aristotelian method, the scientist would first state a self-evident truth (such as that heavy bodies fall faster than light ones) and then prove it by finding a phenomenon that could be explained by this truth (such as that a pound of lead obviously falls faster than a feather). In the scientific method, the scientist derives a general conclusion from a multitude of observations and experiments (such as dropping a ten-pound lead weight and a five-pound lead weight from the top of a tall building and seeing that they fall at exactly the same speed) and then applies the conclusion to specific cases (such as that a pound of lead and a feather will fall at the same speed in a vacuum).

BETTMANN ARCHIVE

THE LABORATORY of a medieval alchemist.

▪TRUE CHEMISTS.—Following these pioneers, men like Jean Rey (fl. 1630) and John Mayow (1640–1679) became chemists in the modern sense of the word. However, it is Robert Boyle (1627–1691) who deserves to be called the father of modern chemistry. Boyle was the first to distinguish between elements, compounds, and mixtures. In his book, *The Sceptical Chymist,* published in 1661, he stated the goal of modern chemistry—investigation of the composition and properties of substances, to be explained in terms of elements. In addition, Boyle pioneered chemical analysis and the study of chemical reactions, and established the relationship between the pressure and volume of a fixed amount of gas, a relationship known as Boyle's law.

▪PHLOGISTON THEORY.—Among the major questions that the seventeenth century chemists tried to answer was what caused combustion. In an effort to explain this, Johann Joachim Becher (1635–1682) and his student Georg Ernst Stahl (1660–1734) developed the theory of *phlogiston.* Phlogiston was a principle and therefore could not be isolated; it could be known only by its effects. According to the phlogiston theory, only those substances that contained phlogiston could burn. When such a substance burned, the phlogiston was thought to escape into the air. It was not until the eighteenth century and the introduction of quantitative chemistry that the phlogiston theory was finally disproved. Phlogiston's importance to chemistry was that it was the first unified chemical theory.

Rise of Quantitative Chemistry.—The next great advance in chemistry was made in the eighteenth century. The industrial revolution began during this period, and chemists were called upon to make their contributions. As a result, the eighteenth century was a period of remarkable growth and discovery. Among the more noteworthy advances were the studies of chemical affinity by Etienne François Geoffroy (1672–1731) and Torbern Olof Bergman (1735–1784); the independent isolation of oxygen by Joseph Priestley (1733–1804) and Karl Wilhelm Scheele (1742–1786); the discovery of hydrogen by Henry Cavendish (1731–1810); and the work of Joseph Black (1728–1799) on heat and temperature.

The greatest chemist of the eighteenth century, however, was Antoine Laurent Lavoisier (1743–1794). Based on the work of his predecessors, Lavoisier devised a method of chemical nomenclature, developed a theory of acid and base formation,

BETTMANN ARCHIVE

ANTOINE LAVOISIER (*left*) proposed the law of conservation of mass. During the French Revolution, he was arrested (*above*) and executed.

and proposed the law of conservation of mass. His greatest contributions, however, were to propose the theory of oxidation and to disprove the phlogiston principle. In 1756, a Russian chemist, Mikhail Lomonosov (1711–1765), had proved that when iron rusted, it gained weight by combining with something in the air. In 1774, Lavoisier repeated Lomonosov's experiment and concluded that the iron had combined with oxygen. By 1778, utilizing his theory of oxidation, Lavoisier was able to explain quantitatively many of the reactions the phlogistonists had either ignored or been unable to explain.

Atomic Hypothesis.—Having been freed from the phlogiston principle by Lavoisier's work, chemistry made great strides forward. John Dalton (1766–1844), an English schoolmaster, conducted a number of experiments on pure and mixed gases. In an effort to explain his results, and influenced by the writings of the early Greek philosophers Democritus and Leucippus, Dalton proposed his atomic hypothesis in 1808. Dalton's hypothesis stated that matter was composed of discrete particles called atoms, and that all atoms of the same element were alike; but that atoms of different elements were different. As part of his hypothesis, Dalton also stated that compounds were formed by the union of atoms of different elements in simple numerical proportions, and that if elements formed more than one kind of compound, the different weights of each compound were in the ratio of small whole numbers (*the law of multiple proportions*).

Progress in the Nineteenth Century.—Stimulated by Dalton's atomic hypothesis, the progress made during the following one hundred years was astounding. More than half of the elements presently known to man were discovered. Jöns Jakob Berzelius (1779–1848) developed a standardized system of chemical symbols and formulas in 1819. Robert Wilhelm Bunsen (1811–1899) and Gustav Robert Kirchhoff (1824–1887) developed the science of chemical spectroscopy. The studies of electrolysis by Michael Faraday (1791–1867) led to the theories of ionic interactions proposed by Svante August Arrhenius (1859–1927) and Jacobus Hendricus van't Hoff (1852–1911), while the phase rule proposed by Henri Louis Le Châtelier (1850–1936) advanced the field of physical chemistry considerably. Indeed, the number of men who made significant contributions to the development of chemical science during this period is so great it is impossible to list them all here. Nevertheless, two advances were made whose importance necessitates their mention in great detail—the development of organic chemistry and establishment of the periodic system.

■GROWTH OF ORGANIC CHEMISTRY.—Between 1810 and 1815, improved methods of analysis led to the development of methods for the analysis of *organic compounds* (compounds of carbon that were thought to be producible only by living organisms). During this period, several classes of organic compounds were recognized. The greatest advance, however, was not made until 1828, when Friedrich Wöhler (1800–1882) converted an inorganic salt (ammonium cyanate, a compound produced by neither plants nor animals) into the organic compound urea. Justus von Liebig (1803–1873) later collaborated with Wöhler to establish the existence of a number of organic groups that remain unaltered through many chemical changes. The next stride forward was made by Friedrich Kekulé (1829–1896), who developed the concept that carbon had a fixed combining power, or *valence*. This idea that carbon had the ability to form four "bonds" with other atoms led to the use of structural formulas to represent organic compounds and made the organic chemist's task far simpler.

BETTMANN ARCHIVE

JOHN DALTON, famous English chemist.

■PERIODIC SYSTEM.—Once John Dalton established his atomic hypothesis, chemists began to search for some relation between the properties of the elements and their atomic weight (the relative weights of the elements when the weight of carbon is arbitrarily set at 12.000). In 1829, Johann Döbereiner (1780–1849) proposed his rule of triads. Döbereiner noted that in several groupings of three elements of similar properties, called *triads*, the central element had an atomic weight that was almost the arithmetical mean of the other two. (For example, if chlorine, bromine, and iodine are grouped, it is found that the atomic weight of chlorine is 35.453, the atomic weight of bromine is 79.909, and the atomic weight of iodine is 126.904; the arithmetical mean of the atomic weights of chlorine and iodine is 81.179, or slightly more than the atomic weight of bromine.) This regularity led to the idea that the triads were but a part of a regular system that included all the elements. However, atomic weights were not put on an accurate basis until 1858, when Stanislao Cannizzaro (1826–1910) developed a new method for their determination, so no significant advances were made through the use of Döbereiner's triads. In 1863, John A. R. Newlands (1837–1898) discovered another means of categorizing the elements. He found that when the elements were listed in order of increasing atomic weight, each succeeding eighth element had similar properties. Since the eighth element was "a kind of repetition of the first, like the eighth note of an octave in music," Newlands called this relation the *law of octaves*. Unfortunately, the system did not work for elements heavier than chlorine, and the law of octaves fell into disuse. However, in 1869, Dmitri I. Mendeleev (1834–1907) published a periodic table based on the law of octaves, in which he had made adjustments for the difficulty presented by the elements heavier than chlorine. This arrangement was so good, and the concept was so clearly developed, that Mendeleev's name has been connected with the periodic table ever since, even though Julius Lothar Meyer (1830–1895) independently made the same discovery one year after Mendeleev.

Chemistry in the Twentieth Century.—In the twentieth century, the line separating chemistry and physics has blurred to the point where it is now almost indistinguishable. The reason for this is that the chemist has found he can no longer deal solely with the thought of reactions between elements and compounds. He must now do his thinking on the atomic and subatomic scale, and thus his sphere of interest considerably overlaps that of the nuclear physicist. In the early part of the twentieth century, the emphasis was on developing a clear picture of the atom. Great progress in the determination of atomic structure was achieved through the work of Sir Joseph John Thomson (1856–1940), Ernest Rutherford (1871–1937), and Henry G. J. Moseley (1887–1915). Much effort was concentrated on the study of radioactivity, based upon the discovery of radium by Pierre and Marie Curie in 1898. Sir William Henry Bragg (1862–1942) and his son William Lawrence Bragg (1890–) pioneered in the study of crystal structure through the use of X-ray diffraction techniques. As the century progressed, Gilbert Newton Lewis (1875–1946), Irving Langmuir (1881–1957), and Linus Carl Pauling (1901–) contributed to the chemist's understanding of chemical bonding. Others, such as Peter Joseph Wilhelm Debye (1884–1966), clarified the mechanism of chemical reactions. In 1940, three radioactive elements that do not occur naturally on earth (promethium, plutonium, neptunium) were synthesized, as have been nine other elements since.

During the 1950s and 1960s, the most spectacular discoveries came in the area of molecular biology. Such advances as the synthesis of self-replicating DNA (deoxyribonucleic acid) by Arthur Kornberg (1918–) have brought us to an understanding of the chemistry of life. Given such achievements, it is reasonable to assume that the discoveries of the second half of the twentieth century will dwarf those of the first half in the same manner that the contributions of the iatrochemists have dwarfed those of the alchemists.

Branches of Chemistry.—Prior to the seventeenth century, there were no branches of chemistry. Indeed, prior to that time, there were no chemists. Until the early 1600s, it was not necessary for men to specialize in any particular field of science. Instead, men were scientists—they studied and experimented in all fields.

However, as the years passed and the amount of background knowledge an experimenter had to possess increased, men began to specialize. First the biological scientists separated from the physical scientists; then the physical scientists separated to form the branches of chemistry, physics, and geology. Thus, scientists like Robert Boyle (1627–1691), who studied gases, became chemists; those like Isaac Newton (1642–1727), who studied mechanics, became physicists. As the background knowledge an experimenter needed continued to grow, these subdivisions subdivided within themselves. In this way, a number of specialized branches of chemistry came into being before the end of the nineteenth century.

As time passed, it became evident that the sharp boundaries separating one branch of chemistry from another, as well as those separating chemistry from the other branches of science, were practically nonexistent. The branches of chemistry are, therefore, of only limited significance, for they overlap each other and the other branches of science. Indeed, it might be said that the divisions between the sciences are divisions of convenience, and that nature has but small regard for them. Nevertheless, the branches of chemistry are of considerable historical importance and deserve recognition, if only for this reason.

Inorganic chemistry is the study of the properties and reactions of all the elements and their compounds, except most of the compounds of carbon. It is concerned with the occurrence of these elements in nature, methods of obtaining them, and the laboratory techniques, theories, hypotheses, and laws basic to all branches of chemistry.

Organic chemistry is the study of the compounds of carbon, the hydrocarbons, and the fluorocarbons. The majority of compounds found in living organisms, dyes, plastics, petroleum, petroleum products, and many other materials are organic. Thus, there are hundreds of thousands of carbon compounds for the organic chemist to study.

Physical chemistry is the study of the dependence of physical properties of materials upon their chemical composition, and of the physical changes that accompany chemical reactions. Physical chemistry is itself divided into several branches. *Colloid chemistry* is concerned with the behavior of finely divided particles of matter; *surface chemistry*, with the behavior of surfaces; *chemical thermodynamics*, with the role of energy in chemical reactions; *chemical kinetics*, with the speed and mechanism of chemical reactions; *electrochemistry*, with the behavior of chemical substances when treated with an electric current and with the production of electricity by chemical means; and *photochemistry*, with the chemical effects of electromagnetic radiation and with the production of electromagnetic radiation in chemical change.

Analytical chemistry is concerned with the splitting of materials into component parts (or constituents) by chemical methods to determine composition. In *qualitative analysis*, the chemist determines the constituents of the material, irrespective of their amount. In *quantitative analysis*, the chemist must determine the amounts in which the various constituents are present in a material.

Nuclear chemistry is the study of reactions that produce new elements, of the chemical effects produced by high-energy radiation, and of the tracing and behavior of radioactive isotopes.

Biochemistry is the study of the chemical composition and behavior of living organisms and life processes. In many respects it is a combination of organic chemistry, physical chemistry, botany, and zoology.

Geochemistry is concerned with the chemical study of the earth's crust, and the application of chemistry to processes (such as the formation of minerals and the metamorphosis of rocks) within the earth.

Industrial chemistry merges chemistry and chemical engineering. It includes the design, installation, and maintenance of industrial equipment for chemical reactions, the technology and financial practices necessary for collecting raw materials and marketing products, and the development of large-scale, economically feasible industrial processes from research.

—A. D. Levy

BETTMANN ARCHIVE

NIELS BOHR, Danish nuclear physicist, was one of the first to suggest to President Franklin D. Roosevelt that an atomic chain reaction, key to the atomic bomb, was a possibility.

INORGANIC CHEMISTRY

Basic Definitions.—*Chemistry* is the body of knowledge and practices concerned with the properties of atoms, ions, and molecules and with the changes that can be brought about in the arrangement of their electrons. Specifically excluded from chemistry is the study of the changes that occur in the nucleus of the atom. Before atomic structure was understood, chemistry was defined as the study of the properties of substances and of the transformations that substances undergo. *Inorganic chemistry,* one of the primary branches of chemistry, is the study of the specific properties of the atoms of the different elements, the ions and molecules they form (excluding molecules of certain carbon compounds which are the concern of organic chemistry), their occurrence in nature, the methods of obtaining them, their usefulness in our culture, and the laboratory techniques, theories, and laws all chemists must use.

Three concepts basic to all sciences, —and especially relevant to inorganic chemistry—are volume, mass, and energy. The volume of an object is the amount of space it occupies. Mass and energy are more abstract concepts, however. Mass can best be understood in terms of the *law of universal gravitation,* which states that all matter is attracted to all other matter by a force that varies inversely as the square of the distance between them. The closer the bits of matter are to each other, the more strongly they are attracted. The *weight* of an object is a measure of this attraction or force. The weight of an object thus varies with its position relative to other bodies near it, though the actual amount of matter in it is always the same. The word *mass* expresses this unchanging inertial quantity of matter in an object. The concept of *energy* is even more abstract. Many kinds of energy have been distinguished, some of the most familiar being mechanical, chemical, heat, radiant, electrical, nuclear, and kinetic. Each can be converted into any of the other kinds. For example, in an automobile engine the chemical energy of a gasoline and air mixture is converted into the heat energy of the gases produced by the explosion of the mixture; the gases expand and push the piston down, turning the crankshaft and the rear wheels in a transfer of mechanical energy; as the car begins to move, it acquires kinetic energy. Another example is that of the kinetic energy of falling water, which can be transformed through a turbine into mechanical and then into electric energy, which in turn can be changed into heat, radiant, or again to mechanical energy.

Such concepts are useful because they can be expressed quantitatively, that is, in numbers. All the physical properties of materials, such as index of refraction, density, color, conductivity, magnetic strength, tensile strength, coefficient of expansion, heats of fusion and vaporization, compressibility, and many more, are measured by mathematical formulas derived from the basic units of measurement—length, time, and mass, and a fourth unit for electricity. On the other hand, chemical properties cannot be reported quantitatively; they are descriptions of chemical reactions in which one or more substances will participate.

The most important basic principle in science is the law of conservation of matter and of energy, which states that matter can be converted into energy, and vice versa, but the sum of the two before any event is the same as the sum of the two after the event. All experimental evidence supports the principle. It is further assumed, because so far nothing denies it, that all matter and energy obey the identical laws throughout the universe.

Kinetic Molecular Theory.—The basic structure of matter is defined by the *kinetic molecular theory,* which states that all matter consists of ultimate, or finite, particles in constant motion, and that their motion, or kinetic energy, is what we call heat. The greater their heat energy, the faster they move, and thus the greater their kinetic energy. The kinetic energy of an object equals $\frac{1}{2}mv^2$, m being the object's mass and v its speed. The law applies to any atom, ion (electrically charged atom), or molecule (atoms in chemical union); and it applies with qualifications to any subatomic particle, although chemists are interested only in electrons, protons, and neutrons. All these particles are perfectly elastic and lose none of their kinetic energy to friction when they collide, as bulk matter always does. The identification of heat energy with movement of basic particles explains many baffling aspects of nature, including the structure and properties of the three *states,* or *phases,* of matter: gas, liquid, and solid (see also *Properties of Matter,* page 1202).

■GASES.—A gas consists of atoms, ions, or molecules moving at random in space. They travel along individual paths free of the influence of other particles' attraction but, because of collisions with one another, constantly change direction. The volume of the particles themselves is but a minute fraction of the space they can occupy. Since they are perfectly elastic, the sum of the energies of two particles before a collision equals the sum of their energies after the collision, although the energy will be distributed differently between them. At no instant do all the particles in a gas move with the same speed or in the same direction. Therefore, when we speak of the temperature of a gas, we are speaking of the average kinetic energy of all the particles. Slow ones are being speeded up and fast ones are being slowed down by collisions, so that, according to calculations made by the British physicist James Clerk Maxwell (1831–1879), about half of the particles possess speeds within 30 per cent of the average. A few are always traveling very rapidly and a few very slowly, but in any gas or mixture of gases, the kinetic energy of many molecules is close to the average.

BETTMANN ARCHIVE

CUNEIFORM TEXT of a Babylonian chemical formula from the seventeenth century B.C.

The pressure of a gas is understandable in terms of the continual bombardment of the walls of a container by rapidly moving particles. The pressure, which is the same everywhere on every wall of the container, is determined by the number of impacts per unit time and by the force of the impacts. If the gas in a container with a movable top is compressed, the average distance the particles travel between collisions is reduced, and the number of impacts each second on the container walls is increased. Therefore, when a gas is compressed, its pressure increases. If the gas is allowed to expand, there will be fewer impacts per second on the container walls, and the pressure will be less. *Boyle's law*—named for its discoverer, Robert Boyle (1627–1691)—states that a volume (V) of a fixed amount of gas is inversely proportional to the pressure (P), if the temperature (T) is constant. Thus, $V \propto 1/P$ or $V = k/P$ (k is the constant of proportionality) if T is constant.

If the gas in the container is heated from the outside, the kinetic energy of each particle is increased, and the particle will hit with greater impact and greater frequency. Therefore, the pressure of the gas will rise. In order to maintain the original pressure, the gas must be allowed to expand. On the other hand, if the gas is cooled from the outside, the particles will move more slowly, the number of their impacts per unit time will decrease, and the pressure will decrease. In order to exert the same pressure as it did at a higher temperature in a larger space, the gas must be compressed. If this reduction in volume is measured, it is found that for each drop of 1° C. starting at 0° C., the gas will shrink 1/273 of its volume. Theoretically then, at —273° C. the gas will vanish. This means that, since shrinking in volume due to cooling is proportional to the lowering of kinetic energy, at −273° C. all motion of the particles ceases—they will be without any heat energy. This temperature is called *absolute zero*. Thus, on a Kelvin or Rankine thermometer, both of which are calibrated to register absolute temperatures, absolute zero (−273.16° C. or −459.6° F.) is 0°. *Charles's law*, formulated by Jacques Alexandre César Charles (1746–1823), states that the volume of a fixed amount of gas is proportional to the absolute temperature, if the pressure remains the same. Thus $V \propto T_a$ (V is the volume and T_a is the absolute temperature), or $V = kT_a$ (k is the constant of proportionality), when P is constant. Boyle's law and Charles's law can be combined into one equation: $V = 1/P\ (T_a)$, or $PV = kT_a$.

If different gases are mixed, each will exert the pressure that it would in that space if it were by itself. The statement that the total pressure of several different gases in the same containers equals the sum of their individual pressures is *Dalton's law of partial pressures*, named after its discoverer, John Dalton (1766–1844). *Graham's law*, formulated by Thomas Graham (1805–1869), states that a gas diffuses through small openings at a rate inversely proportional to the square root of its density.

In two containers of equal size, containing the same number of molecules at the same temperature, the pressures will be the same. Therefore, equal volumes of all gases at the same temperature and pressure must contain equal numbers of molecules. This was proved in 1811 by Amadeo Avogadro (1776–1856) and is called *Avogadro's hypothesis*. It provides a method for comparing the masses of different molecules. If a known volume of gas *A* is weighed and the same volume of gas *B* is weighed under the same conditions, then the number of *A* molecules equals the number of *B* molecules; if the total weight of all the molecules of gas *B* is four times as great as that of all the molecules of gas *A*, then each molecule of *B* weighs four times as much as each molecule of *A*. Actually, it is not necessary that the volume, the temperature, and the pressure of the gases be the same, because any set of these measurements can be compared to a selected standard of temperature and pressure, which by international agreement is 0° C. and 760 mm. of mercury.

The earth's atmosphere is a mixture of gases whose composition, density, and temperature are constantly changing. Any experiment not sealed from the atmosphere will be affected by it; if the effect is significant, it is necessarily part of the report of that experiment. A gas in a container exerts the same pressure in all directions because of its kinetic energy; the atmosphere, which is not enclosed, does not escape into space because of gravity, which keeps it much more dense at the surface of the earth. The pressure of the atmosphere at a point on the earth's surface is the actual weight of air above that point.

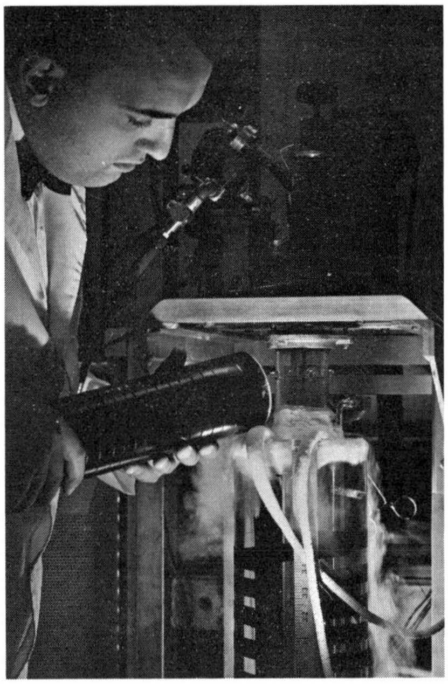

INTERNATIONAL BUSINESS MACHINES CORPORATION

PRECOOLING low-temperature apparatus with liquid nitrogen involves cryogenics.

A common experience with a gas is that it cools as it expands. This phenomenon is called the *Joule-Thomson effect* after its discoverers, James Prescott Joule (1818–1889) and William Thomson (1824–1907). The effect is explained by the fact that there is an attraction between gas molecules during the instant of a collision; this must be overcome as they bounce apart. The energy to overcome the attraction is derived from the molecules' heat. When a gas has a fixed volume, the number of collisions each second does not change, and the speeding up and heating just before each impact takes place equal the cooling just after. If the gas is expanding, the heat lost by molecules working against attraction is not all regained, and the gas cools. The Joule-Thomson effect is the basis of all refrigeration systems.

The gas laws are accurate for an imagined ideal gas whose molecules are considered to be mathematical points. Since real molecules have volume and mass and attract one another when in close proximity, the gas laws are inaccurate for high pressures or low temperatures.

■LIQUIDS.—When a gas is cooled and the average kinetic energy of its molecules is reduced, the molecules will spend more time in each other's vicinity. The forces of attraction between them will make it more difficult for the molecules to bounce apart. Eventually they will lose enough energy, as the cooling is continued, to be unable to bounce apart and will cling together. As the temperature is lowered still further, clusters grow into droplets of liquid. Within the liquid phase the molecules continue to move with the same average kinetic energy that they have in the gas phase at the same temperature, but they cannot escape each other and merely slide about. A liquid maintains its volume at any fixed temperature.

To turn a liquid into a gas, heat must be added to it from the outside. The heat is absorbed by the molecules and increases their kinetic energy; and the increased energy enables them to overcome each other's attraction. Whatever the temperature of the liquid, the vapor directly above it has the same temperature, and all the molecules in both phases have the same average kinetic energy. The energy that a molecule must acquire in order to move from liquid to gas is called its *heat of vaporization*. When a gas molecule gives up its freedom in space and returns to the liquid phase, it gives up to the environment its *heat of condensation*, which is identical to its heat of vaporization. In a closed container partly filled with liquid, fast-moving molecules in the liquid that have absorbed their heat of vaporization through collisions or radiation leave the liquid surface, while slow-moving molecules in the vapor are trapped by the liquid surface when they hit it and give up their heat of condensation. If the temperature is kept constant in the closed container, the exchange of molecules is equal, and the total number of molecules in the gaseous phase does not change. A *dynamic equilibrium* exists between liquid and gas. Any change in temperature changes the kinetic energy of the gas molecules and therefore changes the vapor pressure. If the gas molecules escape from the container, the liquid evaporates; an evaporating liquid is always cooler than its environment because each escaping molecule takes away with it a measure of heat.

In a boiling liquid enough heat is being supplied from outside to cause evaporation throughout the liquid. However, no matter how fast a liquid is boiled (that is, no matter how much heat is applied), its temperature remains constant and is the same as the temperature of its vapor until all the liquid has evaporated. Then more heat will raise the temperature of the vapor.

An average molecule at the surface of a liquid is pulled downward into the liquid by forces of attraction that are greater than the upward attraction exerted by the vapor. This greater attraction of surface molecules pulls drops into a spherical shape. The wetting of surfaces is determined by a complicated set of affinities and repulsions between a liquid and a solid. In a narrow tube the result may be to pull a liquid upward or to depress it, a phenomenon called *capillary action*. Trees lift tons of water from their roots to their foliage, partly by capillary action and partly by osmosis.

A gas can be compressed until its molecules are so close together that they release their heat of condensation and become liquid. Each gas, however, has a *critical temperature* above which it cannot be liquefied by pressure.

■SOLIDS.—When a liquid is cooled, a temperature, called the *freezing point,* is reached at which the slowly moving molecules are caught by new forces of attraction, forces stronger still than those of the liquid. The molecules relinquish their freedom of motion entirely and cling together. As cooling is continued, more are trapped, and a crystal begins to form within which the molecules continue to vibrate around one position. Besides their freedom they have lost a specific amount of energy, called their *heat of fusion.*

Under certain conditions, all three phases of some substances, such as water, can thus exist together at the freezing point. Any molecule in this three-phase system can change its condition to solid, liquid, or vapor, according to whether it gains or loses its heat of fusion and its heat of vaporization. It should be noted, however, that the temperatures of the solid, the liquid, and the vapor in a three-phase system, are the same.

Each substance has a distinctive freezing point and boiling point, but all substances lose all heat and kinetic energy at 0° absolute. At temperatures close to this, matter shows bizarre properties that the kinetic molecular theory cannot explain. For example, at cold near 0° Kelvin, helium creeps out of its container, and metals become superconductors of electricity.

When the molecules in a crystal freeze, they arrange themselves to give the crystal one of half a dozen geometric shapes by which the substance can be identified. Solids without crystal shape are called *amorphous.* Some substances have more than one crystal shape, each melting at a different temperature; and these are called *allotropic forms.* Most solids have vapor pressures far smaller than those of their liquids, but a few have such high vapor pressures that they turn directly into a gas, or *sublimate,* without passing through the liquid phase. Solid carbon dioxide (dry ice), iodine, and naphthalene are examples of substances that undergo sublimation.

Atomic Theory.—The kinetic molecular theory does not try to explain the chemical properties of substances or what happens in a chemical reaction. This is accomplished by the *atomic theory* and by knowledge of atomic structure.

All the matter in the universe consists of substances called *elements;* the smallest part of an element that still has all the properties of that element in bulk is called an *atom.* In chemical combination, the atoms of the elements unite to form molecules. A *molecule* is the smallest particle of a compound that shows all the properties of that compound. There are 91 naturally occurring elements and 13 others that have been created by man in nuclear reactors. The universe consists of these hundred-odd elements and of the compounds they form with one another, a number that is almost unlimited, theoretically and in reality. Over a million compounds are known

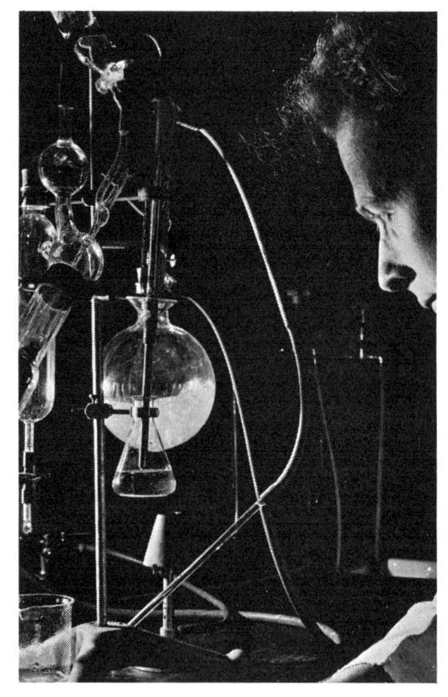

EXPERIMENTAL CHEMIST is determining the nitrogen content in petroleum fractions.

today, a great many of which never existed before they were synthesized in the laboratory.

All the atoms of an element are alike in every way (except in mass: see ISOTOPES) and are unlike the atoms of every other element. All the molecules of a compound are alike in every way; that is, the same kinds of atoms always join together in the same proportion to form the same kind of molecule. The same atoms may also join together in different proportions to form different molecules. For example, carbon atoms may unite with oxygen atoms in three different proportions to form three different compounds: one carbon atom may join with one oxygen atom to form carbon monoxide; one carbon atom may join with two oxygen atoms to form carbon dioxide; three carbon atoms may join with two oxygen atoms to form carbon suboxide. During the process of chemical combination, neither matter nor energy is created or destroyed. However, an energy change is always associated with chemical reactions. Usually heat and/or light will be released to the surroundings or absorbed from it. The quantity of heat and/or light involved is specific for each reaction.

Each element has been assigned a symbol consisting of one or two letters—for example, the symbol for hydrogen is H; for oxygen, O; for carbon, C; for helium, He; for iron, Fe; and for gold, Au. These symbols are used as a form of shorthand by the chemist. To designate a compound, the symbol of each element present in the molecule is written. If more than one atom of any of the elements is present in the molecule, the number of atoms is written as a subscript to the right of the symbol for that element. Thus, the three compounds of carbon and oxygen named above would be written as CO (carbon monoxide), CO_2 (carbon dioxide), and C_3O_2 (carbon suboxide). If the chemist desires to indicate more than one molecule of the compound, the number of molecules is written to the left of the formula of the compound—for example, 2CO, $3CO_2$, $15C_3O_2$, etc.

Atomic Structure.—The modern concept of *atomic structure* is based upon existence of three subatomic particles: the electron, the proton, and the neutron. The *electron* has a negative electrical charge; the *proton,* a positive electrical charge; and the *neutron,* no electrical charge (that is, it is electrically neutral). Like charges repel one another and unlike charges attract each other; all protons repel all other protons and all electrons repel all other electrons, but any proton and any electron attract each other. The positive charge on the proton exactly equals the negative charge on the electron. The atom as a whole is neutral.

The first structure proposed for the atom was that of a sphere of positive electricity in which electrons were embedded to make it electrically neutral. When experiments proved that the atom, tiny though it is, consists largely of space, the electrons were imagined in positions at some distance from a positive nucleus. However, this could not explain why the electrons were not pulled into the nucleus.

To overcome this and several other serious objections to the model, in 1913 the Danish physicist Niels Bohr (1885–1962) postulated a different kind of electron structure around the nucleus. In the Bohr model of the atom, every atom consists of a nucleus that is positively charged and, at a distance from it, a number of electrons that exactly balance the positive charge and are in continuous motion around the nucleus. The simplest pictorialization is a dot for the nucleus and a circle around it for the path of the electron, but this is an extremely symbolic representation of what really exists. An atom with a radius of 10^{-8} centimeters has a nucleus only a thousandth as large. It is impossible to draw a scale model on a sheet of paper because a visible dot representing the nucleus would have to have a dot representing the electron drawn many yards away. To make representation still more difficult, nearly all the mass of the atom is in the nucleus; each proton and neutron weighs almost 1,800 times as much as an electron, yet the electron can be said to have a volume 2½ times that of a proton.

According to the Bohr model of the atom, the electron travels around the nucleus at speeds in the range of the speed of light. In the pictorial representation of the atom, the direction of movement of the electron can be indicated by an arrow on the circle representing the electron's orbit around the nucleus. Although the direction is constant, the electron orbits in all planes around the nucleus, and sometimes its path is circular, sometimes elliptical, sometimes dumbbell-shaped, and sometimes too complicated to describe. When an electron orbiting the nucleus in a certain way receives more energy from the outside through collision or through absorption of radiation, the electron will move into a different orbit farther from the nucleus. Finally, each electron and each nucleus spins on its own axis; and the atom as a whole also spins about, or vibrates around one spot, according to the kinetic molecular theory.

A reference to physics is necessary to understand why the atom retains its shape when all its parts are moving. If a stone on the end of a string is swung fast enough, it will travel around, or orbit, the person swinging it. This phenomenon is explained by Sir Isaac Newton's first law of motion, which states that a body at rest tends to remain at rest and a body in motion tends to remain in motion in the same direction (that is, along a straight-line path) unless an external force acts upon it. The tendency to remain at rest or in straight-line motion is called *inertia*. Thus the stone, having been placed in motion, tends to fly off in a straight line, at right angles to the string. The inertial pull away from the center of the orbit is called *centrifugal force* (although it is not really a force). The string exerts the continuous external force required to overcome the straight-line motion of the stone and change the stone's path into a circular orbit. The force exerted by the string toward the center of the orbit is called *centripetal force*.

In an atom, the electron is attracted to the nucleus because the electron has a negative charge and the nucleus has a positive charge. However, the electron does not fall into the nucleus because the inertial tendency of its motion—its tendency to move away in a straight line—balances the centripetal force of the attraction of the nucleus.

The contemporary model of the atom with all its details is based on information collected by a variety of instruments. Highly heated atoms emit light. If the light is studied in a spectroscope, distinctive lines are found in the spectrum for each element. The lines can be interpreted as specific variations in electron arrangements within the different atoms. In the mass spectrograph, atoms are given an electrical charge and are then lobbed by a magnetic field at a photographic plate; the heavier ones will land at nearer places on the plate than the lighter ones. The point of each atom's impact appears as a spot on the film. By studying the distances traveled by the atoms, the masses of the different atoms can be calculated. X rays projected through a crystal are scattered by the latticework of atomic nuclei in the crystal and, falling on a photographic plate, produce a pattern of dots. From the pattern the three-dimensional lattice of atoms in the crystal can be worked out to give accurate figures for the diameter of the atoms. A system of using electrons as though they were rays of light produces photographs of shadows cast by very large molecular shapes from which a great deal of information about atomic structure can be deduced.

Most vital of all the various sources of information are the *atom smashers*, or accelerators, the first of which, the *cyclotron*, was built in 1933, and the *nuclear reactors*, or *atomic piles*, the first of which was developed in 1941. In all these, the nuclei of atoms are broken up and the results are recorded and studied. For example, the results may be recorded as streaks and spots on photographic plates, as needles moving across instrument faces, as threads of mist, as scintillations on a screen of fluorescent chemical, as intensity of a glow, or as a rate of clicking in earphones.

These records and other kinds of sensory observations are made in quantitative terms, and the figures are then translated by formulas into statements of mathematical relationship between different parts of atoms. Such statements are developed into concepts of structure that, though far from fully understood, are nevertheless used in probing further into that structure. Modern chemistry utilizes a great deal of nuclear knowledge to explain chemical activity.

ELECTRONS.—The most important parts of the atom in chemical reactions are the electrons orbiting around the nucleus. The electron exhibits two sets of properties. One set indicates that the electron is a solid particle. The other set indicates that it is a wave. That is, the electron behaves both as a particle and as a wave, a dual nature for which there is no theory. The wave aspect is defined by *quantum mechanics*, which was developed when units of energy were discovered to exist and to have measureable mass. The units were named *quanta*, and a quantum of light is called a *photon*. Electrons orbiting in an atom can absorb and emit photons. Another aspect of the electron is that no fixed position for it in the atom can be ascertained. Instead of saying that an electron orbits at a certain distance from the nucleus, it is more accurate to say that the probability of finding the electron is greatest at a certain distance from the nucleus.

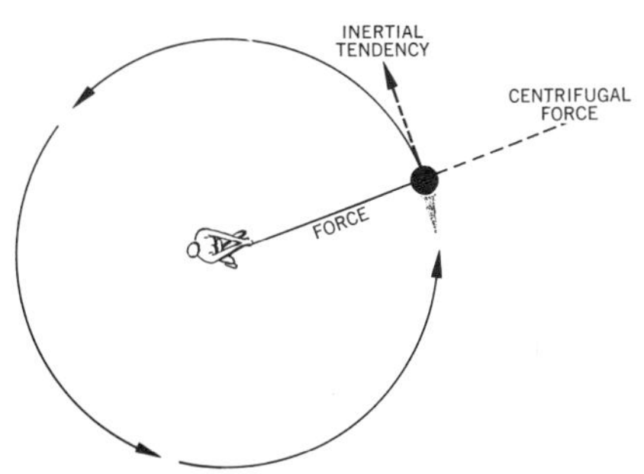

CENTRIPETAL AND CENTRIFUGAL FORCE are opposing actions causing atoms to retain their shape. If a man were to swing a weight, the inertial tendency would be for the weight, like electrons about a nucleus, to move away in a straight line.

Chemists have found that the most useful concept of an electron is that it is a solid particle whose position can always be stated precisely; the probability and wave aspects of the electron are seldom referred to. Nevertheless, both the physical and chemical properties of the elements are being viewed with increasing freedom as expressions of mathematical relationships based on quantum mechanics. The reason for this recent shift from physical models to mathematical definitions lies in the fact that knowledge of the energy distribution within the atom is far more useful for explaining and predicting chemical reactions than knowledge of the geographical distribution of bits of subatomic matter. However, no single view explains everything that is known.

NUCLEUS.—Physicists have identified several hundred different kinds of elementary particles. These particles can be defined in terms of their mass, speed, electrical charge, spin, mathematical definition as either bits of energy or bits of matter, and reaction with each other upon collision.

Chemistry is primarily concerned with only three of these particles—the electron, the neutron, and the proton. The neutron, which has no electrical charge, weighs slightly more than the proton, which has a positive charge. Their mass is so much greater than that of any of the other particles, that the mass of all the others is considered negligible in chemical computations.

Protons and neutrons are joined in the nucleus of the atom in a manner that is not yet clear. Because of their comparatively great mass, the entire mass of the atom is considered as being concentrated in the nucleus.

Electrons, protons, and neutrons can exist singly and the first two can endure, as far as is known, indefinitely. The other elementary particles appear only when a nucleus is shattered. They decay quickly or are annihilated by collisions with other particles, becoming either quanta of energy or protons and electrons. Because the nucleus is never ruptured in chemical reactions, the unstable elementary particles do not come into chemical discussion.

Thus, the basis of atomic structure is a nucleus containing protons and neutrons surrounded by rapidly moving electrons. Since, as has been stated, the atoms of each element are identical (except for mass) but are different from the atoms of every other element, it stands to reason that the difference between atoms must be one of protons, neutrons, and electrons. There must be at least 104 different combinations, one for each of the 104 known elements. Studies of atomic structure have established that elements can be listed in such a way that an atom of each succeeding element on the list has exactly one proton more than the element preceding it. For example, hydrogen, No. 1 on the list, has one proton; helium, No. 2 on the list, has two

TABLE OF ATOMIC WEIGHTS
(Based on Carbon-12)

Name	Symbol	Atomic Number	Atomic Weight	Name	Symbol	Atomic Number	Atomic Weight
Actinium	Ac	89	227	Mercury	Hg	80	200.59
Aluminum	Al	13	26.9815	Molybdenum	Mo	42	95.94
Americium	Am	95	243	Neodymium	Nd	60	144.24
Antimony	Sb	51	121.75	Neon	Ne	10	20.183
Argon	Ar	18	39.948	Neptunium	Np	93	237
Arsenic	As	33	74.9216	Nickel	Ni	28	58.71
Astatine	At	85	210	Niobium	Nb	41	92.906
Barium	Ba	56	137.34	Nitrogen	N	7	14.0067
Berkelium	Bk	97	249	Nobelium	No	102	254
Beryllium	Be	4	9.0122	Osmium	Os	76	190.2
Bismuth	Bi	83	208.980	Oxygen	O	8	15.9994
Boron	B	5	10.811	Palladium	Pd	46	106.4
Bromine	Br	35	79.909	Phosphorus	P	15	30.9738
Cadmium	Cd	48	112.40	Platinum	Pt	78	195.09
Calcium	Ca	20	40.08	Plutonium	Pu	94	242
Californium	Cf	98	249	Polonium	Po	84	210
Carbon	C	6	12.01115	Potassium	K	19	39.102
Cerium	Ce	58	140.12	Praseodymium	Pr	59	140.907
Cesium	Cs	55	132.905	Promethium	Pm	61	147
Chlorine	Cl	17	35.453	Protactinium	Pa	91	231
Chromium	Cr	24	51.996	Radium	Ra	88	226.05
Cobalt	Co	27	58.9332	Radon	Rn	86	222
Copper	Cu	29	63.54	Rhenium	Re	75	186.2
Curium	Cm	96	248	Rhodium	Rh	45	102.905
Dysprosium	Dy	66	162.50	Rubidium	Rb	37	85.47
Einsteinium	Es	99	254	Ruthenium	Ru	44	101.07
Erbium	Er	68	167.26	Samarium	Sm	62	150.35
Europium	Eu	63	151.96	Scandium	Sc	21	44.956
Fermium	Fm	100	252	Selenium	Se	34	78.96
Fluorine	F	9	18.9984	Silicon	Si	14	28.086
Francium	Fr	87	223	Silver	Ag	47	107.870
Gadolinium	Gd	64	157.25	Sodium	Na	11	22.9898
Gallium	Ga	31	69.72	Strontium	Sr	38	87.62
Germanium	Ge	32	72.59	Sulfur	S	16	32.064
Gold	Au	79	196.967	Tantalum	Ta	73	180.948
Hafnium	Hf	72	178.49	Technetium	Tc	43	99
Helium	He	2	4.0026	Tellurium	Te	52	127.60
Holmium	Ho	67	164.930	Terbium	Tb	65	158.924
Hydrogen	H	1	1.00797	Thallium	Tl	81	204.37
Indium	In	49	114.82	Thorium	Th	90	232.038
Iodine	I	53	126.9044	Thulium	Tm	69	168.934
Iridium	Ir	77	192.2	Tin	Sn	50	118.69
Iron	Fe	26	55.847	Titanium	Ti	22	47.90
Krypton	Kr	36	83.80	Tungsten	W	74	183.85
Lanthanum	La	57	138.91	Uranium	U	92	238.03
Lawrencium	Lw	103	257	Vanadium	V	23	50.942
Lead	Pb	82	207.19	Xenon	Xe	54	131.30
Lithium	Li	3	6.939	Ytterbium	Yb	70	173.04
Lutetium	Lu	71	174.97	Yttrium	Y	39	88.905
Magnesium	Mg	12	24.312	Zinc	Zn	30	65.37
Manganese	Mn	25	54.9380	Zirconium	Zr	40	91.22
Mendelevium	Md	101	256				

protons; lithium, No. 3, has three protons; gold, No. 79, has 79 protons; and lawrencium, No. 103 on the list, has 103 protons. The number of an element on the list is called the *atomic number* of that element. It is usually designated by the letter *Z*, and is the same as the number of its protons. Since an atom is electrically neutral, the number of electrons orbiting the nucleus is equal to the number of protons within the nucleus.

While the number of protons in the nucleus is constant for any element, the number of neutrons in the nucleus does not follow any order and can vary in the atoms of an element. For example, oxygen, which is element No. 8 because each of its atoms contains eight protons, has some atoms with eight neutrons, some with nine neutrons, and some with ten neutrons. Each is called an *isotope* of oxygen; almost every element has a number of isotopes. Neutrons do not affect the properties of an element except its mass. The only atom that has no neutrons is protium, the most prevalent isotope of hydrogen, which is element No. 1 on the list because all of its isotopes have a single proton for a nucleus. An isotope of hydrogen called *deuterium* has one neutron with the single proton; water molecules containing deuterium are called *heavy water*. Another hydrogen isotope, called *tritium*, has two neutrons with the single proton.

Atomic Weight.—The chemist is not interested in the actual weight of a single atom, for he deals in bulk matter that contains millions upon millions of atoms. Nevertheless, he must know how the masses of atoms compare with one another. For this purpose the commonest isotope of carbon, which has six protons and six neutrons, has been selected as the standard of *atomic weight* and has been assigned the value of 12. Until 1963 the standard was oxygen, with an atomic weight of 16, but the change affects the old table of atomic weights only very slightly. On the atomic weight scale, the proton has a relative weight of 1.0073; the neutron, 1.0087; and the electron, 0.0005486. Thus, very few elements have atomic weights that are whole numbers. For many calculations in chemistry and physics, however, protons and neutrons are considered to have relative weights of 1 each, and the electrons a mass of zero. Therefore, the *mass number* of an atom is the sum of its neutrons and protons, each counting as 1. It amounts to a workable approximation of the atomic weight. The mass number of hydrogen's most common isotope, protium, is 1; of deutrium, 2; and of tritium, 3. The mixture of hydrogen isotopes found in nature produces a relative atomic weight of 1.008. The atomic weight of uranium's mixture of isotopes is 238.03, but the mass number of the best-known uranium isotope is 235, the sum of 92 protons and 143 neutrons.

The isotopes of an element are chemically indistinguishable, but they can be separated by methods sensitive to the slight differences among their masses. The mass spectrograph is one device that will separate isotopes. Repeated diffusion of gases through porous clays will concentrate the lighter, faster isotopes. The isotopes of an element are so thoroughly mixed by natural forces that, whatever their source, the percentages are always the same. Tin mined in South America has a mixture of isotopes in the same proportion as tin mined in Asia; and refining and use do not vary this proportion. No matter where a sample of water is taken, from Arctic ice or the smoke of burning coal, for every 5,000 atoms of protium there will be 1 atom of deuterium.

MOLECULAR MODEL shows the orbital relationships of electrons within molecules.

Electron Cloud.—Every atom is electrically neutral; for each proton in the nucleus there must be an electron orbiting outside it. Hydrogen, with its single proton nucleus, has a single orbiting electron; and uranium's 92 protons are neutralized by 92 orbiting electrons. If any free electron comes near any free proton, they will affect each other's travel; and if conditions are right, the electron will go into orbit around the proton to create a hydrogen atom. If a naked nucleus with 26 protons and 30 neutrons plows into a cloud of electrons, and the temperature is not over a few thousand degrees, 26 of the electrons will at once arrange themselves around the nucleus to form an atom of iron.

The electrons are both spinning about the nucleus that attracts them and also adjusting against each other's movements and the forces of repulsion that exist among them. At any time the energy of an electron in an atom can be increased, causing it to change its position. But there is a lowest energy level for each electron in an atom; and in the discussion that follows, all the electrons are considered to be at their lowest level of energy, which is also called their *unexcited state* or *ground state*. Several modes of arrangement for electrons can be imagined, but the evidence shows that they are spaced at different distances from the nucleus, as though on the surfaces of concentric spheres, or *shells*. Outer shells hold more electrons than inner ones, since they have a greater surface over which to distribute them. The nucleus has a field of forces around it in space and the electrons, with their own fields, occupy specific positions within it, positions determined by the total balance of energies. The total energy of an electron is defined in four ways, called *quantum numbers;* there is a choice in each category, and every electron can be described by these four quantum numbers. No two electrons in an atom can have all four of their quantum numbers the same, but they can have as many as three alike.

According to an electron's total energy, or quantum number, it will occupy a definable energy area in the atom relative to other electrons. The principal quantum number relates to the number of the shell, or sphere, starting with 1 for the shell nearest the nucleus and proceeding outward to shell 7. Shells are also labeled *K, L, M, N, O, P, Q*. The *K* shell or shell 1, is complete when it has two electrons in it. No more can be accommodated at that distance from the nucleus, whether in a helium atom or a lead atom. The *L* shell is complete when it contains eight electrons. The total number of electrons a shell can contain is given by the term $2n^2$, where n is the number of the shell. Thus the *M* shell accommodates a total of 2×3^2, or 18, electrons. The *N* shell has a maximum of 32; the *O* shell could have 50, the *P* shell could have 72, and the *Q* shell, 98. The three outer shells are never completely filled, although in the heavier atoms all of them contain some electrons and the first four are completely filled.

Each shell consists of a band of subshells, lettered from inner to outer *s, p, d, f* into which the electrons of that shell are distributed. Each shell begins with an *s* subshell, and every *s* subshell has a maximum of two electrons; every *p* subshell has a maximum of six electrons; every *d* subshell has room for ten electrons; and those atoms large enough to have *f* subshells occupied fill them with fourteen electrons. Each subshell imposes a different kind of orbit on its resident electrons. In the *s* subshell, both electrons have circular orbits. All *p* subshell orbits are dumbbell-shaped. The *d* and *f* subshell orbits have more complicated shapes.

As the atomic numbers of the elements increase and protons are added to the nucleus, each balancing electron enters the position that requires the least energy to be maintained against the pull of the nucleus and the repulsion of other electrons. For the first twenty elements, the positions are filled in regular manner, each subshell becoming complete before the next one is begun. After calcium (Ca), the twentieth element, the sequence of energy levels does not coincide with the sequence of subshells. The larger shells overlap each other, and new shells are begun before inner ones are completed. The order in which subshells are filled is shown in the accompanying table.

When the *K* shell's *s* subshell is complete in helium (He) with two electrons, the *L* shell, or shell 2, begins outside it with one electron in the $2s$ subshell, to make lithium (Li). Beryllium (Be), with four protons, adds the fourth electron to the $2s$ subshell. Boron (B) adds its fifth electron to the $2p$ subshell. Carbon (C) also adds its sixth electron to the $2p$ subshell, and so do nitrogen (N), oxygen (O), fluorine (F), and neon (Ne), so that the $2p$ subshell has its full quota of six electrons and shell 2 is complete. It cannot accommodate any more electrons in its other possible subshells $2d$ and $2f$. A new shell, *M*, or shell 3, begins with its $3s$ subshell by taking the eleventh electron in sodium (Na) and the twelfth electron in magnesium (Mg). Then the $3p$ subshell fills until argon (Ar) is reached, which now has eight electrons in shell 3. Theoretically, the $3d$ subshell should begin to fill. Instead, shell 4, or *N*, begins outside the incomplete shell 3. Two electrons go into shell 4, and with scandium (Sc), the 21st element, the 21st electron finds the position of lowest energy in the incomplete *M* shell's $3d$ subshell. The $4p$ subshell completes itself in krypton (Kr) after the $3d$ subshell fills up. With only eight electrons in the *N*, or 4, shell, which has a total capacity of 32, the *O* shell, or shell 5, begins in rubidium (Rb). After the $5s$ subshell has two electrons, and the $4d$ subshell is completed, only after that does the $5p$ subshell resume filling up. When the *O* shell, shell 5, has eight electrons, the sixth, or *P*, shell is begun.

From the above description, it can be seen that the four following rules are always observed:

(1) The outermost shell can contain no more than eight electrons.

(2) The next-to-the-outermost shell can contain no more than 18 electrons.

(3) An outermost shell can contain no more than two electrons if the next-to-the-outermost has not reached its maximum.

(4) A next-to-the-outermost shell cannot contain more than nine electrons unless the second-from-outermost has reached its maximum.

Thus, a new shell begins whenever an outermost shell has filled its *s* and *p* subshells, even though that shell does not have its full quota of electrons and must then complete itself inside the *s* subshell of the new outermost shell. This phenomenon suggests that the configuration of two electrons in the *K* shell and the first eight electrons in any other shell is an extremely stable one, for it creates a recurring pattern in the electron arrangement. When the elements are listed in the order of increasing atomic number and the distribution of their electrons in the various energy levels is noted, the stability of this electron structure becomes evident. As can be seen from the table of electron distribution, each of the so-called inert gases has this arrangement: helium (He), atomic number 2, has two electrons in the *K* shell; neon (Ne), atomic number 10, has two electrons in the *K* shell and eight in the *L* shell; argon (Ar), atomic number 18, has two electrons in the *K* shell, eight in the *L* shell, and eight in the *M* shell; krypton (Kr), atomic number 36, has two electrons in the *K* shell, eight in the *L* shell, eighteen in the *M* shell, and eight in the *N* shell; xenon (Xe), atomic number 54, has two electrons in the *K* shell, eight in the *L* shell, eighteen in the *M* shell, eighteen in the *N* shell, and eight in the *O* shell; and radon (Rn), atomic number 86, has two electrons in the *K* shell, eight in the *L* shell, eighteen in the *M* shell, thirty-two in the *N* shell, eighteen in the *O* shell, and eight in the *P* shell.

The Periodic Law.—As a result of the recurring electron pattern, the elements can be grouped in periods beginning with an element that has one electron in a new shell and ending with an element that has eight in that same shell (or, in the case of the *K* shell, only two electrons). These periods can then be placed under one another in such a way that all elements with a single electron in the outermost shell fall under one another, forming a vertical group, those with two electrons in the outermost shell fall under one another, and so on, ending with a vertical group that contains all the elements with eight electrons in the outermost shell. The facts that allow the elements to be grouped in such a way are summed up in the *periodic law,* which states that the atomic structures of the elements are a periodic function of the number of protons in the nucleus. Since the properties of the elements can be explained only by their atomic structures, the properties are also a periodic function of the number of protons in the nucleus. Based on these facts, a periodic chart, or table, of the elements can be worked out.

All modern chemistry functions on principles based upon the periodic law. For example, without it there would be no theoretical justification for saying that metals can be recognized as a group of elements with certain properties in common. Some similarities, such as those of certain metals, were known even to the alchemists, and early in the nineteenth

A Typical Periodic Arrangement

Group I-A	II-A	III-B	IV-B	V-B	VI-B	VII-B	VIII-B			I-B	II-B	III-A	IV-A	V-A	VI-A	VII-A	
Alkali Metals	Alkaline Earths	←Transition Metals→					←Platinum Metal Triads→			Coinage Metals		Boron–Aluminum Family	Carbon Family	Nitrogen Family	Chalcogens	Halogens	Inert Gases
1 H																	2 He
3 Li	4 Be											5 B	6 C	7 N	8 O	9 F	10 Ne
11 Na	12 Mg											13 Al	14 Si	15 P	16 S	17 Cl	18 A
19 K	20 Ca	21 Sc	22 Ti	23 V	24 Cr	25 Mn	26 Fe	27 Co	28 Ni	29 Cu	30 Zn	31 Ga	32 Ge	33 As	34 Se	35 Br	36 Kr
37 Rb	38 Sr	39 Y	40 Zr	41 Nb	42 Mo	43 Tc	44 Ru	45 Rh	46 Pd	47 Ag	48 Cd	49 In	50 Sn	51 Sb	52 Te	53 I	54 Xe
55 Cs	56 Ba	57 *La	72 Hf	73 Ta	74 W	75 Re	76 Os	77 Ir	78 Pt	79 Au	80 Hg	81 Tl	82 Pb	83 Bi	84 Po	85 At	86 Rn
87 Fr	88 Ra	89 †Ac															

* Lanthanide Series													
58 Ce	59 Pr	60 Nd	61 Pm	62 Sm	63 Eu	64 Gd	65 Tb	66 Dy	67 Ho	68 Er	69 Tm	70 Yb	71 Lu

† Actinide Series													
90 Th	91 Pa	92 U	93 Np	94 Pu	95 Am	96 Cm	97 Bk	98 Cf	99 Es	100 Fm	101 Md	102 No	103 Lw

century chemists began to study these inexplicable groupings with the hope of discovering the cause. In 1869, Dmitri I. Mendeleev announced his periodic law, which stated that the properties of the elements are a periodic function of their atomic weights. He presented the scientific world with the first periodic chart of the 70 or so elements that were then known, arranged in horizontal periods and vertical groups. If his table were brought up to date and rearranged to accommodate new information, it would coincide exactly with the table based on electron structures. In other words, the properties of the elements are a result of the electron structures of the atoms.

The structural reason for the separation into A groups and B groups lies in the fact that the A-group elements are filling their outermost shells, while the B-group elements are filling inner shells and have only a few electrons in the outermost shell. This difference is strikingly reflected in the different properties of the A and B elements.

Large modern wall charts carry a great deal of information in the box devoted to each element: physical properties, such as melting and boiling points, crystal structure, color, density, hardness, conductivity, heat of fusion, solubility; chemical properties, such as heats of reaction and types of compounds formed with oxygen; and radioactive properties, such as half-life and what kind of particle the atom emits. Careful study reveals trends in the properties of a group rather than precise similarities.

Chemical properties cannot be graphed because they are not numerical. In general, the elements of any one group react chemically to form compounds in a way that is very similar. The compounds formed by the elements in any one group have similar properties. There are many exceptions, but the cyclical, or periodic, tendencies are unmistakable.

One of the most revealing graphs is that constructed with the atomic number and the *atomic volume,* which is calculated by dividing the atomic weight of an element by its density, thus obtaining a comparative figure. As the mass of the nucleus increases, it is found that the volume of the whole atom does not always increase—it sometimes shrinks. The largest atomic volumes are found in the group I-A elements; the smallest, in the middle groups. The explanation lies in the force with which the nucleus holds its cloud of electrons. Lithium's (Li) three protons hold two electrons in the *K* shell and one in the *L* shell. Beryllium (Be) has four protons and the greater attractive force of this nucleus holds four electrons, held in the same two shells as those of lithium, a little closer to itself. Boron (B) has five protons that draw its five electrons even closer to the nucleus. Carbon (C), although it has six electrons, is almost the same size as boron. Oxygen (O) and fluorine (F), with six and seven electrons in the *L* shell respectively, are smaller in volume than one would predict. However, helium (He), which ends the period and completes the eight-electron configuration in the outermost shell, suddenly enlarges. Apparently, in order to get that eighth electron into the *L* shell, the nucleus of eight protons has to loosen its grip on the electrons and allow a greater distance between them. Sodium (Na), in whose atoms a new shell begins, cannot shrink itself because of that new shell, and is therefore larger than lithium (Li) above it. But magnesium (Mg), with two electrons in the new shell, is smaller than sodium (Na). Again the increasing attractive power of the nucleus pulls the larger number of electrons, in the same period or shell arrangement, closer to itself until the end of the period. The largest atom, franconium (Fr), following the trends outlined above, is found at the bottom of group I-A; the smallest, except for hydrogen, is boron (B).

ELECTRONEGATIVITIES
Pauling's Scale

Element		Electronegativity
Potassium	(K)	0.8
Sodium	(Na)	0.9
Lithium	(Li)	1.0
Magnesium	(Mg)	1.2
Silicon	(Si)	1.8
Boron	(B)	2.0
Arsenic	(As)	2.0
Hydrogen	(H)	2.1
Phosphorus	(P)	2.1
Selenium	(Se)	2.4
Carbon	(C)	2.5
Sulfur	(S)	2.5
Iodine	(I)	2.5
Bromine	(Br)	2.8
Nitrogen	(N)	3.0
Chlorine	(Cl)	3.0
Oxygen	(O)	3.5
Fluorine	(F)	4.0

The cyclic nature of atomic volume is extremely important when studying the structure of crystals and the formulas of complex compounds. Obviously, a small atom will be arranged in different geometrical positions with a group of considerably larger atoms than it will with a group of atoms its own size.

Ionization. — The key group around which the periodic nature of the elements turns is the one numbered 0, called the *inert gases* or *rare gases:* helium (He), neon (Ne), argon (Ar), krypton (Kr), xenon (Xe), and radon (Rn). The energies in these atoms are so firmly balanced that until 1962 (when compounds of xenon were prepared) none of them was known to form any compounds. They will not combine with other elements under ordinary conditions because their eight outermost electrons (two outermost, in the case of helium) mesh to produce a structure that is more stable than that which any other number of outermost electrons can produce. Inert gases will not even cling to each other to form liquids, except at extremely low temperatures.

The arrangement of eight outermost electrons is so stable that an element immediately following an inert gas tends to lose its new electron quite readily in order to achieve such a structure. Elements with two or three outermost electrons will also give them up to uncover a stable electron arrangement. An atom that has lost electrons to acquire a stable configuration is no longer electrically neutral because the protons outnumber the remaining electrons; the particle as a whole has a positive electrical charge. In such a state, the atom is called a *positive ion,* or a *cation.*

An element that precedes an inert gas in the same period and has seven, six, or five electrons in its outermost shell tends to seize any free electrons it finds, adding one, two, or three to its outermost shell in order to bring the total to eight. An atom with such an arrangement has more electrons than protons and therefore has a negative charge; it is called a *negative ion* or *anion.*

All atoms, except those of the inert gases, form ions easily. The tendency to lose electrons is called *ionization potential,* and is a measure of the energy required to remove the electrons. The lower the ionization potential, the less energy required to remove the electrons and the more easily ions are formed. As a result, the ionization potential is lowest for the elements in the first few groups of the periodic table, for they lose their few outermost electrons easily. The measure of the tendency to gain electrons is called *electron affinity.* The greater an atom's electron affinity, the more strongly the atom tends to attract electrons in an effort to assume a stable electron structure. Electron affinity is highest for the elements in the groups at the right-hand side of the periodic table because these atoms gain electrons. Both ionization potential and electron affinity are useful concepts, but more useful than either is the concept of *relative electronegativity.* The harder it is to remove an electron from a neutral atom or an ion, the higher the electronegativity of that atom or ion is. Elements with seven outermost electrons have the highest electronegativities, and elements with a single outermost electron have the lowest. Since the term is a relative one, fluorine (F), with the highest affinity for electrons of all the elements, was the standard chosen against which the others are compared; and it was given an electronegativity of 4. The inert gases, which would have even higher electronegativities, were omitted because they do not usually form compounds.

Atoms of elements in groups I-A, II-A, and III-A can lose their few outermost electrons as a result of energy changes due to ordinary electric fields, heat, collisions, radiation, or chemical reactions. Considerably more energy than is available in ordinary laboratory work is required to remove the outer shell when it contains four or more electrons. It can be done in strong electric fields or at very high temperatures, but not in chemical reactions. After the outer shell of an atom has been stripped, electrons can be plucked out of the inner shells one by one, either by increasingly powerful positive fields that will overcome the pull of the nucleus or by intense heat. All the electrons of any atom can be boiled off by temperatures equivalent to those at the surface of the sun, which can be duplicated in special laboratory equipment. The heats estimated for the interiors of stars decompose the nucleus itself into its components.

Valence.—Chemical reactions involve only the loss and gain of electrons in the outer shell and, in some cases, a few electrons in the inner shells of atoms. Such electrons are called *valence electrons,* and the outermost shell is called the *valence shell.* The number of electrons an atom can lose or gain in a chemical reaction is the *electrovalence* of that element. All the elements in group I-A have one valence electron, which is lost in chemical reactions to leave a positive ion with a single positive charge; the electrovalence of such an element is +1. Group II-A elements have an electrovalence of +2, for they can give up two electrons to form an ion with a double positive charge. Similarly, group III-A elements have an electrovalence of +3, for they can give up three electrons. The group IV-A elements have an electron affinity high enough to prevent them from losing their electrons easily in chemical reactions; therefore, their electrovalence varies. The group V-A elements do not lose their five valence electrons, but instead pull three electrons into their valence shell to complete the octet; thus, their electrovalence is −3. The group VI-A elements pull two electrons into their valence shell to complete the octet and have an electrovalence of −2. The group VII-A elements need pull only one electron into their valence shell to complete the octet and have an electrovalence of −1. In A groups, the above rule is almost always observed. In B groups, however, there is fluctuation; and the group number does not necessarily indicate the electrovalence. In addition, the B-group elements often have more than one electrovalence; for example, mercury (Hg), which is in group I-B, has electrovalences of +1 and +2. The ion formed when an atom gives up or gains electrons in its valence shell is written with the electrical charge as a suffix to the symbol of the element. For example, an ion of sodium, which has an electrovalence of +1 and whose ions thus have a single positive charge, is represented as Na^+; magnesium, with an electrovalence of +2, forms the ion Mg^{++}; aluminum, which has an electrovalence of +3, forms the ion Al^{+++}; phosphorus, with an electrovalence of −3, forms the ion P^{---}; the ion of sulfur, electrovalence −2, is S^{--}; chlorine, electrovalence −1, forms Cl^-.

All elements that lose electrons in chemical reaction to become cations are called *metals;* elements that gain electrons to become anions are called *nonmetals.* Metals are on the left side of the table in the first three A groups; nonmetals, on the right in the last five A groups. The inert gases, in group 0, are also classed as nonmetals. In all A groups, only the *s* and *p* subshells of the outermost shell are being filled. The B groups, placed between groups II-A and III-A, contain the *transition metals,* whose *s* and *p* subshells are only partly filled while their *d,* and sometimes *f,* subshells are filling up. They have electrovalences of +1, +2, and +3. A few elements on either side of a dividing line between metals and nonmetals are called *metalloids* because they behave either as metals or as nonmetals, depending on the conditions.

Group I-A elements are called the *alkali metals,* group II-A consists of the *alkaline earth metals,* and the elements of group VII-A are called the *halogens.* Into the place occupied by lanthanum (La), element 57, 14 other elements must be squeezed together because they are completing *f* subshells deep within and are chemically very similar. The grouping is called the *lanthamide series,* or the *rare earth metals,* because when the first few were discovered they were thought to be much less abundant in the earth's crust than they have proved to be. Element 89, actinium (Ac), also has 14 other elements with it in its box, all of them filling *f* subshells. The *actinide series* contains 11 of the 12 man-made elements. (The other man-made element is promethium.) Both of these series are placed at the bottom of the periodic table in order not to stretch the table impractically.

Size of Ions.—Since all atoms lose or gain electrons, it is useful to know how this affects volume of the atom. When lithium (Li) loses its single valence electron, the nucleus—with its now unbalanced positive charge—pulls the remaining electrons much closer to itself. The diameter of the lithium ion, Li^+, is less than a third of the lithium atom's. Beryllium (Be) loses two electrons, and the two extra positive charges in the nucleus pull the remaining electrons into an even smaller volume than Li^+. Boron (B) and carbon (C) do not form ions easily, but oxygen (O) adds two electrons to its valence shell. Its ion, O^{--}, has a diameter more than three times as large as the electrically neutral atom, O. Fluorine (F), which adds a single electron, enlarges even more. In any metal group, the ions are always much smaller than the atoms, although the heavier ions are larger than the lighter ones in the same group. Among the nonmetals, the ions are always much larger than the atoms and they, too, increase in size as they get heavier. The largest ion of all is thus that of radon, Rn^-, at the bottom of group VII-A. The smallest, except for hydrogen, H^+, would be that of boron, B^{+++}.

Chemical Bonding.—An atom does not have any of the properties of its ion except mass number. The two particles are completely different in their chemical and physical behavior. Any ion can be forced—through chemical reaction, electrical fields, radiation, or heat—to accept or relinquish electrons and become a neutral atom again. The energy content of an atom can never be that of its ion, so a shift of electrons always involves an energy change in the particle. Such a shift and energy change always occur in a chemical reaction. Two general kinds of chemical reaction are recognized, resulting in either electrovalent or covalent bonds.

The difference in electron affinity, or in relative electronegativity, between metals and nonmetals is great enough so that if their atoms are brought together under the right conditions the metal loses its lightly held electrons to the much more electronegative nonmetal. Indicating valence electrons as dots around the symbol for an element, the following happens in the reaction between sodium (Na) and chlorine (Cl):

$$Na\cdot + \cdot\overset{..}{\underset{..}{Cl}}: \rightarrow Na^+ + :\overset{..}{\underset{..}{Cl}}:^-$$

The transfer produces two ions, Na^+ and Cl^-. Sodium metal (in the atomic state) in bulk is silvery, lighter than water, and reacts with water to produce hydrogen gas and a powerful, poisonous alkali, sodium hydroxide. Chlorine in bulk is a greenish poisonous gas and a strong bleaching agent. The shift of a single electron between the two atoms liberates a great deal of heat energy, and the resulting ions, clinging together because of their opposite charges, crystallize into a white solid that is ordinary table salt, Na^+Cl^-.

A calcium (Ca) atom can lose its two valence electrons to two chlorine (Cl) atoms, each chlorine atom accepting one electron, in the following manner:

$$:\overset{..}{\underset{..}{Cl}}\cdot \nwarrow \cdot Ca\cdot \nearrow \cdot\overset{..}{\underset{..}{Cl}}:$$

This reaction produces the compound calcium chloride, $Ca^{++}Cl_2^-$. Magnesium (Mg), which has an electrovalence of +2, and oxygen (O), which has an electrovalence of −2, can transfer electrons in the following manner to form the compound magnesium oxide, $Mg^{++}O^{--}$:

$$Mg: \rightarrow \overset{..}{\underset{..}{O}}:$$

The shift of electrons from less electronegative to more electronegative atoms is called *electrovalent bonding,* or *ionic bonding,* and it produces one kind of chemical reaction.

Two metals cannot form electrovalent bonds with each other, for while both can donate electrons, neither is sufficiently electronegative to accept them. (Metal atoms do bond together; the mechanism involved is discussed in a later section of this article.) Two nonmetals cannot form electrovalent bonds with each other, either, because neither will give up electrons. However, two nonmetals can share their valence electrons to produce a molecule that has a *covalent bond.* Sharing of electrons, instead of transferring them, is a second kind of mechanism found in chemical combination.

For example, when two chlorine atoms combine to form a chlorine molecule, each shares one of its own electrons with the other; in this manner, both have an outermost shell containing eight electrons, two of which are orbiting around both nuclei. The covalent bond may be pictured as follows:

$$:\ddot{\underset{..}{Cl}}\ :\ \ddot{\underset{..}{Cl}}:$$

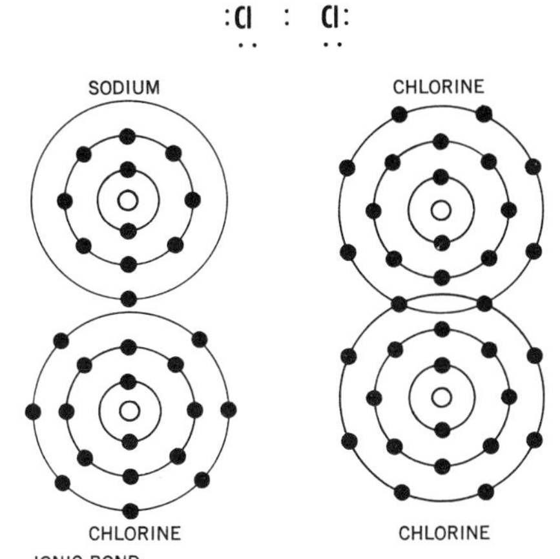

IONIC BOND SODIUM CHLORIDE (TABLE SALT) MOLECULE

COVALENT BOND CHLORINE GAS MOLECULE

The formula for the molecule is Cl_2. Two oxygen atoms may form a covalent bond between them by sharing two electrons from each atom, a total of four. Thus, four electrons in the diatomic oxygen molecule, which has the formula O_2, are orbiting both nuclei. The covalent bond is:

$$:\ddot{O}\ ::\ \ddot{O}:$$

Nitrogen atoms combine into a diatomic molecule with a triple covalent bond, N_2:

$$:N\ :\ :\ :\ N:$$

Hydrogen has only one electron, but this is in the *K* shell, which is complete with two; therefore, hydrogen atoms combine with a covalent bond to form H_2, which is a molecule of hydrogen gas. Sulfur and oxygen form two kinds of compounds with covalently bonded molecules: sulfur dioxide, SO_2, and sulfur trioxide, SO_3. Carbon dioxide, CO_2; carbon monoxide, CO; carbon tetrachloride, CCl_4; and ammonia, NH_3, are all compounds in whose molecules atoms are covalently bonded.

It is possible for one atom to provide both electrons of a covalent bond with another atom and establish what is called a *coordinate covalent bond.* All covalent bonds are an overlapping of *s* and *p* electron orbitals.

If one of the atoms in a covalently bonded molecule has a stronger affinity for electrons than has the other atom, then the shared pair will be held more closely by the stronger nucleus. Such a shared pair will give the molecule an unbalanced configuration of charges, making the more electronegative area of the molecule more negative in character than the other area. The molecule as a whole remains neutral, but one end is negative and the other positive. It is called a *polar molecule,* and the bond is a *polar covalent bond.*

Covalence represents the number of electrons the element shares, and has no positive or negative sign because no ions are produced. Most nonmetals have more than one covalence, being able to share their electrons in several ways. Metalloids have covalence, as well as negative and positive electrovalence. Some transition metals form covalent bonds.

All chemical bonds are electrovalent, covalent, or polar covalent; the latter can be so strongly polar that the molecule has almost an electrovalent configuration, or it can be so little polar that it is almost covalent. The difference in electronegativity between the bonding atoms determines the kind of bond they form.

A distinction must be made between a covalently bonded molecule and the ions of an electrovalent bond. The ions cling together because of their opposite charges, and in the gaseous state the ions of a compound can be considered as a single molecule. In the liquid state the ions are free to move about, and they cling to whatever oppositely charged ion is nearest. A molecule, therefore, cannot exist among ions in the liquid state. In the solid state, each ion is held in its position according to the specific crystal structure of the compound and is surrounded by oppositely charged ions, to each of which it is attracted with equal force. Therefore, in the solid state no molecule exists among ions.

The bonds between atoms in a covalently bonded molecule are far more powerful than any that can exist between ions. When molecules of a covalent compound are liquefied, they continue to be separate entities; whatever attraction exists between the separate molecules, holding them in liquid contact, is very weak compared to the bond between the atoms constituting the molecule. When a covalent compound solidifies, the attraction between molecular surfaces, binding them into the crystal lattice, is still very weak compared to the bond between atoms within such molecules.

Chemical Formulas.—A formula for a compound is a precise representation of the actual number of atoms bonded into a unit molecule, or of ions associated in the simplest combination. In all inorganic formulas, the least electronegative, or most metallic, element is named first, followed by the progressively more electronegative, or less metallic, elements. The last element has an ending formed according to a system that is part of the language of chemistry. Compounds composed of only two elements always have the less metallic element end in *-ide,* whether the compound is electrovalent or covalent. Examples are carbon monoxide, CO; carbon dioxide, CO_2; aluminum bromide, $AlBr_3$; hydrogen oxide (or water), H_2O; and carbon tetrachloride, CCl_4. Formulas for ionic or electrovalent compounds, although they cannot be considered to represent true molecules, are often written as groups of ions, for example, sodium oxide, $Na_2^+O^{--}$, calcium bromide, $Ca^{++}Br_2^-$, and potassium sulfide, $K_2^+S^{--}$. Formulas for compounds with three or more elements use the same sequence, but the endings vary.

Chemical Equations.—A chemical reaction is a shift of electrons in the valence shell of an atom, and the shift is always brought about by a change in the distribution of energies within the atom. All chemical reactions can be represented in the form of a chemical equation. When two atoms of chlorine collide and combine to form covalently bonded chlorine molecules, the equation is:

$$Cl + Cl \rightarrow Cl_2.$$

The plus sign indicates that the masses and energies reacting are added together and also that the two atoms touch under conditions that bring about a shift of electrons. The arrow indicates that the atoms have actually reacted and that their electron configuration has changed; the arrow is also an "equals" sign.

The primary usefulness of a chemical equation lies in the fact that the total number of atoms of each element on one side of the arrow is equal to the total number of atoms of each element on the other side of the arrow. If the total number of atoms is not equal, the equation is said to be *unbalanced;* if the total number of atoms is equal, the equation is said to be *balanced.* A good example of the balancing of a chemical equation is the reaction in which hydrogen (H) and oxygen (O) combine to yield water (H_2O), which could be written:

$$H + O \rightarrow H_2O \text{ (unbalanced).}$$

However, hydrogen and oxygen both exist as diatomic molecules, so this is somewhat more properly written as:

$$H_2 + O_2 \rightarrow H_2O \text{ (unbalanced).}$$

The equation is still unbalanced because there are two oxygen atoms on the left side of the arrow but only one on the right side. To balance the number of oxygen atoms, it reads:

$$H_2 + O_2 \rightarrow 2H_2O \text{ (unbalanced).}$$

Now the number of hydrogen atoms no longer balances, for there are only two on the left but four on the right. Therefore, the equation must be:

$$2H_2 + O_2 \rightarrow 2H_2O \text{ (balanced).}$$

This equation for the reaction is now balanced, with four hydrogen atoms and two oxygen atoms on each side.

Other equations, however, are much more complex and their equations much more difficult to balance. For example, the unbalanced equation for the reaction between phosphorus (P) and nitric acid (HNO_3) in the presence of water, which yields phosphoric acid (H_3PO_4) and nitric oxide (NO), may be written:

$$P + HNO_3 + H_2O \rightarrow H_3PO_4 + NO \text{ (unbalanced).}$$

The balanced equation is:

$$3P + 5HNO_3 + 2H_2O \rightarrow 3H_3PO_4 + 5NO.$$

In addition to the basic method of writing and balancing chemical equations, several conventions are used. Vertical arrows indicate that an element or compound is a gas, as in $H_2\uparrow$ or $CO_2\uparrow$, or an insoluble substance as in $AgCl\downarrow$ or $BaSO_4\downarrow$. The letters *g, s,* and *l* as subscripts indicate whether the reactant or product is in the liquid, gaseous, or solid state, for example: $H_{2(g)}$, $H_2O_{(l)}$, or $BaSO_{4(s)}$. The symbol $+E$ or $-E$ may be added to the right-hand side of an equation to show that energy is either gained from or lost to the surroundings during the reaction. An "equals" sign ($=$) or arrows pointed in opposite directions ($\rightleftarrows$) may replace the single arrow ($\rightarrow$) to show that the reaction is in equilibrium or that the reaction can proceed in either direction. The arrow may also have a triangle placed above ($\overset{\Delta}{\rightarrow}$) or below ($\underset{\Delta}{\rightarrow}$) it to show that the reactants must be heated if the reaction is to proceed.

Calculations. — The determination of correct formulas for compounds is one of the most important aspects of chemical research, and many kinds of analytical procedure have been worked out using relative atomic weight figures in the calculations. For example, suppose the formula is wanted for the gas formed when pure carbon burns in oxygen. By calculating the actual weights of carbon and oxygen used up in the reaction, or by breaking up the gas molecules in various ways and calculating the weights of carbon and oxygen obtained, it is found that 27.29 per cent of the compound formed is carbon and 72.71 per cent is oxygen. In round numbers, the atomic weights of oxygen and carbon are 16 and 12, respectively. Dividing the percentage of each element by its atomic weight gives the ratios of the elements in the compound. In this case the ratio is one atom of carbon to two of oxygen. The empirical formula is therefore CO_2, but the actual molecule could be C_2O_4 or C_3O_6 or any other combination of the atoms in the ratio of one to two. Using Avogadro's hypothesis, or any other method for determining relative molecular weights, the gas formed is found to have a molecular weight of 44. Since the molecular weight of the compound CO_2 would be $12 + (2 \times 16)$ or 44, the molecular weight of C_2O_4 would be $(2 \times 12) + (4 \times 16)$ or 88, the molecular weight of C_3O_6 would be $(3 \times 12) + (6 \times 16)$ or 132, etc. Thus, the correct molecular formula is CO_2, carbon dioxide.

In a balanced equation, symbols stand for single atoms and formulas for single molecules. Symbols also represent their relative atomic weights and formulas their relative molecular weights:

$$\underset{12}{C} + \underset{32}{O_2} \rightarrow \underset{44}{CO_2}$$

As long as these ratios are observed, any weights of carbon and oxygen will combine completely, without leftovers. Thus, 12 grams, 12 pounds, or 12 tons of carbon will need 32 grams, 32 pounds, or 32 tons of oxygen to combine completely and produce 44 grams, 44 pounds, or 44 tons of carbon dioxide. One ton of carbon requires 32/12 or 2⅔ tons of oxygen for complete combustion and will produce 44/12 or 3⅔ tons of carbon dioxide. Thus, the equivalent weight of any unknown in a chemical reaction can be calculated. Without this quantitative use of chemical equations, there could be no chemistry in the modern sense. In fact, the equation was developed precisely to handle the quantitative realities of modern chemistry.

The equation is used to calculate volumes of gases as well. A quantity of oxygen that weighs 15.999 grams, a quantity of hydrogen that weighs 1.008 grams, a quantity of sulfur that weighs 32.064 grams, and a quantity of any element that weighs its atomic weight number in grams will all have the same number of atoms, since atomic weights are ratios obtained when the weights of equal numbers of atoms are compared, according to Avogadro's hypothesis. This amount of an element is called its *gram atomic weight.* The same reasoning holds for molecules, and the weight in grams of the molecular weight number is the *gram molecular weight* of a compound. The *gram equivalent weight* of an element is the weight that will displace one gram atomic weight of hydrogen or will combine with half of one gram atomic weight of oxygen. More practically, the gram equivalent weight of an element is its gram atomic weight divided by its *oxidation number.* The number of atoms or molecules in the gram atomic weight or gram molecular weight of any element or compound has been calculated from various experiments. It is called *Avogadro's number,* or a mole, and is represented by the letter *N.* It is 6.024×10^{23} atoms or molecules. When a mole of any substance is vaporized at standard temperature and pressure (STP), it will occupy 22.4 liters, which is called the *gram molecular volume.* Thus, 32 grams of oxygen gas occupy 22.4 liters at STP, and 44 grams of CO_2 gas occupy 22.4 liters. Therefore, 100 grams of carbon would react completely with $22.4/12 \times 100$ liters of oxygen.

When 1 mole, or 12 grams, of carbon is completely burned, the energy produced has been measured as 94,030 calories. From this, the heat produced by burning any amount of carbon can be calculated, using the same method as above.

If there is not enough oxygen present when carbon burns, carbon monoxide is produced. The balanced equation is

$$2C + O_2 \rightarrow 2CO + E_a.$$

If the CO is then burned, the balanced equation for that reaction is

$$2CO + O_2 \rightarrow 2CO_2 + E_b.$$

The sum of $E_a + E_b$, for one mole of carbon, is 94,030 calories. The reaction illustrates the *law of Hess* (Germain Henri Hess, 1802–1850), which states that no matter how many steps a reaction takes for its completion, the sum of the energies produced at each step exactly equals the energy that would be produced if the reaction took place in one step. It is a special case of the law of conservation of matter and energy.

A reaction that produces energy in the form of heat is called *exothermic,* and a reaction that absorbs heat in order to proceed is called *endothermic.* Most reactions take place in a series of steps, and not all of the steps need be exothermic.

Radicals.—Groups of covalently bonded atoms, which function as a single particle but have an electrical charge and therefore are not molecules, are called *radicals.* SO_4^{--}, NO_3^-, NH_4^+, and CO_3^{--} are some common radicals. They participate in many reactions, exactly as ions do; and they form parts of compounds, just as ions do. Since they function as ions, they are often referred to as ions. Radicals, however, cannot be neutralized into molecules with the same composition. There is no such molecule as SO_4, for example, nor one whose formula is NH_4. If radicals are robbed of their electrical charge, they break up into smaller radicals or into molecules, atoms, and ions.

Mechanism of Reactions. — Obviously, not every collision between atoms and molecules results in a reaction. The wood of a match does not burn, though its molecules are incessantly bombarded by millions of oxygen molecules every second, until the temperature has been raised sufficiently by the flaming match head, which has been forced into reacting by the heat of friction. The temperature at which the energy configuration of atoms allows an electron shift into a new configuration of energies is called the *activation level* of energy in the atoms. Below this excited state, no reaction can take place.

Molecules can be excited, or activated, by radiation, electrical fields, collisions, and—by far the most common—heat. Activation levels of energy are thus usually reported as temperatures below which no appreciable reaction can take place. At no time, according to the kinetic molecular theory, do all the molecules in any system have the same kinetic energy; the temperature of a gas, liquid, or solid is only a measure of the average kinetic energy of the molecules. In many cases where no reaction seems to be taking place, a few molecules with very high energy content may be reacting. An example is iron rusting at ordinary temperature. But even when the bulk of a substance has been raised to activation level, all its molecules do not instantly react, partly because some will have lower energies, partly because most molecules have key spots on which they must be hit, and partly because they cannot all collide at the same instant.

In the case of gases whose molecules are moving very rapidly and colliding billions of times a second, a reaction often becomes an explosion because all the molecules have a chance to react within a second or two. When the mixture of reactants includes a solid or a liquid, collisions are restricted to their surfaces. The number of collisions is increased enormously by powdering and vaporizing the reactants or by dissolving them in a common liquid. Increasing the concentration of one or all of the reactants will increase the number of collisions each second. Raising the temperature will increase the number of collisions, their force, and the number of molecules at activation level.

Finally, a great many reactions will not take place to any appreciable extent unless a *catalyst* is present. A *catalyst* is a substance, usually a solid, that participates in a reaction but emerges from it unchanged. Very small amounts are needed. The theory of catalysis is that reacting molecules will combine on the molecular surface of the catalyst to form a complex—sometimes with the catalyst—that then allows the final step of the reaction to take place at a lower activation level. The molecules are twisted, as it were, by the catalyst into configurations that produce the same readiness to react that an elevation of temperature would. Some catalysts, called *negative catalysts,* work in reverse and inhibit, or retard, reactions. However, chemists are primarily interested in catalysts that accelerate reactions.

The *rate of reaction* can be measured by weighing the reactants before the reaction begins, then isolating and weighing them at regular intervals after the reaction has started. The rate is reported as the number of grams reacting per second. It is of prime importance in industry, where the largest yield is wanted in the least amount of time. Industrial research seeks ways to speed up reactions by altering temperatures and concentrations, which include pressure conditions; by new catalysts; by engineering changes in the pattern of flow in and out of reactors; or by changes in the design of equipment.

The actual steps by which reactions take place are being studied with radioactive isotopes having the same chemical properties as the stable atoms; their presence in intermediate compounds can be traced with radiation detectors. Some of the simplest reactions have proved to be very complicated in the order in which they form intermediate molecules. Every formation of a molecule is attended by an energy change, and no theory can yet predict or follow the involved balancing of forces that determine the reaction sequence.

REVERSIBLE REACTIONS.—The study of reactions is complicated because most reactions are reversible; that is, stable atoms and molecules that react under one set of conditions will be re-formed under another set of conditions because there is no reason why the shift of electrons cannot be reversed. In the reversible reaction

$$A + B \rightleftarrows C + D,$$

consider that all the reactants and products are gases and that the reaction takes place in a closed container. At the start of the reaction only A and B are present, and their concentration is measured. A little later some C and D will have formed, and the concentration of A and B will be less. The number of fruitful collisions, and thus the rate of the reaction of $A + B$, will steadily decrease as they are used up. Eventually there will be no A and no B left, and the container will be completely filled with C and D. Let us assume that the configurations of $A + B$ contain more energy than the configurations of $C + D$. Therefore, energy in the form of heat will be produced when the reaction moves to the right. If the temperature of the pure C and D mixture is raised so that the heat, which is the difference of energy content between the two sets of reactants, is provided from the outside, then C and D will absorb that heat and, with this extra energy, their atoms re-form A and B.

Pure cases of reaction in one direction at one temperature and the reverse at another are extremely rare, however, because at all times there are molecules present with greater and lesser energies than the average. As soon as some C and D are formed, a few of them will have kinetic energies and potentials high enough so that they will re-form into A and B when they collide. As the concentration of C and D increases, the number of such fruitful collisions, and thus the rate of the reaction $C + D$, will increase. The rate $A + B$ decreases. At a certain instant the two rates will be equal, and the reaction is said to have reached a *dynamic equilibrium.* However, the two reactions do not stop. $A + B$ continues at the rate set by the particular conditions in the container, and so does $C + D$ at exactly the same rate. If the concentration of A, B, C, or D is changed, or if the temperature is changed, then the rate of reaction of each side of the equation will change until they are the same and dynamic equilibrium is reestablished.

This phenomenon is defined by the *principle of Le Châtelier* (Henri Louis Le Châtelier, 1850–1936): when a system is in equilibrium and one of the factors producing the equilibrium is altered, the system will always react in such a way as to absorb the change and re-establish a new equilibrium with the changed factors.

LAW OF MASS ACTION.—A bracket around a formula or symbol stands for concentration. $[A]$ means the percentage of all the A molecules in a reaction. Of course, instead of counting molecules, the reactants are weighed. The rate of reaction is always directly proportional to the number of collisions, which is directly proportional to the concentration of molecules. The rate of reaction of $A + B$ is expressed as the product of their concentrations, $[A] \times [B]$. The rate of a reaction at any fixed temperature, no matter how much or how little of the reactants are present, is always the same. That is, in any mixture at that temperature the concentrations will be changed by a shift in one or the other direction in order to achieve the same product of concentrations. $[A]$ and $[B]$ will not always be the same, but their product will; and it is expressed mathematically in the equation

$$[A] \times [B] = k_t.$$

At the same temperature t,

$$[C] \times [D] = k_t.$$

At equilibrium the two equations can be combined into

$$\frac{[C] \times [D]}{[A] \times [B]} = K_t.$$

This equation expresses what is called the *law of mass action,* or *law of chemical equilibrium;* K is the equilibrium constant for that reaction at that temperature. In the reversible reaction

$$mA + nB + \cdots \rightleftarrows pC + qD + \cdots$$

$$\frac{[C]^p\ [D]^q}{[A]^m\ [B]^n} = K_t.$$

Any variation in concentration of any reactant produces a shift in the equilibrium that will change the concentration of all the other reactants and reestablish the same constant k_t. Some well-known reversible reactions are:

$$H_2 + I_2 \rightleftarrows 2HI,$$

$$CaCO_3 \rightleftarrows CaO + CO_2$$

(in the production of quicklime from limestone),

$$2SO_2 + O_2 \rightleftarrows 2SO_3$$

(in the manufacture of sulfuric acid),

$$N_2 + 3H_2 \rightleftarrows 2NH_3$$

(in the manufacture of ammonia by the Haber process).

The total removal of one of the participants in a reversible reaction will bring about a completion of the reaction in the direction of the lost participant. In the equation for the equilibrium constant, if one of the concentrations becomes zero, K becomes zero or infinity. A product can be removed easily if it is a gas or an insoluble solid in a solution. When there is more than one gas or there is a mixture of miscible liquids, or a mixture of solids and liquids dissolved in one another, the problem of bringing about a complete reaction is often too difficult or too expensive, and only a percentage of the theoretical yield can be obtained. For instance, in the Haber process all the reactants are gases at reacting temperatures. By raising the temperature, the concentrations are shifted toward the right, and larger yields of ammonia are obtained. But above a certain temperature the decomposition of ammonia into nitrogen and hydrogen proceeds at an accelerated rate, and the yield of ammonia diminishes. Thus, for every reversible reaction there is an optimum temperature that must be determined by experiment, for no theoretical computation is possible. In industry, the problem of obtaining maximum yields is often solved indirectly by having one of the by-products, or an undesired compound, react with something that will carry it away as a gas or precipitate it. However, no matter how complex a set of multiple reactions may be, in a closed system, for each reaction in that mixture, the law of mass action holds, as does the law of conservation of matter and energy.

If a catalyst is added to a reversible reaction, the rates in both directions are equally affected, and thus the same constant is obtained. Of course, if the system is not in equilibrium when the catalyst is added, the catalyst will seem to speed up the rate only in one direction until equilibrium is reached.

HEAT OF REACTION.—One of the most important aspects of chemical reaction is the heat or some other form of energy, such as light, produced. The *heat of reaction* is the primary goal in the burning of fuels. In other reactions the heat is usually wasted, but often it can be utilized to raise the temperature of fresh reactants to activation levels. The accompanying graph suggests the energy changes involved. The energy, in the form of heat, added to $A + B$ to raise them to activation level, is r-q. The same amount of heat is given back after the formation of C and D and is not part of the heat of reaction. The amount q-p is the energy that has been released by the shift of electrons and is always the same, no matter at what temperature the reaction occurs.

In an endothermic process the graph can be reversed. The heat of reaction q-p must be provided, together with the heat to raise $C + D$ to activation level, in order to fix the configuration of atoms in A and B.

OXIDATION-REDUCTION REACTIONS.—The commonest of all reactions is the combination of elements and compounds with oxygen. It is always an exothermic process. When it occurs rapidly enough, the evolution of heat is sufficient to produce light and flames; and the process of oxidation is then called *burning*. Almost every element combines with oxygen, and some form more than one oxide. If the element is a metal, the general reaction can be written:

$$\mathrm{M}\!: + \ddot{\underset{..}{\mathrm{O}}}\!: \rightarrow \mathrm{M}^{++} \,:\!\ddot{\underset{..}{\mathrm{O}}}\!:^{--}$$

If the element is a nonmetal:

$$:\!\ddot{\underset{..}{\mathrm{A}}}\!: + :\!\ddot{\underset{..}{\mathrm{O}}}\!: \rightarrow :\!\ddot{\underset{..}{\mathrm{A}}}\!:\ :\!\ddot{\underset{..}{\mathrm{O}}}\!:$$

A great many reactions that do not include oxygen can be considered similar to the ones shown above. The meaning of *oxidation* has been expanded to include all reactions in which electrons are actually transferred or in which they can be counted as though they had been transferred. In the reaction between sodium and chlorine,

$$\mathrm{Na}\cdot + \cdot\ddot{\underset{..}{\mathrm{Cl}}}\!: \rightarrow \mathrm{Na}\!:\!\ddot{\underset{..}{\mathrm{Cl}}}\!:$$

sodium loses its electron to chlorine and is therefore oxidized. Chlorine, which gained an electron, was the *oxidizing agent.*

Every element is said to be in a *zero oxidation state* when its atom is neutral; this state can be diminished or increased. In the above equation, sodium's oxidation state is raised from 0 to +1, while chlorine's is lowered from 0 to −1 when they react. Every molecule, including the theoretical union of an ionic compound, is neutral; therefore the + and − oxidation states of the elements in the compounds must cancel out to 0. The *oxidation number* of sodium in table salt is +1, and that of chlorine is −1. The sum of the two is zero.

The process opposite to oxidation is called *reduction.* The word is even older than oxidation, and it originally meant the process of liberating a metal from its ore by making the oxygen of the metallic oxide combine with carbon. In the above reaction, chlorine has been reduced, the *reducing agent* being sodium. In other words, in an oxidation-reduction reaction an oxidizing agent is reduced, a reducing agent, oxidized.

Elaborating the concept, *oxidation numbers* replace the concept of valence and are much more useful in tracing electron shifts through different molecular formulas in an equation. The total increase of oxidation numbers must equal the total decrease, while with valence no such mathematical equating is possible. Every element has several oxidation states that become oxidation numbers when their atoms are in a molecule. Since the majority of inorganic reactions are of the oxidation-reduction type, the count of electrons, as though they were really gained or lost whatever the bond, in effect ties the chemical equation directly into the reality of relative electronegativities.

To illustrate this, the reaction between nitric acid (HNO_3) and iodine (I_2) can be used. They form iodic acid (HIO_3), nitrogen dioxide (NO_2), and water (H_2O). The unbalanced equation is

$$HNO_3 + I_2 \rightarrow HIO_3 + NO_2 + H_2O$$
$$\text{(unbalanced).}$$

The hydrogen atoms do not change their oxidation state, nor do the oxygen atoms. Iodine in the molecular state has an oxidation number of 0. In the HIO_3 molecule, oxygen has a total of −6 and hydrogen has a total of +1. Thus, iodine must have an oxidation number of +5 in order to create a neutral molecule. In HNO_3, nitrogen has an oxidation number of +5 for the same reason, but in NO_2 its oxidation number is only +4. Thus, by scanning the totals, a balanced equation is arrived at:

$$10HNO_3 + I_2 \rightarrow 2HIO_3 + 10NO_2 + 4H_2O$$

Elements, compounds, ions, and radicals are known as either good oxidizing agents or good reducing agents, depending on the energy with which they seize or donate electrons. The general rule follows the shift of electronegativity across the periodic table, nonmetals being good oxidizing agents and metals being good reducing agents. Transition metals vary considerably in this property, acting sometimes as reducing agents and sometimes as oxidizing agents, depending on the other reactant's relative electronegativity. Oxidation-reduction reactions can be expressed in partial electronic equations:

$$2M^0 - 4e^- \rightarrow 2M^{++};$$
$$0^0{}_2 + 4e^- \rightarrow 20^{--}.$$

When the two partials are added:

$$2M^0 + 0^0{}_2 \rightarrow 2M^{++}0^{--}.$$

Bonding in Solids.—The only types of bonding discussed have been ionic (or electrovalent), covalent, polar covalent, and coordinate covalent. When an ionic compound solidifies, each cation (+) is surrounded by anions (−), and each anion is surrounded by cations. The alternating lattice of + and − charges in three dimensions is one in which the bond between an ion and any other oppositely charged ion near it is identical; that is, the bonding force of the solid at any point in it is between oppositely charged particles and is therefore equal in all directions. When covalently bonded molecules solidify, the bonds between the molecules are weak compared to the bonds within the molecule and are the result of what are called *van der Waals forces of attraction,* named after Johannes Diderik van der Waals (1837–1923) who first measured them. The nucleus of each molecule weakly attracts the electrons of nearby molecules, and the molecules are pulled into a pattern in the solid such that the forces acting on individual atoms vary considerably in different directions. Polar covalent molecules bond into the solid state with an orientation of their positive and negative ends. When nonmetallic elements solidify, the atoms bond to one another with covalent arrangements of their electrons, sometimes in very complicated fashion, to form interlocking rings or, like carbon atoms in the diamond structure, to form chains. The entire crystal of such an element can be considered a molecule, in a sense, because there is no variation in the energy of the bonds from atom to atom. No unit can be found consisting of only a few atoms bound differently from the way the unit is bound to other units.

METALLIC BONDING.—When metal atoms solidify, there are not enough valence electrons to share in the usual way, and in any case the low electronegativity of metals sheds electrons. The bonding is imagined as one in which nuclei, having lost their freedom of motion due to cooling, attract the clouds of electrons of adjacent nuclei with a force far stronger than that of electrical charges or of weak van der Waals forces. In fact, the forces of attraction are in the range that binds atoms into covalent molecules. They actually pull the inner shells of metal atoms together, squeezing the few valence electrons into the empty spaces among the atoms. These empty spaces exist not merely as physical holes in the structure but as areas where the liberated valence electrons can repose and expend least energy against the complex forces of attraction and repulsion around them.

When pressure is exerted on a crystal, or when it receives a blow, the lattice particles shear or the structure shatters. When a solid metal is hammered or bent, it gives. The atoms are not held rigidly by their electron distribution. Valence electrons are associated with all the adjacent nuclei and belong to no particular one. Thus, when the structure is bent, these relatively free electrons move from one vacant space to another, and the atoms are allowed to slide past one another into new positions. Along the inside of a bend, the atoms are squeezed out by pressure and slide toward the outside of the bend to take up the space that would otherwise be a crack or tear in the metal. When a metal is hammered, the outer atoms are simply pounded down among the others, the free valence electrons taking up new positions. Thus, a solid metal is also an ionic structure, but there are only positive ions with electrons dispersed among them without regard to any particular nucleus.

■**CONDUCTIVITY.**—The concept of metallic bonding explains electric current as a flow of electrons. If fresh electrons are pressed into one end of a metal wire from a source outside the wire (such as a battery), they will force the nearest valence electrons to shift position. They, in turn, will bump the next layers of free electrons, and the pressure will be passed along to the end of the wire, where the final layer of free electrons will be forced to leave the wire altogether. The speed at which the pressure is transmitted is the speed of an electric current, which is close to the speed of light; and it is conceived of as a wave. The actual movement of the electrons along the wire is a relatively slow affair. If the temperature of a metal is raised, its conductivity becomes less because the atoms are vibrating with greater energy and obstruct flow of the valence electrons.

Electrochemistry.—If a strip of copper is put in water, a few copper (Cu) atoms will be dislodged and will go into solution as cations, Cu^{++}, leaving their valence electrons in the strip and giving it a minute negative charge. The negative charge serves to attract the positive copper ions back to the strip to re-form atoms. However, initially the copper atoms are dislodged to form ions faster than the ions are attracted back to re-form atoms. Therefore, the number of excess electrons on the strip continues to increase and the negative charge slowly grows. When a sufficient number of excess electrons produces a large enough negative charge on the strip, the ions will be attracted back to form atoms at the same rate that atoms are discharged to form ions. Thus, a dynamic equilibrium will exist between the copper ions in solution and the copper atoms of the strip:

$$Cu^0 \rightleftharpoons Cu^{++} + 2e.$$

Any strip of metal will allow some of its atoms to dissolve in water, leaving the strip negatively charged. However, the affinity of each metal for its electrons varies, so that the number of excess electrons that build up on a strip varies with the metal. Zinc metal clings to its electrons with less affinity than does copper; therefore, a larger concentration of electrons will build up on a zinc strip before the equilibrium is established. In other words, the zinc strip will build up a higher electric charge than the copper strip.

If a zinc strip is put in water containing Cu^{++} ions, from such a salt as $Cu^{++}SO_4^{--}$, a copper ion hitting the zinc strip will take up two of the excess electrons left by the dissolving zinc ions because copper has greater affinity for electrons than has zinc. The resulting copper atom will stick to the zinc. Whenever a copper ion, Cu^{++}, removes electrons and becomes a neutral atom, Cu^0, another zinc atom, Zn^0, is freed to escape as an ion, Zn^{++}. Copper ions in solution are blue; zinc metal is grey. Gradually the zinc strip becomes coated with reddish copper while the blue of the solution fades. The reaction is a chemical one and can be written as an equation:

$$Zn^0 + Cu^{++} \rightarrow Zn^{++} + Cu^0.$$

According to the convention of oxidation-reduction reactions, when a metal is ionized, it has also been oxidized; when the ion is turned back into the neutral metal, it has been reduced. Thus, copper has been reduced and zinc oxidized in the above reaction, which has been called an electrochemical reaction.

■**ELECTROCHEMICAL CELLS.**—If a zinc strip and a copper strip are put into the same container of water, the charge on the zinc will become higher than that on the copper, for the oxidation potential of zinc is greater. If the strips are connected by a wire conductor, the difference in charges on them will produce a pressure of electrons from the zinc to the copper. If, at the same time, electrons can be removed or neutralized on the copper strip, electrons will flow from the zinc to the copper through the connecting wire. The neutralizing is done by adding an electrovalent compound, a compound whose ions will act as transporters of electrons to the water. Copper sulfate is suitable for the zinc-copper combination. The copper ions added to the solution by the copper sulfate will be far in excess of the concentration that the copper strip would produce in pure water. The copper ions will therefore deposit on the copper strip in order to bring about the proper dynamic equilibrium between copper atoms and copper ions. As they deposit, using up electrons on the strip, the zinc strip's larger concentration of electrons will flow along the wire toward the copper. This in turn allows more zinc ions to go into solution liberating more electrons. A current of electrons will flow from zinc to copper along the conductor as long as the oxidation-reduction reactions continue at both strips.

The various parts of such a system are put together in what is called a *voltaic cell,* after Alessandro Volta (1745–1827), who first constructed one. There are several variations of voltaic cells, one of the more common being the *Daniell cell,* invented by John F. Daniell in 1836. In the Daniell cell, a zinc electrode is immersed in a zinc sulfate solution and a copper electrode is immersed in a copper sulfate solution. The two solutions may be separated by a porous membrane, or a gravity-type cell may be used in which the less dense zinc sulfate floats on top of the more dense copper sulfate. The ions pass through the porous membrane or through the interface between the two solutions.

Various combinations of electrovalent or ionic compounds in water solution, called *electrolytes,* and of metal strips, called *electrodes,* have been designed for different purposes. Cells can be constructed in which the two solutions contain metal ions that are oxidized and reduced at electrodes made of nonreacting substances, such as carbon. In the familiar dry cell, pastes have been substituted for liquid solutions, and one electrode is nonreactive. Batteries consist of many electrochemical cells arranged in such a way that each one's electromotive force is fed to a single pair of electrodes. In a storage cell, the reaction proceeds until an equilibrium is reached when no more current can flow; by forcing a current from another source through the cell the reactions are reversed and the potentials in the cell are built up again: the cell has been recharged.

■**ELECTROMOTIVE SERIES.**—The chemical reaction that takes place at each electrode of a voltaic cell is half of the complete reaction that takes place in the cell as a whole. Thus, chemical reaction in each half-cell can be seen as a measure of the oxidation potential for the metal in contact with its own ion. When both half-cells have the same oxidation potential, no current can flow. The *electromotive force,* or voltage, is produced between half-cells of different oxidation potential. The oxidation potential for a half-cell cannot be measured directly, but one can be hooked up to a standard half-cell and comparative figures can be obtained. The standard chosen is hydrogen gas in contact with its ion in a molal solution, and this is given the value of 0. Metals, in contact with their ions in molal solution, can be arranged in order of decreasing oxidation potential. Such a list is called an *electromotive series.* Any metal will take electrons from any metal above it on the list and will replace the ion in solution of any metal below it on the list.

■**ELECTROLYTES.**—Electrolytes are ionic compounds. In metallic conductors, the electricity is conducted by electrons. In *electrolytic conductors* or *electrolytes,* the current is carried by ions rather than by electrons. Any compound that has one or more ionic bonds can function as an electrolyte. For example, Na^+Cl^-, $Mg^{++}Cl_2^-$, and $Al^{+++}Cl_3^-$, are all electrolytes. Compounds that contain radicals, such as $Na^+NO_3^-$ or $Ca^{++}SO_4^{--}$, are also electrolytes. The radical's atoms are covalently bonded, and the radical is extremely stable. However, it will accept or donate electrons to form ionic compounds and thus can electrolytically conduct electricity.

■**ELECTROLYSIS.**—If an electrolyte is melted, or fused, in a container and electrodes connected to a source of electrical current are placed into the container, then the cations and positive radicals will move toward the negatively charged electrode, or *cathode,* where they are reduced, while the anions and negatively charged radicals will move toward the positive electrode, or *anode,* where they are oxidized. At the electrodes, also called *poles,* each kind of ion and radical is relieved of its electrical charge. The ions are turned into neutral atoms, and the radicals are broken up into their component atoms. If the electrolyte is table salt, Na^+Cl^-, the Na^+ moves to the negative pole and the Cl^- to the positive pole. The reactions that take place are: $Na^+ + e^- \rightarrow Na^0$ and $Cl^- - e^- \rightarrow Cl^0$ or, more properly, $Cl^- \rightarrow Cl^0 + e^-$. The sodium atoms appear as pure, silvery, solid sodium metal at the cathode, while the chlorine atoms combine into diatomic molecules of greenish-yellow chlorine gas at the anode.

In taking up electrons at the cathode, where the source of the current piles them, sodium allows more electrons to move there. In giving up their electrons at the anode, chlorine ions replenish the source of the current. Thus, a melted electrolyte conducts electricity, though in a completely different way from that of a metal.

An electrolyte dissolved in water also conducts electricity. The radicals and ions move about freely in the water; and when electrodes are introduced, each radical or ion moves to the oppositely charged pole, where it accepts or gives up electrons. However, the neutral metal and nonmetal produced in the water may combine with the water, so that the end products of the electrolysis of a table salt and water solution are chlorine gas and hydrogen gas, never chlorine and sodium metal. A water solution of hydrochloric acid, HCl, on the other hand, results in the decomposition of the water into hydrogen gas and oxygen gas. The explanation makes use of the concept of oxidation-reduction and of relative electronegativities.

Electrolysis is the principle used in *electroplating.* The object to be plated is made the cathode in a solution containing ions of the metal with which the object is to be plated. A current is passed through the solution, the metal ions in the solution are reduced at the cathode, and they bond to the surface of the object.

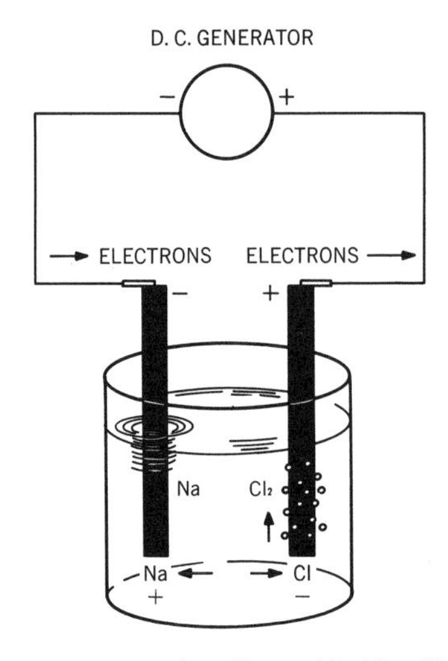

ELECTROLYSIS of sodium chloride, NaCl, liberates sodium ions at the negative electrode, or *cathode,* and chlorine gas at the positive electrode, or *anode.*

Solutions.—Most chemical reactions take place in solutions, which are vital to laboratory work as well as industry, partly because of their general use and partly because they facilitate measuring out accurate quantities of reagents. The world can be thought of as being made of pure elements, pure compounds, and mixtures of elements and compounds in any numbers and any proportions. In heterogeneous mixtures, at least one compound is present in the form of particles or droplets large enough to settle out or to be separated by filtration, centrifuging, or mechanical means. Heterogeneous mixtures, in other words, contain "lumps" large enough to be seen through a microscope. Homogeneous mixtures, on the other hand, are uniform throughout; that is, their components are individual molecules, atoms, or ions that cannot be separated by filtration. They are called *solutions.* There are solutions of a gas in a gas, gas in a liquid, and gas in a solid; of a liquid in a gas, liquid in a liquid, and liquid in a solid; and of a solid in a gas, solid in a liquid, and solid in a solid. The component present in the smaller amount is called the *solute;* the component present in the larger amount is called the *solvent.* A solvent may have many solutes dissolved in it.

Solutions are defined in several ways: *concentration* is the weight of solute in a certain volume of solvent; *percentage composition* is the weight of solute in relation to the total weight of solute plus solvent. A *molar solution* contains 1 mole (1 gram molecular weight) of solute in 1 liter of solvent. Thus, since calcium carbonate, $CaCO_3$, has a gram molecular weight of 100.06, a molar solution of $CaCO_3$ would contain 100.06 grams of $CaCO_3$ in 1 liter of water (50.03 grams in a half-liter, or 200.12 grams in 2 liters), and a 2.5 molar solution would contain 250.15 grams of $CaCO_3$ in 1 liter of water. A *molal solution* contains one mole of solute in 1,000 grams of solvent. Thus, a molal solution of $CaCO_3$ would contain 100.06 grams of $CaCO_3$ in 1,000 grams of water (or 50.03 grams in 500 grams of water, or 200.12 grams in 2,000 grams of water), and a 2.5 molal solution would contain 250.15 grams of $CaCO_3$ in 1,000 grams of water. A *normal solution* contains one gram equivalent weight of solute in 1 liter of solvent. Thus, a 1 normal solution of $CaCO_3$ contains $100.06 \div 2$ or 50.03 grams of $CaCO_3$ in 1 liter of water (25.015 grams in a half-liter, 100.06 grams in 2 liters), and a 2.5 normal solution contains 125.075 grams of $CaCO_3$ in 1 liter of water.

In the laboratory, to avoid weighing out very small amounts of some commonly used reagent many times a day, a solution of the reagent is made up in large quantities and is standardized; that is, its exact composition is determined by analysis and then computed into molarity, molality, or normality. Liquid-measuring apparatus, such as pipettes and burettes, are calibrated to 1/10 cc.; and thus, minute quantities of the desired solute for a reaction can be measured accurately and rapidly.

All gases and some liquids, such as alcohol and water, dissolve to an unlimited extent in each other; but usually there is a limit to the amount of solute that a solvent will absorb. The process of dissolving involves the weakening of the bonds between the solute molecules by the solvent molecules. If both are covalent and similar in other ways, they generally dissolve each other because the bonding is similar. Ionic compounds generally do not dissolve well in covalent solvents but will, often to an unlimited extent, in ionic solvents. Often a compound is formed by the solute and solvent molecules, and this compound then disperses through the solvent. When solids break up and dissolve in a liquid, they must take up some heat from their environment, cooling the solution, just as a solid melts only when it has acquired its heat of fusion. If the solute simultaneously forms a compound with the solvent, heat may be generated.

When a solvent has absorbed as much solute as the various forces of attraction and repulsion permit, the solution is called *saturated.* If a solid solute is being stirred into a liquid solvent, at the saturation point the molecules of the undissolved solid will be in dynamic equilibrium with the dissolved molecules. There will also be a constant exchange between the two states, but the actual number of dissolved molecules will not change. The concentration of a substance in a saturated solution is usually expressed as grams of solute per 100 grams of solvent, and is called the *solubility* of the substance. Solubility varies with temperature and generally increases with a rise in temperature. In the case of a gas dissolved in a liquid or solid, additional heat makes the gas molecules move about more actively and escape. Thus, heating decreases the solubility of gases in liquids. However, an increase in pressure usually increases the solubility of a gas.

With care, it is possible to dissolve more solute than a normally saturated solution will contain. The solution is then called *supersaturated,* and any disturbance of this condition will immediately precipitate out the amount of the solute in excess of the saturation limit.

The freezing point of any pure solvent is lowered by the addition of a solute, and its boiling point is raised, to an extent proportional to the concentration. When a mole of any covalent solute is added to 1,000 grams of water, the freezing point is always depressed 1.86° C. and the boiling point is raised 0.513° C. By measuring the freezing point or boiling point of an unknown solution, its molality can be calculated, and from that the molecular weight of the solute can be computed.

This mathematical change in boiling and freezing points can be explained in terms of the kinetic molecular theory, and relates to the actual number of molecules of solute present and to the vapor pressure they exert. Ions released into solution are more numerous than covalent molecules would be for a solution of the same molality. Ordinary salt fills its solvent with twice as many particles as sugar, which is a covalent compound. Thus, the depression of the freezing point and raising of the boiling point with electrovalent solutes is always greater than with covalent solutes. In addition, the change is not constant for all concentrations of electrovalent solutes because of the varying degrees of ionization that take place as the solution is diluted. Alcohol and glycol are familiar solutes added to water to lower its freezing point, especially in engine coolants. Salt would be better and cheaper, but it corrodes metal parts; it is used in ice.

Colloids.—Intermediate between true solutions and heterogeneous mixtures is *colloidal dispersion.* The *dispersion medium,* analogous to a solvent, contains the *disperse phase,* analogous to a solute. The particles of the disperse phase do not settle out, cannot be filtered out, and cannot be seen in a microscope; these are properties of a true solution. However, the particles scatter light in the *Tyndall effect,* show *Brownian movement,* have electrical charges on them without being ionic, and in other ways reveal that they are not molecules but agglomorations of molecules, hundreds or thousands of them gathered in submicroscopic kernels. The kernels are prevented from coagulating or growing because they have all acquired the same kind of electrical charge, so that each is repelled by the others, or because they have each acquired a covering of solvent molecules that act as buffers.

The size of a colloidal particle has been defined as between 1 mμ and 200 mμ in diameter, 1 mμ being one millionth of a millimeter. If the molecule is small, a great many can cling together to form a colloidal particle, but some organic molecules are themselves so large that a single one can be considered a colloidal particle. Since most reactions begin with a mixing of bulk matter, proceed through molecules, and end with bulk matter, the process must pass through the colloidal stage. The properties of this special state of matter influences just about everything in the universe. They especially affect life organisms, whose cells are put together out of colloidal dispersions.

In industry, colloidal systems are created by grinding down bulk matter, by tearing it apart in an electric arc, or by reactions that produce the desired compound in molecular state in a solution that then arrests its coagulation. Colloidal systems are stabilized by adding substances that will maintain the separation of the particles. For example, oil and water will not stay mixed, but the addition of soap *emulsifies* the mixture; that is, it stabilizes the colloidal dispersion of oil droplets in water.

Colloidal dispersions can be broken up by boiling, by centrifuging, by adding substances that strip the coating off the particles, or by neutralizing the charge on the particles with an electrolyte or in an electrical field. This last method, called *electrophoresis,* is the principle used in Cottrell precipitators. These precipitators are installed in factory chimneys to remove both poisonous and valuable wastes before they are spewed into the atmosphere. The smoke passes through baffles that are electrically, and oppositely, charged. Whatever their charges, particles will stick to the oppositely charged plate, become neutralized, and coagulate rapidly.

Water.—In a vast number of reactions, traces of water must be present to act as a catalyst. Water is the most universal solvent known, and its properties primarily derive from the structure of its molecule. The oxygen atom is covalently bonded to two hydrogen atoms, which are not directly opposite one another but at an angle of 105°. The positive charges of the hydrogen nuclei are concentrated at one side of the molecule, while on the opposite side, the oxygen atom, with its far greater number of electrons, creates a concentration of negative charges. Thus, the water molecule has one end more positively charged than the other negative end. It is a polar molecule and is called a *dipole.* The oxygen end of the dipole, being more negative, repels the oxygen end of all other water molecules but attracts the more positive, or hydrogen, end of all water molecules. This orientation gives water a lower freezing point, a higher boiling point, and a greater heat of fusion and heat of vaporization than similar compounds whose molecules are not polar. The polar nature of water enables it to dissolve other polar compounds more easily, each water molecule prying into the solid with the appropriate positive or negative end.

When hydrogen is covalently bonded to atoms that have high electronegativity, such as oxygen, nitrogen, and fluorine, the proton tends to link into the electron cloud of other molecules containing such atoms. Although the link is weak, it is substantial enough to measure and is called a *hydrogen bond.*

Acids, Bases, and Salts.—Three groups of compounds—acids, bases, and salts—have been known from antiquity. Most acids taste sour, turn a dye called litmus a red color, react with many metals to produce hydrogen gas, and combine with bases to form salts and water. Most bases taste bitter, have a soapy feel, turn litmus a blue color, and combine with acids to form salts and water. Salts are generally crystalline solids at ordinary temperatures. Thus, there are no naturally occurring bases and acids in the Earth's inorganic crust, but a large percentage of the crust consists of hundreds of different kinds of salts. It is from these salts that many acids and bases are commercially manufactured.

Acid-base reactions have been known from antiquity and are as important in chemistry as oxidation-reduction reactions, though there are relatively far fewer acids and bases than oxidizing agents. In the modern theory of acid-base reactions, water plays a key role. By far the most important property of water is the fact that its molecules react with one another to produce ions. The equation expressing this can be written as follows:

$$H_2O + H_2O \rightleftarrows H_3O^+ + OH^-$$

The cation is called the *hydronium ion,* and the anion is the *hydroxide ion.* If the OH^- is somehow removed or suppressed so that the concentration of H_3O^+ is in excess, then the solution will have a sour taste, turn litmus blue, react with many metals to produce hydrogen, and combine with bases to form salts. In other words, a solution with a greater concentration of H_3O^+ than of OH^- will have all the properties of an acid. If the H_3O^+ is suppressed or eliminated, and the concentration of OH^- is increased, the solution will have all the properties of a base. It will taste bitter, have a soapy feel, turn litmus blue, and combine with acids to produce salts. Thus, the ions into which pure water dissociates are the ions that give aqueous solutions of acids and bases their characteristic properties. In pure water, the concentration of H_3O^+ always equals the concentration of OH^- and the two ions cancel out each others' properties, so that pure water is neither sour nor bitter and does not affect litmus. However, it can react with either a base or an acid as an acid or a base.

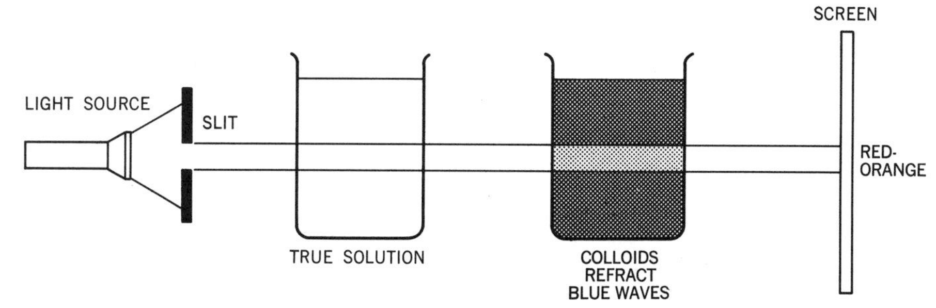

TYNDALL EFFECT can be used to distinguish colloidal suspensions from true solutions. A beam of white light passes through the true solution without change but is refracted when passing through the colloid. Here the particles selectively refract only the blue waves, and the red-orange waves pass through the colloid to illuminate the screen.

Several theories have been used to explain acid-base reactions. One is the *proton transfer theory* (announced independently by Johannes Nicolaus Brønsted and Thomas Martin Lowery in 1923), by which an acid is defined as a substance that has a tendency to lose a proton, and a base as a substance that has a tendency to gain a proton. A proton in any chemical reaction is always the nucleus of a hydrogen atom or is a hydrogen ion, H^+. The strength of an acid is a measure of its tendency to lose a proton, and the strength of a base is a measure of its tendency to take up a proton. It follows that when an acid gives up a proton, it becomes a unit that, in turn, may take up, or accept, a proton to reproduce the acid. It is then possible to write the general equation for an acid as

$$\underset{\text{Acid}}{A} \rightleftarrows \underset{\text{Proton}}{H^+} + \underset{\text{Base}}{B}.$$

An acid and a base related in this manner are said to form a *conjugate pair;* the *conjugate acid* is the proton donor and the *conjugate base* is the proton acceptor; the reaction is called protolysis.

To illustrate that an acid or a base may be either a molecule or an ion, these equations should be noted:

Acid		Proton		Base
HCl	$\rightleftarrows$	H^+	+	Cl^-
NH_4^+	$\rightleftarrows$	H^+	+	NH_3
H_2SO_4	$\rightleftarrows$	H^+	+	HSO_4^-
HSO_4^-	$\rightleftarrows$	H^+	+	SO_4^{--}

The proton transfer theory is also useful in describing the ionization of water. In the ionization of water, one water molecule acts as an acid and donates a hydrogen ion, or proton, H^+, to the other molecule that, acting as a base, attaches the proton to a pair of available valence electrons.

$$H_2\ddot{\underset{..}{O}}: + H_2\ddot{\underset{..}{O}}: \rightleftarrows (H_3\ddot{O}:)^+ + (H\ddot{\underset{..}{O}}:)^-$$

Since the reaction is reversible, the hydronium ion then acts as an acid and donates a proton to the hydroxide ion acting as a base. No theory explains why, in pure water, some molecules accept and others donate protons.

In pure water the concentration of each ion that is formed is always the same and is very small in quantity, which explains why pure water is such a poor conductor of electricity. Using the law of mass action formula,

$$\frac{(OH^-)\ (H_3O^+)}{(H_2O)} = K$$

where K is the equilibrium constant. Since the concentration of water molecules is relatively enormous, and therefore constant compared to the slight concentration of ions in it, the equation for the equilibrium constant can be written:

$$(OH^-)\ (H_3O^+) = K_w.$$

By experiment, K_w has been determined to be 1×10^{-14} at room temperature. The concentration of each ion in pure water is therefore 1×10^{-7}. If the concentration of H_3O^+ is changed, the concentration of OH^- must change proportionately in order to maintain the constant K_w.

A former theory of acid-base reactions, propounded by Svante Arrhenius in 1883, held that the hydrogen ion itself, the H^+, was the acidic particle and that the concentration of H^+ in a solution indicated its strength. A notation was devised in which H^+ was transformed by logarithms into simple numbers and called the *pH factor,* or value. This notation is still the only one in use. The *p*H of a neutral solution is 7; a *p*H of less than 7 indicates an acid solution; and a *p*H of more than 7 indicates a basic, or alkaline, solution. Blood, for example, has a *p*H value of about 7.2, which is slightly alkaline.

Instead of *p*H we should write *p*H_3O, but the change has not been seriously proposed. A strong acid is one that ionizes vigorously and completely even in small amounts of water, thus donating all its protons at a high rate. A weak acid ionizes only slightly in solutions of high concentration, and it must be diluted with large amounts of water before it will yield all its protons. The common strong acids are sulfuric, H_2SO_4; hydrochloric, HCl; and nitric, HNO_3. Among the weak acids are phosphoric, H_3PO_4; carbonic, H_2CO_3; and all the organic acids, such as acetic, $HC_2H_3O_2$, and oxalic, $C_2H_2O_4$. Strong bases produce large concentrations of hydroxide ions in water. They include the hydroxides of the alkali metals, such as sodium hydroxide, NaOH, and of the alkaline earth metals, such as calcium hydroxide, $Ca(OH)_2$. A weak base is ammonium hydroxide, NH_4OH, which must be diluted before it will ionize completely.

Most pure acids are covalently bonded compounds, and it is their aqueous solutions that are used in chemistry. In the mixture produced by the reaction

$$HCl + H_2O \rightleftarrows H_3O^+ + Cl^-,$$

it is the H_3O^+ that acts as the proton donor when a base is added. Most familiar bases already have a hydroxide ion that they yield in solution, for example, Na^+OH^-. Some substances yield protons to strong bases and accept protons from strong acids; these are called *amphoprotic compounds.* One of these is water.

All acids react with all bases to produce water and a salt. For example:

$$HCl + NaOH \rightleftarrows Na^+Cl^- + H_2O,$$
$$H_2SO_4 + Ca(OH)_2 \rightleftarrows Ca^{++}SO_4^{--} + 2H_2O,$$
$$HNO_3 + KOH \rightleftarrows K^+NO_3^- + H_2O.$$

Salts are always electrovalent compounds composed of those parts of acids and bases, whether radicals or ions, that are neither the hydronium nor the hydroxide ion. Thus NaCl, KNO_3, and $CaSO_4$ are salts.

Pure water is neutral; that is, when an acid and a base are allowed to react in exactly the right proportions, so that equal concentrations of H_3O^+ and of OH^- are produced, the solution will be neutral. In other words, it is neither acidic nor basic and has a *p*H of 7. The process of adding acid and base together is called *neutralization.* In the laboratory the technique used in performing neutralization reactions is called *titration.* An acid of known concentration, or molality, is measured slowly from a burette into a sample of base of unknown strength; when the mixture has become neutral, the known amount of acid that was used to neutralize the sample enables calculation of the amount of base present in the sample. Similarly, the strength of the basic solution is standardized by titrating it with an acid solution.

When the solution becomes neutral, the *end point* of the reaction has been reached. It is revealed by an *indicator,* which is a dye that has a different color in acid solutions from its hue in basic solutions. A few drops of the indicator, added to the solution being titrated, colors the solution according to its *p*H value. At the end point, the color changes quite abruptly. There are many indicators, each of which changes color at a specific *p*H value. Two commonly used indicators are litmus, which is red in acid and blue in base, and phenolthalein, colorless in acid and purple in base. Titration is often carried out with a potentiometer, an electrical apparatus that measures the conductivity of a solution and indicates neutrality faster and more precisely than visual indicators.

Salts of strong acids and strong bases (such as NaCl and KNO_3) ionize in water to form neutral solutions. However, the ions of salts of weak acids and strong bases (such as $NaC_2H_3O_2$ and K_2CO_3) will react with water to form alkaline solutions, while the ions of salts of strong acids and weak bases (such as NH_4Cl and $CaSO_4$) will react with water to form acidic solutions. Any process of reaction with water is called *hydrolysis.*

■BUFFER SOLUTIONS. — Both sodium chloride, NaCl, and ammonium carbonate, $(NH_4)_2CO_3$, have a *p*H of 7 in water solution. If a small quantity of dilute hydrochloric acid, HCl, is added to the NaCl solution, it becomes strongly acidic, lowering the *p*H to about 4. When the HCl is added to the $(NH_4)_2CO_3$ solution, its *p*H is hardly changed. Similarly, the addition of a small quantity of sodium hydroxide, NaOH, would raise the *p*H of the NaCl solution to about 10, while the *p*H of the $(NH_4)_2CO_3$ solution would hardly be changed. The ammonium carbonate solution resists a change of *p*H when an acid or a base is added; this is called a *buffer action.* A *buffer solution* is defined as a solution that is resistant to change of *p*H upon the addition of an acid or base. Buffer solutions usually consist of a weak acid and its salt (which is the weak acid's conjugate base), or of a weak base and its salt (which is the weak base's conjugate acid). The salt of a weak acid and a weak base, as shown above, also acts as a buffer. The fluids in a living organism maintain their constant *p*H value because they are buffered solutions.

—Louis Vaczek

ORGANIC CHEMISTRY

Historical Development.—The term "organic" was first associated with chemistry as a convenient classification for substances of plant or animal origin. Although organic substances (alcohols, oils, fats, sugars, and such) had been known for thousands of years, their chemistry received little attention until the eighteenth century. At this time substances or compounds of natural origin were divided arbitrarily into three classes: mineral, vegetable, and animal. It was quickly recognized that compounds obtained from vegetable and animal sources always contained carbon and hydrogen and were more closely related to each other than to compounds of mineral origin. The discovery that in many instances the *same* compound could be isolated from both vegetable and animal sources initiated a reclassification of substances into two groups: those substances produced by living organisms were defined as *organic* and those substances not produced by living organisms were classified as *inorganic.*

This first general definition unfortunately was understood to imply that organic compounds were produced under the influence of a *vital force.* Laboratory synthesis of organic compounds was considered in the realm of fantasy, although inorganic compounds routinely were prepared in the laboratory. This *vitalistic theory* was doomed by the discovery by a German chemist, Friedrich Wöhler (1800–1882), that an organic compound could be produced without the aid of a living organism. In 1828 Wöhler noted that evaporation of an aqueous solution of ammonium cyanate, an inorganic substance, resulted in the formation of urea, an organic compound previously found in animal urine. Stimulated by this observation, nineteenth-century chemists inaugurated a new era in synthetic chemistry by devising methods of forming simple organic compounds from their elements. Today, few organic compounds are considered beyond the scope of laboratory synthesis.

While the terms "inorganic chemistry" and "organic chemistry" apparently have lost their original meanings, the early method of classification remains. All so-called organic compounds contain carbon. This element commands a major branch of chemistry because its compounds far outnumber the known compounds of all of the other elements combined. Thus, *organic chemistry* often is defined as the chemistry of carbon compounds. While not completely accurate (the oxides of carbon, their metal salts, and a few miscellaneous carbon-containing compounds commonly are treated in textbooks of inorganic chemistry), this brief definition adequately distinguishes organic chemistry from other branches of chemistry.

Nature of the Covalent Bond.—The concept of how carbon is linked or bonded to itself and to other atoms is fundamental to understanding organic chemistry. The carbon nucleus contains six positively charged protons that are complemented by six negatively charged electrons distributed in two shells outside the nucleus. The inner shell contains two electrons and does not tend to accept or lose electrons. The outer shell (*valence shell*) contains the remaining four electrons, four fewer than the eight required to fill the second shell. In the periodic table carbon occupies a position between lithium (an electron donor) and fluorine (an electron acceptor). To achieve maximum stability, an atom prefers to take the easiest route to a full valence shell. In the case of lithium, the lone electron in the outer shell tends to be donated to a convenient electron acceptor such as fluorine, which possesses seven electrons in its valence shell. The carbon atom, with four electrons in the outer shell, could acquire a full valence shell either by losing all four of these electrons or by gaining four additional electrons to complete the partially filled shell. However, either process would be energetically unfeasible. This becomes evident when one envisions the state of an atom after one electron is lost. The atom is no longer electrically neutral, but possesses a positive charge. Loss of a second electron is now more difficult because of enhanced attraction by the nucleus. Loss of a third and fourth electron would be even more difficult. By similar reasoning, acquisition of four electrons is prohibited by repulsion of additional incoming electrons through the negative charge established by the first electron accepted.

While carbon does not tend to gain or lose electrons and form a charged ion, it may combine with other elements by "sharing" electrons. Carbon may share its four valence electrons (represented by crosses) with the single valence electrons (represented by dots) of four hydrogen atoms to form the organic compound methane. In effect the carbon atom has attained a full outer shell, and each hydrogen atom has attained a full valence shell by gaining an electron. This sharing of electrons results in a stable molecule. Since there has been no net gain or loss of electrons, methane is electrically neutral. Each electron pair linking hydrogen to carbon constitutes a *covalent bond.*

The ability of carbon to bond in a covalent manner accounts primarily for the distinguishing characteristics of organic compounds. Hydrogen commonly is considered an electron donor, as in hydrogen chloride. In the same sense chlorine is considered an

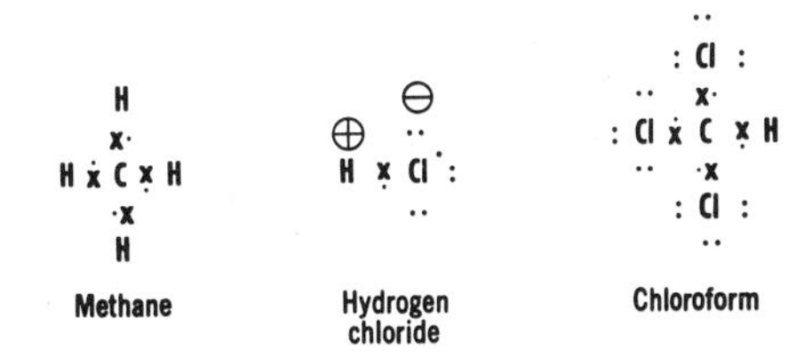

electron acceptor, but both hydrogen and chlorine form covalent bonds with carbon to form stable, electrically neutral compounds. In chloroform the outer shells of all atoms involved possess a complete valence shell without actual electron transfer or formation of ionic charge.

Structure of Organic Molecules.—The *structural formula* for methane, the simplest organic compound, usually is drawn as a planar projection in which the carbon atom is surrounded by four hydrogen atoms linked by solid lines, each line representing an electron pair or covalent bond. Further abbreviation is common in the literature of organic chemistry, and the "stick figure" is shortened to CH_4. Application of these conventions

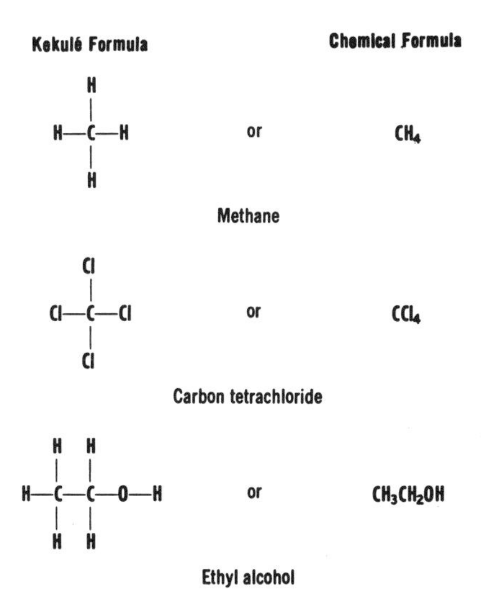

to more complex molecules may be illustrated by the structural formulas for carbon tetrachloride and ethyl alcohol.

Although structural formulas representing organic compounds routinely are depicted as planar, they must be interpreted as three-dimensional entities. Careful investigations

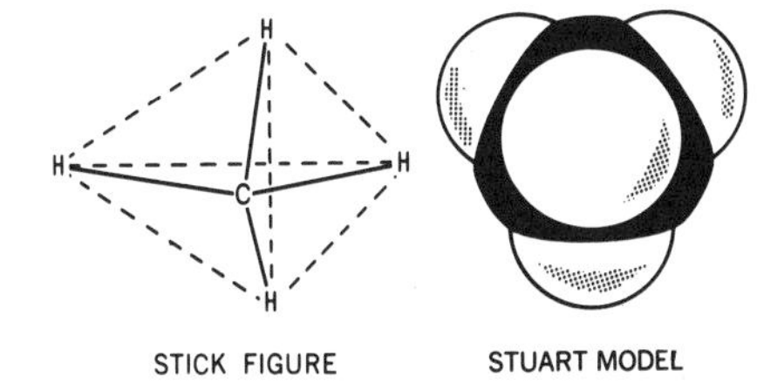

STICK FIGURE   STUART MODEL

have proved that the four covalent bonds of carbon project from the nucleus in such a manner that they are equidistant from each other. Thus, the methane colecule possesses a tetrahedral configuration; that is, the bonds are directed toward the four corners of a regular tetrahedron or pyramid. This three-dimensional concept of the arrangement of bonds about a carbon atom applies not only to methane, but also to the carbon atom in general when it is bonded to four distinct atoms or groups.

While methane may be unambiguously depicted by its molecular formula CH_4, more complex organic compounds require the use of structural formulas. The molecular for-

Organic Compounds

<table>
<tr><th>Name</th><th>Chemical Formula</th><th>Kekulé Formula</th><th>Structural Formula</th></tr>
<tr><td>Ethane . . . .</td><td>C_2H_6</td><td>H H
| |
H—C—C—H
| |
H H</td><td>CH_3CH_3</td></tr>
<tr><td>Propane . . . .</td><td>C_3H_8</td><td>H H H
| | |
H—C—C—C—H
| | |
H H H</td><td>$CH_3CH_2CH_3$</td></tr>
<tr><td>Butane . . . .</td><td>C_4H_{10}</td><td>H H H H
| | | |
H—C—C—C—C—H
| | | |
H H H H</td><td>$CH_3CH_2CH_2CH_3$</td></tr>
<tr><td>Isobutane . . . .</td><td>C_4H_{10}</td><td>H H H
| | |
H—C—C—C—H
| | |
H | H
H—C—H
|
H</td><td>CH_3CHCH_3
|
CH_3</td></tr>
</table>

mula indicates the number of each kind of atom present in the molecule, but does not describe the arrangement. As can be seen from the accompanying table, increasing the length of the carbon chain to four carbons leads to the molecular formula C_4H_{10}, which does not distinguish between two different compounds, butane and isobutane. In organic chemistry there are many cases where a given molecular formula represents two or more compounds that differ in physical and chemical properties. Such compounds, having the same molecular formula but differing in physical and chemical properties, are known as *isomers.* This phenomenon, known as *structural isomerism,* prevails because atoms are arranged in a fixed manner in an organic molecule; and the arrangement differs in each isomer.

Structural isomerism may be manifested when atoms other than carbon and hydrogen are involved. Substitution of a chlorine atom for one of the hydrogen atoms of butane would lead to two possible compounds: 1-chlorobutane and 2-chlorobutane.

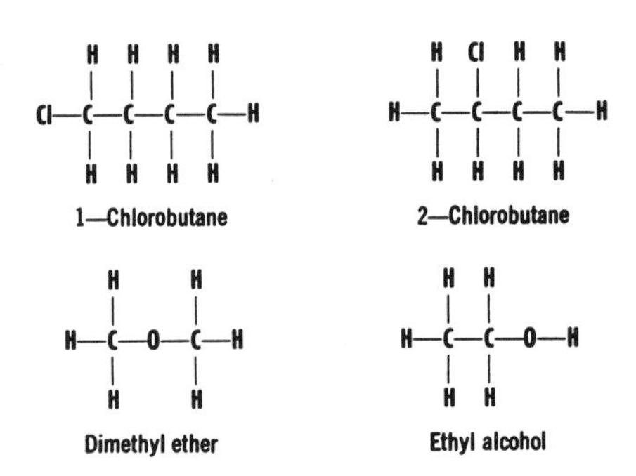

(In naming organic compounds, the carbon atoms are numbered consecutively from one end of the chain to the other: in 1-chlorobutane, the chlorine atom is attached to the first carbon; in 2-chlorobutane, to the second carbon atom.) Even though butane contains ten replaceable hydrogen atoms, inspection of the three-dimensional pictures of all ten possible chlorobutanes (C_4H_9Cl) will demonstrate that they are identical to either 1-chlorobutane or 2-chlorobutane. A third type of structural isomerism is evident with the two compounds of the molecular formula C_2H_6O.

An organic compound of two or more carbon atoms linked solely by carbon-carbon single bonds is described as *saturated.* However, carbon may form multiple bonds by sharing more than one pair of electrons with an adjacent carbon atom. Such compounds are said to be *unsaturated. Alkenes,* organic compounds containing one or more carbon-carbon double bonds, may be represented by their simplest member, ethylene. Both

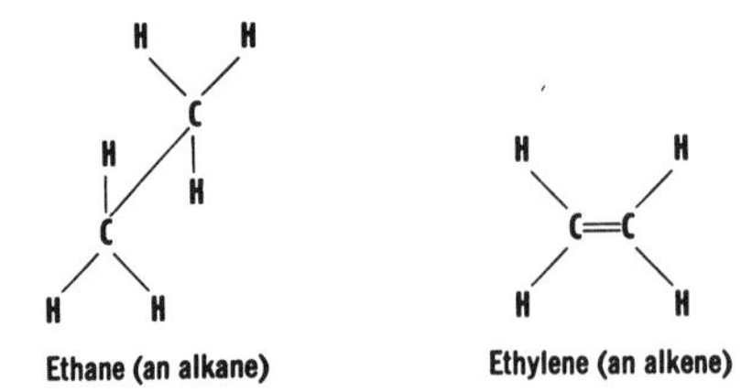

carbon atoms of ethylene are tetravalent, but are not in the tetrahedral configuration of ethylene's saturated counterpart, ethane. Each carbon shares two of its valence electrons with the other, and the result is a *covalent double bond* in which both carbons and the atoms bonded to them lie in one plane. The carbon-carbon double bond is more reactive than the single bond and will undergo *addition reactions* with such reagents as hydrogen, bromine, and hydrogen bromide.

Alkynes are organic compounds that contain one or more *covalent triple bonds.* In the simplest member, acetylene, both carbons share six valence electrons and the molecule is linear; that is, both carbons and the two groups bonded with them lie in a straight line. The triple bond is more reactive than the double bond and under appropriate conditions may undergo a *double addition reaction.*

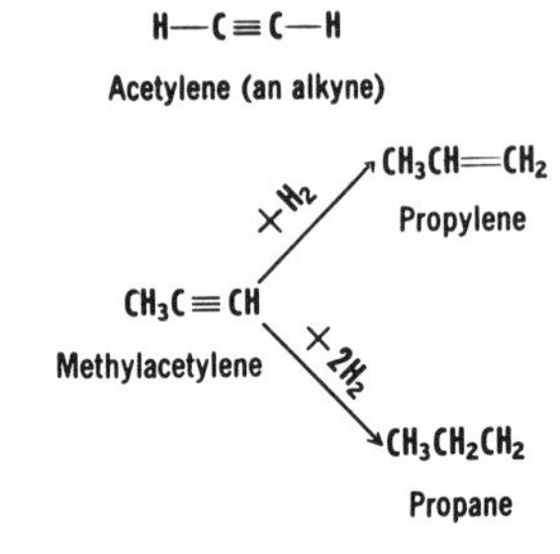

Many organic compounds contain carbon chains closed into rings. They are known collectively as *alicyclic compounds* and behave chemically as do their open chain derivatives. The simplest member of the series is cyclopropane, a gas used as a surgical anesthetic. Cyclobutene and cyclopentadene are but two of the possible variations on this theme that extends to *bicyclic, tricyclic, tetracyclic,* and *polycyclic* systems. A

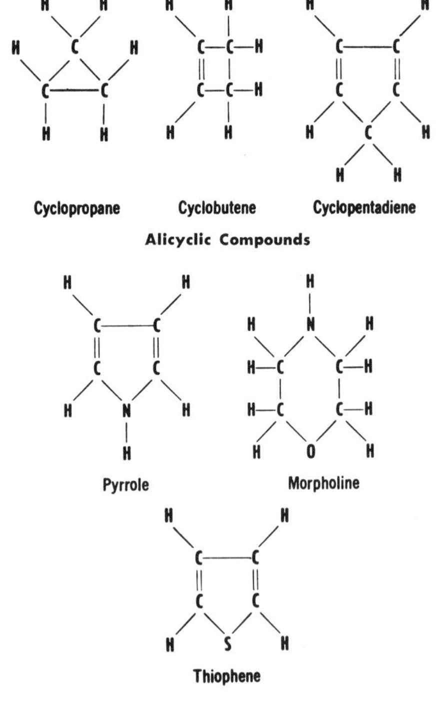

major branch of chemistry deals with *heterocyclic compounds:* that is, cyclic derivatives containing atoms other than carbon (pyrrole, morpholine, thiophene, and so on). Thus, the unique bonding nature of carbon provides a vast number of possible structures.

Aliphatic and Aromatic Compounds.—Compounds containing carbon and hydrogen fall into two broad categories:

aliphatic and aromatic. *Aliphatic compounds* were so named because the first members of this class to be studied were natural fats and fatty acids (the Greek word for fat is *aliphos*). *Aromatic compounds* derive their name from the pleasant odor common to many earlier-known representatives. In general, aromatic compounds are cyclic systems containing two or more double bonds. Benzene, a product of coal tar, is a familiar compound in the aromatic series; other examples, such as phenanthrene, quinoline, and furan, provide some insight into the variety of compounds classified as aromatic. For convenience, detailed representation of carbon and hydrogen atoms commonly is omitted from the structural formula; however, it should be kept in mind that carbon always is tetravalent, and undesignated valences represent carbon-hydrogen covalent bonds.

The modern basis for classification of an organic compound as aliphatic or aromatic involves the chemical properties, that is, the types of reaction, the molecule will undergo. Aromatic compounds, although structurally unsaturated, do not lend themselves to addition reactions as readily as do aliphatic alkenes. Under conditions that transform cyclohexene into cyclohexane by addition of hydrogen, benzene exhibits no tendency to be hydrogenated. When treated with chlorine or bromine, alkenes react by *addition*, whereas aromatic compounds react by *substitution* of a hydrogen atom. The tendency for an unsaturated cyclic compound to undergo substitution rather than addition reactions is a common criterion for assignment to the aromatic classification group.

Benzene

Phenanthrene

Quinoline

Furan

Aromatic Compounds

Functional Groups.—An organic compound may also be classified according to the type or types of functional groups present in the molecule. A *functional group* is a portion of a molecule (either one atom or a group of atoms) that passes through a given reaction unchanged, or that reacts independently of the rest of the molecule. In organic reactions, the functional group frequently is the "handle" by which one organic compound is converted to another, either by undergoing a chemical change itself or by its ability to confer special chemical properties on an adjacent carbon atom or atoms.

The carbon-carbon double bond and the carbon-carbon triple bond are examples of functional groups, since their presence confers unique properties on a molecule. Alkenes generally are denoted in the nomenclature of organic chemistry by replacing the suffix *-ane* of the saturated alkanes with the suffix *-ene*. Similarly, the alkynes utilize the suffix *-yne*. Examples are shown in the accompanying table (common names appear in parentheses). The position of the multiple bond is indicated by numbering the carbon chain with the functional group nearest the end of the chain. Alkanes containing more than five carbon atoms are termed hexane (C_6), heptane (C_7), octane (C_8), nonane (C_9), and decane (C_{10}). The corresponding alkenes and alkynes derive their nomenclature in an analogous manner: 3-hexene, 2-octyne, and so on.

Alkane	Alkene	Alkyne
CH_4 Methane	. . .	. . .
CH_3Ch_3 Ethane	CH_2-CH_2 Ethene (Ethylene)	HC-CH Ethyne (Acetylene)
$CH_3CH_2CH_3$ Propane	CH_2-$CHCH_3$ Propene (Propylene)	HC-CCH_3 Propyne (Methylacetylene)
$CH_3CH_2CH_2CH_3$ Butane	CH_2-$CHCH_2CH_3$ 1-Butene	CH_3C-CCH_3 2-Butyne
$CH_3CH_2CH_2CH_2CH_3$ Pentane	CH_3CH-$CHCH_2CH_3$ 2-Pentene	HC-CCH_2-CH_2CH_3 1-Pentyne

The majority of functional groups contain carbon and/or hydrogen in conjunction with other atoms, while others are composed solely of atoms other than carbon and hydrogen. The halogens (fluorine, chlorine, bromine, and iodine) commonly are present in organic compounds. These compounds are called either *alkyl halides* or *haloalkanes*.

When organic halogen compounds are treated with sodium or potassium hydroxide, they form the corresponding *alcohols* by replacement of the halogen with a *hydroxyl function* (-O-H). Alcohols generally are denoted by the suffix *-ol* or the prefix *hydroxy-*. *Ethers* are obtained by treating two molecules of an alcohol with an acid or by treating the sodium derivative of an alcohol with an alkyl halide. The ethers, characterized by one oxygen bonded to two carbon atoms, may be symmetrical or unsymmetrical, depending upon whether the oxygen-linked groups are the same or different.

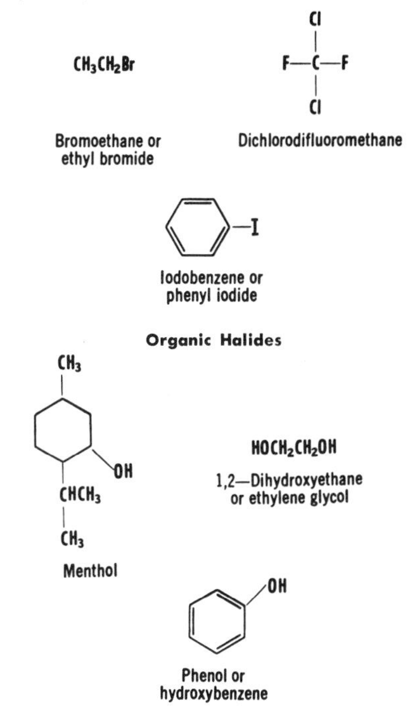

$$2CH_3CH_2OH \xrightarrow{\text{acid}} CH_3CH_2OCH_2CH_3 + H_2O$$

Diethyl ether

Sodium phenoxide + Benzyl bromide ⟶ Phenylbenzyl ether + NaBr

Nitrogen is a common component of functional groups. The *amines* represent an important class of organic compounds that may be regarded as derivatives of ammonia NH_3. A *primary amine* is simply ammonia with one of its hydrogens replaced by carbon. Substitution of carbons for two of the ammonia hydrogens results in a *secondary amine;* a *tertiary amine* is a nitrogen atom completely substituted by carbon. By stepwise oxidation, amines may be converted to *nitroso*-compounds and to *nitro-*

compounds. A well-known explosive, trinitrotoluene (TNT), contains three nitro groups that are bonded in a benzene ring.

In many respects phosphorus behaves like nitrogen in organic reactions. *Phosphines* possess many of the chemical properties of amines. The basic differences between nitrogen and phosphorus—phosphorus is a

CH_3NH_2 Methylamine (primary amine)

Piperidine (secondary amine)

Organic Nitrogen Compounds

Aminobenzene or aniline — Nitrosobenzene

$CH_3CH_2PH_2$ Ethylphosphine — CH_3PHCH_3 Dimethylphosphine

Triphenylphosphine

Organic Phosphorus Compounds

larger atom and forms stronger bonds with oxygen, carbon, and the halogens—have stimulated a great deal of research in *organophospharus chemistry* over the past decade.

Carbon is capable of forming multiple bonds with elements other than itself. A variety of functional groups embody this property, among them

$CH_3\overset{O}{\overset{\|}{C}}H$ Acetaldehyde — Vanillin

Aldehydes

$CH_3\overset{O}{\overset{\|}{C}}CH_3$ Dimethyl ketone or acetone — Camphor

Ketones

the *imine group* ($>C{=}N-$) and the *thiocarbonyl group* ($>C{=}S$), but most important to the organic chemist is the *carbonyl group* ($>C{=}O$). Many alcohols may be oxidized by removing two hydrogen atoms to yield a carbonyl compound. When the carbonyl derivative contains a $C-\overset{O}{\overset{\|}{C}}-H$ group, it is termed an *aldehyde*; when it contains a $C-\overset{O}{\overset{\|}{C}}-C$ group, it is known as a *ketone*. Specific examples of aldehydes are acetaldehyde and vanillin. The latter, a flavoring agent found in vanilla, contains three functional groups: an aldehyde, an ether, and an alcohol. Specific examples of ketones are acetone and camphor. Camphor, which imparts the characteristic odor to camphorated oil, is a bicyclic aliphatic ketone.

$H\overset{O}{\overset{\|}{C}}OH$ Formic acid — $CH_3\overset{O}{\overset{\|}{C}}OH$ Acetic acid

Benzoic acid — Nicotinic acid

$CH_3CH_2\overset{O}{\overset{\|}{C}}OH + NaOH \longrightarrow CH_3CH_2\overset{O}{\overset{\|}{C}}ONa + H_2O$

Propionic acid — Sodium propionate

Carboxylic Acids

$2\,CH_3COOH \xrightarrow{\text{heat}} (CH_3CO)_2O + H_2O$

Acetic anhydride

$CH_3\overset{O}{\overset{\|}{C}}OCH_2CH_3$ Ethyl acetate — $CH_3\overset{O}{\overset{\|}{C}}OCH_2CH_2\overset{CH_3}{\overset{|}{C}}HCH_3$ Isopentyl acetate

Glyceryl tristearate

Procaine or Novocaine

Carboxylic Acid Derivatives

The carbon-oxygen double bond is present in the carboxyl group ($-\overset{O}{\overset{\|}{C}}-O-H$) of *carboxylic acids*. Carboxylic acids, such as formic acid, acetic acid (vinegar), benzoic acid, and nicotinic acid (niacin, a vitamin of the B complex), are commonly found in nature. The term "acid" refers to the ability of the O—H bond to react with alkali to form salts. Salts of aliphatic carboxylic acids with long carbon chains are useful as soaps.

Carboxylic acid anhydrides are prepared by strong heating of carboxylic acids alone or in the presence of a dehydrating agent. As the name suggests, acetic anhydride is the product of two molecules of acetic acid less one molecule of water. *Carboxylic acid halides* are the product of replacement of the hydroxyl portion of the carboxyl function by halogen. These reactive compounds are useful in the preparation of other carboxylic acid derivatives, for they react rapidly with alcohols to yield *carboxylic esters* and with ammonia or amines to yield *carboxylic amides*. The odor of specific carboxylic esters may be detected in many common fruits. Animal fats are triple esters of glycerin (1,2,3-trihydroxypropane) with three long-chain (16 or 18 carbons) aliphatic carboxylic acids. Esters are frequently used in medicinal agents, such as aspirin and Novocain.

$CH_3C{\equiv}N$ Acetonitrile — Benzonitrile

$CH_3CH_2O\overset{O}{\overset{\|}{C}}CH_2C{\equiv}N$ Ethyl cyanoacetate

Nitriles

A *nitrile*, or *organic cyanide*, may be obtained upon dehydration of a carboxylic amide and is characterized by a carbon-nitrogen triple bond. Some examples are acetonitrile, benzonitrile, and ethyl cyanoacetate.

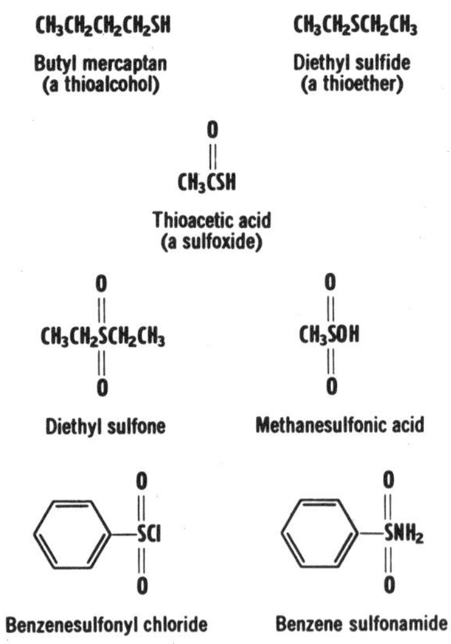

Organic sulfur compounds

Sulfur is chemically similar to oxygen, and most of the oxygen-bearing functional groups have sulfur-bearing counterparts. The nomenclature follows the same pattern, with the prefix *thio-* incorporated into the chemical name. *Thioalcohols,* or *mercaptans,* are responsible for the unpleasant odor of household gas. Other sulfur-containing functional groups include *thioethers, thioketones,* and *thioacids.* In addition, sulfur can combine with oxygen to form sulfoxides ($-\overset{O}{\overset{\|}{S}}-$), sulfones ($-\overset{O}{\overset{\|}{\underset{\underset{O}{\|}}{S}}}-$), *sulfinic acids* ($-\overset{O}{\overset{\|}{S}}-O-H$), and *sulfonic acids* ($-\overset{O}{\overset{\|}{\underset{\underset{O}{\|}}{S}}}-OH$). Sulfonic acids are related to carboxylic acids in that they may be converted to similar derivatives termed *sulfonyl halides, sulfonic esters,* and *sulfonamides.*

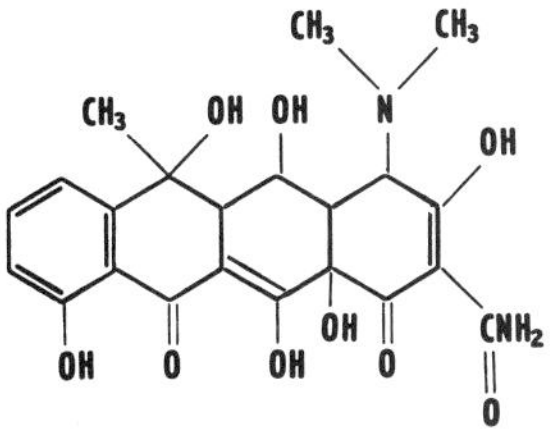

Terramycin

This summary is merely a random sampling of the more common functional groups present in organic molecules. A number of the specific examples demonstrate that more than one group may be present in a given molecule. However, even with the most complex multifunctional molecule, such as Terramycin, it is possible to manipulate a single group selectively without affecting the other groups.

Electron Displacement.—Within a molecule, when a covalent bond links two identical atoms, the two electrons forming the covalent bond may be considered as shared equally by both nuclei. The positive charge of each atomic nucleus is completely neutralized by the valence electrons (represented by dots); the centers of the electron orbits (represented by crosses) coincide with the nuclei (represented by circles). In a covalent bond between two dissimilar atoms, the bonding electrons, together with the other valence electrons, are displaced toward the atom with the greater *electronegativity* (electron affinity, or attraction for electrons). Chlorine is more electronegative than carbon; therefore, in a carbon-chlorine bond the electrical centers of the valence electrons do not coincide with the nuclei, but are *polarized* toward the chlorine atom. This polarization causes chlorine to assume a partial negative charge ($\delta-$) and carbon a partial positive charge ($\delta+$). The polarization of single bonds is termed the *inductive effect,* and it imparts a *polar character* (analogous to a magnet) to the molecule.

Valence electrons may be attracted or repelled by an approaching positive or negative charge. This mode of electron displacement is termed *polarizability* and differs from permanent polarization (inductive effect) in that electrons will resume their previous positions relative to the nucleus when the charge is withdrawn. The ability to displace valence electrons from their resting position is dependent upon the force by which they are attracted to the nucleus. The greater the nuclear attraction, the more difficult it is for the electrons to be displaced by an oncoming ion. As the distance between the valence electrons and the nucleus of a given atom becomes greater, nuclear attraction of the electrons become weaker. The polarizability of a particular bond determines its ability to be cleaved and thus enter into a chemical reaction, such as substitution of a hydroxyl group for a halogen in the conversion of an alkyl halide to an alcohol.

Multiple bonds are readily polarizable and on this basis are prone to undergo addition reactions. Treatment of an alkene, such as propene, with hydrogen bromide results in the formation of an alkyl bromide. The addition may be considered to proceed in a stepwise fashion, initiated by approach of a positively charged hydrogen ion. As the hydrogen ion nears the double bond, electrons are displaced toward the positive charge (curved arrow). Should the hydrogen ion migrate away from the propene molecule, the electrons would return to their original position. However, the polarizability of the double-bond electrons allows displacement to a degree whereby they enter the sphere of influence of the hydrogen nucleus to form a covalent carbon-hydrogen bond. At this stage the adjacent carbon bears a positive charge, but its electron deficiency is quickly relieved by attraction of a bromide ion that supplies two electrons to form a covalent carbon-bromine bond. Thus, the polarizability of a carbon-carbon double bond is instrumental in the progress of an organic reaction that is characterized by the breaking of a C—C bond and the formation of C—H and C—Br bonds.

While the carbon-carbon multiple bonds are theoretically polarizable in either direction, multiple bonds between two dissimilar atoms prefer electron displacement toward the more electronegative atom. The carbonyl group not only is polarized toward the oxygen atom by virtue of its inductive effect, but also readily undergoes a more exaggerated electron displacement when a negative ion is in the vicinity. For example, acetaldehyde may be converted to 2-hydroxypropanenitrile in the presence of hydrogen cyanide. The partial positive charge on carbon induced by oxygen attracts the negatively charged cyanide ion, whose approach repels the polarizable double-bond electrons. When the negative ion is within bonding distance of the carbonyl carbon, the free electron pair of the cyanide ion is capable of forming a covalent carbon-carbon bond. At this point, the electron pair that previously constituted a carbon-oxygen bond resides solely at the oxygen atom, confirming upon it a negative charge. A nearby hydrogen ion is capable of neutralizing the charge by utilizing the free electron pair to form a covalent hydrogen-oxygen bond. This type of addition reaction demonstrates the formation of a C—C bond at the expense of a C—O bond and represents one method of extending the carbon chain of an organic molecule.

Nature of Organic Reactions.—Organic reactions differ from inorganic reactions in a number of interesting respects. Whereas inorganic reactions frequently are instantaneous and quantitative, organic reactions tend to proceed at a measurable rate and rarely lead exclusively to a single product. These differences are reasonable when one considers that inorganic reactions, such as the neutralization of hydrogen chloride by sodium hydroxide, involve small, mobile ions that are attracted to each other by virtue of their opposite electric charges. On the other hand, organic compounds commonly are bulky and possess minimal attraction for each other. Reaction usually occurs as a result of random collision of molecules; however, the rate of collision may be increased by accelerating molecular motion through the application of heat. Unfortunately, sensitive organic compounds tend to decompose upon heating and are lost by conversion to degradation products instead of reacting by the desired route. Frequently the reactants may combine in more than one way or may react with the desired product to yield by-products. It is necessary to design experimental conditions—reactants, solvent, catalyst, temperature, time, and other factors—to achieve a selective conversion to the desired product, but the quantitative organic reaction is rare indeed. Degradation products and by-products not only account for the loss of starting materials but also complicate the isolation of the product. Although a variety of elegant purification techniques are available, separation of a useful product from contaminants usually involves a significant, but unavoidable, waste of material.

To be more specific, consider what takes place during an organic reaction: the process of bond cleavage and formation. The conversion of A—B to C—A involves breaking the covalent bond A—B and forming a new covalent bond, C—A. The *mechanism* of the reaction depends upon the manner in

$$C + A{-}B \rightarrow C{-}A + B$$

which these bonds are broken. Three courses are possible, and the natures of A, B, and C, in conjunction with the experimental conditions, determine which course is operant in a given reaction.

When each atom of A—B retains one electron of the pair, the reaction mechanism is termed *homolytic fission,* or a *free-radical reaction.* Free radicals are electrically neutral, extremely reactive particles that tend to react by addition to multiple bonds. In general, reactions of this type are induced by heat, light, or compounds that themselves generate free radicals, such as peroxides.

A–B cleavage may proceed with A capturing the electron pair. A *hetero-*

$A{:}B \rightarrow A\cdot + B\cdot$

$A\cdot + C \rightarrow AC\cdot$

Homolytic fission

$A{:}B \rightarrow A{:}^{\ominus} + B^{\oplus}$

$A{:}^{\ominus} + C \rightarrow A{:}C^{\ominus}$

$B^{\oplus} + {:}C \rightarrow B{:}C^{\oplus}$

Heterolytic fission

lytic fission, or *ionic reaction,* occurs and C is considered an *electrophilic* (electron-seeking) *reagent* that now shares the electron pair that A wrested from B. An electrophilic reagent is an electron-deficient atom or group that prefers to attack a molecule at the point of greatest negative charge. When A is a group in which carbon bears the free electron pair, the group is termed a *carbanion.*

The remaining generalized reaction mechanism involves heterolytic cleavage in which C is a *nucleophilic* (nucleus-seeking) *reagent* possessing an unshared electron pair available for formation of a covalent bond with B. A nucleophilic reagent prefers to attack a center of electron deficiency. A group, such as B, in which a carbon atom bears the positive charge is termed a *carbonium ion.*

For simplicity, the reaction mechanisms have been depicted as stepwise: first bond cleavage, then bond formation. Although this sequence commonly occurs, the two steps may blend into a concerted transition from A–B to C–A in a manner that requires less energy. When methyl iodide reacts with hydroxide ion, the resultant methyl alcohol has been shown to be formed by Mechanism I rather than Mechanism II. As a hydroxide ion approaches methyl iodide from "behind" carbon in a line with the C–I bond, the bond is polarized and stretched to the point where the C–O bond begins to form and the C–I bond begins to break. When both bonds are of equal strength, neither molecule, CH_3I or CH_3OH, exists as a separate entity. In this state, the system can either form CH_3OH or revert to CH_3I. When the three-dimensional concept of the carbon atom is envisioned, it becomes clear that this *concerted displacement* of the iodide ion has inverted the spatial relationship of the bonds about carbon, similar to turning an umbrella inside out.

The task of discussing the vast number of useful organic reactions becomes less overwhelming when one considers them merely as sequences of a few basic transformations, each usually involving a limited portion of a molecule. A particular sequence, or *reaction mechanism,* may be common to a large number of seemingly different reactions and the transformations involved, that is, the breaking and forming of bonds, by nature must be the same, regardless of the process involved. There are three major types of reaction mechanisms: substitution, addition, and elimination.

■SUBSTITUTION.—*Nucleophilic substitution* reactions in general involve displacement of a group (B) from an atom (A) by a reagent (C) that possesses an unshared pair of electrons. The attacking nucleophilic reagent may or may not bear a negative charge. Conversion of an alkyl halide to an alcohol affords an excellent example in which the reagent, hydroxide ion, bears a negative charge. Other examples of this important group of reactions include conversion of an alcohol to an alkyl halide and conversion of an alkyl halide to an ether.

Electrophilic substitution may be described as displacement of a group

Nucleophilic Substitution

$A{-}B + C^{\oplus} \rightarrow A{-}C + B^{\oplus}$

$CH_3O{-}C_6H_4{-}H + NO_2^{\oplus} \rightarrow$

$CH_3O{-}C_6H_4{-}NO_2 + H^{\oplus}$

Electrophilic substitution

$A{-}B + {:}C \rightarrow A{:}C + B$

$C_6H_5{-}CH_2{-}OH + H{:}Br \rightarrow$

Alcohol

$C_6H_5{-}CH_2{:}Br + H{-}OH$

Alkyl halide

$C_5H_9{-}Br + {}^{\ominus}{:}O{-}C_6H_4{-}Cl \rightarrow$

Alkyl halide

$C_5H_9{:}\,O{-}C_6H_4{-}Cl + Br^{\ominus}$

Ether

(B) from an atom (A) by a reagent (C) that is capable of accommodating two additional electrons. Electrophilic reagents commonly bear a positive charge. Aromatic compounds may undergo substitution by displacement of a hydrogen ion from a carbon ring upon attack of a positively charged group. In this manner, an aromatic system may acquire a variety of such substituents as nitro, halogen, sulfonic acid, and carbonyl. The synthesis of sulfanilamide, an antibacterial drug, illustrates the utility of electrophilic aromatic substitution. Acetanilide is treated with chlorosulfonic acid to yield sulfonyl chloride; conversion to the sulfonamide is effected with ammonia; and sulfanilamide is produced upon cleavage of the carboxylic amide group by hydroxide ion.

■ADDITION. — *Addition reactions* involve the attack of an electrophilic or nucleophilic reagent upon a multiple bond. Alkenes and alkynes may react with a wide variety of reagents, such as hydrogen, halogens, and inorganic acids, to yield saturated products. The mechanism of addition was demonstrated by the conversion of propylene to 2-bromopropane.

Carbonyl groups react by addition of nucleophilic reagents, such as alcohols, amines, mercaptans, and hydrogen cyanide. The addition of hydrogen cyanide to acetaldehyde, discussed earlier, proceeds via a mechanism common to this general type of addition reaction. Carbanion addition to carbonyl groups is an important method of forming a carbon-carbon bond. A carbanion may arise from a compound containing a particular type of functional group, such as nitrile or carbonyl, that permits cleavage of a neighboring C–H bond by a strong base, such as sodamide ($NaNH_2$). The highly reactive carbanion may attack a carbonyl group in another molecule to form an intermediate possessing a negatively charged oxygen atom. Two specific examples below demonstrate the manner in which this intermediate may react to yield products retaining most of the atoms present in the original compounds. Carbanion reactions make possible the construction of the carbon skeleton of a molecule, thereby laying the foundation for the synthesis of complex organic compounds.

■ELIMINATION.—*Elimination reactions* may be viewed as the reverse of addition reactions; that is, multiple bonds are formed by the loss of two groups from adjacent atoms. Specific examples include conversion of 1,2-dibromopropane to propene or propyne, and conversion of cyclohexanol to cyclohexene or cyclohexanone. Obviously, choice of reaction conditions is of prime importance to the course of elimination.

The three major reaction mechanisms do not include many types of reactions in which the carbon skeleton of a molecule may be built up, torn down, or even rearranged. While only a superficial treatment of organic reactions is practical here, it should be remembered that the fundamental processes of bond cleavage and formation are the same whatever course a particular reaction may follow.

Structure Determination. — Knowledge of the structural formula is essential in dealing with an organic compound. Intelligent research and application of the principles of organic chemistry demand that the chemist have an accurate picture of the molecule under study. The development of modern methods of measuring physical properties has facilitated an understanding of the structure of complex molecules.

In order to determine unambiguously the structure of an organic molecule, the compound must be obtained in a highly pure state. Such purification techniques as solvent extraction, distillation, crystallization, and adsorption are employed. Every organic compound possesses distinct physical properties (boiling point, melting point, refractive index, molecular rotation, and so on, and purification is assumed to be complete when these properties become constant upon repeated application of the above techniques.

The pure compound is first subjected to a variety of physical measurements that provide a great deal of information concerning the structure. Elemental analysis and molecular-weight determination define the molecular formula (the types of atoms present and the number of each). The infrared spectrum allows identification of certain functional groups. Information concerning the distribution of multiple bonds is available from the ultraviolet spectrum. Nuclear-magnetic resonance spectroscopy primarily presents a picture of the hydrogen atoms present in the molecule and their spatial relationship to each other. The mass spectrometer cleaves a molecule in relatively predictable fashion; and the number and weight of each fragment, coupled with information from other physical measurements, provide further insight into the structural formula. X-ray crystallography, which yields a diffraction pattern with crystalline solids, is capable of outlining an accurate three-dimensional picture of the molecule; but at present the calculations involved are too tedious for routine use of this technique.

While in some cases the measurement of physical properties is sufficient to define clearly the structure of an organic compound, the majority of structural problems require a study of chemical properties. The knowledge of characteristic reactions of functional groups allows interpretation of the behavior toward various reagents. Reaction of the molecule with acids, alkalis, oxidizing agents, and other substances often leads to identifiable fragments whose structure and possible mode of formation may further clarify the problem. Analysis of the bits of data concerning the physical and chemical properties makes it possible to assign a structure that distinguishes the compound in question from all other possibilities. Confirmation of the structural proposal usually is effected by total synthesis from compounds of known structure.

Complex Organic Molecules.—Due to the complexity of most organic substances isolated from living organisms, little progress was made in structure determination of natural products prior to the twentieth century. Multifunctional compounds of high molecular weight are present throughout plant and animal tissue. The most ubiquitous fall into two categories: carbohydrates and proteins. In general, these compounds are *polymers,* a series of simple molecules bonded to each other in repeating units.

Carbohydrates are distinguished by their high oxygen content. Simple carbohydrates, such as glucose, are termed *monosaccharides;* these may be linked together as ethers to form *polysaccharides.* Glucose, a source of energy for many plants and animals, is stored by animals as glycogen, a polysaccharide composed of 30,000 or more glucose units. Plants synthesize two important glucose polysaccharides: starch, their energy reserve, and cellulose, a supporting tissue.

Proteins, which carry out a multitude of functions important to living organisms, are large molecules composed of a sequence of *amino acids* linked by carboxylic amide groups. Approximately twenty amino acids,

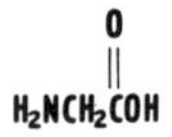

Glycine

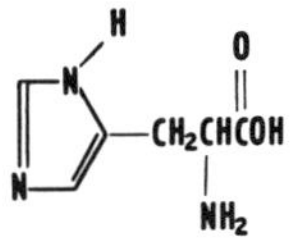

Histidine

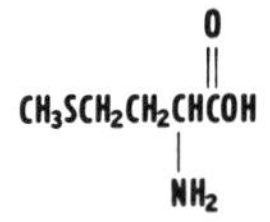

Methionine

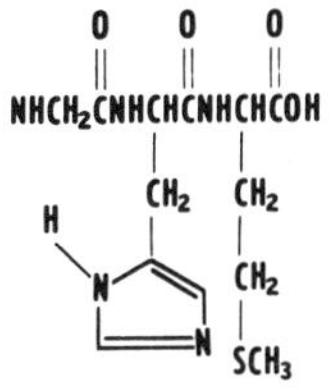

Glycylhistidylmethionine

such as glycine, histidine, and methionine, are common to proteins and may exist in an infinite number of sequences. The glycylhistidylmethionine residue represents a unit typical of those found in proteins. Many familiar substances are natural proteins: fibroin (silk), keratin (hair, skin, and nails), gelatin, insulin, albumin, and globin (the major component of hemoglobin, the oxygen carrier in blood).

Vitamins, while not structurally related to each other, are dietary factors essential to animal growth. The variation in structure is evident

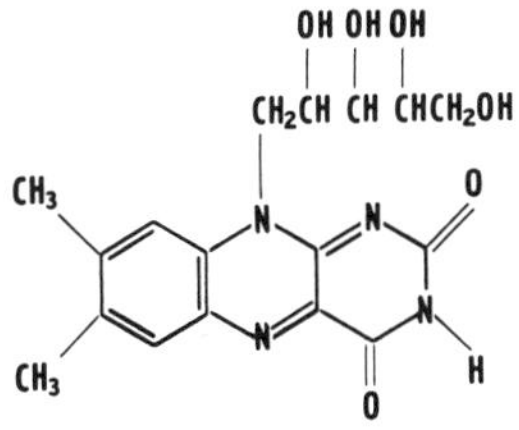

Vitamin B₂

by inspection of the structures of vitamins B_2 (riboflavin), D_2 (calciferol), and nicotinic acid (niacin).

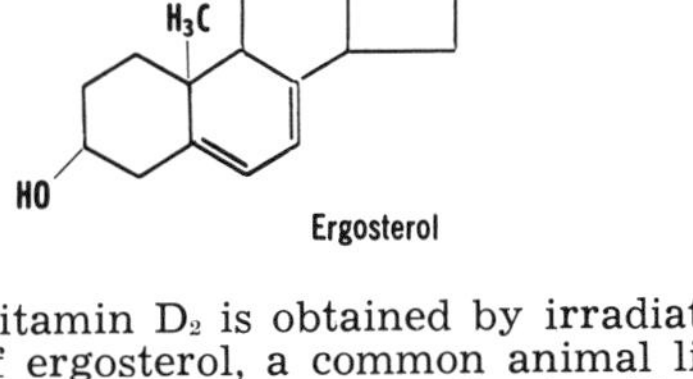

Ergosterol

Vitamin D_2 is obtained by irradiation of ergosterol, a common animal lipid containing the tetracyclic *steroid* ring system. Other naturally occurring steroids include cholesterol (a widely publicized lipid), testosterone (a male hormone), estradiol (a female hormone), and cortisone (a hormone of the adrenal cortex).

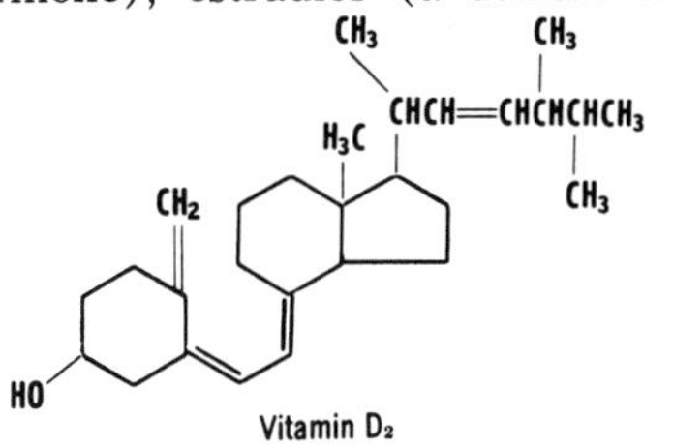

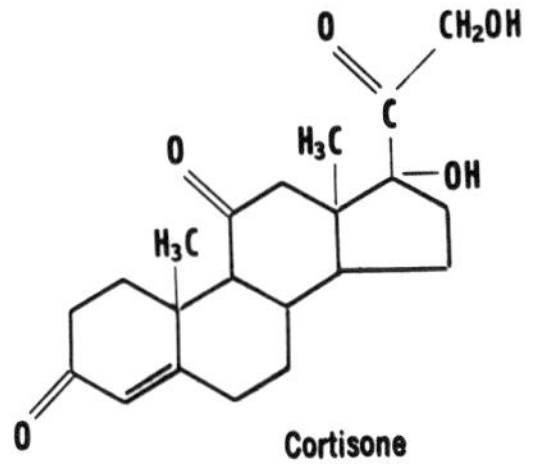

Vitamin D₂

Cortisone

Plants and lower animals are capable of synthesizing complex molecules that are useful to man. Familiar examples in the field of medicine include such alkaloids as morphine (a potent analgesic) and quinine (an antimalarial drug), as well as the antibiotics Terramycin and penicillin. *Macromolecules* (a general term for organic compounds of high molecular weight) produced by plants play an important role in our everyday life. Rubber is a polymer of isoprene ($CH_2=CCH_3-CH=CH_2$); cotton contains 97 to 99 per cent cellulose; and wool is a protein. In recent years organic chemists have been able to prepare synthetic polymers with tensile strength, elasticity, heat stability, and chemical stability that vastly exceed those of natural rubber. These macromolecules are synthesized by joining many small molecules, and contain thousands of repeating units.

Modern methods of polymerization have been responsible to a large degree for our high standard of living. Synthetic fibers and plastics have played no small part in the excellent quality and low cost of consumer goods. Such terms as polyamide, polyester, polyvinyl, and polyacrylate refer to polymers derived from simple organic molecules. A selection of examples includes nylon (a polyamide fiber), Dacron (a polyester fiber noted for its tensile strength and resiliency), Saran (a vinyl polymer useful as a fiber or plastic), Orlon (a fiber, obtained from polyacrylonitrile, known for its resistance to sunlight, weathering, and chemicals), and Lucite (a polyacrylate plastic, useful as an adhesive and protective coating).

Organic Synthesis.—The importance of organic synthesis becomes apparent when one considers that the function of protoplasm, the material that is the basis of life, depends upon the availability of organic substances. Synthesis of the building blocks of life constantly is under way. Although the organic chemist cannot compete with living tissue, he has come a long way since Friedrich Wöhler first discovered that an organic compound could be elaborated from inorganic matter outside a living cell. During the past century man has learned some of the basic principles involved in the construction of carbon-containing substances and has applied this knowledge to duplicate the products of plant and animal chemistry. Delight in his newly discovered ability has prompted him to devote a great deal of effort to the synthesis of natural products readily available to him. Many crowning achievements, such as the syntheses of glucose, morphine, penicillin, and chlorophyll, must be regarded only as academic in view of the fact that these compounds are obtained more economically from nature.

The value of these achievements, however, should not be underestimated. The mere fact that a chemist can produce a complex organic molecule from inorganic sources has demolished the psychological barriers to progress. Nature has provided many useful substances that the chemist has improved through synthesis. The physician has at his disposal drugs that calm agitated patients or induce sleep when natural sleep is impossible. Before the advent of synthetic tranquilizers and sedatives, these results were obtained only with opium, cocaine, or ethyl alcohol. While opium and alcohol still are popular in other areas, synthetic medicinals are used extensively in therapy. Natural fibers are being displaced by more versatile and durable synthetics. Such structural materials as wood and metals are being replaced by plastics.

Biological Chemistry.—The branch of science that deals with the chemistry of all forms of living organisms is termed *biological chemistry,* or *biochemistry.* This relatively recent discipline evolved after organic and physical chemists had developed theories and techniques applicable to biological problems. Only fifty years ago, biochemistry was considered an applied science concerned primarily with problems of the medical and agricultural worlds. Today biochemistry has taken a place among the pure or theoretical sciences, and its primary aim is to investigate the chemical transformations that occur within living cells. The modern term *molecular biology* has been applied to the analysis of the laws that control life on the molecular level.

The chemical constitution of living organisms differs tremendously from one species to another. Within the same organism such various tissues as blood, bone, and muscle are quite different in their chemical makeup. However, a pattern of similarity in the individual cells is evident from tissue to tissue or from species to species. All living cells contain water, inorganic salts, and a myriad of organic compounds. In general, the majority of organic matter falls roughly into three categories: carbohydrate, lipid, and protein. Carbohydrate compounds include monosaccharides and their higher-molecular-weight derivatives. The more common lipids are fatty acids and their esters (fats), long-chain aliphatic alcohols and their esters (waxes), and steroids. Proteins, which are amino-acid polymers, occupy a central position in the construction and function of living matter. They are intimately related to all phases of activity that constitute life in the cell. Collagen is the main structural protein of connective tissue. Enzymes, the substances that catalyze the chemical reactions of the cell, are primarily protein. Other proteinaceous materials function to transport oxygen (hemoglobin), regulate life processes (insulin and thyroid hormone), and protect the organism from infection (gamma globulin). In fact, the type of protein present in a cell is primarily responsible for the obvious anatomical and functional differences between various kinds of cells and tissues.

Living organisms depend upon energy to function. Animals require an intake of organic material; their diet may consist of carbohydrates, fats, and/or proteins, since all may be used more or less interchangeably for the production of energy. Mammals also depend upon their diet for vitamins, some of which are indispensable portions of certain key enzymes. Plants, on the other hand, obtain their energy from sunlight and require no organic nutrients. Their structural materials are synthesized from carbon dioxide, water, inorganic salts, and a source of nitrogen. Light-catalyzed synthesis of carbohydrates in plants (*photosynthesis*) is not only responsible for supplying the plant with its needs, but also indirectly furnishes animal nutritional requirements. The mere fact that plant and animal organisms utilize the same basic organic compounds denotes a great similarity in cellular chemistry.

Higher animals possess a digestive system that functions to degrade their macromolecular diet into small molecules capable of being assimilated by individual cells. Polysaccharides are converted to monosaccharides, and proteins are digested to their component amino acids. Monosaccharides are utilized for the production of energy, while the amino acids are polymerized within the individual cell to form its characteristic proteins. All living cells function in a state of continual degradation and repair. While the structure of the molecules involved may differ, the patterns of breakdown and synthesis are similar in all living organisms.

The most important monosaccharide in animal organisms is glucose. Glucose commonly is stored in the liver as glycogen and released into the blood stream as needed to nourish body tissues. Muscles also store glucose as glycogen to be utilized when energy is needed. When the muscle oxygen supply is adequate, glucose is oxidized completely to carbon dioxide and water; when it is inadequate, glucose is converted to lactic acid. The process of converting glucose (or its polymer, glycogen) to lactic acid is termed *glycolysis* and occurs *anaerobically* (without the utilization of oxygen). Glycolysis occurs in most animal tissues and may be represented by the general equation:

$$\underset{\text{Glucose}}{C_6H_{12}O_6} + 6O_2 + 38ADP + 38P_i \longrightarrow 6CO_2 + 6H_2O + 38ATP$$

The conversion of each glucose molecule to two lactic acid molecules is simultaneous with *phosphorylation* of adenosine diphosphate (ADP) by inorganic phosphate (P_i) to adenosine triphosphate (ATP). Glycolysis actually occurs by a series of nine reactions, each requiring a specific enzyme catalyst. The individual conversions are simple and may involve transfer of hydrogen or phosphate from one molecule to another, but the important feature is the concomitant formation of ATP from ADP. The energy released in the breakdown of glucose is stored as high-energy phosphorus-oxygen bonds in the two molecules of ATP produced. This "stored energy" is available for use by the cell as needed. Cleavage of ATP to ADP and inorganic phosphate allows energy to be released for use in muscle contraction, synthesis of complex molecules, conduction of nerve impulses, and a host of other energy-requiring activities necessary for maintaining life processes.

An important intermediate in the glycolytic pathway, pyruvic acid ($CH_3\text{-}\overset{\overset{\displaystyle O}{\|}}{C}\text{-}\overset{\overset{\displaystyle O}{\|}}{C}\text{-}OH$), may be *aerobically* (by means of oxygen) converted to carbon dioxide and water in a stepwise fashion. This biochemical pathway frequently is termed the *tricarboxylic acid cycle.* Each step in the sequence is catalyzed by a specific enzyme and, as with glycolysis, the energy released is captured by ATP for future use in the tissue. The aerobic phase of glucose metabolism yields a far greater quantity of energy than the anaerobic phase, since the complete oxidation of one molecule of glucose produces 38 molecules of ATP. This simplified picture of carbohydrate metabolism exemplifies the secrets of molecular biology that the biochemist can learn.

Much of the energy released as a result of metabolism is utilized by the organism in synthesizing the materials necessary for life and growth. The nutritive materials consumed by an animal often bear little resemblance to the substances required for normal function. Thus, small molecules are converted to carbohydrates, lipids, and proteins that are native to the particular organism. The uncanny ability of each type of cell to repair itself by reproducing a variety of specialized macromolecules has amazed scientists through the ages. Reproduction by cell division, replication of a complete cell with all its inherent complexities, represents a wonder of nature that was considered beyond the realm of man's intellect. Recently scientists have identified *nucleic acids* as the information-transmitting molecules responsible for the orderly reproduction of proteins, the major cellular constituents. Nucleic acids are macromolecules composed of subunits called *nucleotides.*

IVAN MASSAR, BLACK STAR

CHEMISTRY is concerned with transformation of matter. Within these pipes and distillation towers, chemical reactions are induced to produce high polymers, the basic materials of plastics such as acetate vinyl.

THE BASIC STRUCTURE OF MATTER and the chemical riddle of life itself are subjected to relentless scrutiny. The scientist (*below*), studying cancers produced by urethane in mice thyroids, seeks to learn why cells become malignant. Others search to find (*center, left to right*) what substances change drone larvae into queen wasps; how a simple penicillin mold can destroy bacteria; and what forces inform the crystal patterns of quinidine. Modern chemistry also continues an older search, expressing the amorphous mystery of nature in exact, definable, and communicable symbols; these are more precise than those inherited from the alchemists (*below*).

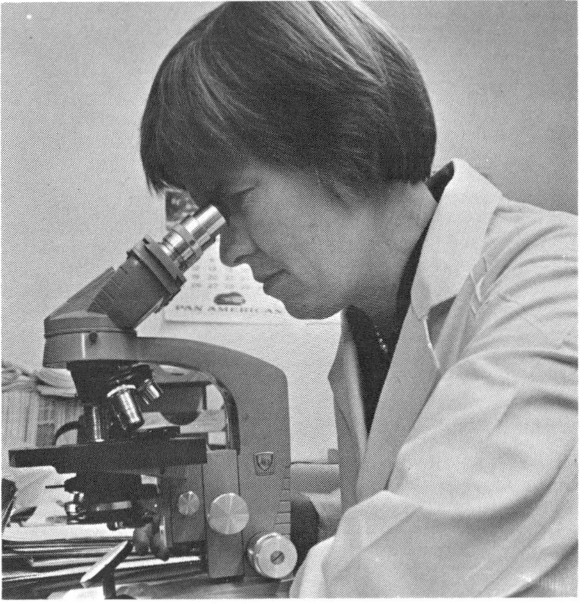

TOM KING, AMERICAN CANCER SOCIETY

U.S. DEPARTMENT OF AGRICULTURE

CHARLES PFIZER AND COMPANY

LOOK MAGAZINE

[illegible]	[illegible]	[illegible]	[illegible]	🜃	[illegible]	[illegible]	*SM	SM*	🜍	☿	♄	♀	☽	♂	[illegible]	▽	*▽	*▽
[illegible]	♃	♂	[illegible]	[illegible]	[illegible]	[illegible]	*[illegible]	[illegible]*	[illegible]	☉	☽	☿	♄	[illegible]	♂	▽	*▽	*▽
[illegible]	[illegible]	♀	[illegible]	[illegible]	[illegible]	[illegible]	*[illegible]	[illegible]*	♂	☽	♀	PC	♀	☽♄ ♀	☽♄ ♀	⊖	*[illegible]	*R
🜃	♀	♄	[illegible]	[illegible]	[illegible]	[illegible]	*[illegible]	[illegible]*	♀	♄								
SM	☽	☿	[illegible]	*[illegible]	[illegible]	*[illegible]	*[illegible]	[illegible]*	♄	♀								
	☿	☽	♂	*[illegible]	*[illegible]	*[illegible]			☽	[illegible]								
			♀	*🜍	*🜍	*🜍			[illegible]	[illegible]								

Nucleotides may be further subdivided into three units: a nitrogenous base (for example, adenine, a component of ATP), a monosaccharide, and a phosphate group. Thus nucleic acids are *polynucleotides* in which the phosphate residues act as bridges in the manner illustrated. All cells contain two types of nucleic acids—ribonucleic acid (RNA) and deoxyribonucleic acid (DNA)—that are differentiated on the basis of whether they contain ribose or deoxyribose as the monosaccharide portion. DNA functions as the storehouse for synthesizing information, while RNA plays a variety of roles in transmitting the information. Even the simplest cell contains thousands of proteins, all recorded in DNA and coordinated and controlled by RNA. DNA determines which protein will be manufactured; RNA picks up the blueprint of DNA's information, collects the individual amino acids needed, arranges them in the proper order, and releases the finished protein after the carboxylic amide linkages are formed. Much of the detail of this process has been uncovered, and research is in progress to determine the role of nucleic acids in memory and learning.

—James Muren

ANALYTICAL CHEMISTRY

Types of Analysis.—*Analytical chemistry* is the branch of chemistry that seeks to determine the composition of matter. It is divided into qualitative analysis and quantitative analysis. When performing a *qualitative analysis,* the chemist seeks to determine what elements or compounds are in a sample. When performing a *quantitative analysis,* he must determine the exact percentage or concentration of the various constituents in the sample. Obviously, a qualitative analysis must be performed before a quantitative analysis can be made. Analytical chemistry is divided into *organic analysis* and *inorganic analysis,* depending on the nature of the sample.

Analysis is also divided according to the size of the sample tested. *Macroanalysis* usually deals with samples weighing from 0.1 to 5 grams. *Microanalysis* deals with samples in the range of 0.001 to 0.01 gram. Microanalysis is not as accurate as macroanalysis, but it is useful in biochemistry, criminal investigation, analysis of paint in valuable paintings, and other situations where the size of the sample is necessarily limited. *Semimicroanalysis* refers to testing done with samples intermediate in size between those characteristic of microanalysis and macroanalysis; *ultramicroanalysis* extends microanalysis to samples smaller than 0.001 gram.

An analysis is said to be *ultimate* if the composition is reported in terms of the elements present; it is *proximate* when larger chemical units, such as molecules, are determined.

Analysis plays an important role in practically all manufacturing processes, controlling the quality of starting materials, intermediates, and finished products. It also is a major part of chemical and medical research.

Methods of Analysis.—The study of the elements has revealed certain characteristic reactions of elements and their compounds. Where other elements do not interfere (or mask these reactions), these reactions provide a shortcut to identifying constituents in a sample. However, because of interference, these characteristic reactions (for example, flame tests and borax bead tests) have limited use.

Analysis of a sample usually takes place in four steps: sampling, preparation for analysis, determination of constituents, and evaluation of results.

■**SAMPLING TECHNIQUES.**—*Sampling* can be as simple as pouring a liquid from a bottle. More often it involves obtaining a representative portion of the material to be studied. For example, how does one find the composition of a boatload of ore? Usually a portion (about a pound) is removed from each ton as it is unloaded, to produce a *gross sample.* This is still too large to work with, so it is crushed, blended, and divided into quarters. The process is repeated until a suitably small sample is obtained. Sampling a single lot of a solid material usually involves collecting a representative portion with a hollow auger that cuts through the various layers. Sampling liquids in several containers of different size is usually done by taking proportionate amounts from each and mixing the liquids.

■**PREPARATION FOR ANALYSIS.**—*Preparation for analysis* consists of putting the sample in aqueous solution. If the sample is water-soluble or already in solution, it is ready for analysis without further treatment. Samples of alloys and minerals insoluble in water can generally be dissolved in an appropriate acid. If no acid can dissolve the sample, the analyst must resort to *fusion,* heating the solid until it melts and then treating it with a suitable reagent (called a *flux*) to make it water-soluble.

■**DETERMINATION.**—*Determination* is the step that identifies and finds the amount of each constituent in a sample. It is usually accomplished by two groups of methods: those depending on a chemical reaction of the constituent and those based on some physical property of the constituent. Chemical methods include titrametric and kinetic. In *titrametric* methods, the analyst determines the amount of a reagent that will react with the sample. In *kinetic* methods, he determines the rate of reaction of the sample with a reagent.

Qualitative analysis of inorganic substances by chemical methods makes use of the fact that most inorganic substances are *ionic;* that is, they are composed of atoms or groups of atoms that have either lost or gained electrons. Atoms or groups that have lost electrons are positively charged and are called *cations;* those that have gained electrons are negatively charged and are called *anions.* There are about 30 common members of each group. Determination consists of treating the sample with a series of reagents to separate it into groups of similar constituents and then further treating these groups until the exact composition is known.

Quantitative analyses of inorganic materials are done by gravimetric or volumetric methods. In *gravimetric analysis* the element or compound is isolated in the form of a definite chemical compound and then weighed. In *volumetric analysis* the amount of reagent needed to react completely with the sample is determined; no isolation of constituents is needed. Volumetric methods are faster but less accurate than gravimetric methods. However, errors can usually be kept within allowable limits by careful calibration of equipment and by the use of controlled experimental conditions.

Most physical methods involve observing the behavior of the sample when it is exposed to some form of energy or to elementary particles of high energy. Classification of these methods is usually based on the type of energy involved.

Methods using some form of electromagnetic energy are called *optical methods* (a somewhat confusing term, since it includes use of short and long wavelengths in addition to those in the normal visual spectrum). Optical methods depend on the fact that under suitable conditions matter can be made to emit, absorb, reflect, or refract the electromagnetic radiation. Typical *emission methods* help identify elements by the characteristic radiation they give off when excited by an electric charge, ultraviolet light, or other intense radiant-energy source. *Spectrophotometry* and *colorimetry* measure the amount of energy absorbed by the sample. *Reflection methods* study the characteristics of energy reflected from the sample—*X-ray diffraction,* for example, studies the interference or reinforcement of reflections from the planes of a crystal. *Refraction methods* make use of the change in the speed of light as it passes through matter—the amount of change being a characteristic of the sample material.

The most widely used *electrical methods* are based on the reaction of materials in solution at electrode surfaces. Current, potential, and time required for a reaction are measured.

Magnetic methods study the behavior of the sample in a magnetic field. The mass spectrograph, for example, studies the movement of charged particles in a magnetic field.

Thermal methods determine the amount of heat generated or absorbed during a chemical reaction. Sometimes the amount of heat gained or lost is noted; other methods note the temperature at which a change occurs.

Analysis of organic samples may be done to determine the elements or compounds present in a sample. To determine the elements present, the sample is decomposed to form inorganic compounds of the constituents; these are then identified and measured by inorganic methods.

Determining compounds present is more difficult. If the sample has a constant boiling or melting point, it probably contains only one compound. The reason for this is that if a sample contains more than one compound, and these compounds have significantly different boiling or melting points, then the component with the lower boiling or melting point will be removed more quickly than the component with the higher boiling or melting point; the boiling or melting point of the sample, instead of being constant, will steadily increase. If only one compound is present in the sample, it can be identified by studying its physical properties and its reactions with various reagents. If the sample has more than one constituent, these must be separated by fractional distillation, extraction, or fractional crystallization. The individual compounds can then be identified.

EVALUATION OF RESULTS.—*Evaluations of results* are usually reported along with the possible amount of error that may be inherent in the method of analysis used. The error is reported in terms of the *relative error* and expressed in parts per thousand, since parts per hundred or percentage error might be confused with the other frequent references to percentage. If an analysis is performed by several chemists, each using different methods, the weighted average of their results is reported as the *most probable value.*

—John Price

PHYSICAL CHEMISTRY

Scope.—*Physical chemistry* is the branch of chemistry that studies the dependence of physical properties on the chemical composition of the different forms of matter and notes the changes these forms undergo. In describing these relationships, three states of matter are considered: gaseous, liquid, and solid.

Gases.—Matter in the gaseous state is distinguished by a tendency to occupy all the space available. This tendency can be counteracted by external forces; for example, the earth's atmosphere is confined within definite limits by the gravitational pull on the gaseous molecules. Gases are particularly susceptible to changes in pressure and temperature.

Robert Boyle in 1622 showed that the volume occupied by gas varies inversely with its pressure if its temperature is kept constant. In 1802 Joseph Louis Gay-Lussac and John Dalton discovered that at constant pressure gases expand when heated and that the change in volume is proportional to the rise in temperature.

Gay-Lussac later discovered that volumes of reacting gases and of their gaseous products can be expressed by simple numerical relationships. Interpretation of these simple volume ratios and of identical behavior of gases toward pressure and temperature is aided by Amadeo Avogadro's hypothesis, which states that equal volumes of different gases at the same temperature and pressure contain the same number of molecules.

The molecules of a gas move about in space at very high velocities. They collide with each other frequently and strike the walls of the containing vessel, creating the pressure exerted by the gas. If the volume of the gas is increased, the number of molecular impacts on a given area is decreased —hence, the pressure drops. The pressure of a gas does not decrease with time, showing that the collisions between molecules are perfectly elastic and that there is no loss of velocity as a result of such collisions.

Liquids.—If the temperature of a gas is brought below a certain value and the pressure is gradually increased, the gas condenses into a *liquid.* Above a *critical temperature,* a gas will not condense into a liquid no matter how much pressure is applied. Substances in the liquid state have greater densities, greater internal friction, greater cohesive pressures, and much smaller compressibilities than they have in the gaseous state. Many of the changes are due to the increase in attractive forces acting between the molecules, which are more closely packed in the liquid state than they are in the gaseous state.

The *vapor pressure* of a pure liquid is the pressure at which the liquid is in equilibrium with its vapor phase. This equilibrium pressure depends only on temperature, not on the relative amounts of liquid and vapor present. The *boiling point* of a liquid is defined as the temperature at which the vapor pressure equals the external pressure.

The *surface tension* of a liquid accounts for the fact that drops of a liquid generally are spherical. A molecule in the bulk of a liquid is usually attracted equally in all directions by surrounding molecules in an area of equal density. But at or near the surface of a liquid, molecules are more strongly attracted by the molecules in the liquid phase than they are by those in the gaseous phase, thus creating an inward force at the interface. This inward force pulls the molecule into a spherical shape.

Different liquids have varying optical properties as a result of differences in their chemical composition. *Refraction,* the change in direction of a homogeneous ray of light when it passes from a medium of one density to a medium of different density, is used to help determine the structure of liquids; different liquids bend a ray of light by different amounts. *Dispersivity,* the difference in refractivity for light of differing wavelengths, also lends itself to the study of structural differences in liquids.

When a beam of polarized light is passed through certain liquids, the plane of polarization is rotated to a new direction. Noting the direction of rotation and the amount of angular displacement helps identify unknown substances. This is important in examining complex organic liquids.

Solids.—When liquids are cooled in such a way as to form regular polyhedral crystals, they are said to have passed into the solid state. For some time it was customary to speak of *amorphous* (uncrystallized) and *crystalline* solids, but analysis has shown that many seemingly amorphous substances are definitely *microcrystalline,* while others are *supercooled liquids* with very high viscosity. The solid state now is generally identified with the crystalline condition; matter in this form resists shearing stresses and has properties that differ in different directions—unlike gases and liquids.

Normal formation of crystals depends on time. Rapid separation of a solid tends to produce granular aggregates whose crystalline structure is difficult to recognize. There is an overlap between the liquid and solid states as far as rigidity is concerned. Highly viscous fluids (such as glass) pass into a fluid condition gradually and continuously when the temperature is raised—unlike crystals, which change properties suddenly at the melting point. On the other hand, most crystals are rigid and fracture when subjected to pressure; but some have such weak forces in the crystal structure that they can be easily distorted and made to flow. Hence, they simulate some of the properties of liquids, yet are solids because they have the optical properties of crystals and possess a definite melting point.

The properties and forms of crystalline solids depend on the arrangement of the atoms or molecules within the crystal. These are no longer free to move from point to point, but occupy a definite mean position in the lattice of the crystal. However, the atoms or molecules do vibrate about this mean position; and if the temperature is raised enough, they will break free of the force holding them in place and move freely, marking the transition to the liquid state.

The onetime supposition that every substance has a unique crystalline form has been disproved by the discovery of *isomorphism* (the tendency of different substances to crystallize in the same form) and *polymorphism* (the existence of the same chemical substance in more than one crystalline form). Sulfur, for example, is polymorphous—it has two interconvertible crystalline forms. Below 96° C. the rhombic form predominates; above 96° C. the stable form is monoclinic sulfur. The temperature at which both forms may exist is known as the *transition temperature.*

—John Price

NUCLEAR CHEMISTRY

Scope.—*Nuclear chemistry* is the branch of chemistry dealing with the effects of radiation and the processes that produce it—fusion, fission, and radioactive decay. Among the important applications of radiation are the tracing of chemical and biological reactions, radiography (X-ray photography), the study of catalysis in chemical reactions, medical therapy, and the supplying of power for orbiting satellites.

FISSION REACTIONS.—The raw material for *fission reactions* is uranium, which has several isotopes. The only significant naturally occurring isotope is that of mass number 235, abbreviated U-235. When an atom of U-235 absorbs a neutron, its nucleus becomes unstable and splits into two *fission fragments,* releasing radiant energy and a number of neutrons. This process is the basis for atomic bombs, nuclear reactors, and the peaceful uses of atomic energy planned under *Project Plowshare:* excavating harbors and canals, creating underground heat supplies for removing oil from shale, and mining.

■FUSION REACTIONS.—The *fusion reaction* involves the combination of atoms to produce heavier atoms, such as the combination of two hydrogen atoms to form helium. As is the case with fission, fusion results in a loss of mass, which is converted to energy. Extremely high temperatures are required for fusion; so far, only the explosion of a fission bomb has produced the needed temperature on a large scale. Research is now going on to perfect a controlled fusion reaction. One big problem is to find a vessel to contain the reaction. The heat of a fusion reaction is so intense that it would melt any solid material; therefore, a strong magnetic field called a *magnetic bottle* is used to contain the reaction. Another problem is to keep out impurities that inhibit the reaction; the use of a high vacuum is providing the answer. The third major problem is to obtain the high temperature and particle density required for the reaction; the solution appears to be the introduction of large amounts of energy to produce a current of ionized gas and use of a magnetic field to compress the gas.

When controlled fusion reactions are perfected on a large scale, they will be able to provide energy for the world for many centuries. So far, however, fusion reactions have been accomplished only on a small scale in laboratories.

■CHAIN REACTIONS.—Large-scale uses of fission depend on the *chain reaction.* In a chain reaction, the splitting of one nucleus by a neutron liberates other neutrons that can trigger further splitting of other fissionable nuclei. Elements that will produce this effect are known as *fertile materials;* the most important ones are U-235, U-238, and thorium-232. Fertile materials produce fissionable isotopes by neutron irradiation and capture.

The first successful chain reaction was carried out at the University of Chicago on December 2, 1942, under the direction of Enrico Fermi. Fermi used an *atomic pile* made up of uranium cubes (as the fissionable material) embedded in graphite, which would absorb neutrons and thereby control the reaction.

Nuclear Reactors.—In recent years, commercial applications of atomic energy have been growing. The nuclear industry has perfected a cycle of operation consisting of mining uranium and thorium ores, chemical processing, fabrication into fuels and other components of nuclear reactors, reprocessing of spent fuel, and disposal of by-product waste. Nuclear reactors are producing electricity that promises, within a few years, to become competitive in cost with conventional power sources in many areas, and they have proved successful in providing long-lasting power supplies for isolated military installations and ships. The possibility of using nuclear power to provide heat for industrial chemical processes is also being investigated.

Nuclear reactors are usually classified as burners, breeders, or converters, depending on the isotope used as fuel. *Burners* use separated U-235 fuel and find application as research, test, portable, and mobile reactors. *Converters* use natural or partially enriched uranium fuel, producing plutonium as a final product. The reactors at Hanford, Washington, and Savannah River, South Carolina, are converters. *Breeders* are a special type of converter in which more fissionable material is produced than is actually used.

Reactors are also classified by the physical state of the fuel: *heterogeneous,* in which fuel is formed into long rods or thin plates, or *homogeneous,* in which fuel is dispersed in an aqueous, molten-salt, or molten-metal solution or *slurry;* by the neutron energy in the fission-capture process: *fast, intermediate,* or *thermal;* and by the coolant used: *gas, pressurized water, boiling water, organic metal,* or *liquid metal.*

Uses of Radioactive Isotopes.—Medical science has made widespread use of radioactive isotopes of various elements in diagnosis and treatment. Some 40 isotopes have been used in diagnosis, relying on the fact that the radioactive isotopes of most elements behave similarly to isotopes that are not radioactive. Radioactive isotopes can easily be traced by a counting device, such as a Geiger counter. Radioactive iodine, for example, is used in studying thyroid conditions and also can be an aid in operations for removal of diseased thyroid glands—before the operation the patient is given a dose of radioactive iodine that settles in the gland; when no excess radiation is left in the area, the surgeon knows the entire organ has been removed.

Other radioisotopes have been used in studying various body functions of humans and animals—as in testing metabolism and in determining the importance of trace elements in the diet. They can also be used to detect various malfunctions and to treat malignancies.

Radioisotope tracers have helped to determine residues of detergents left on food products, to estimate the lean-meat content of animals, and to trace the movement of water. This last process makes use of the fact that the hydrogen in some atoms of rainwater is converted to heavy hydrogen by cosmic-ray bombardment in the atmosphere and by nuclear weapons tests. After twelve years, half of the radioactive hydrogen in the rainwater decays, providing a gauge for measuring the age of underground water.

In addition, the tracing process has helped chemists discover the mechanisms of certain chemical reactions. In some organic reactions, for example, it is difficult to determine which reagent has supplied which atoms in the end product. By selectively "labeling" the reagents with radioisotopes, the chemist can determine the donor of a particular atom or group of atoms by checking the final product for radioactivity. Radiation also has been used to speed up chemical reactions, such as the polymerization of certain plastics, and to vulcanize rubber. Radioactive isotopes are also being investigated as a possible means of preserving foods for long periods of time without refrigeration; in the process, called *irradiation,* the bacteria that cause decay of the food are killed by exposure to radioactivity.

—John Price

BIBLIOGRAPHY

ANDREWS, DONALD HATCH, and RICHARD J. KOKES. *Fundamental Chemistry.* 2nd ed. John Wiley & Sons, Inc., 1965.

AYRES, GILBERT H. *Quantitative Chemical Analysis.* 2nd ed. Harper and Row, 1968.

CLEMENTS, RICHARD. *Modern Chemical Discoveries.* E. P. Dutton & Co., Inc., 1963.

COTTON, F. ALBERT, and LAWRENCE D. LYNCH, JR. *Chemistry: An Investigative Approach.* Houghton Mifflin Company, 1968.

DUFFY, GEORGE H. *Physical Chemistry.* McGraw-Hill, Inc., 1962.

FARBER, EDUARD. *The Evolution of Chemistry.* 2nd ed. The Ronald Press Co., 1968.

FIESER, LOUIS F., and MARY A. FIESER. *Introduction to Organic Chemistry.* D. C. Heath & Co., 1957.

FRIEDLANDER, GERHART and J. W. KENNEDY. *Nuclear and Radiochemistry.* 2nd ed. John Wiley & Sons, Inc., 1964.

GERO, ALEXANDER. *Textbook of Organic Chemistry.* John Wiley & Sons, Inc., 1963.

GLASSTONE, SAMUEL, and D. LEWIS. *Elements of Physical Chemistry.* 2nd ed. D. Van Nostrand Co., Inc., 1960.

HAMPEL, CLIFFORD A., ed. *The Encyclopedia of the Chemical Elements.* Reinhold Publishing Corp., 1968.

MAHLER, HENRY R., and EUGENE H. CORDES. *Biological Chemistry.* Harper and Row, 1968.

MARTIN, ROBERT BRUCE. *Introduction to Biophysical Chemistry.* McGraw-Hill, Inc., 1964.

MCCUE, JOHN JOSEPH GERALD. *An Introduction to Physical Science: The World of Atoms.* 2nd ed. The Ronald Press Co., 1963.

NECHAMKIN, HOWARD. *The Chemistry of the Elements.* McGraw-Hill, Inc., 1968.

NITZ, OTTO WILLIAM JULIUS. *Introductory Chemistry.* D. Van Nostrand Co., Inc., 1961.

REICHEN, CHARLES ALBERT. *History of Chemistry.* Hawthorn Books, Inc., 1963.

TAYLOR, FRANK SHERWOOD. *The Alchemists, Founders of Modern Chemistry,* Abelard-Schuman Ltd., 1949.

TABLE OF ELEMENTS

At. wt., atomic weight (C = 12); At. no., atomic number; Sp. gr., specific gravity; M.P., melting point °C.; B.P., boiling point °C.; C.S., crystal structure: C = cubic, whether body- or face-centered unknown; Cb = cubic, body-centered; Cf = cubic, face-centered; H = hexagonal; T = tetragonal. All temperatures in the text are in degrees Centigrade (°C.).

Element	Occurrence, Preparation, Date of Discovery, and Discoverer	Properties	Chief Compounds and Uses
Actinium (Ac) At. wt. 227 At. no. 89 Valence 3 Sp. gr. 10.1 M.P. 1,050° B.P. 3,200°	In uranium ores from radioactive decay of uranium • Bombardment of bismuth. • 1899; Debierne.	Radioactive; half-life ranges from 20 years to 3.7 seconds. It is chemically similar to lanthanum.	Used in research.
Aluminum (Al) At. wt. 26.9815 At. no. 13 Valence 3 Sp. gr. 2.7 M.P. 660.2° B.P. 2,647° C.S. Cf	Cryolite, bauxite, impure emery, ruby, sapphire (Al_2O_3). • Commercially, by electrolysis of Al_2O_3 from bauxite, dissolved in cryolite; water power usual source of electrical energy. • 1825; Oersted.	Silver-white, ductile metal; malleable at 120°; tensile strength (wrought) 16 tons per square inch. Better conductor of electricity, weight for weight, than copper. Acted upon by dilute hydrochloric acid, slowly by sulfuric acid, but not by nitric acid or the acids in foods. Soluble in alkaline hydroxides.	Used for cooking utensils, boatbuilding, airplanes, small articles requiring lightness and strength, and electric leads. The powdered metal is used as a body for paint; its mixture with ferric oxide, called thermite, is used for producing very high temperatures (up to 2,700°).
Americium (Am) At. wt. 243 At. no. 95 Valences 3, 4, 5, 6 Sp. gr. 11.87 M.P. 850°	Does not occur naturally. • Bombardment of uranium-238 with very high energy electrons. • 1944; Seaborg, James, Morgan, Ghiorso.	Radioactive, transuranium element. Alpha activity amounts to 70 billion alpha disintegratons per minute per milligram.	Used in research.
Antimony (Sb) At. wt. 121.75 At. no. 51 Valences 3, 5 Sp. gr. 6.62 M.P. 630.15° B.P. 1,380° C.S. H	Free and as stibnite (Sb_2S_3). • Roasting stibnite gives Sb_2O_4, which is then reduced by heating with carbon. • 1450; Valentine.	White, brittle, crystalline metal. Its alloys expand on solidification, give very sharp castings for type. Does not tarnish, but may be burned in air; unites directly with the halogens.	Constituent of type metal, Britannia metal, Babbitt metal (used for bearings), and other alloys. Oxide (Sb_2O_3) is both basic and acidic. Trichloride, butter of antimony ($SbCl_3$), is easily hydrolyzed. Tartar emetic is used in medicine and dyeing.
Argon (Ar) At. wt. 39.948 At. no. 18 Valence 0 M.P. − 189.2° B.P. − 185.7°	Present in the air 0.94 per cent by volume. • To isolate from air, carbon dioxide is removed by soda lime, water by phosphorus pentaoxide, oxygen by red-hot copper, nitrogen by magnesium and calcium; fractional distillation of residue yields argon. • 1894; Rayleigh, Ramsay.	Monatomic gas, identified by its characteristic spectrum seen by examining light emitted when the gas is placed in a vacuum tube at low pressure and sparked. More soluble than nitrogen in water; 100 volumes of water dissolves 4 volumes of argon under ordinary conditions.	Forms no compounds, hence its name, argon, meaning 'inert.'
Arsenic (As) At. wt. 74.9216 At. no. 33 Valences 3, 5 Sp. gr. 5.7 M.P. 817° (under pressure) B.P. 613° (sublimes)	Free as arsenical pyrites (FeSAs), orpigment (As_2S_3), and realgar (As_2S_2). • By heating arsenical pyrites ($FeSAs \rightarrow FeS + As$). • 1649; Schröder.	Steel-gray, dull metallic, crystalline element classed as a metalloid because it is between metals and nonmetals. Vapor density corresponds to As_4 at 644°, and to As_2 at 1,700°. Burns in air and unites directly with the halogens, sulfur, and many metals.	Used for hardening lead shot. All compounds are poisonous. White arsenic (As_2O_3) is partly basic, forming a chloride, and partly acidic, forming arsenites. Scheele's green is a dangerous pigment used in wallpaper. Traces of arsenic may be detected by Marsh's test, in which intensely poisonous arsine (AsH_3) is formed.
Astatine (At) At. wt. 210 At. no. 85 Valences 1, 3, 5, 7	Does not occur naturally. • By bombardment of bismuth with alpha particles. • 1940; Segré, Corson, MacKenzie.	Synthetic element similar to polonium. All the known isotopes are short-lived, the longest-lived isotope having a half-life of 8.3 hours.	Used in research.
Barium (Ba) At. wt. 137.34 At. no. 56 Valence 2 Sp. gr. 3.75 M.P. 725° B.P. 1,140° C.S. Cb	As barite, or heavy spar, ($BaSo_4$) and witherite ($BaCO_3$). • By electrolysis of the fused chloride ($BaCl_2$). • 1808; Davy.	Silver-white, lustrous, malleable metal harder than lead. Like calcium, it reacts slowly with water to give barium hydroxide and hydrogen. The vapors of its compounds impart green color to the Bunsen flame.	Peroxide (BaO_2) is used in manufacture of oxygen and hydrogen peroxide; nitrate and chlorate in pyrotechnics to produce green fires; sulfate as the body for permanent white paint and for filling glazed paper. All its soluble compounds are poisonous.
Berkelium (Bk) At. wt. 249 At. no. 97 Valences 3, 4 M.P. 1,278°	Does not occur naturally. • By bombardment of americium-241 with alpha particles. • 1949; Seaborg, Ghiorso, Thompson.	Radioactive, transuranium element. The longest-lived isotope has a half-life of several thousand years.	Used in research.
Beryllium (Be) At. wt. 9.0122 At. no. 4 Valence 2 Sp. gr. 1.8 M.P. 1,278° B.P. 2,970° C.S. H	In beryl [$Be_3Al_2(SiO_3)_6$]. • By electrolysis of the fused double fluoride ($BeF_2 \cdot 2KF$). • 1797; Vauquelin.	Hard, white metal that tarnishes when heated in air; soluble in dilute acids when powdered.	Hydroxide [$Be(OH)_2$] is feebly acidic as well as basic, thus resembling the hydroxide of zinc. Emerald is beryl colored green by chromium oxide. Used as nuclear reactor fuel-rod casing.
Bismuth (Bi) At. wt. 208.980 At. no. 83 Valences 3, 5 Sp. gr. 9.8 M.P. 271.3° B.P. 1,560° C.S. H	Free and as trioxide (Bi_2O_3) and trisulfide (Bi_2S_3). • Ore is roasted, then heated with charcoal and metallic iron to remove traces of sulfur. • 1450; Valentine.	Exceedingly brittle, crystalline, shiny metal, white with tinge of pink. Expands on soldification. Does not tarnish, and can be burned in air. Dissolves in oxygen acids. Most diamagnetic substance known.	Used for making fusible alloys such as Wood's metal (melting point 60.5°), used in plugs of fire sprinklers, boiler safety valves, and for taking casts. The oxynitrate is used in medicine and as a cosmetic.

TABLE OF ELEMENTS (Continued)

Element	Occurrence, Preparation, Date of Discovery, and Discoverer	Properties	Chief Compounds and Uses
Boron (B) At. wt. 10.811 At. no. 5 Valence 3 Sp. gr. amorphous 2.4 crystalline 2.5 M.P. 2,550° B.P. 2,300°	As boric acid (H_3BO_3), borax ($Na_2B_4O_7 \cdot 10H_2O$), and colemanite ($Ca_2B_6O_{11} \cdot 5H_2O$). • Amorphous, by reducing B_2O_3 with magnesium; impure crystalline, by reducing B_2O_3 with excess aluminum. • 1808; Gay-Lussac, Thénard, Davy.	Amorphous form is a greenish-black powder that burns in air at 700°, forming B_2O_3 and BN. Boron is oxidized by adding hot concentrated nitric acid or sulfuric acid to boric acid.	Compounds are analogous to those of silicon. Borax is used as a flux and, in solution, as a mild alkali because of its hydrolysis. Boric acid is used as a weak antiseptic and preservative.
Bromine (Br) At. wt. 79.909 At. no. 35 Valences 1, 3, 5, 7 Sp. gr. 3.12 M.P. −7.2° B.P. 58.78°	In sea water, as alkali bromide; in upper layers of salt deposits as sodium and magnesium bromide. • By treatment of brine with sulfuric acid and manganese dioxide, or with chlorine. • 1826; Balard.	Dark-red liquid; smells like chlorine; vapor irritates eyes, throat, and nose. Dissolves in 30 parts of water (bromine water). Combines with most other elements, but less vigorously than chlorine.	Potassium bromide is used in pharmacy; silver bromide, in photography. Bromine is used in preparation of organic dyes and as a disinfectant and a bleach.
Cadmium (Cd) At. wt. 112.40 At. no. 48 Valence 2 Sp. gr. 8.6 M.P. 320.7° B.P. 765°	With zinc ores, as carbonate and sulfide. • Comes over in first portions in distillation of impure zinc. • 1817; Stromeyer.	Silver-white metal, more ductile and more malleable than zinc. Burns in air; is attacked by dilute acids.	All compounds are poisonous, little ionized. Sulfide is basis of cadmium yellow; iodide is used in pharmacy; metal as protective plating.
Calcium (Ca) At. wt. 40.08 At. no. 20 Valence 2 Sp. gr. 1.55 M.P. 842° B.P. 1,487° C.S. C	As carbonate (Iceland spar, calcite, aragonite, marble, chalk, limestone), sulfate (gypsum), phosphate (apatite), fluoride (fluorspar), complex silicates (feldspars, pyroxines, amphiboles). • By electrolysis of the fused chloride, or heating the iodide with sodium. • 1808; Davy.	White crystalline metal, harder than lead. Can be cut, drawn, rolled, and turned. Reacts with water and burns in air at red heat, forming the oxide (CaO) and the nitride (Ca_3N_2). Unites with hydrogen to form CaH_2, whose reaction with water is source of hydrogen for balloons. Salts color test flame yellowish-red.	Oxide (quicklime) is used for mortar and to remove hair from hides. Hydroxide mixed with sand forms mortar; solution is limewater. Plaster of Paris is less hydrated sulfate; takes up water on setting to form gypsum. Phosphates are fertilizers.
Californium (Cf) At. wt. 249 At. no. 98 Valence 3	Does not occur naturally. • By bombardment of curium-242 with alpha particles. • 1950; Seaborg, Thompson, Ghiorso, Street.	Radioactive; transuranium element.	Used in research.
Carbon (C) At. wt. 12.01115 At. no. 6 Valences 4, 6 Sp. gr. diamond 3.5 graphite 2.3 amorphous 1.9 M.P. Not realized; volatilizes near 3,500° B.P. 4,827°	Free as diamond and graphite; in combination with hydrogen as petroleum; with oxygen as carbon dioxide; with these and other elements as coal and in plant and animal tissues; as many carbonates. • By dry distillation of wood or coal, yielding charcoal and coke, respectively. • Known in antiquity.	Diamond is crystalline, and is the hardest of minerals; dark-colored bort used for cutting and grinding. Graphite has black metallic luster, is crystalline, and may be scratched by the fingernail. Charcoal is amorphous and can absorb gases and coloring matters. All three forms burn in oxygen to form carbon dioxide.	The carbon compounds form the substance of organic chemistry. Carbon dioxide results from burning coal, coke, wood, oil, or illuminating gases, from fermentation and decay (slow burning), and from exhalation. Carbon monoxide is a deadly gas. Graphite is a popular lubricant; diamond is a precious gem; coal, coke, and petroleum are fuels.
Cerium (Ce) At. wt. 140.12 At. no. 58 Valences 3, 4, 6 Sp. gr. 6.78 M.P. 795° B.P. 3,468° C.S. C	As silicate in cerite, along with neodymium, praseodymium, and lanthanum; also in monazite sand. • By electrolysis of the fused chloride. • 1803; Berzelius, Klaproth, Hisinger.	Rare earth metal with the color and luster of iron; like tin in hardness; very ductile and malleable. Burns in air more easily and more brightly than magnesium. Emits sparks when scratched with steel.	Welsbach incandescent gas mantles contain 1 per cent cerium dioxide (CeO_2). Alloys are used for gas and cigar lighters.
Cesium (Cs) At. wt. 132.905 At. no. 55 Valence 1 Sp. gr. 1.9 M.P. 28.5° B.P. 690° C.S. Cb	In certain micas, mineral waters, and the ashes of certain plants. • By heating the hydroxide (CsOH) with magnesium or by electrolysis. • 1860; Bunsen, Kirchhoff.	Silver-white metal resembling rubidium and potassium. The softest of all solid metals; one of the most active metals and the most electropositive. Reacts violently with water. Cesium gives two bright lines in the blue of the spectrum; its name comes from *caesius*, meaning 'sky-blue.'	Used in certain photoelectric cells; in vacuum tubes to eliminate traces of gases. One of its radioactive isotopes is used in radiation therapy.
Chlorine (Cl) At. wt. 35.453 At. no. 17 Valences 1, 3, 5, 7 Sp. gr. (liquid) 1.5 M.P. −100.98° B.P. −34.6°	In sea water as chlorides of the alkalis and alkaline earths; in salt deposits as like compounds. • By electrolysis of alkali chloride, fused or in solution; by the action of manganese dioxide (MnO_2) on hydrochloric acid (HCl). • 1774; Scheele.	Greenish-yellow gas with characteristic odor. Acts violently on respiratory tract. Unites directly with all elements except oxygen, nitrogen, and the argon family. Displaces bromine and iodine from their compounds, substitutes for hydrogen in organic compounds.	Gas is used in extracting gold and in preparing bleaching and disinfecting agents. In presence of water it bleaches many coloring matters. Forms chlorides, such as NaCl, KCl, $CaCl_2$; hypochlorites, as solution of $Ca(OCl)_2$; chlorates, as $KClO_3$, used for matches and in pyrotechnics; and perchlorates, as $KClO_4$. Common table salt is NaCl.
Chromium (Cr) At. wt. 51.996 At. no. 24 Valences 2, 3, 6 Sp. gr. 7.1 M.P. 1,890° B.P. 2,482° C.S. Cb	As chromite [$Fe(CrO_2)_2$]. • By reducing the oxide of chromic acetate (Cr_2O_3) with aluminum filings. • 1797; Vauquelin.	Steel-gray, lustrous, brittle, very hard metal. At high temperatures it burns in air to form green Cr_2O_3. Attacked by dilute sulfuric acid or hydrochloric acid, but not by nitric acid.	Used in alloys of steel and nickel. Chrome green (Cr_2O_3) and chrome yellow ($PbCrO_4$) are pigments. Bichromates (such as $K_2Cr_2O_7$) are used in photo processes, tanning and dyeing, and as oxidizing agents, as in batteries. The metal, like nickel, is used as a protective and decorative plating.
Cobalt (Co) At. wt. 58.9332 At. no. 27 Valences 2, 3 Sp. gr. 8.9 M.P. 1,495° B.P. 3,000° C.S. H (?)	As smaltite ($CoAs_2$) and cobaltite (CoAsS) found with iron and nickel. • By igniting the oxide in hydrogen. • 1735; Brandt.	White, malleable metal, less tenacious than iron. Turns pinkish on exposure to air. Less active chemically than iron.	Intensely blue silicates are used in coloring porcelain and constitute the pigment smalt. Used in commercial dyes; also used in radiation therapy.

TABLE OF ELEMENTS (Continued)

Element	Occurrence, Preparation, Date of Discovery, and Discoverer	Properties	Chief Compounds and Uses
Copper (Cu) At. wt. 63.54 At. no. 29 Valences 1, 2 Sp. gr. 8.92 M.P. 1,083° B.P. 2,595° C.S. Cf	Free as cuprite (Cu_2O), copper glance, chalcopyrite ($CuFeS_2$), and malachite [$Cu_2(OH)_2CO_3$]. • After removal of iron and sulfur, the oxide is reduced by heating with carbon. It is refined electrolytically. • Known in antiquity.	Red, lustrous, very ductile and malleable metal; high tensile strength (14 tons per square inch); second only to silver in electrical conductivity. In ordinary air it gradually becomes coated with basic carbonate. In absence of air, nitric acid alone among the dilute acids attacks it; in the presence of air, even acids in foodstuffs can dissolve it.	Used for coins, ornaments, electrical leads, electroplating, roofing, cooking vessels, and for making such alloys as brass, bell and gun metals, German silver, and the bronzes. The soluble compounds are poisonous and are used as agricultural germicides. Blue vitriol is $CuSO_4 \cdot 5H_2O$; the basic acetate is verdigris.
Curium (Cm) At. wt. 248 At. no. 96 Valence 3 Sp. gr. 7 (?)	Does not occur naturally. • By bombardment of plutonium-239 with alpha particles. • 1944; Seaborg, James, Ghiorso, Morgan.	Radioactive, transuranium element. The longest-lived isotope has a half-life of a half-million years.	Used in research.
Dysprosium (Dy) At. wt. 162.50 At. no. 66 Valence 3 Sp. gr. 8.56 M.P. 1,407° B.P. 2,600°	In monazite, gadolinite, and other rare minerals. • By fractional crystallization of bromates. • 1886; Boisbaudran.	Rare earth. The oxide, dysprosia (Dy_2O_3), is found with three other rare earths.	Salts are green or yellow and show characteristic absorption bands. They are the most magnetic of all salts.
Einsteinium (Es) At. wt. 254 At. no. 99 Valence 3	Does not occur naturally. • Prepared by intensive neutron irradiation of plutonium-239. • 1952; Thompson, Harvey, Choppin, Seaborg, Ghiorso.	Radioactive, transuranium element. Half-life of longest-lived isotope is 280 days.	Used in research.
Erbium (Er) At. wt. 167.26 At. no. 68 Valence 3 Sp. gr. 9.16 M.P. 1,497° B.P. 2,900°	In gadolinite and other rare minerals. • 1843; Mosander.	Rare earth. The oxide erbia (Er_2O_3) is found with holmia, thulia, and dysprosia.	Salts are rose-colored and show characteristic absorption spectra.
Europium (Eu) At. wt. 151.96 At. no. 63 Valences 2, 3 Sp. gr. 5.22 M.P. 826° B.P. 1,439°	In monazite and other rare minerals. • By electrolysis of the chloride. • 1896; Demarçay.	Rare earth. This element so closely resembles samarium that they are difficult to separate analytically.	Salts are pinkish and show a faint absorption spectrum. Sometimes used in control rods of nuclear reactors.
Fermium (Fm) At. wt. 252 At. no. 100 Valence 3	Does not occur naturally. • Prepared by intensive neutron irradiation of plutonium-239. • 1953; Thompson, Harvey, Choppin, Seaborg, Ghiorso.	Radioactive, transuranium element.	Used in research.
Fluorine (F) At. wt. 18.9984 At. no. 9 Valence 1 Sp. gr. (liquid) 1.1 at −187° M.P. −219.62° B.P. −188.14°	As cryolite (Na_3AlF_6), fluorspar (CaF_2), and very widely elsewhere in small quantities. • By electrolysis of dry hydrogen fluoride at −23°. • 1886; Moissan.	Pale yellowish-green gas that unites with every element except oxygen. Rapidly displaces oxygen from water, chlorine from hydrogen chloride.	Hydrogen fluoride is used for etching glass and in silicate analysis. Silver fluoride is soluble and calcium fluoride is insoluble, in contrast with the other halides of these metals.
Francium (Fr) At. wt. 223 At. no. 87 Valence 1	Does not occur naturally. • Disintegrates so rapidly that it is almost impossible to obtain in sufficient quantity for weighing. • 1939; Perey.	Synthetic radioactive element; heaviest of the alkali metals.	Used in research.
Gadolinium (Gd) At. wt. 157.25 At. no. 64 Valence 3 Sp. gr. 7.94 M.P. 1,312° B.P. 3,000°	In gadolinite and samarskite. • By electrolysis of the chloride. • 1886; Marignac.	Rare earth. Closely resembles terbium in its compounds.	Salts are colorless and show absorption bands only in the ultraviolet.
Gallium (Ga) At. wt. 69.72 At. no. 31 Valences 2, 3 Sp. gr. 5.93 M.P. 29.78° B.P. 2,403°	In iron ores, zinc blende (ZnS), and bauxite (Al_2O_3). • By electrolysis of an alkaline solution of its salts secured from zinc. • 1875; Boisbaudran.	Bluish-white, tough metal that can be cut with a knife. Like aluminum, it is soluble in hydrochloric acid and caustic soda, but not in nitric acid.	Forms two chlorides, $GaCl_3$ and $GaCl_2$, that yield very characteristic spark spectra. Alloys with aluminum and cadmium are used for optical mirrors and cathodes.
Germanium (Ge) At. wt. 72.59 At. no. 32 Valences 2, 4 Sp. gr. 5.5 M.P. 937.4° B.P. 2,830°	In the rare mineral argyrodite. • By the reduction of the dioxide (GeO_2) by carbon. • 1886; Winkler.	Grayish-white, brittle, lustrous metal, insoluble in hydrochloric acid. Combines directly with the halogens.	Close relation of this element to carbon and silicon is shown in the compound germanium chloroform. Mendeleev described it before its discovery, calling it ekasilicon. The oxide is used to treat pernicious anemia.
Gold (Au) At. wt. 196.967 At. no. 79 Valences 1, 3 Sp. gr. 19.32 M.P. 1,063° B.P. 2,966°	Chiefly free, but also a telluride; many specimens of iron pyrites are auriferous. • From gold-bearing sands by washing away the lighter materials and dissolving the gold from the residue in mercury, which is subsequently separated from the gold by distillation. • Known in antiquity.	Soft, bright-yellow metal, easily scratched by a knife; most ductile and malleable of the metals; excellent conductor of heat and electricity. Chemically, gold is rather inert and is not attacked by the oxygen of the air, by hydrogen sulfide, or by any single acid.	Pure gold is 24-carat gold. Jewelry is made in 18, 14, and 9-carat gold because alloying it with copper increases hardness and rigidity. Sodium chloraurate is used for toning in photography, and potassium auricyanide is used in electrogilding.

TABLE OF ELEMENTS (Continued)

Element	Occurrence, Preparation, Date of Discovery, and Discoverer	Properties	Chief Compounds and Uses
Hafnium (Hf) At. wt. 178.49 At. no. 72 Valence 4 Sp. gr. 13.3 M.P. 2,150° B.P. 5,400°	Associated with zirconium. • By decomposing the tetraiodide. • 1922; Coster, Hevesy.	Analogous to zirconium.	Similar to zirconium compounds, from which it is separated by fractional crystallization.
Helium (He) At. wt. 4.0026 At. no. 2 Valence 0 Sp. gr. (liquid) 0.15 M.P. −272.2° (at 26 atmospheres) B.P. −268.9°	In the air to the extent of 1 to 2 volumes per million; also occluded in certain minerals. First observed in sun's spectrum (1868; Lockyer, Jannsen). • Neon and helium are boiled off crude argon, and the neon, when cooled, solidifies. • 1895; Ramsay.	Lightest gas except hydrogen, transparent, odorless, and colorless.	Forms no compounds. Used for balloons; not flammable.
Holmium (Ho) At. wt. 164.930 At. no. 67 Valence 3 Sp. gr. 8.76 M.P. 1,461° B.P. 2,600°	Occurs with, and is separated from the erbium subgroup of the rare earths. Has never been isolated. • 1879; Cleve.	Rare earth. Salts are orange-yellow and similar to those of dysprosium.	Used in research.
Hydrogen (H) At. wt. 1.00797 At. no. 1 Valence 1 Sp. gr. (liquid) 0.07 M.P. −259.1° B.P. −252.7°	In the air to the extent of 1 volume per 20,000 volumes of air; combined, in water (11.19 per cent by weight), natural gas, petroleum, and animal and vegetable bodies. • By treating zinc with hydrochloric or sulfuric acid; by electrolysis. • 1766; Cavendish.	Lightest gas, transparent, odorless, and colorless. Soluble in water (2 volumes in 100 volumes of water under average conditions), in platinum, in palladium (502 volumes in 1 volume of palladium). Burns in air and in chlorine and unites with many other elements.	Its two oxides are water and hydrogen peroxide, the latter used in solution as a bleaching agent. Every acid contains hydrogen as an essential constituent. Its compounds with carbon and other elements number over 100,000.
Indium (In) At. wt. 114.82 At. no. 49 Valences 1, 3 Sp. gr. 7.3 M.P. 156° B.P. 2,000°	In zinc blende (ZnS) in small quantities. •Electrolytically from solutions of its salts. •1863; Reich, Richter.	White, malleable metal, softer than lead; about as heavy as tin.	Compounds color the nonluminous gas flame blue and show a characteristic indigo blue line in the spectrum—hence its name. Sometimes used as a coating on bearings.
Iodine (I) At. wt. 126.9044 At. no. 53 Valences 1, 3, 5, 7 Sp. gr. 4.95 M.P. 113.5° B.P. 184.35°	In the ocean and certain seaweeds; always in the combined state. • From iodides by displacement of their iodine by chlorine. • 1811; Courtois.	Dark gray, brittle solid with a metallic luster. Vapor is violet, as are its solutions in chloroform and carbon bisulfide. Requires more than 5,000 parts of water for solution. Combines directly with many elements, but is much less active than chlorine or bromine.	Used in pharmacy as an antiseptic and in prescriptions for the treatment of goiter. Potassium iodide and iodoform likewise find application in medicine. The alkyl iodides are much used in synthetic organic chemistry.
Iridium (Ir) At. wt. 192.2 At. no. 77 Valences 3, 4 Sp. gr. 22.42 M.P. 2,410° B.P. 4,527° C.S. Cf	With platinum and osmium. • From platinum ores by a complex series of operations. • 1804; Tennant.	White metal, brittle when cold, very hard, and one of the densest substances known. Attacked by fused alkalis, but not by aqua regia.	Used for pointing gold pens. Its alloy with 9 parts of platinum is used for standard meter bars because of its unalterability. Used as a black color in china decorations.
Iron (Fe) At. wt. 55.847 At. no. 26 Valences 2, 3 Sp. gr. 7.86 pig 7.03−7.73 M.P. 1,535° wrought 1,600° steel 1,375° gray pig 1,275° white pig 1,075° B.P. 3,000° C.S. Cf, Cb	As magnetite (Fe_3O_4), hematite (Fe_2O_3), limonite ($2Fe_2O_3 \cdot 3H_2O$), siderite ($FeCO_3$), which are important ores; iron pyrites (FeS_2); in rocks as complex silicates; in plants and animals. • Pig iron is prepared in blast furnace by reducation of the ore by means of carbon monoxide in presence of suitable flux. From pig iron, wrought iron is obtained by puddling, and steel by the Bessemer, open-hearth, or other processes. • Known in antiquity.	Malleable, ductile, magnetic metal, that is unchanged in dry air, but rusts in water and moist air. Easily attacked by dilute acids, but not by fused alkalis. Cast iron contains 2 to 5 per cent carbon and other impurities, and is hard and brittle. Wrought iron contains less than 0.2 per cent carbon and is softer and tougher, with tensile strength of 22 to 25 tons per square inch. Steel contains from 0.2 to 1.5 per cent carbon, and is permanently magnetic.	The metal is used as a structural material for rails, machinery, tools, etc. Jeweler's rouge and Venetian red consist of the oxide (Fe_2O_3). Rust is chiefly hydrated oxide. Hammer scale and lodestone have the composition Fe_3O_4. Ferric chloride ($FeCl_3$), ferrous iodide (FeI_2), and other iron compounds are used in medicine. Green vitriol ($FeSO_4 \cdot 7H_2O$) is used in making ink and in dyeing.
Krypton (Kr) At. wt. 83.80 At. no. 36 Valences 0, 2, 4 M.P. −156.6° B.P. −152.3°	In minute quantities in the air. • From crude argon by fractional distillation. • 1898; Ramsay, Travers.	Inert, colorless, and odorless gas resembling, but denser than, argon.	Once thought to be chemically inert; forms series of fluoride compounds, such as KrF_2 and KrF_4.
Lanthanum (La) At. wt. 138.91 At. no. 57 Valence 3 Sp. gr. 6.15	As lanthanite [$La_2(CO_3)_3 \cdot 8H_2O$]. • By electrolysis of the fused chloride ($LaCl_3$). • 1839; Mosander.	Rare earth; iron-gray metal; tarnishes in air to steel-blue; malleable and ductile. Attacked slowly even by cold water.	When heated in air, it forms a strongly basic oxide (La_2O_3) that is diamagnetic, and a nitride (LaN).
Lawrencium (Lw) At. wt. 257 At. no. 103 Valence 3	Does not occur naturally. • By bombardment of californium-252 with boron ions. • 1961; Ghiorso, Sikkeland, Larsh, Latimer.	Radioactive, transuranium element.	Used in research.
Lead (Pb) At. wt. 207.19 At. no. 82 Valences 2, 4 Sp. gr. 11.34 M.P. 327.5° B.P. 1,744° C.S. Cf	End product of certain radioactive decompositions. As galena (PbS) and in silver ores. •By calcination of partially roasted galena. Purification is effected by Parkes process. • Known in antiquity.	Soft, gray metal; malleable and of low tensile strength, relatively impermeable to X rays and atomic radiation. In presence of air, water acts on lead to produce the hydroxide which, being slightly soluble, may cause lead poisoning. When heated in air, lead is oxidized to litharge (PbO) and, under suitable conditions, to minium (Pb_3O_4).	Used for water pipes, roofs and gutters, and storage batteries. For shot it is alloyed with 0.4 per cent arsenic. Type metal contains 80 per cent lead. Babbitt metal, for bearings, contains over 70 per cent lead. Solder and pewter are alloys of lead and tin. The basic carbonate, white lead, is the basis of most oil paints.

TABLE OF ELEMENTS (Continued)

Element	Occurrence, Preparation, Date of Discovery, and Discoverer	Properties	Chief Compounds and Uses
Lithium (Li) At. wt. 6,939 At. no. 3 Valence 1 Sp. gr. 0.53 M.P. 179° B.P. 1,317° C.S. C	In amblygonite [Li(AlF)PO_4]. • By electrolysis of the fused chloride (LiCl). • 1817; Arfvedson.	Lightest metal; silver-white, softer than lead, tarnishes quickly in air, and easily reacts with water. When heated, it unites vigorously with nitrogen.	The carbonate (Li_2CO_3) is used in medicine as a solvent for uric acid, lithium urate being soluble. The salts give a carmine flame coloration.
Lutetium (Lu) At. wt. 174.97 At. no. 71 Valence 3 Sp. gr. 9.749 M.P. 1,652° B.P. 3,327°	In euxenite. • By electrolysis of the chloride. • 1907; Urbain, Welsbach.	Rare earth. Like ytterbium but has lower magnetic susceptibility.	Its compounds resemble those of ytterbium. It was once known as Cassiopium.
Magnesium (Mg) At. wt. 24.312 At. no. 12 Valence 2 Sp. gr. 1.74 M.P. 651° B.P. 1,107° C.S. H	As magnesite ($MgCO_3$), dolomite ($MgCO_3 \cdot CaCO_3$), carnallite ($MgCl_2 \cdot KCl \cdot 6H_2O$), and in very many complex silicates. • By electrolysis of dried, fused carnallite. • 1808; Davy.	Silver-white, very lightweight metal, ductile when hot, and malleable. It tarnishes in air and reacts slowly with water, rapidly with steam. Burns in air to the oxide (MgO), emitting a very bright light. Unites directly with nitrogen.	Used as a reducing agent. Sulfate, known as Epsom salts, is used in medicine, as are the oxide (magnesia), the carbonates, and the citrate. The bright light emitted when the metal is burned in air is used in photography.
Manganese (Mn) At. wt. 54.9380 At. no. 25 Valences 2, 3, 4, 6, 7 Sp. gr. 7.2 M.P. 1,244° B.P. 2,097° C.S. CT	As pyrolusite (MnO_2), braunite (Mn_2O_3), hausmannite (Mn_3O_4), and manganese spar ($MnCO_3$). • By heating Mn_3O_4 with aluminum filings. • 1774; Gahn.	Steel-gray, hard, brittle metal with a pinkish tinge. Rusts in moist air and is attacked by dilute acids.	Ferromanganese and spiegeleisen are alloys with iron, used in making steel tougher. With copper it forms the tough, hard manganese bronzes, with tensile strength up to 30 tons per square inch.
Mendelevium (Md) At. wt. 256 At. no. 101 Valence 3	Does not occur naturally. • By bombardment of einsteinium-253 with alpha particles. • 1955; Thompson, Harvey, Choppin, Ghiorso, Seaborg.	Radioactive, transuranium element.	Used in research.
Mercury (Hg) At. wt. 200.59 At. no. 80 Valence 1, 2 Sp. gr. 13.6 M.P. −38.87° B.P. 356.58° C.S. H	Free and as cinnabar (HgS). • By roasting cinnabar: $HgS + O_2 \rightarrow Hg + SO_2$ • Known in antiquity.	Silver-white, mobile liquid, 20 per cent heavier than lead. Has vapor pressure of 0.0002 millimeter at 0°. Tarnishes slowly in air and is attacked only by nitric among the dilute acids. Vapor is monatomic.	Used in thermometers and barometers. Alloys, some of which are used in dentistry, are called amalgams. Calomel (HgCl) is administered internally in medicine; corrosive sublimate ($HgCl_2$) forms a solution with very powerful germicidal properties.
Molybdenum (Mo) At. wt. 95.94 At. no. 42 Valences 3, 4, 5, 6 Sp. gr. 10.2 M.P. 2,610° B.P. 5,560° C.S. Cb	As molybdenite (MoS_2) and wulfenite ($PbMoO_4$). • By reducing the oxides with aluminum powder. • 1778; Scheele.	White metal as malleable as iron; will not scratch glass. Insoluble in hydrochloric or dilute sulfuric acid.	Ferromolybdenum alloys are used in the manufacture of special steels.
Neodymium (Nd) At. wt. 144.24 At. no. 60 Valences 3, 4 Sp. gr. 6.9 M.P. 1,024° B.P. 3,027°	With cerium and lanthanum. • By electrolysis of the fused chloride. • 1885; Welsbach.	Rare earth; yellowish metal; tarnishes in air.	Salts are rose-violet; solutions show characteristic spectra.
Neon (Ne) At. wt. 20.183 At. no. 10 Valence 0 M.P. −248.67° B.P. −245.92°	Minute quantities in atmosphere. • Neon and helium are boiled out of crude argon and the neon separated from helium by cooling with liquid hydrogen. •1898; Ramsay, Travers.	Colorless, odorless, transparent, monatomic, inert gas.	Forms no compounds; is recognized by its characteristic spectrum. Used in glow tubes for display signs.
Neptunium (Np) At. wt. 237 At. no. 93 Valences 3, 4, 5, 6 Sp. gr. 20.45 M.P. 640°	Does not occur naturally. • By bombardment of uranium with neutrons. • 1940; McMillan, Abelson.	Radioactive element. First transuranium element to be synthesized. Emits alpha particles. Half-life of longest-lived isotope is 2,200,000 years.	Oxide is dark brown. Used in research.
Nickel (Ni) At. wt. 58.71 At. no. 28 Valences 2, 3 Sp. gr. 8.8 M.P. 1,455° B.P. 2,900° C.S. Cf	As nicollite (NiAs) and nickel glance (NiAsS). • By igniting the oxalate in hydrogen. • 1751; Cronstedt.	White, very hard, lustrous metal; malleable, ductile, and tenacious. Rusts slowly in air and is easily attacked only by nitric acid.	Metal furnishes protective coating when plated on iron. German silver is an alloy of nickel, copper, and zinc. Nickel chromium steel is used for armor. Manganin, containing nickel, copper, and manganese, is used for electrical resistors. It is a catalyst, especially in hydrogenation.

TABLE OF ELEMENTS (Continued)

Element	Occurrence, Preparation, Date of Discovery, and Discoverer	Properties	Chief Compounds and Uses
Niobium (Nb) At. wt. 92.906 At. no. 41 Valences 1, 2, 4, 5 Sp. gr. 8.55 M.P. 2,468° B.P. 4,927°	In the mineral columbite ($FeCb_2O_6$). • By reduction of the dioxide (NbO_2) by paraffin. • 1801; Hatchett.	Light gray, malleable, ductile metal, as hard as wrought iron; not affected by acids, even aqua regia. The hydride (NbH) burns in air.	Compounds occur with those of tantalum, which they closely resemble. It was originally called Columbium.
Nitrogen (N) At. wt. 14.0067 At. no. 7 Valences 3, 5 Sp. gr. (liquid) 0.808 M.P. −209.86° B.P. −195.8°	Free nitrogen forms about 79 per cent of the air by volume. Also in Bengal saltpeter (KNO_3), Chile saltpeter ($NaNO_3$). • By fractional distillation of liquid air. • 1772; Rutherford.	Colorless, odorless, transparent gas, rather inactive chemically. At ordinary temperature and pressure, 100 volumes of water dissolve 1.5 volumes of nitrogen.	Nitrous oxide, or laughing gas, is used by dentists. Nitric acid has many applications in analytic and industrial chemistry. Ammonia is a very soluble gas. Many nitrogen compounds are used as fertilizers, explosives, dyes, and drugs.
Nobelium (No) At. wt. 254 At. no. 102 Valence 2	Does not occur naturally. • By bombardment of curium-246 with carbon-12 ions. • 1958; Ghiorso, Sikkeland, Walton, Seaborg.	Radioactive, transuranium element.	Used in research.
Osmium (Os) At. wt. 190.2 At. no. 76 Valences 2, 3, 4, 6, 8 Sp. gr. 22.48 M.P. 3,000° B.P. 5,000°	With platinum and iridium. • By reducing the tetroxide (OsO_4). • 1804; Tennant.	Gray metal, harder than glass; densest of the known elements.	Its alloy with iridium is used in tipping gold pens. Osmium tetroxide is used as a microscope stain for fat.
Oxygen (O) At. wt. 15.9994 At. no. 8 Valence 2 Sp. gr. (liquid) 1.13 M.P. −218.4° B.P. −183°	Free oxygen forms about 20 per cent of the air by volume. Water contains 88.88 per cent oxygen. The rocks of the earth's crust contain about 46 per cent in combination, chiefly as silicates. • In the laboratory, by heating potassium chlorate ($KClO_3$). Commercially, by fractional distillation of air. • 1774; Priestley, Scheele.	Colorless, odorless, tasteless, transparent gas, slightly heavier than air. At ordinary temperature and pressure, 100 volumes of water dissolves 3 volumes of oxygen. Very active chemically, combining directly with all but a few elements to form oxides. Most substances burn more vigorously in oxygen than in air. Liquid oxygen is magnetic.	Gas is sold compressed in mild steel cylinders, and is used for the oxyhydrogen blowpipe, in medicine, and for chemical purposes. Necessary to support animal respiration and ordinary combustion. Enters as a constituent into all oxides, most salts, and many organic compounds. Liquid oxygen (LOX) is an important propellant for rockets.
Palladium (Pd) At. wt. 106.4 At. no. 46 Valences 2, 4, 6 Sp. gr. 12.02 M.P. 1,552° B.P. 2,927° C.S. Cf	With platinum and gold in nickel ores. • By a complex series of processes from platinum ores. •1803; Wollaston.	Silvery, malleable, ductile metal, related to platinum, unlike which, however, it may be attacked by nitric acid. Under suitable conditions it can absorb over 900 volumes of hydrogen.	Since it does not tarnish, it is used for coating silver goods, and by dentists as a substitute for gold. Like platinum, it is used as a catalyst.
Phosphorus (P) At. wt. 30.9738 At. no. 15 Valences 3, 5 Sp. gr. white 1.82 red 2.2 M.P. (white) 44° B.P. (white) 280°	As phosphates, such as apatite [$Ca_5F(PO_4)_3$]; in bones, teeth, and brain; and in seeds of plants. • By reduction of calcium phosphate by carbon with a suitable flux in an electric furnace. • 1669; Brand.	Exists in two allotropic modifications: white phosphorus, which is waxy in consistency, soluble in carbon bisulfide, foul-smelling, and poisonous; and red phosphorus, which is a solid, insoluble in carbon bisulfide, odorless, and not poisonous. White phosphorus has a low ignition temperature, hence its former use in matches.	Red phosphorus is used in the manufacture of matches, as is the compound P_4S_3. In the form of superphosphate of lime, phosphorus is an important artificial fertilizer. The chlorides PCl_5 and PCl_3 are much used in organic chemistry. Compounds are used in medicine. Phosphine, PH_3, is a poison gas.
Platinum (Pt) At. wt. 195.09 At. no. 78 Valences 2, 4 Sp. gr. 21.45 M.P. 1,769° B.P. 3,827° C.S. Cf	Free, alloyed with iridium and osmium, as nuggets in alluvial sands. •Freed from the metals with which it is alloyed by a complex series of processes. • 1557; Scaliger.	Silvery, tenacious, very heavy, ductile, malleable metal, unaltered in moist air and not attacked by any single common acid. Aqua regia, fused alkalis, alkali nitrates, and cyanides, however, do attack it. Platinum sponge and platinum black are finely divided forms.	Because of its resistance to acids, platinum is used for chemical vessels and electrodes. Since its coefficient of expansion is close to that of glass, platinum wires can be fused through glass without danger of breakage on cooling. The salts are used in photography. The metal is more expensive than gold and is used in jewelry. Platinum is used as a catalyst in many industrial reactions.
Plutonium (Pu) At. wt. 242 At. no. 94 Valences 3, 4, 5, 6 Sp. gr. 19.7 M.P. 640°	Present to a small extent in uranium ores. • Produced in nuclear reactors from natural uranium. • 1940; Seaborg, McMillan, Wahl, Kennedy.	Radioactive transuranium element. Emits alpha particles. Half-life of longest-lived isotope is 24,300 years.	Used in nuclear weapons and reactors.
Polonium (Po) At. wt. 210 At. no. 84 Valences 2, 4 Sp. gr. 9.3 M.P. 254° B.P. 962°	With bismuth in uranium minerals. •Metal has been isolated only in minute quantities because almost 13 tons of pitchblende ore yields only about one gram of the element. • 1898; the Curies.	Radioactive element. Half-life is 138.7 days.	Compounds resemble those of tellurium.
Potassium (K) At. wt. 39.102 At. no. 19 Valence 1 Sp. gr. 0.86 M.P. 63.65° B.P. 774° C.S. Cb	As sylvite (KCl), carnallite ($KCl \cdot MgCl_2 \cdot 6H_2O$); in plant and animal ashes, and in many complex silicates. • By reduction or electrolysis of fused potassium hydroxide (KOH). • 1807; Davy.	Silver-white, lustrous, very lightweight metal, as soft as wax; tarnishes instantly in moist air. Chemically very active, decomposing in the cold and uniting violently with the halogens, sulfur, and oxygen.	Alloy (with sodium) is used in high-temperature thermometers. Bengal saltpeter is the nitrate and is used in pyrotechnics, for gunpowders, and as a preservative. Iodide, KI, is used in pharmacy. It is one of the three basic fertilizer elements, nitrogen and phosphorus being the others.

TABLE OF ELEMENTS (Continued)

Element	Occurrence, Preparation, Date of Discovery, and Discoverer	Properties	Chief Compounds and Uses
Praseodymium (Pr) At. wt. 140.907 At. no. 59 Valences 3, 4 Sp. gr. 6.78 M.P. 935° B.P. 3,127°	With cerium and lanthanum. • By electrolysis of the fused chloride. • 1885; Welsbach.	Rare earth; yellowish metal; remains untarnished in air.	Salts are leek-green, and their solutions have characteristic absorption spectra.
Promethium (Pm) At. wt. 147 At. no. 61 Valence 3	Does not occur naturally. • 1947; Marinsky, Glendenin, Coryell.	Rare earth metal, produced artificially. Recognized by its X-ray spectrum and its optical absorption spectrum. Radioisotopes have been identified.	Compounds are similar to those of samarium and neodymium.
Protactinium (Pa) At. wt. 231 At. no. 91 Valence 5 Sp. gr. 15.37 M.P. 1,230° B.P. ?	In uranium ores. • Metal has been isolated; about 70 milligrams may be secured from 1,000 kilograms of pitchblende. • 1917; Hahn, Meitner; Soddy, Cranston.	Radioactive element, emitting alpha particles. Its half-life is 12,000 years.	Compounds resemble those of tantalum.
Radium (Ra) At. wt. 226.05 At. no. 88 Valence 2 Sp. gr. 5 (?) M.P. 700° B.P. <1,737°	In minute quantities in pitchblende and other uranium ores. • Metal has been isolated; bromide is separated from the barium bromide prepared from pitchblende by fractional crystallization. • 1898; the Curies, Bémont.	In all of its compounds, the metal has the power of emitting certain radiations. These can pass through materials that are opaque to light, render air a conductor, affect a photographic plate, and cause a zinc-sulfide screen to fluoresce visibly.	Rays from radium compounds (such as $RaBr_2$, $RaCl_2$, $RaCO_3$) act destructively on living tissues.
Radon (Rn) At. wt. 222 At. no. 86 Valences 0, 2, 4 Sp. gr. (liquid) 4.4 M.P. −71° B.P. −61.8°	Admixed with air. • By passing air through solutions of radium salts. • 1900; Dorn.	Inert gas of the helium family; radioactive, emitting alpha particles; half-life of longest-lived isotope is 3.83 days.	Forms fluoride salts. Used in treatment of cancer; enclosed in minute glass vessels the size of a small match head, it is inserted into the tumor.
Rhenium (Re) At. wt. 186.2 At. no. 75 Valences 1, 4, 6, 7 Sp. gr. 20.53 M.P. 3,180° B.P. 5,627°	In molybdenum and platinum ores. • Hydrogen reduction of NH_4ReO_4. • 1925; Noddack, Tacke, Berg.	Silver-white, hard metal, heavier than gold. Only tungsten is less fusible. Chemical properties are similar to those of manganese.	Used in electronics.
Rhodium (Rh) At. wt. 102.905 At. no. 45 Valences 2, 3, 4 Sp. gr. 12.1 M.P. 1,966° B.P. 3,727° C.S. Cf	In the ores of platinum. • By a complex series of processes from platinum ores. • 1803; Wollaston.	Silvery, malleable, ductile metal; does not tarnish in air; not attacked by aqua regia.	The red chloride ($RhCl_3$) is formed by the action of chlorine on the metal. Rhodium-platinum alloy is used for thermocouples to measure high temperatures.
Rubidium (Rb) At. wt. 85.47 At. no. 37 Valences 1, 3, 5 Sp. gr. 1.53 M.P. 38.89° B.P. 688° C.S. Cb	Found with cesium. Salts are associated with those of potassium. • By heating the hydroxides with magnesium or by the electrolysis of the cyanides or hydroxides. • 1860; Bunsen, Kirchhoff.	Silver-white metal resembling potassium; reacts vigorously with water.	Compounds show characteristic flame spectra with two red lines. Used in photocells and in pharmaceuticals.
Ruthenium (Ru) At. wt. 101.07 At. no. 44 Valences 3, 4, 6, 7, 8 Sp. gr. 12.1 M.P. 2,250° B.P. 3,900°	In the ores of platinum. • By a complex series of processes from platinum ores. • 1844; Klaus.	Hard, white, brittle metal, oxidized when heated in air. Scarcely attacked by aqua regia; very infusible. Chemical properties resemble those of osmium.	The following oxides are known: Ru_2O_3, RuO_2, RuO_4, as well as salts corresponding to RuO_3 and Ru_2O_7. Ruthenium red, an ammoniacal compound, dyes silk a beautiful yellow, but its high price limits its usefulness.
Samarium (Sm) At. wt. 150.35 At. no. 62 Valences 2, 3 Sp. gr. 6.93 M.P. 1,072° B.P. 1,900°	In monazite and samarskite. • By electrolysis of the chloride. • 1879; Boisbaudran.	Rare earth; whitish-gray metal; tarnishes in air.	Salts are topaz-yellow and are similar to those of lanthanum.
Scandium (Sc) At. wt. 44.956 At. no. 21 Valence 3 Sp. gr. 3.02 M.P. 1,539° B.P. 2,727°	In the minerals euxenite and gadolinite. Existence of this element was predicted by Mendeleev in 1869; he called it ekaboron. • Leached from ores with sulfuric acid. • 1879; Nilson.	Forms an oxide and a number of colorless salts.	An alloying element for nickel and nickel steels.
Selenium (Se) At. wt. 78.96 At. no. 34 Valences 2, 4, 6 Sp. gr. amorphous 4.26 monoclinic 4.28 hexagonal 4.8 M.P. amorphous 50° monoclinic 170° hexagonal 217° B.P. 684.9° C.S. THC	Free in some specimens of sulfur and in combination with lead, iron, and other metals, as in pyrites. • Amorphous, by reducing selenious acid (H_2SeO_3) with sulfur dioxide. With tellurium it is obtained from the anode slime of copper refineries. • 1817; Berzelius.	Three varieties are known: (1) red amorphous, soluble in carbon bisulfide from which it is deposited as (2) red translucent monoclinic crystals, soluble in carbon bisulfide; (3) blue-gray metallic selenium, insoluble in carbon bisulfide. This last form conducts electricity much better when exposed to light; conductivity increases with light intensity.	Selenium cells are used as indicators of intensity of illumination. The compounds strongly resemble those of sulfur. Hydrogen selenide is a foul-smelling, flammable gas. Selenic acid (H_2SeO_4) is a more powerful oxidizer than sulfuric acid and dissolves gold. The oxychloride is a valuable solvent for resins, fish oils, etc. Selenium is used in the manufacture of colorless and red-tinted glass, and in electronic rectifiers.

TABLE OF ELEMENTS (Continued)

Element	Occurrence, Preparation, Date of Discovery, and Discoverer	Properties	Chief Compounds and Uses
Silicon (Si) At. wt. 28.086 At. no. 14 Valence 4 Sp. gr. amorphous 2.35 crystalline 2.4 M.P. 1,410° B.P. 2,355°	Silicon dioxide (SiO_2) occurs as flint, quartz, quartz sand, etc. Igneous rocks are composed largely of silicates, and silicon constitutes more than 27 per cent of the earth's crust —more than any other element except oxygen. • By reducing sand with coke in an electric furnace. • 1823; Berzelius.	Amorphous silicon is a brown powder that burns when heated in air. Crystalline silicon forms black needles. It is less active than the amorphous variety and is attacked only slowly by a mixture of hydrofluoric and nitric acids. It unites with fluorine, however, at ordinary temperatures.	Silicon is used in steelmaking. Silicon steel is more magnetic than iron. Ornamental varieties of quartz find uses as gems, as do several natural silicates. Silicon carbide, or carborundum (SiC), is used as an abrasive. Sodium silicate solution is water glass, used to protect sandstone and to preserve eggs. Common glass is a mixture of sodium and calcium silicates.
Silver (Ag) At. wt. 107.870 At. no. 47 Valence 1 Sp. gr. 10.53 M.P. 960.8° B.P. 2,212° C.S. Cf	Native, as sulfide (AgS_2) often associated with galena; as chloride (AgCl). • From lead alloys by the Pattinson process or the Parkes process; from the ores by the Mexican and other processes. • Known in antiquity.	White, highly lustrous, tough, very ductile, malleable metal; best conductor of heat and electricity known. Liquid silver dissolves oxygen. It is unaffected by the oxygen of moist air; its tarnishing is caused by the action of hydrogen sulfide. It dissolves in dilute nitric acid and in hot concentrated sulfuric acid.	Used for tableware, ornaments, coins, etc. U.S. sterling silver contains 90 per cent silver, 10 per cent copper. Lunar caustic is silver nitrate. This salt and the halides of silver are used extensively in photography. For electroplating, a bath of potassium argenticyanide is used.
Sodium (Na) At. wt. 22.9898 At. no. 11 Valence 1 Sp. gr. 0.97 M.P. 97.8° B.P. 892° C.S. Cb	In the sea as chloride (NaCl); in salt deposits as chloride, borate, and nitrate; in many complex silicates in rocks. • By electrolysis of fused sodium hydroxide (NaOH). • 1807; Davy.	Silver-white metal, soft as wax. Immediately tarnishes at ordinary temperatures. Like potassium, it is very active, uniting directly with many other elements and vigorously reacting with cold water.	Used in manufacture of chemicals. Sodium chloride (common salt) is a necessity of life for most animals, and is used in manufacture of hydrochloric acid, chlorine, and sodium compounds. Sodium carbonate, or washing soda, and sodium hydroxide are used for cleaning and for manufacture of soap and chemicals. Sodium bicarbonate is baking soda. The sulfate is known as Glauber's salt; the thiosulfate, by photographers, as "hypo."
Strontium (Sr) At. wt. 87.62 At. no. 38 Valence 2 Sp. gr. 2.55 M.P. 769° B.P. 1,384° C.S. C	As strontianite ($SrCO_3$) and celestite ($SrSO_4$). • By electrolysis of the chloride. • 1790; Crawford.	White metal; harder than sodium, softer than calcium; tarnishes to a yellow tint. Like calcium, it is active enough to react vigorously with cold water.	The nitrate and chlorate are used in fireworks for red color. All volatile compounds color the Bunsen flame red.
Sulfur (S) At. wt. 32.064 At. no. 16 Valences 2, 3, 4, 6 Sp. gr. rhombic 2.07 monoclinic 1.96 M.P. rhombic 112.8° monoclinic 119° B.P. 444.6°	Native, in combination with most metals as sulfides, and with some metals as sulfates. • By melting the free sulfur away from the rocky matrix (Frasch process), and subsequent purification by distillation. •Known in antiquity.	Natural sulfur is rhombic in crystalline form, yellow, brittle, and of vitreous luster. It is a poor conductor of heat and electricity. This and the monoclinic variety are soluble in carbon bisulfide, while amorphous sulfur is not. When heated, sulfur unites directly with most other elements.	Used to prepare sulfur dioxide (SO_2), which is used in making sulfuric acid and sulfites and for bleaching; also for vulcanizing rubber and in manufacture of black gunpowder. Sulfuric acid (H_2SO_4) is to the chemical industry what iron is to engineering. Thiosulfuric acid and its salts are important in the processing of film.
Tantalum (Ta) At. wt. 180.948 At. no. 73 Valences 2, 4, 5 Sp. gr. 16.6 M.P. 2,996° B.P. 5,425°	In tantalite and many other rare minerals. • By the action of sodium tantalofluoride (Na_2TaF_7). • 1802; Ekeberg.	Hard, silver-white metal; ductile and malleable when hot; of very high tensile strength. The hot metal can absorb 740 volumes of hydrogen. Not attacked by aqua regia.	Used for filaments for electric lamps until tungsten replaced it; in surgical instruments and in rectifiers; and as a substitute for platinum.
Technetium (Tc) At. wt. 99 At. no. 43 Valences 6, 7 Sp. gr. 11.50 M.P. 2,140° B.P. ?	Does not occur naturally. • By bombardment of molybdenum with deuterons. • 1937; Perrier, Segré.	The first artifically produced element. Resembles rhenium and manganese.	Used in research.
Tellurium (Te) At. wt. 127.60 At. no. 52 Valences 2, 4, 6 Sp. gr. rhombic 5.93 monoclinic 6.3 M.P. 449.5° B.P. 989.8°	Free and as tellurides. • By reducing tellurous acid (H_2TeO_3) by means of sulfur dioxide. • 1782; Müller von Reichenstein.	Crystalline variety is white, has metallic luster, and conducts heat and electricity. Precipitated variety is black and of lower density. Element is related to sulfur, but is more metallic.	Compounds find few applications; in coloring glass, gives silver a platinum finish. Telluric acid (H_6TeO_6) has basic as well as acidic characteristics, in keeping with the position of the element between the metals and nonmetals.
Terbium (Tb) At. wt. 158.924 At. no. 65 Valence 3 Sp. gr. 8.33 M.P. 1,356° B.P. 2,800°	In gadolinite, samarskite, and other rare minerals. • By electrolysis of the chloride. • 1843; Mosander.	Rare earth element. Closely resembles gadolinium.	Salts are almost colorless; oxide is almost black.
Thallium (Ti) At. wt. 204.37 At. no. 81 Valences 1, 3 Sp. gr. 11.86 M.P. 303.5° B.P. 1,457° C.S. T	In crookesite and in small quantities in many samples of iron pyrites. • Precipitated by zinc from solution obtained by suitable treatment of flue dust from sulfuric acid works. • 1861; Crookes.	Bluish-white, leadlike metal; rather soft and malleable, but of low tensile strength. Decomposes water rapidly at red heat and dissolves in dilute acids.	Forms two sets of salts; thallous and thallic. The salts, used in making optical glass, are poisonous. All the compounds show a characteristic green line in the spectrum.

TABLE OF ELEMENTS (Continued)

Element	Occurrence, Preparation, Date of Discovery, and Discoverer	Properties	Chief Compounds and Uses
Thorium (Th) At. wt. 232.038 At. no. 90 Valence 4 Sp. gr. 11.0 M.P. 1,700° B.P. 4,000° C.S. C	In monazite sand. • By reducing potassium thorium chloride with sodium, or by electrolysis of fused potassium and sodium chlorides. • 1828; Berzelius.	Metal has the color of nickel; can be burned in air. Hydrochloric acid attacks it slowly. Most isotopes are radioactive.	The nitrate $Th(NO_3)_4 \cdot 6H_2O$ is used in making Welsbach incandescent mantles, which consist of 99 per cent thorium dioxide. Alloyed with magnesium for special purposes.
Thulium (Tm) At. wt. 168.934 At. no. 69 Valence 3 Sp. gr. 9.34 M.P. 1,545° B.P. 1,727°	In gadolinite and other yttrium minerals. • 1879; Cleve.	Rare earth metal; has never been isolated.	Salts are a pale green that is destroyed very easily by minute quantities of erbium.
Tin (Sn) At. wt. 118.69 At. no. 50 Valences 2, 4 Sp. gr. 7.3 M.P. 231.89° B.P. 2,270° C.S. TC	As cassiterite (SnO_2). • After roasting, the ore is reduced by heating with carbon. • Known in antiquity.	Silver-white, rather soft, very malleable, ductile metal; practically unchanged in air. When heated, it may be burned in air. Dilute nitric acid is the only dilute acid that attacks it rapidly. When kept long at temperatures below 0°, ordinary tin changes to a brittle, gray, powdery form.	Much tin is used in coating iron as tinplate. A constituent of Britannia metal, pewter, solder, bronze, etc. Forms two sets of salts: stannous and stannic. Pink salt is used in dyeing. Mosaic gold is essentially stannic sulfide.
Titanium (Ti) At. wt. 47.90 At. no. 22 Valences 2, 3, 4 Sp. gr. 4.5 M.P. 1,675° B.P. 3,260° C.S. C	As rutile (TiO_2) and ilmenite ($FeTiO_3$). • By reducing the chloride ($TiCl_4$) by means of sodium. • 1791; Gregor.	Hard, brittle metal, resembling polished steel. May be forged at low red heat. Dissolves in dilute sulfuric acid and decomposes in steam at 800°. Unites easily with nitrogen.	Used in alloys, as a white pigment (paint and paper), and as a coloring for ceramics. Used as structural material in supersonic aircraft.
Tungsten (W) At. wt. 183.85 At. no. 74 Valences 2, 4, 5, 6 Sp. gr. 19.3 M.P. 3,410° B.P. 5,927° C.S. Cb	As wolframite ($FeWO_4$) and as scheelite ($CaWO_4$). • By reducing tungstic acid (H_2WO_4) with carbon at high temperatures. • 1783; the De Elhuyars.	Hard, brittle, gray metal, attacked by chlorine only at 250°, although it can be caused to burn in air. Slowly acted upon by dilute acids and even by water.	Used for filaments of incandescent electric lamps, giving an efficiency of 1.3 watts per candlepower. Tungsten steel has 5 per cent tungsten. Sodium tugnstates are used as mordants in dyeing. Was originally known as wolfram.
Uranium (U) At. wt. 238.03 At. no. 92 Valences 2, 3, 4, 5, 6 Sp. gr. 18.7 M.P. 1,132.3° B.P. 3,818°	As pitchblende, which contains U_3O_8. • By reducing the oxides with aluminum. • 1789; Klaproth.	White, lustrous metal; tarnishes in air and reacts slowly with cold water. Combines directly with many other elements.	All compounds are radioactive in proportion to their radium content. Glass to which uranium compounds have been added shows a greenish-yellow fluorescence. Used in nuclear weapons. Chief fuel for nuclear reactors.
Vanadium (V) At. st. 50.942 At. no. 23 Valences 2, 3, 4, 5 Sp. gr. 5.69 M.P. 1,890° B.P. 3,000° C.S. Cb	In a few rather rare minerals. • By reduction of the dichloride (VCl_2) in hydrogen. • 1830; Sefström.	Silver-white lustrous metal, harder than quartz. Does not tarnish or react with water at ordinary temperatures, but can be burned in oxygen.	Added to steel even in small quantities (0.2 per cent), it increases the tenacity and elastic limit without reducing ductility.
Xenon (Xe) At. wt. 131.30 At. no. 54 Valences 0, 2, 4 Sp. gr. (liquid) 3.52 M.P. −111.9° B.P. −107.1°	In minute quantities in the air, 1 volume in 170 million. • By fractionation of liquid argon. • 1898; Ramsay, Travers.	Transparent, colorless, odorless gas. Densest member of the noble gases.	Once thought to be chemically inert; forms fluoride (XeF_2, XeF_4, XeF_6), oxide (XeO_3), and hexafluoroplatinate (Xe_2PtF_6) compounds.
Ytterbium (Yb) At. wt. 173.04 At. no. 70 Valence 3 Sp. gr. 7.01 M.P. 824° B.P. 1,427°	In gadolinite, euxenite, and other rare minerals. • By electrolysis of the chloride. • 1878; Marignac.	Rare earth. Forms colorless salts.	Compounds show a characteristic spark spectrum.
Yttrium (Y) At. wt. 88.905 At. no. 39 Valence 3 Sp. gr. 5.51 M.P. 1,495° B.P. 2,927°	In gadolinite, euxenite, and other rare minerals. • By electrolysis of sodium yttrium chloride. • 1794; Gadolin.	Gray, lustrous metal.	Chloride yields a characteristic, though complex spectrum.
Zinc (Zn) At. wt. 65.37 At. no. 30 Valence 2 Sp. gr. 7.14 M.P. 419.4° B.P. 907° C.S. H	As zinc blende (ZnS), calamine ($ZnCO_3$), zincite (ZnO), etc. • After roasting, the ore is reduced by coal, the metal distilling off. • Known in antiquity.	Bluish-white, lustrous, brittle metal; malleable and ductile at 120°; tarnishes in moist air. Reacts slowly with cold water, and rapidly when heated in steam. Dissolves in dilute acids and sodium hydroxide solution.	Used for roofs, gutters; galvanic batteries. Iron galvanized with zinc, preventing rust. Zinc alloyed with copper to make brass. In paint, zinc oxide is less toxic than lead oxide. Salts used in medicine, chloride and sulfate in antiseptic solutions.
Zirconium (Zr) At. wt. 91.22 At. no. 40 Valence 4 Sp. gr. 6.4 M.P. 1,852° B.P. 3,578°	As zircon ($ZrSiO_4$). • By reducing the oxide (ZrO_2) with carbon in an electric furnace. • 1789; Klaproth.	Hard, gray metal remaining bright in air; oxidizes slowly at white heat. Dissolves in aqua regia and caustic potash solution.	Oxide is contained in some incandescent gas mantles; is used for furnace linings and as a cleansing agent in metallurgy. Increases tensile strength of armor plate. Carbide is an abrasive.

National Geographic Magazine. Reproduced from the National Geographic Magazine by special permission.

The First Photograph Ever Made Showing the Actual Curvature of the Earth

EARTH SCIENCES

The earth sciences are *geology*, a study of the lithosphere, or solid part of the earth; *oceanography*, a study of the oceans, which are included within the hydrosphere, or liquid part of the earth; and *meteorology*, a study of the atmosphere, or gaseous part of the earth.

The earth sciences are grouped within the natural sciences, or sciences of natural phenomena; other natural sciences include zoology and botany. The earth sciences are largely concerned with inorganic substances, or nonliving matter.

Because the earth sciences are devoted to the study of natural objects, they are, almost by definition, interdisciplinary. The geologist, for example, uses chemistry to determine the composition of rocks and minerals, and biology to classify the fossils found in rocks.

GEOLOGY

Geology is divided into a number of somewhat separate branches of study. *Paleontology* is the study of ancient life—the evolution of animals and plants, their various adaptations and extinctions, and their environments. The need for metals and for fuels other than wood has led to *economic geology*, the study of commercially usable fuel and mineral deposits in the earth's crust. *Mineralogy* explains the origin of ores and minerals, the building blocks of rocks. The search for minerals and petroleum has led to *structural geology*, the study of rock structures and how they are formed. The study of rocks themselves is called *petrology*. The application of physical investigations to geology is known as *geophysics*.

The two general divisions of geology that include the branches mentioned above are physical geology and historical geology. *Physical geology* is the study of the earth as we see it around us; of the physical development of the face of the earth and its associated rocks. *Historical geology* is the study of the development of our planet and its life.

Physical Geology.—To understand the earth's surface we must know what it is made of, how it changes, and the processes that cause it to change. A number of processes can operate at the same time. Some operate quite abruptly with spectacular effect, but these are only occasional. Most act slowly but continuously, and have done so for billions of years.

■WEATHERING.—*Weathering* is one of the earth's slow and continuous processes. The crust of the earth is made up of many kinds of rock, each with a different chemical composition. The elements in rock may be attacked by elements in the atmosphere, with a resulting change in composition that may also alter the durability and size of the particles. On the other hand, physical forces may break up rocks, with little or no chemical change. The first process is called chemical weathering; the second, mechanical weathering.

In *chemical weathering*, or decomposition, certain minerals in rocks are affected by the carbon dioxide, oxygen, or water vapor in the atmosphere. For example, when feldspar —a silicate mineral common in granitic rocks—is attacked by carbon dioxide dissolved in water vapor, it changes to clay and soluble compounds. Oxygen and water vapor attack the iron in rocks, causing both a change in the size and composition of particles and a familiar rust-colored stain. Chemical weathering is more common in hot, humid climates than in cool, dry climates. In some places in the Arctic and Antarctic, unweathered rock is exposed at the surface; but toward the equator there is a thickness of overlying weathered rock. Weathering may extend several hundred feet below the surface in parts of the wet tropics.

Mechanical weathering, or disintegration, does not change the chemical composition of rocks or minerals, but breaks them into smaller particles. If water seeps into a crack in a rock, then freezes and expands, it may break the rock. This is especially common where freezing and thawing alternate rapidly. The roots of plants may push into cracks in a rock and grow, slowly forcing the sides of the cracks apart. Burrowing animals often expose rocks to the attack of mechanical weathering.

Where the earth's surface is relatively level, the layer of weathered rock gradually becomes thicker. The weathering processes become slower with depth, however, until they almost stop. If the surface is relatively vertical, the weathered fragments continually fall off and expose a fresh surface to the effects of weathering, the process continuing until the slope almost disappears. The weathered material and fragments usually accumulate at the bottom of the slope, forming *talus*, or *scree*.

■EROSION.—Weathering softens and loosens rocky material but does not move it. The process of wearing away and transporting weathered particles is called erosion. The agents of erosion are gravity, wind, water, and glacial ice.

Gravity erosion is called *mass wasting*. The most spectacular form of mass wasting is the *landslide*. Here thousands of cubic yards of rock, soil, trees, and other debris move downslope. Most landslides start on steep water-soaked slopes. Mass wasting also takes the form of the *mudflow*. The mudflow is usually a narrow tongue of mud that flows down a steep valley.

Wind erosion can occur only if the surface of the ground is composed of loose particles that are small enough to be moved. These conditions exist in deserts, along lake shores and seashores, and, less commonly, in fields where plant cover is scarce or absent. Fine dust blows entirely out of such areas, leaving only the heavier and coarser sand grains behind. The dust may accumulate elsewhere in a thick deposit called *loess*, which is usually very fertile. As it compacts, the loess develops a vertical, columnar structure, and water can pass downward through it easily. Therefore it does not erode readily, but tends to stand in high banks if cut into by a river or man-made excavation. Loess deposits, thought to have blown out of the Gobi Desert, are common in parts of China. In Europe there are deposits of loess that probably came partly from the Sahara Desert and partly from glacial debris deposited during the Ice Ages. Other loess deposits which are generally considered to be of Ice Age origin, occur in the upper Mississippi valley in the United States.

As desert sand is blown along the ground by the wind, the grains cut and shape the surfaces they move over. This action sometimes carves weird shapes in rocks by wearing away the soft material while leaving the harder rock standing out in relief. Where sand has been blown off the desert floor, there is *desert pavement*, usually covered with sand-blasted stones and bare rock. Elsewhere the sand accumulates in wind-deposited *dunes*, shaped much like snowdrifts. A dune is formed when the velocity of the wind is lessened by some obstacle, and the wind-borne sand is dropped. Oncoming sand is rolled up the windward side of the dune, and the grains drop down the lee side through the force of gravity. In general, the side toward the wind is less steep. Dunes are slowly moved before the wind; this movement often exposes objects previously buried. Where there is little shift in the wind's direction, the dunes tend to be *crescentic* as seen from above; but most dunes have an irregular outline.

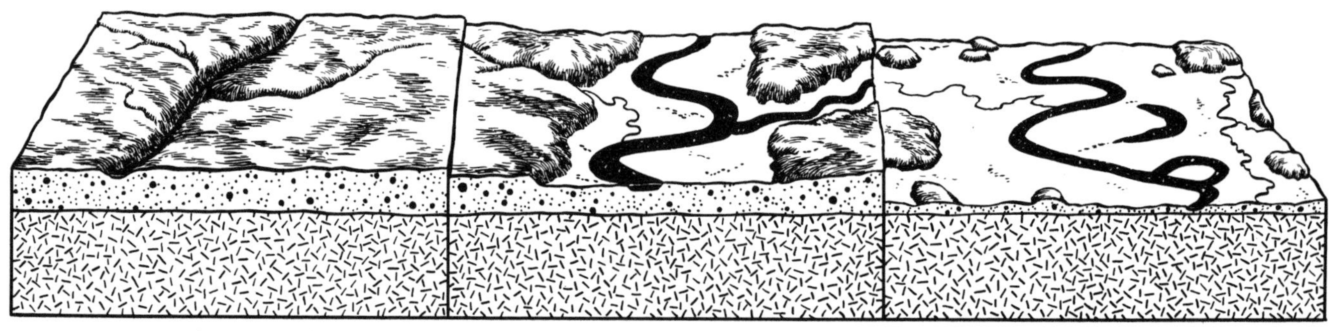

STREAM EROSION. A stream with a steep gradient (left) has a high velocity and cuts downward to form a deep, V-shaped valley. A stream with a low gradient (center and right) has a low velocity and cuts sideways, forming a wide valley with a flat floor, or floodplain.

Stream erosion is among the most active agents in lowering the elevation of the land. Clear water is a poor cutting agent; but armed with fragments of weathered rock, it can cut even the solid rock of a stream bed. Very fine material, like mud, is carried in suspension; coarse material is pushed along the stream floor; and the intermediate sizes are rolled and bounced over the bottom. All of this material is the stream's *load.* Sediment in suspension may travel the river's entire course without being dropped; the largest fragments move only during floods, when the current is exceptionally strong.

A river valley is usually the product of the stream flowing through it. A narrow, V-shaped valley shows that the river has a steep *gradient,* or longitudinal slope. Such a river actively cuts downward as it flows and has waterfalls and many rapids. An example of a downward cutting river is the Colorado River in the Grand Canyon.

A river valley which is wide and flat is produced by a stream with a low gradient. Such a stream is close to its *base level,* the landward projection of the surface of the ocean or lake into which the river flows. The stream does not cut into its bed but into its valley walls. This may cause the river to *meander,* or wind back and forth, forming great meander loops. The river flows across a *floodplain* composed of sediment—sand, silt, and clay particles—deposited by the river in time of flood.

Natural *levees* may build up along the river channel where the initial decrease in velocity occurs when the river overflows its banks. Man may enlarge these levees in an effort to contain the river in its channel and prevent inundation of the floodplain.

The lower Mississippi River is an example of a stream with a broad, flat floodplain. Here one can see *oxbow lakes,* which are formed by parts of meander loops that are isolated when the river cuts across the neck of land at both ends of the curve.

The land surface drained by streams may not be stable, but move upward or downward. An upward moving land mass can cause a lateral cutting stream to cut downward and produce a narrow valley with steep walls. In such a case, great meander loops can be entrenched or cut down below the general level of the land. The incised meanders, or Goosenecks, of the San Juan River in southern Utah were once meanders on a broad, flat floodplain.

Where the land surface sinks downward, the river valley may be drowned by an apparent rise in sea level. Chesapeake Bay and its adjacent tributaries are part of such a drowned river system.

Groundwater.—Almost everywhere from a few feet to several thousand feet below the surface, water is present in the cracks and pore spaces of rocks and unconsolidated sediments. Called *groundwater,* it is the source of all well water. In some places the consumption of groundwater has been so great that the *water table,* or upper surface of the groundwater, has brought sea water into many wells. Inland, however, the salt water that is sometimes pumped from deep oil wells is usually water that was trapped at the time when the rocks were deposited in ancient seas.

Easily soluble rocks, such as limestone, are sometimes dissolved by circulating groundwater to form caves below the ground and *sinkholes,* or *swallow holes,* on the surface. In such areas, the drainage may be entirely underground, with no surface streams in evidence. The subterranean water may come to the surface elsewhere in large springs, like some of those in Florida. Where the rock is less soluble, instead of solution there may be commercial mineral deposits enriched by groundwater.

Oceans and Lakes.—On the shores of oceans, seas, and large lakes, waves constantly attack the land. When a wave recedes, it can cause a *rip tide,* or undertow. This continual bombardment by waves gradually erodes the shore. In areas where there are high tides, a wider vertical range of wave action is possible. Incoming waves sometimes force air into cracks in a rock and, in so doing, burst the rock apart. Waves may also carry rocks and pebbles that shatter the loose shoreline material and thereby cause erosion. The retreating waves then drag the finer debris into deep water and return the coarser material for another attack. This wave action gradually cuts back the shoreline, eventually forming a wavecut platform bordered on the land side by a cliff. The platform may become so wide that the waves hit the cliff only during storms. In general, this kind of erosion is more advanced along the coasts of oceans and seas than on lake shores.

Ocean currents can move sediments produced by wave erosion or deposited at the mouths of rivers. Material eroded from headlands can be deposited nearby, forming a *spit, bar,* or *hook* that extends from the headland. Such deposits are usually built and maintained by currents flowing parallel to the shore. As erosion progresses, the headlands are cut back, forming sea cliffs, and the shoreline becomes straighter.

Evidence of wave action is emphasized where the shoreline moves up or down in relation to the sea level. A shoreline that rises is called an *emergent shoreline,* and one that sinks is called a *submerged shoreline.* In some areas, there are features of both submerged and emergent coasts combined in a *compound shore development,* which may have both drowned valleys and long shore bars. The best harbors are usually found along submerged shorelines.

River water entering the sea is slowed down, and its sediment is deposited to form a large, low, flat, swampy area called a *delta.* A delta is roughly triangular, with the apex of the triangle pointing upstream, thereby splitting the river into a number of channels called *distributaries.* As the water slows down, the heaviest and coarsest sediment is dropped first; the finest, last. This forms a series of inclined layers of sediment, called *fore-set beds,* on the slope of the land margin. Beyond the fore-set beds are roughly horizontal layers of fine material called *bottom-set beds.* On top of the fore-set beds near the land margin are the finest sediments, the *top-set beds.* Deltas are fertile because of the constant addition of fresh soil.

Sea water contains many dissolved minerals. These include, in order of their abundance, sodium chloride, magnesium chloride, magnesium sulfate, calcium sulfate, potassium sulfate, and calcium carbonate. These soluble salts come from the chemical weathering of rocks on the land. They are carried by ground water to the rivers and then to the seas.

River water contains much more calcium in solution than sodium or chlorine. In the ocean, various organisms use calcium in the construction of external shells and internal skeletons. It is estimated that the calcium content of sea water would double in about 100,000 years if it was not precipitated in this way.

In contrast, the amount of sodium chloride, or salt, in sea water has increased through geologic time. Where the rate of evaporation is high, as in parts of the Indian Ocean, the salinity of sea water is higher than the average. Sodium chloride can be precipitated as the mineral halite when salinity exceeds the saturation point. This occurs in a restricted basin, such as a small lagoon or a large bay, where high-salinity sea water caused by evaporation cannot be exchanged for sea water of normal salinity.

Geologically speaking, lakes are temporary inland bodies of standing water. They may contain either fresh or salt water, depending on the presence or absence of outlets. A lake with no outlet eventually grows salty because the water that drains into it tends to evaporate, leaving behind an ever-growing concentration of dissolved minerals. Streams that empty into a lake bring sediment with them, and in the quiet water this sediment settles. This explains why water flowing out of a lake is usually clear. Lakes gradually fill with this sediment and turn first into swamps, then into level plains with streams wandering over them. Most lake deposits are thin-bedded clays and silts, which may contain remains of land-dwelling or fresh-water plants and animals. The world's coal deposits are the remains of lush vegetation that once grew in swamps that were cool enough so that the plant material did not rot, but was preserved in the form of carbon.

In the geologic past, large deposits of sodium chloride were precipitated in restricted basins. These salt deposits are mined in such states as Kansas, Louisiana, Michigan, New York, Ohio, and Texas, and in countries such as England, France, Germany, India, and China. The halite, or native salt, is ground to form rock salt and table salt. It is also used as a source of chlorine, sodium, and compounds of these elements.

Glaciation.—In Antarctica and Greenland, and on the upper slopes of high mountains elsewhere, there are *glaciers*—accumulations of snow and ice that move slowly over the ground. Some 10,000 to 20,000 years ago, during the Ice Age, the northern parts of North America, Europe, Asia, and much of southern South America were covered with ice just as Greenland and Antarctica are today. Such conditions result from a climate so cold that the winter's snows never melt entirely. The snow accumulates, and the lower layers gradually turn into ice under the increased weight.

Glaciers are generally divided into two types: *mountain,* or *alpine, glaciers* and *continental glaciers,* or *ice sheets.* Alpine glaciers have been compared to rivers of ice, but they are more like bulldozers that slowly plow down a valley, straightening it out and deepening it. When an alpine glacier eventually disappears, it leaves a U-shaped valley with steep sides and a *cirque,* or amphitheater-like depression, in the mountainside at its head; the cirque often contains a lake. Alpine glaciers produced the rugged landscapes of the Alps and Himalayas.

Continental glaciers, on the other hand, tend to smooth out the surface over which they move. At the margin of the ice, a continental glacier pushes up a mass of irregular, unsorted rock debris called an *end moraine.* On the side of the moraine away from the ice is a relatively level area, called an *outwash plain,* composed of fine material washed away from the end moraine by meltwater from the glacier. There are irregular *kettle holes* in both the end moraine and the outwash plain, caused by the melting of buried blocks of ice. Behind the end moraine is the *till plain,* or *ground moraine,* a wide, gently undulating area of unsorted debris called *till,* which was carried in and under the ice and was dropped when the glacier receded. Continental glaciers also form *eskers,* or winding ridges of sorted till, and *drumlins,* or long, narrow hills of unsorted till. When it melts back, the ice also deposits *erratics,* or single boulders composed of rock foreign to the area. There may be many lakes in end moraines, but few in ground moraines or outwash plains.

Rocks.—The crust of the earth is made up of three main kinds of rock: igneous, sedimentary, and metamorphic. *Igneous rock* has crystallized and hardened from a hot, liquid mass. *Sedimentary rock* is derived from fragmental material carried as sediment, or in solution by water. *Metamorphic rock* is formed by the alteration of other kinds of rock by pressure or heat. Igneous rock makes up the greatest volume of the crust, but is often hidden under layers of sedimentary or metamorphic rock. Compared with the bulk of the crust, the sedimentary rocks are relatively insignificant, but spread over a greater area. Metamorphic rocks are common in areas of mountain-building.

■SEDIMENTARY ROCKS.—There are three groupings of sedimentary rock, based on origin. Those that were carried in suspension are *clastic sediments;* those that were deposited from solutions are *chemical,* or *precipitated, sediments;* those that form such deposits as coal beds and coral reefs are *organic rocks*—that is, they are composed of the remains of either plant or animal organisms.

Clastic sedimentary rocks are classified largely according to the size of

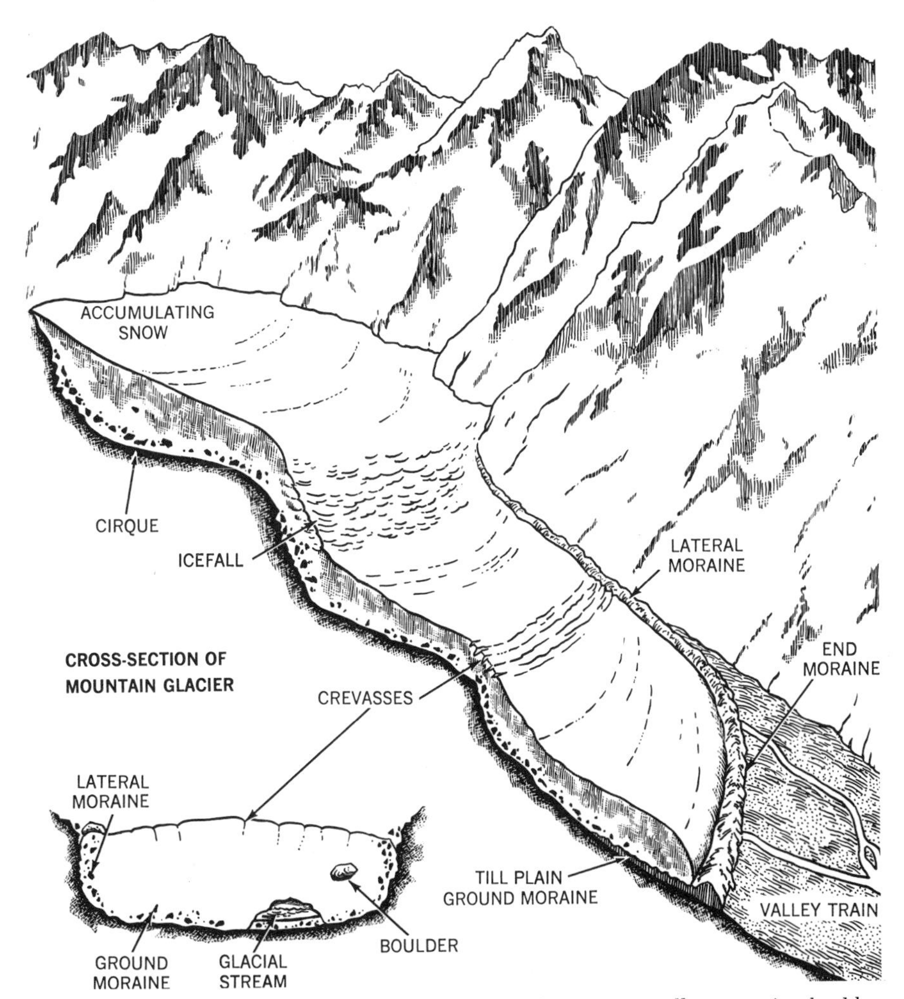

MOUNTAIN GLACIERS flow from high snowfields through mountain valleys, gouging boulders from mountain valleys and walls and changing the configuration of the terrain as they pass.

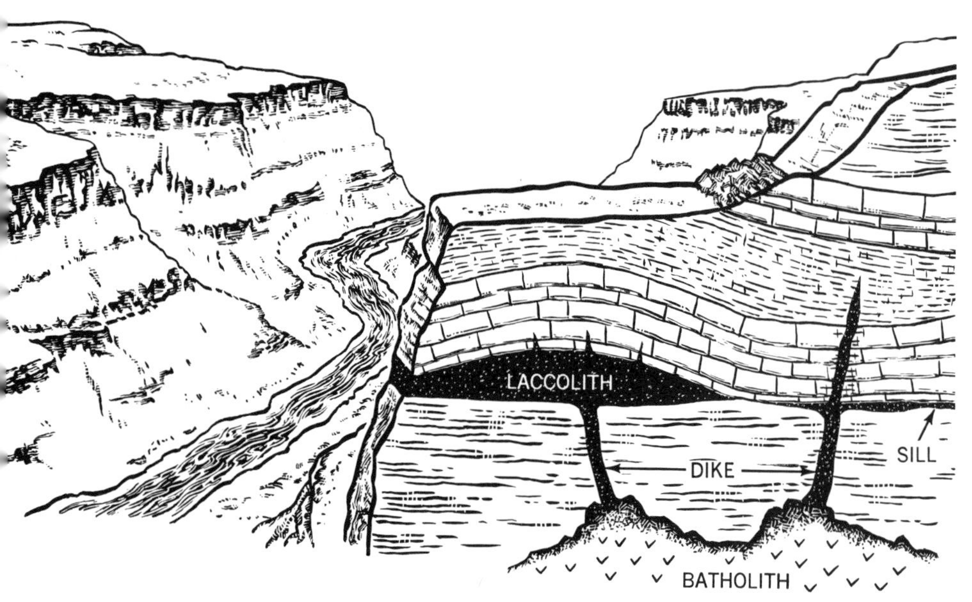

VOLCANIC INTRUSIONS such as the batholith extend deep beneath the earth's crust. From the igneous batholiths emerge lens-shaped laccoliths, vertical dikes, and horizontal sills.

the fragments of which they are composed. The finest sediments are deposited as *clay*—a soft, slippery, plastic, mudlike material that varies greatly in color. As clay becomes more compact, the water between the particles is squeezed out, and the clay becomes *shale*. Clay is used extensively to make brick, china, and similar products. *Sand grains* are much coarser than clay—up to about 2 mm in diameter. Newly deposited sand is loosely packed, but with time compacts into *sandstone*. There may be *cement*, such as calcium carbonate, or iron oxide present, but usually not enough to fill the spaces between sand grains. Therefore sandstone is a good reservoir for groundwater and petroleum. Some sandstone is compact enough for use as building stone and some, composed almost entirely of quartz grains, is pure enough to be used as a source of silica for glassmaking. The coarsest clastic material, composed of gravel, cobbles, and even boulders, may become cemented to form a rock called *conglomerate*. In conglomerates, the particles are rounded from abrasion during transport, usually by running water. In a similar but less common kind of rock known as *breccia*, the particles are angular because they have not traveled far from their source. Many breccias are of volcanic origin. In a few places, glacial till has been compacted into a rock called *tillite*. There are some large deposits of tillite in South Africa, indicating that this region was once glaciated.

Among the precipitated sedimentary rocks are rock salt (halite, or sodium chloride), fertilizer salts (sylvite, or potassium chloride, and associated chlorides), gypsum, anhydrite (calcium sulfate), sedimentary iron ore deposits (hematite), dolomite (calcium magnesium carbonate), and, most abundant, limestone (calcium carbonate). During the past in many parts of the world there have been times when great quantities of sodium chloride were precipitated as rock salt, or halite, in beds hundreds of feet thick. Halite, which has only to be crushed to be used as table salt, is mined in great quantities in North America from New York State to Michigan and along the Gulf Coast, especially in Louisiana. Near Strasbourg, France, and Carlsbad, New Mexico, there are sedimentary basins that are the chief sources of potassium salts. The salts are interbedded, indicating that the composition of the sea water changed with alternating evaporation and flooding. In other parts of the world there are deposits of gypsum, which is used in making plaster. After sandstone and shale, the most abundant sedimentary rocks are limestone and dolomite, which can be of either chemical or organic origin. Limestone is used decoratively on many large buildings. It was used in the past to build the great cathedrals of Europe and the pyramids of Egypt.

The many coral islands in the Pacific Ocean are for the most part composed of organic limestone. Many of these limestones are full of shells or the remains of corals and other lime-secreting organisms, which indicate the kinds of animals that lived when the rocks were deposited. In a few places there are accumulations of the siliceous cases of microscopic plants called diatoms. These surface-dwelling marine plants sink to the bottom when they die, and accumulate by the billions. They form *diatomaceous earth,* or *diatomite,* which is used in sugar refining and oil refining, and for heat insulation. Some deposits, now exposed on land, are up to 100 feet thick. *Phosphorus,* an important fertilizer material, is obtained from deposits of *phosphatic rock* of organic origin. *Coal* is organic sedimentary rock that accumulated in large swamps where the temperature of the water was low enough to keep the material from decaying completely. Coal seams range in thickness from a fraction of an inch to over 400 feet. A few seams as thin as one foot have been mined, but those less than two feet thick are not usually worked by underground mining. If they are not too deep, however, they can be worked by *strip mining,* or surface mining. Other organic materials that originate in sedimentary rocks are *petroleum* and *natural gas* which, together with coal, are known as *fossil fuels.*

■IGNEOUS ROCKS.—By far the most abundant rocks are the igneous rocks that crystallized from hot liquids, such as *lava,* which flows from erupting volcanoes, cools, and hardens to form such rocks as *basalt* and *felsite.* Lava can be any color, and some types contain gas bubbles. The amount of bubbles in lava and its color and fluidity depend on its formation temperature and its composition. Lava that is viscous, dark, and full of large bubbles forms *scoria;* whereas lava that is light gray, relatively fluid, and full of small bubbles forms *pumice.* Scoria and pumice are types of volcanic glass that cooled so rapidly crystals could not form. Pumice is so lightweight that it will float on water. Volcanic glass that does not contain bubbles is called *obsidian.* Other volcanic rocks are crystalline, even though the crystals may be microscopic. These rocks are either *felsic* or *mafic,* depending on their composition. The felsic rocks are usually white to gray or pink; and the mafic ones are dark green to black. Felsic rocks are often called *felsites;* and the mafic rocks, *basalt.* Basalt is a fairly common rock, underlying such large areas as the Columbia Plateau in the northwestern United States and the Deccan Plateau in India. Basaltic rocks also underlie most ocean basins.

Underlying the lava flows and sediments on the continents are large quantities of igneous rock with easily visible crystals. If this coarse-grained rock is light-colored, like felsite, it is called *granite.* Granite is widely used as building stone and for curbstones and cobblestones. It is strong and durable. The corresponding mafic rock is *gabbro,* a coarse, black rock that differs little from basalt in composition. *Porphyry,* an igneous rock that contains relatively large crystals in a fine-grained matrix, may be either felsic or mafic.

Igneous rocks do not occur in beds like sedimentary rocks, since they are intruded, or injected, from below into preexisting rocks, or flow on the surface as lava. Tabular, relatively horizontal intrusions are called *sills.* Tabular intrusions that are more or less vertical are called *dikes.* Sills that have arched up the overlying rocks are known as *laccoliths.* The largest igneous intrusions are *batholiths,* which have no recognized floor and may have incorporated some of the overlying rock. These bodies underlie many mountain ranges.

■METAMORPHIC ROCKS.—Alteration, or *metamorphism,* is caused by heat or pressure. *Gneiss* is one of the most easily recognized metamorphic rocks. It can be altered granite, with the mineral grains so oriented that the rock is banded. *Slate,* or altered shale, is a fine-grained metamorphic rock that can be split into thin layers. This again is caused by orientation of the mineral grains. Slate is used for roofing and blackboards. *Schist* is similar to slate, but the mineral grains are larger and visible to the unaided eye. The grain shows foliation. Sandstone becomes *quartzite* when it is metamorphosed. The *carbonate rocks* are altered to *marble,* which may be fine-grained or coarse-grained. Very fine-grained marble is used in sculpture, and coarse-grained marble is often used as building stone. In marble, the original fragments have recrystallized, destroying any fossils. Partly recrystallized marble, which is sometimes used for interior decoration, usually contains some fossil remains.

At the boundary between many igneous intrusions and the surrounding rock there is a zone of contact metamorphism, where the surrounding rock has been baked or burned. There may also be a zone of chemical alteration caused by the movement of solutions from the igneous mass into the surrounding rock or by the dissolving of some of the preexisting rock by the molten intrusion. Valuable mineral deposits are formed in this manner.

Diastrophism.—The deformation of the earth's crust is called *diastrophism.* Beds of rock are sometimes bent, and rock materials sometimes abruptly change. These changes are caused by the *folding* or *faulting* (breaking) of the rocks in the crust by natural phenomena, such as internal pressure and strain. Uparched folds are called *anticlines,* downfolds are called *synclines,* and simple folds from one level to another are called *monoclines.* When layers of rock are faulted, beds that were once continuous are offset laterally or vertically. Sometimes younger beds are found under older ones without evidence of overturning, indicating that the older deposits have been thrust over the younger ones. In all cases the opposite sides of the fault zone have been moved in different directions. More or less vertical faults often have displacements of up to a few thousand feet, whereas relatively horizontal faults may have displacements measured in miles. Where a fault intersects the surface there may be a definite *scarp,* but this, like all surface features, will eventually disappear through erosion. Some faults continue to be active, but a great many are stationary. Old fault zones may be filled with mineral veins of economic value because they often form channels for mineralizing solutions. The study of faults and folds is important because of the relation of geologic structures to the production of oil, gas, and various minerals. Other aspects of diastrophism are *volcanoes* and *earthquakes.*

VOLCANOES.—Active volcanoes occur mainly in two belts. One follows the shores of the Pacific Ocean; the other extends around the earth from east to west, crossing the Pacific belt in Indonesia and Central America. However, traces of ancient volcanism can be found in many other parts of the world. Volcanoes are relatively quiet most of the time, but now and again they explode violently, ejecting thousands of tons of liquid rock and fragmental material, along with large quantities of water vapor and other gases. Many volcanoes form large, cone-shaped mountains, like Mt. Hood in Oregon. Some build cones of lava that have low, gently sloping sides, like volcanoes in the Hawaiian Islands. Others are combinations of lava and cinders and form steep cones, like Mt. Vesuvius in Italy. Some volcanic cones cave in or blow up to form *calderas,* like Crater Lake in Oregon. If a volcano becomes dormant, the cone starts to wear away, leaving a core standing alone or with dikes radiating from it. Ship Rock in New Mexico had such an origin. The estimate of active volcanoes in the world is 400 to 500.

EARTHQUAKES.—*Earthquakes* are the result of a sudden release of strains in the earth's crust. If a strain is released slowly, there is little noticeable effect; but if the release is sudden, the rocks in the crust move against one another and set up vibrations. When these vibrations reach the surface, loose soil and surface objects are shaken, sometimes violently. The violent earthquake belts of the world are roughly the same as the volcanic belts, but no region is immune to earthquake shocks.

Conclusion.—The information quite literally dug out of the earth by geologists is used in many fields, from exploration for fuels and building materials to the location of power and water-storage dams. Even the disposal of atomic wastes involves geology, because it is essential that these wastes not leak through porous rocks or fault zones and cause harm to life. The earth's geological processes, such as erosion and diastrophism, are slow and continuous. Spectacular events, such as earthquakes and volcanic eruptions, are relatively infrequent, and they constitute only minor aspects of the continuing changes that affect the earth.

—E. Willard Berry;
S. Duncan Heron, Jr.

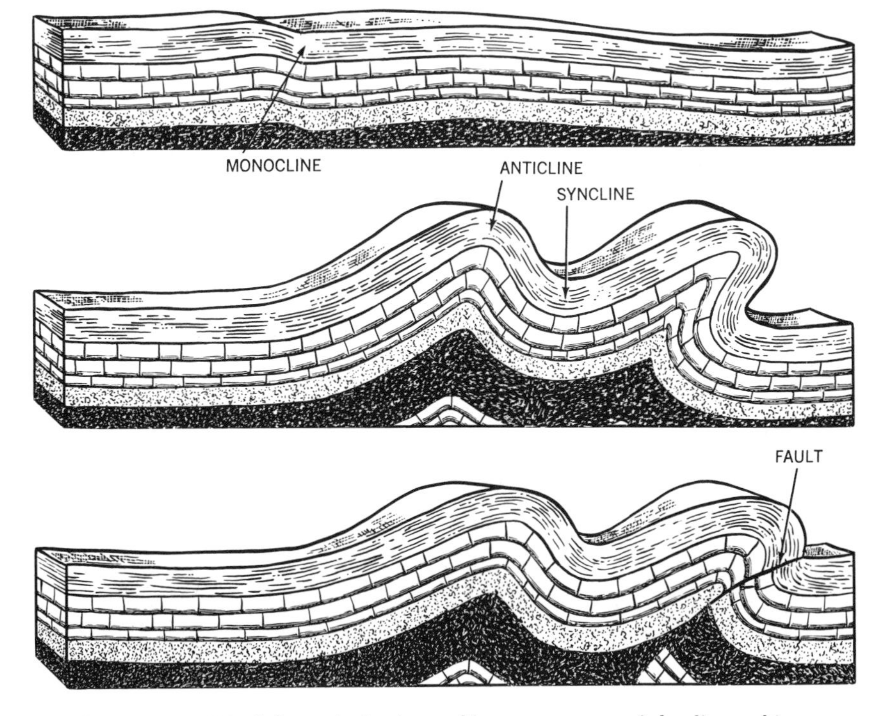

FOLDING AND FAULTING of the rocks in the earth's crust are part of the diastrophic process.

MINERALOGY

History.—Mineralogy, a branch of geology, is a systematic, integrated science intimately related to chemistry and physics. Historically, it is one of the oldest sciences practiced by man. Minerals were known to, and used by, early man throughout the Stone Age, the Bronze Age, and the Iron Age. As far back as 3400 B.C. the inhabitants of the valleys of the Tigris and Euphrates and surrounding areas were searching for, mining, and polishing many-colored gem stones. They were familiar with amethyst, carnelian, agate, beryl, turquoise, lapis lazuli, malachite, jasper, chalcedony, and garnet. In the societies of those days, gems were as much of a status symbol as they are today. They were of special importance to the Egyptians, who used them to adorn the bodies of their dead, which they considered sacred. These gem stones were also buried in ancestral tombs, to be taken along by the deceased and enjoyed in the afterlife. It was from such religious and social practices that mineralogy was born.

The first mineralogy textbook was the *Book of Stones,* written by Theophrastus (c. 372–c. 287 B.C.), a student of Aristotle. In this book he classified 16 minerals under three groupings: metals, stones, and earths. Pliny the Elder (23–79) described minerals and mineral deposits in his books on natural history, and Georg Bauer (1494–1555), better known as Agricola, published an outstanding treatise on economic mineralogy over 400 years ago. Through the years, mineralogy has prospered, and its scope has enlarged to encompass many new areas.

Today mineralogy is divided into a number of branches, including chemical mineralogy, crystallography, descriptive mineralogy, determinative mineralogy, and physical mineralogy. The mineralogist employs geological methods to map rock formations, mineral deposits, and structures of the earth's crust. He collects mineral specimens, tests them by sight, touch, taste, and weight—and then examines them further in the laboratory, using the techniques of the chemist and the physicist.

Minerals.—A *mineral* is defined as a solid, homogeneous, natural substance with definite physical properties and a chemical composition that is fixed within narrow limits—its composition must be such that it can be expressed with a chemical formula.

In order to be classified as a mineral, a substance must be formed by inorganic processes. Thus coal, oil, amber, and pearls are not minerals, for they are produced from plant and animal substances. Also, materials such as the man-made sapphire are not minerals, even though they may be chemically, structurally, and physically identical with the natural substance.

Minerals are the building blocks of the earth's crust. Yet of all the known minerals, only about 50 are rock-making, and only about 20 of the 50 could be said to be essential constituents of rock.

Crystal Structure.—Minerals are crystalline; that is, their atoms or groups of atoms are arranged in a symmetrical, three-dimensional, geometric pattern called a *crystal lattice.* There are a few minerals, known as *mineraloids,* that have a haphazard internal structure. These are said to be *amorphous.*

When minerals are free and uncrowded, they develop as *crystals*—solid bodies having smooth plane surfaces, or *faces.* Crystals give minerals characteristic outward shapes that reflect the internal crystalline arrangement. The angle between corresponding faces of any given mineral is always the same, no matter what the size of the specimen or its origin. An important part of mineralogy consists of measuring these angles. This is done with an instrument called a *goniometer.*

All similar faces on a crystal constitute a *form.* The most common forms are cubes, prisms, and pyramids. Crystal faces on the minerals contained in rocks are seldom distinguishable because they are so closely packed.

■**CRYSTAL FORMS.**—The symmetry in the geometrical form of a crystal is due to regularities in the positions of the corresponding similar faces and edges. Because of this regularity, crystals have planes and axes of symmetry. A *plane of symmetry* divides the crystal into two similar halves, one the mirror image of the other. Crystal forms are divided into six systems: cubic, tetragonal, orthorhombic, monoclinic, triclinic, and hexagonal. The systems are identified by the relative lengths of the axes of symmetry and the angles that the axes make with one another. In *cubic* crystals, the axes are all of the same length and are perpendicular to one another. In *tetragonal* crystals, two axes are of the same length and different from the third; all three axes are perpendicular to one another. In *orthorhombic* crystals, the three axes are of different lengths and perpendicular to one another. In *monoclinic* crystals, the three axes are of different lengths; two of the axes are perpendicular to one another but the third is not perpendicular to either. In *triclinic* crystals, the three axes are of different lengths and none is perpendicular to either of the other two. In *hexagonal* crystals, there are four axes—three of them are of identical lengths and at angles of 60° to one another in one plane; the fourth is of a different length and perpendicular to the plane of the other three.

■**HABIT.**—The crystal form that a mineral characteristically takes in response to rate of growth, heat, and pressure is called its *habit.* In the case of a mineral made up of single and distinct crystals, its habit may be *acicular* (needlelike), *capillary* and *filiform* (hairlike and threadlike), or *bladed* (elongated and flattened).

If the mineral is made up of a group of distinct crystals, its habit may be *dendritic* (branchlike), *reticulated* (latticelike), *divergent* (or *radiated*), or *drusy* (covered with a layer of small crystals).

When a mineral is made up of parallel or radiating groups of single crystals, its habits may be *columnar, bladed, fibrous, stellated* (starlike), *globular, botryoidal* (grapelike), *reniform* (kidney-shaped), or *mammillary* (breastlike). A material consisting of scales could have a habit that is *foliated, micaceous* (capable of being split into very fine sheets), *tabular,* or *plumose* (featherlike). A mineral habit can also be *granular, stalactitic* (with pendant cylinders or cones), *oölitic* (like fish roe), or *pisolitic* (rounded and pea-sized).

Mineral Identification.—When the hand specimen of a mineral is examined, the first thing seen is its outward appearance. It can be granular, compact, or earthy. Other physical aids in identification are the color, luster, hardness, cleavage, tenacity, specific gravity, and magnetism.

■**COLOR.**—Among the rock-forming minerals, the color depends largely upon the presence of iron. Minerals that contain this element are dark: black, brown, deep red, rust, or green. Minerals without iron are usually light-colored or white and are lighter in weight than the dark ones. Some dark minerals can be powdered by rubbing them against a hard, rough surface. In the testing laboratory, a streak plate made of unglazed porcelain is used for this purpose. The

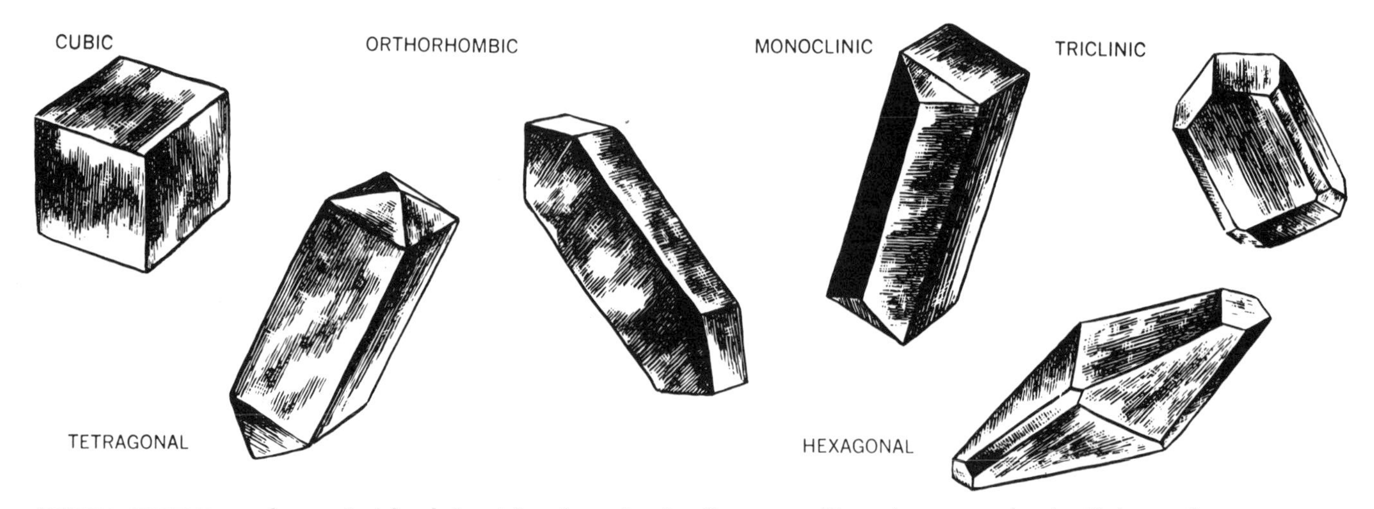

CRYSTAL SYSTEMS are characterized by their axial angles and ratios. Every crystalline substance can be classified according to structure.

color of the streak is, in most cases, characteristic of the mineral.

LUSTER.—Luster is the quality and intensity of the light that a mineral reflects. Luster can be metallic (resembling iron or brass) or nonmetallic—pearly, greasy, silky, resinous, vitreous (glasslike), or adamantine (diamondlike). Some minerals will show a play of colors, or an iridescent effect.

HARDNESS.—The hardness of a mineral is tested by attempting to scratch it with a series of minerals that have been chosen as a standard scale. This scale, called *Mohs' scale,* was proposed by the German mineralogist Friedrich Mohs in 1820. Mohs designated the softest known mineral, talc, as having a hardness of 1; and the hardest, diamond, 10. The ten minerals of the scale, arranged in order of increasing hardness (with their numerical designation on Mohs' scale in parentheses) are: talc (1), gypsum (2), calcite (3), fluorite (4), apatite (5), orthoclase (6), quartz (7), topaz (8), corundum (9), and diamond (10). Each of these minerals will be scratched only by minerals with a higher number on the scale and will scratch only those with a lower number; hence talc will be scratched by all (and will scratch none) and diamond will scratch all (and be scratched by none). Quick approximations of the hardness of minerals may be made by using handy substances to scratch them. These substances, with their numerical designation on the Mohs' scale, are: a fingernail (2½), a copper penny (3), the blade of a penknife or a piece of window glass (5½), and a steel file (6½).

CLEAVAGE.—Many minerals have the tendency to split evenly or break in definite directions along the planes of weakness in their crystal lattice. This is called *cleavage.* The number and arrangement of cleavage planes provide a reliable clue to the identification of minerals. Mica, for instance, has only one direction of cleavage; orthoclase has two, at right angles; calcite has three, mutually oblique. Minerals that break irregularly are said to *fracture.* Fractures can be *conchoidal* (shell-like), as in glass; *hackly* (jagged-edged); *even,* if the break is smooth; *uneven,* if it is irregular; *fibrous* or *splintery,* if like wood; or *earthy.*

TENACITY.—*Tenacity* is the resistance of a mineral to breaking, crushing, bending, or tearing. Some minerals are *sectile* (cutable), *malleable* (capable of being hammered into thin plates), *ductile* (can be drawn into wires), *flexible,* or *elastic.*

SPECIFIC GRAVITY.—The *specific gravity* refers to a mineral's weight, expressed in a number that shows how many times heavier a given volume of that mineral is than an equal volume of water. The specific gravities of minerals range from 1.5 to 20.0, but most fall in the range between 2.0 and 4.0.

MAGNETISM.—A few minerals, such as magnetite, an oxide of iron, will respond to an ordinary pocket magnet and can be identified by this property.

Common Rock-Forming Minerals.—A rock is an aggregate of minerals of different kinds in varying proportions. The following are some of the minerals more commonly found in rocks.

Quartz, whose chemical composition is silicon dioxide, is one of the most widely occurring minerals. It is the most common vein mineral, and makes up the largest part of most sands. Quartz is usually colorless or white; but it can be any color, depending upon its impurities. The colored varieties—amethyst, rose quartz, smoky quartz, citrine, chalcedony, and agate—are used in the manufacture of jewelry. Quartz is also used as an abrasive, in the manufacture of glass and porcelain, in paints, and in scouring soaps. As sand it is used in mortars and cements. Quartzite and sandstone—rocks made up largely of quartz—are used in the building trades.

Orthoclase feldspar, whose chemical composition is sodium-calcium-aluminum silicate, is usually found in various shades of gray, and sometimes white, although the latter is less common. It is a feldspar with a pearly to vitreous luster. Two distinctive subspecies are white *albite* and the dark *labradorite,* which often shows a play of colors when rotated in a good light. These subspecies occur in the same way as orthoclase.

Pyroxene is a silicate of calcium and magnesium, and also contains varying amounts of aluminum, iron, and sodium. It is the name of a group of minerals comprising many varieties that differ slightly in chemical composition. It is light green to dark green or black in color and commonly opaque. The most frequent member of this group is *augite,* which occurs in stubby, irregular crystals. It is a very abundant rock-making mineral that occurs chiefly in dark-colored igneous rocks. It is rarely found in rocks that contain quartz.

Amphibole, like pyroxene the name of a group of slightly differing minerals, is a silicate of calcium and magnesium, with varying amounts of aluminum, iron, and sodium. It is similar to pyroxene in composition, but differs in that it also contains water of constitution. Amphibole has a brighter luster and longer crystals than pyroxene. This mineral is usually opaque. The most familiar member of this group is *hornblende,* which has a luster of silky to dull and a color of black, dark brown, or dark green. It is a common rock-forming mineral that occurs in both igneous and metamorphic rocks.

Mica consists of characteristically shiny, flexible, elastic flakes that are stronger than steel. *Muscovite* (white mica) is a complex silicate containing potassium and aluminum. It occurs in granite, together with quartz and feldspar. Muscovite is typical of mica *schists*—rocks that split in flakes and slabs parallel to the cleavage of mica. It is used in insulation materials and in the manufacture of electrical equipment. *Biotite* (black mica) is a complex silicate containing potassium, magnesium, iron, and aluminum. It is generally dark green, brown, or black. Thin sheets of biotite have a smoky color, which distinguishes it from muscovite. This common rock-making mineral is found in *gneisses*—laminated or foliated rocks—and schists.

Hematite is the most abundant ore of iron. Some specimens are metallic, others earthy and red. More than 90 per cent of iron in the United States comes from ores containing this mineral. Michigan, Wisconsin, and Minnesota are important hematite mining areas.

Limonite is the general name for all hydrous oxides of iron. It is an earthy material, reddish brown, yellow, or orange, that often forms crusts. Limonite, also a valuable source of iron, is formed by the oxidation and hydration of iron in previously existing minerals.

Pyrite, or iron sulfide, is also known as "fool's gold." It is metallic, brassy, and generally granular. It is the most common sulfide ore and an important vein mineral. Pyrite is often a carrier of gold or copper, thus becoming an ore for both of these metals. It is an important source of sulfur in the manufacture of sulfuric acid.

Chalcopyrite is a copper-iron sulfide. It is a golden yellow mineral, although it is generally seen tarnished to iridescent or bronze. It is the most important ore of copper, and is often an ore of gold or silver.

Sphalerite, the most important source of zinc, is a zinc sulfide. It is yellow-brown to dark brown, and has a resinous to submetallic luster. It is widely distributed, generally in veins or irregular bodies in limestone.

Galena, a lead sulfide, is the chief source of lead. Its color is lead-gray, and it has a bright metallic luster. Galena sometimes contains silver; therefore it is also an ore of this metal.

Cassiterite, a tin dioxide, is almost the sole source of tin. It usually occurs in pyramid crystals or rounded pebbles. It is brown to black, with a diamond to metallic luster. Cassiterite is mined on a commercial scale in Malaysia, Bolivia, Indonesia, the Congo republics, and Nigeria.

Bauxite, a mixture of hydrous aluminum oxides of indefinite composition, is the only commercial source of aluminum. Its color varies from white to gray, yellow, or red. Bauxite is translucent, with a dull to earthy luster. The chief deposits of bauxite in the United States are in Georgia, Alabama, Mississippi, and Arkansas. The principal world producers are Jamaica and Surinam.

Uraninite, or *pitchblende,* a uranium dioxide, is usually massive and grapelike in crystal structure. It is black, with a submetallic to pitch-like, dull luster. It is the most valuable source of uranium. The Congo republics and Canada are the most important producers.

Other important rock-forming minerals are *garnet* (a ferromagnesium silicate), used as a gem stone; *calcite* (a calcium carbonate), the chief constituent of marbles and limestones;

and *chlorite,* a complex hydrous magnesium-iron-aluminum silicate with micaceous cleavage. Chlorite is a common rock-forming mineral that gives a green color to many rocks. It is typical of schists and green roofing slates. *Serpentine,* a hydrous magnesium silicate, usually is the altered product of *olivine,* and is the chief constituent of the rock of the same name. *Gypsum,* a hydrous calcium sulfate, is used in the production of plaster of Paris. *Halite,* chemically sodium chloride, is common table salt and is used for seasoning food, as a preservative, and in the chemical industry. *Kaolinite,* chemically hydrous aluminum disilicate, is common clay. It occurs widely and is used in making pottery and brick.

—Doris D. Shaner

HISTORICAL GEOLOGY

Origin of the Earth.—Man's desire to explain the means by which the earth was formed extends to prehistoric times. From the early legends of mythology to the modern theories of cosmogony, numerous explanations of the earth's formation have been offered. Today almost all scientists agree on one point—that the earth and its sister planets are related in their origin. However, no theory yet presented has been generally accepted, for each has left many important questions unanswered.

TWO-STAR HYPOTHESIS.—Probably the most popular theory prior to World War II was that proposed by the American geologist-astronomer team of Thomas Chrowder Chamberlin (1843–1928) and Forest Ray Moulton (1872–1952), called the *two-star,* or *collision, hypothesis.* Chamberlin and Moulton proposed that the planets were born when the sun and a larger star passed so closely that they almost collided. Great bolts of gaseous material were pulled from the sun by the larger star's gravitational attraction. This gaseous material assumed an elliptical orbit around the sun and then condensed to form the planets and their satellites.

ONE-STAR HYPOTHESIS.—In recent years a number of new theories have been suggested. The general trend today is toward the acceptance of a modification of the oldest truly scientific hypothesis ever proposed, the *one-star,* or *nebular, hypothesis.* It was presented in 1755 by the German metaphysician Immanuel Kant (1724–1804), who based his theory on the work of Nikolaus Copernicus (1473–1543), Johannes Kepler (1571–1630), and Sir Isaac Newton (1642–1727). This theory was further refined some 40 years later by Pierre Simon de Laplace (1749–1827), a French mathematician and astronomer. The one-star hypothesis assumes that a hot gaseous *nebula,* or cloud, automatically developed into a solar system as it cooled, without interference by an outside star or other body. This theory was widely accepted during the nineteenth century, then fell into disfavor until it was reconsidered in recent years.

The one-star theory can be summed up in the following way: Eons ago a greatly diffused, spherical gas cloud, or nebula, existed, the radius of which was at least as great as the distance of Pluto, the outermost planet, from the sun today. The cloud rotated slowly, and as it cooled—and therefore contracted—its velocity increased in the same way that a dancer will whirl faster and faster as he draws his arms closer to the body. The gaseous mass developed a disk, or equatorial bulge, around it; indeed, the present appearance of the planet Saturn, with its equatorial rings, resembles this, although on an infinitely smaller scale. At critical points during its rotation, rings of fiery gas are assumed to have been thrown off from the whirling disk by centrifugal force. Each ring then broke up into fragments, which gathered into a sphere. In this way a planet, which began to revolve in the same orbit as the ring from which it had been formed, was produced. A comparable process accounts for the formation of satellites, such as the moon. The planet liquefied as it cooled, and with further cooling acquired a solid crust. The main body of the gas meanwhile condensed and became the sun.

In 1943 a German physicist, Carl von Weizacker, was able to answer the greatest objection to the Kant-Laplace one-star hypothesis. According to mathematical analysis, the forces operating to cause the dispersion of the gaseous nebula revolving around the sun should have been just as strong as those forces acting toward the nebula's formation into planets. Weizacker's addition to the one-star hypothesis was that the materials that went into the formation of a planet such as the earth would have constituted no more than 1 per cent of the entire revolving gaseous mass—which would have been composed mostly of hydrogen and helium. In this milieu, tiny particles of dense material, revolving with the greater part of the gaseous nebula, could have collided. The smaller particles would have been absorbed into the mass of the larger, resulting in the eventual depletion of the supply of particles and the formation of the giant aggregates we know as planets.

Age of the Earth.—Until recently most geologists, astronomers, and cosmogonists (specialists dealing with the origin of the universe), believed the earth to be about two billion years old. But new data seem to indicate that the earth and the solar system are at least four, and perhaps as much as five, billion years old.

DATING METHODS.—One of the most valuable tools available to scientists today for dating the earth is *radioactive decay,* or *disintegration.* Certain radioactive elements, such as uranium, found in minerals and rocks are very unstable. Uranium breaks down into lead at a constant and measurable rate that apparently is unaffected by heat, pressure, or other conditions. If the amount of lead and the amount of uranium are known, the age of the rock can be determined. The ratio is 1/7,600,000,000. This means that the presence in a sample of one gram of lead to 76 grams of uranium indicates that the parent rock is 100 million years old. This method is good exclusively for igneous rocks, since uranium is not likely to be found in either metamorphic or sedimentary rock.

Scientists are experimenting with other ratios, too, such as the ratio between several different isotopes of lead (the isotopes of an element are distinguished by differences in their atomic weights). Scientists are also experimenting with the decay of potassium to argon, and of rubidium to strontium.

GEOLOGIC PROCESSES.—The present is the key to the past—all changes in the earth's crust that occurred billions and billions of years ago are the result of the same physical laws that are in operation today. Thus, mountains have loomed, then have been leveled to nothing by erosion; and arid lands have been flooded by invading seas that later retreated, leaving behind traces of the marine life that inhabited them.

All of the methods of radioactive dating, however, have one disadvantage—they cannot be used to date "recent" events in geologic history; that is, events that occurred less than two million years ago. Only one technique, which uses the radioactive isotope carbon-14, has proved to be an accurate time gauge within this period. And even this technique, known as *radiocarbon dating,* has its limitations, for it can be used to date only organic material that is less than 40,000 years old. The principle of radiocarbon dating is as follows: When cosmic rays bombard nitrogen in the outer atmosphere, the nitrogen may be converted into carbon-14. This carbon-14 combines with oxygen to form a special carbon dioxide. The carbon dioxide circulates through the atmosphere, reaches the earth's surface, and is absorbed by living, or organic, matter. The distribution of this special carbon dioxide has been found to be constant throughout the world. Therefore, there is an identical—although very small—amount of carbon-14 in all living organisms. When death comes to an organism—whether it is an animal or a plant—it ceases to absorb carbon-14. Instead, the carbon-14 present in the organism begins to be converted back to nitrogen at a constant rate. Thus, the longer the organism has been dead, the smaller the amount of carbon-14 that will remain within it. By comparing the amount of carbon-14 present in a no-longer-living organism with the uniform amount of carbon-14 present in all living organisms, the amount of time that has elapsed since death can be calculated.

Geologic Time.—Geologic time is generally taken as the period extending from the end of the earth's formative period to the beginning of the historical period. Thus, geologic time did not begin when the earth was born, but much, much later. The hot gaseous jets thrown off by the sun had to cool into a liquid and then into a solid crust. There was upheaval beneath the earth's crust; the crust

EARTH SCIENCE

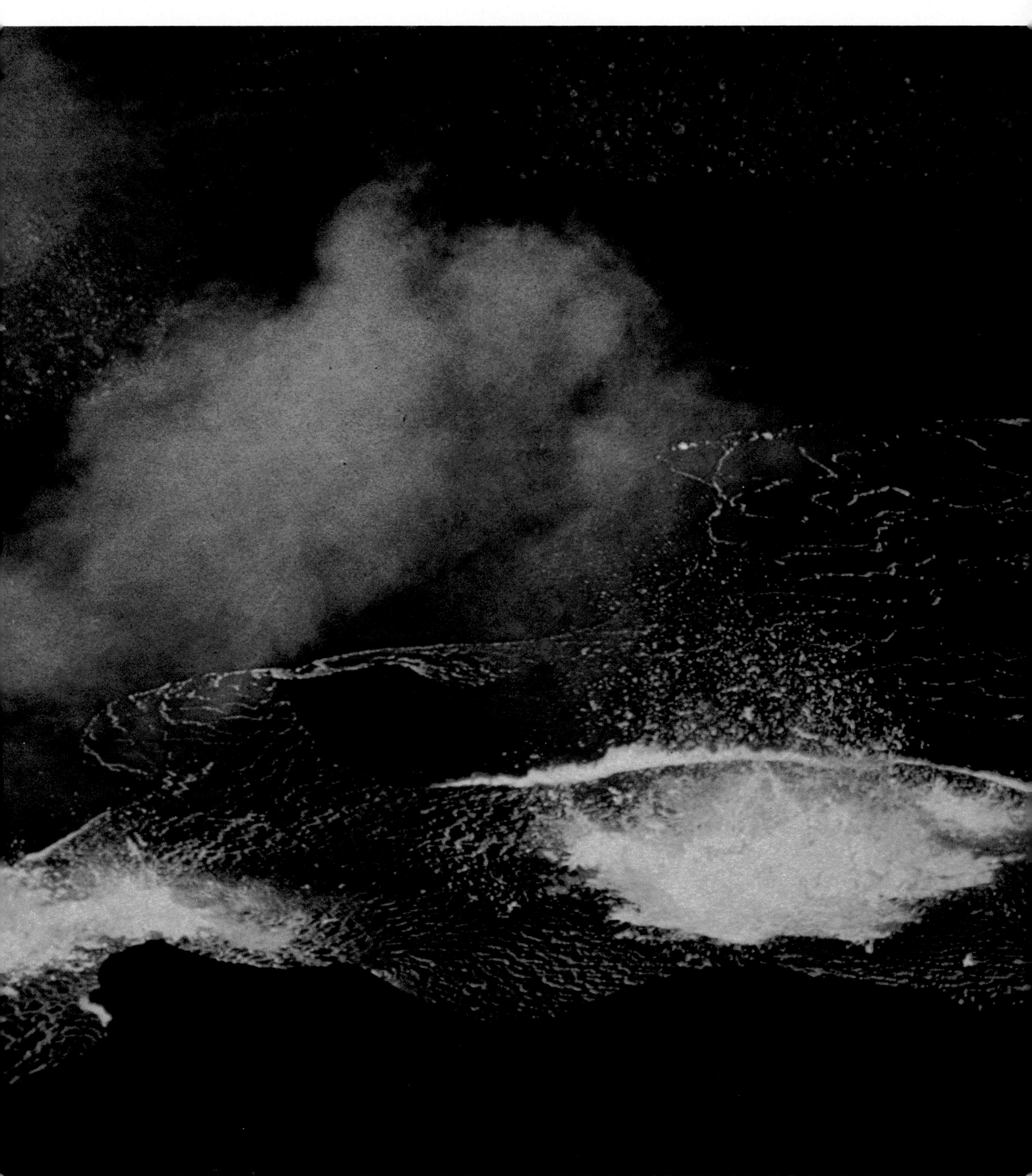

HAWAIIAN VOLCANO ERUPTING

1. RHODENITE with cabochon.
2. TURQUOISE with cabochon.
3. FORTIFICATION AGATE with flat cabochon.
4. MALACHITE with square cabochon.
5. VARISCITE with tumbled specimen, also called Utahite.
6. JASPER with oval cabochon; usually red, also brown and green, sometimes banded.
7. JADE with carved ring.
8. EPIDOTE.
9. PETRIFIED WOOD; minerals are opal, agate, and jasper.
10. CHRYSOCOLLA; soft, but gem value if evenly colored.
11. BERYL EMERALD when dark green and clear.
12. AMAZONITE, also Amazon stone.
13. SULPHUR CRYSTALS; too soft to cut, but sought as beautiful specimens.
14. SMITHSONITE; most valuable in sea green.
15. SODALITE; bright, deep blue most valuable.
16. OBSIDIAN, a volcanic glass, here with white "snowflakes."
17. UNIKITE, a form of epidote with feldspar blotches.
18. MOSS AGATE, another popular agate type; note "madonna" pattern in this piece.
19. GARNET with faceted stone.
20. AQUAMARINE with faceted stone; a popular light green beryl.
21. TOURMALINE with faceted rubellite.
22. ROSE QUARTZ with deep cabochon; popular pinkrose quartz.
23. CITRINE with faceted stone; another popular quartz variety.
24. AMETHYST with faceted stone; another quartz.
25. CLEAR QUARTZ with faceted stone, called rock crystal.

broke, and the fragments sank into the thick, molten rock underneath; then the crust solidified once more. Gradually, the earth cooled enough to allow the water vapor in the envelope of gas surrounding it to condense into rain, and the earth's surface was sufficiently cooled so that the rain could remain as water. It is with erosion that geologic processes, and consequently geologic time, started.

The geologic processes that have left their mark on the face of the earth during the period of geologic time fall into three categories: gradation, volcanism, and diastrophism. The process of *gradation* consists of *erosion,* which is the weathering (wearing away) of rocks and soil by the action of water, ice, and wind, and *deposition,* which is the building up of rock layers through the accumulation of sediments laid down by the action of water, ice, and wind; thus deposition is the converse of erosion. *Volcanism* includes all movements of molten rock, or *magma*—which is assumed to be the earth's inner core—and the formation of solid rock from the molten state, both within the solid crust and on the surface. *Diastrophism* is the process by which the earth's crust is deformed to produce continents, mountains, ocean basins, and plateaus; it therefore includes the processes of *epeirogenesis,* or continent building, and *orogenesis,* or mountain building.

■**GEOLOGIC TIME DIVISIONS.**—A logical way had to be found to divide the vast periods of geologic time. This was done by using the most obvious physical breaks in the biological record. Because the progress of life has been greatly affected by physical disruptions on the earth, a correspondence can be found between the radical changes that occurred on the earth and those that occurred in the development of plants and animals.

During periods of radical change or great diastrophism, tremendous upheavals of the earth's crust occurred, and the forces within the crust caused the rocks to fold like layers of soft modeling clay. The molten interior of the earth pushed into the overlying older rocks of the crust in the form of great *batholiths*—masses of intruded igneous rock—mountains were formed, and parts of the continents were lifted high above sea level. Then shorelines emerged, streams were rejuvenated, and a great amount of erosion took place.

This uplift of the continents caused a change in the climate, which became cold as the lands were removed from the tempering effects of the ocean. *Glaciation,* the formation of large bodies of ice over the land, occurred when the uplift was great enough; and life changed accordingly. Some of the forms of life adapted themselves to colder climates; some died out or migrated to warmer areas. The climate at times destroyed vegetation; and as a result, the animals feeding on it died out. Each new cycle—the uplift of the land, the retreating of the seas, the downwarp of the continents, and the encroaching seas—meant a new phase with new life.

The largest portions of geologic time are called *eras,* and are separated by periods of revolution. These revolutions were most likely worldwide in scope and profoundly affected plant and animal life. The rocks deposited during an era are called a *group.* Eras are divided into *periods,* which are separated by such minor diastrophism as folding, the advance or retreat of the sea, or simply a change in life. A *system* of rocks is deposited during one period. Periods are further subdivided into *epochs,* which are often separated by retreats of the sea on a local scale. A *series* of rocks is deposited during one epoch. Epochs are divided into *stages;* and the rocks deposited then constitute a *stage,* which can be broken down into *substages* and still further into *zones,* named according to the fossils they contain. Basic rock units are called *formations* and are made up of a single layer or several layers in which all sediments have been deposited continuously and under the same conditions.

■**GEOLOGIC COLUMN AND TIME CHART.**—The seas have flooded the land many times since the world began. This has occurred either because the sea level has risen generally or because the continent has warped downward. Consequently, the profile of the land and sea has been vastly different from age to age. All these movements of the continents and the seas have left their telltale marks in the rocks.

To trace the earth's history, records of local regions all over the world have to be painstakingly pieced together, like a jigsaw puzzle, in the proper chronological order. This way, geologists can construct a composite for the world—by superposing the major rock units from different parts of the world in the form of a *geologic column,* representing formations as they would appear in a well core, with the oldest bed at the bottom and the youngest layer on top. The counterpart of the geologic column is the *geologic time chart,* where major units of geologic time are arranged to correspond with the geologic column. Within this framework, geologists are reconstructing the history of the earth.

A complete record of all geologic time cannot be found in any single area. Because of the irregular warping of the earth's surface, the areas of deposition have shifted. However, deposition has always been going on in one place or another. Therefore, while no area contains a complete record, it is only necessary to discover and correlate enough of the scattered fragments to piece together a composite record of all geologic time. Geologists and allied scientists the world over have pooled their knowledge and skill in tagging and timing the earth's rocks. More than 500,000 feet of rock are classified.

Two laws of historic geology form the basis for the construction of the geologic column and time chart. These are the law of superposition and the law of faunal and floral (animal and plant life) succession.

The *law of superposition* assumes that layers of sediments are deposited one at a time, one on top of the other. Therefore, in any normal section—one that has not been deformed—the oldest bed is on the bottom and each bed in turn is younger than the one on which it rests.

The *law of faunal and floral succession* assumes that any grouping of remains of animal and plant life is a collection of organisms that existed together at one time and in one place. In addition, fossil floras and faunas succeed one another in a definite and determinable order.

Fossils.—Fossils are any recognizable organic structures or impressions of organisms preserved from prehistoric time. Referred to in former times as "devices of the Devil placed in rocks to delude men" and "relics of the accursed race that perished with the Flood," fossils were not universally recognized until 1800 A.D., as representatives of life in the geologic past. There were great controversies concerning fossils during the Dark Ages because men took their Scripture so literally. These men, who believed in a special creation, could not accept relics of a life older than 6,000 years. But there have been men through the ages who have recognized the significance of fossils. Herodotus, around 450 B.C., was one of the first to identify fossils. He found fossil seashells in Egypt and the Libyan desert during his African travels, and he came to the accurate conclusion that the Mediterranean Sea must have extended much farther to the south at some past time than in his day.

Fossils can be found in the state of original preservation, such as the woolly mammoth embalmed intact in the Arctic ice. They can also exist as molds, casts, and imprints, as well as footprints and trails. *Coprolites*—prehistoric excrement—are also considered a class of fossils. Although coal and petroleum are referred to as fossil fuels, they are not true fossils despite their age and organic beginnings because they have no recognizable structure. Remains of animals or plants recently dead are not considered fossils even though the species may be extinct because the word "fossil" necessarily implies antiquity. Fossils occur only when and where the environment was favorable for their existence and preservation, and two special conditions are of the utmost importance: the fossils must possess such internal or external hard parts as bones, teeth, scales, shells, or wood, which are left behind when the animal or plant decomposes; and the remains must be buried quickly to protect them against weathering, bacteria, and scavengers. Molds, casts, and imprints also must have quick burial if they are to be preserved. The sea bottom is by far the most important and favorable environment for the preservation of fossils because marine life is particularly prolific; and the shells are quickly, sometimes instantaneously, buried with mud and sand during storms. The beds of prehistoric lakes, bogs, the frozen tundra, asphalt or tar pits, volcanic ash, lava flows, and windblown sediment, such as loess,

CRYPTOZOIC — "Time Of Hidden Life"	PHANEROZOIC—						
PRE-CAMBRIAN	PALEOZOIC "Era Of Old Life"						
KEWEENAWAN, HURONIAN, TIMISKAMING & KEEWATIN	CAMBRIAN	ORDOVICIAN	SILURIAN	DEVONIAN	MISSISSIPPIAN	PENNSYLVANIAN "Age Of Cockroaches"	PERMIAN
?	100	70	15	43	40	45	40
4 ? BILLION	550	450	380	355	310	270	225

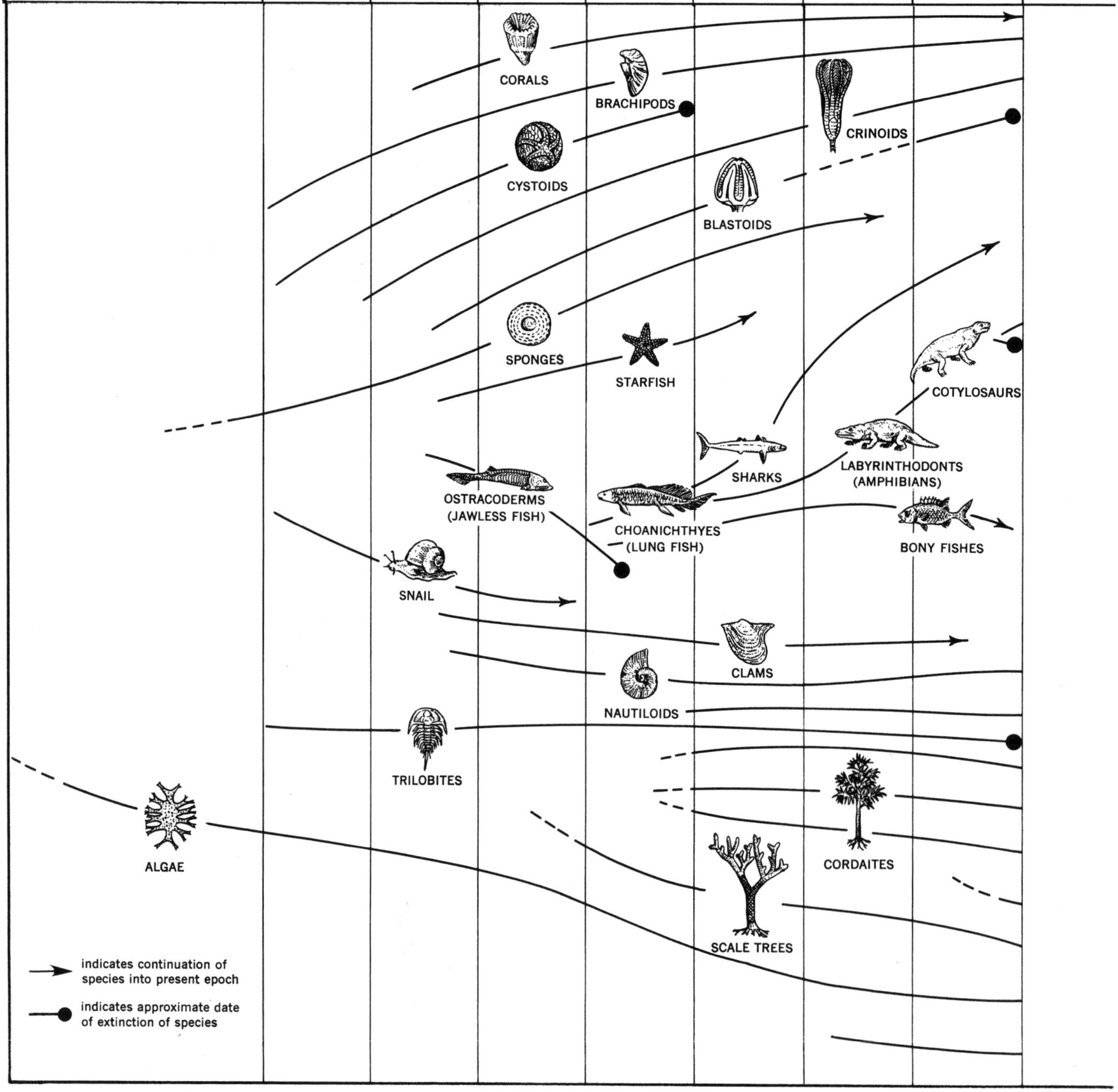

											EON
"Time Of Visible Animal Life"											EON
MESOZOIC "Era Of Middle Life"			CENOZOIC "Era Of Recent Life"								ERA
TRIASSIC	JURASSIC	CRETACEOUS "Time Of Great Dying"	TERTIARY					QUATERNARY			PERIOD
			PALEOCENE	EOCENE	OLIGOCENE	MIDCENE "Golden Age Of Mammals"	PLIOCENE	PLEISTOCENE "The Ice Age"	RECENT		EPOCH
27	33	65	10	10	10	15	14	1	LATE ARCHEOLOGIC AND HISTORIC TIME		DURATION (Millions of Years)
185	158	125	60	50	40	30	15	1			BEGAN Millions of Years Ago

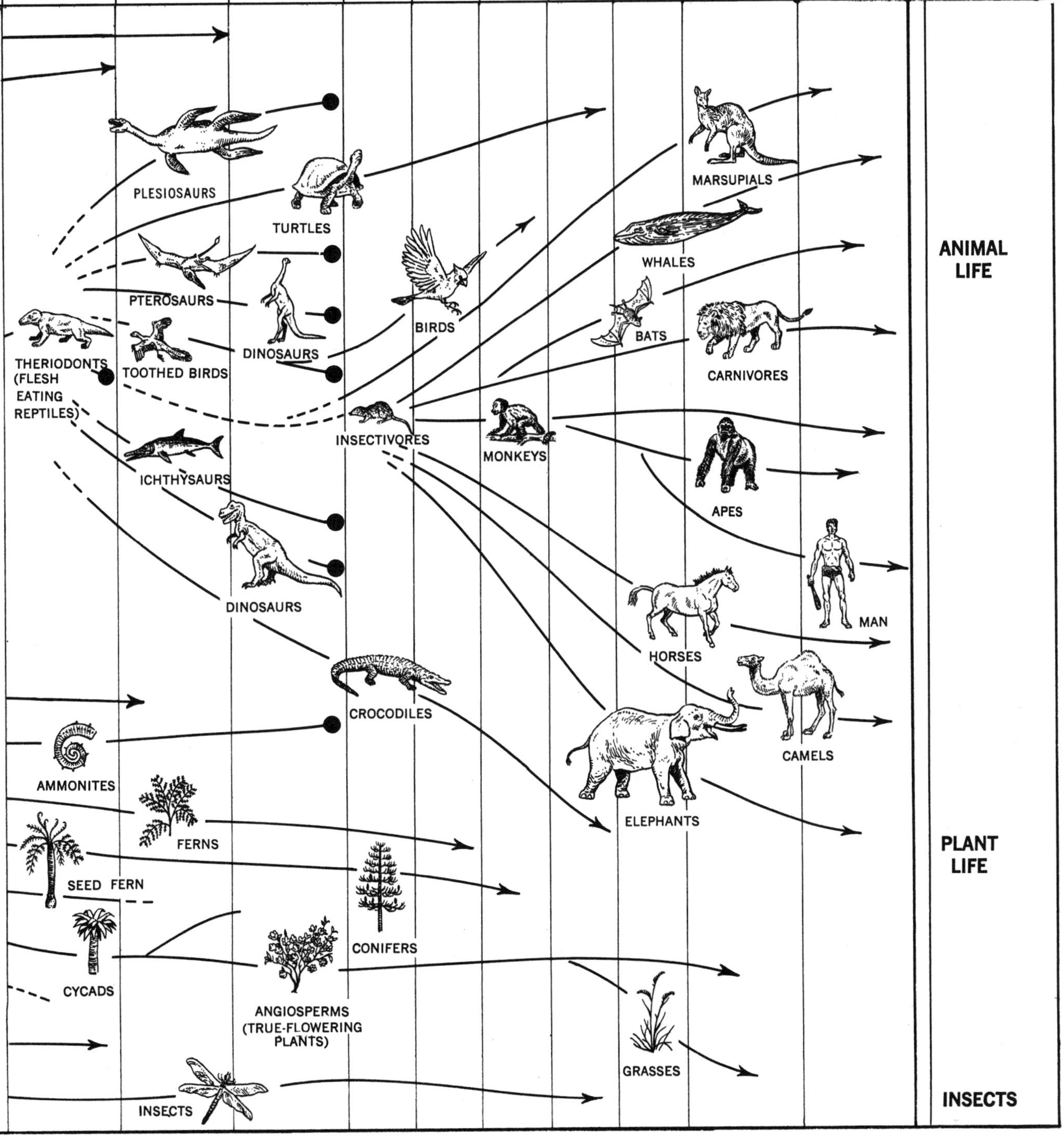

ANIMAL LIFE

PLESIOSAURS
TURTLES
MARSUPIALS
WHALES
PTEROSAURS
BIRDS
DINOSAURS
THERIODONTS (FLESH EATING REPTILES)
TOOTHED BIRDS
BATS
CARNIVORES
INSECTIVORES
MONKEYS
ICHTHYSAURS
APES
DINOSAURS
MAN
HORSES
CROCODILES
CAMELS
AMMONITES
ELEPHANTS

PLANT LIFE

FERNS
SEED FERN
CONIFERS
CYCADS
ANGIOSPERMS (TRUE-FLOWERING PLANTS)
GRASSES

INSECTS

INSECTS

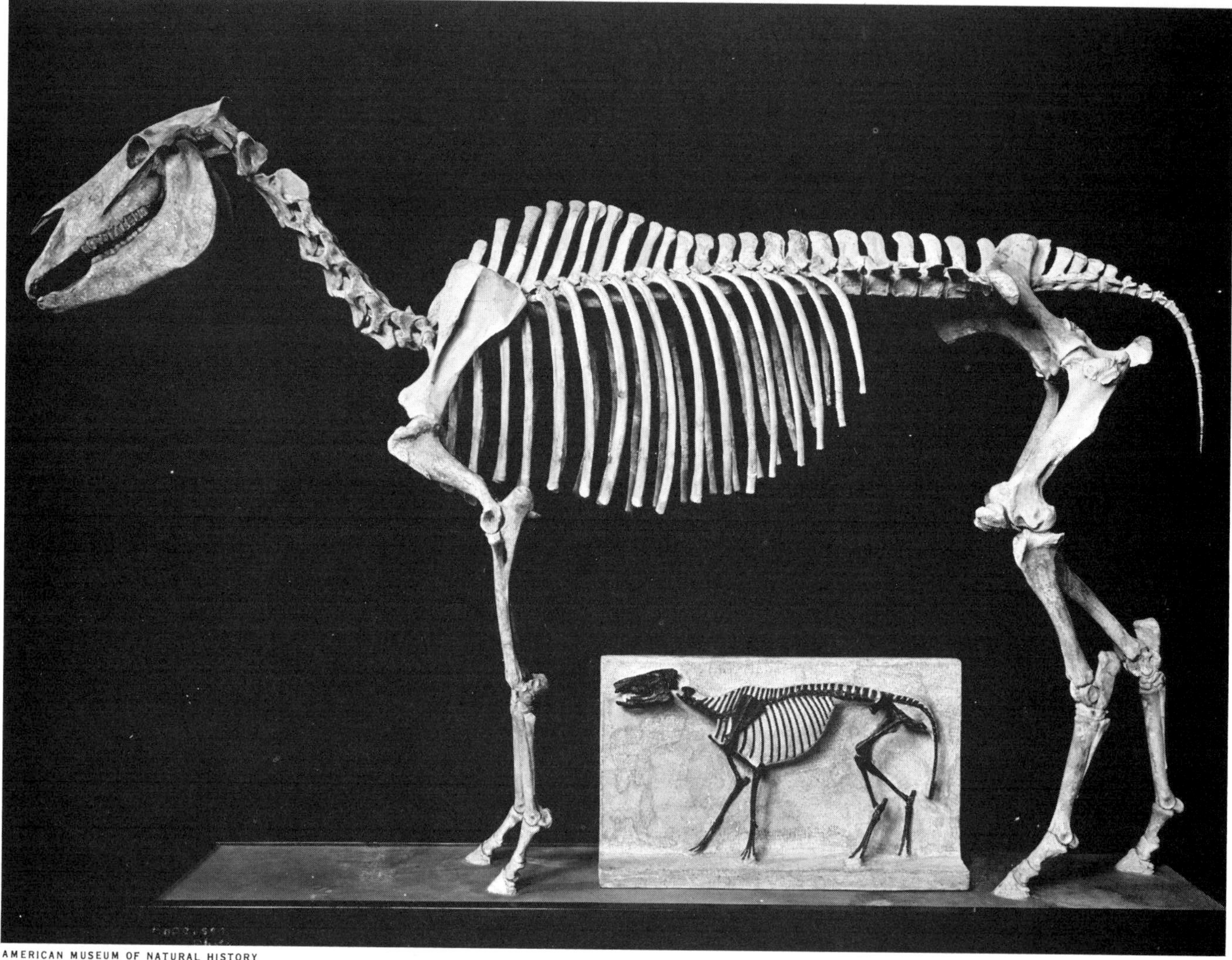

AMERICAN MUSEUM OF NATURAL HISTORY

EQUUS SCOTTI, the one-toed, last stage in the evolution of American horses, compared to the smaller, four-toed *Eohippus venticolus,* the first stage. *Equus scotti,* about 3 feet 9 inches in height, became extinct some 250,000 years ago. *Eohippus venticolus* was about 12 inches in height.

are other environments that encouraged fossil preservation.

Fossils can be the clues to many things: whether the rocks were laid on land or in the sea; whether the climate was warm or cold; and how life has unfolded through the ages. Fossils are the only documentary evidence that life has developed from simple plants and animals to more and more complex forms. They provide the geologist with a clock—a chronology. Rocks of each geologic age contain fossils that are different from those of any other age, and this is the way the geologic record is dated.

Fossils typical of a certain *stratum,* or layer of earth or sediment, are called *index* or *guide fossils.* A guide fossil preferably is an organism that can float or swim, since it will therefore be distributed over a wide area by the sea. It must also have a relatively short life in geologic time.

Paleontology.—*Paleontology* is essentially that branch of historical geology that studies the flora and fauna in past geologic periods. It deals with the succession of life that has been on the earth since the earliest times, and with the environment, evolution, structure, and relationships of that life. Thus, paleontology is the most reliable means available to correlate rock strata when dealing with expansive formations, complex formations and structures, and those strata found beneath the surface of the ground, such as the cores brought to the surface by oil well drillers.

The first of the geologic time divisions is the *Archeozoic era,* during which the oldest exposed rocks that have been found on the continents were formed. This era, of course, does not start with the beginning of the earth, since there was an extensive interval between the time the planet first began to cool and the solidification of the first rocks. The Archeozoic was followed by the *Proterozoic era,* and in the rocks of the Proterozoic, traces of the first living organisms appear. About 550 million years ago the Proterozoic ended and the *Paleozoic era* began. This era, which is divided into seven unequal periods, lasted for over 350 million years. During the Paleozoic the first vertebrate fish, amphibia, reptiles, and the first spore-bearing, conifer, and cycad land plants appeared. The Paleozoic was followed by the *Mesozoic era,* which is divided into three unequal periods. During the first of these, the *Triassic,* the dinosaurs made their appearance; during the second, the *Jurassic,* the first birds appeared; and during the third, the *Cretaceous,* the dinosaurs became extinct and flowering plants appeared. About 60 million years ago, the Mesozoic era gave way to the *Cenozoic era.* The Cenozoic is divided into two periods—the *Tertiary* and the *Quaternary*—and six epochs. During the first of these epochs, the *Paleocene,* the first mammals, the marsupials, appeared; during the second, the *Eocene,* the primates appeared; during the third, the *Oligocene,* elephants made their appearance; during the fourth, the *Miocene,* horses evolved; during the fifth, the *Pliocene,* grasses became more abundant; and during the sixth, the *Pleistocene,* which began about one million years ago (or even earlier according to recent datings), man made his debut.

—Doris D. Shaner

ARABIAN AMERICAN OIL COMPANY

SCIENTIFIC OIL EXPLORATION in Saudi Arabia. A portable corer drills shallow holes in the desert floor and flushes rock fragments to the surface. Petroleum geologists study the rock fragments to learn whether, farther below, there is the type of strata that may contain oil.

ECONOMIC GEOLOGY

Scope.—The economic geologist is concerned with the mineral substances in the earth's crust that are necessary to man's survival and comfort. Only about 200 of the nearly 2,000 minerals that have been identified are of economic interest. The most important of these in today's economy are the mineral or fossil fuels, such as petroleum and coal. (Technically, these fuels are neither true minerals, since their composition cannot be expressed by a chemical formula, nor fossils, since they have no identifiable structure.) Economic geology deals not only with ores and mineral deposits, but also with the rocks that contain them—how the rocks were formed, their nature and structure, and the geological formations developed on them. An economic geologist uses all the principles and techniques of physical and historical geology.

Historical Development.—The early history of economic geology closely parallels that of mineralogy, since both sciences evolved from man's desire for ornamentation. One might say that economic geology, however, began when the first man used a rock to help him survive. The first known economic geologist was probably Haroeris—an Egyptian captain who led an expedition to the Sinai Peninsula around 2000 B.C. After prospecting for several months, Haroeris found and extracted large amounts of the semiprecious stone turquoise. The early Egyptians, the Greeks, and their various neighbors prized gems, gold, and silver highly for the ornamentation of their bodies, their homes, and their dead.

Mining as an industry began with the search for gems and decorative stones during the time when the Egyptian pharaohs were most powerful. Gems and ornaments were so much a part of religious belief in those days that they were considered a necessity of life. However, the gem minerals that were economically important at that time play only a minor role in our life today—they are our luxuries. At the conclusion of the Dark Ages, the chief mineral substances in use were iron, copper, lead, tin, gold, silver, mercury, precious stones, clay, and building stones—compared with more than 75 minerals traded internationally today. The chief minerals now are oil and gas, coal, iron ores, iron-alloy metals (chromium, manganese, molybdenum, nickel, tungsten, vanadium), nonferrous metals (copper, lead, zinc, tin, aluminum), the minor metals, metallurgical minerals, chemical minerals, ceramic materials, and abrasives. Gold is not considered as an industrial mineral because of its monetary value, but it plays an important part in affording means of purchasing needed mineral supplies.

Locating Minerals.—All the easy-to-find mineral deposits have already been exploited, and the search for new deposits has taken scientists to strange and hard-to-reach places, such as deep into the earth and far out over the oceans. One of the main jobs of the economic geologist is to use the principles of geology in searching for and helping to develop valuable mineral deposits. He also works with other scientists and technicians in finding new methods and developing new instruments that will aid in this search and lower production costs. In the future, such new techniques will help develop deposits that are considered too costly to exploit today.

The economic geologist has such powerful prospecting aids as the *seismograph,* a device for recording shock waves that travel through the earth, the airborne *magnetometer,* a device for measuring the strength and direction of magnetic forces, and the *gravity meter,* a device for detecting differences in gravitational attraction. He takes readings from these instruments, interprets them according to his knowledge of the geology of the area, and evaluates the results. From this information, he constructs a map of the geologic structures below the surface of the ground. Thus a geologist can select spots where economically exploitable mineral deposits are most likely to be found.

Mineral Products.—The principal mineral products in the United States, in order of value, are crude petroleum, natural gas, and coal—with petroleum contributing more than half of all the revenue received from the total mineral production. In 1966 more than 3 billion barrels of crude oil were produced; at the wells, this quantity of oil was valued at $8.7 billion. In the same year, 13 million tons of anthracite (hard coal) were produced, with a value of $100.6 million, and 533.9 million tons of bituminous (soft coal) with a value of $2.4 billion. A ready yardstick to indicate the order of importance of mineral fuels in the American economy is the fact that three-fourths of the geologists in the United States are petroleum geologists, employed by oil companies in the search for oil and natural gas. The nation produces about half of the world's oil, and consumes some two-thirds of it. The greatest oil deposits are found in the Middle East. However, significant new discoveries of deposits in North Africa in recent years indicate that this, too, may become a major production region. Besides petroleum and coal, the nonmetallic minerals of economic value include cement materials, ceramic products, building stones, gems, and sulfur. The total value of these nonmetals (excluding fuels) in the United States in 1966 was more than $5 billion.

Metallic minerals include the industrial and precious metals—gold, copper, tin, lead, zinc, uranium, and a host of others. They produced a revenue of $2.6 billion in the United States during 1966.

■ **PETROLEUM.**—Petroleum is a complex mixture of gaseous, liquid, and solid hydrocarbons (compounds containing hydrogen and carbon). The consensus among geologists is that petroleum originated from marine plant and animal life that died and fell to the bottom of the shallow prehistoric seas. Here it decomposed as the result of bacterial action, yielding carbon and hydrogen. The residue was buried by sediments and subjected to further chemical change. Finally, the weight of the sediments squeezed the oil and gas into porous rocks, from which they migrated to suitable reservoirs.

Four conditions are necessary for the formation of an oil deposit. First, a *source rock,* which contains the carbonaceous matter from which oil can be formed, must be present. The most common source rocks are marine bituminous shales. Next, there must be a *reservoir rock* in which the oil can collect. Sandstones are the most common reservoir rocks because they are porous and permeable—that is, they have connected pore spaces large enough to permit the oil to move through the rock. Limestones and dolomites are also important reservoir rocks. The third condition is a *structural* or *stratigraphic trap*—the rock strata must be arranged or deposited in such a way that there is a place where oil can collect in quantity. The simplest structure is an *anticline,* where the rock strata have been folded into the shape of an inverted soup bowl. This was the first type of trap recognized, but today more than a score of different traps are known. Finally, there must be an impervious layer called *cap rock*—generally shale or clay—that overlies the reservoir rock and seals in the oil.

■ **COAL.**—Coal, another vitally important mineral fuel, is a compact mass of carbonized plant debris. It occurs in beds, which are usually sandwiched between layers of sandstone and shale. The great coal-making eras of geological history were the Mississippian and the Pennsylvanian periods, over 230 million years ago. During these periods, tropical climates and lush swamp vegetation encouraged great accumulations of vegetable matter, from which coal was later formed. Coal beds range in thickness from hardly more than a film to hundreds of feet. The grades of coal depend on the concentration of carbon, or how much of the volatile constituents of the carbonaceous mass have been driven off. The stages in coal formation are *peat, lignite* (brown coal), *bituminous* (soft coal), *anthracite* (hard coal), and under favorable conditions, *graphite.*

■ **ORE DEPOSITS.**—Most ores (metal-bearing mineral deposits) are concentrations of metals brought about by igneous activity. *Magma,* the molten material underneath the surface of the earth from which the igneous rocks were formed, supplied the metals in the ore deposits. The metals were released from the magma at the time it solidified. Most deposits of metallic minerals are situated either in border zones of granite *batholiths* or *stocks,* or in the rocks immediately surrounding such intrusive masses.

However, other methods of concentration were also important. Iron ore deposits—iron is by far the most useful and abundant metal—were formed in more ways than deposits of any other metal. But the most important method was *sedimentation,* a process that has produced far larger accumulations of ore than any other process of concentration.

—Doris D. Shaner

BIBLIOGRAPHY

American Geological Institute. *Dictionary of Geological Terms.* Doubleday & Co., Inc., 1960.

Compton, Robert Ross. *Manual of Field Geology.* John Wiley & Sons, Inc., 1962.

Dunbar, Carl Owen. *Historical Geology.* 2nd ed. John Wiley & Sons, Inc., 1960.

Leet, Lewis Don and Florence J. (eds.). *World of Geology.* McGraw-Hill, Inc., 1961.

Longwell, Chester R., and Richard F. Flint. *Introduction to Physical Geology.* 2nd ed. John Wiley & Sons, Inc., 1962.

Moore, Ruth. *The Earth We Live On.* Alfred A. Knopf, Inc., 1956.

Thompson, Henry Dewey. *Fundamentals of Earth Science.* 2nd ed. Appleton-Century-Crofts, 1960.

OCEANOGRAPHY

Scope.—The ocean is the most striking physical feature of our planet, covering over two-thirds of the globe's surface. Without the waters of the seas, there could be no life. Since earliest time, man has used the ocean as a highway, as a great moat to protect him from enemies, as a source of food, and as a final resting place. Only recently, however, has man begun to delve deeply into the sea to learn its secrets.

Oceanography is an environmental science encompassing the study of all processes in the ocean and its boundaries. It includes the study of plant and animal life at all depths (*biological oceanography,* or *marine biology*); the study of the origin of the ocean, its structure, and the stratigraphy and composition of its bottom sediments (*geological oceanography,* or *marine geology*); the study of sea water and its composition (*chemical oceanography*); the study of currents, tides, waves, temperature, salinity, density, and the general circulation of the sea (*physical oceanography*); the study of the food, mineral, and energy sources of the sea and the uses of the ocean for recreation, navigation, communication, and war (*marine technology*).

The Oceans.—Billions of years ago the earth was a lifeless planet plummeting through the darkness of space. As the mass of gases and molten metals cooled, water was squeezed from its interior. The planet became a world unique, as far as we know, in the entire solar system; it had an ocean.

■ORIGIN OF LIFE.—For millions of years the sterile waves lapped against the cold, dead shores. Then, through a series of processes whose nature is still unknown, organic matter developed and became concentrated into a living cell capable of reproducing.

There is no way of knowing what this first life was like, or what the conditions were that produced it. Chances are that some of the earliest organisms were similar to the single-celled plants still found in the surface layers of the ocean. These contain chlorophyll, the substance that enables plants to utilize the energy of sunlight to produce organic material from water and carbon dioxide. Through this process, known as *photosynthesis,* the oxygen of the earth's atmosphere was created from the water in the ocean.

Over the countless generations, spanning millions of years, differences developed among the single-celled organisms. Some preyed on others, so organisms evolved different methods of finding food and of escaping from being eaten; in response to changing conditions they developed different tolerances and sensitivities to light and to chemical variations. Eventually, life in the sea made the jump from simple individual cells to complicated, highly specialized plants and animals.

■CURRENTS.—Only recently has man known anything of the internal movements of the sea. The movement of

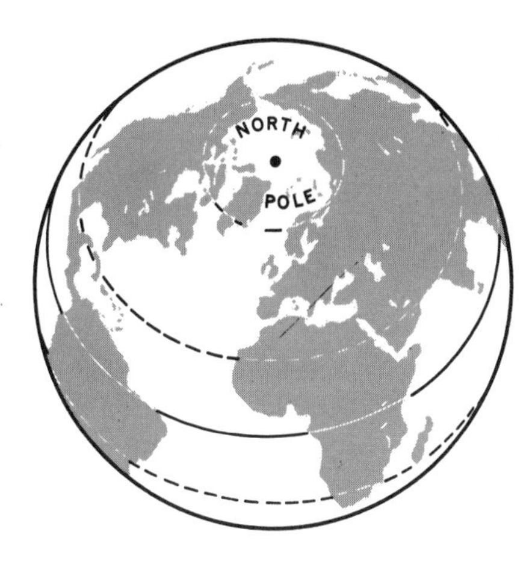

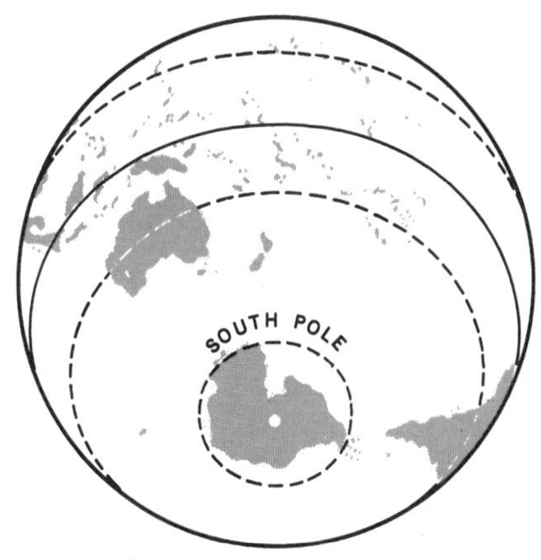

OCEANS, covering two-thirds of the earth, are our most striking physical feature.

water on the surface has been charted, but not the currents that flow beneath the surface. Now, however, through new techniques of measuring natural radioactivity, heat flow, and salt content, and by taking direct measurements of deep currents, a new picture of the ocean's circulation is being revealed.

Along with the horizontal currents in the ocean, there is a constant vertical motion. Practically nothing is known of this vertical movement, but it is vital to life in the sea, since it is the chief process by which the surface waters are constantly supplied with the nutrients required by the *phytoplankton,* the microscopic plants and animals without which no life could exist in the ocean.

Origin of Oceanography. — When man first ventured onto the sea to travel from one place to another, he began to navigate. He found that waves and currents would either help or hinder him in reaching his destination. He noticed that the tidal movements were greatest at certain times of the month, and he saw a correlation between the tidal movements and the moon phases. Going to the sea in search of food, he learned that certain types of bottoms were apt to harbor certain kinds of fishes, mollusks, or crustaceans. As he became conscious of these things and began to search into their causes and relationships, he became, in essence if not in name, a student of oceanography.

The study of oceanography is as important today to navigation and shipping as it was to primitive seagoing man. Today oceanographic knowledge is vital not only to ships traveling on the surface of the sea, but also to craft that travel under the sea and in the air above. Information about currents, tides, and temperatures is necessary in both peacetime and wartime navigation and shipping, as are further studies of the strange behavior of sound and light under water. And today, more than ever before, knowledge of the ocean's circulation is of prime importance—for the sea may be the only place on earth where man can safely store nuclear waste products.

History of Oceanography.—Oceanography as a science began fairly recently. In 1750 the first scientific dredge was invented by Marsigli and Donati; a few years later, in 1769, Benjamin Franklin published the first chart of the Gulf Stream. During the following decade Captain James Cook, in his explorations of the Pacific, took a naturalist along to make observations and to record data. During Cook's second voyage, the first subsurface water temperatures were taken and were found to differ markedly from temperatures on the surface. On this voyage, too, some deep-water soundings were made, and a sample of blue mud was brought up from a depth of 683 fathoms.

■DARWIN.—It later became customary to take a trained naturalist on long survey or exploratory voyages. Charles Darwin's around-the-world voyage on the H.M.S. *Beagle* from 1831 to 1836 was one of the first purely scientific voyages. On this trip Darwin's two great theories were developed: the theory of natural selection and the less revolutionary but oceanographically more important theory of the origin of coral reefs.

■MAURY.—The first textbook on the subject of oceanography, entitled *The Physical Geography of the Sea,* appeared in 1855. The author, Matthew Fontaine Maury (1806–1873), was an American naval officer who was forced by an accident to retire from sea duty. He compiled and analyzed material from ships' logs; and the wind and current charts he drew from them soon became well-known throughout the world and greatly shortened sailing times between the continents. With the aid of a new sounding apparatus that employed a detachable weight, data were obtained from which Maury prepared the first bathymetric chart of the North Atlantic.

■THOMPSON — Until a century ago, most marine scientists believed the depths of the ocean to be utterly devoid of life. The absolute limit of life was thought to be about 300 fathoms. In 1860, however, the transatlantic cable, laid only two years before, was broken. Upon being brought up from a depth of 1,000 fathoms, it was found to be encrusted with living organisms, including a deep-sea coral. Subsequently Wyville Thompson made a series of deep hauls in North Atlantic waters—to a maximum depth of 2,435 fathoms—and in every case the dredge brought up living organisms.

The stage was now set for the first world-wide scientific investigation of the oceans. In 1872 Thompson's book, *The Depths of the Sea,* was published, and it stimulated the scientific world to renewed interest in the deep sea. In the same year the 2,000-ton British corvette H.M.S. *Challenger* embarked on a 70,000-mile voyage through the Atlantic, Pacific, and Indian oceans for the purpose of learning "everything about the sea." During the next three and a half years the *Challenger* scientists, under the direction of Wyville Thompson, and his assistant John Murray, collected animals at great depths, settling for all time the dispute over whether or not the depths are inhabited. Altogether they described a total of 4,417 new species of plants and animals. The *Challenger* expedition occupied 362 oceanographic stations and collected 77 water samples for total chemical analysis. Papers and reports of the voyage filled 50 volumes and required 20 years to complete. This mass of data, and the wide range of the samples collected, were vitally important to the development of modern oceanographic theory.

Geological Oceanography.—The marine geologist is interested in the structure of the ocean basins, the topography of the bottom, and the sediments that have settled on the sea floor.

OCEAN DEPTHS.—Although surfaces of the oceans have been charted and mapped since the time of the earliest navigators, the great bulk of the vast oceanic depths have never been explored or even accurately mapped. In the last twenty years, however, marine geologists have managed to trace rough outlines of some seas through the use of *depth recorders.* These echo-sounding devices measure the time it takes a sound wave to travel to a solid object and back under water. By measuring the time elapsed against the speed at which the sound travels, the distance covered can be computed. Many large areas still exist, though, where no soundings have been made, and precipitous peaks still unknown may rise from great depths nearly to the surface. Several are detected each year by oceanographic research vessels.

The average depth of the oceans is about 13,000 feet, and the greatest depth so far discovered is 35,640 feet, in the Marianas Trench in the Pacific. This exceeds, by over 6,000 feet, the highest elevation above sea level, Mt. Everest, 29,025 feet.

CONTINENTAL SHELF. — Around the edges of all continents is a shallow fringe of submerged land known as the *continental shelf.* On this shelf, throughout the ages, such sedimentary rocks as limestone and sandstone have been and still are being formed. The continental shelf averages about 30 miles in width along most continents, although in some parts of Siberia it extends to 800 miles and along mountainous coasts it diminishes to almost nothing. The shelf is not a smooth, flat surface but is broken up into terraces, ridges, and hills.

Beyond the shelf, at a depth of about 600 feet, a more precipitous drop occurs. This is known as the *continental slope,* and it continues downward to the bottom of the sea—two or three miles, on the average. The deepest spots in the ocean, the ocean *trenches,* are found at the bottoms of the continental slopes.

The continental slopes are explained by the fact that the continental crust consists of rocks less dense than the oceanic crust. The continents can therefore be said to float, like ice floes, on the earth's mantle higher than the oceanic crust. Within the continental shelves and slopes are great canyons similar to deep river valleys, some as large as the Grand Canyon.

OCEAN FLOOR.—The floor of the sea is quite different from the surface of the land. One reason is that the lack of erosion caused by wind, rain, and ice has preserved the submerged peaks, valleys, and canyons—just as the face of the moon has not been changed by weathering. However, there is also a tendency for sediment to very slowly drown these features on the sea floor.

SEAMOUNTS.—All ocean basins have a mid-ocean rise and a range of moun-

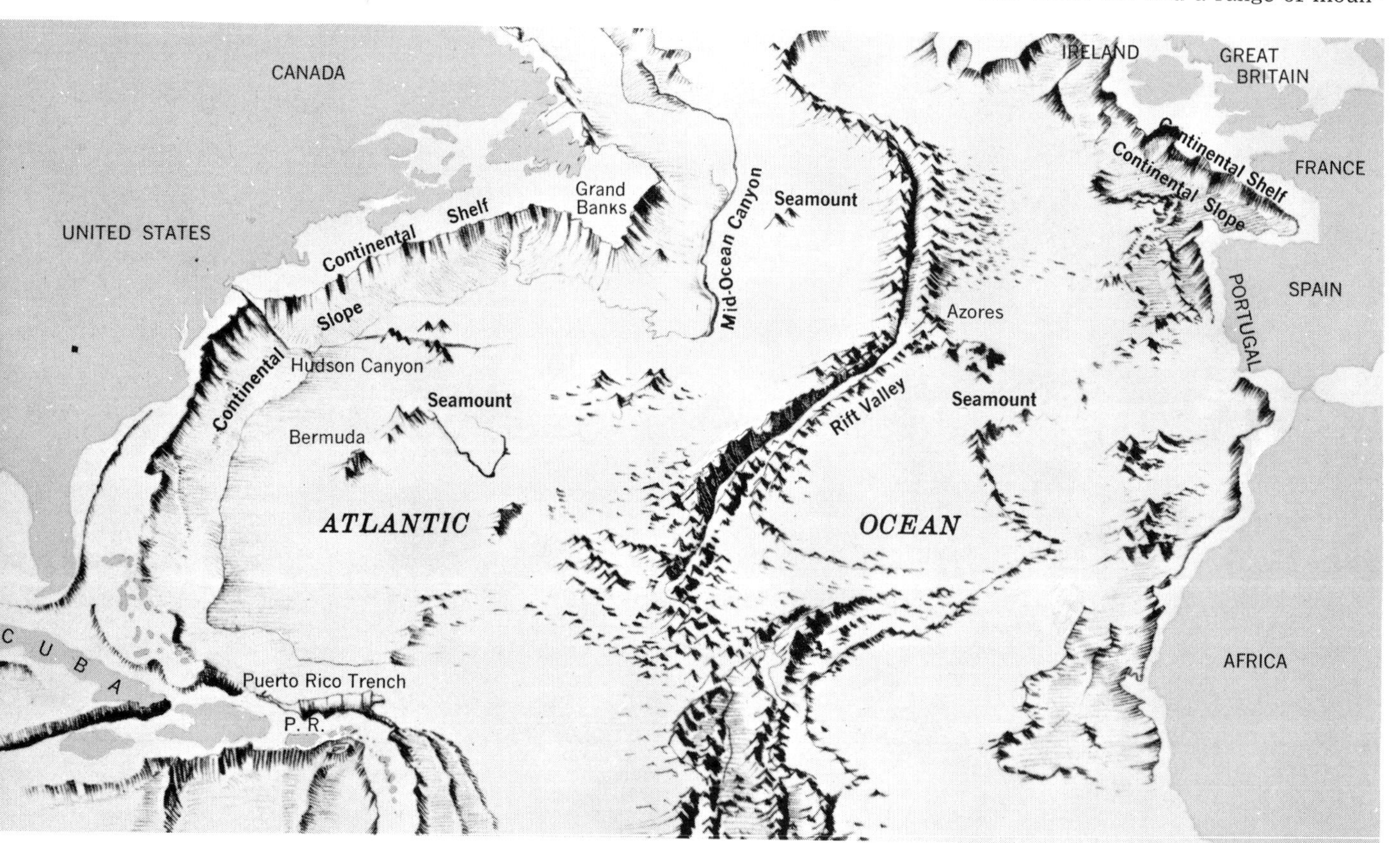

RELIEF OF THE OCEAN FLOOR. The floors of the oceans have their own valleys, mountains, and plains. Until recently, these regions could not be mapped. Today, scientists are discovering new regions at the ocean's bottom. Notice the 10,000-mile-long Mid-Atlantic Ridge.

tains down the middle. Thousands of volcanic peaks called *seamounts* dot the ocean floor. A comparative few of these peaks reach the surface to form islands, such as the Hawaiian chain in the Pacific or the Azores in the Atlantic. Many of the drowned peaks are the foundations of coral atolls, with the dead remains of reef-building corals and calcareous algae extending sometimes thousands of feet downward to the submerged peak. Since reef corals grow only in well-lighted waters (to a maximum depth of about 180 feet), a dead coral cap several thousand feet thick suggests that the atoll's base sank slowly over thousands or millions of years. This was first postulated by Charles Darwin and recently confirmed by deep drillings.

Many seamounts are flat on top, their peaks cut off at depths of 2,000 to 6,000 feet below the ocean's surface. These seamounts are called *guyots*. They were flattened by erosion at some period in the past when they were islands. Since the tops of these seamounts are found at various levels, it is unlikely that they were decapitated by a succession of great rises and falls of sea level. A more likely explanation is that they sank with the collapse of the earth's crust under their tremendous weight.

■SEDIMENT.—Much of the ocean floor is covered with a layer of sediment that has, in undisturbed regions, been accumulating for a hundred million years or more. This sediment contains a record of the earth's early history. Layers of volcanic ash are there, telling of great eruptions; ice-scarred stones from glaciers; and the remains of multitudinous planktonic plants and animals tell of climatic and evolutionary changes. Hollow tubes, or *corers,* are pushed into the sea floor to withdraw cross sections of the sedimentary carpet. An examination of the types of shells and other materials in these cores reveals the climatic variations of the past, and, by deduction, much information about the evolution of early life. Since some species of ancient marine life lived only in cold waters and others flourished in temperate or tropical waters, the sequence of fossils tells a great deal about sea temperatures and productivity millions of years ago.

Of primary importance in geological research on the composition of sediments is the determination of their specific origin. All sediments can be divided into three different groups according to their mode and place of origin. The first group is composed of material having its origin on land, such as soil, clay, and unweathered rock fragments. The second is made up of material formed in the ocean by inorganic precipitation. To this group belong some clays, manganese nodules, and other components of crystalline or gel consistency. The third group consists of organic material formed in the ocean, such as skeletons and other remains of animal and plant life. In addition to these sources, which account for most of the material, outer space contributes very tiny spherules of nickel-iron, remnants of meteorites.

By a careful study of the composition of the sediments, the relationship of the different components to one another, the shape of the particles, their chemical composition, grain size, and color, marine geologists are able to trace their origin and mode of transport before they were buried in the scientist's great treasure trove —the ocean floor.

There are several means by which sediments are transported to the ocean. The finest dust can be blown very long distances from the land (desert sand is found 2,000 miles from Africa in the Atlantic Ocean). Soil particles transported from North Africa across the Mediterranean may settle in rain on Germany, causing the landscape to be colored red. In many places on the ocean floor, the inorganic material consists nearly entirely of wind-transported matter.

The ocean currents also play a part. Their speed is much slower than the wind speed; on the other hand, the particles settle much more slowly in water than in air. Consequently, they may be carried across the ocean before they settle to the bottom. The greatest quantity of sediment, however, settles on the continetal shelf. There is evidence that on occasion large quantities of this unconsolidated material slide off the shelf and down the continental slope to the sea floor, spreading over large areas. Such large-scale movements, known as *turbidity currents,* may have played a part in forming the great canyons found on the continental slopes.

Chemical Oceanography.—The water that makes up the ocean probably had two sources. Some came as rain from the gases surrounding the earth; some from the earth's interior, forced from the rocks as they recrystallized to form the crust. Water is still being released from the interior of the earth through volcanic eruptions.

■SEA WATER.—Analyses of 77 water samples taken during the around-the-world cruise of H.M.S. *Challenger* established two vitally important properties of sea water: the *salinity,* or total content of salts, in sea water varies only slightly throughout the world; and, even where variations do exist, the relative quantity of the major dissolved salts remains constant. This property, known as the *constancy of relative proportions,* helps determine the salinity of water.

The major constituents of sea-water salts are sodium, chlorine, magnesium, sulfur, calcium, and potassium. In addition, sea water contains traces of all natural elements. Unlike the major constituents, these trace elements are found in widely differing proportions in different places and at different times. Studies of the concentrations of trace elements aid in understanding life processes in the ocean. Perhaps the appearance or

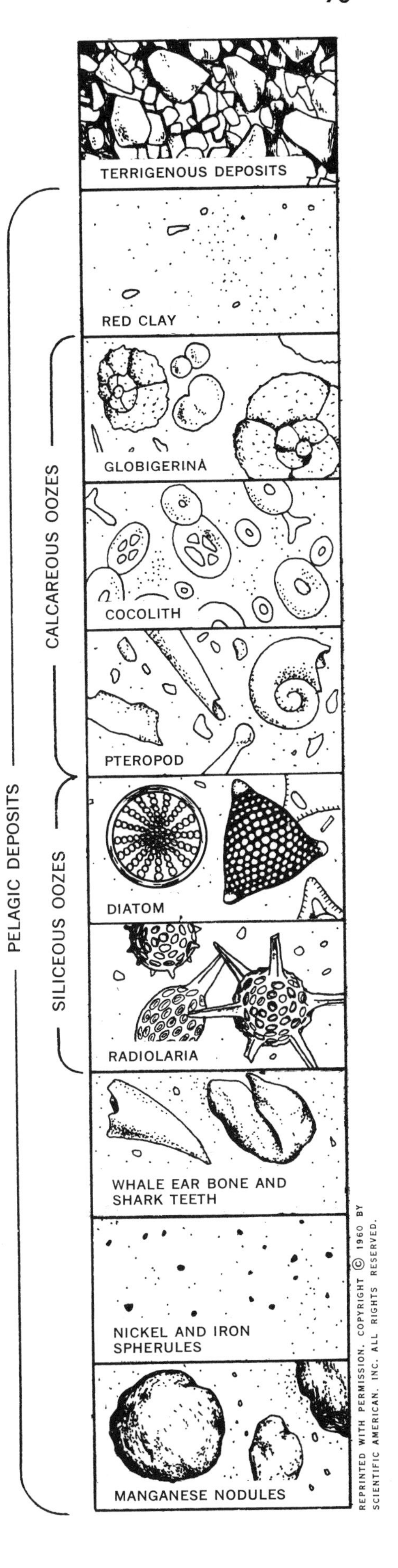

SEDIMENTS found on the ocean's floor are composed of many organic materials. When more than 30 per cent of a sediment consists of dead plant and animal life, it is called an *ooze.* Nearly half of the ocean floor is covered with Globigerina ooze.

CANADIAN NATIONAL RAILWAYS

TIDAL BORE sweeps upstream as a wave.

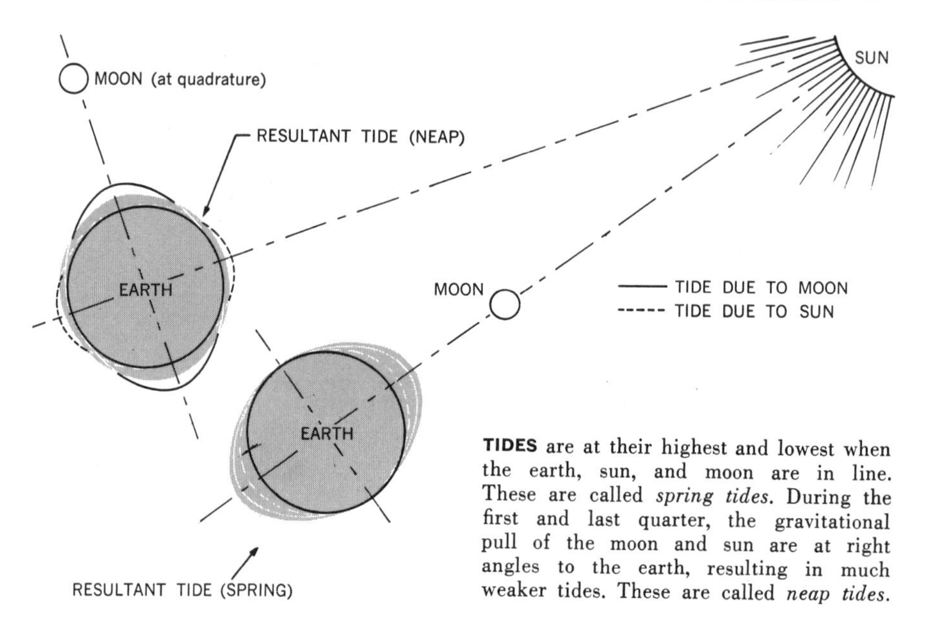

TIDES are at their highest and lowest when the earth, sun, and moon are in line. These are called *spring tides.* During the first and last quarter, the gravitational pull of the moon and sun are at right angles to the earth, resulting in much weaker tides. These are called *neap tides.*

disappearance—the proliferation or decline—of certain kinds of marine life can be attributed to the presence or absence of some substances found only in minute quantities in the sea, such as iron or copper.

The ocean, on the average, contains about 35 parts of dissolved salts per 1,000 parts of water. It is estimated that more than a thousand million tons of salts are being discharged into the sea each year by rivers. Thus, it would appear that the ocean is getting saltier; however, this does not seem to be the case. Apparently the input of new salts is offset by the amount of salts in the materials deposited on the sea floor as sediments and by spray returned to the land.

Physical Oceanography. — The entire ocean is in constant motion. Billions of tons of water course in various patterns throughout their basins, pushed and pulled by winds, currents, and tides, as well as by the motion of the earth itself.

■**TIDES.**—To the oceanographer waves and tides are physically the same. A tide is merely a very long wave moving about the earth as a surface bulge created by the gravitational forces of the moon and the sun.

If all the earth's surface were water, and the depth were the same everywhere, tides would be easy to understand and to predict. But the uneven bottom and the landmasses themselves cause the waves to pile up or to be diverted in their paths, setting up oscillations in enclosed bays. Tides are also influenced by changes in atmospheric pressure and by the action of winds.

When the moon and the sun are in alignment with the earth, as during the full moon and the new moon, the tidal movement is greater than at any other time of the month. Extra-high tides are caused when the time of high tide coincides with storm winds and extraordinarily low atmospheric pressure.

Spectacular tides can also be caused when a large standing wave arrives at the same time as a high tide. When the tide in some areas rises high enough to move against a river, it may sweep upstream as a thundering wave called a *tidal bore.* Tidal bores occur in certain rivers in Europe, North America, South America, and Asia. One of the highest tides is that which occurs in the Bay of Fundy in eastern Canada.

■**WAVES.**—There are two basic kinds of waves: standing waves and progressive waves. A *standing wave* moves back and forth in a confined space, its speed and size determined by the size and depth of the bay or estuary where it moves. A *progressive wave,* on the other hand, moves across an open area. In neither case do the particles of water themselves move very much. They may move in the arc of a pendulum, or they may travel in a circular motion, but they move very little horizontally.

This motion is the same in all waves, from the smallest ripple to the greatest ocean roller. The highest point in a wave is the *crest,* and the lowest is the *trough;* the *height* is measured by the distance between these two points. The *length* is the distance from one crest to that of the following wave, while the *period* of a wave is the time it takes to pass a given point. In the open ocean, waves higher than 25 feet are rare, although waves of 60 and perhaps even 100 feet have been reported in the great unbroken expanses of the Pacific and Antarctic oceans.

The great waves commonly known as *tidal waves,* but more accurately called *tsunamis,* are of seismic, not tidal, origin. Caused by earthquakes on the sea floor or near shore on land, a tsunami wave is hardly noticeable at sea, but it may pile up to form tremendous crests on striking shore. Hawaii and Japan have suffered many devastating tsunamis. The speed of a tsunami wave is very great and is determined by the ocean depth.

Not all ocean waves are on the surface; the sea has internal waves as well. These waves develop on the interfaces between layers of dif-

EWING GALLOWAY

TSUNAMIS, or tidal waves, are caused by earthquakes. They are most common in the Pacific.

ferent density and are usually larger than surface waves, although they move more slowly. The existence of these waves has only recently been determined, largely because of their effect on submarines; and much more is still to be learned about their formation and movements.

CURRENTS.—Surface currents of water are driven mainly by winds. Each hemisphere has three similar wind zones. Along the equator and extending north and south for some 30 degrees of latitude, the wind blows from the east. For the next 30 degrees, the wind is primarily from the west. Nearer the poles, the wind again comes from the east. The prevailing winds—equatorial easterlies and middle latitude westerlies—impart a clockwise circulation to the surface waters of the North Pacific and North Atlantic and counterclockwise circulation to those of the South Pacific and South Atlantic.

Where the wind blows steadily in the same direction the water mass, on the average, tends to move away at an angle 90 degrees to the right of the wind direction in the Northern Hemisphere and 90 degrees to the left in the Southern Hemisphere. The surface moves at 45 degrees. This angle of difference increases with depth. This effect, known as the *Ekman transport,* can produce upwellings when winds blow parallel to coastlines. As warm surface water is pushed away from a coast, colder water from underneath takes its place. The Ekman transport also creates *eddies*, or circular movements of water, such as the Sargasso Sea.

The direction of the ocean currents, as well as the great vertical movements in the sea, are also determined by the differences in the temperature of the water. All parts of the ocean are layered, with the warmer water at the surface. Deep water is always close to the freezing point, even at the equator. In tropical and temperate regions, the upper warmer layers tend to stay on top because they are less dense than the cold, deeper water. Near the poles, however, the surface layers are cooled by the frigid air, become heavy, and tend to sink underneath warmer layers. They then move in the direction of the equator.

Water flows toward areas of lower pressure. Cold water weighs more than warm water, and the higher the salinity, the greater the weight. A horizontal movement of water results as the water flows down a horizontal pressure gradient. As water flows from a high-pressure area to a low-pressure area, it follows a curved path. This curving path is caused by the rotation of the earth and is known as the *Coriolis force.* This phenomenon causes the paths of objects moving in the Northern Hemisphere to turn to the right, and those in the Southern Hemisphere to turn to the left.

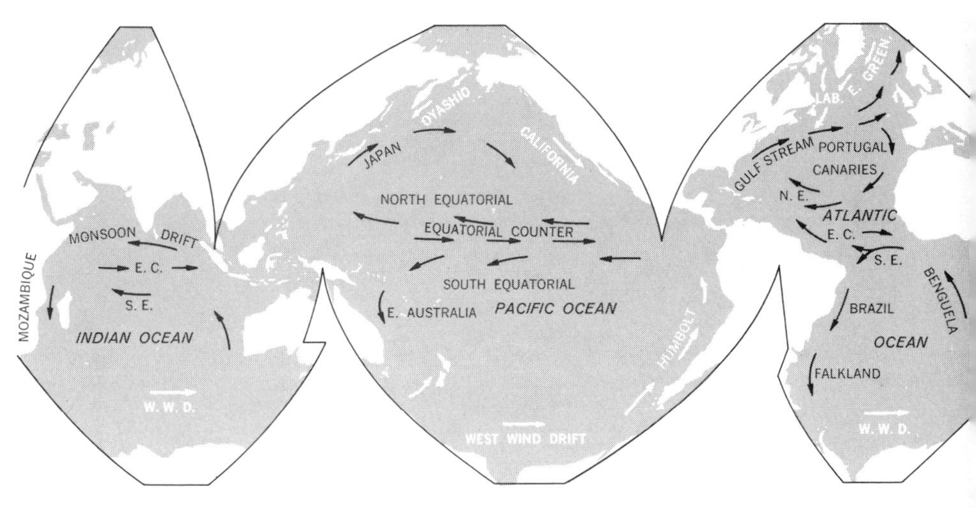

SURFACE CIRCULATION of the world's oceans is controlled mainly by winds. The flow is in a clockwise direction in the Northern Hemisphere and a counterclockwise direction in the Southern Hemisphere. The cooler surface currents are indicated with white arrows.

Biological Oceanography.—It is almost certain that life began in the sea, for all animal phyla on earth have members living in the marine environment. Furthermore, the body fluids of even the land animals are similar to the salty liquid known as sea water. Human blood, for example, is a saline solution that has many of the properties of sea water.

PLANKTON.—All living organisms are either plants or animals. The tiny plants that drift free in the ocean are known as *phytoplankton.* "Phyto" means 'plants,' and "plankton" is from the Greek *planktos,* meaning 'wandering.' Since all plant life depends on sunlight, and sunlight penetrates only to about 600 feet, all chains of life in the sea are linked to the tiny organisms of the upper layers.

Plankton can be defined as those animals and plants that are carried by currents. Since they either drift without independent movement or swim weakly, they cannot move against the current.

The *diatoms* are among the most important forms of life in the sea. These single-celled plants are composed of two shells of silica that fit together like a box with a lid. Other forms of phytoplankton are the *dinoflagellates,* microscopic plants that swim about by beating appendages called flagella.

Mixed with the plants in the upper layers are the planktonic animals that are known collectively as *zooplankton.* Smallest and simplest of these are the *protozoa,* single-celled creatures of a wide diversity of form that feed on the single-celled plants and are themselves eaten by larger and more complex planktonic animals, such as copepods, jellyfish, crustaceans, mollusks, and the larval forms of fishes.

The surface waters of the colder seas hold prodigious quantities of plankton, which furnish food for the great schools of mackerel and herring, the sea birds, and even the herds of baleen whales that subsist on planktonic crustaceans known as krill. This abundance of plant and animal life is made possible by the mineral richness of the colder waters rising from deeper layers.

Although it is evident that cool waters provide more nutrient salts, warmer seas show a much greater diversity of forms. While an area like the Grand Banks of Newfoundland might hold schools of millions of cod and haddock and support a great fishery, a coral atoll might hold thousands of species of marine animals but relatively few of any one species. The reason for this is that life cycles are speeded up in tropical waters; hence a greater number of genetic differences show up.

NEKTON.—The term *nekton* covers the free-swimming creatures that are largely independent of tides, currents, and waves. This group includes most marine fishes, from the tiny sardines to the great whale sharks, which may exceed 50 feet in length. Also among nektonic creatures are the animals of the middle depths, those perpetually black regions thousands of feet down, inhabited by bizarre lantern fish and other grotesque creatures with luminous organs, cavernous mouths, and rapierlike teeth. Here there is no light, and no plants grow. All life must subsist on other living creatures or on the remains of dead animals and plants sifting down from the surface. Most creatures in the middle regions are small, but here are also found the giant squid and the massive sperm whale that pursues the squid in prodigious breath-holding dives of 2,000 feet or more.

BENTHOS.—The *benthic* animals, the dwellers of the sea floor, are, in many cases, almost identical to bottom forms found in shallow water. There are brittle stars, sea cucumbers, sea spiders, crustaceans, flounderlike fishes, and others that would not look greatly out of place on any tidal flat. On the floor of the ocean, the pressure is thousands of pounds per square inch, but it is so evenly distributed throughout the bodies of these creatures that they feel no pressure at all. Except for certain animals with air spaces in their bodies (such as many fishes that have air bladders), the deep-water creatures can be raised from the greatest depths to the surface without injury due to changes in pressure.

AUSTRALIAN NEWS AND INFORMATION BUREAU

MARINE LIFE occurs in many forms, from the smallest one-celled animal to the largest living earth creature, the whale. Very common are the many varieties of coral (*above*) that abound in all oceans. A more familiar sea mammal, the porpoise (*above right*), leaps high above the water. Some sea creatures are as deadly as they are beautiful. The Portuguese man-of-war (*below left*) is also known as the "cobra of the sea" because of its poisonous sting. Fish, such as these mullet (*below center*), are a food source for many people. The tiny rotifers (*below right*) are consumed by the larger sea animals.

MARINE LABORATORY, UNIVERSITY OF MIAMI

AMERICAN MUSEUM OF NATURAL HISTORY

GENERAL ELECTRIC RESEARCH LABORATORY

ARTIFICIAL GILL extracts air from the water but prevents the passage of the liquid.

TAXONOMY.—Biological oceanography takes many forms. Most early students were primarily taxonomists who collected, pickled, classified, and described organisms. Many an outstanding authority on a certain group never saw a living member of that group. Now that the kinds of organisms found in the sea have become better known, biologists have begun to study these animals and plants as living organisms, not as museum specimens. Today studies are made in the undisturbed natural habitats of organisms, wherever possible.

Such environmental studies have been given great impetus by the development of undersea research vehicles and SCUBA (Self-Contained Underwater Breathing Apparatus). With SCUBA, almost any scientist can become a diver and observe marine life in its own habitat. Many shallow-water studies are made with no more than a diving mask, a snorkel, and a pair of flippers.

MARINE ECOLOGY.—With this new ease of penetration into the sea—or at least into the upper sunlit layers—the scientific discipline known as *marine ecology* is coming into its own.

Ecology, or environmental biology, can be defined as a study of organisms in relation to their environment; it is also a study of the interrelationships of individuals and groups. There are many different approaches to ecology, of course. One biologist may be interested in the ecology of a particular taxonomic group—for example, fishes, crustaceans, worms—or even a single species of one of these. Others may be interested in pelagic, benthic, littoral, or coral reef ecology, and yet others in communities or populations. One kind of ecologist may be interested in *function,* the things that animals and plants do. The study of animal *behavior* is still in its infancy, especially as applied to marine animals, but several long-range projects are now under way.

STANDARD OIL COMPANY, NEW JERSEY

WEALTH FROM THE SEA. The vast mineral reserves stored beneath the ocean's floor are only now being tapped. Here, an oil derrick is being set up in the Gulf of Mexico.

Marine Technology.—Marine scientists and technicians are using the tools and techniques of the oceanographer to harvest many of the resources of the sea. Offshore drilling rigs in the Gulf of Mexico tap reservoirs of oil beneath the sea floor, while prospectors with aqualungs and dredges are bringing up diamonds from the continental shelf off Africa. The use of recorded mating sounds to attract food fishes to nets is under experiment by some fishing fleets. Another technological invention transferred from land is undersea television, which is in use as an important oceanographic research aid and also as a tool of the fishing and communications industries. Even the water of the sea and the invisible plankton that inhabit it are being put to man's service. Modern plants now extract magnesium from sea water, and new and cheaper methods are being developed for producing fresh water from the sea. Eventually we will find a way to utilize directly the vast supply of protein-rich plankton.

Researchers are working on ways to reduce the devastation of hurricanes and the perennial damage caused by beach erosion and by shipworms, barnacles, and other ocean pests. Investigators seek safe, clean methods for disposing of industrial, human, and atomic wastes, and for coping with increased oil pollution.

From a military standpoint, our greatest problem of national defense is an oceanographic one. More effective methods of communication under water must be found—from submarine to submarine and from submarine to surface craft and airplanes—and of detecting and destroying enemy submarines in time of war. Radar and radio cannot be used under water, so all detection and communications systems in the sea involve the use of sound.

The unknown regions and untapped resources of the sea present one of the greatest challenges to man's ingenuity and daring. As new methods of oceanic exploration are devised and craft descend for longer periods into the ocean depths, men will finally return to the seas from which they came, to explore, to work, and perhaps even to live in the last frontier regions of the earth.

—Friedrich Frans Koczy;
Cesare Emiliani

BIBLIOGRAPHY

BASCOM, WILLARD. *Waves and Beaches.* Anchor Books, 1964.

COWEN, ROBERT C. *Frontiers of the Sea.* Doubleday & Co., Inc., 1960.

KING, CUCHLAINE AUDREY MURIEL. *Introduction to Oceanography.* McGraw-Hill, Inc., 1963.

MAURY, MATTHEW FONTAINE. *The Physical Geography of the Sea.* Edited by John Leighly. Harvard University Press, 1963.

TUREKIAN, KARL K. *Oceans.* Prentice-Hall, Inc., 1968.

WALFORD, LIONEL ALBERT. *Living Resources of the Sea.* The Ronald Press Co., 1958.

Meteorology is the science of the atmosphere. The *meteorologist* deals with atmospheric processes. Although the meteorologist is often thought of as one who forecasts weather conditions, there is much more to meteorology than weather forecasting. Some meteorologists, for instance, are concerned only with the effects of the atmosphere on the flight of missiles and satellites. Some deal with atmospheric processes as they affect the health and behavior of plants, animals, and human beings. Other meteorologists work with weather conditions as they influence the operations of specific businesses and industries, and still others concern themselves entirely with seeking ways to modify and control the weather. These are only a few of the studies that make up the science of meteorology. In essence, the goal of the meteorologist is to be able to describe and predict the atmosphere's behavior. The first step is to learn everything possible about the properties of the air in which we live.

Historical Background.—The early attempts at weather forecasting related the condition of the surroundings, or the appearance of certain objects, to particular types of future weather. This type of weather prediction dates back to at least 700 B.C., when the Assyrians used the ring around the sun or moon as a sign of coming weather—a sign that is still used. The first book on meteorology was written by Aristotle in about 350 B.C., but no real progress was made in the development of meteorology as a science until the invention of the thermometer by Galileo Galilei in 1640 and of the barometer by Evangelista Torricelli in 1643. The use of these instruments led to recognition of the fact that high-pressure and low-pressure areas can be associated with certain types of weather. In 1743 Benjamin Franklin presented evidence that high-pressure and low-pressure systems move across the earth and carry the weather with them.

Composition of the Atmosphere.—We live at the bottom of a mixture of gases that envelops the earth. This mixture of gases is called *air. Pure, dry air* is made up of 78 per cent nitrogen, 21 per cent oxygen, less than 1 per cent argon, and very small amounts of carbon dioxide, hydrogen, neon, helium, krypton, and xenon. *Ordinary air* contains many other substances, the most important of which is water vapor. Depending upon prevailing conditions, the water vapor can be either liquid or solid, producing fog, clouds, rain, sleet, snow, or hail. Very small sea-salt and dust particles in the air provide nuclei around which rain drops form. Impurities in the air such as smoke and dust sometimes become great enough to cause the condition known as *smog.*

■TROPOSPHERE.—Although the atmosphere extends to great heights, most of it is contained in a six- to ten-mile-high layer that is highest over the equator and lowest over the poles. Since almost all water vapor is distributed throughout this layer, most ordinary weather conditions occur within it. This region of the atmosphere is called the *troposphere.* The troposphere is characterized by a decrease in temperature with each increase in altitude—the greater the height, the lower the temperature. On the average the temperature in the troposphere drops slightly more than 1° F. with each 300-foot increase in elevation.

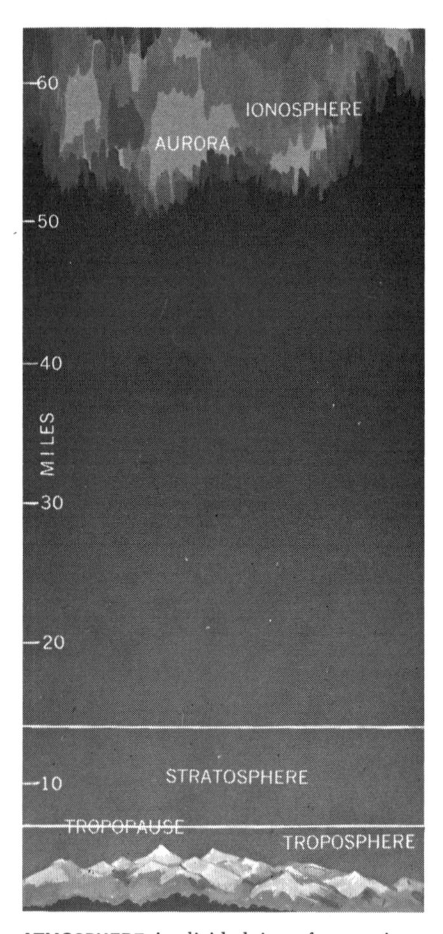

ATMOSPHERE is divided into four regions.

■TROPOPAUSE AND STRATOSPHERE.—The layer separating the troposphere and *stratosphere* is known as the *tropopause.* Within the stratosphere the atmosphere thins with each increase in altitude until, at a height of 60 miles, all but one-millionth of the atmosphere is below.

■IONOSPHERE.—The region above the stratosphere is called the *ionosphere* because it contains several layers of electrically charged particles known as *ions.* These particles reflect radio waves, thereby making world-wide radio communication possible. The colorful *aurora borealis,* or northern lights, produced by charged particles from the sun, also occurs in the ionosphere.

■TEMPERATURE.—The temperature of the atmosphere decreases through the troposphere, becoming as low as —80° F. at the tropopause. It remains constant through the lower stratosphere until, at an elevation of from 12 to 15 miles, it begins to rise again. Above 30 miles the very thin air may become as warm as 70° F. This rise in temperature is due to absorption of ultraviolet rays from the sun by the *ozone,* a form of oxygen found in this region. Above this warm layer the temperature again decreases steadily until, at a height of 45 to 50 miles, it has fallen to as low as —90° F. There are indications that the temperature again begins to rise above the 50-mile level, but the atmosphere is so thin at this height that it is difficult to measure temperature in the usual manner.

Elements of Weather

■ATMOSPHERIC PRESSURE.—The expression "light as air" and the fact that air is invisible to the eye imply that, in a physical sense, there is nothing to the atmosphere. Air is quite heavy, especially near the surface of the earth. The weight of air over a unit area is called *air pressure.*

On the average the weight of the atmosphere over every square inch of the earth's surface at sea level is 14.7 pounds, or *one atmosphere.* Thus the total weight of air on a 100-foot square plot of ground is more than 10,000 tons. Buildings and other objects are not crushed by this weight because the same pressure inside them equalizes the outside pressure and makes the resultant force zero.

Air pressure is an important element of weather because, generally speaking, high pressure can be associated with fair weather and low pressure can be associated with clouds and precipitation. Pressure patterns and pressure changes are important, therefore, as indicators of future weather. In addition, it is the difference between pressure in one area and another that causes the air to move, thereby creating winds.

■PRESSURE MEASUREMENT.—An instrument used to measure air pressure is a *barometer.* The most direct method of measuring the pressure or weight per unit area of air is to balance it against the weight of some other substance. Mercury is usually used for this purpose because it is the heaviest of liquids.

A simple *mercury-in-glass barometer* consists of a glass tube about three feet long that is open at one end and closed at the other. The tube, filled with mercury, stands vertically in a dish of mercury, its open end submerged. Since the weight of the air pressing down on the surface of the mercury in the container will support a column of mercury of equal weight in the glass tube, air pressure is often spoken of in terms of the equivalent height of a mercury column. When barometric pressure is stated as 30.00 inches, it means simply that the weight per unit area of the air is equal to the weight per unit area of a column of mercury 30 inches high. This type of barometer is very accurate, but has numerous disadvantages. It is difficult to transport; the glass tube is easily broken; mercury can be harmful under certain circumstances and is subject to expansion and contraction with temperature variations.

The *aneroid barometer* is portable, liquidless, and automatically corrected for temperature contraction and expansion. Most aneroid barometers have a dial face with the descriptive words "stormy," "rain," "fair," and "dry" printed on it. Mechanically, it consists of a "dry" pressure-sensing element linked by a system of levers and gears to a pointer moving across the dial face.

The barometer is probably the most useful instrument in making a weather forecast, but should not itself be considered as a forecaster of weather. Although it frequently proves true that high pressure means "fair" weather and low pressure means "bad" weather, there are some exceptions. It is, therefore, important to realize that the function of the barometer is to read air pressure; the change in pressure is more important than the actual reading. A pressure change indicates that a change in the weather is imminent; the more rapid the change in pressure, the sooner the change in prevailing weather conditions will occur.

TEMPERATURE.—The earth's atmosphere behaves much like a heat engine operating on energy supplied by the sun. *Temperature* is the measure of the intensity of heat energy supplied. Although as a concept temperature is technically more difficult to define than pressure, it is more easily understood because of the human body's temperature sense. The body can feel temperature differences (that some things are warmer than others), but it cannot ordinarily feel differences in atmospheric pressure. The body sensations of hot and cold are very crude measurements, but accurate values of temperature can be established with instruments.

TEMPERATURE MEASUREMENT.—Most objects change in size as the temperature changes, expanding as the temperature increases and contracting as the temperature decreases. This relationship is used to assign a definite numerical value to a temperature.

In 1724 Gabriel Daniel Fahrenheit, a German physicist, produced a *thermometer* similar to the instrument in use today, consisting of a thin glass tube with a bulb or "bulge" at one end. The bulb is filled with mercury; as the temperature rises, the mercury expands and is forced up into the tube. The level of the mercury in the tube, gauged on a calibrated scale, indicates the temperature—the higher the mercury rises in the tube, the higher the temperature.

Alcohol is frequently used in place of mercury because it has two advantages. Alcohol has a much lower freezing point and can therefore be used to measure air temperatures at which mercury would freeze; and it can be colored, making it readable.

There are many other types of thermometers used for special purposes. The *bimetallic thermometer* consists of two strips of different metals with different expansion characteristics—one expands more than the other as the temperature rises. When the two metals are fastened together, the resulting compound strip bends as the temperature changes because of the difference in rates of the metals' expansion. One end of the compound strip is fixed, and a pointer attached to the free end moves across a temperature scale as the strip changes shape. A pen is sometimes attached to the free end of the bimetal strip instead of a pointer. The pen is positioned so that it touches a piece of graph paper wrapped around a slowly rotating cylinder. Such an instrument, called a *thermograph,* keeps a continuous record of temperature. Thermometers designed to read the highest and lowest temperatures over a given period of time are called *maximum thermometers* and *minimum thermometers,* respectively.

HUMIDITY.—The amount of water in the air in vapor or gaseous form is referred to as *humidity.* Compared to the amount of oxygen and nitrogen in the atmosphere, the amount of water vapor is very small—usually only 1 to 2 per cent. The total volume of water vapor in the atmosphere, however, is substantial. The amount of water, for instance, carried in the atmosphere over the continent of North America is about six times the amount flowing in all the rivers on the continent.

Water vapor in the atmosphere sometimes condenses to form clouds, fog, or rain; sometimes it solidifies to form snow, sleet, or hail. Water vapor is, therefore, responsible for most of the conditions we commonly call "weather." Water vapor also affects the distribution of heat in the atmosphere by absorbing and reflecting solar radiation and by releasing heat during the process of condensation. The amount of water vapor in the air also governs, to a large extent, the degree of human comfort, especially in hot weather.

HUMIDITY MEASUREMENT.—A *hygrometer* measures water vapor (humidity) in the atmosphere through the use of substances that vary in size with humidity. Hair, for instance, expands when the humidity increases and contracts when it decreases. Human hair is especially sensitive to such changes, and is used in an instrument called a *hair hygrometer.*

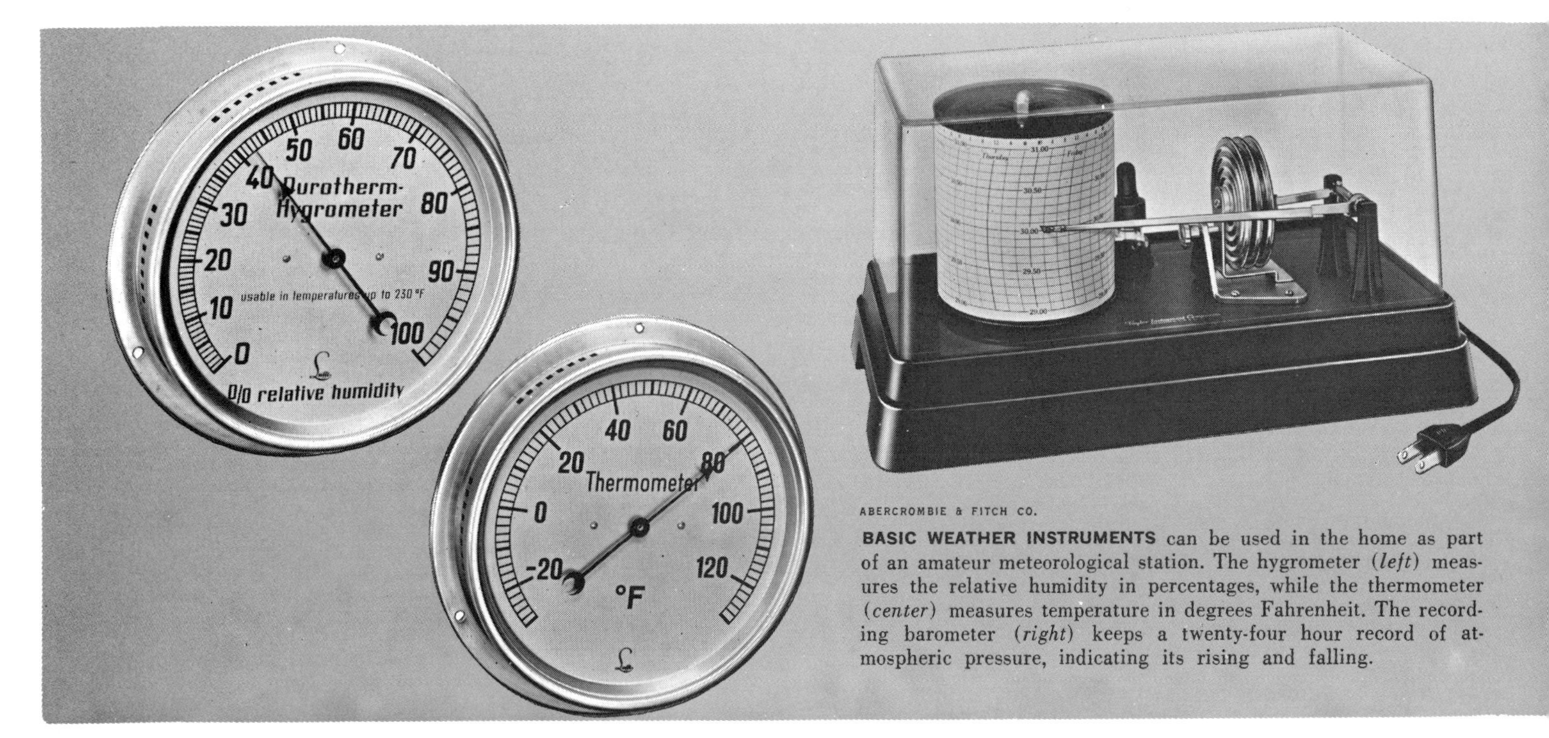

ABERCROMBIE & FITCH CO.

BASIC WEATHER INSTRUMENTS can be used in the home as part of an amateur meteorological station. The hygrometer (*left*) measures the relative humidity in percentages, while the thermometer (*center*) measures temperature in degrees Fahrenheit. The recording barometer (*right*) keeps a twenty-four hour record of atmospheric pressure, indicating its rising and falling.

One end of a bundle of hairs is fixed, and the other end is attached to a pointer that moves across a humidity scale as the hair length changes. An instrument of this type making a constant record of humidity changes, in the same manner that a thermograph records temperature changes, is called a *hygrograph.*

An instrument that measures humidity through the principle of evaporation is called a *psychrometer. Evaporation,* the process of changing a liquid into a gas, is a cooling process. As such it is used to measure atmospheric humidity. The psychrometer consists of two liquid-in-glass thermometers. The bulb of one of these, called the *wet-bulb thermometer,* is covered with a tight-fitting piece of cloth that has been dipped in water. As the water evaporates, the bulb is cooled, thus lowering the wet-bulb thermometer's temperature reading. The other thermometer records the ordinary air temperature, known in this instance as the *dry-bulb temperature.* Due to the evaporative cooling, the temperature of the wet-bulb thermometer is normally lower than that of the dry-bulb thermometer—the lower the relative humidity, the greater the difference between wet-bulb and dry-bulb temperatures. Using these readings, the relative humidity is then determined by reference to *psychrometric tables.*

An instrument that utilizes condensation in measuring the water vapor content of the atmosphere is called a *dew-point hygrometer.* Just as liquid water can be evaporated into the air, water vapor can also be drawn from the air in the form of a liquid. This process is called *condensation.* As air is cooled, a temperature is reached at which condensation occurs. This is known as the *dew-point temperature.* For every value of water vapor content (humidity) there exists a corresponding dew point. The point at which condensation begins and dew forms is the *dew point*—the more water vapor in the air, the higher the temperature at which the vapor will begin to condense. The dew-point hygrometer usually consists of a piece of highly polished metal arranged so that its temperature can be measured as it is cooled to the dew point. An advanced type of this instrument consists of an electronically cooled mirror that is automatically monitored by a photoelectric cell, which signals the point at which condensation occurs.

HUMIDITY UNITS.—The amount of water vapor in the air at a given temperature has a certain maximum value. When that value is reached, the air is said to be *saturated* or to have reached its water-vapor capacity—the point at which the water vapor begins to condense into a liquid or a solid, and fog, clouds, rain, snow, frost, or dew begin to form. It is important, therefore, to know just how close the air is to saturation. The amount of water vapor in a unit volume of air is called the *absolute humidity.*

Relative humidity defines the ratio of absolute humidity to capacity—a measure of how close the air is to saturation at a given temperature.

$$\text{relative humidity (per cent)} = \frac{\text{absolute humidity}}{\text{capacity}} \times 100$$

Since capacity depends upon temperature, relative humidity also depends upon temperature. Thus, even if the absolute humidity remains the same, the relative humidity will change; it will be higher at lower temperatures. In other words, although the actual amount of water vapor in the air remains the same, the relative humidity will fall during the day as the temperature rises and increase at night as the air cools. Atmospheric pressure is another factor to be considered because as the pressure changes, the volume of air will also change, thus affecting both absolute and relative humidity.

A humidity unit that remains constant although pressure or temperature varies is defined as *specific humidity,* the amount of water vapor per unit *mass* of air. Specific humidity is usually expressed as grams of vapor per kilogram of air. One kilogram of air containing 15 grams of water vapor at a given temperature and pressure has a specific humidity of 15 grams per kilogram. If the water vapor content remains the same, this value will likewise remain the same for all temperatures and pressures because the mass of air is not affected by temperature and pressure changes.

Clouds.—Weather conditions on the earth's surface are often governed by conditions at higher altitudes. In the past, cloud observations were the only available means of determining conditions in the upper atmosphere. They are still useful indicators of weather changes.

U.S. DEPARTMENT OF COMMERCE, WEATHER BUREAU

RADIOSONDE is carried aloft by a balloon.

CLOUD TYPES.—The highest clouds are *cirrus clouds.* "Cirrus," which means "curl" in Latin, describes the characteristic hooks or curls these clouds have at their borders (sometimes called "mares' tails"). Cirrus clouds are feathery and white, and range to heights at which temperatures are well below freezing. These clouds are therefore usually composed of ice crystals. When the cirrus clouds remain feathery or slowly disappear, fair weather is indicated. When they grow thicker and blanket the sky, it is likely that lower clouds will form, followed by rain or snow.

Cirrostratus clouds take the form of a continuous white sheet and often give the sky a milky appearance. Enough sunlight penetrates this sheet to cast shadows on the ground. Like cirrus clouds, the cirrostratus are made up of ice crystals, and generally produce a halo or ring around the sun or moon; only clouds composed of ice crystals produce such a ring. When cirrostratus clouds increase and thicken, rain or snow can be expected within 24 hours.

Cirrocumulus clouds appear at high levels as small, white patches. When arranged in rows or waves, they produce what is sometimes called a "mackerel sky." Cirrocumulus, cirrus, and cirrostratus clouds are usually classed together. They all form at altitudes exceeding 20,000 feet, and are sometimes called *high clouds.*

Altostratus clouds form a heavy, gray sheet across the sky. The sun is usually visible as a bright spot, but is not bright enough to cast shadows on the ground. These clouds are composed of water droplets, even at temperatures below freezing, and therefore do not form halos. The appearance of altostratus clouds usually means that rain or snow will follow shortly.

Altocumulus clouds appear in closely spaced patches and are white or gray. They are similar in appearance to the cirrocumulus variety, but are larger and usually at lower levels. When they are in the proper position between the sun and the earth below, beautiful colors often appear around their edges as the sunlight passes through the water droplets. Alto-type clouds usually range from 6,500 to 20,000 feet above the earth and are called the *middle clouds.*

Stratocumulus clouds are irregularly shaped, appearing in rolls or patches that often blend together. They are larger and thicker, and appear at lower levels than altocumulus clouds. Sometimes only a few thousand feet above the ground, stratocumulus clouds often appear just after a storm, but before complete clearing sets in. Light showers of rain or snow may fall from these clouds.

Stratus clouds are low-level gray sheets located from a few hundred to a few thousand feet above the ground. They are relatively thin, and sometimes produce light drizzle.

Nimbostratus clouds are low, dark, gray clouds. They are thicker and darker than the stratus variety and often are the result of a thickening and lowering of altostratus clouds.

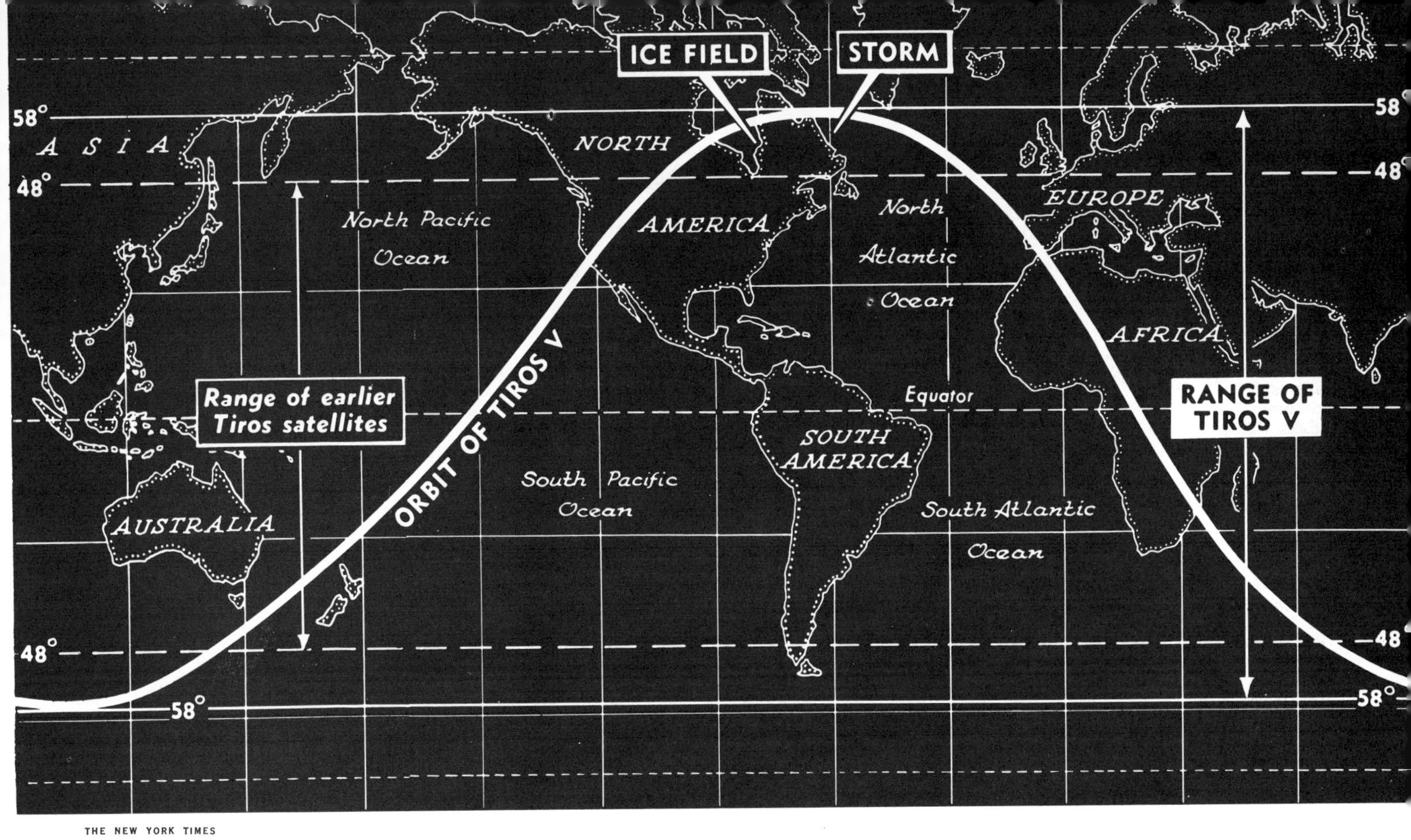

THE NEW YORK TIMES

Weather eyes on the world watch, predict and warn

FROM 400 MILES above the earth, artificial satellites like Tiros V, whose orbital path is shown above, look down on the globe through the eyes of their television cameras. The pictures they transmit enable meteorologists to keep constant track of weather all over the world. Left, a glimpse into the vortex of an Atlantic storm sheds light on the origin, development and movement of storms, to warn against their approach and perhaps, eventually, to tame them. Right, photograph of an ice field in Hudson Bay helps predict future icebergs which menace shipping lanes.

NASA

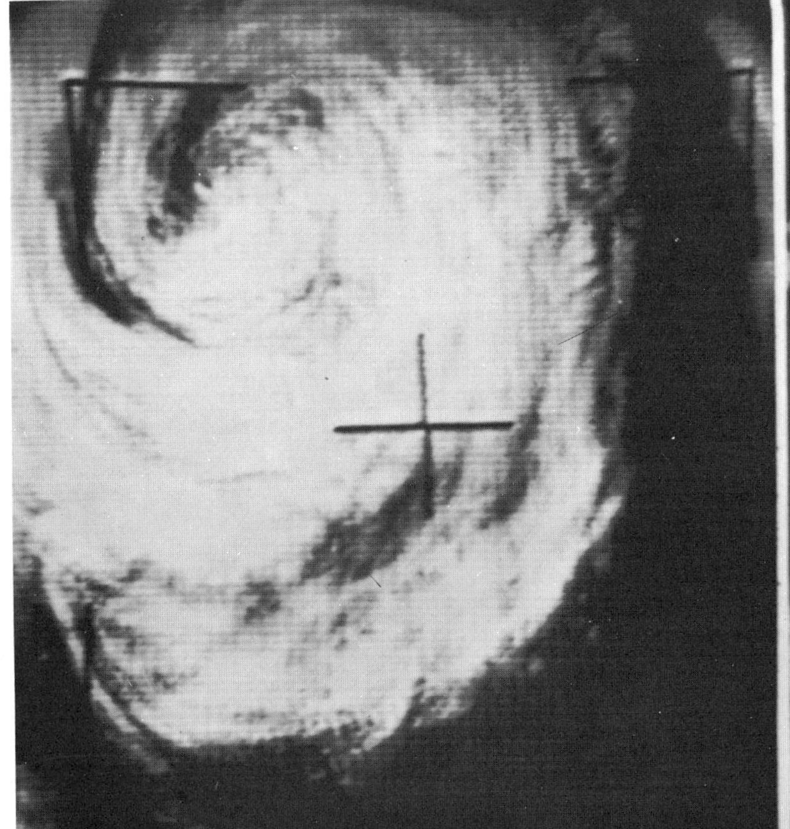

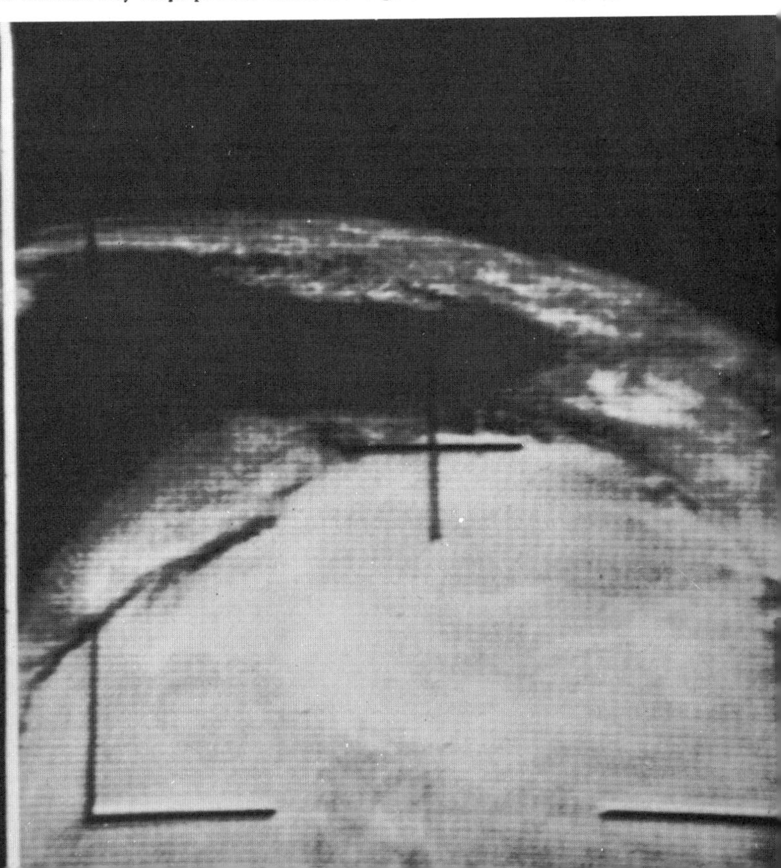

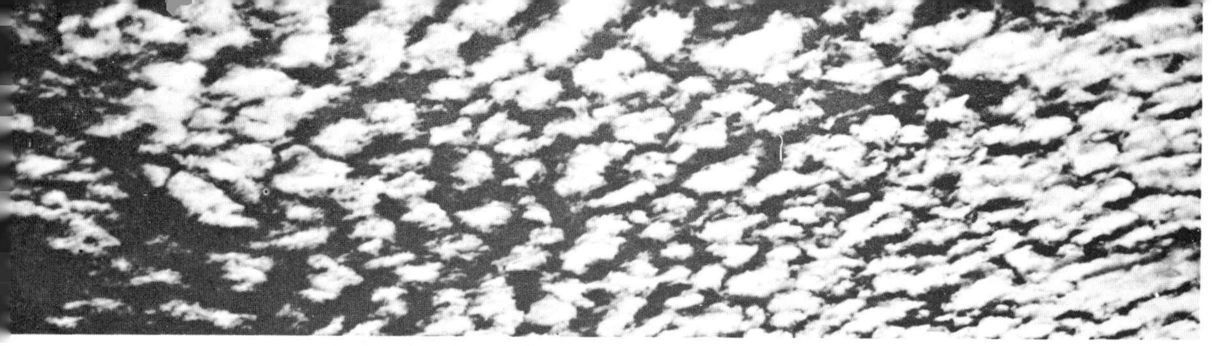

TUFTED PATCHES of Altocumulus water clouds spread across the sky

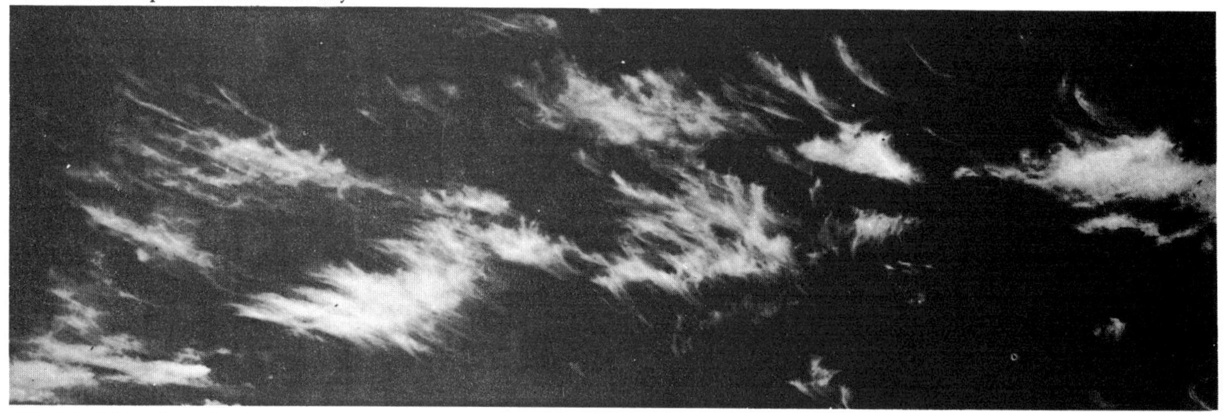

CIRRUS, high ice clouds, may portend rain or snow

CUMULUS, usually seen on sunny days, can develop into rain-clouds

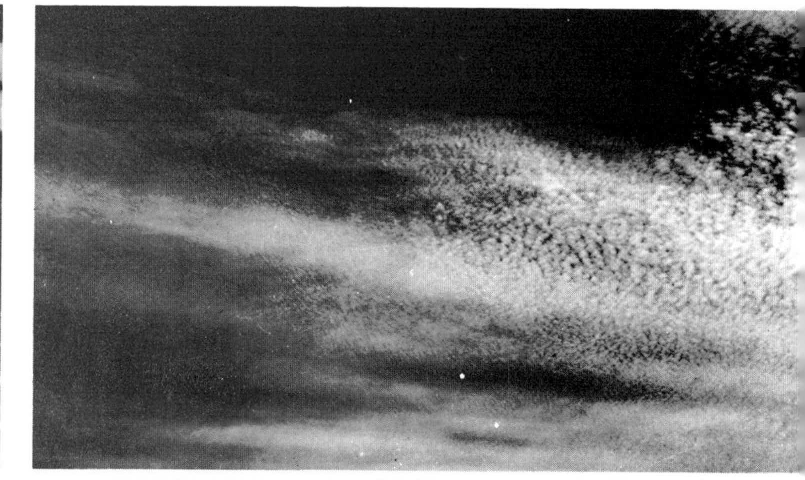

CIRROCUMULUS, merging with Cirrostratus, are harbingers of rain

ALTOCUMULUS LENTICULARIS, named for their strange lens-like shape

CUMULONIMBUS, towering over the landscape in great thunderheads

E. FONTSERÉ

COURTESY OF U.S. DEPT. OF COMMERCE, WEATHER BUREAU

When nimbostratus clouds appear, heavy rain or snow usually follows immediately.

Cumulus clouds are probably the most familiar. They are fluffy, white, flat-based, puff-sided, and round-topped, and usually result from surface heating of the earth. Over land they form most frequently in the late morning and early afternoon. They are essentially fair-weather clouds. If they slowly disappear as the sun begins to set, the weather will probably remain fair throughout the night. The air inside cumulus clouds is in a continuous state of up-and-down motion; so much so, in fact, that airplanes flying through them often experience a bumpy ride. Cumulus clouds change size and shape very rapidly.

Cumulonimbus clouds are the largest clouds in the sky. They are actually overgrown cumulus clouds, often extending to more than five miles above the earth. At the top of these clouds the rounded edges break into a flattened, anvil-shaped layer. Cumulonimbus clouds are the familiar thunderclouds that produce heavy showers, lightning, thunder, and often hail and high winds. As cumulus clouds grow in the process of becoming cumulonimbus clouds, but before the anvil-shaped top appears, they are known as *cumulus congestus*.

The accompanying illustration shows the forms and relative heights of the main cloud types. There are two main divisions based on shape. Cumulus-type clouds have a heaped-up, or bulging, appearance. Stratus-type clouds are spread out in a layer. There are four main divisions based on height. Cirrus-type clouds appear at the greatest heights; alto-type clouds appear at middle levels; stratus and stratocumulus are at low levels; and nimbostratus, cumulus, and cumulonimbus occur in more than one level. The nimbus-type clouds usually produce large amounts of rain.

Condensation and Precipitation

CONDENSATION.—When the air becomes *saturated* (relative humidity reaches 100 per cent), the water vapor in it usually condenses into a liquid. The drops form upon small airborne particles of sea salt, dust, and other matter, known in this instance as *condensation nuclei*. Air may become saturated by the addition of water vapor or through cooling. Condensation in the atmosphere is most often due to cooling, which may take place in a number of ways.

Radiative cooling occurs when, on a clear night, heat from the earth's surface is radiated into space; air near the ground is cooled by contact with the earth. Air may also be cooled by *expansion*. Since atmospheric pressure decreases with each increase in altitude, air rising from a lower level expands. In so doing, its temperature decreases.

An upward air movement can result from the heating of the earth's surface by the sun. Air may also be forced upward by hills or mountains over which it flows, or may be displaced by a heavier mass of air.

SNOWFLAKES are symmetrical hexagons.

Condensation sometimes results in the formation of small drops of water near the ground, called *fog*. These drops are so tiny it would take billions to fill a teaspoon. If the drops formed by this near ground-level condensation grow large enough to fall to the ground, they produce *drizzle* or *mist*, although drizzle most often falls from low clouds. If condensation forms directly on the ground or on some other surface, it is called *dew*. Dew that forms on a surface whose temperature is below freezing will become small ice needles called *frost*. Condensation that is produced by the expansion of rising air and takes place above ground level will result in the formation of *clouds*.

PRECIPITATION.—When cloud drops grow big enough to fall from a cloud and reach the ground, the condition produced is known as *precipitation*. Very few clouds produce precipitation unless the interior temperature first falls below freezing, and ice crystals form. Thus, most precipitation begins as ice. The small cloud drops begin to condense on the ice crystals, forming larger drops. As the drops grow in size, they begin to fall through the cloud, collecting still more cloud drops. The average rain drop equals about one million cloud drops.

TYPES OF PRECIPITATION.—Precipitation that reaches the ground in liquid form is called *rain* or *drizzle*. If the liquid freezes on coming in contact with the ground, it is known as *freezing rain* or *freezing drizzle*. Precipitation that falls through the air as ice is called *sleet*. These ice crystals sometimes have very complex and beautiful shapes and are known as *snow*. Snow crystals are always hexagonal in shape, but no two are alike. *Hail* is formed from raindrops that freeze. Often hail is carried up and down by air currents within a cloud, collecting liquid water drops as it falls and is again carried aloft. This process is sometimes repeated many times, the *hailstones* growing larger with each passage. Hail is almost always associated with thunderstorms.

PRECIPITATION MEASUREMENT.—All that is needed to measure rainfall is a large, open can to catch the rainwater and a ruler to measure the depth of the water collected. A rainfall measurement, therefore, represents the height to which rainwater accumulates on a level surface. The *standard rain gauge* used by the U.S. Weather Bureau is a cylindrical can 24 inches high and 8 inches across. A funnel channels the rain water into a measuring tube with cross-sectional area exactly one-tenth that of the catching can. The depth of rain water measured is, therefore, ten times its actual depth. In other words, one inch of rain in the catching can is measured as ten inches in the tube. The height of rainwater in the measuring tube is divided by ten to find the actual reading. This allows for more accurate measurement—to 1/100 of an inch. The funnel minimizes evaporation because it covers the collected water, except a very small opening.

Snowfall is measured in terms of both depth and weight. The depth is determined by choosing a level area where there has been no drifting and inserting a ruler into the snow until it reaches the ground. Several depth measurements are taken at different points in the area, and the average of these is used. Weight is determined by melting a volume of snow and then measuring the depth of the resultant water. This is known as the *water equivalent* of snow. A standard rain gauge, with the measuring tube and funnel removed, is used to catch the snow. The snow collected in the can is then melted, and the water depth is measured. On the average, ten inches of snow will melt down to approximately one inch of water, but the water equivalent (weight) of snow varies widely.

Wind

PRESSURE GRADIENTS.—*Wind*, which is air in motion, sharply influences the weather by transporting heat and moisture. If the temperature of the atmosphere were the same all over the world, atmospheric pressure would also be the same and air would not move—there would be no wind. Heat from the sun, however, is distributed unevenly over the earth. The variations in temperature produce the differences in atmospheric pressure that, in turn, produce air movement. (The greater the difference in pressure, the faster the air will move.) The force producing this air movement is known as the *pressure-gradient force*. If this were the only force acting on the atmosphere, winds would always blow directly from areas of high pressure to areas of low pressure.

CORIOLIS EFFECT.—The rotation of the earth deflects the wind; instead of flowing in a straight line from high to low, winds flow *around* pressure centers. This *Coriolis force* turns the wind to the right (clockwise) in the Northern Hemisphere and to the left (counterclockwise) in the Southern

Hemisphere. If a person stands with his back to the wind in the Northern Hemisphere high pressure will be on the right; low pressure on the left. In the Southern Hemisphere, it would be just the opposite; low pressure on the right, high on the left. In the Northern Hemisphere, winds blow counterclockwise around a low-pressure center (*cyclone*) and clockwise around a high-pressure center (*anticyclone*). In the Southern Hemisphere, winds blow clockwise around a low-pressure center; counterclockwise around a high-pressure center. Friction with the earth's surface also affects the wind, making it spiral *in* toward the center of a low-pressure area, *out* from a high-pressure area.

■WIND SYSTEMS.—The distribution of heat over the surface of the earth results in world-wide pressure patterns and relatively steady wind systems. These *prevailing winds* are known as *polar easterlies* in polar latitudes, *prevailing westerlies* in middle latitudes, and *trade winds* in tropical regions. Smaller cyclones and anticyclones migrate within these large-scale wind systems.

■MONSOONS.— Localized temperature variations sometimes create small-scale wind systems. The *sea breeze* is one example. During the day the air over the land becomes hot and rises, reducing the air pressure near the ground. The cooler (higher-pressure) air over the water flows toward the land. A wind system produced in this manner is called a *monsoon*. The most famous monsoon circulation occurs over India.

■JET STREAM.—A special, high-altitude wind system, generally found from 30,000 to 35,000 feet above the earth, just below the tropopause, is known as the *jet stream*. Winds in the jet stream average 175 miles per hour, but speeds as high as 400 miles per hour have been recorded.

■WIND MEASUREMENT.—In some localities, the direction from which the wind blows can be associated with a particular kind of weather. For example, in many sections of the United States a south wind signals the approach of warmer weather, while a north wind indicates the opposite. A wind from the west will most often bring fair weather; an easterly wind, rain or snow. A change in wind direction can therefore serve as a useful indicator of a change in the weather.

Wind speed is expressed in either knots or miles per hour and can be estimated on the basis of constant observations. The first wind scale was devised in 1805 by the British admiral, Sir Francis Beaufort. The *Beaufort Scale* is still used in a modified form.

An instrument that shows wind direction is called a *wind vane*. The most common wind vane is an arrow with a large tail. This arrow rotates freely on a fixed base and points *into* the wind—in the direction from which the wind blows.

The instrument used to measure wind speed is called an *anemometer*. The *cup anemometer* is made up of three or four hollow, hemispherical cups attached to horizontal arms extending from a vertical axis. The force of the wind on the cups causes the apparatus to turn on the axis—the higher the wind speed, the faster the cups turn. The spinning apparatus is linked mechanically to a pointer that moves over a scale and indicates speed in knots or miles per hour.

There are some instruments designed to measure wind direction and wind speed simultaneously. One of these looks like a wind vane, but has a hollow tube in the head of the arrow. The wind speed is measured in terms of the pressure of the air blowing into the tube. Another instrument that measures wind speed and direction simultaneously is the *aerovane*, which looks like a miniature airplane without wings. A tail fin keeps the instrument facing into the wind, and the wind speed is determined by a spinning propeller.

■WIND VELOCITY.—*Wind velocity* encompasses both wind speed and wind direction. It should never be used to indicate wind speed alone. *Wind direction* is specified in compass degree points and indicates the direction *from* which the wind is blowing. An east (90°) wind blows from east to west; a south (180°) wind blows from south to north.

Weather Control.—Since all of us participate in outdoor activities of some sort, we must at times be concerned about the weather. Man has therefore always thought about the possibility of exercising some control over it. The *smudge pots* (pots of burning oil) used by citrus growers to prevent their crops from freezing are one example of small-scale weather control. Weather, of course, also affects indoor environment.

■RAINMAKING.—In most cases, precipitation is the kind of weather we want to regulate. In the past, ritual ceremonies were performed to conjure up, or to stop, rainfall. The most famous of these is the *rain dance* of the North American Hopi Indians.

The modern era of rainmaking began in 1946, when it was discovered that dropping small particles of dry ice into a cloud could initiate precipitation by converting some of the small water drops in the cloud to ice crystals. (Natural precipitation usually begins with the formation of ice

U.S. DEPARTMENT OF COMMERCE, WEATHER BUREAU

U.S. NAVY

HURRICANES periodically lash the Florida coast with high winds and heavy rain (*left*). As part of the hurricane-surveillance program, a photographic plane flies above a hurricane (*right*); the clouds circle into the storm's eye, visible above the tail of the aircraft.

crystals.) This process became known as *cloud seeding*. Later it was discovered that if large amounts of dry ice were dropped into thin clouds, the resulting rain drops would be so small they would evaporate before reaching the ground. A hole could thus be cut in a cloud without causing precipitation. Even later it was found that crystals of silver iodide could produce the same effect as the dry ice. This was important because it enabled scientists to "seed" clouds from the ground by burning a solution containing silver iodide and allowing the vapor to float up into the clouds—a much less expensive method than flying above a cloud to drop dry ice into it.

Clouds must be present before rainmaking activities can begin. No one has yet devised a way to form clouds in clear, dry air. Thus, it is very difficult to determine whether cloud seeding really produces rainfall, over and above that which would have occurred naturally. This question will be answered only after years of experiment and observation.

■STORM CONTROL.—Experiments have been performed to determine whether seeding special types of clouds can prevent damaging storms. The idea is that seeding thunderstorm clouds before they develop fully may start premature precipitation and eliminate the clouds before they reach the thunder, lightning, and hail stage. It is believed that such early seeding may prevent tornadoes and that seeding hurricanes early in their development may stop them from maturing.

■REGULATING SOLAR ENERGY.—Future efforts to control the weather will most likely be concerned with discovering a way to influence and control the distribution of solar energy. This will really be going to the heart of the matter because it is the sun's energy that conditions the earth's atmosphere, producing most of the conditions we call "weather." A number of devices have already been suggested to capture or reject the sun's energy, thereby producing very extensive weather modification. If more solar energy, for instance, could be concentrated on the polar regions, it would increase the rate at which the ice and snow in these regions melt. This would result in a rise in the level of all the oceans. Winters would become much milder, and precipitation in the middle latitudes might be drastically reduced.

Special Storms

■HURRICANES.—High-speed winds sometimes occur near the surface of the earth in conjunction with some types of cyclones. One of these is the most destructive of all weather systems, the *hurricane*. Hurricanes are seasonal storms (most prevalent in August and September) that originate over the tropical regions of the Atlantic Ocean. Sometimes more than 300 miles in diameter, these storms move at 10 to 20 miles per hour and have winds of over 75 miles per hour. (A hurricane generates more energy in one hour than all the electric power generated in the United States in one year.) At the center, or *eye*, of a hurricane is an area about five miles wide where the winds are usually calm and the sky above is sometimes clear. Hurricanes usually last from 5 to 10 days. They lose force rapidly when they move over land. On the average, two hurricanes strike the United States each year. No hurricane has ever been observed south of the equator.

The most destructive element of a hurricane is the extraordinarily high tides it drives before it. The hurricane of 1938 caused 500 deaths in New England, most of them drownings. The tragic Galveston hurricane of 1900 swept a 15-foot wall of water out of the Gulf of Mexico and across the city, killing over 5,000 people.

■TYPHOONS.—A *typhoon* is the Pacific version of a hurricane. It occurs mainly during February and March.

■TORNADOES.—Although limited in size and duration, *tornadoes* are the most violent and deadly of all storms. Ranging from 100 to 500 feet wide, they move at from 25 to 40 miles per hour for up to 50 miles. Tornadoes are characterized by a dark funnel shape descending from a low cloud.

A tornado presents a double hazard —high-speed winds and low pressure. Wind speeds have been estimated to be as high as 500 miles per hour. The air pressure at the center of a tornado is very low—perhaps two-thirds that of the surroundings. Thus, closed houses over which a tornado passes usually explode—the pressure within the house pushes against the abnormally low pressure outside. Tornadoes have played many curious tricks, such as driving boards through utility poles and carrying children and animals through the air for miles without harming them. A tornado occurring over water is a *waterspout*.

Most tornadoes occur in Australia and in the southern and the western sections of the United States. More than 100 tornadoes, killing more than 200 people, are recorded every year in the United States.

■THUNDERSTORMS.—One of the most common and spectacular storms is the *thunderstorm*, a composite of high winds, heavy rain, loud noises, and flashing light. *Thunder* is the result of a weather phenomenon called lightning. *Lightning* is untamed electricity that shoots across the sky, causing intense heating along its path. This heating causes the air to expand suddenly. When the air surfaces separated by the expansion come together again, a series of vibrations causes the violent crash known as thunder. Thunder in itself is not dangerous. Many thunderstorms sometimes travel together, forming what is known as a *squall line*.

Sound waves travel about one mile in five seconds, while light travels at more than 186,000 miles per second. Lightning is therefore seen practically at the instant it occurs, while the thunder is heard somewhat after the flash of light appears in the sky. Hence, the distance of a thunderstorm can be calculated by counting the seconds between the time the lightning flash is seen and the time the thunder is heard, allowing five seconds per mile.

U.S. DEPARTMENT OF COMMERCE, WEATHER BUREAU

THUNDERSTORMS, caused by electrical discharges in the atmosphere, are characterized by streaks of lightning and thunder.

More than 1,800 thunderstorms occur in the atmosphere every hour. Every year about 400 people in the United States are killed by lightning. While the chances of being struck by lightning are very small, the following precautions should be taken. Indoors, the center of a room is the safest spot, provided it is not under a light fixture. Radiators and corners near outside rainspouts should be avoided. Outdoors, isolated trees, tops of hills, and metal fences should be avoided. The inside of an automobile is one of the safest places to be during a thunderstorm. If you are caught on open water in a small boat, or on a wide expanse of flat land, the best precaution is to lie down.

Air Masses.—Weather in any area depends greatly upon the characteristics of the air mass over the area. An *air mass* is a large portion of the atmosphere with relatively uniform properties throughout. An air mass develops when a large section of the atmosphere remains relatively stationary over a surface area with

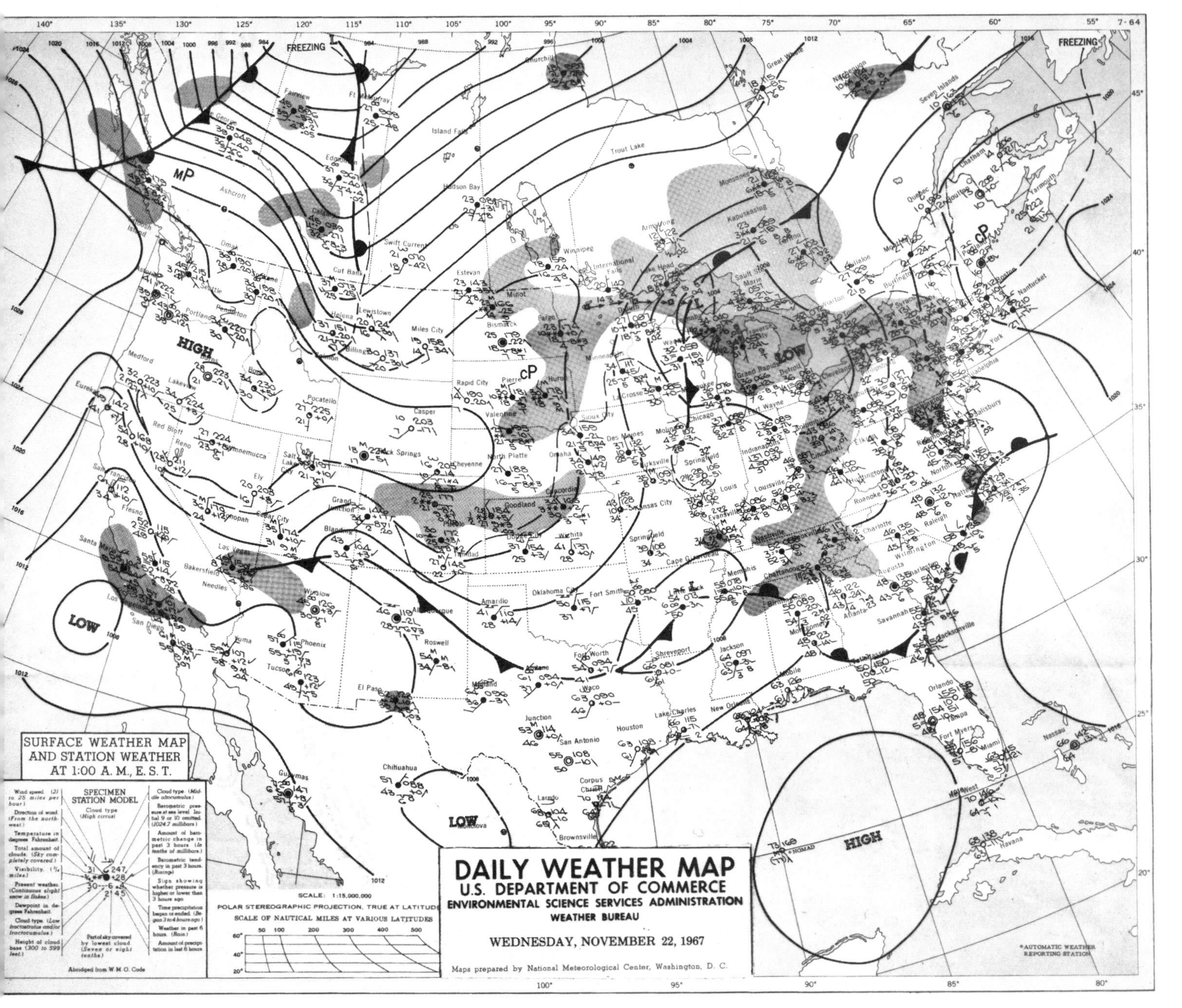

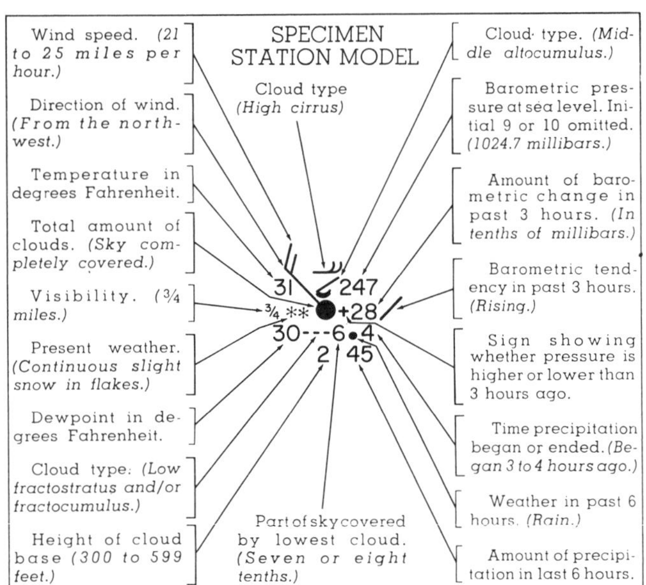

SURFACE WEATHER MAPS, such as the one above, are published daily to show the development and movement of weather systems across the United States. The solid lines across the face of the map, which are called *isobars*, connect points of equal sea-level barometric pressure. The centers of high and low barometric pressure are marked HIGH (as over Oregon and over the Gulf of Mexico) and LOW (as over Lake Huron, off the coast of California, and over Mexico). Air masses are represented by the following symbols: mP, for maritime-polar, over British Columbia in Canada; and cP, for continental-polar, over the states of South Dakota and Maine. The heavy lines with the hemispheres on one side and the triangles on the other represent the boundaries of stationary fronts. The side with the hemispheres is the edge of a warm front, while that with the triangles is the edge of a cold front. The shaded areas of the map are those in which precipitation was taking place at the time the observations were being made. The remainder of the information encoded on the map was collected from a multitude of reporting stations at which a wide variety of meteorological measurements are made at regular intervals, located all over the nation. The location of the reporting station is printed on the map as a small circle, and the information it has reported is arranged in a specified manner, as shown on the left. By international agreement, these symbols are universal.

uniform properties. Such a surface area is known as a *source region.*

Movements and interactions among different air masses also affect the weather.

Air masses are classified primarily by source region or place of origin. The principal air-mass source regions are those areas where high pressure centers tend to develop—over the land in high latitudes and over the ocean in low latitudes. The principal air masses are classified as *polar* and *tropical,* identified P and T on a weather map. Classifications *arctic* (A) and *equatorial* (E) are also used.

A secondary classification distinguishes whether the air mass formed over land or over water, designated *continental* (c) and *maritime* (m) respectively. In general, maritime-tropical air (mT) is warm and humid, continental-tropical (cT) warm and dry, maritime-polar (mP) cold and moist, and continental-polar (cP) cold and dry.

FRONTS.—Major weather changes occur at the boundaries separating air masses, called *fronts.* When cold air is advancing at the boundary, the boundary line is called a *cold front.* Cold fronts are usually accompanied by heavy showers and followed by lower temperatures. When warm air is advancing at the boundary, the boundary line is called a *warm front.* Gentle rains usually precede a warm front, followed by higher temperatures. A boundary line that is not moving is called a *stationary front.*

Frontal boundaries are shown as lines on a weather map, but they really extend upward from the ground. Since cold air is heavier than warm air, the surface slopes; and the cold air forms a wedge underneath the warm air. When cold and warm fronts meet, one of the air masses is lifted from the ground, thus forming an *occluded front.*

Climate.—The condition of the weather at a given time and place is the sum total of temperature, air pressure, relative humidity, wind velocity, precipitation, and cloudiness. Although changes in the weather in one particular area are many and varied, it is possible to arrive at a composite picture of the weather by averaging these variations. Such a generalization is called the *climate* of an area.

The climate of an area, however, is not determined solely by the long-term annual averages of the meteorological elements. Edinburgh, Scotland, and Boston, Massachusetts, for example, have nearly the same annual average temperature (48° F.), but the temperature extremes to which they are subjected during the year are markedly different. At Edinburgh average temperatures range from 38° F. to 58° F., and at Boston, from 27° F. to 70° F. Thus, to characterize climate it is first necessary to consider the regular variations the meteorological elements are subjected to, particularly seasonal changes. But temperature is not the only element to be considered. Cairo, United Arab Republic, and New Orleans, Louisiana, have about the same mean temperature (68° F.) and similar temperature variations, but the annual 1.3-inch rainfall at Cairo and the 56.5-inch rainfall at New Orleans make their climates quite different. Climate thus can be defined as the mean state of the atmosphere at a given place and the variations to which that mean state is subjected.

In general, climates are classified according to the effect they have on animal and plant life. Temperature and precipitation are the two principal elements in most classifications because heat and water are the two factors most profoundly affecting living organisms. They are also the two variables most regularly and generally observed. Evaporation, ground temperature, radiation, and winds are also important, but the distribution of these elements is reflected in temperature and precipitation.

Climate is governed by a number of geographic factors. The most important are latitude, altitude, topography, and proximity to an ocean. Latitude influences average temperatures. Warm climates are generally nearer the equator than cold climates, and vice versa. Altitude and topography influence average temperatures and precipitation—the higher the altitude, the colder the climate. The average midsummer temperature on Pike's Peak in Colorado is more than 30° lower than at Denver, which is 9,000 feet lower. The topographical location of an area in relation to mountains is also important. Air flowing up the windward side of a mountain is cooled, and the water vapor condenses to form clouds and rain. The air flowing down the leeward side of the mountain is heated as it descends and thus is dry. The climate therefore is often cool and wet on the windward side of a mountain range, warm and dry on the lee side. Mountains can also block the flow of air. The Rocky Mountains in the United States, for instance, act as a barrier preventing cold winter air masses from reaching the Pacific Coast.

Finally, oceans are relatively cool in the summer and relatively warm in the winter. Thus they tend to stabilize the climates of nearby land areas, particularly if the prevailing winds are from the ocean to the land. Such climates (*maritime*), on the average, change very little in temperature from season to season. Amid large land masses, temperature variations are usually more pronounced; winters are frequently very cold and summers very hot. This type of climate is called a *continental climate.*

—Francis K. Davis, Jr.

THE WEATHER BUREAU

The *Weather Bureau,* operated by the U.S. Department of Commerce, provides daily weather bulletins, forecasts, and storm warnings for use by public, agricultural, aviation, commercial, industrial, and other interests. It maintains approximately 300 offices inside and outside the continental United States.

History.—The *National Weather Service,* first organized in 1870, was operated by the U.S. Army Signal Corps. In 1891 it was transferred to the newly formed Weather Bureau, an agency put under the jurisdiction of the Department of Commerce in 1940.

Operations.—Today, weather observations are made at stations located in cities, at airports, in the Antarctic and the Arctic, and at fixed points in the Atlantic and Pacific oceans. These observations are supplemented by reports from military land and sea stations, other federal agencies, avia-

U.S. DEPARTMENT OF COMMERCE, WEATHER BUREAU

ELECTRONIC COMPUTER-PLOTTER automatically draws contours of the pressure surface of the atmosphere at specified altitudes above the Northern Hemisphere in under three minutes.

tion interests, merchant vessels of all nationalities, and from foreign countries under international agreement. In addition, there are almost 13,000 private citizens in the United States who assist the Bureau by making daily weather observations.

Weather information collected by this vast network is transmitted to the Weather Bureau's *National Meteorological Center* near Washington, D.C. Meteorologists there use the collected data to analyze the current weather situation and in turn issue comprehensive forecasts covering areas as large as the entire Northern Hemisphere. Much of this work is done with the help of electronic computers and other automatic equipment. One such machine is capable of drawing weather maps automatically. The completed forecasts and maps are then transmitted to area Weather Bureau offices and military field stations.

Using the guidance material provided by the National Meteorological Center, the Weather Bureau's *Area Forecast Centers* issue forecasts and storm warnings covering their specific areas of responsibility. Each local station in turn adapts the forecasts prepared by the Area Forecast Centers to its particular locality. Local weather information is then distributed by the press, radio and television stations, and automatic telephone answering devices.

Modern Observational Instruments

■**SATELLITES.**—Since 1960, eight *Tiros* weather satellites have been launched by the *National Aeronautics and Space Administration.* The cloud pictures taken by cameras in these satellites are used by the Weather Bureau to supplement other reports. The satellites have been of great value to forecasters by providing information about vast areas of the earth where few weather observations are made. Satellite photographs also provide early storm warnings. Since their inception the Tiros satellites have demonstrated their effectiveness by identifying and tracking storms. When significant weather developments, such as hurricane and typhoon formations, are detected by the satellites, the Weather Bureau issues special international bulletins to the nations that may be affected. Several Tiros satellites have carried sensors that measure the radiation balance of the earth and its atmosphere, thereby yielding valuable information that will aid in understanding the atmosphere and predicting its behavior.

■**RADAR.**—The Weather Bureau maintains 32 long-range, and 64 medium-range, radar installations at strategic points in the United States. This equipment is used to track severe storms. The eye of hurricane Carla (1961), for example, first appeared on the scope of the Galveston, Texas, radar when the storm was 220 nautical miles south, over the Gulf of Mexico. This particular hurricane was tracked continuously by the Galveston station for 46 hours.

■**AUTOMATIC STATIONS.**—In addition to manned installations, the Weather Bureau also maintains automatic stations to provide weather information from remote, inaccessible locations. These automatic stations observe cloud height, runway visual range for pilots, air pressure, and other weather conditions, transmitting these observations over teletype circuits. A marine automatic meteorological observation station is being tested in the Gulf of Mexico. For extremely remote locations, an atomic-powered weather station was developed through the efforts of the Atomic Energy Commission and the Weather Bureau. One such station was installed in the Canadian Arctic during the summer of 1961, and has been operating satisfactorily ever since.

■**COMPUTERS.**—To aid the weatherman in assembling and analyzing the wealth of weather data received, high-speed electronic computers have been installed at the *National Meteorological Center,* the *National Weather Satellite Center,* the *National Weather Records Center* at Asheville, North Carolina, and at various other research and forecast centers.

Weather Research.—Much of the progress made in hurricane forecasting is the result of intensive studies made by the Weather Bureau's *National Hurrican Research Project.* Basic information for this research is gathered by a fleet of flying laboratories and networks of upper-air observational stations. In 1961, under the joint sponsorship of the Weather Bureau, the Navy, and the *National Science Foundation,* a series of experiments began to explore the possibilities of modifying hurricanes by releasing silver iodide into the cloud tops near the area of maximum hurricane winds.

The Weather Bureau also initiated the *National Severe Storms Project* in 1961. This project is designed to collect detailed and comprehensive information on the behavior of tornadoes and severe local storms, using aircraft, radar, and observational networks in Oklahoma, Texas, and Kansas—the heart of the tornado belt. The program has received the active cooperation of many government agencies, including the Air Force, the Navy, the *Federal Aviation Agency,* and the National Aeronautics and Space Administration. Many universities and private agencies have also cooperated in the project.

Weather Bureau Publications.—The Weather Bureau publishes the *Daily Weather Map,* the *Weekly Weather and Crop Bulletin, Climatological Data,* the *Mariners Weather Log,* the *Average Monthly Weather Resumé and Outlook,* and the *Monthly Weather Review.* Its research findings are published in the *Research Paper Series,* the *Monthly Weather Review,* and various scientific journals.

—Robert M. White

BIBLIOGRAPHY

BLUMENSTOCK, DAVID I. *Ocean of Air.* Rutgers University Press, 1959.

DAY, JOHN A. *The Science of Weather.* Addison-Wesley, 1966.

LONGSTRETH, THOMAS MORRIS. *Understanding the Weather.* Collier Books, Inc., 1962.

MASON, BASIL JOHN. *Clouds, Rain, and Rainmaking.* Cambridge University Press, 1962.

RIEHL, H. *Introduction to the Atmosphere.* McGraw-Hill, Inc., 1965.

SUTTON, O. G. *The Challenge of the Atmosphere.* Harper and Row, Publishers, 1961.

U.S. DEPARTMENT OF COMMERCE, WEATHER BUREAU

RADAR has become a valuable tool to meteorologists for the tracking of storms. The storm vortex, presented on this radarscope, can be tracked at distances of up to 250 miles.

ORIGIN OF LIFE

Theories dealing with the origin of life attempt to account for the fact that there are living organisms on a planet that contained no life when it was formed. It is difficult to define what is meant by a "living organism" because there are many characteristics associated with living things. However, there are only two essential characteristics necessary for an organism to undergo evolution—the ability to *reproduce* and the ability to *mutate* (undergo change). Therefore, a *living organism* will be defined as an entity that is (a) capable of making a reasonably accurate copy of itself, and (b) subject to a low rate of mutation, with these mutations transmitted to its progeny.

Spontaneous Generation.—The problem of the origin of life in the modern sense was not considered by the ancient thinkers. They held that life could arise spontaneously from organic matter, as well as by sexual and asexual reproduction. The evidence for this belief was the common observation that insects and small animals arose from decaying meat and rotting grain. Thus life was considered to be originating at that time, the process was apparently a simple one, and there was no difficult problem to consider.

In 1668, Francesco Redi performed an experiment to disprove the theory of spontaneous generation. He placed some meat in a flask and covered the flask with muslin so that flies could not lay their eggs on the meat. No maggots, which are the larval form of flies, developed in the meat as long as the flask remained covered. This demonstration was sufficient to disprove the theory of spontaneous generation for higher organisms, but it did not apply to microorganisms. In 1765, Lazzarro Spallanzani conducted similar experiments showing that microorganisms would not appear in various nutrient broths if the flasks were sealed and boiled. Objections were raised that the boiling had destroyed the "vital force" in both the nutrient broth and the air. This "vital force" was thought to be necessary for the spontaneous generation of life. Spallanzani could show that the broth was still suitable for the growth of organisms by readmitting air to the flasks, but he could not prove that the heating process had left the air in the flasks unchanged.

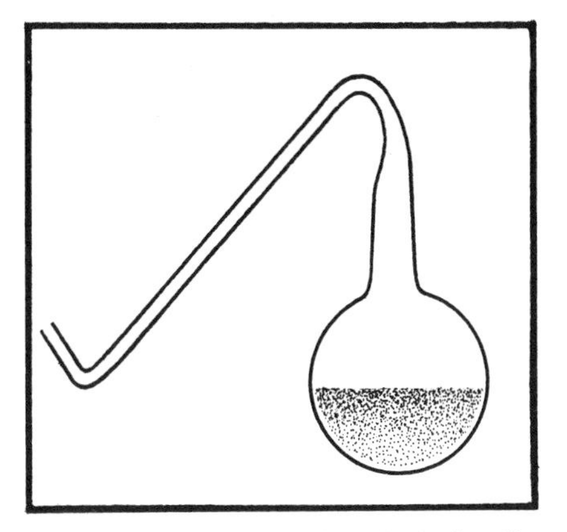

PASTEUR'S apparatus, by which he disproved spontaneous generation of life.

It was not until 1862 that the theory of spontaneous generation was conclusively disproved by Louis Pasteur. He placed a nutrient broth in a flask that had a long, S-shaped tube attached. This S-shaped tube allowed air to pass freely in and out of the flask. However, all dust, molds, and bacteria were caught on the sides of the curved tube The broth was boiled to kill any microorganisms present, and the flask was then allowed to cool. No microorganisms formed in the cooled broth. Since the air could pass freely in and out of the flask, and the broth subsequently could be shown capable of growing microorganisms, this experiment showed that no "vital force" in the broth or the air had been destroyed. Pasteur and John Tyndall were able to extend these experiments to show that the "spontaneous generation" of organisms in nutrient broths is due to contamination by atmospheric microorganisms. These experiments were convincing, and no serious case has since been made in favor of the spontaneous generation of living organisms from nutrient broths.

Other Theories.—Charles Darwin's theory of evolution by natural selection simplified the problem of the origin of life. With this theory he could account for the evolution of the most complex plants and animals from the simplest single-celled organisms. Acceptance of his theory means the problem of origin of life is concerned with how this most primitive organism arose on earth.

It has been proposed that life was created by a supernatural event. This proposal, however, is not a scientific hypothesis since, by its very nature, it is not subject to experimental investigation.

In 1903, Svante Arrhenius offered a theory that life developed on earth as a result of a spore or other stable form of life coming to this planet in a meteorite from outer space or driven to the earth by the pressure of sunlight. One form of this theory assumes that life had no origin but, like matter, has always existed. Analysis of long-lived radioactive elements shows that the elements were formed about five billion years ago. If the elements have not always existed, it is difficult to understand how life could have always existed. Another form of this theory assumes that life was formed on another planet. However, most scientists doubt that any known form of life could survive for very long in outer space

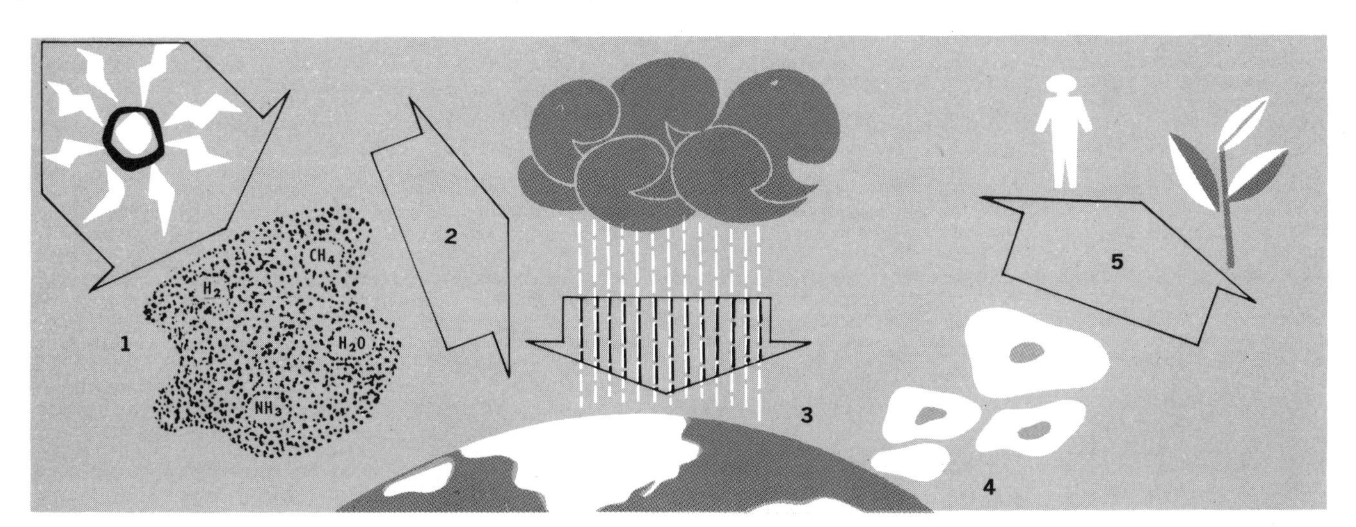

OPARIN'S THEORY maintained that sunlight caused a reaction of the materials in a reducing atmosphere (1), which contains hydrogen (H_2), ammonia (NH_3), methane (CH_4), and water (H_2O), to form simple organic compounds. These collected in clouds (2) and were carried to the earth (3) by precipitation. Continued reaction produced life forms (4) and, finally, more advanced life (5).

then fall through the earth's atmosphere without being destroyed. Therefore, although not disproved, this theory is held to be improbable.

It has been proposed that life developed from inorganic matter by a very improbable event—a spectacular accident. Such an organism would have had to live in an inorganic environment, and it would have had to synthesize all of its cellular components from carbon dioxide, water, and other inorganic materials. The chances for this improbable event are much too small for it to have occurred in the five billion years since the earth was formed.

OPARIN'S THEORY.—The most plausible theory was proposed in 1938 by the Russian biologist, Alexander I. Oparin. He suggested that the first living organism arose spontaneously, not out of inorganic material, but out of the large quantities of organic material that he proposed were present in the oceans of the primitive earth. The simple organic compounds reacted to form structures of greater and greater complexity, until finally something was formed that could be called living. The formation of the first living organism was, then, the product of a series of simple reactions, none highly improbable.

Oparin's hypothesis is not in conflict with the demonstration by Pasteur that spontaneous generation does not take place. Pasteur only showed that spontaneous generation cannot take place at the present time and under present conditions. Such a demonstration does not say anything about spontaneous generation in the past and under different conditions. Two of the conditions necessary for spontaneous generation are that large quantities of organic compounds accumulate and that sufficient time is available for these compounds to organize into a living organism. This implies that spontaneous generation could occur on earth only when there was no life. Wherever there are living organisms, they will devour any organic compounds, thereby preventing their accumulation and reducing the time available for their organization into a living organism.

Oparin proposed that the organic compounds in the primitive oceans could have been formed if the atmosphere was not an *oxidizing atmosphere* as it presently is, but instead was a *reducing atmosphere* of methane, ammonia, water, and hydrogen. In 1952, Harold Urey showed that present theories on formation of the solar system require the earth to have had a reducing atmosphere in its early stages. He also showed that this reducing atmosphere would be present as long as there was molecular hydrogen in it, because methane and ammonia are the stable forms of carbon and nitrogen in the presence of hydrogen. The cosmic dust cloud, from which the solar system was formed, contained a large excess of hydrogen. The planets Jupiter, Saturn, Uranus, and Neptune still have reducing atmospheres. The planets Mercury, Venus, Earth, and Mars have developed oxidizing atmospheres since they were formed. This results from the fact that a water molecule, when exposed to the ultraviolet light from the sun, splits into oxygen and hydrogen; the hydrogen escapes into outer space. The free oxygen does not escape, thus helping to form an oxidizing atmosphere. The atmospheres of Jupiter, Saturn, Uranus, and Neptune have not become oxidizing because the escape of hydrogen is very slow on these planets due to their low temperatures and high gravitational attraction.

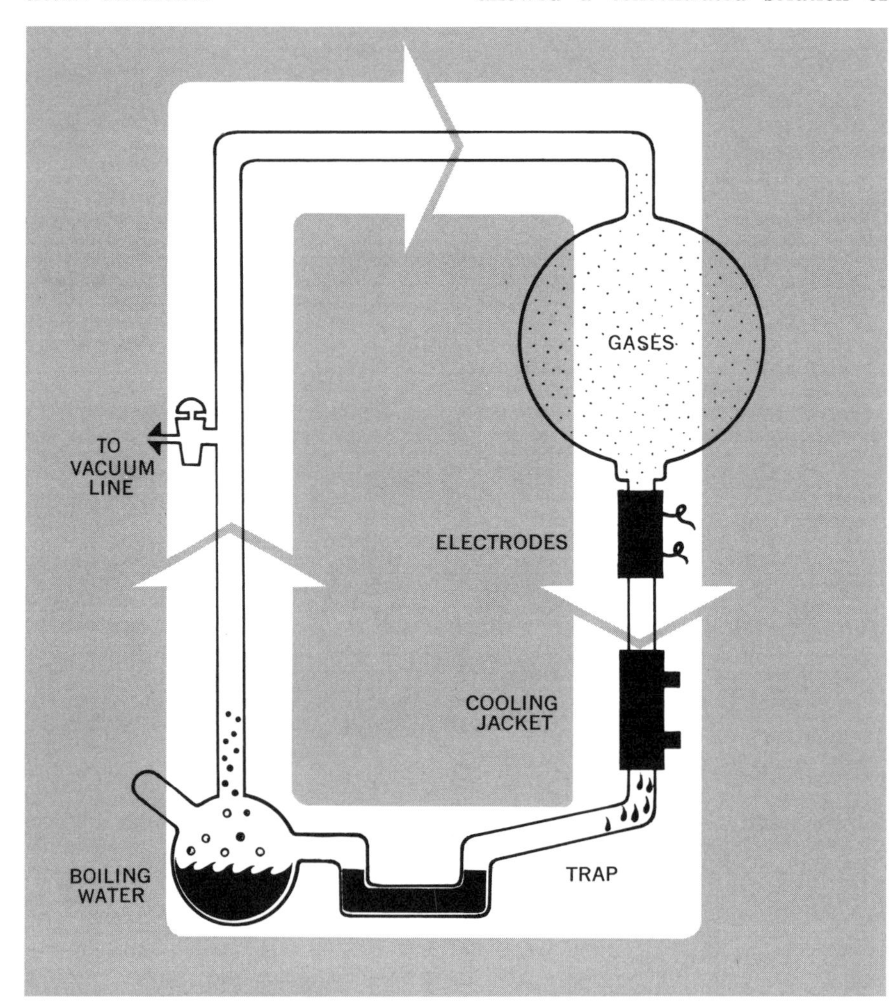

MILLER'S EXPERIMENT, in which the four gases found in an oxidizing atmosphere were passed through an electric discharge to produce amino acids, strengthened Oparin's theory.

Origin of Organic Compounds.—In 1953, Stanley Miller performed a series of experiments that strengthened the theories of Oparin and Urey. He circulated the gases that make up a reducing atmosphere (methane, ammonia, water, and hydrogen) past an electric discharge. Although ultraviolet light was the major source of energy on the primitive earth, electric discharges also were important. In addition, the products of ultraviolet light experiments would be similar to those of electric discharges. The result of Miller's experiments was the production of *amino acids,* the basic building blocks of proteins. In addition to the amino acids, the electric discharge experiments produced hydroxy acids, fatty acids, urea, and a number of other organic compounds.

The amino acids were not produced directly by the electric discharge, but rather by the reaction of smaller molecules produced by it. The smaller molecules included hydrogen cyanide and aldehydes.

In 1961, John Oro showed how hydrogen cyanide could be used to synthesize adenine, one of the purine bases in nucleic acids. He simply allowed a concentrated solution of ammonium cyanide to stand. A large amount of black *polymer* of unknown structure formed, as well as a number of smaller molecules, including adenine. A number of other bases that occur in nucleic acids can be synthesized by similar processes. Since hydrogen cyanide is synthesized by ultraviolet light and especially by electric discharges, such polymerizations were probably important on the primitive earth.

These experiments, as well as related ones, have shown how a number of simple organic compounds may have been synthesized on the primitive earth. This is a small part of the total problem. It is still necessary to understand in more detail how the amino acids, purines, pyrimidine, sugars, and fatty acids were synthesized. An important problem is to show how *peptides,* which are polymers of the amino acids, could have been synthesized under conditions that were present on the primitive earth. The same problem occurs for the synthesis of polymers containing purines, pyrimidines, sugars, and phosphate. These polymers are called *polynucleotides.* There are also the various difficulties of organizing these polymers into structures that are able to perform a primitive "biological" function.

These are difficult problems, but they are all subject to laboratory investigation. In this area of science, what had been thought to be extremely difficult has frequently turned out to be very simple. Therefore, although there are no explanations for these problems at the present time, it is reasonable to believe they eventually will be solved, and with relatively simple answers.

Nature of First Living Organisms.—In present living organisms, reproduction proceeds by duplicating the genes, followed by the synthesis of more enzymes and other cell constituents, and division of the cell into two fragments. The *genes,* which are located in the *chromosomes,* are composed of *deoxyribonucleic acid,* or DNA. Mutations occur when the base composition of the DNA is changed by an imperfect duplication, by ionizing radiation, or by other factors. Since the characteristics required in order to call an organism "living" are the ability to duplicate and the ability to mutate, it has been proposed that the first living organism was simply a strand of DNA, which, with the presence of necessary enzymes, could duplicate.

This organism would be similar to a virus except that it would have the enzymes necessary for its reproduction. A virus consists of DNA—in some viruses, the nucleic acid is *ribonucleic acid* (RNA) instead of DNA—surrounded by a coat of protein. A virus is capable of duplication, but only within another living cell, where the virus makes use of the cell's enzymes and metabolites.

Oparin proposed a different model for the first living organism, a *coacervate particle.* (A coacervate is a type of *colloid*—a substance that consists of very small particles suspended in solution.) These coacervates would accumulate organic material from their environment, grow in size, and then split into two or more fragments. In the course of time, the coacervate particles would develop the ability to form fragments more and more like each other. Later, these coacervate particles would incorporate a genetic apparatus to carry out very accurate duplication.

Our knowledge of present living organisms would speak in favor of the DNA model for the first living organism. DNA carries the biological information for the synthesis of the entire organism. The duplication of DNA appears to be a simpler process than the synthesis of protein and other cell constituents. DNA can be duplicated outside a living cell in a system containing the DNA to be duplicated, the *monomers* of the DNA, and a single enzyme. A system more complex than this would probably be needed to duplicate a strand of DNA on the primitive earth. A mechanism to accumulate the monomers and to hold the system together would probably be needed in addition to the single polymerizing enzyme. It might also be necessary for the first organism to synthesize this enzyme and perhaps several others. It is reasonable to think that such a system may have developed on the primitive earth.

Evolution of Early Organisms.—The first living organism must have obtained all of its small-molecule *metabolites* from the environment and then used these small molecules to build up polymers. There are many bacteria that obtain their metabolites from their environment. These are called *heterotrophic bacteria.* Many other bacteria and all plants synthesize their cell constituents from carbon dioxide, water, and other minerals; they are called *autotrophic organisms.* The first living organisms must have been heterotrophic organisms; it is necessary to explain how a heterotrophic organism could evolve into an autotrophic one, since the first organisms would have used up the available metabolites.

When the supply of a needed metabolite became exhausted, it must have been necessary for the organism to learn to synthesize this metabolite without which it would not have been able to live and to grow. A mechanism by which heterotrophic organisms could acquire various *biosynthetic pathways,* some of which are very long and complicated, was proposed in 1945 by Norman Horowitz. It has been found that the presence of an enzyme in an organism is often dependent on a single gene. This is known as the "one gene—one enzyme" hypothesis. Suppose that the synthesis of A involves the steps

$$D \xrightarrow{c} C \xrightarrow{b} B \xrightarrow{a} A$$

where *a, b,* and *c* are the enzymes, and A, B, and C are compounds that the organism cannot synthesize. If A becomes exhausted from the environment, then the organism must synthesize A in order to survive. It is extremely unlikely that there would be three simultaneous mutations to give the enzymes *a, b,* and *c;* but a single mutation to give enzyme *a* would not be unlikely. If compounds D, C, and B were in the environment when A was exhausted, an organism with enzyme *a* could survive while the others would die out. Similarly, when compound B was exhausted, enzyme *b* could arise by a single mutation, and organisms without this enzyme would die out. By continuing this process, the various steps of a biosynthetic process could be developed, with the last enzyme in the sequence being the first to develop, and the first enzyme developing last.

ENERGY SOURCES.—Every organism must have a source of energy in order to carry out its metabolic functions. Animals obtain their energy from the oxidation of organic compounds. Plants obtain their energy from sunlight. There are many microorganisms that obtain their energy from simple fermentation reactions. For instance, the lactic acid bacteria obtain energy by fermenting glucose:

$$\underset{\text{glucose}}{C_6H_{12}O_6} \rightarrow \underset{\text{lactic acid}}{2CH_3CH(OH)COOH} + \text{energy}$$

The energy appears in the biologically useful form of adenosine triphosphate (ATP). Fermentation reactions do not require the use of molecular oxygen, which was absent from the primitive earth. The first organisms could have obtained their energy supply by fermentation until the supply of fermentable compounds in the environment was exhausted. Then it would have been necessary to develop the more complicated process of photosynthesis. *Photosynthesis* is the process whereby the energy in sunlight is used to make ATP and to reduce carbon dioxide to carbohydrate. With the development of photosynthesis and the pathways for the synthesis of necessary metabolites, organisms would become autotrophic.

The general picture of the origin of life and early evolution presented here is believed by many scientists to be basically correct. However, there is little detailed knowledge of any of the various steps in this process and no knowledge at all of some of the difficult steps.

Although this discussion is directed toward the events that have taken place on our planet, the same process could have taken place on other planets as well. Those planets with the proper temperature and atmosphere could undergo a similar process of chemical evolution. Mars is the nearest planet where life may be present, even though the temperature is barely high enough for life, as we know it, to persist. In a few years devices for the detection of life will be sent to Mars. The finding of life on Mars would confirm our ideas about the origin of life occurring under favorable conditions, and would be one of the greatest achievements of modern science.

It is likely that most stars have planetary systems. Life might also have arisen on many of these planets and may still be there. It follows that some of these planets may have very advanced civilizations that are attempting to communicate with other planets. Some attempts are being made to detect such signals. This is a very difficult technical problem, but the scientific results are of sufficient interest to warrant that some effort be made to detect these signals.

—Stanley Miller

BIBLIOGRAPHY

OPARIN, A. I. *The Origin of Life.* Dover Books, Inc., 1962.

SCHRODINGER, ERWIN. *What Is Life?* Cambridge University Press, 1963.

MICROBIOLOGY

Scope.—*Microbiology* is the study of microscopic living creatures. Generally, it is the study of viruses, *Rickettsiae,* bacteria, protozoa, yeasts, molds, and the small algae, thus including both plant and animal kingdoms. Some authorities do not use these kingdoms for the forms of life that show no tissue differentiation, but instead place them all in the kingdom *Protista,* a group established in 1866 by the German zoologist Ernst Heinrich Haeckel (1834–1919).

Cell Types.—Microorganisms, with the exception of viruses, are organized into two cell types, according to the structure of the nucleus. If there is a visible nucleus, the cell is *eucaryotic.* If the nucleus is not visible, the cell is *procaryotic.*

Eucaryotic cells have a definite outer cell membrane that varies from an unsupported, flexible membrane in animal-like cells to a fragile membrane inside a rigid cell wall in plant-like cells. The cell is filled with a more or less fluid material, *cytoplasm,* which contains a variety of granules (*plastids*) and membranes that function in the metabolism of the cell. The cytoplasm generally is in motion except during cell division.

■**PLASTIDS.**—There are two kinds of plastids. One, the *protoplast,* is a colorless granule that contains many of the cell's respiratory and synthetic functions assembled in a highly organized state. All eucaryotic cells have at least one protoplast and cannot live without it.

The other plastid, the *chloroplast,* occurs in plant cells in addition to the protoplast. The chloroplast contains chlorophyll and all mechanisms for synthesizing chemical compounds through the process of *photosynthesis.* The simplest eucaryotic algae contain one protoplast and one chloroplast. Chloroplasts are not essential for life —since some single-celled plants may lose their chloroplast and still be able to live on dissolved food materials. Once a chloroplast has been lost, there is no evidence of its regeneration from the cytoplasm.

■**NUCLEUS.**—The *nucleus* is suspended in the cytoplasm but is separated from it by a definite membrane. The nucleus contains the *chromosomes,* the hereditary material of the cell, which govern the structure and function of the rest of the cell. The cell usually multiplies by splitting in two. Prior to this, the chromosomes divide, half going into each new nucleus. Chromosomes are composed of *deoxyribonucleic acid* (*DNA*), which is not found in any other structure. Its sole function is to duplicate itself before cell division and to serve as a primary pattern or information center containing all the instructions that a cell or higher organism needs to grow and develop properly. If the DNA makes an error in duplicating itself, a mutation results. The mutant cell is either unable to function properly and therefore dies, or it lives on but is different from the parent cell; its *progeny,* or offspring, will continue to be different in the same way. Since a mutation is always inheritable, evolution is thought to have occurred through a series of mutations. Some mutations add characteristics to the cell or higher forms of life that make the organism better able to compete for food and the other necessities of life. Most mutations, however, cause the loss or damage of a characteristic; such mutations usually are harmful to the species and have no value in its evolution.

■**EUCARYOTIC CELLS.**—Some eucaryotic cells either do not have a cell membrane, or have only an incomplete wall between nuclei. This is true of most of the molds. Such plants look and behave like a large cell with many nuclei. The eucaryotic microorganisms include most algae, protozoans, molds, and yeasts.

■**PROCARYOTIC CELLS.**—Procaryotic cells differ from eucaryotic cells primarily in lacking a definite nucleus surrounded by its own membrane. They do, however, contain nuclear material in a many-folded circular strand of DNA that behaves as a single chromosome. The internal organization of procaryotic cells, such as those of bacteria and blue-green algae, is not well understood. However, the cytoplasm is free of plastids and does not move. The chlorophyll of the blue-green algae and the photosynthetic bacteria is in several special organs and is organized in a variety of fashions. Procaryotic microorganisms include the blue-green algae, bacteria, and *Rickettsiae.*

Viruses.—*Viruses* are the smallest of the microorganisms and lack the structural complexity of a cell. Their unit is a particle that is able to cause a host plant, animal, or bacterium to make more virus. This infective unit, called a *virion,* ranges in size from about 0.025 micron (poliomyelitis virus) up to about 0.3 micron (smallpox virus). A *micron,* μ, is $1/1{,}000$ of a millimeter, or about $1/25{,}000$ of an inch. Most viruses are too small to be seen with the light microscope, which has a maximum useful magnification of about 2,000 diameters. The electron microscope, which can produce magnifications of more than 100,000 diameters, can be used to photograph viruses. Some viruses are called *filterable viruses* because they can pass through a filter of unglazed porcelain that holds back bacteria.

A virus lacks usual cellular structure and instead consists of a nucleic acid associated with protein. The nucleic acid may be DNA, the material from which chromosomes are formed, or *ribonucleic acid* (*RNA*), which serves as a secondary pattern for synthesis of cellular components.

A virus is not capable of independent life and reproduction, and must multiply within the cells of a host. Hence, viruses are *obligately parasitic.* Apparently, viral nucleic acid intrudes into a cell in such a way that the host cell's nucleic acid loses control of the cell's function. The cell then makes more virus particles instead of its own nuclear and cytoplasmic materials. The presence of a virus in a host usually causes changes recognizable as a disease, although animal viruses have been found that cause no recognizable symptoms in the host. These are called *orphan viruses.*

■**VIRAL PLANT DISEASES.**—Plant diseases of viral origin include the mosaics. The tobacco mosaic virus (TMV) was isolated by Wendell Stanley in a crystalline form and has remained infective after years of storage in a bottle. This virus has also been separated into its two components, protein and nucleic acid, and regenerated by their subsequent reunion. The protein, however, is not necessary for infection if it is possible for the nucleic acid to enter the cell without its aid. The protein may act as an enzyme, dissolving part of the cell wall to admit the nucleic acid fraction.

■**VIRAL ANIMAL DISEASES.**—Animal and human diseases caused by viruses

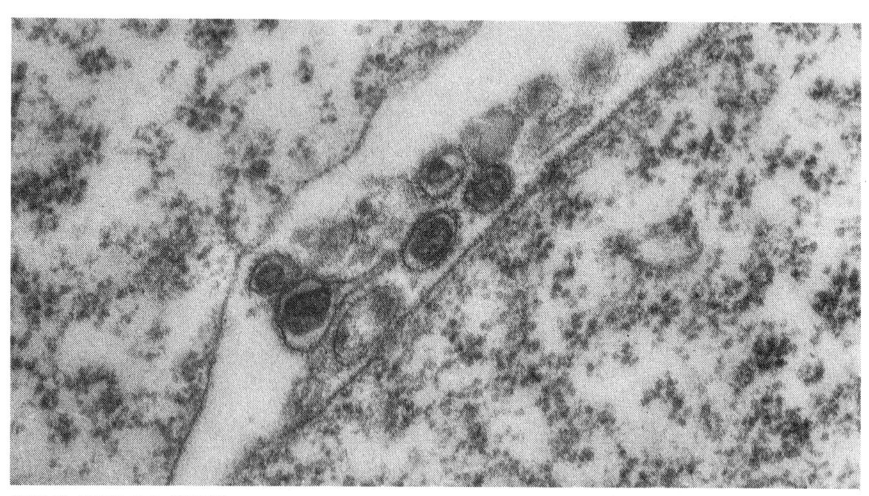

CHARLES PFIZER AND COMPANY

LEUKEMIA VIRUS in mouse tissue as seen through an electron microscope. Clearly visible are the outer shell of the virus, the dense nucleoid, and also the small, tail-like structure.

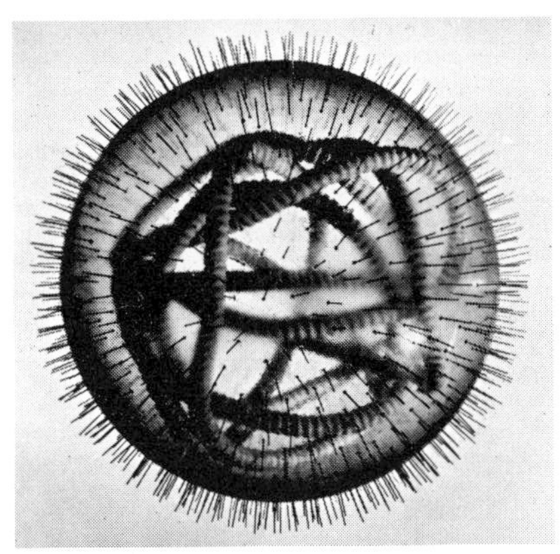

CHARLES PFIZER AND COMPANY

MEASLES VIRUS is classed as a cubic form. This model was designed from micrographs.

include yellow fever, distemper, hoof-and-mouth disease, influenza, fever blisters, measles, mumps, poliomyelitis, chickenpox, small pox, warts, rabies, and the common cold.

Viral diseases are transmitted to human beings and animals by various means. The virus that causes yellow fever is transmitted from person to person by the mosquito *Aedes aegypti*. The mosquito is infected when it bites a diseased person; if the mosquito later bites a healthy person, the virus may be transmitted. In the tropics, reservoirs of viral infection are maintained in monkeys, marmosets, and perhaps other species of animals. Psittacosis (parrot fever) and other ornithoses, or diseases of birds, were once thought to be caused by viruses, but are now known to be caused by Chlamydia, organisms having a cellular structure. Chlamydial diseases, including trachoma, are transferred to man by several routes from infected birds. The contagious hoof-and-mouth disease of livestock is of economic importance. The virus, transmitted by ingestion of contaminated particles, attacks the mouth and hoofs. Man may contract the disease in the same manner. Many viral diseases, such as measles, mumps, and influenza, are transmitted by simple contact with patients or by contact with infected materials. The virus of rabies is usually transmitted by the bite of an infected animal. Recently, an abundance of evidence has been found proving that some forms of animal cancers are caused by viruses. There is good reason to believe that some types of human cancer are also caused by viruses, either directly or indirectly.

Many human and animal viruses can be grown outside their normal hosts if they are inoculated into fertile chicken eggs before they hatch, or if they are inoculated into cultures of animal tissue cells maintained outside of the whole body. This use of chicken embryos and tissue cultures has made possible the laboratory cultivation of large quantities of viruses and has led to new methods of immunization against viral diseases.

A virus that appears to be of little if any importance to man, except as a tool, is the type that infects bacteria, a *bacteriophage*—literally, a "bacteria-eater." (The bacteriophages are treated more fully in the discussion of bacteria.)

Bacteria.—The *bacteria,* larger than viruses, are organized with a cellular structure contained within a rigid cell wall. They are usually divided into three large groups, depending on their shape. The spherical bacteria are *cocci* (singular, *coccus*), the cylindrical are *bacilli* (singular, *bacillus*), and the spiral ones are *spirilla* (singular, *spirillum*). Multiplication of bacterial cells occurs by simple splitting of the cell into two new cells, each of which can then grow and split again. Under ideal conditions this process can occur as often as every twenty minutes. Bacilli and spirilla usually split across the short diameter of the cell, and the newly formed cells seldom hang together for long. However, some species of bacilli form long chains of cells. The cocci have a more complicated system of multiplication that has led to their division into several groups. The *staphylococci* split in a random fashion, producing grapelike clusters. The *streptococci* always divide in the same plane and produce chains of cells. The *tetracocci,* or *gaffkya,* split alternately in two planes at right angles to each other, producing flat sheets of cells. The *sarcina* split successively in three planes at right angles to each other so that they tend to form cubical packets of eight cells.

The smallest bacteria are barely larger than some viruses, about 0.5 μ in diameter. Cocci are generally from 0.5 to 1.0 μ across. Bacilli vary in size, but most are from 0.5 to 1.0 μ across and 1 to 5 μ in length. Spirilla are about the size of bacilli or a little larger. A few giant bacteria will form cells over 50 μ long.

About 2,000 species of bacteria have been found. They are considered to be plantlike because of their rigid cell walls. Since there are so few basic shapes of bacteria, they are classified on the basis of size and shape, the materials they use as food, and the products formed from the food. The products are acids, alcohols, gases, pigments, and toxins. Oxygen relationship is also important. Those bacteria capable of growth only in the presence of air are called *aerobic;* those that cannot grow in the presence of air are called *anaerobic.* Bacteria able to grow in either situation are said to be *facultatively anaerobic.*

A few bacteria contain a special type of chlorophyll and hence are able to live photosynthetically in a manner similar to that of green plants. Bacterial photosynthesis differs from that of green plants mainly in that it is anaerobic and oxygen is not produced. Instead, oxidized organic compounds or sulfur compounds are the end products. Most photosynthetic bacteria contain a high concentration of *carotenoid pigments;* consequently, instead of appearing green, they are red or brown. There are a few green bacteria.

Most bacteria utilize nonliving organic materials for food and for growth, causing the materials' breakdown or decay. In fact, every naturally occurring organic material, including rubber, paraffin, and asphalt, can be used as a source of food by some microorganism. This breakdown is called *saprophytic action.* A few bacterial species are able to oxidize simple inorganic materials and secure their energy for growth from such processes. These processes include oxidations of ammonia, nitrites, sulfur, sulfides, hydrogen, carbon monoxide, iron, and manganese. Few, if any, bacteria are strictly parasitic. Many pathogenic (disease-causing) bacteria are unable to survive for long under natural conditions except in the host animal, but most of them have been cultivated in the laboratory on nonliving materials.

■BACTERIAL MOVEMENT.—Many bacteria cannot move by their own efforts. Some, however, have hairlike projections from the cell that enable them to swim rapidly in a liquid medium. These projections are called *flagella* (singular, *flagellum*). They may occur singly or in clusters at the ends of the cells, or may be scattered over the cell's surface. Bacterial species with a single terminal flagellum seem to swim as well as those having many flagella. How flagella are used to swim is not known. Most bacterial flagella are too small to be seen except when stained by a special method in which the stain accumulates around them so that they become visible under the microscope.

All bacteria have an outer slime layer that generally is thin and may be difficult to detect. Some bacteria, however, have a very thick gelatinous slime layer, a capsule, that can easily be seen. Some disease-causing bacteria, such as the *pneumococcus,* are able to resist defense mechanisms of the host because the latter react with the capsular material and do not

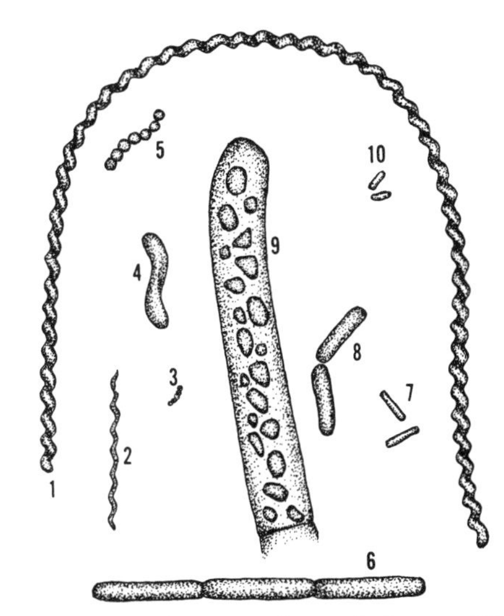

BACTERIA of various sizes and shapes: (1) *Spirochaeta plicatilis;* (2) *Treponema pallidum;* (3) *Peptostreptococcus parvulus;* (4) *Spirillum undula;* (5) *Streptococcus pyogenes;* (6) *Bacillus anthracis;* (7) *Mycobacterium tuberculosis;* (8) *Bacillus megatherium;* (9) *Beggiatoa alba;* and (10) *Bordetella pertussis.*

contact the bacterial cell itself.

A few species of the bacilli produce dormant forms that are more resistant to killing by heat, drying, and chemicals than are the original cells. These forms are called *spores* and are the most heat-tolerant form of life known. Spores of some species will survive boiling-water temperatures for several hours. The bacterial spore is thought to be a means of survival, not reproduction: a single cell usually forms but one spore, and when conditions are favorable for germination, that spore forms only one new growing cell.

The bacterial cell's nucleus is apparently not a clearly defined structure separated from the rest of the cell by its own membrane. The nucleus does, however, contain the cell's hereditary material (DNA), which controls the cell's processes in a manner that is probably similar to that in cells that have a well-defined nucleus. The bacteria that have been studied most completely appear to have only a single chromosome in the nuclear material. Since the chromosomes are not paired (*diploid*), the cell is *haploid;* and there are no dominant and recessive characteristics. Exchanges of nuclear material can occur in bacteria in at least three ways. These are recombination, transformation, and transduction. The latter two processes are apparently confined to bacteria.

NUCLEAR MATERIAL EXCHANGE.— *Recombination* occurs when two bacterial cells conjugate, or mate, and part of a chromosome is transferred from one to the other. Donor (F+) strains always transfer to recipient (F−) strains. Although conjugation is rarely observed, a few F+ strains combine readily with F− strains and are therefore called *high-frequency recombinants* (*HFR*). After conjugation, the F− cell becomes F+ and gains the positive characteristics of the portion of the chromosome transferred. The F+ cell continues to grow because many such cells have more than one center of nuclear material and can use any center that has not been depleted by the conjugation.

Transformation is the incorporation into living bacterial cells of nuclear material (DNA) extracted from living or dead bacteria of the same species which possess a few different characteristics. Some of the transformed cells will acquire some of the characteristics of the other bacteria. The means of transfer of DNA into the cell is not known. Not all bacteria in a culture are competent to receive the DNA.

Transduction is the accidental transfer of bacterial DNA from a cell of one bacterial strain to a cell of another by a bacteriophage and the incorporation of this DNA into the chromosome of the recipient bacterium. The resulting cell has the positive characteristics of both cells for the portion of the chromosome transferred. This transfer requires that the bacteriophage be propagated on a bacterial strain that is destroyed by the virus, and that the recipient strain of bacteria be one that harbors the virus but is not ordinarily destroyed by it. Bacteria of the latter type are called *lysogenic* and may carry the virus with little or no evidence of infection.

Most of the work on transfer of nuclear material among bacteria has been done on the genera *Escherichia, Salmonella,* and *Shigella.* However, these three processes occur in most genera of bacteria examined.

CLASSIFICATION.—Bacteria have been classified in many ways. A completely *phylogenetic system,* one based upon obvious relationships and evolutionary patterns, is not possible with our present knowledge and perhaps may never be satisfactorily achieved. There are several large groups that have the status of orders and appear to be valid subdivisions of bacteria. The current system of classification, presented in the seventh edition of Bergey's *Manual of Determinative Bacteriology,* lists nine orders of bacteria. Four orders are listed below; the other five orders are of doubtful phylogenetic significance and may be listed as variations of true bacteria.

The *Eubacteriales* include the simple, or true, bacteria that have definite shapes and rigid cell walls. They multiply by transverse fission and do not show branching of cells. If they are motile, they move by means of flagella.

The *Myxobacteriales* are the slime bacteria that have no definite cell wall. They do not have flagella but are motile by a gliding motion, the explanation of which is unknown. The whole colony of cells may be motile and move across a surface in a small mound of secreted slime. The colony may organize into a complex fruiting body in which some cells form a base and others, a stalk. A few may become resting stages called *microcysts.* These dry and may be spread by the wind. New colonies of myxobacteria are started if they fall upon a suitable food source. Myxobacterial colonies are often found growing on animal dung deposits.

The *Actinomycetales* include the filamentous bacteria that often show true branching of the filaments. The intergradations between these and the eubacteria are so numerous that there is no clear line of demarcation.

The *Spirochaetales* form the order of the helical bacteria that are flexible, yet retain the coils of the helix. The spiral cell is wrapped around an axial filament that prevents the cell from straightening.

CONTROL OF BACTERIA.—Bacteria can be useful, harmful, or of no known importance to man. They are present in large numbers in most soils and waters and are carried in the air by dust particles. They are abundant on the skin, in the mouth, and in the digestive system of all animals. All ordinary objects have bacteria or microorganisms on their outer surfaces. Since some bacteria cause disease and many contribute to food spoilage, it is important to be able to control their activities. It is necessary to sterilize equipment to be used during surgical treatment of patients and to sterilize foods that are to be preserved for long periods. It is also necessary to sterilize containers for materials used in laboratory work when controlled changes are to be brought about by the use of a pure culture containing only one microbial species. The most common method of sterilization is use of steam under a pressure of about 15 pounds per square inch to give a temperature of 248° F. (121° C.) for 15 to 20 minutes. The apparatus used for this heat treatment is a large pressure cooker, an *autoclave.* Large containers of liquids or cans of dense, viscous material may have to be heated for an hour or more to ensure that all of the contents have been at 248° F. for the necessary length of time. Dry materials that should not be exposed to steam can generally be sterilized by heating them to 338° F. (170° C.) for two hours. If the material to be sterilized is severely altered by the heat treatment, chemical methods, although less convenient than heat, are available for sterilizing solid objects. Some liquids can be sterilized by passing them through a filter that holds back the bacteria. This will not, however, remove viruses; and if it is important that viruses be inactivated, this usually can be done by exposing thin layers of the liquid to high concentrations of ultraviolet light.

A reduction in the number of bacteria or complete sterility of water to be used for drinking, bottling, or industrial processes usually can be achieved by chemical treatment. Chlorine (in a concentration of 0.1 to 1.0 parts per million), bromine, and iodine are generally effective.

Where sterility is not as important as the destruction of disease-causing bacteria or the improvement of keeping quality by reducing the numbers of bacteria, pasteurization is the usual treatment. This relatively mild heat treatment was initiated by Louis Pasteur (1822–1895) to eliminate bacteria that spoil beer and wine. As it is commonly applied today to milk, pasteurization may use a temperature of 143° F. (61.3° C.) for 30 minutes or 161° F. (71.6° C.) for 15 to 20 seconds. Either process kills the bacteria that cause tuberculosis and brucellosis, as well as many of the common milk spoilage bacteria.

Early investigators did not understand that bacteria are everywhere and did not know that many of them can grow in the absence of air. Nor did they know that some species produce heat-resistant spores. Their crude methods of sterilization therefore generally failed, and subsequent growth of bacteria in the inadequately sterilized materials led them to believe that these living forms had originated spontaneously. The classical argument between J. T. Needham (1713–1781) and Lazzaro Spallanzani (1729–1799) in the eighteenth century pointed out the difficulties in establishing that spontaneous generation of life is not a common event under present conditions. The extensive experiments of Pasteur during the 1860's demonstrated that spontaneous generation of microorganisms does not occur—even bacteria must have ancestors.

BACTERIAL DISEASES.—The demonstration that bacteria can cause disease and that a particular bacterial species is always the cause of the same disease was made by the German physician Robert Koch (1843–1910) and his students during the period from 1870 to 1890. The rules for the establishment of this relationship are known as *Koch's postulates.* Four conditions must be achieved: (1) The suspected microorganism must be present in every case of the disease. (2) A pure culture of the microorganism must be obtained from a case of the disease. (3) The pure culture must cause the same disease when inoculated into a suitable experimental animal. (4) The original microorganism must be reisolated from the experimental case of the disease. These criteria often have been used to establish the cause-and-effect relationship between a particular bacterial species and a disease.

Some diseases in man caused by bacteria are scarlet fever, boils, diphtheria, bacillary dysentery, typhoid fever, whooping cough, pneumococcal pneumonia, asiatic cholera, tuberculosis, tetanus, gonorrhea, syphilis, and leprosy.

There are also many bacterial diseases of plants. Among these are fire blight of apple and pear trees, crown gall of trees and flowering plants, bean blight, tomato wilt, and soft rots of many vegetables.

Large quantities of foods are spoiled because of bacterial activities. Both stored fresh foods and canned foods that have been improperly sealed or processed are subject to spoilage. Several types of spoilage are of more than economic importance; poisonous toxins are sometimes produced and cause severe illness or death of persons who eat the affected food. *Staphylococci, streptococci, salmonellae,* and one of the anaerobic spore-forming bacilli, *Clostridium perfringens,* form toxins when they grow in food that has not been adequately refrigerated after preparation. These toxins may cause severe illness, but are seldom fatal. *Clostridium botulinum,* growing in canned or salted foods that have been inadequately processed, produces a toxin commonly fatal to man. Many cases of botulism have been caused by eating home-canned vegetables, but smoked fish have also been implicated. The toxin is destroyed by boiling food for ten minutes before it is eaten.

The manner in which a disease is transmitted depends in a measure on the living habits of the bacterium that causes it. Thus, the typhoid fever organism, *Salmonella typhosa,* which is carried in the feces, can contaminate meat, milk, shellfish, and water. It may infect subclinical carriers, who actually cause more outbreaks than persons with a full-blown case of typhoid. Infected food handlers are an important source of the disease. The gonococcus, *Neisseria gonorrhoeae,* attacks the mucous membranes of the genital tract and is usually transmitted during sexual intercourse. The whooping cough organism, *Bordetella pertussis,* is transmitted by inhalation of droplets from a coughing victim. Cholera is transmitted by contaminated water and food and by flies infected with the cholera organism, *Vibrio comma;* it may also be spread by person-to-person contact.

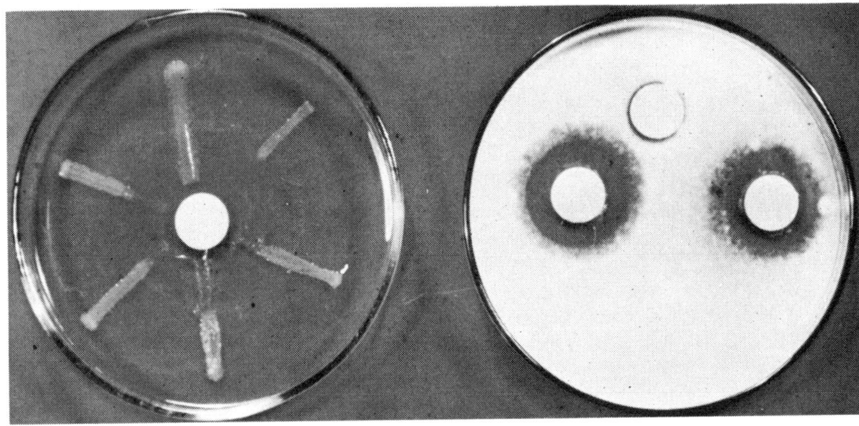

CHARLES PFIZER AND COMPANY

SCREENING PROCESS to test the antibiotic power of mold fluids. The dish on the left contains one mold fluid and six different germs; that on the right, four mold fluids and one germ organism. The dark rings signify fluids that are effective against the organisms.

INDUSTRIAL USES.—Bacteria have important industrial uses. Indeed, many huge industries entirely depend upon bacterial action for their existence.

In the manufacture of cheese, cultures of bacteria are added to pasteurized milk. These bacteria slowly ferment the lactose to lactic acid and break down fats and proteins to produce substances that are responsible for the various flavors, textures, and aromas that characterize the different types of cheese. In fact, the "eyes" in Swiss cheese owe their existence to carbon dioxide made by bacteria.

To make commercial vinegar, the bacterium *Acetobacter* is used to oxidize alcohol to acetic acid, which in a concentration of about 4 per cent in water constitutes commercial vinegar. In the process, the alcohol trickles over a bed of shavings or other finely divided material that has been inoculated with the bacteria.

Bacterial fermentation is important in the manufacture of many other food products, such as sauerkraut, pickles, soy sauce, yogurt, and butter.

Bacteria are also used to make industrial chemicals. Species of the genus *Clostridium* ferment carbohydrates to produce butyl alcohol, isopropyl (rubbing) alcohol, acetone, and numerous other substances used in the manufacture of drugs, paints, synthetic rubber, explosives, and some plastics.

The brewing and related industries utilize bacteria. Species of the genus *Bacillus* produce enzymes from such organic wastes as soybean or peanut cake. These enzymes are used to convert raw starches into materials that can be fermented by yeasts. The textile and paper industries also make wide use of enzymes.

Mixtures of microorganisms are used in separating flax or hemp fiber from the woody plant tissue. Similarly, hides are subjected to bacterial action in leather-making. Bacteria are even used to remove sulfur compounds from petroleum.

Following Sir Alexander Fleming's (1881–1955) discovery in 1929 of the antibacterial action of penicillin (derived from the mold *Penicillium notatum*), Selman Waksman, René Dubos, and others sought other types of antibiotics among bacterial species. They found that the soil bacterium *Streptomyces* is especially antagonistic to pathogenic organisms, and species of this organism have yielded streptomycin, chloramphenicol, erythromycin, neomycin, and many other antibiotics useful in the treatment of a host of infectious diseases. Bacitracin and subtilin derive from species of *Bacillus.* Xerosin comes from *Achromabacter.* Millions of lives have been saved during the past 20 years because of the discovery of antibiotics. Today, drug manufacturers sell more than $100 million worth of antibiotics annually; as a class of medicals their sales are rivaled only by the vitamins.

BACTERIA AND WATER.—Microorganisms are used in one of the methods for the purification of municipal water supplies and also in the purification of sewage in sewage disposal plants. In purifying water supplies by the slow sand filter, the raw water, perhaps from a river, is filtered through sand and gravel beds. As filtration proceeds, a slimy, jelly-like film accumulates around each sand grain, particularly in the upper few inches of sand. This film is composed of billions of bacteria and protozoa; and as it develops, it slows the flow of water through the sand. The water is purified by the action of enzymes, by biological oxidation and reduction processes, and by the ingestion of bacteria by the protozoa in the film. The filter removes about 99 per cent of the impurities from raw water.

The more recent and more commonly used rapid sand filter method of purifying water depends upon a thin layer of a chemical gel on the sand and upon other chemical treatments for removal of organic compounds and bacteria from the water. Both processes are generally followed by the addition of chlorine to complete the sterilization of the water.

BACTERIA AND SEWAGE.—Bacteria are vital to sewage disposal plants. Many different types of organisms abound in sewage; the abundance of any type changes as the sewage proceeds toward purification. One form of organism succeeds others as the environment changes.

Sewage disposal plants perform several functions. First, the sewage, which is about 95 per cent water, is screened to remove such large or inorganic matter as bottles and wooden boxes, and other large refuse. Then it is passed into separation tanks, where the heavier organic matter settles out. The supernatant liquid is then aerated by one of several processes to help the bacteria decompose the dissolved organic materials. This step is generally followed by another sedimentation, and the final water is chlorinated to remove harmful bacteria before it is discharged into a river, lake, or ocean.

The organic sediment from these treatments is pumped into closed tanks, or *sludge digesters,* and heated to about 100° F. Under these conditions the anaerobic bacteria produce a stable product that will not putrify later. The water from the sludge digesters is discharged with the other water; the sediment is dried and sold as fertilizer. The gases from the sludge digester are about 75 per cent methane, the same material as natural gas, which can be used to heat the digestion tanks and can be burned in gas engines to run the pumps in the sewage treatment plant.

Most kinds of organic sewage are decomposed by bacteria. Greases and oils, which decompose slowly, are usually skimmed off and burned. Phosphates and nitrates can be removed by algae, which are fed to livestock.

BACTERIA IN AGRICULTURE.—Agriculture is largely dependent upon microorganisms for decomposition of plant roots and residues in the soil and for conversion of nitrogen-, phosphorus-, and sulfur-containing materials into forms useful to plants. Some soil bacteria can take nitrogen from the atmosphere and build it into their bodies, thus eventually contributing to the fertility of the soil. This process, which is called *nitrogen fixation,* also occurs in root nodules of many plants, where the plants and bacteria of various types grow together in an association beneficial to both. The best example of such a *symbiotic* (living together) association is that found between the bacteria of the genus *Rhizobium* and the legumes, plants such as peas, beans, clover, alfalfa, and locust trees. When the proper strain of *Rhizobium* invades the plant rootlet, a nodule containing the bacteria develops on the root. The nodule is the site of nitrogen fixation, enabling legumes to grow well in soils that are deficient in nitrogen and therefore unable to support other plants.

ANIMALS AND BACTERIA. — Although animals can live without bacteria in their environment, as has been shown by the raising of large colonies of "germ-free" animals, many animals benefit greatly from the presence of bacteria in their digestive systems. In the *rumen* (first stomach) of cattle, sheep, goats, deer, antelope, and other ruminant animals, bacteria and protozoa are responsible for the digestion of much of the food, especially the cellulose and other difficult-to-digest portions of the food. The microorganisms ferment these products and produce a number of acids from them. The acids, chiefly acetic, propionic, and butyric, are absorbed by the animal and metabolized as sources of energy. Microorganisms help the animal by producing vitamins and proteins that can later be digested in the true stomach. A similar situation exists in the caeca of horses and some rodents and birds.

Study of bacteria began as an applied science because of their relationship to disease, agriculture, and industrial processes. In recent years, investigators have attempted to study bacteria to learn more about the bacteria themselves. These studies have given much knowledge about the metabolism and genetics of higher plants and animals, because the basic chemical reactions of living creatures are all fundamentally similar. Because of rapid growth of bacteria, large crops of cells can be grown under carefully controlled conditions in a few hours. Similarly, genetic studies that need large populations or many generations have profited from observations of bacteria. The results of crossing various strains of bacteria can be observed in a few hours or days instead of the months or years that are required when higher plants or animals are used.

Rickettsiae.—*Rickettsiae* are intracellular parasites of the size and shape of small bacteria. They differ from bacteria in that they are dependent upon a host cell for part of their life processes. They have not been cultivated outside of a living host. Most *Rickettsiae* appear to be parasites of insects, ticks, or mites. They may or may not cause recognized illness in their normal host. Some, when transferred to other species, such as man, cause disease. Generally, rickettsial diseases are transmitted to man by the bite of their normal host—typhus fever by the body louse, tsutsugamushi fever by the mite, Rocky Mountain spotted fever by a tick. An exception to this is Q fever, caused by *Coxiella burnetii.* The fever is transmitted through milk from cattle to man, or by one's inhaling dust containing the dried manure of infected cattle. *Rickettsiae* usually can be cultivated in fertile chicken eggs; laboratory cultures are maintained in this manner.

Fungi.—The term *fungus* is poorly defined, but it includes those eucaryotic plantlike organisms that lack chlorophyll and the tissue differentiation of higher plants. Fungi vary in size from the single-celled yeasts, smaller than large bacteria, to mushrooms. Such related forms as puffballs may reach several feet in diameter.

The most commonly observed fungal growth is a cotton-like mass of filaments (*mycelium*) called *mold.* The individual filaments (*hyphae*) grow from the tip and may branch repeatedly. They have a chitin-like rigid wall and many nuclei. The nuclei may or may not be separated by cross walls (*septa*) in the hyphae; but even where septa are present, separation is generally incomplete, and the cytoplasm streams continuously through the mycelium.

Molds and other fungi with mycelia are abundant in the soil, where they are responsible for the decomposition of most dead plant material. Most of the mycelial structure is invisible to the naked eye. We become aware of it only when an unusually large amount of organic matter leads to the production of mycelia and mold fruiting bodies on its surface.

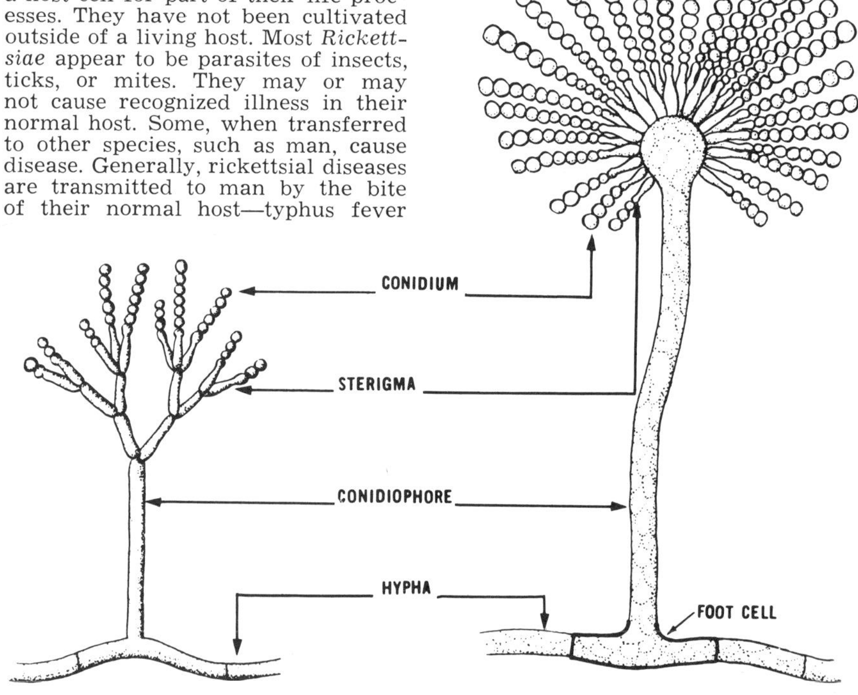

FUNGI occur in many types and forms. These two, *Penicillium* (*left*) and *Aspergillus* (*right*), are typical. Spores are produced at the tip of the branch called a *sterigma.*

REPRODUCTION OF FUNGI.—Fungi reproduce by forming a variety of small bodies called *spores.* Many of these, such as those most often formed by molds of the genus *Penicillium,* are small spheres exuded from the tips of special mycelial cells. They are called asexual spores because no nuclear fusion is required for their formation. Spores of this mold can often be seen on moldy oranges or cheese. They are green or blue-green and give it its characteristic color.

Most fungi have a sexual process for forming spores after the fusion of nuclei from two different mycelia or within multinucleated mycelia. Some of these processes are complex and lead to special spore-bearing organs such as the mushrooms.

FUNGAL DISEASES.—Many fungi cause diseases of plants; a few infect animals. Some fungi cause rust on grain and most other groups of plants. Many require different hosts for different stages of development. For example, black-stem wheat rust needs the common barberry for one phase of its sexual spore development. Another fungus, ergot, causes smuts of cereal crops and may cause death of animals and people who eat too much of the infected grain. However, ergot is also the source of a drug useful in controlling hemorrhage.

Human and animal diseases caused by fungi include athlete's foot, ringworm, and aspergillosis. Fungal diseases spread by contact with skin infected with the offending fungus. Athlete's foot is thought to be spread by the use of common showers and dressing rooms. In the course of this infection, the victim may become hypersensitive to the fungus or its products and develop allergic manifestations on parts of the body other than the feet, often the hands.

USES OF FUNGI.—Fungi are useful in the production of drugs and chemicals. The first of the antibiotics, penicillin, was discovered as a product of *Penicillium notatum. P. chysogenum* is now used for commercial preparation of penicillin. Fungi are also used to produce commercial quantities of citric and gluconic acids and such enzymes as amylase, cellulase, and glucose oxidase.

Yeasts.—A large miscellaneous group of fungi that seldom form mycelia but that generally grow as oval or round cells is known as yeasts. Many have a spore stage that follows fusion of nuclei from two cells. Others show only asexual reproduction, generally by *budding.* A small knob grows from the cell. As it increases in size, the cell nucleus divides and one of the new nuclei migrates into the protuberance. The bud is then walled off from the cell and is capable of forming its own buds. As seen in the figure, many buds may be formed by one cell. If a mass of cells stays attached, it may give the superficial appearance of a mycelium. A few yeasts multiply by fission in a manner similar to that of bacteria.

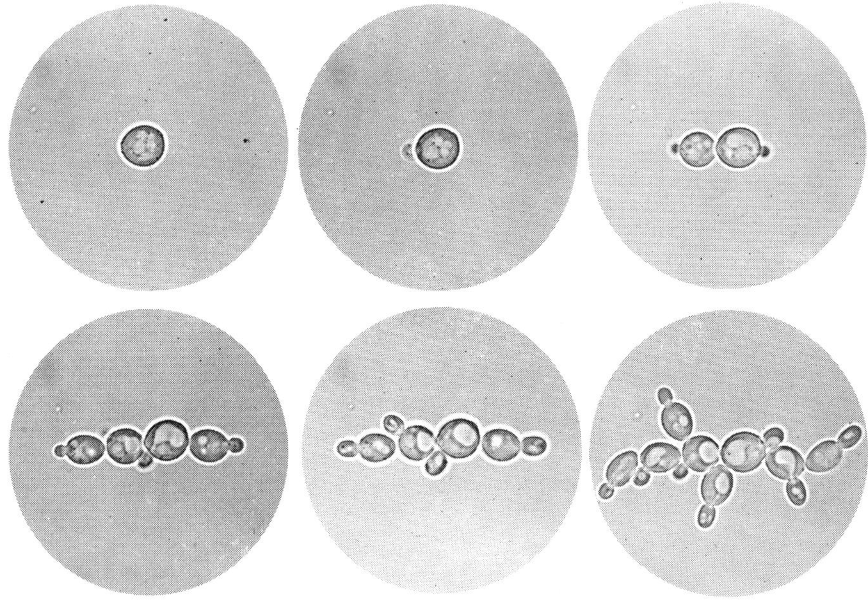

FLEISCHMANN LABORATORIES, STANDARD BRANDS, INC.

REPRODUCTION OF YEASTS by budding. These photomicrographs, taken over a four-hour period, show the firm cell wall. It appears to be formed of a material similar to that found in the cell walls of molds. The yeast cell, however, is usually ovoid or ellipsoidal and produces no flagella. Motility of yeasts has not been observed in any experiment.

USES OF YEASTS.—Some of the yeasts, especially the species *Saccharomyces cerevisiae*, rapidly ferment sugars to carbon dioxide and alcohol. They have been used since antiquity to make beer and wine and to leaven bread. *S. cerevisiae,* variety *ellipsoideus,* is the yeast most often used in wine-making. Special distiller's yeast strains have been selected for making distilled alcoholic beverages and commercial ethyl alcohol. Yeasts also supplement human and animal diets because of their high content of B vitamins and protein. Food yeast may be recovered brewer's yeast, or it may be a special type, usually *Cryptococcus utilis,* grown on waste sugar solutions from a variety of industries.

YEAST INFECTIONS.—A few yeasts, such as *Candida albicans,* cause infections of human mucous membranes. *Candida* infection is not thought to be communicable; most persons unknowingly harbor the organism. Sometimes, when antibiotics are administered, the normal microbial flora of the body are disturbed and a *Candida* flare-up occurs. It is thought that administration of some antibiotics may stimulate its growth. *Cryptococcus neoformans,* a yeast found in the soil, may cause a fatal meningitis.

Algae.—Closely related to fungi and bacteria are the *algae,* which are mainly aquatic and marine plants. The algae vary greatly in size and characteristics; they include the large marine kelps and tiny single-celled plants. All, except for the blue-green algae, are eucaryotic. They all carry out the same photosynthetic reactions as the higher green plants do with the aid of chlorophyll.

The algae are among the most abundant of living things. The mass of microscopic algae in the oceans exceeds that of all green plants on land, making the oceans the principal sites of photosynthesis. Algae constitute a principal part of the diet of the largest whales.

The green algae are the most closely related to higher plants. They have cellulosic cell walls and store starch as a reserve food supply. Some of the unicellular green algae, such as *Chlorella,* are nonmotile while others, *Chlamydomonas,* for example, have flagella and are motile.

Euglena and related algae are quite similar to the green algae. However, they have no rigid cell wall, and hence resemble the protozoans more closely than other algae.

Other algae belong to the groups called *dinoflagellates, brown algae, red algae,* and *diatoms.* The diatoms are of interest because their walls are composed of overlapping halves, as shown in the figure, reinforced with silica. Diatoms are particularly abundant in the oceans and give the brown color to the foam seen on beaches. Their chlorophyll is obscured by carotinoid pigments. Large deposits of diatom shells have been found where prehistoric seas existed. They form a soft, white, powdery stone called *diatomaceous earth,* which is a common base for scouring powders.

Blue-green algae lack the organization of the other algae. They are procaryotic, and their chlorophyll is held in folded membranes instead of being separated from the rest of the cytoplasm in a chloroplast. Many blue-green algae can use gaseous nitrogen from the air and are therefore among the most self-sufficient microorganisms on Earth. They are the first plants to recolonize areas devastated by volcanic action or other catastrophes that kill all forms of life. They also can live in association with certain fungi, producing a complex plant structure called a *lichen.*

Lichens grow in areas of extreme cold and dryness, such as the polar land masses and mountain heights.

Protozoa.—The *protozoa* are generally considered to be unicellular animals, but in some instances there is evidence that a protozoan has arisen from an alga that lost its ability to form chlorophyll. Most protozoans are sufficiently different from the algae so that no readily recognizable relationship can be established. Most protozoa lack a definite cell wall and vary greatly in size and shape. They are much larger than bacteria. *Paramecium,* a commonly studied form, is elliptical, with dimensions of 200 μ to 40 μ. Although they are regarded as the most primitive creatures in the animal kingdom, the protozoa are vastly more complex than the bacteria. The protozoa usually have well-defined portions that perform the functions of specialized organs in the more highly organized animals.

The four main groups of protozoa are *Sarcodina, Mastigophora, Sporozoa,* and *Ciliophora.* There are about 20,000 recognized species.

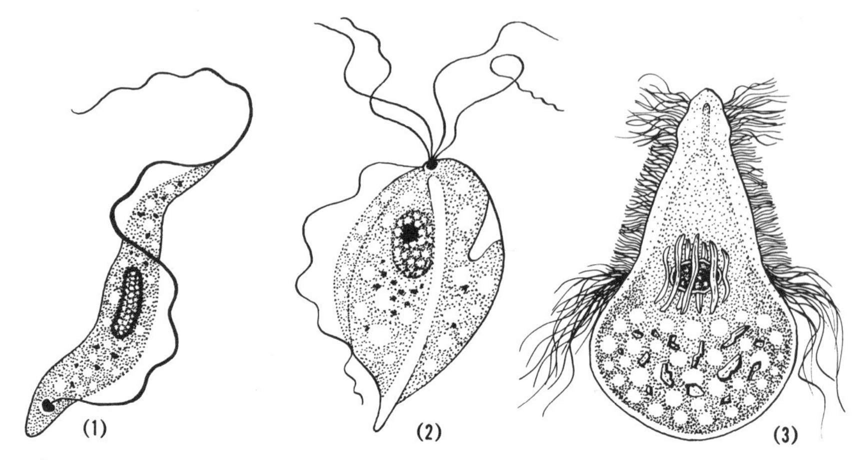

NONPHOTOSYNTHETIC PROTOZOA: (1) *Trypanosoma;* (2) *Trichomonas;* (3) *Trichonympha.*

■**SARCODINA.**—*Sarcodina* include the *amoebae,* the *foraminifera,* and the *radiolaria.* The adult amoebae lack definite shape or form and are bounded by a poorly defined cytoplasmic membrane. They move by extending arms of cytoplasm called *pseudopodia* ("false feet"), followed by movement of the rest of the cytoplasm into the pseudopod. Pseudopodia may completely surround food particles, thus bringing them into the cell. Multiplication is generally by fission of a single cell. This may, however, be preceded by conjugation of two cells. Many amoebae form resting stages, known as *cysts,* that are resistant to drying. When the environment is favorable, the cyst grows, producing an adult cell.

Most of the amoebae are free-living forms, but many are also parasitic on man and other animals. Only one of these, *Endamoeba histolytica,* causes human disease. This organism is carried in the feces and causes amoebic dysentery, a fairly common intestinal disease spread by contaminated water and food.

The foraminifera have a more or less complex outer shell structure, usually formed of calcium carbonate. Pseudopodia may extend through a mouth opening and also through holes in the shell to capture bacteria and other food materials. Geologists use the shells of foraminifera to determine geological age of rocks. Foraminifera occur only in salt water, and their skeletons make up a large part of the ooze of the ocean bottom. Throughout the world there are great chalk deposits, formed in the past geological ages, that consist largely of skeletons of foraminifera. The magnitude of these deposits is apparent when one considers that England alone produces over five million tons of chalk annually. The White Cliffs of Dover are shells of foraminifera.

The radiolaria have internal skeletons with radial spines that extend through and beyond the protoplasm. Their skeletons, which are siliceous, are almost indestructible. Their remains form the principal materials of millions of square miles of the bottoms of the Pacific and Indian oceans. They are thought to have been among the first animals on Earth; their skeletons form the oldest fossil-bearing rocks. Yet living forms closely resemble those of antiquity.

■**MASTIGOPHORA.**— The *Mastigophora* are protozoans that have from one to many long flagella in their principal phase. This group includes very simple organisms that appear to be flagellated amoebae and highly organized cells with definite shapes and internal structures; one finds protozoa that appear to be derived from algae. In fact, the protozoologist usually includes many species containing chlorophyll in the subclass *Phytomastigophora,* which contains the *Euglena* and related microorganisms. In this article all of the chlorophyll-containing protozoa have been classed as algae. Three important groups of flagellated protozoa are the trypanosomes, the trichomonads, and the trichonympha.

The *trypanosomes* and related protozoa have bladelike cells with a single terminal flagellum. An undulating membrane runs lengthwise along one side of the cell. Some of these protozoa cause diseases of insects and plants; others cause serious diseases in man and other animals. Among the latter are kala azar, caused by *Leishmania donovani* and thought to be transmitted by the bite of the sand flea. Another disease caused by a protozoan of this group is African sleeping sickness, caused by *Trypanosoma gambiense* and transmitted from man to man by an intermediate host, the tsetse fly. Most of the diseases of this group are spread by insects or other intermediate hosts.

Trichomonads are pear-shaped protozoans with four flagella at the smaller end. Many species of this group occur in animal intestines and can easily be observed in frogs and other amphibians. *Trichomonas vaginalis* causes an infection of the female genital tract; another form, *T. foetus,* causes abortion in cattle.

Trichonympha and related complex flagellates live in the digestive tracts of many insects, such as roaches and termites. In some instances they are beneficial to the insect, aiding in the digestion of such material as wood.

■**SPOROZOA.**—All *sporozoa* are parasitic protozoans that at some stage of their life cycle form spores. Each species is usually restricted to one or two hosts in which different stages of its life cycle may occur. All groups of animals have sporozoan parasites. The most common sporozoan disease of man is malaria. Part of the sporozoan's life cycle occurs in the Anopheles mosquito, which may become infected from feeding on blood of persons suffering from malaria. After a developmental stage in the mos-

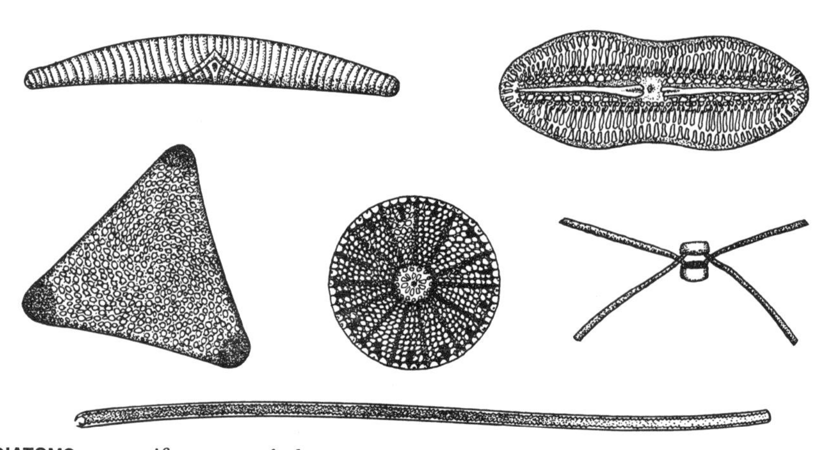

DIATOMS, a specific group of algae, occur in a wide variety of sizes, forms, and shapes.

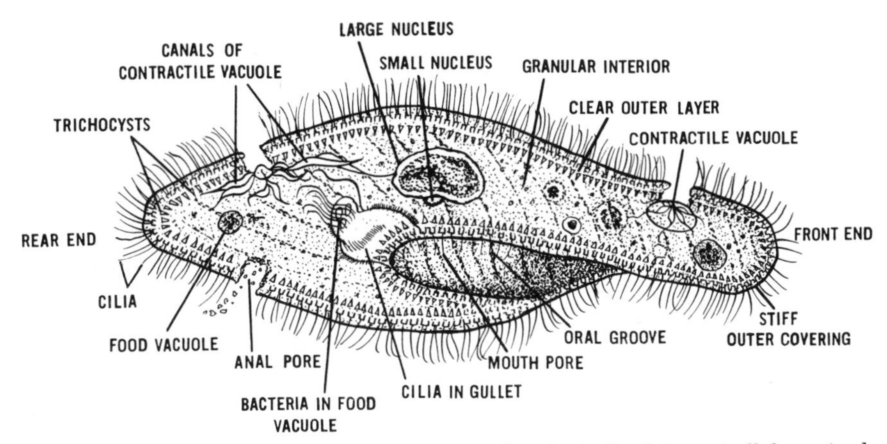

THE PARAMECIUM is one of the most highly developed of all of the unicellular animals.

quito, malaria can be transmitted to man again.

■**CILIOPHORA.**—The *Ciliophora* (*Infusaria*) are protozoa with *cilia*, short protoplasmic threads, on their surface. These cilia may cover nearly the whole surface, acting as organs of locomotion, as in paramecia, or they may be restricted to certain areas, such as around an oral opening where they help collect food. Most ciliates are free-living aquatic animals, but some inhabit the digestive tracts of animals. Many of them, such as the paramecia, readily develop in stagnant water containing dead plant materials. They live largely on other microorganisms, such as bacteria.

Some of the most complex ciliates, belonging to the genus *Diplodinium*, are found in the rumen of cattle and other ruminant animals. They appear to aid in the digestion of food particles and thus are beneficial, although not essential, to the well-being of the animal.

Immunity to Disease.—*Immunology* is the science that is concerned with the study of immunity to disease. The field developed in the course of study of viruses, bacteria, and *Rickettsiae*. Prior to the eighteenth century, people recognized that most persons and animals that recovered from some diseases did not get a second attack of the same disease. They had what we now call acquired immunity. Observation of this fact led to the intentional propagation of a mild form of smallpox, *variola minor*, that killed very few persons. This prevented a larger number of deaths than would otherwise have been the case during epidemics of a more severe form of smallpox, *variola major*. The severe form often killed 50 to 75 per cent of its victims. However, the practice was too risky for wide acceptance.

The English physician Edward Jenner (1749–1823) observed that persons who had had a disease of cattle, cowpox, did not contract smallpox during epidemics. He dramatically demonstrated the value of this knowledge by artificially passing cowpox to a child and then later inoculating the child with material from a smallpox patient. The child did not contract smallpox. This discovery in 1796 was soon put to practical use to control smallpox, although the reasons for its effectiveness were not known for about another century.

Further studies during the period of 1879 to 1900 showed that immunity to several animal and human diseases could be induced by injections of bacteria or viruses treated to make them noninfective. Diseases in which poisonous bacterial products, *toxins*, are involved can be prevented by repeated small injections of toxin. Further, the toxin-neutralizing effect can be passed from an immunized person to a nonimmunized person by a transfer of blood serum. All of these immunities have been found to be specific—they protect the person only against the disease-causing agent used to develop the immunity. Acquired immunity is now known to be caused by the formation of substances called *antibodies*, which are altered blood-serum proteins. The substance inducing antibody formation is called an *antigen*.

The reaction of serum antibodies with the corresponding antigen can be demonstrated in a test tube by one of several means. If the antigen is particulate, such as dead bacterial cells, the antibody will cause it to *agglutinate*, or form clumps, and settle out. If the antigen is soluble, such as egg white or some other protein foreign to the host animal, the antibody will cause a visible precipitate to form. If the antigen is a suspension of living bacterial cells, the antibody will increase the rate at which they can be engulfed by living white blood cells. If the antigen is a toxin, the antibody (*antitoxin*) will neutralize it so that it may safely be injected into an experimental animal that it would otherwise kill. These reactions can be used to identify either the antigen or the antibody if the other half of the system is known.

The exact means whereby antibodies protect an animal from a disease caused by its corresponding pathogenic antigen are not known. They may, however, involve some of the same reactions that can be demonstrated outside the body.

■**IMMUNIZATION.**—Studies of antibody formation have led to many procedures that can be used successfully to protect man and domestic animals from disease. These include: (1) use of living but not virulent bacteria, such as the BCG (*bacille Calmette Guérin*) vaccine for tuberculosis; (2) use of dead bacteria to give immunity, as in vaccination against typhoid fever; (3) use of chemically modified extracts of bacteria, as in whooping cough vaccinations; (4) use of chemically modified toxins, *toxoids*, to form antitoxins against diphtheria and tetanus toxins; (5) use of dead *Rickettsiae*, as in vaccines for typhus fever and Rocky Mountain spotted fever; and (6) use of modified viruses to prevent such diseases as smallpox, measles, and poliomyelitis. Despite these many methods of immunization, present knowledge does not permit us to immunize against all diseases.

All of the above immunization procedures lead to what is called *active immunity;* the patient develops it himself. It is also possible to obtain *passive immunity* by transfer of antibodies, such as antitoxin, from one person to another or from an animal to a human being. Tetanus antitoxin is prepared by immunizing horses to tetanus toxin. The horse serum contains the antitoxin.

Active immunity is generally superior to passive immunity in preventing disease, since it lasts for a time from several months to the remainder of one's life; passive immunity disappears in a few weeks. Passive immunization is superior for treatment of a patient, since the maximum amount of antibody is present at the end of the injection. Active immunity generally requires several days or weeks to reach its maximum level.

The successful use of artificial immunity to prevent disease is dependent on an adequate vaccine, the length of immunity, the relative risk of immunization as compared with having the disease, and whether there are healthy carriers of the disease agent.

In addition to controlling disease, immunology includes studies of allergies and similar reactions and of the rejection of transplanted organs from a genetically different body of the same species.

—Richard H. McBee

BIBLIOGRAPHY

Brock, Thomas D. (ed.). *Milestones in Microbiology*. Prentice-Hall, Inc., 1961.

Dubos, René Jules. *The Unseen World*. The Rockefeller Institute Press, 1962.

Lechevalier, H. H., and M. Solotovsky. *Three Centuries of Microbiology*. McGraw-Hill, Inc., 1965.

Stanier, Roger Yates, and others. *The Microbial World*. 2nd ed. Prentice-Hall, Inc., 1963.

Walter, William G. *Dictionary of Microbiology*. D. Van Nostrand, 1962.

Walter, William G., and Richard H. McBee. *General Microbiology*. 2nd ed. D. Van Nostrand, 1962.

GENETICS

Continuity of Life.—*Genetics* is the study of the inheritance of biological characteristics in living things—characteristics that are passed from one generation to the next. What is inherited is a code message in the genetic material (*genes*) of egg and sperm. The code directs embryonic development and organization of cells into tissues and organs, and the function of each tissue and organ. The development is also influenced by external and internal environment. Thus, the organism is the product of interaction between genetic material and environment.

What is the nature of genetic material? How is it reproduced? What is the material's mechanism of action in the cell's function and the individual's development? How is the material transmitted? What is its role in the process of organic evolution? Each of these questions is complex, but progress has been made in answering them.

Mendelian Genetics.—The first problem to be solved is related to the mechanism of transmission of genetic material. This was the basis of Mendelian, or classical, genetics. Even in the nineteenth century, biologists knew an embryo develops from fusion of egg and sperm. Hence, the new organism is the product of materials from each parent. However, biologists were under the false impression that hereditary traits were transmitted in a bloodlike fluid from each parent and were mixed in the offspring. Hence we have the terms "half-blooded" and "full-blooded."

Gregor Johann Mendel (1822–1884), an Austrian monk, analyzed the basic laws of inheritance in 1866. The results were lost until 1900, when investigators in Holland, Germany, and Austria each independently rediscovered the Mendelian concept.

■MENDEL'S CONTRIBUTION.—Mendel had proved hereditary traits are transmitted by pairs of distinct units, later called *genes,* which reshuffle, segregate, and redistribute, rather than blend, in the offspring.

Mendel used garden peas in his experiments because they hybridize easily. When a purebred tall plant was crossed with a purebred short plant, all hybrid offspring were tall, no matter which type was the mother and which the father. The hybrids self-fertilized. Mendel counted the offspring and found 787 tall and 277 short plants, a ratio of about 3 to 1. When the short plants self-fertilized, they produced only short offspring, but when the tall plants self-fertilized, there were two types of offspring: one-third had only tall offspring, and two-thirds produced both tall and short in a ratio of 3 to 1. Mendel crossed six other characters: round and wrinkled peas, colored and uncolored flowers, and yellow and green peas. He had approximately the same results.

Mendel then formulated the *law of segregation.* Today this principle states that hereditary traits (such as tallness or shortness of peas) are transmitted by *zygotes* (fertilized eggs). One member of the pair of traits comes from the male parent; the other, from the female. In the mature plant, these paired genes segregate during the formation of *gametes* (sperms and eggs) so that just one of the pair is transmitted by a particular gamete. The gamete has only one gene of each pair, and is called *haploid.* When the male and female gametes unite to form the zygote, it is called *double* or *diploid.*

Mendel's studies showed the *principle of dominance.* The trait of tallness is dominant over shortness. (When there is a gene for tallness and one for shortness, all peas are tall.) The opposite, unexpressed factor is *recessive.*

An individual with unlike paired genes can be represented as *Tt. T* represents the dominant gene for tallness, and *t* the recessive gene for shortness. Such an individual is called a *heterozygote.* If both genes are alike (*tt* or *TT*) the individual is a *homozygote.* The genetic makeup is called the *genotype;* the character determined by this genotype and expressed in the individual is the *phenotype.* If the genotype is *TT,* the phenotype is tallness. Another genotype that can give the phenotype tallness is *Tt.* The alternative forms of a gene are called *alleles.* This cross is illustrated in the diagram.

Mendel concluded that dominant and recessive genes do not affect each other; gametes are haploid and have only one of a pair of genes; each type of gamete is produced in equal numbers by a hybrid parent; and combination between gametes depends on chance—the frequency of each class of offspring depends on frequency of the gametes produced by each parent.

Mendel next determined how two or more pairs of genes would behave in crosses. He crossed plants with round, yellow seeds with those with wrinkled, green seeds. He knew a cross between round (*R*) and wrinkled (*r*) seeds produced round seeds in the F_1 generation and three round to one wrinkled seed in the F_2 plants; he knew crossing yellow (*Y*) with green (*y*) produced all yellow in the F_1 and three yellow to one green seed in the F_2 generation. This showed the dominance of roundness and yellowness over their respective contrasting alleles. When Mendel crossed round-yellow with

BROOKHAVEN NATIONAL LABORATORY

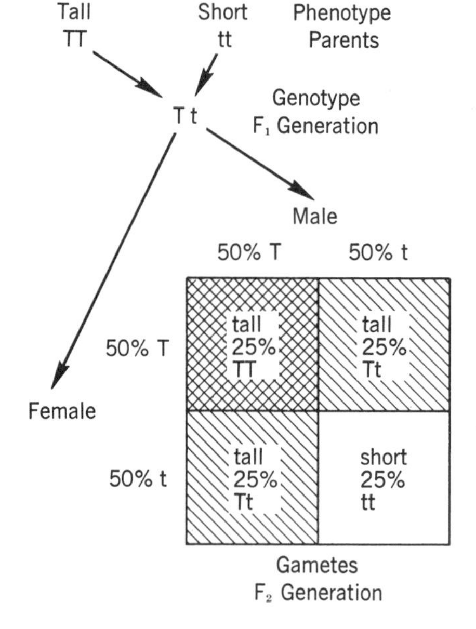

DROSOPHILA (FRUIT FLIES) aid geneticists in studying evolutionary characteristics. The Mendelian square (*right*) illustrates the principle of dominance. Since the trait of tallness is dominant, the combination *TT,tt* yields three tall individuals out of four offspring.

wrinkled-green, the F_1 produced all round-yellow seeds. In the F_2 generation, these seed types were obtained:

Two combinations, round-green and wrinkled-yellow, not present in either parental or the F_1 generation, have appeared. This result can be explained by Mendel's *law of independent assortment,* which states that members of one pair of genes segregate independently of other pairs. The various combinations are illustrated in the accompanying table.

Type	*Proportion*
Round-yellow	9/16
Round-green	3/16
Wrinkled-yellow	3/16
Wrinkled-green	1/16

Mendel also tested his F_2 plants to determine whether all of a single phenotype class, such as round-yellow, were alike in genotype. According to his hypothesis, there should be four different genotypes in this group: *RR, YY; RR, Yy; Rr, YY;* and *Rr, Yy.* When F_2 plants self-fertilized, he found four classes of round-yellow seeded plants; the ratios fitted expectations. The breeding behavior of the F_2 round-green, wrinkled-yellow, and wrinkled-green were in accord with the hypothesis that each pair of genes segregates independently from other pairs of genes and is transmitted independently to the next generation.

Mendel's inheritance rules were later found to apply in other plants and in animals. More information explained the seeming exceptions.

In all characters studied by Mendel, the heterozygote was phenotypically identical with the homozygote dominant. In some cases, however, the heterozygote is intermediate. This is true in the color of the flower known as the "four o'clock." If a red parent is crossed with a white, all F_1 hybrids are intermediate in color, one-quarter of the F_2 offspring are red, one-quarter white, and one-half intermediate. In some cases, both alleles are equally "dominant," for example, those that determine the *MN* factors in blood. If one parent is *M* and the other *N,* all children will be *MN.* If both parents are *M,* or both *N,* all children will be *MM* or *NN,* respectively. If both parents are *MN,* one-quarter of the children are *M,* one-quarter *N,* and one-half *MN.* To understand other exceptions, where independent segregation does not seem to apply, one must consider the chromosomal basis of heredity.

Phenotype Parents: wrinkled-green rryy → ry; Round-Yellow RRYY → RY
Genotype F_1 Generation: Round-Yellow RrYy
Gametes F_2 Generation

MALE \ FEMALE	RY	Ry	rY	ry
ry	RrYy	Rryy	rrYy	rryy
rY	RrYY	RrYy	rrYY	rrYy
Ry	RRYy	RRyy	RrYy	Rryy
RY	RRYY	RRYy	RrYY	RrYy

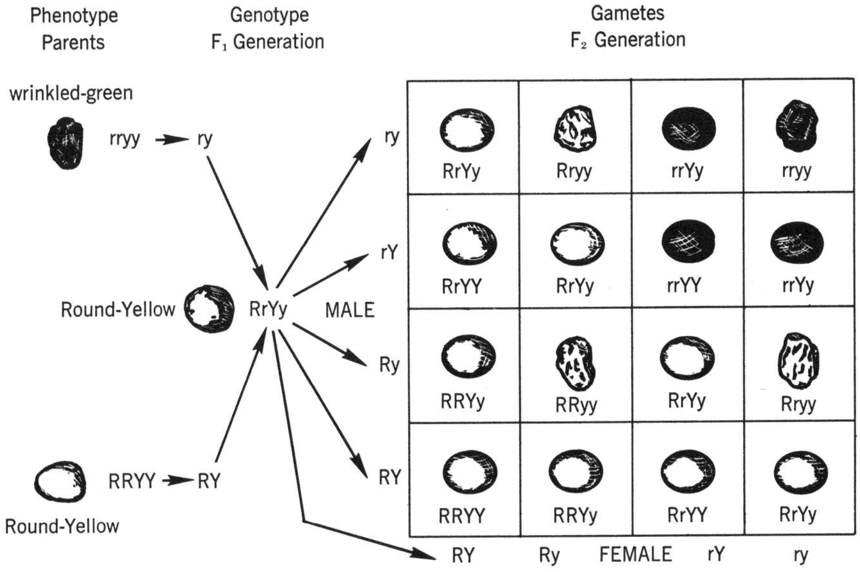

THIS MENDELIAN SQUARE shows the many genetic combinations possible in hybrid peas.

Chromosomal Basis of Heredity.—All living things are composed of cells and begin life as a single cell. In organisms reproducing sexually, the cell is the fertilized zygote, which divides to form all the cells of the body of the organism (*somatic cells*).

Each cell has an inner body (*nucleus*) surrounded by a less dense semifluid material called *cytoplasm,* which is enclosed by a cell membrane. In the cytoplasm are various vital structures. In the nucleus, which is enclosed by its own membrane, are threadlike structures called *chromosomes,* and one or more bodies called *nucleoli,* which are dark when they are stained.

In 1902 a graduate student, W. S. Sutton, and a German cytologist, T. Boveri, decided independently that genes are in the chromosomes. Sutton's arguments were:

(1) Since sperm and egg give continuity from one generation to the next, the hereditary traits must be carried by the sperm and egg. (2) The sperm is almost all nucleus and yet contributes as much to heredity as does the egg, which has both cytoplasm and nucleus. Hence, the hereditary characters must be in the nucleus. (3) The visible nuclear parts that divide during cell division are chromosomes. The genes, then, must be on the chromosomes. (4) Chromosomes occur in pairs, as do genes. (5) Chromosomes segregate during maturation of egg and sperm. Genes segregate during formation of the gametes (law of segregation). (6) Members of one pair of chromosomes segregate independently of other chromosome pairs. (Mendel showed that one pair of genes also segregates independently of other pairs.)

This gives a logical basis for the hypothesis that genes must be on chromosomes. In sixty years, evidence has accumulated proving the truth of this hypothesis and identifying the chemical nature of the hereditary chromosomal material. Chromosomes are equally distributed during cell division (*mitosis*) and each member of the chromosome pair segregates during maturation of egg and sperm (*meiosis*).

■MITOSIS.—Chromosomes are accurately reproduced and transmitted in a precise process that assures that each new cell formed receives one of each chromosome. The number of chromosomes characteristic of each species remains constant. Every somatic cell in the human being has 46 chromosomes; in the fruit fly, 8; in the garden pea, 14.

Although mitosis is a continuous process, it is described in terms of five phases:

Interphase. Chromosomes are usually not individually distinguishable; they are stretched out in long, diffuse threads. In this phase, each chromosome copies itself by a mechanism not yet understood.

Prophase. The chromosomes coil up and become short and thick. The nuclear membrane disappears and spindle fibers, denser than the cytoplasm, appear.

Metaphase. The chromosomes line up in a single plane across the center of the cell.

Anaphase. Each chromosome, at a fixed point on its body, has a minute structure called the *centromere,* attached to a spindle fiber. This fiber apparently contracts, and thereby guides each of the duplicated chromosomes away from the center of the cell in opposite directions.

Telophase. One complete set of chromosomes is in each half of the cell. The entire cell then divides between the two sets, and the chromosomes uncoil and lengthen. A new interphase is then begun in each daughter cell.

The number and make-up of chromosomes remain constant in each cell because successive chromosome duplication is followed by cell division.

■MEIOSIS.—Each cell's chromosomes occur in pairs, one from the mother and one from the father. Each member of the chromosomal pair has similar genes; they are called *homologous chromosomes,* or *homologues.* Human somatic cells have 23 pairs of chromosomes each.

The egg and sperm must each have half the somatic number of chromosomes, or be haploid, so that when they unite, the fertilized zygote will be *diploid,* as will be all cells derived from it by mitosis. The chromosome number of somatic cells remains constant from generation to generation

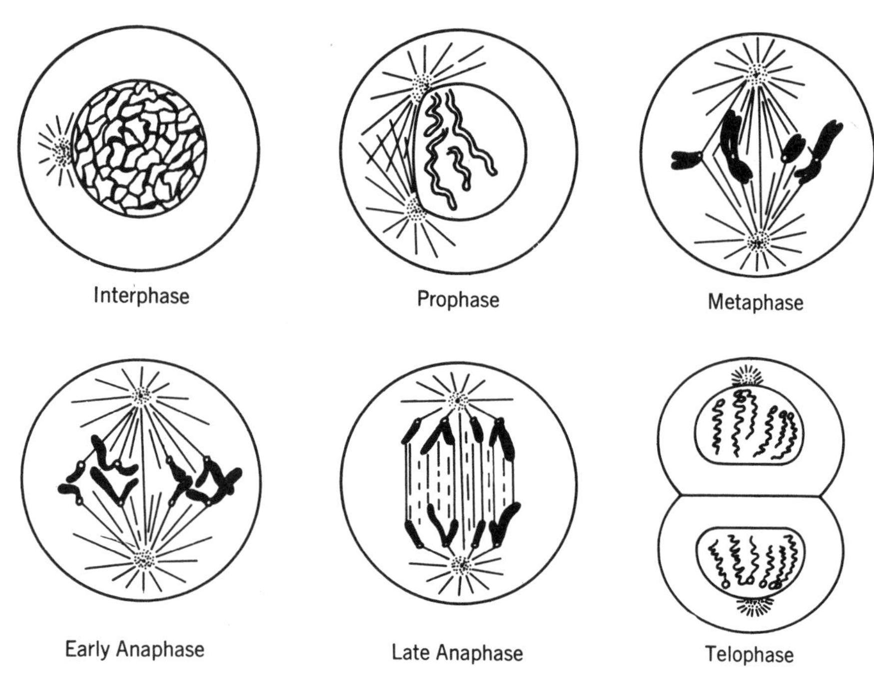

MITOSIS, or cell division, is the process by which a cell splits into identical twins. The number and makeup of chromosomes remain constant in each cell during this process.

of the species and there is a mechanism to ensure that one member of each chromosome pair is in each gamete. The mechanism is known as *meiosis* and has two divisions:

First meiotic division. In this phase, each chromosome copies itself once, creating two *chromatids;* the chromatids are held together by the undivided centromere. The homologous chromosomes move toward each other and pair tightly, so that similar genes, or *alleles,* lie alongside each other. This makes a bundle of four chromatids and two centromeres. During this time the partner, or matching, strands twist around each other, break, and then reunite, thus exchanging homologous sections. As a result, chromosomes consist of parts from both paternal and maternal chromosomes. The centromere of each of the homologous chromosomes, attached to the spindle fiber and to the chromosome, is pulled to an opposite pole of the cell. The cell divides. Each new cell now has the haploid chromosome number, with either a maternally or paternally derived member of each chromosomal pair, depending on chance and on the random attachment of each centromere to the spindle fibers. This is the chromosomal basis for the independent assortment of each member of a pair of chromosomes with respect to the other.

Second meiotic division. In this phase, the centromeres divide so that each chromatid separates from its duplicate. Another division follows that is much like ordinary mitosis—two duplicate cells are formed.

Because of the two meiotic divisions, four cells are formed from the original cell; each has a haploid number of chromosomes, one of each homologous pair.

■SEX-LINKAGE.—Chromosomes occur in pairs, one maternal and one paternal. All but one pair are identical in both sexes. In one sex (usually the male) there is one pair of unidentical chromosomes—*X* and Y. In the human male, there are 22 pairs of identical (nonsex) chromosomes called *autosomes,* and one *X* and one Y chromosome. The female has 22 pairs of autosomes and a pair of *X* chromosomes. The *X* and Y chromosomes are called *sex chromosomes.* If an egg is fertilized by a sperm bearing an *X* chromosome, it becomes female (*XX*). If it is fertilized by a Y-chromosome-bearing sperm, it becomes male (*XY*). Thus, sex determination occurs at the moment of fertilization. Since segregation of the sex chromosomes takes place during meiosis exactly as does segregation of the other chromosomes, and is completely random, the chance is even that any sperm will contain a Y chromosome. Since all eggs contain one *X* chromosome, the probability of the offspring's being either a boy or girl is exactly equal.

The Y chromosomes of organisms contain few or no genes. (None are known to occur on the human Y chromosome.) The *X* contains many genes. Because of this, these genes are segregated differently in the two sexes, resulting in the phenomenon called *sex-linkage.*

Red-green color blindness is the most common sex-linked trait in human beings, occurring in about 8 per cent of men and in about .5 per cent of women. This is explained by the hypothesis that the recessive gene responsible is contained in the *X* chromosome and that there is no corresponding allele in the Y. A woman heterozygous for the trait married to a normal man would have daughters with normal vision, but probably half of her sons would be color-blind. The children of a homozygous (normal) woman married to a color-blind man would all be normal, but probably half the daughters would be heterozygous and would transmit the trait to half their sons.

■X CHROMOSOME.—It has long been a puzzle as to how males can function with only one *X* chromosome, and therefore only one set of sex-linked genes, whereas females have two *X*'s and a double set of sex-linked genes.

Since 1949 geneticists have been able to distinguish between body cells from the male and female of many mammals, including human beings, by a simple staining technique. Cells from females contain a darkly staining body, the *Barr body* or *sex chromatin,* that is missing from male cells.

In 1961 the English geneticist, Mary Lyon, formulated a hypothesis based largely on experiments with mice, which resolved the riddle as to why a single *X* chromosome is enough for a male. She states that one *X* is also enough for a female, since only one of her two *X*'s functions. The other is inactive and condensed, and so it stains deeply. This inactivation occurs early in the embryo's development. In some embryonic cells it may be the maternal *X* that is inactivated; in others, it may be the paternal *X*. All cells descended from the embryonic cells thus would only have the same single active *X* chromosome. The best evidence for this randomness of inactivation is the inheritance of certain sex-linked coat-color genes in mammals. Females heterozygous for any of these always show a mottled or dappled phenotype, with patches of normal and mutant color, as in a tortoise shell cat.

The Barr bodies in female cells fit the theory of the single active *X*. The theory is that all inactive *X*'s show up as Barr bodies. Males (*XY*)

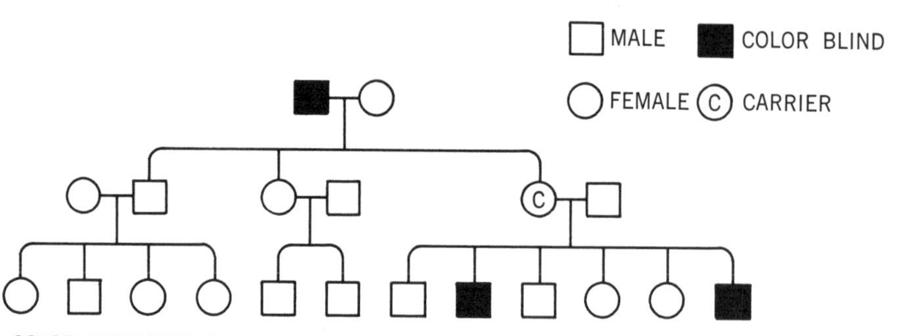

COLOR BLINDNESS is a common sex-linked trait in human beings. A color-blind male passes his recessive trait to a male grandchild through his daughter, who serves as a carrier.

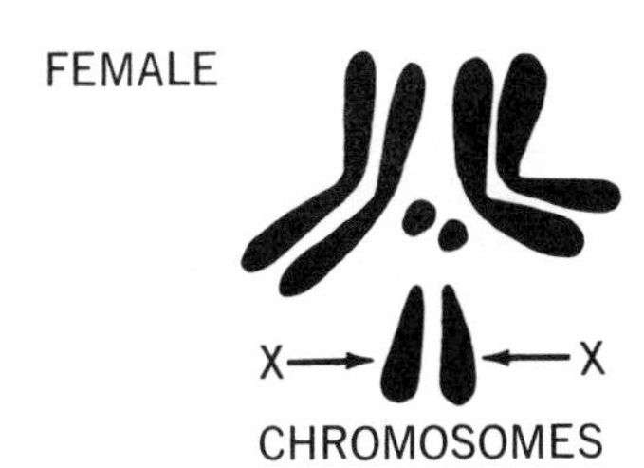

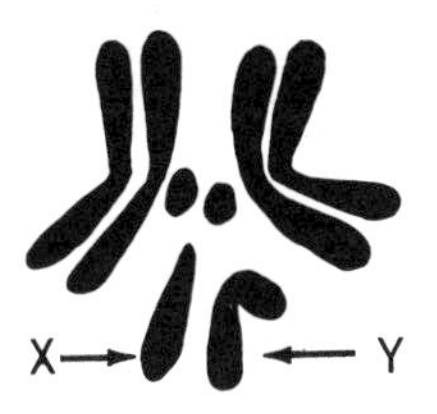

CHROMOSOMES in Drosophila. Two *X*'s combine to make a female; *X* and *Y*, a male.

and females with an abnormal chromosome complement of one *X* and no Y *(XO)* have none; normal females *(XX)* and males with an abnormal chromosome complement (*XXY*) have one. Other abnormalities in sex chromosome number have been found. There are females with three *X* chromosomes and two Barr bodies. Females with four *X*'s and three Barr bodies have also been reported.

The single active *X* chromosome hypothesis is not yet verified, but evidence indicates that it is true.

Y CHROMOSOME.—Experiments with *Drosophila* (small two-winged flies) showed that two *X* chromosomes were necessary to produce a female, and one *X* to produce a male. The Y seemed to be neutral, as proved in certain flies with the *XO* chromosomal constitution (one *X* and no Y) that were sterile males. Female flies were found with two *X* chromosomes and a Y (*XXY*). Until recently, it was thought that sex was determined the same way in the human being. It is now known, however, that the Y chromosome is essential to determine maleness in human beings. Humans with rare chromosomal abnormalities have been studied, and those with the *XO* chromosome constitution (one *X* and no Y) are abnormal females. Some abnormal males have an *XXY* complement. Since the Y is essential in human beings for a male to develop, he will be abnormal if this is unbalanced by two *X*'s. One *X* will produce a female, but two are necessary for normality.

CHROMOSOME MAPPING.—The principle of independent assortment is explained in that the maternal and paternal members of each homologous pair of chromosomes are distributed independently to the gametes at meiosis. The seven factors Mendel studied were carried by different chromosomes and were independently inherited. Each chromosome has many genes; in *Drosophila* there are hundreds of genes known, but only four pairs of chromosomes. Genes located on the same chromosome tend to be inherited together and are said to be *linked*. During meiosis, when homologous chromosomes pair, they twist around each other with resultant breakage and reunion; hence, some parts of a chromosome segregated to the new cell can be either maternal or paternal. This process of chromosomal recombination is called crossover. There is indisputable evidence the crossover occurs at the stage of meiosis when there are four chromatids held together by the undivided centromeres.

Crossover occurs at random sites along the chromosome; the frequency is determined by the distance between the points—the closer the points, the less frequent the crossover; the farther apart, the more frequent.

This is the basis for mapping the locations (*loci*) of the genes; the crossover frequency is the unit of map distance between loci. To determine the order of the gene loci, crosses involving three different pairs of linked factors are used. Analysis of such crosses has proved that the genes within each chromosome are arranged linearly in a definite serial order at fixed loci.

Genetic maps of chromosomes are graphic representations of the relative distances of genes in each linkage group, as determined by the percentage of *recombination* (crossover) among the genes. The four pairs of chromosomes from *Drosophila melanogaster* have been extensively mapped. Corn has also been mapped; each of its ten linkage groups (corresponding to ten pairs of chromosomes) is represented by many genes. Other maps are available for pink bread mold (*Neurospora*), the colon bacillus (*Escherichia coli*), and the mouse. A few gene markers are mapped on the *X* chromosome of man, but the 22 autosomes (the nonsex chromosomes) are largely unmapped. New techniques in tissue culture may make possible mapping of all human chromosomes.

Correspondence between the genetic maps of chromosomes and morphologically detectable defects in the chromosomes has been good. The presence in the salivary glands of *Drosophila* of huge chromosomes has made possible detection of minor structural changes. Sometimes a portion of a chromosome is deleted; this defect is detected in salivary chromosomes. Flies missing a small region near one end of the *X* chromosome also lack the white eye gene; hence, the locus of the gene for white is in this region of the *X* chromosome. Likewise, the precise location of other genes has been ascertained; generally the relative distance as determined by linkage studies has been confirmed by cytological studies.

Nature of Genetic Material.—The chemical nature of genetic material and the chemical basis for its reproduction are important. Generally it is accepted that except for some viruses, genetic information is carried in *deoxyribonucleic acid* (DNA). Principally, the evidence is as follows: DNA is unique to the chromosomes. The amount of DNA is remarkably constant from cell to cell within any organism and within a single species. Only the egg cells and the sperm cells have a different amount; they contain half the normal amount of DNA (and half the somatic chromosome number).

Proteins and *ribonucleic acids* (RNA), other substances found in chromosomes, vary considerably in amount in different tissues within a species. Inheritable changes, *mutations,* are in the genetic material. These seldom occur spontaneously, but the frequency may be greatly increased by exposure to X rays, ultraviolet light, and certain chemicals. DNA and the genes are normally stable. A wavelength of 260 μ ultraviolet light is effective in producing mutations because it is most absorbed by nucleic acids.

The best direct evidence that DNA is the genetic material comes from experiments in transformation of certain bacteria; these show that the so-called transforming principle is pure DNA. Classic work by O. T. Avery, C. M. MacLeod, and M. McCarty in 1944 was based on earlier observations that when an extract from dead cells of one strain of *Pneumococcus* was added to living cells of another strain, it transformed some characters of the living cells so they were identical to the extract strain. The new characters were inherited by progeny of the transformed strain as if the latter had extract genes. Avery's group analyzed the extract and proved that the active part of the transforming extract was pure DNA, so the genetic material must be DNA.

Evidence that DNA transmits genetic information was obtained by the 1952 studies of Hershey and Chase on viral infection of colon bacilli. *Bacterial viruses,* or *bacteriophages,* consist of a protein coat and a DNA core. When viral protein is labeled with a radioactive isotope of

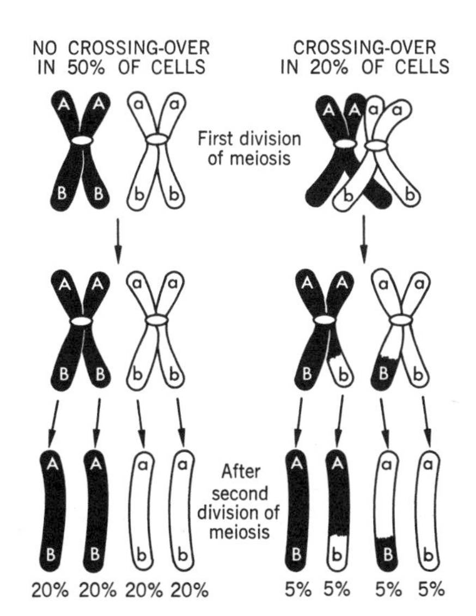

CROSSOVER during meiosis brings about new combinations of genes in the offspring.

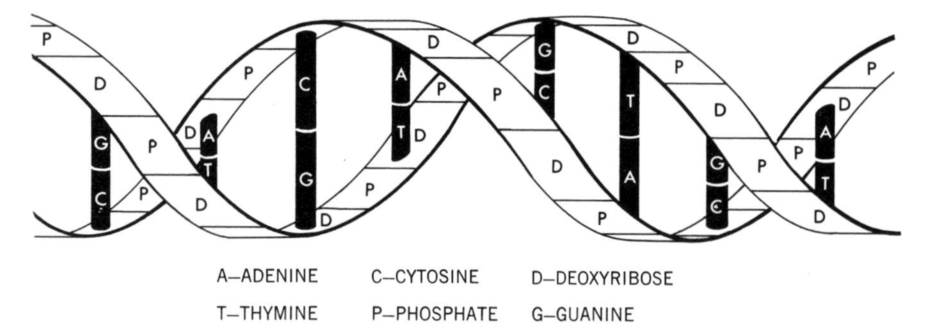

WATSON-CRICK MODEL of the DNA molecule is formed from a few simple molecules repeatedly linked. These various combinations represent a complex genetic code of life.

sulfur (S-35) and the DNA with radioactive phosphorus (P-32) and each is allowed to infect the bacterial host, S-35 remains outside and P-32 inside the host. The P-32 is found in the new viruses released when the bacteria open. Since the viral part (*phage*) within the bacterium contains genetic information that directs its host to make more phage (both DNA and protein), it must be DNA that is the genetic material.

N. Zinder and J. Lederberg discovered bacterial transduction, providing more proof that DNA transmits genetic information. In *bacterial transduction,* a hereditary trait can be transferred from one bacterial cell to another via a virus that infects first one cell and then the other. In transduction, the viral DNA picks up a tiny bit of its host's DNA. When the host cell ruptures and the virus infects a new host, the virus carries the genes of the first host with it. These genes are expressed in the new host and its offspring. A streptomycin-resistant bacterium may be infected with a viral strain of low virulence, a temperate bacteriophage. The virus may multiply within it, and some viral particles may pick up the streptomycin-resistant gene. If the virus is allowed to infect a streptomycin-sensitive bacterial strain, some of these bacteria and progeny are then found to be streptomycin-resistant. If two or more genes are transduced simultaneously, they are always closely linked. Transduction is indirect evidence that DNA is genetic material. Assuming that the part of the virus that enters the host is DNA and that the bacterial material carried by the virus is also DNA, then transduction is like transformation, except that instead of man transferring the DNA from donor to recipient, a temperate bacteriophage does so.

Some viruses, such as tobacco mosaic virus, consist of only RNA and protein. In 1957, Fraenkel-Conrat separated RNA and protein of related strains, recombining them so that strain *A*'s RNA was combined with protein of strain *B,* and vice versa. In each case, genetic properties of the hybrid virus particle, as determined by infection, were always of the strain from which the RNA came.

The Watson-Crick Model.—A high-molecular-weight polymer, DNA is a large molecule formed from a few simple molecules linked repeatedly by chemical bonds. Repeating units are *nucleotides,* each of which consists of a phosphate group; a five-carbon sugar, *deoxyribose;* and one of four different nitrogenous bases. The four bases are two *purines* called *adenine* (A) and *guanine* (G), and two *pyrimidines, cytosine* (C) and *thymine* (T). In polymerized form, as in nucleic acids, nucleotides are connected by a chemical bond between the phosphate group of one nucleotide and the 3-hydroxyl of an adjacent deoxyribose. The DNA has a deoxyribose-phosphate backbone with bases projecting inward and perpendicular to the axis.

Chargaff and co-workers showed that the ratio of adenine to thymine, and of guanine to cytosine, is about 1/1 in any DNA preparation. However, the ratio of adenine (or thymine) to guanine (or cytosine) varies. They thought that since A = T and G = C, each was associated with the other in DNA.

This ratio, together with X-ray diffraction studies of DNA by M. H. F. Wilkins, led James Watson and F. H. C. Crick, in 1953, to propose a DNA structure. Their structure is essentially a twisted ladder or *double helix;* the sides are made of the sugar-phosphate backbone and the rungs are the *bases.* The *base sequence* in one strand determines that in the complementary strand; an A must always be matched with a T, and a C with G. Weak hydrogen bonds hold A to T and G to C, giving the ladder firmness and ability to separate when replication takes place during mitosis.

The *Watson-Crick model* makes it possible to understand how genetic information is duplicated and transmitted. If double helix strands separate, each is a *template,* or mold that specifies the replication of its complementary copy. This would result in two identical DNA molecules, each with an original and a newly synthesized strand.

The model explains how DNA could carry genetic information, or *code,* translated into protein-producing instructions. The code depends on the sequence of the four bases in relation to each other; in effect, a four-letter alphabet. Many configurations are possible, since there are ten base pairs in each complete turn of the double helix; and this is a small portion of the entire molecule. For each turn, the number of configurations possible would be 4^{10}, or 1,048,576. Not all combinations of nucleotides are meaningful, but the storage potential of genetic information is vast.

The model gives a chemical basis for mutation. If DNA is genetic material, then a change in the DNA molecule should change the code and cause mutation. This could happen if an error is made during replication, or if A picks up C instead of T and the error is perpetuated. At one position, base pairs would be C-G instead of A-T, upsetting the code and preventing formation of a normal protein. If a compound similar to a normal base is introduced, a compound that differs so slightly that the replicating DNA could easily mistake it for the normal, mutations might occur. For example, 5-bromouracil can quantitatively replace thymine; in some cases this can produce mutation. Changing the code results in production of abnormal protein or of no protein.

The model fits requirements of linear arrangement of genes in linkage groups; the four bases are arranged linearly along the helix, forming a linear code.

The most direct work supporting the model is that done in 1958 by Meselson and Stahl. They grew many generations of bacteria in heavy nitrogen (N-15) and then transferred cells to ordinary N-14 medium for varying lengths of time. At specified times, they extracted the DNA from the cells and centrifuged it, using a technique that causes molecules to concentrate in bands at definite positions, depending on their density. DNA taken from bacteria grown for a long time in N-15 and allowed to divide once in N-14 formed a hybrid band halfway between that of DNA in bacteria grown in N-15 and in N-14. If the bacteria divided twice in N-14, two bands were visible, one a hybrid band and one the N-14 band. In succeeding divisions, the hybrid band became smaller and the N-14 band larger.

When the two strands of N-15 DNA separate, each is a template for information of a complementary N-14 strand. After one replication, two new DNA helices are formed—each a hybrid of N-14/N-15. At the next replication, each N-14 strand synthesizes another N-14 strand, forming a pure N-14/N-14 helix, and N-15 strands replicate, forming a hybrid N-14/N-15 helix. In succeeding generations, there is but one hybrid strand, but the number of pure N-14/N-14 molecules steadily increases. Facts and theory coincide, suggesting that the DNA helix unwinds into two single strands during replication.

Kornberg proved in a test tube the theoretical model of DNA synthesis. When he mixed a certain enzyme, triphosphates of each of the four nitrogen bases, and a primer, or starter, of a small amount of preformed DNA, more DNA would be synthesized. The DNA formed always had the same ratio of A, G, T, and C as the primer used had, showing that this had provided the template for new DNA.

In late 1967, Kornberg announced the laboratory synthesis of functionally active viral DNA. Spiegelman had earlier synthesized functionally active viral RNA. These two achievements were new milestones in the history of genetics.

Formation of Proteins.—DNA, now accepted as genetic material, is inactive in cellular metabolic processes, which are performed by proteins. Most proteins serve as enzymes and catalysts.

To translate instructions carried within DNA into protein structure, DNA within the chromosomes gives its message to an RNA form present in the nucleus; RNA acts as messenger and transmits the message to the site of protein synthesis within the cytoplasm. This site is another form of RNA found in many particles (*ribosomes*) in the cytoplasm. On the ribosome, an RNA template is formed according to the messenger RNA's code. Proteins are long chains of amino acids hooked together linearly by *peptide bonds*. The chains are *polypeptides*. Twenty amino acids occur in proteins; the smallest protein (*ribonuclease*) has a molecular weight of 13,500 and is a single chain of 124 amino acids. Many amino acids are repeated several times; not all need be present in any individual protein. Proteins have a definite shape, on which their function depends. Polypeptide chains form a helix that folds into a characteristic shape. The amino acid sequence is believed to determine the nature of this folding.

To synthesize proteins on the RNA template on the ribosome, the amino acids must be brought to their proper positions. This is done by a soluble RNA (S-RNA or transfer-RNA), of which there is a specific one for each of the twenty-odd amino acids. S-RNA picks up its own amino acid, which has been activated by its specific enzyme, and brings it to a specific site on the template that it "recognizes," presumably by some kind of complementarity.

Once each amino acid fits into place on the template, sequence is established, peptide bonds are formed, RNA is sloughed off, and protein zips off the template and takes on the shape necessary to carry out its function in the cell.

Summary of RNA Codons (code words)

Amino Acid	RNA Codons*					
Alanine	GCC	GCU		GCA	GCG	
Arginine	AGA	AGC	CGU	CGC	CGA	CGG
Asparagine	AAU	AAC				
Aspartic acid	GAU	GAC				
Cysteine	UGU	UGC				
Glutamic acid	GAA	GAG				
Glutamine	CAA	CAG				
Glycine	GGG	GGA		GGC	GGU	
Histidine	CAU	CAC				
Isoleucine	AUU	AUC		AUA		
Leucine	CUC	CUU		UUG	UUA	
Lysine	AAA	AAG				
Methionine	AUG					
Phenylalanine	UUU	UUC				
Proline	CCU	CCC		CCA		CCG
Serine	UCU	UCC		UCA		UCG
Threonine	ACA	ACU		ACC		ACG
Tryptophan	UGG					
Tyrosine	UAU	UAC				
Valine	GUU	GUC		GUG		GUA

* Codons are read from the 5′ to the 3′ end of the triplet.

Genetic Code.—DNA carries genetic information and gives its message to one form of RNA; this message is translated into protein structure, with the help of other forms of RNA. Although less is known about RNA structure than about either DNA or protein, it is known to be similar to DNA in that it is also a polymer containing four kinds of nucleotides. These nucleotides are similar to those of DNA except that *uracil* (U) replaces thymine as a pyrimidine base, and that the five-carbon sugar is ribose instead of a deoxyribose. The other components are the same as those in DNA. DNA may transmit its information to messenger RNA through complementarity between the bases, which by their alignment in both DNA and RNA contain the key to the code. The problem is how a four-letter code consisting of nucleotides can give a dictionary that is thought to contain the twenty amino acid "words."

Nirenberg and Matthaei, and Severo Ochoa's group, made an RNA, "poly-U" (polyuridylic acid), all of whose bases are uracil (UUUU...). When poly-U was put in a test tube with ingredients needed for protein synthesis, a long chain of repeating units of only one kind of amino acid, phenylalanine, was formed. The RNA code for phenylalanine is thus an unknown number of uracil bases. The favored number is three.

Since then, RNAs have been synthesized containing all possible combinations of A, G, C, and U. These experiments and others by Crick, Brenner, Khorana, and others lead to the following conclusions: (1) The code message is read in groups of three bases (triplets or trinucleotides). Each triplet is called a *codon*. (2) The message is read in nonoverlapping triplets, starting from a fixed point, probably at the 5′ end. (3) Most triplets are meaningful. They allow the gene to function; each triplet probably represents an amino acid. (4) Since four bases combine into groups of three, there is the possibility of $4 \times 4 \times 4$, or 64 triplets, more than enough to code the twenty amino acids found in protein.

In experiments with synthetic RNA, more than one triplet was found to code the same amino acid. A code such as the genetic code, in which more than one word can signify the same object, is called *degenerate*. This does not imply lack of specificity in protein structure; it simply means that more than one codon can direct the same amino acid to its specific site on the forming polypeptide chain. The accompanying table lists the RNA triplets that have been shown to code each of the twenty amino acids. This table is virtually complete, and experiments to refine it are in progress.

Present evidence suggests that the genetic code is universal—all species are now thought to utilize approximately the same code.

Internal Structure of the Gene.—The gene, defined by classical genetics, is the unit of function, recombination, and mutation; the chromosome is pictured as a series of beads (genes) on a string. The genetic code resides in the nucleotide sequence of a long DNA molecule; mutation and recombination can be detected within the borders of the genes of rapidly reproducing organisms. This changes the classical concept of the gene.

In recombination of genetic material, sexual reproduction brings about a great variety in the number of possible genotypes. Crossover results in further recombination of parts of chromosomes. Many microorganisms have alternatives to sexual reproduction that produce new combinations of genes in their progeny. Some bacteria mate, and there even is a form of gene recombination in bacteriophagic viruses.

Because great numbers of progeny can be obtained in bacterial and viral "crosses," it is possible to "dissect" the gene. The following ideas have emerged: (1) The gene locus may have more than one function. The function unit is called a *cistron*, and is perhaps a coded sequence of nucleotide pairs in a DNA molecule. This sequence carries information needed to specify the order of amino acids in a large polypeptide chain. (2) Mutation can occur in a cistron. The mutation unit may be as small as a single nucleotide pair. A *missense mutation* occurs when a codon specific to one amino acid replaces a codon specific to another amino acid. A *nonsense mutation* occurs when a codon such as UAA, which does not code for any amino acid, replaces a normal codon. Single nucleotide insertions or deletions can change the genetic message so that a mutation occurs. A *suppressor mutation* is one that reverses the effect of the original mutation. (3) Recombination can occur almost anywhere along the DNA molecule. The recombination unit is called a *recon*.

■**GENES AND CHROMOSOMES.**—Knowledge obtained from chemical, X-ray, and electron microscope studies of chromosomes is summarized as follows: (1) Chromosomes consist of a complex of DNA and protein, with a variable amount of RNA. (2) The nucleoprotein complex forms individual fibrous particles with a molecular weight of about 18,000,000 each, of which 8,000,000 is attributable to a single DNA molecule. (3) The chromosomes may consist of a *fibril bundle;* each fibril may have either two or four DNA double helices with associated protein or a large strand made up of 1,000 to 100,000 DNA molecules combined with proteins.

Biochemical Genetics.—Genes may control metabolic activity. In 1902 A. E. Garrod studied a human metabolic disease, alcaptonuria, a hereditary defect caused by a block in the normal series of metabolic reactions of the amino acid tyrosine. He called alcaptonuria and similar conditions "inborn errors of metabolism." This was the first example of recessive inheritance recognized in man, and since then other inborn metabolic diseases have been found. Each is controlled by one mutant gene locus.

Human beings are not usually subject to experimental manipulation and breeding. However, George W. Beadle and Edward L. Tatum found that by studying biochemical mutants in a mold, they could determine how genes act in producing such diseases and in normal metabolism. They formulated a hypothesis called the "one gene–one enzyme hypothesis." To paraphrase in terms of DNA: The DNA of each gene carries the information to specify the formation of a single protein—in most cases an enzyme that controls a specific chemical reaction in the organism. Today, based on very recent work, the hypothesis is modified to state that one gene specifies one polypeptide.

Beadle and Tatum chose an organism in which it was possible to prove unequivocally that single genes affect single enzymes—which govern single biochemical reactions. They looked for and found an organism simple enough to show the direct relationship between a gene and its product. The organism was *Neurospora.*

Neurospora is a fungus that usually reproduces asexually by means of spores called *conidia.* All nuclei are haploid; hence, every gene is expressed: no allele interacts or proves dominant. *Neurospora* also undergoes sexual reproduction, which requires fusion of two haploid nuclei to produce a diploid zygote. This occurs only through the union of two strains of opposite mating types. These strains are indistinguishable morphologically, but can be shown to differ. If strains of opposite mating types are grown together, characteristic black sexual spores (*ascospores*) are formed; strains of the same mating type, if grown together, do not form such spores.

Fungi like *Neurospora* grow on a simple synthetic medium, called a *minimal medium,* containing inorganic salts, sugar, and the vitamin biotin. The ordinary wild type of *Neurospora* thrives on this, synthesizing all the organic compounds of which protoplasm is made: amino acids, nucleic acids, fats, vitamins, etc.

Beadle and Tatum irradiated conidia to produce mutations and crossed them with a strain of the opposite mating type. The sexual spores produced were then isolated and each spore grown separately on a medium supplemented with one class of ingredients, either amino acids or vitamins. The germinated spores were tested on minimal medium. If they grew, they were the wild type, still able to synthesize all necessary growth factors. If unable to grow on minimal medium, but able to grow on minimal medium supplemented with, say, amino acids, they were deficient in the ability to synthesize one or more of these acids. Conidia of the new mutant were tested separately on minimal medium supplemented with each amino acid until the one that permitted growth was found. To test that this was truly a genetic defect, each mutant strain was crossed with a wild type and the offspring were tested; a ratio of one mutant to one wild type was found. A large number of mutants were isolated. When a single supplementary compound produced normal growth, it meant that a single gene was affected by the mutagenic treatment. Thus, Beadle and Tatum demonstrated that genes have a single primary function in the process of metabolism. It is, of course, assumed that the normal, or wild type, gene performs the function that is deficient in the mutant.

Essential compounds, the amino acids and vitamins, are synthesized by each cell through a series of reactions. If by mutation each of several strains loses the ability to synthesize a particular amino acid, such as arginine, would all the strains be mutants for the same gene? Would they all be affected at the same step in the synthesis of the amino acid? Could one deduce the order of the processes from a study of mutants and their nutritional requirements?

In the case of arginine, three classes of mutants, designated as *X,* *Y,* and *Z,* were found to grow in media as indicated in the following table. This study suggests that a linear series of reactions results in arginine synthesis: Prior substance X → ornithine → Y → citrulline → Z → arginine. Strain Z grows only on arginine and is blocked at the step lettered Z, conversion of citrulline to arginine; thus Z is unable to grow on any of the earlier compounds in the series, but will grow if arginine, which comes after the block, is supplied.

Mutant Strain	X	Y	Z
Arginine	+	+	+
Citrulline	+	+	–
Ornithine	+	–	–
No supplement	–	–	–

+ indicates growth.
– indicates no growth.

Strain Y can grow on either citrulline or arginine, each of which comes after the block; but it will not grow on ornithine or minimal medium with no supplement. Strain *X* will grow on all three; it is blocked at a step preceding ornithine production, and because of this blocked step is unable to grow on unsupplemented minimal medium. Each step is blocked because an enzyme needed to catalyze that specific step is missing or defective.

Crosses involving mutants of each strain showed that each step involves mutation of one specific gene. Since there are three distinct enzymes, this is an example of the workability of the one gene–one enzyme hypothesis.

The same principle extends to all living organisms, including man. The hereditary disease *galactosemia* is caused by a single recessive gene that controls and prevents production of an enzyme essential for conversion of milk sugar, galactose, into glucose, the sugar that the body can use. A child homozygous for the galactosemia gene develops cataracts, mental retardation, and other defects because of abnormal accumulation of galactose in the body. By eliminating milk from the diet of children with galactosemia, the disease is controlled.

There is a test for the presence of the enzyme needed to convert galactose to glucose in red blood cells, and thus glactosemic infants can be identified before damage is extensive. A homozygote has virtually no enzyme, a normal person has a high level of enzyme, and a heterozygote usually has a level about halfway between. This intermediate level appears to be sufficient for normal function. If galactosemia runs in the family, parents and newborn infants can be tested.

There are many examples of inherited metabolic diseases that show that in man, as in microorganisms, genes control biochemical reactions.

■GENES AND PROTEIN STRUCTURE.—The best evidence for the precise effect of mutation on protein structure comes from a study of the abnormal hemoglobins produced in certain inherited anemias of man, particularly of sickle cell anemia. In *sickle cell anemia,* the red blood cells form a sickle-shaped structure when oxygen concentration is low. The disease is serious, usually fatal in childhood. There is also a mild form of abnormality called *sickle cell trait.*

Both sickle cell trait and sickle cell anemia tend to occur in certain families originating in Central Africa, the central Mediterranean area, the Persian Gulf, and India. Since it is an inherited disease, those with sickle cell anemia are homozygous for a partially dominant gene, *S.* They are designated *SS.* People with sickle cell trait are heterozygotes, *AS,* with *A* the normal gene. Only if both parents are heterozygotes would a child have the disease. The cross would be as follows:

AS × *AS*

SS	*AS*	*AA*
anemia	*trait*	*normal*

Pauling and co-workers reported in 1949 that sickle cell anemia is a "molecular disease," since the large hemoglobin molecule of victims is different from that of normal people. More recent work by Ingram has defined the exact nature of the change. Hemoglobin, a globular protein with the molecular weight 66,200, consists of four polypeptide chains, two of one type (*alpha chains*) and two of another (*beta chains*). Each consists of about 140 amino acids.

Ingram broke the hemoglobin molecule into small fragments and analyzed each fragment for its amino acid sequence. The two alpha chains of normal hemoglobin A fragments were identical with those of hemoglobin S. One fragment from the beta chain showed a difference. This difference resided in only one amino acid in the fragment concerned.

Hemoglobin A	*Hemoglobin S*
Val	Val
His	His
Leu	Leu
Thr	Thr
Pro	Pro
GLU	VAL
Glu	Glu
Lys	Lys

The only difference is the substitution of valine for glutamic acid. This single amino acid substitution also changes other hemoglobin properties, making the red blood cells abnormal.

Mutation, then, can result in an altered protein, as in the hemoglobin, or in no recognizable product at all, as in galactosemia and other diseases caused by metabolic blocks. It is possible that in cases where no enzyme activity is found, an altered, nonfunctioning protein is produced that is undetectable by means available today.

Cytogenetics.—Information can be obtained about the hereditability of traits from the analysis of pedigrees.

Twin studies have been useful in determining the relationship between heredity and environment. There are fraternal twins and identical twins. Fraternal twins result from separate fertilization of two different eggs produced by the mother at the same time. Each zygote develops separately and is no more like the other than any two siblings (brothers and/or sisters). They can be of the same or opposite sex, depending on the sperm that fertilized each egg. Identical twins come from one fertilized egg that divides and separates into two parts at some stage in development. Since both children come from one egg, they have identical genes and must, of course, be of the same sex. Any difference between them is due to environmental influences. Studies of identical twins raised apart have been useful in assessing the relative roles of hereditary and environmental influences. Certain traits—blood groups, fingerprints, eye color—are inherited and not noticeably influenced by environment. Body build, height, weight, and I.Q. are examples of traits that have a very large hereditary component, but are also much influenced by environmental factors.

This fits with the concept of the norm, or range of reaction. Genotypes of living things react with the environment in which they develop. There is a range of potentiality as to what the final phenotypes might be, depending on the interactions. A child potentially tall because of his genotype might be stunted in growth by poor food, disease, etc. But a child whose genotype limits his height to 5½ feet could never be six feet tall, regardless of all the food, vitamins, and good health that might be provided for him.

Another aspect of human genetics has developed because of new cytological techniques for study of human chromosomes. *Cytogenetics* is the study of the role played by cell components, particularly chromosomes, in heredity. Certain congenital abnormalities are associated with abnormal numbers of chromosomes. The normal number of chromosomes is 46. Mongolism, an abnormality accompanied by mental retardation, is associated with 47 chromosomes, the extra chromosome being one of the smallest autosomes.

■GENES AND DEVELOPMENT.—A challenging genetic problem is the gene's role in the development of the embryo. Each cell in an embryo has the same chromosomal complement because it is derived from the original zygote by mitosis. Yet some cells become spindle-shaped muscle cells whose major protein is myosin; some, red blood cells whose major protein is hemoglobin; some, glandular cells that secrete digestive enzymes. Each cell type has the same genes, but is different from other cell types. Genes control formation of proteins found in each cell; but some genes are active in muscle cells and inactive in liver cells, and vice versa. Differentiation involves a continual interaction between the nucleus and the cytoplasm, with substances in the cytoplasm acting on the genes in the nucleus to repress or stimulate certain activities. The nature of this interaction is of interest.

There are two approaches to the study of how gene action is regulated. One may trace the development of an inherited defect back to the earliest stage of the embryo. There is a mutation in chickens known as the *creeper fowl,* which when homozygous (*CpCp*) usually kills the chick in the egg after three days of incubation. The mutant lags behind the normal in growth at as early as 1½ days, and the limb rudiments fail to grow, although in normal chicks they grow rapidly at this time.

Transplantation and tissue culture studies have revealed that most tissue taken from day-old embryos will grow normally for many days. Had they developed within a creeper embryo, they would have died with it. Mutant chick cells can also live normally if the environment supplies something that could be missing in their usual environment.

Heart tissue from early creeper chicks is an exception. It will not grow normally in tissue culture, regardless of what is supplied. Thus, the gene *Cp* in some way produces a substance that affects the development of one embryonic organ, the heart. This leads to a defective circulatory system that cannot distribute food and oxygen to normal cells of the developing chick. Limb buds, which normally grow rapidly at 1½ days, are most in need of food and oxygen and are affected. The defect spreads and the chick dies.

A partial verification of this hypothesis comes from the following experiments. Limb buds from normal embryos grown in tissue culture with too few nutrients show many characteristics of creeper limb rudiments. Also, if certain chemicals that suppress normal metabolism are added to nutritive tissue culture media in which early limb rudiments are growing, development is suppressed in a manner that mimics the creeper phenotype.

Another approach is to start with the genes themselves in a less complex system, usually a microorganism, to determine how gene action is regulated. The method of regulation discovered may provide a model for gene regulation in differentiation. Jacob and Monod in Paris studied the bacillus *Escherichia coli* and its ability to metabolize the sugar lactose. They found two kinds of genes: —those that specify, via messenger RNA, the protein's structure and those that regulate the time and rate of activity of the structural genes. A regulator gene produces a repressor which enters the cytoplasm and inhibits the structural gene. When there are specific molecules present in the cell that combine with the repressor and thus inactivates it, the enzyme can be made. One such molecule is the material on which the enzyme acts, called its *substrate.* This control prevents production of unnecessary protein and thus has a selective advantage.

The particular system the French scientists studied in *E. coli* consists of four genes, three of which are closely linked to form a so-called *operon,* while the fourth is an unlinked regulator gene. The operon has two structural genes, one which controls the synthesis of an enzyme, called a *permease,* which permits the sugar to enter the cell. The second gene controls the synthesis of another enzyme, *beta-galactosidase,* which changes the lactose to simpler sugars. The third, an operator gene, does not control the enzyme structure; it coordinates the activity of the structural genes as follows: The unlinked regulator gene constantly produces a repressor substance (possibly an RNA) that keeps the operator inert. As soon as lactose is present in the cellular environment, the repressor combines with some of it and is no longer able to inhibit the operator gene. This gene starts a chain reaction along the operon, diagramed below; and the two structural genes produce their enzymes, permitting more lactose to enter the cell and to be metabolized to simpler sugars.

operator gene	*beta-galactosidase gene*	*permease gene*
—o—	—o—	—o—

This is an example of coordination of gene action and interaction between genes and substances within the cytoplasm, as well as control of the movement of substances through the cell membrane from the extracellular environment, all influencing one another. Such actions and interactions must take place in a regular sequence of time and space to produce each stage of the embryo until the fully developed organism is formed.

CYTOPLASMIC INHERITANCE.—The primary mechanism of heredity is the self-duplicating gene of the chromosome. There are also cases of hereditary factors carried within the cytoplasm. Self-duplicating cytoplasmic factors are detected by a test in sexually reproducing organisms. Both egg and sperm contribute equal amounts of nuclear material to the zygote. The egg contributes the cytoplasm as well as the other haploid nucleus.

If chromosomal genes are involved, no difference will be observed in offspring from reciprocal crosses. However, if a cytoplasmic factor is involved, it will be inherited through the cytoplasmic donor, the egg. As a result, there will be a difference between offspring from reciprocal crosses. The clearest examples are in plants and involve *plastids,* the small self-reproducing bodies that carry chlorophyll. This type of inheritance accounts for many of the green-and-white spotted leaves on ornamental plants.

Other small bodies in the cytoplasm, including the *centrosomes, mitochondria,* viruses, and virus-like bodies, are thought to be self-duplicating. In addition to green plants, maternal inheritance has been found in mice, *Drosophila* (fruit fly), *Paramecium,* some other one-celled animals, and molds such as *Neurospora* and yeast.

Genes and Evolution.—The theory of natural selection (developed by Charles Darwin) as the mechanism of organic evolution has been the integrating force common to all biology. It is of interest to observe how the mechanism has been applied to large populations. Consider, for example, the blood groups. The most common in the American white population is group O at 45 per cent; next is group A at 38.5 per cent; group B is 13 per cent; and AB is 3.5 per cent.

Blood groups are inherited in a Mendelian fashion. One might question how mutations caused by radiation or chemicals would affect the frequency of the blood group genes, or whether the same frequency occurs in different geographic locations, or whether the frequency is changing with time. These questions belong to the realm of population genetics. Since organic evolution involves the changes in gene frequencies in populations in time and place, the branch of genetics called population genetics strives to explain the mechanism of evolution.

The principle of population genetics was developed independently in 1908 by the Englishman Hardy and the German Weinberg, and hence is known as the Hardy-Weinberg law. It states that relative gene frequencies remain constant from generation to generation in an infinitely large interbreeding population in which mating is at random, and in which there is no selection, migration, or mutation.

How, then, could evolution, which involves change in the genetic composition of populations take place? Obviously, both mutation and selection do occur, as well as some migration and isolation of small populations. These are the factors that influence evolution.

Let us consider mutation first. Mutations occur spontaneously, for unknown reasons, at predictable but low rates. Most mutations are harmful, because the genes already present in the populations have been the most successful survivors over millions of years that life has existed. Dominant lethal mutations are eliminated from the reservoir of genes in the population because the individuals in whom they occur die early in life. An occasional mutation is beneficial because it confers on the recipient a better chance to survive and pass the gene on to offspring. Such a gene is said to have a selective advantage.

Darwin observed that in most species of plants and animals the offspring produced are more numerous than their parents, yet most populations remain relatively stable in size. Also, many variations exist in nature, and most of these are inherited. As a result of the great numbers of offspring, competition exists; those best fitted by virtue of their variations will survive and pass on these variations to the next generation.

Beneficial genes remain in the gene pool, and dominant lethals are driven out. Unless they are disadvantageous to the heterozygote in the competition for food and mate, lethal recessive genes will be passed on, and not eliminated from the gene pool. If they are a disadvantage to the heterozygote, they will be eliminated very slowly over many generations. Many recessive harmful genes persist in populations because they are beneficial to the heterozygote. Sickle cell anemia is an example. The homozygote *SS* dies early, and the heterozygote *AS* has both normal and sickle cell hemoglobin. The trait is common in central Africa where a severe form of malaria exists. People with normal hemoglobin *AA* readily succumb to the disease, whereas heterozygotes *AS* are relatively resistant to this disease and have as much as a 25 per cent better chance of attaining adulthood than do the normal homozygotes.

Thus, a gene is maintained in the population even though individuals homozygous for it die before reaching productive age. This is a case of *balanced polymorphism,* which maintains alternative genotypes in a population by a balance between forces selecting for and against the gene. It is closely related to *heterosis,* or hybrid vigor, exemplified in hybrid corn in which a combination of genes makes it better than any homozygous line.

Although evolution is usually too slow a process for a person to observe in a lifetime, there are some examples of evolutionary change that have been observed recently. One is the development of DDT-resistant strains of insects. No doubt there were always some insects that could have survived DDT, but this was not a selective advantage in a DDT-free world. Once the chemical came into wide use, they were the ones who survived and reproduced, while the DDT-sensitive insects died. Thus, a change in environment (use of DDT) brought about a change in the characteristics of the insect world.

An important question to the population geneticist is: What is the genetic basis of the origin of species? He defines a single species as one in which members can crossbreed and produce fertile hybrids. New species arise through isolation of one group from another. Within each region of different environment, the population over many generations will become unique because of the selection of traits adaptively advantageous in that particular area. After a long time, the individual groups will have diverged to the extent that they no longer can interbreed. They are then separate species, and will remain separate even if they should occupy the same environment. This has happened many times in the course of evolution.

Progress in Genetics.—The significant advances made in the study of genetics are reflected by the number of Nobel Prize awards received by geneticists in recent years. Thomas Hunt Morgan received the award in 1934; Hermann Joseph Muller in 1946; George Wells Beadle, Edward Lawrie Tatum, and Joshua Lederberg in 1958; Arthur Kornberg and Severo Ochoa in 1959; and most recently, Francis Harry Compton Crick, James Watson, and Maurice Hugh Frederick Wilkins in 1962; François Jacob, André Lwoff, and Jacques Monod in 1965; and Marshall W. Nirenberg, H. Gobind Khorana, and Robert W. Holley in 1968.

—Selma Silagi

BIBLIOGRAPHY

CARSON, HAMPTON LAWRENCE. *Heredity and Human Life.* Columbia University Press, 1963.

DOBZHANSKY, THEODOSIUS. *Genetics and the Origin of Species.* Columbia University Press, 1951.

MCKUSICK, V. A. *Human Genetics.* Prentice-Hall, Inc., 1964.

SAGER, RUTH, and FRANCIS JOSEPH RYAN. *Cell Heredity.* John Wiley and Sons, Inc., 1961.

SRB, ADRIAN M., R. OWEN, and R. S. EDGAR. *General Genetics.* 2nd ed. W. H. Freeman and Company, 1965.

SUTTON, HARRY E. *An Introduction to Human Genetics.* Holt, Rinehart, and Winston, 1965.

WATSON, JAMES D. *Molecular Biology of the Gene.* W. A. Benjamin, Inc., 1965.

BIOLOGICAL RELATIONSHIPS

Understanding Living Organisms.

—In a world increasingly oriented toward technology it is sometimes difficult to remember our relationships to the earth and to other living things on it. Although it has become easy to make the false assumption that man is now independent of the old biological world, today we are almost as dependent upon our biological environment as our ancestors were thousands of years ago. We still require the same quantities of oxygen, water, and minerals from the soil that they did. We must still obtain the same quantity of energy from our food. The things that we eat must, as in the past, acquire this energy from sunlight and must still grow from the soil. We have not changed our tolerance to heat or cold, drought or humidity, radiation or pressure. Our advances have not come from any increased ability to adapt to a broader environment, but from learning how to maintain a suitable environment around us. Further technical advancement and survival in a world with an expanding population depend upon an understanding of ourselves as biological organisms and of our relationships with other organisms in the physical world.

In nature no living thing exists by itself. Each is part of an intricate structure composed of other living organisms and of the physical environment that encompasses them.

ECOLOGY.—The study of organisms in relation to each other and to their environment is known as *ecology*. One of the younger fields of biology, ecology has rapidly gained in importance; most of its developments have come since 1900 and more particularly since the 1940's. Ecology has been subdivided into two areas: *autecology*, concerned with the relation of the individual plant or animal to its environment; and *synecology*, concerned with the relation of populations of individuals to other groups and to their total environment.

Ecology has also been divided into the two fields of plant ecology and animal ecology, but one does not have to go very deeply into either field to realize that this separation is artificial. In studying the ecology of plants, an involvement with animals is inevitable. The study of animal ecology almost immediately entails the study of vegetation. There is fundamentally just one ecology, and it is concerned with the study of ecological systems, *ecosystems*, and the plants or animals of which these systems are composed.

Population, Community, Ecosystem.

—Fundamental to the study of ecology are three different concepts: population, community, and ecosystem. A *population* is the total of all the individuals of a given species occupying a particular area. Each individual is necessarily part of a population; everything about it is affected by its place or status in the population; and even as an outcast it exists in relationship to the whole population.

But a population does not exist alone. Each occurs as part of a *biotic community*, a community of living creatures. Any species forms only a part of the community, for it is dependent upon other species for food or shelter, or in turn provides food or shelter for another species. Similarly, the biotic community is not an isolated entity, but bears relationships to other communities and to its physical environment. It cannot exist without the soil or rock, water, atmosphere, and sunlight. Thus each biotic community forms, in combination with its physical environment, an ecosystem.

POPULATION.—Although defined as the sum of individuals, a population is something more than the total of its parts and contains qualities not possessed by any individual member. Important properties of a population that are of interest to both ecologist and census-taker are the number of individuals and their relationship to the area occupied. Knowledge of these characteristics permits the calculation of *density*, which is the number of individuals per unit of area. Also important is the *structure* of the population. This refers to the sex and age composition of the whole: the number of males in relation to females and the age distribution of each sex. Finally, the student of population is interested in population *dynamics*, changes over a period of time and the forces that influence these changes. These are brought about by birth rates, or *natality;* death rates, or *mortality;* and *movements* of individuals into or out of the population.

In order to understand these movements, the factors influencing birth and death rates, and the determinants of population density, the student of animal population may investigate questions similar to those regarding the economic and social status of human populations. In a study of the Norway rat, investigators found that rat populations were essentially self-limiting. Even with abundant food and shelter their numbers were restricted by social factors, particularly an aversion to crowding. Other animals sometimes show no such aversion and, if their numbers are not checked by outside forces, eat themselves out of house and home.

Biotic Potential.—Populations that are not self-limiting are eventually checked in their growth by the pressures of environment. Each species is capable, if unchecked by mortality, of a high rate of increase, known as its *biotic potential*. Small animals with rapid breeding rates have a higher biotic potential than do large animals. But even the largest and slowest breeders could, if entirely unchecked, overrun the earth. Mankind, with a low biotic potential, currently shows signs of doing just that. But no population is long immune to mortality. Inevitably factors in the environment will cause losses or inhibit the birth rate. Predators kill other animals; diseases and parasites decimate species; weather causes loss or checks gains from natality. If all other factors fail, the lack of food, water, or some other essential will limit population growth. The total of all agents in an environment that cause loss or arrest population growth is known as the *environmental resistance*.

A *stable population* is one in which the biotic potential and the environmental resistance are in balance. Interference with such a population can affect this stability and cause severe fluctuations. Some populations normally fluctuate in a regular and predictable manner. These are known as *cyclic populations*. The snowshoe hare and Canada lynx, for example, regularly reach a population peak at nine- to ten-year intervals. This is followed by a marked decline. Other populations are normally stable, but occasionally show a striking increase to "plague" proportions, followed by a major decrease. Such population changes are called *irruptions*. However, most populations are relatively stable. This indicates the presence of constant environmental resistance.

COMMUNITY.—Studies of any population soon lead to questions about the total community of which the population is a part. The community of the ecologist differs from the community of the sociologist in that the biotic community is always composed of more than one species: for example, populations of plants and of the animals that feed upon those plants. Communities vary in complexity. A simple community would be that of hardy lichens growing on an exposed rock surface along with the few associated organisms, mostly small to microscopic, that can find food and shelter among the lichens. The lichen itself is an example of the close ecological relationship between species. It is not a simple plant, but consists of two different kinds of plants living in close association and depending upon one another: a green alga that manufactures food from sunlight, water, and atmospheric gases; and a colorless fungus that shelters and anchors the algae and in turn receives food from the algal cells. Such a close association and mutual dependence between species is an example of a relationship known as *symbiosis* or *mutualism*.

At the other extreme of complexity from the simple lichen community is the tropical rain forest. Here the growing conditions are so nearly ideal for plants that hundreds of different species of trees sometimes occur within a small area. Associated with the trees are an even greater variety of other plants, including *epiphytes* (plants such as the orchids that grow high on tree trunks) and giant vines, or *lianas*, that also depend upon the trees for their support. Finding food and shelter in this mass of vegetation are a greater variety of birds, insects,

SIMPLIFIED ECOSYSTEM with its five basic components: an energy source (the sun), consumer organisms (animals designated *A, B, C,* and *D*), producer organisms (plants *E, F,* and *G*), reducer organisms (remains *H* and *I*), and abiotic, or nonliving, chemicals (*J, K,* and *L*). Reducer organisms return these abiotic chemicals to the soil or the sea.

and other forms of animal life than one can find in any comparable area.

Biotic communities do not spring suddenly into existence, but instead develop through a long process known as *biotic succession.* Succession, in its primary form, occurs in areas that have not previously supported life: bare rocks or newly formed lakes or ponds. The first invaders of such areas are always the more hardy plants and animals. They change the environment so that it can be occupied in turn by more demanding species. A predictable series of changes which lead to greater complexity usually occurs. The soil is further developed and becomes able to support a greater variety of life. Eventually a relatively stable community, in balance with the prevailing climate and adjusted to a mature soil, occupies the area until some disturbance destroys it. When this occurs, the process of succession begins again. This secondary succession may have fewer stages than a primary succession and usually resembles a primary succession's later stages. The relatively stable community resulting from a succession is known as a *climax community.*

■ **ECOSYSTEM.**—No biotic community exists apart from its physical environment. Each depends upon sunlight to provide energy, soil minerals, water, and atmospheric gases. Each is influenced by all the physical and chemical forces that characterize the area in which it is found. Since the interrelationship between the living portion of a community and the nonliving (*abiotic*) environment is so intricate that the two are virtually inseparable, it is necessary to consider them together as an ecological system. The ecosystem therefore becomes the fundamental unit of study for the ecologist. Ecosystems, like the communities that comprise them, can be simple or complex. However, even the most simple artificial system, set up in a laboratory test tube, often reveals complexities that require detailed study. Natural ecosystems may seem to defy any complete analysis or understanding.

Any ecosystem has five basic components: (1) *Energy,* usually derived from sunlight, but rarely and in small quantities from other sources. This moves through the ecosystem along pathways known as food chains, which are described below. (2) *Abiotic chemicals,* including soil minerals, water, and atmospheric gases. (3) *Producer organisms,* usually green plants capable of capturing sunlight energy through the process of photosynthesis. They utilize the energy to construct the organic chemical compounds which form the plant body, or they store it in energy bonds and link the various atoms or molecules in these organic compounds. (4) *Consumer organisms,* such as some colorless, nongreen plants, and all animals in the community. Consumers do not obtain their energy directly, but acquire it secondhand from the sunlight energy originally stored in green plants. All animals are completely dependent upon the producers for energy and for the chemicals that they require for nutrition. Consumers are subdivided into two categories: *primary consumers,* or *herbivores,* that feed directly upon plants; and *secondary consumers,* or *carnivores,* that feed mainly on other animals and thus receive their energy or food chemicals after they have been processed through two other kinds of organisms. (5) *Reducer organisms,* mainly bacteria and fungi that decay and decompose the bodies of dead plants and animals. These organisms feed upon the plants' complex chemical compounds and in turn release simpler compounds. Through this process mineral materials that can be picked up and used once more by the roots of growing plants are eventually returned to the soil or water. Without such organisms an entire community would stagnate, choked by its own debris, and the fertility of the soil would be drained without being restored.

Concept of Niche.—Each of the many species within an ecosystem occupies a particular place in the environment. This place is known as the *ecologic niche* for the species and is not inhabited by any other group. In a broad sense the niche for a herbivore includes suitable green plants on which it can feed, and is influenced by the presence of secondary consumers that will in turn feed upon the herbivore. A herbivore such as the deer requires certain shrubs and herbs which are of a suitable height and contain essential nutritional elements in a palatable form. Drinking water should not be too far from the food supply, and both should be accessible. Shelter is also necessary so that the deer can escape from extreme heat or cold, avoid storms, and evade enemies. Salt licks to provide minerals lacking in food may also be essential. The presence and availability of these elements usually guarantee the deer's occupation of a place in the biotic community.

In turn, the deer's existence, along with other factors, will help create a niche for such other species as the mountain lion which feeds on the deer and the various parasites that depend upon the deer for food and shelter. Deer may also influence the vegetation and prevent the establishment of certain plants on which they feed too heavily, or alter the form or abundance of other plants.

Similar kinds of vegetation usually provide similar niches for animals. A tropical rain forest may harbor a leopard in Africa, a panther in Southeast Asia, and a jaguar in South America, each species occupying a similar niche. Grasslands everywhere provide niches for large, grazing herbivores. But the species of herbivore occupying the niche may vary from one continent to another.

Energy Flow.—The ancients worshipped the sun-god as the giver of life. Today's ecologists could have provided them with a much more complete justification for their religion than their high priest could have imagined. Until recently, all of the energy upon which life depended, and all of the power which made human civilization possible, came directly or indirectly from sunlight. With the discovery of atomic energy and the technology permitting its utilization, man has for the first time established a small degree of independence from solar energy. Complete independence, however, probably cannot be attained. The calories that sustain our work and maintain our bodies were originally sunlight calories. Heat given off by petroleum and coal runs our machinery. This heat was once trapped from solar energy by plants in the swamp forests and ocean waters. The water spinning the turbines in the hydroelectric dams was lifted from the oceans and transported to the streams by solar energy.

■**PHOTOSYNTHESIS.**—In ecosystems the source of energy is sunlight, and only green plants are equipped to utilize it. A few kinds of plants, the iron and sulfur bacteria, can exist without sunlight because they use energy stored in iron or sulfur compounds. But they do not contribute significant amounts of energy or chemicals to the earth's ecosystems. The mechanisms by which green plants use solar energy, known as *photosynthesis,* are extremely complex, and plant physiologists and biochemists have been unable as yet to work out all of the details. Suffice to say that the presence of a complex green compound, *chlorophyll,* permits the capturing of particles of sunlight energy and their storage in chemical bonds in the various parts of the plant. The simple sugar *glucose* is one of the first of these storage compounds. Through further use of sunlight energy, molecules of glucose are broken down and linked with other chemicals. This results in the formation of the various carbohydrates, proteins, vitamins, and other substances that constitute the body of a plant.

During photosynthesis two chemical compounds, carbon dioxide from the air and water from the soil, are combined into simple sugars. In the process, oxygen is released back into the atmosphere. The presence of this gas in the earth's atmosphere is believed to be a contribution from the past generations of plants. Without green plants or some other means of restoring atmospheric oxygen, the continued respiration by animals would eventually exhaust the supply of oxygen on which we all depend.

■**LOSS OF ENERGY.**—Green plants are the only organisms capable of storing large amounts of solar energy. Man has learned various ways of making direct use of this energy, but he has not yet devised effective ways of storing it in quantity. However, photosynthesis is not an efficient process. It has been calculated that only about 1 per cent of the total solar energy reaching the earth is actually fixed and stored by plants. The rest is lost, either because it is in wavelengths of light that plants cannot use; because it is reflected from the surface of plants or from bare soil, rock, or water; or because it is dissipated in the form of heat. Nevertheless, the total quantity of solar energy reaching the earth is so large that the 1 per cent remaining is more than adequate to maintain terrestrial life.

Inefficiency is also apparent in the step from plants to herbivorous animals. The energy stored within plant bodies cannot be transferred to animal tissues without loss. Some remains in the indigestible residue of plants; some is lost as heat generated in the process of digestion; and other fractions are lost during various metabolic processes in the animal's body. At most, 20 per cent of the energy is stored in the body tissues of herbivores. A diminished amount of energy is thus available to the animals which feed on them. Further energy is lost in eating, digesting, and metabolizing the energy stored in a herbivore's body. Of the total herbivore supply of energy (which could be determined by burning tissues in a calorimeter), only a quarter or less will end up as energy stored in the body of a carnivore and available, therefore, to any creature that feeds upon carnivores.

It is obvious that this process cannot go on for long because one cannot have an unending chain of organisms feeding upon one another. Energy follows a one-way path through the ecosystem, with the initial supply rapidly dwindling as it passes from one organism to the next. In order for the system to function, energy must be supplied continually at the green plant end of the chain.

Energy relationships within an ecosystem illustrate the operation of the *second law of thermodynamics.* This law states that in any transfer of energy some energy is lost to the system and dispersed in a degraded form no longer capable of doing work. The various levels through which energy is transferred in a community are known as *trophic levels.* Producers, primary and secondary consumers, and reducers represent trophic levels. Food chains are the pathways over which energy is transferred from one organism to another.

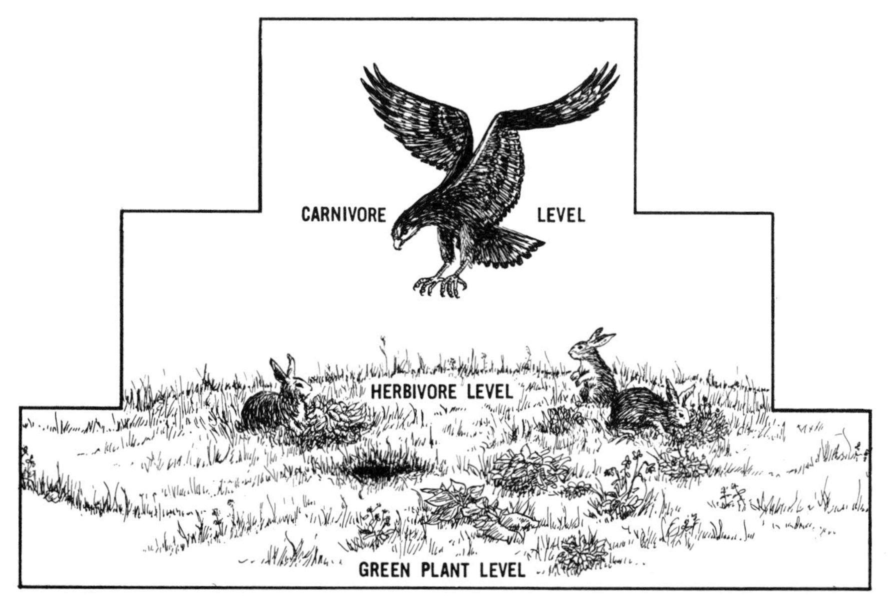

BIOTIC PYRAMID, which is based upon the diminishing quantities at each stage of the food chain, is illustrated by the large number of individuals at the green-plant level, fewer rabbits at the herbivore level, and the solitary eagle at the carnivore level.

Food Chains.—A simple predator food chain can be represented by the grass-steer-man linkage, where grass, steer, and man typify separate trophic levels and links. It is possible to have a longer food chain of this type. In a pond, for example, microscopic green algae are fed upon by small, floating animals (*zooplankton*). These in turn are fed upon by aquatic insects that provide food for small carnivorous fish. These small fish may in turn support a population of such large fish as bass or pike. Because of the energy relationships involved, it is rare to have more than five or six links in such a chain. In addition to predator food chains, other food chains go from large animals down through small. There are also food chains composed of reducer organisms which are involved in the breakdown of dead plant or animal tissues. Food chains are difficult to isolate in natural ecosystems because they are usually intertwined into complex *food webs.* Besides feeding a steer, a green plant furnishes food for a variety of small animals (including insects and microorganisms) which are then eaten by other species. Hence it is difficult to unravel the chains and webs in any complex community.

BIOTIC PYRAMIDS.—The necessary loss of energy between links in each food chain has a direct effect upon the number of organisms that can be supported at any trophic level. Thus the number of green plants upon which deer will feed is always greater than the number of deer that will be supported by them. The number of deer is in turn always greater than the number of mountain lions that feed upon them. These relationships can be diagramed in the form of *biotic pyramids,* which may illustrate either the number of organisms, the total weight of organisms, or the calories of energy stored in each layer of organisms. In a pyramid of numbers there will be more green plants than herbivores supported by them, and more herbivores than carnivores. Therefore the pyramid will show a broad base of plants and a narrow apex of carnivores. The picture would be similar if the relative weights were charted. It would take about 12,000 pounds of range forage to support a 1,000-pound steer for a year, and the steer could be converted into beef and support a 170-pound man.

Man, an omnivorous creature, can support himself upon either a predominantly plant or animal diet. If he acts as a herbivore, or *vegetarian,* more food energy is available to him because energy is not lost in the transfer through another herbivore. Less food, but of higher nutritional quality, is available when man functions as a carnivore. This is not just a matter of theoretical interest. In order to feed the mass of people in an overpopulated country such as China, the waste of energy involved in feeding cereal grains to domestic animals must be eliminated. People must consume food plants directly, and dietary quality must be sacrificed to provide the calories needed to sustain life.

Chemical Cycles.—Besides aiding the flow of energy, food chains provide pathways for the chemical materials required by the body tissues of plants and animals. Some of these chemical materials enter the soil when the rocks in which they originate are decomposed. Others are washed away to ponds and eventually come to rest in the oceans. From any of these substrates—soil, fresh water, or the sea—these minerals can be taken up by green plants and introduced into food chains. Unlike energy flow, however, the flow of chemical materials is not one-way, but circular. The same atom or molecule is used again and again. Moving from plant to animal, it is returned to the soil only to be taken up once more by some other plant. It is likely that the calcium and phosphorus in the bones of a living man were once part of the bones of a now extinct animal. These same elements doubtlessly passed through countless generations of prairie plants, antelopes, buffaloes, and wolves before being taken up by a wheat plant and later ground into flour.

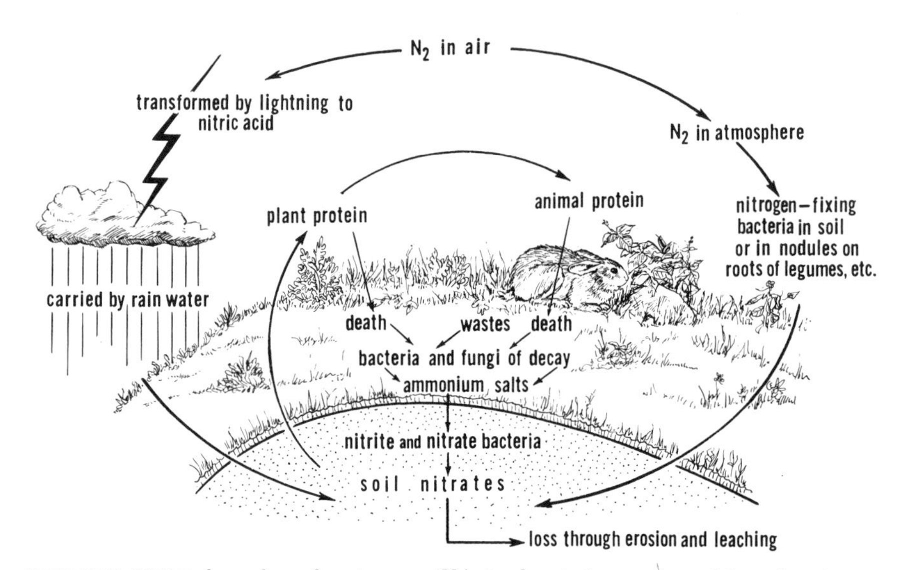

NITROGEN CYCLE shows how the nitrogen (N_2) in the air is transformed into the nitrates utilized by living organisms and then returned to the soil for further use. Nitrogen is essential to life because it is an integral part of the protein material present in living cells.

NITROGEN CYCLE.—One of the mineral pathways which have been thoroughly studied is the cycle through which nitrogen becomes available to living creatures. Nitrogen is essential for life because it forms an integral part of the proteins which must be present in each living cell. It is also one of the more common elements on earth and constitutes nearly 80 per cent of the atmosphere. Atmospheric nitrogen, however, is a relatively inert gas and does not combine readily with other elements. It cannot be used directly by most plants or animals, but must first be oxidized and converted to a nitrate. This conversion occurs when lightning ionizes atmospheric nitrogen and permits the gases to combine. The resultant nitrate dissolves and becomes dilute nitric acid, which enters the soil in rainwater.

Much of the soil nitrogen, however, is formed by the action of certain kinds of bacteria called *nitrogen fixers.* These live either free in the soil or in nodules found on the roots of legumes such as beans, peas, and alfalfa. Such bacteria can take nitrogen from the soil and convert it into nitrates. Plant roots then absorb these nitrates, and the plants combine them with other materials to form plant proteins. These in turn may be eaten by animals and reconverted into animal protein. From the animal they may pass back to the soil in the form of urea or other wastes. When the animal dies, its proteins are attacked by bacteria which break them down into simple nitrogen compounds such as ammonia or ammonium salts. These compounds are used by certain bacteria and oxidized once more into nitrates.

OTHER CYCLES.—Cycles similar to the nitrogen cycle have been traced for various other elements. Since the chemicals required for life are numerous, and the supply in the soil is not inexhaustible, there must be a continuous turnover of these materials if an area is to continue to grow living things and support living animals. In some complex biotic communities, such as dense, luxuriant forests, a high percentage of the nutrients in the ecosystem may be within the bodies of the plants and animals. When soil nutrients are scarce, new growth depends upon the decay of dead plants and animals. Organisms such as earthworms process great amounts of plant litter through their bodies. Their actions accelerate decomposition and make available the materials necessary for new growth. Caterpillars that feed on leaves and add their excrement to the soil similarly hasten the rate of chemical turnover.

WATER CYCLE.—Water, which is essential to life, is stored for the most part in the oceans. Transferred through the atmosphere, it reaches the vegetation and soil as rainwater. Not all of it becomes available to living things, however, for some accumulates on the surface of the ground and returns to the atmosphere through evaporation. Of the water which enters the earth, much moves through the soil and runs off through underground channels. In heavy downpours, or when the soil is soaked, much water may run off the surface and again be lost to the ecosystem. Some that enters the soil becomes closely bound to soil compounds and unavailable to plants. The rest is held as a soil solution which provides not only the water but also the dissolved chemicals necessary for plant life. The solution enters the plant roots and is drawn up by the leaves to be lost through their pores in the process known as *transpiration.* Therefore only some of it enters the plant cells and becomes part of their living protoplasm. From plants the water is transferred to animals, which may also obtain a supply directly from runoff held for a time in streams or pools or from underground sources

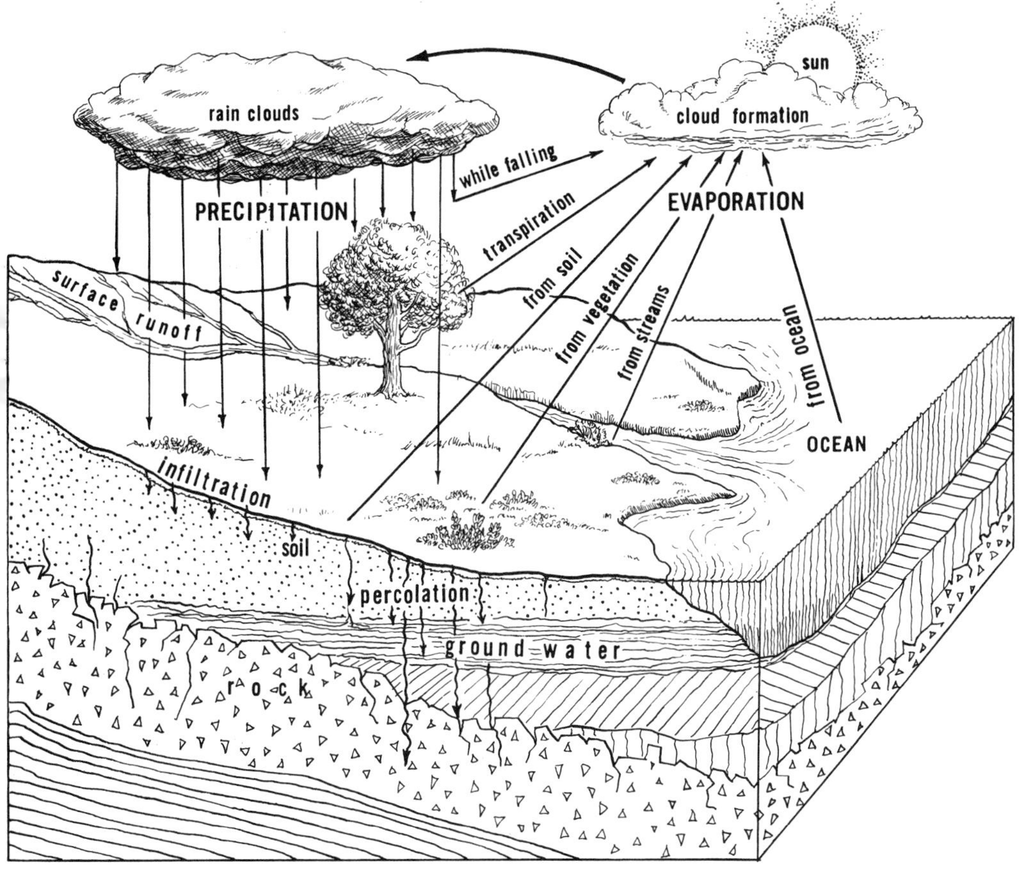

HYDROLOGIC CYCLE is composed of (1) evaporation of water from the surface of the earth to form water-vapor clouds, (2) precipitation of the water in the form of rain and snow back to the surface, and (3) infiltration and percolation of the water into the ground.

that reach the surface as springs. Eventually, however, all of the water used by plants and animals returns to the soil or is lost directly into the atmosphere. In either event it reenters the complex cycle.

Varieties of Ecosystems.—Because their distribution is determined largely by climate and topography, the kinds of ecosystems vary greatly from one part of the world to another. Classifications of ecosystems vary according to the emphasis that ecologists place upon their distinguishing features. Perhaps the most useful broad classification of terrestrial ecosystems is the *biome* system, which recognizes certain major natural communities distributed over the world in accordance with the occurrence of the major types of climate. Climate, vegetation, and animal life are so closely related that in the past, when meteorological records were scarce, geographers mapped the boundaries of climatic regions according to the occurrence of major changes in the vegetation.

■**DISTRIBUTION OF BIOMES.**—If you were to fly down the western coast of North America, starting at Point Barrow, Alaska, you would observe the sequence of biomes that extends between the Arctic and the tropics. First is an extensive region of treeless ground covered with low-growing vegetation. These arctic barren grounds are known as the *tundra*. In the vicinity of the Alaskan peninsula are the first fringes of dark spruce forest, the *taiga*. Next, across the peninsula, beyond the open spruce, are the northward reaches of the *temperate rain forest*. This is a tall, dense forest of Sitka spruce, cedar, and hemlock that continues, with some changes in species, all the way south along the coast to the San Francisco Bay region.

North of San Francisco the forest is broken; and to the south it is replaced by a lower-growing woodland of evergreen oaks or, more frequently, by the dense brush known as *chaparral*. Chaparral and woodland dominate as far south as Baja California, where they are replaced by open desert vegetation. Still farther south, on the mainland of Mexico, the desert vegetation is supplanted by dry tropical scrub and woodland. In Central America this gives way to the dense, luxuriant rain forests of the humid tropics. A continuation of the journey south along the western coast of South America would reveal a similar pattern of vegetation. However, the biomes would appear in a reverse order. Rain forest would give way to woodland and scrub, these regions to desert. In central Chile chaparral would replace the desert, and still farther south would be a dense temperate rain forest not unlike that of the Pacific Northwest. These changes in vegetation represent the major biomes of the continental west coasts of the world and the principal climatic regions of these coasts. Along the eastern coasts of continents some biomes will be different because of different climatic influences.

Ecologists classify the biomes of the regions of the world in a number of diverse ways. All of them, however, recognize the major divisions described below:

■**TUNDRA.**—The tundra characterizes regions of arctic climate with long, cold winters when the sun hardly appears above the horizon, short summer growing seasons of perpetual daylight, and relatively little precipitation. The soil is underlain in most places by permanently frozen ground, the *permafrost*, because summer temperatures are too low to thaw more than the surface layers of ground. Poor drainage causes boggy ground at lower elevations. The vegetation consists of dwarf shrubs and trees, matlike, broad-leaved herbs, grasses, sedges, and, in places, extensive stands of reindeer moss or lichen. This is the home of the caribou, reindeer, and musk ox; of the ptarmigan, white fox, arctic hare, and lemming. North of it lies the barren icefield of the Arctic Ocean. Above it on the mountains are bare rock, snowfields, and glaciers. Composed of those hardy species of plant and animal that have adapted to the extremes of climate, the tundra is the farthest extension of life to the north. It covers the northern fringe of Canada and Alaska and then extends in a band across northern Europe and Asia. Tundra is found also in the higher mountains, above timberline, extending south along mountain ranges into the temperate zone.

■**BOREAL AND MONTANE FORESTS.**—South of the tundra zone, or below it on the mountain ranges, is a forest dominated by needle-leaved evergreen conifers, mainly spruce and fir. This forest biome is the most extensive on the earth and covers much of Canada, Alaska, northern Europe, and Siberia. On its northern border the dark spruce trees are stunted and widely spaced where they merge with tundra. Farther south they grow in taller, denser stands and become mixed with fir, tamarack, or pine. In areas ravaged by fire the conifers are replaced by the broad-leaved birch and aspen. This biome occurs in regions of subarctic climate where winters are severe, but summer growing seasons are longer than in the tundra.

The summer heat is sufficient to prevent development of permafrost. Precipitation is higher than in the arctic region, averaging 15 to 30 inches a year. This biome is the home of the moose, snowshoe hare, northern grouse, goshawk, horned owl, red fox, and Canada lynx. Neither the tundra nor the boreal forest is found in the Southern Hemisphere, although some areas near the tip of South America are similar. This occurs because the southern continents do not have a large enough landmass close enough to the antarctic climatic regions to have the rigorous temperatures characteristic of the northern biomes. The climate of Antarctica is too extreme to support tundra.

TEMPERATE RAIN FOREST.—This forest is dominated in the Northern Hemisphere by a dense, luxuriant stand of tall conifers, usually spruce, cedar, hemlock, Douglas fir, and redwood. In the Southern Hemisphere a forest of similar appearance is dominated by the southern beech (*Nothofagus*) and such southern conifers as *Araucaria* and *Podocarpus*. For sheer volume of wood supported by each acre of land, the forests of coastal North America are unsurpassed. Some of the world's largest trees are found among the redwoods and Douglas firs of California and the Northwest. The reason for this vegetative abundance is the climate, which presents no extremes of cold or drought. Mild temperatures and high rainfall characterize the winters; the summers are cool and seldom without moisture. The growing conditions are thus second only to those of the humid tropical rain forest. The temperate rain forest supports no great mass of animal life, but provides a home for a great variety of smaller species. Characteristic of the North American region are the Roosevelt elk, mountain beaver, and black-tailed deer.

Because of the Gulf Stream influence, the climate necessary for a temperate rain forest does not occur in Europe. Neither Africa nor Australia reaches far enough south to have the necessary weather conditions. Similar climate and vegetation occur, however, on the western coast of New Zealand's South Island.

TEMPERATE DECIDUOUS FOREST.—In the eastern United States, western Europe, and northern China, the original vegetation was a forest of such broad-leaved trees as beech, maple, walnut, hickory, and oak. Man's influence has been felt more readily in these regions than in the others and has radically changed the environment. The climate of this biome is one of warm, wet summers and moderately cold, often snowy winters. A total rainfall of between 40 and 60 inches is adequate to support such a dense forest. Most of the trees adapt to the unfavorable winter growing conditions by shedding their leaves and becoming dormant. Unlike the conifers, which are mostly soft-wooded trees, the broad-leaved species are hardwoods and are considered among the most valued cabinet and furniture woods. In the United States this biome is the home of the white-tailed deer, wild turkey, gray squirrel, and cottontail rabbit.

MEDITERRANEAN FOREST AND SCRUB.—This biome exists in much of California, central Chile, the Cape of Good Hope region in South Africa, and southern Australia. It occurs most widely around the Mediterranean Sea in Europe, Asia, and Africa. The vegetation consists of broad-leaved evergreens and is called *sclerophyll* because the leaves are hard and waxy. Live oak, madroña, and laurel are most widespread in the California woodland, but over much of this biome brush (known as chaparral in California and maquis in Europe) has replaced the forest or woodland. The climate of this biome is one of warm, rainless summers and cool, moderately wet winters. Average rainfall is between 15 and 30 inches per year. In California the mule deer, gray fox, jackrabbit, and California quail are characteristic animals.

TROPICAL DECIDUOUS FOREST.—This is a grouping of similar biomes that includes the savanna forests, monsoon forests, thorn forests, and a variety of other tropical dry forest and scrub. They occur in tropical areas that are seasonally dry and stretch from the equatorial regions, where two wet and two dry seasons are normal, to the marginal tropics, where the summers are wet and the winters dry. Typically the trees and shrubs are leafless and the country barren during the dry season. In the wet season, however, the trees burst into bloom and leaf; and grasses and herbs cover the ground. In Africa these biomes form the big-game country and support great herds of antelope, zebra, buffalo, elephant, and other grazers and browsers, along with a variety of carnivores.

Because of fire, and to a lesser extent differences in soil, much of the area within these biomes is covered with *savanna*. This is an open interspersion of grassland with woody vegetation. Fires sweeping over these grassy regions kill the seedlings of invading shrubs or trees and keep the country open.

TROPICAL RAIN FOREST.—Several layers of trees of different heights characterize the luxuriant biome known

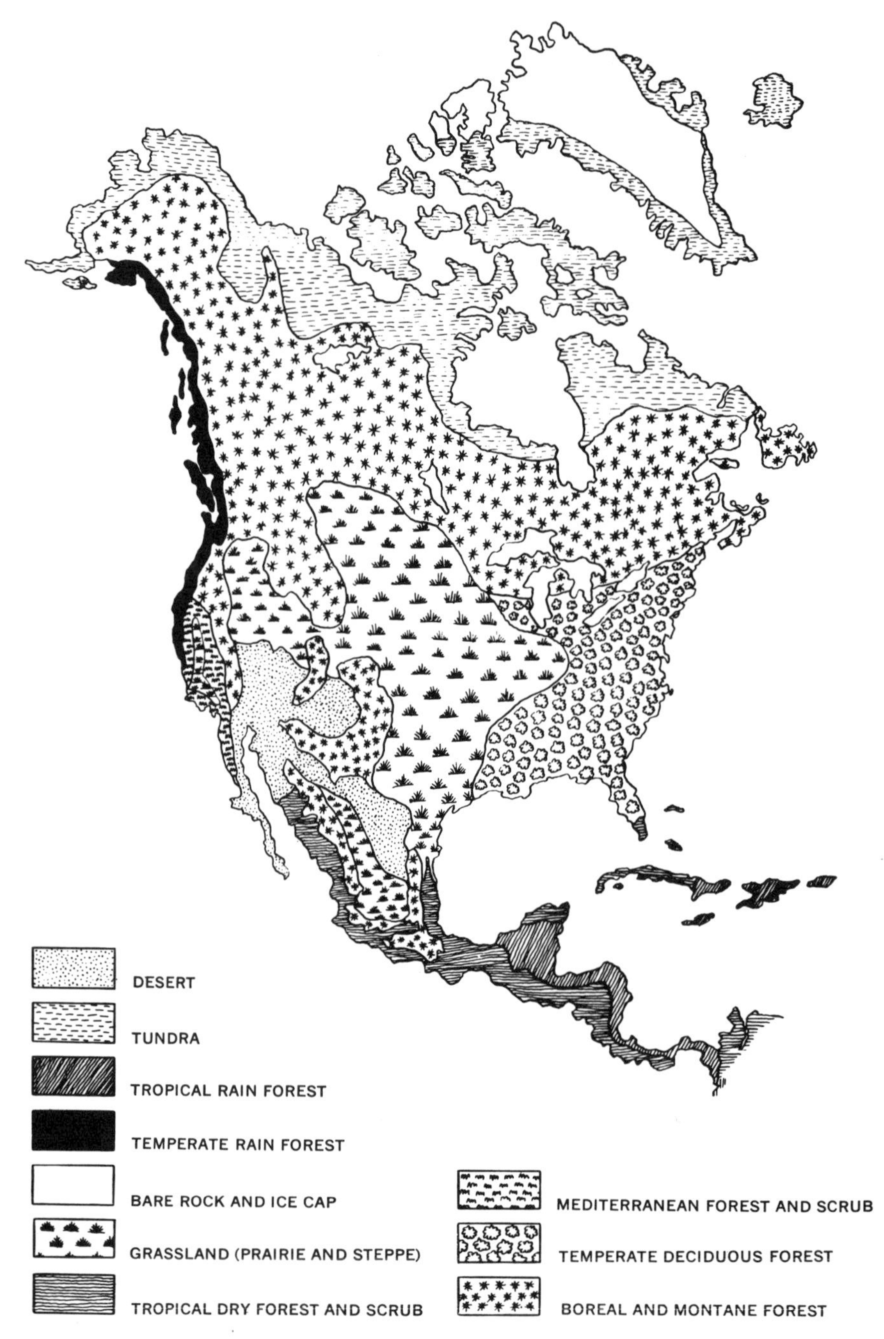

BIOMES OF NORTH AMERICA

as the *tropical rain forest.* The trees are broad-leaved and evergreen, and in the mature forest the floor is relatively open. In some places one can walk on a dense mat of decomposing litter through aisles formed by the trunks of the rain forest trees. Where the mature forest has been cut or otherwise disturbed, however, a dense, almost impenetrable jungle springs up. This in turn is replaced, after many years, by another mature forest. Numerous species of trees occur within this region. However, it is unusual to find more than a few of each type in any one area. This has prevented the development of forest industries to any great extent, since the commercially valuable trees are widely scattered. Along with the great variety of plants, the rain forest supports more birds, insects, and small, tree-dwelling mammals than does any other biome.

This biome exists only in permanently warm and humid tropical areas. It is developed extensively in tropical areas with year-round rainfall; but in modified form, as gallery forest, it follows the banks of the larger permanent streams in the tropics. On tropical mountains, forests derived from the lowland rain forests occur, but these have fewer species and fewer layers of vegetation. The tropical rain forest is not as extensive as was once believed. It is found in the Amazon basin of South America and in other lowland areas of South and Central America. It occurs in the Congo basin and along the western coast of Africa, and is developed also in Southeast Asia.

GRASSLANDS.—*Grasslands* exist on all continents, either as extensive areas dominated exclusively by grasses and other herbaceous plants or as grassy areas interspersed among woodlands and scrub. North American grasslands once extended in an unbroken mass from Illinois to the Rocky Mountains and westward through the intermountain region to the Central Valley of California and the Palouse region of Washington. In Eurasia they stretched from Hungary to the Pacific. Formerly, huge herds of bison, antelope, and elk roamed the plains, and the North American grasslands supported the greatest mass of wild animal life on the continent.

Climatically, these areas are best developed in the zones between moist forests and arid deserts. Since this region is seasonally dry, it is subject to fires which suppress the invasion of woody vegetation. Grasslands can be divided into two general categories: the more humid prairies, dominated by tall grasses; and the dry steppes, where short grasses are the prevailing cover. Partly as a result of overgrazing, small shrubs such as the sagebrush (*Artemisia*) and saltbush (*Atriplex*) invaded widely and have changed the appearance of the dry steppes. Such areas of shrub invasion often look like deserts.

DESERTS.—The warm, dry areas of the earth can be considered together as the desert biome or group of biomes. In these areas rainfall is seldom in excess of 5 inches per year and is so erratic that some places often go without rain for a long time. The most barren deserts are lifeless, but when undisturbed by human influences the desert usually supports an open scrub vegetation. In the American deserts the creosote bush (*Larrea*) covers great areas in uniform, open stands. Elsewhere various cactuses or thorny leguminous shrubs are dominant. Deserts support an interesting and varied fauna including the desert fox, kangaroo rat, and desert jackrabbits in North America; and the oryx, gazelle, and jerboa in Asia and Africa. Many of these animals can live without drinking water and obtain all of their water from their food.

LIFE ZONES.—Biomes occupy broad continental areas corresponding to the major climatic regions. In the higher mountains, however, it is possible to find in one small area the same series of biomes that would ordinarily be encountered in a journey over thousands of miles. The biomes in these mountain areas are distributed in altitudinally arranged belts known as *life zones.* In the western United States one can start at the base of a mountain range in the desert biome and pass successively through zones of grassland, Mediterranean scrub, and pine and fir forest until he finally reaches a tundralike zone above the timberline. Such life zone changes correspond to the decrease in temperature and increase in precipitation resulting from altitude changes.

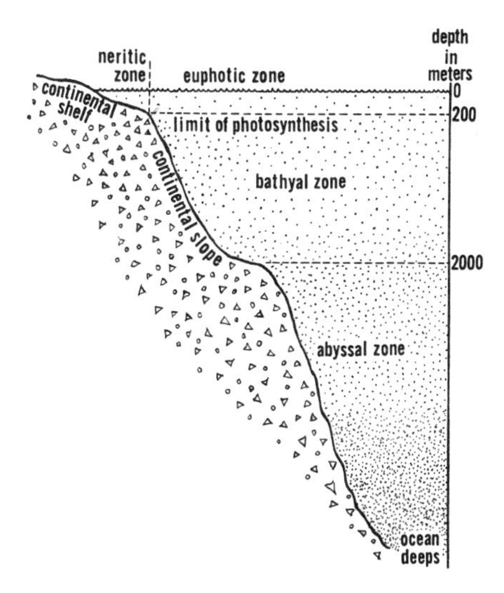

OCEAN is divided into the neritic zone, which lies above the continental shelf; the euphotic zone, in which all food is produced; and the bathyal and abyssal zones.

Aquatic Ecosystems.

Aquatic Ecosystems.—Most of the earth is covered with water, and most of the life on earth still finds its home in an aquatic environment. Life originated in the oceans and from them spread into fresh water, but relatively few of the many sea animals could adapt to the more rigorous conditions. Because the aquatic environment is much more uniform than that of dry land, it is difficult to divide it into separate biomes or easily recognizable ecosystems. Nevertheless, there are marked differences among the life of arctic, temperate, and tropical seas, and between various freshwater environments.

MARINE ENVIRONMENT.—Temperature and moisture are two major factors influencing the distribution of the major continental communities. In the oceans these are less important because of their tendency to be uniform and constant. On land, except beneath the canopies of dense forests, light is seldom a limiting factor. The distribution of life in the marine environment, however, is vitally determined by light. Much of the sunlight that strikes the face of the ocean is reflected back into the atmosphere. Only part of the total solar radiation penetrates the water; and of this, little reaches any great depth. To accomplish photosynthesis the green plants must live in the lighted surface layers of the water. Only here can they successfully produce the food and fix the energy that will support all other ocean life. Six hundred feet is the greatest depth at which this life-giving process can take place, and it usually occurs in more shallow water. Beyond that depth lie miles of water that are forever dark and unproductive. The lighted surface layer, termed the *euphotic zone,* produces all of the food and supports the greatest mass of marine life.

Since the organisms that live beneath the euphotic zone must depend for food upon materials that sink or are carried down from the surface, it was previously assumed that they were few in number. In recent decades, however, it has been found that this assumption is not necessarily true. Great layers of animal life have been located at depths well below the level of light penetration. Some of these layers consist of squid, which move to the surface to feed at night and submerge into the darkness during daylight. A great variety of other fish adjust to life in darkness in strange ways. In order to survive in their dark ocean homes they must feed either upon other organisms that move between the surface and the depths or upon materials that sink from above. Life can be found even in the great oceanic deeps, where animals must scavenge on the organic material that filters down through the upper layers of life. A diagram of life in the oceans, arranged according to mass, would be an inverted pyramid, with a broad base of living material near the surface and a narrow apex in the deeps.

The euphotic zone is not the only major life zone in the ocean. In the open ocean the zone below the euphotic, extending down to about 6,500 feet, is called the *bathyal region.* Still deeper lies the *abyssal region.* The principal way of life for plants and animals in the open ocean is termed *pelagic.* This is a free-swimming or free-floating existence, independent of contact with land. Of the pelagic types of life, those that have limited or no swimming ability are termed *plankton.* Their movements are dependent to a large degree upon the ocean currents, as opposed to those active swimmers that can move against or across currents.

Some animals in the open ocean region, however, live a *benthic* existence. This means that they are attached to, or moving over, the ocean floor. Around the edges of the main oceanic region is an extensive area called the *continental shelf.* This district represents the submerged portions of the continents. Life is usually more fruitful on a continental shelf than in the oceanic region beyond it. Here light can penetrate the water to support attached plants growing on the shelf's floor. These, in addition to the floating plants, constitute a great mass of productive plant material. Living a benthic existence on the continental shelf is also a much greater variety of attached or bottom-dwelling animal life than will be found on the main ocean floor. The water area on the continental shelf is classified as the *neritic zone.*

On its upper edge is the *intertidal zone,* that portion of the ocean with which most land dwellers are familiar. Here are found those plants and animals that can stand exposure to the air during periods of low tide. Although narrow by comparison with the great breadth of the oceans, the productive neritic zone occupies a considerable area. It follows the edges of all the continents, surrounds all of the islands, and occurs wherever there are submerged banks or reefs near the surface.

CHEMICAL NUTRIENTS.—Although light is the major factor limiting the distribution and amount of life in the ocean, chemical nutrients are also of great importance. Just as there are sterile soils on land that support little life, there are relatively sterile waters in the oceans. Those salts needed for plant nutrition are eroded from the land and carried in dissolved form down all of the streams and rivers to the oceans. Some come to rest in bays, in estuaries, or along the continental shelves. The salts that remain in the euphotic zone are available to green plants, but those that sink into the ocean depths are beyond the reach of food chains.

In a general sense there is no shortage of salt in the ocean. But not all of the salts necessary for nutrition are abundant. In particular, nitrates and phosphates are often in short supply. Those that are available are picked up by floating plants, *phytoplankton,* and pass from these to animals. When plants or animals die and their remains sink below the euphotic zone, the chemical nutrients in their bodies are removed from circulation. With a steady supply of nutrients flowing from the land and with an abundance of shallow, lighted water, it follows that the continental shelf areas are the most productive of life. Because mineral sources are unavailable, much of the oceanic euphotic zone is deficient in minerals and cannot support the mass of floating plankton on which larger marine organisms must feed.

The constant motion of the ocean waters encourages life in regions that would not ordinarily support it. Were it not for this motion, the required nutrients would have long ago sunk to the bottom and over the greater part of the seas the water would be lifeless. However, there are both deep and shallow currents in the ocean that keep the waters in constant movement and prevent stagnation. Deep currents can pick up those nutrients that have been lost to the surface waters. Where conditions are favorable, these deep currents can carry nutrient-rich water once more to the surface. Such favorable conditions exist along the western coasts of the major continents. Here the forces generated by the earth's rotation tend to push the warmer surface waters away from the land and to allow an *upwelling* of the cold waters from the depths. These fertile waters are then further distributed by surface currents. It is in these regions that some of the world's major fisheries are located.

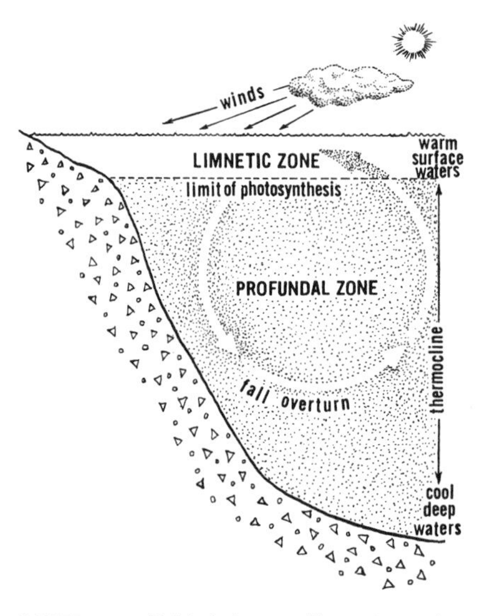

LAKES are divided into a limnetic and a profundal zone. The fall overturn replaces the nutrients that were depleted from the limnetic zone during photosynthesis.

FRESHWATER ENVIRONMENTS.—Since fresh water is less richly supplied with dissolved minerals than is the ocean, the availability of those minerals needed for plant and animal nutrition assumes a more controlling influence on the quantity and distribution of life. A common classification of freshwater environments separates the *oligotrophic* (low nutrient) waters from the *eutrophic* (high nutrient) waters. At one extreme is the glacial lake of the high mountains, fed by melting ice or snow, resting on a sterile substrate of granite; at the other, the farm pond in an area of rich soil, green with algae and teeming with animal life. Along with mineral content, the amount of dissolved oxygen in the water becomes limiting in fresh water more often than in the ocean. While photosynthesis occurs, shallow, eutrophic lakes are rich in oxygen; but in the winter light is screened out by a layer of ice and snow, and the abundant animal populations may exhaust the oxygen supply. Temperature also is much more variable in fresh water than in the ocean. The trout that can survive in a cold, oxygen-rich stream cannot live in the warm waters of sluggish streams or shallow lakes. Passing through several kinds of microorganisms and small invertebrates before ending in some large predatory fish, food chains in fresh water are also frequently complicated.

LAKES.—Deep lakes present many of the same problems for life that are encountered in the oceans. Below the limit of light penetration, called the *limnetic zone* in lakes, is a deeper layer in which no photosynthesis takes place. This is called the *profundal zone.* Fish dwelling in this deep zone depend upon the production of the limnetic zone for their food supply. Periodic mixing of the upper and lower layers of water is essential if their united capability for sustaining life is to be maintained. During the winter, for example, lake waters cool. Since water reaches its maximum density at 4° Centigrade, the water that cools to this temperature will sink to the bottom. As winter progresses, however, the surface layer of water cools below 4° C., expands, and at 0° C. becomes ice. At this stage the bottom waters are cut off from the surface layer; and a sharp temperature gradient, the *thermocline,* exists between the surface and the depths. Since no new oxygen can be added from the surface, animal life dwelling in the profundal zone can exhaust the oxygen supply during the winter.

In the spring, surface waters warm once more and soon reach the same temperature as the bottom waters. At this stage the thermocline disappears. Wind action induces mixing of the waters from the surface and the depths and thereby restores oxygen to the deeper waters. However, as summer approaches, the surface waters continue to warm and a new thermocline develops. A layer of light, warm water occupies the surface of the lake and may not mix with the deeper, cool waters. This condition can continue until declining autumn temperatures encourage another period of mixing. Following the period of mixing in the spring, the *spring overturn,* the surface waters usually experience a blooming of the algae and a consequent increase in animal life. This can exhaust the available nitrates, phosphates, and other essential materials in the surface layer. If this happens, the summer is a period of low productivity because the nutrients in the surface area cannot be replaced until the *fall overturn.*

Although the supply of nutrients controls the abundance of life in fresh water, there are some continental aquatic environments where an excess of salts presents a problem. Great Salt Lake, for example, is so much more saline than the oceans that a swimmer has difficulty in submerging. Mono Lake in California has such a high salt concentration that only two species of organisms, a salt-tolerant fly larva and a brine shrimp, can survive. Because drainage is interior rather than toward the oceans, high levels of salinity or alkalinity are usually found in most lakes in the intermountain region of the American West.

Biology and Conservation.—Since the early days of civilization, when man first domesticated plants and animals, human societies have exploited natural communities. Sometimes man has taken only the few products which he has needed and has left the community to restore itself through natural succession. Often, however, the wild community has been deformed and reshaped into a tamed community intended to serve the human welfare. At times, and in some places, people have managed to strike a balance with the forces of nature and have created stable ecosystems that include civilized man as a component. Too frequently, however, through failure to understand the biological forces involved, the deformation and change wrought in natural communities have initiated chains of consequences that in time have worked against human society. Many old civilizations are buried in jungle or have left their ruins standing windswept in what are now barren wastes. These ruins often reflect the failure to balance human demands against the biological necessities of ecosystems.

Today conservation movements attempt to counterbalance the destructive effects that formerly accompanied the exploitation of nature. *Conservation* is concerned primarily with maintaining a suitable world in which man can live. It is less involved with the quantity of production than with the quality of living. It seeks to preserve at least the remnants of unmodified natural communities and to balance man's demands upon the land and the capabilities of the land to produce; to substitute careful management of resources for unbridled, destructive use. Where it is involved with those resources that form a part of ecological systems, conservation depends upon a knowledge of biological relationships.

It is a recognized tenet of biological conservation that all living things grow, reproduce, and therefore show an annual growth. In a stable community this yearly increase represents a quantity of living material that can be safely harvested by man without destroying the basic resource. If it is not used by man, it will sooner or later die of natural causes, for no living thing can exist forever. Consequently, if the annual production of wood in a forest, brought about by tree growth, is balanced against the amount that is cut and converted into lumber, and if reasonable care is taken in removing this crop, the forest can remain permanently productive. Conversely, if too little is cut, old age, disease, insects, or other natural forces will eventually destroy the surplus. Balancing growth against harvest leads to a *sustained yield* of forest products. Sustained-yield management characterizes all modern systems of conservation-based forest management. The consequences of excessive harvesting and of careless handling of cut-over lands can be seen in the devastated hillsides of many parts of the world or in the scrubby growths of birch and aspen that have replaced the once magnificent pine forests of the Great Lakes states.

■PRESERVATION OF WILD ANIMALS.—A sustained-yield system of harvesting should characterize the commercial use of all living resources, including agricultural soils. Wild animal populations, for example, can be managed in this way. If the annual take by hunters is balanced against the annual replacement of adults by young, a game population can be permanently maintained. If an annual crop is not removed, the animals will die from natural causes; man's protection cannot make them eternal. In the United States game managers try to balance the annual increase of deer, quail, grouse, or rabbits against the annual demand from hunters. If game laws are tailored to allow a breeding stock to remain each year, the game populations will remain abundant. Game managers also attempt, through improving the vegetation or other aspects of the environment, to create a more suitable habitat for game and thus produce larger surpluses for the hunter.

In countries where hunting for sport is not extensive, as in Africa, it has been found that the annual crop of wild game can be taken for commercial purposes. The complex of wild grazers and browsers that forms a part of the biotic communities of Africa has been shown to produce more meat and other products of value, per acre, than can be produced when the same lands are used by domestic livestock. The Soviet Union has had similar experiences with some of its game animals. For example, the saiga antelope, an abundant game animal of the dry steppes and desert margins, is now managed for meat production. A harvest is taken each year by hunters; and the meat, hides, and other by-products enter the domestic economy.

It has become obvious with many kinds of wild animals that excessive protection can do as much harm as excessive hunting. A population of deer or elk, when completely protected, overbrowses and destroys the food plants on which it depends. Before man modified the natural scene, wild predators helped to remove the annual surplus and to keep populations in balance. But in order to protect his domestic animals, man has destroyed most of the wild predators, with the result that the wild herbivores are left without adequate checks on their biotic potential.

■DOMESTIC ANIMALS.—It is in the handling of domestic animals that man has done the most damage to his environment. A sparse human population maintaining great herds of cattle, sheep, or goats can destroy a vast region in a surprisingly short time. Failure to realize that range forage, like timber or wildlife, produces an annual surplus, and that only this annual surplus can safely be cropped, lies behind this devastation. The arid steppe and savanna regions are most susceptible to damage by domestic livestock. Here the plant production during a wet season or good rainfall year must hold the soil during the long, dry periods that follow. Because plant growth is usually slow, the annual surplus that can safely be cropped without injury to the grasses or shrubs is small. Such areas can sustain only light grazing.

If the land is overgrazed, the plants are destroyed; and the bare ground is subject to rapid erosion by wind or rain. When the layer of more fertile topsoil is lost, vegetation is slow to reoccupy the ground, and damage continues. Over wide areas of the globe the desert has expanded

U.S. DEPARTMENT OF AGRICULTURE

MAN'S ACTIVITIES tend to overbalance ecosystems, thus destroying their symbiotic existence. The dust bowl of the American Midwest was the end product of such an overbalancing.

into formerly productive lands following the impact of too many hoofs or grazing mouths of domestic livestock. The Sahara, the Arabian Desert, the deserts of Pakistan and of our American Southwest have all spread into lands that formerly were more productive. *Range management* is a relatively new field of conservation. It attempts to instill a knowledge of how best to harvest the annual forage crop from grasslands without damage to the growing stock of range plants.

Faulty management of farming lands has also resulted from a failure to understand biological necessities. Soils that developed as part of natural ecosystems cannot for long be separated from those forces that contributed to their development and guaranteed their stability. In the virgin prairie soil, structure was maintained by a dense network of plant roots. Erosion was checked by the perennial cover of growing plants. Fertility was held in balance by the annual return of dead plant and animal materials to the ground that had originally produced them.

When these prairie soils were first plowed, they were remarkable in both fertility and stability. With continued cultivation and annual planting of the same kind of crop, and with no effort to protect them from wind or rain, they began to deteriorate. Surface soil was lost to erosion. Structure was destroyed because the roots of wheat or corn failed to provide the mechanical or chemical action that was necessary to maintain it. Fertility was lost through the steady drain of nutrients into crops that were subsequently transported to distant markets.

Recent soil-conservation activities have attempted to repair this damage. The original soil-forming and soil-holding functions of natural vegetation have been replaced by techniques compatible with agricultural production. Crop rotation, cover-cropping, contour cultivation, mulching, and fertilizing all are part of sound, conservation-based agricultural practices today.

■FISHERIES.—The early exploitation of the resources of fresh water and the oceans was based on a misunderstanding of the abundance of aquatic organisms. Too often fishermen have assumed that new sea fisheries can always be found to replace those that have been exhausted. In fact, commercially valuable fisheries are restricted to the few areas where nutrient supplies permit an abundant production of plants. Far from being inexhaustible, such localized fisheries can readily be overfished if sufficient pressure is placed upon them. Once it is reduced to a low population, a fishery may be a long time recovering, for the ocean environment cannot yet be controlled; and natural losses are high. Steady exploitation of depleted fish populations prevents any recovery. For the most part, excluding a few highly valued species of marine life, overfishing has been a localized problem. The ocean still has great resources of living materials suitable for human food. But with the increasing demands of growing human populations, biologically sound management of these marine resources is essential if they are to remain productive.

■PRESERVATION OF NATURAL AREAS.— Since man is as dependent today upon the products of the lands and waters for his food as he was in ages past, it follows that much of the world must be used for the production of those things that people require. However, as more and more land comes into use, the value of the still wild, unmanaged parts of the world increases. Most nations have at least made gestures toward maintaining such natural areas in systems of national parks or wilderness reserves. However, no nation has gone far enough in this direction.

It is now clear that there is an acute need to preserve representative areas of all kinds of natural ecosystems, from humble tracts of bog or moor to vast areas of tropical rain forest. This is because it is important to have ecological check areas against which the progress or loss of the lands under management can be measured. In some parts of the world, devastation has been so complete that we do not know what the land could produce if it were left to recover. Some livestock owners have never seen an undamaged rangeland. They consistently settle for a lower productivity because they know no better. If natural grasslands of high productivity remained in each region, the range manager would better realize his goal.

Natural ecological systems contain the maximum variety and abundance of life that a climate and substrate will support. In development such ecosystems usually tend toward increased complexity. As each species moves in, it creates a niche for some new species that can feed or shelter upon it. Generally speaking, the more complex an ecosystem, the more stable it becomes and the less liable to extreme fluctuations in the abundance of any species. A herbivore that grows too numerous is fed upon by a variety of carnivores until its numbers are reduced. One that grows scarce is spared the pressure of predators that can turn to some other more abundant and readily available prey. Thus the entire system retains a natural balance.

Man, through his efforts at agriculture, pastoralism, logging, hunting, and fishing, tends to simplify complex ecosystems. The more they are simplified, the more likely they are to lose balance. A natural grassland is complex; a wheat field is simple. The wheat field provides an ideal environment for those things that prefer to live on, or feed upon, wheat—the wheat-destroying fungi and insect pests among them. Without natural enemies in this artificial system, these pests can increase to such proportions that they destroy the wheat crop. Similarly, an overgrazed range is simplified and supports fewer species of organisms than does a virgin rangeland. It is therefore out of balance. A species of rodent can increase to plague proportions and do serious damage, whereas such an increase would be prevented by natural checks in an undisturbed grassland community.

■CHEMICAL CONTROL OF INSECTS AND PESTS.—Where agricultural or pastoral peoples have been, they have changed and simplified natural communities. Hence, they have often created an ideal environment for those pests, diseases, parasites, and predators that are the worst enemies of their crops and herds. In their efforts to control these pests that follow them, they have resorted to hunting, trapping, burning, and other techniques. Such attempts at regulation have usually been unsuccessful. With the development of the chemical industry, however, new weapons have been added to the battle against pests in the form of chemical insecticides, fungicides, and herbicides. For a while great progress was made by the use of these materials. Agricultural and pastoral losses were reduced, and production soared. Recently, however, it has become apparent that the use of poisonous chemicals on the land can, unless carefully controlled, wipe out the very species of plants and animals that man most wants to preserve. In the long run, the unrestrained and uninterrupted employment of some of these poisons could pollute the land and its waters to such an extent as to imperil man's very existence.

The pollution of the human environment, not just with pesticides but with all of the wastes and by-products of man's activities, creates the greatest problem for biological conservation today. As human populations increase, it will become even more severe. Perhaps it will finally force us to observe the biological rules of order that should govern our actions on this planet. It is the hope of the ecologist, who has an interest in natural things, that man will yet develop what Aldo Leopold has called an "ecological conscience." This involves a recognition that man is still a part of, not an enemy of, nature, and that other living things have as much right to a place on this earth in the future as they had in the past.

—Raymond F. Dasmann

BIBLIOGRAPHY

Bates, Marston. *Animal Worlds.* Random House, Inc., 1963.

Bonner, John Tyler. *Cells and Societies.* Princeton University Press, 1955.

Carson, Rachel L. *Silent Spring.* Houghton Mifflin Co., 1962.

Clements, Edith S. *Adventures in Ecology: Half a Million Miles, from Mud to Macadam.* Pageant Press, 1960.

Dasmann, Raymond F. *Environmental Conservation.* John Wiley & Sons, Inc., 1959.

Odum, Eugene P., and Howard T. Odum. *Fundamentals of Ecology* 2nd ed., W. B. Saunders Co., 1959.

Wallace, Bruce. *Ecology.* Prentice-Hall, Inc., 1961.

ANIMAL LIFE

The branch of science that deals with animals is called Zoology. Although it deals with the structures and functions of animals, perhaps its best-known aspect is the orderly classification of animal life, from the simplest to the most complex.

Although it is easy to see that a bear is an animal and that a pine tree is a plant, some of the smaller animals and plants are not obviously members of their respective kingdoms. Most animals can be distinguished by movement; yet there are microscopic water plants that swim as freely as animals do.

All animals large enough to be seen with the naked eye obtain energy by eating plants or other animals. Surprisingly, a few microscopic animals are like green plants in that through the process of photosynthesis they can capture energy from sunlight and can use simple chemical compounds dissolved in water as food.

Thus, it is obvious that methods must be found to distinguish animals from plants and to separate one kind of animal from another. One way to do this is an analysis of food habits. Another way is by studying habitat, or where an animal lives. The most important factors, however, are structure and function—how an animal moves, digests its food, breathes, and reproduces.

FOOD HABITS. The food habits of an animal will give information concerning its structure and function, some of the animals related to it, and a general idea of the environment in which it lives. Sometimes a broad category of food habits will cover a wide variety of creatures.

Herbivores. Any animal that eats only vegetable matter is a *herbivore,* or plant-eater. Herbivores eat grasses, leaves, twigs, succulent plants, and other types of vegetation. The classification encompasses such different creatures as caterpillars and cows.

Carnivores. Animals that eat the flesh of other animals are called *carnivores,* or meat-eaters. Animals as different as lions and ladybird beetles are in this category. When a cat pounces on a mouse, kills it, and eats it, the cat becomes a predatory carnivore, or *predator.*

Animals such as the vulture and hyena are also carnivores, although they usually prefer to feed on dead animals; this makes them *scavengers.* Domestic animals also become scavengers at times, such as when they rummage through refuse.

Omnivores. The most familiar *omnivores,* creatures that eat both animal and vegetable matter, are man himself and the domestic pig. There are, however, less familiar omnivores that are far more numerous.

The *aquatic omnivores* subsist on food so small that they must strain it from the water. Clams and oysters do so by producing a current within the shell, filtering out the food, and expelling the water. Worms that burrow in the ocean floor build U-shaped tubes, then wiggle in order to draw water in at one end; after they have strained the food from the water, they force the water from the other end of the tube.

Even the giant whales, which may be as long as 110 feet, filter their food. They swim, mouth opened, until small crustaceans, plankton, and other types of food are caught between thin plates known as whalebone that hang down in the mouth cavity. Then they close their mouths and swallow the contents.

Symbionts. Animals that form a beneficial partnership with animals of some other kind or a similar partnership with a living plant are *symbionts.* If both partners in a symbiotic arrangement benefit equally, the relationship is *mutualistic.* If one benefits without harming the other, it is a *commensal* relationship; if one gains at the expense of the other, it is *parasitic.*

Many termites illustrate mutualism. They chew and swallow wood but cannot digest the wood fibers until they are predigested by minute animals that inhabit the termites' digestive tract. These minute animals could not obtain wood fibers without the termites; the termites could not utilize wood fibers if the minute animals did not first digest them.

The shark sucker, also called pilot fish or remora, has a commensal relation with sharks. It attaches itself to the shark by means of a suction disk and is carried from place to place to share the shark's food. The shark sucker detaches itself while the shark is feeding, eats, and reattaches itself, to be carried elsewhere. Occasionally a shark sucker will be found attached to a sea turtle or a small boat.

Most parasites are harmless unless they have become numerous. One chicken louse will cause a bird mild discomfort; hundreds will cause a bird to scratch itself constantly, weaken, and grow ill. Such parasites as lice, fleas, ticks, and mosquitoes are called *ectoparasites.*

There are also *endoparasites,* which are internal parasites. They include tapeworms, which inhabit the digestive tract; flukes, which inhabit the lungs; and malaria parasites, which attack the red cells in the blood.

Certain minute insects parasitize plants by producing chemicals that irritate the plants into forming unnatural swellings on the leaf or stem, producing deformed terminal buds. These *galls* provide a place for the insects to live while they suck sap from the plant. Each type has a distinctive shape and inner structure.

HABITAT. The oceans are the home of the greatest variety of animals. Near the sea's surface, sunlight penetrates and enables the small, drifting plants called *phytoplankton* to carry on photosynthesis. Swimming weakly through the phytoplankton and feeding on them are *zooplankton.* Larger animals, called *nekton,* swim strongly and feed on the smaller forms of life. The largest of the nekton are the whales, which may migrate from the Arctic to the Antarctic and back again in a period of a single year.

Deep-sea Dwellers. Many inhabitants of the dark ocean depths are scavengers, dependent on material that sinks from the surface; a few are predators. Even in the muddy ooze that covers the sea floor at the greatest depths, there are animals utilizing the organic materials in the ooze. Many of them eat the bacteria that decompose material that sinks from the surface. Some animals at the deepest levels produce their own light; its function is not certain.

Coastal Dwellers. Much animal life is found near the shores, where the seaweeds are larger and more plentiful because the water is rich in minerals washed from the land. Wave action incorporates air bubbles in the water, thus providing more oxygen for the animal life.

The main danger to coastal dwellers is that of being thrown ashore by the waves. Some, like the sea urchins and sea stars, attach themselves to rocks by means of suction disks. Others burrow, or they live only in areas protected from wave action.

Marsh Dwellers. Animals that live in salt marshes and river estuaries must be able to tolerate great fluctuations of the water's salt content—from very low after a heavy rain to very high after a long drought. This is also true of the plants and animals on which they feed.

Freshwater Dwellers. Fresh water, whether flowing or still, supports fewer forms of life than salt water because it contains fewer dissolved minerals. It is often muddy, however, from undissolved particles. These reduce the amount of light penetrating the water and thus reduce the amount of plant food available to animals.

Fresh waters change level rapidly during floods and droughts, thereby altering the habitat of many creatures. They also freeze over in the winter, thereby greatly reducing the available oxygen and forcing some creatures, such as frogs and turtles, to hibernate.

Land Dwellers. There are many types of habitats to be found on the land, and various types of animals live in each. In the soil there are, among others, earthworms, moles, and many insects. Such animals as bears and deer choose forests or their edges.

The "edge" may grade off into willow or alder swamp, or into grassy field or pasture. In either case, there will be plants for herbivores, small herbivores for medium-sized carnivores, and large herbivores, such as deer, for large carnivores, for example, pumas.

Open plains, covered chiefly by grasses, have inhabitants like antelope, prairie dogs, rabbits, prairie chickens, and vast numbers of grasshoppers and ants. Wolves, coyotes, badgers, and snakes formerly were the main carnivores, but the spread of civilization has greatly reduced their numbers. Summer droughts have also served to curb predators, as well as to keep the herbivores from overgrazing the range.

Water is so scarce in deserts that few animals live there; those that do are generally small and have bodies specially adapted to their habitat. Only a few insects can live on ice or in hot springs. Maggots of one or two kinds of flies have been found at the bottoms of shallow petroleum pools in oil fields, feeding on insects that happen to fall in.

Caves also have special animals that exist as external parasites on bats or eat the mold that grows on the bat droppings. Some of these animals prey on other cave creatures.

STRUCTURE AND FUNCTION. Any animal that is to continue to exist must at some time in its life perform the activities common to all animals: movement; food handling, that is, digestion, absorption, and excretion; respiration; coordination, both chemical and nervous; and reproduction, followed by the growth to maturity of new individuals able to carry on the same activities as the parents.

Structure. *Unicellular* (single-celled) marine animals are believed to have been the first form of life. Today there are about 30,000 species of animals that carry on all of their life processes within the one cell.

Most modern animals are *multicellular* (composed of many cells). Among these, the sponges are unique in that any cell can take over the function of any other cell. A sponge may evolve into a different kind of sponge, but it can never become any other kind of organism.

Function. All multicellular animals except sponges have their cells arranged in layers called *tissues;* each tissue is composed of cells with a definite structure and function. In the higher animals, the tissues are connected in *organ systems.*

A man is composed of organ systems, such as the digestive system; this system in turn is composed of such organs as the esophagus, stomach, intestines, and colon. The stomach is composed of a lining layer, muscle layers, and a covering layer; each tissue layer is composed of a definite type of cell. All multicellular animals have the same general types of tissues.

Contractile tissue, composed of tissues that shorten and lengthen, does the work of the body. This type of tissue forms either muscles moving the body or continuous sleeves around cavities, such as the digestive organs or the blood vessels.

Connective and *supporting tissue* is composed of cells that produce nonliving secretions between themselves. The tissue may be a solid mass in cartilage or bone; or tough strands in tendons, which connect bones; or it may be in ligaments, which connect muscle and bone. Connective tissues form the walls of capsules that hold the lubricant in each joint, as well as fine fibers resembling cobwebs that hold organs in place.

Epithelial tissue is composed of thin tilelike, cuboidal, or close-packed columnar cells that are on the surface of the body or are the lining layer of body cavities, ducts, and tubes. Epithelial cells produce such important nonliving substances as shells, hair, antlers, feathers, milk, sweat, and digestive juices. Some epithelial cells have microscopic extensions called *cilia,* which pulsate and propel fluids over the tissues.

Circulating tissue, or *blood,* is a fluid made of blood cells suspended in plasma. It is circulated by the pulsation of the heart muscles and of the blood vessel walls; it may also move incidentally when the body moves.

Endocrine tissue consists of cells that secrete hormones into the blood, to be carried throughout the body in order to coordinate its activities. The pituitary and thyroid glands are endocrine glands, or glands of internal secretion. *Exocrine* glands, or glands of external secretion, such as sweat glands, are epithelial tissue.

Conducting tissue is characteristic of the nervous system. Its individual cells, known as *neurons,* conduct impulses of electrochemical charge and help to coordinate the body by linking *receptors,* such as the light-sensitive cells of the eye, to *effectors,* the muscle cells or glands that respond. Large units of conducting tissue are *ganglia,* which are clusters of neurons, and such centers as the brain and spinal cord, from which many neurons extend in bundles known as *nerves.*

Reproductive tissue consists of egg cells produced in the female's *ovaries* and sperm cells produced in the male's *testes.* An egg is normally fertilized only by union with a sperm cell, becoming a new individual.

CLASSIFICATION

Animals are identified and grouped into a scheme of classes on the basis of their physical structure and the development of the body parts.

BINOMIAL NOMENCLATURE. The system of naming animals by specifying two names was developed in 1758 by Karl von Linné, a Swedish physician and naturalist. He is better known by his Latin signature, Carolus Linnaeus.

Under von Linné's system, a *species* is a group of creatures that can interbreed with complete fertility; several species may be included in a *genus* if they are very similar in structure, but crosses between two species either result in no offspring or in offspring with incomplete fertility. A familiar example is the mule, which is comparatively sterile. It is produced by a mating between a horse *(Equus caballus)* and a donkey *(Equus asinus).*

Von Linné also saw the need for levels of classification between the animal kingdom and the binomial nomenclature of the individual creature. He therefore set within the animal kingdom, in descending order, the *phylum,* the *class,* the *order,* and the *family,* which is composed of the *genus* and the *species.*

For example, the full classification of man is: phylum *Chordata* (vertebrates and near kin), class *Mammalia* (mammals), order *Primates* (mammals with nails and opposable thumbs or big toes or both), family *Hominidae* (man and manlike primates), genus *Homo,* species *sapiens.*

In some cases, additional levels of classification are added to show differences that are considered important. These include *subphyla,* each with one or more classes; *subclasses,* each with one or more orders; *suborders,* each with one or more families; and *subfamilies.*

TRINOMIAL CLASSIFICATION. Geographical divisions of a species are *subspecies,* also called *races.* The distinction among the members of a species is made by using *trinomial nomenclature.* Thus, for example, the white-footed mouse of Vermont is *Peromyscus maniculatus gracilis;* the white-footed mouse found in Canada north of Lake Superior is *Peromyscus maniculatus maniculatus,* the "typical" or "standard" member of the species. These and all other races of the species will interbreed freely if brought together.

EVOLUTION. The idea of evolution received little attention for more than a century after von Linné established the binomial classification. In 1859 Charles Darwin's *On the Origin of Species by Means of Natural Selection* gave evidence and explanation of change in body form over time.

Darwin pointed out that all animals reproduce faster than is necessary merely to maintain a stable population. Competition develops for food and habitat, and the weaker members of the species, unable to compete effectively, are likely to die before being able to reproduce. Only the fittest survive, and they pass on to their offspring any inheritable advantages.

Gradually a species adapts, splits into races, or dies out. Many extinct

FIN-BACKED EDAPHOSAURUS, a giant, plant-eating reptile, roamed through forests of exotic plants more than 300 million years ago.

TRIO OF ORNITHOMIMUSES, *left,* called "ostrich dinosaurs," around a drying water hole as the rain forests give way to desert lands.

GIANT PTERANODONS, *right,* flying lizards with leathery 25-foot wings and three-foot beaks, lived over 150 million years ago.

BIRDS

YOUNG TOUCANS

FRIGATE BIRDS

SANGUE-DE-BOIS

JANDAIA

B. A. LEERBERGER, JR.

PENGUINS

RED-FOOTED BOOBY CHICK

HARPY EAGLE

INSECTS

TARANTULA

WOLF SPIDER

SCORPION

HORNED BEETLE

RICHARD PARKER FROM NATIONAL AUDUBON SOCIETY

RED-BANDED LEAFHOPPER

PRAYING MANTIS

BEE AT FLOWER

DENNIS BROKAW FROM NATIONAL AUDUBON SOCIETY

GRASSHOPPER ON THISTLE

RICHARD PARKER FROM NATIONAL AUDUBON SOCIETY

CICADA

JAGUAR

KOALA

RHINOCEROSES

BEARS

MAMMALS

BIGHORN MOUNTAIN SHEEP

GIRAFFES

RACCOON

ANTEATER

RINGTAIL CAT

MUSTACHED TAMARING

REPTILES

CORAL SNAKE

LAND IGUANA

BRAZILIAN HORNED TOAD

GILA MONSTER

MUSSERANA

RATTLESNAKE

TEJU LIZARD

species, known only through their fossil records, have been classified through their similarities to living animals. All the animals of today are descended from those of the past and are ancestors of all future animals.

PHYLA

Of the more than one million species that have been named, all but 2 percent are classified in 14 phyla. Small additional phyla have been established for the remaining 2 percent, most of them inconspicuous deep-sea dwellers or parasites.

PHYLUM PROTOZOA. There are about 30,000 species of unicellular "first animals," most of them too small to be seen with the naked eye. They are commonly called *protozoans.*

Class Mastigophora. Because they propel themselves through the water by means of one or more whiplike projections from the body, these "whip-carriers" are called *flagellates.* Some, such as *Euglena,* have a single extension, or *flagellum.* Others have two or as many as ten.

Euglena is a large genus, including many species that contain chlorophyll and carry on photosynthesis. Some of these are so numerous in still water that after a period of heat and drought, the water appears bright green. Many colorless flagellates feed on bacteria. Still others are symbionts of termites, living in their digestive tracts and predigesting the wood fibers the insect swallows.

African sleeping sickness is caused by a parasitic flagellate, *Trypanosoma gambiense,* which is transmitted to man through the bite of the tsetse fly. Ordinarily the parasite lives in various wild animals, which seem to be unaffected.

Class Sarcodina. The most famous sarcodinians are the *amoebas,* the *radiolarians,* and the *foraminiferans.* All move about by extending lobes or networks of protoplasm, called *pseudopodia,* or false feet. Amoebas are extraordinary in that they have no definite shape; they flow into one pseudopodium after another as they travel in fresh water or in the digestive tract of an animal.

Radiolarians and foraminiferans, which live on bacteria and microscopic green plants, produce minute skeletons or chambered shells of either silica or lime, from the surface waters of the oceans. When these sarcodinians die, their skeletons sink to the bottom and build up great thicknesses of ooze. Radiolarian ooze becomes an inert powder suitable for making filters and bonding material in dynamite. Foraminiferous ooze gradually becomes a type of limestone.

Class Sporozoa. All sporozoans are parasites of multicellular animals and absorb their food in dissolved form directly from the *host,* the animal to which they attach themselves. They have no means of moving by themselves but must be transferred from one host to another by the activities of *carriers,* such as mosquitoes.

Perhaps the best-known sporozoan is *Plasmodium falciparum,* the cause of the most dangerous type of malaria in man. It penetrates the red blood cells, reproduces, and causes the cells to break open, releasing more parasites to attack other red cells. Plasmodia are spread by mosquitoes of the genus *Anopheles,* which feed on blood. If an *Anopheles* consumes infected blood, the parasites undergo changes and then migrate to its salivary glands. If the mosquito bites a healthy person, the plasmodia, which go with the saliva into the victim's blood stream, start a new infection.

Class Ciliata. Ciliates are named for the many hairlike *cilia* that project from their microscopic bodies, beating in rhythmic waves and driving the cell through the water. The animals are unique in possessing two different kinds of cell nuclei—*macronuclei* and *micronuclei.*

Ciliates live in water and feed on bacteria and small protozoans. Among the best-known are the slipper-shaped *Paramecium,* the bell-shaped *Vorticella,* and the trumpet-shaped *Stentor.* In a strong light, all of these *animalcules* are large enough to be seen without a microscope.

PHYLUM PORIFERA. Phylum *Porifera* contains some 4,500 species of colonial animals that remain attached to the bottom of the sea or to other solid objects, while cells, known as *flagellated collar cells,* draw water and minute particles of food through small holes that lead to a central chamber or system of chambers. After the collar cells have caught the food particles, the water is released through one or more large openings.

Most of these animalcules, commonly known as *sponges,* are marine; a few live in fresh water.

Class Calcarea. *Calcarea* are marine sponges whose cells secrete needle-shaped or branching *spicules* of lime. The spicules usually project from the surface of the colony and mesh, giving it structural support; however, they are regarded as an internal skeleton. Common genera of this class include *Grantia* and *Leucosolenia,* some of whose species grow to an inch in length.

Class Hyalospongiae. *Hyalospongiae* are deep-sea sponges, often of great beauty, that produce a skeleton of silica. *Euplectella* is the Venus' flower basket sponge.

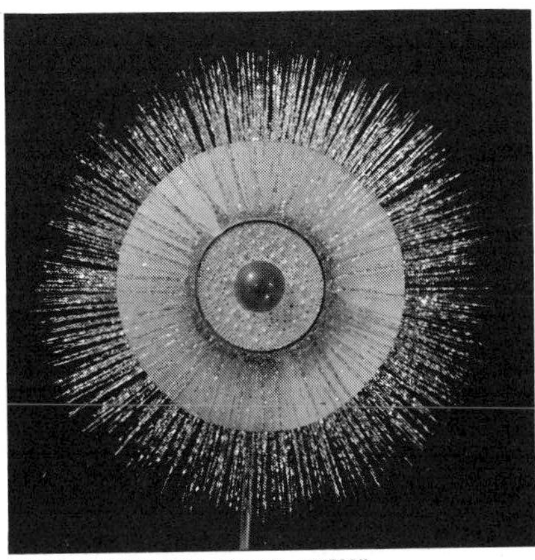

AMERICAN MUSEUM OF NATURAL HISTORY

THE RADIOLARIA has an inner skeleton of silica that houses the spherical nucleus.

Class Demospongiae. *Demospongiae,* the commonest sponges, either lack a skeleton or have one composed of a plastic-like secretion called *spongin.* Most of them live in relatively shallow seas, but one family lives in fresh water. The old-fashioned bath sponge is *Spongia.*

The freshwater sponges belong to the genus *Spongilla;* they are usually bright green or golden-green because they have microscopic plants as mutualistic symbionts.

PHYLUM COELENTERATA. The phylum *Coelenterata* is composed of some 9,600 species of aquatic animals, most of them marine, that have a saclike digestive cavity with a mouth opening at one end. The bodies of the coelenterates are radially or biradially symmetrical, with a ring of tentacles surrounding the mouth. On the tentacles, and often elsewhere, there are unique cells with which smaller animals are stung and paralyzed before they are thrust into the digestive cavity of the coelenterate.

Class Hydrozoa. Hydrozoans are characterized by their method of reproduction. Usually a *polyp* (hydroid) stage reproduces by asexual budding and releases free-swimming *medusae* (jellyfishes), which reproduce sexually; the embryos resulting from this mating settle to the bottom, become attached as polyps, and repeat the cycle.

Colonial hydroids include *Obelia* and *Plumularia,* also known as sea firs, and *Millipora,* or stinging coral. Freshwater hydras have no medusa stage, and some of the larger marine hydrozoan medusae have no known hydroid stage. Freshwater medusae are usually *Craspedacusta.*

Hydrozoans also include such free-floating colonies as the Portuguese man-of-war *(Physalia),* the by-the-wind sailor *(Velella),* and the porpita *(Porpita).*

Class Scyphozoa. Scyphozoans, the larger marine medusae, have armlike tentacles extending from the four corners of the pendant, tubular mouth. *Aurelia,* the moon jelly, is commonly found near shore; it is usually about eight inches in diameter. *Cyanea,* a giant medusa of the open ocean, may be seven feet across.

Class Anthozoa. All anthozoans, which are marine, lack a medusa stage; many are colonial. The most familiar are the sea anemones, the true corals, the sea fans, and the sea whips. They differ from other coelenterates in that they have additional cells in the usually noncellular *mesoglea,* or jelly, that separates the outer epithelium (epidermis) and the inner epithelium (gastrodermis). This additional cellular material makes the anthozoans' bodies firmer than those of other coelenterates.

The reef-forming corals obtain lime from seawater through their symbiotic relationship with microscopic green plants. Since reef-making corals depend on green plants, they occur only where sunlight penetrates warm seas.

PHYLUM CTENOPHORA. About 80 species of free-swimming marine animals with transparent, biradially symmetrical bodies make up the phylum *Ctenophora*. Comb jellies, as they are also known, swim by rhythmically beating eight lengthwise rows of paddle-like comb plates.

Ctenophores feed on small planktonic animals that they usually capture by means of tentacles studded with adhesive cells. Many, such as *Mnemiopsis*, are thimble-shaped and glow in the dark when disturbed. *Cestus*, known as Venus' girdle, is ribbon-shaped and can be three feet long and two inches wide.

PHYLUM PLATYHELMINTHES. *Platyhelminthes* phylum is composed of 15,000 species of flat-bodied animals known as flatworms, which are bilaterally symmetrical and have well-organized muscle bands and muscle sheets. Flatworms also have a distinct nervous system consisting of at least one anterior ring of nerve fibers and lengthwise nerve cords.

Class Turbellaria. Turbellarians are free-living flatworms that glide on a ciliated lower epidermis or swim by bodily undulations. Most of these scavengers have a straight, Y-shaped, or multibranched blind digestive cavity. Although they are chiefly marine, turbellarians are also found in fresh water and in very moist soil. Freshwater turbellarians are known as *planarians*.

Class Trematoda. Trematodes are cilialess, parasitic flatworms that have one or more circular suckers with which to attach themselves to a host animal. The blind digestive tract is Y-shaped; the mouth, anterior.

The trematodes include flukes, such as the destructive liver fluke *(Fasciola)* found in sheep; intestinal and pulmonary parasites; and the dangerous African blood fluke *Schistosoma*. Some of the flukes, including *Schistosoma*, undergo a series of complex bodily changes that requires a sequence of hosts, one of which is usually a freshwater snail.

Class Cestoda. Cestodes, also known as *tapeworms*, are parasitic flatworms without digestive systems. They attach themselves to the intestine or body cavity of a vertebrate and absorb food directly.

Most tapeworms consist of an anterior individual *(scolex)* with suckers and hooks, and a series of posterior individuals. The latter are produced asexually by the anterior individual but can develop sex organs and produce fertilized eggs and embryos before breaking away from the oldest part of the chain and emerging from the host's body with the wastes.

Taenia solium, the tapeworm that attacks man when he eats improperly cooked, infected pork, can reach a length of 25 feet.

PHYLUM NEMATODA. The phylum *Nematoda* consists of about 10,500 species of cylindrical, unsegmented animals called roundworms with a straight digestive tube from anterior mouth to posterior anus. Between the outer body wall, which has only lengthwise muscles, and the digestive tract is a bloodlike fluid that churns back and forth as the animal moves.

Some of these roundworms are free-living in moist soil and other aquatic situations, including hot springs and glaciers. However, parasitic nematodes are the best known. *Necator* and *Ancylostoma* are hookworms that attack man; *Enterobius* is the pinworm; *Trichinella* is a dangerous parasite acquired by eating infected improperly cooked pork. *Filaria* causes elephantiasis.

PHYLUM ROTIFERA. Some 1,500 species of unsegmented aquatic animals, none over 1/16 inch long, compose the phylum *Rotifera*. The head region has a mouth with a muscular grinding mill nearby; two whorls of cilia move food particles toward the mouth and aid in swimming. Posterior to the anus, the wheel animalcule usually has a two-toed foot with cement glands that temporarily anchor the rotifer to a solid object.

Most rotifers are free-living freshwater dwellers. Although they are multicellular, rotifers are approximately the same size as the larger protozoans.

PHYLUM MOLLUSCA. There are about 100,000 living and 40,000 extinct species in the phylum *Mollusca*. All mollusks have a soft muscular, usually unsegmented body with a dorsal mantle that generally secretes a limy shell. Usually the anterior head has a unique rasping instrument, called the *radula*, inside the mouth.

Class Amphineura. The class *Amphineura* consists of the *chitons*, all of which are marine. Most have a dorsal shell consisting of eight transverse overlapping plates. In dangerous situations, chitons curl up to protect the muscular ventral foot, exposing only the hard shell. The class name refers to the two pairs of ventral nerve cords that extend lengthwise from a nerve ring around the chiton's mouth.

Class Scaphopoda. Belonging mainly to the genus *Dentalium* and known as tooth shells, scaphopods are marine mollusks that have a slender, slightly tapered, tubular shell open at both ends. They use their muscular foot, somewhat resembling a horse's foot, to dig themselves into the ocean floor, leaving only the smaller opening of the shell exposed. Through this opening they draw in and expel water and minute particles of food.

Indians on the western coast of North America once used tooth shells as money; natives of New Guinea often wear them as ornaments in pierced ears, noses, and lips.

Class Gastropoda. Some two-thirds of the known species of mollusks are gastropods, known as snails if they have coiled shells, or slugs if they lack shells. All of them creep or cling on a flat, bilaterally symmetrical ventral foot; above the foot is a spiral body covered by the mantle.

Most gastropods are marine herbivores; but some, such as the whelk *Busycon* and the oyster drill *Urosalpinx*, are carnivorous predators.

Physa is a freshwater snail; *Helix pomatea* is the edible garden snail that has been cultivated in Europe for centuries.

Class Pelecypoda. Pelecypods are the *bivalves*, which have two-part limy shells hinged and controlled by muscles between them. These aquatic mollusks are chiefly marine; they are unique in that they lack a head region and a radula. The best-known genera include the scallop *Pecten;* the mussel *Mytilus;* the oyster *Ostraea;* the freshwater clam *Unio;* and *Tridacna*, the huge bear's-paw clam of the South Pacific reefs.

AMERICAN MUSEUM OF NATURAL HISTORY

CLAM WORM, *Nereis virens*, shows an evolutionary link between mollusks and annelids.

Class Cephalopoda. Cephalopods are marine mollusks that are "head-footed" in the sense that from eight or ten to as many as ninety tentacles extend from the part of the foot that contains the mouth; the rest of the animal is almost hidden by a high, conical mantle. Among fossil cephalopods a chambered shell was common; today, however, only the pearly *Nautilus* of the East Indies produces such a shell, in whose outermost chamber it lives.

Other living cephalopods have a greatly reduced shell, such as the "cuttlebone" of the cuttlefish *Sepia*, or none at all, as in the eight-tentacled *Octopus*. Most modern cephalopods have a pair of very large, camera-style eyes, which are much like those of vertebrates, and a highly developed brain.

All cephalopods are predators, most of them grasping their victims with suction disks on the tentacles, then rasping out flesh with the radula. Others are equipped with a special pair of concealed horny jaws, like the beak of a parrot, used for biting their prey.

Class Monoplacophora. Monoplacophorans, discovered alive for the first time in 1957, differ from all other mollusks in that they show signs of segmentation. A low, conical, one-piece shell characterizes and names this class, and is attached to the animal by from eight to twelve pairs of muscles. The sides of the flat foot bear a corresponding number of paired gills and excretory organs; the latter resemble those of the segmented worms.

Living species of Monoplacophorans are members of the genus *Neopilina*. Regarded as "living fossils" because they show an evolutionary link between mollusks and annelids, they are found in the deep waters of the eastern Pacific. Fossils of this class have been found in stratified rocks of the early Paleozoic Era, covering a period from the Ordovician Age to the Devonian Age.

PHYLUM ANNELIDA. Some 7,000 species of cylindrical or flattened segmented worms are included in the phylum *Annelida.* The body cavity of annelids is transversely divided into definite segments, each of which usually contains a portion of the straight digestive tract that extends from anterior mouth to posterior anus.

The body cavity also contains a ganglion of the ventral nerve cord; branches of the closed blood-vessel system; a pair of excretory organs, called *nephridia;* and a set of bristles used in locomotion.

Class Polychaeta. Most members of the class *Polychaeta* are marine worms with a distinct head and a fleshy paddle on each side of most body segments. The paddles are supported and moved by the body bristles embedded in them, and both paddles and bristles are controlled by muscles within the body wall.

Many of these annelids, such as the clam worm *Nereis,* are free-swimming predators and scavengers. Others build U-shaped burrows in the bottom mud and use their paddles to create a current that brings a constant supply of food and oxygen through the burrow. Polychaetes include the lugworm *Arenicola* and the parchment worm *Chaetopterus.*

Class Oligochaeta. The best-known oligochaetes are the terrestrial earthworms, which burrow and scavenge in the soil for decaying plant material. Earthworms have no distinct head and no lateral paddles; they creep or cling by means of bristles that can be extended from each body segment.

The body usually has a swelling, the *clitellum,* about one-third of the way along the body from the head. The clitellum provides a sheathlike case for the eggs.

The most common earthworms are *Lumbricus, Allolobophora,* and *Eisenia,* which can be distinguished by the location of the paired pores that connect to the sex organs, Other smaller oligochaetes live in fresh water, where they burrow into bottom sediments.

Class Hirudinea. The class *Hirudinea* is made up of predatory bloodsuckers whose bodies are composed of exactly 34 segments that are concealed among transverse wrinkles. All these annelids have a large posterior sucker; many also have an anterior one surrounding the mouth, which has three horny jaws to capture prey or to cut through the skin of a victim in order to reach the blood vessels.

Most leeches live in fresh water, but a few are marine; there is also one that lives in the rainy Malayan jungles. The medicinal leech, *Hirudo medicinalis,* has long been used in the bloodletting thought to be a remedy for many diseases in various parts of the world.

PHYLUM ONYCHOPHORA. The onychophores are the "velvet worms" or "walking worms" of humid climates, chiefly the tropics. Their cylindrical bodies, which may be as much as eight inches long, have from 15 to 43 pairs of soft legs, each ending in two claws. The anterior head region is indistinct and bears a pair of simple eyes, a pair of short, flexible tentacles, and a pair of blunt *papillae* through which large salivary glands open near the mouth.

Distinct impressions of onychophorans have been found among the oldest fossils, and those species alive today are referred to as "living fossils" because they show features of both arthropods and annelids.

Like other arthropods, they have the periodically shed exoskeleton of chitin; the reduced body cavity, which is mainly replaced by large *sinuses,* or cavities, through which blood flows in an open circulatory system; the system of fine *tubules* through which air reaches inner organs; and claw-tipped legs.

Like annelids, they have paired excretory organs (nephridia) and simple eyes; they also lack a distinct head (or head plus thorax). The best-known of the Onychophores belong to the genus *Peripatus.*

PHYLUM ARTHROPODA. There are more than 770,000 species of arthropods, of which almost 700,000 are insects. Typically, each arthropod has a segmented body enclosed by an external skeleton containing the polysaccharide *chitin;* this *exoskeleton* is shed periodically. Many of the body segments have a pair of jointed appendages, from which the phylum takes its name.

Almost 80 percent of the known animals are included in this phylum, which includes marine, freshwater, and terrestrial creatures of many types: free-living, *sessile* (attached by the base), commensal, and parasitic.

Class Trilobita. Extinct marine arthropods, of which over 2,000 species are known from Paleozoic times, compose the class *Trilobita.* Each had a flattened, elliptical body marked by lengthwise furrows that separated a central lobe and two lateral lobes (the three lobes gave the class its name).

Transversely, the body was divided into a head with a pair of joined antennae, four pairs of jointed *maxillae* (mouth parts), and a pair of compound eyes; a thorax of 2 to 29 segments, each with a pair of jointed appendages used in swimming and creeping; and an abdomen made up of several segments fused into one plate.

All trilobites, the longest of which were 26 inches long, seem to have been scavengers. Their numbers decreased when fishes with jaws, their natural enemies, proliferated.

Class Crustacea. Although generally marine, crustaceans can also be found in fresh water and on land. A few are parasites that attach themselves to fishes. All crustaceans have a head region with two pairs of *antennae,* a pair of *mandibles* (jaws), and at least two other pairs of maxillae. Both the thorax and the abdomen may have paired appendages for swimming and walking.

Familiar genera of these creatures include *Artemia,* the brine shrimp of alkaline lakes; *Balanus,* the acorn barnacle of seacoasts; *Lepas,* the goose barnacle; *Oniscus* and *Porcellio,* terrestrial pillbugs; *Homarus* the Atlantic lobster; and *Callinectes,* the blue crab.

Class Diplopoda. The diplopods are the *millipedes,* or "thousand-legged worms." Each has a pair of antennae, a pair of jaws, and a pair of maxillae on the head; four segments and three pairs of legs on the thorax; and from nine to more than a hundred segments on the trunk. Each of these segments is really two that have been fused in the course of evolution; thus, as the class name indicates, each segment has two pairs of legs.

Most diplopods are harmless terrestrial scavengers that inhabit moist places.

Class Chilopoda. Terrestrial predators commonly known as *centipedes,* chilopods have flattened bodies and only one pair of legs per segment—there may be from 15 to 181 pairs of legs. The head has a pair of jointed antennae; a pair of jaws; and two pairs of maxillae, the second pair partially joined to form a lower lip that gives the class its name.

The first pair of legs are hooklike and have poison glands that open at the sharp tip of each; these poisonous hooks inflict painful or dangerous wounds.

Class Insecta. Almost 700,000 species of insects, the only flying animals without backbones, are known today; and more are discovered every year. Insects are primarily terrestrial arthropods whose bodies are distinctly divided into head, thorax, and abdomen.

The head has a pair of jointed antennae, a pair of mandibles, a pair of jointed maxillae, and a *labium* (lower lip) that evolved from another pair of maxillae; the maxillae may be modified for chewing, sucking, or lapping. Typically, each of the three thorax segments has a pair of legs, in adult insects the second and third segments may also have a pair of wings apiece.

Insects may be classified by structure and by development. Structurally, the details of the maxillae and the wings are considered. The ancient insects are wingless, and change little in body form from hatching to maturity.

The most advanced of modern insects undergo indirect metamorphosis, a spectacular transformation from a specialized *larva* that spends most of its time eating and growing, to a quiet, non-feeding *pupa* that encases the larva, to an adult that reproduces. The wings develop internally during the pupal stage.

Less advanced modern insects experience direct metamorphosis, an incomplete transformation that lacks a pupal stage; the animal progresses from immature stage to mature stage. The wings of these insects develop as pads on the back.

Of the 16 orders described here, the first two are ancient, the next nine have direct metamorphosis with no pupal stage, and the last five have a pupal stage and indirect metamorphosis.

Order *Collembola* is made up of the *springtails,* minute insects that are rarely more than ⅛ inch long and have chewing mandibles. They leap by flipping a special ventral springing organ on the fourth abdominal

GEORGE A. SMITH

THE MONARCH BUTTERFLY is one of roughly 122,000 species of *Lepidoptera*. The butterfly shown is just emerging from its chrysalis.

segment from under a hook on the third segment. The order includes some 2,000 species that live on land, in soil, and over water.

Order *Thysanura* is composed of about 700 species of *bristletails*, including the silverfish *Lepisma*. They grow to 1¼ inches long, and their bodies are covered with overlapping scales. Thysanurans have chewing mandibles and long, threadlike antennae. There is also a pair of antenna-like structures on the posterior end, and the last body segment of thysanurans may extend as a third antenna-like "tail."

Order *Dermaptera* includes the *earwigs*, which may grow to two inches in length and have chewing mandibles and, at the end of the abdomen, a pair of strong forceps. Some are wingless; others have a short pair of leathery wings that, when the animal is at rest, cover a large pair of membranous, semicircular hind wings. Dermapterans include about 1,100 species, some of them harmful to crops.

Order *Orthoptera* includes some 23,000 species of insects, including cockroaches, stick insects, short-horned grasshoppers (locusts), long-horned grasshoppers (including katydids), and crickets. Some of these insects reach a length of 12 inches.

Most of them have as adults a pair of narrow forewings that cover the hind wings when the animal is at rest. The hind wings are folded fanwise. Certain species cause considerable damage to crops.

Order *Isoptera* is made up of the social insects called *termites* or "white ants." All of the some 1,800 species of this class have chewing mandibles and may reach a length of two inches. Only adult sexual individuals have wings, of which there are two narrow membranous pairs. The wings lie flat on the back when the termite is at rest and are detached after the nuptial flight. The thorax and abdomen are joined broadly; there is no "waist" like that of a true ant.

Termites that eat wood depend upon intestinal flagellates to predigest the fibers. Some tropical termites cultivate fungus plants on chewed vegetation, then eat the fungi.

Order *Odonata* contains about 6,000 species of damselflies and dragonflies, which in their immature stages are freshwater predators. Members of the order have chewing mandibles and may reach a length of six inches, with a wingspread up to one foot—fossil dragonflies had wingspreads of as much as 28 inches. Adults have two pairs of membranous wings crisscrossed with many veins; a head with huge compound eyes; and a long, slender abdomen.

Order *Ephemeroptera* consists of the mayflies, of which there are approximately 1,500 species. They have chewing mandibles in the immature aquatic stages, but these become only vestigial at maturity. Mayflies reach a length of as much as two inches but may look longer because the abdominal tip has two or three filamentous "tails."

Mayflies are unique in that they develop wings and emerge from the water before becoming adult; they must shed their skins once more —even over the wings. The wings consist of a large forepair and a small hind pair, both pairs membranous and crisscrossed with veins. Mayflies rarely survive more than a day as an adult, but may require a year or more to reach this stage of development.

Order *Mallophaga* is composed of the biting lice, which grow to ¼ inch long and have chewing mandibles. Their bodies are flat and wingless, and they have either no eyes or small eyes. There are about 2,700 species of biting lice, all of them external parasites on birds and mammals. The genus *Menopon* includes the hen lice.

Order *Anoplura* includes some 200 species of sucking lice, whose flat, wingless bodies may be up to ¼ inch long. These lice have sucking mouthparts and small eyes—or no eyes. *Pediculus capitis* is the head louse, or "cootie," which transmits typhus fever and other diseases; *Haematopinus suis* is the hog louse.

Order *Heteroptera* is the order of the true bugs, which may grow to a length of four inches and have piercing, sucking mouthparts that arise far forward on the head. Some heteropterans are wingless as adults; those with wings have a forepair that is thick and horny at the base but membranous at the tips, where they overlap when held flat and slightly crossed at rest. The hind wings are membranous and fold slightly below the forewings.

Among the order's approximately 45,000 species are such water striders as *Gerris*, the stinkbugs *Pentatoma*, the milkweed bug *Lygaeus*, and wingless bedbug *Cimex*.

Order *Homoptera* contains the cicadas and their kin, about 25,000 species in all. These insects, many of them injurious to plants, grow to five inches long and have piercing, sucking mouthparts that rise far back on the head. Winged adults have forewings larger than the hind wings; both sets are membranous and at rest are held in tent fashion over the back. Some of the better-known homopterans are the cicadas, aphids (plant lice), scale insects, leafhoppers, and spittle bugs.

Order *Lepidoptera* consists of the moths and butterflies, insects that may have a length of four inches and a wingspread of almost one foot. A caterpillar usually has biting mandibles, three pairs of thoracic legs, up to four pairs of soft abdominal prolegs, and labral openings of silk glands, used in spinning the cocoon. Adults have maxillae joined to form a coiled sucking tube, or *proboscis*. They also usually have two pairs of broad, membranous wings covered by overlapping scales.

There are roughly 122,000 species of lepidopterans, including many that, as caterpillars, eat man's crops or possessions.

Order *Diptera* is composed of the two-winged, or "true," flies. These insects may attain a length of two inches and have a wingspread of three inches. The larval stages are usually legless maggots, some of which have chewing mouthparts. The adults have piercing and sucking, or lapping, mouthparts and one pair of membranous wings; the hind wings are represented by a pair of short, knobbed balancers.

Some of the roughly 90,000 species in the order are the mosquitoes *Anopheles*, *Aedes*, and *Culex*; the

GEORGE A. SMITH

FULLY EMERGED from its pupa case, the butterfly is in the final stage of its metamorphosis from egg to larva and finally to butterfly.

black flies *Simulium;* the beneficial tachinid flies that parasitize caterpillars; the fruit fly *Drosophila;* the housefly *Musca domestica;* and the wingless sheep tick or sheep ked *Melophagus.*

Order *Coleoptera* is made up of the beetles and weevils, which have biting mandibles in both larval and adult stages and may grow to six inches in length. The larvae are usually wormlike creatures with well-developed legs. Adults have a pair of thick, veinless forewings that, at rest, meet along the midline above the membranous hind wings, whose tips are folded when the wings are not in use.

Coleopterans are the largest order of insects, containing some 260,000 species. Many, such as the ladybug beetle *Coccinella,* are beneficial to man; others, such as the boll weevil *Anthonomus grandis,* are destructive.

Order *Siphonaptera* includes the fleas, which in the adult stage are external parasites on birds and mammals. The minute, legless larvae are scavengers with biting mandibles. The adults, about ¼ inch long, have laterally compressed, wingless bodies; their mouthparts are adapted for piercing and sucking.

Of the approximtely 300 species, two of the best known are *Pulex irritans,* which attack rats and humans, and *Xenopsylla cheopis,* the Indian rat flea, which transmits bubonic plague.

Order *Hymenoptera* takes in the ants, bees, wasps, and their kin, some 103,000 species. Some hymenopterans grow to three inches in length and may have a five-inch wingspread. Their larvae may be either legless maggots or caterpillar-like creatures; the latter are distinguished from caterpillars of the order Lepidoptera by having more than four pairs of fleshy prolegs. Some larvae are parasitic, usually attaching themselves to other insects, or living inside them.

Adults usually are solitary, but some species build colonies and organize societies that show distinct castes. Those adults that can fly have membranous forewings longer than the hind wings; the latter are hooked together in flight. The female's *ovipositor* is commonly modified as a saw, drill, or stinger.

Among the better-known species of the order are sawflies, with herbivorous, caterpillar-like larvae; beneficial *ichneumon* flies and *chalcids,* which parasitize harmful insects; gall wasps, ants, such as *Formica;* wasps, such as *Vespa* and *Polistes;* bees, such as the bumblebee *Bombus* and domesticated honeybee *Apis mellifera.*

Class Merostomata. Only four "living fossil" species of horseshoe crabs remain of the ancient "divided mouth" class, Merostomata. All of them dwell in shallow, offshore seas, where they scavenge on seaweeds, sea worms, and young mollusks.

Merostomates are armored creatures with an unsegmented *cephalothorax* joined broadly to an abdomen that ends in a bayonet-like tail spine; they lack antennae and true jaws. Food is chewed between the spiny bases of the four pairs of walking legs that flank the elongated mouth slit. Also near the mouth are two pairs of appendages, the usually pincer-like *chelicerae* and the *pedipalpi.*

Perhaps the best-known species is *Limulus,* of the American eastern coast, which comes to shore each spring to lay its eggs in beaches from Maine to Yucatan.

During the Paleozoic Era, the merostomates included sea scorpions, up to six feet long, whose clearly segmented, flexible, tapering abdomens suggest that they may have been the ancestors of land scorpions. Horseshoe crabs, by contrast, have an abdomen fused into a single unit; however, like the sea scorpions, they have a ventral series of five or six pairs of plates that are used in swimming and as protection for the gills.

Class Arachnida. The class *Arachnida* includes some 30,000 species of spiders and their kin. All species have a cephalothorax and an abdomen, a pair of chelicerae, a pair of pedipalpi, and four pairs of legs, but lacks antennae, jaws, and paired appendages on the abdomen.

The class arachnida includes the scorpion *Scorpio,* the house spider *Theridion,* the orb-web spider *Argiope,* the harvestman *Phalangium,* the tick *Dermacentor,* and the spider mite *Tetranychus.* Scorpions, spiders, and many of the mites are predators; ticks and the rest of the mites are external parasites of animals or plants and may transmit diseases from infected hosts to healthy ones.

PHYLUM BRACHIOPODA. Brachiopods, known as lamp shells because of a resemblance between one type of shell and ancient oil lamps, are marine bivalves that secrete the dorsal half and the ventral half of the shell from the mantle. There are 260 living species and over 5,000 fossil species.

Most adult brachiopods remain permanently attached to their surroundings by means of a short posterior stalk that emerges through an opening in the ventral valve of the shell, near the hinge. The unsegmented body has two *lophophores,* spiral or V-shaped arms, from which the phylum takes its name; the lophophores bear ciliated tentacles that create currents to bring oxygen and microscopic food particles into the shell.

One class of brachiopods has a largely chitinous shell and no hinge teeth; these animals include *Lingula* and *Crania,* genera with the longest fossil histories in the animal kingdom. A second class has a limy shell and teeth that keep the halves of the shell in alignment. These include *Terebratulina* and *Rafinesquina.*

PHYLUM ECHINODERMATA. There are roughly 5,700 species of echinoderms (or "spiny skins"), which begin life as bilaterally symmetrical embryos, then take on a false radial symmetry, and still later may become conspicuously biradial or even bisymmetrical; the symmetry usually has five parts.

If the creature has a skeleton, it is internal, composed of limy spicules or plates. Part of the large body cavity is separated to form a unique water-vascular system with special tube-like feet used in feeding and locomotion.

Echinoderms include the sea lilies and feather stars (class *Crinoidea*); sea cucumbers (class *Holothuroidea*); sea urchins, heart urchins, and sand dollars (class *Echinoidea*); sea stars, or starfishes (class *Asteroidea*); and serpent stars, or brittle stars (class *Ophiuroidea*).

PHYLUM CHAETOGNATHA. Roughly 50 species of arrow worms, all of them very slender, cylindrical, and unsegmented, compose the phylum *Chaetognatha.*

The lateral cranial lobes of these three-inch marine predators have chitinous bristles with which prey is captured and pushed into the mouth (hence the phylum name of "bristle jaws"). The digestive tract is straight, and the anus is just anterior to the tail. A pair of lateral fins, supported by fine chitinous rods, give the creature better stability and allow it to dart after minute crustaceans.

The arrow worms' remarkable transparency often causes them to be overlooked. They are plentiful, however, and provide an important source of food for whalebone whales.

PHYLUM CHORDATA. Chordates, of which there are some 45,000 species, are distinguished from all other animals by the development of a flexible supporting rod, called the *notochord.* This lies immediately below the hollow dorsal nerve cord and gives the phylum its name.

At some stage in development, gill slits connect the pharyngeal area to the outside of the body. Usually the body is bilaterally symmetrical and has a complete digestive tract, a closed circulatory system, and a tail posterior to the anus. Four subphyla are recognized.

Subphylum *Tunicata* is composed of about 1,600 marine species that have a notochord and a nerve cord only during the larval stage; the adult is a degenerate form surrounded by a secreted tunic, usually of cellulose.

The most familiar members of this subphylum are the sea squirts (class *Ascidiacea*). They are small and tadpole-shaped as larvae. The adult is permanently attached to the sea floor, where it draws in water, filters out microscopic food and absorbs oxygen, and expels the water.

Subphylum *Cephalochordata* includes about thirty marine species that retain the notochord and the nerve cord; both extend the full length of the body. Cephalochordates are the lancelets, or *amphioxi,* of the class *Leptocardii* and mainly of the genus *Brachiostoma.* They are slender, pointed at each end, and laterally compressed; they swim freely or make shallow burrows in the sandy sea floor near shore.

Subphylum *Agnatha* originally contained the earliest vertebrates, which are known through the fossilized covering of armor-like, bony scales over the head and much of the body; these members of the class *Ostracodermi* apparently were freshwater bottom dwellers during Ordovician, Silurian, and Devonian times.

Class Cyclostomata. Modern agnathans have a cartilaginous troughlike skull and a series of cartilaginous bars protecting the nerve cord. These smooth-skinned, cylindrical, unarmored creatures have horny teeth in their cup-shaped mouths, but no paired fins. They propel themselves by sinuous swimming movements of the whole body.

The hagfishes or slime eels, such as *Myxine,* are direct-developing marine scavengers that eat dead and dying fishes; the lampreys, such as *Petromyzon,* spend at least their larval stage in fresh water and then may move out to sea, where they attack living fishes. *Petromyzon marinus* spread through the Great Lakes in recent years and almost destroyed commercial fishing.

Subphylum *Gnathostomata* consists of chordates with an upper and lower jaw and blocks of cartilage or bone that serve as *vertebrae* and largely or completely replace the notochord in stiffening and supporting the body.

Usually these creatures have paired appendages—a pectoral pair of fins, legs, wings, or arms, and a pelvic pair of fins or legs. If the appendages are fins, the chordate is a fish; if they are limbs, it is a *tetrapod.* Six of the seven classes in the subphylum are represented by living species, the familiar vertebrates.

Class Placodermi. Placoderms (skin of plates) are extinct, jawed fishes that usually had an armor of bony plates or bony scales and two or more pairs of fins. These ancient fish are known from both freshwater and marine fossils of Upper Silurian to Devonian times; the best-known are *Dinichthys* and *Acanthodes.*

Class Chondrichthyes. About 275 living species, all primarily marine, compose this class of cartilaginous fishes that includes the sharks, skates, rays, and chimeras. All of them have cartilaginous rather than bony skeletons (hence the class name); in some, the cartilage is calcified. The scales are minute and usually have an enamel covering over a dentine base, similar to that of teeth.

Class Osteichthyes. A skeleton at least partly of true bone rather than of cartilage characterizes the approximately 25,000 marine or freshwater bony fishes in this class. Their bony scales either fit together in a diamond pattern or overlap like shingles.

Included in the class are the sturgeon *Acipenser,* whose eggs are caviar; the herring *Clupea;* the eel *Anguilla;* the sea horse *Hippocampus;* the cod *Gadus;* the freshwater perch *Perca;* the lungfish *Protopterus;* and the coelacanth *Latimeria,* sole known survivor of the lobe-fin fishes that are close to the ancestral line from which the tetrapods sprang.

Class Amphibia. Most of the nearly 2,000 species of amphibians transform from a gill-breathing immature stage (such as a tadpole) in fresh water to a lung-breathing, terrestrial adult stage. Most adults have forelegs and hind legs, the latter linked by way of a pelvic girdle to a specialized sacral vertebra. Unlike fishes, which have a two-chambered heart, adult amphibians have a three-chambered heart.

Common genera include the mud puppy *Necturus,* the salamander *Ambystoma,* the frog *Rana,* and the toad *Bufo.*

Class Reptilia. Some 5,000 living species of tetrapod chordates that have a dry skin, usually covered with overlapping scales, belong to the class *Reptilia* (creepers). Their skeletons are completely bony, and the pelvic girdle, if present, is linked to two sacral vertebrae.

These turtles, snakes, lizards, and crocodilians all obliterate the gill slits while developing within the egg; at no stage do they possess gills. Special membranes extend from the embryo to the eggshell and enable the embryo to breathe while surrounded by a watery egg "white" provided by the mother. Presumably the dinosaurs and all other extinct reptiles, including some that flew, were similar in general structure and development to modern reptiles.

Class Aves. More than 8,600 living species of birds are known, all of them warm-blooded and covered with feathers. They have a four-chambered heart and a system of blood vessels that carries all blood from the heart to the lungs for aeration before pumping it through the body again.

Other characteristics of birds are a mouth with a specialized beak; one pair of wings; and one pair of legs, which are linked to several vertebrae by way of a light but strong pelvic girdle. All birds lay eggs. Most can fly, but some—such as the ostrich *Struthio,* the kiwi *Apteryx,* and the penguin *Spheniscus*—are flightless and apparently had completely flightless ancestors.

Some well-known genera of class *Aves* are the domestic duck *Anas,* the fowl *Gallus,* the pigeon *Columba,* the crow *Corvus,* and the sparrow *Passer.*

Class Mammalia. There are more than 4,500 species of living mammals, all of them warm-blooded and at least partly covered with hair. The four-chambered heart pumps the blood through the lungs before it is circulated through the body. The pelvic girdle is fused to five vertebrae. Mothers secrete milk from special *mammary glands,* from which the newborn young gain nourishment. Twelve of the eighteen orders are of special interest.

Order *Monotremata* is composed of the egg-laying mammals. Some five genera belong to this order, all being found in Tasmania, New Guinea, and Australia; they include the duck-bill or platypus *Ornithorhynchus* and the spiny anteater *Tachyglossus.* Only the young have teeth; adults have a horny beak. The large, yolky eggs are unique among mammals, as is the practice of incubating them.

The order is named to draw attention to the single body opening (cloaca) that serves the digestive, urinary, and reproductive tracts.

Order *Marsupialia* is made up of the pouched mammals, of which all but the American opossums inhabit Australasia. Unlike class Monotremata, the adult marsupials have teeth. The females have a pouch on the undersurface of the abdomen; the extremely immature newborn young creep in, attach themselves to a nipple, and remain attached until fully formed.

In all other orders of mammals (except Monotremata), the young are linked to the mother by a special membrane *(placenta)* formed by the embryo and used to transfer food and oxygen from the mother to the embryo, and wastes, including carbon dioxide, from the embryo to the mother. Such mammals are placental mammals.

Order *Insectivora* contains the insect-eating moles and shrews and their kin. These small mammals have pointed snouts and sharp teeth that are less specialized than those of other orders. The upper jaw has six to eight incisors, one pair of canine teeth, and three to four pairs of grinding teeth (molars and premolars); the lower jaw has no canines and often has fewer incisors.

Widely known genera are the mole *Talpa,* the shrews *Sorex* and *Blarina,* and the European hedgehog *Erinaceus.*

Order *Chiroptera* is composed of the bats, the only mammals that are capable of flapping flight. They fly by using their modified forelimbs, whose second to fifth toes are greatly elongated and support a thin, leathery membrane that extends to the hind legs, and usually to the short tail as well. The upper jaw often has one pair of incisors; the lower jaw, three.

Order *Primates* includes the monkeys and apes and their kin. Usually there are fewer incisors on both upper and lower jaws. Nails, rather than claws, are found on at least some fingers and toes. Characteristically, either the thumbs or great toes—or both—are opposable, and the shoulder girdle is linked to the breastbone by a collarbone on each side.

Modern man, *Homo sapiens,* and species of fossil man belong to this order, along with the chimpanzee *Pan,* the gorilla *Gorilla,* the orangutan *Pongo,* the rhesus monkey *Macacus,* the capuchin monkey *Cebus,* and the lemur *Lemur.*

Order *Edentata* includes the anteaters, sloths, and armadillos, about 30 different kinds of which live in tropical and warm-temperate America. The name means "without teeth," but actually only the anteaters are toothless. No kind has incisor or canine teeth, and there is no enamel on the premolar and molar teeth with which sloths and armadillos chew their plant food.

Order *Pholidota* consists of seven different kinds of pangolins, or scaly anteaters, of tropical Africa and Southeast Asia. These mammals are also toothless. They capture insects, particularly ants and termites, with a long, slender, sticky tongue and swallow the prey whole.

Order *Lagomorpha* includes rabbits and their kin. These animals may be distinguished by their teeth: two pairs of incisors in the upper jaw, one pair behind the other; one pair of incisors in the lower jaw; no canine teeth. The lower jaw can move from side to side but not from front to back; the jaws are not opposable. The tail is short. Some common genera are the pika or coney *Ochotona,* the hare *Lepus,* and the cottontail rabbit *Sylvilagus.*

Order *Rodentia* is made up of gnawing mammals that have only two incisors in the upper jaw and two in the lower; they have no canine teeth. Rodents have opposable jaws, and their lower jaws move forward and backward as well as from side to side.

Common rodents include the squirrel *Sciurus,* the marmot *Marmota,* the rat *Rattus,* the mouse *Mus,* the beaver *Castor,* the porcupines *Hystrix* and *Erethizon,* and the South American capybara *Hydrochaerus* (the largest rodent, which grows to four feet in length).

Order *Carnivora* comprises land mammals with well-developed canine teeth used in tearing the flesh of their animal food. They have six small incisors above and below. Included in the order are the dog, wolf, and coyote *Canis,* the bear *Ursus,* the cat *Felis,* and weasel *Mustela.*

Order *Pinnipedia* consists of swimming mammals similar in many ways to carnivores, such as the seal *Phoca,* walrus *Odobenus,* and sea lion *Zalophus.*

Order *Tubulidentata* contains only the aardvark *Orycteropus* of Africa south of the Sahara. It resembles a large-size pig with teeth of a strange, tabular form found only in the sides of the mouth.

Order *Proboscidea* is now represented by only one type of animal, the elephant. These massive, thick-skinned mammals have a nose and upper lip that extend into a trunk tipped with nostrils.

The two upper incisors are elongated as tusks, and only one or two molars at a time are on each side of the upper and lower jaws; there are no canines or premolars. The teeth are large and have many folded rows of enamel.

An elastic pad behind the toes bears the animal's weight; the toes have nail-like hoofs on three to five digits, depending on the species.

Order *Artiodactyla* contains the even-toed hoofed mammals. These animals have two or four toes; on each toe is a horny hoof that reaches the ground. *Sus,* the boar or pig, and *Hippopotamus* have four toes on each foot. They also have a simple stomach.

Other artiodactyls have two toes on each foot and a four-part stomach; they chew regurgitated food in the form of a cud. Included in this second group are the camel, the caribou, the deer, the cow, the giraffe, the antelope , the sheep, the goat, and the musk ox.

Order *Hyracoidea* contains nine kinds of African and Near Eastern animals known as conies, dassies, or hyraxes, whose toes bear flattened nails resembling hoofs—four on the front feet and three on the rear. The

BIRTH OF AN OSTRICH. The egg, which is the largest laid by any living bird, is deposited in the sand and incubated by the heat of the sun. The ostrich, a flightless bird, lives in many parts of Africa.

EARL THEISEN

soles of the feet have special suction cups that help the animals climb on rocks and trees.

Order *Sirenia* is composed of four kinds of sea cows, large animals of the seacoasts that are said to have confused homesick sailors into believing in mermaids.

Order *Perissodactyla* is composed of the odd-toed hoofed animals. They bear their weight on either the middle toe or the three middle toes; there is a hoof on each functional toe. There are no canine teeth, but there are incisors and molars in both jaws. The stomach is simple, so no cud is formed. The best-known genera include the horse and zebra *Equus*, the tapir *Tapirus*, and the rhinoceros *Rhinoceros*.

Order *Cetacea* contains the whales and their kin. Toothed whales, such as the sperm whale *Physeter*, the killer whale *Orcinus*, and the dolphin *Delphinus*, have identical enamel-less teeth and are carnivorous, preying on fishes, squids, and other marine animals.

Toothless (whalebone) whales, such as the great blue whale *Balaenoptera* (which, at 110 feet and 150 tons, is the largest animal of all time) and the right whale *Balaena*, strain food from the sea between parallel fringed plates of whalebone that hang from the inside of the upper jaw.

Both types of whale have a body highly adapted for swimming and diving: the forelimbs are reduced to flippers, the hind limbs have disappeared (although a pelvic girdle remains), and the tail is flattened into a pair of transverse fleshy flukes.

ZOOLOGY IN PERSPECTIVE

The scientific study of animals, with information carefully arranged, began with the works (336–323 BC) of Aristotle—the "Father of Zoology"—a Greek physician and naturalist who studied under Plato and served as tutor to the prince who later became Alexander the Great. Through his works, Aristotle showed himself eager for knowledge for its own sake and ready to relate his knowledge of nonhuman creatures to man.

For some 1,800 years after Aristotle, few people realized that some ideas and information in his works were incomplete and erroneous, and that new discoveries could be important. Among the first to correct Aristotle's mistakes was Andreas Vesalius (1514–1564), a Belgian physician, whose illustrated work on human anatomy appeared in 1543 and earned him the reputation of the "Father of Modern Anatomy." A Swiss contemporary of Vesalius, the naturalist Konrad von Gesner (1516–1565), compiled information on the known kinds of animals in a five-volume encyclopedia he published between 1551 and 1587.

In the 1600s, 1700s, and 1800s, discoveries came so quickly that it was hard for zoologists to fit them all together, and some facts remained unappreciated for many years. Even details visible through the microscope did not lead to immediate understanding.

Milestones in this period included William Harvey's (1578–1657) proof that the human heart circulates blood (1628), Robert Hooke's (1635–1703) discovery and naming of cells in cork (1665), Jean Baptiste de Lamarck's (1744–1829) conclusion that living things are evolving (1801—his theory that evolution came about through use and disuse was disproved), and Mattias Schleiden (1804–1881) and Theodor Schwann's (1810–1882) theory that all living things are either cells or composed of cells. Many zoologists after 1760 were content to improve upon the classification system established by Karl von Linné (1758).

EVOLUTION. Just as Schleiden and Schwann's cell theory provided a unifying concept among living things, so the theory of evolution provided an explanation for Sir Richard Owen's (1804–1892) principles of homology and analogy, set forth in 1843. Overwhelming evidence that evolution had occurred and a theory of its method through natural selection was provided in 1859 when Charles Darwin (1809–1882) published his book *On the Origin of Species*.

Darwin's lack of information on genetics forced him to assume that the visible variations he saw in each species followed an inheritable pattern. This first book and his later works stimulated zoologists all over the world to fresh research, which uncovered a wealth of new evidence supporting the theory of organic evolution.

The first experimental work to support Darwin's work came during his lifetime, in the statistical studies of inheritance in garden peas by the Austrian monk Gregor Mendel (1822–1884). This work, published in 1865, was not "discovered" until 1900, when three different research scientists brought it to the world's attention. In the meantime, W. Kuhne had discovered the nature of enzyme action (1878), W. Flemming had gained a consecutive understanding of the events in the cell division (1882), E. Van Beneden had discovered *meiosis* (1887), and Henry F. Osborne (1857–1935) had recognized the evolutionary principle of adaptive radiation.

MOLECULAR STUDIES. Many recent discoveries and theories have drawn attention to chemical similarities among animals, among plants, and between plants and animals. They have focused attention at the molecular level, leading to a greater appreciation of the steps in organic evolution—particularly those that occurred before the animals that have left fossils.

In 1916, Thomas H. Morgan (1866–1945) presented his theory of the gene; in 1953, the nature of the genetic code, in terms of the molecular structure of the DNA in the chromosomes, was visualized from the work of M. H. F. Wilkins, F. H. C. Crick, and J. D. Watson; today the fine details of the code are being worked out.

In 1929, K. Lohmann discovered ATP, the carrier of energy in living systems; in 1937, Sir Hans Krebs accounted for the citric acid cycle in the mitochondria of each cell as it carries on respiration, using oxygen and producing ATP.

Wendell Stanley (1904–) discovered in 1935 that a virus can be purified until it becomes a nonliving crystal without losing its ability to cause a disease; this reopened the question of the line between the living and the nonliving and led to new considerations of the origin of life and the chemical evolution that preceded the appearance of recognizable plants and animals. In 1953, in Harold Urey's laboratory, Stanley L. Miller demonstrated that organic compounds can form spontaneously under conditions very similar to those that geologists envision for the Earth during the Archeozoic Era.

NEW PERSPECTIVES. With all of this information, the zoologist is able to interpret the range of animal life, both extinct and living, in a new way. He sees the first long period after life began as one during which chemical systems evolved. To survive, each system had to meet the fundamental requirements for life: the ability to absorb from its environment the chemical substances and energy it needed and the ability to reproduce. The zoologist assumes that an almost infinite number of combinations was tried and that each successful one progressed by adding slight variations that improved its chances of survival.

Until a modern form of photosynthesis released quantities of oxygen into the atmosphere, it is believed that there was no basis for *aerobic respiration* on Earth. It is possible that some ancestors of protozoans lived without oxygen, and the same may be true of unicellular ancestors of other phyla.

Predatory animals could not have existed until oxygen was present, for only aerobic respiration allows rapid expenditure of energy for more than a few seconds. Multicellular parasites can thrive without oxygen, but only so long as they have larger, multicellular hosts that carry on aerobic respiration.

Since the Cambrian Age, which began some 550 million years ago, when animals reached a size and firmness of body that made fossils possible, essentially all the phyla—and many of the classes—are represented. Each ancestral line is of about the same length, but some animals have changed more in structure than others. To a great extent, the ones that have changed most have spread from the sea to fresh water and onto the land, into deserts, hot springs, and petroleum pools; today, the seas hold most of the animals that have changed least.

Although all animals have become more specialized in structure and function, some have become too specialized to survive a change in environment. An example is the dinosaurs, which became extinct with the advent of the Ice Age. Thus, the phyla of modern animals are regarded as alternative ways of living, all equally successful in their respective environments and all incorporating general features of life that evolved before the Cambrian Age.

—Lorus J. Milne and Margery Milne

PLANT LIFE

All living organisms on earth are either animals or plants. The main characteristics that distinguish plants are ability to manufacture their own food from raw materials through the process of photosynthesis; general presence of chlorophyll; fixed location in their environment; and cellulose in their body structure.

These differentiating qualities are quite obvious in the higher forms of plant life. The acacia tree, for example, could hardly be confused with the giraffe that browses on its leaves. In the lower forms, however, the differences are not so clear: the molds and fungi, lacking in chlorophyll, cannot manufacture their own food and therefore must live parasitically on other plants; some plants ingest and devour insects; the algae, diatoms, bacteria, and many seaweeds move about freely and often vigorously in water, soil, and air; the slime molds are totally lacking in the characteristic plant building materials, cellulose. Yet all of these are plants.

In the very lowest orders the distinctions between plants and animals cannot be made at all, and in the realm of microbiology, plants and animals are considered as a single group of living organisms. This provides still more evidence of the common ancestry of plant and animal life.

Plants are literally vital to all animal life on earth. Without plants there would be no oxygen; without grass the grazing animals—and man—could not survive. Plants make up the staple food of mankind throughout the world, and only with great difficulty can man adapt himself to land where they do not thrive.

Even in the arctic wastes the chain of life depends on the simplest plants. The caribou and reindeers feed on lichens, a simple combination of algae and fungi; the seal, whose body furnishes the Eskimo with almost all the necessities of life, feeds on fish that feed on other fish, which in turn are all dependent on the simple protozoan plankton—floating sea plants.

METROPOLITAN MUSEUM OF ART: THE CLOISTERS

NATURAL SCIENCE in the Middle Ages was studied with a naive naturalism that often wandered into the fantastic. The unicorn in this fourteenth-century tapestry is a product of the imagination; the plants are not. Most have been identified by genus and species.

BOTANY

The branch of biology dealing with all aspects of plant life is called *botany*. It includes the study of structure, activities, distribution, origin, classification, and uses of plants.

Botany also touches on many other areas of study, some of which are *taxonomy*, or *plant systematics*, the grouping of related forms in a systematic order; *morphology*, the description of physical forms; *anatomy*, the phase of morphology dealing with structure; *histology*, the study of tissues; *physiology*, dealing with the processes, activities, and phenomena incidental to and characteristic of living matter; *cytology*, the study of the structure and physiology of individual cells; *pathology*, the description and investigation of the causes and control of diseases; *genetics*, the study of inheritance and breeding; *ecology*, the relationship of living organisms to their environment; and *paleobotany*, the study of fossilized plants and the evolution of plants.

In addition, there are several botanic specialities dealing with particular groups of plants. Some of these are *microbiology*, the study of microscopic forms of life; *bacteriology*, the study of bacteria; and such other branches of botany as *algology*, *mycology*, and *lichenology*, dealing with algae, fungi, and lichens, respectively.

Other related fields, once considered parts of botany but now regarded as practical sciences are *agronomy*, dealing with field-crop production; *horticulture*, dealing with greenhouse, garden, and orchard plants; and *forestry*, dealing with trees and forests.

HISTORY. The first studies of plants were primarily concerned with their magical and medicinal values; and the early stages of botany are more closely allied to myth, magic, and poetry than to the scientific method. Yet such contemporary drugs as digitalis, quinine, paregoric, and morphine had their origins in medicinal plants that for centuries have been man's pharmacological storehouse.

The first actual botanist was Theophrastus (c. 372–c. 287 BC), a pupil of Aristotle. He described and categorized plants, dividing them arbitrarily into *trees*, *bushes*, and *herbs*. In the first century AD Dioscorides and Pliny the Elder described many plants, stressing medical uses.

It was not until the 1500s that botany, and especially the classification of plants, became systematized. The invention of printing made possible the publication of the first *herbals*, books describing wild and cultivated plants and illustrated with woodcuts. One of the best of these herbals was written by Otto Brunfels (1488–1534), a German botanist.

Contemporary with Brunfels were Hieronymus Bock, author of *Materia Medica*, and Leonhard Fuchs (1501–1566), who wrote a glossary of technical terms, the first terminology of botany. Up to this point, attempts at systematic classification were crude at best; and the main concern of botanists was still with the medicinal virtues of plants.

Taxonomy. The need to distinguish useful plants led gradually to greater accuracy of description and eventually to an interest in *taxonomy*, the organization of the myriad plant forms in a scheme demonstrating their interrelationships.

The Italian botanist Andrea Cesalpino (1519–1603) made the first formal attempt at a methodical classification of plants. In his *De Plantis* (1583), he divided the 1,520 plants then known into 15 classes, basing his divisions on the character of the fruit. John Ray (1627–1705), an English naturalist, developed a system of natural affinities. He separated the flowering from the flowerless plants, calling them *dicotyledons* (having two seed leaves) and *monocotyledons* (having one seed leaf), respectively. The same names are used today.

NEW YORK PUBLIC LIBRARY

LINNAEUS, naturalist, physician, and philosopher, founded systematic botany.

Linnaeus. Karl von Linné (1707–1778), the Swedish botanist who is also known by his Latin name Carolus Linnaeus, founded a system of nomenclature that was based on the characteristics of stamens and pistils; since these are the reproductive organs of the flower, the system is often called the *sexual system.* It was essentially an artificial arrangement, as Linnaeus himself knew; he considered it only a temporary method, to be used until a natural system of classification was developed.

Linnaeus also contributed much to *nomenclature*, or the naming of plants; and his *binomial method*, with one name for the *genus* and a second, qualifying word for the *species*, is universally accepted.

DARWIN was often caricatured and his ideas mocked by an angry and sceptical public.

A notable advance was made by Antoine de Jussieu (1748–1836), professor of botany at the Jardin des Plantes in Paris. The necessity for logical arrangement of the plants caused him to devote considerable time to the problem, and in his *Genera Plantarum* (1789) he outlined a plan that included the best features of Ray's and Linnaeus' systems. It was based on a close study of plant organs, made use of Linnaeus' simple definitions, and showed in general the natural relationships of plants, thus forming the basis for the natural classification predicted by Linnaeus.

Augustin de Candolle (1778–1841) showed that the natural affinities of plants must be found by a study of morphology, not of physiology.

Darwin. The most important influence on taxonomy was Darwin's theory of evolution and the origin of species by natural selection, published in 1859. "Natural" came to mean related by descent, and any classification scheme became a cross section of the course of evolution. This relationship of systematics and evolution, together with the identification and classification of new species, is the main concern of taxonomists now.

CLASSIFICATION OF PLANTS. The ordering of the more than 350,000 known forms of plants into groups of related organisms is the task of taxonomy. In the taxonomic system, every living plant form has its place in relationship to all other forms, both living and extinct.

Plants that seem to be related because of similarities in form and structure are assigned to a definite group; and smaller groups, based on some common characteristics, are formed within larger divisions. Some organisms do not fit well into any group, or perhaps they fit indifferently into several groups, for during the course of evolution all types and degrees of diversity have developed.

Nature is not concerned with the maintenance of groups, and organisms may be shifted from one group to another as taxonomic knowledge increases. The purpose of classification is to present a natural system of relationships; ideally, it aims not to create an artificial ordering but to discover the order inherent in the working of natural forces. Thus, a truly natural system of classification is also an evolutionary system, reflecting in its organization progressive differentiation of plant forms.

Logic of Classification. The logic of the taxonomic system is that of describing the order and connection of the various forms in terms of their relative similarity or, stated conversely, in terms of their progressive differentiation. Starting from the most general or inclusive group, all living organisms belong to the *Organisma;* this expresses the similarities between all animals and plants and their differences from all nonliving forms.

The next step in differentiation is that between plants and animals: all plants belong to the kingdom *Planta*, whose members share some of the properties of animals and are differentiated from them in other commonly shared properties. Differentiation continues in this fashion, with each group sharing certain characteristics with the preceding group, yet also differing from it in other characteristics.

In each case the succeeding group is a function of the preceding one and branches from it like a limb on a tree. Each limb in turn has smaller branches, and these branches give rise to even smaller ones. When a particular group has been so closely defined that there can be found no further characteristics to differentiate its members, the system is closed.

Every property of every plant in the series has a place in the system at some level. Thus, every member of the last group in each series can be placed in relation to every other plant, and organism, in the world.

As just described, the taxonomic system represents a progression from overall similarity to greater and greater differentiation, or *heterogeneity*. Viewed from the other direction, that is starting from the most differentiated unit, the same system presents a picture of greater and greater similarity, or *homogeneity*.

In order to find the position of a particular plant—for example, a white oak tree *(Quercus alba)*—in the whole taxonomic system, it is necessary to start with the particular and proceed through the more and more general groups. From this standpoint, the units of classification are ranged in order of increasing inclusiveness, from the most differentiated to the most undifferentiated.

The basic unit is the *species*, in this case *alba* (the white oak); the next step is the *genus* (plural *genera*), which for the oak is *Quercus*. (In some cases there are also types or *varieties* of species, but these are generally artificial and are not maintained in nature.) A group of closely related genera is a *family*; the oaks belong to the family *Fagaceae*, together with the chestnuts and beeches.

Families are in turn grouped in *orders*, and the *Fagaceae* belong to the order *Fagales*, which includes two other families, the birch and the beech. Groups of related orders are called *classes*, in this case the *Dicotyledoneae*, a large class of 47 orders that includes most flowering trees and many flowering plants.

The class *Dicotyledoneae*, together with the *Monocotyledoneae*, forms the important subdivision *Angiospermae*, to which belong all flowering trees and plants. The largest group within the plant kingdom is the *phylum*, and the oak is a member of the seed-bearing *Spermatophyta*.

The divisions into orders, classes, and phyla, and the groupings of the phyla into subkingdoms, has not been conclusively determined; and a number of systems are current. Many botanists hold that the *Thallophyta* as formerly constituted and the *Pteridophyta* do not represent true natural groupings; some have designated two main subkingdoms as *Thallophyta* and *Embryophyta*. Other systems propose a phylum *Tracheophyta*, which groups together all of the vascular plants.

MANDRAKE, long believed to possess magical powers and credited with human attributes, was thought to shriek if uprooted.

Kingdom Planta

Phylum **Thallophyta** (simple plants without roots, stems, leaves; usually one-celled reproductive organs).
- Subdivision **Algae** (containing chlorophyll, e.g., pond scums, seaweeds).
- Subdivision **Fungi** (lacking Chlorophyll, e.g., molds, mushrooms).

Phylum **Bryophyta** (simple plants without roots, stems, leaves; many-celled reproductive organs).
- Class **Hepaticae** (live worts)
- Class **Musci** (mosses)

Phylum **Pteridophyta** (complex plants with true roots, stems, and leaves, and possessing vascular tissue, but lacking seeds).
- Class **Filicineae** (ferns)
- Class **Equisetineae** (horsetails)
- Class **Lycopodineae** (club mosses)

Phylum **Spermatophyta** (complex plants with true roots, stems, and leaves, vascular tissue, and bearing seeds).
- Subdivision **Gymnospermae** ("naked seed" plants, bearing cones, e.g., pines, spruce).
- Subdivision **Angiospermae** ("covered seed" plants or true flowering plants; e.g., grasses, maples, roses, orchids)
  - Class **Monocotyledoneae** (embryo bearing one cotyledon and flower parts typically in 3's; e.g., tulips, orchids).
  - Class **Dicotyledoneae** (embryo bearing two cotyledons and flower parts in 4's or 5's; e.g., roses, beans).

The classification given above, although not the most recent or the most accurate in terms of evolutionary relationships, is still common.

LIVING PLANTS

THALLOPHYTES. The phylum *Thallophyta* consists of plants possessing neither true roots, stems, nor leaves. They may be unicellular or multicellular; each type may be found in various forms.

Two divisions are recognized, algae and fungi. Algae are usually "independent," possess chlorophyll, and are able to manufacture their own food. Fungi do not possess chlorophyll and are dependent upon an outside source of carbon-furnishing food, such as the carbohydrates.

Algae. The oldest and simplest of all green plants, the *algae* range from simple unicellular organisms to complex, multicellular colonies. They vary in size from diatoms only a fraction of an inch in diameter to seaweeds 150 to 200 feet long. Although primarily water plants, algae grow all over the globe, from the ice and snow of the Arctic regions to the backs of certain turtles living in tropical regions.

While many species are attached, others constitute much of the floating life of aquatic habitats. The smaller forms especially are an important source of food for aquatic animal life. Reproduction may be effected bv simple *fission* (splitting off) or by nonsexual spores that may be motile or nonmotile.

Except for the flowering plants, the algae are the most numerous and widespread of all green plants; but they are a heterogeneous group whose exact interrelationships are not known. They include over 50,000 known species, grouped under different systems in various phyla. Some of the most important types are the following:

Green algae, the *Chlorophyceae*, are those algae in which chlorophyll is conspicuous. They are more numerous than all others combined and include about 10,000 species. The more primitive forms are unicellular, and some possess whiplike motile organs (cilia) in the active state.

Reproduction is primarily by simple fission—a parent cell becoming quiescent and dividing in two. Since death of the parent cell does not occur, any cell may be immortal. Sexual reproduction also occurs when two motile cells (or gametes) fuse to form a new cell.

The simple green algae are of special interest as steps in the chain of evolution. *Euglena*, a single-celled form with one whiplike cilia, may stand at the diverging point of plants and animals. Other green algae may be ancestors of the higher plants.

Filamentous forms of the green algae are numerous; these are the typical pond scums. Green algae of various types also occur in the sea, from the many one-celled forms that constitute large parts of plankton to such complex seaweeds as *Ulva*, the sea lettuce, which consists of colonies of cells forming broad ribbons or leaflike structures.

Diatoms are peculiar unicellular algae whose cell walls are impregnated with silica and fitted together in a boxlike shape. The hard, glass-like walls persist after the death of the cells and settle to the ocean floor, forming large deposits of *diatomaceous earth*, an exceptionally fine abrasive. Much petroleum is also of diatom origin.

Blue-green algae, the *Myxophyceae*, owe their common name to the occurrence of a blue pigment along with the chlorophyll. They may occur as single cells, but colony-forming species are more common. A gelatinous sheath or extensive jelly may enclose the colony, and in stagnant water the blue-green algae may give off a disagreeable odor. There are simple algae, apparently with less organization of the cell than any other algal group. Reproduction is by simple fission.

Brown algae, the *Phaeophyceae,* show great structural complexity and include the largest of all algal forms. They are almost exclusively marine, and include common rockweed and seaweeds. Species of *Laminaria,* or kelp, occur in deeper, colder waters. On the Pacific coast a giant kelp *Nereocystis,* often grows to exceed one hundred feet in a season. Many species break loose and float with the ocean current, often in great quantities; the Sargasso Sea is named after one algae genus, *Sargassum.* Chlorophyll is present in these algae, but is masked by the occurrence of another, golden-brown pigment in the plastid.

Reproduction in the brown algae is by spores or, often, by gamete production. The common rockweeds, *Fucus,* reproduce only by sexual gametes, and their life cycle resembles that of the flowering plants. The brown algae are used for food, especially in the Orient, and supply iodine and considerable fertilizer.

Red algae, the *Rhodophyceae,* are generally red and typically marine. They may be filamentous, massive, and highly differentiated; or they may be membranous. They often frequent deep waters, and in regions of plentiful sunshine have been found at depths exceeding three hundred feet.

FUNGI. The second great division of the thallophytes, the *fungi,* are commonly filamentous in the vegetative condition and are distinguishable from the algae primarily by the absence of chlorophyll.

Practically speaking, fungi occur wherever organic matter exists, since one or more species may inhabit any dead or non-living material and many species attack living tissues, especially those of seed plants. The chief classes of fungi are *Phycomycetes, Ascomycetes,* and *Basidiomycetes;* they also include the bacteria and the slime molds.

Phycomycetes are the alga-like fungi. The vegetative plant body is a *mycelium* and consists of a greatly branched system of threads, or *hyphae,* not clearly divided into distinct cells. There are two large groups.

The first group, the *zygomycetes,* includes the common bread mold. The mold reproduces both nonsexually and sexually. Nonsexual spores are produced in structures known as *sporangia,* which are borne at the ends of specialized filaments; in sexual reproduction, hyphae cut off at their ends form cells that act as gametes, which fuse together and form a thick-walled resting spore known as the *zygospore.*

The second group, the *Oomycetes,* includes the water molds and downy mildews. Some species of water molds are parasitic; one form causes a disease of fish. The downy mildews are parasitic on seed plants, such as the grape.

Ascomycetes are fungi characterized by a saclike reproductive body that produces spores. More than 10,000 species of ascomycetes are known, and they occur in many different situations.

The *saprophytic* species (those inhabiting dead material) are found abundantly upon decaying vegetation and in or on the soil. The blue and green molds of foods, such as *penicillium,* from which the drug penicillin is made, belong to this group. It also includes a few families of fungi with large and fleshy fruit bodies, such as the edible morels and truffles. Another useful group is that of the *yeasts.*

Parasitic species are likewise numerous, and cause such plant diseases as apple scab and rose mildew.

Basidiomycetes are a class of fungi that comprises orders and families varying in both structure and habitat, but the different subclasses are all related through the possession of *basidia* (club-shaped cells) that typically bear four spores. *Rusts* and *smuts* constitute the main parasitic groups; other divisions include the vast majority of the fleshy or woody fungi, such as mushrooms. Smut fungi are most recognizable during the spore stage.

Some rust fungi exhibit an exceedingly complicated life history. The black-stem rust of wheat has one stage on the wheat and related plants and another on the barberry. The apple rust passes from the apple to the red cedar and from the red cedar back to the apple. On the host plant the fungi are ordinarily characterized by the occurrence of rusty spots; these are the beds of the fungal spores.

Woods and fields yield hundreds of fungi variously known as mushrooms, toadstools, and puffballs. They grow in the soil or on decaying logs and vegetation, and a considerable number cause heartwood or sapwood decay of trees. Among fleshy forms there are both edible and poisonous species. They vary in texture from soft, spongy forms to hard, woodlike growths; in size, they range from the microscopic to two feet across.

Bacteria. Among the fungi there are usually included the smallest plants known, the *bacteria,* or *schizomycetes.* They seem, however, to constitute an independent group, perhaps related to some of the lower algae. They are so small that they are visible to the unaided eye only when growing in colonies of many hundreds or thousands. If they were arranged end to end, it would require about 500 of these bacteria of average size to reach across the head of a pin.

Under favorable conditions, bacteria reproduce by fission at an astonishingly rapid rate. Some also produce spores at the rate of a single spore per cell; these enable them to survive hostile conditions.

Bacteria are universally found in air, water, and soil, as well as on and within all living bodies. They have harmful effects as producers of disease in plants, animals, and man, but they are also beneficial as agents of the decay that return nutrients to the soil, and as the *nitrogen bacteria* that are vital in completing the cycle of this important element. Bacteria are also the agents of many useful fermentation processes. They are used commercially in the dairy industry, in wine-making, and in the curing of tobacco.

Slime Molds. *Myxomycetes,* or *slime molds,* are organisms sometime classed as plants and sometimes classed as animals. They form a slimy mass of naked protoplasm in decaying matter in moist, warm places.

The multinucleate protoplasm moves in amoeboid fashion and can ingest solid food, a characteristic that links the slime molds closely to the animal world.

Lichens. A *lichen* consists of a fungus and an algae growing together in a *symbiotic,* or mutually helpful, relationship to form a dual colony so closely associated that the colonial composite acts as one. To a degree the fungus is parasitic upon the alga and

BETTMANN ARCHIVE

BENEATH THE FOREST FLOOR, the mushroom plant extends a network of filaments. The familiar cap, with its spore-producing gills, makes up the "fruit," or reproductive body.

NATIONAL COAL ASSOCIATION

HELEN BUTTFIELD

FERNS have survived almost unchanged from the time 350 million years ago when they left their imprint (*left*) fossilized in the earth.

holds it captive, but the alga is thus enabled to grow in many places where it could not otherwise exist. Within the lichen, the alga multiplies vegetatively; the fungus has its characteristic spore reproduction.

Lichens commonly grow on trunks and branches of trees, on rocks, and on soil. Some are flat and leafy, and some are mosslike. Reindeer moss, really a lichen, is typical of the mosslike group. Lichens are often gray-green and are very resistant to extremes of cold and drought. In the Arctic, they sustain the vast reindeer herds.

MOSSES AND MOSSLIKE PLANTS. Members of the class *Bryophyta,* mosses and mosslike plants, constitute a considerable group of green plants higher in the scale of development than the algae, but less complex than seed plants. In some classifications they are considered the lowest members of the subkingdom *Embryophyta,* plants that form embryos. However, they lack the vascular tissues characteristic of the higher forms.

Mosses. Small, green, flowerless plants, the mosses grow erect but are usually no more than one to two inches high. They occur commonly in moist environments as miniature velvety or feathery growths carpeting the ground, growing on rocks, on the trunks of trees, and even in ponds or running water. The parts of the mosses usually visible look like those of higher plants; they have a stemlike axis with numerous leaves. But there is no true stem or root, such as characterize seed plants.

The mosses are among the most primitive of plants and have persisted almost unchanged since they arose about 300 million years ago. Because they have given rise to no other forms, the mosses are considered a terminal evolutionary group.

Liverworts. Primitive land and fresh-water plants, the liverworts are closely related to the mosses. Their growth pattern is flat and branching, and they resemble some seaweeds in appearance. They are the simplest land plants surviving, and may have evolved from certain algae.

FERNS AND FERN ALLIES. The *ferns,* together with a few related families of the class *Pteridophyta,* are the remnants of a once flourishing form of plant life that dominated the earth's vegetation for years. Geologically one of the oldest groups, the ferns originated in the Paleozoic era, 350 million years ago. Giant *tree ferns* once covered the earth, and the energy they gathered from the sun and stored in their tissues is preserved in the earth's great coal deposits.

Ferns are characterized by large, divided, feather-like leaves, or *fronds,* which usually uncurl from the tip. They have short stems that often grow underground. Unlike the mosses, they have true roots. Ferns, like the higher plants, have vascular tissues that transport nutrients and water from the roots to the leaves.

Ferns are abundant on the moist, shaded, forest floor and along streams; some species, however, thrive on rocky cliffs or slopes. Geographically, these plants range from the Arctic to the equatorial jungles and rain forests, where tree ferns often reach heights of forty feet.

Alternation of Generations. Reproduction in the ferns is marked by a distinct *alternation of generations,* which involves two different forms: the first is the familiar fern plant, which has a root, stem, and leaves; the second is a thin, flat, heart-shaped plant called the *prothallium.* Prothallia produce male and female sex organs in which gametes are developed. A male gamete fuses with an egg, and

HELEN BUTTFIELD

LICHENS, as enduring as the stone they cover, are among the most primitive land plants.

STAR OF BETHLEHEM, BY LEONARDO DA VINCI: ALINARI-ART REFERENCE BUREAU

THE FLOWERING PLANTS are the highest evolutionary form in the plant kingdom and the flower itself is the most effective means of reproduction evolved by any of the plants.

from the fertilized egg develops the fern plant with its roots, stem, and leaves.

On the underside of the fern's leaves there are small spore cases, or *sporangia,* which are often distributed as dots, rows, or larger masses. In these spore cases, large numbers of single-celled spores are produced; when they germinate, they give rise to the prothallia. Thus, the prothallia are *gametophytes* producing gametes, and the leafy fern is a *sporophyte* producing spores.

Club Mosses. Also called *ground pines, club mosses* are rarely more than three feet high. These relatives of the fern are neither mosses nor pines; they are small evergreen plants that have simple leaves resembling pine or hemlock needles. Sometimes they grow upright, but often they trail on the ground, where they propagate by means of *runners,* or elongations of the root stock that send up new sprounts. They also reproduce spores.

Horsetails. Only twenty-five living species of *horsetails* are known, and all of these belong to a single genus. Many extinct forms, however, have been found in fossils. Horsetails are the only surviving representatives of a group containing members that grew to over ninety feet.

Horsetails are composed of underground stems that send up tall, vertical, jointed stalks and of branches covered with scalelike leaves. These plants are also called *scouring rushes,* a name derived from the rough texture imparted by the silica they contain.

Cycads and the Ginkgo. The *cycads* are plants that resemble tree ferns in general appearance, but reproduce by means of seeds. Few in number today, they are the remains of a once dominant group which flourished in the great fern forests of the Permian age.

The *ginkgo,* also called the maidenhair tree, has fan-shaped leaflets, similar in form to those of the maidenhair fern. This leaf-form is found on no other flowering plant and the ginkgo represents a missing link between ferns and flowering plants.

SEED PLANTS. The *spermatophyta,* or seed plants, the highest division in the plant kingdom, contains all plants that reproduce by means of seeds. It includes all living trees and flowering plants. Seed plants form the dominant part of the vegetation of the earth today and thus can be considered the highest, or most successful, plant form.

Seed plants embrace two major divisions: the *gymnosperms,* or "naked seed" plants, in which the seeds lie exposed and unprotected on the cone scales, and the *angiosperms* or "enclosed seed" plants in which the seeds are borne inside a jar-shaped swelling, the ovary, located at the base of the flower pistil.

Gymnosperms. Familiar examples of gymnosperms are the pines, spruces, and other evergreen trees or shrubs with conelike fruits, known as the *Coniferae.* This group forms the great coniferous forests of the temperate zones and includes the largest plants on earth, the giant redwoods.

Vascular or woody tissues reach a high state of development in the stems and roots of conifers. The leaves of mature pines are called *needles*; they are retained for more than a year, giving rise to the name *evergreen.* These needles are highly specialized in structure and are adapted to resist extremes of cold and dryness. Morphologically they are grouped in *fascicles* (clusters) of two, three, or five needles, depending upon the species.

In the conifers the spore-bearing leaves or scales *(sporophylls)* do not constitute a flower. As the scales of the seed-bearing cones separate, the winged seeds may be seen, a pair under each scale.

In the pine, the embryo plant within the seed is surrounded by a considerable food-storage layer, called the *endosperm,* and the whole is enclosed by the seed coat, or *integument.*

Maturity of the seed is accompanied by a certain amount of drying out and by a period of inactivity, or *dormancy.* The seed remains dormant until it absorbs water, and *germination,* or sprouting, ensues. The young seedling, free of the seed coat, consists of root; bud, or *plumule;* and a whorl of long, green seed leaves, or cotyledons.

Angiosperms. The angiosperms, or true *flowering plants,* include the great majority of the familiar flowers and weeds, as well as all the trees and shrubs except the gymnosperms. The angiosperms have seeds and a complex tissue system in which vascular or woody elements attain a further advanced state of differentiation, and they exhibit the highest evolutionary development through the production of flowers.

The angiosperms, which include more than half the known plants—about 200,000 species—are subdivided into two groups. In the *monocotyledons* the young seedling bears a single seed leaf, or *cotyledon. Dicotyledons* bear two seed leaves, sometimes thickened, that provide reserves of organic food. Another characteristic of dicotyledons is apparent in the mature plant: the *venation,* or vein system, of the monocotyledons is parallel, as in corn; in dicotyledons it is netted, or *reticulate,* as in the rose.

There are other characteristics, of both floral parts and inner structure, that are generally distinctive of the two groups, such as the three-part flower structure of the monocotyledons, as opposed to the four or five parts that are found in dicotyledons.

Monocotyledons include such families as the grasses *(Gramineae)* which furnish the cereal grains, pasture grasses, and ornamental grasses;

tropical palms *(Palmaceae)*, including the date and coconut palms; lilies and related plants *(Liliaceae)*; bananas and plantains *(Musaceae)*; and the highly prized orchids *(Orchidaceae)*, famous both for the delicacy and variety of their waxlike flowers and for the remarkable modification in floral parts, the latter having come about in adjustment to insect pollination.

The orchids and the grasses show extreme differences in floral anatomy. The orchid flower's extreme complexity represents a high point in evolutionary development; the grass flower, also an advanced form, shows a high degree of reduction of parts, consisting of a pistil with ovary, style, and two feathery stigmas, and three stamens enclosed in scalelike leaves, or *bracts*.

Dicotyledons are nearly five times as numerous as monocotyledons; they are also regarded as geologically the older group. They have attained far greater diversity in form and have advanced to higher types of development.

Among families exhibiting less complex types of floral structure are the willows and the poplars *(Salicaceae)*, which have simple flowers in spikelike branches called *catkins*. Related to the willows are other families without showy floral parts, such as the walnut-hickory family *(Juglandoceae)* and elms *(Ulmaceae)*.

In a varied group of families that includes the pinks, sweet Williams, carnations *(Caryophyllaceae)*, water lilies *(Nymphaeceae)*, buttercups, columbines, peonies, and clematis *(Ranunculaceae)*, there are many wild and garden plants. All of these have distinct petals (are *polypetalous*), and the carpels are either separate or united.

The rose family *(Rosaceae)* and the pulse family *(Leguminosae)* are closely related, and together they constitute a very considerable part of the separate-petal series of families. As a whole, the rose family is rather heterogeneous; its members include the blackberries, roses, and cherries.

In the pulse family there are several thousand species, among which are the lupines, clovers, beans, peas, and acacias. Generally the *corolla* (floral leaf) is *papilionaceous* (butterfly-like); the fruit is characteristically a pod or legume.

Among the families of dicotyledons in which the petals are united are the mints *(Labiatae)*, fragrant herbs, and shrubs with flowers that are commonly two-lipped. The nightshade family *(Solanaceae)* includes the potato, tomato, eggplant, tobacco, and petunia; the flowers are prevailingly rotate, and the fruit is a *berry* (as in tomato) or a capsule (as in Jimson weed). Melons are the fruit of certain members of the gourd family *(Cucurbitaceae)*.

Botanists generally agree in placing the composite family *(Compositae)* at the pinnacle of development in the plant kingdom. This is the family of sunflowers and goldenrods, asters and thistles, chrysanthemums and dahlias. It also includes lettuce, globe artichoke, and salsify. It is a huge family with perhaps 12,000 species, nearly all of which are annuals or perennial herbs; the woody members are not trees.

The outstanding characteristic of the composite family is the compact head of many flowers—some with ray flowers (as in sunflowers) and some without (as in thistles). The fruits are hard and one-seeded *(achenes)*, and are often provided with a feathery appendage, or *pappus* (as in dandelions), that helps wind distribution of the seeds.

AMERICAN MUSEUM OF NATURAL HISTORY

DECIMATED by the caddis fly, these leaf skeletons reveal the supporting venation.

PLANT STRUCTURE

From the smallest bacteria, 1/50,000 of an inch long, to the largest angiosperm, 372 feet tall, plants are made up of *cells*, the basic units of life. The study of cell structure and function is called *cytology*.

CELLS. A typical plant cell is enclosed by the *cell wall*, which is composed primarily of cellulose and nonliving substances and gives support and form to the cell and the whole plant. The enclosed *protoplasm* is a viscous, transparent, living matter composed of water, proteins, sugars, fats, acids, and salts.

Protoplasm is differentiated into two aspects. The dark, usually round *nucleus* controls the chemical activities of the cell and bears the *chromosomes*, carriers of hereditary traits. Surrounding the nucleus is the *cytoplasm*, in which are the structures that carry on the physiological cell functions: the *plastids*, including chlorophyll-bearing *chloroplasts* that are capable of photosynthesis; the *chromoplasts*, containing red and yellow coloring; the *leucoplasts*, which build sugar into starch grains; *mitochondria*, which produce chemical regulators; *vacuoles*, liquid-filled spaces that serve as storage areas; starch grains and crystals.

TISSUES. In all multicellular plants the individual cells are organized into *tissue*. *Histology* is the study of these groups of structurally similar cells that are organized to perform physiological functions.

Embryonic Tissue. *Meristem*, or young tissue, consists of thin-walled, active cells rich in protoplasm. They are able to divide by cell division and thus to continue the growth of the organ. All other tissues are derived from meristem by differentiation or modification. *Cambium*, the layer of active growth and cell formation in stems and roots of dicotyledons, is a form of meristem.

Permanent Tissue. The new cells produced by division and differentiation of the meristematic cells become the permanent tissue of the plant. As these cells mature, they no longer reproduce by cell division but expand in size. The cell walls stretch and, in many instances, thicken; the cytoplasm thins out and the major portion of the cell interior is taken up by the enlarged vacuole. Further modifications in shape and structure occur as the cell takes its place within one of the various types of permanent tissue.

There are two kinds of permanent tissues: *simple tissues*, in which each tissue is composed of similar cells; and *complex tissues*, in which different types of cells work together as a unit to perform certain physiological functions. Most plant parts are composed of simple tissues. However, the *xylem* and *phloem*, the vital carriers of water and food, are complex tissues consisting of parenchyma, fibrous, sieve, and vascular tissues.

Parenchyma, or soft tissue, consists of mature, thin-walled, usually short cells that have attained their growth, such as the tissue of ripened fruits and the green tissue of leaves.

Collenchyma tissue consists of long cells with thick angles. In many stems it occurs as a strengthening tissue just beneath the epidermis.

Sclerenchyma, or stone tissue, consists of thick-walled usually short cells, so tightly packed that they form a hard mass, as in nutshells and the "stones" of many fruits.

Fibrous and bast tissue consists of thick-walled, elongated cells so tightly packed together that they make up wood fibers (fibrous tissue) and bark (bast tissue) of the stems of most higher plants. Bast fibers are used in ropemaking and linen making.

Sieve tissue consists of elongated, usually large cells, more or less united into tubes and having only slightly thickened walls. The name derives from the perforation of the transverse partitions between the cells in a sievelike pattern; through these perforations the protoplasm connects from cell to cell. Sieve tissue occurs primarily in the young bark and is important in conveying organic food.

Vascular tissue is also tubular, but the continuity of the cavity is usually more complete than in sieve tissue. When they are young, these tubes contain protoplasm; but eventually they contain water or air. Vascular tissue occurs in the woody parts of stems and leaves.

Epidermal tissue, a single layer of cubical cells, constitutes the outer layers of leaves, young stems, and roots.

ANATOMY OF SEED PLANTS

Although the lower plants have developed quite complicated forms and structures and have adapted very successfully to a wide range of environments, the most highly differentiated and evolved structures are found in the angiosperms.

The bodies of typical seed plants are composed of four kinds of parts: *roots, stems, leaves,* and *flowers* (which in turn produce *fruit* and *seeds*). The first three structures are found also in the mosses, ferns, liverworts, and other lower forms; but flowers and their accessory products are exclusive to seed plants.

ROOTS. All living organisms, including the plants, are descended from forms which originated in the sea; and no cells, animal or vegetable, can exist without access to that most important element, water.

In the higher plants, the *root* is the vital link between the plant and its lost aquatic environment. Through its roots a plant absorbs the water that makes up most of its tissues and that it uses to carry on its life processes, as well as some oxygen and most of the mineral nutrients it requires.

Root Structure. All roots are covered at the growing tip with a *root cap* which protects the sensitive tip and serves as a boring point to push into the soil. Behind the root caps, the sides of the roots are covered with many fine *root hairs,* made up of single cells, which penetrate the soil and increase the roots' absorption.

Root Systems. The slender, branching roots of grasses, corn, and wheat are *diffuse systems* spreading over wide but relatively shallow areas. *Tap-root systems,* such as those of the carrot and beet, are made up of a single primary root that probes deeply into the soil and often is thickened to serve in food storage.

Specialized Roots. Adaptations to particular needs are the *aerial roots* of ivy and other *epiphytes* (plants that grow on other plants but are not parasitic); *prop roots* of corn and fig trees, which act as buttresses; *adventitious roots,* which drop down from the stem to lend additional support to many tropical trees; and the *aerating roots* of the cypress and mangrove, which grow above water to obtain oxygen.

STEMS. The *stem* is the part of the plant that supports the leaves and reproductive organs and supplies them with water and mineral nutrients absorbed by the roots. It also serves to carry food back down to roots and other parts for storage.

Most stems grow above the ground *(aerial stems),* but some grow below the surface *(subterranean stems).* Aerial stems are either *herbaceous* or *woody,* an important feature distinguishing groups of plants. All gymnosperms have woody stems, while those of angiosperms can be either woody or herbaceous.

It is generally believed that woody stems are the more primitive and that herbaceous stems have evolved from them. Herbaceous stems are soft and green, covered by an epidermis; they grow in length but little in diameter and are chiefly *annual* (confined to a single growing season). Woody stems are hard and are covered with a tough layer of bark. They grow considerably in diameter and are chiefly *perennial* (surviving through many growing seasons).

HELEN BUTTFIELD

POWERFUL ROOTS anchor the beech tree firmly in the ground while, deep in the soil, their growing tips spread out, searching for the moisture that nourishes the crown.

Woody Stems. Internally, *woody stems* are made up of two kinds of tissue, the outer *bark* and the inner *wood;* these are separated by a single layer of cells, the growing *cambium* layer. As the cambium cells divide the outer portions become differentiated into a layer of cells called *phloem,* and the inner cells form the *xylem.*

Phloem cells transport food, and the xylem cells conduct moisture throughout the plant. This formation of new cells causes a thickening in the diameter of the stem, and it is thus that the trunks of trees continue to increase in girth. The cells produced in the summer are larger and therefore lighter in color; the winter cells are smaller and appear darker.

This color differential creates the *annual rings* marking a year's growth. As a tree grows, the inner layers become clogged and loaded with tannins, resins, and gums. These make up the hard, dark *heartwood,* while the outer, more active layers constitute the *sapwood.*

Herbaceous Stems. Herbaceous stems resemble young woody stems in structure except that the xylem and phloem are arranged in clusters called *vascular bundles,* which are either scattered throughout the stem (monocotyledons) or arranged in a circle (dicotyledons). Herbaceous stems are chiefly annual, and growth is primarily in the length, not the diameter.

Buds. Buds are the growing ends of the stems of both woody and herbaceous plants and are the immature forms of leaves and flowers. *Terminal buds* provide growth at the tip of a stem or twig; *lateral buds* or *axillary buds* form side branches as well as leaves and flowers.

In annuals and tropical woody plants, the buds grow continuously; but in woody plants of temperate climates they become dormant when the plant slows its metabolism during the winter.

Subterranean Stems. *Rhizomes* are horizontal stems growing on or beneath the ground; they serve for storage of food and for reproduction. They are perennial and send up new shoots each year. *Tubers* are the enlarged tips of rhizomes; highly specialized for the storage of food, they are also, as with the potato, important means of propagation.

Bulbs, such as those of the onion and tulip, and *corms,* such as those of the gladiolus and crocus, are enlarged stem buds with overlapping leaves. They serve for storage and reproduction and as a means of carrying the dormant plant through seasons unfavorable to growth.

LEAVES. Leaves are the plant structures specialized primarily for the manufacture of food through photosynthesis. Thus, most leaves are constructed so as to provide the greatest possible surface area and are ar-

PLANT TYPES

IRVIN L. OAKES FROM NATIONAL AUDUBON SOCIETY

MOREL MUSHROOM

© WALT DISNEY PRODUCTIONS

AIRPLANT BLOSSOM

ANTHONY MERCICA FROM NATIONAL AUDUBON SOCIETY

BEAR GRASS

EL AL

BARLEY FIELD

EL AL

DWARF BANANA

© WALT DISNEY PRODUCTIONS

HEDGEHOG CACTUS

© WALT DISNEY PRODUCTIONS

BARCO

© WALT DISNEY PRODUCTIONS

PAPAYA

FLOWERS

ZINNIA

LOTUS

CLEOME (CASPER PLANT)

WATER LILY

SINGLE CHRYSANTHEMUM

COSMOS

PETUNIA

FLOWERS

PHOTOS BAUMAN/LOOK

GLADIOLUS

DAHLIA

HELIOTROPE

PHLOX

JAPANESE CHRYSANTHEMUM

ROSES

TREES

WILLIAM HARLOW FROM NATIONAL AUDUBON SOCIETY

TWO CROWN SHAPES OF AMERICAN ELM

DENNIS BROWN FROM NATIONAL AUDUBON SOCIETY

REDWOOD

ARTHUR W. AMBLER FROM NATIONAL AUDUBON SOCIETY

NORWAY MAPLE

RICHARD PARKER FROM NATIONAL AUDUBON SOCIETY

ASPENS

ranged so as to expose this surface to the maximum amount of sunlight.

Leaf Form. Most leaves consist of a slender stalk, or *petiole,* and a broad, expanded blade, or *lamina.* From the apex of the petiole, *veins* extend into the blade. Veins serve for mechanical support, for the transport of raw materials through the leaf, and for carrying away the products elaborated by the green cells.

Leaves vary in size from a fraction of an inch up to 60 feet, as in the palm tree. In shape they range from the long, thin blades of grass to the circular leaves of the nasturtium and water lily; there are many irregular forms also, as in the maples and oaks.

The *venation* (arrrangement of the vein) is characteristic for different kinds of plants: *net* venation in dicotyledons and *parallel* venation in monocotyledons. In *pinnate* net venation, the petiole extends as a main vein or midrib through the center of the leaf, lateral veins extending off it on each side to give a featherlike appearance; this is found in the apple and elm. In *palmate* net venation, several larger veins radiate from the apex of the petiole, as in the maple.

The leaf blade may be simple, as in the apple, or compound, consisting of *leaflets.* The leaflets may have the pinnate arrangement, as in the pea and tree of heaven; or the leaf may be palmately compound, the leaflets arranged at the end of the petiole like the fingers on a hand.

The leaves of grasses are specially modified in that, instead of a distinct petiole, the basal part of the petiole is a sheath that closely surrounds the stem. At the base of the blade there is a special growing zone, so that grass leaves continue to increase in length for long periods.

Leaves may undergo many modifications. *Bud scales* are greatly reduced leaves that overlap and protect the delicate bud tissues during cold and drought. The thick, fleshy leaves of sedum and other succulent plants store quantities of water.

Some barberry leaves are modified into *spines;* and in the pea, some of the leaflets are modified into *tendrils* that assist the plant in climbing. Very curious modifications occur in the insectivorous plants, such as the pitcher plants and Venus's flytraps; the latter's leaves are modified to attract and capture the insects that are digested by these plants.

Longevity. The leaves of most of the temperate zone plants do not continue to grow as do the grasses; they live for a single season and then fall. Such plants are *deciduous,* as opposed to the evergreens, which retain their leaves for a longer period—normally not for more than four years.

The falling of the leaves is the result of chemical changes in which the substance *auxin* plays a considerable part. In autumn, a special layer of cells, called the *abscission layer,* forms at the base of the petiole; these cells block the flow of water and nutrients. As a result, the chlorophyll decomposes, resulting in the yellow and orange pigments typical of autumn foliage. The red colors are new pigments that develop within the cells. Meanwhile, the middle cells of the abscission layers disintegrate and the leaf breaks off and falls, leaving a *leaf scar* on the stem.

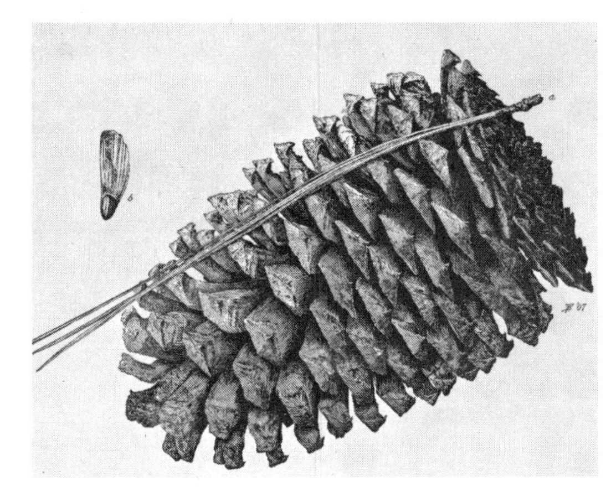

HELEN BUTTFIELD

THE WINGED SEED of the Jeffrey Pine and the feathery seed of the dandelion are two examples of highly efficient seed dispersal; these need merely the air and the wind.

Leaf Arrangement. Leaves usually are arranged on the stem in one of three ways: *spiral,* or *alternate,* one leaf at a node, the leaves forming a continuous ascending spiral on the stem, as in the apple or elm; *opposite,* two leaves that usually are directly opposite each other, as in the maple; *whorled,* three or more leaves at the same node, as in the lily.

Buds are regularly found in the *axils* of the leaves and, under favorable conditions, will grow into branches bearing leaves or flowers.

Internal Structure. In a cross section of a leaf blade, different types of tissues and cells are observed.

Epidermis.—The *upper epidermis* and *lower epidermis* each consists of a single layer of cells that covers the entire leaf surface, protecting the tissues within from mechanical injury and drying out. The epidermis, particularly the upper one, is usually covered with a waxy *cuticle,* which further prevents loss of water.

The epidermal cells are longer and wider than they are deep and, as viewed from the surface, have wavy cell outlines. Among the epidermal cells are special crescent-shaped *guard cells,* which contain chloroplasts. Between the guard cells are openings called *stomata* (singular, *stoma).* The size of these openings is regulated by the movements of the guard cells. It is through the stomata that the gaseous exchange takes place between the interior of the leaf and the air. Water vapor escapes and carbon dioxide enters for the process of photosynthesis; oxygen escapes as a product of photosynthesis or enters in connection with respiration.

Mesophyll.—Between the upper and lower epidermis there are specialized chlorophyll-bearing cells. Beneath the upper epidermis, long, narrow cells are arranged in a palisade fashion, and between these and the lower epidermis are more or less rounded or irregularly shaped cells arranged to form a loose, spongy tissue, with abundant air chambers between. These cells are largely concerned with carbohydrate manufacture.

Vascular bundles.—These are the veins of the leaves, as seen from the surface; the larger ones branch into smaller bundles, which may end in a single cell.

FLOWERS. From the standpoint of the continuation of the species, which is the only *natural* purpose, the flower and its resulting fruit and seed are the most important structures of the plant. It is for the flower's support that the entire plant is designed. The flower is the reproductive organ of the higher plants; its function is to produce the seed containing the male and female gametes that will unite in sexual reproduction and produce the new generation.

Flower Parts. The basic flower parts can be divided into four sets of structures. The outer set of parts (green in some plants) is the *calyx,* the individual leaves of which are the *sepals.* Next there is a set of showy leaves known as the *corolla,* the individual parts of which are *petals;* these are brightly colored and often secrete an aromatic and sweet substance *(nectar)* that attracts insects. Within the petals are a group of small sporophylls, the *stamens.* A stamen consists of a slender stalk *(filament)* with a pollen-bearing *anther* at its tip.

The central and last structure constitutes a united set, termed a *pistil,* that is made up of several parts, or *carpels.* Each group of carpels has an enlarged basal *ovule sac,* a terminal *stigma,* and a connecting shank known as the *style.* In the ovule sac there are differentiated ovules within which the egg gametes are produced.

The sepals and petals are known as *accessory parts* because they are not directly concerned with reproduction. The *essential parts* are the stamens (male parts) and the pistils (female parts).

Pollination. The development of the seed is preceded by the process of *pollination. Pollen* is produced on certain specialized structures and carried by the wind or by insects to the female structures. *Self-pollination* is the transfer of pollen within a single flower or from one flower to another on the same plant; *cross pollination* is the transfer of pollen from one plant to the female structures of another plant. There the pollen germinates, giving rise in a short time to male gametes that fuse with the female gamete. The latter develops into the embryo plant that, surrounded by protective structures, becomes the seed.

SEEDS AND FRUITS. The *seed* develops from the ovule and consists of the young embryo plant with its surrounding nutritive and protective tissues. The seeds are enclosed in the ovary during their development; the fully developed ovary with its adjacent parts constitutes the *fruit.*

Seeds. The seed consists of a *seed coat,* usually tough and partly impervious to the water that is necessary for germination; an *embryo,* the miniature plant that develops from the fertilized egg, or *zygote*; and the *endosperm,* or food-storage tissue that nourishes the germinating plant or seedling.

Effective seed dispersal is vital to the angiosperms because the parent plant is not mobile; seeds must be widely scattered to provide optimum conditions for germination and growth. Most of the striking characteristics of seeds and fruits are adaptations that aid dispersal: by wind, as in the wings of the maple seed and plumes of the dandelion; by spines and burrs that adhere to animals, as in the cockleburs; by floating, as in the coconut; or by the fleshy fruits of such plants as apples and cherries, which are eaten by animals and thus have their seeds distributed.

Fruits. There are several different kinds of fruits.

Fleshy fruits.—Soft and pulpy at maturity, *fleshy fruits* include the *berries* (grape, tomato, orange, squash); the *drupes* (cherry, peach, olive); and the *pomes* (pear, apple).

Dry fruits—Dry and hard at maturity, dry fruits fall into two types. Those that split open at maturity are the *dehiscent* fruits and include *follicle* (milkweed); *legume* (pea); and *capsule* (iris, lily). Fruits not splitting open are *indehiscent: achene* (buckwheat, sunflower); *caryopsis,* or grain (cereals); *samara,* or winged (maple, ash, elm); and *nut* (oak, chestnut).

Structural differences.—Fruits are called *simple* when they develop from a single or a compound ovary within the flower. If there are several separate carpels in the flower, an *aggregate* fruit is formed; in the raspberry there is an aggregation of many drupes. A *multiple* fruit arises when the fruits developed from separate flowers remain united, as in the mulberry and pineapple.

PLANT PHYSIOLOGY

ABSORPTION OF WATER AND NUTRIENTS. Active living plants contain a high percentage of water, and every growing plant must in some way be in contact with a water supply. While the seaweeds and floating algae are surrounded by water, the land plants have had to develop specialized structures, the roots, to penetrate the soil in search of moisture.

In such complex plants the cell walls are in close contact; and thus, although the leaf cells may be many feet from the absorbing roots, the distribution of water is so perfect that there is indirect water contact with the soil through long chains of tissue cells.

Each cell takes its water from some other cell or conducting vessel nearer the constant supply. Unless a plant contains a quantity of dead tissue, as in trees or shrubs, the water content is usually 75 percent to 80 percent of its weight.

Chemical Nutrients. Plants contain the chemical elements carbon, hydrogen, oxygen, nitrogen, phosphorus, sulfur, potassium, calcium, magnesium, and iron. Carbon, hydrogen, and oxygen enter into the composition of the carbohydrates as starch, cellulose, and dextrins. Proteins contain these same three elements, plus nitrogen, and minute traces of phosphorus and sulfur. In addition to these ten elements, minute traces of copper, manganese, boron, and zinc have been found to be necessary for plant growth.

Carbon, in the form of carbon dioxide, and oxygen are obtained mainly from the air; the other elements, usually in the form of nitrates, phosphates, and sulfates, are obtained in solution from the soil.

Absorption and Conduction. All the elements except carbon and oxygen, as well as the vital water molecules in which these elements are dissolved, must somehow be absorbed by the roots and carried, sometimes hundreds of feet, upward through the stem or trunk to the leaves. This task, which is actually an extraordinary and complex chemical and physical process, begins in the single-celled root hairs.

The walls of the root-hair cells are made up of *semipermeable membranes* that permit the passage of liquids from a less dense to a more dense solution, a process known as *osmosis.* Since the protoplasm inside the cells is of greater density than the water outside, the root cells absorb water; the cells become *turgid* from the increasing water content, and *root pressure* then forces the water up into the stem. However, this force is not strong enough to raise water more than a few inches off the ground.

The force that moves the nutrient-laden water up to the leaves is actually a form of suction and results from the plant's loss of water vapor through its leaves *(transpiration).* Thus, the actual force that draws the water upward is the great absorptive capacity of the dry, waterless air surrounding the plant.

Transpiration. The water and mineral nutrients absorbed by the roots are conducted through the xylem to the leaves, where they are utilized in food production. A large proportion of water absorbed is there transpired through millions of small pores, or *stomata,* on the leaf surfaces, which open or close according to the water pressure in surrounding cells.

PHOTOSYNTHESIS. *Photosynthesis* is the unique process of green plants in which they manufacture not only their own food, but also the food for all higher organisms. The raw materials are carbon dioxide from the air and water from the soil; the energy for the process is obtained from sunlight.

Photosynthesis takes place in the chlorophyll-bearing part of the living cells. Since most green cells are found in the leaf, the latter is the principal site of photosynthesis. The chief product synthesized is sugar; oxygen is given off.

Leaves are thin, broad, expanded structures and thus expose a large surface for the absorption of sunlight. The carbon dioxide necessary for photosynthesis enters the leaf through the same pores that function in transpiration, the stomata. The number of stomata varies greatly in different plants. Sometimes they are confined to one surface of the leaf; in other cases they are found on both.

FOOD STORAGE AND DIGESTION. The sugar manufactured in the leaves and other green portions of the plant may be stored temporarily in the form of starch. Most of it is soon transported to other parts of the plant, where it is utilized for building up the plant tissues. Some of it is used in the synthesis of proteins, which are manufactured in any living plant cell; the nitrogen for the proteins is derived from salts absorbed through the roots.

The legumes, such as peas and clover, are able to utilize the free nitrogen of the air through the presence of certain bacteria, called *nitrogen-fixing bacteria,* that develop characteristic nodules on the roots of these plants.

Much of the organic material manufactured during the growing season is stored as carbohydrates, fats, and proteins in the seeds and fruits, tuber, and roots. Upon the return of growing conditions, the plant utilizes this stored material; but, before it can do this, it must change it from an insoluble form to a soluble form for transportation. This is the process of *digestion* and is carried out by means of enzymes.

RESPIRATION. The plant must have a supply of energy in order to carry on the various life processes. This energy

HELEN BUTTFIELD

SUNLIGHT AND AIR surround the growing, green leaf. Within the leaf, the mysterious processes of life are carried on; these include: respiration, transpiration, and the chemical transformations of photosynthesis.

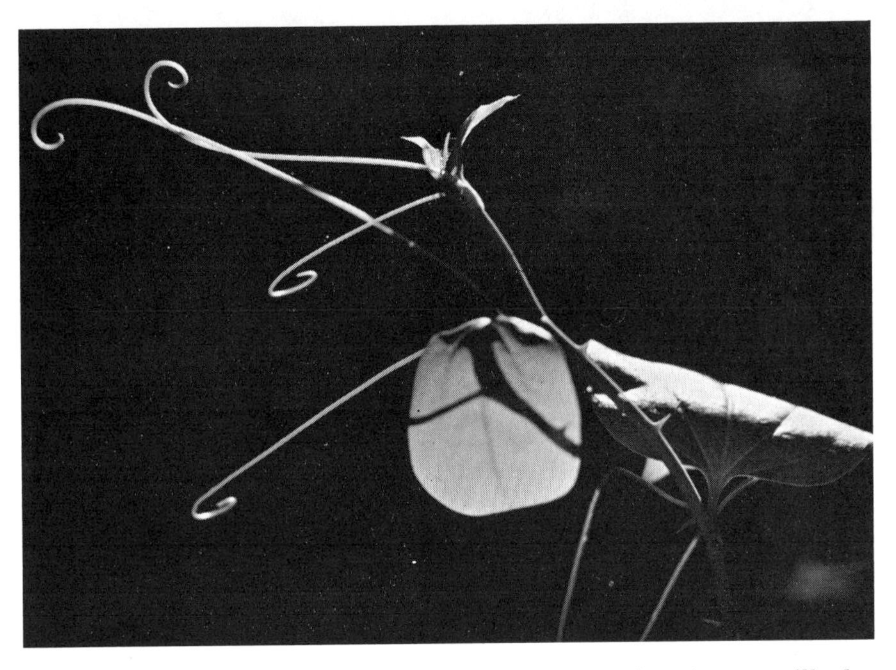

HELEN BUTTFIELD

PUSHING UP through the dead leaves in the forest, the young mayapple will break free, expanding in a delicate umbrella. Tendrils uncoiling toward the light exemplify every plant's constant growth.

comes from the breaking down of the complex organic compounds within the cell; most of these breakdowns are associated with *respiration.*

It is a common mistake to confuse photosynthesis and respiration. Respiration is the same in green plants, nongreen plants, and animals. The chief end product of respiration is carbon dioxide, but one of the end products of photosynthesis is oxygen.

ASSIMILATION. The final conversion of carbohydrates, fats, and proteins into living material (protoplasm) is called *assimilation,* a process whose inner working still eludes the scientists' grasp. Very little is known except that assimilation can take place only where life already exists.

Since assimilation is the process by which living, organic protoplasm is created out of nonliving, inorganic materials, the problem of its mechanics is very close to the mystery of life itself.

STIMULUS AND RESPONSE. One of the basic characteristics of living matter is a property known as *irritability.* It is this that enables living protoplasm to be stimulated by outside forces and to respond to them, either positively or negatively. Without this capacity to respond to outside stimuli, no organism could live and grow. Although plants are less sensitive than animals in some respects, there are many stimuli to which they respond quickly and intensely.

Tropisms. In general, the most important stimulus to almost all plants is light, and their response to it is called *phototropism.* If a sunflower seedling, for example, is illuminated from one side, the stem will bend and the plant grow toward the source of light. This bending is the result of the concentration of auxin, a growth-stimulating hormone, on the side of the stem away from the light. Since the cells on that side grow more rapidly they force the stem to bend gradually toward the light. *Mimosa pudica* and certain other legumes close their leaves when they are placed in darkness and expand them when moved to the light.

Another important though invisible force is the pull of the earth itself. If a seedling is laid horizontally in the dark, the stem will grow upward and the root will bend and grow downward; both growths are in response to the influence of gravity and are, respectively, negative and positive forms of *geotropism.* Roots are also sensitive to the closeness of water and will grow toward it; this is known as *hydrotropism.*

Other Motions. Movements called *turgor movements* are responses of certain plants to light, temperature, and touch, and are usually more rapid than the tropisms. The leaves of clover fold up at night in *sleep movements;* the leaves of the mimosa will fold and droop on being touched; and the specialized leaves of the Venus's flytrap are extremely sensitive to contact, closing to entrap any insect that lights on them.

Photoperiodism. A recent development in physiology is the discovery of *photoperiodism,* the effect of the daily duration of light exposure on reproductive activity. Some plants are *short-day* plants that require brief periods of light in order to reach the flowering stage; others are *long-day* plants and flower only when they are exposed to long periods of light.

PLANT GROWTH. Growth is a familiar characteristic of all living organisms and a vital one for plants, for, unlike animals, when plants cease to grow, they will soon die. Growth is closely related to movement, which is obvious in the slow upward growth of the young seedling, less apparent in the steady spiraling motion made by growing shoots and even by roots.

Germination. Within each seed is a dormant embryo plant, conceived and nurtured on the parent plant, then dispersed and arrested in its growth until it finds the conditions necessary to renew its activity. Primary among these conditions are warmth and water, although other factors will also affect the chances of germination. Most seeds are able to withstand fairly long periods of inactivity before germinating—some up to 2,000 years, as in the case of certain lotus seeds.

When water permeates the hard seed coat, it permeates the cells, expanding their length up to one hundred times. The root of the sprouting plant emerges, the cells growing actively at the tip. Behind the tip, the cells elongate rapidly, continually increasing the length of the root.

As soon as the cells in any zone cease elongating, root hairs develop to establish the vital connection with the soil and water. The same elongation takes place in the stem until the young seedling is formed. Further growth requires new cells produced by cell division, which takes place in the *meristematic zone* at the tip of both stem and root.

Growth Regulation. The hormone *auxin,* present in the growing tips, has been shown to control growth in the young stem and root. This growth is precisely regulated, for each plant has a certain pattern that it must follow. Buds are formed and the leaves set out at exact and regular intervals, and modifications of the various parts are controlled to produce the proper structures when and where they are needed.

Growth rate, as well as form, is a function of the specific inheritance of the plant and of the favorableness of the environment. The bamboo grows with great rapidity, the oak and the bristlecone pine with extreme slowness; the maximum relative-growth rate of each is highest with adequate sunlight, warmth, moisture, and soil nutrients.

Every structure, organ, or individual plant has growth characteristics peculiar to itself and different requirements as to growth conditions. Therefore, some plants grow primarily in the spring or in midsummer; some grow in all seasons.

—Helen Buttfield

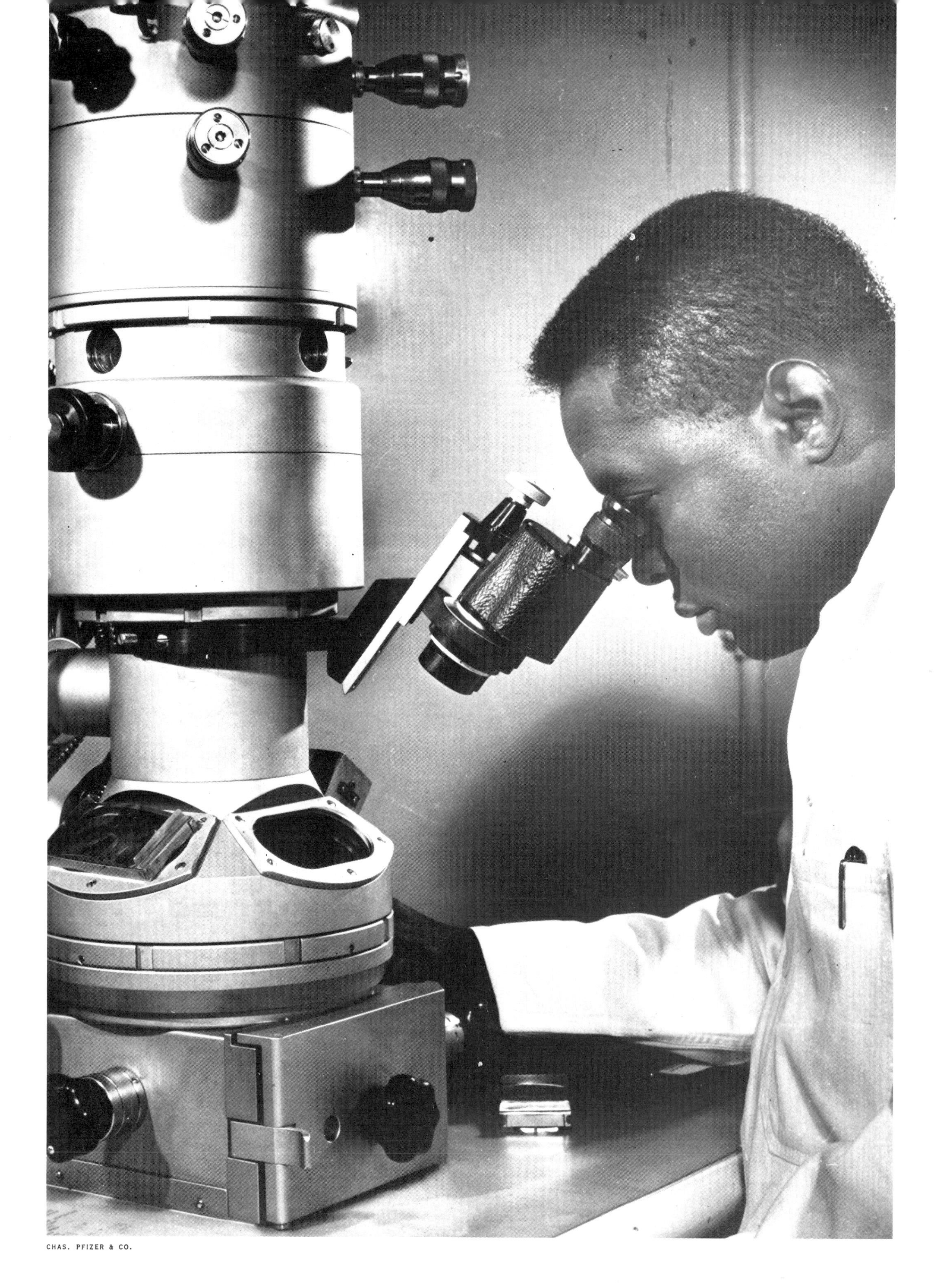

CHAS. PFIZER & CO.

PHYSIOLOGY

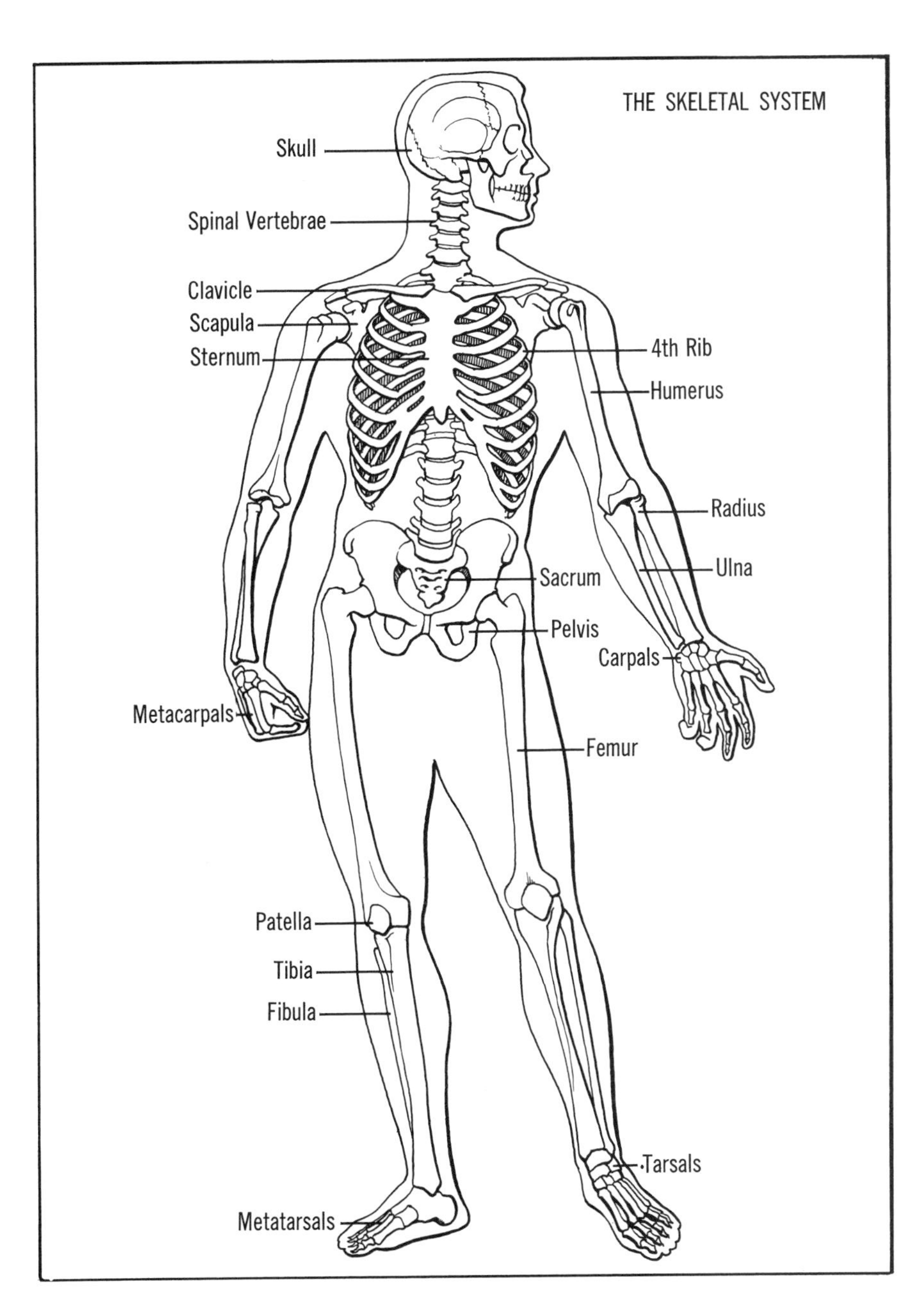

The human body is composed of billions of microscopic units called *cells.* Each cell is a chemical factory capable of producing energy from food substances, of harnessing this energy, and of using it to build the materials it needs to function. Although all cells share these basic activities, each different type performs a unique function—nerve cells conduct impulses; muscle cells contract; kidney cells excrete waste chemicals; red blood cells carry oxygen; and so on. Groups of similar cells are gathered into *tissues,* and tissues of different types are united to form *organs* such as the liver, kidney, eyes, heart, and brain. Organs function as parts of larger *systems* such as the nervous, digestive, and circulatory systems. These systems compose the human body.

Structural Elements of the Body

■ **BONE.**—*Bone* is the hard underpinning on which the soft tissues of the body rest. It is a complex, living tissue, consisting of a matrix of organic fibers with mineral crystals deposited throughout. Bone development begins in the embryo with the formation of a *cartilage* (elastic tissue) model; and as the embryo develops, centers of ossification appear. *Ossification* is the process of bone formation in which the *cartilage,* or elastic tissue, is replaced by hard, bony tissue. During body growth the regions of ossification enlarge until all cartilaginous tissue has been replaced by bone.

The hardness of bone tends to obscure the fact that it is a physiologically active tissue with important functions. Bone is constantly being destroyed and rebuilt, thus enabling the body's framework to reshape its structure according to the stresses it must bear. This aspect of bone activity is reflected by the disorder called *osteoporosis,* or softening of the bones. In this condition bone destruction proceeds at a normal rate, but synthesis of new bone tissue is slow. Osteoporosis may be caused by lack of normal bone stimulation as a result of inactivity, and occurs mainly in elderly or bedridden individuals.

The *calcium* salts deposited in bone are continuously being exchanged with calcium from the blood stream. In the disease called *osteomalacia* the exchange is unbalanced, resulting in the loss of calcium salts from the bone and a decrease in bone hardness. It may be caused by a dietary deficiency of *vitamin D,* a substance that promotes the absorption of calcium from the intestine. Without proper absorption calcium is withdrawn from the bone but is not replaced, leaving the bone soft and easily susceptible to fracture.

Another aspect of bone activity is the production of blood cells in the marrow of such bones as those of the spinal column and the ribs.

The body's framework is the *skeleton,* which consists of 206 bones, ranging in size from the small bones of the hands and feet to the massive *femur,* the long bone of the thigh. The bones are organized into various structures which encase and protect

CROSS SECTION OF SKIN

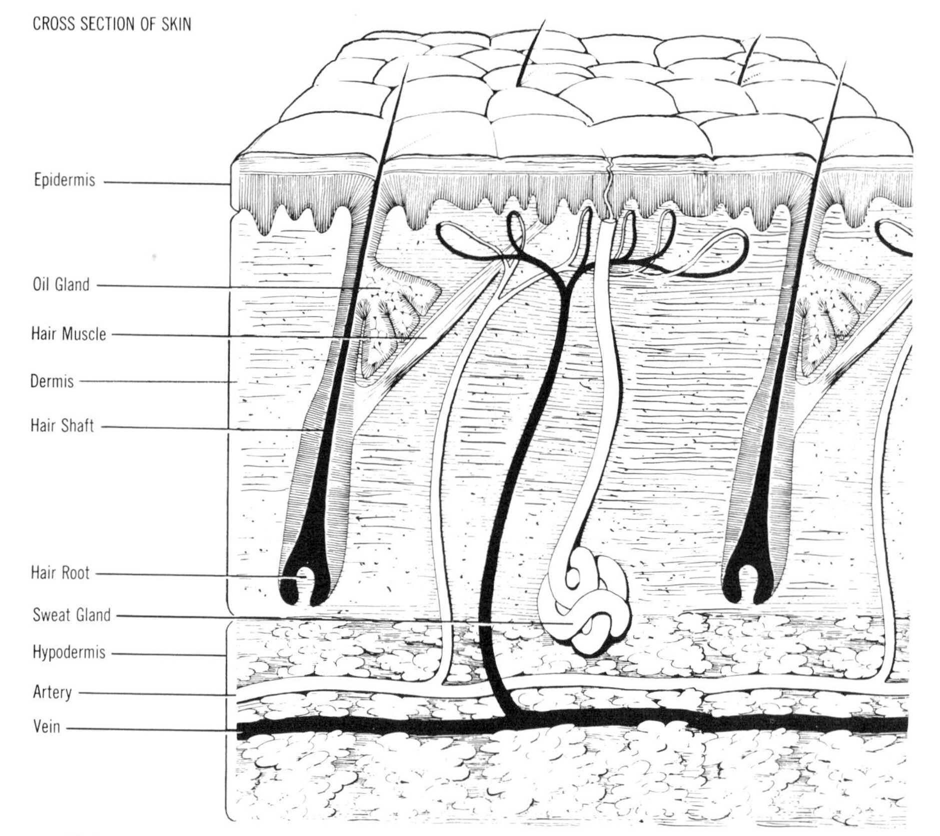

vital organs or act as a hard surface for the attachment of muscles and other soft tissues.

The *skull* is the protective bony structure that shields the brain. It consists of 22 bones—8 bones in the cranium or braincase and 14 in the face. The *rib cage* encloses the heart and lungs. It is composed of that portion of the spinal column at the rear of the chest (*thoracic* section), the ribs, and the *sternum*, or breastbone. The *pelvis* is formed by the hipbones and the lower, or *sacral*, portion of the spinal column. It holds the bladder, the reproductive organs, and the lower parts of the intestines. The *shoulder girdle* consists of the *scapula* (shoulder blade) in the back and the *clavicle*, or collarbone, in the front. It serves as an attachment for the heavy muscles of the chest, back, and upper arms.

Fractures, or breaks, are common bone injuries; and a confusing variety of descriptive terms has been applied to them. *Spiral* and *greenstick* describe the appearance of certain fractured bones. A *Pott's fracture* is a specific type of break involving several bones in the ankle region. A *pathological fracture* is a fracture of diseased bone that has yielded to stress that would not affect normal, healthy bone. A *closed* or *simple fracture* describes a fracture in which the skin overlying the break is intact. In an *open* or *compound fracture* the skin is broken, and the injured tissues are exposed to air and dirt. Such fractures are very likely to become infected.

The basic principle of treating fractures is to *reduce* them—to bring the parts of the bone into normal alignment. The physician attempts to do this as quickly as possible, before too much swelling has occurred. Anesthesia is given to permit manipulation of the fracture. In a *closed reduction* the parts can be brought into alignment by external manipulation. In an *open reduction* surgery is necessary. Once the parts are in correct position they are held in place by a plaster cast. In some cases weights are applied by pulley devices to maintain tension, thus holding the parts in proper relation. This is called *traction*. Nails or screws may also be used for this purpose. The limb is kept immobile until new bone has grown into the fracture region. After the plaster cast or other fixation device has been removed, the patient is guided in a program of exercise and physical therapy to restore the normal functions of muscles and joints.

Skin.—The *skin*, the living protective covering of the body, constantly replenishes itself as it is worn away. This multipurpose covering aids in the regulation of body temperature and shields the body from harmful microorganisms. It also contains numerous sense receptors for touch, pain, and temperature to provide the brain with information about the outside world.

The skin consists of two layers, the *epidermis* (upper skin) and the *dermis* (true skin), which overlie a layer of loose connective tissue. When the epidermis is subjected to irritation, it produces corns and calluses. It also contains the pigment-producing cells that generate a tan when exposed to the sun's ultraviolet rays. The epidermis has no blood vessels of its own, but receives nutrition from the dermis, which is divided into an upper *papillary* layer and a lower *reticular* layer. The papillary layer is convoluted; the overlying epidermis follows these convolutions, which are seen by the naked eye as fine lines and whorls. This is particularly evident on the fingertips, where the ridged papillary layer forms distinctive patterns, or fingerprints, on the undersurfaces. The dermis contains sweat glands, sebaceous (oil) glands, nerve endings, and small blood vessels.

Through their secretion the sweat glands help regulate the body temperature—body heat is expended when sweat is evaporated from the skin. The warm-weather disorders of *heat cramp* and *heat stroke* are associated with the functioning of the glands. Heat cramps are caused by loss of large amounts of salt through the sweat. This disturbs the electrochemical balance of the body, resulting in painful muscular contractions. Heat stroke develops when the sweat glands cease to function. In this extremely dangerous condition the body temperature may rise above 105° F.

■HAIR.—*Hair* is a skin appendage formed by an infolding of the epidermis. The hair grows outward from the *papilla*, an egg-shaped mass of cells at the base of the hair *follicle* (sac). Hair color varies with the amount of pigment, the greatest amount being present in black hair. Hairs on different parts of the body are constantly being replaced. The hairs of the scalp have a life span of from two to four years, while the hairs of the eyelashes may be replaced every three to five months.

Baldness, or loss of scalp hair, is common in men over twenty. It may begin as a receding hairline or a patch of hair lost at the top of the head. While most baldness is inherited, specific diseases, such as tuberculosis, and certain endocrine disorders or scalp injury may be responsible. Balding is usually permanent when the scalp is scarred by injury—by radiation exposure, for example. When scarring does not occur, hair will regrow if the cause is corrected. Treatment of baldness is feasible only when a specific disorder is responsible; inherited baldness cannot be successfully treated.

■SEBACEOUS GLANDS.—The natural oiliness of the hair and skin is maintained by the *sebaceous glands*, which pour their oily secretion, called *sebum*, around the hair shaft. In *acne*, a skin condition particularly common in adolescence, the ducts of the sebaceous glands become blocked by accumulations of dirt and sebum.

■NAILS.—*Nails* are horny plates that grow from a matrix at the rear of the nail plate; the growing nail glides forward over the nail bed. Disorders of the nail may be caused by damage to the matrix, by infectious diseases such as fungus, or by general body disorders which secondarily affect nail growth.

■TEMPERATURE CONTROL.—The blood vessels of the skin are a principal means of body temperature regulation. When these vessels expand (*vasodilatation*), more blood is brought to the skin where it is cooled; when the vessels contract (*vasoconstriction*), blood is retained in the interior of the body, and heat is conserved. This contraction and expansion is largely under the control of the *autonomic* nervous system (see *Nervous System*) and responds to emotional factors and as to changes in the outside temperature.

■SUBCUTANEOUS TISSUE.—Beneath the skin is a layer of loose tissue called the *hypodermis,* consisting of connective tissue and fat. This is one of the main storage sites for body fat; in obese individuals it may be several inches thick. The connective tissue is a supporting network composed of cells, tough and flexible collagen fibers, elastic fibers, and a ground substance. The connective tissue of the skin and hypodermis is composed of loosely woven fibers; other types of connective tissue are elsewhere. *Dense connective* tissue, consisting of tightly packed fibers, is found in the intestines; *regular connective* tissue, made up of fibers aligned in definite patterns, is found in the *tendons*—the bands of tissue connecting muscle to bone.

Nutrition.—The human body has often been compared to a machine. This analogy, although useful to some extent, should be modified by the realization that the body is a machine which continually rebuilds itself. For this the body needs three kinds of materials: substances that provide energy, substances that can be used to build tissue, and substances which can initiate and facilitate the chemical reactions that produce energy and synthesize tissue.

The principal energy-rich foods in human nutrition are *carbohydrates* and *fats.*

■CARBOHYDRATES.—*Carbohydrates* are produced by green plants using the energy of sunlight. These substances, commonly called *starches* and *sugars,* are the ultimate source of energy for animals. Only plants possess the chlorophyll and similar light-sensitive pigments capable of trapping solar energy and storing it in the form of carbohydrates. Animals obtain their carbohydrates from cereals, vegetables, and fruits. Among the carbohydrate-rich foods in our diet are candies, ice cream, bread, spaghetti, and potatoes.

■FATS.—*Fats* are obtained from both plant and animal foods. Vegetable fats are found in olive oil, wheat germ oil, and various nuts. Animal fats are contained in milk, eggs, cheese, and meats. Fats are the principal form in which animals store energy-rich material in the tissues. In addition to being a source of energy, fats evidently are essential for other body activities. It has been demonstrated that skin disorders may develop if fats are not in the diet.

■PROTEINS.—Principal sources of tissue-building material are *proteins,* which are found in meat, milk, eggs, and fish. Proteins are the most complex chemicals in nature. They are composed of hundreds of thousands of building blocks called *amino acids.* Some 20 different amino acids are linked in various combinations to form the vast number of proteins found in nature. The importance of proteins in plant and animal life cannot be overestimated. Every body cell, whether muscle, nerve, skin, blood, bone, or any other type, is composed of protein. Each plant and animal is composed of proteins that are unique in structure. In addition to the *structural* function which proteins have in the cells of living things, they also perform a vital role as *enzymes,* chemical agents that promote the thousands of chemical reactions occurring within the tissues.

■MINERALS.—For normal health the body requires certain *minerals* that may be used in structural tissue, in enzyme systems, in the synthesis of particular chemicals, or in various other body reactions. Among these essential minerals are calcium, phosphorus, iodine, iron, sodium, chlorine, and potassium. *Calcium* is contained in milk and cheese. In the body it is found in the bones and teeth. It is essential minerals are calcium, phosand for muscular activity. *Phosphorus* is also found in bones and teeth and is particularly important as a part of certain chemicals, most notably ATP (adenosine triphosphate) which makes possible energy storage and exchange in cells. *Iodine* is used in the synthesis of the thyroid hormone, which regulates the pace of body activity. *Iron* is essential for the manufacture of hemoglobin, the pigment of the red blood cells which carries oxygen to the tissues. *Sodium, chlorine,* and *potassium* are involved in the regulation of water balance and in nerve and muscle activity. Other minerals found

A CHART OF THE VITAMINS

Vitamin
1. ***Food Sources***
2. ***Functions***
3. ***Daily Adult Requirement***
4. ***Effects of Deficiency***

Vitamin A
1. Fish-liver oils, yellow and leafy green vegetables
2. Preserves night vision, integrity of skin
3. 5,000 I.U.
4. Night blindness, skin lesions

Vitamin B_1 (Thiamine)
1. Bran (coats of grain), yeast, meat
2. Helps body process foodstuffs for energy
3. 1.5 mg.
4. Beriberi, marked by loss of weight, body swelling, muscle wasting and weakness, nervous disturbances, heart damage

Vitamin B_2 (Riboflavin)
1. Yeast, liver, eggs, milk
2. Participates in energy-producing reactions and promotes growth
3. 2–3 mg.
4. Inflammation of tongue, lesions at corners of mouth and nostrils

Niacin (Nicotinic Acid)
1. Yeasts, meat, fish, peanuts
2. Various chemical reactions in cells of body
3. 20 mg.
4. Pellagra, marked by skin, digestive, and nervous system disturbances

Vitamin B_6 (Pyridoxine)
1. Liver, yeast, whole cereal grains
2. Functions in metabolism of amino acids
3. No minimum requirements set
4. Deficiency has been known to cause convulsions in infants

Pantothenic Acid
1. Yeast, liver, eggs, milk
2. Functions in energy-producing reactions and in synthesis of various chemicals
3. No minimum requirements set
4. No deficiency symptoms observed

Biotin
1. Liver, yeast, eggs, peanuts, milk
2. Participates in chemical reactions involved in tissue synthesis
3. No minimum requirements set
4. No symptoms of deficiency observed

Folic Acid
1. Widely distributed in animal and plant foods; also synthesized by intestinal bacteria
2. Synthesis of nucleic acids, cell formation
3. No minimum requirements set
4. Anemia

Vitamin B_{12}
1. Liver, lean meat, fish, milk
2. Manufacture of nucleic acids
3. No minimum requirements set
4. Pernicious anemia

Choline
1. Widely distributed in animal and plant tissues
2. Transportation of fat and formation of animal cells
3. No minimum requirements set
4. Accumulation of fat in liver; may aid in onset of cirrhosis

Inositol
1. Widely distributed in animal and plant tissues
2. Lipotropic agent
3. No minimum requirements set
4. Accumulation of fat in liver

Vitamin C (Ascorbic Acid)
1. Citrus fruits, raw leafy vegetables
2. Protects other vitamins from destruction in body
3. 75 mg.
4. Scurvy, marked by loosening of teeth, joint pains, hemorrhage

Vitamin D
1. Fish-liver oils, butter, egg yolk; also formed in body by action of sunlight on skin
2. Regulates absorption of calcium and phosphorus
3. 400 U.S.P. units
4. Bone disorders (rickets in children; osteomalacia in adults)

Vitamin E
1. Plant oils
2. Protects vitamin A from destruction in body
3. No minimum requirements set
4. Infertility, muscular dystrophy, or vascular diseases

Vitamin K
1. Synthesized by intestinal bacteria
2. Important in blood clotting
3. No minimum requirements set
4. Tendency to bleed easily

in the body include magnesium, sulfur, zinc, molybdenum, manganese, and cobalt.

VITAMINS.—*Vitamins* are important in the enzyme systems which carry out essential chemical reactions. They are needed only in small quantities and generally must be supplied by the diet because they are not synthesized in the body. Vitamin K, which is synthesized by intestinal bacteria, is an exception.

Vitamins may be grouped in two categories: fat-soluble vitamins and water-soluble vitamins. The *fat-soluble vitamins*—A, D, E, and K—can be stored in the body and therefore do not have to be continuously supplied. *Water-soluble vitamins*—the B-complex vitamins and vitamin C—are not effectively stored in the body and must be regularly supplied through the diet.

NUTRITIONAL DISEASES.—Nutritional disorders are of three types: those caused by a dietary deficiency of a necessary substance; those caused by the body's inability to absorb a substance; those caused by an excess of a nutrient.

The vitamin deficiency diseases are historically interesting for the role they have played in certain populations that have not had access to a sufficiently varied diet. *Scurvy,* for example, was a notorious scourge of sailors on long sea voyages where they had no fresh foods containing vitamin C. *Beriberi,* a disease caused by a deficiency of vitamin B_1, has long been prevalent in areas of the Far East where polished rice is the dietary staple. The protein deficiency disease *kwashiorkor* is found in certain areas of Africa where children are weaned on starchy foods.

In the disease called *sprue,* a disorder of absorption, the small intestine does not properly absorb nutrients. This may result in multiple vitamin deficiencies, *anemia* (a deficiency of red blood cells), and loss of weight. In *pernicious anemia* vitamin B_{12} is not absorbed normally from the digestive tract because absorption is hindered by the absence of an intrinsic factor from the gastric juice. Red blood cell formation declines in the absence of the vitamin.

Vitamins A and D may be harmful if given in large doses over long periods of time. Excess vitamin A may cause bone and skin abnormalities. An oversupply of vitamin D may result in pathological deposition of calcium in soft tissues of the body.

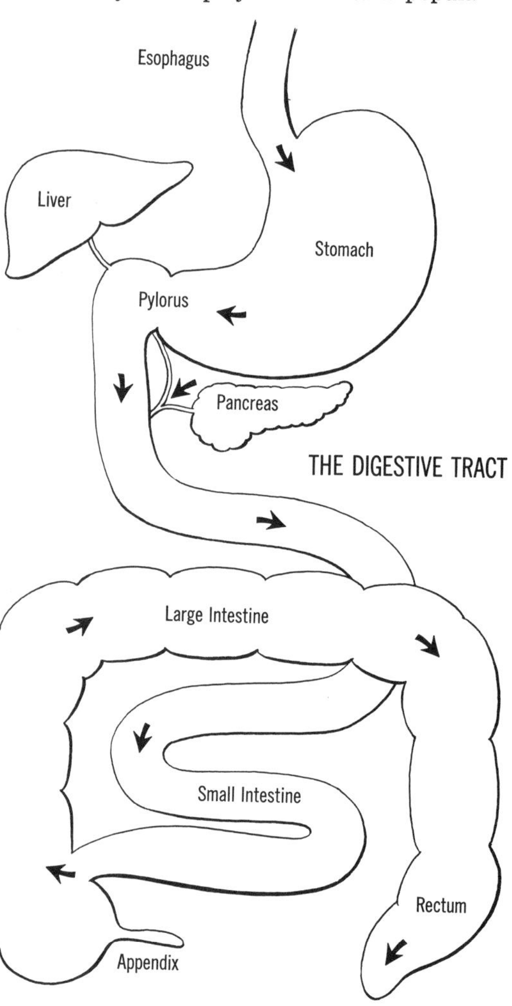

DIGESTIVE TRACT. Food is kneaded in the stomach and mixed with digestive enzymes; most digestion occurs in the small intestine.

Digestive System

TEETH.—The first set of teeth—the *milk,* or *deciduous,* teeth—appear between the ages of six months and two years. These 20 deciduous teeth are replaced, beginning about the sixth year, with the 32 permanent teeth. Each tooth is capped by smooth, dense *enamel,* the hardest substance in the body. Underneath the enamel is the softer *dentine,* which is sensitive to heat and cold and is also less resistant to decay than the enamel is. The *pulp cavity* under the dentine contains the nerves and blood vessels of the tooth.

Decay, the most common dental disorder, is more common in children and adolescents than in adults. It begins on the outer surface of the tooth and spreads rapidly once it penetrates the hard enamel. If it is left untreated, decay causes inflammation of the pulp cavity, with excruciating pain, and eventual loss of the tooth. Another frequent tooth disorder is *malocclusion,* an abnormal positioning of the teeth. *Pyorrhea* is an inflammation of the gums which may be caused by irritation from deposits of tartar or by malocclusion.

SALIVARY GLANDS.—There are three pairs of *salivary glands:* the *parotid glands,* which lie in front of the ears; the *submaxillary glands* of the jaw; and the *sublingual glands,* which lie under the floor of the mouth.

DIGESTION IN THE MOUTH.—The first stage of digestion is initiated by the teeth, which grind the food and mix it with saliva containing the enzyme *salivary amylase.* This enzyme breaks down starches into simpler compounds. The food mass is then swallowed and propelled to the stomach via the *esophagus,* the long, tubular passageway connecting the mouth and the stomach.

DIGESTION IN THE STOMACH.—In the stomach the food is kneaded and mixed with gastric juices containing the enzyme *pepsin* and *hydrochloric acid.* Pepsin decomposes the long protein chains into smaller units called *peptides* and *proteoses.*

DIGESTION IN THE INTESTINES.—The *intestines* are divided into two parts: the *small intestine,* a coiled tube about 21 feet long, and the *large intestine,* about 5 feet long. These terms "small" and "large" refer to the diameter of the intestines.

The small intestine is the major site of digestive activity. Here foods are broken down into simpler substances which can be absorbed by the body, thus completing the process which began in the mouth. In the small intestine fats are decomposed into *fatty acids* and *glycerol;* carbohydrates are decomposed into simple sugars; and proteins are decomposed into amino acids. This intense digestive activity is made possible by digestive enzymes secreted by the pancreas and by the intestinal lining itself. The liver contributes bile salts, which break fat globules into smaller particles that can be acted upon by fat-splitting enzymes. The products of digestion are absorbed through the intestinal wall by the blood and lymph streams and are distributed to the body cells. The undigested residue from the small intestine is moved by wavelike contractions of the intestinal wall (*peristaltic waves*), towards the large intestine. There water is absorbed from the residues, and the remainder passes to the *rectum* to be excreted.

PHASES OF DIGESTION.—In the description of digestion given above, it is noted that at various points enzyme-containing secretions are poured into the digestive tract. An important aspect of digestion is the way that the flow of these secretions is controlled. With respect to enzyme flow, digestion is divided into three phases.

In the *psychic phase,* before the food is eaten, the flow of secretions from the salivary glands, stomach, and pancreas is stimulated by the sight or smell of food. This is effected by nerve pathways from the brain. The presence of food in the stomach during the *gastric phase* of digestion liberates hormones from the stomach lining that travel through the blood to stimulate glandular cells in the stomach. Then, in the *intestinal phase,* when food is in the small intestine, hormones are liberated from the intestinal lining to stimulate the stomach and the pancreas. Fatty substances in the small intestine stimulate the release of a hormone that inhibits digestive activity in the stomach.

Metabolism

Metabolism.—The process through which the body makes use of the end products of digestion is called *metabolism.* This process includes what happens to digested material after it enters the blood stream and the way in which the body uses food to build tissues and to produce energy.

LIVER.—The *liver,* a major metabolic organ, has often been compared to a chemical factory because of its many functions. It is the largest gland in the body, comprising approximately

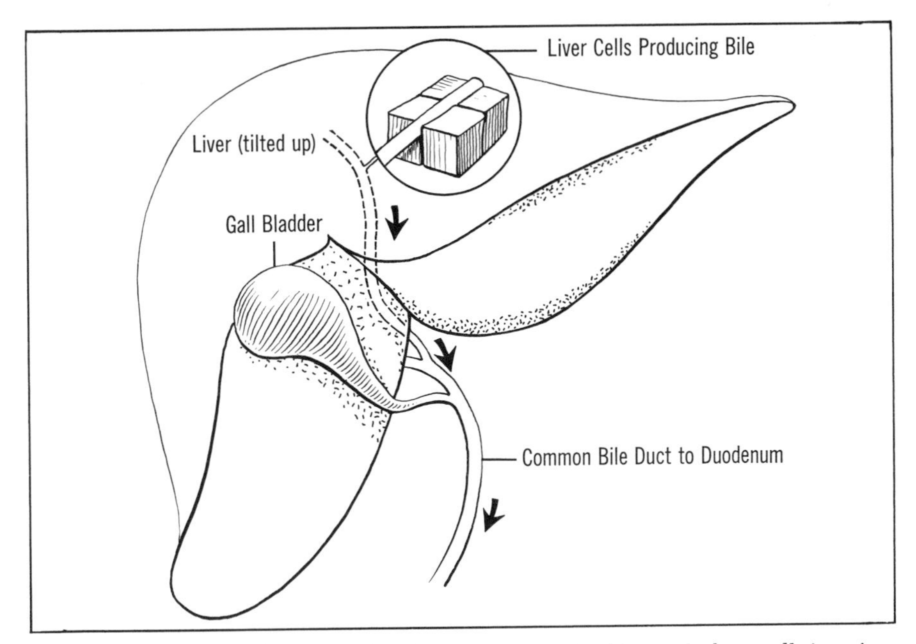

LIVER. Bile secretions flow from the liver to the gallbladder and the small intestine.

one-thirtieth of the body weight, or about five pounds in the average person. It lies on the right side of the body, immediately under the *diaphragm* (the disclike muscle which stretches horizontally across the body, separating the abdominal cavity from the chest cavity).

The liver is composed of two distinct types of cells—the parenchymal cells and the Kupffer cells. The *parenchymal* cells are the glandular, or secreting, cells, which perform most of the liver's metabolic activity. The *Kupffer* cells belong to the widely distributed *reticuloendothelial system* of cells, a system which plays an important role in the body's defense against infection. One of the main functions of this system is to remove bacteria and other foreign particles from the blood.

The liver plays a key role in the metabolism of carbohydrates, proteins, and fats. It stores vitamins and minerals and produces, among other things, the bile which is so important in the digestive process. Blood rich in digestive products from the intestine enters the liver by way of the portal vein. Sugars, the products of carbohydrate digestion, are stored in the liver in the form of the starch *glycogen*. Liver glycogen can be reconverted into other sugars and released into the blood as needed. If enough carbohydrates are not supplied in the diet, the liver can manufacture glycogen from other substances. *Amino acids,* the end products of protein digestion, are used by the liver as energy-producing substances. In this process a portion of the amino acid molecule is detached and converted into *urea,* which is excreted by the kidneys. Fatty acids which enter the liver are broken down into still simpler substances that can be more easily used for energy production. The liver also stores fats.

Bile is a liver product that contains both secretions and excretions. The secretions are manufactured by the parenchymal cells and contain bile salts, which aid in the digestion of fats. Bile also enables fats to be absorbed from the intestine. The excretions include the bile pigments, waste products derived from the breakdown of the hemoglobin of red blood cells. Bile pigments are produced by the Kupffer cells and also by other cells of the reticuloendothelial system. Bile is stored in the *gallbladder*—a small, pear-shaped pouch partially embedded in the undersurface of the liver—and is emptied into the small intestine through the *bile ducts*.

The liver manufactures certain essential *blood proteins*. Two of these, *prothrombin* and *fibrinogen,* are important in blood clotting; others (principally the *albumins*) contribute to the osmotic pressure of the blood. In this context *osmotic pressure* refers to the ability of the blood to retain water. This is attributed largely to the presence of large protein molecules that tend to draw water from the tissues into the blood, thus compensating for the leakage of fluid from the blood into the tissue spaces.

■ENERGY PRODUCTION IN THE CELL.—The energy-rich substances leaving the liver are principally sugars (*glucose*) and the products of fat breakdown. These are carried by the blood stream to all the body tissues, where they enter the individual cells. Here the energy-rich material is processed through a complex series of chemical reactions. The overall equation describing the process is: food substances + oxygen → carbon dioxide + water + energy.

In some popular treatments of physiology the statement is made that the body "burns" food as a fuel, thereby liberating energy. This suggests the inaccurate image of a stove-like device somewhere in the body being stoked with fuel and giving off heat. Actually each body cell burns its fuel in microscopic structures called *mitochondria*. The burning occurs in a series of reactions (the *Krebs cycle*), with small amounts of energy being liberated in the course of the series. This explains why the cells are not consumed by the violent reaction that would result if fuel combined directly with oxygen and ignited, as in the gasoline engine.

Respiration

■RESPIRATORY PROCESS.—Energy production in the cells requires oxygen. The process by which oxygen is taken

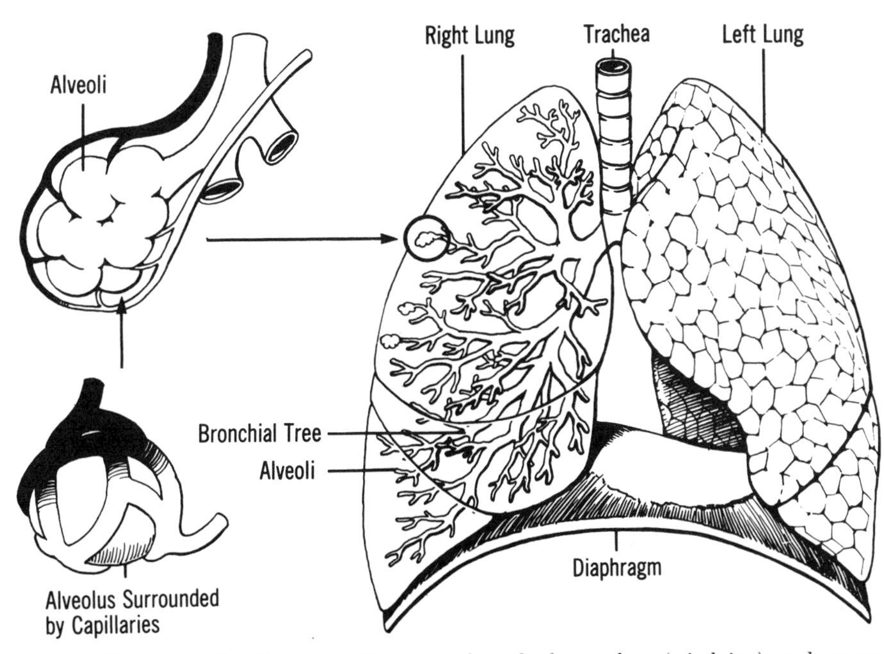

RESPIRATORY SYSTEM. Air enters the lungs through the trachea (windpipe) and passes through the successively smaller passages of the bronchial tree, ultimately reaching the alveoli, where there is an exchange of gases between the air and the blood stream.

into the body and used by the cells is called *respiration*. The mechanical work of breathing is accomplished principally by the *diaphragm*, a sheet of muscle which stretches across the lower boundary of the chest cavity. The size of the chest cavity increases as the diaphragm contracts, thus drawing air into the *lungs*. Air is expelled as the diaphragm relaxes. The muscles attached to the ribs also aid in breathing by slightly increasing and decreasing the size of the chest cavity.

Air is taken in through the mouth and nose and enters the *trachea*, or windpipe, a long, tubular structure which branches off at its terminal point into two main *bronchi*. In the lungs the bronchi divide into smaller and smaller branches, terminating in tiny saclike structures called *alveoli*, where the actual exchange of gases between the air and blood occurs. The surfaces of the alveoli are covered with thousands of minute blood vessels. The arrangement is such that only two fragile membranes separate a large sheet of moving blood from the air in the alveoli. At this point the blood surrenders some of its carbon dioxide to the air and absorbs oxygen from the air. This oxygen is bound in a loose chemical combination to *hemoglobin*, the pigment of the red blood cells.

The oxygen-bearing red blood cells are then carried to the farthest reaches of the body, finally arriving at the microscopic capillaries of the tissues, where a second exchange of gases occurs; oxygen leaves the capillaries to enter the tissue cells, and carbon dioxide enters the blood. Once it is inside the cells, oxygen is used in the burning of food substances to produce energy. The carbon dioxide entering the blood stream is a waste product of respiration and is carried in the venous blood to the lungs, where it passes into the air in the *alveolar sacs* and is exhaled as part of the breathing process.

REGULATION OF BREATHING.—Breathing proceeds automatically, without awareness or conscious direction. This is made possible by breathing centers in the base of the brain which receive nerve impulses from receptors in the lungs. When these receptors are stretched during *inspiration* (breathing in), they generate nerve impulses to the breathing centers to halt the inspiratory movement. It is believed that this mechanism, called the *Hering-Breur reflex*, is the principal control applied in normal rhythmic breathing.

Aside from this normal regulation of the breathing cycle, there is the question of how the body adjusts the rate of breathing to changing physiological conditions. For example, how is the breathing rate increased during ascents to high altitudes or during vigorous exercise? The answer to the question of high altitude seems to lie in the acidity of the blood and in its carbon dioxide content. When either or both of these factors increase, as they are likely to do in an oxygen-thin atmosphere, special *chemoreceptors* in certain arteries and in the brain relay impulses to the breathing centers of the brain, which in turn increase the breathing rate. The explanation of the increase in breathing during exercise is less well understood because the carbon dioxide concentration of the blood does not change at this time. It has been suggested that some aspect of muscle activity (such as the formation of waste products during muscle metabolism) may indirectly stimulate the breathing centers of the brain.

Circulatory System

BLOOD FUNCTIONS.—The blood has two major functions: the maintenance of life in the individual cells and the regulation and coordination of activities affecting tissues, organs, and organ systems. Oxygen and food substances in the form of sugars, fats, and amino acids are brought to the cells by the blood, and waste products, such as carbon dioxide, uric acid, and urea are carried away. The blood stream is important in the distribution of heat throughout the body. The body is cooled when greater volumes of blood flow to the blood vessels of the skin; heat is conserved when these vessels constrict, shunting blood to the deeper tissues.

The blood is a major line of defense against disease. It brings bacteria-fighting white blood cells to sites of infection. It also contains special proteins (*antibodies*) which neutralize disease-producing microorganisms. In addition to this, blood serves as a transportation network by carrying *hormones*, the secretions of the endocrine glands. These substances regulate certain important activities, such as metabolism and menstruation, and help the body adjust to sudden stress.

BLOOD COMPONENTS.—The fluid part of the blood, or *plasma*, is about 90 per cent water; but it also contains dissolved minerals, hormones, and such organic substances as fats, sugars, and enzymes.

BLOOD CELLS.—The red blood cells, or *erythrocytes*, are disc-shaped structures which contain the oxygen-carrying pigment *hemoglobin*. There are several types of white blood cells, or *leukocytes*; they can be distinguished by their appearance. These are *lymphocytes, monocytes, neutrophils, eosinophils*, and *basophils*. Some of the white blood cells possess the property of *phagocytosis*, which means they are able to engulf and destroy bacteria and other foreign particles. This property comes into play when a part of the body is attacked by bacteria. The invasion gives rise to an *inflammatory response*. The local blood vessels expand, bringing more blood to the infected site; white blood cells pass through the walls of the capillaries and attack the bacteria in the tissues. This warfare between the white blood cells and the bacteria results in the formation of *pus*, debris of dead cells.

BLOOD PLATELETS.—The tiny fragments of protoplasm called platelets are important in blood clotting. It is believed that they rupture during clotting, releasing enzymes that promote the formation of the blood clot.

BLOOD PROTEINS.—Blood proteins are contained in the plasma portion of the blood. *Albumins* are manufactured in the liver and aid the blood in maintaining the water balance of the body. *Fibrinogen* is essential for blood clotting. The clotting reaction can be described as occurring in roughly three stages, beginning when blood comes into contact with injured tissue: The blood platelets rupture, releasing an enzyme called *thromboplastin*. Thromboplastin, in combination with the plasma protein *prothrombin*, produces the enzyme *thrombin*. The thrombin plus fibrinogen produces the *fibrin* threads which form the core of the blood clot.

The *gamma globulins* are a group of blood proteins which include antibodies, the specialized substances which react with and neutralize, or destroy, invading microorganisms. The body produces antibodies to combat a specific disease after having been exposed to it. This accounts for the immunity which develops following exposure to, and recovery from, certain infectious diseases. This is also the principle on which *vaccination* works: a preparation is administered to cause the formation of disease-combating antibodies.

BLOOD GROUPS AND TRANSFUSIONS.—The blood of different individuals may differ in respect to the presence or absence of certain substances called *blood factors*. The most widely known factor system concerns the *antigens A* and *B* and their antibodies, *anti-A* and *anti-B*. The antigens and their respective antibodies are incompatible. For instance, if blood containing A is mixed with blood containing anti-A, the blood cells will *agglutinate*, or clump together. Antigens are contained in the cells; antibodies are found in the plasma. The blood of any human being will be in one of four *groups—O, B, A*, or *AB*—according to the presence or absence of these two antigens and two antibodies.

	Antigen	*Antibodies*
Group O	None	Anti-A; Anti-B
Group B	B	Anti-A
Group A	A	Anti-B
Group AB	AB	None

A fatal reaction may occur if a transfusion of the wrong type of blood is given to a patient. If group A blood is transfused into a group B patient, the anti-B antibodies of the transfused blood could agglutinate the recipient's blood cells (which contain the B antigen). This can result in chills, rapid pulse, and possibly kidney failure and death. For this reason the blood of both the donor and the patient are tested before a transfusion.

Rh FACTOR.—An additional and quite complicated blood factor is called the *Rh factor* (after the Rhesus monkey, which was used for the experimental work). The Rh factor is present in a majority of the population; such persons are called *Rh positive*. Individuals lacking the factor are called *Rh negative*. The occurrence of the Rh factor is associated with *erythroblastosis fetalis*, a disease of newborn infants of an Rh negative woman and an Rh positive man. If the child is

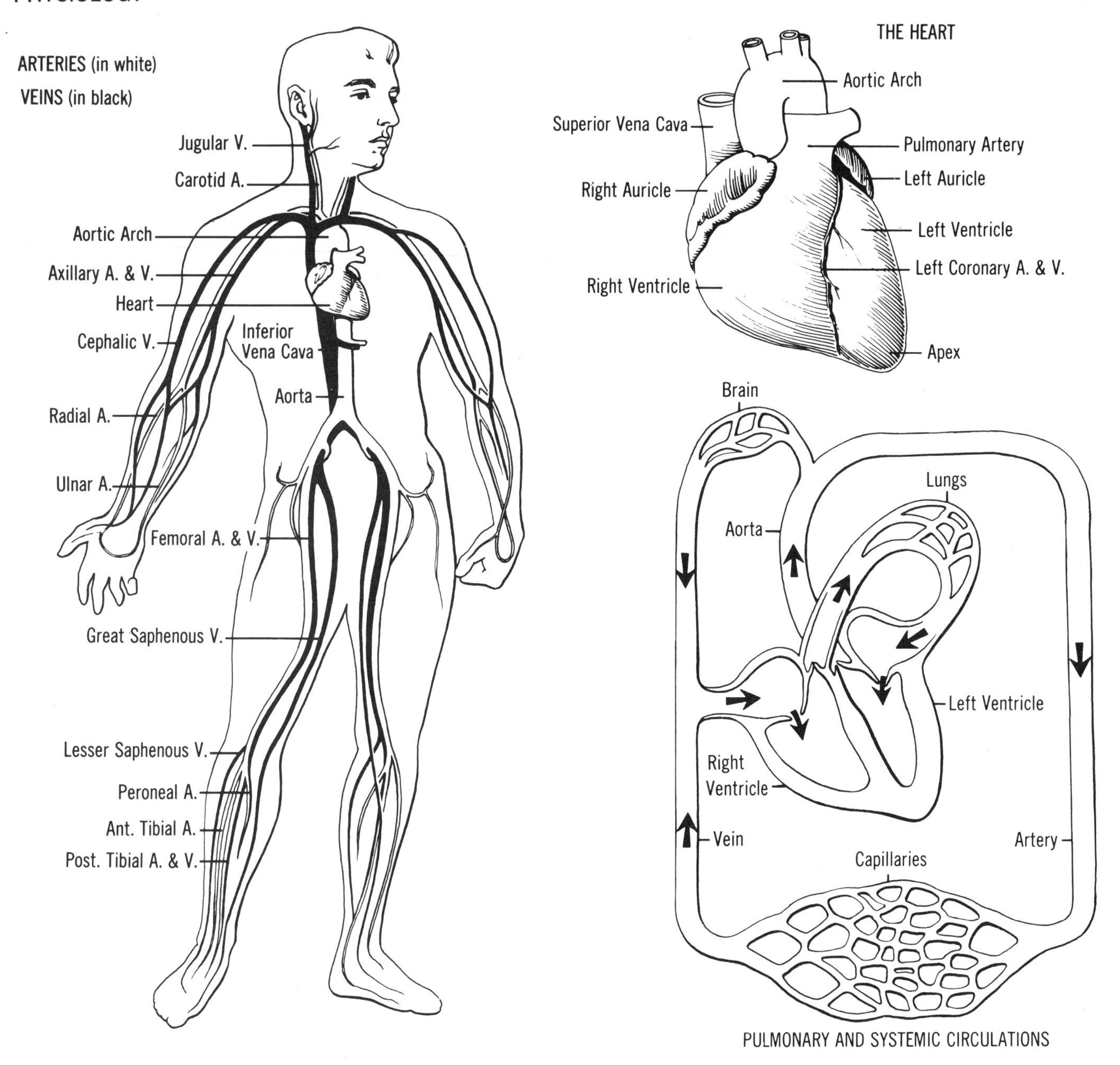

PULMONARY AND SYSTEMIC CIRCULATIONS

Rh positive, the mother's body may produce antibodies to destroy the fetus's red blood cells; the child is born with a severe anemia. Such children can be treated after birth with transfusions of Rh negative blood.

BLOOD CIRCULATION.—The blood flows throughout the body in a closed circuit of vessels of varying diameter and structure. It is propelled through this circuit by the pumping action of the *heart*, a muscular organ about the size of the fist. The human heart consists of four chambers—two thin-walled *atria* (*auricles*) and two powerfully muscled *ventricles*. The heart can be considered as two synchronized pumps. The *right heart*, consisting of the right atrium and the right ventricle, receives blood from the great veins known as the *inferior* and *superior vena cavae*, and pumps this blood from the right ventricle to the lungs. The *left heart* receives oxygenated blood from the lungs and pumps it into the arteries. The path of blood through the heart can thus be plotted as follows: the great veins —right atrium—right ventricle—lungs (via the *pulmonary artery*)—left atrium (via the *pulmonary vein*)—left ventricle—*aorta* (the largest artery of the body). As the blood passes through the lungs, it absorbs oxygen and surrenders carbon dioxide.

DIASTOLE AND SYSTOLE.—Diastole and systole are the terms used to describe the principal phases of the heartbeat. *Diastole* is the relaxed phase during which the blood passes from the atria to the ventricles. During *systole* the ventricles contract vigorously, propelling blood to the lungs and the arteries. In these two phases the action of both halves of the heart are synchronized. During diastole both the left and right ventricles relax; during systole both contract.

HEARTBEAT REGULATION.—The heart beats regularly and automatically throughout life, adjusting its output to meet the changing needs of the body. This is possible because the normal heartbeat rhythm is controlled by bundles of specialized conductive tissue in the heart which act as a built-in timing system. Along the upper portion of the right atrium is the "pacemaker" of the heart (the *SA*, or *sinoauricular node*), which generates impulses at the rate of 60 to 90 per minute. These impulses spread along the walls of the left and right atria, stimulating contraction. At the *interatrial septum*, the boundary of the left and right atria, the impulse stimulates another structure, the *AV*, or *atrioventricular node*. This generates another wave of impulses that travels along the walls of both ventricles, causing them to contract.

However, since the system is purely a clocklike internal system, it is not sensitive to physiological conditions which require changes in the heart's output. For this purpose the heart is supplied with two sets of nerves from the brain and spinal cord. One set, the *sympathetic pathways,* increases heart action. The other set, the *parasympathetic pathways,* decreases heart action. The impulses that flow along these pathways are triggered by special reflexes and by emotion, excitement, and other factors which stimulate the brain control centers that regulate heart action.

■ARTERIES.—Blood leaves the heart through the *aorta,* the largest artery of the body, and then flows through successively smaller branches of the arterial tree as it moves away from the heart. The systolic contraction of the ventricles, which propels the blood, can be felt as the *pulse* in the arteries near the skin's surface.

Arteries are not merely passive carriers of blood, however. They also regulate its flow and distribution. Elastic tissue in the arterial walls, particularly in the larger arteries, expands under the force of the wave of blood propelled by the systolic contraction of the heart, and recoils during diastole.

A second important arterial function involves the smooth muscles of the smallest arteries, the *arterioles.* These muscles are supplied with fine nerve endings from the *autonomic branch* of the nervous system, the branch concerned with the unconscious coordination of internal body activities. Depending upon the type of nerve impulse, the arterioles either expand or contract, decreasing or increasing blood flow to a particular part of the body. During digestion, for example, the pattern of nerve stimulation contracts arterioles in the skin and muscles while expanding those in the digestive tract, with the net effect of increasing the blood supply to the digestive organs.

Another effect occurs during vigorous exercise. Nerve stimulation to the *adrenal glands,* located above the kidneys, releases hormones that act directly on the arterioles, *constricting* (contracting) those of the skin and *dilating* (expanding) those in the muscles of the trunk and limbs. This causes blood to be diverted to those muscles being used.

■BLOOD PRESSURE.—The terms used to describe blood pressure are *systolic* and *diastolic.* The first refers to the force, or pressure, of the blood during heart contraction. The second refers to the pressure of the blood when the heart is relaxed. Units of blood pressure are given in terms of millimeters of mercury, or the ability of the blood pressure to support a given amount of mercury in a glass tube.

Blood pressure generally increases with age. The average blood pressure of infants is usually about 80 millimeters of mercury systolic and 55 millimeters diastolic, or 80/55; the average figure for young adults is about 120/80.

The significance of blood pressure may be understood by an analogy between the circulatory system of the body and the plumbing system of a house. If the water pressure in a house is too low, the upper floors do not receive an adequate supply. Similarly, if the blood pressure in the body is too low, the brain does not receive enough blood. Low blood pressure can result in dizziness, fainting, and possible brain damage. On the other hand, if the water pressure in a house is excessive, there is danger of the pipes' breaking, especially if they are old and fragile. Similarly, when the human pipes, or arteries, are fragile and the blood pressure is high, a rupture may occur. Such a rupture in the brain is called a *stroke.*

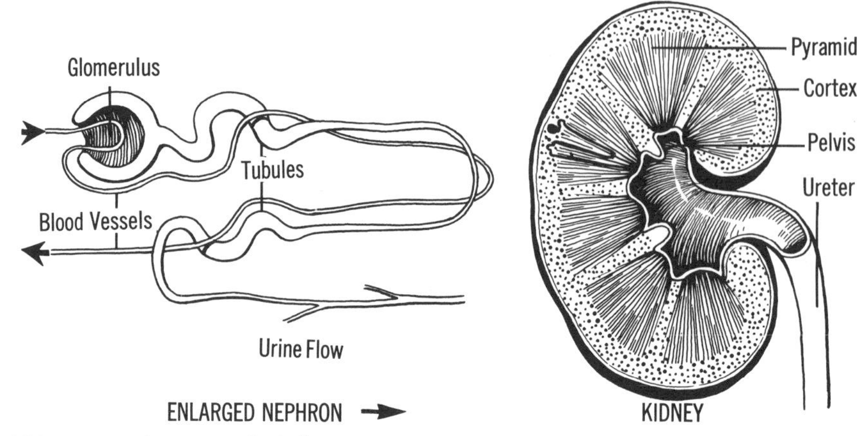

EACH KIDNEY is composed of about one million microscopic, urine-producing nephrons.

■CAPILLARIES.—The basic functions of the blood are carried out in the *capillaries,* which are situated between the arterial and venous systems and consist of microscopic tubes whose walls are often only one cell thick. These walls permit fluid, oxygen, sugar, amino acids, and even blood cells to enter the tissue fluids, while *carbon dioxide, lactic acid,* and other by-products of cell metabolism pass into the capillary to be carried away by the blood stream.

■VEINS.—Blood leaving the capillaries enters the *venules,* minute veins connecting the capillary beds to larger veins. Venous blood flow differs from arterial blood flow in several important respects. Blood in the veins is under considerably less pressure than blood in the arteries because the force of the heart's pumping action is not transmitted to the veins. For this reason arterial bleeding is much more dangerous than venous bleeding. When an artery is cut, blood spurts out with great force; and enough blood may be lost in only a few minutes to cause death. When a vein is cut, however, blood flows out slowly and steadily.

The low pressure of venous blood flow creates a problem in the return of blood to the heart. This is especially true with respect to the return of blood—against the force of gravity—from the lower extremities. To facilitate this movement, the veins are equipped with valves that permit the blood to flow in only one direction—toward the heart. In *varicose veins* these valves fail, permitting backflow of blood and the consequent appearance of saclike expansions of the venous walls.

Another difference between venous and arterial blood flow is oxygen and carbon dioxide content. The arterial blood, having just made the circuit through the lungs, has more oxygen and less carbon dioxide than does venous blood. This explains the observable difference in color. Arterial blood is bright red; venous blood is darker and bluish.

■SPECIAL CIRCULATORY SYSTEMS.—In the circulatory system arterial blood normally flows to an organ, and venous blood flows away from it. However, this is not true in two cases—the pulmonary and portal circulations. The *pulmonary circulatory system* serves the lungs. The blood which flows to the lungs is venous blood (oxygen-poor); but the vessel that carries it is an artery, the *pulmonary artery.* Blood flowing away from the lungs is arterial blood (oxygen-rich), but it is carried by a vein, the *pulmonary vein.*

The *portal circulatory system* consists entirely of veins. Blood rich in nutrients is brought to the liver from the digestive tract by the *portal vein.* Blood is carried away from the liver by the *hepatic vein.*

Lymphatic System.—The *lymphatic system* is a second circulatory system which returns fluids and proteins from the tissue spaces to the blood. As blood flows through the tiny capillaries of the tissues, fluids and proteins leak out. Without the lymph channels this material would collect in the tissues, producing fluid swellings known as *edema.*

The *lymph vessels* begin as small closed tubes in the tissues and join into larger vessels, eventually emptying into large veins in the chest. Lymph vessels are found in the skin, the muscle, and the linings of various organs. The *lymph nodes* distributed along the system contain cells that consume foreign particles and bacteria in the lymph fluid. This is an important defense mechanism against disease. Swelling of the lymph nodes is often a sign that the region which they drain is infected.

Reticuloendothelial System.—The cells of the *reticuloendothelial system* (*RE cells*) are widely distributed in the body. They are found in the liver, in connective tissue, in the spleen, in the lymph nodes, and in certain glands. In addition to this "fixed" type, "wandering" RE cells travel in the blood stream to various parts of the body. All RE cells are scavengers, and as such play an important role in the body's defense against infection. They are able to *phagocytize,* to consume such foreign substances as bacteria. There is also evidence that RE cells participate in the production of *antibodies,* special substances that can destroy or neutralize invading microorganisms. In the spleen, liver, and bone marrow, RE cells consume old red blood cells and convert the hemoglobin into bile pigments, which are excreted with the bile.

Spleen.—The *spleen* is located on the left side of the abdomen underneath the diaphragm. This organ contains a rich network of blood channels, thus exposing a large volume of blood to the action of its cells.

The spleen acts upon the blood system in several ways. It manufactures white blood cells known as *lymphocytes* and destroys old red blood cells, white cells, and platelets circulating in the blood stream. Since it is part of the reticuloendothelial system, the spleen can also produce disease-fighting antibodies and is capable of phagocytizing bacteria and other foreign substances in the blood.

Kidneys.—Waste products are carried away from the body cells by the circulatory system. One of these wastes, carbon dioxide, is carried to the lungs, from which it is expelled. Other chemical wastes are disposed of by the *kidneys,* two bean-shaped organs embedded in the posterior abdominal wall. The kidneys are located one on each side of the vertebral column.

The basic functional unit of the kidney is the *nephron,* of which there are approximately one million in each kidney. The microscopic nephrons contain a filtering mechanism called the *glomerulus,* which leads into a long, twisted tubule. Blood enters the glomerulus, where virtually everything, except blood cells and large protein molecules, passes through to the long tubule. The material which enters the tubule is known as the *glomerular filtrate.* This is the first step of *urine* formation, which is essentially an indirect process based on the principle of selective reabsorption. Varying quantities of such substances as urea, sugar, salts, and most of the water are reabsorbed into the blood through the tubule wall. After a secretion of the cells lining the tubule has been added, the resulting urine leaves the tubule and passes through the *ureters* to the *bladder,* where it is stored until it is finally excreted.

Endocrine Glands.—The *endocrine glands* (as distinguished from the *exocrine glands,* such as the sweat glands of the skin) pour secretions directly into the blood stream. These secretions, known as *hormones,* serve as a means of long-distance control, enabling the body to coordinate its internal activities.

■**PITUITARY GLAND.**—The *pituitary gland,* or *hypophysis,* is a pea-sized structure situated under the brain. It is often called the "master" endocrine gland because of the control it exercises over other endocrine glands. The pituitary gland is divided into three parts: the anterior, intermediate, and posterior portions.

The anterior portion, or *adenohypophysis,* secretes hormones that stimulate the activity of other endocrine glands. Among these are the *thyrotrophic hormone* (*TSH*), which stimulates the thyroid gland, and the *adrenocorticotrophic hormone* (*ACTH*), which stimulates the outer layer of the adrenal glands. The anterior pituitary also secretes hormones that stimulate sexual development and regulate the menstrual cycle.

Secretion of many of these hormones is controlled by a chemical "feedback" principle. (This is a term borrowed from engineering, where similar principles are used in devices such as thermostats and governors.) Feedback control is illustrated in the secretion of the thyrotrophic hormone, which acts upon the thyroid gland. As the concentration of thyroid hormone in the blood declines, the pituitary secretion of thyrotrophic hormone increases. This stimulates the thyroid gland and elevates the blood concentration of thyroid hormone, which in turn inhibits further secretion of TSH by the pituitary gland.

Secretion of some pituitary hormones is controlled by stimulation from the brain. In emergency reactions, for example, the hypothalmus of the brain may stimulate the secretion of ACTH from the pituitary. ACTH in turn stimulates the secretion of *cortisone* and related hormones from the *adrenal cortex,* the outer portion of the adrenal gland. These *adrenocortical hormones* help prepare the body to meet the impending stress.

The functions of the intermediate portion of the human pituitary gland are not yet clearly understood. In lower animals it secretes a hormone that regulates the distribution of pigment in the skin (melohocyte-stimulating hormone, *MSH*).

The posterior portion of the pituitary, or *neurohypophysis,* secretes the *antidiuretic hormone* (*ADH*), which acts upon the kidneys to regulate the passage of water in the urine. It also secretes the hormone *oxytocin,* which stimulates contractions of the uterus during childbirth and milk flow from the breasts during nursing.

■**THYROID GLAND.**—The *thyroid gland,* located in the throat adjacent to the windpipe, or *trachea,* secretes hormones which regulate the rate at which the cells consume oxygen in the process of energy production.

Adjacent to the thyroid are the *parathyroid glands,* small brownish-red bodies that regulate the concentration of calcium in the blood. The action of the *parathyroid hormone* closely resembles that of vitamin D.

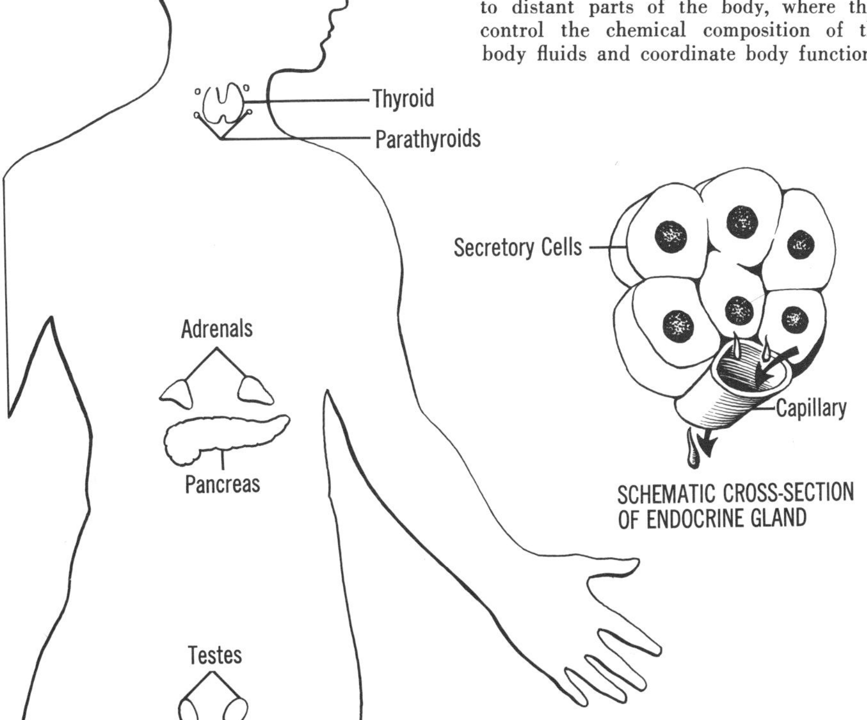

ENDOCRINE GLANDS, important chemical regulators of body activity, produce *hormones;* these travel via the blood stream to distant parts of the body, where they control the chemical composition of the body fluids and coordinate body functions.

Both substances regulate the calcium metabolism of the body.

ADRENAL GLANDS.—The two *adrenal glands* are located one on each kidney. These small structures actually consist of two distinct glands.

The outer part of the adrenal gland, or *cortex*, produces three types of hormones: those involved in regulating the energy-producing activities of the body (including *cortisone*), those which regulate the mineral composition of the body fluids (principally *aldosterone*), and those concerned with the regulation of sexual characteristics (*androgens*). The group of adrenal cortex hormones is called *corticosteroids*. The release of these substances is controlled by ACTH, which is secreted by the anterior portion of the pituitary gland.

The inner portion of the adrenal gland, the *medulla*, secretes the hormones *epinephrine* (*adrenalin*) and *norepinephrine* (*noradrenalin*), which constrict small arteries and thus raise the blood pressure to prepare the body to meet impending stress. The secretion of these substances is controlled by direct nervous stimulation from the brain.

PANCREAS.—The *pancreas* is both an endocrine gland and an exocrine gland because it secretes substances through ducts and also pours them directly into the blood stream. The exocrine portion of the pancreas is involved with digestion: the *acinar cells* secrete digestive enzymes into the *pancreatic duct*, which empties into the small intestine. The endocrine portion of the gland is involved in regulating the sugar level of the blood. For this purpose the *islet cells* (*islets of Langerhans*) secrete the hormones *insulin* and *glucagon*. Insulin lowers the blood sugar; glucagon apparently raises it. A partial or complete failure of the insulin supply results in *diabetes mellitus*, a disease which impairs the body's ability to utilize carbohydrates.

SEX GLANDS OR GONADS.—The human sex glands include the *testes* of the male and the *ovaries* of the female. The testes produce *testosterone*, a hormone which promotes development of certain male physical characteristics. The ovaries secrete *estrogen*, the female sex hormone that influences the secondary sexual characteristics in women.

PLACENTA.—The endocrine structure which appears only during pregnancy is called the *placenta*. It indirectly connects the blood stream of the mother with that of the fetus and also secretes the hormone *progesterone*, which regulates and maintains pregnancy. The placenta is passed from the body as the *afterbirth*.

THYMUS.—The *thymus* is an enigmatic gland located in the upper chest. It reaches its maximum size at puberty and then begins to shrink. Evidence suggests that the thymus may play a role in the body's immunity mechanisms. Several hormones affecting growth have also been isolated from the gland.

MENSTRUATION.—A simple listing of the names and locations of the endocrine glands and their hormones fails to give a full picture of their delicate interaction in the affairs of the body. An understanding of how hormones coordinate physiological activities can best be obtained from a description of the part they play in one body function, *menstruation*.

The pituitary gland initiates the menstrual cycle by secreting the *follicle stimulating hormone* (*FSH*), which causes the ripening of one of the egg-bearing follicles of the ovary. As this occurs, the ovary gradually increases its production of *estrogen*, the female sex hormone. Estrogen in turn stimulates development of the uterus lining, preparing it to receive the egg if it should be fertilized. Approximately halfway through the cycle (about the fourteenth day) the egg ruptures from the follicle, leaves the ovary, and travels down the fallopian tubes to the uterus. At this time the pituitary gland has already begun to secrete another hormone, the *luteotrophic hormone*. This causes the ruptured follicle, which is left behind in the ovary, to develop into the *corpus luteum* (yellow body). This in turn secretes the hormone progesterone, which joins estrogen in stimulating the uterine lining. Progesterone also acts upon the pituitary, inhibiting further secretion of the follicle-stimulating hormone. This prevents new eggs from ripening in the ovary.

The final stage of hormone control in the menstrual cycle presumably occurs when the concentration of progesterone in the blood reaches a critical level and inhibits secretion of the luteotrophic hormone. This sharply drops the progesterone level of the blood because, as the luteotrophic hormone is withdrawn, the corpus luteum degenerates and ceases to produce progesterone. The decrease in progesterone concentration causes the lining of the uterus to slough off, producing the menstrual flow.

Reproductive System

MALE SYSTEM.—The male reproductive apparatus consists of the testes, seminal ducts, seminal vesicles, prostate and bulbourethral glands, urethra, and penis. The *testes* are the primary male sex organs. They are contained in a cutaneous pouch called the *scrotum*, which is suspended outside the abdominal cavity. This is important because the male sex cells need a lower temperature than that inside the body in order to develop. In the testes are the *seminiferous tubules*. These tubules are lined with *spermatogenic cells*, specialized cells from which arise the male sex cells known as *spermatozoa*. The spermatozoa travel from the testes via the *vas deferens* (*seminal ducts*) to the *urethra*, a membranous tube that carries both urine and semen outside the body. *Semen*, the whitish fluid ejaculated by the male, is composed of spermatozoa suspended in nutrient secretions contributed by the *prostate gland*, the *seminal vesicles*, and the *bulbourethral*, or *Cowper's, gland*. The *penis* is the organ used in the act of copulation.

FEMALE SYSTEM.—The female reproductive apparatus consists of the ovaries, fallopian or uterine tubes, uterus, and vagina. The *ovaries*, or *female gonads*, are two small almond-shaped bodies located on each side of the body in a shallow depression on the lateral wall of the pelvis. An ovary is composed of about 50,000 ova-containing follicles. After the age of puberty the follicle-stimulating hormone (FSH) is produced, and one of these follicles matures during each menstrual cycle. It discharges its *ovum*, or egg, which then travels to the uterus by way of the *fallopian tubes*, two long, slender tubes connecting the ovaries to the uterus. The ovum is propelled through the fallopian tube by the action of *cilia* (tiny hairs) and smooth muscle. The *uterus* receives the egg and houses the fetus during pregnancy. It is a hollow, muscular, pear-shaped organ about three inches long, with a broad, flattened body above and a narrow, cylindrical part known as the *cervix* below. The *vagina*, a curved musculomembranous canal, leads from the outside of the body to the cervix and receives the male penis during copulation.

Pregnancy begins when the female egg is fertilized by the male sperm. This usually occurs in the fallopian tubes. The fertilized egg then implants itself in the uterine lining and develops.

PREGNANCY TESTS.—The most common sign of pregnancy is failure to menstruate. Several tests are available, however, to detect pregnancy in its early stages.

The most widely used pregnancy tests are the *Aschheim-Zondek test* and its modifications. The woman's urine is injected into an immature female rat, mouse, or rabbit; after one to five days the animal's ovaries are examined. If the woman tested is pregnant, her urine will contain hormones from the tissues surrounding the embryo. These hormones induce specific changes in the ovaries of the test animal.

In the *frog test*, the urine is injected into a male frog. If the woman tested is pregnant, the frog will eject spermatozoa within several hours.

Another test utilizes progesterone, the hormone involved in the menstrual cycle. The progesterone is administered over a brief period and then promptly withdrawn. The woman should then menstruate if she is not pregnant.

COURSE OF PREGNANCY.—During the first three months of pregnancy the embryo grows to a length of three to four inches. By the end of the sixth month the fetus will be approximately 14 inches long. At this stage sexual features will be discernible, and the major portion of muscle, kidney, and nervous system development will have taken place. By the thirty-second week of pregnancy the fetus will have developed to the point where it could survive if it were born prematurely.

Most *miscarriages*, or spontaneous abortions, occur in the first three months. They are usually caused by imperfections of the fertilized egg, a condition known as *blighted ovum*. Spontaneous abortions after the first three months of pregnancy can be

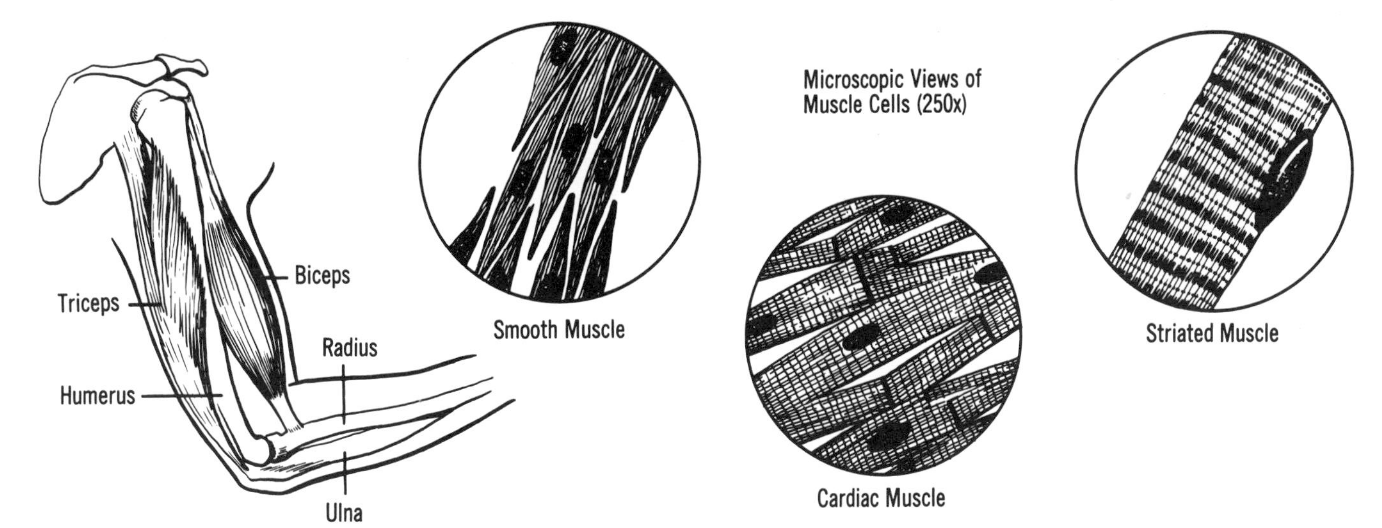

MUSCLES. *Striated muscle* comprises the consciously controlled muscles of the trunk and limbs; *smooth muscle,* which is found in the walls of the digestive tract, in blood vessels, and in air passages, is unconsciously controlled by the autonomic nervous system; *cardiac muscle,* found only in the heart, is automatically controlled by the internal excitatory system of that organ.

caused by such disorders of the mother as syphilis, kidney disease, or a structural weakness of the neck of the womb, *incompetent cervix.*

An *ectopic pregnancy* occurs when the fertilized ovum becomes lodged in the fallopian tube, or some other site, instead of being normally implanted in the womb. Surgery is sometimes necessary to prevent serious injury to the mother.

Muscles.—There are three types of muscle in the body: skeletal muscle, attached to the bones of the arms, legs, trunk, and head; smooth muscle, found in blood vessels and the lining of the digestive tract; and heart muscle. The following discussion is limited to skeletal muscle.

This type of muscle has been described by three different adjectives —skeletal, striated, and voluntary. *Skeletal* emphasizes the fact that these muscles are attached to the bones. *Striated* refers to the regular markings, or striations, of the muscle fiber that are visible under a microscope. *Voluntary* refers to the fact that these muscles are under the conscious control of the brain; they can be moved at will.

■**MUSCLE STRUCTURE.**—Muscle tissue can be described as fibers within fibers within fibers—rather akin to an electrical cable made up of smaller wires which are in turn composed of still smaller wires. Muscle is composed of parallel muscle fibers bound together in groups by bands of connective tissue. Each fiber is a separate cell with several nuclei distributed along its length. Each cell contains numerous *myofibrils* which extend the length of the cell. The myofibrils consist of short segments of the protein fibers *actin* and *myosin.* The clue to muscle contraction is believed to lie in the spatial arrangement of these protein fibers.

The functioning of muscle can be considered in relation to the questions of how muscles produce energy, how this energy is used for contraction, and how muscle is controlled and stimulated by the nervous system.

■**ENERGY PRODUCTION IN MUSCLE.**—Energy production within the muscle occurs in two phases. One proceeds *anaerobically,* or without oxygen; the other requires oxygen. The principal source of muscle energy is *glucose.* During the processes of digestion and metabolism, carbohydrates (starches and sugars) are broken down into sugar (glucose) and absorbed into the blood. In the liver this sugar is stored in the form of the starch *glycogen,* which consists of thousands of sugar units. Liver glycogen is decomposed into sugar as needed and released into the blood stream, where it circulates to the muscles.

In the muscle cell, sugar is stored as *muscle glycogen.* When the muscle requires energy, this glycogen is broken down into its constituent sugar units. The muscle cells produce energy from sugar by breaking it down through a series of chemical reactions into *lactic acid.* In the course of these reactions energy is produced in the form of *adenosine triphosphate,* or *ATP.* This first phase of energy production is a "fast reaction" which does not require oxygen. The advantage of this is that muscles may be called upon to produce energy before the body is able to increase its oxygen intake.

The second phase of energy production involves the "burning" of lactic acid with oxygen. This phase occurs for the most part in the liver and requires another series of reactions, known as the *Krebs,* or *citric acid, cycle.* In the course of these reactions considerably more ATP is produced, along with the waste products, carbon dioxide and water.

It is helpful to place these chemical reactions in context, as they actually occur in conjunction with muscular activity. The case of a sprinter is a good example. A runner needs quick energy in order to race. While he is running, his muscles produce energy by breaking down muscle glycogen into sugar, which is then broken down into lactic acid. During this time he is said to acquire an "oxygen debt," referring to the extra oxygen he will have to breathe to rebuild the glycogen stores he has consumed. He repays this debt as his rate of breathing increases. At this time the second phase of energy production proceeds: lactic acid is transported from the muscle to the liver by the blood and is burned in the citric acid cycle, thus producing energy-rich ATP and more glycogen. This is generally described as the "second wind," the shifting of energy production to the *aerobic* or oxygen-consuming phase, once the rate of breathing has been increased.

■**MUSCULAR CONTRACTION.**—In order to understand muscular activity, it is necessary to know how chemical energy stored in ATP and in *phosphocreatine* (a high-energy phosphate that, like ATP, acts as the cell's immediate energy reserve) is converted into the mechanical energy of a muscular contraction. At one time it was believed that this resulted from the shortening of the actin and myosin fibers, but this hypothesis has been generally discarded in favor of what might be called the "sliding filament" theory of muscular contraction, which likens the muscle to a folding telescope opened to full length, with a spring inside tending to "telescope" it closed. The parts of the telescope lock in the open position by a rachet-like device. As the lock is released, the parts of the telescope are pulled together by spring action. This is roughly what is thought to occur in the muscle. The actin and myosin fibers are thought to lie in a telescoping arrangement, locked in position by bridges that cross between them. ATP may release these bridges, at which time the actin and myosin fibers slide together and lock in a new position.

■**MUSCLE AND THE NERVOUS SYSTEM.**—The large muscles of the trunk, limbs, and head are under voluntary control. It is important to know how this is achieved and how the muscle is stimulated to contract.

The voluntary control of muscles is under the direction of cells in the *cerebral cortex,* the upper part of the brain. Each muscle is represented by cells in the motor area of the cerebrum. The number of cells representing a specific muscle is proportional to the degree of control the brain exerts over the muscle. Thus the muscles of the hands have more representative controlling cells in the brain than do muscles of the back. This reflects the exquisite control a person has over the small muscles that enable him to move his fingers in such a great variety of skilled movements.

Nerve fibers from the brain pass down the spinal column, where they connect with the motor cells of the spinal cord. These fibers then leave the spinal cord through the spinal nerves, ultimately arriving at a *motor end plate* on a muscle cell. The nerve ending secretes a substance called *acetylcholine,* which alters the electrical properties of the membrane of the muscle cell. This triggers the muscle contraction, in which the energy contained in ATP somehow alters the spatial alignment of the actin and the myosin fibers.

Sense Organs.—It may be said that all the systems of the body exist for the sake of the nervous system. This incredibly intricate structure directs both the internal activities of the body and its interactions with the outside world. The *sensory receptors* are specialized structures that feed information into the nervous system. They include the *distance receptors* (eyes and ears), the *chemical receptors* (nose and taste buds), the *sensory receptors of the skin* that react to heat, cold, and pressure, and the *internal receptors* that maintain muscle position and balance. Finally, there is the poorly understood phenomenon of pain, which is frequently considered to be a sensation.

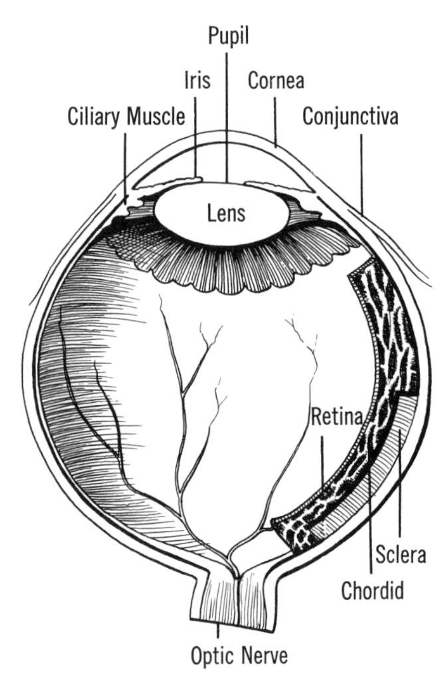

THE EYE IN CROSS-SECTION

■EYE.—Light enters the eye through a transparent outer covering, the *cornea.* The amount of light that passes through the cornea is adjusted by the size of the *pupil,* a circular opening in the *iris* (a membrane covering the lens). The size of the pupil is automatically regulated by reflex mechanisms. It narrows in bright light and widens as the light grows dim. The discomfort experienced in passing from a dark movie theater into bright afternoon sunlight results from the fact that the pupil has opened wide in the dim light of the theater and requires a few moments to adjust to the sudden change in illumination.

Behind the iris is the transparent *lens,* an elastic body that is convex on both its front and back surfaces. The shape of the lens is adjusted by surrounding muscle that focuses light on the retina by altering the lens curvature.

The *retina* is the light-sensitive layer in the rear of the eye, where the image registers. Its structure is such that the actual light-sensitive cells (the *rods* and *cones*) are covered by two additional cell layers through which light must pass before it can activate impulses. The retina contains approximately 120 million rods, which serve for black-and-white perception, and 6 million cones, which distinguish color.

■OPTIC NERVE.—An image is transmitted to the brain by the *optic nerve,* which consists of about 1 million fibers. Before it reaches the brain, the optic nerve divides at the *optic chiasma.* This is a crossover point where the images from the right halves of both retinas travel to the right side of the brain, and images from the left halves travel to the left side of the brain. The visual impulses finally reach the *visual cortex* (the outer layer of gray matter of the cerebrum), located in the *occipital* (back) region of the cerebrum.

■NATURE OF VISION.—A simple description of the visual apparatus would make it seem as though the process of perception is well understood, whereas actually there are many unsolved problems in this field.

Vision is an active process. One point where the familiar analogy between the eye and a camera fails is in the idea that the visual image simply registers on the retina, from which it is communicated to the brain via the optic nerve. There is much evidence to indicate that the eye does not merely fix an object in focus, but moves about in definite patterns, "scanning" an object much in the manner of a photoelectric tube scanning an image broadcast on television. Much work in the field of vision concerns the nature of eye activity in the process of perception.

Still unanswered, too, is the question of depth perception—how things are seen in three dimensions. Some researchers feel that this is an inborn ability of the visual mechanism,

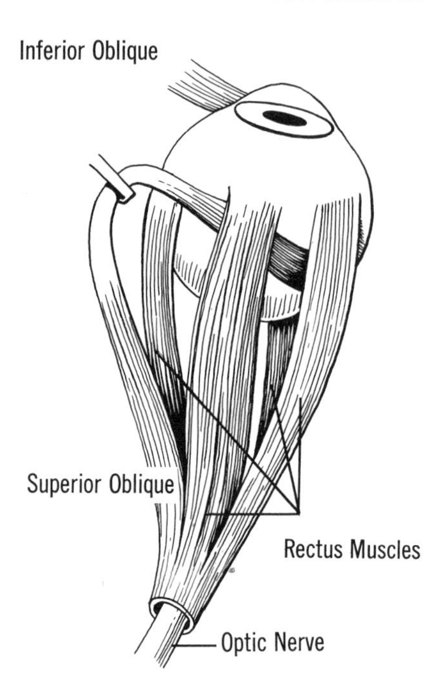

while others consider that it is learned in the course of development. One feature of depth perception, the fact that each eye sees a slightly different image, raises the additional question of *binocular fusion*—how the images from the two eyes are combined in the brain to form a single image.

■COLOR VISION.—The nature of color perception is another perplexing problem in the study of vision. Most theories of color vision propose the existence of specialized cone cells for the various colors. One of the best-known hypotheses, the *Young-Helmholtz theory,* proposes that there are three types of cones—one for blue, one for red, and one for green. All other colors would be created by stimulation in varying degrees and combinations of these three basic receptors. For example, yellow light would stimulate a specific number of red and green receptors. Impulses from these would somehow be mixed in the visual pathway or the brain to produce the experience of yellow. However, the Young-Helmholtz theory has been challenged by others. The *Ladd-Franklin theory,* for instance, proposes four types of receptors. As yet no completely satisfactory color theory has evolved.

■EYE DISORDERS AND DISEASES.—*Color blindness* is the inability to see a full range of colors. It is an inherited disorder which, according to the Young-Helmholtz theory, results from the absence, or failure to function, of one of the three basic cone types.

Blindness may be caused by disturbances to the visual mechanism at any point along the visual pathway. *Trachoma* is a viral disease which may cause a growth of tissue over the cornea; scarring of the cornea may result in blindness. *Cataracts* cause clouding of the lens and may be severe enough to blind the victim. A buildup of internal eye pressure occurs in *glaucoma,* at times

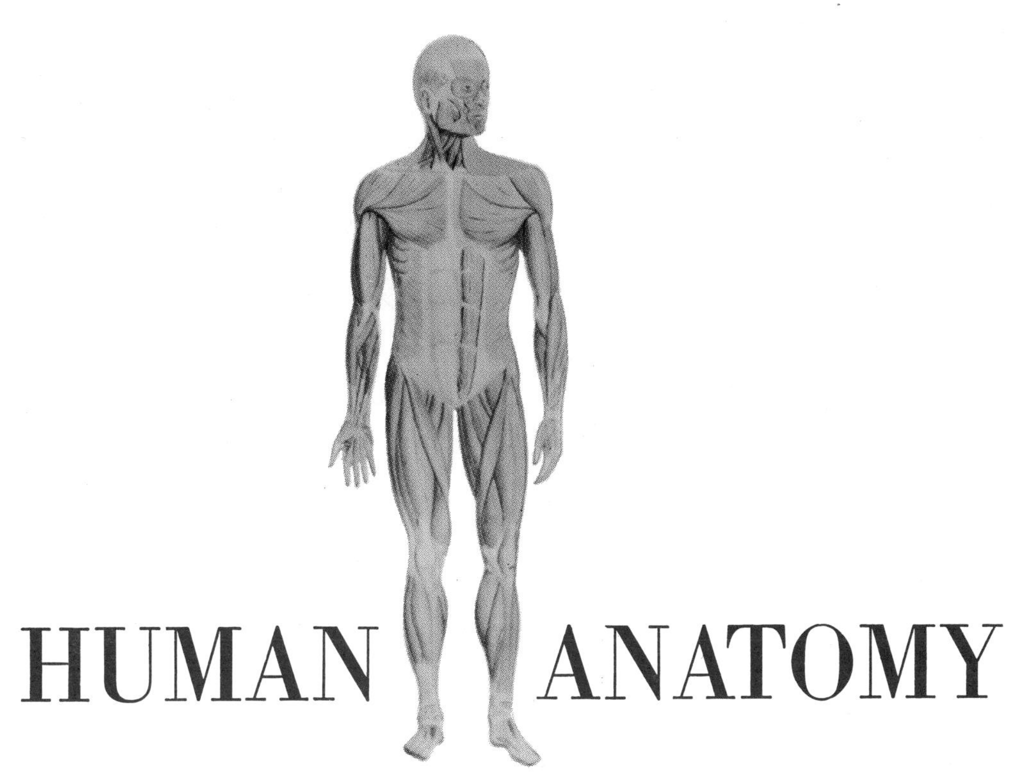

HUMAN ANATOMY

ILLUSTRATIONS BY RONALD KELLER

Prepared in consultation with
Charles N. Berry, Ph.D.
Seton Hall College of Medicine and Dentistry

This section on Human Anatomy is presented to help you to form a conception of the structure of the human body. The various structures of the body may be seen in their exact locations and in relation to the other structures. By the use of the transparencies one can determine these relationships in three dimensions—horizontally, vertically, and also in depth. In this way an understanding can be developed about the systems of the body and, therefore, about the total human organism.

Plate A shows the inside of the rib cage looking toward the front, while Plate F shows the inside of the skeleton looking toward the back. On the front of the first transparency (Plate B) most of the organs of the respiratory and digestive systems can be seen as viewed from the front. These organs include the trachea, the lungs, the intestines, pancreas, liver, and gall bladder. The thyroid glands (a part of the endocrine system) and the domelike diaphragm muscle are also seen. The back view of these organs is found on the reverse side of the transparency (Plate C). In Plate D are shown the principal parts of the circulatory system, plus the kidneys, suprarenals, and parts of the excretory system as seen from the front. In Plate E a back view of the same systems is shown.

By turning the transparencies and studying the organs and the systems that are contained from both the front and the back views, a more meaningful interpretation can be derived from the discussions in the text of the structure and functioning of the human body.

PLATE A

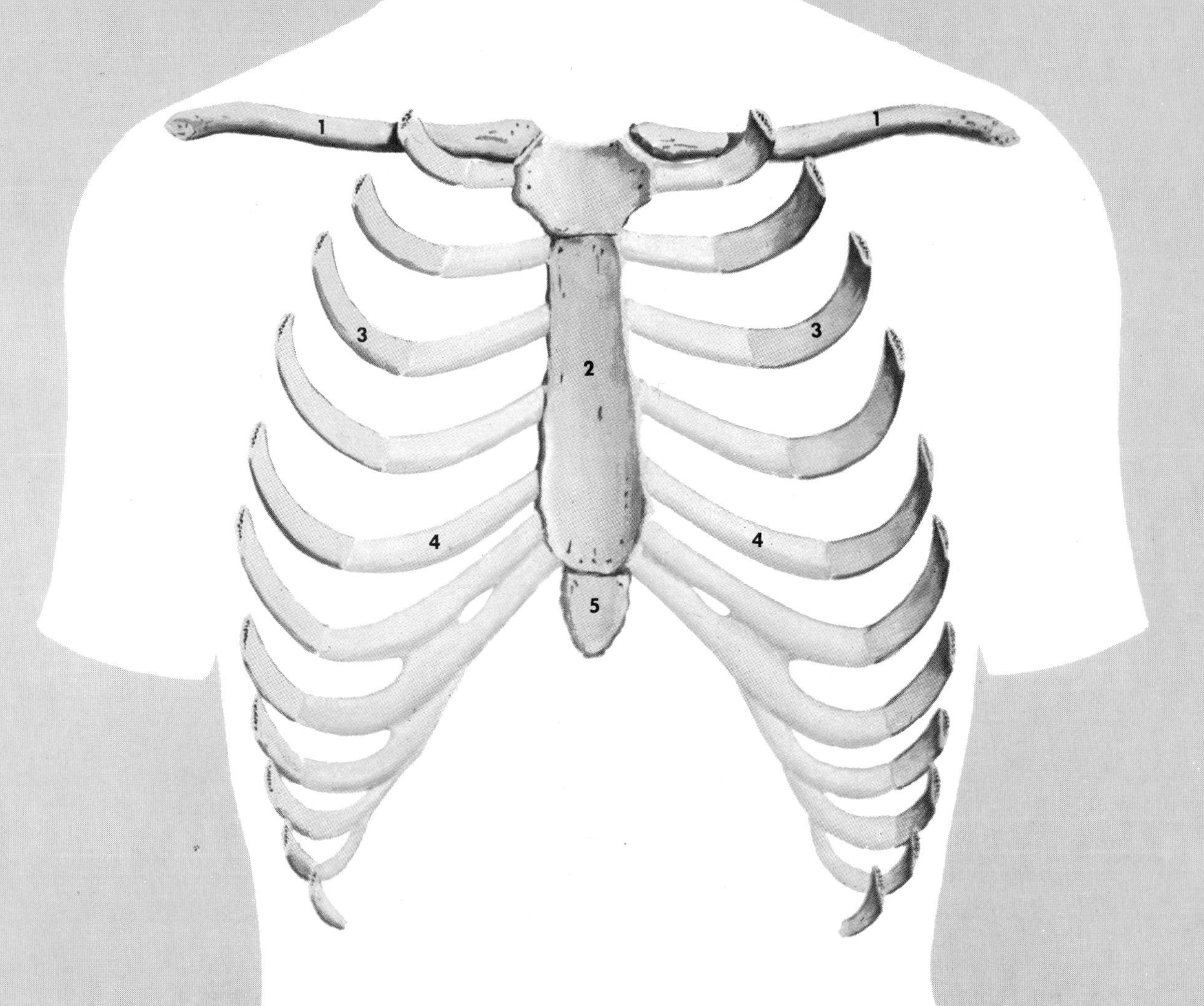

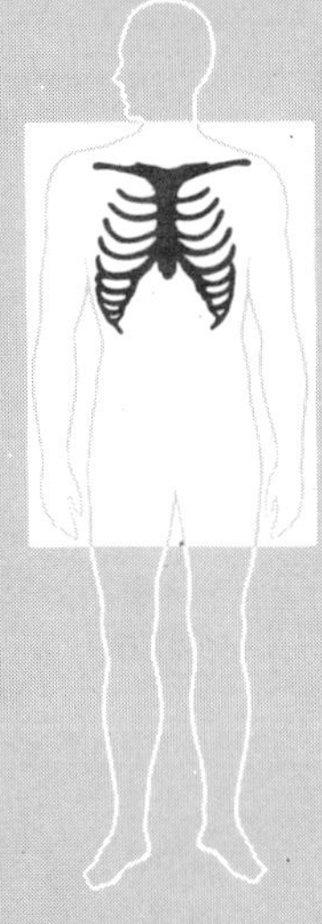

1. Collarbone (clavicle)
2. Sternum
3. Rib
4. Cartilage of the rib
5. Xiphoid process

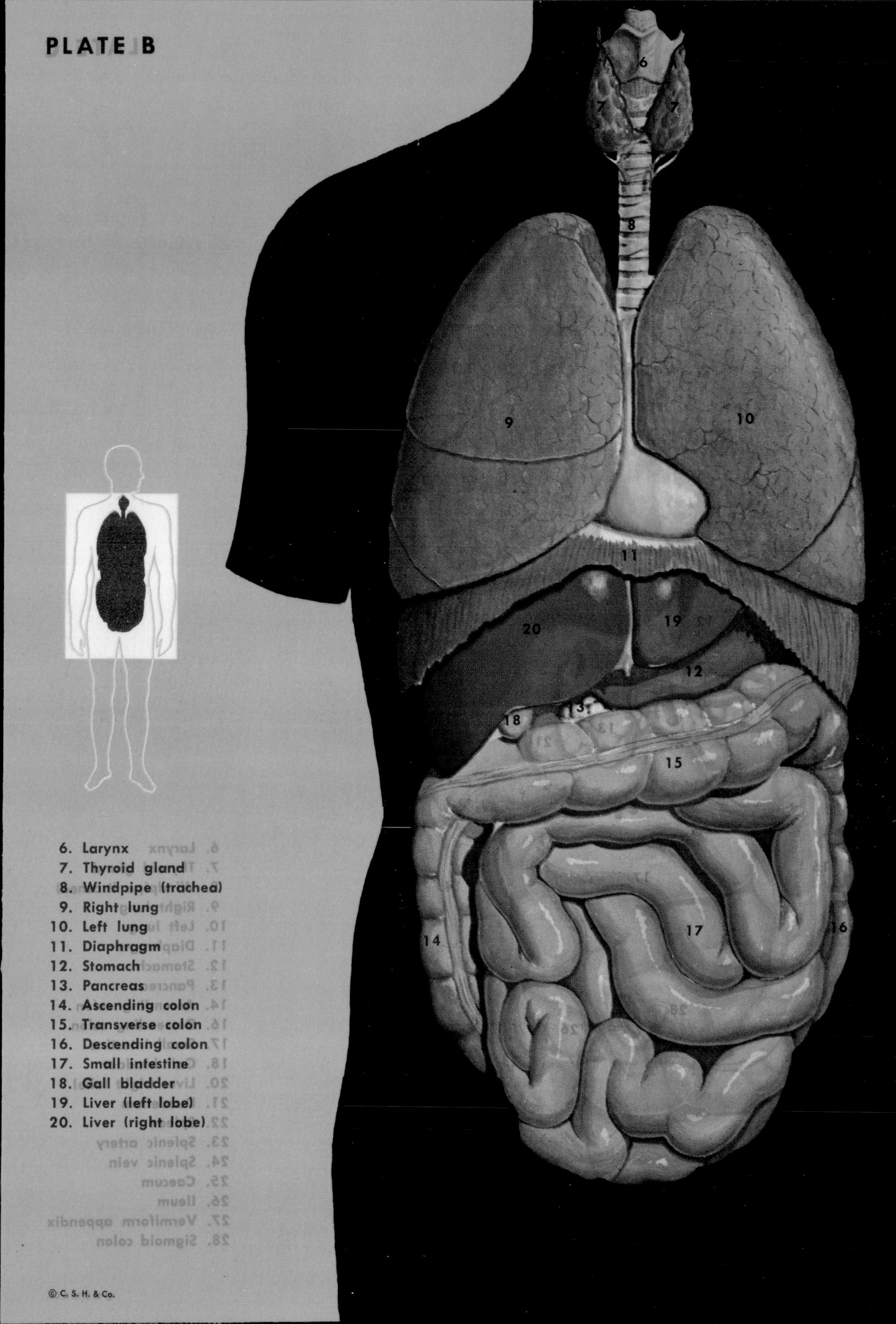
PLATE B
6
7
7
8
9
10
11
19
20
12
13
18
15
17
16
14
6. Larynx
7. Thyroid gland
8. Windpipe (trachea)
9. Right lung
10. Left lung
11. Diaphragm
12. Stomach
13. Pancreas
14. Ascending colon
15. Transverse colon
16. Descending colon
17. Small intestine
18. Gall bladder
19. Liver (left lobe)
20. Liver (right lobe)

PLATE C

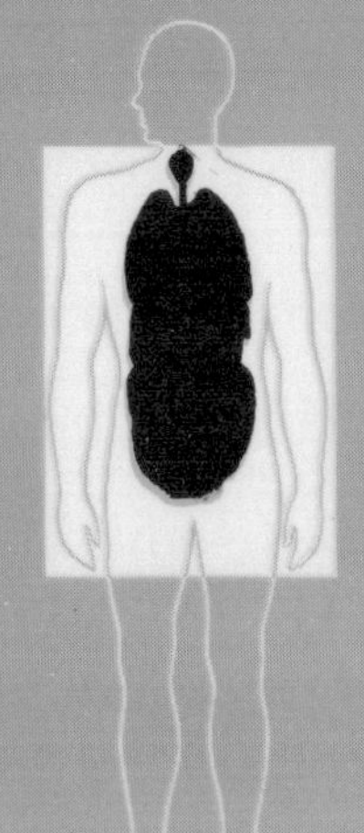

6. Larynx
7. Thyroid gland
8. Windpipe (trachea)
9. Right lung
10. Left lung
11. Diaphragm
12. Stomach
13. Pancreas
14. Ascending colon
16. Descending colon
17. Small intestine
18. Gall bladder
20. Liver (right lobe)
21. Duodenum
22. Spleen
23. Splenic artery
24. Splenic vein
25. Caecum
26. Ileum
27. Vermiform appendix
28. Sigmoid colon

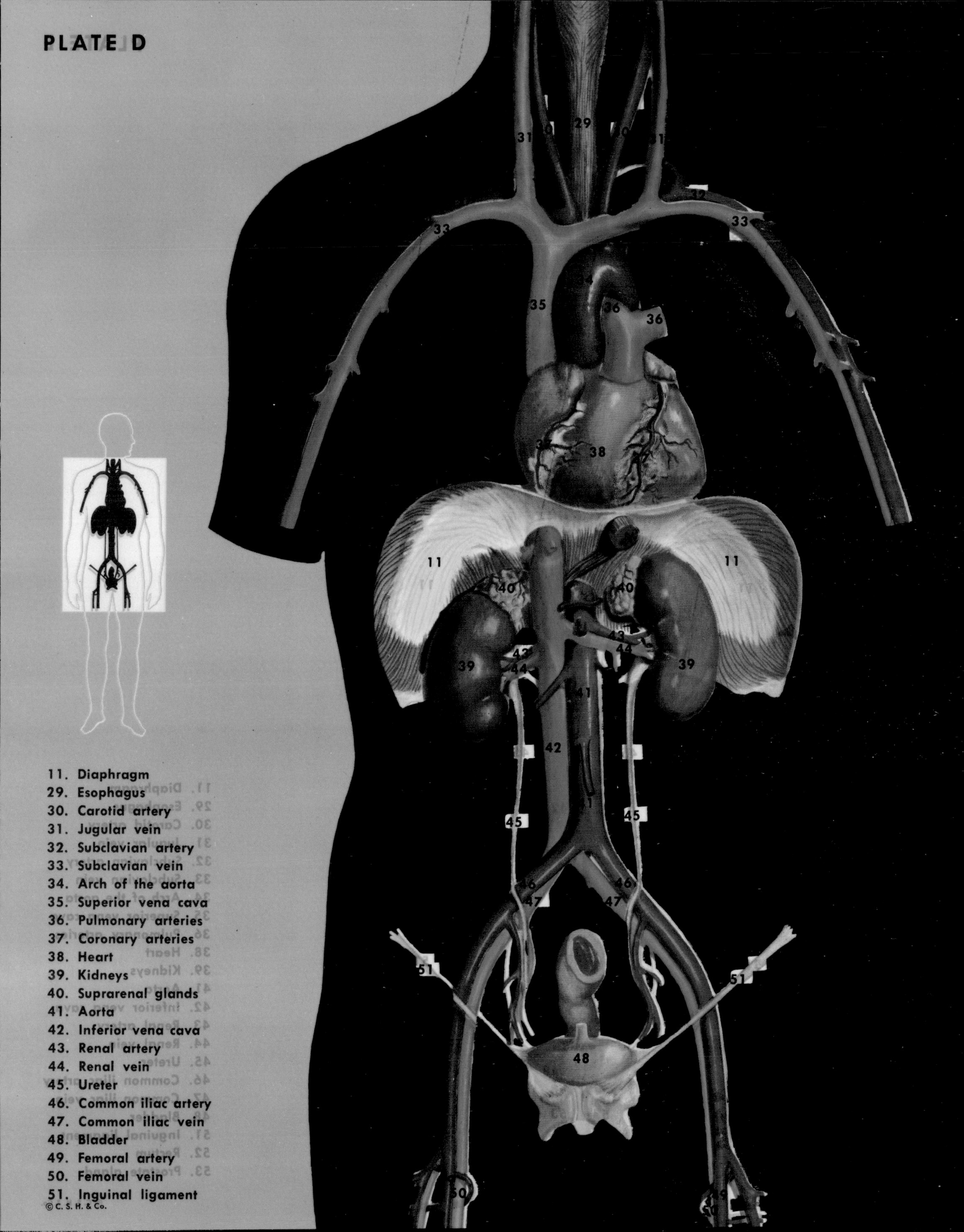
PLATE D
11. Diaphragm
29. Esophagus
30. Carotid artery
31. Jugular vein
32. Subclavian artery
33. Subclavian vein
34. Arch of the aorta
35. Superior vena cava
36. Pulmonary arteries
37. Coronary arteries
38. Heart
39. Kidneys
40. Suprarenal glands
41. Aorta
42. Inferior vena cava
43. Renal artery
44. Renal vein
45. Ureter
46. Common iliac artery
47. Common iliac vein
48. Bladder
49. Femoral artery
50. Femoral vein
51. Inguinal ligament
© C. S. H. & Co.

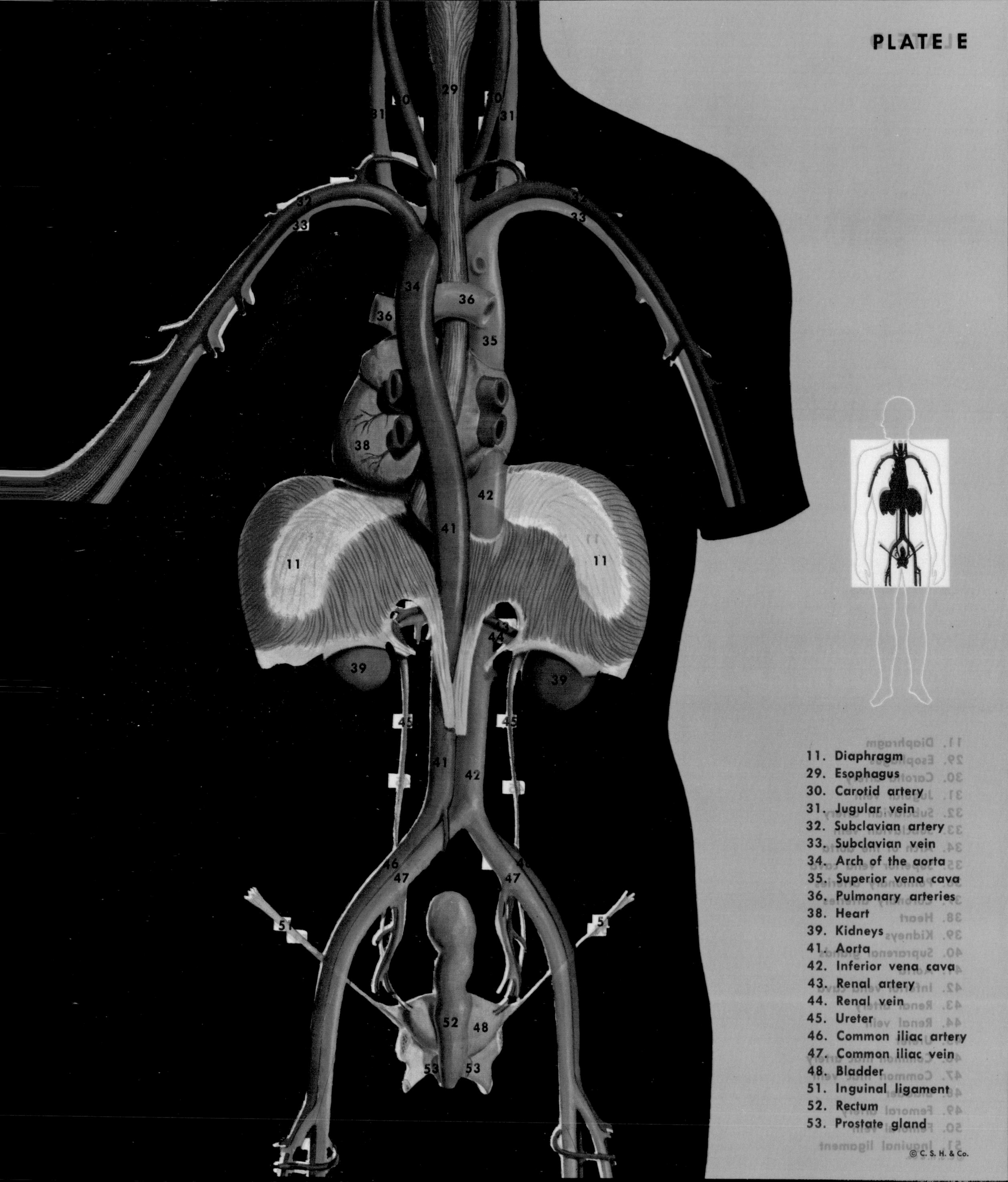
PLATE E
29
30
30
31
31
32
33
32
33
34
36
36
35
38
42
41
11
11
43
44
39
39
45
45
41
42
46
47
47
51
51
52
48
53
53
11. Diaphragm
29. Esophagus
30. Carotid artery
31. Jugular vein
32. Subclavian artery
33. Subclavian vein
34. Arch of the aorta
35. Superior vena cava
36. Pulmonary arteries
38. Heart
39. Kidneys
41. Aorta
42. Inferior vena cava
43. Renal artery
44. Renal vein
45. Ureter
46. Common iliac artery
47. Common iliac vein
48. Bladder
51. Inguinal ligament
52. Rectum
53. Prostate gland

PLATE F

3. Rib
54. Cervical vertebrae
55. Thoracic vertebrae
56. Lumbar vertebrae
57. Intervertebral discs
58. Sacrum
59. Coccyx
60. Scapula
61. Humerus
62. Head of humerus
63. Ilium
64. Femur
65. Head of femur

INDEX

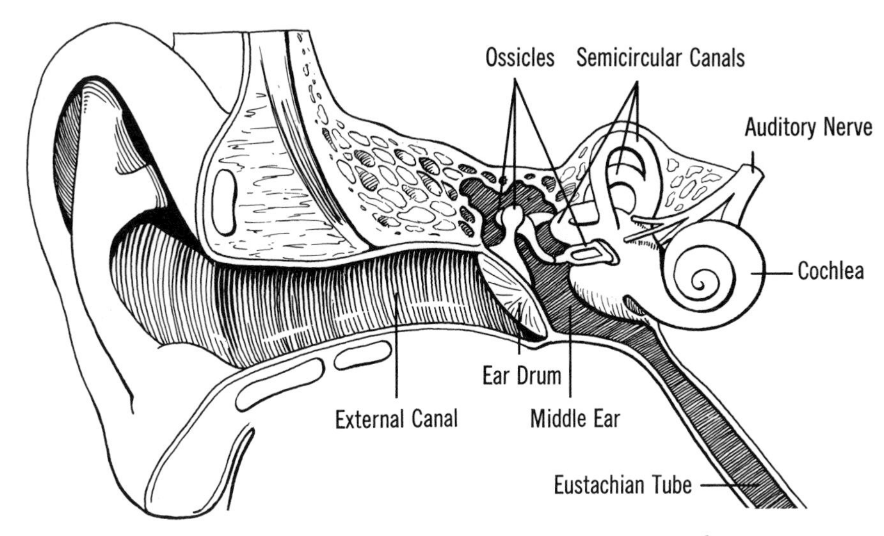

THE EAR. Sound waves are ultimately converted into nerve impulses in the inner ear.

severe enough to blind by damaging the optic nerve. Glaucoma may be caused by blockage of the normal channels through which the eye fluid drains. Detachment of the retina caused by injury may produce blindness. Growing tumors may cut off sight by pressure on the optic nerve.

Treatment of blindness depends upon the cause and the structures involved. Damaged corneas can be replaced by transplanting new ones, and detached retinas can often be sewn back in place. Damage to the optic nerve and visual regions of the brain, however, is permanent.

■EAR.—The *ear* is an organ of hearing and balance. The visible structure called the ear is actually only part of the *external ear.* It is joined to the *middle ear* by the *external ear canal,* which ends at the *eardrum.*

The middle ear contains three small bones—the *malleus, incus,* and *stapes* (also called the *hammer, anvil,* and *stirrup*)—so arranged that vibrations of the eardrum are transmitted through them in succession. The stapes transmits the vibrations to the inner ear at the oval window of the cochlea.

The inner ear consists of a spiral, snail-shaped structure called the *cochlea,* which is divided into three chambers, each of which extends the length of the structure. The middle chamber contains the *organ of Corti,* a delicate arrangement of hair cells and nerve fibers that converts the vibrations of the *cochlear fluid* into electrical impulses which travel along the *auditory nerve* to the brain.

The process of hearing can be understood as a translation of the physical properties of a sound wave into coded electrochemical nerve impulses. In brief, sound waves, which are vibrations of molecules in air, are transformed into vibrations of the eardrum. The movements of the eardrum are changed into vibrations of the bones of the middle ear. These are converted into movements of the cochlear fluid, exerting tension on the hair cells of the organ of Corti and finally generating nerve impulses in the auditory nerve.

■DEAFNESS.—*Deafness* results from interference with the hearing process at any point. In *otosclerosis,* bony deposits fix the stapes in place, preventing them from transmitting vibrations to the inner ear. This condition can be corrected by freeing the stapes or by cutting a window between the middle and inner ear to permit vibrations to pass through. Damage to the cochlea or the auditory nerve, however, produces incurable deafness.

■BALANCE.—Close to the cochlea in the inner ear are the *semicircular canals,* the organs of balance, consisting of three fluid-filled semicircular structures, each lying in a different plane. Within the open ends of the canals are the cone-shaped *cristae.* As the head is moved, the fluid in the canals drags against the cristae, bending them and generating nerve impulses to the brain. This stimulation, however, occurs only when motion begins or ends. When the body is spun at a fairly constant speed, dizziness results because the fluid and the cristae move together. Thus no stimulation is produced, and the brain receives no balancing signals. Ballet dancers avoid such dizziness by alternately increasing and decreasing their speed of rotation when executing pirouettes.

In addition to the aforementioned receptors there are also "gravity" receptors that are stimulated by changes in the position of the head.

■BODY SENSE AND BALANCE.— The brain receives other types of information that enable it to maintain body balance. One source is the eyes, which perceive horizontal and vertical structures in the surroundings. Other information comes from sensory receptors located in the muscles and tendons. These are called *proprioceptors* and are activated when muscles contract and tendons stretch. Their signals inform the brain of the position and degree of muscular tension of the various parts of the body. This information aids in balance and also permits smooth, coordinated muscle action.

■SENSE OF SMELL.—*Smell,* also called *olfaction,* is one of the most primitive sensory mechanisms. In lower animals it affects behavior considerably. In man the sense of smell as a means of locating food and avoiding danger has yielded to the senses of sight and hearing.

The olfactory tissue is located at the upper rear portion of the nasal cavity. The olfactory cells are covered by a fluid layer that seems to dissolve odorous materials. However, the exact manner in which these materials excite nerve impulses is still not known.

The property of adaptation, common to all the senses, is particularly prominent in the sense of smell. When a person is first exposed to an odor (upon entering a room, for example) he senses it sharply. After a few moments, however, he ceases to smell it—the sense receptors have adapted to the stimulation and no longer report its presence.

■SENSE OF TASTE.—The taste receptors are located on the surface of the tongue. There are four different types of taste cells corresponding to the four basic tastes—sweet, sour (acid), salty, and bitter. More complex taste sensations occur when two or more of these receptors are stimulated in various combinations. There is evidence that the different types of receptors are found in specific areas of the tongue—sweet and salt receptors on the tip, acid cells on the sides, and bitter receptors toward the back.

Flavor is a compound of the sensations of taste, smell, and also of touch, since there are also touch receptors in the tongue.

■TOUCH AND TEMPERATURE SENSATIONS.—The skin has receptors that respond to mechanical pressure and temperature change. Impulses from these receptors travel to the brain, which has a type of map enabling it to identify stimulation as coming from a specific part of the body. The distribution of the various receptors over the body varies considerably. For example, more touch receptors are found in the tips of the fingers than in the skin of the back.

■PAIN.—*Pain* is called a sensation since it involves receptors that transmit impulses to the brain. Pain differs from other sensations, however, in some important respects. It dominates consciousness, driving all other sensations into the background, and is not experienced in direct proportion to the stimulus. A sound grows louder as more energy beats upon the eardrums, but pain does not increase step by step in intensity as the stimulus is increased. For example, a severe razor cut may go unnoticed, whereas the pain from an injection can be heightened out of proportion to the injury. Pain is excited in several different ways. Other sensations are aroused by specific forms of stimulation, such as sound waves falling on the ear; but pain may be caused by tension and pressure in swollen tissues, by mechanical injury, or by chemical factors, such as are believed to be associated with the pain generated by strenuous muscle exercise.

There are other peculiarities associated with pain, such as the difficulty sometimes experienced in localizing it and the fact that the perception of pain is so strongly conditioned by social and cultural factors. From a strictly physiological point of view, it has been established that there are specific pain pathways in the brain and spinal cord.

Nervous System

PERIPHERAL NERVOUS SYSTEM.—The sensations discussed above generally have their receptors in the outlying, or peripheral, regions of the body. The nerve impulses from these receptors enter the brain through either of two channels—the 12 cranial nerves or the spinal cord. The cranial nerves supply the structures of the head and neck. (An exception to this is the *vagus nerve,* which supplies the abdominal organs and the heart.) Entrance to the spinal cord is made through the spinal nerves. There are two spinal nerves for each segment or level of the spinal cord.

The cranial nerves and the spinal nerves, with their branches, constitute the *peripheral nervous system,* the system of nerve pathways linking the brain and spinal cord to the organs of the body. The peripheral nerves consist of *sensory* (*afferent*) *fibers,* which carry impulses to the brain, and *motor* (*efferent*) *fibers,* which carry impulses in the opposite direction. At various points along the peripheral pathways are structures called *ganglions,* collections of nerve cell bodies that act as relay stations. An incoming fiber ends on the body of a ganglion cell; another fiber conducts the impulse away from the ganglion.

CENTRAL NERVOUS SYSTEM.—The *spinal cord* is a great nerve highway linking the brain to the rest of the body. Within its central *gray matter* are nerve cell bodies through which various linkages in the nerve pathways are made. Around the gray matter is *white matter,* consisting of bundles of nerve fibers grouped into ascending and descending tracts.

The *brain* is the most intricate structure known in nature. Although we can list the parts of the brain and attempt to describe their functions, this should not imply that man understands even a small fraction of what actually occurs in the brain.

The *medulla,* located at the base of the brain, is the point of entrance for the spinal cord and the location of the control centers that regulate heartbeat and respiration. The *cerebellum* is a rounded structure to the rear of the midbrain. It is a coordinating center for muscular activity. Here impulses from the spinal cord and cerebral cortex are integrated.

The upper part of the brain, the convoluted *cerebrum,* is divided into two *cerebral hemispheres.* The surface of the cerebrum, called the *cerebral cortex,* consists of several layers of cells that dip and twist in intricate folds. It is here that the centers for vision, hearing, touch, pain, and muscle control are located. It also contains the centers for speech, which are unique in that they are located in only one half of the brain; other activities are generally represented in both hemispheres of the cerebrum. The frontal lobes of the cerebrum are thought to be especially important in abstract thinking, although it should be emphasized that the brain's activities can hardly be compartmentalized and simplified in this manner. Evidence suggests that the brain to a great extent functions as a whole; events occurring in one portion greatly influence what is simultaneously occurring in other portions.

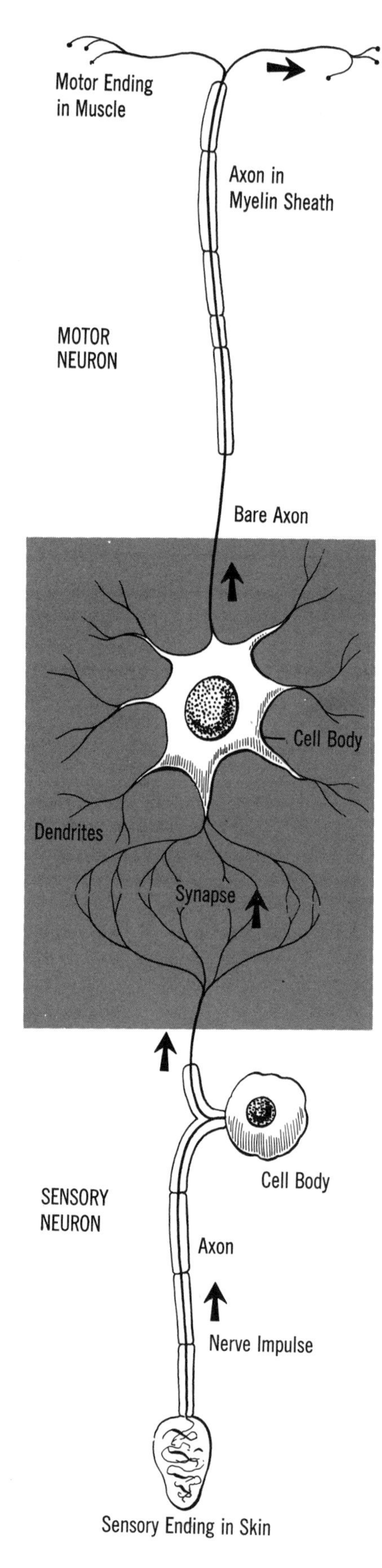

A NERVE CIRCUIT. Impulses from the skin travel to the spinal cord, then to muscle.

Running through the lower brain and the midbrain is a core of tissue known as the *reticular formation,* or the *nonspecific activating system.* This network receives impulses from all the sensory systems that feed into the brain and activates other portions of the brain and spinal cord. For this reason it is believed to play an important role in attention and consciousness. The reticular formation also feeds impulses into the *hypothalamus,* a key structure through which the brain controls the activities of the internal organs. The hypothalamus provides the connection to the autonomic nervous system.

AUTONOMIC NERVOUS SYSTEM.—The nervous system not only regulates the activities of the body in relation to the outside world but also coordinates the activities of the internal organs. This second type of control proceeds without conscious direction and is made possible by special nerve pathways and cells organized into the *autonomic nervous system.* The system has two major divisions: the sympathetic nervous system and the parasympathetic nervous system.

The *sympathetic nervous system* includes centers in the middle of the brain and at its base. Its fibers extend to the various organs through the spinal nerves. These fibers end on smooth muscles in blood vessels and on glandular cells in the various tissues. When it is stimulated, the sympathetic nervous system helps to mobilize the body's resources for emergency action. This includes constriction of blood vessels in the skin and abdominal region and expansion of blood vessels in the muscles. Also, the heartbeat increases; and the movements of the digestive tract and the secretions of the digestive organs are reduced.

The *parasympathetic nervous system* generally produces effects which are the opposite of those associated with sympathetic activity. The parasympathetic fibers reach the organs largely through the vagus nerve and through the lower part of the spinal cord in the *sacral region.* When they are stimulated, the parasympathetic fibers slow the heartbeat and stimulate the movements and secretions of the digestive organs.

The autonomic nervous system exerts considerable control over the endocrine glands, and through this control the brain and nervous system achieve complete integration of body activity. One aspect of this control is the relationship between the hypothalamus and the pituitary gland, the "master" endocrine gland; pituitary secretions are regulated by the nervous system through the hypothalamus. The pituitary, in turn, regulates the secretions of the thyroid, adrenal, and sex glands. Thus the brain controls the endocrine system by regulation of pituitary secretions.

A second aspect of the nervous system's control over endocrine function is the direct nerve connection between the autonomic nervous system and the *adrenal medulla,* the inner section of the adrenal gland. This pathway is activated during times of stress, when nerve stimulation to the adrenal medulla liberates epinephrine and norepinephrine. These hormones increase heart action and otherwise prepare the body for vigorous exertion.

■NERVE CELL AND NERVE FIBER.—The structural element of the nervous system is a highly specialized nerve cell, the *neuron.* Knowledge of neuron function is still meager, but certain facts have been established.

An impulse passes from one nerve cell (A) to another (B) by way of a nerve fiber from A which ends on the cell body of B, or on a dendrite of B. (A *dendrite* is a branched structure attached to the cell body.) The gap which separates the nerve fiber of A from the dendrite or cell body of B is called the *synapse.* It has been established that at many synapses the fiber of cell A secretes a chemical transmitter substance which crosses this gap to act on the membrane of cell B, causing a series of electrochemical reactions in the membrane which generate a new nerve impulse in cell B. The impulse is called a *spike potential* and travels along the cell body of B like a spark along a trail of gunpowder, passing out of the cell body along the *axon,* the nerve fiber which conducts the impulse to the next cell. In some cells the axon may be several feet long, as in the large *pyramidal cells* of the brain that direct muscle movement. Axons from the body of the pyramidal cells extend down the spinal cord to motor cells in the anterior horn of the spinal cord. Nerves are thus collections of axons, each carrying its nerve messages from one cell to the next.

Facts on nerve cell function are: (1) The *all-or-none law* describes the fact that the nerve cell either generates a full nerve impulse or none at all, just as a bullet is either propelled from the barrel of a gun at its maximum speed or does not leave it at all. (2) All nerve impulses are alike. Nerve impulses generated by cells of varying type and size may differ from one another in the speed at which they travel or in their strength, but they all have the same electrochemical nature. This means that a nerve impulse communicating an odor to the brain along the olfactory pathway is essentially the same as a nerve impulse racing down the vagus nerve to stimulate the pancreas to secrete digestive juices. (3) *Inhibition.* An important class of nerve impulses in the nervous system are those which retard or inhibit nerve activity. Thus, an inhibitory impulse arriving at a synapse would lessen the likelihood that the next cell would be stimulated.

■REFLEXES.—*Reflexes* are comparatively simple patterns of behavior which are built into the nervous system. They do not have to be learned and require no conscious effort. They are often adaptive, and aid in functioning of the body or its protection.

The elements of a reflex are: the *receptor,* a sensory element such as the eye, ear, or the stretch-sensitive fibers which are coiled around certain muscle fibers; the nerve pathway between the receptor and the brain or the spinal cord; the connecting nerve pathways in the central nervous system; and the fibers that carry impulses away from the central nervous system to the muscles that put the reflex into action.

Although its importance is not generally appreciated, one of the most common reflexes operates every day of our lives. This is the *stretch reflex.* Distributed among the voluntary muscles of the body are stretch receptors that generate nerve impulses when they are put under tension. As a muscle relaxes, these receptors are stretched and give off impulses which enter the spinal cord, resulting in motor impulses that stimulate muscle contraction. This reflex maintains the tension of the muscles which hold the body upright. These receptors also temper muscular action, creating a smooth movement out of what might otherwise be a series of spasmodic jerks.

Other familiar reflexes include coughing, sneezing, blinking, and the diagnostic aid, the knee jerk.

—Martin Spencer and Daniel Monroe

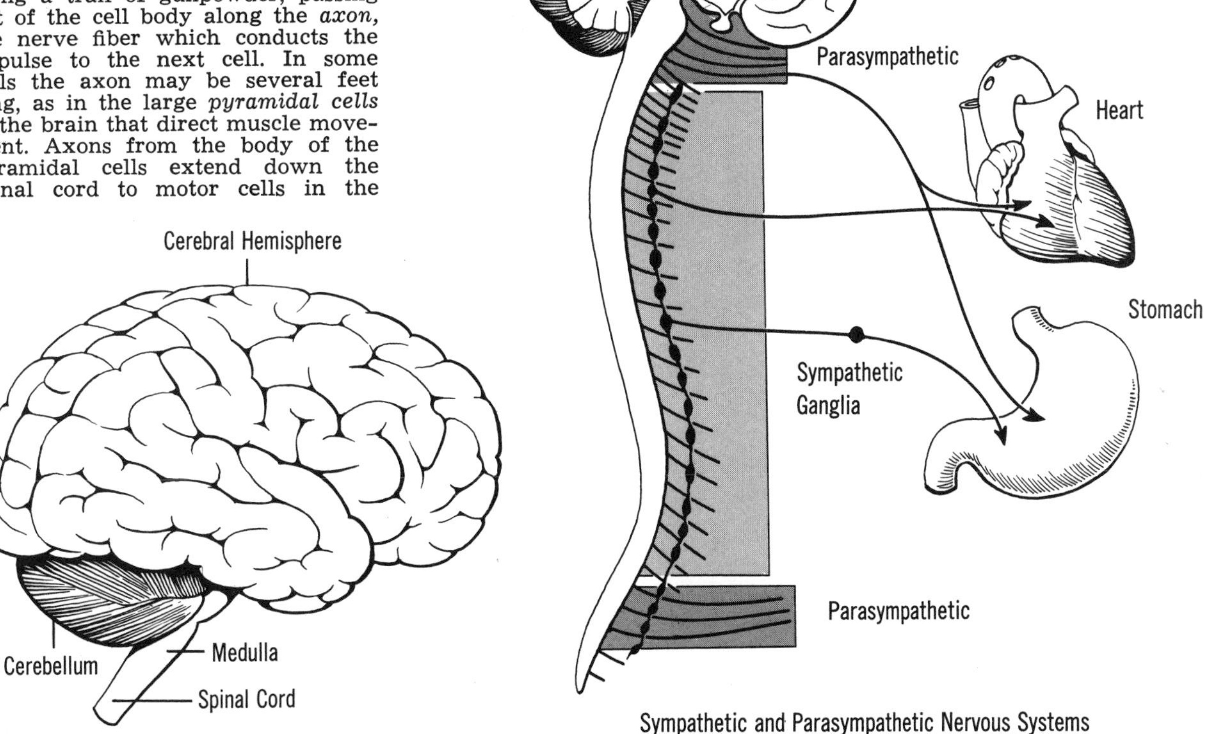

THE AUTONOMIC NERVOUS SYSTEM comprises the *parasympathetic system,* which slows the heartbeat, and the *sympathetic system,* which raises the sugar concentration in the blood, quickens the heartbeat, and mobilizes the body's resources for vigorous activity.

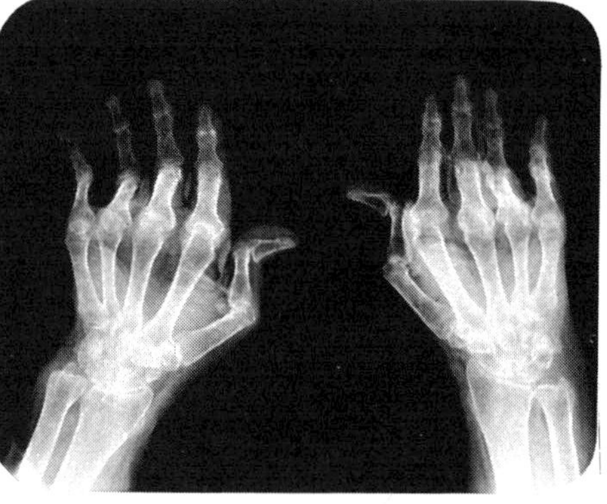

AMERICAN HEART ASSOCIATION AND WNEW-TV, CHANNEL 5

MEDICAL CARE requires the combined efforts of nurses, technicians and doctors. A technician examining a slide may be able to provide valuable information about the blood chemistry of a patient. The X-ray might aid the specialist in determining the extent of bone damage caused by arthritis. The surgeon operating on the young child (*right*) will require assistance from many individuals in order to perform a complex operation with relative safety.

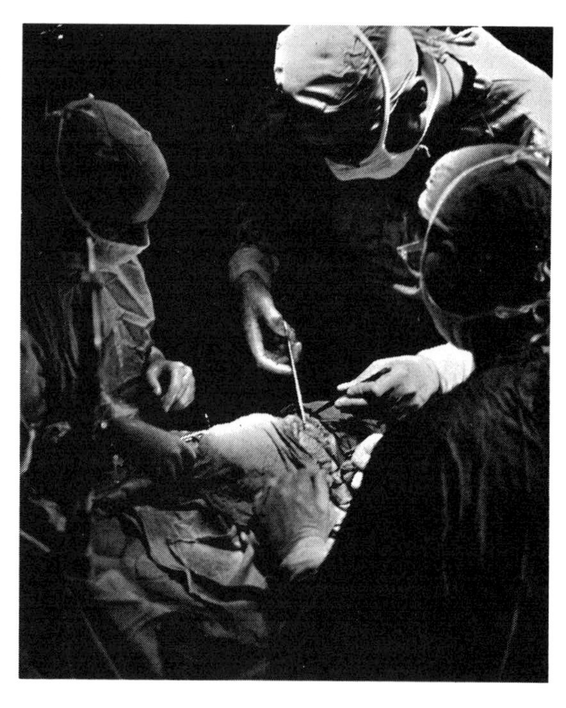

GLOSSARY OF DISEASES

(*See also chart of infectious diseases.*)

Acromegaly.—A pituitary gland disorder in which excessive secretion of growth hormone causes overgrowth of bone and tissues in face, hands, and feet.

Addison's disease.—A disturbance of the cortex (outer layer) of the adrenal glands. Undersecretion of adrenocortical hormones causes weakness, skin pigmentation, low blood pressure, weight loss, and other symptoms. May result from tuberculosis of the adrenal gland or from unknown causes.

Agranulocytosis.—A disorder marked by a decline in the number of certain white blood cells, usually caused by drug sensitivity. Victims are abnormally susceptible to infection.

Air sickness.—See *Motion sickness.*

Albinism.—The congenital absence of pigmentation, either complete or partial. In albinos, the hair color is white, the pupils of the eyes are red, and the skin is pink.

Allergy.—An abnormal sensitivity of the body to certain substances. In allergic reactions the allergy-producing substance (*allergen*) reacts with special substances in the body (*antibodies*), resulting in fluid swellings in various parts of the body. Asthma and hay fever are allergic disorders.

Altitude sickness.—A disorder seen in mountain climbing and in flights in unpressurized airplanes at high altitudes. Oxygen deficiency causes dizziness and disturbances of speech and coordination.

Anemia.—A decline in the oxygen-carrying capacity of the blood, caused by a decrease in the number of red blood cells or in the quantity of hemoglobin they carry. It may result from hemorrhage, excessive blood destruction, or damage to the blood-forming tissues. Anemia causes dizziness and shortness of breath.

Aneurysm.—A bulging or outpocketing in the wall of an artery, caused by structural weakness. It is seen in the later stages of syphilis and in arteriosclerosis (hardening of the arteries).

Aphasia.—A disorder of language functions, caused by brain injury or disease. The aphasic individual may have difficulty in naming objects, communicating ideas, or understanding either spoken or written language.

Appendicitis.—An inflammation of the appendix, a blind, wormlike pouch that opens into the large intestine in the lower-right region of the abdominal cavity. Appendicitis usually causes pain in the lower-right abdomen and an increase in the number of white blood cells.

Arteriosclerosis (*hardening of the arteries*).—An arterial disease marked by narrowing of the arterial passageway and by rigidity and loss of elasticity in the arterial wall. Symptoms depend upon the part of the body affected. Arteriosclerosis of the arteries in the heart is one of the most common forms of heart disease. In the arteries of the brain it may result in rupture and hemorrhage (stroke). Arteriosclerosis of the arteries of the legs produces numbness, tingling of the feet, and muscle cramps when its victims walk.

Arthritis.—A disease of the joints. The major types of arthritis include: *rheumatoid arthritis*—most common in the 30–40 age group, producing inflammation of the joint, with possible deformity in advanced stages; *gout*—a disorder of body chemistry marked by deposition of uric acid crystals in the joints; *osteoarthritis*—destruction of the joints, seen in elderly persons and possibly resulting from mechanical "wear and tear."

Asthma.—A breathing disorder. The most common type is *allergic bronchial asthma,* caused by sensitivity to such substances as pollen, wheat, chocolate, and various drugs. Asthmatic attacks exhibit shortness of breath, wheezing, and coughing.

Astigmatism.—A disorder of vision marked by an inability to focus properly all parts of the visual field. It is caused by differences in the refracting, or "light-bending," power of different parts of the visual apparatus.

Bedsores (*decubitus ulcers*).—Skin ulcers caused by prolonged pressure on such parts of the body as the heels, elbows, and back. Bedsores occur in bedridden patients who are kept in the same position for long periods of time.

Bell's palsy.—Paralysis of one side of the face, causing muffled speech, difficulty in eating, and tearing of the eye on the affected side.

Beriberi.—A nutritional disease caused by a deficiency of vitamin B_1.

Berylliosis.—Beryllium poisoning, caused by inhalation of, or contact with, beryllium fumes.

Blue baby.—A newborn infant suffering from congenital heart defects that interfere with the normal oxygenation of blood in the lungs. In such cases large amounts of oxygen-poor blood give the skin a bluish cast.

Botulism.—An often fatal type of food poisoning, caused by the consumption of improperly sterilized canned food contaminated with the anaerobic bacterium *Clostridium botulinum.*

Bronchiectasis.—A disease of the *bronchi* (the air passages to the lungs) in which the walls of the small bronchi are weakened and dilated. The disease may develop as a complication of lung infections. It produces coughing accompanied by bloody sputum.

Bronchitis.—An inflammation of the bronchi. *Acute bronchitis* may be caused by infection or by irritation from tobacco, chemicals, dust, and similar substances. Chronic bronchitis may be associated with lung diseases. Symptoms include coughing, wheezing, and shortness of breath.

Buerger's disease (*thromboangitis obliterans*).—A circulatory disease marked by progressive obstruction of the blood supply to the extremities. Occurs principally in smokers.

Bunion (*hallux valgus*).—A foot disorder in which the great toe points away from the midline of the body. The joint at the base of the toe is enlarged and deformed. The overlying skin is usually calloused. It is caused mainly by wearing narrow, pointed shoes.

Bursitis.—An inflammation of a *bursa* (fluid-filled sacs usually found at points of friction in the body, such as where tendons pass over bony prominences). *Subdeltoid bursitis* involves inflammation of the bursa in the shoulder region.

Cancer.—An abnormal, uncontrolled growth of cells. Occurs in plants, in human beings, and in other animals.

Car sickness.—See *Motion sickness*.

Cataract.—A clouding of the normally transparent lens of the eye, occurring mainly in the aged. Also a complication of *diabetes mellitus*.

Celiac disease.—A children's disease marked by interference with the absorption of nutrients from the intestine. Victims are undernourished and fail to grow normally.

Cerebral palsy.—Disturbances in muscular function and coordination caused by brain damage. It appears mainly in children and is associated with prenatal brain damage. In *spastic palsy*, the muscles contract vigorously, producing jerky, uncontrolled movements. *Athetoid palsy* is marked by constant, irregular, writhing movements.

Chilblain (*pernio*).—A redness, burning, itching, and blistering of parts of the skin, caused by repeated exposure to cold.

Cholangitis.—An inflammation of the bile ducts.

Chorea (*St. Vitus' dance*).—A brain disorder, marked by involuntary movements of the lips, cheeks, arms, and other parts of the body. It occurs as a complication of rheumatic fever (in children), pneumonia, and scarlet fever, and is occasionally seen in pregnancy.

Cirrhosis of the liver.—A scarring and degeneration of the liver that may be associated with alcoholism or liver infections. One serious complication is the weakening of certain large veins of the chest (*esophageal varices*), caused by obstruction of the blood flow through the liver.

Cleft palate.—Congenital openings in the *palate* (roof of the mouth) that join the nasal and oral cavities. Causes trouble in eating and speaking.

Clubfoot.—A congenital deformity of the foot and ankle. In most cases the toes point down and the foot is twisted inward at the ankle.

Colitis.—An intestinal inflammation that may be caused by infection, poisoning, or unknown causes. A major form of colitis is *nonspecific ulcerative colitis*, which occurs principally in young adults and produces pain, fever, diarrhea, and ulcers of the intestinal wall.

Cretinism.—A disorder of children in which a deficiency of thyroid hormone causes physical and mental retardation.

Cryptorchism.—Undescended testicle or testicles. The testicles usually descend during the seventh to the ninth month of fetal life. Failure of both testicles to descend usually causes sterility.

Cystic fibrosis.—An inherited disorder of the *exocrine* (duct) glands of the body (sweat glands, acinar cells of the pancreas, glands lining the bronchi). Characteristic symptoms include digestive difficulties, malnutrition, and chronic lung disease.

Cystocele.—A female disorder, marked by a rupture of the urinary bladder through the vagina. May cause backache and a burning sensation during urination.

Decompression sickness (*caisson disease, or the "bends"*).—Paralysis, loss of equilibrium, and pains in the muscles and joints caused by liberation of nitrogen bubbles from the blood. It occurs in individuals undergoing rapid reduction in air pressure, such as caisson workers, deep-sea divers, and others who work under increased air pressure.

Dermatitis.—An inflammation of the skin. May be caused by allergies, harsh chemicals, radiation, or overexposure to the sun.

Deviated septum.—A displacement of the partition (*septum*) that divides the nasal cavity. May interfere with breathing and speech.

Diabetes insipidus.—An endocrine disease marked by the passage of large amounts of urine. Caused by undersecretion of a pituitary hormone that regulates the reabsorption of water in the kidney.

Diabetes mellitus.—Disease caused by inadequate secretion of the pancreatic hormone insulin, with a consequent impairment of the body's ability to burn *carbohydrates* (starches and sugars) as fuel.

Diaper rash.—An inflammation and maceration of the diaper region in infants, caused by the action of urine on the skin.

Dwarfism.—Stunted growth, usually caused by disorders of the bones or of the endocrine glands. Types include *pituitary dwarfism*, in which body proportions are abnormal, and cretinism, in which both mental development and physical growth are impaired.

Eclampsia.—Convulsions and coma that may appear in certain disorders of pregnancy, called *toxemias*. Associated symptoms include swelling of the body, increased blood pressure, and abdominal pains.

Eczema.—A skin disorder marked by scaling, redness, itching, and blistering. In chronic cases, patches of skin become thickened and roughened.

Edema (*dropsy*).—A swelling of the body caused by accumulations of fluid in the tissues. May be caused by kidney disorders and circulatory disorders.

Elephantiasis.—A thickening of skin and underlying tissues. May be caused by interference in the circulation to a part of the body as a result of parasitic disease (*filariasis*), or by surgery.

Embolus.—A fragment of a blood clot or other substance that travels freely through the blood stream, eventually lodging in and obstructing a blood vessel. It may appear following blood clots in the vessels of the heart or after surgery. Emboli lodging in the blood vessels of the heart, brain, or lungs are highly dangerous.

Emphysema, pulmonary.—A lung disorder marked by expansion of the tiny sacs of the lungs (*alveoli*) where the exchange of gases occurs. It occurs in cases of bronchitis and interferes with respiration. It may result in coughing, shortness of breath, and, possibly, heart failure.

Epilepsy.—A brain disorder marked by convulsive seizures (*grand mal*) or by momentary faints or spells (*petit mal*), possibly caused by brain tumors, injury, or infection. In cases of *idiopathic epilepsy*, no observable cause can be ascertained. Such cases tend to appear in family groups.

Eunuchoidism.—A glandular disorder of males, characterized by female fat distribution, high-pitched voice, and other feminine changes in both the body and the psyche. It is caused by inadequate production of male hormones by the testes.

Exophthalmos.—A protrusion of the eyeballs, seen frequently in cases of hyperthyroidism.

Farsightedness (*hypermetropia*).—A visual disturbance in which the eye cannot focus properly. Distant objects can be focused by contractions of the muscle that controls the shape of the lens of the eye. Close objects can also be focused by this method, but at the cost of strain, eye fatigue, and headaches.

Fibroma.—A tumor of fibrous tissue, seen most often in the skin and subcutaneous tissue.

Fistula.—An abnormal opening between two body cavities (*internal fistula*) or between the body and the outside (*external fistula*).

Flat feet.—A depression of the long arch of the foot. It may be present congenitally, in which case the foot is usually strong and does not require treatment. Other cases develop as a result of mechanical strain to the normal foot.

Food poisoning.—A condition that usually results from consumption of food contaminated by pathogenic bacteria or by their secretions. The commonest type is caused by the *staphylococcus* bacteria, which may contaminate improperly refrigerated salads and pastries. Inadequately sterilized canned foods may contain the deadly *botulinum* bacteria.

Frostbite.—Tissue destruction caused by freezing. Frostbite depends not only upon the outside temperature, but also on how rapidly heat is conducted from the tissues by the wind. Thus, if the wind is sharp, frostbite can occur at temperatures as high as 23° F. The skin first becomes white and numb. Later, redness, blistering, and ulceration develop.

Gangrene.—Tissue destruction caused

by interference with the circulation to a part of the body. It may occur in arteriosclerosis, or through mechanical interference with the circulation, as when a portion of the bowel becomes twisted upon itself.

Gigantism.—Abnormal stature that may be caused by undersecretion of the sex hormones or by oversecretion of the growth hormones of the pituitary gland. In the first case, bone growth continues beyond the age at which it is usually halted by the sex hormones. In the second case, all the tissues of the body are stimulated to continued growth.

Glaucoma. — A visual disorder in which obstruction of fluid drainage from the eye causes an increase in internal pressure that may lead to blindness. Glaucoma occurs principally in older persons. Early symptoms are blurred vision, colored halos seen around lights, and headaches.

Glomerulonephritis.—A kidney inflammation seen primarily in young adults and children. Symptoms include puffiness of the face, swelling of the ankles, and high blood pressure.

Glossitis.—A redness and swelling of the tongue, arising from vitamin deficiencies, infections, anemia, or irritation.

Goiter.—An enlargement of the thyroid gland, often associated with a deficiency of iodine in the diet (*endemic goiter*). *Toxic goiters* cause oversecretion of thyroid hormone, resulting in weakness, tremors, and increased heart activity.

Gout.—A disorder of body chemistry in which crystals of uric acid are deposited in various tissues, particularly in the joint at the base of the great toe and in the earlobes.

Harelip (*cleft lip*).—A congenital defect of the face, characterized by a single or double fissure of the upper lip under the nostrils. Often seen with cleft palate.

Hay fever (*allergic rhinitis*).—An allergic reaction to certain air-borne pollens, or fungus spores. It often occurs in family groups, and causes sneezing, congestion of nasal passages, tearing, and itching of the eyes. Some cases improve spontaneously over a period of time, while others may develop into asthma.

Heartburn.—A burning sensation in the abdomen and chest, caused by leakage of stomach fluids into the *esophagus* (tubular passageway between the mouth and stomach).

Hematoma.—Swelling of a part of the body with blood that has escaped from the circulatory system. They are usually caused by injuries that rupture blood vessels. Hematomas disappear spontaneously, although hematomas on the surface of the brain (*subdural hematomas*) may persist and may eventually require surgery.

Hemophilia.—An inherited disorder of the blood-clotting function caused by the absence of certain essential proteins from the blood. The hemophiliac may suffer severe blood loss from trivial injuries, and at such times require transfusions of normal blood.

Hemorrhoids.—Enlarged veins in the anus or lower rectum, causing bleeding and pain during defecation.

Hernia.—A rupture of tissues through the walls of a body cavity. The commonest type involves the protrusion of the abdominal viscera into the inguinal canal, which normally connects the abdominal cavity and the scrotum.

Hirsutism.—An abnormal growth of hair, referring especially to the growth of a beard and the masculine distribution of body hair occurring in women suffering from certain glandular disorders.

Hives (*urticaria*).—Itchy swellings of the skin caused by allergic reactions to food, medicine, or infection.

Hodgkin's disease.—A disorder characterized by progressive enlargement of the lymph glands. Causes weight loss, itching, fatigue, anemia, and other symptoms attributable to pressure exerted on various tissues by the enlarged glands.

Hydrocephalus.—A disorder in which excessive quantities of cerebrospinal fluid accumulate within the cavities of the brain (*ventricles*). It may be caused by obstruction of the flow of cerebrospinal fluid by congenital malformations, tumors, or infections. In infants, hydrocephalus causes a characteristic enlargement of the head, accompanied by mental retardation.

Hydronephrosis.—A distention of parts of the kidney due to back-pressure of urine, resulting from obstruction of urine flow. It may be caused by tumors, congenital abnormalities, or stones. Severe cases may result in kidney failure.

Hyperthyroidism.—A disorder caused by excessive secretion of thyroid hormone. It produces rapid pulse, weight loss, diarrhea, and, in many cases, heart complications.

Hypogonadism.—Inadequate functioning of the sex glands, usually occurring in the male. In young boys, male sexual characteristics fail to develop. Instead, the body acquires a female fat distribution, weak musculature, and small genitalia.

Intestinal obstruction.—A disturbance involving interference with the passage of materials along the intestinal channel. It may be caused by a mechanical obstruction (tumors, scar tissue), irritation of the nerve supply to the intestines, or through interference with the blood supply to the intestinal wall.

Jaundice (*icterus*).—A yellowish tinting of the skin and eyeballs, resulting from an accumulation of bile pigments in body fluids. Bile pigments are produced from the hemoglobin of broken-down red blood cells and are normally excreted into the gallbladder and small intestine. Jaundice may be caused by obstruction of the bile ducts (*obstructive jaundice*), by liver disease (*hepatogenous jaundice*), or by massive destruction of red blood cells, leading to excessive production of bile pigments (*hemolytic jaundice*).

Ketosis.—A disorder of body chemistry, developing when the body uses fat as a principal source of energy. It may be caused by starvation or *diabetes mellitus.*

Leukemia.—A serious blood disorder involving overproduction of white blood cells. Principal changes are anemia, tendency to bleed, and susceptibility to infection.

Lichen planus.—Small, flat, purplish skin eruptions, usually seen on the wrists and above the ankles. It may spread slowly and persist for several months, but it is usually limited in extent.

Lipoma.—A tumor of fat tissue, usually seen on the back, neck, shoulders, or extremities. Lipomas are painless and grow slowly, sometimes achieving such size that they interfere with normal functioning by pressing on surrounding tissues.

Lupus erythematosus.—A disease of the connective, or supporting, tissues, marked by skin eruptions, joint pains, swelling of the lymph glands, and possible involvement of the kidneys, heart, lungs, digestive system, and nervous system. It occurs primarily in young women.

Lymphosarcoma.—A disease in which the lymph glands enlarge and produce massive numbers of certain types of white blood cells.

Melanoma.—A tumor containing the pigment *melanin,* which is normally found in the skin, hair, and eyes. Melanomas of the face and feet frequently become cancerous.

Ménière's disease.—A disorder of the balancing structures of the inner ear, resulting in attacks of vertigo, nausea, and hissing or ringing noises in the ear. It occurs primarily in elderly persons and is believed to be caused by fluid accumulations in the inner ear.

Morning sickness.—Nausea and vomiting that may occur in early pregnancy. It usually disappears by the end of the third month.

Motion sickness (*air sickness, sea sickness, car sickness*).—Nausea, dizziness, headache, and pallor, caused by continued rocking or up-and-down motions, and believed to result from interference with the body balance mechanisms.

Multiple sclerosis.—A nervous disease involving destruction of the fatty insulation that envelops many nerve fibers. The disease is marked by attacks of paralysis, disturbances of coordination, double vision, and other nervous symptoms that usually subside in days or weeks, only to recur later, usually with greater severity. The cause is unknown.

Muscular dystrophy. — An inherited disorder marked by a progressive deterioration of the voluntary muscles. Some forms of the disease attack children only.

Myasthenia gravis.—A disorder in which muscle weakness produces drooping eyelids, easy fatigability, and difficulties in eating and speaking. It is apparently caused by an interruption in the transmission of impulses from the nerves to the muscles.

Myxedema.—A disease caused by undersecretion of thyroid hormone in adults. Signs and symptoms include joint and muscle pains, voice changes, puffiness of the eyes, constipation, anemia, dizziness, and, occasionally, psychosis. (See also *Cretinism.*)

Nearsightedness (*myopia*).—A visual

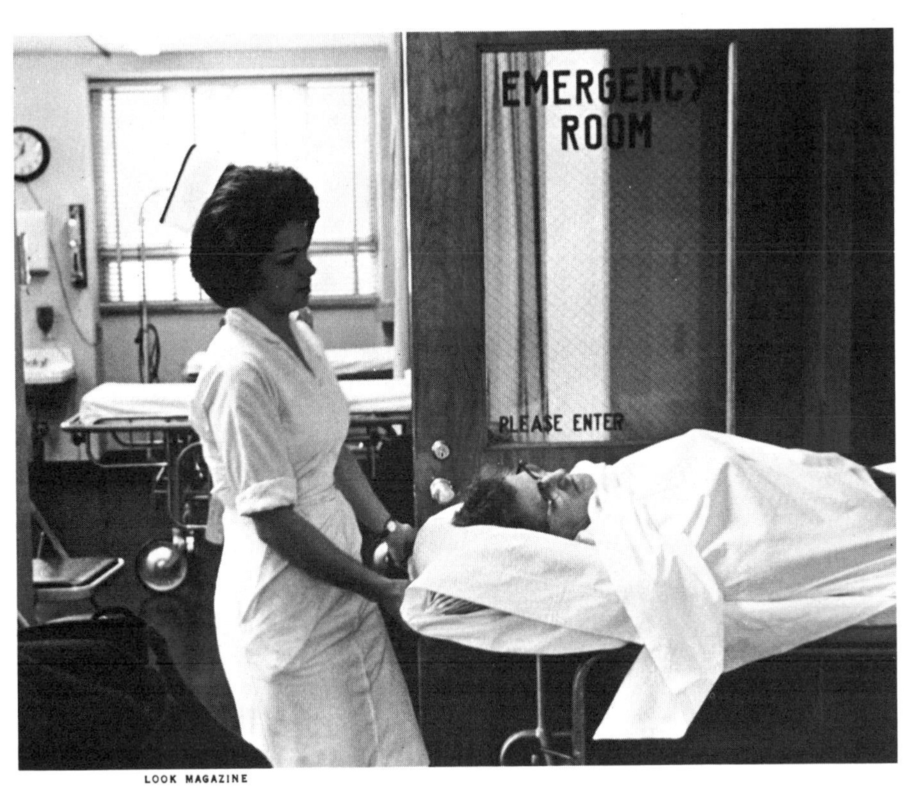

LOOK MAGAZINE

AMERICAN CANCER SOCIETY

HOSPITAL TREATMENT includes emergency care for new arrivals (*above left*), radiation therapy (*above right*), and surgery (*below*).

AMERICAN CANCER SOCIETY

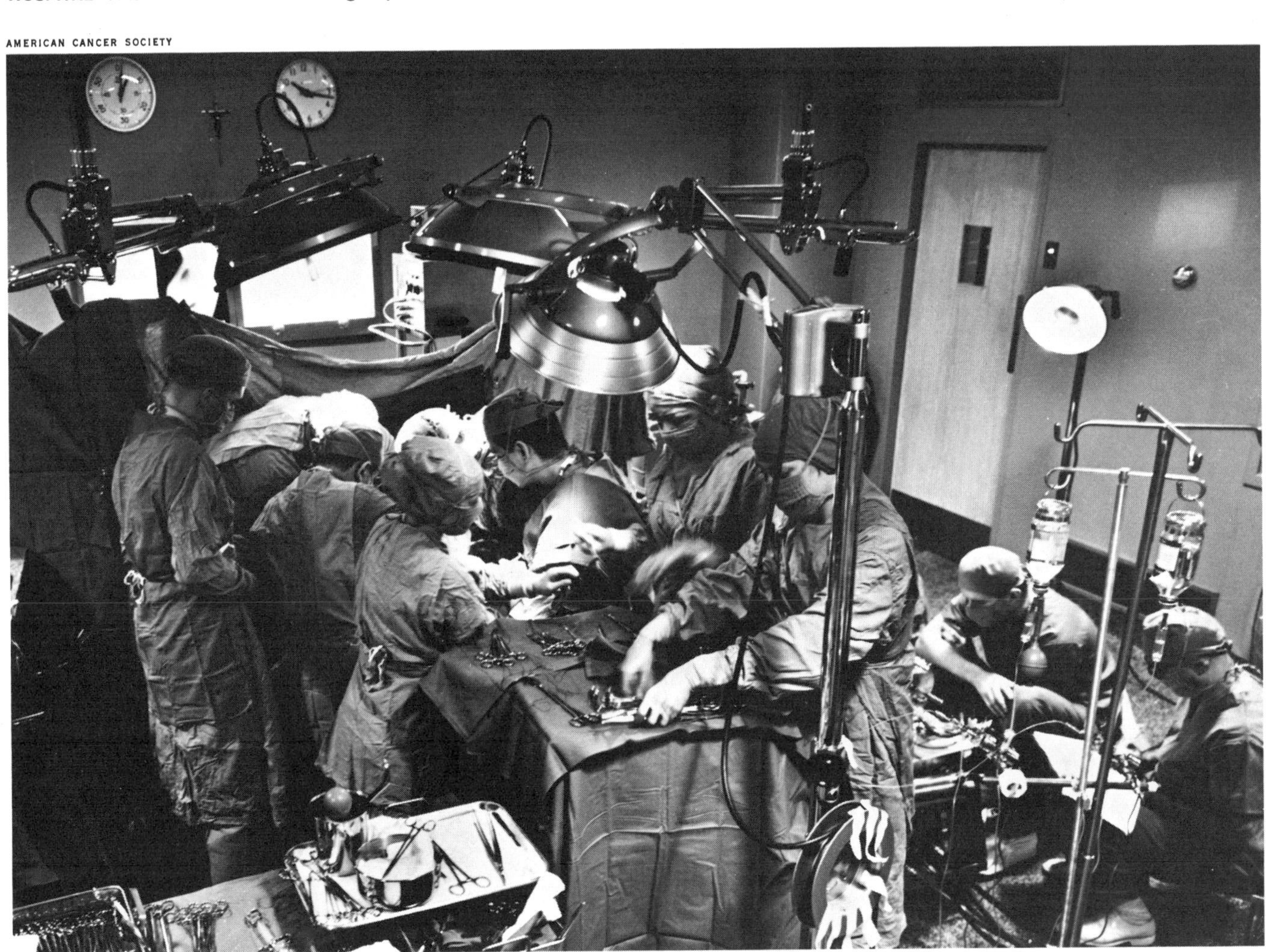

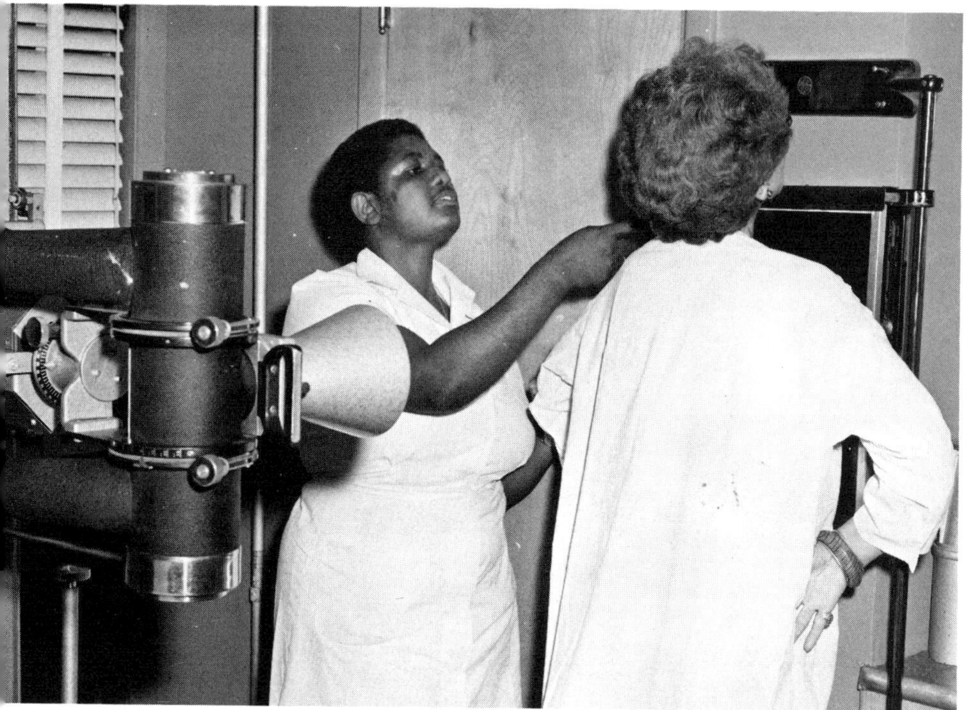

HEALTH INSURANCE PLAN OF GREATER NEW YORK

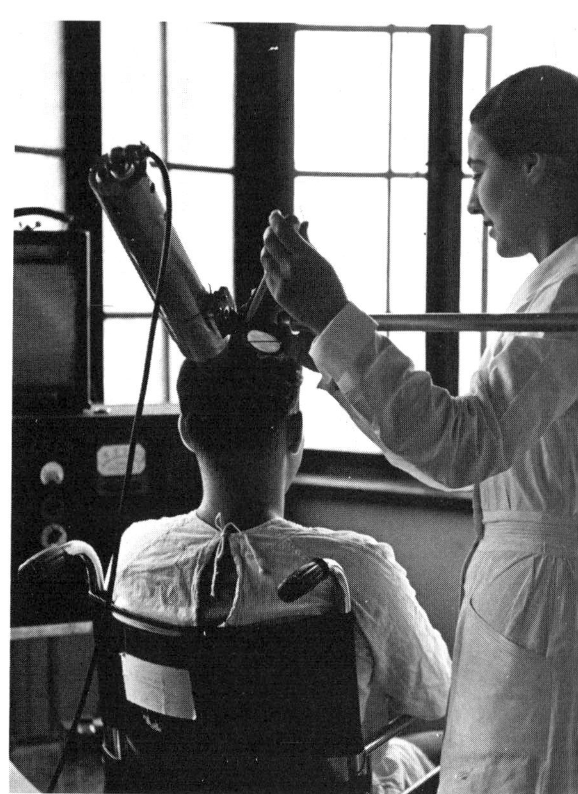

AMERICAN CANCER SOCIETY

DISEASE PREVENTION. The x-ray machine (*above left*) and the scintillation detector (*right*) can give early warning. Vaccination (*below*) affords immunity.

UNITED NATIONS

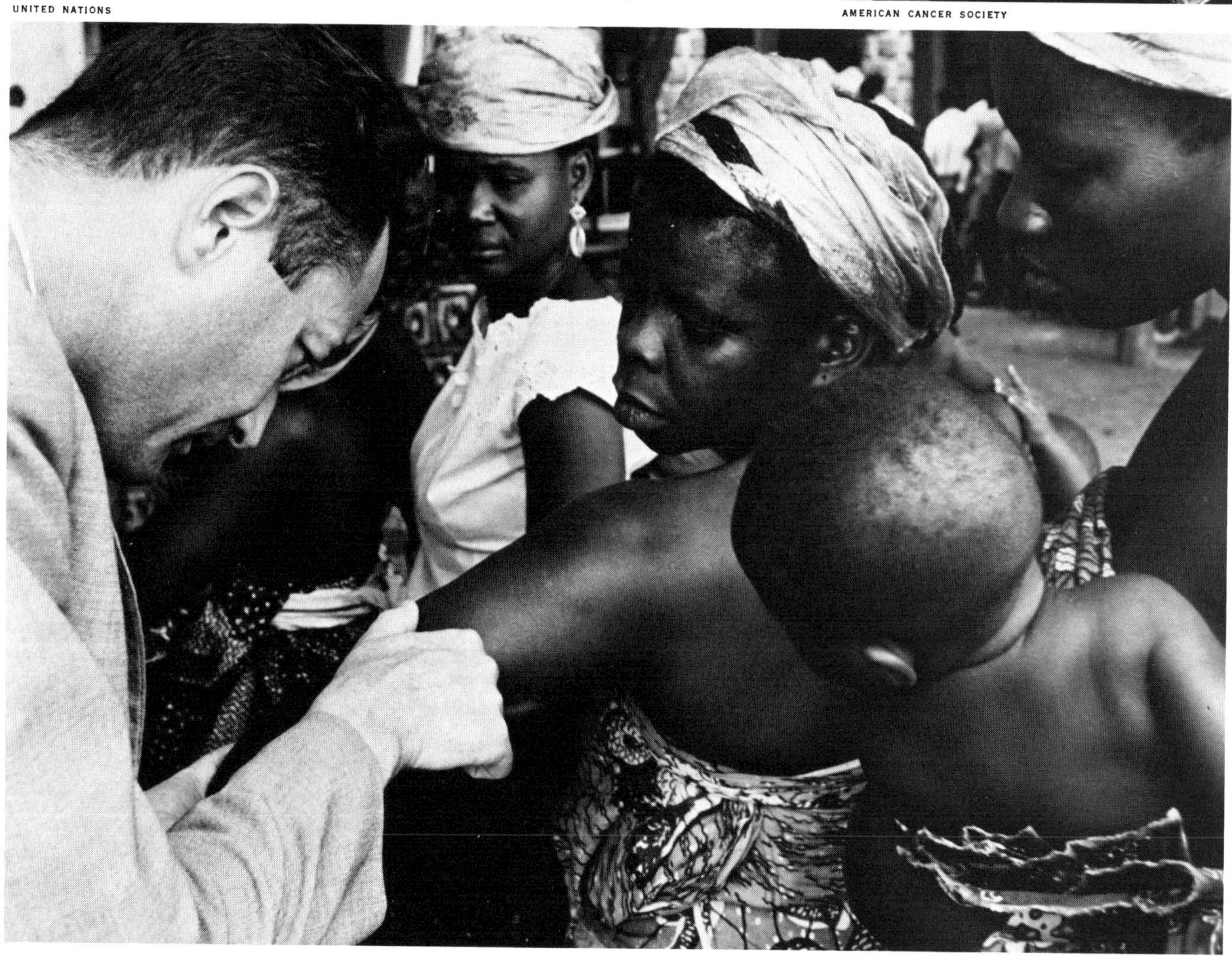

disturbance marked by difficulty in focusing distant objects. It usually develops between the ages of 9 and 13, becoming worse during puberty and then tending to stabilize.

Nephrosis.—A kidney disease marked by the excretion of proteins in the urine and by fluid swellings of the face, ankles, and other parts of the body.

Neuralgia.—A term describing intense shooting pain along the course of various nerves.

Neuritis.—An inflammation or degeneration of the *peripheral nerves* (those outside the brain and spinal cord). It may be caused by infection, poisoning, metabolic diseases, tumors, or by unknown causes. Symptoms include pain, tingling, numbness, muscle weakness, and, possibly, paralysis.

Night blindness (*nyctalopia*).—A disturbance of night vision, often associated with a deficiency of vitamin A in the diet. It may also occur as a complication of other eye disorders, such as nearsightedness and glaucoma.

Osteochondritis.—A bone disease of unknown cause, characterized by destruction of a limited section of bone. It occurs primarily in children and usually affects bones of the lower extremities.

Osteomalacia.—A loss of minerals from the bone, resulting in bone tenderness and *pseudofractures,* or splits in the bone without actual separation. Osteomalacia may be caused by deficiencies of calcium and vitamin D or by kidney disease.

Paget's disease (*osteitis deformans*).—A bone disease seen principally in older persons and marked by painful thickening and deformity of the bones of the skull, spine, pelvis, and thighs.

Parkinson's disease (*paralysis agitans*).—A nervous disorder marked by characteristic tremors and disturbances of movement. Occurs primarily in older persons.

Pellagra.—A disease caused by a deficiency of niacin in the diet and marked by disturbances of the skin, gastrointestinal tract, and nervous system.

Pheochromocytoma.—A tumor in the adrenal glands, or in various nerve centers in the body. The tumor secretes the adrenal hormones *epinephrine* (*adrenalin*) and *norepinephrine,* which raise the blood pressure and increase heart action.

Pityriasis rosea.—A skin disease, characterized by red, scaly, itching patches with lighter centers.

Pleurisy.—Inflammations of the membrances lining the chest cavity and covering the lungs (*pleura*). May occur as the result of a number of such infectious and noninfectious diseases as tuberculosis, rheumatoid arthritis, and rheumatic fever. Symptoms include chest pain and shortness of breath.

Pneumothorax.—The presence of air between the layers of the pleura. This may result from penetrating chest wounds or from various diseases of the lungs and may collapse the lung. May be induced to treat tuberculosis.

Polycythemia vera.—A blood disease, characterized by an increase in the number of red blood cells.

Prickly heat (*milaria*).—A rash caused by blocking of sweat gland ducts.

Psoriasis.—A chronic skin disease characterized by thickened red patches of skin covered by a silvery scale. It usually affects the elbows, scalp, and knees.

Pulmonary embolism.—An obstruction of a lung artery by a blood clot or other substance carried by the blood from another part of the body.

Purpura.—The appearance of hemorrhages in the skin and mucous membranes, associated with fragility of the capillaries or with a decline in the number of blood platelets.

Raynaud's disease.—A circulatory disorder, seen primarily in young women. It is marked by attacks of pallor, coldness, tingling, and pain in the fingers and toes.

Rectocele.—A female disorder in which the rectum presses against, or ruptures through, the wall of the vagina. It may occur after childbirth.

Regional enteritis.—An intestinal disease marked by scarring, thickening, and, possibly, by perforation of the intestinal wall. Usually appears in individuals under age 40.

Rheumatic fever.—A disorder seen most often in children and characterized by wandering joint pains, skin eruptions, abdominal pains, and, possibly, heart damage. Usually occurs some weeks following a sore throat caused by *streptococcus* bacteria.

Rickets.—A children's disease caused by a deficiency of vitamin D. Bones soften and fail to grow normally.

Sarcoidosis.—A disease in which nodules and scar tissue appear in the skin, lungs, bones, and other parts of the body. It may cause chest pain, shortness of breath, heart failure, and other changes, depending upon the organs affected.

Sciatica.—A radiating pain along the course of the sciatic nerve, which runs along the buttock and back of the thigh. It may be caused by spinal disorders, injuries, or tumors.

Scleroderma.—A disease of the connective tissue, marked by skin changes and, possibly, by disturbances of the lungs, blood vessels, intestines, and other organs.

Scoliosis.—A lateral deviation of the spine that may be caused by tumors, injury, paralysis, or a shortened leg.

Sea sickness.—See *Motion sickness.*

Scurvy.—A disease caused by a deficiency of vitamin C in the diet.

Serum sickness.—An allergic reaction to vaccines containing animal serums. It usually appears several days after the injection, and is characterized by skin eruptions and swellings of the face, hands, or feet.

Shock.—A circulatory collapse in which the effective blood volume is insufficient to carry out the normal functions of the circulatory system. It may be caused by extensive bleeding, by sudden expansion of the blood vessels, or by heart damage. Changes observed in shock are lowered blood pressure, pale, cold, and clammy skin, and weak pulse.

Sickle cell anemia. — An inherited blood disorder, marked by crescent-shaped red blood cells that tend to be destroyed easily, thereby causing clots to form in small blood vessels.

Silicosis.—An occupational lung disease, caused by inhalation of dust containing silica.

Slipped disc.—The displacement of an *intervertebral disc* (one of the cartilaginous structures that lie between the bones of the spine). It may be caused by strain or injury.

Spina bifida.—A congenital fissure of the back of the spinal column. In some bases the spinal membranes protrude through the cleft. In severe cases, disturbances of the nerve supply below the fissures may cause muscle weakness and neurological disturbances.

Sprue.—A disease in which food material cannot be properly absorbed through the intestinal wall.

Stroke (*apoplexy*).—Brain damage caused by a disturbance in the blood supply to the brain. May occur in arteriosclerosis and in some types of heart disease. Symptoms include paralysis and disturbances of speech and thought.

Tetany.—A state of extreme muscle excitability, usually accompanied by muscle spasms in various parts of the body. Associated with a decline in the concentration of calcium ions in the blood, it occurs in rickets and some kidney and intestinal disorders.

Thrombophlebitis.—A blood clot in an inflamed vein, usually in a limb.

Thrombosis.—A blood clot occurring in an artery or vein. It is often serious when the blood vessels of the heart, brain, or lungs are involved.

Torticollis.—A twisting of the neck to one side so that the ear approaches the shoulder. It may be either congenital or caused by injury to, or irritation of, the muscles and tendons of the neck.

Trigeminal neuralgia (*tic douloureux*).—A disorder marked by recurring attacks of severe pain along the trigeminal nerve of the face.

Tumor.—A swelling or growth in the body. *Benign* tumors do not invade healthy tissues. *Malignant* (cancerous) tumors invade adjacent tissues, causing cell destruction, and may spread to other parts of the body.

Ulcers.—Areas of destroyed tissue, affecting the skin or mucous membranes of the body. Skin ulcers may be caused by infection or by circulatory disorders. Ulcers of the gastrointestinal tract may be associated with oversecretion of gastric juice or may appear from unknown causes.

Uremia.—Changes in the concentration of urea, potassium, sodium, and other substances in the blood, seen in kidney disorders.

Varicose veins.—Veins that have become weakened and distended because of "incompetence" of the valves that normally prevent back-flow of blood. They usually occur in the legs or as hemorrhoids or varicoceles.

Vertigo.—Dizziness and loss of balance that may be caused by disturbances of the body's balancing mechanisms. Vertigo occurs in certain ear and brain disorders and also in motion sickness, where the usual visual balancing cues shift position.

—Martin Spencer and Daniel Monroe

A CHART OF INFECTIOUS DISEASES

Name of Disease
1. ***Nature of Infecting Organisms***
2. ***Special Geographic Distribution***
3. ***Mode of Transmission***
4. ***Groups Primarily Affected***
5. ***Incubation Period***
6. ***Signs and Symptoms***
7. ***Prevention and Control***
8. ***Remarks***

Actinomycosis
1. Fungi
3. Chewing, swallowing, or inhaling contaminated materials
4. Young males
5. Few weeks to a year
6. Draining wounds on head and neck, cough and abdominal pains if organisms attack lungs and viscera
8. Also occurs in cattle and other animals

Anthrax
1. Bacteria
2. Especially prevalent in Mediterranean region, Africa, and Asia
3. Skin contact with infected cattle hides, breathing infected dust, consuming contaminated milk or meat
4. Butchers, farmers, tannery workers, woolsorters
6. Skin—pustules
   Lungs—cough, difficulty in breathing, prostration
   Viscera—vomiting, pain, constipation
7. Vaccination of domestic animals; disinfection of animal products

Blastomycosis (Gilchrist's Disease)
1. Fungi
2. North America
4. Men between the ages of 20 and 40
6. Skin—abscesses, pustules
   Lungs—cough, chest pain, fever
   Bones—lesions
8. Also occurs in dogs, but no dog-to-man transmission has been established

Brucellosis (Undulant Fever)
1. Bacteria
3. Contaminated milk, direct contact with infected animals
4. Farmers and others who come into contact with infected animals
5. 10–30 days
6. Chills, backache, fever, headache, joint pains, loss of appetite and weight
7. Pasteurization of milk, elimination of disease in animals, human immunization
8. Disease occurs in goats, sheep, cattle, and other domestic animals

Candidiasis (Moniliasis)
1. Fungi
6. Attacks skin between fingers and toes, underarms, also mouth, nails, vagina, and lungs
8. Organisms may be present in mouth, skin, and intestinal tract without producing any signs of candidiasis

Chancroid
1. Bacteria
3. Sexual contact
5. 3–7 days
6. Ulcers on genitalia

Chicken Pox (Varicella)
1. Viruses
3. Through inhalation of infected airborne droplets
4. Children
5. 14–20 days
6. Rash, fever, headache, loss of appetite
8. Varicella virus is the same as that which causes herpes zoster (shingles)

Cholera
1. Bacteria
2. Far East
3. Consumption of contaminated water, food, milk
5. 1–6 days
6. Dehydration, diarrhea, weakness, possible circulatory failure
7. Purification of water supply, vaccination
8. Organisms produce a toxin which irritates intestinal wall; high death rate

Coccidioidomycosis
1. Fungi
2. Western and southwestern U.S., also parts of Central and South America
3. Organisms are carried by dust
4. 25–55 age group
5. 8–14 days
6. Usually attacks lungs; may also involve skin, nervous system, and viscera
7. Paving of roads and other dust-control measures
8. Disease is usually mild in Caucasians, but may be fatal in dark-skinned races

Cold, Common
1. Viruses
3. Probably transmitted by airborne droplets from nose and throat of carriers
5. 2–3 days
6. Sneezing, coughing, sore throat, possibly fever
7. Vaccines have been developed but their effectiveness is questionable

Cold Sores (Herpes Simplex)
1. Viruses
6. Single or multiple blisters on lips, eyes, or other parts of the body
7. Smallpox vaccinations sometimes prevent recurrences
8. Cold sores on eyes may produce scarring of cornea

Colorado Tick Fever
1. Viruses
2. Western U.S.
3. Ticks
5. 4–5 days
6. Fever, chills, headache, back and eye pain
7. Preventive measures against ticks
8. Only known tick-transmitted virus disease of Northern Hemisphere

Cryptococcosis (Torulosis)
1. Fungi
3. Contaminated soil
6. Attacks brain meninges, lungs, skin

Dengue Fever (Breakbone Fever)
1. Viruses
2. Tropical and subtropical regions
3. Mosquitoes (Aedes)
5. 5–8 days
6. Fever, headache, muscle soreness, pain behind the eyes
7. Mosquito control. A vaccine is available

Diphtheria
1. Bacilli
3. Airborne droplets from throat of carrier
4. Children
5. 2–4 days
6. Difficulty in breathing caused by diphtheritic membrane across the throat. Diphtheria toxin may attack heart and nervous system
7. Vaccine is available. Incidence of diphtheria has been greatly reduced since introduction of immunization
8. An antitoxin is available for treatment but must be used early if it is to be effective

Dysentery, Amebic
1. Amebae
3. Contaminated food
6. Attacks digestive tract causing diarrhea, weakness, nausea, vomiting; may produce liver abscesses
7. Sanitary disposal of wastes; hygienic handling of food

Dysentery, Bacillary
1. Bacteria
2. Most dangerous types occur in Orient
3. Contaminated food, water
4. Often occurs in epidemic outbreaks in institutions, military camps, and the like
5. Varies from 24 hours to several days
6. Violent diarrhea, fever, chills, cramps, nausea, vomiting. Develops more rapidly than amebic dysentery
7. Prevention of spread by treatment of those infected
8. Disease has been of historical importance in military campaigns: defeat of Persian army in 380 B.C. was in part ascribed to dysentery

Filariasis
1. Roundworms
2. Tropics and subtropics
3. Insects
6. Attacks skin, lymph vessels, eyes; massive swellings of arms or legs may develop
7. Sanitary measures to destroy insects
8. Some forms may cause blindness

German Measles (Rubella)
1. Viruses
3. Airborne droplets from respiratory passages of carriers
4. Children
5. 21 days
6. Skin rash, swollen lymph glands
8. German measles in pregnant women may cause serious damage to the fetus

Gonorrhea
1. Bacteria
3. Sexual contact
4. Young adults
5. 2–8 days
6. Attacks reproductive organs causing pain, fever, swellings. May result in sterility or permanent joint damage
7. Especially hard to control because of the difficulty of diagnosing the disease in women
8. Infants of gonorrheal mothers may develop eye infection while passing through the birth canal. To avoid this, drops of an antibiotic solution are routinely placed in the eyes of newborn infants

Hepatitis, Infectious
1. Viruses (type A)
2. Prevalent in Mediterranean region
3. Contaminated food or water; close personal contact with carriers
4. Children and young adults
5. 20–40 days
6. Fever, skin rash, weakness, fatigue, loss of appetite, chills, nausea, jaundice
7. Hygienic handling of food
8. Epidemics may occur in boarding schools and similar institutions

Hepatitis, Serum
1. Viruses (type B)
3. Transfusions of infected blood; contaminated needles used for injection
5. 60–160 days
6. Same as in infectious hepatitis (above)
7. Proper handling and source of blood supplies; adequate sterilization of needles

Histoplasmosis
1. Fungi
2. Especially prevalent in midwestern U.S.
3. Inhalation of fungus spores. Fungus is found in soil
6. Attacks lungs, causing fever, coughing, sweating, loss of weight

Hookworm
1. Roundworms
2. Tropical regions *(Necator americanus)* Europe, North Africa, Far East *(Ancylostoma duodenale)*
3. Walking barefoot in infected soil, larvae penetrate skin of feet
6. Bronchitis, abdominal pain, anemia, weakness, pallor. In children causes physical and mental retardation
7. Sanitary disposal of excreta. Wearing of shoes

A CHART OF INFECTIOUS DISEASES (Continued)

Influenza
1. Viruses
2. Worldwide epidemics
3. Airborne droplets from carriers
5. 1–2 days
6. Chills, fever, pain behind eyes, muscle pains, sneezing
7. Vaccines give temporary immunity
8. Dangerous to old and debilitated persons, who may develop pneumonia

Leishmaniasis
1. Protozoa
2. Asia, Africa, Central and South America
3. Sandflies
6. *Leishmania donovani*—anemia, fever, chills, sweats, dizziness
   Leishmania tropica—sores on skin, face, ears, neck, hands
   Leishmania braziliensis — attacks mucous membranes of the mouth, nose, throat; may cause ulcerations
7. Elimination of sandflies; treatment of humans and dogs who carry the disease

Leprosy
1. Bacteria
2. Primarily in the tropics
3. Probably through infection of superficial skin abrasions
5. 2–4 years or more
6. Ulcers of skin, nerve damage, hair loss
7. Treatment of carriers
8. Sulfones now control most cases

Leptospirosis
1. Bacteria
3. Contact with field and swamp water contaminated by urine of infected animals
5. 7–10 days
6. Fever, vomiting, headache, congestion of eyes, muscle pains, small skin hemorrhages; in some types, jaundice
7. Eliminating rats and other carriers. Protective clothing to prevent organisms from entering the body

Lymphogranuloma Venereum
1. Viruslike microorganisms
2. In tropics and Mediterranean region
3. Sexual contact
5. 3–20 days
6. Genital ulcer, abscess, fever, joint pain, swelling, scarring of genitalia
7. Identification and treatment of carriers

Malaria
1. Protozoa
2. Temperate and tropical regions
3. Mosquitoes
5. Varies with type
6. Periodic attacks of chills, fever
   Chronic malaria—listlessness, headache, fatigue, enlargement of spleen
7. Mosquito control, treatment of carriers
8. Malaria is the major parasitic disease of mankind

Mumps (Epidemic Parotitis)
1. Viruses
3. Direct contact; airborne droplets from mouths of infected persons
4. Children
5. 8–21 days
6. Attacks salivary glands, particularly parotid glands adjacent to ear. Causes pain, swelling, fever
7. A vaccine is available
8. Frequently produces inflammation of the testicles in adult males. In some cases sterility may result

Parrot Fever (Psittacosis, Ornithosis)
1. Viruses
3. Contaminated air droplets, feather dust from infected birds
5. 7–14 days
6. Fever, chills, headache, muscle pains. Eyes may become sensitive to light. Also pneumonia, insomnia, apathy
7. Treatment of infected birds; chemical prophylaxis of birds
8. Disease affects parrots, parakeets, ducks, pigeons, turkeys

Pinworm (Seatworm)
1. Intestinal roundworms
3. Contaminated food and drink. Pinworm eggs are carried in the air
6. Itching in anal region, poor appetite, weight loss
7. Worms may be removed by drugs
8. In some cases no symptoms observed

Plague (Black Death)
1. Bacteria
2. Asia
3. Rat flea transmits bacteria to man
   Pneumonic plague—bacteria are carried by airborne droplets from victims
5. 2–10 days
6. Swelling of lymph glands in groins and armpits. Chills, fever, delirium, headache, thirst. Hemorrhages into the skin and other parts of the body *Pneumonic plague*—attacks the lungs, producing a bloody sputum
7. Rat and insect control. Vaccines give temporary limited immunity
8. Disease is of historic significance, having caused devastating pandemics in Europe and the Orient

Pneumonia
1. Usually bacteria, but also viruses and fungi
3. Bacteria may be normally present in the body and attack when resistance is weakened by colds, malnutrition, exhaustion, or lung ailments
6. Coughing, fever, lung abscesses, aches, chills, chest pains, bloody or pus-laden sputum, loss of appetite
8. Drug-resistant staphylococcus bacteria cause many hospital cases of pneumonia

Poliomyelitis (Infantile Paralysis)
1. Viruses
2. Industrialized countries of temperate zones
3. Airborne droplets from carriers; contaminated food or other objects
4. Children and young adults
5. 3–35 days
6. May produce only fever and a mild feeling of illness. In other cases, sore throat, aches, headaches, paralysis. Viruses attack nerve cells controlling muscles
7. Vaccines are available
8. In underdeveloped countries where sanitary facilities are poor, infection usually occurs early in infancy when children are protected by antibodies received from the mother. This enables them to withstand the disease and develop permanent immunity

Q Fever
1. Rickettsiae
2. Western U.S., Australia, Italy, Greece
3. Inhalation of infected-tick excreta from animal hides and fleece
4. Animal handlers, woolsorters
5. 14–28 days
6. Chills, fever, headache, muscle pains, weight loss
7. Immunization; pasteurization of milk

Rabies
1. Viruses
3. Bites of infected animals
5. 10 days to several months
6. Fever, headache, burning and tingling sensations around the bite, difficulty in swallowing, breathing spasms, convulsions, fits. Invariably fatal unless injections of vaccine are given at once
7. Immunization of animals
8. Affects foxes, coyotes, jackals, wolves, skunks, and bats

Relapsing Fever
1. Bacteria
3. Ticks, body lice
5. About 7 days
6. Chills, fever, headache, vomiting, muscle and joint pain. Symptoms vanish and recur in cycles of decreasing severity
7. Control of parasites; spraying of homes and clothing with DDT
8. May occur in epidemic form

Rheumatic Fever
1. Bacteria
4. Children between the ages of 5 and 15
5. Develops some weeks after streptococcal infection of upper respiratory tract
6. Joint pains which move, or "migrate," from joint to joint in an unpredictable pattern, nosebleeds, abdominal pains, skin eruptions. Heart damage may be a serious complication
7. Antibiotics used to prevent recurrence
8. Disease is considered to be different from ordinary infectious diseases in that it develops *after* infection by bacteria. Apparently the signs and symptoms are not *directly* related to the presence of the bacteria in the tissues

Rickettsial Pox
1. Rickettsiae
2. Eastern U.S.
3. Mites
5. 1–2 weeks
6. Red, black-scabbed pimple in region of bite, fever, chills, headache, muscle pains, skin rash

Rocky Mountain Spotted Fever
1. Rickettsiae
2. Western Hemisphere
3. Ticks
5. 3–12 days
6. Headache, chills, fever, muscle-bone-joint pains, rash
7. Tick control; immunization
8. Occurs principally in spring and summer

Scarlet Fever (Scarlatina)
1. Bacteria
3. Airborne droplets from carriers, contaminated food and other objects
4. Children (rare in infants)
5. 2–5 days
6. Fever, headache, vomiting, sore throat, rash
8. Rheumatic fever or kidney inflammation may follow 2–3 weeks after recovery

Schistosomiasis
(Snail Fever, Bilharziasis)
1. Flatworms
2. Far East, Middle East
3. Infected water—organisms are carried by snails
6. *Intestinal schistosomiasis*—fever, abdominal pains, weight loss. In chronic stage anemia and irregular fever may develop
   Vesical schistosomiasis—bloody urine, ulcers of urinary tract, painful urination

Scrub Typhus
(Tsutsugamushi Disease)
1. Rickettsiae
2. Japan and South Pacific
3. Mites
5. 10–12 days
6. Skin lesion at bite; fever, chills, headache, skin rash
7. Control of mites

Shingles (Herpes Zoster)
1. Viruses
4. Adults
6. Attacks outlying (peripheral) nerves, causing inflammation, blistering, pain and tenderness along skin overlying nerve pathways, particularly of head and trunk. Serious eye complications if nerves in region of eye are involved
8. Virus of herpes zoster is the same virus that causes chicken pox. It is believed that the organism may be latent in the tissues and be reactivated by injury, medication, or tumors

(continued on next page)

A CHART OF INFECTIOUS DISEASES (Continued)

Sleeping Sickness
(African Trypanosomiasis)
1. Trypanosomes
2. Tropical Africa
3. Tsetse flies
6. Lesion at fly bite; fever, headache, enlarged lymph glands, weakness, muscle tenderness, physical and mental depression. Disease is most serious when central nervous system is attacked
7. Use of drugs to ward off infection in dangerous areas

Smallpox (Variola)
1. Viruses
3. Airborne droplets from carrier; contaminated clothing, eating utensils
5. 8–14 days
6. Fever, vomiting, headache, backache, skin rash
7. Vaccination
8. Scarring usually permanent. Vaccination has eradicated the disease in many countries, but pockets of smallpox still exist

Syphilis
1. Bacteria
2. Worldwide
3. Sexual contact, inheritance
5. *Primary stage*—3–4 weeks
Secondary stage—4–6 weeks later
Tertiary stage—20 or more years after original infection
6. *Primary stage*—lesion on genitals, mouth
Secondary stage—skin eruptions, fever, sore throat, enlarged lymph nodes
Tertiary stage—mental changes, heart disease, spinal cord damage
7. Educational program, early detection and treatment with antibiotics, prevention of congenital syphilis by treatment of pregnant mother
8. Decline in death rates of adults and infants, but disease not under control

Tapeworm
1. Flatworms
3. Contaminated beef, pork, fish
6. In many cases no symptoms. In others abdominal pains, digestive disturbances, nausea, anemia. Non-intestinal variety (hydatid disease) may attack liver, lungs, muscles; is contracted from infected soil or from contact with infected dogs
7. Proper cooking of meat, use of specific drugs, effective sewerage systems

Tetanus (Lockjaw)
1. Bacteria
3. Contamination of wounds with dirt containing tetanus spores
5. Varies from days to weeks to years
6. Violent muscle spasms. Death may result from interference with breathing. Tetanus toxin attacks nerve tissue
7. Immunization
8. Tetanus bacteria are frequently normal inhabitants of the intestinal tracts of man and various animals

Trachoma
1. Viruses
2. Near and Far East, southern Europe
3. Direct contact with carriers; contaminated personal articles
4. Children
5. Approximately one week
6. Tearing of eyes, sensitivity to light, scarring and ulceration of cornea and eyelids. May cause blindness
7. Treatment of carriers
8. Victims are susceptible to repeated attacks since no immunity develops

Trichinosis
1. Roundworms
2. Temperate zones
3. Contaminated pork
5. 1–4 days
6. No symptoms in mild infestations. In other cases nausea, diarrhea, vomiting, muscle pains, weakness, fever. Eyelids become puffy and hemorrhages appear under nails
7. Pork should be thoroughly cooked—30 minutes at 140°F. for each pound
8. Larvae become permanently encysted in muscles

Tuberculosis
1. Bacteria
3. Airborne droplets from carriers, contaminated milk
6. In early stage fever, fatigue, loss of weight. In later chronic stage also night sweating, bloody sputum, chest pains
7. Early detection and treatment of carriers; pasteurization of milk
8. Although a vaccine is available (BCG vaccine) and has been used for mass vaccinations its value for this purpose is not universally accepted

Tularemia (Rabbit Fever)
1. Bacteria
3. Contact with infected rodents, bloodsucking flies, and ticks; contaminated meat
4. Hunters, butchers, campers
6. Skin ulcers, eye inflammation or mouth ulcers, depending upon where the organisms enter the body; headache, fever, vomiting, chills
7. Immunization. Rabbit meat should be thoroughly cooked. Water from streams in infected areas should be avoided

Typhoid Fever
1. Bacteria
3. Contaminated food or water
5. 10–12 days
6. Headache, remittent fever, chills, nausea, cough, constipation, nosebleeds, "rose spots" on skin of trunk
7. Vaccination; detection and treatment of carriers; purification of drinking water

Typhus (Epidemic Typhus)
1. Rickettsiae
3. Body lice
5. 10–14 days
6. Chills, headache, aches, pains, fever, delirium, pink spots on skin later developing into hemorrhagic spots. Possibly gangrene of toes, fingers, or earlobes
7. Immunization, delousing procedures (using DDT)

Whipworm
1. Intestinal roundworms
3. Contaminated food or water
6. Mild infections go unnoticed. In other cases abdominal pain, nausea, vomiting, flatulence, headache, bloody stools, anemia, weight loss
7. Sanitary disposal of human wastes

Whooping Cough (Pertussis)
1. Bacteria
3. Airborne droplets from carriers
4. Children; females more often than males
5. 7–14 days
6. Explosive cough, convulsions, possible lung complications
7. Immunization
8. Epidemics have occurred in large cities at intervals of two to four years

Yaws
1. Bacteria
2. Tropics
3. Direct contact, organisms enter through the skin
4. Children living in crowded, unhygienic conditions
5. 3–4 weeks
6. Skin lesions. May attack bones in later stages
7. Treatment of carriers
8. The *Treponema pertenue*, which causes yaws, is identical in appearance to the *Treponema pallidum*, which causes syphilis

Yellow Fever
1. Viruses
2. Central and South America
3. Mosquitoes
5. 3–6 days
6. Fever, headache, backache, congestion of eyes, slow pulse, nausea, vomiting, hemorrhage from mucous membranes, jaundice; kidney failure in severe cases
7. Immunization; mosquito control

BIBLIOGRAPHY

BEST, CHARLES H. and TAYLOR, NORMAN B. *The Living Body* (4th ed.). Holt, Rinehart and Winston, Inc., 1958.

BURNET, MACFARLANE. *Natural History of Infectious Diseases*. Cambridge University Press, 1953.

CARLSON, ANTON J. and JOHNSON, VICTOR. *The Machinery of the Body*. The University of Chicago Press, 1953.

CLARK, RANDOLPH LEE and CUMLEY, RUSSELL W. *The Book of Health: An Encyclopedia for Everyone* (2nd ed.). D. Van Nostrand Co., Inc., 1962.

EDWARDS, LINDEN F. *Concise Anatomy*. McGraw-Hill, Inc., 1956.

EMERSON, CHARLES P., JR. and BRAGDON, JANE S. *Essentials of Medicine*. J. B. Lippincott Co., 1959.

GALLAGHER, J. R., GOLDBERGER, I. H., and HALLOCK, G. T. *Health for Life*. Ginn & Co., 1963.

GERARD, RALPH W. *The Body Functions*. John Wiley & Sons, Inc., 1941.

GOSS, CHARLES M. *Gray's Anatomy of the Human Body* (27th ed.). Lea & Febiger, 1959.

HICKMAN, CLEVELAND P. *Health for College Students* (2nd ed.). Prentice-Hall, Inc., 1963.

MITCHELL, PHILIP H. *A Textbook of General Physiology*. McGraw-Hill, Inc., 1956.

SCHIFFERES, JUSTUS J. *Healthier Living*. John Wiley & Sons, Inc., 1955.

SEXTON, W. A. *Chemical Constitution and Biological Activity* (2nd ed.). D. Van Nostrand Co., Inc., 1953.

STANLEY, W. M. and VALENS, E. G. *Viruses and the Nature of Life*. E. P. Dutton & Co., Inc., 1961.

SUNDGAARD, ARNOLD. *The Miracle of Growth*. University of Illinois Press, 1950.

YOUNG, CLARENCE and others. *The Human Organism and the World of Life*. Harper & Bros., 1951.

PUBLIC HEALTH

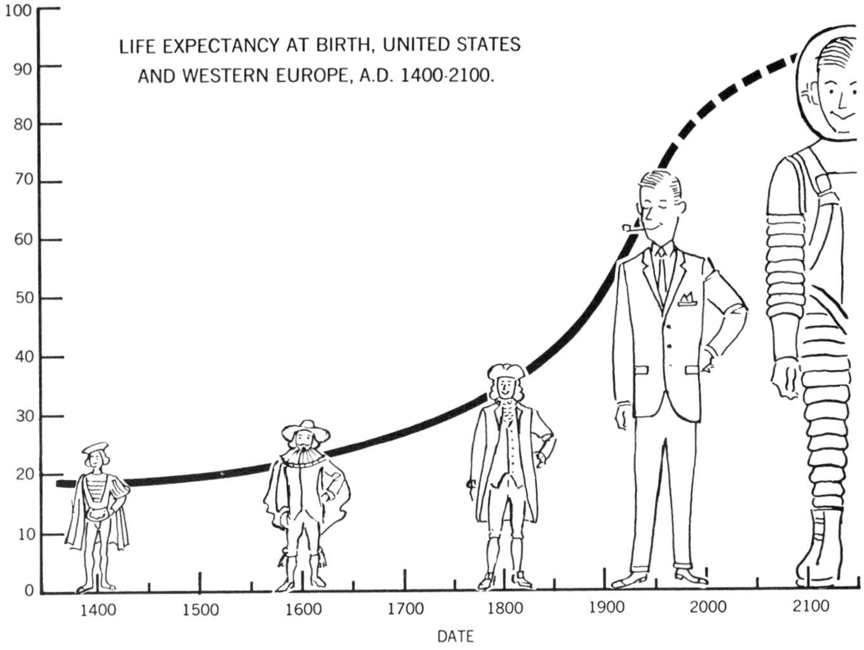

Public health, as defined by C. E. A. Winslow (1877–1957), is "the art and science of preventing disease, prolonging life, and promoting physical and mental efficiency through organized community effort for the sanitation of the environment, the control of community infections, the education of the individual in principles of personal hygiene, the organization of medical and nursing services for the early diagnosis and preventive treatment of disease, and the development of social machinery which will ensure to every individual a standard of living adequate for the maintenance of health."

LIFE EXPECTANCY AT BIRTH, UNITED STATES AND WESTERN EUROPE, A.D. 1400-2100.

Life Expectancy.—Accomplishments in the field of public health are reflected by the increase in life expectancy. In the United States the *average life expectancy* at birth is now 67 years for males and 73 for females. In contrast, it averaged only 50 years at the beginning of the twentieth century; a century ago it averaged only 40 years. Looking ahead, it is estimated that the average American born in the year 2000 will live to the age of 82, and one born in the year 2400 can expect to approach the age of 100.

Although life expectancy has increased conspicuously in recent years, the natural span, or limit, of life has not changed appreciably. Barring "unnatural" death from disease or injury, man's lifetime is now, and always has been, about 100 years. This is the *biological limit* of human life. Although there have been reports of people living for 150 years and more, these cases lack scientific confirmation.

Traumatic Injuries. — Accidents are among the greatest public health problems in the world today, in terms of both disabilities and deaths. In the United States, accidents kill nearly 112,000 people each year and injure another 50 million. More than a third of these deaths and injuries are caused by motor vehicles, and another third occur in the home. Altogether, accidents are fourth on the list of the leading causes of death in the United States, being exceeded only by heart disease, cancer, and strokes.

LEADING CAUSES OF DEATH IN THE UNITED STATES
(rates per 100,000)

HEART DISEASES	371.2
CANCERS	155.1
STROKES	104.6
ACCIDENTS	58.0
INFLUENZA AND PNEUMONIA [1]	32.5
DISEASES OF INFANCY	26.4
GENERAL ARTERIOSCLEROSIS	19.9
DIABETES MELLITUS	17.7
OTHER CIRCULATORY DISEASES	14.6
OTHER BRONCHOPULMONIC DISEASES	14.5

[1] Except pneumonia of newborn.

Source: U.S. Statistical Abstract, 1968

Among the other forms of violent death, suicides accounted for about 22,000 deaths in 1966 and homicides for 11,000 more. Disasters, including fires, floods, windstorms, earthquakes, explosions, droughts, famines, epidemics, and extremes of heat and cold, account for about 1,000 deaths annually in the United States.

Wound Infections.—Accidental injuries are often complicated by invading microorganisms; these wound infections may prove more lethal than the traumatic injury itself. Such infections were responsible for about 8,724 deaths in 1965 in the United States.

Staphylococci are the most common bacterial invaders but generally give rise to only localized lesions, described as *ordinary abscesses*. These specific microorganisms have proved susceptible to some of the newer antibiotics. However drug-resistant strains are emerging that are more difficult to control.

Streptococci are also fairly common secondary invaders. These bacteria tend to diffuse deeply into the tissues, leading to a generalized *sepsis* or *septicemia* (blood poisoning). Such complications of wounds were once so common that a minor scratch or simple surgical procedure would frequently prove fatal. At present, however, these infections are comparatively rare and can usually be controlled by the sulfonamides and antibiotics.

Tetanus (lockjaw) is still another type of wound infection. Although uncommon, it is a most excruciating affliction and ultimately kills half of its victims (181 deaths in the United States in 1965). Tetanus can easily be prevented by active immunization, provided it is initiated a month or more before injury. The causative agent of tetanus is a microorganism present in ordinary soil, particularly soil fertilized with animal manure. When they are introduced into a puncture wound, in which most of the air is excluded, the organisms multiply and produce an extremely virulent toxin that then diffuses through the body, causing tetanic muscle spasms and other symptoms of the disease. Formerly, treatment with specific antitoxins (passive immunization) used the serum of actively immunized animals. Now, a specific antitoxin against tetanus is available which is derived from actively immunized humans. This product is more costly, but eliminates the danger of serious horse-serum reaction. The *tetanus toxoid* used to produce active immunity is harmless because it is not an animal-serum preparation. Ideally, everyone should be immunized against tetanus as a child and thereafter should receive occasional booster shots. Such immunization will reduce the risk of this dread malady, but proper surgical management of wounds remains a necessity.

Infectious Diseases.—Progress in the extension of life has been due chiefly to the discovery of the specific causes and controls of infectious diseases. Recognition of microorganisms as a cause of disease and the application of control measures in the fields of environmental sanitation and immunization have advanced human welfare further during the twentieth century than has anything in the previous thirty centuries. Since those most susceptible to infectious diseases are infants and children, the life expectancy of the young has been increased about twenty years since the beginning of the twentieth century; that of adults over fifty, only about two years.

The conquest of disease has not only prolonged life but has also improved the general health of the individual during his entire life. An appraisal of the extent and degree of illness (*morbidity statistics*) is not as simple as counting the number of deaths (*mortality statistics*); nevertheless, there is abundant evidence of a tremendous improvement in man's general health. Even with this improvement, however, many health problems still remain to be solved. The World Health Organization lists the six major health problems as tuberculosis, malaria, venereal disease, malnutrition, mental illness, and environmental sanitation.

Airborne Infections.—Most of the common acute communicable diseases are classified as *contact infections* and *airborne infections,* as indicated by their modes of transmission. As a group they may also be described as *respiratory diseases,* because they involve various parts of the respiratory tract and are generally spread by discharges from the nose and mouth. They account for the largest proportion of disabling illness among persons of all ages, particularly among children. In the United States this group of diseases takes 100,000 lives annually. Control measures depend largely upon whether the causative agent is virus, fungus, or bacterium.

INCUBATION PERIOD.—The *incubation period* of an infection is the time between its entry into the body and the appearance of the first symptoms. As a rule, *viral infections* have an incubation period of from 10 to 14 days (with the exception of the common cold, influenza, and cowpox, which have an incubation period of from one to three days). Most *bacterial infections* have an incubation period of from three to five days, with the exception of rheumatic fever, tuberculosis, and leprosy, which may evolve insidiously (without symptoms) over a period of many days. *Fungal infections* have an incubation period extending over a period of weeks.

TREATMENT. — There is no specific treatment for most viral infections, but the majority of the bacterial infections and some of the fungal infections yield to the newer sulfonamides and antibiotics. Most viral infections produce a lasting immunity so that a second attack is unusual. But most bacterial and fungal infections do not produce a lasting immunity; repeated attacks can occur.

CARRIERS.—The ultimate source of almost all infections is man himself, although animals may share a few of the bacterial and fungal infections. Viral infections usually arise from *active cases*—patients with obvious symptoms. Most bacterial and fungal infections arise either from active cases or from *carriers*—people who are infected but appear to be healthy.

Disease Control.—The control of infectious diseases involves an understanding of them and the application of the following measures.

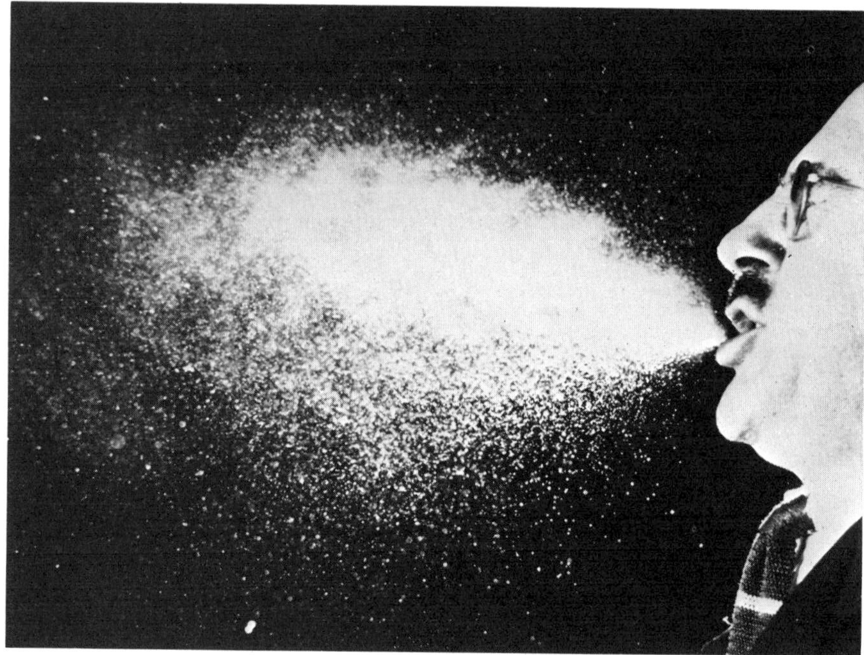

M. W. JENNISON

SNEEZING AND COUGHING spread infectious germs of the respiratory tract through the air.

ISOLATION. — Patients with obvious symptoms should be isolated and precautions taken so that objects contaminated by them (*fomites*) are not contacted by others. Isolation rather than strict quarantine is now common practice, with a possible exception in cases of smallpox. Detection of healthy carriers (as in diphtheria) and *case-finding* (as in tuberculosis) are considered good public health practices. Likewise, the reporting of all cases of infectious diseases to the local health department and surveillance by authorities to prevent epidemics are important control factors.

DECONTAMINATION.—Contamination of the environment should be reduced. The practice of spitting must be condemned because *droplet nuclei* disseminated into the air by coughing and even by ordinary breathing are important factors in the spread of disease. Efforts should be made to provide proper ventilation, dust control, and the sterilization of indoor air by means of sunlight, ultraviolet light, and chemical aerosols. Avoidance of crowds, particularly indoors, is encouraged; avoidance of known patients and carriers is essential.

RESISTANCE.—Resistance to infection should be maintained. Specific immunization for smallpox, diphtheria, poliomyelitis, whooping cough, influenza, and measles should be a routine precaution. Vaccination against mumps, German measles, and chickenpox is now experimental, and against tuberculosis is controversial, but should be considered. Vaccination is compulsory in many countries, but it is urged rather than forced in the United States. Some other measures that reduce susceptibility to infections are adequate rest, exercise, recreation, and a proper diet. At the same time, allergies and other conditions that predispose a person to infections should be kept under control.

Problems of Nutrition.—Nutrition may seem to be an unimportant consideration to most Americans, but for two-thirds of the world's population the primary objective in life is getting enough food to maintain life. The world has never had enough to eat, nor will it in the immediate future. Even the modern miracles of science that have advanced food technology at a rate comparable to nuclear and space explorations have not kept abreast of the population explosion. Recent studies have revealed that at least 800 million people around the world are constantly faced with the prospect of starvation.

MALNUTRITION.—The United States, in contrast, has such an abundance of food that millions of tons can be exported annually—and yet surveys indicate that the diets of more than half of the population are inadequate in some respect. Many people in the United States, through whim, fad, custom, or ignorance, become the victims of malnutrition. Nutritional deficiency diseases are very common, and even overt clinical cases are frequently observed; a surprising number of deaths are directly attributable to malnutrition.

DIET.—An optimal diet must fulfill many criteria. It must, first of all, satisfy the appetite. Second, it must meet the caloric needs of the body in order to supply sufficient heat and energy for movement, thought, growth, and replacement of tissue. An average adult needs about 3,000 calories a day. (If this amount of calories was converted into heat energy, it would be enough to raise the temperature of eight gallons of water from 32° to 212° F.; if converted into work energy, it would be enough to lift a 150-pound person 60,000 feet.) An optimal diet must also contain the correct amounts and combinations of carbohydrates, proteins, fats, minerals, and vitamins. Most of these essentials may be obtained by eating a combination of dairy foods (including milk, cheese, ice cream, and butter), meats (including fish, poultry, and eggs), vegetables and fruits, and breads and cereals (particularly the whole-grain and enriched varieties).

VITAMINS.—Although a diet made up of the above foods could fulfill the body's needs, it would not necessarily provide adequate amounts of certain minerals and vitamins—notably iron, iodine, calcium, and vitamins B_1, B_2, and C. A simple solution is to supplement natural foods with manufactured concentrates. Natural foods, however, are still the best source of vitamins and minerals; too often people take vitamins as a substitute for a balanced diet instead of using them only as a supplement.

EDUCATION.—On the whole, proper nutrition depends largely upon the application of exact knowledge of food substances rather than upon the dictates of appetite, custom, habit, or fad. Although many public health agencies disseminate diet information, more and better education is needed.

Digestive Disorders.—An average, nutritionally correct diet may be inadequate for an individual who has an imperfect digestive system or defective endocrine system. As a rule, digestive and endocrine disorders are not readily controlled by the same type of organized efforts that are so effective in reducing infectious diseases and nutritional deficiencies. The control of digestive and endocrine disorders depends largely upon early diagnosis and appropriate treatment rather than upon specific preventive measures. Exceptions to this generalization are numerous, as may be illustrated by the many inflammatory diseases of the intestinal tract that result from the ingestion of irritants, such as foods that are too hot, too cold, too highly seasoned, or too coarse. Likewise, it is possible to control uses of excessive alcohol, laxatives, or cathartics. Emotional stress plays an important role in some digestive disorders; in others, heredity may be a significant factor.

Food Poisoning.—Afflictions associated with the ingestion of food can be serious and sometimes fatal. In recent years such afflictions have become less common, yet almost every person will experience at least one attack of food poisoning in his life.

STAPHYLOCOCCUS.—At present, most cases of food poisoning are caused by a toxin that forms in foods contaminated by certain strains of *staphylococcus* bacteria. Nausea, vomiting, and abdominal cramps soon follow the ingestion of such contaminated foods. "Staph" food poisoning is associated mainly with prepared foods containing meat, eggs, or milk that have been seeded by the discharges from skin lesions, nose, or throat and then have been allowed to incubate at room temperature. There is no specific treatment for this type of food poisoning and no immunization against it, although the hazard may be reduced by carefully protecting prepared foods from contamination, especially by refrigerating them.

BOTULISM.—One type of food poisoning that has received much publicity because of its high fatality rate is *botulism.* Fortunately this disease is very rare. Botulism results from an *enterotoxin* produced by germs that remain viable in improperly processed canned foods. These germs are found in soil (particularly soils fertilized with animal manure) and are therefore usually associated with certain garden vegetables. The bacteria are *spore-forming* and cannot be destroyed by ordinary cooking; but the enterotoxin that they produce is *heat-labile* and can be destroyed by reboiling home-canned foods immediately before serving.

CHEMICALS.—Another type of food poisoning can result from chemical contamination. Chemicals are often introduced into food unwittingly; for example, zinc poisoning can result from the ingestion of an acid food that was packed in a galvanized (zinc-coated) metal container. Chemical poisoning also occurs when a toxic chemical is confused with an edible one, as when cockroach powder is mistaken for baking soda.

The public has long been concerned with the detrimental effects of insecticides, preservatives, and other chemicals used in the production and processing of food products. In the United States this concern led to the enactment in 1906 of the Pure Food and Drug Act, and subsequently to its numerous amendments and to new laws. Enforcement of these regulations by the Food and Drug Administration (F.D.A.) and other federal, state, and local authorities, combined with the voluntary efforts of the food manufacturers themselves, has virtually eliminated any major hazard of food poisoning by chemicals.

PREVENTION.—Extreme care in the procurement, processing, preservation, and preparation of food products is essential to the prevention of the various types of food poisoning and food-borne infections. Milk and meats deserve special attention because they deteriorate very rapidly. The *pasteurization* of milk by heating (143° F. for 30 minutes) is a practical means of eliminating disease-producing microorganisms that may have originated with the cow or have been introduced during later processing. Most meats are inspected by the U.S. Department of Agriculture's Bureau of Animal Industry, acting under the Meat Inspection Act of 1907. But even the most meticulous inspection of meat cannot guarantee its purity, and it is therefore imperative that all meats be cooked adequately before being eaten. The cooking of meats and other foods has probably prevented more illness than any other single public health measure. Proper refrigeration of foods is also important. However, refrigeration merely inhibits the growth of microorganisms, and even frozen foods cannot be considered free of *pathogens* (disease-causing bacteria or viruses). In contrast, canned foods that are properly processed are sterile because they have been subjected to temperatures sufficient to destroy microorganisms. Even the spore-forming bacteria that cause botulism can be eliminated through the pressure-cooking routine in commercial canning.

Fecal-borne Infections.—Typhoid fever and similar infectious diseases are spread from person to person, and occasionally from animal to man, by the contamination of food and drink by intestinal wastes. These diseases are among the most common and universally distributed of all afflictions, but prevail mainly in areas where sewage is improperly treated. Throughout history these diseases have plagued armies in the field and have often played an important part in determining the outcome of wars. **Today the degree of civilization of an area can be measured by the incidence of fecal-borne infections.**

ANNUAL DEATH RATE PER 1,000 FROM DISEASE, INJURY, AND BATTLES; U.S. ARMY.
(From data furnished by the Surgeon General.)

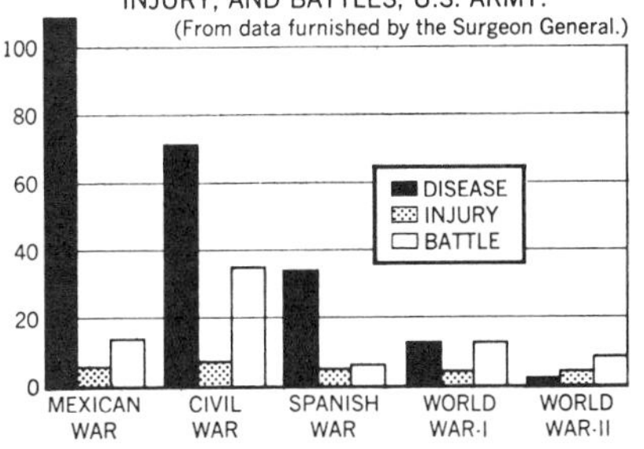

AGENTS.—Specific agents responsible for filth-borne diseases include the typhoid bacillus; the *Salmonella* that cause paratyphoid fever; the *Shigella* that cause the bacillary dysenteries; the *Vibrio* that causes Asiatic cholera; a number of viruses that cause polio, infectious hepatitis, and some forms of enteritis; many intestinal protozoa that cause the amoebic dysenteries; a formidable list of roundworms, including the common pinworm and hookworm; and flatworms, including the flukes and tapeworms.

CONTROL.—Although immunization is of some value in the prevention of certain of the bacterial and viral filth-borne diseases, control measures are largely dependent upon the proper isolation of infected persons, sanitary disposal of body wastes, protection of food and drink from contamination, pasteurization of milk, and purification of water supplies. Treatment with the newer sulfonamides and antibiotics has some limited success, and some *anthelmintic drugs* are effective against certain of the worms. Prevention, rather than treatment, however, is the great challenge.

Water Supplies.—Communal water supplies are generally derived from lakes, rivers, and streams that are exposed to contamination by all types of filth, including the fecal wastes of animals and humans. Thus, all surface water must be considered as potentially dangerous and should be properly treated before ingestion. Procurement of an adequate supply of potable water is largely an engineering problem, but safety of the water is of immediate concern to health authorities.

TREATMENT.—The usual treatment of a public water supply consists of these steps: aeration, chemical treatment, sedimentation, sand filtration, and chlorination. First, the water is aerated by spraying it into the air, where it loses objectionable tastes and odors. Chemical treatment follows, involving the addition of alum, lime, or other agents that cause the aggregation of suspended particles; these are then heavy enough to settle out. Next, the supernatant fluid is passed through a bed of sand and gravel to remove any remaining particles. Finally, a small amount of chlorine is added to purify the water as it flows to the consumer.

Sewage Disposal.—Proper sewage disposal is a major public health problem closely akin to that of a pure water supply. Here the problem is basically one of avoiding direct contamination of oneself, food, and drink with fecal wastes. This problem is largely controlled by the chemical treatment and removal of the solids from sewage, and the terminal disinfection of the effluent waters. While the water carriage system of waste disposal is commonplace today, the method is only about 100 years old. At present, about 25 per cent of the urban communities in the United States still have inadequate sewage treatment facilities.

Arthropod-borne Diseases.—More than half of all human misery and death may be attributed to insects and other arthropods.

FLIES.—It is difficult to conceive of a better transient mechanical disseminator for disease germs, especially those of filth-borne diseases, than the ordinary fly. The fly's intestinal tract can carry a four-day food supply and still have enough room left for 20 million disease-causing pathogens. The bristled hair on the fly's body and its sticky foot pads can transport just as many pathogens. While many of the 100,000 known species of flies are transmitters of fecal-borne diseases, the common housefly is the greatest menace to human health because of its close association with man. In summer the housefly's life expectancy is about two months, but those that hibernate during cold weather may survive nine months or more. During her lifetime, a female fly will lay from six to ten batches of a hundred or more eggs, usually in the feces of man and animals or in decaying organic matter. The cycle from egg to larva to pupa to adult fly is about a week.

INFESTATION.—Larvae, or maggots, of many species of flies are dangerous to human health because they invade living tissue, causing a condition known as *myiasis,* or fly infestation. Some eggs or larvae may be ingested with food or drink and develop within the human intestine, although

FLIES spread cholera and typhoid fever.

myiasis of the genital and urinary tracts is also possible. Many species of maggots invade the nose, sinuses, eyes, ears, and open wounds. Myiasis in any of these locations may produce festering, deep, disfiguring wounds that exude a bloody, foul pus.

FLY CONTROL.—Control should begin early in the breeding season. Common methods within the home include the use of screening, flypaper, fly swatters, and insecticides. Ordinary cleanliness, however, is the most important control measure. Elimination of feeding and breeding grounds requires the proper disposal of garbage and body wastes. Insecticides, such as DDT, are useful in controlling flies, but environmental sanitation is more important.

MOSQUITOES.—While most flies act as transient mechanical disseminators of fecal-borne diseases, other arthropods also transmit disease. The blood-sucking *Anopheles mosquito,* for example, first infects itself and then infects a human host. Anophelene mosquitoes are *vectors* of malaria, a dread fever that leaves its victims frail, anemic, and wasted. Malaria used to attack 100 million people yearly and kill more than one million. Efforts of the World Health Organization have reduced these figures to about one-third the amount. Even temporary neglect of control measures would result in a resurgence of the widespread malaria epidemics

PHILIP GENDREAU

RODENTS transmit at least 18 diseases.

that once plagued the United States. Other mosquitoes are vectors of yellow fever, dengue, filariasis, and some viral encephalitides (brain fevers). Of these, only the viral encephalitides are now of any great significance in the United States.

OTHER ARTHROPODS.—Other arthropod-borne diseases are the flea-borne endemic typhus and bubonic plague. In addition, certain ticks are responsible for some spotted fevers, *Sarcoptes* mites are responsible for scabies (the so-called seven-year itch), and chiggers (immature harvest mites) are a common cause of skin irritation.

RANGE.—Most arthropod-borne diseases are common in the tropics because the majority of insects and other arthropods flourish in warm regions and prevail in direct proportion to man's lack of control over them. (Louse-borne diseases are a notable exception. Lice flourish in the colder climes where more clothing is worn, thus providing the ideal circumstance for their propagation.)

VECTOR CONTROL.—Control of all arthropod-borne diseases depends upon an exact knowledge of the habitat and life cycle of the vectors; the most effective control depends upon interrupting the vector's life cycle. Insecticides, fumigants, and repellents are useful. Of the insecticides, the *pyrethrums* are noteworthy because of their rapid "knock-down" effect, but DDT, although slower acting, is even more lethal. These chemicals, however, are dangerous and must be used very carefully.

TREATMENT. — Specific treatment is available for a few of the arthropod-borne diseases. For example, quinine-like drugs are highly effective against malaria; and some antibiotics are useful against plague, yaws, and certain other diseases. For many of these diseases, however, there is no specific therapy; and for only a few, such as yellow fever, is immunization practical.

Animal-borne Diseases.—Animals are directly or indirectly responsible for transmitting more than 100 diseases; these are often transferred from the animal to a susceptible person by blood-sucking arthropods.

RODENTS.—Rats and other rodents have always been a scourge of mankind, spreading terrible epidemics of disease and destroying or contaminating man's food. Rodents are responsible for the transmission of at least 18 diseases and, in the United States alone, destroy more than $2 billion worth of food and products yearly.

Rodent populations are limited primarily by the availability of food, so starvation should be a major part of any antirodent campaign. Rodent-proofing of buildings also limits the rodents' access to food and shelter, but poisoning is the simplest and most efficient method of control. The complete extermination of the species seems impossible, however, because wholesale killing gives the few survivors an easier life.

LIVESTOCK.—One of the most significant animal diseases directly transmissible to man is *brucellosis,* a prolonged illness usually referred to as *undulant fever* and associated with cattle, goats, and swine. More than 5,000 human cases are reported each year in the United States; of these, about 80 are fatal. Transmission may occur by direct contact with infected animals or infected animal tissue, or by the consumption of unpasteurized milk and milk products. *Trichinosis* is an animal-borne disease that can be contracted from the ingestion of insufficiently cooked pork. *Tularemia,* a plaguelike disease, is usually communicated by direct contact with the tissue of infected rabbits. *Rabies,* or *hydrophobia,* is associated with the bites of certain infected animals, often dogs or bats.

Environmental Health Hazards.—*Radiation* is an important environmental health problem. There has always been some natural radiation in the environment, including that from radioactive substances found in rocks and soils, such as uranium and radium, as well as cosmic rays and other types of radiation from outer space. But man-made radiation, consisting of the radiation from medical and dental X-ray and fluoroscopic equipment and the much-publicized fallout from nuclear explosions, has become the major cause for concern in recent years. While overexposure to these types of radiation is potentially dangerous, the average person presently receives no more radiation from these man-made sources than from the natural sources. A number of research groups are now studying the effects of radiation on humans and animals. These studies center on the genetic effects (cell mutations) and long-term influences of both the natural and man-made forms of radiation.

■PESTICIDES. — Agricultural chemicals are another environmental threat to human health. Recent reports have asserted that pesticides are polluting rivers, destroying wildlife, and generally upsetting the balance of nature; however, many of these reports tend to give insufficient consideration to the benefits gained from the use of pesticides. Actually, modern agriculture depends heavily on the use of pesticides, and millions of people are alive today only because of the chemicals used in suppressing the vectors of malaria, yellow fever, and typhus. At the same time, there are nearly 100 deaths a year in the United States that are caused by pesticides; most of these are the result of improper use. While pesticides at present constitute no great hazard, there still is need for continued government regulation of their production and use.

■AIR POLLUTION.—A major public health problem that increases with industrialization is *air pollution.* In the eastern sections of the United States, coal smoke and sulphur dioxide are the most common pollutants; in the western part of the country, pollutants arising from petroleum refining, chemical and metal manufacturing, and rubbish incineration are the major contributors. Automobiles are a ubiquitous source of impurities. Much thought is being given to the problem of air pollution, and most large cities now have active control programs.

■CHEMICALS. — Acute poisoning from ingested chemicals is a common medical emergency. During the first six months of 1961, 6,414 cases were reported in New York City alone. Internal medicines are the agents most frequently responsible; but topical medicines, cosmetics, detergents, insecticides, bleaches, solvents, disinfectants, and polishes are also involved. The number of potentially toxic substances in the home increases daily with the continuing advances in the synthesis and distribution of new chemical compounds. A *poison-control center* was opened in Chicago in 1953, and since then almost 500 more officially recognized centers have been established.

The first step in the treatment of poisoning cases is generally to empty the stomach by inducing vomiting. If the poisonous agent can be identified, the correct antidote can be given; for this reason, physicians usually request that the container with the remains of the poison be brought with the patient. It has been suggested that *ipecac* or a similar natural emetic be added to potentially dangerous preparations so that ingestion will induce vomiting at once.

■GASES. — Carbon monoxide is the most common poisonous gas, causing more fatalities than any other form of chemical poisoning (although many of these deaths are suicides rather than accidents). It is colorless, tasteless, and odorless; this imperceptibility, combined with the fact that the gas is found in the exhaust fumes of virtually every furnace, stove, and engine, makes carbon monoxide particularly dangerous.

Sociosexual Problems. — The venereal diseases are a major threat to health in every country of the world.

■SYPHILIS.—The most serious venereal disease is syphilis, an unremitting, progressive disease that caused 2,850 deaths in the United States in 1964.

The first symptom of syphilis is a painless, ulcerated sore, or *chancre,* on or near the genitals. Prompt treatment with penicillin at this stage will render the individual noninfectious and will usually result in a complete cure; if neglected, the disease lapses into a secondary stage within a month. This stage is characterized by disseminated eruptions that may occur anywhere on the body and resemble many other skin conditions. The symptoms of secondary syphilis are often mild, but can be distressing. After about a year the disease lapses into a late stage characterized by destructive lesions, or *gummata,* which may affect any tissue or organ system of the body. Some complications of late syphilis include heart damage, blindness, paralysis, and brain damage. The late stages of syphilis are usually noninfectious, except in the case of a pregnant woman, who may transmit *congenital syphilis* to her child at this stage. Both the congenital and late forms of syphilis have, through the use of penicillin and other antibiotics, become uncommon in recent years in the United States.

■GONORRHEA.—A more common but less dangerous venereal disease than syphilis is *gonorrhea,* characterized by an inflammation of the mucous membranes of the genital organs. Over 1½ million cases are estimated in the United States annually; for every reported case, supposedly four are unreported. Fatalities are low, but gonorrhea can produce complications, including infertility. In its early stages, gonorrhea is easily controlled by penicillin and other antibiotics and by sulfonamides.

■MINOR DISEASES.—The three minor venereal diseases—*chancroid, lymphogranuloma venereum,* and *granuloma inguinal*—are "minor" only in contrast to syphilis and gonorrhea.

■CONTROL.—The control of venereal diseases centers on detecting and treating existing cases and carriers. Since infected persons are the only known source, or reservoir, early recognition and treatment of these cases offers the best chance of curbing the spread of infection. There is no active means of immunization, nor are convalescent cases immune to reinfection. The spread of venereal diseases can be lessened by the use of prophylaxes but depends upon reducing promiscuity.

Maternity.—In addition to the prevention, control, and treatment of traumatic injuries and specific diseases, public health is concerned with the general daily welfare, physical as well as emotional, of the individual from the time of birth. Public health progress in the United States is reflected in the decrease in maternal and infant mortality.

■MATERNAL MORTALITY.—Before 1930 more than six mothers died per 1,000 live births. In 1966 the rate was .291 per 1,000 live births. (Maternal deaths in the United States totaled 1,049 in 1965.) The chief causes of maternal deaths, in descending order of magnitude, are hemorrhage, toxemia, abortion, infection, and ectopic pregnancy (a pregnancy outside of the uterus as, for example, in the Fallopian tube).

■INFANT MORTALITY.—There has also

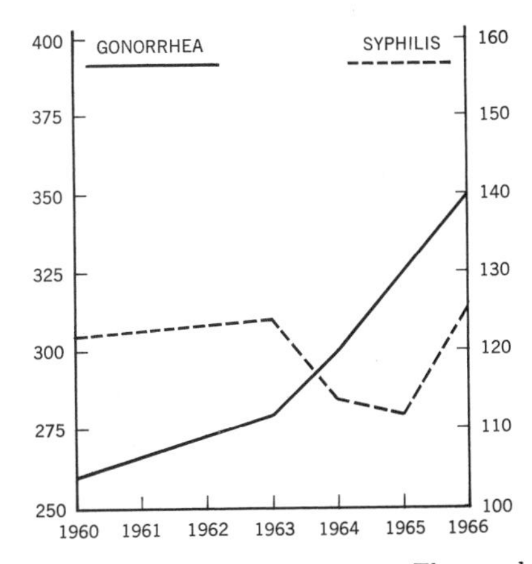

RISE IN U.S. VENEREAL DISEASE. The graph above illustrates the number of *reported* cases from 1960 to 1966 in thousands.

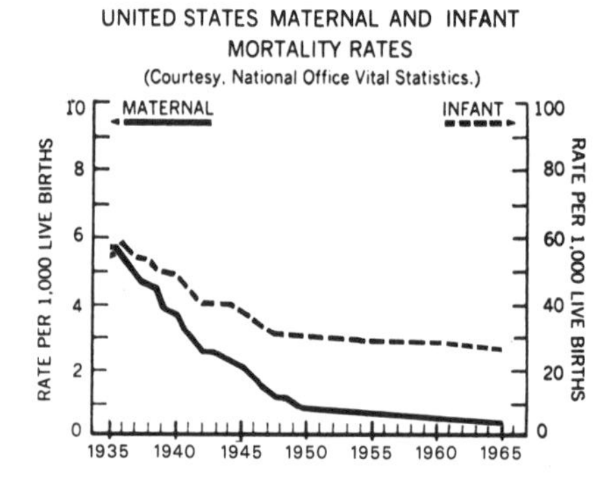

UNITED STATES MATERNAL AND INFANT MORTALITY RATES
(Courtesy, National Office Vital Statistics.)

been a consistent reduction of infant mortality. In 1915 the death rate of infants under one year of age in the United States was about 100 per 1,000 live births; it is now less than 25 per 1,000. The majority of infant deaths occur within the first month of life as the result of (in descending order of magnitude) premature birth, congenital malformations, infections, and birth injuries. The extent of the existing infant mortality problem, however, may be judged from the 85,516 infant deaths in the United States in 1965 and from the fact that the toll is even higher in less developed countries.

▪PREGNANCY.—A further reduction of infant and maternal mortality rates, as well as of birth abnormalities, can be insured if, as soon as pregnancy is suspected, the mother seeks competent medical supervision. Diet, rest, and moderate exercise are important during pregnancy, and an adequate intake of calcium, iron, protein, and selected vitamins should be carefully maintained for the benefit of both mother and developing child. This is particularly important, because during pregnancy eccentricities of the appetite may occur, resulting in vitamin deficiency. Extra precaution should be taken to avoid infectious diseases, particularly German measles, a disease that, when contracted during the first three months of pregnancy, may cause congenital abnormalities in the unborn child.

▪CHILDBIRTH.—Birth complications affecting the infant are far more common than those affecting the mother. Of the 4 million births annually in this country, 80,000 (or 2 per cent) are stillbirths (babies born dead whose period of gestation was over 20 weeks). Since the hours of labor and childbirth and the days following delivery are very critical for the baby, modern facilities and expert medical attention are imperative.

▪FEEDING.—Milk from the mother's breast has many advantages over a prepared formula, but breast feeding is not always possible, due to an insufficient milk flow, infections, or other complications involving the mother's breasts. When natural feeding is possible, it should be continued until the infant is six months of age. If a cow's milk formula is substituted, boiling the milk makes it safe and more digestible. Formulas usually consist of milk, water, sugar, and vitamin C and D supplements.

▪EDUCATION.—Programs that aid in insuring the proper care of babies deserve a high priority in community public health efforts. Those not able to afford the services of a private practitioner should be able to obtain competent help from public clinics and health departments.

Childhood Health Problems. — Dental health is of particular concern to a growing child, but diseases associated with the teeth are not limited to any one age group. Although diseases of the teeth are rarely a direct cause of death, they do contribute to ill health. The health of the entire body depends upon the maintenance of sound teeth and, conversely, the soundness of the teeth depends largely upon the health of the rest of the body. Although at least 90 per cent of the people in the United States have significant dental disease, only 25 per cent of these are getting reasonably adequate dental care.

Dental caries (cavities), the most common dental disease, are a curse of modern civilization. Primitive people, sustained by relatively crude diets, suffered less tooth decay than does modern man. Some factors involved in dental caries are diet, oral hygiene, and mechanical defects.

▪FLUORIDES.—The amount of fluorides ingested with food and drink has recently been shown to be of great significance; drinking water containing one part per million of fluoride dramatically reduces the incidence of dental caries, as has been demonstrated in numerous communities where the fluoride content of the water has been controlled at that level. Excessive fluorides found in some natural waters may give rise to a harmless condition known as *mottled enamel,* characterized by teeth that appear to be stained brown.

▪EYES AND EARS.—The eyes and ears, vital to lifelong welfare, are very susceptible to various afflictions during childhood. It is important, therefore, that abnormalities be detected and treated at an early age, when chances for correction are greatest.

Loss of sight or hearing means personal tragedy to the individual and a burden to society. There are at present some 260,000 blind and 57,000 totally deaf persons in the United States; some impairment of vision is present in nearly half the population, and at least 3 million persons are handicapped by impaired hearing. Diseases of the eyes account for about 50 deaths in the United States each year, and diseases of the ears for about 600. Disorders of the nervous system can adversely affect the sensations of sight, sound, taste, smell, and touch. Protection of every part of the nervous system is one of the foremost objectives of preventive medicine and public health.

Mental Health.—Of every 100 American children of school age, 13 will fail to reach emotional maturity, 8 will be shattered by emotional breakdowns, 4 will be confined in mental hospitals for a time, and 1 will turn to crime. It is estimated that half of all the persons seeking medical treatment are suffering emotional upsets, while nervous disorders alone or in combination with organic disease account for about 70 per cent of the medical cases being treated.

Of the 15 million United States draftees examined for military service during World War II, nearly 2 million (37 per cent) were rejected on the grounds of neuropsychiatric disorders; even after this initial screening, 39 per cent of all servicemen receiving medical discharges were classified as psychoneurotics. At present, more than 625,000 patients with nervous disorders occupy half the available hospital beds in the United States, and approximately 140,000 more need institutional care. Nearly 300,000 persons are admitted to mental hospitals annually; two-thirds of these are first admissions. This figure is expected to increase by 20,000 each year. Since the duration of treatment is usually so long, the burden of hospitalization would soon bankrupt most patients and their families. Therefore, the increasing responsibility for the care of the mentally ill has fallen to hospitals supported by public funds; at present, about 87 per cent of mental patients in the United States are cared for in state-maintained institutions. Although expenditures connected with disorders of the nervous system now exceed $200 million annually, another $500 million is needed to provide for a reasonable standard of treatment and care. Diseases affecting the nervous system account for 13 per cent of the deaths in the United States, collectively ranking second as the leading cause of death and being exceeded only by heart disease. Other deaths attributable to mental disorders during 1966 were 21,281 suicides and 11,606 homicides. In 1965 there were 2,665 fatal cases of alcoholism.

Problems of Later Life.—In 1900, 18 per cent of the people in the United States were over 45 years old; today 30 per cent are in this group, and estimates indicate that by 1980, 40 per cent of the population will fall into this category. At present, 10 million persons in the United States are over 65 years of age. Accompanying this rapid increase in the number

NATIONAL ASSOCIATION FOR MENTAL HEALTH

POSTER urges all to fight mental illness.

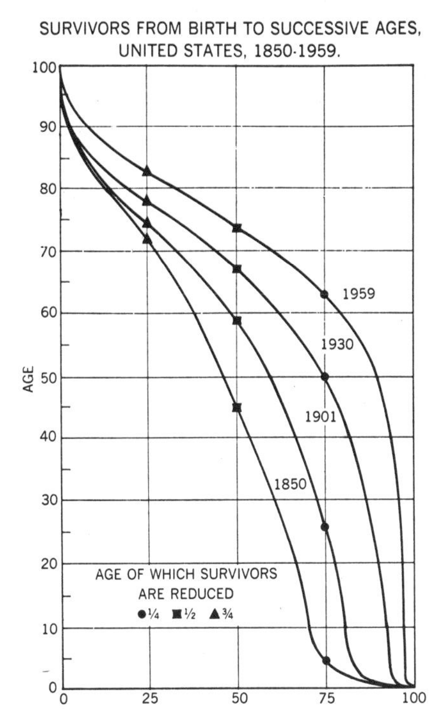

SURVIVAL CHART shows survivors from birth to successive ages. People are living longer today than at any time in history. In 1901, 18 per cent of the people in the United States were over 45 years old; in 1964, 30 per cent of the people were in this group.

of older people has been a corresponding increase of chronic illness, disability, degenerative diseases, and physical and mental impairment. Seventy per cent of those permanently disabled and 50 per cent of the chronically ill are over 45 years old. The most important illnesses of this age group are heart disease, arteriosclerosis, high blood pressure, nervous and mental disorders, cancer, arthritis, diabetes, kidney disease, asthma, and tuberculosis.

The intensity of the problems of later life has awakened an interest in *gerontology,* the study of the aging process, and *geriatrics,* medical care of the aged. Numerous voluntary and official health organizations have concerned themselves with various facets of these problems. Notable among these are the American Heart Association, the American Cancer Society, the National Tuberculosis Association, and the National Institutes of Health—the research arm of the U.S. Public Health Service.

Promoting Health.—Although the people of the United States are among the healthiest in the world, sickness produces widespread suffering and a heavy drain on the nation's productivity. Numerous surveys indicate that on any given day approximately 1 per cent of America's people are disabled in hospitals, 4 per cent are disabled at home, 20 per cent are suffering from nondisabling ailments, and 65 per cent are semiwell or subwell. Only 10 per cent of the population enjoy really good health.

During 1962 a patient entered some hospital in the country every 1.2 seconds; 7,028 hospitals registered with the American Medical Association admitted 27 million patients who spent a total of 512 million days in the hospital. This meant an average daily hospital census of over 1.4 million persons. One-third of these hospitals are tax-supported; their average daily census represents almost three-fourths of the total.

The present proportion of physicians to the general population is greater in the United States, with its 1 physician for every 750 persons, than in any other country except Israel. The distribution of physicians in the various states differs from 1 per 496 persons in New York to 1 per 1,459 in Mississippi. As a rule there is a greater concentration of physicians in large urban areas, paralleling the hospital monetary resources of the community.

■COST.—Part of the problem of getting satisfactory medical care is the practical aspect of affording it. Those who need medical care the most are usually least able to pay for it because the ability to earn money to pay for the necessities of life, including medical care, depends upon the individual's health. Each year the costs of illness must be borne by the 5 per cent of the population unfortunate enough to need medical care, and not every family is able to set aside money for the time of sickness.

Not only has the present system of fee-for-service failed to provide adequate medical care for all persons in the low-income group, but it has also failed to distribute good services to rural dwellers and to draw the public away from quack doctors and patent medicines. Moreover, it has failed to develop preventive medical services in proportion to their importance, as judged by the fact that less than $2 per capita is spent annually for public health measures while the average person spends nearly $135 for curative measures.

In spite of these serious defects in the present system of medical care, progress during the last century of American history has been spectacular. It is debatable whether such a record could have been accomplished under any other system.

■FUNCTIONS.—Functions of the various state health departments are very much alike. Whether carried out by separate departments or bureaus, their activities generally include statistical tabulation and analysis, control of contagious diseases through laboratory diagnosis and public health nursing in homes and clinics, promotion of maternal and child hygiene, environmental sanitation, food and drug control, industrial hygiene, and health education.

Local health departments are official agencies of local governments—city, county, or district. Their power is embodied in the charter of incorporation, which establishes the local governmental unit and is supplemented by local ordinances and the state health code; this is done through delegation of authority from the state health officer to the local health officer. The local department is officially responsible to the taxpayers, through their elected officials, for protection of the health of the community. Its functions are similar to those of the state health department, but modified to meet the outstanding needs of the community.

Most voluntary (nonofficial) health agencies were created by persons of vision when circumstances dictated a real need. Except for the Red Cross and a few tuberculosis societies, voluntary health agencies were almost unknown before the beginning of the twentieth century. Now there are more than 50,000 such organizations.

Public Health Agencies.—Development of state and local public health agencies in the United States has been gradual. The first local board of health was authorized in 1797 in Boston, Massachusetts, with Paul Revere as its chairman. The first state board of health was that of Massachusetts in 1869; the last state health department to be organized was that of New Mexico in 1919. Baltimore, Maryland, was the first large city to organize a full-time health department (1798), although several smaller cities had established part-time services earlier.

At present, no single agency of the United States government is responsible for public health. Instead, many agencies are individually concerned with the health and welfare of the people. The primary objective of most of these is to provide services to the various state governments in order to aid them in carrying out their individual programs. Notable among these are the Department of Health, Education, and Welfare, the Department of Agriculture, the Department of Defense, the Atomic Energy Commission, and the many branches of the Veterans Administration.

The U.S. Public Health Service, a major division of the Department of Health, Education, and Welfare and one of the oldest governmental agencies, was organized in 1798 as the Marine Hospital Service. Over the years its responsibilities have been extended to practically all phases of medical care, preventive services, and related research.

The first international public health agency, the Egyptian Quarantine Board, was established in 1831 to control plague and other infectious diseases then prevalent in the Mediterranean area. Its scope was enlarged with the opening of the Suez Canal in 1869. The Pan-American Sanitary Bureau, established in 1902, was the first international health agency to involve many nations of the western world. International conferences held in Europe at intervals beginning in 1851 culminated in the establishment of the International Public Health Office in 1909. The Health Organization of the League of Nations, organized after World War I, was succeeded in 1948 by the World Health Organization, a specialized agency of the United Nations.

—William Stiles

THE STUDY OF MAN, his evolution and early history, his customs, his beliefs, and the sometimes puzzling patterns of his existence, is the province of the social sciences. Anthropology and sociology deal respectively with the physical and social aspects of culture—from the way a South African Xhosa woman paints her face with clay (*below*), to the living conditions in a Turkish slum (*right*). Psychology and psychiatry examine the inner life of man, the structure of the self, and the nature of human behavior. Together, these fields comprise a picture of man's past and present, seen through the eyes of modern science.

UNITED NATIONS

SATOUR

HELEN BUTTFIELD

LEON DELLER, MONKMEYER

SYMBOLS of the search for meaning, the churches which men build are the physical counterparts of philosophical and religious ideas. The Blue Mosque in Istanbul (*left*) embodies submission to the will of Allah; Le Corbusier's Chapel (*above*), a renewed Christianity.

ANTHROPOLOGY

Scope.—The term *anthropology* is derived from two Greek words: *anthropos,* meaning 'human being,' and *logos,* meaning 'ordered knowledge.' It is, in effect, *the science of man.*

Physical anthropology is the study of human beings as biological organisms varying in both time and space. It thus includes human evolution, various forms of adaptation to climate and culture, and racial variations in anatomy and physiology.

Prehistoric archaeology is the study of the caves, rock shelters, camp sites, village mounds, and other deposits in which human beings have lived and where they have left such remains as stone tools, broken animal bones, and pottery. A thin line separates this subject from the study of literate urban cultures, which fall under such separate headings as classical archaeology, Egyptology, and Assyriology, depending on the region, time span, and people concerned.

Cultural anthropology is the study of the habitual behavior patterns, or "cultures," of peoples before the impact of modern western civilization. A specific culture, as studied by anthropologists, consists of the sum total of the ways in which a certain people lives, transmitted from generation to generation. Most of these cultures are preliterate and considered primitive.

Relationship to Other Fields.—Anthropology stands midway between the natural and social sciences and includes aspects of many branches of learning. To many of the physical and social sciences it has contributed breadth of understanding and masses of comparative data, while it has also helped to prepare the leading nations of the world for the close interrelation of races and cultures created by the modern age.

At the same time, other disciplines have enriched anthropology. Anatomical studies have demonstrated some racial differences in certain physical characteristics, such as the positions of muscles, sizes of endocrine glands, and adaptation to heat and cold. Geneticists have found racial variations in blood groups, as well as in fingerprints and hand prints; they have also explained the principle of balanced polymorphism in which the "mixed" form of a gene may be more advantageous than either "pure" form. Population geneticists, working mostly with domestic animals, have thrown light on the structure of breeding units; zoogeographers have clarified the roles of marginal and nuclear populations in evolution; and animal taxonomists have helped to straighten out the tangled classification of human races, both living and extinct.

Systematic, scientific archaeological research has been made possible by studies in geology. Paleobotany has contributed knowledge of the flora of bygone times, and paleontology, that of the animals that men hunted; both have provided techniques of establishing the dates of fossil remains. Physicists have contributed the techniques of long-range dating by carbon-14, argon-potassium, the quantities of fluorine and nitrogen in ancient bone, and the amount of surface erosion in obsidian tools. Mine detectors and other electronic devices and skin divers are bringing new evidence to light.

In the field of cultural anthropology, modern transportation and communications, tape recorders, motion picture cameras, aerial photographs, and other mechanical devices have greatly helped the field worker. Year-round residence in a community enables the anthropologist to witness all seasonal aspects of a culture; the techniques of statistical analysis, using sorting machines and even computers, and the detailed mapping of villages have replaced earlier practices of making simple generalizations from hasty observation or the employment of single informants.

Psychology has contributed almost as much to anthropology as it has received in return. Some field workers have had themselves psychoanalyzed before going into the field in order to eradicate the biases that stem from their own cultures and personalities. The concept of the unconscious has taught field workers not to ask primitive people *why* they do this or that; the people themselves do not know, but in order to satisfy the inquirer, produce a variety of rationalizations and misleading answers.

Personality studies have been conducted on individuals in other cultures and on the whole cultures themselves, to determine, among other factors, what role infant training and child-rearing practices have on the cultural attitudes of the adults. Primate studies made on many species of monkeys and apes have also revealed the common behavioral background of man and his zoological kin, thus contributing to an understanding of human behavior; experiments in "handling" other animals have been made to supplement studies of infant training in human beings.

PHYSICAL ANTHROPOLOGY

The search for fossil man is one of the most exciting aspects of anthropology, and almost every year there are new finds that supplement or even change the human family tree. This tree, or *genealogy,* of man tells the story of his gradual evolution and his differentiation from the other descendants of a common primate ancestor. Man's evolution was not the direct, uninterrupted path it may seem in retrospect, but a series of staggering steps with many sideways leaps—and some disastrous dives down blind alleys. Many species that branched off, following one or more specializations and adaptations to specific environments, failed and slowly died out. Many of the fossil remains belong to these offshoots; the others, which stand in the line of the descent of man, constitute the story of the evolution of *Homo sapiens.*

Primates.—Man belongs to the order *Primates,* suborder *Catarrhinae,* superfamily *Hominoidea,* family *Hominidae,* subfamily *Homininae,* genus *Homo,* and species *sapiens.* Our nearest living relatives are the great apes, the *Pongidae,* which includes *Gorilla; Pan,* the chimpanzee; and *Pongo,* the orangutan. The gibbons are also closely related, monkeys, more distantly.

■ MAN'S EARLIEST ANCESTORS.—The oldest known possible common ancestor of the great apes and man is *Propliopithecus,* who lived in Egypt during the mid-Oligocene period, or about 30 million years ago. The *Dryopithecines,* a family of apes found in Africa, Europe, and India, are probable descendants of Propliopithecus. From other branches of Propliopithecus came the gorilla, the chimpanzee, and *Ramapithecus* (originally called *Kenyapithecus*) of India and East Africa, who lived about 14 million years ago. Unlike the living apes, Ramapithecus had short canine teeth similar to those of modern man, and is probably man's ancestor.

After Ramapithecus there is a gap of between 10 to 12 million years before the Pleistocene, the last major period of geologic time. Until recently the Pleistocene was thought to have begun about 1 million years ago. Argon-potassium datings of new finds in Africa and elsewhere have set this date back to about 3 million years ago. The Lower Pleistocene, or Villafranchian, took up about the first 2 million years of this period.

The Pleistocene was a period of extreme climatic changes, with four great ice ages being separated by periods of warm temperatures in some areas and by periods of alternating moisture and drought in others, ending about 10,000 years ago. It is possible that changing climatic and environmental conditions induced the anatomical and social modifications that led to the emergence of man.

The earliest fossil specimen of a man-like creature is a piece of upper arm bone, including its elbow joint, found in Kenya and dated at about 2.4 million years ago. Its shape shows that it could only have belonged to an individual that walked on his hind legs.

AUSTRALOPITHECINES.—This man-like creature may have been a member of the sub-family Australopithecinae, also called the Southern Apes, which is represented by two species, *Australopithecus africanus* and *Australopithecus robustus,* located in East and South Africa. The ages and zoological names of many of their specimens, as well as their relationships to each other and to man, are uncertain at present. In East Africa the two species were contemporary from about 1.75 million to about 500,000 years ago; in South Africa *A. africanus* preceded *A. robustus,* and their combined time span probably began and ended later.

Both species were under five feet tall and walked more or less upright. Both had brains within the upper size range of apes, under 650 cc in volume. Their teeth were more specialized than those of modern man; *A. robustus* had particularly enlarged molar and premolar teeth and small canines and incisors, which indicates heavy chewing. In East Africa crude stone tools have been found with bones of both species.

HOMO ERECTUS.—Between 500,000 and 400,000 years ago in both East and South Africa, the Australopithecines were replaced by *Homo erectus,* the ancestor of *Homo sapiens,* modern man.

Java. That the Australopithecines evolved into *Homo erectus* in Africa has not been demonstrated. However, a fossil jaw found in Java contains teeth comparable to those of *A. africanus* (or, as proposed, *Homo habilis*) and has been dated at about 700,000 years ago, long before *A. africanus* in Africa became extinct. A cranium of the same age as the jaw and probably of the same population has a cranial capacity of about 900 cc. It is called *Pithecanthropus 4,* or *Homo erectus,* because it walked upright. Four similar specimens are 200,000 years younger. All five skulls have heavy brow ridges, receding foreheads, and crests for neck-muscle attachments. They range from about 750 to about 1100 cc in cranial capacity.

About 100,000 years ago (the date is approximate), Solo Man appeared. He is represented by a group of 11 skulls with cranial capacities ranging from about 1035 to about 1255 cc. An adolescent skull attributed to the grade of *Homo sapiens* was found in Borneo and dated at about 40,000 years ago; two skulls of the same or lesser age were also found in Java. These skulls show a gradual development of what was probably a single evolutionary line, the Australoid, culminating in the living Australian aborigines and in certain peoples of Southeast Asia and India.

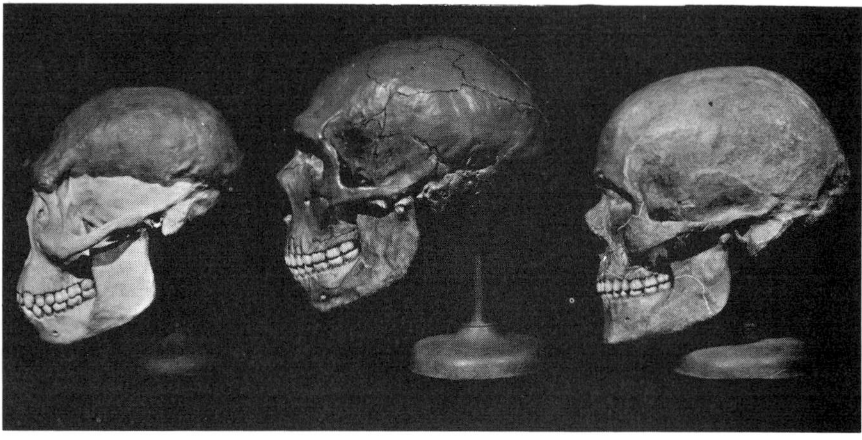

AMERICAN MUSEUM OF NATURAL HISTORY

SKULLS OF EARLY MAN show the evolutionary development of our species. The *Pithecanthropus* (*left*) dates back about 500,000 years. The *Neanderthal* (*center*) existed about 90,000 years ago, while the *Cro-Magnon* (*right*) may have lived only 30,000 years ago.

China. The oldest human specimens from China are a skull and a lower jaw between 500,000 and 600,000 years old, provisionally called *Sinanthropus lantianensis.* The skull has a cranial capacity of about 750 cc and otherwise resembles the better-known *Sinanthropus pekinensis* specimens, now called *Homo erectus pekinensis,* or simply Peking Man. These specimens are fragmentary remains of about 40 individuals quarried out of former caves near Peking and dated at about 400,000 years ago. They are of about the same evolutionary stage as Solo Man. Heavily built and with a slightly larger cranial capacity, they show certain Mongoloid characteristics in the teeth and especially in the high cheekbones typical of modern Mongoloids. Remains in cave deposits where they were found indicate that they not only chipped rough stone tools but also used fire for warmth, protection against animals, and possibly for cooking.

These finds are followed by a late Pleistocene skull from Mapa (Kwangtung) in South China, dated at about 150,000 to 100,000 years ago; like Peking Man, it is still essentially a type of *Homo erectus.* A skull from Tze Yang in West China, of less than half that age, is much closer to the form of *Homo sapiens.* Finally, from the end of the Pleistocene, about 10,000 years ago, are three skeletons also found at Choukoutien; called the Upper Cave people, they are virtually modern Mongoloids.

Africa. A skullcap of *Homo erectus* dated at about 500,000 years ago was found immediately above the Australopithecine specimens (widely known as Zinjanthropus, but doubtless an early form of *A. robustus*) at Olduvai in Tanzania. The skullcap is similar to the *H. erectus* skulls of Java and China, except that the brow ridges are curved rather than straight, giving it a closer resemblance to a Caucasoid or Negroid type. Another *H. erectus* skull, about 30,000 years old, was found at Saldanha Bay in South Africa. It is morphologically similar but has a cranial capacity of 1200 to 1250 cc. Still another, some 23,000 years old, was removed from a mine at Broken Hill, in Zambia (formerly Northern Rhodesia). It is the only truly primitive fossil skull yet found that is complete, and its facial structure is of a very primitive Negroid type. The cranial capacity is about 1280 cc. A series of four fragmentary skulls from Kanjera, Kenya, of uncertain date are also Negroid in type but more evolved than the Olduvai-Broken Hill examples. Aside from these specimens, we have nothing substantial with which to trace the origin of the African Negroes and their dwarf relatives, the Pygmies.

In North Africa the earliest *Homo erectus* remains are two lower jaws from Ternefine, Algeria, of about the same age as *Sinanthropus* and resembling him in form and tooth structure. A series of lower jaws from three sites in Morocco suggests that this local race survived until the early Late Pleistocene. Two skulls from Jebel Ighoud, Morocco, are heavy-browed, flat-faced and prognathous, similar to *Sinanthropus* and probably prototypes of the modern South African Bushmen. The upper jaw of a child, found in a cave of Tangier, carries this series into the Late Pleistocene, the time of modern Caucasoid men in Europe and the Near East.

Europe. Very few examples of man have been discovered in Europe dating from the Early or Middle Pleistocene, but the record from the later eras is more detailed than on any other continent. The European (and associated Middle Eastern) sequence begins with a large occipital bone discovered in 1965 at Vertesszollos, Hungary. The skull of which it formed a part must have had a cranial capacity of more than 1400 cc, a figure which is above the mean for *Homo sapiens.* Although the bone is closer in form to *Homo sapiens* than to *Homo erectus,* it has been provisionally labelled *Homo* (*erectus* seu *sapiens*) *palaeohungaricus.* Such a skull could have matched the famous

Heidelberg jaw found in Germany, which is dated at 400,000 to 360,000 years ago.

There are two skulls from about 250,000 years ago, one from Steinheim, Germany, and the other from Swanscombe, England. Except for the heavy brow ridges of the Steinheim specimen (the Swanscombe specimen has no frontal bone), they are both remarkably similar in size and form to skulls of modern man. Between 250,000 years ago and the final, or Würm, glaciation, Europeans changed little morphologically. The remains of 27 individuals have been discovered, who are probably Caucasoid.

NEANDERTHALS.—The Würm glacial period was the age of the Neanderthals, an enigmatic people concentrated in France and the Near East; the remains of 83 individuals had been unearthed by 1963. They were people of normal body build with deep chests and short, square-toed feet. They did *not* walk stooped over or with a shambling gait, as is commonly represented, but stood straight and held their heads erect. Like most fossil men, the Neanderthals had heavy brow ridges, long faces, and prominent noses; but they were not the brutish churls they are often made out to be, and their brain capacity was equal to that of many living Europeans. The teeth of the individual skulls studied are considerably worn down, possibly from using them to dress animal skins worn as protection against the cold.

Both the origin and the destiny of the Neanderthals is greatly disputed. Some scholars believe the Neanderthals were a separate species that migrated to Europe and replaced the *Homo sapiens* living there, while others believe they evolved from their *sapiens* predecessors. They survived in both Europe and the Middle East throughout the first Würm glaciation, but with the temporary retreat of the ice they gradually disappeared, either replaced by the Caucasoid ancestors of living Europeans and Middle Easterners or genetically absorbed by the Caucasoid race.

HOMO SAPIENS.—The strain *Homo sapiens* inhabited western Europe during the rest of the Pleistocene. Their practice of burying their dead has left many complete skeletons for physical anthropologists, and the rich and highly developed artistic tradition of their cave paintings provides both evidence to their culture and further clues to their appearance. A skeleton from Combe Capelle, a skull from Chancelade, the buried remains of a Cro-Magnon man found near Les Eyzies in France are among the more than fifty skeletons recovered from this period. They range in height generally from five to six feet, and their skulls are almost identical to modern man's, with high foreheads and prominent chins; in general appearance they could nearly pass for contemporary Europeans. Their origin is still uncertain, although recent finds from Israel (a skull from Mt. Carmel and the Skhül skeletons) show a mixture of Neanderthal and Caucasoid characteristics that indicate a Middle Eastern orgin.

MIGRATIONS.—In Asia, by the end of the Pleistocene, and probably as early as 18,000 years ago, the Mongoloids had expanded from their home in China, crossed the Bering Strait, and entered the previously uninhabited New World, becoming the ancestors of the American Indian. Between 12,000 and 10,000 years ago, other Mongoloids spread southward into Southeast Asia and Indonesia, displacing the original Australoids. Some of these Australoids lived on as dwarfs in impenetrable forests, mountain refuges, and fringing islands; others moved out to sea across strings of small islands (exposed by the falling sea level during the glacial periods) from Indonesia to New Guinea and Australia.

At the time of this second Mongoloid expansion, Caucasoids coming from Europe or the Near East invaded North Africa, expelling many of the earlier inhabitants of the Jebel Ighoud race. The latter migrated southward as far as the Cape of Good Hope, where they became the ancestors of the modern Bushmen, now a partially dwarfed, partially infantile race. The Caucasoids who replaced them in North Africa were the ancestors of the living Berbers.

Living Races.—All living men belong to a single genus and species, *Homo sapiens*. Within this classification they fall into groups that differ in certain physical characteristics; these groups are called *races*. The differences between races are purely biological and are marked by the hereditary transmission of anatomical and physiological characteristics, including blood groups. Races differ in such things as adaptation to extremes of temperature, solar radiation, atmospheric moisture, water vapor pressure, and their resistance to certain diseases. Intermediate, or clinal, populations are found in regions where geographical barriers such as mountains do not impede mixing.

Whether or not races differ, as individuals do, in such behavioral characteristics as temperament and intellectual capacity is a matter of dispute, but the basic drives that lead to the rise of familial and other social institutions are undoubtedly the same. The term "race" cannot be applied to national, cultural, or linguistic groups.

CLASSIFICATION.—Many attempts have been made to classify the living races of man on the basis of their physical characteristics, but no complete agreement has been reached. Other classifications based strictly on genetics have stressed such factors as blood groups. The various classifications have ranged from the commonly held, threefold, Caucasoid, Negroid, and Mongoloid to as many as ten or more divisions. Most recently, C. S. Coon has proposed a fivefold classification, based on the fossil history of man as outlined above. He recognizes five subspecies, or geographical races, as follows: Mongoloid, Australoid, Caucasoid, Capoid, and Congoid. The first three are clearly differentiated and generally accepted; there are less obvious differences between the last two races, which have the same hair form and similar blood groups. Mixture of Capoids and Congoids is probably the result of the translocation of the Capoids from North Africa to South Africa. The two races, however, have definitely separate histories.

MONGOLOIDS.—The principal characteristics of the Mongoloids are a flattish face, often with an internal eye fold, subnasal prognathism, and shovel-shaped incisors and canine teeth. They also have a minimum of beard and body hair; coarse, straight, black head hair; small wrists, ankles, hands and feet; and skin color varying from brunet white to dark brown. Many are of stocky build, with relatively short limbs.

AUSTRALOIDS.—The Australoids also have flattish faces, but their eyes are sunken like those of Caucasoids rather than flush like those of the Mongoloids. They also lack the characteristic Mongoloid eye folds. The jaws of Australoids are prognathous, their nasal tips broad and fleshy, and their teeth are large and unspecialized. They are bearded like Caucasoids; some have heavy body hair, and their hair form is straight, wavy, or curly. Australoid skin color varies from black to almost brunet white; dark skin is the result of one genetic factor, whereas in African Negroes at least two such factors are responsible for the characteristic pigmentation.

As a rule, Australoids are of slender build, with limb proportions similar to those of Caucasoids. Unlike the Mongoloids, who show little *sexual dimorphism* (difference in body form between the sexes), the Australoid women are much smaller than the men. Both men and women also tend to become gray-haired early in life, and some of the men are bald; among Mongoloids baldness is almost unknown, except among the Japanese.

The "hairy" Ainu of northern Japan, Sakhalin Island, and the Kuriles, resemble Caucasoids in skin color, hair form, hair distribution, and, to a certain extent, facial features. In other respects they resemble the aborigines of southeastern Australia. But their blood groups closely relate the Ainu to neither of these peoples. The cultural evidence suggests that they migrated long ago from Europe, West Asia, or both. Like other peripheral peoples, the Ainu may be of composite (mostly archaic Caucasoid) origin.

CAUCASOIDS.—The Caucasoids are essentially the peoples of Europe, North Africa, the Arab countries of the Near East, Iran, Afghanistan, West Pakistan, much of India, and Ceylon. They also include settlers in the Americas, Australia, New Zealand, and South Africa. Their physical characteristics are well-known. Their range in skin color is the greatest found in any subspecies, extending from the complete depigmentation of the Baltic peoples to the almost black skin color of the natives of southern India.

Essentially, the Caucasoids gravitate toward two morphological poles: the narrow-faced, thin-nosed extreme as seen in Europe among the Basques,

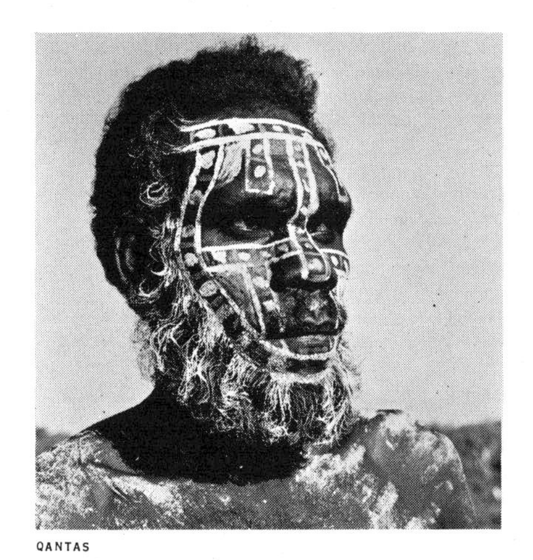

QANTAS

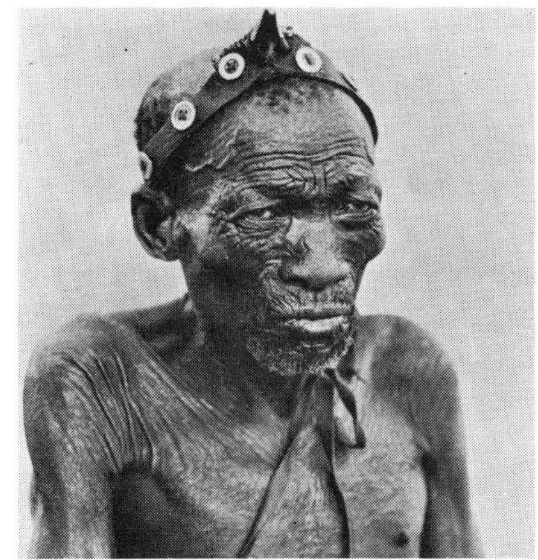

SATOUR

SATOUR

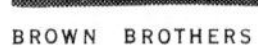

BROWN BROTHERS

THE RACES OF MAN cover the world with a wide variation on a single theme: *Homo sapiens.* The Australoids (*top left*) are the original inhabitants of the Australian subcontinent; the Congoid peoples of Africa are the Negroes and Pygmies; the Capoids are represented by a Bushman and a Zulu woman is another Congoid (*top center and right*). The Caucasoids shown here are five Russian immigrants; the Mongoloids (*below, left to right*) include the pure Mongoloid Chinese and two offshoots, the Eskimo and the American Indian, who were the earliest immigrants to America.

NEW YORK PUBLIC LIBRARY

AMERICAN MUSEUM OF NATURAL HISTORY

NEW YORK PUBLIC LIBRARY

and a broader, flatter-faced form with a low nasal bridge and widely separated eyes, as represented by Finns and Slavs. Although many anthropologists have seen in the latter form evidence of Mongoloid admixture, modern blood-group research has disproved this theory. The Lapps, who stand at the extreme in this respect, are not Mongoloid at all but purely European.

CONGOIDS.—The classification of African peoples is confused by a scarcity of paleontological material and by the number of migrations along the highroads of the Sudan and the East African highlands. The Congoids include both the Negroes and the Pygmies, who apparently became dwarfs independently of the Asiatic and Philippine Negritos. We have absolutely no evidence of their beginnings. The historic center of the Negroes has extended from West Africa and the northern fringes of the Congo into the marsh country of the Sudan. From this point, some of them migrated in various directions, eventually reaching South Africa.

Their physical features are also well-known. They have less sexual dimorphism than either Australoids or Caucasoids, the same kind of hair as the Papuans and Melanesians (with whom they have no relation either historically or as shown by blood groups), long lower-leg and arm segments, and narrow pelves. The Pygmies, aside from being dwarfs, differ from the Negroes in having less everted lips, more body hair, and lighter skins, sometimes mahogany-colored or reddish.

CAPOIDS.—The Capoids are represented today by the Bushmen of the Kalahari and a few relict tribes in Tanganyika; they are yellow-skinned, with flat faces, Mongoloid eye folds, and scanty body hair. Their local peculiarities include excessive steatopygia (fattening of the buttocks), a horizontal stance of the penis, and a great enlargement of the *labiae minorae*. Von Eickstedt, a modern classifier of races, places them with the Mongoloids, which may be justified if they are traced back far enough in time. Blood-group classifiers place them with Congoids, and Coon gives them separate status as a subspecies. In any case, they are the most enigmatic of the major living races of *Homo sapiens*.

INTERMEDIATE PEOPLES. — Between these five subspecies live intermediate populations with mixtures of various features. They form what physical anthropologists call *"clines"*: graded series of variations in bodily structure or function forming a kind of bridge or inclined slope between different races. Between the Mongoloids and Caucasoids in central Asia are various intermediate Turkish-speaking peoples, such as the Turkmen and Uzbeks. On the southern slope of the Himalayas, a steep Mongoloid-Caucasoid cline is seen, particularly in Nepal. The Melanesians, Micronesians, and Polynesians probably are the product of a similar ancient cline in southern China and Indonesia, with the Polynesians being the most Mongoloid and the Melanesians the most Australoid.

AMERICAN MUSEUM OF NATURAL HISTORY

MAORI WOMAN of New Zealand. The tattooed chin is believed to be a sign of beauty.

Much of Africa is clinal country, with mixed Caucasoid-Congoid peoples in the Sudan and East Africa, and a special cline in Ethiopia. The Hottentots and Korana, cattle people who speak languages similar to those of the Bushmen, stand between the latter and Negroids.

RACIAL MIXTURE.—If the existence of interracial clines indicates a more or less constant flow of genetic material between centers of racial differentiation, why do races continue to exist? Why did not the constant mixing of genetic material homogenize the human species long ago?

One answer to this question is that man is a cultural animal. Other species communicate by using symbols that are mostly inherent, or common to the species as a whole. Man, on the other hand, communicates by using symbols which are mostly arbitrary in nature and acquired by learning. Man's enormous variety of culturally determined sets of symbols —including languages, ethical codes, religions, political ideologies, and ways of doing things in general—set human groups apart from each other and impede mixture and homogenization.

ADAPTATION.—The races of man also resist mixture because some of them are adapted to specific kinds of environmental stress. The Australian aborigines can sleep naked on the ground at 32° F. because the veins and arteries in their arms and legs exchange heat, reducing the caloric requirements of the body. The same kind of cold adaptation has been found among the Reindeer Lapps. The Indians of Tierra del Fuego and northern Canada, the Eskimo, and peoples of Siberia and Manchuria resist cold by having a high basal metabolism that burns a high-caloric diet. The Mongoloid Tibetans and Andean Indians can tolerate the thin air of their highland homes because their hearts are large and their blood thick. Such adaptations account for the persistence of races in their homelands despite intermingling with other less adaptable racial types.

Regardless of race, certain other rules follow climatic zones. Within each race, stature and weight increase with isotherms of winter cold and decrease with heat. Pigment increases with the amount of sunlight, and tolerance to sunlight with pigmentation. Another kind of adaptation is the capacity to tolerate crowding, which is at its height in long-settled urban populations and at its minimum among those in isolated, thinly populated areas.

PREHISTORIC ARCHAEOLOGY

While physical anthropology deals with the changing body structure and biological adaptations of early men, prehistoric archaeology is concerned with man's developing *technology* and his first steps toward civilization. It is the record of what men made: the tools and weapons they fashioned first from stone, then ivory and bone, and later from metals.

During the Stone Age, the Bronze Age, and the Iron Age, men changed less in physical form than they had in previous eras. After the evolution of *Homo sapiens* in the Upper Paleolithic Age, man's adaptation became centered on the ways in which he added to his meager physical equipment by the use of clothing, tools, and weapons and by the application of his brain to work with other men in elementary forms of society, through speech and later through magic and art. This *cultural evolution,* which started with the manufacture of the first stone tool artifact, continued through the Pleistocene. Archaeologically, this prehistoric period is divided into the Old Stone Age (Paleolithic), Middle Stone Age (Mesolithic), and Late Stone Age (Neolithic). The Paleolithic is further subdivided into Lower (the earliest), Middle, and Upper. In Europe, the time span covered is more than 500,000 years, ending about 10,000 years ago. In some remote areas of the world it has continued.

STONE IMPLEMENTS of prehistoric man.

Tool Making.—The earliest known tools are simple *pebble tools,* made by breaking a rounded pebble in such a way that a sharp edge will be formed. These have been found principally in Africa and Asia, and some exist in Australopithecine remains.

CORE TOOLS.—The next step is the manufacture of *choppers,* which are shaped stones with a cutting edge produced by flaking on only one side, and *chopping tools,* made by flaking both sides. Both of these are called *core tools* because the central portion that remains after flakes are chipped off becomes the tool. They have been found in North Africa, East Africa, Israel, India, Southeast Asia, Java, China, Japan, and even in North America and South America. In modern times they were still made and used by Indians of the Great Basin of the American West and by some of the Australian aborigines.

While in the beginning choppers and chopping tools were as widely distributed as man himself, by the time of the first interglacial period, or roughly 500,000 years ago, the inhabited parts of the world diverged in the manufacture of tool types. In Europe, Africa, and western Asia as far as India, people began to make almond-shaped tools chipped on both sides, known as *hand axes,* and similar tools without points, known as cleavers. The manufacture of these so-called *bifacial* tools improved as time went on, reaching perfection by the first interglacial period, roughly 100,000 years ago. From India to China, choppers and chopping tools continued to be made for an even longer period.

FLAKE TOOLS.—In both archaeological regions *flake tools* were also used, made by chipping flakes off the stone, then retouching the edges of these flakes to make many sharp-edged tools. From 150,000 to 75,000 years ago most of the world's tools were made of flakes, although in remote places such as South Africa, bifacial tools continued to be used until less than 25,000 years ago.

Some time between 75,000 and 30,000 years ago, men slowly perfected the technique of making *blades*—flat, parallel-sided strips of flint suitable for use as knives. Blades were made by carefully preparing a tubular piece of flint with a flat end so that a carefully aimed blow with a piece of horn or bone would break off long, thin slivers of stone. The first blades were probably made in the Near East, but by 30,000 B.C., they were in common use in such widely separated regions as France and Afghanistan. Blade making did not reach Africa or the Far East until after the Pleistocene. At the time they were discovered, the Australian aborigines were still in a technological Stone Age, making tools in many fashions: choppers on the island fringes and in Tasmania, simple flakes in parts of the desert, and blade tools in the very southeast portion of the continent. The American Indians made both chopping tool artifacts and elaborate flake tools (the finely pointed arrowhead); they also made blades with beautifully fluted points.

Paleolithic.—By far most of the work on the Paleolithic, or Old Stone Age, has been done in Europe, where detailed sequences have been established. The *Lower Paleolithic* is the period of bifacial tools, which served the rudimentary hunting needs of men whose primary diet was still roots, nuts, and fruits. The *Middle Paleolithic* is characterized by flake tools of several industries and by the gradual elaboration of the flaking technique. This is the period of the culture of the various Neanderthals, more skillful hunters than their predecessors, who used fire and were beginning in some areas to bury their dead. From this latter practice we can infer the beginning of a rudimentary society and some dim concept of an after-life.

The *Upper* or *Late Paleolithic* began about 30,000 B.C., the time of a relatively warm interval in the last glaciation in Europe. It was essentially a time of blade making in both Europe and western Asia. The blade-makers, such as *Cro-Magnon* man, also made tools and weapons, including harpoons and needles, out of bone, ivory, and antler. They also clothed themselves with the skins of the beasts they killed, sewn together with bone and ivory needles, which, together with crude buttons, have been found in some sites. Along with their developing technology they produced one of the most sophisticated representational art forms the world has known: the *Magdalenian cave paintings.* Figures of bison, mammoth, and reindeer were incised and painted with earth colors in the remote recesses of caves in southern France and in Spain; they undoubtedly served a magical purpose. The lifelike representation of the animals they hunted, often shown pierced by arrows, became incantations to insure the success of the hunt. Likewise, the *Aurignacian Venuses,* similar to female figures found in existing primitive societies, were probably charms to increase human fertility. The practice of these arts, and of magic, together with the collective hunting required to kill the giant mammoths, show an increasing complexity of thought and activity that can for the first time be called *culture.*

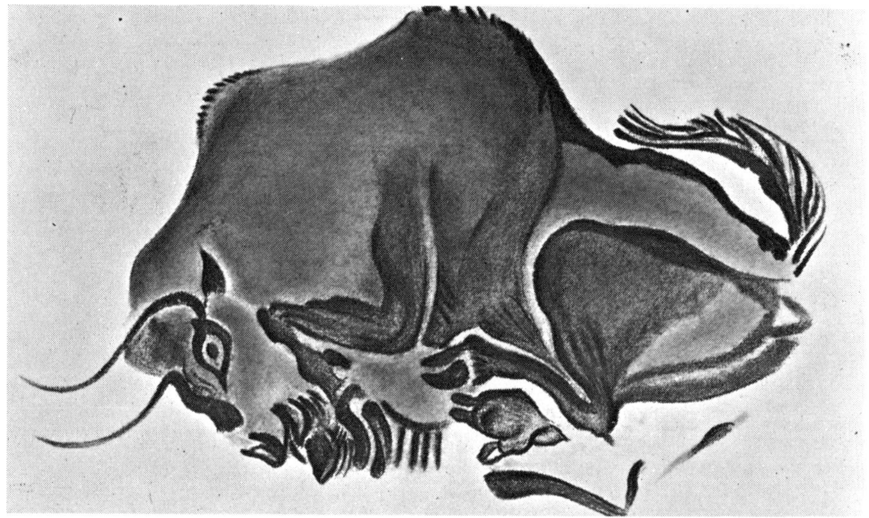

AMERICAN MUSEUM OF NATURAL HISTORY

CAVE DRAWING of a female bison was probably done about 50,000 years ago. This drawing, found near Altamira, Spain, gives some idea of the life that was led by the cave dwellers.

Mesolithic.—The *Middle Stone Age* is the period of the European industries in the time span immediately following the Pleistocene. With the shrinking of the ice caps, forests invaded central and northern Europe, the sea level rose, and vast migrations of salmon and other fish filled the rivers and estuaries. The giant animals of the Pleistocene—mammoths, hairy rhinoceroses, cave bears, huge wild oxen—as well as the reindeer, were replaced by smaller game, particularly red deer. Bows and arrows and domestic dogs had been introduced, and fish weirs, nets, hooks, and spears were used. The stone industry was an offshoot of the Upper Paleolithic blade industry, with an emphasis on *microliths*—tiny blades that can be hafted in a row to produce a blade of any length. These tools later helped prepare the way for the practice of agriculture, since they could easily be adapted to use as simple sickles. One important new tool was invented, the axe hafted onto a wooden or bone handle; it was both a product of, and the means of coping with, the new forest environment. These dense forests, which had replaced the open tundra of the Paleolithic, limited migration and communication between groups; thus the Mesolithic is not a single homogeneous culture but consists of a number of isolated separate cultures.

Neolithic.—The New Stone Age marks a change in way of life so decisive for man that it has been called the *Neolithic,* or *Agricultural Revolution.* For the first time men planted seeds and collected the harvest of their crops instead of gathering them where they could be found growing wild; they domesticated animals, spun and wove fibers into cloth, and made pottery. With polished stone axes they felled trees to clear the land, and built boats and houses. Villages grew up, industries and the arts flourished, and slowly the fabric of urban society evolved.

The initial date for the European

SATOUR
ANDRE EMMERICH GALLERY

WEAPONS used by the South African Bushmen are very similar to the ancient bow and arrow pictured in this sixth-century B.C. drawing.

Neolithic is at least 5200 B.C.; in the Middle East it is much earlier. In Turkey it goes back to 7000 B.C., and the fortress of Jericho was once a Neolithic trading post. The Neolithic began in the Middle East, in the warm and well-watered areas of the great river valleys, and from there it spread to India and China, as well as to Europe and North Africa. In the New World, an apparently independent Neolithic culture arose in one or more centers of Central America and South America at about the same time. It spread northward as far as the St. Lawrence valley and south to Bolivia and Chile. Well into modern times the Neolithic was still the way of life of many of the earth's peoples, including the Polynesians, Micronesians, Melanesians, Papuans, and all of the agricultural Indians of North America and the forested regions of South America.

Ages of Bronze and Iron.—The discovery of the technique of extracting metal from ore and forging or casting metal tools marks the end of the Stone Age. Flint and bone were replaced by the more malleable and durable materials bronze and iron. The rise of metallurgy was soon followed by the invention of writing, with which man passed from prehistory into history.

While the Old World Bronze Age and Iron Age cultures were evolving, many peoples on the peripheries of the centers of origin of these techniques remained untouched by that development. They adopted metal work late in history, without passing through the regular sequences of cultural stages found in the Middle East and Europe. The Indians of Peru invented bronze metallurgy independently, and were in the Bronze Age when the Spanish *conquistadores* arrived. In Africa there was no proper Bronze Age outside of Egypt and a few sites along the Mediterranean coast. Iron followed stone tools in North Africa in Carthaginian and Roman times, having been introduced to the interior mostly by Jewish traders and craftsmen. In sub-Saharan Africa during the middle of the first millennium B.C., ironworking was set up on a large scale at Meroë by the Egyptians under Hittite influence. From that center, and possibly also from Indonesian contacts on the Indian Ocean coast, techniques of smelting and forging iron spread over Negro Africa during the first millennium A.D.

CULTURAL ANTHROPOLOGY

Methods of Study.—*Cultural anthropology* can be studied in terms of individual cultures, such as those of the Hopi Indians of New Mexico or the Zulu of South Africa, or by subject matter such as marriage or religion. The first method is known as *ethnography*. Within the scope of ethnography is the study of groups of related cultures that occupy a geographical region and follow a common way of life. This is called the *culture area* method and includes such relatively small and homogeneous groups as the buffalo-hunting Indians of the American plains or the cattle people of East Africa; it also encompasses such a widespread and complex culture as that of the Islamic Middle East.

The study of cultures by subject matter is known as *ethnology*. It is broken down into studies of material culture, or *primitive technology*, and *social anthropology*, which emphasizes the human relationships binding men in a social fabric of behavior, custom, and institutions.

In studying the behavior patterns, beliefs, and customs of primitive cultures, anthropologists are hampered by the changes wrought on primitive peoples by the impact of modern civilization. Only rarely can the field worker find unspoiled cultures to examine and record; often he must reconstruct many of the details from elderly informants and early records. The *ethnographic present* is the term used to describe the period when a culture was still unspoiled, as yet unaffected by the various stages of acculturation.

Primitive Technology.—One of the working definitions of man is "the toolmaker," for from the beginning of his history man has used tools. There is no people on Earth who do not use some kind of tool, no matter how primitive. Toolmaking is in the material sense the beginning of civilization, for it is the root of technology, the means by which modern man so efficiently controls his environment.

The first tools used by men were stones thrown at small game, or sticks used in the same way a chimpanzee will use one to reach fruit beyond his grasp. The stones, were shaped and chipped to provide a cutting surface, as seen in the development of the Stone Age cultures, or tied in groups of two or three to form the *bolas*, a kind of thrown snare still used in South America. The stick pointed at one end became the digging stick, used to unearth edible roots and later, to plant and harvest cultivated crops. With a stone point at one end, it became the spear, the javelin, or the harpoon; later, the arrow was combined with the ingenious combination of the curved stick and tensed string that is the bow. A curved piece of wood alone, deftly shaped and accurately thrown, is a boomerang, which will return to the hand of the thrower if it fails to hit the target.

Where wood and stone were not available, man used whatever materials lay at hand: fish and animal teeth, coral, bamboo, clam shells, ivory, and whalebone. Without tools there would be no culture.

■**FIREMAKING.**—Another vital necessity to man is *fire*, which keeps him warm, protects him from predatory animals, and makes his food tender and easy to digest. It gives light and warmth at night to a group of people who can sit, talk, and dance together. Fire, which brings men together around a common center, is perhaps the beginning of social life as living peoples know it.

Fire can be obtained from nature —from burning forests or smoldering volcanoes—or it can be made by man himself. The two commonest methods are by striking flint against iron pyrites to make a spark, or by friction of one piece of wood against another. Although almost all primitive peoples use a variation of one of these methods, there are a few who cannot make fire. When first discovered, neither the Andaman Islanders nor the Tasmanians knew how to make fire. Instead they carried burning or smoldering torches with them wherever they went.

PROVIDING SHELTER.—All peoples know how to make some kind of *shelter*. The most elementary protection is a *windbreak* of trees and brushwood, or a rectangle of skins or matting set at a slant with its back to the wind, the only shelter known to the Ona of Tierra del Fuego and the Great Andaman Andamanese. The simplest type of *hut* is made by bending over branches like croquet wickets, tying them together, and covering them with leaves or pieces of bark. In a cold climate, like that of the southern coast of Chile, skins are placed over such frames, and fires are built inside. In the northern regions of the Old World and New World, the commonest house type is a conical *tipi* made by leaning poles together and covering the surface with bark or skins. Desert nomads and the pastoral Tibetans dwell in tents of skin or goat hair, which are portable, durable, cool in summer, and warm in winter. But they are not as warm as the *yurt,* a demountable, portable house made by fitting sticks into a circular frame and covering the frame with sheets of felt. This is the traditional house of the Mongols and nomadic Turks.

Permanent houses everywhere are made of the best available materials, such as wood, straw, coconut fronds, stone, and earth, either poured and pounded between frames (pisé) or in the form of unburned (adobe) or burned bricks. Desert peoples have learned to build houses with thick walls and hidden windows; some tropical islanders make houses without walls. Northern fisherfolk and hunters in Siberia construct houses underground to conserve heat. The inland Chuckchi of Bering Straits, who breed reindeer, sleep in fur boxes suspended inside skin tents.

Some houses are single-family shelters, and others are communal dwellings in which each family has a cubicle or section. In primitive communities houses are built according to a single traditional plan. Usually a person in authority takes the first step, such as killing a cock and sprinkling its blood on the roof poles before building the roof, a practice in rural Morocco. Wherever permanent structures are to be erected, as many people as are available join in and help build each other's houses. This requires a certain amount of organization and is important in maintaining the mutual relationships of the persons concerned.

BODY COVERING.—In most climates, people need some kind of portable protection, although they can become physiologically adapted to the extremes of their climate.

However, *body covering* serves another purpose besides protection: it is an important means of communication. By painting designs on their faces or wearing a special kind of clothing, people inform each other about their status, role, or condition. Even naked people use some kind of body paint, which may well be the oldest kind of body covering. Earth colors mixed with water, fat, or marrow offer a certain amount of incidental protection from wind and sun. In the Amazonian forests, where such mineral pigments are lacking and no such protection is needed, the Indians paint themselves with vegetable juices instead. Lacking mirrors, people who use body paint, paint each other, thus making this exercise an extension of grooming, one of the oldest forms of interpersonal activity among primates.

The habit of going naked is common among people who live in hot places, where extreme mobility is needed and nothing unnecessary to survival can be carried. It is also prevalent in chilly and wet climates, where clothing cannot be kept dry.

The simplest garments are the breech clout and the robe. A breech clout protects and conceals the genitals, and the robe keeps the body warm, although it impedes the use of the arms. These two are universal where clothes are worn at all and where tailored garments have not been introduced.

METICULOUS GROOMING and concern for personal appearance occurs in most cultures.

Going barefoot is normal in most climates, at least in summer. In many regions, sandals are worn on hot and stony ground, and buskins, moccasins, and boots are worn in cold and wet climates. Many peoples go bareheaded. A simple and widespread kind of headgear is the fillet, a cord or band wrapped around the head, leaving the top bare. This may be combined with a dustcloth, as among desert Arabs. Visored caps, to protect the eyes from glare, are worn in Tibet, among the Aleuts of Alaska, and by Europeans. The Eskimo wear slitted wooden snow glasses as protection against snow blindness.

BODY MUTILATION.—Many cultures habitually mutilate their bodies for symbolic purposes. Simple mutilations include tying bands around arms, legs, and waists so tightly that the trunk or limbs are deformed on either side, a practice found among the most primitive peoples. Those with black skins may decorate themselves with raised scars (Australians, Africans), and those with lighter skins with *tattooing* (Polynesians, Berbers, Ainu, modern Europeans, Japanese, and Malaysians).

Circumcision, a symbol of reaching manhood, is widespread in the Old World, but among Muslims and Jews it has been moved from puberty to infancy. Many kinds of head and facial *deformation* are practiced. Skulls are intentionally deformed with bands and boards in Africa and among many American Indian tribes. Among the Ubangi in Africa and one Indian tribe in Brazil, lips are pierced, stretched, and held out with discs like duck bills. Many peoples such as the Eskimo, wear labrets through holes in the lower lip, and some Indians of Brazil use cheek holes as receptacles for feathers. The piercing and stretching of nasal septa, nostrils, and earlobes is common. In Upper Burma, some tribesmen stretch the necks of women with brass rings.

In sum, people put themselves to

SMITHSONIAN INSTITUTION

AMERICAN INDIANS OF THE PLAINS lived a nomadic life, as shown in this rare 1892 photograph of Kiowa Indians. The tepee was the typical shelter used by these tribes.

SAM MINSKOFF AND SONS

PAN AMERICAN AIRWAYS

MAN'S SHELTER can be a modern apartment building (*above left*), a regal castle (*above right*), or a nomad tent of cloth and hides (*below*).

UNITED NATIONS

every conceivable pain and discomfort, even mutilation, in order to transform their bodies into permanent symbols. Profiles of human hands with chopped-off fingers, found in Upper Paleolithic caves in France, show that this has been going on for a long time; and early Mesolithic skulls found in North Africa reveal the removal of incisor teeth in adolescence. Tooth-knocking is still practiced in Australia and Africa, and finger-chopping, until recently among the Plains Indians.

CONTAINERS.—One of the conspicuous differences between man and other primates is that human beings habitually carry water and food, while the other primates (with few exceptions) drink and eat on the spot. Man alone uses containers. The Tasmanians carry water in seaweed bubbles; the Bushmen of South Africa use ostrich-egg shells. Human ingenuity is well demonstrated in the use of local materials: paper bark for baskets in Australia, birch bark in North America and northern Asia, bamboo nodes in Malaysia. The use of fibers, in the form of string bags, plaited baskets, and coiled baskets, is nearly universal, as is the craft of *pottery*. The invention of the potter's wheel, a time-saver in shaping pottery, in the Near East may have led to the invention of the wheel as a means of locomotion.

FOOD PROCESSING.—Like most other primates, man eats a mixed diet of fruits, vegetables, and animal proteins. No other primate, however, eats as much meat as do human beings. The shift to a meat diet hundreds of thousands of years ago must have exposed our remote ancestors to many animal parasites and caused a rigorous selection of men with the greatest immunity. The invention of *cooking* reduced the hazards of infection. The simplest and probably the oldest method of cooking is simply to drop whole animals and roots on a fire and roast them. This is the usual system among such simple hunters and gatherers as the Australian aborigines, the Bushmen, and the Fuegian Indians. Refinements are grilling chunks of meat on spits, common among pastoral nomads, or grilling fish over fires of green sticks as do South American Indians. Some Australians encase a bird in clay and throw it on the fire; when they break off the clay, the feathers and skin come with it. Polynesians bake with hot stones in pits; Indians of the northern forests of North America boiled food in birch-bark kettles held off the flame. Malaysians boil food in green bamboo nodes; Eskimo, in soapstone kettles over blubber lamps; and Plains Indians, like the ancient Scythians, boil meat by dropping hot stones into a skin lining a hole in the ground.

Food is processed both for immediate consumption and for future use. Peoples who live in regions where there is little seasonal change store little food; where seasonal changes are marked, *food storage* is necessary for survival. Cereals can be stored without special attention. Many peoples dry meat, fish, and fruits, particularly in arid climates where they will not spoil. Smoking meat and fish, as is done with salmon in the Pacific Northwest, is more suitable in cool, wet climates. Where it is cold enough, as in the circumpolar region, whole carcasses can be frozen for the winter.

HUNTING FOR FOOD is one of man's earliest technologies. Shown in this early print are American Indians draped in deer hide, trying to decoy deer that are close enough to kill.

A common way to treat food materials, whether meat, fish, vegetables, or fruit, is to pulverize them in a mortar and pestle, on a hand grindstone (*metate*), or on a rotary *quern,* a primitive hand mill invented in Roman times.

Fermentation is another form of food preservation, for grapes will keep as well in the form of wine as in that of raisins; it is also a means of producing alcohol for ceremonial intoxication. This process, however, is virtually limited to agricultural and pastoral peoples, the only ones who have anything in bulk to ferment. *Distillation,* a relatively modern refinement of fermentation, is limited to the high cultures of the Old World and to those of Mexico.

COLLECTING.—The simple collection of natural materials is common to animal life as a whole. Natural materials include roots, fruits, edible leaves, stalks, tubers, shellfish (which can be gathered at low tide), insects in various forms, honey, eggs, and slow game (small, slow-moving mammals, snakes, lizards, and fledglings, which a woman or child can catch by hand or with a stick). Collecting as so described is as old as man—and older, for it was apparently done by the Australopithecines.

HUNTING.—Hunting requires great endurance, keen observation, and a knowledge of animal behavior. Some hunters wear disguises that imitate the animal they are stalking; others lure animals to ambushes by imitating their mating calls.

Agricultural people, for whom hunting is only part-time work, are very apt to employ *traps* in order to capture animals. These traps generally depend upon the ingenuity of the hunter in creating delicate trigger-release mechanisms. Some people catch birds by smearing sticky gums on favorite perches; another curiosity is the Eskimo's wolf trap—a piece of sharp-edged whalebone coiled and frozen in a piece of fat. The heat of the wolf's stomach melts the fat, and the whalebone uncoils.

Solo hunting is rare. Most successful hunters operate in small groups, in which coordination and direction are essential. A group of men can drive a herd of animals into an opening between two converging fences, leading to a butchering corral, and they can carry home and share the spoils of a successful hunt. There are elaborate rules governing both men's preparation for the hunt and the behavior of women while their hunter husbands are away. A common belief is that a wife's infidelity will spoil the men's luck.

FISHING.—Primitive fishermen sometimes catch fish by hand, but more often by lines with or without hooks; a simple device is the fish gorge—a double-pointed spindle of bone or wood. In still waters, fishermen sometimes shoot fish with bows and arrows, and many use fish spears with barbed or pronged heads. More sophisticated fishermen, as in Melanesia and the Pacific Northwest, use elaborate traps and nets.

Such fishing gear is made by hand, and most fishing also requires only hand work, unless boats are used to reach the fishing grounds. In Fiji many people get together to operate nets, but the pattern of interaction needed for fishing is simple except when extensive voyages are made in large boats.

HUSBANDRY.—*Husbandry* began when people first grew plants and bred animals in captivity. By so doing they sheltered the organisms, protecting them from predators and from the rigors of the environment. They thereby preserved random mutations, which would otherwise have been quickly pruned off by natural selection. Some of these mutations, being useful to man, were selected deliberately, and the genetic structures of individual plant and animal species were changed. Grain that remains on the stalk after ripening, instead of scattering on the ground, is a mutation that provided primitive farmers with grain to reap and store.

The most useful and widely grown plants are *cereals,* which provide mainly starches; the *legumes,* including peas, beans, and lentils, grown principally for proteins; and the *oil seeds,* such as sesame, flax, and sunflower seeds. Seeds can be sown broadcast, harvested by reaping, threshed with flails or oxen, and easily stored—they can be handled by mass production methods. Only maize and wet rice require individual handling, but in compensation they have high yields. Also, the stalks and stems of seed plants can be used for fodder, bedding for animals, building materials, and fuel. A Neolithic family with a mixed crop of annuals, legumes, and oil seeds could be fed on a fraction of the land needed by seed collectors.

Primitive agriculture, however, is not limited to seed plants. Roots and tubers, such as yams, sweet potatoes, and manioc, grow best in the wet tropics, while white potatoes can be grown in moist, temperate climates. Temperate-land farmers get much of their sugar from fruits, such as apples, pears, figs, and peaches, while tropical gardeners get sugar from citrus fruits, persimmons, and papayas. Trees, particularly the coconut and the date palm, provide building and clothing materials; the coconut also yields drinking water. Without coconuts and other domestic plants, the Polynesians and Micronesians could not have settled their islands.

Compared to the number of plant species, very few species of animals have been domesticated, because the number of useful species is limited and because the requirements of catching, taming, feeding, and protecting animals is greater. The total is only 13 useful kinds of mammals, 11 of birds, 2 of insects, and 1 type of fish; and most of these were domesticated in the Old World. In the New World, only in the Andes did man tame economically valuable animals, the llama and the alpaca.

It is often stated that the dog was the first animal to be domesticated, but ancient dog bones are hard to tell from wolf bones. In the Neolithic, dogs became useful not only for hunting but also for guarding flocks. Some people ate them, and others sheared them for blanket material. Still others, including the Eskimo and Maritime Chuckchi, depend on dogs for winter transportation.

Domestic animals provide people with many substances—meat, skin, horn, bone, ivory, wool, hair, blood, guts, dung—but that is not their chief importance. They also provide a source of energy, as when oxen pull a plough across a field, camels draw water from deep wells, or donkeys turn grain mills.

WALKING, the simplest form of land transportation, is predominant today in India.

■TRANSPORTATION.—The simplest form of land transport is *walking.* However, several ancient inventions facilitate walking under special conditions; the simplest are sandals and boots. Snowshoes and skis make traveling possible in deep snow; western Europeans and Chinese skate on ice, and Eskimos walk on it with ivory creepers. In Polynesia, southern France, and the Canary Islands, some people walk on stilts.

In different regions special ways of *carrying* things are characteristic. African Negroes prefer to carry loads on their heads; American Indians, on their backs. Many special carrying devices have been invented, such as the tumpline over the forehead (American Indies, Formosa) and the shoulder pole or shoulder yoke with loads hanging from each end (China, Japan, Europe, colonial United States).

The use of animals for transport involved *packing, riding,* and *traction.* Nearly every domestic animal is packed, and many ingenious packing devices have been invented. Those animals large and strong enough to hold a man have all been ridden, but the most widely used riding animals are the donkey, camel, and horse. Without the camel the great deserts could not have been efficiently crossed, and without the riding horse the great empires of the Old World could not have arisen.

Long before the camel and horse were ridden, people used animals for traction. The simplest such device is the *travois,* an A-frame attached to an animal's back with the two poles dragging on the ground and the load fastened across the poles behind the animal's rump. The Plains Indians first used the travois with dogs, later with horses.

In the Old World, the first known traction devices were land sleds, used long before the invention of the *wheel.* The use of wheeled vehicles was long limited to the Old World, and reached Oceania and Africa south of the Sahara only in modern times.

■WATER TRANSPORT. — Simple water craft are used by some of the world's most primitive peoples. The Tasmanians make rafts out of bundles of reeds that they tie together; the Andean Indians sail similar craft on Lake Titicaca. Some of the Australian aborigines make bark canoes with very crude tools; without such boats and simple rafts, the ancestors of these aborigines could not have reached Australia from Indonesia. Birch bark is the familiar canoe material of the Indians of northern North America, while the nomads of southern Iran make rafts of inflated skins and swim across swift streams on single skins.

People with Neolithic or metal tools hollow out large tree trunks, usually by alternately burning, quenching, and cutting the wood. Some very large hulls, carrying up to fifty people, have been made in this way. The next step is to raise the sides with plank gunwales (Polynesia), and the final one is to build the whole hull of planks on a single frame (China, India, Arabia, Europe).

The simplest way to propel a craft is to pole it in shallow water or to paddle it in deep water. When fifty men are paddling in stroke, a boat can move very fast. Rowing and sculling are European, West Asian, and Chinese techniques diffused to other peoples in recent times. Short of machine power, the only other ways of propelling boats are towing in canals and rivers, and sailing. Sails were first known in ancient Egypt and Mesopotamia.

Water transport is cheaper and more efficient than land transport. The ancient Egyptians sailed up and floated down the Nile; the Chinese long ago perfected a system of inland waterways; the Aztecs built their city on a large lake. Thus, the great civilizations of the Old World, and one of the greatest in the New World, arose in regions of efficient water transport. The ethnographic present, the time of autonomous native cultures, approached its end when European navigators crossed the great oceans and circumnavigated the world.

Techniques of transport on land and sea have influenced the growth of complex cultures in yet another way. A caravan crossing a desert needs careful organization and planning, a firm chain of command, and strict discipline, not only because of the rigors of travel but also because of hostile natives. The same is true of a ship at sea. Pastoral nomads, too, need organization and discipline in their annual migrations and are therefore more powerful than the sedentary peoples whom they sometimes conquer. Thus, the disciplines learned through techniques of transport may be transferred to those of government and warfare.

COMMUNICATION.—Human beings communicate by means of mutually understood symbols. While gestures, body paint, clothing, religious objects, ritual procedures, and the like are all vehicles of communication, the prime medium of communication is language, and the study of language is *linguistics.*

LANGUAGE.—Human *speech* is produced by the coordination of several organs that originally evolved in response to different needs. Neurological coordination of these organs for speech required the further evolution of the brain to a size and degree of complexity found among all normal human beings and probably all fossil men thus far discovered. Because the number of sounds capable of being produced and distinguished by the human ear far exceeds the number needed for a single language, no two languages have exactly the same sounds. And because languages are learned and people live in more or less isolated groups, different peoples have devoloped their own systems of speech.

Thus, the number of languages spoken in a given area is a function of the degree of isolation of the peoples concerned and the length of time they have lived together as cultural units. In aboriginal California and in Australia, where local tribes were small, self-sufficient, and relatively isolated, dozens of languages were spoken. The same is true of refuge areas, such as the Caucasus and the Himalayas. In contrast, English, Arabic, Spanish, and Chinese are spoken by millions, over vast regions.

Some linguists arrange languages in families and compare the basic vocabularies—words for family relationships and such universals as water and fire—to find out how long the languages have been separated. A comparative study of whole vocabularies and of types of language also reveals borrowings and, hence, early cultural contacts. This is important in dealing with cultures without written history. The ways in which different languages classify concepts—by number, gender, size, shape, texture, chronological age, social classes—also reveal some of the psychological differences between cultures in their outlook on human relations and the outer environment.

EXTENSIONS OF LANGUAGE.—Although many peoples habitually speak more than one language, they also come in contact with others with whom they cannot converse and therefore learn *sign languages,* as among the central Australians and the Plains Indians. Other extensions of language permit communication beyond the range of the spoken human voice. Examples are smoke signals, the West African drum languages, and the whistling language of the Canary Islands.

But the prime extension of language is *writing,* which not only permits sending messages but also provides an extension of the limits of memory, the accumulation of knowledge, and the accurate transmission of knowledge from generation to generation. Thus, the use of writing facilitates the growth and spatial expansion of cultures.

Mnemonic devices short of writing are used by many illiterate peoples. Australian aborigines send message sticks from camp to camp, inviting each other to meetings and ceremonies, with the meaning conveyed by simple carved designs and notches that indicate the number of intervening days. The Plains Indians painted pictures on skins to record events. In true writing, however, the symbols represent individual words or word components.

HELEN BUTTFIELD

WATER TRANSPORTATION has been one of man's oldest means of trade and travel. These boats along Egypt's Nile River look today just as they did during the time of Christ.

INDIA TOURIST OFFICE

SHEPHERD boy of 14 from Saurashtra, India.

Organization of Society.—Since the acquisition and the preparation of food is the basic occupation of primitive societies, the division of labor is one of the most elementary forms of social organization.

DIVISION OF LABOR.—The most fundamental *division of labor* is that based on *age.* In no society is a child expected to become fully self-reliant until after *puberty.* In any camp or village, children sort themselves out by age into natural grades and teach each other play activities, including games. As they reach puberty, they participate in the work of the society, performing different actions at successive periods; for instance, among the Galla of Ethiopia the men are first cattleherds, then warriors, and finally settle down as married men.

Because men specialize in hunting and the more exacting kinds of manufacturing, most social systems also have a clear division of labor based on *sex.* Among the simplest food gatherers, whose men neither hunt nor make anything very complicated, men and women do more or less the same things. Among hunting peoples, however, women collect wild vegetables and slow game, and fetch firewood and water. They are also usually responsible for tending the fire, while the men usually build shelters and handle meat. In agricultural societies the men fell and burn trees, and the women plant the crops; the men may go hunting or raiding while the crops are growing, and then come home to help the women with the harvest. Where the plow is used, however, planting becomes the province of the male.

In manufacturing, work relegated to women on a simple level becomes men's work when complex techniques require the skill of specialists. Thus, women make pottery by hand but men manufacture it on the potter's wheel. Women weave cloth on simple, usually vertical, looms and men, on complex horizontal looms.

In some countries, notably India, the contact between peoples of different racial and cultural origins and levels of cultural complexity has given rise to still another kind of divi-

AMERICAN MUSEUM OF NATURAL HISTORY

TRIBAL GAMES are an ancient device used to teach the methods of war. Often, neighboring tribes will compete against each other until the first man is killed. Both sides will then stop fighting, hold his funeral, and then continue the games until both sides have had enough.

sion of labor, the *ethnic*. Classes of people perform special tasks, which are inherited. This is the well-known *caste system*, also prevalent in many parts of Africa. It is characteristic in regions of interracial clines and in domestic and institutional slavery.

Many peoples keep household *slaves*, who are either war captives, as among Pacific Northwest Indians and Homeric Greeks; poor people who have sold themselves for debt, as described in the Old Testament; or children sold by poor parents, as in China. While *domestic slavery* is common to many relatively simple cultures, *institutional slavery* is confined to a few complex ones. The Romans kept large numbers of slaves at work on plantations, in grain mills, in mines, and in galleys; African slaves once worked on farms and plantations in Arabia, Brazil, and the United States.

INSTITUTIONS.—An *institution* (in the sociological sense) is a group of persons who perform common activities regularly enough to develop a pattern of behavior and have their own rules and *esprit de corps*.

Anthropologists recognize several kinds of institutions based on the activities involved; and Coon has suggested that most, if not all, institutions can be traced to instincts held in common, not only with other primates but also with many other kinds of animals. The family is, in this sense, concerned with reproduction and child care; the political institution, with territoriality; the religious, with fear, deprivation, and ornament; the economic, with the food quest and possibly sharing and storing; and voluntary associations, with the pecking order, a basic status system.

KINSHIP AND THE FAMILY.—In its simplest form the *family* consists of father, mother, and their children. But other relationships are recognized in all human societies, and each involves special *roles*, or patterns of behavior. Every language contains terms to indicate these relationships, vertically in age and laterally in *sibship* (a *sib* is brothers and sisters). The number of kin recognized by special names varies with societies. In the simplest cultures everyone is recognized by a *kinship* term; and if the actual relationship cannot be traced, as with a stranger, an appropriate one is given.

Every person need not be given a separate kinship term, because whole categories of kin are supposed to behave toward him in the same way—and vice versa—and they will be lumped under a single term. This operates through the principle of *the equivalence of brothers*. An individual's father's brother may also be called father, and his paternal grandfather's brother is also grandfather. Thus, he may expect the same kind of treatment from his paternal uncle that he received from his father, particularly if his father has died and, as is usual in some cultures, his paternal uncle has married his mother (*levirate*). Similarly, his mother's sister may be called mother, and his father may marry his deceased wife's sister (*sororate*) or be married to both of them at once.

Some cultures permit only monogamous marriages, but by far the greatest number permit *polygyny* (plural wives), which is usual among other primates and nearly all the hunters and food-gatherers. *Polyandry*, the marriage of one woman to two or more husbands, is usually confined to impoverished people.

Kinship terminology serves the function of specifying who should avoid whom, who may joke with whom, who must feed whom under certain circumstances, and particularly who may or may not marry whom. When these rules are studied carefully in terms of observed behavior, it is seen that they automatically perform two functions. Except in rare cases involving royalty, they forbid genetically deleterious *inbreeding*, as in brother-sister, father-daughter, and mother-son matings, and often specify the greatest possible outbreeding in small gene pools. But these *eugenic* aspects are no more deliberate than is the other function, that of so regulating the behavior of the community so that disturbances are kept at a minimum. Both have evolved through subconscious trial and error.

The family institution often reaches beyond the bounds of the local community to extend the range of the gene pool. In many cases it is the rule that a man is forbidden to marry within his own group, which may be designated as a *clan* or simply as a *marriage class*. This system is called *exogamy*. If, as sometimes happens in more elaborate cultures, social classes, and religious communities, he is obliged to marry within his group, this is called *endogamy*. Exogamy tends to foster amicable intergroup relations; endogamy, to strengthen the local unit and to accentuate its genetic peculiarities. Both systems can be useful.

POLITICAL INSTITUTIONS.—The *political institution* differs from others in that it posseses the ultimate authority and is expected to use force in both internal and external crises. In its simplest form the *state* is no more than a band of a few related families living, hunting, and sharing food together. Its leadership is informal. The best hunter may divide the meat or feed the greatest number of people; the wisest and ablest man may settle disputes. Many such communities are really extended families, and the leader is simply the head of the family; further extensions of the family are clans and tribes.

In simple agricultural communities some form of *chieftainship* is usual in order to keep the peace in the village and to defend it against enemies. Such a chief may also act as judge. In some societies chieftainship is inherited, in others, normally acquired through competition.

In one Micronesian atoll early navigators found a small community ruled by a king. Although a monarch seemed unnecessary on so small an island, when a typhoon roared across the lagoon, shaking down the coconuts—the islanders' only source of drinking water—the king took charge of rationing, and order prevailed.

Pastoral nomads also require a firm political institution to regulate the times of migration, the routes followed, the position of each family and its flocks on the march, the assignment of pasture, and the use of water. So well-disciplined are nomads that they often defy established governments and even conquer cities and take over the reins of government. Many dynasties have been founded in this fashion, disappearing when a new group of nomads replaced them.

Whenever a sovereign state becomes so big that all its members do not know each other, formality creeps in; kings wear crowns or other insignia and follow a stereotyped pattern. Heralds run before them to clear the road, secretaries note conversations and prompt their master, jesters amuse them to break tension. The king may become so sacred that not even his shadow may be touched, as in Hawaii, or that he is not seen, as among Nilotic tribes.

■GAMES AND WARFARE. — Anthropologists sometimes find a village or tribe divided into two or more rival units that compete with each other in formal games, such as football, wrestling matches, or track events. These games are likely to be held at some relatively slack time of year, thus keeping idle men out of trouble.

Some games are rougher than others, and sometimes they are lethal. From games to formal warfare is a small but critical step, because primitive people usually fight according to gamelike rules. In New Caledonia, for example, rival armies used to meet at a prearranged place and time. The two forces would line up facing each other, hold an oratorical contest, and finally fight with clubs. After the first man was killed, they stopped and held his funeral, then continued to fight until they agreed both sides had had enough.

■PRIMITIVE LAW.—Every culture has its own rules for behavior in all aspects of human relations and in every kind of institution, but only those rules that are enforced by the political institution constitute *law*. The function of law is to prevent disturbances to the group. In modern nations, laws are made and repealed by governments; but in simple, mostly preliterate societies, laws are based on precedent and custom.

The usual explanation of a law, often related in the form of a myth, is that it was handed down by the tribal ancestors. During puberty ceremonies, when youthful minds are impressionable, the older men teach the boys rules of conduct toward women, children, older people, and each other. In more complex societies, where the body of law is great, men with particularly good memories memorize them, and recite them when needed.

In small, simply organized communities, everyone is under close observation by everyone else, and few breaches of custom pass unnoticed. In larger communities, however, the culprit may go undetected; and the chief or king may call in sorcerers and diviners to discover his identity. If the accused denies his guilt, trial by ordeal may follow. In West Africa, for example, he may be forced to drink a potion made from poisonous tree sap; if he is able to vomit it and thus to survive, he is considered innocent. Whether guilty or not, the victim is usually someone who disturbs the group. Among primitive peoples there are usually no facilities for imprisonment; punishment is death, exile, or enslavement. In the simplest societies, as in central Australia, the elders secretly meet and kill an offender under cover of darkness, disguising their tracks with emu-feather shoes.

■PRIMITIVE ECONOMICS AND TRADE.—In its most primitive form, *trade* is simply the practice of exchanging identical or similar goods between neighboring bands, in order to maintain friendly relations. Body paint and tool materials, owing to their uneven distribution, are particularly subject to trading. Obsidian was widely traded in both western Asia and Mexico; and Aztec traders, traveling through hostile territory in disguise, carried obsidian blades among many Central American tribes. Polished axe materials were traded even more extensively in the Pacific.

■PRIMITIVE RELIGION. — *Religion* is a broad category of phenomena involving symbols, beliefs, practices, ritual specialists, and organizations of various degrees of complexity. A principal function of religion is to restore equilibrium to individuals and groups after a disturbance. The only animal who knows death to be inevitable, man lives under constant if usually unconscious tension that increases at critical times of the individual's life: at puberty, marriage, childbearing, illness, and death. However, the disturbance affects the family and associates of the individual as much as, or even more than, it affects him. Among all peoples living in normal cultural situations, rituals are performed on at least some of these occasions, and they are called *rites of passage*. Changes in the work schedules of groups of people caused by the progression of the seasons occasion ceremonies of another kind, called *rites of intensification*, such as Thanksgiving and Hopi Snake Dance.

In most primitive societies there are religious specialists called *shamans, medicine men*, or *witchdoctors*. One of their functions to heal the sick; they sometimes succeed by use of herbs and other medicines, by massage, bloodletting, and the like, but their principal role is relieving the psychosomatic element, if any, of the patient's ailment.

A shaman may, for example, dance himself into a trance and talk in strange voices, which indicate that his spirit travels in the sky, and consult another spirit, who reveals the cause of the illness. The shaman's spirit now returns; the trance is over, and the shaman approaches the patient dramatically. He massages the affected part, sucks it, and spits out some strange object that, he states, was inside the patient's body, causing the disease. The audience, at least, feels better.

Nearly all peoples perform magic of some kind. Much more is done to protect from harm than to cause it; more is suspected than is performed.

As a culture becomes more complex, shamans begin to specialize; some are healers, while others try to influence the weather. The Maya priests, for example, kept a precise calendar, could predict the movement of the celestial bodies, including eclipses, and told the people the right time—before the rains were to begin—to plant crops.

■PRIMITIVE ART.—The creation and enjoyment of *art* is a universal human attribute that may be even older than our species. In its simplest forms it requires no artifacts, for even chimpanzees can make rhythmic sounds and can dance. Art's most primitive function is to communicate emotion, more basic than the communication of ideas and facts. Art is an element in all ritual—political, familial, and economic as well as religious. Each group of people is conditioned to respond agreeably to its own symbols and with distaste to the symbols of others; we revere our flag, but Muslims detest the sound of Christian church bells. Thus, art serves both to unite and to divide peoples at all cultural levels.

Art exists in space in the form of surface decoration and sculpture; in time as dance, music, and literature; and in space and time combined in the dance and drama. All of the most primitive peoples decorate their bodies and usually decorate their implements. Pottery is a natural art medium, as are textiles.

Since some individuals are more gifted in artistic expression than others, specialization in these media arises in relatively simple cultural levels, as for example, among the Australian aborigines of Arnhemland, where the Didjeridoo Man, who blows a hollow-log trumpet, goes from band to band to perform and is fed by his audiences. In Polynesia there were whole companies of itinerant actors, and in North Africa troupes of religious students wander about as entertainers. Besides priests, ministers, and jesters, royal courts in many countries have their own staffs of musicians, just as Scottish chiefs used to walk preceded by pipers.

—Carleton S. Coon

BIBLIOGRAPHY

CHILDE, V. GORDON. *Man Makes Himself*. Penguin Books, 1964.

COON, CARLETON S. *A Reader in General Anthropology*. Holt, Rinehart & Winston, 1948.

COON, CARLETON S. *The Living Races of Man*. Alfred A. Knopf, 1965.

COON, CARLETON S. *The Story of Man*. 2nd edition, revised, Alfred A. Knopf, 1962.

KROEBER, ALFRED L. *Anthropology*. Revised edition. Harcourt, Brace & World, 1948.

OAKLEY, K. P. *Frameworks for Dating Fossil Man*. Aldine Publishing Co., 1964.

PSYCHOLOGY AND PSYCHIATRY

Introduction

Psychology is the science which studies the mental life of human beings and animals. More technically, psychology is the science that deals with the study of *behavior,* that is, with the facts and events arising from the interaction of a living creature and its environment by means of sense organs, nervous systems (including the brain), and such effector organs as glands and muscles.

Older writers defined psychology simply as the study of the mind or of the psychical processes, but the more modern definition given above includes all the mental phenomena embraced in the older definition, as well as all the processes seen in the adaptive and nonadaptive behavioral responses of the individual throughout his entire life span.

Psychiatry is the medical specialty dealing with the diagnosis, treatment, and care of those who are mentally ill or defective, as well as with the prevention of mental illness. Sometimes psychiatry is considered to include the study of the dynamic basis of personality adjustments of all human beings, both normal and abnormal.

■**PSYCHOTHERAPY.**—This is a method of treatment of mental and psychosomatic disorders by the use of various nonchemical and nonphysical psychological techniques. The term is usually reserved for types of treatment given by professional workers, such as psychiatrists, psychologists, or psychiatric social workers.

■**PSYCHOANALYSIS.**—This is the study and treatment of processes in the abnormal and the normal mental life of human beings as developed by Sigmund Freud (1856–1939) and his associates. This term is at times used somewhat loosely to include all dynamic psychiatric points of view. In this sense it includes *analytical psychology* as developed by Carl G. Jung (1875–1961), *individual psychology* as developed by Alfred Adler (1870–1937), and many other modern approaches to psychiatry.

■**PARAPSYCHOLOGY.**—This name covers the study of reported psychical phenomena that appear to fall beyond the known limits of the sciences of physics, chemistry, biology, or experimental psychology. Its subject matter is not easy to define, but it includes the consideration of alleged supranormal and extrasensory mental processes. It studies telepathy and the activities of self-styled "mediums." *Psychic research* is the investigation of the phenomena that make up the field of parapsychology.

PSYCHOLOGY

History.—The word "mind" was once used to designate all conscious human processes, including mental faculties, powers, aptitudes, and dispositions, both acquired and innate. Some philosophers considered mind and body as two ultimately distinct aspects of life. They gave much time to speculation about the body-mind problem and how consciousness is related to the physical matter of the brain and body.

Today "mind" is regarded by scientific psychologists as a collective noun used to describe all the aspects of what is commonly called *mental life.* These mental phenomena include memory, perception, and thought as well as all the behavior, unlearned or learned, related to the individual's adaptation to his environment.

Mental life, the subject matter of psychology, has long interested and puzzled human beings. Some primitive men thought of breath as having a special significance in connection with mental processes, a concept described by anthropologists as belief in a thin, unsubstantial vapor, film, or shadow which was regarded as basic to the actions and thought of the individual. Breath, in this special sense, was sometimes considered the animating principle of the body. In dreams it might journey from the body and return, but at death it left the body forever. This concept is known as *primitive animism.* Today, this animistic or vitalistic view is not generally accepted, principally because scientists have never been able to determine how a nonmeasurable or not directly observable mental force of any kind can act upon any part of the living body, such as the cells of the brain, and thereby initiate observable changes in, or modify, behavior. This is further dealt with under parapsychology.

LOOK MAGAZINE

THE MIND, when withdrawn from the external world, has the power to create its own realm—which may become a prison.

The term "consciousness" is not used today in the writings of many scientific psychologists because the word has proved difficult to define. Those who still wish to use this term tend to think of it as a name for the sum of those private processes, such as sensations, feelings, thoughts, and impulses to action, that an individual himself knows and is able to report upon, but that are never directly observable by others.

Experimental Psychology.—Scientific experimental psychology began in the laboratory studies of a number of physicians and physiologists who explored the structure and function of the human nervous system. Sir Charles Bell (1774–1842) was the first to demonstrate the existence of two anatomically distinct types of nerves: those which transmit sensory impulses from the outside world to the brain and those which send motor impulses out from the brain to the muscles. His studies of retinal sensitivity led to the realization that the sense organs, or receptors, of the body are each specialized to be stimulated by one sort of energy rather than by others. This means that human beings are aware of, and respond to, the activities set up in their own nerves (the optic nerve, the auditory nerve, and so on) and not to the physical or chemical characteristics of the stimuli. For example, an individual who has congenital red-green blindness cannot discriminate these colors, even though the stimuli for red and for green are acting on his eyes.

■**WUNDT.**—Although many physiologists, physicists, doctors of medicine, and philosophers made early contributions to the science of psychology, the founder of experimental or scientific psychology is considered to be Wilhelm Wundt (1832–1920), who founded the first formal psychological laboratory in 1878 at the University of Leipzig in Germany.

■**JAMES.**—The great American psychologist William James (1842–1910) also started his scientific work as a physiologist, and as early as 1875, he organized at Harvard an informal laboratory to demonstrate phenomena in the field of psychology to his students. This was not, however, a formal scientific research laboratory of the sort that Wundt founded, one dedicated to the discovery of new facts. Following these early experimental and demonstrational laboratories, many similar centers were established in other universities. Today, scientific psychology is largely made up of the facts and theories developed in such laboratories.

Schools of Psychology.—The organized teachings of philosophers on the nature of the physical and mental worlds and of man's place in nature have long been called "systems" or "schools" of philosophy. Toward the end of the nineteenth century there was a tendency also to use the term "school" to describe different systems of organizing the observed facts and the general theories of psychology. Today, many scientific students of psychology tend to be much less concerned with schools than was formerly the case. Nevertheless, it is worthwhile to mention briefly some of these systems because they are so frequently referred to.

STRUCTURAL SCHOOL. — Psychologists who emphasize the study of conscious experience are called *introspective psychologists.* Those who attempt to analyze the content of immediately given conscious experience (or experience in the phenomenal field) into mental elements are said to belong to the *structural school.* Structure in this sense does not refer to the make-up of the cells of the brain or of the rest of the body, but rather to the assumed structure of conscious experience. The units resulting from the analysis are called by such names as sensations, images, and feelings. The leading American exponent of the structural school was Edward B. Titchener (1867–1927), an Englishman who had studied with Wundt.

NEW YORK PUBLIC LIBRARY

WILLIAM JAMES, an American psychologist.

ACT SCHOOL.—Certain psychologists soon came to feel that man's conscious life is not made up exclusively of mental building blocks, as described by the structuralists. They became concerned with the active processes of motivation and "set," or a readiness to perceive or to respond in a certain way, that steer and direct human experience and thought. These psychologists, represented by O. Külpe in Germany and R. M. Ogden in America, were classed as members of the *act school* of psychology.

FUNCTIONAL SCHOOL.— Some American psychologists rebelled against the idea that psychology was exclusively concerned with an analysis of conscious experiences. They were very much influenced in this view by the study of Darwinian evolutionary biology. They investigated the role of intelligence, emotion, learning, and so-called instinct in the adaptive and maladaptive behavior of animals and man. This point of view came to be called the *American functional school.* William James, whose great and stylistically elegant two-volume *Principles of Psychology* was published in 1890, did much to make this approach dominant in the United States, although as a "school" it may be better to associate it with the names of John Dewey (1859–1952) and J. R. Angell (1869–1949).

BEHAVIORIST SCHOOL.—Gradually laboratory studies of human reactions and the intensive scientific investigation of animal behavior led to the study of bodily responses for their own sake. Ultimately some students came to think of this objective study of behavior as the main, or indeed the only proper, subject matter of psychology. Holders of an extreme form of this belief, the *behaviorists,* came to hold that conscious phenomena cannot really be studied scientifically and are thus irrelevant in a scientific study of psychology. Some behaviorists are greatly interested in the neural and other bodily structures that make responses possible, while others are mainly concerned with a quantitative correlation between measured stimuli and measured responses, without reference either to intervening conscious processes or the nervous system.

GESTALT SCHOOL.— The German word for form, shape, or pattern is *Gestalt,* and this has been used to name a school of psychology. Gestalt psychology emphasizes the fact that behavior and experience must be studied as organized wholes and cannot be analyzed into independent units. Responses are not a sum of independent reflexes, and perceptions are not a sum of independent sensations or other elementary mental processes. Gestalt psychologists see relationships, patterns, and grouping as primary elements of perception. Wolfgang Köhler is one of the greatest exponents of this view. Another position that has developed, at least in part, out of Gestalt psychology is called *psychological field theory.* This is a way of looking at some parts of psychology on the basis of a description of a so-called *life space,* which includes the individual and the conceptualized forces of the environment in which he exists. This school, which owes much to K. Lewin, uses terms such as goals, barriers, and boundaries and is interested in the mathematical concepts of topology as applied to psychology.

General Scientific Psychology.—The name *general psychology* is often given to the consideration of all the psychological processes of normal adult humans without special reference to any limited school. Once, it is true, the study of general psychology was almost wholly concerned with an analysis of human conscious experience; but now this conscious experience is discussed under such headings as sensation, perception, feeling, emotion, and thought. *Perception,* for example, is recognized as primarily controlled by the present excitation of an individual's sense organs. Yet a full study of every perception shows that all such processes are also influenced by the inborn makeup of the organism and by previous behavior and learning. In any perceptual experience, such as seeing, hearing or touching, it is difficult to distinguish between the raw sensory data and the recognition of these data as an object or a person. Part of the experience is immediately and clearly given; but a large part is derived from previous experience and learning. A consideration of geometrical visual illusions may make this clear. Because of the way their eyes and brains work, all normal people "see what is not there" if measurement is taken as a test of reality in such illusions. Also, the influence of other people, past and present, is recognized as playing a special part in much perception. It has even been shown that children of poor families tend to judge coins as larger than do rich children similarly tested.

SENSATION.—In the older "structural" experimental psychology the basic unit of perceptual experience was considered to be the sensation. A *sensation* was defined as applying to an elementary and generally not further analyzable item or fixed unit of experience that could be shown to be related to present receptor activity. Specific colors, sounds, odors, tastes, temperature experiences, pains, and pressures are examples of types of sensations. Each sensory experience, such as that of a particular color, can be further described in terms of its conscious dimensions or attributes. Every seen color has a particular hue (red or green, for instance), a particular brightness (relation of the color to a point on the black-white continuum), and a particular saturation (purity of color, freedom from admixture with gray). Other visual attributes have been isolated, and auditory and other types of sensory experience described.

AFFECTION.—The word *affection* is used by some psychologists to distinguish emotions and their attendant feelings from the cognitive, perceptual or sensory experiences just considered. Feeling tones are typically pleasant or unpleasant. Emotional experiences are thought of as involving conscious experiences and ways of acting commonly called fear, an-

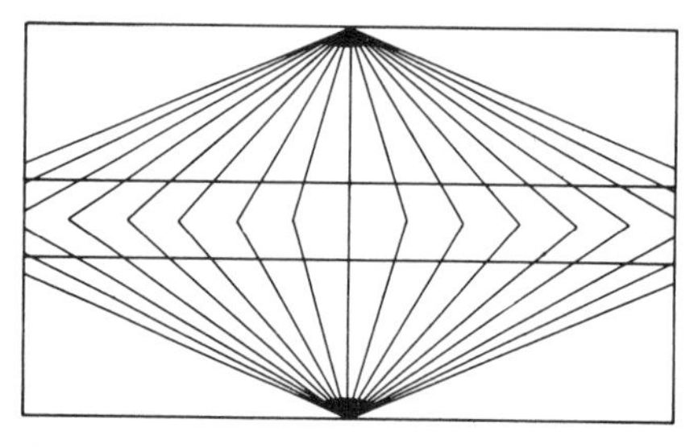

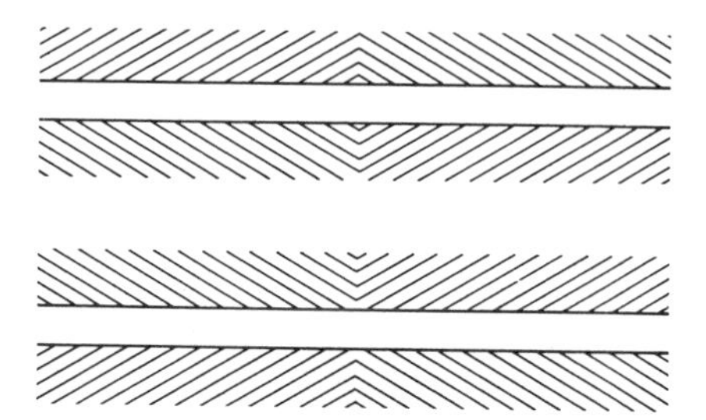

VISUAL ILLUSIONS show how our sensory organs may perceive things differently from the way they actually are. The horizontal lines are parallel though they do not appear so.

ger, joy, disgust, pity, and so forth. It is known that special parts of the brain, such as the hypothalamus, are active in emotions and that the complex autonomic nervous system is involved in bringing about the bodily activities characteristic of emotional behavior.

CONATION.—In the older psychology, *conation* was used to describe those aspects of mental life which are related to action and striving. In some writings on psychology the terms cognition, affection, and conation are spoken of as basic mental processes.

General psychology traditionally considers all the topics just mentioned and also gives special emphasis to a study of behavior changes resulting from growth, that is, maturation, or from learning. When one attempts to analyze learning, it becomes clear that only an active or motivated organism learns. Today, as a result of this understanding, much of general psychology is devoted to a consideration of the factors that explain the basis of human responses and patterns of action and the way in which organisms are modified by the processes of habit formation.

MOTIVATION.—The term *motivation* is frequently used to describe a continuing physiological condition or drive of an organism which impels it to activity, persisting until the organism reacts in such a way as temporarily or permanently to satiate or eliminate the drive. Physiological drives of this sort provide the most basic motivation. In an organism that has had an opportunity to acquire habits, drive-satiation behavior may be seen as having an end in view, or as *goal-seeking*. Sometimes an internal drive must reach a certain level before other stimuli, often in the external world, are able to start or direct behavior. For example, an organism must often be in a certain definite state of hunger before the presence of food in its environment will cause it to eat.

Among the basic drives are those connected with the need for food, water, sexual partners, oxygen, the elimination of waste materials, and the securing of an external temperature that is neither too warm nor too cold. In the adult human, however, motivation is by no means limited to such physiological drives. All those aspects of human mental life that are related to acquired desires, preferences, and purposes act as motives. Many of these motives take their form from the normal social experience of the group to which the individual belongs.

Learning.—Motives are basic in producing the behavior which can be modified by the processes called *learning,* a word used in psychology as a name for any change in an individual's experience or behavior that can be shown to be brought about by previous responses or activities that the individual organism has carried out, and that would not have developed by maturation alone. The physiological process of learning in the higher animals is generally considered to consist of some alteration in the brain structure. This change takes place in the cortex, or outer part of the cerebral hemispheres. When a motivated organism reacts, the brain is involved, and under certain conditions such brain activity produces more or less permanent changes in the brain's living cells. What these changes are is not at present well understood, but their effect is to modify the subsequent reaction of the organism. When the stimuli are repeated later the organism that has "learned" reacts as it would not have reacted if the brain changes had not taken place.

During the twentieth century much experimental study of learning has involved investigation of what are called *conditioned reflexes,* or, to use the more common term, *conditioned responses*. This approach to the study of learning developed out of the investigations of a Russian physiologist, Ivan P. Pavlov (1849–1936), in the physiology of digestion. He discovered that a dog that salivated when given food could later be made to salivate when another stimulus, such as a bell, was sounded. This took place, however, only when the bell had been sounded at the same time or just before the food was given to the dog. When the bell-food sequence was repeated a number of times, the bell alone could elicit the salivary response, even when food was not present. Food, in this example, which causes salivation without previous laboratory experience, is called the *unconditioned stimulus*. The bell that elicits the response which is acquired after the two stimuli have been presented to the organism together is called a *conditioned stimulus*. The response elicited by this stimulus is termed the *conditioned response*.

A second type of conditioning, called *instrumental* or *operant conditioning,* describes a modification of behavior that is different in certain respects from the classical conditioning described above. If a situation is established in which a hungry animal may respond in any way that is natural to it, and if one of these responses leads to the reward of securing food, it often turns out that the specific response which leads to the securing of food is repeated and thus learned. The behavior that has been instrumental in obtaining the reward is thus especially important to the organism. The second type of conditioning includes the acquisition of the new behavior patterns formed when the response in question allows the organism to escape from painful or noxious stimuli. Operant conditioning of all types has been studied in great detail by B. F. Skinner and his students and has been made the basis of a general theory that is seen to have much relevance to an understanding of normal and even abnormal human and animal behavior.

When some previously conditioned responses can no longer be elicited, they are said to be *extinguished*. The establishment and the extinction of responses can be determined by the way *reinforcers,* or rewards, are given. This pattern of reward-giving, both in time and frequency, is a *schedule of reinforcement*. For example, in the case of certain animal experiments, the presentation or withholding of a pellet of food after an animal has pressed a lever which sometimes releases the food will determine the rate of learning and, to a degree, the rate of extinction. One rate will follow if each lever press produces food; another, if every other press produces food; and still another if the reinforcement is irregular.

HARVARD UNIVERSITY

CONDITIONED RESPONSES have enabled these pigeons to play a game. If the ball falls into the slot, the defending pigeon receives a mild shock; the other pigeon receives food.

The idea of conditioning is a modern, objective behavioristic extension of what the older writers called the *association of ideas*. This phrase was used to describe a functional relationship that made possible the recall of words or ideas. In a typical case of association, a sight, sound, odor, or word may recall a specific memory. Laws of association were developed to explain differences in recall. For example, it was noted that items were associated if they were similar or contrasting, or if they were presented at the same time or in immediate succession. In general, associationistic psychologists concluded, as do students of conditioned responses, that if two presentations occur simultaneously or in close succession, recall is most apt to take place. If such presentations are related to intense stimulation, recall is also especially apt to occur.

MEMORY.—The word *memory* is employed not only in cases of association, but also to describe some types of conditioning. Classically, memory has been recognized as involving four logically separate stages: *impression* or *presentation,* sometimes called *memorization; retention,* probably involving the maintenance of an established brain trace; *recall* or *retrieval,* hypothetically the reactivation of the brain trace; and *recognition,* that is, placing and dating, or a knowledge that "this has been experienced before."

Much experimental study of learning and memory has been conducted by requiring subjects to memorize lists of words or of "nonsense" syllables such as zug and yarp. The factors that influence the subjects' ability to recall these associations can thus be measured or "quantified" and curves of recall can be drawn to show the effect of time and other factors on forgetting. For example, it is found that the normal effects of learning may be impaired if the first presentation is followed closely by another activity, especially if it is very similar to the first. This has been called *retroactive inhibition,* and is important in much active forgetting. Forgetting is known to be a complex process which depends on many factors besides the mere passage of time. In the older psychology, memories were thought to be composed of images, that is, impressions which remain after an external stimulus has been removed. Images were seen as fundamental also in thought, including reasoning and constructive imagination, and in dreams. When images are confused with perceptions the individual is said to have hallucinations. The pink elephants of alcoholic intoxication are examples.

The term *trial-and-error learning* describes a situation in which a human being or an animal that has not established a way of satisfying a drive responds with a series of varied actions until one response proves rewarding. This is *accidental success.* As the situation is repeated, the adaptive act appears more and more quickly until, when it is learned, it is the only response that appears. Certain Gestalt psychologists have emphasized the fact that some learning is not based on many repetitions of blind trial and error, but may involve only one immediate response that is correct the first time. This is said to result from insight into what must be done in order to meet the situation adaptively. *Insight* is thus learning guided by all relevant existing relations. Sometimes it seems as though the solution to the problem were already prepared, and the adding of one item leads to "closure." The study of thought and the psychological investigation of language, both spoken aloud and said to oneself (sometimes called *subvocal*), are closely related to the study of learning. The psychology of language is a specialized and important topic. Reasoning and thought are also fields now much studied by psychologists. Reasoning has been seen sometimes to be like trial-and-error learning and sometimes like insight learning. In reasoning, symbolic representations such as images, rather than overt acts, are manipulated.

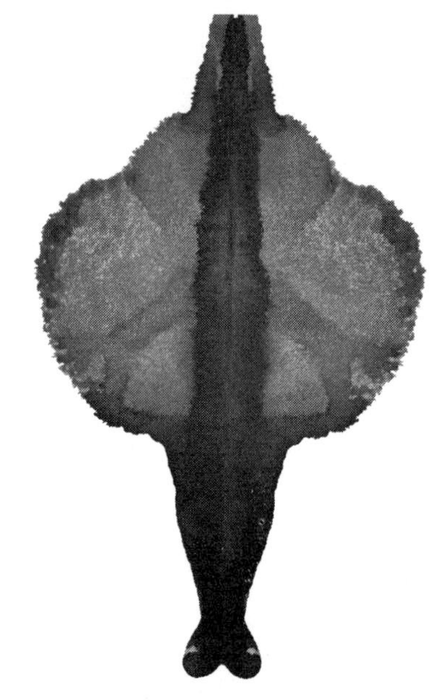

SYMMETRICAL INKBLOT is similar to the type used in the Rorschach personality test.

Much recent work has been devoted to the chemistry of learning. It is assumed that learning and memory will some day be explained in terms of a molecular "code" of some kind, just as the genetic code was found to be the basis of organic heredity.

Intelligence.—The term "intelligence" has been used in many different senses in technical psychology. As applied to man, *intelligence* is often used to describe differences in ability to solve problems or to deal with new life situations. The rate at which the individual develops the capacity to learn the solutions to new problems and his ability to deal effectively with linguistic, numerical, or other abstractions when solving problems are both spoken of as depending on intelligence or as constituting intelligence.

■ **INTELLIGENCE TESTS.**—Sets of problems have been devised that can be given to persons of different ages to determine the individual's capacity to solve them. When such tests are given to large numbers of individuals and the results are carefully tabulated, it is possible to establish norms against which an individual can be tested or judged, or, in a sense, measured. Errors resulting from different linguistic and social backgrounds and prior opportunities for learning must be guarded against in evaluating such test scores. It is known that an individual's scores on intelligence tests or specialized verbal or mathematical aptitude tests are a result of both his inborn capacity and his experiences, including his formal education. The most famous of all intelligence tests is the Binet or Binet-Simon scale, so named because of the psychologists who developed it and first used it to test French school children in 1905. In all its forms this test battery consists of a series of problems that are so arranged that they can be solved by average children of different chronological ages. By the use of this scale a mental age may be computed for the child being tested as compared with those on whom the test was standardized. The *I.Q.,* or *intelligence quotient,* is calculated by dividing the mental age secured on the test by the chronological age of the subject and multiplying by 100. If the chronological age and the mental age are the same, the I.Q. is 100. If the child's mental age is 12 and his chronological age is 10, he has an I.Q. of 120. If his mental age is 8 and his chronological age is 10, he has an I.Q. of 80.

Personality.—"Personality" is a term used in modern psychology in many different ways. Human personalities are, it is agreed, always a result of inborn characteristics and of the learning that has molded the individual during his formative years. The word "personality" derives its meaning from the Latin *persona,* meaning 'theatrical mask.' Thus, the term is often used to describe what may be spoken of as the "mask" that an individual wears for others and, indeed, for himself. Each personality, when studied, is seen to be an organization and, to a degree, a unification, of motives, dynamic tendencies, perceptual sensitivities, intellectual capacities, and aesthetic and other characteristics. Many people who have investigated the topic of personality have found that it is illuminating to study the specific types of differences that exist among individuals. Students of abnormal psychology have attempted to define the limits of what may be called socially normal personalities.

■ **PERSONALITY TESTS.**—Psychiatrists and clinical psychologists are interested in what are called *projective tests,* in which personality characteristics are revealed in relatively free, constructive, and imaginative situations. The Rorschach inkblot test is a well-known projective test. In this test, subjects are shown ten inkblots, one at a time; their replies to such questions as "What could that be?" are recorded. These replies are then scored in a standard and elaborate way, and the results provide clues to some of the basic personality mechanisms of the individual.

Physiological Psychology.—The name *physiological psychology* is given to the study of the correlation between anatomical and physiological processes of the body and the behavior and reported experience of individuals. Some authors prefer the term *psychophysiology* for this study. Physiological psychologists have learned much about mental life and have described in detail how the sense organs work. Much of this study, as carried on by psychologists today, involves the use of very complex electronic devices, both to control and measure stimulation and to record the electrical reactions of active nerve and other cells. In their studies of emotion they have described the activities of the smooth muscles (typical of the viscera) and the glands in relation to responses seen in different psychological states. The effect of inherited physique has also been examined by some physiological psychologists. Above all, such students have been interested in the brain in relation to conscious experience and behavior, particularly language and defects of speech such as aphasia, or specific difficulties in reading, such as alexia. Studies suggest that the human brain is unlike that of any other animal. Only the human brain is capable of developing real language and manipulating complex symbols, as in solving mathematical problems. Certain large areas of the cortex of the brain are equipotential (one part is as important as another part) in facilitating certain types of learning and behavior. However, some processes, such as those dependent on the brain mechanisms fundamental to vision, are precisely localized. The electrical waves accompanying brain activity can be detected by placing electrodes on the scalp of subjects and recording the patterns on an *electroencephalograph.* These tracings of brain waves, or *electroencephalograms,* have been found to have many clinical uses, such as in the identification of types of epilepsy.

The study of physiological psychology is closely related to *psychophysics,* which is the study of the relation between the measurable characteristics of the physical energies of stimuli and the quantitative attributes of sensory or other experiences or of behavior. *Weber's law* or, in its development, the *Weber-Fechner law* holds that in a general way and in certain ranges of the strength of stimulation, the intensity of a sensory experience or the magnitude of a response has a definite mathematical relation to the intensity of the stimulus.

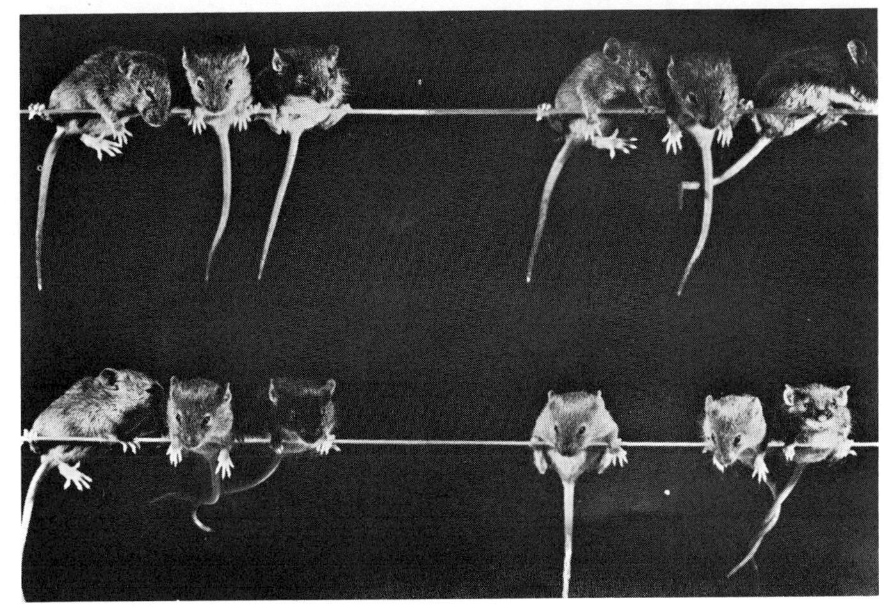

WORLD HEALTH ORGANIZATION

SOCIAL BEHAVIOR OF ANIMALS can be studied under a variety of controlled laboratory conditions. These two groups of mice are living in various radioactive environments.

Comparative and Animal Psychology.—The term *comparative psychology* is given to the branch of psychology which studies and compares the behavior and mental life of different animal species, including man. This is actually *phylogenetic psychology. Comparative psychology* is also used, but less frequently, in the comparison of typical patterns of mental life of different human races or of different development stages in the growth of the individual. The latter, however, is better called *child psychology, developmental psychology,* or *ontogenetic psychology.* The term *genetic psychology* is used by some writers as a synonym for both phylogenetic and ontogenetic psychology.

As we have seen, behavior is the main field of study of many present-day scientific psychologists. Behavior also is a subject studied by some modern zoologists who are interested in *taxonomy*—the classification of animal species—and in the natural history of living organisms. Students of behavior who are primarily trained in zoology tend to be especially interested in what has sometimes been called the *ecology of behavior;* they consider the all-inclusive environment that makes an animal's life what it is as a basis for understanding the natural responses of organisms. The term *ethology* is generally used for zoologically oriented study of the complete behavior of animals, especially in their natural habitats. Modern ethologists tend to avoid terms referring to consciousness in their scientific reports, but they do speak of instincts and of the characteristic responses of different types of organisms. Responses of this type are referred to frequently as categories of *species-specific* behavior. Ethologists refer to the quick learning or lasting change of behavior under specific conditions as *imprinting.*

■ **INSTINCT.**—The word "instinct" has had a checkered career in scientific psychological work since the 1860's. At first, after Darwin's concepts of evolution had become important in psychology, it was assumed that all animals, including man and the other mammals, displayed many specific and identifiable tendencies or dispositions to action which were inherited. Some of these were called *instincts.* The older psychologists often gave long lists of instincts, such as pugnacity and maternal behavior. But the difficulty of proving that such classes of response, particularly in human beings, are inborn, rather than learned, led to a strong opposition to the use of the term "instinct" in psychology. Today, it is once again recognized that a concept such as "instinct" may be useful in describing many forms of behavior in insects and vertebrates, even in man. In Freudian psychiatry "instinct" is used to characterize the basic tendencies of the individual, the *life* (or *love* or *sex*) *instinct* and the *death instinct.*

Other Fields of Psychology

Differential psychology is the name given to the branch of psychology in which tests to measure or define intelligence and personality are used, as well as tests that measure differences in sensory thresholds, motor skills, musical ability, and many other capacities. In general, such tests give the person being tested a rating so that he can be compared to others being tested. Personality tests, vocational aptitude tests, and various measures used by psychologists interested in selecting individuals to play a part in complex man-machine systems are all part of this branch.

Human engineering is the field concerned with man-machine organizations. The work of human engineers is of special importance in highly automated industries and in the production of vehicles in which man penetrates hostile environments, such as aircraft, space capsules, and submarines.

Child psychology is the branch of psychology which considers the psychological processes of the developing human being from the first responses of prenatal life until he has reached adult life. Psychologists also study the behavior characteristics of senescence which result, in part, from the brain and body changes due to tissue deterioration in old age.

Social psychology is the branch of psychology which considers the reactions of the individual that depend on his inborn character and the learning that has taken place during his growth in a world made up of other people. Social psychology emphasizes the importance of the interstimulation and response between the individual and the small, as well as the large, groups of people of which he is a part. It studies the reactions found in regimented authoritarian situations as contrasted with those that take place in congenial and permissive environments. Special social situations, such as the essentially pathological reactions brought out in individuals when they become part of a mob, have been studied.

Cognitive psychology is the branch of psychology that considers the processes by means of which organisms become aware of or obtain knowledge about their environments. *Cognitive dissonance* is a term used to describe an individual's awareness of a conflict between his attitudes and his behavior.

Abnormal psychology is the branch of psychology given to a scientific consideration of the mental life and the reactions of atypical or abnormal individuals. To some degree, practically every person may be considered as not fully normal in all respects, but the term "abnormal" is most often used to describe persons with personality characteristics which lead them into obvious social maladjustments and, in extreme cases, require their confinement in mental hospitals.

Educational psychology occupies itself with the study of the human individual in learning situations, especially in formal school life. It is concerned with the nature of the academic learning of such skills as those required in reading and arithmetic and with the use of various differential tests, which assist in assessing the educational capacity of each pupil. In education, scholastic aptitude is tested for by examinations very similar to those for measuring intelligence. Modern educational psychologists are interested in the social climate of the classroom and in the proper and constructive use of rewards, punishments, and discipline in connection with education. They are also concerned with the overall mental hygiene of pupils. Educational psychologists have given much attention to visual aids to education, such as television, and to programmed learning, made possible by teaching machines.

Clinical psychology is directly concerned with the study and treatment of individuals who have psychological problems of various kinds. The term is often used today to describe the procedures used to diagnose and correct maladjustments of behavior which interfere with the individual's educational or social life. An important modern technique of clinical psychology is *nondirective counseling,* as developed by C. R. Rogers and his students. By the use of this technique individuals are helped to "talk out" and solve their own, often deep-seated problems, enabling them to develop better-adjusted personalities and thus to live more effective lives.

Vocational psychology includes all those techniques by means of which the psychologist assists in selection, classification, training, and effective use of the aptitudes and skills of individuals in their vocations. Certain specific vocational tests, such as those for clerical aptitude, musical aptitude, and for the performance of various military tasks, such as those of aircraft pilots, have been used with marked success.

Parapsychology, also called *psychical research,* is the study of seemingly supranormal phenomena. It is not, as yet, an accepted part of a recognized field of science. It includes such phenomena as the *transmittance* of impressions from one person to another by channels outside the known means of sense perception; *psychometry,* the supposed capacity to receive supranormal knowledge of an object's history or that of the present owner while holding it; *precognition,* the ability to forecast specific occurrences without scientific explanations of such predictions; and *mediumship,* the reported ability of persons to mediate communications between the living and the deceased; and phenomena such as the movement of objects without visible or other known contacts.

NEW YORK PUBLIC LIBRARY

NINETEENTH-CENTURY MENTAL HOSPITALS were often no better than penal institutions. In this hospital in Paris the patients were free to roam the grounds under supervision.

PSYCHIATRY

Introduction.—*Psychiatry* is the branch of medicine dealing with the prevention, treatment, and cure of mental diseases and emotional disorders. Both psychiatry and the specific method of treatment developed by Freud, psychoanalysis, are deeply involved in the study of the nature and development of the mental and emotional life. Although studies in these fields are drawn from the examination of disturbed and mentally ill individuals, the findings have contributed greatly to man's understanding of himself and of his often mystifying behavior.

Psychiatry emerged as a branch of *neurology,* the study of the anatomy and physiology of the nervous system, including the brain. Most of the early pioneers were themselves neurologists. Medically, the *neurologist* is a physician who treats organic diseases of the nervous system. *Neurosurgeons* specialize in operations on the nervous system and brain, and often treat such illnesses as the convulsive disorders of the epilepsies which have various psychological symptoms. Since some mental illness is related to pathology of the brain, neurologists are sometimes thought of as physicians who specialize in the treatment of those aspects of mental illness that are directly related to brain disease.

■CLASSIFICATION.—*Mental illness* can roughly be classified according to two general categories: the organic and the functional. Basically, the *organic* illnesses are those which derive from a physical disorder, deficiency, or malfunctioning of the nervous system, especially the brain. This may be caused by a failure to develop normally (*amentia*), or by some later damage to or deterioration of the brain through disease, accident, or the process of aging (*senescence*). This category involves mental deficiencies and many of the severe psychoses. The *functional* diseases are those in which there is no apparent physical change in the brain or other organs of the body; these include some psychoses and most neuroses and so-called character disorders.

■MENTAL DEFICIENCY.—The term *mental deficiency* is used to describe all levels of subnormal intellectual development, the range of which is measured in terms of intelligence quotient (*I.Q.*). The term *idiot* is applied to an individual with an I.Q. generally below 25. Typically, idiots are not able to learn effective speech and cannot guard themselves from the common physical or social dangers. An *imbecile* is an individual with an I.Q. of roughly between 25 and 49. Imbeciles can be taught simple habits, and may learn to protect themselves against the dangers of the environment; however, they cannot become self-supporting members of society. The *moron* is an individual whose I.Q. is approximately between 50 and 69. Under favorable and protected conditions, morons can be taught to earn a living; but in general they cannot compete on equal terms with normal individuals or demonstrate necessary prudence in managing their lives. *Borderline deficiency* describes the mental state of individuals with I.Q.'s of approximately 70 to 80. Such individuals, in general, may be considered as legally competent, but they are slightly subnormal in their ability to deal with intellectual problems and to adjust themselves in social situations. An individual whose test performance is *normal* has an I.Q. of between 90 and 110. Individuals with I.Q.'s between 80 and 90 are sometimes spoken of as having *low normal* intelligence. Individuals with I.Q.'s above 120 are generally classified as *superior*, and in much earlier writing those with an I.Q. of 140 or more were spoken of as having the ability of a *genius*. All these definitions are subject to the qualification that the determination of the I.Q. is far from fixed or absolute.

Mental deficiency is generally considered a result of an inborn failure of the brain to develop normally. It is believed to depend on atypical chromosomes or upon disease or other environmental factors affecting the individual in prenatal or early postnatal life. Some types of mental retardation are related to brain injuries sustained during the birth process, but such injuries characteristically show themselves in motor disabilities rather than in intellectual defects. The term *mongolism* is given to one type of inborn failure to develop in a normal way. Good evidence is available that this condition is a result of an extra chromosome. The condition of a child who characteristically has a dwarfed physique and limited intelligence seldom rising above the imbecile level is called cretinism. If diagnosed early, and if the hormone of the thyroid gland, thyroxin, is regularly administered, the child may be normal.

Psychoses.—The most serious types of mental illness are *psychoses,* which severely hamper the individual's ability to function socially, to express himself, or to deal with reality. The *psychotic* often creates his own special environment in which his perceptions are grossly distorted, as in delusions and hallucinations. Such a condition is known legally as *insanity* and requires very specialized treatment, often in a mental hospital.

■SCHIZOPHRENIA.—One of the most severe and widespread of all psychoses is *schizophrenia;* it is also one of the most baffling to psychiatrists, for although it has been prevalent since early times, its causes and cure are still largely unknown. It was formerly called *dementia praecox,* the "precocious disease," because it often occurs in middle or late adolescence; the term now used is *schizophrenia,* meaning a "splitting of the mind" or personality. In the early stages it is characterized by a *dissociation of affect,* a general decline in intelligence, emotional blunting, and withdrawal from normal social relations. His feelings become separated from and often at variance with his thoughts; he may laugh when his normal feeling would be one of sadness, or cry at an instance of great humor. His powers of association deteriorate and his speech frequently takes on a characteristic disorderly, incoherent, and utterly irrelevant quality. This endless chain of discourse may not, as it so often does in less severe mental states, represent a hidden or private meaning; it is so totally disconnected and meaningless that it has been likened to a tangled, interminable string which, no matter how long it is pulled, has no end. In its extreme form schizophrenia is marked by severe mental deterioration and increasing withdrawal from reality ending in virtually complete separation from the world. It is the most serious of all psychotic disorders and accounts for the great majority of patients in mental hospitals. Medical treatments include the early use of tranquilizers, with effectiveness sometimes increased by electroshock therapy.

In *manic-depressive psychosis,* another grave form of mental illness, the victim typically suffers alternate periods of great excitement and severe depression with or without intervals of normality. In the *manic phase,* the individual is in an excited state in which restlessness, feelings of elation, flights of ideas, and sometimes a destructive violence appear. In the *depressive phase,* the person shows deep feelings of sadness, anxiety, and hopelessness, often with the conviction that the body and even the person are disappearing; sometimes these symptoms are accompanied by attempted or successful suicide or by a deep stupor.

Paranoia is the term used to describe a severe mental illness which, in its pure form, is relatively rare and is characterized by delusions such as an unsubstantiated belief in the person's own grandeur ("I am Napoleon, you know") or in fancied persecution ("Wall Street is trying to get me"), but with little or no accompanying dementia. Sometimes individuals with strong paranoiac tendencies have personalities which in other ways are relatively unaffected. On the other hand, paranoia is often associated with schizophrenia, in which case the most prominent symptom is a well-developed delusional system with hallucinations and feelings of persecution. People who are relatively normal, but who tend to be unduly pompous, sensitive to criticism, and suspicious of those around them, are sometimes half-jokingly spoken of as paranoid.

Involutional melancholia or *involutional psychotic reaction* is a form of mental illness which occurs most frequently at the time of the climacteric. The symptoms of this illness are often complex, involving deep worry, feelings of guilt, delusional ideas, and a prolonged and intense general depression. Special psychotic conditions of short or very long duration sometimes follow childbirth or severe infectious diseases.

Senile psychosis refers to a chronic condition of the aged which is almost certainly related to an impairment of the structure and function of the brain and a change in the endocrine balance of the body fluids. This condition is characterized by marked loss of memory, stubbornness, irritability, and irresponsibility. The term *paresis* is given to a psychosis in which a progressive loss of intellectual capacity is accompanied by various types of motor paralysis and specific psychotic manifestations. General paresis, or general paralysis of the insane (*dementia paralytica*), is typically a result of deterioration of certain parts of the brain brought about by syphilitic infection.

THAMES & HUDSON, LONDON

DRAWINGS BY MENTAL PATIENTS both enable psychiatrists to learn more about the patient and allow the patient the opportunity to express all of his inner feelings in visual terms.

Neuroses.—The *psychoneuroses,* or *neuroses,* are functional behavior disorders of a generally less severe order. While deeply troublesome to the individual and often to his close associates, they do not exhibit the flagrant disturbances of external behavior characteristic of the psychoses, and are usually not severe enough to require commitment to a mental hospital. The neurotic usually maintains contact with his environment; both his thinking and feeling are distorted, but he is not typically delusional.

Character Disorders.—A third category used by some psychiatrists is that of *character disorders* or *personality disorders.* These patterns of reaction are closely associated with neuroses, in which the basic character structure of the individual is distorted or deformed. This condition is characterized by patterns of inadequate or antisocial behavior, emotional instability, excessive reaction to stress, extreme dependence or aggressiveness, and, above all, anxiety. In this category are many individuals called *sociopaths,* who act out their hostility on society; many criminals, delinquents, and sexual deviates are often included in this category of mental illness.

The distinctions between these categories are not rigidly defined and are interpreted differently by various schools of psychiatry and psychoanalysis. The distinction between the psychoses and neuroses is not as clearly defined as it once was, especially in the view of some psychoanalytically oriented psychiatrists.

The term *psychosomatic disorders* is given to the physical illnesses which are considered to have full or partial base in the abnormal mental life of the individual. Many experts in psychosomatic medicine rely largely on the techniques of psychoanalysis in their clinical procedures. There is evidence that certain allergies at times have a psychosomatic origin, and some types of stomach ulcer have been treated successfully by psychosomatic techniques. The term *conversion hysteria* is sometimes used to name a condition, such as paralysis, which results from emotional conflict and which may be treated by psychosomatic procedures.

The terminology of mental illness is not as fixed as those of many other classes of disease, due to the very complex nature of the symptoms of mental illness. Indeed, some psychiatrists no longer use the word "disease" in speaking of any atypical mental reaction. The present terminology owes much to the work of Emil Kraepelin, a student of Wundt, whose categorization of mental symptoms, as early as 1883, helped bring a scientific viewpoint to psychiatry.

The Growth of Psychiatry.—Modern psychiatry and psychoanalysis have had a complex and fascinating history. In primitive society, in classical times, and almost down to the modern era, mental illness was believed to result from the activities of a malevolent spirit inhabiting the human body. This concept was an extension of the primitive animism prevalent in almost all early societies. The causes and cure of this "possession" were the province of magic, and the first professional psychiatrists were the witch doctors and charmers. Later, the clergy performed rites of exorcism which, if successful, drove out the evil spirits. The term *lunacy* is now only a legal term denoting a deviation from normal mental life that makes it necessary to appoint a guardian for the afflicted individual. The term comes from the Latin word *luna,* 'the moon,' whose changing phases were believed to cause the waxing and waning of symptoms in some types of mental disease. Many of the witches who were burned in medieval times, or executed in colonial America, were undoubtedly mentally ill individuals. *Bedlam,* the earliest insane asylum in England, was really a prison for the insane. Londoners, even in the eighteenth century, considered it an amusement to visit Bedlam and to poke fun at and excite the unfortunate inmates. Humanitarianism, reinforced by religious feelings, led to a gradual transformation of the old inhuman custodial madhouses of Europe and America into much more humane asylums for the insane in the latter part of the eighteenth century and first half of the nineteenth century. The great French physician Philippe Pinel gave impetus to this movement when, in 1795, he removed the shackles from the inmates of Paris' notorious Bicêtre, the hospital for the insane, and initiated their treatment as patients instead of as prisoners. In America progress in reforms owes much to Dorothea Lynde Dix.

The modern scientific study of mental illness at first placed great emphasis on the investigation of the forms of brain pathology seen at the autopsies of the mentally ill. Much new and important information was secured by such studies, for instance, the base of some forms of mental illness, such as those caused by syphilis and by the brain deterioration resulting from old age.

CLARK UNIVERSITY

PSYCHIATRY PIONEERS. (*Left to right, standing*) A. A. Brill, Ernest Jones, Sandor Ferenczi; (*seated*) Sigmund Freud, G. Stanley Hall, and Carl G. Jung. Photo was taken in 1909.

■ NEUROSIS.—The diagnosis and treatment of *neurotic disorders,* as distinguished from the psychoses, have been especially important in the history of psychiatry and psychology. Such illnesses also have not been shown to be related to easily demonstrable changes in the brain. They are sometimes described as functional, although it is recognized by scientists that all human functions have an organic base.

The condition known as *hysteria* early attracted medical attention. In extreme cases such neurotics may suddenly become blind or deaf, lose feeling in their limbs, become wholly or partially paralyzed, or even forget who they are. As early as 1766, Friedrich Mesmer began the use of *hypnosis* in the treatment of hysteria, and the term *mesmerism* has often been used as a synonym for hypnosis. It is a trancelike state induced by the hypnotist through suggestions. The subject becomes sleepy, relaxes his muscles, and performs various acts as directed by the hypnotist. Josef Breuer in 1880–1882 relieved a patient's hysterical symptoms by causing her to relive, under hypnosis, certain scenes associated with the onset of her illness. At the same time in Paris the French neurologist J. M. Charcot was using hypnosis in the study and cure of hysterical patients. One of Charcot's students, Theodore Janet, developed the teachings of his master into a new psychological view of the importance of the integration of the personality and insisted on psychological treatment for hysteria. He pointed out that neurotics tend to have divided personalities, whereas those of normal individuals are more fully integrated. Another student of Charcot was Morton Prince, a distinguished American neurologist and psychiatrist who did much to advance the understanding of dissociated and multiple personalities. *Dissociative reactions* such as fugue and amnesia are attempts to escape from tension by unconsciously blocking off part of one's conscious recognition.

NATIONAL ASSOCIATION FOR MENTAL HEALTH

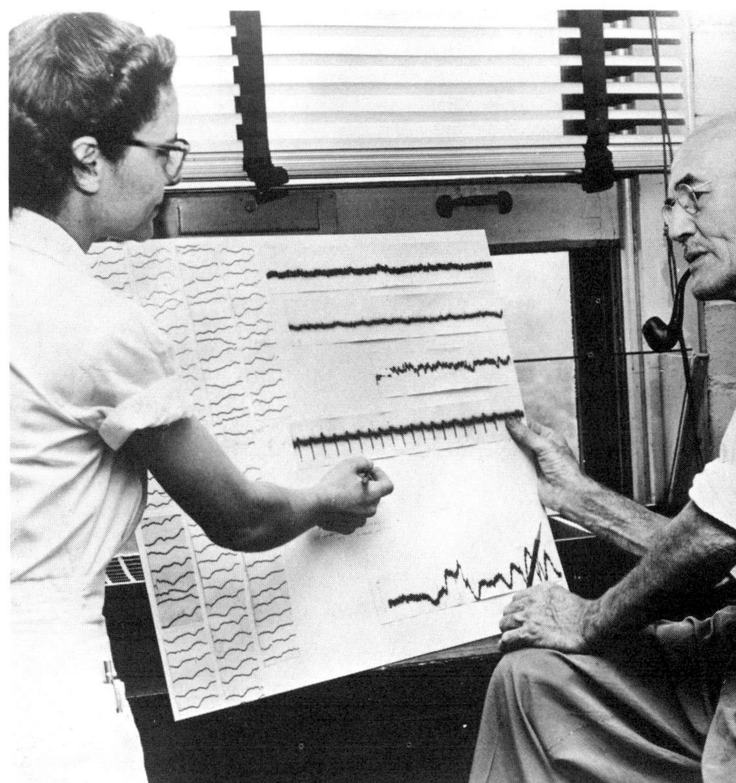

NATIONAL ASSOCIATION FOR MENTAL HEALTH

RESEARCH. Scientists study the relationship of body chemicals (*above left*) and brain waves (*above right*) to behavior disorders. Machines can stimulate and record electrical activity in the nervous system. (*below*).

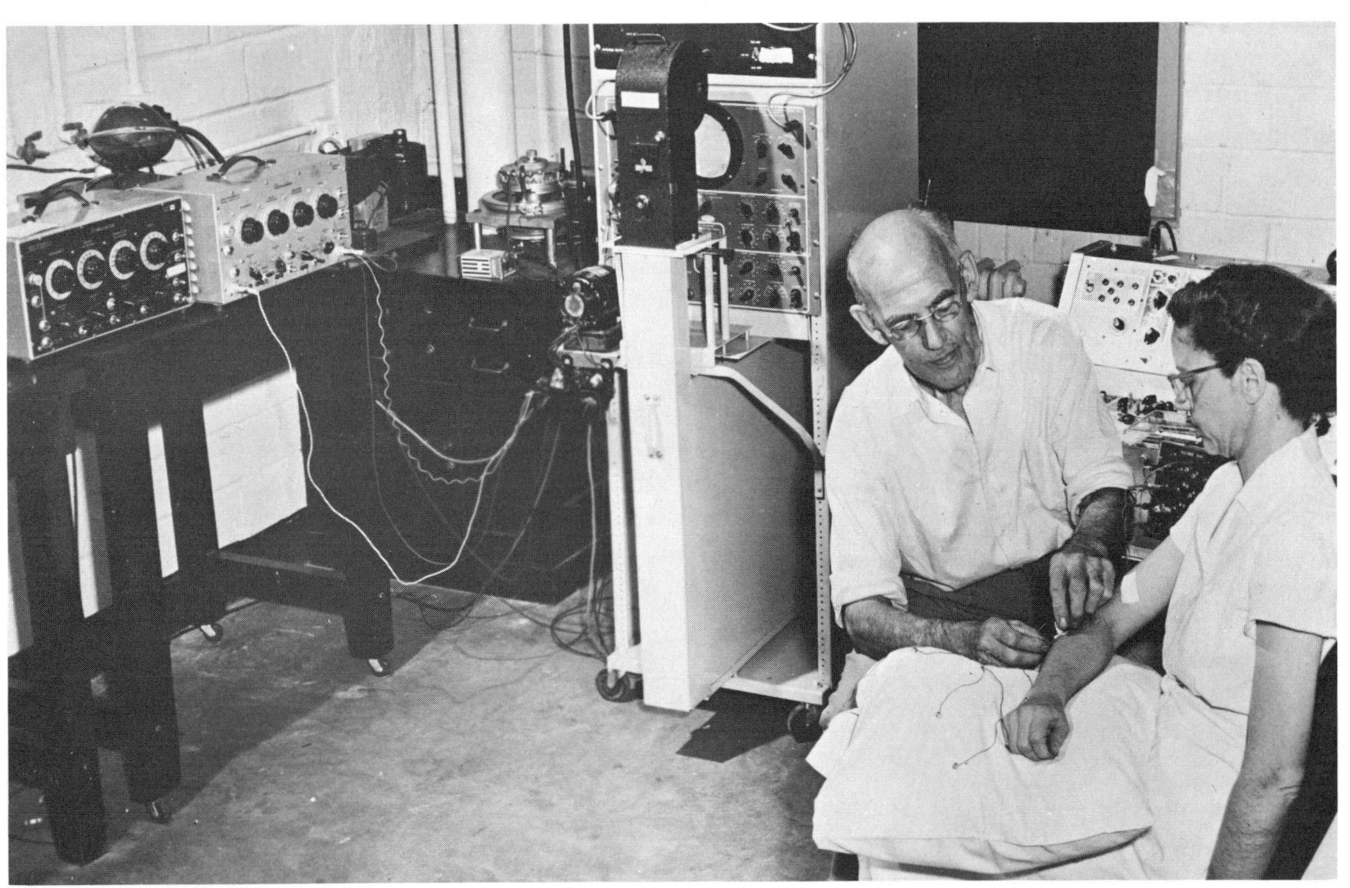

NATIONAL ASSOCIATION FOR MENTAL HEALTH

NATIONAL ASSOCIATION FOR MENTAL HEALTH

TREATMENT. Group discussions (*above*) help to increase understanding. Crafts (*below left*) and games (*below right*) afford relaxation.

NATIONAL ASSOCIATION FOR MENTAL HEALTH

NATIONAL ASSOCIATION FOR MENTAL HEALTH

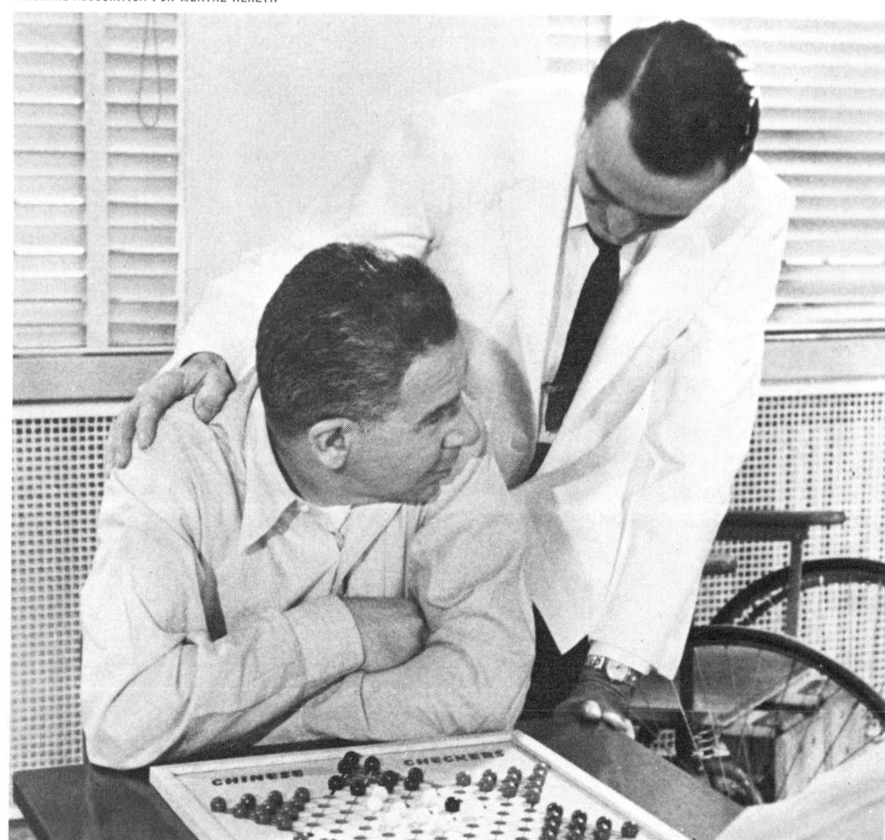

The term *multiple personality* describes a condition in which two or more personalities inhabit an individual, alternating with one another, appearing and disappearing, with no awareness on the individual's part.

FREUD.—By far the most important student to work with Charcot was Sigmund Freud (1856–1939), considered by many to be one of the greatest figures of our time. Born in Czechoslovakia, he lived almost all his life in Vienna, where he made his studies and his far-reaching discoveries concerning the mental life. As a young student, he began the microscopic study of nerve cells in connection with the physiology and pathology of the brain. In his study of mental disease, he became especially puzzled by the neuroses, and in 1885 went to Paris to study with Charcot. For a time Freud adopted Charcot's technique of hypnosis, and on returning to Vienna he became associated with Breuer in the treatment of neurotic patients by this method. They observed that certain patients were relieved of their symptoms by the recalling of an otherwise forgotten event, as had first been noted, but not pursued, by Breuer. This *mental catharsis,* or "talking-out treatment," as one patient named it, was dropped by Breuer; but Freud continued in the work that led to the development of psychoanalysis.

PSYCHOANALYSIS.—Noting that patients talked freely even when not under hypnosis, Freud soon abandoned that technique, which he found too constricting, and urged his patients to say whatever came into their minds without concern for relevance or propriety. By this method of *free association* and by the study of the dreams which they reported, he assisted his patients in recovering previously forgotten memories. At certain times, however, patients had great difficulty in speaking freely, and Freud concluded that this occurred in instances where an experience had been so painful and disturbing that it was *repressed* from awareness and memory by the conscious self. Such experiences are called *traumatic* because they appear to cause a mental wound or shock. Freud also noted that many of these repressed experiences were sexual in nature and occurred very early in childhood; and he therefore concluded that in many cases these sexual experiences were the primary cause of the neurosis. This discovery, on which the theory of the *sexual origin* of neuroses is based, is one of the most fundamental and controversial of all Freud's discoveries. It led to the theory of *infantile sexuality*—that the first sexual feelings appear not at puberty, but in early childhood. Freud believed that the child's sexual development passes through three stages: the *oral,* associated with sucking and mouth activity; the *anal,* in which pleasure is derived from the excretory functions; and finally the localization of sexuality in the *genital* region, leading to full adult sexuality and finally to normal reproductive behavior. If development is arrested, or *fixated,* at one of the earlier stages, it may cause a serious *regression* of the adult to that stage, thus determining the form that his neurosis takes.

In general, Freudian psychoanalysis may be described as a form of *depth psychiatry* or *depth psychology* in which an attempt is made to bring "into consciousness" the impulses and mental processes that have been active in ordinarily hidden parts of the individual's mental life. These deep mechanisms are spoken of as existing in the patient's *unconscious,* whose existence was one of Freud's most important postulations. In treating patients, psychoanalysts believe that it is necessary to break through *resistances* in order to bring some of these hidden, previously unconscious, but important and powerful mental processes into awareness. The physician who conducts the psychoanalysis therefore helps the patient to break down various forms of resistance and later assists him in reorganizing his life by facing reality in a constructive way. The patient comes to avoid the emotional conflicts that had previously led to obstructive and damaging behavior.

What is known as *motivation* in psychological terms is a fundamental concept in psychoanalysis. Freud believed that the basis of all human motivation was to be found in the *libido,* the manifestation of the sexual instinct, or Eros, a term which he closely identified with the *life instinct.* Mental and emotional life results from the interplay of libidinal energy with other factors in the individual and in his environment.

One of the most important of these factors is the *pleasure-pain principle,* which describes the individual's striving toward pleasure and withdrawal from or avoidance of pain. During growth and development, the original pleasure principle is modified by its contact with the *reality principle,* the sum total of outside forces leading to new developments in the individual's mental life.

Toward the end of his life, Freud posited a second great principle which he believed acted in opposition to *Eros,* the "life-instinct"; he called it *Thanatos,* the "death instinct" or wish for death. Thus the process of life represents a struggle of counterbalancing forces, the constructive, or life-creating, force, versus the destructive, or life-destroying, force.

In psychoanalysis the mental life of the individual is often considered to consist of three interacting parts: the id, the ego, and the superego. The *id* is the division of the psyche from which arise blind, impersonal, instinctual impulses related to the gratification of primitive needs. These impulses dominate the earliest life of the individual, but during later life are forced into various forms of submission and conformity. The *ego* is the aspect of the personality which is in contact with the external world through perception and *reality-regulated striving,* the struggle between the pleasure principle and the reality principle. As the individual develops, the ego is further differentiated and a superego develops. The ego comes to mediate between the superego and the id by building up what have been called *ego defenses.* The *superego* represents the inhibitions of instinct in man, the system within the individual's total mental life developed by incorporating parental standards as perceived by the ego. Thus, moral standards as perceived by the ego become part of the personality. Sometimes the superego is considered as having two parts, the *ego ideal* and the *conscience.*

It can be seen in a dynamic theory of depth psychiatry of the sort developed by Freud that conflict and repression are inevitable as an individual develops. Various complexes are thus characteristic of the growth of the normal, as well as of the neurotic, individual. In the neurotic, however, this conflict leads to an arousal of *anxiety* against which the individual erects *defense mechanisms* which severely impede his personal and social life. Basically, the differences between normal and abnormal are in degree or quality, and not in kind.

MUSEUM OF MODERN ART

FREE ASSOCIATION in Chagall's *I and the Village* resembles psychoanalysis.

Psychoanalysis is a clinical method used by properly trained physicians or other qualified professional persons in treating many neurotic conditions and personality disorders. In America virtually all qualified psychoanalysts are physicians. Today psychoanalysis is held by many psychiatrists to have an important place also in dealing with aspects of the graver psychoses.

JUNG.—Carl G. Jung, a psychiatrist in Zurich, developed a school described as *analytic psychology.* In his early years, Jung was personally associated with Freud in the development of psychoanalysis but later broke with Freud over the question of the sexual basis of neurosis. Jung gives less emphasis to sex in his treatment of the basic dynamics of personality than does Freud and instead sees a general life urge as fundamental in motivation. The concepts of *introversion* and *extroversion* were developed by Jung to describe the characteristics of individuals whose attention and interest are centered internally upon themselves or externally upon the social and physical world. For Jung, depth psychology and the concept of the unconscious are just as important as they are for Freud, but Jung deals not only with an individual's unconscious but also with the concept of a *collective* or *racial unconscious*, which he considered to be inherited by all human beings. Primitive notions related to birth, death, and other magical phenomena are described by Jung as natural ways of thinking and as inherited in the unconscious. The system developed by Jung, like that of Freud, is very complicated and important for the modern student. Jung emphasizes the deep significance of religion in human life, and his system is of great importance to the *psychology of religion*. Those who have studied Jung's work in detail and who know of his clinical procedures are convinced that he brought great wisdom to the assessment and cure of individual personality disorders—disorders arising from the complex demands of today's industrial world. The human body and the basic patterns of human behavior evolved for survival in a simpler and more "natural" environment than that which modern civilization provides.

■**ADLER.**—Alfred Adler, a Viennese physician who was an early associate of Freud, developed another system which is characterized as *individual psychology*. Adler believed that the fundamental fact in a neurosis is a feeling of inferiority and that the most important striving is for power. He did not deny the importance of sex, but like Jung placed less emphasis on it than did Freud. Adler taught that individuals who have fancied inferiorities try to compensate or sometimes to overcompensate for these intensely felt self-deficiencies and thus develop an urge for power and domination over others. The commonly used term *inferiority complex* was coined by Adler to describe the style of life of weak individuals which allows them to avoid at least certain of the realistic demands of their environments. Adler held up as the human ideal a balanced emotional life in which each person thinks of those with whom he is associated not as superiors to whom he should be subservient nor as inferiors who should be dominated by him, but rather as equals. Adler's teachings, although less complex than those of Freud, have had a strong effect in reinterpreting some of the fundamental meanings of democracy for today's generation.

■**RANK.**—Another psychoanalyst whose work has had wide influence, Otto Rank stressed what is called the *birth trauma*, which is defined as the effect upon the individual's psyche of the strain of being born. Emphasis is given not so much to the actual pain of delivery as to the fact that the newborn child must begin his difficult adaptation to a hostile environment. The child is, as it were, removed by birth from the security of the mother's body and from the first object of its libido, the mother. Adult anxiety neuroses and other neurotic symptoms, it is held, may at times be traced back to this initial trauma. Rank emphasized the importance of observation of the full environment, the complete present state of life of his patients. This point of view has been found helpful by many clinicians and social workers who must help individuals to adjust in difficult family and social situations.

■**SULLIVAN.**—The American psychiatrist Harry Stack Sullivan emphasized the interaction between the growing child and his parents, and between the adult and other individuals, or society. He believed that these *interpersonal relations* were of greater importance in the origin and treatment of mental illness than were developments and relationships within the individual psyche.

The preceding paragraphs have given in barest outline some suggestions of modern dynamic psychoanalytic approaches to an understanding of mental life. These psychoanalytic theories and others related to them have had an important influence upon modern thought. Sociology, legal history, and the criticism of literature and art have all been affected by different schools of psychoanalysis. Today, as the pioneer workers in these specialized schools, who were trained by Freud, Jung, or Adler, or the other pioneers in depth psychology, are being replaced by men and women fully trained in modern medicine and experimental and clinical psychology, there is an inevitable reorientation of the basic concepts of these dynamic approaches to an understanding of mental life and mental illness.

For example, a new biological orientation in dynamic psychology seems to be developing. New psychopharmacologic knowledge, based on an understanding of the effects of tranquilizing drugs, must be considered in connection with basic psychological and analytical approaches to the personality. Drugs that produce hallucinations and motivational changes must be assessed in the same context. In general, drugs are seen as aids in enabling a patient to obtain greater benefits from psychotherapy, and not as a substitute for professional care.

Mental Hygiene.—All these facts and theories are considered in the new approach to mental hygiene, that is, the study of techniques that may be used in the prevention of mental illness. Indeed, growing knowledge of the anatomy of the brain, the physiology of the nervous system, experimental psychology, general psychiatry, psychoanalysis, and psychopharmacology are all making contributions to the techniques that may be used in preventing, insofar as possible, incapacitating mental illness and in its cure. The dream of a world without mental hospitals may not be a realizable one, but there can be no doubt that modern scientific approaches to the understanding of mental life and of mental illness will not only reduce the number of patients who must be sent to such institutions, but will also cut down the duration of confinement. Part of the present-day improvement in the care of mental illness is related to new drugs, a wide variety of techniques of psychotherapy, and the use of imagination to improve the lives of patients while they are in mental hospitals. Above all, many types of mental patients must be made to feel that they are important to those treating them, no matter by what "system." Useful treatment techniques are occupational therapy and the creative use of drama, dance, music, and painting. All these approaches help to bring about the readjustment of the patient to the real world and many emphasize an understanding of learning in the reeducation of patients.

A generation ago Adolf Meyer, of Johns Hopkins, who originated the term "mental hygiene," taught that understanding of mental illness depended upon greater scientific knowledge in psychology and in all aspects of medicine. This approach, called *psychobiology*, gives full attention to the findings of all schools of psychology, psychoanalysis, pharmacology, and physiology. It promises great strides in mental medicine. Today advances are being made that will enable man to achieve a better understanding of one of the most important and complex of all subjects —the human mind.

—Leonard Carmichael

BIBLIOGRAPHY

CAMERON, NORMAN. *Personality and Psychopathology: A Dynamic Approach.* Houghton Mifflin Company, 1963.

CARMICHAEL, LEONARD. *Basic Psychology: A Study of the Modern Healthy Mind.* Random House of Canada, Ltd., 1957.

FREUD, SIGMUND. *Collected Papers.* 10 vols. Collier Books, 1963.

HEBB, DONALD O. *A Textbook of Psychology.* 2nd ed. W. B. Saunders Company, 1966.

HENDERSON, DAVID, and R. D. GILLESPIE. *A Text-Book of Psychiatry for Students and Practitioners.* David Henderson and Ivor R. Batchelor, eds. 9th ed. Oxford University Press, 1962.

JONES, ERNEST. *The Life and Work of Sigmund Freud.* Edited and abridged by Lionel Trilling and Steven Marcus. Basic Books, 1961.

MARLER, PETER, and W. J. HAMILTON. *Mechanisms of Animal Behavior.* John Wiley and Sons, 1966.

MUNN, NORMAN L. *Psychology, The Fundamentals of Human Adjustment.* 5th ed. Houghton Mifflin Company, 1966.

PIAGET, JEAN. *The Language and Thought of the Child.* The World Publishing Company, 1963.

WATSON, ROBERT I. *The Great Psychologists.* 2nd ed. J. B. Lippincott Company, 1968.

WHYTE, LANCELOT LAW. *The Unconscious Before Freud.* Basic Books, Inc., 1963.

MATHEMATICS

HISTORY

The Origins of Mathematics.—When prehistoric man found a way to answer the questions "how much?" and "how many?" he laid the foundations of mathematics.

Stones from the Paleolithic period, with notches and geometric designs carved on them, indicate that a low level of mathematical thinking existed in the early part of the Stone Age. In the fifth and fourth millennia B.C., large groups of people inhabiting the fertile regions on the banks of long, navigable rivers—the Nile in Egypt, the Euphrates and Tigris in Mesopotamia, the Indus in Pakistan, the Ganges in India, and the Hwang Ho (Yellow River) in China—developed a utilitarian form of mathematics that made it possible for them to keep pace with their advances in engineering, agriculture, and trade. At this time the notches formerly used in recording a count were replaced by symbols (*numerals*), and calendars were devised to let the farmer know when to plant, to harvest, and to expect heavy rains. A crude geometry used to survey land also existed in this period.

Egyptian Contributions.—Although all these river civilizations can claim great antiquity for their achievements in mathematics, only the Egyptians and Mesopotamians left tangible evidence of a systematic development of the subject prior to 1100 B.C. From the Egypt of 3500 B.C., for example, there is a royal mace inscribed with a record of the amount of loot seized by the king in a great battle. There are drawings on the walls of Egyptian temples and pyramids indicating that a method of taxation involving extensive bookkeeping was in use as long ago as 3000 B.C. Several papyri dealing with pharmacology prove that a system of weights and measures as sophisticated as any in use up to the nineteenth century was well established 5,000 years ago among the people of the Nile valley.

Knowledge of Egyptian mathematics is derived mainly from a study of 110 problems and their solutions, which are to be found in two well-preserved papyri—the *Moscow Papyrus*, written about 1850 B.C., and the *Rhind* (*Ahmes*) *Papyrus*, which was copied in 1650 B.C. by Ahmes from an older manuscript. These papyri, in a cursive writing called *hieratic*, show that the Egyptians had complete mastery of the fundamental operations of arithmetic (addition, subtraction, multiplication, and division) for both whole numbers and fractions; that they could solve simple equations; and that their interest in mathematics was not confined to its utilitarian aspects, as had been the case in an earlier day—problems of a recreational nature such as puzzles were beginning to appear. The geometry mentioned in these papyri was of a strictly mensurational type, limited to finding areas and volumes of the common geometric figures and solids. The solutions to these measurement problems were probably based on good guesses backed up by observation of sample figures. As a result, some formulae are indicative of great ingenuity on the part of their discoverers, while others fall so far short of the correct result that they bear witness to the lack of the power to generalize and to a complete absence of a deductive basis. Thus, Egyptian mathematics might be classed as trial and error rather than as a true science.

Babylonian Contributions.—Although the failure to develop a deductive system of geometry also characterized the efforts of the Babylonians, they towered over their Egyptian neighbors in the other branches of mathematics, algebra and the theory of numbers. These talented people wrote by pressing a stylus on a tablet of wet clay. Since the clay was later fired to harden it, many thousands of these tablets, most of them dating from 2000 B.C. to 200 B.C., have been preserved. Recent studies of these tablets indicate that the pre-eminence of the Babylonians in mathematics was due in no small part to their ability to represent whole or fractional numbers of any size by using a place system of numeration with the base 60, called a *sexagesimal system*. Their system was used in the same manner in which we use the base 10, with one exception—the Babylonians did not have a symbol for zero. Late in their history they employed a punctuation mark, the period, to indicate the absence of a symbol between two other symbols in a numeral. Since this mark was never used alone or at the end of a numeral to indicate the absence of units (ones) in the number, it cannot be called a "zero." Moreover, the Babylonians did not have a counterpart of the decimal point, a symbol that serves to separate the whole number from the fraction. With all its shortcomings, however, the Babylonian system of writing fractions was so superior to other methods that European mathematicians and astronomers used it until the late sixteenth century, when the decimal fraction was invented. The base 60, moreover, has survived a span of more than 4,000 years, for it is still used in the ordinary units for telling time and measuring angles (60 seconds = 1 minute; 60 minutes = 1 degree or 1 hour). In the solution of equations and sets of equations, of both first and second degrees, the Babylonians had no peers among the ancient peoples. They also had a knowledge of number theory, such as the method of finding sets of three whole numbers that can be measures of the sides of a right triangle.

Greek Mathematics.—While observation and intuition, fortified by a certain amount of reasoning, play a role in the mathematical invention, the major discoveries in the subject have been made through a formal system of deductive thinking. Such a system was completely foreign to the methods of the Egyptians and Babylonians. It is this formalization of deductive thinking, which has given rise to present-day *axiomatics*, that must be considered the foremost Greek contribution to mathematics. In the hands of the Greeks, the method consisted of the adoption of a list of geometrical statements (*postulates* or *axioms*) that were considered so self-evident as to be acceptable without proof, followed by logical deduction of a whole body of statements, each of which was based on a postulate or on some previously proved statements. Moreover, an attempt was made to define all terms used. Although the Greek concept of a deductive science has undergone a significant revision in modern times, the Greeks provided the world with the first example of a procedure that is now characteristic of all branches of mathematics.

The Greeks also were the first to mention mathematicians by name and to take pride in their exploits. Thales of Miletus (c. 640–546 B.C.), who is credited with the invention of (*demonstrative*) *geometry*, was the earliest

Preclarissimus liber elementorum Euclidis perspi/
cacissimi: in artem Geometrie incipit quã foelicissime:

Punctus est cuius ps nõ est. Linea est
lõgitudo sine latitudine cui⁹ quidẽ ex/
tremitates st duo pũcta. Linea recta
ẽ ab vno pũcto ad aliũ breuissima extẽ/
sio i extremitates suas vtrũqz eorum reci
piens. Supficies ẽ q̃ lõgitudinẽ ⁊ lati
tudinẽ tñ hz: cui⁹termi quidẽ sũt linee.
Supficies plana ẽ ab vna linea ad a/
liã extẽsio i extremitates suas recipiẽs
Angulus planus ẽ duarũ linearũ al/
ternus ꝯtactus: quarum expãsio ẽ sup sup/
ficiẽ applicatioqz nõ directa. Quãdo aũt angulum ꝯtinẽt due
linee recte rectiline⁹ angulus noiat. Qñ recta linea sup rectã
steterit duoqz anguli vtrobiqz fuerit eq̃les: eorum vterqz rect⁹ erit
Lineaqz linee supstãs ei cui supstat ppendicularis vocat. An
gulus vo qui recto maior ẽ obtusus dicit. Angul⁹ vo minor re
cto acut⁹ appellat. Termin⁹ ẽ qd vniuscuiusqz finis ẽ. Figura
ẽ q̃ tmino vl termis ꝯtinet. Circul⁹ ẽ figura plana vna qdem li/
nea ꝯtẽta: q̃ circũferentia noiat: in cui⁹ medio pũct⁹ ẽ: a quo oẽs
linee recte ad circũferẽtiã exeũtes sibiĩuicez sũt equales. Et hic
quidẽ pũct⁹ cẽtrũ circuli dr. Diameter circuli ẽ linea recta que
sup ci⁹ centrum trãsiens extremitatesqz suas circũferẽtie applicans
circulũ i duo media diuidit. Semicirculus ẽ figura plana dia/
metro circuli ⁊ medietate circũferentie ꝯtenta. Portio circu/
li ẽ figura plana recta linea ⁊ parte circũferẽtie ꝯtẽta: semicircu/
lo quidẽ aut maior aut minor. Rectilinee figure sũt q̃ rectis li/
neis cõtinent quarũ quedã trilatere q̃ trib⁹ rectis lineis: quedã
quadrilatere q̃ q̃tuor rectis lineis. q̃dã mltilatere que pluribus
q̃z quatuor rectis lineis continẽt. Figurarũ trilaterarũ: alia
est triangulus hñs tria latera equalia. Alia triangulus duo hñs
eq̃lia latera. Alia triangulus triũ inequalium laterũ. Harum iterũ
alia est orthogoniũ: vnũ .s. rectum angulum habens. Alia ẽ am/
bligonium aliquem obtusum angulum habens. Alia est oxigoni
um: in qua tres anguli sunt acuti. Figurarũ autẽ quadrilaterarum
Alia est q̃dratum quod est equilaterũ atqz rectangulũ. Alia est
tetragon⁹ long⁹: q̃ est figura rectangula: sed equilatera non est.
Alia est helmuaym: que est equilatera: sed rectangula non est.

NEW YORK PUBLIC LIBRARY

THE "ELEMENTS" by Euclid was written about 300 B.C. This book has had more editions and translations than any other on mathematics. Shown is a page in Latin printed in 1482.

mathematician whose name has come down to us. Thales set up a list of postulates for geometry and, on the basis of these, proved about a half-dozen *theorems*, one of which gives a set of conditions under which two triangles are congruent.

The work begun by Thales was carried forward by Pythagoras (c. 582–507 B.C.), who organized geometry into a deductive science and proved many theorems. The most famous of these theorems, which states that the sum of the squares of the sides of a right triangle equals the square of the hypotenuse, bears his name. Pythagoras founded a secret society that had for one of its tenets the belief that "number" is the basis of all creation. A member of the group, or possibly Pythagoras himself, made one of the most far-reaching discoveries in the history of mathematics—the recognition of the existence of segments of lines with measures that could not be represented by whole numbers or quotients of whole numbers —the *irrationals*. (In the decimal system, irrational numbers are represented by the nonrepeating, infinite decimals.) Pythagoras is also credited with the discovery of the mathematical laws of harmony, which deal with a certain progression of numbers. Because of their connection with music, the system is called "harmonic."

In the two centuries following Pythagoras, the Greek school made notable contributions to the philosophical aspects of mathematics. Among the problems that engaged mathematicians of the period were the paradoxes of Zeno of Elea (c. 475 B.C.). One of the simplest of these is the argument that it is impossible for a moving body to reach its destination, because before the body in motion can traverse an entire distance, it must reach the halfway mark; and before getting to the halfway mark, it must get to the quarter mark, the eighth, the sixteenth, and so on *ad infinitum*. Movement is therefore impossible, claimed Zeno, since the moving body would have to traverse an infinite number of segments in finite time.

Another problem of this period that left its mark on mathematics is that of "squaring the circle"; that is, constructing a square equal in area to a particular circle. The proof that this construction cannot be made with the two instruments—the unmarked straight edge and the compasses to which geometricians of the Greek school were limited—had to wait until 1882 and the work of the German mathematician Ferdinand Lindemann. In the centuries that intervened between the framing of the problem and Lindemann's work, attempts to solve the problem yielded many new branches of mathematics and revealed the nature of number.

With the founding of Alexandria (named for Alexander the Great) in 332 B.C., the center of mathematical studies moved from Athens back to Egypt. The first of the Hellenistic (Alexandrian) mathematicians of note was Euclid, who wrote the *Elements*, a book on geometry that has gone through more editions and translations than any other on mathematics. It might be said that with Euclid the Greek development of axiomatics reached its zenith. Until the mid-twentieth century, the *Elements* was studied in simplified and abridged versions by secondary-school children the world over. Geometry that employs the axioms set forth by Euclid is called Euclidean geometry.

Although Euclid is the most widely known of the Hellenistic mathematicians, greatest (perhaps of all antiquity) was Archimedes (c. 287–212 B.C.), who lived in the Greek colony of Sicily after a period of schooling in Alexandria. Archimedes devised a way of representing large numbers with a few symbols, in order, as he said in his *Sand-Reckoner*, to be able to represent the number of grains of sand in a volume equal to that of the earth, a volume that he had correctly calculated by using as diameter a measure obtained by Eratosthenes (c. 275–195 B.C.). Besides finding a formula for the volume of the sphere, Archimedes discovered the formula for its surface area and for areas of figures bounded by various curves, such as parabolas and spirals. His method foreshadowed the invention of the integral calculus 18 centuries later. However, Archimedes' discoveries in the realm of physics (such as the lever, the pulley, and the screw) overshadow his less utilitarian work in mathematics, and he is better known as the world's first mechanical engineer than as an outstanding mathematician.

A contemporary of Archimedes, Apollonius of Perga (fl. 247–205 B.C.) carried almost to completion the study by geometric means of the sections of a right circular cone made by cutting the cone at varying angles. He subjected the sections (ellipse, hyperbola, parabola), which had been known to mathematicians since about 335 B.C., to detailed analysis. Apollonius discussed the basic properties of asymptotes, chords, axes, tangents, and so on, for these curves and thereby paved the way for their application to modern mechanics and astronomy. The significance of the work of Apollonius in this field was not realized until the development of projective geometry, which did not occur until the seventeenth century.

Archimedes and Apollonius were the last of the great theoretical mathematicians until the Renaissance. With the exception of the work of these two—and even some of their work was motivated by the possibility of application—beginning in the third century B.C., practical arithmetic and astronomy provided the impetus for all the major advances in mathematics until the sixteenth century. Thus, the Alexandrian school of mathematics became a partner of the budding science of astronomy, and from it came the opposing concepts of the motion of the earth in relation to the sun—the *heliocentric theory* of Aristarchus of Samos (third century B.C.), the Copernicus of antiquity—and the *geocentric theory* of Apollonius of Perga, which had the support of most of the ancient astronomers who considered the topic. The greatest of these astronomers, Hipparchus (fl. 130 B.C.), the father of trigonometry, adopted the geocentric theory, as did Claudius Ptolemy (fl. 150 A.D.), whose *Almagest* became the astronomical Bible of medieval and Renaissance Europe.

NEW YORK PUBLIC LIBRARY

THE ABACUS, an early form of the digital computer, was used to perform mathematical computations prior to the Christian Era. This sixteenth-century woodcut shows two men calculating sums on their abaci while a scribe records their answers.

Alexandria also fathered the beginning of algebra as a subject studied without recourse to geometry, through the works of Diophantus (third century B.C.). He invented a symbolism that was to be a transition between the primitive algebra (called *rhetorical algebra*), in which the statement of problems is given solely in words, and that of modern times, which is completely symbolic. The symbolism invented by Diophantus is called "syncopated" because it used the first letters of words to replace the words. Diophantus also made a systematic study of equations in which the number of quantitative conditions put on the unknowns is smaller than the number of unknowns. Diophantus showed how to solve these if a qualitative condition, such as "the answer must be a whole number," is placed on the variables; this type of solution comes under the heading of *Diophantine analysis*.

The Middle Ages.—Beginning with the fall of Rome, 476 A.D., mathematics entered a stagnation that lasted nearly 800 years. During this sterile era, very little work of consequence can be attributed to European mathematicians. One of the few exceptions during this period was Anicius Manlius Severinus Boethius (c. 480–c. 524), who managed to reconstruct a fragment of Euclid's geometry and some of the Greek theory of numbers. However, between 400 and 1200, the development of mathematics advanced mainly through the efforts of the Hindus and the Arabs, who have contributed some of the most important of mathematical concepts.

Hindu Influence.—After the burning of the library of Alexandria by the Arabs in 642 A.D. the center of mathematical activity moved to Persia (Iran and Afghanistan) and to India, countries that had previously benefited by an influx of scholars from Egypt, Mesopotamia, and Greece during the decline of Roman power in the East. With the advantages gained by their invention of zero and a positional decimal system of numeration (the Hindu-Arabic system we use today), the Indians managed to surpass all their predecessors in the development of methods of calculation (*algorithms*) and in the solution of algebraic problems. They devised a formula for solving quadratic (second-degree) equations, and managed to solve some cubic (third-degree) equations. They gave rules for simplifying complicated radicals, for operating with zero and with negative numbers, and discovered "imaginary" numbers (numbers involving $\sqrt{-1}$). Their algebra was rhetorical.

Arab Influence.—When they conquered Persia and culturally penetrated India, the Arabs became acquainted with the mathematics of the vanquished people and adapted it to the mathematics they had found in Alexandria. The great importance of the Arab school of mathematics stems from the fact that their scholars translated the Greek works into Arabic, thus preserving the manuscripts that they had not destroyed. Moreover, they added to this store of Greek mathematics the arithmetic and algebra that had been developed in India. The only fields in which they made major contributions of their own were astronomy and spherical trigonometry. From the ninth century to the twelfth century, the Arab universities in Spain attracted European mathematicians, among whom were Gerbert, who later became Pope Sylvester II (d. 1003), and Adelard of Bath (c. 1120), who made a translation into Latin of an Arabic version of Euclid's *Elements* and commented on an arithmetic prepared from earlier works by Al-Khowarizmi (fl. 820). A corruption of the latter's name resulted in the word *algorism* (an efficient written method of calculation), while the Arabic word *al-jabr*, in the title of his work on the subject, accounts for the word "algebra."

Revival of European Mathematics.—The meager writings of Boethius constituted the only source of mathematical knowledge in the medieval church schools until the appearance in Ireland, England, and continental Europe of translations of Arabic works. The first book by a European explaining the Hindu-Arabic numerals, and methods of computing with them, that was more than a translation of an Arabic source was *Liber Abaci* ("Book of the Abacus") written in 1202 by Leonardo da Pisa, also known by his self-given nickname of Fibonacci ("Son of a Simpleton").

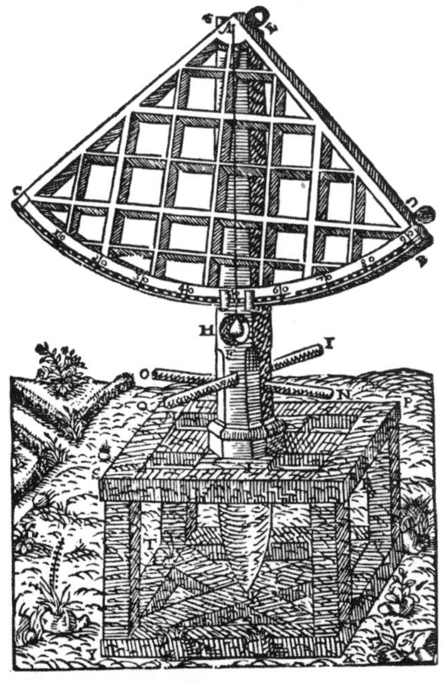

NEW YORK PUBLIC LIBRARY

PRE-GALILEAN ASTRONOMY relied on such nontelescopic instruments as this quadrant reproduced from Tycho Brahe's *Astronomiae Instauratae Mechanica*, published in 1598. Despite the crudeness of these instruments, astronomers were quite accurate.

BETTMANN ARCHIVE

GOTTFRIED WILHELM VON LEIBNIZ, noted for development of differential calculus.

The Fibonacci sequence 1, 1, 2, 3, 5, 8, . . . , in which each term (after the first two) is the sum of the two preceding terms, made its first appearance in this book. Mathematicians have connected this sequence with the "divine proportion" or "golden section," a proportion for line segments that is pleasing to the eye, prized by artists of the Renaissance. The advent of Fibonacci marked the beginning of three and a half centuries of Italian domination in the field of mathematics. Students from all over the Western world flocked to Italian universities to learn to multiply and divide with the new Arabic numerals and to study algebra. To spur discoveries in methods of solving equations, public contests for honors and a purse were held between two contenders to see which one could solve third-degree equations proposed by the other. One such contest led to the formula for the solution of all cubic equations by Niccolò Tartaglia (c. 1500–1557) and to the subsequent publication of Tartaglia's solution in the majestic treatise on algebra, the *Ars Magna* ("Great Art") of Jerome Cardan in 1545. It was a student of Cardan, Lodovico Ferrari, who solved the general fourth-degree (quartic) equation. For the three centuries that followed, activity in algebra was directed toward solving the general fifth-degree (quintic) equation. Attempts to unravel this knotty problem produced much new mathematics, but no solution.

The interest in algebra that was aroused by the solution of the cubic and quartic equations led to a great advance in the development of symbolic notation. This intermittent but agelong struggle started with the use by the Egyptians of the word "heap" for the unknown and hieroglyphs of legs walking left or right to represent addition and subtraction, respectively. By the middle of the eighteenth century, all the symbols of classical algebra had been invented, with mathematicians of many European countries taking part in the development. The question of notation, however, still occupies mathematicians.

Seventeenth-Century Developments.— Although every development in mathematics has its impact, some are so revolutionary that they alter the subject's course of development. One such advance came in 1637, when the French mathematician René Descartes published his *Discourse on Method.* In his essay on geometry, Descartes made use of a pair of numbers to locate a point in reference to a pair of conveniently drawn intersecting lines. This correspondence between number and point made it possible to study geometric relations by means of equations. The union of algebra and geometry in this manner is known as *analytic geometry,* and was a necessary step toward the invention of the calculus.

Most inventions in mathematics have been preceded by an extended period of preparation and by a particular need for the innovation. This was the case in the formulation of the *integral calculus,* which had its beginnings in Archimedes' great work on quadrature (finding areas bounded by curves). When Johannes Kepler (1571–1630) tried to explain his observations on the motion of the earth in relation to the sun, he made use of the conic sections invented in one form by the Greek mathematician Menaechmus (c. 350 B.C.) and in another form by Apollonius of Perga. Kepler also made use of a crude type of calculus to formulate one of his laws on planetary motion. It remained for Sir Isaac Newton (1642–1727) and Gottfried Wilhelm von Leibniz (1646–1716) to invent a *differential calculus*

BETTMANN ARCHIVE

BLAISE PASCAL, French mathematician, developed a theory of projective geometry.

and to discover that the processes of differentiation and integration are related operations. Although Leibniz published his results before Newton and used a notation superior to that of the English mathematician, the fact that the two had met and had exchanged correspondence led to one of the most bitter priority quarrels in the annals of science. Today both are granted credit for independent discovery.

The great astronomical developments that followed the use of the telescope by Galileo Galilei (1564–1642) in making celestial observations required quick methods of calculation. The invention of *logarithms* by John Napier in 1614 filled this need and gave great impetus to the development of *trigonometry.*

ERIC POLLITZER

PROJECTIVE GEOMETRY was one of the scientific principles utilized by Renaissance artists as they brought realism into painting. The dotted lines in this seventeenth-century woodcut illustrate the perspective the artists used to achieve a three-dimensional effect.

Moreover, the application of mathematics to the art of gambling, which had small beginnings among Italian algebraists in the fifteenth and sixteenth centuries, came to the mathematical foreground with the formulation of the laws of probability in the work of the French mathematician Pierre de Fermat (1601–1665) and Blaise Pascal (1623–1662) in 1653. The latter was also responsible for the first mechanical adding machine based on the use of gears. This device followed closely the invention of the slide rule by William Oughtred (1574–1660) in 1632.

The seventeenth century, which provided the setting for the invention of logarithms, analytic (coordinate) geometry, the theory of probability, and the calculus, also witnessed another important development—the invention by Gérard Desargues (1593–1662) and Blaise Pascal of *projective geometry*, which treats of geometric relations that remain invariant under central projection, a subject related to perspective.

Eighteenth-Century Developments.—During the seventeenth and eighteenth centuries, the attention of mathematicians was focused on the extension of the scope of probability of the calculus to physics and astronomy. The eighteenth century can be called the century of the great analysts, as those who have made signal contributions to the calculus are known. The Swiss mathematician Leonhard Euler (1707–1783) was the most prolific of these. The years he spent in Russia at the invitation of Catherine the Great laid the foundation for the excellence of Russian mathematicians in modern times. He and Joseph Louis Lagrange (1736–1813) invented the *calculus of variations*, a subject of great importance in applied mathematics.

NEW YORK PUBLIC LIBRARY

JOSEPH LOUIS LAGRANGE

NEW YORK PUBLIC LIBRARY

CHINESE OBSERVATORY at Peking was equipped with a large number of surprisingly accurate astronomical instruments, including the equatorial armillary (3), celestial sphere (4), zodiacal armillary (5), azimuth indicator (6), quadrant (7), and sextant (8). These instruments were indicative of a correspondingly high degree of mathematical sophistication.

■ **CHINESE CONTRIBUTIONS.**—The eighteenth century also witnessed the introduction into Europe, via the returning Jesuit missionaries, of Chinese mathematics. Although little is known of the early development of Chinese mathematics, because their early writings—written on perishable substances such as bark and bamboo—have not survived, it has been ascertained that the Chinese had developed a body of mathematical knowledge that was already ancient in 1100 B.C. In the years following the Arab conquest of Alexandria, many scholars moved eastward into India, which had trade and diplomatic relations with China. It is therefore assumed by historians that a very limited exchange of mathematical knowledge did take place between Europeans and Chinese and may have been responsible for the Chinese influence on the development in Europe of such areas as *determinants* (square arrays of numbers that are related to the coefficients of the unknowns in a set of equations that can be used to solve the equations) and of a set of numbers giving the coefficients of the expansion of a binomial raised to any positive integral power (the *Pascal triangle of numbers*).

It is definitely known, moreover, that Jesuit missionaries to China brought back to Europe, as early as the seventeenth century, knowledge of an equivalent of the *binary system* of numeration. This place system of numeration, which requires only two symbols—0 and 1—was resurrected by mathematicians building the first electronic computers; the "0" was indicative of the absence of a current and the "1" of the presence of a current. In the more recently developed computers, the binary system has been replaced by other methods of registering a number.

Rise of Non-Euclidean Geometry.—One of the topics that was intensively investigated during the eighteenth century was Euclid's *parallel postulate*, the modern equivalent of which is the statement that through a point outside a straight line, one, and only one, straight line may be drawn parallel to the given line. Mathematicians over the centuries had tried to prove this statement; that is, to derive this statement as a consequence of the other postulates and definitions of Euclid. In 1733, a book giving the results of an investigation into the consequences of denying this postulate and of substituting for it two postulates, each contradicting Euclid's, was published by an Italian Jesuit, Girolamo Saccheri (1667–1733). The two types of geometry that stemmed from the postulates that Saccheri substituted for Euclid's were so strange that even a hundred years after the publication of Saccheri's work, when Europe's most renowned mathematician discovered one of the two geometries of Saccheri, he was too timid to make

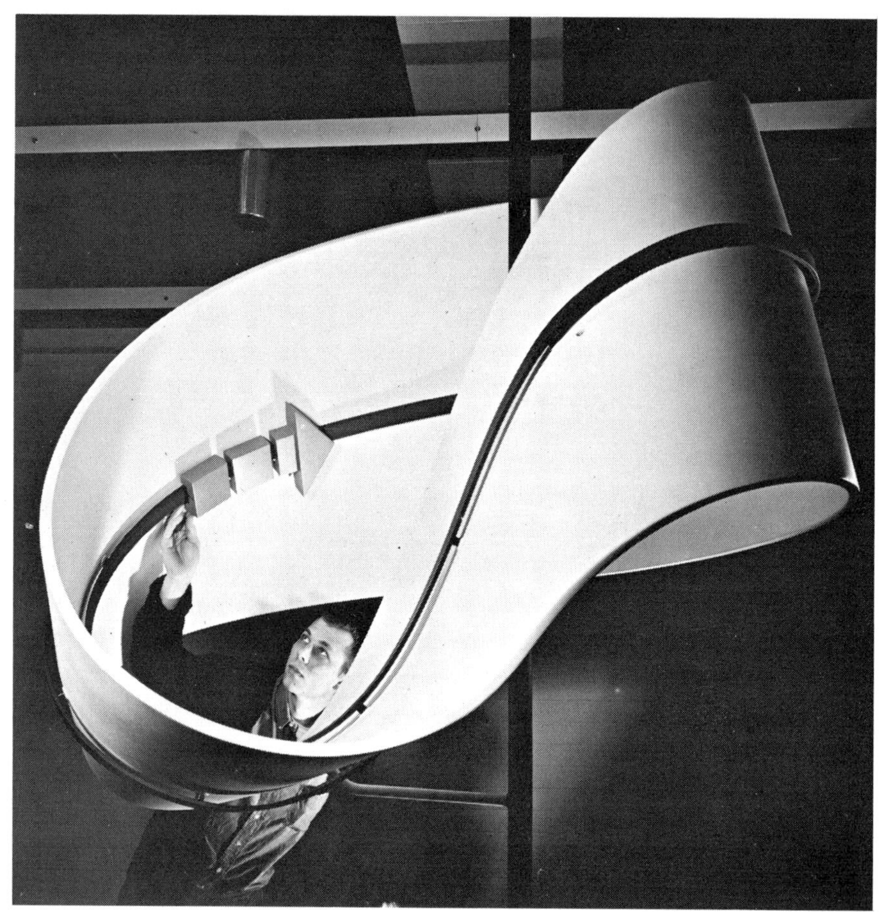

INTERNATIONAL BUSINESS MACHINES CORPORATION

MÖBIUS STRIP, discovered by A. F. Möbius in 1858, is a topological curiosity that has only one surface and only one edge. For this reason, when pushed completely around the strip along its track, the arrow will return to its starting point but will be inverted.

public his investigations. The mathematician in question was Karl Friedrich Gauss (1777–1855), the founder of the modern German school of mathematics, who is today credited with having been the first mathematician of note to have reached the conclusion that Euclid's *parallel postulate* was independent of the other postulates of Euclid, and to have recognized that it is possible to create another type of geometry that is as valid as that of Euclid. Gauss coined the word "non-Euclidean" to describe the new geometry. The name most widely associated with the discovery of the non-Euclidean geometry of the type studied by Gauss, however, is that of Nikolai I. Lobachevsky (1793–1856), whose first publication on the subject appeared in 1829. The Hungarian János Bolyai (1802–1860) is also given credit for the independent discovery of a non-Euclidean geometry, although his work did not make its appearance until 1832 and resembled that of Gauss and Lobachevsky. In 1854 a student of Gauss, G. F. B. Riemann (1826–1866), lectured at Göttingen University on another type of non-Euclidean geometry—the *elliptic geometry* that was first mentioned in Saccheri's work.

The importance to mathematics of the discovery of non-Euclidean geometries lies in the fact that it freed mathematics from the notion that a postulate, or axiom, had to meet with acceptance because it was a "self-evident" truth; that is, that it was consistent with our experience in the physical world. With the advent of the non-Euclidean geometries and the subsequent change in the concept of the function of a set of postulates, it became possible to develop mathematical systems entirely divorced from what we consider to be reality. Mathematics became abstract—it could no longer be conceived of as a subject that expounds absolute truths, but instead had to be thought of as one that draws necessary inferences from a set of postulates that may or may not be true.

Modern Mathematics.—While geometry was undergoing profound changes of aspect in the early part of the nineteenth century, the study of algebra was also being subjected to sweeping revision. The problem that sparked the revolution in algebra concerned the search for a solution to the equation of fifth degree, a search that had been going on in mathematical circles for the three centuries following Lodovico Ferrari's solution of the fourth-degree equation. Two young men were involved, and both were tragic figures in that they died without having enjoyed the public acclaim their work earned for them. Niels Henrik Abel (1802–1829), who died of tuberculosis caused by deprivation, proved the impossibility of solving the general equation of fifth degree in terms of radicals. This brought to an abrupt close the three-century-old search for a solution to the general quintic. (Unknown to Abel, however, the theorem had been proved inadequate by Paolo Ruffini [1765–1822] some 25 years earlier. It is now called the Abel-Ruffini theorem.) Evariste Galois (1811–1832), the other young genius, spent the night before a fatal duel writing a memoir on his theory of groups as they applied to the problem of solving algebraic equations. Although he had been anticipated by both Lagrange and Ruffini in this work, Galois was the first mathematician to show the structure of the *transformation group* associated with the roots of an algebraic equation and to point the way to the development of an abstract algebra. Today the Galois theory of equations has been superseded by more powerful methods, but his was the work that initiated a new school of mathematics—one in which the study of both algebra and geometry became a matter of studying transformation groups.

■ **STATISTICS.**—The study of *statistics* emerged as a separate discipline when mathematicians began to apply the theory of probability to fields other than gambling. Abraham De Moivre (1667–1754), known for his contributions to the theory of complex numbers, investigated the distributions of errors in a large collection of data and came upon a bell-shaped curve, now known as the *normal curve*. The first applications of statistics were related to astronomical data, and it was in this area that major contributions were made to the theory by several members of the Bernoulli family (seventeenth-eighteenth centuries), Pierre Simon de Laplace (1749–1827), Adrien Marie Legendre (1752–1833), Karl Friedrich Gauss, and a host of other mathematicians. The application of statistics to the analysis of social problems began with L. A. J. Quételet (1796–1874), a Belgian mathematician and astronomer who conceived of studying the incidence of various types of crime in a given population. This type of application was furthered by the work of Sir Francis Galton (1822–1911), who widened the scope of statistics to many fields, including education. The use of statistics in solving problems related to business and manufacturing is comparatively new, having received great impetus during World War II. It should be noted that the Russians have been active in this area and have made important contributions.

■ **TOPOLOGY.**—One of the subjects that has been largely responsible for changing the whole aspect of mathematics is *topology*. Generally classified as a branch of geometry, topology is concerned with the transformations that can be applied to a surface by twisting, stretching, and bending—but not cutting or breaking into—the surface. It is also applied to the study of nets of lines connecting points and to a variety of problems involving the location of geometric entities without regard to size. Although conceived originally as an oddity for mathematical recreation (and called *analysis situs* by the first to mention it, Gottfried Wilhelm von Leibniz), it has at the present time many applications to such diverse studies as economics and electrical circuitry. A pure mathematician now studies it by algebraic means.

■ **BOOLEAN ALGEBRA.**—When Leibniz searched for a "universal characteristic" that would make it possible to study logic from the point of view of the form of a statement divorced from the meaning of the words used in framing it, he anticipated the modern development of *symbolic logic*, which had its real beginnings about 1850 in the work of the Englishman George Boole (1815–1864). In his books *Symbolic Logic* and *Investigation into the Laws of Thought*, Boole used algebraic symbolism and the operations of arithmetic, as well as the symbols 0 and 1, to indicate "falsity" or "truth" of a statement. He also used the concept of classes of entities, the finite sets of modern mathematics. The modern Boolean algebra is an outgrowth of this effort and has become one of the tools of applied mathematics.

■ **TRANSFINITE NUMBERS.**—Whereas Boole developed an algebra of finite sets, the German mathematician Georg Cantor (1845–1918) dared attack the problem of formulating an algebra of infinite sets, the measures of which are called *transfinite numbers*. This work met with fierce denunciation by many prominent contemporaries, and it has only recently come to be fully appreciated as one of the most original and useful concepts of modern mathematics, forming the foundation of function theory and topology.

INTERNATIONAL BUSINESS MACHINES CORPORATION

CUBE OF LIGHTS visually illustrates the answer when any number from one to eight is multiplied, squared, or cubed by lighting the appropriate number of bulbs.

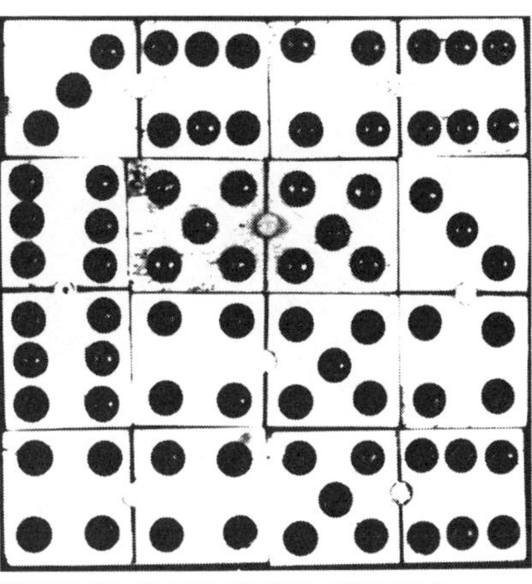

INTERNATIONAL BUSINESS MACHINES CORPORATION

MAGIC SQUARES are arrays of numbers whose columns, rows, and diagonals yield the same sum upon addition. The square on the left, whose sum is 19, is composed of dominoes. The array on the right, which was devised by the German painter Albrecht Dürer in 1514, uses each of the consecutive numbers from 1 to 16 only once and has a sum of 34.

■ **MATHEMATICAL ANALYSIS.**—Based on the work of Joseph Louis Lagrange, *mathematical analysis* began its development during the first third of the nineteenth century. Among the pioneers in mathematical analysis, besides Lagrange, were Sir William Rowan Hamilton (1805–1865) and K. G. J. Jacobi (1804–1851) who formulated some of the basic equations of *analytical dynamics*, which has proved to be of great value to engineers and physicists. In the late nineteenth century, the American physicist Josiah Willard Gibbs (1839–1903) was attracted to the work of Hamilton, as well as to that of Hermann Günther Grassmann, who had made formidable contributions to *geometric algebra*. By modifying their theories and combining them with his own, Gibbs formulated *vector analysis*, one of the physicist's most powerful mathematical tools.

Modern Mathematical Progress.—The theory of sets was evolved by Cantor at a time when mathematicians were occupied with the foundations of the subject. Since most of mathematics depends on (or grows out of) arithmetic, the first concern was with a definition of number that, together with certain postulates, could be used to evolve the whole body of mathematics. Advances in arithmetization were made by the German mathematician Karl T. Weierstrass (1815–1897), Georg Cantor, and Richard Dedekind (1831–1916), and the French mathematician Charles Meray (1835–1911). The work was carried forward by Giuseppe Peano (1858–1932) who, in collaboration with a large group of mathematicians, undertook to rewrite the whole system of mathematics in symbolic form. In the course of this work the group, publishing under the name of Peano, invented many of the symbols now used in set theory and logic. Peano's system of postulates makes it possible to derive the entire arithmetic of numbers from a small number of statements.

Another group of mathematicians is rewriting the whole body of mathematics in the rigor available to mathematicians in the mid-1960's. This group, which is composed of a varying number of France's best mathematicians (with a few nationals of other countries), writes under the name of the nonexistent "General Nicholas Bourbaki." They have produced about three dozen volumes on mathematics that have been used as resource materials by other writers and present the ultimate in the axiomatization of mathematics. It is this group that was instrumental in causing the revision of the curricula and methods of teaching mathematics that has recently been under way in most countries of the world.

The theory of sets left two questions that have only recently been answered. At the International Congress of Mathematicians in 1900, the German mathematician David Hilbert (1862–1943) gave an address in the course of which he suggested 15 problems that required solution. Among these problems was that of determining whether or not there was a transfinite number between that of the measure of Cantor's smallest transfinite, which corresponds to the size of the set of counting numbers, and the larger one, which corresponds to the size of the set of rational numbers. Professor Paul Cohen of Stanford University proved that the existence of an in-between number could neither be proved nor disproved.

The outcome of Cohen's work is that the theory of sets is placed in exactly the same position in which geometry found itself when the non-Euclidean geometries were proved to have the same degree of validity as that of Euclid. It is quite possible, therefore, that his work may have as profound an impact on the development of mathematics as the invention of the non-Euclidean geometry had more than a century ago.

—Linda Allegri

FLAGS, SYMBOLIZING ENEMY PLANES shot down, were used by pilots. This is an example of one-to-one correspondence, or *equivalent sets.*

Numbers.—The study of arithmetic is concerned with the notion of number and the fundamental concepts and relations that pertain to the use of numbers. As is often the case, the basic terms are not easily definable, if, indeed, they can be defined at all without circular reasoning. Although the sense of quantity is undoubtedly as old as historical man, records of the earliest development of numbers are lost. Man could probably keep "count" long before he had developed a written language. Records do indicate, however, that work with a system of numbers dates back to approximately 3500 B.C.

A number is a complete abstraction—something that cannot be seen. In order to better understand the nature of number, the concept of *set* is introduced. Basically, the term "set" is also undefined, although it does have several synonyms. A set may be thought of as a collection, a class, an aggregate, or an ensemble of anything whatsoever that one might wish to include in it. The items, objects, people, numbers, or ideas that may be used in constructing a particular set are referred to as the *members* or *elements* of the given set. Ordinarily, the elements that belong to a given set are identified by certain distinguishing characteristics. Thus, it is always possible to ascertain which elements belong to a given set and which elements do not. The set containing no elements is referred to as the *null*, or *empty, set*. A few common illustrations of the set concept are: a set of dishes, a herd of sheep, a class of students, a collection of stamps, a set of numbers. Since sets can be created mentally, there is no end to the kinds or quantity of sets that can be defined. Man's mental processes are continually organizing data into sets, for sets enable him to understand the nature of number. The two sets below include as elements only those markings (which have no particular meaning) seen within the closed curves.

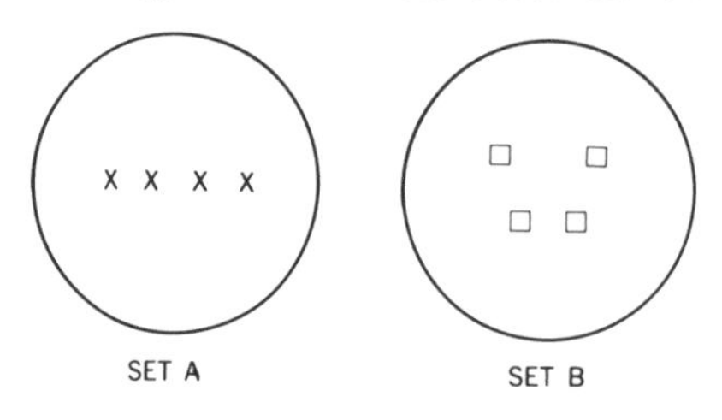

Each of these sets has different elements, yet one's mind is immediately aware of a property that is common to both of them. That property can be demonstrated in the diagram below.

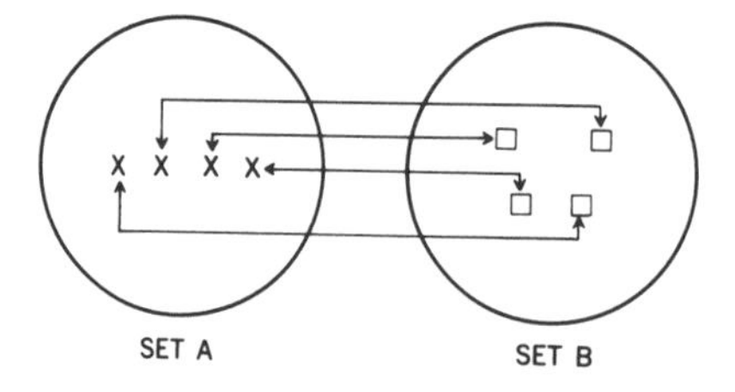

When the elements from set A are paired with the elements of set B, there are no elements left over in either set. Thus, each element of set A is said to correspond to one, and only one, element of set B, and each element of set B corresponds to one, and only one, element of set A. This pairing, or matching, of elements in two (or more) sets is called a *one-to-one correspondence.*

All of the sets whose elements can be placed in a one-to-one correspondence with each other are said to be *equivalent sets,* regardless of what the individual elements in each set may be. All equivalent sets are said to possess the same *cardinal* property. This property of sets answers the question "how many?" and is one of the fundamental aspects of the number concept. Equivalent sets are said to have the same cardinal number.

It is common practice to use capital letters to denote particular sets, and to enclose within a pair of braces the elements or statements describing the elements of the set. For example, to indicate the set of letters that constitute the English alphabet, one could use either: set $A = \{a,b,c,d,e,f,g,h,i,j,k,l,m,n,o,p,q,r,s,t,u,v,w,x,y,z\}$, or set $A = \{$all the letters of the English alphabet$\}$. To express the number associated with the set of letters from the English alphabet, one could write $n(A) = 26$. If set $A = \{\square\square\square\}$, then $n(A) = 3$; if set $A = \{1,2,3,4\}$, then $n(A) = 4$. In general, if A represents a particular set, then $n(A)$ is a number. However, it should be noted that a number is not the same as a set.

It is conjectured that, prior to the development of a number vocabulary, ancient man used bags of stones, notches in a stick, and similar devices to keep track of how many animals he had in a given year. For example, for each sheep or goat in a herd he would have one, and only one, stone. He had thus established a one-to-one correspondence between the elements in his herd and the elements in his bag of stones. During recent wars, airplane crews painted a picture on their planes for each enemy plane they destroyed. The familiar practice of making penciled tally marks when keeping a count of certain quantities of items is still another illustration of the use of the basic one-to-one correspondence idea, which is similar to the techniques used by primitive man.

■ MODEL COLLECTIONS.—An extension of the one-to-one correspondence concept occurred with the development of model collections that served to represent the cardinality (the "how many") of certain sets. For example, the set of wings on birds was used to symbolize the concept of "twoness"; the legs on certain animals, "fourness"; the fingers on a man's hand, "fiveness"; and so on. The names of these model collections could then be used to express the same information that had previously required the use of bags of stones or notches in a stick. The words for the model collections eventually underwent changes; they eventually lost their concrete and descriptive origins and became abstract words. Vestiges of these model collections are still apparent in some of the number-words of various cultures that exist today.

A second aspect of the concept of number can be illustrated in the following manner. In any racing competition there may be many contestants who begin; however, the final results are indicated by the order in which the contestants finish, that is, first, second, third, and so on. Here the concept of number tells the order of events. This is called the *ordinal* property of numbers. Common illustrations include house numbers, telephone numbers, and license numbers.

The process of counting could not be developed until the model collections had been arranged in a definite, ordered sequence. The very essence of counting implies that it must be possible to establish a natural progression of increasing magnitude among the model collections; that is, the model collections that are used to denote cardinality cannot be used in counting until they have been ordered (arranged) so that it is possible to move from one model to its successor that has one more member in it. Names and symbols can now be assigned to the cardinal numbers of these sets. To count a set of objects involves

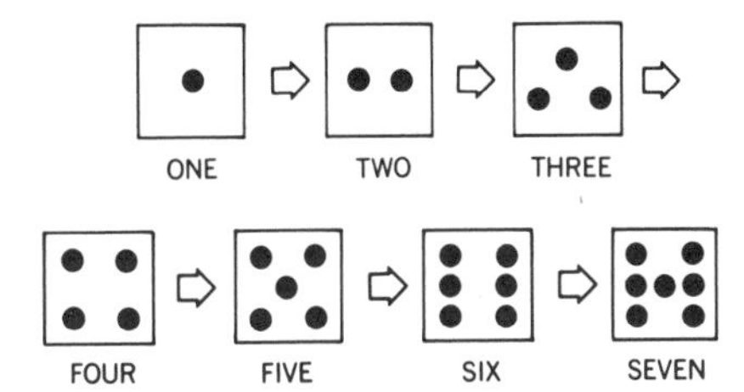

assigning to every element in the set a name from the ordered model collection beginning with the smallest and continuing until the last element in the set to be counted has been assigned a name from the model collection. The last name to be assigned is the ordinal number of the set. By this matching, two questions can be answered: "how many?" and "which one?" These questions correspond to the terms "cardinal" and "ordinal," respectively. The property of immediate succession that all counting numbers have is basic to the scientific process.

Systems of Numeration.—Knowledge of the historical development of a set of symbols (model collections) and of a language for conveying number sense is largely a matter of extrapolation and conjecture. The dates of man's earliest attempts at symbolization are lost. However, it is known from various artifacts that the Egyptians by 3500 B.C., and the Greeks and Chinese at about the same time, had used the tally mark (either a simple horizontal or vertical stroke) to refer to their beginning numbers. As the quantity of strokes increased, other symbols were introduced to summarize collections of the simple strokes. The various sets of symbols that were the inventions of men and have differed between cultures are called the *numerals* or *number symbols*. These numerals are the pictorial representations that man uses to refer to numbers which are complete abstractions. A set of numerals and the rules for their usage constitute a *numeration system*.

The set of symbols used today in the Western world is called *Hindu-Arabic* because it was probably first developed by the Hindus by 250 B.C.; it was later transported to Europe by the Arab traders. The earliest set of Hindu symbols did not contain the symbol zero; this was not introduced until about 600 A.D. Similarly, the general appearance of some of the original symbols has undergone modifications. Until the invention of the printing press in the fifteenth century, standardization of all written material was difficult.

■ EGYPTIAN.—In studying systems of numeration, there are certain underlying principles that are evident. For example, an *additive principle* is involved in the systems where different symbols are used to indicate collections of other symbols. Thus, to find the number that is being represented by a set of symbols (numerals), the individual numbers represented are simply added. An early illustration of this principle is the ancient Egyptian hieroglyphic system of numeration. (See the accompanying chart.) Note that considerable repetition of basic symbols was necessary in order to

WIDE WORLD

ORDINAL PROPERTY of numbers can be illustrated by a race. This aspect of number tells the order of events, or how the runners finish.

Egyptian System of Numeration

Modern Hindu-Arabic Numerals	Egyptian Numerals	
1	|	a simple stroke
2-9	||–|||||||||	series of simple strokes
10	∩	a heel bone
20	∩∩	two heel bones
55	∩∩∩∩∩|||||	five heel bones and five simple strokes
100	9	a coiled rope
1000	𓆼	a lotus flower
1,000,000	𓁨	a man in astonishment

Roman System of Numeration

Modern Hindu-Arabic Numerals	Roman Numerals
1	I
2–3	II–III
4	IV
5	V
6–8	VI–VIII
9	IX
10	X
20	XX
50	L
100	C
500	D
1,000	M

Babylonian System of Numeration

Modern Hindu-Arabic Numerals	Babylonian Numerals
1	Y
2-9	YY–YYY YYY YYY
10	<
60	Y
600	<
3602	<YY (early)
3602	<⁝YY (later; use of a symbol for the empty place)

represent large numbers. Also, the Egyptian system did not consider as important the order or position in which the symbol occurred in the representation of a particular number. Thus, a number such as 13 could be represented either as III ∩ or ∩ III. It is interesting to see that their system was based on ten, but that they did not have a symbol for zero and they did not use place value. Despite these limitations in their system of numeration, they did build the great pyramids between 3000 and 2000 B.C., which certainly required considerable mathematical sophistication.

■ **ROMAN.**—The Romans, whose influence was at its peak at about 100 A.D., had developed a system of numeration that was also essentially additive at first. However, later developments introduced both *subtractive* and *multiplicative principles*. The Romans used an *abacus* to perform their computations and then recorded the results using the Roman numerals. Subtraction was later introduced and was used when a smaller number preceded a larger one; the two quantities were considered together and the smaller was subtracted from the larger.

To express the number nine, the Romans used IX, which meant subtracting the quantity one from the quantity ten. This technique is believed to have been introduced in order to save space in writing or printing and therefore was probably a relatively recent innovation to the system. To represent large numbers, the Roman system used a bar over particular numerals. One bar meant that the symbol underneath was to be multiplied by one thousand. For example: $\overline{XX}IV$ represented the quantity 20,004. Using two bars above a numeral meant that the symbols underneath were to be multiplied by one million. Thus $\overline{\overline{XX}}IV$ represented 20,000,004. Whereas the Egyptian numeration system was a simple additive system, the Romans employed some subtraction and multiplication. The Romans also made some use of place value based on ten; that is, L = 5 × 10, C = 10 × 10, D = 5 × 10 × 10, and M = 10 × 10 × 10. While addition and subtraction may be performed quite easily in the Roman system, multiplication and division are difficult.

■ **BABYLONIAN.**—From ancient Babylonian (about 3000 B.C.) recordings on cuneiform tablets (cuneiform was the written language of the Sumerians, or Babylonians) there are indications that they were the first to use the principle of position, or place value. They performed their writing on soft pieces of clay with a stylus; the clay tablets were left to dry in the sun or were baked to form hard tablets that have been preserved to the present time. The oldest of these tablets, dated at about 3000 B.C., is from the city of Ur. The recognition and subsequent use of place value has been referred to as a most important step in the development of an efficient numeration system.

Greek System of Numeration

Modern Hindu-Arabic Numerals	Greek Symbol	Greek Name
1	A	alpha
2	B	beta
3	Γ	gamma
4	Δ	delta
5	E	epsilon
6	F	digamma
7	Z	zeta
8	H	eta
9	Θ	theta
10	I	iota
20	K	kappa
30	Λ	lambda
40	M	mu
50	N	nu
60	Ξ	xi
70	O	omicron
80	Π	pi
90	ϙ	koppa
100	P	rho
200	Σ	sigma
300	T	tau
400	Υ	upsilon
500	Φ	phi
600	X	chi
700	Ψ	psi
800	Ω	omega
900	ϡ	sampi

The Babylonians used an essential base of sixty; and since they wrote with a stylus on their clay tablets, the symbol for unity (one) appears as a wedge, Y. This wedge symbol was used to indicate one, sixty, sixty-sixties, or higher multiples of sixty, depending upon its position within the numerical representation. To denote the quantity ten, they put two wedge marks together at a slight angle to each other, <. In some of the records it appears that they may have used different-size wedges to distinguish between one and sixty. Their early use of place value consisted primarily of leaving a vacant space between wedges. Later, they put two double wedges together to represent zero. Some examples of their numeration are shown in the accompanying chart.

Today units of time are used that reflect the Babylonian base sixty (*sexagesimal*): sixty seconds in a minute, sixty minutes in an hour. Angle measurements also reflect the base sixty. Babylonian astronomers kept very accurate records of sun and moon eclipses. These and other records that have been well preserved on clay tablets have provided a complete legacy of Babylonian mathematical and scientific accomplishments.

■ **GREEK.**—The Greeks, whose civilization lasted roughly from 600 B.C. to 400 A.D., used a very simple system of numeration. It was based on ten, but neither had a symbol for zero nor used place value. They used letters from their alphabet to represent numbers. (See the accompanying chart.)

The Greeks made many contributions to the area of mathematics, in particular the area of proofs. Their greatest work was in geometry; they even did their algebra problems geometrically whenever possible because their numeration system was poorly developed.

■ **HINDU-ARABIC.**—The numeration system that is presently used in the United States and most other countries is basically Hindu in origin. However, through the 2,000 years since its first known usage, it has undergone several major modifications. The earliest Hindu numbers were called Brahmi numbers and did not have a symbol for zero nor did they make use of place value. By 500 A.D., the Indian symbols did include a zero and made use of place value. It was the Arab traders of this period that brought the system from India to Europe, and thus today it is referred to as the *Hindu-Arabic numeration system.* It is called a *base,* or *scale, ten (decimal) system,* because it uses only ten basic symbols or digits: 0, 1, 2, 3, 4, 5, 6, 7, 8, 9. These basic symbols are said to have face value. In addition to their face value, the location of a particular digit in a number representation is significant. This gives the digits positional, or place, value (similar to the technique introduced earlier by the Babylonians) in addition to their face value. Finally, the system uses the additive principle. Consider the Hindu-Arabic numeral, 5,965. This expresses the number five thousand nine hundred and sixty-five. The representation may be rewritten in the following form to illustrate the positional and additive principles: 5,000 + 900 + 60 + 5. By using powers of ten (exponential representation) this can be written in the form: $(5 \cdot 10^3) + (9 \cdot 10^2) + (6 \cdot 10^1) + (5 \cdot 10^0)$. Note that $10^0 = 1$. In this system of numeration, place value is in terms of powers of ten. The English language has definite words to describe some of these, as shown below.

Words Used to Describe Powers of Ten in U.S.*

Word	Numeral	Power of Ten
Units	1	10^0
Tens	10	10^1
Hundreds	100	10^2
Thousands	1,000	10^3
Ten thousands	10,000	10^4
Hundred thousands	100,000	10^5
Millions	1,000,000	10^6
Ten millions	10,000,000	10^7
Hundred millions	100,000,000	10^8
Billions	1,000,000,000	10^9
Ten billions	10,000,000,000	10^{10}
Hundred billions	100,000,000,000	10^{11}
Trillions	1,000,000,000,000	10^{12}

* In the United States and France, the system shown in this table is used to name the powers of ten; in England and Germany, a slightly different system is used.

In a system of numeration (such as the Hindu-Arabic) that makes use of a definite base and possesses the important symbol zero, uses positional notation, and is additive, any whole number can be represented by the general expression:

$$N = a_n x^n + a_{n-1} x^{n-1} + \ldots + a_2 x^2 + a_1 x^1 + a_0 x^0,$$

where x represents the particular base and a represents the respective coefficient. For example, in the Hindu-Arabic decimal system, the numeral 5,965 has 10 as its base, or x, value, and the coefficients are 5, 9, 6, and 5, reading from right to left.

$$\begin{aligned} N &= 5{,}965 \\ &= (5 \cdot 1000) + (9 \cdot 100) + (6 \cdot 10) + (5) \\ &= (5 \cdot 10^3) + (9 \cdot 10^2) + (6 \cdot 10^1) + (5 \cdot 10^0) \end{aligned}$$

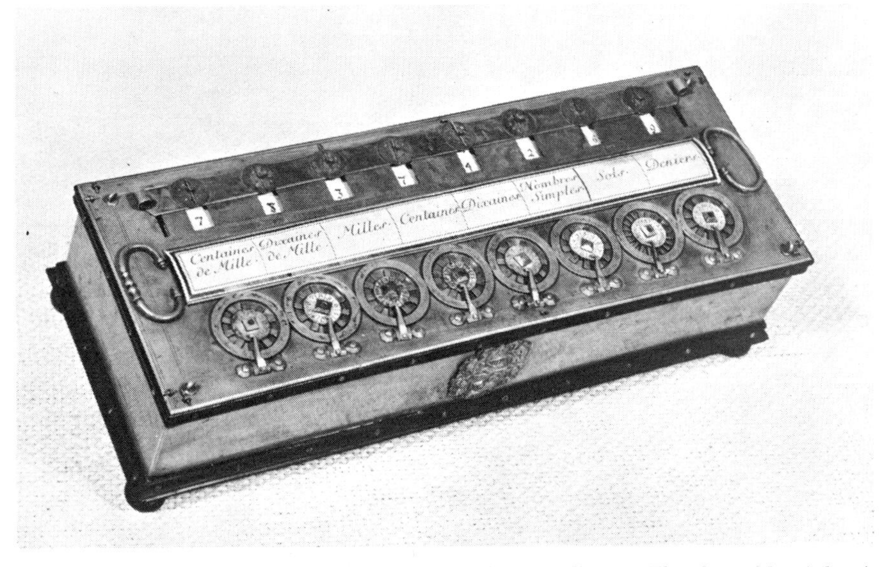

MATHEMATICAL CALCULATING DEVICES have existed for millennia. The clay tablet (*above*) was used in Babylon about 1600 B.C. The Roman abacus (*center*) dates back more than two thousand years and is not unlike the abacus used by the Chinese. The calculator (*below*), developed by Blaise Pascal in 1647, led the way to the computer of today.

QUINARY SYSTEM.—The use of other base values has received extensive consideration during recent years. For example, consider a system of numeration that has all the properties of the Hindu-Arabic, but which, instead of having ten different digits, has only five digits, 0, 1, 2, 3, 4. This would constitute a *base*, or *scale*, *five* (*quinary*) *system*. Having reached the number represented by 4, and thereby exhausting the individual digits, "fiveness" would be represented by the numeral 10, "sixness" by 11, "sevenness" by 12, and so on.

Note that the numerals 10, 11, and 12 are not read as ten, eleven, and twelve. They are read as either one-zero (one collection of five and no ones), one-one (one collection of five and one), one-two (one collection of five and two) or as five, six, and seven, respectively. Whereas our base ten system is referred to as a *decimal system* because it has ten different digits, the base five system is called a *quinary system*. In base ten, there is a *decimal* point to indicate the separation of a whole number and a fraction, whereas in base five this separation point would be called the *quinary* point. The accompanying table shows a comparison of base ten and base five in terms of the symbols, basic collections, and possible word names.

BINARY SYSTEM.—Another base that has found extensive usage in recent years is the *binary*, or *base two*, *system*. This is the simplest significant base that can be used. It has only two symbols and again it is convenient to use the symbols from base ten, 0 and 1. Records indicate that Thomas Harriot used a binary system during the early seventeenth century. Later in the same century, Gottfried Wilhelm Leibniz attempted to use the binary system for a religious purpose. He had hoped to convince the Chinese emperor that God (whose symbol was unity, or one) created everything from the void, which was, of course, represented by the zero. However, until the recent development of electronic data processing and computers, usage of base two was negligible. The reason that base two has become so vital to computers is that there are many electrical components, such as switches, relays, magnetic particles, and electronic tubes, that are capable of only two different positions, "On" and "Off." To represent the number one, a switch or relay may be closed so that electrical current will flow; to represent zero, the switch or relay may be open so that no current will flow.

In the binary system, the place values are in multiples of two. Thus, the numeral (1 or 0) in the rightmost place represents either 1 or 0, the next place represents the two's place, the next the four's, etc. For example, 110101 may be written in base two as follows:

$$(1\cdot 2^5)+(1\cdot 2^4)+(0\cdot 2^3)+(1\cdot 2^2)+(0\cdot 2^1)+(1\cdot 2^0)$$

The accompanying table compares base ten and base two in terms of their symbolization, basic collections, and possible word names.

DUODECIMAL SYSTEM.—Another base that has received considerable attention recently is the *duodecimal*, or *base twelve*, *system*. The utilization of this base necessitates the introduction of two new symbols; often *t* and *e* are used to represent the concept of "tenness" and "elevenness." Thus, a complete set of digits for base twelve is: 0, 1, 2, 3, 4, 5, 6, 7, 8, 9, *t*, and *e*. Twelve is represented by 10 (one-zero). The advantages of this base are to be found in the area of division. Whereas the base ten has only 2 and 5 as factors, a base of twelve has 2, 3, 4, and 6. This provides a considerable advantage in computations that involve fractions. Many products are handled in lots, or collections, of twelve. For example, eggs are sold by the dozen, or dozen-dozen (gross). In 1944, the Duodecimal Society of America was founded and has served as the center of activity for those persons interested in the promotion of this base. The accompanying table compares base ten and base twelve in terms of symbolization, basic collections, and possible word names.

To summarize the discussion of numeration systems, the accompanying table compares the four bases which have been presented.

In order to perform the basic operations of arithmetic (addition, subtraction, multiplication, and division) in the various bases, the basic addition and multiplication facts must be known. The following tables present this information, which will be referred to in the next section.

Comparison of Base Ten with Base Five

Symbol	Base Ten Collections	Names	Symbol	Base Five Collections	Names
0		zero	0	X	zero
1	X	one	1	X	one
2	XX	two	2	XX	two
3	XXX	three	3	XXX	three
4	XXXX	four	4	XXXX	four
5	XXXXX	five	10	XXXXX	one five and zero
6	XXXXXX	six	11	XXXXX X	one five and one
7	XXXXXXX	seven	12	XXXXX XX	one five and two
8	XXXXXXXX	eight	13	XXXXX XXX	one five and three
9	XXXXXXXXX	nine	14	XXXXX XXXX	one five and four
10	XXXXXXXXXX	ten	20	XXXXX XXXXX	two fives and zero
11	XXXXXXXXXX X	eleven	21	XXXXX XXXXX X	two fives and one
25	XXXXXXXXXX XXXXXXXXXX XXXXX	twenty-five	100	XXXXX XXXXX XXXXX XXXXX XXXXX	one five-fives, zero fives and zero
43	XXXXXXXXXX XXXXXXXXXX XXXXXXXXXX XXXXXXXXXX XXX	forty-three	133	XXXXX XXXXX XXXXX XXXXX XXXXX XXXXX XXXXX XXXXX XXX	one five-fives three fives and three $(1\cdot 5^2+3\cdot 5^1+3\cdot 5^0)$

Comparison of Base Ten with Base Two

Symbol	Base Ten Collections	Names	Symbol	Base Two Collections	Names
0		zero	0		zero
1	X	one	1	X	one
2	XX	two	10	XX	one two and zero
3	XXX	three	11	XX X	one two and one
4	XXXX	four	100	XX XX	one two-twos and zero
5	XXXXX	five	101	XX XX X	one two-twos and one
10	XXXXXXXXXX	ten	1010	XX XX XX XX XX	(names become very cumbersome)
16	XXXXXXXXXX XXXXXX	sixteen	10000	XX XX XX XX XX XX XX XX	

Comparison of Base Ten with Base Twelve

Symbol	Base Ten Collections	Names	Symbol	Base Twelve Collections	Names
0		zero	0		zero
1	X	one	1	X	one
9	XXXXXXXXX	nine	9	XXXXXXXXX	nine
10	XXXXXXXXXX	ten	*t*	XXXXXXXXXX	ten
11	XXXXXXXXXX X	eleven	*e*	XXXXXXXXXXX	eleven
12	XXXXXXXXXX XX	twelve	10	XXXXXXXXXXXX	one twelve and zero
18	XXXXXXXXXX XXXXXXXX	eighteen	16	XXXXXXXXXXXX XXXXXX	one twelve and six
24	XXXXXXXXXX XXXXXXXXXX XXXX	twenty-four	20	XXXXXXXXXXXX XXXXXXXXXXXX	two twelves and no ones

Addition Facts

Base Ten

+	0	1	2	3	4	5	6	7	8	9	10
0	0	1	2	3	4	5	6	7	8	9	10
1	1	2	3	4	5	6	7	8	9	10	11
2	2	3	4	5	6	7	8	9	10	11	12
3	3	4	5	6	7	8	9	10	11	12	13
4	4	5	6	7	8	9	10	11	12	13	14
5	5	6	7	8	9	10	11	12	13	14	15
6	6	7	8	9	10	11	12	13	14	15	16
7	7	8	9	10	11	12	13	14	15	16	17
8	8	9	10	11	12	13	14	15	16	17	18
9	9	10	11	12	13	14	15	16	17	18	19
10	10	11	12	13	14	15	16	17	18	19	20

Base Twelve

+	0	1	2	3	4	5	6	7	8	9	*t*	e	10
0	0	1	2	3	4	5	6	7	8	9	*t*	e	10
1	1	2	3	4	5	6	7	8	9	*t*	*e*	10	11
2	2	3	4	5	6	7	8	9	*t*	*e*	10	11	12
3	3	4	5	6	7	8	9	*t*	*e*	10	11	12	13
4	4	5	6	7	8	9	*t*	*e*	10	11	12	13	14
5	5	6	7	8	9	*t*	*e*	10	11	12	13	14	15
6	6	7	8	9	*t*	*e*	10	11	12	13	14	15	16
7	7	8	9	*t*	*e*	10	11	12	13	14	15	16	17
8	8	9	*t*	*e*	10	11	12	13	14	15	16	17	18
9	9	*t*	*e*	10	11	12	13	14	15	16	17	18	19
t	*t*	*e*	10	11	12	13	14	15	16	17	18	19	1*t*
e	*e*	10	11	12	13	14	15	16	17	18	19	1*t*	1*e*
10	10	11	12	13	14	15	16	17	18	19	1*t*	1*e*	20

Base Five

+	0	1	2	3	4	10
0	0	1	2	3	4	10
1	1	2	3	4	10	11
2	2	3	4	10	11	12
3	3	4	10	11	12	13
4	4	10	11	12	13	14
10	10	11	12	13	14	20

Base Two

+	0	1
0	0	1
1	1	10

Multiplication Facts

Base Ten

×	0	1	2	3	4	5	6	7	8	9	10
0	0	0	0	0	0	0	0	0	0	0	0
1	0	1	2	3	4	5	6	7	8	9	10
2	0	2	4	6	8	10	12	14	16	18	20
3	0	3	6	9	12	15	18	21	24	27	30
4	0	4	8	12	16	20	24	28	32	36	40
5	0	5	10	15	20	25	30	35	40	45	50
6	0	6	12	18	24	30	36	42	48	54	60
7	0	7	14	21	28	35	42	49	56	63	70
8	0	8	16	24	32	40	48	56	64	72	80
9	0	9	18	27	36	45	54	63	72	81	90
10	0	10	20	30	40	50	60	70	80	90	100

Base Twelve

×	0	1	2	3	4	5	6	7	8	9	*t*	e	10
0	0	0	0	0	0	0	0	0	0	0	0	0	0
1	0	1	2	3	4	5	6	7	8	9	*t*	*e*	10
2	0	2	4	6	8	*t*	10	12	14	16	18	1*t*	20
3	0	3	6	9	10	13	16	19	20	23	26	29	30
4	0	4	8	10	14	18	20	24	28	30	34	48	40
5	0	5	*t*	13	18	21	26	2*e*	34	39	42	47	50
6	0	6	10	16	20	26	30	36	40	46	50	56	60
7	0	7	12	19	24	2*e*	36	41	48	53	5*t*	65	70
8	0	8	14	20	28	34	40	48	54	60	68	74	80
9	0	9	16	23	30	39	46	53	60	69	76	83	90
t	0	*t*	18	26	34	42	50	5t	68	76	84	92	*t*0
e	0	*e*	1*t*	29	38	47	56	65	74	83	92	*t*1	*e*0

Base Five

×	0	1	2	3	4	10
0	0	0	0	0	0	0
1	0	1	2	3	4	10
2	0	2	4	11	13	20
3	0	3	11	14	22	30
4	0	4	13	22	31	40
10	0	10	20	30	40	100

Base Two

×	0	1
0	0	0
1	0	1

■ **CONVERSIONS.**—Numerals written in various bases (that is, numeration systems that employ the same general principles as the Hindu-Arabic system) can be converted to base ten notation.

Examples:

Change 324_{five} to base ten notation.

$= (3 \times \text{five} \times \text{five}) + (2 \times \text{five}) + 4$
$= (3 \times 25) + (10) + 4$
$= 75 + 10 + 4$
$= 89_{ten}$

Change 1101_{two} to base ten notation.

$= (1 \times \text{two} \times \text{two} \times \text{two}) + (1 \times \text{two} \times \text{two}) + (0 \times 2) + 1$
$= (1 \times 2 \times 2 \times 2) + (1 \times 2 \times 2) + (0 \times 2) + 1$
$= 8 + 4 + 0 + 1$
$= 13_{ten}$

Change $2e5_{twelve}$ to base ten notation.

$= (2 \times \text{twelve} \times \text{twelve}) + (e \times \text{twelve}) + 5$
$= (2 \times 12 \times 12) + (11 \times 12) + 5$
$= 288 + 132 + 5$
$= 425_{ten}$

To change numbers written in base ten numerals to other base numerals we again make use of the fact that the size (or quantity) of the base collections has to be altered. Whether the collections are larger or smaller than ten will depend upon whether the new base is larger or smaller than ten.

Example:

Change 97_{ten} to base five notation. In base five the values of the places are one, five, five × five, five × five × five, and so on. In base ten these place values

Comparison of the Four Bases Presented

Base Ten (Decimal)	Base Five (Quinary)	Base Two (Binary)	Base Twelve (Duodecimal)
1	1	1	1
2	2	10	2
3	3	11	3
4	4	100	4
5	10	101	5
6	11	110	6
7	12	111	7
8	13	1000	8
9	14	1001	9
10	20	1010	*t*
11	21	1011	*e*
12	22	1100	10
50	200	110010	42
100	400	1101100	84
1000	13000	1111101000	6e4

are 1, 5, 25, and 125. Thus 97_{ten}

$= 75 + 20 + 2$
$= (3 \times 25) + (4 \times 5) + 2$
$= (3 \times \text{five} \times \text{five}) + (4 \times \text{five}) + 2.$

This last expression is written as 342 in base five, since it represents three collections of five-fives, four collections of five, plus two ones. The following procedure of a step by step breakdown into multiples of five is equivalent to repeated division by five.

5)97 = 19	with a remainder of 2	This indicates there are 19 fives and 2 ones.
5)19 = 3	with a remainder of 4	This indicates there are 3 five-fives and 4 fives.
5)3 = 0	with a remainder of 3	This indicates there are no collections of five × five × five; however, there are 3 five-fives.

INTERNATIONAL BUSINESS MACHINES CORPORATION

ELECTRONIC COMPUTERS rely almost entirely upon the binary system because electronic tubes and transistors are capable of only two different positions: "On" and "Off."

By writing the remainders in the reverse order of the repeated division, the correct result of 342 is obtained. Therefore, $97_{ten} = 342_{five}$.

Example:

Change 475_{ten} to base two notation using the repeated division procedure as shown above.

2)475 — 237 with a remainder of 1
2)237 — 118 with a remainder of 1
2)118 — 59 with a remainder of 0
2)59 — 29 with a remainder of 1
2)29 — 14 with a remainder of 1
2)14 — 7 with a remainder of 0
2)7 — 3 with a remainder of 1
2)3 — 1 with a remainder of 1
2)1 — 0 with a remainder of 1

By writing the remainders in the reverse order of the repeated division as in the previous example, the result $475_{ten} = 111011011_{two}$ is obtained.

Example:

Change 1862_{ten} to base twelve notation using the repeated division procedure as shown above.

12)1862 — 155 with a remainder of 2
12)155 — 12 with a remainder of *e*
12)12 — 1 with a remainder of 0
12)1 — 0 with a remainder of 1

By writing the remainders in the reverse order of the repeated division as in the previous examples, the result $1862_{ten} = 10e2_{twelve}$ is obtained.

OPERATIONS.—To perform the operation of addition in base ten, one has had to learn the 100 fundamental addition facts or combinations. To add in base five, the 25 fundamental combinations must be known; in base twelve there are 144, while in base two there are only 4 basic combinations. The tables previously given illustrate what the fundamental facts are for each of the bases mentioned.

The following example is given in order to review the steps that are taken in adding, or finding the sum of, two numbers in base ten.

Example:

Find the sum of 28_{ten} and 45_{ten}.

2 tens + 8 ones
+ 4 tens + 5 ones
6 tens + 13 ones
= 7 tens + 3 ones
= 73_{ten}

The same general procedure is used to find the sum of two numbers in other bases. The following examples illustrate this for the bases five, twelve, and two.

Examples:

Find the sum of 22_{five} and 14_{five}.

2 fives + 2 ones
+ 1 five + 4 ones
3 fives + 11 ones*
= 4 fives + 1 one
= 41_{five}

(*Note: 2+4 = 11 in base five.)

Find the sum of 36_{twelve} and 75_{twelve}.

3 twelves + 6 ones
+ 7 twelves + 5 ones
t twelves + *e* ones*
= te_{twelve}

(*Note: 3 + 7 = *t* and 6 + 5 = *e* in base twelve.)

Find the sum of 101_{two} and 011_{two}.

1(two × two) + 0(two) + 1 one
+ 0(two × two) + 1(two) + 1 one
1(two × two) + 1(two) + 10 ones
= 1(two × two) + 10(two) + 0 one
= 10(two × two) + 0(two) + 0 one
= 1000_{two}

To perform the operation of subtraction (inverse operation of addition) the fundamental combinations in the addition table must again be referred to. To find the difference of two numbers, that is, $a - b$, a number c must be found which, when added to b, will equal a.

For example, in base ten $8 - 3 = 5$ since $8 = 3 + 5$. The following examples are given to show the steps performed in subtraction.

Examples:

Find the difference of 861_{ten} minus 284_{ten}.

8 hundreds + 6 tens + 1 one
− 2 hundreds + 8 tens + 4 ones

7 hundreds + 15 tens + 11 ones
= − 2 hundreds + 8 tens + 4 ones
5 hundreds + 7 tens + 7 ones
= 577_{ten}

(Check: 861 − 284 = 577 since 861 = 284 + 577 in base ten.)

Find the difference of 321_{five} minus 142_{five}.

3 five-fives + 2 fives + 1 one
− 1 five-fives + 4 fives + 2 ones

2 five-fives + 11 fives + 11 ones
= − 1 five-fives + 4 fives + 2 ones
1 five-fives + 2 fives + 4 ones
= 124_{five}

(Check: 321 − 142 = 124 since 321 = 142 + 124 in base five.)

Find the difference of $4e2_{twelve}$ minus 212_{twelve}.

4 twelve-twelves + *e* twelves + 2 ones
− 2 twelve-twelves + 1 twelve + 2 ones
2 twelve-twelves + *t* twelves + 0 ones

(Check: 4*e*2 − 212 = 2*t*0 since 4*e*2 = 212 + 2*t*0 in base twelve.)

To multiply and divide in the various bases, the fundamental multiplication facts for each base are necessary. The operation of multiplication can be performed by either a horizontal or a vertical procedure.

Examples:

Multiply 4563_{ten} by 18_{ten} using the horizontal procedure.

= 18(4 thousands + 5 hundreds + 6 tens + 3 ones)
= $(18 \times 4) \times 10^3 + (18 \times 5) \times 10^2 + (18 \times 6) \times 10^1 + (18 \times 3)$
= $(72 \times 10^3) + (90 \times 10^2) + (108 \times 10^1) + 54$
= 82134_{ten}

Multiply 4563_{ten} by 18_{ten} using the vertical procedure.

4563_{ten}
$\times\ 18_{ten}$
36504
4563
82134_{ten}

These procedures are equally applicable to multiplication in other bases, but the vertical procedure is more frequently used.

Examples:

Multiply 34_{five} by 4_{five}.

34_{five}
$\times\ 4_{five}$
301_{five}

(Note: 4 × 4 = 31 in base five. Write the 1 and carry 3 fives. The second step involves multiplying 4 × 3, which equals 22 in base five, and then adding the carry of 3; 22 + 3 = 30 in base five.)

Multiply 243_{five} by 24_{five}.

243_{five}
$\times\ 24_{five}$
2132
1041
13042_{five}

Multiply 39_{twelve} by 8_{twelve}.

39_{twelve}
$\times\ 8_{twelve}$
260_{twelve}

(Note: 8 × 9 = 60 in base twelve. Write the 0 and carry 6 twelves. The second step involves multiplying 8 × 3, which equals 20 in base twelve, and then adding the carry of 6; 20 + 6 = 26 in base twelve.)

Multiply 11011_{two} by 111_{two}.

11011_{two}
$\times\ 111_{two}$
11011
11011
11011
10111101_{two}

(Note: In this base, multiplication is very easy; the only problem is in adding the various products.)

To perform the operation of division (inverse operation of multiplication), the facts from the multiplication table are used. In dividing we are "undoing" multiplication. In base ten, for example, $8 \times 6 = 48$, while $48 \div 8 = 6$. This can be stated in a general way which holds true for all bases: If $a = b \times c$, then $a \div b = c$ and $a \div c = b$. There are several forms used in performing the operation of division, which are discussed more fully in the section on operations on whole numbers. For the purpose here, the more commonly used method, as illustrated for the base ten below, will be employed.

Example:

Divide 6270_{ten} by 38_{ten}.

```
           165ten
    38ten)6270teu
          38
          247
          228
           190
           190
          (no remainder)
```

(Note: It is assumed that the reader is familiar with this procedure. If not, see the section on division of whole numbers, page 1091.)

In this example there was no remainder after the last step. In practice, however, there frequently is a remainder that may be expressed as a fraction or a whole number of units; or further steps may be taken in the division process to express the remainder as a decimal fraction.

Example:

Divide 422_{ten} by 4_{five}.

```
          103five
    4five)422five
          4
           22
           22
```

Divide 434_{five} by 23_{five}.

```
           14five
    23five)434five
           23
           204
           202
             2  (with a remainder
                   of 2)
```

Divide 11_{two} by 10_{two}.

```
            1two
    10two)11two
          10
           1  (with a remainder
                 of 1)
```

Divide 11101_{two} by 101_{two}.

```
              101two
    101two)11101two
           101
            1001
             101
             100  (with a remainder
                      of 100)
```

Divide $e3_{twelve}$ by 5_{twelve}.

```
              23twelve
    5twelve)e3twelve
            t
            13
            13
```

Divide $9et_{twelve}$ by $5e_{twelve}$.

```
               18twelve
    5etwelve)9ettwelve
             5e
             40t
             3e4
              16  (with a remainder
                      of 16)
```

The operation of division is relatively more difficult than the other operations since it involves addition, subtraction, and multiplication.

It is possible to use fractions in these other bases, even though their appearance is somewhat confusing at first. For example, in base ten, one-half is indicated by the symbol $\frac{1}{2}$. Since base two has no symbol 2, one-half would be expressed as $\frac{1}{10}_{two}$. The fraction eleven-twelfths from base ten ($\frac{11}{12}$) would be $\frac{e}{10}$ in base twelve. If it is desired to express these fractions as decimal fractions, the need for naming the decimal point must be recognized, since it by name refers to the decimal, or base ten, system. For base five, the term *quinary point* could be used, for base two, the *binary point*, and for base twelve, the *duodecimal point*.

It should be noted that after having performed any one of the four operations in a base other than ten, it is always possible to check the work by converting the numbers involved into base ten and then reperforming the operation.

The study of the growth and development of numeration systems closely parallels the growth and scientific accomplishments of civilizations. This study also permits a better understanding and appreciation of the present base ten system. The fact that base ten is used does not necessarily mean that it is the best of all possible numeration systems. It has been said that, had Cleopatra's nose been an inch longer, the course of history might have been different. So too, the course of numeration systems might have been appreciably different had man been born with more or less than ten fingers. It is important to realize, however, that without an efficient numeration system advances in other areas of knowledge would be greatly impeded, if not made impossible at some points. It can also be speculated that if man had a better system (whatever that might be), his civilization would now be further advanced.

The Whole Numbers.—In the previous section the set of natural numbers was defined. For instance, the natural number 2 is the number of the set $A = \{a, b\}$ and is also the number of any set which may be placed in one-to-one correspondence with the set A. The set of all such numbers is $N = \{1, 2, 3, 4, 5, 6, \ldots\}$. The set of *whole numbers* will be thought of as the natural numbers together with the number zero. Zero (0) is defined as the number of an empty set, that is, a set with no elements. An example of an empty set would be the set of all living men over 12 feet tall. For convenience, the set of whole numbers will be denoted by the capital letter $W = \{0, 1, 2, 3, 4, 5, 6, 7, \ldots\}$.

In studying the system of whole numbers, we shall be interested in the usual operations on whole numbers and the laws which govern the behavior of whole numbers with respect to these operations. Let us consider what is meant by the term *operation*. By an operation on the whole numbers we mean any process which associates a single unique whole number with one, two, three, or more whole numbers. For example, the operation which associates 1 with 1, 4 with 2, 9 with 3, 16 with 4, and so on, is called the *squaring* operation. Since this operation associates a single element of W with a single element of W it is called a *unary* operation. Most of the usual operations with whole numbers are *binary* operations, since they associate a single unique element of W with a *pair* of elements of W. Addition associates the single element 5 with the pair of elements (2, 3) and multiplication associates the element 6 with the pair (2, 3). As an example of a *ternary* operation, consider the operation of taking the maximum of three whole numbers. Under this operation max $(2, 4, 7) = 7$.

Basic Operations.—The operations of addition, multiplication, subtraction, and division are commonly called the four fundamental operations of arithmetic. Of these operations, addition and multiplication are basic, since subtraction and division may be defined from them. In the operation of addition, two whole numbers are associated with a third whole number called the *sum*. Each of the two given numbers is called an *addend*. In the operation of multiplication, two whole numbers are associated with a third whole number called the *product*. Each of the two given whole numbers is called a *factor*. Thus, (a, b, c elements of W) $a + b = c$ is written for addition and $a \times b = c$, or $a \cdot b = c$, or $ab = c$ for multiplication.

■ **ADDITION.**—The operation of addition is very closely allied with the counting process and the experience of combining sets of objects. For example, suppose that there are two separate bookshelves and it is desired to determine the total number of books on them. There are two alternatives: the two collections of books can be mentally combined into one set and the total number of books can be added; or, the number of books on each shelf can be counted and then "added" to find the total number of books. It is in the comparison of these two techniques that the germ of the meaning of addition is found.

Now consider the problem of finding the sum of 3 and 4 $(3 + 4)$, assuming no previous knowledge of such sums. To determine the sum, choose any set A which has 3 elements and any other set B which has no elements in common with A and contains 4 elements. For example, let $A = \{r, s, t\}$ and $B = \{a, b, c, d\}$. As in the bookshelf example, the total number of letters of the alphabet that are in the two sets can be found by combining them into a single set and counting. Combining two sets into a single set in this way is called taking their union and is denoted by the symbol $A \cup B$. (In general the union of two sets is the set which contains all elements that belong to either of them.) Here $A \cup B = \{r, s, t, a, b, c, d\}$ and the number of elements in $A \cup B$ is 7, by counting. It is on this basis that it can be concluded that $3 + 4 = 7$.

Of course it is not necessary or practical to carry out such an involved procedure in every situation that requires addition. Rather, it is used as an illustration of the basic meaning of the operation from which the definition and further properties are deduced. Any process which leads to the sum of two whole numbers in the above sense will be called an addition.

■ **DEFINITION.**—The definition of addition of whole numbers can now be easily framed in terms of the above example and the language of sets. If m and w are any two whole numbers, and A and B are two sets with no elements in common such that $n(A) = m$ and $n(B) = w$, then $m + w = n(A) + n(B) = n(A \cup B)$. Note that the restriction that A and B have no elements in common is a necessary one. If the two sets were to have elements in common, these would be counted only once in the union of the two sets, whereas they would be counted once in each of the individual sets. This is made clear if one considers the set A to be members of the Rotary Club in a given community and B to be the members of the Chamber of Commerce in the same community. Presumably, there will be some overlap in the membership of these two organizations. It is clear that the total number of persons in the two organizations is not the sum of the numbers of persons in each organization.

On the basis of this definition there are several important and readily apparent properties possessed by the operation of addition with whole numbers: (1) For any two whole numbers m and w there is always a whole number v associated with them as their sum; this is called the *closure property of addition* for whole numbers. (2) For any two whole numbers m and w the sum is the same regardless of the order in which taken, that is, $m + w = w + m$; this principle is called the *commutative law of addition*. (3) For any three whole numbers m, w, and v, $m + w + v = (m + w) + v = m + (w + v)$. Since the definition of addition is as a binary operation it does not directly provide a technique for adding more than two whole numbers. The law just stated, called the *associative law of addition*, gives a technique for adding three whole numbers in two different ways. By this law $2 + 3 + 5 = (2 + 3) + 5 = 5 + 5 = 10$, or $2 + 3 + 5 = 2 + (3 + 5) = 2 + 8 = 10$. In practice, the *associative* and *commutative laws* allow the addition of whole numbers in any order or grouping. (4) The whole number zero (0) has the special property that when added to any whole number the result is that same whole number. Thus $m + 0 = m$ for any whole number m. 0 is called the *identity* element for the operation of addition. No other whole number has this property. This property of zero, though seemingly trivial, will be seen to be very important.

■ **MULTIPLICATION.**—There are at least two possible alternative interpretations for the operation of multiplication. The first of these is essentially in terms of the operation of addition. For example, $3 \cdot 4$

Justification of the Process of Addition

Steps	Justification
$21 + 65 = (20 + 1) + (60 + 5)$	Hindu-Arabic system of numeration.
$= (2 \cdot 10^1 + 1 \cdot 10^0) + (6 \cdot 10^1 + 5 \cdot 10^0)$	Hindu-Arabic system of numeration.
$= 2 \cdot 10^1 + (1 \cdot 10^0 + 6 \cdot 10^1) + 5 \cdot 10^0$	Associative law for addition.
$= 2 \cdot 10^1 + (6 \cdot 10^1 + 1 \cdot 10^0) + 5 \cdot 10^0$	Commutative law for addition.
$= (2 \cdot 10^1 + 6 \cdot 10^1) + (1 \cdot 10^0 + 5 \cdot 10^0)$	Associative law for addition.
$= (2 + 6) \cdot 10^1 + (1 + 5) \cdot 10^0$	Distributive law.
$= 8 \cdot 10^1 + 6 \cdot 10^0$	Basic addition facts.
$= 80 + 6$	Hindu-Arabic system of numeration.
$= 86$	Hindu-Arabic system of numeration.

may be thought of as finding the total number of elements in three sets of 4 elements each, or $4 + 4 + 4$. Using the *associative law of addition*, $4 + 4 + 4 = 8 + 4 = 12$. One difficulty with the above process is that from this interpretation it is not readily evident that $3 \cdot 4 = 4 \cdot 3$. In this case it would appear that $3 \cdot 4 = 4 + 4 + 4$ and $4 \cdot 3 = 3 + 3 + 3 + 3$. These two results are entirely different in form, and in the general case would not appear alike at all. However, returning to the original example of three sets of four elements, a reasonable interpretation can be worked out. Place the three sets of four elements each in an array where the elements of each set appear in a row.

a	b	c	d
e	f	g	h
i	j	k	l

From this it is apparent that the elements in each column make up four sets of three elements each and that the total number of elements in each case must be the same. The fact that the elements of three sets are displayed in a rectangular array suggests that it may be possible to arrive at the definition of multiplication directly from notions about sets and counting, as was done with addition. First, suppose that there are two sets A and B such that $n(A) = 3$ and $n(B) = 4$. To establish an operation or process on these sets that will yield a set containing a number of elements equal to $3 \cdot 4$, or 12, which is completely general. As model sets let $A = \{a, d, f\}$ and $B = \{w, r, t, y\}$. Let an array of the elements in B appear as follows.

y	y	y
t	t	t
r	r	r
w	w	w
a	d	f

For each element of the set A, the elements of B will be written in a column over that element. Then, to determine $3 \cdot 4$, the number of squares in the array containing an element of B are counted and this is seen to be 12. As a further refinement of this procedure, one might construct an array as shown below. Here

y	x	x	x
t	x	x	x
r	x	x	x
w	x	x	x
	a	d	f

instead of actually inserting the letters w, r, t, y in each square of the columns, the positions are identified by placing the letters in a reference column at the left, just as the elements of A appear at the bottom of the array. Note also that the symbol x appearing in each square actually identifies a pair of letters, one each from set A and set B, in that order. As a further refinement of this process, construct a *lattice* of points as shown.

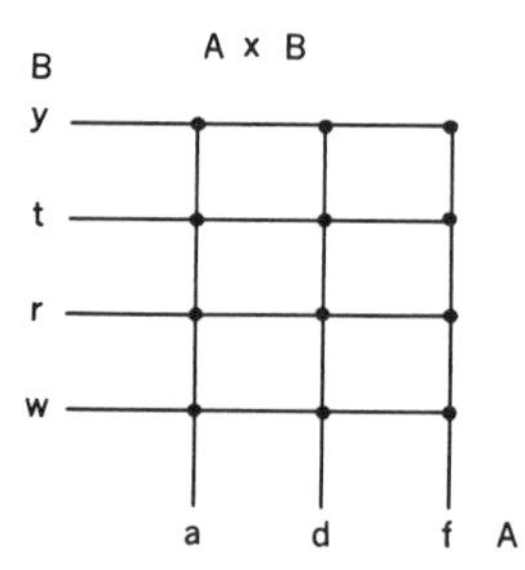

Each point of the lattice represents a pairing of an element of the set A with an element of the set B, for example, the points in the first column represent the pairs (a, w), (a, r), (a, t), (a, y). From the lattice, the product $3 \cdot 4$ may be determined by counting the total number of points in the lattice or by counting the number of elements in each column and then adding. Since there are 4 elements in each column, $3 \cdot 4 = 4 + 4 + 4 = 12$ and the two interpretations of multiplication are related. The process illustrated of forming ordered pairs of elements from two sets, A and B, where the first element of each pair is selected from A and the second from B, is called taking the *set product*, and is written $A \times B$. These set products clearly give a model for defining the operation of multiplication of whole numbers in set language.

■ **DEFINITION.**—If r and s are two whole numbers and if A and B are two sets such that $n(A) = r$ and $n(B) = s$, then

$r \cdot s = n(A) \cdot n(B) = n(A \times B)$. In terms of this definition and the previous illustration, it is evident that $4 \cdot 3 = n(B \times A)$. The lattice for $B \times A$ as constructed clearly indicates that the

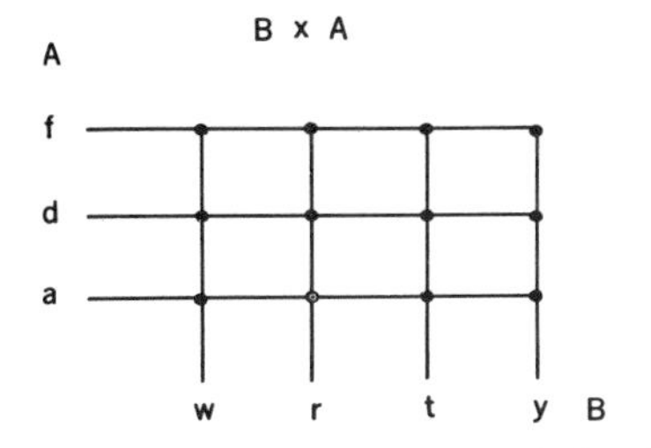

number of elements in $B \times A$ is the same as the number of elements in $A \times B$. In general, a one-to-one correspondence could be set up by associating the ordered pair (y, x) of $B \times A$ with the ordered pair (x, y) of $A \times B$. Thus, from this definition, it follows that $3 \cdot 4 = 4 \cdot 3$ and generally $r \cdot s = s \cdot r$. On the basis of this definition, the operation of multiplication with whole numbers is seen to possess many properties analogous to those of addition:

(1) *Closure property:* for any two whole numbers r and s, there is a third whole number such that $r \cdot s = t$.

(2) *Commutative law:* for any two whole numbers r and s, $r \cdot s = s \cdot r$.

(3) *Associative law:* for any three whole numbers r, s, and t, $r \cdot s \cdot t = (r \cdot s) \cdot t = r \cdot (s \cdot t)$. Since multiplication, like addition, is a binary operation, this law gives a means of multiplying three whole numbers in two different ways. This law can be illustrated as follows: $2 \cdot 3 \cdot 5 = (2 \cdot 3) \cdot 5 = 6 \cdot 5 = 30$, or $2 \cdot 3 \cdot 5 = 2 \cdot (3 \cdot 5) = 2 \cdot 15 = 30$.

(4) *Identity element:* the whole number 1 has the property that if a is any whole number $a \cdot 1 = a$. No other whole number has this property, which is a very important one in carrying out the operations of arithmetic.

(5) The number zero (0) plays a special role in multiplication in that for any whole number a, $a \cdot 0 = 0$.

Each of the principles previously discussed has referred to addition and multiplication as an individual process. In much of the work with the operations of arithmetic, the relationship of the two operations is the primary concern. This relationship is called the *distributive law*, which is stated as follows: for any three whole numbers a, b, and c, $a \cdot (b + c) = (a \cdot b) + (a \cdot c)$. By this law, the product $3(4 + 7)$ may be calculated in two different ways: $3(4 + 7) = 3 \cdot 11 = 33$ or $3(4 + 7) = (3 \cdot 4) + (3 \cdot 7) = 12 + 21 = 33$. The *distributive law* follows as a consequence of the definitions of addition and multiplication.

Inverse Operations.—The two operations of subtraction and division may be defined in terms of set language, but may also be defined in terms of the operations of addition and multiplication. Both approaches are useful and will be illustrated in the following paragraphs.

■ **SUBTRACTION.**—Suppose that two whole numbers u and v are given and it is desired to determine what the difference of v and u, or $v - u$, is. The process of determining the whole number associated with v and u, $v - u$, is called subtraction. The whole number v is called the *minuend*, and the whole number u, the *subtrahend*.

As a specific example, let $u = 4$ and $v = 7$. Then take as model sets $A = \{a, d, s, f\}$ and $B = \{z, x, c, b, j, g, h\}$. Clearly $n(A) = 4$ and $n(B) = 7$. The operation of subtraction can then be thought of as placing the set A in one-to-one correspondence with some subset of B and then counting the number of elements in the remaining subset of B. Placing A in one-to-one correspondence with $\{x, b, j, h\}$ the remaining subset of B is $\{z, c, g\}$ and its number is 3. Therefore, it can be concluded that $7 - 4 = 3$. This is equivalent to the "take away" interpretation of subtraction.

A clear restriction on this operation is that u is less than or equal to v. In general, given two whole numbers u and v, u may be less than v, v may be less than u, or u may be equal to v. Therefore, it is not always possible to subtract two whole numbers. The operation of subtraction must be carried out in a *specific order*. The fact that $7 - 4 = 3$ does not tell anything about the difference $4 - 7$, which does not exist in the set of whole numbers. In order for such differences to exist the scope of the number system must be expanded.

A desirable alternative to the preceding interpretation of the operation of subtraction relates subtraction to the operation of addition. In the discussion of the meaning of subtraction, note that it was necessary to partition the set B into two sets with no elements in common so that the sum of the numbers of these two sets is $n(B)$, or 7. The definition of subtraction in terms of addition is then that the difference of two whole numbers v and u is a whole number w (if it exists) such that the sum of u and w is v. In the example $7 - 4 = 3$ since $4 + 3 = 7$. This relationship of addition and subtraction is called *inverse*, or subtraction is the *inverse operation* to addition. This is equivalent to the "additive" interpretation of subtraction.

The properties of subtraction are much more restricted than those of addition. It is evident from the discussion above that the whole numbers are not closed under the operation of subtraction and that subtraction does not obey a commutative law. The operation also does not have an associative law. With respect to the whole number 0 and any whole number a, $a - a = 0$, $a - 0 = a$, but $0 - a$ is not defined unless $a = 0$. These properties follow from the definition of subtraction and properties of addition since $a = a + 0$, $a = 0 + a$ and $0 = 0 + 0$. The *distributive law* can be extended to subtraction in a restricted sense. For example, $3(7 - 4) = 3 \cdot 3 = 9$ or $3(7 - 4) = 21 - 12 = 9$. In general, $a \cdot (b - c) = (a \cdot b) - (a \cdot c)$ provided c is less than or equal to b.

■ **DIVISION.**—In the operation of division one is given two whole numbers, m and n, and asked to associate them with a third whole number, p, called the *quotient* of m and n; this is written $m \div n = p$, $m/n = p$, or $n\overline{)m}$ with p above. In this operation, m is called the *dividend* and n the *divisor*. The operation of division may be interpreted in more than one way. For instance, 12 divided by 4 may be thought of as partitioning a set of 12 elements into 4 equal sets and then counting the number of elements in each set as the quotient. The operation can also be thought of as partitioning the 12 elements into distinct sets each of which contains 4 elements and then counting the number of such sets as the quotient. Division can also be thought of in terms of repeated subtraction. That is, subtract 4 from 12, then subtract 4 from the difference and continue until the difference is 0. The number of times 4 has been subtracted before reaching this difference of 0 is the quotient of 12 and 4.

All of these interpretations are mathematically similar in that the quotient is a number such that when it is multiplied by the divisor the product is the dividend. In the example, $3 \cdot 4 = 12$ in each case. Division is thus related to multiplication as an inverse operation. The definition of division in terms of multiplication is then that the quotient of two whole numbers m and n is a whole number p (if it exists) such that

Justification of the Process of Multiplication

Steps	Justification
$34 \times 5 = (30 + 4)(5)$	Hindu-Arabic system of numeration.
$= (3 \cdot 10^1 + 4 \cdot 10^0)(5 \cdot 10^0)$	Hindu-Arabic system of numeration.
$= (3 \cdot 10^1 \times 5 \cdot 10^0) + (4 \cdot 10^0 \times 5 \cdot 10^0)$	Distributive law.
$= 3(10^1 \times 5)10^0 + 4(10^0 \times 5)10^0$	Associative law for multiplication.
$= 3(5 \cdot 10^1)10^0 + 4(5 \cdot 10^0)10^0$	Commutative law for multiplication.
$= (3 \times 5)(10^1 \times 10^0) + (4 \times 5)(10^0 \times 10^0)$	Associative law for multiplication.
$= (3 \times 5)(10^1) + (4 \times 5)(10^0)$	Law of exponents.
$= 15 \cdot 10^1 + 20 \cdot 10^0$	Basic multiplication facts.
$= (1 \cdot 10^1 + 5 \cdot 10^0) 10^1 + (2 \cdot 10^1 + 0 \cdot 10^0)10^0$	Hindu-Arabic system of numeration.
$= (1 \cdot 10^1 \cdot 10^1 + 5 \cdot 10^0 \cdot 10^1) + (2 \cdot 10^1 \cdot 10^0 + 0 \cdot 10^0 \cdot 10^0)$	Distributive law.
$= 1(10^1 \cdot 10^1) + 5(10^0 \cdot 10^1) + 2(10^1 \cdot 10^0) + 0(10^0 \cdot 10^0)$	Associative law for multiplication.
$= 1 \cdot 10^2 + 5 \cdot 10^1 + 2 \cdot 10^1 + 0 \cdot 10^0$	Law of exponents.
$= 1 \cdot 10^2 + (5 + 2)10^1 + 0 \cdot 10^0$	Distributive law.
$= 1 \cdot 10^2 + 7 \cdot 10^1 + 0 \cdot 10^0$	Basic addition facts.
$= 100 + 70 + 0$	Hindu-Arabic system of numeration.
$= 170$	Hindu-Arabic system of numeration.

the product of n and p is m, that is, $m = n \cdot p$.

Consideration of the conditions under which the quotient of two whole numbers exists leads to some interesting properties of the division operation: (1) The quotient of a nonzero number and zero must be *undefined*. As an example consider $6 \div 0$. Suppose that quotient is some whole number, k. Then by definition $0 \cdot k = 6$. But this is impossible since $0 \cdot k = 0$. Thus, there is no whole number which is the quotient of 6 and 0. (2) The quotient of 0 and 0 is *indeterminate*. Suppose $0 \div 0 = t$. Then $0 = 0 \cdot t = 0$ and the definition is satisfied. But suppose also that $0 \div 0 = v$, where $v \neq t$. Here again, $0 = 0 \cdot v = 0$ and the definition is again satisfied. Therefore, $0 \div 0$ may be any whole number whatsoever and is said to be *indeterminate*. On the basis of (1) and (2), *0 is ruled out as a divisor*. Hence, in succeeding examples the *divisor will always be nonzero*. (3) $a \div 1 = a$, since $a = 1 \cdot a$. (4) $a \div a = 1$, since $a \cdot 1 = a$. (5) $0 \div c = 0$, since $c \cdot 0 = 0$. (6) In general, $a \div b$ exists if and only if b is a factor of a. This means that there is a whole number c such that $b \cdot c = a$. Then by definition, c is the quotient of a and b. If b is not a factor of a, as in $13 \div 4$, the quotient does not exist in the set of whole numbers. For such quotients to exist, it is necessary to extend the scope of the number system.

The whole numbers do not have closure for the operation of division but quotients when they exist are *unique*. Furthermore, neither the *commutative* nor the *associative laws* hold for division. There is a restricted distributive law for division with respect to addition and subtraction. If a, b, and c are whole numbers and $c \neq 0$, $(a + b) \div c = (a \div c) + (b \div c)$, provided that c is a factor of both a and b. Under the same restrictions, and $a > b$, $(a - b) \div c = (a \div c) - (b \div c)$.

Examples:

$$(12 + 9) \div 3 = 4 + 3 = 7$$
$$(12 - 9) \div 3 = 4 - 3 = 1$$

The Number Line.—By marking an initial point on a straight line as 0 and establishing a certain length as a unit distance, it is possible to establish a correspondence between the whole numbers and a set of points on a line. Using this geometric representation of the whole numbers, the four operations of addition, subtraction, multiplication, and division can be illustrated.

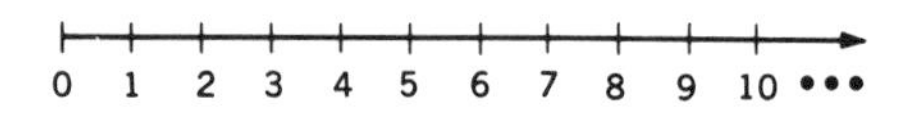

To add two whole numbers, for example, 3 and 4, on the number line, mark off 3 units on the number line from 0 to 3, and then mark off 4 units from 3. The terminal point is found to be 7. Thus $3 + 4 = 7$.

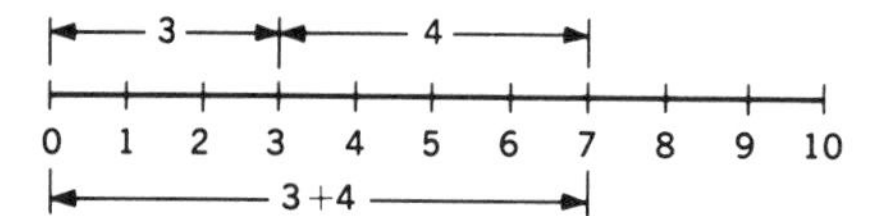

In like fashion 4 and 3 are added and the result is again 7; $3 + 4 = 4 + 3 = 7$.

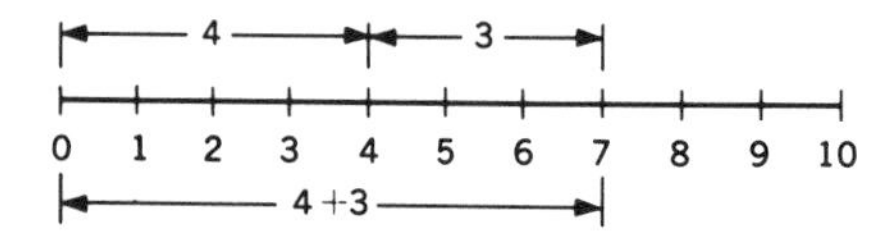

To multiply two whole numbers, say 2 and 3, the additive interpretation is used and 3 units and 3 units are added, again obtaining $2 \cdot 3 = 6$. Again, $3 \cdot 2 = 6$ by a similar process, as illustrated.

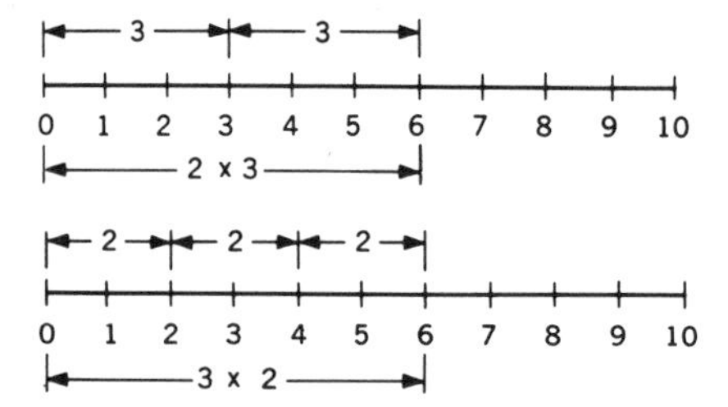

Subtraction on the number line may be thought of in two ways. To find the difference $8 - 3$, mark off 8 units from 0 to the right and then mark off 3 units from 8 to the left, arriving at 5 as the result. This is the "take away" interpretation of subtraction.

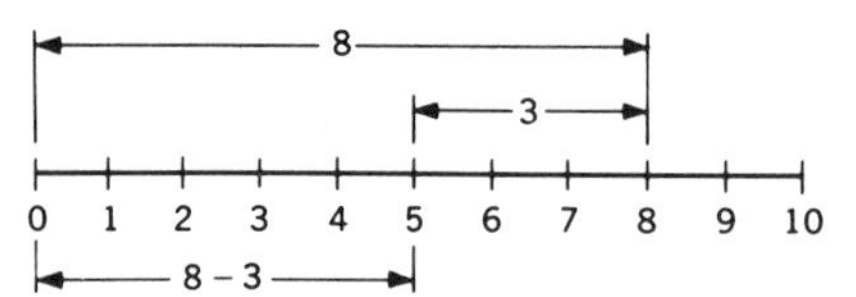

The second illustration is in terms of the definition of the operation of subtraction. Here 3 units are marked off to the right of 0 and 8 units to the right of zero and then count the number of units from 3 to 8; that is, we find the number which, when added to 3, yields 8 as the sum. This is the "additive" interpretation of subtraction.

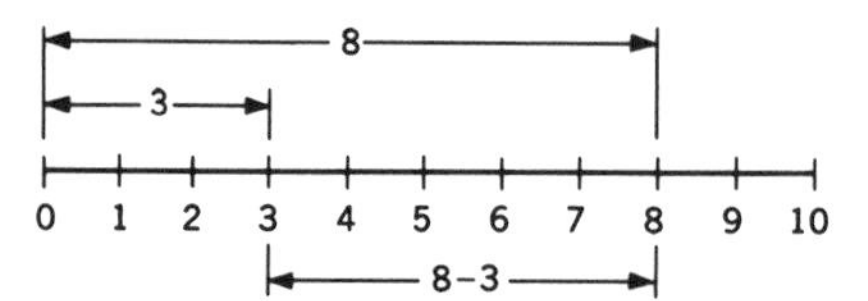

The operation of division can also be carried out in two ways on the number line. To find $6 \div 3$, for example, mark off 6 units to the right of 0 and then divide this segment into distinct segments of 3 units each. Finding that there are two such segments, it is concluded that $6 \div 3 = 2$.

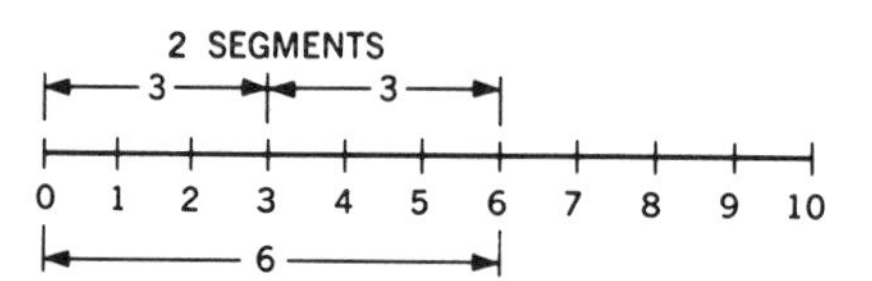

In the second interpretation, divide the segment from 0 to 6 into 3 equal segments. The number of units in each segment is 2. Again, $6 \div 3 = 2$.

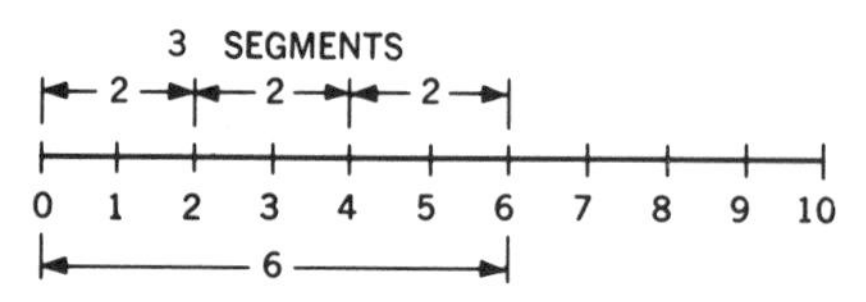

Calculation with Whole Numbers.—In dealing with single-digit whole numbers, it is a relatively simple matter to develop their sums and products directly from the definitions. This information for the digits 0 to 9 is usually collected in the addition and multiplication tables. From these tables, using the definitions of subtraction and division, it is possible to obtain many differences and quotients. The difference $9 - 4$ would be found by identifying the number which, when added to 4, gives 9 as a sum, or by completing the sentence $4 + \square = 9$. The quotient $6 \div 3$ would be found in a similar manner by completing the sentence $3 \times \square = 6$.

To find sums and products for numbers having a multiple-digit numeral, such as 288 and 24, and to find differences and quotients, these elementary methods do not suffice. The definitions guarantee that $288 + 24$, $288 \cdot 24$, $288 - 24$ and $288 \div 24$ all exist but do not give any practical technique for finding their numerical representation. In order to calculate efficiently with whole numbers, it is necessary to develop systematic processes that consist of a series of easy steps for finding the numeral representing such sums, products, differences, and quotients. To be efficient, these processes should make each step in the process as automatic as possible and should require as little writing as possible. Such processes for carrying out the operations of arithmetic are called *algorithms*.

Historically there have been many different algorithms used in calculations with the operations of arithmetic. The algorithms presently used are the result of centuries of refinement. Each of these algorithms compresses a large number of concepts into a relatively short process. These algorithms make it possible for elementary school children today to perform calculations that were exceedingly difficult if not impossible for scholars of a few centuries ago.

■ **ADDITION.**—This is the way the addition of 34 and 23 appears using the customary algorithm for addition:

$$\begin{array}{r} 34 \\ +23 \\ \hline 57 \end{array}$$

The numbers to be added are arranged in a vertical fashion so that the units digits fall in a column. This automatically ensures that the other digits will fall in the proper place-value position. The columns are then added individually and, if no column totals more than 9, these column totals yield the digits of the sum. To show how this process is derived from the basic properties of whole numbers, represent the sum as $34 + 23$. Then, using the numerals in expanded form and the commutative and associative laws for addition: $34 + 23 = (30 + 4) + (20 + 3) = (30 + 20) + (4 + 3) = 50 + 7 = 57$. Adding 30 and 20 strictly in terms of the fundamental combinations is accomplished by writing $30 + 20 = 3 \cdot 10 + 2 \cdot 10 = (3 + 2) \cdot 10 = 5 \cdot 10$ (by the distributive law) $= 50$. This notion operates generally whenever digits are added in a given place-value column. Now consider the following addition:

$$\begin{array}{r} 326 \\ 275 \\ 142 \\ \hline 743 \end{array}$$

Here addition of the digits in the units column results in a number greater than 9, in this case 13 units. Ten units are changed to 1 ten and shifted to the tens column. Again, adding ten digits in the tens column results in 14 tens. Changing 10 tens to 1 hundred and shifting it to the hundreds column, the digits in the hundreds column are added, and the final result thus obtained. Further insight into this process is obtained by writing each numeral in expanded form:

$$\begin{aligned} &(300 + 20 + 6) + (200 + 70 + 5) \\ &\quad + (100 + 40 + 2) \\ &= (300 + 200 + 100) + (20 + 70 \\ &\quad + 40) + (6 + 5 + 2). \end{aligned}$$

This rearrangement is accomplished through repeated applications of the commutative and associative laws and is identical with the column arrangement of digits by place value. Adding the quantities in each parentheses $600 + 130 + 13$ is obtained. Each of these additions involves the basic combinations and the distributive law. To complete the process:

$$\begin{aligned} &600 + (100 + 30) + (10 + 3) \\ &\quad = (600 + 100) + (30 + 10) + 3 \\ &\quad = 700 + 40 + 3 = 743. \end{aligned}$$

This latter process is equivalent to what is called "carrying" in arithmetic.

In performing such additions the columns may be added up or down. Often the columns are added one way and then added the other way as a check on the calculation. Although the present algorithm adds the units column first, the reason is purely one of economy in writing. However, it is possible to begin with the column farthest to the left:

$$\begin{array}{r} 457 \\ 391 \\ 729 \\ \hline 1400 \\ 160 \\ 17 \\ \hline 1577 \end{array} \qquad \begin{array}{r} 457 \\ 391 \\ 729 \\ \hline 1\not{4}\not{6}7 \\ 57 \\ \\ 1577 \end{array}$$

The second method given is one of a number of ancient algorithms known as scratch methods. These originated from the practice of performing calculations on a slate or dust board. Rather than scratching out the digit as was done above, the corrected digits would be wiped out and replaced. The advent of printing and pen-and-pencil work made this process impractical. Theoretically, there are some real advantages in the method, since the digits that make the least difference in the answer are added last, when the mind may have fatigued and error is more likely.

▪ **MULTIPLICATION.**—The operation of multiplication involves more steps and is slightly more complex than the operation of addition. The process of multiplication will first be illustrated with several examples and then the reasoning that lies behind it will be examined.

Example:

$$3 \times 72$$

To carry out such multiplications, the two numerals are first arranged vertically with the place values aligned as in addition. The digit 3 of the multiplier then multiplies in turn each digit of the multiplicand from right to left, the products being obtained from the basic multiplication table and written in order from right to left below the line.

$$\begin{array}{r} 72 \\ \times\ 3 \\ \hline 216 \end{array}$$

In this case, $3 \times 2 = 6$ and $3 \times 7 = 21$, the digits 21 being written to the left of 6 in the product.

Example:

$$6 \times 247$$

Again, the factors are arranged in vertical order with the place values aligned. In this example, however, when multiplying 6×7, the product is 42, a two-digit numeral. Therefore, do not write 42 below the line, but just the 2 in the units place, and hold the 4. Multiply 6×4, obtaining 24. Now add the 4 as a carry to 24, obtaining 28. Again, write only the 8 below the line and to the left of 2, holding the 2 as a carry to the next place. Then take the product of 6 and 2, which is 12, and, after adding the carry, 2, write the sum 14 to the left of 8 and 2, obtaining the product, 1482.

$$\begin{array}{r} 247 \\ \times\ 6 \\ \hline 1482 \end{array}$$

Example:

$$47 \times 597$$

Proceeding as before, write the two factors in vertical order with place values aligned. In this example there are two digits in the multiplier.

$$\begin{array}{r} 597 \\ \times\ 47 \\ \hline 4179 \\ 23880 \\ \hline 28059 \end{array}$$

The multiplication is carried out in two steps. First step, using 7, the units digit of 47, is precisely as in the preceding example. This product is called a partial product. Then obtain a second partial product by multiplying the second digit, 4: 4×597. Since 4 really represents 40, or 4 tens, the units digit in this partial product is 0. Otherwise, this partial product is obtained in the same way as was the first. Finally, the two partial products are added to obtain the result, 28,059. In practice, the 0 at the right of the second partial product is not written. If the multiplier in examples similar to this one contains more than two digits, the process is continued until as many partial products as digits in the multiplier have been obtained.

Example:

$$30 \times 342$$

This is a special case where the units digit of the multiplier is zero.

$$\begin{array}{r} 342 \\ \times\ 30 \\ \hline 000 \\ 10260 \\ \hline 10260 \end{array} \qquad \begin{array}{r} 342 \\ \times\ 30 \\ \hline 10260 \end{array}$$

In the first process, follow through the example as before. The second process is a convenient shortcut that may also be extended to multipliers ending in any number of zeros. Note that this is a special case of multiplication by 10, 100, 1000, etc., for $10 \times 36 = 360$, $100 \times 36 = 3600$, $1000 \times 36 = 36{,}000$, etc.

Example:

$$206 \times 3152$$

This example illustrates the handling of a zero digit in the multiplier when it is not a terminal digit.

$$\begin{array}{r} 3152 \\ \times\ 206 \\ \hline 18912 \\ 00000 \\ 630400 \\ \hline 649312 \end{array} \qquad \begin{array}{r} 3152 \\ \times\ 206 \\ \hline 18912 \\ 63040 \\ \hline 649312 \end{array}$$

The example on the left has been worked out completely, as in preceding examples; the example on the right appears in the abbreviated form commonly used.

The rationale behind these processes in multiplication lies mainly in the use of the distributive law. For instance, in multiplying 3×72, first write 72 in expanded form as $70 + 2$. Then $3 \cdot (70 + 2)$ is $3 \cdot 70 + 3 \cdot 2$, by the distributive law, and $3 \cdot 70 + 3 \cdot 2 = 210 + 6 = 216$. In similar fashion, $6 \cdot (247) = 6 \cdot (200 + 40 + 7) = 6 \cdot 200 + 6 \cdot 40 + 6 \cdot 7 = 1200 + 240 + 42 = 1200 + 282 = 1482$. Where the multiplier has more than a single digit, the process can still be explained in terms of the distributive law. For the product 47×597, first write the multiplier in expanded form, obtaining $(40 + 7) \cdot 597$. Then, using the distributive law once again, this equals $(40 \cdot 597) + (7 \cdot 597)$. Now, expanding 597 gives $40 \cdot (500 + 90 + 7) + 7 \cdot (500 + 90 + 7)$. To arrange this in the usual order of the algorithm, use the commutative law of addition to write $7 \cdot (500 + 90 + 7) + 40 \cdot (500 + 90 + 7)$. Again, using the distributive law, this then becomes $(7 \cdot 500 + 7 \cdot 90 + 7 \cdot 7) + (40 \cdot 500 + 40 \cdot 90 + 40 \cdot 7)$. Carrying out these operations gives $(3500 + 630 + 49) + (20{,}000 + 3600 + 280) = (3500 + 679) + (20{,}000 + 3880) = 4179 + 23{,}880$. These last two numbers are the partial products which are added to obtain the product, 28,059.

As with addition, the multiplication process can be carried out from left to right, and scratch methods were used in the past. First two methods illustrated below are left-to-right methods.

$$\begin{array}{r} 637 \\ \times\ 24 \\ \hline 12 \\ 6 \\ 14 \\ 24 \\ 12 \\ 28 \\ \hline 15288 \end{array} \qquad \begin{array}{r} 637 \\ 24 \\ \hline 1274 \\ 2548 \\ \hline 15288 \end{array} \qquad \begin{array}{r} 28 \\ 5\not{1}\not{6} \\ \not{4}7\not{4}8 \\ 12\not{6}\boxed{24} \\ \not{6}\not{3}7\not{7} \\ \not{6}\not{3} \\ 15288 \end{array}$$

In the first method, the digit 2 is first used as a multiplier, place values arranged in order from left to right. Then 4 is used as a multiplier, place values again in order from left to right but shifted one place to the right. This method can be used letting the multiplier digits occur in any order if the place values are properly accounted for. By using a little mental arithmetic for carries, the process can be further condensed. In the second method the multiplier digits are used from left to right but the digits of the multiplicand are in the usual order.

The third method is the ancient scratch method. The multiplicand is first placed underneath the multiplier so that the units digit falls under the first digit of the multiplier. The digit 2 is then used to multiply each digit of 637 from left to right, carries taken care of by scratching out a previously written digit and writing the corrected digit above it. The multiplicand is then shifted one place to the right and multiplied by 4, repeating the previous steps. The product then appears around the top of the work.

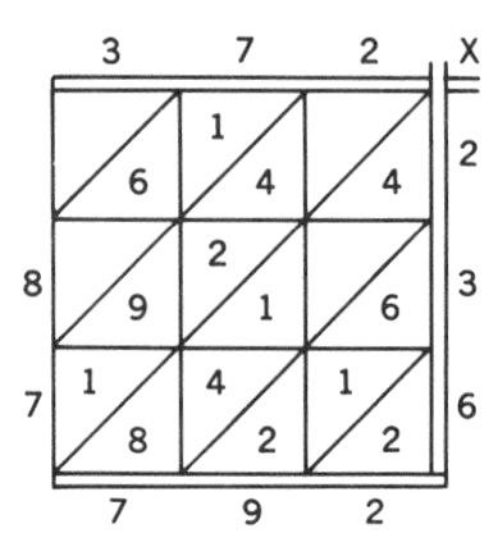

Another ancient algorithm for multiplication, which is very easy to use, is the *lattice*, or *grating, method*. To illustrate $236 \cdot 372$: in this method a grating having as many columns as the multiplicand and as many rows as the multiplier is set up. The digits of the two factors are then arranged as shown. Diagonal lines are drawn from upper right to lower left. The numbers in each row represent the digital products, the first digit of a two-digit product being placed above the diagonal. Then, the digits in each diagonal path are summed up to obtain the product, 87,792.

■ SUBTRACTION.—There are several algorithms presently in use for the operation of subtraction. These may be classified in terms of the way that the numerals are transformed when digital differences do not exist. Consider $379 - 36$, usually written in vertical form:

$$\begin{array}{r} 379 \\ -\ 36 \\ \hline 343 \end{array}$$

The two numerals are placed so that the units digits lie in a column thus aligning place values. Then each digit is subtracted from the one above ($9 - 6 = 3$, $7 - 3 = 4$, and $3 - 0 = 3$). By definition these would be found by completing the sentence $6 + \square = 9$, $3 + \square = 7$, and $0 + \square = 3$. It is quite common, however, for the take-away combinations to be taught as distinct from sums previously learned. In this case the subtraction of the digits would be carried out as: 9 take away 6 is 3, and so on. The second basic procedure in subtraction is illustrated in the difference $263 - 137$. Here, if the digits are aligned in the proper way, as below, it is found that the difference $3 - 7$ in the units column does not exist. Thus, some transformation in the structure of the numeral must be made in order to apply the same procedure as before. The two usual ones are illustrated:

$$\begin{array}{r} 263 \\ -\ 137 \\ \hline \end{array} \qquad \begin{array}{r} 25^{1}3 \\ -\ 13\ 7 \\ \hline 12\ 6 \end{array}$$

$$\begin{array}{r} 263 \\ -\ 137 \\ \hline \end{array} \qquad \begin{array}{r} 26^{1}3 \\ -\ 14\ 7 \\ \hline 12\ 6 \end{array}$$

In the first example, ten is borrowed from the tens place of 263, reducing the tens digit to 5. The 10 is then added to the 3 in the units place making a total of 13 units. Then $13 - 7 = 6$, $5 - 3 = 2$, $2 - 1 = 1$ and the subtraction is complete. In the second example, 10 is added to the 3 in the units place, making a total of 13 units and 1 is added to the tens place of 137, making this number 147. Then $13 - 7 = 6$, $6 - 4 = 2$ and $2 - 1 = 1$, completing the subtraction. To further illustrate these two processes, each numeral is written in expanded form.

$$\begin{array}{rl} 263 & = 200 + 60 + 3 \\ & = 200 + (50 + 10) + 3 = 200 + 50 + 13 \\ 137 & = 100 + 30 + 7 \qquad\quad\ = 100 + 30 + 7 \\ \hline & \qquad\qquad\qquad\qquad\quad\ \ 100 + 20 + 6 = 126 \end{array}$$

Here the use of the numeration system and the associative law of addition enables the exchange of place values so that the 7 may be subtracted from 13, eliminating the need to perform the impossible subtraction $3 - 7$.

$$\begin{array}{rl} 263 & = (200 + 60 + 3) + 10 = 200 + 60 + 13 \\ 137 & = (100 + 30 + 7) + 10 = 100 + 40 + 7 \\ \hline & \qquad\qquad\qquad\qquad\quad\ 100 + 20 + 6 \\ & \qquad\qquad\qquad\qquad\quad\ = 126 \end{array}$$

Here the principle operating is what might be called "compensation." Add 10 to the 3 in the units place to obtain a number from which 7 may be subtracted and then add ten to the subtrahend to compensate for this change in the minuend. This 10 cannot be added to the 7 in the units place here, for it would again give an impossible subtraction. Therefore, add it to the 30 in the tens place, leaving the 7 in the units place undisturbed. Since 40 is less than 60, the subtraction can be completed.

The two processes illustrated apply not only when the digital differences do not exist in the units column, but also when such differences fail to exist in any place-value column. An example is given using each method.

$$\begin{array}{r} 53004 \\ -\ 14765 \\ \hline 38239 \end{array}$$

$$\begin{array}{rl} 53004 & = 50000 + 3000 + 000 + 00 + 4 \\ & = 50000 + 2000 + 1000 + 00 + 4 \\ & = 50000 + 2000 + 900 + 100 + 4 \\ & = 50000 + 2000 + 900 + 90 + 14 \\ & = 40000 + 12000 + 900 + 90 + 14 \end{array}$$

Then,

$$\begin{array}{rl} & 40000 + 12000 + 900 + 90 + 14 \\ - & 10000 + 4000 + 700 + 60 + 5 \\ \hline & 30000 + 8000 + 200 + 30 + 9 = 38239 \end{array}$$

or,

$$\begin{array}{r} 4\ 2\ 9\ 9 \\ \not{5}^{1}\not{3}^{1}\not{0}^{1}\not{0}^{1}4 \\ -1\ 4\ 7\ 6\ 5 \\ \hline 3\ 8\ 2\ 3\ 9 \end{array}$$

Using the first process, it is necessary to make many exchanges in place value to arrive at a form in which the subtraction may finally be completed. This example also indicates what is done when an exchange cannot be made with the place value immediately at the left. To obtain a 10 to add to 4 in the units place, it was necessary to exchange 10 hundreds for 1 thousand, then 10 tens for 1 hundred, and finally 10 units for one ten. It was also necessary to exchange 10 thousands for 1 ten-thousand. The problem as it might appear with all place-value exchanges indicated is also given above. Usually these exchanges are done mentally so that the result would appear as in the initial statement of the problem. Another approach is:

$$\begin{array}{l} 50000 + 13000 + 1000 + 100 + 14 \\ 20000 + 5000 + 800 + 70 + 5 \\ \hline 30000 + 8000 + 200 + 30 + 9 = 38239 \end{array}$$

or

$$\begin{array}{l} 5^{1}3^{1}0^{1}0^{1}4 \\ 2\ 5\ 8\ 7 \\ \not{1}\ \not{4}\ \not{7}\ \not{6}\ 5 \\ \hline 3\ 8\ 2\ 3\ 9 \end{array}$$

In this process, we have added 10, 100, 1000, and 10,000 to the minuend in the units, tens, hundreds, and thousands places and compensated by adding the same amounts to the subtrahend, but in the corresponding place-value columns. The problem as it might appear with the changes indicated in the two numerals is given above. Here, again, the changes are made mentally so that such a subtraction would actually appear as in the original problem. The principles in operation in the first process illustrated are the principles of the system of numeration and the laws governing the operation of addition and subtraction. The second process involves a rule of law, not previously discussed, relating to the operation of subtraction: if the difference $a - b$ of any two whole numbers exists and is equal to c, then for any whole number d, the difference $(a + d) - (b + d)$ exists and is also equal to c. To establish this principle, we note that if $a - b = c$ then by definition $a = b + c$. Adding d to both sides of this equality $(a + d) = (b + d) + c$ is then obtained. Therefore, $(a + d) - (b + d) = c$ by definition. Thus the steps used in the second process are justified in terms of the basic properties of whole numbers.

■ DIVISION.—The division algorithm is perhaps the most difficult of the algorithms for the four fundamental operations with whole numbers. For students it is often the least understood; it is apt to be performed in a mechanical way and with little understanding. Thus, it is also the operation that is the most subject to error on the student's part. Although division, as previously indicated, is defined only for the case in which the divisor is an exact factor of the dividend (that is, even division), we can make certain statements about pairs of whole numbers, m and n, even though m is not an even divisor of n. For instance, given the pair of whole numbers, 63 and 5, we can express 63 in terms of the largest multiple of 5 which is less than 63, or $60 = 12 \cdot 5$, and the excess, or 3: $63 = 12 \cdot 5 + 3$. Here 12 is called the quotient and 3 the remainder. The whole numbers have the property that, given any pair of them, a quotient and remainder can be found as above, provided that the second number given is not zero. Thus, the pair 27, 7 yields 3 as a quotient and 6 as a remainder, since $27 = 7 \cdot 3 + 6$. Note that the remainder is a number less than the second number of the pair of numbers given. This relation is a basic property of the whole numbers, usually stated as follows: If m and n are two whole numbers, and $n \neq 0$, there is a unique pair of whole numbers, q and r, such that $m = n \cdot q + r$ and r is less than n. The process by which the numbers q and r are actually determined is the familiar process of long (or short) division. Either of these, since they are basically the same, can be referred to as the *division algorithm*. In this context it is usual to call the number n the *divisor*, the number m the *dividend*, the number q the *quotient*, and the number r the *remainder*. Using this terminology, the *basic division relation* given above may be restated as *dividend* = *divisor* × *quotient* + *remainder*, where the remainder is less than the divisor and the divisor is nonzero.

With this information, the usual division algorithm used to find the numbers q and r, that is, a quotient and remainder for a given dividend and divisor, will be considered. In the examples given, the steps necessary to find the result will be noted, deferring detailed explanation of the process at present.

Example:

$$838 \div 3.$$

It is customary to arrange the work in the form $3\overline{)838}$. (1) In the stepwise procedure only the first digit, 8, of the dividend is considered at first and one must find the highest multiple of 3 contained in 8, which is 2. Since 8 is in the hundreds place of the dividend, 2 is in the hundreds place of the quotient. To indicate this, the 2 is usually written directly above the eight, as shown below. Then the product $2 \cdot 3 = 6$ is written below the 8 and subtracted from it, giving a difference of 2. (2) The next step then is to bring down (adjoin) the tens digit of the dividend, 3, to 2, yielding 23. Next consider the highest multiple of the divisor, 3, contained in 23, which is 7; 7 is thus written in the tens place of the quotient and the product $7 \cdot 3 = 21$ is written below 23 and subtracted from it, yielding a difference of 2. (3) Then the units digit of the dividend, 8, is adjoined to this, yielding 28. Since the highest multiple of 3 contained in 28 is 9, the final digit of the quotient is 9. Multiplying 9 times 3 and subtracting the result, 27, from 28, the difference is 1. Since 1 is less than 3, the process is complete giving the quotient as 279 and the remainder as 1. The result is checked by verifying that $838 = 3 \cdot 279 + 1$ as it does to satisfy the basic division relation.

```
        2            27            279
(1) 3)838     (2) 3)838     (3) 3)838
      6             6             6
      2             23            23
                    21            21
                     2             28
                                   27
                                    1
```

If there are more digits in the dividend, the process is simply repeated in the same way until all the digits are exhausted and the final remainder is less than the divisor.

Example:

$$2457 \div 5.$$

In the preceding example the divisor was contained in the first digit of the dividend. In this example this is no longer true. Therefore, in the first step of the process it is necessary to consider not just the first digit of the dividend but the first two digits, 24. Then the highest multiple of 5 contained in 24 is 4, which is then the first digit of the quotient. This digit is not written over the 2 of the dividend, but over the 4 in the hundreds place. Otherwise the division is carried out exactly as in the first example. Note that in the second step the difference obtained is zero. Thus, when the units digit is adjoined, it becomes 07, which is the same as the single digit 7.

```
   491
5)2457
  20
   45
   45
    07
     5
     2
```

Example:

$$3219 \div 8.$$

```
      402              402
(1) 8)3219       (2) 8)3219
      32               32
       01               019
        0                16
       19                 3
       16
        3
```

This example is given to illustrate the handling of a situation such as occurs in the second step where the divisor, 8, is not contained properly in 1. Thus the highest multiple of 8 that is less than or equal to 1 is 0, since $0 \cdot 8 = 0$. We therefore write 0 as the second digit of the quotient, subtract 0 from 1, and continue as before. This step is sometimes abbreviated as shown in (2).

Example:

$$2345 \div 9.$$

```
      260              260
(1) 9)2345       (2) 9)2345
      18               18
       54               54
       54               54
        05               05
         0
         5
```

This example illustrates the occurrence of the same type of situation as in the final step of the division in the previous example, where the final digit of the quotient is 0 and the remainder is 5. The abbreviated form is shown in (2).

Example:

Each of the preceding examples may be carried out by what is known as *short division*, where the work is performed as shown below. The intermediate steps are done mentally. (R denotes the remainder.)

$$3\underline{)8^{2}3^{2}8} \qquad 5\underline{)24^{4}57}$$
$$\;\;2\;7\;9R1 \qquad\quad 4\;91R2$$

$$8\underline{)3219} \qquad 9\underline{)23^{5}45}$$
$$\;\;402R3 \qquad\quad 2\;60R5$$

The preceding section has outlined the basic processes involved in the division algorithm. The extension of these processes to more complex examples will now be illustrated.

Example:

$475 \div 13.$

```
     36
13)475
   39
    85
    78
     7
```

Since the divisor contains two digits we must take at least two digits of the dividend for the first step of the division. In this case the number formed by the first two digits of the dividend contains 13 and the highest multiple of 13 contained in 47 is 3. Thus the first digit of the quotient is 3. This digit will be in the tens place of the quotient, since the 7 of 47 is in the tens place of the dividend. Subtracting $3 \cdot 13 = 39$ from 47 the difference is 8. Adjoining the last digit of the dividend gives 85. Since the largest multiple of 13 contained in 85 is 6, the last digit of the quotient is 6. Subtracting $6 \cdot 13 = 78$ from 85, the difference is 7. Since 7 is less than the divisor, 13, the process is complete, yielding a quotient of 36 and a remainder of 7.

Example:

$2351 \div 46.$

```
      51
46)2351
   230
     51
     46
      5
```

In the event that the number formed by the first two digits of the dividend does not contain the divisor we take the number formed by the first three digits of the divisor in the first step of the process and place the first digit of the quotient accordingly. Otherwise the process is completed in the same way.

If the divisor contains more than two digits the procedures shown for two-digit divisors apply.

Example:

```
         557
416)231745
    2080
     2374
     2080
      2945
      2912
        33
```

The first step entails taking at least the number formed by the first three digits in the dividend. Since this number, 231, does not contain the divisor, add an additional digit to obtain 2317. Then obtain the largest multiple of 416 contained in 2317. By trial, this is seen to be 5. Then $5 \cdot 416 = 2080$ is subtracted from 2317 and the process continued as before. The first digit, 5, of the quotient is the same place value as the 7 of 2317 is in the dividend.

In general, then, in dividing a multiple-digit number, the process is initiated by taking the number formed by the successive digits of the dividend, either as many digits or one more digit than contained in the divisor. The highest multiple of the divisor contained in this number is the first digit of the quotient. The place value of this digit in the quotient is the same as the place value of the last digit used in the dividend. The process is then completed as in previous examples.

In the examples given so far, the process of finding the highest multiple of the divisor contained in a given number was found by simple trial. In some cases this may involve several trials before the highest such multiple is found. In the following examples, a simple procedure for estimating quotient digits or obtaining trial quotient digits is given.

Example:

$10{,}769 \div 423.$

```
          25
423)10769
     846
     2309
     2115
      194
```

According to the rule, the number formed by the first three digits of the dividend must be considered first. Since 423 is not contained in 107, it is necessary to take 1076. The procedure to be illustrated for obtaining trial quotient digits is sometimes called the one-step rule. This involves taking just the first digit, 4, of the divisor, and obtaining the highest multiple of 4 contained in the number formed by the first two digits of the dividend, 10. This number is 2. Then try 2 as the highest multiple of 423 contained in 1076. Since $2 \cdot 423 = 846$, the choice is correct. Therefore, the first digit of the quotient is 2. Continuing the process, the highest multiple of 423 contained in 2309 must be found next. Again, take 4 and the number formed by the first two digits of 2309, or 23. The highest multiple of 4 contained in 23 is 5; $5 \cdot 423 = 2115$ which is less than 2309. Hence, the second quotient digit is 5. Subtracting, a remainder of 194 is obtained and the process is complete.

In the above example the estimated quotient digit turned out to be the correct quotient digit in each step. This, however, is not always the case.

Example:

$9627 \div 36.$

```
     267
36)9627
   72
   242
   216
    267
    252
     15
```

Since 36 is contained in the number formed by the first two digits of the dividend, the first quotient digit can be placed in the same place value as the 6 of 9627. To estimate this quotient digit, use only the first digit, 3, of 36 and the first digit, 9, of the dividend. This yields a trial quotient digit of 3. Since $3 \cdot 36 = 108$, this is too large. But $2 \cdot 36 = 72$. Hence, the first quotient digit is 2. Continuing, the next step is to find the highest multiple of 36 contained in 242. It is seen that the highest multiple of 3 contained in 24 is 8, but $8 \cdot 36 = 288$, which is too large. Reducing the digit to 7, $7 \cdot 36 = 52$ which is still too large. Reducing again, $6 \cdot 36 = 216$. The second quotient digit is then 6. In the final step, it is necessary to find the highest multiple of 36 contained in 267. The one-step rule gives 8 as a trial quotient digit, but it is known from the previous step that this has to be reduced to 7. Thus the final quotient of 267 and a remainder of 15 is obtained.

In this second example, the one-step rule did not give the correct quotient digit in any step, whereas in the first example it yielded the correct quotient digit every time. In general, the rule will give the correct quotient digit about 65 per cent of the time. Fortunately, whenever the quotient digit is incorrect it is invariably too large. For this reason, it is known that whenever the product of the trial quotient digit and the divisor is less than the dividend, the quotient digit is correct. The one-step rule may be made more useful if it is observed that divisors such as 38, 176, and 567 are actually closer to 40, 200, and 600 than they are to 30, 100, and 500 and that, therefore, with such divisors the one-step rule is apt to yield too high a quotient digit.

To gain an understanding of the reasoning involved in the division algorithm, it is necessary to examine the processes used more closely. The processes previously illustrated represent a considerable refinement of properties associated with the basic division relation.

Let us examine a typical division by a single-digit divisor, as illustrated in a previous example: 838 divided by 3, or $3\overline{)838}$. First, rewrite the dividend in expanded form, so that the problem is then written $3\overline{)800 + 30 + 8}$. The first step in the algorithm, finding the largest multiple of 3 contained in 8, is then seen to actually represent finding the largest multiple of 3 in terms of 100's contained in 800, which is 200. Multiplying $200 \cdot 3$ and subtracting the product from 800 the difference is 200.

```
  200
3)800 + 30 + 8
  600
  200 + 30 = 230
```

Adding the 30 to this leaves a remainder of $230 + 8$, or 23 tens and 8 units. The next step in the algorithm is to find the largest multiple of 3 in terms of tens. This will be 7 tens, or 70. Multiplying

70 · 3 and subtracting the product from

```
      200 + 70
   3)800 + 30 + 8
      600
      200 + 30 = 230
                 210
                  20 + 8 = 28
```

230 the difference is 20. Adding 8 to this the remainder is 28. In the final step, the

```
     200 + 70 + 9 = 279
   3)800 + 30 + 8
     600
     200 + 30 = 230
                210
                 20 + 8 = 28
                          27
                           1
```

largest multiple of 3 in terms of units contained in 28, which is 9, is found. Multiplying 9 · 3 and subtracting the product from 28 a final remainder of 1 is obtained. This process may be summarized in a more familiar form:

```
      9
     70
    200
  3)838
    600
    238
    210
     28
     27
      1
```

One of the greatest difficulties with the process of division previously illustrated is that it is imperative that the quotient digits be correct at each step of the process. A method for finding the quotient and remainder for two given numbers which avoids this difficulty has seen increasing acceptance in recent years. This process is illustrated below.

To find the quotient and remainder for 475 divided by 27, for example, the numbers are set up as 27)475. The work is then carried out as displayed. In step

```
27)475 |   8   (1)
   216 |
   259 |   7   (2)
   189 |
    70 |   2   (3)
    54 |
    16 |   0   (4)
       |  17   (5)
```

(1) it is estimated that 8 · 27 is contained in 475. 8 is thus written in the first line at the right and 8 · 27 = 216 is subtracted from 475. (2) Since the remainder 259 is larger than 27, the process is repeated. This time it is estimated that 7 · 27 is contained in 259 and 7 is written on a line to the right of 259; 7 · 27 = 189 is then subtracted from 259. Since the remainder, 70, is larger than 27, the process is again repeated. This time the estimate that 2 · 27 is contained in 70 is made and 2 is written on a line at the right of 70 (3). Subtracting 2 · 27 = 54 from 70, the remainder is 16, which is less than 27. Write a 0 on the line at the right (4) to indicate this and then total the multiples of 27 contained in 475. This total is 17 (5) and thus the quotient is 17 and the remainder is 16. This process will work no matter what multiples of 27 are used at each step, except that they must not be greater than the number involved; for example, in the first step, 20 could not be used since 20 · 27 = 540, which is larger than 475. Below two other solutions to the problem are given, using this method.

```
27)475 |  8        27)475 | 10
   216 |              270 |
   259 |  8           205 |  7
   216 |              189 |
    43 |               16 |  0
    27 |  1               | 17
    16 |  0
       | 17
```

In the second solution above, the steps are identical with those in the usual algorithm for long division. The advantages of this alternate method of performing long division lie in its clarity and its direct relation to the fundamental processes involved. The usual method still remains probably the most efficient for the skilled calculator.

Prime and Composite Numbers.—The entire set of whole numbers can be separated into two categories called the primes and composites. A *prime* number is any whole number, with the exception of one, that is divisible only by one and itself. One is not usually considered to be a prime number. Numbers that are not prime are called *composite* numbers. For example, 5 is a prime number, since only 1 and 5 are the divisors, or factors, of 5; whereas 6 is a composite number, since it has the factors 2 and 3 in addition to 1 and 6. More specifically, if three whole numbers, a, b, and c, are related in the following manner, $ab = c$, then both a and b are referred to as the divisors or *factors* of c. The number c is called a multiple of the numbers a and b. When c can be expressed as a product of a and b, ab, where both a and b are greater than 1, then c is called a composite number. This may be observed from the two examples given above: 5 · 1 = 5 (prime); 2 · 3 = 6 (composite).

The *fundamental theorem of arithmetic* (also called the *unique factorization theorem*) states that every composite number is capable of being factored uniquely (that is, one and only one way, except for the order) into a product of prime numbers. Note the following examples:

$$30 = 2 \cdot 3 \cdot 5$$
$$252 = 2^2 \cdot 3^2 \cdot 7$$
$$646 = 2 \cdot 17 \cdot 19$$

Thus, every whole number greater than 1 is either a prime or is expressible as a product of primes.

Mathematicians have devoted considerable effort to investigate the nature of prime numbers. Euclid presented a proof that there is an infinitude of prime numbers, thus simultaneously demonstrating that there is no largest prime. Although others have reached the same conclusion in different manners, no one has found a technique or formula which will produce all of the prime numbers. Given a particular large number (odd number, since all even numbers with the exception of 2 are obviously composite), it is usually not easy to determine whether it is a prime or composite number. Tables have been made which list prime numbers up to a certain value; however, to extend these tables becomes a formidable task. The task of finding and determining which numbers are primes has been greatly eased by the use of high-speed electronic computers, even though these efforts are limited. An early technique for determining all prime numbers less than a given number is credited to the Greek scholar, Eratosthenes (276–194 B.C.) and is called the "Sieve of Eratosthenes." This method may be illustrated as follows: To find all of the prime numbers less than 100, begin by writing down all of the numbers up to 100. To begin the sieve cross out all of the numbers that contain 2 as a

Sieve of Eratosthenes

1	2	3	~~4~~	5	~~6~~	7	~~8~~	~~9~~	~~10~~
11	~~12~~	13	~~14~~	~~15~~	~~16~~	17	~~18~~	19	~~20~~
~~21~~	~~22~~	23	~~24~~	~~25~~	~~26~~	~~27~~	~~28~~	29	~~30~~
31	~~32~~	~~33~~	~~34~~	~~35~~	~~36~~	37	~~38~~	~~39~~	~~40~~
41	~~42~~	43	~~44~~	~~45~~	~~46~~	47	~~48~~	~~49~~	~~50~~
~~51~~	~~52~~	53	~~54~~	~~55~~	~~56~~	~~57~~	~~58~~	59	~~60~~
61	~~62~~	~~63~~	~~64~~	~~65~~	~~66~~	67	~~68~~	~~69~~	~~70~~
71	~~72~~	73	~~74~~	~~75~~	~~76~~	~~77~~	~~78~~	79	~~80~~
~~81~~	~~82~~	83	~~84~~	~~85~~	~~86~~	~~87~~	~~88~~	89	~~90~~
~~91~~	~~92~~	~~93~~	~~94~~	~~95~~	~~96~~	97	~~98~~	~~99~~	~~100~~

factor, with the exception of 2. This then removes all of the even numbers greater than 2 which are obviously not primes. The next step is to cross out all of the numbers that contain 3 as a factor, with the exception of 3. This will remove numbers such as 9, 15, and 21. The next number to consider is 4—however, it has already been crossed out since it contained 2 as a factor. The next number to consider is 5. Again cross out the numbers above 5 which contain 5 as a factor. This includes such numbers as 25, 35, 40. Continue this general procedure of moving up from one number to the next until all numbers have been crossed out which contain the smaller numbers as factors. The accompanying table indicates the final result of this procedure. The numbers that are not crossed out are the prime numbers less than 100. It is to be observed that in this example it was only necessary to cross out multiples of prime numbers less than the square root of 100, namely 2, 3, 5, and 7. In finding all of the primes less than

1,000, it is necessary to consider only primes less than $\sqrt{1{,}000}$, or the primes up to and including 31. To determine whether a given number is prime or composite without going through the sieve completely, simply try one prime after another until all of the primes up to the square root of the number under consideration have been tried. For example, to determine if the number 161 is prime or composite, first find the $\sqrt{161}$, which is 12 plus a fraction. If the number 161 is composite it must have as a divisor one of the following primes: 2, 3, 5, 7, 11. These are the only divisors that need to be tried. Since 7 divides 161 ($7 \cdot 23$), the number is composite. This technique can of course be extended to any number desirable and since it is very mechanical, it is easily adaptable to devices such as punched cards. Computers can be programmed to determine primes quite easily.

In looking at the prime numbers up to 100—2, 3, 5, 7, 11, 13, 17, 19, 23, 29, 31, 37, 41, 43, 47, 53, 59, 61, 67, 71, 73, 79, 83, 89, 97—no pattern is easily discernible. The primes do not seem to occur at regular intervals; in fact they are very erratically distributed. This apparent lack of pattern continues as larger numbers are investigated and is the reason that no formula or technique has been developed for determining what numbers are prime.

There have been many attempts made at finding a formula; one of these is $n^2 - n + 41$, where n is any whole number. It is interesting to note that when n is 1, 2, 3, . . . the expression will give primes—for a while. When $n = 1$, the answer is 41, which is prime; when $n = 2$, it is 43, which is the next prime, etc. But when $n = 41$, the answer is $41^2 - 41 + 41$, which is obviously not prime, since 41^2 is a composite number.

When large-scale distributions of primes are considered, the following values appear. The first five entries are the result of actual counts, while the last two are the result of calculations.

Number of Primes

up to	is
1,000	168
10,000	1,229
100,000	9,592
1,000,000	78,498
10,000,000	664,579
100,000,000	5,761,455
1,000,000,000	50,847,478

Advanced number theory seems to indicate that this pattern has meaning in terms of the ultimate distribution of primes. Until 1951, the largest known prime number was $2^{127} - 1$, which is written out as 170,141,183,460,469,231,731,687,303,715,884,105,727. In 1951, with the help of electronic computers, a much larger prime was determined. This number is $180 \times (2^{127} - 1)^2 + 1$.

One of the many unsolved problems involving prime numbers is called "Goldbach's conjecture." C. Goldbach (1690–1764) noted that every even number can be written as the sum of two primes, that is, it worked for all the even numbers he tested. However this conjecture has never been proven, nor have any cases been found where it is not true.

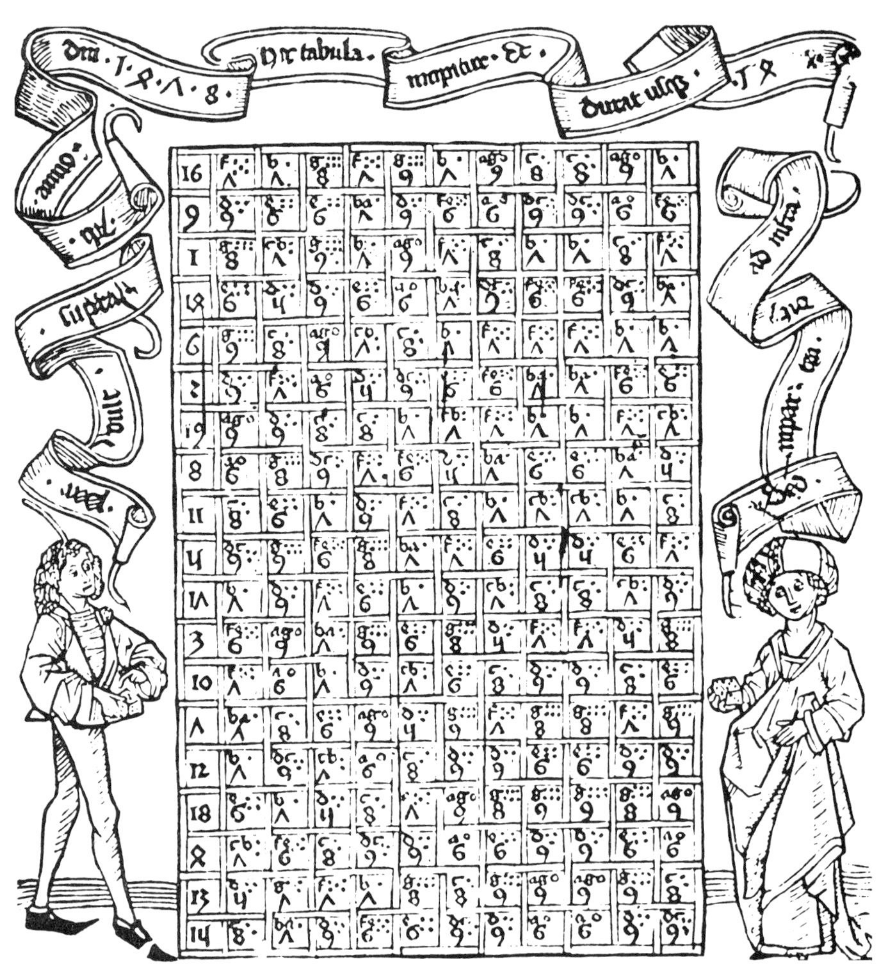

BETTMANN ARCHIVE

THIS MEDIEVAL CALENDAR with mystic figures relied upon both astronomy and mathematics.

There are some prime numbers that differ from each other by only 2, such as 3 and 5, 5 and 7, 11 and 13, 17 and 19, 29 and 31, 41 and 43, . . . , 10,006,427 and 10,006,429 . . . Such pairs of prime numbers are called prime twins and these, too, seem to occur less frequently as the numbers get larger. However it is believed that there is an infinite number of these, though again no proof of this conjecture has been found.

■ **PERFECT NUMBERS.**—The ancient Greeks had developed an extensive belief in what we today call numerology. They were intrigued with numbers such as 6 and 28, since these numbers are equal to the sum of their proper divisors. That is, the proper divisors of 6 are 1, 2, and 3. Adding $1 + 2 + 3$ gives 6. The number 28 has as proper divisors 1, 2, 4, 7, and 14. The sum of these is 28. Numbers which possess this characteristic were called the *perfect numbers*. These perfect numbers are indeed scarce; 6 and 28 are the first two, while the fifth one is 33,550,336. Even though Euclid found an expression which will find all of the perfect numbers, there are still only 17 known perfect numbers, even with the help of high-speed computers. It is not known how many there are, nor have any odd perfect numbers been found.

■ **NUMEROLOGY.**—These same Greeks that found the first perfect numbers also developed other mystic meanings for certain numbers. They considered 1 as the source of all other numbers. Since even numbers are divisible by 2, they are considered to be weak and therefore feminine. The odd numbers were considered to be strong and masculine. The number 5 was used to represent marriage, since it is the sum, or union, of 2 and 3, male and female. Death was represented by the number 8, while 9 represented immortality.

The Greeks also used the number 7 extensively in their culture; for example, the seven wise men, the seven wonders of the world, and the seven liberal arts that constituted the curriculum in their schools. The number 7 is still very much in vogue; it is seen in the seven days in a week and citizens' vote upon reaching 21, which is equal to 3 times 7. There are many other examples that reflect ancient numerology.

■ GEOMETRY.—Another interesting classification of numbers is based on the geometric figures that can be made by using objects to represent numbers. These figurate numbers are called rectangular, triangular, or square numbers.

The rectangular numbers may be illustrated in the following manner:

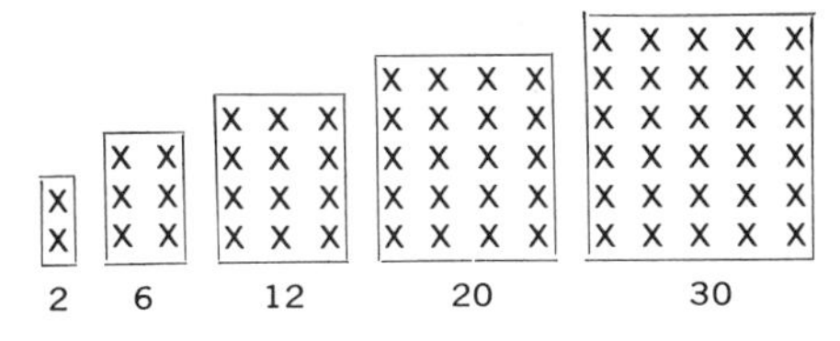

These numbers may also be represented in this manner:

$1+1=2$
$1+1+2+2=6$
$1+1+2+2+3+3=12$
$1+1+2+2+3+3+4+4=20$
$1+1+2+2+3+3+4+4+5+5=30$

The triangular numbers may be illustrated in this manner:

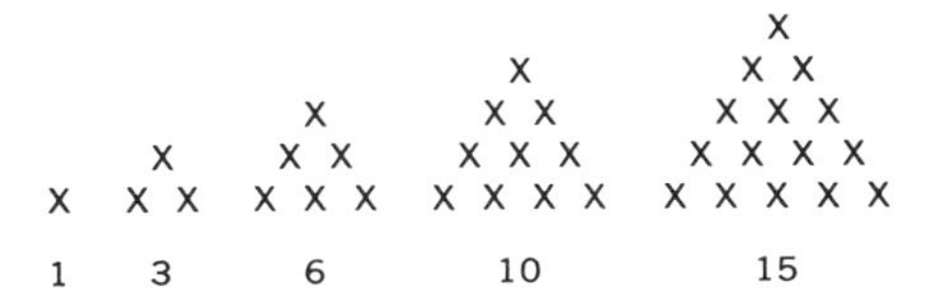

These numbers are represented in the following manner:

1
$1+2=3$
$1+2+3=6$
$1+2+3+4=10$
$1+2+3+4+5=15$

The square numbers are illustrated by making square configurations:

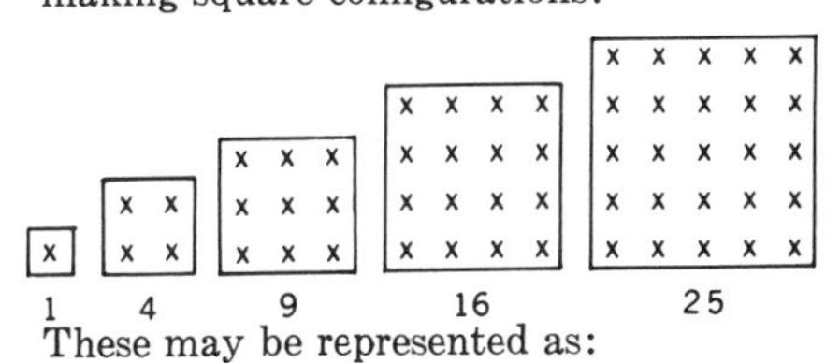

These may be represented as:

1
$1+3=4$
$1+3+5=9$
$1+3+5+7=16$
$1+3+5+7+9=25$

There are many other possible configurations that can be made to represent various kinds of numbers. It is interesting to note that cubes can be represented by three-dimensional arrangements of objects.

Greatest Common Factor.—The problem of finding the factorization of a whole number is important to the arithmetic of fractions. Specifically, in order to "reduce" a particular fraction it is necessary to make use of the *greatest common factor* (also called the *greatest common divisor*) of the numerator and the denominator. Consider the factors or divisors of the two numbers 48 and 36.

The number 48 has the factors: 1, 2, 3, 4, 6, 8, 12, 16, 24, 48.

The number 36 has the factors: 1, 2, 3, 4, 6, 9, 12, 18, 36.

Note that these two numbers have the following factors (divisors) in common: 1, 2, 3, 4, 6, 12. Since 12 is the largest of these, it is called the greatest common factor (g.c.f.) of 48 and 36.

Additional illustrations are:

$$(14, 21)\text{ g.c.f.} = 7$$
$$(3, 12)\text{ g.c.f.} = 3$$

It is always possible to find the g.c.f. after the numbers have been factored into all possible factors. A simple visual inspection is all that is necessary to select the one that is the largest among the factors that are common. (Two numbers which have no factors in common other than 1, such as 5 and 19, are said to be *relatively prime* to each other.)

A second method for finding the g.c.f. of two numbers involves the use of the division algorithm and is called the *Euclidean Algorithm* after Euclid, who is credited with having developed it. To understand its use, first observe that the g.c.f. of any two numbers must divide the difference between the two numbers. Note the examples above. In each case the difference between the two numbers is divisible by the g.c.f. Using this fact, it is possible to perform repeated division until the g.c.f. is found. The rule for carrying out this procedure is as follows: Divide the larger of the two numbers by the smaller. Then divide the divisor by the remainder from the first division. Continue dividing the last divisor by the last remainder until there is no remainder except zero. The last divisor prior to attaining the zero remainder is the g.c.f. of the two numbers. When this number turns out to be one, then the numbers are relatively prime. Note the process in the following examples.

Example:

Find the g.c.f. of the numbers in the box below by using the Euclidean Algorithm.

A. (36, 48)	36)48 = 1, 36, remainder 12; 12)36 = 3, 36, remainder 0	(Since 12 is the last divisor used to obtain the zero remainder, it is the g.c.f. of 36 and 48. This checks with the previous method.)
B. (5, 19)	5)19 = 3, 15, remainder 4; 4)5 = 1, 4, remainder 1; 1)4 = 4, 4, remainder 0	(Since the last divisor here is 1, the two numbers are relatively prime.)
C. (362, 1832)	362)1832 = 5, 1810, remainder 22; 22)362 = 16, 352, remainder 10; 10)22 = 2, 20, remainder 2; 2)10 = 5, 10, remainder 0	

The steps in C may be rewritten in the following fashion:

$$1832 = (5 \times 362) + 22$$
$$362 = (16 \times 22) + 10$$
$$22 = (2 \times 10) + 2$$
$$10 = (5 \times 2) + 0$$

$$1832 - (5 \times 362) = 22$$
$$362 - (16 \times 22) = 10$$
$$22 - (2 \times 10) = 2$$

It is possible to find the greatest common factor (divisor) of more than two numbers by simply considering the numbers in pairs. For example if there are three numbers, a, b, and c, the first step is to find the g.c.f. of a and b. The second step involves finding the g.c.f. of the g.c.f. of a and b and the number c. As an illustration consider finding the g.c.f. of (30, 168, 231).

First step: Find the g.c.f. of (30, 168).

30)168 = 5, 150, remainder 18
18)30 = 1, 18, remainder 12
12)18 = 1, 12, remainder 6
6)12 = 2, 12 g.c.f. = 6

Second step: Find the g.c.f. of (6, 231).

6)231 = 38, 228, remainder 3
3)6 = 2, 6 g.c.f. = 3

The g.c.f. of (30, 168, 231) is therefore 3.

Least Common Multiple.—Another term that is used in conjunction with the arithmetic of fractions is the *least common multiple* (*l.c.m.*). The least common multiple of two numbers is the smallest number which is a multiple of both of them. For example, the least common multiple of 5 and 6 is 30, while the least common multiple of 21 and 30 is 210. Observe that if the two numbers are relatively prime, then the l.c.m. is their product. The least common multiple and the greatest common factor are related in the following manner. The least common multiple is equal to the product of the two numbers under consideration divided by the greatest common factor of the two numbers, as shown below.

$$\text{l.c.m.} = \frac{21 \times 30}{\text{g.c.f.}}$$

Since the g.c.f. of (21, 30) is 3, the product of 21 × 30 is divided by 3.

$$\text{l.c.m.} = \frac{21 \times 30}{3} = \frac{630}{3} = 210$$

The least common multiple can also be found by inspecting the prime factorization of the two numbers. Again using the above example,

$$21 = 3 \times 7$$

The least common multiple must have factors 2, 3, 5, and 7, producing 210.

$$30 = 2 \times 3 \times 5$$

The greatest common factor (divisor) and the least common multiple are used primarily in the arithmetic operations involving the rational numbers (commonly called the fractions).

Fractions.—The numbers that are usually referred to as the fractions are represented by symbols such as 2/3, 8/6, 7/1, 5/6, etc. Such symbols may have several possible interpretations. Consider, for instance, 2/3. This may be interpreted as dividing a quantity into three equal parts and then taking *two* of them to represent 2/3. Or, 2/3 may be thought of as the result of dividing two whole quantities into three equal parts.

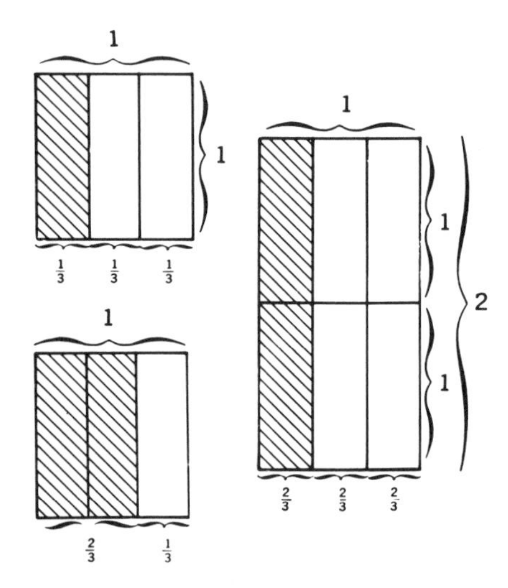

The accompanying diagrams represent these processes geometrically. In this sense, 2/3 is thought of as 2 times 1/3 in the first instance and as 1/3 times 2 in the second. Also, 2/3 may be thought of as 2 ÷ 3. That is, $2 \div 3 = 2/3 = k$, where k is a number such that $2 = 3k$, satisfying the definition of division. Still yet another interpretation of the symbol is that of ratio, which is sometimes written 2 : 3. This interpretation results from comparing sets, or magnitudes. If set A has 2 elements and set B has 3 elements, then the ratio of the number of elements in A to the number of elements in B is written as 2/3.

All of these interpretations are important and have had a long development historically. Since all are identical mathematically, the following may be chosen as the basic interpretations of the symbol a/b: (1) Interpret a/b as dividing a whole quantity into "b" parts and then taking "a" of them, or (2) interpret a/b as taking "a" whole quantities and dividing the aggregate into "b" equal parts, then taking one of these equal parts.

In each interpretation of fractions the essential meaning is conveyed in terms of a pair of whole numbers. Furthermore, the pair is ordered since 2/3 is distinct from 3/2 in any of the interpretations. These considerations allow us to state the following definition: A fraction is an ordered pair of whole numbers a and b, usually written in the form a/b or $\frac{a}{b}$, where b is not 0. In the usual terminology, a is the *numerator* and b is the *denominator* of the given fraction. In the case that a is less than b, a/b is a proper fraction, for example, 5/8. In the case that a is greater than or equal to b, a/b is called an improper fraction, for example, 9/5. This does not imply that there is anything wrong with writing such a fraction, but the words "proper" and "improper" merely distinguish two particular types of fractions.

Since by the basic interpretation, the fraction 3/1 would mean beginning with three whole quantities, dividing the aggregate into one equal part, and then taking one of these parts, it is evident that $3/1 = 3$ or, in general, $n/1 = n$. Also, the two diagrams below illustrate the geometric interpretation of the improper fraction 4/3. In the smaller diagram, a unit area has been divided into thirds and a fourth third has been added. In the larger diagram, one third of each of four unit areas has been shaded. In both diagrams, the total shaded area is *4 thirds*. From these, it is again evident that $4/3 = 1 + 1/3$. Hence, we should always interpret such symbols as $2\frac{3}{4}$, $5\frac{3}{5}$, and $7\frac{5}{8}$ as a short way of writing $2 + 3/4$, $5 + 3/5$, and $7 + 5/8$.

In similar fashion, 4/4, 3/3, and, in general, n/n (n is any nonzero whole number) are all equal to 1. This is illustrated by the accompanying diagram.

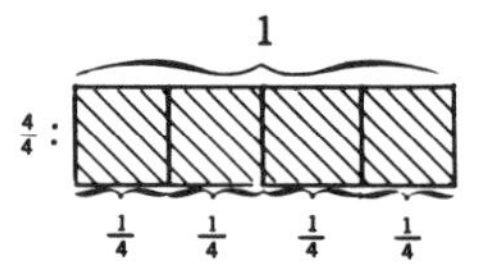

■ EQUALITY.—The two fractions 2/3 and 4/6 are easily seen to be equivalent, in the sense that they represent the same quantity. The solid lines in the rectangle represent its division into three equal parts. The dotted and solid lines represent its division into six equal parts. Thus, taking 4 of the 1/6 parts is the same as taking 2 of the 1/3 parts and hence it agrees that $2/3 = 4/6$.

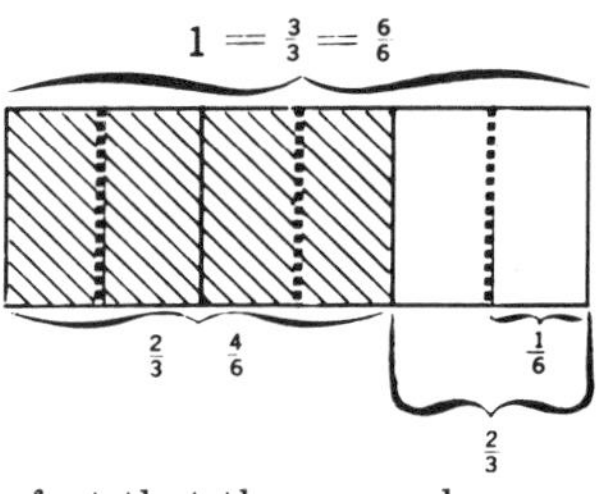

The fact that there may be many different fractions that are equivalent, in the sense that they represent the same quantity in terms of the basic interpretation, suggests that operating with fractions may not be quite the same as operating with whole numbers. For instance, the fraction 2/3 is one member of a set containing an unlimited number of fractions equivalent to it: $2/3 = 4/6 = 6/9 = 8/12 = 10/15 = \ldots$ However, the fact that all of these fractions represent the same quantity suggests that there may be only *one number* involved here which has many representations, or names, as a fraction. This single number associated with the set of all fractions equal to 2/3 is called a *rational number*. (A fraction has been defined in terms of whole numbers and therefore these fractions represent only that subset of the rational numbers known as the nonnegative rational numbers. Henceforth in this article, when rational numbers are spoken of they will mean a nonnegative rational number.) Since the same is true for any fraction a/b, any particular fraction is a name for a unique rational number, but there may be many fractions which name the same rational number. With the agreement that all fractions equivalent to a given fraction are merely names for the same number, rules can be developed for operating with them just as with whole numbers.

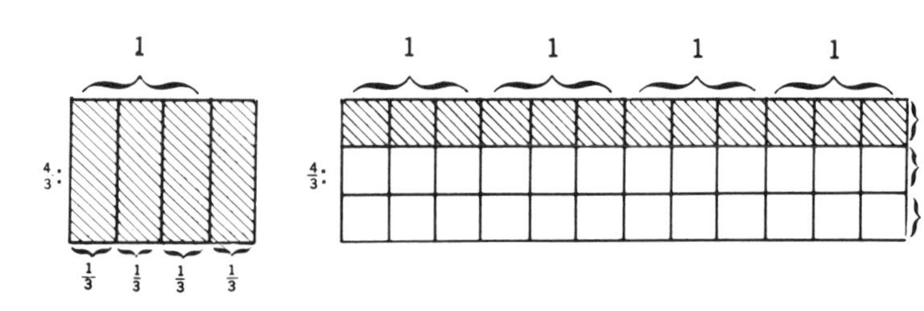

A simple test for equivalence of fractions may be formulated as follows: If a/b and c/d are fractions then $a/b = c/d$ if, and only if, $ad = bc$. To see that the test really works, consider the example given above, 2/3 and 4/6. Here $a = 2$, $b = 3$, $c = 4$ and $d = 6$; $a \cdot d = 2 \cdot 6 = 12$, and $b \cdot c = 3 \cdot 4 = 12$, verifying that $2/3 = 4/6$. If the problem of determining whether or not the fractions 6/9 and 8/12 are equivalent by the basic interpretation were considered, it would be found to be a considerable task; but by the rule given it is easy, since $6 \cdot 12 = 8 \cdot 9 = 72$. It is important to note that a, b, and c are whole numbers and $b \neq 0$, $c \neq 0$; then $ac/bc = a/b$ follows, using the above basic rule for equivalence of fractions, since $(ac)b = (bc)a$. The latter equality holds, since the properties of whole numbers assures us that the product of three whole numbers is the same regardless of the order or grouping used in the multiplication. Thus, using this principle, $6/9 = (2 \cdot 3)/(3 \cdot 3) = 2/3$ and $8/12 = (2 \cdot 4)/(3 \cdot 4) = 2/3$, verifying again that $6/9 = 8/12$. This latter principle is the basic rule for changing a given fraction to an equivalent fraction, which is another name for the same unique rational number.

■ ADDITION.—The addition of whole numbers was defined in terms of properties of sets. A similar approach will be used to motivate the definition for addition of fractions. Consider $3/12 + 5/12$.

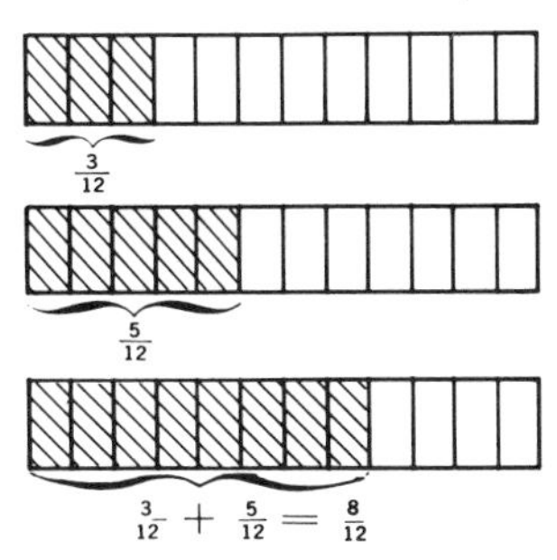

Thus, the sum must be 8/12, and it is evident that whenever two fractions have the same whole number in the denominator the sum is found by taking the sum of the whole numbers in their numerators as the numerator of the sum and the *common denominator* as the denominator of the sum. Hence, for two fractions, a/b and c/b, the definition is that $a/b + c/b = (a + c)/b$. However, this definition does not immediately show how to add two fractions which do not have the same whole number as a denominator. To see how to proceed, consider $1/4 + 2/3$. These two fractions do not have the same denominator, but by the rule developed for changing a fraction to an equivalent fraction the denominator of 2/3, or 3, may be used to change 1/4 to $(1 \cdot 3)/(4 \cdot 3) = 3/12$ and the denominator of 1/4, or 4, to change 2/3 to $(2 \cdot 4)/(3 \cdot 4) = 8/12$. The fractions then become $1/4 + 2/3 = 3/12 + 8/12 = 11/12$. On this basis any two fractions may be added regardless of whether or not they have the same denominator. Thus, in general,

$$a/b + c/d = ad/bd + bc/bd = (ad + bc)/bd$$

for any two fractions a/b and c/d. It should be noted that this process is not the simplest possible one in that the denominator bd may be larger than is absolutely necessary for the two fractions involved. It is true, nevertheless, that this denominator will produce the desired sum, but perhaps not in what is usually called the simplified, or reduced, form. For example, if 3/8 and 5/12 are added by the above rule the sum is

$$(3 \cdot 12) + (8 \cdot 5)/(8 \cdot 12) = 76/96.$$

The reduced or simplest form of any fraction is a fraction equivalent to the given fraction such that the numerator and denominator have no factor in common. Thus, since

$$76/96 = (19 \cdot 4)/(24 \cdot 4),$$

it is not the simplest form but is equivalent to 19/24, which is in simplest form.

In practice the addition of fractions is carried out by means of the rule which applies to fractions having the same denominator. When the two fractions to be added do not have the same denominator, the notion of the *least common denominator* is used to obtain two equivalent fractions having the same denominator. Consider the example $3/8 + 5/12$ again. These two fractions must be changed to equivalent fractions having the same denominator. Certainly both fractions may be changed to fractions with a denominator $8 \cdot 12$, or 96, as shown above. But is this the smallest number that will serve as a common denominator? What is desired, of course, is to change each fraction to one having a denominator which is the smallest whole-number multiple of each given denominator. Thus, it is desired to change each denominator to a number which is the *least common multiple* of the given denominators. In this case the least common multiple of 8 and 12 is 24. This number is then referred to as the *least common denominator* of the two given fractions. Then the fractions may be changed to $(3 \cdot 3)/(8 \cdot 3) = 9/24$, $(5 \cdot 2)/(12 \cdot 2) = 10/24$. Then the sum is 19/24, as before. Note that the result is now in simplest, or reduced, form.

In some cases it is not readily evident what the least common denominator of two given fractions would be by inspection. In this case the techniques used for finding the least common multiple of two whole numbers may be helpful. Consider $13/36 + 28/45$. To find the least common multiple of 36 and 45, write each in factored form: $36 = 2 \cdot 2 \cdot 3 \cdot 3$, and $45 = 3 \cdot 3 \cdot 5$; then write the product of all prime factors that appear in either 36 or 45, repeating each of these as a factor the greatest number of times it is repeated in either 36 or 45. This gives the product $2 \cdot 2 \cdot 3 \cdot 3 \cdot 5$, or 180—the least common multiple of 36 and 45 and the least common denominator of the given fractions. When the numbers are written in factored form it is easy to see what multiples of 36 and 45 must be taken in changing the two fractions to ones having denominator $2 \cdot 2 \cdot 3 \cdot 3 \cdot 5 = 180$. The denominator 36 would have to be multiplied by 5, and 45 would have to be multiplied by $2 \cdot 2$. These multiples could also be found by dividing 180 by 36 and by 45. Thus it follows that

$$\begin{aligned} 13/36 + 28/45 &= (13 \cdot 5)/(36 \cdot 5) \\ &\quad + (28 \cdot 4)/(45 \cdot 4) \\ &= 65/180 + 112/180 = 177/180 \end{aligned}$$

Several special cases in the addition of fractions arise when considering improper fractions, addition of whole number and fractions, and the treatment of symbols such as 4-2/3. The following examples illustrate procedures for dealing with these special cases.

Example:

$$\begin{aligned} 5\text{-}3/4 &= 5 + 3/4 \\ &= 5/1 + 3/4 \\ &= 20/4 + 3/4 \\ &= 23/4 \end{aligned}$$

Example:

$$\begin{aligned} 2\text{-}3/8 + 3\text{-}7/8 &= (2 + 3/8) + (3 + 7/8) \\ &= (2 + 3) + (3/8 + 7/8) \\ &= 5 + 10/8 \\ &= 5 + (8/8 + 2/8) \\ &= 5 + (1 + 2/8) \\ &= 6 + 2/8 \\ &= 6 + 1/4 \\ &= 6\text{-}1/4 \end{aligned}$$

The addition in this example can also be carried out by converting 2-3/8 to 19/8 and 3-7/8 to 31/8 as follows:

$$\begin{aligned} 2\text{-}3/8 + 3\text{-}7/8 &= 19/8 + 31/8 \\ &= 50/8 \\ &= 48/8 + 2/8 \\ &= 6 + 2/8 \\ &= 6 + 1/4 \\ &= 6\text{-}1/4 \end{aligned}$$

The addition can be further altered by converting 50/8 to 25/4 and then continuing as above.

The question also arises in connection with the above as to which of the symbols, 5-3/4 or 23/4, is preferable, since they both represent the same number. There is no absolute answer to this question. In the mathematical sense, 23/4 is possibly the preferred form because of the consistency of symbolism, but one would certainly not go to a store and ask for 23/4 yards of material. The preferable symbol would then be the one that is most appropriate to the situation. In any case, to work successfully with fractions, the meaning of both symbols must be understood.

■ MULTIPLICATION.—To illustrate the operation of multiplication of fractions, consider the following series of examples; base the conclusions on the initial interpretations given to the symbol a/b, where a and b are whole numbers, and b is not zero.

Example:

$$3 \times \frac{2}{5}$$

Interpreting multiplication by a whole number as continued addition (as was illustrated with multiplication of a whole number by a whole number) this result would be $2/5 + 2/5 + 2/5 = 6/5$. Also, $2/5$ could be interpreted as $2 \times 1/5$. Then $3 \times (2 \times 1/5) = (3 \times 2) \times 1/5 = 6 \times 1/5 = 6/5$. This has the geometric interpretation shown below.

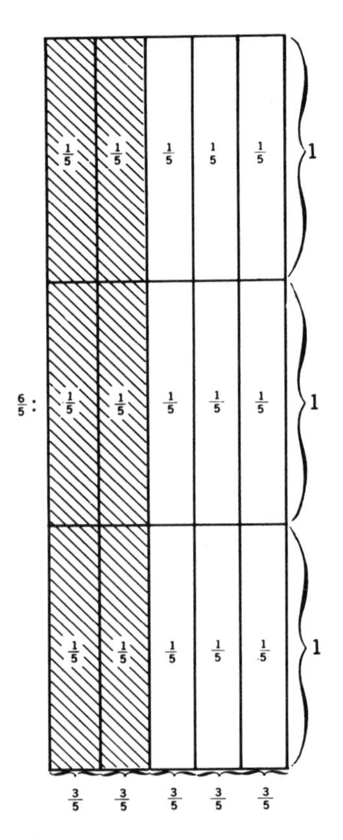

Example:

$$\frac{2}{5} \times 3$$

Here $2/5$ is interpreted as $2 \times 1/5$ again and then $2 \times 1/5 \times 3 = 3 \times 1/5 \times 2 = 2 \times 3/5$. Geometrically, this may be thought of as dividing the aggregate of three whole quantities into five equal parts and then taking two of them.

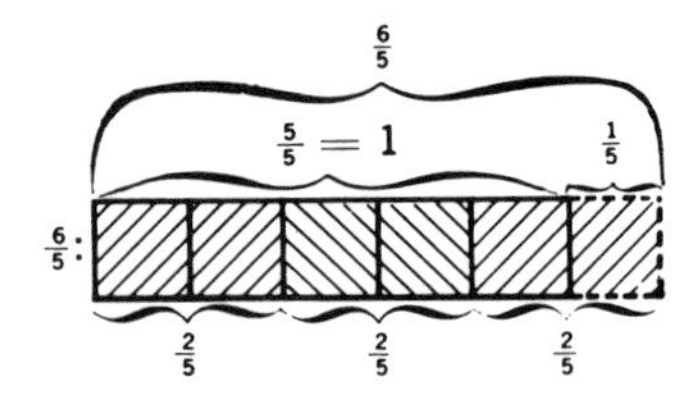

Example:

$$\frac{2}{3} \times \frac{3}{4}$$

In terms of the basic interpretation of symbols, this may be interpreted as 2 units of $3/4$ divided into three equal parts, or as dividing $3/4$ into three equal parts and then taking two of them. In either case $2/3 \times 3/4$ will be $2/4 = 1/2$. The result is illustrated geometrically.

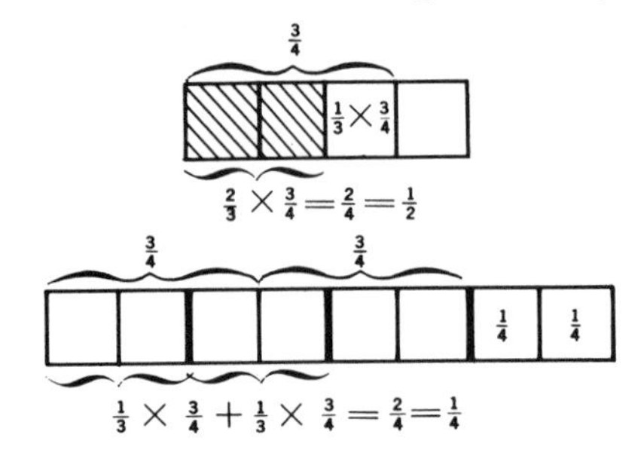

Example:

$$\frac{2}{5} \times \frac{3}{4}$$

This could be interpreted as dividing $3/4$ into five equal parts and then taking two of them. To divide $3/4$ into five equal parts, it is easier if $3/4$ is first changed to $(3 \cdot 5)/(4 \cdot 5) = 15/20$. Then, if $15/20$ is divided into five equal parts, each equal part will be $3/20$ and then taking two of them gives the result $6/20$, or $3/10$.

In each of these examples it is easily verifiable that the same result could be obtained by writing each factor in fractional form and multiplying the whole numbers that appear respectively in the numerators and denominators to obtain the fraction that represents the product. Thus, $3 \times 2/5 = 3/1 \times 2/5 = 6/5$, $2/3 \times 3/4 = 6/12$, and $2/5 \times 3/4 = 6/20 = 3/10$. With the suggestion of these examples, multiplication can be defined as follows: If a/b and c/d are fractions, then $a/b \times c/d = ac/bd$.

The examples below illustrate several general procedures that are helpful in carrying out the multiplication of fractions in varying situations.

Example:

$$\frac{12}{17} \times \frac{2}{3} = \frac{12 \cdot 2}{17 \cdot 3} = \frac{4 \cdot 3 \cdot 2}{17 \cdot 3} = \frac{(4 \cdot 2) \cdot 3}{17 \cdot 3} = \frac{8}{17}$$

Here it is shown that in multiplying fractions it is helpful to try to identify common factors in the numerator and denominator before carrying out the actual multiplication, using the rule $ac/bc = a/b$ to obtain the simplest form of the product.

Example:

$$12 \times \frac{11}{24} = \frac{12}{1} \times \frac{11}{24} = \frac{12 \cdot 11}{12 \cdot 2} = \frac{11}{2}$$

Example:

$$6 \times 3\frac{4}{7} = 6\left(3 + \frac{4}{7}\right) = (6 \cdot 3) + \left(6 \cdot \frac{4}{7}\right) = 18 + \frac{24}{7} = 18 + \left(3 + \frac{3}{7}\right) = (18 + 3) + \frac{3}{7} = 21 + \frac{3}{7} = 21\frac{3}{7}$$

Alternatively, since 3-4/7 = 25/7, $6 \times 25/7 = 6/1 \times 25/7 = 150/7$ = 21-3/7.

Example:

$$2\frac{2}{3} \times 3\frac{5}{8} = \left(2 + \frac{2}{3}\right) \times \left(3 + \frac{5}{8}\right) = 2\left(3 + \frac{5}{8}\right) + \frac{2}{3}\left(3 + \frac{5}{8}\right)$$
$$= 6 + \left(2 \times \frac{5}{8}\right) + \left(\frac{2}{3} \times 3\right) \times \left(\frac{2}{3} \times \frac{5}{8}\right) = 6 + \frac{5}{4} + 2 + \frac{5}{12} = 6 + 2 + \left(\frac{15}{12} + \frac{5}{12}\right)$$
$$= 8 + \frac{20}{12} = 8 + 1 + \frac{2}{3} = 9 + \frac{2}{3} = 9\frac{2}{3}$$

Since 2-2/3 = 8/3, and 3-5/8 = 29/8, the product may also be found by finding $8/3 \times 29/8 = (8 \cdot 29)/(3 \cdot 8) = (29 \cdot 8)/(3 \cdot 8) = 29/3$ = 9-2/3. Usually in cases such as the two previous examples, it is easier to change the symbols to fractional form before carrying out the multiplication.

Note that in the above examples it was necessary to assume commutative and associative laws for multiplication and addition and also that the distributive law holds for fractions.

■ **SUBTRACTION.**—The operation of subtraction for fractions is defined in a way that is analogous to the definition of subtraction of whole numbers. That is, the difference of two fractions, a/b and c/d, is e/f if, and only if, e/f is a fraction such that $a/b = c/d + e/f$. Thus to find $7/10 - 3/10$, note that $3/10 + 4/10 = 7/10$, and hence the answer is $4/10$. Note also that the result could be obtained by writing $7/10 - 3/10 = (7 - 3)/10 = 4/10$, which is precisely the usual procedure for obtaining the difference of two fractions with the same denominator. As with whole numbers, this process can only be carried out if the minuend is greater than the subtrahend. This is an obvious result of the observation that in obtaining the numerator of the difference fraction it is necessary to find the difference of two whole numbers.

The question then arises as to how one determines which of two given fractions is the greater. This question did not arise with the whole numbers, since the system of numeration renders the result immediately. Of course, if the two fractions have the same denominator, as in the example, then the result follows from the fact that $7 > 4$. But what about $7/12$ and $11/20$? One procedure would be to change both fractions so that they have the same denominator. The most direct way to do this is to write $7/12 = (7 \cdot 20)/(12 \cdot 20) = 140/240$, and $11/20 = (11 \cdot 12)/(20 \cdot 12) = 132/240$. It then is immediately evident that $7/12$ is greater than $11/20$. Thus, for any two fractions a/b and c/d, the first would be changed to ad/bd and the second to bc/bd; the first is greater than the second if, and only if, ad is greater than bc.

In the same way, if a/b and c/d are any two fractions, the first may be changed to ad/bd and the second to bc/bd; their difference will then be the fraction $(ad - bc)/bd$, provided $ad > bc$.

This is similar to the general rule for the addition of fractions.

In subtraction, as in addition, the least common multiple of denominators (*least common denominator*) is used to make the process as efficient as possible. Also, similar comments apply where symbols such as 3-5/8 appear.

Example:

$$\frac{11}{21} - \frac{3}{7} = \frac{11}{7 \cdot 3} - \frac{3}{7} = \frac{11}{7 \cdot 3} - \frac{3 \cdot 3}{7 \cdot 3} = \frac{(11 - 9)}{21} = \frac{2}{21}$$

Example:

$$7\frac{3}{8} - 2\frac{5}{12} = \left(7 + \frac{3}{8}\right) - \left(2 + \frac{5}{12}\right) = (7 - 2) + \left(\frac{3}{8} - \frac{5}{12}\right)$$

But $3 \cdot 12$ is less than $8 \cdot 5$. Hence, the last subtraction cannot be performed. Therefore, it is necessary to go back and change $7 + 3/8$ to $6 + (1 + 3/8) = 6 + 11/8$. This gives $(6 - 2) + (11/8 - 5/12)$. However, $6 - 2 = 4$ and the least common multiple of 8 and 12 is 24. Then $11/8 - 5/12 = (11 \cdot 3)/(8 \cdot 3) - (5 \cdot 2)/(12 \cdot 2) = 33/24 - 10/24 = (33 - 10)/24 = 23/24$. Therefore, the difference is $4 + 23/24 =$ 4-23/24. The problem might also be carried out by changing 7-3/8 to 59/8 and 2-5/12 to 29/12, then subtracting.

DIVISION.—As has been seen, the operations of addition and subtraction are quite readily explained in terms of the basic interpretations. Multiplication is a little more difficult, but is fairly easily understood in terms of these interpretations. Division, however, seems to be more difficult, and an attempt to provide an adequate rationale will start with the most elementary considerations.

Division of a whole number by a fraction: It seems reasonable to interpret the symbol $3 \div 1/4$ as asking the question: how many 1/4ths are contained in 3? Since there are four 1/4ths in 1, there must be three times as many, or 12, in 3. Thus the result of the division is 12. Note that $12 = 3 \times 4/1$. Again, what is $3 \div 2/3$? First the question, how many 2/3rds are contained in 1, must be answered. Now $1 = 2/3 + 1/3$ and 1/3 is 1/2 of 2/3. Hence the number of 2/3rds in 1 is 1-1/2. The number of 2/3rds in 3 is then $3 \times$ 1-1/2 = 4-1/2. Note that 4-1/2 $= 9/2 = 3 \times 3/2$.

Division of a fraction by a whole number: Here it is necessary to give a slightly different interpretation of the operation of division, namely, that of dividing 1/4 into three equal parts. Thus, 1/4 is changed to the equivalent fraction 3/12. Taking 1/3 of 3/12, the result of the division is then 1/12. Again, note that $1/12 = 1/4 \times 1/3$.

Division of a fraction by a fraction: The symbol $2/3 \div 3/5$ may be regarded as asking: How many 3/5 in 2/3? And again, it is first necessary to find how many units of 3/5 are contained in 1. Since $1 = 5/5 = 3/5 + 2/5$, there is one unit of 3/5 in 1 with 2/5 remaining. But $2/3 \times 3/5 = 2/5$. Hence, there are 1-2/3 units of 3/5 in 1. Therefore, there will be 2/3 as many in 2/3. Now $2/3 \times$ 1-2/3 $= 2/3 \times 5/3 = 10/9 =$ 1-1/9, which is $2/3 \times 5/3$.

In every case it is evident that the final result could have been obtained by means of the usual rule for division, which is to invert the divisor and multiply. However, the arguments used are not complete and perhaps not sufficient to justify this rule for division. A fresh look at the operation of division from the standpoint of the properties of fractions themselves is therefore in order.

First note that any fraction of the form a/a, where a is not zero, has the identity property for multiplication. That is, for any given fraction c/d,

$$c/d \times a/a = ca/da = c/d.$$

Thus 2/2 has this property, as does any fraction equivalent to it. Recalling the earlier discussion, 1/1, 2/2, 3/3, and all fractions of this form are names for the same rational number. It is customary to give this rational number the name 1, since it behaves as an identity for multiplication. Thus, it is seen that any fraction, which is a particular name for some rational number, multiplied by any fraction of the form a/a, yields a fraction which is a name for the same rational number.

Also, it is easily seen that for any nonzero fraction, say 2/3, there is a fraction formed by inverting the given fraction, 3/2, which has the property that $2/3 \times 3/2 = 6/6 = 1$, the identity for multiplication. It must be remembered that 3/2 is not the only fraction which has this property in relation to 2/3, since any fraction equivalent to 3/2 has the same property. But if the fraction 2/3 is interpreted as naming a particular *rational number*, then there is only *one* rational number, whether it be named by 3/2 or by some equivalent, that yields 1 as a product. This rational number is called the *inverse*, or *reciprocal*, of the given rational number. Therefore, the rational number named by 3/2 is called the *inverse* of the rational number named by 2/3. In this sense every fraction $a/b \neq 0$ has an inverse, b/a.

It is now possible to frame a definition of division of fractions similar to that framed for division of whole numbers (note: reference is made to the symbols as fractions with the understanding that they be thought of as a name for a rational number): The quotient of a/b and c/d is e/f if, and only if, $a/b = c/d \times e/f$. For example, $2/3 \div 3/5 = 10/9$, since $2/3 = 3/5 \times 10/9$.

Now let us look once again at the usual rule for division, that $a/b \div c/d = a/b \times d/c$. Or, stated in words, to divide two fractions, invert the divisor and multiply. For example, consider $3/4 \div 5/7$. Let x/y name the quotient. Then, by the definition of division, it must follow that $3/4 = 5/7 \times x/y$. Now if these two expressions are equal they will certainly remain equal if they are both multiplied by the 7/5, which is the inverse of 5/7. Thus $7/5 \times 3/4 = 7/5 \times 5/7 \times x/y = 1 \times x/y = x/y$. Hence, the quotient x/y is $7/5 \times 3/4 = 3/4 \times 7/5$, justifying the usual rule on the basis of the definition of division and the existence of inverses. (Note that the associative law and the commutative law for multiplication are assumed.)

Another procedure for performing the operation of division with fractions is illustrated below.

$$6/7 \div 2/3 = (6/7 \times 21) \div (2/3 \times 21) = 18 \div 14 = 18/14 = 9/7$$

This procedure is based on the principle that the quotient remains unchanged when the dividend and divisor are multiplied by the same or equal fractions. In the illustration, the multiplier used is the least common denominator of the fractions involved in the division. Also, the interpretation of the fractional symbol as representing an indicated division is utilized; that is, $a/b = a \div b$. Further utilization of this idea would enable the solution to be written as follows:

$$\frac{6/7}{2/3} = \frac{6/7 \times 21}{2/3 \times 21} = 18/14 = 9/7$$

In still another variation on this theme, the reciprocal of the divisor, 3/2, is used as a multiplier and the result is obtained as follows:

$$\frac{6/7}{2/3} = \frac{6/7 \times 3/2}{2/3 \times 3/2} = \frac{6/7 \times 3/2}{1} = 6/7 \times 3/2 = 18/14 = 9/7$$

Neither of these procedures involves learning the rule given for division usually stated, yet each gives the required result.

In concluding the section on fractions, note once more that a fraction names a particular number, called a rational number, but that there may be many fractions which name the same rational number. A given rational number is usually named by a fraction in its reduced, or simplest, form.

The set of rational numbers has many properties analogous to the whole numbers plus some additional ones. This set obeys the commutative and associative laws of addition and multiplication. It is closed for the operations of addition and multiplication and also for division. This means that given any two rational numbers, their sum, product, and quotient are also rational numbers. The set contains identity elements for addition and multiplication, the rational numbers named by 0/1 and 1/1. It has been seen that each rational number, a/b, has a multiplicative inverse, b/a, such that $a/b \times b/a = 1$. The distributive law of multiplication with respect to addition also holds, that is,

$$a/b \times (c/d + e/f) = (a/b \times c/d) + (a/b \times e/f)$$

The rational numbers have an additional property that distinguishes them from the whole numbers. This is that the rational numbers are said to constitute a *dense* set. This means that between any two rational numbers, no matter how close together, there still remains another rational number. In fact, between any two rational numbers there are an unlimited number of rational numbers. If the rational numbers are interpreted as points on a line, this

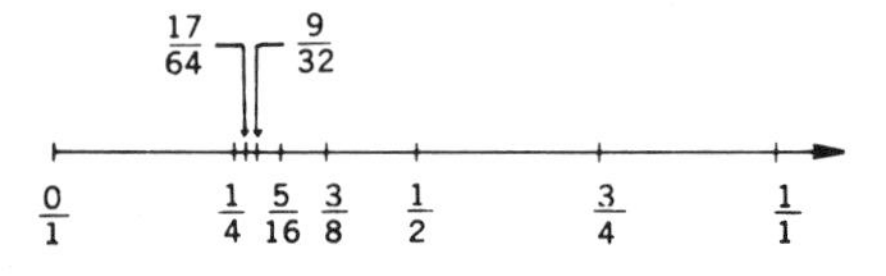

DENSE SET OF RATIONAL NUMBERS

means that the points representing rational numbers are closely packed together. The property of denseness makes a rational number completely adequate for all practical measurement purposes, whereas the whole numbers are adequate only for counting purposes. This is illustrated in practice when rulers of increasing precision are obtained simply by increasing the number of subdivisions in the unit of measure.

Ratio, Proportion, and Per Cent.—Many everyday problems involve the comparison of two numbers. Some examples are: the number of games a team wins compared with the total number of games played; the circumference of a circle compared to its diameter ($c/d = \pi$); the number of boys compared to the number of girls in a class; and, from trigonometry, there are the comparisons of the measures of the sides of a right triangle. Comparisons may be done in two ways. One may compare by subtraction or by division. Comparison by division is referred to as a *ratio* and is expressed as a/b (or $a : b$), where a and b are natural numbers. Notice that they look like the rational numbers—indeed the rational numbers derived their name from ratios. Ratios imply a quotient. To illustrate: suppose that Kathy is 3 years old and Anne is 5 years old. The ratio of Kathy's age to Anne's age is 3/5; that is, Kathy is three-fifths as old as Anne. Records indicate that about 2000 years ago the Greeks made extensive use of ratios in comparing lengths of line segments, sizes of angles, and areas of regions.

Proportion is a statement that shows that two ratios are equal. Thus, if the ratios a/b and c/d represent the same comparison they may be set equal to each other as: $a/b = c/d$. The values a, b, c, and d are called the four members of the proportion. One of the useful facts about proportions is that if $a/b = c/d$, then $ad = bc$, as proven below.

Proof:

Given $a/b = c/d$. Multiply both numbers by bd:

$$(bd)(a/b) = (bd)(c/d)$$

Simplify by dividing both the numerator and denominator by b on the left and by d on the right side of the equality, with the result:

$$ad = bc.$$

The importance of this is that if any three members of a proportion are known, the fourth can always be found. For example, what number bears the same ratio to 20 as 3 does to 4? The ratio x to 20 is equal to the ratio 3 to 4.

$$x/20 = 3/4$$
$$4x = 60$$
$$x = 15$$

A recipe calls for 6 eggs to make 24 cookies. How many eggs are necessary to make 36 cookies? Again a proportion:

$$6/24 = x/36$$
$$24x = 216$$
$$x = 9$$

When comparing denominative quantities by division, the values must be expressed in common units. Thus 6 inches to 2 feet would give a ratio 1/2:2 which is equal to 1/4 when both are expressed in feet. If inches are used, 6/24 is obtained, which is also equal to 1/4. The ratio itself is a pure number and has no units attached to it.

Per cent is closely related to the concept of ratio. The term "per cent" is derived from the Latin *per centum*, which means "by the hundred." Thus, 5 per cent means 5 out of one hundred. As a ratio, this would be written 5/100. The per cent symbol, %, was introduced as early as the fifteenth century, although it was not used extensively until recently. Today the meaning of per cent is somewhat broader than was the original interpretation. For example, expressions such as 700%, 5000%, and so on, are often used. Per cent is viewed as a ratio a/b where b is always 100. The three ways in which a per cent may be expressed are: $a/100$, $0.01 \times a$, and $a\%$.

Examples:

An investment of $5000.00 yields an annual return of $200.00. What is the per cent of return?

$$\frac{200}{5000} = \frac{x}{100}$$
$$5000x = 20000$$
$$x = 4\%$$

A bank pays interest at the rate of 5 per cent per year. How much interest will an investment of $250.00 yield annually?

$$\frac{5}{100} = \frac{x}{250}$$
$$x = \$12.50$$

How many dollars should be invested at 6 per cent to yield an annual income of $300.00?

$$\frac{6}{100} = \frac{300}{x}$$
$$x = \$5{,}000$$

In general this can be formulated as: $i/100 = p/b$, where $i/100$ is the rate of interest, p is the percentage of return, and b is the base, or amount invested.

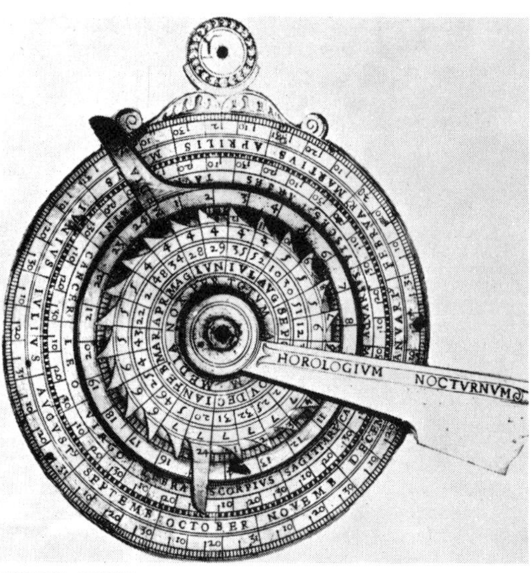

BETTMANN ARCHIVE

CIRCULAR ITALIAN CALENDAR, used during the sixteenth century, was designed to be used in the evening as well as daytime.

Number Congruence.—The concept of number congruence and the symbolism related to this idea were introduced by Karl Friedrich Gauss (1777–1855). To illustrate the concept of number congruence, look at the face of a watch or clock. If it is 10 P.M. now, what time will it be in 7 hours? The problem is to add 10 and 7 on the face of a clock. As is seen, one gets 5 as the answer rather than 17. The reason for getting 5 o'clock is that the common clock repeats itself every 12 hours. The odometers in automobiles repeat every 100,000 miles. The days of the months repeat in approximately 30-day cycles. The days of the year repeat in $365\frac{1}{4}$ days. There are many additional examples, among them such common devices as gas meters, water meters, and electric meters.

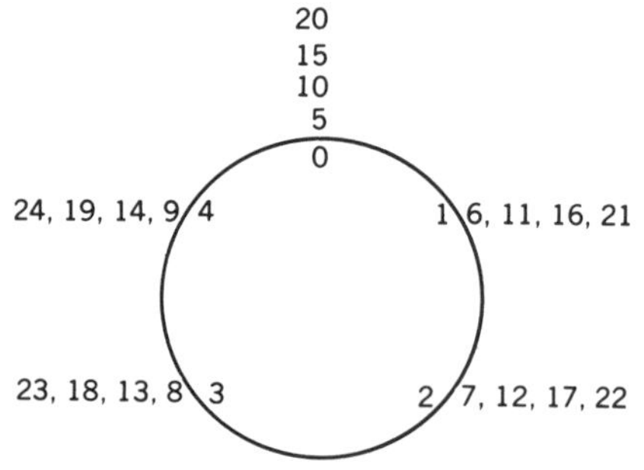

Suppose one were to look at the face of a clock that has only 5 positions on it. With a little reflection, it will be seen that these basic positions are the remainders one can obtain when dividing any whole number by 5. In this example, one can say that all the whole numbers that have the same remainder when divided by 5 are congruent to each other. Thus 1, 6, 11, 16, . . . belong to the same set; that is, they are congruent because they each have a remainder of 1 when divided by 5. The numbers 4, 9, 14, 19, . . . each have 4 as a remainder when divided

BETTMANN ARCHIVE

TRAVEL DESKCLOCK with a sun motif on the cover was made in Sweden in 1760. Early timepieces such as these contained a rather complex system of gears powered by a central mainspring. Most of these clocks had to be wound and adjusted at least once a day.

BETTMANN ARCHIVE

CALENDAR sculptured in clay was used about 3,500 years ago in Crete. Agricultural symbols often represented time periods.

by 5 and these are called congruent.

Thus, *two whole numbers "a" and "b," which have the same remainder when divided by a number "m," are said to be congruent in modulo "m."* The number m is called the modulus, which is the Latin term meaning "little measure." The symbolization of this expression is: $a = b$ mod m (mod is used as an abbreviation for modulo) if, and only if, $a - b = k \cdot m$, where k is an integer. This statement is read: a is congruent to b in modulo m if, and only if, the difference between a and b can be expressed as a multiple of m. Another way of saying this is that the difference between a and b is exactly divisible by m. Using modulo 5, it is seen that the numbers 22 and 7 are congruent since $22 - 7 = 3 \cdot 5$. However, 31 and 8 are not congruent in modulo 5, for $31 - 8$ cannot be expressed as a multiple of 5. Associated with each modulus is a table of addition and multiplication facts. For modulo 5 the following facts are presented.

Modulo 5

+	0	1	2	3	4
0	0	1	2	3	4
1	1	2	3	4	0
2	2	3	4	0	1
3	3	4	0	1	2
4	4	0	1	2	3

×	0	1	2	3	4
0	0	0	0	0	0
1	0	1	2	3	4
2	0	2	4	1	3
3	0	3	1	4	2
4	0	4	3	2	1

For modulo 2, it is seen that there would be two sets of congruent numbers, the even and the odd. The addition and multiplication facts for modulo 2 are presented below:

Modulo 2

+	0	1
0	0	1
1	1	0

×	0	1
0	0	0
1	0	1

The common clock is thus modulo 12.

Shortcuts.—As man learned more about the numbers he had created and their uses he began to find timesaving techniques for doing ordinary calculations. Most of these techniques can very simply be explained in terms of the basic structure of the base ten numeration system. For example, the use of exponents has made it possible to multiply and divide very large and very small numbers easily, for example, $10^{15} \times 10^5 = 10^{20}$, and $10^{-16}/10^5 = 10^{-21}$. Some of the powers of 10 are shown below.

$$1000 = 10 \cdot 10 \cdot 10 = 10^3$$
$$100 = 10 \cdot 10 = 10^2$$
$$10 = 10 = 10^1$$
$$1 = \frac{10}{10} = 10^0$$
$$.10 = \frac{1}{10} = 10^{-1}$$
$$.01 = \frac{1}{100} = 10^{-2}$$
$$.001 = \frac{1}{1000} = 10^{-3}$$

More specialized techniques have also been devised; for example, to multiply a number by 5, divide the number by 2 and multiply by 10, for example, $46 \times 5 = 46/2 \times 10 = 230$. To multiply a number by 25, divide the number by 4 and multiply by 100; for example, $84 \times 25 = 84/4 \times 100 = 2100$.

Notice that both of these techniques depend on the number being multiplied also being divisible by 2 and 4, respectively. If this is not the case, then very little time is saved by using this technique.

To multiply a number by 11 is very simple. Notice the usual procedure:

$$\begin{array}{r} 3542 \\ \times\ 11 \\ \hline 3542 \\ 3542 \\ \hline 38962 \end{array}$$

Careful examination of this process reveals it can be performed as follows:

```
 3 5 4 2
3 8 9 6 2
```

Begin by writing down 2.
Then add $4 + 2 = 6$.
Then add $5 + 4 = 9$.
Then add $3 + 5 = 8$.
Finally write down 3.

There are many rules for testing divisibility of numbers. Some of the common ones are:

Divisibility by 2: For a number to be divisible by 2 it must end in 0, 2, 4, 6, or 8.

Divisibility by 3: A number is divisible by 3 only if 3 divides the sum of the digits in the number; for example, 435 is divisible by 3, because $4 + 3 + 5 = 12$, which is divisible by 3.

Divisibility by 4: If the last two digits of a number form a number that is divisible by 4 then the entire number is divisible by 4; for example, 2536 is divisible by 4, since 36 is divisible by 4.

Divisibility by 5: Only numbers ending in 0 or 5 are divisible by 5.

Divisibility by 6: Only numbers divisible both by 2 and by 3 are divisible by 6, for example, 156.

Divisibility by 7: A number is divisible by 7 if the difference between twice the units digit and the number formed by the remaining digits is exactly divisible by 7, for example, 161, since $16 - 2 = 14$ which is divisible by 7.

Divisibility by 8: If 8 divides the number formed by the last three digits, then 8 divides the number; for example, 5144 is divisible by 8 since 144 is.

Divisibility by 9: A number is divisible by 9 if the sum of the digits is divisible by 9; for example, 3987 is divisible by 9, since $3 + 9 + 8 + 7 = 27$, a number which is divisible by 9. This fact, which was used rather extensively in the past in checking the results of basic operations, is called "casting out nines."

—H. Laverne Thomas and Viggo P. Hansen

Scope.—Algebra is the study of numbers and the relations among them. It is more generalized than arithmetic, since it deals with symbols that can represent many different numbers. However, these symbols are manipulated by the same four basic operations—addition, subtraction, multiplication, and division—as are numbers. Indeed, algebra might be thought of as generalized arithmetic.

Types of Numbers.—The natural numbers (counting numbers, such as 1, 2, 3, . . .) behave well in addition and multiplication. The addition of two natural numbers results in a sum that is also a natural number, and the same holds true for the result of the multiplication of two natural numbers. But subtraction and division raise problems that are solved only by expanding the system of natural numbers.

Consider the expression $a - b$. If a and b are both natural numbers, and if a is larger than b, there is no problem; the result is a natural number. But if b is larger than a, no natural number can express the result. Hence, *negative numbers* were added to the "vocabulary" of numbers. (The expanded system, which includes negative whole numbers and also the number 0, is called the *integers*.) Negative numbers make subtraction possible for all numbers. For example, $6 - 8$ is said to be negative 2, written as -2. Positive numbers are written with a plus sign. (Numbers without signs are assumed to be positive.)

Division, however, presents a problem that even the integers cannot always handle. Sometimes division of two integers results in another integer ($6 \div 3 = 2$), but often it does not ($2 \div 3$ and $5 \div 2$ cannot be expressed as integers). Hence, fractions were included, producing the system of *rational numbers*. Fractions make division of one number by any other number (except zero) always possible. Other extensions of the number system will be discussed.

Number Line.—The positive numbers, negative numbers, and fractions are shown on a number line in Figure 1.

Figure 1.

$-3\frac{1}{2}$ $+3\frac{1}{2}$

-5 -4 -3 -2 -1 0 $+1$ $+2$ $+3$ $+4$ $+5$

Beginning at a fixed point on the line designated as 0, the positive whole numbers are marked off in order at points that lie at equal intervals to the right of 0; the negative numbers are similarly marked off at points to the left of 0. Fractions are marked off at points between the whole numbers. Thus $-3\frac{1}{2}$ lies midway between -3 and -4. Every number, positive or negative, is thus assigned to one, and only one, point on the number line.

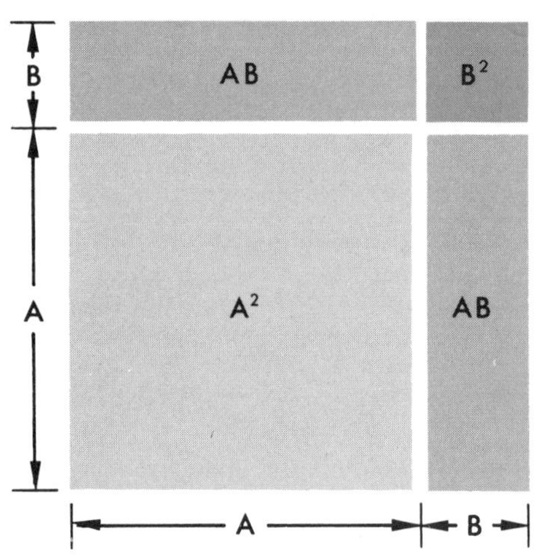

ALGEBRAIC EXPANSION was one of the many mathematical principles first understood by the Greeks over two thousand years ago. Although they lacked the algebraic shorthand with which to express such problems, they knew how to solve them geometrically. The diagram above illustrates the principle of *squaring*, in which a square of a known area is increased proportionately on all sides. If the original area, A^2, whose sides are A feet long, is increased on each side by the length B, the area of the resulting square can be determined by adding the areas of the small rectangles (AB, AB, and B^2) to the area of the original square (A^2). Expressed algebraically, this would read: $2AB + B^2 + A^2$. The Greeks, however, did not formulate algebra as a branch of mathematics distinct from geometry until much later. Diophantus (c. 275) was the first to devise an algebraic symbolism to express such concepts without geometry.

Basic Postulates.—Algebra, like all branches of mathematics, is a postulational system. Certain basic statements are assumed to be true about numbers and the relations among them; other truths are then evolved by logical reasoning. The basic statements that are assumed to be true are called *postulates*, or *axioms*.

The basic postulates about the operations of addition and multiplication on the numbers of arithmetic are assumed to hold true for the expanded system of positive and negative numbers and fractions of algebra. These postulates make it possible to determine the meaning of addition, subtraction, multiplication, and division on the signed numbers of algebra, for they govern the ways numbers can be combined.

The *commutative postulate for addition* states that if two numbers are added, they may be added in any order without affecting the result. That is, $a + b = b + a$ if a and b represent any of the signed numbers of algebra.

The *associative postulate for addition* states that if three numbers are to be added, any two may be added first and their sum then added to the third without affecting the result. That is, $(a + b) + c = a + (b + c)$. (Parentheses are used as a symbol of grouping; the parentheses around $a + b$ in the statement above signify that the sum of a and b is to be treated as a whole.)

The *commutative postulate for multiplication* states that if two numbers are to be multiplied, they may be multiplied in either order without affecting the result; that is, ab is equal to ba. (In algebra the multiplication of two letters, or of a letter and a number, is indicated by writing them adjacent to each other without a sign between: xy means "x times y," and $5c$ means "5 times c.")

The *associative postulate for multiplication* states that if three numbers are to be multiplied, any two may be multiplied first and their product then multiplied by the third without affecting the result: $(ab)c$ is equal to $a(bc)$.

The *distributive postulate* states that if the sum of two numbers is to be multiplied by a third number, the result will be the same if the sum is taken first and then multiplied, or if each of the two numbers is multiplied by the third number and the two resulting products are then added; that is, $a(b + c)$ is equal to $ab + ac$.

Operations.—Each of the fundamental operations of arithmetic can be performed on the signed numbers in such a way that the five basic postulates above will be true.

■ **ADDITION.**—The expression $(+3) + (+2)$ means positive 3 plus positive 2. This may be done on the number line. A positive number is added by moving to the right (starting at 0) a number of spaces equivalent to this number's distance; a negative number is added by a similar movement to the left. Thus, to add $+2$ to $+3$, start at $+3$ and move two spaces to the right, arriving at $+5$ (Fig. 2). Therefore, $(+3) + (+2) = +5$.

Similarly, $(+3) + (-2) = +1$; move two spaces to the left from $+3$ to add -2 to it. Reversing the order does not

Figure 2.

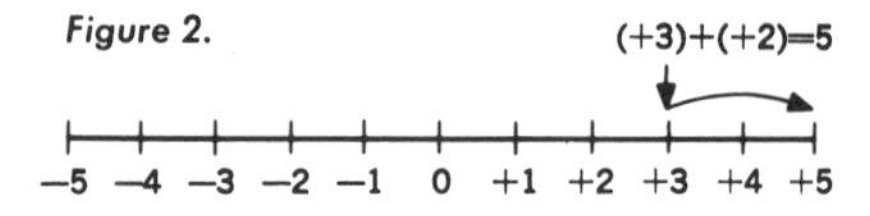

change the result; thus $(-2) + (+3)$ also gives $+1$ under the procedure used to interpret addition. These examples illustrate that the commutative postulate for addition holds true.

The number line should be used to add groups of three signed numbers, using various orders of addition, to show that this procedure also satisfies the associative postulate for addition.

It is possible, of course, to add signed numbers without using the number line.

To do so requires use of the concept of the absolute value of a number. The *absolute value* of a number is its magnitude without regard to sign. The absolute value of both +2.3 and −2.3 is 2.3; the absolute value of −5 is 5.

Thus, there are two principles for adding signed numbers that will give the same results as were obtained by the use of the number line:

(1) The sum of two numbers that have the same sign is the sum of their absolute values with the common sign written in front of it.

Examples:

$$\begin{array}{r} +3 \\ (+)\ \underline{+2} \\ +5 \end{array} \qquad \begin{array}{r} -4 \\ (+)\ \underline{-5} \\ -9 \end{array} \qquad \begin{array}{r} +2\frac{1}{2} \\ (+)\ \underline{+3\frac{1}{4}} \\ +5\frac{3}{4} \end{array}$$

(2) The sum of two numbers that have different signs is the difference of their absolute values with the sign of the number having the larger absolute value written in front of it.

Examples:

$$\begin{array}{r} +3 \\ (+)\underline{-2} \\ +1 \end{array} \qquad \begin{array}{r} -2 \\ (+)\underline{+3} \\ +1 \end{array} \qquad \begin{array}{r} -8 \\ (+)\underline{+5} \\ -3 \end{array} \qquad \begin{array}{r} -9 \\ (+)\underline{+9} \\ 0 \end{array}$$

■ **SUBTRACTION.**—Subtraction is the inverse operation to addition. This means that it is possible to subtract +3 from +8 by asking what must be added to +3 to get +8. On the number line, one must move five spaces to the right (+5) to get from +3 to +8. Therefore if +3 is subtracted from +8, the result is +5.

Similarly, one must move three spaces to the left to get from +7 to +4, and therefore $(+4) - (+7) = -3$. Again, $(-5) - (-3) = +2$, since one must move two spaces to the right to get from −5 to −3. Finally, $(+4) - (-7) = +11$, for going from −7 to +4 is a move of eleven spaces to the right.

To be able to subtract signed numbers without referring to the number line requires the concept of an additive inverse.

The *additive inverse* of a number is the number with the same absolute value but opposite sign. Thus, the additive inverse of +3 is −3, and that of −8 is +8.

Rewriting the subtraction examples above:

$$\begin{array}{r} +8 \\ (-)\underline{+3} \\ +5 \end{array} \qquad \begin{array}{r} +4 \\ (-)\underline{+7} \\ -3 \end{array} \qquad \begin{array}{r} -3 \\ (-)\underline{-5} \\ +2 \end{array} \qquad \begin{array}{r} +4 \\ (-)\underline{-7} \\ +11 \end{array}$$

and comparing them with the following addition examples:

$$\begin{array}{r} +8 \\ (+)\underline{-3} \\ +5 \end{array} \qquad \begin{array}{r} +4 \\ (+)\underline{-7} \\ -3 \end{array} \qquad \begin{array}{r} -3 \\ (+)\underline{+5} \\ +2 \end{array} \qquad \begin{array}{r} +4 \\ (+)\underline{+7} \\ +11 \end{array}$$

a principle for subtracting signed numbers is evident: A number may be subtracted from another number by adding its additive inverse to the other number. Thus,

$$\begin{array}{r} -7 \\ (-)\underline{+5} \end{array}$$

becomes

$$\begin{array}{r} -7 \\ (+)\underline{-5} \\ -12 \end{array}$$

(since −5 is the additive inverse of +5). Similarly:

$$\begin{array}{r} -3 \\ (-)\underline{-4} \\ +1 \end{array} \qquad \begin{array}{r} -7 \\ (-)\underline{+5} \\ -12 \end{array} \qquad \begin{array}{r} +8 \\ (-)\underline{-3} \\ +11 \end{array} \qquad \begin{array}{r} -3 \\ (-)\underline{-3} \\ 0 \end{array}$$

In algebra, the plus sign and the minus sign are used in more than one way. A minus sign may indicate the operation of subtraction, as in $(+3) - (+5)$, or it may indicate a negative

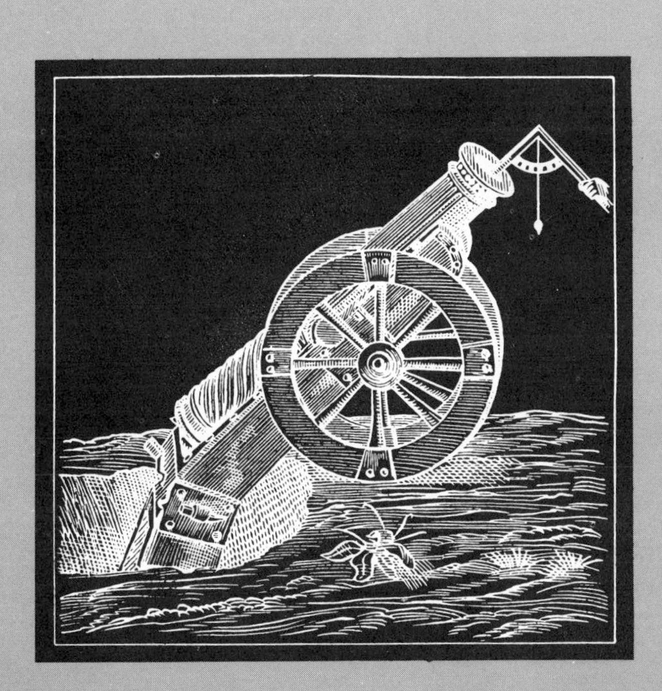

ITALIAN ALGEBRAISTS of the sixteenth century matched wits in many problem-solving contests, but the most brilliant feat of all was Niccolò Tartaglia's solution of cubic equations. Between bouts of pure mathematics, Tartaglia (1500–1557) solved such military problems as the angle of a cannon's trajectory (*above*). Jerome Cardan (1501–1576) stole his cubic solution, added his own solution of quartic equations, and published both in 1545 in his *Ars Magna*, covering all known algebra.

number, as in -3. The plus sign is similarly used to indicate addition and also to indicate a positive number. Confusion may arise as to which meaning of the minus sign is intended in an expression such as $5 - 3$. It is customary to think of this expression as meaning "positive 5 combined with (added to) negative 3," which is equal to positive 2. This way of interpreting the sign is convenient for handling expressions involving more than two terms. Thus, $4 - 6 + 3 + 2$ means "positive 4 combined with negative 6, then combined with positive 3, and then with positive 2"; the result is positive 3. Try combining $7 - 11 - 9 + 3$ in this manner. It is advantageous to make use of the associative postulate by adding all positive numbers and all negative numbers first, then combining the two results (answer is -10).

■ MULTIPLICATION.—The expression $(+3)(+4)$ represents the product of positive 3 and positive 4 and is evaluated as $+12$ since the positive numbers are the counterparts of the natural numbers of arithmetic. Similarly, $(+3)(-4)$ can be interpreted as the sum of three negative fours $(-4 \; -4 \; -4)$ and, hence, is equal to -12. Also, $(-3)(+4)$ can be changed to $(+4)(-3)$ by the commutative postulate for multiplication and is, therefore, also -12 by an interpretation similar to that of the previous example. On the other hand, $(-3)(-4)$ cannot be assigned a value using an interpretation of multiplication as repeated addition, as was done in the previous examples, because both factors are negative. However, it has already been shown that $(-3)(+2)$ is -6. But $(-3)(+2)$ is the same as $(-3)(6 - 4)$ and therefore $(-3)(6 - 4)$ must equal -6. By the distributive postulate, $(-3)(6 - 4) = -18 + (-3)(-4)$. In order to make this last result, $-18 + (-3)(-4)$, equal -6, the expression $(-3)(-4)$ must be equal to $+12$.

Rewriting the examples above, we obtain:

$+3$	$+3$	-3	-3
$(\times)+4$	$(\times)-4$	$(\times)+4$	$(\times)-4$
$+12$	-12	-12	$+12$

Thus, the principles for multiplying signed numbers are:

(1) The product of two numbers having the same sign is the product of their absolute values with a positive sign written in front of it.

(2) The product of two numbers having different signs is the product of their absolute values with a negative sign in front of it.

Examples:

$$(-5)\,(-6) = +30$$
$$(-7)\,(+2) = -14$$
$$(+9)\,(-3) = -27$$
$$(+8\tfrac{1}{2})(+2) = +17$$

■ DIVISION.—Division is the inverse operation to multiplication. The *multiplicative inverse* of a number is the number which when multiplied by it gives a product of 1. Thus, $\frac{1}{3}$ is the multiplicative inverse of 3, since $3 \times \frac{1}{3} = 1$, and $\frac{2}{5}$ is the multiplicative inverse of $\frac{5}{2}$. The multiplicative inverse of a number is also called its *reciprocal*.

It must be kept in mind that each division example may be transformed into a multiplication example by changing the divisor to its multiplicative inverse and then multiplying by it. For example, the division example $8 \div 2$ is the same as the multiplication example $8 \times \frac{1}{2}$, since $\frac{1}{2}$ is the multiplicative inverse of 2.

Since any division example can be changed into a multiplication example, the method of obtaining the sign of the answer in division (quotient) must be the same as that used in multiplication. Therefore, these principles for the division of signed numbers logically follow:

(1) The quotient of two numbers having the same sign is the quotient of their absolute values with a positive sign written before it.

(2) The quotient of two numbers having different signs is the quotient of their absolute values with a negative sign written before it.

Examples:

$$(+12) \div (+3) = +4$$
$$(-18) \div (-2) = +9$$
$$(-27) \div (+3) = -9$$
$$(+15) \div (-30) = -\tfrac{1}{2}$$

Use of Letters.—It has already been pointed out that algebra achieves a generalization not possible in arithmetic, because in algebra letters are employed to represent unspecified numbers. Each letter represents any one of a certain set of numbers called the *domain* or *replacement set* for that letter. Suppose that in the expression $b + 7$ the domain of b is the set $\{1, 2, 3, 4, 5\}$. Then b may be replaced by 1, 2, 3, 4, or 5. Hence, b is termed a *variable*, and its possible replacements are 1, 2, 3, 4, or 5.

In contrast to a variable, a number such as 75, which has one fixed value, is called a *constant*. If b is replaced by 1, the expression $b + 7$ equals 8; if b is 2, the expression equals 9, and so on. To take another example, suppose that the domain of x in the expression

$$\frac{x+3}{2}$$

is the set $\{-2, -1, 0, 1, 2\}$. If the variable x is replaced by -2, the expression equals $\frac{1}{2}$; if by -1, the expression equals 1, and so on. If no domain is specified for a particular letter, the domain is considered to be the largest possible set of numbers for that letter. (At this point, the largest possible set of numbers can be thought of as including all positive and negative integers and fractions, and zero.)

Evaluating Expressions.—An algebraic expression is a number expressed by means of letters and numbers connected by signs of operation, for example, $5a + 3b^2 - \frac{1}{2}bc$. To *evaluate* an algebraic expression means to find the number it represents. This is done by replacing the variables in the expression by the numbers they represent.

Example:

To evaluate $2a + 3b$ when $a = 4$ and $b = -2$ means to replace a by 4 and b by -2. After substitution, the expression yields $2(4) + 3(-2)$, or $8 - 6$, which is 2.

It has already been noted that writing two letters or a number and a letter adjacent to one another without any sign between them implies that they are to be multiplied together. Thus xy means x is to be multiplied by y and $3x$ means x is to be multiplied by 3.

When a number represented by a symbol is to be multiplied by itself, it is customary to write x^2 instead of xx. Similarly, x^3 stands for xxx, y^7 stands for the product of seven y's, and 3^2 means $3(3)$, or 9. When a number is multiplied by itself, the product is called a *power* of that number. Thus, x^2 is called the second power of x, or x squared; x^3 is the third power of x, or x cubed; x^4 is the fourth power of x; x^5 is the fifth power of x, and so on. The number that is multiplied by itself is the *base*, and the superscript, which indicates the power, is called the *exponent*. Thus, in the example $5^2 = 25$, 5 is the base, 2 is the exponent, and 25 is the second power of 5, or 5 squared.

If s represents the number of inches in the side of a square, s^2 represents the number of square inches in its area; if $s = 9$, $s^2 = 81$. If e represents the number of inches in the edge of a cube, e^3 represents the number of cubic inches in its volume; if $e = 2$, $e^3 = 8$, that is, $2 \times 2 \times 2$.

An exponent applies only to the base that precedes it. Thus, xy^2 means xyy and $2x^3y$ means $2xxxy$. If two or more bases are to be raised to the same power, they are placed in parentheses and the superscript is placed outside. Thus, $a(bc)^2$ means $a(bc)(bc)$, or ab^2c^2.

Examples:

To evaluate $8a^2$ when $a = -3$, replace a by -3:

$$8(-3)^2 = 8(9) = 72$$

To evaluate $-2a^3b$ when $a = -1$ and $b = 3$, replace a by -1 and b by 3:

$$-2(-1)^3(3) = -2(-1)(3) = 6$$

To evaluate $5x(yz)^2$ when $x = -2$, $y = 4$, and $z = 2$, replace x by -2, y by 4, and z by 2:

$$5(-2)(4 \times 2)^2$$
$$5(-2)(8)^2$$
$$5(-2)(64)$$
$$-640$$

Order of Operations.—The commutative and associative postulates for addition imply that the order in which numbers are added does not affect the result. Thus, $2 + 3$ is the same as $3 + 2$, and $2 + (3 + 5)$ is the same as $(2 + 3) + 5$. Similarly, the order in which numbers are multiplied does not affect the result according to the commutative and associative postulates for multiplication. Subtraction and division, however, are not commutative: $5 - 2$ is not the same as $2 - 5$, and $8 \div 2$ is different from $2 \div 8$. Furthermore, when more than one operation is involved in evaluating an expression, the order in which the operations are performed does affect the result. In evaluating $3a^2$ when $a = 5$, if 5 is squared first ($5^2 = 25$) and then multiplied by 3, the result is 75; but if 3 is multiplied by 5 first ($3 \times 5 = 15$) and the result is then squared, the final result is 225. Similarly, in evaluating $a + 2b$ when $a = 5$ and $b = 6$, if 5 is added to 2 first ($5 + 2 = 7$) and the answer is then multiplied by 6, the result is 42; but if 6 is first multiplied by 2 ($6 \times 2 = 12$) and the answer is then added to 5, the result is 17.

To prevent such ambiguities in evaluating expressions, the following order is always used in performing operations:

(1) Operations included within signs of grouping, such as parentheses or fraction lines, are performed first.

(2) Numbers are then raised to powers.

(3) Multiplications and divisions are performed next.

(4) Additions and subtractions are performed last.

Following this order in evaluating $3a^2$ when $a = 5$, the power (25) is obtained first and then multiplied by 3, producing the correct result, 75. In evaluating $a + 2b$ when $a = 5$ and $b = 6$, the multiplication of 2 by 6 is performed before adding the 5, producing the correct result, 17.

Example:

To evaluate $2\pi r(r + h) + 3h$, when $\pi = 22/7$, $r = 7$, and $h = 2$, substitute the numbers for the letters:

$$2\left(\frac{22}{7}\right)(7)(7 + 2) + 3(2);$$

perform the operations in parentheses first:

$$2\left(\frac{22}{7}\right)(7)(9) + 3(2);$$

multiplications and divisions next:

$$396 + 6;$$

and add and subtract last:

$$402.$$

Parts of Algebraic Expressions.—The quantities added together algebraically in an expression are called the *terms* of the expression. Thus, in $2a^2 - 3bc + 5c$, there are three terms: $2a^2$, $-3bc$, and $+5c$. Terms can be recognized by the fact that they are separated from one another by plus or minus signs.

An expression containing only one term is called a *monomial*. Thus, $-17x^3y$ is a monomial. An expression containing two terms, such as $2a + 3$, is called a *binomial*. An expression containing three terms, such as $y^2 - 3 + 4x$, is called a *trinomial*. Any expression having more than one term is called a *polynomial*. For example, $4z - 7$ and $7a + 15b^2c - 36a^2 + 17c$ are polynomials.

The numbers that are multiplied together to form an expression are called the *factors* of the expression. The factors of the term $7x^2y$ are 7, x^2, y, $7x$, $7x^2$, xy, x^2y, $7y$, and so on. (These may be considered as exact divisors of elements contained in the expression.)

The numerical factor of a term is called the *numerical coefficient* of that term. In $-7x^2y$, -7 is the numerical coefficient. In c^3d^2, the numerical coefficient is not explicitly written but is understood to be 1.

The factors of a term that consist of letters form the *literal factor* of the term. In $-7x^2y$, x^2y is the literal factor.

Like terms are terms that have the same literal factors. Thus, $5x^2y$, $-3x^2y$, and $+\frac{1}{2}x^2y$ are like terms.

Unlike terms are terms that do not have the same literal factors. Thus, $-2a$, $5b$, 5, $7x^2$, $6x$, $2x^2y$, and $2xy^2$ are unlike terms. Note that literal factors are unlike unless both the letters and the exponents associated with each of them are exactly the same.

Operations on Monomials.—Since algebraic expressions represent numbers, the question arises as to how the operations of addition, subtraction, multiplication, and division can be performed on them. This question will be examined first with respect to monomials.

COMBINING TERMS.—Like terms may be combined by making use of the postulates for operations on numbers. To combine $-3xy$ and $+5xy$, the commutative postulate for multiplication permits changing them to $xy(-3)$ and $xy(+5)$. By the distributive postulate, this sum is the same as $xy(-3 + 5)$ or $xy(+2)$, and is equal to $2xy$. Comparing this result with the two original terms shows that like terms may be combined by combining their numerical coefficients and multiplying the result by their common literal factor. To add $+7d$ and $-9d$, we combine their numerical coefficients ($+7 - 9 = -2$) and multiply this by the common literal factor, getting $-2d$.

Examples:

$$\begin{array}{r} +3a \\ (+)\underline{-7a} \\ -4a \end{array} \qquad \begin{array}{r} +7x^2y \\ -2x^2y \\ (+)\underline{-3x^2y} \\ +2x^2y \end{array} \qquad \begin{array}{r} -4c^2 \\ +7c^2 \\ (+)\underline{-2c^2} \\ c^2 \end{array}$$

Combining like terms in effect makes it possible to add or subtract any like terms.

MULTIPLYING MONOMIALS.—To multiply $+13a$ by $-2b$, the indicated product $(+13a)(-2b)$ may be rewritten as $(+13)(-2)(ab)$ by the application of the commutative and associative postulates for multiplication. Hence, the product is $-26ab$.

Comparing the last result with the original factors shows that a short way to get the product of two monomials is to multiply their coefficients and then to multiply this product by the product of their literal factors. Thus, $(-3x^2)(-2y)$ would be $+6x^2y$.

Sometimes the two monomials to be multiplied contain powers of the same base, as in the case of $(3a^2)(5a^4)$. Since $(3a^2)(5a^4)$ means $(3aa)(5aaaa)$, which is $15aaaaaa$, or $15a^6$, it can be seen that powers of the same base can be multiplied by adding the exponents. Thus, $(x^3)(x^4) = x^7$.

Examples:

$$(+2a^3b^2)(-7a^2b^7) = -14a^5b^9$$
$$c^3(5c^2d) = 5c^5d$$
$$3xy^3(-4x^2y^2) = -12x^3y^5$$
$$2a(3b^2)(-5b) = -30ab^3$$

In the last example above, the literal factor b in the last parentheses is understood to have the exponent 1. The literal factor b^3 in the result is obtained by adding the exponent 1 to the exponent 2 of b^2 to produce the exponent 3 in the product b^3.

Repeated multiplication of monomials may occur when raising a monomial to a power. Thus, $(-2x^2)^4$ means $(-2x^2)(-2x^2)(-2x^2)(-2x^2)$, or $16x^8$.

DIVIDING MONOMIALS.—Since division is the inverse operation to multiplication, to divide $+16a^5$ by $-2a^3$ meano to find the expression which when multiplied by $-2a^3$ will give $+16a^5$. This expression is $-8a^2$. Division of monomials is thus accomplished by dividing the numerical coefficients and multiplying this result by the result obtained by dividing the literal factors. Literal factors with the same base are divided by keeping that base and subtracting the exponents.

Examples:

$$(+6x^5) \div (-2x^2) = -3x^3$$
$$\frac{-12x^3y^2}{4xy^2} = -3x^2$$
$$\frac{-14ab^3c}{-2ab} = 7b^2c$$

Operations on Polynomials.—The same operations that were shown to be workable with monomials can also be worked with polynomials.

ADDING POLYNOMIALS.—Since the like terms can be combined, two or more polynomials can be added by arranging them so that their like terms are in vertical columns; the like terms in each column can then be combined.

Example:

To add the polynomials $3x^2 - 2xy + 4y^2$, $5y^2 - x^2$, and $-5xy + 3x^2 + 3$, ar-

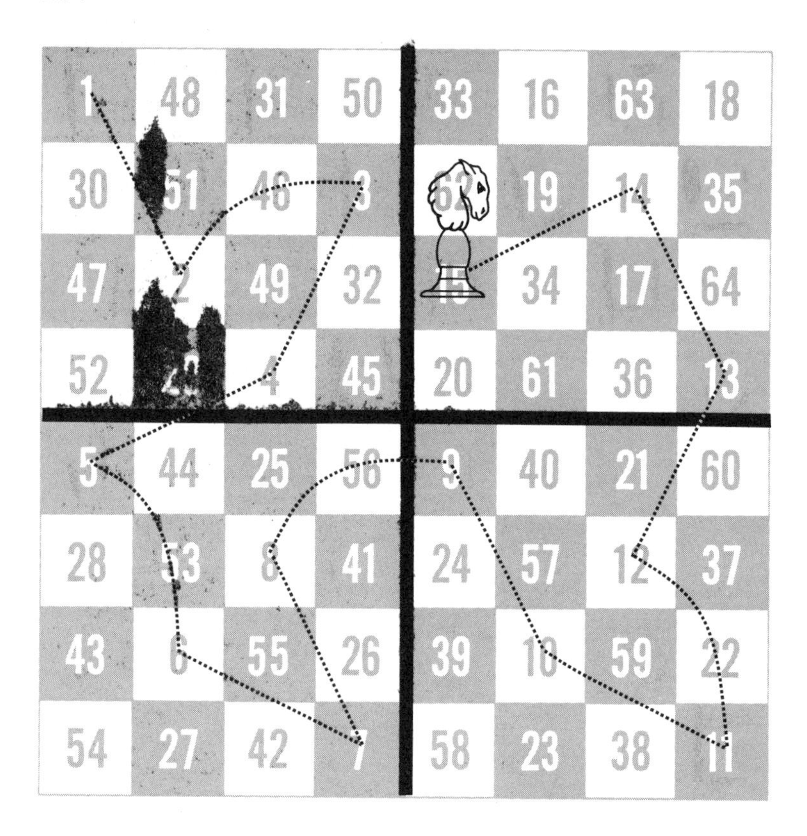

SCRIPTA MATHEMATICA

IN THIS MAGIC SQUARE devised by Leonhard Euler (1707–1783) each horizontal or vertical row totals 260, while half of each row totals 130. Its most fascinating feature is that a chess knight starting from box 1 and moving in its pattern of two squares up and one over will land on all 64 boxes in numerical sequence.

range like terms in vertical columns and add:

$$\begin{array}{l} 3x^2 - 2xy + 4y^2 \\ -x^2 \qquad\quad + 5y^2 \\ \underline{3x^2 - 5xy \qquad\quad + 3} \\ 5x^2 - 7xy + 9y^2 + 3 \end{array}$$

■ SUBTRACTING POLYNOMIALS.—It was shown that subtracting a quantity is equivalent to adding the additive inverse of that quantity. Thus, polynomials may be subtracted by arranging them so that like terms are in vertical columns, changing the signs of the terms in the polynomial to be subtracted (thus obtaining its additive inverse), and then combining like terms.

Example:

To subtract $-5b + 2a + 4c$ from $3a + 7b$, arrange like terms in columns, write the additive inverse of the polynomial to be subtracted, then combine like terms:

$$\begin{array}{r} 3a + 7b \qquad \\ \underline{-2a + 5b - 4c} \\ a + 12b - 4c \end{array}$$

■ MULTIPLYING POLYNOMIALS.—To multiply a polynomial, such as $2x(3x^2 - y + 5)$, the distributive postulate requires that each term of the polynomial be multiplied, in turn, by the monomial. The product is the sum of all the resulting terms.

Examples:

The expression is then multiplied:

$$2x(3x^2 - y + 5)$$
$$6x^3 - 2xy + 10x$$

Another illustration:

$$-2a^2(2 + 4ab + 7b^2)$$
$$-4a^2 - 8a^3b - 14a^2b^2$$

If an expression involves a set of parentheses preceded by a plus sign, the parentheses may be removed by multiplying the polynomial enclosed within it by $+1$. Thus, $2 + (x - 5y)$ becomes $2 + x - 5y$.

To remove a set of parentheses preceded by a minus sign, multiply the enclosed polynomial by -1. Thus, $5 - (2a + 3b - 7c)$ becomes $5 - 2a - 3b + 7c$.

Suppose it is desired to multiply $3x - 5y$ by $2x + 3y$. By the distributive postulate, $(3x - 5y)(2x + 3y)$ equals $(3x - 5y)(2x) + (3x - 5y)(3y)$. Using the commutative postulate, this becomes $2x(3x - 5y) + 3y(3x - 5y)$, and a second application of the distributive postulate shows it to be equivalent to $2x(3x) + 2x(-5y) + 3y(3x) + 3y(-5y)$. Thus, the multiplication of two polynomials has been reduced to a series of multiplications of pairs of monomials: $6x^2 - 10xy + 9xy - 15y^2$. By combining the two like terms, $6x^2 - xy - 15y^2$ is the final product. In other words, two polynomials are multiplied by each other by multiplying each term of the first polynomial by each term of the second and adding the resulting terms.

In practice, this operation is carried out by arranging the two polynomials to be multiplied in such a way that one is under the other as in multiplication in arithmetic. Partial products are obtained by multiplying the first polynomial by each term of the second much as the top number is multiplied by each digit of the lower one in multiplication in arithmetic. Finally, the partial products are added.

Example:

To multiply $3x - 5y$ by $2x + 3y$, multiply each term of $3x - 5y$ by $2x$, and each term of $3x - 5y$ by $3y$. Then place like terms in columns and add:

$$\begin{array}{l} 3x \ - 5y \\ \underline{2x \ + 3y} \\ 6x^2 - 10xy \\ \underline{\qquad + \ 9xy - 15y^2} \\ 6x^2 - \quad xy - 15y^2 \end{array}$$

Similarly,

$$\begin{array}{r} 2a - 5 \qquad \\ \underline{6a + 7} \qquad \\ 12a^2 - 30a \qquad \\ \underline{+ 14a - 35} \\ 12a^2 - 16a - 35 \end{array}$$

■ DIVIDING POLYNOMIALS.—Since the operation of dividing a polynomial by a monomial is the inverse of multiplying, the division operation may be accomplished by dividing each term of the polynomial, in turn, by the monomial and writing the quotient as the sum of the resulting terms.

Example:

To divide $5x^3 - 10x^2 + 15x$ by $5x$, divide each term of the numerator by $5x$:

$$x^2 - 2x + 3.$$

Factoring.—The *factors* of an expression are the expressions that when multiplied together give the original expression. Thus, 6 and 2 are factors of 12; so are 4 and 3. In the expression, $3x^2 + 6x$, the factors are $3x$ and $(x + 2)$, because $3x(x + 2) = 3x^2 + 6x$.

To *factor* an expression means to express it as the product of two or more quantities. Thus, $3x^2 + 6x$ is factored when it is written as $3x(x + 2)$. In factoring, it is agreed that only integers are to be used as coefficients and that the factors are not to include the original expression and 1.

If an expression cannot be written as the product of any factors other than itself and 1, the expression is said to be *prime*.

Highest Common Factors.—Factoring and multiplication are opposite processes. The expression $2a(x - 3y)$ is changed to $2ax - 6ay$ by multiplication; the expression $2ax - 6ay$ is changed back to $2a(x - 3y)$ by factoring. Whenever a polynomial is multiplied by a monomial, an expression results that can be factored back into a polynomial multiplied by the monomial. The monomial is called a *common factor* because it is a factor of each term of the polynomial. It is customary to factor in such a way as to get the largest such common monomial factor, and hence this type of factor is known as the *highest common monomial factor*. The highest common monomial factor has as its coefficient the largest integer that will divide into every term of the polynomial; its literal factor consists of the highest power of each literal factor that will divide into every term of the polynomial.

To factor $2x^3 - 8x^2y + 6x^2$, note that 2 is the largest integer that will divide into every term of the polynomial, and x^2 is the highest power of x that will divide into every term. Since y will not divide into every term, it is not in the common factor. The highest common monomial factor is therefore $2x^2$. The other factor is obtained by dividing $2x^2$ into the polynomial. Thus, $2x^3 - 8x^2y + 6x^2$ can be factored into $2x^2(x - 4y + 3)$. Note that multiplication should produce the original expression again and therefore will serve as a check on factoring.

For example, $24a^2b^3 - 18a^2b^4$ factors into $6a^2b^3(4 - 3b)$; $\pi R^2 - \pi r^2$ factors into $\pi(R^2 - r^2)$; and $2x^2y - 3xy - 5y^2$ factors into $y(2x^2 - 3x - 5y)$.

Multiplying Binomials.—As explained earlier, two polynomials such as $5a - 3b$ and $2a - 4b$ may be multiplied by a method similar to that used for multiplying large numbers in arithmetic:

$$\begin{array}{r} 5a - 3b \\ 2a - 4b \\ \hline 10a^2 - 6ab \qquad \\ - 20ab + 12b^2 \\ \hline 10a^2 - 26ab + 12b^2 \end{array}$$

However, when both polynomials are binomials, as is the case here, it is possible to write down the product almost at sight.

Note that the first term of the answer $10a^2$ is the product of the first terms of the two binomials $(5a)(2a)$. The last term of the answer $12b^2$ is the product of the last term of the two binomials $(-3b)(-4b)$. The middle term is the sum of the two cross products, $(5a)(-4b) + (2a)(-3b)$, or $(-20ab) + (-6ab)$, or $-26ab$.

The entire work for the multiplication can be conveniently arranged in the following manner:

$$\begin{array}{c} -20ab \\ -6ab \\ (5a - 3b)(2a - 4b) \\ 10a^2 - 26ab + 12b^2 \end{array}$$

The inner and outer cross products are generally listed as shown before they are combined algebraically to give the middle term.

Examples:

Multiply $(2x - 1)(3x + 2)$ at sight.

$$\begin{array}{c} +4x \\ -3x \\ (2x - 1)(3x + 2) \\ 6x^2 + x - 2 \end{array}$$

Multiply $(x + 7)^2$, which is the same as $(x + 7)(x + 7)$.

$$\begin{array}{c} +7x \\ +7x \\ (x + 7)(x + 7) \\ x^2 + 14x + 49 \end{array}$$

The product in each of these examples is a trinomial which is quadratic; that is, the variable appears to the second power and to no higher power. Such a trinomial is called a *quadratic trinomial*.

Factoring Quadratic Trinomials.—A quadratic trinomial is an expression of the form $ax^2 + bx + c$ where x stands for some variable and a, b, and c are constants not equal to ($\neq$) 0. Since such an expression results when two binomials are multiplied, factoring such a quadratic trinomial should result in two binomial factors.

To factor a quadratic trinomial, such as $5x^2 + 28x - 12$, look for two binomial factors such that the product of their first terms is $5x^2$: $(5x \quad)(x \quad)$. The product of the last terms of the binomials must be -12; but -12 has many factors: 2 and -6, -2 and 6, 3 and -4, -3 and 4, 12 and -1, and -12 and 1. The factors to select for the last term of the binomial must be such that the resulting cross products combine algebraically to give $28x$, the middle term of the quadratic trinomial. By trial and error it can be determined that 6 and -2 will do this provided -2 is placed in the same binomial factor with $5x$ and 6 placed with x:

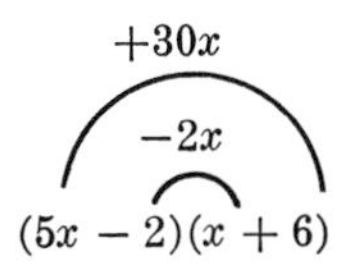

It may happen that the first term of the quadratic trinomial has more than one possible set of factors, as occurred in the previous example with the last term. If this is the case, the correct factors can only be determined by trial and error, making sure that the combination of the cross products of the binomials gives the correct middle term. Thus, in factoring $12a^2 + a - 6$, $12a^2$ has several pairs of factors and so does -6. The correct factors are

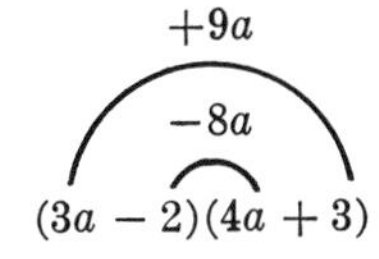

Any other combination will not result in a product equivalent to the original expression.

If the terms of a quadratic trinomial are not in descending order of exponents, it is necessary to arrange them in this order before factoring.

Example:

To factor $4a^2 + 3 + 8a$, rearrange in descending order of exponents and factor as follows:

$$4a^2 + 8a + 3$$

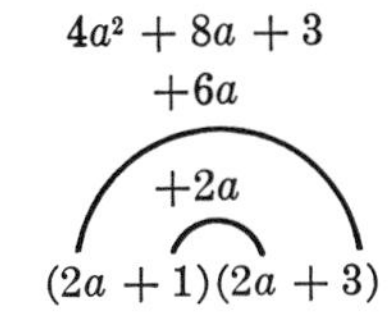

If the squared term of a quadratic trinomial is negative and the constant is positive, it is more convenient to arrange the terms in ascending order of exponents although it is possible to factor if they are arranged in descending order.

Example:

To factor $56 - b^2 - b$, rearrange in ascending order of exponents and factor as follows:

$$\begin{array}{c} 56 - b - b^2 \\ -8b \\ +7b \\ (8 + b)(7 - b) \end{array}$$

Sometimes a quadratic trinomial has two equal factors: $x^2 - 6x + 9$ factors as $(x - 3)(x - 3)$. This may also be written as $(x - 3)^2$.

Not all quadratic trinomials are factorable into the type of factors specified. For example, $x^2 + 4x + 2$ cannot be factored to give two factors with coefficients that are integers. The first term, x^2, requires that x be the first term of each binomial. The last term, 2, can be written only as the product of 2 and 1. But $(x+2)(x+1)$ does not equal x^2+4x+2, since its middle term is $3x$ instead of $4x$. (Actually the factors involve decimals, so the solution cannot be found by sight.)

Product of Sum and Difference.—If the binomials $(x+3)$ and $(x-3)$, which represent the sum and difference of the same two terms x and 3, are multiplied together, a quadratic without a middle term results because the cross products add to zero:

$$-3x \quad +3x$$
$$(x+3)(x-3)$$
$$x^2-9$$

The resulting quadratic is the difference of the squares of the two original terms. When binomials representing the sum and difference of two terms are multiplied, the product is always the difference of their squares.

Examples:

$$(x-5)(x+5) = x^2-25$$
$$(2x+3y)(2x-3y) = 4x^2-9y^2$$
$$(2-x)(2+x) = 4-x^2$$

Since multiplication of the sum and difference of two terms will produce an expression that is the difference of their squares, factoring an expression that is the difference of two squares should change it back to the product of two binomials that are the sum and difference of the squared terms.

Examples:

The term x^2-9 factors as $(x+3)(x-3)$; $4y^2-1$ factors as $(2y-1)(2y+1)$; $49a^2-36b^2$ factors as $(7a-6b)(7a+6b)$.

Complete Factoring.—In factoring an expression, the highest common monomial factor should be removed first. The resulting factor may be a factorable quadratic trinomial or a difference of two squares. If it is one of these, it may be replaced by two binomial factors by factoring again. In other words, factoring should be continued until *prime factors* (factors that cannot be factored further) are obtained.

In factoring $12x^2+14x-10$, the highest common monomial factor is 2, giving $2(6x^2+7x-5)$. But $6x^2+7x-5$ is also factorable, giving

$$-3x \quad +10x$$
$$2(3x+5)(2x-1)$$

Examples:

To factor $4a^3+48a-28a^2$, rearrange in descending order of exponents, factor the highest common monomial, and then factor the quadratic trinomial as follows:

$$4a^3-28a^2+48a$$
$$4a(a^2-7a+12)$$
$$-3a \quad -4a$$
$$4a(a-4)(a-3)$$

To factor $x-25x^3$, factor the highest common monomial and then factor the difference of two squares:

$$x(1-25x^2)$$
$$x(1-5x)(1+5x)$$

When a difference of two squares is factored, it is possible for one of the factors to be another difference of two squares.

To factor $16x^4-y^4$, factor as a difference of two squares and then factor $4x^2-y^2$:

$$(4x^2-y^2)(4x^2+y^2)$$
$$(2x-y)(2x+y)(4x^2+y^2)$$

Fractions.—In arithmetic, fractions are reduced to lowest terms by making use of the principle that the numerator and denominator of a fraction may be divided by any nonzero quantity without changing the value of the fraction. This principle is a consequence of the fact that when the numerator and denominator are divided by the same nonzero quantity, a, the fraction is really divided by a/a, or 1, and hence remains the same. Using this principle, $\frac{6}{8}$ is reduced to the equivalent fraction $\frac{3}{4}$ by dividing both numerator and denominator by 2. The quantity by which the numerator and denominator are divided (in this case 2) must be a factor of both.

Algebraic fractions are reduced by making use of the same principle as is used in arithmetic. The value of any algebraic fraction is unchanged if the numerator and denominator are both divided by the same nonzero quantity; in order to divide, the quantity must be a factor of both numerator and denominator.

Example:

To reduce

$$\frac{-4xy^3}{10xy^2}$$

to lowest terms, divide numerator and denominator by $2xy^2$. The answer is

$$\frac{-2y}{5}.$$

Division by zero is undefined, but the above solution need not be qualified by adding that $x \neq 0$ and $y \neq 0$. The original fraction contains x and y as factors of its denominator, hence it may be assumed they do not equal zero.

If the numerator or denominator of an algebraic fraction is a polynomial, it is usually necessary to factor it before reducing to determine what factors may be divided into both numerator and denominator.

Example:

To reduce

$$\frac{3x-6}{9x-18},$$

factor numerator and denominator,

$$\frac{3(x-2)}{9(x-2)},$$

and divide both numerator and denominator by 3 and $(x-2)$. The answer is $\frac{1}{3}$. Notice that x cannot equal 2, since this would result in division by 0; but $x \neq 2$, because the original fraction would be undefined if it did.

Example:

To reduce

$$\frac{x^2+2x}{x^2-3x-10},$$

factor numerator and denominator:

$$\frac{x(x+2)}{(x-5)(x+2)};$$

and divide numerator and denominator by $(x+2)$:

$$\frac{x}{x-5}.$$

Signs of Fractions.—Every fraction may be considered to have three signs: the sign of the numerator, the sign of the denominator, and the sign of the whole fraction. All three are shown in the fraction

$$+\frac{-a}{+b}.$$

Consider the effect of these signs on the value of a fraction:

The fraction $+(+4/+2)$ has the value $+2$, because it may be interpreted as the positive value of the result of dividing $+4$ by $+2$.

The fraction $+(-4/-2)$ also has the value $+2$, because it means the positive value of the result of dividing -4 by -2.

The fraction $-(-4/+2)$ has the value $+2$, because it means the negative value (additive inverse) of the result of dividing -4 by $+2$.

The fraction $-(+4/-2)$ has the value $+2$, because it means the negative value (additive inverse) of the result of dividing $+4$ by -2.

Each of the four fractions above can be obtained from any one of the other three by changing exactly two signs, and all four have the value $+2$. These fractions illustrate the principle that any two of the signs of a fraction may be changed without changing the value of the fraction.

It is sometimes necessary to employ the above principle to change the signs of a fraction in order to facilitate reducing the fraction or performing other operations with it.

Accordingly, in attempting to reduce the expression

$$\frac{25-5x}{x^2-25},$$

factoring produces the form

$$\frac{5(5-x)}{(x-5)(x+5)}.$$

In this form, there is no common factor to divide both numerator and denominator. By changing two signs, such as the signs of the fraction and of its numerator, this fraction becomes

$$-\frac{5(x-5)}{(x-5)(x+5)},$$

which reduces to

$$-\frac{5}{x+5}.$$

It should be emphasized at this point that in a case such as

$$\frac{3(x+2)}{2(2+x)},$$

it is not necessary to make any sign changes in order to reduce to $\frac{3}{2}$, since $x+2$ and $2+x$ are the same except for the fact that their terms happen to be written down in different orders; they are equal by the commutative postulate for addition.

MULTIPLICATION.—In arithmetic, fractions are multiplied by multiplying their numerators together to form the numerator of the product and multiplying their denominators together to form the denominator of the product. For example,

$$\frac{2}{5}\cdot\frac{3}{7}=\frac{6}{35}.$$

Algebraic fractions are multiplied in exactly the same way. As in arithmetic, *like factors* in the numerator and denominator may be "canceled," that is, divided out by applying the principle that the numerator and denominator may be divided by the same nonzero number without changing the value of the fraction. (The dot [$\cdot$] between the fractions is used to denote multiplication since the usual times sign [$\times$] might be confused with x.)

Example:

To multiply

$$\frac{10x^2y}{3}\cdot\frac{2y}{5x},$$

divide numerator and denominator by $5x$, with the result:

$$\frac{2xy}{3}\cdot\frac{2y}{1};$$

and multiply numerators together and denominators together for the answer:

$$\frac{4xy^2}{3}.$$

If an integral expression is to be multiplied by a fraction, it is helpful to think of it as having the denominator 1.

Example:

To multiply

$$2x\cdot\frac{3y}{x^2},$$

write $2x$ with the denominator 1:

$$\frac{2x}{1}\cdot\frac{3y}{x^2};$$

divide numerator and denominator by x:

$$\frac{2}{1}\cdot\frac{3y}{x};$$

and multiply numerators together and denominators together with the result:

$$\frac{6y}{x}.$$

If in fractions to be multiplied, the numerator or denominator or both are polynomials, it is advisable to factor them first. This makes it easy to reduce by dividing numerator and denominator by the factors that appear in both.

Example:

To multiply

$$\frac{x^2-16}{x}\cdot\frac{2x}{3x-12},$$

factor:

$$\frac{(x-4)(x+4)}{x}\cdot\frac{2x}{3(x-4)};$$

divide numerator and denominator by x and by $(x-4)$:

$$\frac{x+4}{1}\cdot\frac{2}{3};$$

and multiply numerators together and denominators together:

$$\frac{2(x+4)}{3}.$$

DIVISION.—In arithmetic, the division example $\frac{3}{4}\div\frac{7}{5}$ is performed by changing the divisor to its multiplicative inverse and then multiplying. Thus $\frac{3}{4}\div\frac{7}{5}$ becomes $\frac{3}{4}\cdot\frac{5}{7}$, or $\frac{15}{28}$. The multiplicative inverse, or reciprocal, of a number is formed by inverting the number. Thus, the multiplicative inverse, or reciprocal, of $\frac{4}{3}$ is $\frac{3}{4}$; and $\frac{3}{1}$ is the multiplicative inverse, or reciprocal, of $\frac{1}{3}$.

Algebraic fractions can also be divided by multiplying by the multiplicative inverse, or reciprocal, of the fraction that is the divisor.

Example:

To divide

$$\frac{3x^3}{6}\div\frac{x^2}{9},$$

change to multiplication by replacing the divisor with its multiplicative inverse:

$$\frac{3x^3}{6}\cdot\frac{9}{x^2};$$

divide numerator and denominator by 3 and by x^2:

$$\frac{x}{2}\cdot\frac{9}{1};$$

and multiply numerators together and denominators together:

$$\frac{9x}{2}.$$

Numerators and denominators that are polynomials should be factored to make it possible to reduce the answer.

Example:

To divide

$$\frac{x^2-4x-21}{3x+6}\div\frac{x-3}{x+2},$$

factor and change to multiplication by replacing the divisor with its multiplicative inverse:

$$\frac{(x-7)(x+3)}{3(x+2)}\cdot\frac{x+2}{x-3};$$

divide numerator and denominator by $(x+2)$:

$$\frac{(x-7)(x+3)}{3}\cdot\frac{1}{x-3};$$

and write the products of numerators and denominators to give the respective numerator and denominator of the answer:

$$\frac{(x-7)(x+3)}{3(x-3)}.$$

ADDITION AND SUBTRACTION.—In arithmetic, if fractions have the same denominator, they are added or subtracted by adding or subtracting their numerators and placing the result over their common denominator. Thus, $\frac{2}{5}+\frac{1}{5}=\frac{3}{5}$.

Algebraic fractions that have the same denominator are added and subtracted (combined) in the same way.

Example:

To combine

$$\frac{2}{x-2}+\frac{1}{x-2}-\frac{5x}{x-2},$$

add and subtract the numerators and place the sum over the common denominator:

$$\frac{2+1-5x}{x-2};$$

and combine like terms:

$$\frac{3-5x}{x-2}.$$

In arithmetic, when fractions do not have the same denominators, they must first be changed to equivalent fractions having the same denominators before they can be added or subtracted. Such equivalent fractions are obtained by multiplying the numerator and denominator of each fraction by some nonzero number so selected that the multiplication will produce the common denominator in each. To add $\frac{1}{3}$ and $\frac{2}{7}$, $\frac{1}{3}$ is first changed to $\frac{7}{21}$ by multiplying its numerator and denominator by 7. In thus employing the principle that the value of a fraction is unchanged if its numerator and denominator are both multiplied by the same nonzero number, in effect the fraction is multiplied by 1. The second fraction, $\frac{2}{7}$, is changed to the equivalent fraction, $\frac{6}{21}$, by multiplying both its numerator and denominator by 3. In their new equivalent forms, $\frac{7}{21}$ and $\frac{6}{21}$ can be added to give $\frac{13}{21}$. The combination of the fractions was made possible by obtaining a common denominator (21 in this case) that was divisible by each of the original denominators.

Algebraic fractions can also be added or subtracted by converting them to equivalent fractions with a common denominator that is divisible by each of the original denominators.

Example:

To combine

$$\frac{5}{2x^2y} - \frac{2}{3xy} + \frac{4}{x^2},$$

the common denominator must be divisible by $2x^2y$, $3xy$, and x^2. The lowest common denominator is $6x^2y$. Note that each literal factor appears in the lowest common denominator raised to the highest power to which it appears in any individual denominator. The denominator of the first fraction, $2x^2y$, must be multiplied by 3 to convert it to the common denominator, $6x^2y$. Therefore, its numerator, 5, must also be multiplied by 3 to keep the value of the fraction unchanged. In a similar manner, the second fraction must be multiplied by $2x/2x$, and the third by $6y/6y$. The example then becomes

$$\frac{5(3)}{2x^2y(3)} - \frac{2(2x)}{3xy(2x)} + \frac{4(6y)}{x^2(6y)};$$

and after multiplying:

$$\frac{15}{6x^2y} - \frac{4x}{6x^2y} + \frac{24y}{6x^2y};$$

and combining the numerators over the common denominator, we obtain

$$\frac{15 - 4x + 24y}{6x^2y}.$$

It is sometimes necessary to change two of the signs in one fraction to convert it to a form with the same denominator as another.

Example:

To combine

$$\frac{2}{x-2} + \frac{3}{2-x},$$

change the sign of the second fraction and the sign of its denominator:

$$\frac{2}{x-2} - \frac{3}{-2+x} \quad \text{or} \quad \frac{2}{x-2} - \frac{3}{x-2};$$

combine the numerators over the common denominator:

$$\frac{2-3}{x-2}$$

and combine like terms:

$$\frac{-1}{x-2}.$$

This answer may also be written as

$$\frac{1}{2-x}$$

by changing the signs of the numerator and denominator; this form has the advantage of having fewer minus signs.

In combining fractions with polynomial denominators, it is necessary to factor the denominators in order to discover the lowest common denominator.

Example:

To combine

$$\frac{3}{y^2 + 7y + 10} - \frac{2}{y^2 - 25},$$

first factor the denominators:

$$\frac{3}{(y+5)(y+2)} - \frac{2}{(y+5)(y-5)}.$$

Since the lowest common denominator must be divisible by each of the original denominators, it must contain as factors all the factors in any one of the original denominators. In this case the lowest common denominator is $(y+5)(y-5)(y+2)$.

Next change each fraction to an equivalent fraction with the common denominator:

$$\frac{3(y-5)}{(y+5)(y+2)(y-5)} - \frac{2(y+2)}{(y+5)(y-5)(y+2)};$$

multiply in the numerators:

$$\frac{3y-15}{(y+5)(y+2)(y-5)} - \frac{2y+4}{(y+5)(y-5)(y+2)};$$

and combine the numerators over the common denominator:

$$\frac{3y - 15 - (2y+4)}{(y+5)(y-5)(y+2)}.$$

Note how the parentheses preceded by the minus sign are used to ensure that the second fraction is subtracted from the first. To complete the operation, remove the parentheses:

$$\frac{3y - 15 - 2y - 4}{(y+5)(y-5)(y+2)};$$

and combine like terms:

$$\frac{y-19}{(y+5)(y-5)(y+2)}.$$

Sentences.—In algebra, statements such as $2 + 3 = 5$ are called *sentences.* This particular type of sentence is called an *equation,* since it states that two quantities are equal. Other types of sentences such as $8 - 11 \neq 5$ are called *inequalities,* because they state that two quantities are unequal; the sign $\neq$ is read "is not equal to." The expression $8 > 5$ is also an inequality which states that 8 is greater than 5. The symbol $>$ is read "is greater than" and the symbol $<$ is read "is less than." It should be noted that the larger end of the symbols $>$ and $<$ faces the larger number. Thus, $-3 < -1$ is an inequality stating that -3 is less than -1.

All of the sentences above, both equations and inequalities, are true statements. The expressions on each side of the equals or inequality signs are called the sides or members of the equation or inequality. In true sentences, both members actually bear the relation to each other which is expressed by the sign connecting them.

Sentences may also be false. For example, $-8 + 5 = 5$, $-13 > -2$, $8 < 2$, and $4 + 3 \neq 7$ are all false statements, because their members are not related in the manner expressed by the sign connecting them.

■ OPEN SENTENCES.—Some sentences contain variables denoted by letter symbols. Such sentences may be either true or false, depending on the value assigned to the variable; hence they are called *open sentences.* For example, $x + 3 = 7$ is true if, and only if, $x = 4$; $y + 2 > 5$ if and only if $y > 3$. Some open sentences are combinations of equations and inequalities. For example, $x + 2 \geqq 5$ states that the number x added to 2 is either greater than or equal to 5 (this will be true if, and only if, $x \geqq 3$).

■ SOLVING OPEN SENTENCES.—Solving an open sentence is the process of finding the numbers in the domain of the variable that make the sentence true. The set of numbers that make an open sentence true is called the *solution set* of the sentence. The numbers that form the solution set of an equation are also known as the *roots* or *solutions* of the equation. In the equation $x + 2 = 8$ the solution set is $\{6\}$. This solution set has only one element; thus the equation has only one root, 6. In contrast to this, the open sentence $2x + 3 = x + x + 3$ is an equation whose solution set consists of all possible values of x. This equation has an infinite number of roots if the domain of x is infinite. On the other hand, the sentence $x - 3 = x$ will not be true for any replacement for x. This equation has no root; its solution set is the empty set, which can be written as θ. The solution set for the inequality $x + 3 > 7$ consists of all numbers greater than 4; there are an infinite number of replacements for the variable that will make this inequality true if the domain of x includes all the positive integers and fractions greater than 4.

Set Notation.—One method of indicating a set is by enclosing the members in braces. Thus, $\{2, 3, 5\}$ stands for the set consisting of 2, 3, and 5. However, since it is impossible to list all the members of a solution set that has an infinite number of members, mathematicians also use another notation for showing sets that is especially suitable in such a case: $\{x \mid x > 4\}$ is read "the set of all values of x such that x is greater than 4." This notation, called the *set-builder notation,* may also be used for sets that are not infinite. Thus, $\{x \mid -2 < x \leqq 5, x \text{ is an integer}\}$ is read "the set of x such that x is greater than -2 and less than or equal to 5 and x is an integer," and is the same as the set $\{-1, 0, 1, 2, 3, 4, 5\}$.

Checking Solutions.—The process of showing that a number is a member of the solution set of an equation or inequality is called *checking* the solution. An equation is checked by replacing the variable with the number that is the supposed root, and showing that the resulting values of both members of the equation are equal, that is, showing that the expressions on both sides of the equals sign are the same. To check that -10 is a root of $2x + 3 = x - 7$, replace x by -10 and evaluate both members of the equation. The check would appear as follows:

$$\begin{aligned} 2x + 3 &= x - 7 \\ 2(-10) + 3 &\stackrel{?}{=} -10 - 7 \\ -20 + 3 &\stackrel{?}{=} -10 - 7 \\ -17 &= -17 \end{aligned}$$

A solution to an inequality may similarly be checked by replacing the variable with the number being checked and seeing whether the resulting values of both members make the sentence true. To check $-1\frac{1}{3}$ as a solution for $x - 3 < 7$, replace x by $-1\frac{1}{3}$:

$$\begin{aligned} x - 3 &< 7 \\ -1\tfrac{1}{3} - 3 &\stackrel{?}{<} 7 \\ -4\tfrac{1}{3} &< 7 \end{aligned}$$

Equivalent Expressions.—Two equations or two inequalities that have the same solution set are said to be *equivalent*. The equations $x + 3 = 5$ and $2x + 7 = 11$ are equivalent equations; both have $\{2\}$ as their solution set, as may be seen by checking $x = 2$ in each of them. However, $x + 3 = 5$ and $x + 3 = 6$ are not equivalent equations; the solution set of the first is $\{2\}$, but the solution set of the second is $\{3\}$. The inequalities $x + 3 > 5$ and $x > 2$ are equivalent, since the solution set for both consists of all numbers greater than 2, that is, the solution set of both is $\{x \mid x > 2\}$.

The concept of equivalent equations is important in solving or finding the solution set of an equation. It is very obvious that the solution set of the simple equation $x = 3$ is $\{3\}$, but it is not so easy to discover that the solution set of the more complicated equation $5x - 7 = 2x + 2$ is also $\{3\}$. However, both equations are equivalent, since 3 is the only root of each. In algebra, solving equations consists of changing complicated equations into equivalent equations that are simpler in the sense that their solution sets are more obvious.

Solving First-Degree Equations.—*First-degree equations* are equations whose variables are raised to the first power but no higher. Thus, $2x + 3 = 5x - 7$ is a first-degree equation, but $x^2 + x + 3 = 0$ is not, because the variable appears to the second power. First-degree equations are also called *linear equations* because they can be represented as straight lines; the section on the graphs of equations makes this clear.

As stated above, the method for solving first-degree equations consists of transforming them into equivalent equations in which the variable is alone on one side and a number is alone on the other. The four postulates that follow make it possible to solve various forms of first-degree equations. (Actually these postulates apply to an equation of any degree, but they are used here to solve only first-degree equations.)

FIRST POSTULATE.—If the same number is added to both members of an equation, an equivalent equation results.

This postulate makes it possible to get the variable alone on one side of an equation by removing a number that was subtracted from it originally.

Example:

To solve $x - 5 = 2$, add 5 to each member:

$$\begin{aligned} x - 5 &= 2 \\ +5 &\quad +5 \\ \hline x &= 7 \end{aligned}$$

Check:

$$\begin{aligned} x - 5 &= 2 \\ 7 - 5 &\stackrel{?}{=} 2 \\ 2 &= 2 \end{aligned}$$

Since the variable represents a number, this postulate may be used to add an expression containing the variable to each member of the equation. Thus, it is possible to remove a negative term containing the variable from the side of the equation on which this term is not wanted.

Example:

To solve $-x = 7 - 2x$, add $2x$ to each member:

$$\begin{aligned} -x &= 7 - 2x \\ +2x &\quad +2x \\ \hline x &= 7 \end{aligned}$$

Check:

$$\begin{aligned} -x &= 7 - 2x \\ -7 &\stackrel{?}{=} 7 - 2(7) \\ -7 &\stackrel{?}{=} 7 - 14 \\ -7 &= -7 \end{aligned}$$

In order to have a simple equation with an obvious solution set, it is not necessary to have the variable on the left side of the equation; it is sufficient that the variable be alone on one side and that the other side consist of a single number. If it is convenient, the variable may appear on the right side.

Example:

To solve $13 = x - 5$, add 5 to each member:

$$\begin{aligned} 13 &= x - 5 \\ +5 &= \quad +5 \\ \hline 18 &= x \end{aligned}$$

Check:

$$\begin{aligned} 13 &= x - 5 \\ 13 &\stackrel{?}{=} 18 - 5 \\ 13 &= 13 \end{aligned}$$

SECOND POSTULATE.—If the same number is subtracted from both members of an equation, an equivalent equation results.

This postulate makes it possible to get the variable alone on one side of the equation by removing any number that was added to it originally. It can also be used to remove a term containing the variable from a side of the equation on which it is not wanted.

Example:

To solve $x + 5 = 9$, subtract 5 from each member:

$$\begin{aligned} x + 5 &= 9 \\ -5 &\quad -5 \\ \hline x &= 4 \end{aligned}$$

Check:

$$\begin{aligned} x + 5 &= 9 \\ 4 + 5 &\stackrel{?}{=} 9 \\ 9 &= 9 \end{aligned}$$

Example:

To solve $x + 7 = 2$, subtract 7 from each member:

$$\begin{aligned} x + 7 &= 2 \\ -7 &\quad -7 \\ \hline x &= -5 \end{aligned}$$

Check:

$$\begin{aligned} x + 7 &= 2 \\ -5 + 7 &\stackrel{?}{=} 2 \\ 2 &= 2 \end{aligned}$$

The number subtracted from both members of an equation may be represented by an expression containing the variable.

Example:

To solve $3x = 2x - 8$, subtract $2x$ from each member:

$$\begin{aligned} 3x &= 2x - 8 \\ -2x &\quad -2x \\ \hline x &= \quad -8 \end{aligned}$$

Check:

$$\begin{aligned} 3x &= 2x - 8 \\ 3(-8) &\stackrel{?}{=} 2(-8) - 8 \\ -24 &\stackrel{?}{=} -16 - 8 \\ -24 &= -24 \end{aligned}$$

The first two postulates for solving equations, in effect, permit removing a term from one side of an equation and transferring it to the other side with its sign changed. If 6 is added to both members of $x - 6 = 5$, the -6 term disappears from the left member and appears as $+6$ on the right side of the equation. If 4 is subtracted from both members of the equation $x + 4 = 12$, the $+4$ term disappears from the left member and appears as -4 on the right side of the equation.

The situation illustrated in the two examples above makes it possible to write the result of adding or subtracting the same quantity from both members of an equation without actually showing the intermediate step of adding or subtracting. Thus, $x - 7 = 5$ becomes $x = 5 + 7$, by adding 7 to both members. To solve, combine like terms:

$$x = 12.$$

Similarly,

$$x + 15 = 2$$

becomes

$$x = 2 - 15$$

by subtracting 15 from both members. Therefore,

$$x = -13.$$

The short-cut method for adding or subtracting the same quantity from both members of an equation by moving a term from one side to the other and changing its sign is called *transposition*.

■ THIRD POSTULATE.—If both members of an equation are multiplied by the same nonzero number, an equivalent equation results.

This postulate makes it possible to get a variable alone by removing a number by which the variable was originally divided. The reason for requiring a nonzero multiplier is that if both members of an equation are multiplied by 0, the result is always 0 = 0. Since 0 = 0 no matter what x is, the resulting equation is not truly equivalent to the original.

Example:

To solve $x/5 = 7$, multiply both members by 5:

$$5\left(\frac{x}{5}\right) = 5(7)$$
$$x = 35$$

Check:

$$\frac{x}{5} = 7$$
$$\frac{35}{5} \stackrel{?}{=} 7$$
$$7 = 7$$

Example:

To solve $-\frac{3}{4}x = 12$, multiply both members by $-\frac{4}{3}$:

$$-\frac{4}{3}\left(-\frac{3}{4}x\right) = -\frac{4}{3}(12)$$
$$x = -16$$

Check:

$$-\frac{3}{4}x = 12$$
$$-\frac{3}{4}(-16) \stackrel{?}{=} 12$$
$$12 = 12$$

■ FOURTH POSTULATE.—If both members of an equation are divided by the same nonzero number, an equivalent equation results.

This postulate makes it possible to get a variable alone by removing the number by which it was multiplied originally (that is, its numerical coefficient).

Example:

To solve $4x = 12$, divide both members of the equation by 4:

$$\frac{4x}{4} = \frac{12}{4}$$
$$x = 3$$

Check:

$$4x = 12$$
$$4(3) \stackrel{?}{=} 12$$
$$12 = 12$$

Example:

To solve $18 = -2x$, divide both members by -2:

$$\frac{18}{-2} = \frac{-2x}{-2}$$
$$-9 = x$$

Check:

$$18 = -2x$$
$$18 \stackrel{?}{=} -2(-9)$$
$$18 = 18$$

It is frequently necessary to employ more than one of the postulates to find the solution of an equation.

Example:

Consider the equation $3y + 9 = 15$. In order to get the variable alone on one side of the equation it is necessary to remove both the 9 which was added to it and the coefficient 3 which was multiplied by it. To solve, subtract 9 from both members:

$$\begin{array}{rcr} 3y + 9 & = & 15 \\ -\ 9 & & -9 \\ \hline 3y \quad & = & 6 \end{array}$$

and divide both members by 3:

$$\frac{3y}{3} = \frac{6}{3}$$
$$y = 2$$

Check:

$$3y + 9 = 15$$
$$3(2) + 9 \stackrel{?}{=} 15$$
$$6 + 9 \stackrel{?}{=} 15$$
$$15 = 15$$

Example:

To solve $y + 16 = 3y + 2$, subtract y and 2 from both members:

$$\begin{array}{rcl} y + 16 & = & 3y + 2 \\ -y - \ 2 & = & -y - 2 \\ \hline 14 & = & 2y \end{array}$$

and divide both members by 2:

$$7 = y.$$

Check:

$$y + 16 = 3y + 2$$
$$7 + 16 \stackrel{?}{=} 3(7) + 2$$
$$7 + 16 \stackrel{?}{=} 21 + 2$$
$$23 = 23$$

Note that the check of the solution to an equation must always be made by substituting the value being checked in the *original* equation. If an error has been made at any point in the solution, the equations appearing after the error will not be equivalent to the original equation, and the fact that a solution checks in one of these later equations will not guarantee that it is also a root of the original.

If an equation contains several like terms, it is advisable to combine them before using the postulates that transform the equation into a simpler equivalent equation.

Example:

To solve $14x + 35 - 6x - 48 = 5 + 2$, combine like terms so that $8x - 13 = 7$, and add 13 to both members:

$$\begin{array}{rcr} 8x - 13 & = & 7 \\ +13 & & +13 \\ \hline 8x \quad & = & 20 \end{array}$$

and divide both members of the equation by 8:

$$\frac{8x}{8} = \frac{20}{8}$$
$$x = 2\frac{1}{2}$$

Check:

$$14x + 35 - 6x - 48 = 5 + 2$$
$$14\left(2\frac{1}{2}\right) + 35 - 6\left(2\frac{1}{2}\right) - 48 \stackrel{?}{=} 5 + 2$$
$$35 + 35 - 15 - 48 \stackrel{?}{=} 5 + 2$$
$$7 = 7$$

Example:

To solve $8x + 5 - 2x = 7 + 2x - 10$, combine like terms:

$$6x + 5 = 2x - 3;$$

subtract $2x$ and 5 from both members by the method of transposition:

$$6x - 2x = -3 - 5;$$

combine like terms:

$$4x = -8$$

and divide both members by 4:

$$\frac{4x}{4} = \frac{-8}{4}$$
$$x = -2$$

Check:

$$8x + 5 - 2x = 7 + 2x - 10$$
$$8(-2) + 5 - 2(-2) \stackrel{?}{=} 7 + 2(-2) - 10$$
$$-16 + 5 + 4 \stackrel{?}{=} 7 - 4 - 10$$
$$-7 = -7$$

If an equation contains parentheses, they should be removed before proceeding with the rest of the solution.

Example:

To solve $2 - 5(x - 3) = 1 + (x - 8)$, remove parentheses:

$$2 - 5x + 15 = 1 + x - 8;$$

combine like terms:

$$17 - 5x = x - 7;$$

add 7 and $5x$ to both members by transposing:

$$17 + 7 = x + 5x;$$

combine like terms:

$$24 = 6x;$$

and divide both members of the equation by 6:

$$\frac{24}{6} = \frac{6x}{6}$$
$$4 = x$$

Check:

$$2 - 5(x - 3) \stackrel{?}{=} 1 + (x - 8)$$
$$2 - 5(4 - 3) \stackrel{?}{=} 1 + (4 - 8)$$
$$2 - 5(1) \stackrel{?}{=} 1 + (-4)$$
$$2 - 5 \stackrel{?}{=} 1 - 4$$
$$-3 = -3$$

Solving Problems.—One of the important applications of algebra is solving problems through the use of equations.

Consider this problem: If a certain number is subtracted from 40, the result is the same as 1 more than twice the number.

The problem is solved by choosing a variable x to represent the number to be found. Then $40 - x$ is the result when the number is subtracted from 40, and $2x + 1$ represents 1 more than twice the number. The statement of the problem is an open sentence which, when translated into algebraic symbols, is the equation $40 - x = 2x + 1$.

To solve the problem add x and subtract 1 from both members:

$$40 - 1 = 2x + x;$$

combine like terms:

$$39 = 3x;$$

and divide both members by 3:

$$13 = x.$$

As a check of the solution to a problem, the supposed answer must satisfy the conditions prescribed by the problem. Substitution in the equation would not constitute a satisfactory check for the solution to a problem, since the problem may have been translated incorrectly in writing the equation. Checking the conditions of the problem solved here, if the number is 13, then subtracting it from 40 would give 27. One more than twice 13 is also 27, thus showing that the answer is correct.

Sometimes the identification and representation of the quantities involved in a problem are more subtle than in the previous one. Consider this problem: Find three consecutive odd integers whose sum is 249. Begin by choosing n to represent the first odd integer. Consideration of two consecutive odd integers (such as 7 and 9) will show that the second integer is 2 more than the first. Therefore, $n + 2$ will represent the second odd integer. Adding another 2 will produce the third odd integer, $n + 4$. It is customary to list and identify the variable and any other quantities that must be represented in terms of it in order to write the equation.

In this problem let n equal the first odd integer; $n + 2$, the second consecutive odd integer; and $n + 4$, the third consecutive odd integer.

The problem states that the sum of the three consecutive odd integers is 249. Translating this into an algebraic equation gives

$$n + (n + 2) + (n + 4) = 249.$$

To solve the problem, remove parentheses:

$$n + n + 2 + n + 4 = 249;$$

combine like terms:

$$3n + 6 = 249;$$

subtract 6 from both members:

$$3n = 249 - 6$$
$$3n = 243$$

and divide both members by 3:

$$n = 81$$
$$n + 2 = 83$$
$$n + 4 = 85$$

Check:

Inspection shows that the answers are three consecutive odd integers. And since $81 + 83 + 85 = 249$, the answers are correct.

Sometimes it is helpful to represent the information in a problem by using a tabular form.

Example:

How many pounds of candy worth 63 cents per pound should be mixed with candy worth 81 cents per pound to produce a mixture of 60 pounds to sell at 75 cents per pound?

Let x be the number of pounds of 63-cent candy to be used. The remainder of the 60-pound mixture, $60 - x$, must be taken from the 81-cent candy. All of the quantities used in the problem are shown in the following table:

Price per Pound (cents)	*Weight (pounds)*	*Total Value (cents)*
63	x	$63x$
81	$60 - x$	$81(60 - x)$
75	60	$75(60)$

The equation in this problem cannot be obtained by a direct translation of any explicit statement in the problem. However, it is implied that the sum of the values of the two ingredients must equal the value of the mixture. Writing this relationship in algebraic symbols:

$$63x + 81(60 - x) = 75(60).$$

Solution:

$$63x + 4860 - 81x = 4500$$
$$63x - 81x = 4500 - 4860$$
$$-18x = -360$$
$$x = 20$$

Thus, the amount of 60-cent candy required is 20 pounds.

Check:

60 − 20 = 40 lb. of 81¢ candy
20 lb. at 63¢ = \$12.60
40 lb. at 81¢ = \$32.40
60 lb. at 75¢ = \$45.00

In some problems, such as those involving uniform motion, it may be helpful to draw a diagram in order to discover what the equation is.

Example:

A boat left port traveling at 30 miles per hour. Four hours later a helicopter was sent after the boat from the same port traveling at 90 miles per hour over the same course. How long will it take the helicopter to catch the boat?

Draw a diagram to represent the distances traveled by the boat and by the helicopter (Fig. 3). It is evident that

Figure 3.

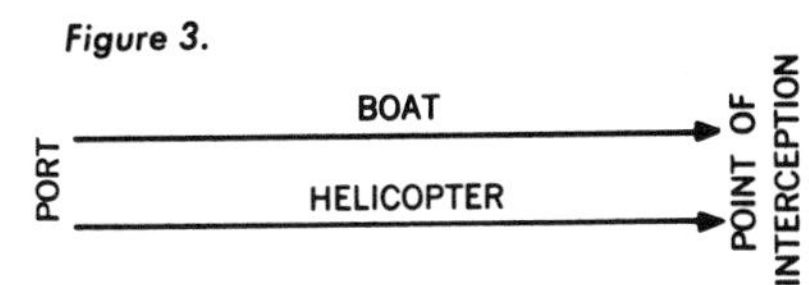

these distances are equal, since both started from the same point and ended at the same point. Therefore, expressions are needed for the distances traveled by each; these are obtained by multiplying the rate of each by the time of travel.

Letting x equal the number of hours for the helicopter to catch the boat, and noting that the boat will therefore travel $4 + x$ hours before being intercepted, the data for the problem can be represented as follows:

	Rate (miles per hour)	× *Time (hours)*	= *Distance (miles)*
Boat	30	$4 + x$	$30(4 + x)$
Helicopter	90	x	$90x$

Solution:

$$30(4 + x) = 90x$$
$$120 + 30x = 90x$$
$$120 = 90x - 30x$$
$$120 = 60x$$
$$2 = x$$

Thus, the helicopter would overtake the boat in 2 hours.

Check:

In 2 hours the helicopter would go 90×2, or 180 miles. The boat would travel $4 + 2$, or 6 hours, and in 6 hours, it would go 30×6, or 180 miles, and the helicopter would overtake the boat at this point.

A second example will show the importance of drawing a distance diagram to get the equation.

Example:

Two trains are scheduled to leave stations 284 miles apart and travel toward each other at 33 mph and 38 mph, respectively. How long will it take them to reach the point where they are supposed to pass each other?

Figure 4.

284 miles

Drawing the diagram to show the distances traveled by the two trains up to the point where they meet, it is evident that the sum of the distances moved is equal to the 284 miles between the two stations (Fig. 4).

Letting x equal the time each train travels until they meet, the facts are:

	Rate (miles per hour)	× Time (hours)	= Distance (miles)
First train	33	x	$33x$
Second train	38	x	$38x$

Solution:

$$\begin{aligned} 33x + 38x &= 284 \\ 71x &= 284 \\ x &= 4 \end{aligned}$$

Thus, the trains will meet in 4 hours.

Check:

In 4 hours the first train will travel 33×4, or 132 miles; in 4 hours the second train will travel 38×4, or 152 miles, and $132 + 152 = 284$ miles.

Solving Inequalities.—Just as first-degree equations are solved by transforming them into equivalent equations, so inequalities are solved by transforming them into inequalities in which it is easy to see the solution set.

The inequality $13x - 5 > 6 + 2x$ and the inequality $x > 1$ are equivalent, for in both the solution set consists of all values of the variable x that are greater than 1; that is, the solution set is $\{x \mid x > 1\}$. In the second inequality it is easy to see what the solution set is; in the first it is not (the reader should try substituting a value of x greater than 1, say 2, to satisfy himself that it actually makes the first sentence true).

To find the solution set of a complicated inequality, it is transformed into a simpler but equivalent inequality by using four postulates similar to those used in solving equations.

■ **FIRST POSTULATE.**—If the same number is added to both members of an inequality, an equivalent inequality results.

Example:

To solve $x - 10 < 28$, add 10 to each member:

$$\begin{array}{rcl} x - 10 & < & 28 \\ +10 & & +10 \\ \hline x & < & 38 \end{array}$$

Example:

To solve $-x < 4 - 2x$, add $2x$ to each member:

$$\begin{array}{rcl} -x & < & 4 - 2x \\ +2x & & +2x \\ \hline x & < & 4 \end{array}$$

■ **SECOND POSTULATE.**—If the same number is subtracted from both members of an inequality, an equivalent inequality results.

Example:

To solve $x + 9 < 2$, subtract 9 from each member of the inequality, thus yielding an equivalent inequality:

$$\begin{array}{rcl} x + 9 & < & 2 \\ -9 & & -9 \\ \hline x & < & -7 \end{array}$$

Example:

To solve $7x > 6x + 3$, subtract $6x$ from each member:

$$\begin{array}{rcl} 7x & > & 6x + 3 \\ -6x & & -6x \\ \hline x & > & 3 \end{array}$$

■ **THIRD POSTULATE.**—If both members of an inequality are multiplied by the same positive number, an equivalent inequality results; if both members of an inequality are multiplied by the same negative number and the direction of the inequality is reversed, an equivalent inequality results.

The need for making a distinction between what happens when inequalities are multiplied by positive numbers and what happens when they are multiplied by negative numbers will be clear if an example is considered. Suppose we take the true sentence $5 > 3$. Multiply both members by positive 2 and the result is

$$+2(5) > +2(3)$$
$$10 > 6$$

Take the original true sentence but multiply both members by negative 2 and reverse the direction of the inequality:

$$5 > 3$$
$$-2(5) < -2(3)$$

Then the true sentence results:

$$-10 < -6.$$

The application of the third postulate to the solution of an inequality requiring a positive multiplier will be illustrated first in solving

$$\frac{x}{4} > 5.$$

Multiply both members by positive 4:

$$4\left(\frac{x}{4}\right) > 4(5)$$
$$x > 20$$

An application of the third postulate to the solution of an inequality requiring a negative multiplier is shown next in solving

$$\frac{x}{-2} < 7.$$

Multiply both members by negative 2 and reverse the direction of the inequality:

$$-2\left(\frac{x}{-2}\right) > -2(7)$$
$$x > -14$$

Example:

To solve $5 < x/4$, multiply both members by positive 4:

$$4(5) < 4\left(\frac{x}{4}\right)$$
$$20 < x$$

The answer to this inequality can also be written as $x > 20$.

Example:

To solve $x/-8 > 1/2$, multiply both members by -8 and reverse the direction of the inequality:

$$-8\left(\frac{x}{-8}\right) < -8\left(\frac{1}{2}\right)$$
$$x < -4$$

■ **FOURTH POSTULATE.**—If both members of an inequality are divided by the same positive number, an equivalent inequality results; if both members of an inequality are divided by the same negative number and the direction of the inequality is reversed, an equivalent inequality results.

The need to distinguish between what happens when the members of an inequality are divided by a positive number and when they are divided by a negative number is similar to the reasoning for the third postulate.

Example:

To solve the inequality $3x > 1/2$, divide both members by $+3$:

$$\frac{3x}{3} > \frac{1}{2} \div 3$$
$$x > \frac{1}{6}$$

Example:

To solve the inequality $26 < -2x$, divide both members by -2 and reverse the direction of the inequality:

$$\frac{26}{-2} > \frac{-2x}{-2}$$
$$-13 > x$$

This answer can also be written as $x < -13$.

More than one of the postulates for transforming an inequality into a simpler equivalent inequality may have to be used to discover the solution set.

Example:

To solve $x - 2 > 3x + 6$, add 2 to, and subtract $3x$ from, each member of the inequality (using transposition method):

$$x - 3x > 6 + 2;$$

combine like terms:

$$-2x > 8;$$

divide both members by -2, and reverse the direction of the inequality:

$$x < -4.$$

Quadratic Equations.—A *quadratic equation* is an equation in which the variable appears raised to the second power and to no higher power than the second. Thus, a quadratic equation is also called a *second-degree equation.* Any quadratic equation can be put in the form $ax^2 + bx + c = 0$ where $a \neq 0$. In this form, x stands for the variable, and a, b, and c stand for the coefficients. Note that b and c may be 0 but that a cannot be 0, since the equation would then have no term with the second power of the variable. The equations $x^2 + 2x - 8 = 0$, $3x^2 - 2 = 5x$, $7 - x^2 = 3$, $-3x^2 = 22$, and $x^2 + x = 0$ are all quadratic, but not $2x + 3 = 4$ or $x^3 - 3x^2 + 2 = 0$.

A quadratic equation such as $x^2 + 2x - 8 = 0$ cannot be solved solely by the four postulates for first-degree equations. These postulates provide no means for disposing of the x^2 term so that an equivalent equation with a single x on one side and a single number on the other can be obtained.

■ **SOLUTION BY FACTORING.**—The quadratic equation $x^2 + 2x - 8 = 0$ is in a form that suggests factoring the left member: $(x + 4)(x - 2) = 0$. In factored form, the equation represents a situation in which the product of two numbers is equal to zero: $(\ \ )(\ \ ) = 0$. The product of two numbers can be zero if, and only if, one or both of the two numbers is zero. Thus, if $(x + 4)(x - 2) = 0$, either the factor $x + 4$ must equal 0, or the factor $x - 2$ must equal 0, or both must equal 0.

Thus, if the first factor, $x + 4$, equals 0, we have a first-degree equation $x + 4 = 0$, whose solution is $x = -4$. Setting the second factor, $x - 2$, equal to 0 gives another first-degree equation, $x - 2 = 0$, whose solution is $x = 2$. It happens that in this case both factors cannot equal 0 at the same time. Thus the solution set of the original quadratic equation is $\{-4, 2\}$. Both of these roots may be checked by substitution, one at a time, in the quadratic equation.

The complete solution and check for $x^2 + 2x - 8 = 0$ would be as follows:

$$x^2 + 2x - 8 = 0;$$

factor:

$$(x + 4)(x - 2) = 0;$$

set each factor equal to 0:

$$x + 4 = 0 \qquad x - 2 = 0$$

and solve the resulting first-degree equations:

$$x = -4 \qquad x = +2$$

Check:

$$\begin{aligned} &\text{For } x = -4: \\ x^2 + 2x - 8 &= 0 \\ (-4)^2 + 2(-4) - 8 &\stackrel{?}{=} 0 \\ 16 - 8 - 8 &\stackrel{?}{=} 0 \\ 0 &= 0 \end{aligned}$$

$$\begin{aligned} &\text{For } x = 2: \\ x^2 + 2x - 8 &= 0 \\ (2)^2 + 2(2) - 8 &\stackrel{?}{=} 0 \\ 4 + 4 - 8 &\stackrel{?}{=} 0 \\ 0 &= 0 \end{aligned}$$

It is extremely important to realize that the procedure for solving quadratic equations by factoring hinges on the fact that the product of the factors must be zero. If the product of two factors is anything but zero, it is impossible to reach any conclusion concerning the value of the individual factors. For example, if the product of two factors is +12, one factor may be +6 and the other +2; one may be −6 and the other −2; or they may be 4 and 3, 12 and 1, 24 and $\frac{1}{2}$, 36 and $\frac{1}{3}$, or 18 and $\frac{2}{3}$, and so on. For this reason, all the nonzero terms of a quadratic equation must be brought together on one side of the equation with a single zero on the other side before one can proceed to the solution of the equation through factoring.

Example:

To solve $2x^2 + 5x = 3$, subtract 3 from both members:

$$2x^2 + 5x - 3 = 0;$$

factor:

$$(2x - 1)(x + 3) = 0;$$

set each factor equal to 0:

$$2x - 1 = 0 \qquad x + 3 = 0$$

and solve the resulting first-degree equations:

$$2x = 1$$

$$x = \frac{1}{2} \qquad x = -3$$

Check:

$$\begin{aligned} &\text{For } x = \tfrac{1}{2}: \\ 2x^2 + 5x &= 3 \\ 2\left(\frac{1}{2}\right)^2 + 5\left(\frac{1}{2}\right) &\stackrel{?}{=} 3 \\ \frac{1}{2} + \frac{5}{2} &\stackrel{?}{=} 3 \\ 3 &= 3 \end{aligned}$$

$$\begin{aligned} &\text{For } x = -3: \\ 2x^2 + 5x &= 3 \\ 2(-3)^2 + 5(-3) &\stackrel{?}{=} 3 \\ 2(9) + 5(-3) &\stackrel{?}{=} 3 \\ 18 - 15 &\stackrel{?}{=} 3 \\ 3 &= 3 \end{aligned}$$

The procedure for solving a quadratic equation by factoring may make use of any one of the types of factoring. It may be necessary, for example, to factor the difference of two squares in order to solve.

Example:

To solve $x^2 = 64$, subtract 64 from both members:

$$x^2 - 64 = 0;$$

factor:

$$(x - 8)(x + 8) = 0;$$

set each factor equal to 0:

$$x - 8 = 0 \qquad x + 8 = 0$$

and solve these two equations:

$$x = 8 \qquad x = -8$$

Check:

$$\begin{aligned} &\text{For } x = 8: \\ x^2 &= 64 \\ (8)^2 &\stackrel{?}{=} 64 \\ 64 &= 64 \end{aligned} \qquad \begin{aligned} &\text{For } x = -8: \\ x^2 &= 64 \\ (-8)^2 &\stackrel{?}{=} 64 \\ 64 &= 64 \end{aligned}$$

It may be necessary to take out the highest common monomial factor.

Example:

To solve $2y^2 - y = 0$, factor:

$$y(2y - 1) = 0;$$

set each factor equal to 0:

$$y = 0 \qquad 2y - 1 = 0$$

and solve:

$$2y = 1$$

$$y = 0 \qquad y = \frac{1}{2}$$

Check:

$$\begin{aligned} &\text{For } y = 0: \\ 2y^2 - y &= 0 \\ 2(0)^2 - 0 &\stackrel{?}{=} 0 \\ 0 - 0 &\stackrel{?}{=} 0 \\ 0 &= 0 \end{aligned} \qquad \begin{aligned} &\text{For } y = \tfrac{1}{2}: \\ 2y^2 - y &= 0 \\ 2\left(\frac{1}{2}\right)^2 - \frac{1}{2} &\stackrel{?}{=} 0 \\ 2\left(\frac{1}{4}\right) - \frac{1}{2} &\stackrel{?}{=} 0 \\ \frac{1}{2} - \frac{1}{2} &\stackrel{?}{=} 0 \\ 0 &= 0 \end{aligned}$$

An equation of this type is often incorrectly solved because of a misapplication of one of the postulates concerning equivalent equations. It would be incorrect to change the equation $2y^2 - y = 0$ to $2y - 1 = 0$ by dividing both members by y. The postulate governing division of both members of an equation by a number states that an equivalent equation is produced only when the division is by a nonzero number. The variable y may possibly equal 0, hence it cannot be used as a divisor. The equation $2y - 1 = 0$ is not equivalent to the original equation, $2y^2 - y = 0$, because $2y - 1 = 0$ has only one root, $\frac{1}{2}$, while $2y^2 - y = 0$ has two roots, 0 and $\frac{1}{2}$. In general, one may not divide by an expression containing the variable without considering the possibility that the expression may equal zero for some value or values of the variable. Note, on the other hand, that in the equation $2y - 2 = 0$, it is perfectly correct to divide both members by 2, yielding the equivalent equation $y - 1 = 0$; in this case, the division is by a nonzero number as required by the postulate. The equations $2y - 2 = 0$ and $y - 1 = 0$ are equivalent; both have the root 1 as their only solution.

It is sometimes advantageous to divide both members of a quadratic equation by a nonzero number even before factoring in order to simplify the solution of the equation.

Example:

To solve $15x^2 + 30 = 45x$, subtract $45x$ from each member:

$$15x^2 - 45x + 30 = 0;$$

divide both members by 15:

$$x^2 - 3x + 2 = 0;$$

factor:

$$(x - 2)(x - 1) = 0;$$

set each factor equal to 0:

$$x - 2 = 0 \qquad x - 1 = 0$$

and solve these two equations:

$$x = 2 \qquad x = 1$$

Check:

For $x = 2$:
$$15x^2 + 30 = 45x$$
$$15(2)^2 + 30 \overset{?}{=} 45(2)$$
$$15(4) + 30 \overset{?}{=} 45(2)$$
$$60 + 30 \overset{?}{=} 90$$
$$90 = 90$$

For $x = 1$:
$$15x^2 + 30 = 45x$$
$$15(1)^2 + 30 \overset{?}{=} 45(1)$$
$$15(1) + 30 \overset{?}{=} 45(1)$$
$$15 + 30 \overset{?}{=} 45$$
$$45 = 45$$

In all of the quadratic equations solved so far, two roots have been obtained. It is true, in general, that quadratic equations have two roots. These roots, however, may be equal to each other.

Example:

To solve $x^2 = 6x - 9$, subtract $6x$ from each member and add 9 to each member:

$$x^2 - 6x + 9 = 0;$$

factor:

$$(x - 3)(x - 3) = 0;$$

and set each factor equal to 0:

$$x - 3 = 0 \qquad x - 3 = 0$$
$$x = 3 \qquad x = 3$$

Check:

For $x = 3$:
$$x^2 = 6x - 9$$
$$(3)^2 \overset{?}{=} 6(3) - 9$$
$$9 \overset{?}{=} 18 - 9$$
$$9 = 9$$

The quadratic equation in the above example is considered to have two roots, although both roots are 3. Mathematicians prefer to establish general statements which have as few exceptions as possible. In higher mathematics, it is shown that an equation of the nth degree, where n is a positive integer whose members are polynomials, has exactly n roots, although some or all of these roots may be equal. Thus, a first-degree equation has exactly one root, a second-degree equation has exactly two roots; a third-degree equation has exactly three roots, and so on. When the graphs of second-degree equations are studied, other reasons also appear for considering them to have two roots, even though these roots may be equal.

■ INCOMPLETE QUADRATIC EQUATIONS.—Any quadratic equation can be put in the form $ax^2 + bx + c = 0$, where $a \neq 0$. When $b = 0$, the equation is a special case of the quadratic of the form $ax^2 + c = 0$ and is known as an *incomplete quadratic equation.* For example, $6x^2 - 24 = 0$ is an incomplete quadratic equation. It may be solved by dividing both members by 6 to yield $x^2 - 4 = 0$. This can be factored into $(x - 2)(x + 2) = 0$; when each factor is set equal to 0, two roots, $+2$ and -2, are obtained.

However, $6x^2 - 24 = 0$, like other incomplete quadratic equations, can also be solved by a method that depends on transforming it into a form in which the square of the variable is equal to a number: $6x^2 = 24$, $x^2 = 4$, so $x = +2$ or $x = -2$. Finding the number whose square is the given number is called finding the square root of the given number. The square root of 4 was found to be $+2$ or -2. The equation $x^2 = 4$ was really solved by taking the square root of both members and assuming that the square roots of equal quantities are equal. To solve $x^2 = 36$, the square roots of both members can be taken, obtaining $x = +6$ and $x = -6$ (this result is usually written in the shorter form, $x = \pm 6$).

The method for solving an incomplete quadratic equation without factoring is illustrated below.

Example:

To solve $3x^2 - 75 = 0$, divide both members by 3:

$$x^2 - 25 = 0;$$

add 25 to both members:

$$x^2 = 25;$$

and take the square roots of both members:

$$x = \pm 5.$$

Check:

For $x = +5$:
$$3x^2 - 75 = 0$$
$$3(5)^2 - 75 \overset{?}{=} 0$$
$$3(25) - 75 \overset{?}{=} 0$$
$$75 - 75 \overset{?}{=} 0$$
$$0 = 0$$

For $x = -5$:
$$3x^2 - 75 = 0$$
$$3(-5)^2 - 75 \overset{?}{=} 0$$
$$3(25) - 75 \overset{?}{=} 0$$
$$75 - 75 \overset{?}{=} 0$$
$$0 = 0$$

Square Roots.—In solving $x^2 = 4$, it was pointed out that this open sentence speaks of a number whose square is 4. A number whose square is a given number is said to be a *square root* of the given number. In other words, the square root of a number is one of the two equal factors of the number.

Every perfect square has two square roots which are opposite in sign. The positive one is called the *principal square root* of the number. By conventional notation the symbol $\sqrt{\quad}$ (called a *radical sign*) is used to stand for the principal square root only. Thus $\sqrt{25} = 5$, $\sqrt{100} = 10$, $\sqrt{36} = 6$. Note that $\sqrt{2}(\sqrt{2}) = 2$. The negative square root is denoted by placing a minus sign before the radical sign: $-\sqrt{9} = -3$.

In the examples considered so far, only the square roots of perfect squares have been mentioned. The question may arise as to whether there is a solution to the equation $x^2 = 3$. Certainly there is no integer or fraction whose square is 3. In order to make it possible to take the square root of any positive number, the number system must be extended by inventing new numbers that are the square roots of *nonsquare* positive numbers. This is analogous to the extensions of the number system that were made earlier to make other operations possible in general. The present extension makes it possible to take the square root of any positive number. The reason for limiting the square root to positive numbers in the present extension of the number system is that the product of two equal numbers of the kinds which have so far been used will always be a positive number, whether the numbers being multiplied are both positive or both negative. Therefore, the square root of a negative number would have to be a number possessing different properties from the numbers used up to this point. However, another extension of the number system is made later to provide numbers that are the square roots of negative numbers, called *imaginary numbers.*

■ IRRATIONAL NUMBERS.—The integers and fractions used so far can all be expressed as ratios of two integers. Thus, $\frac{1}{2}$ is the ratio 1 : 2, 3 is 3 : 1, $2\frac{1}{3}$ is 7 : 3, and so on. The set of integers and fractions is therefore called the set of *rational numbers* because of their relation to ratios.

The square roots of nonsquare numbers, however, cannot be expressed as the ratio of two integers; therefore, they are included in the set of numbers called the irrational numbers. An *irrational number* is one that cannot be expressed as the ratio of two integers. Thus $\sqrt{3}$ is irrational, but $\sqrt{4}$ is rational, since $\sqrt{4} = 2$.

Square roots of nonsquare numbers are not the only irrational numbers; for example, π is irrational. Just as the square root of a number is one of its two equal factors, so the *cube root* is one of its three equal factors; and the cube roots of noncube numbers are also irrational. The expression $\sqrt[3]{8}$ represents the principal cube root of 8, which is 2. Therefore $\sqrt[3]{8}$ is rational, but $\sqrt[3]{9}$ is irrational. The small 3 used with the radical sign to indicate the cube root is called the *index* of the radical; when no index is shown, it is understood to be 2, as in $\sqrt{3}$. Note also that it is possible to have a cube root of a negative number: $\sqrt[3]{-27}$ is -3.

When rational numbers are expressed as decimal fractions, the decimal fractions either terminate or continue indefinitely, repeating a certain sequence of digits. For example, $5\frac{1}{2} = 5.5$ and $\frac{1}{4} = 0.25$ both terminate, but $\frac{1}{3} = 0.33333\ldots$ and $\frac{1}{7} = 0.142857142857\ldots$ do not—they continue onward (the three dots at the end indicate a continuing repetition of the pattern of digits). Irrational numbers, when expressed as decimal fractions, neither terminate nor repeat a pattern. For example, $\sqrt{2}$ is approximately equal to 1.41421, but its exact value cannot be represented as a finite decimal. Although there is no exact decimal representation for an irrational number, there is a precise location on the number line for every irrational number. For example, $\sqrt{2}$, since it is approximately 1.41421, is located between 1 and 2 (Fig. 5). Its precise position can be found by marking off a distance to the right of the zero point that is exactly equal to $\sqrt{2}$ units. This

Figure 5.

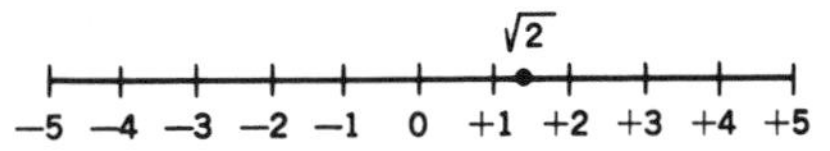

exact length can be found as the hypotenuse of a right triangle whose legs are each 1 unit. If such a triangle is constructed using the same distance to represent a unit as is used on the number line (Fig. 6), the hypotenuse will be $\sqrt{2}$, since the square of the hypotenuse must equal the sum of the squares of the legs by the Pythagorean theorem: $(\text{hypotenuse})^2 = 1^2 + 1^2 = 2$.

Figure 6.

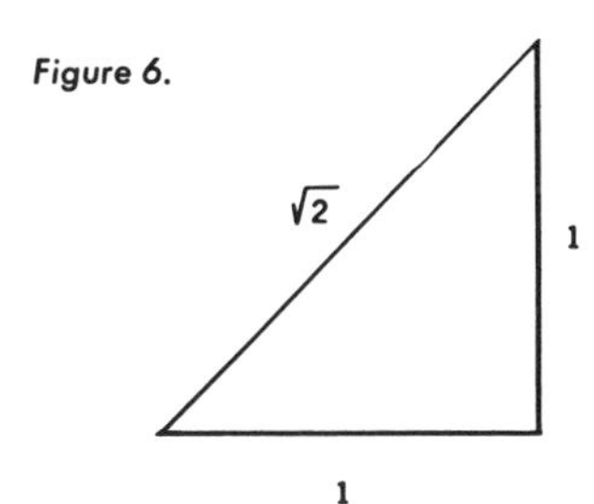

The rational and irrational numbers together form the set of numbers called the *real numbers*. Mathematicians have shown that every point on the number line represents a real number, and that every real number represents a point on the number line.

■ COMPUTING SQUARE ROOTS.—The square root of a number, even if it is a perfect square, may not be obvious by inspection. The process of finding a square root in such a case will be illustrated with the square root of 6.76: The digits are first grouped in pairs beginning at the decimal point and pairing in both directions:

$$\sqrt{6.76}$$

The largest square less than or equal to the first pair (4 in this case) is written below it and its square root, 2, is placed above the pair in the answer:

$$\begin{array}{l} \ \ 2 \\ \sqrt{6.76} \\ \ \ 4 \end{array}$$

Subtraction is performed as in long division and the next pair, 76, is brought down:

$$\begin{array}{l} \ \ 2 \\ \sqrt{6.76} \\ \ \ 4 \\ \ \ 2\ 76 \end{array}$$

The answer, 2, is doubled and used as a trial divisor:

$$\begin{array}{r|l} & \ \ 2. \\ & \sqrt{6.76} \\ & \ \ 4 \\ 4 & \ \ 2\ 76 \end{array}$$

A digit, in this case 6, is placed above the next pair in the answer and also written at the right end of the trial divisor, the digit being chosen in such a way that the product of the divisor and itself will be as large as possible without exceeding the remainder, 276:

$$\begin{array}{r|l} & \ \ 2.\ 6 \\ & \sqrt{6.76} \\ & \ \ 4 \\ 46 & \ \ 2\ 76 \\ & \ \ 2\ 76 \end{array}$$

Continue this until all number pairs have been brought down and used.

In the example above, 2.6 is the square root of 6.76, and this may be checked by multiplying 2.6 by 2.6 to show that the product is 6.76.

■ APPROXIMATING IRRATIONALS.—The approximate decimal values of square roots of numbers that are not perfect squares may be calculated by the same process as that used to get the square root of perfect square numbers.

Suppose it is desired to find the square root of 19 to the nearest tenth. Pairs of zeros are added after the decimal point, enough pairs being added to calculate one more decimal place than required in the answer.

$$\begin{array}{r|l} & \ \ 4.\ 3\ \ 5\ + \\ & \sqrt{19.00\ 00} \\ & \ \ 16 \\ 83 & \ \ 3\ 00 \\ & \ \ 2\ 49 \\ 865 & \ \ \ \ 51\ 00 \\ & \ \ \ \ 43\ 25 \end{array}$$

The answer to the nearest tenth is 4.4 since 4.35+ is more than the halfway point (4.35) between 4.3 and 4.4. Had the calculated answer been 4.34 or less, the answer to the nearest tenth would have been given as 4.3.

Operations on Radicals.—Irrational numbers involving radical signs, such as $2\sqrt{5}$, are called *radicals*. Since they are numbers, it can be assumed that the results of adding, subtracting, multiplying, and dividing them must be consistent with the postulates that have been assumed to govern these operations on the other sets of numbers, such as the integers and fractions.

■ MULTIPLICATION.—Suppose it is desired to multiply two radicals with the same index, such as $\sqrt{25}$ and $\sqrt{4}$. The product, $(\sqrt{25})(\sqrt{4})$, is the same as (5)(2), the answer to which is 10. The same answer would result if the numbers under the radical sign, called the *radicands*, were multiplied together to form the radicand of the product: $(\sqrt{25})(\sqrt{4}) = \sqrt{100}$, which is 10. If the radicands are not perfect squares as they are in the above example, consideration of their decimal approximations will show that this is a suitable procedure for producing their product, provided they have the same index. Thus, $(\sqrt{3})(\sqrt{7}) = \sqrt{21}$ and $(\sqrt[3]{4})(\sqrt[3]{5}) = \sqrt[3]{20}$.

If the radicals to be multiplied have numerical coefficients, for example, $2\sqrt{5}$ and $3\sqrt{6}$, they may be multiplied by multiplying the coefficients and making this product the coefficient of the answer. By the commutative and associative postulates for multiplication, $(2\sqrt{5})(3\sqrt{6})$ is the same as $(2)(3)(\sqrt{5})(\sqrt{6})$, or $6\sqrt{30}$.

Multiplication of radicals of the same index is thus performed by multiplying their coefficients to form the coefficient of the answer and multiplying their radicands to produce the radicand of the answer.

Examples:

$$(-4\sqrt{5})(3\sqrt{7}) = -12\sqrt{35}$$
$$(3\sqrt{2})(\sqrt{5}) = 3\sqrt{10}$$

■ DIVISION.—Since division is the inverse of multiplication, solving $21\sqrt{18} \div 3\sqrt{6}$ is equivalent to asking the question, "By what number must $3\sqrt{6}$ be multiplied to give $21\sqrt{18}$?" The answer, $7\sqrt{3}$, shows that division of radicals of the same index may be performed by dividing their coefficients to form the coefficient of the answer and dividing their radicands to form the radicand of the answer.

Examples:

$$36\sqrt{75} \div 9\sqrt{3} = 4\sqrt{25}$$
$$\frac{28\sqrt{30}}{7\sqrt{10}} = 4\sqrt{3}$$

■ SIMPLIFICATION.—Since $(\sqrt{4})(\sqrt{3}) = \sqrt{12}$, in turn $\sqrt{12}$ can be transformed into $(\sqrt{4})(\sqrt{3})$. But $\sqrt{4} = 2$, and therefore $\sqrt{12} = 2\sqrt{3}$. The form $2\sqrt{3}$ is considered simpler than $\sqrt{12}$ because its radicand is a smaller number. Radicals that are square roots are simplified by "removing" any factors that are perfect squares in order to make the radicand as small as possible. When $\sqrt{18}$ is broken up into $(\sqrt{9})(\sqrt{2})$ and then written as $3\sqrt{2}$, the perfect square, 9, is, in effect, "removed" from the radicand.

Examples:

$$\sqrt{50} = (\sqrt{25})(\sqrt{2}) = 5\sqrt{2}$$
$$\sqrt{27} = (\sqrt{9})(\sqrt{3}) = 3\sqrt{3}$$
$$3\sqrt{12} = 3(\sqrt{4})(\sqrt{3}) = (3)(2)\sqrt{3} = 6\sqrt{3}$$

It is customary to simplify radicals before listing them as final answers. In multiplying $3\sqrt{6}$ by $2\sqrt{15}$, $6\sqrt{90}$ is first obtained. Upon simplifying, this becomes $(6\sqrt{9})(\sqrt{10})$, or $(6)(3)\sqrt{10}$; hence, $18\sqrt{10}$.

It is not necessary to remove the largest perfect square factor in simplifying a radical as long as the process is continued until there are no more perfect square factors to be removed. Thus, in simplifying $\sqrt{72}$, if one had noticed that $\sqrt{72} = (\sqrt{36})(\sqrt{2})$, one would have obtained $6\sqrt{2}$ by removing the 36. Yet one might have written $\sqrt{72}$ as $(\sqrt{9})(\sqrt{8})$ instead, and gotten $3\sqrt{8}$, which, in turn, is $(3\sqrt{4})(\sqrt{2})$, or $(3)(2)\sqrt{2}$, and finally $6\sqrt{2}$.

■ ADDITION AND SUBTRACTION.—Just as $5x - 3x + 7x$ can be combined into one term, $9x$, by the application of the distributive postulate, so the same postulate requires that $5\sqrt{3} - 3\sqrt{3} + 7\sqrt{3}$ become $9\sqrt{3}$ when combined. Just as

$2x + 3y$ cannot be combined into one term, neither can $2\sqrt{5} + 3\sqrt{7}$ be combined. Radicands must be the same in order to combine radicals. But $5\sqrt[3]{7} + 2\sqrt{7}$ cannot be combined into one term although both radicands are the same, for in order to combine radicals their indices must be the same as well as their radicands. Radicals having the same index and same radicand may be combined by algebraically combining their coefficients and writing the sum before their common radical factor.

Examples:

$$7\sqrt{2} - 4\sqrt{2} + 9\sqrt{2} = 12\sqrt{2}$$
$$8\sqrt{7} - 18\sqrt{7} = -10\sqrt{7}$$
$$9\sqrt[3]{5} - 8\sqrt[3]{5} = \sqrt[3]{5}$$

Radicals cannot be combined into a single term if their radicands differ, but it is sometimes possible to change a radicand by simplifying it to make it the same as that in another radical to which it is to be added.

Example:

To combine $7\sqrt{50} - 4\sqrt{98}$, simplify as follows:

$$(7)(\sqrt{25})(\sqrt{2}) - (4)(\sqrt{49})(\sqrt{2})$$
$$(7)(5)\sqrt{2} - (4)(7)\sqrt{2}$$
$$35\sqrt{2} - 28\sqrt{2}$$
$$7\sqrt{2}$$

Imaginary Numbers.—The extension of the number system to include irrational numbers makes it possible to have a square root for every positive number in the number system. However, the rational and irrational numbers do not include the square roots of negative numbers. In fact, the property of multiplication that makes the product of two positive numbers a positive number and the product of two negative numbers also a positive number makes it impossible to get a negative number by squaring any of the real numbers.

To compute the square roots of negative numbers, the number system is extended to include another class of numbers whose squares are negative. The symbol i is used for the number which is $\sqrt{-1}$. The number i is called an *imaginary number*. This symbol, i, was chosen to represent this number because it is the initial letter of the word "imaginary."

Notice that $i^2 = -1$. Since $\sqrt{-4}$ can be simplified as $(\sqrt{4})(\sqrt{-1})$, or $2\sqrt{-1}$, $\sqrt{-4}$ is $2i$; $\sqrt{-49}$ would be $7i$, and so on. It is possible to simplify $\sqrt{-3}$ as $(\sqrt{3})(\sqrt{-1})$; therefore, $\sqrt{-3}$ is $i\sqrt{3}$. If $2i$ is squared, it equals $4i^2$, which in turn is $4(-1)$, or -4.

If $-3i$ is squared, $(-3i)^2$, it equals $9i^2$, which, in turn, is $9(-1)$, or -9. The square of an imaginary number is a negative real number.

Complex Numbers.—The imaginary numbers, when combined with the real numbers, form an extended number system called the *complex number system*. Complex numbers include all the real numbers, such as the rational numbers 5 and $\frac{2}{3}$; all the irrational numbers, such as $\sqrt{2}$ or $\sqrt[3]{73}$; all the imaginary numbers, such as $-2i$ and $i\sqrt{5}$; and combinations of both real and imaginary terms, such as $2 - 3i$ or $\sqrt{2} + 7i$. The complex numbers obey the commutative and associative postulates for addition and multiplication and the distributive postulate. But most important, any first-degree or second-degree equation whose coefficients are complex numbers will have solutions which are also complex numbers.

Completing the Square.—Since factoring is restricted to the use of integral coefficients, a quadratic equation whose roots are irrational cannot be solved by factoring. A method for solving a quadratic equation, called *completing the square*, does permit solutions whether the roots are rational or irrational.

The method of completing the square requires transforming the equation into a form in which the square root of both members can be taken, as in the method for solving an incomplete quadratic equation. In order to take the square root, a perfect-square trinomial is needed on the side containing the variable. Therefore, a method is required for getting a perfect square there. Examination of quadratic trinomials that are perfect squares, such as $x^2 + 6x + 9$ which is $(x + 3)^2$, or $x^2 - 8x + 16$ which is $(x - 4)^2$, shows that the constant term (9 in the first example) must be the square of one-half the middle term's coefficient (in the first example, one-half of 6 is 3). In the method of solving a quadratic equation by completing the square this fact is used to create, or "complete," a perfect square on one side.

Example:

To solve $x^2 + 4x - 21 = 0$, add 21 to both members:

$$x^2 + 4x = 21;$$

complete the square by adding to both members the number found by taking one-half of 4 and squaring it ($4 \div 2 = 2$ and $2^2 = 4$):

$$x^2 + 4x + 4 = 21 + 4;$$

factor:

$$(x + 2)^2 = 25;$$

take the square root of both members:

$$x + 2 = \pm 5;$$

and subtract 2 from both members:

$$x = \pm 5 - 2,$$

combine, and solve:

$$x = 3 \qquad x = -7$$

This example illustrates the use of the method of completing the square in an equation whose roots are rational. This equation could have been solved by factoring. The next example illustrates the use of the method for an equation whose roots are irrational and which, therefore, cannot be solved by factoring.

Example:

To solve $x^2 + 6x - 4 = 0$, add 4 to both members:

$$x^2 + 6x = 4;$$

complete the square by taking one-half of 6, squaring it, and adding the result to both members:

$$x^2 + 6x + 9 = 4 + 9;$$

factor:

$$(x + 3)^2 = 13;$$

take the square root of both members:

$$x + 3 = \pm\sqrt{13};$$

and subtract 3 from both members:

$$x = -3 + \sqrt{13}$$
$$x = -3 - \sqrt{13}$$

Irrational roots may be checked by substitution in the original equation just as rational roots are checked. The next example illustrates this procedure by checking one root in this equation.

Check:

For $x = -3 + \sqrt{13}$:

$$x^2 + 6x - 4 = 0$$
$$(-3 + \sqrt{13})^2 + 6(-3 + \sqrt{13}) - 4 \stackrel{?}{=} 0$$
$$9 - 6\sqrt{13} + 13 - 18 + 6\sqrt{13} - 4 \stackrel{?}{=} 0$$
$$0 = 0$$

Quadratic Formula.—Any quadratic equation in one variable may be put into the form $ax^2 + bx + c = 0$ where a, b, and c stand for the coefficients of x^2, x, and the constant term, respectively, and $a \neq 0$. By solving the equation $ax^2 + bx + c = 0$, which stands for any quadratic equation, mathematicians evolved a formula that gives the roots of any quadratic equation when the particular values of a, b, and c for that equation are substituted in it. The method of completing the square was used to solve $ax^2 + bx + c = 0$. Starting with $ax^2 + bx + c = 0$, divide both members by a:

$$x^2 + \frac{b}{a}x + \frac{c}{a} = 0;$$

subtract c/a from both members:

$$x^2 + \frac{b}{a}x = -\frac{c}{a};$$

complete the square (one-half of b/a is $b/2a$, and squaring gives $b^2/4a^2$):

$$x^2 + \frac{b}{a}x + \frac{b^2}{4a^2} = -\frac{c}{a} + \frac{b^2}{4a^2};$$

factor:

$$\left(x + \frac{b}{2a}\right)^2 = \frac{-4ac + b^2}{4a^2};$$

take the square root of both members:

$$x + \frac{b}{2a} = \frac{\pm\sqrt{b^2 - 4ac}}{2a};$$

and subtract $b/2a$ from both members:

$$x = \frac{-b \pm\sqrt{b^2 - 4ac}}{2a}.$$

Since $ax^2 + bx + c = 0$ represents any quadratic equation, its solution,

$$x = \frac{-b \pm \sqrt{b^2 - 4ac}}{2a},$$

is a formula for the roots of any quadratic equation.

The use of the formula is illustrated below in solving a quadratic equation that has rational roots.

Example:

To solve $x^2 = 10 - 3x$, put the equation in $ax^2 + bx + c = 0$ form:

$$x^2 + 3x - 10 = 0;$$

identify a, b, and c:

$$a = 1 \quad b = 3 \quad c = -10$$

use the formula:

$$x = \frac{-b \pm \sqrt{b^2 - 4ac}}{2a};$$

substitute values of a, b, and c:

$$x = \frac{-3 \pm \sqrt{9 - 4(1)(-10)}}{2(1)};$$

and simplify:

$$x = \frac{-3 \pm \sqrt{9 + 40}}{2}$$

$$x = \frac{-3 \pm \sqrt{49}}{2}$$

$$x = \frac{-3 \pm 7}{2}$$

$$x = 2 \qquad x = -5$$

The use of the quadratic formula to obtain irrational roots is shown next.

Example:

To solve $4x^2 - 3x = 2$, put the equation in the $ax^2 + bx + c = 0$ form:

$$4x^2 - 3x - 2 = 0;$$

identify a, b, and c:

$$a = 4 \quad b = -3 \quad c = -2$$

use the formula:

$$x = \frac{-b \pm \sqrt{b^2 - 4ac}}{2a};$$

substitute:

$$x = \frac{3 \pm \sqrt{(-3)^2 - 4(4)(-2)}}{2(4)};$$

and simplify:

$$x = \frac{3 \pm \sqrt{9 + 32}}{8},$$

$$x = \frac{3 \pm \sqrt{41}}{8},$$

$$x = \frac{3 + \sqrt{41}}{8} \qquad x = \frac{3 - \sqrt{41}}{8}$$

■ SOLVING PROBLEMS.—Some problems result in quadratic equations.

Example:

The length of a rectangular plot of ground exceeds three times its width by 1 foot. Its area is 52 square feet. Find its length and width.

Let x equal the number of feet in width, $3x + 1$ the length. As the area of a rectangle is width times length, $x(3x + 1)$, the area is $x(3x + 1) = 52$. Remove parentheses:

$$3x^2 + x = 52;$$

subtract 52 from both members:

$$3x^2 + x - 52 = 0;$$

factor:

$$(3x + 13)(x - 4) = 0;$$

set each factor equal to 0:

$$3x + 13 = 0 \qquad x - 4 = 0$$

and solve:

$$3x = -13$$

$$x = \frac{-13}{3} \qquad x = 4$$

The root $-13/3$ must be rejected as an answer to the problem, since the domain of the variable, x, must be restricted to positive numbers if it is to represent the number of feet in the width of a rectangle. If $x = 4$, $3x + 1 = 13$. Thus the plot is 4 feet wide and 13 feet long.

Check:

For length:	For area:
4	13
× 3	× 4
12	52
+ 1	
13	

Some problems may result in roots that are not rational.

Example:

A man has 30 feet of fencing to enclose a rectangular area of 40 square feet. Find the length and width of the rectangle to the nearest tenth of a foot.

Let x equal the number of feet in the length of the rectangle. Since 30 represents two lengths plus two widths, one length plus one width is 15 feet. Therefore, $15 - x$ is the number of feet in the width. Area is length times width:

$$x(15 - x) = 40$$

$$15x - x^2 = 40$$

Add x^2 and subtract $15x$ from both members:

$$0 = x^2 - 15x + 40.$$

The right member of this equation is not factorable. However, it may be solved by using the formula

$$x = \frac{-b \pm \sqrt{b^2 - 4ac}}{2a}.$$

Identify a, b, and c:

$$a = 1 \quad b = -15 \quad c = 40$$

substitute:

$$x = \frac{15 \pm \sqrt{225 - 4(1)(40)}}{2(1)}$$

$$x = \frac{15 \pm \sqrt{225 - 160}}{2}$$

$$x = \frac{15 \pm \sqrt{65}}{2}$$

and to get the answer to the nearest tenth, calculate $\sqrt{65}$ to the hundredths place:

```
          8. 0  6+
        √65.00 00
         64
  160   | 1 00
        |    0
  1606  | 1 00 00
        |   96 36
```

The hundredths place in the approximate square root of 65 is kept through the rest of the calculation and the rounding off to the nearest tenth is done last.

$$x = \frac{15 \pm 8.06}{2}$$

$$x = \frac{23.06}{2} \qquad x = \frac{6.94}{2}$$

$$x = 11.53 \qquad x = 3.47$$

$$x = 11.5 \qquad x = 3.5$$

The length could not be 3.5 feet because it would then be smaller than the width of the rectangle. Therefore, the length is 11.5 ft.

If $x = 11.5$, $15 - x = 3.5$. The rectangle is 11.5 feet long and 3.5 feet wide.

A check of these answers above will not produce the exact area specified in the problem. But since the dimensions are approximate, their product will approximate the area:

```
  11.5
 × 3.5
  575
 345
 40.25
```

(The exact area is 40.)

Sentences in Two Variables.—The open sentence $2x + y = 7$ contains two variables, x and y, each of which may be regarded as having a set of possible replacements or a domain. The open sentence will be true or false when a pair of numbers—one for x and one for y—is substituted in the equation. Thus, if $x = 2$ and $y = 3$, the sentence is true. It is also true if $x = 3$ and $y = 1$, if $x = 5$ and $y = -3$, or if $x = 2\frac{1}{2}$ and $y = 2$. There are an infinite number of pairs of values of x and y that will make the sentence true. However, not all pairs of values of x and y make it true. If $x = 2$ and $y = 1$, for example, the sentence will be false.

When an open sentence contains two variables, a solution consists of a pair of numbers—one number for each variable. Furthermore, the pair of numbers causes the sentence to be true only if the numbers are substituted for the correct variables. That is, the pair $x = 2$ and $y = 3$ makes $2x + y = 7$ a true sentence, but $x = 3$ and $y = 2$ does not. The pair of numbers in which $x = 2$ and $y = 3$ is represented by the notation (2,3), with the number representing x listed before that representing y (note that they are in alphabetical order). Such number pairs as (2,3) are said to be *ordered pairs*, since any pair is different from the pair consisting of the same numbers with their order reversed. That is, $(-5,2)$ represents the ordered pair in which $x = -5$ and $y = 2$, whereas $(2,-5)$ represents the ordered pair in which $x = 2$ and $y = -5$.

The solution set of an open sentence in two variables is a set of ordered pairs of numbers. Thus, in $2x + y = 7$, some of the ordered pairs in the solution set are $(2,3)$, $(1,5)$, $(5,-3)$, $(2\frac{1}{2},2)$.

Example:

Some of the ordered pairs in the solution set of $3x - 2y = 8$ are:

$$(2,-1),\ (3,\tfrac{1}{2}),\ (4,2),\ (0,-4).$$

Finding Number Pairs.—A good way to obtain some of the ordered pairs included in the solution set of an equation in two variables is to solve the equation for one variable in terms of the other. Then, by substituting arbitrarily selected values for the other variable, the corresponding values of the one solved for can be discovered. Thus, to get some of the ordered pairs in the solution set for $3x + y = 5$, transform the equation into $y = -3x + 5$ by subtracting $3x$ from each member. In this form, y is expressed in terms of x. Arbitrarily picking a value of x, say $x = 2$, gives $y = -3(2) + 5$, or $y = -1$, as the corresponding y value that will make the sentence true. If $x = 3$, $y = -3(3) + 5$, or $y = -4$. Therefore, $(3,-4)$ is another ordered pair in the solution set of $3x + y = 5$. For convenience, the ordered pairs are often arranged in a table:

x	2	3	4	5
y	-1	-4	-7	-10

A more difficult example might involve more than one step in solving for one variable in terms of the other. To get ordered pairs for the solution set of $5x - 2y = 3$, add $2y - 3$ to each member:

$$5x - 3 = 2y;$$

and divide both members by 2:

$$\frac{5x - 3}{2} = y.$$

The equation is now transformed so that it is solved for y in terms of x. Substituting arbitrarily selected values of x, the corresponding values of y are:

x	0	1	2	-7
y	$-\frac{3}{2}$	1	$\frac{7}{2}$	-19

Solution Sets for Inequalities.—A procedure similar to that for obtaining ordered number pairs in the solution set of an equation in two variables is useful for getting the same information for an inequality in two variables. The inequality is transformed so that one of the variables alone constitutes one member and the other member is an expression in terms of the other variable. This form permits discovery of possible values of the first variable when arbitrary values of the second are substituted for it.

Consider the inequality $2x - y > 5$. Subtracting $2x$ from each member gives $-y > -2x + 5$; dividing by -1 and reversing the direction of the inequality gives $y < 2x - 5$. If a value is now chosen for x, say $x = 1$, y can be any number less than $2(1) - 5$ that is less than -3, for instance, $-3\frac{1}{2}$. Thus the ordered pair $(1,-3\frac{1}{2})$ is in the solution set; so are $(1,-4)$, $(1,-10)$, and so on.

Simultaneous Equations.—Suppose two equations, for example, $x + y = 1$ and $2x + y = 9$, involve the same two variables, each variable having the same domain in both equations. Such equations are said to be *simultaneous* and the pair of equations is called a *system of equations*.

A certain set of ordered pairs will constitute the solution set of the first equation and another set of ordered pairs will constitute the solution set of the second equation. In the example considered here, the table

x	0	1	5	8
y	1	0	-4	-7

shows a small part of the solution set of $x + y = 1$, and the table

x	0	1	5	8
y	9	7	-1	-7

shows a small part of the solution set of $2x + y = 9$. It will be noted that the ordered pairs in the two solution sets differ in general, but that one pair, $(8,-7)$, is in both sets.

If the respective members of the two equations are added:

$$\begin{aligned} x + y &= 1 \\ 2x + y &= 9 \\ \hline 3x + 2y &= 10 \end{aligned}$$

a new equation is obtained. This new equation, also in two variables, has a solution set that consists of ordered pairs of numbers, and among them is the pair $(8,-7)$, which was in the solution sets of both of the two equations that were added. The reason for this will be apparent if each equation is thought of as being transformed so that its right member is zero. The first equation will then be in the form $A = 0$, where A stands for all the terms in the left member. The second equation will be in the form $B = 0$. If these are added, the sum will be in the form $A + B = 0$. Replacing x and y by any ordered pair from the solution set of the first equation will make A zero, and replacing x and y by any ordered number pair from the second equation will make B zero. The ordered number pair that makes both A and B zero will also make $A + B$ zero. Furthermore, any solution of the first equation that is not a solution of the second will make A zero but not B, and therefore will not make $A + B$ zero; that is, it will not be a solution to $A + B = 0$. Consequently, the only number pairs from the solution sets of two equations being added that will be in the solution sets of the resulting equation are those number pairs that are in the solution sets of both.

Adding two equations thus gives an equation that has their common solution in its own solution set. If, before adding two equations, the members of either or both are multiplied by nonzero constants, their common solutions will also be solutions of the resulting equation. For if the constants used to multiply each are m and n, respectively, then the resulting equation will take the form $mA + nB = 0$. Whenever A and B are zero, mA and nB also will be zero and so will their sum. Whenever either A or B is not zero, mA or nB will not be zero and neither will their sum.

■ **SOLVING BY ADDITION.**—Solving a pair of simultaneous equations means finding their common solutions, that is, the ordered number pairs that make both of them true.

The principle of the preceding section can be used to find the common solution of two simultaneous equations. This is done by adding them in such a way as to eliminate one of the variables.

Example:

To solve the system

$$\begin{aligned} 3x + y &= 5 \\ x - y &= 7 \\ \hline 4x \quad &= 12 \\ x \quad &= 3 \end{aligned}$$

substitute this value of x in either of the two original equations (in this case the second, $x - y$, is easier to use):

$$\begin{aligned} 3 - y &= 7 \\ -y &= 7 - 3 \\ y &= -4 \end{aligned}$$

The common solution is $x = 3$, $y = -4$, or the number pair $(3,-4)$.

To check this solution, the number pair must be substituted in *both* original equations, since the fact that it checks in one does not guarantee that it will also check in the other.

Check:

$$\begin{aligned} 3x + y &= 5 \\ 3(3) - 4 &\stackrel{?}{=} 5 \\ 9 - 4 &\stackrel{?}{=} 5 \\ 5 &= 5 \end{aligned} \qquad \begin{aligned} x - y &= 7 \\ 3 - (-4) &\stackrel{?}{=} 7 \\ 3 + 4 &\stackrel{?}{=} 7 \\ 7 &= 7 \end{aligned}$$

The method of solving simultaneous equations by addition depends on eliminating one of the variables through adding. Adding two equations will eliminate one variable when the coefficients of that variable are additive inverses of one another, that is, are equal in absolute value but opposite in sign, as in the above equations ($+y$ and $-y$).

Example:

In the system

$$\begin{aligned} 2a + b &= 3 \\ 2a - 3b &= -41 \end{aligned}$$

neither variable has coefficients that are additive inverses of each other in the two equations. However, this situation can be achieved by multiplying both members of the second equation by -1. The preceding section has shown that such a procedure will not affect the fact that the common solution will be a solution of the equation formed by adding.

Multiply the second equation by -1 and add:

$$\begin{aligned} 2a + b &= 3 \\ -2a + 3b &= 41 \\ \hline 4b &= 44 \end{aligned}$$

divide both members by 4:

$$b = 11;$$

and substitute 11 for b in the first equation and solve for a:

$$\begin{aligned} 2a + 11 &= 3 \\ 2a &= 3 - 11 \\ 2a &= -8 \\ a &= -4 \end{aligned}$$

$$a = -4 \qquad b = 11$$

Check:

$$\begin{aligned} 2a + b &= 3 & 2a - 3b &= -41 \\ 2(-4) + 11 &\stackrel{?}{=} 3 & 2(-4) - 3(11) &\stackrel{?}{=} -41 \\ -8 + 11 &\stackrel{?}{=} 3 & -8 - 33 &\stackrel{?}{=} -41 \\ 3 &= 3 & -41 &= -41 \end{aligned}$$

In the system

$$\begin{aligned} 3x + 4y &= 18 \\ 5x - y &= 7 \end{aligned}$$

adding will not eliminate either variable. However, if the second equation is first multiplied by 4, adding will eliminate y, since its coefficients will then be additive inverses of each other: $+4y$ and $-4y$.

Multiply both members of the second equation by 4 and add:

$$\begin{aligned} 3x + 4y &= 18 \\ 20x - 4y &= 28 \\ \hline 23x \qquad &= 46 \end{aligned}$$

divide both members by 23:

$$x = 2;$$

substitute 2 for x in the second equation and solve for y:

$$\begin{aligned} 5x - y &= 7 \\ 5(2) - y &= 7 \\ 10 - y &= 7 \\ -y &= 7 - 10 \\ -y &= -3 \\ y &= 3 \end{aligned}$$

$$x = 2 \qquad y = 3$$

Check:

$$\begin{aligned} 3x + 4y &= 18 & 5x - y &= 7 \\ 3(2) + 4(3) &\stackrel{?}{=} 18 & 5(2) - 3 &\stackrel{?}{=} 7 \\ 6 + 12 &\stackrel{?}{=} 18 & 10 - 3 &\stackrel{?}{=} 7 \\ 18 &= 18 & 7 &= 7 \end{aligned}$$

At times it is convenient to multiply both equations in a system by different nonzero constants in order to get the coefficient of one variable to be the additive inverse of the other.

Example:

Solve:

$$\begin{aligned} 2x - 3y &= 2 \\ 5x + 4y &= 51 \end{aligned}$$

This system will be solved by first multiplying the first equation by 4 and the second by 3 to make it possible to eliminate y by adding. The first equation might have been multiplied by 5 and the second by -2 to cause the elimination of x by adding. Multiply the first equation by 4, the second by 3, then add:

$$\begin{aligned} 8x - 12y &= 8 \\ 15x + 12y &= 153 \\ \hline 23x \qquad &= 161 \end{aligned}$$

divide both members by 23:

$$x = 7;$$

substitute 7 for x in the first equation and solve for y:

$$\begin{aligned} 2x - 3y &= 2 \\ 2(7) - 3y &= 2 \\ 14 - 3y &= 2 \\ -3y &= 2 - 14 \\ -3y &= -12 \\ y &= 4 \end{aligned}$$

$$x = 7 \qquad y = 4$$

Check:

$$\begin{aligned} 2x - 3y &= 2 & 5x + 4y &= 51 \\ 2(7) - 3(4) &\stackrel{?}{=} 2 & 5(7) + 4(4) &\stackrel{?}{=} 51 \\ 14 - 12 &\stackrel{?}{=} 2 & 35 + 16 &\stackrel{?}{=} 51 \\ 2 &= 2 & 51 &= 51 \end{aligned}$$

Before adding the equations in a system, the equations should first be transformed so that the terms containing the variables are on one side of the equation and the constant terms are on the other side.

Example:

To solve

$$\begin{aligned} 5x &= 5 - 2y \\ 3y &= 15 - 9x \end{aligned}$$

change the equations:

$$\begin{aligned} 5x + 2y &= 5 \\ 9x + 3y &= 15 \end{aligned}$$

multiply the first equation by -3, the second by 2, then add:

$$\begin{aligned} -15x - 6y &= -15 \\ 18x + 6y &= 30 \\ \hline 3x \qquad &= 15 \end{aligned}$$

divide both members by 3:

$$x = 5;$$

substitute 5 for x in the second equation and solve for y:

$$\begin{aligned} 3y &= 15 - 9x \\ 3y &= 15 - 9(5) \\ 3y &= 15 - 45 \\ 3y &= -30 \\ y &= -10 \end{aligned}$$

$$x = 5 \qquad y = -10$$

Check:

$$\begin{aligned} 5x &= 5 - 2y & 3y &= 15 - 9x \\ 5(5) &\stackrel{?}{=} 5 - 2(-10) & 3(-10) &\stackrel{?}{=} 15 - 9(5) \\ 25 &\stackrel{?}{=} 5 + 20 & -30 &\stackrel{?}{=} 15 - 45 \\ 25 &= 25 & -30 &= -30 \end{aligned}$$

■ **SOLVING BY SUBSTITUTION.**—In the system

$$\begin{aligned} x + 3y &= 9 \\ 4x + 5y &= 22 \end{aligned}$$

it is very easy to transform the first equation in order to have one variable—x in this case—expressed in terms of the other:

$$x = 9 - 3y.$$

If this expression for x is substituted for x in the other equation, the resulting equation will contain only the variable y and will have in its solution set the ordered number pairs common to the solution sets of both original equations. Substituting $9 - 3y$ for x in $4x + 5y = 22$ gives

$$4(9 - 3y) + 5y = 22.$$

To solve the problem remove parentheses and combine like terms:

$$\begin{aligned} 36 - 12y + 5y &= 22 \\ 36 - 7y &= 22 \end{aligned}$$

subtract 36 from both members:

$$\begin{aligned} -7y &= 22 - 36 \\ -7y &= -14 \end{aligned}$$

divide both members by -7:

$$y = 2;$$

and substitute 2 for y and solve for x as follows:

$$\begin{aligned} x &= 9 - 3y \\ x &= 9 - 3(2) \\ x &= 3 \end{aligned}$$

$$x = 3 \qquad y = 2$$

Check:

$$\begin{aligned} x + 3y &= 9 & 4x + 5y &= 22 \\ 3 + 3(2) &\stackrel{?}{=} 9 & 4(3) + 5(2) &\stackrel{?}{=} 22 \\ 3 + 6 &\stackrel{?}{=} 9 & 12 + 10 &\stackrel{?}{=} 22 \\ 9 &= 9 & 22 &= 22 \end{aligned}$$

This method of solving a system of simultaneous equations is known as the *method of substitution*, because one variable was eliminated by substituting for it an expression in terms of the other variable.

The method of substitution may be used on any pair of simultaneous first-degree equations, but it is particularly convenient to use when one of the equations is in a form that expresses one variable in terms of the other.

Example:

Thus, in the system

$$\begin{aligned} a - 14 &= 2b \\ b &= 4a \end{aligned}$$

the second equation gives b in terms of a. Substituting this expression for b in the first equation gives

$$\begin{aligned} a - 14 &= 2(4a) \\ a - 14 &= 8a \end{aligned}$$

To solve the problem, subtract a from each member and combine like terms:

$$\begin{aligned} -14 &= 8a - a \\ -14 &= 7a \end{aligned}$$

divide both members by 7:

$$-2 = a;$$

and substitute -2 for a and solve for b:

$$b = 4a$$
$$b = 4(-2)$$
$$b = -8$$
$$a = -2 \qquad b = -8$$

Check:

$$a - 14 = 2b \qquad b = 4a$$
$$-2 - 14 \stackrel{?}{=} 2(-8) \qquad -8 \stackrel{?}{=} 4(-2)$$
$$-2 - 14 \stackrel{?}{=} -16 \qquad -8 = -8$$
$$-16 = -16$$

If one of the variables appears with a coefficient of 1 in one of the equations in a system, this equation can be easily transformed so that it is solved for this variable in terms of the other. This, then, is the same situation as existed in the last illustration. Therefore, the method of substitution is convenient to use whenever one of the variables appears with a coefficient of 1 in one of the equations.

Example:

Solve:

$$6x - 4y = 2$$
$$y - 4x = -3$$

In the second equation, the coefficient of y is 1. Solving this equation for y gives

$$y = 4x - 3.$$

Substitute $4x - 3$ for y in the first equation:

$$6x - 4(4x - 3) = 2;$$

remove parentheses and combine like terms:

$$6x - 16x + 12 = 2$$
$$-10x + 12 = 2$$

subtract 12 from both members and combine like terms:

$$-10x = 2 - 12$$
$$-10x = -10$$

divide by -10:

$$x = 1;$$

and substitute 1 for x and solve for y:

$$y = 4x - 3$$
$$y = 4(1) - 3$$
$$y = 1$$
$$x = 1 \qquad y = 1$$

Check:

$$6x - 4y = 2 \qquad y - 4x = -3$$
$$6(1) - 4(1) \stackrel{?}{=} 2 \qquad 1 - 4(1) \stackrel{?}{=} -3$$
$$6 - 4 \stackrel{?}{=} 2 \qquad 1 - 4 \stackrel{?}{=} -3$$
$$2 = 2 \qquad -3 = -3$$

The next example illustrates an application of the method of substitution in a case in which no variable has a coefficient of 1. This will show that the method of substitution is applicable even in this situation, although elimination by addition is probably easier to use.

Example:

In the system

$$2x - 3y = 2$$
$$5x + 4y = 51$$

solve the first equation for x in terms of y:

$$2x = 2 + 3y$$
$$x = \frac{2 + 3y}{2}$$

substitute for x in the second equation:

$$5\left(\frac{2 + 3y}{2}\right) + 4y = 51;$$

remove parentheses:

$$\frac{10 + 15y}{2} + 4y = 51;$$

multiply by 2 and combine like terms:

$$10 + 15y + 8y = 102$$
$$10 + 23y = 102$$

subtract 10 from both members and combine like terms:

$$23y = 102 - 10$$
$$23y = 92$$

divide by 23:

$$y = 4;$$

and substitute 4 for y and solve for x:

$$x = \frac{2 + 3y}{2}$$
$$x = \frac{2 + 3(4)}{2}$$
$$x = \frac{2 + 12}{2}$$
$$x = 7$$
$$x = 7 \qquad y = 4$$

Check:

$$2x - 3y = 2 \qquad 5x + 4y = 51$$
$$2(7) - 3(4) \stackrel{?}{=} 2 \qquad 5(7) + 4(4) \stackrel{?}{=} 51$$
$$14 - 12 \stackrel{?}{=} 2 \qquad 35 + 16 \stackrel{?}{=} 51$$
$$2 = 2 \qquad 51 = 51$$

Simultaneous Equation Problems.— When problems involve two unknown quantities it is often convenient to represent each of them by a separate variable. To solve the problem, values of the two variables must then be found. If one equation is written to express a relationship between these two variables, it will have an infinite number of pairs in its solution set. If a second equation expressing another relationship between the same two variables can be written, it will also have an infinite number of pairs in its solution set. However, the common solution, or solutions, to the two equations will give answers for the variables that satisfy the relationships between them expressed in both equations. The number of such common solutions is generally small. In fact, all of the simultaneous equations (all of which were first degree) solved so far have had only one common solution. The question of the possible number of common solutions will be discussed in detail in the section on graphs.

Example:

A motorist paid \$3.23 for 10 gallons of gasoline and 1 quart of oil. At the same time, another motorist paid \$3.31 for 8 gallons of the same grade of gasoline and 3 quarts of the same type of oil. Find the price of 1 gallon of gasoline and 1 quart of oil.

Let x represent the price in cents of 1 gallon of gasoline and y represent the price in cents of 1 quart of oil.

Since the cost of an item is obtained by multiplying the price per unit by the number of units bought, the purchase of the first motorist gives the equation

$$10x + y = 323;$$

that of the second motorist:

$$8x + 3y = 331.$$

(Note that \$3.23 is 323 cents; expressing the equation in cents avoids decimals.)

It is convenient to solve the system of equations above by using the method of substitution, since the first equation can be solved for y in terms of x by transforming it into

$$y = 323 - 10x.$$

Substitute for y in the second equation:

$$8x + 3y = 331$$
$$8x + 3(323 - 10x) = 331$$

remove parentheses and combine like terms:

$$8x + 969 - 30x = 331$$
$$-22x + 969 = 331$$

subtract 969 from both members and combine like terms:

$$-22x = 331 - 969$$
$$-22x = -638$$

divide by -22:

$$x = 29;$$

and substitute 29 for x and solve for y:

$$y = 323 - 10x$$
$$y = 323 - 10(29)$$
$$y = 323 - 290$$
$$y = 33$$

$$x = 29 \qquad y = 33$$
Gas : 29¢ gal. Oil : 33¢ gal.

Check:

First motorist's purchase:

Gas : 10 gal. at 29¢ = \$2.90
Oil : 1 qt. at 33¢ = .33
\$3.23

Second motorist's purchase:

Gas : 8 gal. at 29¢ = \$2.32
Oil : 3 qt. at 33¢ = .99
\$3.31

The use of two variables and two simultaneous equations is particularly valuable in solving problems involving aircraft flying with or against the wind or ships sailing with or against the current.

Example:

A plane can travel 1,080 miles in 6 hours when flying with the wind but it takes 5 hours to go only 600 miles when flying against the wind. Find the speed of the plane in still air and the speed of the wind.

If we let x equal the speed of the plane

in miles per hour when flying in still air and y equal the speed of the wind in miles per hour, then the speed of the plane flying with the wind will be $x + y$, and speed of the plane flying against the wind will be $x - y$.

	Rate (miles per hour) ×	*Time (hours)* =	*Distance (miles)*
With wind	$x + y$	6	$6(x + y)$
Against wind	$x - y$	5	$5(x - y)$

The equation representing the flight with the wind is

$$6(x + y) = 1{,}080;$$

and the flight against the wind:

$$5(x - y) = 600.$$

Removing parentheses gives the system:

$$\begin{aligned} 6x + 6y &= 1{,}080 \\ 5x - 5y &= 600 \end{aligned}$$

It is convenient to solve this system by dividing both members of the first equation by 6 and both members of the second equation by 5 and then adding:

$$\begin{aligned} x + y &= 180 \\ x - y &= 120 \\ \hline 2x \quad &= 300 \end{aligned}$$

Then divide both members by 2:

$$x = 150;$$

substitute 150 for x and solve for y:

$$\begin{aligned} x + y &= 180 \\ 150 + y &= 180 \end{aligned}$$

subtract 150 from both members:

$$y = 30$$

$x = 150$ $\qquad$ $y = 30$

Plane : 150 mph $\qquad$ Wind : 30 mph

Check:

The plane's speed with the wind is 150 + 30, or 180 mph. In 6 hours it will go 6 × 180, or 1,080 miles. The plane's speed against the wind is 150 − 30, or 120 mph. In 5 hours it will go 5 × 120, or 600 miles.

Graphs.—It was pointed out previously that the real numbers may be represented on a number line. A fixed point on this line is chosen to represent zero, and the positive numbers are arranged to the right of the zero point at points appropriately spaced according to the scale used. The negative numbers are similarly arranged at points to the left.

■ **ONE VARIABLE.**—Consider the equation $x + 5 = 7$. It has one solution, 2, which may be pictured as a darkened point on the number line as in Figure 7.

Figure 7.

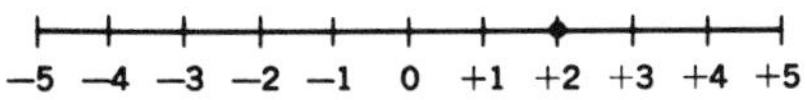

−5 −4 −3 −2 −1 0 +1 +2 +3 +4 +5

This illustration represents the solution set of $x + 5 = 7$ pictorially, and the darkened point on the number line is therefore called the *graph* of the solution set of $x + 5 = 7$.

For the inequality $x + 5 > 7$, if the domain of x is all the real numbers, the solution set consists of all the real numbers greater than 2. This solution set can be pictured on the number line as a darkened line beginning at 2 and extending indefinitely to the right, a circle around 2 being used to indicate that 2 itself is not included in the solution set (Fig. 8). Note, however, that

Figure 8.

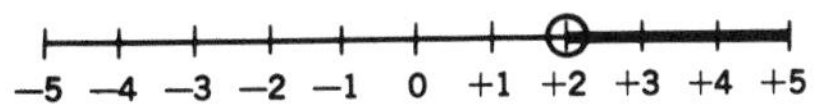

the solution set contains numbers on the darkened line that are very close to 2, for example, 2.00001. The darkened line without the point 2 is the graph of the solution set of $x + 5 > 7$.

The solution set of any sentence consists of all those numbers *in the domain of the variable* that make the sentence true. If the inequality $x + 5 > 7$, whose graph was considered in the previous example, is again examined, but if the domain of the variable x is now the set of integers instead of the set of all real numbers, a different graph will result. The solution set with the new domain will consist of integers only, and the graph will appear as a series of discrete points, as shown in Figure 9.

Figure 9.

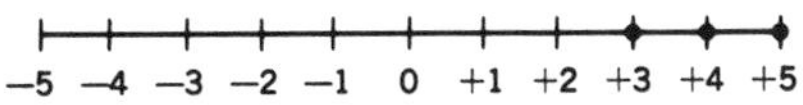

−5 −4 −3 −2 −1 0 +1 +2 +3 +4 +5

The graph of the open sentence $x + 5 \geqq 7$ is a combination of the graph of the equation $x + 5 = 7$ and the graph of the inequality $x + 5 > 7$. If the domain of x is the set of all real numbers, the graph of $x + 5 \geqq 7$ is a line (Fig. 10). There is no circle at 2, since 2 is a member of the solution set.

Figure 10.

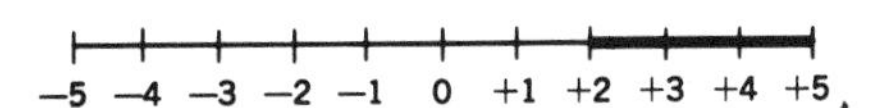

The quadratic equation $x^2 - 2x - 3 = 0$ has two roots, 3 and −1, which are obtained after factoring it into the form $(x - 3)(x + 1) = 0$ and setting each factor equal to 0. The graph of the solution set of this equation appears as two points on the number line (Fig. 11).

Figure 11.

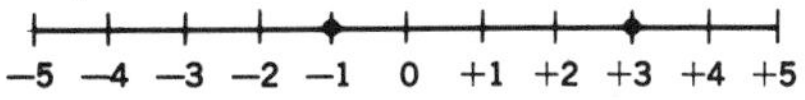

−5 −4 −3 −2 −1 0 +1 +2 +3 +4 +5

■ **TWO VARIABLES.**—The solution sets of sentences in two variables consist of ordered number pairs. These cannot be pictured on the number line, since the number line can show only numbers, and not number pairs.

In order to picture, or graph, ordered number pairs, they are represented by points on a plane, such as the surface of a page of this book, instead of by points along a line. Each point of a plane representing a number pair is located with reference to two axes perpendicular to each other, as in Figure 13. The horizontal axis—the x axis—is a number line used to locate the first number in a number pair. The vertical axis—the y axis—is another number line that is used to locate the second number in the number pair. The two axes, or number lines, intersect at their zero points, and this point of intersection, therefore, represents the point (0,0), which is called the *origin*. To locate a point representing some number pair, such as (3,2), count three units along the x axis to the right of the origin (this is the positive direction for x), and then up two units in a direction parallel to the y axis (the positive direction for y). In Figure 12, P is the point (3,2).

Negative values of x are counted off along the x axis to the left of the origin and negative values of y are counted off in the descending direction parallel to the y axis. Q in Figure 12 is the point (−1,4), while R is the point (−3,−5), and S is the point $(2, -1\frac{1}{2})$.

The procedure for locating a point that represents an ordered number pair is similar to the method of locating a street corner by giving the number of the avenue and the number of the street which intersect there. It may also be compared to the method a navigator uses for locating a position by giving the longitude and latitude of that position.

Every ordered number pair represents one, and only one, point in the plane, and every point in the plane has one, and only one, ordered number pair corresponding to it. The pair of numbers that corresponds to a point is known as the *coordinates* of the point. Thus, (3,2) are the coordinates of P in Figure 12. The x coordinate is called the *abscissa* of the point; the y coordinate is its *ordinate*. Hence, the abscissa of P is 3 and the ordinate is 2.

The method of setting up a correspondence between ordered number pairs and the points in a plane which is explained here is known as the *rectangular coordinate system* (because the axes are at right angles to each other). It is also called the *Cartesian coordinate system* in honor of the French mathematician René Descartes (1596–1650), who invented it. Other coordinate systems are used in more advanced mathematics.

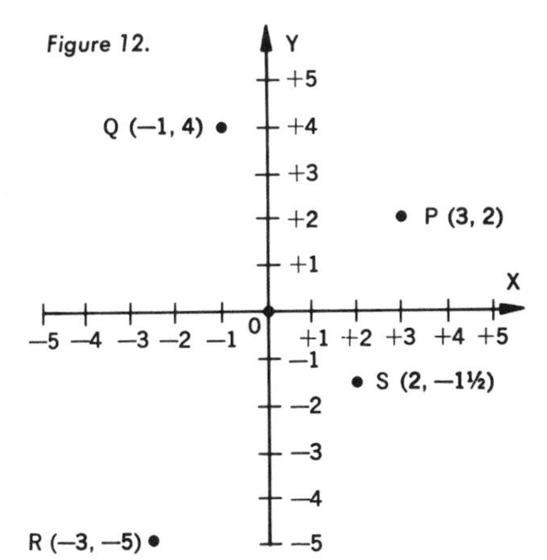

Figure 13.

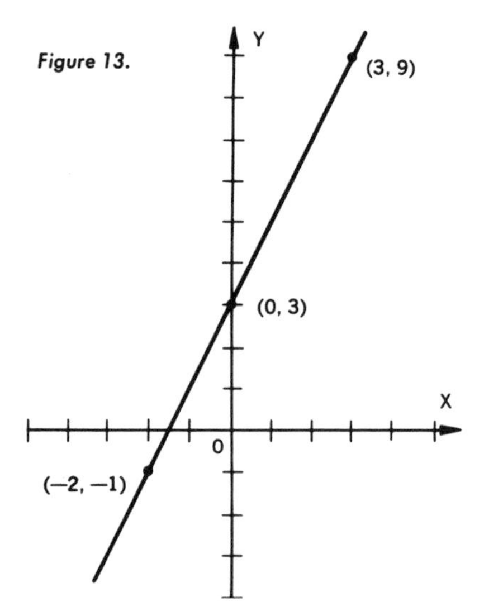

Graphing First-Degree Equations.—The solution set of the equation $y = 2x + 3$ is a set of ordered pairs. Some of them are shown in the following table:

x	-2	0	3
y	-1	3	9

If the points corresponding to these number pairs are located on a rectangular coordinate system they will lie in a straight line (Fig. 13).

In advanced mathematics it is proved that the points representing the solution set of any first-degree equation in two variables must lie in a straight line. Furthermore, all the points along the line will represent the ordered number pairs in the solution set. The graph of the solution set of $y = 2x + 3$ is the line (Fig. 13) that passes through the three points whose coordinates are shown in the table. The graph includes all the points on the line—each representing an ordered pair in the solution set. The graph is usually referred to as the graph of the equation $y = 2x + 3$. Notice that if any point whatsoever is now taken on the line, for example, $(\frac{1}{2},4)$, its coordinates will make the open sentence true: $4 = 2(\frac{1}{2}) + 3$.

To draw the graph of a first-degree equation, it is customary to plot at least three points on the Cartesian coordinate system; two points will enable the line to be drawn, and the third point serves as a check on the accuracy of the first two. Accuracy in drawing the graph is improved if the points are not taken too close together. The coordinates of the three needed points are most easily obtained by solving the equation for one letter in terms of the other and then choosing arbitrary values for that other letter. (This was done in the section covering sentences in two variables.)

Example:

For the graph of $4y - 3x + 4 = 0$, solve for y in terms of x:

$$4y = 3x - 4$$
$$y = \tfrac{3}{4}x - 1$$

Then let x take the arbitrary values -4,

Figure 14.

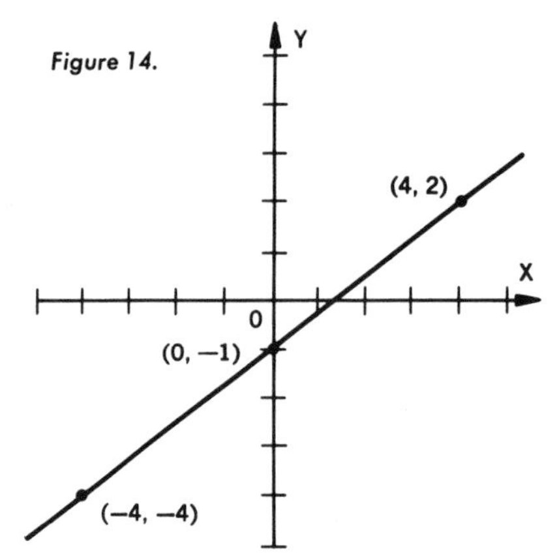

0, and 4. The corresponding values for y are:

x	-4	0	4
y	-4	-1	2

Plotting these points and drawing the straight line determined by them gives the graph of this equation (Fig. 14).

A first-degree equation in one variable, such as $y = 3$, can be regarded as a first-degree equation in two variables by considering the missing variable to have a coefficient of 0. The equation $y = 3$ could be thought of as $y + 0x = 3$. The solution set of $y = 3$ will then consist of ordered number pairs (the second number will always be 3) and the graph of $y = 3$ can be represented on the Cartesian coordinate system. Some of the ordered pairs in the solution set of $y = 3$ are:

x	-2	0	4
y	3	3	3

What this really means is that $y = 3$ no matter what value x assumes; hence the graph is a line parallel to the x axis and three units above it, as in Figure 15.

A first-degree equation that contains only the variable x, such as $x = -2$, may be regarded as the first-degree equation $x + 0y = -2$. Since in this form it has two variables, its solution set will consist of ordered number pairs, some of which are shown in the following table:

x	-2	-2	-2
y	3	0	-2

Here $x = -2$ no matter what value y assumes, hence the graph of $x = -2$ is a line parallel to the y axis and two units to the left of it, as shown in Figure 16.

Inequalities in Two Variables.—To draw the graph of the inequality $y > 2x + 3$, note that this open sentence requires the y value of each ordered pair in its solution set to be larger than the corresponding y in the equation $y = 2x + 3$. The graph of the inequality $y > 2x + 3$, therefore, consists of all points in the plane above the line that represents the graph of $y = 2x + 3$. These points are shown in the shaded half-plane above the line in Figure 17. The shaded area is referred to as a half-plane since it extends indefinitely upward and to the left.

Figure 15.

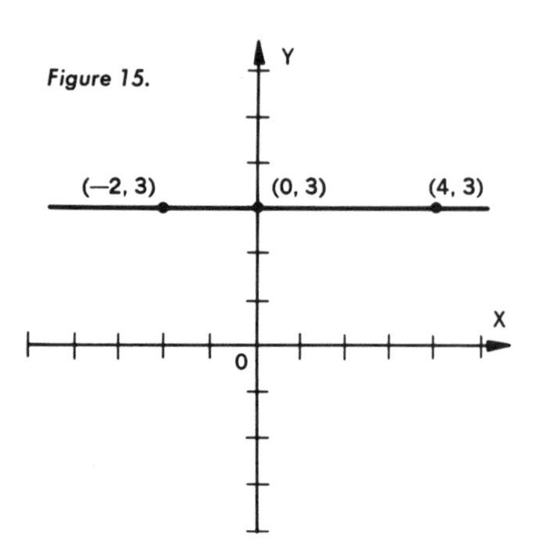

Figure 16.

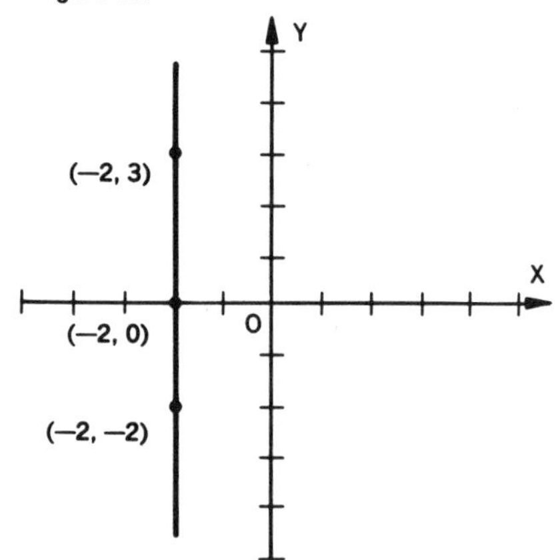

Figure 17.

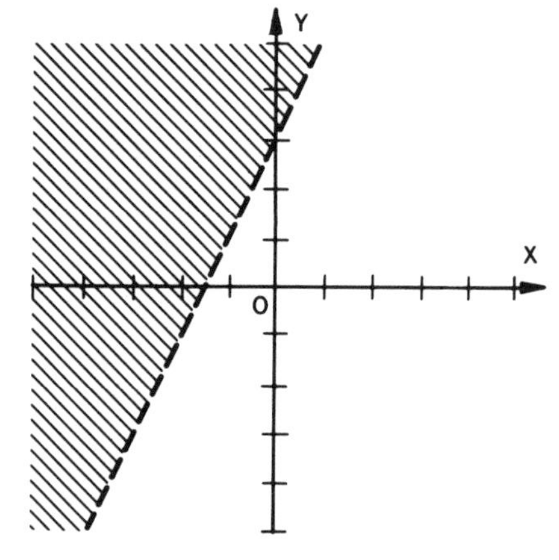

Figure 18.

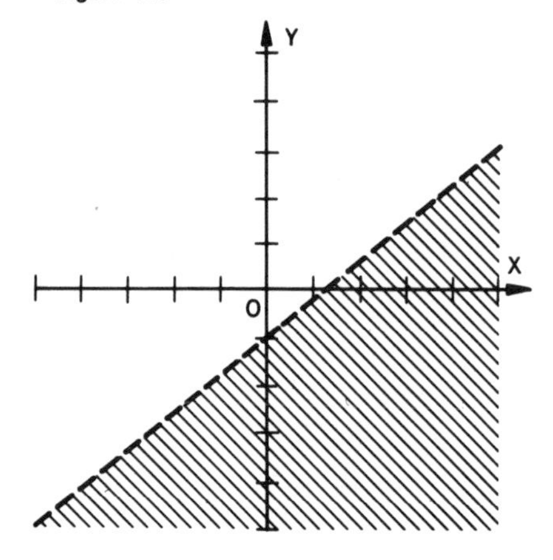

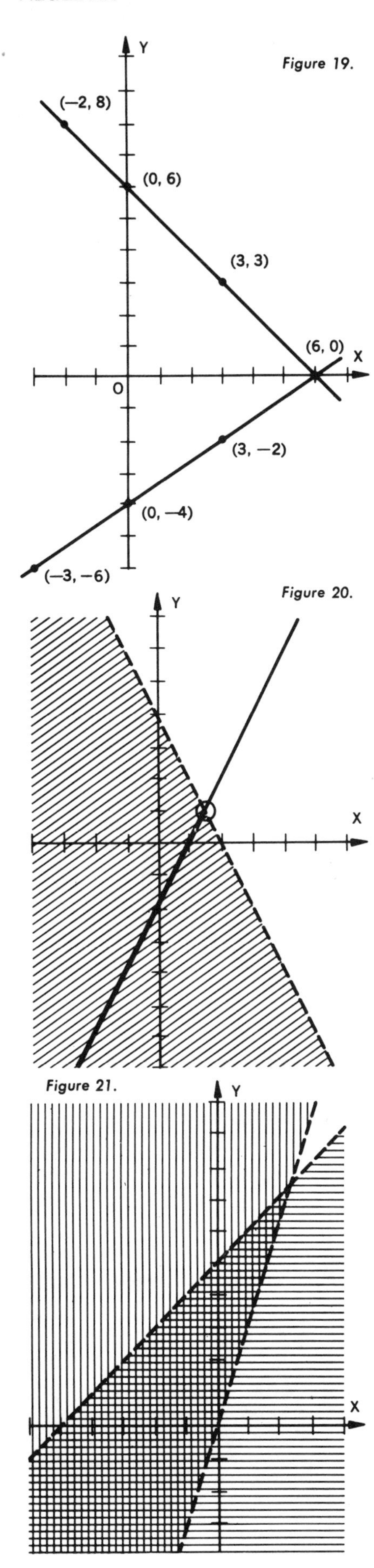

Figure 19.

Figure 20.

Figure 21.

However, the line representing $y = 2x + 3$ is not included in the graph of $y > 2x + 3$. If the open sentence had been $y \geqq 2x + 3$, the solution set would consist of the shaded portion including the line that forms its lower boundary.

To obtain the graph of the solution set for the inequality $y < \frac{3}{4}x - 1$, draw the graph of $y = \frac{3}{4}x - 1$. This graph has already been worked out in the preceding section. The graph of $y < \frac{3}{4}x - 1$ consists of all points in the plane below the line representing $y = \frac{3}{4}x - 1$ and not including it. This is the shaded portion in Figure 18.

Graphing Two Variables.—The graph of an equation in two variables shows the points whose coordinates are the solutions to the equation. If the graphs of two such equations are drawn on the same axes, then their common solution can be obtained by reading the coordinates of any points where the graphs intersect. This is illustrated on two first-degree equations in two variables.

Example:

To solve graphically the system

$$3y = 2x - 12$$
$$x + y = 6$$

solve each equation for y:

$$y = \frac{2x - 12}{3}$$
$$y = 6 - x$$

For $y = \dfrac{2x - 12}{3}$:

x	−3	0	3	6
y	−6	−4	−2	0

For $y = 6 - x$:

x	−2	0	3	6
y	8	6	3	0

Both graphs are plotted on the same axes in Figure 19. The common solution is read off the graph as the point (6,0) where the two graphs intersect. Thus the common solution is $x = 6$ and $y = 0$.

Check:

$$3y = 2x - 12 \qquad x + y = 6$$
$$3(0) \stackrel{?}{=} 2(6) - 12 \qquad 6 + 0 \stackrel{?}{=} 6$$
$$0 \stackrel{?}{=} 12 - 12 \qquad 6 = 6$$
$$0 = 0$$

The last example helps to explain why there is generally one common solution to two simultaneous first-degree equations. A first-degree equation has a graph that is a straight line. If two straight lines intersect at all, they intersect in exactly one point. Of course, two straight lines may not intersect. If a graphic solution of $y = 2x + 3$ and $y = 2x - 2$ is attempted, it will be discovered that they are parallel lines and their common solution is, therefore, the empty set. The reader should attempt to solve this system by one of the two algebraic methods developed earlier to see that these procedures will not give a common solution either.

In addition to the possibility that the graphs of two equations may not meet at all, there is also the possibility that two graphs may coincide, so that all the solutions of one are solutions of the other. The equations $y = 3x - 1$ and $2y = 6x - 2$ have the same graph and all the number pairs that make one true are also solutions to the other. The reader should notice that one of the postulates concerning transforming equations indicates that these two equations must be equivalent and, therefore, must have the same solutions.

In the discussion so far, only first-degree equations have been considered, and first-degree equations have graphs that are straight lines. Equations of a higher degree than first have graphs that are curved lines. The graphs of such equations may intersect at more than one point even when the graphs do not coincide. For example, two circles can intersect at two points. As a result, there is usually more than one common solution for a system of equations if at least one of them is of a higher degree than the first.

Graphing Equalities and Inequalities.—A system consisting of an inequality and an equality may also be solved graphically.

Example:

Solve graphically the system

$$y < 4 - 2x$$
$$y = 2x - 2$$

The graph of $y < 4 - 2x$ consists of all points below the line representing $y = 4 - 2x$ (the shaded part in Figure 20). The line representing $y = 2x - 2$ lies partly within this shaded portion and partly outside of it. The darkened lower part of the line, which is within the shaded portion, represents the common solution set of the inequality and the equality. Note that the encircled point $(1\frac{1}{2},1)$ is not one of the number pairs in the common solution, because the shaded solution set does not include the dotted line that forms its boundary.

The graphic method may be used to get the solution set of a system of inequalities.

Example:

Show graphically the solution set of

$$y > 3x$$
$$y < x + 5$$

The graph of the first sentence consists of all points above the line $y = 3x$, as shown by vertical cross-hatching in Figure 21. The graph of the second sentence consists of all points below the line $y = x + 5$, shown by horizontal cross-hatching on the graph. The common solutions to the two inequalities are represented by all points within the area covered by both horizontal and vertical cross-hatching.

—Lester W. Schlumpf

GEOMETRY

EGYPTIAN STATE TOURIST ADMINISTRATION

STONE ARCHES at Egypt's Temple of Deir el-Bahri reveal an early knowledge of geometry.

SECTION ONE

Practical Geometry.—Geometry has its origins in antiquity. Archeological discoveries show that ancient peoples were very much aware of geometric shapes and relations. The designs on early Egyptian pottery, 4000–3500 B.C., use groups of parallel lines, and ancient pottery from all parts of the world uses concentric circles, squares, rectangles, and triangles.

Daily problems of farmers and artisans of long ago motivated the search for methods to measure lengths, areas, and volumes. Astronomers learned how to measure the size of an angle. Both ancient Babylonians and Egyptians used a knowledge of practical geometry in their irrigation and construction projects.

Thus, it must have been recognized early in man's development that the shortest distance between two points is along a straight line between those points.

Early use of the plumb line indicates a knowledge of some of the properties of right angles and perpendiculars to planes. Similarly, the use of circular shapes such as wheels in ancient Babylonia clearly indicates a knowledge of some of the properties of circles.

A famous example of a geometric proposition was used by the rope-stretchers (*harpedonaptae*) of ancient Egypt. They made a triangle of rope with sides such that their lengths had the ratios of 3:4:5. This gave them a right angle between the two shorter sides; and they used this device in squaring corners, much as a carpenter's square is used today. When rivers such as the Nile overflowed and obliterated boundary markers, men were forced to recognize those geometric principles that led to the art of surveying.

A growing body of evidence supports the fact that practical geometry also was known in ancient China and India. In all areas of the world, when men learned how to produce their food rather than gather it, when they learned to make houses rather than seek shelter in caves, it was natural that knowledge of practical geometry should ensue. It may safely be asserted that wherever pottery was made, farms cultivated, or some form of building constructed, there was also a knowledge of practical geometry.

With the advent of writing, such knowledge was inscribed in manuals. The first known record was an Egyptian papyrus, called the *Rhind papyrus*. This was written by Ahmes, an Egyptian priest, sometime between 2000 and 1700 B.C. It was described by Samuel Birch (1813–1885) in 1868, and translated by August Eisenlohr (1832–1902) in 1877. Ahmes wrote of the mathematical knowledge of his time, but did not concern himself with mathematical theory. His handbook tells how to solve particular problems and how to calculate areas and volumes. It is interesting to note that in his calculations for areas of triangles, instead of an altitude Ahmes used a side of a triangle.

In the Rhind papyrus and in other documents found in Babylonia, China, and India, there are passages that tell how to do things, but not why the instructions produce correct results. They tell, for instance, how to draw a square, or say that base angles of an isosceles

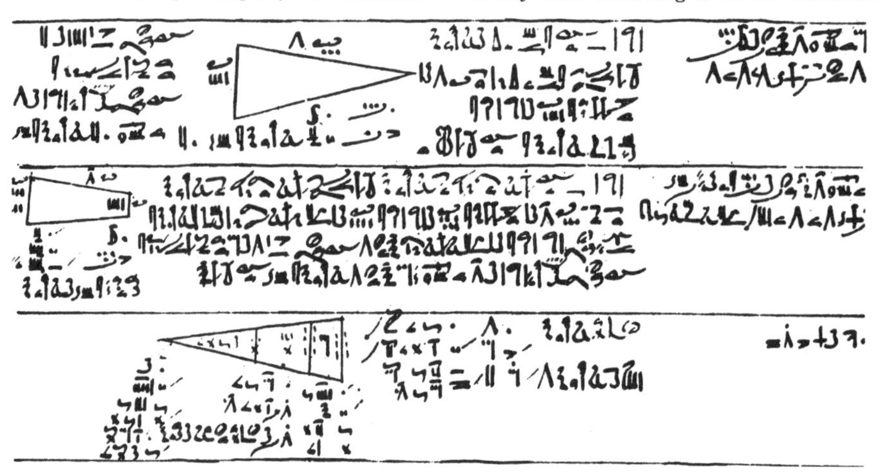

PREBIBLICAL MATHEMATICS written on Rhind papyrus indicate that the early Near Eastern civilizations were familiar with fractions. Translated, the document reads: "What equal areas should be taken from 5 fields if the sum of these areas is to be 3 setat?"

ENGELHARD, MONKMEYER

GREEK ARCHITECTURE, with its logic and balance, represents, in stone, the characteristics of early Greek philosopher-mathematicians.

triangle have equal measures, but no effort is made to explain why the figure will be a square, or why in a triangle the equality of the measures of the two base angles follows logically from the equality of the lengths of the two sides opposite these angles. There were no known attempts to demonstrate logical geometric relations before 600 B.C.

The geometry that is taught today in the elementary grades is *practical geometry*, sometimes called *informal geometry*. Students learn it by drawing figures with rulers, and by measuring with rulers and protractors. The facts they learn usually are not logically related to each other. Generally in the tenth grade, these logical relations are studied as *theoretical geometry*, sometimes called *formal geometry*.

SECTION TWO

Theoretical Geometry.—As far as is known, Thales (c. 640–546 B.C.), a Greek philosopher of Miletus, was the first to use logic in geometry. His travels took him to Babylon and Egypt, and there he increased his knowledge of practical geometry. There are no primary sources of the development of early Greek mathematics, only references in documents dating from the fourth century B.C. But it is safe to say that Thales and his successors were not content to know that a proposition about geometric figures was true; they wanted to know why it was true.

The essential characteristic of this school of thinking is the use of logic, that is, the demonstration that one statement follows inevitably from another statement or group of other statements. The impulse to demonstrate such a relation is not motivated by utility, but by the desire to understand. It is the difference between knowing how to do something (the *know-how*), and knowing why it succeeds (the *know-why*).

From 600 B.C. to 300 B.C., there was a veritable explosion of geometric discovery that can be accounted for by the *method of logic*, or, as it is sometimes called, the *method of deduction*. This method produced understanding and insight, which in turn led to new discoveries. Thales found a deductive proof for about six geometric facts, and his success inspired others to follow his example. A partial list of famous Greek mathematicians who followed Thales includes Pythagoras (c. 582–507 B.C.), Anaxagoras (c. 500–428 B.C.), Hippocrates (c. 460–c. 377 B.C.), Democritus (c. 460–357 B.C.), and Eudoxus of Cnidus (fl. 300 B.C.).

The deductive method is vital in the study of formal geometry as well as in all mathematics and theoretical science. It is illustrated by the following example.

Suppose it is known from practical geometry that the sum of the measures of the three angles of any triangle is 180°. It would then follow that in a right triangle there are two angles whose measures have the sum of 90°. The proof might be:

Let the measures of the angles of the right triangle be A, B, and C, with $C = 90°$. It then follows that:

$$(1)\quad A + B + C = 180°$$
$$(2)\quad C = 90°$$
$$(3)\quad A + B = 90°$$

The first statement is assumed to be true through a knowledge of practical geometry. The second statement is true because it is characteristic of all right triangles. The third statement, which is the desired conclusion, follows from a knowledge of *number facts*. (This number fact will be discussed later along with other number principles.)

About 300 B.C., Euclid (fl. 300 B.C.) made a compilation of the geometry that was known at the time. His most famous work is the thirteen-book *Stoicheia* ("Elements"). The organization of Euclid's geometry follows the form that is known today as a *postulational system*. Euclid selected a few geometric facts as a basis and demonstrated how all the others could be deduced from them as logical consequences flowing from these few.

The impact of the *Elements* on the development of mathematics, science, and philosophy cannot be overrated. It demonstrated how a large body of knowledge could be organized in such a manner that it is a coherent structure rather than a collection of unrelated facts. As such, it serves as a model for scientists even to this day.

Mathematicians scrutinized the *Elements* for imperfections, and these finally were corrected by David Hilbert (1862–1943). The *Elements* also stimulated the discovery of new geometric facts. Mathematicians devised alternate postulational systems that began with different assumptions. This was particularly true of Nikolai Lobachevsky (1793–1856), János Bolyai (1802–1860), Karl Gauss (1777–1855), and G. F. B. Riemann (1826–1866).

The origin and development of theoretical geometry is rooted in the fact that Thales' logical demonstrations worked with relations or properties of geometric figures rather than with the figures themselves. In the example of a logical proof concerning a right triangle, such a triangle was not drawn, nor were the sizes of its angles examined. Three statements or propositions were employed: (1) about the sum of the measures of the three angles of any triangle; (2) about a characteristic property of all right triangles; (3) about a property of numbers.

Consider some properties of such a simple object as a line. Perhaps it is a taut string or a line drawn with ruler and pencil. The four properties of a physical line are: (1) It can be seen. (2) It has a beginning and an end. (3) It consists of molecules that have spaces between them; therefore it is discontinuous. (4) It has a measurable width, even if it is necessary to use a microscope to determine this.

On the other hand, mathematicians think of a line in geometry as having the following properties: (1) It cannot be seen. (2) It has neither beginning nor end. (3) It is continuous, made up of points that have no space between them. (4) It has no width.

In other words, the mathematician's line is an abstraction. It is not the same as the carpenter's line or the draftsman's line. Similarly, mathematical points and planes are abstractions, not to be confused with physical dots and physical flat surfaces. One is prompted to ask why the mathematician has conceived of a line that differs from a physical line.

THE ELEMENTS
OF GEOMETRIE
of the moſt auncient Philoſopher
EVCLIDE
of Megara.

Faithfully (now firſt) tranſlated into the Engliſhe toung, by H. Billingſley, *Citizen of London. Whereunto are annexed certaine Scholies, Annotations, and Inuentions, of the beſt Mathematicians, both of time paſt, and in this our age.*

With a very fruitfull Præface made by M. I. Dee, *ſpecifying the chiefe Mathematicall Sciēces, what they are, and wherunto commodious: where, alſo, are diſcloſed certaine new Secrets Mathematicall and Mechanicall, vntill theſe our daies, greatly miſſed.*

NEW YORK PUBLIC LIBRARY

THE ELEMENTS, a book written by Euclid about 300 B.C., has influenced all mathematics.

The answer lies in the simplicity that ensues when he constructs logical proofs. One might then ask whether the logical consequences bear any resemblance to the facts that are true of physical lines. This is answered only by finding the logical consequences and then noting how well they conform to reality. The success of postulational thinking lies in such continual verifications. Then one might ask why geometry books have diagrams, since these are physical objects. The diagrams serve to suggest (not to prove) relations among geometric abstractions.

We might say that practical geometry studies physical geometric objects and results in a list of assorted facts. On the other hand, theoretical geometries study geometric abstractions and result in a list of propositions that are organized as a *postulational system*.

There are four steps in constructing a postulational system. The first is to list the concepts or terms that are to be *primitive*, or undefined. It is impossible to define all terms of the system because any definition of a term contains other terms. In the case of the first definition, these other terms thus are undefined.

This procedure is not the method of a dictionary, which attempts to define all terms and therefore must indulge in some circular definitions.

There are four initial, undefined terms: "set," "point," "line," and "plane." Also accepted as undefined are such common words of daily discourse as "and," "or," "of," and "all." The first definition is that of *space*, which is defined as the set of all points. Other definitions are postponed until the "betweenness" relation has been established. One might ask how objects with no definitions can be discussed. This is partly remedied in the next step.

The second step in constructing a postulational system is to list all the statements about the undefined terms for which no proof is offered. This helps to identify what the undefined terms represent. These statements are called *axioms*, *postulates*, or *assumptions*. Euclid seemed to reserve the term "postulate" for those statements that apply to geometric objects, and the term "axiom" for all other statements. Many contemporary mathematicians call all assumptions "postulates." We cannot tell whether postulates are true until a meaning is given to each of their undefined terms. When a meaning is assigned to each undefined term and each postulate then becomes true, a model of the postulational system is formed. In that case, the postulates are *consistent*. A postulational system may have more than one model if more than one meaning can be given to undefined terms. Mathematicians find it convenient to start with a few postulates, but this is not a logical necessity.

The third step is to decide on the rules of logic by means of which new statements are deduced from the postulates. Generally the classical Aristotelian logic, in which a statement is either true or false, is used.

The fourth step is to use the postulates and the rules of logic to deduce new statements. These are called *theorems*. A *corollary* is a theorem that is an immediate logical consequence of another theorem. A *lemma* is a theorem whose only importance is to prepare the way for another theorem that is considered significant in the system. This fourth step occupies practically all of the student's effort and time.

The choice of postulates in a system is a mathematician's prerogative. This became clear in the early part of the nineteenth century, when non-Euclidean geometries were created in which at least one of the postulates differed from one in Euclid's set. Since then, a variety of geometries have been created. A brief account of some of these geometries appears later.

A postulational system is therefore a system of propositions based on a set of undefined terms; definitions using these terms or already defined terms; a set of postulates; a set of rules of logic; and, using these as a basis, the theorems that can be deduced from them.

SECTION THREE

Principles of Logic.—The following are principles of logic often used in geometry.

1. A *proposition* is a statement capable of being true or false.

Notation. Propositions are denoted by p, q, r, etc.

Example. p: Some living trees grow leaves. q: The earth is flat.

2. A *conjunction* of two or more propositions uses the connective "and." It is also a proposition. A conjunction is true if, and only if, each of its component propositions is true.

Example. In the above example, "p and q" is false because "q" is false.

3. A *disjunction* of two propositions uses the connective "or." It is also a proposition. A disjunction is true if either or both parts are true.

Example. In the example above, "p or q" is true because "p" is true.

4. The *negation* of "p" is "p is false," or "not p."

Example. The negation of "it is raining" is the statement "it is not raining."

5. An *implication* has the form "p implies q," or "if p then q." The proposition following "if" is called the *antecedent*, or *hypothesis*. The proposition following "then" is called the *consequent*, or *conclusion*. An implication is considered true if the antecedent is false or the consequent is true.

6. The implication "if p then q" may be expressed, "the fact that p is true is a *sufficient condition* that q is true" or "the fact that q is true is a *necessary condition* that p is true."

7. The proposition "p is a *necessary* and *sufficient* condition for q" means, "if p then q and if q then p." In this case, p and q are *equivalent conditions* and must agree in their true values. This may also be stated, "q is true if, and only if, p is true."

8. The *converse* of "if p then q" is "if q then p." The converse of a true implication need not be true.

9. The *inverse* of "if p then q" is "if not p then not q." The inverse of a true implication need not be true.

10. The *contrapositive* of "if p then q" is "if not q then not p." The contrapositive of a true implication is necessarily true.

Comment. The converse of the inverse of an implication is the contrapositive of the implication.

11. "If p implies q and q implies r, then p implies r" is called the *law of syllogism*. Most proofs in geometry follow this pattern. They consist in showing that a implies b, that b implies c . . . , and finally that r implies s. The conclusion is that a implies s. It is sometimes known as the *direct method*.

12. The truth of "if p then q" does not guarantee that q is true. It also must be known that p is true. The principle "if p is true and if p implies q, then q is true" is called the *rule of detachment*, or *modus ponens*.

13. In the *indirect method* used to prove "if p then q," "p and not q" is assumed, and by reasoning with these propositions, postulates, and theorems, it is deduced either that p is false or that some postulate or theorem already accepted as true is false.

14. A *definition* has the "if, and only if" meaning.

Example. The definition "a right triangle is a triangle one of whose angles is a right angle" yields the following two propositions: (1) If a triangle is a right triangle, one of its angles is a right angle. (2) If one of the angles of a triangle is a right angle, it is a right triangle. The definition may also be stated: A triangle is a right triangle if, and only if, one of its angles is a right angle.

The purpose of a definition is to replace a larger set of words with a smaller set.

SECTION FOUR

Basic Number Principles.—Properties of points and point sets, such as lines and planes, can be examined by relating them to properties of numbers. Following are the basic principles and properties of real numbers that are used in elementary algebra.

Definition of Equality.—In the equality $a = b$, the symbol $=$ means that a and b are names of the same object. In this article the object may be a number, a point, or a set of points.

Properties of Number Equalities.—In the following sentences, a, b, and c are real numbers.

Substitution Principle. In any equal-

BETTMANN ARCHIVE

ARISTOTLE, Greek philosopher and logician, taught Alexander the Great of Macedon.

ity, one name of an object may be replaced by another.

Reflexive Property. For any a, $a = a$.

Symmetric Property. If $a = b$, then $b = a$.

Transitive Property. If $a = b$ and $b = c$, then $a = c$.

Addition Property. If $a = b$, then $a + c = b + c$.

Multiplication Property. If $a = b$, then $ac = bc$. Also, if $ac = bc$ and $c \neq 0$, then $a = b$.

Properties of Order (Inequalities).—If a and b are distinct numbers, then either $a > b$ (read a is greater than b) or $b > a$, but not both.

Transitive Property. If $a > b$ and $b > c$, then $a > c$.

Addition Property. If $a > b$, then $a + c > b + c$.

Multiplication Properties. If $c > 0$ and $a > b$, then $ac > bc$. If $0 > c$ and $a > b$, then $bc > ac$.

The statement $a > b$ may also be written $b < a$ (b is less than a). The order properties may be rephrased by using $<$ in place of $>$, except for the multiplication properties.

The statement $a < b$ and $b < c$ may be abbreviated as $a < b < c$. A number x that is between a and b may be referred to by writing either $a < x < b$, or $b < x < a$, depending on the respective values of a and b.

It is now possible to develop a postulational system that is a revision of Euclid's geometry. The choice of postulates is guided by the contributions of several groups who have made valuable suggestions concerning the revision of school geometry, particularly the *School Mathematics Study Group.*

SECTION FIVE

INCIDENCE RELATIONS

Postulate 1.—Space contains at least two *distinct points.* Points generally are identified by capital letters.

Postulate 2.—Every line is a set of points and contains at least two points.

Postulate 3.—If P and Q are two distinct points, there is exactly one line that contains them.

Notation. The symbol $\overleftrightarrow{PQ}$ is used to denote the *unique line* that contains P and Q. P and Q *determine* one *line.* Lower-case letters are used to name a line.

Postulate 4.—No line contains all points of space.

■ THEOREM 5-1.–Space contains at least three points that are not in one line.

Proof. Postulate 1 states that space contains at least two points that may be called A and B, respectively. By Postulate 3, they determine $\overleftrightarrow{AB}$. Postulate 4 supplies a third point not in $\overleftrightarrow{AB}$, Q.E.D. (*quod erat demonstrandum,* meaning "which was to be proved").

■ THEOREM 5-2.–Two distinct lines cannot intersect in more than one point.

Proof. Let the two lines be r and s. If r and s intersect in two distinct points, A and B, then by Postulate 3, $AB = s$ and $AB = r$. Therefore, $r = s$. But this contradicts the information that r and s are distinct. Thus it must be concluded that r and s cannot intersect in two points, Q.E.D.

Definition. If three or more points are contained in one line, they are *collinear* (lie on or pass through the line). If three points are not in one line, they are *noncollinear.*

Postulate 5.—Every plane is a set of points and contains at least three noncollinear points.

Postulate 6.—If P, Q, and R are three distinct noncollinear points, then there is exactly one plane that contains them.

Notation. A plane is named by listing three noncollinear points of the plane, for example, plane PQR.

■ THEOREM 5-3.–Space contains at least one plane.

Proof. Theorem 5-1 supplies three noncollinear points, and Postulate 6 states that there is one plane that contains them, Q.E.D.

Comment. Some aspects of geometry are confined to one plane. When limited to one plane, the subject is called *plane geometry.*

Postulate 7.—No plane contains all points of space.

Notation. A *set* is denoted by $\{\ \}$. For instance, the set consisting of the pair A,B is written $\{A,B\}$.

■ THEOREM 5-4.–Space contains at least two distinct planes.

Proof. Theorem 5-3 supplies one plane, which contains three noncollinear points, A, B, and C. Postulate 7 supplies a fourth point, D, not in plane ABC. The sets $\{D,A,B\}$ and $\{D,A,C\}$ cannot both contain collinear points since, if they did, $\overleftrightarrow{DA}$ would contain B and C and therefore A, B, C would be collinear. But this would contradict the given information that they are not collinear. Therefore, either $\{D,A,B\}$ or $\{D,A,C\}$ determines a plane different from plane ABC, which does not contain D. Finally, this plane is a set of points, and, by Postulate 5, it is in space, Q.E.D.

Postulate 8.—If two distinct points of a line belong to a plane, then every point of the line belongs to that plane.

Postulate 9.—If two distinct planes intersect, then their intersection is a line.

Comment. Postulates 8 and 9 express for theoretical geometry what is felt intuitively for physical flat surfaces. The study of properties of point sets in more than one plane is called *solid geometry.*

NEW YORK PUBLIC LIBRARY

EIGHTEENTH-CENTURY FRENCH TEXTBOOK concentrated entirely on the geometry of Euclid.

SECTION SIX

Distances and Coordinates.—Practical geometry uses a ruler to measure how far one dot is from another. Theoretical geometry uses an "abstract ruler." The notion of a correspondence helps in creating this tool. A *correspondence* is a matching scheme by which each member of one set is associated with a unique member of a second set. In Postulate 10 two sets are used. The first consists of pairs of points in space; the second consists of the positive numbers.

Postulate 10.—If A and B are distinct points, there exists a correspondence that associates with each pair of distinct points in space a *unique positive number*, such that the number assigned to the pair $\{A,B\}$ is one.

Definition. The pair $\{A,B\}$ in Postulate 1 is a *unit pair*. The number associated with a pair of distinct points is the distance between them relative to $\{A,B\}$. In Postulate 1 it is stated that any pair of points may be used as a unit pair, and that the distance between two points relative to this unit pair is a positive number.

Definition. The distance between a point and itself relative to any unit pair is zero.

Notation. The distance between a point P and a point Q relative to a given unit pair is written PQ. The phrase "relative to a unit pair" corresponds in the physical world or practical geometry to a linear unit of measure, such as an inch, mile, or meter.

Postulate 11.—Let $\{A,B\}$ and $\{C,D\}$ be two unit pairs. Then for every pair of distinct points P and Q in space,

$$\frac{PQ \text{ (relative to } \{A,B\})}{PQ \text{ (relative to } \{C,D\})}$$

is a *constant.*

An interpretation of this postulate may take the following form. If any distance is measured in inches and also in feet, the quotient of the two measures is constant. In this case the quotient is 12, since there are always 12 inches in every foot.

Definition. A *one-to-one correspondence* between two sets is a correspondence that matches every member of the first set with a unique member of the second set, and also matches every member of the second set with a unique member of the first set.

Definition. Let $\{A,B\}$ be any unit pair and let l be any line. A coordinate system on l relative to $\{A,B\}$ is a one-to-one correspondence between the set of points on l and the set of all real numbers if for any two points of l, P and Q, associated respectively with numbers p and q, and $p > q$, then PQ (relative to $\{A,B\}$) $= p - q$. The point on l that is associated with zero is the *origin*, and the point on l associated with l is the *unit point*. The number that a coordinate system on a line associates with a point is the *coordinate* of that point in the coordinate system.

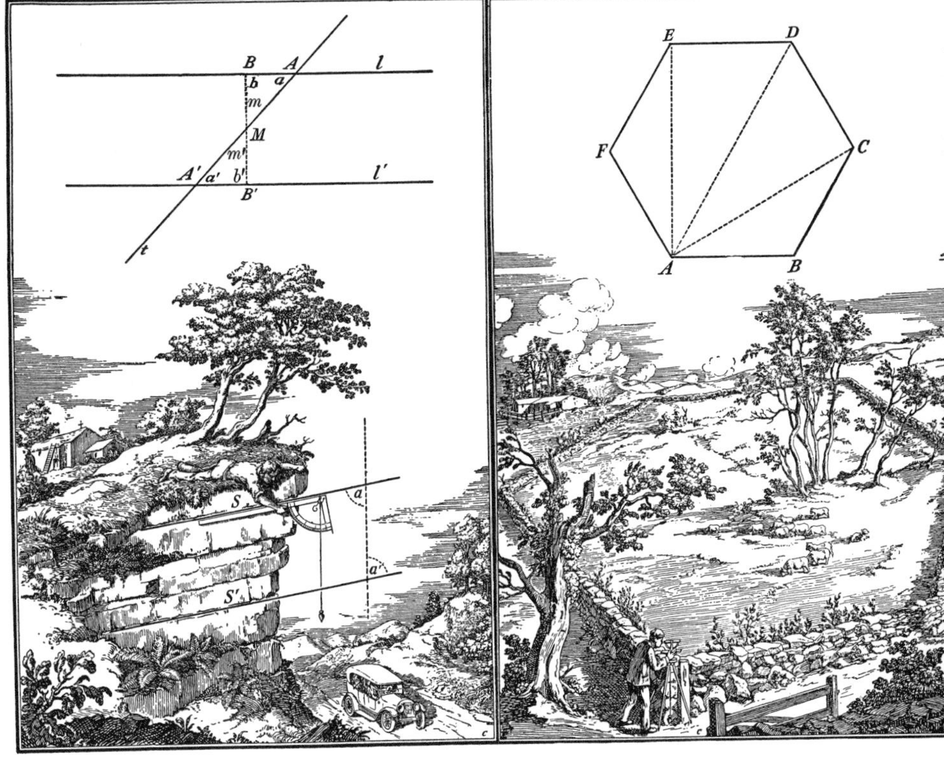

NEW YORK PUBLIC LIBRARY

EARLY TWENTIETH-CENTURY TEXTBOOKS in America taught mathematics by presenting practical problems. The problems illustrated above involve the use of trigonometry and geometry in the measurement of distances in various practical applications.

Postulate 12 (Ruler Postulate).—If $\{A,B\}$ is a unit pair and l is any line, and if P and Q are any two distinct points on l, then there is a unique coordinate system on l relative to $\{A,B\}$, such that its origin is P and the coordinate of Q is positive.

Comment. This postulate implies that a line contains as many points as there are real numbers.

■ THEOREM 6-1 (Origin and Unit-Point Theorem).—If P and Q are any two distinct points, then there is a unique coordinate system on $\overleftrightarrow{PQ}$ relative to $\{P,Q\}$, such that P is the origin and Q is the unit point of the system. The proof of this theorem follows directly from Postulate 12 (Ruler Postulate) by taking $\{A,B\} = \{P,Q\}$.

■ THEOREM 6-2 (Two-Coordinate-System Theorem).—Let a line, l, and two coordinate systems on l be given. Then there exist two numbers, a and b $(a \neq 0)$, such that for any point on l with coordinate x in one system and y in the other, $y = ax + b$.

Proof. Let distinct points P and Q on l have coordinates x_1 and x_2 in the first coordinate system, and y_1 and y_2 in the second coordinate system. Let X be any other point of l with coordinate x in the first coordinate system and y in the second.

$$\begin{array}{ccc} P & Q & X \\ x_1 & x_2 & x \\ y_1 & y_2 & y \end{array} \quad l$$

A number of cases need to be considered.

To begin with, it may be assumed (without loss of generality) that $x_1 < x_2 < x$. There are six possible inequalities for the y coordinates. Postulate 11 says that if $y_1 < y_2 < y$, then

$$\frac{x - x_2}{y - y_2} = \frac{x - x_1}{y - y_1} = \frac{x_2 - x_1}{y_2 - y_1}.$$

The equality of any two of the three fractions and $x_1 \neq x_2$ imply that

$$y = \frac{y_2 - y_1}{x_2 - x_1},$$

$$x = \frac{x_1y_2 - x_2y_1}{x_2 - x_1}.$$

Let $a = (y_2 - y_1)/(x_2 - x_1)$ and $b = (x_1y_2 - x_2y_1)/(x_2 - x_1)$, and the theorem for the case $y_1 < y_2 < y$, is proved.

If $y_1 < y < y_2$, then

$$\frac{x - x_2}{y_2 - y} = \frac{x_2 - x_1}{y - y_1} = \frac{x_2 - x_1}{y_2 - y_1}.$$

These equations lead to the result $x = x_2$, which contradicts the assumption $x_2 < x$.

In all the cases to be considered, the outcome, therefore, will be either $y = ax + b$, or a contradiction, such as $x = x_2$, $x = x_1$, or $y = y_1$.

Finally, it is verified that if $x = x_1$, then $y = y_1$; and if $x = x_2$, then $y = y_2$. Therefore, X can be any point on the line l.

Definition. Let P, Q, and R be points on line l and let there be a coordinate system on l that gives P, Q, and R coordinates p, q, and r, respectively. Q is between P and R if, and only if, q is between p and r.

■ THEOREM 6-3.–If P, Q, and R are points on line l and Q is between P and R in one coordinate system on l, then Q will be between P and R in any other coordinate system on l.

Proof. Suppose that the coordinates of P, Q, and R are p, q, and r in the given coordinate system on l. Then, by definition, either $p < q < r$ or $r < q < p$. Let a be any nonzero number. Then by either of the multiplication properties of order, $ap < aq < ar$ or $ar < aq < ap$. Let b be any number. Then, $ap + b < aq + b < ar + b$ or $ar + b < aq + b < ap + b$, proving that Q is between P and R in any other coordinate system, Q.E.D.

The significance of this theorem is that the betweenness relation among these collinear points is independent of the coordinate system on the line and is a property of the points themselves.

Definition. Given two distinct points A and B, the set of points consisting of A, B, X, and Y (such that for all X, X is between A and B, and B is between A and Y) is the ray AB and is denoted $\overrightarrow{AB}$. All other points of $\overleftrightarrow{AB}$ and A form the ray opposite $\overrightarrow{AB}$. A is the *vertex* or *end point* of each of these rays. If the vertex of a ray is deleted from the ray, what is left is a half line with end point A. Rays that have the same vertex are *concurrent rays*.

Definition. The set of points consisting of A, B, and all points between A and B is *segment AB* and is denoted $\overline{AB}$. A and B are the *end points* of $\overline{AB}$ and all its other points are *interior points*.

The length of a segment joining two points is the distance between the two points.

Two segments are congruent to each other if they have the same length. "$\overline{AB}$ is congruent to $\overline{CD}$" is abbreviated as follows: $\overline{AB} \cong \overline{CD}$.

The following are easily proved properties of congruence of line segments.

Reflexive property. For all $\overline{AB}$, $\overline{AB} \cong \overline{AB}$.

Symmetric property. If $\overline{AB} \cong \overline{CD}$, then $\overline{CD} \cong \overline{AB}$.

Transitive property. If $\overline{AB} \cong \overline{CD}$ and $\overline{CD} \cong \overline{EF}$, then $\overline{AB} \cong \overline{EF}$.

■ THEOREM 6-4 (Point-Plotting Theorem).— Given a unit pair $\{A,B\}$, point P, and positive number p, then on any ray with vertex P there is a unique point R such that $PR = p$.

Proof. Let $\overrightarrow{PS}$ be any ray. By Theorem 6-1, there is a unique coordinate system on $\overleftrightarrow{PS}$ such that P is the origin and S is the unit point. By Postulate 11,

$$\frac{PS \text{ (relative to } \{P,S\})}{PS \text{ (relative to } \{A,B\})} = k.$$

In this coordinate system kp determines point R such that PR (relative to $\{A,B\}$) $= p$, Q.E.D.

■ THEOREM 6-5 (Two-Point Theorem).— In a given coordinate system on line l, let x_1 and x_2 be the respective coordinates of given distinct points x_1 and x_2 on l. Then if x is the coordinate of any point X on l, the number k can be found such that $x = x_1 + k(x_2 - x_1)$.

X_1	X_2	X	l
x_1	x_2	x	
0	1	k	

Proof. Consider the coordinate system on l that assigns 0 to X_1 and 1 to X_2, and let k be the number assigned to X in this system. By Theorem 6–2 (Two-Coordinate-System Theorem), $x = ak + b$, where a and b are to be found. At point X_1, $x_1 = a \cdot 0 + b$, or $b = x_1$. At point X_2, $x_2 = a \cdot 1 + x_1$, or $a = x_2 - x_1$. Therefore, at X, $x = (x_2 - x_1)k + x_1$, or $x_1 + k(x_2 - x_1)$, Q.E.D.

Definition. The midpoint of a segment $\overline{AB}$ is a point X such that it is a point on $\overline{AB}$ and such that $AX = XB$.

■ THEOREM 6-6.–The midpoint of a segment is unique. This can be proved by taking $k = l/2$ in Theorem 6–5 (Two-Point Theorem). If the endpoints of a segment are x_1 and x_2, the coordinate of the midpoint is $(x_1 + x_2)/2$.

■ THEOREM 6-7 (Betweenness-Distance Theorem).—Let A, B, and C be three points such that B is between A and C. Then $AB + BC = AC$ for any given unit pair. Or, $AB = AC - BC$ or $BC = AC - AB$.

Proof. By Postulate 12 (Ruler Postulate), take the coordinate system with A as the origin and to C assign a positive number, c. Let the number assigned to B be b. Then $0 < b < c$. Since $AB = b$, $BC = c - b$, and $AC = c$, it follows that $AB + BC = AC$. However, no reference was made to a specific unit pair, so the proof applies to any unit pair, Q.E.D.

Comment. This theorem served as part of the basis for David Hilbert's (1862–1943) development of *betweenness relations for points.*

SECTION SEVEN

Angles.—To prepare for a study of angles, *separation principles* must be considered.

Definition. A set containing more than one point is a *convex set* if, and only if, the segment joining any two of its points contains only points of the set.

Example. A segment is a convex set, but the set consisting of one exterior point of a segment and the segment is not.

■ THEOREM 7-1.–The intersection of any two convex sets is a convex set.

Proof. The intersection of any two sets contains only elements that belong to both sets. Consider any two points, A and B, in the intersection. Because A and B are members of both convex sets, $\overline{AB}$ also belongs to both and hence is contained in the intersection, Q.E.D.

Postulate 13 (Plane-Separation Postulate).—For any plane and any line in that plane, the points of the plane that are not contained in the line form two sets, each of which is convex; and every segment that joins a point of one set to a point of the other intersects the given line.

Definition. Each of the convex sets in Postulate 13 is a *half plane.* The line separates the plane into two opposite half planes, and is the *edge of each half plane.* If a half plane has edge l and contains point P, it is the *P side of l.*

■ THEOREM 7-2.–If a ray intersects a line only in its end point, then the interior of the ray is contained in one of the half planes whose edge is the given line.

Postulate 14.—For any plane, the points of space that do not lie in that plane form two convex sets such that every segment that joins a point of one set to a point of the other intersects the given plane.

Definition. An *angle* is the set of points contained in two concurrent, noncollinear rays. (Two collinear rays form either a *zero angle* or a *straight angle,* neither of which is included in this definition. They are omitted in order to simplify the mathematical notion of the interior of an angle.) The rays are the *sides of the angle,* and the common vertex of the rays is the *vertex of the angle.*

Notation. The angle formed by $\overrightarrow{VA}$ and $\overrightarrow{VB}$ is named $\angle AVB$. The vertex of the angle appears in the middle. If it is not ambiguous, the angle may also be called $\angle V$.

Comment. $\angle AVB = \angle BVA$ because they consist of the same set of points and hence name the same subject.

Comment. To some extent, the method of constructing a measure for abstract (mathematical) angles is similar to that of constructing a measure for the distance between two points.

Postulate 15.—There exists a correspondence that associates each angle in space with a unique number between 0 and 180.

Definition. The number in Postulate 15 is the *measure of the angle.*

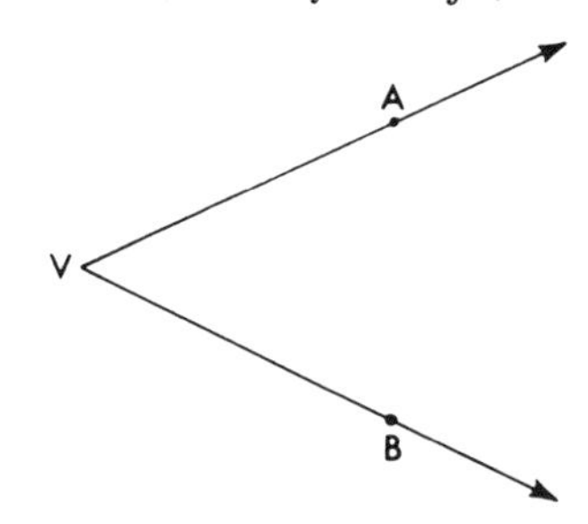

Notation. The measure of $\angle ABC$ is denoted: $m\angle ABC$.

Definition. Let V be any point in a plane. A *ray coordinate system* in that plane, relative to V, is the one-to-one correspondence between all the rays in the plane with vertex V, and the set of all numbers x such that $0 < x < 360$ with the following properties: If numbers r and s correspond to $\overrightarrow{VR}$ and $\overrightarrow{VS}$ in the plane, and if $r > s$ then, $m\angle RVS = r - s$, if $r - s < 180$. $m\angle RVS = 360 - (r - s)$, if $r - s > 180$. $\overrightarrow{VR}$ and $\overrightarrow{VS}$ are opposite rays if, and only if, $r - s = 180$.

Postulate 16 (Protractor Postulate).—If $\overrightarrow{VA}$ and $\overrightarrow{VS}$ are noncollinear rays in a plane, then there is a unique ray coordinate system in the plane relative to V such that $\overrightarrow{VA}$ corresponds to zero, and every ray in the B side of $\overleftrightarrow{VA}$ corresponds to a number less than 180.

■ THEOREM 7-3 (Angle Construction Theorem).—Given a half plane with edge $\overleftrightarrow{VA}$ and a number r, such that $0 < r < 180$, there is then a unique ray $\overrightarrow{VR}$ such that $\overrightarrow{VR}$ is in the given half plane and $m\angle AVR = r$.

Definition. Given three concurrent rays in a plane—$\overrightarrow{VA}$, $\overrightarrow{VB}$, and $\overrightarrow{VC}$—$\overrightarrow{VB}$ is said to be *between* $\overrightarrow{VA}$ and $\overrightarrow{VC}$ if, and only if, there is a ray coordinate system in the plane relative to V, such that the respective ray coordinates, 0, b, and c of $\overrightarrow{VA}$, $\overrightarrow{VB}$, and $\overrightarrow{VC}$, are such that $b < c$.

■ THEOREM 7-4 (Betweenness-Angles Theorem).—If $\overrightarrow{VB}$ is between $\overrightarrow{VA}$ and $\overrightarrow{VC}$, then $m\angle AVB + m\angle BVC = m\angle AVC$. This may also be expressed: $m\angle AVC - m\angle AVB = m\angle BVC$.

Definition. A ray is the *midray of an angle* if the ray is between the sides of the angle and forms with them two angles of equal measure.

■ THEOREM 7-5.–Every angle has a unique midray.

Definition. The midray of an angle bisects the angle and is called the *angle bisector*.

The *interior of an angle* may be described in any one of the following ways:

(1) The interior of $\angle AVB$ is the set of rays between $\overrightarrow{VA}$ and $\overrightarrow{VB}$ if V is deleted from the set. This is set R.

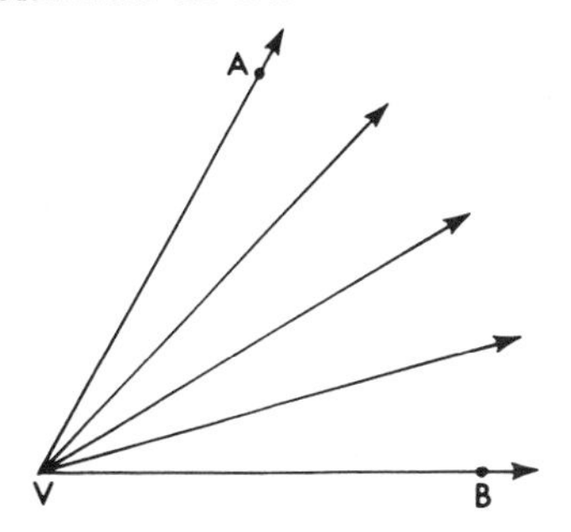

(2) The interior of $\angle AVB$ is the intersection of two half planes: the A side of $\overleftrightarrow{VB}$ and the B side of $\overleftrightarrow{VA}$. This is set I.

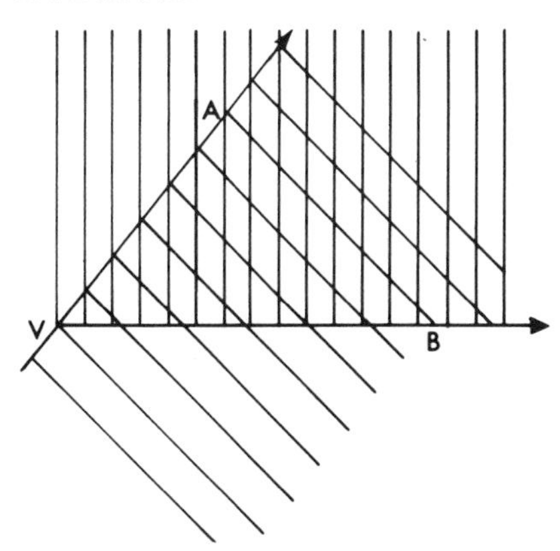

(3) The set of interior points of all segments that have one end point in $\overrightarrow{VA}$ and the other in $\overrightarrow{VB}$, with V deleted from this set, is set S.

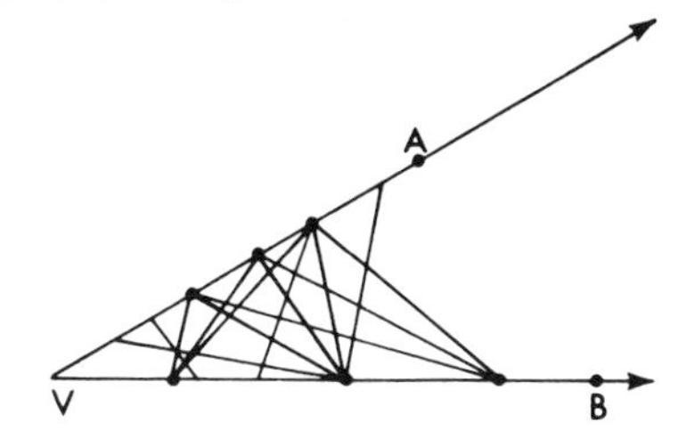

Postulate 17.—Sets R and I described above are equal and contain set S.

■ THEOREM 7-6.–The interior of any angle is a convex set.

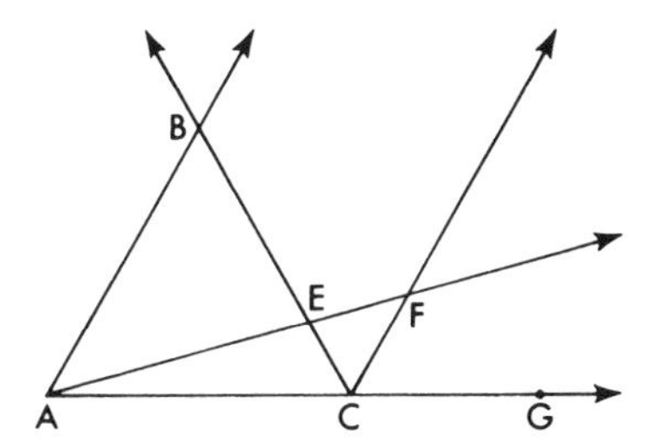

■ THEOREM 7-7.–Let A, B, C, E, F, and G be *coplanar points* such that A, B, and C are noncollinear, E is between B and C, $\overrightarrow{EF}$ and $\overrightarrow{EA}$ are opposite rays, and $\overrightarrow{CA}$ and $\overrightarrow{CG}$ are opposite rays. Then $\overrightarrow{CF}$ is between $\overrightarrow{CB}$ and $\overrightarrow{CG}$.

Comment. The proof depends upon establishing that F is on the B side of $\overleftrightarrow{CG}$ and also on the G side of $\overleftrightarrow{BC}$, and is therefore an interior point of $\angle BCG$. Then $\overrightarrow{CF}$ is between $\overrightarrow{CB}$ and $\overrightarrow{CG}$.

Definition. The *exterior of an angle* is the set of all points in the plane of the angle that are contained neither in the angle nor in its interior.

Definition. The two angles formed by three concurrent rays, two of which are opposite rays, are a *linear pair of angles.*

■ THEOREM 7-8.–The sum of the measures of the angles in a linear pair is 180.

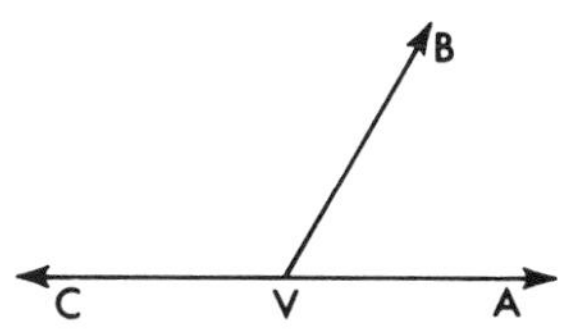

Proof. Suppose the opposite rays are $\overrightarrow{VA}$ and $\overrightarrow{VC}$, and the third ray is $\overrightarrow{VB}$. By Postulate 16 (Protractor Postulate), there is a ray-coordinate system in which $\overrightarrow{VA}$ corresponds to 0, $\overrightarrow{VC}$ corresponds to 180, and $\overrightarrow{VB}$ corresponds to a number, b, between 0 and 180. Therefore, $m\angle AVB = b$, $m\angle BVC = 180 - b$, and $m\angle AVB + m\angle BVC = 180$, Q.E.D.

■ COROLLARY 7-8-1.–If $\overrightarrow{OA}$ and $\overrightarrow{OB}$ are opposite rays, and $\overrightarrow{OC}$ and $\overrightarrow{OD}$ are in the same half plane with edge $\overleftrightarrow{AB}$ such that $\overrightarrow{OC}$ is between $\overrightarrow{OD}$ and $\overrightarrow{OA}$, then $m\angle AOC + m\angle COD + m\angle DOB = 180$.

Definition. Two coplanar angles are a pair of *adjacent angles* if, and only if, they have one side in common and their interiors have no points in common.

Comment. A *linear pair* of angles is also a pair of adjacent angles.

■ THEOREM 7-9.–If the sum of the measures of two adjacent angles is 180, they are a linear pair of angles.

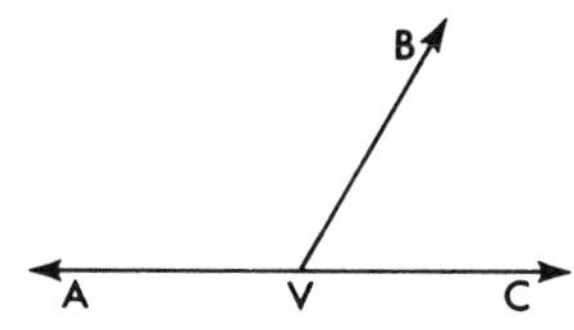

Proof. Let the two adjacent angles be $\angle AVB$ and $\angle BVC$. By Postulate 16 (Protractor Postulate), there is a ray coordinate system that assigns 0 to $\overrightarrow{VB}$, a to $\overrightarrow{VA}$ and c to $\overrightarrow{VC}$. Since $\overrightarrow{VB}$ intersects $\overleftrightarrow{AC}$, A and C are in opposite half planes with edge $\overleftrightarrow{VB}$; either a or c is less than 180, and the other is more than 180. If $a < 180$, then $m\angle BVA = a$, and $m\angle BVC = 360 - c$. By hypothesis, $a + 360 - c = 180$, or $c - a = 180$. Therefore, $\overrightarrow{VA}$ and $\overrightarrow{VC}$ are opposite rays, Q.E.D.

Definitions. An angle whose measure is 90 is a *right angle.* An angle whose measure is less than 90 is an *acute angle.* An angle whose measure is greater than 90 is an *obtuse angle.*

■ THEOREM 7-10.–If the measures of two angles in a linear pair are equal, then each angle is a right angle.

Definition. The lines determined by two rays that are sides of a right angle are *perpendicular.* If two rays or two segments determine perpendicular lines, they are *perpendicular rays,* or *perpendicular segments.*

Notation. $\overleftrightarrow{AB}$ is perpendicular to $\overleftrightarrow{BC}$ is written: $\overleftrightarrow{AB} \perp \overleftrightarrow{CD}$.

Definition. Two angles that have the same measure are *congruent* to each other.

Notation. $\angle ABC$ is congruent to $\angle DEF$ is written: $\angle ABC \cong \angle DEF$.

The properties of congruence of angles are as follows:

Reflexive property. For all $\angle A$, $\angle A \cong \angle A$.

Symmetric property. If $\angle A \cong \angle B$, then $\angle B \cong \angle A$.

Transitive property. If $\angle A \cong \angle B$ and $\angle B \cong \angle C$, then $\angle A \cong \angle C$.

■ THEOREM 7-11.–If two angles are right angles, they are congruent to each other.

Definitions. If the sum of the measures of two angles is 180, they are a pair of *supplementary angles,* and each is said to be the *supplement* of the other. If the sum of the measures of two angles is 90, they are a pair of *complementary angles,* and each is said to be the *complement* of the other.

■ THEOREM 7-12.–The two angles in a linear pair are supplementary angles.

■ THEOREM 7-13.–If two angles are congruent and supplementary, then each is a right angle.

■ THEOREM 7-14.–Supplements of congruent angles are congruent to each other.

■ THEOREM 7-15.–Complements of congruent angles are congruent to each other.

Definition. If the sides of two angles form two pairs of opposite rays, they are a pair of *vertical angles.*

■ THEOREM 7-16.–If two angles are vertical, they are congruent.

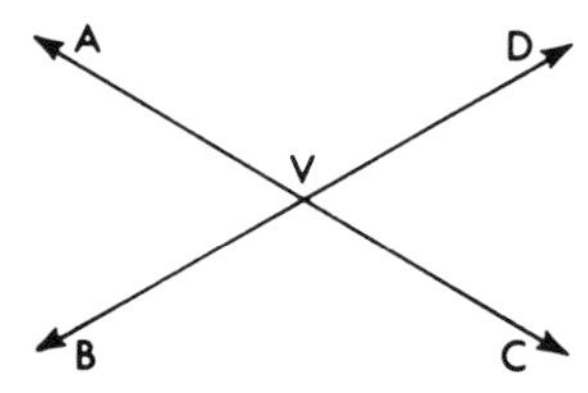

Proof. Let the angles be $\angle AVB$ and $\angle DVC$, with opposite rays $\overrightarrow{VA}$ and $\overrightarrow{VC}$, and $\overrightarrow{VB}$ and $\overrightarrow{VD}$. Since $\angle BVA$ and $\angle AVD$ are a linear pair, they are supplementary. In like manner $\angle AVD$ and $\angle DVC$ are supplementary. Therefore, $\angle BVA$ and $\angle DVC$ are congruent, Q.E.D.

Comment. If two lines intersect, they form two pairs of vertical angles.

■ THEOREM 7-17.–If two intersecting lines are perpendicular, they form four right angles.

■ THEOREM 7-18.–Given a line in a plane and a point on the line, then there is one, and only one, line in the plane perpendicular to the given line containing the given point.

Comment. The proof of this theorem depends on Theorem 7-3 (Angle-Construction Theorem) and Theorem 7-17.

Definition. If three points are noncollinear, the set of points in the three segments determined by them is a *triangle.* Each of the three given points is a *vertex of the triangle,* and each segment is a *side of the triangle.*

Notation. If the three given points are A, B, and C, the triangle is denoted $\triangle ABC$; the sides are denoted $\overline{AB}$, $\overline{BC}$, $\overline{CA}$.

Definition. In $\triangle ABC$, $\angle ABC$, $\angle BCA$, and $\angle CAB$ are angles of $\triangle ABC$. A side of a triangle and the angle whose vertex is not a point of that side are *opposite* each other.

Definition. The set of points common to the interiors of the three angles of a triangle is the *interior of the triangle.*

■ THEOREM 7-19.–The interior of a triangle is a convex set.

Definition. Let A, B, C, and D be given coplanar points such that no three are collinear and $\overline{AB}$, $\overline{BC}$, $\overline{CD}$, and $\overline{DA}$ do not intersect each other in interior points. The set of points contained in the four segments is a *quadrilateral;* each of the given points is a *vertex of the quadrilateral;* each segment is a *side of the quadrilateral;* and a segment joining two nonconsecutive vertices is a *diagonal of the quadrilateral.*

Notation. The quadrilateral defined above is denoted, quadrilateral $ABCD$.

Definition. A *polygon* is a *convex polygon* if, and only if, each side lies in the edge of a half plane that contains the rest of the polygon.

Definition. The *interior of a convex polygon* is the set of points common to all the half planes of the previous definition.

■ THEOREM 7-20.–The interior of a convex polygon is a convex set.

■ COROLLARY 7-20-1.–The interior of each diagonal of a convex polygon is in the interior of the polygon.

Definition. Any angle determined by a pair of consecutive sides of a convex polygon is an *angle of the polygon.* Two angles having consecutive vertices of the polygon are *consecutive angles of the polygon.*

Definition. Two nonconsecutive angles of a quadrilateral are *opposite angles.* Two nonconsecutive sides of a quadrilateral are *opposite sides.*

Definition. Two nonplanar half planes having a common edge form a *dihedral angle.* The common edge is the *edge of the half plane,* and all points in the edge and in one half plane are a *face of the dihedral angle.*

SECTION EIGHT

Congruences.—In the physical world, two objects are congruent if they have the same shape and size. The congruence of two physical objects can sometimes be demonstrated by noting how corresponding parts fit each other.

In theoretical geometry there is also a concern with one-to-one correspondences, and congruence between two triangles is considered to be a property of a correspondence between their vertices.

Definition. If the vertices of $\triangle ABC$ and the vertices of $\triangle DEF$ are made to correspond A to D, B to E, and C to F, then the *corresponding parts* of the triangles are, in pairs, $\angle A$ and $\angle D$, $\angle B$ and $\angle E$, $\angle C$ and $\angle F$, $\overline{AB}$ and $\overline{DE}$, $\overline{BC}$ and $\overline{EF}$, and $\overline{CA}$ and $\overline{FD}$.

Definition. A one-to-one correspondence between the vertices of two triangles, in which corresponding parts are congruent, is a *congruence* between the two triangles.

Definition. Two triangles are *congruent* if, and only if, there is a congruence between the two triangles.

Notation. $\triangle ABC \cong \triangle DEF$ means that the one-to-one correspondence, A to D, B to E, and C to F, is a congruence.

Properties of Triangle Congruence.—These are consequences of the definition of congruent triangles:

Reflexive property. For all $\triangle ABC$, $\triangle ABC \cong \triangle ABC$.

Symmetric property. If $\triangle ABC \cong \triangle DEF$, then $\triangle DEF \cong \triangle ABC$.

Transitive property. If $\triangle ABC \cong \triangle DEF$ and $\triangle DEF \cong \triangle GHI$, then $\triangle ABC \cong \triangle GHI$.

Definition. An angle of a triangle is said to be *included* between the two sides of the triangle that are contained in the rays of the angle. A side of a triangle is said to be *included* between two angles of the triangle if its end points are vertices of the angles.

Notation. "Two sides and their included angle of a triangle" is abbreviated *S.A.S.* "Two angles and their included side" is abbreviated *A.S.A.* "The three sides of a triangle" is abbreviated *S.S.S.* "The three angles of a triangle" is abbreviated *A.A.A.*

Postulate 18 (S.A.S. Postulate).—Given a one-to-one correspondence between the vertices of two triangles (not necessarily distinct), if two sides and the included angle of one triangle are congruent to the corresponding parts of the other triangle, then the correspondence is a congruence.

Postulate 19 (A.S.A. Postulate).—Given a one-to-one correspondence between the vertices of two triangles (not necessarily distinct), if two angles and the included side of one triangle are congruent to the corresponding parts of the other triangle, then the correspondence is a congruence.

Postulate 20 (S.S.S. Postulate).—Given a one-to-one correspondence between the vertices of two triangles (not necessarily distinct), if the three sides of one triangle are congruent to the three sides of the other triangle, then the correspondence is a congruence.

Comment. Postulates 19 and 20 can be deduced as a logical sequence from Postulate 18 and previous postulates. Some schools postulate them.

■ THEOREM 8-1 (Betweenness-Addition Theorem for Points).—If points B and C are between A and D, and if $\overline{AB} \cong \overline{CD}$, then $\overline{AC} \cong \overline{BD}$.

Proof. The order of points might be A, B, C, D, or A, C, B, D. Suppose that the order is A, B, C, D. Then $\overline{AB} \cong \overline{CD}$, by definition of congruent segments, implies that $AB = CD$. By the addition property of equality, it follows that $AB + BC = CD + BC$. By Theorem 6-7 (Betweenness-Distance Theorem), $AB + BC = AC$, and $CD + BC = BD$. By the substitution principle of equality, $AC = BD$ and, using the definition of congruent segments, again the conclusion is that $\overline{AC} \cong \overline{BD}$.

If the order is A, C, B, D, the proof is the same, with the modification that $AB - BC = CD - BD$; but all reasons remain unchanged.

■ THEOREM 8-2 (Betweenness-Addition Theorem for Rays).—If $\overrightarrow{OB}$ and $\overrightarrow{OC}$ are between $\overrightarrow{OA}$ and $\overrightarrow{OD}$, and $\angle AOB \cong \angle COD$, then $\angle AOC \cong \angle BOD$. The proof is similar to that of Theorem 8-1.

The accompanying examples illustrate how a two-column proof is written and also how triangle congruences are used in formal proofs.

Comment 1. In Proof 2, some details are expressed that ordinarily are taken for granted. It is a common practice, for instance, to condense steps 7-10 inclusive by first proving the theorem: If two angles are congruent, their midrays form four congruent angles with the sides of the angles.

Two-Column Proofs

Proof 1.

Hypothesis: P, Q, R, S are collinear in that order, with $PQ \cong RS$. A and B are on opposite sides of PS. $PA \cong SB$ and $RA \cong QB$.
Conclusion:
$\angle PAR \cong \angle SBQ$.

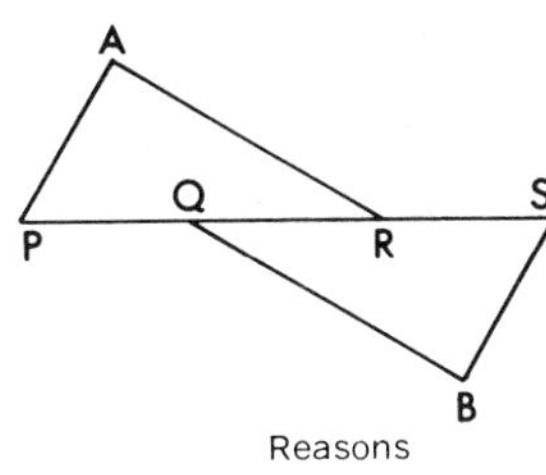

Statements	Reasons
1. Q and R are between P and S.	Hypothesis.
2. $PQ \cong RS$.	Hypothesis.
3. $PR \cong QS$.	Betweenness-Addition Theorem for points.
4. $PA \cong SB$ and $RA \cong QB$.	Hypothesis.
5. $\triangle PAR \cong \triangle SBQ$.	S.S.S. Postulate.
6. $\angle PAR \cong \angle SBQ$, Q.E.D.	Definition of congruent triangles.

Proof 2.

Hypothesis: A, B, C, D are collinear points, in that order. E is not in AD. F is between E and B, and G is between E and C, such that CF is the midray of $\angle ECB$, and BG is the midray of $\angle EBC$. $\angle ABE \cong \angle DCE$.
Conclusion: $CF \cong BG$.

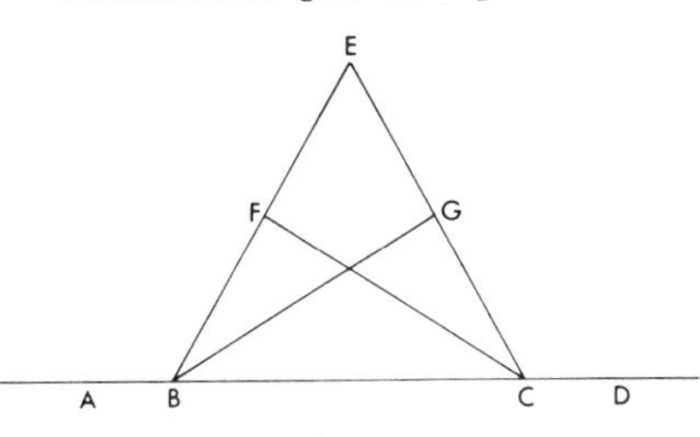

Statements	Reasons
1. A, B, C, D are collinear, in that order.	Hypothesis.
2. $\angle ABE$ and $\angle EBC$ are a linear pair. Also, $\angle DCE$ and $\angle ECB$ are a linear pair.	Definition of a linear pair.
3. $\angle EBC$ is the supplement of $\angle ABE$; $\angle ECB$ is the supplement of $\angle DCE$.	If two angles form a linear pair, they are supplementary.
4. $\angle ABE \cong \angle DEC$.	Hypothesis.
5. $\angle EBC \cong \angle ECB$.	If two angles are congruent, their supplements are congruent.
6. $BC \cong CB$.	Reflexive property of congruence of segments.
7. CF is the midray of $\angle ECB$; BG is the midray of $\angle EBC$.	Hypothesis.
8. $m\angle FCB = \frac{1}{2}m\angle ECB$; $m\angle GBC = \frac{1}{2}m\angle EBC$.	Definition of a midray.
9. $m\angle FCB = m\angle GBC$.	Multiplication property of equality of numbers.
10. $\angle FCB \cong \angle GBC$.	If two angles have equal measures, they are congruent.
11. $\triangle FBC \cong \triangle GCB$.	A.S.A. Postulate.
12. $CF \cong BG$, Q.E.D.	If two triangles are congruent, their corresponding parts are congruent.

Proof 3.

Hypothesis: In $\triangle ABC$, $AB \cong AC$.
Conclusion: $\angle B \cong \angle C$.

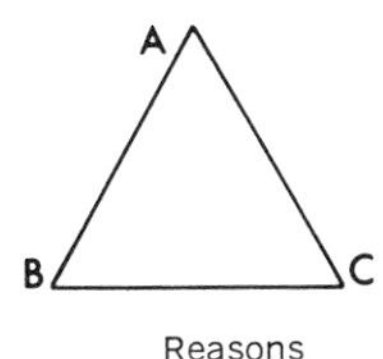

Statements	Reasons
1. $AB \cong AC$.	Hypothesis.
2. $BC \cong CB$.	Reflexive property of congruence.
3. $CA \cong BA$.	Symmetric property of congruence.
4. $\triangle ABC \cong \triangle ACB$.	S.S.S. Postulate.
5. $\angle B \cong \angle C$, Q.E.D.	If triangles are congruent, corresponding parts are congruent.

Comment 2. It is possible to read this proof without referring to the figure, if it is kept in mind that A corresponds with D, B with C, F with G, and E with itself. The beginning student of geometry will find this a valuable exercise.

Definition. A triangle is called *isosceles* if two of its sides are congruent. In an isosceles triangle, the congruent sides are *arms* (or *legs*); the third side is the *base*. The angle included between the arms is the *vertex angle*. The angles that include the base are the *base angles*.

■ THEOREM 8-3.–If a triangle is isosceles, then its base angles are congruent.

Proof. Let the triangle be $\triangle ABC$, with $\overline{AB} \cong \overline{AC}$. Considering $\triangle ABC$ and $\triangle ACB$, refer to Proof 3.

Definition. A triangle is *equilateral* if its three sides are congruent to each other. A triangle is *equiangular* if its three angles are congruent to each other.

■ COROLLARY 8-3-1.–If a triangle is equilateral, it is equiangular.

■ THEOREM 8-4 (Converse of Theorem 8-3).—If two angles of a triangle are congruent, then the sides opposite these angles are congruent.

Comment. The proof of this theorem is similar to that of Theorem 8-3.

■ COROLLARY 8-4-1.–If a triangle is equiangular, then it is equilateral.

Example. How Theorem 8-3 and Theorem 8-4 can be used.

To Prove. If in $\triangle ABC$, $\overline{AB} \cong \overline{AC}$, E is between A and B, F is between A and C, $\overrightarrow{CE}$ is the midray of $\angle ACB$, $\overrightarrow{BF}$ is the midray of $\angle ABC$, and CE and BF intersect in G, then $\overline{CG} \cong \overline{BG}$.

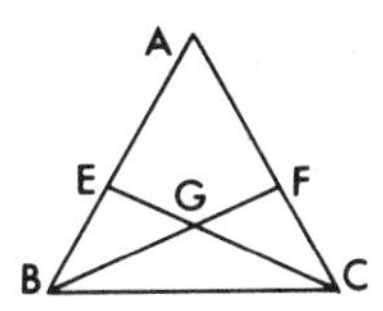

Proof. By Theorem 8-3 and the hypothesis that $\overline{AB} \cong \overline{AC}$, it follows that $\angle ACB \cong \angle ABC$. From the hypothesis that $\overrightarrow{CE}$ is the midray of $\angle ACB$ and that $\overrightarrow{BF}$ is the midray of $\angle ABC$, and from the multiplication property of equality, it follows that $\angle ECB \cong \angle FBC$. By Theorem 8-4, it follows that $\overline{BG} \cong \overline{CG}$, Q.E.D.

Comment. Little attention was given to the statement that if $\angle ACB \cong \angle ABC$, then $m\angle ACB = m\angle ABC$; also, if $m\angle ECB = m\angle FBC$, then $\angle ECB \cong \angle FBC$. Both of these follow from the definition of congruent angles. It is common practice to make this kind of omission.

Definition. A *median of a triangle* is a segment whose end points are a vertex of the triangle and the midpoint of the opposite side.

■ THEOREM 8-5.–The median to the base of an isosceles triangle bisects the vertex angle and is perpendicular to the base.

Proof. If in $\triangle ABC$, $\overline{AB} = \overline{AC}$, and D is the midpoint of $\overline{BC}$, it follows from Postulate 20 (S.S.S. Postulate) that $\triangle ADB \cong \triangle ADC$. This leads to the conclusion that $\angle BAD \cong \angle CAD$, Q.E.D. It also follows that $\angle ADB \cong \angle ADC$.

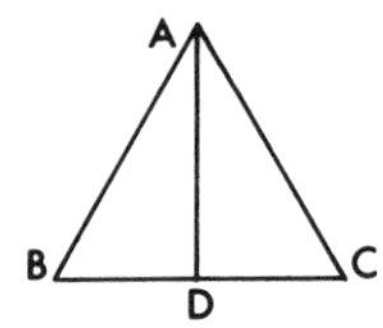

Since the latter pair of angles is a linear pair, they are right angles and $\overline{AD} \perp \overline{BC}$, Q.E.D.

■ THEOREM 8-6.–The bisector of the vertex angle of an isosceles triangle bisects the base and is perpendicular to the base.

Definitions. Each angle of a triangle is an *interior angle of the triangle.* An angle that forms a linear pair with an interior angle of a triangle is an *exterior angle of the triangle.* Each exterior angle is *adjacent* to the interior angle with which it forms a linear pair, and *remote* to the other interior angles of the triangle.

■ THEOREM 8-7.–The measure of an exterior angle is greater than the measure of either of its remote interior angles.

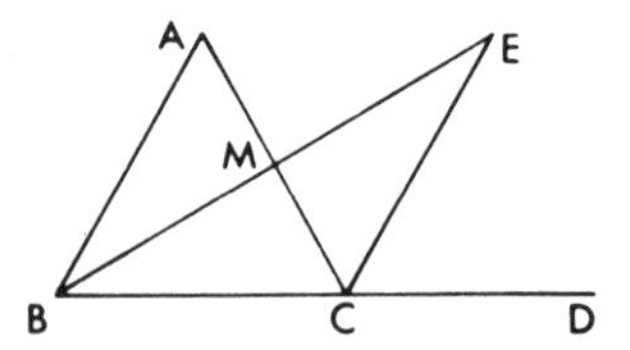

Proof. Let the triangle be $\triangle ABC$. Consider one case in which it is to be proved that the measure of exterior $\angle ACD > m\angle A$. $\overline{AC}$ has a unique midpoint, M, and by Theorem 6-4 (Point-Plotting Theorem), there is a point on the ray opposite $\overrightarrow{MB}$, $\overrightarrow{ME}$, such that $\overrightarrow{ME} = \overrightarrow{MB}$. $\triangle AMB \cong \triangle CME$ (S.A.S.). By Theorem 7-7, $\overrightarrow{CE}$ is between $\overrightarrow{CA}$ and $\overrightarrow{CD}$. This implies that $m\angle ACD > m\angle ACE$. Since $m\angle ACE = m\angle A$, $m\angle ACD > m\angle A$. Other cases are similarly proved, Q.E.D.

■ THEOREM 8-8.–Given a line and a point not on the line, then there is only one line that contains the point and is perpendicular to the given line.

Proof. If there were two perpendiculars, there would be a contradiction of Theorem 8-7, Q.E.D.

SECTION NINE

Parallel Lines.—A formal treatment of parallel lines requires a postulate concerning their existence. Euclid used one known today as the *Fifth Postulate,* or *Euclid's Parallel Postulate:*

"If a straight line falling on two straight lines makes the interior angles on the same side less than two right angles, the two straight lines, if produced indefinitely, meet on that side on which are the angles less than two right angles."

The complexity of this statement suggested to many mathematicians that it could be shown to be a logical consequence of Euclid's other postulates. This effort, over many centuries, led to the discovery of geometries that were based on postulates, one of which was contradictory to the Fifth Postulate. These are known as *non-Euclidean geometries.*

Other mathematicians used simpler alternative statements that are equivalent to the Fifth Postulate:

Proclus (c. 410–485): "Parallels remain at a finite distance from each other."

John Wallis (1616–1703): "Given a triangle, it is possible to construct another triangle similar to it."

Adrien Legendre (1752–c. 1833): "The sum of the three angles of a triangle is two right angles."

John Playfair (1748–1819): "Through a given point, not on a given line, only one parallel can be drawn to the given line." (This is also attributed to Proclus.)

Karl Gauss (1777–1855): "If k is any integer, there exists a triangle whose area exceeds k."

János Bolyai (1802–1860): "Given any three points not on a straight line, there is a circle that passes through them."

The *Playfair Postulate* has been the most favored in text books. A slight modification of it is used here.

Definition. Two lines are *parallel* if they are coplanar and do not intersect.

Notation. The statement "$\overleftrightarrow{AB}$ is parallel to $\overleftrightarrow{CD}$" is abbreviated $\overleftrightarrow{AB} \parallel \overleftrightarrow{CD}$.

Definition. Given two distinct coplanar lines, a line that intersects them in two distinct points is a *transversal* of the two lines.

■ THEOREM 9-1.–If a transversal is a perpendicular to each of two lines, then the lines are parallel.

Proof. Let the lines be p and q; the transversal, s. If p and q are not parallel, they meet at a point from which two perpendiculars to s exist. This contradicts Theorem 8-8, Q.E.D.

Comment. The significance of this theorem lies in the conclusion that through a given point not on a line, there exists at least one line containing the point and parallel to the line. But it does not permit the conclusion that this parallel is unique. To accomplish this, a postulate is introduced.

Postulate 21 (Parallel Postulate).—There is at most one line parallel to a given line and containing a given point not on the given line.

Definition. Let p and q be two distinct lines cut by transversal t in points P and Q, respectively. Let A and B be points of p and q, respectively, in opposite half planes having edge t. Then $\angle APQ$ and $\angle BQP$ are a *pair of alternate interior angles.*

If two angles are a pair of alternate interior angles, then each of these angles and the vertical angle of the other are a *pair of corresponding angles.*

Let p and q be two distinct lines cut by a transversal t in points P and Q, respectively. Let C and B be points of p and q, respectively, in the same half plane with edge t. Then $\angle CPQ$ and $\angle BQP$ are a *pair of consecutive interior angles.*

BETTMANN ARCHIVE

KARL FRIEDRICH GAUSS, German mathematician, founded the modern theory of numbers and also worked in geometry.

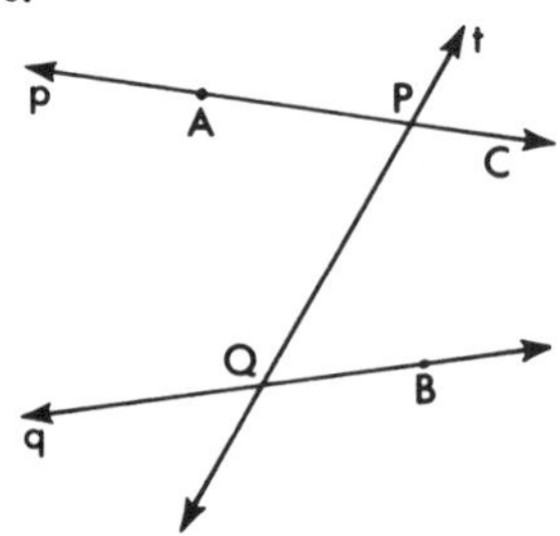

■ THEOREM 9-2.–If a transversal of two coplanar lines forms a pair of congruent alternate interior angles, then the lines are parallel.

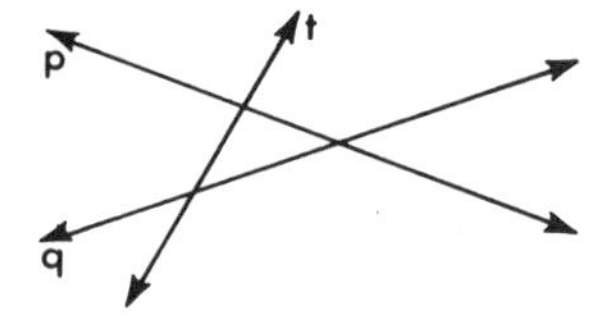

Proof. Let the lines be p and q and the transversal t. Using the indirect method of proof, start with the assumption that p and q intersect, thus determining a triangle with t. But by Theorem 8-7, this implies that the alternate interior angles have unequal measures. Therefore, $p \parallel q$, Q.E.D.

■ COROLLARY 9-2-1.–If a transversal of two coplanar lines forms a pair of congruent corresponding angles, then the lines are parallel.

■ COROLLARY 9-2-2.–If a transversal of two coplanar lines forms a pair of consecutive interior angles that are supplementary, then the lines are parallel.

■ THEOREM 9-3.–If two lines are parallel, then any transversal of those lines forms with them a *pair of congruent alternate interior angles.*

Proof. Let the parallel lines be p and q, and let the transversal t intersect p and q in P and Q, respectively. Take A

NEW YORK PUBLIC LIBRARY

EARLY EUROPEAN MATHEMATICIANS. John Wallis (1616–1703) (*left*), English geometry professor, laid the groundwork for calculus. Adrien Legendre (1752–c. 1833) (*center*) contributed to the theory of ellipsoids. John Playfair (1748–1819) (*right*) wrote *Elements of Geometry.*

and B on opposite sides of t so that $\angle APQ$ and $\angle BQP$ are a pair of alternate interior angles. By Postulate 16

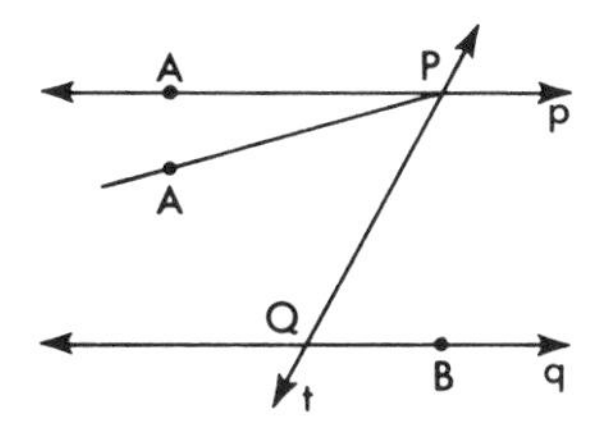

(Protractor Postulate), there is a point A' on the A side of t such that $m\angle A'PQ = m\angle BQP$. Then $\overleftrightarrow{PA'} \parallel q$, by Theorem 9–2. But Postulate 21 (Parallel Postulate) states that there is at most one parallel to q containing P. Therefore $\overrightarrow{PA'} = \overrightarrow{PA}$, and $m\angle APQ = m\angle A'PQ = m\angle BQP$, Q.E.D.

■ COROLLARY 9–3–1.–If two lines are parallel, then any transversal of these lines forms with them a *pair of congruent corresponding angles.*

■ COROLLARY 9–3–2.–If two lines are parallel, then any transversal of these lines forms with them a *pair of consecutive interior angles* that are supplementary.

■ COROLLARY 9–3–3.–If two lines are parallel, any transversal of these lines that is perpendicular to one is perpendicular to the others.

■ THEOREM 9–4.–If each of two coplanar lines is parallel to the same third line, then they are parallel to each other.

Proof. If the two coplanar lines intersect, there will be two intersecting lines parallel to the third line—a contradiction of Postulate 21 (Parallel Postulate). Therefore, they do not intersect and, being coplanar, they are parallel, Q.E.D.

Comment. This theorem includes a case of three lines not in one plane.

Comment. The relationship of parallelism for lines in a plane has the symmetric and transitive properties. If the definition of parallel lines were to include the case of a line being parallel to itself, the relationship would have the reflexive property as well.

■ COROLLARY 9–4–1.–If a line that is coplanar with two parallel lines intersects one, it also intersects the other.

■ COROLLARY 9–4–2.–Given two sets of parallel lines in a plane such that a line of one set is perpendicular to a line of the second set, then every line in one set is perpendicular to every line in the second set.

Definition. Two segments are *parallel segments* if, and only if, they lie in parallel lines.

Definition. A quadrilateral is a *parallelogram* if each of its sides is parallel to the side opposite it.

Notation. Parallelogram $ABCD$ is written: $\square ABCD$.

Comment. In $\square ABCD$, $\overline{AB} \parallel \overline{CD}$, $\overline{BC} \parallel \overline{DA}$.

■ THEOREM 9–5.–In any parallelogram, each side is congruent to the side opposite it.

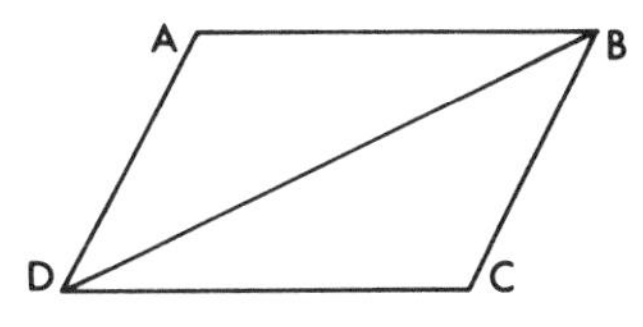

Proof. By Theorem 9–3, $\angle ABD \cong \angle CDB$, $\angle ADB \cong \angle CBD$. Since $\overline{BD} \cong \overline{DB}$, $\triangle ABD \cong \triangle CDB$, leading to the conclusion that $\overline{AB} \cong \overline{CD}$, and $\overline{AD} \cong \overline{CB}$, Q.E.D.

■ COROLLARY 9–5–1.–The diagonal of a parallelogram forms two congruent triangles with the sides of the parallelogram.

■ COROLLARY 9–5–2.–In any parallelogram, each angle is congruent to its opposite and supplementary to its adjacent angle.

■ COROLLARY 9–5–3.–Given two parallel lines, then segments that have their end points in the given lines and are perpendicular to them have the same length.

Definition. The length of any segment in Collary 9–5–3 is the distance between the two parallel lines.

Comment. Corollary 9–5–3 is sometimes phrased, "Parallel lines are everywhere equidistant."

■ THEOREM 9–6 (Converse of Theorem 9–5).—If each side of a quadrilateral is congruent to the side opposite it, then the quadrilateral is a parallelogram.

Proof. Let the quadrilateral be $ABCD$. Then by Postulate 20 (S.S.S. Postulate) $\triangle ABD \cong \triangle CDB$, $\angle ABD \cong \angle CDB$, and $\overline{AB} \parallel \overline{CD}$. Similarly, $\overline{AD} \parallel \overline{CB}$, Q.E.D.

Comment. Theorems 9–5 and 9–6 can be stated as one theorem as follows: A quadrilateral is a parallelogram if, and only if, each side is congruent to the side opposite.

■ COROLLARY 9–6–1.–If each angle of a quadrilateral is congruent to its opposite angle, then the quadrilateral is a parallelogram.

■ THEOREM 9–7.–If two sides of a quadrilateral are parallel and congruent, then the quadrilateral is a parallelogram.

■ THEOREM 9–8.–The diagonals of a quadrilateral bisect each other if, and only if, the quadrilateral is a parallelogram.

Comment. If the quadrilateral is $ABCD$, then two statements are to be proved: If $ABCD$ is a parallelogram, $\overline{AC}$ and $\overline{BD}$ bisect each other. If $\overline{AC}$ and $\overline{BD}$ bisect each other, then $ABCD$ is a parallelogram.

Definition. A parallelogram is a *rhombus* if, and only if, it is equilateral.

■ THEOREM 9–9.–A parallelogram is a rhombus if, and only if, each diagonal bisects one of its angles.

■ THEOREM 9–10.–A parallelogram is a rhombus if, and only if, its diagonals are perpendicular to each other.

Definition. A parallelogram is a *rectangle* if, and only if, one of its angles is a right angle.

■ THEOREM 9–11.–A parallelogram is a rectangle if, and only if, it is equiangular.

■ THEOREM 9–12.–A parallelogram is a rectangle if, and only if, its diagonals are congruent.

Definition. A parallelogram is a *square*

if, and only if, it is a rhombus and a rectangle.

Comment. The square has all the properties of rectangles and rhombuses.

Definition. A quadrilateral is a *trapezoid* if, and only if, exactly two of its sides are parallel. The sides that are parallel are its *bases*, and the nonparallel sides are its *arms* (or *legs*). The angles containing a base are a *pair of base angles*.

Definition. A trapezoid is an *isosceles trapezoid* if, and only if, its arms are congruent.

■ THEOREM 9-13.–A trapezoid is an isosceles trapezoid if, and only if, a pair of base angles are congruent.

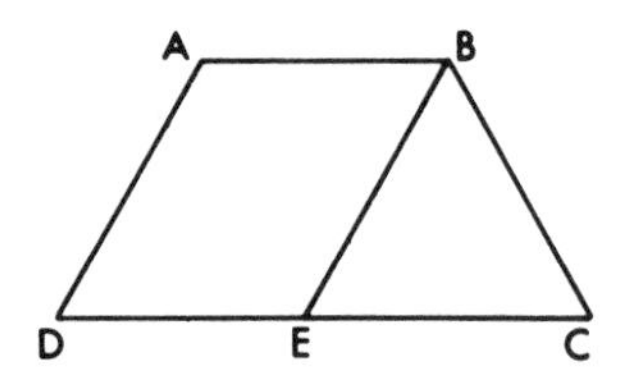

Proof. Let $ABCD$ be the trapezoid with $\overline{AB} \parallel \overline{CD}$. Let E be a point in $\overleftrightarrow{DC}$, such that $\overline{BE} \parallel \overline{AD}$. Then $\overline{AD} \cong \overline{BE}$. If $\overline{AD} \cong \overline{BC}$, then $\overline{BE} \cong \overline{BC}$ and $\angle C \cong \angle BEC \cong \angle D$. If $\angle C \cong \angle D$, then $\angle BEC \cong \angle C$ and $\overline{BC} \cong \overline{BE} \cong \overline{AD}$, Q.E.D.

■ THEOREM 9-14.—A trapezoid is an isosceles trapezoid if, and only if, its diagonals are congruent.

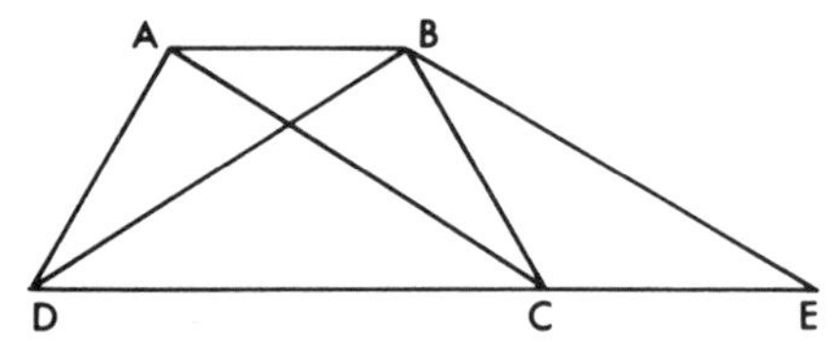

Proof. Let $ABCD$ be the trapezoid with $\overline{AB} \parallel \overline{CD}$. Let E be in $\overrightarrow{DC}$ such that $\overline{BE} \parallel \overline{AC}$. Then $\overline{AC} \cong \overline{BE}$. If $\overline{AC} \cong \overline{BD}$, then $\overline{BD} \cong \overline{BE}$, $\angle BDE \cong \angle BED \cong \angle ACD$, $\triangle ACD \cong \triangle BDC$ (S.A.S.), and $\overline{AD} \cong \overline{BC}$. If $\overline{AD} \cong \overline{BC}$, $\angle ADC \cong \angle BCD$, $\triangle ADC \cong \triangle BCD$, and $\overline{AC} \cong \overline{BD}$, Q.E.D.

■ THEOREM 9-15.–The sum of the measures of the angles of a triangle is 180.

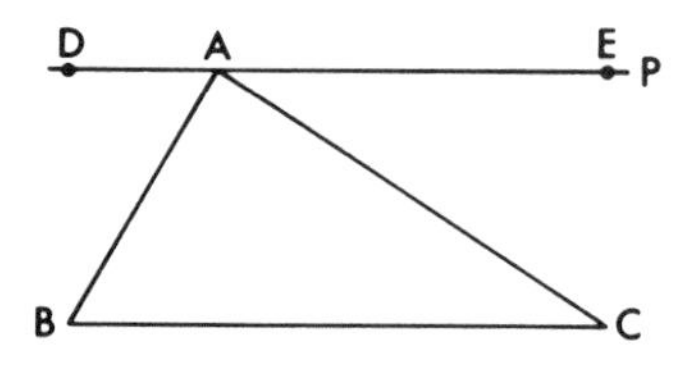

Proof. Let the vertices of the triangle be A, B, and C. There is a unique line p, containing A and parallel to $\overleftrightarrow{BC}$. Let D be a point on p such that D and C are on opposite sides of $\overleftrightarrow{AB}$. Let E be a point on p such that E and B are on opposite sides of $\overleftrightarrow{AC}$. By Corollary 7-8-1, $m\angle DAB + m\angle BAC + m\angle CAE = 180$. By Theorem 9-3, $m\angle DAB = m\angle ABC$, and $m\angle EAC = m\angle ACB$. By the substitution principle, $m\angle ABC + m\angle BAC + m\angle ACB = 180$, Q.E.D.

Comment. In the geometry developed by Nikolai Lobachevsky, this sum is less than 180; in the geometry developed by Bernhard Riemann, it is greater than 180.

■ COROLLARY 9-15-1.–The measure of an exterior angle of a triangle is equal to the sum of the measures of the remote interior angles.

■ COROLLARY 9-15-2.–Given a correspondence between the vertices of two triangles, if two pairs of corresponding angles are congruent, then the third pair of corresponding angles is congruent.

■ COROLLARY 9-15-3.–(S.A.A. Theorem.) Given a correspondence between the vertices of two triangles, if two angles and a side opposite one of them in one triangle are congruent to the corresponding parts of the second triangle, then the correspondence is a congruence.

■ COROLLARY 9-15-4.–The sum of the measures of the angles of a quadrilateral is 360.

Comment. If $ABCD$ is the quadrilateral, then the convexity of the quadrilateral permits the conclusion that $\overrightarrow{AC}$ is between $\overrightarrow{AB}$ and $\overrightarrow{AD}$. This allows the conclusion that $m\angle BAC + m\angle CAD = m\angle BAD$, a key step in the proof of this theorem.

Definition. If one of the angles of a triangle is a right angle, the triangle is a *right triangle*. The side opposite the right angle is the *hypotenuse of the right triangle*. Each of the other two sides is a *leg of the right triangle*.

■ COROLLARY 9-15-5.–The angles of a right triangle that are opposite the legs are complementary.

■ THEOREM 9-16 (Hypotenuse-Leg Theorem).—Given a correspondence between the vertices of two right triangles in which the vertices of the two right angles correspond, if the hypotenuse and leg of one triangle are congruent to the corresponding parts of the other triangle, then the correspondence is a congruence.

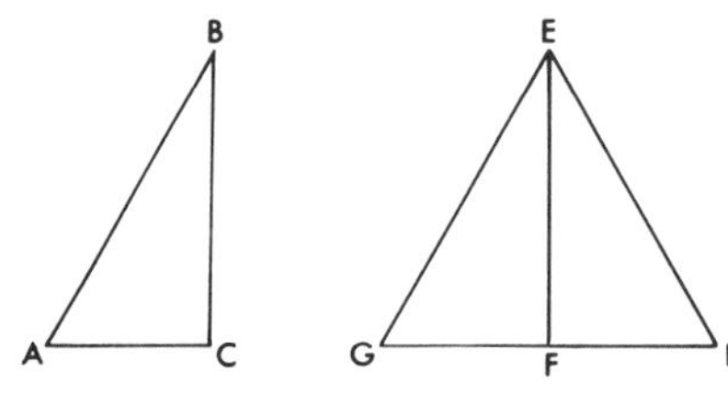

Proof. Let the triangles be $\triangle ABC$ and $\triangle DEF$, with $\angle C$ and $\angle F$ as the right angles. Let $\overline{AB} \cong \overline{DE}$ and $\overline{BC} \cong \overline{EF}$. On the ray opposite $\overrightarrow{FD}$, there is a point G such that $\overline{FG} \cong \overline{CA}$. $\triangle ABC \cong \triangle GEF$ (S.A.S.) and hence, $\angle A \cong \angle G$. But $\overline{GE} \cong \overline{AB} \cong \overline{DE}$. Therefore, $\angle D \cong \angle G \cong \angle A$, and by Corollary 9-15-3 (S.A.A. Theorem), $\triangle ABC \cong \triangle DEF$, Q.E.D.

Definition. If the members of $\{x,y\}$ and $\{x',y'\}$ are numbers such that x corresponds to x', y corresponds to y', and $x > y$, then the numbers of $\{x,y\}$ are unequal in the same order as the corresponding numbers of $\{x',y'\}$ if, and only if, $x' > y'$.

■ THEOREM 9-17.–If the length of two sides of a triangle are unequal, then the measures of the angles opposite these sides are *unequal in the same order.*

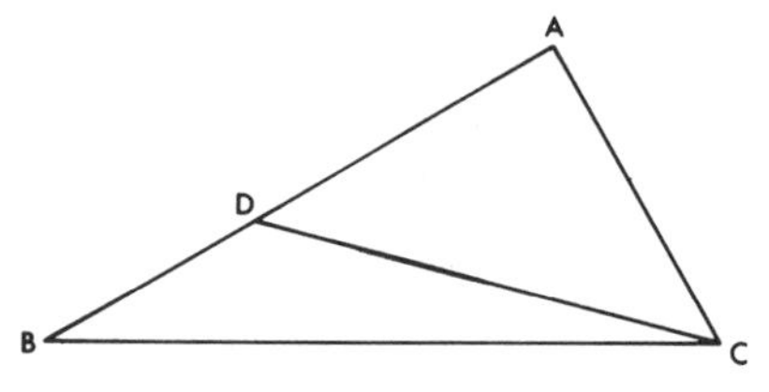

Proof. Let the triangle be $\triangle ABC$ with $\overline{AB} > \overline{AC}$. Then by Theorem 6-4 (Point-Plotting Theorem), there is a point D between A and B such that $\overline{AD} = \overline{AC}$. Therefore, $\overrightarrow{CD}$ is between $\overrightarrow{CA}$ and $\overrightarrow{CB}$ and $m\angle ACB > m\angle ACD = m\angle ADC$. But $m\angle ADC > \angle B$. Therefore, by the transitive property of order, $m\angle ACB > m\angle ABC$.

■ THEOREM 9-18.–If the measures of two angles of a triangle are unequal, the lengths of the sides opposite these angles are unequal in the same order.

■ COROLLARY 9-18-1.–The hypotenuse of a right triangle is the longest side of the triangle.

■ COROLLARY 9-18-2.–Given a line and a point not on this line, the shortest segment joining the given point to the given line is the segment perpendicular to the line.

Definition. The distance between a point and a line not containing this point is the length of the perpendicular segment joining the point to the line. The distance between a line and a point on the line is defined as zero.

■ COROLLARY 9-18-3.–If $\overline{AB}$ is the longest side of $\triangle ABC$, then the perpendicular segment joining C to $\overleftrightarrow{AB}$ intersects $\overline{AB}$ between A and B.

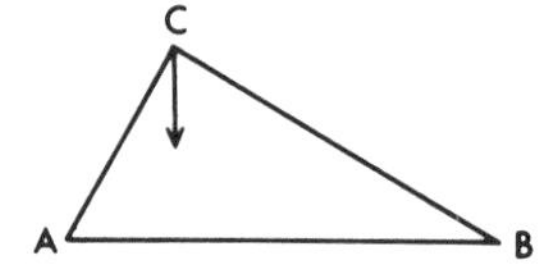

Proof. Let the perpendicular segment from C to $\overleftrightarrow{AB}$ be $\overline{CD}$. If $D = A$, then $\overline{CB} > \overline{AB}$. If $D = B$, then $\overline{CA} > \overline{AB}$. If D is in ray opposite $\overrightarrow{AB}$, then $\overline{CB} > \overline{BD} > \overline{BA}$. If D is in the ray opposite $\overrightarrow{BA}$, then $\overline{AC} > \overline{AD} > \overline{AB}$. Therefore, D is between A and B, Q.E.D.

■ THEOREM 9-19.–The set of points equally distant from the sides of an angle is the midray of the angle.

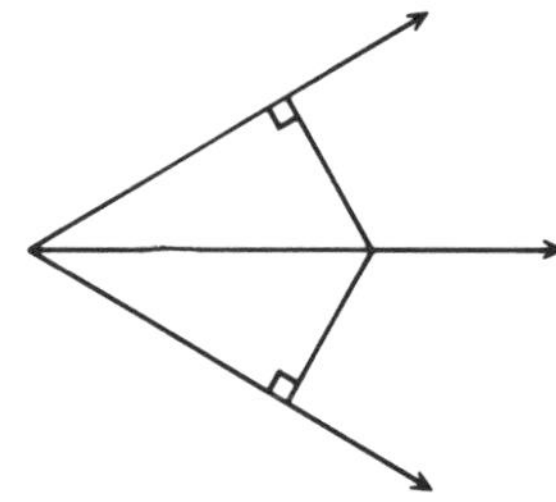

Comment. The proof of this theorem requires two parts: to show that every point on the midray is equally distant from the sides (S.A.A.); to show that if a point is equally distant from the sides, it is on the midray (Hypotenuse-Leg Theorem).

■ THEOREM 9-20.–The set of points in a plane, equally distant from two given points in the plane, is the perpendicular in the plane to the segment joining the given points and containing its midpoint.

■ THEOREM 9-21 (Triangle-Inequality Theorem).—The sum of the lengths of two sides of a triangle is greater than the length of the third side.

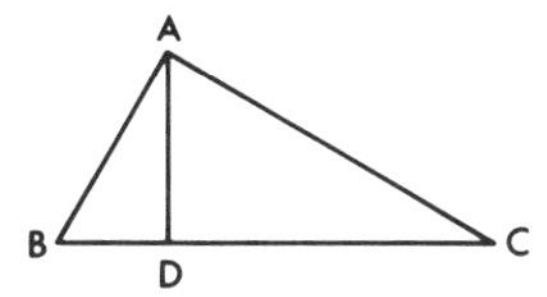

Proof. Let the triangle be $\triangle ABC$, with $\overline{BC}$ the longest side. Then by Corollary 9–18–3, the perpendicular from A to $\overleftrightarrow{BC}$ intersects it in D, between B and C. $\overline{AB} > \overline{BD}$ and $\overline{AC} > \overline{DC}$. Therefore, $\overline{AB} + \overline{AC} > \overline{BD} + \overline{DC} = \overline{BC}$. If $\overline{BC} = \overline{AB} > \overline{AC}$, then $\overline{AB} + \overline{AC} > \overline{BC}$. The proof is obvious, if $\overline{BC}$ is not the longest side, Q.E.D.

SECTION TEN

Similarity.—A formal treatment of similarity is suggested by a knowledge of the practical geometry of objects that have the same shape. Instances of such objects are models, architect's plans, and maps. As with congruences, similarities are based on one-to-one correspondences. In these correspondences, corresponding angles have equal measures and corresponding lengths are proportional. Since a formal treatment of congruent angles already has been presented, it is now time for a formal treatment of proportionality.

Definitions. Suppose the numbers in $\{p,q,r, \ldots\}$ have a one-to-one correspondence with the numbers in $\{a,b,c, \ldots\}$, p corresponding to a, q to b, r to c, and so on, where there are at least two numbers in each set and where the dots (. . .) indicate there may be any amount of numbers. The numbers $p, q, r, \ldots$ are proportional to the numbers $a,b,c, \ldots$ if, and only if, there is a nonzero number k such that $p = ka$, $q = kb$, $r = kc, \ldots$. The number k is the *constant of proportionality.*

Notation. $\bar{p}$ is used to mean "are proportional to." An ordered set of numbers is indicated by writing them in parentheses. Thus, $(3,6,5)\ \bar{p}\ (6,12,10)$ means that $3 = k \cdot 6$, $6 = k \cdot 12$, $5 = k \cdot 10$. It is obvious that the constant in this proportionality is $\frac{1}{2}$.

Properties of Proportionality

Reflexive property. $(a,b,c, \ldots)\ \bar{p}\ (a,b,c, \ldots)$. In this case, $k = 1$.

Symmetric property. If $(p,q,r, \ldots)\ \bar{p}\ (a,b,c, \ldots)$, then $(a,b,c, \ldots)\ \bar{p}\ (p,q,r, \ldots)$. If in the first-mentioned proportionality the constant of proportionality is k, it is $\frac{1}{2}k$ in the second.

Transitive property. If $(a,b,c, \ldots)\ \bar{p}\ (d,e,f, \ldots)$ and $(d,e,f, \ldots)\ \bar{p}\ (h,i,k, \ldots)$, then $(a,b,c, \ldots)\ \bar{p}\ (h,i,k, \ldots)$.

Addition property. If $(a,b,c, \ldots)\ \bar{p}\ (r,s,t, \ldots)$, then $(a + b + c + \ldots, a,b,c, \ldots)\ \bar{p}\ (r + s + t + \ldots, r,s,t, \ldots)$. This is true because $a = kr$, $b = ks$, $c = kt, \ldots$, and $(a + b + c + \ldots) = k(r + s + t + \ldots)$.

Examples. Since $(10,15,25)\ \bar{p}\ (2,3,5)$, then by the addition property of proportionality $(50,10,15,25)\ \bar{p}\ (10,2,3,5)$.

Given: $(a,b,7)\ \bar{p}\ (4,8,3)$. To find: a, b: $7 = 3k$; therefore $k = \frac{7}{3}$. Hence, $a = \frac{7}{3} \cdot 4 = \frac{28}{3}$ and $b = \frac{7}{3} \cdot 8 = \frac{56}{3}$. Given: $(a,b,c)\ \bar{p}\ (d,e,f)$. To write a proportionality starting with (b,c,a), (d,e,f) must be rewritten to maintain the one-to-one correspondence. Hence, $(b,c,a)\ \bar{p}\ (e,f,d)$.

Definition. A proportionality is a *proportion* if, and only if, each of its sets of numbers consist of two numbers.

Properties of Proportions.—Given that a, b, c, and d are positive numbers, then:

Inversion property. If $(a,b)\ \bar{p}\ (c,d)$, then $(b,a)\ \bar{p}\ (d,c)$.

Alternation property. If $(a,b)\ \bar{p}\ (c,d)$, then $(a,c)\ \bar{p}\ (b,d)$.

Product property. $(a,b)\ \bar{p}\ (c,d)$ if, and only if, $ad = bc$.

Comment. $(a,b)\ \bar{p}\ (c,d)$ is also written $a : b = c : d$ or $a/b = c/d$.

Example. Given: $(4,9)\ \bar{p}\ (x,2)$. To find x, use the product property to write $9x = 8$, and $x = \frac{8}{9}$. By the inversion property, $(9,4)\ \bar{p}\ (2,x)$. By the alternation property, $(4,x)\ \bar{p}\ (9,2)$. By the addition property, $(13,4)\ \bar{p}\ (x + 2,x)$ or $(13,9)\ \bar{p}\ (x + 2,2)$.

Definition. A one-to-one correspondence between the vertices of two triangles (not necessarily distinct) such that corresponding angles are congruent and corresponding sides are proportional, is called a *similarity,* and the two triangles are said to be *similar* to one another.

Notation. The statement $\triangle ABC$ is similar to $\triangle DEF$ is abbreviated: $\triangle ABC \sim \triangle DEF$.

Comment. The definition may be generalized for the one-to-one correspondence between two convex polygons having the same number of sides.

In any similarity there are three conditions to be satisfied: (1) There is a correspondence between all the vertices of one polygon with all the vertices of a second polygon. For this correspondence: (2) Corresponding angles are congruent. (3) Corresponding sides are proportional.

Notation. Given $\triangle ABC$, the length of the side opposite A is designated by a; the length of the side opposite B, by b; and so on. In general, the lower-case letter designates the length of the side opposite the vertex with the corresponding capital letter. Thus, if $\triangle ABC \sim \triangle DEF$, it follows that $(a,b,c)\ \bar{p}\ (d,e,f)$.

Definition. If $\triangle ABC \sim \triangle DEF$, the *proportionality constant of the similarity* is k, the proportionality constant in the proportionality $(a,b,c)\ \bar{p}\ (d,e,f)$.

AN ARCHITECT'S MODEL illustrates one-to-one correspondence, or *similarity.* In these correspondences, angles have equal measures and corresponding lengths are proportional.

Comment. Given $\triangle ABC \sim \triangle DEF$, then $\triangle ABC \cong \triangle DEF$ if, and only if, $k = 1$. From this fact it is obvious that a congruence between two triangles is a special case of similarity.

■ THEOREM 10–1.–The similarity relation between convex polygons is reflexive, symmetric, and transitive.

Comment. The proof follows directly from the corresponding properties of the congruence relation between angles and the proportionality relations between two sets of numbers.

Postulate 22 (Proportional-Segments Postulate).—If a line is parallel to one side of a triangle and intersects the other two sides in interior points, then the measures of one of those sides and the two segments into which it is cut are proportional to the measures of the corresponding segments in the other side.

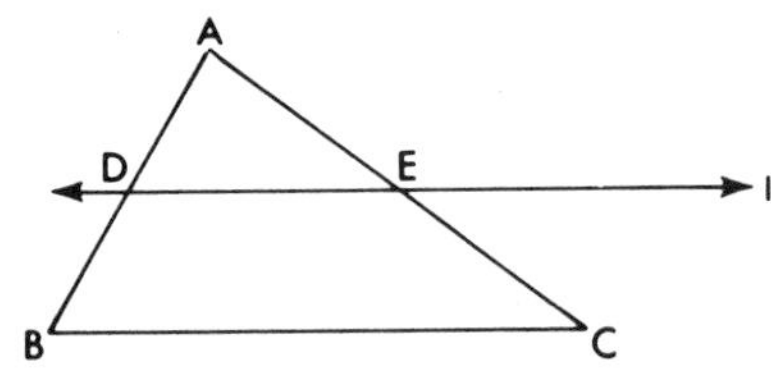

Comment. Postulate 22 is sometimes stated more briefly as follows. If a line is parallel to one side of a triangle, it divides the other two sides proportionally. In terms of the diagram, if $l \parallel \overline{BC}$ and l cuts $\overline{AB}$ in D, between A and B, and cuts $\overline{AC}$ in E, between A and C, then $(AB, AD, DB)\ \bar{p}\ (AC, AE, EC)$.

■ THEOREM 10–2.–Given three coplanar parallel lines and two transversals, then the corresponding segments on the transversals are proportional.

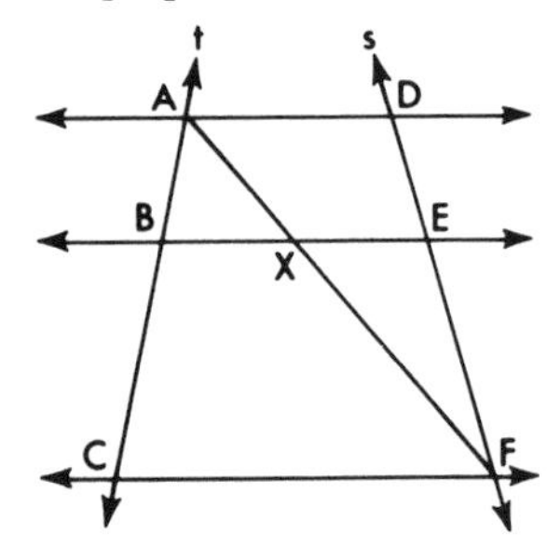

Proof. In the diagram above, let $\overleftrightarrow{AD} \parallel \overleftrightarrow{BE} \parallel \overleftrightarrow{CF}$, where A, B, and C are in transversal t and D, E, and F are in transversal s. By Corollary 9–4–1, $\overleftrightarrow{AF}$ intersects $\overleftrightarrow{BE}$ in X. Then by Postulate 22, $(AB,BC)\ \bar{p}\ (AX,XF)\ \bar{p}\ (DE,EF)$. Therefore, by the transitive property of a proportionality, $(AB,BC)\ \bar{p}\ (DE,EF)$, Q.E.D.

■ THEOREM 10–3.–If a triangle and a positive number k are given, there is a triangle that is similar to the given triangle with proportionality constant k.

Proof. Let the given triangle be $\triangle ABC$. If $k < 1$, there is a point D in $\overline{AB}$ such that $AD = kc$. The parallel to $\overline{BC}$ containing D cuts $\overline{AC}$ in E, such that $AE = kb$. Also, there is a point F in $\overline{AB}$ such that $BF = kc$. The parallel to $\overline{AC}$ containing F cuts $\overline{BC}$ in G, such that $BG = ka$. $\triangle ADE \cong \triangle FBG$ (A.S.A.). Therefore, $DE = ka$. $\triangle ADE \sim \triangle ABC$ because corresponding angles are congruent and corresponding sides are proportional, with k as the proportionality constant.

If $k = 1$, then the given triangle is congruent to itself and hence similar.

If $k > 1$, a similar proof can be constructed with D in $\overrightarrow{AB}$ and F in $\overrightarrow{BA}$ such that $AD = BF = kc$, Q.E.D.

■ THEOREM 10–4 (S.S.S. Similarity Theorem).—A correspondence between two triangles for which corresponding sides are proportional is a similarity.

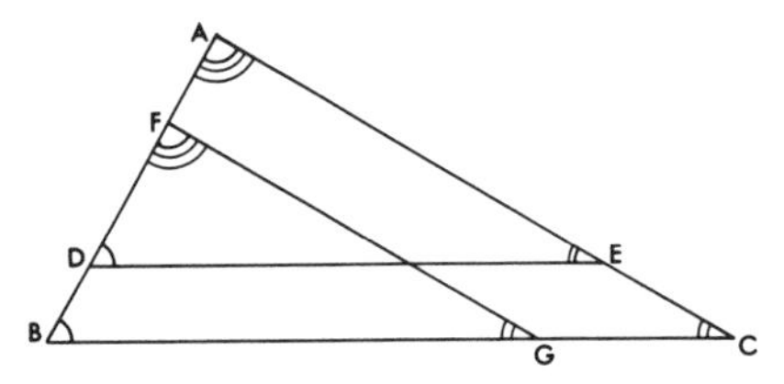

Proof. Let the triangles be $\triangle ABC$ and $\triangle A'B'C'$, with A corresponding to A', B to B', and C to C'. Then $(a,b,c)\ \bar{p}\ (a',b',c')$, with the proportionality factor k. There exists $\triangle A''B''C''$, similar to $\triangle ABC$, with the proportionality factor k, by Theorem 10–2. Its sides have lengths ka, kb, and kc, and hence $\triangle A''B''C'' \cong \triangle A'B'C'$. Therefore, by the transitivity property of similarities, $\triangle ABC \sim \triangle A'B'C'$, Q.E.D.

■ THEOREM 10–5 (S.A.S. Similarity Theorem).—If for a correspondence between two triangles, two sides of one triangle are proportional to the corresponding sides of the other and the included angles are congruent, then the correspondence is a similarity.

Proof. Let the triangles be $\triangle ABC$ and $\triangle A'B'C'$, with $(a,b)\ \bar{p}\ (a',b')$ and $\angle C = \angle C'$. There is a $\triangle A''B''C''$ similar to $\triangle ABC$ with proportionality factor $k = a''/a = b''/b$. Therefore, $a'' = ka = a'$, $b'' = kb = b'$. Moreover, $\angle C'' = \angle C = \angle C'$. Therefore, $\triangle A''B''C'' \cong \triangle A'B'C'$ (S.A.S.), and $\triangle ABC \sim \triangle A'B'C'$, Q.E.D.

■ COROLLARY 10–5–1.–If a line divides two sides of a triangle into proportional segments, it is parallel to the third side.

■ COROLLARY 10–5–2.–The line that bisects one side of a triangle and is parallel to a second side bisects the third side.

■ COROLLARY 10–5–3.–The segment that joins the midpoints of two sides of a triangle is parallel to the third side and is half as long as the third side.

■ THEOREM 10–6 (A.A. Similarity Theorem).— If for a given correspondence between two triangles, two angles of one triangle are congruent to the corresponding angles of the other, then the correspondence is a similarity.

Proof. In $\triangle ABC$ and $\triangle A'B'C'$, let $\angle A \cong \angle A'$, and $\angle B \cong \angle B'$. Take k such that $c' = kc$, and consider $\triangle A''B''C'' \sim \triangle ABC$ with the proportionality constant k. Then $c'' = kc$ and $c' = c''$. Also $\angle A'' \cong \angle A \cong \angle A'$ and $\angle B'' \cong \angle B \cong \angle B'$. Therefore, $\triangle A'B'C' \cong \triangle A''B''C''$ and $\triangle A'B'C' \sim \triangle ABC$, Q.E.D.

Definition. Given a point not in a line, the intersection of the perpendicular to the line containing the given point is called the *projection of the point on the line.* Given a point in a line, its *projection on that line* is defined to be the point itself.

Definition. The segment that joins a vertex of a triangle to the projection of the vertex on the line of the opposite side is called the *altitude of the triangle from that vertex.*

■ COROLLARY 10–6–1.–If two triangles are similar, the lengths of the sides and the altitudes of the first triangle are proportional to the lengths of the corresponding segments of the second triangle.

Definition. The *projection of a segment on a line* is the set of projections of all the points in the segment on the line.

Comment. As here defined, an altitude, or a projection of a segment on a line, is a set of points. The phrases "length of an altitude" or "length of the projection of a segment on a line" are often abbreviated to "altitude" or "projection of a segment on a line," respectively, for convenience. The context indicates which meaning is intended.

■ THEOREM 10–7.–In any right triangle, the altitude to the hypotenuse separates the triangle into two similar triangles, each similar to the original right triangle.

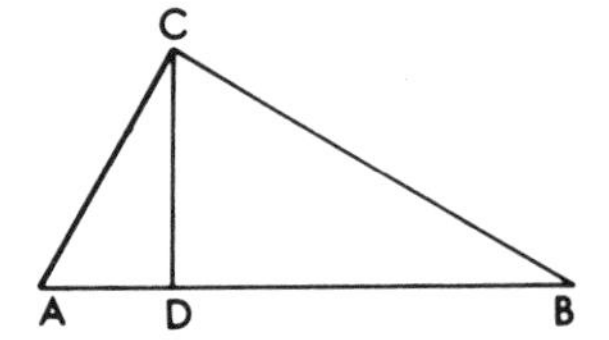

Proof. Since the hypotenuse $\overline{AB}$ is the longest side, the projection of C on $\overline{AB}$, D, is between A and B. $\triangle CAD \sim \triangle BAC$, and $\triangle BCD \sim \triangle BAC$ (A.A. Similarity Theorem 10–6). By the transitive property, $\triangle CAD \sim \triangle BCD$, Q.E.D.

■ COROLLARY 10–7–1.–The square of the altitude to the hypotenuse of a right triangle is equal to the product of the projection of the legs on the hypotenuse.

■ COROLLARY 10–7–2.–The square of the length of either leg of a right triangle is equal to the product of the projection of that leg on the hypotenuse and the length of the hypotenuse.

■ THEOREM 10–8 (Pythagorean Theorem).—In any right triangle, the square of the length of the hypotenuse is equal to the sum of the squares of the lengths of the two legs.

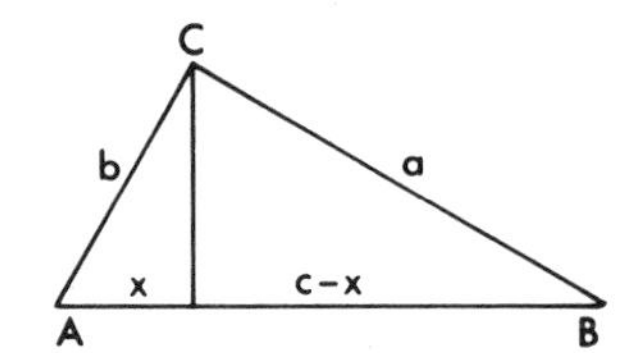

Proof. Let the triangle be $\triangle ABC$, with $\angle C$ the right angle. Let the projection of $\overline{AC}$ on $\overline{AB}$ be x and the projection of $\overline{CB}$ on $\overline{AB}$ be $c - x$. By Corollary 10–7–2, $b^2 = cx$, and $a^2 = c(c - x)$, or $c^2 - cx$. By the addition property of equality, $a^2 + b^2 = c^2$, Q.E.D.

Comment. The Pythagorean Theorem is probably the most famous of all mathematical theorems, and was known to the ancient Egyptians and Indians. However, Pythagoras was the first to prove it formally. A key principle in surveying and construction, the theorem is likewise important in scientific and mathematical studies. It has more formal proofs than any other theorem. (One of these proofs was discovered by President James A. Garfield.)

■ THEOREM 10–9 (Converse of the Pythagorean Theorem).—If the square of the length of one side of a triangle is equal to the sum of the squares of the lengths of the other two sides, then the triangle is a right triangle with the first side as hypotenuse.

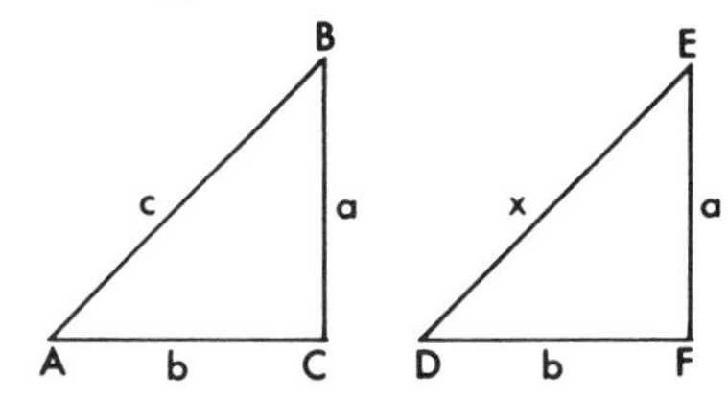

Proof. Let the triangle be $\triangle ABC$, with $c^2 = a^2 + b^2$. There is a right triangle, DEF, such that $\overline{DF} = b$ (Ruler Postulate), $\angle F$ is a right angle (Protractor Postulate) and $\overline{FE} = a$ (Point Plotting Theorem). If $\overline{DE} = x$, then by the Pythagorean Theorem, $x^2 = a^2 + b^2$. Since x and c are positive numbers, $x = c$, and $\triangle ABC \cong \triangle DEF$. Therefore, $\angle C$ is a right angle, Q.E.D.

Comment. Because it is convenient, the Pythagorean Theorem and its converse often are abbreviated, "the square of a side" instead of "the square of the length of a side." If this convenience is used, the Pythagorean Theorem and its converse may be stated: A triangle is a right triangle if, and only if, the square of one side is equal to the sum of the squares of the other two sides.

Comment. Certain families of similar right triangles occur frequently in engineering and mathematical problems, and are listed below.

3, 4, 5 triangles. These include all triangles whose sides have lengths $3k$, $4k$, $5k$, where k is any positive number. They are right triangles, since $(3k)^2 + (4k)^2 = (5k)^2$ for any k.

Example. If the lengths of the legs of a right triangle are 12 inches and 16 inches, the hypotenuse must be 25 inches long ($k = 3$).

5, 12, 13 triangles. These include all triangles whose lengths are $5k$, $12k$, $13k$.

Example. If the hypotenuse of a triangle is $13 \cdot 4$ inches long and one leg is $12 \cdot 4$ inches long, the other leg must be $5 \cdot 4$ inches long.

Right isosceles triangles. In this case, $(a,b,c)\ \bar{p}\ (1,1,\sqrt{2})$ and each acute angle measures 45.

Example. If the hypotenuse of a right isosceles triangle is 20 feet long, and the length of each leg is represented by x, then $(x,x,20)\ \bar{p}\ (1,1,\sqrt{2})$; the constant of proportionality is $20/\sqrt{2}$ and $x = 20/\sqrt{2}$.

30, 60, 90 triangles. In these triangles, if $m\angle A = 30$, $m\angle B = 60$, $m\angle C = 90$, then $(a,b,c)\ \bar{p}\ (1,\sqrt{3},2)$.

Example. If $b = 12$ inches, then $(a,12,c)\ \bar{p}\ (1,\sqrt{3},2)$; the constant of proportionality is $12/\sqrt{3}$ and $a = 12/\sqrt{3}$, $c = 24/\sqrt{3}$.

SECTION ELEVEN

Circles.—A *circle* is the set of all points in a given plane whose distances from a given point in the plane are a given number. The given point is the *center of the circle*. The given number is the *radius of the circle*. A *chord of a circle* is a segment that joins two points of the circle. A *diameter of a circle* is a chord that contains the center. A *radius* (plural, *radii*) *of a circle* is also defined as the segment that joins the center to any of its points. (The context will suggest whether a radius is a number or a set of points.) The *end point* of a radius that is a point of the circle is its *outer end*.

Definition. Circles with congruent radii are *congruent circles*.

■ THEOREM 11–1.—The radii of a circle or of congruent circles are congruent.

■ THEOREM 11–2.—The diameters of a circle or of congruent circles are congruent.

Definition. The *interior of a circle* is the set of all points in the plane of the circle whose distances from the center are less than its radius. The *exterior of a circle* is the set of all points in the plane of the circle whose distances from the center of the circle are greater than its radius.

Comment. "The inside of a circle" is frequently used to mean "the interior of a circle." A point is *on* or *in* a circle if it is a point of the circle.

Definition. A *tangent to a circle* is a line in the plane of the circle that intersects the circle in exactly one point. This point is the *point of tangency*, or the *point of contact*. The circle and the line are *tangent* at this point.

Definition. A *secant to a circle* is a line in the plane of the circle that intersects the circle in exactly two distinct points.

■ THEOREM 11–3.—Given: a line and a circle in the same plane. If the projection of the center of the circle on this line is: (a) outside the circle, then the line is outside the circle; (b) on the circle, then the line is a tangent to the circle; (c) inside the circle, then the line is a secant to the circle.

Proof. Let the center be C, the line l, the projection of C on l, P, and radius r:

(a) If P is outside the circle, $\overline{CP} > r$. Since $\overline{CP}$ is the shortest distance from C to l, all points of C are exterior points.

(b) If P is on C, then $\overline{CP} = r$. $\overline{CP}$ is the shortest distance from C to l, and therefore all other points of l are exterior points. Hence, l is a tangent.

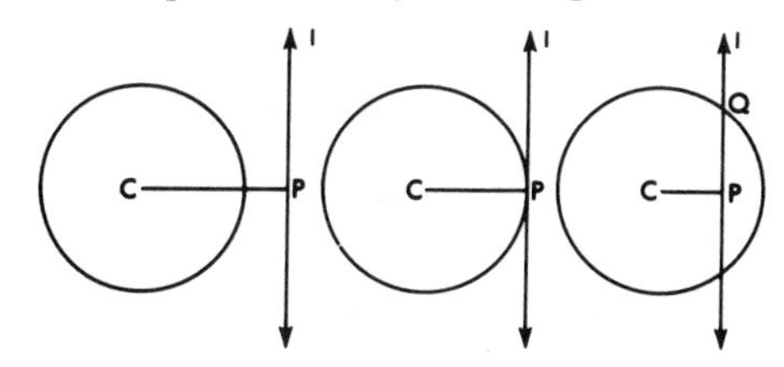

(c) If P is an interior point, then $\overline{CP} < r$. If Q is a point on l and also on the circle, then $(\overline{CP})^2 + (\overline{PQ})^2 = r^2$, or $(\overline{PQ})^2 = r^2 - (\overline{CP})^2$. Since $r > \overline{CP}$, and r and $\overline{CP}$ are positive numbers, $r^2 > (\overline{CP})^2$ and $r^2 - (\overline{CP})^2 > 0$. Therefore, $\overline{PQ}$ exists and is equal to $\sqrt{r^2 - (\overline{CP})^2}$. By Theorem 6–4 (Point-Plotting Theorem), Q can be found in either of the two rays contained in l with endpoint P. Therefore, there are two points that are on l and also on the circle. Hence, l is a secant, Q.E.D.

■ COROLLARY 11–3–1.—Given a circle and a coplanar line, the line is a tangent to the circle if, and only if, it is perpendicular to the radius of the circle at the outer end of the radius.

■ COROLLARY 11–3–2.—A diameter of a circle bisects a nondiameter chord of a circle if, and only if, it is perpendicular to the chord.

■ COROLLARY 11–3–3.—In the plane of a circle, the perpendicular bisector of a chord contains the center of the circle.

■ COROLLARY 11–3–4.—If a line in the plane of a circle contains an interior point of the circle, it intersects the circle in exactly two points.

■ THEOREM 11–4.—Chords of congruent circles are congruent if, and only if, their distances from the center are the same.

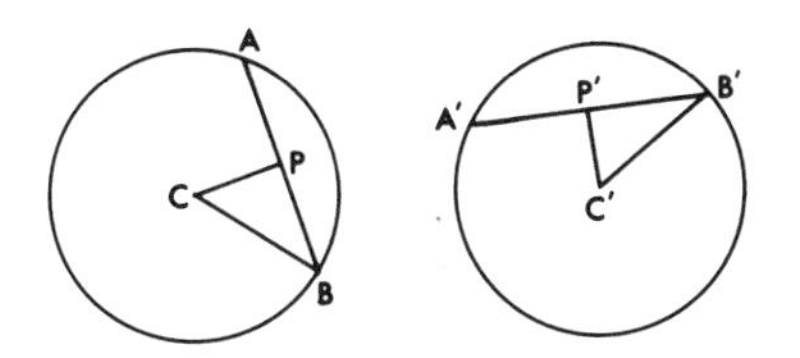

Proof. Let two circles have centers C and C', and let the chords be $\overline{AB}$ and $\overline{A'B'}$ in the respective circles. Let the projection of C on $\overline{AB}$ be P, and that of C' on $\overline{A'B'}$ be P'. By Theorem 10–8 (Pythagorean Theorem), $(\overline{CB})^2 = (\overline{CP})^2 + (\overline{PB})^2$ and $(\overline{C'B'})^2 = (\overline{C'P'})^2 + (\overline{P'B'})^2$. Since $\overline{CB} = \overline{C'B'}$, $(\overline{CP})^2 + (\overline{PB})^2 = (\overline{C'P'})^2 + (\overline{P'B'})^2$. It follows that $\overline{PB} = \overline{P'B'}$ if, and only if, $\overline{CP} = \overline{C'P'}$. By Corollary 11–3–2, $\overline{PB} = \frac{1}{2}\overline{AB}$ and $\overline{P'B'} = \frac{1}{2}\overline{A'B'}$. Therefore, $\overline{AB} \cong \overline{A'B'}$ if, and only if, $\overline{CP} = \overline{C'P'}$.

Definition. Two circles are *tangent* if, and only if, they are coplanar and tangent to the same line at the same point. Tangent circles are *internally tangent* if their centers lie on the same side of the common tangent, and *externally tangent* if their centers are on opposite sides of their common tangent.

■ THEOREM 11–5.—If two circles are tangent, then their centers are collinear with the point of contact.

Definition. A *central angle of a circle* is an angle whose vertex is the center of the circle.

Definition. Given: two distinct points of a circle, not the ends of a diameter of that circle. The set of points consisting of these two points and all the points of the circle in the interior of the central angle that contains the given points is called a *minor arc of the circle.* The set consisting of the two points and all the points of the circle in the exterior of the central angle is a *major arc of the circle.* If the two points are ends of a diameter of the circle, then the set of points consisting of the two points and all points of the circle on the same side of the diameter is a *semicircle.* The two given points in a minor arc, a major arc, or a semicircle are the *end points of the arc.*

Notation. An arc whose end points are A and B is denoted by $\widehat{AB}$. Since there are two arcs of a circle with these end-points, ambiguity can be avoided by using a third point of the arc, such as

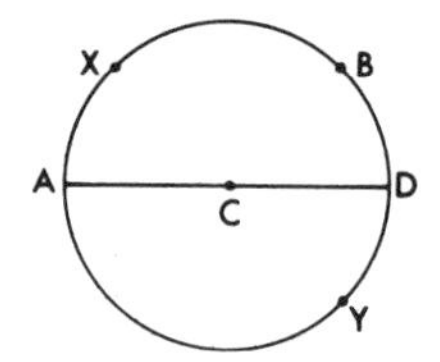

X and then designating the arc $\widehat{AXB}$. In the illustration $\widehat{AXB}$ is a minor arc and $\widehat{AYB}$ is a major arc. $\widehat{ABD}$ or $\widehat{AYD}$ is a diameter if $\overline{AD}$ contains C, the center of the circle.

Definition. If $\widehat{AXB}$ is any arc, then its *degree measure,* designated $m\widehat{AXB}$, is defined as follows:

If $\widehat{AXB}$ is a minor arc, $m\widehat{AXB}$ equals the measure of its associated central angle.

If $\widehat{AYB}$ is a major arc, and $\widehat{AXB}$ is the associated minor arc, then $m\widehat{AYB} = 360 - m\widehat{AXB}$.

If $\widehat{AB}$ is a diameter then $m\widehat{AB} = 180$.

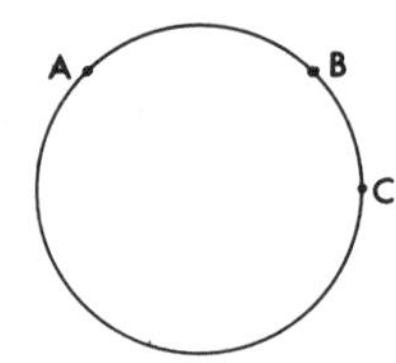

Postulate 23.—If $\widehat{AB}$ and $\widehat{BC}$ of a circle have only B in common and if the set of all points in $\widehat{AB}$ and $\widehat{BC}$ is an arc, $\widehat{AC}$, then $m\widehat{AB} + m\widehat{BC} = m\widehat{AC}$.

Comment. This postulate can be proved easily if $\widehat{AC}$ is a minor arc or a semicircle. If $\widehat{AC}$ is a major arc, the proof is difficult. In this article it is taken as a postulate.

Definition. An angle is *inscribed in an arc* if the angle contains the two end points of the arc and the vertex is a point of the arc, but not an end point. $\angle ABC$ is inscribed in a major arc, $\angle DEF$ is inscribed in a semicircle, and $\angle GHK$ is inscribed in a minor arc.

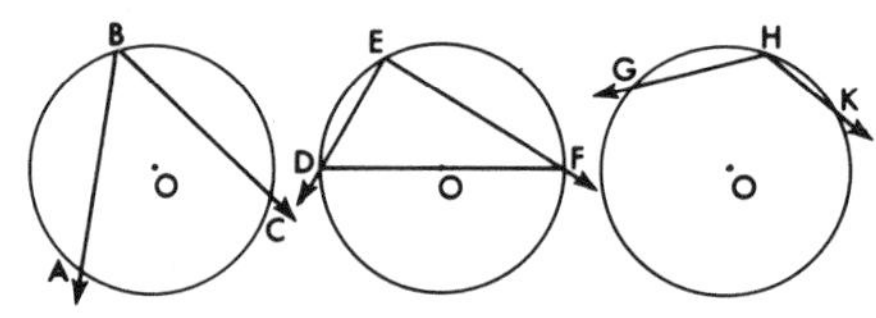

Definition. An angle *intercepts an arc* if each side of the angle contains one of the end points of the arc, and except for its end points, the arc is in the interior of the angle, as shown below.

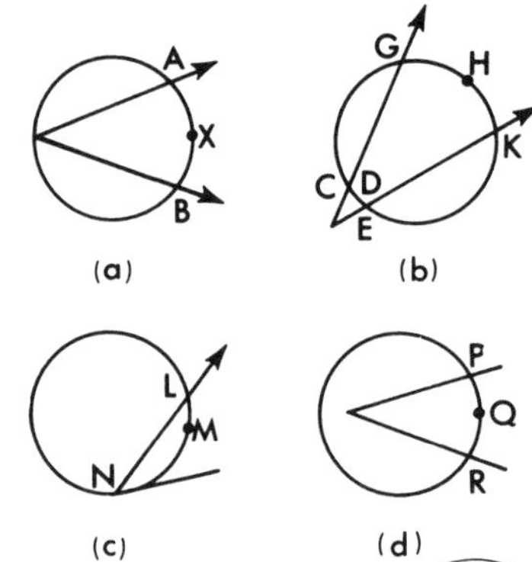

In (a) the angle intercepts $\widehat{AXB}$; in (b) the angle intercepts $\widehat{CDE}$ and $\widehat{GHK}$; in (c) the angle intercepts $\widehat{LMN}$; and in (d) the angle intercepts $\widehat{PQR}$.

■ THEOREM 11-6.—The measure of an inscribed angle is half the measure of the intercepted arc.

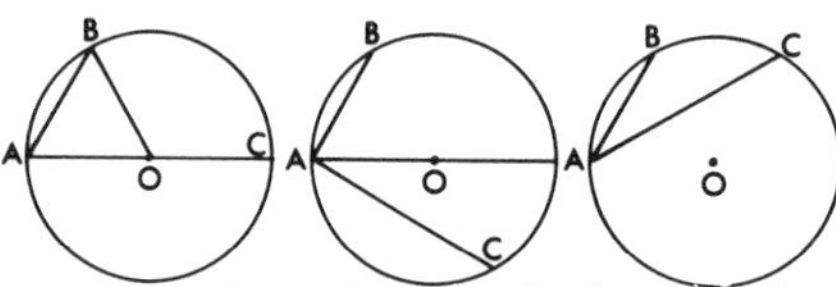

Proof. Let the inscribed angle be $\angle BAC$ intercepting $\widehat{BC}$. Let O be the center of the circle. Consider three cases:

(1) If O is on one side of $\angle A$, say in $\overrightarrow{BC}$, then $\overline{BO} \cong \overline{AO}$ and by Corollary 9-15-1, $m\angle BOC = 2m\angle A$. By the definition of an arc measure, $m\angle BOC = m\widehat{BC}$. Therefore, $2m\angle A = m\widehat{BC}$ and by the multiplication property of equality, $m\angle A = \frac{1}{2}m\widehat{BC}$.

(2) If O is an interior point of $\angle BAC$, then $\overrightarrow{AO}$ is between $\overrightarrow{AB}$ and $\overrightarrow{AC}$. Therefore $m\angle BAC = m\angle BAD + m\angle DAC$. But by example (1), $m\angle BAD = \frac{1}{2}m\widehat{BD}$, $m\angle DAC = \frac{1}{2}m\widehat{DC}$. By postulate 23 $m\widehat{BD} + m\widehat{DC} = m\widehat{BC}$. Therefore, $m\angle BAC = \frac{1}{2}m\widehat{BC}$.

(3) If O is an exterior point of $\angle BAC$, a proof similar to that of example (2) can be written by considering $\overrightarrow{AC}$ to be between $\overrightarrow{AB}$ and $\overrightarrow{AO}$, Q.E.D.

■ COROLLARY 11-6-1.—An angle inscribed in a semicircle is a right angle.

■ COROLLARY 11-6-2.—Angles inscribed in the same arc are congruent.

Definition. In congruent circles (not necessarily distinct) two arcs are congruent if they have the same degree measure.

■ COROLLARY 11-6-3.—Congruent angles inscribed in congruent circles intercept congruent arcs.

■ THEOREM 11-7.—In congruent circles, if two chords, not diameters, are congruent, then the associated minor arcs are congruent and the associated major arcs are congruent.

Comment. The proof can be based on Postulate 20 (S.S.S. Postulate) and the definition of the degree measure of an arc.

■ THEOREM 11-8.—In congruent circles, if two arcs are congruent, then the associated chords are congruent.

Definition. If the vertex of an angle is on a circle, and one side is contained in a tangent and its other side contains a chord, the angle is a *tangent-chord angle.*

Definition. If the vertex of an angle is an exterior point of a circle and its sides are contained in two secants, or two tangents, or a secant and a tangent, then it is, respectively, a *secant-secant angle,* a *tangent-tangent angle,* or a *secant-tangent angle.*

■ THEOREM 11-9.—The measure of a tangent-chord angle is one-half the measure of its intercepted arc.

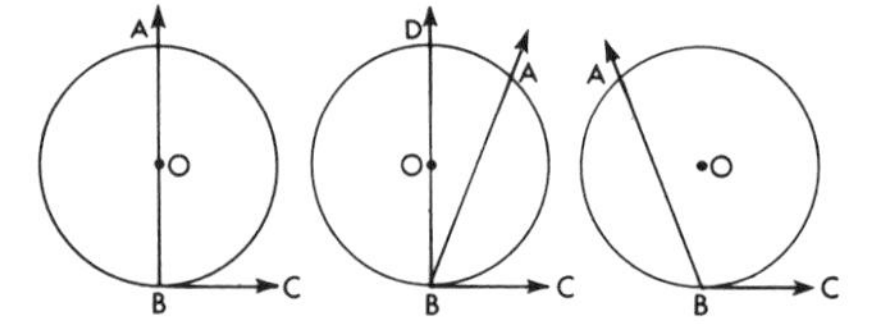

Proof. Let the angle be $\angle ABC$ with $\overrightarrow{BC}$ contained in tangent $\overleftrightarrow{BC}$ at B, and let O be the center of the circle. If O is in $\overrightarrow{BA}$, then $\overline{AB} \perp \overline{BC}$ and $m\angle ABC = 90$ or one-half the measure of a semicircle. If O is an exterior point of $\angle ABC$, and $\overline{BD}$ is a diameter, then $\overrightarrow{BA}$ is between $\overrightarrow{BD}$ and $\overrightarrow{BC}$ or $m\angle ABC = m\angle DBC - m\angle DBA = 90 - \frac{1}{2}\widehat{DA} = \frac{1}{2}\widehat{AB}$. A similar proof can be given if O is an interior point of $\angle ABC$, Q.E.D.

■ THEOREM 11-10.—If the vertex of an angle is an interior point of a circle and its sides are contained in two secants, then its measure is one-half the sum of its intercepted arc and the intercepted arc of its vertical angle.

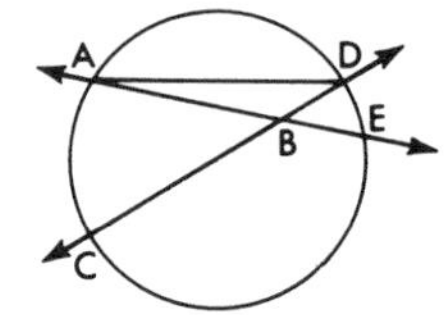

Proof. If $\angle ABC$ is the angle and the secants cut the circle in A, D, E, and C, then $m\angle ABC = m\angle BDA + m\angle BAD = \frac{1}{2}(m\widehat{AC} + m\widehat{DE})$, Q.E.D.

■ THEOREM 11-11.—The measure of a secant-secant angle, or a tangent-tangent angle, or a secant-tangent angle is one-half the difference of the intercepted arcs.

Example. Suppose as shown, $m\widehat{DA} =$

100, $\overline{DE} \parallel \overline{AB}$, $\overleftrightarrow{AB}$ is a tangent at A, and $\widehat{EF} \cong \widehat{FA}$. Then $\angle E \cong \angle A$,

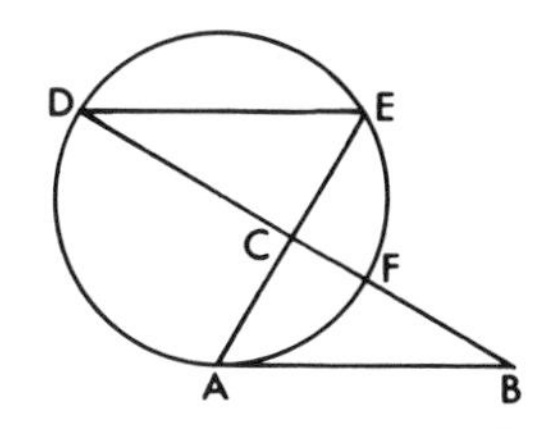

$m\angle E = 50 = m\angle A$, and $m\widehat{EFA} = 100$. $m\widehat{FA} = 50$, $m\angle B = \frac{1}{2}(100 - 50) = 25$, $m\angle DCA = \frac{1}{2}(100 + 50) = 75$.

Comment. The above angle measure theorems may be remembered by noting that when the vertex of the angle is on the circle (inscribed angle or tangent-chord angle), its measure is one-half the *measure* of the intercepted arc; when it is an interior point, its measure is one-half the *sum* of the measures of the intercepted arcs; when it is an exterior point, one-half the *difference* of the intercepted arcs.

Definition. If $\overleftrightarrow{AB}$ is a tangent to a circle at B, then $\overline{AB}$ is a *tangent segment* from A to the circle.

■ THEOREM 11-12.–The two tangent segments from an exterior point to a circle are congruent.

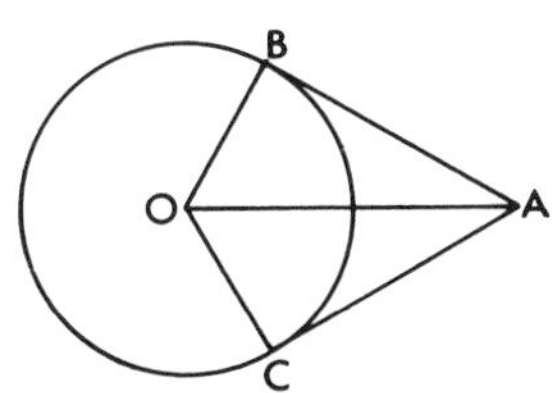

Proof. Let A be the exterior point, $\overline{AB}$ and $\overline{AC}$ the tangent segments, and O the center of the circle. Then $\triangle OBA \cong \triangle OCA$ by Theorem 9–16 (Hypotenuse-Leg Theorem), and $\overline{AB} \cong \overline{AC}$, Q.E.D.

■ COROLLARY 11-12-1.–The line containing an exterior point and the center of a circle bisects the tangent-tangent angle whose vertex is the exterior point.

Definition. If secant $\overleftrightarrow{CD}$ intersects a circle in A and B such that A is between C and B, then $\overline{CB}$ is the *secant segment* from C to the circle, and $\overline{CA}$ is its *external secant segment* from C to the circle.

■ THEOREM 11-13.–Given an exterior point of a circle, the product of the lengths of any secant segment from the point to the circle and its external secant segment is a constant.

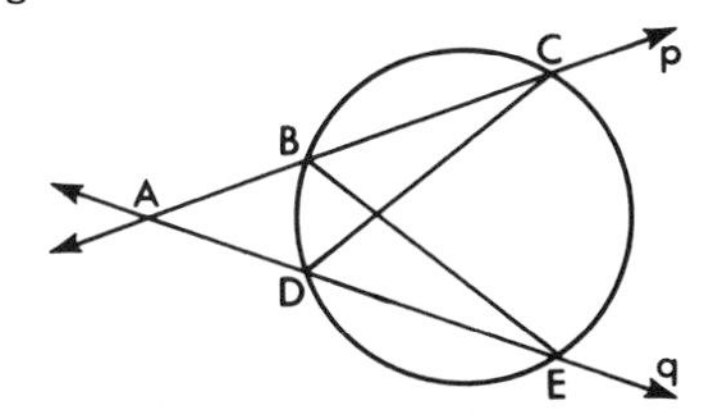

Proof. Let the given point be A and let p and q be any two secants containing A and intersecting the circle as shown. Then by Theorem 10–6 (A.A. Similarity Theorem) for similar triangles, $\triangle ADC \sim \triangle ABE$. It follows that $(\overline{AC},\overline{AD})$ p̄ $(\overline{AE},\overline{AB})$, and by the product property of proportionality, that $AC \cdot AB = AE \cdot AD$. Since p and q are any secants containing A, the proof is complete, Q.E.D.

■ THEOREM 11-14.–Given an exterior point of a circle, and a tangent segment and a secant segment from this point, the product of the lengths of the secant segment and its external secant segment is equal to the square of the length of the tangent segment.

Definition. The square of the length of a tangent segment from a point to a circle is the *power of the point* with respect to the circle.

■ THEOREM 11-15.–If two chords of a circle intersect, the product of the lengths of the segments of one is equal to the product of the lengths of the segments of the other.

Comment. The proof depends on showing that a pair of triangles are similar, as was done for Theorem 11–13.

Comment. For any circle, given an interior point of the circle and any chord containing this point, the product of the lengths of the chord's segments is a constant.

SECTION TWELVE

Areas.—Like the length of a segment, the measure of an angle, and the degree measure of an arc, area is also a measure and therefore a number. First, the mathematical object to which this kind of measure is applied must be defined.

Definition. A *triangular region* is the set of all points of a triangle and its interior. A *polygonal region* consists of a finite number of coplanar triangular regions. Examples are shown in the following diagram.

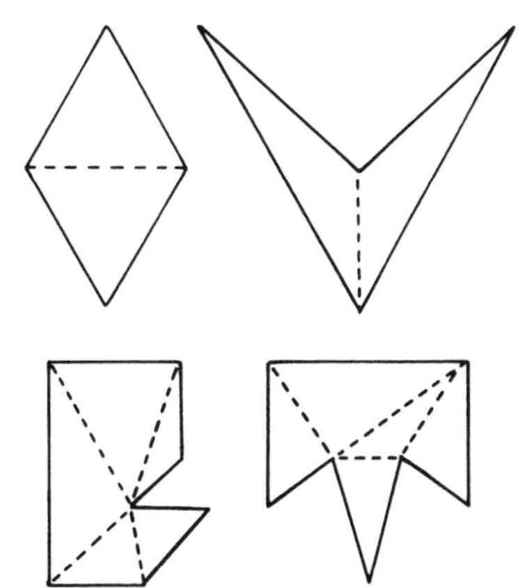

Definition. If a polygonal region consists of a convex polygon and its interior, then the polygon is the *boundary of the polygonal region* and the interior of the polygon is the *interior of the polygonal region.*

Comment. A convex polygon can be considered as a combination of triangular regions in more than one way, as shown below.

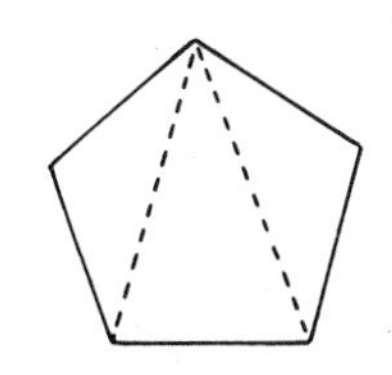
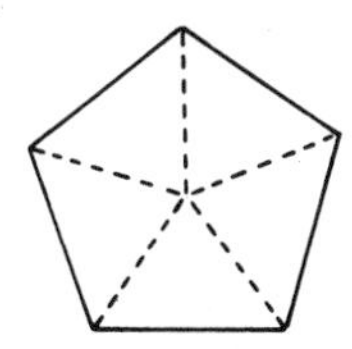

Postulate 24.—If S is any given polygonal region, there is a correspondence that associates to each polygonal region in space a unique positive number, such that the number assigned to S is 1.

Definition. The given polygonal region in this postulate is the *unit area* and, relative to this unit area, the number that is associated with a polygonal region is its *area.*

Comment. Postulate 24 does not tell what the area of a polygonal region is, except that of the unit area. For this information, more postulates are needed.

Postulate 25.—Given a polygonal region R and also given that it consists of two polygonal regions R_1 and R_2 such that R_1 and R_2 have in common only a finite number of segments, then relative to a given unit area, the area of R is the sum of the areas of R_1 and R_2.

Examples. This postulate is illustrated below. Figure (*d*) shows two triangular regions that have a triangular region in common; therefore Postulate 25 does not apply.

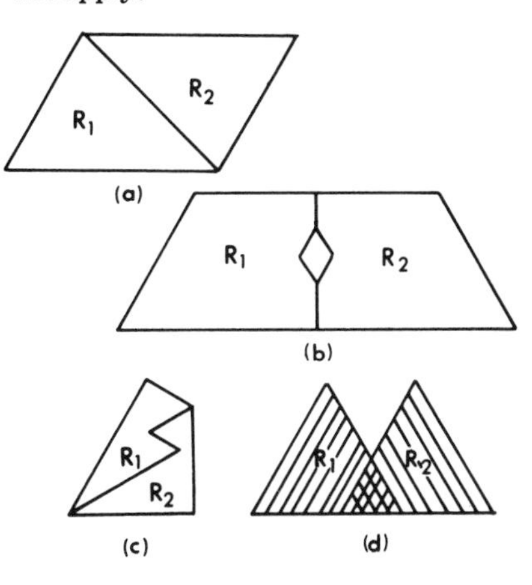

Postulate 26.—If two triangles are congruent, then their triangular regions have the same area relative to any given unit area.

Comment. It is often convenient to abbreviate "area of polygonal region" as "area of a polygon." However, the strict interpretation of "area of a polygon" would refer to the area of the points in a set of line segments, which is not a polygonal region.

Definition. Given a unit pair for measuring distances, a unit area is a *unit square* if, and only if, the unit area is bounded by a square such that the measure of a side of the square is one.

Postulate 27.—Given a unit pair for measuring distances, the area of a rectangle relative to a unit square is the product of the measures of any two consecutive sides.

Definition. Any side of a parallelogram is a *base of the parallelogram.* An *altitude*

of the parallelogram relative to a base is the segment perpendicular to the base whose end points are in the base and the side opposite the base.

Comment. Base and altitude of a parallelogram (and hence of a rectangle) have been defined as segments. When the context makes it clear, "the length of the base" sometimes is abbreviated to "base;" "altitude" is treated similarly. Using these abbreviations, Postulate 27 states that the area of a rectangle is the product of a base and the related altitude. If the area, base, and related altitude of a rectangle are denoted by A, b, and h, respectively, then $A = bh$. The area of a square A, each of whose sides has length S, is then given by $A = s^2$.

■ THEOREM 12-1.–The area of a right triangle is one-half the product of the lengths of two legs.

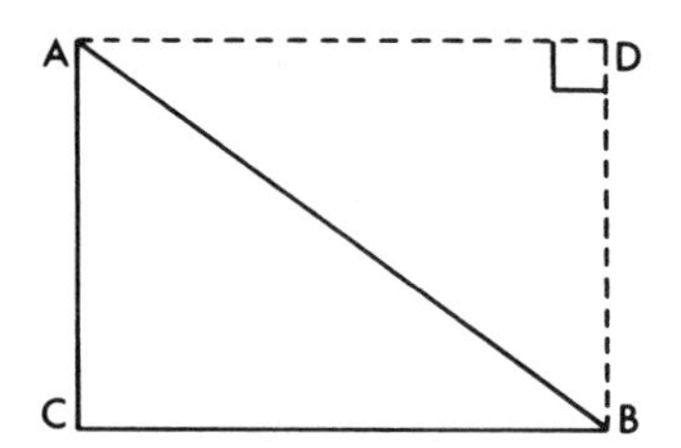

Proof. This theorem can be proved by considering the rectangle having $\overline{AC}$ and $\overline{CB}$, the legs of the right triangle ABC, as consecutive sides and by showing that $\overline{AB}$ divides the rectangle into two congruent triangles. By Postulate 26, their areas are equal; by Postulate 27, the area of $\triangle ABC = \frac{1}{2}(AC \cdot CB)$, Q.E.D.

■ THEOREM 12-2.–The area of a triangle is one-half the product of any base and the altitude to that base.

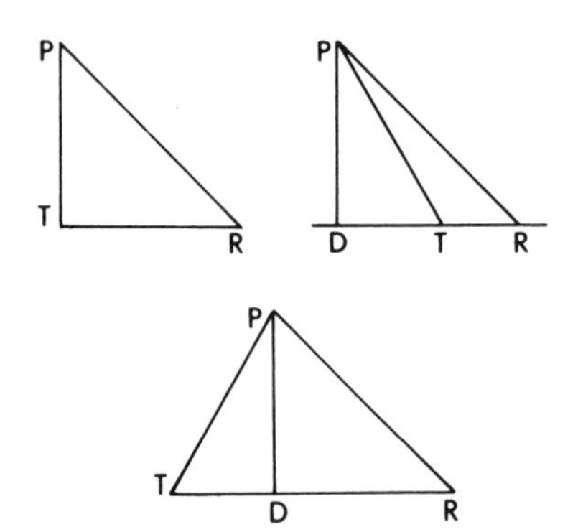

Proof. Let the triangle be $\triangle PTR$ and let the projection of P on $\overleftrightarrow{TR}$ be D. Let $\overleftrightarrow{TR} = b$, $\overleftrightarrow{PD} = a$, and A equal the area of $\triangle PTR$. If $D = T$ or $D = R$, then by Theorem 12–1, $A = \frac{1}{2}ab$. If D is between T and R, then A = area of $\triangle TPD +$ area $\triangle DPR = \frac{1}{2}a \cdot TD + \frac{1}{2}a \cdot DR = \frac{1}{2}ab$. If T is between D and R, then the area of $\triangle DPR$ = area of $\triangle DPT + A$, and $A = \frac{1}{2}a(\overline{DR} - \overline{TR}) = \frac{1}{2}ab$. If R is between T and D, a similar proof follows, Q.E.D.

■ COROLLARY 12-2-1.–The area of an equilateral triangle whose side has length s is $(s^2\sqrt{3})/4$.

■ COROLLARY 12-2-2.–The area of a rhombus is one-half the product of the lengths of the diagonals.

■ THEOREM 12-3.–The area of a parallelogram is the product of a base and the associated altitude.

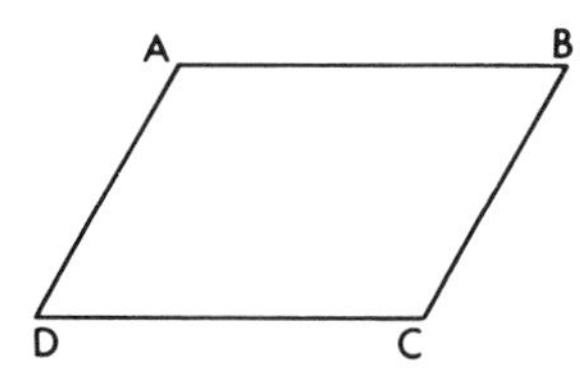

Proof. If $ABCD$ is a parallelogram with base $\overline{DC}$ and associated altitude h, then the area of $\triangle ADC = \frac{1}{2}\overline{DC} \cdot h$. Since $\triangle ADC \cong \triangle CBA$, and the area of $ABCD$ = area of $\triangle ADC$ + area of $\triangle ACB$, it follows that the area of $ABCD = DC \cdot h$, Q.E.D.

■ THEOREM 12-4.–The area of a trapezoid is one-half the product of its altitude and the sum of its bases.

Definition. The *median of a trapezoid* is the segment that joins the midpoints of its two nonparallel sides.

■ COROLLARY 12-4-1.–The area of a trapezoid is equal to the product of its altitude and the length of its median.

Summary of area formulas

Rectangle	$A = bh$
Square	$A = s^2$
Triangle	$A = \frac{1}{2}bh$
Parallelogram	$A = bh$
Equilateral triangle	$A = \frac{s^2\sqrt{3}}{4}$
Rhombus	$A = \frac{1}{2}d \cdot d'$
Trapezoid	$A = \frac{h}{2}(b + b')$; $A = hm$

Definition. Given two ordered sets of n positive numbers $(p,q,r,\ldots)$ and $(a,b,c,\ldots)$, then $p,q,r,\ldots$ are *inversely proportional* to $a,b,c,\ldots$ if, and only if, the products of corresponding numbers are equal; that is, if $pa = qb = rc = \ldots$.

■ THEOREM 12-5.–Given a set of two or more triangles:

If the bases of all triangles are equal, then their areas are proportional to the corresponding associated altitudes.

If their altitudes are equal, then their areas are proportional to the corresponding associated bases.

If their areas are equal, then the bases are inversely proportional to the corresponding associated altitudes.

Comment. A theorem similar to Theorem 12–5 may be written about parallelograms.

■ THEOREM 12-6.–If two triangles are similar, then their areas are proportional to the squares of the lengths of the corresponding sides.

Proof. Let A and A' represent the areas of these triangles. Let b represent the length of one side of the first triangle and a represent the altitude to that side. Let b and a' represent the lengths of the corresponding segments in the second triangle. Then $(b',a') \ \bar{p} \ (b,a)$, $b' = kb$, and $a' = ka$. It follows that $A' = \frac{1}{2}b'a' = \frac{1}{2}k^2b \cdot a = k^2A$. But $k^2 = b'^2/b^2$ or $(A,A') \ \bar{p} \ (b^2,b'^2)$, Q.E.D.

■ THEOREM 12-7.–If two convex polygons are similar, then their areas are proportional to the squares of the lengths of corresponding sides.

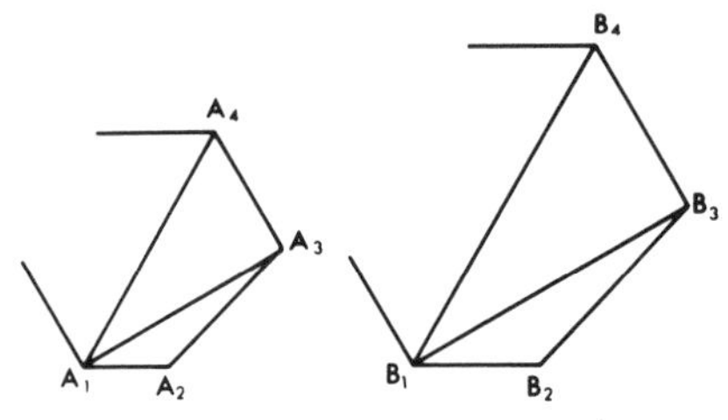

Proof. Since there is a one-to-one correspondence between the vertices of the two polygons, they have an equal number of vertices. Let the polygons be $A_1A_2A_3A_4 \ldots A_n$ and $B_1B_2B_3B_4 \ldots B_n$. By Theorem 10–5 (S.A.S. Similarity Theorem), $\triangle A_1A_2A_3 \sim \triangle B_1B_2B_3$. Since $\overrightarrow{A_3A_1}$ is between $\overrightarrow{A_3A_2}$ and $\overrightarrow{A_3A_4}$, it follows that $\angle A_4A_3A_1 = \angle A_4A_3A_2 - \angle A_1A_3A_2$. A similar conclusion follows for $\angle B_4B_3B_1$. Thus, $\triangle A_1A_3A_4 \sim \triangle B_1B_3B_4$ also can be proved by the S.A.S. Triangle Similarity Theorem as can the area of $\triangle A_1A_2A_3$ equals k^2 times the area of $\triangle B_1B_3B_4$. The process may continue in this manner until it is proved that $\triangle A_1A_{n-1}A_n$ equals k^2 times the area of $\triangle B_1B_{n-1}B_n$. Hence, the area of $A_1A_2A_3 \ldots A_n$ equals k^2 times the area of $B_1B_2B_3 \ldots B_n$. This is equivalent to the conclusion of the theorem, Q.E.D.

Definition. The *perimeter of a polygon* is the sum of the lengths of its sides.

■ COROLLARY 12-7-2.–If two convex polygons are similar, their perimeters are proportional to the lengths of corresponding sides or corresponding diagonals.

SECTION THIRTEEN

Regular Polygons and Circles.—A convex polygon is *regular* if its sides are congruent and its angles are congruent.

■ THEOREM 13-1.–The bisectors of the interior angles of a regular polygon meet at a point.

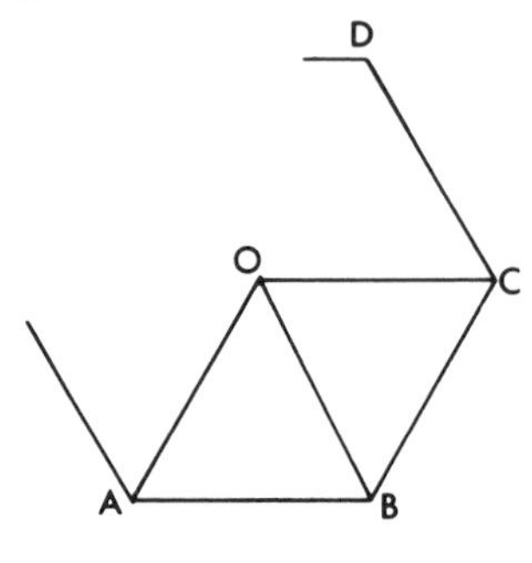

Proof. Let the polygon be $ABCD \ldots$, with $\angle A \cong \angle B \cong \angle C \ldots$ and $\overline{AB} \cong \overline{BC} \cong \overline{CD} \ldots$. Let the bisectors of $\angle A$ and $\angle B$ meet at O. Then, since $\frac{1}{2}m \angle A =$

$\frac{1}{2}m\angle B$, $\overline{OA} = \overline{OB}$. It can be proved that $\triangle OAB \cong \triangle OBC$ ($\overline{OA} = \overline{OB}$, $\angle OAB \cong \angle OBC$, $\overline{AB} \cong \overline{BC}$); hence, $m\angle OAB = m\angle OCB = \frac{1}{2}m\angle B = \frac{1}{2}m\angle C$. Therefore, the bisector of $\angle C$ contains O. Similarly, the bisectors of $\angle D \ldots$ meet in O, Q.E.D.

Definition. The *center of a regular polygon* is the point of intersection of the midrays of the angles of the polygon. The triangle formed by any two consecutive vertices of a regular polygon and its center is a *central triangle of the polygon.*

■ COROLLARY 13-1-1.–The central triangles of a regular polygon are isosceles, and are congruent to each other.

Definition. A *radius of a regular polygon* is the segment from a vertex to its center.

■ COROLLARY 13-1-2.–The radii of a regular polygon are congruent.

Definition. An *apothegm of a regular polygon* is the segment that joins the center to its projection on a side.

■ COROLLARY 13-1-3.–The apothegms of a regular polygon are congruent.

■ THEOREM 13-2.–The sum of the measures of the interior angles of a convex polygon having n sides is $(n - 2) \cdot 180$.

Proof. Let the polygon be $ABCD \ldots N$, and let O be any interior point. Then the sum of the measures of the angles in $\triangle OAB$, $\triangle OBC$, $\ldots$ is $n \cdot 180$. But the sum of the measures of the angles whose vertex is O, is $2 \cdot 180$. Since $\overrightarrow{BO}$ is between $\overrightarrow{BA}$ and $\overrightarrow{BC}$, $m\angle ABO + m\angle OBC = m\angle ABC$.

Similar conclusions can be made of $m\angle BCD$, $m\angle CED \ldots$. Thus, the sum of the measures of the interior angles of the polygon is $n \cdot 180 - 2 \cdot 180 = (n - 2) \cdot 180$, Q.E.D.

■ COROLLARY 13-2-1.–The measure of each interior angle of a regular polygon having n sides is $(n - 2)/n \cdot 180$.

■ COROLLARY 13-2-2.–The measure of each exterior angle of a regular polygon having n sides is $360/n$.

Comment. Each exterior angle forms a linear pair with its adjacent interior angle.

■ COROLLARY 13-2-3.–The sum of the measures of the exterior angles of a polygon, one at each vertex, is 360.

■ COROLLARY 13-2-4.–The measure of each central angle of a regular polygon of n sides is $360/n$.

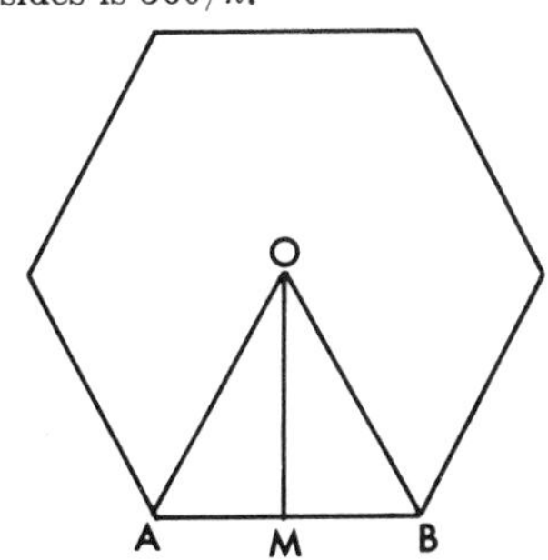

Example 1. Suppose a polygon, as shown, is a regular hexagon with each side having length 10. $m\angle AOB = 60$, $\overline{AM} = 5$, $m\angle AOM = 30$, $m\angle MAO = 60$. $(\overline{AM}, \overline{MO}, \overline{OA})$ $\bar{\bar{p}}$ $(1,\sqrt{3},2)$. Therefore, $\overline{MO} = 5\sqrt{3}$, $\overline{OA} = 10$.

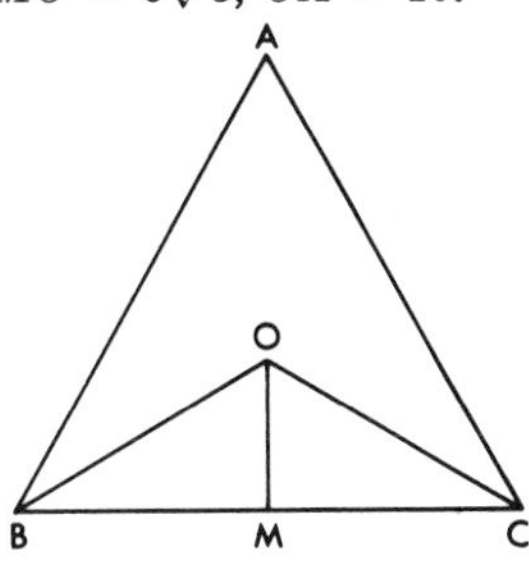

Example 2. Suppose ABC is an equilateral triangle with radius 12. $m\angle BOM = 60$, $(\overline{OM}, \overline{MB}, \overline{BO})$ $\bar{\bar{p}}$ $(1,\sqrt{3},2)$. Therefore, $k = 6$, $\overline{OM} = 6$ and $\overline{MB} = 6\sqrt{3}$.

■ THEOREM 13-3.–If two regular polygons have the same number of sides, they are similar.

Proof. Since they are equilateral, the sides of one are proportional to the sides of the other. That the corresponding angles are congruent can be seen from Corollary 13-2-1. Therefore, the polygons are similar, Q.E.D.

■ COROLLARY 13-3-1.–If two regular polygons have the same number of sides: (a) their perimeters are proportional to the lengths of the sides or radii or apothems; (b) their areas are proportional to the squares of the lengths of the sides or radii or apothegms.

■ THEOREM 13-4.–The area of a regular polygon is one-half the product of an apothegm and the perimeter ($A = \frac{1}{2}ap$).

■ COROLLARY 13-4-1.–If s is the length of each side of a regular polygon having n sides and a is the length of an apothegm, then its area is $\frac{1}{2}ans$.

A rigorous treatment of the circumference of a circle or the area of a circular region depends upon the *theory of limits* and is usually studied in the calculus. Lacking a knowledge of this theory, a plausible treatment is presented here.

■ THEOREM 13-5.–If a circle is divided into n congruent, nonoverlapping arcs, and their end points are joined successively by chords, a regular polygon is formed.

Proof. The chords are congruent and the angles of the polygon are inscribed in congruent arcs, Q.E.D.

Definition. A polygon whose vertices are on a circle is an *inscribed polygon.*

Comment. Let P_n represent the perimeter of a regular polygon inscribed in a given circle and having n sides. Consider the sequence of numbers $p_3, p_4, p_5, \ldots$ for a given circle. It can be shown that this sequence of numbers is increasing; that it is limited even though it is increasing; that by taking n large enough, P_n can be made to be as close to this limit as desired. This limit is defined as the circumference of the circle. This is denoted by $P_n \to C$ (P_n approaches C).

Definition. The *circumference of a circle* is the limit of the perimeters of the inscribed regular polygon.

■ THEOREM 13-6.–The quotient of the circumference of a circle divided by its diameter, or $C/2r$, is the same for all circles.

Proof. (Not rigorous.) Given a circle with center O and radius r and another circle with radius O' and radius r', consider the regular n-gons inscribed in each.

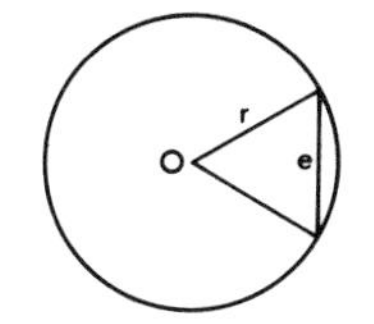

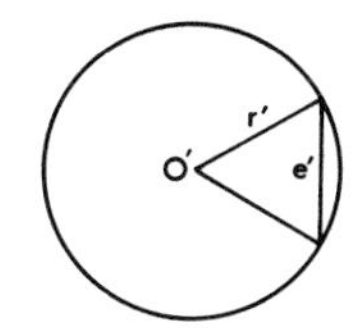

Let e and e' be the lengths of their respective sides. Then (e,e') $\bar{\bar{p}}$ (r,r') and (ne,ne') $\bar{\bar{p}}$ $(2r,2r')$ or (p,p') $\bar{\bar{p}}$ $(2r,2r')$. As n increases, $p \to C$, and $p' \to C'$. Therefore, it is plausible that (C,C') $\bar{\bar{p}}$ $(2r,2r') \cdot C/2r$ is a constant of proportionality, Q.E.D.

Comment. $C/2r$ is denoted by π.

■ COROLLARY 13-6-1.–C equals $2\pi r$.

■ COROLLARY 13-6-2.–Circumferences of circles are proportional to their radii.

If A_n represents the area of an inscribed regular polygon, it is possible for a given circle, to write the sequence of numbers $A_3, A_4, A_5 \ldots$. Again it can be proved that A_n approaches a limit.

Definition. A *circular region* consists of all the points in a circle and in its interior.

Definition. The *area of a circular region* is the limit of the areas of the inscribed regular polygons. It is convenient to write "area of a circle" for "area of a circular region."

■ THEOREM 13-7.–The area of a circle with radius r is πr^2.

Proof. (Not rigorous.) Consider A_n the area of the inscribed regular polygon having n sides ($An = \frac{1}{2}ap$). As n increases, $An \to A$, the area of the circle; $a \to r$; and $p \to C$. It is plausible that $A = \frac{1}{2}r \cdot C = \frac{1}{2}r \cdot 2\pi r$ or $A = \pi r^2$, Q.E.D.

■ COROLLARY 13-7-1.–The area of a circle is proportional to the square of its radius (or diameter).

Definition. If $\widehat{AB}$ is an arc of a circle with center O, and $P_1, P_2, P_3 \ldots P_{n-1}$ are distinct points of $\widehat{AB}$ such that $m\angle AOP_1 \cong \angle P_1OP_2 \cong \ldots \cong \angle P_{n-1}OB$, then the length of $\widehat{AB}$ is the limit of $AP_1 + P_1P_2 + \ldots P_{n-1}B$ as n is taken larger and larger.

Notation. The length of $\widehat{AB}$ is designated $l\widehat{AB}$.

Comment. The degree measure of $\widehat{AB}$ is not to be confused with $l\widehat{AB}$.

Postulate 28.—The lengths of arcs in congruent circles are proportional to their degree measures.

Comment. If $\widehat{AB}$ is a semicircle of a circle with radius r, then $m\widehat{AB} = 180$, but $l\widehat{AB} = \pi r$. It is clear that $(l\widehat{AB}, \pi r)$ $\bar{\bar{p}}$ $(m\widehat{AB}, 180)$ and that $k = \pi r/180$.

■ THEOREM 13-8.–An arc of degree measure q contained in a circle with radius r has length $L = (\pi r/180) \cdot q$.

Proof. By Postulate 28, $(L, \pi r)$ $\bar{\bar{p}}$ $(q,180)$. Therefore, $L = (\pi r/180) \cdot q$, Q.E.D.

Definition. If $\widehat{AB}$ is an arc of a circle with center O and radius r, then all the points of the radii of the circle that are in the interior of $\angle AOB$ and the points in $\overline{OA}$ and $\overline{OB}$ are a *sector.* The *arc of the sector* is $\widehat{AB}$, and the *radius of the sector* is r.

■ THEOREM 13-9.–The area of a sector is half the product of its radius and the length of its arc.

Proof. (Not rigorous.) Consider the set of points in the arc as described in the definition for the lengths of $\widehat{AB}$. Let h be the altitude from O to each segment such as $\overline{P_2P_3}$, and let b be $\overline{P_2P_3}$. The sum of the areas of $\triangle AOP$, $\triangle P_1OP_2, \ldots$ $\triangle P_{n-1}OB = n \cdot \frac{1}{2}h \cdot b$ or $\frac{1}{2}h \cdot nb$. But $np \rightarrow l\widehat{AB}$ and $h \rightarrow r$. Therefore, the area of the sector is $\frac{1}{2}(r \cdot \frac{1}{2}l\widehat{AB})$, Q.E.D.

■ COROLLARY 13-8-1.–The area of a sector of radius r and arc degree measure q is $(\pi r^2/360) \cdot q$.

SECTION FOURTEEN

Parallelism and Perpendicularity

Theorem 5-3 states that space contains at least two planes. This assertion rests primarily on the assumptions that a line contains at least two points and that a plane is determined by three noncollinear points. However, Postulate 12 (Ruler Postulate) states that every line contains as many points as there are real numbers. This implies the existence of a limitless number of planes. In this section there will be an investigation of some relations among these planes and the general relations among points, lines, and planes.

Comment. It can be shown that a unique plane is determined by any of the following: (a) three noncollinear points; (b) two intersecting lines; (c) a line and a point not contained in the line; or, (d) two parallel lines.

Definition. A line and a plane are *perpendicular to each other* if, and only if, they intersect and every line in the plane that contains the point of intersection is perpendicular to the given line.

Postulate 29.—There exists exactly one plane that contains a given point and is perpendicular to a given line.

Comment. The given point may or may not be in the given line.

■ THEOREM 14-1.–The plane that is perpendicular to a given line at a given point in the line contains every line that is perpendicular to the given line at that point.

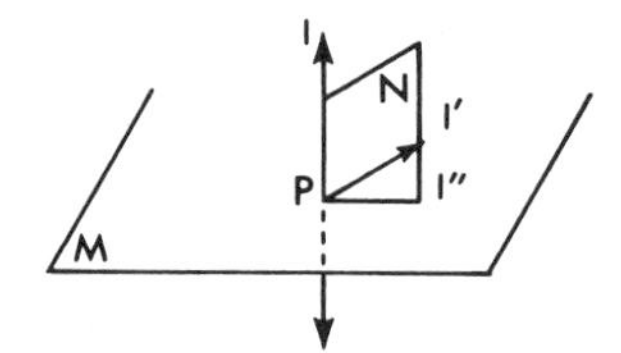

Proof. Let l be the given line, P the given point, and M the plane that is perpendicular to l at P. If $l' \perp M$ at P, then l and l' intersect and determine a plane, N, that intersects plane M in a line, l''. By the definition of perpendicularity between a line and a plane, $l'' \perp l$. But in plane N, there can be only one line perpendicular to l at P. Therefore $l'' = l'$, and l' is in plane M. In this manner it is proved that every line perpendicular to l at P is in M, Q.E.D.

■ THEOREM 14-2.–If a line is perpendicular to each of two intersecting lines at their intersection, it is perpendicular to the plane determined by the two lines.

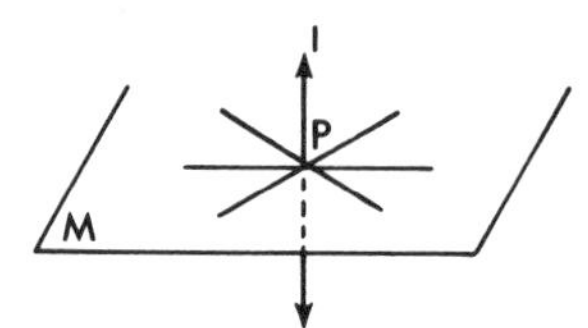

The above figure shows only three of the lines in plane M that contain P. Line l, which is perpendicular to each of these three (and all others in M containing P), is perpendicular to plane M and plane M is perpendicular to line l.

■ THEOREM 14-3.–There is exactly one line that is perpendicular to a given plane at a point in the plane.

Definitions. A line and a plane are *parallel to each other* if, and only if, they have no point in common. Two planes are *parallel to each other* if, and only if, they have no point in common.

■ THEOREM 14-4.–If a plane intersects one of two parallel lines in a point, then it also intersects the other in a point.

Proof. Let the plane be M and the lines be l_1 and l_2, and let M and l_1 intersect in P only. Since $l_1 \parallel l_2$, they determine a plane, N, that is necessarily distinct from M. But M and N have P in common. Hence, they intersect in a line, q, that is distinct from l_1. Therefore, q must intersect l_2 in exactly one point, Q.E.D.

■ COROLLARY 14-4-1.–If a plane is parallel to one of two parallel planes, it is also parallel to the other.

■ THEOREM 14-5.–If a plane intersects each of two parallel planes, the intersections are parallel lines.

Proof. Let the intersections be p and q. If they were to intersect, then the planes that are given parallel would also intersect. Therefore, p and q do not meet, and since they are coplanar, they are parallel, Q.E.D.

■ THEOREM 14-6.–If a line intersects one of two parallel planes, it intersects the other.

■ COROLLARY 14-6-1.–If a line is parallel to one of two parallel planes, it is parallel to the other.

■ THEOREM 14-7.–If two distinct planes are perpendicular to the same line, they are parallel.

Proof. If they meet, then they will form a triangle with two right angles, Q.E.D.

■ THEOREM 14-8.–If a line is perpendicular to one of two parallel planes, it is perpendicular to the other as well.

Postulate 30.—If two lines are perpendicular to the same plane, they are parallel.

Comment. This proposition can be proved. Because the proof is difficult for beginning geometry students, some schools prefer to postulate it.

■ THEOREM 14-9.–If a plane is perpendicular to one of two parallel lines, it is perpendicular to the other also.

■ THEOREM 14-10.–If two lines are each parallel to a third line, they are parallel to each other.

Proof. A plane is considered perpendicular to one of these lines, and it is shown that each line is perpendicular to it. By Postulate 30, the lines are parallel, Q.E.D.

Comment. This theorem extends a similar theorem for two coplanar lines to two lines not necessarily coplanar.

■ THEOREM 14-11.–Given a plane and a point not in the plane, there is exactly one line that contains the point and is perpendicular to the plane.

Comment. This theorem extends Theorem 14-3.

■ THEOREM 14-12.–There is exactly one plane parallel to a given plane and containing a given point not in the given plane.

■ THEOREM 14-13.–If two planes are parallel to a third plane, they are parallel to each other.

Definition. Given a plane and a point not in the plane, the *projection of the point on the plane* is the intersection of the line containing the point and the perpendicular to the plane.

■ THEOREM 14-14.–The shortest segment between a plane and a point not in the plane is the segment that joins the point to its projection on the plane.

Definition. The *distance between a point and a plane* that does not contain the point is the length of the segment that joins the point to its projection on the plane.

■ THEOREM 14-15.–If two planes are parallel, then all points in one plane have the same distance to the other plane.

■ COROLLARY 14-15-1.–If a line is parallel to a plane, then all the points in the line have the same distance to the plane.

■ THEOREM 14-16.–The set of all points that are equidistant from the end points of a given segment is the plane that contains the midpoint of the segment and is perpendicular to the line determined by the segment.

Definition. The intersection of a dihedral angle and a plane perpendicular to the edge of the dihedral angle is a *plane angle of the dihedral angle.*

■ THEOREM 14-17.–Any two plane angles of a dihedral angle are congruent.

Proof. Let A and A' be vertices of two distinct plane angles of the dihedral $\angle X - AA' - Y$. By Theorem 6-4 (Point-Plotting Theorem), B,B' and C,C' can be located as indicated in the diagram, so that $\overline{AB} = \overline{A'B'}$ and $\overline{AC} = \overline{A'C'}$. Since plane ABC is parallel to plane $A'B'C'$, $\overline{AB} \parallel \overline{A'B'}$, and $ABB'A'$ is a parallelogram. Similarly, $ACC'A'$ is a parallelogram. Hence, $\overline{BB'} \parallel \overline{AA'}$, $\overline{BB'} = \overline{AA'}$, $\overline{CC'} \parallel \overline{AA'}$, $\overline{CC'} = \overline{AA'}$. By the transitivity properties of parallelism and equality, $\overline{BB'} \parallel \overline{CC'}$ and $\overline{BB'} = \overline{CC'}$. Hence, $BB'C'C$ is a parallelogram and

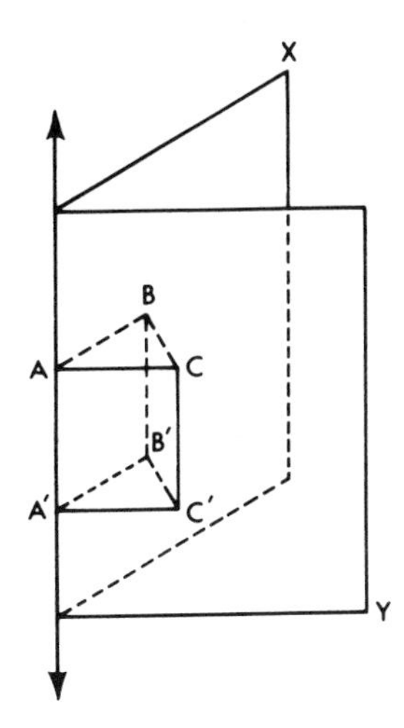

$\overline{CB} \cong \overline{C'B'}$. Therefore, $\triangle ABC \cong \triangle A'B'C'$ (S.S.S.) and $\angle A \cong \angle A'$, Q.E.D.

Definitions. The *measure of a dihedral angle* is the measure of any of its plane angles. A *right dihedral angle* is a dihedral angle whose measure is 90. The planes determined by the faces of a right dihedral angle are *perpendicular to each other.*

■ THEOREM 14–18.–If a line is perpendicular to a plane, then any plane containing this line is perpendicular to the given plane.

■ THEOREM 14–19.–If two planes are perpendicular, then any line in one plane that is perpendicular to their line of intersection is perpendicular to the other plane.

■ THEOREM 14–20.–If two planes are perpendicular, then any line perpendicular to one at a point on their intersection is contained in the other.

■ THEOREM 14–21.–If each of two intersecting planes is perpendicular to a third plane, then their line of intersection is perpendicular to this plane.

Proof. Let the intersection of the two planes intersect the third plane at P. The line perpendicular to the third plane at P must lie in each of the two intersecting planes by Theorem 14–20. It must therefore be the line of intersection of these two planes, Q.E.D.

SECTION FIFTEEN

Polyhedrons.—A *polyhedron* is the set of all the points in a finite number of polygonal regions, each of which is bounded by a convex polygon such that the interiors of no two polygons have points in common, and such that each side of each polygon is also a side of exactly one of the other polygons. Each vertex of each polygon is a *vertex of the polyhedron.* Each side of each polygon is an *edge of the polyhedron.* Each of the polygonal regions is a *face of the polyhedron.*

A polyhedron is named according to the number of faces that it contains.

Polyhedrons

Faces	*Name*
4	tetrahedron
5	pentahedron
6	hexahedron
8	octahedron
10	decahedron
12	dodecahedron
20	icosahedron

Definition. If a plane and a polyhedron intersect, the intersection is a *section of the polyhedron.*

Definition. A polyhedron is a *convex polyhedron* if every section that contains at least three noncollinear points is either a convex polygon or a face of the polyhedron.

Definition. A *regular polyhedron* is a convex polyhedron, all of whose faces are bounded by regular polygons having the same number of sides, and such that all vertices of the polyhedron belong to the same number of faces.

Comment 1. There are exactly five types of regular polyhedrons, as shown below. They are called the *Platonic figures,* or *Platonic solids.*

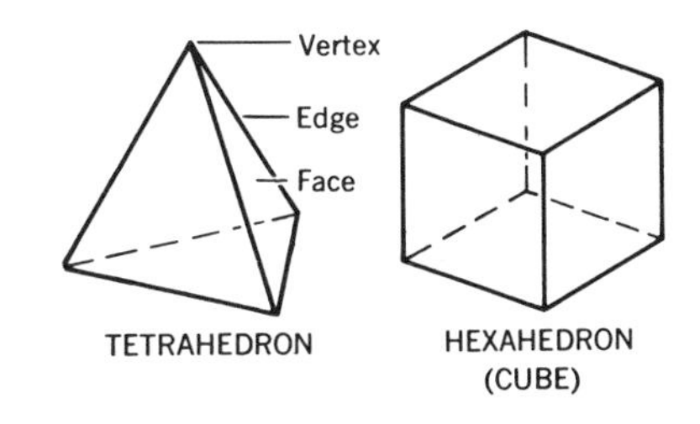

TETRAHEDRON HEXAHEDRON (CUBE)

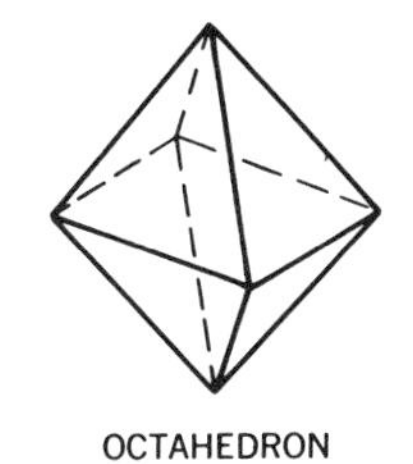

OCTAHEDRON

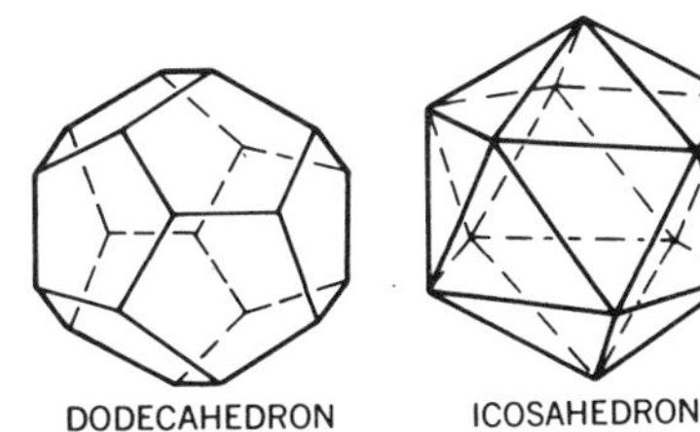

DODECAHEDRON ICOSAHEDRON

Comment 2. It is easily proved that the faces of a regular polyhedron are congruent. This property is often used as part of the definition of a regular polygon.

Definitions. Given: a convex polygon and a point not in the plane of the polygon. The set of all rays, each of which has the given point as end point and a point of the polygon, is called a *polyhedral angle.* The given point is the *vertex of the polyhedral angle,* and each ray containing a vertex of the polygon is an *edge of the polyhedral angle.* An angle with the given point as vertex and containing two consecutive vertices of the polygon is a *face angle of the polyhedral angle.* A *face of a polyhedral angle* is the set of points in a face angle and its interior. A polyhedral angle having three faces is a *trihedral angle.*

■ THEOREM 15–1.–The sum of the measures of any two face angles of a trihedral angle is greater than the measure of the third face of the angle.

■ THEOREM 15–2.–The sum of the measures of all the face angles of a polyhedral angle is less than 360.

Example. If two face angles of a trihedral angle have measures 70 and 120, and x represents the measure of the third face angle, then: by Theorem 15–1, $x + 70 > 120$; by Theorem 15–2, $x + 70 + 120 < 360$.

These imply that $x > 50$ and $x < 170$; that is, x is between 50 and 170.

Comment. Using Theorems 15–1 and 15–2, and the definition of a regular polyhedron, it can be proved that there are exactly five types of regular polyhedrons.

Definition. A *prism* is a polyhedron such that two of its faces are bounded by congruent polygons in parallel planes, and each of the remaining faces is bounded by a parallelogram with two sides in the congruent polygons. The faces bounded by the congruent polygons are *bases.* Prisms are classified according to their bases. One with a triangular-region base is a *triangular prism;* one with a rectangular-region base is a *rectangular prism;* and so on.

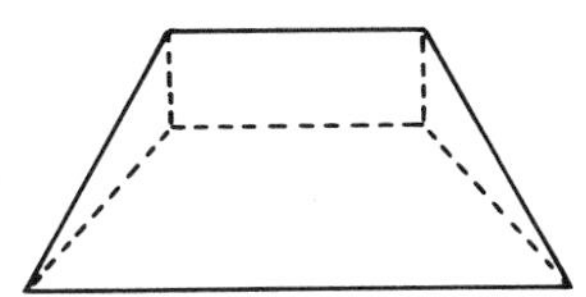

Definitions. A *parallelepiped* is a prism whose base is bounded by a parallelogram. A *rectangular parallelepiped* is a parallelepiped each of whose faces is bounded by a rectangle. A *cube* is a parallelepiped each of whose faces is bounded by a square.

The faces of a prism that are not the bases are *lateral faces.* The *lateral surface of a prism* is the set of all points in the lateral faces. The intersection of two lateral faces is a *lateral edge of the prism.* A prism is a *right prism* if, and only if, a lateral edge is perpendicular to a base.

A *right section of a prism* is the intersection of the prism with a plane that is perpendicular to and intersects the interior of every lateral edge of the prism. The *altitude of a prism* is a segment whose end points are in the planes that contain the bases of the prism and is perpendicular to these planes.

The *lateral area of a prism* is the sum of the areas of all the lateral faces of the prism.

■ THEOREM 15–3.–The lateral area of a prism is equal to the product of the length of a lateral edge and the perimeter of a right section.

■ COROLLARY 15-3-1.–The lateral area of a right prism is the product of the length of a lateral edge and the perimeter of a base.

Definition. A *pyramid* is a convex polyhedron that is contained in a polyhedral angle, except for one of its faces, which is its *base.* The *vertex of the pyramid* is the vertex of the polyhedral angle.

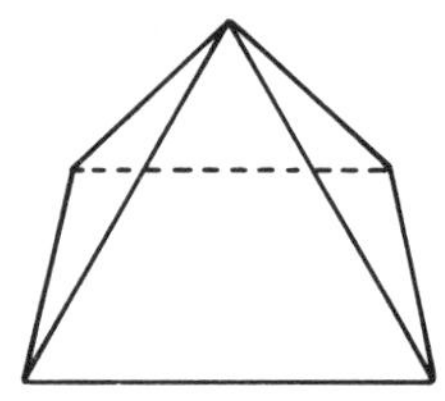

The segment that joins the vertex to its projection on the plane of the base is the *altitude of the pyramid.*

Like prisms, pyramids are classified according to the boundary of the base.

Definitions. A pyramid is a *regular pyramid* if, and only if, its base is bounded by a regular polygon and one end point of the altitude is the center of the base. A *slant height of a pyramid* is the distance between its vertex and an edge in its base.

Definition. The *lateral faces of a pyramid* are the faces that are contained in its polyhedral angle. The *lateral edges of a pyramid* are those of its edges that have the vertex of the pyramid as one end point.

Definition. The *lateral area of a pyramid* is the sum of the areas of its lateral faces.

■ THEOREM 15-4.–The lateral edges of a regular pyramid are congruent.

■ COROLLARY 15-5-1.–The slant heights of a regular pyramid are equal.

■ THEOREM 15-5.–The lateral area of a regular pyramid is one-half the product of a slant height and the perimeter of the base.

■ THEOREM 15-6.–Every polygon section of a pyramid, made by a plane parallel to that of its base, is similar to the boundary of the base. The distances from the vertex to the planes of the intersection and the base are proportional to the corresponding sides of these similar polygons, and the areas of the polygonal regions are proportional to the squares of these distances.

Definition. A *frustum of a pyramid* is a polyhedron whose edges consist of those of the base of the pyramid, the intersection of the pyramid by a plane parallel to that of the base, and the segments contained in the edges of the pyramid between corresponding vertices of the base and the section. The section and its interior is also a *base of the frustum.*

■ THEOREM 15-7.–The lateral faces of a frustum of a regular pyramid are congruent isosceles trapezoids.

■ THEOREM 15-8.–The lateral area of a frustum of a regular pyramid is one-half the product of the height of a lateral face and the sum of the perimeters of the bases: $L = \frac{1}{2}h(p + p')$.

By the method of limits similar to that described for polygons and circles, a cylinder and a cone may be defined.

In a right circular cylinder with base radius r and height h:

$$\text{lateral area} = 2\pi rh,$$
$$\text{total area} = 2\pi rh + 2r^2.$$

In a right circular cone with base radius r, height h, and slant height s:

$$\text{lateral area} = \pi rs,$$
$$\text{total area} = \pi rs + \pi r^2.$$

The area of a sphere with radius r is $4\pi r^2$.

The commonly used volumes are listed below for reference:

B = area of base
h = height
s = slant height
r = radius of the base of a right circular cylinder or cone, or radius of a sphere
V = volume
A = area
V of right prism = Bh
V of pyramid = $\frac{1}{3}Bh$
V of cylinder = πr^2h or Bh
V of cone = $\frac{1}{3}\pi r^2h$ or $\frac{1}{3}Bh$
V of sphere = $\frac{4}{3}\pi r^3$
V of frustum of a pyramid or cone = $\frac{1}{3}h(B + B' + \sqrt{BB'})$ (where B' is the area of the second base)
A of zone = $2\pi rh$

Comment. Cavalieri's principle may be used to prove some of the above volume formulas. It states: If two polyhedrons have their bases in parallel planes and if any intersections with these polyhedrons made by a plane parallel to the bases are boundaries of polygonal regions with equal areas, then the polyhedrons have equal volumes.

The principle also can be used to treat other types of solids, such as spheres, cylinders, and cones.

SECTION SIXTEEN

Constructions.—The drawing with straightedge and compass only of figures that satisfy certain requirements is an application of the theory of geometry. In such problems, abstract points and sets of points are interpreted as physical objects. After finding how to make the drawing, one shows how the theory of geometry explains why the construction is correct.

The word "construct" is used here to indicate a drawing operation employing only a straightedge and one or more compasses.

It is assumed that:

(1) Compasses can be used to draw a circle, or part of a circle, given its center and either its radius or a point in the circle.

(2) The straightedge can be used to draw a line segment that contains one or more given points.

(3) The circles and line segments so drawn are the physical counterparts of mathematical circles, lines, or line segments.

In each of the construction problems that follow, the construction and the related theory that explains why the con-

EGYPTIAN STATE TOURIST ADMINISTRATION

THE PYRAMIDS at Giza, Egypt, were constructed over a period of centuries by people who exhibited an amazingly accurate knowledge of both mathematics and engineering.

struction yields the desired results are indicated.

1. Given a ray $\overrightarrow{AB}$ and a segment $\overline{CD}$, construct a point E on $\overrightarrow{AB}$ such that $\overline{AE} \cong \overline{CD}$.

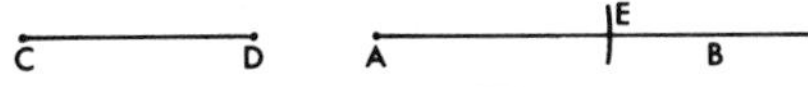

Construction. With $\overline{CD}$ as radius and A as center, draw the arc of the circle that intersects $\overrightarrow{AB}$. The intersection E is the desired point.

Related Theory. Since A is inside the circle, $\overline{AE}$ is a secant and hence cuts the circle in two points, one of them in $\overrightarrow{AB}$. In addition, the radii of a circle are congruent.

2. Given $\overrightarrow{AB}$ and $\overline{CD}$, construct E in $\overrightarrow{AB}$ such that $\overline{AE} = 2 \cdot CD$.

Construction. Locate E' such that $\overline{AE'} = \overline{CD}$, and then in $\overrightarrow{E'B}$ locate E such that $\overline{E'E} = \overline{CD}$ (it is assumed that E is between A and B). Thus $\overline{AE} = 2 \cdot \overline{CD}$.

Related Theory. Theorem 6–4 (Point-Plotting Theorem).

Comment. This construction can be extended to find F such that $\overline{AF} = m \cdot \overline{CD}$, where m is any positive integer.

3. Given $\overline{AB}$, construct its midpoint.

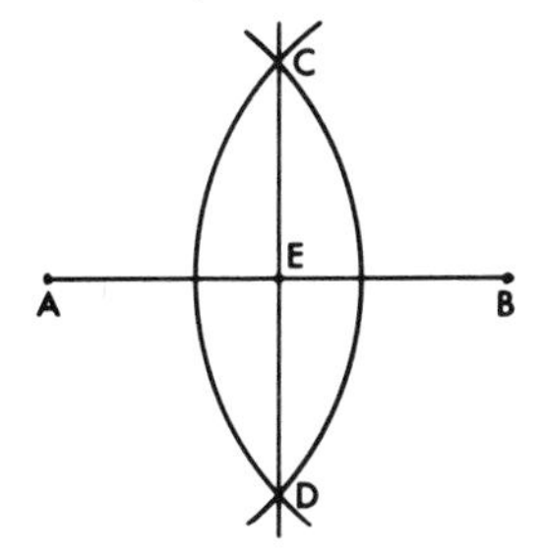

Construction. The arcs shown are drawn with A and B as centers, and with a radius that is greater than half of $\overline{AB} \cdot \overrightarrow{CD}$ intersects $\overline{AB}$ in E, the desired midpoint.

Related Theory. $\triangle ACD \cong \triangle BCD$ (S.S.S.). Therefore, $\overrightarrow{CE}$ is the midray of $\angle ACB$. Since $\triangle ACB$ is an isosceles triangle with $\overline{AC} \cong \overline{BC}$, then $\overline{CE}$ bisects $\overline{AB}$. (Theorem 9–20 may be used for an alternate proof.)

4. Given $\overline{AB}$, construct the perpendicular to $\overline{AB}$ at its midpoint.

Construction. Use Construction 3.

Related Proof. The bisector of the vertex angle of an isosceles triangle is perpendicular to the base and bisects the base.

5. Given $\angle ABC$, construct its midray.

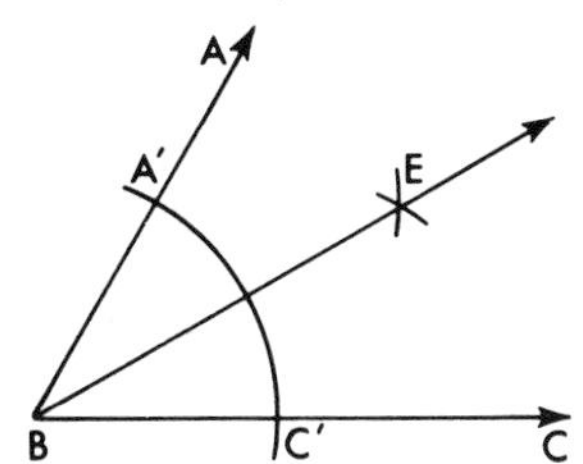

Construction. With B as center and any radius, draw an arc intersecting $\overrightarrow{BA}$ and $\overrightarrow{BC}$ in A' and C', respectively. With A' and C' as centers and a radius greater than half of $A'C'$, draw arcs intersecting in E. Then BE is the desired midray.

Related Theory. $\triangle BA'E \cong \triangle BC'E$ (S.S.S.); therefore, $\angle A'BE \cong \angle C'BE$.

6. Given line l and P in l, construct the perpendicular to l at P.

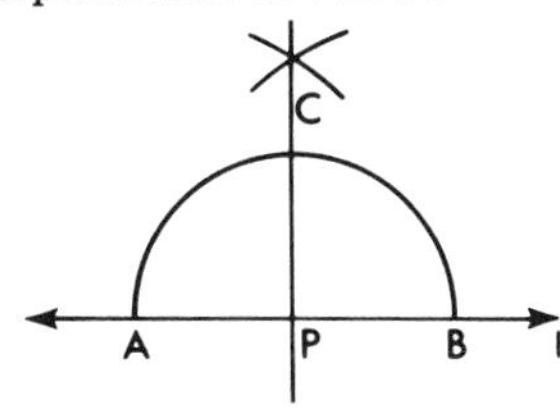

Construction. With P as center and any radius, draw an arc intersecting l in A and B. Then use Construction 4 on $\overline{AB}$, locating C. $\overleftrightarrow{CP}$ is the desired line.

Related Theory. $\triangle ACP \cong \triangle BCP$ (S.S.S.). Then $\angle APC \cong \angle BPC$. Since these angles are a linear pair, each is a right angle.

7. Given line l and P not in l, construct the perpendicular to l containing P.

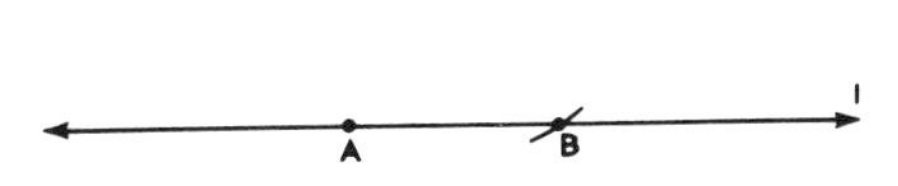

Construction. Take point A in l. If $\overline{PA} \perp l$, the desired construction is completed. If $\overline{PA}$ is not perpendicular to l, then with P as center, and $\overline{PA}$ as radius, draw an arc. Since $\overline{PA}$ is greater than the shortest distance from P to l, this circle intersects l in a second point, B. Complete the construction using Construction 3 on $\overline{AB}$.

Related Theory. If $\overleftrightarrow{PA}$ is not perpendicular to l, then Theorem 11–3 states that l and the circle with P as center and $\overline{PA}$ as radius will intersect in two points.

8. Given $\angle ABC$, $\overrightarrow{DE}$ and a half plane H with edge $\overleftrightarrow{DE}$, construct $\overrightarrow{DF}$ in H, such that $\angle ABC \cong \angle EDF$.

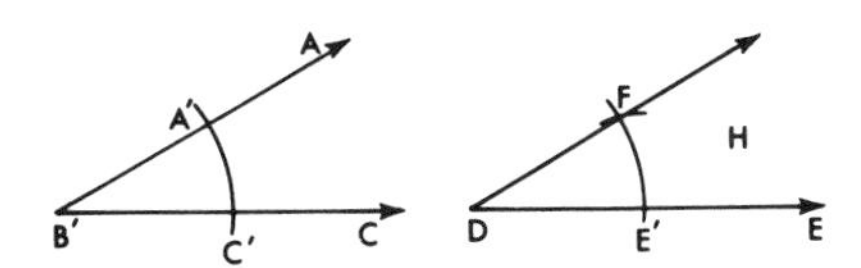

Construction. Construct congruent circles with B and D as centers and any radius, thus locating C' on $\overrightarrow{BC}$, A' on $\overrightarrow{BA}$ and E' on $\overrightarrow{DE}$. With E' as center and $\overline{A'C'}$ as radius, draw an arc in H intersecting the circle with D as center, thus locating F. Then $\overrightarrow{DF}$ is the desired ray.

Related Theory. Since $\overline{E'F}$ and $\overline{A'C'}$ are congruent chords in congruent circles, $\triangle A'BC' \cong \triangle FDE'$ (S.S.S.) and $\angle ABC \cong \angle EDF$.

9. Given line l and point P not in l, construct the line parallel to l, containing P.

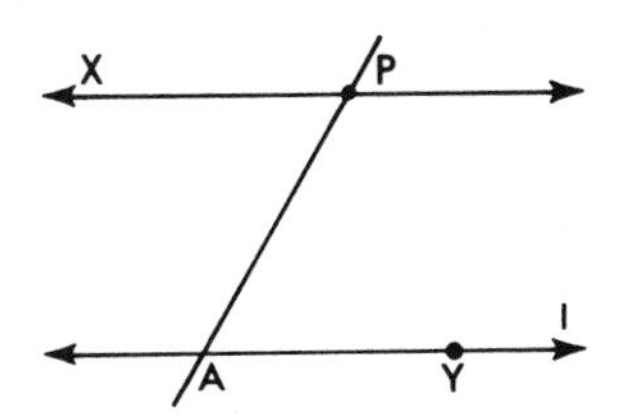

Construction. Choose any point in l, A, and use Construction 8 to make $\angle XPA \cong \angle YAP$, where Y is in l and X and Y are in opposite half planes having $\overleftrightarrow{AP}$ as edge.

Related Theorem. Theorem 9–2.

10. Given three segments with lengths a, b, c, respectively, construct a fourth segment with length d such that $(a,b)\ \bar{\text{p}}\ (c,d)$.

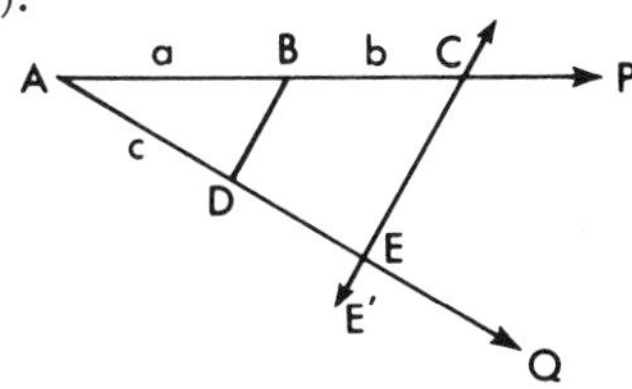

Construction. With any point A as endpoint, draw two distinct rays, $\overrightarrow{AP}$ and $\overrightarrow{AQ}$. With radius a and center A, locate B on $\overrightarrow{AP}$ such that $\overline{AB} = a$. Then with center B and radius b, locate C on $\overrightarrow{BP}$ such that $\overline{BC} = b$. (It is assumed that B is between A and P.) With center A and radius c, locate D on $\overrightarrow{AQ}$ such that $\overline{AD} = c$. Use Construction 9 to construct $\overleftrightarrow{CE'} \parallel \overleftrightarrow{BD}$. Let $\overleftrightarrow{CE'}$ intersect $\overrightarrow{AQ}$ in E. Then $DE = d$.

Related Theory. Postulate 22 (Proportional-Segments Postulate).

11. Given $\overline{AB}$, and two positive numbers, a and b, locate a point C in $\overline{AB}$ such that $(\overline{AC},\overline{CB})\ \bar{\text{p}}\ (a,b)$.

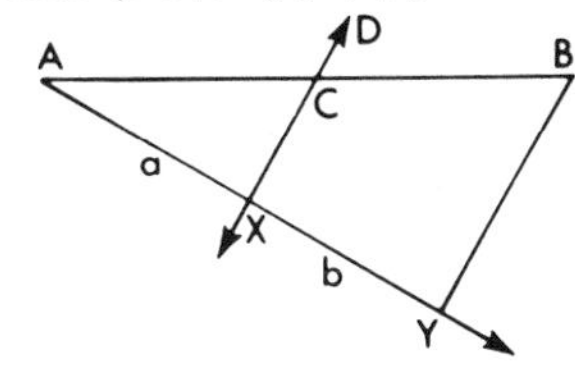

Construction. Draw any ray $\overrightarrow{AP}$ not collinear with $\overline{AB}$. On $\overrightarrow{AP}$ locate X and Y, with X between A and Y such that $\overline{AX} = a$, and $\overline{XY} = b$. Apply Construction 9 to draw $\overleftrightarrow{XD} \parallel \overleftrightarrow{YB}$. Then $\overleftrightarrow{XD}$ intersects $\overline{AB}$ in the desired point C.

Related Theory. Theorem 6–4 (Point-Plotting Theorem) and Postulate 22 (Proportional-Segments Postulate).

SECTION SEVENTEEN

Some Non-Euclidean Geometries.—Euclid's formulation of his parallel postulate is quite complex, as can be seen by referring to the discussion of parallel lines. On the other hand, his other postulates are stated simply. This contrast suggested to many that the parallel postu-

FAMOUS GEOMETRICIANS. *(Top to bottom)* Nikolai Lobachevsky (1793–1856), G. F. B. Riemann (1826–1866), Jules Henri Poincaré (1854–1912), and Arthur Cayley (1821–1895).

late could be deduced from Euclid's other postulates, but all attempts to do this failed. Some mathematicians hoped to show that the parallel postulate is a logical consequence of the other postulates by showing that if a contradictory of the parallel postulate is assumed, a contradiction within the postulational system could eventually be deduced. Such an attempt was made by the Italian priest, Girolamo Saccheri (1667–1733), who taught mathematics at the University of Pavia. His results were published in a book entitled *Euclides ob omni naevo vindicatus* (*Euclid Freed of Every Flaw*).

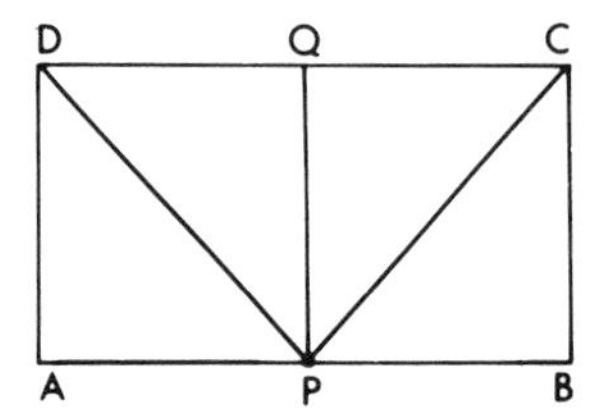

Saccheri studied what is known today as the *Saccheri quadrilateral.* He considered two congruent segments, perpendicular to $\overline{AB}$ at A and B, respectively, on the same side of $\overleftrightarrow{AB}$. If P and Q are midpoints of $\overline{AB}$ and $\overline{DC}$, respectively, then $\triangle ADP \cong \triangle BCP$ (S.A.S.), $\angle ADP \cong \angle BCP$, and $\overline{DP} \cong \overline{CP}$. Also, $\triangle DPQ \cong \triangle CPQ$ (S.S.S.). $\angle PDQ \cong \angle PCQ$ and $\angle ADC \cong \angle BCD$. If the last pair of angles are called the *summit angles* of the Saccheri quadrilateral, three possibilities may be considered: (1) The summit angles are both right angles. (2) The summit angles are both acute angles. (3) The summit angles are both obtuse angles.

Saccheri called these the *right-angle hypothesis*, the *acute-angle hypothesis*, and the *obtuse-angle hypothesis.* He was able to show that the parallel postulate is a consequence of the right-angle hypothesis and (using the infinitude of a straight line) that the obtuse-angle hypothesis produces a contradiction of another postulate. But when he assumed the acute-angle hypothesis, he obtained many seemingly strange results. None of them contradicted theorems that do not depend on the parallel postulate, so he finally took refuge in the conclusion that the "hypothesis of the acute-angle is absolutely false because it is repugnant to the nature of a straight line." Actually, Saccheri had discovered a new non-Euclidean geometry, but failed to realize the importance of the discovery.

It was not until the early nineteenth century that Karl Gauss (1777–1855) stated: "The assumption that the acute sum (of a triangle) is less than 180° leads to a curious geometry, quite different from ours (the Euclidean) but thoroughly consistent, which I have developed to my entire satisfaction. The theorems of this geometry appear to be paradoxical, and, to the uninitiated, absurd, but calm, steady reflection reveals that they contain nothing at all impossible."

Gauss was the greatest mathematician of his time, if not of all time. He started a study of parallels while still in his twenties, and kept up the study for more than thirty years. But he did not publish his results; he was fearful that he would become involved in controversies that would surely follow publication.

The first mathematician to publish a non-Euclidean geometry was the Russian, Nikolai Lobachevsky (1793–1856), a professor at the University of Kazan. The second mathematician (who was not aware of the work of Lobachevsky) was János Bolyai (1802–1860) who wrote a 26-page appendix for his father's two-volume treatise on geometry, which was published in 1832–1833.

Gauss, Lobachevsky, and Bolyai assumed a statement that might be expressed: There are at least two distinct lines containing a given point that are parallel to a given line not containing the given point.

The geometry in which this postulate holds is *hyperbolic geometry.* Some of the theorems in this geometry are as follows: (1) There are an infinite number of lines, containing a given point, that are parallel to a given line not containing the point. (2) The angle sum of any triangle is not constant and is less than 180. (3) The difference between 180 and the angle sum of a triangle can be taken as the measure of the area of the triangle.

A second type of non-Euclidean geometry was developed by Bernhard Riemann (1826–1866). It assumes that there are no lines containing a given point that are parallel to a given line not containing the point.

The geometry on a sphere is an example of this type of geometry, if "line" is interpreted as the "arc of a great circle" and "distance between two points" as "the length of the arc of the great circle." In such a geometry, the following theorems can be proved: (1) The angle sum of any triangle is greater than 180 and less than 540. (2) The excess of the angle sum of a triangle over 180 can be taken as a measure of the area of the triangle. (3) Two lines meet in two points.

There are also other geometries in which the "no parallel" postulate is used. Among them is *elliptic geometry,* in which the first two of the above theorems are valid, but in which two lines meet in exactly one point. The German mathematician Felix Klein (1849–1925) suggested the term "hyperbolic" for the two-parallel geometry, "parabolic" for the one-parallel geometry, and "elliptic" for the no-parallel geometry. These terms arose from the projective approach to non-Euclidean geometries. In this approach, the number of "infinite points" on a straight line is two, one, or none, according to whether the acute-angle, right-angle, or obtuse-angle hypothesis, respectively, is assumed.

—Harry Sitomer

TRIGONOMETRY

Analytic Trigonometry.—The word "trigonometry" literally means "measurement of triangles." For a long time, this classical aspect of trigonometry overshadowed the subject, with impressive applications to such fields as surveying, navigation, and astronomy. However, after the seventeenth century, the development of calculus revealed applications of trigonometry to such phenomena as heat flow, mechanical oscillations, and electronics, which had nothing to do with triangles or angles. The *analytic* rather than the *geometric* aspects of trigonometry became an important tool for mathematicians, physicists, engineers, and other scientists.

Coordinate Systems.—An idea that lies at the very root of many other branches of contemporary mathematics, from elementary plane geometry to advanced mathematical analysis, is the assumption that a *one-to-one correspondence* can be established between the points of any straight line and the members of the ordinary system of numbers—the so-called *real numbers*. Such a one-to-one correspondence is depicted in Figure 1. It is called a *coordinate system* on the line L.

The basic idea of a coordinate system is that once a unit length is defined on any line L by choosing "zero" and "one" as distinct points, then to each point P of that given line L there is associated a *unique* real number r, the coordinate of P. Conversely, to each real number (coordinate) r, there corresponds a unique point P on line L. The point O, which corresponds to the coordinate zero, is called the *origin* of the coordinate system. Positive coordinates correspond to points of L that are on one side of the origin (*positive side* of line L), while negative coordinates correspond to points of L on the other side of the origin (*negative side*). The natural order of the coordinates yields a corresponding arrangement of the points of line L. The *natural order* is defined as follows: Given any pair of real numbers, a and b, only one of three relationships is true—a is less than b ($a < b$), a is equal to b ($a = b$), or a is greater than b ($a > b$). The common convention for the arrangement of points of a line is that the coordinates are in an increasing order from left to right, as shown in Figure 1.

The great advantage of this correspondence of points with numbers is that one has available all properties of the real number system when studying geometrical properties of points and lines.

The idea of a coordinate system is readily extended to the points of any plane by choosing two intersecting lines, X and Y, in this plane and setting up a coordinate system on each of these lines, using the point of intersection O as a common origin for these coordinate systems. The lines X and Y are usually chosen perpendicular to each other, and are called the *X-axis* and *Y-axis*, respectively. The same scale is usually chosen on both axes, especially for the purpose of defining distance. (*Oblique axes*, or axes with different scales, are also possible and useful for certain purposes; however, they will not be discussed in this article.)

As shown in Figure 2, perpendiculars PQ and PR are drawn from any point P to the X-axis and the Y-axis, respectively. On the X-axis there is a unique coordinate x corresponding to Q and, similarly, on the Y-axis there is a unique coordinate y corresponding to R. Any point P of the plane now corresponds to a unique *ordered pair* of numbers (x, y). The value of x, the first coordinate, is called the *abscissa* of point P. The value of y, the second coordinate, is called the *ordinate* of point P.

An important formula from elementary coordinate geometry must be used now. It is the *distance formula*, which expresses the distance between any two points in the plane. By letting the coordinates of these points be x_1, y_1 and x_2, y_2, and applying the Pythagorean theorem to a diagram such as Figure 3, we obtain

$$d = \sqrt{(x_2 - x_1)^2 + (y_2 - y_1)^2}\,.$$

Unit Circle.—An important case of this distance formula arises when one considers the distance of any point (x, y) from the origin, whose coordinates are, of course, (O, O). Denoting this distance by r (Fig. 4), we obtain

$$r = \sqrt{x^2 + y^2}\,.$$

If the distance r is held fixed, while the point (x, y) is permitted to vary, the *locus* of point (x, y)—that is, the set of all points (x, y) with a fixed distance from the origin—is clearly a circle of radius r. This circle is completely specified by this equation for each fixed positive value of r. An equivalent equation is

$$x^2 + y^2 = r^2\,,$$

provided it is understood that the value of r is a fixed number greater than or equal to ($\geqq$) zero; if $r = 1$, the preceding equation becomes a *unit circle* (Fig. 5), and the equation obtained is

$$x^2 + y^2 = 1\,.$$

The study of trigonometry in this article will be based upon the unit circle.

Figure 2.

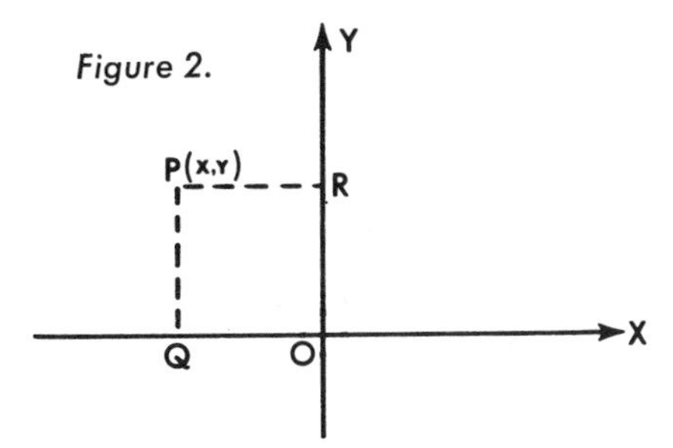

Figure 3.

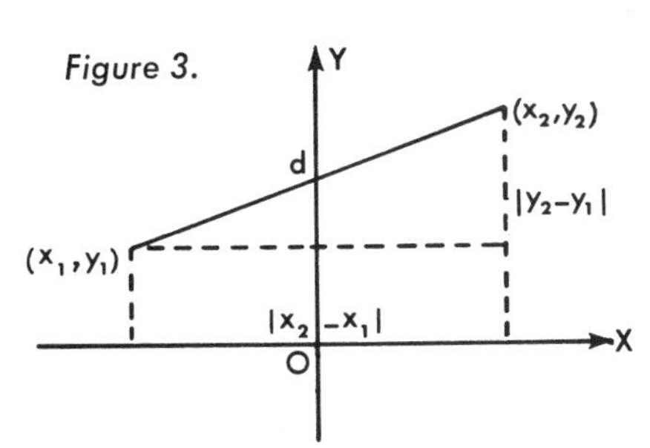

Figure 4.

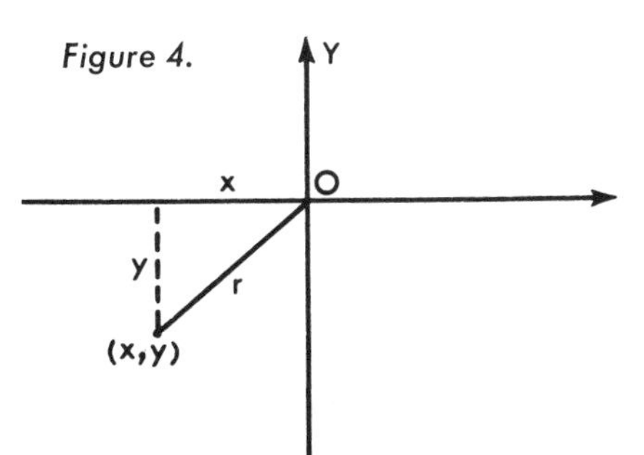

Figure 5.

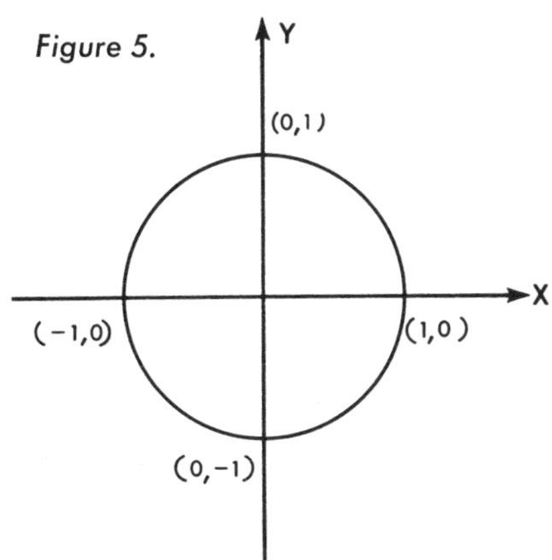

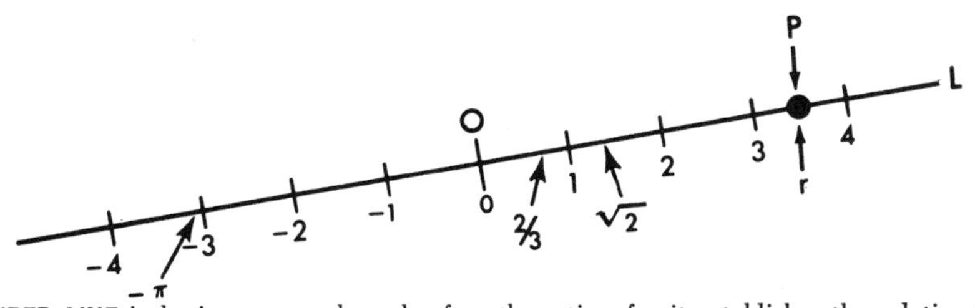

NUMBER LINE is basic to every branch of mathematics, for it establishes the relationship among the elements of the number system that is being used. The close-packed number line shown here contains both the rational and irrational and the positive and negative numbers.

Figure 6.

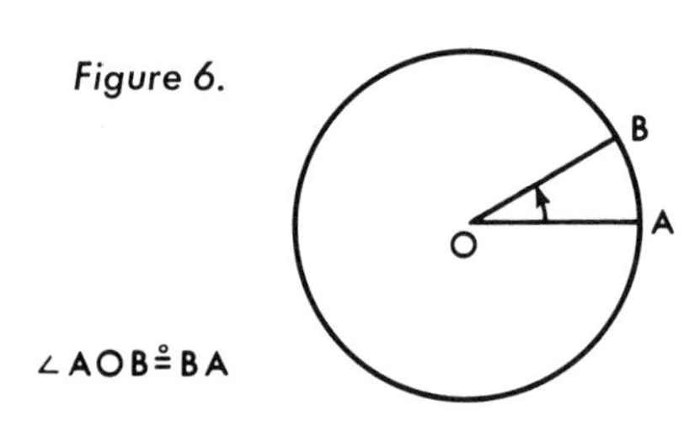

∠AOB ≗ BA

Figure 7.

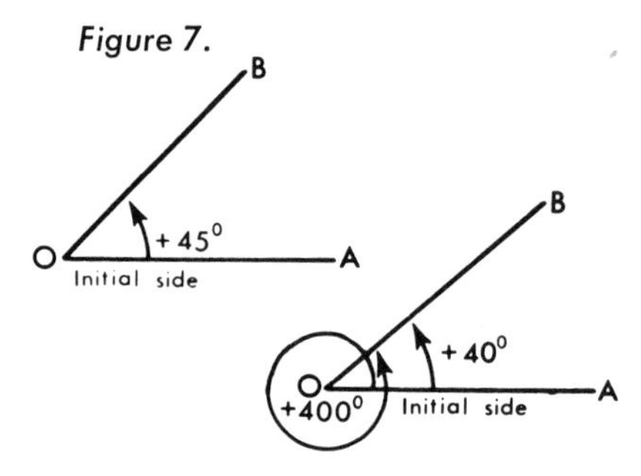

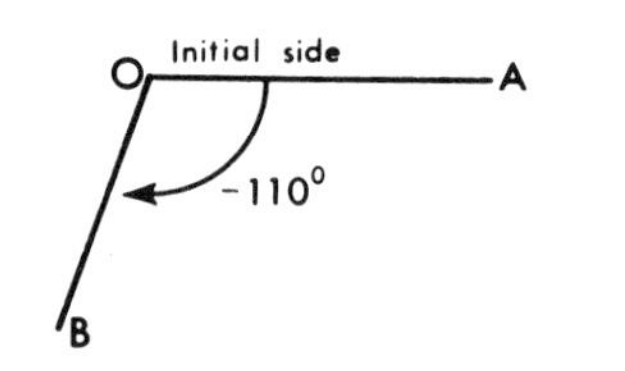

Figure 8.

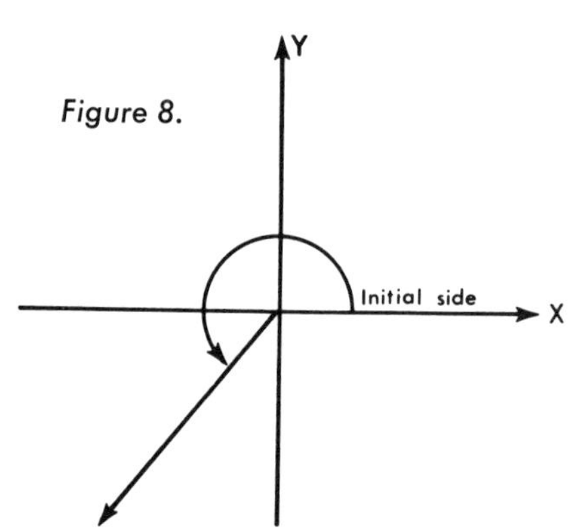

Figure 9.

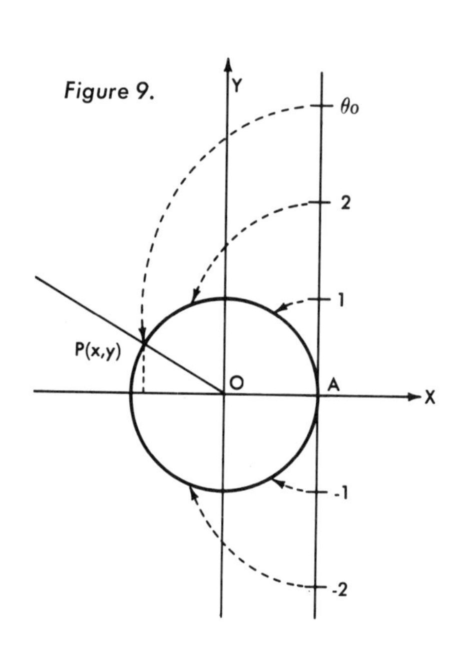

Circular Measure.—The circumference of a circle is divided into 360 equal parts, called *arc degrees*. Also, the number of angle degrees in a central angle equals the number of arc degrees in the intercepted arc, as shown in Figure 6.

In trigonometry, both the amount of rotation and the direction of rotation are important. Thus, an angle can be defined as *positive* if the rotation is counterclockwise, and as *negative* if the rotation is clockwise. The angle may have any magnitude, depending on the amount and direction of the rotation, as shown in Figure 7.

When an angle has its *vertex* at the origin and its *initial side* along the positive X-axis, it is said to be in the *standard position*, as shown in Figure 8.

It is useful in trigonometry to extend the method of measuring angles by introducing *circular measure*. Instead of associating a unique number of degrees with each given angle, in circular measure an *entire set* of real numbers is associated with the angle. Each number in this set measures an amount of rotation from the initial side to the terminal side. The amounts of rotation are expressed precisely in the following way: At point A in Figure 9, a line is drawn tangent to the unit circle, and the coordinate system is assigned to this line as it was assigned to the Y-axis, but starting at point A.

This new line is "wrapped around" the circle as indicated by the dotted arcs in Figure 9; each point of the line will fall upon a definite point of the circle. However, each time the line wraps around the circle, a new point of the line will fall onto the same place as a previous point. Therefore, each point P on the circle will correspond to an infinite number of points of the wrapping line, each new point arising after one complete rotation. These points are therefore spaced along the wrapping line at intervals of 2π, because that is the circumference of the unit circle.

Suppose θ_0 is the coordinate of the first point on the positive side of the wrapping line, which corresponds to point P_1 on the circle. Then

$$0 \leqq \theta_0 < 2\pi\,. \tag{1}$$

If θ is the coordinate of any other point on the wrapping line that corresponds to the same point P on the circle, then θ must differ from θ_0 by an integral multiple of 2π; that is,

$$\theta - \theta_0 = 2n\pi\,,$$

where n is an integer. Hence for any real number θ,

$$\theta = \theta_0 + 2n\pi\,, \tag{2}$$

where $n = 0, \pm 1, \pm 2, \ldots$. Note that positive values of n correspond to points on the positive side of the wrapping line; i.e., to positive (counterclockwise) rotations. Negative values of n correspond to points on the negative side of the wrapping line; i.e., to negative (clockwise) rotations.

The smallest counterclockwise rotation that corresponds to any specific point P on the unit circle is expressed by the value θ_0. All rotations that yield this (same) point P are represented by the set of all values of θ in equation (2) for a given θ_0. It is this set of real numbers that is associated with $\angle AOP$ in Figure 9. This special value θ_0 is often referred to as the *circular residue* of θ.

Although there are many possible values associated with a given angle (in standard position), it should be observed that a real number θ determines a unique point P on the unit circle and hence a unique angle in standard position. Each angle in standard position has many circular measures, but each circular measure determines a unique angle in standard position. For the purposes of this discussion, an angle of circular measure will be referred to as $\angle\theta$.

The relationship between circular measure and degree measure for angles is easily seen in the table below, where a circular measure of 2π corresponds to one complete rotation of the wrapping line (360° in degree measure).

Degree Measure	*Circular Measure*
360°	2π
180°	π
90°	$\frac{\pi}{2}$
60°	$\frac{\pi}{3}$
45°	$\frac{\pi}{4}$
$\frac{180^\circ}{\pi}$	1
1°	$\frac{\pi}{180}$
$\frac{180^\circ}{\pi}$	θ
d°	$\frac{\pi d}{180}$

An angle of circular measure 1 is often called an angle of 1 *radian*. For this reason, circular measure is often called *radian measure*.

From Figure 9 it is clear that an angle of 1 radian intercepts an arc of unit length on a unit circle. It will be seen that an angle of 1 radian is equal to an angle of $180^\circ/\pi$—about 57.3° in degree measure.

Arc Length and Sector Area.—One advantage of circular measure over degree measure is that the former yields simpler formulas for the length of an arc of a circle and the area of a sector. In geometry, in any given circle both arc length and sector area are proportional to the size of the central angle. Thus it is possible to set up the following proportions involving arc length s and sector area A:

$$\frac{s}{2\pi r} = \frac{\theta}{2\pi}\,, \tag{3}$$

$$\frac{A}{r^2} = \frac{\theta}{2\pi}\,, \tag{4}$$

where r is the length of the radius and the central angle θ is expressed in circu-

Figure 10.

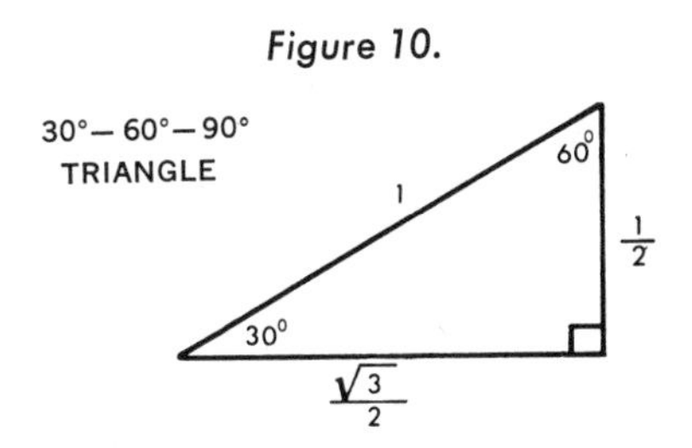

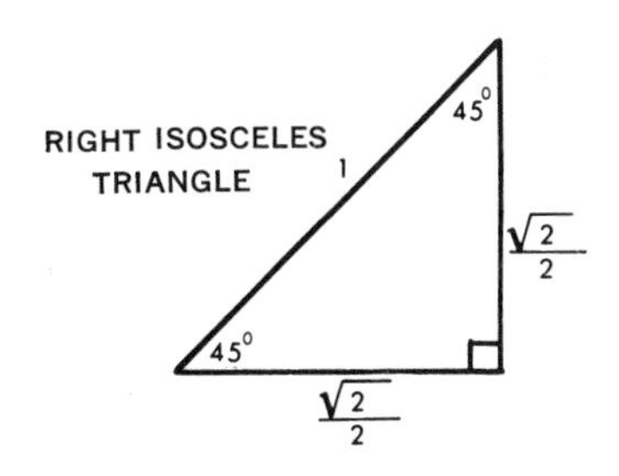

lar measure. From this it follows that

$$s = r\theta \text{ (arc length)}\,, \qquad (5)$$

$$A = \tfrac{1}{2}r^2\theta \text{ (sector area)}\,. \qquad (6)$$

These formulas are most useful in advanced mathematics.

Sine and Cosine Functions.—Take any real number θ. Corresponding to the value θ, there is a unique point P on the unit circle. The point P has a unique pair of coordinates (x, y) and determines a unique $\angle AOP$ in standard position (which will also be called $\angle\theta$). The values of the coordinates x and y, which are determined in this manner from the given real number θ, are called the *cosine of* θ and the *sine of* θ, respectively. They are abbreviated as follows:

$$\cos\theta = x \qquad \text{and} \qquad \sin\theta = y\,. \quad (7)$$

Since

$$x^2 + y^2 = 1\,,$$

we obtain

$$(\cos\theta)^2 + (\sin\theta)^2 = 1\,,$$

for all real θ. This is generally written:

$$\cos^2\theta + \sin^2\theta = 1\,, \qquad (8)$$

for all real θ. However, to each real number θ, there corresponds a unique value of $\cos\theta$ as well as a unique value of $\sin\theta$. Thus, *sine* and *cosine* may properly be referred to as *functions*. The *domain* of each of these functions (the values that θ may assume) consists of all the real numbers. The *range* of each of these functions (the values that $\cos\theta$ or $\sin\theta$ may assume) consists of all the real numbers from -1 to $+1$, inclusive.

The determination of the values of $\cos\theta$ and $\sin\theta$ for every real value of θ requires higher mathematical techniques. However, these values can be tabulated for a large number of particular values of θ, corresponding to special angles such as 30°, 45°, 60°, and 90°, which are readily handled by the methods of elementary plane geometry. In this connection, consider the special right triangles in Figure 10. Using these special triangles and simple coordinate geometry, the accompanying table is constructed.

A more extensive table, listing approximate values of these functions, is included at the end of this article.

The graphs of the cosine and the sine functions can easily be plotted with the aid of the accompanying table. In plotting these graphs, the values of θ are found along the X-axis; the values of cos θ (or sin θ) are then found along the Y-axis.

Values of Cos θ and Sin θ

θ (Degree Measure)	θ (Circular Measure)	Cos θ	Sin θ
0°	0	1	0
30°	$\frac{\pi}{6}$	$\frac{\sqrt{3}}{2}$	$\frac{1}{2}$
45°	$\frac{\pi}{4}$	$\frac{\sqrt{2}}{2}$	$\frac{\sqrt{2}}{2}$
60°	$\frac{\pi}{3}$	$\frac{1}{2}$	$\frac{\sqrt{3}}{2}$
90°	$\frac{\pi}{2}$	0	1
120°	$\frac{2\pi}{3}$	$-\frac{1}{2}$	$\frac{\sqrt{3}}{2}$
135°	$\frac{3\pi}{4}$	$-\frac{\sqrt{2}}{2}$	$\frac{\sqrt{2}}{2}$
150°	$\frac{5\pi}{6}$	$-\frac{\sqrt{3}}{2}$	$\frac{1}{2}$
180°	π	-1	0
210°	$\frac{7\pi}{6}$	$-\frac{\sqrt{3}}{2}$	$-\frac{1}{2}$
225°	$\frac{5\pi}{4}$	$-\frac{\sqrt{2}}{2}$	$-\frac{\sqrt{2}}{2}$
240°	$\frac{4\pi}{3}$	$-\frac{1}{2}$	$-\frac{\sqrt{3}}{2}$
270°	$\frac{3\pi}{2}$	0	-1
300°	$\frac{5\pi}{3}$	$\frac{1}{2}$	$-\frac{\sqrt{3}}{2}$
315°	$\frac{7\pi}{4}$	$\frac{\sqrt{2}}{2}$	$-\frac{\sqrt{2}}{2}$
330°	$\frac{11\pi}{6}$	$\frac{\sqrt{3}}{2}$	$-\frac{1}{2}$
360°	2π	1	0

Considering Figure 9 once again, observe that the coordinates of point P are by definition cos θ, sin θ, because point P is located on a unit circle. Suppose point P is located on the terminal ray of $\angle\theta$ at a distance $OP = r$ from its vertex, as shown in Figure 12.

Let P' be the point where the terminal ray OP intersects the unit circle; i.e., $OP' = 1$. If the coordinates of P are (x, y) while those of P' are (x', y'), then (by similar triangles):

$$\frac{x}{x'} = \frac{y}{y'} = \frac{OP}{OP'} = \frac{r}{1}\,.$$

Hence,

$$x = rx' \qquad \text{and} \qquad y = ry'\,.$$

However since P' is on the unit circle,

$$x' = r\cos\theta \qquad \text{and} \qquad y' = r\sin\theta.$$

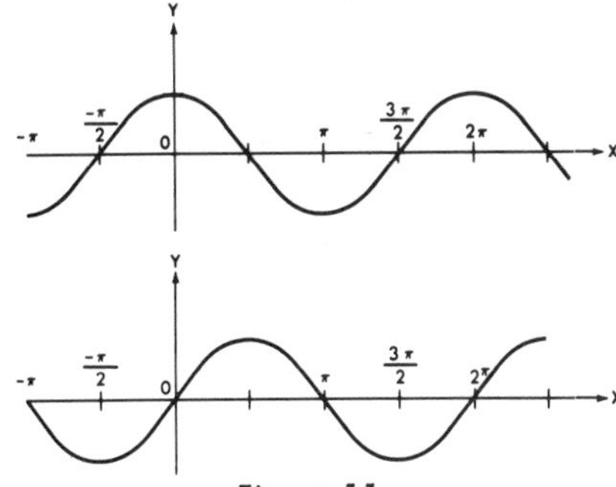

Figure 11.

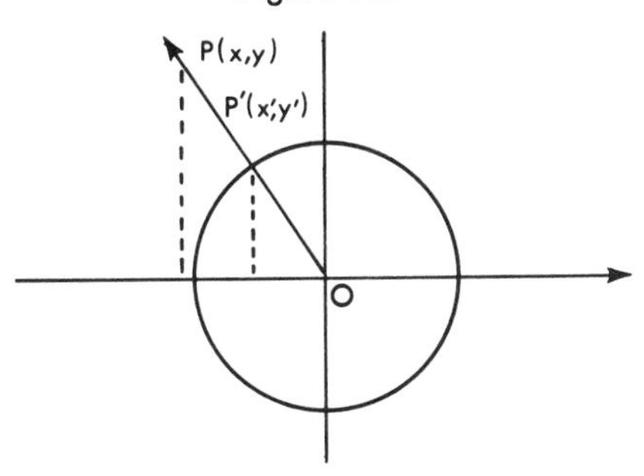

Figure 12.

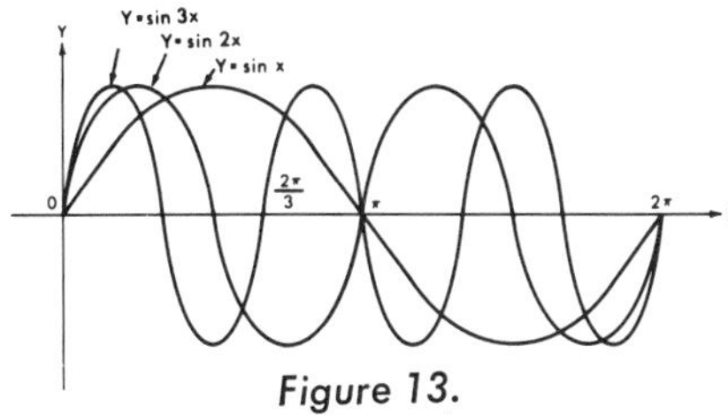

Figure 13.

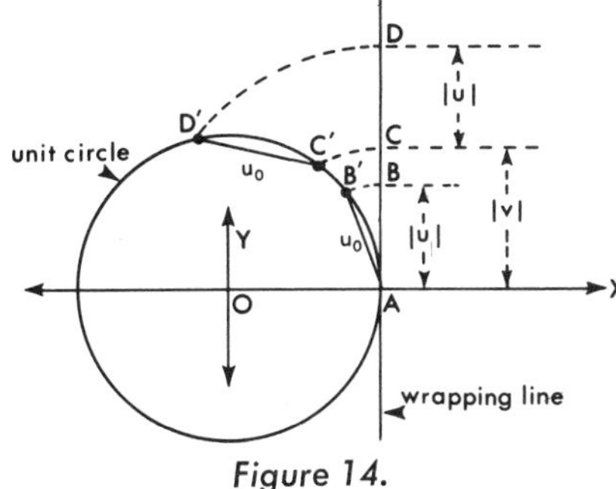

Figure 14.

Hence,

$$x = r\cos\theta \qquad \text{and} \qquad y = r\sin\theta\,. \quad (9)$$

This holds true for any point P' (x', y').

■ **ANALYTIC PROPERTIES.**—From the definition of the sine and cosine functions, it follows that they each possess the important property of *periodicity*. This means that they each repeat the values they assume over certain specified intervals. The smallest interval over which sine and cosine exhibit this repetition is an interval of length, 2π. These functions are therefore called periodic functions with period 2π. The periodicity is conveniently expressed as follows:

$$\cos(\theta + 2\pi) = \cos\theta\,, \qquad (10)$$

$$\sin(\theta + 2\pi) = \sin\theta\,, \qquad (11)$$

for all real θ. The periodicity is also apparent in the graphs of Figure 11.

One can readily construct periodic functions with a period other than 2π by forming *composite functions*. For example, the functions whose values are represented by the expressions

$$\cos 2\theta \quad \text{and} \quad \sin 2\theta$$

have the smaller period π. This follows from equations (10) and (11) because

$$\cos 2(\theta + \pi) = \cos (2\theta + 2\pi) = \cos 2\theta \,,$$
$$\sin 2(\theta + \pi) = \sin (2\theta + 2\pi) = \sin 2\theta \,.$$

The functions defined by

$$\cos 3\theta \quad \text{and} \quad \sin 3\theta$$

similarly are shown to have the period $2\pi/3$ and, in general, each function defined by

$$\cos (A\theta) \quad \text{and} \quad \sin (A\theta)$$

has the period $2\pi/A$. For example, each function defined by

$$\cos \pi\theta \quad \text{and} \quad \sin \pi\theta$$

has the period $2\pi/\pi = 2$. The construction of periodic functions with arbitrary periods is most important for advanced applications of trigonometry.

Figure 13 depicts sine graphs having different periods. They are defined by $y = \sin Ax$, since $A = 1, 2$, and 3. Analogous graphs may be drawn for the cosine function.

An important relation that connects the sine and cosine functions is the *phase difference*. The two curves in Figure 11 look very much alike, but a more careful glance reveals that the cosine curve reaches its peak when $\theta = 0$, while the sine curve attains the same peak value when $\theta = \pi/2$. At every stage this same 90° "lag" is maintained; i.e., the sine curve attains a particular value 90° later than the cosine curve attains this value. This 90° *difference in phase* is conveniently expressed in the following purely analytic form:

$$\sin \left(\theta + \frac{\pi}{2}\right) = \cos \theta \,, \qquad (12)$$

for all real θ.

This relation can be derived purely analytically from an even more general relation connecting the cosine and sine functions. However, before proceeding with this task, note that all the relations mentioned thus far have an important quality: they are true for all real values of θ. For example, equation (8) asserts that

$$\cos^2 \theta + \sin^2 \theta = 1 \,,$$

for all real θ. Whenever an equation containing a variable is true for all permissible values of this variable (such as θ), the equation is called an *identity*. (The permissible values will usually be clear from the context.)

The identity

$$\cos^2 \theta + \sin^2 \theta = 1$$

follows directly from the definitions of the cosine and sine functions. It will be used here to define many important identities.

■ BASIC IDENTITIES.—Let u and v be arbitrary real numbers. Using the diagram of Figure 9, points B, C, and D are marked on the wrapping line so as to correspond respectively to the values u, v, and $u + v$. If this line is now wrapped around the circle, these points fall upon points B', C', and D', as indicated in Figure 14.

A number of wrappings may of course occur before point B falls on B', and the same is true for the points C' and D'. In all cases, however, the length of arc AB' will certainly be the *circular residue* of AB, namely u_0. Hence, chord AB' and chord $C'D'$ have the same length. This length can be determined by applying the distance formula in equation (1) to the pairs of points A, B' and C', D'. By definition, the coordinates of these points are:

A: $(1, 0)$
B': $(\cos u, \sin u)$
C': $(\cos v, \sin v)$
D': $\cos (u + v), \sin (u + v)$.

Applying the distance formula to the equality

$$(AB')^2 = (C'D')^2 \,,$$

we obtain

$$(\cos u - 1)^2 + (\sin u)^2 = [\cos (u + v) - \cos u]^2 + [\sin (u + v) - \sin v]^2 \,.$$

The left side of the above equation then becomes

$$\cos^2 u - 2 \cos u + 1 + \sin^2 u \,.$$

Because

$$\cos^2 u + \sin^2 u = 1 \,,$$

this reduces to

$$2 - 2 \cos u \,.$$

Similarly, the right side of this equation reduces to

$$2 - 2[\cos (u + v) \cos v + \sin (u + v) \sin v] \,.$$

Equating these two expressions yields the following fundamental identity:

$$\cos u = \cos (u + v) \cos v + \sin (u + v) \sin v \,, \qquad (13)$$

for all real values of u and v.

From this identity further important identities can be obtained. First, since equation (13) holds true for all real values of u and v, then if w is any real number, u may be replaced by $w - v$:

$$\cos (w - v) = \cos w \cos v + \sin w \sin v \,, \qquad (14)$$

for all real values of w and v.

If $w = 0$, equation (13) becomes

$$\cos (-v) = \cos 0 \cos v + \sin 0 \sin v \,.$$

But $\cos 0 = 1$ and $\sin 0 = 0$. Hence,

$$\cos (-v) = \cos v \,, \qquad (15)$$

for all real values of v.

Equation (14) shows that cosine is an *even function*. A function f is called an even function only if $f(-x) = f(x)$ for all permissible values of x.

Returning to equation (14), substitute $w = \pi/2$:

$$\cos \left(\frac{\pi}{2} - v\right) = \cos \frac{\pi}{2} \cos v + \sin \frac{\pi}{2} \sin v \,.$$

But $\cos \pi/2 = 0$ and $\sin \pi/2 = 1$. Hence,

$$\cos \left(\frac{\pi}{2} - v\right) = \sin v \,, \qquad (16)$$

for all real values of v.

Replacing v by $(\pi/2 - u)$ in equation (16) yields the following identity as an immediate corollary:

$$\sin \left(\frac{\pi}{2} - u\right) = \cos u \,, \qquad (17)$$

for all real values of u.

Equations (16) and (17) exhibit the important relationship of *complementarity*. Whenever $u + v = \pi/2$, u and v are *complementary*. Either of these values is then the *complement* of the other. Identities (16) and (17) may now be expressed verbally as follows:

The sine of a real number is the cosine of its complement.

The cosine of a real number is the sine of its complement.

Identities (15) and (16) together yield a derivation of the "90° phase difference" property mentioned previously. In equation (16), let $v = \theta + \pi/2$, where θ may be any real number. This gives

$$\cos (-\theta) = \sin \left(\theta + \frac{\pi}{2}\right) .$$

But, by equation (15), $\cos (-\theta) = \cos \theta$. Hence,

$$\sin \left(\theta + \frac{\pi}{2}\right) = \cos \theta \,.$$

for all real θ. This completes the derivation of equation (12).

Return to equation (14) and substitute $v = -(\pi/2)$:

$$\cos \left(w + \frac{\pi}{2}\right) = \cos w \cos \left(-\frac{\pi}{2}\right) + \sin w \sin \left(-\frac{\pi}{2}\right).$$

But from equation (15) $\cos -(\pi/2) = \cos \pi/2 = 0$, and from equation (16), letting $v = \pi$,

$$\sin \left(-\frac{\pi}{2}\right) = \cos \pi = -1 \,.$$

Hence,

$$\cos \left(w + \frac{\pi}{2}\right) = -\sin w \,, \qquad (18)$$

for all real w. In this formula, replace w by $-v$:

$$\cos \left(\frac{\pi}{2} - v\right) = -\sin (-v),$$

and make use of equation (16); thus,

$$\sin (-v) = -\sin v \,, \qquad (19)$$

for all real v. Equation (19) shows that sine is an *odd function*. A function f is called an odd function only if $f(-x) = -f(x)$ for all permissible values of x.

Identities (15) and (19) make it unnecessary to tabulate sines and cosines of negative numbers. For example, to determine the value of $\sin(-\pi/3)$ and $\cos(-\pi/3)$, proceed as follows:

$$\sin\left(-\frac{\pi}{3}\right) = -\sin\frac{\pi}{3} = -\frac{\sqrt{3}}{2},$$

$$\cos\left(-\frac{\pi}{3}\right) = \cos\frac{\pi}{3} = +\frac{1}{2}.$$

With periodicity of sine and cosine, it is unnecessary to tabulate their values outside of the interval $(0, 2\pi)$. Identities (12) and (18) further cut this down to the interval $(0, \pi/2)$. Thus, trigonometry tables need actually be computed only for numbers from 0 to $\pi/4$ (0° to 45° in degree measure).

From equation (14) another very important identity may be obtained by replacing v by $(-v)$:

$$\cos(w+v) = \cos w \cos(-v) + \sin w \sin(-v).$$

Using equations (15) and (19),

$$\cos(w+v) = \cos w \cos v - \sin w \sin v, \quad (20)$$

for all real values of w and v.

There are also formulas for $\sin(w+v)$ and $\sin(w-v)$. Using equation (20) replace w by $w+\pi/2$:

$$\cos\left(w+\frac{\pi}{2}+v\right) = \cos\left(w+\frac{\pi}{2}\right)\cos v - \sin\left(w+\frac{\pi}{2}\right)\sin v.$$

Using equations (18) and (12) we have

$$-\sin(w+v) = -\sin w \cos v - \cos w \sin v,$$

$$\sin(w+v) = \sin w \cos v + \cos w \sin v, \quad (21)$$

for all real values of w and v.

Replacing v by $-v$ in equation (20) and using equations (14) and (18), we obtain

$$\sin(w-v) = \sin w \cos v - \cos w \sin v, \quad (22)$$

for all real values of w and v.

Special cases of identities (22) and (14) are

$$\sin(\pi - v) = \sin v, \quad (23)$$

$$\cos(\pi - v) = -\cos v, \quad (24)$$

for all real v.

The last two identities exhibit the *supplementarity* relationship. Whenever $u + v = \pi$, u and v are supplementary. Either of these values is then the *supplement* of the other. Identities (23) and (24) may thus be expressed verbally:

The sine of a real number is the sine of its supplement.

The cosine of a real number is the negative of the cosine of its supplement.

Double- and Half-Argument Identities.— In identities (20) and (21), let $w = v = \theta$, where θ may be any real number. Thus,

$$\cos 2\theta = \cos^2\theta - \sin^2\theta, \quad (25)$$

$$\sin 2\theta = 2\sin\theta\cos\theta, \quad (26)$$

for all real θ. Identity (24) can be expressed in two other useful ways by using the basic identity (8):

$$\cos^2\theta + \sin^2\theta = 1.$$

From this we obtain

$$\cos^2\theta = 1 - \sin^2\theta,$$
$$\sin^2\theta = 1 - \cos^2\theta.$$

Substituting each of these in turn into equation (24), we obtain

$$\cos 2\theta = 1 - 2\sin^2\theta, \quad (27)$$

$$\cos 2\theta = 2\cos^2\theta - 1, \quad (28)$$

for all real θ. Formulas (25) through (28) are known as the *double-argument* formulas. (They are also called the *double-angle* formulas when θ is interpreted as the measure of an angle.)

Solving equation (27) for $\sin^2\theta$, we obtain

$$\sin^2\theta = \frac{1-\cos 2\theta}{2},$$

for all real θ. As a result, therefore, we obtain

$$\sin\theta = \pm\sqrt{\frac{1-\cos 2\theta}{2}}.$$

However, the ambiguity of sign requires some further discussion. By definition, the values of $\sin\theta$ are positive whenever the circular residue θ_0 is between 0 and π; the values of $\sin\theta$ are negative whenever the circular residue θ_0 is between π and 2π. Hence, to be precise,

$$\sin\theta = \sqrt{\frac{1-\cos 2\theta}{2}},$$

whenever $0 \leq \theta_0 \leq \pi$, and (29)

$$\sin\theta = -\sqrt{\frac{1-\cos 2\theta}{2}},$$

whenever $\pi < \theta_0 < 2\pi$.

These formulas are also useful when considered as *half-argument* rather than double-argument formulas. This is done by letting $\phi = 2\theta$; that is, $\theta = \phi/2$. Then equation (29) becomes

$$\sin\frac{\phi}{2} = \sqrt{\frac{1-\cos\phi}{2}}$$

whenever $0 \leq (\phi/2)_0 \leq \pi$, and (30)

$$\sin\frac{\phi}{2} = -\sqrt{\frac{1-\cos\phi}{2}}$$

whenever $\pi \leq (\phi/2)_0 \leq 2\pi$.

Similarly, by solving equation (28) for $\cos\theta$ and replacing θ by $\phi/2$ we obtain

$$\cos\frac{\phi}{2} = \sqrt{\frac{1+\cos\phi}{2}}$$

whenever $0 \leq (\phi/2)_0 \leq \pi/2$, or $3\pi/2 \leq (\phi/2)_0 \leq 2\pi$, and (31)

$$\cos\frac{\phi}{2} = -\sqrt{\frac{1-\cos\phi}{2}},$$

whenever $\pi/2 \leq (\phi/2)_0 \leq 3\pi/2$.

By way of an application, use these formulas to determine the values of sin 15° and cos 15° ($\sin\pi/12$ and $\cos\pi/12$). Letting $\phi = 30°$ ($\phi = \pi/6$), $\phi/2 = 15°$. Since $(\phi/2)_0 = \pi/12$, use the first formula in each case:

$$\sin 15° = \sin\frac{\pi}{12} = \sqrt{\frac{1-\frac{\sqrt{3}}{2}}{2}},$$

$$\cos 15° = \cos\frac{\pi}{12} = \sqrt{\frac{1+\frac{\sqrt{3}}{2}}{2}}.$$

These expressions simplify to

$$\sin 15° = \frac{1}{2}\sqrt{2-\sqrt{3}},$$

$$\cos 15° = \frac{1}{2}\sqrt{2+\sqrt{3}}.$$

Equivalent expressions are

$$\sin 15° = \frac{\sqrt{6}-\sqrt{2}}{4},$$

$$\cos 15° = \frac{\sqrt{6}+\sqrt{2}}{4},$$

because

$$\left(\frac{\sqrt{6}\pm\sqrt{2}}{2}\right)^2 = \frac{6 \pm 2\sqrt{12}+2}{4} = \frac{8 \pm 4\sqrt{3}}{4} = 2 \pm \sqrt{3}.$$

Hence,

$$\sqrt{2\pm\sqrt{3}} = \frac{\sqrt{6}\pm\sqrt{2}}{2}.$$

Sum and Difference Formulas.— From equations (14), (20), (21), and (22), it is possible to obtain further identities, useful in more advanced work:

$$\begin{aligned}
\sin(w+v) + \sin(w-v) &= 2\sin w\cos v,\\
\sin(w+v) - \sin(w-v) &= 2\cos w\sin v,\\
\cos(w+v) + \cos(w-v) &= 2\cos w\cos v,\\
\cos(w+v) - \cos(w-v) &= -2\sin w\sin v,
\end{aligned} \quad (32)$$

for all real values of w and v.

If $w + v = r$ and $w - v = s$, these identities can be rewritten as follows:

$$\begin{aligned}
\sin r + \sin s &= 2\sin\frac{r+s}{2}\cos\frac{r-s}{2},\\
\sin r - \sin s &= 2\cos\frac{r+s}{2}\sin\frac{r-s}{2},\\
\cos r + \cos s &= 2\cos\frac{r+s}{2}\cos\frac{r-s}{2},\\
\cos r - \cos s &= -2\sin\frac{r+s}{2}\sin\frac{r-s}{2},
\end{aligned} \quad (33)$$

for all real values of r and s.

Tangent and Cotangent Functions.— The functions tangent and cotangent are abbreviated and defined as

$$\tan\theta = \frac{\sin\theta}{\cos\theta} \quad (34)$$

provided $\cos\theta \neq 0$,

$$\cot\theta = \frac{\cos\theta}{\sin\theta} \quad (35)$$

provided $\sin\theta \neq 0$.

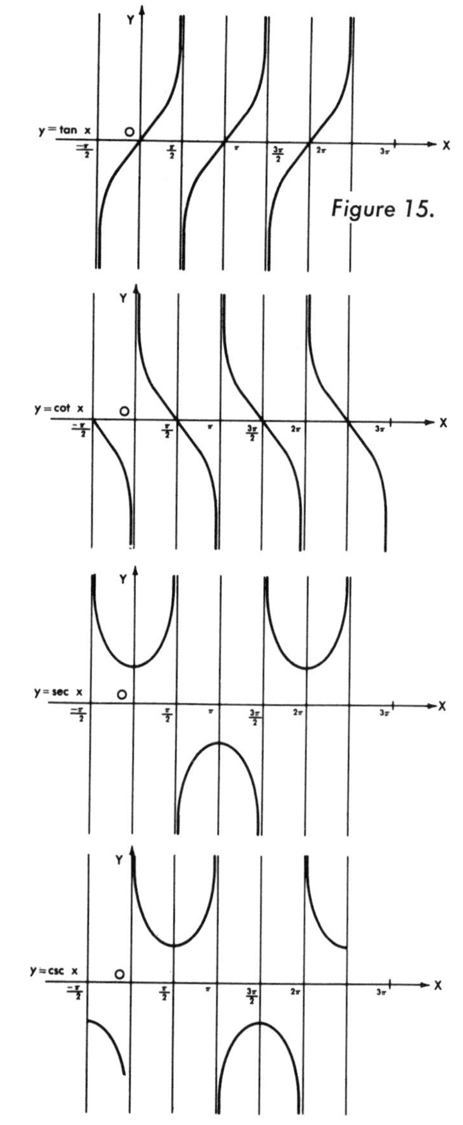

Figure 15.

Secant and Cosecant Functions.—The functions secant and cosecant are abbreviated and defined as follows:

$$\sec\theta = \frac{1}{\cos\theta} \tag{36}$$

provided $\cos\theta \neq 0$,

$$\csc\theta = \frac{1}{\sin\theta} \tag{37}$$

provided $\sin\theta \neq 0$.

Observe that all the functions tangent, cotangent, secant, and cosecant are not defined for certain real values of θ. The tangent and secant are undefined whenever θ_0 (the circular residue of θ) is either $\pi/2$ or $3\pi/2$. The cotangent and cosecant are undefined whenever θ_0 is either 0 or π. For all other values of θ, these functions have well-defined values given by equations (34) through (37). Values of the tangent function are found in the accompanying table. See graphs of these functions in Figure 15.

Further Basic Identities.—Important new identities, readily derived from equations (34) through (37) and the earlier identities, are listed below. The reader should verify these identities. (The proofs for most of them are fairly easy.) Note that these equations are true for all *permissible* values of θ, u, v, w, etc.

$$\tan\theta = \frac{1}{\cot\theta}, \quad \cot\theta = \frac{1}{\tan\theta} \tag{38}$$

$$\tan^2\theta + 1 = \sec^2\theta \tag{39}$$

$$\cot^2\theta + 1 = \csc^2\theta \tag{40}$$

$$\tan(v+w) = \frac{\tan v + \tan w}{1 - \tan v \tan w} \tag{41}$$

$$\tan(v-w) = \frac{\tan v - \tan w}{1 + \tan v \tan w} \tag{42}$$

$$\tan 2\theta = \frac{2\tan\theta}{1-\tan^2\theta} \tag{43}$$

$$\tan\frac{\theta}{2} = \sqrt{\frac{1-\cos\theta}{1+\cos\theta}}$$

whenever $0 \leqq (\theta/2)_0 < \pi/2$, or $\pi \leqq (\theta/2)_0 < 3\pi/2$, and

$$\tan\frac{\theta}{2} = -\sqrt{\frac{1-\cos\theta}{1+\cos\theta}}, \tag{44}$$

whenever $\pi/2 < (\theta/2)_0 \leqq \pi$, or $3\pi/2 < (\theta/2)_0 \leqq 2\pi$.

$$\tan\frac{\theta}{2} = \frac{1-\cos\theta}{\sin\theta} \tag{45}$$

provided $\sin\theta \neq 0$.

Inverses of Functions.—If real numbers u and v are so related that

$$u = \sin v, \tag{46}$$

then every real number is a possible value of v; but not every real number is a possible value of u. In fact, the values of u are confined to the interval from -1 to $+1$, inclusive. This fact has already been expressed in the statement that sine is a function whose domain consists of all real numbers, but whose range consists of all real numbers from -1 to $+1$, inclusive. If the values of v are plotted on a graph along the X-axis and the corresponding values of u (i.e., of $\sin v$) are located on the Y-axis, then the sine curve of Figure 11 is obtained.

Now reverse the graphing procedure and locate the values of u along the X-axis and the corresponding values of v along the Y-axis (Fig. 16).

Plotting the graph in this way amounts to interpreting equation (46) as follows: For each real value u in the interval from -1 to $+1$, inclusive, there are various real values v (infinitely many, in fact) such that $u = \sin v$. Stated differently, *v is a real number whose sine is u*. It has become customary to designate the set of all real numbers whose sine is u by either of the following expressions: $\sin^{-1} u$ (read: "inverse sine of u"), or arc sin u (read: "arc sine u"). The first of these expressions will be used. Equation (46) is therefore seen to be completely equivalent to the sentence, *v is a member of the inverse sine of u*. This may be abbreviated:

$$v \in \sin^{-1} u. \tag{47}$$

(*Note:* Many textbooks still use the incorrect notation "$v = \sin^{-1} u$," when they really mean "$v \sin^{-1} u$.")

Figure 16.

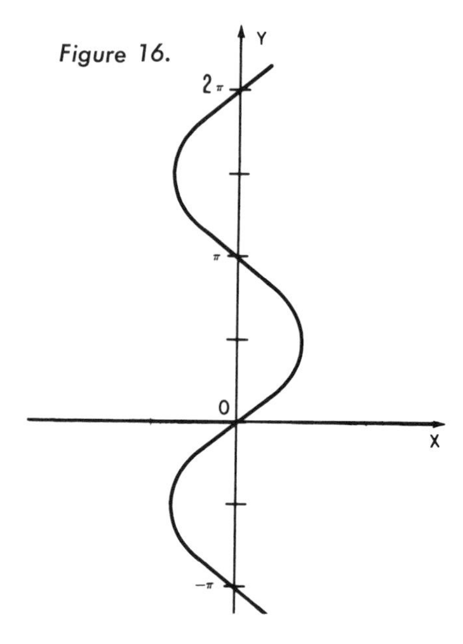

As an example, observe that

$$\sin^{-1} 1 = \left\{\frac{\pi}{2}, \frac{-3\pi}{2}, \frac{5\pi}{2}, \frac{-7\pi}{2}, \ldots\right\}.$$

Hence, the equations

$$\sin\frac{\pi}{2} = 1, \quad \sin\left(\frac{-3\pi}{2}\right) = 1, \ldots$$

may be expressed in "inverse notation" as follows:

$$\frac{\pi}{2} \in (\sin^{-1} 1), \quad \frac{-3\pi}{2} \in (\sin^{-1} 1), \ldots$$

The inverse sine is not a function according to the modern use of this term, because $\sin^{-1} u$ is not a unique real number for each permissible value of u. The inverse sine is an example of a *relation*. (Relations are more general than functions. In fact, a function is a "single-valued relation.")

Although $\sin^{-1}$ is not a function, it can be used to define a function by imposing appropriate restrictions on its range; i.e., on the values it may assume. A restriction that is usually adopted is one that confines the values of $\sin^{-1} u$ to the interval from $-\pi/2$ to $+\pi/2$, inclusive. For each permissible value of u (from -1 to $+1$, inclusive) there is only one member in the set $\sin^{-1} u$ that meets this restriction. This value is called the *principal value* of $\sin^{-1} u$. This principal value is denoted $\text{Sin}^{-1} u$ (read: "principal inverse Sine of u"). Observe that for each value of u in the interval -1 to $+1$, inclusive, there is a unique value $\text{Sin}^{-1} u$, with the properties

$$\text{Sin}^{-1} u \in \sin^{-1} u, \quad -\frac{\pi}{2} \leqq \text{Sin}^{-1} u \leqq \frac{\pi}{2}. \tag{48}$$

Examples are

$$\text{Sin}^{-1} 0 = 0,$$

$$\text{Sin}^{-1} 1 = \frac{\pi}{2},$$

$$\text{Sin}^{-1}(-1) = -\frac{\pi}{2},$$

$$\text{Sin}^{-1}\left(-\frac{1}{2}\right) = -\frac{\pi}{3}.$$

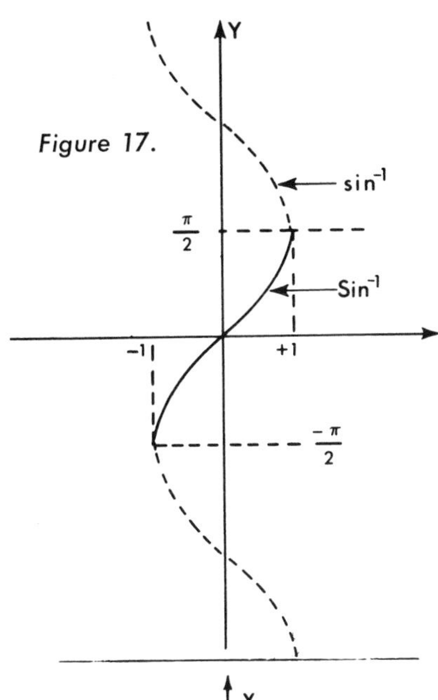

Figure 17.

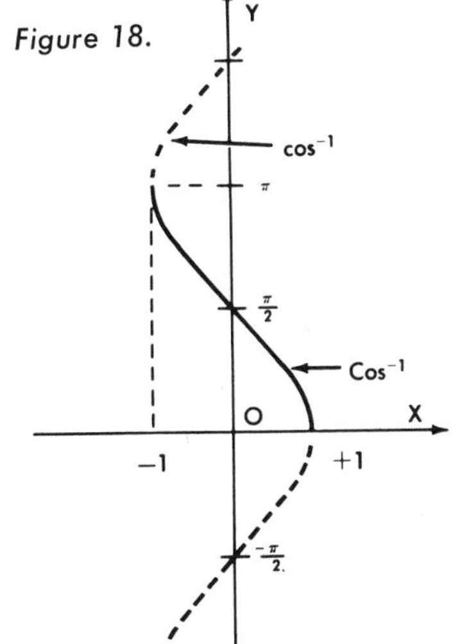

Figure 18.

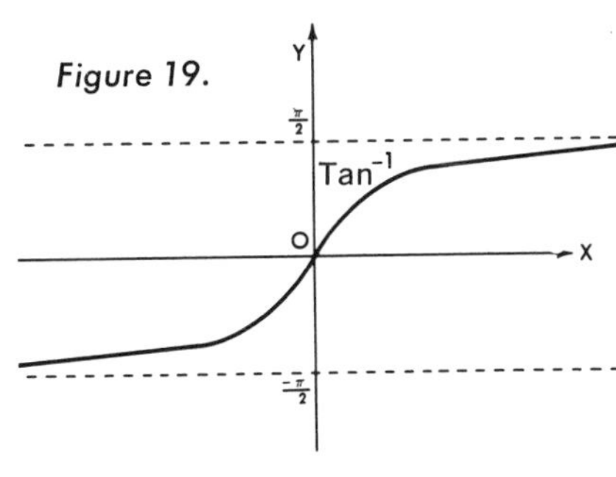

Figure 19.

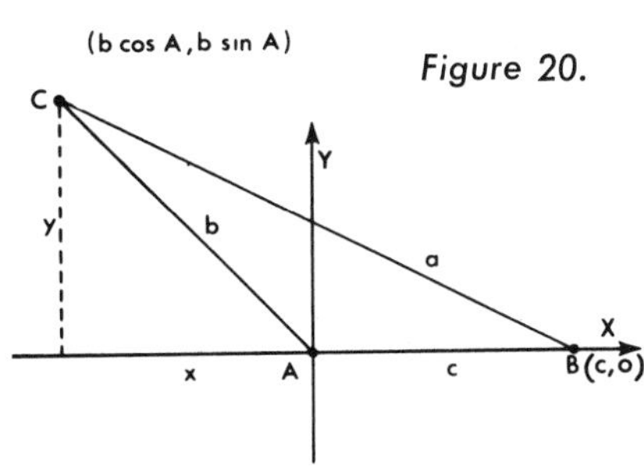

Figure 20.

The principal inverse sine is therefore a function in the modern sense of the term. Its graph is actually a portion of the graph of the inverse sine relation, as shown in Figure 17.

In a completely analogous manner, one may define inverse relations for each of the other five trigonometric functions. Thus the equation

$$u = \cos v \tag{49}$$

is written equivalently as

$$v \in \cos^{-1} u \tag{50}$$

where $\cos^{-1} v = \{\text{real numbers whose cosine is } v\}$.

The *principal inverse cosine* is defined by

$$\text{Cos}^{-1} u \in \cos^{-1} u$$

and (51)

$$0 \leqq \text{Cos}^{-1} u \leqq \pi .$$

Its graph is depicted in Figure 18.

The equation

$$u = \tan v \tag{52}$$

is equivalent to the sentence

$$v \in \tan^{-1} u \tag{53}$$

where $\tan^{-1} u = \{\text{real numbers whose tangent is } u\}$.

The *principal inverse tangent* is defined by

$$\text{Tan}^{-1} u \in \tan^{-1} u$$

and (54)

$$\frac{-\pi}{2} \leqq \text{Tan}^{-1} u \leqq \frac{\pi}{2} .$$

Its graph is given in Figure 19.

The principal values of the remaining three trigonometric functions may conveniently be defined in terms of the three that have already been defined:

$$\text{Cot}^{-1} u = \text{Tan}^{-1} \left(\frac{1}{u}\right), \tag{55}$$

$$\text{Sec}^{-1} u = \text{Cos}^{-1} \left(\frac{1}{u}\right), \tag{56}$$

$$\text{Csc}^{-1} u = \text{Sin}^{-1} \left(\frac{1}{u}\right), \tag{57}$$

for all permissible values of u.

The inverse relations and inverse functions of trigonometry play an important role in calculus and more advanced analysis.

Laws of Cosines and Sines.—Most of the applications of the trigonometric functions to geometry stem from two important laws—the *law of cosines* and the *law of sines*.

■ **LAW OF COSINES.**—Consider any triangle ABC and designate the measures of its sides and angles in the conventional way by a, b, c and A, B, C, respectively. Let a rectangular coordinate system be chosen so that angle A is in standard position, as indicated in Figure 20. Relative to this set of axes, the coordinates of point C are $(b \cos A, b \sin A)$. Applying the distance formula to side BC, we find that

$$\begin{aligned} a^2 &= (b \cos A - c)^2 + (b \sin A - 0)^2 \\ &= b^2 \cos^2 A - 2bc \cos A + c^2 + b^2 \sin^2 A \\ &= b^2 (\cos^2 A + \sin^2 A) + c^2 - 2bc \cos A \end{aligned}$$

$$\therefore a^2 = b^2 + c^2 - 2bc \cos A , \tag{58}$$

for any triangle ABC.

A completely analogous argument, with either angle B or angle C in standard position, establishes the following:

$$b^2 = c^2 + a^2 - 2ac \cos B , \tag{59}$$

$$c^2 = a^2 + b^2 - 2ab \cos C , \tag{60}$$

for any triangle ABC.

Each of the formulas (58), (59), and (60) is referred to as the *law of cosines.* This law is clearly a generalization of the Pythagorean theorem, and is applicable to all triangles, not just to right triangles.

■ **LAW OF SINES.**—This law can be derived directly from the law of cosines in a purely algebraic fashion. Although there are simpler derivations based upon purely geometrical arguments, it is useful to give the purely analytic proof that follows. Solving equation (58) for $\cos A$:

$$\cos A = \frac{b^2 + c^2 - a^2}{2bc} .$$

From this, it follows that

$$\begin{aligned} \sin^2 A &= 1 - \cos^2 A \\ &= 1 - \frac{(b^2 + c^2 - a^2)^2}{4b^2c^2} \\ &= \frac{4b^2c^2 - (b^2 + c^2 - a^2)^2}{4b^2c^2} . \end{aligned}$$

This may be written

$$\begin{aligned} 4b^2c^2 \sin^2 A &= [2bc + (b^2 + c^2 - a^2)][2bc - (b^2 + c^2 - a^2)] \\ &= [(b^2 + 2bc + c^2) - a^2][a^2 - (b^2 - 2bc - c^2)] \\ &= [(b + c)^2 - a^2][a^2 - (b - c)^2] \\ &= (b + c + a)(b + c - a)(a - b + c)(a + b - c) . \end{aligned}$$

Now if s is the *semi-perimeter* of triangle ABC, that is,

$$s = \frac{a + b + c}{2} ,$$

then

$$\begin{aligned} 2s &= a + b + c , \\ 2(s - a) &= b + c - a , \\ 2(s - b) &= a - b + c , \\ 2(s - c) &= a + b - c . \end{aligned}$$

Hence,

$$4b^2c^2 \sin^2 A = 2s \cdot 2(s - a) \cdot 2(s - b) \cdot 2(s - c) ,$$

or

$$b^2c^2 \sin^2 A = 4s(s - a)(s - b)(s - c) ;$$

if

$$K = \sqrt{s(s - a)(s - b)(s - c)} , \tag{61}$$

this becomes

$$b^2c^2 \sin^2 A = 4K^2 .$$

Solving for $\sin A$, and noting that $\sin A > 0$ because $0 < A < \pi$, we then obtain

$$\sin A = \frac{2K}{bc} . \tag{62}$$

In exactly the same manner it can be proved that

$$\sin B = \frac{2K}{ac}, \tag{63}$$

$$\sin C = \frac{2K}{ab}. \tag{64}$$

Further, if we apply equations (62), (63), and (64) it is then possible to obtain the following:

$$\frac{\sin A}{a} = \frac{\sin B}{b} = \frac{\sin C}{c} = \frac{2K}{abc}. \tag{65}$$

This is the *law of sines*.

The geometrical significance of K is evident if equation (62) is written in the following form:

$$K = \frac{1}{2}bc \sin A = \frac{1}{2}c(b \sin A).$$

Referring to Figure 20, it can be seen that this is the same as

$$K = \frac{1}{2}cy,$$

so K is actually the area of triangle ABC.

We have thus succeeded not only in deriving the law of sines from the law of cosines, but we have also obtained incidentally a famous formula (*Heron's formula*) for the area of a triangle, namely

$$K = \sqrt{s(s-a)(s-b)(s-c)}.$$

Another interesting geometric result involving K is obtained by circumscribing a circle around triangle ABC of Figure 20, as indicated in Figure 21. This circle must intersect the Y-axis at another point C', and since angle BAC' is a right angle, the segment joining B to C' is a diameter of the circle. Its length is therefore $2R$, where R is the radius of the circumscribed circle. Furthermore, the inscribed angles C and C' have the same measure because they intercept the same arc AB. Applying the law of sines to triangle ABC', we obtain

$$\frac{\sin C'}{c} = \frac{\sin \angle BAC}{2R} = \frac{\sin \frac{\pi}{2}}{2R} = \frac{1}{2R}.$$

But since $C = C'$, this gives

$$\frac{\sin C}{c} = \frac{1}{2R}. \tag{66}$$

From this we may obtain two interesting results. Inverting both sides, we see that the common value of the ratios $a/\sin A$, $b/\sin B$, and $c/\sin C$ is actually the length of the diameter of the circumscribed circle, namely $2R$. Furthermore, in applying equations (65) and (66) we see that the following result is obtained:

$$\frac{1}{2R} = \frac{2K}{abc},$$

$$\therefore R = \frac{abc}{4K}. \tag{67}$$

This expresses the radius of the circumscribed circle in terms of the three sides of triangle ABC and its area K.

Two other classical formulas that are derivable from the laws of sines and cosines are

$$\frac{a-b}{a+b} = \frac{\tan \frac{1}{2}(A-B)}{\tan \frac{1}{2}(A+B)}, \tag{68}$$

$$\cos \frac{1}{2}A = \sqrt{\frac{s(s-a)}{bc}}. \tag{69}$$

Equation (68) is called the *law of tangents* and equation (69) is known as the *cosine of the half angle* formula. Both are useful in computations involving triangles.

De Moivre's Theorem.—A fundamental interweaving of trigonometry with the theory of complex numbers is seen to be incorporated in the principle that has come to be known as *De Moivre's theorem*, as follows:

$$\begin{aligned}(\cos \theta + i \sin \theta)^2 &= \cos^2 \theta + (2 \sin \theta \cos \theta)i + i^2 \sin^2 \theta \\ &= (\cos^2 \theta - \sin^2 \theta) + i(2 \sin \theta \cos \theta);\end{aligned}$$

that is,

$$(\cos \theta + i \sin \theta)^2 = \cos 2\theta + i \sin 2\theta.$$

Then, multiplying both sides again by the expression

$$(\cos \theta + i \sin \theta),$$

the equation may be expressed in the following manner:

$$\begin{aligned}(\cos \theta + i \sin \theta)^3 &= \cos 2\theta \cos \theta - \sin 2\theta + i(\sin 2\theta \cos \theta + \cos 2\theta \sin 2\theta) \\ &= \cos(2\theta + \theta) + i \sin(2\theta + \theta).\end{aligned}$$

Therefore,

$$(\cos \theta + i \sin \theta)^3 = \cos 3\theta + i \sin 3\theta.$$

The general rule is apparent and can be readily proved by the method of mathematical induction. For example, if n is a positive integer and θ is any real number, then the equation will read as follows:

$$(\cos \theta + i \sin \theta)^n = \cos n\theta + i \sin n\theta. \tag{70}$$

As a matter of fact, this formula can be shown to be true when n is a negative integer or even a fraction, provided a suitable interpretation is given to the fractional power of a complex number when it appears on the left side of equation (70).

Sample Applications.—Following are a few sample applications of the above trigonometrical principles.

■ NAVIGATION.—A lighthouse is located 14 nautical miles from a harbor and its bearing from the harbor is 35° 20′ west of north. A ship sails due west from the harbor for two hours and then observes the bearing of the lighthouse is 41° 10′ east of north. Determine the speed of the ship in knots, that is, nautical miles per hour.

Solution. Draw an appropriate diagram (Fig. 22), from which the angles of triangle ABC can be readily determined: $A = 54° 40'$, $B = 48° 50'$, and $C = 76° 30'$. Furthermore, $b = 14$ miles. Therefore, if we apply the law of sines in the form

$$\frac{c}{\sin C} = \frac{b}{\sin B},$$

we obtain

$$c = \frac{b \sin C}{\sin B}$$

$$c \approx \frac{14(.9724)}{.7528}$$

$$c \approx 18.08.$$

The ship's speed is 18.08/2, or approximately 9 knots.

■ FORCES IN PHYSICS.—Two forces of magnitude 10 lb. and 20 lb. act upon an object along lines of action that form an angle of 60°. Determine the magnitude of the resultant force on the object.

Solution. The resultant force is represented by a vector along the diagonal of a parallelogram, as indicated in Figure 23. From this figure the following facts can be determined about triangle ABC: $a = 10$, $c = 20$, $B = 120°$. Hence,

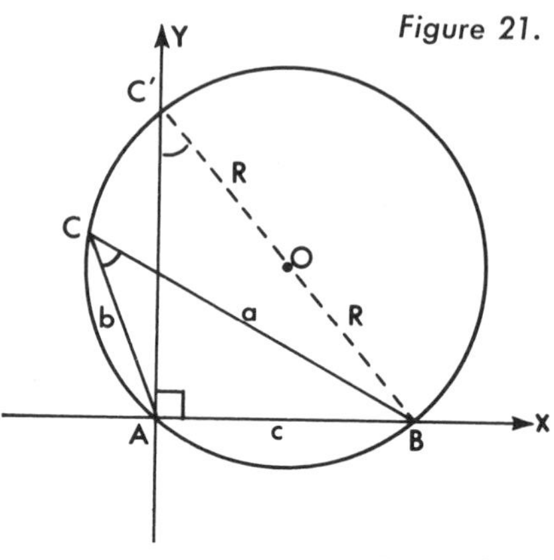

Figure 21.

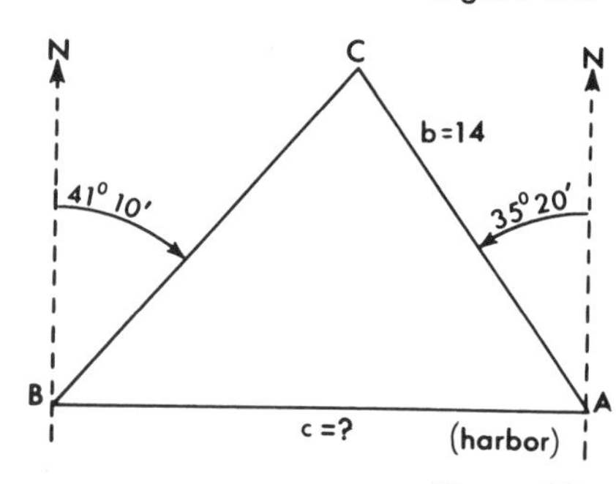

Figure 22.

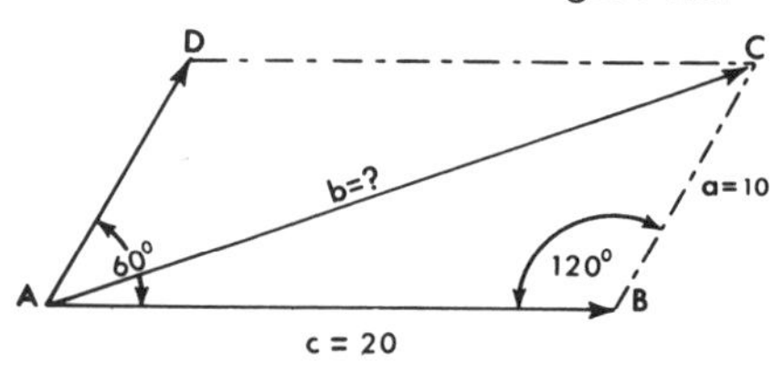

Figure 23.

using the law of cosines in the form

$$b^2 = a^2 + c^2 - 2ac \cos B\,,$$

we obtain

$$b^2 = 100 + 400 - 2(10)(20)(-.5000)$$
$$= 500 + 200$$
$$= 700$$
$$\therefore b = \sqrt{700} \approx 26.5\,.$$

The magnitude of the resultant force is therefore approximately 26.5 lb.

■ **ALTERNATING CURRENT.**—A 60-cycle alternating current has a peak voltage of 100 volts. Determine the drop in voltage .01 second after it reaches its peak.

Solution. The voltage at any time (t seconds) after reaching its peak may be represented by

$$E = 100 \cos 120\pi t\,,$$

because the period of $\cos 120\pi t$ is $2\pi/120\pi$, or 1/60 second. When $t = 0$, this yields the peak voltage

$$E_0 = 100 \cos 0 = 100 \text{ volts}\,.$$

When $t = .01$ this yields the new voltage

$$E_{.01} = 100 \cos 1.2\pi$$
$$= 100 \cos 216^\circ$$
$$= -100 \cos 36^\circ$$
$$\approx -100\,(.5878)$$
$$E_{.01} \approx -59 \text{ volts}\,.$$

Note. The voltage has already reversed itself by this time (.01 second after peak). The drop from peak voltage is thus given by

$$E_0 - E_{.01} \approx 100 - (-59) = 159 \text{ volts}\,.$$

—A. M. Glicksman

Values of Trigonometric Functions (Using Degree Measure)

Angle	Sin	Cos	Tan	Angle	Sin	Cos	Tan
1°	.0175	.9998	.0175	46°	.7193	.6947	1.0355
2°	.0349	.9994	.0349	47°	.7314	.6820	1.0724
3°	.0523	.9986	.0524	48°	.7431	.6691	1.1106
4°	.0698	.9976	.0699	49°	.7547	.6561	1.1504
5°	.0872	.9962	.0875	50°	.7660	.6428	1.1918
6°	.1045	.9945	.1051	51°	.7771	.6293	1.2349
7°	.1219	.9925	.1228	52°	.7880	.6157	1.2799
8°	.1392	.9903	.1405	53°	.7986	.6018	1.3279
9°	.1564	.9877	.1584	54°	.8090	.5878	1.3764
10°	.1736	.9848	.1763	55°	.8192	.5736	1.4281
11°	.1908	.9816	.1944	56°	.8290	.5592	1.4826
12°	.2079	.9781	.2126	57°	.8387	.5446	1.5399
13°	.2250	.9744	.2309	58°	.8480	.5299	1.6003
14°	.2419	.9703	.2493	59°	.8572	.5150	1.6643
15°	.2588	.9659	.2679	60°	.8660	.5000	1.7321
16°	.2756	.9613	.2867	61°	.8746	.4848	1.8040
17°	.2924	.9563	.3057	62°	.8829	.4695	1.8807
18°	.3090	.9511	.3249	63°	.8910	.4540	1.9626
19°	.3256	.9455	.3443	64°	.8988	.4384	2.0503
20°	.3420	.9397	.3640	65°	.9063	.4226	2.1445
21°	.3584	.9336	.3839	66°	.9135	.4067	2.2460
22°	.3746	.9272	.4040	67°	.9205	.3907	2.3559
23°	.3907	.9205	.4245	68°	.9272	.3746	2.4751
24°	.4067	.9135	.4452	69°	.9336	.3584	2.6051
25°	.4226	.9063	.4663	70°	.9397	.3420	2.7475
26°	.4384	.8988	.4877	71°	.9455	.3256	2.9042
27°	.4540	.8910	.5095	72°	.9511	.3090	3.0777
28°	.4695	.8829	.5317	73°	.9563	.2924	3.2709
29°	.4848	.8746	.5543	74°	.9613	.2756	3.4874
30°	.5000	.8660	.5774	75°	.9659	.2588	3.7321
31°	.5150	.8572	.6009	76°	.9703	.2419	4.0108
32°	.5299	.8480	.6249	77°	.9744	.2250	4.3315
33°	.5446	.8387	.6494	78°	.9781	.2079	4.7046
34°	.5592	.8290	.6745	79°	.9816	.1908	5.1446
35°	.5736	.8192	.7002	80°	.9848	.1736	5.6713
36°	.5878	.8090	.7265	81°	.9877	.1564	6.3138
37°	.6018	.7986	.7536	82°	.9903	.1392	7.1154
38°	.6157	.7880	.7813	83°	.9925	.1219	8.1443
39°	.6293	.7771	.8098	84°	.9945	.1045	9.5144
40°	.6428	.7660	.8391	85°	.9962	.0872	11.4301
41°	.6561	.7547	.8693	86°	.9976	.0698	14.3007
42°	.6691	.7431	.9004	87°	.9986	.0523	19.0811
43°	.6820	.7314	.9325	88°	.9994	.0349	28.6363
44°	.6947	.7193	.9657	89°	.9998	.0175	57.2900
45°	.7071	.7071	1.0000	90°	1.0000	.0000	

Values of Sin x and Cos x (Using Circular Measure: for 0 ≦ x ≦ 1.57)

x	Sin x	Cos x	x	Sin x	Cos x	x	Sin x	Cos x	x	Sin x	Cos x
.00	.0000	1.0000	.40	.3894	.9211	.80	.7174	.6967	1.20	.9320	.3624
.01	.0100	1.0000	.41	.3986	.9171	.81	.7243	.6895	1.21	.9356	.3530
.02	.0200	.9998	.42	.4078	.9131	.82	.7311	.6822	1.22	.9391	.3436
.03	.0300	.9996	.43	.4169	.9090	.83	.7379	.6749	1.23	.9425	.3342
.04	.0400	.9992	.44	.4259	.9048	.84	.7446	.6675	1.24	.9458	.3248
.05	.0500	.9988	.45	.4350	.9004	.85	.7513	.6600	1.25	.9490	.3153
.06	.0600	.9982	.46	.4439	.8961	.86	.7578	.6524	1.26	.9521	.3058
.07	.0699	.9976	.47	.4529	.8916	.87	.7643	.6448	1.27	.9551	.2963
.08	.0799	.9968	.48	.4618	.8870	.88	.7707	.6372	1.28	.9580	.2867
.09	.0899	.9960	.49	.4706	.8823	.89	.7771	.6294	1.29	.9608	.2771
.10	.0998	.9950	.50	.4794	.8776	.90	.7833	.6216	1.30	.9636	.2675
.11	.1098	.9940	.51	.4882	.8727	.91	.7895	.6137	1.31	.9662	.2579
.12	.1197	.9928	.52	.4969	.8678	.92	.7956	.6058	1.32	.9687	.2482
.13	.1296	.9916	.53	.5055	.8628	.93	.8016	.5978	1.33	.9711	.2385
.14	.1395	.9902	.54	.5141	.8577	.94	.8076	.5898	1.34	.9735	.2288
.15	.1494	.9888	.55	.5227	.8525	.95	.8134	.5817	1.35	.9757	.2190
.16	.1593	.9872	.56	.5312	.8473	.96	.8192	.5735	1.36	.9779	.2092
.17	.1692	.9856	.57	.5396	.8419	.97	.8249	.5653	1.37	.9799	.1994
.18	.1790	.9838	.58	.5480	.8365	.98	.8305	.5570	1.38	.9819	.1896
.19	.1889	.9820	.59	.5564	.8309	.99	.8360	.5487	1.39	.9837	.1798
.20	.1987	.9801	.60	.5646	.8253	1.00	.8415	.5403	1.40	.9854	.1700
.21	.2085	.9780	.61	.5729	.8196	1.01	.8468	.5319	1.41	.9871	.1601
.22	.2182	.9759	.62	.5810	.8139	1.02	.8521	.5234	1.42	.9887	.1502
.23	.2280	.9737	.63	.5891	.8080	1.03	.8573	.5148	1.43	.9901	.1403
.24	.2377	.9713	.64	.5972	.8021	1.04	.8624	.5062	1.44	.9915	.1304
.25	.2474	.9689	.65	.6052	.7961	1.05	.8674	.4976	1.45	.9927	.1205
.26	.2571	.9664	.66	.6131	.7900	1.06	.8724	.4889	1.46	.9939	.1106
.27	.2667	.9638	.67	.6210	.7838	1.07	.8772	.4801	1.47	.9949	.1006
.28	.2764	.9611	.68	.6288	.7776	1.08	.8820	.4713	1.48	.9959	.0907
.29	.2860	.9582	.69	.6365	.7712	1.09	.8866	.4625	1.49	.9967	.0807
.30	.2955	.9553	.70	.6442	.7648	1.10	.8912	.4536	1.50	.9975	.0707
.31	.3051	.9523	.71	.6518	.7584	1.11	.8957	.4447	1.51	.9982	.0608
.32	.3146	.9492	.72	.6594	.7518	1.12	.9001	.4357	1.52	.9987	.0508
.33	.3240	.9460	.73	.6669	.7452	1.13	.9044	.4267	1.53	.9992	.0408
.34	.3335	.9428	.74	.6743	.7385	1.14	.9086	.4176	1.54	.9995	.0308
.35	.3429	.9394	.75	.6816	.7317	1.15	.9128	.4085	1.55	.9998	.0208
.36	.3523	.9359	.76	.6889	.7248	1.16	.9168	.3993	1.56	.9999	.0108
.37	.3616	.9323	.77	.6961	.7179	1.17	.9208	.3902	1.57	1.0000	.0008
.38	.3709	.9287	.78	.7033	.7109	1.18	.9246	.3809			
.39	.3802	.9249	.79	.7104	.7038	1.19	.9284	.3717			

ANALYTIC GEOMETRY

Introduction.—Historically, geometry began as the study of the measurement and properties of geometric figures. The intuitive findings of the Egyptians and Babylonians were organized into a mathematical science by Euclid around 300 B.C., and for centuries geometry developed through Euclid's "axiomatic method." The growth of arithmetic and algebra paralleled the development of geometry. In the seventeenth century, mathematicians began to employ numerical and algebraic techniques in the study of geometric figures. This method, now called *analytic geometry,* was systematized by René Descartes (1596–1650). Later it was found that the methods of geometry likewise could be applied to problems in algebra.

Coordinate Systems.—The central fact that allows analytic geometry to serve as a link between algebra and geometry is that sets of numbers (or sets of sets of numbers) can be put into a one-to-one correspondence with a set of points—a correspondence called a *coordinate system.* Simplest coordinate system is the correspondence between the set of real numbers and the points on a line.

On a line of infinite extent, below, choose a point P. Label this point 0.

Choose points at equal distances in both directions on the line; label those to the right 1, 2, 3, . . . and those to the left −1, −2, −3, ... It is a basic assumption of mathematics that all of the points on the line will be in a one-to-one correspondence with the set of real numbers.

We can now describe any set of points

BETTMANN ARCHIVE

RENÉ DESCARTES, French philosopher and mathematician, greatly influenced geometry.

on the line by means of *open sentences* (equations, inequalities, and similar relationships). Indicating the point or points suggested by an open sentence is called *graphing the open sentence.*

Examples:

(1) $x = 2$

(2) $|x| = 2$

(3) $x < 4$
(the point +4 is not included)

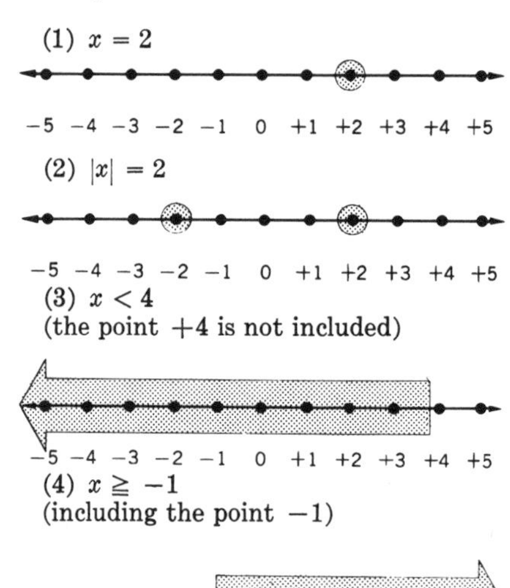

(4) $x \geqq -1$
(including the point −1)

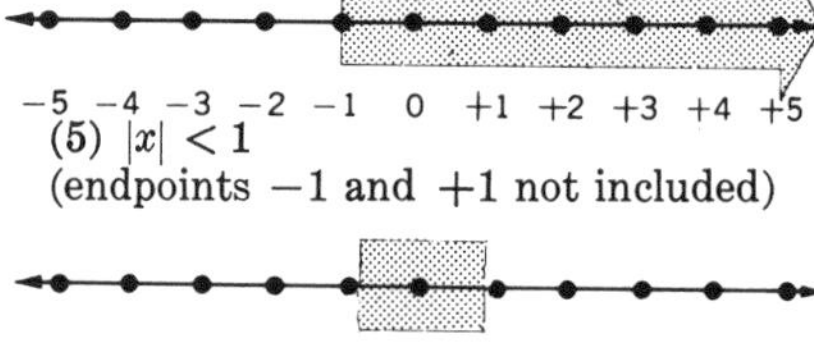

(5) $|x| < 1$
(endpoints −1 and +1 not included)

(6) $-2 \leqq x \leqq 1$
(including endpoints −2 and +1)

(The sign $<$ means "is less than"; $>$ means "is greater than"; $\leqq$ means "is less than or equal to"; $\geqq$ means "is greater than or equal to." The sign $|\ |$ indicates the absolute value of a number, that is, $|x| = x$ if $x > 0$, $|x| = -x$ if $x < 0$, and $|0| = 0$.)

The geometric figure designated by (4) is a *ray;* that by (5) is an *interval open at both ends;* that by (6) is a *segment.* It can be seen in these examples that geometric figures can be designated by algebraic open sentences. The basic idea of analytic geometry is to examine these geometric figures by studying the corresponding open sentences.

Coordinates in a Plane.—If two of the number lines described in the preceding section are drawn so that they are perpendicular to each other and intersect at their zero points, they form a set of *axes* for a plane coordinate system. The location of any point in the plane is given by an *ordered pair of numbers.* The first number in the pair designates the distance (in the positive or negative direction) from the vertical axis, which is called the *Y axis;* the second number designates the distance from the horizontal axis, which is called the *X axis.* The two numbers in the ordered pair are called the *coordinates* of the point they designate. The first number is the x coordinate, or *abscissa;* the second is the y coordinate, or *ordinate.* Below are some points and their coordinates:

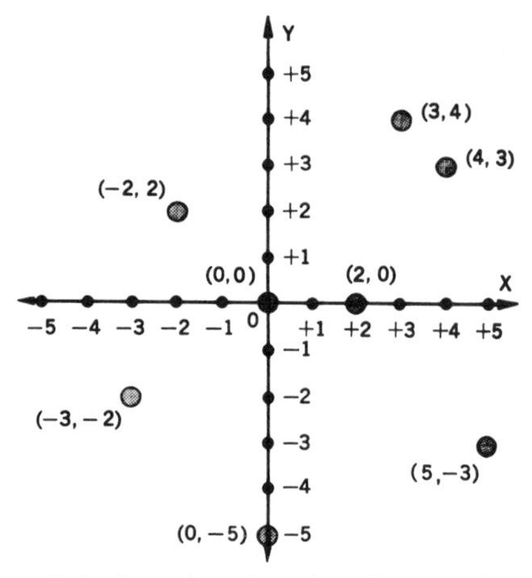

It is clear that there is only one point designated by a given pair of coordinates and that there is only one pair of coordinates corresponding to any given point. As in the case of the coordinate system on a line, open sentences involving the variables x and y may be used to designate sets of points in the plane.

Open Sentences and Their Graphs.—Sets of points can be described by statements, open sentences in two variables, and functional notation.

Examples:

(1) Find the set of points (x, y), such that y has the value of the largest whole number that is less than or equal to x. Thus y will always be a whole number, while x need not be. The ordered pairs $(3\frac{1}{4}, 3)$, $(2.3, 2)$, $(0, 0)$, $(-2, -2)$, and so on, belong to this set. The graph of the set is:

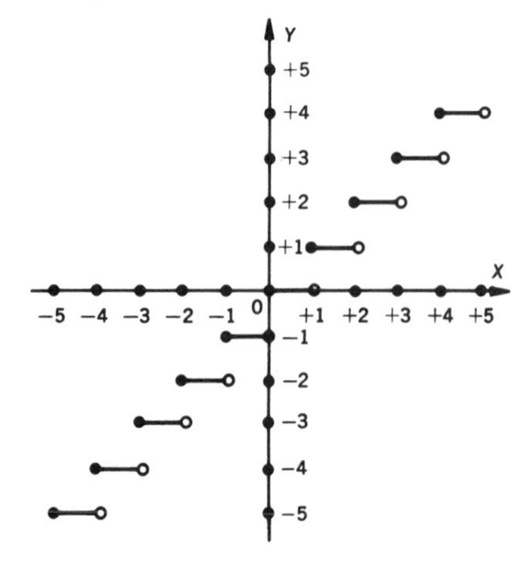

(The solid dots on the graph indicate points included in the segment; hollow dots, points not in the segment.)

(2) To every x, where x is **a real** number, associate its **absolute** value. The graph is:

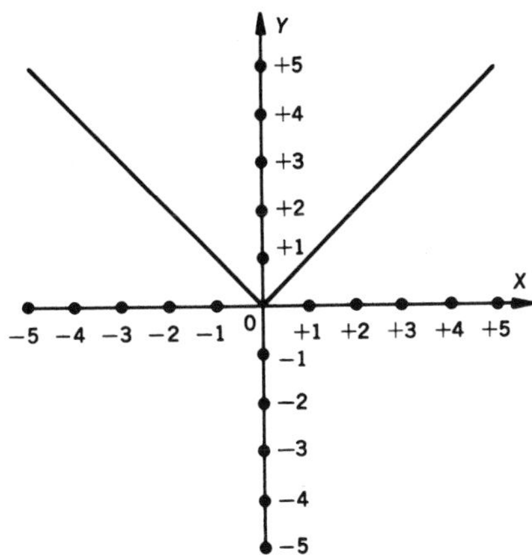

(3) The solution set of an equation or inequality may also be graphed. The *solution set* is the set of numbers that make the open sentence true. The solution sets for the following open sentences are:

(a) $y = x$, a straight line.
(b) $y^2 = x$, a parabola passing through the origin.
(c) $y = x^2 + 1$, a parabola displaced from the origin.
(d) $x^2 + y^2 \leqq 25$, a circle.

(a) $y = x$

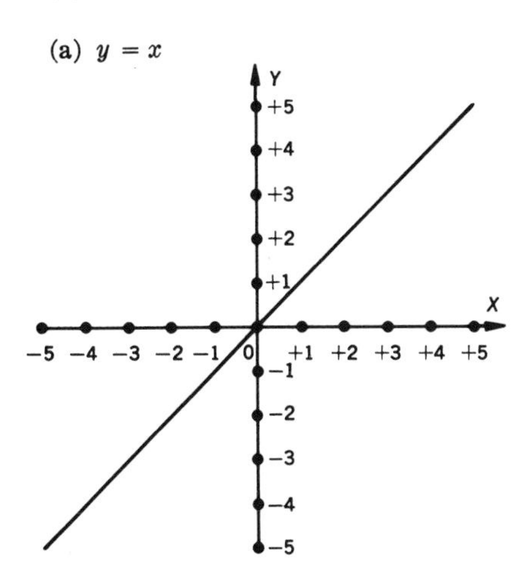

(b) $y^2 = x$

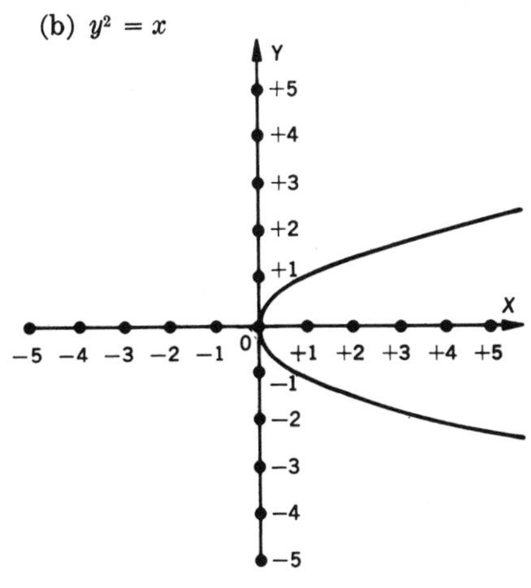

(c) $y = x^2 + 1$

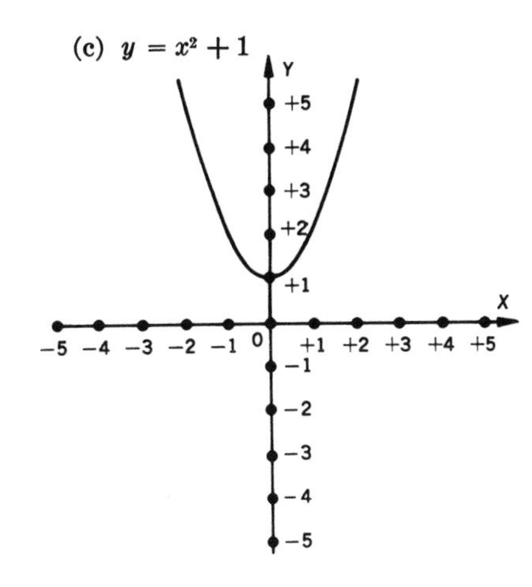

(d) $x^2 + y^2 \leqq 25$

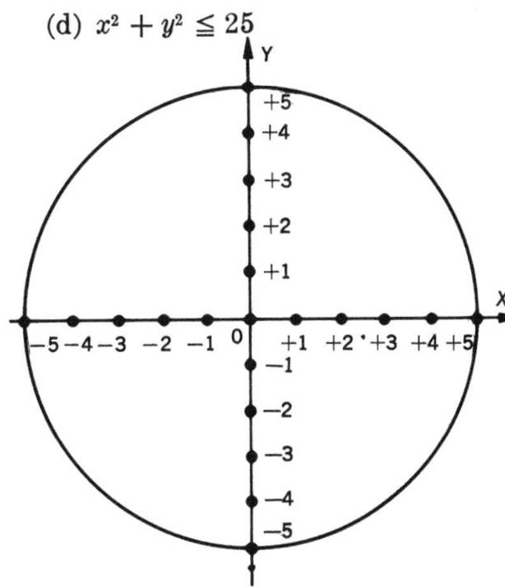

Linear Functions.—Certain subsets of points in a plane form patterns that have the shape of common geometric figures, such as a straight line, circle, ellipse, or cycloid. These figures correspond to certain specific types of equations. In analytic geometry equations of the form

$$Ax + By = C$$

(where A, B, and C are real numbers) have solution sets that form a straight line when graphed. Such equations as $3x - y = 5$ and $y = 2x + 7$ are of this type. It is also true that every straight line can be represented by an equation of this general form. Hence such equations are known as *linear equations.*

Slope-Intercept Form.—Any linear equation can be put into the form

$$y = mx + b.$$

By inspection we see that the number pair $(0, b)$ is a member of the solution set for this equation. The number b shows the graph's distance from the X axis when $x = 0$. Since $x = 0$ on the Y axis, b indicates the point at which the graph crosses this axis. For this reason, b is called the *y intercept of the line.*

Further inspection shows that the number m determines how much of a change there will be in the value of y for every change of one unit in the value of x. For example, if $m = 2$, y will increase by two units whenever x increases by one unit; if $m = -\frac{1}{2}$, y will decrease by one-half unit whenever x increases by one unit. In terms of the graph of the equation $y = mx + b$, m shows how much of a change there will be in the y direction for each change of one unit in the x direction; that is, it indicates what the *slope* of the line will be.

If (x_1, y_1) and (x_2, y_2) are two points on the line, then

$$y_1 = mx_1 + b,$$
$$y_2 = mx_2 + b.$$

Subtracting equations to eliminate b,

$$y_2 - y_1 = m(x_2 - x_1);$$

$$m = \frac{y_2 - y_1}{x_2 - x_1},$$

which is the slope of the line. Note that this expression is not defined if $x_2 = x_1$, since this would make the denominator equal 0. It follows that vertical lines, for which $x_2 = x_1$, do not have slopes. The equations $x = 0$, $x = -7$, and $x = 2$ are

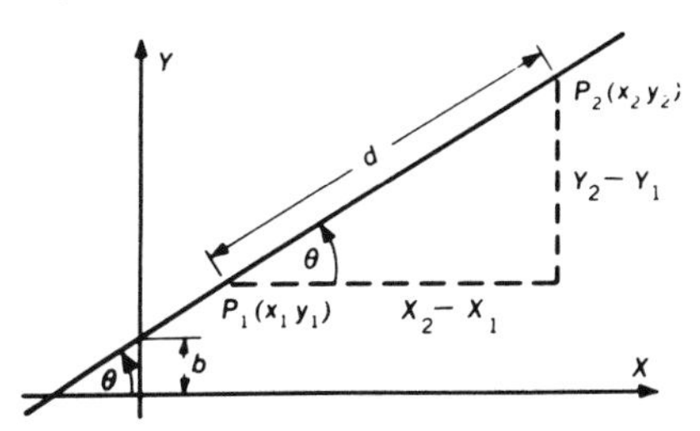

of this type. The diagram above shows that the slope m represents the tangent of the angle the line makes with the positive direction of the X axis. The angle, θ, is the *inclination of the line.* Note that if a linear equation is put into the form $y = mx + b$, it can be determined by inspection where the graph will cross the Y axis and what its slope will be. For this reason it is called the *slope-intercept form.* When an equation is in this form, it is simple to construct its graph by finding the intercept and constructing the slope.

If two lines are parallel, they have equal slopes. The converse is also true. Therefore, two lines with slopes m_1 and m_2 are parallel if, and only if, $m_1 = m_2$. (Two vertical lines that have no slopes are, of course, parallel.) By deduction it can be seen that two lines will be perpendicular if, and only if, $m_1m_2 = -1$. (Here, again, neither line may be vertical; that is, neither m_1 nor m_2 may equal zero.)

The diagram above also suggests a method by which the midpoint of the line P_1P_2 can be found. Consideration of the properties of similar triangles shows that the midpoint will be

$$\left(\frac{x_1 + x_2}{2}, \frac{y_1 + y_2}{2}\right).$$

With the help of the Pythagorean theorem, the diagram can be used to develop a method of finding the distance d between two points on a line:

$$d^2 = (P_1P_2)^2 = (x_2 - x_1)^2 + (y_2 - y_1)^2;$$

$$d = \sqrt{(x_2 - x_1)^2 + (y_2 - y_1)^2}.$$

This is known as the *distance formula.*

Simultaneous Equations.—It is clear that a very simple application of the geometry of straight lines leads to a technique for solving simultaneous linear equations. For example, consider the two equations

$$x + y = 4 \qquad \text{and} \qquad x - y = 2.$$

The graphs of the solution sets of these equations are shown in the diagram below. The coordinates of the point of intersection of the two lines clearly constitute the *intersection* of the solution sets of the two equations.

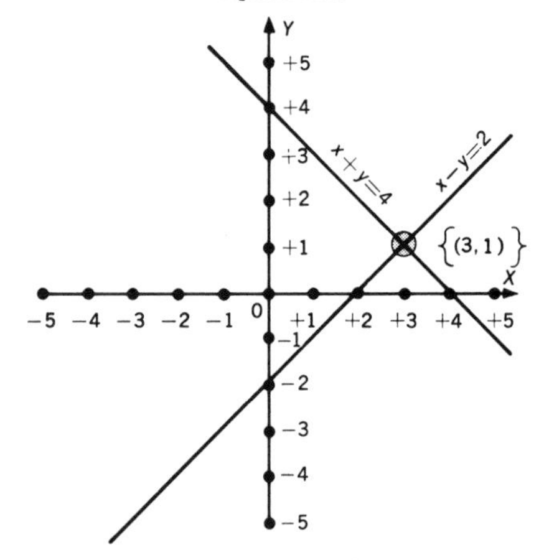

Quadratic Functions.—Consider an equation of the form

$$y = ax^2 + bx + c.$$

The set of ordered pairs of numbers (x, y) that satisfy such an equation—that is, constitute the solution set of such an equation—is the *quadratic function:*

$$f(x) \to ax^2 + bx + c.$$

The graphs of all such functions take the form of a *parabola.*

Examples:

$y = x^2 - 2x + 1$

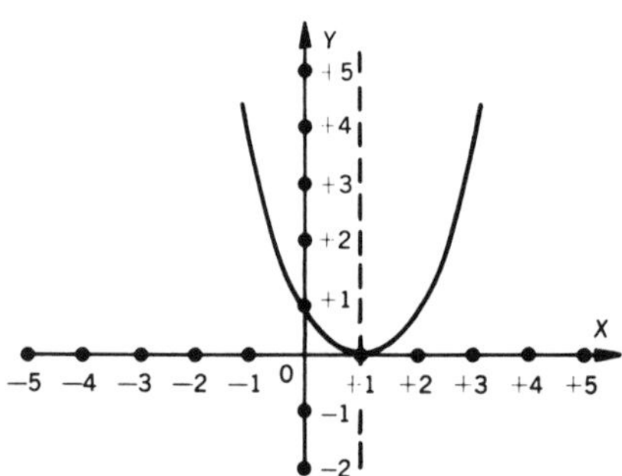

$y = x^2 + 1$

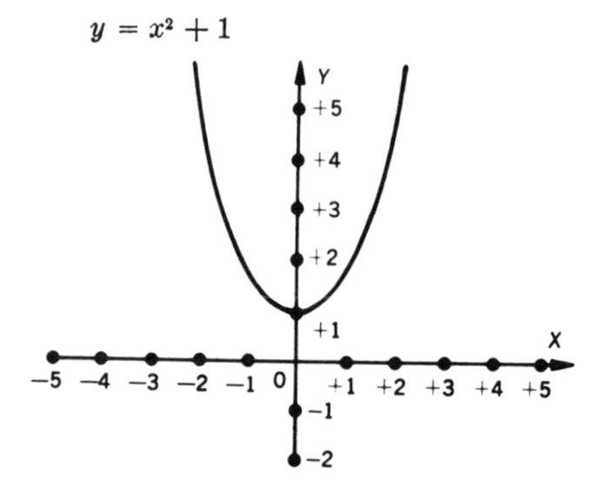

$y = -x^2 + 3x - 1$

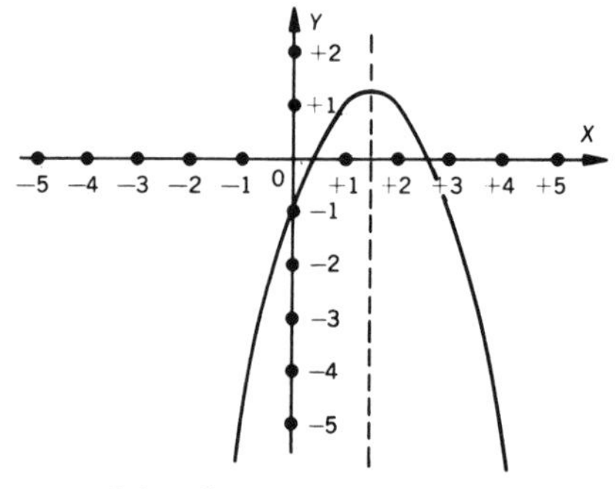

$y = 2x^2 + 7x$

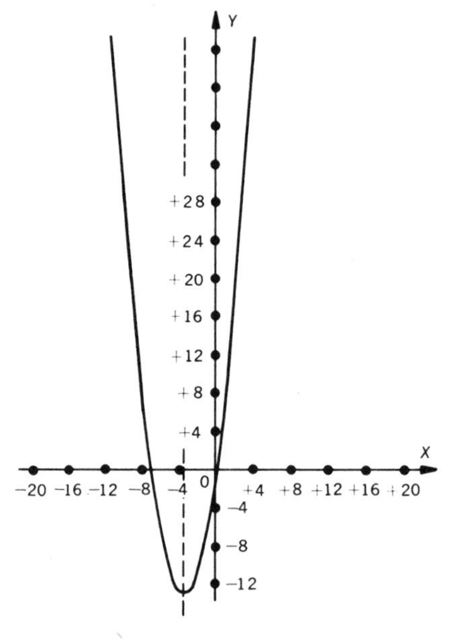

Again, the geometric properties of these figures can be studied by examining the corresponding equations. For example, each of the parabolas is symmetric about a line whose equation is given by $x = -(b/2a)$. In the examples above this yields: $x = 1$, $x = 0$, $x = 3/2$, and $x = -(7/4)$. The parabolas also have *maximum* and *minimum* points. The coordinates of these are given by $[-(b/2a), (4ac - b^2)/4a]$. In the examples, these work out to $(1, 0)$, $(0, 1)$, $(3/2, 5/4)$, and $[-(7/4), -(49/8)]$. The sign of c determines the direction in which the parabola opens. If the sign is positive, the opening is up or to the right; if it is negative, the opening is down or to the left.

Conic Sections.—Just as the graphs of all linear equations prove to be straight lines, so all quadratic equations yield graphs that have a family resemblance.

Consider a conic surface consisting of two infinitely extending cones placed tip to tip. Planes intersecting such a surface can produce circles, parabolas, ellipses, and hyperbolas. (In special cases the intersections can also produce a point, a line, or two intersecting lines.) These figures are known as *conic sections.* They are the graphs of various possible types of quadratic equations.

■ THE CIRCLE.—The equation of a *circle* with its center at the origin and radius r is

$$x^2 + y^2 = r^2.$$

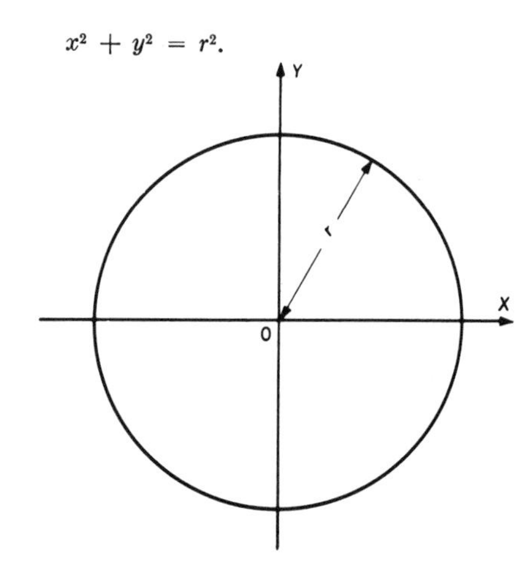

The equation of a circle with its center at point (a, b) and radius r is

$$(x - a)^2 + (y - b)^2 = r^2.$$

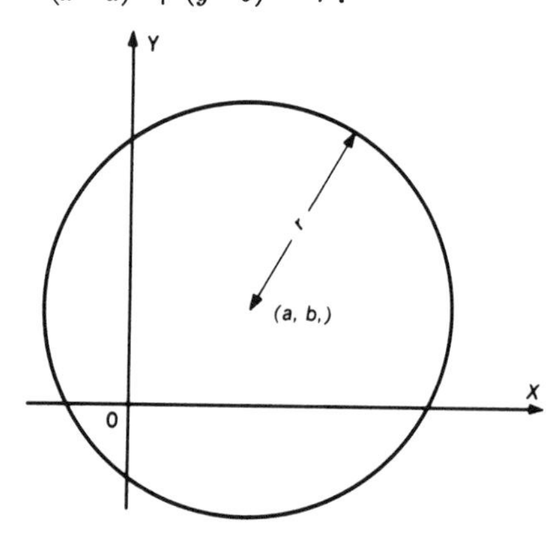

Geometrically, a *circle* is defined as the locus of points in a plane at a given distance from a point in the plane.

■ THE ELLIPSE.—An *ellipse* can be defined as the locus of points in a plane such that the sum of the distances of the points from two given points (called *foci*) is constant.

The algebraic equation for an ellipse with its center at the origin and with major and minor axes equal to a and b, respectively, is

$$\frac{x^2}{a^2} + \frac{y^2}{b^2} = 1.$$

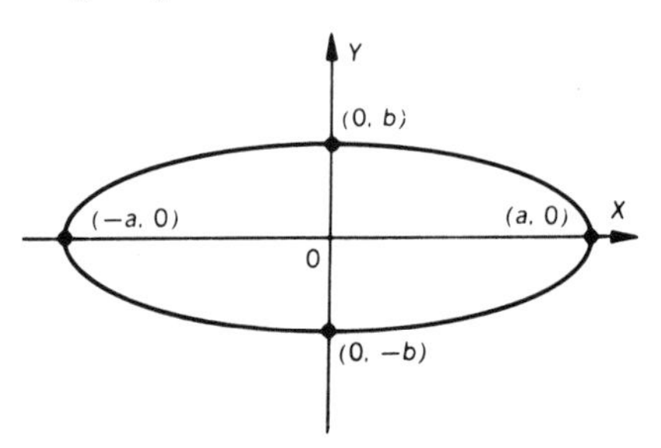

■ THE HYPERBOLA.—The *hyperbola* is the locus of points in a plane such that the difference of the distances from two given points (*foci*) is a constant.

The equation for a hyperbola with its center at the origin is

$$\frac{x^2}{a^2} - \frac{y^2}{b^2} = 1.$$

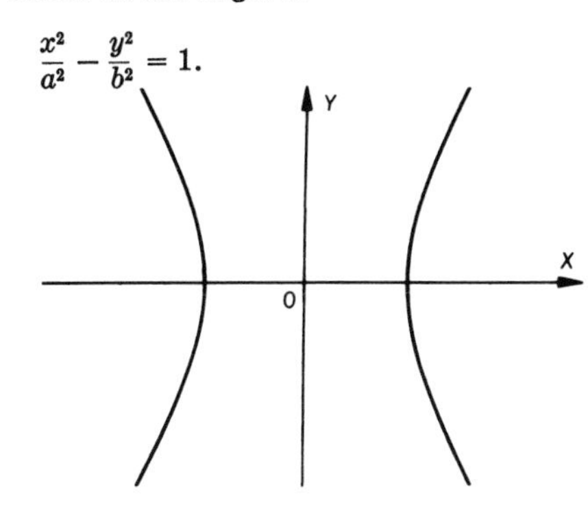

■ **THE PARABOLA.**—It has already been seen that the parabola is the graph of the quadratic equation $y = ax^2 + bx + c$. Geometrically, the *parabola* is defined as the locus of points in a plane equally distant from a given point (*focus*) and a given line (*directrix*). The equation of a parabola that has its vertex at the origin and its focus at $(p, 0)$ is

$$y^2 = 4px.$$

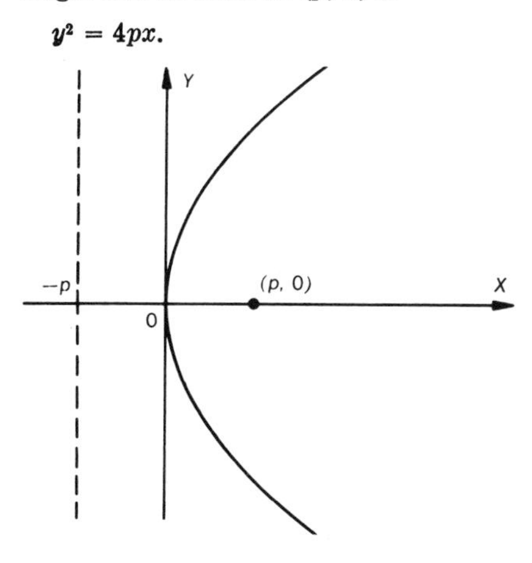

In this case the equation of the directrix is $x = -p$.

Other Coordinate Systems.—The coordinate system used here for the plane is known as the *rectangular* (or *Cartesian*, in honor of Descartes) *coordinate system*. There are several others. Rectangular coordinates can be extended to three-dimensional space by adding a third axis perpendicular to the two used so far. Points, then, are located by ordered sets of triple numbers (x, y, z) which give the point's distance from three mutually perpendicular planes. The graphs of equations in three variables form three-dimensional geometric figures, such as the sphere.

■ **POLAR COORDINATES.**—Points in a plane can be designated by *polar coordinates* as well as by rectangular coordinates. Select a ray AB in the plane. Choose some point P in the plane (assuming for sake of illustration that it does not lie on AB). Draw a line from A to P. Clearly, the location of P may be described in terms of the length of AP and the angle PAB. (For simplicity, the length of AB is called r, and $\angle PAB$ is called θ.) Thus the point P can be located in terms of an ordered pair of numbers (r, θ). As can be seen from the drawing (with the help of trigonometry), polar coordinates and rectangular coordinates are related as follows:

$$y = r \sin \theta,$$
$$x = r \cos \theta,$$

where (x, y) are the rectangular coordinates of P.

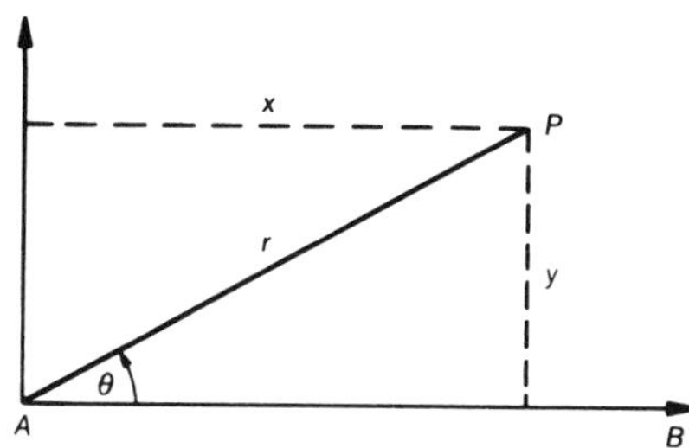

Descriptions of loci that give coordinates of points in terms of other variables are called *parametric equations.*

Equations of Some Loci

Trigonometric Functions

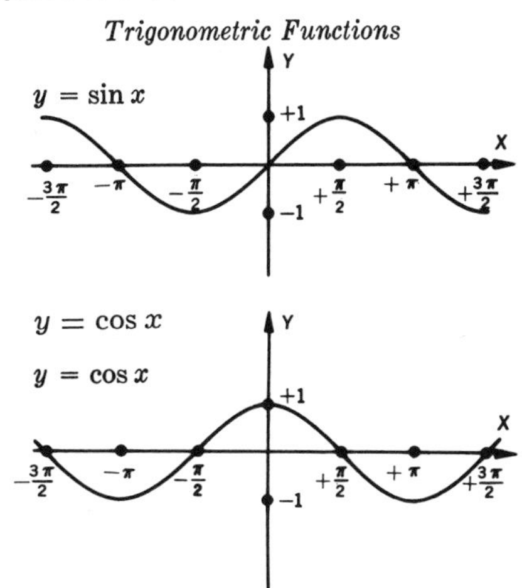

Exponential and Logarithmic Functions

Y +1 X −1 0 +1 +2 +3 +4 +5 −1 −2

$y = \log x$

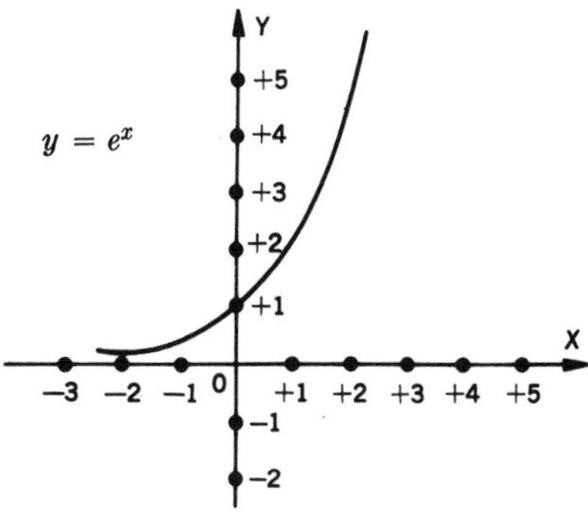

Spirals

$r = e^{a\theta}$ (logarithmic spiral)

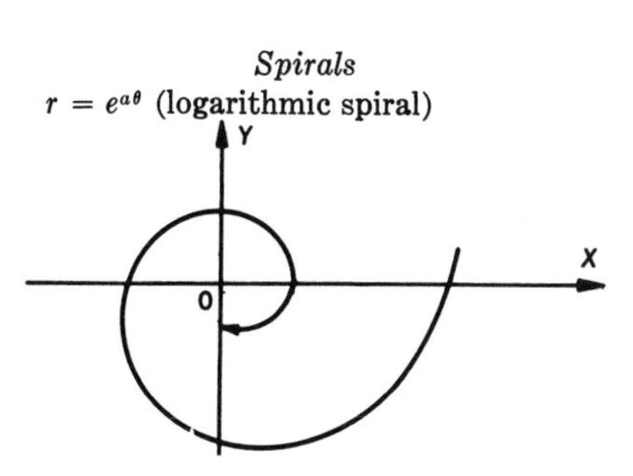

$r = a\theta$ (spiral of Archimedes)

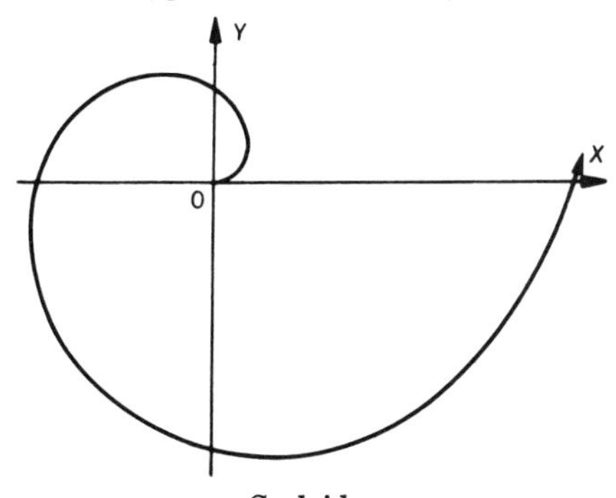

Cycloid

(Path of point on circumference of circle as circle rolls along straight line.)

$$x = a(\theta - \sin \theta)$$
$$y = a(1 - \cos \theta)$$

or

$$x = a \cos^{-1}\left(\frac{a-y}{a}\right) \pm \sqrt{2ay - y^2}$$

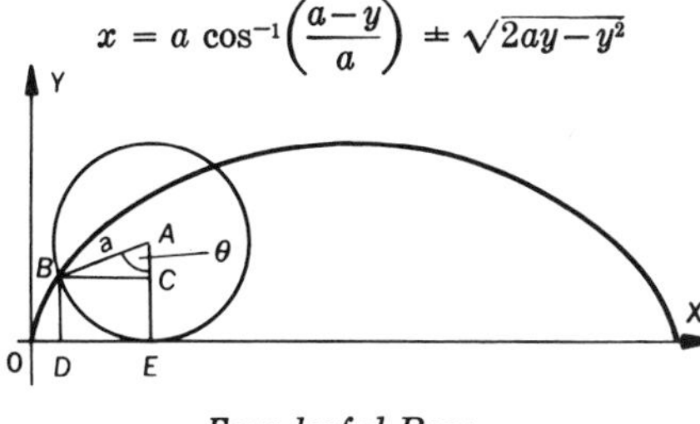

Four-leafed Rose

$r = \sin 2\theta$

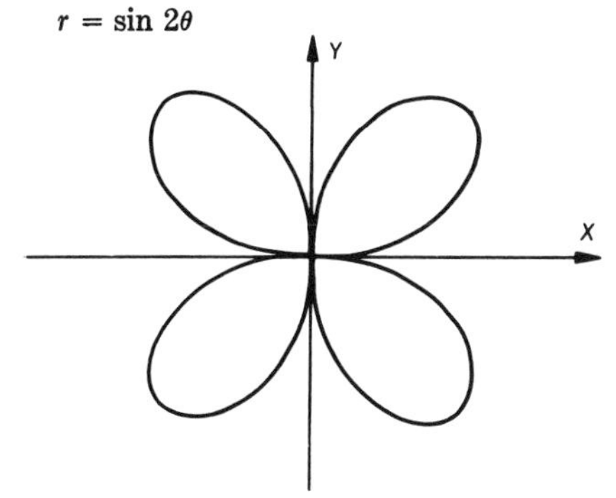

Lemniscate

$r = a^2 \cos 2\theta$

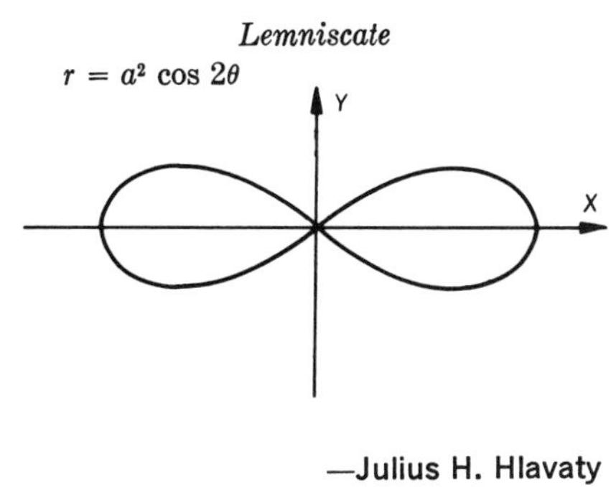

—Julius H. Hlavaty

Basic Concepts.—It has been shown that the properties of geometric figures can be studied by means of their equations. This method gives rise to problems that involve change or motion. These phenomena are the subject matter of *differential calculus* and *integral calculus*.

Consider a curve and a point P on it.

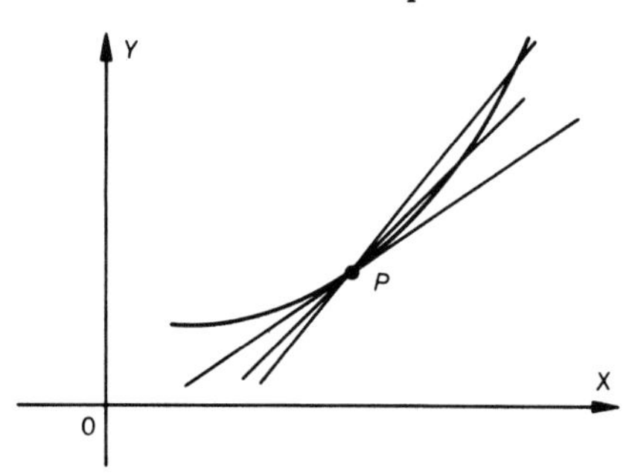

Among the possible lines through P, several will intersect the curve again; these are called *secants*. One line will touch the curve only at the point P; this is the *tangent* to the curve at P. The slope of this tangent can be determined by differential calculus.

Take a curve with points P_1 and P_2.

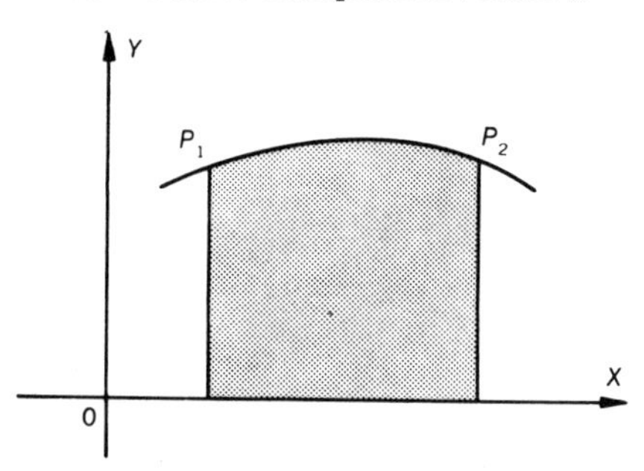

If the equation of the curve is known, what area will be bounded by the curve, the two perpendiculars to the X axis drawn from the points P_1 and P_2 on the curve, and the X axis (shaded portion of the graph)? Such a problem can be solved by integral calculus.

Limit Concept.—Both of the problems mentioned above, and many others, depend heavily on the *limit concept*.

Consider the three sequences:

(1) $1, 2, 3, 4, \ldots$
(2) $1, \frac{1}{2}, \frac{1}{4}, \frac{1}{8}, \ldots$
(3) $1, \frac{1}{2}, \frac{1}{3}, \frac{1}{4}, \ldots$

It is clear that the terms of the first sequence, going far enough along the sequence, increase beyond any finite number. In the other two, no number in the sequence is less than (or equal to) zero, but either one may be extended to a number as close to zero as desired.

Now examine the following sequences:

(4) $1, 3, 6, 10, \ldots$
(5) $1, 1\frac{1}{2}, 1\frac{3}{4}, 1\frac{7}{8}, \ldots$
(6) $1, 1\frac{1}{2}, 1\frac{5}{6}, 2\frac{1}{12}, \ldots$

These sequences were formed by taking the first term, the sum of the first two terms, the sum of the first three terms, and so on, of sequences (1), (2), and (3), respectively.

The terms of sequence (4) increase beyond any given number. In this respect, sequence (4) is similar to sequence (1) from which it was derived.

The terms of sequence (5) can never be 2 or more than 2. In this respect it is similar to sequence (2), in which case the terms could never reach or pass 0.

But the terms of sequence (6) differ. Sequence (3), from which sequence (6) was derived, did not extend beyond a certain point (0); but the terms of sequence (6) can be extended to a number greater than any given number.

Sequences (2), (3), and (5) approach, or have, a *limit*. The limits are, respectively, 0, 0, and 2. The other sequences have no limit. Hence a sequence has a limit if, and only if, no matter how many terms are taken, a number (*limit*) cannot be reached or surpassed.

Slope of Tangent, Derivatives.—Consider the second-degree equation $y = x^2$.

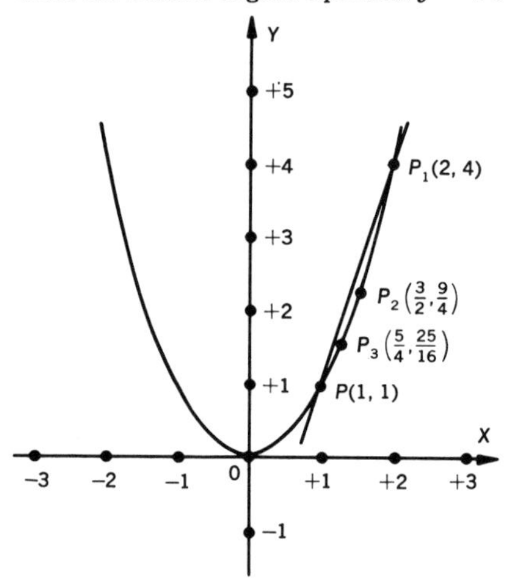

Select points $P, P_1, P_2, P_3, P_4, \ldots$ on the curve so that their x coordinates are $1, 2, 3/2, 5/4, 9/8, \ldots$. Using the equation, corresponding y coordinates are

$P\ (1, 1)$,
$P_1\ (2, 4)$,
$P_2\ (3/2, 9/4)$,
$P_3\ (5/4, 25/16)$,
$P_4\ (9/8, 81/64)$.

Since the slope of a line is given as

$$m = \frac{y_2 - y_1}{x_2 - x_1},$$

the slopes of the lines $PP_1, PP_2, PP_3, PP_4, \ldots$ can be computed:

First Point	*Second Point*	*Slope*
$P\ (1,1)$	$P_1\ (2,4)$	$\frac{4-1}{2-1} = 3$
$P\ (1,1)$	$P_2\ (3/2,9/4)$	$\frac{9/4-1}{3/2-1} = 2\frac{1}{2}$
$P\ (1,1)$	$P_3\ (5/4,25/16)$	$\frac{25/16-1}{5/4-1} = 2\frac{1}{4}$
$P\ (1,1)$	$P_4\ (9/8,81/64)$	$\frac{81/64-1}{9/8-1} = 2\frac{1}{8}$

It can be shown that the sequence $3, 2\frac{1}{2}, 2\frac{1}{4}, 2\frac{1}{8}, \ldots$ approaches 2 as a limit. This limit can be interpreted as the slope of the tangent to the curve $y = x^2$ at the point (1, 1). If the secants for several points in the original sequence are plotted, it will be seen that they approach the position of a line with slope 2 drawn through point P. From the graph, it is clear that no secant drawn from P through some point to the right of P can have a slope less than 2. Thus a graphic illustration of the fact that slope 2 is the limit of the sequence comprising the slopes of the secants is produced.

A more general description can be obtained for this example. Let $y = f(x)$. [The expression $f(x)$, which is read "f of x," simply means "some function of x." In the previous example, $f(x) = x^2$.] Choose a point (x_1, y_1), that is, $[x_1, f(x_1)]$ and a point $[x_1 + h, f(x_1 + h)]$, where $x_1 + h$ is the abscissa of some point near (x_1, y_1). Now consider the quotient

$$\frac{f(x_1 + h) - f(x_1)}{(x_1 + h) - x_1}.$$

For a sequence of decreasing values of h, these quotients give the slopes of secant lines through (x_1, y_1). The limit of this sequence gives the slope of the tangent line. For the special equation $y = x^2$ the following would be obtained:

$$\frac{f(x_1 + h) - f(x_1)}{(x_1 + h) - x_1} = \frac{(x_1 + h)^2 - x_1^2}{(x_1 + h) - x_1}$$
$$= \frac{x_1^2 + 2x_1h + h^2 - x_1^2}{x_1 + h - x_1}$$
$$= \frac{2x_1h + h^2}{h}$$
$$= 2x_1 + h.$$

Intuitively, it is clear that the limit would be $2x_1$ (for h can be made as small as necessary). This limit, $2x$ (where we now wish to consider the x coordinate of *any* point on the curve), is called the *derivative* of x^2. The process of finding derivatives is called *differentiation*.

The derivative of a function $f(x)$ is designated by a variety of symbols in the literature. For example, if $y = f(x)$, the derivative of y with respect to x can be stated as

$$y' = \frac{dy}{dx} = D_xy = \lim_{h \to 0} \frac{f(x + h) - f(x)}{(x + h) - h} = f'(x).$$

Formulas of Differentiation.—The differentiation of $y = x^2$ that was shown is a special case of a general formula. If $y = x^n$,

$$\frac{dy}{dx} = nx^{n-1}.$$

Complete formulas for differentiating various expressions can be found in textbooks on calculus. Following are some of the more common such formulas:

$$\frac{d(C)}{dx} = 0 \qquad (C = \text{constant})$$

If $u = f(x)$ and $v = g(x)$:

$$\frac{d(Cu)}{dx} = C\frac{du}{dx},$$

$$\frac{d}{dx}(u + v + \ldots) = \frac{du}{dx} + \frac{dv}{dx} + \ldots,$$

$$\frac{d}{dx}(uv) = u\frac{dv}{dx} + v\frac{du}{dx},$$

$$\frac{d}{dx}\left(\frac{u}{v}\right) = \frac{v\frac{du}{dx} - u\frac{dv}{dx}}{v^2},$$

$$\frac{d}{dx}(u^n) = nu^{n-1}\frac{du}{dx},$$

$$\frac{d}{dx}(\sin x) = \cos x,$$

$$\frac{d}{dx}(\cos x) = -\sin x,$$

$$\frac{d}{dx}(\ln u) = \frac{1}{u}\frac{du}{dx} \qquad \text{(natural log)},$$

$$\frac{d}{dx}(e^u) = e^u\frac{du}{dx}.$$

The Definite Integral.—Consider the equation $y = x^2$, the graph of which is shown below. Take two points on the curve P_1 (1, 1) and P_2 (2, 4). The problem is to find the area of the shaded portion—

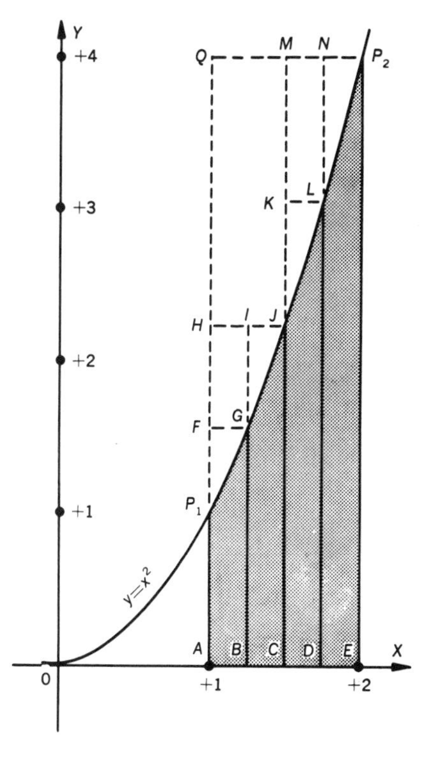

the area bounded by the curve, the X axis, and the two perpendiculars P_1A and P_2E.

P_1A is extended to Q, and QP_2 is drawn; a rectangle is formed the area of which gives a very rough approximation of the shaded area. If this rectangle is divided in half by drawing CM, and lines MP_2 and HJ are added, a better approximation can be had by taking the area of rectangles $AHJC$ and CMP_2E.

Determining a Curve Area Algebraically

Rectangle Approximation	Base	Height	Area
(1) AEP_2Q	AE = 1	EP_2 = 4	4
(2) AHJC	AC = 1/2	CJ = 9/4	9/8
$+CMP_2E$	CE = 1/2	EP_2 = 4	2
			3 1/8
(3) ABGF	AB = 1/4	BG = 25/16	25/64
+BCJI	BC = 1/4	CJ = 9/4	9/16
+CDLM	CD = 1/4	DL = 49/16	49/64
$+DEP_2N$	DE = 1/4	EP_2 = 4	1
			2 23/32

Dividing these two rectangles in half, as shown in the illustration, and repeating the process, the approximation can be improved further. At this point it becomes evident that better and better approximations are produced by repeating this procedure until the actual area under the curve is approached as a limit. The above table on determining a curve area serves to illustrate, algebraically, what is occurring.

It can be shown that the sequence 4, $3\frac{1}{8}$, $2\frac{23}{32}$, . . . approaches $2\frac{1}{3}$ as a limit. This means that if the units of measure on the two axes are 1 inch each, the area bounded by the curve, the X axis, and the perpendiculars drawn at $x = 1$ and $x = 2$ would be $2\frac{1}{3}$ square inches.

The process of finding the limit of this sequence is called *integration*, and the number $2\frac{1}{3}$ is called the *definite integral* of $y = x^2$ from $x = 1$ to $x = 2$.

The Indefinite Integral.—The process of finding the *indefinite integral* is the fundamental theorem of integral calculus. It develops that there is a remarkable relationship between differentiation and integration. In regard to the function $f(x)$, the process of finding a function $F(x)$ such that $f(x)$ is the derivative of $F(x)$, or $F'(x) = f(x)$, is called *integration*. $F(x)$ is called an *integral* of $f(x)$, or an *antiderivative* of $f(x)$. The symbol for integration is $\int$ and is written

$$F(x) = \int f(x),$$

although the more customary form is

$$F(x) = \int f(x)dx.$$

For example,

$$\int x^2dx = \tfrac{1}{3}x^3,$$

since the formula for differentiation shows that

$$\frac{d}{dx}(\tfrac{1}{3}x^3) = 3 \cdot \tfrac{1}{3} \cdot x^2 = x^2.$$

Note that $\frac{1}{3}x^3 + 5$ and $\frac{1}{3}x^3 + C$ (where C is any constant) are also integrals of x^2, for

$$\frac{d}{dx}(\tfrac{1}{3}x^3 + 5) = \frac{d}{dx}(\tfrac{1}{3}x^3) + \frac{d}{dx}(5) = x^2 + 0 = x^2.$$

In general, if $F(x)$ is an integral of $f(x)$, so is $F(x) + C$. For this reason, $\int f(x)dx$ is called the *indefinite integral* of $f(x)$.

The connection between the definite integral and the indefinite integral is illustrated by the following. It develops that if $F(x)$ is an indefinite integral of $f(x)$, then the definite integral of $f(x)$ from $x = a$ to $x = b$ is given by

$$\int_a^b f(x)dx = F(b) - F(a).$$

For the special case in the previous problem the shaded area is

$$\int_1^2 x^2\,dx.$$

Following the general formula

$$\int f(x)dx = \tfrac{1}{3}x^3 = F(x),$$
$$F(2) = 8/3 \qquad F(1) = 1/3.$$

The shaded area is

$$\frac{8}{3} - \frac{1}{3} = \frac{7}{3} = 2\frac{1}{3}.$$

The techniques of integration depend on using tables of differentiation because of the relationship between these two branches of calculus.

Applications.—The problem used to introduce differential calculus—finding the slope of the tangent to a given curve at a given point—is related to the problem of having two variables change their values, when the object is to find the *rate* of change of one with respect to the other. This is exactly the situation in problems of velocity of a moving body where the variables are time and distance. Calculus also relates to the inverse physical problem, for example, when the velocity of a moving body is known, to find the law that describes its motion. Calculus is a basic tool in modern science and engineering, because it supplies a language for expressing physical laws in precise mathematical terms and provides a technique for studying the consequences of these laws.

Derivatives are used in problems of instantaneous rates of change, velocities, and accelerations. Differentiation is used in solving problems of maxima and minima, in the study of related rates of change, and in tracing simple and complex curves.

Integral calculus may be applied to the study of complicated regions, moments of inertia, hydrostatic pressure, work, distance, acceleration, and volume.

—Julius H. Hlavaty

ADVANCED MATHEMATICS

Logarithms.—The simplest arithmetical operation is addition. Next in complexity would probably be subtraction, then multiplication, and last, division. It would be convenient, then, if one could find the product of two numbers by addition rather than by the use of traditional multiplication techniques.

Though it seems impossible, John Napier (1550–1617), a Scottish inventor, discovered a solution which utilizes the peculiar properties of the powers of numbers. To multiply numbers raised to powers, their exponents are added:

$$3^1 \times 3^2 = 3^{1+2} = 3^3$$
$$3 \times 9 = 27 = 3^3$$

The two base numbers (the 3's in this example) must be the same. But with that as the only requirement, Napier had a tool to do exactly what he wanted: a method of multiplying by simple addition. An English geometry professor, Henry Briggs (1561–1630), helped him set up the system. For convenience they chose the number 10 as the base rather than 3. Then they made a table of powers. Just a small part of the table is shown below. The full table is given at the end of this article.

$1 = 10^{0.0000}$	$6 = 10^{0.7782}$
$2 = 10^{0.3010}$	$7 = 10^{0.8451}$
$3 = 10^{0.4771}$	$8 = 10^{0.9031}$
$4 = 10^{0.6021}$	$9 = 10^{0.9542}$
$5 = 10^{0.6990}$	$10 = 10^{1.0000}$

Now, to multiply two numbers, say 2×4, translate them first into their powers of 10, or

$$2 \times 4 = 10^{0.3010} \times 10^{0.6021}.$$

Add the exponents,

$$10^{0.3010} \times 10^{0.6021} = 10^{0.9031},$$

and reconvert to the numerical form:

$$8 = 10^{0.9031}.$$

For convenience, it is possible to eliminate the constant writing of the exponents, which Napier called *logarithms* (abbreviated *log*). Thus the logarithm of 2 is 0.3010. The operation can be stated: *To find the product of two numbers add their logarithms.*

There is no special magic about the base number 10. Any positive number except 1 can serve as base. Today two bases are popular. Natural logarithms use base 2.71828. . . . Common logarithms use the base 10. Briggs and Napier chose 10 for a very practical reason. Note that

$$5 = 10^{0.6990} \text{ (log } 5 = 0.6990)$$
$$50 = 10^{1.6990} \text{ (log } 50 = 1.6990)$$
$$500 = 10^{2.6990} \text{ (log } 500 = 2.6990).$$

The *mantissa* (the four characteristic numbers to the right of the decimal point in the logarithms) is the same whenever the same significant figures appear in the original number. The convenience of 10 as a base therefore comes from the fact that only one table of logarithms for 1,000 numbers (three significant figures) is needed. It doesn't matter if the number to be multiplied is actually in the billions; if there are no more than three significant figures, the one table will serve.

For example, multiply 455 by 81.1. To find the mantissa of 455, look for 45 in the left column of the table and check across the row to the column headed 5 at the top. This gives the mantissa, or 6580. Since 455 lies between 100 and 1,000, its logarithm lies between 2 and 3, or 2 plus a decimal fraction, hence 2.6580. (In other words, the number added in front of the mantissa is equal to the number of places to the right of the first digit before the decimal point.) Find the logarithm of 81.1; then add both logarithms:

$$\log 4.55 = 0.6580, \log 455 = 2.6580$$
$$\log 8.11 = 0.9090, \log 81.1 = 1.9090$$
$$\text{sum} = 4.5670$$

Look up 0.5670 in the body of the table and read up and across to find 369 as its *antilogarithm* (the number whose logarithm was found). Then take care of the number 4 preceding the decimal point by counting off 4 decimal places in the antilogarithm from the right of the first digit, adding zeros as necessary. The answer is 36,900.

To divide, reverse the process and subtract logs; to solve 36,900 ÷ 455,

$$\log 36{,}900 = 4.5670$$
$$\log 455 = 2.6580$$
$$\text{difference} = 1.9090$$
$$\text{antilog } 1.9090 = 81.1.$$

Slide Rule.—The slide rule is merely a mechanical tool for adding logarithms in exactly the way described above. Instead of adding numbers, however, it adds lengths. Take two rulers and place them side by side. By placing the 0 of the upper scale on the 2 of the lower scale, mark off a 2-inch length. Now, to add 3 inches to the 2 inches, read opposite the 3 on the upper scale and find a 5 on the lower scale: $2 + 3 = 5$.

Now suppose instead of equal lengths, positions are marked according to the logarithmic proportions given in the short list above. By adding log 2 and log 3 log 6 is found. The two numbers have been multiplied by adding their logarithms. For simplicity most slide rules are not marked with the mantissas as shown here, but simply with the numbers whose logarithms their lengths represent. Also, decimal points are omitted in slide rules (these must be calculated). Thus, the above setting could just as well represent $20 \times 300 = 6{,}000$ or $2.0 \times 0.3 = 0.6$.

By subtracting logarithmic lengths, division may be performed. This slide rule has been set: $9 \div 3 = 3$.

Statistics.—Statistics is the name given to the study of large sets of numbers which, by their sheer bulk, become unwieldy if not meaningless. For example, suppose a list were made of the exact age of every man, woman, and child in the United States on January 1 of this year. Such facts could be very useful to manufacturers of many products. Educators could use the information to predict teacher loads and classroom needs. The government would be able to estimate social security benefits, taxes, and a thousand other things.

Take all the ages on the list, add them up, divide by the number of people,

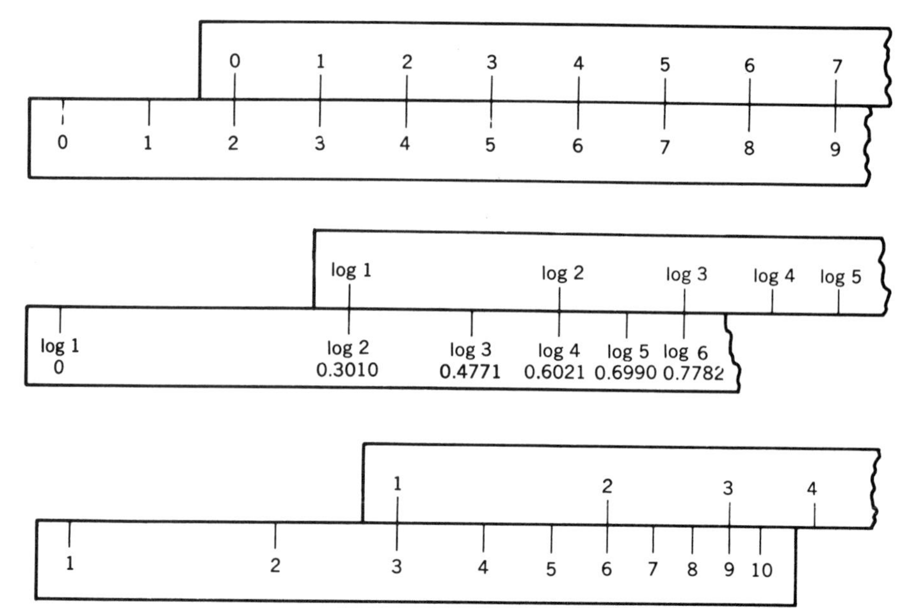

SLIDE RULE is a mechanical tool for adding or subtracting logarithms of numbers.

and an average (statisticians call it a *mean*) is obtained. The mean is significant because it may, for some purposes, represent the whole group. Thus the average weight of the American man has been increasing over the past hundred years, so it can be said that Americans are getting heavier. But there are dangers in giving too much importance to the mean. The mean annual income of U.S. citizens in the year 1960 was $6,900. What does this mean? There are many people who earn less than this and there are people who earn much more. Further, the few people who earn high incomes have an effect which distorts the mean. In this case, it makes more sense to list the incomes of all the people in increasing amounts and then count off from the bottom or from the top until you have gone exactly halfway through this list. The individual at this position earns more than what half the total population does and less than the other half. His income is the *median*, and for the United States in 1960 the median income was $5,600 per year, or a good deal less than the mean. For that reason, perhaps it is more representative.

PROBABILITY.—Statisticians have discovered an interesting thing about large sets of numbers. Certain kinds of data take very regular distributions. Suppose we were to divide a group of 899 men by weight. A typical count might be:

130 to 139 lb.— 2 men
140 to 149 lb.— 30 men
150 to 159 lb.—140 men
160 to 169 lb.—275 men
170 to 179 lb.—270 men
180 to 189 lb.—145 men
190 to 199 lb.— 32 men
200 to 209 lb.— 5 men

These numbers may be plotted in a graph with weight along the horizontal axis and numbers of men along the vertical axis. Note that the plot is symmetrical about the centerline, which in this case is the mean (170 lb.). This characteristic curve occurs so frequently that it is called a *normal* or *probability curve*. The essential shape is always the same. It is best described by the distance from the centerline to the inflection point (where the curve changes from concave to convex), and this distance is always called *sigma* (Greek σ).

It turns out that almost two-thirds of all the things measured (899 men in this example) fall between the two inflection points of any normal curve (between 159 lb. and 181 lb., above).

The probability curve has many interesting and important applications, but only two will be considered here. The first is what the statistician calls a *sample*, and his problem arises in this way: He would like to know whether the distribution of weights in the example above really represents the weights of all male adults in the United States. If it is a good representation, then his job is substantially reduced, since there are over fifty million men he would need to weigh if the limited group were untrustworthy.

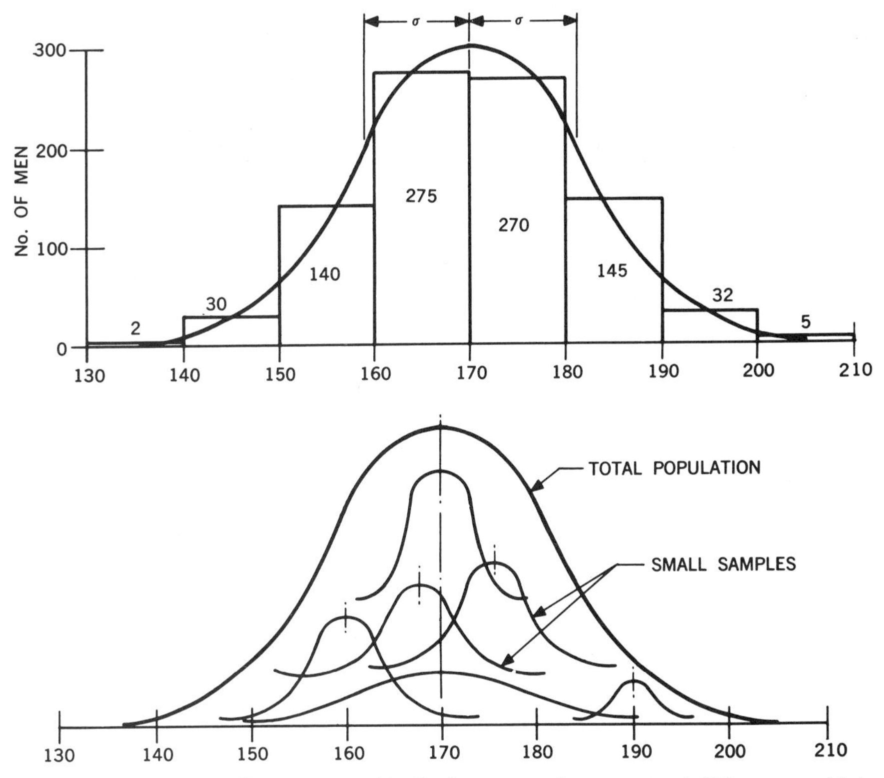

PROBABILITY CURVE illustrates graphically how a random group of 899 men could be represented by weight. The lower curve is a breakdown of several random samples.

Imagine for a moment all the many such limited groups that could be chosen out of the total population of male adults. Some groups would be small, but quite representative of the total; others would be less representative.

Once again, almost two-thirds of the groups would have means that were within one sigma of the mean of the total population. So we could say that any group chosen at random has a two-out-of-three chance of having its mean within one sigma (11 lb.) of the mean of all groups. If these odds are not satisfactory they can be improved by widening the "tolerance." For example, 99.7 per cent of all groups would have means between three sigmas below and three sigmas above the mean of the total population. Thus, it is 99.7 per cent certain that the mean of the sample is within 33 lb. of the average for the whole population.

This calculation is called a finding of the *confidence level* (odds) for any particular range, and the procedures here are exactly the same in principle as those used to figure the odds in any probability calculation.

Probability always starts by finding out how many ways a thing can possibly happen: the number of groups in the discussion above, the number of ways two dice can fall, the number of card hands that can be held. Then how many particular combinations are possible must be determined: the number of ways a seven can be thrown, or the number of royal flushes that can be turned in order from the first five cards at the top of a deck.

There are many rules for figuring out such combinations, but rules never substitute for simple common sense. Consider a pair of dice. Each has six sides. The first die can fall in any of six ways. For each way the first die falls, the second can fall in any of six ways. This makes a total of 36 possible throws—6×6. Now, how many ways can a seven be thrown? Just six: 1–6, 2–5, 3–4, 4–3, 5–2, 6–1. Therefore, what are the odds of throwing a seven? Six out of 36.

What are the odds of throwing a five? This makes four possibilities out of the 36 (1–4, 2–3, 3–2, 4–1). Therefore, it is more likely that a seven will be thrown than a five: 6 out of 36 to 4 out of 36, or 6 to 4.

The straight flush calculation is even simpler. The first card must be a 10. There are only four 10's so the odds are 4 out of 52. The second card must be a jack of the same suit as the 10. Only one card will do and only 51 cards remain. The odds are 1 out of 51 that you will draw the right jack. The same holds true for the other three cards needed:

$$\frac{4}{52} \times \frac{1}{51} \times \frac{1}{50} \times \frac{1}{49} \times \frac{1}{48} = \frac{4}{311{,}875{,}200}$$

Thus, if five cards were drawn an indefinite number of times, only once in

77,968,799 such tries would a straight flush appear. Or, to state it another way, the odds would be overwhelmingly against such a hand being dealt, namely, 77,968,799 to 1.

Note in this calculation that the probability of the first happening (turning up the first 10) is multiplied by the probability of the second (the jack) to get a probability of that particular combination of two cards. Then each succeeding probability is multiplied to get the probability of the five-card hand.

Symbolic Logic.—In ancient times the only respectable way to discover anything about the world around was by thought and reason. No philosopher could soil his hands with an experiment or even a close examination of physical things. He would sit back, close his eyes perhaps, and create propositions for later discussion and analysis with his colleagues.

This analysis of the validity (truth or falseness) of propositions is the heart of all logic. What makes logic so difficult as well as so fascinating is the variability of words and sentences depending on how, exactly, they are used—even on how they are spoken.

In the eighteenth century, when great political as well as industrial revolutions were stimulating the minds of men, several mathematicians wondered whether the clear-cut and precise techniques of mathematics might not be of assistance in logic. They were concerned with the need for a technique of logical analysis, not the content of the propositions, just as algebra is concerned with the correct relationships of the variables x, y, and so forth, not that these are heights in feet, or quantities of water in gallons, or some other measurable thing.

Logic had always been presented in words and sentences whose exact meanings had to be defined carefully. Wouldn't it be simpler if words were left out completely, were switched to symbols—perhaps like those used in mathematics—which could be defined exactly and then manipulated to discover new relationships? The answer is yes; there is such a simple and very powerful tool.

In the most widely used system (obviously many sets of symbols and rules of manipulation are possible) single alphabet letters represent single propositions or sentences. The letters p, q, r, and so on might mean "It is raining," or "All men are mortal," or "My aunt's half brother has a twisted ankle." They are much like the variables in algebra and, as in that branch of mathematics, symbolic logic is not concerned with what propositions they actually represent.

Words and sentences in language are related by certain connectives. The symbols in the new logic must be similarly related. The following is a basic set of connectives.

p	~ p	
T	F	If p is true, p is not true.
F	T	If p is false, p is true.

Denial ($\sim$) is a symbol placed in front of a letter or combination of letters and symbols to mean *not*. For example, $\sim p$ might mean "the sun is not shining."

Conjunction ($\cdot$) is placed between letters to mean *and*, as $p \cdot q$, which is read "p and q."

Alternation ($\vee$) is placed between letters to mean *or*, as $p \vee q$, which is read "p or q or both."

Implication ($\supset$) is placed between letters as $p \supset q$, which is read "if p then q," or, less exactly, "p implies q."

Equivalence ($=$) is placed between letters as $p = q$ and is read "p is equivalent to q."

With one additional idea, enough connectives are present to build an entire logic system. The idea is *truth*. It is assumed that a letter representing a statement can take only two values; it is either true or false. Knowing whether a number of simple statements are true or false means that it can be discovered whether and when a complex combination of these statements is true or false.

Analyses of truth are rather easily made with a *truth table*, a simple mechanical method of considering all possibilities. For example, the following tables clarify the meanings of the symbols defined above.

p	q	p·q	p∨q	p⊃q	p = q
T	T	T	T	T	T
T	F	F	T	F	F
F	T	F	T	T	F
F	F	F	F	T	T

p	q	r	~q	p·(~q)	p∨r	(p∨r)⊃q	~{[p·(~q)] ∨ [(p∨r) ⊃ q]}
T	T	T	F	F	T	T	F
T	T	F	F	F	T	T	F
T	F	T	T	T	T	F	F
T	F	F	T	T	T	F	F
F	T	T	F	F	T	T	F
F	T	F	F	F	F	T	F
F	F	T	T	F	T	F	T
F	F	F	T	F	F	T	F

The first table shows: only when p and q are both true does $p \cdot q$ become true; if either p or q is true then $p \vee q$ is true; the statement $p \supset q$ is false only when the implication fails (that is, when p is true even though q is false); and $p = q$ is true whenever p and q are both false or both true.

Now this set of values can be applied to much more complex statements. Consider this one: "It never happens that either my wineglass is full and I am not thirsty or else I am thirsty if either my wineglass is full or the wine is sour."

Let p represent "My wineglass is full"; q represent "I am thirsty"; and r represent "The wine is sour." In symbolic form the sentence is:

$$\sim\{[p \cdot (\sim q)] \vee [(p \vee r) \supset q]\}$$

The second truth table then sets up all possible combinations of truth and falseness for the three propositions.

The results are rather surprising. The only time the statement is true is when the wineglass is not full, I am not thirsty, and the wine is sour. Or, to think about it in another way, if the statement as a whole is true it can be deduced that the wineglass is not full, I am not thirsty, and the wine is sour.

All this may seem like an interesting exercise which common sense could have performed more quickly. But many practical applications have been made, applications as far apart as can be imagined. The first is in the design of complex electrical control systems. Here the electrical engineer substitutes 1 and 0, or ON and OFF, for true and false. He then substitutes his electric switches for the p, q, and r propositions and sets them up so that for any particular combination of switch settings the entire circuit will be ON or OFF as he needs.

Suppose, in the previous example, instead of a wineglass and thirst you had three switches, p, q, and r, and the only time a certain light was to be ON was when there was no sun (p is 0), the window is covered (q is 0), and someone is in the room (r is 1). The engineer could make a circuit like the symbolic proposition written in the table above and have his answer.

Another application of symbolic logic is for the analysis of the significance of statements in legal documents, in treaties, in newspaper reports, or in propaganda. In English, the statements would be quite confusing, but transformed into symbols and manipulated in these mechanical ways (sometimes by computers) the truth suddenly becomes apparent.

A TRUTH TABLE is a graphical method of evaluating all possible variables. This table evaluates the many combinations of p, q, and r, as discussed in the text. Symbolic logic is the mathematician's way of converting concepts into a form that can be analyzed numerically. Computers can be programmed to handle such concepts.

Topology.—At first glance, topology seems to be simply an extension of plane and solid geometry—an extension that goes beyond the lines and angles of these studies and concerns itself with shape. However, the applications of the theorems of topology are by no means limited to physical objects. They have been applied in logic, philosophy, and even in the strategies of war.

Topology really began with the discovery by René Descartes (1596–1650) in 1640 and Leonhard Euler (1707–1783) in 1752 that a simple polyhedron—a solid made up of plane figures such as triangles, rectangles, and so on—with no through-holes has a consistent relationship between the number of vertices, edges, and faces:

$$\text{vertices} - \text{edges} + \text{faces} = 2\,.$$

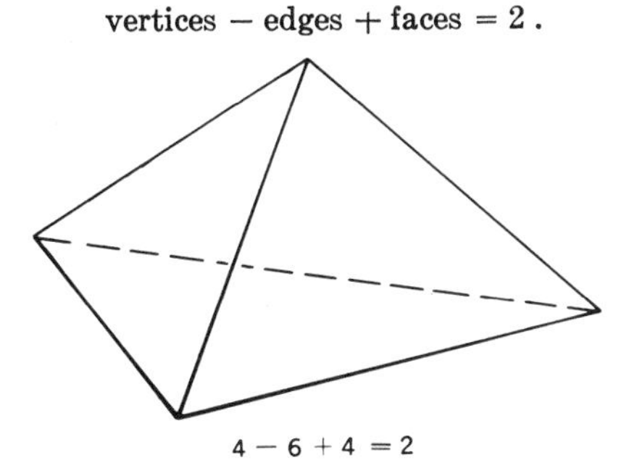

$4 - 6 + 4 = 2$

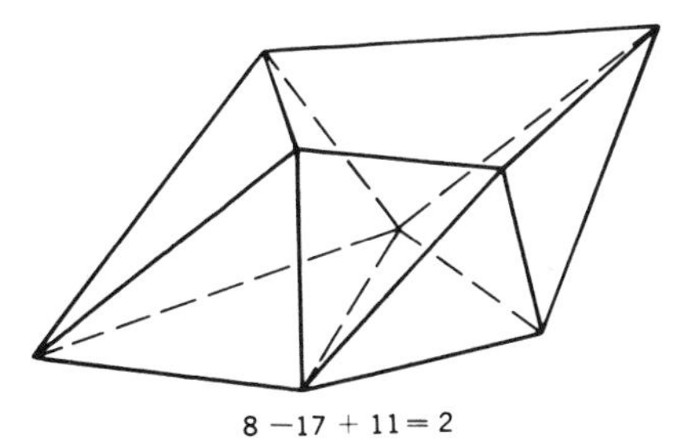

$8 - 17 + 11 = 2$

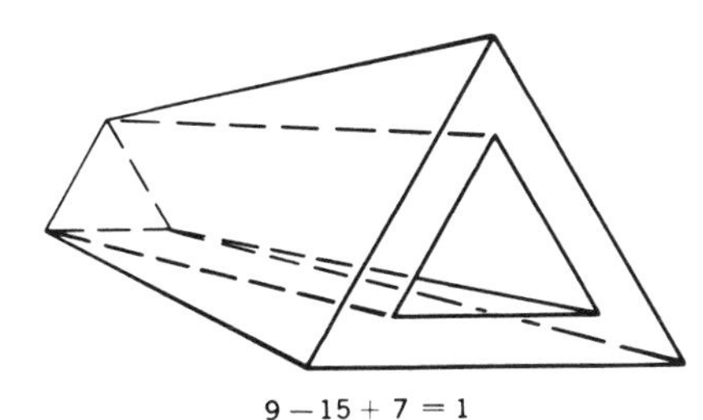

$9 - 15 + 7 = 1$

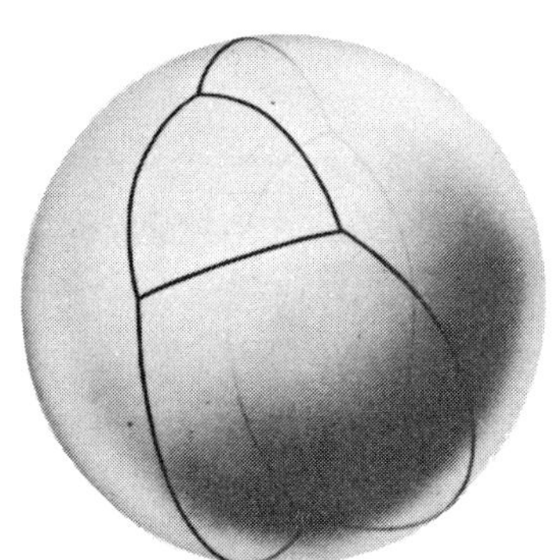

$5 - 8 + 5 = 2$

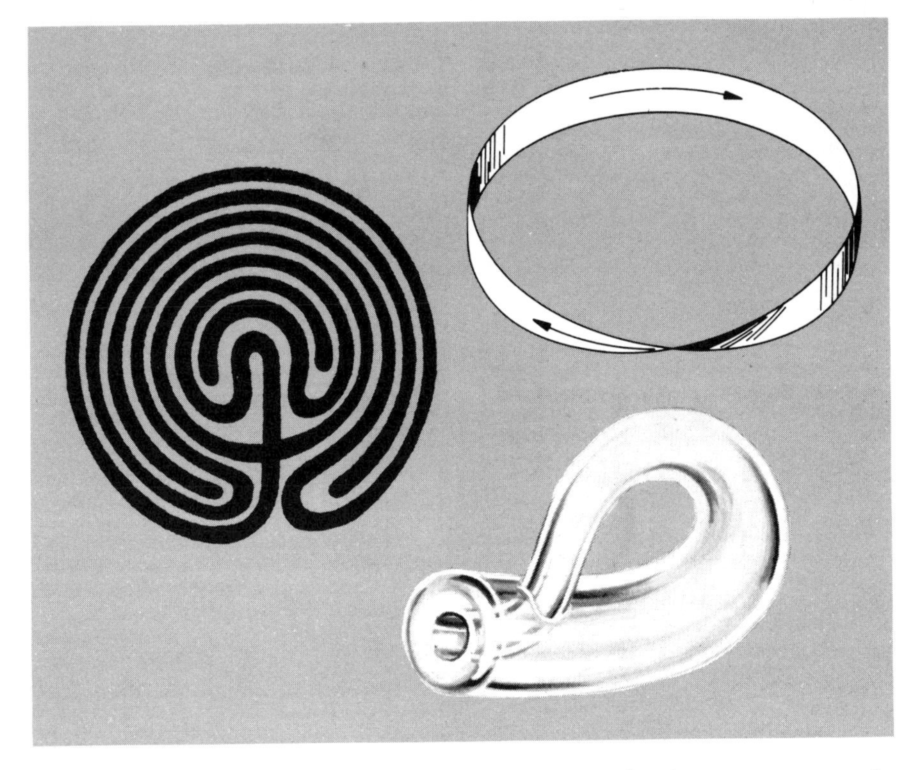

TOPOLOGICAL FIGURES attempt to give dimension to concepts that, in many cases, are only theoretical. The maze (*left*) and the Möbius strip (*right*) are both closed curves. The Klein bottle (*lower right*) is a one-sided figure; it has neither inside nor outside.

Note that the illustration with a hole in it does not follow Descartes's formula because the polyhedron is not simple. However, the formula will apply to all simple polyhedrons and even to solids with curved faces and edges or even to a sphere with areas marked off by curved arcs and vertices indicated by the intersections of these arcs.

This wide applicability suggests that the formula is really independent of the exact dimensions or angles and concerns itself only with shape. Thus, it was the first truly topological formula.

How can this shape, characteristic, or property be precisely defined? First, the topologist looks at any solid or plane figure as an aggregate of an infinite number of points. The body shape is defined by those points which happen to lie on its boundaries (the *perimeter* of a plane figure, the *surface* of a solid). He says that two figures are topologically similar when a point can always be found in the one which will correspond to one, and only one, point in the other.

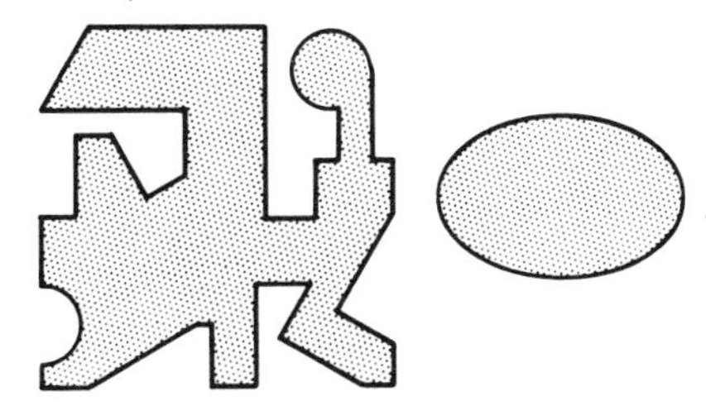

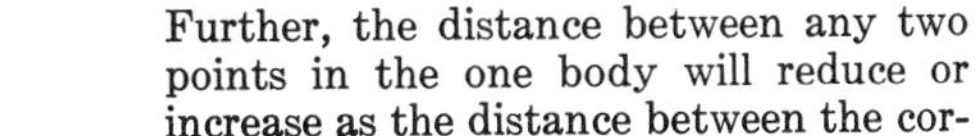

Further, the distance between any two points in the one body will reduce or increase as the distance between the corresponding points in the other is reduced or increased.

Following these rules, solids of quite different shape would have the same topological properties. However, if a hole were punched in the rubber triangle it would change the topological properties because a circle that encloses the hole would not now be able to reduce in size to a point as its corresponding circle can in the unpierced triangle.

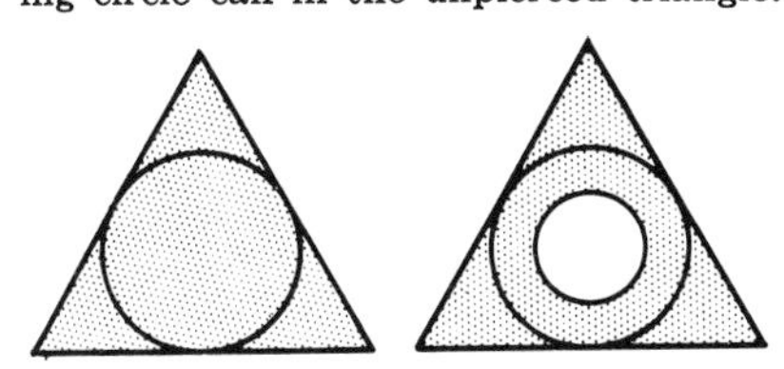

The solid triangle is said to be *simply connected* because any encircling line (starting with the outer boundary) can be reduced in size to a point within the triangle. The triangle with the hole does not permit this. The outer perimeter would have to shrink across the hole to reduce to a point. This figure is said to be *multiply connected*.

It should be noted here that any multiply connected figure may be made into a simply connected figure by making

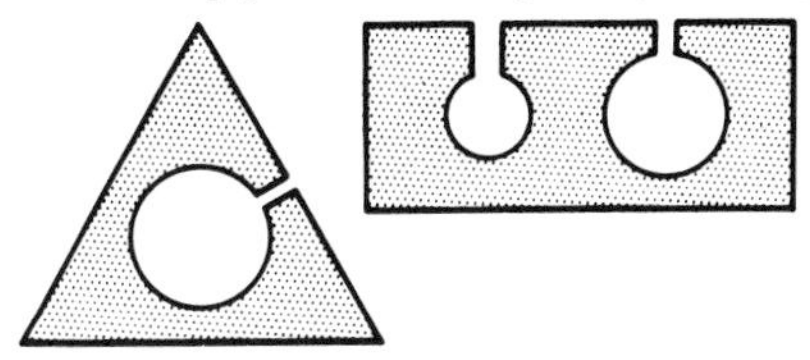

one cut (or more if there is more than one hole) from the hole to the outer perimeter.

One of the interesting discoveries of topology is that of one-sided surfaces. The page on which these words are printed has two sides. If a line were on one side, it could not extend to the other side without turning around the edge. But consider a loop of paper cut, twisted, and reglued as shown in the accompanying illustration. A line started along the inside would wind up on the outside of the strip, for it has only one surface. A. F. Möbius (1790–1868) discovered this strange property; he also noted the strip has only one edge.

Another one-sided surface is the Klein bottle. Here again a line can be drawn starting on the inside surface and traced around the loop and onto the outer surface of the same bottle.

These are some of the peculiarities that topology considers. Sometimes these subjects seem so obvious as to be unworthy of mention. For example, the *Jordan curve theorem*, named after Marie Ennemond Camille Jordan (1838–1922), states that any closed curve drawn on a plane without crossing itself divides the plane into two domains, an inside and an outside. Oddly enough the proof of this seemingly obvious statement is by no means simple. Jordan's own proof was in error and only recently have simple proofs been demonstrated.

Another interesting problem which has not yet been solved is called the *four-color map*. It is well known by the experience of a thousand years of mapmaking that any map of the countries of the world, or of any portion of the world, needs only four different colors for the countries. That is, it can be printed so that no two countries having the same color have boundaries of any finite length in common. But is this always so? It is suspected, of course, that it is, but so far the proof has been made that five colors are always sufficient. No one has yet demonstrated that this can be achieved with four.

Thus far only geometrical figures have been considered, but obviously the "points" that make up the plane or solid figure do not have to be points in space. They could be members of a group (or *class*, to use a logician's term) and in that way something about the general relationships of groups of people or things might be discovered. This is why topology has become the mathematician's mathematics, the philosopher's logic, and the mapmaker's amusement.

—Richard M. Koff

TABLE OF COMMON LOGARITHMS

	0	1	2	3	4	5	6	7	8	9	Proportional Parts								
											1	2	3	4	5	6	7	8	9
10	0000	0043	0086	0128	0170						4	9	13	17	21	25	30	34	38
						0212	0253	0294	0334	0374	4	8	12	16	20	24	28	32	37
11	0414	0453	0492	0531	0569						4	8	12	15	19	23	27	31	35
						0607	0645	0682	0719	0755	4	7	11	15	19	22	26	30	33
12	0792	0828	0864	0899	0934	0969					3	7	11	14	18	21	25	28	32
							1004	1038	1072	1106	3	7	10	14	17	20	24	27	31
13	1139	1173	1206	1239	1271						3	7	10	13	16	20	23	26	30
						1303	1335	1367	1399	1430	3	7	10	12	16	19	22	25	29
14	1461	1492	1523	1553							3	6	9	12	15	18	21	24	28
					1584	1614	1644	1673	1703	1732	3	6	9	12	15	17	20	23	26
15	1761	1790	1818	1847	1875	1903					3	6	9	11	14	17	20	23	26
							1931	1959	1987	2014	3	5	8	11	14	16	19	22	25
16	2041	2068	2095	2122	2148						3	5	8	11	14	16	19	22	24
						2175	2201	2227	2253	2279	3	5	8	10	13	15	18	21	23
17	2304	2330	2355	2380	2405	2430					3	5	8	10	13	15	18	20	23
							2455	2480	2504	2529	2	5	7	10	12	15	17	19	22
18	2553	2577	2601	2625	2648						2	5	7	9	12	14	16	19	21
						2672	2695	2718	2742	2765	2	5	7	9	11	14	16	18	21
19	2788	2810	2833	2856	2878						2	4	7	9	11	13	16	18	20
						2900	2923	2945	2967	2989	2	4	6	8	11	13	15	17	19
20	3010	3032	3054	3075	3096	3118	3139	3160	3181	3201	2	4	6	8	11	13	15	17	19
21	3222	3243	3263	3284	3304	3324	3345	3365	3385	3404	2	4	6	8	10	12	14	16	18
22	3424	3444	3464	3483	3502	3522	3541	3560	3579	3598	2	4	6	8	10	12	14	15	17
23	3617	3636	3655	3674	3692	3711	3729	3747	3766	3784	2	4	6	7	9	11	13	15	17
24	3802	3820	3838	3856	3874	3892	3909	3927	3945	3962	2	4	5	7	9	11	12	14	16
25	3979	3997	4014	4031	4048	4065	4082	4099	4116	4133	2	3	5	7	9	10	12	14	15
26	4150	4166	4183	4200	4216	4232	4249	4265	4281	4298	2	3	5	7	8	10	11	13	15
27	4314	4330	4346	4362	4378	4393	4409	4425	4440	4456	2	3	5	6	8	9	11	13	14
28	4472	4487	4502	4518	4533	4548	4564	4579	4594	4609	2	3	5	6	8	9	11	12	14
29	4624	4639	4654	4669	4683	4698	4713	4728	4742	4757	1	3	4	6	7	9	10	12	13
30	4771	4786	4800	4814	4829	4843	4857	4871	4886	4900	1	3	4	6	7	9	10	11	13
31	4914	4928	4942	4955	4969	4983	4997	5011	5024	5038	1	3	4	6	7	8	10	11	12
32	5051	5065	5079	5092	5105	5119	5132	5145	5159	5172	1	3	4	5	7	8	9	11	12
33	5185	5198	5211	5224	5237	5250	5263	5276	5289	5302	1	3	4	5	6	8	9	10	12
34	5315	5328	5340	5353	5366	6378	5391	5403	5416	5428	1	3	4	5	6	8	9	10	11
35	5441	5453	5465	5478	5490	5502	5514	5527	5539	5551	1	2	4	5	6	7	9	10	11
36	5563	5575	5587	5599	5611	5623	5635	5647	5658	5670	1	2	4	5	6	7	8	10	11
37	5682	5694	5705	5717	5729	5740	5752	5763	5775	5786	1	2	3	5	6	7	8	9	10
38	5798	5809	5821	5832	5843	5855	5866	5877	5888	5899	1	2	3	5	6	7	8	9	10
39	5911	5922	5933	5944	5955	5966	5977	5988	5999	6010	1	2	3	4	5	7	8	9	10
40	6021	6031	6042	6053	6064	6075	6085	6096	6107	6117	1	2	3	4	5	6	8	9	10
41	6128	6138	6149	6160	6170	6180	6191	6201	6212	6222	1	2	3	4	5	6	7	8	9
42	6232	6243	6253	6263	6274	6284	6294	6304	6314	6325	1	2	3	4	5	6	7	8	9
43	6335	6345	6355	6365	6375	6385	6395	6405	6415	6425	1	2	3	4	5	6	7	8	9
44	6435	6444	6454	6464	6474	6484	6493	6503	6513	6522	1	2	3	4	5	6	7	8	9
45	6532	6542	6551	6561	6571	6580	6590	6599	6609	6618	1	2	3	4	5	6	7	8	9
46	6628	6637	6646	6656	6665	6675	6684	6693	6702	6712	1	2	3	4	5	6	7	7	8
47	6721	6730	6739	6749	6758	6767	6776	6785	6794	6803	1	2	3	4	5	5	6	7	8
48	6812	6821	6830	6839	6848	6857	6866	6875	6884	6893	1	2	3	4	4	5	6	7	8
49	6902	6911	6920	6928	6937	6946	6955	6964	6972	6981	1	2	3	4	4	5	6	7	8
50	6990	6998	7007	7016	7024	7033	7042	7050	7059	7067	1	2	3	3	4	5	6	7	8

	0	1	2	3	4	5	6	7	8	9	Proportional Parts								
											1	2	3	4	5	6	7	8	9
51	7076	7084	7093	7101	7110	7118	7126	7135	7143	7152	1	2	3	3	4	5	6	7	8
52	7160	7168	7177	7185	7193	7202	7210	7218	7226	7235	1	2	2	3	4	5	6	7	7
53	7243	7251	7259	7267	7275	7284	7292	7300	7308	7316	1	2	2	3	4	5	6	6	7
54	7324	7332	7340	7348	7356	7364	7372	7380	7388	7396	1	2	2	3	4	5	6	6	7
55	7404	7412	7419	7427	7435	7443	7451	7459	7466	7474	1	2	2	3	4	5	5	6	7
56	7482	7490	7497	7505	7513	7520	7528	7536	7543	7551	1	2	2	3	4	5	5	6	7
57	7559	7566	7574	7582	7589	7597	7604	7612	7619	7627	1	2	2	3	4	5	5	6	7
58	7634	7642	7649	7657	7664	7672	7679	7686	7694	7701	1	1	2	3	4	4	5	6	7
59	7709	7716	7723	7731	7738	7745	7752	7760	7767	7774	1	1	2	3	4	4	5	6	7
60	7782	7789	7796	7803	7810	7818	7825	7832	7839	7846	1	1	2	3	4	4	5	6	6
61	7853	7860	7868	7875	7882	7889	7896	7903	7910	7917	1	1	2	3	4	4	5	6	6
62	7924	7931	7938	7945	7952	7959	7966	7973	7980	7987	1	1	2	3	3	4	5	6	6
63	7993	8000	8007	8014	8021	8028	8035	8041	8048	8055	1	1	2	3	3	4	5	5	6
64	8062	8069	8075	8082	8089	8096	8102	8109	8116	8122	1	1	2	3	3	4	5	5	6
65	8129	8136	8142	8149	8156	8162	8169	8176	8182	8189	1	1	2	3	3	4	5	5	6
66	8195	8202	8209	8215	8222	8228	8335	8241	8248	8254	1	1	2	3	3	4	5	5	6
67	8261	8267	8274	8280	8287	8293	8299	8306	8312	8319	1	1	2	3	3	4	5	5	6
68	8325	8331	8338	8344	8351	8357	8363	8370	8376	8382	1	1	2	3	3	4	4	5	6
69	8388	8395	8401	8407	8414	8420	8426	8432	8439	8445	1	1	2	2	3	4	4	5	6
70	8451	8457	8463	8470	8476	8482	8488	8494	8500	8506	1	1	2	2	3	4	4	5	6
71	8513	8519	8525	8531	8537	8543	8549	8555	8561	8567	1	1	2	2	3	4	4	5	5
72	8573	8579	8585	8591	8597	8603	8609	8615	8621	8627	1	1	2	2	3	4	4	5	5
73	8633	8639	8645	8651	8657	8663	8669	8675	8681	8686	1	1	2	2	3	4	4	5	5
74	8692	8698	8704	8710	8716	8722	8727	8733	8739	8745	1	1	2	2	3	4	4	5	5
75	8751	8756	8762	8768	8774	8779	8785	8791	8797	8802	1	1	2	2	3	3	4	5	5
76	8808	8814	8820	8825	8831	8837	8842	8848	8854	8859	1	1	2	2	3	3	4	5	5
77	8865	8871	8876	8882	8887	8893	8899	8904	8910	8915	1	1	2	2	3	3	4	4	5
78	8921	8927	8932	8938	8943	8949	8954	8960	8965	8971	1	1	2	2	3	3	4	4	5
79	8976	8982	8987	8993	8998	9004	9009	9015	9020	9025	1	1	2	2	3	3	4	4	5
80	9031	9036	9042	9047	9053	9058	9063	9069	9074	9079	1	1	2	2	3	3	4	4	5
81	9085	9090	9096	9101	9106	9112	9117	9122	9128	9133	1	1	2	2	3	3	4	4	5
82	9138	9143	9149	9154	9159	9165	9170	9175	9180	9186	1	1	2	2	3	3	4	4	5
83	9191	9196	9201	9206	9212	9217	9222	9227	9232	9238	1	1	2	2	3	3	4	4	5
84	9243	9248	9253	9258	9263	9269	9274	9279	9284	9289	1	1	2	2	3	3	4	4	5
85	9294	9299	9304	9309	9315	9320	9325	9330	9335	9340	1	1	2	2	3	3	4	4	5
86	9345	9350	9355	9360	9365	9370	9375	9380	9385	9390	1	1	2	2	3	3	4	4	5
87	9395	9400	9405	9410	9415	9420	9425	9430	9435	9440	0	1	1	2	2	3	3	4	4
88	9445	9450	9455	9460	9465	9469	9474	9479	9484	9489	0	1	1	2	2	3	3	4	4
89	9494	9499	9504	9509	9513	9518	9523	9528	9533	9538	0	1	1	2	2	3	3	4	4
90	9542	9547	9552	9557	9562	9566	9571	9576	9581	9586	0	1	1	2	2	3	3	4	4
91	9590	9595	9600	9605	9609	9614	9619	9624	9628	9633	0	1	1	2	2	3	3	4	4
92	9638	9643	9647	9652	9657	9661	9666	9671	9675	9680	0	1	1	2	2	3	3	4	4
93	9685	9689	9694	9699	9703	9708	9713	9717	9722	9727	0	1	1	2	2	3	3	4	4
94	9731	9736	9741	9745	9750	9754	9759	9763	9768	9773	0	1	1	2	2	3	3	4	4
95	9777	9782	9786	9791	9795	9800	9805	9809	9814	9818	0	1	1	2	2	3	3	4	4
96	9823	9827	9832	9836	9841	9845	9850	9854	9859	9863	0	1	1	2	2	3	3	4	4
97	9868	9872	9877	9881	9886	9890	9894	9899	9903	9908	0	1	1	2	2	3	3	4	4
98	9912	9917	9921	9926	9930	9934	9939	9943	9948	9952	0	1	1	2	2	3	3	4	4
99	9956	9961	9965	9969	9974	9978	9983	9987	9991	9996	0	1	1	2	2	3	3	3	4

MATHEMATICS GLOSSARY

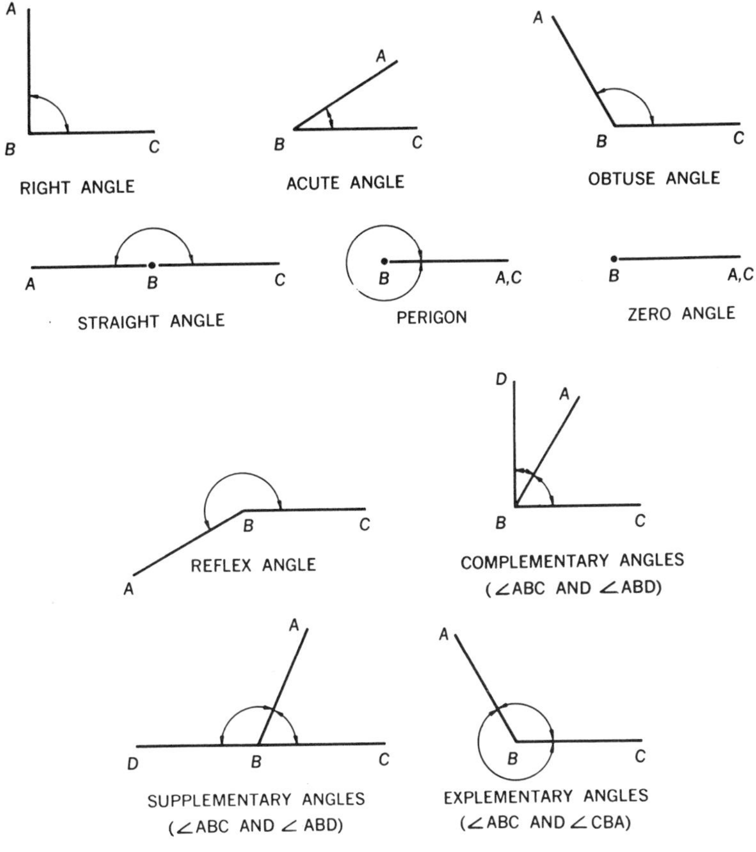

Abscissa.—See *Cartesian coordinates.*

Absolute value (numerical value).—In geometry, the distance of a point from the zero point on an ordinary number scale, regardless of its direction. Hence, the absolute value of 3 is 3; of −7, 7; and of zero, 0. The absolute value of a is denoted by $|a|$; hence, $|3| = 3$, $|-7| = 7$, and $|0| = 0$.

Abstract mathematics (pure mathematics).—Those branches of modern mathematics, such as topology, that have no meaning in terms of "real" or "specific" interpretations. Unlike applied mathematics, which has meaning in terms of physical experiences, these branches are studied solely as a means of extending knowledge.

Addition.—One of the fundamental operations of arithmetic. The combination of two numbers, x and y, by addition is denoted by $x + y$; the result of the combination is the *sum* of the numbers.

Algebra.—The branch of mathematics that involves the investigation of the properties of numbers by means of literal symbols, or variables, such as a, b, x, and y, instead of constants, such as 1, 2, and 100. Since algebraists can work in generalities rather than with specific instances, the scope of their work is not as limited as that of arithmeticians. Typical algebraic problems are the solving of equations and the summation of series.

Algebraic equation.—See *Equation.*

Algorithm (algorism).—Any method of computation whose procedure has been standardized in the form of rules. Familiar examples of algorithms are the processes of long division and of checking addition by "casting out nines."

Alternating series.—See *Series.*

Analysis.—The branch of higher mathematics that deals with the solution of complex physical problems. A powerful tool used by physicists and engineers, analysis grew out of the calculus and was extensively developed by Joseph Louis Lagrange (1736–1813), K. G. J. Jacobi (1804–1851), and Sir William Rowan Hamilton (1805–1865).

Analytic geometry (coordinate geometry).—The branch of mathematics that involves the application of algebraic methods to geometric figures and systems. Developed primarily by René Descartes (1596–1650), it involves the use of coordinate systems that relate the point, which is the fundamental element of geometry, to the number, which is the fundamental element of algebra.

Angle.—The inclination of one line, or *ray*, to another, measured in *degrees* or *radians*. The lines are called the *sides* of the angle; their common point, the *vertex* of the angle. An angle is represented by the symbol ∠, and $\angle ABC$ designates an angle with sides AB and BC and vertex B. The *interior angle* is the smaller of the angles formed by the intersection of the two lines, and the *exterior angle* is the larger angle formed. A *right angle* is one of 90°, or $\pi/2$ radians. An *acute angle* is one of between 0° and 90°, and an *obtuse angle* is one of between 90° and 180°. A *reflex angle* is one between 180° and 360°. A *straight angle* is one of 180°, or π radians; hence, a straight angle is one in which the sides form a straight line. An angle of 360°, or 2π radians, is a *perigon*, while one of 0° is a *zero angle;* in both of these, the sides of the angle coincide. An *oblique angle* is one that is not a multiple of 90°. Two angles are *complementary* if their sum is 90°; each is called the *complement* of the other. Two angles are *supplementary* if their sum is 180°; each is called the *supplement* of the other. Two angles are *explementary*, or *conjugate*, if their sum is 360°; each is called the *explement*, or *conjugate*, of the other. In solid geometry, a *dihedral angle* is formed by the intersection of two planes.

Antilogarithm.—See *Logarithm.*

Applied mathematics.—Those branches of mathematics, such as analysis, that deal with physical, biological, and sociological experiences. While applied mathematics, in its present sense, deals only with mathematical operations, such as computer programming, in its broadest sense it is one of the principal tools of the engineer, the physicist, and the social scientist.

Area.—The two-dimensional measure of the region enclosed within the boundaries of any plane geometric figure.

Arithmetic.—The art of calculation using positive real numbers. Four fundamental operations—addition, division, multiplication, and subtraction—form the basis of all arithmetical calculations.

Arithmetic progression.—See *Progression.*

Arithmetic series.—See *Series.*

Axiom (postulate).—A self-evident proposition that is accepted without further proof; axioms are the basic propositions by which all other propositions of the theory are proved. Although the terms "axiom" and "postulate" were used to differentiate between "common notions" (axioms) and "geometric properties" (postulates) by Euclid, the terms are generally considered synonymous in modern usage.

Axis.—Any one of several lines, usually imaginary, that have importance in a particular connotation. An *axis of symmetry* divides a figure into two mirror images. An *axis of rotation* is the line about which an object turns. An *axis of revolution* is the line about which a plane figure is rotated to generate a surface of revolution.

Binary number system.—See *Number system.*

Binomial.—A polynomial that is the sum of two terms, such as $3x^2 + 2y$, or $ax^3 - bx$.

Boolean algebra.—An algebraic notation used in logic, devised by George Boole (1815–1864). While some of the rules of Boolean algebra resemble those of ordinary algebra, others are quite different.

Calculus.—A branch of mathematics developed independently by Sir Isaac Newton (1642–1727) and Gottfried Wilhelm von Leibniz (1646–1716) that deals with continuously varying quantities. The calculus is divided into two subbranches. *Differential calculus* deals with the rate of change of one quantity in respect to another. *Integral calculus,* which is the inverse of differential calculus, deals with the value of a changing quantity at any particular stage of its variation; it is particularly useful in finding the length, area, and volume of geometric figures.

Cartesian coordinates.—A standard method of locating points. The points can be located in a plane by denoting their distances from two intersecting lines called *axes,* the distance from either axis being measured along a line parallel to the other axis. The axes are said to be *rectangular* when they are mutually perpendicular and *oblique* when they are not mutually perpendicular. Each point is located in terms of its *coordinates.* The coordinate measured from the y-axis parallel to the x-axis is called the *abscissa,* and the coordinate measured from the x-axis parallel to the y-axis is called the *ordinate.* A point can similarly be located in space, or three dimensions, by measuring its distance from three mutually perpendicular axes, usually designated as the x-axis, the y-axis, and the z-axis.

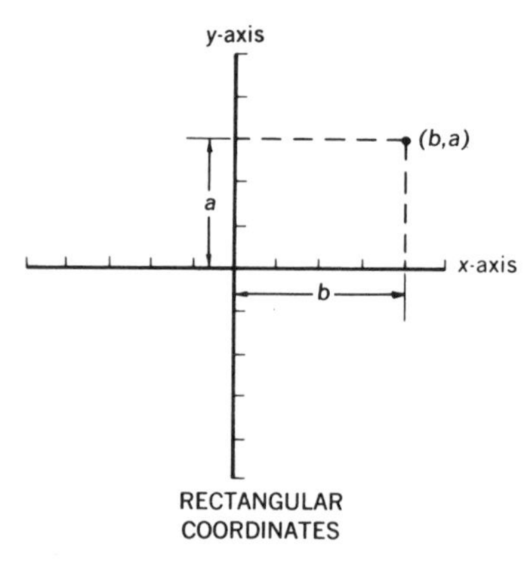

RECTANGULAR COORDINATES

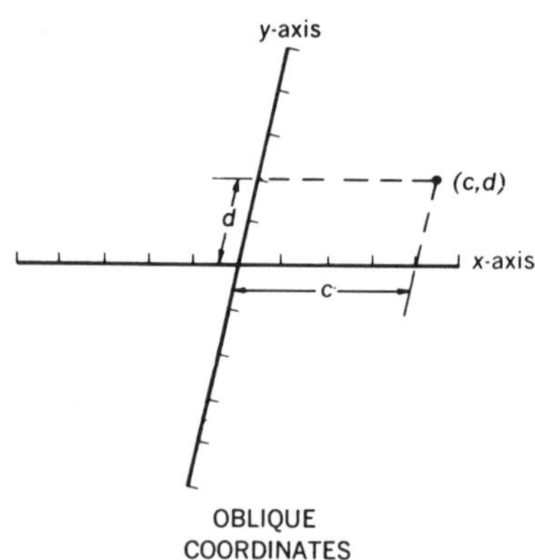

OBLIQUE COORDINATES

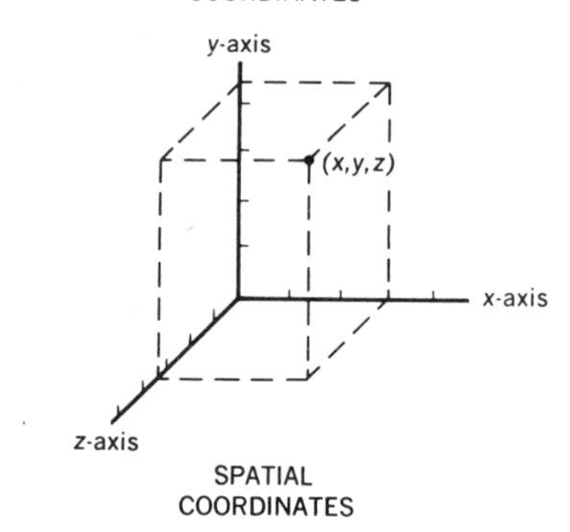

SPATIAL COORDINATES

Circle.—See *Conic.*

Circumference.—The total length of the boundary line of a circle or any other closed curvilinear figure. $C = 2\pi r$.

Coefficient.—In algebra, a constant term written before a variable quantity. Thus in $3x$, $3(x + y)$, and $3xy$, 3 is the coefficient of x, $x + y$, and xy, respectively. In general, the product of all the factors of a term except for one term is the coefficient of that one term. Thus, in $-3axyz$, $-3axy$ is the coefficient of z, $-3axz$ is the coefficient of y, and so on for all terms.

Combinations.—The different sets of one or more objects that can be selected from a given number of objects without regard to order. For example, x,y, y,z, and x,z are all the possible sets, or combinations, of two objects that can be selected from the three objects x, y, and z. The number of combinations of n objects selected r at a time is any collection of r objects selected from a given number of objects, n. This may be expressed by the general formula

$$\frac{n!}{(n-r)!\,r!}$$

In the example, $n = 3$ (x, y, z) and $r = 2$; substituting in the formula, $3!/(3-2)!\,2!$ or $(3 \times 2 \times 1)/(2 \times 1) = 3$.

Common denominator.—Of two or more fractions, an integer or polynomial that is an exact multiple of each denominator. For example, the common denominator of $\frac{1}{2}$, $\frac{1}{4}$, and $\frac{1}{5}$ is 20, 40, 60 and so on. The smallest common denominator (in the above example, 20) is called the *least,* or *lowest, common denominator.*

Common factor (common divisor).—Of two or more integers or polynomials, an integer or polynomial that is a divisor of each. For example, common factors of 12 and 24 are 2, 3, 4, 6, and 12. The greatest common divisor (in the above example, 12) is called the *highest common factor.*

Common logarithm.—See *Logarithm.*

Complementary angle.—See *Angle.*

Concurrent.—In geometry, two or more lines or planes having one point in common.

Cone.—In solid geometry, a figure whose base is a circle, ellipse, or convex curve and whose sides taper up to a *vertex,* or *apex.* If the base is a circle, the figure is called a *circular cone;* if the base is an ellipse, the figure is called an *elliptic cone;* and if the base is a convex curve, the figure is called a *convex cone.* If the axis (the line from the center of the base to the vertex) is perpendicular to the base, the cone is a *right cone;* in any other configuration, it is an *oblique cone.* The *frustrum of a cone* is the figure that

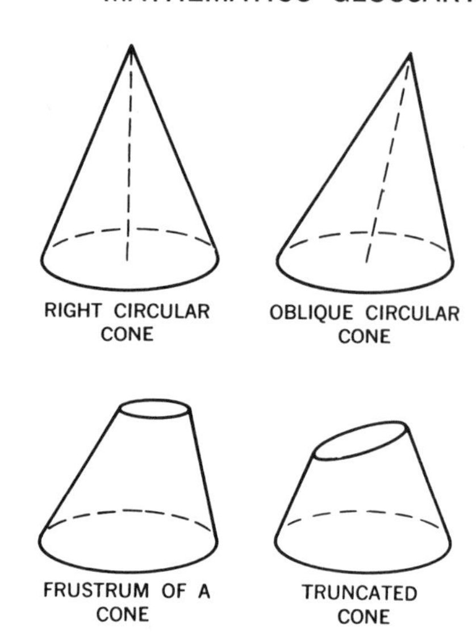
RIGHT CIRCULAR CONE — OBLIQUE CIRCULAR CONE — FRUSTRUM OF A CONE — TRUNCATED CONE

results when a cone is cut by a plane parallel to the base, and a *truncated cone* is the figure that results when a cone is cut by a plane not parallel to the base.

Congruent.—In geometry, figures that coincide when superimposed upon one another. Examples of congruent figures are line segments of equal length and angles of equal measure.

Conic (conic section).—The intersection of a circular conical surface by a plane. The shape of the section depends upon the position of the cutting plane. If the cutting plane is perpendicular to the axis, the section is a *circle.* If the plane is not parallel but does not cut the base, the section is an *ellipse.* If the plane is not parallel and does cut the base, the section is a *parabola.* If the plane cuts both *nappe* of the cone, the section is a *hyperbola.* Conics are sometimes defined in terms of their *eccentricity,* which is the ratio of the distances from a fixed point to a fixed line of

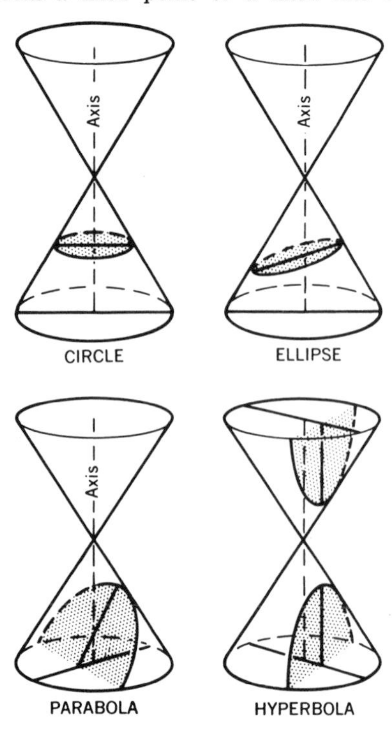

CIRCLE — ELLIPSE — PARABOLA — HYPERBOLA

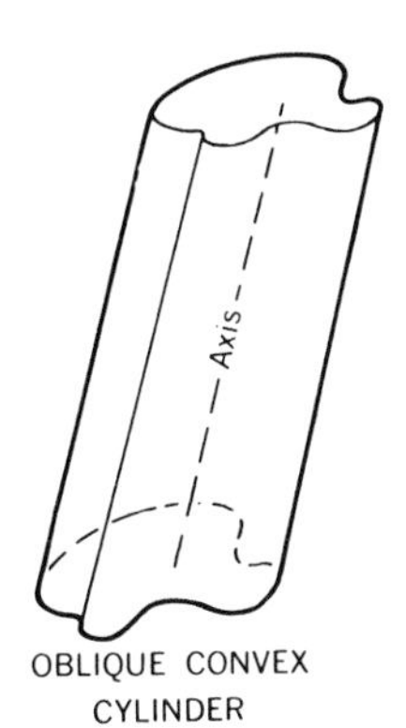

OBLIQUE CONVEX CYLINDER

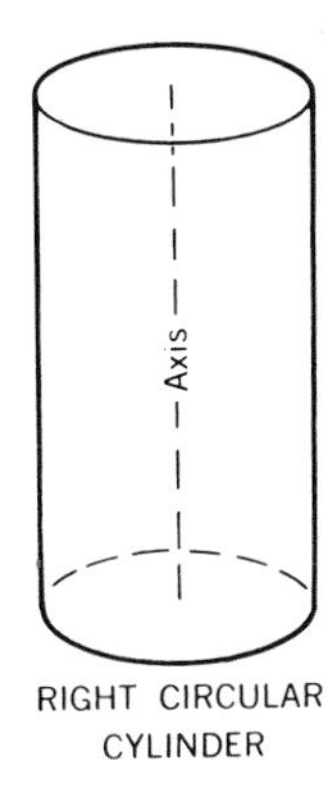

RIGHT CIRCULAR CYLINDER

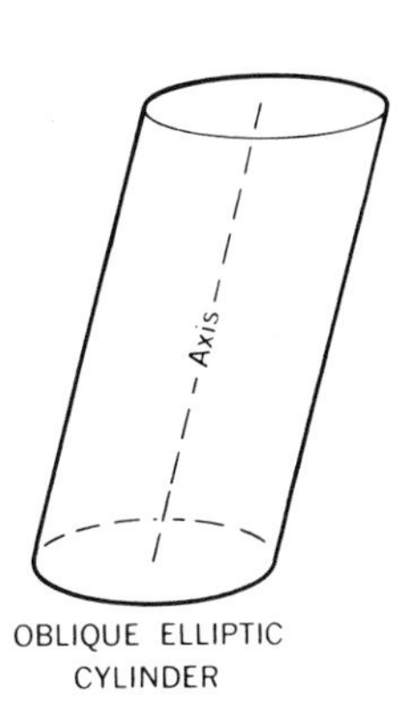

OBLIQUE ELLIPTIC CYLINDER

any point on the conic. If the eccentricity equals 0, the conic is a circle; if the eccentricity is less than 1, the conic is an ellipse; if the eccentricity equals 1, the conic is a parabola; and if the eccentricity is greater than 1, the conic is a hyperbola.

Consistent equation.—See *Equation.*

Constant.—Any quantity, such as 2 or π, whose value does not change, or is regarded as fixed, during any sequence of mathematical operations. A constant number is sometimes called an *absolute number.*

Curve.—A line that can be described in terms of an equation; hence, a straight line is a special case of a curve. An *arc* is any segment of a curve between two of its points.

Cylinder.—In solid geometry, a figure of constant cross section whose base is either a circle, an ellipse, or a convex curve. If the base is a circle, the cylinder is a *circular cylinder;* if the base is an ellipse, the cylinder is an *elliptic cylinder;* and if the base is a convex curve, the cylinder is a *convex cylinder.* If the axis (a line joining the centers of the two bases) is perpendicular to the bases, the cylinder is a *right cylinder;* in any other configuration, it is an *oblique cylinder.*

Decimal number system.—See *Number system.*

Denominator.—See *Fraction.*

Differential calculus.—See *Calculus.*

Diophantine analysis.—A method of finding the solutions of certain algebraic equations in two or more variables, provided the solutions are restricted to whole numbers.

Discriminant.—In algebra, an expression used in determining whether or not the roots of an equation are real or imaginary. For a quadratic equation, $ax^2 + bx + c = 0$, the discriminant is $b^2 - 4ac$: if the discriminant equals zero, the two roots are equal; if the discriminant is negative, the roots are imaginary; and if the discriminant is positive, the roots are real.

Division.—One of the fundamental operations of arithmetic. Basically, division is the determination of how many times one number or quantity is contained in another; thus, it is the inverse operation to multiplication. When one number or quantity, the *dividend,* is divided by another, the *divisor,* the result is called their *quotient.* For example, the quotient x/y of two numbers x and y is the number z (so that $x = yz$), provided that y does not equal zero.

Duodecimal number system.—See *Number system.*

e—The base of hyperbolic or natural logarithms. The value of e, to ten places, is 2.7182818284 . . .; it is a nonrepeating decimal. The term, which ranks with pi (π) in its importance to mathematics, was introduced by the Swiss mathematician Leonhard Euler (1707–1783).

Ellipse.—See *Conic.*

Ellipsoid (ellipsoid of revolution).—The hollow figure whose plane sections are all either ellipses or circles. An ellipsoid is produced by rotating an ellipse about one of its axes: when formed by rotation about the longer, or major, axis of the ellipse, the ellipsoid is *prolate;* when formed by rotation about the shorter, or minor, axis of the ellipse, the ellipsoid is *oblate.* An ellipsoid is sometimes called a *spheroid.*

Equality.—The state of being identical. Equality is expressed by the sign "=." The statement "$1 + 1 = 2$" means that "$1 + 1$" and "2" are different terms for the same quantity. A *continued equality* is three or more quantities set equal by means of two or more equality signs: $x = y = z$. An *inequality,* expressed by the sign "$\neq$," denotes two quantities that are not identical: $1 + 1 \neq 3$.

Equation.—A mathematical statement of equality between two quantities. There are two types of equations, *identities* and *conditional equations,* although the latter are usually referred to simply as equations. An equation is an identity when it is always true for any values of the variables within it; for example, $a(a - b)(a + b) = a^3 + ab^2$. An equation is conditional if it is valid for only certain values of the variables in the equation; for example, $x + y = 3$ is valid when $x = 2$ and $y = 1$ (and for some other values), but is invalid when $x = 5$ and $y = 3$. An *algebraic equation* is one in which each side is an algebraic expression. A *linear equation* is an algebraic equation of the first degree, such as $y = mx + b$. A *binomial equation* is one that has the form $x^n - a = 0$. A *quadratic equation* is one of the second degree, such as $ax^2 + bx + c = y$. A *reciprocal equation* is one in which the variables can be replaced by their reciprocals without changing the value of the equation; for example, if in $x + 1 = 0$ the x is replaced by its reciprocal $1/x$, when the resulting equation is simplified it becomes $1 + x = 0$, which is identical to the original equation. When two or more equations involving two or more unknowns can be solved for one or more values of the unknowns that are common to all equations, the equations are called *consistent.* In order to test the equations for consistency, they must be *simultaneous;* that is, they must represent curves with points that are in conjunction.

Equilateral.—In geometry, having all sides equal. The word is usually used as an adjective to describe a polygon with sides of equal length.

Equilateral triangle.—See *Triangle.*

Even number.—See *Number.*

Exponent.—A term in which a number is placed to the right of and above a symbol. For example, in the expression x^a, x is called the *base* and a the *power.* (It should be noted that some mathematicians use "power" in the same sense that they use "exponent.") When the exponent is a positive integer, the power is a symbol for repeated multiplication; that is, $5^2 = 5 \times 5$ and $5^4 = 5 \times 5 \times 5 \times 5$. When the power is a negative integer, it indicates that the value of the term is the reciprocal of the repeated multiplications, or $x^{-n} = 1/x^n$; for example, $5^{-2} = 1/(5 \times 5) = 1/25$. When the power is 0, regardless of the base, the value of the term is 1. The five laws of exponents are summarized in the table at the end of the article on advanced mathematics.

The *degree* of a term is the power, or sum of the powers. For example, x^2 is of the second degree, $2x^5$ is of the fifth degree, and $2xy^2z^3$ is of the sixth degree. The degree of an equation is that of the highest-degree term in that equation; hence, $3x^3 + 2xy + y = 0$ is a third-degree equation.

Exponential series.—See *Series.*

Factorial.—For a positive integer n, the product of all the positive integers equal to or less than n, denoted by $n!$ (read "n factorial"); hence, $n! = 1 \times 2 \times 3 \times \ldots \times (n - 1) \times n$. For example, $6! = 1 \times 2 \times 3 \times 4 \times 5 \times 6 = 720$. By convention, $0!$ has been given the value 1.

Fermat number.—A whole number of the form $F_n = 2^{2n} + 1$, where $n = 0$, 1, 2, 3, and so on. The equation was devised by Pierre de Fermat (1601–1665) to compute prime numbers. However, Leonhard Euler (1707–1783) showed the $F_5 = 2^{2 \cdot 5} + 1 = 4{,}294{,}967{,}297$ is factorable into (641)(6,700,417) and, therefore, that all Fermat numbers are not necessarily prime.

Fibonacci sequence (Fibonacci numbers).—An unending sequence of integers proposed by Leonardo da Pisa, or Leonardo Fibonacci (thirteenth century), formed according to the rule that each integer (except the first two) is the sum of the preceding two. The first 14 numbers of the sequence are: 1, 1, 2, 3, 5, 8, 13, 21, 34, 55, 89, 144, 233, 377.

Formula.—A fixed rule, general principle, or standard form stated in mathematical terms. For example, the area of a triangle is given by the formula $A = \frac{1}{2}bh$, where b is the length of the base and h is the height of the altitude.

Fraction.—The indicated quotient of two quantities, a/b, where a is called the *numerator* and b is called the *denominator*. A *simple fraction* (sometimes called a *common* or *vulgar fraction*) is one in which both the numerator and the denominator are integers. A *unit fraction* is a simple fraction in which the numerator is unity. A *complex fraction* is one in which the numerator or the denominator, or both, are simple fractions. A *proper fraction* is one in which the numerator is smaller than the denominator. An *improper fraction* is one in which the numerator is equal to or greater than the denominator.

Function.—A relationship of the object or objects of one set with each object in another set. For example, the expression $y = x^2 + x + 2$ defines y as a function of x. If the function $y = x^2 + x + 2$ is denoted by $y = f(x)$, then the value of y when $x = 3$ is $f(3) = 3^2 + 3 + 2 = 14$.

Geometric construction.—In Euclidean geometry, any construction that is made by using only an unmarked straightedge and a compass. A geometric construction is limited to the drawing of straight lines, arcs, and circles.

Geometric mean.—See *Mean.*

Geometric progression.—See *Progression.*

Geometric series.—See *Series.*

Geometry.—The branch of mathematics that deals with the shape and size of objects and the nature of space. *Plane geometry* deals with the properties and relations of plane figures, such as angles, polygons, and conics. *Solid geometry* deals with the properties and relations of three-dimensional figures, such as spheres, polyhedrons, and planes. These geometries were developed by Euclid (c. 300 B.C.) in his 13-volume book *Elements*, and are therefore referred to as *Euclidean geometries. Non-Euclidean geometries* differ in that they consider Euclid's "parallel postulate"—that only one parallel can be drawn to a given line through a point outside that line—to be invalid, thus changing the nature of the space in which the geometries are discussed.

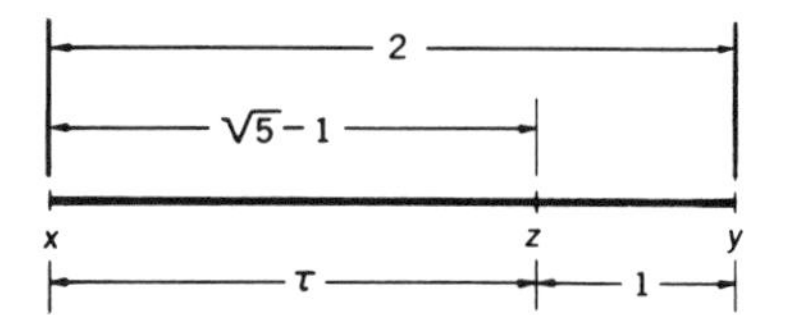

Golden section (golden ratio).—A division of a line segment xy by a point z so that the two parts of the segment are in the ratio τ:1, where τ is the positive root of the expression $x^2 - x - 1 = 0$; that is, $(\sqrt{5} + 1)/2$. This division of a line is very pleasing to the eye and is frequently used by artists; the ratio also has applications in musical theory.

Graph.—A drawing or geometric representation that shows the relation between certain sets of numbers.

Harmonic mean.—See *Mean.*

Hyperbola.—See *Conic.*

Hyperboloid (hyperboloid of revolution).—The hollow figure produced by the rotation of a hyperbola about one of its axes, called its axis of revolution. When the hyperbola is rotated about its conjugate axis, a hyperboloid of *one sheet* is produced. When the hyperbola is rotated about its transverse axis, a hyperboloid of *two sheets* is produced.

Imaginary number.—See *Number.*

Integer.—Any of the numbers . . . , −2, −1, 0, +1, +2, ... *Positive integers* are +1, +2, and so on; *negative integers* are −1, −2, and so on.

Integral calculus.—See *Calculus.*

Intercept.—See *Cartesian coordinates.*

Inverse operation.—The operation that, when performed following a given operation, would cancel the effect of the given operation. Thus, subtraction is the inverse operation to addition, and division is the inverse operation to multiplication.

Inverse trigonometric functions.—See *Trigonometric functions.*

Irrational number.—See *Number.*

Isosceles.—In geometry, having two equal sides. The word is usually used as an adjective to describe a polygon.

Isosceles triangle.—See *Triangle.*

Lemma.—A theorem that is proved for use in the proof of another theorem.

Line (straight line).—A special case of a curve. In Euclidean geometry, the concept of "line" is one of an unswerving path. In Cartesian coordinates, a line has the equation $y = mx + b$, where m is the slope and b is the y-intercept. A *ray* is either of the two portions into which a line is divided by a single point on the line.

Linear equation.—See *Equation.*

Logarithm.—A mathematical device by means of which numerical computation may be facilitated. The logarithm of a number, N, to a given base, x, is the index of the power, a, to which that base must be raised to produce the number; hence, $x^a = N$. For example, since $10^3 = 1{,}000$, 3 is the logarithm of 1,000 to the base 10, written $\log_{10} = 1{,}000 = 3$. Logarithms that use 10 as a base are called *common,* or *Briggsian logarithms,* after Henry Briggs (1561–1630), who introduced them. Logarithms that use e (2.71828 . . .) as a base are called *natural,* or *Napierian logarithms,* after John Napier (1550–1617), who introduced them. Natural logarithms may be converted to common logarithms by multiplying the natural logarithm by 0.434294, a factor called the *modulus of common logarithms;* common logarithms may be converted to natural logarithms by multiplying the common logarithm by 2.302585, a factor called the *modulus of natural logarithms.* Thus, $\log_{10}N = 0.434294 \log_e N$, and $\log_e N = 2.302585 \log_{10} N$. The integer preceding the decimal point of the logarithm is called the *characteristic,* and the decimal fraction that follows is called the *mantissa.* The *antilogarithm,* or *inverse logarithm,* designated antilog a, is the number whose logarithm is the given number; for example, $\text{antilog}_{10} 3 = 1{,}000$.

4	14	15	1
9	7	6	12
5	11	10	8
16	2	3	13

4 x 4 SQUARE

17	24	1	8	15
23	5	7	14	16
4	6	13	20	22
10	12	19	21	3
11	18	25	2	9

5 x 5 SQUARE

Magic square.—A square array of integers in which the sum of the integers in each column, each diagonal, and each row is the same. The *order* of the square is the number of integers in the columns and rows; thus, a 4×4 square is of the order four, and 5×5 square is of the order five.

Mantissa.—See *Logarithm.*

Mathematical signs and symbols.—Letters, marks, and abbreviations representing operations, quantities, or relations used in mathematical expressions, equations, and formulas. Some of the most common signs and symbols are shown in the table on the next page.

Mathematics.—The systematic study of arrangement, number, quantity, and shape. Mathematics is generally described in terms of abstract, or pure, mathematics and applied mathematics. The study of mathematics is usually divided into three major branches: algebra, analysis, and geometry.

Mean (average).—The "central quantity" of a set of quantities. There are three types of means: the *arithmetic mean* (usually referred to as the *mean* or *average*), m; the *geometric mean, g;* and the *harmonic mean, h.* For two numbers, x and y, $m = (x + y)/2$; $g = \sqrt{xy}$; and $1/h = (\frac{1}{2})(1/x + 1/y)$. If x and y are positive, the harmonic mean is less than the geometric mean, which is less than the arithmetic mean. For example, if

Selected Mathematical Signs and Symbols

Symbol	Meaning
$+$	plus; positive
$-$	minus; negative
$\pm$	plus or minus
$\mp$	minus or plus
$\times$ or $\cdot$	multiplied by; times
$/$ or $\div$	divided by
$>$	is greater than
$<$	is less than
$\equiv$	is identical to
$=$	is equal to
$\geqq$ or $\geq$	is greater than or equal to
$\leqq$ or $\leq$	is less than or equal to
$\neq$	is not equal to
$\approx$ or $\doteqdot$	is approximately equal to
$\sim$	is similar to; equivalent
$\cong$	is congruent to; congruent
$\parallel$	is parallel to; parallel
$\perp$	is perpendicular to; perpendicular
$:$	is to; the ratio of
$\therefore$	therefore; hence
$\because$	because; since
$\angle$ or $\measuredangle$	angle
∟ or ⦜	right angle
$\propto$	varies directly as
∞	infinity
a^n	reciprocal ($1/a^n$)
$\lvert a \rvert$	absolute value (of a)
ϵ	is a member of the set of
%	per cent
$f(a)$ or $g(a)$	function of f (or of g) at a
log or $\log_{10}$	common logarithm; logarithm to the base 10
ln or $\log_e$	natural logarithm; logarithm to the base e
antilog	antilogarithm
$\sqrt{}$	root of
i	$\sqrt{-1}$
$\triangle$	triangle
▱	parallelogram
$\square$	square
○ or ⊙	circle
Q.E.D.	which was to be proved
$\doteq$	is measured by
Σ	summation
$\int$	integral
Π	product
a'	a prime
dy/dx or $f'(x)$	the derivative of y with respect to x
$\delta u/\delta x$ or u_x	the partial derivative of u with respect to x

$x = 7$ and $y = 28$, then $h = 11.2$, $g = 14.0$, and $m = 17.5$. For n numbers, $x_1, x_2, \ldots x_n$, the three means may be found by the following formulas:

$$\frac{1}{h} = \left(\frac{1}{n}\right)\left(\frac{1}{x_1} + \frac{1}{x_2} + \ldots + \frac{1}{x_n}\right)$$

$$g = \sqrt[n]{x_1 x_2 \ldots x_n}$$

$$m = \frac{1}{n}\left(x_1 + x_2 + \ldots + x_n\right)$$

Median.—In statistics, of an odd number of increasing numerical values, the middle value; of an even number of increasing numerical values, the arithmetic mean of the two middle values. For example, the median of 15, 20, 50, 60, 133 is 50; of 15, 20, 50, 60, 133, 134 is 55.

Mixed number.—In arithmetic, the sum of an integer and a fraction, such as $3 + \frac{1}{2}$, or $3\frac{1}{2}$; in algebra, the sum of a polynomial and a rational algebraic fraction, such as $3x + 3 + 1/(x - 1)$.

Multiplication.—One of the fundamental operations of arithmetic. Basically, multiplication is the determination of the value of a specified number of repeated additions of the same quantity or integer; thus, $m \times n = m + m + \ldots + m$ (m added to m $n - 1$ times). In $m \times n$, which may also be written $(m)(n)$, $m \cdot n$, or mn, $= p$, m is the *multiplicand*, n is the *multiplier*, and p is the *product*.

Natural logarithm.—See *Logarithm.*

Nomograph (nomogram or alignment chart).—A graphic method of computation in which three lines or curves, usually parallel to one another, are graduated in such a way that a straightedge cutting across them will give the related values of three variables. For example, a nomograph can be constructed, as in the accompanying illustration, in which one scale represents the distance in miles traveled by an automobile, a second scale represents the number of gallons of gasoline used by the automobile, and the third scale represents the number of miles per gallon.

Number.—In general terminology, a positive integer. A *rational number* is either an integer or a fraction of the form a/b, where a and b are both integers but do not equal zero. An *irrational number*, such as e, π, or $\sqrt{2}$, is a number that is not expressible as an integer or a fraction composed of integers. *Real numbers* include all rational and irrational numbers. *Imaginary numbers* include all complex numbers of the form $a + bi$ (where $i = \sqrt{-1}$), a and b are real numbers, and $b \neq 0$. An *even number* is any integer divisible by 2, and an *odd number* is any number that is not so divisible.

Number system.—A method of numeration and the rules for arithmetic computation with the notations thus devised. The most familiar of the number systems is the *decimal number system.* This system uses place values and the base, or *radix*, 10; for example, 3502.16 is the notation for $3 \times 10^3 + 5 \times 10^2 + 0 \times 10^1 + 2 \times 10^0 + 1 \times 10^{-1} + 6 \times 10^{-2}$. The *binary number system*, which is also known as the *dyadic number system*, also uses place values, but the base number is 2 rather than 10. Only 2 digits, 0 and 1, are required, and these are called *bits* (from *bi*nary digi*ts*). The *octonary number system* is also a place-value system and has a radix of 8, while the *duodecimal number system*, which is another place-value system, has a radix of 12.

Numerator.—See *Fraction.*

Oblique.—An adjective meaning "slanted," that is, neither parallel nor perpendicular. An *oblique angle* is any angle that is not a multiple of 90°, an *oblique line* is one that is neither parallel nor perpendicular to another line, and an *oblique triangle* is one that does not contain a right angle.

Obtuse angle.—See *Angle.*

Octonary number system.—See *Number system.*

Odd number.—See *Number.*

Ordinate.—See *Cartesian coordinates.*

Parabola.—See *Conic.*

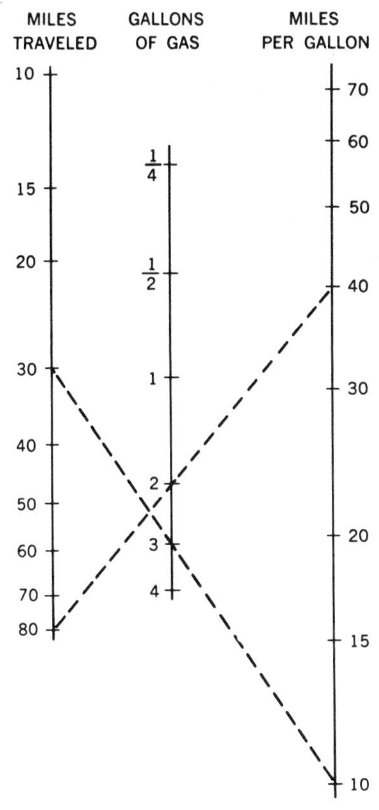

Paraboloid (paraboloid of revolution).—The hollow figure produced by the rotation of a parabola about its axis. An *elliptic paraboloid* is one whose plane sections parallel to the axis of revolution yield parabolas, and whose plane sections perpendicular to these parabolas yield ellipses. A *hyperbolic paraboloid* is one whose plane sections yield hyperbolas and parabolas.

Parallelogram.—A quadrilateral in which the opposite sides are parallel. A *rhombus* is a parallelogram in which all four sides are equal in length; a *square* is a parallelogram in which all four sides are of equal length and in which adjacent sides are perpendicular; and a *rectangle* is a parallelogram in which the adjacent sides are perpendicular and are of different lengths.

Parameter.—An *arbitrary constant* (a variable representing an unspecified constant) in a mathematical expression or formula that distinguishes one member of a group from the others.

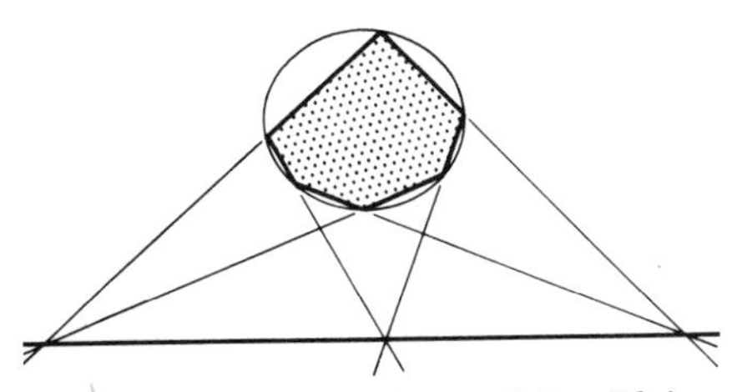

Pascal's theorem.—A proof by Blaise Pascal (1623–1662), showing that when a hexagon is inscribed in a conic, the three points of intersection of the three pairs of opposite sides lie on a straight line.

Pascal's triangle.—A means devised by Blaise Pascal (1623–1662) to determine the coefficients of any binomial expansion, $(x + y)^n$. The triangular array is prepared by writing the number 1 along one vertical leg and along the hypotenuse of an isosceles right triangle. Every other number in the array is the sum of the number directly above it and the number above it and directly to its left. Thus, by using the accompanying array, the coefficients of the expanded form of $(x + y)^5$ can be seen to be $x^5 + 5x^4y + 10x^3y^2 + 10x^2y^3 + 5xy^4 + y^5$.

n	coefficients							
0	1							
1	1	1						
2	1	2	1					
3	1	3	3	1				
4	1	4	6	4	1			
5	1	5	10	10	5	1		
6	1	6	15	20	15	6	1	
7	1	7	21	35	35	21	7	1

Perfect number.—An integer equal to the sum of all its factors, including 1 but excluding itself. For example, 496 is a perfect number because 496 = 1 + 2 + 4 + 8 + 16 + 31 + 62 + 124 + 248. The first five perfect numbers are 6, 28, 496, 8,128, and 33,550,336.

Perfect power.—An integer that is exactly the *n*th power of another integer. For example, 4, 9, 16, and 25 are perfect squares (2^2, 3^2, 4^2, and 5^2, respectively); and 8, 27, 64, and 125 are perfect cubes (2^3, 3^3, 4^3, and 5^3, respectively).

Permutation.—The different arrangements or sequences that can be made of all or part of a given number of objects. All possible permutations of x, y, and z are:

x	x,y	y,x	x,y,z	z,x,y
y	y,z	z,y	x,z,y	z,y,x
z	x,z	z,x	y,x,z	y,z,x

A permutation of n objects taken r at a time, denoted by ${}_nP_r$, is equal to $n!/(n-r)!$ Thus, in the first column (above), ${}_3P_1 = 3!/2! = 3$; in the next two columns, ${}_3P_2 = 3!/1! = 6$. A permutation of n objects taken all at a time is equal to $n!$ Thus, in the last two columns, $3! = 6$.

Perpendicular.—Intersecting at right angles. If two lines in a plane intersect at right angles, each is called a perpendicular, or *normal,* to the other. A line is perpendicular to a curve if it is perpendicular to a tangent to the curve at the point of intersection. Two curves are perpendicular if their tangent lines at the point of intersection are at right angles. A line is perpendicular to a plane if it is perpendicular to every line in the plane that passes through the point of intersection.

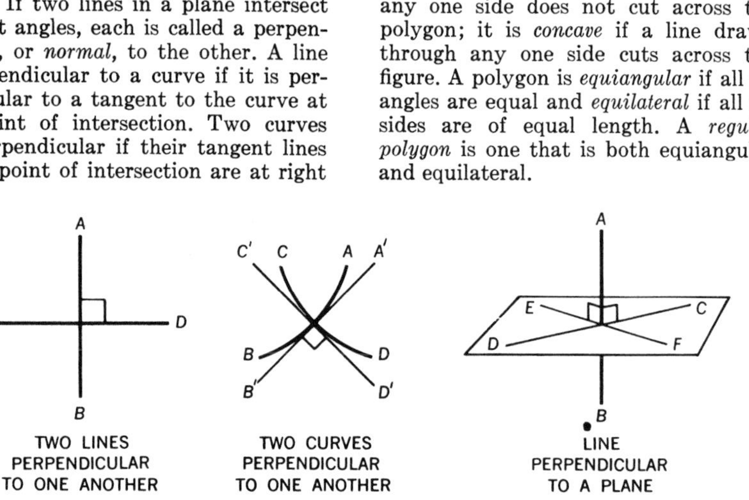

TWO LINES PERPENDICULAR TO ONE ANOTHER

TWO CURVES PERPENDICULAR TO ONE ANOTHER

LINE PERPENDICULAR TO A PLANE

Pi (π).—The ratio of the circumference of a circle to its diameter. The value of π, to ten places, is 3.1415926535 . . .; it is a nonrepeating decimal. A rational approximation of π that is adequate for most calculations is 22/7; a more accurate approximation is 355/113. π ranks with *e* in importance to mathematics.

Plane geometry.—See *Geometry.*

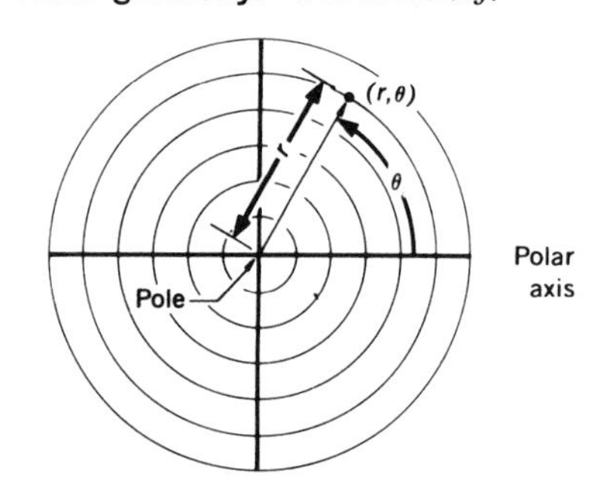

Polar coordinates.—A standard method of locating points. The points can be located in a plane by finding their distances from a fixed point and the angles that the lines from this point to the given points make with a fixed line, which is called the *polar axis.* The line from the fixed point, which is called the *pole,* to a given point is called a *radius vector* and is designated by the letter *r;* the angle of rotation from the polar axis to the radius vector is designated by the Greek letter *theta,* θ. The polar coordinates of a point are thus written (r, θ).

Polygon.—A plane figure in which n points, called *vertices,* are connected by n lines, called *sides,* which may or may not cut across one another. A polygon is named according to its number of sides: one of three sides is a *triangle;* of four sides, a *quadrilateral;* of five sides, a *pentagon;* of six sides, a *hexagon;* of seven sides, a *heptagon;* of eight sides, an *octagon;* of nine sides, a *nonagon;* of ten sides, a *decagon;* of twelve sides, a *dodecagon;* and of n sides, an *n-gon.* A polygon is *convex* if a line drawn through any one side does not cut across the polygon; it is *concave* if a line drawn through any one side cuts across the figure. A polygon is *equiangular* if all its angles are equal and *equilateral* if all its sides are of equal length. A *regular polygon* is one that is both equiangular and equilateral.

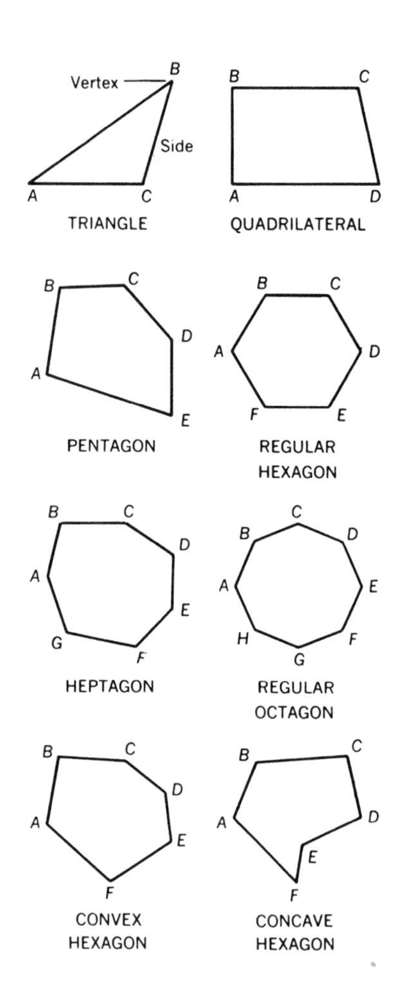

TRIANGLE

QUADRILATERAL

PENTAGON

REGULAR HEXAGON

HEPTAGON

REGULAR OCTAGON

CONVEX HEXAGON

CONCAVE HEXAGON

Polyhedron.—A three-dimensional figure bounded by plane surfaces, called *faces,* that are all polygons. A polyhedron is named according to its number of faces: one of four faces is a *tetrahedron;* of five faces, a *pentahedron;* of six faces, a *hexahedron;* of seven faces, a *heptahedron;* of eight faces, an *octahedron;* of nine faces, a *nonahedron;* of ten faces, a *decahedron;* of twelve faces, a *dodecahedron;* and of twenty faces, an *icosahedron.* A polyhedron is *convex* if a plane containing any one of its faces does not pass through the figure; it is *concave* if the plane cuts through the figure. A *regular polyhedron* is one whose faces are all congruent polygons and all of whose polyhedral angles are congruent. There are only five regular polyhedrons: a tetrahedron, a hexahedron (or cube), an octahedron, a dodecahedron, and an icosahedron.

Postulate.—See *Axiom.*

Power.—See *Exponent.*

Prime number.—A positive integer divisible only by itself and by unity. The first ten prime numbers are 2, 3, 5, 7, 11, 13, 17, 19, 23, and 29 (1 is usually excluded). There is no limit to the number of prime numbers, but no one has yet been able to devise a general formula by which the prime numbers may be determined.

Probability.—The likelihood, or chance, of any event taking place. For example, there are two possible results when a coin is tossed—heads and tails; hence, the likelihood of tossing a head (or of tossing a tail) is one in two, and the probability is 1/2.

Progression.—A sequence of terms that has a first number but no fixed last number, and in which each term is related to every other term by a fixed law. An *arithmetic progression* is a sequence in which each term, except the first, is equal to the preceding term plus a constant; for example, 1, 3, 5, 7, 9, ... A *geometric progression* is a sequence in which the ratio of each term, except the first, to the preceding one is the same throughout the sequence; for example, 1, 4, 16, 64, 256, ... A *harmonic progression* is one in which the reciprocals of the terms form an arithmetic progression, for example, 1, ⅓, ⅕, ⅐, ⅑, ...

Projective geometry.—The study of those properties of plane geometric figures that do not change upon being projected.

Pythagorean theorem.—The relationship of the lengths of the sides of a right triangle, first introduced by the Greek mathematician Pythagoras (died c. 497 B.C.). The theorem states that the sum of the squares of the lengths of the legs, *a* and *b*, of a right triangle equals the square of the length of the hypotenuse, *c*; $a^2 + b^2 = c^2$. Any three positive integers that satisfy this equation are called *Pythagorean numbers*.

Quadratic equation.—See *Equation*.

Quadrilateral.—See *Polygon*.

Radical.—In algebra, the indicated root to be extracted from a quantity, such as $\sqrt{x}$ or $\sqrt[3]{5}$. The *index*, or root to be extracted, is written over the radical sign; for example, $\sqrt[2]{x}$, $\sqrt[3]{x}$, $\sqrt[4]{x}$, $\sqrt[n]{x}$ denote that the square root, cube root, fourth root, and *n*th root of *x* are to be extracted. The quantity under the radical sign (in the above examples, *x*) is called the *radicand*.

Radix.—See *Number system*.

Ratio.—The relative magnitudes of two numbers or quantities. The ratio of *m* to *n* is written *m*/*n* or *m*:*n*. The equality of two ratios is called a *proportion;* for example, $m{:}n = o{:}p$ (read "*m* is to *n* as *o* is to *p*").

Rational number.—See *Number*.

Real number.—See *Number*.

Regular polygon.—See *Polygon*.

Regular polyhedron.—See *Polyhedron*.

Right triangle.—See *Triangle*.

Right angle.—See *Angle, Perpendicular*.

Scalene triangle.—See *Triangle*.

Sequence.—A set of numbers arranged in the same order as are the positive integers. Thus, 1, 2, 3, 4 . . . *n* and $1/x$, $1/2x^2$, $1/3x^3$, $1/4x^4$ . . . $1/nx^n$ are typical sequences. If a sequence continues after the last written term, it is an *infinite sequence;* if the sequence terminates at a finite number, it is a *finite sequence*.

Series.—An expression that indicates the sum of the terms of either a finite sequence (a *finite series*) or an infinite sequence (an *infinite series*). An *alternating series* is an infinite series whose terms are alternately positive and negative, for example, $1 - \frac{1}{2} + \frac{1}{3} - \frac{1}{4} + \ldots + (-1)^n - \frac{1}{n} + \ldots$. An *arithmetic series* is the expression indicating the sum of the terms in an arithmetic progression, for example, $1 + 3 + 5 + \ldots + n + \ldots$. An *exponential series* is an expansion of the exponential function e^x, where $e = 2.71828$; $e^x = x/1! + x^2/2! + x^3/3! + \ldots + x^n/n! + \ldots$. A *geometric series* is the expression indicating the sum of the terms in a geometric progression, for example, $1 + \frac{1}{2} + \frac{1}{4} + \frac{1}{8} + \ldots + 1/2^{n-1} + \ldots$. A *harmonic series* is the expression indicating the sum of a harmonic progression, for example, $1 + \frac{1}{2} + \frac{1}{3} + \frac{1}{4} + \ldots + 1/n + \ldots$.

Set (class).—A collection of particular things, such as a set of points or a set of numbers. The individual items within a set are the *elements* or *members;* each element is said to belong to the set.

Simultaneous equation.—See *Equation*.

Solid geometry.—See *Geometry*.

Statistics.—The branch of mathematics that deals with the methods of obtaining and analyzing quantitative data through the use of probability theory. Statistics finds application in both physical sciences, such as physics, and social sciences, such as economics.

Subtraction.—A fundamental operation of arithmetic. Basically, subtraction is the inverse operation to addition. The difference $m - n$ is the number *a* such that $m + a = n$. In $m - n = a$, *m* is called the *minuend*, *n* is called the *subtrahend*, and *a* is called the *remainder*.

Supplementary angle.—See *Angle*.

Theorem.—A general conclusion, proved through the use of axioms and/or lemmas, of a mathematical theory.

Topology.—The branch of modern geometry that deals with the changes undergone by three-dimensional figures as they are pulled, stretched, bent, twisted, or otherwise distorted without tearing or breaking them and without causing the loss of identity of the points contained within them. One of the fields of abstract, or pure, mathematics, topology has little practical application.

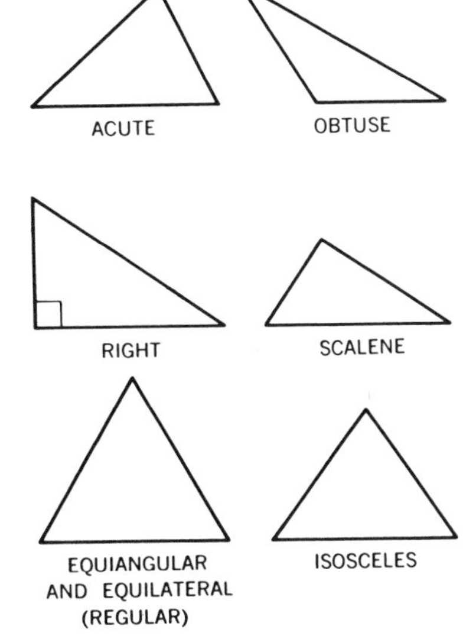

Triangle.—A three-sided polygon formed by joining three points, called *vertices*, that do not fall in a straight line, by three straight lines, called *sides*. A triangle is *acute* if all the interior angles are less than 90°, *obtuse* if one angle is greater than 90°, *right* if one angle is exactly 90°, *oblique* if no angle is 90°, and *equiangular* if all of the angles are equal (60°). A triangle is *scalene* if none of its sides are of equal length, *isosceles* if two of its sides are of equal length, and *equilateral* if all of its sides are of equal length.

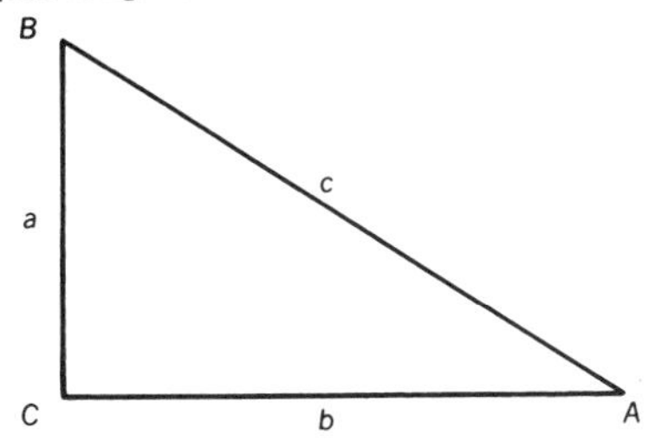

Trigonometric functions.—The six ratios of the sides of a right triangle. The functions are called *sine, cosine, tangent, cotangent, secant,* and *cosecant*, and are abbreviated as *sin, cos, tan, cot* (or *ctn*), *sec*, and *csc*, respectively. The ratios may be defined in relation to the accompanying diagram as follows:

$$\sin A = \frac{\text{opposite side}}{\text{hypotenuse}} = \frac{a}{c} = \frac{1}{\csc A}$$
$$\cos A = \frac{\text{adjacent side}}{\text{hypotenuse}} = \frac{b}{c} = \frac{1}{\sec A}$$
$$\tan A = \frac{\text{opposite side}}{\text{adjacent side}} = \frac{a}{b} = \frac{1}{\cot A}$$
$$\cot A = \frac{\text{adjacent side}}{\text{opposite side}} = \frac{b}{a} = \frac{1}{\tan A}$$
$$\sec A = \frac{\text{hypotenuse}}{\text{adjacent side}} = \frac{c}{b} = \frac{1}{\cos A}$$
$$\csc A = \frac{\text{hypotenuse}}{\text{opposite side}} = \frac{c}{a} = \frac{1}{\sin A}$$

The *inverse trigonometric functions*, or *antitrigonometric functions*, are denoted by *sin*$^{-1}$, *cos*$^{-1}$, *tan*$^{-1}$, *cot*$^{-1}$, and so on, or by *arc sin, arc cos, arc tan, arc cot,* and so on. The relationship between the trigonometric and inverse trigonometric functions may be shown as follows:

Function	*Inverse Function*
$\sin 30° = \frac{1}{2}$	$\sin^{-1} \frac{1}{2} = 30°$
$\cos 30° = \sqrt{3}/2$	$\text{arc cos } \sqrt{3}/2 = 30°$
$\tan 60° = \sqrt{3}$	$\tan^{-1} \sqrt{3} = 60°$
$\cot 45° = 1$	$\text{arc cot } 1 = 45°$
$\sec 60° = 2$	$\sec^{-1} 2 = 60°$
$\csc 30° = 2$	$\text{arc csc } 2 = 30°$

Variable.—An arbitrary symbol, such as *a, b, c . . . x, y,* or *z*, that can represent any one of a set of numbers, quantities, or other entities. The set is called the *range* of the variable, and any element within the set is called a *value* of the variable.

Volume.—The three-dimensional measure of the region enclosed within the boundaries of any solid geometric figure.

Zero.—The integer denoting the absence of any elements in the set under consideration. The introduction of the concept of "zero" by the Hindus made possible the extent of the science of mathematics that is studied today.

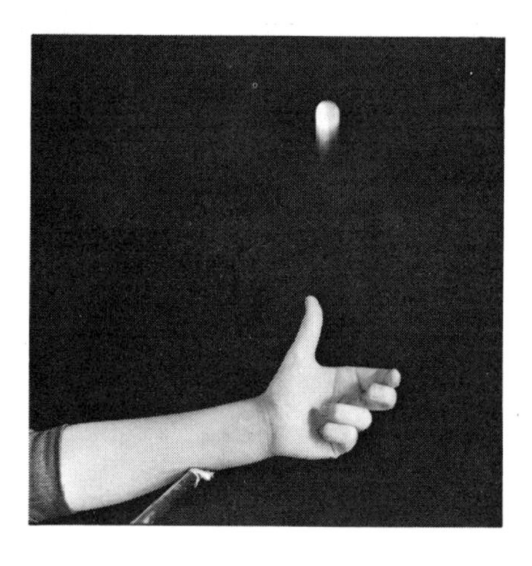

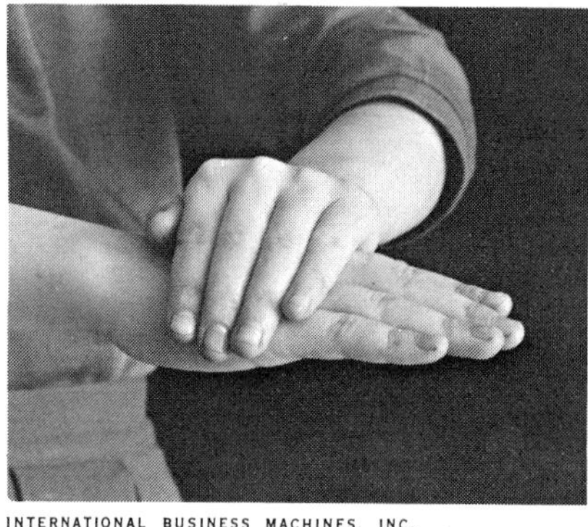

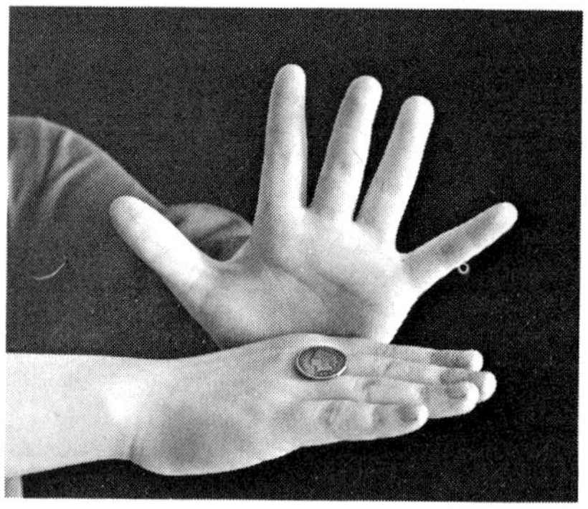

INTERNATIONAL BUSINESS MACHINES, INC.

THE PROBABILITY of an event's occurring can be calculated. For example, when a coin is tossed, it can fall as either heads or tails; the probability of tossing heads is 50–50.

BIBLIOGRAPHY

HISTORY AND GENERAL ASPECTS

Adler, I. *The Magic House of Numbers.* New American Library of World Literature, Inc., 1957.

Bakst, Aaron. *Mathematics: Its Magic and Mastery.* D. Van Nostrand Co., Inc., 1952.

Eves, H. *An Introduction to the History of Mathematics.* Holt, Rinehart & Winston, Inc., 1953.

Fehr, H. F. and Sobel, M. A. *Mathematics for Everybody.* Pocket Books, Inc., 1963.

Felix, L. *The Modern Aspect of Mathematics.* Translated by J. H. Hlavaty and F. H. Hlavaty. Basic Books, Inc., 1960.

Heimer, Ralph T. and Newman, Miriam. *The New Mathematics for Parents.* Holt, Rinehart and Winston, Inc., 1954.

Hogben, L. T. *Mathematics for the Millions* (3rd ed.). W. W. Norton & Co., Inc., 1951.

Johnson, Donovan A. and Glenn, William H. *Exploring Mathematics on Your Own.* Webster Publishing Co., 1960.

Kasner, Edward and Newman, James. *Mathematics and the Imagination.* Simon and Schuster, Inc., 1963.

Kramer, E. E. *The Mainstream of Mathematics.* Fawcett Publications, Inc., 1961.

Land, Frank. *The Language of Mathematics.* Doubleday & Co., Inc., 1963.

Mott-Smith, Geoffrey. *Mathematical Puzzles for Beginners and Enthusiasts.* Dover Publications, Inc., 1955.

Muir, J. *Of Men and Numbers.* Dell Books, 1961.

Newman, James (ed.). *The World of Mathematics.* 4 vols. Simon and Schuster, Inc., 1963.

Reid, Constance. *A Long Way from Euclid.* Thomas Y. Crowell Co., 1963.

Sawyer, W. W. *Prelude to Mathematics.* Penguin Books, Inc., 1955.

Smith, D. E. *History of Mathematics.* 2 vols. Dover Publications, Inc., 1923.

Turnbull, Herbert. *The Great Mathematicians.* New York University Press, 1961.

ELEMENTARY MATHEMATICS

Allendoerfer, C. B. and Oakley, C. O. *Principles of Mathematics.* McGraw-Hill Book Co., Inc., 1955.

Cooley, H. R. and others. *Introduction to Mathematics.* Houghton Mifflin Co., 1949.

Richardson, M. *Fundamentals of Mathematics.* The Macmillan Co., 1958.

Arithmetic

Banks, J. Houston. *Learning and Teaching Arithmetic.* Allyn and Bacon, Inc., 1959.

Bendick, Jeanne. *How Much and How Many?* McGraw-Hill, Inc., 1947.

Brumfiel, Charles F.; Eicholz, Robert E.; and Shanks, Merrill E. *Fundamental Concepts of Elementary Mathematics.* Addison-Wesley Publishing Co., Inc., 1962.

Lay, L. Clark. *Arithmetic: An Introduction to Mathematics.* The Macmillan Co., 1961.

Mueller, F. J. *Arithmetic: Its Structure and Concepts.* Prentice-Hall, Inc., 1955.

Algebra

Asimov, Isaac. *Realm of Algebra.* Houghton Mifflin Co., 1961.

Birkhoff, G. and MacLane, S. *A Survey of Modern Algebra.* The Macmillan Co., 1953.

Brumfiel, Charles F.; Eicholz, Robert E.; and Shanks, Merrill E. *Algebra Book I* and *Book II.* Addison-Wesley Publishing Co., 1961.

Johnson, R. E. *First Course in Abstract Algebra.* Prentice-Hall, Inc., 1953.

Kelley, John L. *Introduction to Modern Algebra.* D. Van Nostrand Co., Inc., 1962.

Maria, May Hickey. *The Structure of Arithmetic and Algebra.* John Wiley & Sons, Inc., 1958.

Richardson, C. H. and Miller, I. L. *Algebra: Commercial and Statistical.* D. Van Nostrand Co., Inc., 1962.

Rosenbach, Joseph B. and others. *College Algebra.* Ginn & Co., 1958.

Stein, Edwin I. *Algebra in Easy Steps.* D. Van Nostrand Co., Inc., 1962.

Geometry

Brumfiel, Charles F.; Eicholz, Robert E.; and Shanks, Merrill E. *Geometry.* Addison-Wesley Publishing Co., Inc., 1960.

Moise, Edwin and Downs, Floyd L. *Geometry.* Addison-Wesley Publishing Co., Inc., 1963.

Skolnick, D. *Dynamic Solid Geometry.* D. Van Nostrand Co., Inc., 1952.

Smith, Rolland R.; Ulrich, James F.; and Clark, John R. *Plane Geometry.* Harcourt, Brace & World, Inc., 1956.

Smith, Rolland R., and Ulrich, James F. *Solid Geometry.* Harcourt, Brace & World, Inc., 1957.

INTERMEDIATE MATHEMATICS

Trigonometry

Fisher, Robert C. and Ziebur, Allen D. *Integrated Algebra and Trigonometry.* Prentice-Hall, Inc., 1958.

Smith, Rolland R., and Hanson, Paul P. *Trigonometry.* Harcourt, Brace & World, Inc., 1957.

HIGHER MATHEMATICS

Analytic Geometry and Calculus

Fisher, Robert C. and Ziebur, Allen D. *Calculus and Analytic Geometry.* Prentice-Hall, Inc., 1961.

Sawyer, W. W. *What is Calculus About?* Random House, New Mathematical Library, 1961.

Smith, E. S. and others. *Analytic Geometry.* John Wiley & Sons, Inc., 1954.

Thomas, George B. *Elements of Calculus and Analytic Geometry.* Addison-Wesley Publishing Co., Inc., 1959.

Wells, Volney H. *Elementary Calculus.* D. Van Nostrand Co., Inc., 1941.

Miscellaneous

Boehm, George A. W. and the editors of *Fortune* (eds.). *The New World of Math.* The Dial Press, Inc., 1959.

Kelley, John L. *An Introduction to Modern Algebra.* D. Van Nostrand Co., Inc., 1962.

Langer, Suzanne. *An Introduction to Symbolic Logic.* Dover Publications, Inc., 1962.

Mosteller, Frederick; Rourke, Robert E. K.; Thomas, George B. *Probability and Statistics.* Addison-Wesley Publishing Co., Inc., 1961.

Weaver, Warren. *Lady Luck: The Theory of Probability.* Doubleday & Co., Inc., 1963.

PHYSICS

Aspects.—Physics may be said to be the story of matter and radiation. This division of physics seems most appropriate at the present time. If it were possible to describe and predict the behavior of matter and radiation, the interaction between them, the energy exchanges among the forms of matter and between matter and radiation, the story would be complete.

There is little point in attempting to define the basic entities that are a part of the physical world. *Matter* appears everywhere in various forms. It has certain properties that characterize it. Everyone instinctively knows what matter is. There is nothing more basic than "stuff," or matter, or whatever one may call it.

In a similar sense, *radiation* is "radiation." It has certain characteristic properties. Man is constantly submerged in radiation, visible or invisible. There is no place from which all electromagnetic radiation is excluded. Even if no other objects are around him, man's own body emits heat radiation. Radiation is not matter, although it has some of the properties of matter. The futility of definition or description in terms of something else is illustrated by the so-called duality of radiation. Radiation has been described both as an electromagnetic wave (in the ether) and as a particle. Neither description alone is adequate, and the combination seems to involve contradictory parts. Thus radiation cannot be thought of as just a wave (in a medium) or as a particle, such as a grain of sand. It is "radiation," and it has its own characteristics.

The situation with respect to the basic concepts, such as time, length, and mass, used in physics is somewhat similar. How can one define *time?* In any attempt at definition, one finds the definition becomes circular—it is possible to describe methods of measuring time precisely, but one cannot define time in terms of something more basic.

Length is defined operationally. It is the number of times a standard length can be placed alongside an object being measured. *Mass* is determined by comparison with a standard mass. Attempts are often made to define mass as the quantity of matter in a body, but how does one determine the quantity of matter? He measures the mass of the object. *Electrical charge* is in the same category as time, length, and mass. The electrical charge in any situation is determined by comparison with a standard charge. *Electric currents* are measured by the magnetic effect of moving charges. There is no other physical entity or "stuff" in terms of which an electrical charge can be described or defined.

The physicist is in somewhat the same situation as a lexicographer who, in compiling a dictionary, defines all words in terms of those previously defined. He has little trouble with the words beginning with "X," "Y," or "Z" at the end of the book, but what can he do with the first half-dozen words? These must be taken as known, for they cannot be defined by words previously defined.

Today, the great majority of physicists adhere to the point of view developed by Percy Williams Bridgman (1882–1961), who wrote, "The concept is synonymous with the corresponding set of operations." The length of a table is the number of times a measuring rod can be placed along the edge of the table. One starts with one end of the rod coinciding with the end of the table and then, by marks on the table, moves the rod along until he arrives at the other end of the table. Time is measured by comparing an interval with the periods of some standard clock—the clock is the earth and the period its time of rotation.

The concepts of time and space upon which much of physics is based have been greatly modified since the early physicists, such as Sir Isaac Newton (1642–1727), first tried to define them. One might say they have been simplified. They are no longer considered abstract and beyond the realm of experiment. The physicist now is in the position Newton assigned to the common people who, as he said, "conceive those quantities (time, space, place, or motion) under no other notions but from the relation they bear to sensible objects."

Today the physicist realizes that his concept and measurement of time are associated with motion. Time is not something that "flows along without relation to sensible objects." Without motion—in a static universe in which nothing moves—the concept of time would not exist. There would be no change in any relationship that would permit a measurement of a time interval. When one attempts to measure an "absolute velocity," or a velocity with respect to an "ether," and then with this absolute velocity set an absolute time scale, his experiments yield no velocity that could be called absolute. All velocities are relative, and indeed a concept such as the simultaneity of two events at different places is found to be dependent on the observer and his motion relative to other observers.

MAX PLANCK (1858–1947), German physicist, winner of the Nobel Prize in Physics.

This does not mean that the physicist no longer allows his imagination to operate or that he no longer makes postulates that have not been tested. Great advances are based on just such excursions of the mind—a notable example being Max Planck's (1858–1947) hypothesis regarding the quantum nature of radiation. But such projections must stand the severe test of experiment. Are they verified by experimental test? Do they predict phenomena that are later observed? If they do, then the physicist may say he has an "explanation," or a theory.

Scope of Physics.—Physics, once called "natural philosophy," is basic to several other sciences. *Astronomy* may be called the physics of the stars; *geology* (at least structural geology, as distinguished from historical geology), the physics of the earth. Indeed, the term "earth sciences" includes geology, *meteorology* (the physics of the air), and *oceanography* (the physics of the ocean). *Chemistry* might be called a branch of atomic physics. The composition of substances and the transformations they undergo are determined by the properties of the substance's atoms. There are well-recognized fields of science called *astrophysics, geophysics, physical chemistry,* and *biophysics.* The practice of medicine and medical research are becoming more and more dependent on physics. Isotopes, radioactivity, and electronic devices are familiar to the physician.

Engineering is now often called *applied physics.* The borderline between engineering and physics is often hard to distinguish. Engineering laboratories and courses in the engineering curricula in universities now resemble those in physics, especially those courses and laboratories that pertain to the properties and structure of matter in bulk.

The commonly recognized divisions of physics are mechanics, thermodynamics, electricity and magnetism, acoustics, optics, and a variety of topics often called "modern physics" to distinguish them from the classical divisions named above. The topics classed as *modern physics* include radioactivity and nuclear physics, atomic and molecular spectra, the quantum theory of matter and radiation, solid-state physics, relativity, high-energy physics, including the study of cosmic radiation, and the physics of fundamental particles.

The various categories are by no means separate and independent. Nature is not divided into nicely compartmentalized units. It is man's attempt to comprehend nature that leads to a division into subjects of areas of phenomena.

Mechanics, the oldest of the subjects dealt with in a quantitative manner, has three branches. The first is *kinematics,* the geometry of motion without regard to forces or energy. *Dynamics,* the second part of mechanics, includes the forces acting and the energy involved when bodies are in motion. The third branch, *statics,* deals with bodies in equilibrium. Mechanics is basic to much of engineering, certainly to structural engineering. Newton developed mechanics to the stage where he could (and did) determine what velocity and energy would be needed to put a satellite into orbit.

Thermodynamics involves heat and the behavior of matter with respect to thermal energy. Generalizations summarized by three laws of thermodynamics include the conservation of energy and the trend found in nature for the distribution and the availability of energy. Thermodynamics also encompasses the study of radiation.

The characteristics of electricity at rest and in motion are the starting topics for *electricity* and *magnetism.* These serve as the bases for the operation of electrical machinery of all kinds. They also lead to electromagnetic radiation and optics, which indicates the interdependence of the various divisions of physics. *Optics* also includes *geometrical optics,* in which the paths of light rays are traced through such devices as lenses.

It was an attempt to measure the difference in the velocity of electromagnetic radiation (light) in different directions as the earth moved about its orbit that led to the theory of relativity propounded by Albert Einstein (1879–1955), whose interpretation of the negative experimental result has had a profound effect on all of physical science and philosophical writings on science.

Acoustics, the phenomena related to sound, is one branch of a topic that might have been called elastic waves in solids, liquids, and gases.

Radioactivity was the first nuclear phenomenon studied by the physicist. In this he discovered that not all atomic nuclei are stable. Today nuclear physics engages the attention of thousands of scientists and large segments of industry. Atomic energy, more appropriately called "nuclear energy," is in the category of *nuclear physics.*

The characteristic radiation, visible and invisible, emitted by atoms and molecules under a variety of conditions is the major source of data on which our notions of atomic and molecular structure are based. The quantum theory of matter and radiation was initiated by Planck's hypothesis regarding the manner in which radiation is emitted from a solid. This was followed by the work of Neils Bohr (1885–1962) and Einstein. Today, quantum phenomena pervade all physics.

Engineers and physicists have recently given much attention to the electrical and mechanical properties of solids. Although solids have always been a part of man's environment, he has not, until recently, been able to relate the behavior and characteristics of solids to the properties of the atoms of which solids are composed.

For half a century the physicist has been aware of a penetrating radiation coming to Earth from outer space. Although called "radiation," the primary component that strikes the upper atmosphere is now known to be made up largely, if not completely, of atomic nuclei. Protons, the nuclei of hydrogen atoms, predominate. These come in with extremely high energies. Their origin and the source of their energy are not known. They are messengers that, with the radio signals detected by the recently constructed large radio telescopes, may yield previously unobtainable information about the stars and galaxies. The high energy of the primary components of cosmic rays has been approached but not equaled by the very large accelerators built in the United States, western Europe, and the Soviet Union.

Role of Physics.—Before an astronaut circles the earth, the energy required for launching, the direction of launching, and the direction of the final boost to his satellite must be accurately determined. Principles of physics supply the answers. The communication the astronaut has with Earth utilizes electronic apparatus and the knowledge of the behavior of electromagnetic radiation developed in the physics laboratories. Man's ability to communicate almost instantly with all parts of the world is the result of discoveries made by scientists in the past. Much of modern engineering is based on the products of the physics laboratory. The X rays, radium, heat therapy, radioactive isotopes, and electrocardigraphs used in hospitals had their origins, or were discovered, in the physics laboratories. Many of the comforts and conveniences in our homes—the temperature controls, the air conditioning, such kitchen devices as the refrigerator, the high-fidelity reproducer, and the television set—were not the objects of research but are by-products of such research.

The theory of relativity, with its requirement that mass and energy must be equivalent, and that the relation between them is $E = mc^2$, where E is energy, m represents mass, and c is the velocity of light, may be called the achievement in physics that will characterize this century. The theoretical work has been confirmed by experimentation (fission and fusion of atomic nuclei) in which a small amount of the mass of the nuclei changes to kinetic energy of the nuclear fragments and photons. The peaceful uses of nuclear energy will, no doubt, free man from fear that the fossil fuels will some day be exhausted.

PROPERTIES OF MATTER

Solids.—A body is said to be a *solid* when it retains its shape and offers resistance to forces tending to deform it. A solid has perceptible strength and does not flow. When described in this way, a solid is characterized by its *elastic properties.* There is no difficulty in differentiating a solid such as ice and a liquid such as water. But it is not always so easy to distinguish between a solid and a liquid. Suppose a piece of asphalt used in paving is broken from a large block. Its behavior while being broken seems to be that of a solid. But if the piece with rough edges and surface is placed in a container and kept at room temperature for a number of days, it will change its shape and appear to be flowing over the bottom of the container. It has not melted as ice melts, but it has not retained its shape. Is it a liquid or a solid? It behaves like very slowly moving molasses and has very high viscosity. Long pieces of glass tubing held in storage racks, with a support near each end of the tubing, will sag after some years. This behavior has led to the designation of glass as a supercooled liquid. Materials such as asphalt and glass, which do not exhibit all the properties of a true solid, are said to be *amorphous* because their molecules are randomly arranged. The molecules of true solids, on the other hand, have a definite pattern, called *crystal structure.* For this reason, solids are described as *crystalline* in structure.

■**CRYSTAL STRUCTURE.**—The molecules in a crystalline solid, such as ice or snow, are not arranged randomly. A definite geometrical spatial arrangement of the molecules is found in all crystals, although different arrangements are found in different crystals. A three-dimensional pattern of some type repeats itself regularly in all crystals. The definite pattern in snow is made evident in the large snow crystals that often can be seen during the first snowfall of the year. The beautiful "Jack Frost" figures formed on a cold window pane, and the crystals of salt formed when a saltwater solution is allowed to evaporate, show the crystalline nature of ice and salt. Metals in the solid state are usually composed of a myriad of tiny crystals, more or less randomly oriented and interlocking with one another. It is possible to grow large single metal crystals, each having dimensions of several centimeters. Such crystals have elastic properties differing markedly from the polycrystalline forms. A single crystal of copper, a centimeter in thickness and several centimeters long, can be bent easily. In copper, the atoms in the crystal are arranged in one of the ways (face-centered cubic) uniform spheres would be packed if the group was to occupy the smallest volume possible.

In common salt (sodium chloride) the atoms are not identical, as in copper; the sodium and chloride atoms are alternately placed in a cubic pattern, or *lattice.* The crystals may be very small or very large, but in every case the faces or surfaces of the crystals make definite angles with each other (90° in the case of sodium chloride).

Sometimes two or more crystalline forms of a substance are found. For example, although both the diamond and graphite are pure carbon, they have different crystal forms. In the diamond, each carbon atom is surrounded by four other carbon atoms at the corners of a regular tetrahedron. In graphite, the carbon atoms lie in planes, or flat sheets, and each atom is attached to three others in this plane to form a series of flat hexagons. The flat sheets of atoms are relatively easily separated or moved over one another. As a result of their respective structures, graphite acts as a lubricant, while diamond is the hardest naturally occurring substance known to man.

■AMORPHOUS SOLIDS.—Carbon is also found in the amorphous (noncrystalline) state as carbon black. Here, if crystals exist, they are microscopic, and the bulk substance does not behave as a crystal. There are numerous solids that are not crystalline in structure. Wood, for example, exhibits a growth pattern, but is not crystalline in the sense that metals are crystalline.

■PROPERTIES.—All solids, both crystalline and amorphous, resist change of shape and of volume. Many crystalline solids are *anisotropic;* that is, many of their properties are not the same in all directions. For example, the heat conductivity and optical properties (such as the speed of light as it passes through) vary with the direction in which these properties are measured. Amorphous solids and solids with a cubic crystal structure, on the other hand, are *isotropic;* that is, their properties are the same in all directions. Crystalline solids melt at definite, clearly defined temperatures. Amorphous solids do not melt at specific temperatures, but instead gradually soften and become more fluid with increasing temperatures.

Liquids.—A *liquid* is generally characterized as being in that state of matter in which the molecules are free to move among themselves without being able to separate from one another (as molecules of a gas do). A liquid will take the shape of any container into which it is placed. (Although a gas will also take the shape of any container into which it is placed, it differs from a liquid in that a gas will expand to fill all the available space in the container.) The volume of a liquid is relatively fixed; that is, it is changed only slightly by variations in temperature or pressure or both. The atoms in a liquid are about as closely spaced as are those of a solid, but there is no long-range order as in a crystal.

■SURFACE TENSION.—This is the property of liquids that causes the surface to behave as if covered by a thin, elastic membrane under tension. This results from the fact that the molecules in the surface layer of the liquid are attracted downward by the molecules within the liquid with a greater force than they are attracted upward by molecules in the atmosphere. The surface of the liquid always tends to contract to the smallest possible area; thus, drops of liquid assume a shape as close to spherical as possible because the surface area is then minimum for a given volume.

■CAPILLARY ACTION.—This causes liquids to rise in tubes of very small diameter, and is a phenomenon related to surface tension. Because of the difference in the magnitude of the forces between the molecules of the liquid themselves, and between these and the molecules of the solid composing the tube, the surface of the liquid is always curved. If the attraction of the liquid's molecules is greater than that of the solid's molecules, the surface curves downward. If the attraction of the liquid's molecules is less than that of the solid's molecules,

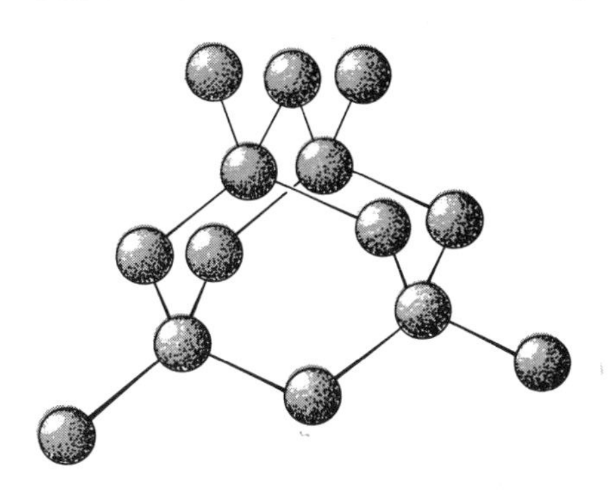

CRYSTAL STRUCTURE of carbon can be either tetrahedral, as in the diamond (*left*), or in plane hexagons arranged in flat sheets bonded to one another, as in graphite (*right*).

the surface curves upward. In the latter case, if the radius of the curvature is sufficiently small, a pressure differential will cause the liquid to rise in the tube.

■VISCOSITY.—The measure of the resistance of a liquid to flow is called *viscosity.* Like surface tension, viscosity is caused by the attraction forces between layers of molecules. The greater the attractive forces, the more difficult it is for one layer to flow over another, and the greater the viscosity. The viscosity of a liquid usually decreases significantly with increasing temperatures.

Gases.—A *gas* is characterized as being that state of matter that has no bounding surface. A gas will take the shape of, and expand to fill, any container in which it is placed. Gases are also characterized by their sensitivity to changes of temperature and pressure. The behavior of most gases on a macroscopic (visible to the naked eye) scale is best described by some relatively simple laws. These are Boyle's law, proposed by Robert Boyle (1627–1691); Charles's law (sometimes called Gay-Lussac's law), proposed by Jacques Alexandre César Charles (1746–1823); Dalton's law of partial pressures, proposed by John Dalton (1766–1844); Joule's law, proposed by James Prescott Joule (1818–1889), and Gay-Lussac's law of combining volumes, proposed by Joseph Gay-Lussac (1778–1850).

Boyle's law.—If a given mass of gas is compressed into a smaller volume and its *pressure* (force on each unit area of the container) increased, or if it is expanded into a larger volume, the product of the pressure and the volume remains constant if the temperature is kept constant.

Charles's law.—If a fixed mass of gas is heated through a given temperature range while the volume is kept constant, the fractional change in pressure will be the same for all gases. Also, if the pressure is kept constant while the temperature is changed over the same range, the fractional change in volume will be the same for all gases. In addition, the fractional change of pressure is equal to the fractional change of volume in the two cases described.

Dalton's law.—If two or more gases are placed in a container, the total pressure on the walls of the container is equal to the sum of the pressures each gas would exert if it were alone in the container.

Joule's law.—The energy associated with a given mass of gas is not a function of its volume or pressure if its temperature is kept constant; that is, the internal energy of a given gas depends only on its temperature.

Gay-Lussac's law.—The volumes of gases required for complete chemical combination to form new products bear a simple relation to each other and to the volume of the product if it is a gas.

The above "laws" are based on experiment. From these an Italian chemist, Amedeo Avogadro (1776–1856), proposed a hypothesis that was compatible with the experimental laws but was not subject to experimental verification by measurement of such macroscopic quantities as pressure or volume. The hypothesis, however, is now given the status of a "law" because it can be indirectly verified in a number of ways.

Avogadro's law.—All gases at a given pressure and temperature contain the same number of molecules per unit volume.

These laws are found valid for all gases if the temperature is not too low and the pressure is not too high. This suggests that under these conditions all gases have some common characteristics. A gas that is described accurately by these laws is called an *ideal gas.* One relation that combines several laws is: Pressure × Volume = Constant × Temperature, or $PV = RT$. The constant R is the same for all gases (at the appropriate temperature and pressure), providing one has a mass of gas numerically equal to its molecular weight. This mass is called a *gram molecular weight*, or *mole*. The relation $PV = RT$ is the *ideal gas law*.

Kinetic Theory of Gases.—One of the earliest important achievements of theoretical physics was the *kinetic theory of gases*. By means of this theory, the laws describing the macroscopic behavior of a gas can be explained. The kinetic theory is based on four assumptions:

(1) The gas is composed of a very large number of molecules.

(2) The size of each molecule is extremely small, and therefore the volume actually occupied by all the molecules is a very small fraction of the space ordinarily considered to be occupied by the gas.

(3) The molecules behave as hard elastic spheres. There are no forces of attraction or repulsion except at collision; the collisions are elastic and of negligible duration. No energy of motion is "lost" through collisions between the molecules.

(4) The molecules are in random motion, and Newton's laws describe their motion.

On the basis of these assumptions, the macroscopic behavior of a gas can be explained. Assume the existence of a closed cubical box, with edges L centimeters long. Its volume is L^3 cubic centimeters. Within the box there are N identical molecules, where N is a very large number. Let M be the mass of each molecule and let u centimeters per second be its speed. For simplicity in calculation, assume that all the molecules have the same speed and that ⅓ of them are moving vertically (along the z axis), another ⅓ are moving horizontally (along the x axis), and the remaining ⅓ are moving horizontally but perpendicular to the second group (that is, along the y axis).

On striking a wall (say the vertical wall), a molecule of mass M will bounce back with the same speed it had before striking the wall. Its momentum before collision is Mu and after collision is $-Mu$. The change in momentum $Mu - (-Mu)$ is $2Mu$. In the actual gas, this molecule may strike another before it crosses the box to the opposite side and, in so doing, communicate momentum to the other molecule. The net result, however, will be the same as if the molecule had traveled across the box without collision with another molecule and had struck the opposite side. It will then rebound and again strike the first side. The time for crossing the box twice is $2L/u$, and the number of collisions per second for one molecule with the side of the box will be $u/2L$. Each collision results in a change of momentum of the molecule of $2Mu$. Since $N/3$ of the molecules are moving across the box in this manner, the total change of momentum per second for the molecules striking one side of the box will be $(N/3)\ (2Mu)\ (u/2L)$, or $NMu^2/3L$. This change in momentum per second of the molecules is, from Newton's laws, the force exerted on the side of the box. The pressure on the side of the box will be the force divided by the area (L^2) of the side.

Or the pressure, P, is given by

$$P = \frac{NMu^2}{3L^3}.$$

But L^3 is the volume of the box (V). Hence,

$$PV = \frac{NMu^2}{3}.$$

The kinetic energy of transitory motion of each molecule is $Mu^2/2$. Thus:

$$PV = \left(\frac{2N}{3}\right)\left(\frac{Mu^2}{2}\right)$$

This equation resembles the general gas law if the kinetic energy of transitory motion of each molecule ($\frac{1}{2}Mu^2$) is related to the temperature. If the temperature is assumed to be proportional to the kinetic energy of the molecules, and N is equal to the numbers of molecules in a mole, the equation can be written

$$PV = RT,$$

where:

$$RT = \left(\frac{2N}{3}\right)\left(\frac{Mu^2}{2}\right)$$

Sometimes the general gas constant R for N molecules is replaced by Nk where $R/N = k$. The constant k is called the gas constant for one molecule, or *Boltzmann's constant*.

The equation $PV = RT$ may be written:

$$PV = NkT = \left(\frac{2N}{3}\right)\left(\frac{Mu^2}{2}\right)$$

The kinetic energy per molecule is:

$$\frac{Mu^2}{2} = \frac{3kT}{2}$$

Thus, if the temperature T is related to the kinetic energy of the molecules, as in the last equation, the general gas law (for ideal gases) has been deduced from the assumptions regarding the nature of a gas. Boyle's law and Charles's law also are included, and Joule's law is predicted by the fact that the energy of the molecules is dependent only on the temperature (as shown by the last equation) and not on the volume occupied or on the pressure.

Since $PV = NkT$, one may write $P = NkT/V$. In this, k is a constant. Hence, if the pressure P and the temperature T are the same for two gases, then N/V, or the number of molecules per unit volume, must also be the same. This is Avogadro's law.

A gas's specific heat can also be predicted if it is assumed that the kinetic energy of motion is equally distributed among the *degrees of freedom* (the number of squared terms in the expression for its energy) of the molecules. For a *monatomic gas*, such as helium, the number of degrees of freedom is three; for a *diatomic gas*, such as oxygen, which may also tumble as a dumbbell, the number of degrees of freedom is five.

Triple Point.—Depending upon the conditions of temperature and pressure, many substances can exist in any of the three states of matter—solid, liquid, or gas. For any particular substance, the physical states in which it can exist are called *phases*. Thus, by varying the conditions, a substance in the solid phase can be converted into the liquid phase (melting); the liquid phase can be converted into the solid phase (freezing); the liquid phase can be converted into the vapor phase (evaporation); the vapor phase can be converted into the liquid phase (condensation); the solid phase can be converted into the vapor phase (sublimation); and the vapor phase can be converted into the solid phase (condensation).

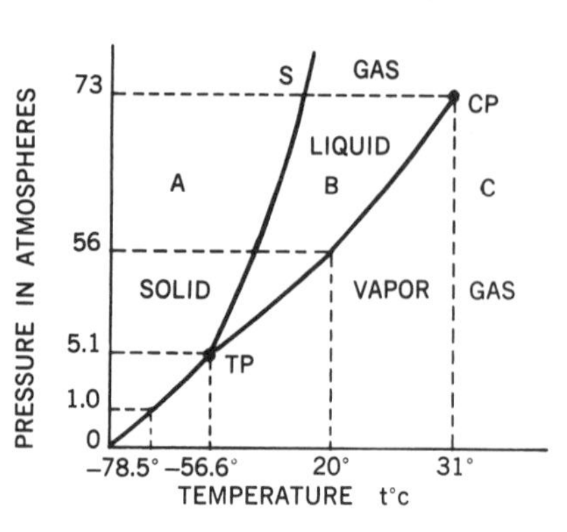

TEMPERATURE-PRESSURE (phase) diagram for carbon dioxide. (Scales not uniform.)

The transition from one phase to another can be illustrated with carbon dioxide. Carbon dioxide (CO_2) is a gaseous constituent of the atmosphere. But it can be condensed, and at normal atmospheric pressure and a temperature of about −80° C. it can be frozen into the solid commonly known as "dry ice." At a pressure of 5.1 atmospheres (75 pounds per square inch) and a temperature of −56.6° C., the solid, liquid, and gaseous states may all exist simultaneously. This is known as the *triple point*. The illustration shows the temperature-pressure relations between the various phases of carbon dioxide.

At the temperature and pressure represented by the point A in the diagram and for any point above the lines O-TP-S, CO_2 is a solid. In the region between the lines TP-S and TP-CP (point B, for example) the substance is a liquid. At the temperature and pressure represented by points to the right of O-TP-CP (point C, for example) CO_2 is a gas for vapor. From this diagram it can be seen that CO_2 can exist as a liquid only when the pressure is above 5.1 atmospheres. This is the pressure at the triple point.

At a pressure of 1 atmosphere CO_2 (dry ice) will be in equilibrium with its vapor at a temperature of −78.5° C.

Along the curve TP-CP (triple point-critical point) the vapor and liquid may exist in contact and in equilibrium. Thus at 20° C (room temperature) carbon dioxide stored in steel cylinders commercially will be in the liquid form with vapor above the liquid surface at a pressure of 56 atmospheres. The curve TP-CP is sometimes called the "boiling point" curve. It shows the temperatures at which the liquid will boil at various pressures. Points along

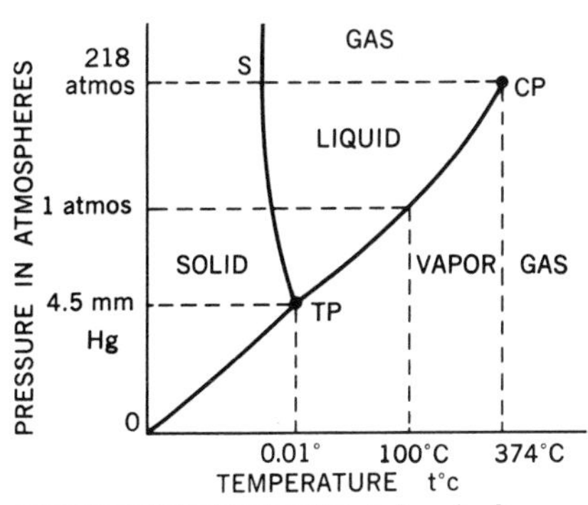

TEMPERATURE-PRESSURE (phase) diagram for water. (Scales not uniform.)

the line TP-S represent the freezing points of carbon dioxide when in contact with the liquid phase. As the pressure increases the freezing temperature rises.

Point CP is called the *critical point* at this point where the line from *TP* ends, the density of the liquid and the vapor are the same. No distinction can be made between them when the temperature is above 31° C. For all temperatures above 31° C. (say, for example, point *A*) the substance is a gas. No increase of pressure will cause it to become a liquid.

Diagrams such as this can be drawn for all substances. That for water is also shown. For water, the triple point temperature is just slightly above 0° C. The critical point is 374° C., and the corresponding pressure is about 218 atmospheres.

It should be noted that the line *TP-S* slopes backward; that is, the freezing point of water is lowered as the pressure is increased. It should also be noted that the line *TP-S* has no definite terminal point; as does *TP-CP*. As the pressure on water goes to very high values, water's behavior becomes more complex.

The terms "gas" and "vapor" refer to the substance in the region to the right of the line *TP-CP*. The term "vapor" is reserved for that state in which high pressure may condense a substance and "gas" is reserved for the condition of the substance when it is above its critical temperature and no increase of pressure is sufficient to condense the substance into the liquid state.

The portion of the curve below the triple point represents equilibrium between the solid and its vapor. All solids will evaporate: some, such as iron, extremely slowly at ordinary temperatures, but ice, even at temperatures well below 0° C., will readily evaporate.

The pressure-temperature diagrams are sometimes called *phase diagrams*.

—J. W. Buchta; Clifford E. Swartz

MECHANICS

Measurement.—*Mechanics* is the division of physics that deals with force, motion, inertia, and energy. Of most interest are the laws that state what effect a *force,* which may be thought of as a push or a pull, will have upon the form and motion of an object. Mowing the lawn, for example, requires a force to push the mower. The engine of a car, truck, or locomotive exerts a pulling force to move the load. The laws of mechanics enable the prediction of the path of a space probe and the trajectory of an electron in a television tube. A study of mechanics is aided by an understanding of the division of mathematics called measurement.

In mechanics only three fundamental measurements are needed: length, mass or force, and time.

■LENGTH.—As it is commonly used in the United States, *length* specifies a distance in *inches, feet, yards,* or *miles.* Most other countries use the *metric system,* with the *meter (m.)* as the basic unit of length. From this base, 1/100 of a meter (0.01 m.) is called a *centimeter (cm.)*; 1/1000 of a meter (0.001 m.) is a *millimeter (mm.)*, and 1,000 meters is a *kilometer (km.)*. The relationship between an inch and a centimeter is such that 1 inch equals 2.54 centimeters. The metric system is commonly used for measurements in mechanics.

■MASS.—The amount of material in an object as shown by its inertia is called *mass. Inertia* is the measure of resistance to change of motion. The fundamental unit of mass in the metric system is the *kilogram (kg.)*, and the unit most used in mechanics is the *gram (gm.)*—1/1000 of the fundamental unit of mass. Other common values are the *milligram* (0.001 gm.), *centigram* (0.01 gm.), and *decigram* (0.1 gm.). In the United States the *pound (lb.)* is the basic unit of mass. The conversion to the metric system is as follows: 1 pound equals 453.6 grams, so that 1 kilogram equals 2.2046 pounds.

But a pound is also a weight. Sir Isaac Newton (1642–1727) found that besides possessing inertia (and thus having mass), all objects have the ability to attract all other objects. This is known as *universal gravitation.* Thus, everything on or near the surface of the earth is attracted to it. For example, a ball thrown into the air returns to earth. The force with which the earth pulls on a mass of 1 pound under standard conditions (at sea level) is called the weight of 1 pound, or 1 pound of force.

■TIME.—Regardless of country or the system of measuring length, or mass or force, *time* is measured in *hours, minutes,* and *seconds.*

As can be seen, there are several systems possible with the three fundamental measurements discussed. In the United States, the *foot, pound, second* (abbreviated as *f.p.s.*) *system* is generally used. This system is common to almost all engineering work except electrical engineering. There is a desire among some engineers and scientists in the United States to adopt the system generally used in scientific work throughout the world—the *c.g.s. system.* Here, the basic unit of length is the *centimeter;* of mass, the *gram;* and of time, the *second.* One problem in the c.g.s. system is the smallness of the values, although its decimal nature makes it simple to convert values to other units.

A third system, which overcomes the small values of the c.g.s. system, is the *m.k.s. system*—based on the *meter, kilogram,* and *second.*

Regardless of which system is used, a few simple equivalents permit easy conversion to any of the other systems. The accompanying table gives some of the equivalents.

1 foot = 0.3048 meter
1 foot = 30.4801 centimeters
1 inch = 2.5400 centimeters
1 ounce = 28.350 grams
1 pound = 453.6 grams
1 kilometer = 3,281.8 feet
1 meter = 39.37 inches
1 centimeter = 0.3937 inch
1 gram = 0.03527 ounce
1 kilogram = 2.2046 pounds

Vectors.—In mechanics, there are two systems of addition that must be used. When adding two numbers (with identical units), an answer is easy. For example, it is easy to see that 8 feet and 2 feet equal 10 feet. But some answers also require a direction. If a direction is added—8 feet north and 2 feet east—the problem is no longer simple arithmetic.

A quantity that is specified by a number and its unit—8 feet, 50 miles per hour—is called a *scalar quantity.* A quantity that also has direction—8 feet north, 50 miles per hour east—is known as a *vector quantity.*

Vector quantities can be handled mathematically, but it is often easier to figure them graphically. For example, an object initially at a point *O* (Fig. 1) is moved 3 feet east and 4 feet north. The final position of the object is then distance *B* from the starting point *O*, or is displaced distance *OB*. *Displacement* of body is the direction and distance from the origin, and is a vector quantity.

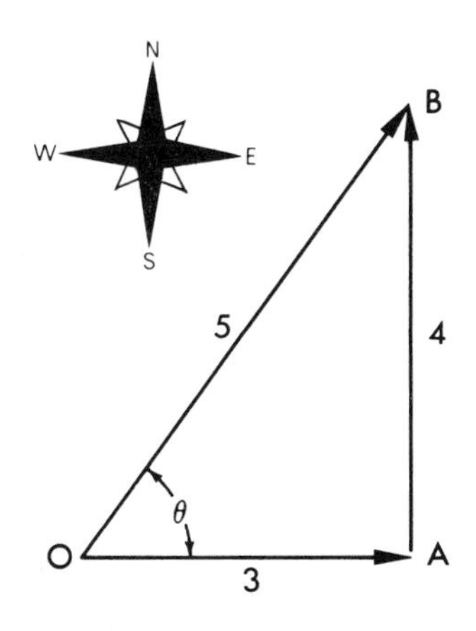

Figure 1.

This is indicated in Figure 1 by the arrow-tipped lines. The horizontal (east) line is 3 units long and shows the magnitude; the arrow head placement shows the direction and sense of the vector quantity. Similarly, the vertical (north) line locates the final point of the object. The resultant displacement from

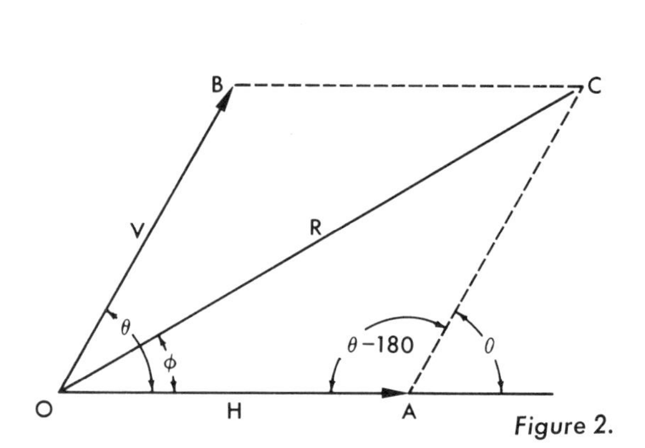

Figure 2.

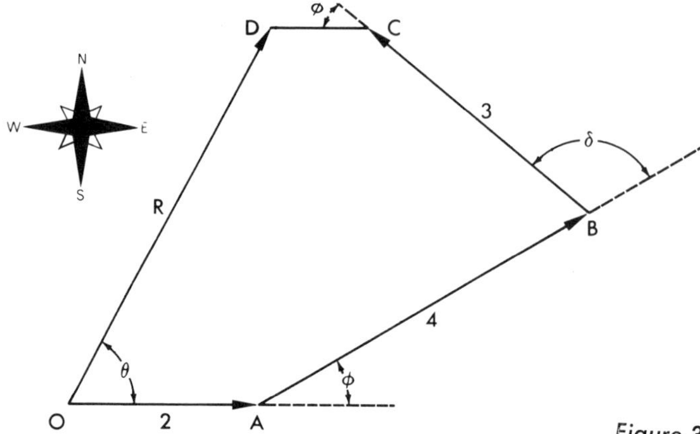

Figure 3.

point O is represented by vector OB, and the magnitude can be measured.

Mathematically, $3 + 4 = 5$ is not true. But vectors are added by a geometrical law that can be written as $\vec{OA} + \vec{AB} = \vec{OB}$, where the arrows show that the vector values are added in a special way.

In Figure 1, the vector OA is at right angles to vector AB. The resultant vector drawn from the origin to the head of the second vector closes the triangle OAB. Since it is a right triangle, and the Pythagorean theorem states that the sum of the squares of the two sides of a right triangle equals the square of the hypotenuse, then $(OA)^2 + (AB)^2 = (OB)^2$, which is $3^2 + 4^2 = (OB)^2$, or $9 + 16 = 25$. Hence, $OB = 5$ units.

Another way in which point B and the direction of vector OB can be described is by specifying the angle θ. The sine of the angle θ is the opposite side divided by the hypotenuse, or sine $\theta = AB/OB = 4/5 = 0.8 = 53°$ (approximately). Therefore, the resultant vector OB is 5 units long in a direction 53° northeast of point O.

Any time two or more vectors are given, the resultant or sum may be found. Three methods most often used are the parallelogram method, the polygon method, and the method of components.

PARALLELOGRAM METHOD.—In the *parallelogram method* any two vectors can be brought to a common point or origin, O, as shown in Figure 2. A parallelogram can then be constructed, with the opposite sides parallel to the original vector and equal in length. Side $BC = OA = OB$ in Figure 2. The resultant, R, is the diagonal drawn from O to the intersection of the parallel sides at C. With a ruler the resultant is measured in the same units as those chosen for vectors V and H. Angle ϕ is measured with a protractor. This graphic solution is only as accurate as the drawing and care used in measuring the lengths of the vectors.

POLYGON METHOD. — The parallelogram method is useful where only two vectors are involved. Where there are more than two vectors, the *polygon method* simplifies the number of lines that need be drawn. Here, the first vector is drawn from an origin, O, and each additional vector and its angle are drawn from the head of the succeeding one. Figure 3 illustrates the method.

For example, vector OA is drawn 2 miles east, and to its head is added vector AB, 4 miles northeast. Then BC is drawn 3 miles northwest, and finally CD, 1 mile west. The resultant OD is then drawn and its length measured in the units chosen for the other vectors. Angle θ is measured with a protractor. As can be seen, this is the parallelogram method but with each vector moved from the origin point parallel to itself until it coincides with the head of the previously drawn vector. Also, the vectors to be added follow each other head to tail. Only the resultant vector touches head to head with the last-drawn vector.

METHOD OF COMPONENTS.—Probably the most useful process in mechanics is the *method of components*. It is used when a number of vectors at various angles to the origin must be combined into a single resultant. The method for achieving this is the reverse of that shown in Figure 1. In Figure 1 two vectors at right angles to each other were given and from these the resultant was found. Conversely, if the resultant (5 units) and its direction (θ) were given, the resultant could be resolved into its x and y components, namely 3 units and 4 units. This resolution into x and y is the method of components.

As an example, in Figure 4A there are four vectors acting at point O. Each vector can be resolved into x and y components (Fig. 4B) by drawing a line at right angles (*perpendicular*) from the vector head to the proper axis. If all the components along the x axis (A_x, B_x, C_x, D_x) are now added, the total is the *summation* of all x components and is written Σ_x. By performing the same operation for the y components, Σ_y is obtained.

Some values will be minus, such as D_x, C_x, A_x, and C_y, since their components point in a direction that, by convention, is negative. The resultant of the four forces is found by the Pythagorean theorem as in Figure 1, and is shown in Figure 4C.

Equilibrium.—Equilibrium is the state achieved when the resultant of all the forces upon a body, which are always represented by vectors, is zero. When equilibrium is attained, there is no change in the motion of the body. The body may be at rest,

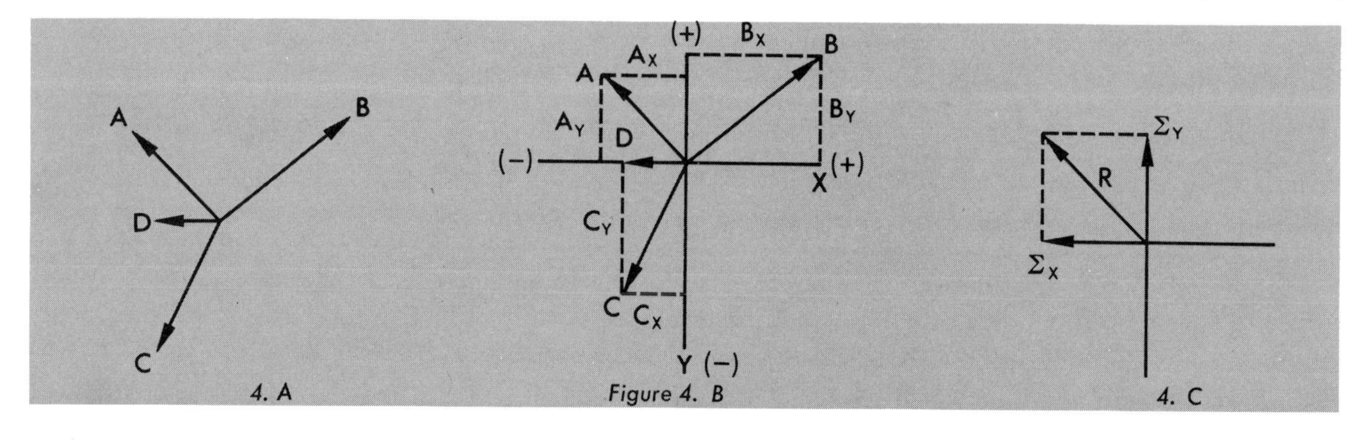

Figure 4. B

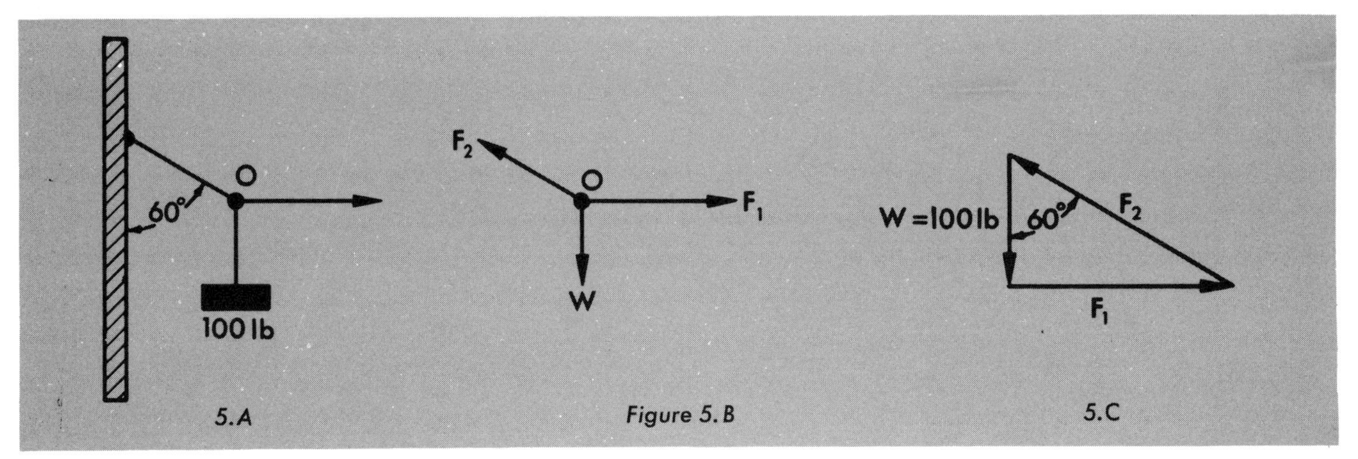

Figure 5.B

moving in a straight line at uniform speed, or rotating about a fixed axis at a uniform rate.

For example, assume a 100-pound object is suspended by a rope and held away from a wall by a second rope (Fig. 5A). If the first rope makes a 60° angle with the wall, what are the tensions in the rope? The system is in equilibrium if the vector sum of the forces acting upon it is zero. If a point O (Fig. 5B) is taken as the point in equilibrium, then the weight W and the forces F_1 and F_2 must equal zero.

The vector diagram (Fig. 5C) can be drawn to form a polygon that closes and the forces F_1 and F_2 can be measured to find the load, or the loads can be computed by trigonometric methods as follows:

$$\tan 60° = \frac{F_1}{100}$$
$$F_1 = 100 \tan 60° = 173.2 \text{ lb.}$$
$$\cos 60° = \frac{100}{F_2}$$
$$F_2 = \frac{100}{\cos 60°} = 200 \text{ lb.}$$

Therefore, to achieve equilibrium (to hold the system as drawn in Figure 5A), a pull of 173.2 pounds is needed on the horizontal rope, and tension in the hanging rope is 200 pounds.

This problem can be solved by the component method as well. If the horizontal component and vertical component of each force are found, the sums must equal zero if the system is in equilibrium. The x and y components of the 100-pound force are 0 and 100 pounds down, respectively. The components of F_1 are F_1 (right) and 0, respectively. The horizontal component of F_2 is $F_2 \cos 60°$ up. For equilibrium, these four forces must equal zero. Therefore

$$F_1 - F_2 \sin 60° = 0 \quad \text{(horizontal)}$$
$$F_2 \cos 60° - 100 \text{ lb.} = 0 \quad \text{(vertical)}$$

Solving the second equation, F_2 equals 200 pounds as previously found. By substituting the F_2 value in the first equation, F_1 equals 173.2 pounds.

Velocity and Acceleration.—The discussion of equilibrium began with the supposition that the resultant, or summation, of the forces acting upon a body was zero. Equilibrium was defined as the state in which there was no change in the motion of a body. However, in mechanics one must also consider bodies upon which the resultant force is *not* zero, that is, upon which the forces do not result in a state of equilibrium. Since when the resultant force on a body is zero the motion of the body does not change, it logically follows that when the resultant force on a body is not zero the motion of the body must change.

The motion of a body is described by its velocity. *Velocity* is the rate of change of position, or rate of displacement; it is usually expressed in feet or centimeters per second. Velocity is a vector quantity. For its complete specification, both its magnitude and its direction must be stated. Thus, when a body is not in equilibrium and its motion is changing, its velocity must also be changing. The rate of change of velocity is *acceleration*. Acceleration is usually expressed in feet or centimeters per second per second, or feet or centimeters per second squared.

The study of mechanics is divided into three areas. One area, *statics*, deals with forces in equilibrium. A second area, *dynamics*, deals with the production and causes of motion. The third area, *kinematics*, treats motion without regard to the forces that produce it. The difference between statics, dynamics, and kinematics can be illustrated by the following discussion.

A box resting on the floor is in equilibrium. The weight of the box, W, is reacted by the ground, R (statics). If a rope is attached to the box and pulled (with a force, F), the box will move along the ground (kinematics). However, the force F must first overcome the weight of the box and the frictional force f between the box and the ground (dynamics). Once the friction is overcome, the box moves along as the force is continually exerted (statics). If F is increased, the box moves faster; if it is decreased, the box slows down and will stop when the force no longer overcomes friction (dynamics).

Motion, then, is velocity and acceleration. *Velocity* may be defined as the time rate of motion in a given direction, *acceleration* as the time rate of change of velocity. When, for example, a car speeds up, it is accelerating; when it slows down, it is decelerating. Acceleration may be expressed algebraically as the final velocity (v_f) minus the original velocity (v_o) divided by the time interval (t), or

$$a = \frac{v_f - v_o}{t}.$$

Newton's Laws of Motion.—The relationship of force to motion was described in three laws formulated by Sir Isaac Newton. Without these fundamental laws the science of mechanics might well be impossible.

■**NEWTON'S FIRST LAW.**—A body at rest remains at rest, and a body in motion continues to move at constant speed along a straight line unless the body is acted upon, in either case, by an unbalanced force.

The first part of the law is simple enough. Forces act on a book placed on a table so that it remains at rest. Gravity pulls it down; the table pushes up. Hence, the forces are in equilibrium and the book remains stationary.

The second part of the law is more difficult, but consider the flights of the astronauts. An initial force accelerates the capsule to a constant speed in a vacuum where there is no opposing force. The capsule follows an orbit until the astronaut fires an unbalancing force (retro-rockets), impelling the capsule toward earth.

In both cases, whether the body is at rest or moving with constant speed along a line, its acceleration is zero. It follows that a body will not accelerate or decelerate unless an unbalanced force acts upon it.

■**NEWTON'S SECOND LAW.**—An unbalanced force acting on a body causes the body to accelerate in the direction of the force, and the acceleration is directly proportional to the unbalanced force and inversely proportional to the mass of the body.

Expressed algebraically, the law states that a varies as F/M, where a is the acceleration, F is the force, and M is the mass. This proportion may be expressed as

$$a = \frac{F}{M} \quad \text{or} \quad F = Ma.$$

To illustrate the law, suppose two identical cars are driven along a road and more force, in the form of push, is applied to the first than to the second. This so-called push is really acceleration. Hence, the force acting on the car causes acceleration proportional to the unbalanced force in the direction of the force.

Now imagine that six persons are added to one of two identical cars and that equal forces are applied to both vehicles. The empty car will have the greater acceleration. Generally, the greater the unbalanced force and the less the mass, the greater the acceleration.

NEWTON'S THIRD LAW.—For every action or force there is an equal and opposite reaction or force.

Of all of Newton's laws this is the easiest to understand. A book on a table presses down; the table pushes up. Action is the book pressing on the table; reaction is the table pushing against the book. Consider a missile being launched. The rocket engines fire hot gases downward. This is the action. The reaction is the upward motion of the missile away from the rocket-engine force.

Note that two bodies are involved in each case and that action and reaction, although equal and opposite, can never balance each other because, to balance each other, they must be exerted on the same body.

There are also forces at work on a body in circular or curvilinear motion. A body will move in a curve only when a lateral force is exerted upon it. The classic example is the stone whirled at the end of a string. The stone pulls outward on the string and, as the cord becomes taut, it pulls inward on the stone. In the same way an astronaut's capsule moves in an orbit around the earth because, as it is drawn inward by the pull of the earth's gravity, it maintains its position along a curved path due to its velocity.

Hence, the motion of a body traveling in a circular path with constant speed is of special interest. In such circular motion the moving object is pulled toward the center of the circle by a force called *centripetal force*. Since, by the first law of motion, an object in motion tends to travel along a straight-line path, the inertial tendency of the object opposes the inward pull. For many years, it was thought that the opposition to centripetal force caused by the body's inertial tendency was actually a reaction force, called *centrifugal force*, as stated in the third law. However, to repeat, action-reaction pairs are never exerted upon the same body.

GRAVITATION.—Each particle of matter attracts every other particle with a force directly proportional to the product of their masses and inversely proportional to the square of the distance between them.

The most familiar example of universal gravitation is an object's falling when released. The amount of the earth's attraction is different for different bodies, varying with their mass; this attraction is known as the *weight* of the body. Weight is proportional to mass; if the same force (gravitational attraction) produces the same acceleration on two bodies, the weights of the two bodies are equal. Since mass is the measure of quantity of matter, a quart of water, for example, must have twice the mass of a pint. By experiment, it can be shown that a quart of water weighs twice as much as a pint.

All bodies would fall with the same velocity if there were no resistance of air. This theory was proved by dropping a feather and a metal pellet in a vacuum; they both reached bottom at the same time.

The laws of falling bodies are based on the fact that their motion is uniformly accelerated. The uniform acceleration does not continue indefinitely, however. If the body falls a sufficient distance, the gravitational force causing acceleration will be equaled by the increasing resistance of the air. At this point, acceleration ceases and the body continues to fall at a uniform maximum velocity called *terminal velocity*.

A relationship exists between velocity, distance covered, acceleration, and time in uniformly accelerated motion. Based on the acceleration formula,

$$a = \frac{v_f - v_o}{t},$$

the formula for the final velocity is

$$v_f = v_o + at.$$

The distance traveled by a body having constant acceleration is found by averaging its velocities during the time interval t. Then distance traveled, s, may be found by the equation

$$s = v_{avg}t,$$
$$= \frac{v_o + v_f}{2}\,t.$$

Replacing v_f by $v_o + at$,

$$s = \frac{v_o + (v_o + at)}{2}\,t,$$
$$= v_ot + \tfrac{1}{2}at^2.$$

The third equation of uniformly accelerated motion is derived by eliminating time:

$$as = \left(\frac{v_f - v_o}{t}\right)\left(\frac{v_o + v_f}{2}\,t\right),$$
$$= \tfrac{1}{2}(v_f - v_o)(v_o + v_f),$$
$$v_f^2 = v_o^2 + 2as.$$

The substitution of *gravitational acceleration, g,* upon these uniformly

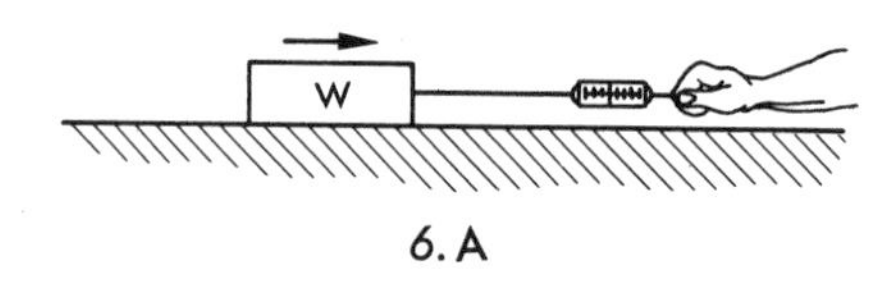

6. A

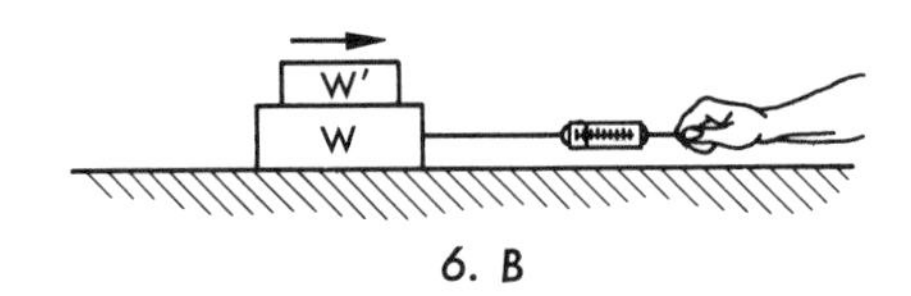

6. B

accelerated motion formulas yields

$$v_f = v_o + gt,$$
$$s = v_ot + \tfrac{1}{2}gt^2,$$
$$v_f^2 = v_o^2 + 2gs.$$

The only change is that acceleration due to gravity is substituted for acceleration. The value of g is 32.2 feet per second per second.

The *angular acceleration* of a body is defined as its time rate of change of angular velocity. This is similar to the definition of linear acceleration, and the equations are also similar:

$$\omega_f = \omega_o + 2\alpha t$$
$$\theta = \omega_o t + \tfrac{1}{2}\alpha t^2$$
$$\omega_f^2 = \omega_o^2 + 2\alpha\theta$$

where θ is the angular displacement in time t; α is the angular acceleration; ω_o is the initial velocity; and ω_f is the final velocity.

Friction.—In the preceding paragraphs, friction has been named several times without being defined. When two bodies in contact are in motion relative to one another, a force—called *frictional force*, or the force of friction—opposes this motion; that is, the frictional force acts in a direction opposite to the direction of motion. There are several theories regarding the cause of friction. One theory contends that the cause of friction is the roughness of the two surfaces in contact. No matter how smooth the surfaces may appear, when they are observed under magnification, irregularities—looking like so many bumps—can be seen. It is thought that these "bumps" interlock to cause opposition to motion. A second theory contends that the same atomic forces that hold the molecules together in each surface also tend to hold the molecules of the two surfaces together. The magnitude of the frictional force depends upon the materials in contact, the condition of the surfaces, and the forces that are pressing the surfaces together.

To illustrate frictional force, imagine a box at rest to which a rope is attached and is pulled upon with a 1-pound force. If the box does not move, the resultant force is zero and the ground must be exerting a frictional force of 1 pound opposite to the force on the rope. If the pull is increased, the box will eventually move; and at the moment it does move and continues to move at constant speed, friction has been overcome. Any increase in the pull will then cause acceleration of the box.

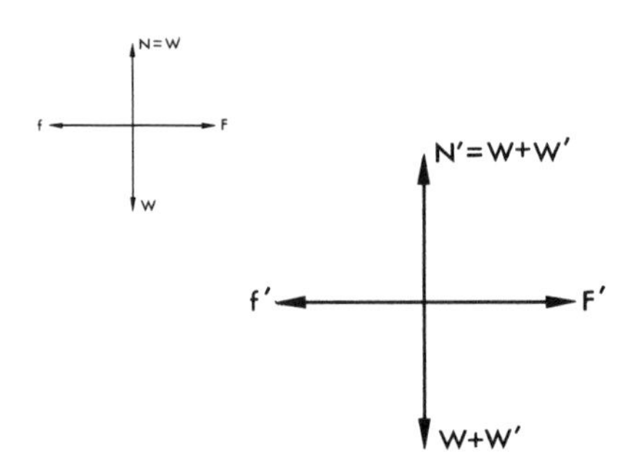

The Coefficient of Friction.—Figure 6A shows a box being moved to the right at a constant velocity by a force, F. The box pushes against the ground with a force equal to its weight, W, and the ground pushes back with an equal force called the normal, N.

Since the box is moving at a constant velocity, the system must be in equilibrium, and frictional force f must therefore equal F. If a second box is placed atop the first (Fig. 6B), to keep the system moving uniformly, the force to the right must be increased to a magnitude of F'. The boxes are pressing against the ground with a force equal to their combined weights, $W + W'$, and the ground is pushing back with the equal force N'. Since the system is still in equilibrium, the frictional force must have increased to f' to remain equal to F'. By experiment, it has been shown that the frictional force increases in the same proportions as the normal force:

$$\frac{f'}{f} = \frac{N'}{N}$$

or

$$\frac{f}{N} = \frac{f'}{N'} = \text{constant.}$$

The constant, which is designated as μ, is called the *coefficient of friction,* and has been determined for a large number of surface pairs. It can be used to determine the frictional forces in a system through the equation

$$f = \mu N.$$

As an example, assume a box weighs 100 pounds and the coefficient of friction is 0.1. How long would it take to move the box 100 feet if it were pulled with a constant force of 20 pounds? (Fig. 7.)

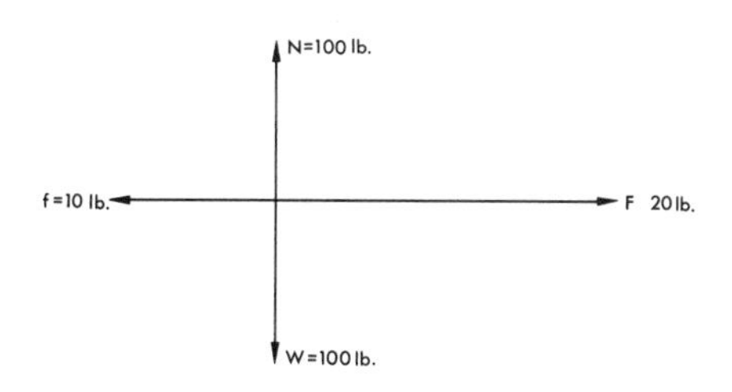

Figure 7.

The frictional force on the box is:

$$\begin{aligned} f &= \mu N \\ &= (0.1)(100) \\ &= 10 \text{ lb.} \end{aligned}$$

The acceleration of the box, where $M = W/g$, is as follows:

$$\begin{aligned} a &= \frac{F}{M} \\ &= \frac{(F)(g)}{W} \\ &= \frac{(20-10)(32.2)}{100} \\ &= \frac{(10)(32.2)}{100} \\ &= \frac{322}{100} \\ &= 3.22 \text{ ft./sec.}^2 \end{aligned}$$

Substituting acceleration in the distance equation,

$$s = v_o t + \tfrac{1}{2}at^2.$$

Since the initial velocity, v_o, is zero,

$$s = \tfrac{1}{2}at^2;$$

and since s = 100 ft. and a = 3.22 ft./sec.², then

$$\begin{aligned} t &= \sqrt{\frac{2s}{a}} \\ &= \sqrt{\frac{200}{3.22}} \\ &= \sqrt{62.1 \text{ sec.}^2} \\ &= 7.88 \text{ sec.} \end{aligned}$$

Work.—In physics, *work* is used to describe a situation where an applied force produces movement in the direction that the force is applied. A truck does not work in holding up a load of bricks, but it does work in moving the load up a hill. Work (W) can be expressed as the product of the applied force (F) times the distance (s) covered by movement in the direction of the force, or

$$W = Fs \cos \theta.$$

With a constant force applied in the direction of the displacement, cos θ equals 1, and the work produced becomes the product of the force times the displacement, or

$$W = Fs.$$

The *foot-pound* is the unit of work in the f.p.s. system. One foot-pound is the work produced by a 1-pound force applied for a distance of 1 foot. In the c.g.s. system, the unit of work is the *erg.* One foot-pound is equal to 13,560,000 ergs. How much work is done while pulling a 650-pound log 20 feet with a chain that makes an angle of 30° with the ground? The coefficient of friction, μ, between the log and ground can be considered to be 0.40.

Thus,

$$\begin{aligned} F &= \mu N \\ &= (0.40)(650) \\ &= 260 \text{ lb.} \end{aligned}$$

and

$$\begin{aligned} W &= Fs \cos \theta \\ &= (260)(20)(\cos 30°) \\ &= (5{,}200)(0.86) \\ &= 4{,}472 \text{ ft.-lb.} \end{aligned}$$

Energy.—The capacity for doing work, *energy,* can exist in many forms and can be converted from one form to another. *Potential energy* is the energy a body has by virtue of its position. *Kinetic energy* is the energy possessed by a body by virtue of its motion. Ignoring friction, the work done on a body equals the change in kinetic and potential energy. The energy change is expressed in the same units as work—foot-pounds in the f.p.s. system and ergs in the c.g.s system. It is converted or transformed by machines to a usable form. Although energy cannot be destroyed, it may be transformed into some unusable form and be dissipated. For example, electrical energy used to heat the coils in a toaster is irretrievably dissipated. Mechanical energy used to stop a moving car is converted into heat in the brakes and is dissipated. The energy used to operate any machine, therefore, can be said to follow a standard pattern: it is converted into a more usable or convenient form and finally dissipated as heat.

■**POTENTIAL ENERGY.**—The energy a body possesses by virtue of its location or configuration is referred to as *potential energy.* Water in a reservoir, or a weight lifted to an elevated position, has potential energy that can be converted by a machine into a more desirable or convenient form. *Gravitational potential energy* is the most common form of potential energy. Since the earth's gravity attracts every body, work is required to elevate a body to a higher level. The work expended on the body (weight of the body times the elevation) represents energy that is stored and can be recovered. This potential energy (P.E.) is the product of the weight (W) and the height (h) to which the body was raised, and can be stated as

$$\text{P.E.} = Wh.$$

If W is given in pounds and h is given in feet, the potential energy is expressed in foot-pounds. When a mass is elevated, its potential energy is increased; when a mass is lowered, its potential energy is decreased. In either case the potential energy is measured from an arbitrary zero point, such as sea level, the floor, or a table top. A stone on a table has no potential energy when the table top is selected as zero point. With the floor as zero point, the potential energy of the stone is equal to its weight times the height of the table.

■**KINETIC ENERGY.** — The energy a body possesses due to its motion is called *kinetic energy.* A bullet in flight, a spinning flywheel, and a speeding automobile all possess kinetic energy. The amount of work a moving body can do while being brought to rest or the work required to produce the velocity at which the body moves is a measure of its kinetic energy (K.E.).

A body at rest acted upon by an unbalanced force (F) through a distance (s) is accelerated (a) to a velocity (v). The work done to accelerate the body is equal to its kinetic energy, or

$$\text{K.E.} = Fs = mas.$$

From the earlier discussion of acceleration and velocity, it can be seen that the following substitutions can be made for a and s:

$$a = \frac{v_f - v_o}{t},$$

$$s = \frac{v_o + v_f}{2}t.$$

Thus,

$$\begin{aligned} \text{K.E.} &= mas \\ &= m\left(\frac{v_f - v_o}{t}\right)\left(\frac{v_o + v_f}{2}t\right). \end{aligned}$$

Starting from rest, when v_o is zero,

$$\text{K.E.} = \frac{mv^2}{2}.$$

If the mass (m) is expressed in *slugs* and the velocity (v) in feet per second, the kinetic energy is expressed in foot-pounds. Using the above formula, the kinetic energy of a 3,200-pound elephant charging at 15 miles per hour (22 feet per second) can be calculated. In slugs the

3,200-pound animal's mass is:

$$m = \frac{W}{g}$$

$$= \frac{3{,}200 \text{ lb.}}{32 \text{ ft./sec.}^2}$$

$$= 100 \text{ slugs.}$$

Therefore,

$$\text{K.E.} = \frac{mv^2}{2}$$

$$= \frac{(100 \text{ slugs})(22 \text{ ft./sec.}^2)}{2}$$

$$= 24{,}200 \text{ ft.-lb.}$$

Accelerating or decelerating a body requires the application of a force that produces a change in the kinetic energy. The change in kinetic energy (ΔK.E.) is equal to the total kinetic energy when the body is accelerated from rest or brought to a standstill.

To stop a moving object, its kinetic energy must be absorbed by work done in opposition to its motion. Since the kinetic energy is proportional to the square of an object's velocity, doubling the speed increases the kinetic energy four times. Therefore four times the amount of work is needed to bring the body to a stop.

Simple Machines. — Machines cannot create or destroy energy. Instead, they simply transform it or apply it in a more convenient or usable form. The energy received by the machine may be converted into useful work, but the work produced cannot exceed the energy received. Machines may be used to convert chemical energy (a gasoline engine), electric energy (an electric motor), or mechanical energy (a hydraulic jack) but only the conversion of mechanical energy will be considered in this discussion. In simple machines, energy is applied by a single force doing useful work against a single resisting force. Regardless of their complexity, all machines can be resolved into combinations of the wedge, or inclined plane, and the lever.

The *actual mechanical advantage* (*A.M.A.*) of any machine is the ratio of the force delivered (F_o) to the force applied (F_i), or

$$\text{A.M.A.} = \frac{\text{force out}}{\text{force in}}$$

$$= \frac{F_o}{F_i}.$$

To lift a 3,000-pound automobile with a 50-pound force requires an actual mechanical advantage of 3,000 lb./50 lb., or 60. Most machines are force multipliers and therefore have an A.M.A. of more than 1. Speed-increasing machines, such as a bicycle chain drive or a hand drill, have an A.M.A. of less than 1.

The *ideal mechanical advantage* (*I.M.A.*) is the ratio of the distance the applied force travels (s_i) to the distance the delivered force travels (S_o):

$$\text{I.M.A.} = \frac{\text{distance in}}{\text{distance out}}$$

$$= \frac{s_i}{s_o}.$$

Due to friction, the work produced is less than the work applied to any machine. Work in ($F_i s_i$) is greater than work out ($F_o s_o$):

$$F_o s_o < F_i s_i$$

When each part of this expression is divided by $F_i s_i$, the result is:

$$\frac{F_o}{F_i} < \frac{s_i}{s_o}.$$

Therefore,

$$\text{A.M.A.} < \text{I.M.A.}$$

The actual mechanical advantage is therefore always less than the ratio of the input distance (s_i) to the output distance (s_o), or the ideal mechanical advantage. The ideal mechanical advantage is frequently referred to as the *velocity ratio,* since the forces are moved these distances in the same time period.

In a theoretical machine (without friction), the work in, $F_i s_i$, equals the work out, $F_o s_o$; thus A.M.A. equals I.M.A. In this case the ratio of forces in and out equals the ratio of distances the forces travel in and out, or

$$\frac{F_o}{F_i} = \frac{s_i}{s_o}.$$

Therefore,

$$\text{A.M.A.} = \text{I.M.A.}$$

The efficiency of a machine is a measure of the useful work produced, and it is usually expressed as a percentage. The useful work produced is always less than the energy input, and the loss, if no energy is stored in the machine, is caused by friction. Expressing this concept in terms of the principle of conservation of energy: the energy in equals the energy out plus the wasted energy.

Efficiency is expressed as the ratio of work output to work input, or

$$\text{efficiency} = \frac{\text{work out}}{\text{work in}}$$

$$= \frac{F_o s_o}{F_i s_i}.$$

The resulting fraction is multiplied by 100 and expressed as a percentage. The efficiency is always less than 1. The efficiency may also be calculated as one of the following ratios multiplied by 100:

$$\text{efficiency} = \frac{F_o/F_i}{s_o/s_i}$$

$$= \frac{\text{A.M.A.}}{\text{I.M.A.}}.$$

Power.—Power introduces a time element into the study of work. The time rate of doing work is *power.* By dividing the amount of work involved by the time required to complete the work, the average power can be determined. Both the work (W) and the time (t) must be measured to determine the average power (P).

$$P = \frac{W}{t} = \frac{Fs}{t}$$

It takes 20 seconds to carry a 3-pound ball to the roof of a building 30 feet tall. The same ball can be thrown to the roof in 2 seconds. The work done in either case is the same, but the power required to throw the ball is 10 times greater:

$$\textit{carrying } P = \frac{Fs}{t}$$

$$= \frac{(3 \text{ lb.})(30 \text{ ft.})}{20 \text{ sec.}}$$

$$= 4.5 \text{ ft.-lb./sec.}$$

$$\textit{throwing } P = \frac{Fs}{t}$$

$$= \frac{(3 \text{ lb.})(30 \text{ ft.})}{2 \text{ sec.}}$$

$$= 45 \text{ ft.-lb./sec.}$$

The units of power in the f.p.s. system are the same as those of work (foot-pounds), divided by time in seconds. Since *1 horsepower* is defined as the energy needed to move a 550-pound weight 1 foot in 1 second (550 ft.-lb./sec.), the average horsepower can be expressed by dividing by this figure.

By absorbing the entire output of a machine with a friction brake, the power output can be measured. The energy absorbed is converted into heat and is dissipated. The band is tightened around the pulley (Fig. 8)

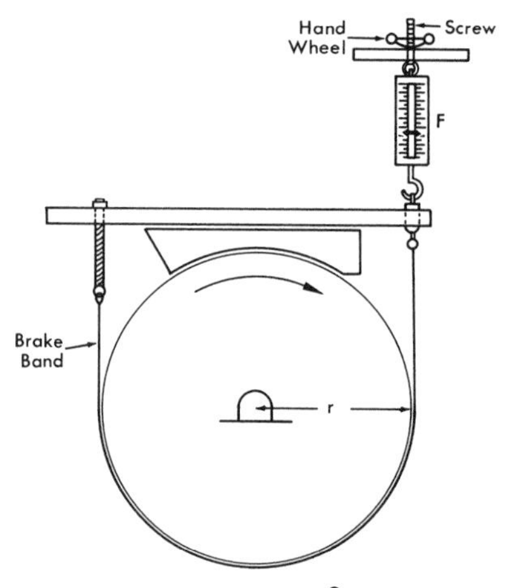

Figure 8.

by rotating the handwheel, and the scale measures the force of the friction produced. The work done by the machine in opposing the friction force F is equal to $F2\pi r$ for each revolution. The work produced in unit time is $F2\pi rn$, where n represents the number of complete revolutions made in unit time.

Torque.—A body at rest or in uniform motion can be said to be in equilibrium. A body will remain in equilibrium as long as all forces acting on the body pass through a common point and the sum of their vectors is zero. When the forces do not pass through a common point, the rotation as well as the linear motion will be changed. *Torque* is the force that produces or tends to produce rotation.

When opposing but equal forces act on a body (Fig. 9A), the resultant movement will be zero because the sum of the vector forces will be zero. If the same forces are applied to the body (Fig. 9B), the vector sum of the forces will again equal zero, but the body will rotate. When the vector sum of the forces is zero, the body will not move in a linear manner; but rotation is possible. Equilibrium is not assured, therefore, unless a second condition is satisfied. The second condition requires that the sum of the torques generated by the applied forces also be zero.

The *torque* (L) about a selected axis is a product of the force (F) and its moment arm (s). (The *moment arm* is the perpendicular distance from a selected axis to the line of the applied force.) Then

$$L = Fs.$$

When combined into the single quantity torque (or moment of force), the magnitude of the force and the length of the moment arm are of equal importance. For a given moment arm, increasing the force increases the torque; and for a given force, increasing the length of the moment arm increases the torque. Torque, being a product of force (pounds) and a measure of length (feet), is expressed in the f.p.s. system of units as *pound-feet (lb.-ft.).*

Since torque is a vector quantity, the selection of the axis is doubly important. First, the selection of the axis determines the algebraic sign of the vector, since altering the location of the axis can change rotation from clockwise (positive) to counterclockwise (negative). Second, the torque produced by a force of unknown magnitude can be reduced to zero by selecting an axis through which the unknown passes, thus simplifying problem solution.

Forces whose vectors intersect at a common point are called *concurrent forces.* The torque produced by each concurrent force, about an axis through their intersection, is zero. Selecting an axis at any other location will usually result in the sum of the torques being greater than zero. Should the sum of the torques around any axis outside the intersection equal zero, they will equal zero around any axis selected. Therefore torque does not enter into calculations involving concurrent forces in equilibrium. With nonconcurrent forces, no single axis exists about which the sum of the torques equals zero. It is essential, therefore, that the resultant torque be considered when analyzing a set of nonconcurrent forces in equilibrium.

The *center of gravity* is that point at which the entire weight of a body may be considered to be concentrated. The sum of all torque produced by the weights of all parts of the body around a horizontal axis through the center of gravity must equal zero. Locating the center of gravity of bodies is necessary in equilibrium problems to find the location of the single vector representing the body's weight.

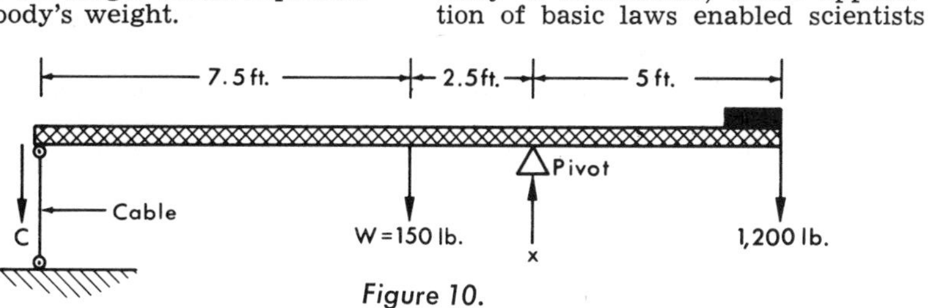

Figure 10.

A steel beam (Fig. 10) 15 feet long and weighing 150 pounds is supported 5 feet from its right end by a pivot. A 1,200-pound weight rests on the right end, and the left end is secured by a cable holding the bar in equilibrium. What is the load on the pivot (the downward force it must resist), and what force must the cable resist?

With a uniform bar such as this, the center of gravity (and its weight) can be considered to be concentrated at its center (W). A force (x) acts upward at the fulcrum, and the 1,200-pound weight acts downward at the right end. Since the system is in equilibrium, two conditions must be satisfied. First, the sum of all vector forces must equal zero:

$$-C - 150 \text{ lb.} + x - 1{,}200 \text{ lb.} = 0,$$

when C is the force on the cable.

With two unknowns this equation cannot be solved, so the second condition for equilibrium must also be considered. The sum of all torques (Σ_L) must equal zero in such a system. Taking the pivot (x) as the axis and starting at the right end, with counterclockwise torques taken as positive (+):

$$\begin{aligned}\Sigma_L = &- (5 \text{ ft.})\ (1{,}200 \text{ lb.}) \\ &+ (x)\ (0 \text{ ft.}) \\ &+ (150 \text{ lb.})\ (2.5 \text{ ft.}) \\ &+ (C)\ (7.5 \text{ ft.}) + (2.5 \text{ ft.}) = 0\end{aligned}$$

$$\begin{aligned}\Sigma_L = &- 6{,}000 \text{ lb.-ft.} \\ &+ 0 + 375 \text{ lb.-ft.} \\ &+ (10 \text{ ft.})\ C = 0\end{aligned}$$

$$(10 \text{ ft.})\ C = 5{,}625 \text{ lb.-ft.}$$

$$C = 562.5 \text{ lb.}$$

Having found one unknown—the force on the cable (C)—this can be substituted in the first equation and a complete solution arrived at:

$$-562.5 \text{ lb.} - 150 \text{ lb.} + x - 1{,}200 \text{ lb.} = 0$$

$$x = 1912.5 \text{ lb.}$$

Momentum.—During the study of basic mechanics, it must not be forgotten that the same laws apply to atomic particles that apply to larger objects. This is particularly true in the study of momentum, where application of basic laws enabled scientists to predict the movement and properties of atomic particles years before their existence was proved.

The *momentum* (P) of a body is defined as the product of its mass (m) times its velocity (v), or

$$P = mv.$$

Since they consist of mass and velocity, the units for momentum can be considered composite units. Following the f.p.s. system, mass (w/g) is given in slugs, velocity in feet per second, and momentum in slug-feet per second. Thus, the momentum of a 5,000-pound truck passing a point at 60 miles per hour (88 feet per second) is:

$$\begin{aligned}m &= \frac{w}{g} \\ &= \frac{5{,}000 \text{ lb.}}{32.2 \text{ ft./sec.}^2} \\ &= 155.3 \text{ slugs}\end{aligned}$$

$$\begin{aligned}P &= mv \\ &= (155.3 \text{ slugs})\ (88 \text{ ft./sec.}) \\ &= 13{,}666.4 \text{ slug-ft./sec.}\end{aligned}$$

A vector quantity, momentum takes its direction from the velocity. When two or more bodies are involved in a system, the momentum of each body must be added vectorially. If two balls of equal mass are rolled toward each other with equal velocity, the momentum of each is equal. Since the vectors are opposite in direction, the vector sum equals zero.

Newton's first law states that there is no change in the motion of a body unless a resistant force acts upon it. A body's mass is constant; therefore, its momentum will remain constant unless an external force is applied. Keeping this basic fact in mind will allow the behavior of everyday objects to be studied and analyzed. When a force is applied to a system of bodies, the system's momentum is altered; but some other set of bodies will gain (or lose) momentum equal to the loss (or gain) produced in the first system. This conservation of momentum can be expressed simply as: momentum lost = momentum gained. If the balls discussed previously collide, they will rebound (if they are elastic). Since the law of conservation of momentum applies to the system, the velocity of the rebounding bodies will be equal to, but not necessarily the same as, the original speed. The system's total momentum is still zero because the vector quantities cancel each other.

A system comprised of unequal masses, such as a projectile leaving

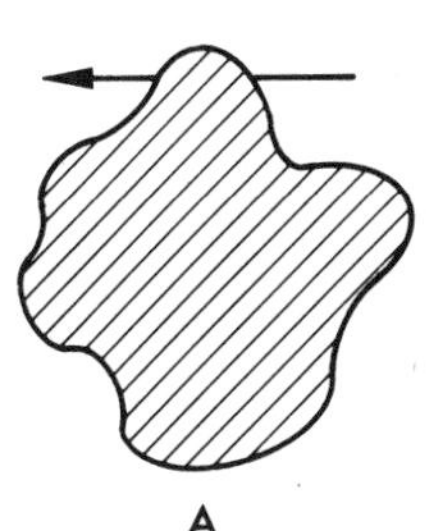

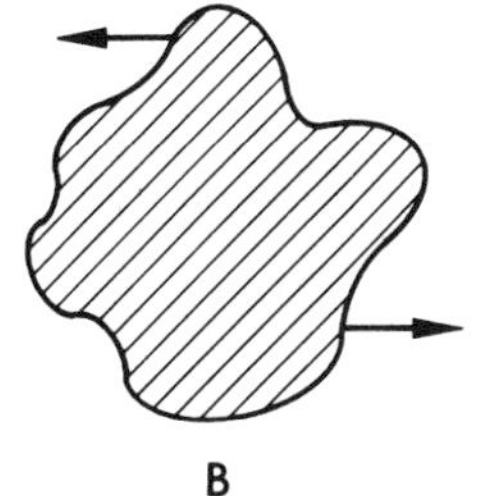

Figure 9.

a cannon, will also follow the same laws. For example, if a 12-pound shell leaves the muzzle at 2,400 feet per second, how fast will the 3-ton cannon recoil? Before firing, the momentum of cannon and shell is zero; therefore the momentum of the cannon must equal that of the shell, but in the opposite direction.

$$m_1v_1 = m_2v_2$$

$$v_1 = \frac{m_2}{m_1} \times v_2$$

$$m_1 = \frac{w_1}{g} = \frac{6{,}000 \text{ lb.}}{32.2 \text{ ft./sec.}^2}$$

$$m_2 = \frac{w_2}{g} = \frac{12 \text{ lb.}}{32.2 \text{ ft./sec.}^2}$$

$$v_1 = \frac{(12 \text{ lb.})/(32.2 \text{ ft./sec.}^2)}{(6{,}000 \text{ lb.})/(32.2 \text{ ft./sec.}^2)} \times 2{,}400 \text{ ft./sec.} = \frac{12}{6{,}000} \times 2{,}400 \text{ ft./sec.} = 4.8 \text{ ft./sec.}$$

Impulse.—Impulse and momentum are related concepts. Indeed, impulse may be defined in terms of momentum: *impulse* is the product of the average value of a force and the time during which it acts, and this product is equal to the change in momentum produced in the force. Although this may sound complex, impulse can be shown to be derived from the second law of motion. Sir Isaac Newton originally stated his second law in terms of force, time, and momentum:

$$F = \frac{mv - mv_o}{t},$$

where F is the force, m is the mass, v is the terminal velocity, v_o is the starting velocity, mv and mv_o are the terminal and initial momentums, respectively, and t is the time. (This equation can be changed to its more familiar form by factoring out the mass:

$$F = \frac{m(v - v_o)}{t}.$$

Since

$$\frac{v - v_o}{t} = a,$$

$$F = ma.$$

By multiplying the above equation by t, it becomes

$$Ft = mv - mv_o.$$

This is called the *impulse equation:* Ft is the impulse, and $mv - mv_o$ is the change in momentum.

As an illustration of impulse, a hammer of mass m accelerating at a rate a to a velocity v strikes a nail with a force F. This force, which lasts for but a fraction of a second, drives the nail into the wood by impulse. The hammer has supplied an impulse equal to its loss of momentum. In the f.p.s. system, the unit of impulse is the *pound-second*. In the c.g.s. system, the unit of impulse is referred to as the *dynesecond.*

Liquids.—Liquids are unlike solid bodies because they have no definite size or shape. A liquid will fill any container to a certain level and assume the shape of the container below that level. One property liquids and solids share is density. A 1-inch block of lead feels (and is) much heavier than a 1-inch block of wood. The difference is their *density* (ρ), or mass per unit volume:

$$\rho = \frac{m}{V}.$$

In units of the f.p.s. system, density is expressed as *slugs per cubic foot.* (In the c.g.s. system, density is expressed in *grams per cubic centimeter.*) The weight per unit volume is called the *weight-density* (D):

$$D = \frac{W}{V}.$$

Since

$$W = mg,$$

$$D = \frac{mg}{V} = \rho g.$$

Density (ρ) is used when problems involving mass are being considered,

NASA

LAUNCHING A MISSILE requires the combined knowledge of chemical, electrical, aeronautical, and mechanical engineers.

while weight-density (D) is used with respect to the effects of force.

The ratio of the density of a substance to that of water at 39.2° F. is called its *specific gravity,* and is expressed as:

$$\text{specific gravity} = \frac{\rho}{\rho_w},$$

where ρ_w is the density of water; and

$$\text{specific gravity} = \frac{D}{D_w},$$

where D_w is the weight-density of water. Specific gravity has no units because the densities of both substances have the same units. Since specific gravity is frequently tabulated, it is convenient to find densities of specific substances by the formula

$$\rho = (\text{specific gravity})(\rho_w).$$

Densities obtained by this formula will be in the units of the system used for the density of water.

A liquid confined in a container exerts a force on the walls and bottom of the container. With a still liquid, the force is normal or perpendicular to the surface. This perpendicular force (F) per unit area (A) is called pressure (P) expressed as:

$$P = \frac{F}{A}.$$

A scalar quantity, pressure is exerted whenever force is applied over an area. Even a very light woman can exert high pressure if her weight is applied over a small area, such as a spike shoe heel. *Pounds per square inch* is the most commonly used unit in the f.p.s. system. (In the c.g.s. system, the unit of pressure is *dynes per square centimeter.*) The pressure concept is essential in work involving liquids and gases. A typical problem is: What is the pressure on a 10-inch by 10-inch plate at the bottom of a swimming pool 10 feet (120 inches) deep? The force perpendicular to the pool bottom is produced by the weight of the column of liquid resting on it. Water weighs 62.4 pounds per cubic foot, or 0.036 pound per cubic inch. Hence,

$$P = \frac{F}{A} = \frac{(0.036 \text{ lb./in.}^3)(120 \text{ in.})(100 \text{ in.}^2)}{100 \text{ in.}^2}$$

$$P = 4.32 \text{ lb./in.}^2$$

When studying liquids at rest, several general statements that always apply should be kept in mind:

Every point within the body of a liquid is subjected to pressure.

Pressure is proportional to the vertical height of the liquid above the point in question.

Since the liquid is at rest, the force on a surface is the same regardless of the orientation of the surface, that is, how the surface is viewed.

Pressure is the same at any point selected on a horizontal plane through the liquid.

Regardless of shape or orientation of the fluid's container, pressure is always exerted perpendicular to the surface.

Force on a container bottom can be more than, less than, or equal to the weight of the liquid in a container. The force on the bottom is equal to the pressure at the bottom times the area of the bottom.

—Joseph J. Kelleher and Robert Abbott

ELECTRICITY AND MAGNETISM

History.—Electricity and magnetism are two very closely associated phenomena that play a vital part in nearly every aspect of modern living: in homes, in transportation, in communications, and in industry. Their impact on civilization has been tremendous and far-reaching. Because their many applications have made them commonplace, it perhaps is difficult to realize that electricity and magnetism have been harnessed for the benefit of mankind only within the last century and a half. This is true despite the fact that there are three natural manifestations of electricity and magnetism that are readily observable. These were known to the ancients, but were considered unrelated and of no apparent use to man.

One of these manifestations is electrical attraction. Records indicate that as early as 600 B.C. it was known that a piece of amber, which was used for ornamental purposes and called "electron," could be rubbed with wool or fur and made to attract small bits of straw. In 1600 Sir William Gilbert discovered that many other substances could be made to exhibit the same effect. He called the effect "electric," from "electron."

A second manifestation of electricity, lightning, has been observed since the dawn of history. However, its electrical nature was not recognized until 1751, when Benjamin Franklin performed his kite experiment.

A third natural manifestation, involving magnetism, was known to early Greek philosophers. A *lodestone* was known to attract iron and to orient itself in a north-south direction if freely suspended. The magnetizing of a steel needle from a lodestone, and use of the needle as a compass, were recorded by the Chinese at the beginning of the twelfth century. The connection between electricity and magnetism was discovered by Hans Christian Oersted in 1819. Other epoch-making discoveries followed rapidly, and the application of electricity and magnetism to the service of mankind soon began.

Static Electricity.—Static electricity is electricity at rest. It is responsible for the attraction between a bit of straw and a piece of amber that has been rubbed with wool or fur. Present theory holds that all matter is composed of tiny *molecules* that are themselves composed of even smaller structures called *atoms*. Atoms contain three principal types of particles—protons, neutrons, and electrons. Each *proton* carries a specific amount of *positive charge,* and each *electron* carries the same amount of *negative charge*. *Neutrons* carry no electric charge and therefore are electrically neutral. The protons and neutrons are located together at the center of the atom, to form the nucleus, while the electrons move about the nucleus in tiny orbits. Each proton or neutron has more than 1,800 times the mass of an electron. The number of protons and neutrons in the nucleus, and the number and arrangement of the orbiting electrons, determine the nature of the substance.

Normally each atom has the same number of electrons as protons; and since the total quantity of positive charge equals the total quantity of negative charge, the atom is electrically neutral. However, electrons can be removed to leave the atom with a net positive charge, or they can be added to give a net negative charge.

One method of altering the number of electrons in the atoms is by friction. Atoms have the ability to attract more electrons than the number normally moving about each nucleus; this attraction is of different strength for different materials. When two substances of different attractive power are rubbed together, some of the electrons in one will leave their orbits and move to new orbits in the other. The first substance will be left with a net positive charge, and the second will acquire a net negative charge. Examples may be tabulated as follows:

Rubbed Material	*Charge*	*Rubbing Material*	*Charge*
Amber	−	Fur or wool	+
Rubber	−	Fur or wool	+
Glass	+	Silk	−

■FORCES OF ATTRACTION AND REPULSION.—The presence of a force acting between two charged bodies a little distance apart can be readily demonstrated by suspending one charged body by a thread and approaching it with another charged body held in the hand (Fig. 1). Experimentation with the materials above will demonstrate the general law that like charges repel one another and unlike charges attract one another.

An uncharged bit of paper or straw is attracted to the charged body because a few of the electrons in the paper or straw are able to move about. Assume, for example, that a positively charged glass rod is brought near a bit of paper. Electrons, drawn toward the rod, will make the surface of the paper negatively charged and attract it to the rod. The more distant surface of the paper will be positively charged but, being farther away, it will be repelled with less force; therefore the paper is attracted. If the paper touches the rod, the movable electrons will go onto the rod, leaving the paper positively charged. The paper will then jump away. This involves another general law: the force between two charges is inversely proportional to the square of the distance between them, and directly proportional to the product of the values of the two charges.

■ELECTRIC FIELDS.—The force that acts between electric charges suggests that there is an electric field associated with every electric charge. For convenience, the *electric field* is considered to consist of electric lines of force, a *line of force* being a line along which a charge tends to move. The positive direction is that direction in which a positive charge would move. Lines of force may be thought of as extending radially outward in all directions from a positive charge, and radially inward from all directions toward a negative charge. The *intensity,* or strength, of the electric field is measured by the force that acts upon a unit electric charge at the point of measurement and is represented pictorially by the density with which the lines of force occur in the vicinity of the point. Electric lines of force never cross one another. They begin at a positive charge and terminate at a negative charge.

The effects of static electricity are of limited use. However, one use of increasing importance is in the electrostatic precipitator, which is used in some air-conditioning systems and in air-pollution control. In this device, dust and smoke particles are attracted and held to charged surfaces.

Current Electricity.—The most familiar uses of electricity involve electricity in motion, a flow of electric charges. A stream of moving electric charges constitutes an *electric current.*

■CONDUCTORS AND INSULATORS.—The electrons orbiting about atomic nuclei are not all unchangeably fixed in

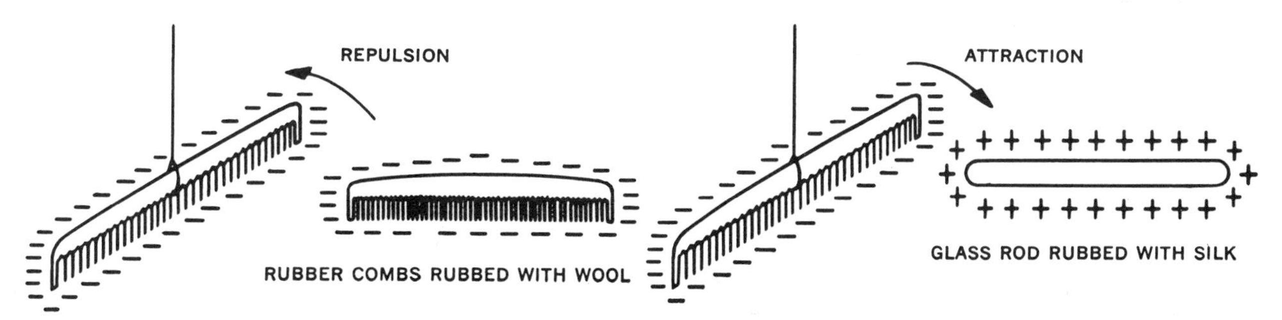

Figure 1.

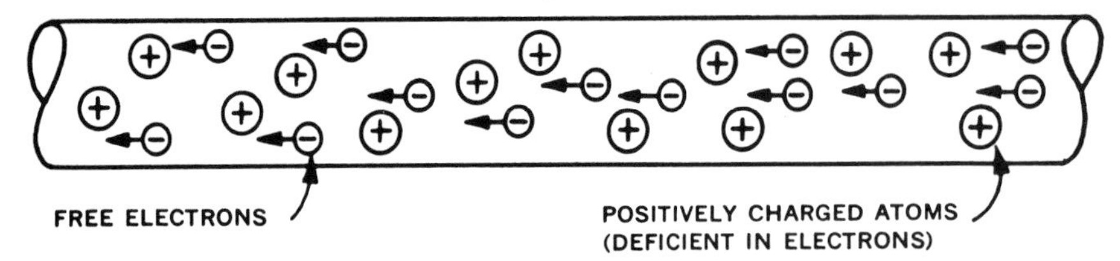

Figure 2.

their orbits. In some materials there are electrons that move readily from atom to atom. Generally speaking, metals contain quantities of such free electrons, and are said to be good *conductors* (Fig. 2). In most nonmetallic substances, all the electrons are tightly bound in their orbits and can move from atom to atom only with great difficulty. These substances are called nonconductors, or *insulators.* Metals differ in the ease with which electrons can move about, and are said to differ in their electrical *conductivity.* The conductivities of several metals, relative to that of copper, are listed in the accompanying table. Of these metals, copper is the most widely used because it is relatively

Metal	*Conductivity*
Copper	100%
Silver	105%
Gold	71%
Aluminum	61%
Iron	18%
Low-carbon steel	8–13%
Some alloy steels	2%

inexpensive and an excellent conductor. Aluminum also is frequently used because it is relatively inexpensive and lightweight. High-power transmission lines frequently are constructed of a stranded steel cable around which aluminum strands are spiraled. This combination provides high strength, light weight, and good conductivity at a low total cost.

Resistivity, the converse (reciprocal) of conductivity, often is used to describe the property of a material as a conductor. Good conductors have low resistivity. Good insulators have very high resistivity. For example, plate glass has approximately 10^{19} times, and porcelain approximately 10^{20} times, the resistivity of copper. It is this enormous difference in resistivity between good conductors and good insulators that makes possible the efficient control and distribution of electricity.

Under certain conditions a gas can become a conductor. The neon light is an example. Under the influence of an electric field, atoms of neon in the glass tube become positively charged ions and are drawn to the negatively charged conductor. At the same time, electrons that have left the neon atoms are drawn to the positive conductor (Fig. 3). A more spectacular example of an electric current through a gas (air) is a lightning flash.

Some liquids, principally water solutions of acids, bases, or salts, also are able to conduct electricity. A portion of the molecules of the solute (the acid, base, or salt) split, or *dissociate,* into two parts. One part has an excess of electrons, the other a deficiency, so that they are negatively and positively charged, respectively. Under the influence of an electric field between two conductors in the solution, these charged particles move, constituting a current. The solution is called an *electrolyte;* the conductors, *electrodes.* Pure water dissociates only to a very slight extent and therefore is a poor electrolyte (Fig. 4).

■TYPES OF ELECTRIC CURRENTS.—There are two basic types of electric currents: direct current and alternating current. *Direct current* is a stream of electrons past any one point in one, and only one, direction. *Alternating current,* on the other hand, alternates its direction, first flowing past a point in one direction and then in the opposite direction. The *frequency* of an alternating current is its number of cycles per second. A *complete cycle* is the interval during which a current starts from zero, pulses to a maximum in one direction, drops back to zero, pulses to a maximum in the opposite direction, and returns to zero once again.

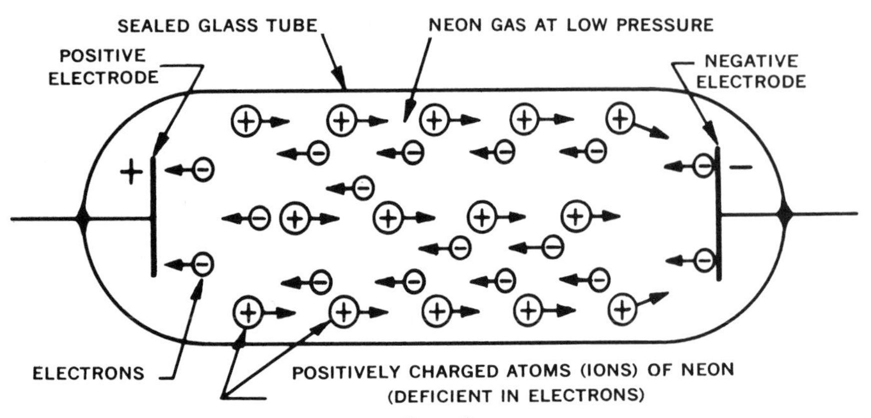

Figure 3.

■UNITS OF MEASURE.—The unit of quantity of electricity is the *coulomb.* One coulomb is equivalent to the charge carried by 6.24×10^{18} electrons. Electricity flowing at the rate of 1 coulomb per second is a current of 1 *ampere.* Thus one ampere is equal to the passage through a given point of 6.24×10^{18} electrons per second.

The electric pressure, or *electromotive force (e.m.f.),* which causes the flow of current, is measured in *volts.* The intensity of an electric field may be expressed as volts per centimeter or volts per inch.

The *resistance* of a conductor is measured in *ohms.* An e.m.f. of 1 volt applied to a resistance of 1 ohm will produce a current of 1 ampere. Ohm's law, named after its propounder Georg Simon Ohm (1787–1854), states that current is proportional to the voltage applied and inversely proportional to the resistance. Stated in the form of an equation, $I = E/R$, where

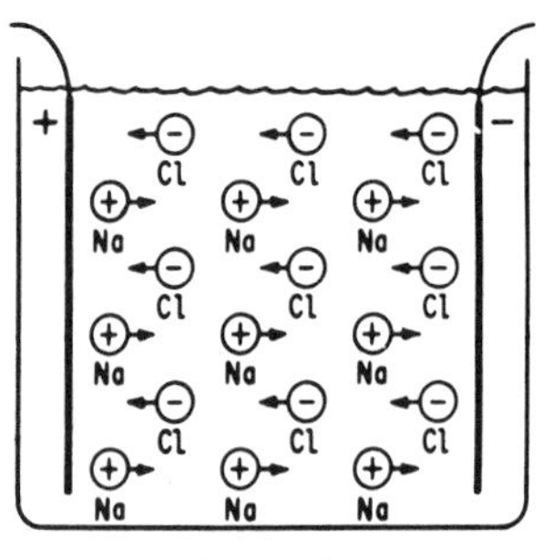

Figure 4.

ELECTROLYSIS consists of salt solution (electrolyte), electrodes, and container.

I is the current in amperes, E is the e.m.f. in volts, and R is the resistance in ohms. The converse of resistance, *conductance,* is measured in *mhos* —which is *ohm(s)* spelled in reverse. Resistance and conductance refer to the properties of a piece of material of a particular shape and size. The corresponding characteristics of the material itself, which do not depend on size or shape, are *resistivity* and *conductivity.* These are the resistance and conductance, respectively, of a cube of the material of unit size. The units mentioned above (the volt, ohm, ampere, and coulomb) are named in honor of the following pioneers in the study of electricity and magnetism: Alessandro Volta (1745–1827); Georg Simon Ohm; André Marie Ampère (1775–1836); and Charles de Coulomb (1736–1806).

Production of Electric Current.—Aside from the transient discharges of static electricity, the first production of electric current was by chemical means. This latter is still a very important source of electric current, as shown by the large number of battery-operated devices. The great bulk of electric power, however, is generated by electromagnetic induction in rotating machinery. The thermocouple and the photocell are sources of current in small amounts.

CHEMICAL CELLS.—If electrodes of two different materials, such as zinc and copper, are immersed in an electrolytic solution, one plate acquires a positive charge and the other a negative charge. If they are then connected externally by a wire, a current will flow through. This is a *voltaic cell,* discovered about 1800. Chemical action between the electrolyte and the electrodes causes atoms of one electode to go into solution, each atom leaving behind one or more electrons. This leaves the electrode with a negative charge. The electrolyte takes electrons from the other electrode, thus giving it a positive charge. The electrons flow through the external wire from the negative electrode to the positive electrode. A group of voltaic cells connected together form a battery.

The *dry cell* (Fig. 5) consists of a zinc container, which serves as the negative electrode; a carbon rod in the center, which serves as a current

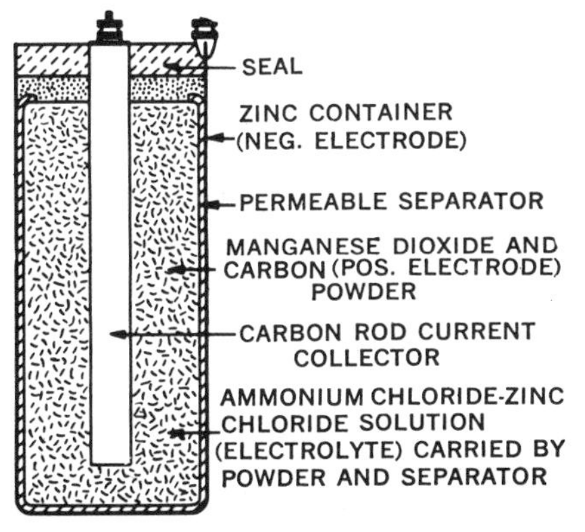

Figure 5.

collector for the positive electrode; and a paste made of ammonium chloride, zinc chloride, and water, which serves as electrolyte. Carbon and manganese dioxide powder are packed around the carbon rod to form the positive electrode. They also react with the hydrogen liberated there, and the zinc chloride reacts with the ammonia gas liberated at the zinc electrode to prevent build-up of gas pressure. These components are thoroughly sealed in, giving the cell its name. A present-day dry cell develops an e.m.f. of about 1.5 volts.

Dry cells and others of the same class are called *primary cells.* Primary cells must be discarded when one of the electrodes is consumed, or the consumed materials must be replaced. *Secondary cells,* such as those used in the *storage battery,* can be recharged by passing current through them in the direction opposite to that in which they produce current. The chemical action is reversed by this process, and the electrodes and electrolyte are restored to their original condition. A common example is the *lead-acid cell* used in automobile batteries and many other applications. The electrolyte is dilute sulfuric acid, and the electrodes are flat lead grids, the positive grid being filled with a porous paste of lead dioxide and the negative grid with spongy lead. A fully charged cell of this type develops an e.m.f. of about 2.2 volts. Another widely used storage cell is the *Edison cell,* which was invented by Thomas Alva Edison (1847–1931); it has electrodes of nickel peroxide and spongy iron and a potassium hydroxide solution as electrolyte.

A more recent development is the *nickel-cadmium storage cell,* which is somewhat lighter than older types and has a longer life. It is capable of being completely sealed, since there is no evolution of gas during operation. Because of this feature it is used in satellites and space vehicles. It also is being used more and more in the development of battery-powered portable tools and appliances. Because there is no evolution of flammable gas or spillage of corrosive liquids, it is not a fire or corrosion hazard.

A development that may have great importance to industry is the *fuel cell.* This is a primary cell in which the chemical reactions are carried on between inexpensive and continually replaceable materials, such as hydrogen or hydrocarbons and oxygen or air. Successfully operating fuel cells have been built and demonstrated, but the development has not yet permitted widespread economic application. The over-all efficiency of converting the energy in the fuel cell to electricity promises to be much greater than that obtained in presently used turbines or diesel engines and generators.

THERMOCOUPLES.—In 1822 Thomas J. Seebeck discovered that if two wires of dissimilar metals are joined at both ends and one junction is kept at a different temperature from the other, an electric current will flow. This is known as the *thermoelectric effect,* and the device is called a *thermocouple.* A number of thermocouples connected in series are a *thermopile.*

Thermocouples are widely used to measure temperature. One junction is maintained at a constant reference temperature, such as that of melting ice; the other, at the temperature to be measured. A sensitive electric instrument is used to indicate the voltage developed. Different combinations of metals are used to cover different temperature ranges. Thermopiles may be used as a source of electric power for some special applications, although they are inefficient. Combined with suitable reflectors, they may be used as sensitive detectors, indicating the presence of a hot object a considerable distance away.

PHOTOCELLS.—The *photovoltaic cell,* or *photoelectric cell,* such as that used in photographic light meters, has been familiar for many years. More recently such cells have been combined into groups to form solar batteries that serve as power sources for the instrumentation of orbiting space satellites. By this means the operating life of the equipment in orbit is extended far beyond that which could be obtained with chemical batteries alone. Although they may become important for isolated applications requiring little power, solar batteries have not been a significant source of power in any quantity for earthbound applications.

A photovoltaic cell consists of a very thin layer of semiconductor material deposited on the surface of another semiconductor material. (*Semiconductor materials* are so called because they lie in a range between good insulators and good conductors. Germanium and silicon are important examples.) When light falls upon the thin film, penetrating to the junction region between the two semiconductors, electrons move across the junction from one material to the other, forming a current.

Magnetic Fields.—The attraction and repulsion between magnets follows a set of principles similar to those described previously for electrostatic forces. Consider the familiar magnetic compass, the needle of which orients itself in a north-south direction. For convenience, the end

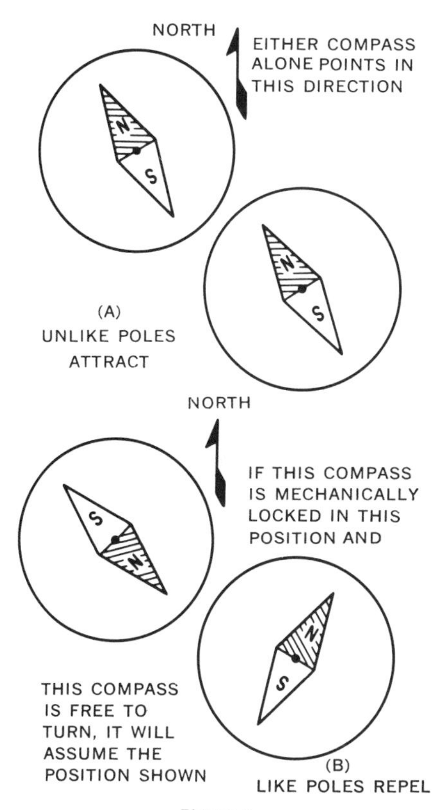

Figure 6.

pointing north is marked N, or "north-seeking," and the other end S, or "south-seeking." When another compass is brought near the first and is moved about it, the two needles deviate from their north-south direction and tend to point toward each other. The attracting and repelling forces reside in the ends of the needles, and ends marked alike repel one another. The ends of the needles (and of all magnets) are called *poles.* Like poles repel one another and unlike poles attract one another, just as like electric charges repel and unlike electric charges attract (Fig. 6).

There is an important difference between a magnetized bar and a charged body, however. The charged body may carry the same polarity of charge over its entire surface, whereas the magnetized bar always has two poles of unlike polarity and of equal strength. A single magnetic pole does not exist by itself.

The forces that act between magnets at a distance suggest that the magnets are surrounded by a magnetic field. This field, like the electric field, is considered to consist of lines of force, a line of force being a line along which a magnet pole tends to move. The positive direction is that in which the N pole tends to move. Thus a compass needle in a magnetic field will orient itself parallel to the lines of force, with the N pole pointing in the positive direction. There is an important difference, however, between the concepts of the magnetic field and the electric field. Whereas the electric lines of force may emerge from a positive charge and terminate on a negative charge, the magnetic lines of force do not terminate at the S pole but continue unbroken through the magnet to emerge at the N pole. Magnetic lines of force exist only as closed loops, and they never cross. They commonly are spoken of as *magnetic flux lines.* A single magnetic flux line is called a *maxwell* (after James C. Maxwell) in the centimeter-gram-second (c.g.s.) system of units, and is known as a *weber* (after Wilhelm E. Weber) in the meter-kilogram-second (m.k.s.) system of measurement. One weber is equivalent to 10^8 maxwells.

Force Between Current and Fields.—In 1819 Hans Christian Oersted found that a compass needle was deflected when a conductor carrying current was held over it. Further experimentation demonstrated that there is a magnetic field about an electric current and that the lines of force form closed loops encircling the conductor in planes perpendicular to the conductor. The lines of force about two parallel wires carrying current produce a force between the two wires. They attract each other if the current flows in the same direction in both, and repel each other if the currents flow in opposite directions (Fig. 7).

If a wire is bent into a closed loop and current is passed through it, magnetic lines of force will form closed loops around the wire (Fig. 8). If a series of wire loops is joined to form a *helix,* or coil, lines of force will pass inside parallel to the axis, curving around outside to close upon themselves. Such a coil is called a *solenoid,* and it exhibits all the characteristics of a bar magnet (Fig. 9).

If a bar of soft iron is put inside the solenoid and current is then passed through the solenoid, the bar will exhibit all the characteristics of a strong magnet. When the current is turned off, the magnetic properties of the bar disappear. This device is called an *electromagnet.*

The discovery of the electromagnet and the development of the chemical cell as a source of electricity opened the way for the invention of the telegraph by Samuel F. B. Morse in 1837, and were essential to the invention of the telephone by Alexander Graham Bell in 1875. Electromagnets are used to open and close electric circuits, to operate valves, and to do a multitude of other mechanical tasks.

Ferromagnetism.—The magnetic characteristics exhibited when a core of soft iron is put into a solenoid are many times greater than those of the solenoid alone. The magnetic properties contributed by the iron are classed as *ferromagnetism,* and the iron is said to be *ferromagnetic.* Two other metals, nickel and cobalt, and a few nonferrous metal alloys also are ferromagnetic, but to a much lesser extent than iron. Some of the most widely used ferromagnetic materials are alloys of iron with nickel and cobalt. No other materials exhibit magnetic properties, except in extremely small amounts.

The increase in magnetic effects produced by putting iron in a solenoid indicates that it is much easier to establish magnetic flux in the iron bar than in the air that previously occupied the space. The iron is said to have a high *relative permeability.* For air and, practically speaking, for all materials other than ferromagnetic materials (and for space itself), the relative permeability is 1. As a comparison, for certain carefully prepared iron alloys it may be over 100,000. Some materials have relative permeabilities slightly greater than 1 and are called *paramagnetic materials.* Others have relative permeabilities slightly less than 1; they are called *diamagnetic materials.*

The source of ferromagnetic properties lies in the electrons orbiting about the atomic nuclei of the material. The electrons spin about their own axis as they orbit, producing a magnetic field along their axis. In all materials other than ferromagnetic materials, the electrons in each atom are so oriented that there is practically no magnetic field produced outside the atom. In ferromagnetic materials, however, there is a net unbalance in each atom so that the atom has a magnetic field about it. The ferromagnetic material thus is composed of a great many tiny magnets. Large groups of these tiny magnets are lined up in one direction, forming *domains.* Normally the domains are grouped together, forming closed loops; and no magnetic field appears outside the material. Under the influence of a fairly small magnetic field, however, these groups are altered until nearly all are oriented parallel to the magnetic field and add their own fields to it. The total field of all the tiny magnets in the material may be many thousand times greater than the field required to line them up. When the external field is removed, the tiny magnets revert approximately to their former orientations, producing no outside field.

In some ferromagnetic materials the magnetic domains are reoriented with great difficulty, and a strong field is required to accomplish this. Moreover, when the field is removed, a large proportion of the domains do not revert to their initial orientations. The bar is said to be magnetized, and is called a *permanent magnet.* Early permanent magnets and compass needles were made of hardened carbon steel. Recent decades have seen the development of materials with vastly improved permanent magnet characteristics, making possible the development of many new instruments and devices. Permanent magnets, when properly prepared and treated, exhibit great stability of their magnetic field and are used widely in electricity-measuring instruments. The watt-hour meters that measure household electrical consumption depend on the constancy of two permanent magnets for their accuracy.

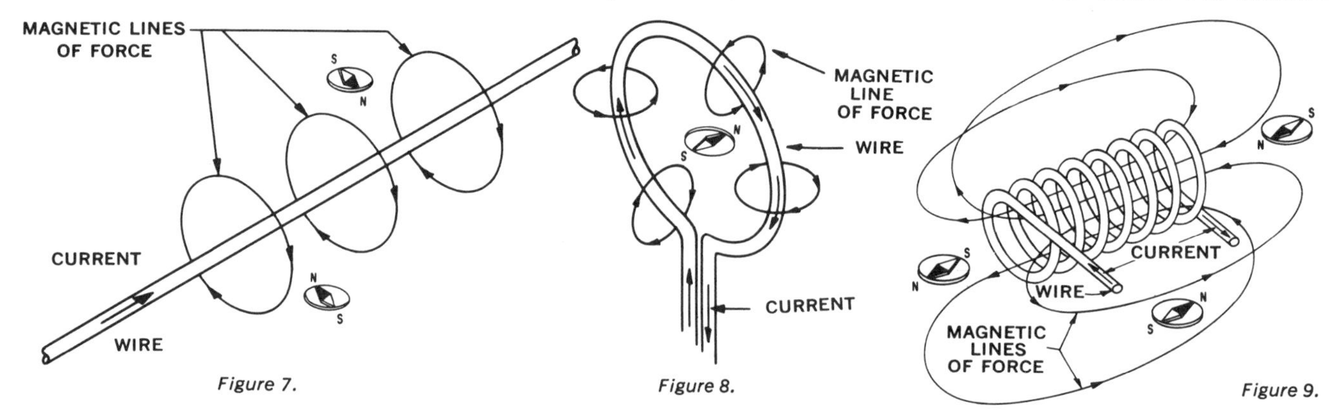

Figure 7. *Figure 8.* *Figure 9.*

Electromagnetic Induction.—In 1831 Michael Faraday discovered that when he plunged a magnet into a coil of wire, a current flowed in the wire, and that the current flowed only while the magnet moved. The principle thus illustrated is commonly stated in two forms. One says that the e.m.f. induced in a loop of wire is proportional to the rate at which the number of magnetic lines of force interlinking the loop increases or decreases. The other says that if a conductor is placed in a magnetic field perpendicular to the lines of force and moved in a direction mutually perpendicular to itself and the lines of force, an e.m.f. will be induced proportional to the rate of movement. The conductor is said to "cut" the lines of force. Actually, of course, they are not severed; they pass through the conductor as it moves.

If the magnetic flux interlinked by a loop of wire changes at the rate of 10^8 maxwells, or 1 weber, per second, the e.m.f. induced is equal to 1 volt. The generation of e.m.f. in this manner is spoken of as *electromagnetic induction.*

Generators.—Almost all the electricity used in homes and in industry is produced by machines, called *electric generators,* that make use of the principles of electromagnetic induction. In all such generators, conductors are moved through a magnetic field to produce an e.m.f. Consider a very simple arrangement in which a magnet is mounted on a shaft through its center so it can be turned end for end rapidly by turning the shaft. Imagine a coil of wire mounted close to the tumbling magnet so that each time a pole of the magnet passes the coil, its flux interlinks the turns of the coil for a brief time. Successive pulses of e.m.f. will be induced in the coil, and an alternating current will flow in a circuit connected to the coil. These are the various elements of an *alternating-current generator,* which produces current that changes direction many times a second. In practice there are several coils, and a laminated iron core is placed inside them to increase the flux from the magnet. In larger generators an electromagnet is used for the rotating magnet, and connection is made to it by means of carbon blocks or brushes that make sliding contact with insulated metal rings mounted on the shaft. The magnet may be arranged with two, four, six, or any even number of poles; the stationary coils must be provided to match. The stationary member is called the *stator;* the rotating one, the *rotating field* or *rotor.*

In order to generate direct current, a rotary switching device called a *commutator* is provided; it reverses the connection to the coils as the current in them reverses. In a *direct-current generator* the magnet is stationary, and the coils are mounted on the shaft with a laminated iron core. The coils are connected to copper bars insulated from one another and arranged to form a drum-type commutator revolving with the shaft. Two or more sets of stationary carbon brushes bear on the outside of the commutator and are located so that they always connect through the bars to coils in which an e.m.f. is being generated in the same direction. When a coil has turned halfway around and is generating e.m.f. in the opposite direction, its commutator bars are in contact with the brushes in the opposite direction. The combination of the rotating coil and its core is called the *armature.* The stationary magnet is called the *field.* Although it may be a permanent magnet, it is usually an electromagnet in larger generators, receiving its current from the armature through the brushes (Fig. 10).

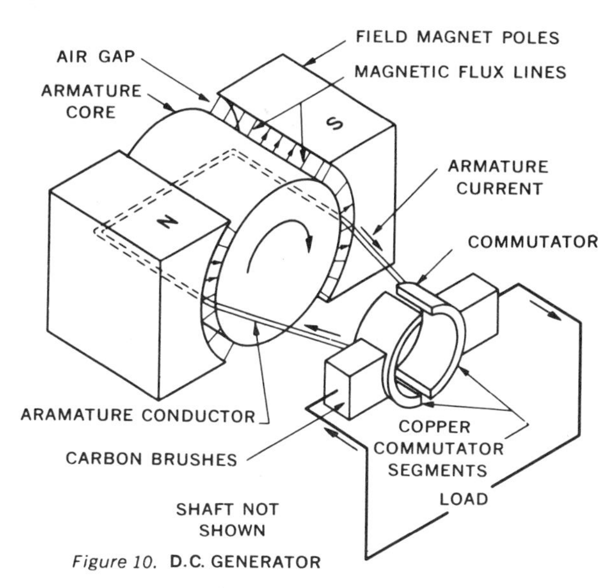

Figure 10. D.C. GENERATOR

Historically, direct-current generators came into use before alternating-current generators, although the latter may seem simpler. Direct-current generators provided a substitute for batteries, producing current in greater quantities, more economically, and in a form with which men were familiar. However, nearly all electric power now is generated as alternating current.

Motors.—Electric motors depend for their operation upon the force developed between electric currents and magnetic fields. Motors are very similar to generators in construction, and many may be used interchangeably as either. Consider the direct-current generator just discussed. If a current is passed through the armature, the brushes and commutator cause the current to pass through coils that are in the magnetic field, and a *turning moment* (torque) is produced. As the armature turns, the brushes and commutator always direct the current into coils that are passing through the magnetic field, and continuous torque is developed.

There are three basic types of alternating-current motors: the series, commutator-type motor; the induction motor; and the synchronous motor. A number of other types of alternating-current motors are used, but they are modifications or combinations of these three basic types.

The *series, commutator-type motor* is similar to the direct-current motor. The field electromagnet has a laminated iron core and is connected in series with the armature so that the same current passes through both. The magnetic field and the armature current alternate together so that the turning moment is always in the same direction. This type of motor is used extensively in appliances and portable tools. It has good starting torque and operates over a wide speed range.

The *induction motor* is the most common type, particularly in motors of fractional horsepower and moderately larger sizes. The rotor consists simply of a soft iron cylinder carrying a short-circuited winding. There is no commutator. The alternating magnetic field produced by the stator current induces alternating current in the rotor coils; the alternating current reacts with the magnetic field to produce torque. These motors are very simple and sturdy.

The *synchronous motor* is similar to the alternating-current generator, with stationary coils and rotating field magnet. Suppose that the rotor is turning at just the right speed so that current in the stationary coils alternates in exact synchronism with the passing of the rotor poles. A turning moment is produced at each pole, always in the same direction. If a load sufficient to slow the rotor is applied, it will stop. Synchronous motors will not start of themselves, but must have auxiliary starting means provided. The familiar electric clock is driven by a synchronous motor. Some of the less expensive ones are not provided with a self-starting mechanism, but must be started by twirling a knob at the back. Synchronous motors are made in all sizes.

■SINGLE-PHASE AND POLYPHASE MACHINES.— Alternating-current generators and motors may be characterized as single-phase or polyphase, depending upon the arrangement of their stator coils. A *single-phase generator* has a single set of coils connected to the load circuit. A *polyphase generator* has two or three sets of coils spaced around the stator so that the voltage induced in the successive coils alternates in sequence rather than simultaneously. Synchronous and induction motors similarly may have single-phase or polyphase stator windings. The series, commutator-type motor is always single phase. Three-phase machines are more common than two-phase machines.

Transformers.—The large-scale, electric-power transmission systems were made possible by the development of the transformer and the use of alternating current. In any transmission line, part of the voltage applied at the generator end is absorbed in the resistance of the wires, so that less voltage is available at the receiving end. A *transformer* is a device that transforms the voltage to a higher level at the generator end while reducing the current. In this way the voltage available to drive the current over the line is multiplied many times, and actual voltage required is reduced. The wires don't need to be as large and heavy, and the cost of the entire line is reduced. At the receiving end, another transformer is used to transform the voltage down to a safe, convenient level, while the current is increased so that nearly the same total power is available.

The transformer works only with alternating current. It consists basically of two coils wound one over the other on an iron core that completely interlinks the coils, forming a closed path in the iron core for the magnetic flux. The voltage to be transformed is applied to one coil, called the *primary coil.* Alternating current flows in the primary coil, causing alternating magnetic flux in the core. This alternating flux, which interlinks the other, the *secondary coil,* induces an alternating voltage in that coil. The same flux induces a voltage in the primary coil that is very nearly equal and is opposite to the applied voltage. The voltage induced in the coils is proportional to the number of turns in the coils, so that if the secondary coil has, say, ten times as many turns as the primary coil, the voltage induced in the secondary coil will be very nearly ten times that applied to the primary coil. The current that flows from the secondary coil over the line produces in the transformer coil a magnetizing force that tries to change the flux there. This magnetizing force is proportional to the current and the number of secondary-coil turns. It is compensated for by an automatic increase in the current that flows in the primary coil, producing an almost equal and opposite magnetizing force. Thus, if the secondary coil has ten times as many turns as the primary coil, the current in the secondary coil will be only one-tenth its counterpart in the primary coil.

Transformer cores are made of thin sheets of a special alloy iron that has high permeability and resistivity. The sheets are insulated from one another by coats of insulating varnish. This reduces to a minimum the current that flows in the core itself as a result of the alternating flux.

Small transformers may consist simply of the coils and core with mounting brackets. Others may be in a case filled with pitch or other compound. Large transformers are mounted in tanks filled with a special grade of mineral oil or a liquid chemical compound. The oil or compound serves to insulate electrically the coils and their leads, and aids in transferring heat from the coils to the outside case.

Transformers are made in a wide range of sizes for a great many applications apart from the use just considered. Radio and television receivers commonly contain one or more, and the home doorbell and the heater-control thermostat operate on a low voltage supplied by a small transformer. Still another use of transformers is in electric-measuring instruments to reduce the voltage at the instrument to a safe and convenient value and still permit accurate measurement.

Electric Circuits.—Electric devices may be connected in a variety of ways with respect to the paths, or circuits, the current takes through them. In a *series circuit* the same current flows in sequence through one device after the other. There is only one path. One type of Christmas tree light string is an example. If one light burns out, the circuit is broken and all lights go out. In a *parallel circuit* there are two or more paths. The current divides, and part flows in each path simultaneously. In another type of Christmas tree lights, the bulbs are connected in parallel. The full supply voltage is applied to each bulb, and the current through each bulb comes directly from the source and returns to it. Thus, if one bulb burns out, only it goes out.

Many street-lighting and highway-lighting circuits are in series. However, most of the circuits normally encountered are parallel circuits. House wiring consists of several parallel circuits. When an appliance is plugged into an outlet, it is put in parallel with other appliances already in use. Circuits may become very complex, including branches of elements in series, forming various series-parallel combinations. The current flowing in parts of the circuit and the voltage across those parts depend upon the resistances of the circuit elements and the circuit complexity. The complexity is further increased in alternating-current circuits in that some of the elements may possess inductance or capacitance as well as resistance. *Inductance* appears when the current is interlinked by a magnetic field, as in an electromagnet. *Capacitance* appears when two parts of a circuit are close together and an alternating voltage establishes an electric field between them, continually building up electric charges of alternating polarity on them. The effect of the inductance is called *inductive reactance,* and that of the capacitance is called *capacitive reactance.* These combine *vectorially* (not directly) with the resistance to form *impedance,* which is the apparent resistance to current flow in an alternating-current circuit—or, to put it another way, impedance is to an alternating-current circuit what resistance is to a direct-current circuit. Both reactances depend upon the frequency of alternation, inductive reactance being directly proportional to frequency, and capacitive reactance inversely proportional to frequency.

Another type of circuit common to electromagnetic devices is the *magnetic circuit.* This describes the path of closed loops of magnetic flux. These circuits also may have series and parallel elements. Because the magnetic flux is not confined to specific paths, as electric current is confined to a conductor, there may be parallel paths, called *leakage paths,* beside or around the desired useful paths. Magnetic circuits may be difficult to calculate with accuracy because the actual extent of the leakage paths frequently is difficult to determine, and they may constitute a large part of the entire circuit.

Electronics.—In a broad sense the term *electronics* may seem relevant to all the applications of electricity, since they all involve the electron, its charge, and its movement. As commonly used, however, electronics applies to the use of electron currents in a vacuum or in a low-pressure gas. More recently it also has come to be used in connection with semiconductor devices.

■DISCOVERY OF THE ELECTRON.—The discovery of the electron in 1897 by Joseph John Thomson was a result of his study of the electric discharges in evacuated glass tubes. As the pressure is reduced in such a tube, the electric discharge first produces a blue glow. This changes to pink, dark areas appear, and the pink discharge disappears. Finally the glass itself glows with a faint greenish light. This is caused by invisible rays coming from the *cathode* (negatively charged electrode). They were first called *cathode rays,* and later were found to be tiny particles torn from the cathode by the strong electric field and drawn toward the *anode* (positively charged electrode). These particles now are known to be electrons.

■ELECTRONIC EMISSION.—A very strong field is required to tear electrons from the atoms in the surface of a metal at room temperature. As the metal is heated, a temperature is reached at which electrons are emitted spontaneously. This is called *thermionic emission.* It may require incandescence, as in the case of tungsten, or only a dull red glow, as with thorium oxide.

Electrons also are emitted at room temperature when light falls upon a metal. This is called the *photoelectric effect.* Most metals exhibit this effect only under ultraviolet light, but such metals as rubidium and cesium emit electrons in visible light.

■THE FLEMING VALVE.—The first important, successful application of thermionic emission was developed by Sir John Ambrose Fleming (1849–1945) in his Fleming valve. This consists of a wire filament surrounded by a metal cylinder in a highly evacuated glass bulb. The filament is heated to incandescence by an electric current, and electrons are emitted from it. If the cylinder is made positive with respect to the filament, the electrons are drawn to it, constituting a current. If the cylinder is made negative, the electrons are repelled and no current flows. The device that permits current to flow in only one direction is called a *rectifier.* It is also called a *diode* because it has two electrodes. In an important modification of this tube the glass bulb is filled with a gas, such as mercury vapor or argon, at low pressure. Tubes of this type, both vacuum and gas-filled, are used in radio and television equipment, battery chargers, and other electronic devices. For example, large gas-filled rectifiers are used to provide direct-current power to electric railways.

THE DE FOREST AUDION.—A most important invention basic to present-day telephone and radio was made by Lee De Forest in 1907 when he placed a wire grid between the filament and plate of a Fleming valve and called the new tube an *audion.* Such a tube also is called a *triode,* indicating the use of three electrodes. A relatively small voltage applied to the grid has a strong controlling effect on the passage of electrons from filament to plate, much greater than if the voltage were applied directly to the plate. Through this effect the triode functions as an amplifier, by means of which very small amounts of power can be used to control large amounts of power. Additional grids may be added, making a *tetrode* or *pentode,* to modify the characteristics of the tube. Another important modification is the use of gas at low pressure in the tube. Such gas-filled tubes can control large amounts of power more efficiently than vacuum tubes.

Many modern tubes use indirectly heated cathodes. The filament is placed in two small holes along the axis of a ceramic rod, heating the rod. A metal cylinder coated with thorium, strontium, or cesium oxide fits snugly over the rod. These oxides are copious emitters of electrons when heated only to a dull red.

TELEPHONY.—The development of vacuum-tube amplifiers greatly extended the range of long-distance telephony. Amplifiers, called *repeaters,* are connected into the lines at intervals so that the strength of the signal is repeatedly restored to its original value. Amplifiers built to operate for twenty years without attention are built into transatlantic telephone cables at twenty-mile intervals. Through their use, telephone calls are transmitted for over 4,000 miles with full strength and clarity.

RADIO COMMUNICATION.—Vacuum tubes play a vital part in both the transmitting and receiving ends of radio communication. At the transmitter, a source of very-high-frequency alternating voltage is necessary to excite the antenna and generate the electromagnetic radiation that provides the means of communication. Early sources were the spark gap, the electric arc, and the high-frequency alternator. These were limited to relatively low frequency and could be used only for code transmission. The triode is capable of working as an oscillator, as well as an amplifier, thus providing a source of power that can cover a much wider range of frequency. The oscillator also is readily adaptable to the modulation of its output by tiny voltages generated by sound waves in a microphone, so that it can be used for transmission of voice as well as of code. Except for extremely high frequencies, where different types of electronic tubes are used, the triode or a modification is used universally in radio transmitters.

At the receiving antenna, the amount of energy is exceedingly small. By use of amplifiers this energy can be received and amplified to give adequate response in headphones or loudspeakers. An outstanding example of long-distance radio transmission and reception is the Mariner space probe, which transmitted signals to earth while passing the planet Venus. The transmitter was of relatively low power, but special high-sensitivity receivers were able to receive the signals reliably.

This transmission of signals from space vehicles in both manned and unmanned flights constitutes *telemetry,* a vitally important use of radio in which large quantities of data gathered by measuring devices on board the vehicle are automatically transmitted to observers on the ground and are recorded for analysis.

TELEVISION.—Television is a familiar example of the application of electronics. The basic principles are the same as those governing the transmission of sound by means of radio. A picture is focused on the photosensitive screen of a television camera. The screen is scanned by a sharply focused exploring beam of electrons moving rapidly across it in a succession of lines. Small electric currents are generated in response to the light and dark areas traversed by the end of the electron beam, and these currents are used to modulate the signal being transmitted. At the receiver a similar beam of electrons moves across a fluorescent screen, forming a tiny scanning spot. The brightness of the scanning spot varies in response to the received signal and reconstructs the picture as seen by the transmitter camera. Network programs may use both wire and radio transmission. Television transmission is tremendously more complex than sound transmission, for the frequencies generated by the scanning process are much higher and cover a broader range. Greater complexity yet is introduced in color television.

RADAR.—An extremely important application of radio came during World War II with the development of radar. A *radar transmitter* is a radio transmitter that emits a succession of short, sharp pulses. These signals are reflected by objects in their paths, and reflected signals return as echoes. A sensitive device receives these signals and measures the very short time they took to get back. This time is a measure of the distance to the object. The transmitting antenna concentrates its output into a beam that is constantly rotated so that the position at the instant the reflected signal is received gives the direction in which the object lies. Radar now is used as a means of navigation through fog by ships at sea, for the control of aircraft near airports, for military warning systems, and for the apprehension of speeding motorists by traffic police.

RECORDING.—Vacuum-tube amplification makes possible the high-fidelity recording and reproduction enjoyed today, and makes practical the slow-speed long-playing record. It also makes practical the tape recorder, both for sound recording and for the recording of television programs.

INDUSTRIAL APPLICATIONS.—Electronics plays a vital part in the rapidly increasing move to automation. Machine tools are automatically operated by electronic control devices. Steel mills are electronically operated to improve the product and reduce the cost. These controls may include computers that almost instantaneously process and analyze data received from measuring devices continually monitoring the operation, and automatically initiate corrections to be applied by controls.

Computers, which really are giant calculating machines, are used in design work to reduce the labor of calculation and to solve complex relationships that otherwise might require months of work. They are widely used in accounting procedures. As another example of their wide application, a flour milling company uses a computer to calculate formulas for cake mix and animal feed.

Semiconductor devices are competing strongly with vacuum tubes in radio and many other applications, particularly for portable use where small size and low battery drain are important. Examples are the transistor radio and the transistor hearing aid. Larger semiconductor devices are competing with gas-filled power rectifier tubes and other gas-filled tubes.

X RAYS.—X-ray machines are electronic devices. X rays are produced when a beam of electrons impinges on a metal target in a highly evacuated chamber. The penetrating power of the X rays is dependent upon the energy of the electron beam, expressed as the number of volts used to accelerate the electrons. Thus, a surgical X-ray machine may use 50,000–100,000 volts, while a therapeutic X-ray machine used for treating disease may require several hundred thousand volts. Machines using 10 million volts are used to X-ray large castings for detection of internal flaws or cracks. Even larger machines, requiring hundreds of millions of volts, are used for scientific investigation and research.

ELECTRON MICROSCOPE.—The electron microscope is a device in which a beam of electrons is used to examine and produce magnified images of objects that are many times smaller than the best light-beam microscope can see. This has been very useful in studying the structure of matter and the nature of viruses.

PARTICLE ACCELERATORS.—*Particle accelerators,* sometimes called "atom smashers," are huge electronic devices in which positively charged particles are accelerated to tremendous energy, measured in hundreds of millions—and billions—of volts. The radiations and new particles that are produced when these energetic particles bombard a target give new insights into the structure of matter.

ELECTRONIC WELDING.—Concentrated electron beams in evacuated chambers are used to heat and weld small parts. The heat applied can be precisely controlled, and materials that might be attacked or contaminated if heated in a gaseous environment can be welded.

ILLUMINATION.—Several of the familiar types of electric lamps are electronic in nature. The neon lamp is a tube containing neon at low pressure through which an electric current passes. Mercury vapor lamps are widely used for highway lighting. Fluorescent lights are glass tubes filled with argon and mercury vapor at low pressure, and coated on the inside surface with a phosphorescent material. An electric current in the gas produces light containing wavelengths that cause the phosphorescent material to glow brightly. Fluorescent lights produce several times more light for the same expenditure of power than do incandescent lights, which must dissipate relatively large amounts of heat.

—Robert Ferguson Edgar

SEMICONDUCTORS

Classification.—Solid substances are classified, on the basis of their electrical behavior, as conductors, insulators, and semiconductors. Such substances as metals, which offer little resistance to the flow of electrons, are classed as *conductors*. Substances such as glass, which offer a great deal of resistance to the flow of electrons, are classed as *insulators*. Materials such as germanium and silicon, which offer more resistance to the flow of electrons than conductors, but less resistance than insulators, are classed as *semiconductors*. All metals show increased resistance as the temperature is increased. Semiconductors and insulators, on the other hand, drastically decrease their resistance as temperature increases.

Basic Principles.—In a good conductor, such as copper, electrons are free to move through the material when an electric field is applied. This flow of electrons is called *electric current*. Semiconductor materials, by contrast, hold their electrons more securely and normally do not have enough free electrons to carry electric current. However, if the temperature of a semiconductor is raised, electrons are freed and become available to conduct current; thus, the resistance in the semiconductor material is reduced.

If a small amount of an impurity, such as antimony, arsenic, or phosphorus, is added to a semiconductor material such as silicon, a small number of electrons are freed. These electrons flow through the material producing an electric current. Semiconductors of this type are called *n-type semiconductors*. The addition of small amounts of other impurities —aluminum, boron, or gallium—to a semiconductor material such as silicon takes electrons away from a few of the atoms of the semiconductor. This lack of an electron in an atom is called a *hole*. Just as an electron can flow through a material, so can a hole, and this flow of holes also produces an electric current. Semiconductors of this type are called *p-type semiconductors*.

Semiconductor Devices.—The first important application of semiconductors was to provide rectifiers for low-frequency alternating currents. A *rectifier* is a device that passes electric current in only one direction.

About 1904, the rectifying properties of semiconductors were used to provide a detector of the high-frequency currents set up in circuits by radio waves. The *crystal rectifier*, consisting of a fine metal wire, or cat-whisker, in contact with a crystalline piece of lead sulfide or silicon, was the best detector for radio waves in the early days of radio. Vacuum tubes replaced such devices, and during World War II metal point-contact silicon devices were used in radar.

P-N JUNCTION DIODE.—More recently, crystal rectifiers have been developed that consist of a single piece of semiconductor containing both *p*-type and *n*-type materials. The junction between the *p*-type and *n*-type semiconducting regions, called a *p-n junction*, acts as a very good rectifier. *Silicon p-n junction diodes* are now widely used as rectifiers and logical elements in electronic equipment, particularly in computer systems.

TRANSISTORS.—The most important development in the application of semiconductors came in 1948 with the discovery of the *germanium transistor* by John Bardeen and Walter H. Brattain at the Bell Telephone Laboratories. The *transistor* is an electronic device that can be used to control or amplify electric current. The earliest type consisted of two metal point contacts placed close together on the surface of a germanium crystal and a third soldered at its base. One of the point contacts is called the *emitter;* the other, the *collector;* the soldered contact is the *base*.

In 1949 William Shockley of the Bell Telephone Laboratories outlined many of the theories that led to a basic understanding of semiconductors, and subsequently, with his co-workers, developed the *junction transistor*. This device consists either of a thin section of *p*-type semiconductor between two parts of *n*-type semiconductor (known as an *n-p-n junction transistor*) or vice versa (called the *p-n-p junction transistor*). Junction transistors and junction rectifiers caused a revolution in the electronics industry. Modern computers could not be built without these devices because of their superior reliability, speed, and useful life, as well as their small size and increased efficiency compared to electronic tubes. John Bardeen, Walter H. Brattain, and William Shockley received the 1956 Nobel prize for their work leading to the discovery of the transistor.

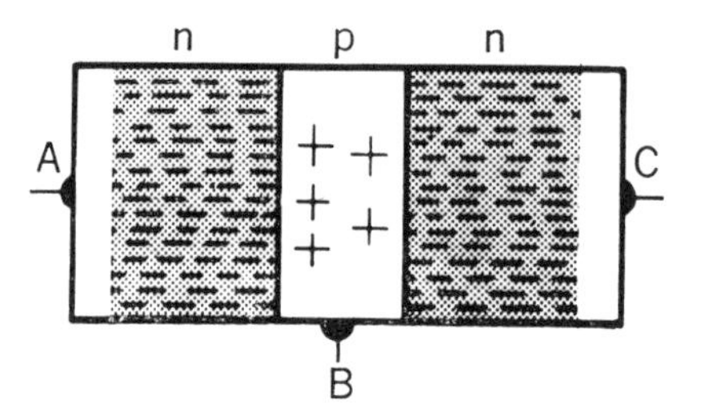

A JUNCTION TRANSISTOR can be used for amplification. Shown in simplified form: emitter (A), collector (C), and base (B).

The transistor, since its invention in 1948, has had more impact on modern electronics than any other device. Modern electronic computers depend for their existence on transistors because tubes are too unreliable, slow-acting, and bulky to be used in a large-scale computer. Radios, hearing aids, military electronic systems, and space communications systems now use transistors as active circuit elements because of the many advantages the transistor has over the vacuum tube.

FIELD EFFECT TRANSISTORS.—In recent years the field effect transistor (FET) has come into use. In the FET a flow of current from a *source* to a *drain* is controlled by a transverse voltage impressed on a gate. There are several kinds of FET. Most common is the MOSFET (metal-oxide-silicon). Its insulating layer of silicon oxide eliminates the need for physical isolation of the components. The MOSFET is favored for use in *integrated circuits*. Experimental versions of a more recent kind of FET, the Schottky variety, need no oxide gatelayer. They have produced oscillations in the microwave range.

INJECTION LASERS.—The *injection laser* was developed simultaneously in 1962 by Marshall I. Nathan and co-workers at IBM and by Robert N. Hall and co-workers at General Electric. It consists of a *p-n* junction in a compound semiconductor, such as gallium arsenide, indium phosphide, or indium arsenide, that emits coherent light (light rays that stay in a tight beam rather than spread) when sufficient electrons are injected into the *p*-region (region of holes). The electrons recombine radiatively —fill the holes—and emit light in the process. (See MASERS AND LASERS.) This device requires a specific geometrical configuration to permit electrical current to be converted directly into a highly directional and pure beam of light through laser action. The injection laser may ultimately be used in communications systems

TUNNEL DIODES.—The tunnel diode was invented by Leo Esaki in 1957. A *tunnel diode* differs from an ordinary *p-n* junction diode in that both the *n*-region and *p*-region are 100 to 1,000 times more impure in the tunnel diode, creating a very thin, abrupt junction between the regions. Under certain conditions, because of the junction, electrons can go from the *n*-side directly to the *p*-side without having to "climb over" the junction barrier. This process is called *tunneling*. Because tunneling is inherently a very rapid process, tunnel diodes are being used in computer switching circuits where high speed is needed.

PHOTOCELLS.—Semiconducting *photocells* are devices that detect light and convert its energy into electricity. There are three main types of semiconducting photocells: *photoconductive, photodiode,* and *photovoltaic*. Photoconductive and photodiode photocells are used as detectors of light, particularly of infrared light, while the photovoltaic type is used as a power source. An example of the latter is the *silicon solar cell* that has been used to power electronic circuits in satellites by direct conversion of solar energy into electrical energy.

■THERMOELECTRIC DEVICES.—Semiconducting devices that can convert thermal (heat) energy into electrical energy and vice versa are called *thermoelectric devices*. In these devices, two dissimilar semiconducting materials, having junctions or joints between the two materials held at different temperatures, develop an electric current due to a heat flow through the systems. A reverse effect is achieved when current is passed through junctions of dissimilar materials; that is, one junction is cooled and the other junction is heated. The first of these effects may have application in the conversion of heat developed in a nuclear reactor into electrical energy; the second may have application in small refrigerators with no moving parts.

Methods of Device Fabrication.—Since the most important semiconductor devices—diodes and transistors—depend upon *p-n* junctions, some of the fabrication techniques for preparing junctions will be mentioned. Some semiconductors are frequently prepared by melting the pure material in a sealed system and pulling a crystal from the melt by slowly dipping and withdrawing a *seed crystal* affixed to a cooled rod. In this fashion it is possible to prepare pure material as a single crystal. If *dopants* (impurity atoms) are added to the melt before crystallization, the result is a *doped* (impure) crystal.

Some *p-n* junctions are *grown junctions* where first one type of impurity and then another is added during the growth process. A modern modification of a grown junction is accomplished by depositing semiconductor material from a vapor (gas) phase onto a semiconductor *substrate* (subsurface). This method is widely used for high-frequency silicon devices.

An *alloy junction diode* or *fused junction diode* is prepared by placing a small dot or pellet of an acceptor impurity, such as indium, on one surface of a wafer of *n*-type germanium. When this combination is heated to the proper temperature, part of the indium dissolves into a portion of the germanium and forms a *p-n* junction between the indium-germanium alloy and the original *n*-type germanium. A *diffused junction* is formed by exposing a piece of *n*-type germanium, silicon, or some other semiconductor to a gaseous *p*-type material. The atoms from the gas diffuse through the crystal lattice of the solid; and where the diffusion stops, a junction is formed. This method is by far the most widely used for preparing junctions.

The integrated circuit, a recent innovation, is a small chip of semiconductor on which are deployed microminiaturized diodes, transistors, resistors, capacitors, and the needed insulation and connections. By using diffusion steps, hundreds of elements can be created on a single chip.

Although germanium is under study, silicon is still preferred for integrated circuits because the oxide of the element is relatively impermeable. This feature makes possible the precise diffusion of junction impurities and of insulation and connection materials on certain areas of the chip surface.

Large-scale integration (LSI), or the incorporation of additional elements on a single chip, has resulted from advanced applications in such areas as reduction photography and electron-beam experiments.

Semiconductor Theory.—The electrical properties of all crystalline solids can be explained in terms of a unified concept called the *energy band theory*. According to atomic theory, electrons in an isolated atom are in stable orbits surrounding a nucleus that is made up of protons and neutrons. The electrons orbiting about the nucleus have exact energies. If a large number of atoms with discrete electronic energy levels are assembled into a three-dimensional array called a *crystal lattice*, the energy levels become perturbed and form bands of energy. These bands are made up of closely spaced electronic energy states that, according to the *Pauli exclusion principle*, can accommodate only two electrons per state. The bands are separated by forbidden gaps, or values of energy, that electrons cannot possess in a perfect lattice. The occupancy of these bands and the width of the forbidden gaps determine the electrical conductivity of solids.

Metals are solids in which the last occupied band (the one of highest energy) is only partially full. This is called the *conduction band*. When an electric field is applied to a piece of metal by a battery, the electrons near the top of the occupied energy states become excited and move into unoccupied energy states within the same band; thus they can move freely through the crystal. As temperature is increased, atoms of the lattice vibrate more due to thermal energy, and electrons are scattered more, thereby decreasing the metal's conductivity.

In insulators the last occupied band, called the *valence band*, is completely filled and the nearest unoccupied energy states are in the conduction band, which is at much higher energies. Transitions of electrons from the valence band across the energy gap to the conduction band cannot occur except under very high electric fields or at extremely high temperatures. The band gap for a good insulator is approximately 10 electron volts, compared to a thermal energy of the electrons at room temperature of 0.025 electron volt. Thus insulators have no electrons in the conduction band to conduct electricity and their conductivity is very low.

An *intrinsic* (pure) *semiconductor* at a temperature of absolute zero (−273° Centigrade) has a filled valence band and an empty conduction band. Therefore, at low temperatures an intrinsic semiconductor is an insulator. The band gap of semiconductors is much smaller than that in insulators. Typical band gaps range between 2.5 and 0.2 electron volts. Therefore, thermal energy available to electrons at room temperature and above is enough to excite a measurable number of electrons from the valence band to the conduction band. Then the negatively charged electrons in the conduction band and the holes in the valence band can be moved through the lattice by an electric field. The hole moves away from the electron, but also has an opposite charge and so adds to the total current. The number of holes is equal to the number of electrons in the conduction band of an intrinsic semiconductor because a hole is created every time an electron is excited to the conduction band. The reason the conductivity of intrinsic semiconductors increases markedly as temperature increases is that the number of electrons excited to the conduction band depends very strongly on temperature. The band gap of a semiconductor is an important property of the material. The table lists the room-temperature energy gaps of a few important semiconductors.

Semiconductor	*Chemical Symbol*	*Energy Gap**
Germanium	Ge	0.66
Silicon	Si	1.09
Indium phophide	InP	1.29
Indium arsenide	InAs	0.36
Indium antimonide	InSb	0.175
Gallium phosphide	GaP	2.25
Gallium arsenide	GaAs	1.43
Gallium antimonide	GaSb	0.72
Cadmium sulfide	CdS	2.4
Cadmium telluride	CdTe	1.5
Lead sulfide	PbS	0.37
Lead telluride	PbTe	0.29

*Electron volts.

In this discussion of the band theory of solids, the electrical behavior of metals, insulators, and intrinsic semiconductors has been covered. *Extrinsic* (impure) *semiconductors* are technologically much more important in the electronics industry than intrinsic material. The effects of impurity atoms in controlling the electrical behavior of semiconductors can also be explained by the band theory. When impurity atoms enter an intrinsic semiconductor lattice, in addition to the conduction band or valence band there are electron energy states in the forbidden gap that are close to either the conduction band or valence band.

Impurities that introduce electronic states that can give up electrons to the conduction band are called *do-*

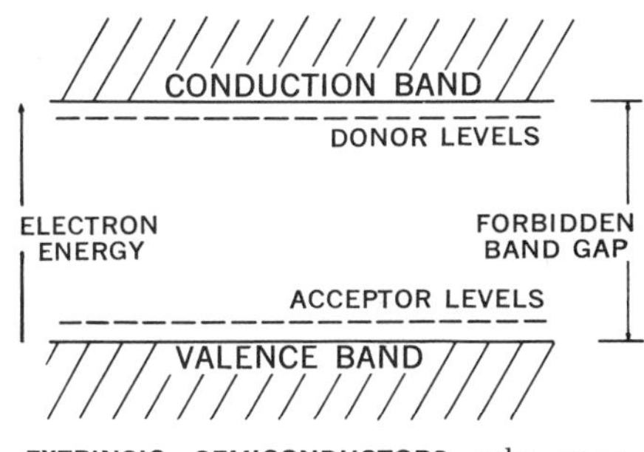

EXTRINSIC SEMICONDUCTORS rely upon impurities to control electrical behavior.

nors, while those that produce electronic levels that can accept electrons from the valence band are known as *acceptors.* Frequently the donor levels are near the conduction band in energy and can donate electrons to the conduction band when excited by a small fraction of the band-gap energy. Similarly, the acceptors are frequently near the valence band and can accept electrons excited from the valence band, leaving holes in the valence band. Because of the closeness of the impurity levels to the bands, small numbers of impurities donate or accept electrons, producing conduction electrons or holes that would not be available if the electrons had to be excited from the valence band to the conduction band. In an extrinsic semiconductor, when the number of donor impurities giving up electrons exceeds the number of acceptor impurities producing holes, the conduction is dominated by electrons and is *n*-type. The inverse situation occurs in *p*-type material, where holes dominate.

In 1963 J. B. Gunn discovered that small samples of gallium arsenide showed regular, high-frequency current oscillations when subjected to a threshhold voltage. Similar oscillations were later produced in germanium samples. This so-called "bulk" behavior of specific lengths of material under fixed conditions may lead to the development of tiny, reliable devices free of complex tuning circuitry. The range of possible applications extends from spacecraft communications systems to hand-held radar devices for the blind.

P-N Junctions.—It is possible to produce both an *n*-type region and a *p*-type region in a single crystal piece of semiconductor material. Such a

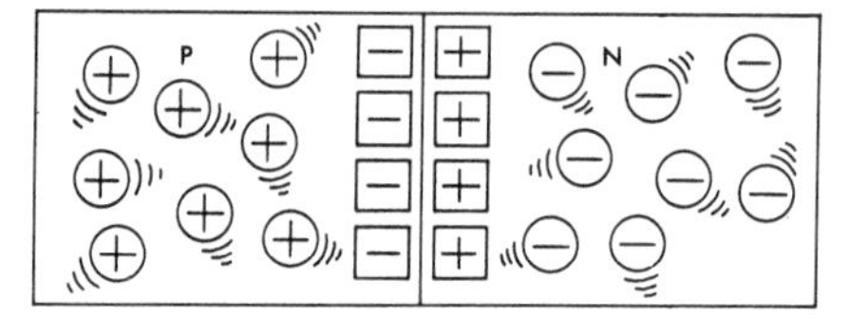

P-N JUNCTION in diagram form. The circles represent mobile electrons (—) and hole (+) carriers. The signs in the squares represent stationary ions at the junction.

combination is called a *p-n junction* and is the basis for many of the semiconductor devices. In the *n*-region, donor impurity atoms give up electrons in greater numbers than acceptor atoms contribute positive holes. In the *p*-region the situation is just the opposite: holes are in excess compared to electrons. Most of the electrons remain in the *n*-region, and most of the holes remain in the *p*-region. However, some electrons from the *n*-region diffuse into the *p*-region, leaving behind positive donor ions. Similarly, some holes from the *p*-region diffuse into the *n*-region, leaving negatively charged acceptor ions. An electric field is established at the junction due to the impurity ions, and this field inhibits additional electrons from diffusing out of the *n*-type material or holes from diffusing out of the *p*-type material.

If a battery is connected with its negative terminal on the *n*-type material and its positive terminal on the *p*-type material, then electrons from the battery neutralize the positive ions on the *n*-side of the junction and electrons enter the positive terminal of the battery from the *p*-region, thereby decreasing the negative charge on the *p*-side. Consequently, the junction field is diminished and electrons can move from the *n*-type material through the junction to recombine with holes. The holes on the *p*-type side carry the current from the positive battery terminal to the junction. A *p-n* junction connected to a voltage source in this fashion is said to be *forward biased,* and the device can carry appreciable current.

If, however, the *p-n* junction is *reverse biased* (with negative voltage applied to the *p*-side and positive voltage to the *n*-side), a field is present that causes the mobile holes and electrons to move away from the junction and each other, toward the *p*-contact and *n*-contact respectively. Electrons are added to the *p*-region and removed from the *n*-region by the battery, leaving a wider region near the junction charged with immobile impurity ions but depleted of mobile charge carriers. The junction field is enhanced by reverse biasing, but very little current flows because of the depletion of mobile charges. A small amount of current flows because of thermal excitation of holes and electrons in the junction region. These carriers contribute a small but finite reverse current. A *p-n* junction passes electrical current when forward biased, but opposes the flow when reverse biased.

Such a device is called a rectifier and can be used to convert alternating current into direct current. Solid-state rectifiers have wide commercial applications as radar or microwave detectors, computer switching diodes, television and radio diodes, and power rectifiers for battery chargers and electronic power supplies. Although vacuum-tube rectifiers were widely used for fifty years, the invention of the *p-n* junction rectifier and the vast needs of computer companies for reliable, small, efficient, and long-lived circuit elements have made semiconductor diodes dominant.

Junction Transistor.—The *junction transistor* is a solid-state device consisting of a thin region of *p*-type or *n*-type semiconductor between two regions of the opposite type semiconductor. Germanium and silicon are the most important semiconductor materials for transistors. While both *n-p-n* and *p-n-p* transistors are used commercially, only the *n-p-n* will be discussed here. Two *p-n* junctions exist that share a common narrow *p*-region. Three contacts called the *emitter, base,* and *collector* are made to the three regions. The *n-p* junction between the emitter and the base contact is the emitter junction and is forward biased. The junction between the collector and the base is the collector junction and is reverse biased. Because the emitter junction is forward biased, electrons flow from the emitter contact through the *n*-region of the emitter to the emitter junction. There they are injected into the *p*-region. These injected electrons diffuse as minority carriers across the thin *p*-base region and are collected by the reverse biased collector junction.

The base region should be thin or narrow so that practically all the injected electrons are drawn across to the collector. They are swept into the collector by the large junction field of the reverse biased collector before they find their way to the base contact or recombine with holes. The emitter current also includes holes injected from the base into the emitter region. By introducing impurities more heavily in the *n*-region of the base, the electron injection can be made to dominate as required for an efficient transistor. The fraction

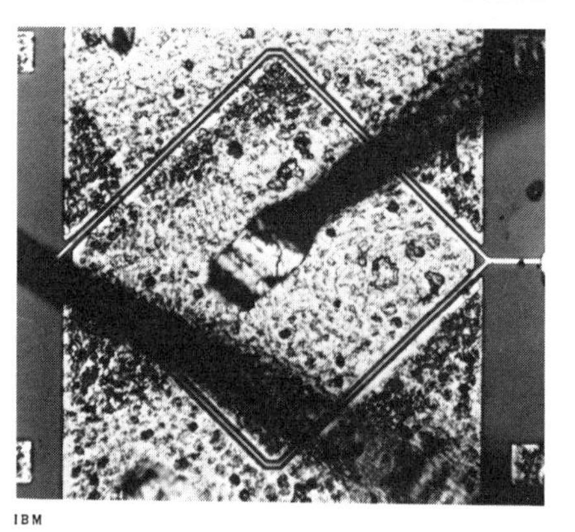

IBM

MINIATURIZED TRANSISTOR. The four "legs" forming the square gate are only 1/250 inch.

of the emitter current that crosses the collector junction is known as the *alpha* (α) of the transistor. For junction transistors α can reach 0.99.

If the emitter current is varied by a signal voltage introduced at the input terminals, then there will be a corresponding variation of the collector current. Since the collector *impedance* (resistance) is high compared to the impedance of the forward biased emitter, a large-load resistor can be used in the output to get voltage amplification or power amplification. For example, if the impedances of the input and output circuits are 100 and 1,000,000 ohms respectively, then there is a power gain and a voltage gain of about 10,000 ohms.

For a more quantitative treatment of the junction transistor, let us assume that the input signal voltage increases in a negative amount of ΔV_i. The forward biased emitter has a low resistance, r_e, so that an increase in the emitter voltage will produce a corresponding increase in emitter current of

$$\Delta I_e = \frac{\Delta V_i}{r_e}.$$

The amount of this increased emitter current that reaches the collector depends on the current gain (α) of the transistor defined by

$$\alpha = \frac{\Delta I_c}{\Delta I_e},$$

where ΔI_c is the change in the collector current. Thus

$$\Delta I_c = \alpha \Delta I_e.$$

If the load resistor in the collector circuit is R_L, then the voltage gain of the transistor is

$$G_v = \frac{\textit{output voltage}}{\textit{input voltage}}$$

$$G_v = \frac{\alpha \Delta I_e R_L}{\Delta I_e r_e} = \alpha \frac{R_L}{r_e}.$$

If $\alpha \cong 1$,

$$G_v = \frac{R_L}{r_e}.$$

The power gain is simply given by

$$G_P = \frac{\textit{power out}}{\textit{power in}} = \frac{\Delta I_c^2 R_L}{\Delta I_e^2 r_e}$$

$$= \frac{\alpha^2 \Delta I_e^2 R_L}{\Delta I_e^2 r_e}$$

or

$$G_P = \alpha^2 \frac{R_L}{r_e},$$

but if $\alpha \sim 1$,

$$G_P \cong \frac{R_L}{r_e}.$$

Because the junction transistor in different circuit configurations can produce voltage gain, power gain, or current gain, the transistor can be used as an amplifier in place of a vacuum tube. Transistors have many advantages over tubes. They are more reliable, more durable, more efficient, much smaller, and can work at lower signal levels with lower noise. They have no heater or cathode to fail and can be flexibly biased because of the existence of both the *n-p-n* and the *p-n-p* structures.

—William J. Turner

SOUND

Waves and Vibrations.—It is very difficult to realize that sound is a wave, yet it is. To visualize a wave, imagine you are holding one end of a rope whose other end is tied to a doorknob. If you make a sharp up-and-down motion with your hand, a bump starts moving down the length of the rope. The rope itself does not move forward—only the bump does. This is one of the very important characteristics of waves: the medium (in this case the rope) through which the wave travels does not move forward or back, it swings up and down across the direction of wave motion.

If a pebble is dropped in the center of a pond, the waves move away from the point of impact in ever-widening circles. The wave in the rope demonstrates a one-dimensional wave—it moves on a plane. Sound, on the other hand, like the waves in the pond, is a three-dimensional wave. The crests and troughs of sound are not bumps, but high-pressure areas (*compressions*) and low-pressure areas (*rarefactions*) of the spongy mass we know as air. Each pressure layer forms the surface of an ever-growing sphere. Each following rarefaction is a slightly smaller sphere just inside the pressure sphere and is itself followed by another sphere of high pressure.

A sound wave moves forward even though the air really only compacts a little and then thins out a little as the sound wave passes. The wave is started by the quick motion (*vibration*) of an object. When you rap the head of a drum, it moves away from the stick, creating a partial vacuum in the layer of air touching the surface of the drum. The surrounding air rushes in to fill this vacuum, leaving a partial vacuum in the second layer just above the drum head, and so on away from the surface. The drum head then returns from its depressed position and bounces to a level higher than normal. This compacts the air immediately above the drum, and the high-pressure area in this space also moves out into the air immediately behind the rarefaction. Thus a wave is created by the back-and-forth motion of the drum head. A radio speaker cone works in exactly the same way. It is pulled sharply back and forth in a quick series of vibrations (*oscillations*) that transmit themselves into the adjacent layers of air.

The sound waves move out on the surfaces of ever-growing spheres. Since the amount of energy carried by a single wave at the start remains with that wave throughout its existence, the amount of compression and rarefaction decreases as the area of the sphere increases. In theory, the wave never disappears entirely; in actuality, the rubbing of air molecules against each other slowly absorbs all of the initial energy and turns it into heat, so that eventually the sound wave is dissipated.

This dissipation of sound energy is even more pronounced in enclosed areas. In a room or auditorium the sound waves initiated on the stage strike the walls, the ceiling, and the floor. While some of the sound is reflected back into the room (causing *reverberation*, or echo), the rest is absorbed by such things as curtains, rugs, and clothes.

VOLUME AND PITCH.—Sound has two important characteristics. The first is *volume* (loudness), which is the intensity of the pressure in the compression part of the wave and the reduction in pressure in the rarefaction. The greater the rise of pressure and the thinner the rarefaction, the louder the sound will seem. The second characteristic is *pitch*, which is the frequency of the waves. The more rapidly the sound waves follow each other, the higher the sound's pitch. A loudspeaker cone that moves back and forth very quickly produces a tone much higher in pitch than one that goes back and forth slowly.

VELOCITY.—Regardless of the frequency of the wave, sound always travels through a particular medium at the same speed. The speed depends on the "springiness," temperature, and density of the medium. Since air is a relatively soft material, sound travels through it at the fairly slow speed of approximately 1,100 feet per second—about 750 miles per hour—at sea level at a temperature of 70° F. Water is stiffer and denser than air, so sound moves faster in the sea—4,800 feet per second. Sound moves fastest in a piece of hard steel —over 16,000 feet per second (more than 10,000 miles per hour).

WAVELENGTH.—In air, sound waves always travel at about 1,100 feet per second, no matter how quickly a loudspeaker cone moves back and forth. Therefore, if a speaker completes one cycle in 1/30 of a second (forward in 1/60 of a second and back in 1/60 of a second), the wave that moves away from the speaker will have traveled about 37 feet by the time the cone is ready to start its second swing; that is, 1/30 (of a second) times 1,100 (feet per second). But if the speaker moves back and forth at the high pitch of 20,000 cycles per second, the wave would only be 1/20,000 times 1,100, or about 6½ inches long.

Engineers have discovered that if a sound wave is to be created efficiently—that is, if a reasonable amount of the energy needed to move the speaker cone back and forth is to be transformed into sound —the diameter of the speaker cone and the length of the sound wave must be approximately equal. This is why small speakers (*tweeters*) work very well for the very-high-frequency (very short) waves, while the deeper sounds require much larger speakers (*woofers*) coupled with large wooden enclosures to move an appreciable amount of sound energy into a room.

Resonance.—*Resonance* is the phenomenon by which sound waves are reinforced so that they sound louder. Resonance may take place in two specific ways, both of which work on the same principle. If the handle of a vibrating tuning fork touches a table, the sound emanating from the tuning fork becomes louder. The reason for this is that the tuning fork causes the particles of the table top to vibrate. Thus the table top also produces sound waves (by causing compressions and rarefactions of the air layers above it) of the same frequency as the sound waves coming from the tuning fork. The two sound waves join to produce a larger and louder sound wave. This type of resonance—one vibrating body causing another to vibrate—is called *sympathetic vibration*.

The other method of producing resonance can be illustrated by holding one end of a long cardboard tube in a bowl of water. If a vibrating tuning fork is held near the open end of the tube and the tube is moved slowly into the water, eventually (when the tube has reached the proper position) the sound will be amplified. This happens because the sound waves emanating from the tuning fork travel down the tube, bounce off the water, and travel back up the tube. When the length of the tube (measured from the open end to the water surface) is equal to the wavelength

DOPPLER EFFECT explains the apparent change in pitch of a train's whistle by the bunching and spreading of sound waves as the train moves toward and away from the listener. This principle is used by submarines to detect the speed and direction of other vessels.

of the sound waves produced by the tuning fork, the reflected waves will be *in phase* (the rarefactions and compressions will coincide) with the waves emanating from the tuning fork. This reinforces the sound waves and makes them sound louder. This resonance is an *echo phenomenon.*

Doppler Effect.—If an observer and a source of sound are moving relative to one another (one or both may be moving, but they cannot both be moving in the same direction at the same speed), the observed frequency or pitch of the sound waves differs from the frequency or pitch emitted. This is called the *Doppler effect,* after its discoverer, Christian Johann Doppler (1803–1853), an Austrian physicist. Picture a series of sound waves emanating from the whistle of a railroad locomotive. When the locomotive is standing still, the waves move out from the whistle in an ever-widening set of spheres. If the locomotive is moving, the sound waves going in front of the train travel at a speed of about 1,100 feet per second (the speed of sound waves in air) plus the speed of the locomotive, while the waves going behind the train are slowed by an amount equal to the speed of the locomotive.

A person standing alongside the track notices this difference. As the train comes toward him, its whistle seems to rise in pitch because the waves reach his ear at a frequency greater than the real one. But as soon as the train passes, the hearer is in the retarded sound area; the pitch suddenly drops and then slowly climbs to normal as the train moves into the distance. The Doppler effect is used to help submarines detect the speed and direction of surface vessels and even of other submarines through a device called *sonar.* (It is also used in astronomy, where the change in the frequency of light waves is measured to determine the velocities of distant stars.)

Hearing.—The human ear is sensitive to sound frequencies from about 30 cycles per second (the very low sounds that can be made by the enormous pipe organs in some old churches) up to 17,000 to 20,000 cycles per second, which are more like barely audible whistles or squeaks than real sound. Dogs are said to be able to hear sounds in the range of 20,000 to 25,000 cycles per second—the reason why whistles producing sounds in that range can be heard by dogs but not by people. Bats use high-pitched squeaks, echoed from trees and walls, to fly between obstacles and even to locate and catch their prey on the darkest nights.

When a sound wave approaches the ear, it is directed down a horn-shaped duct that ends at a thin membrane called the *ear drum.* This membrane reacts to the alternating compression and rarefaction waves by moving back and forth, much as the speaker cone moves back and forth in making a sound wave. Just behind the ear drum, and attached to it at one point, is a tiny linkage composed of three small bones that provide a mechanical transfer of the motion of the ear drum across the middle ear. The *hammer* swings back and forth with the motion of the ear drum and transmits this motion to the *anvil;* the anvil in turn transmits it to the *stirrup,* the foot plate of which covers an oval window at the far side of the middle ear.

This bone linkage hangs in the middle ear from a set of ligaments that support and protect the hearing. When extremely loud sounds are received, causing the ear drum to bulge unduly in either direction, muscles holding the ligaments tense reflexively, limiting the motion of the linkage and thereby reducing the reaction at the sensitive inner ear.

Sound is actually sensed by the *cochlea,* a spiral organ shaped like a small snail shell, the larger end of which is attached to the middle ear. If the cochlea were unwound, it would make a hollow, tapered tube about 1½ inches long. Across the center of the tube, and running its full length, is a thin web carrying hundreds of nerve fibers that end in the roots of tiny hair cells standing crosswise in the web. The entire cochlea is filled with fluid.

When a pressure wave, received by the ear drum and transmitted through the linkage in the middle ear, arrives at the window to the cochlea, the window is driven into the fluid of the cochlea and pushes the dividing web down slightly. The extra pressure is relieved by a bulge in a second window to the cochlea from the middle ear. When the following rarefaction pulls the ear drum and linkage away, the dividing web in the cochlea moves up slightly and the second communicating window bows inward to compensate.

Thus sound waves received at the ear drum are transmitted very efficiently to the sensitive web in the cochlea. In some way not precisely understood, the nerve endings in the web interpret these complex vibrations in terms of pitch and volume. When these are compared to the sounds received at the other ear, the brain can determine the direction from which the sound comes.

The most popular theory explaining how the ear hears is based on sympathetic vibration. Just as the strings of a piano start to vibrate when the same notes are sounded on another instrument so, it is thought, the tiniest hairs in the small end of the cochlea respond to the higher frequencies while the longer ones at the thick end respond to the lower frequencies. The chief objection to this theory is in the fantastic sensitivity of human hearing. The range of 30 to 20,000 cycles per second would require a much greater range in thickness of cochlea membrane and in length of hair cells than actually exists. Recent studies of the electrical signals sent out by ear nerve cells suggest that the frequency of impulses sent out by the nerve fibers themselves may be related to the frequency of sound waves.

Acoustics.—Since ancient times, instrument makers have created magnificent musical instruments with the sole purpose of producing beautiful sounds. Yet their skill was, and to a great extent still is, largely the result of trial and error. The knowledge of how best to shape the resonance chamber, the best woods for construction, the time required for seasoning, and the amount of varnish that will achieve the optimum sound quality is the result of hundreds of years of experimentation.

Over the last hundred years, however, the making, sensing, recording, and reproduction of sound have evolved into a science. Acoustic engineers know what sound is, how to detect it, and how to preserve it on records that can be played again and again. Sound waves are detected by microphones that work much as the ear does—a thin membrane is made to vibrate back and forth as the compressions and rarefactions impinge upon it. The vibrating membrane in the microphone may move a coil of wire in a magnetic field to create an electric voltage. Alternatively, it may move toward and away from a flat metal plate, changing the capacitance between the two and affecting an electric circuit accordingly. In another type it compacts a packet of finely ground carbon particles, thus changing the electric resistance of the packet and altering an electric circuit. Once the sound is transformed into an electric signal (that oscillates in voltage as sound waves do in pressure), it can then be *amplified.* (The voltage is increased while the relative strengths of loud and soft signals are maintained.) Amplification is necessary because the electric signal generated by the microphone is so weak that only the most sensitive electrical measuring instruments can detect any change at all.

SOUND RECORDING.—There are many ways to preserve the pattern of rising and falling voltage. For example, the signal can be made to move a pen back and forth over a moving strip of paper to create a jagged trace; it can be used to swing a light beam back and forth across a moving photographic film, which is then developed to show the changing intensity and frequency by the sweep and density of lines. The signal can drive a pointed stylus back and forth as a soft plastic disk turns under the point, leaving a wavy groove; it can power an electromagnet to create magnetized and reverse-magnetized areas on either a strip of thin plastic tape coated with tiny iron oxide particles or an iron wire. All of these methods are practical. The photographic film is used in movies; the plastic disk is the master from which records are duplicated; the coated tape is used in tape recorders.

SOUND REPRODUCTION.—Sound is recreated by simply reversing the recording process. In the tape recorder an electromagnet senses the changing magnetism of the coated tape passing across its face and reproduces the original electric signal. A stylus rides in the groove of the record disk; and as it moves back and forth, a tiny coil of wire builds up a similarly varying voltage. A light beam shines on the motion picture film and dims and brightens a tube that functions like an electric eye. Once again the weak electric signal must be amplified. It can then drive the electromagnet in the loudspeaker to move a speaker cone back and forth and reproduce a nearly exact duplicate of the sounds originally recorded.

STEREOPHONIC SOUND.—One of the most recent advances in sound recording and reproduction is the achievement of a *stereophonic effect*. As stated before, the brain can recognize the direction from which sound comes by comparing the relative pitch and volume of the sound waves received by each ear. For example, if an orchestra has the brass section on the hearer's left and the percussion section on his right, the sound waves from the trumpets will reach his left ear before reaching the right ear; those from the tympani will reach the right ear before the left. By interpreting the time of arrival and the *intensity* (volume) of the sound reaching each ear, the brain can determine the position of the instruments. In stereophonic recording, two microphones are set some distance apart in front of the orchestra, one recording the sound waves emanating from the left side of the orchestra more strongly, and vice versa. The electric signals produced by each of the microphones are recorded on a separate *channel,* or track. When the sound is reproduced through two separated speakers, each channel is reproduced in only one speaker. Thus the sounds recorded by the left-hand microphone are reproduced by the left-hand speaker; those recorded by the right-hand microphone are reproduced by the right-hand speaker. In this way the listener can sense the direction of the sound, which creates a greater feeling of spaciousness and depth than is possible with *monaural* reproduction.

LINCOLN CENTER FOR THE PERFORMING ARTS

AN ACOUSTICALLY BALANCED AUDITORIUM of the Philharmonic Hall at Lincoln Center.

Room Acoustics.—Sound engineers face enormously difficult problems in designing concert halls. When the Philharmonic Hall at Lincoln Center for the Performing Arts in New York City was planned, studies were made of all the world's great concert halls to see if the best characteristics of each could be duplicated. The ideal hall would distribute music equally to all parts of the auditorium, so that there would be no seats where the music was either overly loud or inaudible. Furthermore, there would be no preferential treatment of some of the instruments at the expense of the others—the base drum would be heard as well as the piccolos. In addition, the *reverberation time*—the time it takes for all the echoes of a sharply struck note to die out—would have to be close to the ideal 1.7 to 2.0 seconds.

All of this can be accomplished by shaping the sides, roof, and floor of the hall so that the sound created on the stage is moved out into the auditorium. The back of the stage is usually shaped like a giant searchlight reflector, to reflect sound rather than light. In Philharmonic Hall carpeted floors, specially treated walls, and thousands of individually positioned reflectors mounted near the ceiling gave the designers an opportunity to change the acoustical characteristics almost at will. Dummy people—blocks of plastic foam with the same acoustic properties as people—were placed in the seats to simulate a filled hall; and engineers studied every area, checking sound levels and characteristics with special meters.

The work done on the Philharmonic Hall is typical of the work of acoustic engineers. They are called upon to design soundproof rooms, to construct rooms that will amplify sounds, to design offices that will minimize the clatter of typewriters, and to construct echo chambers. To meet these wide-ranging demands, a multitude of materials have been designed to meet the specific needs of room acoustics. Sound-absorbing panels for ceilings and walls, thick rugs and drapes, and reflecting panels have become commonplace, but their function in controlling undesirable noise and amplifying pleasant sounds has often gone unnoticed.

Musical Acoustics.—Every impact between two physical bodies, no matter how delicate, starts a pressure wave moving into the surroundings. Two leaves brush against each other, a tree falls in the forest, a moving wind flaps a shutter or whistles through a cracked windowpane—these are all naturally created sounds, but there are many ways to make sound artificially as well. The loudest artificial sound is probably that of the hydrogen bomb, which creates pressure waves so great that buildings are flattened and windows shattered miles from the center of the explosion. Thunder is the sound made when air rushes into the vacuum created by the passage of electric current in a lightning bolt. Sound travels much more slowly than light, so we see the lightning flash first. The sound follows, lagging by five seconds for every mile separating the observer and the lightning.

In an explosion or a thunderclap we are dealing with a single disturbance, possibly followed by echoes

LOOK MAGAZINE

SOUNDS produced by reed (*left*), percussion (*center*), and stringed instruments (*right*) are characterized by the superimposed patterns.

from hills or large buildings nearby. The loudest continuous noise is that produced by sirens. A siren is made by drilling a series of holes in a metal cylinder and then aiming a jet of air at the holes. When the cylinder is turned at high speed, the air jet is broken into a series of *puffs*, or sound waves. The number of puffs per second (the number of holes that pass in front of the air jet in a second) determines the frequency or pitch of the siren.

Musical instruments create a very special type of sound. Music is made up of sounds of clearly defined pitch. Even drums and triangles (percussion instruments) are tuned so that, when struck, a distinct tone is created. The string instruments take advantage of the fact that a taut string will vibrate at a frequency dependent upon its tension, thickness, and length. The shorter, tighter, and thinner the string, the higher the frequency. This is how a violinist can change the notes. He strokes the string with a rosined bow (rosin increases the friction between the string and the bow), which starts the string vibrating and keeps it in motion as long as the bow is in contact with the string. The violinist can shorten the string and thus raise its pitch by clamping it with his finger against the fingerboard.

The wooden box of the violin, which is behind the strings, makes the sound louder and "colors" it as well. The pure note of the string alone in space would be barely audible if there were no wooden box to set larger volumes of air into motion. The differing shapes, sizes, and designs of violins, guitars, mandolins, and pianos create distinctly different sounds, even when strings of the same length and thickness are tightened to the same degree in two or more instruments. This occurs because these shapes do not merely amplify the single note vibrated by the string. They add other, higher notes (*harmonics*) of two, three, four, five, and more times the frequency of the basic tone. The number of these harmonics give all musical instruments their distinctive character. They also cause some instruments to give a richness and mellowness to all the notes within their range, from the very lowest to the very highest.

The third major class of musical instruments is the horns. In these, sound is started by a flow of air moving past a flexible element (a reed in the oboe or clarinet, the musician's lip in the trumpet and tuba) so that the air passage is alternately opened and restricted. The pulsing flow travels down a metal or wooden tube and out a bell-shaped end. The length and diameter of the tube also influence the pitch of the note. When you blow across the open mouths of several bottles of different sizes, the larger bottles make the deeper tones; but a large bottle will give off a higher note than usual if it is almost filled with water. Some horns, such as the oboe or clarinet, have holes drilled along their length so that the musician can adjust the length of the tube by opening and closing different holes. Others, such as the trumpet and tuba, have valves that the musician can open or close to increase or decrease the length of the tube. In this way, combined with control of breath pressure, he changes notes.

The musical scale is no accident; it is a very carefully ordered set of frequencies. The intervals are based on the octave (C to C, for example), which is adjusted so that the higher note is precisely double the frequency, in cycles per second, of the lower note. In 1955 an international agreement set middle A at exactly 440 cycles per second; all other notes are based on this.

For many years the range between any two notes an octave apart has been divided into twelve parts. Originally these were not twelve equal divisions, but intervals determined by ratio. Thus G was always exactly 3/2 of the C below, and E was exactly 81/64 of the C below. This scale was invented by Pythagoras (died c. 497 B.C.) through his experiments with lengths of strings. Each additional note was set by stepping off the so-called *perfect fifths*—the ratio of 3/2 between G and C is a perfect fifth. The next ratio, between D and G, would be found by multiplying 3/2 by 3/2 and then dividing by 2 to reduce the D to the octave under consideration. Thus D is 9/8 above C, and so on.

Later this scale was modified for better harmony. D remained 9/8 of C, but E was made 5/4 (80/64 instead of 81/64) times C. However, all these modifications left the same problem. For example, the F sharp in the G scale is not quite the same frequency as the G flat in the D flat scale. Today all keyboard instruments are tuned to the *equal temperament scale*. In this, the twelve intervals are all equal, so that the frequency of each note is exactly $2^{1/12}$ (1.05946) times the preceding note. Middle A is 440; B flat is $440 \times 2^{1/12} = 466.2$; B is $440 \times 2^{2/12} = 493.9$; C is $440 \times 2^{3/12} = 523.3$; and so on up to A again, which is $440 \times 2^{12/12} = 880$ cycles per second.

The similarity and harmony of all A's, for example, occur because each is a simple multiple of all the others. There is a low A that is 55 cycles per second, another at 110, a third at 220, a fourth at 440, a fifth at 880, a sixth at 1,760, and the highest generally heard in written music at 3,520 cycles per second. Music does not generally take into account the ability of human ears to hear up to 20,000 cycles per second for two reasons. First, these high-pitched squeaks are not appealing to the ear unless they are accompanied by lower tones. Second, these higher frequencies do occur as harmonics (multiples), and they play a large part, as mentioned above, in giving a particular instrument its distinctive character.

CHORDS.—Individual notes sound well when played together (*chords*) because there is a perfect match in the higher harmonics produced when the notes are sounded simultaneously. The similar harmonics of a chord do not appear until well up into the thousands-of-cycles-per-second range, yet this is enough to give them a melodious sound when played together. For example, take a simple major chord: C-E-G. The frequencies of the three basic notes are 261.6, 329.6, and 392.0 cycles per second. The third harmonic of C is 784.8; the second harmonic of G is 784.0—so close as to be undistinguishable. The fifth harmonic of C is 1,308.0; the fourth harmonic of E is 1,308.4—close enough to fool the human ear.

Ultrasonics.—Aviation engineers appropriated the word *supersonic* to define airplane speeds faster than the speed of sound (in air)—over 700 miles per hour. *Ultrasonic* refers to those frequencies of sound waves above the range of human hearing. These high-frequency waves are made with much the same equipment used to reproduce music in the home. However, for these very-high-frequency waves the area of the speaker face need not be so large (a 1,000,000-cycle-per-second ultrasonic signal has a wavelength of approximately the thickness of three pages of this book). Further, these very-high-frequency waves do not move from the sound producer in ever-widening spheres, but tend to stay in a narrow beam.

Certain crystalline materials, such as quartz and Rochelle salts, have the interesting property of changing shape when an electric voltage is applied across their opposite faces. This change in dimension is only a few thousandths of an inch, but that is all that is needed to start a sound wave in adjacent layers of air. These *piezoelectric crystals* are, therefore, the ideal sources of ultrasonic waves (*ultrasound*). The electronics specialist can make an electric signal oscillate through extremes of voltage a million or more times a second with little difficulty so it is quite easy to experiment with ultrasonic waves.

Ultrasound has some very peculiar effects. When it moves through water, it creates billions of tiny bubbles when the rarefaction part of the sound wave tries to thin out a layer of water. The water cannot thin out in this fashion without losing its basic liquid character and therefore turns into a vapor. But the vapor bubble lasts only a fraction of a second before a pressure wave snaps the bubble together with a bang. Scientists think that local pressures in the range of hundreds of thousands of pounds per square inch are generated in these tiny collapsing bubbles. Studies of *cavitation* (the creation and destruction of vapor bubbles) explained why high-speed ship propeller blades wear out faster than lower-speed propeller blades. With their extremely high local pressures, the collapsing bubbles quickly destroy the surfaces of the metal blades. It is exactly like striking the blades with a sharply pointed hammer. The destructiveness of ultrasound is not limited to marine propellers. Fish are killed by the millions when their delicate nerve cells are exposed to a beam of high-frequency sound in the sea.

Ultrasonics also has a number of important uses. High-frequency waves are used to sterilize instruments in hospitals. Narrow pencil beams of ultrasonic energy can be aimed at specific parts of the human body to create warmth—as in the bone marrow—where no other heating device can reach. Ultrasound has been aimed through holes cut in the skull to reduce the pain and torment of the psychoneurotic patient. Ultrasound in air also has the effect of clumping small particles together, so that instead of floating in the air they grow heavy enough to fall. Hence, *ultrasonic precipitators* installed in industrial and incinerator chimneys help clear the air of dust and smoke. Ultrasound in water breaks up dirt and soil, and mixes the tiny particles well into the water. Ultrasonic waves are thus used to clean manufacturing grease and soil from metal parts. Eventually this principle may be used in home dishwashers and clothes washers.

—Richard M. Koff

HEAT

The term "heat" is one of the most misused words in the lexicon of science. An object cannot possess heat because heat is not a property of a material; a hot bar of steel cannot be said to possess more or less heat than a cupful of cold sea water. The unit of comparison between the two should be *energy*, for the concepts "heat" and "energy" are not identical. Energy may be stored in a system, whereas heat may not. The term *heat transfer* describes those processes in which energy is transferred from one object to another owing to a difference in their temperatures. This transfer process is basic to the concept of heat. Thus the definition of heat may be stated: *Heat is that energy which is transferred between two systems by virtue of a temperature differential.*

If a hot steel bar is plunged into cold water, the bar would cool and the water warm. The energies of the bar and water change as a result of heat transfer between them.

In order that we may understand more clearly the relationship between heat and energy, the latter concept must be more fully developed. A basic concept of mechanical energy is that a body in motion possesses *kinetic energy* proportional to the product of its mass and velocity squared. Another type of energy is *potential energy*, which results from movement against a restoring force; for instance, a ball rising against the earth's gravitational force increases in potential energy. Another example is the increase of potential energy of a spring as it is stretched.

All objects consist of molecules in motion, *molecular energy*, and consequently there exist discrete molecular energies. The sum of all the molecular energies of a body is its *energy*. In general, if molecular energies are large, then the *specific energy* (energy per unit of mass) is large; and this condition is reflected quantitatively in the statement that the temperature of the body is high. This concept of temperature is discussed more fully below, but for now, one can consider that the hotter (higher in temperature) a body is, the greater the energies of the molecules making up the body.

If two bodies of unequal temperature, such as a hot steel bar and cold water, are placed in contact with one another, the area of contact has highly energetic molecules (as the iron molecules) vibrating at a large amplitude contacting lower-energy water molecules. By collisions at this interface, the energy of the water molecules increases and the energy of the iron molecules decreases. In this process heat is transferred, and the energies of both systems change.

FIRST LAW OF THERMODYNAMICS.—To express one of the most important and useful basic laws of physics: *The energy of a system can be changed by the transfer of heat.* The only other way in which energy within a system can be changed is by the system's doing work (such as the stretching of a spring or the expansion of a gas enclosed in a cylinder).

If Q is the heat transfer, W the work transfer, and ΔE the change in energy of a system due to these processes, then:

$$\Delta E = Q - W$$

where Q is positive when the heat transfer is such as to increase the energy of the body and W is positive when the body does work on its surroundings. This simple equation, in conjunction with the realization that the energy of any body is determined solely by the energies of its molecules and is independent of the way the molecules obtained these energies (either by work or heat transfer from other bodies), is called the *first law of thermodynamics.*

Historical Background

CALORIC THEORY.—For many decades the manifestations of heat were attributed to an ethereal, invisible fluid called "caloric." This fluid was believed to have the power of penetrating, expanding, solidifying, and dissolving various materials, as well as the power of converting these materials from solid to liquid or from liquid to vapor. The caloric theory pictured heat as a fluid free to flow into a body when the body was heated and out of a body when it was cooled. The expansion of materials upon heating was attributed to the volume of the caloric fluid entering the material. This theory explained many of the known facts, and scientists were able to use it as a basis for the prediction of various phenomena in advance of their experimental discovery.

Although the caloric theory was accepted by many scientists, there remained certain incongruous observations that could not be explained by the postulation of this invisible, weightless, all-pervading fluid. For instance, the generation of heat by the friction arising from the motion of two mechanical objects contacting one another was attributed to a loss of caloric; the caloric was supposed to be ground or squeezed out of the objects. Many scientists of the late eighteenth century found this explanation inadequate.

In a classic experiment conducted in a Bavarian arsenal, Count Benjamin Rumford demonstrated the inadequacy of this explanation and in so doing prepared the way for the downfall of the caloric theory. Rumford's experiments were initiated by his observation that when cannons were bored, a large temperature rise accompanied the boring. In a series of experiments Rumford measured the energy gained by the cannon from the transfer of heat associated with the boring process. He also attempted to measure the weight of the caloric fluid picked up by the hot cannon. Finding none, he concluded that heat was some form of *motion*.

Rumford's experiments alone, however, were not sufficient to disprove the caloric theory. Not until Sir James Joule was able to determine accurately the mechanical equivalent of heat (by melting ice with friction) was heat recognized for what it is—energy in transit. Joule's experiments established beyond all doubt that heat and work are merely different manifestations of the same thing—energy.

Although the caloric theory was disproved more than a century ago, many erroneous concepts of heat that exist today in the mind of the layman are direct results of this theory. The terms associated with the concept of heat in our modern terminology reflect the hold the caloric theory had on scientists in the eighteenth and nineteenth centuries. We speak of heat "flowing" from one body to another, we talk erroneously about the "quantity of heat" in a body, and we measure the quantity in "calories" with a "calorimeter." Nevertheless, it must again be stressed that heat is neither a material object nor an ethereal one and that a body cannot possess heat. Like work, heat is a way to transfer energy from one system to another. Heat energy is transferred by virtue of a temperature difference existing between the systems.

Temperature.—It is common to associate the concept of temperature with the sensations of heat and cold. These physiological sensations can be very misleading. For example, if sheets of copper and paper are cooled by setting them on a dish of crushed ice until their temperatures equal that of the ice, the copper sheet will feel colder to the touch. This is caused by the high thermal conductivity of the copper. Physiological sensations are often deceptive and can lead at most to a qualitative temperature scale. To appreciate the meaning of the concept of temperature, however, we must first discuss quantitative temperature scales that are not associated with human sensations.

Although molecular energies may be used to define a temperature scale, it is extremely inconvenient to measure such energies. Within any given macroscopic system at any given time there exists a variety of molecular energies. It is the average of these molecular energies that we associate with the concept of temperature.

For gases at low pressures, the temperature is defined rather simply as being proportional to the square of the average molecular velocity. In high-pressure gases and in solids and liquids, the same general concept is accepted, although the reasoning is more complicated.

THERMOMETERS.—To translate such molecular concepts into more practical measuring devices, one chooses to measure some other property that is related to molecular energies. It is known that the volume of most liquids and solids increases when they are heated. The same is always true of gases if the pressure is maintained constant. Thus the expansion of mercury, for instance, is used to indicate increases in temperature in the common mercury-in-glass thermometers. Other fluids are also used. It is obvious that there must be some standard for reference so that readings on all thermometers may be compared. These reference points are the freezing point of water (actually the triple point where solid, vapor, and liquid exist in equilibrium), the boiling point of water at one atmosphere of pressure, and the boiling and freezing points of several other materials.

These reference states are assigned numbers in any chosen temperature scale. The two scales with which we are most familiar are the *Centigrade scale* (*C.*) and the *Fahrenheit scale* (*F.*). They are thus related: $\frac{9}{5}$C. + 32 = F. The accompanying chart indicates the reference state temperatures. If the freezing point of water is chosen as 0° C. and the boiling point of water as 100° C., then each degree represents $\frac{1}{100}$ of the scale between 0 and 100. Since the volume change for each degree of temperature change is not constant over the entire scale, the actual distance between degree marks varies slightly on most liquid thermometers. In order that all thermometers may give the same temperature reading when placed in the same environment, they are calibrated against a low-pressure gas thermometer. The latter is used because the volume change of the gas is very nearly the same for each degree change in temperature.

Kelvin and *Rankine scales* are two other temperature scales that deserve mention. It has been previously stated that temperatures are indicative of molecular kinetic energies. As these energies decrease, one might expect some lower limit of temperature where all motion essentially stops. This lower limit is referred to as absolute zero, and the scales that use this point as a basis are the *Kelvin scale* (*K.*) and the *Rankine scale* (*R.*). These absolute temperature scales are related to the Centigrade and Fahrenheit scales: K. = C. + 273.16; R. = F. + 459.69. Thus, −273.16° C. represents the absolute lower limit of temperature on the Centigrade scale.

OTHER MEASURING DEVICES.—In addition to the methods described above, numerous other techniques are used to measure temperature. A *thermocouple* consists of a circuit of two wires of different metals that generate a voltage and current flow when one junction is hotter than the other. The voltage is indicative of the temperature difference, and calibration charts have been prepared for many common thermocouple circuits. This method is generally used to measure temperature in industry.

A *resistance thermometer* measures the change in electrical resistance with change in temperature, and suitable calibration allows the resistance measurement to be converted to a temperature reading. This technique is useful in obtaining very accurate measurements in laboratory experiments.

At very high temperatures, an *optical pyrometer* is often used for temperature measurements. This instrument uses as a basis for comparison the visible light that is emitted by a body at sufficiently

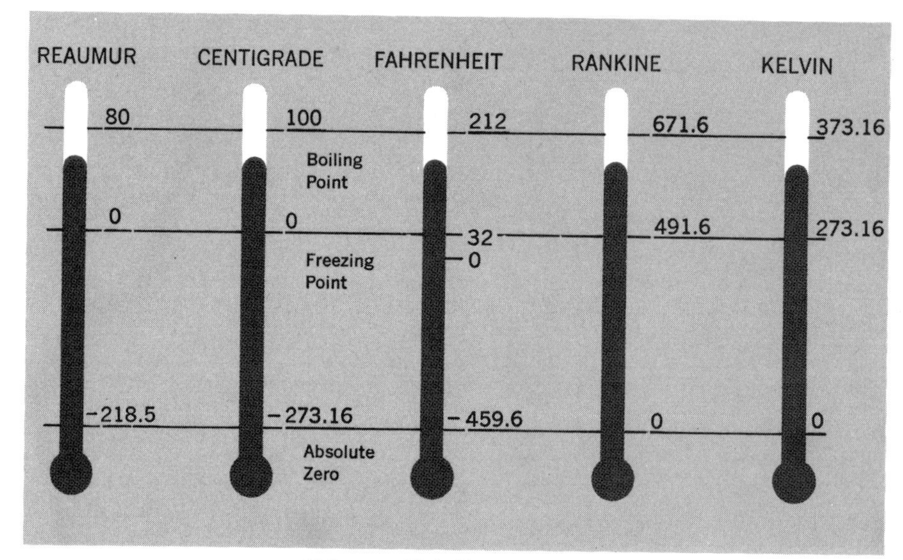

Reference State Temperatures

high temperatures. There are literally hundreds of methods today to measure temperature, all of which measure the outward changes in some physical, optical, magnetic, or other property of a body as the molecular energy changes.

UNITS OF ENERGY.—The energy required to raise the temperature of one gram of water by one degree Centigrade is defined as *one calorie* of energy. If the units are one pound of water and one degree Fahrenheit, then the unit is *one British thermal unit* (Btu). One Btu is the equivalent of 252 calories. Calories are used by most scientists and by the general public in those countries employing the metric system. British thermal units are used by engineers in many English-speaking countries, including the United States. Other units of energy are used when desired (such as kilowatt-hour, joule, electron-volt, erg), but they are all proportional to the calorie or Btu.

HEAT CAPACITY.—The *heat capacity* of a substance is the change in energy necessary for a one-degree rise in its temperature. Values of heat capacities are needed whenever one desires to determine how much heat transfer is necessary to heat or cool a substance through a given temperature interval. Simply stated, the heat transfer requirement per unit of mass is the product of the heat capacity times the temperature change.

A calorie and a Btu were so defined that the heat capacity for water was approximately 1 in the units of calories/gram (C.) or Btu/lb. (F.). Heat capacities for other materials may be greatly different from that of water but are usually smaller. The term *specific heat* is the ratio of the heat capacity of a material to that of water. Since heat capacity is near unity, the two terms are often confused and used interchangeably. However, these terms are not synonymous, inasmuch as specific heats have no units and are simply numbers.

Heat capacities are related to the ability of a body to store energy from the transfer of heat. The more possible ways in which molecules can store this energy, the larger the heat capacity. For example, consider helium. The molecular energies of helium are related to how rapidly the molecules are moving—their velocity. If heat is transferred to helium, then the molecules move faster; this is the only way heat energy may be stored. The average molecular kinetic energy can be shown to be proportional to the absolute temperature (°K. or °R.); thus, if we double the average molecular energy, we double the absolute temperature.

As the heat capacity is the ratio of the energy increase to the temperature increase, it is a constant for helium, independent of temperature. The value is about 0.75 calories/gm for helium heated in a constant volume container. Instead of expressing heat capacities for a gram of matter, the unit of mole is used. A *mole* of any material is the mass that is equal to its molecular weight; since the molecular weight of helium is 4, one mole of helium has a mass of 4 grams. Each mole of any substance has the same number of molecules, so that when the heat capacity is given per mole, the number is simply related to that which would be given on a molecular basis. For helium, then, the heat capacity is about 3 calories/mole; therefore, if the temperature of 4 grams of helium in a closed container were raised one degree Centigrade, then the energy change of the helium would be 3 calories.

Consider next a more complicated molecule, such as ammonia gas, NH_3. There are four atoms in each molecule, the hydrogen atoms being chemically bonded to the nitrogen. Ammonia molecules can move, as could the helium, with various velocities. In this case, however, there are other ways in which energy may be stored. The molecule may rotate as a solid (as the earth does), or the bonds between the hydrogen and nitrogen may vibrate with increasing amplitude, as though the atoms were connected with springs. Each of these possibilities allows energy to be stored. Temperature, however, is still associated with the velocity of the molecule as a whole. Inasmuch as the rotational and vibrational energies increase with increasing temperature, any energy that is supplied to the system must be shared between these two forms of energy and the kinetic energy associated with translational molecular velocities.

Consequently, for a given input of energy, less energy is available to increase the velocities of the ammonia molecules than was the case with helium. The result is a smaller temperature rise for ammonia than for helium. The heat capacity of ammonia per mole is thus considerably greater than that of helium. At room temperature the value is about 8.9 calories/mole. Although the true situation is somewhat more complicated, because some of the ways energy may be stored are not activated except at high temperatures, the essence of the above statements is correct.

Molecules in liquids and solids can store energy in various ways, and liquids and solids have much higher heat capacities than helium when expressed on the basis of a mole. No general rules may be given for liquids, but it can be shown that the heat capacities of all crystalline solid elements are approximately 6.2 ± 0.4 calories/mole. Thus iron, with a molecular weight of 56, has a heat capacity of $6.2 \times 1/56$ calories/gram.

PHASE CHANGES.—In most cases, materials expand when heated. For this expansion to occur, the molecules constituting the material must move farther apart. Just as work is required to stretch a spring, so in separating molecules there must be work done against the forces with which they attract one another. If one stretches a spring too far, the spring may lose its ability to return to its original position; in fact, the spring may break. Similarly, if the molecules in our analogy are moved far enough apart, a point will be reached at which the influence of the attractive forces is almost completely overcome, and the molecules will no longer be constrained. In this condition the molecules can move about freely at random and completely fill the volume available.

The random kinetic energies corresponding to the high temperature are much greater than the potential energy of attraction between the molecules, and the material is a gas. Conversely, as a gas is cooled, the random kinetic energies of molecules decrease. Consequently, when a gas

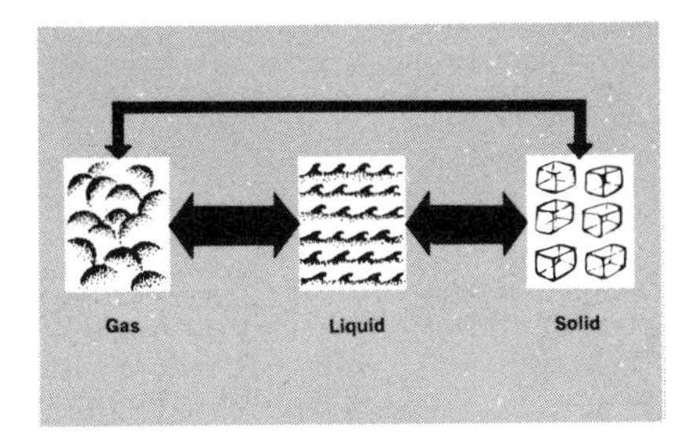

PHASE CHANGES as they occur among the three states of matter.

is cooling, a point is reached at which the attractive forces predominate and the gas collapses, with the molecules losing much of their potential energy of separation. This process, when the substance changes from a gas to a liquid, is called *liquefaction.*

In passing from a gas to a liquid, a substance loses energy. (When a stretched spring is released, it also loses energy). The energy changes accompanying phase changes are known as *latent heat effects* and are quite large. For example, it was seen that the heat capacity of water is approximately 1 Btu/lb. The latent heat associated with the condensation of one pound of water is almost 1,000 Btu/lb. Clearly, a drastic change on the molecular level has occurred.

To extend this picture, if the liquid is further cooled, a temperature will be reached at which the attractive forces so dominate that another phase change occurs, this time one of *solidification.* The molecules are now almost completely prevented from movement because they have been locked in a semirigid structure. This change from a liquid to a solid is accompanied by another energy change, called the *latent heat of fusion.* Heats of fusion are usually much less than heats of vaporization; for instance, for water 133.5 Btu of energy must be removed to freeze one pound of water.

To complete the picture, if the original pressure had been sufficiently low, the cooling process would have resulted in a phase change directly from the gas to a solid. The energy change associated with this process is termed the *heat of sublimation.* The heat of sublimation is approximately equal to the sum of the heats of fusion and vaporization. Carbon dioxide, at one atmosphere of pressure, cannot exist as a liquid; and as the gas is cooled, solid carbon dioxide ("dry ice") is formed.

When certain liquids are rapidly cooled, solidification does not occur at a definite temperature, as occurs on slow cooling; that is, there is no assignable freezing-point temperature. Instead, the ability of the liquid to flow freely decreases steadily until a glassy substance is obtained. This glassy state is characterized by the absence of a regular crystalline structure and by *optical isotropy* (optical transparency) when not in a state of strain. On long standing, especially at high temperatures, glasses may undergo crystallization. This state in which certain substances can exist may be viewed as a condition lying between the liquid and solid states.

Vapor pressures of liquids represent a dynamic state of equilibrium with a continual interchange of molecules between the gaseous and the condensed phase. The higher the temperature, the higher the average kinetic energy of the liquid molecules and the more molecules that can break away from the strong attractive forces existing in the liquid. The vapor pressure thus increases with temperature, and vice versa. Thus, by increasing the pressure above a liquid, one increases the temperature at which a liquid will boil. The modern housewife takes advantage of this fact by using a pressure cooker that permits her to cook foods faster than ordinary pots and pans would permit.

CRITICAL TEMPERATURE.—There is a limiting temperature, called the *critical temperature*, at which the kinetic energy of the liquid molecules exceeds the attractive forces, and no liquid phase can exist regardless of the pressure. At the critical point, the molecular energies of the liquid and gas are equal and the latent heat effects are zero.

It is obvious that in any operation in which a fluid is used to transfer energy, the amount of such fluid required is greatly decreased if a phase change can be utilized. This principle is used in most steam power plants and refrigeration cycles.

Heat Transfer Mechanisms. — Before a discussion of some practical uses of heat transfer can be opened, it is imperative to understand something about the underlying mechanism of heat transfer. Heat transfer will take place when two bodies of unequal temperature are brought together, the hot one becoming colder and the cold one hotter. This result is actually a statement of the *second law of thermodynamics*. However, even though no mention has been made of the rates of heat transfer that result, the size of all heat exchange equipment is based almost entirely on these rates. There are three general mechanisms through which heat transfer may take place.

CONDUCTION.—The temperature of a solid is related to the amplitude of vibration of the molecules around some equilibrium position in the solid lattice. When two solid bodies of unequal temperature are placed in contact, some of the excess vibrational energy of the surface molecules of the hot body is communicated to the surface molecules of the cold body by collisions of the molecules at the points of contact. As a result of this exchange of energy, the surface molecules of the cold body acquire a higher average energy level (higher temperature) than those molecules in the underlying layers. Similarly, the surface molecules of the hot body are now at a lower temperature than those in the underlying solid. This energy exchange process is now repeated between the surface molecules and those just below the surface in both solids as well as between the original interacting surface molecules.

In this manner the heat transfer process is propagated through both solids until the transfer of vibrational energy occurs in both. After some time a condition is achieved in which the temperature of each body varies from point to point within the body and is a steadily increasing or decreasing function of the distance from the point of contact. Later, the energy transfer process is damped out and both bodies achieve a common temperature (approximately equal molecular energies), with no net heat transfer taking place. This type of energy transfer is termed *heat conduction;* no gross movement of any part of the materials occurs.

The rate of heat transfer by conduction has been found to be proportional to the area of contact and *temperature gradient*, the rate of change of temperature with distance. The constant of proportionality is called the *thermal conductivity*. The larger the thermal conductivity, the higher the heat conduction rate. Copper, for example, has a very high conductivity, whereas paper has a very low one. Thus, a piece of copper feels much colder to the touch than paper at the same temperature, since heat can be conducted through copper at a much higher rate than through paper. The rate of heat loss from one's body is greater in the case of copper, and thus copper will feel colder. It is obvious, then, why handles of cooking utensils are made of a material of low thermal conductivity, such as plastic or wood, while the pan bottoms are made of a metal with a high conductivity. Cold metal objects can actually freeze the moisture on a hand so rapidly that the hand will stick to them; nonmetallic objects, which have a low thermal conductivity, do not conduct body heat away at a sufficiently rapid rate to result in the freezing of the surface moisture.

Conduction processes may occur in liquids and gases by an identical mechanism; however, for these "fluids," it is difficult to prevent motion of parts of the material, a process that leads to the second form of heat transfer.

CONVECTION.—Instead of contacting two solids as in the case of conduction, suppose a cold liquid is placed over a warm solid. The solid is cooled by the conduction process, which begins in the liquid phase. However, since liquids expand upon heating, the warmer liquid at the solid interface becomes buoyant and tries to rise over the colder, dense layers of liquid covering it. The liquid-solid surface is renewed with more cold liquid, and the process is repeated. This large-scale mixing of the heated fragments of liquid with the cooler bulk prevents any large temperature gradients from becoming established, and conduction processes (which depend upon such gradients) are small. Heat is transferred by actual movement of warm fluid being carried into and mixing with colder portions. The rate of heat transfer seems to be proportional to the temperature difference between the solid-liquid interface and the bulk liquid, and depends markedly on the degree of agitation near the surface. The convection process described above depends entirely upon the buoyant force of gravity to accomplish the mixing. Such a process is called *natural convection*.

If one were to accentuate the mixing by stirring, shaking, or flowing the liquid across the hot surface, then a similar result would be achieved; this process is called *forced convection*. Forced convection transfer is usually more rapid than natural convection, especially if the fluid has a high *viscosity* (is very syrupy). A thick pudding heated on a stove must be stirred to prevent burning, as natural convection and conduction rates are too low to cause sufficient heat to be transferred from the hot burner to the fluid.

To prevent convection, one inhibits the flow of fluid by packing material of low conductivity somewhat tightly on the interface. For example, for home insulation, wads of a low conductivity material such as fiber glass are tacked to the outer walls. The spacing of the fiber glass filaments prevents easy flow of convection air currents near the outer walls and thus decreases the heat loss due to convection.

RADIATION.—Whereas convection and conduction heat transfer processes depend on molecules contacting molecules, radiation does not. Molecules by their very vibration can emit electromagnetic waves that, when absorbed by other materials, result in an energy transfer mechanism that is termed *radiation*. In a vacuum, radiation is the most efficient means of heat transfer. Materials must be at a rather high temperature before much emission of energy occurs; in fact, the intensity of the radiation (I) is proportional to the fourth power of the absolute temperature ($I \propto T^4$). In general, then, radiation is important only at very high temperatures, or in those cases where heat is transferred between surfaces separated by a vacuum.

Electric heaters transfer heat primarily by radiation from the glowing coils, although some natural convection occurs simultaneously. All the sun's energy is transferred by radiation through the intervening vacuum of space. When radiation strikes a body, the radiation may be reflected, absorbed, or transmitted, depending upon the frequency of the radiation and the properties of the body's surface. Bright metallic surfaces, such as polished aluminum, are considered good reflectors. Black, rough surfaces are usually good absorbers. Glass may or may not be a good absorber or transmitter. The greenhouse effect results from the fact that the glass roof of a greenhouse transmits most of the radiation from the sun. However, this radiation is absorbed by the plants and re-emitted at a different frequency, which is not transmitted by the glass. Thus winter sunlight can be used rather efficiently to heat a greenhouse.

Applications of Heat Transfer.—Having developed the concept of heat as energy flowing from one system to another by virtue of a temperature difference, let us examine how man has managed to utilize this concept of energy transfer in his everyday life. Although such appliances as stoves, refrigerators, and furnaces come immediately to mind as examples of devices that utilize heat transfer processes, electric appliances are also the indirect result of man's ability to convert heat energy to work. Most of the electricity used in the United States is obtained from installations powered by heat engines. Since these devices play such a major role in our industrial society, it is important that we understand the principles by which they operate.

■**HEAT ENGINES.**—A *heat engine* is a machine that absorbs heat at a high temperature, uses some of the energy to perform work, and ejects a portion of this energy as heat at a lower temperature. Although the modern steam turbines used to generate electricity and the rocket engines used to place man in orbit are radically different in outward appearance from the early steam engines of Thomas Savery, Thomas Newcomen, and James Watt, all of these devices operate on the same basic principles. In order to understand these principles, it is instructive to consider in some detail a very simple heat engine, as, for instance, a heat engine operating on a steam cycle.

The *steam power cycle* has continued in basically the same form from the days of Watt to the present. This cycle supplies the world with more power than any other man-made energy device.

The four basic components of a typical steam engine are shown in the figure below.

This basic arrangement is called a *Rankine cycle.* One of the simplest possible means by which thermal energy is converted into mechanical energy by means of this cycle is described below:

Starting at point (1), liquid at a temperature below its boiling point is injected at a high pressure into a boiler and heated by a combustion process or atomic reactor until it is vaporized. The high-temperature, high-pressure steam is piped out of the boiler into a prime mover (2), where it does work by being allowed to expand to a lower pressure against a piston or turbine wheel. The low-pressure exit steam is then condensed in a heat exchanger (3). The liquid water leaving the condenser is pumped (4) back to the boiler, and the cycle is repeated.

Since each pound of water has been returned to its original state, it undergoes no net energy change ($\Delta E = 0$), and the net result of the cycle may be stated as follows:

The system takes in heat (Q_1) at a high temperature in the boiler; some of this energy leaves the system as work (W) in the prime mover, and the remainder leaves the system as heat (Q_2) in the condenser. Thus this heat engine operates as shown in the below right figure, thus

$$0 = Q_1 - Q_2 - W.$$

Although this description has greatly oversimplified the actual situation as far as modern engines are concerned, it contains the elementary features used as a basis for the modern designs.

Steam cycles are often used for the production of electricity, but the *internal-combustion engine* is the type of heat engine that finds widespread application in the transportation industry. Automobiles, trucks, airplanes, and ships are powered by these engines.

In internal-combustion engines, the heat input is supplied by combustion of the working fluid itself—an air-gasoline mixture in the case of the automobile engine. The high-pressure, high-temperature mixture generated by the combustion processes does work by expanding against a piston or turbine. The residual high-temperature gases are then exhausted to the atmosphere. The internal-combustion engine thus operates on the same fundamental principle as the steam turbine. The corresponding portions of each cycle are as follows:

The *efficiency* (ζ) of a heat engine is the ratio of the mechanical work output (W) of the engine to the quantity of heat absorbed at the high temperature (Q_1). Thus:

$$\zeta = \frac{W}{Q_1}.$$

In general, heat engines are quite inefficient devices. A typical efficiency for a modern steam turbine power plant used in generating electricity is 35 per cent. Some of this inefficiency is due to irreversible losses of energy in the engine through friction of one type or another. The major reason for the low efficiencies of these devices, however, is the fact that the engine receives a quantity of heat at one temperature and ejects a smaller quantity at a lower temperature. By virtue of their nature, heat engines cannot be 100 per cent effi-

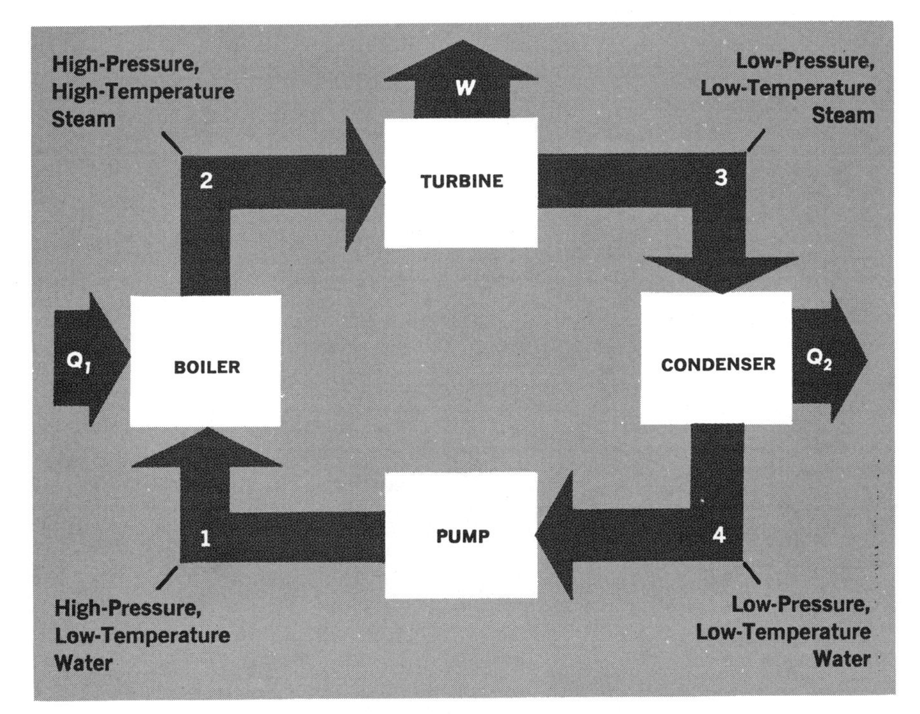

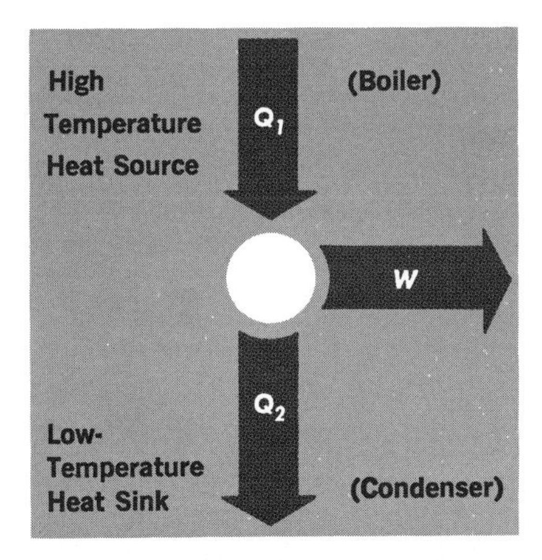

RANKINE CYCLE for the transfer of heat in a steam turbine is shown diagrammatically (*left*). The schematic diagram (*above*) shows that the heat input to the system, Q_1, which is added by combustion or atomic reactor at the boiler, is equal to the sum of the work performed by the turbine, W, and the heat removed from the system during the condensation of steam to water, Q_2.

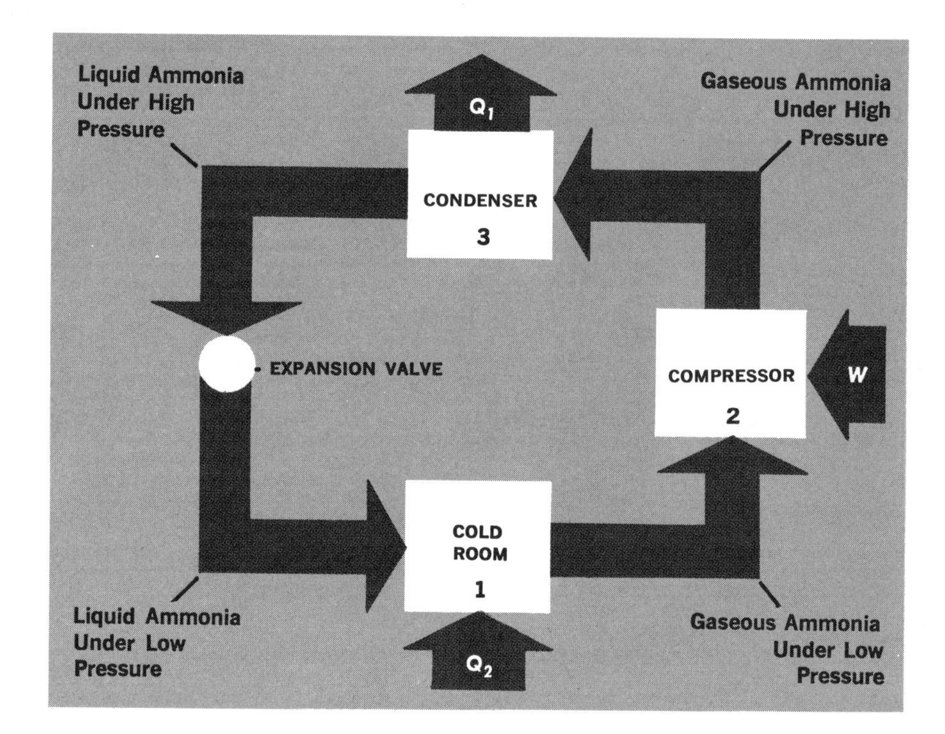

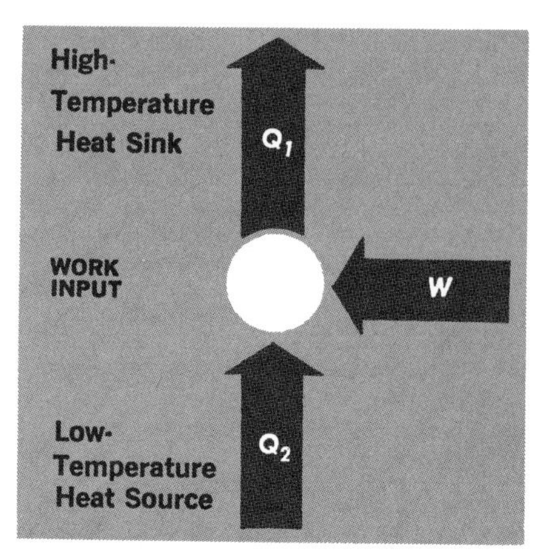

HEAT-TRANSFER CYCLE for an ammonia refrigeration unit (*left*). The schematic diagram (*above*) shows that the heat input, Q_2, which is abstracted by the liquid ammonia from the cold room during the vaporization process, plus the work input of the compressor, W, are equal to the heat liberated to the surroundings by the condensation of the gaseous ammonia, Q_1.

cient. There is a theoretical maximum efficiency at which these machines may operate. This theoretical maximum efficiency is uniquely determined by the temperatures at which the heat engine receives and ejects heat. For the steam cycle discussed earlier, this theoretical efficiency is given by:

$$\zeta \text{ theory} = \frac{T_{boiler} - T_{condenser}}{T_{boiler}}$$

where the temperatures of the boiler and condenser are measured in degrees absolute (Kelvin or Rankine). From this relationship it is easily seen that the theoretical efficiency increases as the temperature at which the boiler is operated increases. In order to operate a boiler at temperatures above 212° F., boiler pressures greater than one atmosphere (14.7 psi) must be maintained. As the operating pressure of the boiler increases, the temperature at which the water is vaporized increases and with it the theoretical maximum efficiency of the steam engine. This is one of the reasons for operating steam turbines and engines at the highest possible pressures.

Refrigeration is the act of producing low temperatures. Most refrigerators are merely heat engines that are operated in reverse. In this sense, the refrigerator is a heat pump that takes in an amount of heat (Q_2) at a relatively low temperature, receives work from an external source, and ejects an amount of heat (Q_1) at a higher temperature. The relationship of this device to a heat engine is apparent from the figure above.

When a drop of ether, acetone, or other highly volatile substance is placed on one's skin, the skin in the immediate vicinity of the drop is cooled as the drop evaporates. The latent heat of vaporization is taken from the skin, causing a decrease in skin temperature in the vicinity of the drop. In the same manner, liquids with low boiling points can be used on a much larger scale to cool large rooms or closed boxes, such as the common household refrigerator. The most common working fluids used in modern refrigeration are ammonia, sulfur dioxide, and various *freons* (halogenated hydrocarbons). Heat from the area being refrigerated is transferred to the working fluid, and this energy is carried away by the vaporized liquid.

A simplified ammonia refrigeration cycle is shown above. The cycle shown is used to cool a room.

1. Cold liquid ammonia is vaporized by abstraction of heat from the room. With good heat transfer, the vaporized ammonia gas leaving the cold room is heated almost to room temperature.

2. The ammonia vapor is compressed to a higher pressure. At this higher pressure, the temperature at which liquid ammonia exists in equilibrium with its vapor is above the temperature of the surroundings, some easily attainable temperature outside the cold room. Consequently, if this vapor is allowed to exchange heat with the surroundings, it will again be liquefied.

3. This heat exchange process is carried out in the condenser, where the heat of liquefaction is removed at a higher temperature than that existing in the cold room. The liquid ammonia passes through an expansion valve separating the higher-pressure area from the lower-pressure portions of the system. Liquid ammonia in the low-pressure side will boil at a lower temperature and may, when evaporated, be used again to remove heat from the cold room.

In actual practice, the ammonia itself is usually not pumped into the cold room, but instead is pumped to a heat exchanger, where it is vaporized by contact with a cold brine solution that is continuously circulated between the cold room and the heat exchanger. The brine thus serves as a heat exchange medium for the liquid ammonia and the cold room.

The performance of refrigerating equipment is measured in terms of the ratio of the heat removed at the lower temperature to the work required to effect this removal. This ratio is termed the *coefficient of performance* (*COP*). Thus:

$$COP = \frac{Q_2}{W}.$$

In order to specify the capacity of a refrigeration unit, the standard ton is used in the United States. This quantity corresponds to a rate of heat removal of 288,000 Btu per day and is approximately equal to the latent heat of fusion of one ton of ice at 32° F.

■CRYOGENICS.—The term *cryogenics* refers to phenomena occurring at temperatures less than about 150° K., some 125° C. below the freezing point of water. At these temperatures most substances are liquids or solids. Those few materials that boil below 150° K. at one atmosphere are often referred to as *cryogenic fluids*. These comprise some of the following materials:

Fluid	*Normal Boiling Point (at 1 atmosphere)*	
	(°*F.*)	(°*K.*)
Helium	−452.1	4.2
Hydrogen	−423.0	20.4
Nitrogen	−320.5	77.3
Oxygen	−297.3	90.1
Methane	−258.6	111.7

When the temperature decreases, molecular velocities decrease, and liquefaction and/or solidification occur. At a temperature of absolute zero, molecular motion has essentially ceased. Now it is apparent, from observations of slow-motion movies, that one may study a particular process or event in more detail and thus gain more information from that event if one is able somehow to slow down the rate at which this event is occurring. Thus, by carrying out experiments at very low temperatures, scientists are able to gain information about the nature of the particular molecular species under investigation that would not otherwise be obtainable. Several unusual changes in the thermodynamic and electric properties of various materials also occur at these temperatures owing to the decrease in molecular motion. For example, the electrical resistance of some metals becomes zero below a certain temperature, a phenomenon known as *superconductivity*.

Because of the difficulties associated with the production and handling of cryogenic fluids, most of their early applications were in the area of basic research. However, as these difficulties have been surmounted by the development of modern techniques, cryogenic fluids have found increasing use in other areas.

The fields of practical application range from medicine to space research to transportation. Cryogenic fluids are used to obtain low temperatures in space-simulation chambers where components and instruments for space vehicles undergo testing. Liquid oxygen is often used as oxidizer in rocket propellant systems.

While man's conquest of space at present involves the most publicized application of cryogenics, some of the most promising applications for the future lie in the field of medicine. Medical researchers have in recent years successfully used cryogenic fluids in the treatment of stomach ulcers and Parkinson's disease. In the treatment of the latter, liquid nitrogen (77° K.) is used to destroy, by freezing, an area of the brain that is responsible for the shaking palsy characteristic of the disease. Cryogenic fluids have also been used in the preservation and storage of blood and various body organs.

In addition to the somewhat esoteric uses described above, cryogenic techniques are now being used in several basic industries. Liquid oxygen is often used in modern blast furnaces for the manufacture of steel. Liquid methane tankers have been developed to decrease long-distance transportation costs of this product. The quick freezing of foods is still another area in which cryogenic fluids are used today. As the techniques for handling these fluids are improved and as more people become aware of their commercial potential, these fluids will undoubtedly play an increasing role in our industrial society.

—Robert C. Reid and Charles Hill

LIGHT

Light in general may be simply defined as the form of radiant energy that produces visual sense impressions by stimulating the retina of the eye. Ultraviolet and infrared light are physically the same as visible light but cannot be seen by the human eye. *Ultraviolet light* is sometimes called black light; *infrared light* is often called heat radiation. The latter, however, is a misnomer.

Light is dualistic in nature; its properties must be explained by two separate and distinct theories—the wave theory and the quantum theory. These two theories are alike in only one respect; in both light has an *associated frequency*.

■WAVE THEORY.—The *wave theory* considers light as both electric and magnetic in character—and as transmitted through space as an *electromagnetic transverse wave*. This transverse wave motion may be compared to the motion of water molecules in a pond stirred by a stone thrown into it. The stone, striking the water, causes a series of ever-widening circles to ripple outward from the center of movement. Each circle consists of water molecules oscillating up and down, at right angles (transverse) to the direction of the wave's movement. These waves form a series of ridges (*crests*) and depressions (*troughs*). The distance between any two wave crests is the *wavelength;* and the *frequency* is the number of wavelengths passing a given point in a second. In a luminous body such as an incandescent bulb, the light waves travel out spherically (three-dimensionally) from the bulb, whereas the water waves travel out in circles (two-dimensionally) from the source of disturbance. However, an electromagnetic wave is transverse, like a water wave, and can be depicted as follows:

A represents the *amplitude* of the transverse wave (in the case of the electromagnetic wave, the maximum electric or magnetic intensity). The *period* (*T*) is the time it takes the wave to travel one wavelength (λ), or $\lambda = VT$. This can be written $V = \frac{\lambda}{T}$. Since the *frequency* (number of waves passing a point in one second) is equal to $\frac{1}{T}$, *V* (velocity) can be equated to frequency (*f*) times wavelength (λ).

Electromagnetic waves are classified by wavelength in the *electromagnetic spectrum,* which is divided into various regions. These regions, from long wavelengths to short, are radio waves, Hertzian waves or microwaves, infrared, visible light, ultraviolet rays, X rays, and gamma rays. In the visible-light region, the waves are classified by the colors that the human eye senses on receiving certain wavelengths—this region is called the *visible spectrum.* The wavelength is usually given in millimicrons (abbreviated Mμ). One millimicron is equal to 10^{-7} (.0000001) centimeters. Wavelengths, especially in the visible region, are also denoted in angstrom units. One *angstrom unit* (1 Å) is equal to 10^{-8} centimeters, or 1 Mμ = 10 angstrom units. The visible region extends only from 760 Mμ (7,600 Å)—the red region—to 380 Mμ (3,800 Å)—the blue region. It is obvious, therefore, that the eye responds to a rather small portion of the over-all spectrum.

■QUANTUM THEORY.—The *quantum theory* explains effects that the electromagnetic theory cannot. It regards

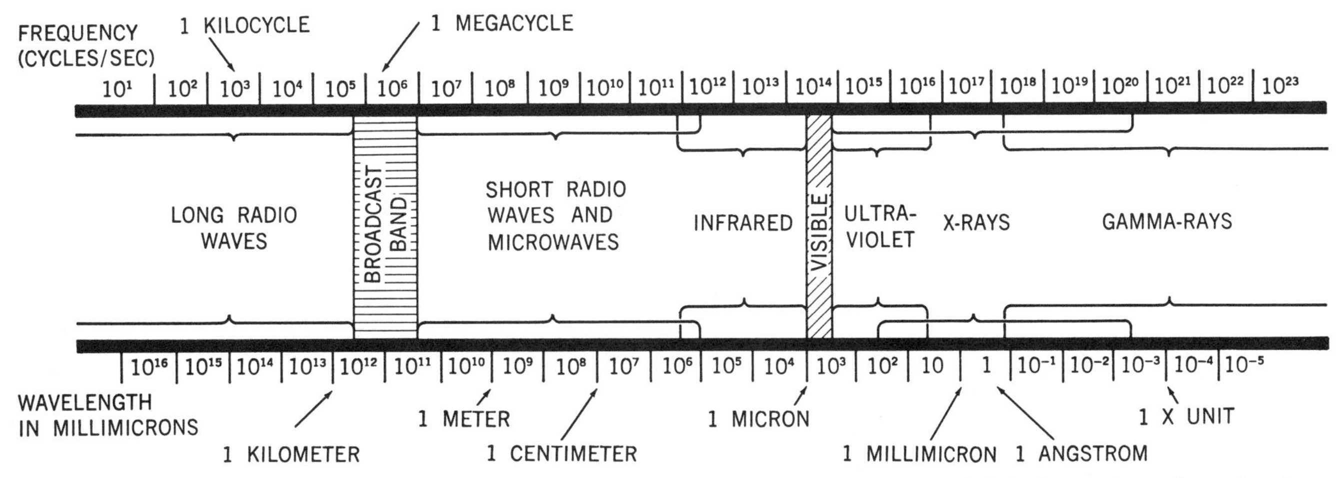

ELECTROMAGNETIC SPECTRUM categorizes the varied forms of electromagnetic radiation in terms of their frequencies and wavelengths.

electromagnetic energy as traveling in small packets. A single packet is called a *quantum* or *photon*. The energy in a quantum is given by the expression $e = hf$, where e is the energy in the photon, h is a constant (called *Planck's constant*), and f is the frequency. The existence of these discrete photons has been shown by the work of Niels Bohr (1885–1962), Albert Einstein (1879–1955), and Max Planck (1858–1947).

Velocity of Light.—Light travels at a *velocity* of approximately 186,283 miles per second in a vacuum and at practically the same speed in the atmosphere. The velocity, however, is less in all other transparent media. For example, in water the velocity is approximately 140,000 miles per second. The distance light travels in one second is roughly equivalent to seven times around the earth at the equator. The American physicist Albert A. Michelson (1852–1931) made the first precise measurement of the speed of light by using a mechanical device. Today, the speed of light in a vacuum is listed as 299,792,500 meters per second.

Reflection and Transmission.—When light strikes a material, it may be reflected, absorbed, or transmitted. An object appears blue because it reflects blue light and absorbs the other wavelengths. Objects appear red through a piece of red glass because the glass transmits red light and absorbs all other colors. A black object is black because it absorbs almost all the light that falls on it and reflects only a very small percentage. When light energy is absorbed by an object, it is imparted to the molecules of the object and consequently increases its temperature.

A substance therefore has a reflection coefficient, an absorption coefficient, and a transmission coefficient. These coefficients vary according to the wavelength of the incident light. For example, there are filters that transmit light in the visible spectrum and absorb light outside the visible spectrum. Such filters have a high transmission coefficient in the visible region, but a coefficient of practically zero elsewhere in the spectrum.

The ratio of the intensity of the light reflected from an object to the light incident on an object (wavelength for wavelength) is called the *reflection coefficient*. The ratio of the intensity of the light converted into heat divided by the incident light intensity (wavelength for wavelength) is called the *absorption coefficient*. The intensity of the light transmitted divided by the intensity of the incident light on an object is termed the *transmission coefficient*.

The *law of reflection* states that when a ray of light is reflected from a surface, the incident ray, the reflected ray, and the normal (a line drawn perpendicular to the surface at the point where the incident ray strikes the surface) all lie in the same plane. It also states that the *angle of incidence* (the angle between the incident ray and the normal) is equal to the *angle of reflection* (the angle between the reflected ray and the normal). In geometrical optics, light rays are thought of as traveling in a straight line and as originating at the object.

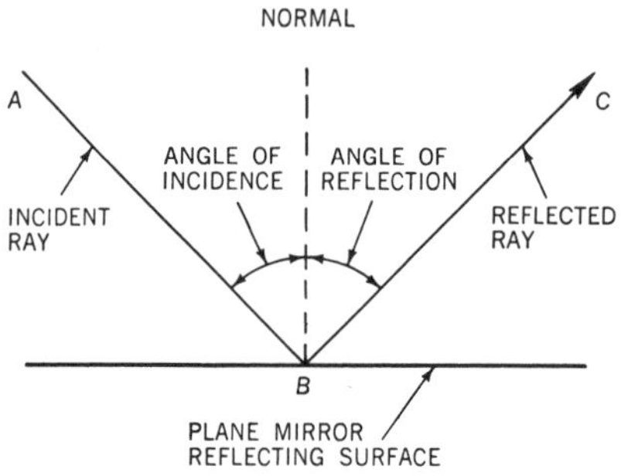

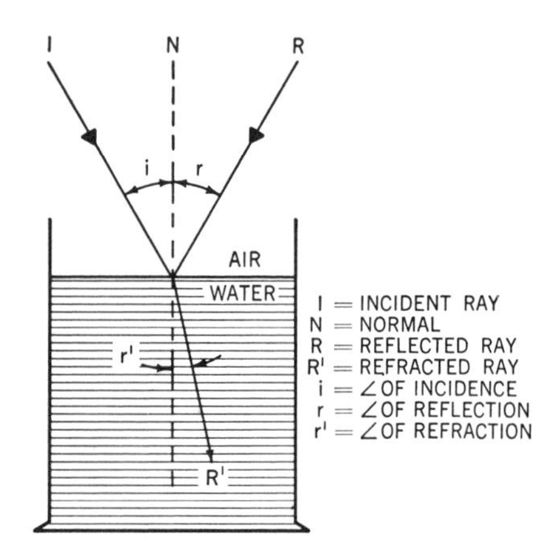

REFLECTION AND REFRACTION are changes in the path of an incident light ray. The angle of incidence equals the angle of reflection, but not the angle of refraction.

When a ray of light from an object strikes a plane mirror, it is unchanged; it merely travels in a new direction. Therefore, when looking at a plane mirror, the eye sees an image that is identical to the object and that appears to be the same distance behind the mirror as the object is in front of it. A line joining the object and the image would be perpendicular to the mirror. In addition, the image is *virtual;* that is, the rays appear to be coming directly from the object, although they actually are not.

Of course, reflecting surfaces do not necessarily have to be plane mirrors. There are convex and concave spherical reflectors, convex and concave parabolic reflectors, convex and concave elliptical reflectors, and many others. Reflection takes place on these curved surfaces according to laws identical to those for plane surfaces, the normal now being the normal to the tangent plane of the surface at the point of reflection.

Refraction and Dispersion.—When light passes from a less dense medium into a denser medium, its path is altered. This change in the direction of a ray of light is called *refraction*. The velocity of light varies in different media; it is greater in a less dense medium than in a more dense medium. For example, the velocity of light in air (a less dense medium) is greater than that in glass (a more dense medium).

Every material that transmits light has an *index of refraction,* which is defined as the velocity of light in a vacuum for a particular wavelength, divided by the velocity of light at that wavelength in the material. This index of refraction is symbolized by n, so that mathematically,

$$n = \frac{\text{velocity of light in vacuum}}{\text{velocity of light in material}}$$

when the two velocities are for the same wavelength of light. The index of refraction depends, therefore, upon the kind of material and upon the wavelength.

The *law of refraction* is stated in two parts: When a ray of light passes from a less dense medium into a more dense medium, the ray is bent towards the normal (to the surface of the medium). Conversely, when the ray passes from a more dense medium into a less dense medium, it is bent away from the normal. Also, the incident ray, the normal, and the refracted ray all lie in the same plane. For any given wavelength, the index of refraction of the first medium multiplied by the sine of the angle of incidence is equal to the index of refraction of the second medium multiplied by the sine of the angle of refraction. This is known as *Snell's law*.

Since the index of refraction of a medium is different for different wavelengths of light, the component wavelengths that make up white light are bent by varying amounts when they pass through a prism. The light is then said to be *dispersed;* that is, it is separated into its component colors. These components are red, orange, yellow, green, blue, indigo, and violet. Note that red light

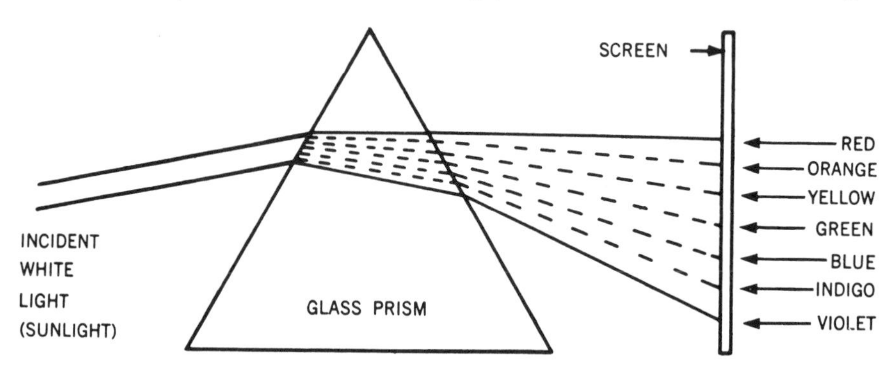

DISPERSION breaks multiwavelength white light into its component colors, or spectrum.

(with the longest wavelength) bends the least when passing through a prism, while violet light (with the shortest wavelength) bends the most.

Interference.—At any point where two or more light waves cross one another, they are said to *interfere*. This is not to be taken to mean that the waves impede one another, but refers to the combined effect at the point in question.

The accompanying illustration shows two transverse light waves emanating from the same source but having different path lengths. The top wave has traveled 1½ wavelengths while the bottom wave has traveled exactly one wavelength. If the amplitudes of the two waves are added, one will cancel the other and the resultant amplitude will be zero.

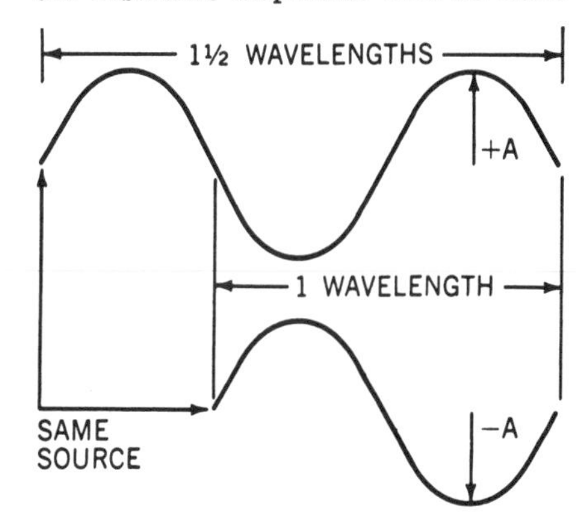

DESTRUCTIVE INTERFERENCE occurs when the crests and troughs of two light waves one-half wavelength apart cancel each other.

Physically speaking, the two waves have interfered; and complete *destructive interference* has taken place, because the two waves, added algebraically, are numerically equal but have opposite signs.

It is also possible to have *reinforcement* of the waves occur. This is not illustrated, but will occur when the path lengths differ by one wavelength or any multiple thereof.

The drawing shows the details of the experiment performed by an English physicist, Thomas Young, in 1804. The experiment was of great importance, for it verified the wave theory and the interference of light. Light from a monochromatic (single-wavelength) source illuminates slit s_1 and the light from this slit in turn illuminates slits s_2 and s_3. Since slits s_2 and s_3 are equidistant from s_1, it is as though s_2 and s_3 are identical sources but are displaced from one another by a distance *d*. The light from both s_2 and s_3 now falls on the screen. The light reaching *P* from both sources has traveled the same distance; therefore, at this point reinforcement will occur and a bright line will be observed. However, at a point such as *E*, the light from s_3 has traveled a distance greater than that from s_2 and interference will take place. If this path difference is one-half wavelength, *destructive interference* takes place; and a black line appears on the screen. The pattern occurring on the screen thus consists of alternate light and dark lines called *interference fringes*.

The various colors observed when looking at a thin film of oil or a soap bubble are caused by interference effects at the surface of the medium. Since the light is reflected from the top and bottom surfaces of a film, interference takes place at the top surface and depends upon the wavelengths of the light, the index of refraction of the material, and the film thickness.

Diffraction.—If light travels in straight lines in a homogeneous material (a material whose composition is identical throughout), then one would expect that a beam of light passing through a slit would form an illuminated image of the slit. However, if the slit width is comparable to the wavelength of the light passing through the slit, a number of alternate light and dark areas will appear on the screen, similar to the light and dark areas produced by interference. The light is spread out in a *diffraction pattern*, consisting of a central bright band that may be much wider than the width of the slit, bordered by alternating dark and bright bands of decreasing intensity. In a sense, the light has bent around the corners of the slit. (This same diffraction effect takes place in the case of sound waves. Sound will travel around obstacles; but since sound waves are longer than light waves, the effect is greater.)

To increase the number of slits through which the beam of light must pass, a *diffraction grating* may be produced by ruling very narrow parallel slits on a piece of glass with a diamond point. When a beam of parallel rays of white light is incident on such a grating, the various wavelengths are spread out by the diffraction, producing a spectrum. This is similar to the color spectrum produced by a prism, except that in the case of diffraction the longest wavelength is bent the most. This is exactly opposite to the case of prism refraction, where red light is bent the least.

Gratings are produced with many thousands of lines per inch. The surface does not necessarily have to be transmitting; lines can also be ruled on reflecting surfaces, which become known as *reflection gratings*. Such ruled surfaces can be either plane or concave. Many gratings of the latter type are used in spectographic systems. A grating system has less light loss and greater dispersion (colors are spread out more) than a prism.

Polarization.—Whereas interference and diffraction can occur with any sort of waves, polarization cannot. *Polarization* is a phenomenon that depends not only on the fact that light travels as a wave but also on the fact that light waves are transverse. Longitudinal waves, such as sound waves, cannot undergo polarization.

According to the wave theory, light is an electromagnetic transverse wave. Ordinary light consists of waves whose vibrations take place in all possible directions (*planes*) perpendicular to the direction of propagation. When light is polarized, all paths of vibration—except those in one direction—are eliminated. Therefore, in polarized light, the vibrations take place in only one plane perpendicular to the direction of propagation. Polarized light of this type, called *linearly polarized* or *plane-polarized* light, can be produced in a number of ways. For example, tourmaline, a transparent mineral containing aluminum, boron, silicon, and oxygen, possesses the property of transmitting vibrations in only one plane. Polaroid, a man-made material developed in 1934 and consisting of transparent sheets resembling cellophane, possesses the same property. Materials such as these are called *dichroic*.

Other crystalline materials, such as calcite and quartz, will split an ordinary light beam into two beams. These two beams will be plane-polarized light beams and the planes of polarization in the two beams will be at right angles to one another. Such crystalline materials are called *birefringent* or *double-refracting*.

Ordinary light can also be plane polarized by reflection. At a certain angle of incidence, vibrations in only one plane predominate in the reflected beam. The angle of incidence at which this occurs depends upon the index of refraction of the reflecting glass; the angle is called *Brewster's angle*. For ordinary glass this angle is approximately 57 degrees. Since light reflected from a road surface on a sunny day is partly polarized, polaroid glasses will decrease the glare.

If a source of light is observed

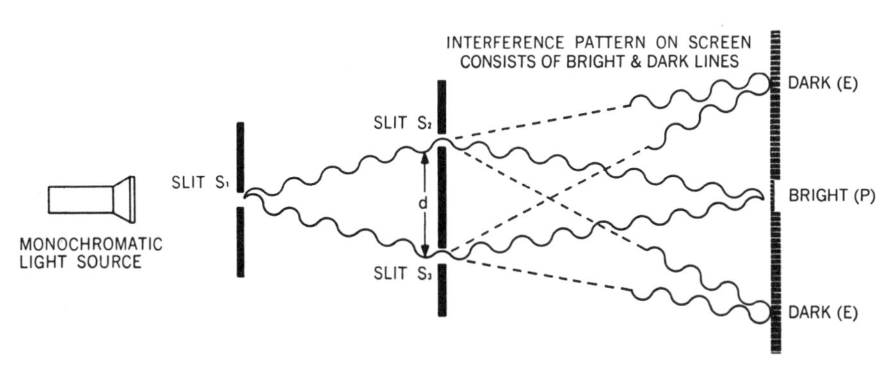

INTERFERENCE EXPERIMENT performed by the English physicist Thomas Young in 1804.

through a sheet of polaroid, the eye cannot detect polarization; and the source appears as it would to the naked eye. However, if the source of light is seen through two pieces of polaroid whose transmitting planes of vibration are perpendicular to one another, the light source is extinguished. For any other angle between the transmission planes of the two sheets of polaroid, some light from the source will reach the eye; and the amount of light will be maximum when the transmission planes are parallel to one another.

Much use is made of polarized light. In chemistry, for example, it is used to analyze sugar solutions because these solutions rotate the plane of polarization according to the concentration of sugar. Polarized light is also used in engineering. A transparent mechanical model may be constructed that, when placed between crossed polaroids and examined under stress, will show the points of greatest stress.

In addition to plane polarization, light can also be *circularly polarized* and *elliptically polarized.* In the former, the plane of the vibration rotates while the maximum amplitude of the wave remains the same as the wave progresses. In the latter, the plane of vibration rotates and the maximum amplitude of the wave varies as the wave progresses.

Luminescence and Incandescence.— Light may be radiated by atoms and molecules if their electrons, excited by an increase in energy, suddenly lose that energy in the form of photons. When the energy is increased by heating, the resulting radiation is called *incandescence.* When the energy is increased by any means other than heating, the radiation of light is called *luminescence.*

When an electrical discharge travels through a gas at low pressure, or when a salt is volatized in a flame, light of a specific wavelength—called the *characteristic color*—is emitted. In the case of the electrical discharge passing through a vapor, familiar examples are the neon tube, the sodium vapor lamp, and the fluorescent lamp; the characteristic color of neon is reddish-orange; of sodium, yellow; and of the fluorescent lamp, bluish-white. If cadmium vapor were present in the tube, a red light would be emitted; and if other gaseous elements were present, colors characteristic of those elements would be emitted. The colors emitted when the atoms of an element are involved are few and widely separated in the spectrum, appearing as isolated lines. These *line spectra* are characteristic of energized atoms of an element.

If a compound is excited instead of an element, the emitted radiation, instead of being discrete and widely separated wavelengths or colors, will be a number of different wavelengths or colors (a series of line spectra, each representing one of the elements making up the compound) covering a span or band of wavelengths. These spectra are called *band spectra.*

If a solid or liquid is heated, all wavelengths of electromagnetic energy are produced. A spectrum of this type is called a *continuous spectrum.* (This phenomenon is described in THERMAL RADIATION.)

Luminescence occurs when the electrons in an atom are raised to a higher energy state by absorbing the energy in the photons of the light illuminating the material. The electrons, when raised to the higher energy states, are unstable and tend to return to their original state. When they do so, light of a frequency dependent upon the difference in energy states is emitted. Usually the light emitted is at a longer wavelength; that is, when materials luminesce, they usually absorb light of a shorter wavelength and emit light of a longer wavelength. For example, vegetation containing chlorophyll will emit orange light when illuminated by short-wavelength ultraviolet light.

Luminescence is categorized into *fluorescence* and *phosphorescence.* A fluorescent material emits light only while being illuminated and for a very short time afterward. A phosphorescent material emits light while being illuminated and may phosphoresce (continue to emit light) for hours and even weeks after the illumination is removed.

There are many ways to excite luminescence other than illuminating a material with light waves. A television screen is made to fluoresce by bombarding the front of the tube with electrons. This is called *cathode luminescence.* Luminescence can also be caused by the application of an electrical field to a material. This is called *electro-luminescence.* Luminescence may even be caused by chemical or physiological processes, as in the case of the firefly. It may even be spontaneous, as in the case of radium.

Elements emit their characteristic light waves when caused to luminesce. The science of *spectroscopy* is based on this fact. By determining the wavelengths emitted by a luminescing substance, it is possible to determine the elements of which the substance is composed.

Light will also be absorbed when it passes through a gaseous material; the wavelength absorbed will be the same wavelength that would be emitted if the gas luminesced. Before reaching the earth, sunlight must pass through the outer gaseous layers of the sun. When the spectrum of sunlight is studied, dark lines are observed in it, indicating that certain elements must exist in these gaseous layers to absorb the wavelengths where the lines appear. These lines are called *Fraunhofer lines,* after Joseph von Fraunhofer (1787–1826), who first explained them.

■THERMAL RADIATION.—A continuous spectrum (electromagnetic energy of all wavelengths) may be obtained by heating a solid or liquid. When a solid body such as iron is heated, it turns red; as the temperature is further increased, it turns white and then blue. Perhaps the most common and useful application of thermal radiation is the ordinary incandescent lamp. A tungsten filament is heated by an electrical current flowing through it; the filament then emits visible light. (The filament also emits light that is not in the visible region. If all the light were in the visible region, the luminous efficiency of the lamp would be greater.)

In the latter part of the eighteenth century, Pierre Prévost (1751–1839) investigated the radiation interchange from one body to another. He concluded that when radiant energy strikes a body, various portions of the energy are absorbed, transmitted, or reflected by that body. In 1792 he announced his *theory of interchanges,* which states that all bodies are continuously absorbing energy from surrounding bodies and in turn radiating energy to all surrounding bodies.

Gustav R. Kirchhoff (1824–1887) investigated absorption and radiation of energy by materials, and in 1859 he stated that a good absorber of radiant energy is also a good radiator of radiant energy. This led to the postulation of the so-called "black body"—an opaque body that absorbs all the energy incident upon it. This would also be an ideal radiator. Such a body exists only in theory. The actual radiation from physical bodies is interpreted in terms of "black body" radiation by defining a ratio called *emissivity.* This is the radiation emitted by an object, divided by that of the theoretical "black body," which is at the same temperature as the object. *Kirchhoff's law* states that for all bodies at the same temperature, the radiant energy from that body divided by the absorption of radiant energy of that body is a constant and equals the radiant energy of a "black body" at that temperature.

Units.—The definition of light is based on our visual awareness of radiation in a very narrow band of the electromagnetic spectrum called the visible region. The eye is not equally sensitive to all visual light waves. The variation of sensitivity to each wavelength may be graphed as a *luminosity curve.* The luminosity curve for the average eye shows how the eye responds to equal amounts of energy at the various wavelengths within the visible region. Since the eye is more sensitive to wavelengths in the yellow region, yellows appear very bright as compared to the reds and blues. The most sensitive wavelength is at 555 millimicrons, and the curve is normalized by making the maximum ordinate at this point equal to one. Above and below, the maximum curve falls off sharply and drops to practically zero at 400 and 700 $M\mu$ (the limits of the visible region).

Units based on the visual interpretation of radiant energy are called *photometric units* or *luminous units;* those based on the physical interpretation of radiation are called *radiometric units.* There would be no need for two sets of units if the eye responded equally to all the visible wavelengths; if this were the case, only the purely physical radiometric units would be needed.

The unit of luminous flux or power is called the *lumen;* that for radiant flux is the *watt.* The two are related by means of the luminosity curve.

At 555 Mμ, one watt equals 685 lumens; at any other visible wavelength one watt expressed in lumens will be less than 685 and will equal 685 times the luminosity curve ordinate at the particular wavelength. For example, at 600 Mμ the ordinate of the luminosity curve is 0.6; therefore, one watt of radiant energy of wavelength 600 Mμ is equal to 685 times 0.6, or 411 lumens.

The unit of luminous intensity is the *candle.* One candle equals one *lumen per steradian.* The *steradian* is the unit of solid angle and is so defined that the total number of steradians subtended at the center of a sphere is 4π. A point source of light having a uniform intensity of one candle in all directions therefore emits 4π lumens of luminous flux. This is true because the point source can be thought of as being at the center of an imaginary sphere and there are 4π steradians in the sphere. The analogous unit of radiant intensity is the *watt per steradian.*

The designation "candle" for luminous intensity came about because the first photometric standard for light intensity was a sperm-wax candle constructed in a specific way.

When light is incident on a surface, the surface is *illuminated;* and the unit of *illuminance* is the luminous flux per unit area. If the area is one square foot and the flux is one lumen, then there is one lumen per square foot; this is known as a *foot-candle.* The analogous radiant term is *irradiance,* and the unit is the watt per square foot.

A source of light does not necessarily have to be a point source; it can also be an extended source. For such sources, the larger area of the source is thought of as divided into smaller unit areas, each of which is pictured as emitting light. The total luminous flux emitted per unit area is termed the *luminous emittance* of the extended source, and the units become lumens per square foot, lumens per square meter, lumens per square centimeter, lumens per square inch, and so on. The corresponding radiant quantity is called *radiant emittance,* and the units are watts per square foot, watts per square meter, and so on.

The last unit of significance correspondingly associated with an extended light source is termed *brightness,* or *luminance.* This is the luminous intensity per unit area of the source (candles per square foot, candles per square meter, and such). Many surfaces appear equally bright no matter from which direction they are viewed; these obey *Lambert's law* and are called *perfectly diffuse emitters.* The sun is a good example, since the edge appears exactly as bright as the center. The analogous radiant quantity is *radiance,* radiant intensity per unit area of an extended surface.

The *inverse square law* states that the illuminance on a surface varies inversely as the square of the distance from a point source. For example, if the illuminance on a surface is one lumen per square foot when this surface is at a distance of one foot from a point source, the illuminance on the surface will be one-quarter lumen per square foot when the object is placed at a distance of two feet from the same source. The intensity of two point sources of light can be compared, or if one is known the other can be determined, by making use of this inverse square law. In essence, the sources are so placed that the illuminance on the screen is the same from either source. By measuring the distance of the screen from the sources, the relative intensities, or the unknown intensity, can be calculated. Instruments for making these measurements are called *photometers.*

Color.—The word *color* is commonly used in several different ways. It is used with reference to the sensation received in the brain when the retina of the eye is stimulated by light of a particular wavelength. It is used to describe a property of an object, for example, a "red" barn. Indeed, by definition, everything that is seen has a sensation of color associated with it. The only truly colorless things are those that are invisible, such as air.

■ADDITIVE PROCESS.—In passing through a prism, sunlight is dispersed and broken down into its constituent colors—the spectral colors red, orange, yellow, green, blue, indigo, and violet. Sunlight, which is essentially white light, is therefore an *additive color mixture* of all the spectral colors. If all these spectral colors are added in exactly the same amounts as they appear in the sunlight spectrum, sunlight or white light will be produced.

When colors are projected on a screen simultaneously by two or more projectors, different colors are obtained in regions where the colored images overlap. Similarly, if a disc made up of different-colored sectors is rotated rapidly in front of the eye, the eye will see a color different from that of either of the colored sectors. If half the disk is red and the other green, the disk will appear to be yellow. If a third of the disc is red, a third blue, and a third green, the disk will appear to be grayish white.

A color other than a pure spectral color can be reproduced by adding three spectral colors together in the correct luminous amounts. All colors cannot be produced by an additive mixture of the same three spectral colors, but it is always possible to find three spectral colors that, when added together in the right amounts, will produce the color desired. Since the greatest number of color variations can be produced by additively mixing red, yellow, and blue, these three are called the *primary colors.* In some cases it is possible to produce white light by adding two colored lights instead of three. Such colors are called *complementary colors.* Purple and green, for example, are complementary colors.

■SUBTRACTIVE PROCESS.—If white light is passed through an optical filter, the filter will transmit a percentage of the incident light at each particular wavelength in the visible region. If the filter is yellow, it will transmit a greater percentage of yellow light than light of other wavelengths. A blue filter will transmit a larger percentage of light in the blue region, and so on. If the incident light is passed through a blue filter and yellow filter in combination, the resultant light transmitted will be those wavelengths where the transmission curves of the two filters overlap. In the case of the blue and yellow filters, the greatest percentage of light transmitted is in the green region, and thus green light is observed through the two filters. Since each filter subtracts a certain amount of energy from the incident light, this method is called the *subtractive method* of color mixing.

■COLORS OF PAINTS AND INKS.—By mixing paints and inks of different colors, it is possible to obtain other colors through the subtractive process. If pigments (transparent and dyed) are suspended in a transparent, colorless base liquid such as linseed oil, a colored paint is produced. The color reflected from the paint is produced by the incident light passing through millions of suspended colored filters (pigments) in the base liquid. For example, if blue and yellow paint are mixed, the reflected light from the mixed paints will be green; and the paint produced will therefore be green. This is the same result produced by the blue and yellow filters in the subtractive method.

In the three-color printing process, three transparent, colored inks are used. The colored picture is produced by printing the three colors one on top of the other. The incident

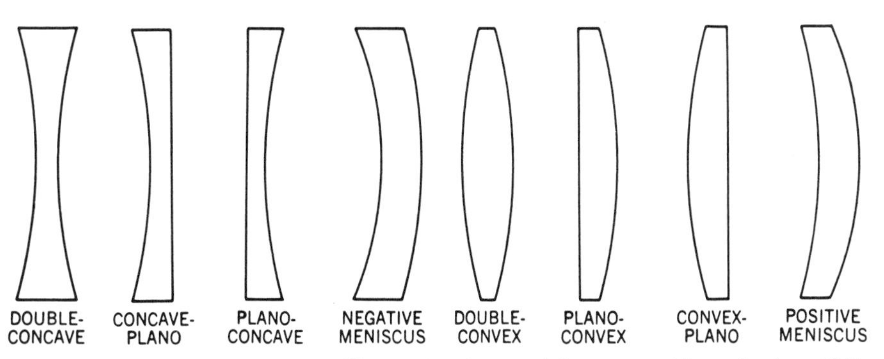

POSITIVE AND NEGATIVE LENSES. All negative (concave) lenses are thinner in the middle than at the edges, while all positive (convex) lenses are thinner at the edges.

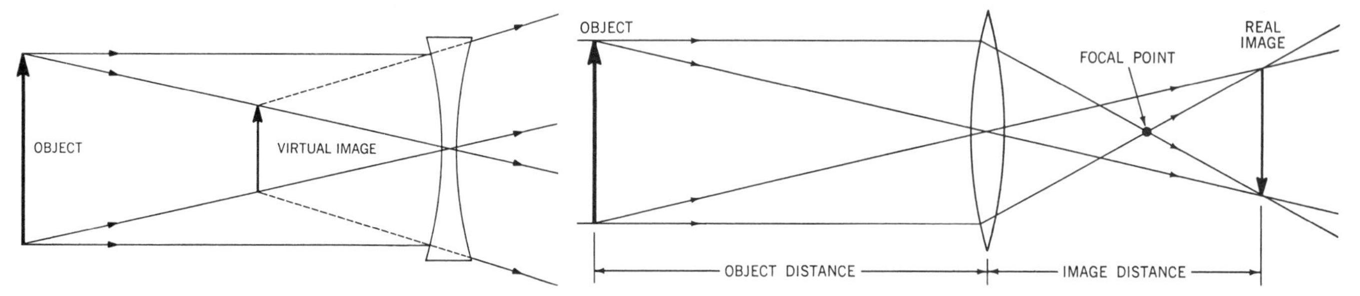

IMAGES formed by a negative lens (*left*) are always virtual, erect, and smaller than the object. Images formed by a positive lens (*right*) are always real, inverted, and smaller than the object when the object is more than twice the lens' focal length from the lens.

light on the picture thus passes through the three inks (filters, in effect), is reflected from the white paper on which the picture is printed, and then passes again through the same three inks to produce a colored picture by the subtractive method of color mixing.

Lenses.—A lens is made of a material that will transmit light. Its surfaces are shaped so that when light passes through it, the light will be refracted or redirected. Lenses are usually made to form images. In general, there are two types of lenses—*positive* (*convex*) and *negative* (*concave*). A positive lens is thicker in the middle than at the edges. A negative lens is thinner in the middle than at the edges.

Positive (*converging*) lenses and negative (*diverging*) lenses are described by their optical axes, positions of their focal points, and their local lengths. (This is true if the lenses are considered to be thin lenses, meaning that the thicknesses of the lenses can be neglected in any computation involving them.)

The *optical axis* is an imaginary line that passes through the centers of the spherical lens surfaces. The *focal point* is that point through which all incident rays parallel to the optical axis and entering the lens will pass. (The focal point for the positive lens is real—the rays actually pass through this point. For the negative lens, the focal point is imaginary—the light rays appear to be passing through this focal point but actually do not.) The *focal length* is the distance of the focal point from the thin lens. The *linear magnification* of a lens is the image height divided by the object height.

In general, it may be said that an image is a *real image* if it is on the opposite side of the lens from the object; that is, the image is real in the sense that, if a screen is placed at the image distance, the image will be formed on the screen. For example, in the case of a slide projector, the slide is the object, the lens becomes the projection lens, and on the screen a real image of whatever is on the slide is produced. The real image produced by a single lens is *inverted*, that is, upside down and turned left for right (the right side of the object becomes, in effect, its left side in the image).

Virtual images only appear to be where they are seen. If a screen is placed at the point where the virtual image appears to be, no image is formed on the screen. However, to an eye looking through the lens, the image appears to be at that point.

With a positive lens, if the object is located between the first focus and the lens, the image is virtual, erect, and magnified. This is the principle of the simple magnifying glass. For an object located at a distance greater than two times the focal length of a positive lens, the image will be real, inverted, and reduced. Magnified, inverted, real images will be produced by a positive lens when the object is at a distance from the lens greater than one time, but less than two times, its focal length. At an object distance equal to two times the focal length of a positive lens, the image will be real, inverted, and the same size as the object.

Most lenses are made with spherical surfaces because these are the cheapest to produce. However, surfaces that are not spherical (*aspherical surfaces*) are becoming available. Such surfaces are used to overcome certain image defects of spherical surfaces. These image defects, called *aberrations*, include coma, spherical aberration, astigmatism, curvature of field, and distortion.

Coma and *spherical aberration* are the result of rays passing through the zones of the lens farther from the center being brought to a focus before those passing through the zones closer to the center. Coma results when the rays originate from a point not on the axis of the lens; spherical aberration, when they originate from a point on the axis. The effect of coma is the spreading of light from a point source off the optical axis into a small, cometlike image because of the displaced, out-of-focus image rays from the outer zones of the lens. The effect of spherical aberration or coma is that no sharp image can be formed; instead, the image is always out-of-focus.

Astigmatism results when the rays originating from a point source and striking the lens along a vertical diameter come to focus at a distance different from those rays striking along a horizontal diameter. The consequence of this is the formation of a line image (*primary image*) that progressively becomes an ellipse, then a circle (called the *circle of least confusion*), and then an ellipse, and finally a line perpendicular to the first one (*secondary image*) as the beam moves outward from the lens. *Curvature of field* is a result of the rays crossing in the same manner as in astigmatism; however, in curvature of field the distance between the primary and secondary images is negligible. The images are curved instead of being in a plane. This is of special importance in photography, where the film is flat and the image must be made to fall upon it in a plane.

Distortion, unlike the other aberrations, does not refer to the inability of a lens to form a point image of a point source. Instead, distortion results from a variation of magnification with distance from the axis. In *pincushion distortion,* the magnification increases with increasing distance from the axis; and the outer parts of the object appear larger than they should. The result is that the sides of a square appear to be caving in toward the center. In *barrel distortion,* the magnification decreases with increasing distance from the axis; and the outer parts of the object appear smaller than they should. The result is that the sides of a square appear to bulge away from the center.

In general, aberrations are corrected by choosing particular radii of curvature for the lenses, by making the lens of two or more parts that have different refractive indices, by using aspherical surfaces, or by controlling the distances between the various lens elements. It is impossible to produce a perfect image, but by careful lens design it is possible to obtain images within the limits desired.

While all of the above-mentioned aberrations are *monochromatic* (occur for any one wavelength), there is also a form of aberration that takes place with only nonmonochromatic light; it is called *chromatic aberration* and occurs only when more than one wavelength is present. Chromatic aberration is explained by the fact that the index of refraction of any substance varies with the wavelength of the light. Therefore, when nonmonochromatic light passes through a lens, some wavelengths are bent more than others. The result is that each of the various colors or wavelengths has a different focal point. If a lens not corrected for chromatic aberration is used to form an image, the image will have a border of spectral colors. All single lenses, when used in nonmonochromatic light, exhibit chromatic aberration.

An *achromatic lens* is one designed to correct chromatic aberration. Such a lens actually consists of two lenses—a convex lens of crown glass and a concave lens of flint glass. These lenses are constructed so that the chromatic aberration produced by one is neutralized by the other.

Prisms.—Reflecting prisms are used in optical instruments for a number of reasons. They may be used to displace a beam of light through a certain distance, to deviate a beam of light through a known angle, to rotate an image, or to erect the image formed by a lens before the image enters another lens. Although there are many types of prisms, only those commonly used are described.

To fully understand prisms, one must know the meanings of reversion and inversion. It was mentioned in the section on lenses that a real image formed by a positive lens is inverted. An *inverted image* is one that is upside down with respect to the object and also is turned left for right. A *reverted image,* on the other hand, is an image that is reversed about only one dimension when compared to its object. When a person looks into a plane mirror or looking glass, for example, he sees himself right side up, but his left side becomes the right side and vice versa. Some prisms are capable of reverting the object, and others are capable of inverting it.

Another aspect of prisms (and other reflecting surfaces) that must be understood is the critical angle. If a ray of light passes from glass into air, and if that ray exceeds a particular angle (called the *critical angle*) at the glass-air interface, none of the light passes from the glass into the air. What happens is that all the rays are reflected back into the glass at the glass-air interface. At such a reflecting surface, it is not necessary to silver the glass because all the rays are reflected. One therefore sees many prisms with no silvered reflecting surfaces simply because the rays of light in the glass exceed the critical angle at the reflecting surface. However, in some prisms this is not the case; there are also prisms with silvered reflecting surfaces.

A *90° prism,* or *right-angle prism,* is used for deviating or bending the light beam through 90°. This prism reverts the image. The *Amici prism* accomplishes the same thing as the right-angle prism, except that the reflecting surface becomes a roof (two planes perpendicular to each other) and inverts the image. The *Porro prism* may cause some lateral displacement of the image, depending on how it is used. The rays of light entering from the object are reflected by the two reflecting surfaces, and the reflected rays come back out of the prism parallel to the entering rays. This is true even if the prism is rotated about its horizontal axis. It should be noted that the image formed by a Porro prism is reverted. The *rhomboidal prism* merely causes a lateral displacement of the image, depending on the length of the prism. The *dove prism* causes an inline reversion of the image. "Inline" means that the rays leaving the prism are parallel to the rays that enter the prism. If one looks at an object through a dove prism and then rotates it, the image of the object will rotate through twice the angle that the prism is rotated through. This is also true if one looks at his reflection in a Porro prism and likewise rotates the Porro. The *Penta prism* bends or deviates the light beam through 90°. Penta prisms are sometimes made with a roof on one of the reflecting surfaces. If there is a roof, the image is reverted. If the Penta prism is rotated about its vertical axis, the entering and exiting rays will still make an angle of 90° with each other. This is a very important characteristic of the Penta prism. Penta prisms are used a great deal in large optical range finders to bend the rays of light entering the ends so that they pass along the axis of the instrument. They are also used to a great extent in optical alignment work. Penta prisms are sometimes called *optical squares.*

Often prisms are combined to make up a *prism system.* For example, two Porro prisms may be combined in such a way that the entering and exiting rays are parallel to one another although laterally displaced. The image formed by this system is inverted. A system such as this is commonly used in binoculars to reinvert the inverted image formed by the objective lens and eyepiece. This also permits higher magnification without increasing the length of the binocular housing.

—Edward F. Mackey and J. A. Mauro

MASERS AND LASERS

Masers.—Maser, an acronym for Microwave Amplification by Stimulated Emission of Radiation, is an amplifier or generator of electromagnetic waves. "Radiation" refers to electromagnetic radiation, a term that encompasses both visible light and invisible light. The latter includes frequencies higher than visible (ultraviolet and x-ray) as well as frequencies lower than visible (infrared, microwave, television, radio, and audio). Amplifiers from audio, radio, and up to microwave frequencies (about 10,000 million cycles per second) use condensers and inductances to make resonant circuits in which amplification or oscillation of electromagnetic radiation is possible. In order to design these resonant circuits to respond only over a particular band of frequencies, the circuits must be made smaller than a wavelength of the radiation, an increasingly difficult task at high frequencies. Shortly after World War II a number of scientists became convinced that the only practical solution to this problem was to use the natural molecular oscillations as the resonant circuits at the very highest frequencies.

Every system made of atoms—molecules and crystals, for example—has its characteristic energy levels. Although the system is almost always in the lowest state, it has the ability to absorb certain amounts of energy, and raise itself to a higher level. A molecule can be raised from a *lower state* to an *upper state* by absorbing radiation in the form of energy packets, or *quanta.* If E_u and E_l are two molecular states, upper and lower respectively, the frequency of emitted or absorbed radiation is given by $E_u - E_l$, divided by the constant h, known as *Planck's constant.*

■STIMULATED EMISSION OF RADIATION.—One of the most important rules of quantum mechanics was first stated by Albert Einstein in 1917. He said that the ability of molecules to absorb radiation depends not only on the number of molecules in the system but also on whether the molecules are in their upper or lower energy states. If more molecules are in their upper states than in their lower states, more molecules are stimulated to emit rather than to absorb radiation. Hence, the intensity of the initial radiation is increased as it passes through the molecules. This is known as *stimulated emission of radiation* and is the principle on which all maser and laser amplifiers and oscillators work.

Of course, to produce useful devices, the molecular amplification must exceed various microwave or optical losses in the molecules through which the radiation is passed. The first molecular amplifier which solved these problems was designed and built in 1954 by Professor Charles H. Townes and a team of graduate students at Columbia University. The high-energy molecules were separated from the low-energy ones in a molecular beam of ammonia gas (NH_3) by passing the beam through appropriately designed electric fields. The frequency of the molecular transition was 23,870 megacycles (in the microwave region). The energy was removed from the molecules by passing them through a *microwave cavity* vibrating at the molecular frequency. This metal-walled "resonant cavity" for electromagnetic waves is like a resonant organ pipe for sound waves. In each case the dimensions of the structure determine the wavelength at which it will vibrate, or be *resonant.* For the simplest vibrations the cavity or pipe contains along its length an integral number of half wavelengths of light or sound, as the case may be. A stream of air excites the organ pipe acoustically; the stream of molecules excites the microwave cavity electromagnetically, building up a density of radiation which stimulates succeeding molecules in the molecular beam to emit more radiation, thus amplifying the radiation.

■MASERS AS A STANDARD.—In succeeding years many more molecular systems were made to oscillate or amplify in the microwave region, and means other than molecular beams

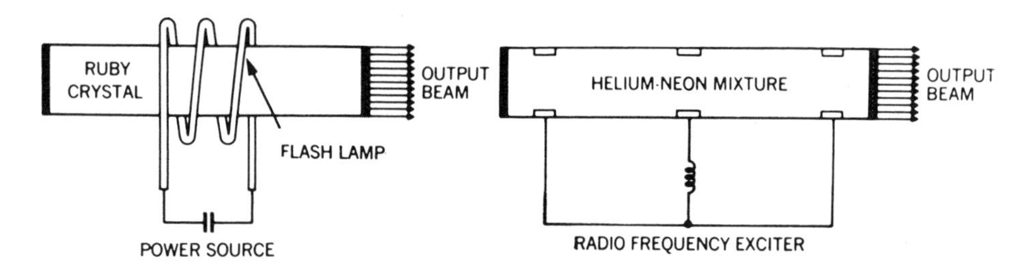

LASERS can use a ruby crystal (*above left*) or a mixture of helium and neon (*above right*) as the material for the molecular system. The ruby, or solid, laser produces high power, whereas the helium-neon, or gas, laser produces high coherence.

PHOTON CASCADE

TOTALLY SILVERED REFLECTING SURFACE
PARTIALLY SILVERED REFLECTING SURFACE
a
b
c
d
e

PHOTON CASCADE amplifies the excitation light in a laser (*left*). In *a*, the atoms (black dots) are in their ground state. The stimulating photons (arrows) have raised most of the atoms to their excited state (open dots) in *b*. The first photon spontaneously emitted parallel to the long axis of the crystal by an excited atom stimulates a second excited atom to emit its photon (*c*). (Photons emitted in other directions pass out of the crystal.) The stimulation continues as the photons are reflected between the ends of the crystal (*d*). When the amplification is great enough, some of the laser beam passes through the partially silvered end (*e*).

were found to selectively raise molecules to their upper states. Microwave masers possess two important electronic properties: as amplifiers they have very low noise figures, and as oscillators the molecular-beam masers in particular have very sharp, stable, and reproducible output frequencies because of collision-free oscillations. Hence the beam oscillators are used as frequency standards at the National Bureau of Standards, where frequency stabilities as great as one part in a million million have been achieved. Since all molecules of a given kind (such as ammonia) have the same oscillation frequency, which is independent of temperature, molecules serve as conveniently reproducible frequency standards.

■MASERS AS LOW NOISE AMPLIFIERS.—Despite their low noise features, microwave maser amplifiers have come into only limited use because they are relatively large and expensive. They are used, however, as sensitive low-noise receivers for radio astronomy experiments and in the Bell Telephone Laboratory's Telstar television relay link. In these applications a form of solid-state maser invented by N. Bloembergen of Harvard is used. Microwave energy of one frequency is absorbed by the material to "pump" the molecules or ions to the desired upper energy levels. These masers work best at temperatures of near −269°, the boiling point of liquid helium. A frequently used maser material is the trivalent chromium ion in Al_2O_3 (ruby). Some energy levels of this ion are shifted by a magnetic field so that the transition frequencies lie in the microwave region.

Lasers.—Soon after the achievement of the microwave maser, scientists sought to extend this principle of operation to the visible region of the electromagnetic spectrum, where no true electronic amplifiers existed. The problems seemed formidable because optical wavelengths (around 5,000 angstrom units, or 0.02 thousandths of an inch) are much smaller than any resonant cavity in which the molecules are located. (As is evident from the discussion of the ammonia maser, it is often desirable to couple the molecular oscillations to some larger resonant structure.) At these optical frequencies, moreover, a new problem is encountered: when molecules are raised to high energy states, there is a great probability that they will radiate energy spontaneously, without requiring a stimulating input signal. Since this spontaneously radiated energy does not change when an input light signal is added, it does not contribute to an amplification of the light, but is wasted energy. The loss is proportional to the fourth power of the frequency, and hence is important at optical frequencies but negligible for most purposes at microwave frequencies.

Again the solution to the problem was proposed by Professor Townes, this time, in 1958, in collaboration with Dr. A. L. Schawlow of the Bell Telephone Laboratories. Their suggestion: mirrors. They noted that if the molecular oscillators of a gas, for instance, are contained between two parallel mirrors, only the radiation bouncing back and forth between the mirrors passes through the molecules often enough to be amplified significantly. In the absence of mirrors, molecules can emit radiation spontaneously over a range of many millions of cycles near each resonant frequency. The mirrors are resonant to some of these frequencies, specifically those for which an integral number of half wavelengths of the radiation can be fitted into the space between the mirrors. (The product of light frequency times wavelength is the velocity of light, 3×10^{10} centimeters per second, or 186,000 miles per second.) Thus the combined mirror-plus-molecules system oscillates most strongly at one or more of these selected frequencies.

The first successful optical maser, or *laser*, an acronym for Light Amplification by Stimulated Emission of Radiation, was built in 1960 by T. H. Maiman of the Hughes Aircraft Company. Again the material was ruby, with mirror coatings of silver on the surfaces of accurately polished, parallel plane faces of the gem. For the ruby laser no magnetic field is required because different energy levels are used from maser levels. Operation is usually at room temperature, although lower temperature improves efficiency.

The laser oscillator emits a beam of light, with a very small angular spread, at right angles to the faces of the parallel mirrors. As the light energy bounces back and forth between the mirrored surfaces, energy is built up or amplified by stimulated emission. (One of the mirror's faces is partly transparent, allowing the amplified light energy to pass through.) As with the maser, this emitted radiation maintains an exact frequency, or is *monochromatic*.

In lasers using a gas (a mixture of helium and neon) as the molecular system, radiation monochromatic to 20 cycles out of 260 million million cycles (one part in 10^{13}) has been achieved, with an extremely small angular beam spread. These properties of narrow beam spread and sharp monochromaticity are the strong *coherence* features of a laser beam. Parts of the beam many wavelengths apart keep in step and thus maintain a flat wave front and a long wave path, or *train*. A very flat wave front possesses strong *space coherence*, and a long wave train has a long *coherence time*.

■LASER APPLICATIONS.—The proposed technical uses of lasers are many. The directional features suggest sending information over laser beams. Furthermore, since the frequency is very high, it should be possible to transmit many messages simultaneously. Theoretically, all the simultaneous telephone calls in the United States could be transmitted over a single laser beam. In practice, however, there are severe limitations in attaining this ideal—the laser must be *modulated*, that is, its light output must be changed to correspond to the message being sent.

Clouds in the atmosphere absorb and scatter the light, so signals on earth might have to be sent through pipes, although no such limitations are imposed on space communications. Laser radar echoes have been bounced off the moon and received on the earth. The high coherence features of laser beams, together with the fact that some lasers can be pulsed to emit their light in extremely short bursts, make them suitable for very precise measurements of distance and the velocity of moving targets (determined from the *Doppler shift* in frequency of the return echo from the target).

Lasers have recently been used to record three dimensional images on film by an interferometric technique known as holography.

Other laser applications result from the focusing properties of very coherent light beams: they can be focused to a spot about a wavelength in diameter and hence can be used to generate intense light, heat, and pressure in one spot. The pulsed output of a ruby laser has been used to punch holes through razor blades, diamonds, and other materials. Thus micromachining, welding, and photographic recording applications for lasers exist. Closely analagous medical applications include welding detached retinas in eyes.

Some examples of all the phases of matter—gas, solid, and liquid—have now been made into lasers. Some feature direct conversion of energy from electrical to optical; and some, like the ruby, conversion from incoherent pump light to coherent laser light.

Although high power (ruby) and high coherence (gas) have been emphasized, for some purposes small size, high electrical efficiency, and ease of modulation are important. For these the *semiconductor injection laser*, developed in 1962 by Marshall I. Nathan and co-workers at International Business Machines and Robert N. Hall and co-workers at the General Electric Company, is a promising development.

—W. V. Smith

NUCLEAR PHYSICS

Scope.—*Nuclear physics* is a concerted study undertaken to determine what the nucleus of the atom consists of, how it behaves, and what holds it together. Attainment of a proper perspective requires an understanding both of the modern picture of the nucleus and of its properties.

Characteristics of the Nucleus. — An atom is about 10^{-8}, or 0.00000001 centimeter (cm.) in diameter. One can better imagine how small this is when it is realized that there are six sextillion (6 followed by 21 zeros) atoms in a drop of water. The nucleus of the atom, which is composed of neutrons and protons (except in the commonest form of hydrogen, in which it consists of a single proton), is much smaller—about 10^{-13} cm. in diameter. Outside the nucleus there are only electrons. The neutron and the proton have almost identical masses, 1.67×10^{-24} gram. This is approximately 1,840 times the mass of an electron. Therefore, in the commonest form of hydrogen, which is composed of one electron and one proton, 1,840 times as much mass is found in the nucleus as is found in the region outside the nucleus. Since the region outside the nucleus is 100,000 times as big, it can be said that typically an atom is composed mostly of empty space, with only a very small region where all the mass is found. Although they have virtually the same mass, the neutron and proton differ in one important characteristic: the proton carries an electrical charge, and the neutron does not. The positive charge on a proton is the smallest electrical charge known in nature. In no experiments have charges of, say, half that on a proton been found. It is as if the charge on a proton were similar to the U.S. penny—no halfpennies are in circulation. Pairs of pennies can be found, but this would be similar to the case where two positive charges are found linked together in some way. The electron has an electrical charge of the same magnitude as that of the proton, but it is negative. Again, no experiment has detected a charge one-half or one-quarter the size of the electron's charge. There have been recent suggestions that "particles" called *quarks*, with one-third or two-thirds the electron's charge, do exist, but none have yet been found.

■ATOMIC STRUCTURE.—The structure of the hydrogen atom has been outlined above. The nucleus consists of one proton and therefore has a positive

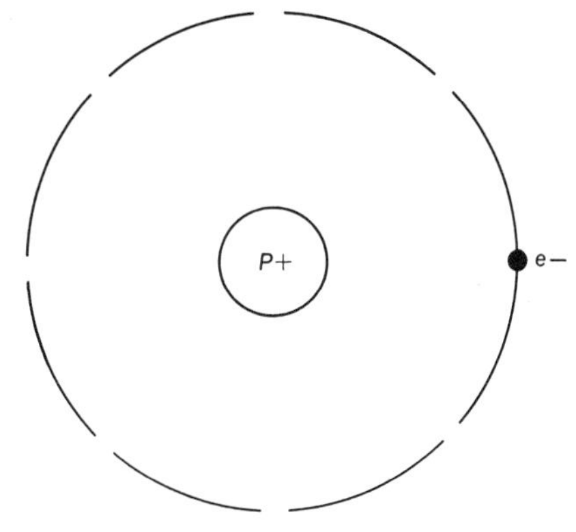

HYDROGEN ATOM consists of an electron, e^-, and a nucleus containing one proton, P^+.

charge. The electron has a negative charge, so the net charge on the hydrogen atom is zero, because the positive charge is equal to, but the opposite of, the negative charge. The force that holds the atom together, that binds the electron to the proton in some way, is the *coulomb force*, the attraction between positive and negative charges. The next smallest atom after hydrogen is that of helium, which has two protons and two neutrons in the nucleus; this results in its having a net positive charge of two units. Therefore, in order for helium to be electrically neutral, two electrons must orbit the nucleus. The forces holding the electrons in the atom are once again the coulomb forces. At this point, however, a major problem of nuclear physics appears. There are two protons in the nucleus of helium. Since these two protons have identical positive charges, they would be expected to repulse each other, causing the nucleus to fly apart. What prevents this from occurring is a major concern of nuclear physics.

■BUILDING ELEMENTS.—The building of the atoms of elements by addition of protons and neutrons in the nucleus and of electrons outside the nucleus continues in orderly fashion until the heaviest naturally occurring element, uranium, is reached. The uranium nucleus contains 92 protons and 146 neutrons. Actually, there are several different forms of uranium, called *isotopes*, which differ in the number of neutrons in the nucleus. It should be noted that the isotopes of every element have been found. The number of electrons present in uranium is 92, again preserving the zero net charge of the atom. About a dozen elements having more than 92 protons in the nuclei have been created, but they are relatively unstable and do not exist in nature. These elements are called *transuranic elements*. Two of the more familiar are plutonium and neptunium.

Scattering Experiment.—A considerable amount has been said about what atoms look like, but nothing has been said about how the knowledge of the structure of the nucleus and the *nucleons* (the neutrons and protons) is obtained. The basic experiment of nuclear physics is the *scattering experiment*, in which projectiles (neutrons, protons, electrons, gamma rays, or various other nuclear particles) are accelerated to high energies and directed at a target consisting of atoms of the material to be studied. At positions around the target, *collectors* (or *detectors*) of various types are set up. This setup and experiment may be compared, in a simple way, to the game of pool or pocket billiards. The cue ball serves as the projectile, the racked balls or an individual ball as the target, the pockets as the collectors, and the pool cue as the accelerator. Some features of scattering immediately emerge. As anyone familiar with the game of pool knows, the effect of "scattering" the cue ball off several other balls is different from that of "scattering" it off a single ball. It is also clear that scattering off a cubic object, rather than a pool ball, would lead to a markedly different scattering pattern. The scattering experiment in nuclear physics differs from pool in that scientists do not know what the scatterers look like. They know what kind of "pool cue" they have (the various types of accelerators will be discussed later); they know what the "cue ball" looks like (the protons, neutrons, or other particles, which come out the end of the accelerator); they know what the pockets do because these collectors are built for a specific perform-

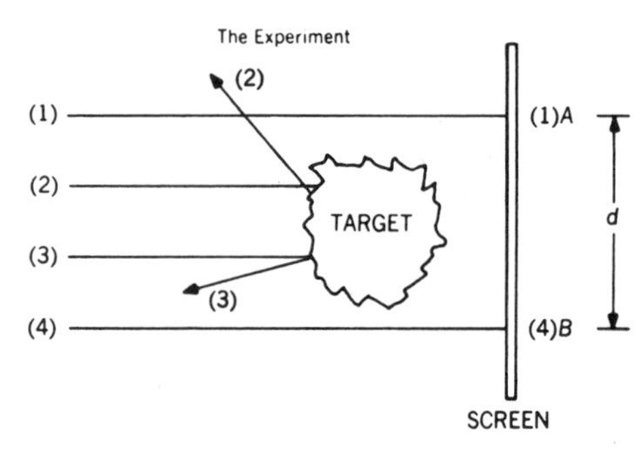

SIZE AND SHAPE OF AN UNKNOWN can be found by noting the way it deflects particles.

ance—but they do not know what the target looks like. However, from their knowledge of these three factors they can deduce the fourth. To see how this might be done in a simple case is shown in the accompanying illustration.

Thinking about a possible shape for the target leads to the following conclusions. The dimensions of the target have to be about the distance from point *A* to point *B* because the two balls passed straight through the target area without deflection. (This is an important concept in nuclear physics, and the dimension—the distance *d*—so obtained is called the *cross section of the target*. Obviously, nuclear physics deals with three dimensions and therefore the dimension, instead of being a length, is an area.) Balls (2) and (3) were scattered to their right. A possible means of accounting for the scattering of all four balls might be a target shaped like the one shown. The scattered paths are indicated. Thus, by observing the scattered particles, the shape and dimensions of targets can, at least in principle, be deduced.

One other phenomenon can be explained in terms of pool balls. All pool players know it is possible to put "English" on the ball. This causes the ball to rotate in one direction while moving in another. The effect of a ball being scattered with spin on it and without spin on it is different, and so indeed is the effect of scattering nuclear projectiles with this sort of spin off targets. Nuclear particles possess spin (and so do the targets), so it is possible that spin effects enter into the interpretation of scattering data. This useful result, although it somewhat complicates the interpretation of the scattering experiment data, does enable nuclear physicists to obtain more information than would otherwise be possible.

Accelerators.—One of the first accelerators used in a nuclear physics experiment was a naturally radioactive element that emitted *alpha particles.* (An *alpha particle* consists of two protons and two neutrons bound together, and thus is a helium atom stripped of electrons.) Ernest Rutherford (1871–1937) used radium as his "accelerator" to bombard thin foils of materials. The results indicated that some of the alpha particles were scattered backward, although most went through relatively undeviated. Backscattering could be explained only if the alpha particles were striking something more massive than themselves. It was in these experiments that the picture of an atom as composed of a small but massive nucleus surrounded by a relatively vacant space was first deduced and then confirmed by Rutherford.

Other types of accelerators include the Cockroft-Walton accelerator, invented by J. Douglas Cockroft and Ernest Thomas Sinton Walton in 1932; the Van de Graaff accelerator, proposed by Robert Jemison Van de Graaff in 1931; the cyclotron, built by Ernest O. Lawrence in 1930; the synchrocyclotron; the synchroton; the bevatron; the betatron; the linear accelerator; and other special types of devices. In all these accelerators, the effect of electric and magnetic fields on charged particles is utilized to give the particles high velocities.

■CYCLOTRONS.—The effect of electric and magnetic fields can best be explained by reference to the *cyclotron,* where the effects of both types of fields are utilized to produce an energetic beam of particles. As shown in the accompanying diagram, charged particles (protons) are introduced at point *A; B* and *C* are two magnets, called "Dees" because of their shape, that have a hollow space between them. An electrically charged particle is pushed on by the electric field, say, in the direction shown by the arrow. The particle then enters the magnetic field of *C*. While in the magnet it is turned in a circular path, as shown, and finally is ready to emerge from the magnet. During the period of time the particle is completing the half circle, the voltage on the magnets has been reversed. Once again the particle is accelerated, but now toward magnet *B*. Once the particle enters magnet *B* it again is turned in a circular path. Upon emerging from the magnet *B*, the particle finds the direction of the electric field has again been reversed to cause acceleration toward *C*. This process continues until the particle emerges from the cyclotron.

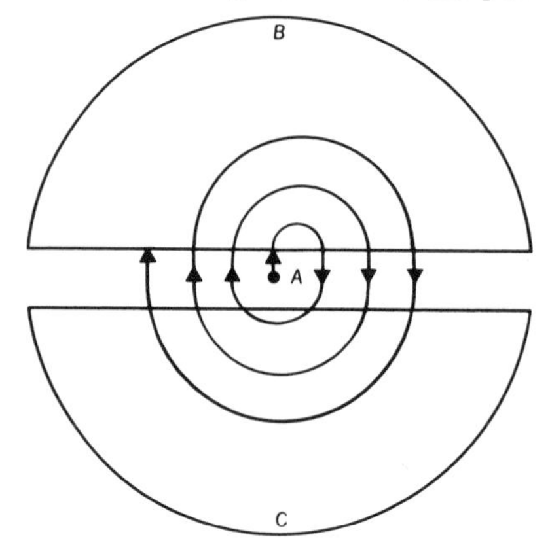

CYCLOTRON is used to accelerate particles.

■SYNCHROCYCLOTRON.—One characteristic of the cyclotron is that the rate at which the voltage on the magnets has to be reversed is independent of the radius of the accelerated particles' path and of their velocity, provided the mass of the accelerated particle remains the same. Relativity predicts that the mass of a particle will increase with its velocity, however, and this effect has been noted in the cyclotron. Such a mass change means that the rate at which the electric accelerating potential is changed must change as the accelerated particle increases in velocity. An accelerator that does this is the *synchrocyclotron,* a device put into operation after World War II. Its principles and those of the cyclotron are basically the same: the electric field accelerates the particles while the magnetic field steers them.

■SYNCHROTRON.—The basic operation of the *synchrotron* is also similar to the cyclotron and the synchrocyclotron. In the synchrotron the particles travel in nearly circular orbits whose radius does not change. The acceleration is performed between steering magnets, again by the use of electric fields. The largest synchrotron is located at the Brookhaven National Laboratory, Long Island, New York. It produces a beam of particles with an energy of 33 billion electron volts. (An *electron volt* is the energy equal to that acquired by an electron or protron in passing through an electric potential of 1 volt. Thus, it is as if the protons accelerated had passed through an electric potential of 30 billion volts.)

■OTHER ACCELERATORS.—A basically different type of accelerator is represented by the Van de Graaff, Cockroft-Walton, and linear accelerators. In these, magnetic fields are not used to bend the beam in a circular or almost circular path. Instead, the charged particles are accelerated in a straight line by the action of an electric field. In the case of the Van de Graaff and Cockroft-Walton accelerators, the electric field is static, while in the linear accelerator it is traveling. The linear accelerator now under construction at Stanford University is about two miles long.

■NEUTRON PRODUCTION.—After accelerated particles emerge from the accelerators, they are allowed to hit targets directly in order to study the targets themselves, or to hit targets in order to produce beams of particles of other than the original type. For example, neutrons, since they have no charge, cannot be accelerated or steered by electric and magnetic fields. Beams of neutrons must be produced by collisions of the primary beam with other targets.

■ACCELERATOR RESEARCH. — Attempts are being made to develop accelerators that can produce primary beams of higher and higher energies, perhaps as much as 1,000 billion electron volts, in the next generation of accelerators. These accelerators will be used further to explore the nucleus, particularly in attempts to unravel the mystery of what holds the nucleus—and hence matter—together. The higher energies are needed because, in order to study the particles that scientists think hold the nucleus together, enough energy must be available to free these particles from the nucleus. Other workers are putting their efforts into developing accelerators that can produce primary beams of greater and greater intensity. The process of accelerating charged particles is a fairly wasteful one: the particles escape, collide, and so on, and do not emerge from the machines. The accuracy of the data obtained from scattering experiments depends to a great extent on the number of projectiles available. (In the example of the pool balls given above, eight balls would give more data and accuracy than would four.) For this reason, many scientists are concentrating on the intensity of the available beams rather than the absolute energy of emerging particles, although they also need relatively high energies. These machines would be used to find out more about the nucleus's structure and behavior than about what holds it together.

Collectors.—In the early days of nuclear physics, nuclear events were observed on photographic plates. In fact, radioactivity was discovered by Antoine Henri Becquerel in 1895 through the darkening of photographic film—charged particles expose film in passing through it. The use of photographic plates and their descendants, the *nuclear track plates,* continues to this day. The data obtained is accurate and useful. The methods, however, are slow; and the plates require developing and processing, steps that make obtaining and processing data more complex and difficult.

Electronic techniques have also been used. The *Geiger counter,* named after its inventor, Hans Geiger, uses the effect of energetic radiation in making an enclosed gas electrically conducting in order to measure the nuclear processes. Details of the events are not visible in the Geiger and similar counters; only the total number of events taking place is accessible to the user.

Scintillation counters make use of the fact that radiation produces flashes of light when it strikes certain crystals such as those of sodium iodide. Electronic circuitry, in conjunction with photomultiplier tubes, detects the flashes and counts them. Again, these devices, although extremely useful, measure only the quantity of a certain type of event rather than giving detailed information about the event. In some experiments in nuclear physics this type of information is extremely important, particularly in a case where accurate measurement of large numbers of events is desired. This is true, for example, in observations of the behavior and structure of the nucleus.

■CLOUD CHAMBER.—In some types of research involving the structure of the nucleus and the forces that hold it together, it is very important to observe individual nuclear events. To do this, a group of detectors has been developed. The first of these was the *Wilson cloud chamber,* designed by Charles Thomas Rees Wilson in 1911. A gas-filled chamber containing supersaturated vapor is exposed to radiation. The incoming particles cause the excess moisture to precipitate along the trail of the particle, creating a visible trail. Thus, the actual path of a particle can be seen. When the cloud chamber is used in conjunction with electric and magnetic fields and photographic films, the charge and momentum of the incident particles can be deduced. The basic difficulty with the cloud chamber is that relatively few events can be observed in a given time, for the chamber is sensitive only when the vapor is supersaturated. The process of supersaturation is a relatively long one, so the cloud chamber is "dead" a good part of the time.

■BUBBLE CHAMBER.—A device whose action is basically similar to the cloud chamber is the *bubble chamber,* which was designed in the 1950's. This

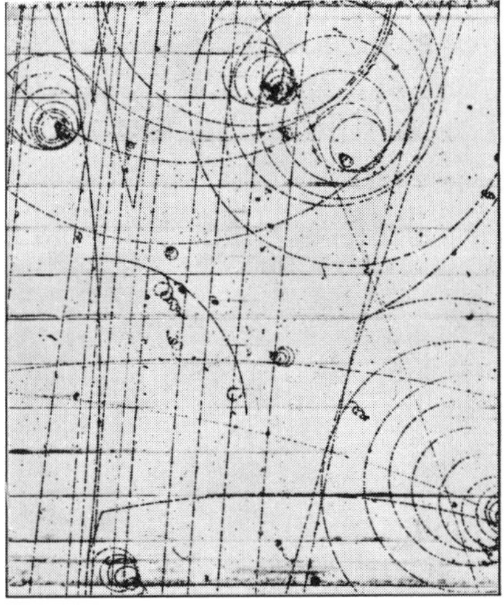

BROOKHAVEN NATIONAL LABORATORIES

VAPOR TRAILS of radioactive particles are seen in this bubble-chamber photograph.

chamber usually contains supersaturated liquid hydrogen; others possible include liquid xenon and helium. The action of the incoming radiation causes a row of bubbles along the trail of the incoming particle, provided the particle has an electrical charge. After supersaturation of the liquid is achieved, a light is flashed so that events of interest can be recorded on photographic film. However, the bubble chamber suffers from relative slowness and, since the picture does not necessarily coincide with a particular type of event, it is not very selective. Because the liquid hydrogen is dense compared to the gases used in the cloud chamber, many more events are photographed in a given volume of liquid than in an equal volume of gas in the cloud chamber. Furthermore, since the nucleus of hydrogen is just a proton, scattering experiments are performed with the detector simultaneously being a very excellent target. Much information concerning scattering of various primary beams off protons has been obtained in the last several years.

■SPARK CHAMBER.—The *spark chamber* is another collector device used extensively to study individual nuclear events. In this device a series of thin metal plates are stacked, with each plate separated from its neighbor. The entire chamber contains neon gas. A high electric potential is maintained between each plate. When radiation enters the chamber, the neon gas along the trail is ionized—one or more electrons are separated from the neon atom. Therefore, the electrons and the neon atom, which now has a net positive charge, are free to be acted upon by the electric potential. Along the trail, neon gas makes visible the path of the incoming radiation through brief discharges. When the sparks occur, photographs are taken and later analyzed. Since the action of the spark chamber is electronically controlled, its action is relatively fast; supersaturation does not have to be achieved, although ions and electrons do have to be swept away. In various ways, through the use of auxiliary electronic circuits, the spark chamber can also be made somewhat selective about what events are recorded.

Nuclear Fission Reactions.—In *nuclear fission reactions* the nucleus of a heavy isotope of an element is ruptured into two segments of almost equal mass, with the resultant release of large quantities of energy and of neutrons. The two segments into which the nucleus splits are the nuclei of other, lighter atoms. The rupture occurs when the nucleus absorbs a free neutron. The neutrons released during the fission process may, in turn, cause other atoms of the heavy isotopes to undergo fission in a *chain reaction.* When the chain reaction is not inhibited in any way, an uncontrolled fission reaction takes place, releasing vast amounts of energy over a very short period; this is what happened in the atomic bombs exploded over Hiroshima and Nagasaki, Japan, in 1945. When the chain reaction is inhibited, a controlled reaction takes place and the same quantity of energy is liberated over a much longer period; this is what takes place in a nuclear reactor, such as in the Yankee power reactor at Rowe, Massachusetts, where electricity is produced, or in the one aboard the nuclear-powered merchant ship *Savannah.*

A typical fission reaction is the one undergone by uranium-235 to produce atoms of barium-141 and krypton-92, as shown in the following equation:

$$ {}_{92}U^{235} + {}_{0}n^{1} \rightarrow {}_{56}Ba^{141} + {}_{36}Kr^{92} + 3\,{}_{0}n^{1} + \text{energy}. $$

Nuclear Fusion Reactions.—The *fusion reaction* is virtually the exact opposite of the fission reaction. In a fusion reaction, the nuclei of two light

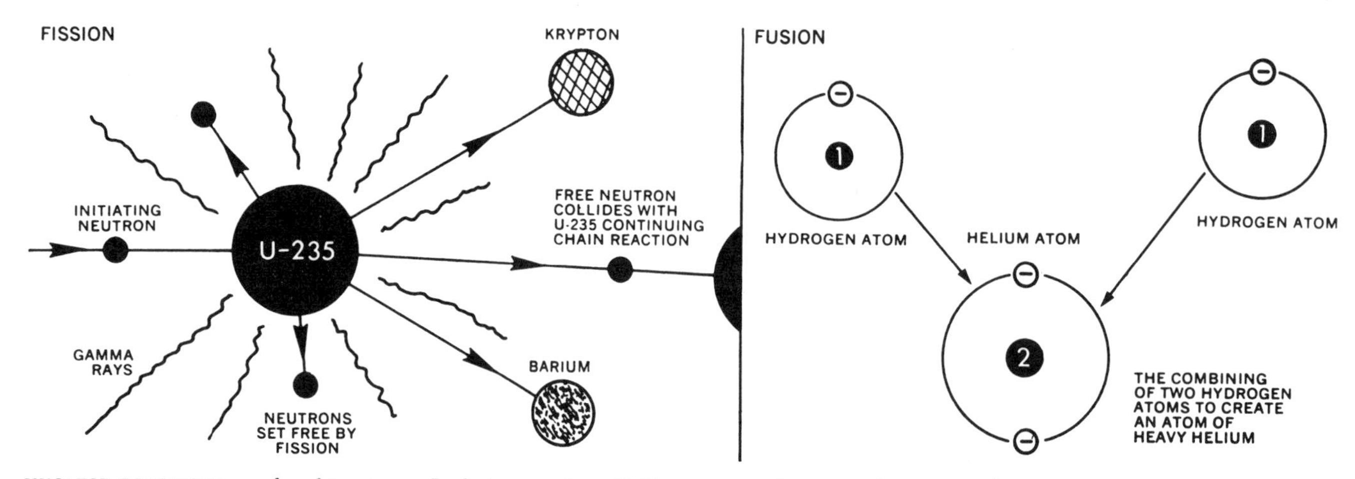

NUCLEAR REACTIONS can be of two types. In fission reactions (*left*), an atom of a heavy element—such as uranium-235—is split into atoms of lighter elements, while in fusion reactions (*right*), atoms of light elements combine to form an atom of a heavier element.

isotopes join to form a heavier isotope; here, too, the reaction results in the release of large quantities of energy. A typical uncontrolled fusion reaction is the one that takes place within the sun, where hydrogen nuclei combine to form a helium nucleus and release solar energy, a process called the *hydrogen-helium cycle*. The equation for this reaction is:

$$4\,{}_1H^1 \rightarrow 2\,{}_2H^4 + W2\,{}_1\beta^0 + \text{energy},$$

where ${}_1\beta^0$ is a beta particle, or electron. Experiments are currently being conducted in an effort to produce a controlled fusion reaction. However, at the present time only limited success has been achieved.

Status of Knowledge.—With the accelerators and collectors that have become available in recent years, progress in understanding nuclear structure phenomena and forces has been rapid. A model of the nucleus has been developed that adequately explains many of the properties and the behavior of atoms. This is, of course, not to say that much more does not need to be done. More accurate measurements, improved theories, and greater variety, intensity, and energy of primary beams is still necessary.

The status of nuclear knowledge can be considered in three important areas that, while not the only areas of interest in nuclear physics, are typical of some of the most important: the structure and properties of the nuclei of atoms; the properties of *nucleons* (the protons and neutrons that make up the nucleus); and the forces that cause the nucleus to stay together (these are related to the "elementary" particles).

Properties of the Nuclei.—As a result of the research so far performed, it is now known that when a nucleus is bombarded by a primary beam from an accelerator, the resulting nuclear reaction can occur in two possible ways. In both of these the bombarding particle is essentially swallowed by the nucleus. The incoming particle and the nucleus stay in this joint state for about 10^{-22} second, after which time one of two things may happen. The incoming particle may escape, or the incoming particle may literally combine with the bombarding nuclei. The processes have been described as one in which the intermediate state immediately proceeds to the final state by direct interaction without disturbing most of the nucleons in the system, and as one in which the intermediate state is transformed into a true compound that lasts for only about 10^{-13} second. The compound nucleus eventually decays by the emission of gamma rays or by evaporation of one or more nucleons to form a new product.

Structure of the Nucleons.—It was mentioned earlier that the proton and the neutron are the basic residents in the nucleus. For many years they, along with several others, were thought to be elementary particles, the simplest particles that could exist. The proton had a positive charge on it, and that was that; the neutron had no charge. However, some disconcerting facts eventually emerged. One of the most disconcerting was that the neutron was found to have a magnetic moment. A *magnetic moment* indicates the presence of charges in motion, and for a long time it was difficult to see how a particle with no charge could have a circulating charge. Robert Hofstadter, using a high-energy electron accelerator, was able to determine in scattering experiments that the neutron, although it has a net charge of zero, does have a *charge structure*. As a matter of fact, it possesses a fringe of positive charge and enough negative charges to make the net charge zero. While Hofstadter's findings cleared up one problem, they might have created others, in that the proton's charge was also found to have a structure. In other words, the charge is not concentrated all at one point. Thus, even though the charge on the proton is the smallest unit of charge so far found, it is "smeared" around to some extent. Hofstadter's findings also confirmed the basic similarity in the behavior of neutron and proton—except, of course, in terms of charge.

Nuclear Forces.—Finally, the investigations of the fundamental forces holding the nucleus together are, at this moment, arranging themselves in order. The quandary of the "elementary" particles is being resolved. As has been mentioned, the search for elementary particles is an old one. These particles, which cannot be further subdivided, have been sought for many years. At first it was thought that they consisted of the electron, the proton, the neutron, the photon, and the neutrino. Then, in 1937, Hideki Yukawa predicted the existence of the *meson* as a carrier of the nuclear force. A *meson* is a particle having a mass approximately 250 times that of the electron. To show how a particle can be the carrier of a force, the following simple analogy may be made. Imagine two boys, each with a pillow. One of the boys throws the pillow to the other, who catches it. In catching it, he is pushed backward. The first boy is also pushed backward, and so the exchange of the pillow leads to a repulsive force. An attractive force occurs if each boy grabs the other's pillow and tugs on it. It is precisely in this way, using particles (mesons) rather than pillows, that Yukawa thought of nuclear forces. One other fact must be borne in mind, however. Creation of a particle to exchange requires energy. Einstein's familiar formula,

$$E = mc^2,$$

is well-known. E is the energy necessary to create a particle, m is the mass of the particle created, and c is a constant equal to the velocity of light. (The mass of an electron is equivalent to about 500,000 electron volts of energy.) The uncertainty principle of Werner Karl von Heisenberg (1901–   ) makes a prediction about how long particles of this type can survive unless enough energy is supplied literally to create them or to free them permanently. This principle states that the more massive the particle, the less time it can live if not freed. This has implications concerning over what distances nuclear forces can extend, for if the forces really arise due to an exchange, then the distance a force can extend depends roughly on how far a particle can move in its lifetime. The shorter the lifetime, the shorter the distance; thus, the relatively large mass of the nuclear particles means the forces are short-range.

Nuclear physics's approach to studying these particles has been straightforward; supply the energy necessary to free the exchanged particles. It is easily seen that the energy necessary is very large because the mass of the particles is very large; thus, the development of the very-high-energy accelerators has been vital. What have physicists found about these particles in their recent studies? At first they thought there were very many particles. As the accelerator energy increased, they found more and more particles. Finally, though, some order began to emerge out of the chaos. The situation is very similar to that which existed in atomic physics and spectra in the early days of research in that area. The spectrum of mercury, represented by lines in an illustration, can be taken as an example. Although a great many lines appear, the spectrum is only from mercury; each line represents transitions between excited energy states.

In the same way, current thinking is that the photons, the neutrino, the electron, the pi-mesons, the protons, the neutrons, and other particles are all related in simpler fashion, much as the line spectrum of mercury (or any atom) has its origin in the basic structure of a single atom. In addition, something should be said about the anti-particles. An *anti-particle* is merely another state of a known particle. For example, the anti-proton and the proton differ only in that their charges are opposite: the mass and other properties remain the same. For this reason, many physicists object to regarding the anti-particles as new particles. Their reasoning is that calling each anti-particle of a given particle a new particle is like doubling the number of animal species by calling the mirror image of each a new species.

A number of physicists maintain that there are only two elementary particles, the *baryon* and the *lepton.* The most familiar baryons are protons and neutrons; they are thought of as being merely different states of the same particles. Other particles, such as the *lambda-zero* and the *sigma-minus,* are excited states (states of higher energy) of the baryon. They can decay to a proton or neutron by emitting pi-mesons or K-mesons (kaons). The most familiar leptons are neutrinos, electrons, and mu-mesons. Again, decay is possible by emission of mesons.

Recently, Yuval Ne'eman and Murray Gell-Mann have independently devised an approach known as the "eight-fold way" to predict the differences in energies between certain excited states. To a large extent these relationships are correct, but physicists are still far from an understanding of the theory of why the relationships exist.

—Donald E. Cunningham

PHYSICS GLOSSARY

Absolute zero.—The temperature at which the molecules of a substance would cease moving and at which, therefore, a perfect gas (if kept at constant volume) would exert no pressure. The temperature has never been reached and can be shown to be unattainable. It is equal to −273.15° C. or −459.67° F. (See also *Temperature scales.*)

Absorption.—The conversion of radiant light energy into other forms of energy, such as heat, as the light energy passes through a particular medium. The *absorptivity,* or *absorptive power,* of a medium is the ratio of the amount of light energy it converts to the total amount of light energy passing through it.

Acceleration.—The change of velocity (speeding up or slowing down) with time. This change may be a change in speed, such as when a ball is dropped from a height, or a change in direction, such as when a runner moves around a curved track at a constant speed, or a combination of the two. *Angular acceleration* is the speeding up or slowing down of a rotating body. The *acceleration of gravity,* usually designated by the letter *g,* is the change in velocity caused by gravitational attraction; on Earth, it is equal to 32.2 feet per second per second, or 980.6 centimeters per second per second.

Acoustics.—The branch of physics that concerns itself with the production, conduction, measurement, perception, reproduction, and control of sound waves.

Adiabatic process.—Any change in matter that takes place without a loss or gain in the heat content of the system.

Alpha particle.—The nucleus of a helium atom. An alpha particle is composed of two protons and two neutrons and therefore has a positive charge. Alpha particles are one of the products of the radioactive decay of the elements radium and uranium.

Ammeter.—An indicating device used to measure the quantity of current in an electric or electronic circuit. Ammeters are always connected in series in the portion of the circuit in which they are used.

Ampere.—The unit of electrical current named after the French physicist André Marie Ampère (1775–1836). A *milliampere* is equal to 0.001 ampere, and a *microampere* is equal to 0.000001 ampere.

Amplitude.—The value measured from the mean, or interim, position to the highest, or extreme, position in any periodic or vibratory function, such as the swinging of a pendulum, the vibration of an alternating current, or the movement of a sound wave. It is sometimes called the *peak position.*

Avogadro's number.—(*Avogadro's constant.*) The total number of molecules present in 22.4 liters of any gas at standard temperature and pressure, named after the Italian physicist Amedeo Avogadro (1776–1856). Avogadro's number, usually designated as *N,* is equal to 6.0254×10^{23}.

Beat.—In acoustics, the tone resulting from the simultaneous arrival of two sound waves of nearly equal frequency. For example, if two tones having frequencies of 200 and 250 cycles per second arrive at the ear simultaneously, the listener will also hear a *beat frequency* tone equal to the difference between the other two tones (50 cycles per second).

Beta particle.—An electron emitted during the radioactive decay of such substances as thorium and uranium. *Positive beta particles* are *positrons,* and these are also emitted during the radioactive decay of certain elements.

Betatron.—See *Particle accelerator.*

Boiling point.—The temperature at which the vapor pressure of a liquid is fractionally greater than the pressure of its surroundings. At this point molecules of the liquid enter the gaseous state at a greater rate than they return to the liquid state, and the liquid boils off.

Calorie.—A unit quantity of heat energy. A *gram-calorie* is the quantity of heat that must be supplied to 1 gram of water to raise its temperature by 1° C. A *kilogram-calorie* is the quantity of heat that must be supplied to 1 kilogram of water to raise its temperature by 1° C.

Capacitance.—The ratio of the electrostatic charge on two bodies to the potential between them. A *capacitor,* or *condenser,* is a device consisting of two conducting plates separated by an insulating material, or *dielectric,* and thus having a high capacitance; it is used to store electric charges.

Carnot cycle.—The series of operations gone through by an ideal heat engine (called the *Carnot engine*), an engine that operates at maximum efficiency. Proposed by Nicolas Léonard Sadi Carnot in 1824, the four phases of the cycle are *isothermal expansion, adiabatic expansion, isothermal compression,* and *adiabatic compression.* Work is done only during the two expansion phases of the cycle, which accounts for the high efficiency of the engine; energy is expended only when maximum energy (highest temperature) is attained in the expansion phases.

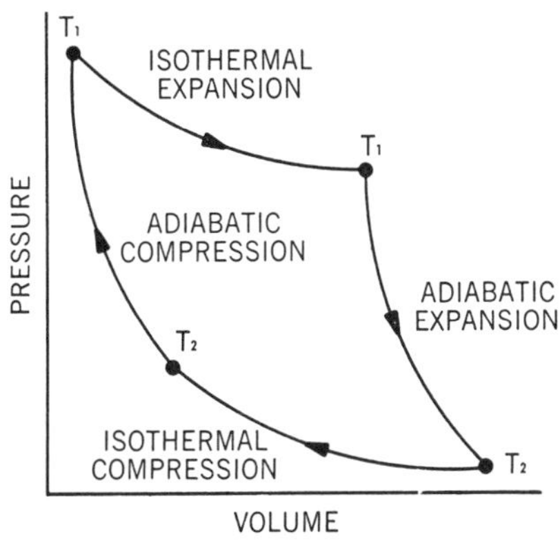

CARNOT CYCLE for an ideal heat engine. Since heat is evolved and taken up only during the two adiabatic parts of the cycle, the engine's efficiency is equal to $1 - (T_2/T_1)$; the greater the starting temperature, T_1, and the lower the interim temperature, T_2, the greater the efficiency.

Cavitation.—The process of forming vapor bubbles in a liquid by the reduction of the pressure at a point within the liquid. A common example of this is the bubbles formed by the rotation of high-speed ships' propellers; cavitation results in the wearing of holes in the metal of the propeller, which are known as *cavitation damage.* When the vapor bubbles collapse in the liquid, the sound produced is called *cavitation noise.*
Celsius scale.—See *Temperature scales.*

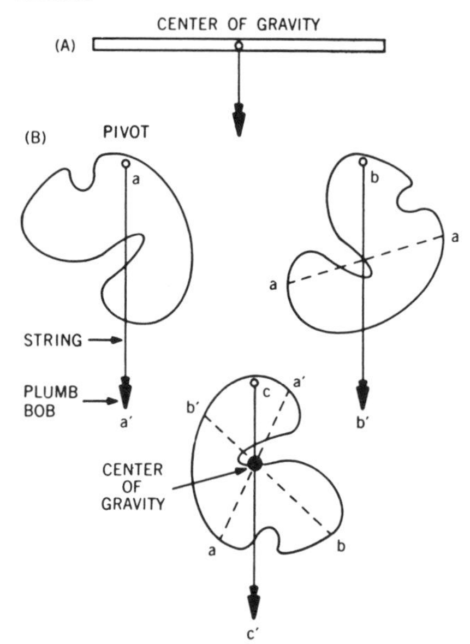

CENTER OF GRAVITY is the point of intersection of three plumb lines dropped from any body in three different orientations.

Center of gravity.—The point at which the weight of a body may be considered as being concentrated and at which the body may be supported by a single force to produce equilibrium. The center of gravity of a uniform beam is at its geometric center. The center of gravity of a nonuniform body may be determined by suspending a plumb line from it while it is in different positions. The center of gravity is the point of intersection of lines drawn through the plumb lines while the body is in any three different positions.
Centigrade scale.—See *Temperature scales.*
Circuit.—Any number of electrical conductors and components connected in such a way that current can flow. An *open circuit* is one through which current can normally flow but with a temporary break, such as an open switch, in it. A *closed circuit* is one through which current is flowing. A *series circuit* is one in which the total current, but only a fraction of the voltage, passes through every component of the circuit. A *parallel circuit* is one in which only a fraction of the current, but the total voltage, passes through every component. A *series-parallel circuit* is a parallel circuit in which some components are connected in series.
Cloud chamber.—A device in which rapidly moving charged particles, such as alpha and beta particles, leave a trail of condensed water-vapor particles. The trails may be photographed and studied to learn more about the particles that made them. The cloud chamber was invented by Charles Thomson Rees Wilson, a Scottish physicist, in 1911.
Color.—An inherent property of light apart from its form. Color is a sensation evoked in the eye in response to the wavelength of the light reaching it. In viewing a normal spectrum of light, going from long wavelength to short wavelength, the eye perceives the colors red, orange, yellow, green, blue, indigo, and violet; these are known as the spectral colors.
Condensation.—The change of a material from the gaseous to the liquid state, such as the condensation of water vapor to form rain.
Condenser.—See *Capacitance.*
Cosmic rays.—Radiation originating in interstellar space and reaching the earth's surface. The rays' exact origin is unknown. Cosmic rays consist primarily of protons, although electrons, mesons, and positrons have also been detected. Cosmic rays often rain upon the earth's surface in *bursts* or *showers.*
Cryogenics.—The branch of physics that deals with the properties of matter at temperatures approaching absolute zero.
Cybernetics.—The branch of physics that deals with the control and internal communication of information within automatic data-processing machines.
Cycle.—Any sequence of events at the end of which conditions have returned to the state that existed at the beginning. A cycle is generally recurrent; that is, the sequence generally repeats over and over. A familiar cycle is that of the four-stroke internal-combustion engine, in which the sequence of events is intake, compression, explosion, and exhaust. Other familiar cycles are found in electricity and in electronic communications: alternating electric current is usually designated in terms of both voltage and cycle rate, such as 120 volts, 60 cycles; and the frequency of radio waves is usually given in cycles per second (*c.p.s.*), thousands of cycles per second (kilocycles per second, *kc.p.s.*), or millions of cycles per second (megacycles per second, *mc.p.s.*).
Cyclotron.—See *Particle accelerators.*
Density.—The mass per unit volume of a material. Density is usually expressed in terms of pounds per cubic foot or grams per cubic centimeter.
Dewar vessel (Dewar flask).—A container used for the storage of liquid air. The vessel, named after its discoverer, Sir James Dewar, consists of a flask-within-a-flask (an arrangement called "double-walled"). The space between the inner and outer containers is a vacuum, and the walls enclosing the space are silvered. Thus, the construction is like that of the familiar vacuum bottle. The vacuum prevents the transfer of heat by conduction or by convection currents. The silvered surface minimizes the transfer of heat by radiation. Dewar vessels are used in cryogenics.
Diffraction.—In optics, the process by which "fuzziness" at the edge of a shadow is caused by the change in direction of light rays from a point source as they interact with the edge of an object. In acoustics, diffraction is the change in direction of a sound wave caused by its passing through layers of different densities or by the varying velocity of wind with altitude.
Diffusion.—The scattering of light in all directions during transmission or reflection. During transmission, it is caused by the light waves' striking minute particles. During reflection, diffusion is caused by irregularities in the reflecting surface.
Dispersion.—The separation of light into its component wavelengths. Dispersion may be accomplished through the use of a prism or a diffraction grating. The breaking of white light into the seven spectral colors (red, orange, yellow, green, blue, indigo, and violet) is a dispersion phenomenon.
Doping.—The introduction of a specified quantity of impurities, called *dopants,* into a solid material in order to produce certain properties. This process is very important in the production of semiconductors, such as transistors.
Echo.—A sound wave that has been reflected in such a manner as to arrive at the listener at a later time than a sound wave coming directly from the source. An *echo chamber* is a room used to produce intentionally a certain amount of reverberation, or echo, of the sounds on a phonograph record or tape recording.
Electroluminescence.—The direct conversion of electrical energy into light in a liquid or solid material. This process has been adapted by illumination engineers to provide a very efficient means of lighting large areas. It is also the basis of operation of the injection laser.
Electromagnetic spectrum.—See *Spectrum.*
Electromotive force (e.m.f.).—The pressure that causes the movement of electrons around an electric circuit. Electromotive force is produced by reversible processes of energy conversion, such as in a storage battery; and the device in which it is produced is called the *seat of electromotive force.* The unit of measurement of this force is the volt.
Electron. — An electrically charged particle that is one of the basic building blocks of the atom. The electron possesses the smallest amount of negative electric charge found thus far in nature. (However, three elementary particles, called *quarks,* with smaller charges have been postulated.) An electron has a mass only $\frac{1}{1,800}$ that of the proton, which is the basic particle of positive electric charge.
Electron-volt (e.v.).—A small unit of energy. It is defined as the kinetic energy possessed by an electron that has been subjected to an electromotive force of one volt. The energy of nuclear particles in particle accelerators is usually expressed in millions of electron volts (*M.e.v.*) or billions of electron volts (*B.e.v.*). An electron acted upon by one billion volts to

give it a kinetic energy of 1 B.e.v. will travel at a speed close to that of light (186,300 miles per second).

Energy. — The ability to do work. There are many forms of energy, such as heat, light, chemical, radiant, mechanical, and nuclear energy. Any form of energy can be converted into any other form of energy, and matter and energy may be converted into one another. However, the total mass and energy of a system must always remain the same. There are two kinds of mechanical energy: a body possesses *kinetic energy* by virtue of its motion, and it possesses *potential energy* by virtue of its position in relation to any specified reference point.

Enthalpy.—The heat possessed by a body or system per unit mass. Enthalpy is an important concept in thermodynamics.

Entropy.—A thermodynamic property of a system that is a measure of the energy that the system has "used" and will, therefore, never be able to use again. Since all thermodynamic processes "use" energy in this way and the energy can never be recovered, the entropy of the universe is constantly increasing.

Fahrenheit scale.—See *Temperature scales.*

Fluorescence.—The absorption of energy of short wavelength and its prompt emission in the form of visible light. The energy absorbed may be from a chemical reaction, heating, ion bombardment, or exposure to light rays of short wavelength. If the energy is stored for a long period before the emission of the visible light, the process is called *phosphorescence.*

Force.—Any action that will impart an acceleration to any mass that is free to move.

Freezing point.—The temperature at which a liquid solidifies. Pure substances generally solidify at a constant temperature, while impure substances solidify over a small range of temperatures. The *melting point,* which is the temperature at which a solid turns into a liquid, is the same temperature as the freezing point.

Frequency.—The number of times an event will take place in a given unit of time. Thus, frequency may be measured in cycles per second, revolutions per minute, pulses per second, and so on. In acoustics, the *audiofrequency* is any sound-wave frequency that can normally be heard; audiofrequencies range from 15 to 20,000 cycles per second. An *infrasonic frequency* is one below the audiofrequency range; an *ultrasonic frequency* is one lying above the audiofrequency range.

Friction.—The inherent resistance to motion when an attempt is made to move one surface over another. The energy used to overcome friction is generally dissipated as heat.

Galvanometer.—An electric measuring instrument. It is generally used for the detection of electric currents. When properly connected to resistance elements and calibrated, galvanometers are called *ammeters* and *voltmeters.*

Gamma rays.—Very high energy X rays. Gamma rays are emitted by radioactive substances in decaying.

Half-life.—The period of time it takes for a radioactive material to lose one-half of its strength. Half-lives vary from trillions of years to fractions of seconds. For example, the half-life of samarium-152 is 1,000,000,000,000 years, while that of polonium-212 is 0.0000003 second.

Harmonic.—A frequency that is a multiple of another frequency to which it is related. A harmonic having a frequency four times that of the fundamental frequency, for example, is called the fourth harmonic. In acoustics, the harmonics of a sound that are heard simultaneously with the fundamental frequency are called *overtones.*

Heat.—A form of energy that is associated with the motions of the atoms and molecules within a substance. Thus, heat is a form of mechanical energy. The more rapid the motion of the atoms and molecules, the greater the heat content of the substance.

Hole.—An empty space in the crystal lattice of a semiconductor material, into which an electron may move. Current flows through a semiconductor by a process called *hole conduction,* in which an electron moves into a "hole," leaving a "hole" behind it; another electron moves to fill the new "hole," and so on.

Impedance.—In electronics, the apparent resistance to the flow of an alternating current that can be equated to the resistance to the flow of direct current through a circuit.

Image.—In optics, the likeness of an object caused by the viewing of the light rays proceeding from it. A *real image* can be shown on a screen, while a *virtual image* cannot.

Inductance.—An electrical property common to all circuits but primarily noticeable in A.C. (alternating current) circuits. Inductance can be seen in two separate effects. In one effect, a varying current tends to produce an electromotive force in its own circuit (called *self-inductance*) or in an adjoining circuit (called *mutual inductance*). In the other effect, inductance tends to even out the alternations in the current by causing a lag in building up to the maximum value and a lag once again in falling to the minimum value. The more rapid the alternations, the greater the inductance. An *inductor* is any electrical apparatus, usually an *inductance coil,* introduced into a circuit with the intention of producing inductance.

Inertia.—The inherent property of a body, as defined by Sir Isaac Newton's first law of motion, that tends to resist any change in the body's motion.

Infinity.—In optics, a distance sufficiently great so that light rays coming from a source located there will —for all practical purposes—be parallel to one another. Conversely, if an object is located at the focal point of a lens, the light rays proceeding from it will emerge from the lens parallel to one another, and the image is said to be formed at infinity.

Ion.—A charged particle that results when an atom or molecule gains or loses one or more electrons. A positive ion is an atom or molecule that has lost an electron; a negative ion is an atom or molecule that has gained an electron.

Isothermal process.—Any change in matter that takes place at a constant temperature; thus, it is a change in the pressure and/or volume of the substance.

Junction diode.—A semiconductor device that can carry current more easily in one direction than in the other; it is therefore widely used as a rectifier. At one end of a junction diode current flows by "hole" conduction, while at the other end it flows by electron conduction. The device is usually a single crystal, and it is a basic component of the *junction transistor.*

Kelvin scale.—See *Temperature scales.*

Laser.—An electronic device that converts incoherent light of various frequencies into a powerful beam of coherent light. *Coherent light* is light in which all the light waves are in phase, or "in step," with one another. The name of the device is an acronym for "*L*ight *A*mplification by *S*timulated *E*mission of *R*adiation." The laser functions in the same manner as its forerunner, the maser, and therefore is sometimes referred to as an "optical maser." The intensity of laser light is one million times that of sunlight at the sun's surface.

Laue pattern (Laue photography).—A pattern of dots produced on a photographic plate by X rays that have been scattered while passing through a crystalline substance. An anaysis of the dot pattern allows the crystal structure of the substance to be determined. The technique is named for Max von Laue, who pioneered in the field in 1912.

Lens.—A transparent optical device with at least one curved surface that bends, or refracts, light rays. A *convex lens* is one in which the curved surface bulges outward; it converges the light rays that pass through it. A *concave lens* is one in which the curved surface bulges inward; it diverges light rays that pass through it.

Mach number.—Specifically, the ratio of the speed of a fluid to the speed of sound at that location. More generally, it is used to designate the ratio of the speed of an aircraft in flight to the speed of sound at the altitude at which the aircraft is flying. Thus, an aircraft flying at Mach 2 would have a speed twice that of sound, or about 1,400 miles per hour. If the Mach number is less than 1.0, the speed is called *subsonic;* if the Mach number is greater than 1.0 the speed is called *supersonic.*

Magnetohydrodynamics (M.H.D.).—The branch of physics that deals with the interaction of a magnetic field with an electrically conducting gas. The study of magnetohydrodynamics is closely related to many of the experiments with nuclear fusion.

Maser. — An electronic device that amplifies low-frequency electromagnetic waves (*microwaves*) by using these waves to stimulate the emission of similar waves from atoms and molecules. Its function is described by its name, which is an acronym for "*M*icrowave *A*mplification by *S*timulated *E*mission of *R*adiation." The inner workings of the maser are almost identical to those of its offshoot, the laser. In a simplified form, it is this:

The atoms of any material have a characteristic energy level, which is called the *ground state*. The atoms can absorb a quantum of energy, which raises them to a higher energy level. Under normal conditions, the atom will eventually emit the quantum it has absorbed and return to the ground state. However, in certain materials, the atom can be stimulated to give up the quantum of energy it has absorbed by exposing it to a second quantum. The unique property of maser and laser materials is that the quantum of energy emitted is in phase with and parallel to the quantum that stimulated its emission. It is this property that has made the maser and the laser the most significant breakthroughs in electronics since the transistor.

Mass.—An inherent property of matter that is a measure of the amount of matter present in a body. Masses are generally measured by comparing the force of gravitation acting upon them; this is called their *weight*. Thus, the weight of a body is proportional to its mass; in the English system, the mass of a body is 1/32 of its weight.

Melting point.—See *Freezing point*.

Meson.—Any one of about a dozen subatomic particles whose mass lies between that of the electron and that of the proton, and that have electrical charges of 0, +1, or −1. Mesons are unstable, and their lifetime is only about 0.000001 second. Mesons are found in cosmic rays and are produced artificially in nuclear reactors. The two most prominent species are called *mu mesons* and *pi mesons*.

Momentum.—An inherent property of a body in motion that determines the amount of time it will take to bring the body to rest when a constant force is applied to the body. Momentum is the product of the body's mass multiplied by its velocity.

Neutron.—A neutral particle that is one of the basic building blocks of the atom. The mass of the neutron is equal to that of the proton.

Nuclear fission.—The splitting of the nucleus of an atom, caused by bombarding it with neutrons, into two parts; the splitting is accompanied by the release of more neutrons and large quantities of energy in the form of light and heat. Only the heavy isotopes, such as uranium-235 and plutonium-239, readily undergo fission, as in military uses.

Nuclear fusion.—The joining of the nuclei of two atoms to form the single nucleus of another heavier element, such as the fusion of two hydrogen nuclei in the sun to form a helium nucleus. When the nuclei of two light atoms, such as hydrogen or lithium, fuse, a large amount of energy is released, according to the formula proposed by Albert Einstein to equate mass and energy: $E = mc^2$, where E is the energy released, m is the mass, and c is a constant equal to the velocity of light.

Ohm.—The unit of electrical resistance named after the German physicist Georg Simon Ohm (1787–1854). The reciprocal of the ohm, the *mho*, is the unit of electrical conductance.

Optics.—The branch of mathematical physics that deals with light. *Physical optics* studies are based on the wave properties of light; studies in *geometric optics* are based on quantum properties.

Particle accelerator.—Any of several devices, sometimes referred to as "atom smashers," used to impart high velocities to such charged particles as alpha particles, electrons, and protons. The *betatron* is used to accelerate electrons. The *bevatron* is a proton accelerator. The *cosmotron* and the *cyclotron* are both used as proton accelerators. The *linear accelerator* is different from the others in that the particles within it always move in straight lines rather than in circles or spirals; it is used to accelerate electrons. The *synchrotron* is similar to the cyclotron, but it can accelerate protons to even higher energies. The *Van de Graaff accelerator* is the only electrostatic accelerator of the group; it is used to accelerate electrons.

Photon.—See *Quantum*.

Piezoelectricity.—An electric current developed in some crystalline materials when they are subjected to a strain in an appropriate direction. This property is utilized commercially in ceramic phonograph cartridges.

Pitch.—The subjective tone of a sound wave. The pitch of a sound wave is a function of its frequency.

Plasma.—The fourth state of matter. Plasmas are similar to gases in that the particles of which they are composed are at great distances from one another. The particles of a plasma have a very high internal energy. Plasmas are mixtures of neutral particles, ions (charged particles), and free electrons. Plasmas are found in the sun, in lightning bolts, and in a number of man-made devices, such as fluorescent lights.

Positron.—A positively charged subatomic particle that has a mass and a magnitude of charge equal to that of the electron.

Potential difference.—The difference in electrical potential at two points that causes an electrical current to flow between them. Potential difference is called *voltage*.

Potentiometer.—An electrical measuring device used to determine the potential difference between two points. It is more accurate than a voltmeter.

Printed circuit.—An electrically conducting wiring diagram deposited, etched, or otherwise superimposed on a nonconducting surface. Electrical components may be soldered to the conducting wiring diagram, as in most inexpensive radios.

Proton.—An electrically charged particle that is one of the basic building blocks of the atom. The proton has a positive electrical charge equal in magnitude to that of the electron. The proton has a mass of 1.66×10^{-24} grams, about 1,800 times larger than that of the electron and equal to that of the neutron.

Quantum.—The smallest indivisible quantity of energy. In an electromagnetic wave, such as light, a quantum is carried in a discrete packet called a *photon*.

Radioactive decay.—The breakdown of an unstable isotope through the emission of alpha and beta particles and gamma rays to form a stable isotope.

Réaumur scale.—See *Temperature scales*.

Rectification.—The conversion of an alternating electrical current into a direct electrical current. Any device that accomplishes this, such as a diode, is called a *rectifier*.

Refraction.—The change in direction of a light ray as it passes between substances of different densities. It results from the fact that the speed of light varies inversely with the density of the medium through which it is traveling.

Resistance.—The opposition offered to the flow of an electrical current by a substance through which it is passing. Those substances that have a low resistance, such as aluminum and copper, are called *conductors*. Those substances that have a high resistance, such as rubber and silk, are called *nonconductors*, or *insulators*. A *resistor* is a device placed in an electric circuit for the express purpose of providing a specified amount of resistance to the current flowing in it.

Semiconductor.—A material whose resistance to the flow of electrical current is greater than that of a conductor, such as aluminum or copper, but less than that of an insulator, such as rubber. Typical semiconductors are germanium and silicon.

Semimetal.—A material whose resistance to the flow of electrical current is greater than that of a conductor, such as aluminum or copper, but less than that of a semiconductor, such as germanium or silicon. Typical semimetals are antimony and bismuth.

Specific gravity.—The ratio of the weight of any substance to that of an equal volume of fresh water. A specific gravity of less than 1.0 indicates a density less than that of water; a specific gravity greater than 1.0, a density greater than that of water.

Spectrum.—The arrangement of any sequence of wave forms in order of increasing or decreasing wavelength or frequency. For example, in acoustics a study is made of the *audio-frequency spectrum*, which includes all sound waves of 15 cycles to 20,000 cycles per second. The term "spectrum," however, is generally taken as a reference to the visible portion of the *electromagnetic spectrum*. The electromagnetic spectrum consists of electromagnetic waves with wavelengths ranging from about 10^{-14} meter (cosmic rays) to about 10^{7} meters (radio waves); the visible portion of the spectrum covers wavelengths from about 4×10^{-7} (violet light) to 7.5×10^{-7} (red light).

Speed.—See *Velocity*.

Temperature.—The average kinetic energy (energy of motion) of the atoms and molecules in a substance. Heat will always flow from a region of higher temperature to one of lower temperature.

Temperature scales.—Arbitrary calibrations of thermometers in order to establish convenient references for the determination of temperatures. There are three thermometric scales based on the boiling and freezing points of water: the Fahrenheit scale; the Celsius, or Centigrade, scale; and the Réaumur scale. The *Fahrenheit scale* (*F.*) was proposed by Gabriel D. Fahrenheit (1686–1736). He set the zero point as the temperature of a mixture of salt and ice water and the temperature of the human body as 96° (this latter measurement has since been proved erroneous, and the temperature of the human body on the Fahrenheit scale is 98.6°); on this basis, the *ice point* (the freezing point of water at a pressure of 1 atmosphere, or 980 millimeters) was 32° and the *steam point* (the boiling point of water at a pressure of 1 atmosphere, or 980 millimeters) was 212°. The *Celsius,* or *Centigrade, scale* (*C.*) was proposed by Anders Celsius (1701–1744). Celsius set the ice point of his scale at 0° and the steam point at 100°. The *Réaumur scale* (*R.*) was proposed by René A. F. Réaumur (1683–1757). Réaumur set the ice point of his scale at 0° and the steam point at 80°. The *Absolute* (*A.*), or *Kelvin* (*K.*), *scale* was proposed by Lord Kelvin (William Thomson, 1824–1907). The zero point is that temperature at which all molecular motion theoretically ceases. The ice point on the Absolute scale is 273° and the steam point is 373°. The scales are related to one another by the following formulas:

$$°F. = \frac{9}{5}°C. + 32$$

$$°F. = \frac{9}{4}°R. + 32$$

$$°C. = \frac{5}{9}(°F. - 32)$$

$$°C. = \frac{5}{4}°R.$$

$$°R. = \frac{4}{5}°C.$$

$$°R. = \frac{4}{9}(°F. - 32)$$

$$°K. = °A.$$

$$°K. = °C. + 273$$

$$°K. = \frac{5}{9}°F. + 255$$

$$°K. = \frac{5}{4}°R. + 273$$

Thermodynamics.—The branch of mathematical physics that deals with the relationship of heat energy to other forms of energy, primarily mechanical energy.

Transistor. — An electronic device, made of a doped semiconductor crystal, that is used to amplify an electrical current. Transistors are replacing vacuum tubes, such as the diode and the triode, in many applications.

Velocity.—The rate of displacement, or rate of change of position, per unit time. The difference between velocity and *speed* is that velocity includes direction, but speed does not. Thus, an airplane that has a velocity of 600 miles per hour east will have a speed of 600 miles per hour. The distinction between velocity and speed is not always made, especially in nontechnical usage, however. For example, the velocity of light is generally given as 186,300 miles per second, but this is really its speed.

Volt.—The unit of electrical potential difference, named after the Italian physicist Alessandro Volta (1745–1827).

Voltmeter.—An indicating device used to measure the difference in potential between two points in an electric circuit. Voltmeters, which are not as accurate as potentiometers, are always connected in parallel between the two points.

Watt.—The unit of power named after the Scottish physicist James Watt (1736–1819). A *kilowatt* is equal to 1,000 watts; 1 *megawatt* is equal to 1,000 kilowatts, or 1,000,000 watts; and 1 *gigawatt* is equal to 1,000 megawatts, 1,000,000 kilowatts, or 1,000,000,000 watts. Electric power (measured in watts) is the product of the current (measured in amperes) and the potential difference (measured in volts).

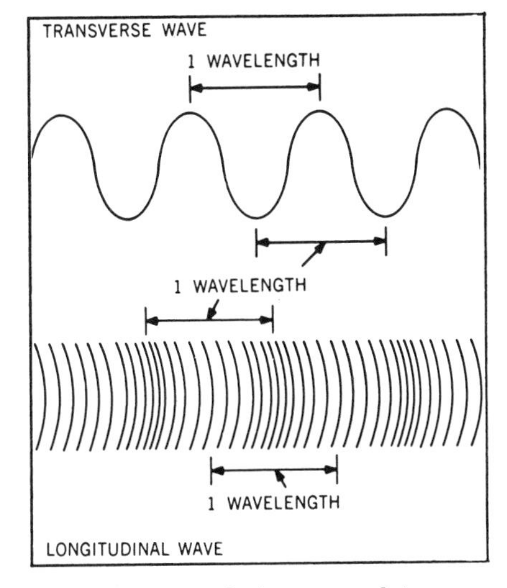

WAVELENGTH as it is measured in transverse and in longitudinal waves.

Wavelength.—The distance between similar, consecutive points along any wave form. Thus, in a *longitudinal wave,* such as a compressional sound wave, the wavelength is the distance between successive condensations and rarefactions; in a *transverse wave,* such as a sinusoidal light wave, the wavelength is the distance between successive crests and troughs. The different colors the eye senses when receiving light waves are due to the different wavelengths of the waves.

Weight.—See *Mass.*

Wheatstone bridge.—A device for determining the resistance in an electrical circuit invented by the English physicist Sir Charles Wheatstone in 1843.

Work.—In mechanics, the product of a force acting through a distance. Thus, work is done when a 5-pound weight is raised 10 feet above the ground, the amount of work being 50 (5 × 10) foot-pounds. In electricity, work is the electric power expended over a period of time. It is usually measured in watt-hours or kilowatt-hours, one kilowatt-hour equaling 1,000 watt-hours.

BIBLIOGRAPHY

GENERAL

ASIMOV, ISAAC. *Inside the Atom.* Rev. ed. Abelard-Schuman, Ltd., 1961.

BLACKWOOD, OSWALD HANCE, and others. *General Physics.* 3rd ed. John Wiley & Sons, Inc., 1963.

COHEN, I. BERNARD. *The Birth of a New Physics.* Anchor Books, 1960.

DULL, CHARLES EDWARD, and others. *Modern Physics.* Holt, Rinehart & Winston, Inc., 1963.

EINSTEIN, ALBERT, and LEOPOLD INFELD. *The Evolution of Physics.* Simon & Schuster, Inc., 1938.

GARDNER, MARTIN. *Relativity for the Millions.* The Macmillan Co., 1962.

REICHEN, CHARLES ALBERT. *History of Physics.* Hawthorn Books, Inc., 1963.

RIPIN, EDWIN M. *How to Solve Physics Problems.* John F. Rider Publisher, Inc., 1961.

WHITE, HARVEY E. *Modern College Physics.* 5th ed. D. Van Nostrand Co., Inc., 1966.

MECHANICS

BEER, FERDINAND PIERRE. *Mechanics for Engineers: Statics and Dynamics.* McGraw-Hill, Inc., 1962.

PHELAN, RICHARD. *Fundamentals of Mechanical Design.* 2nd ed. McGraw-Hill, Inc., 1962.

LIGHT

COLLIS, JOHN STEWART. *World of Light.* Horizon Press, Inc., 1960.

MONK, GEORGE SPENCER. *Light: Principles and Experiments.* Dover Publications, Inc., 1963.

OTIS, ARTHUR SINTON. *Light Velocity and Relativity.* Christian E. Burckel & Associates, 1963.

HEAT

BERGMANN, PETER GABRIEL. *Basic Theories of Physics: Heat and Quanta.* Peter Smith, 1951.

EFRON, ALEXANDER. *Heat.* John F. Rider, Publisher, Inc., 1957.

MOTT-SMITH, MORTON CHURCHILL. *Heat and Its Workings.* Dover Publications, Inc., 1962.

SOUND

RICHARDSON, EDWARD GICK. *Sound.* St. Martin's Press, Inc. 1953.

SEARS, FRANCIS WESTON. *Mechanics, Heat and Sound.* 2nd ed. Addison-Wesley Publishing Co., Inc., 1950.

ELECTRICITY

ATKIN, RONALD HARRY. *Theoretical Electromagnetism.* John Wiley & Sons, 1962.

SUFFERN, MAURICE GRAYLE. *Basic Electrical and Electronic Principles.* 3rd ed. McGraw-Hill, Inc., 1962.

ATOMIC ENERGY

HARNWELL, GAYLORD P. and WILLIAM E. STEPHENS. *Atomic Physics.* McGraw-Hill, Inc., 1955.

HARRISON, GEORGE RUSSELL. *Atoms in Action.* William Morrow & Co., Inc., 1949.

JUKES, JOHN D. *Man-made Sun.* The Viking Press, Inc., 1961.

SPACE SCIENCES

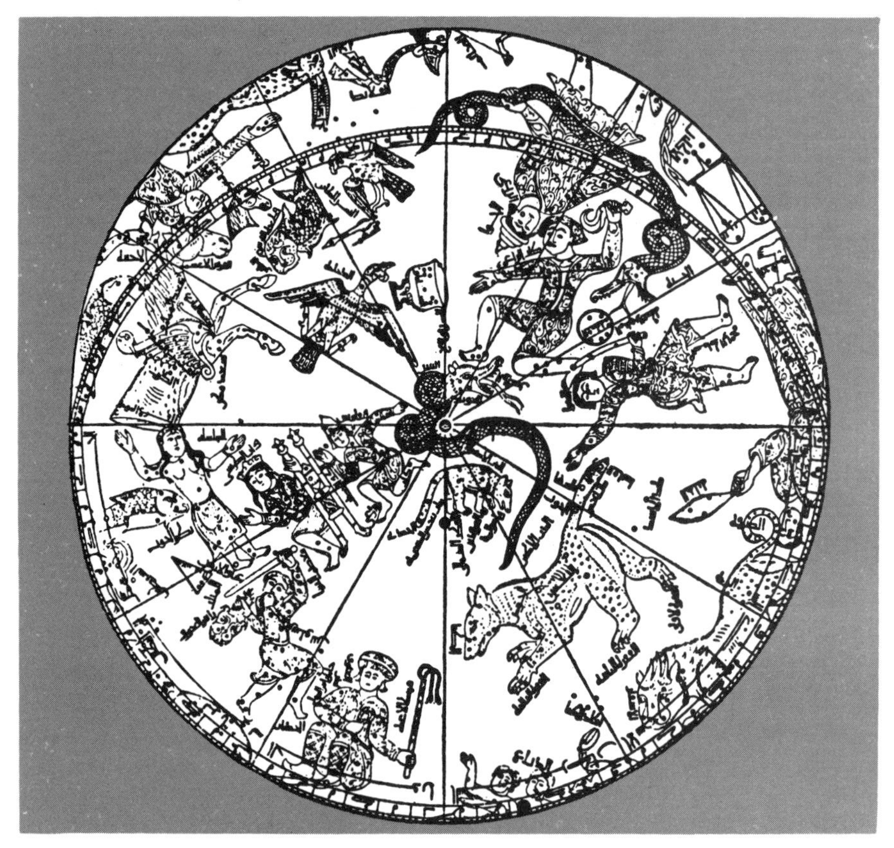

ANCIENT ARABIC celestial sphere showing the constellations of the Northern Hemisphere.

ASTRONOMY

Astronomy is the study of all heavenly bodies, their positions, distances, motions, dimensions, compositions, and physical conditions. It is said to be the oldest science.

Solar System.—The *solar system* is the total system of the sun. Our solar system is made up of nine principal planets, thirty-one known natural satellites or moons that circle some of the planets, thousands of tiny planetoids or asteroids, millions of comets, innumerable meteoroids, and vast quantities of interplanetary dust and gas. The magnetic and radiation fields around the sun and the planets are also important parts of the system.

The volume of space our solar system occupies can be visualized as a sphere more than ten billion miles across, with the sun at its center. Only minor parts of the system extend to the extreme borders of this sphere. The planets are located relatively close to the sun; they lie in a plane, and all revolve around the sun in the same direction.

■**SOLAR GRAVITY.**—All material particles in the solar system, from the giant planet Jupiter to those no bigger than a grain of sand, pursue individual *orbits,* or paths around the sun. Our moon revolves around Earth, but it also revolves with Earth around the sun. Viewed from Earth, the moon's orbit appears as a near circle. Viewed from the sun, the path of the moon would appear much like Earth's but with "wiggles" in it.

Material particles orbit the sun because of the gravitational pull that the sun's large mass exerts over them. This force is continuous and would, theoretically, pull all members of the solar system into the sun if they themselves were not moving. A planet can be visualized as a stone tied to a string; the sun, as a boy swinging the stone in a circle over his head. The pull of the string on the stone keeps the stone from flying off its orbit. If the stone were not moving, however, the same amount of pull would quickly bring the stone to the boy in the same way that a planet, if not moving, would be pulled into the sun.

■**DISTRIBUTION OF MATTER.**—Two striking features of the solar system that are often overlooked are its isolation in space and its relative emptiness. The system is a near vacuum separated by immense distances from other near vacuums. The sun possesses over 99 per cent of all matter in our solar system. If the remaining 1 per cent were spread evenly over the volume of space represented by the sun's sphere of influence, there would still exist a vacuum far better than can be produced in a laboratory.

■**DISTANCE RELATIONSHIPS.**—Most matter outside the sun is concentrated in nine relatively dense pellets, the planets. Relationships in the planetary system can be illustrated as follows. In a 250-foot square representing the space occupied by the solar system, the sun would be represented by a sphere $\frac{1}{3}$ inch in diameter at the center of the field. The planets would appear as nine tiny spheres, the largest of which would be $\frac{1}{30}$ inch in diameter; they would be situated in random directions about the sun at distances of approximately 1, 2, 3, $4\frac{1}{2}$, 15, 30, 60, 90, and 120 feet. Earth would be represented by a $\frac{3}{1000}$-inch sphere about 3 feet from the sun; Pluto, by a sphere approximately the same size as Earth, but nearly 120 feet away. The star closest to the sun would be a sphere similar in size to the sun placed in another field about 150 miles away. This illustration could be expanded to include planetary spheres surrounding the other star to represent its planetary system, but their existence cannot be definitely proved because they cannot be seen. There are, however, many theoretical reasons supporting belief in the existence of other solar systems.

Dynamics of the Solar System.—In the model described above, the relative distances of the planets from the sun (1, 2, 3, $4\frac{1}{2}$, 15, 30, 60, 90, 120 feet) suggests a regular mathematical series. The arrangement seems to be far too orderly to have happened purely by chance.

■**BODE'S LAW.**—This orderliness of division was noted years ago; and an empirical relation, known as *Bode's law,* was formulated by Johann Elert Bode (1747–1826). Using this law, it is possible to make a rough estimate of the distances from all known planets (except Neptune) to the sun in terms of Earth's distance from the sun, or *astronomical units (A.U.).* (Earth is approximately 93 million miles from the sun; therefore, one astronomical unit is equivalent to 93 million miles.)

The law works in this manner: Set down a series of 4's. Add the number 3 to the second 4; then 6 to the third; 12 to the fourth, and so on, doubling the number added each time. Then divide the sums by 10 to obtain the approximate distance from each planet (except Neptune) to the sun in astronomical units. The accompanying chart of Bode's law clearly illustrates this method.

Here Bode's law holds quite well out to Uranus. If Neptune did not exist, the law would also give the distance from the sun to Pluto fairly well. The 38.8 A.U. distance the law predicts for Neptune is close to Pluto's actual 39.5 A.U. distance.

No theoretical basis for Bode's law has ever been found. It does not follow, as do other relationships, from either the laws of motion or the law of gravitation.

In 1781, when the planet Uranus was discovered in an orbit predicted by Bode's law, a great search was begun for a "missing planet" at 2.8 A.U., between the orbits of Mars and

Bode's Law

Mercury	Venus	Earth	Mars	Asteroids	Jupiter	Saturn	Uranus	Neptune	Pluto
4	4	4	4	4	4	4	4	4	4
+0	+3	+6	+12	+24	+48	+96	+192	+384	+768
4	7	10	16	28	52	100	196	388	772
divided by 10 =									
.4	.7	1.0	1.6	2.8	5.2	10.0	19.6	38.8	77.2
actual distances in astronomical units (A.U.) =									
.4	.7	1.0	1.5	2.8	5.2	9.5	19.2	30.0	39.5

Jupiter. This search resulted in the discovery of not one, but hundreds of tiny planets, or *asteroids,* orbiting in the region. It is believed that these asteroids are the remnants of one or more larger bodies that originally orbited in the region.

■KEPLER'S LAWS.—Although no theoretical explanation appears to exist for Bode's law, there is a physical explanation for the exact relation between the distances and periods of planets discovered by Johannes Kepler (1571–1630) and later elaborated upon by Isaac Newton (1642–1727). The *period* of a planet is the time it takes to make one complete *revolution* around the sun. (*Rotation,* on the other hand, describes the turning of a planet on its axis. One complete revolution of the earth takes approximately 365 days; one rotation, one day.) The accompanying chart lists distance from the sun, diameter, periods, mass, density, and number of satellites of each planet.

Kepler showed that the cube of a planet's distance from the sun in astronomical units is equal to the square of that planet's period ($D^3 = P^2$). For example, Mars is 1.52369 A.U.'s from the sun. This figure cubed is 3.5374. The square root of 3.5374 is 1.8808, which is exactly the period of Mars. Therefore, when either a planet's distance from the sun or its period is known, the other quantity can be found by use of this formula. This law applies not only to the planets, but also to each and every member of the solar system orbiting the sun, even the tiny meteoroids. Since, however, the minor members of the system often travel in highly elliptical orbits, their average distance from the sun must be computed before applying Kepler's law to determine the period. The law can also be used to describe the distance and period relationship of a satellite to its parent, or *primary.*

This distance and period relationship was actually the last of three laws discovered by Kepler, known collectively as *Kepler's three laws of planetary motion.* The first law describes the type, or "shape," orbit that every member of the solar system follows—a curve that looks like an elongated circle, called an *ellipse.* The second law describes the rate at which a planet moves on its elliptical path. In order, the three laws are:

1. The orbit of every planet is an ellipse, with the sun at one focus.
2. The *radius vector* (line connecting the sun and planet) sweeps over equal areas in equal times.
3. The cubes of the mean distances of the planets from the sun are proportional to the squares of their sidereal periods.

■CELESTIAL MECHANICS.—More than a half-century after Kepler discovered his descriptive laws, Isaac Newton proved them to be a natural consequence of the universal law of gravitation and of the laws of motion of material objects.

The law of gravitation states:

$$F \propto \frac{m_1 m_2}{d^2}$$

In other words, every material particle (m_1) attracts any other material particle (m_2) with a force (F) that is proportional ($\propto$) to the product of the masses ($m_1 m_2$) divided by the square of the distance between them (d^2).

The law of motion of material objects states:

$$F = ma$$

The force (F) necessary to impart a certain acceleration (a) to a mass (m) is the product of mass and acceleration (ma).

It has been stated that as the planets move about the sun, they must be constantly accelerated, or else they would fly off into space. The force producing this acceleration is gravity. The combination of the two equations thus forms the basis for the complex subject of celestial mechanics. *Celestial mechanics* proves that, under the force of gravitation, the planets move in the way described by Kepler's laws—and in no other.

In addition to the sun's gravitational force, every member of the solar system pulls on every other member to varying degrees. Because of this, calculating the exact orbit of the lighter and more easily influenced members, such as comets, is an exacting task. If, for instance, a comet passes fairly close to the giant planet Jupiter, whose gravitational pull can rival the sun's at close range, the comet's orbit will change.

Celestial mechanics also makes possible the precision with which the occurrence of eclipses of the sun and moon can be predicted many years in advance, and is now employed in calculating the orbits of artificial satellites and rockets. In this respect there is no difference between a natural celestial object and one placed in space by man; they act the same.

The solar system, looked at as a sort of clockwork or mechanical device, can be described as a star (the sun) around which the planets course with mathematical precision, and a great host of minor bodies course in a much less regular fashion.

Physical Nature of the Solar System.—Members of the solar system form a family of individuals with widely different physical characteristics—so different, in fact, that one well might wonder whether they really do belong to one physical family. They differ not only in size, but also in chemical composition and density. In spite of these disparities, they do appear to have had a common origin. The variations in size, composition, and density are probably attributable to the manner in which the planets condensed out of the original solar *nebula* (cloud). Although it is not known exactly how the solar system originated, it is commonly held that there was once an extensive gas and dust nebula from which our system evolved. Such clouds have been detected in other regions of space; and it is thought that they may develop into other solar systems, as ours probably did billions of years ago.

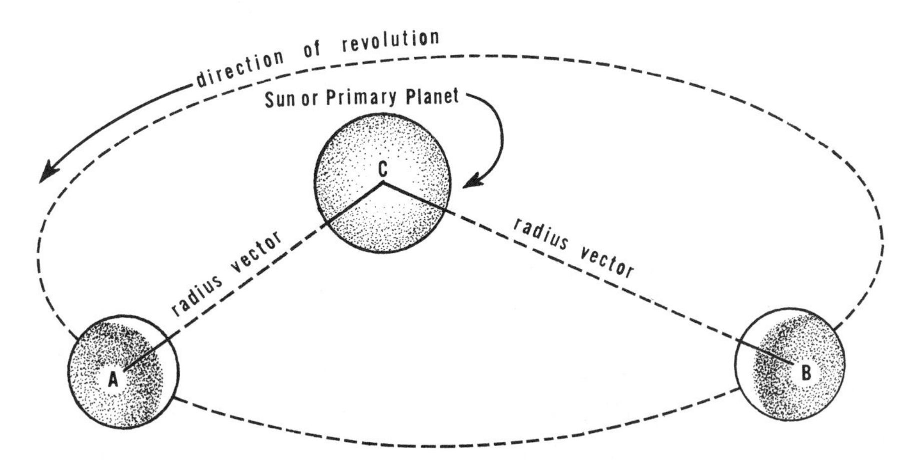

KEPLER'S SECOND LAW states that the radius vector between a body and its primary will pass through equal areas in equal times. Thus, the area of triangle *ABC* is equal to the area formed when the body passes between any other points in its orbit in the same time.

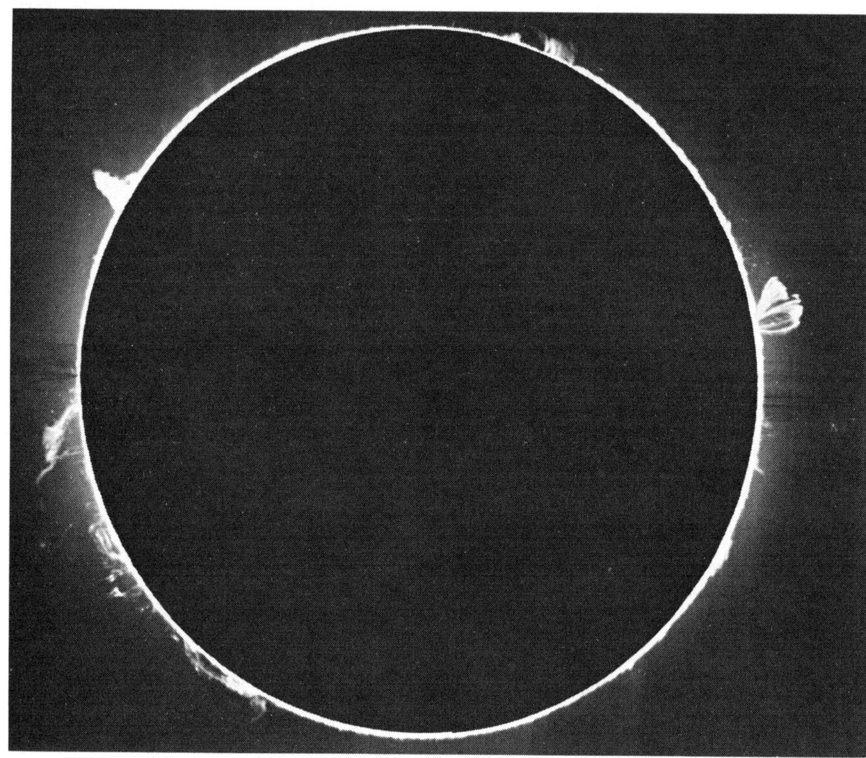

MOUNT WILSON AND PALOMAR OBSERVATORIES

SOLAR PROMINENCES flare out from the sun's surface thousands of miles into space.

The original nebula (see *Origin of the Solar System*) from which our system evolved probably contained much more material than the combined mass of all members of the present solar system. It is suspected that as the central portion condensed and became hot, much of the original nebula literally evaporated into space. The central portion eventually became the sun, and whirlpools of celestial material relatively close to the developing sun formed the planets. Similarly, smaller whirlpools of material in the immediate neighborhood of the planets became satellites or moons. Satellite formation was, therefore, determined by two factors: the existence of whirlpools of celestial material in the immediate area of the developing planet; and the ability of the planet, by virtue of its gravitational strength, to hold this material within its sphere of influence. The giant planets Jupiter and Saturn, because of their strong gravities, were able to hold many developing satellites. Jupiter has twelve known moons, and Saturn has nine; but dwarfish Mars has two tiny satellites only five and ten miles in diameter.

Gravitational strength, or lack of it, probably also accounts for the present difference in chemical composition of the planets. The giant planets, with strong gravities, were able to retain large quantities of such lighter gases as hydrogen, which easily escaped from the atmospheres of the smaller planets. The great abundance of hydrogen in the atmospheres of the giants allowed the formation of many hydrogen compounds not found in the atmospheres of the smaller planets. Location is another factor. Distance from the sun would determine the amount of light and heat received from the sun and would, in time, affect chemical evolution.

The original mass of a planet, therefore, appears to account for the present differences in chemical makeup of the planets rather than any basic differences in the materials from which they had been formed.

Major Members of the Solar System

THE SUN.—Our central star—the sun—is a glowing sphere of gas about 864,000 miles in diameter with a mass of 332,000 Earths. Its temperature varies from a comparatively cool 10,000° F. at the surface to about 28,000,000° F. near the center.

The sun shines because of nuclear reactions that are constantly taking place deep in its interior under intense heat and pressure. The basic reaction is the transformation of hydrogen into helium, in the course of which matter is transformed into *radiant energy,* ultimately resulting in the sunlight we see and the radiated heat we feel.

The sun is the "powerhouse" of the solar system. Its mighty gravitational power holds the planets on their courses; its radiation warms the solar system; its visible light, reflected from the surfaces and atmospheres of the planets, makes them visible to us. (No planet shines by its own light—only by reflected sunlight.) The sun is the source of energy that originally made life on Earth possible, and the one that constantly sustains it. Our power sources (coal, oil, and the like), food, and even weather conditions can be directly attributed to the sun. We, too, are a form of "canned solar energy."

MERCURY.—Mercury is the planet closest to the sun. It is also the smallest, the hottest, and the fastest-moving of the planets. One-third larger than Earth's moon, but smaller than the two largest moons of Jupiter, Mercury is approximately 3,000 miles in diameter. Radio telescopes have shown that the rotation period of Mercury is not the same as its period of revolution around the sun, as was believed. Mercury rotates 1½ times per revolution, a result presumably, of a strong solar gravitational pull on an equatorial tidal bulge which reaches maximum strength when Mercury is closest to the sun. Because of its small size, Mercury has no atmosphere and no satellites. It is a dead planet.

VENUS.—Venus is about 200 miles smaller in diameter than Earth and almost as massive. Because of this similarity in size, Earth and Venus are often considered "twin planets." Like Earth, Venus holds an appreciable atmosphere; but unlike Earth's atmosphere, which is transparent, the atmosphere of Venus is clouded and opaque. Radio waves, however, can "see" through the cloud layer. Signals transmitted from Earth and collected by radio telescopes on Earth and, most recently, by the Mariner spacecraft, indicate the surface temperature of Venus to be about 800° F.

The atmosphere of Venus is rich in carbon dioxide and lacking in free oxygen; only small traces of water vapor have been detected. The surface of the planet therefore must be baked dry, and it may be this condition that causes the Cytherean cloud layer. Continuous "dust storms" ranging over the hot, dry surface may be

Facts About the Planets

Planet	Mean Dist. from Sun (Millions of Miles)	Mean Diameter (Thousands of Miles)	Sidereal Revolution Period (Years)	Axial Rotation Period* (At Equator)	Mass (Earth = 1)	Density (Water = 1)	Natural Satellites
Mercury	36	3.0	0.24	59.3d	0.05	5.2	0
Venus	67	7.6	0.62	247.1d	0.82	5.1	0
Earth	93	7.9	1.00	23h 56m	1.00	5.5	1
Mars	142	4.1	1.88	24h 37m	0.11	4.0	2
Jupiter	483	86.8	11.86	9h 50m	318.20	1.3	12
Saturn	886	71.5	29.46	10h 14m	95.20	0.7	9
Uranus	1,783	29.4	84.02	10h 49m	14.60	1.6	5
Neptune	2,794	27.0	164.78	15h 48m	17.20	2.2	2
Pluto	3,670	3.6	248.42	6.4d	0.91	4.0?	0

*Figures given represent days (d), hours (h), and minutes (m).

largely responsible for the clouds that keep us from seeing the surface. Unlike Earth, Venus has no satellites.

EARTH.—If Earth could be seen from Venus, it might be described as follows: "Earth is a blue-green planet, similar in size to Venus, or about 8,000 miles in diameter. It is noted for its great oceans of water and for its polar ice caps, which remain relatively stable throughout its seasons. Its atmosphere contains oxygen. Earth has one moon, about one-fourth its own size."

MARS.—With about one-half the diameter and one-tenth the mass of Earth, Mars has a thin, transparent atmosphere that allows direct observation of its surface. The distance separating Mars and Earth ranges from 35 million to 245 million miles. When the small planet is closest, it is usually in a good position for viewing. It is our own atmosphere that hinders a truly clear view of Mars.

It was once considered a strong possibility that life as we know it—even a higher form of life—existed on Mars. This idea was prompted by the observation of markings on the surface of Mars that change with the Martian seasons. Certain areas that are bluish or greenish during the Martian spring turn yellowish and then grayish during the Martian summer and autumn. This led to the belief that at least simple botanical forms that grow here on Earth also grow on Mars. Then some astronomers claimed to have observed a series of fine markings, which they called "canals," on the surface of Mars. Hypothesizing that such a network of canals implied intelligent construction, an impressive case was made for the existence of intelligent life on the planet.

Other scientific evidence, however, is overwhelmingly opposed to this idea. The very rare atmosphere, the low temperature, and the lack of appreciable water vapor—along with the absence of oxygen—all argue against the possibility of any life as we know it in the Martian environment (except perhaps lichens or other simple botanical forms).

Mars rotates completely once every 24 hours 37 minutes, and makes one complete circuit around the sun in 687 days. It has two tiny satellites, about five and ten miles in diameter.

JUPITER.—The diameter of the planet Jupiter is eleven times that of Earth; its mass, equal to 318 Earths. It is a giant planet, larger and heavier than all the other planets combined. Jupiter has a strong surface gravity (about three times that of Earth), enabling it to hold a deep atmosphere. Because of this many-hundred-mile-deep layer, the solid surface of Jupiter is visually obscured; we see only the outer atmosphere. Jupiter is also the fastest-rotating of the planets; despite its size, it spins very rapidly, making one complete rotation in just under ten hours. It takes Jupiter twelve years to make one complete revolution around the sun.

Jupiter also has the most satellites. The twelve moons that circle the giant create a veritable miniature solar system. Four of these moons are giants—two of them larger than Earth's. These four—Io, Europa, Ganymede, Callesto—were discovered by Galileo in 1610 with the help of a new invention, the telescope.

SATURN.—The planet Saturn is second only to Jupiter in size, mass, speed of rotation, and number of satellites. Like Jupiter, it possesses an extensive atmosphere of hydrogen compounds, making it impossible for us to see its solid surface with our present astronomical instruments.

Saturn has nine moons, but its most outstanding feature is its system of rings. These three thin concentric rings of tiny celestial particles that encircle the planet are probably the remnants of a crushed moon. What leads to this belief is that the rings lie within a gravitational danger zone known as *Roche's limit,* within which the gravitational grip of the planet is great enough theoretically to crumble a large object. It is thought that a satellite came within the danger zone, was crushed by Saturn's force of gravity, and its remnants circle the planet, forming these rings.

URANUS AND NEPTUNE.—Just as Earth and Venus are considered the "dwarf twins," the planets Uranus and Neptune are the "giant twins." Only Jupiter and Saturn exceed the two in diameter and mass.

Uranus was the first planet discovered by means of a telescope. Neptune was the first planet discovered as the result of an orbital prediction. Uranus has five satellites; Neptune, two. The "giant twins" are so distant from the heat of the sun that many chemical compounds existing as gases on Earth are frozen solids.

PLUTO.—The tiny planet Pluto is the most remote and was the last to be discovered (1930). It is out so far in space, in fact, that the sun, viewed from Pluto's surface, would appear as only a very bright star. With a probable diameter of 3,600 miles, Pluto is only slightly larger than Mercury, the smallest and nearest planet to the sun. Even more than Uranus and Neptune, its surface is probably frozen solid and only dimly lit by the distant sun. It has no known satellites. Pluto also has the longest period of revolution of all nine planets—248 years.

The Minor Members

ASTEROIDS.—Asteroids are small solid bodies found mainly between the orbits of Mars and Jupiter. They are too small to be seen without a telescope, and are thought to be remnants of a collision of larger objects that once orbited in the region they now occupy. Their discovery was the result of a search begun for a "missing planet" predicted by Bode's law (see pages 1308 and 1309).

Since the first asteroid, Ceres, was discovered in 1801, the orbits of more than 1,500 have been computed; many more are known to exist. Some meteorites found on Earth are believed to be closely related to the asteroids.

COMETS.—Beyond the orbit of Pluto, far out in the distant reaches of the solar system, is believed to be the domain of the *comets*—loose configurations of ice and celestial dust formed from scattered material left at the outer edges of the original solar nebula. In this hinterland of the solar system, the formation of large bodies is impossible. Only loose aggregates of material, held together by their own gravities, can exist.

It is believed that the feeble gravitational forces exerted by the distant stars and planets will occasionally alter a comet's orbit, in effect withdrawing it from the distant storehouse and sending it toward the sun on a highly elliptical orbit. (The periods of comets' revolutions vary widely and can range from three years to millions of years.) As the comet approaches the sun, the icy particles of which it is composed are warmed, causing them to dissociate and move away from the main body, thereby forming the luminous and

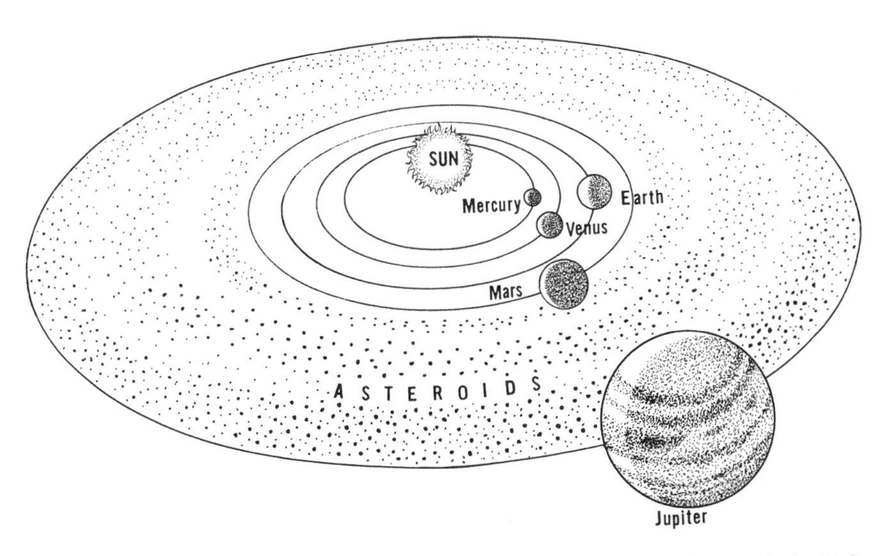

ASTEROID BELT, which is filled with small, solid chunks of rock—almost all of which are too small to be seen without the aid of a telescope—lies between the orbits of Mars and Jupiter. The asteroids, or planetoids, are believed to be the remains of large bodies that once traveled around the sun but were subsequently destroyed in a cosmic disaster.

ARIZONA METEOR CRATER, near Winslow, is about 4,000 feet wide and 600 feet deep.

extensive tail that is regarded as an identifying feature. (The tail shines partly by directly reflected sunlight and partly by reradiated sunlight absorbed by gases in the comet's atmosphere.)

Light also exerts pressure, much too slight to affect more substantial celestial bodies, but strong enough to affect the microscopic particles that make up a comet's tail. This, plus the additional driving force provided by a solar wind of charged particles, explains why comet tails generally point away from the sun. Even when a comet is on its outward journey into space, the tail precedes the comet body—just as smoke from a ship precedes the ship when a strong tail wind is blowing.

With each new approach to the sun, some of the comet's material is driven away from the main body and left strewn along its orbital path. Because of this, any comet that orbits close to the sun will eventually—after many return trips—waste away entirely. When Earth crosses a comet's orbit, even though the comet may have passed many years before, the bits of cometary material left behind collide with Earth's atmosphere, and we experience a meteor shower. Long before these particles can reach the surface of our planet, they burn up by the heat of friction in the upper atmosphere, causing the short, swift streaks of light called *meteors.*

One of the challenging tasks of future space research will be the capture of a portion of a comet—the original material from which the solar system evolved.

▪METEORS.—A meteoric particle traveling in space is known as a *meteoroid.* The same particle, after it enters Earth's atmosphere and is made luminous by friction with the upper air, is called a *meteor.* Any part of the meteor that reaches Earth's surface is known as a *meteorite.*

Meteors arising from a collision of Earth with the cometary debris are neither large enough nor dense enough to withstand the rough ride through Earth's atmosphere. These exceedingly fragile particles are, therefore, destroyed by friction long before reaching Earth's surface.

The very bright meteors that do sometimes penetrate the atmosphere and survive to hit Earth have a different origin. They are really tiny asteroids composed largely of iron, nickel, combinations of iron and nickel, or stone, and are capable of withstanding the enormous heat of entry. The great pit in Arizona, known as the Arizona Meteor Crater, is the result of just such a prehistoric collision of Earth and a small asteroid. The crater is about 4,000 feet across and 600 feet deep. It can easily be seen that if such a meteor ever landed on a modern city, it would cause great destruction.

Many of the lunar craters are believed to have resulted from large meteoroids colliding with the moon. Since the moon has no atmosphere, little frictional resistance is offered entering meteoroids. Therefore, many more survive to hit the surface of the moon than survive to hit Earth. Because it has no atmosphere, the moon is also devoid of wind and rain. Lunar craters therefore do not erode in the manner they would on Earth, but remain as long-lasting records of the many collisions that have occurred during the moon's history.

Origin of the Solar System.—It is not known how the solar system actually originated. Theories that once were generally accepted have been found to possess fatal faults.

Two main divisions of knowledge are presently being applied to the problem of origin. The first is a knowledge of the crucial properties of the solar system, explained in terms of physical processes; the second, a knowledge of the physics of gases, magnetic fields, and gravitational dynamics. All factors that constitute these two divisions of knowledge must be considered before a coherent theory, explaining how they originated, can be formulated. In short, it is first necessary to know exactly what is to be explained before an adequate theory can be deduced; *what is* must be clearly described before we can attempt to explain *what was.*

▪PROPERTIES OF THE SOLAR SYSTEM.—The outstanding properties of the solar system as it exists today are:

1. The system is isolated in space.
2. The sun possesses the preponderant mass of the total system (more than 99 per cent).
3. The planets possess the preponderant angular momentum of the system. (The sun rotates slowly in relation to the revolution and rotation of the planets.)
4. There are two distinct "solar families" in the system. The first family is made up of the solid, spherical planets that revolve about the sun in circular orbits essentially in one plane, rotating on their axes. The second family is made up of smaller objects—the host of comets and meteoroids—that often are of low density, generally with elongated orbits and a great variety of inclination to the planets' central plane.
5. There are satellite systems around some of the planets, with almost all satellites revolving around their primaries in the same direction as the primaries rotate on their axes.
6. Earth's moon is out of proportion to its primary when compared with other satellites in the solar system, which are small compared with their primaries.
7. Small, dense planets are closest to the sun (Mercury, Venus, Earth, Mars) and farthest from the sun (Pluto), with large, massive planets at intermediate distances. The asteroids occupy the Bode gap between one of the smallest planets (Mars) and the largest planet (Jupiter).

These major properties of the solar system must be considered in relation to the estimated age of the sun (six billion years) and the planets (four and one-half billion years).

▪PAST THEORIES.—Incomplete consideration of these outstanding properties of the solar system caused some of the early theories to prove inadequate. One of the earliest theories, the *nebular hypothesis,* proposed by Immanuel Kant (1724–1804) and Pierre Simon Laplace (1749–1827), stated that the sun condensed from a primordial gas cloud and that as it shrank in size, it spun faster and faster, throwing off rings of material from its equatorial regions; these rings condensed into the planets. This hypothesis is untenable for two reasons. First, such rings would disperse rather than condense; second, the sun would still be spinning very rapidly, but this is not true.

The later *close-encounter theories* were based on the proposition that the sun once had a close encounter with another star. One explanation following from this was that the encounter caused huge tidal effects on the surface of the sun; these effects pulled long filaments of material from it, and the material later condensed to form the planets.

Another theory based on the close-encounter idea was that the encounter caused quantities of solar material to be ejected, and that this material quickly cooled to form small, solid *planetesimals* that later grew into the planets through slow accretion.

These close-encounter theories are inadequate because they cannot account for the great angular momentum, or "quantity of motion," of the planets. For example, a passing star could not exert a pull strong enough to bring the mass of Uranus and Neptune out to their present distances from the sun and also, at the same time, impart to them the orbital angular momentum they now possess.

PRESENT THEORIES.—Most current explanations of the origin of the solar system revert to the original nebular hypothesis, but with several very important differences.

One of these theories postulates a large and relatively massive nebula containing much more material than now constitutes the solar system. This nebula is assumed to have possessed a highly turbulent motion, as has been observed in other nebulae in space. As the major portion condensed, it formed a slowly rotating but still dark sun; turbulent eddies in the outer nebula became interlocked gravitationally to form the relatively massive nuclei of the present planets. Lesser eddies around the larger nuclei became the nuclei of the satellites. A preferential direction of revolution and rotation was soon established by the mutual cancellation of nonadditive motions. As the primeval sun condensed, it became hot, heating the embryonic planetary system and evaporating and driving off into space much of the original material of the nebula. New planets were thus freed of their cocoon of primordial material.

Although such a theory overcomes the inadequacies of previous theories, it almost entirely lacks the mathematical precision that is required of a physical explanation. This, however, may be impossible to accomplish because what is being asked is, in effect, a mathematically precise account of an event that occurred billions of years ago under physical conditions that may have been markedly different from those of today.

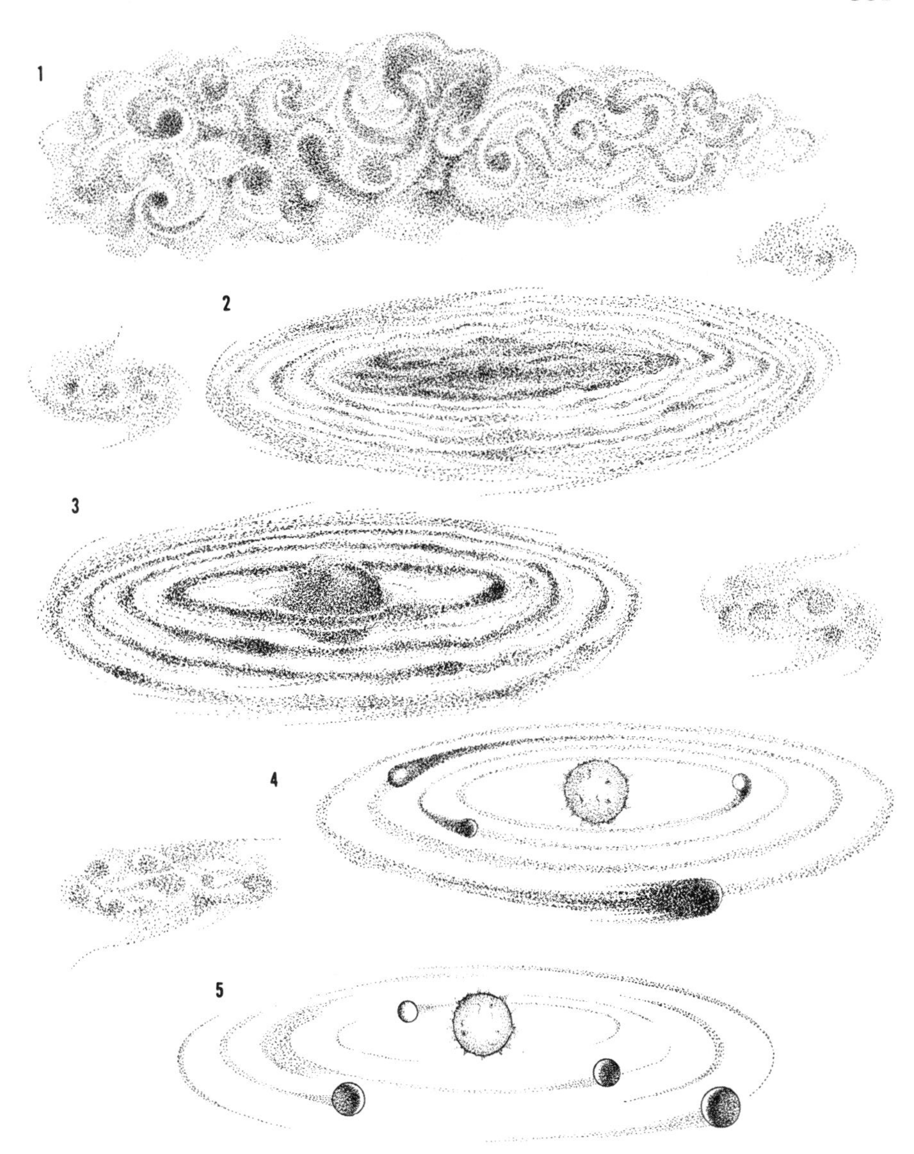

FORMATION OF THE SOLAR SYSTEM theoretically took place when a nebula (1) contracted and began to rotate (2). The major portion of the nebula condensed to form the sun (3), while lesser eddies formed the planets (4), eventually cooling to form the solar system (5).

Facts About the Galaxies

Galaxy	Apparent Magnitude	Distance (Millions of Light Years)	Absolute Magnitude
Andromeda (M 31)	+4.3	2.5	–20
M 101	+8.2	12	–20
M 81	+7.9	6.8	–19
Virgo Cluster	...	30	...
Coma Cluster	...	150	...
Hercules Cluster	...	300	...
Boötes Cluster	...	700	...
Hydra Cluster	...	1,100	...

The Universe of Galaxies.

—The sun is but one of a hundred billion stars that make up an isolated system of stars in space. This system is one among millions of other star systems scattered throughout the universe. Each of these huge systems of stars is called a *galaxy*.

MILKY WAY.—The galaxy to which Earth belongs is called the *Milky Way*. The name describes what we see when we look along the circumference of the galaxy's distant rim. The millions of stars along our line of sight converge, appearing as a milky-white haze.

Our galaxy is a highly flattened, rotating, spiral aggregate of stars, dust, and gas. It has a rim because, unlike some galaxies that are roughly spherical or ellipsoidal in shape, the Milky Way belongs to that large group of galaxies shaped like a watch. Our star—the sun—is located far from the galactic center, near one of the spiral arms (see illustration). Consequently, if we look outward—perpendicular to the page—we see relatively few stars, with many open spaces through which we can peer out toward the realm of the myriad other galaxies. If, however, we look along the plane of the galaxy—along the page—our view of the universe outside our own galaxy is completely

MOUNT WILSON AND PALOMAR OBSERVATORIES

MILKY WAY is a spiral galaxy. It consists of billions of stars, one of which, appearing toward the outer rim of this galaxy, is our sun.

blocked by the concentration of stars, cosmic dust, and gas that lies largely in the plane of our galaxy. This obstruction is so great that before light from stars near the central regions of our galaxy reaches us, it is absorbed and scattered by the intervening material. Radio waves, however, can penetrate these clouded and obscured regions far better than light waves can. Thus the radio telescope has become a major tool used in explorations of the depths of our own galaxy as well as the structure and content of the universe of galaxies beyond.

■DIMENSIONS OF THE MILKY WAY.—Because of the enormous and cumbersome figures that would be involved if we were to measure distances to stars and galaxies in miles, stellar and galactic distances are expressed in terms of *light-years*—the distance light travels in one year. Since light travels at 186,000 miles per second, one light-year is equivalent to about six trillion miles.

Although the term "Milky Way" describes the crowded visible part of our galaxy, it applies to the galaxy as a whole. The length of the Milky Way is nearly 100,000 light-years. Its thickness is variable and difficult to measure accurately. It will suffice to say the galaxy is about five to six times longer than it is thick.

Virtually the only way of estimating distances in the galaxy is by noting the difference between apparent and absolute magnitudes of its various members. The *apparent magnitude* is how bright the member appears to be; the *absolute magnitude* is how bright the member would appear to be from a standard distance of ten *parsecs,* or 32.6 light-years. The absolute magnitudes are assumed from previous knowledge, which generally was determined for similar objects much closer to the sun. Thus the true brightness of blue, giant stars may be regarded as known, no matter where in the galaxy they are found. This is also true of other standard objects. This method of measurement would be nearly perfect were it not for the existence, especially near the plane of our galaxy, of variable amounts of obscuring clouds of fine particles. The apparent magnitude of an object is thus altered; it appears fainter than it would if there were no obstructions. This leads to distance estimates much greater than they actually are, if the obstructions are not taken into consideration in the calculations.

■CEPHEID VARIABLE.—There is one special class of star—*the Cepheid variable*—that is especially important in the determination not only of distances within our own galaxy, but also of the distances to the nearer galaxies. It is a type of variable star whose light varies in a definite manner in periods of from two to forty or more days. The Cepheid has one very special property: its period and brightness are closely related. The longer the period of variation, the greater the star's absolute brightness. Consequently one need only observe the time it takes the Cepheid star to pulsate, and one immediately knows its absolute magnitude. It was by means of these "lighthouses in space" that the distance to the Andromeda nebula was determined, finally settling the great controversy among astronomers in the early part of this century as to whether objects such as the Andromeda galaxy were single peculiar stars, relatively close by, or distant systems of stars. Not until 1924 was mankind presented with the grand picture of a universe of galaxies, rather than a universe of isolated stars.

■POPULATION OF A GALAXY.—A galaxy contains many varieties of individual members. It is incorrect to consider a galaxy as merely a relatively uniform distribution of stars of assorted spectral classes. The population within a galaxy varies with position in the galaxy, much as the type of population and number of people vary with location within a city.

The study of the structure and composition of the galaxy forms a major branch of astronomy, and a brief survey cannot do the subject justice. Yet a summary can be made. A galaxy is composed of stars in various stages of evolution and of dust and gas not in stellar form.

The nonstellar matter is frequently in turbulent motion and appears to be the principal source of the continuous galactic radio radiation observed with radio telescopes over a wide band of frequencies. One particular frequency received from galactic sources is the famous 21-cm. line that arises from neutral hydrogen. By monitoring it, the extent of hydrogen gas clouds throughout the galaxy and beyond can be observed.

Much of the cosmic dust in the galaxy can be detected only because it

Facts About the Stars

Star	Apparent Magnitude	Distance (Light Years)	Absolute Magnitude	Spectral Class
Sirius	−1.6	8.7	+1.4	Ao
Alpha Centauri	+0.3	4.3	+4.7	Go
Vega	+0.1	27	+0.5	Ao
Capella	+0.2	46	−0.5	Go
Arcturus	+0.2	37	0.0	Ko
Rigel	+0.3	650	−6.0	B8
Altair	+0.9	16	+2.4	A5
Aldebaran	+1.1	16	−0.5	K5
Autares	+1.2	170	−3.0	MO
Deneb	+1.3	540	−5.0	A2
61 Cygni	+5.6	11	+7.9	K5

obstructs the passage of light from stars beyond it. The *Great Rift* and other dark lanes in the Milky Way are caused by clouds of dark obscuring matter lying close to the galactic plane and revolving, with the stars, about the central mass of the galaxy.

When, however, such interstellar matter lies in the vicinity of one or more hot, bright stars, it can reflect their light or fluoresce from their ultraviolet radiation. What would otherwise be dark obscuring matter becomes luminous, thereby furnishing some of the more beautiful spectacles on the celestial sphere. The famous Orion, Trifid, and Lagoon nebulae are examples.

New stars are still forming out of the dust and gas in the galaxies. The newly minted stars and those formed within the last several billion years are known as *Population I* stars. Older stars whose histories go back to the early stages of the galaxy are called, for historical reasons, *Population II* stars.

Because of their greater age, Population II stars differ somewhat in chemical composition from the Population I stars; the latter have a greater concentration of metals than their forerunners. It is thought that metals were formed in the interiors of the "first generation" giant stars and later diffused into galactic space. The metal deficiency in the Population II stars causes the *H–R diagrams,* which will be explained later on, of the two main stellar populations to differ sensibly. Population II red giants attain considerably greater brightness than do Population I giants.

There are many subsystems of stars within our galaxy, notably the open clusters and globular clusters. *Open clusters,* such as the Pleiades and the Hyades, are Population I stars. They represent clusters of several hundred stars apparently formed within one large gas cloud in the recent astronomical past. *Globular clusters* are spherical collections of Population II stars, each cluster having thousands of stars. Several hundred globular clusters are known. They do not exhibit the general flattening of a galaxy, but form a spherical halo around its center. All of the globular clusters are situated closer to the center of our galaxy than is the sun.

The sun itself lies about two-thirds of the way out from the center of the galaxy and slightly out of its central plane, close to one of the main spiral arms of our galaxy.

The Milky Way is considered a giant member of the local group of galaxies that includes the Andromeda, which may be even larger.

■GALAXIES.—Galaxies usually occur in large groups. The *Coma cluster* of galaxies, for instance, contains approximately 10,000 separate galaxies. Generally speaking, such clusters include galaxies of all types. Some clusters, however, favor spiral galaxies; others, the elliptical variety.

Even before the true nature of galaxies was known (in the past they were called nebulae, and were included with the true gaseous nebulae), attempts were made to classify them as to types and forms. The *Hubble classification* predominated for many years and is still of importance because of its simplicity and broad coverage, even though more detailed classifications have since been made.

Galaxies have been found in undiminishing numbers to the limits of modern optical and radio telescopic "sight." The distance to the most remote known galaxy is uncertain, but it may be as much as six billion light-years or more away.

The Expanding Universe.—The spectroscopic analysis of galactic light confirms its stellar nature. The spectra of galaxies conform to what could be expected of the combined light of millions of stars. The speed at which a galaxy moves toward or away from us can, therefore, be determined in the same manner that the radial velocities of the individual stars can be determined, namely, by observation of what is known as the *Doppler shift* of the absorption lines in the spectrum.

A most surprising discovery was made as a result of such measurements. The farther away a galaxy is, the faster it appears to be receding from us. This outstanding observation was recently confirmed by radio telescope observations of distant galaxies that emit strong radio waves. The most distant ones seem to be moving away from us at incredible speeds—up to 90,000 miles per second.

Astronomers quickly realized that even if these observations are correct, it does not necessarily imply that our galaxy is at the center of the universe. If an ordinary classroom were expanding, for example, like a balloon being blown up, then no matter where one was sitting, all other seats in the classroom would appear to be receding—the more distant seats at a greater rate of speed.

If the universe is expanding, there is still the question of whether this is a "real" expansion or merely the effect of the geometry of huge distances. In any event, the close correlation of velocity and distance has given us a fine indicator for the determination of the distances to galaxies. Once their spectral speed is determined, their distance also is known.

The speed of recession increases by about 125 kilometers per second for every million parsecs of distance. This "constant" of proportionality is known as the *Hubble parameter.* It is one of the most important proportions in the science of astronomy.

There are two major opposing theories of the expanding universe. One, usually associated with George Gamow (1904–1968), asserts that the universe is evolving and is at present rapidly expanding. The other, associated with the English astronomer Fred Hoyle (1915–), holds that the universe is in a "steady state"; that the galaxies disappearing beyond our observational horizon are replaced by new ones formed from spontaneously generated hydrogen. Observation favors the former theory.

Stars.—Stars are large, generally spherical masses of gas that have become self-luminous through nuclear reactions occurring deep within their interiors. The laws of physics show that once a quantity of interstellar gas and dust, great enough to hold itself together by self-gravitation, is gathered together, a star will result. The gas will slowly contract to form a dark, cold, spherical body. As gravitational contraction continues, the temperature and pressure at the center of the gaseous mass will initiate the nuclear reactions that transform hydrogen into helium and eventually into heavier elements.

In this natural process of element formation, some of the mass of the star is continuously transformed into energy, according to the famous re-

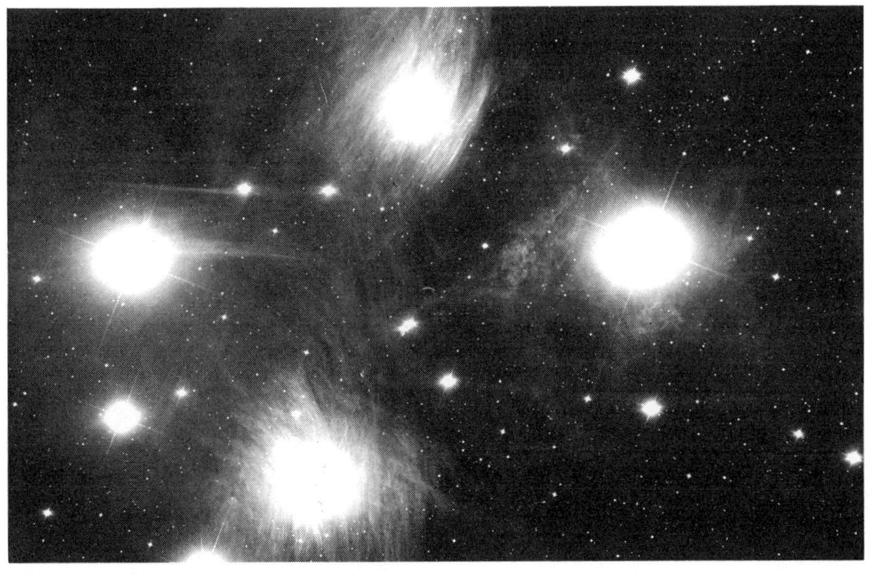

MOUNT WILSON AND PALOMAR OBSERVATORIES

PLEIADES, also known as the Seven Sisters and the Little Dipper, are fourth-magnitude stars. The six bright spots are images of stars brighter than the fourth magnitude.

lation of Einstein: $E = mc^2$. This fundamental law states that a mass (m) of one gram is the equivalent of the square of the velocity of light (c^2, or 9×10^{20}) ergs of energy (E). The star maintains its outflow of luminous energy until the fuel supply is exhausted and the star eventually "dies." The laws of physics also show that the type of star that develops—large or small, brilliant or faint, very hot or relatively cool—depends primarily on the amount of material from which it originally condensed. The twinkling points of light we see in the night sky are, in reality, huge atomic energy plants whose powers immeasurably exceed man's relatively feeble attempts to release the energy locked inside the nuclei of atoms.

The stars appear as twinkling points of light for two reasons: their great distance and the instability of our own atmosphere. Their distance makes them appear to us as true points of light rather than disks (as the planets, moon, and sun appear to us), and the unstable atmosphere causes the feeble rays of light from these seeming points to zigzag as they pass through the atmosphere on their way to our eyes. If Earth had no atmosphere, the stars would not seem to twinkle.

There are many different types of stars. They vary greatly in true brightness, size, temperature, and spectral appearance. The sun is in some ways an average star—brighter than many, not as bright as others.

These individual differences can be explained in relatively simple physical terms. It is first important, however, to distinguish the true properties of stars from those apparent properties that result from the distance between us and them.

STELLAR DISTANCES.—The star (other than the sun) closest to us is about 4½ light-years away. It is the closest of the approximately 100 billion other stars in the gigantic complex of stars we call the Milky Way.

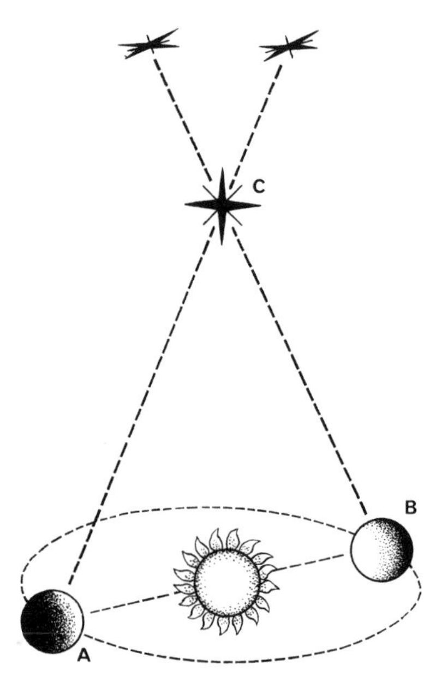

PARALLAX of star at point *C* is one-half its apparent movement as viewed when Earth is at points *A* and *B* in its orbit.

The most distant stars in our own galaxy are more than 50,000 light-years away; other galaxies of stars are millions and even billions of light-years away.

MEASUREMENT OF DISTANCE.—The fundamental method of measuring the distance to a star is known as the *trigonometric parallax method*. As Earth moves around the sun, a relatively nearby star will appear to shift its position in the sky relative to more distant stars. By measuring these minute angular shifts, the distance of the star can be determined trigonometrically. One-half the total apparent shift of the star with respect to the background is called the star's *trigonometric parallax*. It also follows that the parallax of a star is the angle subtended at the star by the radius of Earth's orbit.

No star has a parallax of more than one second of arc (approximately the angle subtended by a penny at a distance of 2½ miles). The nearest star, Alpha Centauri, has a parallax of three-quarters of a second of arc. By definition, a parallax of one second corresponds to a distance of one *parsec* (*par*allax of one *sec*ond). The distance of a star in parsecs is thus 1/parallax. One parsec is equal to 3.26 light-years.

The trigonometric method works only for the relatively nearby stars because the distant stars are used as a background reference. The trigonometric shifts of the latter are virtually negligible, and specialized methods must be used to determine their distances and the distances of the extremely remote galaxies.

There are many highly specialized methods that can be used for certain groups of stars, but the more general are based on the fact that *the intensity of light varies inversely as the square of the distance*. In other words, if two stars of the same intrinsic, actual brightness are placed one and two units of distance away from Earth, the more distant star will appear four times fainter than the closer one. If, then, the actual brightness of a star is known, its distance can be derived simply by noting how bright or faint it appears.

MAGNITUDES.—The brightness of a star as it appears to us is expressed technically as its *apparent magnitude*. The scale of stellar magnitudes is logarithmic and takes into account the fact that visual sensation varies logarithmically; that is, the eye senses the ratios of brightness of sources, rather than the arithmetical differences in brightness. For example, a two candlepower light will look proportionately as bright to the eye when compared with a one candlepower light, as will a 100 candlepower light compared with a 50 candlepower light, because the ratio in both cases is 2:1. The difference in magnitude (Δm) between two stars is expressed mathematically as follows:

$$\Delta m = 2.5 \log_{10} \frac{I_1}{I_2}$$

Thus:

Difference in Magnitude (Δm)	*Ratio of Brightness*
1	2.51
2	6.30
3	15.84
4	39.80
5	100.
10	10,000.
20	100,000,000.

We shall see shortly that the absolute magnitude of a star can be judged fairly accurately from the appearance of its spectrum and, hence, the star's distance can be found.

Distances so determined are called *spectroscopic parallaxes*. This method is an extension of the fundamental, but less accurate, method of trigonometric parallaxes. The latter, however, in addition to methods applicable in special cases to certain groups of stars, serves to establish a range of actual luminosities of stars.

As defined, the absolute magnitudes of stars range from about +20 for the very faintest to −10 for the very brightest. Since every step of five magnitudes corresponds to an advance of 100 in brightness, it can be readily seen that the range in true luminosity of stars is enormous—a total range of thirty magnitudes, or a ratio of 100^6, or 10^{12}—one trillion times.

The truly bright, giant stars can be seen at enormous distances, while the fainter ones must be relatively close to be seen at all. The great majority of stars are fainter than the sun and can be seen only with the aid of a telescope—only when they are in our general part of the galaxy. The stars we see when we glance up at the sky at night are, predominantly, intrinsically bright, faraway stars—beacons far out in space. For every star that can be seen with the naked eye, there are 100 or more closer to us that cannot be seen, even at moderate distances, because they are intrinsically faint.

LUMINOSITY AND TEMPERATURE.—The luminosity of a star depends directly on its size (surface area) and on how brightly each unit area shines (luminosity = surface area × luminosity of unit area). A star can appear bright even though it is relatively faint per square yard. A small, intensely brilliant star will appear as bright as an enormous but faint star.

A star's surface brightness and color depend on its surface temperature. The radiation laws of physics show that the energy flow through the surface of a star varies as the fourth power of the temperature (*Stefan's law*). A star with twice the surface temperature of another will, therefore, emit 16 times as much energy per unit area.

The total luminosity of a star, then, reveals nothing of the star's size unless we also know the star's surface brightness. To find this, the star's temperature must be known.

WIEN'S LAW.—The temperature of a star can be determined fairly accurately by noting its color. Stars behave much like "black bodies," that is, like bodies that obey the laws of

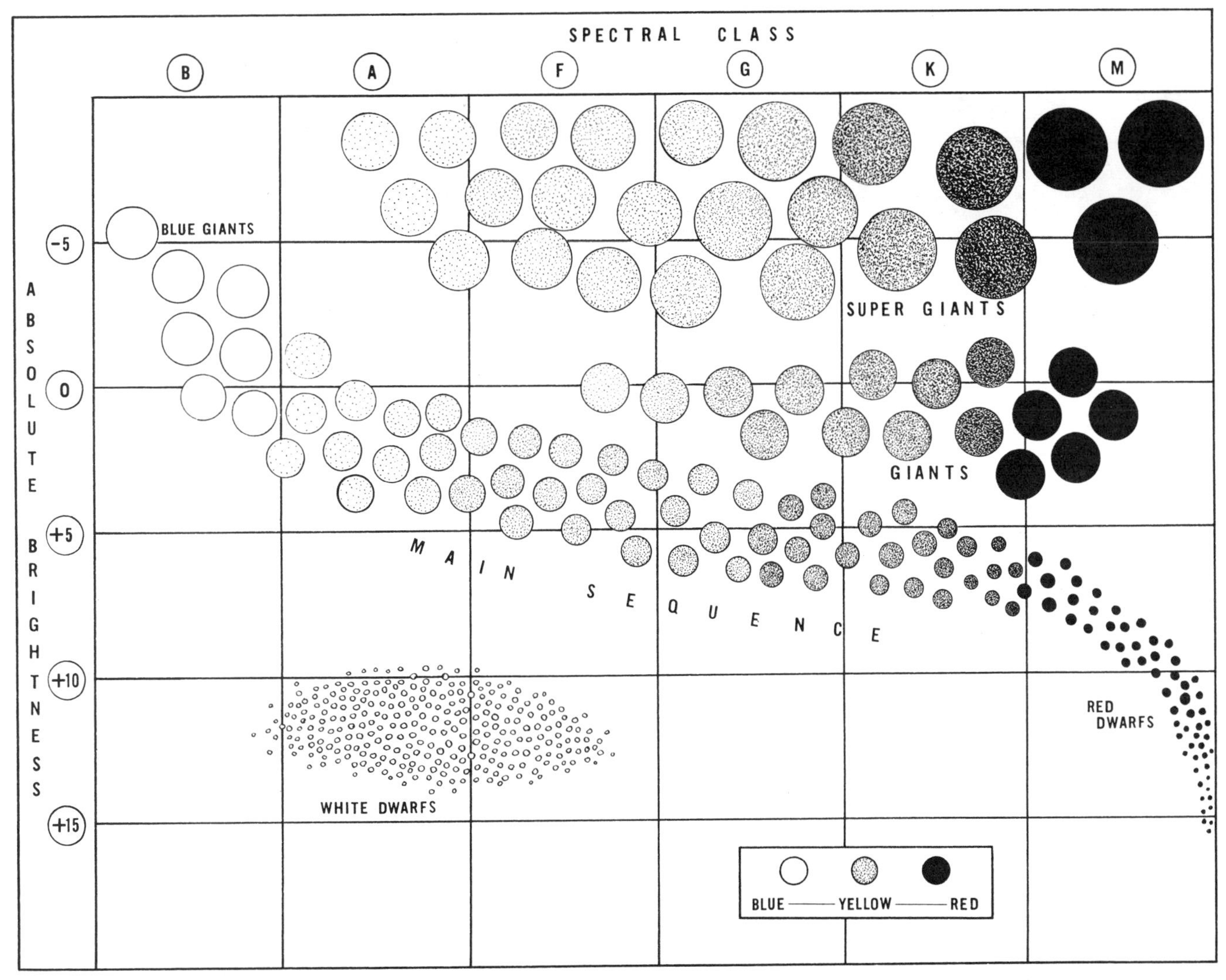

H-R DIAGRAM, which provides a key to stellar evolution, plots the spectral class (based on temperature) of a star against its absolute brightness. The main sequence, which consists of yellow stars bounded by blue giants at one end and red dwarfs at the other, contains most of the known stars. Above the main sequence are the yellow and red giants and supergiants, and below it are the white dwarfs. It is believed that evolutionary changes will cause a star to collapse and eventually to become a white dwarf.

radiation. Of these laws, *Wien's law* states that the most predominant color in the spectrum of a star is a function of the temperature—the hotter the temperature, the bluer the color. Like a piece of metal as it is heated, first it glows a dull red; then, as its temperature rises, it becomes brighter (as stated in Stefan's law), and also becomes, in turn, orange and bright yellow (as in Wien's law). If the metal did not melt and vaporize, it would become green and then blue. Stars of course are completely vaporized but still conform with radiation laws. An orange-yellow star such as Arcturus has a temperature of about 4,500° C.; our yellow sun has a temperature of about 6,000° C.; the bluish star, Rigel, has a surface temperature of some 18,000° C. There is, however, a more accurate method of determining stellar temperatures.

STELLAR SPECTRA.—The surface temperature of a star greatly influences its tenuous atmosphere. Temperature is a direct expression of thermal motion in the material whose temperature is being measured. It is an index of the energy of motion of atoms and molecules; the higher the temperature, the greater the motion of atoms and molecules.

Such motions "thermally excite" the atoms to various energy states. Again, the higher the temperature, the higher the energy states attained by the atoms of the various chemical elements in the star's atmosphere.

The ability of an atom to absorb energy from the starlight pouring out from the lower levels of a stellar atmosphere (thus producing the *Fraunhofer lines,* or familiar absorption lines, in the spectrum of a star) depends on the particular energy state the atom is in. This, as we have seen, depends on the temperature of the stellar surface. It follows, therefore, that the kind of absorption spectrum a star exhibits depends on both the chemical elements present in its atmosphere and the temperature of the atmosphere.

Thus the spectrum of a star is an accurate index of temperature. Long before this fact became known, astronomers had classified stellar spectra on an arbitrary scale of *A, B, C, D,* and so on. It soon developed, however, that certain assigned letters were either superfluous or misplaced. The principal standard spectral classes of stars used today follows, with the approximate temperature corresponding to each spectral class listed.

O	50,000° C.	G	5,500° C.
B	20,000° C.	K	4,500° C.
A	10,000° C.	M	3,000° C.
F	7,500° C.		

In use, each major spectral class is further subdivided into decimal parts. The spectral class of the sun is *G*2, a temperature of 6,000° C.

H-R DIAGRAM.—Although it might seem that for a given stellar temperature there might be a large range of sizes (and thus a large range of luminosity) it does not happen that way in nature for one very good reason: the very great majority of stars fall upon what is termed the *main sequence.*

If the luminosity (absolute magnitude) of a star is plotted against its spectral class (temperature), it will be found that the preponderance of stars are on a nearly straight line, running from highly luminous, hot, bluish stars down to the faint, relatively cool, reddish stars. A given temperature and size go together. Thus, along the main sequence there are no other stars having the sun's temperature (6,000° C.) but a different size. Proper identification of a star's spectrum, indicating where the star belongs along the main sequence, immediately reveals the star's absolute luminosity. Once this is known, the star's distance can be derived.

Not all stars, however, lie along the main sequence. One out of every few hundred stars lies well off it, generally in a band called the *giant sequence.* Such a star may have the temperature of the sun, but be 100 times as luminous. This indicates that the star must be giant in size and in brightness. A few other stars off the main sequence are the *supergiants* and *white dwarfs.*

The *H–R diagram,* named after its formulators, Ejnar Hertzsprung (1873–1967) and Henry Norris Russell (1877–1957), provides a basic key to the problem of stellar evolution. It is generally held that when a star first develops from a primordial gas cloud it "joins" the main sequence at a point appropriate to its mass (the mass-luminosity relationship long noted empirically from a study of double stars and later explained theoretically). A star spends the major portion of its life on the main sequence, but when its hydrogen supply becomes depleted, its internal structure changes to a different "model"; it becomes more luminous and expands greatly in size, lowering its surface temperature. An *A*-type, main-sequence star will thus become a *G*-type or *K*-type giant star.

It is believed that later evolutionary changes cause the star to collapse and eventually to become a white dwarf. White dwarfs might thus be regarded as "celestial clinkers."

DOUBLE AND MULTIPLE STARS.—Nearly one-half of the stars have companion stars with which they revolve about a common center of mass. Apparently, when the original gas cloud from which a star forms is large, several condensation nuclei can develop, resulting in a *multiple star.* Some astronomers favor the idea that in star formation a star becomes either a multiple star or a single star with a family of minor, nonluminous companion planets. This view favors the concept of a multiplicity of solar systems throughout the universe, with all that such a concept may imply.

The study of double stars is a separate branch of astronomy. Those stars whose companions are at relatively great distances are visible in the telescope as two separate stars and are called *visual double stars.* Because of the great separations, the periods of revolution of these stars are often measured in hundreds of years. On the other hand, some double stars are so close together that they appear to be almost in contact. In such cases, the stars spin about each other every few hours.

Such *close doubles* are much too close to be perceived as separate stars. Their existence is revealed by the spectrograph. When one component, in its orbital motion, approaches Earth, its spectrum lines are shifted, in accordance with the *Doppler effect,* toward the blue end of the spectrum. When, half a cycle later, this component recedes from Earth, the lines shift back toward the red. This *Doppler shift* can easily be measured with instruments and has led to the discovery of many such *spectroscopic binaries.*

The manner of motion of one star around another depends on the masses of the bodies involved. Binary stars —particularly visual double stars— thereby offer the only means of direct determination of the masses of stars. The mass of a star is of crucial importance in determining the place of a star in the H–R diagram and also in determining its life cycle.

THE SUN.—As a star, the sun is of spectral type *G*2. It has an absolute magnitude of +4.84, corresponding to an energy outflow of 4×10^{33} ergs/sec, a surface temperature of nearly 6,000° C., a diameter of 864,000 miles, and a mass of 2×10^{33} grams. The sun is an average star. There are stars many times as bright, and others many times fainter. Numerically, it is well above average; there are relatively few stars that are brighter, but a great many that are fainter.

The estimated age of the sun is six billion years. It is further estimated that it has a total useful life of about 12 billion years. The sun appears, therefore, to be at the midpoint of its career as a star.

—J. Allen Hynek

INTERNATIONAL BUSINESS MACHINES CORPORATION

NINETEENTH-CENTURY TELESCOPE opened additional areas to astronomical research.

ASTRONOMICAL INSTRUMENTS

Purpose.—Astronomy is often called an observational science rather than an experimental science because an astronomer must be satisfied with what he can learn from observing what already exists in the universe and how it behaves; there is no opportunity to experiment with celestial bodies. Therefore, astronomical laboratories are called observatories, and astronomical instruments are almost all observing instruments. These instruments are designed to give as much information as possible about objects that are observed.

Only four things can actually be observed about any celestial body: the present position of the body and how its position changes with time; the features of its surface—these are observable on only a few bodies (no more than half a dozen); the brightness of the body (how much light is received from it); and the color of its light (the amount of light received in wavelengths of energy).

All of this information can be obtained by using the eye alone, but only to a limited extent. The eye, guided by such accurate angle-measuring instruments as quadrants or sextants, can determine the positions of stars to a few minutes of arc. The eye alone can see some features on the moon's surface and can distinguish the brightness of the stars and their colors to a surprising degree.

But for very accurate measurements of all these things, the human eye is aided in astronomy by an instrument called a *telescope,* supplemented by several different kinds of accessory instruments.

Optical Telescopes.—An *optical telescope* is a device for collecting a large amount of light, greater than the eye alone can collect, and concentrating it onto a small area where it can be observed by the eye, recorded on a photographic film, or measured in some way. There are two basic kinds of telescopes: the refracting (lens-type) telescope, and the reflecting (mirror-type) telescope.

■**REFRACTING TELESCOPES.**—In a *refracting telescope* a large glass lens, shaped in a convex (outward-bulging) curve on each surface, is called the *telescope objective.* When light from a star reaches the lens, it bends slightly as it passes through. Where the light enters the lens at the greatest angle with the surface, it bends most. The curvature of the lens is designed so that all of the light passing through the glass is bent toward a single point behind the lens called the *focal point,* or *focus.* An observer examines the light at the focus through another small lens, called an *eyepiece,* which works very much like a magnifying glass.

■**REFLECTING TELESCOPES.**—In a *reflecting telescope,* the light from a celestial body passes down a tube to the surface of a mirror mounted at the bottom of the tube. The surface of the mirror, which is the objective, is ground to a very precisely curved shape; on large mirrors, the curve is part of a parabola. The curved surface of the mirror reflects the light back up the tube toward a focus, where all the light passes through a single point. Since the focus is directly in the path of the incoming light, it usually cannot be observed directly, so the light is reflected from a secondary mirror inside the tube to bring the focus to a more convenient location. If the secondary mirror reflects the light to one side of the tube, the telescope is a *Newtonian,* after Sir Isaac Newton, the English astronomer who devised this type of instrument about 1670. If the secondary mirror reflects the light back down the tube through a hole in the center of the primary mirror, the telescope is a *Cassegrain,* after the astronomer Guillaume N. Cassegrain, who invented it in 1672. In very large reflecting telescopes, the focus can be seen directly by an observer in a cage inside the tube itself. The observer's cage blocks off a small amount of the light going down the tube toward the primary mirror (as do the secondary mirrors), but this does not interfere with the quality of the telescope in any way.

■**SIZE.**—The largest telescopes are all of the reflecting variety for a number of reasons. It is easier to cast mirrors than large lenses because only the surface of the mirror must be perfect, while all the glass and both surfaces of a lens must be perfect. It is also easier to mount and support a large mirror because it can be held by a frame over its entire bottom surface, while a lens must be supported around its edge. For the same size telescope, a reflector is much less expensive than a refractor because the glass of the mirror does not have to be perfect and only one surface has to be ground and polished. The largest reflecting telescopes are the 200-inch Hale telescope at Mt. Palomar, California, the 120-inch telescope of the Lick Observatory, Mt. Hamilton, California, and the 107-inch telescope at the University of Texas. The largest lens-type telescope is the 40-inch refractor of the Yerkes Observatory, Williams Bay, Wisconsin. The size of a telescope is always the diameter of the objective lens or mirror, as the case may be.

MOUNT WILSON AND PALOMAR OBSERVATORIES

200-INCH HALE TELESCOPE points skyward to photograph a specific portion of the moon.

■**FUNCTIONS.**—A telescope performs four important functions for an astronomer. First, it increases his *pointing accuracy;* that is, the telescope allows him to point more accurately in a required direction and also to measure more accurately the direction in which he is pointing his instrument. Second, a telescope provides *magnifying power;* that is, it makes distant and small objects appear larger than they appear to the naked eye. Third, a telescope provides *light-gathering power;* that is, it makes faint objects appear brighter because it gathers all the light falling on a large objective and concentrates it for the astronomer to see. Fourth, a telescope gives *resolving power;* that is, it enables the astronomer to separate and to see distinctly objects so close together that they appear as one to the naked eye.

The pointing accuracy of a telescope depends somewhat on the focal length of the telescope. The *focal length* is the distance from the objective to the focal point. Long-focus telescopes can be pointed more accurately than short-focus telescopes. More important than focal length, however, are the size and accuracy of the measuring circles built into the telescope mounting and the sturdiness and accuracy of the mounting itself. Specially designed telescopes with rigid and accurate mountings and with large, finely divided measuring circles are used to determine accurately the positions of stars and changes in their positions. These telescopes are usually mounted so that they are free to move only along the meridian, the north-south line in the sky. They are called *meridian telescopes* or *transits.* The accuracy of positions toward which other telescopes are pointed is checked against the positions of certain stars that have been carefully measured by meridian telescopes.

The magnifying power of a telescope depends upon its focal length. When a telescope is pointed at an extended object, such as the sun or moon, the size of the image at the focus depends upon the distance from the lens or mirror to the focus. If the focal length of a telescope is 10 feet, for example, the image of the moon at the focal plane will be about 1 inch across, but it would be about 10 inches across if the focal length of the telescope were 100 feet. If a photographic film were placed at the focus of such a telescope, it would record a picture of the moon one foot in diameter. Telescopes that are designed to observe the surface features of the sun, moon, and planets are therefore always long-focal-length telescopes.

The magnification of a telescope that is used for visual observation depends upon another factor, the focal length of the eyepiece. The *visual magnifying power* of a telescope is the ratio of the focal length of the objective to the focal length of the eyepiece. Visual magnification can be increased in any telescope by using an eyepiece of shorter focal length, thus increasing the ratio governing the magnification. However, there is a practical limit to the magnifying power that can be used with any given telescope, usually about 30 to 40 times the diameter of the objective. For a 4-inch telescope (diameter of objective), the practical limit of magnifying power is therefore about 150. Using an eyepiece of a focal length that gives a magnifying power higher than 150 for such a telescope results in a loss of quality and light in the image.

For use with stars, faint nebulae, and galaxies, the remaining two functions of a telescope, light-gathering power and resolving power, are the more important. The magnifying power of a telescope is of no value in observing stars, since they are essentially point sources of light without any area to be magnified. Stars are so far away that, in spite of their great size, nothing but a point of light can be seen in even the largest telescopes. As a matter of fact, very large telescopes of good quality produce extremely small images of stars —the smaller the better. Thus, stars that are close together can be better resolved into individual images.

The light-gathering power and the resolving power of a telescope both depend upon the size of the objective lens or mirror. The amount of light that a telescope receives varies with the square of the objective's diameter. Thus, a 100-inch telescope, with a diameter 10 times as great, would receive 10 squared (10^2), or 100, times as much light as a 10-inch telescope. Theoretically, this 100-inch telescope could detect objects 100 times fainter than the faintest object detectable in the smaller instrument. This is the real purpose of building exceptionally large telescopes with objective diameters of 100 or 200 inches. Such telescopes are used primarily to observe stars, nebulae, and galaxies that are too faint to be observed by smaller instruments. In so doing, they can observe objects at very great distances, objects that may be as bright as nearby objects, but have been dimmed by the effect of distance. These very large telescopes, therefore, are effective in pushing the limits of the observable universe to ever-greater distances.

Photographic Telescopes.—Large telescopes are seldom used today for visual observations. They are almost always used with one of several accessory devices that record the images they produce or that measure or analyze the light they gather. For recording the appearance or position of stars, nebulae, and galaxies, telescopes are used as photographic instruments. Photographic film holders are placed in the focal plane of the telescope, and the images are recorded on the film. The resulting picture is a permanent record of the field of the sky being examined.

Photographic telescopes have another tremendous advantage. The image seen by the eye represents the light that reaches the eye in a very brief moment. But the image recorded on a film can represent all of the light accumulated over several hours of exposure. Since the intensity of the image on the film continues to build up as long as it is exposed to light, a photograph can record stars or galaxies that could never be observed visually. Throughout the exposure of the film, of course, the telescope must be pointed in precisely the same direction.

■ **MOUNTING.**—Telescopes are mounted in two ways, alt-azimuth and equatorial. In an *alt-azimuth mount,* the telescope is free to move up or down (in altitude) or horizontally around the sky (in azimuth). This type of mounting is usually found only on smaller telescopes. In an *equatorial mount,* the telescope is free to move around an axis parallel to the axis of the earth's rotation. The telescope is first set to a position corresponding to the star's distance above or below the plane of the earth's equator. Then, when the telescope moves around the polar axis, it moves in a path that corresponds to the star's path of motion across the sky. In a photographic telescope, this motion is provided with motors geared to the speed of the earth's rotation, so that the telescope tracks the stars being photographed as they move. Throughout the exposure of the film, the drive of the telescope compensates exactly for the rotation of the earth.

Photometers and Spectrographs.—Two important accessories used with modern telescopes are photometers and spectrographs. A *photometer* is similar in principle to a photoelectric cell. When light from a star falls on a tiny photosensitive cell, the cell regulates the amount of electric energy sent to a meter according to the intensity of the star's light. The meter then indicates the brightness of the star. A *spectrograph* is an instrument that produces a photograph of a star's spectrum. The star's light is allowed to pass through a device, such as a prism, which spreads the light out into all of its colors, or its *spectrum.* The spectrum is recorded on film; this allows the spectra of faint objects to be observed. The features of the star's spectrum—the nature and intensity of its dark or bright lines—can then be examined.

HARVARD COLLEGE OBSERVATORY

RADIO TELESCOPE hears sounds from space.

Electronic Equipment.—In addition to telescopes and their accessories, astronomers use a variety of other instruments to assist them in analyzing the records provided by their telescopes. They use special instruments to measure the position and intensity of images recorded on film. In today's astronomy, the use of electronic equipment is becoming increasingly important. Electronic image intensifiers have been developed to suppress the background brightness of the sky and to enhance the images of very faint stars and galaxies. Many complex theoretical problems in astronomy, such as the nature of stellar interiors, can be solved only by the use of high-speed electronic computers. And electronic equipment is essential in the radio telescopes that today explore the radio universe.

Radio Telescopes.—The *radio universe* is observed in the radiation that is naturally produced in frequencies of energy called *radio waves.* Celestial bodies produce radio waves just as they produce light waves. *Radio telescopes* collect and focus these radio waves just as optical telescopes collect and focus light. But since radio waves cannot be seen or recorded on films, they must be handled electronically. They are amplified, and the frequency or intensity is recorded on a meter or chart. In this way, the sky is looked at through a new kind of "light." It has been found that many things look different in this "radio light" and that some things can be observed only with the use of radio telescopes.

Space Techniques.—The next developments in astronomical instruments will be those designed for use in space. Some have already been developed and used, such as the *stratoscope telescopes,* which were flown by balloons to altitudes of more than 80,000 feet, and the *orbiting solar observatories* (OSO satellites) and *orbiting astronomical observatories* (OAO satellites). These projects involve instruments for many purposes, but perhaps most important are those for exploring the sky in ultraviolet energy. This short-wave form of energy is blocked from the earthbound instruments by the atmosphere, but can be observed in space. Astronomers expect that the future exploration of the sky in ultraviolet energy will be as valuable as radio observations have been.

—Thomas D. Nicholson

ASTRONOMY GLOSSARY

Aberration of light. — The apparent change of direction (or bending) of light from a celestial body, causing an apparent change in the position of that body, that results from the motion of the earth in its orbit. By analogy, if a person stands in a rainstorm, and there is no wind, the raindrops will come down on him vertically; hence, he must hold an umbrella directly overhead to keep dry. However, if he begins running, he will have to hold the umbrella so that it points ahead of him, at an angle to the vertical, because the raindrops will seem to be coming at a slant. The faster the person runs, the greater the apparent deviation of the raindrops from the vertical.

The earth travels in its orbit at an average speed of 18½ miles per second. Therefore, in order for a particular star to be seen, a telescope must be aimed at an angle just slightly ahead of the true direction, just as a person running in a rainstorm must hold his umbrella at a slant.

If the earth were traveling along a straight-line path, there would be no way to ascertain the true position of a star. However, since the earth travels in a particular direction at one time of the year and in the opposite direction six months later, all the stars seem to shift slightly in position. The *constant of aberration* is the angular displacement thus observed. It is equal to 20.47 seconds of arc and is the same for all stars. In calculating the true position of a star, it is necessary to take note of this aberrational displacement.

Although all stars show the same total displacement in the course of a year, the amount of displacement on a given observation is not constant, except for a star situated precisely at the ecliptic pole. For the actual aberration to be equal to 20.47 seconds of arc, it is necessary that the earth's direction of motion be exactly perpendicular to the direction of the star. This generally happens only twice a year for a given star. Consider, for example, a star in the plane of the ecliptic. Twice a year the earth will cross the imaginary line passing through the star and the center of the earth's orbit; at each of these crossings the star's aberration will be the maximum, 20.47 seconds. At two other points in its orbit the earth will be moving directly away from the star; at these times aberration will be zero.

Albedo.—The fraction of light reflected from the surface of a celestial body that is not self-luminous. The magnitude of the fraction depends upon the atmosphere and the surface of the body. For example, the moon, which has no atmosphere and a very rough and broken surface of dark-colored rock, has a low albedo; Venus, on the other hand, whose atmosphere is filled with dense clouds, has a high albedo. The albedo of each of the planets in the solar system, in order of decreasing magnitude, is: Venus, 0.76; Uranus, 0.66; Jupiter, 0.51; Saturn, 0.50; Earth, 0.39; Neptune, 0.26; Pluto, 0.16; Mars, 0.148; Mercury, 0.058—the moon's albedo, at 0.072, is slightly greater than Mercury's.

Altitude.—See *Horizon system.*

Apex of the sun's way.—The point on the celestial system toward which the solar system is moving at a speed of about 12 miles per second. This point, located in the constellation Hercules, is also known as the *solar apex.*

Aphelion.—The point on the orbit of a planet, or any other member of the solar system, that is farthest from the sun. (See *Orbit.*)

Apogee.—The point on the orbit of the moon or any artificial satellite that is farthest from the earth.

Apsides.—The two points in the orbit of a celestial body that are respectively the farthest from and the closest to the attracting primary. Each of the points is called an *apsis* or *apse.* A line drawn between them is called the *line of apsides* and is identical with the orbit's major axis (longest diameter). (See *Orbit.*)

Aries, first point of.—See *Equinox.*

Asteroids.—Small planetary bodies of varying dimensions. The orbits of most of these lie between the orbits of Mars and Jupiter. They are also called *minor planets* or *planetoids.*

Astronomical unit (a.u.).—The mean distance from the earth to the sun, equal to the semimajor axis (one-half the longest diameter) of the earth's orbit. It is the unit of measurement for distances within the solar system. One astronomical unit is equal to 92,887,000 ± 20,000 miles.

Aurora.—An electrical display seen above the earth's poles as arcs, rays, and streamers of green, red, and yellow. It is the result of the discharge of electricity at heights of 50 to 150 miles caused by the action on the rarefied atmospheric gases of high-speed particles (alpha particles and electrons) shot from the sun to the earth. In the Northern Hemisphere the display is called the *aurora borealis, aurora polaris,* or *northern lights;* in the Southern Hemisphere it is called the *aurora australis.*

Azimuth.—See *Horizon system.*

Baily's beads.—A phenomenon, discovered by Francis Baily (1774–1844), in which, during the last seconds before the solar eclipse becomes total, bright spots are seen on the dark edge of the moon. These spots, or beads, are produced by sunlight shining between the mountains on the surface of the moon.

Barnard's star.—A faint star in the constellation Ophiuchus, discovered by Edward Emerson Barnard in 1916, that has the largest proper motion (the angle through which a star appears to move against the background of stars on the celestial sphere) measured to date. Only about 200 stars are known with a proper motion as great as 1/10 that of Barnard's star, and most stars have a proper motion about 1/100 that of Barnard's star. In 1963, it was established that Barnard's star is circled by a planet, called *Barnard's planet.*

Celestial sphere.—An imaginary sphere, of indeterminate radius, around the earth on whose surface all celestial objects, regardless of their real distance, are assumed to be projected. If the plane of the earth's equator were to be extended indefinitely, it would cut the celestial sphere at the *celestial equator.* The celestial equator may be thought of as a great circle passing completely around the celestial sphere, midway between the celestial poles. The *celestial poles* may be found by extending a line through the earth's poles until it touches the celestial sphere. The point on the celestial sphere directly over the head of the observer is the *zenith;* the point directly opposite the zenith, on the opposite side of the celestial sphere, is the *nadir.* The celestial poles are therefore fixed reference points on the celestial sphere while the zenith and nadir vary with the position of the observer. Another great circle on the celestial sphere is the ecliptic, which is the apparent path of the sun among the stars. Because the earth's axis is not perpendicular to its orbit, the ecliptic cuts the celestial equator at an angle of 23½°. This inclination makes the sun seem to rise and set at a slightly different point each morning and evening and to cause the seasons. The points of intersection of the celestial equator with the ecliptic are called *equinoxes.*

Chromosphere.—The inner atmosphere of the sun, extending from 300 to 6,000 miles above the sun's surface, or photosphere, and consisting of permanent gases. The lower part of the chromosphere is called the *reversing layer;* the name is derived from a phenomenon that takes place in this region, in which the dark absorption lines of the solar spectrum are formed by reversal from bright emission lines. *Prominences,* vast flame-like eruptions of gas reaching hundreds of thousands of miles from the sun, arise from the chromosphere.

Circumpolar star.—A star that does not rise and set but is always above the horizon. The angle between the star and the celestial pole must be less than or equal to the observer's latitude; thus, the number of such stars varies with the places of observation on the earth.

Comet.—A member of the solar system having a very small mass and usually forming a long, gaseous tail as it approaches the sun. The tail of a comet always points away from the sun because it is blown away from the sun by the solar winds. The *coma,* or head of the comet, is probably composed of ice and dust.

Conjunction.—See *Planetary configurations.*

Constellation.—A group of stars, not necessarily in the same system, that seems to form the outline of a figure. The names of constellations are usually of mythological origin. Constellations are much used in astrology, but have little significance in astronomy. (See *Zodiac.*)

Copernican system.—See *Heliocentric system.*
Counterglow.—See *Gegenschein.*
Day.—Any one of several divisions of time. It is the period of the earth's rotation about its axis. An *apparent solar day* is the time elapsing between two successive passages of the true sun across the same meridian; it is not of constant length because of the earth's elliptic orbit. A *mean solar day* is the time elapsing between two successive passages of the mean sun across the same meridian; it is of constant length. A *sidereal day* is the time elapsing between two successive passages of the vernal equinox across the same meridian; it is of constant length.
Declination.—See *Equator system.*
Doppler effect.—A phenomenon, explained by Christian Johann Doppler (1803–1853), in which the observed wavelength of the light received from an approaching or receding body is changed. The amount of change depends upon the speed of approach or recession. When a star or galaxy recedes from the earth, its spectral lines are shifted toward the red end of the spectrum—this is called the *red shift.* When a star or galaxy is approaching the earth, its spectral lines shift toward the violet end of the spectrum—this is called the *violet shift.* The Doppler effect is utilized in many fields; here it is used to determine the motion of a celestial body relative to the earth.
Earthlight (*earthshine*).—The sunlight reflected from the earth's surface that produces a faint illumination of the dark side of the moon.
Eclipse.—The obscuring of the light from one celestial body by another. In a *solar eclipse,* the moon moves between the sun and the earth so as to obscure the light of the sun as seen from the earth. In a *lunar eclipse,* the earth moves between the sun and the moon so that the moon is in the earth's shadow.

The region of partial shadow surrounding the dark cone in an eclipse is called the *penumbra.* The dark cone, or region of complete shadow, is called the *umbra.*

It would seem that there would be an eclipse of the moon at every full moon and an eclipse of the sun at every new moon. But this is not the case because the moon's orbit is inclined to the orbit of the earth. If the full moon is not near the ascending or descending node of its orbit, it will be so far from the plane of the ecliptic that it will pass above or below the earth's shadow and will not be eclipsed. If the new moon is not near one of the nodes, its shadow will pass above or below the earth and the sun will not be eclipsed. If the shadow of the earth and the full moon are near enough to one of the nodes so that the moon passes entirely into the shadow, a *total lunar eclipse* results. If only part of the moon enters the shadow, a *partial lunar eclipse* occurs. Most persons have witnessed a lunar eclipse, for the phenomenon can be seen by all the people on the night side of the earth at the same instant. In fact, an eclipse of the moon is visible to considerably more than a hemisphere because in the four hours that may pass from the time the moon first touches the earth's shadow until it completely leaves it, the earth has turned on its axis 60 degrees. There are between 15 and 16 lunar eclipses in 10 years. The eastern side of the moon becomes darker some time before it actually reaches the shadow; then a "bite" is taken out of the moon's disk, and this "bite" increases in size until the whole moon is darkened. Presently, a small, bright arc emerges on the opposite side and grows until the whole lunar surface is again in sunlight.

If the new moon's cone-shaped shadow does not pass above or below the earth and is long enough to reach it, a *total solar eclipse* occurs. The sun is totally obscured from those parts of the earth that the moon's shadow touches. If the new moon's shadow is in line with, but falls short of reaching, the earth, as it often does, a ring of sunlight shows around the disk of the moon; this is called an *annular eclipse.* A solar eclipse is visible as total only within the narrow shadow path. For some hundreds of miles on either side of the shadow path, the sun may be observed in *partial eclipse.* The shadow travels at high speed, and for any one place on the earth the average duration of total eclipse is three minutes. At any one station a total eclipse of the sun will be visible only once in 300 or 400 years.

Eclipses follow a regular cycle. An eclipse, whether solar or lunar, repeats itself every 6,585 days, or 18 years 11⅓ days. This interval between eclipses is called the *saros.*
Ecliptic.—The apparent path of the sun around the celestial sphere. (See *Celestial sphere.*)
Ephemeris.—A table, published at regular intervals, in which the daily position of the sun, moon, planets, artificial satellites, selected stars, and other data necessary for astronomical and navigational observations have been computed.
Epicycle.—In geocentric or Ptolemaic astronomy, a term used to describe the circular orbit traveled by the sun, the moon, and all the planets other than the earth. The center of the epicycle was on a larger circle, called the *deferent.* The sun, the moon, and the planets revolved in their epicycles, and the epicycles themselves revolved around the earth on the deferent circle. The concept of epicycles was ultimately disproved by Johannes Kepler (1571–1630), when he showed that an ellipse could be substituted for a series of epicycles to explain the motion of each planet around the sun.
Equator system.—A means of fixing the position of a celestial body upon the celestial sphere. The lines of latitude and longitude must be imagined to extend from the earth until they cut the celestial sphere. The equator of the earth will cut the celestial sphere at the *celestial equator.* The parallels of latitude will cut the celestial sphere in great circles, called *parallels of declination,* that are north and south of the celestial equator and parallel to it. The *declination* of a celestial body is its angular distance north or south of the celestial equator. A star with a declination of 30° north, or +30, as this is generally represented, lies directly above the parallel of latitude that is 30° north of the equator. The meridians of longitude intersect the celestial sphere in great circles, called *hour circles,* extending from pole to pole. The angular distance to one of these hour circles, measured in an easterly direction from the vernal equinox, is the *right ascension.* Thus, a star may have a right ascension of 6 hours 30 minutes, which means it is 97°30′ east of the vernal equinox.
Equinox.—A point of intersection of the celestial equator with the ecliptic. There are two equinoxes: the sun crosses the celestial equator at a point called the *vernal equinox,* or *first point of Aries,* on March 20 or 21 (the first day of spring); the sun crosses the celestial equator at a point called the *autumnal equinox* on September 22 or 23 (the first day of autumn).
Faculae.—Large, bright areas of hot gases, at a higher temperature than the average for the sun's surface, located near the top of the photosphere. They can most easily be seen near sunspots and at the edge of the sun's disk.
Fireball.—See *Meteor.*
Flocculi.—Small, bright or dark markings on the chromosphere of the sun—irregular clouds of gas, usually containing calcium or hydrogen.
Gegenschein. — Sometimes called *counterglow,* a faint illumination of the sky, sometimes seen at night in the ecliptic opposite to the position of the sun. It is produced by the scattering of sunlight by dust particles in space, and is related to the zodiacal light. (See *Zodiacal light.*)
Geocentric system (*Ptolemaic system*).—A system of planetary motion, as described in a treatise by Claudius Ptolemy (second century), in which the sun, moon, and planets revolved around a stationary earth; the fixed stars were supposedly attached to an outer sphere surrounding the earth. It was eventually replaced by the idea of the heliocentric system. (See *Heliocentric system.*)
Harvest moon.—In the Northern Hemisphere, the full moon that occurs each year nearest the date of the autumnal equinox, September 22 or 23. Its remarkable feature is that for several successive nights it rises at much more nearly the same hour than at other times of the year. The phenomenon may be understood by considering that the path of the moon through the stars, roughly the same as the ecliptic, is inclined to the celestial equator; and on the date of the autumnal equinox, the full moon is in the vernal equinox, directly opposite the sun.
Heliocentric system (*Copernican system*).—A system of planetary motion, as described by Nikolaus Copernicus (1473–1543), in which the earth and the other planets move in paths around a fixed sun. This system, which is the basis of the present-day concept of the solar sys-

tem, superseded the geocentric system of Claudius Ptolemy. (See *Geocentric system.*)

Horizon system.—A means of fixing the position of a celestial body upon the celestial sphere. The observer projects the visible horizon onto the celestial sphere to form a great circle, called the *astronomical horizon.* Starting from the zenith, an arc, called a *vertical circle,* is drawn so that it intersects the astronomical horizon. The *altitude* of a celestial body is the angular distance from the horizon to the celestial body, measured along the vertical circle passing through the celestial body. The *azimuth* is the direction angle of the celestial body, measured in the plane of the astronomical horizon; the zero point of the azimuth is generally taken as being due south and rotating to the right. Thus, in the horizon system, the position of a star could be given as altitude 45°, azimuth 30°.

Hour circle.—See *Equator system.*

Inferior planet.—See *Planetary configurations.*

Latitude.—On the earth, the angle measured along a meridian to any particular point; this is often called *terrestrial latitude. Celestial latitude,* the angular distance of a celestial body above or below the ecliptic, is one coordinate in the ecliptic system for fixing the position of any celestial body on the celestial sphere (the other coordinate is celestial longitude). *Galactic latitude* is the angular distance of a celestial body above or below a great circle chosen to represent the plane of the Milky Way.

Libration (from the Latin *libra,* 'a balance').—An apparent oscillation, or rocking, in the motion of a celestial body as viewed from its primary, similar to the oscillations of a balance scale before it comes to rest. The term is most commonly applied to the motion of the moon. Libration causes slightly different portions of the moon's surface to be visible from the earth at different times of the month, although on the average the moon keeps the same face always turned to the earth. Libration arises from the fact that the axial rotation of the moon is uniform, while the speed of revolution (orbital motion) is variable. The compounding, or resultant, of the two motions gives the moon a small oscillating rotation relative to the earth. The moon's libration makes about 41 per cent of its surface always visible (neglecting its phases), and 18 per cent alternately visible and invisible; thus, about 41 per cent of the moon's surface is never seen from the earth.

Light-year.—A unit of measure of stellar distances. One light-year is equal to the distance light travels in one year: $5{,}878 \times 10^9$ miles, or 63,290 astronomical units.

Longitude.—On the earth, the angle measured at the center of the earth between the points at which the meridian through Greenwich, England, and the meridian through any particular location cross the equator; this is often called *terrestrial longitude. Celestial longitude,* the angular distance measured eastward along the ecliptic from the vernal equinox to the meridian passing through the poles of the ecliptic and a particular celestial body, is one coordinate in the ecliptic system for fixing the position of a celestial body on the celestial sphere (the other coordinate is celestial latitude). *Galactic longitude* is the angular distance measured eastward along a great circle chosen to represent the plane of the Milky Way from a point designated as the galactic center to the meridian passing through the poles of the galaxy and a particular celestial body.

Magellanic clouds.—Two irregular galaxies in the Southern Hemisphere. Known as the *Greater Magellanic Cloud* and *Lesser Magellanic Cloud,* respectively, their names are derived from their sizes and the fact that as seen from the Straits of Magellan they pass not far from the zenith. They are the two nearest galaxies to the Milky Way. The Greater Magellanic Cloud is 30,000 light-years in diameter and 144,000 light-years from the earth; the Lesser Magellanic Cloud is 23,400 light-years in diameter and 164,000 light-years from the earth.

Mean sun.—A fictitious body that moves around the celestial equator at a uniform rate, postulated for convenience in measuring time. Because the earth does not move around the sun at a constant rate, and because the ecliptic is inclined to the celestial equator, the sun appears to travel around the earth at different speeds at different times of the year. By measuring time through the use of a mean, or average, sun instead of the real sun, clocks do not have to be corrected to compensate for changes in the speed of the sun throughout the year.

Meteor.—A class of celestial bodies that are seen as streaks of light in the night sky as they burn because of frictional heating while falling through the earth's atmosphere. They are sometimes called *shooting stars* (although they are not stars) when they are not brighter than the planet Venus, and *fireballs* when they are brighter than Venus. If the meteor does not vaporize completely as it passes through the atmosphere, the portion that survives to strike the earth's surface is called a *meteorite.*

Midnight sun.—The sun when it shines throughout the night in the Arctic or Antarctic regions during the summer months.

Minor planets.—See *Asteroids.*

Month.—Any one of several divisions of time. A *calendar month* is an arbitrary division of the year. An *anomalistic month* is the time elapsing between two successive passages of the moon through perigee in its orbit; it is equal to 27.5455 days. A *sidereal month* is the time elapsing between two successive passages of the moon through a point in its orbit obtained by drawing a line from the center of the sun to a fixed point on the celestial sphere; it is equal to 27.32166 days. A *synodic month* is the time elapsing between two successive passages of the moon through either conjunction or opposition, or the time elapsing while the moon completes going through all of its phases; it is equal to 29.53059 days. A *tropical month* is the time elapsing between two successive passages of the moon through the ascending node of the ecliptic, equal to 27.32158 days.

Nadir.—The point on the celestial sphere directly below the observer and directly opposite the zenith. (See *Celestial sphere.*)

Nebula.—Any luminous patch seen among the stars. More specifically, a cloud of dust or gas that can be seen because of the absorption, emission, or reflection of light.

Neutron star.—A star composed solely of neutrons packed very closely together. The existence of these stars has been postulated by astrophysicists, and in 1963 the discovery of several through the use of orbiting satellites was claimed, although these claims have not yet been verified. A cubic inch of a neutron star would weigh about one billion tons, its surface temperature would be about 20 million degrees Fahrenheit, and it would emit light only in the X-ray region of the spectrum.

Nodes.—The points on the celestial sphere at which the orbit of any body of the solar system crosses the ecliptic. The term is most commonly applied to the points at which the moon crosses the ecliptic. The *ascending node* is the point at which the body passes from south of the ecliptic to north of it. The *descending node* is the point at which the body passes from north to south. The nodes of an orbit do not remain at the same points on the ecliptic; over long periods of time they move around the ecliptic in the retrograde (west to east) direction. The nodes of the earth's orbit are the vernal and autumnal equinoxes.

Northern lights.—See *Aurora.*

Nova.—A star that flares from obscurity to great brilliance and then sinks slowly back to obscurity. A nova is sometimes called a *new star* or a *temporary star.* A *supernova* is an exceptionally bright nova. A supernova that appeared in Taurus in 1054 was visible in broad daylight for a time, and can now be seen as the Crab nebula. A *recurrent nova* is a star that has increased significantly in brightness more than once. A *permanent nova* is a star that has streams of gas flaring from its surface, as if it were in a continual state of eruption.

Nutation (from the Latin *nutatio,* meaning 'to nod').—A movement of the earth's axis like the nodding of a top, which causes the celestial pole to trace a wavy path as it moves among the stars. Nutation is caused by a variation in the attraction of the sun and moon as their movement alters their positions relative to the earth. (See *Precession.*)

Occultation.—The hiding of one celestial body by another, as eclipse of a star or planet by the moon.

Orbit.—The path traveled by one celestial body as it moves about another under the influence of its gravitational attraction. There are six factors that must be known in order to determine the orbit of a celestial body and the body's position in it: the inclination of the orbit to the plane of the primary, the semimajor axis, the date on which the body passes perihelion, the eccentricity of the orbit, the longitude of perihelion, and the longitude of the ascending node.

Parallax.—The apparent change of position of a celestial body against the celestial sphere when viewed from two different points, called the *ends of the baseline. Geocentric,* or *diurnal, parallax* is the angular distance a celestial body appears to move when the baseline is the diameter of the earth. *Heliocentric,* or *annual, parallax* is the angular distance a celestial body appears to move when the baseline is the diameter of the earth's orbit, or one astronomical unit. *Lunar parallax* is the geocentric parallax of the moon, and *solar parallax* is the geocentric parallax of the sun.

Parsec.—A unit of measure of stellar distance. One parsec equals 1.916×10^{12} miles, 206,205 astronomical units, or 3.26 light-years. It is the distance at which the radius of the earth's orbit subtends an angle of 1 second. Therefore, if a celestial body were exactly 1 parsec from the earth, it would have a parallax of 1 second of arc. The convenience of the parsec as a unit of stellar measurement is that a star's distance in parsecs is equal to the reciprocal of its parallax in seconds of arc.

Perigee.—The point on the orbit of the moon or of any artificial satellite that is closest to the earth.

Perihelion.—The point on the orbit of a planet, or of any other member of the solar system, that is closest to the sun. (See *Orbit.*)

Period.—The time that elapses while a celestial body makes one complete revolution in its orbit. A *sidereal period* is the time that elapses between two successive crossings of a particular line from the sun to a fixed point on the celestial sphere. A *synodic period* is the time that elapses between two successive crossings of a line passing through the center of the earth and of the sun.

Perturbations. — Disturbances caused by the gravitational force of a celestial body that result in the deviation of another celestial body from its regular orbit. An observation of the perturbations caused by a body can help in the calculation of its mass.

Phases of the moon.—The apparent changes seen in the shape of the moon in the course of a month. When the moon is in conjunction—between the earth and the sun—it is called a *new moon.* The next day, a *waxing crescent* appears as the moon begins moving to the east of the sun. A quarter of a lunar month later, the moon is at *first quarter* and is on the meridian, or due south, at sunset; at first quarter, one-half the face of the moon is illuminated. As the moon continues moving eastward, more and more of its face continues to become illuminated, and the phase is called *waxing gibbous.* After another quarter of a lunar month, the entire face of the moon can be seen and it rises at sunset; this is called a *full moon.* As the moon continues moving around the earth, less and less of its face shows each day, and the phase is called *waning gibbous.* After another quarter of a lunar month the moon is at *last quarter;* once again half the face is illuminated, but now the moon rises about midnight and is due south at sunrise. As the moon continues to move in its eastward path, still less and less of the surface is illuminated each day, and the phase is said to be *waning crescent.* Finally, at the completion of the lunar month, the moon fades from sight and is a *new moon* again.

Photosphere.—The visible surface of the sun; the envelope of gas surrounding the sun.

Planetarium.—Both an instrument for artificially duplicating the appearance of the heavens and the large, specially constructed auditorium in which it is housed. By means of a complex projector, the representation of one half of the celestial sphere may be thrown on the interior of a large, concave hemispherical dome. The instrument may be set to depict the celestial sphere as seen from any terrestrial latitude and for any time; it may be operated to show apparent celestial motions, such as the diurnal motion and the changes due to precession. The moon, sun, and planets are shown by means of special projectors, and their apparently complex motions are faithfully reproduced. Various speeds may be used, so that the diurnal motions of the heavens, and even the annual motions, may be compressed into a few seconds. This acceleration gives a realization of the annual course of the sun among the stars, the planetary orbits, and the intricate apparent motions of the planets in them.

Planetary configurations.—The positions of a planet relative to the earth and some other body, usually the sun. For a *superior planet*—one whose orbit is outside that of the earth—there are four critical positions relative to the sun: *conjunction, opposition, east quadrature,* and *west quadrature.* The planets are shown in color on pages 1298 and 1299. For an *inferior planet*—one whose orbit is inside that of the earth—there is no position of opposition or quadrature. There are, however, two conjunctions: *superior conjunction,* when the sun is between the earth and the planet, and *inferior conjunction,* when the planet is between the earth and the sun. There is a third critical position, called *greatest elongation,* when the planet's angular distance from the sun as observed from the earth is a maximum; this occurs when the line joining the earth and the planet is tangent to the planet's orbit. The term *conjunction* is also used to indicate the time when two superior planets have the same celestial longitude; that is, when they are closest together. Jupiter and Mars, for example, are in conjunction when the plane passing through the earth, Jupiter, and Mars is perpendicular to the ecliptic, or when the three planets are most nearly in a straight line.

Planetoids.—See *Asteroids.*

Pole.—One of the ends of the axis of a sphere. The earth's axis passes through the earth's surface at the North and South poles and touches the celestial sphere at the north and south celestial poles.

Precession.—The movement of the equinoxes around the ecliptic and the conical motion of the earth's axis. Discovered by Hipparchus (c. 160–125 B.C.) precession was mathematically explained by Isaac Newton as being caused by the unequal attraction of the sun and moon on the earth's equatorial bulge. The equinoxes take about 215,800 years to complete their revolution about the ecliptic in a retrograde (west to east) direction. Because of precession, in the year 15,000 the north celestial pole will no longer be marked by Polaris, but will instead be near the boundary line between the constellations Lyra and Hercules, about halfway between the bright stars Vega and Gamma Draconis. The angular rate of precession is 50.26 seconds.

Prominences.—See *Chromosphere.*

Protoplanet.—Original ball of matter from which a planet was formed.

Ptolemaic system. — See *Geocentric system.*

Quadrature.—See *Planetary configurations.*

Quasar (*quasi-stellar radio source*).—A brightly shining body that resembles a star, but is so distant that it must be hundreds of billions of times brighter than the brightest star. The first quasars were identified early in 1963 by means of the radio waves they emit; since then, about a dozen others have been located, some of which pulsate like variable stars.

Radiant energy.—Electromagnetic radiation, such as heat, light, radio waves, X rays, and gamma rays, given off by the sun and other stars.

Radio star.—A particular region in the Milky Way or the space beyond from which radio waves are emanating. Only a few of these regions have been identified as containing celestial bodies, as the Crab nebula.

Radio telescope.—A device for detecting radio waves coming to the earth from outer space. The usual structure for the device is that of a large parabolic reflector (a bowl shape, like that of a radar antenna) that can be rotated so as to gather radio signals from any section of the heavens, and that focuses the incoming signals on an aerial at its center. The largest radio telescope in the world, located at Jodrell Bank, England, and operated by the University of Manchester, has a parabolic reflector with a 250-foot diameter.

Regression.—The clockwise (west to east) motion of the vernal and autumnal equinoxes.

Retrograde motion.—The apparent motion of a planet from west to east among the stars, caused by a combination of its true motion with that of the earth. It was this retrograde motion that necessitated the postulation of epicycles in the geocentric, or Ptolemaic, system.

Reversing layer.—See *Chromosphere.*

Revolution.—Orbital motion of a planet or satellite about its primary.

Right ascension.—See *Equator system.*

Roche's limit.—The closest point to which a satellite can approach its primary without the tidal effects produced by the planet's gravitational field pulling the satellite apart. Roche's value is 2.44 times the planet's diameter.

Rotation.—The turning of a celestial body on its axis; for example, the turning of the earth on its polar axis.

Saros.—See *Eclipse.*

Scintillation (*twinkling*).—The irregular variation in the brightness of a star, caused by variations in the density of different layers of the earth's atmosphere. As a ray of light passes through the air, it is bent or refracted. The amount of this refraction depends upon the density of the air. Because the air at different levels is at different temperatures, and the successive layers of air are being continuously shifted by winds, the slight distortions of a ray of light are different from instant to instant.

Seasons.—The four divisions of the tropical year that begin when the sun reaches specific points on the ecliptic. *Spring* begins when the sun reaches the vernal equinox and lasts for 92 days 20 hours 12 minutes. *Summer* begins when the sun reaches the summer solstice, and lasts for 93 days 14 hours 24 minutes. *Autumn* begins when the sun reaches the autumnal equinox, and lasts for 89 days 18 hours 42 minutes. *Winter* begins when the sun reaches the winter solstice, and lasts 89 days 30 minutes.

Shooting star.—See *Meteor.*

Solar apex.—See *Apex of the sun's way.*

Solar wind.—A stream of hot gas, caused by the expanding corona of the sun, that travels at speeds of about 999,000 miles per hour. It is composed of hydrogen and is responsible for the outer portions of the Van Allen radiation belts around the earth, for auroras in the earth's atmosphere, and for terrestrial magnetic storms. It also blows the tails of comets away from the sun.

Solstices.—The two points on the ecliptic when the sun is at its maximum distance from the celestial equator. At the summer solstice (for the Northern Hemisphere) the sun's declination is 23½° south of the celestial equator.

Sunspot.—A dark area representing a cool region in the sun's photosphere. Sunspots vary greatly in size and shape, the largest ones covering more than 100,000 miles of the sun's surface. Each spot has a dark center, called the *umbra,* that has a less dark region, called the *penumbra,* around it. Sunspots start as small points, develop rapidly, and last for periods ranging from a few hours to several months. A *sunspot cycle,* which lasts for an average of 11.1 years, is the period during which the number of sunspots builds up from a minimum to a maximum and then returns to a minimum. The time from a maximum to a minimum, about 6.5 years, is longer than the time from a minimum to a maximum—4.6 years.

Superior planet.—See *Planetary configurations.*

Telescope.—An optical instrument for making distant objects appear closer. An astronomical telescope consists of a system of lenses and mirrors that brings the light to a focus, and the image formed at the focus is then magnified. Almost all astronomical telescopes are equipped with cameras so that pictures may be taken to provide permanent records of observations. In a *reflecting telescope,* such as a *Newtonian* or *Cassegrain reflector,* the light gathered by the telescope is reflected from a polished surface or mirror to form an image that is then magnified by an eyepiece. In a *refracting telescope,* which was the first form of the telescope, no mirror is used. A *Schmidt telescope* is a special type of reflecting telescope that records large areas of the sky in one wide-angle photograph.

Terminator.—Line separating the illumined side of a planet or satellite.

Tides.—The alternate rises and falls of the earth's waters, caused by the gravitational attractions of the moon and, to a lesser extent, of the sun.

Time measurement.—A means of determining the extent of the duration of an event or a series of events. Time is measured by the hour angle (15° equaling one hour) of a particular point on the celestial sphere with respect to the observer. *Apparent solar time* is the westward hour angle of the sun measured from the observer's celestial meridian. *Apparent noon* is the moment when the sun is on the observer's celestial meridian. *Mean solar time* is the westward hour angle of the mean sun measured from the observer's celestial meridian. *Sidereal time* is the westward hour angle of the vernal equinox measured from the observer's celestial meridian. *Standard time* is the mean solar time legally adopted by a city or country for the major portion of the year. The meridian of longitude through Greenwich, England, is the *standard meridian,* and each meridian west of Greenwich that is a multiple of 15° has a standard time the same multiple of one hour earlier than that of Greenwich; thus, the meridian 60° west of Greenwich has a standard time 4 hours (4 × 15°) earlier than that of Greenwich. *Universal time* is the mean solar time of the Greenwich meridian. The *equation of time* is the difference at any instant between apparent and mean solar time.

Transit.—The apparent passage of a celestial body across a line on the celestial sphere or across any other celestial body as seen from the earth. A star is in transit when it crosses a celestial meridian. Mercury and Venus are in transit when they pass across the face of the sun. A satellite, such as one of Jupiter's moons, is in transit when it passes across the face of its primary.

Vernal equinox.—See *Equinox.*

Year.—Any one of several divisions of time. A *tropical year* is the time elapsing between two successive passages of the earth through the vernal equinox; it is equal to 365.242196 days, and is the year on which the calendar is based. A *calendar year* has 365 mean solar days. A *leap year* has 366 mean solar days, and is effected by adding an extra day to the calendar year when the year number is divisible by 4, such as 1968; a leap year is necessary to prevent the calendar from falling behind by one full day every fourth year. An *anomalistic year* is the time elapsing between two successive passages of the earth through the point of perihelion in its orbit; it is equal to 365.25964 days. An *ecliptic year* is the time elapsing between two successive passages of the sun through the ascending node of the moon's orbit; it is equal to 346.62003 days. A *sidereal year* is the time elapsing between two successive passages of the earth through a fixed point in its orbit; it is obtained by drawing a line from the center of the sun to a fixed point on the celestial sphere; it is equal to 365.25636 days.

Zenith.—A point on the celestial sphere directly above the observer and directly opposite the nadir. (See *Celestial sphere.*)

Zodiac.—A zone extending about 9° on either side of the ecliptic that contains the orbits of the sun, moon, and all the planets except Pluto. The ancient Greeks divided the zodiac into 12 equal divisions of 30°, called the *signs of the zodiac.* Each of these signs was named for the principal constellation it contained. The vernal equinox was a convenient reference point for dividing the zodiac. Therefore, the Greek astronomer Hipparchus (second century B.C.), who named the signs of the zodiac systematically, laid off 12 segments of 30°, counting eastward from the vernal equinox. The segments coincided with the following 12 constellations:

♈ Aries (Ram)	♎ Libra (Balance)
♉ Taurus (Bull)	♏ Scorpio (Scorpion)
♊ Gemini (Twins)	♐ Sagittarius (Archer)
♋ Cancer (Crab)	♑ Capricornus (Goat)
♌ Leo (Lion)	♒ Aquarius (Water Bearer)
♍ Virgo (Virgin)	♓ Pisces (Fishes)

These 12 constellations gave the names to the 12 signs of the zodiac.

Zodiacal light.—A faint illumination of the evening sky extending along the ecliptic. It is caused by the scattering of sunlight by meteoric dust in the plane of the solar system.

BIBLIOGRAPHY

Baker, Robert Horace. *Astronomy.* D. Van Nostrand Co., Inc., 1964.

Chamberlain, Joseph Miles. *Planets, Stars and Space.* Creative Educational Society, 1962.

DeGani, Meir H. *Astronomy Made Simple.* Doubleday & Co., Inc., 1963.

Gamow, George. *The Birth and Death of the Sun.* The Viking Press, Inc., 1964.

Hoyle, Fred. *Frontiers of Astronomy.* Harper & Bros., 1955.

Hynch, Allen. *Challenge of the Universe.* Scholastic Press, 1962.

Ley, Willy. *Watchers of the Skies: An Informal History of Astronomy from Babylon to the Space Age.* The Viking Press, Inc., 1963.

Moore, Patrick. *The Picture History of Astronomy.* Grosset & Dunlap, Inc., Publishers, 1962.

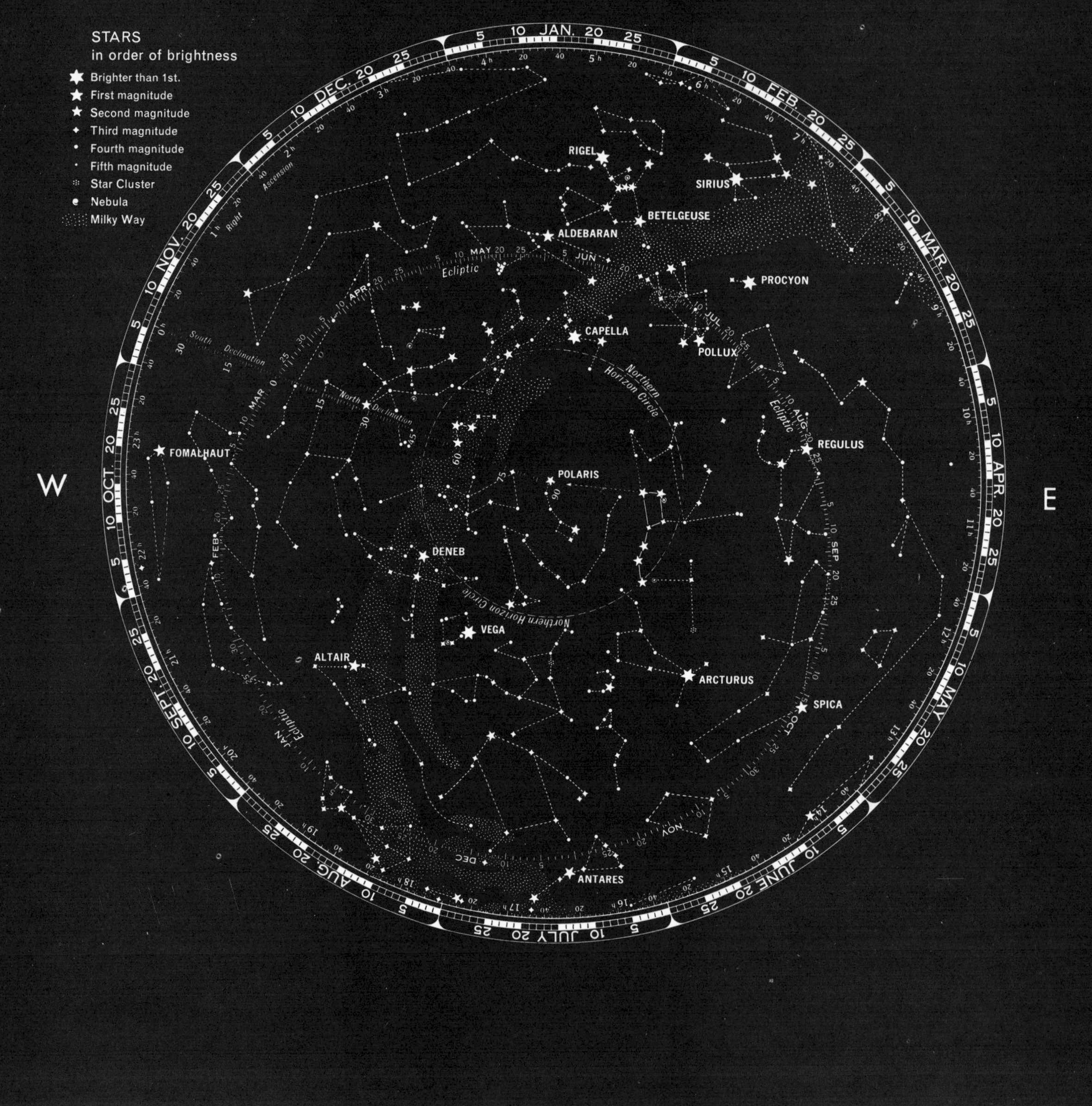

STAR CHART of the Northern skies can be used by anyone living in North America. The approximate position of any star at any time can be obtained from the chart. Face north and rotate the book until the current date is at the top of the chart. The stars and constellations in the upper two-thirds of the chart are those visible in the sky at about 9 P.M. The star nearly at the center of the chart is Polaris, the North Star. The stars within the Northern Horizon Circle correspond to the stars that rotate counter-clockwise around the North Star, and are always above the horizon for anyone living at 40° North Latitude. The stars to the right of center will be visible in the eastern sky; those to the left of center will be visible in the western sky. Stars close to the upper edges of the chart will be close to the southern horizon. To visualize the positions of the stars directly overhead and in the southern sky, hold the star chart with the current date pointing south. Some stars will appear brighter than others. The Order of Brightness guide in the upper left hand corner will aid in identification. Stars close to the horizon may be obscured by the surface haze.

Scope.—*Astronautics* encompasses all human activities in space—unmanned satellites circling the earth, space probes sent to another planet, manned space capsules, and the piloted spaceships to come. The word was coined by the French aviation pioneer Robert Esnault-Pelterie as a title for his book *l'Astronautique,* printed in Paris in 1930. It is derived from the Greek roots *aster,* meaning 'star,' and *nautes,* meaning 'sailor.' An astronaut, therefore, is literally a "sailor to the stars." The Russian term *cosmonaut* is derived from the Greek word *kosmos,* which actually means 'well-ordered' but is used to mean the universe—the well-ordered universe as distinct from chaos.

Early History.—Although astronautics has developed as a science only recently, it has a considerable history; the early part consists of philosophical and scientific speculations on the nature of other worlds.

These speculations began well before the birth of Christ. The classical philosophical school of the Pythagoreans believed the moon to be a solid mass, like the earth, and inhabited. While other classical philosophers disagreed with the Pythagoreans on practically every point, they did agree that the moon was a solid body with mountains and valleys.

LITERARY BACKGROUND.—The first man to devote an entire book to the moon was the Greek biographer Plutarch (c. 46–c. 120). In his work *De facie in orbe lunae* ("On the Face That Can Be Seen in the Orb of the Moon") Plutarch stated that the moon was inhabited, although not by people; he had either "demons" or the souls of the dead in mind. Since the philosophers were in general agreement that the moon was a solid body, and since it did not occur to anyone that a habitable body might not be inhabited, it was only natural to speculate how such a body might be reached. The first speculation known was written forty years after Plutarch's death by the Greek satirist Lucian of Samosata. While Lucian used the opportunity to poke fun at everything in reach of his stylus, he did express a wish to reach the moon.

Some 1,400 years later an English bishop, Francis Godwin, wrote a book in which a number of huge birds, called *gansas,* carry the hero to the moon. In the same year, 1638, another English bishop, John Wilkins, published a serious philosophical discussion on the possibility of life on the moon. Ten years later, Wilkins published a second edition in which he enumerated the possibilities of reaching the moon.

While these early space-flight fantasies are properly catalogued under the heading of literature rather than science, they did keep the idea of traveling into space—specifically, to the moon—alive. A number of ideas now in the process of being realized first appeared in works of fiction.

SCIENTIFIC BACKGROUND.—One of the most important contributions in the field of astronautics was made by the German astronomer Johannes Kepler (1571–1630). It is interesting that Kepler had also written a moon-travel fantasy, *Somnium* ("Sleep"), in which Plutarch's "demons" carry astronomers across the bridge of the earth's shadow during a lunar eclipse.

In his work *Astronomia nova* ("The New Astronomy," 1609), Kepler stated that the true shape of the orbit of a planet around the sun is an ellipse and that the position of the sun is not at the center of the ellipse but at one of its two focal points. In this same work, he also stated that a planet moves more rapidly when it is in the section of its orbit that is closer to the sun, and that it moves more slowly when in the section of its orbit that is farthest from the sun. Looking for a mathematical relationship governing this change in orbital velocity, Kepler imagined a line from the sun to the planet that he called the *radius vector.*

He then saw that "the radius vector sweeps over equal areas in equal times." This law permitted the calculation of both past and future positions of a planet once the size of its orbit and its degree of elongation (*eccentricity*) were known. Kepler called the orbit point closest to the sun the *perihelion* (from the Greek words *helios,* meaning 'sun,' and *peri,* meaning in this case 'to go around'). Of course, on each orbit there also was a point where the planet was farthest from the sun; Kepler (using the Greek word *apo,* which means 'to go away' or 'to be away') called this the *aphelion.*

Kepler also coined the term *satellite.* Before his time, the only known satellite was the moon. However, during Kepler's lifetime, Galileo Galilei (1564–1642) discovered the four major moons of Jupiter. Kepler, in a letter to Galileo, suggested a general term for planetary attendants, coining "satellite" from the Greek *satellos,* 'attendant.'

Kepler had shown *how* the planets move. Sir Isaac Newton (1642–1727) showed *why* they move. By introducing the two concepts of *universal gravitation* (any body in the universe attracts every other, although very weakly if the distance is great) and *inertia* (any body remains in its state of rest or of motion unless influenced by an external force), Newton gave astronomers the laws that had escaped scientists for centuries. In explaining how the moon moves around the earth, in his *Mathematical Principles of Natural Philosophy,* Newton even developed the theory of artificial satellites.

In this work, Newton used a diagram showing the earth with a high imaginary mountain that had a horizontally mounted gun on its top. The principle of satellite motion is really quite simple: if the gun first fires its projectile with a certain velocity, and impact on the ground takes place a certain distance from the mountain then a projectile fired with a higher muzzle velocity will have its impact point a greater distance away. Newton explained this by showing that the earth's gravitational pull re-

NEW YORK PUBLIC LIBRARY

JULES VERNE envisioned a moon trip in 1865. Capsules made up the projectile.

BROWN BROTHERS

JOHANNES KEPLER proved that planets travel in an elliptical orbit around the sun.

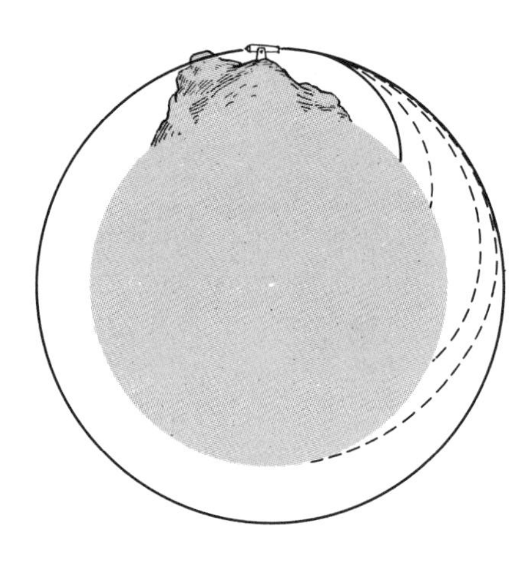

NEWTON'S DIAGRAM showing an earth orbit.

mains constant. The faster projectile will cover a greater horizontal distance per second. The result is that the curvature of the trajectory grows shallower as the muzzle velocity increases. If the velocity of the projectile is great enough, the curvature of its trajectory will become so shallow that it corresponds to the curvature of the ground below. It thus orbits the earth.

However, Kepler and Newton developed only the basic principles of astronautics. Over two centuries of scientific detail had to accumulate before any astronautical applications could come from their work.

It is not known who was the first to calculate the velocity required for an object to stay in orbit around the earth. However, any mathematician could have done it after Karl Friedrich Gauss (1777–1855) had published his work on the calculation of orbits early in the nineteenth century. Whoever performed the calculation first must have jumped to the conclusion that an orbit could never be achieved. If a satellite could be orbited at the earth's surface, it would have to move at the rate of almost 5 miles per second to compensate for the earth's gravitational pull. Even at a point a thousand miles from the earth, the required velocity is 4.4 miles per second.

Early Rocketry.—The name of the inventor of the first rocket is not known. However, it is certain that he was a Chinese, for the first mention of rockets appears in ancient Chinese writings. In a Chinese work, in which the siege of the city of Kai-fung-fu (1232) is described, "arrows of flying fire" are mentioned. The invention of these rockets, similar to those now called *fireworks rockets,* must have taken place sometime between 1200 and 1230. The chronicles of the city of Cologne mention a rocket in 1258, and Arab scientists of the same period referred to rockets by the name of *alsichem alkhatai,* which means "Chinese arrows." For about two centuries rockets were used as weapons of war, but by 1400 the recently invented guns came into general use and replaced them. However, there was one military target against which rockets still found an application—the large, flammable sails and tarred rigging of ships. For this reason rockets continued to be used as a weapon on the high seas. On land the rocket assumed a far more peaceful role in the mid-sixteenth century; it then became the mainstay of large, often expensive fireworks displays.

Nevertheless, a number of military men and private experimenters questioned whether or not improved rockets could perform military duties in competition with artillery. Yet, in spite of some interesting demonstrations, they failed to convince their respective governments.

■**CONGREVE ROCKETS.**—One man, the English army captain (later general) William Congreve (1772–1828), did succeed in returning the rocket to the battlefield. Reading about British defeats caused by rocket fire during the Indian campaigns—especially during the two battles of Seringapatam in 1792 and 1799—he decided to investigate the existing fireworks rockets for their possible usefulness. Within two years he had improved the accuracy and range of rockets, primarily by increasing their size. The largest fireworks rockets he had been able to buy in London had had a diameter of 1½ inches and a length of not much more than a foot. The early *Congreve rockets* had a diameter of 3½ inches and a length of 40½ inches. The front of the sheet-iron rocket case held the warhead, which contained an incendiary charge. The maximum range of the Congreve rockets was 3,000 yards—1,000 yards more than the maximum range of the heavy 10-inch mortar then used by the English army. Congreve rockets were used in various battles during the reign of Napoleon. They also were employed when the British bombarded Fort McHenry during the War of 1812 and are referred to in the American national anthem. Congreve rockets were also used when British troops burned the White House. The rockets were not fired into the building, but were wedged between the beams, then ignited.

Rockets were not used to a great extent because there was no production machinery for them; they had to be made individually by hand. Also, the propulsive charge was unreliable: rockets that had been stored for more than six months were likely to explode on ignition or to fly erratically if they worked at all.

Beginnings of Scientific Astronautics.—Although a great many rockets were produced during the nineteenth century, for the most part they were considered toys and were neglected by science. No mathematician sat down to evolve a theory of rocket flight; no physicist took the trouble to investigate the way in which they operated; and no engineer thought of improving them.

When rockets finally were investigated scientifically, it was because Sir Isaac Newton implied that they would operate in space. His third law stated that reaction is equal (in power) but opposite (in direction) to action, or the actions of two bodies are equal (in power) but point in opposite directions. This law was valid without regard to the surroundings—a rocket would operate whether it was surrounded by water or air, or was in a vacuum. Thus, it could be a means for traveling through space.

This concept occurred independently to several people during the last decade of the nineteenth century. In 1891, an inventor in Berlin, Hermann Ganswindt, lectured on the possibility of space travel and even drew a design for a spaceship.

Another German, mathematics professor Kurd Lasswitz, used his spare time to develop a theory of space travel. He suggested placing space stations near the earth from which a spaceship could be directed into an orbit that would intercept the orbit of another planet. He embodied his ideas in a best-selling novel *Auf zwei Planeten* ("On Two Planets"), which he published in 1897. Although Lasswitz even gave examples of how to perform the calculations necessary to attain the proper orbits, his readers thought his book only a very interesting novel; no one believed that he had propounded valid theories.

Another early space prophet was the Russian high school teacher Konstantin Eduardovitch Ziolkovsky (1857–1935). He also began by writing a space-travel story (*On the Moon,* 1893), but then started work on a nonfictional treatise on space travel. It was finished in 1898 but not published until 1903, under the title *Exploration of Planetary Space by Means of Reaction-powered Equipment.* Although outdated, a few observations from this treatise are worth mentioning. Ziolkovsky introduced the Russian word *sputnik* ('travel companion') in the sense of artificial satellite. He also listed possible fuels for the superrockets he advocated. One fuel listed was kerosene, which is used today for

ESTHER C. GODDARD

large rockets in both the Soviet Union and the United States.

The next trio of space-travel prophets, an American and two Germans, came to the fore two decades later. The American was Robert H. Goddard (1882–1945), a professor of physics. In 1914, he obtained two U.S. patents, one mentioning the possibility of using liquid fuels for rocket propulsion, the other dealing mainly with the concept of the step rocket —a smaller rocket carried by a larger one. During World War I, Goddard measured the thrust of various types of propellant powders. He also invented a device for demonstrating that a rocket will work in a vacuum. In addition, he calculated a shot to the moon.

The two German authors were Hermann Oberth (1894–), whose book *Die Rakete zu den Planetenräumen* ("The Rocket into Interplanetary Space") was published in 1923, and Walter Hohmann (1880–1944), whose book *Die Erreichbarkeit der Himmelskörper* ("The Attainability of the Heavenly Bodies") was published in 1925. Oberth, like Goddard, mentioned high-altitude research rockets, although he went further than Goddard and predicted piloted rockets capable of orbiting the earth. Hohmann, however, assumed the existence of a rocket-powered spaceship and calculated which orbits would have to be traveled in order to reach the nearer planets; he also calculated the amount of fuel required for each maneuver, the duration of the trips, and the amount of food, water, and oxygen required for the crew. Thus, Hohmann's work might be called the most advanced of the three.

■FIRST LIQUID-FUEL ROCKETS.—Although Goddard wrote about solid fuels, mentioning liquid fuels only casually, and Oberth stressed the use of liquid fuels for rocket propulsion, the first man actually to build a liquid-fuel rocket was Goddard. His first rocket made its flight from a farm near Auburn, Massachusetts, on March 16, 1926. Quoting from Goddard's report, "The rocket traveled a distance of 184 feet in 2.5 seconds, as timed by stop watch." The Soviet Union's first liquid-fuel rocket, constructed by a group of young enthusiasts led by Anatoli Blagonravov, made its initial test flight during the spring of 1932. Apparently it was erratic, for no details have ever been revealed.

Mass Ratio.—The theory of rocket propulsion, as worked out by Goddard and Oberth, showed, among other things, that it is important for the rocket to keep burning for a long period of time. Fireworks, the only rockets in existence when the theory was developed, had burning times of half a second or less and consequently could not be propelled very high. It was shown by comparatively simple mathematics that if rocket *A* had a burning time of 25 seconds and could reach a certain altitude, rocket *B* would go twice as high if it had enough fuel to burn for another 5 seconds. It became obvious that a rocket should carry as much fuel as possible yet be as light as possible.

If the mass of the fuel carried is designated as M_F, the mass of the rocket itself as M_S, and the payload (a package of scientific instruments, for example) as M_P, the take-off mass of the rocket is $M_F + M_S + M_P$. After all the fuel is consumed, there is a remaining mass of $M_S + M_P$. Oberth called the take-off mass M_o, the remaining mass M_1, and the ratio M_o/M_1 the *mass ratio* of the rocket.

Oberth also showed that if the mass ratio is equal to *e* (2.71828 . . . , the basis of natural logarithms) the rocket's velocity will equal its *exhaust velocity,* provided that the rocket is not restricted by either air resistance or a gravitational field. For example, if a rocket of the mass ratio *e* were in orbit around the sun at an appreciable distance from the earth, and if the exhaust velocity of the fuel used were equal to 6,000 feet per second, the rocket, after consuming its fuel, would move 6,000 feet per second faster than its velocity prior to firing.

If the rocket had to overcome air resistance and to climb against the gravitational pull of the earth, the mass ratio would have to be higher to produce the same result. The German V-2 rocket had a take-off weight of 28,200 pounds, which was distributed as follows: M_F = 19,400 pounds, M_P = 2,200 pounds, and M_s = 6,600 pounds. The remaining mass, M_1, therefore was 8,800 pounds, and the mass ratio, M_o/M_1, equaled 3.2. The exhaust velocity at sea level was 6,560 feet per second, but the maximum velocity was 5,575 feet per second, almost 1,000 feet per second less than the exhaust velocity. In space, far from the earth, the velocity would have been about 7,700 feet per second. The difference between this theoretical velocity and the actual velocity—2,125 feet per second—was due to the earth's gravitational pull and to air resistance. If the same rocket could have been built with the structure weighing only half of what it did, the mass ratio would have been about 5 and the theoretical velocity would have been nearly 11,000 feet per second. In reality, however, a mass ratio of 5 is most difficult to realize and could be surpassed only in exceptional cases. Even if a rocket with a mass ratio of 6 could be built, it could not reach orbital velocity with the fuel that is now available.

Staging.—One method of attaining an orbit is the step principle, which was stressed by both Goddard and Oberth. To illustrate the principle, assume that the payload of a rocket is another, smaller rocket. When the fuel supply of the first rocket has been consumed, the second stage is ignited. The second rocket can achieve a velocity that no single rocket using the same fuel could ever attain. To use the example of the V-2 once more, if the payload (warhead) of 2,200 pounds had been a 2,200-pound rocket with the same mass ratio as the first stage, then the final velocity of the second stage would have been twice that of the first stage (actually

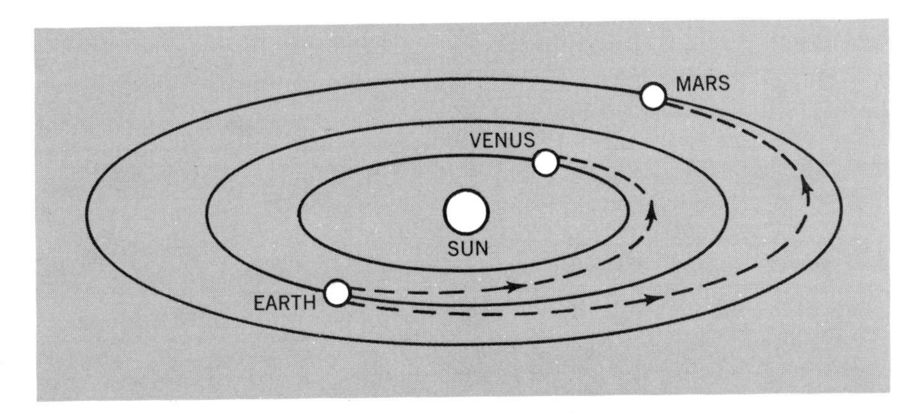

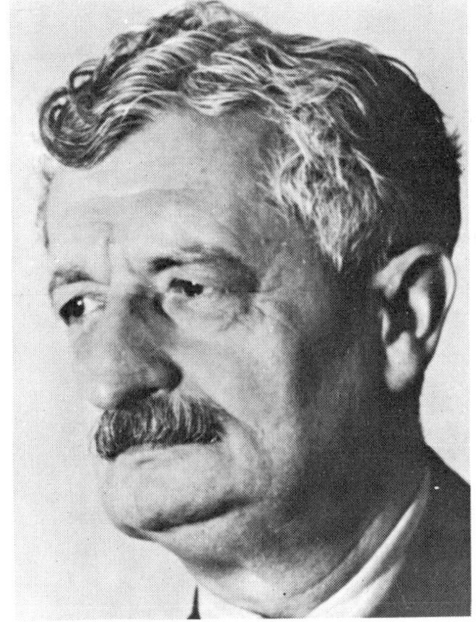

SPACE PIONEERS of the twentieth century used the scientific foundation developed by Newton, Kepler, and Gauss to build the vehicles and perfect a technology to reach into space. Robert Goddard (*left*) developed rocket fuels and flight-tested several successful vehicles; and Hermann Oberth (*right*) calculated the amount of fuel required for a space flight as well as the velocities needed to attain a proper orbit. In 1925, a German, Walter Hohmann, calculated the orbits required to reach the nearest planets. He drew the trajectories (*above*) for journeys to Venus or Mars.

AMERICAN INSTITUTE OF AERONAUTICS AND ASTRONAUTICS

a little more, because the second stage would not be ignited until it attained an altitude of 20 miles, where air resistance is negligible).

The final velocity of the second stage of a two-stage rocket is equal to that of another rocket whose mass ratio is equal to the product of the mass ratios of both stages. That is, if the first stage has a mass ratio of 3 and the upper stage has a mass ratio of 4, then the velocity attained by the upper stage is the same as that which would be attained by a single rocket with a mass ratio of 12. While it is impossible to build a single rocket with a mass ratio of 12, it is entirely feasible to build rockets with mass ratios of 3 and 4. To reach the high velocities necessary to escape the earth, rockets that have two or three stages are needed. Rockets with more than three stages are, as a rule, impractical, although they are sometimes built.

Specific Impulse.—There are two possible yardsticks for measuring the power of a fuel: the theoretical exhaust velocity and the specific impulse. As an example of the former, one could state that ethyl alcohol burned with pure oxygen produces an exhaust velocity of 13,800 feet per second; however, in reality it is impossible for a fuel to reach the theoretical exhaust velocity because there are always losses of some kind. *Specific impulse,* therefore, provides a better yardstick: It measures the amount of thrust (in pounds) produced by burning a unit amount of fuel for one second. To calculate the specific impulse, the thrust of rocket engine, expressed in pounds, is divided by the fuel consumption per second. If a rocket engine produces 80,000 pounds of thrust and consumes 280 pounds of fuel per second, it has a specific impulse of 80,000/280, or 285.7 seconds. If the same rocket engine were designed for a less powerful fuel and had to consume 400 pounds of fuel per second to produce a thrust of 80,000 pounds, the specific impulse would be 200 seconds. The specific impulse of modern rockets is higher than 200 seconds, and for the upper stages, surpasses 300 seconds.

Fuel Systems.—There are three basic types of fuel systems: all-liquid, all-solid, and hybrid. In addition, each type has subdivisions.

■ **LIQUID FUELS.**—In a liquid-fuel rocket, the *bipropellant system* is the rule. In this subdivision, the rocket uses two different liquids housed in separate tanks. One of the liquids is the fuel proper; the other, usually liquefied oxygen (called LOX), is the oxidizer. The two most common fuels burned with oxygen are refined kerosene, which is used in the Atlas rocket, and ethyl alcohol, which was used in the V-2, the Viking, and the first stage of the Vanguard rocket. Bipropellant systems using liquid oxygen have the highest specific impulse. Unfortunately, they have a disadvantage in that liquid oxygen has a very low boiling point and therefore evaporates quickly. Thus, once the rocket has been fueled, it must be fired within half an hour.

The most powerful fuel combination of the all-liquid bipropellant systems is hydrogen burned with oxygen. A hydrogen-oxygen rocket, called *Centaur,* was first tested in 1963, and is now a common top stage.

Another variety of the bipropellant system uses RFNA (red fuming nitric acid) as the oxidizer. RFNA is rich in oxygen and does not evaporate easily. However, it is a very corrosive liquid, and special precautions must be taken when handling it. During early experimentation with RFNA, it was discovered that some fuels, when brought in contact with it, spontaneously burst into flame. This type of fuel-oxidizer combination is known as a *hypergolic* fuel. The first hypergolic fuel to be used was a combination of nitric acid and aniline; it was this fuel that powered the small American WAC-Corporal rocket. Unsymmetrical dimethyl hydrazine, which has proved more powerful and reliable than aniline, was used in combination with RFNA in the second stage of the Vanguard rocket.

Hypergolic combinations are advantageous in that they can be left on the launching pad for some time without loss of fuel and are easy to ignite. Many rocket engineers, however, shy away from hypergols, for if a small leak develops somewhere, it can lead to a disastrous fire.

Another liquid-fuel system is the *monopropellant system.* Here, one fuel tank contains a liquid mixture of fuel and oxidizer. A monopropellant needs ignition only when it enters the combustion chamber. The prime danger is that the flame may travel back through the fuel pipes and ignite the fuel supply in the tank, which would then explode. To date, there is no safe monopropellant.

■ **SOLID PROPELLANTS.**—Solid propellants consist of a mixture of fuel and an oxidizer; thus, they are similar to the propulsive charges of the earliest rockets. The first modern solid rocket fuels, known as *double-base powders,* were developed by Alfred Nobel (1833–1896). Nobel began with nitroglycerine, a powerful but highly unreliable liquid explosive. By soaking up the nitroglycerine in *diatomaceous earth* (a special kind of earth composed of the remains of certain microorganisms, also called *diatomite* and *kieselguhr*), he obtained a solid explosive that was nearly as powerful as nitroglycerine but relatively safe to handle. Nobel called it *dynamite.* Then he began experimenting with nitroglycerine soaked up by guncotton, obtaining a gelatinous substance even more powerful than nitroglycerine itself. He called this *blasting gelatin;* despite its power, it never achieved great use because of the dangers involved in handling it.

Nobel experimented with a number of additives to make his blasting gelatin safer to handle and more solid. The resulting explosives received such names as *Ballistite* and *Cordite.* The overall term, *double-base powders,* referred to the fact that both basic ingredients, nitroglycerine and guncotton, are explosives. Virtually all the military bombard-

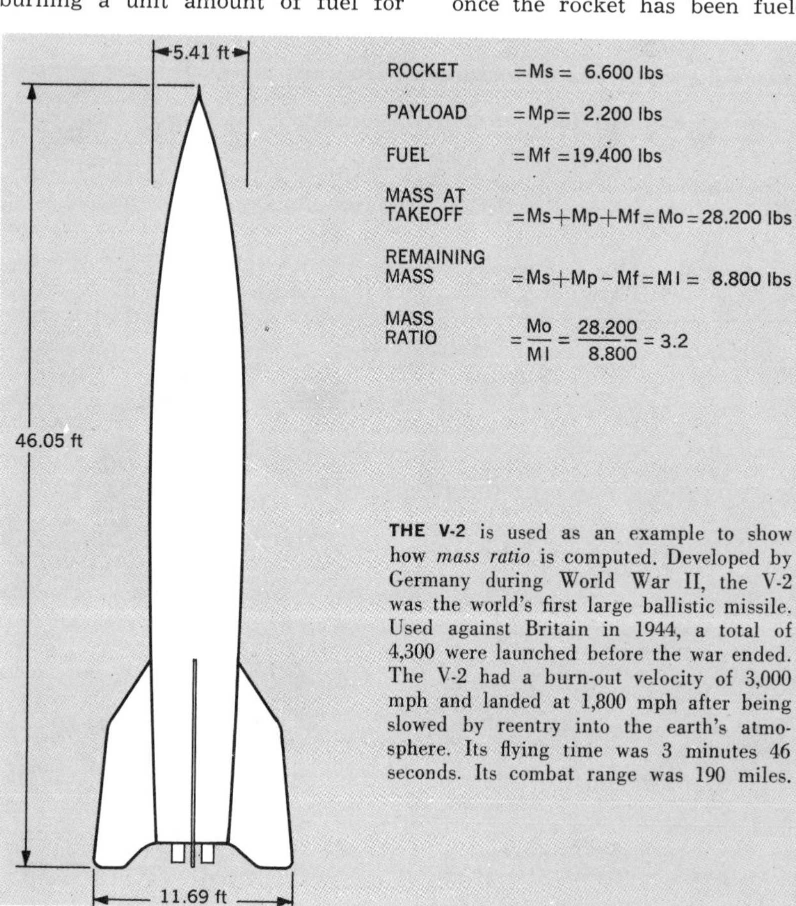

THE V-2 is used as an example to show how *mass ratio* is computed. Developed by Germany during World War II, the V-2 was the world's first large ballistic missile. Used against Britain in 1944, a total of 4,300 were launched before the war ended. The V-2 had a burn-out velocity of 3,000 mph and landed at 1,800 mph after being slowed by reentry into the earth's atmosphere. Its flying time was 3 minutes 46 seconds. Its combat range was 190 miles.

U.S. ARMY

U.S. AIR FORCE

NASA

U.S. MISSILES. The Jupiter (*left*) and the Atlas (*center*) are operational. The Saturn (*right*) will carry a manned capsule to the moon.

ment rockets used by both sides in World War II had some form of Cordite or Ballistite propelling them.

But while the double-base powders performed military functions satisfactorily, it was soon found that they were not ideal rocket propellants. They were sensitive to temperature —on very cold days they would burn more slowly than on warm days—so obviously they could not be used where a high degree of accuracy was required. In addition, it also became clear that double-base grains could not be made very large, and thus would always burn too rapidly.

Double-base powders are still used for comparatively small military rockets, but new solid fuels have been developed for the larger missiles and research rockets. Details of composition and manufacture are, of course, secret; but the main ingredients of these modern solid fuels are synthetic rubber and an oxidizer (for example, ammonium perchlorate) kneaded into the rubber during the manufacturing process. Various additives, such as powdered metals, improve the reliability and sometimes increase the power of this mixture.

There are three main types of modern solid propellants. One is the *suspended grain,* which burns along its whole length; this is used in small military rockets. Because the grain burns along its whole length, one speaks of an *unrestricted burning charge* in these fuels. The other two types of fuels have propelling charges that fit tightly into the casing. They are called *wall-fitting charges,* or *restricted burning charges.* In one type the charge is just a tightly fitting cylinder of propellant that is ignited at one end and slowly burns until it is completely consumed. This type of construction, which has been nicknamed "cigarette burner," provides a comparatively low thrust for a fairly long time. It is rarely used. The common type of wall-fitting charge has a center hole that runs the length of the charge. Ignition takes place in this center hole; the charge burns from the inside outward.

■HYBRID SYSTEMS.—If a wall-fitting charge with a center hole consisted solely of the fuel part of the solid fuel, the fuel could burn only if oxygen were supplied to it. This could be done by blowing gaseous oxygen through the center hole from the front end. Such a system is called a *hybrid system,* and flight tests of such rockets were successful in 1968.

Atomic Rocket.—The *atomic rocket,* now under development as a part of a larger development project called Project Rover, differs from all other rockets in that no fuel is actually burned. In a normal rocket—whether the propellant is liquid, solid, or hybrid—fuel is burned to produce large quantities of combustion gases. These gases are then expelled through the exhaust nozzle to produce a reaction. In the atomic rocket, heat is supplied by an atomic reactor. Hydrogen gas from a storage tank, where it is stored in liquefied form, is then led through the atomic reactor, where it is heated. When heated, the gas expands and is expelled through the exhaust nozzle; however, the exhaust stream is unburned hydrogen.

The atomic rocket, when completed, is expected to show a specific impulse of about 325 seconds, which would make it far superior to any rocket that burns fuel. The ultimate goal in the development of atomic rockets is one that does not need liquid hydrogen but can operate on water. If that can be accomplished, a long step forward will have been taken. It is known, for example, that there is water on Mars. If atomic rockets that operate with water are developed, it would only be necessary to provide enough water for the trip to Mars; the water for the return trip could be obtained on Mars. Thus, the weight required of the rocket could be reduced.

Development of Rocket Engines. — In 1930 more than half a dozen groups of researchers were investigating the problems of the liquid-fuel rocket. There were, in the United States, Robert H. Goddard, who was then in New England but was soon to move to New Mexico; and two societies, the Cleveland Rocket Society in Cleveland and the American Rocket Society in New York. Both of these groups failed to accomplish a great deal, mainly for lack of resources. The Cleveland Rocket Society collapsed, but the American Rocket Society has grown into the largest rocket society in the world.

The remaining researchers were in Europe. In France, one of the pioneers of aviation, Robert Esnault-Pelterie, had established a research laboratory to try to build a high-altitude rocket. In Austria, Eugen Sänger, then professor of engineering at the University of Vienna, carried out a long series of rocket-motor tests. In Russia, an organization called GIRD (from the initial letters of the Russian words for "Group for the Investigation of Motion by Reaction") prepared the first Russian liquid-fuel rockets. In Germany the Society for Space Travel (*Verein für Raumschiffahrt*) actively conducted research.

All these researchers had one main problem on their hands: the building of the device that came to be called the rocket motor. In outline, this was a metal container with inlets for the fuel and the oxidizer at one end and an exhaust nozzle at the other. Even though the rocket motor was a completely new device, some knowledge from other fields of engineering could be utilized. The fuel inlets and the injection nozzles already existed and were even being mass-produced for diesel engines; unfortunately, these were the wrong size, but the detail problem had been solved.

It was also known how the exhaust nozzles had to look: they had to be of the so-called *convergent-divergent* type, which means that the nozzle first narrows to a point of smallest cross section (called the *throat*) and then flares out again. This shape is necessary because a gas flowing through a nozzle at a speed less than that of sound will move faster as the nozzle gets narrower. A gas moving at a speed greater than that of sound will slow down in a narrowing nozzle, but a supersonic gas stream will speed up in a flaring nozzle. The design of the rocket's exhaust nozzle therefore had to take these phenomena into consideration. The combustion gases coming from the chamber where the fuel had been burned would move fast, but not as fast as sound. Hence they would speed up as the nozzle narrowed and, if everything was of the right size, they would reach the speed of sound at the throat. They would then enter the flaring section of the nozzle and speed up more, so that they would leave the nozzle opening (called the *muzzle*) with a speed exceeding that of sound.

The main problem encountered in the construction of the rocket motor, however, was that the heat of burning was much greater than the melting point of any metal that could be used to build the motor. Hence, rocket motors had a tendency to develop two holes while burning: one near the center of the combustion chamber and one near the throat of the nozzle. For short burning times, say up to 1½ seconds, this could be avoided by making the metal quite thick. But for longer burning times, a cooling system of some kind was mandatory. At first a cooling jacket with cold water in it was tried, but the weight of the water was prohibitive. Then it was realized that a cold liquid was available: the fuel from the tank. Therefore, the fuel was no longer injected directly into the rocket, but instead went through the cooling jacket before entering the combustion chamber. The early rocket motors consisted of only a combustion chamber and an exhaust nozzle with a formfitting cooling jacket. As the motors grew larger, however, more than one fuel pipe and one oxygen pipe were needed, and at a still later date the fuel pump was attached to the rocket motor. The entire system —rocket motor, piping, pumps, and whatever auxiliary devices are needed—is called the *rocket engine*.

Progress in Germany.—By 1935 the number of research groups had shrunk. The Cleveland Rocket Society and the German Society for Space Travel had ceased to exist, Eugen Sänger was working in the field of aerodynamics, and the American Rocket Society, while still in existence, had stopped experimenting. There were now only one individual experimenter, Robert H. Goddard, and three active groups: a Russian group under Anatoli Blagonravov, a German army group under Captain (later General) Walter Dornberger and Wernher von Braun, and the small French group around Robert Esnault-Pelterie. Esnault-Pelterie never finished his high-altitude rocket because of the advent of World War II, but the other three had initial successes. In December 1934 the German army group fired two rockets, of a type they called *A-2*, to altitudes of about 6,500 feet. During the summer of 1935 Goddard fired several rockets, one of which went as high as 7,500 feet; and the Russian group made about a dozen shots, of which the highest came close to six miles.

Seven years later the picture had changed again. The Russians had stopped rocket experimentation altogether. Goddard, after a hiatus of a few years, worked on rockets as take-off aid for Navy flying boats. The German army group had progressed to the first large rocket.

After the initial success with the A-2 rockets, the Germans had built a larger rocket called *A-3* and had begun to make sketches for a much bigger one called *A-4*. It then became apparent that work on the A-4 could not be started without more experiments with the A-3. Therefore, before starting the A-4, they built still another type, called the *A-5*. In 1942 the first few A-4's were built. The new rocket was quite troublesome for some time—indeed, there were only two successful shots among the first fourteen. But the two successes proved that the A-4 could be made to work, and a somewhat lighter version of the A-4 became the rocket known as the *V-2*.

U.S. ARMY

A JUPITER intermediate-range ballistic missile is being assembled on a production line.

Postwar Developments.—Since all of the other nations had dropped out of the race, the field of large rockets was a German monopoly for nearly a decade. The V-2 had been such a long step forward that everybody else's plans had become obsolete. Hence it was only logical that, after the war, the V-2 should become the basis of the programs of other countries. The United States had developed only one medium-size liquid-fuel rocket named the *WAC-Corporal;* the work was done by a California group that owed its existence to Theodore von Kármán. Therefore, the rocket program of the United States began with about 70 V-2 rockets that had been captured in parts. The German research and planning staff, including Dornberger and Von Braun, was also brought to the United States.

The Russians had captured two V-2 rockets in working order—they had been earmarked for troop training—and, of course, the manuals that accompanied them. Contrary to popular impression, the Russians did not capture any of the important rocket scientists; a number of men from the original production line were held in Russia only temporarily. The Russians, therefore, began with a production line for V-2 rockets but proceeded to their own designs.

In the United States, the captured V-2 rockets were used up in a program that combined high-altitude research and the familiarization of American military men and scientists with large rockets. The first United States design, about half the size of the V-2, was a rocket called the *774,* in which a new idea for steering a rocket was tried. Instead of having graphite vanes inserted into the exhaust blast, as the Germans had done, the whole rocket motor of the 774 was suspended on gimbals so that it could be swiveled for balancing and steering. This idea was then incorporated into the Navy's Viking rockets, the first American design to become well known. In the meantime, *Project Bumper* had been carried out, consisting of eight V-2 rockets that had been equipped with second stages consisting of specially adapted WAC-Corporal rockets. On February 24, 1949, the second stage of Bumper 5 rose to an altitude of 250 miles, the first man-made device to climb beyond the atmosphere of the earth.

Since the Viking rockets were a Navy project, the Army asked Wernher von Braun to produce a new but bigger V-2-type rocket. This became known as the *Redstone,* after the Redstone Arsenal near Huntsville, Alabama, where it was developed. A number of Redstone rockets were set aside to test components for the next larger Army rocket, the Jupiter, which was to have a range of 1,500 miles. For this reason the Redstone rockets used in this test program were called *Jupiter-C,* the *C* standing for "components." At the same time the Air Force was developing still another 1,500-mile rocket, the *Thor.* Jupiter and Thor were to be intermediate-range missiles. The intercontinental missiles, *Atlas* and *Titan,* were developed later.

U.S. Space Programs.—When rocket development had reached the Redstone rocket, it became clear to the experts in the field that the experiment that Sir Isaac Newton had been able to perform only on paper could now be carried out in actuality. The Redstone itself did not have a great range, but it could carry a heavy payload. If staging were used, the Redstone could carry the upper stages needed to attain orbital velocity.

In 1954 *Project Orbiter* was formulated. It was to be a joint Army-Navy project, with the Army furnishing the rockets and the Navy supplying everything else. The rocket was to be a Redstone, with clusters of small solid-fuel rockets forming upper stages. The uppermost stage, a single, small, solid-fuel rocket, was to be placed into orbit. Project Orbiter could have put a small artificial satellite into orbit in the spring of 1956. However, in 1955 it was decided to discontinue Project Orbiter and proceed with a new project, *Project Vanguard.* The plan was to put a satellite into orbit in the autumn of 1957.

Vanguard was an entirely new rocket of three stages. The first-stage propellant was a liquid-fuel (alcohol and oxygen) rocket and was based on the experience gained with Project Viking. The second stage was to be a new rocket using hypergolic liquid fuels, and the third stage was to be a new solid-fuel rocket.

In 1956 the Russians announced that they, too, had a satellite project. Since Project Vanguard was already falling behind schedule—there were too many new things to be developed and not enough money had been appropriated for the test—the Army proposed to launch an artificial satellite by means of its new Jupiter-C rocket. However, the Secretary of Defense, Charles Wilson, forbade this.

Because of his order, the first artificial satellite to be orbited was the Russian *Sputnik I,* on October 4, 1957. This was followed by *Sputnik II* on November 3, 1957. At this point Neil McElroy, Wilson's successor as Secretary of Defense, ordered the Army to go ahead with the Jupiter-C program. *Explorer I,* first American satellite to orbit, was launched by a Jupiter-C on February 1, 1958.

Artificial Satellites. — The orbits into which artificial satellites can be and have been placed can be divided into two main categories, stable (permanent) and unstable (temporary). A *stable orbit* is one that is well outside the earth's atmosphere along its entire length. The first satellite to attain such an orbit was Vanguard I. Vanguard I will, therefore, continue to orbit the earth until it is artificially removed from orbit, possibly by a manned spaceship.

An *unstable orbit* is one whose perigee section is inside the earth's atmosphere. The process that takes place in this case is technically known as *orbital decay,* the end product of which is reentry and burn-up of the satellite. An explanation of orbital decay is that the satellite, in passing through the upper atmosphere, loses a little of its momentum because of *atmospheric drag* (the resistance of air molecules to the satellite's motion). Since it has lost momentum, the satellite cannot travel as far away from the earth as it did previously. Hence, the next apogee reached is somewhat closer to the earth than the preceding apogee. With each succeeding orbit, the apogee point slowly approaches the earth, thereby making the orbit more circular. Finally the apogee will be as close as the perigee, which means, of course, that the whole orbit is inside the fringes of the earth's atmosphere. Then the nearly circular orbit becomes a spiral; and when the satellite reaches denser layers of the atmosphere, it will burn up in the heat caused by compression of the air in its path.

The majority of the artificial satellites put into orbit were research satellites that were expected to be useful for only a limited period of time, say six months or less. It was desirable to put these satellites into unstable orbits so that they would not clutter up space when their useful life was over. There were some slipups in the early years of the space age—for example, Vanguard II and Vanguard III unnecessarily assumed stable orbits—but in general the early research satellites were placed into unstable orbits and have since reentered and burned up.

However, some satellites should be in stable orbits, namely those that are called the *working satellites.* The U.S. space program includes several types: satellites to aid navigation, such as *Transit;* satellites to aid weather forecasters, such as *Tiros* and *Nimbus;* and finally, communications satellites, such as *Echo, Telstar, Relay,* and *Syncom.* All these satellites must be in stable orbits to perform their functions.

Of the orbits around the earth, there is one that deserves special mention. A satellite in an orbit 22,300 miles above sea level will need 24 hours to circle the earth once. If this satellite is moving eastward over the equator, it will keep pace with the turning of the earth beneath it. It will, therefore, seem to hang motionless over one point of the equator. This type is a *synchronous* satellite, which means that it is moving "in time with" the turning earth.

NASA

A COMMUNICATIONS SATELLITE, such as the Relay (*right*), if put in the orbit shown above, would have an angular velocity identical to that of the earth. Thus, if satellite S_1 were above point A on the earth's surface, it would travel to point S'_1 while point A moved to A'. Three such synchronous satellites could cover the earth's surface, with the exception of a small area at the North and South poles.

NOVOSTI PRESS AGENCY

NASA

NASA

Manned Space Flights

Name of Astronaut	Name of Spacecraft	Date of Launching	Apogee (Miles)	Perigee (Miles)	Earth Orbits	Duration of Flight
Yuri A. Gagarin	Vostok I	Apr. 12, 1961	188	110	1	89 min.
Alan B. Shepard, Jr.	Freedom 7	May 5, 1961	117	....	......	15 min.
Virgil I. Grissom	Liberty Bell 7	July 21, 1961	118	....	......	15 min.
Gherman S. Titov	Vostok II	Aug. 6, 1961	160	110	17½	25 h. 18 min.
John H. Glenn, Jr.	Friendship 7	Feb. 20, 1962	162	92	3	4 h. 50 min.
M. Scott Carpenter	Aurora 7	May 24, 1962	167	99	3	4 h. 50 min.
Andrian G. Nikolayev	Vostok III	Aug. 11, 1962	156	113	64	94 h. 25 min.
Pavel R. Popovich	Vostok IV	Aug. 12, 1962	157	112	48	71 h. 3 min.
Walter M. Schirra, Jr.	Sigma 7	Oct. 3, 1962	176	100	5¾	9 h. 14 min.
L. Gordon Cooper, Jr.	Faith 7	May 15, 1963	166	100	22	34 h. 20 min.
Valery Bykovsky	Vostok V	June 14, 1963	140	109	81	119 h. 6 min.
Valentina A. Tereshkova	Vostok VI	June 16, 1963	144	112	48	70 h. 50 min.
Konstantin Feoktistov, Boris B. Yegorov, Vladimir Komarov	Voskhod	Oct. 12, 1964	254	111	16	24 h. 17 min.
Pavel I. Belyayev, Aleksei Leonov	Voskhod II	Mar. 18, 1965	308	108	17	26 h. 2 min.
Virgil Grissom, John W. Young	Gemini 3	Mar. 23, 1965	140	100	3	4 h. 53 min.
James McDivitt, Edward H. White	Gemini 4	June 3, 1965	182	100	62	97 h. 48 min.
Gordon Cooper, Charles Conrad	Gemini 5	Aug. 21, 1965	219	100	120	190 h. 56 min.
James A. Lovell, Frank Borman	Gemini 7	Dec. 4, 1965	203	100	206	329 h. 30 min.
Walter M. Schirra, Jr., Thomas Stafford	Gemini 6	Dec. 15, 1965	167	100	16	22 h. 55 min.
David R. Scott, Neal Armstrong	Gemini 8	Mar. 16, 1966	147	87	7	10 h. 42 min.
T. P. Stafford, E. A. Cernan	Gemini 9A	June 3, 1966	144	86	44	72 h. 21 min.
John W. Young, Michael Collins	Gemini 10	July 18, 1966	476	86	43	70 h. 46 min.
Charles Conrad, Richard Gordon	Gemini 11	Sept. 12, 1966	850	100	44	71 h. 17 min.
James A. Lovell, Jr., Edwin E. Aldrin, Jr.	Gemini 12	Nov. 11, 1966	180	155	59	94 h. 33 min.
Vladimir M. Komarov	Soyuz 1	Apr. 23, 1967	223	122	17	25 h. 12 min.
Walter M. Schirra, Jr., Don F. Eisele, R. Walter Cunningham	Apollo 7	Oct. 11, 1967	282	102	163	260 h. 14 min.
Frank Borman, James A. Lovell, William A. Anders	Apollo 8	Dec. 21 1968	....	....	(moon orbit)	147 h. 10 min.

SPACE TRAVELERS. Yuri Gagarin (*top*), the first man in space, is a Soviet cosmonaut. John Glenn (*bottom*) was the first American to orbit the earth. American astronauts John Young and Michael Collins (*center*) achieved rendezvous and docking missions in Gemini 10.

Manned Flight.—Soon after the first satellites had been successfully orbited, the Soviet Union and the United States began to plan manned orbits. However, the two countries approached the problem differently.

■**U.S. EFFORTS.**—The United States launched two sub-orbital flights in May and July 1961. This was followed September 13, 1961 by Mercury-Atlas IV, containing a dummy that simulated breathing and perspiring. After one orbit it was deorbited by remote control and recovered. November 29, 1961 Mercury-Atlas V carried aloft the chimpanzee Enos. During the second orbit, the interior of the capsule grew too hot and the test was terminated. The capsule was recovered and the animal found to have endured the trip well. After these tests were successfully concluded, a manned orbit was attempted. On February 20, 1962, Lt. John H. Glenn became the first American to orbit the earth. *Project Mercury* consisted of four orbital flights—Friendship 7, Aurora 7, Sigma 7 and Faith 7—and was concluded with Maj. Gordon Cooper's 34-hour flight in Faith 7.

Project Gemini, a two-man capsule, began with two unmanned capsules and developed the technique of docking. Gemini 7 astronauts Borman and Lovell remained aloft more than 330 hours and completed 206 revolutions.

Project Apollo, a three-man capsule, is designed specifically for a moon-landing. It was to have been tested in February, 1967. However, a flash fire January 27th during a full-scale simulated launching took the lives of Maj. Virgil I. Grissom, Lt. Col. Edward H. White, Jr., and Lt. Comdr. Roger B. Chaffee and compelled the space agency to reappraise the use of pure oxygen as an atmospheric environment in space cabins. The disaster meant a 12-month setback to the program and a possible 2-year delay in a U.S. moon-landing.

■**SOVIET EFFORTS.**—The Russians made no manned sub-orbital flights, but orbited large capsules containing animals, beginning with Sputnik V. After recovering the animal capsules repeatedly, Maj. Yuri Gagarin became the first man in space by completing one orbit on April 12, 1961. In March 1965 Lt. Col. Leonov left Voskhod II for the first "walk in space." On March 24, 1967 Soyuz I had completed 18 orbits when dangerous tumbling forced Col. Komarov to de-orbit. On re-entry, parachute lines snarled, plunging the craft to earth and killing Komarov.

Planetary Probes.—Recalling the early books on space travel, it can be said that everything Robert H. Goddard wrote had been turned into reality by —to pick an arbitrary date—January 1, 1963. Likewise, everything Hermann Oberth set down in his work was reality on that date, with the exception of the manned orbiting satellite, the "space station." The orbits Walter Hohmann investigated are still untraveled by man, but they have been traversed by unmanned devices called *planetary probes.*

Going into orbit around the earth, going to the moon, and traveling to another planet are three entirely different problems. Not only are the distances to be coped with growing larger, but there are also several fundamental differences involved. If a rocket is put into an orbit around the earth, the calculations can be made as if the earth were the only body in the universe. The orbit of an artificial satellite is not influenced by the fact that the earth is moving around the sun at the same time. The orbit is not even disturbed by a relatively close body like the moon—it has been calculated that the path of an artificial satellite would differ from reality by only one yard if the moon did not exist, a difference far too small even to be detected.

Similarly, in sending a rocket to the moon, the calculations can be made as if the earth and its moon were the only two bodies in the universe. Both travel around the sun, at the same distance from the sun (compared to the earth's distance of 93,000,000 miles from the sun, the distance of 240,000 miles to the moon is negligible), and at very nearly the same rate of speed. Hence, in theory, a shot to the moon could be made at any moment. In practice, there are certain periods at which a moon shot is easier to accomplish than at others, but these periods are preferred for one reason only: the problem of guidance is less complicated.

But when it comes to reaching another planet, the problem has to be faced that the various planets move in their orbits with greatly varying velocities. Mercury races around the sun at the rate of nearly 30 miles per second, while Saturn moves at only 6 miles per second.

When Walter Hohmann began his investigations, he first tried to calculate what type of flight path between the orbit of the earth and the orbit of another planet would involve the smallest expenditure of fuel. He found that the flight path that just touched the orbits of the earth and, say, Venus involved the least expenditure of fuel. Hohmann had compared five possible flight paths, which he labeled *A, B, C, D,* and *E*. The one that just touched the two planetary orbits, but did not cross either one of them, happened to be the one he had labeled *A*. Therefore, scientists now refer to it as the *Hohmann-A orbit.* Hohmann calculated the time needed for a ship to travel along the A-orbit from the earth to Venus to be 146 days. In reality, there are minor deviations from this figure, partly because the earth's orbit is slightly elliptical, but mainly because only an approximation of the A-orbit is actually traveled. Since Venus moves at the rate of 21.7 miles per second, and since its average distance from the sun is 67 million miles, it takes 224.7 earth days to complete one orbit. The earth, with an orbital velocity of 18.5 miles per second, is moving at an average distance of 93 million miles from the sun and needs 365¼ days to complete one orbit. Since the rocket needs 146 days for its trip, a Venus probe must be launched at a time when Venus is 146 days of orbital travel away from the point in its orbit that the rocket will reach. Obviously, one cannot launch a probe to Venus (or to Mars, for that matter) at any time one happens to be ready. There are short periods—of about ten days each—when a launching can be made. Of course, one can aim for the orbit of the other planet at any time; but if the target planet is to be at that point of its orbit, attention must be paid to the *periods of possibility,* which are now referred to as *windows.* The time between two windows is a little less than two years, although it varies for each planet. Therefore, missing one such period means a considerable delay before the next time for launching.

Sending a probe to another planet also differs in method from the simple orbit around the earth or the moon shot. For an orbit around the earth, the only requirement is that the satellite have enough velocity to stay in orbit and that the direction of its movement be parallel to the ground. In the case of a moon shot, the principle is to aim at that point of the sky where the moon will be when the rocket arrives there—the transit time of the Russian moon shot was 35 hours—and the motion of the

MOON PROBE. After launching (1) a corrective maneuver must be made (2) to assure the vehicle's proper attitude. If this adjustment is successful, the probe will be on target (3) to hit the moon. Retro-rockets (4) must decelerate the vehicle to ensure a safe impact (5).

Moon Shots and Planetary Probes

Spacecraft	Date	Weight (lb.)	Results
Metcha (Lunik I)	Jan. 2, 1959	3,245	Missed moon and went into orbit around the sun between the earth and Mars. Orbital period is 444 days.
Cosmic Rocket II (Lunik II)	Sept. 12, 1959	860	Hit moon. Transit time was 35 hours; impact velocity was about 6,000 mph.
Pioneer V	Mar. 11, 1960	95	Fired into orbit between orbits of Venus and the earth. Orbital period is 311 days.
Sputnik VIII (Venus probes)	Feb. 12, 1961		Fired from orbiting Sputnik VIII. Transmitter failure; instrument capsule in orbit similar to that of Pioneer V.
Ranger III	Jan. 26, 1962	727	Missed moon and went into orbit around sun; orbit similar to that of Pioneer IV.
Mariner II	Aug. 27, 1962	449	Passed Venus at a distance of 21,648 miles and reported on planetary conditions. First successful planetary probe. Now orbiting the sun.
Ranger V	Oct. 18, 1962	730	Passed 300 miles from the moon. Now in orbit around the sun.
Mars I (Russian Mars probe)	Nov. 1, 1962	1,196	Must have passed Mars in June 1963. Radio contact was lost in March 1963. In orbit around sun between the earth and Mars. Orbital period unknown, must exceed 500 days.
Moon IV	Apr. 2, 1963	3,135	Russian lunar probe, possibly intended for soft landing, missed moon by 5,300 miles.
Ranger VI	Jan. 30, 1964	730	Perfect shot to the moon, impacted near crater Arago, but cameras failed to work.
Zond I	Apr. 2, 1964	? ? ?	Russian probe launched from parking orbit, probably for Venus. In orbit around sun; no data due to transmitter failure.
Ranger VII	July 31, 1964	806	First successful moon shot. Within a distance of from 1,300 miles to 1,000 miles from the moon's surface, 4,316 pictures of the moon were relayed to earth. Crashed into Sea of Clouds.
Mariner IV	Nov. 28, 1964	575	Passed Mars at closest distance of 5,400 miles on July 14, 1965, took pictures showing impact craters on Mars. Now orbiting the sun.
Zond II	Nov. 30, 1964	? ? ?	Russian Mars probe, lost electrical power after a few weeks, must have passed Mars within a day of Mariner IV passage. Orbiting the sun.
Ranger VIII	Feb. 17, 1965	809	U.S. lunar probe, took 64.9 hours to reach the moon, crashed into Sea of Tranquility. Transmitted 7,137 excellent pictures.
Ranger IX	Mar. 21, 1965	809	Reached the moon in 64.5 hours, impacted in crater Arago, transmitted 5,814 pictures.
Zond III	July 18, 1965		In interplanetary orbit, took and transmitted pictures of the moon's far side.
Luna VII	Oct. 4, 1965	3,180	Russian probe was supposed to land on the moon's surface. Missed target and is now orbiting the sun.
Luna VIII	Dec. 3, 1965	3,414	Russian probe designed for soft landing, crashed on moon's surface.
Luna IX	Feb. 3, 1966	? ? ?	220-pound payload with camera landed on the moon, and took approximately nine pictures.
Luna X	Apr. 7, 1966	? ? ?	540-pound payload was put into orbit around the moon by the Russians.
Surveyor I	May 30, 1966	2,194	Made a soft landing on the moon on June 2, more than 10,300 pictures were taken in the first lunar day.
Lunar Orbiter I	Aug. 10, 1966	850	Assumed orbit around moon on Aug. 14, transmitted 215 good pictures. Was finally crashed on the moon to make room for the next orbiter.
Pioneer VII	Aug. 17, 1966	140	Put into orbit around the sun, orbital period 400 days.
Luna XI	Aug. 24, 1966	3,608	Achieved orbit around the moon on Aug. 27, transmitted information but no pictures.
Surveyor II	Sept. 20, 1966	2,100	Tumbled in flight, crashed on the moon on Sept. 23.
Luna XII	Oct. 22, 1966	? ? ?	Achieved orbit around the moon on Oct. 25, transmission difficulties, unsuccessful.
Lunar Orbiter II	Nov. 7, 1966	850	Achieved lunar orbit, lowest altitude 31.3 miles from lunar surface. Many excellent pictures, was later crashed.
Luna XIII	Dec. 21, 1966	? ? ?	Soft-landed on the moon after 80-hour flight. Transmitted good pictures.
Lunar Orbiter III	Feb. 4, 1967	850	First spacecraft to have an orbit going over moon's poles. Fine scientific results and pictures. Was later crashed.
Surveyor III	Apr. 17, 1967	2,283	Soft-landed on the moon after 65-hour flight. Had small scoop digging trenches in lunar soil to test its strength.
Lunar Orbiter IV	May 4, 1967	850	Achieved orbit around the moon May 7, orbit was later changed. Finally crashed into moon.
Venera IV	June 12, 1967	? ? ?	Russian Venus probe, instrument capsule entered atmosphere of Venus, transmitted temperature measurements and particular chemical analysis.
Mariner V	June 14, 1967	550	Made flyby near Venus on Oct. 19, confirming earlier findings.
Surveyor IV	July 14, 1967	2,800	Retro rocket exploded just before landing on the moon.
Explorer 35	July 19, 1967	150	Large orbit around the moon achieved July 22, makes radiation measurements in the vicinity of the moon.
Lunar Orbiter V (last of series)	Aug. 2, 1967	850	Orbit around the moon achieved Aug. 5. Many fine pictures; crashed on moon after film was used up.
Surveyor V	Sept. 8, 1967	2,800	Landed in Sea of Tranquility, made soil analysis.
Surveyor VI	Nov. 7, 1967	2,800	Landed on *Sinus medii,* dark area in center of visible hemisphere. Performed eight-foot jump to photograph its own footprints.
Surveyor VII (last of series)	Jan. 6, 1968		Landed near crater Tycho, made soil analysis.
Luna XIV	Apr. 7, 1968	? ? ?	Achieved orbit around the moon April 10, no other information available.
Zond V	Sept. 14, 1968		Orbited moon and was then steered into return path to earth. Splashed in Indian Ocean; recovered by Russian ships.

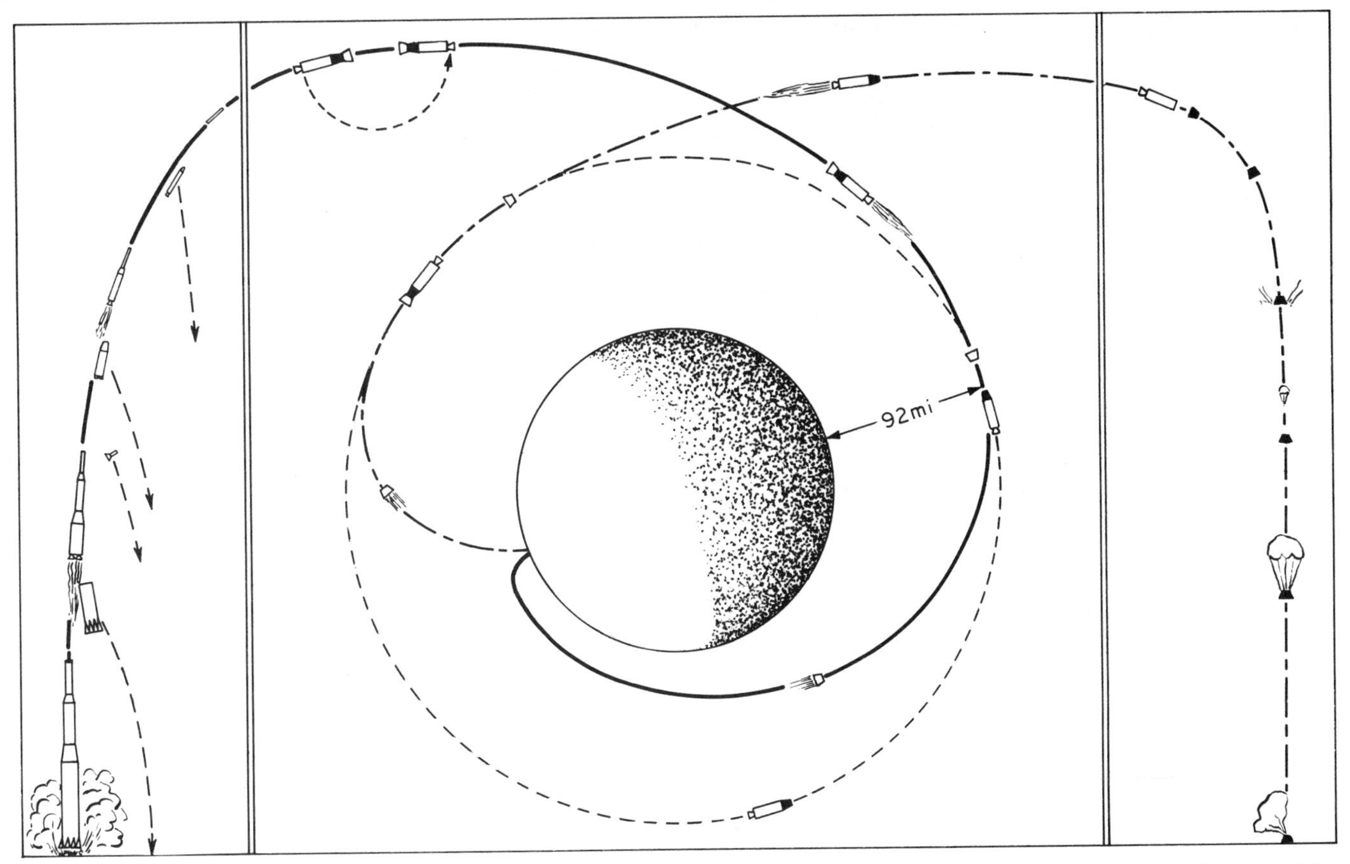

APOLLO FLIGHTPLAN FOR A MANNED LUNAR LANDING. After the booster rockets put the three-part mooncraft into the flightpath to the moon, the craft turns around so that it can slow down, making a lunar orbit possible. Two astronauts descend to the moon in the Lunar Excursion Module, while the third stays in the orbiting Command Module. The lower part of the LEM is the launch pad for the upper part which returns the astronauts to the Command Module. The lower part stays on the moon; the upper part orbits the moon. The Service Module powers the return trip. It is jettisoned before re-entry; the Command Module re-enters and lands.

rocket consists of climbing more or less vertically out of the earth's gravitational field until the gravitational field of the moon becomes more powerful. Then the rocket simply falls into the moon's gravitational field.

The motion of a planetary probe, however, is in the sun's gravitational field. Since, before firing, the rocket is a part of the earth, it moves around the sun in the earth's orbit at the rate of 18½ miles per second. After firing, the rocket has a velocity that is, as seen from the sun, greater or less than 18½ miles per second. If the velocity is greater, the sun will not be able tc hold the rocket in the earth's orbit; and the rocket will drift outward in the solar system in the direction of the orbit of Mars. If the rocket's velocity is less than the orbital velocity of the earth, the sun's gravitational pull will make it drift inward in the solar system, in the direction of the orbit of Venus.

The principle of an interplanetary shot is to add the rocket's velocity to the orbital velocity of the earth, or else to subtract the rocket's velocity from the orbital velocity of the earth —just the right amount of velocity, and at just the right time, while a window exists. Whether the rocket's velocity is added or subtracted depends on the time of day. If the rocket is fired vertically, it will add its velocity to the earth's velocity if the shot is made at dawn. The section of the earth that experiences dawn is its *leading edge;* the section that experiences dusk is the *traveling edge.* Hence, a rocket launched at dawn will move faster than the earth, as seen from the sun, and as a consequence will drift outward in the solar system; the rocket launched at dusk will move more slowly than the earth and will be drawn inward in the solar system.

■**PARKING ORBITS.**—In the foregoing explanation, it has been assumed that the launching is made from the ground. However, it can be done in a different manner, a manner that is preferred by the Russians even though they have suffered quite a number of failures with their planetary probes. This method consists of first putting the rocket, or rather, its upper stages and the payload, into an orbit around the earth. Such an orbit is called a *parking orbit.* The upper stages are then fired by radio command from the ground when their motion has put them in the proper position.

Future Expectations.—Communication satellites, weather research satellites, and navigational satellites were all operational for some time in 1968. In addition, space radiation monitors and surveillance satellites doing work for military reasons were in service. Among the projects planned for the future is a heat-detecting earth-orbiting satellite that may assist in forecasting volcanic eruptions. Another space probe to Mars has definitely been planned, to be followed by a "Mars lander." A less definite project is an exploratory probe to a comet to examine and determine its chemical composition.

The largest American project is Project Apollo, designed to land two astronauts on the moon in 1969. (For a detailed explanation of the proposed lunar landing, see the drawing and caption on the preceding page.) The United States Air Force is working on a project called MOL, (Manned Orbiting Laboratory) which will be a large cylindrical container, 18 to 20 feet in diameter and a minimum of 40 feet long. The Manned Orbiting Laboratory has an airlock where a Gemini capsule will be able to dock and astronauts will then spend weeks and even months inside the Manned Orbiting Laboratory for medical and other research purposes.

ASTRONAUTICAL GLOSSARY

Abort.—A rocket shot that cannot be carried to its conclusion. An abort may be due to a malfunction preventing take-off, or it may be the intentional destruction of the rocket

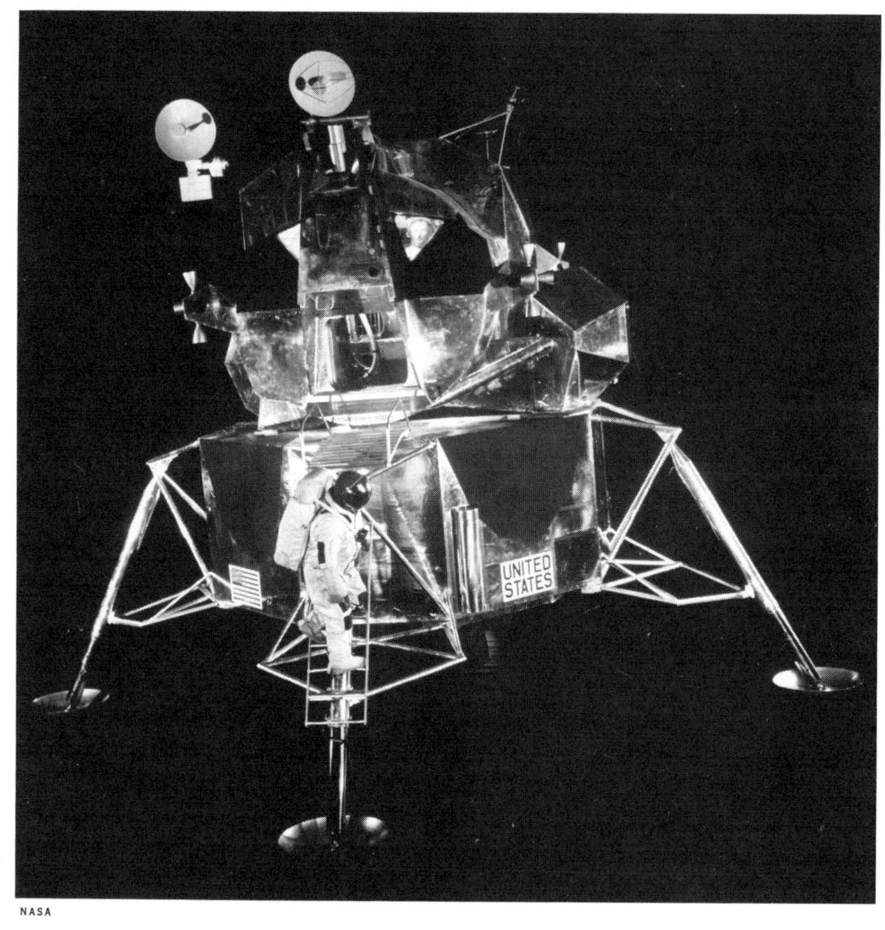

NASA

SCALE MODEL OF THE LUNAR EXCURSION MODULE, expected to land two American astronauts on the moon in 1969. The spider-like capsule is one part of the three-part Apollo craft.

because of its deviation from its trajectory, thereby endangering life or property. (See *Destruct.*)

Acquisition.—A term meaning that a missile or satellite has been "caught" in a radar beam for tracking. Also, the name given to the radar equipment used in tracking.

Airglow.—A very faint luminescence of the night sky, caused by the release of energy stored by molecules of the upper atmosphere during daylight. It can be detected only with very sensitive photographic plates that are exposed for several hours.

Apogee and aphelion.—Both of these terms mean the farthest point of an orbit. Apogee applies to an orbit around the earth, and aphelion applies to an orbit around the sun. The opposites are *perigee* and *perihelion,* orbital points nearest the primary.

Attitude.—The angle formed by the longitudinal axis of the rocket (or space capsule) with the direction of motion. Normally this angle is zero, as is the case when the rocket travels nose first.

Back-up vehicle.—A rocket identical with one being prepared for launching that is held in reserve in case the planned launch goes awry.

Booster.—One or several powerful rocket units that supply a great amount of thrust at take-off. Normally the burning time of a booster is short, sometimes less than thirty seconds. The bottom stage of a multi-stage rocket is sometimes referred to as the booster, but it is more properly called the first stage.

Chemical fuel.—A propellant that depends upon an oxidizer for combustion. Liquid and solid rocket propellants are chemical fuels.

Circum.—A prefix meaning 'around'; a *circumsolar* orbit is an orbit around the sun; a *circumlunar* orbit is an orbit around the moon; and a *circumpolar* orbit is an orbit around the earth that passes over the North and South poles.

Coriolis effect.—Named after its propounder, G. G. Coriolis (1792–1842). This term describes the influence of the earth's rotation on the trajectory of a long-range missile. In the Northern Hemisphere, missiles traveling parallel to the earth's surface are deflected to the right; and in the Southern Hemisphere they are deflected to the left.

Countdown.—The countdown is not just the last ten seconds before take-off counted backward, as commonly believed; it is a complete list of all the things to be done—from placing the rocket on the launching pad to firing. A full countdown usually lasts ten hours; the counting of seconds after lift-off is commonly referred to as the *countup.*

Destruct.—A term coined to signify the premeditated destruction of a rocket in flight by the detonation of packages of high explosives carried for this purpose. The destruct is carried out by the range safety officer when a missile deviates from its course. (See *Abort.*)

Double-base powders. — High explosives, used for the propulsion of military bombardment rockets, that are compounded from nitroglycerine and guncotton, with various stabilizing substances added. The name was given because the two main ingredients are both explosives. Specific double-base powders have such trade names as Cordite and Ballistite.

Escape velocity.—The speed needed to overcome a planet's gravitational attraction so as to escape permanently into space; for the earth, the escape velocity is approximately 7 miles per second, or 25,200 miles per hour.

Galcit fuels.—Solid rocket fuels prepared from asphalt and oil, with potassium perchlorate as the oxidizer. This type of fuel was developed for take-off rockets for military aircraft. The name is a contraction of the Guggenheim Aeronautical Laboratory of the California Institute of Technology, where the fuels were prepared.

Hohmann transfer ellipse.—Orbit, also known as *Hohmann-A* orbit, named after the German engineer Walter Hohmann, who proved in 1925 that the most efficient spaceship trajectory from one planet to another is an ellipse that just touches both planetary orbits.

Impact area.—The area where a long-range missile or space capsule is expected to land when it returns to the earth.

Jettison.—To cast off something no longer needed; for example, the heat shield of a manned capsule is jettisoned after the capsule has re-entered the earth's atmosphere.

Launch pad.—Sometimes referred to only as the pad, it is the platform or other supporting structure from which a rocket vehicle is launched.

Life-support system.—In manned spacecraft, the mechanisms and de-

NASA

ASTRONAUTS James Lovell and Elliot See are studying pictures of the lunar surface.

vices needed to provide food, to control the composition of the cabin air and the dehumidifying apparatus, and to maintain the temperature at a comfortable level.

Liquid propellants. — Those chemical fuels and their oxidizers that are stored in the liquid state aboard rockets. Typical liquid fuels and their oxidizers are: kerosene and liquid oxygen (LOX); ethyl alcohol and liquid oxygen; aniline and red fuming nitric acid (RFNA); and unsymmetrical dimethyl hydrazine (often called "dimazine") and RFNA. (See *Solid propellants.*)

Missile.—A device whose sole purpose is to reach a target and cause destruction. Missiles are classified by their means of propulsion: *rocket-propelled missiles* are powered by rocket motors; *air-breathing missiles* are powered by internal-combustion engines or turbines. (See *Rocket.*)

Nose cone.—The front, or a covering over the front, of a rocket or satellite. Its purpose is to withstand the heat caused by friction as the projectile passes through the atmosphere.

Orbit.—The path described by a rocket, satellite, spacecraft, or planet as it travels through space. A *geocentric orbit* is one that has the earth at its approximate center; a *heliocentric orbit* is one that has the sun at its approximate center.

Outer space.—That region, extending outward indefinitely, beyond the upper limit of the earth's atmosphere. Outer space is subdivided into four parts: *cislunar space* is the region enclosed by the orbit of the moon; *interplanetary space* is the region between any two planets of the solar system (the earth is, naturally, one of the planets); *interstellar space* is the region between any two stars (the sun is the star nearest to us); and *intergalactic space* is the region between any two galaxies (the Milky Way is the galaxy of which our sun is a member).

Oxidizer.—A chemical element or compound, either oxygen or containing oxygen, that supports the burning of a fuel or propellant. The term is also used when fluorine is used in lieu of oxygen. (See *Galcit fuels, Liquid propellants,* and *Solid propellants.*)

Parking orbit.—A geocentric orbit into which the upper stages of a rocket and its payload (usually a planetary probe) may be placed. The payload may then be launched at the proper moment, by radio command from the ground, from this orbit.

Payload.—The useful cargo, such as a manned capsule, planetary probe, or satellite, carried by a launch vehicle.

Perigee and perihelion.—See *Apogee and aphelion.*

Permanent orbit.—A satellite is in a permanent orbit around the earth when every point of its orbit is so far from the earth that no atmospheric resistance is encountered anywhere along the orbit. Such a satellite will never reenter the atmosphere.

Perturbation.—A minor change in orbit caused by a force other than the sun's gravitational attraction, usually the gravitational influence of another planet. Very light and large satellites, such as Echo, are also perturbed by the sun's radiation pressure.

Primary.—The body that is being orbited by another, smaller body. The primary of a satellite is the planet of which it is a satellite; the primary of a planet is the sun.

Retro-rocket.—Generally, a rocket firing in the direction of movement to reduce the velocity, usually to slow down an object in orbit so it will reenter the earth's atmosphere.

Rocket.—Any vehicle that uses internally stored oxygen (or other oxidizer) instead of atmospheric oxygen for the combustion of its fuel. More specifically, in astronautics, the launch vehicle for a manned capsule, planetary probe, or satellite. (See *Missile.*)

Satellite.—Any body, natural or artificial, which is in orbit around a planet. The word is derived from the Greek *satellos,* meaning 'attendant.'

Scrub. — Postponement of a rocket shot because of a malfunction detected during the countdown.

Solid propellants. — Those chemical fuels and their oxidizers that are stored in the solid state aboard rockets. Solid propellants combine a synthetic rubber as the fuel, oxidizers, such as ammonium perchlorate, and small amounts of powdered metals, such as aluminum. (See *Liquid propellants.*)

Sustainer engine.—In a rocket with a booster, the sustainer is the rocket engine that keeps burning, to sustain thrust and acceleration, after the booster has ceased to function.

Ullage rockets.—Small rockets, usually solid-fuel units, needed for the operation of large liquid-fuel units in space. After a liquid-fuel rocket has ceased burning, the fuel remaining in the tanks, being weightless, will distribute itself at random inside the tank. The purpose of the ullage rockets is to provide a short push to "shake down" the fuel into the proper place so that the liquid-fuel motors can be started again.

Umbilical cord.—An external cable attached to the rocket, through which electric current is carried to the devices in the rocket; this saves the batteries in the rocket until they are required. The umbilical cord drops off just before lift-off.

Vernier rockets.—Small rockets used to adjust the velocity of a large rocket—for example, an intercontinental missile—to the precise value needed for the particular mission.

Zero G.—Also called *weightlessness* or *free fall.* Weightlessness occurs when a body freely follows the pull of gravity without resisting it, as is the case when a satellite is in orbit.

—Willy Ley

NASA

MANNED ORBITING RESEARCH LABORATORY (MORL) is presently still in the research stage. MORL, with three Gemini spacecraft attached, would be capable of year-long missions.

BIBLIOGRAPHY

BERGAUST, ERIK. *The Next 50 Years in Space.* The Macmillan Co., 1964.

CHESTER, MICHAEL. *Rockets and Space Craft of the World.* W. W. Norton & Co., Inc., 1964.

CLARKE, ARTHUR C. *Interplanetary Flight.* Harper & Bros., 1960.

DE LEEUW, HENDRIK. *From Flying Horse to Man in the Moon.* St. Martin's Press, Inc., 1963.

DIAMOND, EDWIN. *The Rise and Fall of the Space Age.* Doubleday and Co., Inc., 1964.

LEVITT, ISRAEL M. and COLE, D. M. *Exploring the Secrets of Space.* Prentice-Hall, Inc., 1963.

LEY, WILLY. *Our Work in Space.* The Macmillan Co., 1963.

LEY, WILLY. *Rockets, Missiles, and Men in Space* (rev. ed.). The Viking Press, Inc., 1967.

THOMAS, SHIRLEY. *Men of Space.* Chilton Co., 1960.

VAN ALLEN, JAMES A. (ed.). *Scientific Uses of Earth Satellites.* The University of Michigan Press, 1956.

WEISER, WILLIAM J. *Space Guidebook.* Coward-McCann, Inc., 1960.

Areas of Study.—The new science of *space biology,* which is sometimes also referred to as *bioastronautics,* is concerned with the biological effects and the possible hazards that confront astronauts setting out on long missions through space. This science requires an ultra-careful study of the human body, as well as a physiological and psychological study of those people selected as astronauts. For this reason, the more specific term *space medicine* is often used, since animal studies, so far, are only incidental. Space biology must not be confused with *exobiology,* which is the study of life forms indigenous to celestial bodies other than the earth. Exobiology is only a theoretical science so far, concerned mainly with laying down its own foundations.

The problems encountered in space biology fall into two classes. One consists of those problems produced by the launching and operation of the sapcecraft. The other is comprised of those problems resulting from the environment of space, which differs fundamentally from the normal environment at or near sea level to which our bodies are accustomed. The problems associated with the operation of the space vehicle can further be divided into those associated with the *high-acceleration* phase that accompanies blast-off and reentry (or landing) and those accompanying the *zero-gravity,* or *weightless,* phase that is characteristic of orbital flight, whether the orbit leads around the earth or from the earth to the moon or another planet.

Gravitational Problems.—The letter *g,* used to describe the intensity of acceleration during blast-off, is an abbreviation of the word "gravity." It was chosen because the force the astronaut is subjected to is compared with the force due to gravitational acceleration at sea level. The normal force, designated as 1 g, to which we are subjected during our daily lives is the result of the earth's gravitational attraction. If the gravitational attraction were to be doubled, we would be subjected to twice the force, or 2 g; that is, we would weigh twice as much. The g-force experienced at any moment is the result of resisting the gravitational acceleration of the earth, since we are supported by the ground. If we were falling freely, without resisting the earth's pull, this g-force would disappear.

■CAUSE OF HIGH G.—If a rocket rises vertically with an acceleration equal to that produced by the earth's gravitational field—that is, if it travels 32 feet per second faster at the end of every second—the astronauts in the space capsule atop that rocket will be subjected to a force of 2 g. One component of the 2 g force is caused by the acceleration of the climbing rocket, while the other is the result of the earth's gravitational field. The g-force acting on the astronaut's body is, therefore, always 1 g higher than is accounted for by the rocket's acceleration. All rockets are designed to operate with a constant thrust, which means that the fuel consumption per second of powered flight remains constant. But since fuel consumption and thrust are constant, the acceleration must increase. For example, if the rocket lifted off with an acceleration of 1 g because the thrust of the rocket engine was precisely twice the weight of the rocket, the acceleration will be 3 g when the weight of the rocket, because of fuel consumption, has dropped to one half of the take-off figure. The astronaut will then experience a force of 4 g, 3 g of which is due to the rocket's acceleration and 1 g to the earth's gravity. Since the acceleration due to the earth's gravity decreases with increasing altitude, the g-force exerted by the earth is slightly less than 1 g when the rocket has climbed to an altitude of, for example, 50 miles. The difference for the first hundred miles is so small that it can be disregarded for practical purposes.

It is obvious, therefore, that the g forces on the body of the astronaut reach their peak during the last second of burning of the rocket fuel. At that moment the rocket is close to its minimum weight (the structural weight plus the payload weight) while the thrust is even slightly higher than it was at lift-off, because a rocket's thrust rises somewhat as external air pressure diminishes. For preliminary calculations it is assumed that the thrust of a given rocket engine in a vaccum is 16 per cent higher than it is at sea level. However, the actual increase is slightly different (but rarely by more than 1 per cent) because it is influenced by the design of the exhaust nozzle.

In the case of a two-stage rocket, the g-forces drop to zero when the lower stage ceases burning and remain at zero until the second stage begins burning. Then another cycle of mounting g-forces, beginning at about 2 g, begins. The g-forces become zero during the interval between the burning of the first stage and that of the second stage because during this interval the rocket is effected by the gravitational pull of the earth and loses velocity.

■EFFECTS OF HIGH G.—Before the first launching of a manned space vehicle, it had been calculated that the time from lift-off to insertion into an orbit around the earth would be on the order of five minutes. It was also calculated that the g-forces on the astronaut's body would average about 5 g, with a short-duration peak of over 8 g at the end of the first stage burning and a second short-duration peak of about 7 g at the end of the burning of the second stage. The question was, of course, how the human body would react to an acceleration of 5 g for a period of five minutes. No situation occurs on the ground that imposes a high acceleration for any length of time. The only periods of acceleration substantially higher than 1 g and lasting for more than a few seconds were experienced in making sharp banking turns in high-speed aircraft. Such turns were not made in the course of ordinary flying, but took place mainly in aerial combat during World War II. A rule of thumb arrived at during these experiences was that if the turn did not produce g-forces higher than 3 g they had no problem, but at 4 g or over they lost consciousness, or "blacked out." Fortunately, the blackout period usually lasted less than twenty seconds.

Although previous experience did

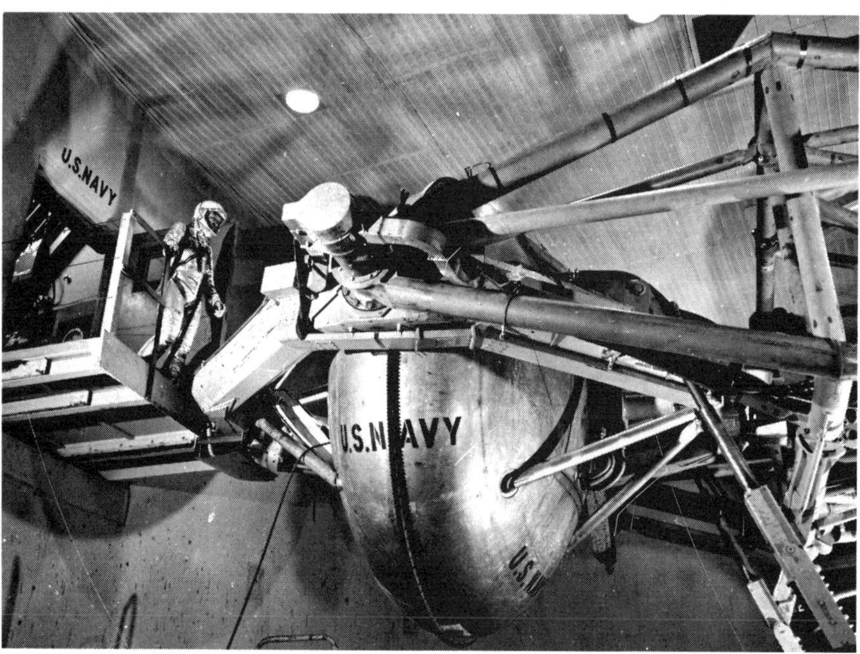

NASA

GIANT CENTRIFUGE is being used by astronaut Walter M. Schirra to test the effects of acceleration. Astronauts will experience acceleration during both launching and reentry.

NASA

WEIGHTLESSNESS is experienced in test flights prior to the astronauts' actual space flight.

not have more than a hint to offer, there was a way of investigating the problem. A rapidly spinning centrifuge could produce g-forces of the proper magnitude and duration. Preliminary experiments with small centrifuges showed that mice were usually killed by 25 g, but that hard-shelled insects (such as medium sized beetles) and very light insects (such as the common housefly) seemed not to be affected at all by the same acceleration. For the purpose of designing a centrifuge large enough to test people, a short table was drawn up. Although it was anticipated that a man would be put only into orbit, with an orbital velocity of 4.6 to 4.7 miles per second, the table was drawn up for a moon-flight, where a velocity of 7 miles per second is required. If it was assumed that the human body could not stand a higher acceleration than 3 g, the time required to reach a velocity of 7 miles per second was 9 minutes and 31 seconds. If 4 g could be used, the time shrank to 6 minutes and 21 seconds. If 5 g was possible, the time would be reduced to 4 minutes and 45 seconds; at 6 g it would be only 3 minutes and 48 seconds; and at 7 g it would be 3 minutes and 10 seconds. If a superman who could survive 10 g were found, only 2 minutes and 6 seconds would be needed.

While the first such centrifuge was being built, some thought was expended on the best position for the volunteer to be tested. The wartime pilots had, of course, been in a sitting position so that the g-forces acted along their spine, from head to buttocks. Theoretically there was a better position. The man should be placed on his back, with his head and shoulders slightly raised and his knees raised to about the level of the head. Then the g-forces would act from front to back and it seemed likely that higher accelerations could be endured in that position.

The actual tests, when they were run, surprised everyone, from the volunteers who permitted themselves to be tested to the engineers and doctors who ran the tests. The series began with 3 g for 9½ minutes; no harmful effects were found, but the volunteers thought they had been left in the centrifuge for 20 minutes. In spite of these complaints a 4 g run was tried. This was easier on the volunteers; the shorter duration more than made up for the extra g. The 5 g and 6 g runs also proved to be easier on the volunteers than the 3 g run had been. When the 10 g run was tried, it was found to cause a physical hardship, but there was no "black-out."

These tests were repeated many times in a variety of ways. Interrupted tests were run to simulate the blast-off of a two- or three-stage rocketship. Tests of only 2 g for 25 minutes were run for the purpose of simulating a reentry into the atmosphere. The actual reentry imposes loads as high as those experienced in take-off, although for a slightly shorter duration. In the end it was clear that a healthy man could endure the acceleration required for going into space without any harm to his health. The centrifuge tests are now a standard item in the training of astronauts, and the position conceived as being the best for enduring high acceleration has been successfully used in the Mercury and Gemini manned-space-flights.

■ZERO G PROBLEMS.—Since the g-forces are the result of resisting acceleration (either gravitational or other forces) a spacecraft in orbit around the earth does not experience any g-forces at all. A capsule in orbit follows a path compounded of its own inertia and of the earth's gravitational attraction. One might say that it falls around the earth and, of course, every one of its parts, including loose objects in its interior, "fall" in the same direction with the same velocity.

While the absence of g-forces, called to "zero-g condition," was perfectly explained from the point of view of the physicist, the physician could think of a large number of questions. What would zero-g do to the astronaut's heart? How would his lungs, stomach and intestines, inner ear, and brain and nervous system react? Would an astronaut be able to tell what was above and what below? Would he become completely confused—"disoriented" was the technical term—and be unable to work? And would he be able to sleep?

At first there seemed to be no answers to these questions and, again, previous experience offered no information. An approximation of zero-g (but only an approximation, because of air resistance) is experienced by a parachute jumper for the short time between leaving the aircraft and the opening of the parachute. It is experienced by a swimmer diving from a high board, and by a ski jumper during a long jump. But in each case the duration is measured in seconds, and the parachute jumper, the ski jumper, and the high diver always have their attention on other things. However, when the medical men concerned with the problem examined the operation of each organ, they found that gravity has very little to do with the functioning of the human body.

Inhaling, for example, is accomplished by expanding the chest so that the pressure inside the lungs is lower than the pressure outside the body. Exhaling is done by making the inside pressure greater than the outside pressure. Could an astronaut swallow without gravity? It was remembered that little boys drink a glass of water, or even eat food, while hanging by their knees from a bar, swallowing *against* gravity. And while there was no doubt that the sense of equilibrium was located in the ear, it was also known that it was not needed provided the man could see. As for the heart, it did have to lift the blood out of the legs and the abdominal cavity. It was soon realized, however, that the main work of the heart consists in overcoming the friction in the blood vessels; and that the friction would not be changed by zero-g conditions. There remained the two questions of how the brain and nervous system would function and whether an astronaut in orbit would be able to sleep.

Then a method of producing zero-g for a short time was devised. If an airplane went into a shallow dive, pulled out of it, and had the power shut down at the same time, it would fly through an arc on inertia only, like a projectile. While it was flying through that arc, the zero-g condition would prevail. After a few cautious experiments, such zero-g flights—which are now also a part of the training of an astronaut—were carried out in a long series of tests. First the ability of the test subject to stick a pencil accurately into the center of a paper target was tested. Then a few cautious swallows of water and then eating were tried. Everything worked well, although some practice was required for certain things. Just to be thorough, animal experiments were also carried out. These were done mainly with cats, for cats jump, are

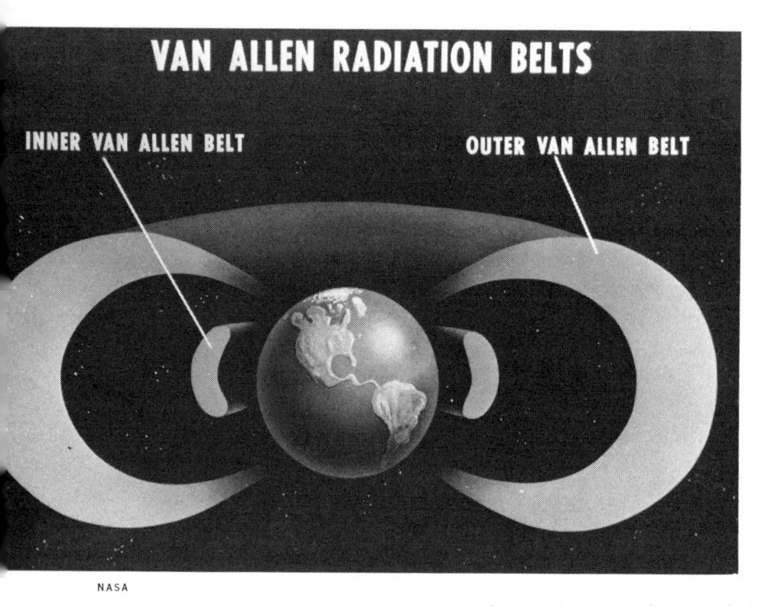

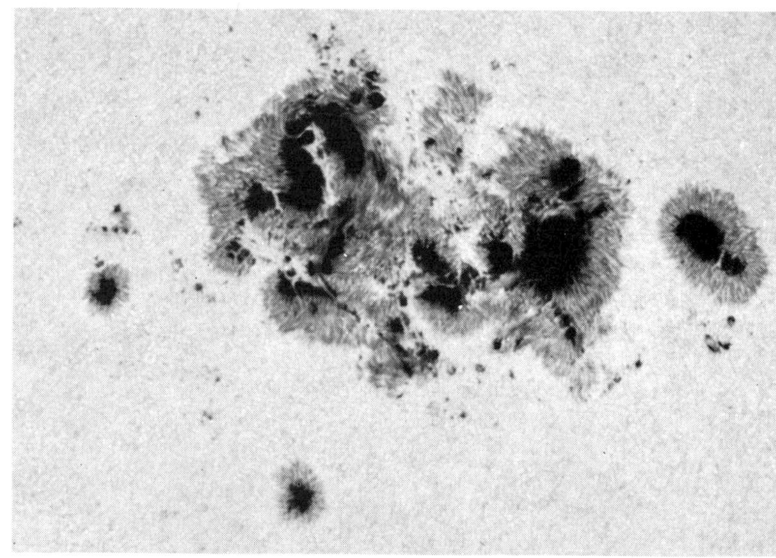

NASA MT. WILSON AND PALOMAR OBSERVATORIES

VAN ALLEN RADIATION BELT presents a danger in manned space flights. The belt may be caused by radiation from the sun, trapped by the earth's magnetic field. During sunspot activity (*right*), great amounts of solar flare radiation are emitted from the interior of the sun.

versatile, and are not easily distracted from what they want to do. Only the question of whether they would be able to sleep could not be answered by these flights, for the condition of weightlessness in the initial tests lasted for only 30 seconds. The overall outcome of these weightless flights was that people could adapt well to the zero-g condition.

■ANIMALS IN ORBIT.—While the high-g and zero-g tests were being conducted, animals were sent up in the nose cones of rockets. A vertically rising rocket that is too slow to go into orbit will produce a high acceleration for about a minute. From then on, while it climbs to its peak by inertia only and while it is falling back, zero-g is obtained. It lasts from the moment the rocket engine cuts off to the moment the parachute attached to the animal capsule opens, which can be a period of six minutes and more. American scientists used rhesus monkeys, while the Russians used dogs (and later an assortment of animals, including hares, mice, and a few insects) for control purposes. Whenever the animals were recovered, they were in good shape, with the exception of one test animal that died of sunstroke because it was several hours before the capsule was found in the desert.

The first animal experiment in orbit was carried out by the Russians with their second artificial satellite, *Sputnik II,* which was launched on November 3, 1957. In addition to an artificial satellite identical to *Sputnik I,* the rocket carried an animal capsule holding a medium-sized female dog named Laika ("barker"), which lived in orbit for more than 150 hours. The dog's breathing, heartbeat, etc., were telemetered to the ground, and it was found that Laika's heartbeat was high above normal after the strain of the acceleration and stayed at a higher level for some time. Precisely the same observation was later made about human astronauts. However, it is interesting to note that during the flight of *Gemini 5,* which lasted eight days, the heartbeat of astronaut Charles Conrad reached a greater frequency and stayed high for a longer time than that of astronaut L. Gordon Cooper, who had made a 1½-day orbital flight before.

Sputnik IV, which was orbited on May 15, 1960, may have carried an animal experiment too, but the capsule disintegrated upon reentry. *Sputnik V,* which was orbited on August 19, 1960, carried an animal capsule housing two dogs named Balka and Strelka, a few smaller animals, and microscopic life forms of various types. It was successfully de-orbited one day after firing and the animals were recovered. *Sputnik VI,* which was orbited on December 1, 1960, also carried two dogs, named Pchelka and Mushka, and smaller life forms; but this capsule burned up during reentry. *Sputnik IX,* launched on March 9, 1961, and *Sputnik X,* orbited on March 25, 1961, carried the dogs Chernushka and Zvezdochkar, respectively. Both were successfully de-orbited and the animals recovered.

■MEN IN ORBIT.—The Russians must have felt by that time that they had mastered the technique of de-orbiting, because their next space shot was a manned capsule, called *Vostok* ("east"). It went into orbit on April 12, 1961, with cosmonaut Yuri Gagarin aboard. The capsule was brought in for a landing a few hundred miles to the west of the lift-off area, so that Gagarin did not quite complete one orbit.

All these flights, however, had not answered the question whether a man could sleep when weightless. Gagarin's flight had lasted for only 108 minutes and it would be hard to tell whether an otherwise healthy dog had been deprived of sleep for twenty-four hours. Theoretical assertions ranged from the statement that weightlessness would make no difference at all, which turned out to be true, to the belief that a man might need only two or three hours of sleep per twenty-four-hour period when weightless. The answer was supplied by the second Russian cosmonaut, Gherman Titov, who went into orbit in *Vostok II* on August 6, 1961. Titov completed 17½ orbits within 25 hours 18 minutes. He reported that he had eaten and slept but had felt "faintly nauseous" for the duration of the flight.

The flight of *Gemini 7* has shown that specially selected and trained men can not only live through two weeks of zero-g without visible harm, but can also perform a multitude of tasks accurately and efficiently. For still longer missions, however, such as a mission to Mars that would involve about 260 days in space for a one-way trip, special measures must be taken.

Environment of Space.—Man, when at home on the ground, is accustomed to what might be called a *particle environment.* The particles in question are the molecules of nitrogen, oxygen, and other gases that compose the atmosphere. Virtually everything we feel but cannot see—heat or cold, calmness or wind, and dryness or moisture—we are made aware of through the collision of these molecules with our skin. Only a strong sun in the sky reminds us that radiation is also present.

Space is not a particle environment, although microscopic and near-microscopic meteoric dust is found there. The roles of radiation and particles are more or less reversed in space when compared to the bottom of the atmosphere. At the bottom of the atmosphere man is in a particle environment with some radiation, while in space there is a radiation environment with some particles. Virtually all of this radiation, for a vehicle in the vicinity of the earth's orbit, originates in the sun.

There are two types of radiation. The first is the kind that comprises the electromagnetic spectrum, and which is usually labeled by its wavelength. The ordinary radio waves used for broadcasting are the longest, and next are the short radio waves used, for example, in radar sets. Infrared (or heat) rays come next, followed by the visible spectrum from red to violet, the ultraviolet, and finally the X rays. Since the sun is a weak radio star as well as a weak X ray star, the quantity of very long or very short waves it emits is small. Other stars produce a much higher percentage of very long waves—radar wavelengths and longer —and it may be assumed that there are also stars that produce an unusual percentage of X rays. This type of radiation from the sun is completely stopped by a very thin metal skin, comparable in thickness to the metal of an ordinary tin can. Since the skin of a space vehicle has to be thicker than that for structural reasons, it can be counted on to stop all wavelengths, including the X rays that naturally occur in space.

It is the second kind of radiation that poses problems. This type consists of *subatomic particles* that move through space at high velocities. Most of these particles are electrons, with a large admixture of protons (hydrogen nuclei). Of course, the energy of a proton depends not only upon its mass, but also upon the velocity with which it travels. The *cosmic rays* that puzzled scientists forty years ago turned out to be, in the main, very fast, and therefore very energetic, protons. Occasionally the nuclei of heavier elements are detected. These are collectively referred to as *heavy primaries;* it is believed that most of them did not originate in the sun but have come to the solar system from far distant places in space where a stellar catastrophe, such as a nova explosion, has taken place.

Since the cosmic rays can penetrate many feet of concrete, there is no way shielding the astronauts in a space capsule against them. For a long time this was considered a major hazard to space flight. Both the United States and the Soviet Union have orbited numerous artificial satellites just for the purpose of detecting and measuring radiation in space. It is now known that the hazards lie elsewhere. The number of very energetic cosmic rays is small and every person sustains a number of hits of such particles in the natural course of events even when on the ground. The actual radiation hazards of space are twofold. One of these hazards, a solar flare, is a temporary event, while the other, the Van Allen belts, is permanent and fixed in space.

■SOLAR FLARES.—A *solar flare* is an especially bright spot on the sun, usually appearing in the vicinity of a sunspot group. The true nature of the sunspots is still not known, but a solar flare marks the spot where a large cloud of subatomic particles is injected into space. These clouds of protons and electrons, which often have a diameter several times that of the earth, constitute a truly major hazard to the life of astronauts. Fortunately, solar flares do not occur without any warning at all. In the first place there are years when the sun is relatively quiet and other years when it is more active. These years of activity and inactivity constitute the so-called "sunspot cycle." There is, therefore, the possibility of a statistical prediction.

Even an individual and unexpected solar flare still has a built-in safety factor. The light emitted by a solar flare needs only about eight minutes to travel from the sun to the orbit of earth. The cloud of protons and electrons, however, travels more slowly than light, needing hours and sometimes days to reach us. There is, therefore, a warning period of at least a few hours. It may be mentioned that at some times a cloud of subatomic particles that was expected to strike the earth was never detected at all. This compels us to conclude that these clouds may not always travel in straight lines. Constant observation of solar phenomena, assisted in the near future by special sun-orbiting satellites, is expected to lead to a better understanding of all the factors involved, with the final result that reliable predictions will become possible.

■VAN ALLEN BELTS.—The first artificial satellite orbited by the United States, *Explorer I,* which was launched on February 1, 1958, discovered the then unknown fact that the earth is surrounded by a radiation belt that begins at an altitude of about 500 miles above sea level. Almost a year later, on December 6, 1958, an attempt to send *Pioneer III* to the moon was unsuccessful in its main objective, but resulted in the discovery of another such radiation belt, about 12,000 miles out. They are now referred to as the inner and outer belt and have been named the *Van Allen belts* in honor of their investigator, James Alfred Van Allen. The inner belt forms a ring around the earth that reaches from about 40° North latitude to 40° South latitude. The inner belt is assumed to be more dangerous than the outer belt and could be easily avoided. All manned space flights performed so far have stayed in the clear space above the atmosphere and below the inner Van Allen belt, although Van Allen himself has stated that he does not consider a quick penetration of the inner belt particularly dangerous. The outer belt is larger, more strongly curved, and apparently more diffuse, and it cannot be avoided by a spacecraft bound for the moon.

Since both belts are a mixture of electrons and protons that have been trapped by the earth's magnetic field, it is logical to assume that the trapped subatomic particles were originally members of solar flare clouds. The curved edges of the outer belt seem, from time to time, to touch the fringes of our atmosphere in the areas of the magnetic poles, and it is thought that this causes the auroras. Therefore, a particle that has been trapped in the outer belt does not stay there indefinitely, but the average time spent in the belt still has to be determined. Of the four bodies in space that have been investigated either directly or by space probe—the earth, the moon, Venus, and Mars—only the earth has been found to have a radiation belt. The conclusion is that of these four only the earth has a powerful magnetic field, but it is not yet known what factor is responsible for this condition.

Life Support.—All missions require certain elements of life support: Water and food; cabin temperature of about 70° F.; removal of excess moisture and carbon dioxide; breathable atmosphere. Until the tragic Apollo fire in January, 1967, the atmosphere was pure oxygen which increased an astronaut's tolerance to stress, simplified leaving the cabin in outer space and allowed a relatively simple system requiring only 5.1 lbs. per square inch pressurization — important in weightlessness. The tragedy virtually ruled out pure oxygen as an atmospheric environment. A multigas system (e.g., helium-oxygen) means more plumbing, greater pressurization thus heavier cabin walls, and an altogether heavier capsule.

■SHORT MISSIONS.—It is elementary to calculate the amount of oxygen, water, and food consumed per person during a twenty-four-hour period. This comes to about three pounds of food (pro-

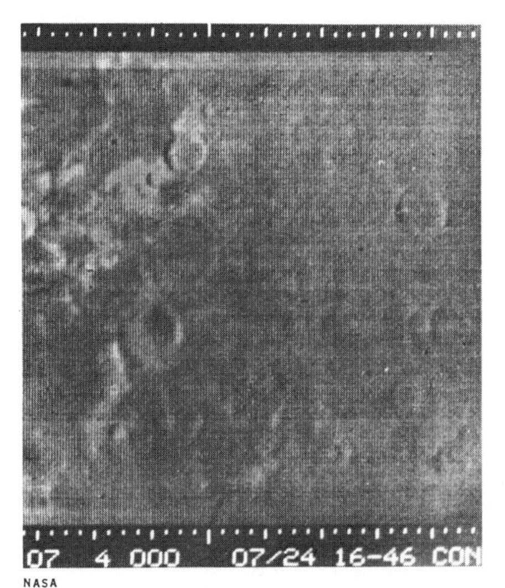

NASA

FIRST CLOSEUP PHOTOGRAPHS of Mars transmitted from the *Mariner IV* spacecraft. The photographs and the vast amount of data relayed from *Mariner IV* answered many questions about the distant planet; the data was invaluable for preparing the future manned trip to Mars.

cessed to eliminate all unnecessary weight and brought to the cheese-like consistency that is the most practical under zero-g), not quite a gallon of liquid, and less than a gallon of oxygen (if it is carried in liquid form). The main problem is the removal of waste matter. Here the ordinary body waste, which is caught and stored in plastic bags, is the simplest aspect. What is more difficult is the removal of the waste that contaminates the cabin atmosphere. Six-tenths of every quart of water a man drinks appears in the cabin atmosphere, partly through skin evaporation, but mainly by exhalation. Everyone knows how annoying high humidity is, so this water vapor must be removed, condensed into water, and stored. It is an interesting fact that a space capsule always ends up with more water than it started out with. Part of the food eaten by the men appears as water, the water being a type of by-product. The other waste material that contaminates the cabin atmosphere is carbon dioxide, which, while not poisonous in the strict meaning of the word, is quite insidious. Ground-based experiments with space-cabin simulators have shown that carbon dioxide, if not removed, results in lethargy and drowsiness. Test subjects simply insisted that they were going to sleep and refused to obey orders. Hence, the carbon dioxide content of the cabin atmosphere must be kept quite low (less than 1/10 of 1 per cent), and this is accomplished by absorbing the carbon dioxide with suitable chemicals.

For a short mission any attempt to reclaim something is meaningless, although it is, of course, likely that tests will be carried out on short missions *as experiments* in order to find good and efficient ways to do it. On a medium-short mission, one lasting more than one day but less than ten days, it is advisable to perform regular exercises in order to keep up the muscle tone. The stretching of rubber bands or steel springs is the simplest exercise, and adds the least weight. The astronauts could perform isometrics, the type of exercises that operate by playing one set of muscles against another set, but psychologically the stretching of springs or rubber bands, where the astronauts can see that something is going on, is a superior exercise.

The characteristic of the life support system for a short mission, then, is that it provides the astronauts with oxygen, food, and liquid in a *clean* atmosphere of a suitable temperature. But all the waste products are merely removed from circulation and stored to be brought back.

■ LONG MISSIONS.—The life support system for astronauts on a long mission has to do everything the short mission support system has to do, but it also has a few additional jobs. For example on a 250-day mission to Mars, boredom will be a very important factor. Even with the most rigorous system of inspection and testing, the working time of an astronaut will hardly be more than four hours per twenty-four-hour period. This means that entertainment, in the form of reading matter, music, and, if possible, movies, must be provided, all of it reduced to the smallest possible size and least possible weight. Another thing that has to be done is to provide what, for want of a better term, is referred to as *artificial gravity*. This could be done by separating the return stage of the rocket system, which would still be full of fuel, from the personnel capsule, connecting the two by a wire rope of adequate length, and then setting the two bodies to spinning around their common center of gravity. In this way centrifugal force would provide pseudo-gravity, which would make living more convenient. It is not necessary that this pseudogravity reach the value of 1 g, for one-third of 1 g would probably be sufficient.

For long missions it will also pay to do some reclaiming of waste matter, particularly of the carbon dioxide. There are chemical means of breaking down the carbon dioxide molecule (CO_2) into its components so that the oxygen can be re-released into the cabin atmosphere. The present methods are fairly clumsy, and it will be necessary either to refine them or to find a new one that can be used inside a spaceship. If external power is available—for example, solar energy collected by a paraboloid mirror mounted outside the ship—it would be comparatively easy to produce oxygen and hydrogen from the surplus water. The oxygen would be retained for breathing, while the hydrogen could be vented into space. A life support system for long missions would thus be characterized by reclaiming some of the waste matter and by jettisoning some of the other.

The ideal situation is, of course, what is known as a *closed ecological system*. Green plants take in carbon dioxide and give off oxygen during photosynthesis. Thus, to green plants carbon dioxide is food, just as is the more solid human waste. It is theoretically possible, therefore, to establish just the right balance, with plants absorbing the carbon dioxide, releasing fresh oxygen, and also utilizing the solid waste as fertilizer for growing edible fruits or parts. Such a completely closed system, however, is too heavy and requires too much room to be used in a moving vehicle. It is a good possibility, however, for a stationary establishment, such as a base on the moon or an outpost on Mars.

—Willy Ley

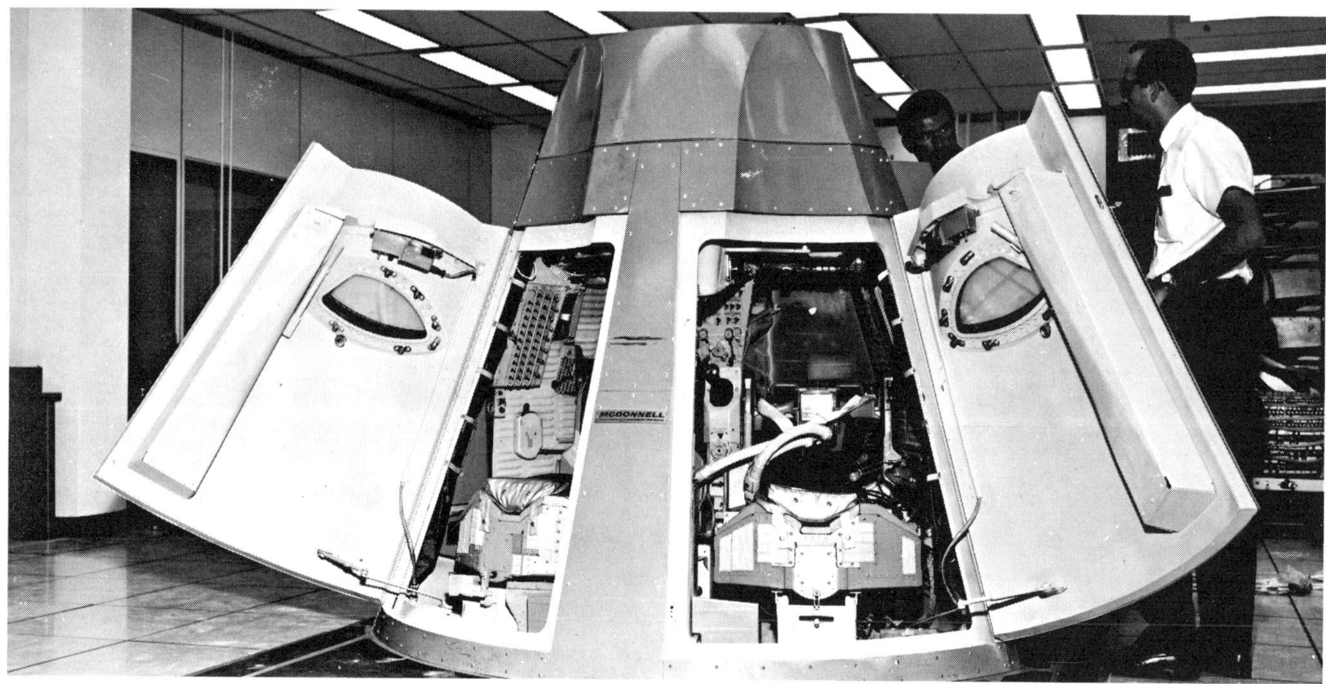

NASA

A GEMINI SIMULATOR, an essential instrument in the space program, provided training for the Gemini flight crew prior to the actual flight.

PART 2

Industry and Technology

INTRODUCTION

Definition.—"The systematic interpretation of nature into a framework of law, we call science; the effort to convert experience and understanding to useful account is engineering," said Julius Stratton, president of Massachusetts Institute of Technology. Elmer Engstrom, former president of the Radio Corporation of America, has added: "Technology is the fruit of this process. It involves both science and engineering, and it depends on their free interaction."

These definitions, as we shall see, are extremely modern in character and primarily applicable only since the mid-twentieth century. In their light, the usual dictionary definition of technology as "industrial science" or the "systematic knowledge of the industrial arts" seems narrow. Historically, technology has been the sum of the tools, methods, experience, and comprehension of a society applied to its improvement, especially by increasing its total productivity. Engineering is the professional application of the tools, skills, and methods that comprise a society's technology.

Quite clearly, technology originated with early man's improvisation of simple stone tools and aids, which enabled him better to cope with his environment. Indeed, the conscious adoption of any form of technology to improve productivity, and to advance toward a more civilized state, is a significant distinction between human beings and animals.

Historically, the definition of technology tends to be influenced by its predominant form at any given time. Thus, the dictionary definition has been influenced by the Industrial Revolution (1750-1850) and the definitions of Stratton and Engstrom by the science-oriented technological revolution of our own time. For an appreciation of its total social significance, however, technology should be defined, in its broadest and deepest terms, as the human employment of any aid—physical or intellectual—in generating structures, products, or services that can increase man's productivity through better understanding, adaptation to, and control of, his environment.

Technology in Society.—The extraordinary development of technology in the twentieth century provides a remarkable perspective on the technological character of a society. The form and management of a society's technology at any time is intimately related to its goals. Indeed, the potentialities of its technology determine, in major measure, the goals to which a society can reasonably aspire. Thus, recent achievement provides an illuminating contrast to the technological levels of society only a few generations ago.

As a solitary human being acting alone, man can enjoy only the barest existence. Alone, he could live only as a wild animal, with a minimum of food, shelter, and protection. Only as he forms a social group can the productivity of the group as a whole be increased to provide more adequately for the needs of all. Through an organized society, each individual can do one or a few tasks more effectively; through learning, through mastery of tools and skills, and through articulation of the jobs done by each individual, the assembled group can provide most effectively for the society that it forms. Thus, perhaps one basic purpose of society is to increase the effectiveness of each individual—enable him to make a greater contribution to, and, thereby obtain greater benefit from, society. Technology is the means by which man's effectiveness can be significantly enhanced; it provides the methods and tools whereby he can multiply his skills and energies.

Of course, material well-being is not the sole purpose of a society. Certainly, government and philosophy, and art and literature are also among its aspirations. Yet man's spiritual and intellectual side cannot help but be blunted when the great mass of people have to struggle for bare existence in an uncontrolled and primitive environment.

Through most of history, improvements in individual productivity were relatively minor—the evolution from stone to bronze to iron for tools; a water wheel or a windmill for energy; a horse-drawn plow for turning the soil. None of these offered society much hope of dramatic improvement in the general standard of living. Under such primitive technologies, with their restricted levels of productivity, most men were destined to poverty with its inevitable brutality of the spirit. No ideology could substitute for inadequate technologies.

■**THE NEW SCIENCE.**—But in recent times, a radical change has taken place. Four centuries ago, dominated by the philosophies of René Descartes and Francis Bacon, science took a new turn. Speculation about nature was deemed useless, unless it could be verified by experiment. Gradually out of man's imagination, rigidly controlled by observation and experiment, a deeper comprehension of the behavior of nature was acquired. The process at first was very slow, but it quickened until, in the twentieth century, the detailed comprehension of natural phenomena has come with a rush. By his ingenuity man has penetrated deep into the vastness of space as well as into the inconceivably tiny atom, and by exploring the chemical structure of the living cell, he is beginning to expose the very secrets of life itself. Through his intuition and his artistic sense, man is connecting seemingly unrelated events of nature in a synthesis of the basic laws that govern these events—laws that, in turn, generate deeper insights and enable man to predict nature's most probable behavior in complex situations. Out of this more comprehensive science have arisen powerful new technologies, new ways to do new things far more effectively and often at a greater economy than ever before.

■**THE NEW TECHNOLOGY.**—In developing and learning to manage the new science-based technology, man has doubtless introduced a vast step-function in his level of civilization. The Stone Age, Bronze Age, and Iron Age represented distinct advances in man's social level as a consequence of new technologies, based on the discovery of new materials and tools that raised his level of opportunity and productivity. Yet each of these social developments was minor compared with the vast opportunity offered by today's science-based technologies.

Man has finally learned that the elimination of poverty can be achieved only through the tremendous increases in productivity that these new technologies can confer. The old engineering, derived from the total of day-to-day experience in the mechanical arts, simply had no hope of satisfying such goals for society. No idealistic philosophy could enrich a society of limited productivity, whose members derived their wealth from elementary technologies. Only by understanding nature in scientific and technological depth could man arrive at the high level of adaptation to his environment at which great advances in productivity could be realized.

The development of the remarkable science-based technologies in our time has enabled man to adopt new social goals with confidence. First, man can now visualize a level of productivity, conferred by the new science-based technologies, in which poverty can be totally obliterated.

Second, the development of new medical technologies represents the opportunity of healthful participation in society for a vast majority of its members. Since the American Revolution, life expectancy has risen from a mere 30 years to the Biblical goal of three score and ten, the major gain being since the turn of the century.

Third, with relief from the compulsion of satisfying daily and immediate needs, society can now establish new and major goals derived from man's age-old yearning for comprehension of the universe around him—goals such as the exploration of space, the moon, and the planets; the elucidation of the character of matter, or of the nature of life itself. As these goals are achieved and enlarged, they extend man's horizons and react in turn by enlarging still further the capability of his technological systems.

No nation can rise toward goals and aspirations which lie beyond its access to technological aids. The quality of its art, its literature, its music—the whole of its human attitudes—is dependent upon a sufficient measure of productivity to release men for such pursuits. But we must understand quite clearly that even a superb technology, suitably employed, cannot alone ensure a high level of cultural development. To employ a technology successfully, a people must be guided by a sense of purpose, a sense of morals and ethics, suitable to govern that technology for society's optimum benefit. However, without a suitable technological base, high cultural goals are beyond the reach of the great mass of people. With only primitive technologies at hand, man had little choice in selection of social goals. Only with the technological revolution of the twentieth century has the opportunity for choice of social goals been opened up. Unfortunately, the experience of the past offers few social guides to the successful management of the future.

Development of Technology.—The earliest record of technology is found in archaeology, where its tortuous development can be traced to the beginning of time. Indeed, the earliest suspected human remains—*Zinjanthropus,* found in 1959 by Louis S. B. Leakey 40 miles east of the eastern tip of Lake Victoria, Africa, and dated as 1,300,000 to 1,750,000 years old—are noted "as probably qualifying as man because his remains were found among stone tools." Thus, the acquisition of a technology of stone tools is considered a critical factor in the identification of early man as distinct from animals.

Archaeology usually divides man's development according to the character of his tools and their technological employment, following the suggestion of Christian Jurgensen Thomsen in 1836. Thus human history can be divided into three main stages: the *Stone Age,* when tools and weapons were made of stone, wood, bone, etc.; the *Bronze Age,* beginning about 2000 B.C., when tools of copper alloyed with tin came to dominate technology in advanced European, Asian, and African societies; the *Iron Age,* when iron began to be used as a basic element of technology, its significant employment dating from about 800 B.C., through earlier uses have been identified.

Thus, archaeology implicitly recognizes that the state of artistic, spiritual, and philosophical development of early civilizations is directly correlated to the technologies accessible to each civilization. While each of these human periods—stone, bronze, and iron—has been finely subdivided in terms of the intellectual development of each relatively isolated civilization, that development was essentially restricted by the character of its technology.

The development of technology down to the present century has been chronicled in many works. From the extraordinarily complex interactions in this development, certain broad generalizations can be perceived.

SMITHSONIAN INSTITUTION

TECHNOLOGY grew from pure science with such inventions as the 1654 von Guericke air pump, which showed the power of a vacuum. Air was pumped out of two hemispheres and the vacuum created withstood the strength of horses pulling in opposite directions.

EARLY TECHNOLOGY.—Man's original technologies were born of necessity by pure discovery and invention. These arose out of the need for shelter and clothing, more efficient and dependable supplies of food, and more adequate means of defense. As cities were formed, man needed water and sanitation, and roads and bridges for communication and commerce. Scientific knowledge was neither sufficiently advanced nor in the form necessary to contribute significantly to early technological progress. Indeed, it might be said that until technology could provide the means and materials for precise measurement and experimentation, modern science was inconceivable.

In its early stage, each technology developed as a "mechanic's-art" out of the gradually accumulated knowledge, skills, and experience of the ages. Technological processes and skills were transmitted by apprenticeship and through guilds (or their equivalent) until relatively recent times. Today's advanced engineering still bears many similarities derived from these early roots. But in the absence of broad and systematized knowledge of nature, provided by an advanced science, the value of these early technologies was slight.

Moreover, there was little stimulus for technological innovation. Scientific speculation was largely the domain of a leisured and cultivated elite, whereas the making and doing of things was left to the lower classes. There was little incentive to devise labor-saving machinery, for mechanical problems involving the application of more power could be resolved simply by the use of more human muscle power in the form of more slaves. Thus, for many centuries slavery retarded technical advances in the mechanical arts. The Hellenistic scientist, Heron of Alexandria (first century A.D.), knew of steam power, but he invented frivolous devices to amuse, delude, and impress worshippers in the temples, rather than to perform work done by humans.

THE MIDDLE AGES.—Although the Middle Ages was a period of economic disintegration, political fragmentation, and intellectual decline in the Western World, it witnessed many advances in technology. The decline of slavery necessitated new sources of power. Consequently, the waterwheel and the windmill were introduced into Europe near the close of the 1100s.

Two major technical advances, as well as several agricultural inventions, helped bring the Middle Ages to a close and establish the basis for modern times. Gunpowder, introduced into Europe in the 1300s, contributed to the demise of feudalism and stimulated the development of the mining and metallurgical industries. Printing, made possible by the invention of movable metal type in the mid-1400s, revolutionized every aspect of man's culture. It helped disseminate the "new" science, which was to be one of the roots of modern technology.

THE SIXTEENTH CENTURY.—Science as we know it today—the rigorous exercise of human thought by observation, and more particularly by critical experiment—can be traced back only four centuries among the 70 centuries of recorded history. This new science seems to have arisen from the convergence of three major events toward the end of the sixteenth century. The first was the conception of the precise description of natural events by the application and development of the mathematics of the ancients. The second was the use of precise measurement in experiment to supply the data from which a mathematical description could be formulated. The third was the invention and use of the printing press for the rapid and widespread communication of scientific data. The scientific conceptions arising out of this convergence were guided by the recognition that the most simple and general synthesis of observed facts best approximated the truth.

Prior to this convergence, science was largely built out of the imagination, and technology out of crude empirical experience. Precise astronomical measurements made by Tycho Brahe in the 1500s laid the basis for discoveries and developments in mechanics by Johannes Kepler, Galileo, and Isaac Newton. The great advances of science awaited the thought of Bacon and Descartes, both of whom championed experimental science. For them theory was only a means to direct experiment and observation, not an end in itself. But experimenting and observing require means of measurement. Thus experimental science grew as measuring instruments became more precise.

THE SEVENTEENTH CENTURY.—The barometer of Evangelista Torricelli (1643); the telescope of Galileo (1609); the early thermometers of Galileo (1593), of Jan Baptista van Helmont (1630), and of Jean Rey (1632), which were followed by the precise instruments of Gabriel Daniel Fahrenheit (1724); the air pump of Otto von Guericke (1654); the pendulum clock of Christian Huygens (1656)—these and related developments represent the beginnings of the precise instrumentation out of which the new science was to grow. Although reasonably precise measurements of mass and length had been made by the ancients, even these awaited the science of the sixteenth and seventeenth centuries for significant refinement.

As instrumentation grew more sophisticated, the interaction between science and technology became greater. Developments of the thirteenth and fourteenth centuries, including convex lenses by the glass workers of Venice, and the box compass, an invention of early navigators, prompted the dreams of Columbus. Throughout the subsequent growth of science, instrument makers drew on the skills of trade to supply the special materials, means of fabrication, knowledge, and manipulative capability derived from technological experience. In turn, the new means of measurement began to lift technology from the morass of empirical art to the solid foundations of science.

THE INDUSTRIAL REVOLUTION.—In tracing the beginnings of the industrial revolution (1750), one historian has observed that "the industrial state, as now understood, did not exist, although the foundations for it had been well laid. . . . The wide use of power-driven machinery, although a basic factor, was not all-important. The application of science to technology played an increasingly effective part." Pure science, however, did not lay at the basis of the social, economic, and technical changes that sparked the Industrial Revolution in Britain. Other major factors were the expansion of trade and agriculture, social and political systems favorable to change and business, the availability of raw materials, and the growth of an extensive transportation network.

Scientific discoveries encouraged technology but technological advances also encouraged the development of science. Even before Watt's steam engine, the Industrial Revolution had begun with the advent of textile factories and powered spinning and weaving machinery.

The men who guided initial industrial growth in Britain were practical, little concerned with purely scientific studies. But the growth of the new industrial technology itself also required scientific growth. The steam engine's efficiency, for example, could not be increased without thermodynamic studies. A close link developed between scientific investigations and practical technical applications. Much of technology was becoming "applied science."

BEGINNINGS OF MODERN ERA.—Before the twentieth century, however, the scientific roots of technology were slow to take hold. Consider the development of electric power. Only after electricity was lifted from the realm of parlor curiosity a mere century ago by the scientific theories and experiments of Henry Cavendish, Michael Faraday, Joseph Henry, Charles Augustin de Coulomb, Georg Simon Ohm, James Clerk Maxwell, and their successors, could man conceive of unlimited energy transmitted from a distant source to await release at his command. Today living standards are directly proportional to the per-capita consumption of electric energy.

Basic technological inventions came directly out of the work of Faraday and Henry, and without much delay. Yet society's acceptance of controlled electric energy was painfully slow and labored. As one searches for the explanation, he finds many factors. Electrical technologists were few, for university training in engineering was in its infancy. Means of employing controlled energy awaited further research, invention, and development. Efficiency of application awaited the elaboration of nature by science. Above all, the public was not ready to buy on a scale sufficient to stimulate and enjoy the tremendous social benefits that electric power offered. Even as the twentieth century unfolded, electricity was viewed by most people as a kind of plaything only distantly related to needs and aspirations.

So any analysis of scientific innovation must take into account that a society has a time-constant during which it exhibits severe resistance. How to shorten this time-constant would be worth some study. One suspects that constant and broad public exposure to the most advanced fronts of technology, such as the space program, may have a powerful influence in reducing public resistance to advanced technology as a whole. If this is so, perhaps the space program will be repaid many times over, merely by the advantages of earlier acceptance of new products and services that can mightily benefit society.

THE MODERN ERA.—The turn of the twentieth century brought full development of the scientific roots of technology—a development that heralded the technological revolution whose full impact was to be felt by the midcentury.

The great inventors—Thomas Alva Edison, Henry Ford, the Wright brothers, Alexander Graham Bell, and Guglielmo Marconi—tapped the most elementary manifestations of science to create new industrial potentials, which, in turn, created altogether different kinds of industry and opened human potentials inconceivable before.

In the ensuing half-century, man's comprehension of the physical character of nature was revised completely and expanded enormously. He became able to deal with the minuscule—the atoms and nuclei from which matter is formed. As a result, all gross products of reactives could be predicted on the basis of a few equations.

With scientific groundwork laid, the full impact of the technological revolution became inevitable. Moreover, World War II had forced consummation of the marriage of science and technology so that the roots of technology by the midcentury were, for the first time, predominantly in science.

The Technological Revolution.—No social revolution is ever instantaneous, yet the rise of technological revolution at the mid-twentieth century was remarkably sudden when measured in the scale of human affairs. Perhaps its basic technical characteristic has been the development of electronics, whereby one can use sensors and amplifiers to detect and measure phenomena far beyond the capability of any human sense. This includes the modern computer, which can solve problems with incredible speed, can order the information, and produce results that could not otherwise be obtained in a lifetime.

Perhaps the most profound social consequence of the technological revolution has been the change in the character of the economy. Prior to and throughout the Industrial Revolution, man's principal effort, with the then primitive and empirical technologies, was to supply the basic necessities for living—food, clothing, shelter, and a minimum of protection and transportation. Even well into the twentieth century, the economy was primarily one of minimum necessity. Such an industrial economy is sharply population-limited to supplying the food that could be eaten, the clothes that could be worn, and so on. But the new science-derived industry of the present century enables man to adapt better to his environment. This new industry might primarily be called the "adaptive" industry, which, significantly, is not population-limited, for there seems to be no foreseeable limit to man's adaptation to, and command of, his environment. Already it forms more than half the economy, with no limit on its applicability in sight. To paraphrase Elmer Engstrom, in the 40 years from 1923 to 1963 the country has grown in population from 108 to 192 million, while its gross national product has multiplied nearly eight times from $80 billion to over $600 billion annually. The bulk of this greater product consists of goods and services that were entirely absent or relatively insignificant in 1923.

The actual process of transition from science to a suitable technology is not completely clear. One widely accepted public misconception should, however, be laid to rest. This is the idea that there is anything like a one-to-one correspondence between a scientific discovery and a consequent technology. A new technology depends on thousands of scientific advances articulated in the most suitable way. Jet aircraft technology, for example, depends on solid-state physics and physical metallurgy for materials of suitable properties; on progress in thermodynamics, fluid dynamics, and fuel technology; and on a wide variety of advances in communications and control. When it is said that a man can be sent to the moon in 1970, what is meant is that scientific progress on a thousand fronts now permits the development of a myriad of related technological processes, all of which must be suitably woven and articulated to form an adequate manned-space technology.

THE RISKS.—The development of a new technology requires the combined efforts of the scientist, the engineer, and the entrepreneur who is willing to take risk. Adopting a new technological strategy presents many risks. Will the estimated combination of scientific resources produce the technological result anticipated? How can these be combined to produce the creative result visualized? How can the resulting product or service be made usable and marketable? How can the product or service be introduced so as to be accepted in areas in which it has either been unknown or previously offered in an inferior manner? The whole process involves a very special kind of imagination, faith, and creative genius, supplemented by a high order of education and manipulative dexterity. Access to a broad range of up-to-date scientific knowledge is imperative.

ABOVE: U.S. AIR FORCE RIGHT: AMERICAN TELEPHONE AND TELEGRAPH COMPANY

GOVERNMENT AND INDUSTRY frequently join forces in technological ventures. Air Force research that led to the delta-winged bomber (*above*) has since been applied to the design of a planned commercial supersonic transport. The Telstar Communications Satellite (*right*), privately built and owned, was launched and is used by the government and industry.

For any technological strategy undertaken by an organization to be successful, the interval from its inception to its profitable exploitation should be about five years, certainly no longer than seven. To adopt longer strategies means the end is too fuzzy, the danger from alternative competing strategies too great, the cost of development too uncertain.

Thus, to initiate new technological strategies, an organization is usually staffed and equipped according to certain criteria. It must have intimate knowledge of the literature and practice of science, so that potential innovations can be recognized as they emerge from whatever combination of scientific sources. Government and industry must have access to groups of men so close to all working aspects of science that potential innovations fitting their general mission can be immediately recognized and interpreted. A central purpose of basic research in industry and government is to acquire a vanguard of men who are sufficiently informed over broad areas of science to recognize and to initiate new technological strategies out of the whole of scientific advance. From alternative strategies, selection can be exercised and decision made. This process appears to be most effective if that vanguard of strategy-makers actually participates in basic scientific research. Thus, most organizations in today's technology support a considerable effort in basic research.

Since innovation may arise from a synthesis of a wide variety of relatively unrelated scientific developments, the task of industrial or government laboratories is to articulate quite independent scientific discoveries or advances into a mission-oriented body of science and technology. This is possible only if the laboratory has full access to the total of basic scientific advance by intimate participation in it. It seems no accident that the formulation of Claude Shannon's information theory, which has revolutionized communications and data handling, should come out of the Bell Telephone Laboratories, though the scientific pieces from which it was constructed were diverse. Radar also originated from work done in the laboratories of several countries, with its ideas having various scientific origins.

Finally, through its own facilities or through contact, the organization must have access to the capability for development and engineering that can bring the idea to fruition.

ROLE OF THE GOVERNMENT.—Experience over the past two decades shows that the really great technological strategies, such as space exploration, Antarctic exploration, atomic energy, radar, aircraft navigation, and supersonic jet travel, must be initiated by government. It is of interest to observe that even in such a highly private business as rail transportation, no significant technological progress was made until the recent experimental program of the Japanese government and the more recent program of the United States. As the outstanding exception, the communications strategies of the telephone and telegraph companies represent great contributions by private initiative.

ROLE OF PRIVATE ENTERPRISE.—On the other hand, smaller yet considerable technological strategies have come from private enterprise and have been produced in the most suitable form out of the competition of the marketplace. In a sense, the greater technological developments await the perfection by private entrepreneurs of the dozens of preceding technologies which make these possible. Experience indicates that the greatest success occurs in an open society with decentralized control, offering competitive opportunity for independent approaches to most suitable solutions.

Underlying the whole process of technological innovation is assumed a steadily expanding body of scientific knowledge to which it has access. Since, in basic science, the end result cannot be foreseen; since, in any event, the lead time from basic research to a product or service may often be too long, there is no broad

economic motivation for basic research other than the need for coupling science to technology, as previously described. Consequently, the support and encouragement for the exploration and elaboration of basic science must come primarily from government.

■ROLE OF THE UNIVERSITY.—In the Western system—for example, in the field of agriculture under the Morrill Act of 1862—basic research has been conducted at the universities, and at other advanced research institutions, largely with government support. This is a particularly efficient method, since the research can be coupled with the training of graduate and postdoctoral students and the advancement of the teachers.

The whole operation of systematically assimilating the resultant science into the technological process is still too new to be highly developed. Although the university cannot be expected to assume direct responsibility for innovation—which are properly assigned to industry and government—it can be expected to organize scientific knowledge in a form most readily digestible by society in its research for new industrial strategies.

Technology and Education.—The universities and professions were slow to recognize the new science as a legitimate branch of learning. Paracelsus' (1493–1541) attempt to introduce chemical science into medicine was instantly resisted. In England, the quarrel between the "new" scientists and the Oxford dons led to the formation of the independent Royal Society (1662). Not until the mid-nineteenth century did science and engineering achieve wider acceptance as legitimate subjects for university training.

As shown in the accompanying table, the onset of the technological revolution is graphically illustrated by the explosively rising demand for advanced education. This is doubtless the most remarkable social development in human history. Not only does it reflect the urgency for advanced education in a technological age, but it also represents a marked change in the tastes and desires of the marketplace for the quality and character of its products.

To meet this educational need, undergraduate institutions have been planted and have flowered in every state of the United States. There are between 3,000 and 4,000 such institutions, depending upon which categories are included. That these must double in their total capacity in the next two decades is no longer questioned.

However, as one views the national capability for graduate education, the picture is quite different. To advance its science base, and to transform that base into new and useful technologies, society quite suddenly has become aware that advanced graduate education is needed on a broad scale. If new technologies are to be made effective, highly educated leaders are needed in sufficient numbers—men whose insights can perceive the opportunities that nature offers and who can organize the environmental patterns to bring advantages to the great mass of our people. This requires training to the level of the doctoral degree, and continuing education beyond.

The first American to receive a doctoral degree in engineering was Josiah Willard Gibbs (Yale, 1863). Since then, the development of graduate training in engineering has resulted in the growth of specialized higher engineering degree programs in the United States.

Specialized technological institutions, such as the Massachusetts Institute of Technology and the California Institute of Technology, which combine teaching and research, have made engineering increasingly scientific and have encouraged technological innovation. In the newer and more sophisticated technologies—for example, nuclear energy and rocket propulsion—the dividing line between science and engineering has virtually disappeared.

Since World War II new fields of study have emerged in such areas as material sciences, operations research, and systems research. Emphasis is increasingly being placed on the optimum use of organizational resources, as well as mechanical devices and material input, in solving problems in the production and distribution of goods, services, and information. Despite the need for technological leadership, only about 2.5 per cent of those intellectually qualified achieve doctoral degrees. This is a great loss of potential experts in technology.

Graduates in the United States

Year	Baccalaureate Graduates (per annum)	Doctoral Graduates (per annum)
1900	28,000	400
1920	50,000	700
1940	200,000	3,500
1960	450,000	10,000
1980	(estimated) 1,000,000	(?)

Problems of the Technological Age.—The achievement of virutally unlimited productivity, made possible by the new science-based technologies, presents society with a plethora of new problems induced by this wholly new situation. These include:

■POPULATION.—With the science-based medical technologies of the present century, populations now double in 35 to 40 years. The hypothesis of Thomas R. Malthus (1798) has become a reality, the arithmetic of which cannot be refuted. Man must quickly find a solution to uncontrolled growth of populations, or he must surrender to widespread hunger, brutality, and death. Our basic resources of atmosphere, water, and materials must soon reach their limits if populations are uncontrolled. With reasonable control, these resources, manipulated with advancing technologies, can be made to support society indefinitely.

■WAR.—The new technology has not eliminated the political problem of war. Man has, in fact, used technology to create weapons of ultimate destruction. The survival of life itself depends on man's ability and willingness to control the arms he has produced.

■UNDERDEVELOPED NATIONS.—The pressures for technological advance derived not only from population growth but also from the "rising expectations" of the new nations that came into being after World War II from the former European colonial empires. Filled with pride at their emerging nationhood and desirous of acquiring the material benefits of Western technology, the less-developed countries in Africa, Asia, and the older states of Latin America, have sought to improve the standard of living of their people and to increase their national power.

The problem of transferring Western technology to non-Western peoples, some of whom are still in the early stages of industrial development, is one which must be resolved lest the differences between the standards of living of the industrial nations and the "have-nots" reach a point where international disorder may result. Despite large-scale technical and financial assistance, the hopes of the "revolution of rising expectations" have not been fulfilled. In many cases, the new nations have sought only "prestige" projects, such as large-scale steel mills, even though their economies will not enable these to run efficiently because of a lack of resources, capital, and skilled labor; or they have sought only to advance their military powers while ignoring the living needs of their people. Furthermore, the aspirations of these people have sometimes been sacrificed for the sake of power by their leaders and by the larger nations of the world in the East-West power struggle.

It is evident that the industrialization of the underdeveloped areas is no simple task. It involves much more than the mere transfer of machinery or the application of large amounts of capital. Industrialization requires cultural adjustments.

and changes in the entire political and social structure of the less-developed areas.

■**METROPOLITAN DEVELOPMENT.**—The impact of the new technology in individual communities has had important consequences. Agriculture has become automated and industrialized to the extent that the open areas of the nation are being depleted of population. In the last decade, in Texas, for example, a half-million people have left the land for the cities. Texas' repidly growing population is centered in nine or ten metropolitan areas.

The same tendency is manifested everywhere in the United States. In whatever way one may interpret this change in light of primitive and historic norms, the U.S. population will soon be centered in some 100 to 125 great and affluent metropolitan areas. And within the framework of present social concepts, employment for this concentrating population will have to be found in the new science-based technologies.

To measure the task ahead, it can be foreseen that each of these metropolitan centers must have a great graduate university at its hub. The university will be required to adopt new forms in its service to the whole community. As Clark Kerr has put it, in the new science-based society every metropolis will have to be a "city of intellect." With this must come viable city planning so that the metropolis remains both competitive and livable in the most artistic sense.

■**POLLUTION.**—Spreading to all nations, and with rising populations, the new technology can cause deadly pollution of the atmosphere, soil, and water, unless it is severely controlled. Strict pollution control must be accepted as a fully justifiable cost if national resources are to be preserved. With new levels of productivity, this cost can easily be absorbed. Moreover, with the new technology, no form of pollution need be tolerated, since solutions to every problem are conceivable.

■**AUTOMATION AND INNOVATION.**—The very basis for increased productivity of the individual is the increase in his control over whatever mechanism or process he may be directing. This is often called *automation*. As the degree of control of the individual is enhanced, the requisite level of training, skill, perception, and education to exercise that control is raised. At the same time, the number of persons previously occupied in the process is diminished.

To provide continued opportunity for employment of those displaced, further *innovation* from science is required. The day of the old-fashioned inventor is largely past; the source of innovation today is the highly educated scientist or engineer with the laboratory facilities to devise and develop new technological ideas and methods.

Thus, sustained employment involves the race between unemployment arising from automation and new opportunity offered by innovation. Success at every level in this race requires ever-enlarging opportunity for more adequate education and higher manipulative skills. This represents a serious problem for those who are displaced. To achieve the benefits of higher productivity offered by the new technologies, the community must provide opportunity for more advanced education and retraining in more advanced skills at every level of employment.

These are but a few examples of the utterly new problems that the new science-based technologies pose to society. As man finds the means to achieve his social goals, at the same time he acquires altogether new problems. But it is of critical importance that he see these problems in their full perspective so that in searching for the solution he does not, metaphorically, kill the goose that lays the golden egg. The power of the new technologies creates a more delicate balance of social forces, in which ancient moralities and ethics must be evaluated and reevaluated in light of their new and far-reaching consequences.

Man or Machine.—Man's most basic aspiration is equality of opportunity for relief from want, disease, and brutality—affluence, if you like. He has now acquired the means to achieve that aspiration—a technology with which he can realize equality of opportunity. But this new technology can also be misused, particularly if it is employed in accordance with the norms of a more primitive civilization. In such a social framework, population increases, war, uncontrolled contaminants of living, unplanned metropolitan growth, and retention of unproductive and primitive pursuits can lead to serious and even destructive consequences. The same technological base at once can free man and destroy him.

If man is to employ the new technology to achieve his goals, he must at the same time adjust his habits to control its potential abuses. He cannot destroy the technological base itself without at the same time destroying his hope of achievement. The focus, therefore, should be on the specific social adjustments that enable man to acquire the benefits and, at the same time, remedy the abuses.

The basic human problem ahead is whether man shall control the machine or fall slave to it. It is of little profit to try to destroy the machine and, with it, man's opportunity to rise to new heights of civilized opportunity. Equally, it would be of little profit to subordinate man to the very machine that can provide him his hope. To use the machine to lift man to new heights in the command of his environment and in the free exercise of his creative spirit is the great human problem of our time. It transcends all others. It cannot be solved by the oversimplified and time-worn clichés of the humanist, the scientist, or the politician. As man moves into the new technological age, with its enormous potentials for individual study and expression, creativity and freedom, he is confronted with problems which, for the most part, have not been encountered before. At the same time, there are at hand altogether new opportunities for their solution. In this delicate social atmosphere, the solutions of the demagogue conceived "off the top of his head" or of the reactionary who relies too heavily on the past are unsuitable, even dangerous. The solutions will require deep and scholarly study, perception, and comprehension of both the machine—its uses and potentials—and of man—his hopes and aspirations. The humanist of the future must comprehend and encompass both. —Lloyd V. Berkner; Melvin Kranzberg

LOS ANGELES AIR POLLUTION CONTROL DISTRICT

AIR POLLUTION in cities such as smcg-bound Los Angeles jeopardizes the public health. Automotive exhaust fumes and industrial wastes are primary contributors to air pollution.

COMMUNICATIONS AND TRANSPORTATION

Early Communications.—Ever since the dawn of history, every step in the growth of civilization has depended on man's ability to improve his techniques of communication in terms of: the amount of information conveyed; the distance over which it is sent; the time taken; the certainty of its arrival; and the number of people who could participate in the exchange simultaneously.

Improvement in response to each of these five challenges over the centuries makes up the history of communication—and the very building blocks for the growth of civilization. Today it is possible in one second to move thousands of words thousands of miles to millions of people. This is a long way from the limitations imposed by face-to-face communications at the dawn of history.

Progress has been achieved through a series of discoveries that continue to come at a fantastically accelerating rate. Each of these has brought new forces to bear on the development of civilization itself.

■**FIRE CODE.**—The signal fire, demonstrated in sending the news of victory back to Greece from Troy in 1084 B.C., was perhaps the first communication breakthrough. A single coded word, which could span land and water at a speed equivalent to 100 miles an hour, replaced the messenger who averaged 10 miles an hour. In the next 2,800 years, the signal-fire approach was perfected. First, scientific coding techniques were developed. The classical historian Polybius, in the fourth century B.C., devised a system of two sets of five lights each to send twenty-five individual letters and numbers, thereby increasing the amount of information that could be rapidly relayed by day or night over long distance.

■**EARLY ADVANCES.**—Then other modes of "transporting" information were developed, such as signal flags at sea, horns in the Swiss Alps, tom-toms in the jungle, and smoke signals on the plains. Even the ancient art of training homing pigeons was revived on grand scale by Napoleon's armies in the early nineteenth century. The latter made it possible to move more information—even a 1,000-word message from a moving army or ship—but the speed was not improved.

■**SEMAPHORE FLAGS.**—It was not until the close of the eighteenth century that another significant advance was made. By 1794 the mechanics of signal towers or flags had reached their ultimate effectiveness through the development by the French of the techniques of semaphore flags. Then, for the first time, *large* amounts of information, not just single words, could be relayed rapidly over long distances at an equivalent speed of 100 miles per hour.

The Nineteenth Century.—By 1800 the art of communication was beginning to be recognized as a significant element in the expansion of civilization. Books and journals were being regularly published; more and more people began learning more and more about others who lived far away. Increased communication among people created new problems in education and understanding, and demanded still more improvement in both the speed and volume of communications. In effect, "brainpower" was being developed, and man's muscles were soon to be multiplied through the Industrial Revolution, but the "nerve system" we call communication was still thin and weak. Society was beginning to need a "nerve system" capable of handling considerably more information and doing it faster.

GENERAL ELECTRIC

MULTI-CHANNEL TRANSMISSION networks sort out and route millions of transmitted signals, providing for efficient handling of voice calls, teleprinter information, and long distance computer/data links.

■**TELEGRAPH.**—In 1844 Samuel F. B. Morse demonstrated the magnetic telegraph by transmitting the message "What hath God wrought?" from Baltimore to Washington, D.C. At that moment the world witnessed the first of what turned out to be a series of communications innovations. Not only could news be sent at a rate equivalent to 1,000 miles per hour (counting the time it took to encode and decode the messages), but a sort of slow-motion two-way conversation was now possible.

Obtaining the acceptance or refusal of a potential political candidate while a convention remained in session, calling for rescue from attacking Indians or rampant floods, issuing crucial business orders to buy or sell on the market at a certain time—these were but a few examples of the new role of communications made possible by the magnetic telegraph. During the U.S. Civil War, the telegraph network was extended to span the nation. In 1868, ten years after a short-lived initial test, telegraph messages could be sent across the Atlantic by cable, thereby linking nations at speeds 100 times faster than before.

■**TELEPHONE.**—Then, in 1875, another discovery presaged a technological explosion in communications. While experimenting to devise a means of carrying several telegraph signals over one wire at the same time, Alexander Graham Bell happened to notice what seemed to be the transmission of sound by way of the reeds that he was trying to use to receive the separate electrical signals. He quickly diverted his efforts to this discovery; within a year he devised and demonstrated the telephone.

Bell's basic invention was the conversion of sound pressure waves into electrical energy and back again. After trials at short distances, telephone lines lengthened steadily, and by 1915 lines reach coast-to-coast.

Twentieth-Century: Radio.—New means for carrying electrical messages where wires could not be run, such as to ships at sea, were also being developed. The real impact of the "wireless" on communications was first demonstrated in 1899, when a lightship 12 miles off England used its radio to save all those aboard after being struck by another ship. Just two years later, in 1901, Guglielmo Marconi showed that the radio waves would follow the curve of the earth when he received the letter "S" transmitted in code across the Atlantic. At first radio, like the telegraph, could be used only with coded signals, but by 1906 the combined effect of Bell's invention and radio was achieved experimentally: Christmas music was played over radio and heard by receivers in the Boston area. By 1915 voice could be carried across the Atlantic, and around the world over a combination of radio and cable by 1935.

Radio also made it possible for one person to communicate with many at the same time on a one-way basis. Broadcasting, as we have come to call this application, started in 1920 with stations KDKA in East Pittsburgh and WWJ in Detroit. Music and news were offered to those experimenters who had receivers.

The invention of the triode vacuum tube by Lee De Forest in 1906 and development of the superheterodyne receiver by Edwin H. Armstrong in 1918 were important steppingstones in the rapid development of radio. By 1940, only 41 years after the first practical use of radio, long-distance two-way telephone and broadcast voice communications were a basic part of civilized society. They reached not only homes but also ships, planes, trains, and automobiles.

Within a span of less than a hundred years, communication had advanced from relaying a letter or a word at a time to the point where any two people could talk as in normal two-way conversation, though separated by thousands of miles. Not only could much more information be conveyed via the telephone—as much as one can generate in his speech—but the fine points of emotion—concern, love, or anger—and even personal recognition of the addressee's voice could enhance the communication. Presidents could now persuade kings directly when all else failed. Business contracts, subject to confirmation in writing, could be achieved on a worldwide basis as fast as political or economic circumstances changed. Vast audiences and whole nations could be entertained, informed, educated, and persuaded.

Television.—Such progress stimulated greater progress, and attention naturally centered on transmitting visual information. The basic truth in the saying that "a picture is worth 10,000 words" is certainly evident in the field of communications, whether measured in terms of effort required or the impact achieved.

In 1924 it first became possible to send a photograph via radio across the Atlantic, but it took 20 minutes. Only 40 years later a new picture could be sent over the same distance 60 times a second (the television rate)—an improvement of almost 5 orders of magnitude in speed, along with better quality and reliability. What happened in just four decades was the technological explosion of television. As in the case of voice transmission, the key lay in the invention and development of the device that would convert the "information" into an electrical signal and *vice-versa.* Vladimir Zworykin developed the "camera" tube (iconoscope) and Philo Farnsworth developed the "projector." In the 1930's television was still in the experimental stages, and it had to be put on the shelf during World War II. However, the tremendous technological strides made in electronics during the war rapidly advanced the television art between 1946 and 1950. Coaxial cables and microwave radio, products of World War II, became the work horses for carrying the high information content in broadcasting pictures coast-to-coast by 1951. And in 1962 the Telstar communications satellite provided an adequately broad communication highway for carrying television signals across the Atlantic. In 1964 commercial videophone service (telephone with picture of caller) was established between Washington, D.C., New York, and Chicago. A year later, pictures of Mars were transmitted 150,000,000 miles to be seen within a few hours by millions of people.

Future Communications.—In the year 2000, history will consider these marvels of our day as but crude beginnings of whole new eras of communication experience. The explosion continues, and within another 40 years it is quite likely that *all* information will be distributed over a single cable to every home, in the same way that we now transport energy by a few wires for every need in the home. Over this "information umbilical cord" will come: visual contact with friends and salesmen; entertainment and cultural programs of a wide variety; and reference information and news in temporary display or printed form.

Perhaps the communication milestones of the future will involve experiments in receiving the information in three-dimensional form, including sensations of touch, smell, and taste. The ultimate in communications, in effect, enables one to become an "on-the-spot" witness—resulting in the experiencing of sensations without the actual movement of matter.

Such thinking is perhaps beyond comprehension today, but the recent trends must lead us to expect continuous achievement. The greater challenge is sociological and political, rather than technical. With the communications techniques of this era, we are playing with a powerful tool and dangerous weapon.

Communications assuredly can bring people to better understanding, but the power of broadcast radio over men's minds for the cause of hate was vividly demonstrated in the 1930's during the rise to power of a man who had once been an obscure house painter. Adolf Hitler made tremendous use of communications in rallying a super-nationalistic force in Germany, and he later used radio in controlling subjugated nations.

One could reasonably compare the development of voice radio communications to the first unleashing of atomic energy—the power of each for good or evil is truly comparable. The parallel can be continued through the development of the H-bomb and worldwide television. The latter, giving access to the minds of millions in different nations, has further multiplied human energy. So far, it has worked for peace, being kept under control of reasonable people. Whether we can keep it this way will depend upon our ability to discern that communications, now or in the future, is like many other of men's inventions in containing the potential to be either a powerful tool for good or a weapon for evil. —R. P. Gifford

Evolution of Transportation.—Transportation technology represents one of man's most spectacular achievements—his conquest of space and time. From its earliest beginnings, transportation has been the servant of man, not only for the movement of people and the exchange of goods but also for the transmission of ideas.

Our prehistoric ancestors relied upon their own leg muscles to move about and upon their own backs to transport goods from one place to another. The beginnings of civilization are concomitant with the domestication of animals, used as beasts of burden as well as for food. Furthermore, the first great civilizations—those in Mesopotamia, Egypt, and China—developed along waterways that served as an adjunct to transportation.

■**THE WHEEL.**—Every step upward in the level of civilization was linked with advances in transportation. One of the earliest—and certainly the greatest and most basic—of all transportation advances was the wheel. Prior to its invention, men pushed or dragged heavy objects on sledges, occasionally over lubricated boards to lessen the friction and ease the task. The inventor or inventors of the wheel, which first came into use about 3500 B.C., are unknown. However, without it men could scarcely have developed civilization beyond a certain point, for the difficulties of transportation would necessarily have limited the exchange of goods and the intercourse between and within communities.

■**ROMAN ROADS.**—Transportation is a system, for it includes not only the vehicle but the way over which it travels. The Romans built great roads, such as the Appian Way, that enabled them to conquer and rule a great empire. But the vehicles of classical antiquity were chariots, rude carts, and primitive wagons, drawn by men, oxen, or horses, whose pulling power was limited by inefficient harnesses. With the introduction of the horseshoe, rigid horsecollar, and tandem harness in the Middle Ages, the horse's tractive power could be more efficiently utilized even though the roads had deteriorated since Roman times.

■**WATER TRANSPORT.**—For centuries thereafter, few major advances were made in land transport, but water transportation improved greatly. The sternpost rudder, the fore-and-aft sailing rig, and navigational aids such as the mariner's compass and better charts enabled men to venture into the open seas in early modern times to discover new continents, to found great overseas empires, and to revolutionize commerce.

■**NEW ROADS AND CANALS.**—In the eighteenth and early nineteenth centuries, overland transportation again made great strides. Roadbuilding, which had languished since the decline of the Roman Empire, began to move forward again with the formation of turnpike companies and with new methods of road construction, associated with the work of Thomas Telford and John Loudon McAdam (Macadam roads) in Britain, and

Pierre Marie Jérôme Trésaguet in France. More immediately important for the beginnings of the Industrial Revolution was the creation of a network of canals, which provided the linkage between raw materials, factories, and consumers. The economic advantages of canals were shown in the Erie Canal. Before this canal was built, it cost $400 and took 20 days to move a ton of freight from Buffalo to New York City; once the canal was opened, it cost only $10 and took only 8 days. The Erie Canal opened the western lands of the United States by connecting them to the eastern seaboard, and it made New York the country's greatest port and largest city.

■STEAM POWER.—Although water transportation still remains cheaper for bulk commodities than any other means, the canals soon fell victim to a new form of transportation—the railroad. Railroads were faster and could be built where canals were impossible. The railroad became the symbol of the new industrialized age of the nineteenth century. The steam engine was also applied to water transport, and the steamship with its increased speed, reliability, and cargo-carrying capacity drove sailing vessels into oblivion by the end of the nineteenth century.

■INTERNAL-COMBUSTION ENGINE.—Yet, at virtually the same time, a new form of engine—the internal-combustion engine—appeared. This light, efficient, and portable power plant made private transportation possible in the form of the automobile. The internal-combustion engine revolutionized the transport of commodities, the movement of peoples, and the social life of the nation.

The internal-combustion engine also made possible the fulfillment of an age-old dream of mankind going back to the myth of Icarus—flight. Culminating the dreams and unsuccessful experiments of many generations, the Wright brothers' demonstration of flight by a heavier-than-air machine marked the beginnings of the aviation age. Yet for half a century man was atmosphere-bound if not still earthbound. Then, at the close of the 1950's, the development of rocket propulsion made it possible for men to venture into outer space. Having quickened transportation on the surface of the land and water, under water, and in the air, men could now think of travel throughout the cosmos.

Transportation and Technology.—Transportation developments are closely connected with those in other areas of technology. The locomotive would not have been possible without the steam engine, which was first developed to pump water from mines. Railroad development then provided a stimulus to other technological fields by vastly expanding the need for coal and iron in order to produce the locomotives and rails. Similarly, the internal-combustion engine, first designed as a stationary power source, proved applicable to automobiles and aircraft; these stimulated the rise of whole new industries, such as petroleum and rubber. In turn, these new industries increased the demand for transportation and, as in the case of petroleum, developed new types of transportation, such as pipelines for the transmission of petroleum products and natural gas. Similarly, new materials, new power sources, the miniaturization of components, and other innovations have resulted from the beginnings of aerospace travel.

Transportation also requires the development of auxiliary technologies. Bridges and roads are essential for land transport, ocean shipping demands ports and harbors, and airplanes must have airports and other auxiliary facilities. Furthermore, mass-production techniques, employed in all industries, derived from Henry Ford's assembly-line technique first applied to the automobile.

Transportation and Society.—More important than transportation's interrelations with other technologies has been its impact upon man, his physical environment, and his society. Because the major function of transportation is to move persons and goods from one place to another, men can utilize natural resources from all over the world and extend Western culture throughout the globe. Transportation makes it possible for each region of the world to specialize in what it can produce best, thereby enabling man to increase his productive capacity and his control over the natural environment. Furthermore, quick transportation covering the entire globe has enriched human life in the twentieth century.

The political configuration of a nation is also affected by its transportation. The transcontinental railroad served to unite the United States near the close of the nineteenth century. The railroad shrank the United States, in terms of travel time, and jet travel has similarly shrunk the entire world. International politics are also determined by transportation resources. Britannia's rule of the waves throughout the nineteenth century was dependent in large measure on Britain's superior naval transportation, in both mercantile shipping and military vessels. At a later date, the agelong rivalry between powers was to be affected by the airplane. And now, in the mid-twentieth-century, a still more advanced form of transportation—space flight—has become involved in the competition for world political prestige.

■URBAN DEVELOPMENT.—Cities, the most populous environment for living in the twentieth century, are completely dependent upon transportation. For one thing, food has to be transported from the countryside to the urban population to keep it alive. For another, raw materials must be brought in to keep the city's factories going, while the finished products of the factories must be shipped to the consumers, many of whom live far away.

Not only did transportation make the city possible, but developments in intra-city transportation—which have been almost as revolutionary as the advances in long-distance transportation—enabled cities to become larger in area and population, to expand to the suburbs and exurbs, to develop into metropolises and now into megalopolises. From the beginning of recorded history, men had lived close to their work. The distances a man could walk in a reasonable time set the limits of city growth. Once street cars and other forms of urban transit came into use, working men could move into residential areas farther away from their work; the automobile expanded the area of the city even further.

■PROBLEM AREAS.—Man's ability to travel quickly to all parts of the earth has had pronounced social and psychological repercussions. Transportation banishes isolation and increases the mobility of society. At an earlier period in history, a man's position in the social structure depended largely upon the status into which he was born and from which he could not easily escape. However, with economical and quick transportation, a man could strike out for himself in a new area and make his own place in society. This was

CULVER

ORVILLE WRIGHT became the first man to fly in a heavier-than-air craft during a test, on December 17, 1903, of the airplane that he and his brother, Wilbur, had designed and built.

certainly the case on the American frontier, and the ease with which Americans have moved about from one place to another has helped to mold a more democratic society.

That most depended-upon technological artifact of American society—the automobile—has influenced every aspect of life. It has done away with the traditional distinction between town and country inhabitants, helped to emancipate women, and made American society mobile and restless, always "on the go." Furthermore, the automobile has become one of the foundations of the American economy—it is estimated that one out of every six American workers makes his living through the automotive industry, either directly or indirectly.

While transportation has led man to new triumphs and new opportunities, it has also presented him with new problems. Thus automobiles, which at first increased the speed with which persons could travel from home to factory, eventually became so numerous and created such great traffic jams as to decrease the mobility of the population. Parking difficulties, air pollution, and danger to the safety of pedestrians and of other drivers have become the concomitants of automobile transport.

Perhaps the cure for these problems is the application of more and better transportation technology. Efforts are being made to develop mass transportation for intra-city and inter-city use—coordinated transportation systems, involving the use of buses, elevated railways (monorail), and rapid transit and subway lines, are being considered for cities. For intermediate distances, one of the most publicized proposals is that of high-speed tube transportation which, for example, would carry passengers from Boston to Washington in 90 minutes in vehicles traveling at speeds up to 500 miles an hour through dual evacuated tubes. Helicopters, vertical takeoff and landing planes, and individual rocket belts are among the devices being investigated to increase still further the mobility of the population and the speed of transportation. At the same time, efforts are under way to increase automobile safety—by making the vehicles crash-proof, by improving road conditions, and by educating drivers.

Despite the problems, there is no doubt that transportation is going to develop further. In terms of the vast span of human history, transportation technology is still in its infancy. If the earth's age is taken as a 24-hour day, the time from the invention of the steam engine to the present would make up only 0.003 seconds of the 24-hour scale. Within the limited time of 200 years, man has evolved from reliance upon natural forces, such as the wind or animals, for transportation to dependence upon machines that carry him under the surface of the ocean, on the surface of the earth, above the earth, and outside the earth's atmosphere. Thus, transportation today provides both opportunities and challenges for the future.

—Melvin Kranzberg

COMMUNICATIONS AND TRANSPORTATION GLOSSARY

Aeronautical Engineering, the branch of engineering that deals with solving the problems of flight and with the design of aircraft. The principal flight vehicles are heavier than air—gliders, propeller-driven aircraft, helicopters, and jet aircraft. Lighter-than-air vehicles, such as balloons and dirigibles, play little part in modern aeronautics.

The physical basis for an airplane's ability to fly is known as *Bernoulli's principle.* According to this principle, a wing experiences a lifting force if its cross-sectional shape is designed to provide a higher air velocity above the wing than exists below it.

A second important factor in aircraft design is *drag,* the frictional force of air operating against the surfaces of the aircraft. Drag is overcome by the thrust of the propeller or jet to cause the wing or other airfoil to move through the air.

With the advent of jet and rocket propulsion, flight has become possible at supersonic speeds, greatly increasing the problems of aeronautical engineering. An aircraft flying at close to the speed of sound, about 760 miles per hour at sea level, begins to catch up with its own pressure or sound waves, producing a shock wave, or *sonic boom.* This considerably increases drag, while reducing the effectiveness of the plane's control surfaces. At supersonic speeds, the skin of the aircraft is heated greatly by friction with the air. At a speed of three times the speed of sound, the temperature of the skin can rise as high as 600° F.

The major tools of aeronautical engineers are the wind tunnel, in which flight conditions can be simulated, and the electronic computer, which solves the difficult mathematical problems that arise in their work.

—William C. Vergara

Airship, any powered, steerable, lighter-than-air craft. The two main types are the rigid frame type and the balloon, or "blimp," type.

Manned flight began in 1783 when the Montgolfier brothers of Annonay, France, discovered the lifting power of a bag full of heated air and devised the first man-carrying balloon. Jacques A. C. Charles' application of hydrogen as a lifting gas in 1784 greatly extended the capability of the free balloon. In the next hundred years, balloon flights became commonplace in Europe and America.

Free ballooning as a sport still has many devotees. It also has certain limited scientific applications, such as high-altitude meteorological and aerological research, and for high altitude astrophysical observations outside the Earth's atmosphere. For practical navigation, however, its utility is near zero. A free balloon drifts with the winds, and only by careful selection of predictable meteorological conditions will it travel in a desired direction.

POWERED FLIGHT.—Prints dating from the late eighteenth and early nineteenth centuries show many ideas for aerial navigation. Among these are the application of sails or great sweeps (oars) to spherical balloons in an effort to guide them in desired directions independent of wind and weather. None of these arrangements had much success, for a truly steerable airship could not be contrived until lightweight power plants became available.

In September 1852, Henri Giffard, a French inventor, put up an elongated, hydrogen-filled balloon that carried a "car" containing a boiler, a steam engine, and a propeller. This demonstrated a certain controllability in a light wind, though it could not be called a complete success. The

NASA

AERONAUTICAL scientists use a wind tunnel to test a model of an Apollo Launch Escape Vehicle at supersonic speeds by launching the device from a special gun. Air passing over the model produces progressive shock waves, which are recorded on a "shadowgraph" and appear as diagonal lines leading from the rocket's shadow in the photograph above.

first fully controlled airship flight occurred in August, 1884, when two French officers, Charles Renard and Arthur Krebs, flew the electric-powered dirigible *La France* on a circular course of five miles at Chalais-Meudon, near Paris.

In the historical development of the airship, two principal approaches were used: the *rigid, dirigible,* or *zeppelin* type of design, and the *non-rigid,* or *pressure,* ships. An intermediate, or *semirigid,* class also appeared briefly.

RIGID AIRSHIPS.—Rigid airships are characterized by lightweight, fabric-covered external skeletons of metal that surround a series of independent lifting gas cells. To these hull structures are attached the control and passenger cabins, the power plant units, and the handling and docking devices. A rigid airship retains its external configuration whether or not the gas cells are inflated.

Count Ferdinand von Zeppelin built the first airship of this type, flown over Lake Constance near Friedrichshafen, Germany, in July 1900. Germany brought the rigid dirigible to a high state of development before and during World War I, when nearly 100 aircraft of this type were built.

SEMIRIGID AIRSHIPS.—Semirigid airships are the least important of the three categories. None survived long in service. Such ships are characterized by an external structural keel from bow to stern that contains the crew and passenger quarters and usually the power plant, and to which is attached a single, large, streamlined bag containing the lifting gas. If the gas escapes, the keel retains its shape, but the bag collapses.

France, Italy, and possibly Russia built several such ships in the 1920's and early 1930's. In 1921 the U.S. Army purchased one from Italy. Christened the *Roma,* it crashed and burned at Langley Field in February, 1922, ending American interest in the type. The most famous semirigid airships were the Italian-built *Norge* and *Italia*. In May, 1936, the *Norge* flew successfully from Norway to Alaska via the North Pole. An attempt to repeat the polar flight two years later ended in disaster.

NONRIGID AIRSHIPS.—Nonrigid airships, or *blimps,* have been the most conspicuously successful type, and have been built and used in relatively large numbers. They have no rigid structural elements in the hull. The gas bag itself, under the internal pressure of the lifting gas, maintains proper aerodynamic shape and provides support for the attached cabin and power plant.

During World War I blimps enjoyed wide military and naval use by most major powers. By the end of the 1920's, however, active development had tapered off to near the vanishing point, except in the United States. Here, the U.S. Navy fostered a design and development program that came to a peak in World War II, then finally came to a halt twenty years later. —S. Paul Johnston

Automobile. see *Motor Vehicle* and *Automobile Manufacturing.*

RYAN AIRCRAFT CORPORATION

SPIRIT OF ST. LOUIS was piloted across the Atlantic by Charles Lindbergh in May, 1927.

Aviation, the operation and use of flight vehicles within the earth's atmosphere. Where "aviation" ends and "space flight" begins is still subject to exact definition. For the present article, however, all forms of heavier-than-air craft designed to operate at altitudes to about 150,000 feet are considered to fall with the scope of aviation.

FIXED WING AIRCRAFT.—The balloon flights of the Montgolfier brothers, Jacques A. C. Charles, and others in the late eighteenth century had freed men forever of being earthbound. Once the experimenters of the early nineteenth century had given up trying to emulate the bird (that is, by *ornithopters*) as a practical means to manned flight, the way was cleared for the accelerated development of the form of aircraft in most common use today—the *fixed-wing airplane.*

Sir George Cayley, the grandfather of aviation, established the basic theory and laid the experimental groundwork for modern aerodynamics in the period between 1799 and 1810. First he concentrated on *gliders,* experiments undoubtedly prompted by the observation that many birds soar and glide with rigid, outstretched wings for long periods. Cayley built gliders of all sizes, up to those with man-carrying dimensions. He postulated powered flight. He might, in fact, have evolved a successful airplane, had a lightweight engine then been available. Almost 100 years passed, however, before a practical power plant for aircraft was available.

Meanwhile, many experimenters, including John Stringfellow, Félix du Temple, Alphonse Penaud, Clément Ader, Laurence Hargrave, and Sir Hiram Maxim, built models and constructed larger craft in many weird and wonderful forms. All of these, however, were totally unsuccessful as practical flying machines. Although most designers attempted to use steam as power for propulsion, it soon became clear that steam engine and boiler designs could not produce power/weight ratios compatible with flight requirements. The light and efficient internal-combustion engine had to be developed before practical flight was to be achieved.

Between 1896 and 1903 Samuel Pierpont Langley constructed several experimental small-scale aircraft, which made extended power flights over the Potomac River below Washington, D.C. After years of such experimental aerodynamic research, coupled with a remarkable adaptation of a Balzer-designed gasoline automobile engine by Charles Manly, his assistant, Langley developed a man-carrying "aerodrome" that was ready for testing late in 1903. With Manly as pilot, two unsuccessful attempts were made to launch the machine from a catapult atop a houseboat on the Potomac River. Nine days after the second attempt, Orville Wright made the world's first power-driven, sustained, controllable airplane flight near Kitty Hawk, North Carolina.

The Wright Brothers. The achievement of the Wright brothers, Orville and Wilbur, was no happenstance. It was the logical result of painstaking study, research, and development over many years. Their basic interest in flight was stimulated by reports of gliding experiments by Otto Lilienthal in Germany. Their natural ingenuity and sound mechanical instincts led them step by step, from experimental kite and glider flying, through development of their own lightweight engine, to ultimate success on December 17, 1903. The longest of four flights made that day lasted 59 seconds and covered 852 feet from the takeoff point.

During the following two years, unnoticed by the press but never in secret, the Wrights built and flew improved machines, and developed their flying techniques at Huffman Prairie near Dayton. By the end of 1905 they were making fully-controlled flights (circles and figure eights) of as much as 38 minutes' duration.

EARLY AIRCRAFT.—European experimenters were also hard at work, but it was not until September 1906 that the first "hop" (21 seconds) was made by Alberto Santos-Dumont at Bagatelle, near Paris. In November, 1907, Henri Farman succeeded in staying in the air for a full minute in a Voisin biplane. During the following year progress accelerated rapidly. Many aircraft appeared in Europe and in America. Sponsored by Alexander Graham Bell's Aerial Experiment Association, Glenn Curtiss first flew the *June Bug* (1 minute, 43 seconds) at Hammondsport, N.Y., on July 4, 1908. On December 31, 1908, Wilbur Wright kept his *Model*

A biplane in the air over France for 2 hours, 23 minutes, 23 seconds.

At the beginning of World War I, the airplane was looked upon as an uncertain and unreliable substitute for battlefield scouts and observers. By the end of 1918 it had become a full-fledged tactical and/or strategic weapon. All major powers were engaged in extensive development programs, and the advance in performance was enormous. Heavy bombers were carrying substantial tonnages for hundreds of miles, and agile well-armed fighters were engaging in aerial combat at two-mile altitudes at speeds up to 130 miles per hour.

POST-WORLD WAR I DEVELOPMENT.—With the end of hostilities in 1918, other tasks for the airplane were quickly found. People began to realize its potential as a vehicle. A period of practical development ensued (1919–1940); it focused largely on civilian applications, and was based on expanding programs of scientific research in aerodynamics, propulsion, structures, materials, and devices for navigation.

The decade 1920–1930 was marked by many spectacular feats of pilots and planes, all laying the groundwork—both technically and with the public—for the widespread acceptance of the airplane as a practical vehicle in the 1930's. Among the more significant feats were:

May 8–31, 1919: the first eastbound Atlantic crossing, via Newfoundland and the Azores, by Lt. Com. A. C. Read in a U.S. Navy NC-4.

June 14–15, 1919: the first non-stop transatlantic crossing, Newfoundland to Ireland, by Sir John W. Alcock and Arthur W. Brown in a Vickers bomber.

May 2–3, 1923: the first non-stop U.S. transcontinental flight, by Oakley Kelly and John A. Macready in a Fokker monoplane.

April 6–Sept. 28, 1924: the first round-the-world flight, by two planes of the U.S. Army Air Corps.

May 8–9, 1926: the first round trip over the North Pole, by Richard A. Byrd and Floyd Bennett.

May 20–21, 1927: the first non-stop flight from New York to Paris, by Charles Lindbergh in *The Spirit of St. Louis*.

July 28–29, 1927: first non-stop flight from California to Hawaii, by Lester J. Maitland and Albert F. Hegenberger.

June 17–18, 1928: the first non-stop transatlantic flight by a woman, Amelia Earhart, with Wilmer Stultz and Lou Gordon aboard the Fokker *Friendship*.

Nov. 28–29, 1929: the first flight to circle the South Pole, by Byrd flying out of Little America.

COMMERCIAL SERVICE.—Although a few embryonic and intermittent passenger-carrying operations appeared in Europe shortly after the Armistice of 1918, precedence in the United States was given to the carriage of mail. Military airplanes were converted for this purpose simply by covering over the forward cockpit to form a mail compartment. During the 1920's a network of contract-operated airmail routes spread across the United States. Airway aids such as light beacons, radio-communication links, and improved weather-reporting services came into being under the aegis of the U.S. Department of Commerce.

By 1930 scheduled passenger services were available between a number of cities, and a combined air and rail transcontinental service had begun. Encouraged by the growing demand for passenger accommodations, several manufacturers designed and put into production fast and efficient commercial aircraft. Notable among these were the Boeing *247,* Curtiss *Condor,* Lockheed *10,* and Douglas *DC-2* and *DC-3.*

What appeared to be a setback for commercial air transport in America occurred early in 1933 when President Franklin D. Roosevelt suddenly cancelled all airmail contracts on the grounds that certain frauds and collusions had been perpetrated by certain operators. This effectively shut down most commercial air services for a time. In the interim, the U.S. Army Air Corps was given the task of flying the mail. It was ill-equipped, under-staffed, and inadequately trained to cope with scheduled flying in the midst of severe winter weather, and a number of Army planes and pilots were lost. While shocking the nation, this tragic series of crashes pointed up the inadequacies of our air arm. During the following years, Congress granted increased funds for the procurement of planes and the training of pilots.

Airports, airways, and navigational aids kept pace with the expanding requirements of commercial air transport. By the mid-1930's, night-and-day air transportation for passengers and mail was commonplace in America. Parallel development, on a smaller scale, took place in Europe and in other parts of the world. Only the oceans stood in the way of a world-wide transportation system.

By 1937–1938 these last barriers began to yield. Exploratory operations by large four-engine flying boats over both the Atlantic and Pacific oceans demonstrated the feasibility of intercontinental cargo and passenger travel. By 1939 several limited commercial overseas services were in actual operation. The outbreak of World War II interrupted most civilian development in this direction. However, the tremendous wartime demands for overseas military air transport so accelerated the development of new long-range cargo aircraft and operational techniques, that a truly world-wide system of commercial air transport was to spring, almost full-grown, into being at the end of hostilities.

WORLD WAR II AIRPLANES.—The performance capabilities of the military aircraft of all countries improved considerably under the impetus of war, but in the end it was the sheer weight of numbers that predominated. In this area, Germany led the field by a wide margin in 1939–1940. By 1945, however, the United States had surpassed the rest of the world combined. Between 1943 and 1945, U.S. factories turned out almost 30,000 bombers, 40,000 fighters, 20,000 trainers, and 13,000 transport and miscellaneous types.

Most of the planes used in World War II were developments of types designed in the late 1930's. Improvement had been continuous through research. Almost without exception, airplanes were powered by gasoline engines equipped with air propellers. In 1943, however, another breakthrough was made with the advent of jet and rocket engines.

The application of direct thermal thrust to aircraft propulsion, either in the form of chemical rocket motors or of turbine-driven jets, immediately opened the way to vast improvement in performance. The so-called *sonic barrier* (Mach 1, which is equal to the speed of sound in air at any given altitude) no longer imposed a physical limitation on airplane speed. Furthermore, since reaction propulsion functions equally well in or beyond the earth's atmosphere, altitudes inaccessible to propeller-driven craft now came within reach.

The Germans flew the first experimental jet engine in August 1939. A Whittle-engined experimental *Gloster* flew in England two years later. The first U.S. jet flight occurred on Oct. 1, 1942, in a Bell *XP59A* aircraft powered by two General Electric turbojet engines. But the Germans made the first military application in production quantities. Some 1,300 Messerschmitt *ME-262* twin-jet fighters were built, and many saw service against American and British attacks in the last months of the war. Their performance was far superior, but their impact came too late to decide the outcome of the war.

POST-WORLD WAR II DEVELOPMENT.—By 1960, practically all military aircraft—except for primary trainers and certain special purpose slow-speed types—were powered by turbine engines, either by pure jets, for highest performance in speed and altitude, or by turbo-propeller combination, for long-range, heavy transport. Fighters tended to become almost missile-like in configuration, with small, almost embryonic wing and tail surfaces. Radar-controlled guided missiles supplemented machine guns and cannon as armament. Mach 2 speeds became routine, and in some cases (for example, YF-12) Mach 3 has been attained. The ultimate in aircraft performance to date has been reached by the experimental X-15 rocket-powered airplane. This craft, which is air-launched from a B-52 bomber, has reached flight speeds of 4,534 miles per hour and altitudes (as a ballistic missile) of 350,000 feet. It has no immediate military application, but the lessons learned from its development will have a bearing on the future.

Although by 1965, with the advent of intercontinental guided missiles, the development of manned super-bombers had practically ceased, large fleets of jet-powered heavy bombers —*B-47*'s, *B-52*'s, and *B-58*'s in the United States, British *Vulcans, Victors,* and *Valiants,* etc.—came into inventories in the late 1950's. The range of such aircraft has been greatly extended by in-flight refueling from special tanker airplanes. A version of the Boeing *707* jet transport, the *K-135*, is in wide use as a tanker for the *B-52* bombers of the Strategic Air Command.

The last of the large bombers (at least in the U.S.) is the experimental *Valkyrie* (XB-70). Based on an Air Force requirement promulgated in 1954, the first of three prototype machines was flown on Sept. 21, 1964. This machine, with a take-off weight of some 530,000 pounds, is designed to fly at Mach 3 and to cover over 7,500 miles nonstop without refueling.

NAVAL AVIATION.—Meanwhile, naval aviation was also developing according to its own special needs. In November 1910, Eugene Ely, a civilian pilot, took off in a Curtiss biplane from the deck of the *Birmingham*. Two months later, Ely landed on and took off from a wooden platform erected on the *Pennsylvania*. In 1911, Glenn Curtiss won the Collier Trophy for the development of the first practical float-type seaplane, the *A-1*.

Stemming from a World War I requirement for offshore patrol, initial naval interest (1918–19) centered on flying boats. Curtiss had built a prototype for Rodman Wanamaker for a transatlantic attempt in 1914, a project abandoned because of the outbreak of the war.

The development of flying boats, for both military and civilian purposes, followed a sporadic course throughout the 1920's and 1940's. Large two- and four-engine boats were used in great numbers for overseas and interisland transport in World War II. With the increased efficiency of large land-type transports and the availability of landing fields everywhere around the world, however, development of flying-boat types had been largely abandoned by the early 1960's.

Shortly after World War I, naval interest centered on the development of aircraft carriers, ships with long, clear decks for the launching and recovery of land-type aircraft. The record of U.S. naval carrier-based aviation in the Atlantic and Pacific during World War II is well known. It was a decisive factor in the defeat of Japan and made major contributions towards the outcome of the Korean War and subsequent peacekeeping operation around the world. The modern high-speed carrier with its complement of catapult-launched, high-performance jet aircraft carrying nuclear weapons and rocket missiles is one of the most powerful weapons systems available in the world today.

Attempts have been made in recent years to design a multipurpose aircraft to be used almost interchangeably over land and at sea—for example, the *F111* or *TFX*. By certain modification of equipment, it was planned to be used by both services. Its most interesting feature, however, is the incorporation of a variable-sweep wing. For slow flight speeds (takeoff and landing), its wings can be extended to an angle of about 90° to the fuselage, to give maximum span. Once the plane is in flight, the wings can be swung rearward to make an almost "delta" configuration for less drag and higher wing loading, the optimum condition for high-speed flight.

In passing, it may be pointed out that the variable-wing configuration described above has been given consideration in at least one of the *supersonic transport* (SST) design studies. The projected Boeing *733*, a 430,000-pound craft designed to carry 150 passengers over a 4,000-mile range at 1,800 miles per hour (Mach 2.7) at 60,000 feet, originally incorporated the variable-geometry wing. Structural weight appeared to be excessive, however, and the swing wing was abandoned in 1968.

ARMY AIRCRAFT.—Apart from Air Force and Navy, the ground forces have developed many requirements for fixed-wing aircraft. Hundreds of small, two-seat aircraft have been incorporated in regiments and divisions of ground troops for battlefield surveillance, artillery spotting, etc. Larger airplanes (single- and multi-engined) are in widespread use by the Army for rapid and efficient command and light-cargo transport. Rugged, slow-landing and steep take-off types are in service in rough terrain, where airfields are small or nonexistent. Also, ground-support and large-scale troop-transport requirements have fostered development of larger and faster cargo and personnel carriers by air and ground forces. Generally speaking, with modifications, the aircraft used are similar to those of commercial airlines. Modern civilian jet transports handle from 80 to 150 passengers at a time. Certain projected designs, primarily for the military, call for huge machines to carry upward of 500 troops. These, however, will probably not appear in service until the 1970's.

An interesting example of a highly specialized form of air-cargo carrier is the so-called *Super Guppy*, the *B-377PG*, which appeared in mid-1965. Its single purpose is to ferry very large rocket and space components from factories to firing sites. In reality, the craft is a *Strato-cruiser*, modified by the addition of a huge bubblelike superstructure over the length of the fuselage. This housing is large enough to enclose a Saturn V rocket, which is too large for transport by rail or over any U.S. highway. Four turbo-prop engines give this bulky and awkward-appearing flying machine a cruising speed of 250 miles per hour.

BUSINESS AIRPLANES.—On the civilian side, a very wide variety of airplanes has evolved for both business and pleasure. Over 100,000 are in service in the United States alone. Here again, a detailed accounting is impossible. Types in current service range from single-seat motorless sailplanes to the large jet-powered transports which carry 100 to 150 passengers, in intercontinental service. The most prominent commercial transports are listed in the following table:

BAC VC10-1100
BAC VC10-1150
BAC 111-400
Boeing 707-120
Boeing 707-320B
Boeing 707-420
Boeing 720
Boeing 727
Boeing 737
Convair 440
Convair 880
Convair 990
DeHavilland Comet 4C
DeHavilland Trident IE
Douglas DC-7C
Douglas DC-8-50
Douglas DC-9
Lockheed Electra
Sud Caravelle 6R
Sud Caravelle Super
Vickers Viscount 812

Since the end of World War II there has been a considerable upsurge in flying purely for fun, or for utilitarian purposes such as surveillance of large farm holdings. For this kind of flying, a number of excellent single-engined, two- to four-seat airplanes are available, either open cockpit or cabin types. Most are high-wing monoplanes (the biplanes of the 1920-30 period have practically disappeared) with tricycle landing gear (frequently retractable) and with automobile-like interiors. Prices range from $5,000 to $50,000, depending partly on the amount of navigation and communication equipment installed.

On an increasing scale, American business organizations are making use of fleets of airplanes for executive travel. Twin-engined light transports (propeller driven) with comfortable accommodations for five to eight passengers and a crew of two are available, fitted with the necessary radio and navigation instrumentation to permit their use on all U.S. and overseas airways. Such aircraft, depending upon the equipment installed,

LOCKHEED AIRCRAFT CORPORATION

SUPERSONIC TRANSPORT design shows the sleek lines that permit its huge size. When built in 1970, the aircraft will carry up to 266 passengers 1,800 miles an hour at low cost.

MANEUVERS, shown here as a pilot sees them on an artificial horizon (*insets*), are accomplished with air control devices. Ailerons alter the flow of air above or below each wing to bank the plane, while the rudder gives vertical stability. When air flows past a turned-down elevator, it causes a dive: a plane with elevator turned up climbs.

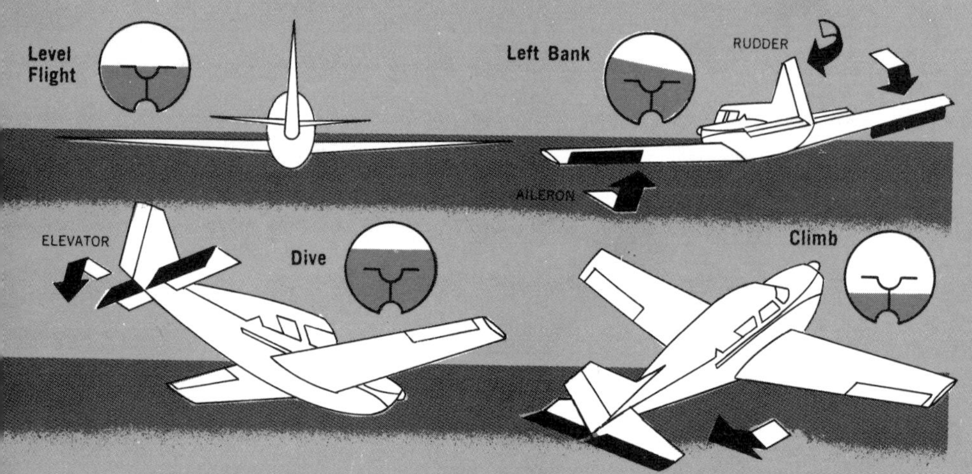

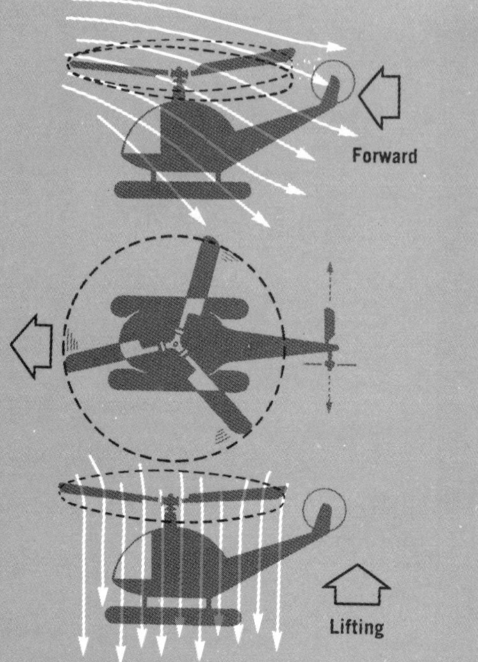

WINGS, or "air foils," create a partial vacuum above the wing's surface. Air pressure below the wing provides enough lift to overcome the force of gravity. Thrust, from jets or propellors, counteracts drag, or air friction, to move the plane forward through the air.

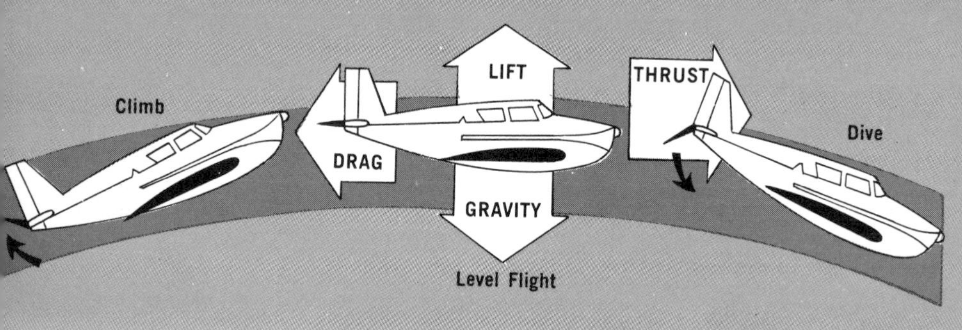

HELICOPTER rotor, with blades shaped as air foils, is both wing and propellor. In vertical flight (*bottom*) the blades' pitch, the angle at which they pass through the air, is increased to push air beneath the craft, causing it to rise; when pitch decreases, less air is pushed down, causing descent. A small tail rotor acts as a rudder to guide the craft and counteract torque from the main rotor (*center*). In horizontal flight the pitch of each blade varies according to the direction desired. Thus in forward flight (*top*) low front blades and high rear ones force air back and down.

range in cost from $30,000 to $100,000.

More recently (1960–65), a number of high-performance jet-powered executive types have come on the market. Typical is the four-engined Lockheed *Jetstar,* which furnishes comfortable accommodations for ten passengers and a crew of two, and can cruise for about 2,000 miles nonstop at almost 500 miles per hour. Twin-jet types for similar service have been produced. Such aircraft are extremely expensive but can provide private executive transport at the same levels of comfort, speed, and safety available on modern airliners.

■VTOL CRAFT. —The next stage in air-transport progress—still a few years away—will be the development of a VTOL (vertical take-off and landing) craft, which can convert in flight to a high-speed configuration, then reconvert to a zero-speed landing mode. Such aircraft, by combining the characteristics of the helicopter and the high-speed airplane, would eliminate the requirement for remote airports and all the difficulties and time delays now attendant upon ground transportation for passengers and cargo. Researchers have been working on the problem for many years, and a number of manufacturers in this country and abroad are developing prototypes. These are all small, experimental aircraft, but enough work has been done to indicate that some such configuration is feasible.

British and American designers (and doubtless those of the Soviet Union, although direct evidence is lacking) have long since been intrigued by the idea. The largest known example is the LTV-Hiller-Ryan *VHR-447* tiltwing transport, in which a relatively conventional wing and engine (four 2,800-horsepower turboprops) combination can be tilted vertically for a direct-thrust vertical takeoff, and, once in the air, the whole returned to "normal" flight position. To land, the procedure is reversed. The wing and engines are turned to the vertical configuration, and engine power progressively reduced to permit a vertical descent. This machine was test-flown late in 1964 and early in 1965. Other designers rotate only the power plants for vertical lift. Still other American and British types depend on the controlled deflection of the exhaust jet of a combination of turbojet engines for vertical lift and forward thrust. Although the concept has considerable promise for the future, the applications to date have not yet proven entirely satisfactory.

It is possible that such aircraft, by making city-center-to-city-center transportation available, and eliminating practically all ground time, might reduce the requirement for supersonic transports.

■ROTARY WING AIRCRAFT.—One of the earliest conceptions of mechanical flight was sketched by Leonardo da Vinci in the late fifteenth century.

Many years later, Sir George Cayley experimented with helicopter-like toys made of corks, feathers, and a spring. About 1842, he drew up a design for a machine that incorporated many of the features of today's vertical risers. As with his work on fixed-wing machines, the lack of a suitable lightweight engine prevented him from designing an aircraft light enough to overcome the weight problem.

The advent of the lightweight gasoline engine and reports of the early airplane flights of the Wright brothers and others in the America and Europe stimulated renewed interest in the helicopter. It occurred to a number of contemporary experimenters that the inherent limitations of the airplane (long take-off and landing runs, high forward speed to maintain flight, relative instability, etc.) might be avoided by building movable-wing systems (that is, large propellers) rotating around vertical axes to develop vertical lift without forward motion of the entire machine. Eventually their theories proved tenable, but many years were to go by, and many unforeseen problems had to be solved, before a useful *helicopter* was to appear. In fact, it was not until the years immediately prior to World War II that practical flying machines of this type came into use.

Unaware of the complexities of the problem and of the difficulties of solution, however, many people in the first quarter of the twentieth century tried to build rotary-wing machines. Louis Breguet, Étienne Oehmichen, Paul Cornu, and Raoul Pescara in France, Igor Sikorsky in Russia, and Henry Berliner and George de Bothezat in the United States designed, built, and tested an extraordinary variety of complicated machines intended for vertical flight. Some of them managed to get a few feet off the ground before shaking themselves to pieces or going out of control. How to cope with gyroscopic forces and the differential lift produced in rotating-wing systems in forward flight was beyond the state of the aeronautical art of the period.

THE AUTOGIRO.—It was a Spanish inventor, Juan de la Cierva, working in a somewhat different direction, who pointed the way to eventual solution. Instead of attempting true vertical flight capability by applying power to a rotating-wing or propeller system (as in a helicopter), Cierva combined a "free-wheeling" lifting rotor system with a conventional forward thrusting engine and propeller. No power is applied to the rotor. Under aerodynamic forces alone, the wing system produces lift by "autorotation" hence, his descriptive name, the *autogiro*.

THE HELICOPTER.—After many years as a designer and builder of large multiengined aircraft, Igor Sikorsky returned to his first love, the helicopter, in the late 1930's. In 1941 Sikorsky built and personally tested the VS-300 helicopter. This experimental machine had a single, articulated-blade lifting rotor, and three smaller rotors to provide stability and control. A year later, a more sophisticated design, the *R-4B*, with a single, large lifting rotor appeared and went into immediate production to supply a military demand. Although under active development by a number of manufacturers during World War II, helicopters played a relatively minor part. It was during the Korean War that this type of aircraft came into its own. The rescue of downed pilots and the evacuation of wounded under fire pointed up dramatically the possibilities of true vertical-lift machines.

Although their basic principles remain the same, modern helicopters have little resemblance to the fragile, open-framework Sikorsky *VS-300* of 1941. A detailed review of the stages of development is beyond our scope, but a few examples of modern helicopter trends will serve to indicate the progress in the past 25 years.

There are many variations of the theme, ranging from the giant *YCH 54-A* Skycrane, designed to "straddle" and lift large and bulky loads of up to 20,000 pounds, down to small, two-place armed and armored attack helicopters to support troop movements.

The one- and two-place machines are generally powered by a single gasoline engine. Larger sizes including both military and commercial types are now almost universally fitted with single or multiple gas-turbine engines. Most designs depend on one lifting rotor (2, 3, or 5 bladed); some machines use twin-rotor systems, arranged fore and aft, and one design is built around twin intermeshing rotors, as in an eggbeater. Although the examples cited here are of U.S. origin, the same patterns can be found in the helicopters of other countries. The Soviet Union in particular has a very active helicopter program, matching the United States type for type, up to and including the huge *Flying Crane*.

—S. Paul Johnston

MECHANICS OF FLIGHT.—It is a fact that the automobile is more complicated in operation than the airplane. Although an airplane is physically easier to operate, a higher degree of skill is required. The airplane pilot is not troubled with a brake, clutch, or gears to change; in these respects he has an advantage over the automobile driver. In clear weather the pilot has only to watch the performance of his motor, keep his ship on an even keel, and follow his compass direction. He governs speed by feeding more or less gasoline to the engine with the throttle. He has two flying controls: a control stick (or wheel) for the hands and the rudder bar in small planes operated by the feet. For a finer adjustments of attitude *trim tabs* are used on control surfaces.

The stick controls the *elevator* at the tail surface and the *ailerons*, movable flaps at the ends of the wings. Moving the stick forward causes the nose of the plane to dip; moving it back causes the nose to rise. When the stick is moved sideways, it operates the ailerons. When one aileron is depressed, it presents a flat surface to the rushing air and acts as a sort of brake, causing that wing to lift. The two ailerons work in opposite directions; when one is depressed, the other is lifted, so that one wing is pushed down and the other up. It is the action of the ailerons that causes a ship to heel over sharply in making a turn. Without the aileron action, a ship making a turn would simply skid sidewise through the air; a sharp turn could not be made and the plane could easily get out of control. But because the ailerons tip the wing so as to present a broad, flat surface to the air, the plane can "bank" around a very sharp curve, just as an automobile can turn a corner faster when the road is inclined than when it is flat.

The rudder bar controls only the rudder at the tail surface for right and left motion. When the pilot makes a turn to the right, he pushes down on the right rudder bar to turn the plane in that direction and at the same time moves the stick to the right to bring the ailerons into play and dip the right wing. Then, to *level out* he must use left rudder and move the stick a little to the left to bring the wings up to level again.

VARIABLE SWEEP WINGS on F-111 plane fold (*left* to *right*) from takeoff position to the "full sweep" for supersonic flight.

GENERAL DYNAMICS CORPORATION

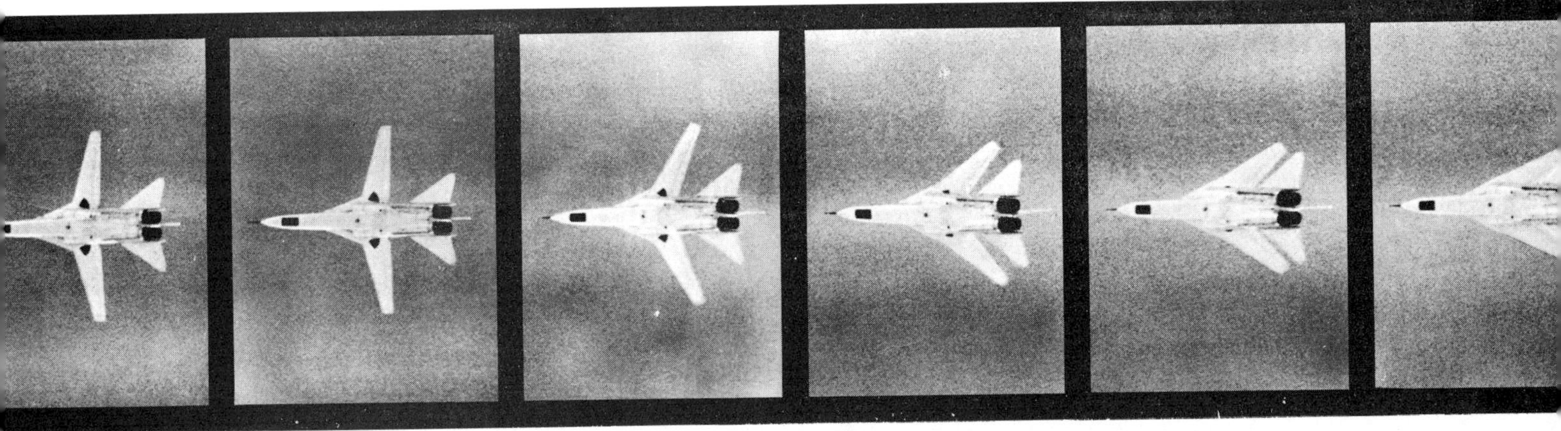

Best-Selling Hard Cover Books*

Year	Fiction	Non-Fiction
1967	**The Arrangement,** Elia Kazan	**Death of a President,** William Manchester
1966	**Valley of the Dolls,** Jaqueline Susann	**How to Avoid Probate,** Norman F. Dacey
1965	**The Source,** James A. Michener	**How to be a Jewish Mother,** Dan Greenberg
1964	**The Spy Who Came In From the Cold,** John Le Carré	**Four Days,** American Heritage & United Press International
1963	**The Shoes of the Fisherman,** Morris West	**Happiness is a Warm Puppy,** Charles Schultz
1962	**Ship of Fools,** Katherine Anne Porter	**Calories Don't Count,** Herman Taller, M.D.
1961	**The Agony and the Ecstasy,** Irving Stone	**The New English Bible: The New Testament**
1960	**Advise and Consent,** Allen Drury	**Folk Medicine,** D. C. Jarvis
1959	**Exodus,** Leon Uris	**Twixt Twelve and Twenty,** Pat Boone
1958	**Doctor Zhivago,** Boris Pasternak	**Kids Say The Darndest Things!** Art Linkletter
1957	**By Love Possessed,** James Gould Cozzens	**Kids Say The Darndest Things!** Art Linkletter

*According to retail sales. Courtesy, *Publishers' Weekly*.

Book Publishing, the procurement and processing of manuscripts and illustrations for publication in either hard-cover or soft-cover volumes. Book publishing in this century has become clearly differentiated from book printing and binding and from retail bookselling, all originally carried on as part of a single enterprise. The publisher's function is to finance the book, to publicize it, and to promote its widest possible distribution.

The publisher usually contracts with an author for the exclusive right to publish his book, in consideration of the payment of a percentage of the selling price as a *royalty*. In the case of textbooks, encyclopedias, and some other kinds of works, the authors may be commissioned by the publisher. The publisher may have, or the author may retain, the right to license translations and foreign editions of the work and dramatic, film, or broadcast versions. Income from such permissions and from the licensing of reprints and book-club editions is shared between the author and the original publisher.

Although a few publishers print their own works, usually the manufacture of a book is done under contract by a printing and binding firm, according to a design and specifications provided by the publisher and on paper supplied by him. (See *Printing* and *Bookbinding*.) Recent developments in manufacture, including the use of photo-offset lithography, of glued rather than sewn bindings, of high-speed presses, and of improvements in color printing have made possible the production of inexpensive paperbound editions and of popular art books and others with illustration in color.

The publisher seeks to promote the sale of books through visits of salesmen to wholesalers, bookstores, schools, and libraries, through advertising, and through the distribution of review copies and other means of publicity. An increasing proportion of sales is to schools and libraries.

Approximately 100 *book clubs,* a few very large clubs of a general nature and many smaller specialized ones, offer books to their members on a subscription basis at a reduced price. Several thousand books annually are reprinted in inexpensive paperbound editions and given mass distribution through newsstands, drugstores, supermarkets, and similar outlets. About 300 million copies a year are distributed in this form.

About 8 to 10 per cent of the total American book output is exported, of which scientific, technical, and medical books, college textbooks, and paperbacks are especially important. Canada is the largest single export market, but there is an increasing worldwide demand for American books.

In addition to the publishing of general *trade books* (fiction, poetry, children's books, and general works of broad interest), there are many specialized sectors of publishing producing school and college textbooks; dictionaries and encyclopedias; scientific, technical, medical, law, and business books; Bibles, missals, and other religious books; and other specialized works. The industry as a whole has expanded considerably in recent years, primarily as a result of the greatly increased enrollment at all levels of education, the growth of library service, and the general increase in reading as a part of the remarkable postwar cultural growth. Total sales of books in the United States rose from $750 million in 1954 to nearly $2.5 billion in 1967. Of this sum, about 30 per cent represents textbook sales and about 20 per cent encyclopedia sales. Of the more than one billion copies of books distributed annually (including textbooks), more than 300 million are paperbound and nearly 120 million are sold through book clubs. About 28,750 different books were published in 1967, as compared with 12,589 in 1955. There are several hundred active book publishers, including church and university presses as well as commercial firms. They vary in size from small houses with only five or six employees to giant firms doing an annual business of more than $100 million.

—Dan Lacy

Canal, any open channel filled with water, used to convey water or shipping from one place to another. *Drainage canals* are built to carry off rainwater and waterborne wastes of all types, including domestic sewage. However, open sewage canals are now considered health hazards. *Water supply canals* bring water long distances, often passing through mountains in big tunnels. Large cities such as Los Angeles receive much of their drinking water via large canals. The canals that supply southern California also make possible the irrigated agriculture that thrives in the warm, sunny, semiarid climate.

■NAVIGATION CANALS.—Navigation canals for ships and barges represent man's first real large-scale transportation system, a natural outgrowth of using the rivers of the world to move passengers and freight from one place to another.

The Chinese started building the 1,000-mile Grand Canal about 500 B.C. The Babylonians linked the Euphrates and Tigris rivers with a canal at about the same time, but historians think this canal was originally built 1,000 years earlier and the Babylonians merely reconstructed the waterway. Late in the fifteenth century the *navigation lock* was invented, and engineers were no longer limited to building canals on level ground. The lock made it possible to lift shipping from a low level to a higher section. A *lock* is a large rectangular chamber, now usually built of concrete. Watertight doors, called gates, seal off each end of the lock chamber. To raise a ship, the gates on the high-level end are closed and big valves drain the water from the chamber so the water level in the chamber coincides with the water level in the low-level canal. The downstream gates are much higher than the upstream gates since they must extend below the surface of the low level to above the surface of the high level when the lock is full. When these big gates open, the ship enters the lock and the doors close behind it. Then different valves open, admitting water from the high-level canal to flow into and fill the chamber. The ship merely floats up as the water rises. When water levels coincide, the upstream gates open and the ship sails out. All this is done without pumping any water. It all flows by gravity and the only energy expended is that needed to open and close the lock gates and the valves. Using a series of locks, such as the Panama Canal does, huge ships are raised and lowered hundreds of feet.

Many countries of Europe have developed elaborate systems of canals, particularly in the Lowlands where they serve the dual purpose of drainage and navigation. Belgium and the Netherlands alone have about 7,000 miles of canals and canalized rivers. Nations with relatively limited seacoast, such as the Soviet Union, also have elaborate inland waterway systems. Much of this canal construction took place in the early 1800's before the advent of the railroads, which ultimately gave the waterways stiff competition in the transport of bulk materials. This same pattern developed in the United States, where the first canal was built in Massachusetts in 1793—a relatively short canal around some rapids in the Connecticut River. The Erie Canal linking the Hudson River with the Great Lakes at Buffalo was finished in 1825. In America, most of the canals were displaced by the railroads.

The Panama and Suez Canals, which handle ocean-going ships, both cut thousands of miles off the ship lanes.

—William W. Jacobus, Jr.

Cinematography, the science and art of taking and projecting motion pictures. This is a two-part process which involves a motion-picture camera and a projector. The camera records the motion by taking a series of still pictures on a strip of film at precise intervals, usually 24 per second. The motion-picture projector flashes the still pictures on a screen at the same frequency at which they were recorded to produce the illusion of motion.

SET IN MOTION, this series of *frames,* or still pictures, shows actress Julie Andrews levitating in the movie *Mary Poppins.*

■**CAMERAS.**—A motion-picture camera is simply a still camera which incorporates a mechanism to automatically advance the film past the lens and actuate the shutter. Every motion-picture camera incorporates a lens, shutter, gate, film-advance mechanism, and magazine. If sound is also being recorded, a *sound head* is added.

As in all cameras, the *lens system* gathers the light from the subjects and focuses it on the film. Many cameras mount several lenses on a rotating turret, enabling the camera operator to change lenses easily and rapidly.

Mounted between the lens and the film, *shutters* for motion-picture cameras consist of a pair of rotating disks with pie-shaped cutouts. To vary exposure time, the angular relationship between the disks is varied to change the size of the opening, while the rotational speed is held constant.

The film is held in the focal plane of the lens by the *gate,* which consists of the edge guides, aperture plate, and pressure plate. Lateral movement of the film is prevented by the edge guides while the pressure plate holds the film against the aperture plate positioned at the focal plane. The opening, or aperture, in the aperture plate defines the size and shape of the exposed area on the film for each picture.

Film movement through the camera is controlled by the *film-advance,* or *pulldown, mechanism.* Since the camera records a series of still pictures, the filmstrip moves intermittently. To advance the film, the pulldown mechanism uses a claw which engages perforations along each edge of the film. The claw pulls the film down, stops, withdraws from the perforation, and rises to repeat the cycle. As the claw rises, the shutter is opened, exposing the stationary film.

Magazines for motion-picture film consist of a pair of light-tight enclosures. Unexposed film is stored in one enclosure while the second collects the exposed film. The magazines are held in place by mechanical latches which allow fresh magazines to be attached to the camera as needed. At the usual exposure rate of 24 frames per second, a 1,000-foot film magazine can record approximately 11 minutes of action.

For *sound recording,* a microphone is used to convert the sound into electric signals which are amplified and applied to a lamp filament. As the sound and the resultant electric signal strength vary, the intensity of the light produced by the lamp varies. This varying light intensity is focused by a lens system onto a narrow sound track on the light-sensitive film. After development, the sound track appears alongside the picture frames as a ribbon with transverse stripes of varying width and intensity.

■**PROJECTORS.**—The prime function of a motion-picture projector is to create the optical illusion of movement called *motion pictures.* The projector accomplishes this by flashing a series of still pictures on a screen. The interval between flashes is so brief that the eye retains the picture image during the periods the screen is dark. The viewer's eye is fooled into believing it sees motion while the viewer actually sees 24 still pictures every second.

To accomplish the desired effect, the projector incorporates a lens system, film-advance mechanism, light source, and, if sound pictures are to be shown, a *sound head.*

The *lens system* inverts the image on the film and focuses the picture on the screen. Moving the lens toward or away from the film focuses the picture for different projector-to-screen distances.

Motion of the film through the projector is controlled by the *film-advance mechanism.* The film is advanced in intermittent steps, frame-by-frame, at a rate of 24 frames per second. Inexpensive projectors of the amateur movie variety usually employ a clawlike device that engages the perforations on the film's edge. The pulldown claw pulls the film down one frame, disengages, and rises to repeat the cycle. More expensive projectors advance the film with a pair of sprockets driven by a *Geneva* mechanism. The sprocket drive has the advantage of distributing the load on the film over numerous perforations, reducing wear on the film and extending its useful life.

After the film has advanced the *shutter* opens, projecting the picture onto the screen. The shutter's function is to shut off light from the *lamphouse* as the film advances. This prevents the image from moving during exposure. Projector shutters usually consist of a rotating disk with cutouts or a pair of fanlike blades mounted on a shaft. Motion of the shutter is synchronized with the film-advance mechanism to allow light to pass only while the film is held stationary.

A powerful *light source* is required in the lamphouse to project the image on the film through the lens to the screen. Portable projectors for short distances rely on a powerful electric bulb or lamp. For motion-picture theaters, where the distance to the screen may be several hundred feet, a carbon arc or high-intensity mercury-vapor light source is required.

Sound for motion pictures is produced by the projector's *sound head,* which scans an optical track running along the film beside the picture frame. An *exciter* lamp projects a beam of light through the sound track to a photoelectric cell behind the film. Variations in the optical density of the sound track are converted into electric signals by the photoelectric cell to reproduce the sound recorded with the film. Silent-film projectors for home or amateur cinematography eliminate the sound head entirely.

Standard projectors are manufactured for 8mm, 16mm, and 35mm film. The 8mm film is really 16mm film which is split longitudinally during processing. Until recently, 8mm projectors were limited to silent pictures, but super-8 film now available provides space for a sound track. The 16mm projectors are widely used for scientific films, television news coverage, and instructional films.

Professional projection over distances exceeding about 100 feet require a 35mm projector, which is therefore used in all commercial motion-picture theaters. A wide range of other film widths has been used for special projection effects such as *Cinerama, CinemaScope, VistaVision, Todd-AO,* and many others. These films require either special projectors or multiple projectors synchronized to throw a single picture on very wide or curved screens. (See *Motion Picture Industry,* page 1602.)

—Joseph J. Kelleher

Clocks and Watches, instruments to indicate time and its passage, usually by means of a dial. Among the earliest of such devices was the *sundial,* whose origins are lost in history. Next came the *water clock,* which measured time lapse by a changing water level in relation to a scale. Its chief advantage was independence of of the sun. Some later water clocks operated mechanical almanacs. The most remarkable of these, made in China about 1100 A.D., embodied a water-activated escapement that foreshadowed the mechanical clock escapement of Western civilization. Another early method involved the burning of such substances as incense, wax, or oil, indicating time by measurement of the amount consumed.

The *sandglass* or *hourglass* of the Middle Ages consisted of two glass containers united at a narrow orifice through which the granular contents flowed from one container to the other. It had to be reset promptly at the end of a cycle to assure accuracy but was long popular at sea because ship motion did not affect its timekeeping. Today it is familiarly used as an egg timer.

The true mechanical clock appeared near the end of the thirteenth century. We know the details of an elaborate clock made by Giovanni de Dondi in 1364, which indicated many astronomical occurrences according to the Ptolemaic system. The timekeeping element consisted of a train of wheels driven by a weight and controlled by an escapement. This and subsequent escapements transferred the driving power to a balance wheel (later also to a pendulum). Teeth on the escape wheel—last of a series in the clock movement—acted on pallets, causing the balance wheel to oscillate and a hand to indicate elapsed time. To make clocks portable, a driving spring instead of a weight was applied during the sixteenth century. All clocks of this era kept time poorly and required frequent setting.

ACCURATE TIMEPIECES.—In 1657, Christian Huygens of Holland successfully applied the pendulum, with its natural period of swing, to timekeeping, making possible the construction of fairly constant and predictable clocks. Soon a hairspring was added to the balance wheels, giving them a similar natural period of oscillation.

Application of these improved timekeepers to scientific tasks revealed many shortcomings. The greatest of these were overcome by improving the escapement and thermometrically compensating for gross errors caused by temperature variations.

Timekeepers of truly scientific utility were then possible, as demonstrated by John Harrison with his invention of chronometers for marine navigation. Practical instruments, with which the navigator could compute accurate longitude, date from the work of Thomas Earnshaw and John Arnold in the latter half of the eighteenth century.

The nineteenth century was notable for the refinement of manufacturing techniques and for the invention of designs for mass production. In clock manufacture the United States took the lead, beginning about 1816 with the work of Eli Terry. Clocks with wooden works were sold by the hundreds of thousands in the United States and surrounding areas; many are still in use. Their manufacture ceased, however, with the introduction of the stamped-brass clock about 1837, and the market for American clocks rapidly expanded overseas. The European industry was strongly affected by these developments and soon adopted American production techniques.

The nineteenth century also brought the introduction of precision *gravity escapements.* The most far-reaching scientific timekeeping advance of the century began in the 1890's with Charles Édouard Guillaume's alloys, which permitted the elimination of temperature compensators.

Further developments were made possible when electricity was applied to the old reliable pendulum or balance wheel. With the advent of electronics, additional refinement led to the quartz-crystal oscillator in which a resonant frequency was obtained and fed to a synchronous motor operating the hands. The *atomic clocks* of today's precision laboratories are similar but use a stream of particles as their resonant element. In watch design, the most radical advance has been the replacement of the balance wheel by a tuning fork.

—Edwin A. Battison

Communications Satellite, an artificial earth satellite used to relay telephone, television, and similar signals between two earthbound stations. *Early Bird,* the world's first commercial communications satellite, was launched from Cape Kennedy, Florida, on April 6, 1965, opening a new era in international communications. After a series of tests and demonstrations, it went into commercial operation on June 28, 1965, linking North America and Europe. The event marked the first step toward the establishment of a worldwide network of satellites by INTELSAT, the International Telecommunication Satellite Consortium, which will provide new channels of communication to many nations.

During 1967, a new two-ocean commercial network using improved synchronous satellites of the INTELSAT II series, larger and more powerful than Early Bird, was established. Two of the new satellites were placed over the Pacific area, and a third was placed over the Atlantic, supplementing and expanding service provided by Early Bird. The satellites serve the communications needs of NASA's Apollo moon-program, and also fulfill other commercial uses.

During the latter part of 1968, plans were underway to launch and emplace even more advanced synchronous satellites over the Atlantic, Pacific, and Indian oceans to establish an initial global system.

International communications have been growing at a rate of 18 to 20 per cent a year. Radio and cable facilities have been constantly expanded to meet demands for more lines of communication to promote commerce and to exchange knowledge between peoples. But ever greater demands are being made.

The vast expanses of space offer one of the greatest potentials for satisfying this demand. Satellites can supplement, and greatly expand, the number of channels available for increased global communications. In addition, satellites can be used for different types of communication not now technically possible by existing transmission systems.

The INTELSAT satellites have proven their versatility for all types of traffic: transmitting telephone calls, high-quality color and two-way black-and-white television, photographs, teletype, facsimile data, and other communications between continents. The INTELSAT IV satellites planned for 1971 will vastly expand this capability to all parts of the world.

MICROWAVES.—Advances in rocketry, in electronics, and in other technologies made possible the successful development of active communications satellites. Such spacecraft carry electronic equipment that can receive signals beamed to them by earth stations, amplify them, and repeat them to other earth stations.

Such satellites provide line-of-sight contact with the earth, receiving and repeating signals between earth sta-

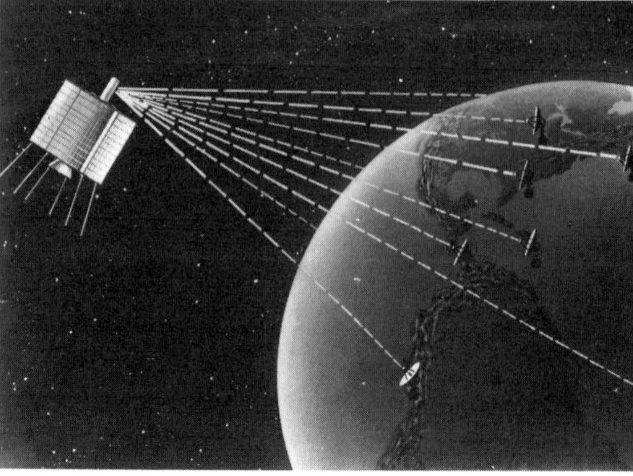

COMSAT

SATELLITES in synchronous equatorial orbit relay all forms of communication simultaneously, via microwave, between all earth stations within their lines of sight.

tions over tremendous distances. They act essentially as radio towers in space. Because of their altitude, satellites can relay signals that otherwise would be blocked by the curvature of the earth. The signals used are microwaves which, like a beam of light, travel in a straight line. They can be transmitted through the air without being affected by lightning, bursts of radio activity associated with sun spots, or other atmospheric conditions that often disrupt ordinary radio signals.

TWO TYPES.—Communications satellites are of two types, passive and active. *Passive* satellites contain no relaying signals; they merely serve to reflect signals between two points. *Active* satellites, which contain intricate electronic equipment, rebroadcast signals that are beamed to the satellite. The accompanying table lists the communications satellites that had been orbited by Jan. 1, 1968.

Communications spacecraft may be placed in medium-altitude orbits ranging from several thousand to about 12,000 miles, or synchronous altitude of 22,300 miles, Twelve or more medium-altitude satellites would be required to provide global communications coverage. Three synchronous satellites could provide coverage for nearly all the earth.

The INTELSAT satellites are in a synchronous orbit; that is, the orbital speed keeps pace, or is synchronized, with the earth's rotation. Thus the satellite appears fixed in one spot in space. From over the equator at an altitude of 22,300 miles, the Atlantic satellites link North and South America with Europe, and the Pacific satellites link North America with the Far East and Australia.

COMSAT.—The successful orbiting of Early Bird launched the Communications Satellite Corporation (COMSAT) on its first commercial space venture in partnership with many nations. Comsat is not a U.S. agency but a shareholder-owned private company whose functions and responsibilities were specified by Congress in the Communications Satellite Act of 1962.

In conformity with this law, Comsat's stock is widely distributed. Ten million shares were issued and sold on June 2, 1964, to provide the initial financing for the corporation: half of the common stock to the general public and half to the U. S. companies authorized by law as communications common carriers.

To implement the concept of a worldwide commercial satellite system, COMSAT represents the United States in the International Telecommunications Satellite Consortium (INTELSAT), a joint venture in which 63 nations share ownership and use of the space system. Comsat acts as manager of the system on behalf of the other members, and retains a majority interest. —Matthew Gordon

Compass, an instrument that indicates the north-south direction. The *magnetic compass* consists of a magnetized needle mounted so that it can rotate freely in the horizontal plane. The needle aligns itself with the earth's magnetic field, and a scale, marked in degrees, is used to indicate bearings relative to the earth's north magnetic pole. Another type, the *gyrocompass,* obtains its directional properties from a spinning gyroscope.

MAGNETIC COMPASS.—The earth acts as if it were a large permanent magnet, with magnetic lines of force extending between the north and south magnetic poles. These lines of force cause the magnetized needle in a compass to rotate until it points toward the magnetic poles. The magnetic compass does not point exactly toward the earth's geographic poles, however, because each magnetic pole is displaced about 700 miles from the corresponding geographic pole. The error, or *variation of the compass,* varies from year to year and from place to place.

The *mariner's magnetic compass* consists of a round compass card, 7½ inches in diameter on which the magnetic material is mounted. The card, graduated in 360 degrees, is attached to an inverted cup. This, in turn, floats on a mixture of alcohol and water in a bowl. The compass bowl is mounted on gimbals for complete freedom of motion.

GYROCOMPASS.—Unlike the magnetic compass, the gyrocompass points toward the geographic north pole. The heart of the gyrocompass is a spinning rotor, which is so affected by gravity that, as its axis tilts because of the earth's rotation, it experiences a continual force causing the axis to align itself with the north-south direction. A typical gyrocompass has a rotor several inches in diameter that spins between 6,000 and 20,000 revolutions per minute. The gyrocompass is used widely as an aid to ship navigation, with the more reliable magnetic compass backing it up.

An instrument called a *directional gyro* is used on many aircraft. It is not north-seeking but holds any preset heading with little error for periods up to half an hour. It must be reset periodically with the help of a magnetic compass. The *Gyrosyn compass* combines the functions of the directional gyro and the magnetic compass. It points toward magnetic north but does not oscillate during the maneuvers of the airplane as a magnetic compass would.

RADIO COMPASS.—A radio compass uses radio waves to indicate the direction of a transmitting station of known location. It consists of a loop antenna, a radio receiver, and a direction indicator. When the loop is rotated so that it points toward the transmitter, it picks up the strongest signal. When the plane of the loop is at right angles to the station's direction, the received signal strength is a minimum. A scale attached to the loop then gives the station's direction with considerable accuracy. Modern designs eliminate the need for manual loop rotation in order to obtain a bearing. Electronic circuits monitor the broadcast signal and give continuous bearings of the station automatically. By reading the directions of two transmitting stations, the pilot can pinpoint his location on a navigational chart. —William C. Vergara

Data Processing Systems, devices for the mathematical manipulation, storage, and retrieval of information. Data processing systems perform a variety of commercial recordkeeping and scientific tasks formerly done by hand. The type of equipment making up a data processing system can vary widely—from a single key-driven calculator to a powerful electronic digital computer linking up a score or more of individual units.

In all cases, the system involves three basic considerations. The first is input of source data, usually in the form of holes in cards or paper tape, magnetized spots on magnetic tape

Communications Satellites

Name	Nationality	Launch Date	Comment
Score	U.S.	Dec. 18, 1958	Broadcast Christmas message from President Eisenhower to the world
Echo 1	U.S.	Aug. 12, 1960	First passive comsat orbited
Courier 1B	U.S.	Oct. 4, 1960	First active repeater comsat orbited
Oscar 1	U.S.	Dec. 12, 1961	First amateur comsat; built by American Radio Relay League
Oscar 2	U.S.	June 2,1962	Amateur comsat; broadcast for 18 days
Telstar 1	U.S.	July 10, 1962	Active repeater comsat
Relay 1	U.S.	Dec. 13, 1962	Active repeater comsat
Syncom 1	U.S.	Feb. 14, 1963	Failed to function after orbiting
Telstar 2	U.S.	May 7, 1963	Active repeater comsat
Syncom 2	U.S.	July 26, 1963	First successful synchronous active repeater comsat
Relay 2	U.S.	Jan. 21, 1964	Active repeater comsat
Echo 2	U.S.	Jan. 25, 1964	Passive comsat; participated in first cooperative program with Soviet Union
Syncom 3	U.S.	Aug. 19, 1964	Synchronous active repeater comsat; used to telecast 1964 Olympic Games from Japan to the United States
Oscar 3	U.S.	Mar. 9, 1965	Amateur comsat; broadcast for 16 days
Early Bird	U.S.	Apr. 6, 1965	First commercial comsat; service initiated on June 28, 1965
Molniya 1	Soviet Union	Apr. 23, 1965	First Soviet comsat; active-repeater type
Molniya 2	Soviet Union	Oct. 13, 1965	Active repeater comsat
INTELSAT II-1	INTELSAT	Oct. 26, 1966	Active repeater; failed to achieve synchronous orbit.
INTELSAT II-2	INTELSAT	Jan. 11, 1967	Synchronous active repeater. Pacific Ocean 174° EL.
INTELSAT II-3	INTELSAT	Mar. 22, 1967	Synchronous active repeater. Atlantic Ocean 6° WL.
INTELSAT II-4	INTELSAT	Sep. 27, 1967	Synchronous active repeater. Pacific Ocean 176° EL.

or disks, or characters printed on documents. The second is the planned processing of this data within the system—adding, subtracting, multiplying, dividing, comparing, storing, etc. The third is the output of an end result, usually in the form of punched cards, paper tape, magnetic tape or disks, visual display, or printed information on paper.

The most widely used data processing method today is the *punched-card system,* which involves the recording of data on cards in the form of punched holes or perforations. A perforation in a particular location, for example, may indicate the number "2" to the machine, while a perforation in another location will stand for the letter "A". Punched-card machines "read" these holes by means of small copper brushes or similar conducting devices. As a card passes under these brushes, an electrical circuit is created wherever there is a hole. This completed circuit then permits an electrical impulse to be routed through the machine where it is used in a predetermined, or programmed way.

■DATA TRANSMISSION.—The electronic movement of data from one location to another for the use of man or machine is not a new concept. Samuel F. B. Morse started it in May 1844 with his telegraph. In June 1876, at the Philadelphia Centennial, the public telephone made its debut in the United States. Since then, both the Western Union and American Telephone & Telegraph companies have transmitted untold billions of business messages.

In recent years the concept of data transmission has taken on new meaning. With the advent of electronic computers, and their voracious appetite for data, the need to maintain a flow of business information between remote offices and plants and a centralized data-processing installation has become acute. As a result, new input/output devices, along with higher-speed methods of communication, have been developed.

Input/output terminals today range from the teleprinter unit to devices that can accept punched-card or magnetic-tape data. For example, a wide range of teleprocessing devices have been introduced that are capable of "reading" punched holes in cards or magnetized spots on tape. These coded numbers and letters are translated into pulse signals on a device called a *Data-Phone.* The signals are then transmitted to a receiving station via regular telephone lines or private wire circuits.

Several types of transmission circuits are used today to move data from one point to another: low-speed wires, used mostly for teleprinter communications; voice-grade wires (regular telephone lines), capable of serving a wide range of input/output devices; and new wide-band, or high-speed, circuits that transmit by means of coaxial cable or microwave channels. This last category includes laser devices producing beams of coherent light that may be an important communications medium in the future. —Stanley Englebardt

Printing has performed a role of achievement unparalleled in the r evelation of new horizons, and in emphasizing the po $1234567890
PRINTING HAS PERFORMED A ROLE OF ACHIEVEMENT UNPA

Printing has performed a role of achievement unparalleled in the r evelation of new horizons, and in emphasizing the po $1234567890
PRINTING HAS PERFORMED A ROLE OF ACHIEVEMENT UNPA

Printing has performed a role of achievement unparalleled in the revelation of new horizons, and in empha $1234567890
PRINTING HAS PERFORMED A ROLE OF ACHIEVEMENT

Printing has performed a role of achievement unparalleled in the revelation of new horizons, and in empha $1234567890
PRINTING HAS PERFORMED A ROLE OF ACHIEVEMENT

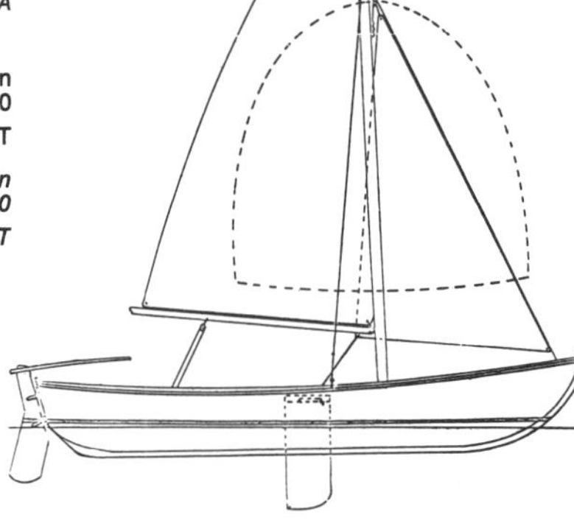

XEROX CORPORATION

XEROGRAPHY, using heat and electrically-charged paper and powder, reproduces in detail print, drawings and three-dimensional objects, as this photograph of a Xerocopy shows.

Duplicating Machines, copying machines used to reproduce or duplicate documents and drawings. The specific type of equipment employed usually depends on the material to be reproduced and the volume of copies desired.

The most familiar copying method, aside from use of carbon paper, is the stencil duplicating or mimeograph process. The word *mimeograph* was originally the trademark of the first manufacturer, but has since become a generic term to describe a particular duplicating process. Mimeograph utilizes stencil sheets made of porous paper covered with an ink-impervious coating. Typing or drawing on the paper with a sharp pointed object displaces the coating, exposing the porous paper backing. The stencil is then wrapped around an inked cylinder and rolled against paper. In this way, ink is forced through the openings onto the paper, and a duplicate of the stencil is printed. Mimeographing is an inexpensive technique, but it is limited to bulletins, forms, letters, and similar purposes.

An offshoot of the mimeograph process is the *spirit-duplicating method.* In this process, a special aniline-dye carbon paper is placed face-up beneath a sheet of coated paper. Typing or drawing on the coated paper produce a reversed copy on the back by the carbon paper. This master copy is then affixed to the drum of a duplicating machine, and the aniline dye is transferred to slightly moistened blank paper, which comes into contact with the master after entering the duplicator. The copies are extremely clear but limited in number.

Offset duplicating is also similar to the mimeograph process, but the equipment is generally larger, sturdier, and capable of turning out a much greater number of clear copies from a single master. The most important feature of offset duplicating is that the master copy can be reproduced from an original directly (without typing or drawing) by a photographic or electrostatic-transfer process. In this way there is no chance of human error.

The *electrostatic-transfer,* or *xerographic, process* uses an electrostatic force to deposit dry powder on copy paper ("xero" means "dry"). This involves exposing an original to an electrically charged plate. Where light strikes this plate, the charge is dissipated; but where opaque material, such as lines or letters, prevents light from coming through, the positive charge remains. The plate is then dusted with negatively charged particles, which are attracted to the positive ones. Positively charged paper is then placed over the plate and the negatively charged image transferred to it. The ink particles are fused to the paper by a heat process. The technique is extremely popular because almost any kind of original can be copied on almost any kind of paper.

An *infrared process* uses heat-sensitive paper that is placed on top of the original to be reproduced. Infrared light in the machine causes the carbon or metallic particles in the print on the original to heat up and transfer their image to the sensitized paper. No further chemicals or developers are needed.

The dry and moist *diazo processes* require translucent originals to be placed on top of azo-dyed paper. The pair are exposed to ultraviolet light. The paper is then developed either in ammonia vapor (dry process) or a special liquid (wet process).

The *dye-transfer process* utilizes a matrix sheet that is exposed to an original and then placed in an activator solution. After activation, the matrix is placed in contact with copy paper, and the dye in it is transferred. Up to five copies can be made from one matrix at low cost.

In the *silver-transfer* and *photographic copying processes,* the technique closely approximates standard photoprocessing methods. In both cases, the original is exposed to sensitized paper, which is then developed and printed. A more complex variation of this process is used to duplicate blueprints. —Stanley Englebardt

Electric Transit, mass transit facilities whose primary source of power is derived from electric generators. Electric-traction motors, powered by electric current from trolley wires and "third rails," have been the most desirable form of propulsion for mass-transit vehicles since the 1880's. From then until the great depression of the 1930's, electric streetcar lines and interurban lines were the major means of transportation, along with railroads. Trains running on elevated structures, originally steam-powered but later electrified, marked the beginnings of separated, right-of-way rapid transit in some of our largest cities.

The mass use of automobiles, starting after World War I, gradually brought on the demise of street railway and interurban lines. But electric transit systems in the form of rapid transit survived and lately have been flourishing due to their superior ability to furnish efficient access to congested urban areas. A single rail track can move as many people as 12 to 16 lanes of automobile highway in a given time.

Several types of electric transit are in use today or are being demonstrated for future use. All utilize electric motors to propel the vehicles and, in the more modern equipment, to slow them down by dynamic motor braking. Either direct-current or alternating-current power is obtained from the trolley wire or third rail, feeding the motors through speed-regulating controls.

■STREETCARS AND TRAMS.—Many cities of the world utilize modern versions of the old electric streetcar. But in the Western Hemisphere, they have virtually been abandoned because it is difficult to operate them efficiently in congested city streets. In Europe, there has been more interest in retaining a version of the streetcar, the *tram,* and converting it to private right-of-way operation.

■TROLLEY BUSES.—Also called "trackless" trolleys, these rubber-tired bus-like vehicles collect electric power from a double trolley wire overhead, but can maneuver independently of tracks. Now in limited use in some cities of the world, they too are slowly being phased out of service because of urban traffic congestion.

■RAIL RAPID TRANSIT.—Characterized by the *subway* and the *elevated,* this form of mass transit has maintained its usefulness over the years. Private right-of-way, fast schedule speeds, and the ability to run underground in subways through the heart of congested urban areas have enabled these rapid transit lines to survive despite increasing use of the automobile, and the tremendous public highway building programs. In this day of perpetual traffic jams and high parking rates, they have been increasing their ridership.

Rail rapid transit provides for the efficient movement of large numbers of people going in the same direction. For example, a single track in the Toronto subways carries 35,000 passengers per hour, the equivalent of 16 lanes of urban expressway.

Seven metropolitan areas in the Western Hemisphere have rail rapid transit in operation: New York, Chicago, Boston, Philadelphia, and Cleveland in the United States, and Toronto and Montreal in Canada. In the Eastern Hemisphere, major systems are operating in London, Paris, Madrid, Milan, Rome, Warsaw, Moscow, and Tokyo.

Completely new rail rapid-transit systems are also being built. In 1971 San Francisco will have the first section of its 75-mile long, 80-mile-per-hour system in operation. The Delaware River Port Authority is completing an 11-mile system for the Philadelphia area, and Washington, D. C., will have a subway by 1972 or 1973.

■ELECTRIFIED COMMUTER RAILROAD SERVICE.—Electrical MU (multiple unit) cars are used in certain major cities in the United States to provide rapid transit service for commuters. The largest commuter service railroad is the Long Island Railroad, connecting Long Island points with New York City.

■MONORAIL.—This class of electrical transit system is characterized by its single running rail or beam, in contrast with the two rails of conventional rail rapid transit. Two general types have been demonstrated: the *suspended monorail,* in which the passenger cab hangs down from the overhead beamway, and the *supported monorail,* in which the vehicle straddles the beamway. One of the earliest suspended monorail systems, built in 1901, is still in operation in Wuppertal, Germany. The monorail operating during the 1964–65 World's Fair in New York was also of the suspended type. The last supported monorail put in service was the line from Haneda International Airport to downtown Tokyo, approximately 8 miles long.

■TRANSIT EXPRESSWAY.—A developmental rapid-transit system was constructed in 1965 for demonstration in South Park, Pittsburgh. On this prototype system, three unmanned vehicles run as single car units or coupled together into trains demonstrating multiple-unit operation completely controlled by computer. A test layout for speeds up to 50 miles per hour is provided by 9,340 feet of roadway in aerial, at-grade, single-track, and double-track configurations. Transit expressway is not a monorail but a dual rail system, with the vehicles running on top of the roadway. It is adaptable to subway construction. —Philip R. Gillespie

ALWEG-SEATTLE MONORAIL INSTALLATION

MONORAIL car moves on horizontal rubber tires that grip the supporting beam, or rail.

Electronic Engineering, the branch of engineering that deals with designing equipment, such as radio and television sets, by exploiting the existing knowledge of the electron. Electronic engineering deals with the emission of electrons from various substances and the control of the emitted electrons, by electrical or magnetic means, to perform useful functions. These include amplification of weak signals, counting operations in computers, modulation of broadcast signals, the generation of alternating-current signals, switching, and rectification. Electronic engineering differs, therefore, from the older field of electrical engineering, which is concerned primarily with the conduction of electricity in wires or solutions.

Radio and television are two everyday applications of electronic engineering. Equally important are microwave radio relay stations, which carry television signals and thousands of telephone conversations across the continents. Communications satellites make it possible to relay such signals around the world.

Electronic engineers design a wide variety of navigation equipment, such as loran and air traffic control systems, to help guide planes and ships to their destinations. The electronic automatic pilot enables a plane to fly along a preset course without help from the pilot.

Knowledge gained in the fields of communications and radar led directly to the electronic computer, which is having a tremendous impact on business, science, and industry. Computers process enormous amounts of data and make decisions more quickly, accurately, and economically than human beings. They also automatically control and optimize industrial operations.

Electronic engineering is vital to space programs, providing communication and remote-control equipment, as well as telemetry for storing and transmitting data to waiting ground stations.

The invention of the transistor has greatly reduced the size and power requirements of electronic equipment. It has also opened the way for microelectronic techniques that promise further to reduce the cost and size of electronic systems.

The field of electronic engineering is expanding dramatically both in importance and complexity. Its contributions to all phases of human endeavor should continue to increase greatly. —William C. Vergara

Elevator, an enclosed conveyance used to vertically raise or lower persons, merchandise, or vehicles from one level to another. The application of elevators goes back to the days of ancient Greece and Egypt, when they were used primarily to lift material. Today, elevators are still used to carry freight, but by far the major use is to carry passengers. The invention of the elevator was one of the factors that made the multistory building feasible, and the development of high-speed elevators contributed to the solution of the skyscraper. It is often said that more Americans ride each day in elevators than in any other form of public transportation.

The development of the modern elevator began in the middle of the nineteenth century. Most of the early elevators were hydraulic- or steam-powered. By the beginning of the twentieth century, however, electric elevators were being built, and this type proved capable of the high speeds that modern design calls for.

Both hydraulic and electric elevators are being manufactured today. The *hydraulic elevator* usually is restricted to buildings having no more than five or six stories and where speeds no greater than 200 feet per minute are required. *Electric elevators* are used for buildings of any height at speeds that may approach 2,000 feet per minute. Normal elevator speeds, however, are between 100 and 800 feet per minute.

■**HYDRAULIC ELEVATOR.**—In a hydraulic elevator, the car is lifted by a hydraulic jack placed underneath the elevator in a hole approximately as deep as the height of the building. The grease rack that raises cars in the local gasoline station is a simple and familiar form of hydraulic lift.

■**ELECTRIC ELEVATOR.**—An electric elevator is lifted by a drive powered by an electric motor. This machine is usually situated at the top of the elevator shafts. Several cables attached to the car are passed over the machine drive sheave (grooved drum) and then down to the counterweight which counterbalances the weight of the car. Friction between the ropes and drive sheave causes the elevator to move vertically as the machine rotates. Both the elevator car and the counterweight are guided in a vertical path through the building by rails or tracks.

■**SAFETY DEVICES.**—Many safety devices are provided to protect the passengers. Among these are the speed governor, which stops the elevator in case of excessive speed by actuating a device on the car that clamps the guide rail; interlocks on the doors to prevent motion of the elevator unless the doors are closed and locked; and buffers or shock absorbers situated under the elevator to stop the car gently should it pass below the bottom landing. Many other electrical and mechanical devices are supplied to assure safe operation.

Originally, an attendant was required to operate an elevator, but since the 1950's almost all new elevators are automatic. Each passenger directs the elevator to the desired floor by pressing a button in the car. The operation of a group of elevators is regulated by a computer type of control to schedule their movement.

—P. L. Fosburg

Escalator, a moving stairway in which the steps move as a unit upward or downward under power at an angle of 30° to carry passengers. The escalator consists of an endless loop of steps attached to chains, all contained within a steel truss. The chains pass over sprocket-wheel assemblies at the top and bottom. An electric motor geared to the top sprocket wheels produces a continuous, even movement of the steps, which are supported by rollers traveling on inclined tracks. Enclosures, or *balustrades,* are provided on each side to form a railing and to conceal the mechanism. Along the top of each balustrade is a handrail that moves in synchronism with the steps. The step treads are grooved, and they mesh with comb-fingers at each end of the travel to protect the passenger as he enters and leaves the moving steps. The maximum allowable speed is 125 feet per minute. Most escalators, however, travel at a speed of 90 feet per minute.

The first true escalator made its appearance in 1900 at the Paris Exposition. Escalators are now used where a large number of people must be transported vertically for only a few floor levels, as in department stores, transportation terminals, and banks.

—P. L. Fosburg

Harbor, any sheltered body of water deep enough for ships and boats to anchor. Most harbors are situated where rivers empty into the sea in what is known as the river's *estuary.* Major harbors have facilities for loading and unloading both cargo and passengers; technically, this makes them *ports.* Harbors and ports often have shipyards where ships are built, launched, and repaired. Many harbors are designed for specific purposes, such as naval bases, small craft marinas, fishing ports, and commercial shipping terminals.

The most common type of harbor is a natural body of water, mostly surrounded by land, with access to the sea via a narrow inlet. Some harbors having the same characteristics are manmade. Usually, a harbor is constructed because the site offers convenient rail and highway connections. Most often, however, the natural harbor already existed and, because of this, the land transportation and port facilities followed. The majority of the world's major cities enjoy that status because of natural harbors. New York City, Rio de Janeiro, and Liverpool are typical harbor cities. But proximity to the sea is not absolutely necessary. New Orleans, Bremen, London, and Bordeaux, all situated far from the sea, are major ports because of their direct, deepwater channels leading to the sea.

In some parts of the world there are very few natural harbors, and ships must dock at piers and wharves which are unprotected from ocean waves, tides, and storms. Breakwaters, jetties, and seawalls are sometimes built to deflect the force of waves, thus creating an artificial sheltered area. However, these provide only limited protection for a limited period of time compared with that of a natural harbor.

The ideal harbor is deep enough to handle the largest ships, has space enough to provide all the docking and unloading facilities for both cargo and passengers, and provides a direct transfer to rail and truck transport. Efficiency is all-important at a commercial port. Like a railroad car, a ship standing idle while it is being loaded or unloaded does not make money for its owner.

—William W. Jacobus, Jr.

Highway Engineering, the branch of civil engineering concerned with the design, construction, and maintenance of roads. The Romans were probably the first to develop engineered highways—roadways built with foundations to sustain heavy traffic without deteriorating. But the art was lost for many years with the decline of the Roman Empire. For centuries, roads were made of dirt and gravel, until the advent of the automobile in the early 1900's.

Divided highways, an important safety development of the late 1930's, have evolved into today's freeways, superhighways, and expressways. By the separation of traffic moving in opposite directions, many accidents were prevented. But the early divided highway allowed traffic to enter or leave the road at almost any point, and service stations and other commercial establishments were permitted along the roadside without limitations. These conditions proved to be a major source of accidents, and engineers developed the *limited-access highway.*

Currently, the Federal government is sponsoring a national interstate highway program involving a vast network of limited-access roads, now under construction throughout the United States. The states, which actually handle construction, generally contribute only 10 per cent of the cost, while the Federal government provides the remainder. The U.S. Bureau of Public Roads oversees the design and construction of the interstate highways, which are regarded as the safest ever built. Access and egress are limited and curves and grades are very gradual. Interchanges are well lighted and provided with deceleration and acceleration lanes. Signs giving directions are large and clearly visible. Billboards and service stations are prohibited within the right-of-way. When completed, the interstate system will enable motorists to drive across the country on a single highway without encountering a single traffic light.

In cities, where traffic control is a major problem, traffic lights are a necessity and the computer has made it possible to completely integrate traffic systems, changing the timing of lights automatically in accordance with changes in traffic patterns.

—William W. Jacobus, Jr.

Lighthouse, a structure surmounted by a powerful light, used in marine navigation. Since the earliest days, the concept of a bright fire on a hill as a warning to ships at sea has been an accepted navigation and warning device. Whether it was recognized or not, the principle that the higher the light is situated the farther it will be visible, was an early proof that the earth was round. A light 10 feet high can be seen 8 miles away. A 50-foot light can be seen 12½ miles away. A lighthouse 1,000 feet high can be seen 40 miles off.

The earliest harbor planners knew that two lights were highly desirable for a safe entrance, for with two lights on which to base his position, the mariner by triangulation could know definitely where he was. Harbor lighthouses are known as *making lights* because they are the ones the ship's officer sees when he is "making" land.

Another type of light is the *warning light,* such as the famous Eddystone Lighthouse, which is used to mark an especially dangerous spot. *Coasting lights,* which lead the sailor along a coast, are still another type. *Leading lights* lead a ship up a channel or into a harbor.

■LIGHT SOURCES.—The earliest lighthouses were probably wooden towers from which were hung crude metal baskets loaded with burning wood or coal. As far back as 300 B.C., one of the Seven Wonders of the World was the famed Pharos of Alexandria, a gigantic lighthouse structure said to have been over 100 feet high. It was built on a 400-foot cliff. Roman-built stone lighthouses were used through the Middle Ages.

After wood and coal, candlepower was the next source of light. Tallow candles were first used, then wax candles. The oil lamp that followed was a great improvement. Sperm oil was used at first, and then various vegetable oils. Petroleum came into use for lighthouses in the latter part of the nineteenth century. The electric light is today's standard where power sources are available, and the acetylene gas light is used in stations out of reach of electricity.

The lantern through which the light is projected is a vital part of the lighthouse. Ordinary glass came first, followed by cut glass of many varieties. The parabolic reflectors introduced at the end of the eighteenth century were an important step in the projection of light. Highly complicated and scientifically designed lenses have been developed to extend the range of the modern lighthouse.

Lighthouses are distinguished from one other by the intervals between the flashing of their lights. These are created by having a revolving lens, certain portions of which are blacked off to create the dark intervals.

The development of radar, loran, and other modern electronic safety and communications devices for ships has reduced the over-all importance of lighthouses. Many more lighthouses are being abandoned than are being constructed, but it will be a long time before the lighthouse is obsolete. —Frank O. Braynard

America's Leading Magazines In 1967*

Rank		Circulation
1	**Reader's Digest**	17,336,168
2	**TV Guide**	12,718,141
3	**McCall's**	8,545,839
4	**Look**	7,756,351
5	**Family Circle**	7,386,700
6	**Life**	7,354,615
7	**Better Homes & Gardens**	7,274,726
8	**Woman's Day**	7,225,073
9	**Saturday Evening Post**	6,811,418
10	**Ladies' Home Journal**	6,779,059

* Based on average circulation per issue July–December 1967. Courtesy, Audit Bureau of Circulation.

Magazines, periodical publications that constitute a major medium of communication throughout the world. There are two major categories, general and special-interest, within which are many subcategories.

Magazines originated in the seventeenth century as book catalogues. Book publishers began adding brief descriptive material to certain books in their lists, and from these descriptions was developed the editorial content of modern magazines.

The *Journal des Savants,* published in France in 1665, has been considered the parent of all magazines, although the *Philosophical Transactions* of the British Royal Society shares the same date. *Weekly Memorials for the Ingenious,* published in London in 1681, may have been the first to accept contributions. As the editorial content broadened in scope and interest grew, periodicals such as *Athenian Mercury* (to which Daniel Defoe was a contributor) and Edward Ward's *London Spy* made valuable additions to the magazine concept.

In the eighteenth century, writers such as Joseph Addison and Sir Richard Steele wrote violent political essays for periodicals such as *Spectator* (1711) which attained a circulation of 4,000; *Tatler* (1709), *Guardian* (1712), and *Examiner* (1710).

The *Gentleman's Magazine,* published in 1731 in England, was the model for Benjamin Franklin's first magazine, which he called *The General Magazine.* The eighteenth and nineteenth centuries saw a rapid growth of periodicals. Charles Dickens, William M. Thackeray, and many other writers were magazine editors. In this same period many special-interest periodicals were introduced, including children's magazines such as *The Young Misses* (1806), *St. Nicholas* (1873), and *Youth's Companion* (1877). Other publications were devoted to art, science, and history. *The Pennsylvania Magazine* (1775–76) has both Robert Aitken and Thomas Paine as authors.

Today, in the United States, magazines are published as weeklies, biweeklies, monthlies, bimonthlies, quarterlies, and annuals. Some are general in format, offering articles, fiction, cartoons, and poetry, while others are noted for their treatment of the news, for photography, or for their coverage of specialized fields. Trade journals cover business, industry, and the professions.

Magazines could not be made available to the general public today at modest prices without advertising revenues. Production costs for any single magazine copy are usually far in excess of its subscription or newsstand sale price; for that reason, advertising revenue must be sought. In 1967, more than 7,500 companies spent almost $1.2 billion to advertise in general magazines. An additional $707 million was spent by advertisers in business publications.

A characteristic of United States magazines that distinguishes them from the majority of European periodicals has been their readiness to change their formats with the times. High-speed printing, color photography, and even three-dimensional printing were adopted first in this country; while British magazines, for example, have retained the same general appearance they had many years ago. —Howard Watson

Marine Engineering, the branch of mechanical engineering concerned with the design and production of propulsion machinery. For almost all of recorded history, the science of marine engineering was limited to devices on ships having to do with cargo handling, hotel facilities for passengers, and engines of war. Propulsion was left to the winds. In the past 200 years, however, marine engineering has come to mean the science of ship propulsion, with all other applications of the term set aside into specialized fields under different categories. In two centuries, marine-propulsion machinery has moved through the inventive stage of the steam engine to today's wide assortment of power sources topped by the atom. Atomic power for ships has been pioneered by American science, as evidenced by the U.S. Navy's *Nautilus,* the first nuclear-powered submarine, and the *Savannah,* the world's first merchant ship to be driven by the atom. The Soviet Union has also constructed a nuclear-powered vessel, the ice breaker *Lenin.*

The use of nuclear power does not mean the end of the steamship, but rather its freedom from the restrictions of old-style fuels such as wood, coal, and oil. Now, once again as with the wind, the ship may sail indefinitely instead of being limited to the distance between coaling stations. To this degree, the atomic era is certain to be viewed as one of the major developments in marine engineering, although nuclear ships technically are still steamships.

The coming of automation to marine engineering was increasingly apparent in 1965, particularly in the American merchant marine. Direct bridge control of the ship's machinery was extended from tugboats, which have had it for some years, to large cargo ships and oil tankers.

Another evolutionary change that might be noted in marine engineering design was the placing of ship engines aft in large passenger ships and freight vessels, as in the British liners *North Star, Southern Cross,* and *Canberra,* and the twin Italian superliners *Michelangelo* and *Raffaello.*

—Frank O. Braynard

Marine Signaling, the method of communicating from one ship to another (or to shore), and the device or system of marking a channel or calling attention to a dangerous area. In one instance, the ship is the transmitter of information; in the other, the ship receives a communication or situation warning.

Marine signaling from ship to ship, or ship to shore, has become so mechanized as to virtually eliminate the need for knowing the traditional old-style manual methods. The day of blinker light or semaphore signaling is fast passing. These systems are still taught in maritime academies, but their use on shipboard has been all but abandoned. The various forms of radio communication have taken over, except in certain unmechanized craft, such as fishing boats. With many types of craft, the use of radio telephone has brought as much as a 50 per cent increase in efficiency of use and productivity.

Automation is also coming to the stationary markings that guide the mariner into a harbor or warn him of dangerous rocks. Although Coast Guard and similar craft are still required to maintain the vast network of *can* and *nun* buoys marking channels, the lighthouse and lightship are becoming automated to a surprising degree. Entirely mechanical lighthouses are commonplace today up and down the coasts of the United States.

Mechanical fog-warning systems and automated lightships are rapidly taking the place of man-operated equipment. The latest in mechanical signaling proposals is the turning-direction-signal system proposed recently for ships. One Navy transport has been so equipped. Another modern device is the signal light that would automatically glow when the ship's whistle is blown.

—Frank O. Braynard

Motion Picture Industry, the production of still pictures which, when shown in rapid succession, produce an illusion of continuous action. The production of motion pictures, in all but the socialist countries, is unique in that it is both an art and an industry. As an industry, it gives employment to more than 325,000 people in the United States alone (production and exhibition) and represents a capital investment of close to $3 billion in theaters, studios, and distribution facilities. As an art form, it reaches—and influences—upwards of 46 million people a week in the United States via theaters, untold millions more via television, and additional millions throughout the world.

The necessary interaction between patron and producer has profoundly affected the nature of the medium. Historically, the producer has always been close to his audience. Such industry veterans as Barney Balaban, Samuel Goldwyn, Spyros Skouras, Jack L. Warner, and Adolph Zukor were originally showmen. Their concept was to "give the public what it wanted." But rarely is the producer himself a film maker. To get a picture on the screen, he must rely upon the services of writers, directors, actors, art directors, composers, and a vast corps of technicians, all of whom take pride in their artistry and craft. These men are less inclined to accept the theory that "the public is always right." As artists, they often feel called upon to offer their own point of view.

■INDEPENDENT PRODUCTION.—Such are the tensions behind most film making. With the rise of independent production and the competition of television since 1950, traditional concepts of making movies to please everybody have not only been reevaluated but revised. When control of production rested in the hands of eight major studios, most of them turning out 50 new films a year, there was little opportunity for the creative artist to experiment with themes or techniques. Increasingly, however, the new breed of independent producers has been seeking out directors and writers who have proved themselves at the box office, and offering them a virtual *carte blanche* to make the films of their choice. The studios, which have come to act more as distributors than as initiators, are quite willing to finance such projects if they have faith in the artists involved.

But perhaps the greatest change has come in the audience itself. Where once movie-going was a habit, accounting for upwards of 90 million weekly customers in the American market, the rise of television—and of box-office prices—has made for increased selectivity. People choose the films they want to see. Actually, with the sole exception of *Gone With the Wind* (1939), all of the all-time top-grossing pictures have been made since 1950, with a world's record gross foreseen for the 1965 production of *The Sound of Music.* While producers dream of a picture that will please everybody, realistically they are willing to settle for a film that will attract some sizable segment of the mass audience. And with the advent of television, increasingly they are searching out themes unacceptable to the even wider mass audiences of that competitive medium.

■FREEDOM OF SCREEN.—As a result, and much to their chagrin, the film makers have found themselves in constant difficulties with censors, both legal and self-appointed. Censors function to protect the *status quo,* to preserve the moral standards of the past. Because of the tremendous immediacy of the film medium, it has always been particularly vulnerable to censorial attack. When films were produced in great quantities for the widest possible audience, this factor was taken into consideration. Not only did the studios support their Production Code Administration, which forced their films to hew to a safe, predetermined line, but they also had advisory staffs to apprise them of regulations abroad. In the changed situation, particularly after 1953 when Otto Preminger's *The Moon Is Blue* emphasized how outmoded such thinking was, the leading independent producers openly flouted their own industry's self-regulatory Code. In 1965, the Motion Picture Association announced plans to overhaul both the Code and its administration, bringing it more in keeping with contemporary standards and practices.

More important, beginning with *The Miracle* case in 1952, the U.S. Supreme Court has issued a series of opinions that have struck out every legal basis for state or community censorship of motion pictures except in cases of pure, hard-core pornography. Late in 1965 even the Catholic Legion of Decency, newly named the Catholic Office for Motion Pictures, announced a softening of policy. Henceforth, it was intimated, instead of condemning "indecent films," the Legion would put its emphasis upon support of "artistically and morally good" entertainment.

Less an endorsement than an acceptance of the realities of picture production, all such moves seemed to clear the way toward an acceptance of film classification, long practiced in Europe as an effective means of keeping children from seeing movies intended for adults. In October 1968 the MPAA officially announced a new voluntary film rating plan that set up four categories for viewers—including an X rating designed to bar children under 16 from theaters showing adult fare.

The result has been a marked upsurge in films dealing with controversial, unconventional, or unsavory subject matter. Issues of the day—politics, nuclear warfare, integration—are handled openly and maturely. Similarly, sexual themes previously barred from the screen, including homosexuality and perversion, have been appearing with increasing frequency. Freed of the imperative to appeal to everyone, the film makers have discovered that they can speak forcefully and effectively to mass minorities.

In this, American directors and writers are approaching more closely the situation that prevails in most European countries. Because of spiraling production costs, the Americans will probably never be as free or as personal as Antonioni, Bergman, Fellini, Godard, De Sica, or Truffaut, to name but a few. But, like them, such directors as Blake Edwards, David Lean, Stanley Kramer, Stanley Kubrick, Otto Preminger, and George Stevens can make pretty much the films they want in the way they want to make them—a situation that obtains so long as they can demonstrate to the studios that finance them that audiences want to see such pictures in the first place.

In the socialist countries, this link between film audience and film maker is even stronger. Replacing the profit motive is social purpose, and preproduction committees (often made up of nonprofessionals) recommend or pass upon scripts on the basis of morality, practicality, and utility. Communication, always a vital concern for the film maker, under these circumstances becomes the primary consideration. (See also *Cinematography,* page 1595.)

—Arthur Knight

Motor Vehicle, a self-propelled conveyance used for passenger transportation or for hauling freight, and generally driven by an internal-combustion engine. The motor vehicle, while a comparatively recent development, had its beginnings in the eighteenth century. During the 1760's, the French military engineer Nicolas Cugnot constructed several workable steam-propelled vehicles. At about the same time, Francis Moore of London was also experimenting with steam propulsion. Richard Trevithick had several passenger-carrying steam vehicles running in England in the first decade of the nineteenth century; in the 1820's and 1830's, a number of heavy steam coaches were in use in that country. The use of these was discouraged during the 1830's by high tolls, and those remaining were practically legislated out of existence in 1865.

In the United States, Oliver Evans operated his *Orukter Amphibolos,* an amphibious dredge, in Philadelphia in 1805. Throughout the remainder of the nineteenth century, numerous Americans experimented with steam-powered carriages. Notable among them were built by Thomas Blanchard, Richard Dudgeon, and Sylvester Roper.

The internal combustion engine was first applied to a vehicle by the Frenchman Jean Joseph Étienne Lenoir, in 1863, in the form of a non-compression engine operating on ordinary illuminating gas. In Vienna the following year, Siegfried Marcus constructed a vehicle using an engine of the Lenoir type, but having a carburetor that would permit the use of liquid fuel. By 1876, the German Nikolaus Otto had developed his four-cycle engine, which was to become the power unit for nearly all the automobiles produced to this day. This type of engine was applied to vehicles in 1885, simultaneously yet independently, by the Germans Gottlieb Daimler and Karl Benz. In the United States, George Selden applied for a patent on a motor vehicle using a Brayton-type engine in 1879.

■**AUTOMOBILE DEVELOPMENT.**—By 1890 the times were nearly ready for the automobile. Practical steam cars could have been developed much earlier, but the lack of public acceptance and interest delayed the movement. Many historians agree with automotive pioneer Hiram P. Maxim, who felt that the bicycle was first necessary to give the public its first taste of private, independent transportation. Many American pioneers produced experimental cars during the 1890's, among them the Duryea brothers, Elwood Haynes, A. L. Riker, R. E. Olds, Alexander Winton, Henry Ford, Louis S. Clarke, and the Stanley twins.

In this early period, steam, gasoline, and electricity were all used to propel vehicles. While steam and electricity were to see limited use for several decades to come, it became apparent very early in the twentieth century that the gasoline engine, through rapid improvements, was to be the ultimate choice as a source of power.

The automotive industry in the United States began early in the twentieth century to gain a small momentum. In 1901, the Olds Motor Works produced 425 curved-dash Oldsmobiles, the first notable instance of mass production. By 1903, more than 125 automobile manufacturers were able to produce and market more than 11,000 motor vehicles a year. A typical car of the period had a one- or two-cylinder engine located somewhere underneath the body, frequently under the seat. Horsepower varied from 4 to 10, which with a two-speed transmission generated speeds up to 20 miles per hour. Final drive was usually

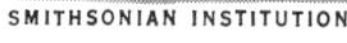

SMITHSONIAN INSTITUTION

FORD MOTOR COMPANY

CHRYSLER CORPORATION

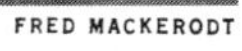

FRED MACKERODT

FORD MOTOR COMPANY

AUTOMOBILES, originally little more than self-propelled buggies like the Stanley Steamer (*upper left*), created a social revolution when Ford's Model T (*upper right*) put the American family on wheels. Today, sleek, luxurious cars like the Plymouth Barracuda (*above*), Chevrolet Camaro (*center right*), and Ford Thunderbird (*right*) fill the superhighways, parking lots, and "drive-ins" constructed for them.

by chain. The body frequently had only one seat, and steering was generally by means of a large, bow-shaped tiller.

By 1905, production had more than doubled; the pattern for the modern automobile was already apparent. A vertical engine, usually four-cylinder and rated between 15 and 30 horsepower, was placed in front, under a hood that was immediately behind the radiator. Bevel gears replaced the chains for the final drive. Equipped with two seats, wheel steering, and a three-speed transmission, some of the cars were capable of speeds up to 35 miles per hour. Production in 1909 went up to 127,000 vehicles, nearly double the 1908 figure. When the famous Model T Ford became available for $850, a trend was begun that put a serviceable automobile within the reach of all.

A few years earlier, in 1906, six-cylinder engines began to appear on the market. By 1910 the sixes were becoming common, though the four-cylinder models were to retain a substantial lead for some years to come. Selective transmissions had now taken a decisive lead over the progressive type. Steam and electric vehicles were being made only in token numbers, and it was obvious that they were about to pass from the scene. During the second decade of the century, some of the more significant advances were the electric starter, demountable rims to facilitate tire repairs, V-8 engines introduced by Cadillac in 1915, forced feed lubrication, and the widespread use of alloy steels.

The 1920's brought several innovations that made automobiles safer and more comfortable. The sedan, previously very expensive, now cost only slightly more than the open car. Balloon tires and hydraulic four-wheel brakes were introduced, and the end of the decade brought both safety glass and the synchromesh transmission. Horsepower varied between 25 and 35, and both four- and six-cylinder engines were common.

The six-cylinder engine gained a lead over the four during the 1930's and by the middle of the decade, streamlining was noticeable in body design. Power brakes became available on some models, and by the end of the decade, automatic transmissions were available in a few models. In the 1940's, advancement in automotive design was impeded by World War II, so the development and use of the automatic transmission did not become widespread until about 1950.

In the 1950's, tubeless tires and seat belts became common, and in 1953 the number of new cars equipped with V-8 engines exceeded the number with six-cylinder engines. Automobiles continued to grow larger and more powerful, causing some buyers to turn to the smaller foreign cars. This demand for smaller cars led the industry to offer a number of compact cars in 1960.

■COMMERCIAL VEHICLES.—In the commercial field, the development and use of motor vehicles was retarded during the early period, partly because of an attempt to adapt the passenger-car chassis to this heavier use. The general prejudice against motor trucks, for many years prevalent among business houses, was only deepened by the weaknesses resulting from this practice. In 1904, only 700 motor trucks were marketed in the United States.

The second decade of the twentieth century brought more substantial trucks. Tractors for street use were offered early in the decade in an effort to interest users who had a heavy investment in horse-drawn vehicles, for the tractors could replace the horses and front-axle assemblies and allow the vehicles to remain in service. In 1914, nearly 25,000 trucks were sold and the number of sales almost tripled in the following year. Except for a few very light models, trucks required solid rubber tires and, because of the condition of the nation's roads, were confined to use on city streets only.

The possibilities of cross-country motor transport were demonstrated soon after the United States entered World War I, when extreme congestion on the railroads forced trucks onto the highways. The war focused the attention of both designers and users on trucks. After the war, the nation's roads and the quality of pneumatic tires were improved, furthering the cause of motor freighting. During the depression years of the 1930's, merchants wanted to replace large inventories and bulk shipments by small, frequent deliveries. Thus began the trend to truck shipment on a large scale.

The development of the motorbus closely paralleled that of the truck. At first a novelty, open buses came into general use for sightseeing in cities. During the few years preceding World War I, a number of closed buses were introduced in cities on regular service lines. Like the truck, the bus moved out onto the open road following the war.

■MOTORCYCLES.—The invention of the motorcycle in 1885 is generally attributed to Gottlieb Daimler. It was not until the early 1900's, however, that the motorcycle became a practical means of conveyance. Used for private transport and traffic control, as well as for sport, the motorcycle today usually has one or two cylinders, an internal combustion engine that is air-cooled and gas-propelled, an electric ignition system and a three-speed transmission.

■VEHICLE STATISTICS.—Of the more than 190 million motor vehicles throughout the world, 94 million are in the United States. In 1962, the 200 millionth U.S. motor vehicle was produced, and in 1967 U.S. production totaled 7,412,659 passenger cars and 1,611,077 commercial vehicles. The U.S. imported 1,195,849 vehicles in 1967, and exported close to 500,000. A total of 13,500,000 wage earners, or one of every seven employed persons in the United States, depends on the motor-vehicle industry and related industries for his livelihood. American motorists drove 967 billion miles in 1967 and paid over 13 billion dollars in motor vehicle taxes.

—Don H. Berkebile

Moving Walk, a powered conveyance for moving passengers between two points either horizontally or along a slight incline. The moving walk has a high carrying capacity comparable to that of the escalator. The construction of a moving walk is also similar, except that the tread surface is flat and does not form steps. The tread surface is grooved and meshes with a combplate at either end of the walk. Moving walks can be used either on a horizontal plane, or at an angle of incline. Normally the maximum angle of incline is 10°, although under certain conditions the building code permits an angle as high as 15°. Normal operating speeds are the same as those for escalators, with the maximum speed usually being 120 feet per minute.

The first moving walk to gain attention was installed at the Columbian Exposition in 1892, but it did not become popular until the early 1960's. Moving walks are found most useful in buildings where large groups of people must move long distances horizontally and go up or down a short rise, as in transportation terminals, stadiums, and shopping centers. (See also *Escalator*.)

—P. L. Fosburg

Newspaper, a publication issued at regular intervals, usually daily or weekly, for the primary purpose of reporting current events. The first newspaper in the American Colonies, called *Publick Occurrences,* appeared on Sept. 25, 1690, in Boston. It was immediately suppressed by the authorities after the first issue because it had not been licensed and had dealt with political and military matters. No further attempt to publish a newspaper was made in America until 1704.

In 1968—278 years after the first publication—there were 1,763 daily newspapers in the United States, selling 60,412,000 copies every day, and almost 9,000 weekly or semiweekly papers, with circulations totaling 24,000,000.

After the *Boston News Letter* appeared in 1704 (the first regularly issued publication in North America), other weeklies were brought out, many under the name *The Gazette.* The first newspaper in New York City, the New York *Gazette,* appeared in 1725.

Another early arrival, the New York *Weekly Journal,* established by John Peter Zenger on Nov. 5, 1733, was to become the subject of the first libel suit brought in the Colonies, known as the "Zenger Trial." Zenger's arrest and imprisonment on

Largest Non-U.S. Newspapers

Newspaper	World Circulation (Millions)
Asahi Shimbun, Tokyo (AM-PM)	9.0
Yomiuri Shimbun, Tokyo (AM-PM)	8.0
Mainichi Shimbun, Tokyo (AM-PM)	7.0
Pravda	6.7
Izvestia	6.0
London Daily Mirror (AM)	5.1
London Daily Express (AM)	4.0
London Daily Mail (AM)	2.2

Source: *Editor & Publisher*, 1968.

Largest American Newspapers

1967		1960	
Newspaper	Daily Circulation	Newspaper	Daily Circulation
New York News (A.M.)	2,112,244	**New York News** (A.M.)	2,021,395
Los Angeles Times (A.M.)	861,350	**Chicago Tribune** (A.M.)	869,958
New York Times (A.M.)	840,495	**New York Mirror** (A.M.)*	836,760
Chicago Tribune (A.M.)	805,851	**Philadelphia Bulletin** (P.M.)	705,599
Los Angeles Herald Examiner (P.M.)	731,473	**New York Times** (A.M.)	644,175
Detroit News (A.M.)	700,321	**Philadelphia Inquirer** (A.M.)	618,902
Philadelphia Bulletin (P.M.)	671,525	**Chicago Sun-Times** (A.M.)	566,219
New York Post (P.M.)	628,146	**Chicago News** (P.M.)	539,448
Detroit Free Press (P.M.)	590,546	**Los Angeles Times** (A.M.)	532,078
Philadelphia Inquirer (A.M.)	516,640	**New York Post** (P.M.)	317,264

Source: *Editor & Publisher.*
* Ceased publication in October, 1963.

Nov. 17, 1734, was ordered by Governor William Cosby for defaming His Majesty's Government. Defended by Alexander Hamilton, Zenger was acquitted by a jury on Aug. 4, 1735, marking the dawn of that liberty which afterward revolutionized America and specifically established the tradition of freedom of the press, later guaranteed in the Bill of Rights.

The first daily newspaper published in the United States was the *American Daily Advertiser*, established in Philadelphia in 1784. New York City's first daily, known as *The Daily Advertiser*, was launched a year later.

Up until 1833, newspapers were mainly political organs espousing the cause of certain political parties and financed by them. In that year, James Gordon Bennett founded the New York *Herald* on purely journalistic principles, supporting it by revenues from readers and advertisers, and not beholden to any political party. Thus began the "independent" press as we know it today.

When the Revolutionary War began in 1775, there were 37 weekly newspapers in the American colonies. By 1835 there were 574 dailies publishing 2,601,000 copies per day, and about 4,500 weeklies and semiweeklies with a total circulation of about 11,000,000.

■MECHANICAL PROGRESS.—The introduction of the Linotype in the late nineteenth century enabled a printer to set a line of type by machine instead of by hand from individual characters. This brought about an industrial revolution in the printing field and a tremendous expansion in the number of newspapers, accelerated by the invention of the high-speed rotary press that replaced the slow and cumbersome method of printing directly from type on a flatbed press.

At the end of World War I, there were 2,078 daily newspapers with circulations averaging 26,443,000 copies per day, plus more than 10,000 weekly newspapers. Since that time, rising production costs have brought consolidations and mergers, so that there are about 300 fewer dailies. But these are larger and stronger than their predecessors, and have twice as many readers.

The 1960's marked the beginning of another revolution in printing techniques and perhaps another expansion in the newspaper business. The offset press, used by several hundred small daily newspapers and many more weekly papers, was adapted for use in conjunction with photocomposition, thus eliminating the use of all hot metal. Larger presses were built for the large newspapers, and the trend promises to continue.

In addition, automation by computers was introduced into newspaper production. Some 50 large newspapers utilized computers in their composing room to set type. The device accepts a perforated tape from an ordinary typist and within seconds produces a second tape with lines justified (filled out to the full measure) and words hyphenated where necessary. The second tape, in turn, is used to operate automatic high-speed typesetting machines. One large computer can handle the requirements for several newspapers connected by wire. The number of newspapers using them for this purpose have increased steadily. The same computers can also be used for bookkeeping, payroll, and billing requirements, depending on the capacity of their memory storage units.

—Robert U. Brown

Phonograph, a device for recording and/or reproducing speech, music, or other sounds by means of lines mechanically introduced on a cylinder or disc. The idea of recording and reproducing sound by mechanics is more than a century old. Many inventors, among them a French scientist named Léon Scott, tried repeatedly but unsuccessfully to achieve it. Scott's device, which he called the *phonautograph*, came to public attention in 1857. It recorded sound in the form of an undulating line on a cylinder coated with lampblack. The shortcoming that doomed the experiment was the inability to reproduce the recorded material.

Twenty years later, in America, Thomas A. Edison independently discovered how to reproduce sound, including the human voice. In so doing, he showed the way to the modern phonograph. The Edison method consisted of a grooved cylinder wrapped with tinfoil, stiffened with antimony; a diaphragm and needle, which rested on the foil; a mouthpiece to introduce sound; a funnel for outcoming sounds; and a crank to turn the cylinder.

Notable among the others who contributed to the development of the phonograph is Emile Berliner, who invented the microphone, the disk record, and the gramaphone. (See also *Sound Recording and Reproduction.*)

—John O'Brine

Photography, the recording of an image through the photochemical reaction of light on a sensitized surface. Photography is a versatile and useful means of communication and does not require the skill of an artist to capture a likeness. It can make a moment of history—of glory, or disaster, or achievement—a part of the lives of millions through its transfer to the printed page, the motion picture, or television screen.

Photography is a combination of chemistry, optics, and mechanics. It existed as an achievable reality in the minds of men for a hundred or more years before its actual birth. Strangely, the chemistry of the film and the optics of the camera were both known separately for many years before they were brought together.

■EXPERIMENTS WITH LIGHT.—The first recorded knowledge of the effect of light upon a chemical substance came in 1727 when Johann Heinrich Schulze, a German physician, discovered that chalk treated with a solution of silver and nitric acid would turn black when exposed to sunlight. He placed the mixture in bottles around which he wrapped stencils. Sunlight passing through the openings in the stencils turned that portion of the mixture black while the rest remained white.

Thomas Wedgwood, the son of Josiah Wedgwood, the famous English potter, made the first actual prints on paper, using sheets of paper coated with silver chloride which Karl Wilhelm Scheele and William Lewis had discovered had the same light-sensitive properties as Schulze's chalk. As negatives, he used silhouettes painted upon glass. He placed the glass plates upon the paper and made the exposure, the result of which was a black-and-white print that was the reverse of the image painted on the glass. In 1802, Wedgwood published a paper, in cooperation with Sir Humphry Davy, entitled *An Account of a Method of Copying Paintings on Glass, and Making Profiles, by the Agency of Light Upon Nitrate of Silver.* The pictures, however, were very fleeting; as soon as the "negative" was removed, the whole paper started turning black.

In 1839, William Henry Fox Talbot, at the suggestion of Sir John Herschel, developed a method of "fixing" the picture with the use of sodium thiosulfate, the *hypo* that is still used today for fixing photographic films and prints.

■EARLY CAMERAS.—The history of the camera goes back many years prior to the experiments in light-chemistry. Probably the earliest mention of a camera was made by Roger Bacon in 1267. Leonardo da Vinci, who died early in the sixteenth century, also described a *camera obscura.*

The camera obscura was a simplified camera. The first models, which were probably constructed in the sixteenth century, consisted of a lens at the front of a box that had a translucent screen at the focal point of the lens. When the box was aimed at a scene, an inverted image was projected on

FAIRCHILD SPACE AND DEFENSE SYSTEMS

SPECTACULAR picture of New York City and surroundings, above, was taken by a new aerial-photography camera that has a rotating prism covering a 180-degree panorama. The curvature is not the earth's, but is a geometric projection caused by the 180-degree span. The chief feature of panoramic cameras is their ability to take detailed photos at high speeds and low altitudes. The picture at left was taken outside an English castle with a fisheye lens and gives the effect of being at the bottom of a well. The photo at right was taken with a high-speed action camera with fast lens.

the screen. A thin sheet of paper was placed over the screen, and the screen was traced. Later, other lenses and mirrors were added to correct the image. More ambitious cameras obscura were made by using tents with a lens and mirror or prism at the apex.

BEGINNINGS OF PHOTOGRAPHY.—It wasn't until the early part of the nineteenth century that light-sensitive chemicals and the camera obscura were brought together to create the first real photographs. Joseph Nicéphore Niepce and Louis Jacques Mandé Daguerre, both Frenchmen, together discovered that when a silver plate of a silver-coated copper plate that had been exposed to iodine fumes, forming the light-sensitive silver iodide, was exposed in a camera obscura and the plate subsequently developed in mercury fumes, a positive picture of the subject was formed. When fixed with sodium thiosulfate, the picture became permanent. Many of these *Daguerrotypes* are still in existence, and even by present photographic standards are considered to be amazingly sharp.

The principal drawback to this method of photography was that there was no way of duplicating the pictures. There were no negatives. Each picture had to be separately exposed.

Talbot, who had discovered the method for "fixing" the light-sensitive chemicals, was also the first to create a negative process from which any number of positive pictures could be made with one exposure. This was called the *Calotype* (or Talbotype) process, In this process he used paper coated with silver nitrate which, after exposure, was developed in gallic acid and silver nitrate. This not only cut exposure from more than an hour to less than a minute, but also yielded a negative print. Positive prints could be made by repeating the process but making the exposure for the final print through the first, or negative, print.

COLLODION PLATES.—To overcome the inadequacies of the paper negative, Claude Niepce de Saint-Victor in 1847 coated a sheet of glass with silver nitrate, using albumen to hold it in place. Collodion was later substituted for the albumen by Frederick Scott Archer of London.

The difficulty with this process was that a fresh wet solution had to be applied to the glass immediately before taking the picture. As a result, the photographer going into the field had to carry his darkroom with him —usually in the form of a tent— as well as all of his chemicals, glass plates, and processing trays. Mathew B. Brady first brought home to people the horrors of war in his Civil War photographs, traveling with a horse-drawn carriage equipped as a darkroom.

The first dry plates were made in England by Frank Charles Luther Wratten in 1877. These revolutionized photography by eliminating the need for the portable darkroom.

WORK OF EASTMAN.—Three years later, George Eastman, a book clerk of Rochester, N.Y., who had become interested in photography, began the manufacture of dry plates, developing faster mechanical methods of coating both plates and paper.

Up to this time, virtually all photographs had been made by men who were seriously engaged in photography either as portrait photographers or who saw in it other financial opportunities. Eastman felt that picture taking, if it could be made simple enough, would appeal to everyone. His belief was justified when, in 1888, he brought out the first simple amateur camera, which was advertised for years with the slogan, "You Push the Button, We do the Rest". The first camera contained a roll of sensitized paper upon which 100 circular pictures could be made. When the last exposure was made, the entire camera was returned to the manufacturer for the processing of the pictures and the reloading of the camera. It was with this camera that amateur picture taking—snapshooting—was born.

It wasn't until 11 years later that film itself came into being, when Eastman first produced a photographic emulsion that was coated on a flexible base. In 1895, he rolled strips of this film with an opaque paper to create the first daylight-loading film and made cameras available that would utilize it. This simplified and further popularized picture taking by making the film available almost anywhere and by eliminating the necessity for returning the camera to the manufacturer. It also brought into being an entirely new business—photofinishing.

COLOR PHOTOGRAPHY.—There still remained one major barrier to be hurdled if photography was to complete its role as a means of communication—color. James Clerk Maxwell, in 1861, had proved that color pictures were possible. At a Royal Institution lecture in London, he had shown lantern slides in color. The lantern slides had previously been tinted with transparent colors, but these were the first photographed and projected in color. They were pictures of colored ribbons of which he had made three exposures, one through a red solution, one through a green solution, and the third through a blue solution. Using three projectors, the slides were projected through the same colors and exactly superimposed on the screen to recreate the original colors of the ribbons.

Every color process since then has utilized the same principle. Early color photographs were made on three separate films exposed through appropriate colored glass filters, either one at a time or in *one-shot-cameras* in which, through an arrangement of prisms, all three films were exposed at once. Prints were usually made by making separate exposures from each of the three negatives on *stripping films,* which were then dyed, carefully stripped from their backing, and superimposed in register on each other on a white backing paper.

The Lumière process used dyed red, green, and blue starch grains over which the emulsion was coated. During development the film was exposed to white light, reversing the image and creating a positive transparency. Each of the tiny grains of starch had acted as a filter creating, when viewed unmagnified, the same effect as Maxwell's three slides had on the screen. The *Dufaycolor process* was similar, utilizing a red and green screen.

The first color film suitable for amateur use was made by two musicians, Leopold D. Mannes and Leopold Godowsky, Jr. This was Kodachrome film, which is made up of three layers of emulsion with dyes incorporated into them. The top layer is sensitive to blue light; yellow filter layer is beneath it, with a green-sensitive layer of emulsion under it; beneath that is a red-sensitive layer. Critically controlled processing is necessary to develop the three layers of emulsion, removing the silver and leaving only the three positive dye images. Placed on the market in 1935, this film was first available only for use in amateur movie cameras. Later, it was made available for the amateur in the candid camera size of 35mm and in sheet film sizes for professional and commercial photographers.

Since then, other color films have been developed that are less critical in the processing stages and can be processed by the professional in his studio or by the amateur at home.

■ APPLICATIONS.—Early photographs, both those taken by commercial and professional photographers, and those taken by the snapshooting amateur, were generally confined to people and places. It was not until later that science, industry, medicine, and education began to realize the full potential of photography as a means of communication and as a tool for their efforts.

Photography has invaded every field of human endeavor. Metals and chemical compounds are analyzed, and their analysis recorded on spectrographs and metallographs. The human body and giant heat exchangers for nuclear power plants are examined by radiography. Whole libraries are recorded on microfilm that can be stored in a desk drawer. Missiles and rockets are tracked across the sky by huge cameras. The planets and the far galaxies of space are explored by astrophotography. The authenticity of documents, postage stamps, and old masters is checked by infrared and ultraviolet photography. Motions that are too fast for the human eye to see are brought to a slow-turning halt by high-speed motion pictures. Photography is used to catch the bank bandit and the kitten at play. It has become an inseparable part of our life. As a means of communication, it speaks all languages with equal facility. —C. Grantly Wallington

HARRIS INTERTYPE CORPORATION

OFFSET presses use photographically treated cylinders to reproduce type and pictures. This web-fed, remote-controlled machine can print as much as 1,000 feet of paper a minute.

Printing, the reproduction on paper or other substance of an image from an inked printing surface. Reproduction may be from a raised surface, as in *letterpress* printing; a depressed surface, an in *intaglio* or *gravure printing;* or from a level surface, as in *lithography* or *planography.* Although popularly associated with paper, printing is done on a variety of other materials—metal, metal laminated to paper, cloth, and plastic—and on objects of various shapes.

■ LETTERPRESS PRINTING.—By this method, the portion of the printing surface which carries the image is raised above the nonprinting areas and only that surface is inked. Letterpresses take various forms and are constructed in sizes from small hand-operated presses to high-speed giants.

A *flat-bed cylinder press* also has a flat printing surface, or *bed,* but the sheet of paper is wrapped around an impression cylinder which rotates to press the paper against the inked printing form. By the addition of a second unit, a flat-bed cylinder press may be designed to print two colors on one side of the paper, or it may be constructed so that it turns the paper and prints both sides, in which case it is called a *perfecting press.*

A *rotary letterpress* consists of a cylindrical printing surface and a cylindrical impression cylinder between which the paper passes. If the rotary press prints on single sheets of paper it is called a *sheet-fed press,* which may run at speeds of up to 7,500 impressions per hour. If it prints on paper that comes from rolls it is known as a *web-fed press,* which may operate at speeds of more than 2,000 feet per minute. Both sheet-fed and web-fed rotary presses may print up to six colors during one pass through the press. Web-fed presses print both sides of the paper. Sheet-fed presses usually print only one side, but they also may be built to print both sides.

Letterpress printing may be done from metal type and engravings or it may be done from page plates, or *electros,* made from the original type and engravings. Rotary letterpresses always print from plates which are curved to fit the printing cylinder.

Letterpress illustrations may be reproduced from line or halftone engravings. A *line engraving* is made from artwork, such as an ink drawing, which is solid color with no shading. If an illustration contains continuous shading, such as a photograph or wash drawing, a *halftone engraving* is made. The artwork is photographed through a screen, or *grid,* of opaque lines at right angles to one another usually 60 to 200 per inch in each direction. The grid breaks up the artwork into fine dots of varying sizes, depending upon the intensity of the tone. In light areas the dots are almost nonexistent, while in extremely dark areas they are so large as to make an almost solid surface. The screened negative thus produced is etched on copper to furnish a printing plate which will reproduce the shading of the original artwork.

Artwork or photographs in color are filtered to produce a plate for each of the primary colors—red, yellow, and blue—plus one for black for definition. When these four plates print together in perfect register, they produce an illustration in full color.

GRAVURE PRINTING.—Gravure, or *intaglio* printing, is the opposite of letterpress. Instead of a raised surface, minute wells etched into a copper cylinder carry the ink. The cylinder revolves in a trough of ink and is completely covered. A flexible metal scraper, called a *doctor blade,* wipes all the ink from the surface, leaving it only in the depressed wells. As the paper comes in contact with the cylinder, it draws the ink from the wells and deposits it in the areas to be printed. Gravure cylinders may be covered with a thin plating of chrome to give added strength for long runs. Like letterpress, gravure presses may be sheet-fed or web-fed. A web-fed gravure press is known as a *rotogravure press* and operates at speeds of more than 1,800 feet per minute.

Gravure cylinders are prepared photographically from type matter and illustrations. Photographs and wash drawings are handled in a manner similar to that used in making halftone illustrations for letterpress, except that all of the dots are the same size and shading is produced by varying the depth of the wells, thus controlling the amount of ink which each one will transfer to the paper.

LITHOGRAPHY.—Lithography, or planography, is based on the principle that oil and water do not mix. The perfectly level printing surface is treated photographically, so that the area which is to print will accept oil but not water. The entire plate is dampened, then inked. The printing image areas accept the ink, while the dampened nonprinting areas repel it. The most widely used form of lithography in use today is *offset,* in which a roller covered with a rubber *blanket* picks up the ink image from the lithographic plate and transfers, or *offsets,* it to the paper.

Offset presses may be either sheet-fed or web-fed. Like letterpresses, offset presses can be built to print up to six colors on both sides of the sheet with one pass through the press. A web-fed offset press can print more than 1,500 feet per minute. The offset process reproduces illustrations much as does letterpress except that no individual plates for illustrations are made. Screened halftones or line illustrations are put on film and transferred photographically to the printing plate.

OTHER PROCESSES.—Other, less common printing processes include *silk screen,* in which the ink is forced through an open-weave mesh stencil; *flexography,* which is similar to letterpress except that it uses rubber plates; *collotype,* a photogelatin process which reproduces artwork in continuous tones; and *dry offset,* which uses no water and has a slightly raised image, but retains the step which transfers the image from the plate to a rubber roll, from which it is offset to the paper. Other recent developments are *three-dimensional printing* (see *Xograph,* page 1188); and *electrostatic printing* which, instead of pressure, rely on opposing electrical charges to attract the dry powdered ink to the paper.

TYPE COMPOSITION.—Since the invention of movable type in the fifteenth century, type has been set by hand. Most type now, however, is set by machine or photographically.

Intertype and *Linotype* machines cast a complete line in a single *slug.* Many of these machines are operated by punched paper-type tape which greatly increases their speed. The most recent development is computerized control.

Monotype casts individual type characters. The casting machine is controlled by a punched tape produced frequently with the help of computers.

Ludlow is a combination of hand and machine composition. The compositor sets individual matrices by hand and then places them in a Ludlow caster which produces a slug.

Photocomposition is similar in operation to hot-metal typecasting, but the product, instead of metal type, is a film negative that can be used to make plates for any of the commercial printing processes. A significant advance in photocomposition has made it possible to form type at the rate of many thousands of characters per second by using a cathode ray tube controlled by a sophisticated electronic computer.

—Robert McGuire

Radar, the use of radio waves to detect a distant target and to determine its precise location. The term *radar* is an acronym for *RA*dio *D*etection *A*nd *R*anging.

A radar set detects an object by means of radio echoes. A short burst, or *pulse,* of radio waves is sent out in a given direction. If these radio waves encounter an airplane, ship, or other solid object, a radio echo returns and is picked up by the radar set. These returning echoes show up as *blips* of light on the radar screen —a cathode-ray tube similar to the picture tube of a television set. Because of its motion, the blip of a target can be distinguished easily from those of fixed objects on the ground.

The target's direction is determined by sending out pulses in one direction at a time. The radar antenna rotates slowly, examining each segment of the search area in turn. When a moving target is detected, the antenna can be stopped "on target"—pointing directly at the object.

To measure the distance, or *range,* of the target, the radar set uses the fact that radio waves travel at a known speed. The farther away the target happens to be, the longer the length of time needed by the pulse to travel to the target and return. A radar set measures this time interval very accurately and converts it to miles or feet for the operator. By combining detection, direction, and range, a radar set gathers all the information needed to pinpoint moving targets.

—William C. Vergara

Radio, the use of radiated electromagnetic energy for various classes of communication, such as telegraphy, telephony, sound broadcasting, television, artificial satellites, navigation guidance, radar, telemetry, and remote control.

The common aspect of radio systems is the controlled radiation of energy into space from a transmitting antenna energized by the radio transmitter. This energy is propagated, often in a directed beam, to a radio receiver. Electromagnetic radiation is a natural phenomenon from substances at temperatures above absolute zero. Being random in nature, it is present in radio apparatus as "noise," which sets a limit to receiver sensitivity. The receiving antenna extracts energy from the passing wave, and, if enough is received to override the receiver's own noise and that of external ambient sources, natural or manmade, the signal can be received intelligibly.

Radio transmissions take place in the *radio spectrum,* which extends roughly from 10,000 to 150 billion cycles per second, or, in terms of electromagnetic wavelengths in free space, from 30 kilometers down to millimeters. The adjustment for a particular transmission on an assigned frequency is called *tuning,* which is familiar to every owner of a radio or television receiver. Many professional radio systems are pretuned to a fixed operating frequency.

All nations of the world follow a frequency allocation plan worked out by the International Telecommunication Union (ITU), organized in 1932, which superseded the International Telegraph Union, organized in 1865. Broad technical principles underlie the designation of different portions of the radio spectrum for various classes of service, namely: wave propagation characteristics; distances to be covered; whether directive or nondirective transmission; the amount of spectrum (bandwidth) needed for a particular type of transmission; time intermittency or continuity of transmission; and the seriousness of interference to a particular service. There are three ITU frequency allocation plans for three geographical regions of the world but, because of their nature, some bands of frequencies are uniformly allocated on a worldwide basis. Within the broad ITU plans, individual countries subdivide bands of radio frequencies in much greater detail to suit individual needs.

In the United States, allocations for nongovernment users are made by the Federal Communications Commission (FCC). Allocations for government radio frequency assignments are made by the Interdepartmental Radio Advisory Committee (IRAC).

All radio transmitters in the world have to be licensed by government authority, and license assignments are registered with ITU. There are rules concerning constancy of operating frequency, the class of transmission and the bandwidth employed, the avoidance of harmful interference, and also rules relating to many technical characteristics required for dis-

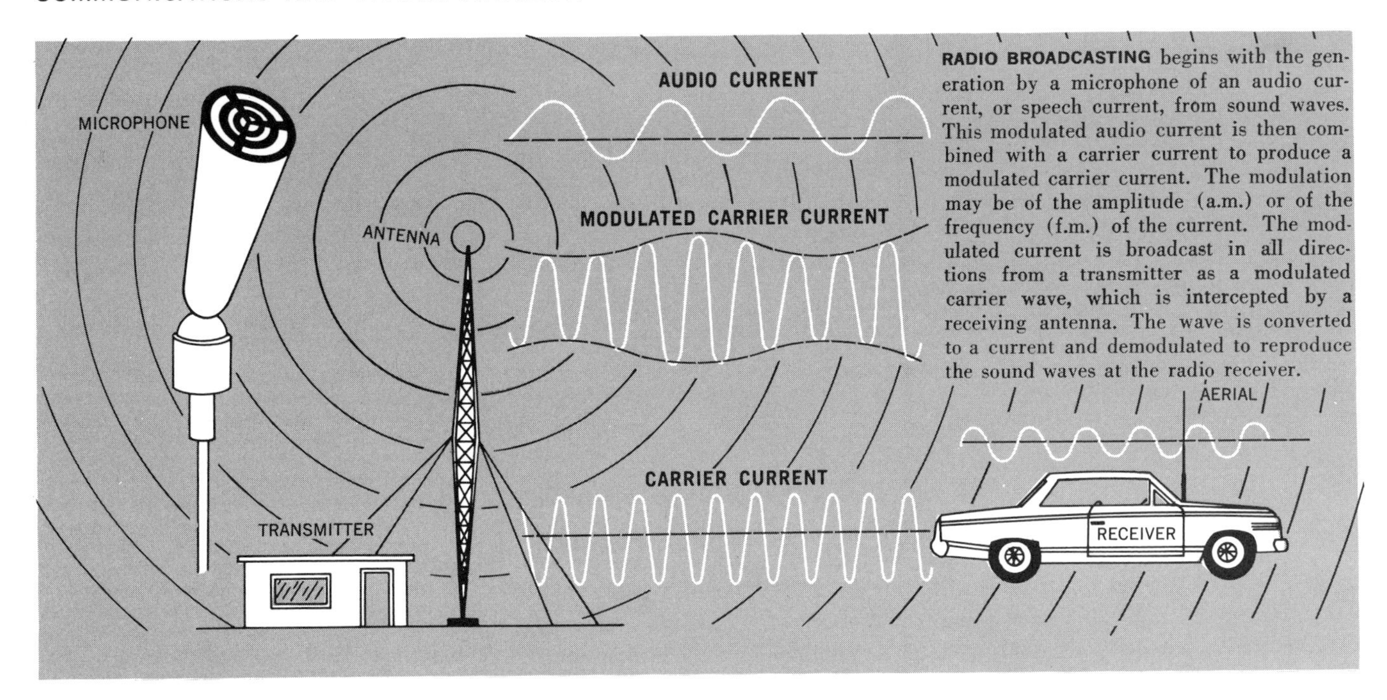

RADIO BROADCASTING begins with the generation by a microphone of an audio current, or speech current, from sound waves. This modulated audio current is then combined with a carrier current to produce a modulated carrier current. The modulation may be of the amplitude (a.m.) or of the frequency (f.m.) of the current. The modulated current is broadcast in all directions from a transmitter as a modulated carrier wave, which is intercepted by a receiving antenna. The wave is converted to a current and demodulated to reproduce the sound waves at the radio receiver.

ciplined use of radio. Though unbelievably complicated, world use of radio is actually quite orderly.

Through its technical consultative committees, ITU continually promotes studies and tests that eventually lead to specific recommendations concerning new rules or arrangements. After formal adoption, they have the status of treaties, which in the United States require Senate ratification. Following this, FCC rules of compliance are promulgated essentially as law.

■ **RADIO WAVE PROPAGATION.**—There are four basic modes of radio wave propagation: *ground wave, skywave,* quasi-optical, and optical. As shown in the accompanying illustration, different parts of the frequency spectrum give best results in different modes, with overlap between them.

Ground waves. Frequencies below two to five million cycles per second (2 to 5 mc) can be propagated over the earth's surface as ground waves, which in some cases cover great distances if sufficient power is radiated. This is the only way the lowest frequencies are usable. Ordinary broadcasting amplitude modulation (AM) depends on ground-wave propagation for primary coverage. However, ground-wave distances decrease with frequency because of earth curvature, so that above five million cycles per second, the losses limit distance so severely that it is more economical to employ other frequencies and propagation modes.

Skywaves. The upper atmosphere of the earth has various strata of ionized gas that bend and reflect some radio waves back to earth. Below one million cycles per second (1 mc), reflected *skywaves* are more bothersome than useful. But from 3 to 30 or 40 mc, the ionized strata, called *ionospheres,* act as efficient mirrors from which long-distance radio waves can be bounced by single or multiple reflection.

This frequency range, commonly known as *short wave,* is used for telegraphy and telephony over long distances, including international short-wave broadcasting. Variations in the ionosphere due to solar influences, and to the daylight-darkness changes, affect choice of best working frequencies and transmission reliability. Every short-wave broadcast listener knows how much variation in reception can occur from time to time. Nevertheless, this frequency range is of utmost utility for aviation, marine, transoceanic telephony and telegraphy, sound broadcasting, navigation, and many other services that could not be realized economically by any other mode of wave propagation. However, all such services are confined to those employing narrow-band emissions, or bandwidths of less than 5-10 kilocycles, which is barely enough for acceptable telephony. Many long-distance circuits are limited to one or two hundred cycles' bandwidth, suitable only for slow telegraphy.

Quasi-optical Propagation, When the wave frequency is too high to be reflected from the ionosphere, the range of a station is limited to a relatively small distance beyond the horizon. Propagation into the trans-horizon region is due to the phenomena of atmospheric refraction and diffraction over the curvature of the earth.

To best utilize these effects, as is done in typical frequency modulation and television broadcasting, the transmitting antenna is raised as high as feasible, and the radiant energy is concentrated toward the horizon. Hills and clusters of tall buildings in the wave path impair transmission by this mode. Reception is improved by using directional antennas (most

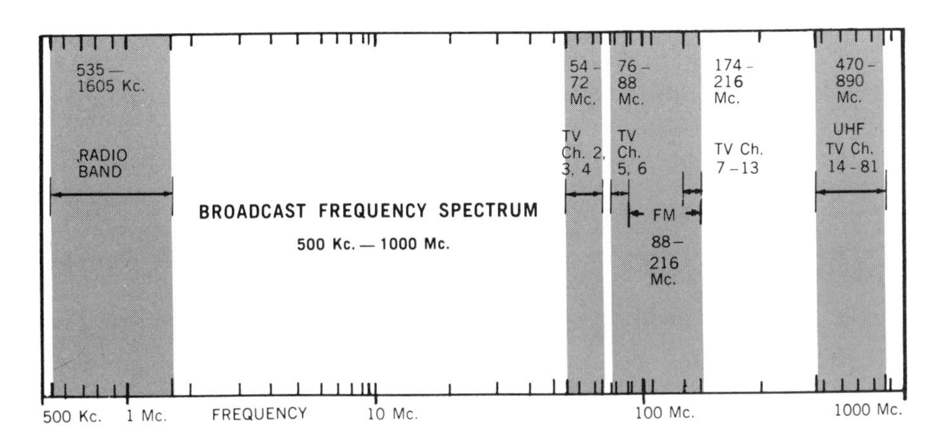

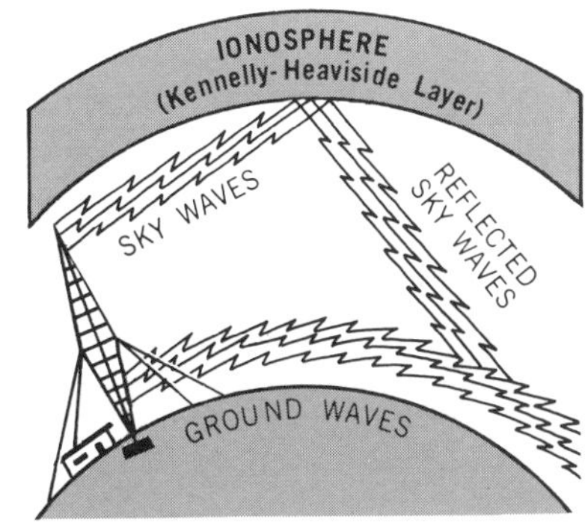

RADIO WAVES follow the curvature of the earth and are reflected by the ionosphere.

TV antennas are directional to some extent) and raising their heights. In television *fringe areas,* reception varies with atmospheric conditions—mainly the vertical distribution of humidity which causes the waves to bend downward and to penetrate further into transhorizon areas. It also varies when aircraft reflections interfere with direct reception, producing flutter in television images.

Exceptional distances are sometimes covered by superrefraction, when vertical humidity gradients exceed certain normal values. The "Venetian-blind" effect in a television picture is due to this condition. It is produced by occasional long-distance interference from another television station on the same channel. Another case of this may be a *surface duct* existing occasionally between the earth and an elevated boundary layer in the lower atmosphere. Such a duct often *traps* the wave and causes it to be propagated by successive reflections between the ground and the boundary layer—frequently a well-defined region between clear and hazy air. Cloud strata indicate similar conditions.

Optical Propagation. This term applies to propagation at still higher frequencies (500 mc or higher) where transhorizon propagation by diffraction almost disappears. These frequencies are useful only when applied to locations within line-of-sight. Though this is a serious limitation, it nevertheless provides a very wide usefulness. A number of extremely important services use optical transmissions, such as microwave radar, radio relaying of all kinds of telecommunications by means of a series of radio repeaters (most transcontinental television programs are carried this way), and satellite and space communications. For example, intercontinental satellite communication takes advantage of the great height of the satellite to give line-of-sight paths from the satellite to Europe and to North America. The satellite carries a repeater, which consists of "receive" and "retransmit" equipment that relays both ways.

Marginal Propagation Modes. It suffices merely to mention certain marginal and intermittent modes of wave propagation other than those discussed above, because they have some utility. They include tropospheric scatter, ionospheric scatter, meteor-trail reflection, auroral reflection, and a few more obscure modes.

■ANTENNAS.—Radio waves are launched into space from structures (usually metallic) called antennas, or aerials. The geometry of the structure, and the distribution of the electric currents in it when it is energized by a transmitter, determine the distribution of radiant energy at a particular frequency. At the lowest radio frequencies, they consist of steel towers, or towers supporting wires, with the active parts of the antenna formed between the earth and the wires. Others, of which television antennas are examples, are complete structures held up in space on suitable supports, and connected to apparatus by a transmission line.

There are endless configurations possible for many special purposes, varying from simple auto "whips" to electromagnetic lenses, horns, paraboloids, etc., all dependent on desired vertical and horizontal directivity. Beams of energy can be shaped as needed for aiming at one location, fanned to cover a certain sector, or distributed uniformly in all directions. Beaming the power in one direction has the same effect as a searchlight, increasing the effective radiated power on the beam. At microwave frequencies, antennas may have power gains as high as a million times and are used for radio relaying, radar, and satellite communications.

Any transmitting antenna can function also as a receiving antenna with the same directional properties. In most microwave systems, the same antenna is used for transmitting and receiving. At the lowest frequencies (longest wavelengths), including ordinary radio broadcasting, the receiving antenna can be extremely simple if the field strength is so great that the antenna efficiency is unimportant. Modern broadcast receivers may only use a part of the power cord as an antenna. *Rabbit-ear* antennas often suffice for television reception. But in systems where economics determine the need for highest operating efficiency, the design of the receiving antenna must be elaborated to extract a maximum of energy from the wave to exclude electrical noise and interference as much as possible.

■RADIO TRANSMITTER.—A transmitter consists of a source of electric oscillations of a required constancy at a specified frequency; amplifiers to attain a specified power; means for modulating the oscillations in such a manner as to carry the desired intelligence; and power supplies to energize the various portions of the transmitter.

The unmodulated radio frequency is called the *carrier wave,* or carrier. Modulation consists of interrupting or modifying its amplitude (AM), phase (PM), or frequency (FM) in accordance with telegraphy, telephony, television, or other form of communication. The information transmitted is contained wholly within the variations imposed on the carrier. It is even possible to transmit this information without the carrier in what is known as *sideband* transmission. Any form of modulation introduces other frequencies clustered around the carrier that consist of the sums and differences between the frequencies composing the modulating signal and the carrier frequency. Once these sidebands are generated, the carrier wave need not be transmitted, provided it can be correctly restored at the receiver to permit detection of the information. Most well-known forms of radio transmit the carrier with the sidebands, thus permitting inexpensive receivers to be used. Suppressed carrier techniques are used for many professional systems. Radio transmitters in use today vary in power output from milliwatts to megawatts.

There are systems that transmit the carrier wave in the form of short pulses. *Pulse radar* uses the silent spacing intervals to receive echoes of the pulse from reflecting objects illuminated by the waves. Distance is measured by timing the return echo. Pulses also can be modulated in amplitude, width, or timing for transmission of all kinds of information, as with a continuous carrier. These are called, respectively, pulse-amplitude, pulse-width, and pulse-position modulation.

■RADIO RECEIVER.—A radio receiver consists of circuitry that accepts the very small energy from the receiving antenna (when tuned to the desired frequency) and amplifies it to a sufficient power to actuate a detector efficiently. The detector separates the carrier wave or carrier pulses from the modulated variations that contain the transmitted information. The recovered information (signal) is then amplified and delivered to an output device, such as a loudspeaker, picture tube, display, teleprinter, or recorder, according to the nature of the system. Additionally, the receiver may have auxiliary functions such as automatic amplification control to compensate for fading, filters to exclude extraneous disturbances, impulse noise suppressors, and many others. The receiver must be adapted to the kind of transmission and the frequencies to be received, since it must work as a coordinated system with the radio transmitter. —E. A. Laport

Railroad, a system of land transportation with a permanent way, or track, of two parallel rails that support and guide vehicles fitted with flanged wheels. The chief advantage of this system is its ability to move heavy loads with a minimum of power because of the smooth running surface offered by the rails.

The rutways of ancient Greece and Rome can be claimed as the forerunner of the railway. Although primitive wooden railways were used in Central European mines during the sixteenth century—and probably earlier—the modern railway is essentially a nineteenth-century development. The Stockton and Darlington Railway in England was the first public railway. Opened in 1825, it ushered in the modern concept of the railway as a common carrier. Other similar enterprises followed, and England became the center of railway construction and technology.

■BEGINNINGS OF U.S. RAIL SYSTEM.—The United States, an aspiring industrial nation at the beginning of the nineteenth century, urgently needed an inland transportation system. Compared to canals, railways were relatively cheap and could be constructed quickly. Furthermore, they had greater capacity than roads and were not subject to the seasonal disruptions (floods, freezing, and drought) of inland waterways. Although a few horse-drawn industrial lines were built earlier in the nineteenth century, no major railroad construction was completed in America until 1830.

Numerous rail lines were projected and built in all settled parts of the country during the 1830's. This first railroad boom, which produced nearly

UNION PACIFIC RAILROAD

A POWERFUL STEAM LOCOMOTIVE, the 4000, was one of the three largest ever built. It was first used to haul freight cars in 1941.

GENERAL MOTORS CORPORATION

AEROTRAIN, an American-built, lightweight diesel-powered passenger train, has ten coaches capable of carrying 400 passengers.

1,500 miles of track, did not end until the panic of 1837. Dull times slowed construction during the early 1840's, after which the system enjoyed another rapid growth. The New England states were united by a network of rail lines by 1850; the Midwestern system was virtually completed during the next seven years. Several lines—the Erie, Baltimore and Ohio, and Pennsylvania Railroads—connected the interior to the Eastern seacoast. The second boom was ended by the panic of 1857, a financial collapse directly associated with the overbuilding of railroads.

■**THE TRANSCONTINENTAL RAILROAD.**—Interest in a railroad to the Pacific was heightened by the admission of California to the Union in 1850. While the first railroad bridge over the Mississippi River was completed in 1856 at Davenport, Iowa, the transcontinental railroad project did not proceed until the federal government agreed in 1862 to subsidize its construction. Work was started the next year by the Union Pacific in Nebraska and by the Central Pacific in California. The two companies, building toward each other, met at Promontory, Utah, on May 10, 1869, and the meeting was symbolized by the driving of a *golden spike*. The completion of the transcontinental road spurred the construction of other Western lines, and by 1890 a nationwide rail system was complete. Railroad mileage continued to expand until 1915.

■**STANDARDIZATION.**—While the national system was being completed, the question of a uniform gauge plagued the industry. The English *standard gauge* of 4 feet 8½ inches was used by many Eastern lines from the 1830's, but a variety of gauges was adopted througout the country. The Erie chose the broad gauge of 6 feet; the Southern railroads, 5 feet; and the New Jersey lines, 4 feet 10 inches. In the 1870's and 1880's the narrow-gauge mania developed, but few major roads adopted the 3-foot gauge. The selection of 4 feet 8½ inches by the transcontinental railway in 1862 prompted other new roads to build on this width. The rebuilding of existing odd-gauge railroads was slow and costly, but by 1886 when the Southern railroads changed from 5-foot to standard gauge, nationwide conversion was virtually completed.

■**ADVANCES IN THE SYSTEM.**—Between about 1870 and World War I, the railroads dominated land transportation in the United States. Long-haul road transport was nonexistent, river traffic languished, and canals were abandoned or reduced to low-tariff haulage. This concentration of traffic on one carrier proved too much for the rail system when demands for service exploded during World War I. In the crisis that developed, the industry was nationalized in 1918. Two years later the railroads were returned to private owners. In addition to being weakened by this experience, the railroads then also met aggressive competition for traffic. The motor truck began intra-city haulage not long after the war. Because of door-to-door delivery, trucks attracted much of the high-tariff freight. Automobiles, buses, and airlines cut deeply into passenger traffic. Pipelines and inland waterways, which developed during the same period, siphoned off much of the bulk traffic. The emergence of these competing forms of transportation, together with the great depression of the 1930's, struck a severe blow at the railroad industry. Many major roads were forced into receivership. Numerous secondary and short lines were entirely abandoned. Greatly weakened by these events, the railroads were none-

BALTIMORE & OHIO RAILROAD

GENERAL MOTORS CORPORATION

"TOM THUMB" (left) was an 1830 experimental steam locomotive for passenger use. At right is a 3600 horsepower Diesel electric locomotive.

theless required to transport the enormous traffic of World War II.

After the war, particularly in the 1960's, railroads have shown greater interest in revising their operations to suit the needs of shippers. Freight-train speeds have been markedly increased, with some "hotshot" freights offering overnight delivery between major cities. Special high-capacity cars, piggyback cars for truck trailers, multilevel automobile cars, and unit trains for efficient coal movements are among the more important reforms of recent years. Although much new equipment has been purchased, a less aggressive program has been pursued to attract passenger traffic, and its decline has been severe—down nearly 50 per cent since 1945. Heavy losses from operations, no longer compensated by equivalent freight business, have forced some railroads to abandon such service.

The railroad industry has made many efforts to streamline performance aside from new rolling stock. Centralized traffic control has speeded trains and eliminated unnecessary tracks. Dieselization has improved service and reduced fuel, service, and labor costs. Mergers between parallel roads eliminated duplicate facilities and personnel.

The railroad industry lost its monopoly in the 1920's, but its prospects are not all dark. Because of highway congestion and the inability of the airlines to carry freight at an economical rate, the railroad will undoubtedly continue as an efficient carrier for many generations.

▪LOCOMOTIVES.—The steam locomotive, like the railway, originated in Great Britain. Richard Trevithick built the first steam locomotive in 1804 for the Pen-y-darran iron works in Wales. The machine proved too heavy for light track, intended only for horse-drawn wagons, but it encouraged other English mechanics to build locomotives. By 1820 the steam locomotive was recognized as a practical and economical form of motive power. Three years later, the world's first locomotive works was established at Newcastle by Robert Stephenson and Company. In 1829 this company completed the famous *Rocket,* a light, well-proportioned mechanism that incorporated all the basic features of subsequent steam locomotives-multitubular boiler, separate firebox, outside connection, and blast pipe exhaust.

The first locomotive in America, the *Stourbridge Lion,* was imported from England in 1829. Importation of British engines continued until 1841. American mechanics, however, quickly entered the field and produced a locomotive that was better suited to the uneven tracks of the cheaply built railroads. The West Point Foundry Association (New York City) built the first locomotive in the United States, the *Best Friend of Charleston,* in 1830. During succeeding years nearly 150 companies entered the field, but by 1915 the industry was dominated by three major companies—Baldwin, Lima, and American Locomotive.

▪AMERICAN STEAM LOCOMOTIVES.—Because of inadequate capital, poor track, and steep grade, the pioneer railroads required flexible, cheap, and powerful locomotives. Simple, direct designs kept costs low—small wheels and high boiler pressures provided power, while leading trucks, bar frames, and equalizing levers offered a flexible running gear. All of these admirable features were incorporated in the eight-wheel, or *American,* locomotive. This machine was used for both freight and passenger service and proved so effective that it remained the standard wheel arrangement almost to the end of the nineteenth century. The development of heavier freight trains during the 1870's caused the decline of the *American* type for that service. More powerful freight locomotives, such as the *Mogul* and *Consolidation,* became common. During the same period, passenger train speeds were increased, and again heavier, more specialized locomotives replaced the *American* type.

The basic concepts of locomotive design were revised during the early years of the twentieth century. Faster schedules, heavier tracks, and more ample capital required a more sophisticated and scientific design. The most obvious change was an enormous increase in size: whereas during most of the nineteenth century, few locomotives weighed over 30 tons, 60-ton locomotives were not uncommon by 1910. A wider variety of wheel arrangements and more specialized designs for each class of service were introduced. Locomotives became increasingly complex as more auxiliary equipment, such as stokers, superheaters, and feedwater heaters, was added to increase the machine's capacity and efficiency.

Locomotives and Tracks in U.S.

Year	Engines	Track
1830	15	23 miles
1850	3,000	9,021 miles
1870	12,500	52,922 miles
1890	31,812	163,597 miles
1910	60,019	240,293 miles
1930	59,553	249,052 miles
1950	42,951	223,779 miles
1960	31,178	217,552 miles

By the late 1940's steam locomotives were huge mechanisms, cluttered with numerous auxiliaries and intricate piping systems. The Union Pacific Railroad had a giant, 600-ton, articulated freight locomotive—the *Big Boy*—which was recognized as the largest steam locomotive ever constructed. Nevertheless, the last steam locomotives built in the United States (1953) continued to follow the basic design of the *Rocket* (1829). As a result of conservative design, poor thermal efficiency, together with several nonmechanical problems, such as labor and terminal costs, steam locomotives on U.S. mainline railroads were abandoned by 1960.

▪DIESEL-ELECTRIC LOCOMOTIVES.—The diesel-electric locomotive succeeded steam as the principal railroad motive power by the early 1950's. First used commercially by the Central Railroad of New Jersey in 1925, diesel-electrics were at first considered only for switching service. During the 1930's, several lightweight streamlined trains were built, and in 1935 the first freight diesel-electric was used on the Baltimore and Ohio Railroad. World War II interrupted the more rapid introduction of diesel locomotives. However, many major railroads began to convert entirely to this form of power at the war's end. By 1954, diesel-electrics were moving 80 per cent of the railroad traffic in the United States. The complete conversion took place in 1960.

The diesel engine drives a generator that supplies electricity to motors geared to the locomotive's wheels. Because of the electric transmission, it is possible to connect several diesel-electric locomotives, or units, thus forming a powerful locomotive under the control of a single crew. In recent years, heavy-duty hydraulic transmissions have been successfully employed for diesels.

▪CARS.—The earliest railroad cars were small, four-wheeled and fabricated almost entirely from wood. Eight-wheel, or *double truck,* cars were introduced in 1831, and within a few years became the accepted model for both freight and passenger service. The basic form of American passenger car, with a center aisle, side seats, and an entrance at each end of the car was introduced at the same time. During most of the nineteenth century, such cars weighed about 15 tons and seated 60 passengers. The clerestory roof, a raised center section with windows for air and light, introduced in the 1860's, was a major innovation in car design. Wood construction continued until 1907, when all-steel coaches were introduced by the Pennsylvania Railroad. Steel passenger cars rapidly replaced weak, flammable wood cars. Lightweight streamlined cars appeared in the 1930's. Air conditioning soon became common.

A wide variety of freight cars—box, flat, hopper, gondola, tank, and others—was developed to handle the diverse products that must be transported. Of these classes, the boxcar is the most numerous, accounting for over a third of the 1,512,306 freight cars now in service in the United States. Wooden freight cars prevailed—although several thousand iron hopper cars were built from the 1840's through the 1880's—until the introduction of all-steel cars in 1897. The greater capacity of all-steel cars won rapid acceptance of this style of construction.

▪RAILS.—A wide variety of rail types came into use during the first half of the nineteenth century. Strap rail, a thin bar of iron spiked to a longitudinal timber, was favored by most early roads because of its cheapness. This form of construction was not used by public railroads after the 1850's. More costly but more durable rolled rails were used on many of the better constructed pioneer railroads. For example, the Camden and Amboy was built with *T rails* in 1831, and the Philadelphia and Columbia with a form of chair rail known as *Clarence rail* in 1833. *Pear rail,* a squat form of T rail, and *U rail* were popular through the 1860's. After that time, however, "T" became the universal type of rail used on American railroads. Steel "T" rails were introduced during the 1860's and by 1890 were generally adopted by all major lines. Rail weight increased from 60 pounds per yard in the 1850's to about 120 pounds per yard at the present time; the heaviest rail in use weighs 155 pounds. Welded rails 1,500 feet long (standard rail length is 39 feet), now being widely installed, are free from troublesome rail joints and provide a smoother track. —John H. White, Jr.

Railroad Engineering, the design, construction, and maintenance of all physical properties needed for railroad operations. These include roadbeds, tracks, bridges, tunnels, stations, repair shops, locomotives, cars, signals, and rails.

A well-constructed track is fundamental to a successful railroad. Such a track is characterized by good alignment, proper drainage, easy grades, and generous curves. Grades rarely exceed two per cent, and curves are generally not sharper than a 1,000-foot radius on mainline railroads. Immense cuts, fills, trestles, and tunnels are required to maintain such a straight and level line. Railroad construction costs, accordingly, are high, but such roads are able to move heavy traffic with a minimum of power.

The track must be resilient to absorb and cushion the weight of passing trains. The wooden crosstie has been retained for this purpose although steel and cast concrete ties fitted with rubber cushions are being considered as the cost of wooden ties increases. Chemically treated wooden ties last as long as 40 years and remain the most popular type of railroad tie. Crushed stone is used by all major railroads for ballast.

At first, American civil engineers believed that the massive stone viaducts favored by British railways were the only design proper for railroad bridges. These structures were difficult and expensive to build. Wood was soon adopted because it was abundant, and a less costly bridge quick and easy to fabricate. Iron bridges of several patterns were used in the 1840's and 1850's, and by 1870 iron was the accepted material for railroad bridge construction. Late in the nineteenth century, iron was supplanted by steel.

Tunnels are avoided wherever possible because of their great cost. There are only about 1,400 tunnels on U.S. railroads. The first railroad tunnel in the United States was built by the Allegheny Portage Railroad, near Johnstown, Pa., in 1833. The Baltimore and Ohio Railroad built a short tunnel near Harpers Ferry, W. Va., in 1839–40. Many other short tunnels were then constructed. The first tunnel of major length was the 4.7-mile Hoosac Tunnel completed in 1875 by the Boston and Maine Railroad. The longest American railroad tunnel is the 7.7-mile Cascade Tunnel of the Great Northern Railroad. Today many railroads are eliminating tunnels by rerouting lines or by replacing tunnels with open cuts.
—John H. White, Jr.

Railroad Signaling, the system of semaphores and lights used to control railroad operation. There was little need for elaborate signals on early American railroads; traffic was light and trains rarely operated at night before 1850. Instead, the timetable, or the *time-interval system,* was the basis for train movements. By strict adherence to the schedules, each train on the line could be accounted for and would be in its proper place. This simple system worked well, even on single track line, as long as there were only a few trains. However, long delays often resulted; if one train was off schedule, other trains on the line could not proceed until it was accounted for. The introduction of telegraph dispatching in 1851 on the Erie Railroad greatly improved the efficiency of the timetable system.

The *space interval* or *block system,* where fixed signals are used, is a more positive method of traffic control. It was first applied on the Newcastle and Frenchtown Railroad (Delaware) in 1832. Manually operated ball signals at three-mile intervals governed train movements. This plan was not generally adopted until traffic increased to a point where the timetable system was inadequate for safe or efficient handling of train movements. The first important installation of the block system was on the United Railways of New Jersey, between Philadelphia and Trenton, in 1863. By 1876, the entire main line of the Pennsylvania Railroad was fitted with block signals, and within a few years most other major railroads had adopted the system.

All signals before Thomas W. Hall's *electric disc signal* of 1866–67 were manually operated. The principal manual signals were the *disc, ball, banner,* and *semaphore.* Automatic electric signaling enjoyed a rapid development after William Robinson's invention of the closed-track circuit in 1872. Because of inadequate electric motors and a dependence on batteries, the first automatic signals were mechanically driven by clockworks, whose action was in turn governed by an electric control. By 1893, more powerful yet compact electric motors permitted the automation of the long-favored semaphore signal. The semaphore, of ancient origin, remained the standard railway signal until about 1940, when it was superseded on many roads by light signals.

Centralized traffic control, or *CTC,* by which more than 150 miles of railroad can be controlled by a single operator, is the most important recent development in railroad signaling. It was first tested by the New York Central Railroad in 1927. By 1967, nearly 36,000 miles of road were operated by centralized traffic control. —John H. White, Jr.

River Engineering, a specialized science combining geology and engineering and devoted to the control and use of rivers, mainly for transportation purposes. It may also deal with the prevention of floods and of water pollution. The river engineer's task is twofold: first he must make the river navigable, deepening and widening its channel, bypassing its rapids, and straightening its curves; then he must try to maintain his handiwork, by far the more challenging job since rivers persistently resist man's efforts to harness them.

Despite dredging, revetments on river banks, and various control structures designed to keep river channels stable, natural forces, not completely understood, cause most rivers to *meander.* Straight reaches develop gradual curves. Winding sections become even more so, with the river always eroding its banks on the outside of the curve and depositing sediment on the inside. Floods often cause pronounced and sudden changes in river channels.

River engineers work to halt or at least slow down the natural river actions because they make navigation both difficult and dangerous. Some rivers are completely canalized, thus preventing many of the above problems. Relatively low dams, built at various points along the river's course, create what are called navigation pools, really reservoirs which together create a watery stairway. River flow spills over these dams and boats and barges go up and down level-by-level through navigation locks. Other navigation dams and locks are built to eliminate rapids and shallow stretches. For example, a dam just below the rapids forms a pool, inundating the rapids so river traffic may pass.
—William W. Jacobus, Jr.

Roads and Highways, thoroughfares for pedestrian, animal, and vehicular traffic. While paths and trails are as old as man, the first substantial, carefully engineered roads were built by the Romans. The influence of their heavy construction was felt down to the eighteenth century, when the modern science of roadbuilding originated. European road surfaces were frequently 18 inches thick until about 1775 when Pierre Trésaguet introduced a lighter construction into France. In England in the early years of the nineteenth century, two experienced road builders, Thomas Telford and John Loudon McAdam, improved the methods for building roads of broken stone. Telford's system, similar to that of Trésaguet, consisted of building a foundation of larger stones, which were then covered with several layers of smaller broken stones. McAdam felt that the foundation of larger stones had a detrimental effect, and built his roads to a depth of 10 inches with three layers of small stones, broken to a size which would pass through a 2-inch ring. Each layer was compacted by traffic before the next layer was applied, and in later years horse-drawn rollers, and eventually, steamrollers, performed the work of com-

UNITED STATES LINES

OCEAN MONARCHS, S.S. *America* (*foreground*) and S.S. *United States,* pass in the Hudson.

pacting. These men, particularly McAdam, also stressed the importance of proper grading and drainage, notably by raising the road above the surrounding terrain.

In the United States, the first long-distance section of broken stone and gravel surfacing was laid in Pennsylvania, on the Philadelphia and Lancaster Turnpike Road, between 1793 and 1795. In 1823 the first surface constructed according to the McAdam system was laid on the Boonsborough Turnpike Road between Boonsboro and Hagerstown, Maryland. During the mid-nineteenth century there was a general neglect of roads, particularly in rural areas. There were new developments, however, such as rock-breaking machinery and steam road rollers, yet the use of these did not become widespread until the end of the century.

In the 1890's many states began to enact road-aid laws, due in part to the efforts of the League of American Wheelmen. In 1893 the Office of Road Inquiry was established under the Department of Agriculture to begin studies on the management and construction of roads. During the first decade of the twentieth century several new constructions, used some years earlier for surfacing city streets, were introduced on rural roads. These were the various types of bituminous surfaces, such as coal tar, crude oil, and asphalt, and Portland cement concrete. The bituminous materials were more frequently used during the earlier years since they could be applied to old macadam roads without the necessity of building an entirely new road surface. The Federal government intensified its interest in roads during the 1920's; since that time, roads have continued to spread across the United States until they totaled 3,697,950 miles at the beginning of 1967. At the present time there is under construction a 41,000-mile National System of Interstate and Defense Highways. This system, scheduled for completion in 1972, will serve all of America's large cities. In the future, it is expected that national road mileage will not increase greatly, but emphasis will be directed toward better and safer roads.

—Don H. Berkebile

Ship, a seagoing vessel used for cargo and passengers or military applications. Nuclear power is opening vast new areas for maritime expansion. Atomic engines, by virtually eliminating the problem of air intake and exhaust, make possible the development of submarine ships of all types, with limitless size potentials. The entire area of underwater exploration and exploitation is becoming a billion-dollar industry. No longer must ships be tied to routes dictated by the location of bunkering stations. The commercial and military potential of atomic power in shipping is tremendous.

■HYDROFOILS.—Equally exciting are the new types of ships that have been developed in the postwar era. The hydrofoil has been perfected for short-haul coastal and lake use. With speeds of 50 to 70 knots, the effect of such ships on old schedules and transportation patterns is clear. In 1965, the U.S. Navy completed the Maritime Administration's *Dennison,* a 50-knot hydrofoil, which has been successfully tested. A hydrofoil vessel is equipped with stiltlike legs; it can rise out of the water and plane on these so-called *foils.* Speeds up to 100 knots are predicted, with possible application to major ocean vessels.

■AIR-CUSHION VESSELS.—Another major new ship type is the *surface-effect craft,* also known as the *hover-craft.* This vessel rides inches off the surface of the ocean, supported by an air cushion maintained by ducted fans. British ship architects developed this type of craft, two of which have been ordered for cross-Channel passenger and auto ferry service. The potential of the hovercraft is great, since it can ride not only over water but over flat land areas, suggesting door-to-door freight delivery to replace expensive cargo-handling steps at each seaport.

■CARGO SHIPS.—Ordinary cargo ships today are no longer so ordinary. Speeds, which for generations remained about 10 or 12 knots for the typical freighter, have now more than doubled. New liners vie for transatlantic freight speed records, breaking one another's marks as often as did the great clippers of a hundred years ago. Wider hatches for better cargo handling, mechanical hatch covers, and heavier cargo-boom capacities allow quicker freight loading. Specialized types of cargo ships are another new phase of shipping. Liquefied gas carriers and tankers built to transport dozens of utterly different, often highly dangerous, cargoes are being built throughout the world.

In size, the oil tanker has attracted the most attention. As recently as World War II, a 16,000-ton vessel was considered a very large tanker. Today, orders for ships of 300,000 tons each have been given.

■PASSENGER SHIPS.—Modern transatlantic passenger ships are marvels of speed and comfort, largely because of the keen competition among the steamship companies and the maritime nations. Among British ships, the *Queen Mary* and the *Queen Elizabeth,* and among the French, the *France,* were built as the utmost in luxury liners. The *Queen Mary,* launched in 1934, has a gross tonnage of 81,237 tons. The *Queen Elizabeth,* launched in 1940, is the world's largest passenger ship, with an over-all length of 1,031 feet and a gross tonnage of 83,673 tons. Her engines generate 246,000 horsepower. The 83,000-ton *Normandie,* launched in 1935, was destroyed in a fire in New York Harbor in 1942. Her recent successor, the 66,348-ton *France* is the world's longest passenger ship; four feet longer than the *Queen Elizabeth.* Faster than either of the Queens, both retired, is the *United States,* launched in 1952; she has 12 decks, two theaters, and two large swimming pools. With an overall length of 990 feet and a gross tonnage of 53,329, the *United States* can carry 1,982 passengers in addition to her 1,000-man crew. Demonstrating the modern passenger ship's adaptability to wartime needs, the *United States* can be converted to a military transport ship capable of carrying 12,000 to 14,000 troops.

—Frank O. Braynard

U.S. NAVY

WORLD'S LARGEST aircraft carrier, the 86,000-ton nuclear-powered U.S.S. Enterprise is shown during construction in 1960.

Sound Recording and Reproduction, the processes whereby sound is converted, using microphones, to electric signals which are stored (recorded) in such a manner that they can be subsequently converted back to sound (reproduced), using loudspeakers. The difference in quality between the reproduced sound and the original sound determines the degree of fidelity, that is, high or low fidelity.

Sound results from vibrations in the air. It encompasses a range of audio frequencies that a human being is capable of hearing—normally from 20 to 20,000 cycles per second. A bass drum produces a low-frequency sound of about 100 cycles per second, while a violin can produce high frequency sounds up to 12,000 cycles per second. A male voice produces a range of frequencies between 100 and 10,000 cycles per second, while a female voice produces a range between 200 and 10,000 cycles per second.

There are three common methods of recording sound, each using a different medium for recording and storing sound information. These are modulated groove recording on phonograph discs, magnetic recording on audio tape, and modulated light recording on a photographic film.

The method of modulated groove recording on phonograph discs was invented by Thomas A. Edison in 1877. The medium originally used was a wax-covered cylinder approximately three inches in diameter. The magnetic recording system was invented by Valdemar Poulsen, a Danish inventor, in 1898. The medium used then was steel wire approximately 20 mils in diameter. No single person is credited with the discovery of the process of modulated light recording on film. Both the Radio Corporation of America and the Western Electric Company developed their own methods of sound-on-film recording in 1910.

MAGNETIC TAPE RECORDING.—Most original performances are recorded by the magnetic process. The sounds produced by an actor or a singer or an orchestra's instruments are sensed by a microphone, or group of microphones, in which sound is converted to an electric signal. The electric signal is controlled, amplified, and directed to an audio magnetic tape-recording machine. In the tape machine, the electric signal acts upon an electromagnetic recording head, causing magnetization of a ¼-inch wide, 1½-mil thick magnetic audio tape as it passes over the head at a speed of 7½ inches per second. The magnetization varies in accordance with the original sound signal.

Audio magnetic recording is used because the tape can be more easily handled by the recording engineer. It can be edited (cut and spliced) to make corrections. Editing is not practical in either of the other recording methods.

Once the original recording is completed and stored on a reel of magnetic audio tape, it can then be dubbed (transferred) to the other media or duplicated on other reels of audio magnetic tape. The latter process requires many tape recording machines to be fed from a playback machine on which the originally recorded tape is placed.

COLUMBIA RECORDS

RECORDING session in a studio is taped and monitored by engineers in the control room.

PHONOGRAPH RECORDS.—The production of phonograph discs (or records) involves the following steps. First, a master disc recording is made, using the originally recorded tape as the source. That tape is placed on a playback machine similar to the one used in the production of audio tapes. The electric signal is picked up from the moving magnetic audio tape by an electromagnetic playback head, amplified, and directed to a phonograph-disc recording system. Such a system includes a powerful amplifier that feeds an electromagnetic cartridge, which in turn actuates a diamond-tipped cutting stylus. As the master phonograph disc rotates at a speed of 33 1/3, or 78 revolutions per minute, the stylus is moved from side to side in accordance with the original sound signal, creating a modulated groove (a groove of varying width). The groove is started from the outside and spirals toward the center. This master disc is used to produce a stamping mold *mother* from which copies of phonograph records, made of a vinyl plastic, are produced.

HIGH-FIDELITY REPRODUCTION.—Audio tape and phonograph discs are the media most commonly used for playback or reproduction by high-fidelity enthusiasts. For reproduction purposes an *audio tape recorder* (capable of both recording and playback) is connected to an amplifier and loudspeaker system. The audio tape is placed (threaded) on the tape machine. The tape passes over an electromagnetic *playback head* located along the tape's path. The magnetization is picked up by the playback head and fed to a *loudspeaker* through a powerful *amplifier*. In order to reproduce a phonograph record, a *turntable*, upon which the record is placed, and a *tone arm* with a phonograph *pickup cartridge*, on which a *stylus* is mounted, are required. The stylus follows the groove and is moved from side to side in accordance with the recorded modulations, producing an electric signal at the output terminals of the cartridge. The signal is fed to an *amplifier* and *loudspeaker* combination.

The phonograph pickup cartridge may be either an *electromagnetic* or a *crystal* type. In the crystal type, the material that responds in the cartridge is Rochelle salt or Barium Titanate (ceramic). In either type, the stylus is tipped with either sapphire or diamond, the latter being the more costly.

In a complete home high-fidelity reproduction system, a *preamplifier* is commonly used. This device can accommodate many playback devices, including a radio tuner.

MODULATED LIGHT RECORDING.—Modulated light recording on a system of photographic film is used with motion pictures. A light source is modulated (varied) in accordance with the sound signal. The light is directed to the unexposed film in the motion-picture camera while the camera is taking pictures. This sound signal is located on the film between the picture and the sprocket holes. After the exposed film is processed, it is ready for reproduction. In reproduction, light is passed through the film at the sound track position. The light intensity is varied by the modulation on the sound track and is directed to a photocell. The output of the photocell (an electric signal) is fed to a powerful amplifier and then to a loudspeaker.

STEREOPHONIC SYSTEMS.—Stereophonic (dual channel) recording and reproduction differs from monaural (single channel) high-fidelity systems in that there are two channels, left side and right side. An original stereophonic recording, using left side and right side microphones, provides two separate but simultaneous electrical signals. On audio tape they are recorded on two parallel tracks, while on phonograph discs they are recorded on each side of the groove. In reproduction, both systems provide left and right side electrical signals that are fed to a pair of amplifiers and loudspeakers. Because the loudspeakers are physically separated, the listener is presented with a more realistic reproduction of the original performance.

—Emil P. Vincent; Frank L. Fleming

Stock Ticker, a telegraph machine that prints the quotations of stock or commodity sales. Each important stock or commodity exchange transmits reports of transactions on its *floor* over a separate ticker network provided and maintained by the telegraph company. To economize on transmission time, standardized abbreviations are used, such as *X* for United States Steel Corporation or *J* for Standard Oil Company of New Jersey.

At an exchange, reports of transactions are speeded to operators who use typewriterlike keyboards. The quotations typed by the operators are automatically combined in a stream of electric impulses flashing over the network, causing tickers in brokers' and other offices to print the quotations on a narrow tape flowing from the tickers.

The nation's exchanges have been provided with stock ticker systems by the Western Union Telegraph Company for about a century. At first, the tickers were slow and could not be brought into unison by central control, but were improved by Thomas A. Edison, a telegraph operator. He was rewarded with $40,000, and set up a shop to make tickers. This enabled Edison to start his career as an inventor.

In 1928 the old glass-domed ticker, initially developed by Edison, proved too slow to handle the volume of quotations on the New York Stock Exchange, and a self-winding ticker was substituted with a speed of 285 characters per minute. A new ticker, with a speed of 500, was installed in 1931. By 1964 that also had become too slow, and another ticker was put into operation with a speed of 900 characters a minute, increasing the rate at which information could be transmitted.

The American Stock Exchange began using the 900-speed ticker in 1968. The older tickers were still used by the Midwest Exchange in Chicago, the Pacific Coast Stock Exchange, the Philadelphia-Baltimore-Washington Exchange, and the commodity exchanges. —George P. Oslin

Submarine, a ship capable of running on or below the surface of the sea. Diesel engines are used for power while the ship is surfaced, and electric motors while it is submerged. Power for the electric motors is provided by banks of storage batteries. The batteries are charged by Diesel generators while the submarine is surfaced. Since the batteries can store only a limited amount of power, the submarines can operate submerged for only limited periods.

The invention of the nuclear power plant and its application to submarines led to the development of the first true submersible. Nuclear submarines can operate submerged indefinitely because their power plants do not require air to be mixed with the fuel. Nuclear submarines are driven by steam turbines connected to the propellers by reduction gearing. When the electric motor is used for propulsion, the turbines and reduction gearing are disconnected. As in conventional submarines, electric power is stored in banks of storage batteries. The batteries are charged by turbine driven generators or Diesel generators when the nuclear reactor is shut down.

Submarines have two distinct *hulls,* or bodies. The outer hull encloses the inner, or pressure, hull, which resists water pressure when the ship is submerged. The pressure hull is built up of cylinders and truncated cones reinforced with bulkheads and frames. Within the pressure hull are the crew's quarters, propulsion machinery, and battery storage areas. Watertight bulkheads divide the inner spaces to restrict flooding to local areas if the pressure hull is pierced.

Built into the space between the inner and outer hulls are the ballast and fuel-oil tanks. To prevent them from being crushed by water pressure, these tanks must be kept filled with fluid or compressed air.

A conning tower, or superstructure, mounted atop the submarine contains a bridge for operating on the surface. Protruding from the top of the conning tower are the periscopes, which provide visibility when the submarine runs just beneath the surface. Because the conning tower is outside the pressure hull, it is flooded during submerged operations. Pipes, fittings and valves atop the hull are enclosed in a free flooding enclosure.

A submarine dives by opening vent valves in the top of the main ballast tanks, allowing water to enter flood holes at the bottom. To surface, the vents are closed and compressed air is admitted to the tanks. This forces water out of the flood holes and enables the submarine to rise. Compressed air to blow the tanks is provided by compressors and stored in high-pressure bottles. Trim tanks are provided fore and aft to compensate for changes in weight as fuel oil and armament stores are consumed. The operation of leveling the submarine with the trim tanks is called *trimming.*

When running submerged, the submarine uses diving planes to control the angle of its motion. These planes are hydrofoils which can be tilted to develop vertical forces on the submarine. Conventional or Diesel-powered submarines have their diving planes mounted on the hull. Nuclear submarines have their forward diving planes mounted high on the sail or conning tower. When running on the surface, the planes of nuclear submarines are well above the water and look like aircraft wings.

Primitive submarines were built by Cornelis Drebbel in England and Le Son in France during the seventeenth century. A muscle-powered submarine, the *Turtle,* built by David Bushnell, was used against the British during the Revolutionary War. Robert Fulton's *Nautilus,* built about 1800, was similar in many respects to Bushnell's craft. After Charles Brun, Thorsten Nordenfeldt, Gustave Zédé, and others had constructed relatively unsuccessful steam- and electric-powered submarines, John P. Holland succeeded in building the first modern submarine. Named the *Holland* and purchased by the U.S. Navy on April 11, 1900, the craft was powered by a gasoline engine when surfaced and by electric motors when submerged.

By the end of World War II, the submarine had evolved into a large warship whose range and underwater capability were limited only by its need for Diesel fuel and air. This problem was solved by the application of nuclear power. The first nuclear-powered submarine, the U.S.S. *Nautilus,* was commissioned on Sept. 30, 1954. Since then, the United States has commissioned more than two dozen other nuclear submarines, and Great Britain and the Soviet Union have also launched vessels of this type. —Joseph J. Kelleher

Telegraph, an electrical method by which information is transmitted to distant places and recorded. At one time, the term *telegraph* was applied to any system for sending coded messages. Thus, over the centuries, tele-

U.S. NAVY

U.S.S. LAFAYETTE, which is a nuclear-powered submarine, capable of cruising for extended periods without surfacing or refueling, undergoes sea try-outs in the Atlantic Ocean in 1963.

graph systems have transmitted signals by means of fire, smoke, colored sails, flags, drums, carrier pigeons, flashing shields in the sun, and semaphore towers with movable arms.

While a means of providing electricity was being developed, hundreds of men devised methods of sending electrical signals over a wire. Some used a wire for each letter of the alphabet; others used disks revolving in unison, with the same character visible when an impulse was sent; others used an electromagnetic needle that deflected to the right or left upon the arrival of electrical impulses.

In 1800, Alessandro Volta of Italy invented the electric cell, which for the first time provided a steady source of current. In 1829 Joseph Henry, a school teacher of Albany, N. Y., devised an electromagnet suitable for telegraph use. Sir Humphry Davy, Sir Charles Wheatstone and William F. Cooke, and others produced early electric telegraph systems.

Samuel F. B. Morse, an American artist, produced the first practical telegraph system to attain general use in the United States. The first line was placed in operation between Washington, D. C., and Baltimore, Md., on May 24, 1844, with the transmission of the words "What Hath God Wrought!" Licensed by Morse and his partners, 50 telegraph lines were built throughout the eastern states by 1851.

Most early lines used the Morse system, in which an operator opened or closed a key for short or long intervals. The combinations of these dots and dashes, the *Morse code,* indicated the letters of the alphabet and figures. Operators *read* messages by listening to the time intervals between clicks of a small iron bar in a *sounder* as it was attracted by an electromagnet, which was energized by the arriving electrical impulses.

In 1853 Wilhelm Gintl of Vienna developed the *duplex,* by which two messages could be sent simultaneously in opposite directions over a single wire. Thomas A. Edison produced the *quadruplex* in 1874, by which two messages in each direction could be sent simultaneously over one wire. Various printing telegraph systems, in which messages are received in printed form, were introduced in the 1890's. From this evolved the *multiplex system,* installed on most trunk lines in 1914. This permitted the simultaneous transmission of eight messages over one wire.

In the 1920's *teleprinters* were installed in the offices of some 30,000 companies, enabling them to send and receive their messages over a direct wire to the nearest main telegraph office. Many thousands of teleprinters also have been installed in branch telegraph offices and in private wire systems leased to large companies and the government.

In the 1940's fifteen high-speed message centers, each serving two or more states, were installed to mechanize the nationwide network. This eliminated manual handling at relay points. Each message is typed at the point of origin, and is switched through to its destination by the pressing of a button at only one point.

Facsimile telegraph systems came into use in the 1930s. An electric eye scans the handwritten or typed copy and flashes impulses over a circuit, causing a recorder to produce a facsimile of what was sent. The arriving current decomposes, or *burns away,* minute portions of the coating of an electrosensitive dry recording paper, and the message is ready for delivery. Facsimile systems are used by the U. S. Weather Bureau and the U.S. Air Force to send weather maps nationwide, and in other networks. Facsimile is also used by over 40,000 firms to send and receive 50,000,000 telegrams a year.

Installation of *frequency-modulation* (FM) *carrier systems* in the 1920's produced a vast increase in telegraphic capacity. With them, as many as 288 messages can be sent simultaneously over a single pair of wires. A number of frequencies are generated, and communications are sent over each frequency.

The first commercial microwave beam system was constructed in 1945, and such systems are now in nationwide use. A series of towers is built on high places, such as tall buildings or mountains, about 30 miles apart. Microwaves travel in a straight line. Reflectors on each tower catch the beam; the signals are strengthened and improved, then flashed on to the next tower. The beam is divided into voice-frequency channels, and an FM carrier system is placed on each channel, permitting 2,000 telegrams to be sent simultaneously over a single beam system. A 7,500-mile beam system was placed in operation in 1964, which added 80 million telegraph channel miles to the nationwide network.

Rapid growth in the use of computers in business management, with data transmitted from distant plants and offices to data centers, has required increasingly automatic and complex private-wire telegraph systems. This is the major growth trend of the telegraph industry in the 1960's.

The best known telegraph services are the *fast telegram, day letter, night letter,* and *telegraph money order.* Introduced in the 1960's was a dial-direct, subscriber-to-subscriber teleprinter exchange service called *Telex,* which produces instant direct connections with other subscribers by the dialing of a number.

—George P. Oslin

Telemetry, the process of making measurements at a distance with the help of sensing devices and radio, wire, or other means of communication. Telemetry is used to make measurements that would ordinarily be extremely difficult or impossible because of danger to human life, inaccessibility, or inconvenience. Telemetering systems range from the most simple (an outdoor-indoor thermometer with adjacent indicating scales) to complex instruments used to measure and transmit data from the upper atmosphere or from outer space.

Telemetry is used in space vehicles to sense environmental quantities such as temperature, pressure, and radiation, and to transmit the data by radio to the earth, where it is decoded and recorded. Sensing elements, or transducers, have been developed to measure: weight, time interval, angle, electric field strength, light intensity, temperature, velocity, atomic radiation, blood pressure, many other measurable aspects of the environment. The sensing elements invariably produce an electrical output signal, because telemetry transmissions are usually made by radio or wire. In practice, the signal from the transducer is applied to a wire line or radio transmitter that forwards the information to a distant receiver.

In addition to its role in space science, telemetry has been adopted for inventory control systems, airline reservation systems, stock quotation boards, and teleprinters used in telegraphy. Industry makes use of telemetry in the control of industrial processes, such as paper manufacture, pipeline flow, and the refining of crude oil. Telemetry is also used widely by meteorologists to gather weather data from the upper atmosphere.

—William C. Vergara

Telephone, a combination of devices by which sound is converted into electric impulses, transmitted, and reconverted into sound waves at the receiving instrument. The simple device Alexander Graham Bell invented in 1876 consisted of only a dozen parts. Although today's telephone contains more than 450 parts, the fundamental principle he discovered still applies.

The principle concerns the relationship between electric current and sound waves. Electric energy, supplied by a battery in a telephone central office, is carried to the telephone by a series of wires. When the handset is lifted, the battery sends a steady current through the *transmitter* (mouthpiece). When a person speaks, sound waves enter the transmitter and strike a diaphragm, which vibrates at various speeds, depending upon the variations in air pressure caused by the varying tones and loudness of the voice. As the diaphragm vibrates, it acts against a chamber filled with carbon granules, which move closer together or farther apart, depending on which way the diaphragm is moving. An increased air pressure momentarily packs the granules of carbon more tightly. This reduces the resistance of the carbon to the electric current flowing through it, and a stronger current goes out over the phone circuit. When a reduction in the air pressure allows the diaphragm to move outward, the granules separate slightly, so a weaker current is sent through the circuit. Thus the variation in the current takes on the pattern of the original sound waves.

The varying electric currents are

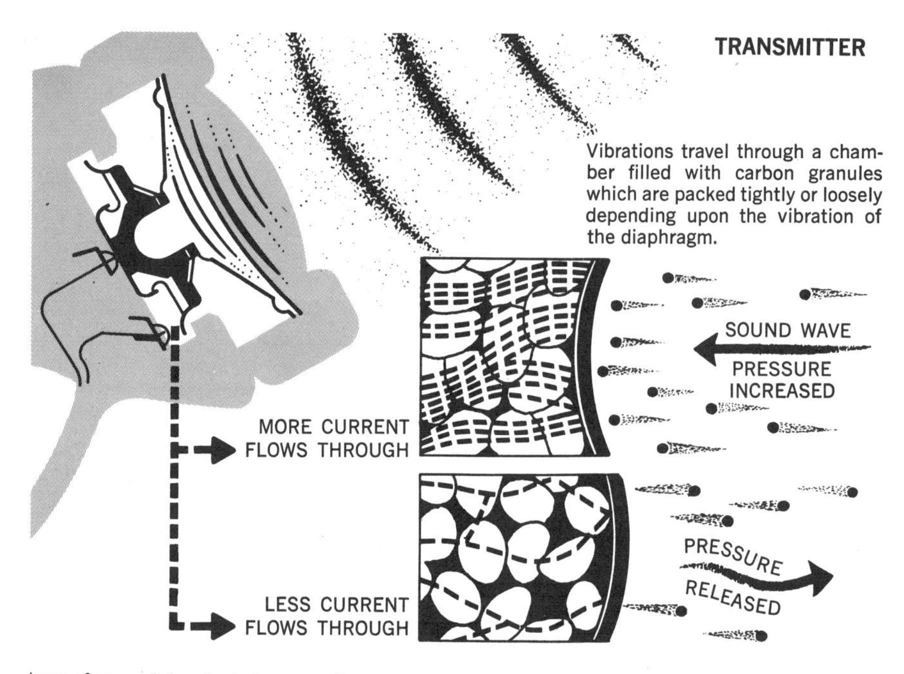

transformed back into speech sounds in the telephone *receiver* (earpiece), where two magnets regulate the movement of a thin circular metal diaphragm. One magnet is a permanent magnet that constantly pulls the diaphragm toward it. The other is an electromagnet, a piece of iron that becomes magnetized when electric current flows through a coil of wire wound around it. When the electromagnet is energized, it pulls the diaphragm away from the permanent magnet. The strength of the pull varies according to the pattern of the current set up by the distant transmitter. As the diaphragm moves in and out, it sets up new sound waves that are replicas of the original voice.

SWITCHING.—The usefulness of the telephone depends on the ability of users to be connected quickly and easily with one another. To accomplish this, each telephone is linked with a central office where connections to other telephones are made by switching equipment. Almost all telephones are dial-operated. Calls are switched automatically by four types of switching systems: step-by-step, panel, crossbar, and electronic.

Step-by-step equipment operates in direct response to the dial pulses generated by the user. *Panel* systems use the "common control" principle, in which the switches are operated by a common set of equipment instead of being under direct control of the dial. *Crossbar* switching involves crosspoint switches operated by a common group of equipment using the common control principle. Common equipment is used only long enough to set up the connection and is then released for other calls. *Electronic* switching, the newest system, is basically a digital data processor (computer) utilizing solid-state electronic devices. Each action takes only microseconds to complete, and information is stored in magnetic "memories" rather than in arrangements of wires.

TRANSMISSION.—To get to and from a switching center, calls travel over wire, cable, and/or radio relay microwave systems. All long-distance channels use the carrier technique, by which many telephone conversations are carried simultaneously over a single pair of wires, coaxial cable, or radio relay. It is possible to carry a number of simultaneous conversations over one pair of wires. Messages are transmitted by modulating alternating currents of different frequencies, *carrier frequencies*.

Coaxial cable, first introduced in 1941, consists of a copper wire held at the center of a metallic tube by insulating washers. Using the carrier technique, a pair of coaxial tubes—one for each direction of transmission—can handle as many as 1,860 simultaneous conversations. These cables may have 8, 12, or even 20 tubes.

Radio relay systems use superhigh frequencies, called *microwaves,* to transmit intelligence from point to point. Because microwaves travel in straight lines, they are usually aimed like a searchlight from one microwave tower to another. At these towers, situated about 30 miles apart, the signals are amplified. Generally, a clear line-of-sight path must exist between amplifying stations. A microwave signal can be aimed so precisely that the same frequency can be used for transmission in different directions from the same station. One microwave carrier system, called *TH,* can carry as many as 11,160 telephone conversations or 12 television programs on one route. *Communications satellites,* such as *Intelsal,* are actually microwave towers in the sky.

Tropospheric scattter propagation, or "over-the-horizon" systems, use that portion of microwave energy that goes beyond the horizon and scatters downward to earth. A *feed horn* sprays the signal against a reflection screen, which beams it over the horizon in the direction of the receiving antenna situated several hundred miles away. The small portion scattered downward is received by an antenna that resembles a huge drive-in movie screen. Here the signal is amplified and sent on to the next station or concentrated into a line-of-sight microwave system or other facility to reach its destination.

The first *undersea telephone cable* was laid between Miami, Florida, and Havana, Cuba, in 1921. The development of submarine repeater devices to amplify the signals made possible the first transatlantic telephone cable—between Scotland and Nova Scotia, via Clarenville, Newfoundland—in 1956. Later, cables were laid between Seattle, Washington, and Ketchikan, Alaska; Oakland, California, and Hawaii; Penmarch, France, and Nova Scotia; Miami and Puerto Rico; and Miami and Bermuda.

In 1963, new armorless cable was placed in service between Florida, Jamaica, and the Canal Zone, and between Tuckerton, New Jersey, and Cornwall, England. This type of cable was also used between the U.S. mainland, Hawaii, and Japan, via Guam, and between Guam and the Philippines. A fourth transatlantic cable, between New Jersey and France, was completed in 1965. These newly designed cables provide two-way communications over a single cable. This cable, with repeaters already spliced in, is placed by the 17,000-ton *C.S. Long Lines,* a specially constructed cable-laying vessel that lays cable while traveling at 8 knots.

SERVICES AND PRODUCTS.—In 1878, when the first telephone exchange opened in New Haven, Conn., 21 people signed up for service. Seven years later, the American Telephone & Telegraph Company (AT&T)—today the world's largest telephone

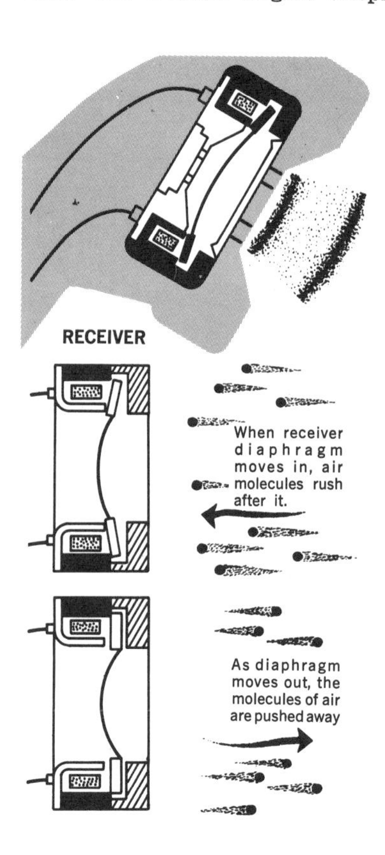

system—was established as a long-distance company. In 1900, AT&T took over the property of the American Bell Telephone Company, becoming the parent organization of the Bell System. In 1968 Bell served about 83 per cent or 90 million phones in the United States and interconnected with 2,000 independent telephone companies. Interstate and international services are regulated by the Federal Communications Commission and intrastate business by state commissions.

Telephone Sets. There are several different telephone sets. Telephones with internal microphones and wall-mounted loudspeakers permit room-to-room communication. A microphone-speaker arrangement for door-answering and a telephone with the dial in the handset are also available.

The *Speakerphone* permits "hand-free" telephoning and also enables a group to participate in a conversation. A sensitive microphone situated near the telephone picks up the speaker's voice. The answering voice is heard over a loudspeaker that can be adjusted for volume.

Among handset accessories is the *automatic card dialer,* which saves time by "reading" prerecorded telephone numbers and dialing calls. By the press of a telephone button, more than one outside line or extension may be held on one line, or transferred, while there is a conversation on another, and one may signal for a transfer or another extension. *Call director* telephones offer up to 30 outside lines—signaling, code dialing of extensions, transfer of calls, or by dialing a code, arranging a conference call.

Teletypewriter Exchange Service, which provides swift, written communications, is dial-operated and sends page and/or punched paper tape at speeds of 60 and 100 words per minute.

Data-Phone service transmits data in any medium—punched cards, punched paper tape, magnetic tape, and facsimile material (such as maps and drawings, as well as handwriting)—over regular telephone lines. Data-Phone *data sets* accept information from various business machines, convert the information to electric signals for transmission over telephone lines and, at the receiving end, reconvert it to its original form. Speeds vary, although they usually range from about 100 to 2,500 words per minute.

Mobile telephone service provides communications between cars, trucks, trains, and ships and the nationwide telephone network.

Recent Developments. One example of improved service is *Touch-Tone* calling, which permits customers to push buttons instead of rotating the dial. These phones can also be used as data input devices. Picturephone see-while-you-talk service was introduced in 1965 between centers in New York, Chicago, and Washington, D.C. The Bell System plans to offer on-premises service to a limited number of customers, probably businesses in the early 1970s.

—James M. Freeman

Teletypewriter, a device for transmitting typewritten messages electrically from one station to another where it is automatically typewritten again. The receiving instrument is especially interesting, as it appears to be operated by an invisible typist. All the operations of typing—including return of the carriage, spacing of lines, and operation of the type box—are performed without the attention of an operator except for removing typed messages from the machine and providing it with new paper.

A keyboard printer is a particular type of teleprinter equipment. It is similar to a normal typewriter in that it has a keyboard consisting of alphabetic, numerical, and function keys and a printing mechanism. In addition, however, it has a device that generates electric signals. These signals are coded to conform with the character represented on the keyboard. When an operator types a character, a coded signal is generated and transmitted over a communication channel to a receiving teleprinter. At the receiving end, the coded signal is interpreted and the character is printed. Two or more operators at teletypewriters on the same circuit can converse simply by typing back and forth. The message appears as a page on the sending teletypewriter and is received as a page.

At the same time, punched paper tape can be created and used for future transmission, filing, or computer input. Using teletypewriter tape permits editing of errors, and it also provides for automatic transmission at speeds much greater than manual typing by an operator. As the operator types a character, a coded signal activates the punch which puts holes in the paper tape. These holes represent the coded form of the particular characters that were typed. Several codes are used, which provide five-, six-, seven-, and eight-level punched paper tape.

Punched paper tape transmission terminals consist of two units, a sender and a receiver. The sender consists of a paper tape reader and a signal generator. As the tape passes through the reader, the holes are read photoelectrically or by sensing pins that are connected to the signal generator. When the presence of a hole in the tape is sensed, an electric signal is generated. The operation is reversed at the receiver, which consists of a sigal interpreter and a paper tape punch. As the electric signals are received from the communications channel, the pins in the tape punch are activated to punch holes accordingly to the paper tape.

Several types of teletypewriter are available: automatic send and receive page machine (ASR); automatic transmitter (ATR); keyboard send and receive page machine (KSR); keyboard typing reperforator (KTR); perforator (PERF); receive-only page machine (RO); receive-only typing perforator (ROTR); and reperforator transmitter (RT). The ROTR provides tape printed down the middle of the coded paper tape. Several models of each type are also available.

Teletypewriters are used to transmit messages, including data, over private-line facilities that connect two or more locations, as well as over the exchange network.

—James M. Freeman

Television, the electrical transmission and reception of transient visual images. In broadcast television, the picture is always accompanied by a sound transmission in the same channel.

Television transmission and reception form a "lock-and-key" system in that the transmitter and receiver must operate by a common set of rules. For example, television sets made for the United States signal are worthless in Great Britain, where transmission standards are different. This is in contrast to radio receivers, which can pick up any transmission to which they are tuned.

Television transmission in the United States is governed by the Federal Communications Commission, which issues a set of *Rules of Good Engineering Practice.* In the United States, television transmission occurs in three bands: the low very-high-frequency (VHF) band, the high VHF band, and the ultrahigh frequency (UHF) band. Each channel has a bandwidth of 6 megacycles (mc.) per second (sec.) disposed as shown in the accompanying table.

Disposition of Television Bands

Band	Channels	Location in Radio Spectrum
Low VHF band	2, 3, and 4	54– 72 mc./sec.
	5 and 6	76– 88 mc./sec.
High VHF band	7 to 13	174–216 mc./sec.
UHF band	14 to 83	470–890 mc./sec.

Other countries have allocated bandwidths of 6, 7, or 8 mc./sec. for their television channels.

VHF and UHF transmissions are normally limited in range to the horizon. This is why it is important that both the transmitting and receiving antennas be situated as high as possible.

For simplicity, an image can be considered as made up of points having different brightnesses. A television camera picks up this image, analyzes the brightness point by point, and converts the varying brightness into a picture signal, whose value varies accordingly. A *synchronizing signal* is also generated which, loosely speaking, establishes the location of each point in the picture. The receiver picks up both the picture and synchronizing signals. It then reproduces the brightness of each point in its proper location, thus reconstructing the image.

The output of the camera and the synchronizing signals occur in the frequency range zero to about 4.0 mc., which are called *video frequencies.* The sound is *frequency-modulated* (FM) on a separate carrier radio wave higher in frequency by 4.5 mc. than the picture carrier. The complete signal is then transmitted on one of the channels listed above.

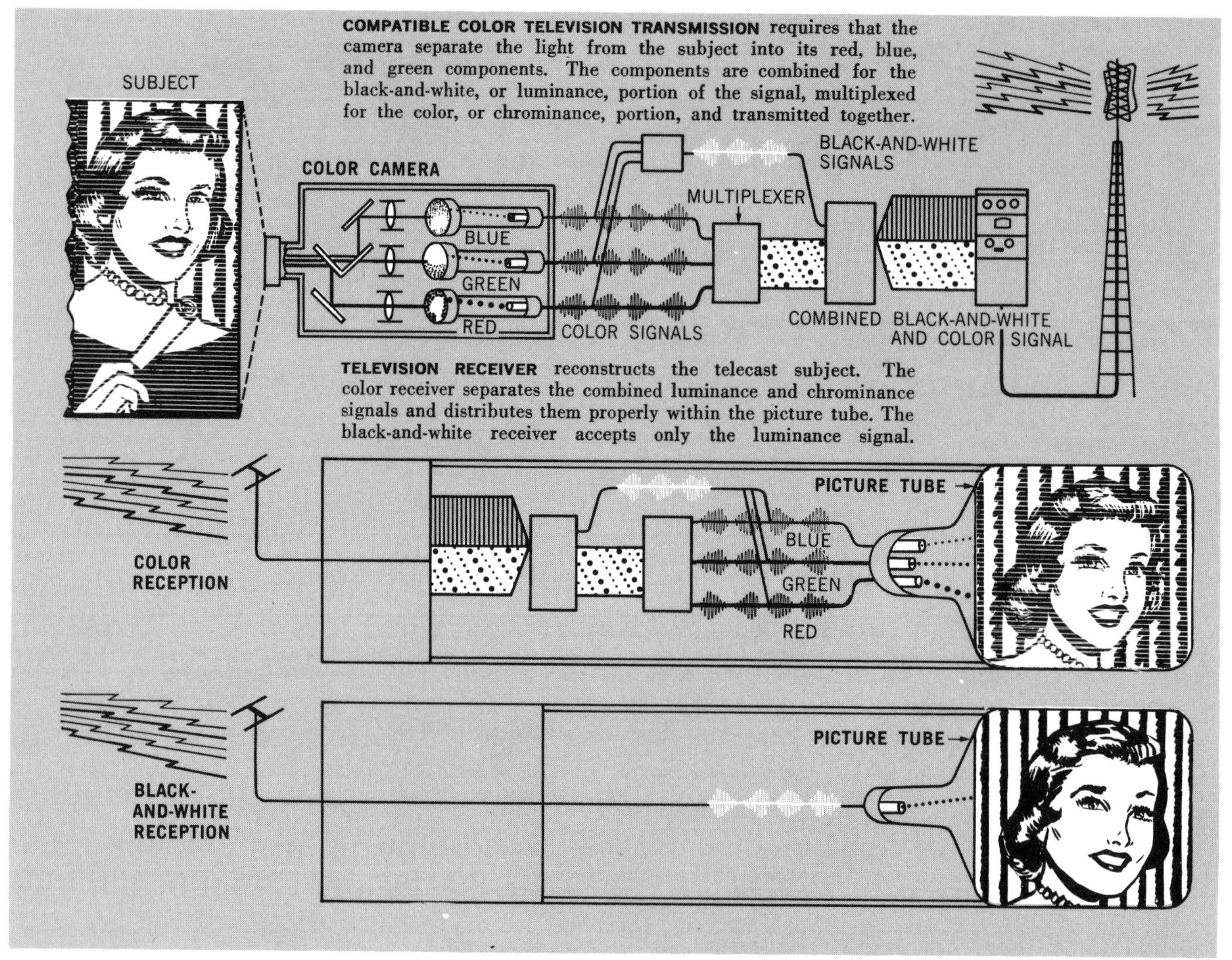

COMPATIBLE COLOR TELEVISION TRANSMISSION requires that the camera separate the light from the subject into its red, blue, and green components. The components are combined for the black-and-white, or luminance, portion of the signal, multiplexed for the color, or chrominance, portion, and transmitted together.

TELEVISION RECEIVER reconstructs the telecast subject. The color receiver separates the combined luminance and chrominance signals and distributes them properly within the picture tube. The black-and-white receiver accepts only the luminance signal.

The receiver *antenna* picks up this signal and delivers it to the receiver which, in the United States, is of the superheterodyne type. The receiver selects the proper channel, detects and amplifies the signal, and separates the video, synchronizing, and sound signals from one another. The video picture signal is impressed on a cathode-ray tube. Within the tube it produces picture elements whose brightness is proportional to the strength of the picture signal and whose location in the picture is established by the synchronizing signal. The sound is treated in the usual manner of FM receivers.

■SCANNING.—At the camera, the picture is analyzed left to right and top to bottom, along a series of vertically displaced horizontal parallel lines, much as a person reads a book. This process, called *scanning,* is performed by a moving electron beam. When the beam has reached the right-hand edge of the scanning line, it quickly retraces to the left for the next line. When it has reached the bottom of the picture, it quickly retraces to the top for the next scan. The picture is blanked during retrace so that the retrace lines are not visible. In the United States, television pictures are resolved into 525 lines repeated 30 times per second; thus, $525 \times 30 = 15{,}750$ lines per second. In Great Britain, the corresponding number is 405 lines repeated 25 times per second (10,125 lines per second); in Europe, it is 625 lines repeated 25 times per second (15,625 lines per second).

To reduce flicker, the number of *images per second* is doubled by an artifice known as *interlaced scanning.* The odd-numbered lines, of which there are $525 \div 2 = 262.5$, are scanned in 1/60 second, and the even-numbered lines are then scanned in the next 1/60 second. Each scan, taking 1/60 second, is called a *field.* Two successive fields comprise a *frame,* or a *picture.* The half-line left at the end of each field causes the lines of the next field to locate themselves halfway between those of the preceding field.

The power-line frequency, 60 cycles per second in the United States and 50 cycles per second in Europe, is the main factor determining the field frequency. Transmitting pictures at a rate of 60 fields per second on a 50-cycle power system, or vice versa, would produce a very annoying 10 cycles per second flicker unless expensive additional filtering were used. This difference in field frequency poses the greatest single problem in the international exchange of television programs.

■MODULATION.—The start of each line and the start of each field are initiated by distinctive synchronizing signals. In the United States, picture-black and peak-white are established by specific signal levels. The synchronizing signals and the pedestals on which they are placed occur at *blacker-than-black* levels, so that they, as well as the retrace lines, are invisible in a receiver whose average brightness is well adjusted.

As applied to the picture tube, black corresponds to zero beam current and increasingly positive signals on its control grid produce increasing brightness. However, the radio transmission makes use of negative modulation so that peak-white, the brightest part of the picture, represents the lowest modulation level.

If a TV set is tuned to a channel on which no signal is being transmitted, the screen will be fully illuminated, or white, indicating a zero signal. A weak signal may produce "snow" indicating that the signal is not strong enough to effect the receiver's scanning beam.

Great Britain and France use positive modulation. Advantages can be cited for both types. The advantages of negative modulation are that noise cannot drive the signal below zero and, therefore, is less bright than for positive modulation, where it can produce brilliant white snow; it also provides fixed levels for black and for peak-white. The advantages of positive modulation are that noise cannot drive the synchronizing peaks below their level of zero volts and that reliable synchronization can be achieved more economically.

IMAGE RESOLUTION.—In the United States, a bandwidth of 4.0 mc. is allocated for picture information. Including positive and negative half-cycles, this permits the transmission of $2 \times 4{,}000{,}000 = 8{,}000{,}000$ picture elements per second. Since there are 15,750 lines per picture, this corresponds to $8{,}000{,}000 \div 15{,}750 = 508$ horizontal picture elements per line. But since 85 per cent of the line period is reserved for picture and 15 per cent is reserved for synchronization, the number of possible picture elements is $508 \times 0.85 = 430$ per line. The picture face of a 23-inch tube is about 18 inches wide and 13.8 inches high. Therefore, the system is capable of resolving a picture element that is $18/430 = 0.042$ inch wide.

This means that a line can have any number of picture elements from 0 to 430, each 0.042 inch disposed in any order along the line. Assuming that 500 of the 525 lines are available for the picture and 25 lines for vertical retrace, the picture element is $13.8 \div 500 = 0.027$ inch high. While the vertical resolution is a function only of the tube size and the number of lines, the horizontal resolution depends on the bandwidth actually passed by the receiver.

As stated above, the sound is transmitted by FM on a separate carrier located 4.5 mc. above the picture carrier. The FM has a maximum deviation of ± 25 kc. for television broadcasting. Great Britain and France use AM for sound.

If the picture signal were transmitted by normal amplitude modulalation, it would require two sidebands and occupy a total bandwidth of 8.0 mc., not including the sound. To save bandwidth, only part of the lower sideband is transmitted.

CAMERA TUBES.—The camera tubes most used currently are the *image orthicon* and the *vidicon*. In both types, the image is focussed on a photosensitive surface, where it is stored as a potential image in the form of a surface distribution of electrical charges. This potential image is then discharged by a scanning electron beam that translates the charge distribution into a current proportional to the brightness of the consecutive points on the surface.

A synchronizing generator produces the synchronizing waveform that is added to the video signal for transmission to the receiver.

The locally generated horizontal and vertical synchronizing signals generate two sawtooth waveforms which respectively occur 15,750 and 60 times per second. The fast one is used for horizontal scanning and the slow one for vertical scanning of the camera in a manner identical to that described below for the cathode-ray tube display, called a *kinescope*.

PICTURE TUBE.—The cathode emits a beam of electrons that is focused into a spot where it strikes the screen. The screen is coated with a phosphor, which emits light when it is struck by electrons. The brightness of the light spot depends on the energy with which the beam strikes the phosphor. This energy is controlled by the strength of the video signal, which in turn depends on the brightness of the corresponding spot at the camera.

The electron beam is made to scan in accordance with the synchronizing signals, which generate two sawtooth waves. One of these occurs 15,750 times a second and is applied to the horizontal deflection coils; the other occurs 60 times per second and is applied to the vertical reflection coils.

It is a law of electricity that a current (a moving electron is a current) going through a magnetic field is deflected at right angles to both the direction of the current and of the magnetic field by an amount that depends on the strength of both. Therefore, a set of coils that produces a vertical magnetic field of linearly increasing intensity (sawtooth) will cause the beam to move horizontally parallel to itself at a uniform rate. Another set of coils produces a horizontal field for vertical deflection. In this manner, the spot on the screen of the cathode-ray tube is made to scan horizontally and vertically, line-by-line and field-by-field, until the whole picture is covered.

The sawtooth wave is initiated by its synchronizing signal. To keep the initiation of the sawtooth, and therefore the start of the line, from being disturbed by noise or interference, a *fly-wheel circuit* is used that is unaffected by transient disturbances. The sawtooth is generated from a free-running oscillator at the normal synchronizing frequency. Its time relation (phase) is adjusted to correspond to the average phase of synchronizing pulse. Thus, the deflection is made essentially noise-free.

To prevent the average picture brightness from being affected by automobile ignition and similar noise, or by airplane flutter, the gain of the receiver is controlled by the average value of many synchronizing pulses, with the rest of the signal shut out. Because the synchronizing pulses occur only 15 per cent of the time, instead of 100 per cent, it is less affected by noise.

Black-and-white television requires the transmission of the "brightness" of successive points in an image. Color television requires that their *hue* and *color saturation* be transmitted as well. This is accomplished by adding a *color subcarrier*, within the black-and-white signal, which is doubly modulated, in *phase* and *amplitude* to transmit *hue* and *saturation* respectively. —Charles J. Hirsch

Typewriter, a mechanical device that produces printed copy a single character at a time as it is operated. Typewriters are either operated entirely manually or with the aid of an electric drive motor. Manual typewriters rely on the force of the operator's hand and fingers applied to the keys to operate the linkage and create the type impression. Electric typewriters are actuated by the motor which supplies the force necessary while the keys are simply control elements. Operators of manual machines must develop a uniform touch or pressure on the keys to produce letters of equal density. The electric typewriter has the advantage of producing uniform letter quality despite variations in the operator's touch or the force applied.

When the operator depresses a key, the printed character is produced by a block of raised type that strikes a carbon-impregnated ribbon, transferring the raised impression to the paper. The roller, or *platen,* around which the paper is wrapped, advances one character each time a key is struck. Normally, a cloth ribbon impregnated with ink is used. While the cloth ribbon is reusable, paper ribbons are good for only a single impression.

The average typewriter has approximately 43 keys, each with an upper and lower case letter selected by means of a shift control. Standard features also include adjustable margins, tabulators, and touch control.

Many special typewriters are available for specific jobs. Long carriage machines are made for the preparation of special documents such as accounting ledgers. The Varityper is a machine with easily interchangeable type faces, enabling a typist to produce copy with numerous type sizes, foreign language characters, and scientific symbols.

Though the typewriter has become an essential tool of industry and is found in many homes, it was not widely used until the 1880's. More than 150 years earlier, in 1714, a patent was granted by Queen Anne of England to Henry Mill for "an artificial machine or method for the impressing or transcribing of letters singly or progressively one after another, as in writing, whereby all writings whatsoever may be engrossed in paper or parchment so neat and exact as not to be distinguished from print." Though no model or drawing of Mill's design

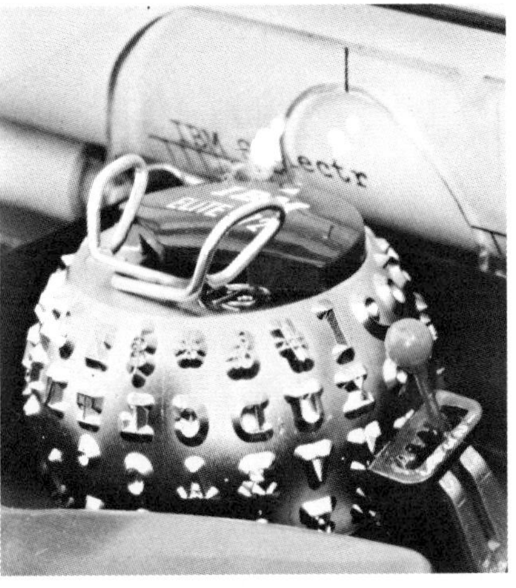

IBM

TYPE sphere aids typing speed, accuracy.

exists, it is clear that he had attempted to produce a typewriter.

The first American patent for a typewriter was granted in 1829 to William Burt of Detroit. Burt's machine, called a *Typographer*, was never put into production. In 1843 Charles Thurber of Worcester, Mass., patented a typewriter that incorporated the first roller platen, a feature of almost all modern typewriters.

Serious development of a practical, commercially successful typing machine was started in 1867 in Milwaukee, Wis., by C. L. Sholes. His initial inspiration was an article in the *Scientific American* that described the invention of the *Pterotype* by John Pratt. The article pointed out the benefits of such a machine and the fame and fortune that its inventor could expect.

Sholes and two friends spent seven years building numerous prototypes until they had a practical working model. Sholes and his financial backer, James Densmore, then made an arrangement with E. Remington and Sons at Ilion, N.Y., in 1873 to put the typewriter into production. Remington, a well-known arms and sewing machine manufacturer, had the experience required to produce equipment consisting of numerous small parts.

A team of expert mechanics was put to work perfecting the details and readying the device for production. One problem was how to arrange the keyboard. Originally the letters were arranged alphabetically, but this proved impractical because the most frequently used letters were not the most accessible and the type bars frequently collided. Rearranging the keyboard in the same sequence as a printer's type case, in which the type is arranged for convenience, proved to be successful and this arrangement is still in use. The first commercial typewriter, Remington No. 1, was produced in September 1873.

This early machine was built into its own stand and used a foot-pedal carriage return. By 1876 a manual return replaced the foot pedal and the typewriter could be used on any table top. Remington Model No. 2 was designed and produced in 1878 as the first machine with both capital and lower-case letters.

In 1925 Remington produced the first electric typewriter, which was designed to produce copies of invoices and other multicarbon documents. By the end of World War II the potential of the electric typewriter to produce superior correspondence had become evident and numerous machines were being sold.

The most unusual electric model now available is the International Business Machine *Selectric*, introduced in 1961. This typewriter features a stationary roll platen and has the raised type characters on the surface of a plastic ball. When a key is depressed the ball rotates to the proper position and is depressed against the ribbon and paper. Character spacing is provided by the ball traveling along the platen and typefaces can be changed quickly by replacing the ball. —Joseph J. Kelleher

Xograph, the first mass-production process for achieving three-dimensional reproduction of a photograph that could be seen without viewing aids. The first photograph to be reproduced in this way was a black-and-white illustration for an article about Thomas A. Edison that appeared in *Look* Magazine on Feb. 25, 1964. A later issue of *Look* on April 7, 1964, contained a four-color advertisement produced by the same process.

To make an Xograph, a screen is placed in front of the film. The image is divided into hundreds of vertical parallel strips. After the film is processed, press plates are made in the conventional manner. In printing, a viewing screen is applied by coating the printed surface. The screen focuses on the vertical strips in the picture and provides the viewer with the illusion of depth.

The Xograph process was cooperatively developed by the *Look* Magazine division of Cowles Communications, Inc., and Eastman Chemical Products, Inc. A major technical breakthrough was the synthesis of the plastic used for coating the viewing screen. The plastic has the required optical properties, is water-clear, adheres to paper, and has an optimum melting point for use on a high-speed press.

The original camera designed for this process was a 6-foot cube weighing half a ton. Later models became more mobile, consisting of three segmented parts: the camera proper, the control panel, and the electronic console. Their lenses are interchangeable from wide-angle to telephoto with a range of focus from 3 feet to infinity. The exposure may vary from 5 to 20 seconds. Although film size is 11 x 14 inches, the maximum of an Xograph is 8½ x 11 inches. The printing accuracy required is ten times normal.

Cowles Communications, Inc., publishers of *Look*, have organized a subsidiary, Visual Panographics, Inc., to make the process available commercially. It is being used by various companies for magazine covers and advertisements, books, counter displays, greeting cards, direct mail, post cards, and calendars.

—Marvin C. Whatmore

VISUAL PANOGRAPHICS, INC.

THREE-DIMENSIONAL CAMERA was used to take the picture facing the title page. The special photographic process, called Xograph, was pioneered by *Look* Magazine. The unique optical equipment produces a print that is coated with a plastic to give a 3-D effect.

BIBLIOGRAPHY

BUCHSBAUM, WALTER H. *Fundamentals of Television.* John F. Rider Publisher, Inc., 1964.

CROWHURST, NORMAN H. *Stereophonic Sound.* John F. Rider Publisher, Inc., 1961.

FOSTER, LEROY E. *Telemetry Systems.* John Wiley & Sons, Inc., 1965.

GLASSTONE, SAMUEL. *Sourcebook on the Space Sciences.* D. Van Nostrand Co., Inc., 1965.

HAY, WILLIAM W. *An Introduction to Transportation Engineering.* John Wiley & Sons, Inc., 1961.

JANE'S ALL THE WORLD'S AIRCRAFT. S. Low, Marston & Co., Ltd., 1965/66.

LEE, MARSHALL. *Bookmaking; the Illustrated Guide to Design and Production.* R. R. Bowker Co., 1965.

MANDL, MATTHEW. *Fundamentals of Electronics.* 2d ed. Prentice-Hall, Inc., 1965.

MARCUS, ABRAHAM AND MARCUS, WILLIAM. *Elements of Radio.* 5th ed. Prentice-Hall, Inc., 1965.

RHODE, ROBERT B. AND MCCALL, FLOYD H. *Introduction to Photography.* The Macmillan Co., 1965.

SCHNEIDER, HERMAN AND SCHNEIDER, NINA. *Your Telephone and How It Works.* 3rd ed. McGraw-Hill, Inc., 1965.

SIMON, IRVING B. *The Story of Printing; From Wood Blocks to Electronics.* Harvey House, Inc., 1965.

WOODS, ALLAN. *Modern Newspaper Production.* Harper & Row, Publishers, Inc., 1963.

ENERGY AND POWER SOURCES

Manpower.—Man's control over nature and his environment took a major step forward when he first domesticated certain animals. This added enormously to his own limited energy. In time, however, man himself was displaced as a principal source of energy.

Man is a limited producer of energy. Thus, as long as man was dependent upon his own energy, supplemented by the contributions of his animate servants, his progress in meeting his requirements for food, clothing, and shelter was limited.

The Industrial Revolution.—The great expansion in man's productive power came with the development of his ability to harness mechanical or inanimate energy. This marked the start of the Industrial Revolution in England approximately 200 years ago. Only then did it become possible for more than a relatively small segment of the population to aspire to and achieve standards of material welfare above the bare subsistence level. This development has continued almost without interruption, so that the world as we know it today would be utterly inconceivable without the ubiquitous use of very large quantities of inanimate energy. This is likely to prevail even more in the world of tomorrow.

Sources of Energy.—There are currently four principal sources of primary energy; falling water (hydro), coal, oil, and gas. Rising in importance on the horizon is a fifth source—nuclear fuel, or atomic power. Minor primary energy sources, more or less exotic and negligible in their over-all economic value, are wind power, solar energy, and tidal power.

Perhaps the most versatile form of energy in use today is electric energy, but electric energy is not a primary energy source. It is a converted, highly refined form of energy. In the United States at the present time, approximately 20 per cent of the total energy used is converted to the electric form. Electric energy is particularly important in technologically and economically advanced societies and has contributed to their progress.

The great technological and economic progress that has taken place in the last two centuries can be viewed in terms of advancing technology and the development of energy-using machines and equipment.

In the United States, the commercial exploitation of mineral energy is roughly 200 years old. The bicentennial of the first commercial production of coal in Virginia was celebrated in 1959. The Titusville, Pa., centennial of the petroleum industry was celebrated a year or so before that.

At first, progress in the use of mineral energy was extremely slow. Even by 1850, in the United States, bituminous and anthracite coal occupied a relatively minor position in an already heavy energy-using country. Although coal was known and in use prior to the Revolutionary War, approximately 90 per cent of the total energy, except that supplied by human or animal power, was still supplied by wood. The period before the Civil War was primarily a wood fuel economy.

Coal was not consumed in significant quantities until about 1860. Thereafter, it quickly began to replace wood as a railroad locomotive fuel and as a source of coke for the growing steel industry. The appearance of the automobile and the expanding automotive use of petroleum provided the major impetus for the growth in petroleum consumption; until then, petroleum had served primarily as a source of light and as a lubricant.

The increased use of natural gas over the past several decades received its greatest momentum with the development of welded seamless pipe. This made possible the extensive long-distance pipeline network that has been constructed largely since the end of World War II, and in turn the expansion of gas markets at locations remote from the source.

■**ELECTRIC ENERGY.**—Electric energy has many unique qualities, notably, ease of transportation and distribution, complete flexibility, cleanliness, safety, sensitive susceptibility to control, potential applicability to almost all energy-using processes, and the ability to be produced from every primary energy source. For these reasons, it has had an especially significant impact on our society since 1882, when Thomas Edison placed in commercial operation the first central-station power supply in the city of New York.

Since electric energy is a converted form of energy, limitations on primary energy supplies bear the seeds of limitations on electric-energy production. However, because electric energy can be produced from any primary energy source and because it is uniquely capable of potential application to any energy-using process, electrification offers an excellent route for solving the problems of the potential exhaustion of our mineral fuels. This prospect has been enhanced by the development of nuclear power—the first new primary energy source in almost 100 years. Its contribution to compensating for the depletion of the other primary fuels is limited primarily to its conversion to electric energy. The extent to which it can supplement older primary-energy resources depends, therefore, on the degree to which energy-using processes can be electrified. It is important to note, in this regard, that at the present time the largest single consumer of energy is transportation, and transportation is based almost entirely on petroleum.

Energy Consumption.—The growth in energy use in the United States reached the following balance for the year 1960: total energy consumption amounted to the equivalent of about 1,700 million tons of bituminous coal,

ANSEL ADAMS

WATER was early man's prime source of power. The energy in a majestically free-flowing waterfall today can be harnessed to provide many kilowatts of electrical power.

Energy Conversion Chart

To \ From	Chemical	Elastic	Electrical	Gravitational	Heat	Kinetic	Nuclear	Radiation
Chemical			Electroplating Storage Battery Electrolysis		Blast Furnace Refining			Photography Photosynthesis
Elastic			Piezoelectric Crystal	Auto Suspension (Springs)	Thermostat Expansion of Heated Metals	Trampoline Diving Board Tire Pump		
Electrical	Fuel Cell Storage Battery Primary Battery	Piezoelectric Crystal	Transformer Motor-Generator Rectifier	Hydroelectric Generators	Thermoelectric Generators	Phonograph Pickup Generator	Nuclear Battery	Solar Cell Photoelectric Cell
Gravitational		Auto Suspension (Springs)			Convection of Fluids and Gases	Clock Pendulum Artillery Shell		
Heat	Combustion of Fuel Heat of Chemical Reactions	Compression Ignition of Gases Gas Cooling by Expansion	Heating Coils Spark Plug Thermoelectric Cooler			Friction Impact	Fission Fusion Reactor	Infra-red Rays High Frequency Cooker Radiant Heating
Kinetic	Explosives Rocket Fuels	Clock Spring Slingshot Trampoline	Electric Motor Loudspeaker Solenoid	Clock Weights Falling Stone Waterfall Tides	Steam Engine Turbine Internal-Combustion Engine	Collision Propeller		Radiation Pressure from Sun
Nuclear	No known method of conversion							
Radiation	Cold Light Emission (Firefly) Chemical Laser		Fluorescent Light Laser Radar Radio-Television	Gravitational Collapse	Light Bulb Heated Surfaces		Radiation from the Sun Nuclear Explosion	Fluorescence Laser

of which coal itself accounted for about 23 per cent, petroleum 42 per cent, and natural gas 31 per cent. Hydroelectric energy provided the remaining 4 per cent.

By 1975, total energy consumption in the United States is expected to grow more than 60 per cent, to the equivalent of 2,750 million tons of bituminous coal. In that growth, coal is likely to increase its share to about 30 per cent of the total, petroleum should show a slight drop to 40 per cent, gas is likely to experience a larger decline to 25 per cent, hydroelectric power will provide about 3 per cent, and for the first time there will be a significant amount of nuclear power, providing the remaining 2 per cent.

In the year 2000, it is expected that changes in the relative importance of alternative sources of energy will still be gradual and relatively modest. At that time, of the projected energy equivalent of 4,000 million tons of bituminous coal, coal itself is likely to provide about 30 per cent; petroleum about 32 per cent; natural gas will have continued to fall, providing only 15 per cent; and hydroelectric power will provide 2 per cent. The major change expected for the year 2000 is the growth in nuclear power, which should provide the remaining 21 per cent.

The most significant aspect of these projections is that, despite the growth in total energy requirements by the year 2000 to almost 2.4 times the 1960 level, and despite the very rapid growth indicated for nuclear energy after 1975, the principal sources of energy for the United States will continue to be the fossil fuels, supplying over 75 per cent of the total. The nuclear-energy figure shown for the year 2000 is based on the most optimistic assumptions about the rate at which nuclear energy can take its place in the economy. The nuclear-energy figure for the year 2000 may be lower, depending on developments within the fossil fuel industries.

In 1960, approximately 19.5 per cent of the total energy consumed in the United States was utilized in the production of the 753 billion kilowatt-hours (kwh) generated by the electric-utility industry. By 1975, even with expected improvements in the efficiency of converting mineral energy into electric energy, about 26 per cent of the total energy will be utilized for the production of about 2 trillion kwh. In the next quarter-century after that, electric-energy generation may be expected to triple to 6 trillion kwh and to account for 40 per cent of the total energy use. In this growth of electric energy, the role of coal will first expand, then gradually diminish as nuclear power increases in importance.

Energy Conversion.—The process of converting any of the primary sources of energy, particularly coal, oil, gas, or nuclear fuel, into electric energy involves a highly advanced and sophisticated technology. The preliminary process of heat release is followed by the conversion of the resulting mechanical energy into electric energy. It is a process that employs complicated machines and equipment to achieve a conversion efficiency of 40 per cent. This high degree of efficiency is the result of slow, painstaking research and development that have been going on since the very inception of central-station electric power more than 80 years ago, when conversion efficiency was only about 5 per cent.

In essence, however, the process of converting the energy in mineral fuel to electric energy employs principles that date back to Michael Faraday's discovery of electromagnetic induction in 1831. The steam turbine that converts the thermal energy resulting from fuel combustion to mechanical energy dates back in its essential principles almost two centuries to Thomas Newcomen and James Watt. At the present time, more exotic ways of converting primary energy into electric energy are coming under increasingly intensive investigation. But these processes, too, are based on principles that have been known for more than 100 years. Four of these—thermionic generation, thermal-electric generation, fuel cells, and magnetohydrodynamic generation (MHD)—show particular promise of leading to advances in energy conversion. Each of these highly intriguing prospects depends on a different principle, but all have in common the aim of directly converting primary energy into electrical energy, without any intermediate mechanical device. While considerable progress has been made in the first three of these technologies, there has been no sign that any can be developed into a competitive, large-scale commercial energy converter in the near future.

Magnetohydrodynamics, the fourth of these new approaches, appears to offer more exciting possibilities. It offers the promise of great simplification of the electric generation process and the possible achievement of very high thermal efficiency—as much as a one-third improvement over the present highest efficiency of 40 per cent. Because of this promise, a considerable amount of research effort is going on today looking toward the development of MHD to commercial practicality.

Looking to the Future.—Whatever the results of present research into these more exotic energy-conversion devices, it is clear that total energy requirements by the year 2000 and during the intervening period will continue to necessitate very large amounts of fossil fuel. Thus, it will be essential to foster the continued development of an adequate supply of coal, oil, and gas for the long period during which nuclear power will gradually assume an increasing part of the total energy burden.

It is in the light of such an over-all outlook for energy requirements that the atomic program needs to be viewed, if an adequate supply of energy in the required forms is to be assured. The long-term promise of nuclear energy cannot be permitted to obscure the continuing importance of the fossil fuels. They must be able to provide, for some time to come, the far larger share of the total energy needs that nuclear power will not be able to satisfy even under the most favorable conditions.

Regardless of the progress made in new conversion technologies, these are unlikely to play a leading part in the supply of energy for the remainder of this century. Furthermore, they are only conversion devices and not new sources of energy. They will still require a supply of primary energy with which to carry out the process of conversion.

Thus, there is clearly a need for continuing concern for the supply and efficient utilization of the fossil fuels. For some time, they will continue to provide the greatest share of the total need for inanimate energy. —Philip Sporn

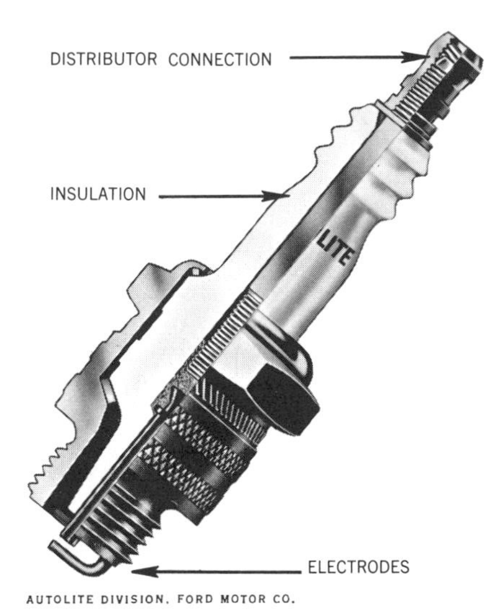

SPARK PLUG is shown in quarter section.

ENERGY AND POWER GLOSSARY

Automotive Electrical System, a complete power plant that produces and stores electricity, and delivers it as it is needed. Electric power is required for lights, radio, horns, windshield wipers, power seats and windows, heater and air conditioning blowers, and to turn the engine during starting. In addition, thousands of pulses of high-voltage electricity must be generated every minute to fire the engine's spark plugs. The electrical system used in most American automobiles operates on 12 volts. The electrical system also contains transformers, switches, contacts, relays, and fuses that distribute and regulate the flow of electricity. The complete electrical system actually consists of five interconnected systems—the generating, or charging system, the ignition system, the cranking system, the lighting system, and the accessory system.

■**GENERATING SYSTEM.**—The generating, or charging, system includes an alternator or generator, ammeter or warning light, voltage regulator, and battery. The generator converts mechanical energy from the engine into electrical energy. Either a *direct-current* (D.C.) *generator* or an *alternator* producing alternating current (A.C.) is used. Alternators are equipped with rectifiers to convert the A.C. to D.C. before it is used in the system.

Current from the generator is routed to the *voltage regulator,* which controls the generator's output. Located between the battery and the generator, the regulator operates to reduce generator output when the battery is fully charged and increase output when the battery is low. When the regulator is set correctly, just enough current is produced to keep the battery fully charged at all times. The *ammeter* (gauge), or warning light, provides a visual indication of whether the generator is charging or discharging.

Aside from the generator, the *battery* is the most important element of the generating system. The battery does not store electricity but converts electricity energy into chemical energy and stores it. When electricity is required, the chemical reaction within the battery is reversed to produce a flow of current at the terminals. A storage battery's capacity to store energy is limited; if current is not supplied continuously, the battery runs down or falls rapidly.

■**IGNITION SYSTEM.**—The ignition system, which consists of the ignition switch, coil, distributor, and spark plugs, receives electricity from the battery and creates a high-voltage spark at the electrodes of the spark plug. To bridge the gap at the plugs, a high voltage surge (20,000 to 30,000 volts) is applied across the electrodes. An additional function of the ignition system is to time and distribute the high-voltage surges exactly, so that the proper spark plug is energized at precisely the right moment.

■**CRANKING SYSTEM.**—The cranking motor is a heavy-duty electric motor that rotates the engine crankshaft during starting. Electric power for the cranking motor is supplied by the battery when the starter switch is turned on. Though similar to other electric motors, the cranking motor is specifically designed to operate for brief periods while heavily overloaded.

■**LIGHTING SYSTEM.**—The lighting system consists of the headlights, parking and tail lights, turn signals, emergency flasher, light switch, foot dimmer switch, instrument lights, and interior lights. The light switch is usually a two-position pull switch. The first position, or notch, illuminates the parking lights while the second notch turns on the headlights. The dimmer switch, which is located on the floor, switches the headlights to low beam to prevent blinding of oncoming drivers.

—Joseph J. Kelleher

Battery, Electric, a device that produces or stores electricity by chemical means. Typically, an electric battery has two electrodes: a positively charged *cathode* and a negatively charged *anode.* Both are immersed in an *electrolyte,* a chemical compound that dissociates into positive and negative particles called *ions.* Migration of the ions within the battery stimulates a flow of current in the external circuit connected to the battery. A battery often consists of a number of identical cells connected in series.

Batteries are of two basic types, primary and secondary. In the *primary cell,* the chemical substances are consumed during their useful life, after which the cell is discarded. The common flashlight battery is a typical primary cell that delivers only a predetermined amount of energy. *Secondary cells,* also called *storage batteries,* can be recharged many times before their components deteriorate.

Primary cells are further divided into wet and dry. In *wet cells,* the electrolyte is a liquid; in *dry cells,* it is a paste. The first battery, invented in 1800 by Alessandro Volta, was a primary wet cell consisting of copper and zinc disks immersed in a salt or acid solution. Most primary cells in use today are dry cells having a zinc anode in the shape of a cup that contains the electrolyte, ammonium chloride. The cathode is a carbon rod immersed in the zinc cup. A *depolarizer,* manganese dioxide, is added to the electrolyte to prevent accumulation of unwanted chemical substances on the electrodes that would stop the reaction.

A newer primary dry cell is the *mercury cell,* which was developed during World War II. It consists of a mercuric oxide cathode and a zinc anode in a potassium hydroxide electrolyte. It has longer shelf life and higher electrical capacity than the zinc ammonium chloride cell.

The *lead-acid cell* is the most common secondary cell in use today. Its

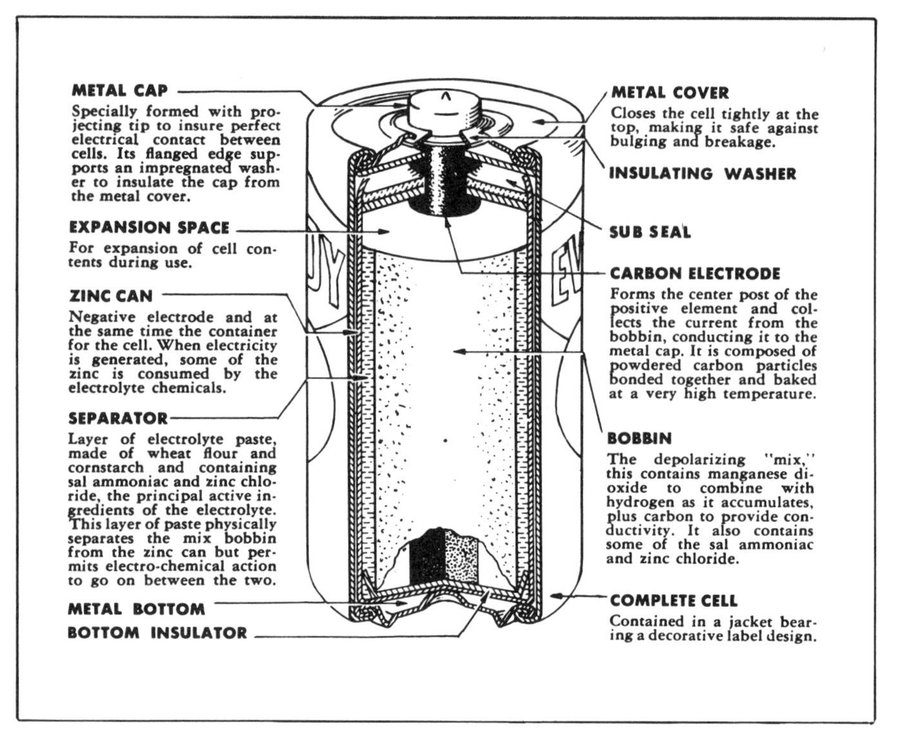

EVEREADY

DRY CELL, rated at one and a half volts, generates electric energy through the electrochemical reaction between the zinc can and the pastelike depolarizing mix, called a bobbin.

positive plate is made of lead peroxide and its negative plate of lead. The electrolyte is dilute sulfuric acid. Like all secondary cells, the lead-acid battery is recharged by the passage of current through it in a direction opposite to the current flow on discharge. The lead-acid battery is low in cost, and its ability to supply high current for short periods of time makes it suitable for starting automobile engines.

The *nickel-cadmium secondary cell* can operate at low temperatures and, because no gases are given off during normal operation, it can be hermetically sealed. The positive plates are made of nickel oxide. The negative plates are made of cadmium containing a small amount of iron. The electrolyte is a solution of potassium hydroxide. Nickel-cadium cells are used in space vehicles and in specialized portable devices, but are too expensive for general automotive use.

Silver-zinc and *silver-cadmium cells* are small in size, light in weight, and perform well at high discharge rates. Despite their relatively short life, they are used in military and lightweight commercial equipment.

Another secondary cell in use today is the *nickel-iron cell,* which uses potassium hydroxide as the electrolyte. Called the *Edison cell,* it is used primarily in industry, where its high efficiency is of particular value.

—William C. Vergara

Diesel Engine, invented by Rudolf Diesel in 1892, is a particular type of internal-combustion engine. It differs from the gasoline internal-combustion engine (Otto cycle) in that the fuel injected into the cylinder burns rather than explodes. No spark is required for ignition of the fuel of the diesel engine. Instead, the air in the cylinder is so highly compressed by the stroke of the piston that it reaches a sufficiently high temperature to ignite the fuel without a spark.

Only a slow-speed engine will follow the classic cycle conceived by Rudolf Diesel, and true diesels are large, heavy engines generally used to drive generators in electric power plants, railroad locomotives, or ships. Small, higher speed semi-diesel engines operate on a modified diesel cycle and are used to drive lighter equipment, including automobiles.

Although the original diesel engine increased mechanical efficiency, and enabled coal dust instead of liquid fuel to be burned, it was short-lived due to an Augsburg factory explosion. Required to revise it, Diesel then created a superior engine.

Diesels do not require high octane gasoline but generally use kerosene or heavier petroleum distillates as fuel. (See *Engine, Internal-Combustion,* page 1568.) —Hunter Hughes

Power-Handling Capacity of EHV Lines

Voltage Rating	Typical Loads Carried in Megawatts (Millions of Watts)	Corresponds to the Average Consumption of:
115,000	120	Wichita, Kan.
230,000	300	Hartford, Conn.
345,000	550	Cincinnati, Ohio
500,000	1200	Montreal, Quebec
765,000	2900	Chicago, Ill.

Electric Power Transmission, the movement of electric energy from its point of generation to its point of use. The scope of electric power transmission has grown from the earliest direct-current (D.C.) transmission lines of several hundred volts, emanating from central community power stations of the 1890s, to today's vast alternating-current (A.C.) transmission networks, with some lines operating at hundreds of thousands of volts. These networks, or grids, extend hundreds of miles between terminals so as to interconnect almost all the major geographic areas in the United States and Canada.

The economic and technical forces that have encouraged higher voltages and greater transmission distances differ mainly in magnitude compared with the early stages of power-system evolution. The utility industry has always found that the most reliable electric service can be achieved by building large central power stations and then interconnecting them with the utilization areas and with one another by power transmission lines. Strong interconnections of this sort have several advantages over independent locally-supplied power generation. They reduce the number of spare generators that are required by allowing spares on one system to serve neighboring systems, and they permit use of the largest, most efficient power plants by allowing one plant to serve a large geographic area. They also allow location of those plants at or near sources of water, which is needed for cooling purposes. In addition they permit neighboring power systems to interchange power when, as is often the case, the time of maximum power usage on one system differs from that of another.

■CARRYING CAPACITY.—In general, the ability of a transmission line to carry power is proportional to the square of its voltage rating; for example, doubling the voltage increases the power-handling capacity by four. For a given voltage rating, the amount of power that can be transmitted is inversely proportional to the distance of transmission.

Power lines, in general, are three-phase. This means that three separate conductors are required, each operating at the same voltage with respect to ground, but each reaching the crest of its alternating voltage at a different time. They are rated in accordance with voltage between conductors or phases. Actual voltage ratings of power lines range from several thousand volts, typically used to distribute power directly to residences, to 500,000 volts, typical of the most modern interconnections between adjacent power systems. Some inter-connections in the United States operate at 765,000 volts. A gauge of the power-handling capacity of high-voltage transmission lines can be seen from the accompanying table.

Higher voltage lines are often limited in their capacity by the need to keep the power systems at both ends of the line in electrical synchronism. If too much power is carried, an electrical disturbance (such as a lightning stroke to the line) will cause the two systems to become unstable, that is, to lose synchronism. This is analogous to a long mechanical shaft coupling two motors. As long as the shaft is intact (even though twisted by the force applied), the motors must run at the same speed. If one machine tries to transmit too much torque to the other, the shaft will break, and the two machines will run independently.

APPEARANCE.—Regardless of their essential role in supplying power to the consumer, the appearance of transmission lines, especially in scenic areas, has become an increasing cause of concern. Utility companies are seeking solutions in several ways. Among them are the use of higher voltages, enabling more power to be carried by each line, improved design of tower and lines, and research into lower cost underground transmission methods.

ECONOMIC FACTOR.—Underground cable has been the only practical means of power transmission in metropolitan areas for many years. Early in the 1900s, engineers found that the spacings of 25 feet or so needed between wires in open air could be reduced to several inches in an underground cable where oil and paper were used as insulation between wires. But this gain in capacity is achieved at an increase of from 10 to 40 times the cost of overhead lines. Conversion of all overhead lines in the United States to underground, using presently available methods and technology, would more than double the cost of power to the consumer.

Some of the possibilities being explored for reduced-cost underground transmission are: the use of synthetic insulation, the use of forced cooling of cables to permit them to carry more power, the use of gas as an insulating material, and the use of cryogenics, that is, super-cooling of the cable to almost absolute zero (−273° Kelvin), where virtually all losses disappear and very high power levels can be transmitted.

GENERAL ELECTRIC

A 345,000 VOLT SUBSTATION serves as a step-down point from a multi-state transmission network to 138,000 volt lines which provide electric power for local cities and industries.

In the case of low-voltage distribution lines, the extra cost for underground cable is not nearly so great since the cable itself is much simpler and can be directly buried without specialized techniques or equipment.

In new residential subdivisions, the cost of an underground system may be almost as low as that of an overhead system.

D.C. TRANSMISSION.—On D.C. transmission lines, the power is (1) received as conventional A.C.; (2) converted to D.C. by huge rectifying valves, of either mercury ore or solid state types; (3) transmitted as D.C.; and (4) reconstituted into three-phase A.C. by an identical array of valves for delivery to the receiving system.

High voltage D.C. lines can carry much more power, for the same size line, than A.C. The power flow over a D.C. line can be regulated independently of the state of synchronism of the systems at the two ends. This is analogous to a fluid coupling in a mechanical system. Two high-voltage D.C. lines, each rated plus and minus 375,000 volts (750,000 between conductors), will link the Pacific Northwest with Los Angeles and with the Hoover Dam in Arizona. High voltage D.C. transmission is attractive for use where long distances and high power levels are required.

COMPETITIVE POWER SOURCES.—In the business world, high-voltage power transmission must be looked on as only one of several competing means of transporting energy. The energy

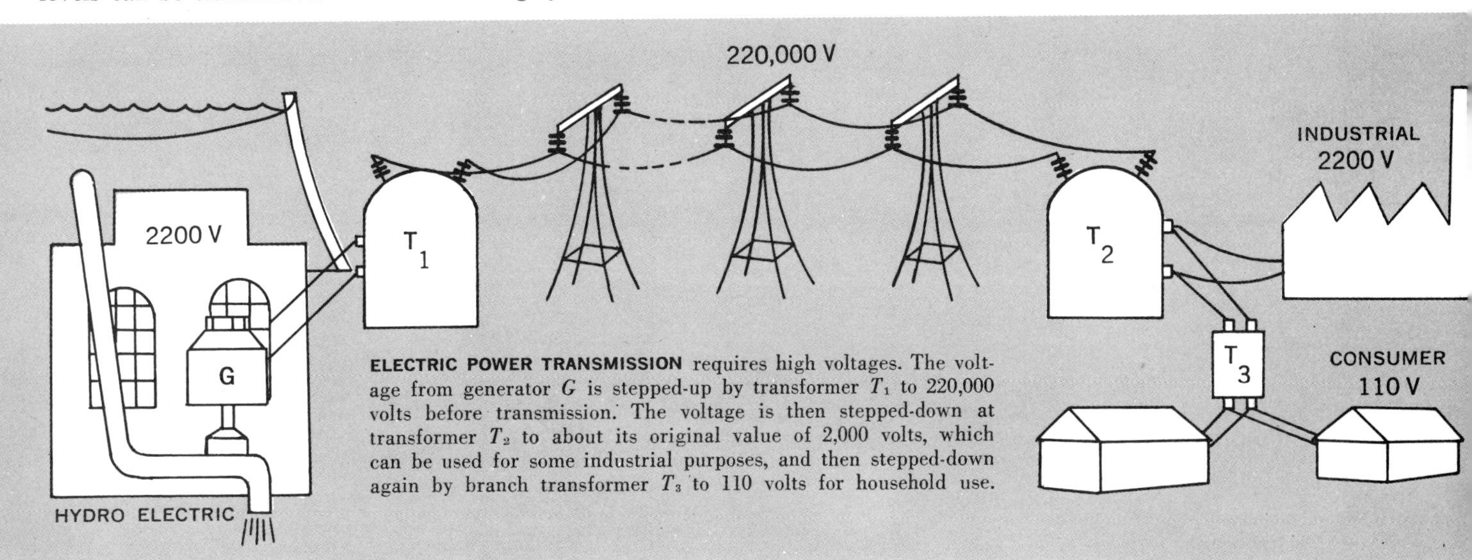

ELECTRIC POWER TRANSMISSION requires high voltages. The voltage from generator G is stepped-up by transformer T_1 to 220,000 volts before transmission. The voltage is then stepped-down at transformer T_2 to about its original value of 2,000 volts, which can be used for some industrial purposes, and then stepped-down again by branch transformer T_3 to 110 volts for household use.

that is ultimately delivered to a city can be converted from fuel resources hundreds of miles remote from the city and transmitted electrically to the urban load, or the fuel itself (coal, gas, oil, etc.) can be transported by rail or pipeline for use in generating the power locally.

Engineers sometimes talk of a day when each home will have its own power source (perhaps using atomic energy), thereby eliminating the need for power transmission altogether. But all the progress in making power more economical has so far led to its generation in larger rather than smaller amounts. Secondly, the total-generating capacity in the United States is now only one-tenth as great as the sum of all individual maximum power demands of U.S. consumers. Extensive power transmission networks allow present-day systems to take statistical advantage of the fact that each consumer's maximum demand occurs at a different time. Individual home power units could not take advantage of this ten-to-one diversity.

—L.O. Barthold

Electrical Engineering, the branch of engineering dealing with the design and application of electrical equipment and electronic systems. The nineteenth century was a mechanical age, but with the invention of the electromagnet and the vacuum tube, electric power and electronic systems replaced many mechanical devices and made possible entirely new practical applications of the forces of electricity.

At the start of the twentieth century, electrical engineering was in its infancy, and electrical engineers dealt primarily with the design and operation of electric generating stations, transmission and distribution lines that delivered the electric energy to its point of use, and motors and electric lights that made use of the energy. This is still an important part of electrical engineering, but communication and electronics have become major aspects of electrical engineering. Telephone, radio, television, radar, X-ray, and many associated instruments and controls are all within the scope of the electrical engineers. He is responsible for the design and application of thousands of different pieces of electrical equipment, from complicated systems in huge research laboratories to electrical equipment in the home.

Electrical engineering is taught in most engineering colleges and is usually a four- or five-year course. In pursuing their careers, graduate electrical engineers may select research, design, application, or operation. They may choose to work in industry, in government, or in private practice as consulting engineers.

Numerically, electrical engineering is the largest of the several branches of engineering; electrical engineers considerably outnumber mechanical, civil, or chemical engineers.

—Hunter Hughes

Electroluminescence. See *Lighting,* page 467.

Engine, Internal-Combustion, a prime mover in which the combustion of the fuel takes place within the engine rather than in an external furnace. Gasoline engines, gas engines, diesel engines, and semi-diesel engines, are all internal-combustion engines.

The internal-combustion engine considered here is the gasoline engine operating on an *Otto cycle,* named for the German inventor Nikolaus Otto, who patented and built (1877) the first practical engines of this type. It is today the commonest of all prime movers, powering all kinds of equipment from toys to aircraft, including the most obvious example, the automobile.

The gasoline engine derives its power from the explosion of a mixture of air and gasoline, as opposed to the diesel engine, in which the fuel burns rather than explodes. The air-fuel mixture, when ignited, expands rapidly in a cylinder, forcing a piston from the top of the cylinder to the bottom. The piston is attached to a crankshaft by means of a piston rod, and the crankshaft translates the lineal movement of the piston into rotary motion.

■**FOUR-STROKE CYCLE.**—The gasoline engine is designed to make use of either a two-stroke or a four-stroke cycle. In the four-stroke cycle, the piston strokes are: (1) *intake*—the piston moves down the cylinder drawing in, through an open intake valve, an explosive mixture of fuel and air; (2) *compression*—all valves are closed, and the piston moves toward the top of the cylinder, compressing the explosive mixture; (3) *power*—while all valves are closed, the mixture is ignited by an electric spark when the piston is near the top of the cylinder toward the end of the compression stroke; the resulting explosion drives the piston downward; (4) *exhaust*—as the piston reaches the end of the power stroke an exhaust valve opens, and on the return stroke the piston drives all exhaust gases from the cylinder to complete the series. It takes two full revolutions of the crankshaft to complete the four strokes.

It is customary to design gasoline engines with four, six, eight, or more cylinders. Four-cylinder engines, for example, provide a power stroke from one of the four pistons on every half revolution of the crankshaft, while the other pistons are going through intake, compression, or exhaust. It can be seen that a multiplicity of cylinders permits a smooth operation of the engine. An engine with more than four cylinders permits a partial overlapping of power strokes in two or more pistons.

■**TWO-STROKE CYCLE.**—The two-stroke-cycle gasoline engine is designed to eliminate the intake and exhaust strokes of the four-stroke cycle. At the bottom of its power stroke, the piston uncovers or permits the opening of both the exhaust and the intake valves. The air-fuel mixture, which has been precompressed in the crankcase or in an outside compressor, enters through the intake valve, *scavenges* the cylinder by driving out the exhaust gases, and is then further compressed by the upward stroke of the piston. As with the four-stroke cycle, explosion results from an electrical spark, and the piston is driven downward on its power stroke.

Gasoline engines operating on the four-stroke cycle greatly outnumber those using the two-stroke, since automobile manufacturers have concentrated on four-stroke designs. There are, however, a few European manufacturers producing cars powered by two-cycle engines. Two-cycle engines have been used extensively in the United States for powering lawnmowers and similar light equipment. The average user is conscious of the difference only in that the two-cycle engine requires the addition of lubricating oil to the gasoline, whereas the four-cycle engine does not. In the United States, the trend is to the four-stroke cycle, even for small engines. However, the amazingly miniaturized, single-cylinder, model aircraft engine, with a cylinder smaller than a thimble, is a two-stroke design familiar to nearly every boy.

■**STROKE AND BORE.**—The number of cylinders has no direct relationship to the power or speed of an internal-combustion engine. A four-cylinder engine with a long piston *stroke* and a large *bore* (cylinder diameter) may have much more power than an eight- or twelve-cylinder engine with a short stroke and a small bore. Many of the great racing automobiles have four-cylinder engines. Equally great winners have six, eight, or twelve. It is a matter of the designer's preference. *Displacement* is a much more useful measure of

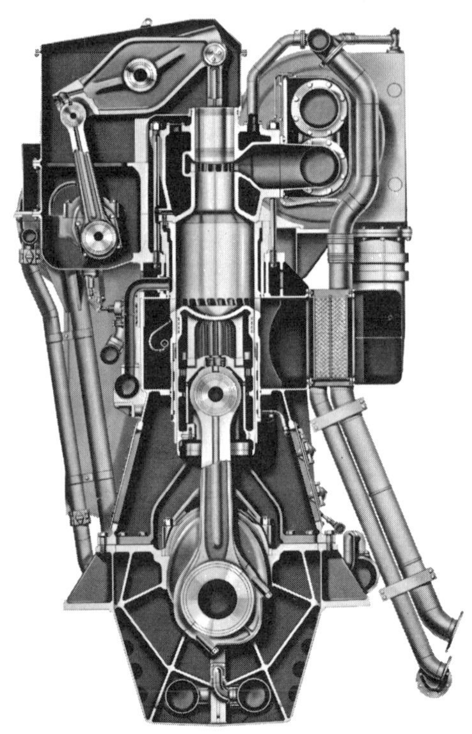

FAIRBANKS MORSE

DIESEL ENGINE in cross-section is a heavy-duty model delivering 1,000 horsepower per cylinder for marine or heavy industrial use.

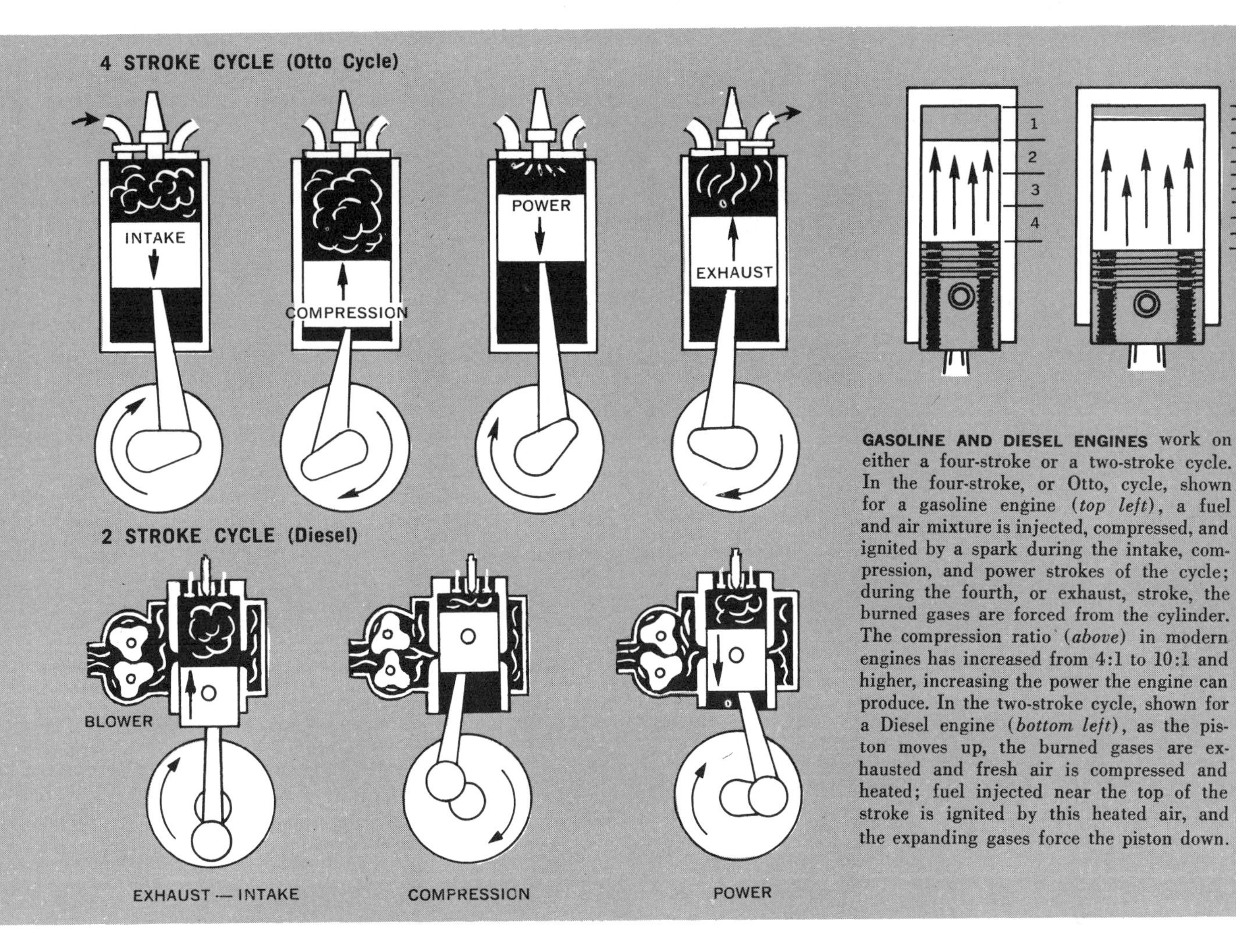

GASOLINE AND DIESEL ENGINES work on either a four-stroke or a two-stroke cycle. In the four-stroke, or Otto, cycle, shown for a gasoline engine (*top left*), a fuel and air mixture is injected, compressed, and ignited by a spark during the intake, compression, and power strokes of the cycle; during the fourth, or exhaust, stroke, the burned gases are forced from the cylinder. The compression ratio (*above*) in modern engines has increased from 4:1 to 10:1 and higher, increasing the power the engine can produce. In the two-stroke cycle, shown for a Diesel engine (*bottom left*), as the piston moves up, the burned gases are exhausted and fresh air is compressed and heated; fuel injected near the top of the stroke is ignited by this heated air, and the expanding gases force the piston down.

power. When a piston moves in the cylinder from the bottom of its stroke to the top, its movement brings about a change in volume in the cylinder. The change in volume can be measured in cubic inches, as in U.S. practice, or in liters, as in European practice (a liter is about 61 cubic inches). This "working volume" of all the engine's cylinders combined is called the displacement of the engine. It can be seen that a four-cylinder engine in which there is a displacement of 100 cubic inches in each cylinder ($4\times100=400$) is larger, and all else being equal, more powerful than an eight-cylinder engine with a displacement of 40 cubic inches in each cylinder ($8\times40=320$).

■COMPRESSION RATIO.—The compression ratio is another factor influencing the power output and efficiency of a gasoline engine. The compression ratio is the ratio of the volume of the cylinder at the bottom of the piston stroke to the volume at the top. In modern automobile engines this ratio may be as high as ten or eleven to one, meaning that the maximum volume of the cylinder is ten or eleven times as great as the minimum volume, as measured with the piston at opposite ends of its stroke. These *high-compression engines* use the higher octane fuels.

■CYLINDER ARRANGEMENTS.—Internal combustion engines are designed with a wide variety of cylinder arrangements. The *in-line,* or straight, engine has all cylinders arranged in a line straight down the engine block. This is a common arrangement for four-, six-, and occasionally eight-cylinder engines. The *V* arrangement is more popular for eight-cylinder engines. It consists of two banks of four cylinders, each set at an angle with the crankshaft at the bottom. This is a compact design permitting the use of a shorter crankshaft. This *V* design is also used for twelve-cylinder engines and for the few sixteen-cylinder engines that have been manufactured.

In recent years the *horizontal design* has become popular for smaller (usually four-cylinder) automobile engines. This is a flat engine with half the cylinders on one side of the crankshaft and half directly on the other. This is a special type of *V* design in which the angle of the *V* has been increased to 180°.

The *radial engine* has been used largely for aircraft and is only now being replaced by the gas turbine and the turbojet. For many years it was the standard engine design for larger aircraft. In the radial engine the cylinders are arranged in a circle with the crankshaft in the center. There is but one crank on the crankshaft to which a master rod from one piston is attached. The rods from the other pistons are connected by wrist pins to the large end of the master rod. There is always an odd number of cylinders.

During the last 75 years of intensive development of the internal-combustion engine, many engines with unusual cylinder arrangements have been built, but few offered any distinct advantages. One of the more interesting is the *rotary engine,* similar in appearance to the radial engine except that the crankchaft is stationary and the cylinders rotate about it. This was used for a few aircraft engines during World War I, but the rotation of the heavy engine in one direction made it difficult to keep the entire aircraft from rotating in the other.

The *W* engine has enjoyed some popularity as an aircraft engine, it being similar to the V but with an additional bank of cylinders between. The *X* is simply two V engines joined together, one above and the other upside-down below. None of these designs is popular today.

There are a few other internal combustion engine designs worthy of mention. The *opposed-piston engine* uses a sleeve, open at both ends, as a cylinder, and this cylinder contains two facing pistons, coming together on one stroke and moving apart on the next. Hugo Junkers, a German aircraft designer, had considerable success with this engine which operated on a two-stroke diesel cycle.

Another type of internal-combustion engine has achieved some success in recent years for small power plants and as an air compressor. This is the *free-piston engine*, which makes use of a cylinder containing two opposed pistons. The power stroke, which starts when the two pistons are close together at the center of the cylinder, drives the pistons apart. These are, however, "free" pistons, meaning that they have no piston rods attached. The cylinder is closed at both ends, and as the pistons spring apart in the power stroke they compress air in the space behind them. This air, or part of it, is used to scavenge the combustion chamber. A small volume of air left in the cylinder behind the pistons is compressed to such a high pressure that it bounces the pistons back together for the compression stroke. The power output of the free-piston engine is in the form of compressed air or compressed air combined with exhaust gases. The compressed air and gas mixture can be used to drive a gas turbine.

■**COOLING.**—While the basic principle of operation of the internal-combustion engine is quite simple, many refinements and auxiliaries are required to achieve an efficient, reliable engine. Since the gasoline engine requires an explosion of fuel for operation, it is obvious that some means must be used to keep it relatively cool, cool enough to permit the lubricating oil to function efficiently. Engines are either *air-cooled* or *liquid-cooled*. For many years liquid-cooled engines were considered more satisfactory, but improvements in design have made the air-cooled design popular for smaller automobiles.

The *liquid-cooled engine* provides jackets around the cylinders through which the liquid flows, carrying off heat. The hot liquid flows or is pumped to a radiator consisting of finned tubes. A fan forces the air through the radiator. The cooling liquid is usually water to which alcohol or some antifreeze is added in cold weather. Occasionally other liquids, such as thin oils, are used as coolants. The principal advantage is the prevention of rust and scale in the cooling system.

The *aircooled engine*, provided with large cooling fins directly outside the cylinders, and with larger fans, manages to keep the cylinder temperature sufficiently low for efficient operation. This type of design is simple and ideal for propeller-driven aircraft, where the rapid flow of air over the cylinders provides excellent cooling

■**IGNITION SYSTEM.**—An ignition system is also essential to a gasoline engine. The *spark* igniting the air-fuel mixture is provided by a *spark plug*. This plug, fitted into the top of the cylinder, has two points of copper with a small gap between. When an electrical voltage is applied to one point, a spark jumps the gap to the other point. It is the electric spark that ignites the explosive mixture. The electric voltage is provided from a *storage battery* or small generator or alternator driven by the engine. Since the generator runs only when the engine is running, a battery must be used to provide the spark as the engine is being started. Once the engine is running under its own power, the required electrical potential is provided by the generator or alternator. The generator or alternator also supplies current to the battery to keep it charged.

The spark must occur in each cylinder at exactly the proper time in the cycle, just at or near the arrival of the piston at the top of the compression stroke when the air-fuel mixture is ready for ignition. This precise timing is provided by a *distributor*, a revolving electric switch mechanically linked to the engine, so that the electric current is switched to each spark plug at exactly the proper time. The distributor may be manually or automatically adjusted to permit the timing of the spark to be slightly retarded or advanced to achieve the best engine performance.

■**CARBURETION.**—An internal-combustion engine must also have some means for the fuel and air to enter each cylinder at the proper time in the cycle. Most engines accomplish this with a *combustor*, a mechanical device that mixes the fuel and air in proper proportions to provide an explosive mixture. The ratio of fuel to air can be changed by *carburetor* adjustment to achieve the optimum mixture for the particular fuel and the particular engine design. The air-fuel mixture from the carburetor is drawn into each cylinder on the intake stroke of the piston in a four-cycle engine or enters under pressure at the start of the compression stroke in the two-cycle engine.

■**FUEL INJECTION.**—Some engines are designed to use *fuel injectors*, rather than carburetors. There is no premixing of air and fuel. Instead, the fuel is injected into the cylinder as air is being compressed on the upward stroke of the piston. This would appear to resemble the diesel engine, but with a high volatile fuel, gasoline, and with a spark ignition, there is still the characteristic explosion of the Otto cycle rather than the relatively slow burning of fuel characteristic of the true diesel.

■**LUBRICATION.**—Lubrication is essential to the operation of the internal-combustion engine. Many types of lubricating systems have been designed, but all manage in some way to supply lubricating oil to the bearings of the crankshaft and to the walls of the pistons and cylinders, thereby reducing friction, and consequent overheating and wear.

Many useful accessories designed to improve engine operation or reduce engine wear have been developed. The *air filter* for the carburetor, the *supercharger*, and the *oil filter* are typical examples.

—Hunter Hughes

Fuel, a solid, liquid, or gas that will burn and give off heat.

■**SOLID FUELS.**—The most common solid fuel is coal, the product of decayed plant material subjected to chemical action and pressure over millions of years. It contains carbon, hydrogen, oxygen, nitrogen, sulfur, and small amounts of various impurities. When coal is analyzed according to its *proximate analysis*, a determination is made of the percentage of carbon, ash, volatile matter (gases), and moisture. Coals vary widely in type and analysis, ranging from lignite, a low grade coal, through subbituminous, and bituminous, to anthracite. *Lignite*, for example, is about 30 per cent carbon, 25 per cent volatile matter, and about 40 per cent moisture. *Anthracite* coal, at the other extreme, contains about 92 per cent carbon, 5 per cent volatile matter, and only 3 per cent moisture. *Bituminous* coal is between the two and is by far the most important as a fuel. Nearly 80 per cent of the coal reserve in the United States is bituminous or subbituminous.

The use of coal as a fuel declined from a peak of 631 million tons in 1947 to a low of 403 million tons in 1961. This decrease was brought about by the change of railroads from coal-burning locomotives to Diesels, and by the increased use of fuel oil and gas for energy in industry and for heating homes and larger buildings.

There has been, however, an increase in the use of coal since 1961, with 553 million tons mined in 1967. Most of this increase resulted from higher generation of electricity in coal-burning thermal power plants, but the trend toward nuclear power may again decrease the use of coal.

Lignite, also, is used primarily as a fuel for electric utility plants and its consumption may be expected to hold fairly steady. The outlook is less bright for anthracite, which has been burned primarily for space heating. Oil and gas have largely replaced it.

There are many other less important solid fuels. *Peat*, which is plant matter that has partially decomposed under water in a bog or swamp, is used as a fuel in many parts of the world. It is an important fuel in the Soviet Union and Ireland, where it is used in power stations as well as for heating. In most parts of the world it is produced as an agricultural additive to improve soil.

Wood is still an important fuel, since many homes have fireplaces, and a wood fire is pleasant if inefficient. Only in sawmills and the wood industries is it burned commercially.

Charcoal, generally produced by driving off the volatile matter in wood, leaving the carbon, is a minor solid fuel now formed into briquettes and used for outdoor cooking. It has some limited industrial uses.

Straw, tan bark, and bagasse (sugar cane from which the juice has been extracted) are burned in industrial boilers where they otherwise would be waste products, but they are not distributed and sold commercially as fuels. *Petroleum coke* is also burned

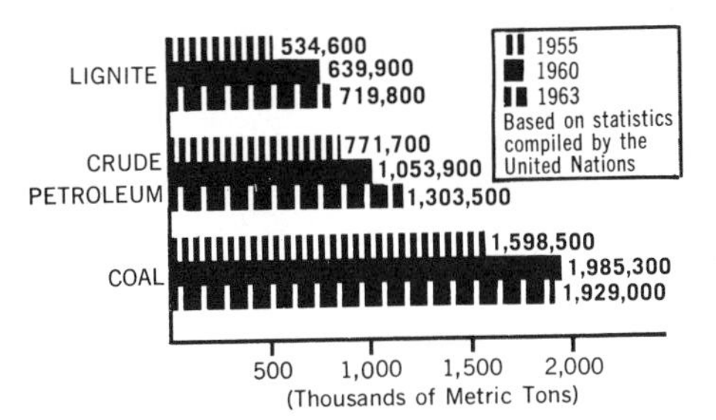

in a few thermal power plants, but it, too, is a minor solid fuel.

LIQUID FUELS.—Liquid fuels have increased enormously in importance during the twentieth-century. In 1900, liquid petroleum fuels were used to produce about 5 per cent of the energy output of the United States; by 1965 this had increased to about 40 per cent, largely at the expense of coal.

The most important liquid fuels are hydrocarbons produced by refining crude petroleum pumped from wells. A typical crude oil from a well contains about 85 per cent carbon, 13 per cent hydrogen, and some minor impurities. This kind of oil is refined by various processes to produce gasoline, kerosene, jet fuels, fuel oil, and many other products, such as lubricants and asphalt.

Gasoline is the most important of the liquid fuels, since it is used in most automobiles, trucks, and aircraft. Over 1.9 billion barrels of gasoline were produced in the United States in 1967. Today, natural gasoline refined from crude petroleum is combined with naphthas, natural gas liquids, and various other compounds and additives to produce a fuel suited to the modern internal-combustion engines.

Most *kerosene,* which is less highly refined than gasoline, is used as a fuel for jet aircraft, though there are other minor fuel uses. Military-grade jet fuel is a high-grade kerosene or a mixture of gasoline and kerosene.

Fuel oil is another product of crude oil refining. It is available in many grades and viscosities from thin Diesel fuel to heavy residual oils that are more solid than liquid and must be warmed before they will flow. The lighter *distilate fuel oils* are used as Diesel fuels and for smaller heating units. The heavier *residual fuel oils* are used in large heating units and as industrial and electric utility fuels.

Much recent research has been devoted to development of fuels for rocket engines. The United States has concentrated primarily on liquid fuels while continuing to investigate solid fuels. One important rocket fuel is kerosene and liquid oxygen (Lox). Liquid hydrogen-oxygen and hydrogen-fluoride serve as fuel for the upper stages of lunar launch vehicles now being developed. Nitric acid and alkyl-hydrazine mixtures are useful as fuels for some special requirements of spacecraft.

There are, of course, many liquid fuels other than those derived from crude petroleum. Considerable liquid fuel is extracted from natural gas. Oil can be extracted also from certain sands and shales, but neither of these sources has yet been exploited to any great extent. Alcohol is a good fuel but expensive. Liquid by-products of some process industries are used as fuels rather than being discarded as waste. So-called *black liquor,* a by-product of paper manufacture, is burned in papermill boilers, for example.

GASEOUS FUELS.—The use of *natural gas* as a fuel is growing rapidly—more rapidly than any other fuel. It is, like liquid petroleum products, composed primarily of hydrogen and carbon, and it, too, is taken from wells. After being processed, it is piped to the point of use. Natural gas, as a fuel, is used in industrial furnaces and utility boilers, but it has also become more and more popular for home heating and cooking and for heating commercial buildings.

The great advantage of gas is its simplicity of use. Solid fuels require large storage facilities and are difficult to handle. Coal varies greatly in burning characteristics, and when used in large utility boilers must be cleaned and pulverized. There is also the problem of ash removal and smoke. Coal's primary advantage is price, and even this advantage does not exist in all areas. Liquid fuels are ideal for moving vehicles and present less of a handling problem than solids, but they, too, must be stored in tanks prior to use. Gas, supplied to the user under pressure, requires no storage, no handling, and there is no ash or smoke. This accounts for the rapid increase in the use of natural gas as a fuel for heating homes and larger buildings.

Where natural gas is not available, *liquid petroleum gas* (LPG), consisting primarily of propane and butane, which are gas products of petroleum refineries, are sold in pressure tanks and used mostly as cooking fuels.

Gases other than natural gas are useful fuels. Before natural gas became so generally available, gas manufactured from coal and coke served as an important industrial and domestic fuel. Experiments have been conducted during the past few years for the gasification of coal in the mines so that the output of the mine is an industrial fuel gas rather than coal. A method for producing from coal gaseous hydrocarbon similar to natural gas, by a process known as hydrogenation, is under investigation. Gas fuels such as hydrogen and acetylene have specialized but limited use.

The term *fuel* now generally used with reference to uranium and plutonium in atomic piles producing heat, is technically incorrect, for there is no combustion involved in the nuclear reaction.

—Hunter Hughes

Fuel Cell, a versatile power source of great simplicity and high efficiency that converts energy from the reaction of a conventional fuel and air (oxygen) into low-voltage direct-current electric energy. Theoretically, a fuel cell can be built in almost any size and capacity. For practical purposes, however, individual cells are stacked in small modules, or *batteries,* and connected electrically in parallel. The major difference between fuel cells and conventional batteries is that the former operate continuously as long as fuel and an oxidizer are supplied and their electrolyte remains chemically unchanged, eliminating the need for recharging. Most fuel cells have no moving parts and require little or no maintenance. The noxious or toxic exhaust products associated with a combustion reaction—heat, smoke, and noise—are minimized. Water and/or carbon dioxide are the usual by-products.

The principle of the fuel cell is not new. Sir Humphry Davy first suggested it in 1802 and the first laboratory demonstration was by Sir William Grove in 1839. However, the fuel cell remained an electric power curiosity, by-passed by the steam engine, the internal-combustion engine, and nuclear energy, until the late 1950's. One of the first practical applications of fuel cells was on the

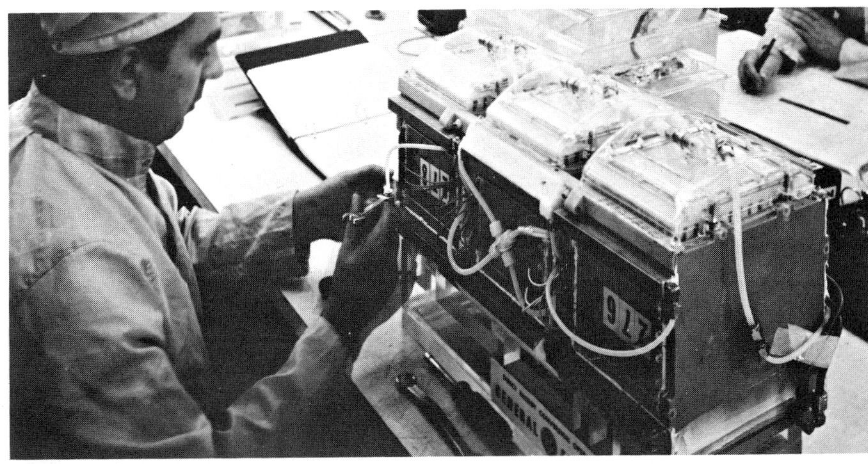

GENERAL ELECTRIC COMPANY

FUEL CELL MODULES (*foreground*) are connected with control equipment and enclosed in cylindrical cases during assembly of batteries that are used in the Gemini space program.

eight-day flight of the Gemini 5 spacecraft in 1965, when two fuel cell batteries operating on hydrogen and nitrogen provided prime electric power for the spacecraft.

A typical fuel cell contains an anode and cathode, which are in contact with an electrolyte (acid or alkali) that acts as an ion-transfer medium. Fuel (normally hydrogen) is introduced at one electrode, and an oxidizer at the other. In an *acid electrolyte fuel cell,* the hydrogen reacts with the electrolyte at the anode, giving up electrons to the electrode and ions to the electrolyte. The ions migrate through the electrolyte to the surface of the other electrode, where they combine with electrons that have traveled through the external circuit and with the oxygen diffused through the electrode, to form water as a by-product. Individual cells are connected in series to produce the desired voltage, and are sized to give the needed current. Scaling up or down does not significantly change performance or efficiency.

The efficiency of fuel cells is far greater than that of conventional engines. By converting chemical energy directly into electric energy, the fuel cell is not subject to the thermal cycle limitation of conventional generating equipment. Unlike conventional engines, the fuel cell uses fuel only on demand and produces energy more efficiently at partial or no-load. Practical fuel cells operating on hydrogen and nitrogen can convert 60 to 70 per cent of chemical energy into electricity, an efficiency about twice that of a gas turbine.

Although recent development activity has brought forth a great variety of fuel cells, cells using hydrogen and oxygen are in the most advanced state of development. For general use, efficient fuel cells that consume a commonly available hydrocarbon fuel—such as natural gas, propane, or such liquid fuels as kerosene and gasoline—are being developed. Some use the hydrocarbon fuels directly; others use hydrogen extracted from the hydrocarbon by means of a reformer. The development of more widescale use of these devices makes it necessary that scientists achieve an increase in operating life and a reduction in manufacturing costs.

Eventually, fuel cells are expected to be used for a wide variety of ground and marine uses. Present applications are in military and space programs where the advantages of small size, low weight, and high efficiency outweigh high initial costs. Fuel cells have been successfully used in the Gemini and Apollo spacecraft programs. The achievement of increased operating life and reduced initial cost, would make it possible to use fuel cell systems ranging in output from 10 watts to 15 kilowatts for a variety of military power sources. Silent and able to use military fuels and air at high efficiency, they could be used to power communications equipment, radar, and sonar, or to charge batteries.

—Roy S. Mushrush

U.S. DEPARTMENT OF THE INTERIOR

GIANT GENERATORS use electromagnets to convert the mechanical energy produced by the Hoover Dam's water supply into electrical power for many areas of the West Coast.

Generator, Electric, a machine that converts mechanical energy into electric energy. It operates by providing relative motion between electric conductors and a magnetic field. Such relative motion, or "cutting," of lines of magnetic force generates electric voltage in the conductors. The conductors are joined in a systematic winding whose ends are terminated on the external frame. When these terminals are connected to an electrical load, the generated voltage causes electric current to flow through the completed circuit.

If the current from the generator flows continuously in one direction, it is a *direct-current* or D.C. *generator.* If the current rapidly reverses direction, it is an *alternating-current* or A.C. *generator.* The 60-cycle power systems in common use are supplied by A.C. generators, in which this cycle from positive to negative and back to positive occurs 60 times every second.

■**D.C. GENERATORS.**—Direct-current generators are of two basic types, homopolar and heteropolar. The *homopolar* or *acyclic* D.C. generator is constructed with a uniform magnetic field perpendicular to its rotor surface. Conductors on the rotor cut magnetic lines uniformly in a single relative direction, thus generating a D.C. voltage. Connection to external load can be made by stationary contacts, such as carbon brushes sliding against collector surfaces on the rotor. Often the rotor is a solid metal piece that serves as the conductor. The acyclic D.C. generator, demonstrated by Michael Faraday in 1831, was the first type of generator.

The *heteropolar* D.C. generator, developed somewhat later than the acyclic generator, usually has several pairs of stationary magnetic poles arranged around the inner periphery of a magnetic yoke. The poles have electromagnet coils connected so that the poles are alternately north and south. The magnetic field passes from the north poles into the magnetic material of the rotor and then back to the stationary south poles. The rotor or armature has insulated conductors near the surface that pass alternately under north and south poles, cutting magnetic lines in two opposite directions. Thus is an A.C. voltage generated; if the conductors were connected directly to an electric load, A.C. voltage would flow. However, the heteropolar D.C. generator has a means of converting the A.C. to D.C. The usual conversion is by a *commutator* rotating with the armature. This is a cylinder comprised of wedge-shaped copper bars separated, one from the next, by thin strips of insulation. The individual copper bars are electrically connected to the armature conductors. Electrical contact is made with the commutator surface by means of *brushes,* which are carbon blocks held against the moving surface by spring pressure. These brushes are spaced around the commutator so that half of them are always in contact with bars connected to conductors generating north voltage, while the other half are similarly associated with south voltage conductors. Therefore, the switching effect of commutator bars sliding past brushes converts alternating-current voltage in the conductors to direct-current voltage at the stationary terminals.

Some small D.C. generators convert the A.C. to D.C. by means of rectifier elements mounted on the rotating armature. These generators usually supply magnetic field current to large synchronous machines.

Direct-current generators are classified as *shunt, series,* or *compound,* depending upon how the magnetic field is produced. The application determines which class should be

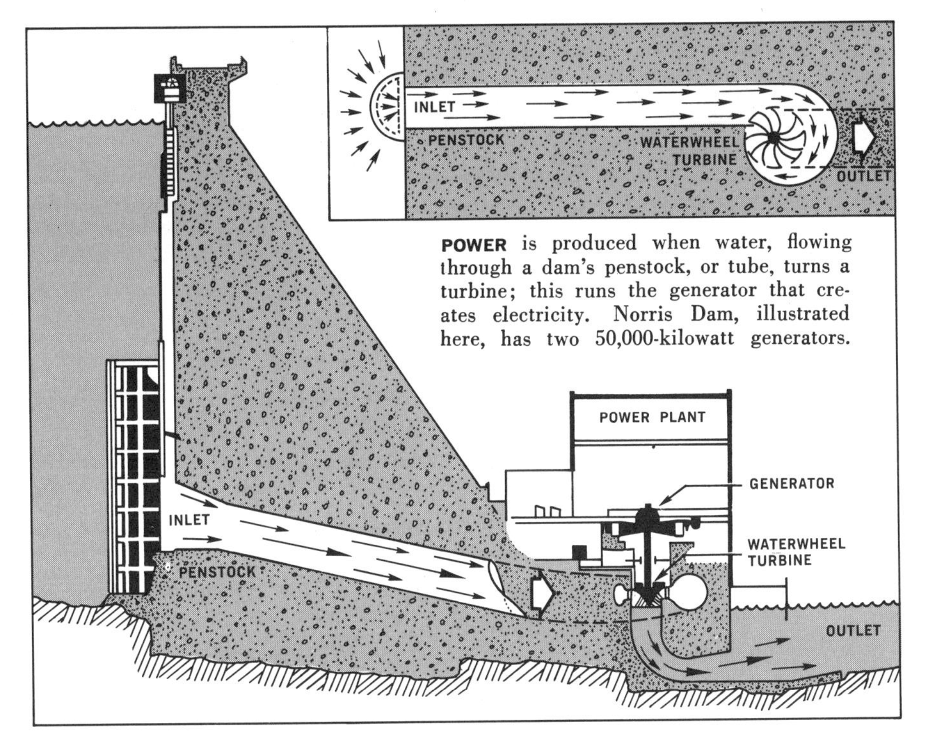

POWER is produced when water, flowing through a dam's penstock, or tube, turns a turbine; this runs the generator that creates electricity. Norris Dam, illustrated here, has two 50,000-kilowatt generators.

used. D.C. generators can be driven by any rotating prime mover, although engines, small turbines, and electric motors are most common. D.C. generators have industrial, utility, and transportation applications. The largest use is to supply power for D.C. motors, but electroplating, chemical refining, and others require bulk D.C. energy. Sizes range up to 6,000 kilowatts and 800 volts.

■**A.C. GENERATORS.**—Alternating-current generators operate on the same principle of voltage generation as do the heteropolar D.C. generators. Indeed, the heteropolar D.C. generator is an A.C. generator until the commutator or other device changes the A.C. to D.C. The usual form of A.C. generator, however, has its electromagnetic poles on the rotor and its conductors for voltage generation on the stationary armature. This permits current to be taken off without large collector rings. However, a small collector is still needed to carry D.C. to the rotating electromagnets.

The type of A.C. generator just described is the most common; it is often called a *synchronous generator* or an *alternator*. Two other forms of A.C. generators, having lesser application, are the *inductor alternator* and the *induction generator*.

World Production of ELECTRICAL ENERGY in %, 1966

Country	%
UNITED STATES	34.6
SOVIET UNION	15.1
JAPAN	6.0
UNITED KINGDOM	5.6
WEST GERMANY	4.8
CANADA	4.4
FRANCE	2.9
ITALY	2.5

Total production: 3,602 million kwh (Source: UN Statistical Yearbook, 1967)

The inductor alternator has a toothed steel rotor that provides magnetic field variations simulating north and south poles. It is inherently limited to small power ratings and is best suited for generating high frequencies, that is, 180 to 10,000 cycles per second (cps).

The induction generator is simply an induction motor (see *Motor, Electric*) with its rotor driven faster than synchronous speed by a prime mover, thus feeding electric energy back into the power system to which it is connected. Because the induction generator requires a connected system for successful operation, it is limited to a few special applications.

The A.C. frequency generated by any alternator is determined by the formula: cycles per second =

$$\frac{(\text{number of poles}) \times (\text{revolutions per minute})}{120}$$

Power systems in the United States and Canada are usually 60 cps. In other countries, however, 50 cps is often the standard frequency.

The principal use of A.C. generators is power generation for residential and industrial consumers. Large generators of two types are generally employed: *steam turbine-generators* and *hydroelectric generators*.

■**STEAM TURBINE-GENERATORS.**—Steam turbine-generators, also called turbo-generators, are of the synchronous-generator type, and are directly connected to high-speed steam turbines. Because of the resulting large centrifugal forces, the rotor is a single-piece, high-strength forging slotted for the magnetic field coils. The result is a so-called *round-rotor field* instead of the *salient-pole field* previously described for synchronous generators. Progress in materials and in mechanical and electrical design has been tremendous over the years, permitting larger and larger ratings of generators to be built. Units rated 1.4 million kilovolt-amperes and higher have been installed. Voltages for the larger machines range up to 26,000 volts.

■**HYDROELECTRIC GENERATORS.**—The hydroelectric generator, sometimes called *water-wheel generator*, is the other major type of large generator. It is a salient-pole synchronous generator, and the usual installation has a vertical shaft with the generator mounted above the prime-mover hydraulic turbine. The combined weight of the generator and turbine rotor plus the downward thrust of the water passing through the turbine is supported by a thrust bearing located near the generator. The speed of the generator depends upon the *head*, or height of fall, of the water and upon the type of turbine. Speeds range from about 50 rpm to 600 rpm for large hydroelectric generators, and ratings above 200,000 kilovolt-amperes have been furnished. The most common voltage rating is 13,800. The larger units may be 45 feet in over-all diameter when installed. Gas turbines and engines also are used to drive A.C. generators, but not to the same extent as steam and hydraulic turbines. —H. D. Snively

Hydroelectric Power, the use of the force of falling water to generate electricity. There are two basic types of water power. The first utilizes the natural flow of rivers and must depend on the particular flow of a river at a given time. The second obtains water from a reservoir. The flowing water passes through turbines which power *hydroelectric generators* thus producing electricity.

The first hydroelectric plant was built in 1882 in Appleton, Wis. Several years later, George Westinghouse (1846–1914) developed a system of transmitting power across long distances by using alternating current. In 1896 current was generated at Niagara Falls, N.Y. and transmitted to Buffalo, a distance of 20 miles. (See also *Electric Power Transmission*, page 1566.) —Joseph J. Kelleher

Lighting, the use of sources of illumination to facilitate seeing. The earliest form of artificial lighting was the open campfire, and this was succeeded by more sophisticated flame sources, such as candles, oil lamps, and gas lights. It was not until the last quarter of the nineteenth century that electric lighting systems became practical. Today, except in emergencies or in areas where electric power is not available, electric lighting is the universal source of artificial illumination. Electric lamps are relied on to supply the illumination needs of a highly complex technology. Electric lamps deliver lighting at standards of quality and quantity far beyond the possibilities of flame sources, and they do this safely and economically. Electric lamps are produced in thousands of shapes and sizes, with widely different performance characteristics.

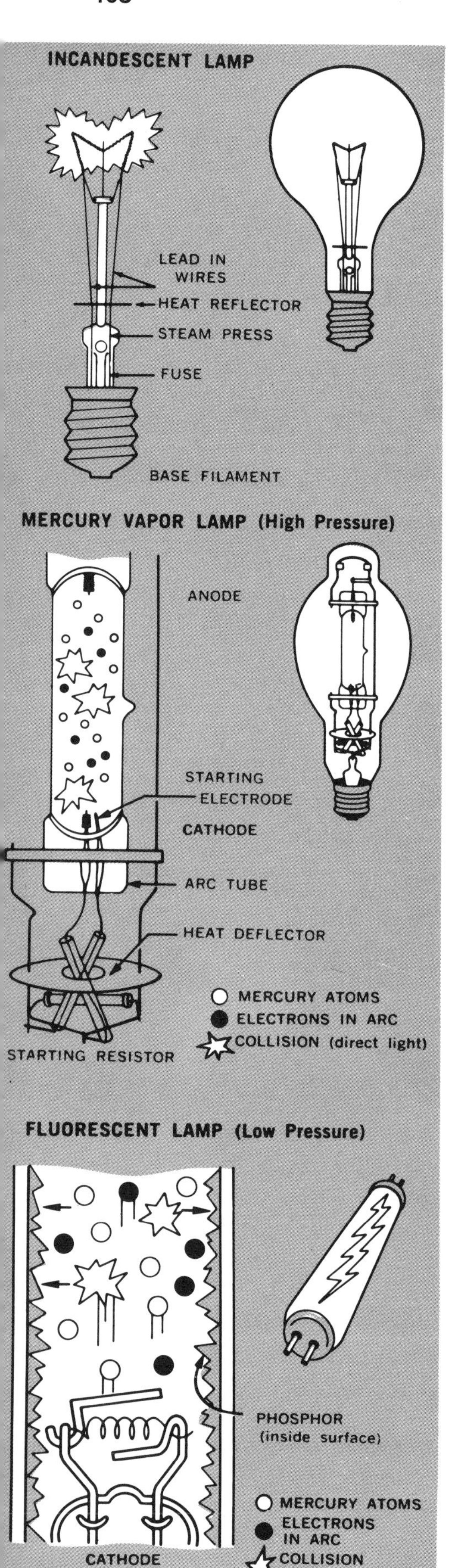

ELECTRIC LIGHTING SOURCES include incandescent, mercury vapor, and fluorescent lamps. In the incandescent lamp, the electric current heats the filament until it glows, or incandesces. In the mercury vapor and fluorescent lamps, the current is carried by an arc through the gas in the lamp. In the former, visible light is produced by the collision of electrons in the arc with mercury ions under high pressure in the gas, while in the latter ultraviolet light emanating from the collision of electrons with mercury ions under low pressure causes a mixture of phosphors coated inside the lamp to fluoresce visible light.

■**EARLY LIGHTING.**—*Candles* have been used throughout the Christian era. The first candles were merely pieces of wood dipped in tallow. Later, a mixture of beeswax and tallow was found to be better, and the wood splinter was replaced with a cotton-fiber wick. For many years the favorite candle material, called *spermaceti,* was a waxy substance obtained from the heads of whales. In the American colonies, candles were often made from the wax of the bayberry fruit. Modern candles are usually made of paraffin wax and stearin, a hardening agent derived from animal fats.

Primitive *oil lamps* were unsatisfactory because of their ineffectual lighting and smoky flame. In 1784 the French mathematician Jean Robert Argand equipped the oil lamp with a glass chimney, which provided for a better, steadier, less smoky light. During the period of oil lamp use, the chief fuel was whale oil.

For many years, gas was one of the most important illuminants used in cities. The early discovery of natural gas was made in England by Thomas Shirley in 1659, when he found gas bubbling through the water in a well. In the second half of the seventeenth century, John Clayton discovered that he could produce a gas that would burn by heating coal in a retort. It was another century, however, before gas came into practical use for illumination.

The first gas lights were merely open jets in which the flame burned with a deficiency of air. The light from the flame came from the incandescent particles of carbon that had failed to ignite. Later other substances were used with burning gas to give a whiter light. The *limelight,* for example, utilized lime (calcium oxide), but the materials found to be the best producers of light when heated to very high temperatures by a gas flame were the oxides of thorium, cerium, and zirconium. In 1885 the gas lamp was improved by Karl Auer von Welsbach's invention of a device, called the *Welsbach mantle,* by which a combination of these oxides could be suspended in the gas flame. The Welsbach mantle produced a brilliant white light and was responsible to a great extent for the adoption of gas lighting in towns and cities.

■**ELECTRIC LIGHTING.**—Although there had been experimentation with *electric* lighting since the early nineteenth century, it was not until 1879 that Thomas Alva Edison produced the first successful incandescent lamp, a long thin carbon filament of high resistance sealed in an evacuated glass container. For 30 years the design of the incandescent lamp remained unchanged, until in 1907 Alexander A. Just and Franz Hanaman devised the pressed tungsten filament. Three years later, William D. Coolidge developed a process for making drawn tungsten wire, which was much stronger than either pressed tungsten or carbon. The next major advance was the introduction of a gas into the evacuated lamp. This improvement, suggested by Coolidge and Irving Langmuir, provided for greatly increased efficiency.

Lamp manufacture is a very large business today, as it must supply the tremendous demand for lighting. Lamp making is organized along market lines. Thus, there are millions of lamps made for photography and projection. Many photographic lamps satisfy their purpose in a fraction of a second, just long enough to take a picture. Millions of other lamps, classified as miniature types, are made to meet the demand for light sources that can be used in limited space for extended service; an example is the automobile lamp. Large lamps, both incandescent and fluorescent, used for general lighting service, account for the largest markets in terms of both numbers and dollars. Mercury lamps, widely used in industry and in outdoor area lighting, belong in the large lamp category.

■**TYPES OF LAMPS.**— *Incandescent lamps,* since Edison's day, have been the dominant source for lighting. The highly efficient fluorescent lamps, introduced in 1938, have by now created a strong and growing market for these sources. Nevertheless, incandescent lamps still account for well over half the money spent for large lamps. They are popular because of low installation cost and simplicity of operation. While they are made in many shapes, the typical incandescent lamp consists of a glass bulb enclosing a filament that is mounted on a glass stem; in the stem there are wires to carry current from the contact on the base to the filament. The instant the current enters the lamp the filament becomes incandescent, and thus produces light. Compared with the fluorescent lamp this process of light production is simple.

Tungsten-halogen lamps. In a conventional incandescent bulb tungsten particles gradually boil off the filament and deposit themselves on the wall of the bulb. This blackening causes depreciation in the amount of light escaping at a steadily increasing rate as the operating life of the bulb increases. In the tungsten-halogen lamp, the tungsten atoms also leave the burning filament, but now, instead of depositing on the tube wall, combine with iodine or bromine atoms (members of the halogen family of chemicals) to form gaseous tungsten-iodide. The gas circulates through the lamp and contacts the filament, whose intense heat causes the iodide to break down. Tungsten atoms are redeposited on the filament, and the iodine atoms are freed again to react with other tungsten atoms. The tungsten particles are not necessarily returned to the same part of the filament from which they departed. Some parts of the filament get thicker and others get thinner until a portion of the wire gets too thin to carry the current, and the filament breaks. But up to that time the inner bulb wall stays clean, and light output remains high. Lamp life is usually doubled.

Fluorescent lamps are in the category of *electric discharge* sources, which includes the mercury types. Electric discharge refers to the arc that is created within the light source when current is applied. The arc excites the atoms of a small amount of mercury within the tube or lamp. In the fluorescent tube the mercury is vaporized to a very low pressure. This is sufficient to create electron collisions in the vapor with resultant radiation that is absorbed by the phosphor lining of the bulb. The tube glows with the absorbed radiation, emitting a very pleasant diffuse illumination. The steps mentioned require several seconds in some fluorescent lamps to produce light. With the "rapid start" and "instant start" types the steps are telescoped by special design of the circuits.

Fluorescent lamps have greatly increased the scope of lighting service. Advanced standards of illumination, widely sought for stores, schools, offices, and factories, require economical fluorescent systems.

Incandescent and fluorescent lamps are often combined in general lighting installations. The two sources complement each other. Incandescent lamps are concentrated "point" sources, rich in red energy. Fluorescent lamps are diffuse, low-brightness sources, cool in effect and color quality, highly efficient and long-lived. In combination they provide versatile, satisfying illumination.

Mercury lamps, inherently producers of discontinuous spectrum concentrated in blue energy, offer economy of light production and long life. They are in demand for industrial and outdoor service where color quality may be less important than low-cost maintenance. Mercury lamps are high-pressure mercury vapor cousins of fluorescent lamps, and like them, are noted for efficiency. Researchers improved the color and increased the efficiency of the basic mercury lamp by injecting selected iodides into the arc chamber. An iodide is a compound of iodine with another element. The metal halides most often used, and the spectral enhancement produced, include the iodides of sodium (yellow), thallium (green), and indium (blue and red). *Neon lamps* simply substitute neon gas for mercury vapor.

High pressure sodium discharge lamps, which were made available in 1966, are the first general-purpose white light sources with efficiency in excess of 100 lumens per watt. These lamps belong to the family of high-intensity electric discharge sources that include the mercury types. Intense, white light, produced in a slender, elliptical-shaped bulb, is made possible with the cigarette-sized high pressure arc tube within the bulb. The ceramic arc tube permits the use of alkali metal vapors at much higher temperatures than have ever before been practical. These lamps are commonly used as street lights and in industrial plants. They are popular too in sports arenas and parking lots. Research has increased the efficiency of these lamps, making them very desirable.

—James L. Tugman

Lightning, a high-voltage electric discharge in the form of a luminous, multibranched spark from cloud to ground, or within or between clouds. A typical lightning flash consists of from 2 to 40 separate strokes, at intervals of a few hundredths of a second, each consisting of tens of thousands of amperes of electric current. To the human eye, which cannot resolve the individual strokes, lightning appears as a single flash lasting about one second. The stroke is usually a mile or two in length, but may extend 50 miles or more on rare occasions. The electric power dissipated in a single stroke is in the range of several billion horsepower.

Lightning is produced by intense electric charges that occur during thunderstorm weather conditions. The central region of a thundercloud contains an intense negative charge, while the top and bottom contain intense positive charges. Lightning can occur within the cloud, or from cloud to cloud, when the insulation of air breaks down in response to these strong electric fields. Cloud-to-ground lightning occurs when the central negative portion of the cloud discharges to the ground below. Lightning is thought to be nature's way of returning positive charges from the earth to the upper atmosphere against the current that flows continuously from air to earth.

The several strokes along the lightning path heat the air, causing it to expand. Repeated expansion and contractions caused by the strokes generate intense sound waves, which we hear as thunder. *Heat lightning,* a reddish, luminous globe about a foot in diameter, is most mysterious of all. It may hover, explode, or quietly disappear.

—William C. Vergara

Magnetohydrodynamics, or *MHD,* a method of generating electricity in which an electrically conducting gas is forced to move past a magnetic field. The gas becomes conductive when it is heated to temperatures greater than 2,500°C, producing a mixture of electrons, ions, and molecules. This mixture, called a *plasma,* is forced past a magnetic field, thereby taking the place of the wires in a conventional electric generator. The major problem in *MHD* power generation is finding an electrical insulating material that can contain the white hot plasma over a period of years without breaking down. No such material has yet been developed.

—William C. Vergara

Motor, Electric, a machine that converts electric energy into mechanical energy. According to the "motor" principle, an electrical conductor carrying current in the presence of a magnetic field experiences a force that tends to move the conductor at a right angle to the field. An electric motor is constructed so that either the magnetic field member or the conductor member is stationary, while the other element becomes the rotor, free to rotate in bearings and coupled to the rotating mechanical load.

In construction, electric motors resemble electric generators. (See *Generator, Electric,* page 1572.) Most motors can operate as generators and vice versa; many practical applications require a machine to perform either function. A motor may be either D.C. or A.C., depending on which type is called for by the main power supply.

Direct-current motors are generally similar in construction to the heteropolar D.C. generator; they both feature a revolving armature and commutator and a stationary magnetic field. Incoming D.C. at the motor terminals flows into the armature conductors through the brushes and the commutator. The current-carrying conductors react with the stationary magnetic field to cause the rotor to revolve. The conductors thus contact positive and negative commutator brushes alternately. The sliding action of commutator bars past brushes performs the function of converting D.C. at the terminals to A.C. in the moving conductors, the exact reverse of its function in a D.C. generator.

A D.C. motor, like most other electrical motors, has an inherent ability to adjust itself to changing load de-

U.S. DEPARTMENT OF COMMERCE

FORKED LIGHTNING splits the sky during a summer electrical storm. Only about one electrical discharge in ten strikes the earth.

mands, as distinguished from many other mechanical drives. Since the motor has all the elements of a generator, it follows that the conductors are generating a so-called *back voltage* at the same time as they are experiencing the force that moves the rotor and drives the load. This back voltage, which always opposes the stationary terminal voltage in direction, is lower, and the difference between them is proportional to the load current. If the mechanical load increases, a motor will tend to slow down, thus reducing its back voltage, which is proportional to the speed. This will increase the difference between back voltage and applied voltage, and the motor will draw an increased current from the power line to match the new load demand.

D.C. motors, like D.C. generators, are classified as *shunt, series,* or *compound* according to how the electromagnetic field coils are supplied with current. In a *shunt* motor, the field winding is parallel to the armature circuit, so the magnetic field force does not vary appreciably with load-current changes. A *series* motor has its field winding in series with the armature, so its magnetic field strength changes directly with the load current. A *compound* motor has both a shunt and a series winding.

Shunt D.C. motors are widely used where it is desirable to have fairly constant speed, regardless of changes in mechanical load. Changes in armature current do not affect the magnetic field strength appreciably, and the motor will change speed only slightly with load changes. However, a big advantage of a shunt motor is that its speed is readily controllable by a *field rheostat,* an adjustable resistance in series with the field winding circuit. Increasing the resistance will reduce the field current and raise the motor speed; reducing the resistance will raise the field current and lower the speed. These characteristics of good speed regulation under load and of simple speed adjustment make D.C. shunt motors desirable for many industrial applications. For example, large D.C. motors drive the main rolls of reversing and continuous metal rolling mills, where wide ranges of loads and of speeds are needed as the metal is processed from the hot ingot to the final product of thin strip or structural shape.

Series D.C. motors undergo a large drop in speed as the load current increases, due to the increased field strength. The driving torque increases rapidly, however, because the conductor current and the magnetic field strength increase together. This makes series motors particularly applicable in land transportation where this load characteristic is desirable; for example, a locomotive climbing a grade requires considerably more torque at a lower speed than it needs on the level run. The series motor also has excellent starting ability. However, the speed of a series motor is not adjustable by simple control methods.

Compound D.C. motors have characteristics ranging between those of shunt and series motors, depending upon the relative strength of the two field winding elements.

D.C. motors are usually powered from D.C. generators or from static rectifiers that convert A.C. to D.C. Small D.C. motors, such as those used in an automobile, can be powered from batteries. Ratings range from tiny units rotating at very high speeds to large, low-speed motors—for example, 7,000 horsepower (hp) at 50 revolutions per minute (rpm) for steel mill or marine applications. Voltages of less than 800 are customary.

Alternating-current motors are either *synchronous* or *asynchronous,* depending upon whether they operate at exactly synchronous speed under all load conditions. Synchronous speed is determined by the formula:

$$\text{synchronous speed (in rpm)} = \frac{120 \times \text{frequency (in cycle per second)}}{\text{number of poles}}$$

A *synchronous motor* has the same construction as a synchronous generator, with a stationary armature and a rotating electromagnetic field, usually salient-pole. An induction-motor type of winding on the pole surface permits self-starting of a synchronous motor having a three-phase or other polyphase armature winding. The motor locks into synchronous speed when field current is applied after it has been accelerated almost to speed by the induction-motor winding, which then ceases to carry current. Changes in load on the motor do not reduce the speed as on a D.C. motor, but do cause changes in the current and its so-called power factor drawn from the A.C. lines.

Synchronous motors are used where their constant speed, high efficiency, or other attributes are important. They are used as drive motors on A.C. to D.C. motor generator sets, and as drives for compressors, blowers, printing presses, centrifugal pumps, and many other devices. An increasing application for very large synchronous motors is in pumped storage hydroelectric units, where the unit acts as an electric generator during peak power periods and as a motor to pump water to a storage reservoir during off-peak periods. Units as large as 200,000 horsepower have been supplied for this purpose. A separate starting motor is sometimes used on such large machines instead of an induction-motor starting winding.

The most common *asynchronous motor* is the *induction* motor. The stationary member of an induction motor is like that of a synchronous

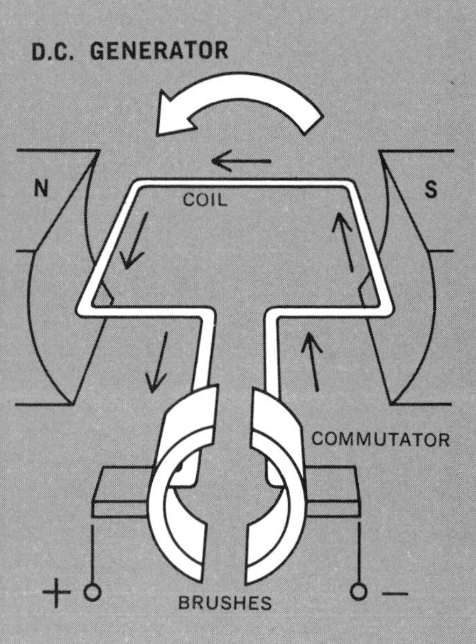

ELECTRIC GENERATORS produce an alternating current when a coil rotated in a magnetic field between the poles of a magnet (*below*), distorts the field. A direct-current generator (*left*) has a commutator, or split ring, to carry the current in only one direction; in an alternating-current generator (*right*) two slip rings change the flow alternately in opposite directions.

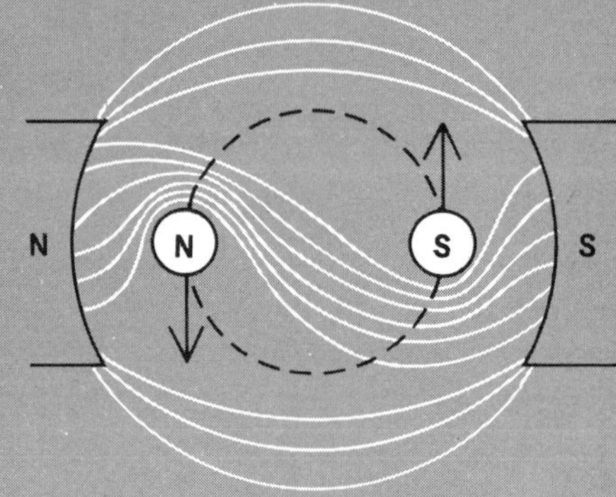

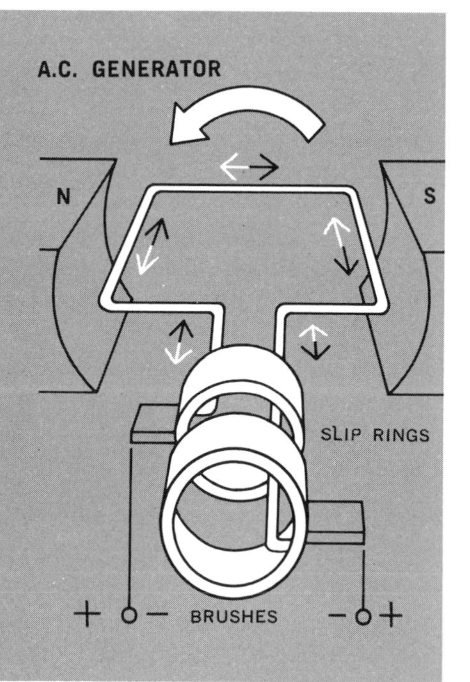

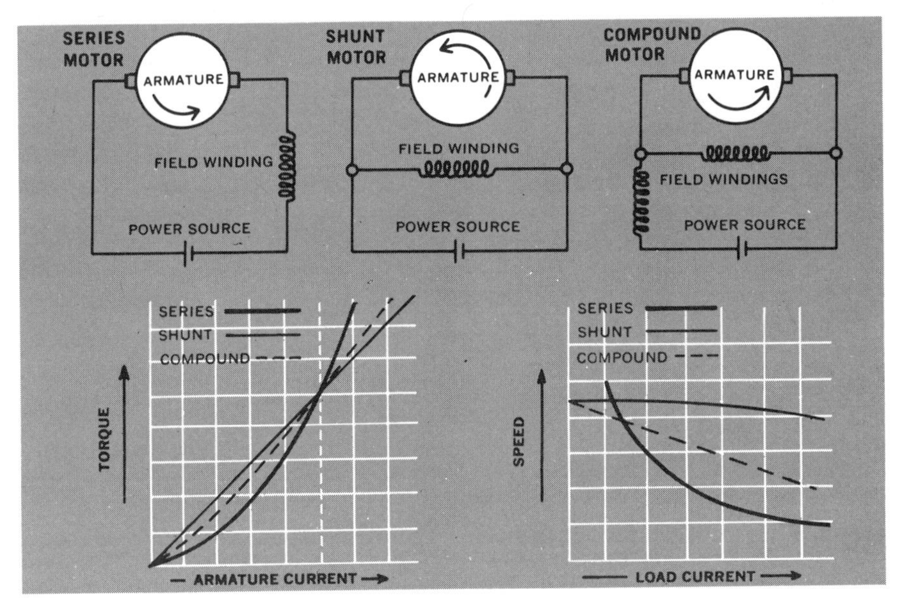

DIRECT-CURRENT MOTORS are classed by the manner in which their field coil is connected in the circuit. In a series motor (*top left*) the armature and field windings are connected in series; in a shunt motor (*top center*) they are connected in parallel; and in a compound motor (*top right*) the field winding is in two parts—one in series and the other in parallel with the armature winding. The increase in torque and decrease in speed of each type with increasing load are compared graphically at bottom left and bottom right, respectively.

motor. The rotor is either a *squirrel-cage* type or a *wound-rotor* type. The *squirrel-cage,* so named for its appearance, has an assembly of bare axial conducting bars embedded in the rotor surface and joined together at either end of the rotor by heavy conducting end rings. The *wound-rotor* has insulated conductors arranged in a polyphase winding, the ends of which are brought to collector rings on the shaft for external control purpose.

In either case, when power is applied to the stationary winding of a three-phase or other polyphase motor, a rotating field is induced that in turn cuts the rotor conductors, causes rotor currents to flow, and causes forces to be exerted on the rotor bars to turn the rotor. The rotor accelerates almost to synchronous speed but never quite attains it. If it did, no rotor current would flow and there would be no force to turn the rotor. The difference between actual speed and synchronous speed is called the *slip.* Slip increases with load and, depending upon the type of induction motor, is from 0.5 to 5 per cent of synchronous speed at rated load.

Wound-rotor motors with external control of their rotor windings are used where large speed adjustment is necessary. *Squirrel-cage* motors, rugged in construction and low in cost, are used in a variety of applications where approximately constant speed is needed but where speed adjustment is not required. Ratings up to 10,000 horsepower have been supplied. Wound-rotor motors over 40,000 horsepower are in operation.

Another large class of asynchronous motor is the *single-phase induction motor.* It is usually small in power rating and designed for use where polyphase power is not readily available. A single-phase induction motor is not self-starting in the manner previously described, because a single-phase armature winding produces a pulsating field rather than a rotating field. It is customary to simulate the effect of a polyphase winding by a special starting arrangement that brings the motor up to speed, whereupon a single-phase motor will behave much as a polyphase motor. *Split-phase, capacitor, shaded-pole,* and *repulsion* motors are several types commonly used. Each name describes a starting method.

Series A.C. motors are asynchronous motors identical with series D.C. motors except that the field magnetic structure is laminated instead of solid. Since the magnetic field and the conductor current change direction simultaneously at the A.C. frequency, this motor type behaves like a D.C. series motor. In fact, it is sometimes called a *universal motor* because it can be used on either A.C. or D.C. Applications are similar to those of D.C. series motors, in addition to considerable use in A.C. household appliances.

Fractional-horsepower motors are a class including all types of D.C. and A.C. motors less than one horsepower in rating. Most modern homes have from 10 to 50 fractional-horsepower motors performing various jobs.

Timing motors for electric clocks and other accurate timing purposes are A.C. motors of special construction, either *hysteresis motors* or *synchronous-inductor motors.*

—H. D. Snively

Neon Light. See *Lighting,* page 467.

Nuclear Engineering, the branch of engineering that deals with the control and utilization of energy and radiations from nuclear sources. Perhaps the most common example of the work of a nuclear engineer is the nuclear reactor. In a nuclear reactor, the heat energy given off in the fission of uranium is converted into useful electric power. Examples of other things a nuclear engineer might design are food sterilization plants and thickness gauges. In a food sterilization plant, nuclear radiations (usually gamma rays from radioactive cobalt) are used to kill insects or bacteria in grain, bacon, and other food products, thus retarding spoilage. A thickness gauge can be built making use of the fact that the intensity of a beam of nuclear radiation is decreased in passing through matter. The amount of the decrease in intensity of a beam of beta rays, for example, can be used to measure the thickness of aluminum foil or the amount of tobacco in a cigarette.

The distinguishing feature of the formal education of a nuclear engineer is that it includes courses in reactor physics (the study of controlled nuclear chain reactions), atomic and nuclear physics, the handling and use of radioactive materials, radiation shielding, and the effects of radiation on materials. The nuclear engineer must also be skilled in other fields of engineering (e.g., electrical, mechanical, metallurgical, and chemical), in order to control, utilize, and understand problems associated with nuclear energy sources. —R. O. Wooton

Nuclear Power, energy that is obtained from nuclear fission and fusion reactions and from radioactive disintegrations. In a *fission reaction,* the nucleus of a heavy atom, such as uranium, splits into smaller fragments. In a *fusion reaction,* two nuclei of a light atom, such as hydrogen, combine or fuse to form a single, heavier nucleus. In a *radioactive disintegration,* the nucleus of an atom emits nuclear radiations in the form of gamma rays, beta particles, and alpha particles.

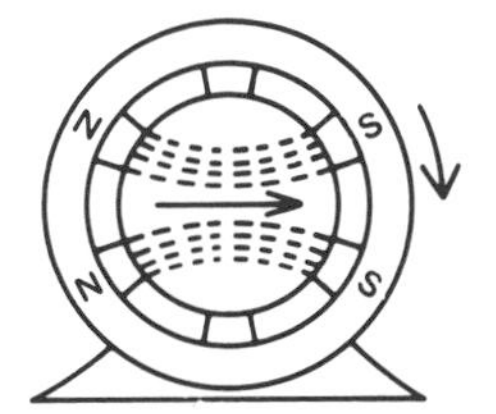

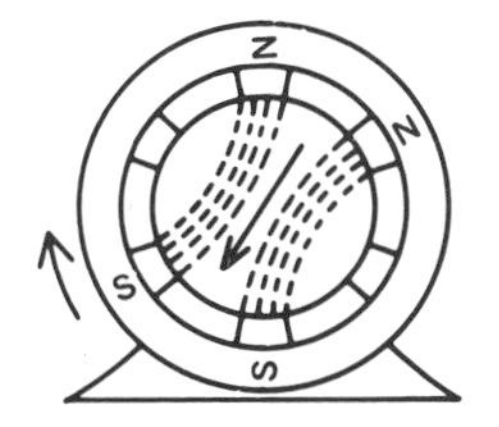

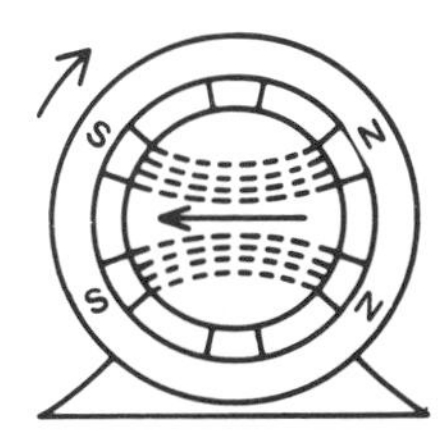

THREE-PHASE induction motor is a fixed rotor surrounded by a rotating magnetic field made by stator windings on three sets of poles. When the three-phase alternating current reaches the motor it moves through each winding, revolving the magnetic field.

The fission of uranium-235 is accomplished by bombarding the nucleus with a neutron, which splits the nucleus into two fragments of about equal mass. In the fission process a few neutrons are emitted, and energy is liberated. The energy liberated is the source of heat in nuclear reactors. The neutrons emitted can cause fissions in other uranium atoms, and a self-sustaining, or nuclear, chain reaction is possible. In addition to uranium-235, other isotopes of uranium and plutonium are readily fissionable by neutrons.

Fusion reactions are responsible for the heat generated by the sun and for the explosive power of the hydrogen bomb. Since about 1952, scientists around the world have been working intensively to attain a controlled fusion reaction, which could be used to generate electric power. Of primary importance for controlled reactions is the fusion of combinations of deuterium and tritium, which are isotopes of hydrogen. As of 1968, a controlled fusion reactor had not been built, and it is not known with certainty whether it is possible. It is a worthwhile effort for scientific research, however, since there is sufficient deuterium fuel in the oceans of the world to provide mankind with an essentially unlimited source of energy.

The nuclear radiations emitted in radioactive disintegrations possess energy of motion that can be converted into heat. This heat energy can be converted into electric power. Devices of this type have been built to provide power for marine life buoys and for space satellites.

—R. O. Wooton

Nuclear Reactor, a device that generates heat from the controlled fission of the atomic fuel. Since fission heat is accompanied by radiation, the reactor is a source of radioisotopes for research as well as for power. The reactor is divided into six basic sections: the *reactor core,* which contains the fuel; the *coolant;* the *shielding,* the *moderator;* the *reflector;* and several *control rods.*

Reactors vary greatly in size, fuel, structural and moderator material, and coolant. Reactors are used to produce electric power, to provide radiation for experimental purposes (research and test reactors), to produce radioisotopes, to produce fissionable material (breeder reactors), to provide propulsion for ships, and to purify water.

Fuel. The atomic fuel is the basic component of the nuclear reactor. The fuel material must undergo fission when it captures a neutron, and this fission process must produce neutrons, in addition to energy, to continue the reaction. The fuel is usually mixed with a metal such as aluminum, zirconium, beryllium, molybdenum, tungsten, niobium, or certain steels to give it added strength and corrosion resistance. It may also be used directly as oxides or carbides of uranium in some ceramic forms with a minimal cladding (coating) for protection. The fuel and metal mixture is formed into spheres, pellets, or thin plates and is usually clad with the same metal that is used to mix with the fuel. The type of fuel and metal used depends upon the reactor temperature, the use of the reactor, or the reactor type. Fissionable materials used in nuclear reactors are uranium-233, uranium-235, and plutonium.

Coolant. A liquid or a gas is circulated through the reactor core as a coolant to transport the fission heat to energy conversion equipment, such as turbines. The most common coolant is high-purity water, but many other materials can also be used as a coolant. In order to increase reactor efficiency, the reactor is operated at maximum temperatures. The more recent high-efficiency reactors are using liquid metals such as mercury, sodium, or potassium as coolants. Organic liquids and high-pressure gases are also being used.

Shielding. The emission of high-intensity radiation accompanying fission necessitates the use of heavy radiation shields. The most common shielding material is high-density concrete, which contains iron or other heavy aggregate. Where size and weight are important factors, other materials are used.

Moderator. A moderator is a material that slows the neutrons so that they are more likely to cause fission. Water is the most common moderating material, but graphite (carbon), beryllium, and heavy water (deuterium oxide) can also be used. Some reactors do not have moderators, for they are so designed that the fast neutrons will cause sufficient fissioning of the fuel. This type is referred to as a "fast reactor."

Reflector. A reflector is a material, placed around the core, that scatters (reflects) some of the escaping neutrons back into the reactor core to cause fissioning. Water is a very good reflector, as are graphite and beryllium. Almost all materials will scatter some of the neutrons back into the reactor core, but the elements listed above are the most efficient.

Control Rod. The reactor control rod is a movable section of the reactor core that contains a neutron-absorbing material such as boron, cadmium, and other elements. Since neutrons cause fission, the insertion of a neutron-absorbing material reduces the fission rate and, therefore, the reactor power. The control rods are moved by electric motors and drive mechanisms.

■ OPERATION.—To help describe the construction and operation of nuclear reactors, an example of a simple reactor system is given below.

One of the most common types of reactors in the United States is the pool-type research reactor. The reactor core is located near the bottom of a large pool of water that has been highly purified by distillation or by an ion-exchange process. All the water impurities are removed because most impurities become radioactive when bombarded by the neutrons

CONSOLIDATED EDISON

ATOMIC ENERGY produces electricity at a New York power station. Below this operating floor, a uranium oxide core is loaded into a reactor covered by water. By 1973 seventy plants in the U.S. will use nuclear energy as an inexpensive, clean power source.

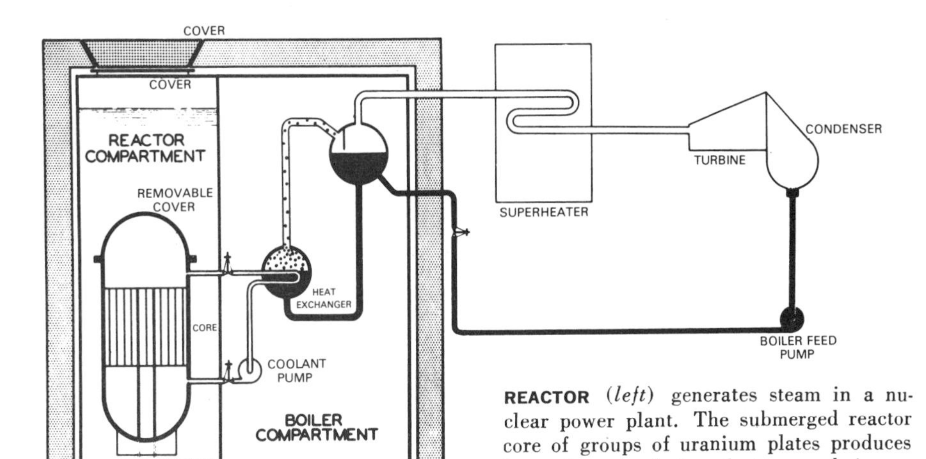

REACTOR (*left*) generates steam in a nuclear power plant. The submerged reactor core of groups of uranium plates produces heat through fission during a chain reaction which is set off by the partial removal of one neutron-absorbing control rod.

CONSOLIDATED EDISON

from the reactor. The pool water serves (1) as the reactor coolant, (2) as a neutron moderator, and (3) as a shield against the high-intensity radiation emitted from the core.

The pool water is circulated through the reactor to remove the heat generated. The water is raised in temperature as it passes through the core and must be cooled before it passes through the core again. The water is cooled by passing it through a "heat exchanger," which operates on a principle similar to that of an automobile radiator. Instead of passing air over the coils to cool the water, as in an automobile radiator, water is passed over the coils.

The reactor core of most of the pool-type reactors uses the plate-type fuel element. The uranium-235 is mixed with aluminum in thin plates and then clad with aluminum to form a plate about 1/16-inch thick. Ten to eighteen plates, spaced about 1/8-inch apart to allow water to flow between them for cooling, are grouped together to form a fuel element. Several of the fuel elements are brought together to make up the reactor core.

The reactor is controlled by placing a neutron-absorbing material (such as boron or cadmium) in the reactor core. If part of the neutron-absorbing material is removed, the reactor will begin to produce a self-sustaining chain reaction, which produces the heat. The reactor can be controlled by the neutron-absorbing material because neutrons, which are produced by fissioning, also cause the fissioning of the uranium. Therefore, absorbing some of the neutrons reduces the number of fissions.

A chamber situated in the reactor core measures the neutron intensity, which is proportional to the reactor power. The neutron chamber is connected to an automatic control device. When the reactor power goes up, the automatic control channel actuates the rod-drive motor and drives the control rod into the core, thereby absorbing more neutrons than before and, therefore, reducing the reactor power. If the reactor power drops below the pre-set power level, the automatic control system actuates the control-rod drive motor in reverse and it pulls the rod further out of the core, increasing the reactor power.

■ **REACTOR TYPES.**—Listed below are several reactor types with brief descriptions of the most common design features.

Research and Test Reactor. This type of reactor has a small core, usually measuring about two feet on a side. The core is fueled with highly enriched uranium (90 per cent or more uranium-235) and the structural material is aluminum or steel. The reactor, normally, is water-cooled and the high-power test reactor is pressurized to prevent boiling. The reactor is designed for experimentation on the effects of radiation on materials.

Power Reactor. The power reactor usually has very large cores, some measuring up to 20 feet or more in diameter. The fuel is normally uranium, slightly enriched (by a few per cent) in the uranium-235 isotope. The structural material is usually steel, and the core is contained in a steel pressure vessel. Pressurized water is the normal coolant, although sodium or organic coolants are sometimes used.

Submarine Reactor. This is a pressurized water reactor that uses highly enriched uranium in the metallic phase or as a ceramic (uranium oxide). The structural material is zirconium alloy or steel, and the core is small—only a few feet square.

Space Reactor. This type of reactor is very small, approximately a foot high and a foot in diameter, with a beryllium reflector. Liquid metals such as sodium or sodium-potassium alloy are used as the coolant to permit very high temperature operation.

Breeder Reactor. This type is a power reactor that has a uranium-238 blanket around it. The fission process in the core is by fast (high-energy) neutrons, as opposed to the usual thermal reactor that uses low-energy neutrons for fission. The fast neutrons collide with the uranium-238 atoms to create uranium-239, which decays to neptunium-239 and then to plutonium-239. The plutonium-239 can then be used to fuel a nuclear reactor. The breeder reactor thus produces fuel even as it consumes fuel to produce power.

—L. O. Gunnels

Prime Mover, a mechanism that changes a natural energy source into mechanical energy in order to power a machine. The term is commonly applied to gasoline or diesel engines, steam turbines, steam engines, water wheels, windmills, and jet engines. The output of a prime mover may be used to rotate a shaft, move a machine element back and forth, or provide a jet for propulsion.

In a diesel-driven electric generator, the generator is turned by the diesel engine, which converts combustible fuel into mechanical energy. The diesel engine is the prime mover, while the generator uses the mechanical energy that is produced. Electric motors are not prime movers because their energy source, electricity, is not a natural source. (See *Engine, Internal-Combustion.*

—Joseph J. Kelleher

Solar Cell, a semiconductor device that generates electricity in significant quantities directly from sunlight. It differs from the older photovoltaic cell mainly in that it can provide a larger amount of electric power. The photovoltaic cell is used primarily as a control device, as in automatic exposure mechanisms for cameras. The solar cell finds its major field of application in power supplies for space vehicles, telephone stations, and portable radio receivers. A large panel of silicon solar cells has even been used experimentally to power an electric automobile.

A silicon solar cell is a thin wafer about twenty-thousandths of an inch thick. It is manufactured in sizes from 3/32-inch square to 1 1/4-inch in diameter. A solar cell weighs less than an equivalent volume of aluminum.

Silicon solar cells are useful for power applications because of their relatively high efficiency. Commercial solar cells are now available having conversion efficiencies between 10 and 13 per cent. This compares with about 11 per cent for the gasoline engine. A 26-square foot panel of solar cells has an electric output power of about 200 watts under solar illumination.

■ **PRINCIPLE.**—The solar cell operates on the principle that when light energy strikes certain kinds of atoms, it will dislodge electrons from the atoms. In practice, the solar cell consists of a thin silicon wafer containing a small amount of a material such as arsenic. The surface layer has an additional admixture of boron or a similar material. The junction of the surface layer (boron-silicon) and the rest of the wafer (arsenic-silicon) forms a kind of electrical barrier. As light penetrates both layers, it creates negatively charged electrons and an equivalent number of positive charges (called *holes*) in each layer. The electric force at the junction of the layers drives the electrons to the arsenic side and the positive charges to the boron side. Conducting wires are connected to each layer, and electrons flow from the arsenic layer through the external circuit to recombine with the positive *holes* in the boron layer. This current provides the useful power of the cell.

A typical silicon solar cell having a useful area of one square inch will provide a voltage of about 0.4 volts and a current of 0.175 amperes under solar illumination. This amounts to about 0.07 watts. If higher voltages are required, a number of solar cells can be connected in series. The output voltage is then equal to the sum of the voltages of the individual cells. Additional current capacity can be obtained by using larger cells, or by connecting additional cells in parallel.

■IN SPACE VEHICLES.—The most dramatic application of solar cells has been in satellites and space vehicles, where they convert sunlight into electricity to power the vehicle's electronic equipment. The Telstar communications satellite carries 3,600 solar cells for that purpose. Other solar-cell powered space vehicles include the Tiros weather satellite, the Mariner Venus probe, the Orbiting Solar Observatory, Explorer satellite, and many others. When a satellite is in the shadow of the earth, it uses electricity from storage batteries that have been charged from solar cells. Solar cells are particularly useful in space vehicles because of their light weight and their ability to provide electric power indefinitely without using fuel. —William C. Vergara

Steam Engine, a simple mechanical prime mover. Steam is produced, at the desired temperature and pressure, in an external boiler fired by coal, oil, gas, or some other fuel. Steam from the boiler is piped to the engine, which in its simplest form consists of a cylinder, valve, piston, piston rod, and flywheel.

At the start of a cycle, the steam enters one end of the cylinder through an intake port. The steam expands in the cylinder, forcing the piston ahead of it. When the piston reaches the end of its stroke, the valve is moved to direct steam to the other end of the cylinder, where it powers the return stroke. As the valve opens the intake at one end of the cylinder, it simultaneously opens the exhaust port at the other end to permit escape of the expanded steam.

The piston rod drives a flywheel, and the sliding valve is activated by a mechanical linkage attached to them.

Reciprocating steam engines were extensively used in the nineteenth century, but by now have been replaced by more efficient steam and gas turbines and internal-combustion engines. —Hunter Hughes

Thermoelectricity, the generation of electricity by heating the junction of two dissimilar materials. When two wires made of different metals are joined at their ends, and the junctions are maintained at different temperatures, an electric current flows around the loop of wire. Such a device is called a *thermocouple.* The basic phenomenon is called the *Seebeck effect,* after Thomas Johann Seebeck who discovered it in 1821.

The opposite effect also takes place. If both junctions have the same initial temperature, and an electric current is passed around the loop of wire, heat is given off at one junction and absorbed at the other. This makes it possible to transfer heat from a cold body to a hot one, as in a conventional refrigerator. This phenomenon is called the *Peltier effect,* after Jean C. A. Peltier, who first demonstrated it in 1834. Both the Seebeck and Peltier effects are manifestations of thermoelectricity.

The Seebeck effect, in the form of thermocouples, has long been used to measure temperatures up to 3,000° F. This is possible because the voltage produced by a thermocouple depends in a known way on the temperature difference between the two junctions. Newly developed semiconductor materials with improved thermoelectric properties have enabled thermocouples to act as sources of electric power for radios in remote areas. Heat is provided by the sun.

The Peltier effect is used today in refrigerators to remove heat from small electronic devices and to keep microscope slides cold during examination. —William C. Vergara

Transducer, a device that transforms energy from one form into another. The pickup cartridge of a phonograph is a transducer that transforms the physical motion of the needle (mechanical energy) into a corresponding variation of an electric current (electrical energy). After amplification, this electric current is applied to another transducer, the loudspeaker. The loudspeaker transforms the electrical signal into mechanical oscillations that produce sound energy vibrations in the air. In a well designed transducer, the output signal must be proportional to the input signal. Otherwise, the output will be a distorted indication of the input information.

A power generator, which changes mechanical energy into electrical energy, is a transducer, as are a photoelectric cell, which transforms light into electricity, and a thermionic converter, which transforms heat into electricity. An important class of transducers is made of piezoelectric crystals, such as quartz, which transform mechanical stress into electricity. Another class is made of magnetostrictive materials, such as nickel, which are often used to transform electrical oscillations into mechanical vibrations.

Transducers are widely used in telemetry, especially in space vehicles, to obtain data for transmission to earth. They are also used in sonar systems to send sound waves into the water, and in the control of industrial processes for economy and safety. In such applications, transducers are designed to have electrical output signals because of the convenience with which they can be transmitted (by radio or wire) and then recorded and translated into accurate, numerical data. —William C. Vergara

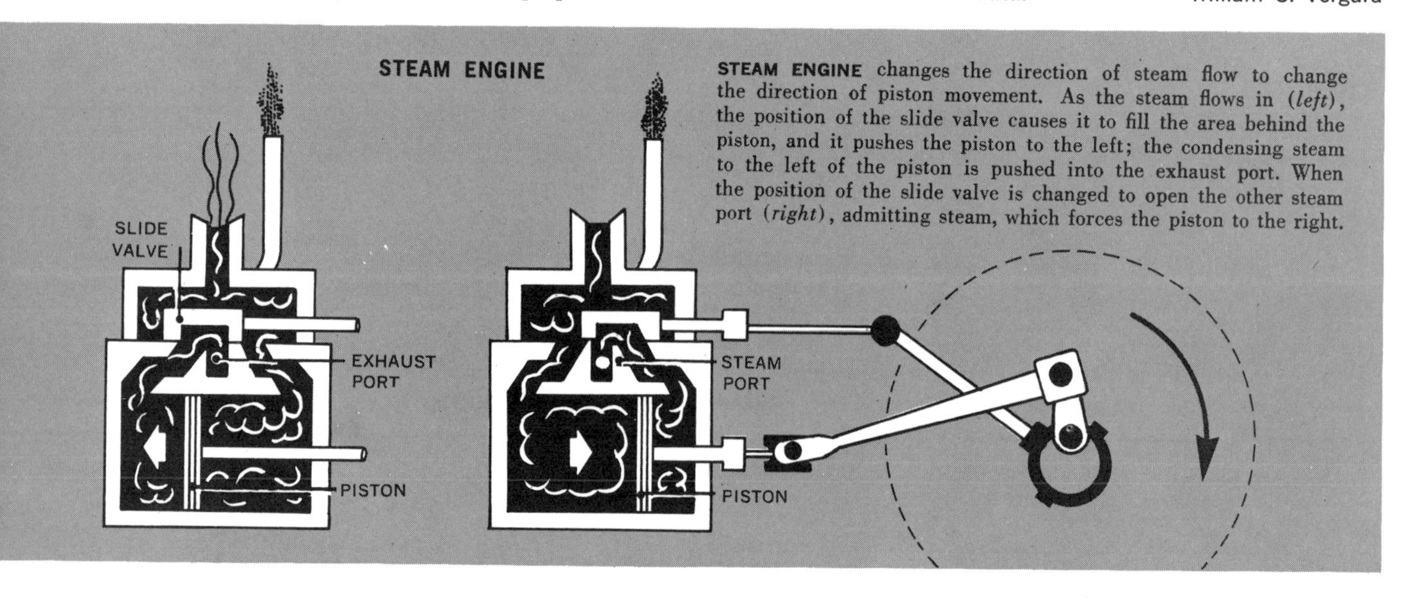

STEAM ENGINE changes the direction of steam flow to change the direction of piston movement. As the steam flows in (*left*), the position of the slide valve causes it to fill the area behind the piston, and it pushes the piston to the left; the condensing steam to the left of the piston is pushed into the exhaust port. When the position of the slide valve is changed to open the other steam port (*right*), admitting steam, which forces the piston to the right.

GENERAL ELECTRIC

STEP-UP TRANSFORMER is assembled, as its iron core and voltage-changing coils are lowered into place. Equipment for cooling the working mechanism circles the center tank.

Transformer, a device that transforms alternating-current (A.C.) voltage to a larger or smaller value.

A transformer that reduces a higher voltage to a lower one is called a *step-down transformer*; one that increases a lower voltage to a higher one is called a *step-up transformer*. The invention of the transformer in the late 1880's made possible the transmission and distribution of energy by electricity. Today, the transformer is utilized in almost every A.C. electrical circuit or system.

The most efficient type of transformer consists of a steel core around which are two or more windings of an electrical conductor. The input winding is designated the *primary*, and the output winding is designated the *secondary*. The secondary winding is normally located next to the core, while the primary winding is wound over the secondary but isolated from it. The core, which is laminated to reduce the electrical losses in the steel, provides a low-reluctance path for effcient magnetic coupling between the primary and secondary windings.

■HOW IT WORKS.—Since the transformer is a static device, an alternating-current voltage must be applied to produce the condition necessary to induce a voltage in the secondary winding. The ever-changing applied voltage (E_p) causes a current to flow through the primary turns (N_p), producing an electrical field. The flux of this field continually expands and compresses, and the primary windings convey this effect to the magnetic core. Since the secondary winding is located next to the core, this same flux causes the same volts per turn of windings to be produced in the secondary winding. Thus the desired secondary voltage (E_S) may be obtained by supplying the required number of secondary turns (N_S). The terminal voltages are therefore proportioned by the ratio of primary to secondary winding turns. This may be expressed mathematically as $E_p/E_S = N_p/N_S =$ turns ratio.

A transformer cannot create energy, and the output is limited to input minus losses. However, a transformer is usually very efficient, with losses of 0.5 to 1.0 per cent at full output, so that the input is almost equal to the output. This may also be expressed as $E_p \times I_p = E_S \times I_S$, where I is the current in amperes and the subscripts denote the primary and secondary currents respectively.

■TWO MAIN TYPES.—A transformer that has one primary winding, one secondary winding, and one core is classified as a *single-phase unit*. Many applications require more than a single-phase supply, because of the amount of power to be transformed or because of a need for a multiphase supply. In such cases, the single-phase unit may also be used by the proper connection of three units.

The *three-phase system* is almost universally used in the transmission of electric power. It carries three equal voltages situated 120 electrical degrees apart and normally varying at a rate of 60 cycles per second. In a three-phase system, a transformer must have the proper primary and secondary phase relationship as well as the proper voltage ratings.

The three-phase transformer may be made by winding three separate secondaries and three separate primaries on each of three legs of a single laminated core. However, a very popular type of transformer, called an *autotransformer*, has only one winding per phase. The primary voltage is applied to the winding, and the secondary terminal is tapped into the winding at the appropriate place to give the correct secondary voltage.

A transformer may also regulate voltage by using an automatic tap changer in one set of windings. A potential signal from the output is compared with a specified reference, and the load tap changer is automatically switched to the correct turns ratio to give constant voltage output. Typical tap ranges allow constant voltage output with an input of ± 10 per cent of rated voltage. Electric utility systems use many of these units to give constant-voltage service to their customers.

There are also transformers that use tap-changing mechanisms to shift the phase angle between the input and output terminals. This change in angle forces more or less power to flow over a given circuit, since power is the function of the sine of the angle between the system's input and output voltages. The phase-shifting transformer may require its circuit to carry more load, or it may limit the amount of power being trans-

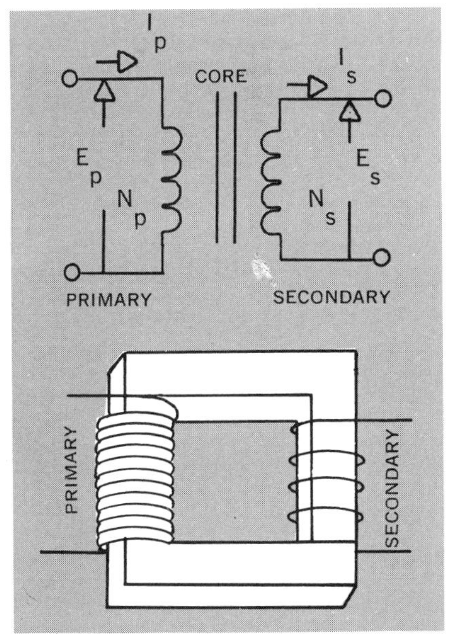

STEP-DOWN TRANSFORMER has more turns of wire on its input winding, or *primary*, than on its output winding, or *secondary* (*bottom*). As labeled in the schematic drawing (*top*), the primary voltage, E_P, primary current, I_P, and number of turns of wire in the primary, N_P, are related to the secondary voltage, E_S, secondary current, I_S, and number of turns in the secondary, N_S, as: $E_P/E_S = I_S/I_P = N_P/N_S$.

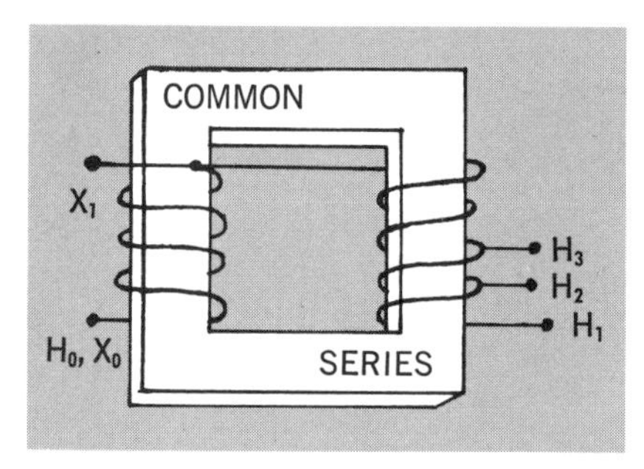

A TRANSFORMER WHOSE WINDING IS TAPPED to obtain a second voltage (as X_1 above) is called an autotransformer. It may be a step-up, step-down or a combination step-up–step-down. (Primary X_1X_0, secondary H_1H_0, etc.) The voltage rating is determined by the ratio of series and common winding turns. The range may be varied by taps such as H_2. On three-phase autotransformers, a third, or tertiary, winding is used to assure balanced voltages.

mitted over a current-limited circuit such as an underground cable.

■HOW THEY ARE USED.—Transformers have a variety of uses. Some of the more spectacular ones are found in the electric power industry. A generator step-up transformer might transform the output of a 1,000 megavolt-amperes (mva) generator from 25 kilovolts (kv) to 345 kv, 500 kv, or 700 kv for transmission over extrahigh-voltage (ehv) lines. Then autotransformers would be used to step down to 115 or 230 kv for transmission of the power into load areas. Here 25–200 mva bulk distribution transformers would again transform the voltage to 4 through 34.5 kv for distribution to homes or industry.

The distribution transformers are usually single-phase units for homes and three-phase for industry. They transform energy from distribution voltage to secondary voltages of 120/230 volts for the home and usually three-phase 277/480 volts for industry, commercial districts, or large office or apartment buildings.

Other uses range from the doorbell transformer in the home to multiphase transformers for electrochemical industries, furnace transformers for steel mills, welding transformers, reduced voltage starters for large motors, or transformers for neon signs.

A special type of transformer, called an instrument transformer, is a low-output, high-accuracy device that steps down high voltages or large currents on high-voltage circuits to 120 volts or 5 amperes respectively for use in relays, meters, or instruments.

The modern power transformer is a sophisticated apparatus. Trends indicate that banks larger than 1,800 mva will be required by 1975. At the present time, most large 500 kv and 700 kv banks must be composed of single-phase units, due to shipping limitations on size and to voltage clearances. Shipping limitations might require future large extrahigh-voltage banks to be made up of single-phase units. —G. W. Alexander

Turbine, a machine that makes use of the pressure or velocity of a fluid to produce rotary motion. Turbines may be classified as hydraulic, steam, or gas, depending upon the fluid employed, and further classified as impulse or reaction types depending upon the manner in which the pressure energy of the fluid is converted to mechanical work.

Hydraulic turbines are large rotating machines used to drive electric generators for the production of power. The simplest type is the *impulse turbine,* or *Pelton wheel,* in which a nozzle directs a stream of water against buckets or cups attached to the periphery of a wheel. The velocity of the water from the jet provides the force to turn the wheel. This type of turbine works best with a high head (pressure) and is generally installed only where water is available at the top of a high hill or mountain for delivery through pipes to a turbine in a valley from five hundred to several thousand feet below. It is a high-speed machine.

The *reaction-type hydraulic turbine* is a slow-speed machine using a relatively greater volume of water than does the impulse turbine, but at lower pressure. The level of the water source may range from only a few feet above the turbine to as much as a thousand feet. A dam provides the water storage and the head for the turbine.

The reaction turbine makes use of a guide case and stationary vanes through which the water passes into vanes of a revolving wheel, or runner. Only part of the water pressure is transformed into velocity in the stationary vanes, its velocity continuing to increase in the vanes of the runner, providing a reaction pressure from which the turbine gets its name.

The *propeller turbine* is a variation of the reaction turbine, but it has fewer blades on the runner, and operates with a partial vacuum at discharge. It is best adapted to low heads and large quantities of water.

Steam turbines are also classified as impulse or reaction types, but most combine the two principles. Steam, in contrast to water, expands as it passes through a turbine. Therefore, an *impulse steam turbine* is one in which expansion of steam, with consequent increase in velocity, takes place only in the stationary blades or nozzles, while in the *reaction steam turbine* expansion occurs in both the stationary and the moving blades. The steam enters the turbine at a high pressure and temperature, and exhausts at a lower temperature and at a negative pressure, the vacuum being produced by a condenser.

Steam turbines vary in size from small machines with a single rotor producing a few horsepower to huge multistage machines driving power plant generators rated at several hundred thousand kilowatts. By far the greater part of the electric energy of the world is provided by steam-turbine driven generators.

In the past twenty years there has been a rapid increase in the use of *gas turbines.* These are somewhat similar in design to steam turbines, but they are powered by the expansion of compressed air and gases from burning kerosene or other hydrocarbon fuels. Part of the power output is used to drive an air compressor which provides the air required for combustion of the fuel within a combustion chamber. Gas turbines drive small and medium-sized electric generators, and are also used as aircraft (turboprop), railroad, and ship engines.

The first airplane generator was the *turbojet,* consisting of an axial or centrifugal air compressor and a gas turbine on one shaft, with a combustion chamber between them. With the innovation of this airborne power plant in 1939, the propeller, crankshaft, cylinder and piston became obsolete. Although many commercial and private airlines continue to use

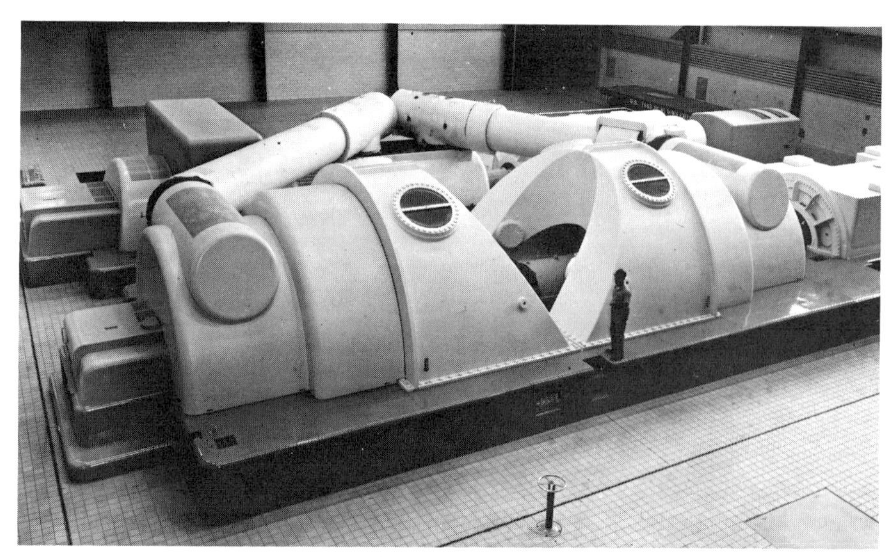

MODERN STEAM TURBINE at the Widow's Creek steam plant of the Tennessee Valley Authority is capable of producing 500,000 kilowatts of electric power. Seven giant turbines at this plant are capable of producing more than 1,250,000 kilowatts of power, which is used in homes, farms, commercial establishments, and industries within the T.V.A. region.

propeller and piston-driven planes for maximum efficiency at low speeds, turbojets are invaluable for providing adequate thrust at transonic and supersonic speeds. —Hunter Hughes

Ultrasonic Motor, a device for converting electric energy into vibrational mechanical energy, with the output exceeding 10,000 cycles per second. An ultrasonic motor is at the heart of every piece of ultrasonic equipment. Just as an electric motor takes electric energy and converts it to rotary motion, an ultrasonic motor converts electric energy to a reciprocating, or vibrating, motion. Ultrasonic vibraatory motors are well suited to the age of automation and robot-controlled complexes, such as are found in space exploration devices. This is because ultrasonic motors have no gross moving parts, require no lubrication, and have the ruggedness needed to operate perfectly either under high stress or in a vacuum.

A typical ultrasonic motor has for its driving agency what is usually known as an electromechanical transducer. This transducer is generally nothing more than a specially fabricated piece of ceramic that is capable of generating vibrations from alternating-current electric energy by means of the piezoelectric effect. Or, instead, a stack of nickel laminations inside a coil of wire may be used. In this case, the magnetostrictive effect is used to convert the electric energy into mechanical vibrations. The transducer is coupled to a transmission member to convey the vibrations to the place where they are required to work.

Broadly speaking, ultrasonic motors are used for observing, testing, and analyzing, on the one hand, and for processing or manufacturing, on the other. The actual use of the motor determines its design very much as electric motor design is influenced by the use intended for it. In analytical applications, the ultrasonic motor vibrations are generally in the megacycle (million cycle) per second range; in processing, the motor frequency will be in the range from about 10,000 to 50,000 cycles per second.

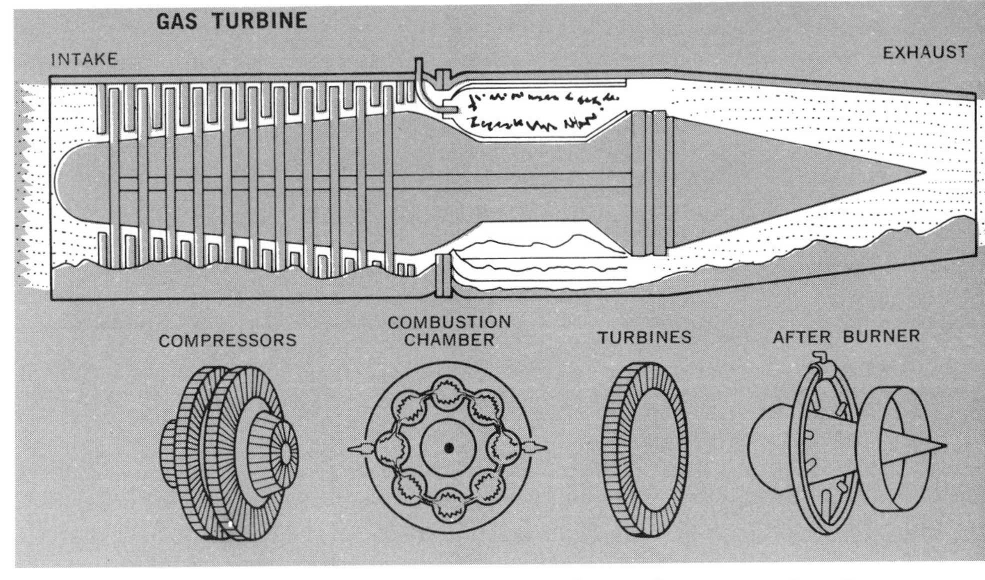

GAS TURBINES work on the principle of the expansion of air and gases occurring after fuel is burned and heat is generated in the combustion chamber. The heated gases strike a turbine wheel—which operates the compressor via a shaft—before escaping to propel the craft. The afterburner consumes any unburned fuel, furnishing additional thrust power.

■**ANALYTICAL USES.**—In the first area, often conceived generally as that of instrumentation, ultrasonic motors are used in sonar, with its various underwater uses; for flaw detection in large castings and forgings; for inspecting rails; for determining liquid levels in tanks; and for many other systems. With the advent of radar, the sonar idea was adapted to "seeing" what goes on inside the human body. As a result, new methods for detecting various kinds of cancer have developed, along with spectacular advances in the fields of gynecology, eye and brain surgery, and the study of the action of the heart and other soft organs of the body.

■**PROCESSING APPLICATIONS.**—The second, larger area of use of ultrasonic motors is also proliferating rapidly. In the 1940's an ultrasonic machine tool was produced for cutting hard, brittle materials such as glass, ceramic, precious stones, and cements. This machine is used for die-making and in the fabrication of transistors, as well as in the field of glass and ceramic technology. In the 1950's, the use of ultrasonic motors attached to tanks filled with liquid was developed to a high point for all sorts of industrial cleaning problems. Such problems range from the cleaning of a tiny watch mechanism, to the in-process cleaning of rapidly moving sheets of metal on the production line. Dentists, too, employ ultrasonic instruments for cleaning teeth.

The 1950's and the 1960's have also brought new uses for the ultrasonic motor in wire-drawing, metal extrusion, metal and plastic welding, and other material-forming and assembly processes. To give one example, it is possible to push a metal screw into a solid plastic at room temperature with an ultrasonic motor. Immediately afterward, the screw may be unscrewed from the plastic, leaving behind a perfect thread.

Today the ultrasonic motor is where the electric motor was about 80 years ago. The 1970's and 1980's should see the rapid spread of the distribution of high-frequency, alternating-current power, just as 60-cycle, alternating-current power is now commonplace. As this expansion occurs, there will be an enormous expansion in the use of ultrasonic motors, not only in industry and medicine, but in the home as well where soon ovens may cook ultrasonically. —Lewis Balamuth

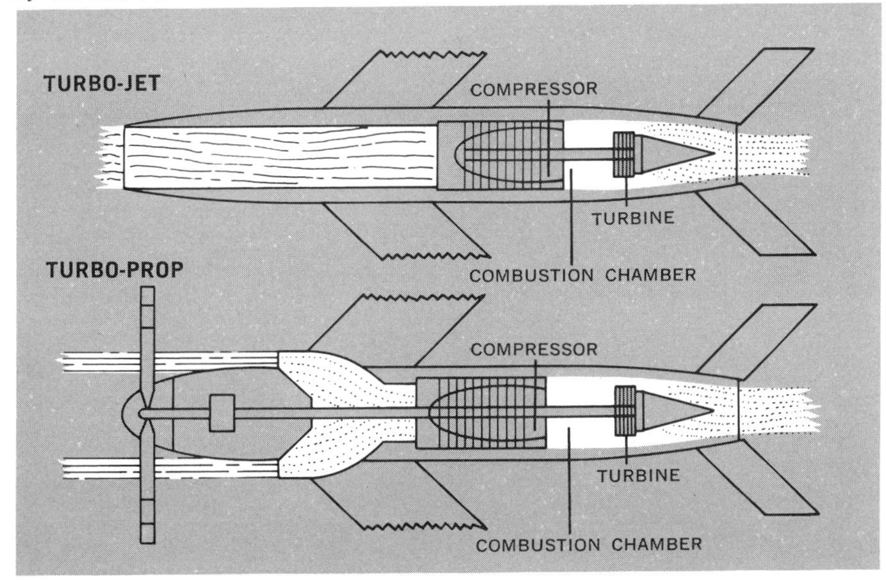

COMMERCIAL AIRCRAFT TURBINES are of two types. In a turbojet (*above*), the escaping gas shoots out the rear of the engine, and the reaction to this force propels the plane. In a turbo-prop (*below*), the power of the escaping gas is used by a larger turbine to turn a propellor by means of a shaft, while the exhaust adds some thrust.

Water Wheel, a device that converts the gravitational energy of falling water into rotary motion (kinetic energy) to do work or to provide power for other machinery. Water wheels have been in use since antiquity and are still being used today. Initially used to grind grain or pump water, water wheels have also been used to power sawmills, drive textile weaving machines, turn metalworking equipment, and generate electricity.

The simplest form of water wheel is the *undershot wheel,* which has a series of paddles mounted on its rim. When the paddles at the bottom of the wheel are immersed in a swiftly running stream, the moving water strikes the paddles and rotates the wheel. The *overshot wheel,* a more efficient design, has buckets attached to or built into the wheel rim. All of the buckets are arranged in the same direction so that, as the wheel rotates, the buckets on the descending side open upward while those on the ascending side open downward. Water delivered to the top of the wheel by a pipe or sluice from a dam fills the buckets. As the wheel rotates, the descending buckets gradually empty as they reach the bottom of the wheel. The weight of the water in the full buckets on the descending side causes the wheel to rotate. The simple undershot and overshot wheels were frequently constructed of wood and some were as large as 70 feet in diameter.

Modern water wheels are of the impulse or reaction turbine types. The *impulse turbine,* or *Pelton wheel,* has buckets or blades around its rim that are struck by high-pressure jets of water. Water pressure in the jets is increased by positioning the impulse wheel as far below the supply reservoir as possible. Some hydroelectric plants have reservoirs located several thousand feet above the impulse wheel that drives the generator.

The *reaction turbine,* mounted horizontally, consists of a circular upper plate and lower ring connected by vertical vanes. Water from a reservoir at higher elevation is fed to the chamber surrounding the turbine, passes through the vanes, and is discharged through the lower ring. The reaction turbine is more efficient than the Pelton wheel because the water pressure is applied to all the vanes at the same time. —Joseph J. Kelleher

Windmill, a device that converts the energy of moving air into rotary motion to do work or to provide power for other machinery. Windmills were used in China and Japan as early as 2000 B.C. These early wind-powered devices had vertical axes, and they powered mills previously driven by animals or men. Although the primary application for these windmills was in pumping water for irrigation, the name "windmill" is derived from the use of wind-driven devices to mill, or grind, corn.

Horizontal-axis mills of the Dutch type were in use in the eleventh century in Europe and in England by the twelfth century. Early horizontal-shaft windmills had simple canvas sails, stretched over wooden poles. These mills could not turn but were built facing the direction of the prevailing winds. Later, *post mills* were developed that allowed the operator to rotate the mill to face the wind.

The body of a post mill, containing the machinery and sails on a horizontal shaft, rests on a single heavy oak post mounted on a brick or stone foundation. A long tailpole projecting from the back of the body is used by the miller to turn the mill into the wind.

Tower mills, a later development, consist of a brick or wooden tower with a rotatable cap or roof. The sails and shaft are mounted in the cap, which turns as the wind turns. In 1745, Edmund Lee invented the fantail to keep the mill turned into the wind automatically. The fantail consists of a small windmill mounted at right angles to the main sails. When the wind shifts, it strikes the fantail, causing it to rotate. Linked to the mill by gears, the fantail drives the mill cap until the sails face the wind.

Experimental work to develop windmills for electric-power generation started in the 1880's and led to the development of the two-bladed, or propeller, windmill. Though many large aerogenerators have been built and tested, commercial applications are limited to smaller units, for isolated homes, with a capacity of about 3 kilowatts. The largest wind-driven generator ever constructed was a 1250-kilowatt unit built in the 1940's in the United States. Though research continues, it is unlikely that wind-powered devices will replace conventional power sources to any significant extent.

—Joseph J. Kelleher

BIBLIOGRAPHY

BLACK, PERRY O. *Audels Diesel Engine Manual.* The Howard W. Sams Co., Inc., 1964.

HALACY, D. S., JR. *The Coming Age of Solar Energy.* Harper & Row, Publishers, Inc., 1963.

HOGERTON, JOHN F. *The Atomic Energy Deskbook.* Reinhold Publication Corp., 1963.

KLEIN, H. ARTHUR. *Fuel Cells: An Introduction to Electrochemistry.* J. B. Lippincott Co., 1966.

LYTEL, ALLEN H. *ABC's of Electric Motors & Generators.* The Howard W. Sams Co., Inc., 1964.

MANDL, MATTHEW. *Fundamentals of Electric and Electronic Circuits.* Prentice-Hall, Inc., 1964.

MANN, MARTIN. *Peacetime Uses of Atomic Energy.* The Viking Press, Inc., 1964.

PURVIS, JUD. *All About Small Gas Engines.* The Goodheart-Willcox Co., Inc., 1960.

ROBERTSON, EDWIN C. AND HERBERT, ROY. *Fuel; the Conquest of Man's Environment.* Harper & Row, Publishers, Inc., 1963.

SHARLIN, HAROLD I. *The Making of the Electrical Age; from the Telegraph to Automation.* Abelard-Schuman Ltd., 1964.

ALLIS-CHALMERS MANUFACTURING COMPANY

WATER WHEELS are made in many sizes: the impulse, or Pelton, wheel (*left*) can produce 150,000 horsepower; the reaction, or Francis, type-turbine (*center*), can produce 35,400 horsepower; and the propeller-type runner (*right*) is rated at a lesser 44,000 horsepower.

FOOD AND AGRICULTURE

Agriculture and Civilization.—Civilization began when man started farming some ten thousand years ago. Before that, for perhaps a million years, people lived a wandering, precarious existence, hunting seeds and nuts and killing small wild animals for food. Agriculture permitted men to live settled lives and encouraged the development of industries and cities, necessary attributes for modern civilization.

Very likely one of mankind's greatest achievements—planting and harvesting crops—resulted from chance observation. A primitive woman may have noticed that grain-bearing plants grew up where grain had been spilled or stored. She then took the vital step of planting seeds, protecting the growing plants, and harvesting the crop.

Keeping animals probably began when primitive man tamed wounded or trapped animals, or when women saved and tamed young animals. Farming and animal husbandry developed together for a long period of time. The nomadic herding of livestock was a later development.

■ **ORIGINS.**—Agriculture originated first in the Middle East, probably in the grassy uplands where the wild grains and the wild animals first to be domesticated were found. By 5000 B.C., crops included wheat and barley, while sheep, goats, pigs, horses, and cattle had all been domesticated. The first farmers used tools of polished or chipped flint and obsidian.

Agriculture spread from the Middle East to the Danubian Basin, the shores of the Black Sea, the fertile crescent bordering the Arabian desert, and the valleys of the Indus River in India and the Hwang Ho in China. The crops, animals, and tools were much the same except in America, where agriculture probably was discovered independently.

Prehistoric man, drawing upon wild stock, developed all the major food plants and animals used today. Wheat and barley were domesticated in the Middle East. Rice and bananas were developed later in southeast Asia, and sorghum and millets in Africa. The New World saw the domestication of maize (corn) and potatoes. Food animals were domesticated first in the Middle East. Chickens were developed in southeast Asia, and turkeys domesticated in the New World.

■ **EARLY TOOLS.**—The first farming tool was a pointed stick, the digging stick. The food gatherers had used it to dig roots; the first farmers used it to dig holes for seeds. The spade was invented by the man who added a crossbar to his digging stick so that he could use his foot to drive it deeper into the soil. A stick with a sharp branch at one end was the first hoe. Later, a sharp stone or shell was tied to the branch to give it a more effective cutting edge. Sharp stones set along one edge converted a stick into a sickle.

ABOUT TEN PER CENT of the world's land is under cultivation today. Much of the growing area is north of the equator.

After animals were domesticated for food, they were soon trained to become beasts of burden. The next step, never taken by the American Indians, was to train animals to pull a heavy hoe through the earth—the beginning of plowing and cultivating by animal power. The Indians, perhaps because they had no animals for plowing, never adopted a *clean field* agriculture. Instead, they planted seeds in hills a few feet apart and cut the weeds only around the hills.

Nearly all prehistoric farmers in arid regions developed some method of irrigating their crops. Many also discovered the advantages of various types of fertilizer, from using farmyard manure to planting fish in hills of maize. They stored grain, seeds, and nuts for winter and also preserved meat and fish by salting and drying.

The discovery of metal and its uses, marking the end of the Neolithic period, enabled farmers to have sharper, stronger blades for hoes, plow points, and sickles. The change came slowly and in some areas, particularly in America, did not take place until Europeans introduced the new tools.

When agriculture first appeared in recorded history, it already was well developed—all the basic advances in domesticating major crops and animals had been made. The rules men followed were based upon longtime observations and trial and error. For centuries, few changes took place.

Advances in Agriculture.—The transformation in English agriculture which took place during the eighteenth century paved the way for modern farming. The key to the change was the enclosure of former open-field farms and the conversion of much arable land into pasture. The rights of villagers to use certain lands in common were largely revoked, and many small landowners and laborers were forced out of farming. At the same time, many improvements in farming were made. Improved methods of cultivation were adopted and machines were more widely used. New crops and more productive varieties were introduced, and controlled animal breeding became possible.

Not all the improvements originated in England. Clover was introduced in Spain, turnip cultivation in Flanders, and new grasses in France. The Rotherham plow, with a coulter and share made of iron, may have originated in Holland.

■ **AGRICULTURAL LEADERS.**—The adoption of advanced practices was encouraged by a number of English agricultural leaders. For 150 years, British farmers learned from such men as Jethro Tull (1674-1740), Charles Townshend (1674-1738), Robert Bakewell (1725-1795), Arthur Young (1741-1820), Sir John Sinclair (1745-1835), and Thomas Coke (1752-1842).

Tull invented a grain drill and advocated more intensive cultivation and the use of animal power. Townshend improved crop rotation and stressed the field cultivation of turnips and clover. Bakewell developed better breeds of livestock. Young and Sinclair were influential writers, and corresponded with American leaders such as George Washington and Thomas Jefferson. Coke developed a model agricultural estate, working particularly with wheat and sheep.

■ **THE NEW WORLD.**—British technological improvements in agriculture gradually spread through western Europe and to the United States. Members of such groups as the Philadelphia Society for Promoting Agriculture and the South Carolina Society for Promoting and Improving Agriculture, both founded in 1785, were familiar with the new English practices and followed some of them. Longhorn cattle, developed by Robert Blakewell, were imported into the United States in 1783, while Henry Clay imported Hereford cattle in 1817. The first agricultural journal appeared in 1810, and John Stuart Skinner began publishing the influential *American Farmer* in 1819. Such periodicals brought some of the English advances to the knowledge of a wider group of American farmers.

The most important technological advance in American agriculture dur-

WORK-SAVING INVENTIONS that raised farm production are the McCormick reaper (*left*), Whitney cotton gin (*right*), and Deere steel plow (*insert*).

ing this period, however, had no relation to English agricultural change. It was, instead, English demand for cotton that led Eli Whitney in 1793 to invent the cotton gin. This practical device for separating the seed from the lint of short-staple cotton revolutionized Southern agriculture. Production of cotton increased from an estimated 10,500 bales in 1793 to 4,486,000 bales in 1861.

Extensive commercial production of cotton, made possible by the cotton gin, led to the expansion of the plantation system, with its use of slave labor. It also led to dependence upon the single staple crop. Cotton cultivation led to the rapid settlement of the region and returned large sums of money to the planters. It also encouraged the economic development of the entire nation by providing large sums for use in foreign exchange.

■IMPROVEMENT OF PLOWS.—Technical ingenuity was not confined to cotton. Many Americans were attempting to build better plows, fundamental to an improved agriculture. In 1793, Thomas Jefferson developed a moldboard, made according to a mathematical plan, that would offer little soil resistance.

The first patent for a plow was issued to Charles Newbold of New Jersey in 1797. The plow, except for the handles and beam, was to be of solid cast iron. Farmers distrusted the new plow, believing that the iron poisoned the soil and made weeds grow.

The next great improvement in the plow was Jethro Wood's cast-iron model, first patented in 1814 but greatly improved in 1819. The moldboard, share, and landside were cast in three parts. The interchangeability of the parts was one of Wood's major contributions to the development of the modern plow.

The cast-iron plow was successful in New England and the Middle Atlantic states. But it would not scour in the prairies; the heavy, sticky soil would cling to the moldboard instead of sliding by and turning over. The steel plow was the answer to this problem. In 1833, John Lane, a blacksmith of Lockport, Ill., began covering moldboards with strips of saw steel. These plows succeeded in turning the prairie soil. In 1837, John Deere, a blacksmith of Grand Detour, Ill., began making a one-piece share and moldboard of saw steel. Deere became a successful manufacturer of steel and wrought-iron plows, which by 1860 had largely displaced the cast-iron plow in the prairies.

■DEVELOPMENT OF THE REAPER.—The mechanical reaper was probably the most significant single invention introduced into American farming between 1800 and 1860. It replaced much human power at a critical point in grain production where the work must be completed quickly to ensure saving the crop. By the American Revolution, the cradle had generally replaced the sickle and ordinary scythe for cutting grain. The cradle was a scythe with a light framework which gathered the stems and laid the grain down evenly.

Many inventors worked on animal-powered machines for harvesting grain. The first such machines in America efficient enough to find a market were patented by Obed Hussey in 1833 and Cyrus H. McCormick in 1834. In the struggle for business that followed, McCormick emerged dominant.

The Marsh harvester, patented in 1858, used a traveling apron to lift the cut grain into a receiving box where men riding on the machine bound it into bundles. Early in the 1870's, an automatic wire binder was perfected, but it was superseded by a twine binder late in the decade.

Other horse-drawn machines followed the improved plows and the grain reapers. Threshing machines were first brought from Scotland, then a practical thresher was patented in the United States in 1837. Other American patents were granted in the 1840's and 1850's for an improved grain drill, a mowing machine, a disk harrow, a corn planter, and a straddle row cultivator. At the same time, improved crop varieties were being introduced from abroad, and the commercial fertilizer industry was just beginning.

American Agricultural Revolution.—The Civil War led to the first American agricultural revolution. Farmers found that the demand for farm products was so great and the labor shortage so pronounced that it seemed both possible and profitable to adopt the new machines and techniques developed in the preceding decades. The establishment of the land-grant colleges to teach agriculture and mechanical arts and of the Department of Agriculture, the provisions for Western settlement in the Homestead Act, and the chartering of the Union Pacific Railroad, all in 1862, were also made immediately possible by the war. Together, these changes marked a transition from subsistence to commercial agriculture and from hand power to animal power.

The first state agricultural experiment station was established in Connecticut in 1875. Congress provided in 1887 for a yearly grant to each state for the support of an agricultural experiment station. These stations, the state colleges, and the Department of Agriculture together brought science to farming. As a result, farmers obtained improved or new varieties of plants and breeds of animals, learned how better to fertilize their crops, feed their animals, and control many plant and animal diseases and pests. This became more prevalent after 1914, when Congress, by the Smith-Lever Act, provided for a county agent in each agricultural county. The county agent, college-trained, carried scientific knowledge directly to the farmers.

The farm machinery widely adopted after the Civil War was mostly horse-drawn, while various devices were used to transmit horse-power to such stationary machines as threshers. A considerable number of steam engines were built for farm use. These usually were mounted on wheels and could be moved from place to place by horses. By 1900, self-propelled steam tractors were being sold for use in agriculture, but their weight made them unwieldy.

The internal-combustion engine, which had been invented in Europe, was a more practical answer to the search for mechanical power. By 1890, a number of American companies were manufacturing stationary engines, some of which were mounted on wheels. John Froelich of Iowa built the first gasoline tractor on record that was an operating success. In 1892 he mounted a gasoline engine on a running gear equipped with a traction arrangement of his own manufacture. The tractor completed a 50-day threshing run. The first company in the United States devoted exclusively to the manufacture of tractors was established in Iowa City, Iowa, about 1903.

The change from animal power to mechanical power came slowly. Farmers already had horses and could grow the feed necessary to maintain them. It took another crisis, World War II with its manpower shortage and seemingly unlimited demand for farm products, to give impetus to the change.

Advances in Technology.—Technological changes, in most instances, resulted from the application of scientific theories to practical problems. This was true of hybrid corn, one of the greatest agricultural innovations in modern times. The theories of two European scientists, Gregor Mendel and Charles Darwin, were applied to the problem. In 1865, Mendel discovered and announced the basic laws of inheritance of specific characteristics, but the importance of his work was not recognized until about 1900. Darwin pointed out in 1876 that inbreeding usually reduced plant vigor while crossbreeding restored it.

■**IMPROVEMENTS IN GRAINS.**—A number of American scientists built upon the work of Darwin and Mendel. Those who contributed directly to the development of hybrid corn included William James Beal, George Shull, Edward Murray East, H. K. Hayes, and Henry A. Wallace. In 1914, Donald F. Jones, who had been appointed to the Connecticut Agricultural Experiment Station, devoted himself to developing a method for making hybrid corn practical. Within three years, drawing upon the work of his predecessors, he developed the technique called double-crossing. It has been used by commercial firms since. The first seed company devoted to the commercial production of hybrid corn was organized in 1926. Mainly because of the use of hybrid seed, the per-acre yield of corn rose from an average 23 bushels in 1933 to 82 bushels in 1968.

The techniques used in breeding hybrid corn were successfully applied to grain sorghum in the 1950's and to semidwarf winter wheat for the Pacific Northwest in the 1960's. There have been gains, less spectacular, for other crops.

■**ADVANCES IN ANIMAL HUSBANDRY.**—Similar breeding techniques have been successfully applied to chickens, both as broilers and egg producers. Productive work has also been done with developing hybrid hogs. Crossbreeding has resulted in the development of new, useful breeds of beef cattle and sheep.

The dairy industry in the United States underwent a major change in the 1940's and 1950's. The changes have been due to several factors, such as the influence of markets, prices of milk and feeds, better knowledge of feeding, and the conquest of many pests and diseases. However, one of the most important has been the increased use of artificial insemination.

Artificial insemination as the best means of speeding up livestock improvement was demonstrated in Russia in the 1920's. The first dairy artificial-breeding cooperative in the United States was organized in New Jersey in 1938. It was found that instead of breeding only about 40 cows a year, artificial insemination made it possible for one bull to impregnate thousands of cows a year. Sires of proven ability to get high-producing offspring could be used to improve a dairy herd quickly. In 1943, the average American dairy cow produced 4,598 pounds of milk and 183 pounds of butterfat. In 1968, the figures were 9,000 and 330 pounds, respectively.

■**FERTILIZERS.**—The growth of the worldwide chemical fertilizer industry, able to supply farmers with the elements needed for more effective production at reasonable cost, has helped bring about major changes in agriculture in many nations. It has been an important factor in what might be called the second American agricultural revolution. The Department of Agriculture has estimated that the increased use of fertilizer was responsible for 55 per cent of the increase in productivity per crop-acre from 1940 to 1955.

■**INCREASES IN PRODUCTIVITY.**—The unparalleled demand for farm products, a doubling of prices, and the shortage of manpower during World War II were basically responsible for the tremendous upsurge in production. The specific changes making up this technological revolution included the displacement of animal power by mechanical power, widespread progress in mechanization, greater use of lime and fertilizer, irrigation, adoption of conservation practices, use of improved varieties, better balanced feeding of livestock and poultry, and more effective control of insects and disease. The widespread use of agricultural chemicals has led some to label the contemporary period in agriculture the *age of chemurgy*.

The major key to progress in farming, with increases in productivity per acre and man-hour, was that technological innovations were adopted in combination. The farmer became a skilled manager and businessman, constantly seeking to find the most efficient and profitable combinations of technology. His success can be measured by the fact that in 1860 the American farmer produced enough food and fiber for 4.5 people; in 1940, for 10.7 people; and in 1968, for 45 people. In the 100 years from 1860 to 1960, the proportion of America's working population engaged in agriculture declined from 58 per cent to 8.3 per cent, and by 1968 to 5.1 per cent, but that percentage was providing abundantly for all Americans and millions of people overseas.

—Wayne D. Rasmussen

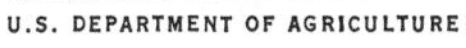
U.S. DEPARTMENT OF AGRICULTURE

HENRY AGARD WALLACE (*left*) developed hybrid corn and, as secretary of agriculture, aided other advancements in farming methods. Through research into new products and processes, George W. Carver (*right*) made contributions to the field of scientific agriculture.

FOOD AND AGRICULTURE GLOSSARY

Agricultural Chemistry, the application of chemistry to the efficient production and utilization of plants and animals of economic importance. While the definition refers mainly to farm crops and livestock, it may be extended to include forest and marine products, and indeed, to anything else that grows and is useful to mankind.

Together with genetics, improved farm equipment, and improved cultural practices, chemistry has played an important part in advancing modern agriculture to the status of big business. Various agencies of the Federal government are engaged in fundamental and applied chemical research, and in analytical and regulatory activities, that are of service to producers, handlers, processors, and consumers of agricultural commodities. Likewise engaged in this field are various state departments of agriculture, universities, and agricultural experiment stations; private institutions that conduct or foster chemical research and development; industrial processors of agricultural products; and manufacturers of agricultural chemicals.

Among the fields of application of agricultural chemistry are crop production; livestock and poultry production; storage, transportation, and marketing; food utilization; and industrial utilization of by-products.

—S. B. Detwiler, Jr.

Agricultural Engineering, the application of engineering principles to agricultural problems. The divisions are power and machinery, structures, electric power, processing, and soil and water.

The strong position of the United States in today's world is due to a prospering mechanized agriculture—an agriculture that produces food and fiber in abundance with only about 6 per cent of the gainfully employed labor force. The millions of workers thus released to other industries, services, and professions contribute to the remarkable industrial expansion and high standard of living.

Crops once thought next to impossible to be harvested mechanically have been mechanized recently. Typical examples are sugar beets, cotton, tomatoes, figs, peaches, prunes, and nuts. Research will extend mechanization to many other crops.

Production structures contribute to increased efficiency. A well-designed milking parlor permits one machine milker to handle up to 60 cows. A crew of 10 can feed 20,000 beef animals in a mechanized feed lot. Dehydrators and fruit and vegetable packing houses prepare products for final consumption.

Fully 95 per cent of the farms in the United States are electrified. In addition to the general improvement in living conditions, electricity operates much of the farmstead equipment. Pumping and heating water, brooding chickens and pigs, milking, refrigerating, and feed processing are now done electrically.

Other engineering contributions are in irrigation and drainage and soil conservation. The western third of the United States depends upon an irrigated agriculture. Drainage is as important in irrigated areas as in regions with abundant rainfall. Soil erosion, a menace to the land, can be controlled to a large extent by agricultural engineers.

—Roy Bainer

Agricultural Science, an interdisciplinary science that provides information on biological, physical, and socio-economic problems in the production, marketing, processing, and use of crops and livestock for food and clothing. Generally, it is divided into several separate but related sciences, each with its own professional groups. These include: plant sciences, animal sciences, engineering sciences, soil sciences, entomology, parasitology, and social and behavioral sciences.

Many persons who are agricultural scientists are also biologists, chemists, physicists, mathematicians, engineers, or economists. Agricultural science consists of portions of these and other basic sciences applied to crop and livestock production, marketing, and use.

Agricultural science includes *molecular biology,* the study of how viruses, bacteria, and molds change virulence due to changes in RNA or DNA, the chemical bases of living materials. Such changes must be understood in order to combat these organisms. It also includes *radio biology,* the use of isotopes to follow the physiology of nutrient assimilation, and of radiation to induce genetic change or insect sterility.

Systems engineering and *electronic data processing* have become a part of agricultural science. The combination of soil moisture, fertility, probability of rainfall, hours of sunshine, and known plant characteristics may be used to determine the best time to plant peas in Wisconsin or the amount, formula, and time of fertilizer application for Indiana corn. Automated equipment places the feed formulated to provide all nutrient requirements at least cost before Delaware broilers. The least-cost formulated feed may have been selected by means of linear programming, a method also used by economists to determine the best use of resources by an individual farmer to increase his income.

EDUCATION AND RESEARCH.—Agricultural science results from research in the laboratories of national governments, of universities, and of private institutions in many countries. In the United States, about 5,000 agricultural research scientists in the U.S. Department of Agriculture conduct research in many locations; about 10,000 do research in state agricultural experiment stations in the 50 states and Puerto Rico; about 10,000 do research in universities and col-

U.S. DEPARTMENT OF AGRICULTURE

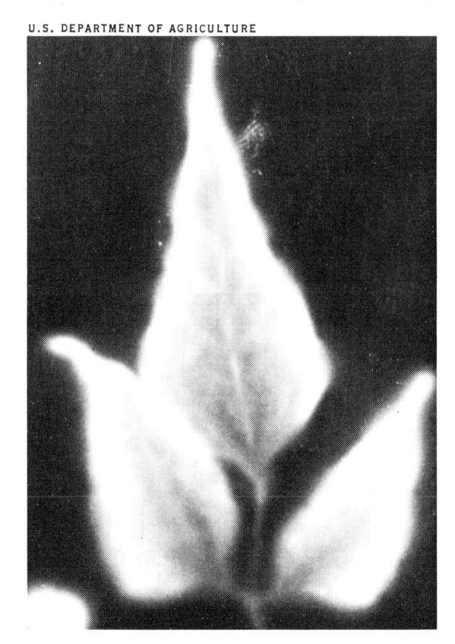

E. I. DU PONT DE NEMOURS CO.

RESEARCH brings agricultural knowledge as a scientist (*above*) studies chemicals' effects on plant development. At *left,* radioactivity records and controls plant growth.

SEAGRAM AND SONS, INC.

A SAMPLE of whiskey is drawn for quality checking by a tester using a "whiskey thief," which reaches to the center of the barrel.

JOSEPH SCHLITZ BREWING COMPANY

MALTING, initial step in brewing beer, begins with "steeping." *Above,* oxygen bubbles through barley and water in "steep tanks."

leges and in private research institutions, including the laboratories of many manufacturers of chemicals, food products, and textiles.

Many more agricultural scientists are engaged in teaching, extension service, regulatory work, and production. They take the technical information developed in research to the farms and factories where it is put to use. This has resulted in yields, per acre and per animal unit, double those of a hundred years ago. It has made fresh meat, milk, and eggs available the year round.

■PEST CONTROL.—*Entomologists, plant pathologists,* and *weed scientists* have developed means of controlling many insect pests, diseases, and weeds that limit the productivity of crop plants, trees, grasses, flowers, and shrubs. The wheat crop depends on the continued development of wheat stocks resistant to old virulent rusts and new virulent ones as they occur. Weeds are controlled on 75,000,000 acres in the United States by effective chemicals. Used according to prescription, these chemicals are innocuous to man and animals. Dramatic control of an animal pest, the screwworm fly, has been achieved by the production and distribution of radiation-sterilized males. Development of power machinery for planting, cultivation, and harvest has led to the displacement of the horse as a source of power, greatly reducing the amount of manual labor required for crop and forest production and livestock care.

Research in animal nutrition has led to the formulation of efficient diets that have cut in half the feed cost of producing broilers. *Nutrition research* has contributed to the improved health and production efficiency of all our livestock.

Research in animal disease has developed diagnostic tests and immunization methods that have made possible the control of such diseases as *pullorum* in poultry, *brucellosis* and *tuberculosis* in cattle, *cholera* in swine. Research has developed effective parasiticides that protect our livestock against protozoan and worm parasites. —T. C. Byerly

Alcoholic Beverages, potable liquids containing ethyl alcohol produced by enzyme action on sugars. There are three kinds of drink containing alcohol: wine, spirits, and beer. *Wine* is fermented from grapes or other berries and plants; *spirits* are distilled; *beer* is malted. Wine was first made by the ancient Egyptians. Grapes were later carried by the Romans to France, Spain, parts of Germany, and even to Britain. More recently, grapes have been grown in the United States, South Africa, and Australia. The cereals used to make spirits, except those made from wine, have spread over the world with the Caucasian population. Beer, known in the ancient East, was also made by the ancient inhabitants of northern Europe.

■WINE.—Grapes are used to make wine in commercial quantities. Other materials are generally used for home winemaking only. The grapes are crushed to express the juice, which ferments naturally because of the presence of skin fungi on the fruit, producing a chemical reaction. Two broad differences result from the color of the fruit—white and red grapes each produce wine of their own color with many intermediate and varying shades. The flavor varies from dry—that is, rather sour—to very sweet. The alcoholic content ranges between 9 and 18 per cent by volume and up to 25 per cent in fortified wines.

Champagne is a sparkling or bubbling wine, the sparkle being caused by bottling before fermentation is finished. Like many other wines, it is named after the place where it is made: Champagne, Marne, France. *Asti spumante* is a sparkling wine made in Italy. Other French wines are *claret* from Bordeaux and burgundy from the province of the same name. Some light white wines are also made in France, but the German white wines, *moselle* and *rhenish,* from the grapes grown along the banks of these rivers, are perhaps the best known of this kind. *Tokay* is a famous Hungarian wine. *Chianti* and *lacrima christi* are well-known Italian products. Sherry comes from Spain, and *port* from Oporto, Portugal. The last is a fortified wine, that is, one to which brandy or alcohol is added. There are many other fortified wines. Wines of all these types are now made in the United States, principally in California and New York; in Australia, and in South Africa. Light white wines of the German type are now exported from Yugoslavia, as are Greek, Cyprian, and other wines formerly consumed locally.

■SPIRITS.—Brandy, whisky, rum, gin, and vodka are the best known spirits. Besides these, there is a variety of liqueurs, usually consumed after a meal. These drinks have a higher alcoholic content than wine, 25 to 50 per cent or more. Spirits are made by distillation; brandy from fermented grape juice, and the others from grain or potatoes.

Brandy, said to be first known in the East, was carried to Italy in the thirteenth century and known as *aqua di vita.* After this, it was made in the Netherlands, where it was known as *brandewijn,* the origin of its modern name. *Cognac* and *Armagnac* are made in those parts of France from which they take their names. Distilled drinks are made from the juice of other fruits, such as apples and cherries.

Whisky (or whiskey) originated in Scotland as *usquebaugh.* It has a high alcoholic content, ranging from about 40 to occasionally 75 per cent. It is made from cereals (rye, barley, maize, etc.) or potatoes. The grain is boiled to mash; then malt is added. Malt is grain sprouted in water, a process that changes the starch content to sugar. When the two are mixed, a chemical change from sugar to alcohol takes place, and this is separated by distilling.

Rum, which originated in the West Indies, is distilled from the fermented juice of sugarcane or from fermented molasses. All rums are colorless when distilled, but acquire color from the casks in which they are aged or from the addition of a coloring agent, such as caramel.

Gin is also distilled from cereals, chiefly barley. Various flavorings are added, the most favored being juniper oil. Gin contains from 25 to 50 per cent alcohol. It is the easiest spirit to fake, and substitutes (the renowned bathtub gin) are often sold.

Vodka is a Russian drink that is now being exported and becoming popular in western Europe and the United States. The materials used are roughly the same as those for other spirits, but the process and mixing are rather different.

■BEER, ALE, AND STOUT.—These beverages are brewed, using barley or other cereal grain for malting, and then boiled with hops, after which yeast is added to the liquor to cause it to ferment. *Beer* is sometimes made from maize, and there is a story that Columbus was given maize beer by American Indians. Maize is also used by the natives of South Africa for making beer. *Lager* beer, a light beer, introduced to the United States by German immigrants in the nineteenth century, is the most popular type. A *dark lager* is made in southern Germany. *Ale* made in England is stronger and more bitter than lager. *Porter* is dark ale, and *stout* is a stronger ale. The color of the liquor is said to depend upon the tint of the barley used, though varied grain is often mixed before malting.

—G. E. Fussell

Alfalfa, a perennial *legume* sometimes referred to as the "queen of forage crops." It is the most nutritious of the commonly-grown hay crops, and richer in protein, minerals, and vitamins than either clover or timothy. It is utilized as pasture, hay, or meal.

Alfalfa is a particularly hardy plant, being drought-resistant as well as resistant to extreme heat and cold. It grows best in deep loam soils with porous subsoils that contain lime. If lime is not present in the soil, as is the case in most areas east of the Mississippi River, it must be added.

Alfalfa grows 2 to 3 feet high, with trifoliate leaves, purple or yellow flowers, and a long tap root that may extend 25 or 30 feet into the ground. It is subject to attack by *bacterial wilt, spotted alfalfa aphids,* and *alfalfa weevil.*

The center of production in the United States is the Middle West, with California, South Dakota, and Nebraska excelling in seed production.

—Robert G. Dunbar

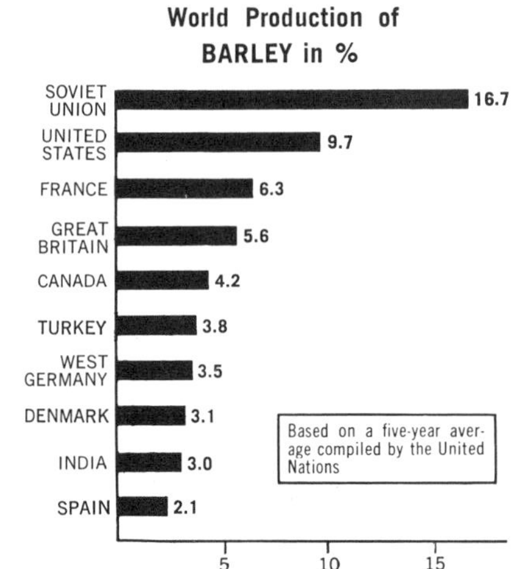

Barley, first brought into cultivation by prehistoric man, one of the two *cereals* that supplied the food necessary to allow civilization to develop, the other being wheat. There are two kinds of barley, two-row and six-row. Another, *Bere,* is called four-row, but its appearance is misleading. Today, barley is grown all over the world in subtropical and temperate climates. It will flourish in high altitudes, in poor soils, and in areas of low rainfall. Owing to its chemical composition, it does not make a satisfactory loaf, but can be made into a hard bread, which English consumers used to call *Barley bangers.*

The principal modern use of this cereal is malting for brewing. Pot and pearl barley are prepared for use in cooking, thickening soups, etc. The protein dissolves in water, and the solution (barley water) is often used in infant and convalescent diets. Barley grain and straw are used for feeding livestock, especially pigs.

There are several different strains useful for growing in particular places, some of which have been bred from a single ear. A good deal of work on breeding strains suitable for cultivation in the widely varied conditions of the United States has been done by the Bureau of Plant Breeding, Soils, and Agricultural Engineering.

—G. E. Fussell

Beekeeping or **apiculture,** the industry and hobby of raising bees and collecting the honey and wax they produce. Just how long man has kept bees is not known, but a cave painting in Spain, estimated to be at least 8,000 years old, shows a man robbing a bee cave. Not until the twentieth century, however, did man learn to control *swarming,* the bees' natural method of increase. Swarm prevention helps him harvest more honey from the bees.

When swarming occurs, the queen, a few drones, and 5,000 to 40,000 worker bees leave their home in a mass flight. They establish a new home in a hollow tree, a cave, the wall cavity of a building some distance away, or a man-made beehive.

The first step in the preparation of the new home is wax construction. Hundreds of worker bees gorge themselves with nectar from flowers, then wax glands on the underside of their abdomens convert the nectar into *white wax.* This is the structural material of the cells of the *comb,* which constitutes the framework of the hive.

Honey is also prepared from nectar. Ripe honey, which is sealed in the wax cells, consists largely of a mixture of sugars. The honey serves as food for the bees.

Other cells are filled with pollen, or *beebread,* which is also collected from flowers and carried in pellets packed ingeniously by the bee on its hind legs.

Propolis, a resinous material on buds and scars of trees, is also collected and carried like pollen on the bees' hind legs into the hive. It is used to seal cracks and holes too small for the bee to pass through, and to keep out rain and wind.

Within the hive, the queen lays about a thousand eggs a day, one to a cell. After three days, the eggs hatch into *larvae,* after five days of feeding, these are sealed in their cells to pupate, or change into adult form. The entire process takes about three weeks.

Eventually, the queen ages sufficiently so that she no longer can continue her egg-laying at the required pace. The worker bees then select a few tiny larvae, enlarge their cells, and feed them a special glandular food called *royal jelly.* These larvae, which would have become worker bees after their pupal stage, instead develop into virgin queens. The first one to emerge, however, seeks out and kills the other virgins before they complete their pupation.

When the surviving virgin is about ten days old, she goes on her mating

U.S. DEPARTMENT OF AGRICULTURE

BEES SWARM by the thousands on a branch, before going to a new hive.

flight, and in the air mates with one to six drones, who exist primarily for this duty. She returns home to begin her lifetime task of egg-laying, never to leave the hive again except with a swarm. The now-ignored mother queen may remain several weeks before she finally disappears.

During all this time, the bees have been storing honey and pollen for winter use. The amount stored depends on the flowers in the area, the weather, and many other factors, but there should be at least 50 pounds of honey and several pounds of pollen.

With approaching winter, the flowers disappear, the drones die, and the queen reduces her egg-laying or ceases altogether. When the temperature goes below about 57° F, the bees form a tight cluster to conserve heat, maintaining a cluster temperature of about 92° F, even when outside temperature falls below zero.

This is the natural way of increase for bees, but it may not always be the best way from the beekeeper's point of view. The swarm may escape. It may leave at the beginning of a good period for storing honey, and before the colony is strong with bees again. Strong colonies are essential if much honey is to be stored.

The beekeeper has found that bees can be kept in multiunit hives with movable frames. These can be taken apart for examination, and if the bees are crowded, more room can be provided. If more colonies are desired, the hive can be divided after the main honey crop has been stored. The part of the hive that lacks a queen will promptly rear another.

When frames containing sheets of wax with embedded cell bases are placed in the hive, the bees will construct straight combs in them. When these combs are filled with honey, they can be cut out, or the honey can be centrifuged from the cells and the empty comb replaced in the hive to be filled again. The average amount of honey to be expected from a colony is about 50 pounds, although trained beekeepers often harvest many times that amount.

Persons unfamiliar with bees should always wear a veil or screen over head and face. When one is working with bees, smoke should be blown into the entrance hole and over the frames as they are exposed. Although the reason is unknown, smoke tends to keep bees away from the worker and discourage stinging.

With movable-frame hives, smoker, and veil, the beekeeper can regulate the colony so it will produce the maximum amount of honey for him, and he can increase the number of colonies as he desires.

—Samuel E. McGregor

Breeding, Animal, the improvement of the quality of domestic animals through breeding, observation, and selection. Little effective progress was made until the rediscovery (1900) of Mendel's laws of heredity and their application in animal breeding. Knowledge of genetics has rapidly increased since then. With this has improved our knowledge of *cytology* (the study of chromosome behavior), statistics, and interdisciplinary science, providing tools for describing populations and developing experimental methods and chemistry. Applied to the fundamental laws of genetics, such knowledge provides the basis for today's science of animal breeding.

PIONEER HI-BREED CORN COMPANY

COMPUTER REPORTS help cattle raisers to select the best animals for breeding purposes.

■ **SELECTION.**—In animal breeding, the basic problem is to identify those animals that have desirable characteristics and to use these in producing superior animals. The application of statistical theory to the fundamental laws of heredity has resulted in the development of techniques to establish the best procedures for selecting animals having the greatest genetic value for specific objectives. These procedures provide bases for evaluating the probability that differences observed will be repeated.

■ **GENETIC IMPORTANCE.**—The expression of most traits of economic importance is determined by the combined efforts of *genotypic* (hereditary determiners) and environmental effects. Therefore, the degree to which the expression of a trait is determined by genetic versus environmental factors must be considered in deciding how to design the most efficient animal-breeding programs. Furthermore, procedures must be devised to estimate the genetic value of groups selected for reproduction.

■ **MATING SYSTEM.**—Individuals selected in an animal-breeding program can be mated together in many different ways to reproduce a population. Thus, *mating system* refers to any set procedure used to mate selected individuals to reproduce a population. Where relationships have been established within a population, matings based upon these relationships may be carried out. Examples of mating systems based upon matings between individuals more closely related than the average of a population are *full-sib* or *half-sib* matings. Mating systems between individuals that appear more like each other or less like each other than the average of the population, ignoring relationships, are called *assortative* and *disassortative mating systems*, respectively.

■ **ARTIFICIAL INSEMINATION.**—The practice of placing sperm cells in the female genitalia of animals by instruments rather than by natural service is known as artificial insemination. This method has been used widely in dairy breeding for animal improvement. Here it is possible to collect large amounts of semen from proven sires and to use this semen in producing calves from a very large number of females. Where rare bulls can be naturally mated to only about 100 cows a year, use of artificial insemination makes it possible for individual bulls to sire more than 10,000 calves a year. Management conditions have made this practice especially worthwhile for dairy cattle in the United States. In some situations this technique is worth considering for beef cattle, sheep, swine, horses, poultry, and fur-bearing animals. It may, someday, also be considered important from a eugenic standpoint for application to human populations.

■ **PROBLEMS.**—It should be realized that there are problems in the application of breeding procedures in animal populations that are not apparent. Where breeding procedures are used to improve animals of economic significance, all groups (germ-plasm sources) which *may* have, but do not now have, significance must be reproduced each generation. If this is not done, sources that may be invaluable for adaptation to conditions of the future may be lost. Artificial breeding, specifically, storage of sperm cells and ova, may in the future provide cheaper ways of overcoming this problem.

—Laurence Baker

U.S. DEPARTMENT OF AGRICULTURE

A HONEY BEE leaves a cotton flower coated with the plant's pollen. Flying to other plants, the bee cross-pollinates the cotton.

Breeding, Plant, the improvement of the quality of crop plants through reproduction, observation, and selection. Little progress was made until the rediscovery (1900) of Mendel's laws of heredity and their application in plant breeding. Since then, knowledge of genetics and related sciences has increased rapidly.

CROSS-POLLINATING SPECIES.—The most extensive plant-breeding efforts to cross-pollinate crops of economic significance have been applied to corn, or maize. The basic procedures that have been applied to corn include the application of various mating systems, such as *self-pollination,* the development of inbred lines from open pollinated varieties; mass selection in improving germ-plasm sources prior to inbreeding; and the production and evaluation of test crosses.

Due to differences in response to different environmental conditions, corn must be developed for each area that differs in length of day. Due to the increased yield obtained by crossing different sources of material, hybrid corn provides the main source of improved seed corn available today. With few exceptions, the primary objective in corn breeding has been to develop new crosses (hybrids) that yield more grain. Though disease and insect resistance is important and sometimes constitutes a problem, it has usually been provided through selection among indigenous sources of material.

SELF-POLLINATING SPECIES.—Until recently, breeding techniques for improving self-pollinating species were limited. The main sources of variability for traits requiring improvement were exotic materials or crosses of related species with existing species. Most self-fertilizing crops are not completely self-fertilized; therefore, limited crossing is possible to provide sources of new material.

Self-pollinating crops that received attention were bread wheats, durum wheat, barley, rice, oats, and grain sorghum. Until recently, the potential value of first-generation hybrids of self-fertilized cereals was given little consideration. However, evidence now exists that increases of 30 per cent in yield may be possible through the development of such crosses. Other advantages, such as increased ease in improving quality, could be more important than increases in yield. A major obstacle still to be overcome in all small grains except sorghum is that cross-fertilization under field conditions must be possible before crosses can be produced in seed for wheat, rice, and barley. The problems are solvable, and first-generation hybrids of many more normally self-fertilized cereals will be available. Increased emphasis is being placed on breeding for nutritive quality in economically important feed grains.

—Laurence Baker

Cacao, the dried and fermented seeds of a green-leafed evergreen tree used in making cocoa and chocolate. Cacao, or cocoa, beans *(Theobroma cacao)* grow along the trunks and some branches of the cacao tree. The tree thrives in hot, rainy climates and is generally cultivated in lands 20 degrees north to 20 degrees south of the equator, where temperatures average between 65° and 75° F.

The tree is grown in shady areas, sometimes as an intercrop sheltered by leguminous trees or, as in some areas in Africa, by banana or rubber trees. In its early growth, the cacao tree is sometimes protected from winds by windbreaks. Most strains bear fruit from 3 to 5 years and produce for an average of 40 years, although some varieties have lived 200 years. The cacao tree bears its fruit, or pods, throughout the year, but the harvest is seasonal. The pods are cut by hand from the tree, and are then split. A machette wielder can open up to 500 pods an hour.

Of the many varieties of cacao, the main classifications are the *Criollo,* which produces a light-colored, thin-skinned bean used for fine chocolate; the *Forastero,* which is easier to cultivate on varied soils; and the *Trinitario,* a cross of the other two types. On plantations in the Western Hemisphere, it is rare to find only a single species.

COCOA AND CHOCOLATE.—The most important products of the cacao bean are chocolate and cocoa. Before processing, the beans are weighed and blended, then roasted for one to two hours in large rotary cylinders. The beans turn a rich, brown shade and lose about 20 per cent of their weight. The remaining cracked *nibs* are conveyed to the mills, where they are crushed between large, heated steel discs. The remaining viscous mass is called *chocolate liquor.* This liquor in a solidified state is used for cooking and is familiar as *bitter chocolate.* About 53 per cent of the liquor is the practical vegetable fat, *cocoa butter.* It lasts for years and melts at low temperatures, but remains solid at room temperatures. The 47 per cent residual is cocoa powder, used in chocolate-flavored foods and beverages. In manufacturing confectioneries, cocoa butter is added as an enriching agent. Breakfast cocoa contains 22 per cent fat; fine-quality chocolate requires a higher proportion of butter.

Cocoa, which is native to the American continent, was grown and used by the Aztecs, Incas, Mayans, and Toltecs. Spanish explorers popularized it in Europe, and by the 1600s cocoa had spread throughout African colonial possessions. In about 1828, a Dutchman, C. T. van Houten, developed a rich, palatable chocolate powder by removing some cocoa butter and adding an alkali to the mixture. About 50 years later, a Swiss invented milk chocolate.

Ghana, the largest producer of cocoa, exports over one-third of the world total. About one-quarter to one-third of all U.S. cocoa imports in the mid-1960s were from Ghana. The United States consumes nearly 30 per cent of all the raw cocoa produced. The Common Market countries of western Europe import about one-third of the world total.

In the early 1960s, effective disease control programs in cocoa-producing countries led to a sharp expansion in world production, and cocoa prices reached low levels by 1965. Low prices and higher incomes stimulated consumption, and a price recovery occurred after 1965. World consumption of cocoa remains high, however, and from 1966 to 1968 exceeded production. Liberalization of trade controls in some European countries and relaxation of internal taxes on cocoa contributed to increased demand. Stock excesses, common in the mid-1960s, were absorbed by the late 1960s, and supply and consumption of cocoa were nearly balanced in 1969.

—Marshall H. Cohen

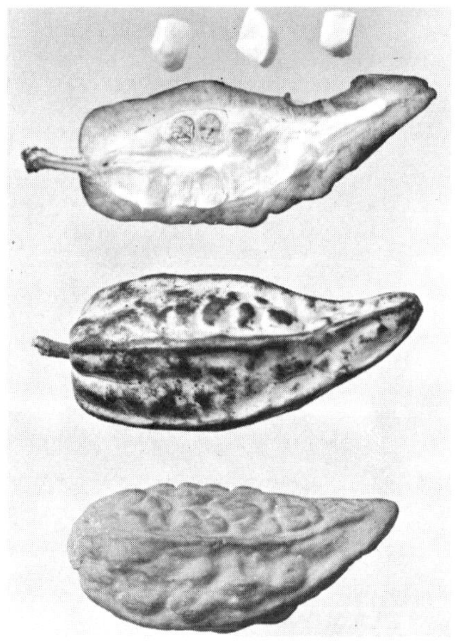

U.S. DEPARTMENT OF AGRICULTURE

CACAO was known to ancient Mayans, as the fossil below two modern beans shows.

Camel, a large herbivorous mammal used for riding, transport, meat, hides, and hair. The *Bactrian,* a two-humped camel, has heavy hair in the winter, hard foot pads, and short legs which fit it for cold and mountainous country. It is found in central Asia from the Black Sea to Manchuria, but cannot stand prolonged spells of great heat. The *Dromedary* has one hump, short hair, a long neck, long legs, and foot pads that hold the animal up in sand. It travels best on level sandy land and cannot move in mud. Dromedaries are found in North Africa, Egypt, Arabia, and India. They have also been successfully introduced into Spain and the Canary Islands, but unsuccessfully into Australia and the Americas.

Camels can metabolize the fat in their hump to get some water, but store most of their emergency water supply in one of their stomachs. They can go up to a week without water if they move slowly. Thick skin inside the mouth allows camels to eat nearly anything, including thorns, although they prefer brush, tree leaves, and fruit. Camels mate at any season and have a gestation period of about 12 months. The female usually produces one foal every other year. The working life of a camel is around 20 years. Camels are susceptible to *rabies, anthrax, tuberculosis, tetanus,* and *foot-and-mouth disease.* All these diseases can be prevented with vaccines. In addition, camels are particularly susceptible to the incurable and frequently fatal *trypanosomiasis* (sleeping sickness).

Camels are ruminants of the genus *Camelus,* which has two species, *C. Bactrianus* and *C. Dromedarius.*

—John T. Schlebecker

Casein, a heterogeneous compound protein derived from amino acids and containing primarily calcium, nitrogen, and phosphorus. Although the lack of uniformity limits its industrial applications, the casein compound is used in a wide variety of processes ranging from cheese-making to plastic manufacture. Casein constitutes the bulk of the curd of milk, and makes up about 3.15 per cent of the whole fluid cow's milk. It can be transformed into cheese, or *paracasein,* by the action of the enzyme *rennin.* For industrial uses, skim milk is treated with sulfuric acid, hydrochloric acid, lactic acid, and the casein is then washed, dried, and pressed. Dried casein is essentially a granulated jelly. It can be used to make a water-resistant wood glue, or a binder for various oil and latex paints. Lacquers sometimes also use a casein binder. Formerly large amounts of casein were used as a binder of paper coatings, but recently the use of casein for high-gloss paper has been declining in the United States. Casein can be hardened with formaldehyde to make plastics, but lately other chemicals have largely replaced casein. Since 1957, United States production of industrial casein has virtually ceased, although casein has been imported, chiefly from Argentina.

—John T. Schlebecker

AMERICAN ANGUS ASSOCIATION

BEEF CATTLE, such as the Angus breed *(above)* are, like dairy cattle, grazers. Angus cattle are highly adaptable to various climatic conditions and they produce high quality beef. They are popular especially in the United States, Argentina, and in Scotland.

Cattle, herbivorous, horned, bovine mammals. The male is called a *bull,* the female a *cow,* a young cow a *heifer,* a castrated male a *steer,* and a young animal a *calf.* In the United States, cattle are raised for milk, meat, and hides. Cattle furnish the bulk of dairy products in the United States, as well as cattle meat, called *beef* or *veal.* Beef, the most popular of all meats in the United States, can be readily preserved, although most of it is sold as red (fresh) meat. Livestock and livestock products rank third as a source of farm income in the United States.

Most cattle are now sold at country selling points; central markets, such as Chicago and Omaha, handle only about 30 per cent of the total marketed. Beef slaughtering and processing have also been decentralized. Chain stores have been directly buying and slaughtering large percentages of cattle. Dairy processing has concentrated in fewer companies, both private and cooperative.

In the United States, the chief beef cattle have derived from British and Indian breeds, such as the English *Devon, Hereford, Red Poll, Shorthorn,* and *Sussex,* and the Scottish *Aberdeen Angus, Highland,* and *Galloway.* The Indian cattle include several *Zebu* breeds which Americans lump together under *Brahman.* Zebus have also been crossed with other cattle to produce various new fixed breeds, including the *Santa Gertrudis, Beefmaster,* and *Brangus.* The French *Charolaise,* and crosses from it, are becoming more important as beef cattle in the United States. The several breeds are valued for their adaptability to climatic and regional differences within the United States.

The dairy breeds of the United States all originated in Europe. The *Guernsey* and *Jersey* came from the Channel Islands, and the most popular, the *Holstein-Friesian,* came from the Netherlands. Dual-purpose cattle provide both beef and milk in commercial quantities. The *Brown Swiss,* the Scottish *Ayrshire,* the *Holsteins,* and the *Milking Shorthorns* are often considered to be dual-purpose.

Cattle are grass grazers, but high milk and meat yields are achieved by feeding silage. Dairymen and beef raisers increasingly supplement feeds with minerals, vitamins, and antibiotics. Antibiotics, however, should not be used as a feed supplement for lactating dairy cows for the drugs may enter the milk. Cows reach maturity in a year and can be bred about every three weeks thereafter, although breeding should not be permitted before the cow is 18 to 24 months old. The period of gestation averages 283 days, and the cow produces one calf, or infrequently two. The cow comes in heat 30 to 60 days after calving.

Some fatal cattle diseases, such as *anthrax* and *blackleg,* can be prevented by vaccination. *Tick fever,* once highly virulent in some places, has been controlled by dipping or spraying to kill the ticks. Dips and sprays have also been effective against arthropod-borne *anaplasmosis.* Vaccines can control *contagious abortion* (brucellosis) and *tuberculosis,* although the slight incidence of tuberculosis in the United States makes slaughter of diseased animals more practical. *Pneumonia* and *shipping fever,* although very common, can be cured with antibiotics. Dairy cattle particularly suffer from *mastitis,* which antibiotics can cure. Cattle succumb to a large variety of poisonous plants, for which the only prevention is poison-free pastures.

Cattle are ruminants of the genus *Bos,* with two living species, *B. longifrons* and *B. indicus.*

—John T. Schlebecker

AMERICAN JERSEY CATTLE CLUB

DAIRY CATTLE BREEDS originated in Europe and were brought to the U.S. by early settlers. The Jersey breed (*above*) gives the richest milk but the quantity produced is only half that of the Holstein, the first-ranking milk producer in America.

Chicken, an edible bird that is probably the most widely domesticated fowl in the world. Almost all *broilers,* or young meat chickens, have white plumage; they are crossbreds, produced from matings of special meat males and females. Usually one breeder produces the male line and another breeder, the female line. All chicks are produced in large incubators, some of which can incubate more than 50,000 eggs at one time. Many chicken producers have an integrated system, controlling the entire process from producing the hatching eggs to marketing the ready-to-cook broilers, TV dinner, or chicken pie. The margin of profit per bird is very small, but integration and large volume make it possible for the producer to market a product of excellent quality at a low cost to the consumer.

Poultry processing and marketing have benefited from technological development. Live birds are placed on a moving conveyor at one end of a plant, and when they reach the other end the birds are in ready-to-cook form. In a short period of time, the poultry is slaughtered, picked, eviscerated, washed, and chilled. Most of the operations are completely automatic and it is not unusual for plants to process 40,000 to 50,000 birds per day. Poultry is offered to today's consumer at an attractive price and in a variety of forms—chilled, frozen, canned, parts, boneless, cooked, and in combinations with other ingredients. (*See also,* Egg Production, page 311.) —Carl W. Hess

Chicory, an annual that grows from seed planted in the spring and best known for the use of its root as a coffee substitute or supplement. Although often regarded as a roadside weed in the eastern United States and Canada, chicory has long been cultivated in America. The base of the chicory plant resembles the dandelion, but the stalk is longer, sometimes rising to a height of 5 feet, and its narrow, flat blossom is blue. Its spreading branches develop coarse-toothed and lobed leaves. These are valued as greens, used raw in salads or boiled. But the plant's long, fleshy root constitutes its most prized portion. Roasted and ground, it gives body, color, and long-lasting flavor to coffee. The plant originated in Europe and spread to other continents. It has been cultivated in the United States since the latter part of the nineteenth century. —Charles E. Rogers

Chocolate. See *Cacao.*

Clover, an important group of annual and perennial plants of the *pea family* having trifoliate leaves and dense flower heads. There are several species, some of which can be recognized by the color of their flowers—red, crimson, or white. All are used as forage crops. Growing clover for forage brought about great progress in farming: livestock were fed better, especially in winter, and became more productive; the extra manure they produced was used to fertilize the land, increasing grain yield; clover root bacteria add nitrogen to the soil, an aid to fertility. This crop is very ancient; its uses were known in southern Europe in Roman times. However, it was neglected until the sixteenth century, when it began to be grown again in Spain. From there it was carried to Holland, and then to Germany and England about 1650. European settlers took the crop to the United States, where it became a grazing and hay crop. —G. E. Fussell

Cocoa. See *Cacao.*

Coconut Palm, a tall tree with feather-like leaves, whose fruit—the coconut—has made it nature's greatest provider to mankind in the tropics. Second only to the grasses among plants useful to man, it thrives in low-lying areas near the coast, 20° to 25° from the equator. The stem of the coconut palm rises to a height of 60 to 100 feet, with leaves growing only at its upper extremity. Within the graceful crown of leaves, the fruit ripens. A single tree may produce as many as 100 coconuts. Inside the outer husk of the fruit is found the familiar hard-shelled *nut.* It contains firm white meat and a white liquid, or *milk.* Dried coconut meat, which is known as *copra,* yields one of the world's most important vegetable oils.

Coconut oil is a product of importance in the modern technology of advanced nations. It is used in making candles, soap, shampoos, and detergents; as an element in synthetic rubber and in brake fluid for airplanes; and in the manufacture of tin cans, roofing plate, margarine, and shortenings. Shredded coconut meat supplies the coconut ingredient in cakes, pies, and confections. The husk surrounding the nut yields *coir,* a tough fiber that has many uses. Finally, young stalks produce a sweetish sap that is a source of sugar, alcohol, and alcoholic beverages. —Charles E. Rogers

Coffee, a large, broad-leafed evergreen shrub, its seeds, and a beverage brewed from the roasted, ground seeds. The shrub belongs to the genus *Coffea* of the madder family. In plantations, its height is generally kept at 6 feet, although it can reach 14 to 20 feet. Coffee grows best in the temperate, tropical highlands. The fruit of the plant, the *cherry,* which ripens about six months after the plant's blossoming, is dried and depulped, and its green seed or bean is transported to consumer countries, where it is roasted and distributed.

The word *coffee* is from the Arabic *Qahwah,* and legend depicts the first user of coffee as an Arab physician named Rhazes around the tenth century. However, mention is made of its cultivation in Ethiopia as early as the sixth century. Its use as a stimulating beverage was popularized by Muslim priests, who found it beneficial during prolonged prayer ritual.

Coffee use was confined to the Middle East and Turkey until the seventeenth century, when traders, explorers, and patrons to the court of Louis XIV introduced it on the Continent. *Coffee houses* became a social institution in the seventeeth century, especially in England and Austria. Until then, all coffee was cultivated in Yemen and Ethiopia. About 1690, Dutch and French explorers introduced planting methods to Java and to the Western Hemisphere from Martinique to Brazil, which today is the world's leading coffee-producing country.

■**GROWING.**—Of about 25 species of coffee, the principal one is *Coffea arabica,* which grows best at altitudes between 2,000 and 6,500 feet in a rich terra cotta soil. Arabica coffees are grown in 58 of the 70 major coffee producing countries, but the most sought-after is the Brazilian *Santos.* Arabicas are subclassified into *Brazils*

PAN-AMERICAN COFFEE BUREAU

COFFEE BEANS must be sorted and graded by hand after they are dried and processed. Once this operation has been completed, the coffee is then packed for export and roasting.

and *Milds,* an example of the latter being the high-quality Colombian bean. Other species are the hardy *C. robusta* and the *C. liberica,* the former indigenous to Congolese Africa and the latter to West Africa. They are grown at low altitudes, from sea level to 2,000 feet. Robustas are primarily used in producing instant coffee due to their lower cost and high caffeine content.

The coffee seed is frequently planted in nurseries under careful temperature control. The young tree is transplanted and shaded by corn plants or banana trees. The tree is pruned after two years and matures in six years. Average trees produce for 15 to 20 years, and yield about 2,000 cherries. The cherries are handpicked, dried, sorted, and bagged with little mechanization. However, Colombian Milds are depulped by a *wet process* in mechanically operated tanks. The *dry process* (sun-drying) is extensively used elsewhere.

PROCESSING.—Coffee should always be commercially roasted shortly before it is marketed and distributed. The green beans are placed in large, revolving cylinders for 15 minutes at 390°–422° F (some new machinery can do this operation at 500° F for 5 minutes). One process, called *wet roast,* includes a water-spraying operation. Cooling and further cleaning, or *stoning,* follow, and occasionally the rich brown beans are preserved with a light coat of molasses. Grinding and packing are the final operation. *Instant* coffee requires a dehydration of the ground coffee, leaving the tiny soluble crystals. A pound of instant coffee requires three times as many beans as regular coffee. Coffee is finally bagged or canned in vacuum tins.

The economic importance of coffee in world trade is obvious: it constitutes 50 per cent of the export earnings of six Latin American countries and provides a livelihood for an estimated 13 million Latin Americans. About 45 per cent of Brazil's total exports was made up of coffee in 1968. Brazil, the major coffee exporting country, exports about one-third of the world's coffee. Demand has increased for African coffee, and in 1967–1968, African countries produced 18 million bags, about one-quarter of the world's exportable production.

—Marshall H. Cohen

Corn or **maize,** a cross-fertilizing *cereal grain* that originated in Central and South America. The common hybrid *dent* corns of the United States Corn Belt were developed first by American farmers and then by plant breeders during the nineteenth century. However, they differ in morphology and performance from forms prevalent in other parts of the world. It is of interest to note that corn still furnishes 80 per cent of the calories and 70 per cent of the protein of certain Central American indians.

CULTURAL PRACTICES.—Until recently, corn was planted for the purpose of harvesting the grain primarily for animal feeding. Under these conditions, corn was usually planted in rows spaced 40 inches apart, and 12,000 to 14,000 plants were planted per acre. With changes in the importance of various forms of livestock raising and crop production in the United States, the extent of the corn-growing area has changed. At the same time, innovations in growing conditions and harvesting methods have been applied. Today it is not uncommon to find corn planted in rows spaced 20 to 30 inches apart and planted in concentrations of 25,000 to 30,000 plants per acre.

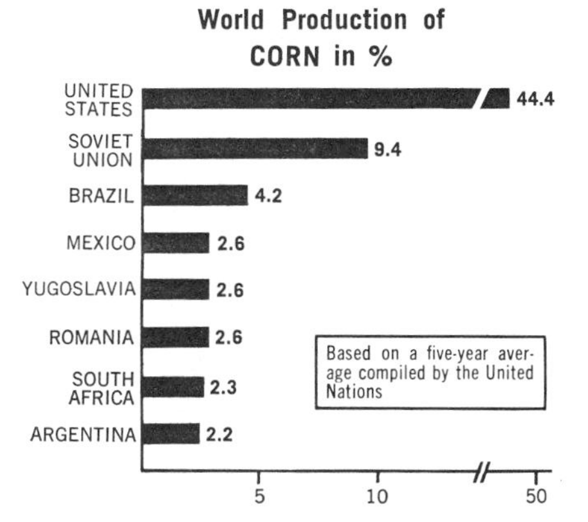

PHYSICAL CONDITIONS.—Hybrid corn is now available to fit a range of maturity requirements. The growing season can range from 120 to 170 days. Warm humid weather is desirable, with 30 to 50 inches of rainfall evenly spaced during the growing season. However, although warm weather is desirable, consistent temperatures higher than 90° F can cause more damage than good to growing corn when large amounts of moisture are not available. Though rich loam is the best soil, any well-drained piece of land can be used to produce corn if it is in a good climate and is properly fertilized to provide the needed soil nutrients.

VARIETIES.—Several hundred varieties of yellow *dent* corns are now in use in the U.S. Corn Belt. The majority (80 to 90 per cent) is used for animal feeding, with a small amount (4 to 5 per cent) for starch and alcohol production.

Flint corns are harder and not as desirable for animal feeding. Even so, many varieties are still produced in the United States for special purposes and even more by many countries in South America, Africa, and the Middle East.

Pod corns, which are not cultivated extensively, have grains (kernels) enclosed in individual husks.

Sweet corn has a higher sugar content than other varieties and is more palatable for human consumption. It is produced throughout the U.S. Corn Belt and is marketed fresh, frozen, or canned.

Flour corns, which are soft and starchy, are grown chiefly in South America.

Popcorn is a type of corn with a hard surface and high moisture content in the endosperm. Steam generated in each kernel by quick heating causes it to explode, thus making it pop. It is also produced throughout the Corn Belt for human consumption.

DISTRIBUTION.—Most of the corn grown in the United States results from planting hybrid seed produced and distributed by commercial companies or released by universities. Such seed is made available throughout the corn-growing area of the United States through seedsmen. Corn is now grown in most areas of North America from Georgia to central Canada and from Colorado to the east coast. In addition, it is an important crop in Central and South America, South Africa, Europe, and the Middle East. Except for the United States, most countries of the world import some form of corn for animal and their livestock.

TRADE.—Corn has taken on a new level of importance in the world market. Though second to wheat for human consumption, it is a cereal feed grain of primary importance. The United States, Argentina, and South Africa are the major exporters of corn. The major importers are countries in western Europe and the Middle East, and Japan.

—Laurence Baker

Cotton, a plant of the genus *Gossypium,* its fiber, and the fabric produced from the fiber. The plant belongs to the mallow family, which includes the okra, hisbiscus, and the rose of Sharon. Although cotton is a perennial shrub in climates where it is not killed by frost, its commercial production in nearly all areas necessitates planting a new crop annually for optimum quality and yield.

Boll segments and fiber found in caves of the Tehuacán Valley of southern Mexico indicate cotton was grown at least 7,000 years ago. And in India, seat of the ancient cotton industry, artisans many centuries ago achieved an unsurpassed degree of skill in spinning and weaving the fiber.

The comfort, launderability, and durability of cotton fabrics are due to the absorbency and strength of the fiber. It is formed when a hollow tube emerges from a cell in the seed wall—one tube from each of thousands of cells. Layers of cellulose fibrils build up in an orderly series of spirals within the tube. The fiber matures and, after the cotton boll opens, it dries into a flat, twisted, ribbonlike shape. The fibers interlock to form a fluffy white mass ideal for spinning into yarn.

While it is estimated that cotton has more than 10,000 uses, apparel and household items account for approximately 81 per cent of the total U.S. consumption. Large quantities of cotton are used also for making thread, bags, machinery belts, and medical supplies, and in automobiles.

Seed separated from the lint during the ginning process was once a waste product. Now all components—*linters* (short fibers clinging to the seed after ginning), *hulls,* and *meats*—are important by-products. Linters are a major source of cellulose for plastics and artificial fibers and are used for cushioning in furniture and mattresses. Hulls are used primarily for cattle feed. Oil is crushed from the meats. Refined cottonseed oil products include salad and cooking oils, mayonnaise, salad dressing, margarine, and shortening.

■ **GROWING.**—There are many types of cotton, but cultivated varieties fall into two general categories—*upland* and *barbadense.* Upland cottons account for a major share of the world's production. Their fiber is shorter and coarser than that of barbadense cottons, such as Egyptian and Peruvian, and clings tightly to the seed.

Cotton is a commercial crop within limits of approximately 37° north latitude and about 32° south latitude in the New World. In the Old World, these extremes range from 47° north in the Ukraine to 30° south in Africa and Australia. With these wide variations, it is possible to find cotton being planted, cultivated, or harvested somewhere in the world practically every day of the year.

A growing season of at least 180 frost-free days is required for cotton. In the United States, planting dates range from early February in the Rio Grande Valley to late May or early June in North Carolina, Oklahoma, and Missouri, which are the upper limits of the Cotton Belt.

Within a week or two after planting, depending on temperature and moisture conditions, young seedlings emerge from the soil. A month to six weeks later, squares (flower buds) appear. In another three weeks the cotton blossom appears. After three days, the blossom withers and falls, leaving the young ovary attached to the plant. The ovary ripens, enlarges, and forms a pod called a *boll.* Inside the boll, the moist fibers grow and push out from the coating of the newly formed seed. Although the boll enlarges rapidly up to maturity, some time elapses before it opens and the fluffy cotton bursts forth. The interval from bloom to open boll is 45 to 60 days, depending on variety, soil fertility, moisture, and climate.

Cotton production has usually required a very large labor force, particularly during the peak seasons of weeding and harvesting. An exodus of workers from the farms of the U.S. Cotton Belt after World War II speeded mechanization. Man-hours required to produce a bale of cotton (500 pounds of lint) were reduced from 145 in 1947 to less than 32 in 1968. And in the 1965–1966 season, yield per acre reached a record 527 pounds, compared with 252 pounds in 1940.

■ **HARVEST.**—In the 1967–1968 crop year, it is estimated that tractor power was used for virtually all of the land preparation, planting, and cultivation of cotton in the United States. Chemicals for weed control were used on more than 80 per cent of the acreage, and 94 per cent of the crop was harvested mechanically.

World cotton production increased from a pre-World War II average of about 32 million bales to about 53 million bales for the 1968–1969 growing season. The United States harvests about 14 million bales of cotton annually. The Soviet Union, the second largest producer, harvests a little more than 9 million bales per year. Significant quantities of cotton are also grown in mainland China, India, Mexico, Brazil, Egypt, Pakistan, and Turkey.

Cotton is grown in 19 states and is a major crop in 14 of them. Texas leads in cotton production, averaging about 4 million bales annually. Mississippi and California each produce an average of almost 2 million bales annually. Other major producing states, in the order named, are Arkansas, Alabama, Arizona, Tennessee, Georgia, Louisiana, South Carolina, Missouri, Oklahoma, New Mexico, and North Carolina.

Although textile consumption has been rising throughout the world since World War II, cotton has not shared proportionately; its percentage of the market declined from 72 to 51 between 1946 and 1967. Cotton's share of the U.S. market declined from 51 per cent of the total in 1947 to 40 per cent in 1967. Cotton's inability to compete effectively is attributed by industry leaders to rising costs, and greater research and promotion expenditures by competitors.

—Willmer L. Foreman

Cover Crops, grains grown to cover the soil and protect it against erosion. Their cultivation is an ancient farm practice, mentioned in the literature of pre-Christian Greece and Rome, as well as in ancient China. Buckwheat, oats, and rye were used as cover crops by colonial farmers in America.

INTERNATIONAL HARVESTER COMPANY

COTTON PRODUCTION, an ancient process, is performed almost entirely by tractor in the U.S. Other significant sources of this versatile fiber are Russia and Egypt.

Some crops among a score or more now widely used for the purpose are crimson clover, rye, and vetch.

Cover crops are often plowed under or disked into the soil as *green manure.* They add organic matter and often put nitrogen into the soil as well as improve its physical conditions. Millions of acres of cover crops are grown for green manure annually in southeast United States. Cover crops commonly utilized as green manure are alfalfa, soybeans, cowpeas, vetch, red clover, sweet clover, rye, buckwheat, and lespedeza. Colorful and exotic names of other cover crops often give us a clue to their origin, use, or appearance. Some of these are kudzu, sudan grass, Australian winter peas, hairy indigo, beggarweed, and the lupines—blue, yellow, and white.

While the main purpose of a cover crop is to prevent or reduce erosion, it may provide temporary grazing or a supply of grain. Cover crops are not as widely grown in summer as in winter, for land is more often used to grow cash crops. Annual rainfall of 20 inches or more is considered essential to the growth of a cover crop.

Agricultural experiment stations in all sections of the United States have supplied evidence of the effectiveness of cover crops. The U.S. Department of Agriculture offers cost-sharing payments through the Agricultural Conservation Program to encourage farmers to use cover crops. Also, in some surplus-reduction programs, farmers replace grains with such crops. Funds of the Commodity Credit Corporation are used to maintain reserve stocks of cover crop seed and to help the seed industry increase newly developed seed varieties for sale to farmers. —Charles E. Rogers

SCHAFFER AND SEAWELL, BLACK STAR

CAMEMBERT CHEESE is exposed to air during the many-day curing, or ripening, stage.

Dairy Products, any products derived from milk, chiefly cow's milk, but including the milk of goats, sheep, horses, reindeer, camels, and yaks. Organized dairying dates back to the third millenium B.C. in the Mesopotamian city-states, but it was unknown to the American Indian until 1607, when cows arrived with the Jamestown colonists. However, the growth of cities and the Industrial Revolution contributed to making it a full-fledged industry by 1900. In the United States, the leading dairy products—other than fluid milk—by volume of milk used in production are: 1) butter; 2) nonfat dry milk; 3) cheese; 4) ice cream; and 5) evaporated and condensed milk. In the United States during the past decade, per capita consumption of butter, evaporated and condensed milk, fluid milk and cream has decreased, while consumption of ice cream, cheese, and nonfat dry milk has increased. The consumption of cottage cheese has increased impressively because of its inclusion in many special diets. Although the per capita consumption of several products has declined, the total amount of milk used has risen steadily for many years. For the farmers of America, cattle and calves alone have produced more total income than have the various dairy products taken together. Dairy products brought substantially more income than any other major commodity, including bread grains, feed grains, and other livestock products.

Dairy-product processing was among the first of the industries to employ automation. Automation not only allowed economies, but more importantly, it allowed a marked increase in sanitary handling of an easily contaminated food product. Nearly all milk, whatever its ultimate processing destination, is drawn from the cow by machine, pumped into stainless-steel tanks, cooled, and then pumped into bulk tank trucks with no human contact in the process and nearly no exposure to air. Tank trucks deliver the milk to automated processing plants where the milk is pasteurized and either delivered to the consumer as fluid milk or cream or changed into some other of the dairy products. Many creameries (butter factories) use a continuous process of butter manufacture wherein the milk, cream, skim milk, and buttermilk are untouched by humans except for testing and quality control. Many of the continuous processes no longer use churns in the strict sense of that term. Many cheese factories pump milk, curd, and whey from one process to another, and move the solid product about by conveyor belts. To some extent, however, cheese making still requires human handling at various stages of manufacture. In all manufacturing most of the equipment is made of stainless steel, and pasteurization is an important element. Ice cream and evaporated milk manufacture can be and often is fully automated, even to quality control by electronic devices.

In addition to the leading dairy foods, a variety of other products are made from milk. Some of these are: buttermilk, the liquid and solids left over after the butter has formed; casein, most of the solid element in milk, used in industry chiefly for glues and paints; whey, the liquid and solids left over after cheese has formed from curd; lactic acid, produced by the fermentation of milk sugar and used in food and industrial processes; and yogurt, a fermented milk with a consistency much like sour cream. —John T. Schlebecker

Principal Varieties of Cheese

Name	Place of Origin	Type of Milk
Brick, Muenster	Germany	Cow
Brie	France	Cow
Caciovallo e.g. Provolone	Italy	Cow
Camembert	France	Cow
Cheddar	England	Cow
Cottage	America	Cow
Cream	America	Cow
Edam	Holland, Denmark, France	Cow
Emmentaler, Swiss	Switzerland	Cow
Gouda	Holland	Cow
Hand	Germany, Austria	Cow
Limburger	Germany, Belgium	Cow
Neufchâtel	France	Cow
Parmesan	Italy	Cow
Process	America, Denmark	Cow
Romano	Italy	Sheep
Roquefort, Gorgonzola (blue cheeses)	France, Italy	Sheep
Sapsago	Switzerland	Cow
Trappist	Canada, America	Cow
Whey-albumin cheeses e.g. Primost, Ricotta	Netherlands, Italy	Goat, Cow

Source: Cheese Unlimited, N.Y.C.

Duck, a web-footed swimming bird related to the goose and swan. Ducks are raised primarily for meat production in the United States, although in some European countries they are kept for egg production. The Khaki-Campbell duck reportedly lays as many as 350 to 360 eggs per year. The Pekin is the most popular breed in the United States. Others include the Rouen, Aylesbury, and Muscovy. Twenty-eight days of incubation are required for *ducklings* to hatch, except for Muscovy eggs which require 35 to 37 days. With proper care, ducklings grow very rapidly, reaching about 7 pounds at 8 weeks of age. A mature duck, or female, weighs about 8 pounds; the *drake,* or male, 9 pounds. At maturity and when fully feathered the drake, but not the female, has a few curled feathers at the base of the tail. The duck also has a much louder voice than the drake, except in the Muscovy breed. Ducks are not as susceptible to disease as most other types of poultry. Ducks are difficult to defeather in the home. They may be scalded by immersing them for 3 minutes in water heated to 140° F. The ratio of meat to bone is lower in ducks than in other types of poultry. —Carl W. Hess

Egg Production, the growing of fowl, principally chickens, for the primary purpose of laying eggs. The great majority of hens used are crosses of strains, of breeds, or of inbred lines. Under proper management, a hen can lay 240 or more eggs per year. The hens are usually kept in artificially lighted and ventilated houses on a litter floor or slatted floor or in cages holding from one to ten or more birds.

While many chicken breeds were used 20 to 30 years ago; today, only a few are used for commercial egg production. Leghorn or Leghorn-type chickens lay white-shelled eggs. Heavies, such as Rhode Island Reds and Plymouth Rocks, lay brown-shelled eggs. Although some consumers prefer one or the other color, there is no nutritional or other difference between these eggs. The color of the yolk will also vary, depending on the feed used. For example, although complete, well-balanced commercial feeds are commonly used, large amounts of yellow corn or alfalfa in the diet will result in a deeper yellow.

Eggs are marketed by grade and size. Grade refers to quality (AA, A, B) and size refers to weight per dozen. Eggs weigh approximately 2 ounces each. Top quality eggs have a large amount of thick albumen (white) and high upstanding yolks, while low quality eggs spread when broken. In many modern egg-packing plants, eggs are automatically washed, graded, sized, and packed.

Eggs are also marketed as frozen or solid (dried) whole eggs, albumen, yolks, and various blends. These products are used in large quantities by institutions, bakers, and other food manufacturers. —Carl W. Hess

U.S. DEPARTMENT OF AGRICULTURE

EGG PRODUCTION for one carefully bred and housed hen may total 240 a year.

Fats and Oils, organic chemical compounds found in plants and animals and, therefore, in foods. As discussed here, fats and oils are *glyceride oils*, in contrast with mineral or petroleum oils. *Fat* is often used to signify a solid or semisolid product, such as that obtained from steam rendering of the fatty tissue of slaughtered animals, mainly cattle and pigs. *Oil* is often used to signify the liquid product obtained from vegetable seeds—such as soybeans, cottonseed, corn, peanut, safflower, linseed, or coconut—by heating and pressing or by extraction with a solvent, which is then evaporated. Purification by treatment with alkalai and bleaching earths and deodorizing under vacuum are usually applied to the crude oil to produce a bland edible oil. Since a semisolid fat will melt when heated, and a liquid oil will solidify when cooled, fats and oils are often considered together as glycerides. This is because they are both made up of chemical combinations known as esters of *glycerine* and *fatty acids*, or as *glycerides* of fatty acids.

Glycerine, or *glycerol*, is a viscous, water-soluble, colorless liquid containing three carbon atoms. Each carbon atom has a hydroxyl group attached to it that can combine with one fatty acid. Glycerine can therefore combine with three fatty acids to make a *triglyceride ester* which is insoluble in water. The three fatty acids may be the same, but usually are different.

These fatty acids are made up of many carbon atoms joined together in a chain, with an acid group at one end. Each acid group is united with one hydroxyl group of the glycerol to make the ester group, as found in fats and oils.

The unsaturated fatty-acid glycerides of liquid vegetable oils, such as cottonseed or soybean oil, can be changed to saturated glycerides by adding hydrogen to the double bonds—a process known as *hydrogenation*. The resulting glyceride has more saturated glycerides, is semisolid, and is also more stable in resisting oxidation and rancidity. Such plastic shortenings are widely used commercially and domestically in baked and fried foods. Certain chemicals known as *antioxidants* may be added to fats and oils in minute amounts to retard oxidation and rancidity.

The liquid unsaturated vegetable oils are used for making salad dressings and mayonnaise, which are emulsions of the oil and water with vinegar, spices, and flavorings. A semisolid or plastic shortening, however, would not give the desired texture.

Fats and oils have the highest caloric value as a food, 9.0 calories per gram, carbohydrates and proteins each having only 4.0 calories. In addition, fats and oils are highly digestible. The body can synthesize fats from carbohydrates, but the fats that are eaten as such also become a part of the body fat.

Certain polyunsaturated acids are considered essential to proper nutrition, and the body cannot synthesize these essential fatty acids from carbohydrates. Linoleic acid, the most common essential fatty acid, is present in sufficient amounts in ordinary diets.

Water can react with oil to break the ester group to form glycerine and fatty acids. *Soaps* are made by reacting the fatty acids with an alkali, such as sodium or potassium hydroxide, or by having it present during the reaction with water. Certain soaps, when mixed with mineral lubricating oils, form greases.

Fatty acids differ as to the number of carbons joined in the chain (usually 12 to 18) and the way they are joined. Shorter chains make the acids and the oils lower melting. If each carbon in the fatty acid chain is joined to the next one in the chain by a single chemical bond, the fatty acid is said to be *saturated*—this makes the acid and the oils higher melting. If certain adjacent carbons are joined with two chemical bonds (a double bond), the acid is *unsaturated*—this makes the acid and the oil lower melting. For example, linoleic acid, which has two double bonds and 18 carbons, is lower melting than oleic acid, which has one double bond and 18 carbons. Oleic acid, in turn, is lower melting than the saturated stearic acid (18 carbons) or palmitic acid (16 carbons).

The animal fats have more saturated fatty-acid glycerides (mostly 18 and 16 carbons), which makes them semisolid at room temperatures, while the vegetable oils usually have more unsaturated acid glycerides, which makes them liquid.

The unsaturated acid glycerides are much more readily oxidized by the oxygen in air than the saturated glycerides. This is especially true if there are fatty-acid glycerides with two or more double bonds (*polyunsaturated acids*). Oxidation in edible fats and oils causes unpleasant tastes and odors.

Many authorities believe that a diet in which the unsaturated vegetable oils replace the more saturated animal fats is desirable in lowering blood cholesterol and preventing atherosclerosis and coronary thrombosis. Others feel that this effect has not been adequately proven for humans.

Unsaturated oils—particularly linseed, soybean, and tung oil—find extensive use in paints, enamels, and varnishes. These oils all contain glycerides of fatty acids with two or three double bonds. The oils, often combined with a resin, and with pigments ground into them, dry on contact with the oxygen of the air due to oxidation and polymerization of the unsaturated acid groups. The polymerized oil becomes solid and holds the pigments as a strong paint film, which is used to protect and decorate wood and metal surfaces.

—Donald H. Wheeler

Fertilizer, a substance used in the cultivation of plants to enhance their growth. In addition to carbon dioxide and water, which green plants convert to carbohydrates, plants need a number of chemical elements for their growth. The plant obtains these essential elements through its roots from the soil. But a particular soil seldom contains as much of each of these essential elements as the plant requires. Even if a soil is extremely fertile, continuous harvesting of crops removes essential elements that have been taken up by the plants. Placing additional amounts of the essential elements in the soil in the form of fertilizer will thus improve the fertility of soil, or maintain it in fertile crop lands.

Organic fertilizers, such as animal manures, dead fish, and guano, have been used as fertilizers since ancient times. These materials are still used, but today most commercial fertilizers are synthetic products.

Nitrogen, phosphorus, and potassium are the three most important plant food elements. The percentage of each is usually specified for each lot of fertilizer sold. For example, a 4-10-6 grade fertilizer contains 4 per cent nitrogen, 10 per cent phosphorus pentoxide equivalent, and 6 per cent potash (potassium oxide equivalent). The designation of phosphorus and potassium by their oxide equivalents is a long established practice.

Three other elements are sometimes called secondary plant nutrients: calcium, magnesium, and sulfur. Boron, copper, iron manganese, molybdenum, zinc, and several other elements needed by plants in very small quantities are frequently added to fertilizers as trace elements.

Nitrogen is essential to plants, since it is a component of many chemical compounds made by the plant, including amino acids, the materials of which proteins are composed. Uncombined, or free, nitrogen makes up about 78 per cent of the volume of air, but in its free state nitrogen cannot be used by plants. It must first be combined with other elements to form compounds which can be used. Certain bacteria living in the soil are able to take nitrogen from the air and "fix" it into compounds available to higher plants.

Nearly all industrial processes for fixing nitrogen from the air involve making ammonia. This is done by combining nitrogen with another gas, hydrogen, at pressures as high as 1,000 atmospheres, in the presence of a catalyst. The ammonia formed is a gas at room temperature, but it can be compressed to form a liquid that can be stored if kept refrigerated or under pressure. With proper equipment, ammonia can be used directly as a fertilizer by injection into the soil. Much ammonia is converted into solution materials to make solid fertilizers, such as *ammonium nitrate, ammonium sulfate, ammonium phosphate,* or *urea.*

Phosphorus is a component of some of the key chemical compounds in all living cells. The chief source of phosphorus for fertilizers is phosphate rock, mined in Florida, Tennessee, Idaho, Montana, Wyoming, and North Carolina. Outside the United States, the most important phosphate mining operations are in North Africa, West Africa, the Soviet Union, and various islands in the Pacific and Caribbean. Phosphorus in phosphate rock is in the form of *tricalcium phosphate,* which is often combined with calcium fluoride. Ground phosphate rock can be used as a fertilizer, but tricalcium phosphate is almost completely insoluble in water, so that phosphorus is not readily available to plants.

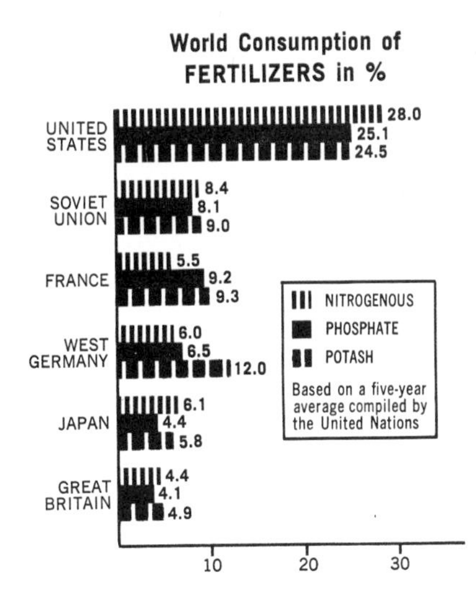

ALLIED CHEMICAL CORPORATION

NITROGEN compounds raise corn yield. Unfertilized plot (*left*) produced a small crop.

Treating ground phosphate rock with sulfuric acid to form a slurry, allowing it to harden, and then storing the mixture for a few weeks yields a solid product called *normal superphosphate.* In this mixture, the tricalcium phosphate has been converted into more soluble forms of phosphate that are readily available to the plant. *Triple superphospate* is a more concentrated phosphatic fertilizer, as is *diammonium phosphate.*

Potassium, essential to a number of chemical activities in the growing plant, is obtained from potassium compounds (collectively known as potash) mined in New Mexico, Canada, Germany, France, and elsewhere. It is also obtained from brines in California and Utah. *Potassium chloride,* often called *muriate of potash,* is the most common potassium material in fertilizers.

Today an efficient farmer has his soil analyzed and calculates exactly how much of each of the primary plant nutrients he needs. He can then order the exact type of fertilizer he wants, often made up specifically for his needs. The companies producing fertilizers tend to make more and more concentrated materials in order to save the cost of transporting inert matter. Special attention is also given to the form in which material is produced. For example, in solid fertilizers the trend is toward material composed of granules of uniform size, so that they will flow easily through application equipment with a minimum of dust.

The use of fertilizer is increasing at a tremendous rate in the United States and in the rest of the world. As the world population increases it is necessary to use more and more fertilizer to produce crops to feed it. A great deal of effort is being expended by the United States, the highly developed countries of Europe, and Japan to increase the use of fertilizers in underdeveloped countries so that they may feed themselves.

—Albert S. Hester

Fiber Crops, providers of the basic materials for the manufacture of clothing, household goods such as sheets and towels, and commercial products such as bags, rope, and twine. Cotton is the leading fiber crop grown in the United States. Other fibers that are significant in world trade are flax, hemp, henequen, and sisal.

About 30 per cent of the world's cotton is grown in the United States. It ranks second among all U.S. cash crops. The fiber is used to make thread and cloth, and the kernel of the seed is crushed for oil, which is used for salad dressing, in cooking, and in making margarine, soap, paints, and lubricants. Cottonseed cake is a cattle feed, and cottonseed hull is utilized for roughage, fertilizer, and making fiberboard. Flax and hemp are also grown in the United States. The leading product of flax is linen thread. Linseed oil is made from the crushed seed of flax, and its straw is employed in papermaking. Hemp, a coarse, strong, and durable fiber, was used in pioneering times to make "homespun" cloth; today it is commonly used to make twine, rope, bagging, rugs, and sailcloth.

—Charles E. Rogers

Fish and Seafood, edible and industrially useful marine life. From a commercial standpoint, fish can be grouped into food fish, including shellfish, and industrial fish. Food fish provides about 10 per cent of the world's animal protein used directly for human consumption. Industrial fish are utilized in the manufacture of fish meal, fish oils, pet food, and fertilizers.

Tuna, salmon, and sardines are

among the important food fish consumed in canned form throughout the world. In the more developed countries, such food fish as ocean perch, haddock, flounder, redfish, herring, and cod are marketed either in fresh or frozen form as fillets, steaks, or dressed whole fish. In countries where refrigeration and other modern preservation methods are not as widely used, food fish in smoked, dried, or pickled, and cured forms are more prevalent.

Shrimp, lobsters, crabs, oysters, and clams are the principal shellfish species. These are distributed mainly in fresh or frozen form.

Industrial, or trash, fish include such species as menhaden, alewives, and anchovies. These, along with minor nonedible species of trash fish and the trimmings and waste from the preparation of food fish, provide the raw product for this growing segment of the fishing industry.

In 1964, U.S. commercial fishermen caught an estimated 4.5 billion pounds of all species. Of this amount, 35 per cent was landed on the northeastern and mid-Atlantic coasts, primarily in New England, 26 per cent in Alaska, Hawaii, and the Pacific coast states, and 36 per cent in the South Atlantic and Gulf states, with the remainder coming from inland lakes and rivers.

The world catch has doubled every ten years for the past several decades. The 102 billion pounds caught in 1963, however, is far short of the 440 billion pounds estimated to be available on an annual sustainable yield basis throughout the waters of the world.

In the United States, direct use of seafood for human consumption is considerably less important than the use of meats and dairy products. The annual per capita consumption is only about 10 to 11 pounds, compared with over 150 pounds of meat products. Nevertheless, the seafood industry is an important segment of the total food industry, both because of the present volume of production and because of its potential for satisfying future needs.

Because the production of food fish in the United States has remained relatively stable over the last 10 years, an increasing part of the U.S. supply of edible fish products is being supplied by imports. The population growth and purchasing power of the United States has provided a lucrative market for high-priced species sold by so many other fishing countries. Imports amounted to about 34 percent in 1964.

Commercial fishermen in the United States employ a variety of gear, methods, and seagoing vessels to catch fish and other marine species. Fish which tend to live in surface waters, such as salmon, tuna, and mackerel, are caught by means of nets, or hooks and lines. Bottom-dwelling species are caught with a variety of gear including nets, dredges, lines, and traps. The ships and boats used in the commercial fisheries also vary in design, tonnage, and size of crew.

The major problems in handling fishery products prior to consumption are to minimize undesirable changes in quality and to prevent spoilage. Ideally, therefore, the catch should be held for the shortest possible period prior to processing. Some fisheries, including some of the Soviet and the Japanese fleets, solve this problem by processing the fish on factory ships equipped for the heat-processing or freezing of seafood. Freezing at sea is used in the United States only to a very limited extent in some Alaskan fisheries.

The usual method of preservation is by icing, that is, surrounding the fish with sufficient ice to keep them at low temperatures for the period needed to bring the catch in. Some marine products, especially crabs and lobsters, must be delivered alive to shore plants. For this purpose, fishing vessels are equipped with holds flooded with circulating fresh sea water in which the lobsters or crabs may be kept alive for several days.

■PROCESSING OF SEAFOODS.—Seafoods used for human consumption are sold in one of the following forms: fresh-refrigerated, frozen, precooked frozen, canned, cured or dehydrated. Fresh products are usually sold by the pound, without prepackaging, in locations close to the port of entry. Lobsters must be sold alive, and are therefore shipped in refrigerated containers filled with seaweed, and kept at a low temperature. Transportation by air has made it possible to market lobsters at great distances from where they are caught.

Much seafood is marketed as frozen products. These may be frozen fresh, as fillets, steaks, and individually quick-frozen shellfish, or precooked prior to freezing as fish sticks, various fish dishes, and precooked frozen shrimp.

Canning is the method of preservation applied to most of the fishery products processed in the United States. Products which are preserved in this manner include tuna meat, salmon, sardines, oysters, crab, and chowders based on fishery products.

TUNA RESEARCH FOUNDATION

TUNA, hauled by net onto fishing boats, is an important commercial fish. Most processed tuna goes to consumers in cans.

Curing of fish by addition of salt and smoking is an old method of preservation, but is used in the United States only for a few specialty products. Dehydration by conventional drying methods is also used to a very limited extent, but recently there has been developed a considerable interest in freeze-dehydration, which is capable of producing high-quality dehydrated products. Shrimp, various precooked seafoods, and fish sticks and fillets are preserved by this method. Another new development is the use of atomic radiation to pasteurize fish and shellfish. Foods pasteurized in this manner retain their natural flavor and appearance, and may be kept at refrigerated temperatures for weeks whereas unpasteurized fisheries products are stable for only a few days. —Marcus Karel

Flax. See *Rope,* page 564.

Food Additives, chemicals that do not occur naturally in foods but are introduced in the course of food production, processing, or packaging. *Intentional additives* are chemicals added in controlled amounts with the purpose of achieving some desirable results. *Nonintentional additives* are contaminants introduced into foods as a result of manufacturing operations.

■INTENTIONAL ADDITIVES.—These are introduced in order to improve the safety, palatability, or nutritional value of foods. Additives inhibiting microbial spoilage are called *chemical preservatives.* Salt, organic acids, and certain other chemicals are used for this purpose.

Chemical reactions producing undesirable changes in foods are often controlled by addition of chemicals. Reactions with atmospheric oxygen are controlled by addition of *antioxidants,* notably ascorbic acid (vitamin C), tocopherol (vitamin E), certain phenolic compounds, and other reducing chemicals. Darkening of fruits and vegetables may be controlled by sulfites, and by ascorbic acid. Reactions involving metals present in foods may be minimized by the use of chemicals that tie up the metals in a nonreactive form. Chemicals effective for this purpose, called *chelating agents,* include salts of citric and phosphoric acids. Undesirable textural changes in certain types of foods may be prevented by addition of colloidal materials, such as carboxymethyl cellulose, which are known as *stabilizers.*

Additives may be used not only to prevent changes, but also to improve the initial quality of foods. Texture of meats, for instance, may be improved by adding plant enzymes such as papain (derived from the papaya plant), or bromellin (derived from pineapple). These enzymes tenderize meat, that is, they partially hydrolyze or digest it.

Taste and flavor may be adjusted to suit consumer preferences by means of synthetic sweetening agents, synthetic or natural flavor mixtures, and by flavor enhancers or potentiators. The last-mentioned category of additives has no characteristic flavor of its own, but is capable of enhancing natural flavors.

Color of foods may be improved by agents that stabilize or desirably alter natural pigments as the addition of sodium nitrite to meats or by addition of synthetic dyes.

Additives are often used to supplement the nutritional value of foods deficient in some vitamins, minerals, or other nutrients. For instance, bread, cereals, milk, and some baby foods are enriched with selected vitamins and minerals. Table salt is enriched by addition of iodides, which not only improve its handling characteristics but are also an important source of iodine, an element needed in the synthesis of the hormone thyroxin (produced by the thyroid gland).

■NONINTENTIONAL ADDITIVES.—These may arise from agricultural practices—residues from pesticides, antibiotics for treating cattle, and similar agents; from manufacturing practices—as, metallic substances acquired from equipment used in processing foods; or packaging procedures—as, plasticizers or adhesives absorbed by foods from packaging materials. These additives serve no useful function, and their occurrence must be prevented or substantially inhibited to avoid possible health hazards.

—Marcus Karel

Food Engineering, the branch of engineering that deals with the development and operation of industrial processes aimed at the production of predictable and controlled changes in the chemical composition, physical characteristics, and biological properties of foods. These industrial processes and the associated controlled transformation of raw materials form the basis of food manufacturing and food preservation.

In designing and analyzing food processes, food engineering makes use of the concept of *unit operations.* A unit operation is an operation that may occur in any of a number of processes, often using different kinds of equipment, but that has a single physical basis and a single set of scientific principles, which apply independently of the process in which the operation is applied.

The purpose of engineering analysis of food processes is the quantitative determination of data needed for equipment design, for control of process variables, and for accurate prediction of process costs. The data obtained by the engineer may be the time needed to achieve a given effect under a given set of conditions, the size of equipment needed, or the specification of conditions such as temperature and pressure for a given process.

The complexity of the composition and structure of foods, and the stringent requirements with respect to wholesomeness and appeal of processed foods, present the food engineer with problems that call for extensive knowledge of scientific principles and the properties of foods. Food engineering must be concerned at all times not only with the efficiency of operations, but also with the effects of these operations on the foods and on the people who consume these foods. For this reason, graduate food engineers are not only trained in mathematical, physical, and chemical subjects, but are also prepared to consider the biological and biochemical problems that may arise in food processing.

—Marcus Karel

Food Manufacturing, the process of preparing farm produce for the consumer. This includes the slaughter of meat, fowl, and seafood, and the packaging of eatable foodstuffs.

■MEAT-PACKING.—The conversion of livestock to meat, as well as to edible and nonedible by-products.

Most of the livestock used in the industry is produced west of the Mississippi River, and the major industry centers are located in Midwestern cities. The industry originated, however, on the East coast. At that time, the locally raised animals were slaughtered in nearby market cities, and surplus meat was salted and packed in barrels for export to the West Indies. This practice gave the industry its name, even though this method of "packing" is no longer practiced to any significant extent. During the nineteenth century, livestock raising moved progressively westward, and so did the industry. By mid-century the major meat-packing center was Cincinnati, Ohio. In the second half of the nineteenth century, with the rapid development of railroads, meat-packing plants were relocated at rail centers. The preeminence of Chicago as the meat-packing capital dates from that period. Chicago, Kansas City, Kans., Kansas City, Mo., and Omaha, Neb., are among the most important industry centers today.

Operations of the meat-packing industry include buying the livestock, slaughtering the animals, converting the carcasses to meat cuts and by-products, and selling these products. The animals are shipped to stockyards by railroad or truck and are held in pens. Prior to slaughter they are stunned. After slaughter the animal is eviscerated, and the carcass divided into two *sides,* which are then moved into a refrigerated room to allow rapid chilling. The viscera and other organs from each animal are collected in a separate container for inspection by a qualified representative of the Meat Inspection Division of the Bureau of Animal Industry of the United States Department of Agriculture. All meat that is moved in interstate commerce must be inspected, and the government inspection stamp guarantees that the animal was free from disease, was slaughtered under sanitary conditions, and that the carcass was wholesome at the time it left the packing plant. In addition, trained government graders may stamp carcasses with one of the official grades, which are based on the eating quality of the meat. The government grading is done only at the packer's request.

After chilling, the sides are divided into wholesale cuts. These are various portions of the carcass from which the retail cuts are prepared. The wholesale cuts for beef include round, sirloin, short loin, flank, rib, plate, chuck, brisket, and foreshank. For pork, they include hind foot, ham, side pork, loin, spareribs, shoulder, and jowl. The wholesale cuts are then shipped, leaving the final preparation of retail portions to the stores. Sometimes the sides are shipped, and all the cutting is done at the retail level. Most fresh meat is shipped at refrigerated temperatures; freezing is also used to some extent. Refrigeration at temperatures slightly above freezing is used also for preservation of meat products such as fresh ground meats, sausages, and a variety of cured products.

Other methods of preservation include heat processing, used for a variety of canned meats; freezing, used for many types of sausages and for certain precooked meat products; chemical preservation by salt, which is usually used in combination with refrigeration; and, more recently, irradiation and freeze dehydration.

A large variety of products forms the category of cured meats, that is, meats to which certain materials known as curing agents are added, frequently over a period of time, to alter the meat products with respect to their keeping qualities, flavor, and appearance. Originally the main purpose of curing was to allow meat to be stored at room temperature. With the advent of refrigeration, this purpose became secondary to the production of characteristic flavor and color. Curing processes vary greatly from product to product, and the curing agents vary according to the different processes. In most of the processes, however, the major components of curing mixtures are salt, nitrite, sugar, and flavoring agents. Some cured products are smoked to impart a characteristic and desirable flavor. Recently it has become possible to produce in meats a smokelike flavor without subjecting the products to actual smokehouse conditions.

In addition to the production of meats and meat products, the meat-packing industry is engaged in the production of a large variety of by-products. These include industrial raw materials, such as hides; chemicals derived from various nonedible portions of the animals, such as gelatin; and pharmaceuticals derived from the glandular organs of the slaughtered animals, such as dried thyroid glands of sheep which are used in treatment of goiter and other thyroid disorders in humans. Another important by-product of meat processing is animal fat, which is used for production of edible fats, such as lard, as well as for industrial greases.

■POULTRY PRODUCTS.—Among meat products, poultry ranks third in quantity, surpassed by beef and pork products.

Poultry production centers are located in every major region of the United States. Modern poultry farming is based on scientific principles, utilizing technological advances in housing, disease control, sanitation, and feeding. The feeding practices are based on knowledge of nutritional requirements of birds, and the feed formulas include nutritional supplements, such as high-protein meals and vitamins.

Poultry is classified in accordance with the weight and age of the bird at the time of slaughter. Thus chickens may be classified as broilers or fryers, roasters, fowl, and cocks or old roosters. Turkeys are classified as fryer-roasters, young hens and toms, yearlings, and mature turkeys. The quality of individual birds is expressed by standards, ranging from *A* to *C* quality, established by the U. S. Department of Agriculture.

At a modern processing plant, the birds are put through a series of operations conducted in production-line fashion, with the birds carried on overhead conveyors. The major steps in processing include weighing, slaughter and bleeding, scalding (dipping in hot water to facilitate defeathering), defeathering, usually by automatic equipment, eviscerating, chilling, and grading. Additional steps depend on the manner in which the products are to be sold—fresh, frozen, cooked, canned, or dehydrated.

A large proportion of poultry is shipped fresh to retail channels. Fresh poultry is stored and distributed under refrigeration, packaged in boxes or in plastic bags. The storage life of refrigerated fresh poultry is limited, but recent research on the application of ionizing radiations and on the use of antibiotic dips indicates that the shelf life may be considerably extended.

Poultry to be stored for longer periods of time than are permitted by refrigeration is usually frozen. Birds to be frozen are usually placed in tightly fitting plastic bags, for proper packaging is necessary to maintain high quality in frozen storage. Freezing is usually done in air or by immersion in liquids.

In addition to uncooked poultry, processed as described above, large quantities are sold cooked. Cooked poultry products include a variety of canned items, such as chicken in broth, various stews and soups, and many others. A smaller quantity of cooked poultry items is available frozen; recently, interest has developed in the production of precooked freeze-dehydrated poultry dishes.

■ FRUIT AND VEGETABLE PRODUCTS.—Most of the fruits and vegetables grown in the United States are harvested within a relatively short season. In order to assure a year-round supply, therefore, a large proportion of these crops is processed before delivery to consumers. Canning, freezing, pickling, and dehydration are the common methods of processing.

Modern production methods are aimed at maximizing the output per man-hour, which requires the introduction of mechanical devices for harvesting, and the mechanization of much of the processing operation. The mechanization of the agricultural practices, in turn, requires the use of herbicides, insecticides, fertilizers, and growth promoters. The efficiency of agricultural operations depends also, to a large extent, on selective breeding of suitable varieties of commercial crops. In recent years, the selection of varieties has been based not only on high yields and resistance to environmental hardships but also on suitability for mechanical harvesting and for processing. Suitability for mechanical harvesting requires a degree of sturdiness and, most important, uniform maturation times. Scheduling of the planting and harvesting operations is often based on the *heat unit system* in which each day of plant growth is credited with a certain number of units depending on the mean temperature of the day.

Generally, vegetables grown for processing are processed soon after harvest, thus minimizing post-harvest deterioration. Whenever delays occur, steps are taken to minimize deterioration, principally by rapid cooling of the crop and storage at low temperature. Some fruits, such as strawberries, are very susceptible to deterioration after harvest and must be cooled and processed rapidly. Other fruits, such as apples, pears, and oranges, may be stored for long periods of time without processing. In order to maximize their storage life, however, they are stored under strictly controlled conditions of temperature, relative humidity, and gas composition in the storage chambers. The preparation for processing of vegetables includes the following operations, most of which are also used in processing of fruits:

Cleaning: This involves washing with water containing disinfectants, brushing, and rinsing with water of potable quality.

Conveying: In most operations, conveying is mechanized. Depending on the nature of the material, this may include belt conveyors, vibrating conveyors, and many other devices.

Grading: Grading for size, appearance, and lack of defects requires some degree of manual labor. Many mechanical sorting devices for size grading, however, are used, and equipment has recently been developed for automatic color sorting and grading.

Peeling and Shelling: Many different methods may be used for this purpose. The simplest is manual peeling. Mechanical devices operate on the abrasion principle, chemical peeling methods (lye peeling), and loosening of the peels by steam with subsequent removal of the peel. Certain vegetables are flame-peeled. In this process, the product is exposed briefly to high-temperature combustion gases, which puffs and loosens the skins so they can then be removed by high-pressure water sprays.

Size Reduction: The fruits and vegetables are reduced to the desired form (slices, halves, purėe) by various manual or mechanized methods.

In juice preparation, a number of different processes are available for pressing the raw material to release the juice, and for removing the undesirable components extracted with the juice. In some cases, the juices may be concentrated by the removal of a portion of the water. Vacuum evaporators of high heat-transfer efficiency are usually used, but some newer processes are based on removal of water as ice crystals in a process known as freeze concentration. The most modern process, not as yet completely developed, involves removal of water by diffusion through a membrane permeable to it but not to the dissolved sugars, flavor compounds, and other juice components.

The final processing and packaging of the fruit and vegetable products varies with the method of preservation. Products preserved by heat processing are usually packaged in cans.

Internal linings of the cans are chosen to assure best taste and appearance of the products and to prevent or minimize corrosion. Some fruit products may be packaged in aluminum cans or in plastic packages. Usually, the heat processing is conducted on the packaged product but, in the case of fluid products, it is often possible to heat the product in heat exchangers, and then to fill presterilized containers while the liquid is hot. This last method is known as *aseptic canning*. The amount of heat required to assure stability and safety varies with the product. Nonacid products, such as peas and many other vegetables, require the most heat since they are capable of allowing the growth of potentially deadly microorganisms, such as *Clostridium botulinum*, the causative agent of botulism—a form of food intoxication with high mortality rates. Most fruits, and some vegetables with high acidity, require only heating to temperatures close to boiling.

Other methods of preservation include freezing, dehydration, and fermentation. Fermentation is one of the oldest methods of preservation, and is still important in the preservation of cucumbers (pickles), tomatoes, peppers, and cabbage (sauerkraut). The fermented products are usually pasteurized in glass jars or cans.

Distribution of fresh fruits and vegetables without processing requires a degree of control of temperature and suitable packaging. Packages for fresh fruits and vegetables must allow gas exchange, since the plant materials are still "alive"—that is, they possess an active enzyme system that produces a high rate of respiration. When the normal respiration is inhibited by interference with gas exchange, the plants are subject to rapid spoilage. In the case of very actively respiring plants, the packages are perforated with holes to allow adequate gas exchange. —Marcus Karel

Food Preservation, the application of a process or processes that permit a normally perishable food to be stored for long periods without spoilage. Food spoilage means the loss of edibility due to formation of offensive changes in taste, flavor, texture, or appearance; loss of wholesomeness due to formation of toxins, extensive loss of nutritive value, or development of high concentrations of microorganisms. Spoilage may occur through chemical and physical processes, or through the growth and activity of microorganisms, such as bacteria, molds, and yeasts. Microbial spoilage is of greatest concern because it usually is rapid, and is most likely to result in conditions adversely affecting the health of consumers. Food preservation, therefore, is aimed primarily at the inhibition of microbial spoilage. Additional measures, such as exclusion of air and addition of chemical agents, may inhibit chemical deterioration.

Microbial spoilage may be prevented by *sterilization* of foods, that is, complete destruction of microorganisms present in foods, or by *inhibition* of growth of microorganisms by producing conditions unfavorable to their growth. Food may be sterilized by heating (thermal sterilization), by exposure to radiation (radiation sterilization), or by treatment with chemical agents (chemical sterilization).

Thermal sterilization is usually performed by exposing foods, packaged in hermetically sealed metallic containers, to temperatures of 240 to 250° F in pressurized vessels (canning) or by heating foods prior to packaging, and then packaging them under conditions preventing recontamination with microorganisms (aseptic canning). Newer methods of canning are aimed at improving the quality of foods, often by heating at very high temperatures for short periods of time.

Radiation sterilization is achieved by exposing packaged foods to radiations such as gamma rays from radioactive isotopes, X-rays, or high-energy electrons produced by electron accelerators. The type and amount of radiation used for food preservation are controlled to achieve sterilization without the production of any radioactivity in the food, and without impairing the safety of food. Radiation preservation, an outgrowth of postwar research interest in atomic energy, has recently received full recognition as a safe method of food preservation by the U. S. Food and Drug Administration.

Chemical sterilization is only rarely applied to food products. It is usually reserved for such items as clothing.

Food preservation without sterilization may be achieved by producing conditions unfavorable to microbial growth. Processes based on this principle include freezing, refrigeration, dehydration, chemical preservation, and fermentation.

Refrigeration, or storage of foods at temperatures slightly above freezing, is effective in slowing down microbial growth, but does not prevent such growth entirely. It is used, therefore, in combination with chemical preservation, or partial destruction of microbes (pasteurization) for products expected to have only a short storage life. Freezing prevents bacterial growth by crystallization of water and by low temperatures of storage. Although freezing can preserve foods with little impairment of their quality, it requires facilities for low temperature distribution and storage.

Dehydration reduces the water content of foods to low levels at which microbial activity ceases and chemical processes are greatly slowed down. It may be performed by a number of processes, ranging from the ancient technique of sun-drying to the most modern freezing dehydration.

Chemical preservation is based on addition to the food, or production in the food, of chemical agents that prevent microbial growth. The most commonly used agents include salt, acetic acid (vinegar), lactic acid, and other organic acids, and in some types of foods, high concentrations of sugar.

Fermentation, which in principle is identical to chemical preservation, is based on production of organic acid (usually lactic acid) by bacteria or molds either naturally present in the food, or introduced to it as a fermentation starter. Production of cheese, various sour milk products, pickles, and numerous other food products is based on fermentation, which is also vitally important in production of antibiotics.

■FROZEN FOODS.—The crystallization of water and the low temperatures of storage produce conditions that prevent spoilage of the food by microorganisms. These conditions are also effective in greatly slowing down most of the chemical reactions and physical changes that adversely affect the eating quality and nutritional value of stored foods. Best results are obtained when high-quality foods are selected for freezing, the food is frozen rapidly, and packaged in containers that prevent loss of water and of volatile flavors, and when the food is kept at temperatures below 0° F throughout its distribution and storage cycle. The food-spoiling microbes are not destroyed by freezing. It is essential, therefore, that after thawing, such foods be stored for only brief periods, preferably at refrigerator temperatures.

Freezing of foods by exposure to natural low temperatures is an old art, but modern food preservation by freezing began in the 1920's with the development of equipment for rapid freezing. At the present time, foods are frozen by one of the following methods:

Blast freezing in tunnels in which the food is exposed to air at temperatures below —20° F and air velocities of several thousand feet per minute; *plate freezing,* in which the food is cooled by contact with plates maintained at low temperatures; *immersion freezing,* in which foods are immersed in liquids such as brine or water-glycol mixtures maintained at temperatures of —20° to —40° F. Superior quality is claimed for foods frozen at these low temperatures.

Chief among foods preserved by freezing are concentrated orange juice, fruits and berries, vegetables, poultry, seafood, and a variety of precooked dinners. The growth in volume has been most spectacular for frozen juices, which increased in consumption from approximately two million pounds in 1946 to close to a billion pounds in 1960, and for precooked frozen dinners, which were virtually unknown in the 1940's but now account for a billion pounds annually. —Marcus Karel

Forests. See *Lumber Industry,* page 552.

Fruit, any one of a large number of cultivated edible flower parts. Botanically, a fruit is usually considered the ripened ovary (or ovaries) of a flower (or flowers), with or without closely related parts. Some botanists prefer to designate it as basically the enlarged pistil of a flower. These definitions, however, include some plants that are classified as vegetables in economic terms (as, tomatoes and melons). We shall consider fruit to include perennial tree fruits and nuts, berries, and grapes. The main types of tree fruits are *citrus* (as, oranges), *pome* (as, apples), and *stone* (as, peaches).

The fruit industry in the United States is perhaps the most highly developed in the world. No other country produces such a quantity and variety of fruit so efficiently, and few, if any, have such an efficient fruit marketing system.

Fruit production has become highly specialized in the United States. Whereas at the turn of the century nearly every farm produced some fruit, today production is generally concentrated on specialized farms in fruit districts. These districts are usually identified by a temperate climate, good soils, and adequate rainfall or access to water. Mechanization is increasing; labor problems have led to special interest in mechanical harvesting.

In recent years, the value of fruit crop at farm level has been between

FLORIDA CITRUS COMMISSION

ORANGE PRODUCTION, a major part of the large, highly-developed U.S. fruit industry.

$1.4 billion and $1.8 billion. The retail value (including marketing charges and processing) is several times this figure. The leading fruit-producing states in terms of farm value are California (with about 40 per cent of the total), Florida (about 20 per cent), Washington, New York, and Michigan. Other important states include Hawaii, Oregon, Pennsylvania, New Jersey, and Virginia. The five most important U.S. fruits, in decreasing order of farm value, are oranges, apples, grapes, peaches, and strawberries. Nuts rate below peaches. Other leading fruits include grapefruit, pears, prunes, lemons, and pineapples.

The fruit crop is utilized in two ways, fresh consumption and processing. A major portion, about 60 per cent in recent years, is processed; this enables us to have nearly all types of fruit throughout the year. The major form of processing is canning, but freezing and drying are also important, particularly for certain crops, notably, frozen concentrated orange juice. A number of new processing techniques, such as freeze-drying, dehydro-freezing, and radiation treatments may become more significant in the future.

Total fruit consumption or use has averaged nearly 200 pounds per capita annually (this figure is in farm weight terms; it excludes nuts but allows for imports and exports). Overall, a slight downward trend is evident since World War II. This decline, however, may be more apparent than real because spoilage has been reduced by improved handling and increased processing. Even so, there has been increasing consumption of processed fruit as a substitute for fresh fruit. In either form, fruits add variety and enjoyment to the diet and are important sources of vitamins and minerals. —Dana G. Dalrymple

Fungicide. See *Pesticides.*

Fur Farming, the raising of fur-bearing animals in captivity under conditions of controlled breeding, feeding, and care. Fur farming, sometimes called *fur ranching,* started in Canada at the end of the nineteenth century. Today most of the pelts from North America come from fur farms.

Fur farming is possible only with animals that breed well in captivity and present minimum problems in care and feeding. Beaver and muskrat, for example, require too much water to be profitably raised in captivity. Silver foxes and mink were the first animals to be raised on farms. Other fur farm animals are the fisher, marten, sable, nutria, skunk, raccoon, and chinchilla. The best chinchilla fur comes from wild Andean animals, but generally the fur of ranch or farm animals is superior to that of wild animals. Rabbits (conies) have long been raised for food, and more recently for fur.

Animals in captivity can be selectively bred to develop special colors of furs, such as the various mink mutations of pinks and even lavenders. In addition, furs from animals kept in captivity are more uniform in size and quality and can be used more easily in the garment industry.

Generally, the animals are kept in cages with wire mesh bottoms for ease of cleaning. The animals are usually separated to keep them from injuring one another. Some animals, notably mink, will kill their young if excited, so great care must be taken not to alarm them. All animals, if excited, are likely to damage their fur.

In captivity the animals are fed raw meat, poultry, and fish, with added supplements of cereals, citrus fruits, vitamins, and minerals. Fish of the carp family contain thiamine-destroying elements. Therefore, carp should be fed sparingly, if at all, for supplements cannot make up the thiamine deficiency, which results in *Chastek's paralysis.* Diseases such as distemper, abortion (*salmonellosis*), pneumonia, tularemia, and streptococcic septicemia either respond to antibiotics or can be prevented by vaccination. Some diseases, such as tularemia in mink, are difficult to treat because antibiotics can kill the animals. Anthrax is too dangerous to cure or treat and is too rare to require vaccination; it should be controlled by using great care in feeding and tending the animals. Avian tuberculosis has become increasingly damaging among rodents, but it can be controlled by making sure that the animals are fed uninfected pork or poultry. Most fur-bearing animals mate in the winter, although some, such as the muskrat and rabbit, produce several large litters a year.

When their fur is of best quality and size, the animals are killed painlessly by injection or electricity. The pelts are then removed, stretched, scraped, dried, and stored until sold. —John T. Schlebecker

NATIONAL FUR NEWS

MINK farming began in the nineteenth century. Today, a good brown pelt may bring $75; a pastel mutation as much as $125.

Gelatin, a purified form of glue, produced principally by the acid treatment of pork skin and the lime or alkaline treatment of calf hides and demineralized cattle bones. It is an easily digestible but nutritionally incomplete protein that is not coagulated by heat. In water, it forms gels that are easily dispersed by the addition of heat.

The properties of gelatin in water render it useful in gelation, emulsion stabilization, water-binding, foaming, solution clarification, and inhibition of crystal formation. These functions are utilized in various applications. In foods, gelatin is used in the preparation of desserts, confections, ice cream, whips, and jellied meats. Pharmaceutically, it is used in the formulation of emulsions, capsules, lozenges, suppositories, and cosmetic preparations, and as a treatment for nail defects.

Gelatin has a number of applications in the photographic, printing, electroplating, and tanning industries. It is also used to size (stiffen) paper and textiles, to fine (clarify) beverages, and to provide culture media for microorganisms. —Isaac J. Wahba

Gibberellic Acid, one of an expanding group of natural plant growth-regulating substances known as gibberellins. The nine closely related compounds that originally constituted this group are designated A_1 through A_9, the most widely known being A_3, or gibberellic acid. Gibberellins, produced by the fungus *Gibberella fujikuroi,* are obtained commercially by growing the fungus in a liquid culture medium and then extracting the gibberellins in a manner similar to that used in the production of antibiotics. In its natural habitat, the fungus attacks rice plants, causing abnormally long stems to develop, a disease known as *bakanae* or *foolish seedling.* The growth-accelerating factors were isolated in 1938 from infected rice plants in Japan. Responses to gibberellic acid have been studied widely in research with plants, since the acid accelerates stem elongation and induces other growth responses. It is used in crop production mainly to increase the size of some varieties of grapes, to improve the quality of navel oranges, and to increase the production of the enzyme alpha-amylase in malt. —John W. Mitchell

Goat, a hollow-horned, hoofed mammal that is similar to sheep. Males are called *bucks,* females *does,* and the young *kids.* Goats are kept for milk, meat, hides, hair, and sometimes work. In the United States, goats are kept primarily for milk, although some *angora* goats are kept for their hair. Goat's milk is easy to digest and is generally safe from tuberculosis, since goats rarely get this disease. The milk breeds used in the United States include the *Toggenburg, Nubian, Saanen, British Alpine,* and *French Alpine.* Goats usually mate in the fall or winter, but can mate anytime, and produce one or two young, with a gestation period of five months. The doe may be bred at eight months. Goats are browsers rather than grazers and will eat almost anything.

They can be especially destructive to young plants and trees because they eat bark. Goats suffer from *milk fever* and *mastitis,* both of which may be controlled by antibiotics. Pneumonia is usually cured with antibiotics, and anthrax can be prevented by vaccination. *Goat pox* and *brucellosis* are especially contagious and incurable, but they can be prevented by sanitation and the removal of diseased animals. Sulfa drugs work well on *actinomycosis,* an inflammation caused by a fungus. Worm parasites respond to treatment by vermifuges, such as copper sulfate, nicotine sulfate, and phenothiazine.

Goats are ruminants of the genus *Capra,* with four wild species, three domesticated species, and one species both wild and domesticated. In addition, *Capra ibex,* with several species, has never been domesticated.

—John T. Schlebecker

Goose, a web-footed swimming bird, related to swans and ducks, raised mainly for its meat. There is a limited demand for geese, probably due to the fatness of the meat. Geese are often kept by some to alert against intruders, or to weed cotton and berry fields. If not more than 5 to 7 geese per acre are used, and are removed before the berries ripen, they will not damage the crop. Of the numerous breeds, the Toulouse and Emden geese are the most popular. It is generally difficult to distinguish between the male and female. However, in Pilgrim geese the *gander,* or male, is white and the female, or goose, light gray. Goose eggs require 28 to 31 days of incubation for hatching. *Goslings* grow rapidly and reach 10–12 pounds at 10 weeks of age, and up to 25 pounds at maturity, depending on the sex and breed. The gander is usually several pounds heavier than the goose. Geese are long-lived—some up to 100 years—but females are normally not useful after 8 or 10 years. Ganders may have two or more mates but they usually remain faithful to the same one for life. Geese are costly, mainly because of low egg production and hatchability.

—Carl W. Hess

Grape, a vine-grown *fruit* ranging in color from greenish white to black. Grapes are the most important fruit produced in the world in terms of tonnage. Most of the production is concentrated in the Mediterranean area. On the basis of individual nations, Italy and France are the leaders, followed by Spain and the United States.

The U.S. produces 3 to 4 million tons, most of which is concentrated in California. In recent years, California has accounted for approximately 90 per cent of the crop, followed by New York, Michigan, Washington, Pennsylvania, and Ohio. Botanically, California grapes are of European varieties; others are generally American.

About one-sixth of production is sold for fresh consumption and about five-sixths for processing. The main forms of processing are: crushing for wine or juice; drying into raisins; and canning of white grapes. Crushing takes a little over half of production, drying about a third, while canning is of relatively minor importance.

—Dana G. Dalrymple

Guinea Fowl, an edible bird characterized by a small, partially unfeathered head and a curved body. The Pearl guinea, with grayish plumage dotted with white, is the most popular variety of domestic guinea. Other varieties are the White and the Lavender. Guinea eggs hatch after 26 to 28 days of incubation. Young guineas, or *keets,* reach about 2½ pounds at 14 weeks of age, and at maturity weigh 3 to 3½ pounds. Guineas are used frequently in rural areas as guardians of the farmstead, for they produce harsh shrieks whenever strangers appear. They are considered a delicacy and are used as a specialty food.

—Carl W. Hess

Herbicide. See *Pesticides.*

Horse, a domesticated, four-legged, herbivorous mammal best known for its use as a beast of burden, although its flesh is eaten in some cultures. Horses range in size from a few hundred pounds to more than a ton. Although a type of prehistoric horse once existed in North America, all horse stocks today are descended from animals that originated on the high plateaus of eastern Russia and Siberia. The currently accepted date for the domestication of the horse is approximately 5,000 B.C. From their native habitat they spread to the south and west. Successive invasions of the Arabs and the Moors brought these native strains into Spain, from which was developed the *Andalusian horse.* This was the horse brought to North and South America by the Spanish conquistadores and from which the South American *Criollo* and the North American *Mustang* are descended. English and Dutch settlers on the Atlantic seaboard brought horses from England and Holland to complete the basic stocks for the North American breeds, while a number of French horses were brought to eastern Canada.

Horses played a major part in the advancing American frontier, U.S. cavalry regiments eventually subduing the mounted Indian. In the wake of these battles came horse-drawn covered wagons, which in turn were followed by the horse-drawn plows that broke the plains to facilitate the sowing of food crops.

The peak of North American horse population was reached before World War I. After that, trucks began to replace horsepower in the cities. By the end of World War II, tractors had virtually driven horses off the farms. Today, the great majority of horses are used for sport—*thoroughbreds* for running and racing, *standardbreds* for harness racing, and *quarter horses* for working cattle, rodeo performances, trail riding, and pleasure riding. The standardbred horse is the only breed exported extensively by the United States; these horses are sent all over the world to provide breeding stock for harness racing. Other American breeds include the *American saddle horse,* both three- and five-gaited, and the *Tennessee walking horse,* both of which are primarily horse show breeds. American breeds characterized by their color are the golden-colored *palomino* and the spotted *appaloosa.* Descended from the horse of conquistadores are horses of the *pinto* breed, which are marked in patches rather than in spots. One of the oldest American breeds is the *Morgan horse,* descended from Justin Morgan, a stallion foaled in the late eighteenth century and belonging to a Vermont school teacher of the same name. These horses are noted for their versatility in being able to perform all types of work.

Ponies are also bred extensively in this country, but the majority of them are imported from England, particularly the *Welsh* and the *Shetland;* the *Connemara* comes from Ireland. A native American breed is the *Pony of the Americas,* a small type of Appaloosa.

The gestation period of the horse is approximately 11 months. Given reasonable care, horses are generally free from disease, although they are subject to a number of respiratory ailments, most of them not fatal. Horses are also subject to sleeping sickness.

There has been a strong revival of interest in riding of all kinds in North America, and in fact throughout the Occident. —Alexander Mackay-Smith

Hydroponics, the science of growing plants in water or sand culture. In *water culture,* the plants grow with their roots suspended in a dilute solution of the essential mineral elements, contained in shallow tanks. They are supported above the water by wire netting or hardware cloth, which is covered with wood shavings, peat moss, or similar material in order to exclude light from the solution and maintain a high humidity around the upper roots. The solution must be renewed at intervals as water and elements are absorbed. It must be aerated to supply oxygen to the roots. In *sand culture,* the nutrient solution is applied frequently to containers of sand, or other inert media (vermiculite, perlite, haydite), in which the plants are growing. This method has been widely used in experimental studies of plant nutrition for many years.

A variation of the sand culture method, known as *sub-irrigation culture* is the only commercial application of hydroponics. Watertight beds are filled with a coarse medium, such as washed gravel. The nutrient solution is pumped into the beds or allowed to flow into them from overhead reservoirs, until the beds are filled to within an inch of the surface. The solution then flows out of the gravel to a storage tank. Since the only source of water and nutrients for the plants is from the films of moisture around the gravel particles, the cycle of filling and draining the beds must be repeated frequently.

The major handicap to economic food production by hydroponics is the cost of the installation. Some technical training and considerable experience are necessary for the efficient management of soilless culture crop production. Soilless culture was used to produce vegetables at certain United States Army bases overseas during World War II. Hydroponics continues to be an absorbing hobby for the home gardener. —Neil Stuart

Irrigation, the practice of making available to crops, pasture, lawns, and turfs a greater quantity of water than that retained from natural rainfall. It thus includes various flooding or ditching techniques for direct water application, called *gravity* or *low-pressure irrigation; overhead sprinkling* or *high-pressure systems* of varying design; *diking* of fields to obstruct the natural draining away of rainfall, as in rice paddies; and situations where groundwater tables are or can be made sufficiently shallow to permit the wetting of root zones from below.

Irrigation has played an important role in agricultural production since the dawn of history, as evidenced by the early civilizations in the Tigris and Euphrates valleys and the dynasties of ancient Egypt supported by the floodwaters of the Nile. Today about 415 million acres—about one-eighth of the world's arable agricultural land—are irrigated. Three-fourths of the total is in Asia, especially China, India, and Pakistan. North and Central America account for about 45 million acres, led by the United States with about 42 million acres of irrigated land. About 35 million acres of this is in crops, and 7 million acres in pasture. It has been estimated that the U.S. total eventually could reach about 75 million acres.

About 92 per cent of the irrigated land in the United States lies in the 17 western states, where the practice originated with aboriginal Indians, was taken up by the Spanish missionaries and by colonists from both Spain and Mexico, and was put on a sustained basis by the Mormon settlers who arrived in Utah in 1847. Rapid expansion through the West began about 1870 from a 32,000-acre project supported by Horace Greeley, near present-day Greeley, Colorado. Other large projects were started in about the same period near Riverside and Anaheim, California. But irrigation is also important in the Mississippi Delta states, in Hawaiian sugarcane areas, in Florida citrus groves, and on truck farms in New Jersey. Also, it is increasing in importance through most of the eastern states.

BUREAU OF RECLAMATION, U.S. DEPARTMENT OF THE INTERIOR

FLOOD CONTROL in the Missouri River Basin Project of Montana performs the secondary function of aiding sugar beet production by diversion of water through irrigation systems.

■**GRAVITY IRRIGATION.**—Methods of irrigation vary considerably from area to area and are closely related to the nature and proximity of water supplies. Gravity methods utilizing ditches, furrows, border dikes, and flooding are still employed on 88 per cent of the acreage in the older irrigated areas—the West, the Mississippi Delta, and Florida. About 52 per cent of this acreage is served by diversions from streams or reservoirs. Most of the remainder is supplied by pumping water from wells. While limited to nearly flat terrain and requiring considerable expense for land-shaping as well as considerable volumes of water, gravity irrigation has the special advantage of requiring a comparatively small investment for irrigation equipment. Operating costs are largely for labor. Most of the needed reservoirs, canals, and land preparation were arranged through mutual irrigation companies or in recent years by the Department of the Interior's Bureau of Reclamation.

■**OVERHEAD IRRIGATION.**—Overhead or sprinkler irrigation is gaining rapidly in importance in all areas of the United States and in other countries too. More than 15 per cent of the irrigated acreage in the United States is now sprinkled, compared with 2 per cent in 1950. The most common system lifts water from wells or adjacent streams and forces it at high pressure into a main pipeline having several portable lateral lines, each with its own series of nozzles or sprinklers. The lateral lines are periodically disconnected and moved to a new valve point on the main line until the field is irrigated completely. To avoid uncoupling of lateral sections, the complete lateral may be mounted on large wheels and rolled to its new position. Or there may be a combination main and delivery line on lugged wheels in which some of the water pressure itself is used to inch the system along automatically. This is called a *hydraulic-move system.* Still another sprinkler system consists of a single large boom, or perhaps several booms, mounted on a trailer with giant nozzles along the booms and at the ends, operating much like an ordinary rotating lawn sprinkler. Some of these rigs can irrigate up to 5.5 acres per setting. Sprinkler systems are well adapted to both level and moderately sloping land but require a considerable investment in equipment. Their special advantages are complete control over rates of water application to match absorptive capacity of soils, uniform application to all areas, and portability.

■**SUBIRRIGATION.**—Subirrigation is possible with little expense under conditions where ground water tables are naturally shallow, as in the cranberry bogs of Massachusetts and Michigan. Favorable conditions also can be created if the subsoil is impervious at a depth of 6 feet or more and if the surface soil is relatively permeable. A common technique is to surround large blocks of land with levees to prevent flooding, to install drainage systems, and then to pump excess water in wet seasons over the levees into the streams. In dry periods, the river water is siphoned or pumped back over the levees into ditches about one foot wide, 2 or 3 feet deep, and up to several hundred feet apart. The water is diffused through the root zone by capillary action, with downward percolation restricted by the impervious subsoil. Subirrigation can be very efficient as expenses for labor and operation are relatively small.

The role of irrigation in the U.S. farm economy is indicated by the fact that irrigated crops account for about one-fifth of the value of all harvested crops but only 8.5 per cent of the total crop acreage. In the order of their market value, the leading irrigated crops are cotton, vegetables, tame hay, potatoes, sugarbeets, corn, sorghums, barley, and wheat. Citrus and other orchard and vineyard crops also rank high. The value of the 7 million acres of irrigated pasture is difficult to estimate, but this acreage is an essential part of the feed base on many dairy farms and cattle or sheep ranches. —George A. Pavelis

Jute. See *Rope,* page 1710.

Meat Packing. See *Food Manufacturing.*

Oats, a grain of the grass family, widely used as a source of human and animal food and an important source of straw. Oats rank fourth among the grain crops, next below wheat, corn, and grain sorghum. Despite the inroads of motor-propelled machines on farms, displacing the horses and mules that formerly were the main consumers of oats, this grain retains a prominent place in the farm economy. Now the oat crop is fed largely to dairy cattle, poultry, hogs, and sheep.

About 4 per cent of the billion-bushel annual United States crop goes into the manufacture of oatmeal for the breakfast table. Some oats are pastured and cut for hay or livestock bedding.

The oat originated in eastern Europe or western Asia. Among its principal growers today are the United States, the Soviet Union, Germany, Canada, France, and the United Kingdom. The type that is generally grown, known as the *common oat,* develops an upright pinnacle with branches falling about equally to all sides. Another type, known as *side oats,* has branches that fall to one side. Oat hulls vary in color according to variety. Some are white or gray, others are yellow, red, or black. The hulls are most often white in the sections of predominant U.S. production—the Corn Belt, the Great Lakes states, and the northern Plains. These areas taken together grow about three-quarters of the 25 to 30 million acres seeded annually in the United States. In the South, oats are usually red or gray. Most oats in cool climates are sown in the spring, but a fall-seeded crop will winter over in the warmer Southern states.

—Charles E. Rogers

Peanuts, a legume of the pea family, deriving the "nut" part of the name from the fact that the plant ripens its product within a shell. Though the peanut originated in South America, probably in Brazil, it has migrated to many parts of the world. China, India, West Africa, and the United States are leading producers. Except in the United States, peanuts are mainly crushed for oil, but here about 50 per cent of the crop goes into peanut butter. Most of the remaining nuts are roasted for direct consumption, used in candy and bakers' goods, or left in the soil to be rooted out by swine. *Peanut vines* produce hay equal in feeding value to clover. In the United States, the peanut is an annual plant grown mainly in the Southern states. Planted from the hulled seed in the spring, it develops *pegs* after blossoming. The pegs elongate and go into the soil where they produce the *groundnut,* as the peanut is sometimes called.

—Charles E. Rogers

Pesticides, any of various substances used by the agriculturist to destroy plant or animal pests. Most pesticides are manufactured chemicals that, by the nature of their purpose, must be lethal to some living organism. The utility of pesticides depends on the fact that there is a wide difference in reaction from species to species.

Specific terms are applied to pesticides according to the organism controlled. The Federal Insecticide, Fungicide, and Rodenticide Act, which regulates the marketing of pesticides or economic poisons shipped in interstate commerce, defines several such terms.

Insecticides are substances or mixtures of substances intended for preventing, destroying, repelling, or mitigating any insects that may be present in any environment. In the same manner, *fungicides* and *herbicides* are defined as substances intended for the control of fungi and weeds, respectively.

■**INSECTICIDES.**—Prior to 1940, agricultural pesticides were rather simple substances such as lime sulfur, arsenic acid, and Paris green (lead arsenate), as well as a few substances of plant origin such as rotenone and pyrethrum dusts. As a by-product of World War II, chemicals from a group developed for the purpose of destroying mankind were found useful in destroying insect pests. During this same period, it was discovered that the chemical *dichlorodiphenyl trichloroethane* (DDT) was very toxic to insects but relatively safe to man. These developments set the stage for a new era in the agricultural chemical industry. Today, there are four important groups of insecticides including chlorinated hydrocarbons, organic phosphates, carbamates, and botanicals.

The *chlorinated hydrocarbons* are organic chemicals that commonly contain chlorine, although they may contain fluorine, bromine, or iodine. There is a wide variation in the complexity of these structures, but the effects on the body are somewhat similar. Little is known of how these effects take place.

The fact that these compounds are soluble in animal fat accounts for scientists' concern regarding their storage in the body fat. Although the storage of DDT in human tissues has been studied for over 20 years, little significance has been associated with the storage except the obvious fact that it is proof of exposure and absorption of the compound.

The basic action of *organic phosphates* is similar within this group, but the degree of toxicity varies widely. In a very general way, these compounds disrupt transmission of nerve impulses, functions of the body over which we have little or no conscious control.

Carbamates affect the same systems of the body as do the organic phosphates. The major difference is that the effects of organic phosphates on the regulatory components of nerve transmission are either nonreversible, or only very slowly reversible, whereas carbamates are rapidly reversible in character. Carbamate compounds also vary widely in their toxicity both to man and insects.

Botanicals have a wide range of actions and toxicities. Nicotine is one of the most toxic to man, and pyrethrum is about the least toxic. *Rotenone,* which by laboratory evaluations is very toxic, is not a problem under practical conditions.

■**HERBICIDES.**—Chemicals of a nonspecific type, such as arsenic acid, had been used as herbicides prior to 1940. About this time, a number of relatively selective herbicides became available. These chemicals are, in general, much less toxic to man than the nonspecific types. *Dinitrophenols* are among the important herbicides toxicologically. Repeated exposure increases the chances of intoxication. Basically, these compounds increase the metabolic rate. Another important group of herbicides is the *chlorophenoxy acetic acid derivatives,* which include 2,4-dichloro phenoxy acetic acid (2,4-D). In plants these compounds function as hormones. The action in animals is thought not to be hormonal in nature, but the exact mechanism is not known. They are among the least toxic pesticides to man.

■**FUNGICIDES.**—There are a number of important agricultural fungicides. However, the two most frequently used classes are the dithiocarbamates and the organic mercurials. While many of the *dithiocarbamates* are irritating to the skin and eyes, they are not sufficiently toxic to be considered hazardous to use. For this reason, they have not been extensively studied as to mode of action. *Organic mercurial compounds,* such as phenyl mercuric acetate, are moderately to very highly toxic to man, since there is a general tendency to accumulate mercury in the tissue of the liver and kidney. Continued exposure to high concentrations of mercury from any source can produce serious injury.

—J. S. Leary, Jr.

U.S. DEPARTMENT OF AGRICULTURE

SPRAYING soft-jet pesticides prevents or destroys various tree diseases.

Pheasant, an edible bird having a compact body, short unfeathered legs, and short rounded wings. The Ringneck is the most common of the pheasants. There are other popular breeds, such as Mongolian and Chinese, and ornamental breeds including Golden, Silver, Lady Amherst, and Reeves. In many sections of the country pheasants are popular as game birds and are raised for sports clubs. They are also used as a specialty food. Pheasant eggs hatch after 22 to 24 days of incubation. Pheasants are highly nervous and may injure their heads when raised in confinement. This is true even when the wing feathers of one wing are clipped or the last joint of one wing is removed to prevent flying. Most states in the United States restrict the raising of game birds, and require a permit or license for the raising or hunting of pheasants.

—Carl W. Hess

Pigeon, a bird with dense fluffy plumage, a stout body and short neck, a small head, and a square or rounded tail. Young pigeons are called *Squab.* Of the many pigeon breeds, the King, Carneau, and Homer are the ones used for squab production. Squabs grow very rapidly and reach market age in about 4 weeks, depending on the size of the breed. *Homing pigeons* are raised by pigeon fanciers for racing. About 17 days of incubation are required for pigeon eggs to hatch. —Carl W. Hess

Poultry Products. See *Food Manufacturing.*

Rice, a native grass of tropical Asia and the cereal food mainstay of the peoples of China, Japan, the mainland of southeast Asia, and the islands of the southwest Pacific. This is the *rice bowl,* where 95 per cent of the world's supply of the cereal is produced and consumed. There are thousands of known varieties, more than of any other crop.

Alone among the world's great crops, rice grows typically in a field of standing water, but the land must slope enough to allow a slight movement of the water. Irrigation is the prime requisite. Upland rice, grown without irrigation, constitutes a negligible part of the total harvest.

Most rice is grown in coastal plains and tidal deltas in tropical, semitropical, and temperate climates. Relatively high humidity and an average temperature of 70° F during the growing season of 4 to 6 months are necessary for best results. Heavy soil is desirable, and the subsoil must be impervious to water. Except in technically advanced countries like the United States, rice cultivation is done almost entirely by hand.

Seed is sown broadcast, and the seedlings that emerge are transplanted when they are 6 to 8 inches in height. They are planted in water 2 to 4 inches deep, in rows about one foot apart, to permit intensive cultivation. From the time the seedlings are transplanted until the harvest, they are supplied with heavy fertilization. Harvested rice is threshed by flailing or beating. It is then winnowed and the hulls removed. Afterward the clean rice is polished.

Communist China leads the world in rice production—about 80 million metric tons annually, according to an estimate made by the Food and Agri-

World Production of RICE in %

Country	%
MAINLAND CHINA	32.9
INDIA	20.8
JAPAN	7.0
PAKISTAN	6.5
INDONESIA	5.2
THAILAND	3.5

Based on a five-year average compiled by the United Nations

culture Organization of the United Nations. The second largest producer is India with 55 million metric tons,

RICE CULTIVATION remains unchanged despite technological improvements in most of Asia, and many farmers still employ the primitive methods used by their ancestors for centuries.

and Japan falls into third place with 17 million. Some rice is grown on every continent. North America produced 4.5 million metric tons in 1964–65, the largest part—3.25 million tons—in the United States. The leading rice-producing states are Louisiana, Texas, Arkansas, and California.

About two-thirds of the world's exports of rice in 1964–65 were from six Asian countries: Thailand, Burma, South Vietnam, Cambodia, Taiwan, and South Korea. The principal importing countries also were Asian—Indonesia, India, Japan, Malaya, the Philippines, and Pakistan. The United States exported a record 2.1 million tons in 1964–65. India was the largest importer of U.S. rice. Increases in both acreage and yield among the principal importing countries of Asia have failed to bring about sufficient production to satisfy growing domestic needs. —Charles E. Rogers

Rye, a cereal of the grass family. Like many others, it probably originated from a wild species growing in the Near East. It is used largely as a bread grain in the Soviet Union and Germany. It flourishes in a variety of conditions, but when grown on good soil yields less profit than other crops. It is therefore cultivated mainly on the poor, light, dry land un-

World Production of RYE in %

Country	%
SOVIET UNION	46.4
POLAND	21.1
EAST GERMANY	9.1
WEST GERMANY	5.1

Based on a five-year average compiled by the United Nations

suited to producing high yields of wheat, characteristic of northeastern Europe. It is extremely winter-hardy and drought-resistant.

In Great Britain, rye is grown as a forage crop, sometimes as a spring crop to be consumed at an early stage of growth. If allowed to reach maturity, it is hard, dry, and unpalatable to stock. The straw is used for thatching, packing, and similar purposes. The crop is not important in the United States, where its annual production is comparatively small and is used chiefly to give flavor to the so-called rye bread. The grain is malted to make rye whiskey, and the straw is used for packing.

Rye is subject to a fungus disease, *ergot,* and if infected grain is made into bread, it can be a dangerous poison, sometimes causing death. Ergot is used, however, as an ingredient of drugs for medical purposes. —G. E. Fussell

Seafood. See *Fish and Seafood,* page 493.

Sheep, horned, woolly mammals that are similar to goats. Males are called *rams,* females *ewes,* castrated males *wethers,* and the young *lambs.* Domesticated sheep are kept for wool, meat, fur, and hides. Outside the United States, ewes are sometimes kept for milk. Some sheep breeds produce higher-quality wool than others, and some produce meat more abundantly and rapidly. Meat animals must be raised comparatively near the point of consumption, because mutton cannot be pickled or smoked. Lambs are preferred for meat in the United States. Various breeds differ in their gregariousness. Sheep that are range-grazed should be gregarious, while sheep kept in farm pastures need not flock. The mutton types are best kept separate and need not be very gregarious.

Various crosses of breeds can be achieved to secure other desired characteristics, such as hardiness, rapid lamb growth, large lamb crops, and dual production of wool and meat. Wool breeds, sufficiently gregarious for herding, include the Rambouillet, Merino, Corriedale, Romeldale, Panama, Columbia, Targhee and the long-wooled Lincolns, Cotswolds, and Romneys. The meat breeds include the Hampshires, Suffolks, Oxfords, and Southdowns.

Sheep usually mate in the fall, and the lambs are born in the spring. The gestation is about five months, and the breeding life of a ewe averages seven years. Several of the most deadly sheep diseases, such as anthrax, blackleg, tetanus, and rabies, have effective vaccines. Crippling or fatal diseases, such as foot-and-mouth, listerellosis (circling disease), mastitis, and shipping fever respond well to antibiotics. Black disease, caused by liver flukes, can be prevented by chemical destruction of the intermediary snail hosts. Brucellosis and tuberculosis in sheep are very rare, and slaughter eradication works well. Liver flukes, long worms, and tapeworms all succumb to phenothiazine or other vermicides. Dips and sprays are effective against disease-carrying ticks and insects.

Sheep are ruminants of the genus *Ovis,* with four wild species-groups, one domesticated species-group, and many subspecies.

—John T. Schlebecker

Sisal. See *Rope,* page 564.

Soils, earth composed of inorganic particles, organic matter, water, and air. They are formed by the weathering of rocks and the decay of organic material.

The formation of soils determines their classification into five broad soil belts, which in turn are divided into groups and subgroups. The *tundra soils* constitute the most northern of these belts; they are frozen much of the year and are available for cultivation only in the summertime. *Podzolic soils* have been formed mainly in the forested areas of eastern Canada, northeastern United States, and northern Eurasia. *Latosolic soils,* tropical and subtropical soils formed by the decay of forest and savanna vegetation, are extremely leached; the soils of the southeastern United States belong to this classification. *Chernozemic soils* are the product of prairie and steppe grasslands; they are relatively unleached and fertile. *Desertic soils* are those of the deserts; they are usually productive when irrigated.

CHEMICAL CONTENT.—Plants need at least 16 chemical elements for growth and seed production: carbon, hydrogen, oxygen, calcium, nitrogen, phosphorus, potassium, sulfur, magnesium, iron, manganese, zinc, copper, molybdenum, boron, and chlorine. Of these, the first nine are the most essential. Carbon, hydrogen, and oxygen are obtainable from the water and the air, but the other elements or nutrients must be obtained from the soil.

The podzolic and latosolic soils of the eastern United States are usually deficient in calcium and consequently are more acid than alkaline. Since many plants are sensitive to acidity, these soils need the addition of calcium in the form of lime. To designate the soil condition, a pH scale from 0 to 14 is used. A pH factor below 7 indicates acidity; above it, alkalinity.

Nitrogen and sulfur are needed in the production of proteins, while phosphorus is necessary for photosynthesis and seed formation. Potassium is essential for such physiological processes as cell growth and the formation of starch and sugars.

These and other nutrients are added to soils deficient in them by means of crop rotation, green manuring, and the application of animal manures and commercial fertilizers. The inclusion of clover or alfalfa in a *rotation cycle* enriches the soil because of the action of nitrogen-fixing bacteria, which live in nodules on the roots of these plants. *Green manuring* is the plowing under of a growing, leafy crop, usually leguminous. Farmers have long used *animal manures,* but recently they have increased greatly their use of *commercial fertilizers.* In the years between 1942 and 1962, the consumption of commercial fertilizers in the United States more than tripled.

MOISTURE.—The *moisture content* of the soil is of major concern to farmers. If there is too much, they must provide surface drainage by grading, terracing, and ditching or subsurface drainage by the construction of *tile* or *mole drains.* If the rainfall is normally sufficient, they must try to retain it by cultivation and by the destruction of weeds. If there is not enough, they must either conserve it or supplement it by irrigation.

In the United States, on the Great Plains and in the intermountain valleys farther west where annual rainfall is less than 20 inches, farmers and agronomists have developed a type of moisture-conserving agriculture known as *dry-land farming.* It consists of the culture of drought-resistant crop varieties and of a system of alternate cropping known as *summer fallowing* in which every other year the land lies fallow in order to store moisture.

In arid regions, water is diverted from streams into ditches or pumped from wells and applied to the land by methods called *flooding, furrow, border, basin,* and *sprinkler irrigation.*

TILLAGE.—To produce crops, the soil must be *tilled.* The first step in tillage is plowing, which loosens, granulates, and turns under organic materials. The seedbed is usually further pulverized by means of spike-tooth and disk harrows. Then crops are planted by means of planters and drills and, in the case of intertilled crops, cultivated during the growing season to reduce competition from weeds.

Loose soil, however, is subject to *erosion* by wind and water. On the Great Plains, where wind erosion is a problem, it has been checked by strip cropping and the creation of a stubble or trashy mulch. *Strip cropping* is the planting of grain in strips alternating with fallow, while *stubble mulch* is created by chisels and sweeps, loosening and granulating the soil but retaining the stubble from the previous crops on the surface. Erosion by water is retarded by *terracing, contour plowing,* and the planting of cover crops.

—Robert G. Dunbar

Soybeans, an Asian legume or its seed. Soybeans originated in the Orient; in China they have been a staple food product for thousands of years. Introduced to America and grown in a small way in the nineteenth century, they were rediscovered as a valuable legume crop during the past half-century. Their rise to third place among cash crops in the United States took place after World War II. Adapted to temperate climate, with warm, humid growing seasons, this annual farm crop does well in tropical and subtropical areas as a hay and cover crop, though not for the development of beans. As a bean crop it is best adapted to the Corn Belt and the Great Lakes and Delta states.

Among the oilseed crops of the Western Hemisphere, soybeans hold first place. The annual value of the soybean crop in recent years has been nearly $2 billion. In the early 1930's, American farmers grew only enough soybeans to sell $10 million worth annually. About 700 million bushels were grown in the United States in 1965. Fifty per cent of the soybean crop in the United States before 1941 was grown for green manure, for grazing use, or for hay. Two-thirds of the crop is used to produce soybean oil and soybean meal. Over 25 per cent of the bean is exported.

The upsurge of soybeans during and since the 1940's is due to several circumstances. During World War II, supplies of fats and oils from the Far East were cut off, and margarine and shortening makers used soybean oil in place of imported oils. After the war, consumers were able and willing to

U.S. DEPARTMENT OF AGRICULTURE

PRIMARY SHEEP PRODUCTS are wool and mutton. The Hampshire breed (*right*) is more suitable for mutton, while the fine-fleeced Merino (*left*) is best used for wool.

M. FELDMAN

SUGAR CANE being dropped into a waiting wagon by a loader that crops a quarter-ton of cane. The new tractor, dubbed "Sugar Babe," has the capacity to pull four loaded trucks, or 65,000 pounds, to a sugar mill. This harvest is part of Florida's expanding sugar industry.

buy more livestock products, thus opening up a market for soybeans in mixed feeds. Recent rapid expansion of soybean crops in the United States has been aided by variety adaptation, and by advances in farm mechanization, marketing, and technology. Of the soybeans grown in the United States, 90 per cent is for food—in the form of oil for human consumption and soybean meal in mixed feeds for animals—and 10 per cent for nonfood products. In food industries, margarine and shortening take about one-third of the soybean oil produced annually; salad oil and mayonnaise, one-fourth. Nonfood articles include paints, varnishes, and lubricants; soaps, sprays, and cosmetics; oilcloth and linoleum.

Soybeans are eaten in many forms in China. *Bean curd* is the most common form, in addition to bean *milk,* which is pressed from the curd. The green leaves of the soybean find their way to the Chinese dinner table boiled as greens. The stalks are fed to pigs and chickens, or dried for fuel.

—Charles E. Rogers

Spices, aromatic vegetable products used primarily for seasoning and preserving food, and to a lesser extent for making perfumes, soaps, and lotions. The "true" spices, so designated by spice specialists, are pepper, vanilla, cloves, cinnamon, nutmeg, and ginger. Some spices come from the seeds, buds, flowers, and fruits of plants; others, from the bark, leaves, or roots. The harvested substance is usually ground, as is pepper, the most widely-consumed spice.

All true spices grow in the tropics, particularly in Asia and Africa. Cinnamon, cloves, nutmeg, pepper, and opium have been cultivated since medieval times in Ceylon, Malacca, and the Malabar Coast. For centuries these areas were exploited by Arabia, Venice, and Portugal, who amassed a great wealth by importing flavoring, meat preservative, and luxury spices to western Europe.

Measured by use and world trade, *pepper* is the leading spice and the United States is the leading consumer. Both black and white pepper are derived from a berry, known as *peppercorn,* that grows on a vine. India and Indonesia produce two-thirds of the world's supply.

Vanilla comes from an *orchid,* which also grows on a vine. About two-thirds of the world production originates on islands off the southeast coast of Africa. Besides its well-known use as a flavoring, it is employed in the manufacture of chocolate, perfumes, and soap.

Cloves grow on trees that are native to Indonesia, though the bulk of world production now comes from Tanzania and the Malagasy Republic. While cloves have been used for centuries to flavor and decorate foods, two-thirds of the world's crop is now ground and mixed with cigarette tobacco.

Cinnamon, a popular flavoring, is derived from the bark of a tree grown in Asia. Its cousin *cassia* is similar, but is usually regarded as somewhat inferior. Trees that supply these spices belong to different but related species.

Nutmeg and *mace* grow on an evergreen tree—nutmeg is the seed, and mace is the membrane around it. Indonesia and the West Indies are the chief sources. Ground nutmeg has many uses in the food and beverage trade, and the oils of nutmeg and mace find their way into soaps, cosmetics, perfumes, confections, and pharmaceutical products.

Ginger, a pungent spice used widely for making ginger ale and for medicinal preparations, is derived from a root native to Asia. About one-half of the world's supply comes from India. Other sources are the west coast of Africa and Jamaica.

In addition to the true spices, there are many herbs and other plants that are commonly regarded as spices. Included among these are oregano, anise, caraway, cardamon, coriander, cumin, fennel, and mustard seed.

—Charles E. Rogers

Sugar, one of the carbohydrates, an important source of energy. Technically known as *sucrose,* it is produced commercially from sugarcane and sugarbeets. There are many types of sugar, notably *dextrose,* produced from corn, and *lactose,* produced from milk. Other sources of sugars are *honey, sugar maple trees, palm trees,* and *sorghum.* Sugars are produced in the leaves of green plants by photosynthesis. Sugarcane and sugarbeets, which store sugar abundantly, have become the primary sources of sugar.

Sugarcane is a large perennial grass, which is produced in tropical and semitropical climates. Cuttings of the cane stalk, rather than seed, are used to propagate the crop. The cane attains heights of 10 to 20 feet. It is normally harvested 12 to 24 months after planting. For centuries, sugarcane was laboriously harvested by hand; today, the crop is often harvested by machine. The stalk is cut near the ground, and the top and the leaves are removed. The stalk or cane is shipped to a nearby sugar mill for processing. Normally, a number of crops may be cut from the same roots. It is not necessary, therefore, to replant every year.

The sugarbeet, a biennial plant, is produced in temperate climates. Normally, the crop is planted in the spring and harvested in the fall. Sugar is produced in its leaves and stored in its root. The crop is harvested by lifting the beet from the ground and removing the tops or leaves. Sugarbeet tops are fed to livestock. The root, or beet, is shipped to a nearby factory for processing.

Sugarcane is believed to have originated in the Orient, where its juice was valued hundreds of years before the birth of Christ. People are believed to have learned to produce sugar from sugarcane sometime about the fifth century. Much early trade revolved around sugar. Columbus introduced sugarcane to the New World on his second voyage.

The production of sugar from sugarbeets is a relatively recent development. In 1747, Andreas Marggraf, a German chemist, proved that sucrose could be extracted from sugarbeets. The world's first beet sugar factory began operating in Europe in 1802. The first successful sugarbeet operation was started in America in 1879.

Approximately 60 per cent of the sugar consumed in the United States is processed from sugarcane and sugarbeets produced by American farmers. The remaining 40 per cent is imported and then refined in the United States.

Per capita consumption of sugar in the United States is approximately 100 pounds a year. One-third of the sugar consumed is distributed through retail stores. The other two-thirds is distributed to food processors—bakers, bottlers, canners, confectioners, ice-cream manufacturers, and others —who use sugar as an ingredient in their products.

Sugar is an essential ingredient in many processed foods. In addition to its contribution as a sweetener, sugar can enhance the flavor, appearance, and texture of many foods. It performs many other purposes, one of which is that of a preservative.

—Nicholas Kominus

Sweeteners, Artificial, chemical compounds that are used for sweetening. They are also known as *synthetic, nonnutritive,* or *noncaloric sweeteners.* Artifiicial sweeteners have no nutritional value. They are primarily produced for people suffering from diabetes and other diseases that prohibit the consumption of sugar, and for dieting.

Saccharin, the first commercial artificial sweetener, is produced from coal tar. It was discovered by Ira Remson, an American chemist, in 1879. Its sweet taste, however, was discovered by Constantin Fahlberg, a German chemist working in Remson's laboratory, who obtained a patent on a process to manufacture saccharin. Commercial production began in 1901 in the United States.

Sodium cyclamate and *calcium cyclamate,* known as the cyclamates, are the major artificial sweeteners produced in the United States. Commercial producation of the cyclamates began in 1950 in the United States.

Mixtures of saccharin and the cyclamates have become increasingly important in recent years.

Saccharin is generally considered to be 300 to 500 times sweeter than sugar. The cyclamates are considered to be 30 times as sweet as sugar.

Artificial sweeteners are produced in a number of plants located throughout the United States and many other countries. —Nicholas Kominus

Swine, omnivorous hoofed mammals that are a domesticated form of wild hog. The male is called a *boar,* the female a *sow,* the young female a *gilt,* the castrated male a *barrow,* the young a *pig.* In the United States, swine are kept primarily for meat and lard, with hides and bristles as side products. Hogs are especially desirable meat animals because pork can readily be preserved by pickling or by smoking. Furthermore, swine eat almost anything and can be raised under a variety of conditions. The animals also mature quickly and reproduce frequently with large litters,

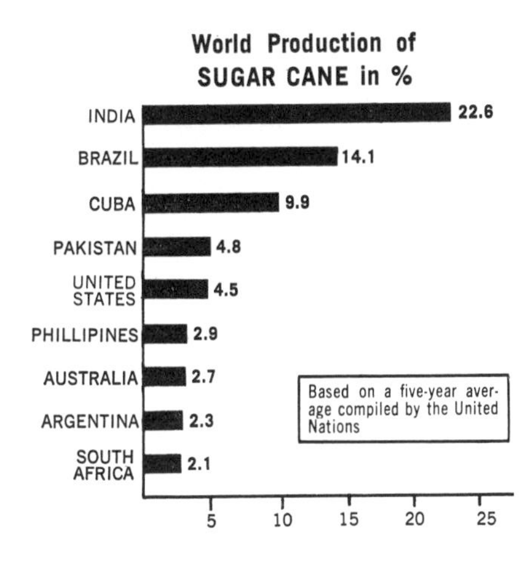

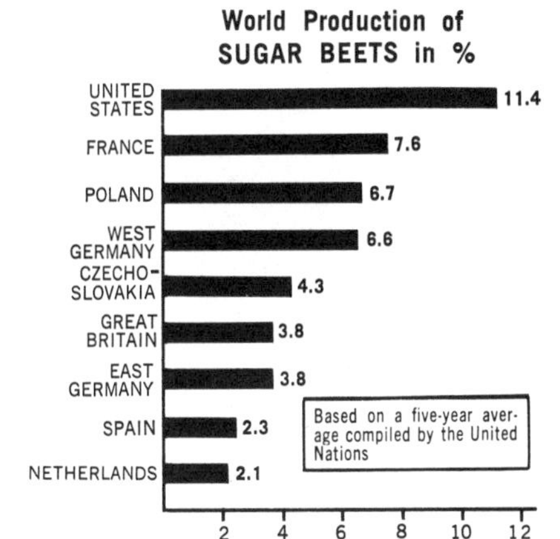

so that they provide a large and steady supply of meat. Currently, hogs reach market weight in five or six months, and with recent advances in pig weaning and feeding, sows can be bred three times a year. The gestation period averages 114 days, and sows come in heat about once every three weeks.

Various breeds of swine have different characteristics, although crosses are frequently preferred. Some breeds tend to produce more meat than others, but there are no distinctive meat or lard breeds. In general, the breeds which produce the most meat are the Yorkshire, Poland-China, Duroc, Tamworth, and Danish Landrace, and the recently developed American Landrace, Maryland #1, Minnesota #1, Minnesota #2, Beltsville #1, Beltsville #2, Montana #1, and Palouse. Other swine with varying advantages in size and hardiness are the Chester White, Berkshire, Hereford, and Spotted-Swine.

Swine gain best on formula rations containing corn, barley, sorghum, and alfalfa, with heavy supplements of vitamins, proteins, minerals, and antibiotics. Supplements should be added even when the swine are on pasture. Swine are susceptible to a wide range of diseases which, even when not fatal, sharply reduce rates of gain. Some diseases, such as brucellosis (abortion), atrophic rhinitis, influenza, and virus pneumonia, must be controlled by sanitation and sometimes by the slaughter of infected animals. Other diseases, such as erysipelas, leptospirosis, and hog cholera, can be prevented with vaccines. Swine parasites are mostly worms such as flukes, tapeworms, and roundworms. Roundworms cause the most trouble in the United States. Group treatment of swine produces the best results. Treatment requires the use of dangerous chemicals. Effective vermifuges include phenothiazine, sodium fluoride, piperazine, and hydrocyin. *Trichina* are not serious in swine because recovery occurs rapidly as the worms enter the cyst state.

Swine belong to the genus *Sus,* with three species groups and several species subgroups.

—John T. Schlebecker

HAMPSHIRE SWINE REGISTRY

AMERICAN YORKSHIRE CLUB

SWINE provide a variety of pork and pork products. The Hampshire sow (*left*) is recognized as an excellent lard producer. The Yorkshire hog (*right*) is prized for its quality bacon.

Tea, an evergreen shrub, its leaves, and the beverage brewed from these leaves. The tea plant, *Camellia sinensis,* whose leaves contain caffeine and tannin, grows in the humid tropics. Frequent rainfall, plentiful sunshine, and good soil fertility—conditions natural to Ceylon and India—produce the best yields and quality. In Japan, rapid yield increases have resulted from the use of organic fertilizers as plant nutrients.

Tea is grown at varied elevations and on different terrains. Indian tea grows at elevations from 200 to 6,000 feet, mainly in the Darjeeling hill district. It is commonly grown on flat foothills in Latin America. In China, tea has been cultivated for centuries on almost vertical terraced banks, where frequent rainfall results in natural irrigation. Tea is regarded as a plantation crop, but it is often cultivated on small holdings and estates in Africa, Asia, and Latin America.

The three major varieties of tea are green, black, and oolong. Green, an unfermented leaf, is immediately toasted, while the black is fermented before being machine-toasted (Ceylon black dominates U.S. tea imports). Oolong is a semifermented leaf.

The tea plant reaches maturity 4 to 6 years after planting. Replanting is customary after about 60 yields per plant. The leaves are graded by size: the fine leaves near the end of the twig produce a high-quality tea, in contrast to coarse plucking.

India and Ceylon export about 75 per cent of the world's tea, although 23 countries are rated as major producers. In 1967 India produced approximately 840 million pounds of tea, or about 38 per cent of world production (excluding Mainland China), and Ceylon produced 487 million pounds, or 20 per cent. In 1967 about 90 per cent of the world's tea was produced in Asia and 10 per cent in Africa and South America.

U.S. DEPARTMENT OF AGRICULTURE

VEGETABLES are always available in U.S. markets. Cabbages (*right*) are grown in a suitable climate year-round, and snap beans (*center*) are bred for adaptability to fresh, frozen, or canned sales. Potato fields (*left*) produce the largest-selling crop.

The United Kingdom, the world's largest tea-drinking area, at just below 10 pounds per person a year, consumed 503 million pounds in 1967, mostly from India and Ceylon. Although the United States is the second largest tea importer, Americans consumed only 0.67 pounds per capita in 1967. U.S. tea imports, mostly from Ceylon, were one-quarter of the United Kingdom's total.

From the 1930s to 1955, tea export quotas were controlled by the International Tea Arrangement. Prices of tea have been falling recently, and world tea production has been setting annual records, due to expanded acreage and use of higher yielding varieties. The 1968 crop should reach another record high, but it is anticipated that expanding world consumption will absorb the larger supplies with little change in global stock positions.

—Marshall H. Cohen

Tobacco, a herbaceous plant of the nightshade family whose leaves are used extensively for smoking and chewing and as snuff. The origin of tobacco was well symbolized by the cigar-store Indian, a hallmark of tobacco shops in earlier generations. The American Indian introduced tobacco to the first white colonists in North America in the fifteenth and sixteenth centuries. He taught the Europeans how to use the leaf in forms common today—for smoking, chewing, and snuff. The fundamentals of cultivation were known by the Indian and passed on to the white man.

Today tobacco is grown in 22 states of the United States. Three-quarters of the crop comes from five states: North Carolina, Kentucky, Virginia, Tennessee, and South Carolina. There are six main classifications of tobacco leaf, according to the U.S. Department of Agriculture: flue-cured, fire-cured, air-cured, cigar-filler, cigar-binder, and cigar-wrapper. The United States leads the world in tobacco production and export, followed by China, India, the Soviet Union, and Japan. The world's annual harvest amounts to 9.5 billion pounds.

Labor input to grow an acre of tobacco exceeds 400 man-hours, contrasted with only 8 for wheat. Seeds are planted in seedbeds protected by cloth covering, and the seedlings that emerge are transplanted to the field in the spring, from March to May. After 90 to 100 days, the crop is harvested, then cured and sold at auction. It takes 1.25 billion pounds of domestic tobacco to manufacture the number of cigarettes produced annually in the United States—a total of 565 billion in all. —Charles E. Rogers

Turkey, an edible game bird native to North America, now almost entirely raised on farms. Of the many turkey varieties, only the Broad-Breasted Bronze, the Broad-Breasted Large White, and the Beltsville Small White are raised in commercial quantities. Much emphasis has been given by turkey breeders to wide breasts and overall meatiness. The Beltsville Small White is a relatively small, broad-breasted bird raised for use by small families. Marketed at about 5 months, the live *hens,* or females, weigh about 9 pounds; the *toms,* or males, about 16. The large broad-breasted varieties weigh about twice as much or more when marketed. Turkeys were traditionally popular, chiefly for holiday consumption, although they are now available and used throughout the year. Reproduction in turkeys is relatively low due to poor egg production, fertility, and hatchability. Eggs hatch after 28 days of incubation. *Blackhead,* a major disease in turkeys, especially when kept with chickens, is caused by a microscopic parasite which attacks the cecum and liver. Pullorum and fowl typhoid used to cause heavy losses but these diseases now have been largely eliminated through blood testing. To help prevent losses from diseases and parasites, sanitary practices should be followed and chickens and turkeys should not be raised together. —Carl W. Hess

U.S. DEPARTMENT OF AGRICULTURE

BELTSVILLE TURKEYS, bred for small ovens, have extra meat on their bodies.

Vegetable, any one of a large number of herbaceous plants cultivated for food. The precise definition of a vegetable is troublesome. No single botanic or economic description is adequate or clear-cut. We shall consider vegetables as annual or semiannual crops (excepting asparagus or artichokes) that have an edible fleshy portion. Thus, some plants that botanically are fruit (as tomatoes and melons) are included.

As in few other nations of the world, fresh domestic vegetables are available the year-round in the United States. This is due to two factors: a wide range of climate, which permits vegetables to be harvested every day of the year somewhere in the country, and an efficient marketing system.

Vegetable production, like other phases of agriculture, is becoming more specialized. Therefore, production is becoming concentrated in fewer but larger farms. These generally are found in areas of temperate climate (excepting potatoes), good soil, and ample supplies of rain or other sources of water. Production is also becoming more mechanized; labor problems have led to special interest in mechanical harvesting.

The farm value of the U.S. commercial vegetable crop ranges widely from year to year but tends to average between $1.5 and $2 billion. The five most important states in terms of value of production are, in the order named, California (about 25 per cent of the total), Florida (about 10 per cent), New York, Idaho, and Maine. The latter two states are preeminent in potatoes. Other leading vegetable-producing states are Texas, Wisconsin, Arizona, Michigan, New Jersey, Oregon, and Washington. The most important single crop is the white potato, followed by tomatoes and lettuce. Other vegetables in approximate order of importance include snap beans, string beans, sweet corn, onions, cantaloupes, celery, cucumbers, carrots, cabbage, green peas, sweet potatoes, and watermelons.

Vegetables are either marketed fresh or shipped for processing. Some potatoes are also used for feed and seed. While more than half of the vegetable crop has been sold fresh in the past, processing is becoming more important. The most common forms of processing are canning and freezing; dehydration is of lesser importance. Potato products such as chips, shoestrings, and frozen French fries are of special significance. New techniques, such as freeze-drying, may play a greater role in the future.

Total vegetable consumption, or apparent use, is about 340 pounds per person per year; nearly one-third of this figure is represented by potatoes (these figures are in farm-weight terms, but allow for imports and exports). Consumption of fresh vegetables is decreasing while that of processed vegetables is increasing. Trends for the many individual vegetables vary: consumption of cabbage, for example, appears to be decreasing while that of salad vegetables, such as lettuce, has been increasing. Potatoes had been going through a long downward trend until a few years ago, when consumption leveled off.

—Dana G. Dalrymple

Wheat, the most widely cultivated *cereal.* It is grown around the world, with the Soviet Union, United States, China, Canada, and France leading in production. From 1955 to 1959 the average world production was 9.3 billion bushels, grown on 493,010,000 acres. Since wheat is a cool-season crop, most of these acres are situated

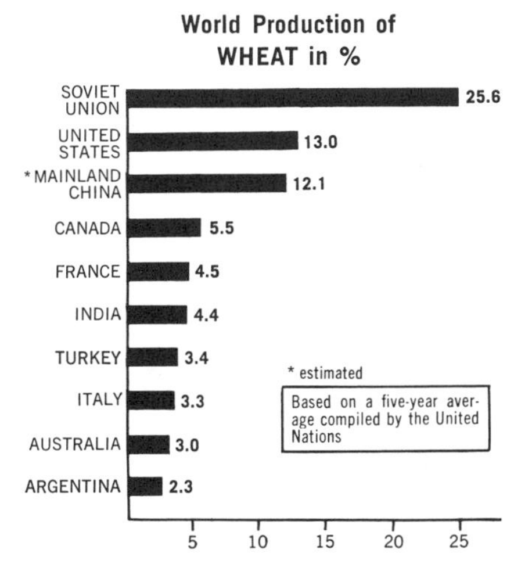

LOOK MAGAZINE

WHEAT, the most widely cultivated cereal, is grown primarily in the climates of the Northern Hemisphere. Here, combines completely process a Kansas wheat crop.

in the Northern Hemisphere. The leading producers in the United States are Kansas, North Dakota, Montana, Oklahoma, and Washington.

Botanists count 15 species of wheat, of which three are generally grown in the United States—*common, durum,* and *club.* However, 95 per cent of the wheat produced in the United States belongs to the first-named species.

There are two principal types of wheat: *winter wheat,* planted in the fall, and *spring wheat,* planted in the spring. Another classification divides wheat into soft and hard varieties.

The soft varieties are grown in the eastern half of the United States and in the states of Washington, Oregon, and Idaho; the hard winter wheats account for most of the production on the central plains, while most of the wheat grown on the northern plains is of the hard spring varieties. When it was realized in the late nineteenth century that soft wheats could not withstand the temperatures and droughts of the plains, plant breeders sought adaptive varieties. Cerealists such as Mark A. Carleton of the U.S. Department of Agriculture learned that the hard winter wheats grown on the steppes of southern Russia were adaptable. He visited Russia and brought back a winter variety known as *Kharkov,* as well as a spring-grown durum wheat. In Canada, William Saunders and his sons bred a hard spring wheat known as *Marquis.* Using these varieties and others as parent stocks, wheat breeders have developed the wheats that now account for much of the nation's production.

Whatever the variety, the wheat plant consists of a root system, leaves, stem, and heads or spikes. The roots ordinarily reach to a depth of 5–6 feet, while the stems vary from 2 to 5 feet. The heads bearing the kernels are usually 2 to 4 inches in length; the kernels are small and oval, consisting of a protective coating, a starchy endosperm, and the germ.

Wheat grows best on well-drained medium-to-heavy soils, especially silt and clay loams. Farmers living in humid areas prepare a seedbed by means of moldboard plows, spiketooth harrows, and disks; in the more arid Western states, they use adapted cultivators and blades. Farmers in both areas sow wheat by means of tractor-drawn drills.

Winter wheat, which is planted in September or October, is harvested in June and July, while spring wheat, planted in April, is harvested in July and August. Whereas 40 years ago most of the wheat crop was cut with a binder and threshed by a thresher, now more than 95 per cent of the crop is cut and threshed by combines.

Wheat is subject to attack by rusts and smuts. Stem rust, which has been particularly destructive on the Great Plains, has been checked by the development of rust-resistant varieties, such as *Thatcher,* which was developed by the Minnesota Agricultural Experiment Station. Insect enemies include the Hessian fly, wheat jointworm, grasshopper, and the wheat-stem sawfly. To combat the attacks of the latter, Canadian cerealists developed the *Rescue* variety with solid rather than hollow stems.

Wheat is used principally for human consumption. The hard varieties, richer in protein than the others, produce flour that is used in the making of bread. Flour from the soft wheats is suitable for pastries, crackers, biscuits, and cakes. Durum wheat is used for the manufacture of macaroni, spaghetti, and vermicelli. Some wheat is converted into breakfast foods.

Wheat is nutritious. The average kernel contains about 70 per cent carbohydrates, 12 per cent protein, 2 per cent fat, 12 per cent water, 1.8 per cent mineral matter, and 2.2 per cent cellulose. The protein content of the hard wheats of the Great Plains may be as high as 15 per cent. A kernel also contains vitamins of the B-group, such as thiamine, riboflavin, and niacin.

Wheat is an important article of international trade. The United States exports more than half of its crop; in 1964 this amounted to approximately 675 million bushels. The other leading exporters are Canada, Australia, and the Soviet Union. Major importers are the United Kingdom, India, Japan, West Germany, and Brazil. (See also *Flour Milling.*)

—Robert G. Dunbar

Wool, a major textile fiber derived from the soft coat of a domesticated sheep. Its chief use is in outer apparel and household articles, such as blankets, with the manufacture of carpets and rugs utilizing a lesser amount. Wool is broadly classed as apparel or carpet quality. It is the seventh largest commodity in world trade.

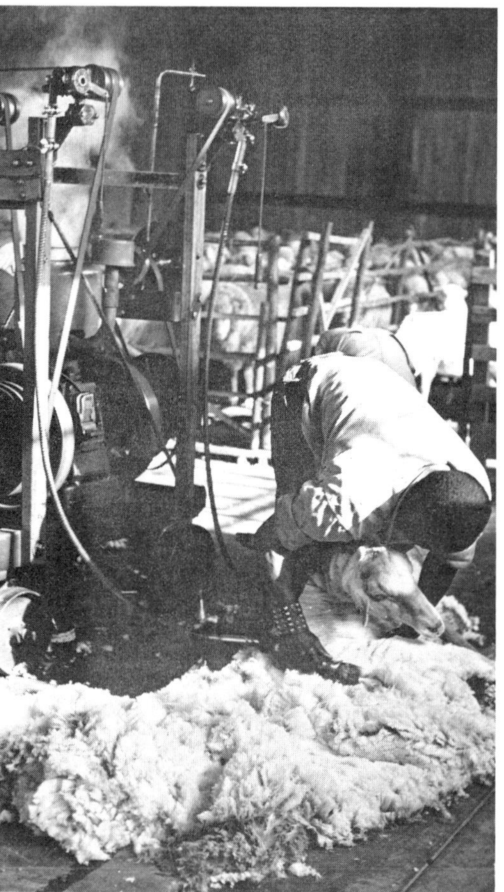

SOUTH AFRICAN INFORMATION SERVICE

SHEARING SHEEP BY MACHINE enables sheep raisers to shear up to 250 sheep a day.

In the decade after 1955, world production of wool gained about 15 per cent to approximately 5.75 billion pounds (grease basis). Consumption rose by about one-fourth. Wool's share of total world fiber consumption dipped slightly, however, as the overall total rose and the use of newly developed man-made fibers increased.

Australia, which grows more than 1.7 billion pounds (grease basis) annually (almost entirely of apparel wool), is the world's largest producer, and has nearly 160 million sheep. Other leading producers of apparel-class wool, in order of importance, are New Zealand, Argentina, South Africa, the United States, Uruguay, and Great Britain. The Soviet Union is considered the second largest producer, and while its wool has been coarse, the output of apparel types is believed to be growing. Others in the coarse-wool category are The People's Republic of China, India, and Pakistan.

Wool has natural crimp and elasticity. It absorbs water vapor readily, but many wool articles resist wetting by liquid water. Wool dyes easily in a wide variety of fast colors. Wool garments impart a feeling of warmth, yet lightweight wool apparel is comfortable in hot weather. It has been said that if a fiber with wool's attributes had just been invented, it would lead the list of so-called miracle fibers. Also regarded as wool-type fibers are the specialty hairs such as mohair, cashmere, camel's hair, and vicuña.

Sorting of fleeces and scouring (washing) of the wool are the first steps in processing that leads to spinning. There are two spinning systems: woolen and worsted. The woolen system makes use of the shorter fibers. Woolen yarns are usually soft and lofty with relatively little twist and the individual fibers lie in all directions. In the worsted system, the aim is to lay the fibers as parallel as possible. The resulting yarn is firmer or harder and is given more twist than woolen yarn. Fabrics made of these two types of yarns are called woolens and worsteds.

Woolens, often used in sportswear and coats, lend themselves to a variety of colors and often have a woolly or hairy surface, which mutes and diffuses the colors and patterns. Fabrics usually made with woolen yarns include tweeds, meltons, coverts, and fleeces. Worsted fabrics, which usually are crisp and springy with little or no surface fiber, include serge, gabardine, and whipcord.

The United States, Great Britain, Italy, and Japan are among the largest manufacturers of wool. Though there has been a severe decrease in the number of mills and amount of wool machinery and a decline in cloth production since World War II, the American industry is regarded as the most efficient in the world, and it leads in the development of products made of wool blended with the new noncellulosic man-made fibers. For example, polyester/wool tropical cloth is the basic quality fabric for summer clothing. Producers make man-made fibers especially for the various systems of textile processing, so the trend to blends is expected to continue. (See also, *Sheep.*)

—Gordon F. Graham

Yeast, a fungus growth consisting of tiny cells of vegetable matter that collect in a frothy, yellowish cluster. There are hundreds of species, widely distributed in nature, each strain possessing distinctive characteristics, properties, and uses. The most familiar species, *Saccharomyces cerevisiae*, is used to prepare cultures adapted specifically for the baker, the brewer, and the manufacturer of primary food yeast. Yeast has been utilized since ancient times as a leavening property in baking, as a fermenting agent in alcoholic beverages, and as a medicine and food. The ability of yeast to change sugar to alcohol makes it indispensable to brewers and distillers.

The yeast organism is grown in a suitable medium and harvested when a sufficient crop of cells has appeared. In former times it was grown from *wort* (an Old English word meaning herb, plant, or root) prepared from grains mashed in water. More recently, primary yeast has been cultivated from refuse material.

—Charles E. Rogers

BIBLIOGRAPHY

ANDERSON, ARTHUR L. and KISER, JAMES J. *Introductory Animal Science.* The Macmillan Co., 1963.

BEAR, FIRMAN E. editor. *Chemistry of the Soil.* 2nd ed. Reinhold Publishing Corp., 1964.

BREWBAKER, JAMES L. *Agricultural Genetics.* Prentice-Hall, Inc., 1964.

FREEMAN, ORVILLE L. *World Without Hunger.* Frederick A. Praeger, Inc., 1968.

LAVERTON, SYLVIA. *Irrigation, its Profitable Use for Agricultural and Horticultural Crops.* Oxford University Press, 1964.

LEONARD, WARREN H. and MARTIN, JOHN H. *Cereal Crops.* The Macmillan Co., 1963.

MALLIS, ARNOLD. *Handbook of Pest Control; the Behavior, Life History and Control of Household Pests.* 4th ed. MacNair-Dorland Co., Inc., 1964.

PORTER, A. R., SIMS, J. A., and FOREMAN, C. F. *Dairy Cattle in American Agriculture.* Iowa State University Press, 1965.

RICHEY, C. B. editor-in-chief. *Agricultural Engineers' Handbook.* McGraw-Hill, 1961.

TAYLOR, NORMAN. *Taylor's Encyclopedia of Gardening, Horticulture and Landscape Design.* 4th ed. Houghton Mifflin Co., 1961.

U. S. DEPARTMENT OF AGRICULTURE. *Consumers All; Yearbook of Agriculture, 1965.* U. S. Government Printing Office, 1965.

U. S. DEPARTMENT OF AGRICULTURE. *The Farmer's World; Yearbook of Agriculture, 1964.* U. S. Government Printing Office, 1964.

U. S. DEPARTMENT OF AGRICULTURE. *Science for Better Living; Yearbook of Agriculture, 1968.* U.S. Government Printing Office, 1968.

Definition.—Agronomy is defined by the American Society of Agronomy as "the theory and practice of field-crop production and soil management." In several countries of Europe and South America, however, agronomy takes on a much broader meaning. In these countries, agriculture is divided into agronomy and animal science, with all phases of plant science, including horticulture, plant pathology, entomology, crop and soil science, combined into agronomy.

Agronomy is a specialized phase of several sciences. It includes soil physics (structural properties of the soil, aeration, and water movement), soil chemistry (physical and biochemical reactions that take place in the soil), soil microbiology (activities and decomposition products of microorganisms), soil fertility (factors affecting the availability of nutrients and the use of fertilizers), soil morphology and genesis (soil formation, classification, and survey), and soil conservation and management (controlling erosion and improving productivity).

History.—Technologically speaking, man has developed greatly; but he is still dependent upon the soil for many of his requirements. Although agronomy is very old, the greatest progress in it has been made only since the 1920's. There are myriads of problems yet to be solved.

The dawn of civilization is frequently recognized as the point at which man recognized that seeds could be planted and a crop grown. He quickly observed that plants grew better in some soils than in others. This was the beginning of soil science. Many references are made to soils and problems of crop production in very early literature. In the Greek epic the *Odyssey*, Homer (who is supposed to have lived at some time between 900 and 700 B.C.) tells how Odysseus, the far wanderer, was recognized at his homecoming by Argos, his faithful hound, who was "lying on a heap of dung with which the thralls were wont to manure the land."

Acid soil conditions (sour soils), a major problem in many areas of the world today, were recognized by the Greeks between 800 and 600 B.C. They recommended application of *marl* (a mixture of clay and calcium carbonate) and shells to overcome this condition, a practice that was adopted by the Romans.

Theophrastus (c. 372–287 B.C.) recommended the abundant manuring of thin soils, but suggested that manure be used sparingly on rich soils. He was one of the first men to recognize the fertilizing value of saltpeter (potassium nitrate); he also recognized differences in the fertilizing values of manures, listing them in the following order of decreasing value: human, swine, goat, sheep, cow, ox, and horse. The importance of growing legumes to improve the soil condition for crops was advocated in the writings of Vergil (70–19 B.C.).

About 1500 interest began to develop in the study of the factors influencing plant growth. Francis Bacon (1561–1626) suggested that water was the principal nourishment of plants, with soil serving only as a support. This concept was substantiated by the famous Flemish physician and chemist Jan Baptista van Helmont (1577–1644), who conducted an experiment by placing 200 pounds of soil in an earthen container. In this soil he planted a five-pound willow tree. He shielded the container from dust and added only water. Five years later, the tree had increased in weight to almost 170 pounds, but the soil had decreased only about two ounces in weight. Van Helmont therefore concluded that water was the only nutrient needed for plant growth.

The discovery of oxygen by Joseph Priestley in 1774 stimulated Nicolas Théodore de Saussure (1767–1845), a Swiss naturalist, to begin studying the effects of air on plant growth. As a result of this study, he was able to demonstrate that plants absorb oxygen and emit carbon dioxide (the process of respiration). He also showed that plants kept in an atmosphere lacking carbon dioxide will die, because they need the carbon dioxide for the process of photosynthesis. De Saussure also wondered about the origin of mineral salts found within plants.

Great advances toward answering this question were made by the brilliant German chemist Justus von Liebig and the two British scientists Sir John Bennet Lawes and Sir Joseph Henry Gilbert during the period 1840 to 1850. Through their efforts, information was gained about the elements required for plant growth. Liebig demonstrated that the plant obtains oxygen, hydrogen, and carbon dioxide from water and air. He also proposed the "law of the minimum," which states that plant growth (and subsequent yield of crops) is limited by the least plentiful nutrient element, providing all other nutrients are present in adequate quantities. Lawes and Gilbert showed that all plants require phosphorus and potassium and that all nonlegumes require nitrogen. They also established the famous Rothamsted Experiment Station—the first agricultural experiment station—which is still one of the outstanding research centers.

Soils.—There are so many definitions of soil that it is difficult to select any one as best. The following definition, however, is widely accepted by soil scientists: *Soil* is a naturally occurring body, three-dimensional in nature, formed at the earth's surface through the action of weathering processes on soil-forming materials (rocks and minerals), under the influence of climatic and biotic factors (plant and animal life). Soil is sometimes described as having the same relation to the earth that an orange peel has to an orange. The orange peel, however, is uniform in compo-

U.S. DEPARTMENT OF AGRICULTURE

SOIL SCIENTISTS at the turn of the century (*left*) lacked the tools and equipment to do as complete a soil study as today's specialists.

sition and thickness, which is not true of soil. Since the parent materials vary widely in composition and since the effects of climate, geologic age, *topography* (geographic features), and biotic factors are variable, soils vary. They may be reddish (because of iron deposits), such as those found in many tropical and subtropical areas (Brazil, India, Hawaii, and the southeastern United States), or they may be black, such as those found in many temperate areas (the northern part of the United States, Canada, and Russia). Soils vary in texture from coarse sand to fine clay; they may be shallow or deep. However, every soil consists of mineral and organic matter, water, and air. It is the link between the rock core of the earth and life on the surface.

The composition of soils varies with depth. A cross section of a soil reveals a series of zones, each somewhat different from the one above and the one beneath. Each of these zones is called a *horizon,* which is defined as a layer of soil, approximately parallel to the land surface, with observable characteristics. A typical soil has three major horizons: the A-horizon, the B-horizon, and the C-horizon. The *A*-horizon is the zone closest to the surface of the earth. Since water-soluble materials are carried downward through the A-horizon by soil water (a process called "leaching," the A-horizon is also called the "zone of leaching." The *B*-horizon lies directly below the A-horizon. Since soil water drains into the B-horizon from above and also rises into the B-horizon from below (high rates of evaporation cause the soil water to be drawn upward) to deposit water-soluble materials, this horizon is often called the "zone of accumulation." The *C*-horizon is a zone of partially broken and decomposed rock or mineral material. Some of the original *bedrock* (parent material) minerals may still be present, but most have been converted into other forms. The C-horizon continues downward to blend into the unweathered bedrock.

All three soil horizons develop from the underlying parent material. When the material is first exposed at the surface, weathering and decomposition proceed rapidly. While the decomposed material increases, downward-percolating water leaches out some of the minerals and begins to deposit them farther down. In this way, the A-horizon and the B-horizon are built up. As the weathering process continues, the bedrock material gives rise to the C-horizon; as time passes, the C-horizon reaches deeper into the bedrock and the B-horizon and A-horizon grow deeper.

Soil Classification.—The classification of soils is very important in identifying and associating their significant characteristics. For example, the thickness and characteristics of the horizons (depth of profile) will determine the amounts of nutrients and water that will be available to plants. Strongly acid or compact subsoils may not allow plant roots to penetrate deeply. Restricting the roots to the surface horizon greatly reduces the volume of soil available to plants as a storehouse of moisture in dry periods.

Soils with similar characteristics, such as parent material, topography, thickness of horizons, and presence of compact layers, may be grouped into a *series.* For easy identification, soil series are given names, usually a geographical term relating to the place where the soil is first defined. The soil series is, in turn, divided into soil *types* and *phases.* The *soil type* identifies the texture of the surface soil. The full name of a soil type includes both the name of the soil series and the textural class of the surface soil down to "plow depth," which is usually the upper six inches of soil. A *phase* indicates a specific condition important in the use and management of the soil. For example, in the soil classed as *Cecil clay loam, sloping eroded phase, Cecil* is the series name, *clay loam* denotes the soil type, and *sloping eroded* indicates that the surface horizon is not as deep as usual for this soil type and is on sloping land.

■SOIL SURVEYS.—There are several thousand kinds of soil, many of them surveyed and mapped. In the United States, soil surveys are made cooperatively by the Soil Conservation Service of the Department of Agriculture, agricultural experiment stations, and other state and federal agencies. The United Nations Food and Agriculture Organization (FAO) supports soil surveys in many countries as a step toward increasing food production. This group is also working on a world soil-survey program.

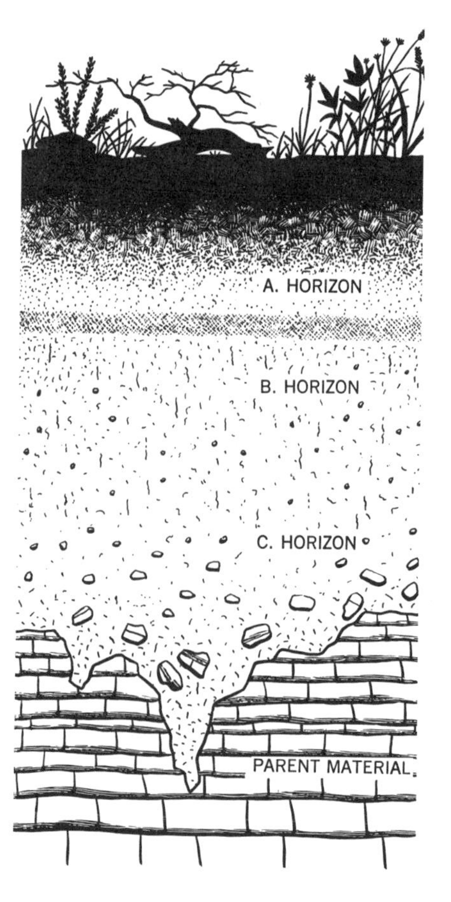

SOIL HORIZONS fall into three categories.

A land capability system has been developed by the Soil Conservation Service to aid farmers in planning their cropping programs. Land conditions, such as slope, degree of erosion, depth of topsoil, and drainage, greatly influence soil productivity and management requirements. As an aid in the selection of management practices, the soil series are grouped into capability classes, subclasses, and units. *Land capability classes* denote magnitude of hazards and limitations; *land capability subclasses* refer to kinds of hazards; and *land capability units* are management groupings for specific purposes.

There are eight land capability classes: I through IV are suitable for cultivation; V through VIII are better adapted to pasture, woodland, or wildlife uses. Each of the subclasses is designated by a letter to indicate the major kind of hazard. The letter *e* indicates an erosion problem; *w,* a water problem; and *s,* drought or low fertility. The capability unit is a division of the subclass and is designated by a number.

Soil Fertility.—Sixteen chemical elements are presently recognized as essential for plant growth. The test of essentiality is that a plant must have the element in order to complete its life cycle. Three of the essential elements—carbon, hydrogen, and oxygen—make up over 90 per cent of a plant. All three are obtained by the plant from water and air. The remaining thirteen essential elements are obtained from the soil.

Soil fertility includes the supply of the thirteen nutrient elements in the soil, plus the capacity of the soil to provide and transmit these elements to the plant over a period of time. Of the elements obtained from the soil, the primary elements—nitrogen, phosphorus, and potassium—are consumed in the largest quantities. The secondary elements are calcium, magnesium, and sulfur; the *micronutrients* (referred to as *minor elements* or *trace elements*), manganese, iron, copper, zinc, boron, molybdenum, and chlorine. The micronutrients are required only in very small quantities. Large amounts of several of these, including boron, manganese, and copper, will create a *toxic* soil condition.

■FERTILITY MANAGEMENT.—Soil-fertility management includes several objectives. One of these is to manage the soil in such a manner that during the growing season it will release amounts of nutrient elements adequate for the desired yield. A second objective is to avoid losses of nutrient elements through downward leaching by water, through excess uptake of the elements by plants, and through fixation by the soil into forms only slightly available to plants. A third objective is to overcome nutrient deficiencies and acidity in order that plant roots may utilize the maximum volume of soil.

LOUDON COUNTY, TENNESSEE - SHEET NUMBER 15

U.S. DEPARTMENT OF AGRICULTURE

COLOR CHARTS are used to identify the various soil layers. Once the soil scientist has classified the many soil layers in a region, he then prepares a special area map for use.

When soil fertility limits the yield obtained from a particular soil, one or more essential elements are deficient. The total amount of an element present in the soil does not indicate the amount available to a plant, for there are numerous factors that determine how much of a given element is able to be absorbed. These factors include moisture, aeration, acidity or alkalinity, presence of competing elements, and nature of the plant itself.

Since soils are developed through the weathering of rocks and minerals, geologic age influences soil fertility. Young soils usually contain mineral fragments that, with continued weathering, release nutrient elements for plant use. Very old, highly weathered soils, on the other hand, have more or less reached a state in which the rate of release of nutrient elements is slow. The number of years a soil has been cultivated also influences the fertility status. Poorly managed soil that has been permitted to erode may have very low fertility, even though it is geologically young. Well-managed soils, on which commercial fertilizers have been used correctly, may actually be much more fertile than they were in their virgin state, even though they are geologically old.

Contrasting soil situations are those in India and in the southeastern United States. The soils of India are young from a geologic viewpoint. They have, however, been farmed for centuries or more without any appreciable addition of commercial fertilizers. This has created widespread nutrient deficiencies. In contrast, the Piedmont region of the southeastern United States has some of the oldest soils, geologically speaking, in the world. These soils were highly weathered when the settlers first came to America—deficient in nitrogen, phosphorus, and potassium. Nevertheless, through judicious management and the proper application of various commercial fertilizers, many of these soils have been to some extent enriched and are presently relatively fertile.

Organic Content.—Organic-matter content frequently shows direct correlation with the fertility level of the soil. Organic matter accumulates in the soil through the centuries from plant and animal residues. Conditions conducive to good plant growth in virgin soil usually carry over into cultivated conditions. The breakdown of the organic matter releases nutrients accumulated by the native plants for use by the cultivated plants. Decomposition of organic matter is usually rapid during the first few years the land is under cultivation, then levels off. Erosion, of course, rapidly removes topsoil and organic matter, upsetting the equilibrium.

Soils developed under poor drainage conditions, such as *peat* and *muck*, are frequently very high in organic matter content. Such soils, however, are usually very low in minerals and also deficient in nitrogen. Draining such areas does not result in fertile soils, although good management and judicious use of fertilizer and lime can lead to good production.

Sandy soils do not usually contain as much organic matter as finer-textured soils do. Since nitrogen is released into soil through the decomposition of organic matter, sandy soils are frequently deficient in nitrogen. Pores are larger in sandy soils than in finer-textured soils, but the total pore space is less. This means that water will penetrate a sandy soil to a greater depth than it would a finer-textured soil. Thus, leaching of soluble nutrient elements, such as nitrogen in the form of nitrates, is more common in sandy soils than in silt loams or clay loams.

Toxic Soils.—Many soil conditions are toxic to plant growth. The more widespread natural toxic conditions (rather than those caused by the application of chemicals by man for the purpose of controlling such pests as weeds, insects, and nematodes) include acid, saline, and alkali soils.

Acid soils, old and highly weathered, are usually found in humid regions. They are low in calcium and magnesium and frequently are high in iron and aluminum, which form complex compounds with phosphorus. Some plants tolerate more acidity than others do. Blueberries, camellias, and azaleas grow better in acid soils, since iron and manganese are more readily available. Legumes, on the other hand, grow better in soils adequately supplied with calcium and magnesium and with a *p*H near neutrality. (A *p*H of 7 indicates neutrality; a *p*H above 7 shows alkalinity; and one below 7 shows acidity—thus, the lower the *p*H, the greater the acidity.) Acidity, as such, is not important to the immediate growth of plants unless the aluminum content is high enough to be toxic. If the aluminum content is low and sufficient nutrients, such as calcium and magnesium, are supplied, most plants can tolerate considerable acidity (to a *p*H of at least 4.0). Most plants, however, prefer a soil condition that is only slightly acid, where adequate quantities of calcium and magnesium are available (*p*H 6.0–7.0). The majority of nutrient elements are most available in a slightly acid soil.

A *saline soil* is one in which there has been such an accumulation of soluble salts that plant growth is reduced or stopped. Saline soils are most common in the drier regions of the world, where the rainfall is not sufficient to wash the soluble salts through the soil profile and into the groundwater. In the United States, most saline soils are found west of the Missouri River. Local areas in the more humid regions may also become saline. Severe windstorms, such as hurricanes, may cover tidewater areas with salt water. Salt accumulations in soils can also result from seepage from drainage ditches in which salt water has "backed up." Irrigation water may contain enough soluble salts to cause a saline condition if the salts accumulate in a particular area of the soil profile, such as the immediate surface. One of the most common localized saline conditions is caused by the improper application of fertilizer; the soluble nitrogen and potassium salts in the fertilizer may reduce growth or kill seedlings.

Good drainage through the soil profile is the first requirement for reclamation of saline soil. To reclaim such a soil, the soluble salts must be drained and washed from the profile to a depth of at least three feet. If the salts are predominately sodium, gypsum must be applied at the rate of two or three tons per acre before leaching. If the soil is *calcareous* (has excess calcium carbonate), it is likely to be deficient in phosphorus because calcium carbonate and phosphorus form tri-calcium phosphate, which is not readily available to plants.

An *alkali soil* has either a high degree of alkalinity (*p*H 8.5 or above) or a high percentage of exchangeable sodium (15 per cent or higher), or both, acting to reduce or stop growth of most crop plants. The exchangeable sodium, when present in sufficient quantities, causes the soil to run together when wet and to become hard when dry. Alkali soils usually evolve from saline conditions in which the soluble salts are predominately sodium.

To improve alkali conditions, the sodium must be replaced with calcium. Gypsum (calcium sulfate) is best for this purpose unless the soil is calcareous, in which case sulfur is used. As with saline soils, adequate drainage is necessary for reclamation of alkali soils.

Soil Testing.—Ever since Justus von Liebig (1803–1873) demonstrated that mineral elements are required for plant growth, there has been much interest in soil analysis as a method of evaluating soil fertility and ascertaining the nutrient deficiencies likely to occur. Early attempts at soil analysis were generally unsatisfactory, but much progress has been made since the 1930's. Today there is strong evidence that competent use of soil tests can make a valuable contribution to intelligent soil management. Soil tests are being conducted on all continents, and hundreds of thousands of soil samples are analyzed annually.

Information gained from soil testing is used in many ways. One of the more important objectives is to group soils into classes in order to formulate the best practices when adding fertilizer and lime. The amount of plant nutrients needed is related to the crop grown and the levels of available nutrients in the soil. A second important objective is to predict the probability of a profitable response to the application of plant nutrients. A third objective is to help evaluate soil productivity. The organic-matter content, level of nutrient elements, and *p*H of the soil have been found to be good guides in estimating the potential productivity of soils. Still another valuable objective is to determine specific soil conditions that may be improved by the addition of soil amendments, such as manures, chemical fertilizers, and lime.

In conducting a soil testing program, there are six major phases that should be carefully considered. These phases are: securing representative soil samples (a poorly taken sample may be worse than none at all); adequate research to correlate the testing procedure with crop response on the type of soils to be tested; satisfactory chemical procedures (the technique used must be accurate and reproducible); interpretation of the results; recommendations based on the interpretation of the test results and other information that is available; and follow-up to see that the reports are understood and that the recommendations can be followed.

A wide range of values has been assigned to soil testing, from merely psychological to cure-all. The value of a soil test is as a good source of information that must be properly interpreted for use in soil management.

Soil Improvement.—Soil consists of inorganic (mineral) and organic matter, water, and air. The inorganic fraction of the soil usually constitutes one-half of the total volume of most surface soils. The organic portion is usually less than 5 per cent (it is higher in peat or muck soils). The remainder consists of water and air. The actual proportions of these four major components vary greatly with depth and type of soil. The proportion of air to water also fluctuates greatly within the same soil from season to season and even from day to day.

The inorganic fraction is frequently divided into size groups referred to as *separates*. The coarser material (larger than 2.0 millimeters) is classed as *gravel* or *stones*, and is not usually included in the discussion of particle size, except to designate a soil as "stony phase" if there are enough stones to warrant recognition. Particles less than 2.0 millimeters in diameter are grouped into *sand* (2.0 to 0.05 millimeters in diameter), *silt* (0.05 to 0.002 millimeters), and *clay* (less than 0.002 millimeters). Sand particles may be observed with the naked eye, but an electron microscope must be used to observe the fine clay particles.

Except for sand grains, soil particles do not usually occur singly. Instead, many fine particles may be grouped into a *cluster*, which is a secondary structural unit referred to as an *aggregate*. The aggregates may be arranged in *platy* blocks or cubes (often referred to as *granular* or *crumb*) or in *columns* (*prisms*). Arrangement of the individual grains or aggregates in the soil is referred to as *soil structure*. The size and shape of the pore space of the soil is largely determined by the structure. So are the movement of air and water through the soil and the capacity of the soil to hold water and air.

Improving the soil structure is a slow and tedious process. It is, therefore, better to maintain soil in a good physical condition than to try to improve the structure later. Freezing and thawing, wetting (swelling) and drying (shrinkage) are important forces in determining soil structure. Aggregates, once formed, must be stable when wet. Soil organic matter, with products formed through activities of microorganisms, is very important in the stabilization of soil structure.

Soil structure is frequently destroyed by *compaction* with heavy machinery, such as rubber-tired tractors. Plowing or cultivating the soil when it is too wet will destroy good structure and result in a puddled condition. A driving rain on a freshly plowed field will have the same effect. One of the big problems in maintaining golf greens and tees is the compaction and puddling caused by avid golfers playing when the soil is wet. Working the soil excessively when it is dry, as with a disc, creates a loose, powdery condition that also destroys desirable structure.

Tilling the soil at the correct moisture level is very important in maintaining the desired physical condition. Rotation of sod with row crops gives the soil a "rest" and, through wetting, drying, freezing, and thawing, will improve the structure.

Natural Fertilizers.—Most "natural" fertilizers are organic, but there are also a few inorganic materials, such as sodium nitrate, that may also be classified as natural. Natural fertilizers include animal manures, bird or bat *guano*, fish meal, sewage sludge, and composts. Many other by-products may also be grouped as "natural," such as meat meal, bone tankage, dried blood, hair, wool, feather wastes, ground leather, horn meal, rice and peanut hulls, and tobacco stems. Seed meals, resulting from the removal of oil from seeds, are frequently used in fertilizers. Some of these meals are castor pomace, cottonseed meal, linseed meal, rapeseed meal, soybean meal, and peanut meal.

Typical Analyses of Animal Manure

Animal	Pounds per Ton		
	Nitrogen (N)	Phosphate (P_2O_5)	Potash (K_2O)
Average barnyard (cattle and horse)	10	5	10
Sheep	21	6	20
Poultry	23	18	9

The mineral elements contained in the organic material were originally obtained from the soil. If the organic material is from an animal source, then the animal ate a plant or another animal that had eaten a plant. The plant obtained the minerals from the soil, and returning all of the plant remains to the soil from which it grew would not increase the mineral content of the soil. Placing organic matter grown on one soil upon another would increase the latter's fertility, providing none had been removed by other means.

Although animal manures are a good source of fertilizer, the only nutrients found in them are those present in the animal's feed. As the feed passes through the animal's digestive tract, certain portions are removed; no nutrients are added. The composition of manure is variable and depends upon several factors. One of these is the kind and amount of feed given the animal; feeds high in protein will result in manure relatively high in nitrogen. Another factor is the kind of animal: poultry manure is much higher in nitrogen and phosphorus than that of most animals. A third factor is the age of the animal; a young, growing animal removes more nutrients than does an old or mature animal. Likewise, a dairy cow producing milk will remove a different fraction of the feed than will a beef cow. Still a fourth factor is the amount and kind of litter used, such as straw or wood shavings.

Although the composition of manure varies greatly, as indicated above, the accompanying table shows the approximate composition of various types of manures. All organic fertilizers are decomposed by microorganisms in the soil, releasing the nutrient elements in a form that growing plants can assimilate. These forms are usually the same as those found in inorganic commercial fertilizers. For example, plants assimilate nitrogen as ammonium or nitrate ions, not as the proteins or amino acids found in the organic material. If temperature and moisture conditions are satisfactory (above 50° F.), decomposition takes place relatively rapidly, and the nutrient elements are released in sizable quantities within three weeks.

Chemical Additives.—The addition of chemical materials to soils is a very old practice. The application of *marl* (largely calcium carbonate), burned lime (largely calcium oxide), and saltpeter (potassium nitrate) was practiced several centuries before the Christian Era. No concentrated effort was made to utilize chemical fertilizers, however, until after the research of Justus von Liebig, Sir John Bennet Lawes (1814–1900), and Sir Joseph Henry Gilbert (1817–1901) was made known to the world. The amounts of chemical fertilizers consumed gradually increased, so that by 1905 the world consumption was about two million tons. A sharper trend did not begin until after World War I. The economic depression in the 1920's, followed by the severe depression and drought of the 1930's, created conditions that retarded the expansion of commercial fertilizer use. The most important chemical fertilizers used from 1905 to 1939 were sodium nitrate, ammonium sulfate, superphosphate (monocalcium phosphate), basic slag (a phosphorus-rich residue produced during the Bessemer process for making steel), and the potassium salts, potassium chloride and potassium sulfate.

During World War II several synthetic-nitrogen plants were built to produce material for explosives. Following the war, these plants began to produce fertilizers. New beds of phosphate rock and potassium salts also were developed. Economic conditions encouraged high crop production, and the use of chemical fertilizers increased at a fantastic rate. Since 1945 the increase in the consumption of chemical fertilizers has been at the rate of approximately 1,400,000 tons a year. The tonnage of chemical fertilizers manufactured has increased rapidly since 1945, as have the kinds of fertilizers available. The technology of fertilizer manufacture also has changed markedly, and even more significant developments are anticipated in the future.

All states of the United States and most countries of the world have laws governing the sale of commercial (chemical) fertilizers. The laws differ from state to state and from country to country, but they are similar in many ways. Almost all of the laws require that the product be labeled to show the percentage of the primary nutrient elements—nitrogen, phosphorus, and potassium—present. In most countries, including the United States, the order of listing the primary elements is nitrogen (N), phosphorus (as phosphorus pentoxide, P_2O_5), and potassium (as potassium oxide, K_2O). The phosphorus and potassium oxides are usually referred to as phosphate and potash respectively.

The *fertilizer grade* refers to a guarantee of the nutrient element content found in a fertilizer in terms of nitrogen, available phosphate, and water-soluble potash. A grade 5–10–15 fertilizer contains 5 per cent nitrogen, 10 per cent available phosphate, and 15 per cent water-soluble potash. The *fertilizer ratio* refers to the relative percentages of N, P_2O_5, and K_2O in the fertilizer. A grade 5–10–15 fertilizer has a ratio of 1–2–3; a grade of 5–10–10, a ratio of 1–2–2; and a grade of 10–10–10, a ratio of 1–1–1.

Understanding the fertilizer grade and ratio is very important in determining the proper fertilizer to purchase, the price to pay, and the rate of application. For example, a soil low in phosphorus requires a fertilizer grade in which the ratio of P_2O_5 to N and K is high. Many fertilizers have the same ratio but different nutrient content. For example, a 4–8–12 fertilizer has the same ratio as a 5–10–15; an 8–8–8, the same ratio as a 10–10–10. The cost per 100 pounds, however, should be different. Likewise, the rate of application should be different. Both the 4–8–12 and the 8–8–8 should cost only 80 percent as much as the 5–10–15 or the 10–10–10 respectively. The rate of application of the 5–10–15 or the 10–10–10 should be only 80 per cent that of the 4–8–12 or the 8–8–8. In other words, an application of 80 pounds of 10–10–10 contains as much nutrient material as 100 pounds of 8–8–8.

There is considerable variation in laws governing the labeling and guaranteeing of essential elements besides primary nutrients. Some states in the United States require labeling and a guarantee of all essential elements, which include the primary, secondary, and micronutrients. Other states require a guarantee only if the essential element is listed on the label.

There are many grades of fertilizers manufactured. In 1960, in the United States alone, over 1,600 grades of fertilizer were sold. The difference among the majority of these grades is not great; only 10 grades account for about 50 per cent of the fertilizer sold in the United States.

Only a portion of the fertilizer applied to the soil is used by plants in any one year. Nitrogen and potash are lost through leaching; some is assimilated by weeds; some, by microorganisms; and a portion reacts with the soil. The changes that the fertilizer undergoes after it is placed in the soil depend upon the condition of the soil. Soil condition also determines the residual effect of the fertilizer. In a broad sense, soils compete with plants for the fertilizer applied: the extent of competition is determined by materials in the soil that react with the fertilizer to change its availability. Iron and aluminum in the soil, for example, react with applied phosphate and greatly reduce its availability to plants. Thus, fertilizer applications should be based upon soil types as well as the economic returns that may be obtained from a crop.

Lime (calcium carbonate and magnesium carbonate) is not usually considered a fertilizer, although it supplies calcium and magnesium for plant use. It is generally applied to reduce soil acidity. Under normal conditions, soils tend to become acid through the formation of organic acids by the decomposition of organic matter, through leaching losses of calcium and magnesium, and other factors. Soils in semihumid or humid

Typical Analyses of Chemical Nitrogen and Natural Organic Fertilizers

Material	Nitrogen Content (Per Cent)
Ammonium nitrate	33.5
Ammonium nitrate-limestone mixtures	20.5
Ammonium phosphate*	11.0–21.0
Ammonium sulfate	20.5
Calcium cyanamide	21.0
Calcium nitrate	15.5
Potassium nitrate	13.0
Sodium nitrate	16.0
Urea	45.0
Anhydrous ammonia	82.0
Nitrogen solutions†	16.0–49.0

Material	Nitrogen Content (Per Cent)	Phosphate Content (Per Cent)	Potash Content (Per Cent)
Animal tankage	5.7–10.0	1.8–3.6	0.1–1.6
Dried blood	12.0–14.0	0.5–2.0	0.1–0.9
Castor pomace	4.0– 6.5	1.0–2.0	1.0–1.5
Cocoa tankage	2.0– 2.5	1.0–1.3	0.6–3.0
Cottonseed meal	6.5– 7.5	2.0–3.0	1.5–2.0
Fish tankage	6.5–10.0	4.0–8.0	0.1–1.1
Garbage tankage	2.5– 3.3	2.0–5.0	0.5–1.0
Tung meal	3.8– 4.4	1.0–1.5	1.0–1.5
Sewage sludge	1.5– 6.3	1.0–4.0	0.1–0.5

* Available phosphate (P_2O_5) content, 20.0%–53.0%.
† Aqueous solutions of ammonia with ammonium nitrate or urea, or both.

regions are likely to require an application of from one to two tons of lime per acre every three to five years.

For greatest efficiency, fertilizer and lime applications should be considered for a cropping system over a period of four to six years. This permits the residual effects of fertilizers and lime to be appraised and more fully utilized. Soil tests and soil survey reports showing the soil type are the best sources of information upon which to base a fertilizer-and-lime program.

Although the use of commercial fertilizers has increased many times since 1945, the amount presently used is only a small portion of that needed. Most soils, even virgin ones, cannot supply the amount of nutrients required for a large crop. The great increase in the use of commercial fertilizers has not been caused by "worn-out" soils, but by a greater yield potential than has ever before been attained. Improving plant nutrition is essential in improving human nutrition, since man's food is obtained directly or indirectly from plants. In order to attain ideal nutrition for the world as a whole, especially for the people in the underdeveloped countries, additional nutrients from chemical fertilizers must become available.

Crop Production.—In the production of crops, the principal interest is in the quantity and quality of yield in relation to the costs (input) of producing the yield. *Crop yield* can be expressed as a function of crop, soil, climate, and management. If an index value or numerical figure can be obtained for each of these variables, then the yield for a given set of conditions can be predicted. This is difficult, if not impossible, since there are so many variables involved. For example, the crop includes such variables as kind, variety, and thickness of stand (closeness of planting). For soil, consideration must be given to the nutrients, water, air, acidity, alkalinity, toxic elements, microorganisms, texture, depth of profile, type of clay, minerals, and many other factors. Precipitation, temperature, day length, light intensity, and wind are some of the climatic factors to be reckoned with. Management, as considered here, includes such factors as insect control, disease control, and weed control, and such seasonal cultural practices as planting, cultivating, and harvesting.

A low crop yield may be caused by defects in any one of the four factors listed above, in a portion thereof, or in a combination of them. The yield actually obtained is the result of the composite of all the factors. As yields increase, the composite of the factors involved must be more favorable. Production of 150 bushels of corn per acre requires the selection of a good corn variety (usually an adapted hybrid), a good stand, properly limed and fertilized soil, ample seedbed preparation, adequate pest control, and favorable climate. On the other hand, production of only ten bushels of corn per acre may occur with a poor variety of corn, a thin stand, little or no fertilizer, poor pest control, and adverse climate.

U.S. DEPARTMENT OF AGRICULTURE

WHEAT YIELD varies from county to county, depending on the condition of the terrain.

Man cannot control all of the factors involved in crop production, but research has given him access to the information and tools with which to make relatively high yields possible.

Unsolved Problems.—Although much progress has been made since 1945, prospects for the future are even greater. The solution of one problem through research usually results in the unfolding of many more; those not anticipated today probably will be the most important tomorrow. There are several problems at present, however, toward which more research should be directed.

One of these is the question of how to raise the productivity of soils to the level where high yields can be sustained. The concept that a virgin soil is in its highest state of productivity is erroneous; most virgin soils cannot support the productivity level that will be required in the future. Soil conservation will change from maintaining soils at the virgin level of productivity to improving the soil so that productivity may be maintained at a higher level. This will require improvement of both physical and chemical properties. Attention also will have to be paid to the microorganism population, which is a very important part of the soil.

More attention must also be given to higher-quality products. The amount of grass produced per acre is not as important as the amount of beef or milk that can be produced from the grass. Balancing of essential elements in the soil is important in obtaining good quality, but availability of the nutrients during various stages of plant development must also be considered. Likewise, nutrients cannot be considered alone, since moisture and other climatic factors are equally important.

Another problem that must be solved is how to achieve greater use of the soil for retaining water. With the rapidly increasing urban demand for water, both domestic and industrial, steps must be taken to utilize a much higher proportion of the annual precipitation. Construction of dams is helpful, but a bigger problem is the management of the watershed on which the precipitation falls. This includes not only farms, wooded areas, and rangeland, but roadsides, yards, and gardens as well.

As civilization develops, many "unnatural" products that come into use may contaminate the soil. Radioactive fallout from nuclear bomb testing is a good example. There are many others, however, such as the widely used insecticides, nematocides, herbicides, and other pest-controlling products. All of these may be very effective for the purpose for which they were originally developed, but their residue in the soil must be reckoned with—or serious damage may result. Reactions that take place in the soil must be ascertained and remedial practices developed where needed.

Waste disposal from cities and manufacturing plants is a big problem at present, and is likely to increase. Questions are being raised about the possibility of utilizing the soil for waste disposal without ruining it for other uses, particularly for crop production.

Land reclamation has been and is a very important phase of agriculture. It is likely to be even more important in the future. Not only will it be necessary to drain swampy areas and improve alkali conditions, as is being done now; man will also have to restore areas formerly used for roads or building sites. With a greater demand for food in the future, the reclamation of such areas is likely to be important.

—J. Walter Fitts

BIBLIOGRAPHY

AMERICAN SOCIETY OF AGRONOMY. *Advances in Agronomy.* Vol. 15. Academic Press, Inc., 1964.

BROMFIELD, LEWIS. *Out of the Earth.* Harper & Bros., 1950.

HIGBEE, EDWARD COUNSELMAN. *The American Oasis: The Land and Its Uses.* Alfred A. Knopf, Inc., 1957.

ISRAELSON, ORSON WINSO. *Irrigation Principles and Practices.* John Wiley & Sons, Inc., 1950.

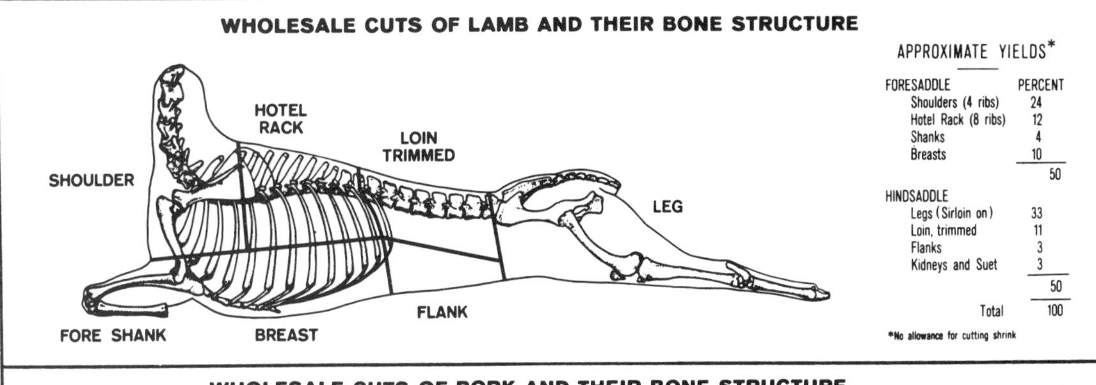

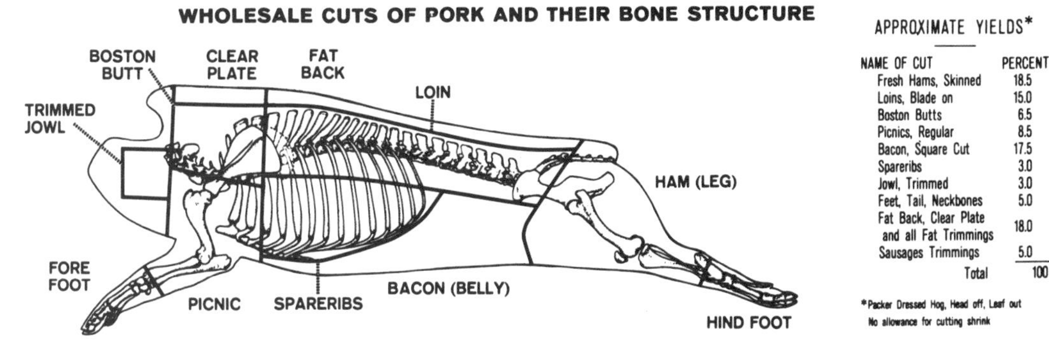

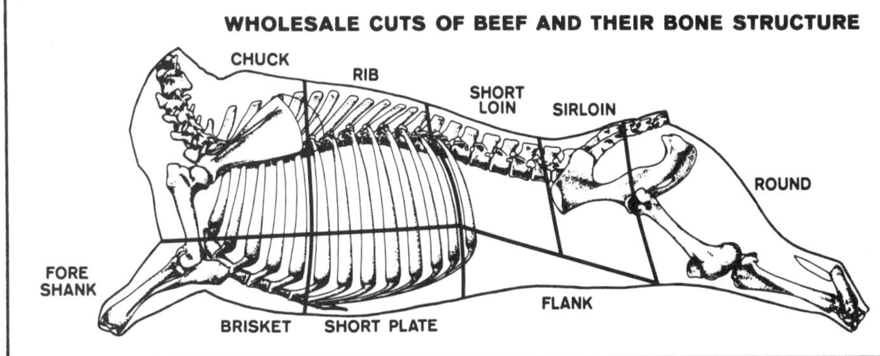

LEFT: U.S. DEPARTMENT OF AGRICULTURE RIGHT: NATIONAL LIVE STOCK AND MEAT BOARD

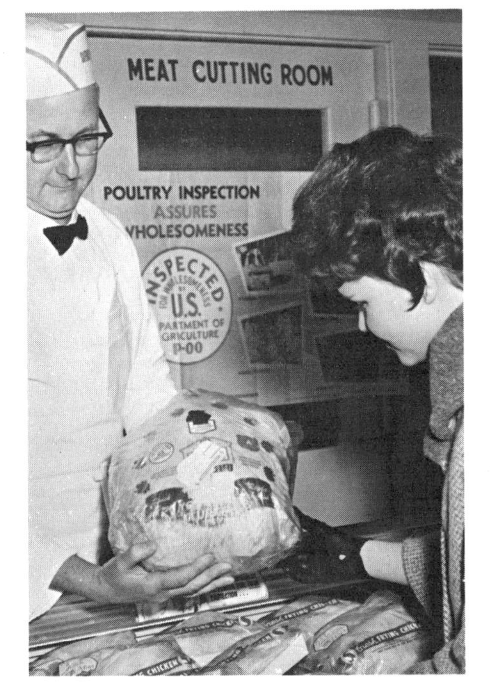

LOWER LEFT: U.S. DEPARTMENT OF AGRICULTURE LOWER RIGHT: SWIFT AND COMPANY

MEATS GRADED into U.S. Department of Agriculture classes from "Choice" to "Utility" (*above left*) are marked for consumers' information. At *left*, a shopper picks a "Grade A" turkey. *Above*, "U.S. Choice" beef ages in the cold storage room of a packing house.

MACHINES AND PROCESSES

Evolution of Machines.—The history of machines may be said to begin with the introduction of complex machines in classical antiquity. In the Greek and Roman period it was human muscle that supplied the power to drive machines. In the Middle Ages human beings began to be replaced by nonhuman power sources, and machines became larger and more elaborate. This innovation continued to develop until the eighteenth century, when a new and more flexible power source—the steam engine—appeared. The steam engine was rapidly followed by the mechanization of the textile industry and by the introduction of machine tools. The Machine Age was further accelerated by the advent of electric power toward the end of the nineteenth century, and progress in invention and development of power-driven machinery has continued.

However, with the early 1950's came a remarkable advance, the *Age of Automation,* characterized by the automatic operation and direction of the machine and the computerized supervision of complex manufacturing processes. The Age of Automation, increasingly freeing man from production, will continue for the foreseeable future.

INTERNATIONAL BUSINESS MACHINES CORPORATION

A RENAISSANCE MACHINE designed by Leonardo da Vinci was a car with a set of springs and gears for each wheel to provide for the difference in rotation when turning a curve.

Role of Machines.—Machines contribute more to modern living than any other single feature of twentieth century society, yet there is no simple definition of machine. As mechanization encompasses a host of activities, it is impossible to formulate a definition that would apply equally, say, to the essential nature of both a drill press and a computer. A general understanding of the function of machines can be gained from the fact that they have long been adjuncts to man's feeble and fallible hands, and that more recently they have begun to add power to some of his similarly slow and erring thought processes.

The machines of antiquity served to modify or concentrate human effort to accomplish a desired end, such as lifting a heavy block of stone up to its position in a building wall. As mentioned in the section on energy and power sources, men of the Middle Ages developed remarkable sources of nonhuman power, and these were linked with machines to extend still further the effectiveness of hand tools.

With the advent of machinery that did not depend on human muscles, man became the guide or director of machine operation, in both a physical and an intellectual sense. He could channel the newly invented power sources into work that man alone could never accomplish. For example, it is impossible to conceive of a system that would directly use the million human beings that would be necessary to supply the power used by a modern jet aircraft.

Just as the Middle Ages began to see men freed from the drudgery of supplying power, such as walking all day and every day on a treadmill, so in the second half of the twentieth century have men begun to be removed almost entirely from productive operations. The present era of automation is characterized by the freeing of men even from their tasks of guiding machines. The development of automatic controls, thus, is at least as important as the medieval invention of nonhuman power sources in terms of freeing men for the pursuit of more intellectual forms of activity.

■ANCIENT MACHINES.—The beginnings of the evolution that led to modern automation lay in the mist of prehistory, for early man had invented two of the classic five simple machines—the lever and the wedge, or inclined plane. Neolithic man then produced the wheel and axle. In historic times, the Greeks invented the pulley and the screw.

Antiquity also saw the invention of the first of the complex machines, all of which were constructed of wood. These included three complex devices employed in food manufacturing: pumps for irrigation, grain mills to produce flour, and presses used in making wine and olive oil. Looms were developed to weave fabric to clothe people, and remarkable man-powered cranes were constructed to build large buildings for shelter and social activity. Other interesting complex machines of antiquity, which had their origins in prehistory, were the potter's wheel, which may have preceded the wheel and axle on carts, and the bow-string drill.

The Greeks also invented heavy missile-throwing machines that were the ancient forerunners of modern ballistic missiles. By the close of the classical period, man had developed complex machines to produce food, to clothe and house himself, and to help him enforce his political aspirations on other groups.

Medieval Industry.—Up until the tenth century, medieval industry was largely domestic. There was little or no improvement of ancient machines or invention of new machines. However, from the tenth to the twelfth centuries there were improvements in ancient machines, accompanied by the introduction of sources of nonhuman power. Then again, from the tenth to the fourteenth centuries, few advances were made in food production and textiles, although the spinning wheel appeared at this time.

Most machines invented and developed in the later Middle Ages were associated with mining, metallurgy, and construction. These included mechanically operated bellows, saws, the pole lathe, toothed wheels for gearing, and the development of many applications for that all-important device, the crank, first used in the ninth century.

Improvements, made possible largely by the new power sources, were made in ancient cranes. Similarly, larger boring machines were developed, which were used at the end of the Middle Ages to produce cannon. At the same time, there were further developments in missile-throwing war machines, exemplified by the introduction of the counterpoise-powered trebuchet — a beam with a sling holding a rock at one end and a heavy counterpoise weight at the other.

One highly significant achievement of medieval technology was the mechanical clock, a spring-driven series of gears that kept time more accurately than earlier devices. A mechanical clock was built in Milan in 1335, and the Italian Jacopo de Dondi improved the design in 1344. As a large mechanism was easy to construct, most early timepieces were tower clocks like that made by Henry de Vick in 1379 for the Paris Palais de Justice.

FORD MOTOR COMPANY

EARLY MASS PRODUCTION in the automobile industry employed devices like this "body drop," used on a 1914 Ford assembly line to speed the manufacture of Model T cars.

Between 1400 and 1700 no important new power source was developed. However, looms, pumps, and mills were extensively improved, and new machines were invented, including knitting machines, clocks, screw-cutting lathes, and printing presses.

Without a doubt the most important of the new machines was the printing press, for it facilitated a remarkable spread of knowledge, beginning in the last half of the fifteenth century. Printing was also important from the mechanical point of view because it was the first process to mass-produce a product having interchangeable parts. A book was printed on sheets of paper that were folded into signatures—a signature usually contained 4, 8, 16, or perhaps 32 pages. Each signature was labeled, the first often being *A*, the second *B*. However, each *A* signature was like every other *A* signature, so that a book could be assembled by using any of the *A* signatures as the first in the book. The basic requirement for interchangeable manufacture is accurate production so that each part is like every other. Since each sheet was printed from the same type, all were exactly alike; therefore, it made no difference in which copy of the book a given sheet was used. However, it was not until the nineteenth century that manufacture using interchangeable parts was extended to other industrial operations.

The Industrial Revolution.—The Industrial Revolution, which began in the eighteenth century and has extended to modern times, was initially associated with James Watt's invention of an efficient steam engine and many new types of power-driven machinery. New power sources—improved water wheels, water turbines, and steam turbines—followed the Watt steam engine, and new machines were developed.

There was little use for complex machines in agriculture until the eighteenth century, when there was an increasing demand for improved agricultural production resulting from an increase in population that has continued through the present day. New machines in the eighteenth century were varieties of seed drills, which mechanized planting. In the nineteenth century, reapers and threshers were developed to mechanize the harvest. More recently, mechanical harvesting techniques have been applied, particularly to fruit but also to field crops.

In the early decades of the Industrial Revolution, the introduction of new textile machinery spurred the revolution forward more than any other type of development. Powered spinning machines led to the invention of powered machines for preparing textile fibers for spinning. Not only were power looms invented, but also figure-weaving looms on which the design could be mechanically woven. These figure-weaving looms used large punched cards, the direct forebears of the punched cards so widely used in modern computers.

Associated with early inventions of the Industrial Revolution was the extensive introduction of machine tools, which are metal-cutting tools. Until the middle of the eighteenth century, machinery was still being built by the carpenter. With the advent of the Industrial Revolution, metal began to be used in construction of machines. To work metal, new machine tools were required. Machine tools were powerful, and they were accurate—without them, the Industrial Revolution would have slowed down in the first half of the nineteenth century.

Typical machine tools are the drill press, metal screw-cutting lathe, planing machines, turret lathes, gear-cutting machines, precision grinders, and milling machines. Most of these devices were invented in the first half of the nineteenth century, and most of them operate with a shearing action of a sharp tool. Metals continued to be machined in this manner until the 1950's, when nontraditional machining techniques began to come into use, including laser machining, ultrasonic machining, plasma-torch cutting, and electrical-discharge machining.

Mass Production.—Aside from book printing, mass-production processes did not come into wide use until the early nineteenth century. Among the first persons to use such procedures were Eli Whitney in the manufacture of muskets and Eli Terry in the production of wooden clocks. In Whitney's musket factory, uniform parts were produced by machine, and these parts were assembled into individual muskets by filing and other fitting procedures. However, it was not possible to interchange parts among these finished muskets. With the advent of machine tools, which worked accurately and produced finished parts, it was possible in 1819 for John H. Hall of the Harpers Ferry Armory to start mass production of a breech-loading rifle employing the system of interchangeable parts. Mass production with interchangeable parts soon came to be known as the *American System* and spread to other areas of manufacture. Today it is almost universally employed, except that the computer is now making it possible to employ mass-production techniques that are entirely mechanized but that yield nonidentical products.

Assembly-line processing was an-

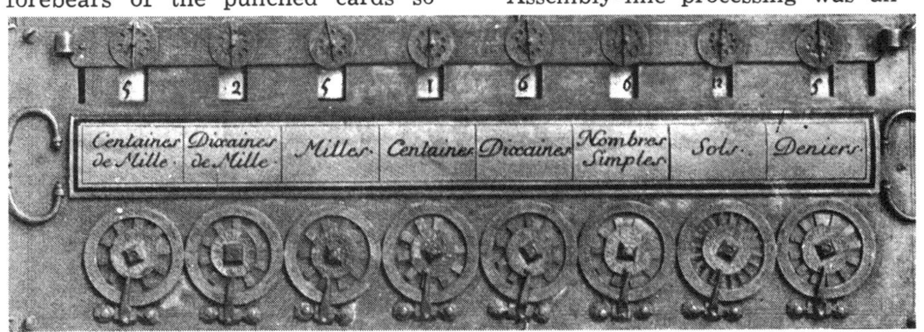

INTERNATIONAL BUSINESS MACHINES CORPORATION

MECHANICAL CALCULATING MACHINE, devised by Blaise Pascal in 1642, was the first one known capable of adding, subtracting, multiplying, and dividing with accuracy.

other important innovation in nineteenth century manufacturing. Assembly-line procedures were probably first introduced in pork packing houses in Cincinnati, Ohio, in the 1860's, and these slaughterhouse procedures, in turn, helped to influence Henry Ford to introduce his famous assembly-line automobile manufacture in 1913. Since Ford's successful employment of the assembly line, it has been widely used in manufacture and, recently, has been extensively automated.

■**AUTOMATION.**—Automation has been characterized by the introduction of automatic control and the electronic digital computer. Automatic controls employing the feedback principle first appeared in the eighteenth century on steam engines and windmills. An example was the fly-ball governor which James Watt adapted from windmill-driven flour mills to control the speed of a steam engine. Watt attached the device to the engine throttle; as engine speed increased, the device closed the throttle, thereby reducing speed. Thus, there was a *feedback* of information from engine speed to control to throttle to engine speed. Controls that use such feedback in "closed loops" are true automatic controls.

Although automatic controls were first known in the eighteenth century, it was only in the twentieth century that they began to be used extensively in continuous processes, for example, in chemical plants. Subsequently, self-regulating automatic control of machine-tool operation was introduced. In this technique, known as *numerical control,* numerically coded instructions on punched paper tape or punched cards are fed into a reading device in the machine, and the information in the tape or cards guides the machine through its task without the aid of human operators. Numerical control has eliminated human error, freed many workers from tedious and repetitious tasks, and has further improved the speed, safety, and accuracy of machine production.

The automation of business methods has been accomplished largely through the electronic digital computer. Although automatic controls had been little used here, data-processing computers have been extensively introduced since 1955 into business procedures, replacing human workers to an amazing extent in such operations as invoicing, accounting, paying, and inventory maintenance.

The computer, also, has been recently introduced into manufacturing processes. Numerical control directs a machine to produce a given product. Computers are employed to direct the several complex machines that will yield the desired product. Here, as the circumstances of machining may vary due to the condition of the tool or of the metal being machined, the computer corrects the process so that the final product is uniform. Numerical control had already replaced the machine operator. The computer is now replacing the supervisor of the operators.

—Frederick G. Kilgour

MACHINES AND PROCESSES GLOSSARY

Air Pollution, the contamination of the atmosphere with injurious substances. The effects of air pollution are serious. Through injury to vegetation and livestock, corrosion and soiling of materials and structures, lowering of property values, and interference with air and surface travel, air pollution costs billions of dollars each year in economic damage in the United States alone. Of even greater concern are its adverse effects on human health. Episodes of extreme pollution have contributed directly to sickness and death. More importantly, studies conducted over the past decade have produced a growing body of evidence that the long-term effects of exposure to ordinary levels of air pollution adversely affects the health of many and may result in chronic disease and premature death. Among the major illnesses which have been linked with air pollution are bronchitis, asthma, lung cancer, emphysema, and even the common cold. The unmistakable upward trends in the factors which contribute to the contemporary air-pollution problem—increasing urban population, increasing industrialization, greater demands for power produced by the burning of fossil fuels, more and more dependence on automobiles, trucks, and buses to meet rising transportation needs—these and other trends of growth leave no doubt that the potential scope and severity of the problem will intensify unless the nation substantially and rapidly augments its efforts to control the sources of air pollution.

The technical means for controlling, or limiting, the discharge of most pollutants to the air are available today. Through such devices as filters, scrubbers, centrifugal separators, electrostatic precipitators, or such means as industrial process modification or fuel substitution, the majority of sources can be adequately controlled. On the other hand, means for controlling the discharge from fossil-fuel burning processes or from motor vehicles, are either not presently available or are only partly effective. To bring such sources under fully adequate control will require substantial augmentation of research and development efforts by both industry and government as well as public awareness.

The determination of the U.S. federal government to control air pollution is clearly expressed in the Clean Air Act of 1963 and the Clean Air Act Amendments of 1965. These new federal authorities include strengthened research and technical assistance activities, a matching-grants program which offers substantial financial stimulation for the creation or improvement of state and local regulatory control agencies, a federal abatement program focused primarily on interstate pollution problems, and the regulation, nationally, of auto exhaust emissions.

If the United States expects to fully meet the challenge of air pollution, great strides must be taken in the application of controls to pollution sources. The forces that shape the contemporary environment make certain that the problem will otherwise continue to worsen.

—Vernon G. MacKenzie

Assaying, the procedure used to determine the portion of pure or precious metal present in an ore sample. Unlike complete analysis, assaying usually determines only the proportion of certain valuable constituents to the whole sample. In essence, assaying procedures are adaptations of ore-treatment procedures designed to separate the valuable and waste portions of a mineral specimen.

Fire assaying, which is used for precious metal ores, makes use of the fact that the gold, silver, or platinum in a sample will dissolve in lead when submitted to high heat in a reducing atmosphere. In this process, litharge (a lead oxide), silica, and other materials are added to an *assay ton* (29.166 grams) of the sample. In the reducing atmosphere the litharge is reduced to lead, and the gold, silver, or platinum in the sample combine with it and sink to the bottom. The silica, borax, and non-valuable portions of the ore become liquid at this high temperature. The lead button containing the precious metals is weighed and then submitted to a high temperature in an oxidizing atmosphere. This drives off the lead, leaving as residue a bead

U.S. PUBLIC HEALTH SERVICE

AIR POLLUTION, a serious problem in many cities today, is caused by vehicle exhaust fumes and by smoke that factories, like those above, pour over congested city areas.

containing only the gold, silver, or platinum. This is weighed and treated with hot nitric acid. The nitric acid dissolves the silver in the button. The difference in weight of the bead, before and after treatment by nitric acid, is the weight in grams of silver in the original sample. The remainder is gold and platinum. The weight in grams of gold or silver in an assay ton of sample represents the number of ounces of gold in a ton of the material of which the sample is supposed to be representative.

There are quicker methods than fire assaying for analysis of any ore material. The assay office can determine everything present in the ore material by chemical means or through the use of equipment such as the spectroscope, in which characteristic colors given by different elements are compared to calibrated standard samples. In assaying radioactive material, a scintillometer may be used to compare the radioactivity of the sample under analysis to a calibrated series of samples to determine the percentage of radioactive material in the sample. —J. C. Fox

Automatic Control Systems, assemblies of devices to maintain a desired physical state, such as temperature or speed, without the need for human intervention. Such systems make it possible for a rocket to travel a precise path to a distant planet or for a home to be automatically and safely kept at a comfortable, even temperature. Automatic control systems play an important role in technology, and give people more leisure time while increasing their comfort and safety.

Automatic control systems for space rockets or even for industrial processes can be extremely complicated, but they are based on the same principles as simple systems. Almost any measurable thing that can be controlled by hand can also be automatically controlled.

An elementary automatic control system is made of three main elements: a *sensor,* which measures whatever is to be controlled; a device to determine what must be done if the measured value does not meet specifications; and a *servo,* which causes a corrective action to be taken. A typical *sensor* is an electric thermostat that turns on a switch connected to a heater when the room temperature falls below a certain point. In this case, the device that determines what is to be done is the wiring of the system itself, causing the heater to be turned on. The *servo* in this case could be a valve that permits fuel to flow to the heater. Heat continues to warm the room until the thermostat signals a shutoff of the heaters. Such a control system is called an *on-off system,* and the difference between the turn-on and turn-off temperatures is called the *differential.*

A fundamental concept is *feedback,* a communication link between the servo, which brings about the corrective action, and the sensor, which commands the servo. In the simple example above, the feedback is through the air in the room. The thermostat commands the heater to come on; the heater warms the air; and the heated air, in turn, feeds this information back to the thermostat—hence, feedback. In some complex systems, feedback can be achieved in ways other than through the controlled medium.

An automatic control system of a different type is illustrated by one of the oldest control systems, the speed governor of a steam engine. It permits gradual increases or decreases in control rather than all or none, as with an on-off control. This is called *proportional control.*

Special problems in automatic control arise if great precision is required. If, for instance, it should be necessary to control the temperature of a room to within a fraction of one degree, the simple arrangement described above would not be adequate, no matter how sensitive or accurate the thermostat. There would be too much time delay in response of the system. When the thermostat turned the heater on, some time would elapse before the temperature rose in the room. By the time the thermostat turned the heater off, the temperature would have continued to rise for a while. As the temperature again fell, the thermostat would turn the heater on again, but too late to prevent a continuing dip in temperature. Any simple control system, no matter what it is controlling, has this tendency to swing rhythmically. This is called *hunting.*

BURROUGHS CORPORATION

AUTOMATED devices process data from tape.

Hunting can be eliminated by adding a device that will respond to the rate at which the controlled medium is changing. Such a control arrangement, called a *rate* system, has many applications. Airplanes and rockets have *rate gyros* that generate electrical signals according to how fast the vehicle is changing direction. These signals go to a computer, which in turn operates servos for steering. Such systems are called *rate damping* systems, and they permit smooth, even flight.

The ultimate in a control system is achieved when the sensors measure those things that would cause the change in what is controlled, rather than measure the controlled medium itself. The reason is that a sensor which is sensing the controlled medium cannot do anything until a change occurs, and that change is the very thing the control system is supposed to prevent. This fundamental is the basis for very elaborate automatic control systems. —John V. Sigford

Automation, a term denoting a wide variety of automatic procedures. To some, it is "the automatic handling of parts between progressive production processes"; to others, it is "an evolutionary extension of mechanization" or "a new technological revolution (that is, a second industrial revolution) with far-reaching social and economic effects."

Actually, a good case can be made for each of these definitions. The earliest so-called automated factories were those which used mechanical methods to move parts from one production machine to another. An automated refinery operation, for example, was one in which remotely controlled pipelines were used to move petroleum products between distillation towers. An automated factory was one that utilized conveyor belts or similar devices to pick up parts at one machine and move them along to the next production operation.

In the 1950's and early 1960's the term "automation" was applied to the growing family of numerically controlled production machines that operate from instructions punched into tapes or cards. Although an operator is still required to set up the tapes and supervise the machine, the actual milling, cutting, or drilling operations are performed automatically.

In a sense, the trend to numerically controlled machines is an extension of the Industrial Revolution of the eighteenth and nineteenth centuries. Originally, all machines and tools were operated and controlled directly by man or animals. The basic change effected by the Industrial Revolution was to replace muscle power with steam and, later, electricity. But the machines themselves were still operated and controlled by man.

Today we find man stepping further and further into the background as tapes or cards, often produced automatically by computers, control and run production machines while the operator functions primarily as an overseer.

All of this has contributed to the "technological revolution" that is sweeping industry today. In the office, we have the increased use of data-processing systems to take over many of the tedious clerical chores formerly handled by clerks; in the factory, we have the growing use of mechanical and electronic devices to manufacture, test, and transport products; and in science, we have a growing dependence on computers to perform computation tasks formerly done by hand—or not done at all. Among the results of these changes has been an increase in leisure time and an improvement in our standard of living. —Stanley L. Englebardt

Automobile Manufacturing, the production of motor vehicles, primarily those powered by an internal-combustion engine and designed to transport two to nine passengers. Although precise details are clouded by conflicting claims, it is generally agreed that production of automobiles in volume began in the early 1890's in western Europe. In the United States, series production of both electric and gasoline automobiles was under way by 1896, when the Pope Manufacturing Company of Hartford, Conn., produced 500 electric and 40 gasoline carriages in one year.

By 1897 Ransom E. Olds and Alexander Winton founded companies to build gasoline-powered cars and the Stanley brothers were building steamers, but only Pope could qualify as a volume producer.

Two important events in 1899 changed the picture radically. The Olds Motor Works moved to Detroit, and then the factory burned down. The only thing saved was the prototype of a low-priced car designed by Olds. By concentrating on the single-cylinder buggylike car, the company achieved instant success. In 1901, about 600 Oldsmobiles were produced; in 1902, production jumped to 2,500; in 1903, the number reached 4,000; and in 1905 a total of 5,000 units rolled from the plant. At that time Olds and his backers severed relations, and production of the economy model was halted.

■MASS PRODUCTION.—Although the Olds company had achieved substantial volume, its manufacturing methods were still based on skilled craftsmen producing one car at a time. The big change started in 1903 when Henry Ford, after several false starts, founded the Ford Motor Company and put his personal stamp on automobile production. He concentrated on developing a car suitable for the mass market, and then on cutting manufacturing costs. By 1907 the prototype of the *Model T* had been developed, and the most famous car ever built was offered to the public in 1908. In 1909, Ford stopped production of all other models and concentrated all his efforts on the Model T. By 1914 a complete assembly-line had been perfected that reduced chassis assembly time from 12½ hours to 1½ hours. The success of the effort to cut manufacturing costs was reflected in the rapid drop of the retail price. In 1912, a Model T sold for $600, and as production increased, costs decreased so the retail price could be dropped even further. Ford had produced 6,000 units in 1908, but by 1916 production soared to 577,000 Model T's and the price was down from $850 in 1908 to $360. A total of 15,007,003 Model T's had been built when production finally stopped on May 31, 1927.

■ASSEMBLY-LINE METHODS.—By the time Ford shut down to retool for a new model in 1927, the manufacturing processes developed to cut the cost of the Model T had been adopted by the entire industry. Assembly methods employed in the auto industry today can be traced to the basic concepts that were in use at Ford as early as World War I. Work is moved past the operator on an assembly line while parts are stacked near at hand or delivered as needed by conveyors. The basic rule—never stop the assembly line—must be followed to get the most from continuous production. If a mistake is made, an operation missed, or a defect discovered, it is bypassed and corrected at the end of the line. Stopping the line would immediately idle every worker on it and result in complete loss of production.

GENERAL MOTORS CORPORATION

MASS PRODUCTION of automobiles requires quick repetition of highly specialized tasks.

Each operator performs one relatively simple task that is repeated on each piece of work. This reduces the need for skilled operators to assemble even complex equipment, since each worker need only know how to perform his simple, repetitive task.

The final assembly line in an auto plant, where thousands of parts are joined to produce a complete automobile, represents the culmination of several years' effort. Hundreds of cars are produced on the final assembly line each day, yet the dramatic assembly operation represents only a small portion of the total effort that goes into making a new car. The uninterrupted flow of vehicles from the assembly line reflects years of planning, engineering, and scheduling by specialists who have worked out thousands of details.

■PRODUCT PLANNING.—All work starts with a product-planning group that evaluates market research data and then, working with other departments, proposes a new model program to company management. New models proposed by product planning must provide what the customer wants, be manufactured at a competitive cost, be competitively priced, and return a profit for the capital invested.

A proposed new model is described in a *paper program,* or *package,* specifying in detail the targets and limits for the new model. The limits include the physical dimensions, total weight, carrying capacity, and power requirements. Targets include the total manufacturing cost and delivered price of the new car.

Production experts then examine the plan in detail to make sure that every part of it is practical from a manufacturing standpoint. Cost analysts and cost estimators examine the proposed product to determine manufacturing costs.

The final product plan, or package, is then presented to the product planning committee for approval. This is the signal to begin detailed styling and engineering work.

■STYLING.—Hundreds of sketches are prepared to develop new and interesting styling ideas, for the overall automobile as well as its different parts. From these sketches, coherent styling ideas gradually emerge. The most promising designs are developed into full-size clay models to get a three-dimensional basis for more effective selection of the final appearance.

Finished to a glossy smoothness, the clay is covered with plastic paint, and exterior trim is added. When the model is complete, it is difficult to tell the difference between the clay *mock-up* and a real automobile.

As the exterior is developed, interior styling is worked on by specialists in textiles, color, fashions, and plastics. Development of the car's interior must be integrated with exterior styling so that the two will harmonize. Design details are developed by artists who sketch ideas for seats, instrument panels, and other interior parts.

The most promising interior designs are blown up to full-size drawings and then developed into full-size models called *trim bucks,* before a final selection is made. Trim bucks are full-size models of the interior seating arrangement, with actual paints and fabrics used to show the proposed interior exactly as it will appear in the finished automobile.

■ENGINEERING.—While the design is being developed, engineers draw plans for each of the more than 13,000 operating parts that will go into the car. In designing a specific automobile and its many parts, engineers refer continually to the basic package document as approved by the product planning committee. This

GENERAL MOTORS CORPORATION

ON ASSEMBLY LINE front end parts are attached to a car body *(left)* and rear suspension is readied for placement *(right)*.

is to be sure that the car will end up within the original objectives that govern construction, weight, cost, performance, and ease of manufacture.

The development of an individual part or assembly begins with an engineer's ideas in the form of sketches. While working out the basic design he keeps in mind the other parts with which it must be assembled, the work it will have to do, and the amount of space in the car available for it. The engineer's job is to make sure that all the pieces of the jigsaw puzzle fit in the spaces allotted to them—and that they all work when joined in final assembly.

Whenever a new model is mass-produced, production men frequently encounter delays, adjustments, and problems that are caused by the production process itself. In 1957, the Ford Motor Company adopted an idea that enables production engineers to discover and correct troubles in advance. The idea was a *pilot plant* that produces a few sample cars. A complete assembly line is set up in the pilot plant, with the same tools, templates, and forming devices, the same gauges, and the same skills that will be used to mass-produce the car on actual production lines. Ford's pilot plant has become a standard operation where the "bugs" are worked out of new car assembly methods well before production begins.

■**FINAL ASSEMBLY.**—Hundreds of suppliers now begin to deliver raw materials and subassemblies, ordered well in advance, to the final assembly plants. Steel is shaped into car bodies from dies that took years to produce, while engines are machined and built from iron castings.

The first step in assembling a car is building the body. The various sheet-steel parts of the car body—*floor, roof,* and *side panels*—are stamped into shape on giant presses and then welded together on the longest feeder line in the assembly plant. Framing fixtures hold the panels in place while electric welding machines spot-weld parts together.

The welded body is mounted on a conveyor or wheeled dolly for further processing. Seams in the body metal are filled with molten solder, which is ground down to blend with the steel, forming smooth surfaces. The *doors* and *deck,* or *trunk lid,* are added and fitted; then all metal surfaces are ground smooth and the body thoroughly cleaned.

Moving along the conveyor system, the body is phosphate-coated in a dip tank to prepare the metal surfaces for painting, joints are sealed with vinyl and asphalt sealers, and protective layers of primer are sprayed on and baked in an oven.

On a parallel assembly line the *front end,* though a part of the body, is built up separately. This subassembly consists of the *front fenders, radiator, grille, headlights,* and *fender aprons* at the sides of the engine compartment. After being painted and trimmed, the front end is moved to the final assembly area, where it is added after the main body structure is secured to the *frame,* or *chassis.*

Trimming an automobile's body is the installation of the upholstery lining, instrument panel, electrical wiring, glass, interior hardware, heater, radio, and accessories, and the application of ornamental chrome. Particular care must be exercised during this subassembly because paint, trim, and upholstery are easily scratched, torn, or soiled.

Actual assembly of the complete automobile begins when a frame is lowered onto the moving assembly line. Subassemblies fed from conveyors or stockpiles close to the line are added in a carefully programmed sequence. The *front suspension, springs,* and *shock absorbers* are among the first parts to be added as the frame advances along the line. Then the *rear springs* and *rear axle* are bolted on. These subassemblies are usually manufactured at other plants and delivered to the final assembly plant. At the *engine drop,* completely assembled and test-run *engines* are lowered onto the frame—in the industry this operation is referred to as *decking the engine.*

With *fuel feed lines, brake lines,* and other connections installed, tightened, and inspected, the chassis enters a booth where *transmission oil* and *hydraulic brake fluids* are added. Beyond the booth, *wheels* with *tires* already mounted and inflated roll down inclined chutes from overhead assembly areas, and are fastened by power wrenches that tighten all five wheel bolts at once. Next a huge *body shim gauge* is lowered onto the chassis to check attachment points on the frame for possible misalignment before the body is added. The *body drop* refers to the operation where painted and trimmed bodies are lowered onto the frame.

Then the complete *front-end subassembly* is swung into place and bolted to the body and chassis. With the front end in place, the *hood* is installed and fitted. Assemblers check the operation of both hinges and latches, and adjustments are made to guarantee smooth operation and a good fit. Finally seats, built up on separate subassembly lines, are installed as the car moves toward completion. After an inspector has checked all previous inspection reports and given a final approval, an employee enters the car, turns the ignition key, and moves the car under its own power for the first time.

The car is driven a few feet to an inspection station where headlight beams are adjusted and front-wheel alignment is checked and set. At the next station, the car is checked on a dynamometer: with the engine running and the car in gear, the rear wheels rotate at varying speeds on rollers set into the floor, while the front wheels remain motionless. This test is a functional check of the complete drive line.

Checked throughout all the assembly operations, the car is then subjected to one final inspection. First a high-pressure water spray test checks for leaks, then a paint check under bright lights, and final examination of operating parts is made before the finished car is driven to a storage yard. Within 48 hours the car will be on the way to a dealer.

—Joseph J. Kelleher

Bearing, a friction-reducing support for a rotating shaft or spindle. Bearings vary in size and type from tiny jewels in watches and miniaturized instruments to huge roller bearings used on heavy machinery.

The simplest bearing is the *plain cylindrical journal bearing* exemplified by a wheel turning on the axle of a wagon or by a machine shaft turning in a cylindrical support. For ideal operation, there should be no metal-to-metal contact between the shaft and sleeve. A lubricant, such as oil or grease, prevents metallic contact by building up a wedge as the shaft turns in the sleeve. This wedge of lubricant separates the shaft from the sleeve and greatly reduces friction and heat. The heavier the bearing load and the slower the speed of rotation, the heavier the lubricant required.

Ball bearings, in their simplest form, consist of an *outer race* fitted into the sleeve and an *inner race* fitted onto the shaft. Carefully ground steel balls occupy the space between the outer and the inner races. There is point contact of the metal, since the bearing load is actually transferred to the steel balls, each of which touches the outer and the inner races. Ball bearings require some lubrication, but they depend upon the free rolling of the balls for their effectiveness rather than upon the lubricant. It is particularly important that they be lubricated lightly, for over-lubrication will cause heating and bearing damage.

The *roller bearing* is a variation of the ball bearing and is particularly suited to heavy loads. It uses cylindrical rollers rather than balls, and thereby provides a line contact of metal to metal rather than a point contact. *Needle bearings* are similar to roller bearings, but the individual rollers are of much smaller diameter.

The plain cylindrical journal bearings, ball bearings, and roller bearings described are all designed to reduce the friction of a shaft turning on its support or a wheel or lever turning on its axis. Another type of bearing is the *thrust bearing,* which takes a load directly along the line of the shaft. Minute jewels may be used as *thrust bearings,* while at the other extreme, thrust bearings support the main shafts of huge hydroelectric turbines. There are thrust bearings to handle various types of loads between these extremes.

Often it is necessary to combine both types of loads, and some bearings are designed to handle a thrust as well as a radial load. The *tapered roller bearing* is a good example. In addition, there is a wide variety of special bearings designed to handle unusual conditions.

A number of materials are used in the manufacture of bearings. Steel, cast iron, and bronze are common bearing materials, and Babbitt metal and lead alloys are frequently used as bearing linings. Nylon and other plastics can be used for both sleeve and thrust bearings, while special applications call for the use of special materials. —Hunter Hughes

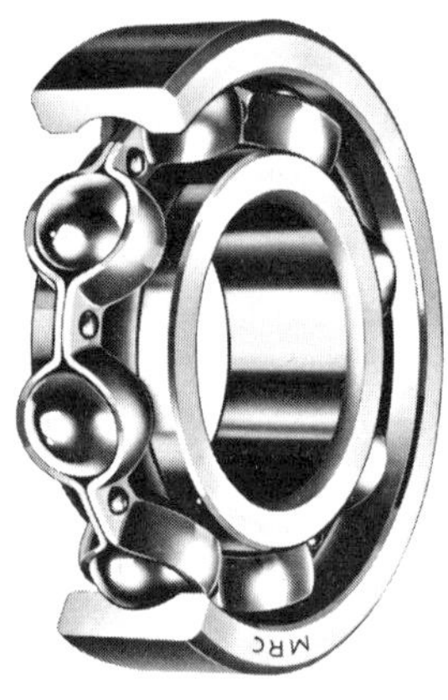

MARLIN-ROCKWELL CORPORATION

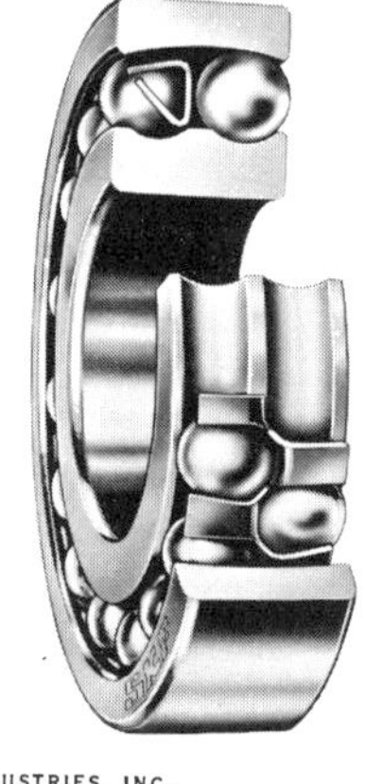
SKF INDUSTRIES, INC.

FEDERAL-MOGUL CORPORATION

BALL BEARINGS (*left; center*) and plain roller bearings (*right*) are used mainly for radial loads, to reduce friction between an axle and its wheel or a shaft and its support.

Boiler, a closed vessel in which water under pressure is converted to steam by the absorption of heat from the combustion of fuel. Boilers are designed to absorb as much heat as possible from the fuel burned. The ratio of the heat absorbed by the water and steam to the heat input of the fuel per unit of time is known as *boiler efficiency.*

There are two general types of boiler: the *fire-tube boiler,* consisting of many tubes through which hot gases pass to heat the surrounding water; the *water-tube boiler,* in which water passes through the tubes and is heated by surrounding gases. Although used extensively, the fire-tube boiler is limited by its design to low pressure and low capacity. Water-tube boilers are used more frequently because they may be designed for either low or high operating temperatures and pressures, and for higher capacities.

Records indicate that certain properties of steam were recognized earlier than 100 A.D. At about that time, Hero of Alexandria invented a primitive boiler-turbine known as *Hero's engine.* The earliest form of water-tube boiler was patented in 1766 by William Blakey, a contemporary of James Watt, but the first successful water-tube boiler as we now know it was developed in 1788 by James Rumsey, an American. Stephen Wilcox, in 1856, was the first to use inclined tubes—the forerunner of modern water-tube boilers.

The first boilers were fired with coal by hand and, until the early 1900's, operated at pressures only up to 100 or 200 pounds per square inch. Now, practically all boilers are fired by mechanical means.

Boilers are designed for many purposes: small package boilers, shipped ready to be installed for heating or process; large package or field-erected boilers for industrial plants; and huge high-pressure boilers for electric utility companies. These last may be 20 or more stories high and produce up to 9,300,000 pounds of steam per hour, at temperatures of 1,050° F and above, and at pressures as high as 5,000 pounds per square inch.

During the mid-1900s, the *once-through boiler,* which can be operated above or below *critical pressure* (3,206 pounds per square inch), has been used successfully by many large utility companies. The once-through boiler is, in essence, a number of parallel tubes through which fluid is pumped and to which heat is applied along their entire length. Water goes through once and comes out as superheated steam. Operation above critical pressure results in a higher efficiency. Improved efficiency is the boiler designer's constant aim.

—Margot Valentine

Bookbinding, the folding and gathering of printed sheets, sewing or gluing them together in proper order, and enclosing them in covers of paper, cloth, leather, or other material. Books usually are printed in sections consisting of multiples of four pages up to as many as 128 pages per section. The individual pages are arranged on the press so that when the printed page is folded, the pages fall into consecutive order. These folded sections are called *signatures.* Flat sheets go to folding machines to be made into signatures, while presses that print from rolls usually have a built-in folder and deliver folded signatures.

When all the sections for a book are printed and folded, endsheets are pasted to the first and last sections. Endsheets are four page signatures of strong paper that will eventually attach the book to its cover. In certain books, added strength is obtained by reinforcing the endsheets with a cloth strip at the fold.

The sections then are ready to be put together. A gathering machine picks up the sections in sequence and assembles them into a complete book. The gathered sections go into a sewing machine that fastens them together with strong thread. Books that are expected to receive hard use may be reinforced by sewing the sections to a cloth tape.

From the sewing machine, the book goes into a machine known as the *smasher.* This machine reduces the bulk at the back caused by the folds and the sewing. Next the book is trimmed on three sides. A rounding and backing machine then gives the backbone and fore edge their familiar rounded shape. The next machine glues a paper liner and a muslin reinforcing strip, if required, to the backbone.

While the body of the book is being put together, other machines are making and decorating the cover or case. The most common cases are made of strong boards (a type of cardboard material) covered with cloth or paper, and sometimes both.

The cases and books are brought together on a *casing-in* machine that glues the endsheets firmly and accurately to the cases. A *building-in* machine next forms the cover around the book under heat and pressure.

Paperback books are bound by gathering the signatures and gluing them to a paper backbone.

—Robert L. McGuire

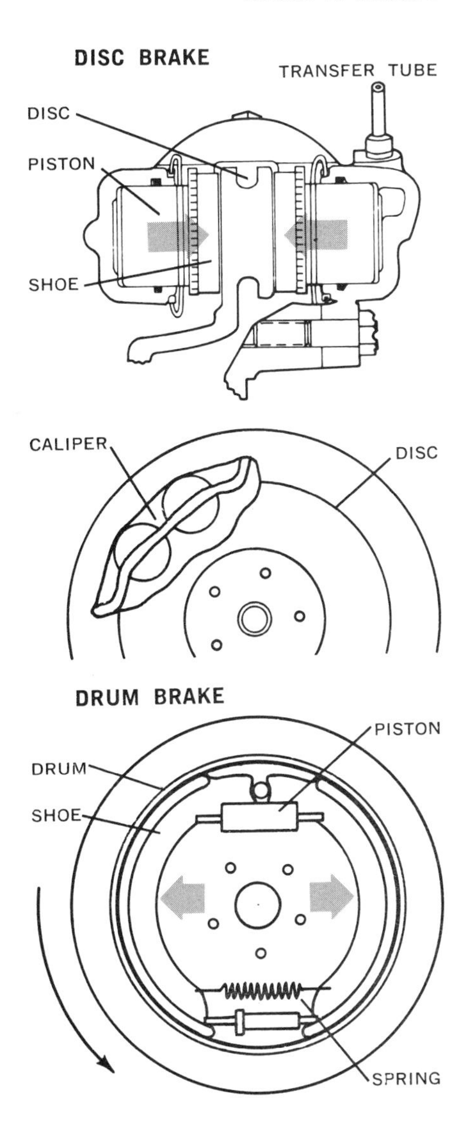

AUTOMOBILE BRAKES may be of two types, disk or drum. In disk brakes (*center*) a caliper straddles the disk above and behind the hub; when the brakes are activated, pistons within the caliper (*top*) force the shoes against the disk and friction slows the wheel. In drum brakes (*bottom*) pistons force the shoes evenly against the drums to create friction inside of the wheel; a spring releases the brake-shoes.

Brake, a device to slow or stop a machine or control its speed. This is generally accomplished by friction between a stationary and a moving part, which converts the energy of motion into heat.

The *block brake* is nothing more than a block, or shoe, bearing against a rotating wheel. Pressure is applied with a lever.

The *band brake* makes use of a band of asbestos fabric or other heat-resistant material passed around a wheel or drum. One or both ends of the band are attached to a lever in such a manner that a movement of the lever will tighten the band and the resulting friction will slow or stop the wheel.

Internal brakes, extensively used on automobiles, consist of brake shoes, covered with asbestos fabric linings, which can be brought to bear upon the inside of a metal drum. Automobile brakes are activated hydraulically. Foot pressure on the brake pedal moves a piston in a master cylinder. This transmits hydraulic pressure through brake fluid (a light oil) to wheel cylinders located inside the brake drum of each wheel. Increased fluid pressure in each wheel cylinder moves a small piston outward, and the piston forces the asbestos-covered brake shoes against the inside of the brake drum, stopping the wheels and the car. The use of hydraulic pressure to activate the brakes assures that an equal force will be applied to each wheel when the brakes are applied.

Cone and *disk brakes* depend for their action on the friction created by contact between a stationary cone or flat disk and a rotating member against which it is forced. Disk brakes have long been used on aircraft and recently have been adapted for automobiles in a somewhat modified design. The automobile disk brake consists of a steel disk secured to and rotating with the wheel. The rectangular friction pads, one on each side of the disk, are actuated by hydraulic pressure so that they press against the disk, creating friction and stopping the car. Disk brakes are usually applied to the front wheels, but some cars have disk brakes on all four wheels.

Many automobiles are now equipped with *power brakes* in which a vacuum from the engine intake manifold supplements the foot pressure of the driver on the brake pedal, making braking easier.

Lifting equipment, such as cranes, hoists, or elevators, require special braking systems. They are frequently equipped with cone or disk brakes designed so that they will engage if the load should start to descend when it should be rising or stopped, or if it descends more rapidly than desired.

A completely different principle is applied in *eddy current brakes,* sometimes used for stopping heavy machinery quickly. A magnetic flux created by electrically excited poles causes eddy currents in the rotating rim. These currents convert the flywheel energy into heat, stopping the wheel.

—Hunter Hughes

Brazing, the process by which metal parts are joined by heating and the space between them filled with a molten, nonferrous filler metal. If the parts are closely fitted, the filler metal will flow by capillary attraction between the base metal surfaces. A chemical flux or cleaning agent is used to assure that the filler alloys, *wets* the base metal, which is heated but not melted. Brazing is distinguished from soldering by the higher temperature required (over 800° F) and the greater strength of the brazing filler. Many different brazing methods are used to join a wide variety of metals and, in some cases, to join metals to specially coated ceramics or glass. The most common method is *torch brazing,* in which a gas torch is used to heat the base metal and melt the filler. Among the most useful filler metals are alloys of copper and silver with added traces of cadmium, nickel, phosphorus, and tin. Fillers of such alloys are used to braze mild steel, stainless steel, precious metals, nickel, and copper alloys. Aluminum brazing requires an aluminum filler alloy with 4 per cent copper and up to 13 per cent silicon. Liquid, powder, or paste fluxes are used to protect the joint from oxidation and to remove oxides formed during heating.

—Joseph J. Kelleher

Cable, or *wire rope,* is made up of iron or steel wire twisted into strands, which are then twisted around a core of hemp or other suitable material. Normally the wires are twisted in one direction to form the strands and the strands are twisted in the other direction around the core. The wire strands may be twisted either to the right (*right lay*) or to the left (*left lay*). If the wires and strands are twisted in the same direction, the wire rope is known as *Lang lay rope.* It is possible to combine the two, using a Lang lay for strands around the core and a regular lay for an additional outer layer of strands. This diminishes the tendency for the cable to spin or lash when used as a hoisting rope.

Wire rope is available in many varieties, varying in size, strength, and design. Standard wire rope is made up of six strands twisted about a hemp center with 19 wires in each strand. The more wires used in each strand, the more flexible the rope. Extra special flexible hoisting rope has six 61-wire twisted strands to attain maximum flexibility.

Iron wire rope is tough and flexible, but when more strength is required, the wires are made of steel. *Steel cables* for suspension bridges or tramways are usually made with six strands of high-strength steel and have wire cores.

Wire rope is used in many types of elevators, hoists, derricks, dredges, scrapers, or other material-handling equipment. It is also used for guying poles, stacks, and tall antenna.

Without wire rope, engineers could not have designed the long-span suspension bridges that are such a vital part of transportation systems.

—Hunter Hughes

Calculating Machines, mechanical, electric, or electronic devices used to speed computation. The *abacus,* the oldest and most widely used calculating machine on earth, was devised some 2,000 years ago. In the United States, Canada, and Europe, the abacus is little more than a curiosity, but in vast areas of Asia it is the only known counting device.

The abacus (from *abax,* an ancient Greek word for slab) was a direct result of early methods of counting. When primitive man satisfied his needs for food and shelter, he began seeking ways of expressing himself. He wanted to tell his family and neighbors "how many" animals he had killed on a hunt, "how many" children he had, and so forth. Thus, symbols were developed to indicate "one," "several," and "many." The next step was a big one—devising symbols to express specific quantities. The first two symbols were, quite naturally, a "two" and a "five" —"two" because man had two hands, and "five" because he had five fingers on each hand. By combining the symbols for hand and fingers, he could express many different specific quantities.

The abacus makes use of this two-five, or biquinary, notation system. The Chinese abacus, or *suan-pan,* for example, consists of a series of rods and wires on which beads are strung. There are seven beads on each wire, separated by a divider into a set of five on the bottom and two on top. Thus, the number "seven" can be expressed by one bead from the top (equalling five) and two from the bottom.

But the abacus had its shortcomings. It couldn't carry over tens from one line to the next. As man expanded his mathematical horizons, this became a problem.

■ADDING MACHINE.—In the seventeenth century, Blaise Pascal, as a young man working in his father's tax office, in Rouen, France, invented a gear-driven machine the size of a shoe box on which sums could be added by means of a series of notched wheels. The machine could perform addition and subtraction and was capable of carrying tens automatically.

Pascal's adding machine was primitive by today's standards, but the principles on which it was based have not changed. Modern adding machines and desk calculators all owe their origin to Pascal's device of three centuries ago.

■SLIDE RULE.—Another form of calculating device is the slide rule. This, however, is an analog device, in contrast to digital machines. The analog calculator simulates an actual problem by using a model that operates according to equivalent physical quantities. The slide rule, for example, doesn't actually add numbers. Instead, it adds the lengths proportional to them. The scales printed on the rule, however, permit reading the numbers themselves.

The slide rule, which was developed by the Englishman William Oughtred in 1632, was the forerunner of many analog devices designed to measure (and calculate) specific quantities. Among these devices are the *planimeter* for measuring areas, the *speedometer* for measuring speeds, and the *differential analyzer* for solving intricate problems in calculus.

■CALCULATOR.—Digital machines, on the other hand, deal with basic arithmetic and can be used for any problem that involves these terms. This was the reasoning behind the attempt of the Englishman Charles Babbage to build a "difference engine" in 1822. In effect, Babbage drew up plans for the first digital computer. His projected machine could do complex calculations and print out results. There was to be a "memory" made up of punched cards, and the machine was to have an arithmetic unit, called a "mill," in which to store this data. Output, according to Babbage's plans, was to be set up automatically in type, thus avoiding transcription errors.

Babbage was 130 years ahead of his time. Unfortunately, the device was to be set up automatically in type, technology, and the machine just couldn't be constructed.

In 1887 Herman Hollerith, of the U.S. Bureau of the Census, crystallized Pascal's and Babbage's ideas in his punched-card system. By working out an electromechanical method for recording, compiling, and tabulating census facts, he initiated a paperwork revolution. Within a few years a wide range of industries were using his punched-card techniques for accounting and record-keeping tasks.

■COMPUTER.—During the 1940's, several electronic machines were constructed. The earliest simply linked up banks of vacuum tubes to existing punched-card machines in order to speed up the calculating process. In 1948, however, International Business Machines Corporation introduced a machine that could store its operating instructions internally, and this machine led to the beginnings of the "computer era."

Since that time, scores of computers have been introduced to perform a large number of functions. While these digital machines are infinitely faster and more flexible than the original abacus, they operate according to the same age-old principles.

—Stanley Englebardt

INTERNATIONAL BUSINESS MACHINES CORPORATION

TABULATOR devised by Herman Hollerith in 1887 for the Census Bureau was based on the Jacquard automatic loom and used metal needles to sort punched data cards. The needles touched a charged surface through the holes, creating a current that activated an instrument that totaled the figures and made a record of the final data.

Cam, a disk, plate, surface, or roller which, when rotated or moved, controls the motion of a *follower* that is in contact with it. The most familiar application of cams is to operate the valves in internal-combustion engines. In these engines, a series of cams is formed on a shaft. Turned by the crankshaft, the cams on the shaft displace *lifters,* or followers, that open the valves. Springs are used to close the valves and keep the follower in contact with the rotating cam.

Design of a cam *begins* with a description of the motions to be produced in the cam follower; this description is usually given in terms of *displacement* and *time.* For a rotating surface cam, displacement of the follower depends on the distance of the cam's surface from the center of rotation, while time depends on how fast the cam rotates. A simple diagram with displacement as the vertical axis and time, expressed in the number of degrees through which the cam has rotated, as the horizontal axis, can be used to show graphically the motion of the follower. The cam *profile* or displacement diagram shown indicates that the cam follower will *rise*—or be displaced—10 units when the cam rotates from 0° to 40°, *dwell*—or stay displaced a constant distance—during rotation from 40° to 100°, and return to starting position while the cam rotates from 100° to 150°. A larger displacement, dwell, and return occur during rotation from 200° to 360°. The slope of the curve indicating a rise or return must be selected carefully to avoid exceeding the maximum acceleration that can be tolerated by the follower and its linkage. See diagram on opposite page.

The actual cam profile is converted to a cam shape by transposing displacements into radii, at the appropriate angular positions, around the cam's center of rotation. After the radii are laid out, a smooth curve is drawn between their ends to describe the shape of the cam. Where the displacement-time diagram indicates a rise, the cam's surface will be a curve of increasing radius. A dwell period results in a curve of constant radius, while a return curve is one of decreasing radius.

Cams are widely used for vending-machine mechanisms, machine tools, power transmission, and control mechanisms.

—Joseph J. Kelleher

Camera, an instrument for taking photographs. All cameras, from the most complicated to the simplest, are basically alike. All are light-tight boxes, with a front opening for light to enter (the aperture), a lens, a shutter, and a view finder.

For a good picture, the film must be properly exposed. Too much light will cause an excessive buildup of the silver in the film, making it too dense to print properly. Resulting prints or transparencies will be too light. If not enough light reaches the film, the result will be a print or transparency that is too dark.

The *shutter* may be likened to a window shade that can be raised and lowered; the speed at which this takes place determines the amount of light it will admit. In a camera, a fast shutter speed is also used to "stop" the motion of objects being photographed, by not allowing a long enough exposure for movement to blur the image.

The *aperture,* which usually consists of an *iris diaphragm* or, in simple cameras, holes of varying size, can be compared to the iris of the eye. If the light is too brilliant, the aperture closes; in dim light, it opens wider to admit more light.

The *lens* focuses an image on the film much as the lens of the eye focuses it on the retina. The larger the lens, the more light it will admit, thus permitting higher shutter speeds and at wider apertures.

In *simple* or *box* cameras, the aperture is fixed, and the lens and shutter are preset by the manufacturer to produce good pictures under ordinary picture-taking conditions. Unless an auxiliary lens is added, such a camera is unable to make a sharp picture at distances of less than six feet. It cannot make pictures under adverse lighting conditions, nor is the shutter speed fast enough to stop motion. When flash is added, however, the picture-taking range is considerably broadened.

Roll-film cameras, which take cartridges of film 35 mm wide, make up the vast majority of cameras on the market today. *Still-film cameras,* recently introduced, take a preloaded plastic cartridge that needs only to be dropped into place in the back of the camera. The cartridge has a notch on one edge which, in some models, engages a finger in the camera and thus adjusts the camera to films of different speeds. The simpler models of this type accept films of only one speed.

These cameras, like many other types on the market today, incorporate built-in exposure meters that in many cases are linked to the diaphragm (or aperture control) and shutter to provide automatic exposure control.

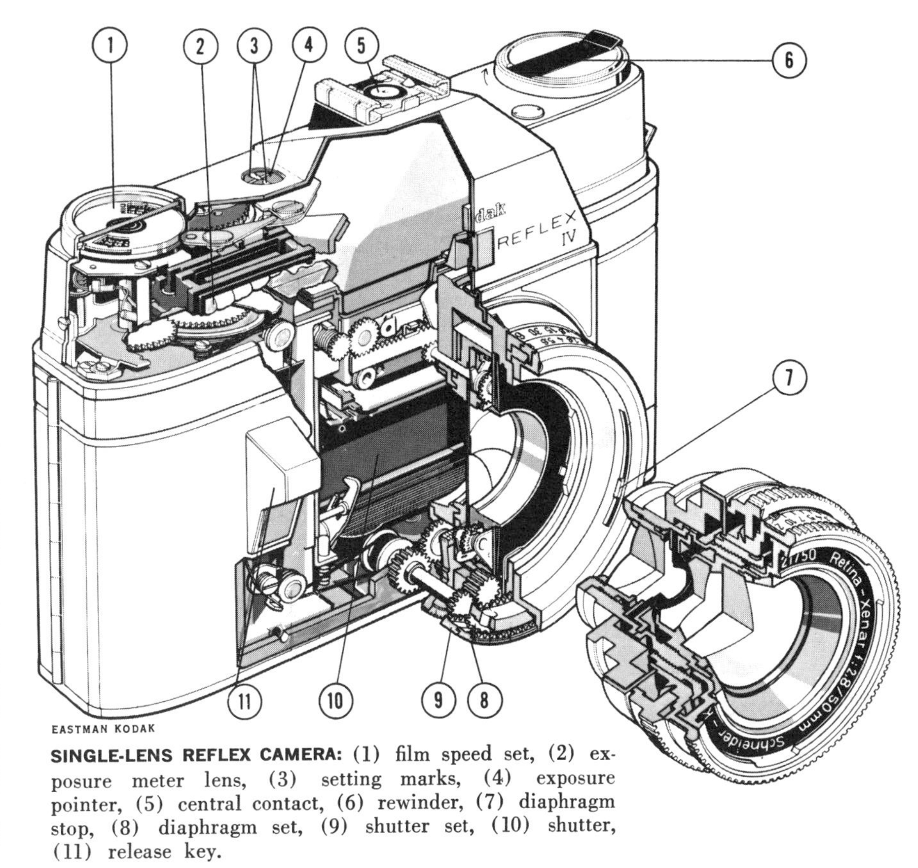

EASTMAN KODAK

SINGLE-LENS REFLEX CAMERA: (1) film speed set, (2) exposure meter lens, (3) setting marks, (4) exposure pointer, (5) central contact, (6) rewinder, (7) diaphragm stop, (8) diaphragm set, (9) shutter set, (10) shutter, (11) release key.

Studio cameras and *press-type cameras* generally use sheet film which is inserted in holders. (This must be done in a dark room, since the film is unprotected.) The holder is clamped to the back of the camera by spring clamps; the film in the holder is protected by a slide which is removed just prior to the exposure. The slide is replaced after the exposure and before the holder is detached from the camera.

Reflex cameras, which in recent years have been finding increasing favor with newspaper and magazine photographers, are of two types: twin-lens and single-lens.

The *twin-lens reflex* has a viewing lens with mirror and ground glass located directly above the taking lens. The ground glass shows the image to be photographed usually in full negative size. The image is clearly visible because the viewing lens is never stopped down to reduce the light. The image is also visible before, during, and after the exposure. Except at very close range, the image shown on the ground glass is exactly what the camera will see. Most of these cameras usually use roll film and provide pictures approximately 2¼ inches square. The larger picture, particularly when color is used, is preferred by many magazine editors.

The *single-lens reflex* usually uses 35 mm film. The ground glass in these cameras shows the image exactly as the camera will see it, even in ultra close-ups or when the camera lenses have been changed. This is accomplished through the use of a mirror, angled at 45° behind the lens, which reflects the image upward onto a ground glass. When the shutter release is pressed, the mirror springs out of the way, allowing the light image to reach the film. These cameras also use a focal-plane shutter in most cases, which prevents any light from reaching the film even though the taking lens is open.

The *focal-plane shutter* is a cloth or metal screen or shade that moves horizontally or vertically directly in front of the film. The amount of light reaching the film is controlled by the width of a slot in the curtain.

The *automatic* or *electric-eye cameras,* which have become so popular in recent years, operate on the same principle as any other camera. The single major difference is that they incorporate a sensitive photoelectric cell. The cell averages the amount of light on the subject at which the camera is aimed and converts that light into electric energy. This power actuates the lens and shutter settings in direct proportion to the amount of

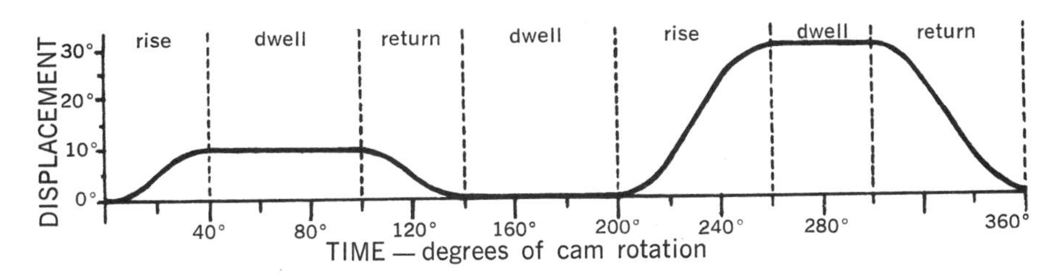

CAM PROFILE describes the motion of a cam "follower," a valve lifter, for example.

energy created by the light reflected from the subject. Such automatic cameras enable almost anyone, regardless of his knowledge of photography, to enjoy the advantages of an adjustable camera while maintaining much of the simplicity of the box camera.

The more advanced cameras usually have some provision for changing the lenses, so that the user can switch from a normal lens to a wide-angle or telephoto lens at will. Many of the newer single-lens reflex cameras incorporate a photoelectric cell within the camera that measures the amount of light striking the mirror. This offers an advantage: when lenses are changed, as when a telephoto lens is added, the light measured is only that which the camera sees through that lens and which will reach the film.

Another feature, which has been added to some cameras in recent years, is the *motorized film advance.* Whether it be an electric motor or a spring-wound motor, it advances the film automatically and cocks the shutter at the same time. With such cameras, whole series of pictures may be made in a few seconds.

Flashholders are built into many modern cameras, including some of the least expensive ones. They are synchronized to the shutter so that the bulb does not fire until the shutter is fully open. The latest type includes provision for a four-sided, sealed-in plastic "cube" containing four miniature bulbs. The cube turns as each exposure is made, placing a new "side" in position. When combined with a motorized film advance, the photographer is enabled to make four fully illuminated flash shots in rapid sequence.

Motion-picture cameras function on the same principles as still cameras which are described above. (See *Cinematography.*) The film, which is usually 8 mm or 16 mm, is drawn past the lens by a spring- or battery-powered motor. Recently, amateur cameras have been made with *zoom lenses,* which allow the focal length to be altered from telephoto to wide-angle—or vice versa—while pictures are being taken.

—C. Grantly Wallington

Camera Accessories, devices that facilitate the taking of photographs with a camera. Many items that were listed as accessories with older-type cameras are now built directly into the instrument, although they will still be required separately with certain types of camera. Included among such accessories are flashholders, range finders, and photoelectric exposure meters. Most professional photographers will continue to use them as separate accessories because of the greater flexibility and accuracy they offer.

The *flashholder,* as its name implies, is made up of a case for the batteries, a reflector, and a receptacle for the flashbulb. It is connected to the camera's shutter through a cord and a plug on the face of the camera or the side of the shutter. Its advantages over the built-in flashholder are that a number of such units can be connected together for more intense light or to cover a wide area; the reflector can be removed or aimed at the ceiling to provide diffuse illumination, and it may be held away from the camera for a better lighting angle. Some units are of the focusing type for photographing distant subjects. *Electronic flash* provides intense light with maximum convenience. The flashholder itself may be attached directly to the camera or held in the photographer's hand. Instead of a flashbulb, it contains a high-intensity tube that is flashed by a battery or power source frequently carried on a strap over the photographer's shoulder. It eliminates the need for carrying flashbulbs and provides an extremely intense light for a very short period of time, the flash of which is fast enough to stop rapid motion. Many studios are equipped with permanent high-powered units.

Separate *range finders* provide an accurate measure of distance by bringing two views of the subject from different angles into coincidence. This is accomplished with lenses and prisms, one of which is movable.

The separate *photoelectric exposure meter* offers several advantages. It is usually larger, thus will pick up more light rays reflected from the subject and give a more accurate measurement. It can be aimed to exclude some parts of the picture, such as large sky areas which, when the light is averaged, might indicate more light than is actually on the subject of the photograph. By aiming it at a similar nearby object with the same selectivity as the subject, a reading can be obtained for the distant object. One type of meter is also designed to read incident light, and is most frequently used by professional photographers. Held in front of the subject, it measures the amount of light reaching it.

Other useful accessories include *sun shades* to protect the camera lens from the direct rays of the sun that otherwise might spoil the picture; *tripods* to keep the camera steady on longer exposures and to hold it in position while the subject is being arranged; and *field cases* to protect the camera while traveling or for carrying it while keeping it convenient for use.

For black-and-white photography, a set of *filters* can add dramatic value to many pictures. A yellow or amber filter, generally designated as a *K2* filter, will darken blue skies and bring out clouds. A red filter will provide skies that are almost black with very dramatic white clouds. Filters are not used to any great extent in amateur color photography, although a *skylight,* or ultraviolet filter, can be useful in reducing the blue of distant haze in the mountains or at the seashore. It is also useful to protect the lens from sand or salt spray at the seashore, since it does not require any increase in exposure.

Probably one of the most useful accessories to tne amateur photographer is the *close-up lens* that permits him to make close-ups of flowers, pets, and people. These lenses are available in different powers covering medium close-ups and extreme close-ups. They are easy to use and add to the camera's versatility.

Other accessories for specialized photographic work include *microscope adapters, copying stands, polarizing filters, ultraviolet* and *infrared filters,* and *light-balancing filters* for extreme accuracy in color rendition.

—C. Grantly Wallington

E. LEITZ, INC.

LENSES of different focal lengths attached to the same camera were used to take these photographs from the same spot. Lenses with greater focal lengths close in on distant objects and enlarge them. The extreme closeup of the sliding board at far *right,* taken with a 400 mm telephoto lens, is an 8X enlargement of the sliding board in the background of the far *left* picture, which was taken with a 21 mm lens.

GENERAL FOODS CORPORATION

FREEZING is an important method of food preservation today. *Above*, green beans are blanched to destroy decay-causing enzymes, cooled, then packaged (*right*) for freezing.

Canning and Preserving, processes for protecting food from deterioration and decay and making it available for future consumption. Such food should retain desirable flavor, color, and texture, as well as its original nutritional value. *Canning* protects food by sterilizing and excluding air: *Preserving* protects food by adding substances such as salt or sugar, by removing moisture, or by fermentation.

Man probably first preserved food by *drying*, even before he learned to farm. Seeds and nuts, dried by the rays of the sun, would keep through the winter. Meat and fish could be preserved in the same way.

Cooking, after man discovered fire, made food more appetizing and was also a means of preservation, since heating killed some of the microorganisms and enzymes that caused spoilage. Preserving meat by *smoking* was an outgrowth of cooking.

Salt was used for flavoring before man learned that meat soaked in salt brine or rubbed with salt would keep for weeks or months. *Brining*, later called *pickling*, became a favorite way of keeping fruits and vegetables for winter use. The preserving properties of sugar were also known in ancient times, and the making of jam and marmalade was widely practiced.

Fermentation, the natural process of chemical change in food, was observed and used thousands of years ago. Fruit juices, when fermented, resulted in wine, a safe beverage in areas of uncertain water supply. Vinegar, a product of fermentation, was useful for pickling meats, fish, fruits, and vegetables. Fermented cabbage, or sauerkraut, was widely used. The fermentation of milk, which does not keep well, resulted in cheese, which does keep well. Bread—that is, the fermented sourdough bread developed by the Egyptians 5,000 years ago—came to be called the staff of life. It was nutritious, easily carried from place to place, kept for months without spoiling, and, with water, could sustain life for long periods.

■HISTORY OF CANNING.—For centuries, these outgrowths of natural processes were the only means that man had for preserving food. Then, in 1810, a French chemist and confectioner, Nicolas Appert, developed canning. Although the theoretical basis for canning was not known until Louis Pasteur observed the relationships between microorganisms and food spoilage some 50 years later, Appert's ideas are still valid. He placed wholesome food in clean, metal containers, which were then sealed and boiled for a sufficiently long time to prevent spoilage.

Canning spread rapidly. In 1810 an Englishman, Peter Durand, patented a can of iron coated with tin. Today's cans are primarily steel, with a thin coating of tin.

Commercial canning began in the United States with the production of pickles, ketchup, and sauces in Boston in 1819 and the canning of seafood in Baltimore in 1820. The first canners found that it was necessary to keep the cans in boiling water for five or six hours. In 1860, a canner found that adding calcium chloride to water raised its boiling point from 212° to 240° Fahrenheit, sharply reducing required cooking times. The invention of the *retort*, or *pressure cooker*, in 1874 was an even more important step. It gave the canner accurate control of temperature during the processing operation.

■MECHANIZATION.—Reduced cooking times meant that many more cans could be processed in a given period. This in turn led to the development of machines that would do many of the tasks formerly done by hand, such as shelling peas, cutting corn from the cob, and cleaning salmon. At the same time, can making became entirely mechanized. The modern open-top can, closed by crimping with a rubber sealing compound, speeded mechanical filling and sealing.

Glass jars have been most widely used for home canning, although tin cans have also been adapted for home use. Home canning of all types of food was greatly encouraged by home economists, state extension-service workers, and others from about 1900 to World War II as a means of utilizing home garden products, providing better diets, and reducing the cost of living on farms. Many farm women still can and preserve at home for winter use. However, the increasingly commercial character of farms, the low cost of commercially canned foods, and the widespread use of freezers have made home canning less important than it was before World War II.

Since World War II, commercial canning has been challenged by other methods of processing, particularly freezing and new types of drying. However, the production of canned foods has continued to grow, although with shifts among particular foods and in the geographical areas in which the canning is done.

The quality of raw materials is of primary concern to the modern canner. Canners work closely with farmers to secure a uniform product of a particular quality. This may take the form of contract farming, in which price, grade, and tonnage are agreed upon before the crop is planted. The canner may specify that particular seeds are to be used and particular cultivating practices are to be followed.

■PROCESSING.—Speed in processing preserves some of the flavor, color, and nutrition that would otherwise be lost and, at the same time, tends to cut unit cost.

After the product to be canned reaches the plant, it is inspected and graded. Instruments measure such qualities as firmness, maturity, and color by mechanical, chemical, or electronic means, although much grading and sorting is still controlled by skilled human labor. Washing, peeling, trimming, grinding, and cutting machines have replaced virtually all hand labor in these processes.

The raw material may undergo preliminary cooking or blanching before the cans are automatically filled and sealed. Some large canneries then place the cans on moving belts that move through the cooking and cooling processes in one operation. Most canning is done with the retort or pressure cooker. The time required for processing depends mainly on the chemical and physical makeup of the food, as well as on the size of the container. After cooling, the cans are labeled and boxed for shipment.

Special techniques have been developed for some products. For example, certain soups are pumped through high-temperature heat exchangers into separately sterilized cans, which are then sealed under aseptic conditions. The goal of such procedures is to reduce processing times.

Some canners have adopted standards developed by the U.S. Department of Agriculture and pay for grading by the department. Others maintain their own grading systems. However, most canned goods do not carry any public indication of grade.

A new method of food preservation, called *irradiation, cold sterilization,* or *radiation sterilization,* is being tested by industrial and government laboratories. The food is exposed to nuclear radiation from a radioactive material or is bombarded with high-energy electrons. Although these techniques destroy microorganisms, they do not destroy enzymes, so objectionable flavors tend to develop in the food. However, bacon prepared by irradiation was purchased by the federal government in 1966 for military use.

—Wayne D. Rasmussen

Carburetor, a device that mixes air with gasoline or other internal-combustion-engine fuels in the correct ratio for complete combustion at all engine speeds. A secondary function of most carburetors is to control the engine's power output by throttling, or metering, the air-fuel mixture admitted to the cylinders.

The simplest type of carburetor consists of a tube with a *venturi,* or tapered restriction, in which a *jet,* or fuel-spray nozzle, is mounted. Fuel is supplied to the jet from a constant-level fuel chamber at atmospheric pressure. When the engine piston moves down, a partial vacuum is created in the cylinder, sucking air in through the carburetor. The speed of the air is increased as it passes through the venturi. Reduced pressure in the carburetor then allows atmospheric pressure on the fuel to force it through the nozzle into the airstream. The air-fuel mixture continues to flow into the cylinder until the piston starts to move upward in the compression stroke.

Automobile carburetors are far more complex, since their engines operate through a wide range of speeds and loads. To function properly, these carburetors incorporate a number of additional features.

For starting, a *choke,* or butterfly valve, is used to restrict airflow to the carburetor, thus increasing the suction on the fuel flow. A *throttle valve* between the carburetor and cylinder controls the volume of air-fuel mixture reaching the cylinders. Linked to the accelerator pedal, the throttle valve adjusts the power output and, thus, the speed of the vehicle in response to the driver's foot pressure. An acceleration pump in the carburetor provides momentary fuel enrichment when the accelerator pedal is depressed rapidly. The pump, linked to the throttle, increases the responsiveness of the engine during acceleration by maintaining the correct fuel-air ratio when the throttle is opened suddenly.

Carburetors for large engines may incorporate several tubes or throats in a single housing. Each tube may be equipped with identical jets, plus choke and throttle valve operating in unison, or the accelerator linkage can be constructed to operate each throttle valve in sequence as additional power is required.

—Joseph J. Kelleher

Cathode-Ray Tube, a device in which a beam of electrons is used to provide a pictorial representation of the current or voltage in a circuit. It is familiarly known as a television *picture tube,* or *kinescope.* It makes use of electric or magnetic fields to deflect a beam of electrons. These originally were called *cathode rays* because they are emitted by a negatively charged electrode, or *cathode.* After leaving the cathode, the narrow electron beam strikes a fluorescent screen on the face of the tube, where it forms a bright spot of light, called the *scanning spot.*

In a television set, the scanning spot moves from side to side and from top to bottom, quickly covering the entire area of the picture tube. This is similar to the way a typewriter covers a page with print. The picture is formed by changing the number of electrons in the beam as it moves from point to point on the screen. This causes the brightness of the scanning spot to change correspondingly, thereby producing a picture.

The cathode-ray tube was invented in 1897 by Karl Ferdinand Braun for the study of alternating voltages. In the modern *oscilloscope,* the electron beam scans from the viewer's left to right, producing a line of light on the screen. The voltage under study causes the beam to deflect vertically, in the course of time, in precise agreement with the magnitude of the voltage. In this way, the track of light represents the variation with time of the voltage under study. The cathode-ray tube is also an indispensable part of all radar sets, where it is used as a visual indication of the presence and location of targets.

The cathode-ray tube had its beginning with the Crookes tube, a gas discharge tube devised by Sir William Crookes about 1897. It consists of a completely enclosed vessel having two metal electrodes sealed through the walls. When a voltage is applied to the electrodes, and the tube contains a partial vacuum, a glow discharge takes place in the tube. This discharge is produced by collisions between cathode rays and gas particles in the tube.

■**X-RAY.**—In 1895, Wilhelm Roentgen discovered that X-rays are produced in the Crookes tube when the electron beam strikes a positively charged electrode, called the *anode.* The invisible X-rays, unlike cathode rays, are able to leave the tube and darken a photographic plate some distance away. They are produced whenever high-speed electrons strike the atoms of an anode. The energy of impact causes the atoms to vibrate violently, and to emit a high-frequency, short-wavelength light. This invisible radiation is electromagnetic in nature and, therefore, similar to ordinary light, infrared light, and ultraviolet light. X-rays differ only in having much shorter wavelengths,

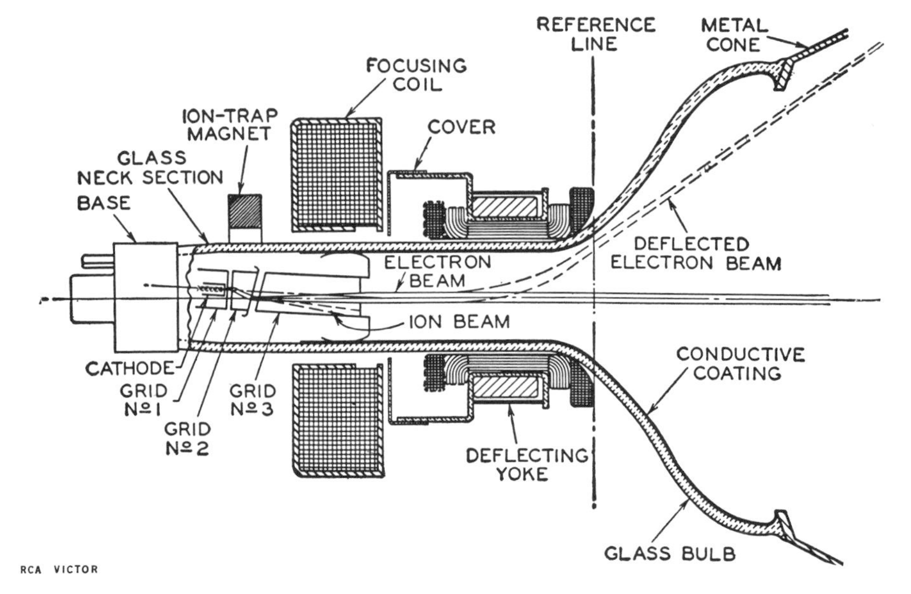

CATHODE-RAY TUBE, better known as a television picture tube, uses an electron beam in order to produce a picture. In the process, the electrons are deflected by magnetic fields.

JERRY SOALT, ILGWU "JUSTICE"

GARMENT pattern is placed by marker (*above*) to utilize maximum fabric, and the cloth is cut along the pattern markings (*lower right*). After a seamstress (*upper right*) has sewn these pieces together, detail work and careful pressing will finish the garment.

which permit them to penetrate, to a greater or lesser degree, various substances that are opaque to ordinary light.

The modern *X-ray tube*, or *Coolidge tube*, invented by William D. Coolidge in 1913, has a hot tungsten filament, which gives off many electrons, and an anode, or target, of tungsten or another metal. Many thousands of volts are used to accelerate the electrons from filament to target. The filament temperature determines the number of electrons in the beam, and the voltage between target and filament determines their speed. The greater the speed, the more penetrating the X-rays. X-rays are used to produce shadowgraphs of the human body and to discover flaws within metal castings and other structures that cannot be studied by ordinary light.

■**TV CAMERA TUBES.**—Cathode rays are also an indispensable part of the *iconoscope*, an early television camera tube. This tube consists of a plate—about 4 by 5 inches in size and holding millions of tiny photoelectric cells—and a narrow beam of electrons that scans the cells. An image of the scene is focused by a lens on the light-sensitive plate, where each tiny cell generates a voltage proportional to the amount of light falling on it. The mosaic of cells creates an overall pattern of voltages that corresponds precisely to the pattern of light in the image. As the electron beam scans the mosaic of cells, it neutralizes the voltages and generates a picture signal in the form of a varying current of electricity.

The *image orthicon* is a generally superior television camera tube. It is more sensitive than the iconoscope and produces less distortion in the brightness of the image. It differs from the iconoscope in that the image is amplified electronically within the tube. A satisfactory picture signal can be produced, therefore, with less illumination on the scene.

—William C. Vergara

Clothing Industry, the manufacture of ready-to-wear apparel. The clothing industry traces its modern beginnings to the sewing machine, invented by Elias Howe in 1846 and improved by Isaac Singer in 1851, and to cutting equipment introduced in the 1870's and 1880's. Before that time, clothing worn by working-class people was made chiefly of coarse homespun by the women of the family. In the United States, only the rich could afford the services of custom tailors or dressmakers or imported garments from England and France.

Actually, manufacture of men's apparel preceded that of women's by almost a half-century. Production of men's clothing in quantity started in the early years of the nineteenth century, when firms made up hand-sewn but inferior *slop clothes* for sailors visiting waterfront shops in East coast ports.

The census of 1860, the first to take notice of the industry, reported 188 factories doing an annual volume of about $7 million and employing fewer than 6,000 "male and female hands." Ninety-six manufacturers turned out cloaks and mantillas. Among other major manufactures were hoop skirts and corsets.

In the 1880's and 1890's, immigration of craftsmen from central and eastern Europe hastened the pace of the industry's growth.

It was not until 1890, however, that female apparel increased its share of total garment production to 50 per cent. Wholesale volume in the women's wear section of the industry had jumped to $174 million by 1900. Factories had sprung up in Chicago, Cleveland, and Baltimore, as well as in the East coast ports of New York, Philadelphia, and Boston. By 1914, manufacture of women's wear had far exceeded that of men's wear.

It was not until 1908 that the separate dress was created by joining a skirt to a bodice. Sportswear and other casual clothes, born in the 1920's, came into their own in the 1950's, partly as a result of suburban living.

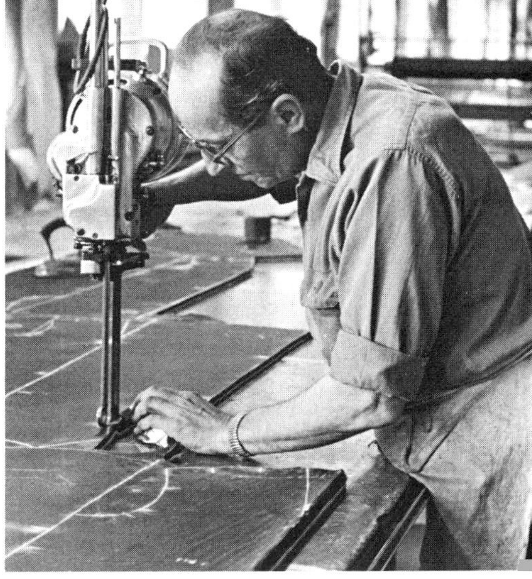

American manufacturers became aware of the Parisian couture just before World War I. Since then, except for the World War II period when the United States industry was thrown on its own resources, France has held a premier position in originating fashions. However, a number of American designers are also in the forefront among fashion leaders.

Newly important as a contributor to the fashion world, and always significant as a producer of attractive clothing for all social levels, the garment industry today is big business as well. In 1967 some 22,000 clothing manufacturers employed 1.4 million workers with a total payroll of $5.3 billion and an estimated value of shipments of $20 billion.

During the past few years, the women's and men's apparel trades, traditionally among the last refuges of small business, have tended to join the ranks of big business by means of diversification of products, mergers, acquisitions, and public stock issues. More and more firms are cutting across women's ready-to-wear lines into accessories and even into men's wear, and vice versa.

—Samuel Feinberg

Clutches, mechanical devices used to engage and disengage rotating machine elements. They can be classified as either positive clutches or friction clutches.

Positive clutches cannot slip and have no friction elements. They cannot be engaged until both shafts are stopped or are rotating at the same speed. The simplest form of positive clutch consists of a pair of mating *dogs,* or *jaws,* that mesh. A shift lever or pedal is used to slide one of the jaws along its shaft, engaging or disengaging the opposite member. *Splines,* or *keys,* are used to transmit rotary motion between the sliding jaw and its shaft.

Friction clutches are widely used in automobile transmissions or wherever the speeds of the rotating elements differ. Motion is transmitted through a friction surface that permits slippage until the speeds of the rotating elements are equal. In the automobile, the clutch is mounted on the flywheel and transmits power through a flat friction disk splined to the driven shaft. The disk is located between a spring-loaded pressure plate and the flywheel of the engine. In the engaged position, axial force is applied by the springs to clamp the disk between the pressure plate and the flywheel. The disk is faced with specially compounded friction material riveted or bonded in place. These clutches are operated dry in most applications. Wet clutches, filled with a fluid, are used for heavy equipment, such as earth movers or caterpillar tractors; the fluid helps dissipate the heat produced by slippage.

Hydraulic clutches, or *fluid couplings,* use fluid to transmit motion between rotating shafts, end to end. These clutches have finned rotors enclosed in a fluid-tight housing filled with liquid. The vanes on the driving rotor impart a rotational motion to the fluid that is transmitted to and drives the vanes of the driven rotor. Smoothness and torque conversion are the advantages of the hydraulic clutch, which can tolerate wide differences in speed between the input and output shafts. These clutches are widely used in automatic transmissions for automobiles, trucks, buses, and military vehicles.

One-way, or *overrunning, clutches* permit transmission of rotary motion in one direction only. This freewheeling action allows the driven shaft to rotate faster than the driving shaft when a second source of power is applied or when the input slows down or stops. Spring-loaded *sprags,* or *balls,* are used to produce a wedging action between the input and output. When the overrunning clutch's input shaft rotates, the balls climb a ramp or the sprags tilt, wedging themselves between the driving hub and a driven sleeve. There is no slippage as long as input-shaft speed exceeds output-shaft speed. Once the speed of the output shaft exceeds that of the inputs, the balls are driven down the ramp to disengage or the sprags are tilted to free the sleeve. These one-way clutches operate automatically and are frequently used to convert reciprocating motion to intermittent unidirectional rotation.

—Joseph J. Kelleher

Cofferdam, any temporary structure built to exclude earth and water from an excavation so that work may be done in the dry. The term, coined by civil engineers, stems from the structure's resemblance to a *coffer,* a little-used name for the rectangular chamber of a navigation lock, which in turn is shaped like a coffin. Cofferdams are commonly associated with marine construction such as bridge piers, canal locks, and dams.

A coffer formed by dams, hence cofferdam, is not to be confused with a *caisson,* although they often look alike and basically accomplish the same thing. A caisson is a chamber, sometimes very large, that is usually sunk by excavating within it, in order to gain access to the bed of a stream or other body of water. If the chamber is closed on top and the water is excluded by air pressure, it is called a *pneumatic caisson.*

Cofferdams are built in many different ways. The simplest version is an earthen dike built from the shoreline of a body of water to enclose a specific area. Pumps remove the water within the enclosure and work proceeds as though on dry land. Since earth dikes are often too temporary, especially in moving water such as a river, engineers have developed cofferdams with protective timber or steel sheeting on at least the riverward side of the structure. Early cofferdams often were formed of a series of *timber cribs* filled with rock and earth. Another common type of cofferdam consisted of two parallel rows of timber or steel sheeting filled with earth and linked by steel rods.

The most common cofferdam used today is the cellular steel sheetpile structure. Pile drivers form large circular cells by driving steel sheetpiles (long narrow pieces of steel with interlocks on their long edges so they can be locked together) into the river bottom. When filled with a granular material such as sand or gravel, each cell stands as an independent structure. When dozens of these are linked together and the construction site is enclosed with what looks like a row of giant tin cans, water can be pumped out. After the construction is completed the cells are removed.

—William W. Jacobus, Jr.

Compressors, devices that compress air or gas by decreasing its volume. This can be accomplished by machines referred to as either positive-displacement or dynamic compressors. *Positive-displacement compressors* can be classified as either *piston* or *rotary machines. Dynamic compressors* fall into the general classification of *centrifugal* or *axial flow machines,* and *jet blowers* or *air ejectors.*

■**PISTON COMPRESSORS.**—Piston compressors admit a quantity of the gas to a closed space where the pressure is increased by reducing the volume as the piston reciprocates. They work from slight vacuum (negative pressures) to several thousand pounds per square inch (psi) pressure. The capacity in cubic feet per minute pumped depends upon the bore and stroke of the piston and the speed at which it is reciprocated. For pressures from 1 to 100 psi, single-cylinder compressors are adequate, while multiple-stage compressors—two, three, or four cylinder machines—can produce pressures up to 5,000 psi or higher. When more than one stage of compression is required, *intercoolers* are used between stages. Intercoolers, or *heat exchangers,* cool the gas as it travels from the discharge part of one cylinder to the intake of the next.

The simplest forms of piston compressor are the hand-operated insecticide sprayer and tire pump. As the operator pumps the handle back and forth, air is admitted to the cylinder and then is compressed as the moving piston reduces the volume of the cylinder. Piston compressors are commonly employed to supply air for paint sprayers or pneumatic tools and in household refrigerators.

■**ROTARY COMPRESSORS.**—Rotary compressors consist of a casing enclosing either a pair of impellers with intermeshed lobes or a single rotor with a series of sliding vanes arranged radially around the center of rotation. The rotors of *dual-impeller units* are symmetrical and interconnected by gears to rotate in opposite directions. As the impellers rotate, air trapped between the lobes and the housing is forced from intake to the exhaust port. *Sliding-vane compressors* consist of a slotted rotor inside a cylindrical housing. Vanes in the slots of the eccentrically mounted rotor are spring-loaded to contact the inner surface of the housing. As the rotor is turned, air trapped between the vanes is compressed as the space between the eccentrically mounted rotor and the housing walls decreases. Rotary compressors are used to supercharge internal combustion engines, for general blower service, or wherever high volume of gas must be moved at low pressure.

■**DYNAMIC COMPRESSORS.**—Dynamic compressors operate by accelerating and diffusing the gas passing through the machine. These compressors consist of a bladed rotor enclosed in a stationary housing. The blades on the rotor accelerate the gas inside the housing in the same way a fan blade accelerates air. The housing guides the gas or air to the rotor, changes the kinetic energy of the gas leaving the rotor blades into pressure, and directs the gas to the outlet.

In *centrifugal compressors,* the gas enters the machine at the center or hub of the rotor and moves radially toward the discharge ports. Gas flowing through *axial compressors* follows a relatively straight path parallel to the rotor shaft. Depending upon the pressure and volume required, either a single-stage or multistage axial compressor is used. The best-known application for these compressors is the aircraft gas turbine or jet engine. In these engines, ten or more sets of blades may be used to achieve the desired compression.

—Joseph J. Kelleher

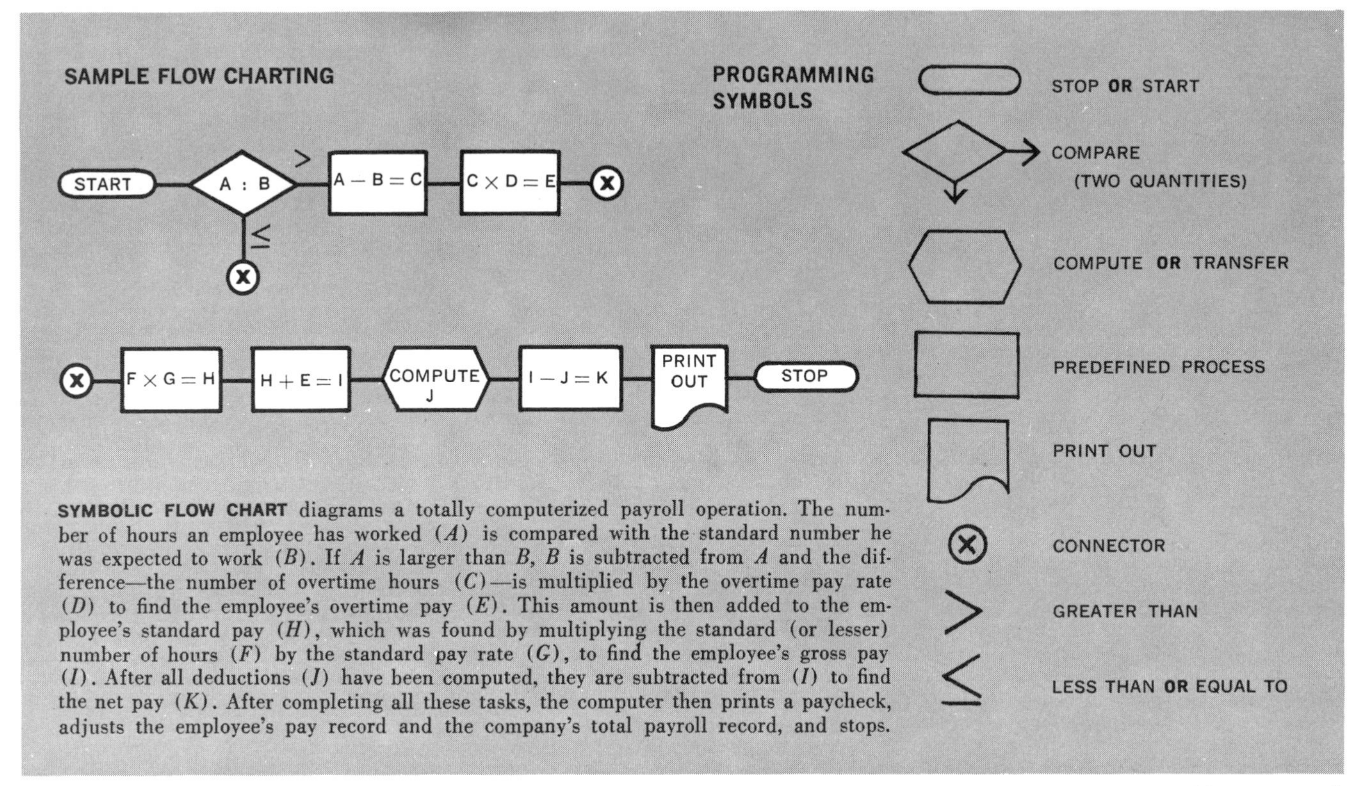

SYMBOLIC FLOW CHART diagrams a totally computerized payroll operation. The number of hours an employee has worked (*A*) is compared with the standard number he was expected to work (*B*). If *A* is larger than *B*, *B* is subtracted from *A* and the difference—the number of overtime hours (*C*)—is multiplied by the overtime pay rate (*D*) to find the employee's overtime pay (*E*). This amount is then added to the employee's standard pay (*H*), which was found by multiplying the standard (or lesser) number of hours (*F*) by the standard pay rate (*G*), to find the employee's gross pay (*I*). After all deductions (*J*) have been computed, they are subtracted from (*I*) to find the net pay (*K*). After completing all these tasks, the computer then prints a paycheck, adjusts the employee's pay record and the company's total payroll record, and stops.

Computer, an electronic or mechanical device that accepts information, performs a mathematical or logical operation on it, and then reports the results in either a visual or machine-readable manner.

■ANALOG COMPUTERS.—The most common form of computer—often not recognized as such—is the analog device, which accepts continuous measurements to produce continuous solutions to mathematical equations. A classic example is the automobile speedometer, which converts the turning of a cylindrical shaft (the axle) into a numerical approximation of speed in terms of miles per hour. Another is the slide rule, which is based on the fact that two numbers can be multiplied by adding their logarithms. On the slide rule, however, we don't actually add numbers—instead we add the lengths proportional to them.

■DIGITAL COMPUTERS.—While the analog computer performs its calculations by measuring, the digital computer actually counts. The principle behind the digital computer evolved through the ages: first, in the form of the abacus, developed some 2000 years ago and still widely used; later, as a crude adding machine invented (1642) by the Frenchman Blaise Pascal; and, still later (c. 1670), as a primitive desk calculator built by the German mathematician and philosopher, Gottfried Wilhelm von Leibniz.

These earlier counting devices were dependent on manual effort—fingers moved beads or punched keys that turned notched wheels. In 1880, Herman Hollerith invented the punched-card technique, which substituted electric power for manual effort. While these electromechanical devices weren't true computers, they were capable of performing a wide range of calculating tasks.

In 1947, J. Presper Eckert, Jr., and John W. Mauchly of the Moore School of Engineering, University of Pennsylvania, built a counting device called the Electronic Numerical Integrator and Computer (ENIAC). Instead of using electricity to move mechanical components, ENIAC employed thousands of switches, which were turned on or off by vacuum-tube-generated electronic impulses. This proved to be the first electronic digital computer.

The term "digital" is derived from counting on our fingers, or "digits." Because we have ten fingers, most computation is based on the familiar decimal system. Computers, on the other hand, use switches or other components which can be only in a state of "on" or "off," "yes" or "no," "0" or "1." For this reason, most digital computers are based on the binary counting system, which utilizes only two digits, 0 and 1.

Although ENIAC could perform 5,000 additions per second (by using electronic impulses that move at the speed of light) its *program*—the series of instructions that tell the machine how to perform a particular job—was preset by means of wires. Thus, it did not have a high degree of flexibility.

■PROGRAMMING.—In 1948, International Business Machines Corporation introduced the Selective Sequence Electronic Calculator. The most important feature of this machine was its ability to store instructions magnetically and thus modify them as dictated by the developing stages of work.

Today there is a wide range of stored-program computers performing virtually every type of scientific and business task. Instead of vacuum tubes they use transistors or, in the latest models, microelectronic components. Although the specific method of data handling varies from model to model, all operate in basically the same manner.

A set of instructions tells the machine exactly how to perform a particular job, for example, process a payroll, update a file, or perform an engineering calculation. Then raw data are recorded in one of several media: punched cards, magnetic tape, direct access disk files, or simply in typed or magnetic ink characters. This information is fed into the system through an *input* device. There it comes under the control of the stored program and is processed accordingly. This may involve addition, subtraction, multiplication, division, or some similar mathematical or logical operation. The system's *output* then reports the results in the form of punched cards, magnetic tape, or printed media, or by display on a cathode-ray tube or some similar visual device.

Most digital computers today are used for routine office jobs as payroll, accounts receivable, or the like. Many, however, are applied to extremely sophisticated tasks, such as simulating real-life systems (human blood factors, war games, business operations) or continually monitoring complex physical systems (petroleum refineries, manufacturing processes). In these latter cases, the input to the digital computer often comes directly from an analog device by means of analog-to-digital converters. —Stanley Englebardt

Condenser, Vapor, a device used in chemical processes and steam power plants to condense vapors to liquids by heat transfer. The most common use for condensers is to convert steam into water. To effect the change, the heat of the vapor is transferred to a cooler fluid. The cooler fluid is most commonly water, but other liquids or gases may be used.

In process systems, condensers are used to extract one specific liquid from the vapor of a mixture of liquids. This is accomplished by holding the temperature of the cooling fluid just below the boiling point of the liquid to be extracted. In power plants, steam is condensed to reduce the back pressure at the discharge end of the power turbine and to recover the condensate for boiler feed water.

According to their construction, condensers are classified as surface or contact condensers. *Surface condensers* separate the vapor and cooling fluid by a surface through which the heat is transferred. Pipes, plates, and partitions are used to prevent mixing of the condensed vapor and the cooling fluid, which are then extracted separately. *Contact condensers* mix the fluid and vapor; the condensed vapor and cooling fluid are extracted through a common outlet.

—Joseph J. Kelleher

Construction Equipment, machinery used in the building of structures and roads. Until the advent of steam power, construction equipment was limited almost entirely to man-operated hoists, pumps, and similar devices. Labor was the major resource, works were long in building, and limitations—such as founding bridge piers at sufficient depth to prevent scour during floods—led to many failures.

The ancient Egyptians had developed no important equipment but, with plentiful labor, built huge temporary earth inclines in erecting their great stone constructions, such as the pyramids. The man-powered hoists and pumps of the Greeks and Romans remained little changed through the eighteenth century.

During the nineteenth century, James Watt's steam engine was applied not only in meeting the needs of industry and transportation but also in replacing earlier construction equipment with new steam-powered devices. An ancient, almost completely manual art was transformed into a highly specialized engineering activity.

A number of nineteenth century advances were British in origin. The early use of the diving bell led to the development of the compressed-air process, which was patented in 1830 and later applied in shaft-sinking and subaqueous tunneling. Underwater foundations, earlier limited to 8 or 10 feet below water level, reached the great depth of 136 feet in James B. Eads' St. Louis arch bridge of 1869. The steam-powered pile driver was likewise a British device following the invention of the steam forging hammer in 1838. The steam shovel, "the great American railroad builder," originated in 1838. The steam percussion rock drill appeared in the United States in 1849. It was followed by the compressed-air drill and, combined with Alfred Nobel's new explosive, dynamite, was used in the building of famous tunnels of the period, such as the Hoosac Tunnel (Massachusetts) of 1866. Nevertheless, other major advances were to come primarily in the United States in the earlier years of the twentieth century. The construction of the Chicago Drainage Canal (1892–1900) and the first New York subway of 1904 were largely pick-and-shovel undertakings.

Various factors contributed to this movement. One was that the increasing cost of labor emphasized labor-saving equipment. Another, at the turn of the century in 1900, was the introduction of the gasoline engine, which was to replace the horse-and-cart and to find increasing application in many construction devices, including pumps, scrapers, rollers, and loading and other equipment. Belt-conveyors and other mechanical means have similarly been developed to expedite construction, while the advent of Diesel engines has further reduced costs and extended the use of mobile power equipment. Modern dipper and hydraulic dredges of great capacity are available; the largest steam shovels—picking up 15 tons in a single scoop—are used in open-pit mining operations, notably coal stripping. In seeking the most favorable balance between labor and capital expenditures or rental costs for machines, the construction expert thus has a wide choice of modern equipment.

—James Kip Finch

Couplings, mechanical devices used to join rotating or reciprocating shafts. Common applications for couplings include the flexible link in the driveshaft of automobiles, the mechanical joint between an electric motor and the equipment it drives, and the universal joint used to transmit power between shafts intersecting at an angle. Rigid, or plain, couplings can be used only to connect shafts whose axes are perfectly aligned. Flexible couplings are used to connect rotating shafts that are slightly misaligned. Universal, or Hooke, joints are used for shafts with large angular misalignment or changing misalignments.

Rigid couplings consist of a sleeve or collar which slips over the joint between two shafts and is clamped or keyed to each. Flanged, rigid couplings have a hub, which is keyed or pressed on the shaft, and a larger diameter flange with a circle of bolt holes. When being mounted on each shaft, the flanged couplings are aligned and then bolts are installed in the holes.

Flexible couplings transmit power between shafts whose axes are not precisely aligned. The misalignment may be offset, angular, or a combination of both. An additional benefit of flexible joints is their ability to absorb or *damp* vibrations produced by speed variation or unbalanced rotating machinery. As the name implies, flexible couplings frequently consist of rigid hubs or spiders

R. G. LE TOURNEAU, INC.

CONSTRUCTION of highways, canals, dams, and other massive projects proceeds quickly with the help of huge earth-movers such as this, which can dig and load many tons of dirt, transport it to a desired location, and spread it into an evenly graded surface.

AQUA-CHEM., INC.

DESALINIZATION PLANTS make salt water potable. The "flash evaporator" (*bottom*) converts 800,000 gallons a day. A desalter provides water for ship *Michelangelo* at sea.

clamped to each shaft and linked together by flexible material. The flexible material may be thin steel disks, steel springs, rubber impregnated canvas, or solid rubber.

The simplest form of flexible coupling consists of a pair of steel hubs joined by a solid-rubber sleeve. The rubber element is molded or clamped onto the hubs to form a solid, one-piece coupling.

One of the oldest couplings known is *Oldham's double slider coupling*, which incorporates the basic features from which most flexible couplings have been developed. These couplings consist of two hubs and a central plate. Ribs on opposing faces of the central disk mate with slots in the hubs. Lubrication is essential in Oldham's couplings because the ribs slide in the slots as the coupling rotates. For low-power, low-speed applications, a plastic such as nylon may be used for the central plate. This assures silent operation and eliminates the lubrication problem.

Universal joints are used to transmit power between shafts with large angular misalignment. These joints consist of a pair of yokes connected by a cross, or spider. Universal joints are widely used in automobile power trains. Common practice calls for the installation of two universal joints in the driveshaft that transmits power to the rear axle; this permits uninterrupted power flow to the rear wheels when the axle bounces up and down over bumps.

—Joseph J. Kelleher

Cybernetics, the science of communication and control. Cybernetics seeks a general theory of the way in which systems control themselves in the performance of a useful task. Such theories should apply equally well to any field of human activity or study: engineering, biology, physics, economics, business, or even baseball.

From the point of view of cybernetics, a system is any collection of items that influence one another as the system performs its function. A system usually consists of smaller subsystems, and is, in turn, often a part of a larger system. For example, a machinist and his tools form a system that produces parts by machine. The machinist's system, in turn, is a part of a larger system, the machine shop. The machine shop, in turn, is part of a still larger system, the factory. In the field of engineering, a system might consist of a jet aircraft. In business, it might be a company that operates a fleet of such aircraft. In politics it might consist of the internal government of the United States, or a collection of all the world's nations and the way they interact to produce good will or war.

The most successful systems known are those found in nature's living things—vision, muscular control, life itself. The secret of this success seems to lie in the way such systems control themselves and communicate among their various parts. By studying such systems, *cyberneticians* hope to discover nature's basic principles of control so they can be applied to manmade systems of all kinds. A major contribution might then be made in the field of international relations. The system of nations normally accepts only force as a means of control. Perhaps the more subtle methods of control used in nature may one day offer a more peaceful and successful kind of regulation of international affairs.

—William C. Vergara

Desalinization, also known as desalination, methods of making seawater, and other polluted sources of water, fit for human and industrial uses. Desalinization has gained added momentum as the world faces an increasingly critical water shortage. This shortage is caused by four major factors: a fast-growing population; an even faster rise in the industrial use of water; the growing pollution of waterways; and maldistribution of water with respect to concentrations of population and industry.

Although millions of dollars are being spent to find new approaches, the long-proven method of *distillation* continues to be dominant. Virtually all seawater desalinization facilities in practical operation today are based on the principle of evaporating water and leaving the brine (or other pollutants) behind.

More recently, significant advances have been made in *multistage flash distillation*. In this process, the pure distillate is obtained by evaporating the seawater at progressively higher vacuums and lower temperatures. This method has steadily brought down the cost of producing potable water to where engineers now talk of its soon approaching the cost of water purified by conventional chemical methods—which in turn are becoming higher as levels of pollution rise and the cost of treatment and distribution increase.

For many years the highest efficiency achieved in multistage flash distillation was about 10 to 1—that is, 10 pounds of pure water were produced for every pound of steam required to raise the temperature of the seawater to the flash point. However, in 1968, the Office of Saline Water, a part of the U.S. Department of the Interior, announced the completion of a contract for the construction of a new high-ratio multistage, multi-effect distillation plant at San Diego, Calif. The plant achieves a ratio of 20 to 1—thus doubling the efficiency of its forerunners. This is expected to permit a substantial reduction in the cost of output water, and to hasten the day when desalinization becomes an even more practical solution to the world's growing need for water.

Research into other methods, such as *reverse osmosis* continues. Furthermore, this process is being constantly improved as the method of choice for situations where the water supply has a relatively low concentration of pollutants—on the order of 5,000 parts per million or lower.

—Gordon F. Leitner

Dies, tools used to form or cut parts or shapes from a wide range of materials. To form parts, the material is stressed beyond its yield point until it takes a permanent set. To cut parts or punch openings, the material must be stressed in shear until it fails. *Forming dies* are used to bend, emboss, forge, draw, or form rigid material that can be permanently deformed. *Cutting dies* are used to cut out blanks for forming, to punch holes, or to cut either rigid or flexible stock to shape.

For high-volume production of complex shapes, *progressive dies* are used. A progressive die consists of as many as ten or more individual dies combined and mounted in a single press. As the workpiece passes from die to die, successive forming operations take place until the part is completed. The brass base of a light bulb, for example, is formed in a seven-step, high-speed progressive die.

In operation, dies are mounted in a press that exerts the force needed to form or cut the workpiece. Mechanical, hydraulic, or pneumatic presses may be used, depending on the force needed and the speed of operation. The basic method of feeding material to the dies is by hand. To increase the speed of operation, coil stock and automatic feeders are frequently used.

To cut soft sheet stock, steel rule dies are used in a press called a *clicker machine.* The machine consists of a wooden table over which a hydraulically operated arm is suspended. The sharpened steel rule die is placed on the material by hand, and the arm is lowered by hydraulic power to cut out a piece of stock. Clicker machines and steel rule dies are widely used to cut leather, canvas, sheet rubber, and gasket materials. —Joseph J. Kelleher

Diving, entering the water and descending to a great depth, to accomplish a specific purpose, such as salvage. Diving was practiced in ancient times, at first without equipment and then with the aid of a *diving bell.* Diving bells originally consisted of a bell-shaped metal housing which retained the air inside when it was lowered into the sea. Divers could enter the open bottom of the bell to replenish their air without returning to the surface.

Drawings from the fourth century A.D. show Roman divers with watertight helmets attached to leather tubes leading to the surface. In 1240 Roger Bacon wrote of "instruments" which allowed men to walk on the bottom of the sea. The first practical diving suit was devised in 1819 by Augustus Siebe, who attached an air pump to a metal helmet. The complete diving apparatus consisted of the metal helmet attached to a leather jacket. Air from the pump was fed to the helmet through flexible tubes and kept the water below the diver's chin. One drawback of the original Siebe suit was that the diver had to remain standing or water would fill his helmet and he would drown.

After many experiments, Siebe developed a closed diving dress in 1830 with a helmet equipped with air inlet and regulating outlet valves. The principles developed by Siebe are still in use today, though numerous improvements have been made.

Modern diving suits with helmets require an air pump, metal helmet and breastplate, flexible air tube, weighted boots, lead weights, and a lifeline. When fully dressed before submerging, the diver is encumbered by several hundred pounds of equipment and cannot move without aid.

When submerged, the diver adjusts the regulating valve on his helmet, which controls the air entering the suit, until the buoyancy created by the air in the suit equals the weight of the equipment; this allows the diver to move freely. Plate glass windows in the helmet allow the diver to see while he is submerged. Divers generally breathe only compressed air. However, for dives deeper than 300 feet, a mixture of gases is necessary for greater resistance against water pressure.

The conventional diving suit and helmet are being rapidly replaced by self-contained breathing apparatus, known as *scuba,* devised by Jean Costeau, a French oceanographer. The diver is supplied with air from tanks strapped on his back and is free to swim or work unencumbered by air hoses or lifelines. Breathing tubes connect the compressed air tanks to the diver's mouthpiece. A demand valve on the pressure tanks admits air to the diver's mouthpiece each time inhalation drops the pressure. When the diver exhales, the air is discharged into the water.

Scuba diving has become a popular sport in Europe and the United States, where amateur divers use the equipment for spear fishing and exploring. Professional divers use the equipment for underwater repair, salvage, and rescue work. After much experimentation with air-gas mixtures replacing ordinary air, divers have descended successfully to depths of 750 feet. Because of its simplicity and mobility, scuba equipment has almost completely replaced the older hard-helmet diving suits. —Joseph J. Kelleher

Dredging, the process of deepening, widening, and maintaining harbors, rivers, and canals. One of the first machines designed specifically to dredge a river channel was a sort of continuous chain with buckets at-

U.S. NAVY SILVER SPRINGS, FLORIDA

DIVING EQUIPMENT (*left*), including metal helmet, suit, shoes, air hose, and air supply, weighs several hundred pounds; the less cumbersome self-contained scuba unit (*right*) consists of a watertight glass face-mask, rubber foot fins, and an aqualung.

ELLICOTT MACHINE CORPORATION

DIPPER DREDGES are used for the removal of heavy earth materials from shipping channels. The dredge shown here can reach to a depth of 47 feet below the water level.

tached at intervals. As the chain took inverted buckets to the bottom, full ones came to the surface and emptied into a barge at the top of the cycle. This dredge was invented in the early 1700's. As large excavating machines were developed for dry-land excavations, marine engineers adapted them for dredging from barges.

The *dipper dredge,* a sort of underwater steam shovel, was among the earliest types of large dredging devices. The booms and rigging were adopted so the shovel could dip well below the surface. The buckets were enlarged and designed so some water leaked out as they emerged.

Huge dipper dredges still do much dredging, but they are being displaced by *hydraulic suction dredges.* These work like mammoth vacuum cleaners sucking the silt and sand from below the surface with pumps. For work in relatively firm material, hydraulic dredges are equipped with rotating cutterheads on the end of the suction pipe.

Hydraulic dredges range in size from small portable models that can be disassembled and moved overland, to the U.S. Army Corps of Engineers' big *hopper dredges.* These are full-scale seagoing vessels which—although they dredge hydraulically—are called hopper dredges because of their huge holds. The holds are filled with dredged material, taken to sea, and dumped through the bottom of the hold, as in hoppers.

Clamshell buckets mounted on cranes are also used for dredging. Another common method used for small-scale work is the *dragline, a* large steel scoop attached to a crane by two cables. One, running to the top of the crane boom, is used to cast the bucket into the water. The other, attached to the mouth of the scoop, is tied directly to the crane's winch; as it is reeled in, it pulls the scoop or dragline toward the crane, and thus fills it. —William W. Jacobus, Jr.

Dyeing, fixing colors uniformly in textile materials. Although man has dyed textiles since prehistoric times, he was forced until 1856 to rely on a mere handful of colors derived from plant and animal sources. In that year, the synthesis of the first coal-tar derived dye by William H. Perkin, a British scientist, marked the beginning of the synthetic dyestuff industry. Today, thousands of different chemical compounds of varying brilliance, fastness, and depth are available as dyes. None is effective on all textile fibers. Some work well on animal fibers, such as wool; others on vegetable fibers, such as cotton; and still others on one or more of the many synthetic fibers, such as nylon.

Dyes are classified into four general categories. *Direct dyes* dissolve in the dyebath and are taken up directly by the fiber. *Developed dyes* are taken up by the fiber but need further chemical treatment to yield their final shade. *Mordant dyes* have an affinity not for the fiber itself but for another substance (usually metallic) that is first applied to the fiber. *Pigment colors* are very fine colored powders dispersed into an emulsion that is linked to the fiber by an adhesive substance.

Within these four categories are many specific classes and subclasses of dyes. *Vat dyes* have maximum fastness to light and laundering and are usually applied to cotton and rayon. *Basic dyes* are used on cotton, silk, and wool where bright shades are desired, but they usually have poor fastness to light and washing. *Sulfur dyes,* another class used on cotton, possess very good fastness to washing but only moderate to good lightfastness. *Acid dyes* are used extensively on wool, silk, and nylon; they offer low cost, brightness, and easy application, and fastness to light and washing ranges from good to very good. *Disperse dyes* are applicable to manmade fibers, such as polyester, nylon, and acetate, and have enjoyed a steady growth along with that of the fibers themselves during the past decade; while fastness properties are good to very good, these dyes often require special treatment to ensure fastness and uniformity on polyester and nylon. *Reactive dyes* are the newest class of dyes; these actually form a chemical union with either cellulosic or wool, depending on the structure of the dye; as a result, they feature very good fastness and bright shades.

Textile materials can be dyed in four basic ways, each related to a stage of textile manufacture.

In *raw stock* or *sliver dyeing,* fibers are dyed in loose form before they are drawn and twisted into yarn. This method, used mainly for wool, results in uniform dye absorption and colorfastness. The well-known expression *dyed in the wool* originally referred to fabric made of fibers dyed this way.

Yarn dyeing includes *skein dyeing,* in which loose hanks of yarn are dyed in a bath; *beam dyeing,* a method of dyeing warp yarns that are later woven with filling; and *package dyeing,* in which yarn is dyed while on tightly-woven spools. In each case, the operation is designed so that the dye penetrates to the innermost section of the mass of yarn to ensure a uniform color throughout. Yarns dyed by these techniques can be used for knitting or as components of yarn-dyed woven fabrics, such as plaids.

Piece dyeing involves the application of dyes to woven or knitted fabrics. (In some cases, complete garments, such as hosiery or sweaters, are dyed in piece form.) Piece dyeing is the most flexible and most economical method of dyeing, especially since most textile products are subject to the whims of fashion. The availability of piece dyeing permits the fabric merchant to maintain large stocks of undyed fabric, portions of which can be dyed on order to the specifications of the end-product manufacturer, and in whatever shades may be popular at a given time. Piece dyeing is carried out continuously or semicontinuously on various types of machines, either in open-width or rope form. Recent developments in machinery for piece dyeing have included faster speeds for economy, higher temperatures and pressures for better dye fixation, and the use of automated devices for greater uniformity and lower labor costs.

In *solution dyeing,* pigments are added to the liquid polymeric solution, from which a synthetic fiber is spun. The color, therefore, is well-fixed throughout the fiber. An economic disadvantage is that mills which make the fiber into fabric or yarn must maintain a very large inventory of differently colored fibers to satisfy customer demands for various colors. Thus, this method is used chiefly for synthetic fibers that are difficult to dye in yarn or piece form, or in cases where the color of the end-use item is of minor importance from the standpoint of fashion.

—Francis A. McNeirney

Electric Furnace, a heating chamber in which electric energy is the "fuel." Electric furnaces may be classified into two large groups, resistance furnaces and arc furnaces, according to the principles of heating employed.

Resistance furnaces may operate on the principle of indirect heating or direct heating. With *indirect heating,* electric current is passed through special resistors in the form of coils, rods, or grids or baths of molten salt that have high electrical resistance and are heated by their resistance to flow of the current. The heat generated is used to heat the furnace and the material in it by radiation, convection, and/or conduction. In the case of *direct heating,* current is made to flow through the material to be heated, and heat is generated in the material by its own electrical resistance. This may be accomplished by applying contacts to the piece to be heated, making it part of a low-voltage circuit, or by using the induction principle. *Induction heating* is accomplished by passing a high- or medium-frequency alternating current through a coil surrounding (but not in contact with) the material to be heated.

Arc furnaces are of three general types: indirect, direct, and a combination of the two.

Indirect-arc furnaces heat solely by radiation from the arc, using alternating current. One type employs two horizontal electrodes in a cylindrical furnace that rotates about the horizontal axis of the cylinder. An arc is maintained between the ends of the electrodes within the furnace above the heated material.

Direct-arc furnaces of the type employed principally for melting and refining ferrous metals use three vertical electrodes and three-phase alternating current. Current passes through an arc from one electrode to the bath of molten metal, passes through the bath, and arcs from the bath to another electrode to complete the electrical circuit. These are known as *series-arc furnaces.* Another type of direct-arc furnace is the *single-arc furnace,* employing alternating current in which the current arcs from one electrode to the bath, passes through the bath, and out through an electrode in the bottom of the furnace.

Consumable-electrode furnaces are direct-arc furnaces for remelting and refining metals under vacuum; they use direct current. In these furnaces, current passes through an electrode made of the metal to be melted, and an arc is maintained between the end of the electrode and a small pool of molten metal in the bottom of a water-cooled mold. Heat generated by the arc causes continuous melting of the end of the electrode. The melted metal drops into the pool and forms an ingot as the metal cools and solidifies in the mold.

The third type of arc furnace, the *combination arc and resistance furnace,* uses both arc radiation and the heat generated by current passing through the refractory bottom of the furnace to heat the charge.

—Harold McGannon

Electrical Measuring Instruments, devices that measure electrical quantities such as voltage, current, power, electric charge, or energy. They are used to obtain quantitative information about the status or performance of an electric circuit. Common measuring instruments include ammeters, voltmeters, and wattmeters.

The *d'Arsonval galvanometer* is a basic electromechanical device for measuring or detecting weak electric currents, and is the sensing element of many electrical measuring instruments. It consists of a small coil of fine wire suspended on bearings between the poles of a permanent magnet. If a current flows through the coil, it produces a magnetic field that interacts with the magnetism of the permanent magnet, causing the coil to rotate. A pointer attached to the coil indicates the amount of current flowing. In the absence of a deflecting current, a light spring returns the pointer to zero.

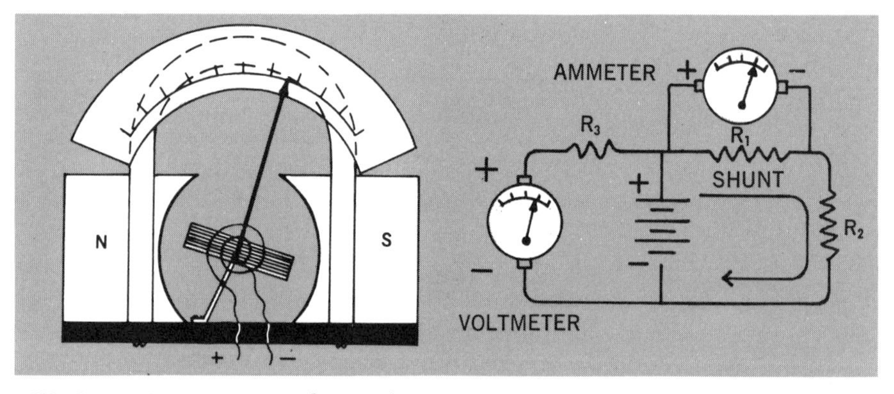

MOVING-COIL GALVANOMETER detects electrical currents with the use of a permanent magnet, whose field fills the gap between its north and south poles, and a coil attached to a spring in that gap. As current passes through the coil, it induces a magnetic field that is repelled by that of the magnet, and the needle is deflected. The spring returns the pointer to the zero position when the current ceases.

AMMETER, a modified galvanometer, measures the electricity flowing through a circuit. Only part of the current enters its magnetic field, but the magnet controlling the indicator is balanced to record the actual strength. Thus, even a current strong enough to destroy the mechanism can be measured. A *voltmeter* is a galvanometer connected to a transistor (R_3) that limits the current through the moving coil.

The ordinary *direct-current ammeter* is merely a sensitive galvanometer that has a shunt of very low resistance connected across the terminals of the coil. The shunt allows a predetermined fraction of the current to bypass the coil. The galvanometer's scale is calibrated in *amperes* to indicate the true current flowing in the circuit, even though only a small fraction of that current actually flows through its coil.

The galvanometer is also used as a *direct-current voltmeter* by connecting a large resistance in series with its coil. The resistance limits the current through the coil to a safe amount, and the scale is calibrated to read in *volts.*

The galvanometer is also the heart of the *ohmmeter,* a useful instrument designed to measure, in units of the *ohm,* the electrical resistance in a circuit. In operation, the unknown resistance is connected in series with a galvanometer and a battery of known voltage. The current that flows through the galvanometer, therefore, depends upon the magnitude of the unknown resistance. A scale, calibrated in ohms, gives that magnitude directly.

The galvanometer finds another application as the indicating element of the *potentiometer,* a voltage-measuring instrument that draws no current from the voltage source being measured. The unknown voltage source is balanced against a known, variable voltage, and any difference between the two will cause the galvanometer to deflect. Equality between the two voltages is achieved when the galvanometer reads zero.

The *Wheatstone bridge,* named for Sir Charles Wheatstone, is used to measure electrical resistance independently of any variations of the voltage source. It consists of three known resistances, an unknown resistance, a battery, and a galvanometer. The four resistances are connected in series, the four connections forming a square. The battery is connected to a pair of opposite points of the square, and the galvanometer to the other pair of opposite points. The bridge is balanced by adjusting the known resistances until the deflection of the galvanometer reaches zero. It is then possible to calculate the magnitude of the unknown resistance. The same fundamental principle is used to measure electrical inductance or capacitance in alternating-current circuits.

The *electrometer* is an extremely sensitive voltmeter that draws much less current in its operation than do moving-coil galvanometers. The first electrometer consisted of a sealed jar containing a metal rod to which were attached two small strips of thin gold foil. An electric charge causes the gold leaves to repel one another and their movement, measured against a calibrated scale by a microscope, gives the magnitude of the unknown voltage. An *electroscope* is merely an electrometer without a calibrated scale. Modern electrometers use electronic techniques to achieve even greater sensitivity. The latter are 100 million times as sensitive as the best galvanometers.

The *Geiger-Müller counter* contains an electronic electrometer that indicates radioactivity levels when radiations from radioactive material enter the ionization chamber of the instrument.

The *fluxmeter* measures the magnetic strength of a magnet, electromagnet, or other source of magnetism. It consists of a search coil connected to a sensitive, moving-coil galvanometer. The coil of the galvanometer is suspended by a fine quartz or silk fiber that minimizes the forces tending to return the pointer to zero. In operation, the search coil is brought close to the magnet under test, and the deflection of the galvanometer indicates the change in magnetic flux, which is a measure of the strength of the magnet.

The direct current *wattmeter* indicates the product of voltage and current in a circuit, hence the electric power. The unit of measurement is the *watt*. The wattmeter is, essentially, a combination of a voltmeter and an ammeter, so arranged that the pointer indicates the product of the number of volts applied to a circuit and the number of amperes flowing through the circuit. The *watt-hour meter* indicates the total amount of energy consumed by a circuit. The usual alternating-current residential type uses a rotating disk whose speed of rotation depends on the power passing through the meter. A counting mechanism records the energy consumed in units of *kilowatt-hours*.

The *oscilloscope* is a versatile measuring instrument that presents a visual indication of the voltage existing in a circuit. It is used primarily for alternating-current measurements to show how the voltage varies in the course of time. This information is presented in visual form on a cathode-ray tube, similar to the picture tube in a television set.

—William C. Vergara

Electron Microscope, a device using electrons to form an image. Its development is encompassed by that branch of physics known as *electron optics*.

The electron microscope is an evacuated tube. At one end is the electron gun, which accelerates freed electrons. The electron beam is directed by electromagnetic and electrostatic fields (lenses) onto the specimen. After interaction with atomic nucleii of the specimen, the electrons enter other lenses, finally hitting a fluorescent screen or photographic emulsion where the image is formed. Magnifications up to 600,000X and resolutions down to a few Angstroms may be obtained.

Electron microscopy (and the co-technique of electron diffraction on the same instrument) has wide ap-

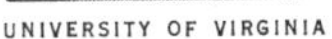
UNIVERSITY OF VIRGINIA

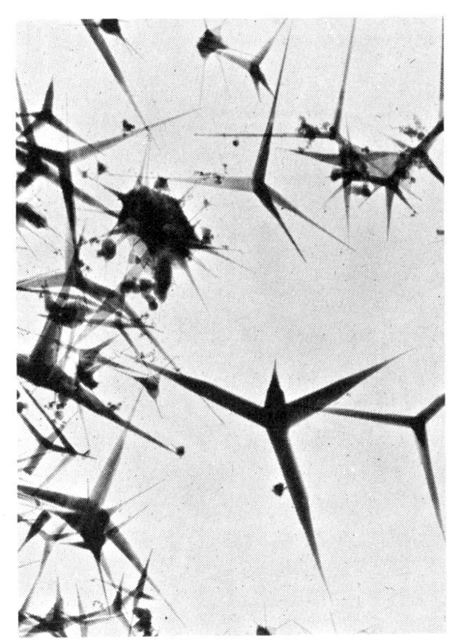
RADIO CORPORATION OF AMERICA

ELECTRON MICROSCOPE (*left*) is an instrument capable of magnifying an object up to 400,000 times. A complex structure weighing some ten tons, it contains about 15,000 parts. The electron micrograph (*right*) shows zinc oxide crystals magnified 65,600 times.

plication. It is an indispensable and ultimate tool in such varied fields as biological ultrastructure, polymer chemistry, and research in solid-state physics.

Of increasing use and interest is the very high voltage instrument, accelerating electrons through one million electron volts (1 MEV) and more. More than 4,000 electron microscopes are employed throughout the world in research institutions, universities, hospitals, and private industry, and they are in increasing demand. (See also *Microscope*).

—John W. Coleman

Electronic Industry, the design, manufacture, and sale of electronic equipment. This equipment includes television sets, radio receivers, radar sets, communication equipment, and countless other electronic marvels made possible by modern engineering.

Progress in the electronic industry has been particularly rapid as a result of automation and the impact of electronics on production methods. In 1966 it ranked fifth among the leading U.S. industries. Almost every area of the industrial and national economy has been affected by the growth of electronics. In 1965, the industry had sales of approximately $18 billion with an annual growth rate of about $1 billion per year. This was about 2.6 per cent of the gross national product. By contrast, the electronic output of the rest of the world was about $10 billion in 1964.

The largest user of electronic equipment is the U. S. government, which purchased about $9.6 billion worth in 1965. Most of this money was spent on electronic equipment for aircraft, missiles, rockets, and space vehicles. This includes radar sets for tracking, automatic flight-control systems, telemetry equipment to relay data to ground stations, and general communication and navigational systems. Because of the high speed of modern aircraft, increasingly complex electronic equipment is needed to aid the pilot in the safe and efficient control of the plane. For unmanned missiles and rockets, even more sophisticated systems are required to provide the necessary guidance and control.

Consumer products accounted for sales of $3.1 billion in 1965. In addition to radio and television sets, consumers purchased phonographs, tape recorders, and a wide variety of hi-fi components.

American industry is another large user of electronic equipment, with purchase of some $5.1 billion in 1965. The largest part, $1.9 billion, went for computing and data-processing equipment. These electronic systems assist in almost all phases of business activity, from inventory control to payroll calculations. Electronic computers are even used to design other pieces of electronic equipment. Electronic equipment is used to control industrial processes, to measure and test a variety of products, and has many applications in the fields of communications and data processing.

Though relatively small in volume, about $212 million in 1966, *microelectronics* is probably the fastest-growing electronic field. Microelectronics replaces a radio tube or transistor and its usual complement of other parts with a tiny microelectronic package smaller than a pencil eraser. Microelectronic circuits have also proven to be more reliable than the parts they replace. As their cost comes down, microelectronic circuits are expected to find their way into all areas of electronics. Microelectronics will then be as important to the electronics industry as the transistor itself.

—William C. Vergara

Electroplating, the process of coating an object made of metal or other electrically conductive material with a layer of another metal or alloy through the use of an electric current. In modern practice, electroplating has either an aesthetic or a more functional aim, such as to improve the corrosion resistance of a product or to ensure low resistivity through electrical contacts. Often the functional and the aesthetic purposes are combined, as in the "chrome" plating of automotive bumpers and other components.

Fundamentally, the process is quite simple and requires a minimum of equipment: an object to be plated; a piece of the metal with which the object is to be plated; a special bath or solution in which both the object and the piece of metal are immersed; and a source of direct current. Plating begins when the negative terminal of the current source is connected to the object to be plated, and the positive terminal is connected to the piece of plating metal.

The *bath* is a solution of a chemical compound of the plating material. At the negative terminal, which is also the part to be plated, the metal in solution becomes deposited on the surface of the object. Simultaneously, more metal is dissolved from the metal at the positive terminal, replenishing the bath. The speed at which these reactions occur varies in direct proportion to the magnitude of the current. In addition, some baths have been found to be more efficient than others, so that a number of proprietary mixtures for baths have been developed.

Metals commonly used for plating are gold, silver, copper, nickel, chromium, tin, zinc, and cadmium. It is not uncommon to cover the same part with successive electroplated deposits of the same or of different metals. For instance, some of the better chromium-plated parts are coated first with copper, then with nickel, and finally with chromium. Others may receive two different deposits of nickel, and so forth.

Integrity of the coated layer is important, as it must act as a barrier to prevent the basis metal from coming into contact with the atmosphere or with other sources of oxidation and corrosion. Cleanliness of the part being plated is essential, as it permits good adhesion of the deposit and eliminates the source of holes in the plated metal. Even so, the deposit later almost always contains minute pores that are likely, sooner or later, to establish a connection between the corroding atmosphere and the basis metal.

To eliminate this possibility, high-quality electroplating includes one or two buffed layers. After each layer has been deposited, the part is washed and dried, and then vigorously buffed so as to smear the deposit. This smearing action bridges over the pores, effectively sealing them and ensuring the protection of the basis metal.

Certain metals adhere more tenaciously to some basis metals than do others. As the total protection provided by electroplating is in part dependent on the thickness of the total plate, there is little lost in dividing this layer into a number of sublayers —as in the copper-nickel-chromium sequence. On the other hand, there is much to be gained, since the various sublayers are chosen for high adhesion to each other and to the basis metal and for their ability to resist different sources and types of corrosion. —Felix Giordano

WESTERN ELECTRIC COMPANY

ETCHING creates electronic circuits (*left*) as well as works of art. Circuit pattern at *right* is photographically etched on plastic film before size reduction and printing.

Engraving, the art of removing metal, especially from plates and cylindrical or other continuous surfaces, by means of shallow cuts. The purpose of engraving is to alter the original surface for decorative or utilitarian purposes.

A mechanical method of engraving is used to inscribe messages on the inside of rings, on watches and other jewelry, on plaques and trophies, and on similar articles. Engraving is also used to produce plates for printing paper currency, postage stamps, bonds, deeds, and other legal documents. In industry, the process is used to produce identification lettering on tools and dies and for the production of master rolls required in various types of embossing.

Much engraving is performed by hand, with tools which the master engraver himself fashions as he needs them. Where the purpose of engraving is to reproduce a master pattern of letters or other line work, a pantograph machine is used. The master pattern is fixed in place, and the lines in the master are traced with the machine's follower. A graver on the machine reproduces the master pattern in smaller size onto the metal surface. In some instances the graver is replaced with a rotating tool, which facilitates cutting through some harder metals but reduces the engraving's edge sharpness.

Engraving is sometimes used as a finishing operation to sharpen the edges of cuts produced in metal surfaces by mechanical or chemical action. Such a finishing operation produces a plate with all the working characteristics of one produced by engraving alone, and at a significant reduction in cost. (See also *Graphic Arts.*) —Felix Giordano

Etching, an engraving process by which metal or glass is removed by chemical action. In some instances, as when etching is to prepare a metal surface for microexamination, the etching action is required to be very shallow. In most instances, a deeper etching action is needed, as in the production of printing plates, and the chemical milling of relatively large surfaces, especially of magnesium, aluminum, and similar metals.

The chemical, or *etchant,* used varies with the material to be etched. Etching time is regulated by the activity of the etchant and the desired depth of etch.

Parts to be etched are prepared by masking out the areas in which no etch is desired, then either immersing the parts in the etching solution or spraying the solution on them. The spraying process is often used in the production of printing plates, as it reduces the amount of undercut at the edges of the masked-out area. When the etching is completed, the masking material is stripped from the surface of the part, which is thoroughly cleaned and dried.

Masking material may be applied locally by hand. More often, masking-out is achieved through a photomechanical process by applying the masking material over the entire surface, then exposing it through a plate negative of the desired design. This exposure activates the sections of the masking coat that are exposed to light, so that subsequent immersion in water or special solutions washes it off, leaving the coat only where no etching is desired. —Felix Giordano

Farm Machinery, or *agricultural implements,* the tools used on the farmstead for crop and animal production. Farm machinery includes tractors, plows, planters, harvesters, and similar equipment. Today these implements and motor vehicles amount to more than 20 per cent of the value of physical assets on farms. Over $4 billion of farm machines and equipment is shipped each year by manufacturers. Farm machines are primarily responsible for the high level of agricultural production in the United States today. Less than 6 per cent of our population live on farms today, compared with 40 per cent only 35 years ago.

The *tractor* is the power unit that replaced the horse on farms. Tractors and electric motors today supply the power that makes it possible for one man to produce enough food for himself and 40 other persons. This ratio is still 1 to 3 or 4 in most countries of the world.

About 4.8 million tractors were used on U.S. farms in 1967. Each farm worker has 50 or more horsepower at his disposal. Tractors sold about the time of World War II were mostly less than 35 horsepower (hp). In the middle 1950's, only 30 per cent of the tractors sold were less than 35 hp, and 20 per cent were over 50 hp. In the middle 1960's, only 12 per cent were less than 35 hp, and 40 per cent were over 50 hp. In fact, many present-day agricultural tractors are in the range of 90 to 130 hp.

Hydraulic steering, comfortable seats, improved safety features, air-conditioned or heated cabs, remote-controlled hydraulic cylinders for lifting implements, and power-shift transmissions are now commonly found on farm tractors. The modern three-point hitch, (a device at the rear of the tractor for attaching implements) has hydraulic lift control and automatic load and depth control; that is, the implement depth is automatically adjusted to a load the tractor can pull through the earth. Lights for night work and safety lights for highway travel are standard equipment on most tractors today.

Plows and other ground-working implements are generally attached to and lifted by the three-point hitch. A rear wheel raised and lowered by a hydraulic cylinder, controlled from the driver's seat, is sometimes placed on the implement to assist in lifting a portion of the weight for turning or transport. Modern tillage equipment is pulled at speeds up to 5 or 6 miles per hour (mph) and will cover 10 to 50 acres a day, depending on the size of tractor and width of implement.

Modern *planters* range from 2- to 12-row units. They may be pull-type with wheels or mounted on a tool bar attached to a three-point hitch. Planters may be equipped with fertilizer, insecticide, fungicide, and herbicide applicators in addition to the seeding mechanism. Hoppers or spray tanks are used, depending on whether the chemical is granular or liquid.

In the Corn Belt, planters are often pulled behind a disk or spring-tooth harrow. This is referred to as a minimum-tillage operation. Another minimum-tillage practice is referred to as "mulch planting"; planters are designed to plant in last year's stalks (mulch) without first plowing or disking. This practice is found mostly in the Great Plains, where moisture must be conserved in the soil.

Cultivators equipped either with sweeps or with rotary hoe wheels are used to till crops at speeds from 2 to 12 mph. However, the number of cultivations is often decreased today by the use of herbicides sprayed on or mixed in the soil at planting time. Selective herbicides are available for some crops that completely eliminate the need of cultivation for weed control. On sloping land, however, cultivation still may be needed to prevent water runoff. After crops are up, chemical herbicides may also be sprayed on weeds and grass. Special applicators are used in order to protect the crop.

Harvesters for most major and many minor crops are highly perfected. Self-propelled models now account for 85 per cent of sales of grain combines, compared with 25 per cent in the early 1950's. Today's larger combines cut, thresh, and deliver grain to transport trucks as they travel through the field, and often harvest as much as 75 acres in a single day. There were approximately 880,000 combines in use on farms in the United States in 1967.

Corn is harvested with both combines and conventional *pickers.* The corn head for combines has been available since 1956, yet already more corn heads than conventional pickers are sold each year. Corn heads on combines pour the shelled corn in the bin mounted on the machine. Pickers put corn ears (generally shucked) into trailers pulled behind the tractor and picker. Shelled corn harvested with combines often has to be dried with heated air before it is put in barns or silos for storage.

Hay balers are used to put up about 75 per cent of the annual 100 million tons of forage (excluding silage). Modern balers pick the hay up after it has dried in the sun and compress it into neatly formed packages approximately 14×16×30 inches. Attachments are available to throw the bales into trailers pulled behind the machines. Bales can be dumped or mechanically conveyed into storage with a minimum of handling.

In 1967 it was estimated that approximately 775,000 hay balers were in use on farms in the United States. The average life of the machine is about seven years or less, depending upon the amount of use and type of terrain.

In the irrigated desert areas of California, Arizona, Nevada, and New Mexico, a new method of packaging hay into small cubes has come into use. Present technology limits this method of harvest to these dry weather areas. The cubes are 1¼×1¼×2½ inches (so-called "bite-size"). The machine picks up dry alfalfa, sprays a small amount of water on it, and compresses it into these cubes at a rate of about 5 tons per hour. These cubes are easier to handle, transport, and feed than the larger conventional bales.

INTERNATIONAL HARVESTER COMPANY

HARVESTING TECHNIQUES have advanced considerably. In the nineteenth century, men labored behind a horse-drawn reaper (*left*) to bind and stack grain. Today, one diesel-powered combine (*right*) not only cuts, but also threshes the grain in one continuous operation.

INTERNATIONAL HARVESTER COMPANY

INTERNATIONAL HARVESTER COMPANY

MODERN MACHINERY has increased productivity on the farm. Automatic planting machinery (*left*) results in a mechanical precision in the sowing of seeds that could never be achieved by men. A pick-up baler (*right*) collects, binds, and deposits hay in a continuous operation.

Cotton has been conquered by mechanical picker and stripper-type harvesters. *Cotton pickers* have several hundred rotating spindles that engage the cotton lint and twist it from the burs that hold it on the plant. Another mechanism unwraps (doffs) the cotton from the teeth on the spindle, then air-blows it up to the carrying basket. In the drier areas of Texas and Oklahoma, *stripper-type harvesters* use long rotating nylon brushes to remove both cotton and bur from the plant. The burs are later removed at the cotton gin prior to the cleaning and removing of the lint from the seed.

Vegetables now harvested with machines include radishes, carrots, beets, green peas, and spinach. However, melons, celery, cauliflower, lettuce, and broccoli are proving more difficult to conquer mechanically.

Tree fruits such as almonds, walnuts, pecans, filberts, prunes, and cherries may be removed from trees by *boom* or *cable shakers*. Some are caught on canvas conveyors, while others are swept up off the ground.

Farmstead buildings and equipment are undergoing a rather rapid change to facilitate mechanization and automation. Electric service on farms and the availability of numerous types of controls and servomechanisms make automatic operations for conveying, processing, and feeding a reality. Automated poultry, dairy, and livestock feeding operations are leading the way.

—H. F. Miller

Fasteners, mechanical elements used to hold two or more parts together. Thousands of different types of fasteners are available, ranging from the common wire paper staple to large structural bolts several feet long. In general, fasteners can be classified as *threaded* or *plain.* Threaded fasteners include machine screws, cap screws, wood screws, nuts, bolts, and studs. Plain fasteners include nails, rivets, pins, and staples.

Wood screws with recessed or slotted heads are available in standard sizes. These screws are produced in lengths from ¼ inch to 5 inches and with flat, round, or oval heads.

Self-tapping screws are used in metal assembly work, and tap or form a thread in a drilled hole. In addition to slotted and recessed heads, self-tapping screws are available with hexagonal heads. These screws are widely used in joining plastic and metal assemblies for household appliances, such as refrigerators and washing machines.

Set screws fit into tapped holes and are available in a wide range of sizes, in either square head or headless types and with flat, cone, cup, oval, or dog points. Either a slot or socket is used to drive the headless types. The most common applications of set screws are lock or position sheaves or impellers on rotating shafts.

Machine screws are classified according to the style of head and/or the type of drive used to install them. Thus, round, flat, fillister, oval, truss, binding, and pan-head machine screws have slotted or recessed heads and are turned by a flat-blade or Phillips-head screwdriver. Hexagonal-head machine screws are turned by conventional wrenches or external sockets.

Bolts are most easily differentiated from screws by the addition of a nut. Bolts are installed in drilled holes and hold two or more parts together by the clamping force created as the nut is tightened. Produced with either round, square, or hexagonal heads, bolts are available in a wide range of diameters, lengths, and styles. The more common styles include machine bolts, stovebolts, and stud bolts.

Used as permanently installed fasteners in castings, stud bolts, or *studs,* are threaded at both ends. One end of the stud is screwed into a tapped hole while a nut is installed on the other end. The nut may be removed and assembled many times without damaging the thread in the soft casting.

Nails are the most common type of plain, or unthreaded, fastener and have been in use for thousands of years. Most nails fall into two categories: *wire nails,* as the name implies, are cold-formed from lengths of wire and are of standard circular cross section; *cut nails* have a rectangular cross section and taper from head to point. Though hundreds of different types of nails are available, house carpenters use only *flat,* or *common, nails* and *finishing nails.* Flat-head nails are used for fastening the structural members of the house, while finishing nails are used on the trim work.

The length of nails is given in terms of pennies, which is a holdover from the days when nails were sold at so many pennies per hundred. The penny system, with *d* used as the symbol for pennies, now designates only the length. For example, 2d nails are 1 inch long; 3d, 1¼ inch; 4d, 1½ inch; up to 10d for 3-inch nails.

Rivets are made in a range of head styles and sizes for joining everything from pot handles to structural steel. They are fitted into a drilled or punched hole and headed by hammering or pressing. Solid rivets of aluminum alloy are used by the millions to fasten the components of a modern jet aircraft. Rivets are also widely used in high-speed assembly operations where they are inserted into the part and headed by automatic equipment.

When only one side of a part is accessible, a *blind rivet* is frequently used. Blind rivets are usually hollow and are set by pulling an oversize mandrel through the hole. When a fluid-tight joint is required, the mandrel is broken off in the rivet, plugging the hole.

Pins are inserted in pre-drilled holes for a great number of fastening operations. Plain round pins are pressed into slightly undersized holes to produce an interference fit. Tapered pins, with a taper of ¼-inch per foot, are used where disassembly of the part may be required.

Roll pins and the *Spirol pin* are made by rolling a strip of steel into a C-shaped sleeve or coil. Compressed and inserted into an undersize hole, the coiled pin expands against the walls to hold the pin securely. These pins are used as fasteners, hinges, shafts, or dowels.

Stapling, or *wire stitching,* is an outgrowth of the staples used to fasten paper together. Equipment is now available to staple sheet steel up to 1/16-inch thick. The process can be used to fasten metal to metal, or metal to rubber, cloth, wood, or plastics. The staples are preformed or supplied in wire reels, cut to length, and formed in the gun. Stapling or stitching is widely used in the automotive industry to attach trim materials to body panels.

—Joseph J. Kelleher

Fire Detection, methods of sensing fire in its early stages to alert people of its occurrence, so it may be quickly controlled and damage held to a minimum. Mechanical, electrical, and electronic equipment are customarily used for the purpose, but in its broadest sense the term would include the employment of watchmen and guards to signal discovery of fire.

A typical fire-detection system includes fire detectors placed at prescribed ceiling locations throughout the structure to be protected. These are connected to a control unit, which in electrical systems energizes the detection circuits. When a detector senses a fire, an audible or visible alarm is automatically given, either on the premises or at an alarm center operated by the fire department or privately. Detectors may be actuated by heat or by smoke or other gaseous product of combustion.

Heat detectors of the fixed-temperature type most commonly are thermostats utilizing the different coefficients of expansion of two metals under heat to close electrical contacts. There is also thermostatic cable employing two tensioned steel wires separated by a covering which melts at the rated temperature. Other heat detectors operate on the rate-of-rise principle, functioning when the rate of temperature increase at the detector exceeds a stated number of degrees a minute. One form uses pneumatic tubing in which pressure builds up as heat reaches the tubing, and the pressure is applied to a diaphragm. Another form operates on the thermoelectric principle, employing two sets of thermocouples so arranged that one set is exposed to convection and radiation while the other is shielded. There are also detectors combining the fixed-temperature and rate-of-rise principles.

The melting of a fusible element by heat is used to actuate a detector employing compressed gas as the energy source for the alarm mechanism, and a mechanical unit using a spring-wound motor to sound the alarm.

Customarily heat detectors are set to operate at 135 to 165 degrees (Fahrenheit) temperatures, but where normal ceiling temperatures exceed 100 degrees higher settings are used.

Smoke detectors most commonly employ photoelectric cells in which the change in current resulting from partial obscuring of a photoelectric beam by smoke is measured and an alarm tripped when this obscuration reaches a critical value. A flame detector also uses a photoelectric circuit which is responsive to changes of light intensity resulting from the flickering of flames. Ionization and resistance-bridge types of detectors are responsive to both smoke and gaseous products of combustion.

FIRE TRUCKS are equipped with high-powered pumps and hoses, chemical foam, and rotating extension ladders to allow firemen to attack a fire effectively from a variety of angles.

■FIRE EXTINGUISHER.—The extinguisher is a device containing a liquid or powder to be discharged on a fire, and capable of extinguishing a fire in its early stages. Basically a fire extinguisher consists of a container, an extinguishing agent, a pressure-producing device or agent, and a discharge orifice or hose and nozzle. A good extinguisher can be put into operation quickly and with reliable effectiveness.

Many types have been developed. To designate their suitability, fires have been divided into four principal classes: A, B, C, and D.

Class A fires, which involve ordinary combustible material, are extinguished by the cooling action of water or water-based liquids or by certain dry chemicals. Extinguishers for this purpose include manually operated *pump tanks; water* or *antifreeze water solutions,* stored under pressure; or *soda-acid extinguishers,* actuated by mixing sulfuric acid with a sodium bicarbonate water solution. Special dry chemicals for Class A fires have a monammonium phosphate base.

Class B fires, which involve flammable and combustible liquids, need a blanketing-smothering or flame-interrupting effect for extinguishment. Extinguishers for this purpose employ *dry chemicals* (having a base of sodium bicarbonate, potassium bicarbonate, or monammonium phosphate) discharged by an expellent gas; *carbon dioxide* stored under pressure as a liquid and discharged as a gas: *foam* generated by mixing aluminum sulfate and a sodium bicarbonate water-based solution; or *bromotrifluoromethane,* a liquefied gas.

Class C fires, which involve "live" electrical equipment, must be extinguished with a nonconductive agent to avoid shock hazard to the user. *Dry chemical, carbon dioxide,* and *bromotrifluoromethane* extinguishers employ such agents and are useful for these fires as well as for Class B fires.

Class D fires, which involve combustible metals, require special extinguishing agents that will not react with the particular metal involved. There are a number of commercial extinguishers suitable for use on magnesium, zirconium, titanium, or sodium, and some special powders, applied by scoops, and some liquids which are effective on metal fires.

Extinguishers employing *carbon tetrachloride* or *chlorobromomethane* have also been used on Class B and C fires, but these are no longer officially recognized because they are not as efficient and produce irritating and toxic vapors. Also available, but not currently in widespread usage, are *loaded-stream* (alkali-metal salts in water) and *wetting-agent extinguishers.*

Portable extinguishers are tested and labeled by Underwriters' Laboratories (UL), Factory Mutual Association (FM), and Underwriters' Laboratories of Canada (ULC). The labels certify that the device meets exacting requirements of construction and performance. Standards on the installation, maintenance, and use of extinguishers are issued by the National Fire Protection Association.

—Deuel Richardson

Fire Prevention, measures directed towards preventing the occurrence of fire. Fire prevention differs from fire protection, which refers to the methods of providing for fire control or fire extinguishment. Prevention methods take three general forms: laws and regulations, inspection programs, and public education.

Fire-prevention laws and regulations control the types of materials, wiring, and equipment—such as those for heating and air conditioning—used in buildings; the storage and handling of flammable liquids and gases; the use of explosives and fireworks; and other common fire hazards. Arson is also a subject of such laws. The laws and regulations may originate with the state or the locality, and administration is customarily in the hands of a state or local fire marshal or similar officer. The great majority of fire prevention laws and regulations derive from standards and codes developed by the National Fire Protection Association.

Inspection programs are carried on, usually by fire departments, in order to discover and correct hazardous conditions before they cause fires. Schools, hospitals, and other public buildings may be inspected monthly for this purpose, and commercial and industrial structures at least once or twice yearly. An increasing number of fire departments inspect multiple residential properties on a regular schedule, and also are instituting inspections of private dwellings, which can be entered only by consent of the occupant. There is also a great deal of valuable self-inspection by occupants of buildings of all types.

Fire-prevention education is based on the premise that people are the principal causes of fire, and that public awareness of and interest in reducing fire hazards is a necessary adjunct to laws and inspections.

U.S. FOREST SERVICE

"SMOKEY" is a symbol of fire prevention.

Many private and public agencies engage in this activity, and a substantial number of commercial and industrial concerns provide employee education in avoiding and correcting fire hazards. Such activity usually climaxes during Fire Prevention Week, observed annually in October, but most programs now have year-round emphasis. —Deuel Richardson

Firearms, weapons that discharge a projectile by means of an explosion. The first firearms were developed in Europe about 1300. The first of these were *cannon,* large tubes closed at the breech end and pierced with a little hole called a *vent* from the top of the tube to the bore near the breech. To fire one of these weapons, a glowing wire or a burning coal was thrust into the vent to ignite the powder inside the tube. By 1350, smaller versions of these guns designed to be held in the hand had appeared. Known as *hand cannon,* they were fastened to a wooden pole, the forerunner of the modern stock, so that they could be held more easily. Soon the vent was moved from the top of the barrel to the right side, and a ledge to hold a supply of priming powder was added beneath it. To fire these improved guns, the shooter used a glowing wick called a *match,* which he held in his right hand. Shortly after 1400, a pivoted arm was added to the gun to hold this match. Before 1475, the first true gunlock, the *matchlock,* appeared, complete with a trigger to operate the pivoted arm.

About 1500, a new era in the history of firearms opened with the invention of the *wheel lock.* This lock produced a spark by holding a piece of iron pyrites in contact with a revolving rough-edged wheel. Guns could now be used with one hand, and *pistols* became practical for the first time. At almost the same time, another form of lock developed that produced a spark by striking a piece of flint against steel. Several versions of this mechanism appeared in various parts of Europe during the next century. They included the *snaphaunce,* the *Scandinavian snaplock,* the *miquelet,* and terminated with the invention of the true *flintlock* in France about 1610.

In 1807, Alexander Forsyth developed a system for igniting the powder charge in firearms by using a compound that exploded when struck a sharp blow. This invention paved the way for the *percussion cap* and the later *metal-cased cartridge,* both of which appeared before 1825. Successful *rimfire cartridges* are usually considered to date from the Smith & Wesson design in 1858, and the *centerfire type* was perfected in the middle 1860's.

Rifling, the system of spiral grooves in a gun barrel that causes the bullet

GUN MANUFACTURING has come a long way since John H. Hall built breechloaders with interchangeable parts. The entire group of parts making up the firing mechanism and chamber of the M-16 rifle, now in combat use (*top*), is interchangeable with the same system on the assault rifle (*center*) and on the submachine gun (*bottom*). To ease military supply problems further, all three weapons can fire the same, standardized ammunition.

to rotate and thus travel straighter, first appeared in central Europe about 1500. Muzzle-loading rifles were slow to load because the bullet had to fit tightly and could not be dropped in loosely as in the *smoothbore*, which used a cartridge consisting of a charge of powder and a bullet wrapped in paper. Finally, in the 1840's, an elongated bullet with a hollow base was developed that could be dropped loosely down the barrel but would expand when fired so that it fit tightly. It was called the *minié ball* after Claude Étienne Minié, one of the men who helped perfect it. The rifle had been used for military purposes from the beginning, but with the minié ball it became a practical arm for all troops, and the smoothbore disappeared except for shotguns and some highly specialized weapons.

From the very beginning, some guns were made to load at the breech as well as at the muzzle. Some used threaded holes with screw plugs; others used moving breechlocks. The problem was to develop a tight gas seal that would not stick and jam. Over the years there were a number of more or less successful solutions of this problem, including the pattern of John H. Hall, which was issued to American troops in 1819 and became the first breechloading military weapon issued on a large scale. The real solution, however, came with the development of the metallic cartridge that not only held the load but also sealed the breech when it was fired.

Repeating firearms also appeared early, shortly after 1500. Some piled shots on top of each other in one barrel; some used multiple barrels that were either stationary or revolving. The true *revolver* appeared before 1600, and there were even two magazine types using loose powder and ball in the early 1600's. Successful revolvers were developed in the early 1800's with such percussion cap weapons as the Colt, but magazine repeaters had to wait for the metallic cartridge to become practical. About 1860 *lever-action rifles*, such as the Henry and the Spencer, paved the way for the later Winchester. The *bolt-action rifles* with box magazine was invented by James P. Lee in 1879. Automatic weapons came next. The first practical *auto-loading* pistol, the Schonberger, was manufactured in Austria in 1892. Rifles followed, then machine guns. Now almost all military and some sporting arms are automatic or at least semiautomatic. —Harold L. Peterson

Flood Control, the erection of dams, walls, and levees to prevent rivers from overflowing their banks. To protect himself from floods, the flood plain dweller builds walls and levees, at least around the cities and towns, and sometimes for many miles along rural riverbanks. Hydrologists measure winter snowpack in the mountains to get some idea of how much water the spring freshet will bring. Nevertheless, floods are unpredictable, for it is not possible to predict with certainty how much rain will fall and how fast snow will melt.

Still, man has made progress toward harnessing the rivers. As a general rule, floods are seasonal and rivers do not leave their banks all year long. Hence, the principle of a flood-control dam.

Engineers build dams in the upper reach of a river and its tributaries to create reservoirs. The flood-control reservoir is unique in that much of the time, especially during the flood season, its water level is kept low by letting normal river flow pass the dam. When a flood develops as a result of heavy rains or rapidly melting snow, gates in the dam are closed and the reservoir traps the floodwaters. On a large river, a series of reservoirs may be needed to handle the volume of water. Most of the extra water must be stored in these reservoirs. When the high river flow returns to normal, the stored floodwaters are released gradually or saved for dry-weather use.

This method can be called *true flood control*. The waters are trapped upstream and prevented from surging downriver, out of control, and from causing serious damage. Although the flood is not really prevented, damage is minimized by confining the high water upstream. —William W. Jacobus, Jr.

Flour Milling, the process of converting wheat and other cereals into meal. Finely ground meal is called *flour* especially when made from wheat. Milling is one of the world's oldest industries, going back to Neolithic culture in Europe.

Milling is almost completely automated; modern plants require the attention of very few skilled millers. The white, inner portion of the kernel, called the *endosperm*, is mechanically separated from the *bran*, or outer layers, and from the *germ*, which is the embryo of the new plant, in the middle. The process consists of continual grinding, called *gradual reduction*, and sifting which produces flour and a by-product called *millfeed*, made up of the coarser particles. Millfeed is used in the livestock industry. About 73 pounds of white flour are extracted from 100 pounds of wheat, although more can be obtained if coarser flour is desired.

Millers make flour from different types of wheat, depending upon ultimate usage. Strong, high-protein wheat flours are used by bread bakers; soft wheat flours by cake and cookie bakers; and extrahard durum wheat flours by spaghetti and macaroni manufacturers. Specialty flours are made from rye, corn, and barley. American flour is exported to more than 100 countries.

White flour is *enriched* with vitamins and minerals, to government specifications, to replace those lost in the milling process. Enrichment, introduced in 1941, adds to the nutritional value of the product.

The larger millers make cake mixes, breakfast foods, and snacks, all having a processed cereal base.

■**HISTORY.**—In early times, grain was milled by rubbing it between two stones. Next came *millstones*, operated in pairs. They were large, flat, and corrugated to provide an abrasive action. The bottom, or *nether*, millstone remained stationary while the top, or *runner*, stone rotated on a vertical axis. The grain was fed into an opening in the center of the upper millstone. As the stone turned, the grain gradually worked its way to the outer edges to emerge in ground form. The product was then sifted to remove the coarser particles, which sometimes were ground. The millstones were powered by hand labor, later by animals, and subsequently by water, wind, and steam power. Today most are electrically driven.

In the 1870's, the milling process was revolutionized. The Minneapolis millers, the leaders of the industry, introduced a new system of milling, developed in Hungary, involving the use of special cast-iron rolls, cylindrical in shape and corrugated.

By 1900, there were about 8,000 flour mills in the United States. Gradually, centralized operation was found to be more economical, with the result that today there are about 400 mills in the United States capable of producing a total of 95 millions pounds of flour every 24 hours. About 200 of these mills produce 85 per cent of the flour.

Buffalo is the largest milling center in the United States, with Kansas City, Mo., ranking second. Minneapolis, once the largest producer, is the headquarters of five of the nation's largest milling firms.

Though some of the larger mills have been closed, new ones have opened elsewhere. The trend is toward the erection of plants in areas of growing population, such as California, or in areas where transportation rates are more favorable to the markets served. Most remaining older mills have been completely modernized and re-equipped.

■**PROCESS.**—Essentially, the milling process still involves the use of three main machines: the roller mill, the purifier, and the sifter. In recent years, other machines have been developed for *impact milling*, and the separation and fractionation of flour by *air classification*. All are refinements and improvements of the basic process. Air conveying of stock, using pneumatic systems, is a major feature of modern mills.

The flour production process is divided into four main parts. First, the wheat is received and stored in elevators alongside the mill. Next, it is cleaned and conditioned, that is, it is prepared so that the separation of the endosperm from the bran can be efficiently performed. Some wheats are too hard and dry; others too soft and wet. Then comes the milling operation itself, after which the flour is moved to the warehouse where it is packed in sacks or placed in huge bins preparatory to conveying in bulk by trucks and railcars.

Throughout the process, chemists are continually checking the product for quality, and sanitarians inspect for purity and cleanliness. —George Swarbreck

Frozen Foods. See *Food Preservation*.

Furnace, an enclosure in which heat is generated by burning a fuel, or in some other manner. The heat produced in a furnace must be removed in some way for external use or made use of within the furnace itself.

The burning of fuel in a home furnace produces heat, which is then distributed through ducts as hot air or through pipes as hot water or steam. A furnace used to produce steam is known as a *boiler furnace*. The largest of these, which are designed for steam power plants, have walls lined with pipes through which water flows as it is being converted to steam. Smaller furnaces are frequently lined with firebrick or other refractories that will withstand high temperature.

Furnaces burning gas or liquid fuels are fired through burners or atomizers, and the exhaust gases are removed through a chimney or stack. Coal-fired boilers may be fed by various types of stokers or through pulverized fuel burners. If coal is to be burned, provision must be made for the removal of ashes.

Industrial furnaces of various types are used in iron and steel and other industries as part of the production process. A *blast furnace*, for example, is a tall column in which coke is burned on the hearth and hot gases pass upward through limestone, coke, and iron ore. This produces pig iron. *Open-hearth furnaces* and *electric furnaces* are also used in the steel industry.

Electric furnaces are available in many different designs. Some of them, particularly the *arc furnace*, operate at extremely high temperatures. —Hunter Hughes

Furniture Manufacturing, the production of home and office furnishings. The furniture manufacturing industry originated with the early cabinetmakers and carpenters who hand-fashioned chests, cupboards, tables, benches, bedframes, and chairs. Early American designs were copies from European sources, principally English, Dutch, French, and Swedish, with certain differences due to the skill and facilities of the makers and the tastes and needs of the users.

Markets, manpower, and materials have influenced the formation and relocation of furniture manufacturing centers. Grand Rapids, Mich., no longer dominates in this respect. Jamestown, N.Y., and Gardner, Mass., are of lesser importance. Chicago and Rockford, Ill., are no longer significant furniture manufacturing centers. New York City is still important, particularly in upholstered and custom furniture. The Los Angeles area has grown extensively, especially in the manufacture of upholstered goods. Fort Smith, Ark., and northern Mississippi have expanded in industry importance, and Indiana accounts for substantial volume, being especially noted for cabinets and desks. The dynamic growth has been in the Southeast, commonly known as the High Point, N.C., area. Big, modern factories are situated in southern Virginia, central and western North Carolina, and eastern Tennessee.

The industry's products are diverse: wood and metal household and office furniture, both upholstered and unupholstered; mattresses and bed springs; public building furniture; partitions and fixtures. Hence, manufacturing practices vary. Depending on the product, operations may include cutting, machining, or fabricating wood or metal parts; making plywood or plastic laminated panels; assembly and fitting of parts; sanding and finishing of surfaces; cutting and sewing of upholstery fabrics; springing up of seating units and installation of padding and upholstery covers; handling of incoming materials and supplies; and expediting production and shipping to conform with sales requirements. Transportation poses unusual problems because furniture is rather bulky and easily damaged.

Production techniques also vary with furniture type and style. Upholstered furniture production lines are seldom conveyorized, but can be. Finishing departments of wood and metal furniture factories usually are equipped with conveyors. Precision, high-production woodworking machinery and powered hand tools are widely used.

Marketing requirements determine the kind, style, and price of furniture to be made. Designers are often employed to do the styling. In a larger company, product engineering is performed by a separate department, while in a smaller firm the plant superintendent may decide how the furniture is to be made.

To improve efficiency, furniture plants usually establish time standards for performing certain jobs. A per-piece rate is then paid the worker, or he is given a bonus for producing quantities above the standard, provided he meets quality requirements.

Most furniture manufacturers depend on vendors for metal and plastic parts, hardware, springs, turnings, carvings, fasteners, upholstery filling and covers. Metal furniture manufacturers do their own fabricating, at least in part.

New materials have brought about changes in furniture construction and processes. Particle board, a pressed board composed of wood flakes and fibers glued together, has almost completely replaced edge-glued lumber for cores of panels with veneer or plastic laminate surfaces. Hardboard, a thinner, dense panel of pressed wood fibers, has replaced plywood in many case and mirror backs as well as on drawer bottoms. It is also used for exterior surfaces to be painted or printed to resemble wood grain.

Plastic materials are of increasing importance in the furniture industry. It is estimated that some 700 million pounds of plastic will be consumed in furniture production by 1975.

Although the furniture industry consists of a large number of small manufacturers, its overall volume constitutes big business. In the mid-1960s the industry employed some 428,000 people and had a payroll of \$2.2 billion. The value of shipments was about \$7.5 billion. —Raymond A. Helmers

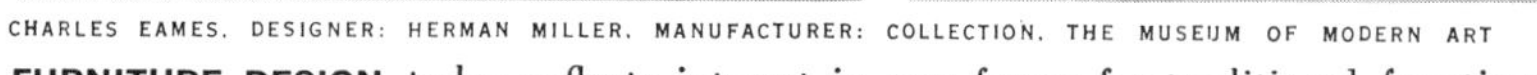

CHARLES EAMES, DESIGNER: HERMAN MILLER, MANUFACTURER: COLLECTION, THE MUSEUM OF MODERN ART

KNOLL ASSOCIATES, INC.: EERO SAARINEN, DESIGNER

FURNITURE DESIGN today reflects interest in new forms for traditional functions. Some architects design furniture for modern structures.

ILLINOIS TOOLWORKS, INC. "PRODUCT ENGINEERING"

GEAR and pinion (*above*) have skewed axis for greater power. Bevel gear is at *right*.

Gears, toothed wheels used to connect rotating shafts and provide positive transmission of rotational motion. Teeth on mating gears mesh to maintain a constant velocity ratio between the driven and driving elements.

Spur gears have their teeth cut parallel to the gear's axis of rotation and can only be used to connect parallel shafts. *External spur gears* have teeth pointing away from the center of the gear, while the teeth of *internal spur gears* point toward the axis. A *rack and pinion* is a pair of spur gears with one gear, the rack, considered to be a circle of infinite radius. The rack and pinion converts rotary motion into linear motion, or vice versa.

Although most gears are circular, noncircular gears—which produce velocity ratios that change in a precise manner—are employed for special applications.

Helical gears have their teeth running on parallel axes but cut obliquely to the gear axis. Cutting the teeth obliquely increases the length of each tooth, and the contact between mating teeth proceeds from one edge across the tooth face to the opposite edge. Helical gears run more quietly than spur gears, especially at high speeds. *Herringbone gears* have double helical teeth of opposing twist, eliminating the axial thrust produced by single helical gears.

When shafts cross at an angle of less than 90°, *crossed helical gears* are used to transmit motion. Crossed helical gears have less load-carrying ability, because the teeth make point contact rather than line contact, as in conventional helical gears.

When shafts cross at right angles or intersect, *bevel gears* are used to transmit motion between them. *Straight bevel gears* have teeth which, if extended inward, would come together at the center line of the intersecting shafts. *Spiral bevel gears* with curved and oblique teeth overcome the limitations of straight bevel gears and provide gradual engagement, eliminating impact. These gears are employed for machine tools, sewing machines, motion picture projector and camera drives, or wherever quiet, smooth operation is required.

A further refinement of bevel gears is the *hypoid gear,* which is used to transmit motion between intersecting shafts that are neither parallel nor intersecting. These gears run smoothly and quietly and are widely used in automobile differentials. Reduction ratios of up to 60 to 1 are possible with hypoid gear sets, but due to the sliding tooth action, their efficiency is lower than bevel gears. The sliding action of these gears requires a special hypoid fluid for their lubrication. —Joseph J. Kelleher

Guided Missile, a class of weapons characterized by the fact that they are self-propelled and that their flightpath can be changed, either by radio commands from a guidance center or by built-in (on-board) devices. While propulsion by rocket is the most common, it can also be by pulsejet, turbojet, or ramjet. However, not every rocket is a guided missile, and unguided bombardment rockets must be considered a special type of artillery. Missiles are classified according to their purpose: for example, an *air-to-air* missile is fired by one aircraft against another, and a *ground-to-air* missile is fired from the ground or from shipboard against aircraft. The most numerous group, the *ground-to-ground* missiles, are subdivided according to their range: *tactical missiles,* with ranges from 50 to 150 miles; *intermediate range ballistic missiles* (IRBM), with ranges from 500 to about 2,500 miles; and *intercontinental ballistic missiles* (ICMB), with ranges of more than 4,000 miles. Naturally, the longer its range, the larger the size of a missile, for most of the take-off weight is fuel. The trend in military missiles is in the direction of solid fuels. Though much more expensive than liquid fuels, a solid-fuel missile is not only more convenient to handle, but also can be launched on very short notice. (See also *Astronautics.*)

—Willy Ley

Gyroscope, a wheel universally mounted on bearings so that its axis can be made to point in any desired direction. The gyroscope is a useful device because, when spinning rapidly, its axis of rotation tends to maintain its initial direction despite the presence of forces that would otherwise change that direction.

Imagine a spinning gyroscope (or top) mounted as in the following illustration. Because of gravity, the spinning wheel tends to fall. But because of its rotation, the wheel revolves instead about the vertical direction Y-Y′. This slow rotation about Y-Y′ is called *precession.*

In a practical gyroscope, the wheel (heavy for its size) and its axle are mounted within three gimbal rings. Three sets of low-friction bearings minimize the external forces that can affect the spinning wheel. A small motor produces the wheel's rotation, and the axis of the wheel constantly maintains any preset direction despite the motion of the airplane or ship to which the gyroscope's frame is attached. The rotors, or wheels, of directional gyroscopes vary from about one inch in diameter to several inches, and they spin at velocities up to 6,000 revolutions per minute. The tendency of a gyroscope to maintain a constant direction is called *gyroscopic inertia.*

■**AUTOMATIC PILOT.**—Gyroscopes are the heart of automatic-pilot systems used on modern aircraft. In practice, a computer, a number of servomechanisms, and other electronic equipment are used to take advantage of the gyroscope's directional properties without disturbing their delicate balance. In the automatic pilot, a vertical gyroscope controls ailerons and elevators, while a directional gyroscope is used for rudder control. With the help of these gyroscopes, which resist deflective forces, the automatic pilot makes it possible for a plane to fly a preset course without human assistance.

■**GYROCOMPASS.**—A gyrocompass is a more complicated device than the directional gyroscope. It makes use of the earth's rotational motion to cause

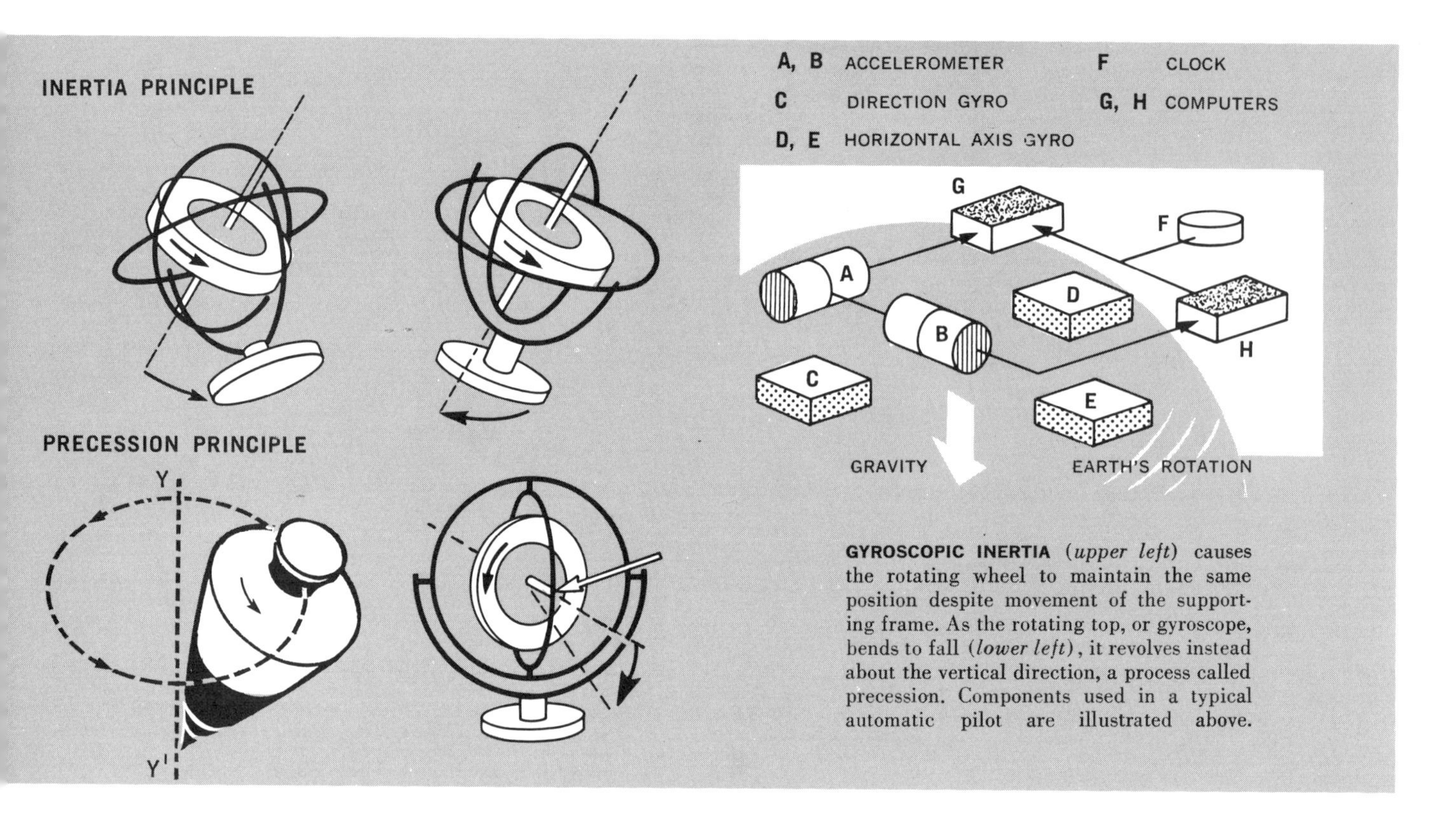

GYROSCOPIC INERTIA (*upper left*) causes the rotating wheel to maintain the same position despite movement of the supporting frame. As the rotating top, or gyroscope, bends to fall (*lower left*), it revolves instead about the vertical direction, a process called precession. Components used in a typical automatic pilot are illustrated above.

precession of the gyroscope until it points in the north-south direction. The gyrocompass is widely used on planes and on ships, where the presence of steel renders the magnetic compass unreliable.

■**INERTIAL GUIDANCE.**—Gyroscopes make it possible for rockets, missiles, and submerged submarines to navigate accurately by means of inertial guidance or dead reckoning. Devices called *accelerometers* are used to measure the acceleration of the vehicle. With this information and the elapsed time, computers can calculate the distance traveled.

In order to operate accurately, the accelerometers must be kept at right angles to the force of gravity at all times. Otherwise, serious errors would be introduced because the devices cannot tell the difference between gravitational effects, which they should ignore, and vehicle accelerations, which they should measure. To eliminate this source of error, accelerometers are mounted on a platform which is stabilized in the horizontal plane by a pair of gyroscopes. This keeps the accelerometers horizontal, or perpendicular to the force of gravity. A third directional gyroscope keeps the stable platform pointing in the desired direction, regardless of the motion of the vehicle. Similar methods are used in the inertial guidance of torpedos.

■**GYROSCOPIC STABILIZER.**—The marine gyroscopic stabilizer is a device that reduces the rolling of a ship by conteracting the forces produced by waves against the hull. As each wave strikes the ship, the stabilizer goes into operation, thereby preventing large oscillations from building up. The heart of the stabilizer is a massive gyroscope fastened to the structural members of the ship. The rotors of such gyroscopes are often over 10 feet in diameter. Gyroscopes and accelerometers are also used in the sensitive apparatus of Gyrofin ship stabilizers. This method uses the speed of the ship through the water to prevent rolling.

—William C. Vergara

Heating Systems, equipment for the maintenance of a closed space—office, home, or shop at a comfortable temperature during cold weather. The amount of heat supplied and the method of application depends on the outside temperature and the heat loss through the enclosure.

When planning a heating installation or plant, a heat balance must be calculated. This involves estimates of the heat lost through the walls of the structure to be heated and average temperature outside the structure. Finally the temperature to be maintained in the structure is selected depending on its occupancy and use. Heat lost through the walls depends on the type of construction, insulation, number of windows, and leakage and ventilation losses.

Once a heat balance has been cal-calculated. This involves estimates heating units can be selected.

■**STEAM HEATING.**—In steam heating, a boiler burning coal, natural gas, or oil converts water into steam. A system of pipes carries the hot steam from the boiler to cast-iron radiators or nonferrous convectors in the specific locations to be heated. The water that condenses within the radiators or convectors is returned to the boiler to be reused. Steam heating systems are relatively inefficient and require considerable maintenance, but are less expensive to install than hot-water systems.

■**HOT-WATER HEATING.**—Hot-water heating systems are similar to steam systems except that the heat-transfer medium is a liquid rather than a gas. Water, heated in a boiler, circulates through the closed system to the radiators and convectors and, after giving up most of its heat, returns to the boiler. Hot-water heating systems are more expensive than steam systems to install, but are more efficient.

■**HOT-AIR HEATING.**—A hot-air heating system requires ducts and blowers to carry warmed air to individual grills, from which it passes into the space to be heated. In some systems, the room air is returned into the system for reheating, for a system of this type is less expensive to operate—although more costly to install—than one requiring the introduction of outside air. Hot-air systems require considerable space for duct work and blowers. However, since the same ducts can also be used for central air conditioning, these systems are becoming more popular.

■**ELECTRIC HEATING.**—An electric heating system relies on the generation of heat in resisting elements at each location to be warmed. Electric systems are the least expensive to install, have quick response and quiet operation, and permit a different temperature to be easily maintained at each location. However, the cost of operation is prohibitive except in localities where electric power is very cheap.

—Joseph J. Kelleher

Household Appliances, electric- or gas-powered devices used primarily in the home to ease the workload of homemakers. Although electricity became widely used by 1910, it served chiefly for lighting and power. By 1918, electric home refrigerators and ranges were developed and marketed, and electrical household appliances gained general acceptance. America's gas industry was born in 1816; a vast network of pipelines now crisscrosses the country. It is believed that the first gas range in the United States was introduced in 1840; 100 families were using gas ranges by 1859. In 1895, an instantaneous gas water heater with thermostatic control, the first automatic home appliance, was invented.

Since the introduction of the first electrical and gas household appliances there has been a flow of new products to the marketplate, interrupted only by World War II. Among the electrical appliances now in use are those for food preparation, such as the blender, can opener, ice crusher, food mixer, drink mixer, food chopper, juicer, and slicing knife. Cooking appliances include gas and electric ranges, gas-fired grills, and portable electric cooking utensils, such as the fry pan, griddle, toaster, coffee maker, waffle baker, egg cooker, broiler, table oven, saucepan, pressure cooker, corn popper, baby-food warmer, and bottle warmer. There are electric refrigerators, freezers, and combination refrigerator-freezers, as well as laundry appliances to wash, dry, and iron clothes.

Appliances for housecleaning include, in addition to several styles of vacuum cleaner, the floor polisher, rug shampooer, floor scrubber, upholstery shampooer, and furniture buffer. For other household chores there are electric and gas dishwashers, electric waste food disposers and gas incinerators, electric starters for charcoal fires, and defrosters for refrigerators. Household appliances now also include beauty and personal care products. The electric shaver was among the first of these; now there are hair dryers, toothbrushes, manicure sets, vibrators, and massagers. One of the newest gas appliances is a toilet with a gas-fired unit for complete disposal of human waste without the use of sewers or plumbing.

■ **NEW DESIGNS.**—There have been many refinements in appliances since their introduction. Extensive use of thermostats and other controls that sense a condition and respond to it has made appliances automatic to varying degrees. In some cases, appliances are programmed to permit the operator to control several factors with the push of one button.

The development of new materials for appliance construction has affected design. For example, seamless plastic door liners for refrigerators can now be formed with shelves and storage compartments for butter and eggs. More efficient insulating materials, such as foamed-in-place plastic, have made possible the larger-capacity refrigerators that fit the floor space of older thick-walled models. Anodized aluminum, stain-resistant porcelain enamels, glass, ceramics, stainless steel, and copper are among the materials used, in addition to steel, in the manufacture of household appliances.

Colored appliances have been on the market for many years. The number of colored major appliances sold annually has increased steadily since the early 1950's. Pink, turquoise, and yellow have been popular; wood tones, brushed stainless steel, and copper finishes have also been used extensively, although white still accounts for a sizable part of the market.

In the years following World War II, people were eager to buy any appliances that were available, and the builder market, with its demands for low prices and simple, low-cost installation, became an important factor in the appliance business. Consumers had many more appliances in their homes than ever before, but there were not enough trained appliance servicemen. The result was a rising wave of complaints about appliance breakdowns, the high cost of repairs, and the delays and other annoyances involved in getting repairs. Consumers became more selective in their purchases, seeking more reliable performance. Manufacturers responded with greater effort than ever before to design appliances with less probability of breakdown and with more readily accessible, more easily replaced or repaired working parts. Manufacturers, distributors, and dealers also increased their efforts to train service personnel and to reduce the time needed to obtain parts.

■ **RANGES.**—Electric ranges require 230-volt, 3-wire service. Gas ranges may be fueled with natural or liquefied petroleum (bottled) gas, and many also use 115-volt electric service for clocks, timers, and lights. Oven thermostats are standard equipment; thermostatically controlled burners for surface cooking are available. Various oven control systems are also available; some use a delay, cook, and stop sequence; some cook and hold; some delay, cook, and hold.

Electronic ranges, which use microwave energy to cook food, require 230-volt, 3-wire electric service. They have been well accepted commercially but have gained only a small share of the domestic market. Electronic cooking is much faster than conventional cooking by heat, but cooking time increases with the amount of food to be cooked. The oven in such a device stays cool unless a heating element is used for browning. Any nonmetallic plates, containers, or wrappings may be used.

A range may be free-standing with one or two ovens, or oven and broiler, beneath the range cooking surface, or it may consist of one or two wall ovens, or oven and broiler, and a separate drop-in range top. The range may be wall-hung or stacked on a base cabinet with an oven at eye level and the cooking units or burners below it at counter level, or it may be a free-standing console with an eye-level oven over the range top and another oven below it.

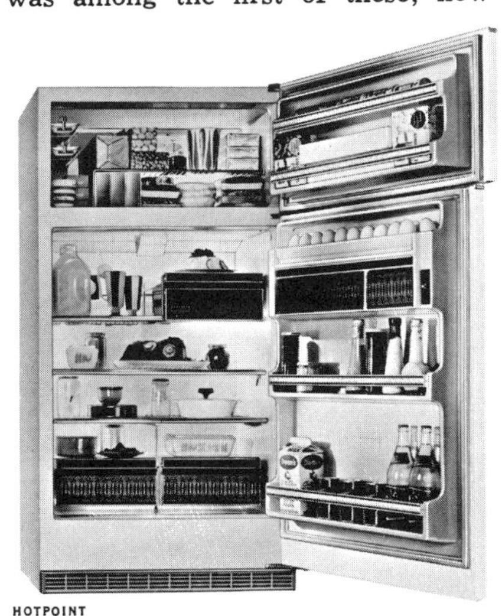

HOTPOINT

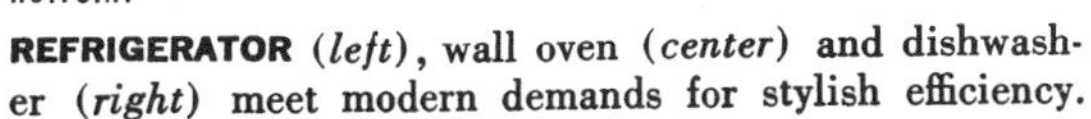

REFRIGERATOR (*left*), wall oven (*center*) and dishwasher (*right*) meet modern demands for stylish efficiency.

REFRIGERATORS AND FREEZERS.—All of the refrigerators, freezers, and combination refrigerator-freezers now on the market are electric. Gas refrigerators did not prove competitive with electric models and are no longer being promoted in the United States.

A refrigerator may have a compartment for relatively short-term storage of frozen food. Such a compartment is separated from the fresh-food storage compartment by a chiller tray or drawer. The freezer compartment of a combination refrigerator-freezer is completely insulated and positioned separate from the fresh-food compartment. The freezer may be above, below, or on the side of the refrigerator section; in some models, the refrigerator is at eye level, with the freezer below it, under a counter-height work surface. Freezers, whether separate chest or upright models or in combination with refrigerators, maintain zero-tone temperatures that are suitable for freezing and long-term storage of frozen foods. There are manually defrosted refrigerators and freezers, automatically defrosted refrigerators, and refrigerators and freezers that are free of frost within the food compartments.

Thermoelectric refrigerators, unlike conventional refrigerators, do not use compressors but rely on the cooling effect produced by the flow of electric current through the junction of two dissimilar metals, the so-called *Peltier effect.* They are in limited production. Thermoelectric refrigerators are quiet and durable. At present, the high cost of materials used in them makes it economically unsound to produce the large sizes generally required for home use.

LAUNDRY APPLIANCES.—A *washer* is a power-driven machine for washing fabrics in water. If automatic, as are the majority sold in the United States, it fills with water at a selected temperature (or as near to it as the water delivered to the washer permits), washes, rinses, extracts water, and stops—all by one setting of the controls without further attention by the user. A semiautomatic washer requires that controls be reset for one or more operations after the original setting, although the fabrics need not be removed. *Spinner washers,* in which items are washed in one container and removed to another for rinsing and water extraction, and *wringer washers* equipped with power-driven rollers for extracting water by squeezing, are still being made.

A *dryer* is a power-driven device for drying fabrics by evaporation through the use of various combinations of heat, air flow, and tumbling. Electricity or gas may be used as the heat source. Dryers may be time-controlled, with the drying time preselected by the operator, or they may be automatically controlled, with the drying time determined by measurements made by the machine.

Automatic ironers are being manufactured, but annual sales are low. *Hand irons* may be dry or combination steam-dry models; some steam-dry irons can also spray fabrics. Thermostatic control of irons is standard; in some models, steam can be produced at the low sole plate temperatures required for ironing thermoplastic synthetic fabrics.

DISHWASHERS.—Dishwashers may be built in or portable, serve to wash, rinse, and dry dishes. Most models are electric; there is a gas dishwasher which so heats the water for the final rinse that the dishes dry without additional heat. The trend has been toward larger capacity, better washability due to improved water distribution and more effective detergents, and choice of cycles.

VACUUM CLEANERS.—Vacuum cleaners utilize straight suction or a combination of suction and power-driven beaters and brushers to loosen and pick up dirt. The trend has been toward uprights with more easily used attachments for above-floor cleaning and toward suction cleaners (tanks and canisters) with better rug-cleaning performance, often achieved by power-driven brushes. Lightweight vacuum cleaners for quick pickup cleaning are often used as second cleaners in the home. Centrally installed vacuum cleaning systems for home use are gaining in importance.

SPECIAL FEATURES.—Detachable thermostatic controls on many types of cookware make them both automatic and immersible. Nonstick coatings and other easily cleaned finishes are widely used. Cordless electric appliances, with rechargeable cells or batteries of cells, offer increased mobility and, in some cases, safety, although the amount of power that can be stored is limited. Available cordless appliances include toothbrushes, shavers, food mixers, ice crushers, and vacuum cleaners.

—Rose Marie Burnley

AMERICAN SUGAR REFINING COMPANY

ELECTRICAL CONTROL DEVICES that represent automation, are the heart and nervous system of many industrial processing centers. A technician in the control room can observe and regulate both the material flow and the product quality at each stage of the process.

Industrial Control, the automatic regulation of machinery for safety, consistency, speed, or to enable operators of very complicated machinery to keep track of all the motions. Electric switches or relays, hydraulic devices, electric circuits, and/or pneumatic devices, are combined to create the control systems. For example, if a large press has two operators and there are four widely-spaced start buttons so connected that the ram of the press will not come down against the die unless all four buttons are pushed at once, it will take four hands pushing start buttons to make the press operate. This is a good safety device, which prevents one operator from starting the press before the other operator is ready.

In the same way, *limit switches* can keep machine tables from traveling too far; can keep two parts from coming together in the same place at the same time; and can stop a machine from cutting threads in a hole that has not yet been drilled.

Tracer controls, which can be either electric or hydraulic in operation, will cause the cutting tool to follow the same path that a tracer finger follows along a pattern or template.

In modern machine tools with numerical control systems, each of the machine motions is equipped with devices that can determine precisely where the machine is and feed this information back to the control. Instructions fed to the control in numerical form by punched tape can then be used to move each of the machine motions in the manner required to do a particular job, and the feedback will keep the control informed of what is happening.

—Anderson Ashburn

Industrial Engineering, the branch of engineering concerned with the planning and control of industrial production operations and the measurement and reduction of their cost. The rise of industrial engineering has been largely an American development, growing out of the struggle to secure increased production in view of ever-increasing labor costs. Some of the basic ideas may be traced far back to ancient times, however. The Greeks recognized that specialization —for example, having one worker make only the soles and another the uppers of sandals—led to increased output. Adam Smith in his *Wealth of Nations* (1776) emphasized the importance of such specialization. More recently, the development of mass production and interchangeable manufacture has also contributed to the movement toward what has been characterized as *engineered production.*

Frederick Winslow Taylor (1856–1915) is usually regarded as the father of scientific management, or industrial engineering. In the 1880's, he undertook studies for the Midvale Steel Company of Philadelphia. Taylor advised that special attention be given to selecting workers with special skills, and that they be directed and guided in their work to secure the more effective operation of their tools and machines and greater production. He also suggested that similar studies be made of management. Frank B. Gilbreth (1868–1924) pioneered in what were known as *time and motion studies,* leading to the reduction of movements on the part of the worker in carrying out his operations.

Opposed at first by both labor and management, efforts have continued to reduce costs and increase production through studies not only of manufacturing operations but also of the selection of tools and machines and all phases of management and operation. It is becoming more widely recognized that higher pay for workers can result only in the inflation of prices unless increased pay is accompanied by increased production. These methods and ideas are also receiving increased attention in foreign lands where labor conditions earlier had discouraged such advances. —James Kip Finch

Instrumentation, the use of instruments for the measurement and/or control of complex systems and processes. A system might consist of a jet aircraft or, a satellite-tracking radar system, and a process might range from a municipal water-purification plant to a steel mill. Instrumentation is a basic factor in automation, where it substitutes the precision of a machine for the judgment or performance of a human operator.

The instrument panel of a car is an example of instrumentation in its simplest form. The operator can observe the speed, water temperature, oil pressure, and fuel level of the vehicle on suitable meters or indicators. With this information, he can control the car efficiently. A more complex system would include recording and control instruments in a central place so that the operator could supervise an entire industrial process. In its most advanced form, an instrumentation system would be completely automatic, requiring little or no human attention. Any failure of the system to function correctly would be brought to the attention of an operator by a signal.

Instruments used in industrial processes vary in complexity from simple pressure gauges to sophisticated interferometers that measure dimensions by means of light waves. In chemical plants, where production is on a continuous flow basis, instruments measure and control flow rates and chemical composition. In modern factories, computers control the operation of machines from the raw material stage to the finished product.

In modern technology, instrumentation is most advanced in the field of space exploration. Complex instrumentation is used to control the velocity and trajectory of a spacecraft and to maintain communication between the spacecraft and the ground. Precise measurements are made of conditions in the spacecraft and automatic corrections are made to the vehicle's temperature, attitude, and orbit. The future will see greater use of instrumentation as man achieves greater efficiency in his daily work. —William C. Vergara

Iron and Steel Production, the processes by which iron is purified and alloyed to produce steel. The *Bessemer process* was the first to answer the Industrial Revolution's need for inexpensive, mass-produced steel. William Kelly, an American, and Henry Bessemer, an Englishman, a few years apart independently conceived of refining molten pig iron into steel by subjecting it to a strong air blast. They correctly theorized that oxygen in the air blast would burn off the excess carbon in the iron and also burn out most of the other undesirable elements formed during combustion.

The process developed by Bessemer gained wider acceptance and became identified with his name. The Bessemer converter is a pear-shaped vessel with many small holes in the bottom for the admission of an air blast. In 20 minutes it can make a heat of steel, ranging from 5 to 20 tons.

Bessemer steel was first produced in the United States in 1864. The Bessemer process quickly became the major steel production technique in this country, a position it held until it was overtaken by the open-hearth process in 1908. The Bessemer process is little used in the United States today and is now chiefly of historical interest.

■OPEN-HEARTH PROCESS.—The open-hearth process, developed by the Siemens brothers in England and the Martin brothers in France, became known abroad as the *Siemens-Martin process* but is more commonly called the open-hearth process in the United States. It is so called because the elongated, saucer-shaped hearth is open to the sweep of the flames that refine the steel. The flames issue from burners at each end of the furnace. Among the various fuels used in producing steel by the open-hearth process are natural gas, coke oven gas, fuel oil, and tar.

The furnace is regenerative in that the hot gases from combustion of the fuel are passed through regenerative

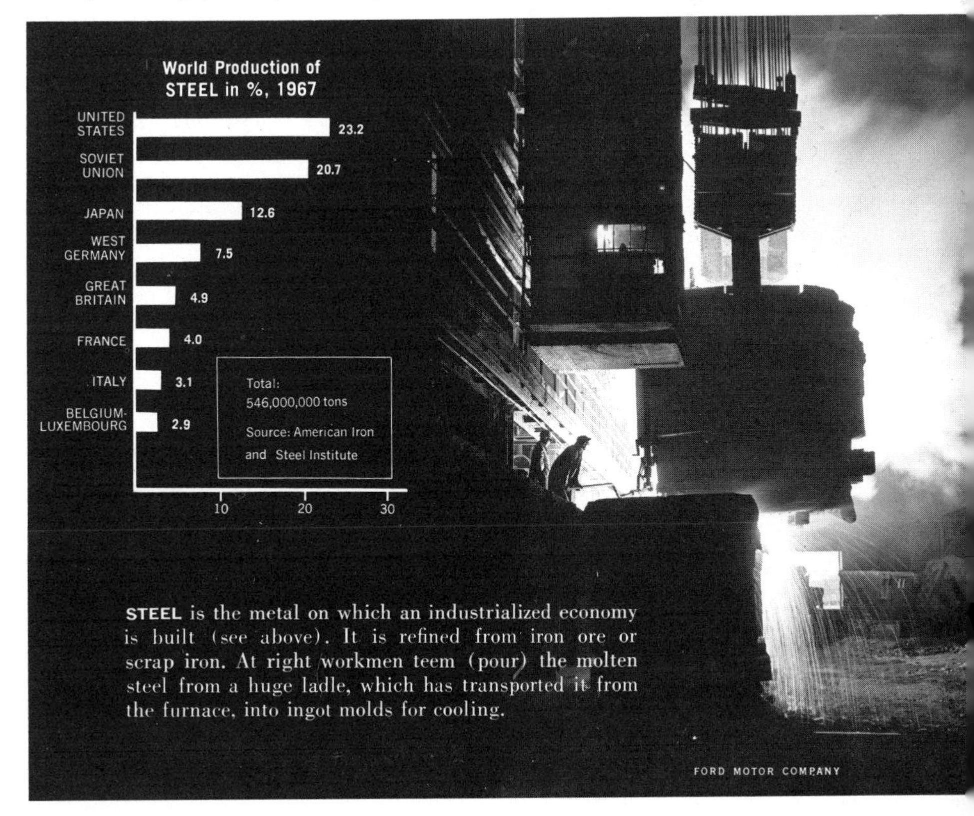

STEEL is the metal on which an industrialized economy is built (see above). It is refined from iron ore or scrap iron. At right workmen teem (pour) the molten steel from a huge ladle, which has transported it from the furnace, into ingot molds for cooling.

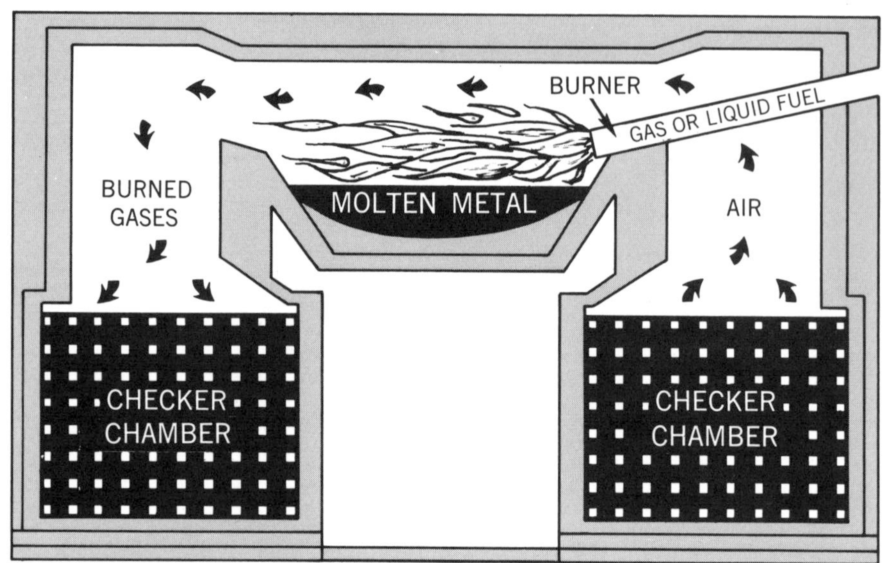

OPEN HEARTH FURNACE (*above*) is charged with scrap steel, iron ore, and limestone. An open flame refines the molten metal. Burned gases are passed through the checker chambers which are then used to heat incoming air.

STEEL is made in an electric furnace (*right*) with a charge of scrap steel, iron, and iron ore (*top*). Lowered electrodes oxidize impurities (*center*), then the slag is poured from the tipped furnace (*bottom*). Molten steel is removed by tipping the furnace to the other side. In a Bessemer converter (*far right*) the charge is molten iron (*top*). Air is blasted through to remove impurities (*center*) and the converter is tipped to remove the steel.

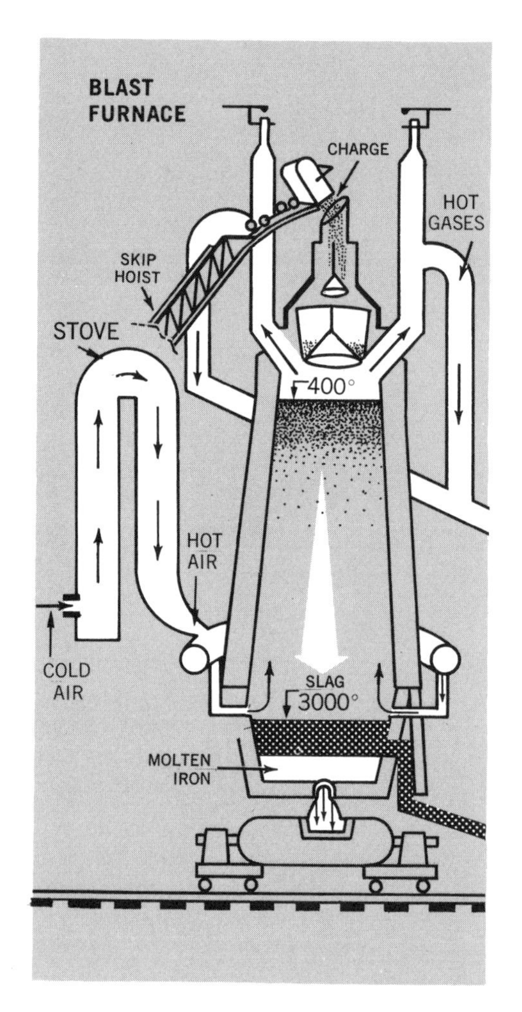

BLAST FURNACE (*left*) is charged by cars moving along a hoist with iron ore, coke, and limestone. The charge melts, reacts, and settles to the bottom, where the slag floats atop the molten iron.

brick chambers where they give up part of their heat. The direction of the gas flow is reversed periodically, and the incoming cold air for combustion is led through the heated brick chambers, absorbing some of their heat. The heated air then joins the flames, raising their temperature higher than could be obtained with cold air.

The open-hearth furnace is generally charged with about equal parts of molten pig iron and steel scrap. Limestone is added to draw off waste matter in the form of slag. The open-hearth furnace produces a heat of steel in about 8 to 10 hours. The time may be shortened by the injection of oxygen. The capacity of the largest furnaces is 600 tons per heat, but most range from 200 to 300 tons.

The open-hearth process accounted at one time for 90 per cent of U.S. steel production. In recent years, however, it has been steadily losing ground to the new basic oxygen process, and accounts for little more than half of American steel production.

■ELECTRIC ARC PROCESS.—Historically the third major steel producer, the electric arc furnace refines steel by the heat of an electric arc and the electrical resistance of the steel bath itself. It owes its development chiefly to Dr. Paul Héroult of France, who produced steel by the process in 1899.

The electric arc furnace resembles a huge teakettle. Three large graphite electrodes, entering through the roof, convey the current to the charge, which consists almost entirely of steel scrap with small amounts of slag-forming material. The chief advantage of the electric arc furnace is the high degree of precision with which the temperature and composition of the bath can be regulated.

Originally its use was reserved for a limited range of high-quality alloy steels, but the electric arc process has become a versatile instrument for a much wider range, including carbon steels, which constitute over half of its production. The electric arc furnace produces practically all the stainless, tool, and special alloy steels used in the chemical, automotive, aviation, machine tool, transportation, food processing, and many other industries. Electric arc production in the late 1960s was 15 million tons annually, or about 12 per cent of total steel production. Capacities range generally from 150 to 200 tons of steel per heat.

■BASIC OXYGEN PROCESS.—A relatively new process rapidly being adopted around the world was developed in Linz-Donawitz, Austria, and is known as the basic oxygen, or L-D, process. It was introduced in Canada and the United States in 1954. The vessel used somewhat resembles a Bessemer converter, but without the air holes in the bottom, since the oxygen is injected from above through a lance into the charge, which consists of about 70 per cent molten pig iron and 30 per cent scrap. The oxygen readily dissolves in the metal and rapidly oxidizes the silicon, manganese, phosphorus, sulfur, and carbon in the ore. The carbon is oxidized to form carbon monoxide gas, which causes vigorous boiling of the entire charge. Phosphorus, sulfur, and nitrogen content in the finished steel are generally lower than in open-hearth steel.

Advantages of the basic oxygen process are speed, lower installation costs than for an open-hearth furnace of equal productivity, and close control of the refining process. It can produce open-hearth quality steel for a wide range of products at a rate of nearly 500 tons per hour, compared to 50 tons per hour in an open-hearth furnace. It can produce carbon, alloy, and stainless steels.

There were basic oxygen plants in 34 nations on 6 continents in the late 1960s with 165 million tons annual capacity. Plants under construction or planned will raise world capacity to 260 million tons by 1970. World production is then expected to equal

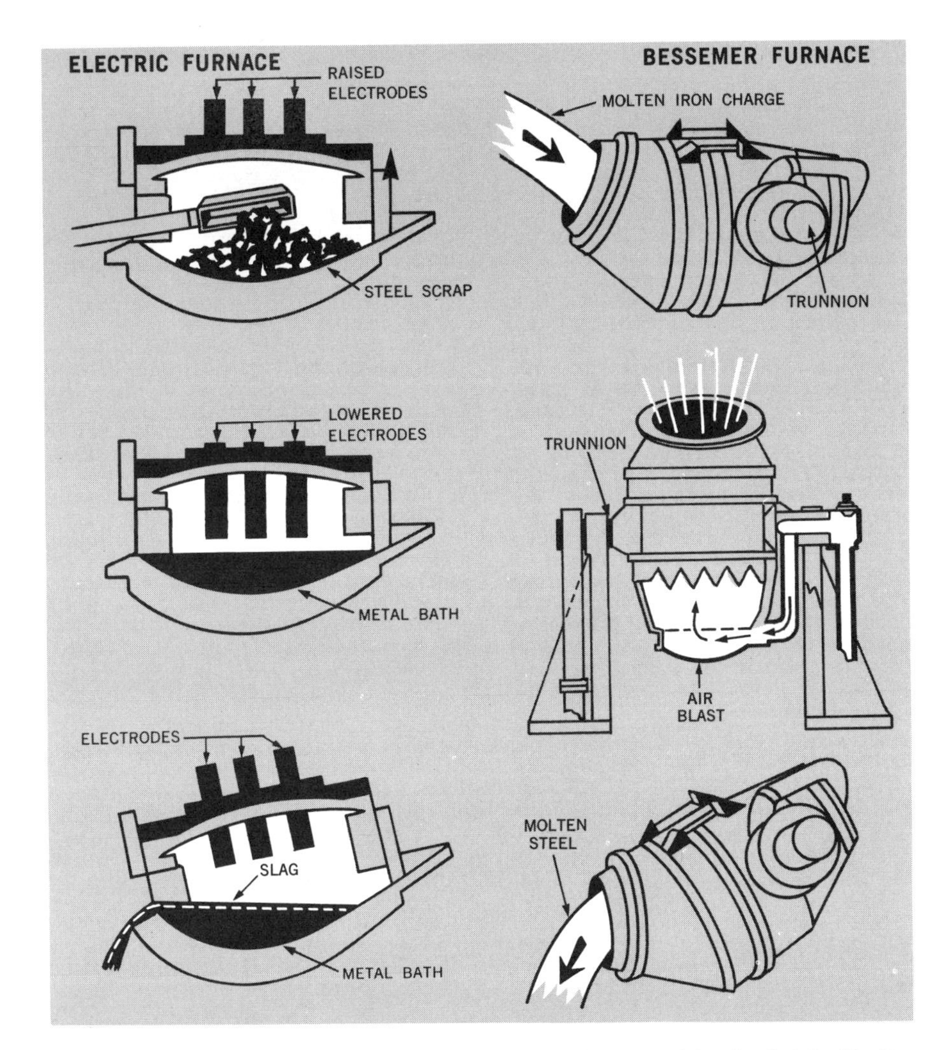

or surpass open-hearth output. In the United States basic oxygen production in the late 1960's was over 40 million tons, or 33 per cent of total U.S. steel production. Rapidly expanding capacity will soon make the basic oxygen process America's major steel producer.

■ VACUUM REFINING PROCESS.—This relatively new process does not manufacture steel but refines steel that is already made. Vacuum refining was developed in response to the insistent demands of the space age for steels of greater strength and heat resistance than are produced by present methods. Minute amounts of certain compounds or elements, not injurious to steel under ordinary conditions may have deleterious effects in steels destined for the severest service. The worst offender is hydrogen. Minute quantities of this gas dissolved in steel—4 to 8 parts per million—may later cause microcracks that can lead to failure of a vital part when subjected to great stresses, such as those resulting from high speeds at elevated temperatures. The residual hydrogen, as well as oxygen and nitrogen, can be reduced in liquid steel by exposing it, under certain conditions, to the suction action of a state of vacuum.

Vacuum refining of steel was long known but little used for lack of a compelling need for it. Originally the principle was applied to melting already highly refined steel within a vacuum. The constant state of vacuum, maintained by powerful pumps, sucks the gases from the melted steel. The need for removing gases from large masses of steel, particularly from ingots intended for forging generator shafts and turbine rotors, led to techniques for degassing steel fresh from the furnace. Some of the degassed ingots weigh as much as 250 tons.

Vacuum degassing, or *vacuum casting*, as the process is called, is being done on an increasing scale. Its main purpose is to reduce the hydrogen content of the steel, although nitrogen and oxygen are simultaneously removed. The operation begins with a ladle of molten steel, freshly tapped from an open-hearth or electric furnace. The ladle is placed over a tank in which a vacuum has been created. The molten steel is allowed to pour slowly from the ladle through the vacuum chamber into the ingot mold below. In its passage through the chamber, the liquid steel separates into droplets, exposing the maximum surface of the steel to the sucking action of the vacuum pumps, thus facilitating the withdrawal of the gases.

Steels melted or cast in a vacuum are "cleaner" and stronger than those produced in the atmosphere. They have improved mechanical properties at high temperatures, greater ductility, and a high degree of uniformity in quality. The electrical industry, for example, now specifies vacuum-cast steel for all critical applications.

■ CONTINUOUS CASTING.—This new process illustrates the trend in the steel industry to simplify processes, aided by computer systems. In a steel plant a long series of steps are taken, from smelting the iron to producing finished steel products. Moving the steel from one stage to the next considerably slows down progress through the mill. Now certain steps in the production of semifinished steel, such as blooms, slabs, and billets, are being eliminated through the process of continuous casting. These three forms are the basic stock from which most finished steel products are derived.

Blooms are chunky blocks of steel, which are rolled chiefly into rails, and structural shapes. *Slabs* are thick, flat sections, and are rolled into flat products such as plates and sheets. *Billets* are long and narrow, the stock for rolling bars, rods, pipes and tubes. All three are rolled from ingots, the first solid form in which most steel is made. From ingot to semi-finished steel, by conventional methods, involves five steps, which can be eliminated by continuous casting. Although the equipment varies somewhat, the basic principle remains the same.

In one type of machine for the production of billets, a ladle of molten steel, fresh from the furnace, is lifted to the top of a tall, vertical structure and placed in position. The elevation and vertical position are needed for the steel to flow downward. A hole in the bottom of the ladle is opened and the steel flows into a receptacle which serves as a reservoir. The steel passes from the reservoir into a long, vertical mold directly below it. The mold is made of copper and is water-cooled. On entering the mold, the steel is chilled, forming a hard outer shell, from ¼ to ½ inch thick. The outwardly hardened casting continues down through the mold, and as it emerges, it is gripped by rollers which pull it downward. Still held by the rollers, the casting is passed through water sprays which do most of the cooling. After the casting is completely solidified, it is cut into lengths of required size by oxyacetylene torches.

Continuous casting represents a major breakthrough in the shaping of steel. As in all simplified operations, it means a reduction in manpower. What formerly took many hours to perform can be done in less than one hour, at lower cost, and the steel is of higher quality than steel produced by the older method. In various countries, basic oxygen converters and continuous casting machines are being combined as a unit, and such combinations should increase rapidly in the future.

—Douglas Alan Fisher

Jigs and Fixtures, devices used to hold the work during machining operations. The basic function of a jig or fixture is to reduce *set-up time,* that is, the time a machinist needs to position and clamp the workpiece on the machine. The additional function of a jig is to guide the tools precisely during the machining so that each workpiece is identical. Fixtures, however, merely hold the part without guiding the tools. The cost of jigs and fixtures is offset by the savings made in labor and machine time.

Both jigs and fixtures usually have a base plate with hardened stops and buttons that locate the part precisely. Buttons are used to locate horizontal surfaces—three buttons are needed to define and determine a plane surface—while steel pins or stops locate vertical surfaces. Jigs are usually equipped with a hinged upper plate into which hardened steel drill bushings are inserted. The bushings guide the drills, reamers, and cutters used to machine the production parts.

Depending on the application and the number of parts to be machined, jigs and fixtures are constructed of cast iron or fabricated of steel weldments with hardened steel inserts at wear points. Aluminum and magnesium are also widely used because they can be easily machined and are lighter to handle. The aircraft industry, which frequently requires large jigs for wings or airframes, uses fiberglass reinforced with steel tubes as the base for the steel wear points and bushings. —Joseph J. Kelleher

Kiln, in general usage, any large chamber in which heat is produced for baking, drying, melting, or firing. In ceramic practice, however, a distinction is often drawn between a *furnace,* in which actual melting occurs (as in a glass furnace), and a *kiln,* which may reach very high temperatures but does not actually melt the ceramic body. All ceramics, except glasses, must be fired in a kiln so that physical and chemical reactions will occur and give the ware strength and hardness.

Kilns are constructed from insulating and refractory brick. They are heated by gas, fuel oil, or electricity. In *periodic kilns* the ware is set in place, brought to the maturing temperature, cooled, and withdrawn. This inefficient method required days or sometimes weeks to complete. Modern *continuous kilns* are usually tunnel-shaped. The ware rides through on metal and refractory cars, starting in a low-temperature zone, passing into a zone of maximum temperature, and finally entering a zone of decreasing temperature before it is removed. The tunnel kiln reduces firing time as much as 75 per cent and controls temperature changes more efficiently. As a result, it can provide a higher quality product at a much lower cost.

Many ceramics require two firings. The *bisque,* or *biscuit,* firing fixes the size and properties of the body. The *glost* firing is performed after a glaze has been applied and fuses this glaze permanently to the ware.
—Charles J. Phillips

Lock, a mechanical device for fastening doors and lids, protecting machinery, or controlling electrical contacts. It may be operated by key or by dial. The earliest known mechanical locks were made by the Egyptians about 2000 B.C. These locks, made of wood, contained pegs that dropped into holes in the lock bolt. Keys with pegs arranged in a pattern similar to the holes in the lock bolt were used to raise the pegs and move the bolt. The Greeks used a simple lock consisting of a notched bolt that was moved with a key the size and shape of a farmer's sickle. The Romans improved upon the Egyptian lock and later developed *warded locks* in which the mechanism was protected by fixed projections, or wards. In the Middle Ages, warded locks became more complicated and ornate, but they were not very secure.

During the Renaissance, the *lever-tumbler* appeared in some warded locks as an added safety measure. The lever dropped into one of two notches in the lock bolt, preventing the bolt from being moved in either direction. However, it was easy for a thief to raise the lever and operate the lock.

In 1778, Robert Barron developed a lock with several lever-tumblers. Each tumbler had a double action; there were projections above and below the notch into which the lock bolt slid, and if any tumbler was raised too high or not high enough, the lock could not be opened. Thus, a very accurate key was needed. A further improvement in this type of lock was made by Jeremiah Chubb in 1818. Chubb's lock utilized six regular levers and, in addition, had a special "detector" lever that indicated to the owner if the lock had been tampered with.

■**MODERN LOCKS.**—The *pin-tumbler cylinder lock* was invented by Linus Yale, Jr., in 1865. With modern improvements, it is still considered the most secure key-operated locking device. This lock consists of an inner plug and outer shell. The plug is connected to the lock bolt or latch, which moves when the plug is rotated. In the locked position, metal pins are pressed into holes in the plug by springs and other pins called drivers, preventing the plug from being turned. There may be from three to seven pins, and these pins are made in from six to ten lengths. Keys are cut with notches to correspond with the number and size of the pins. The correct key raises the pins until their top ends form a separation, or shear line, between the plug and the shell, allowing the plug to be rotated. The slightest variation will prevent the plug from rotating. The pin-tumbler cylinder permits an almost unlimited number of possible key changes. It is used for nearly all types of lock where high security is desired.

The *disk-tumbler lock* is similar to the pin-tumbler cylinder lock but is not as secure. Flat metal discs are used in place of pins, but because of its design and construction, fewer key changes are available in the disk-tumbler lock.

Combination locks provide very high security and are used on safes, bank vaults, lockers, and padlocks. The mechanism consists of three or more rings, each with a slot. The rings are connected to a dial and by dialing the correct series of positions, or *combination,* the slots are aligned, allowing the lock to be operated.
—Eaton Yale & Towne Inc.

Lumber Industry, the growth and processing of wood. *Lumber* is the product of the sawmill and the planing mill, and is not further manufactured except by resawing or planing. The unit of measurement of lumber in the United States and Canada is the *board foot,* which is represented by a piece of wood one foot long, one foot wide, and one inch thick. Originating in the United States lumber industry, the board foot measure is used only in North America and the Philippines. Elsewhere in the world, the common units of measurement are the *cubic meter,* the *cubic foot,* and the *standard* (or *Petersburg standard*), which is equal to 1,980 board feet.

Timber is a term loosely applied to standing trees or certain products made from them, such as large pieces used in construction of buildings, bridges, and ships.

The lumber and timber most important in construction and other industrial uses are *softwoods,* such as pine, spruce, hemlock, cedar, cypress, fir, and redwood. The term *softwood* bears no relation to the hardness of the wood; it refers to the coniferous,

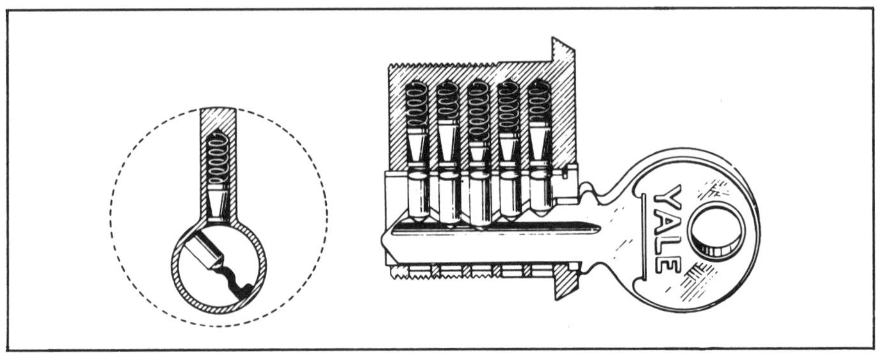

EATON YALE & TOWNE INC.

CYLINDER LOCK will turn only when the five steel pins are raised to a position in which the division between the two halves coincides with rim of cylinder. Then (*left*), it opens.

WEYERHAEUSER COMPANY

LUMBER is "engineered"-molded (*right*) into arches and trusses that are capable of supporting heavy roof loads. At *left*, a lathe "peels" a sheet of veneer off of specially selected log sections. The veneers are then cut to desired size and bonded together to produce plywood.

or needle-bearing trees. The *hardwoods* include the broadleaf trees that usually shed their foliage in the fall; they are used principally in fine veneers, furniture, flooring, and interior paneling in houses. Oak, birch, gum, maple, mahogany, and poplar are among the hardwoods.

■**LUMBER PRODUCTION.**—The United States and Canada together account for 33 per cent of the world's lumber production. The Soviet Union and its satellites account for 32 per cent, while western European countries produce 20 per cent. In 1959, the chief lumber-producing countries of Europe were France, Sweden, West

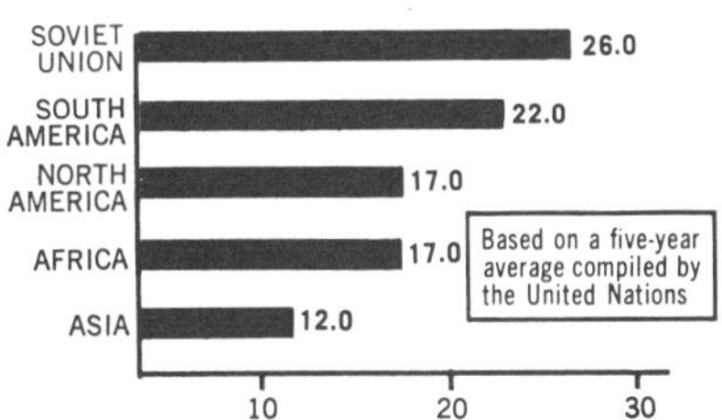

Germany, Poland, and Finland, in that order. Peak lumber production in the United States was reached in 1909, when U.S. mills manufactured 45 billion board feet. Current annual production is approximately 35 billion board feet.

The United States lumber industry originated in the northeastern states about 1631, and then shifted to the mid-Atlantic states about 1860. From there it moved to the Great Lakes states, and, near the end of the century, to the south. After 1920, it moved to the western states, which lead in production today. *Second-growth,* timber that has grown in the regions first cut over, is reviving the lumber industry in the northern and southern states of the eastern United States.

■**FORESTS AND FORESTRY.**—The United States has 758,865,000 acres of forest land, which is about one-third of the total land area of the 50 states. About 509 million of these acres are classed as *commercial forest land,* land capable of growing repeated crops of commercial timber and available for that purpose. The other 250 million acres are made up of unproductive land, as well as productive forest land withdrawn from commercial use for parks, game preserves, and the like.

Forestry, the scientific management of forests for continuous production of high-quality trees, is further advanced in Europe than anywhere else in the world. In Germany, France, and the Scandinavian countries, forestry has been practiced for centuries.

Economic conditions have combined with effective nationwide forest-fire prevention and control measures to render the practice of forestry attractive to private owners in the United States only within the past two decades. Generally the government-owned forest lands are well managed, as are the industrial ownerships, while farmers and other small landowners have been slow to adopt forestry measures.

An important development in United States forestry was the inauguration in 1941 of the *American tree farm system* by the forest products industry. The tree farm program gives public recognition to forest owners who dedicate their woodlands to the growing of repeated crops of trees for commercial use. Since the tree farm system was started, more than 32,800 tree farms containing nearly 72 million acres of well-managed, tax-paying private woodlands have been certified.

—Charles A. Gillett

Machine Tools, power-driven devices that cut or shape metal and other materials. Machine tools are essential to the manufacture of a host of products ranging from safety pins to automobiles, from watches to elevators. In the United States, for example, there are more than three million machine tools operating in thousands of factories. About 75 per cent of these are for metal cutting and 25 per cent are for metal forming.

Metal-cutting machines function in different ways. *Lathes* rotate the work, while cutting tools, which are stationary except for a feeding motion, cut the metal away in the form of chips. Lathes come in many different sizes and variations of design and function to produce parts that are shaped like shafts, cups, cones, or disks. When the cut is made on the outside of the part, the machining operation is called *turning;* when it is made on the inside, it is called *boring.*

Metal-forming machines consist principally of a variety of presses for shaping metal, but also include machines for bending, twisting, rolling, hammering, punching, and shearing.

Milling machines rotate the tool, a cutter that usually has a number of cutting points or teeth. The work is mounted on a reciprocating table and fed past this rotating cutter. By milling, it is easy to machine flat surfaces. Controlling the cutter position with a template and trace makes it possible to mill intricately shaped surfaces of dies and molds.

Planers cut flat surfaces by moving the work past a fixed tool. A succession of strokes can machine a large, flat surface. The same operation, on a smaller scale, with the tool moving instead of the work, is done in a machine called a *shaper.* If the planing or shaping operation is done with a tool that has many teeth, each making a small additional bite into the work so that the entire operation is performed in one stroke of the cutter, the tool is called a *broach;* the machine itself, which must have extreme rigidity and power, is called a *broaching machine.*

Drilling machines cut holes in solid material by feeding a rotating drill into the work. Drilling machines range from simple single-spindle machines, in which the drill is rotated by power but fed by hand, to complex automatic machines which drill hundreds of holes simultaneously in a prescribed pattern.

Grinding machines use an abrasive wheel, disk, or belt to cut material from the surface of the metal. Various types of grinding machines are built to duplicate all the actions of other types of cutting machines. The abrasive wheel, however, must always be moving at high speed. In cylindrical grinding, for example, the work rotates just as it does in a lathe, but the grinding wheel also rotates at the same time that it is fed across the work. Grinding can produce extremely fine, mirrorlike finishes. It can also make deep, rough cuts which are described as *abrasive machining.*

Other types of metal-cutting machine tools are built for special purposes such as sawing, gear cutting, honing, lapping, and polishing.

Some machine tools are loaded and controlled manually by a skilled machinist. These are often used in the manufacture of tools for other machines, and are called *toolroom machines.* Production machines often have automatic cycles, but may still be loaded and unloaded by hand. The operator may also make adjustments to correct for tool wear, changes in the material, or other conditions.

A number of successive operations can be performed by a line of machines working in tandem. These can be separate machines linked together by handling equipment, or they can be a single *transfer machine* in which the different stations are built around a rotating table or along a transfer mechanism that takes the work from station to station.

Some transfer machines perform just two related operations, such as drilling a hole and then cutting threads in it. Others perform hundreds of operations on a complex part, such as an automobile engine block or an electric motor housing.

Conventional automatic machines are generally inflexible—they will only make one part or a few closely related parts. A transfer machine for an engine block may have to be scrapped when the particular design it makes is no longer in demand.

In recent years, a new kind of automatic control has been developed that is more flexible. Called *numerical control,* or *tape control,* it performs a sequence of operations in response to instructions fed to the machine in numerical form, usually by punched or magnetic tape, sometimes directly from a computer. With these machines, automatic operation can be used in the production of small quantities of parts.

—Anderson Ashburn

Material Handling, the science of moving, packaging, and storing substances in any form. Frederick Winslow Taylor, the pioneer industrial engineer, called it "the blood stream of the factory." Material handling, in civilian terms, corresponds roughly to the logistics of military science. The techniques of modern material handling have replaced the "hernia methods" of lifting, carrying, and placing.

By proper handling, the smallest particle of a bulk material (such as coal) up to large heavy units (such as machine tools) can be moved in the most economical manner. It is an engineering science governed by axioms, laws, and economics—just like any other engineering specialty. Inasmuch as the packages and containers in which articles are shipped have an important bearing on the manner of handling before and after shipment, packaging occupies a large place in material-handling science. Material-handling engineering aims at the development of integrated methods, which move raw materials from the receiving department right through the processing plant to the boxcar, ship, airplane, or truck in the most economical manner.

Material handling offers important opportunities to reduce costs, increase production, save time and materials, reduce losses from breakage and pilferage, reduce injuries to personnel, and increase the utilization of cubic space. It has been called industry's chief area of unexplored cost savings.

■**EQUIPMENT.**—The *fork-lift truck* is the most familiar material-handling device, but it is not the only one. A wide variety of equipment is used in modern material-handling systems, notably the various *industrial tractors,* used to pull long trains of *trailers. Powered hand trucks* come under this general heading. Simple, manually-operated *floor trucks* on wheels or casters are also used to great advantage. An important auxiliary to fork-lift trucks is the *pallet,* which may be of wood, aluminum, steel, magnesium, plastic, cardboard, wood and steel, or wire. There is a trend toward use of unit loads, whereby a pallet is loaded with a given number and arrangement of packages at the manufacturer's plant, secured to the pallet with steel strapping, and delivered to its destination as a unit.

Conveyors of all types are important material-handling tools. The *rubber conveyor belt* is the wheel horse of the bulk-handling industries. The *roller skate wheel conveyor* is available in a variety of lengths that can be joined together. In the field of vertical material handling, the *electric hoist* finds many uses. The *overhead traveling bridge crane* has long been a fixture in plants where heavy lifts are common. Although there is a trend toward the single-story plant, *electric and hydraulic elevators* form an important part of the material-handling system in multistoried plants.

■**SYSTEMS.**—Material-handling systems are not new. A specialized mechanical-handling system was used to unload bags of cement from ships during the construction of the Panama Canal in 1912. Other specialized systems have been in existence for a long time. The industrial truck had its start early in the twentieth century when an electric motor powered by batteries was bolted onto an ordinary baggage carrier in a railroad terminal. Subsequent refinements have led today to the fork-lift truck capable of handling almost anything with one of many special attachments. It can be said that material handling only began to come into its own during World War II.

It is the opinion of specialists in this field that great opportunities lie ahead. Already promising development work is being done to apply atomic energy to material handling. Automatic warehouses are being developed in which a key would be punched in a central office to release a part from a bin to fall by gravity onto a conveyor belt, which would take it with other items on an order to a central packaging point.

These are examples from the most progressive segments of industry. There is still much time and money to be saved by the application of standard equipment used more conventionally. —Eaton Yale & Towne Inc.

CLARK EQUIPMENT COMPANY

FORK-LIFT TRUCK with triple-stage upright that extends on rollers from 12¾ feet to 15¾ feet and can lift up to 4,000 pounds.

Metal Coating, the art of coating a metallic surface with a layer of another metal, primarily to protect it from oxidation and corrosion. The coating metals are chosen from among those that are electropositive with respect to the basis metal.

A number of processes have been in use for many years to coat iron, steel sheet for roofing and other outdoor uses, and for the production of tinplate used in canning.

The basic process, still most often used, is *hot-dip galvanizing,* in which the iron or steel sheet is coated with a layer of zinc. In this process, the sheets are thoroughly cleaned by pickling, then washed and passed through sal ammoniac on their way to the pot of molten zinc. The process is continous, with the zinc-coated sheets passing through rollers as they come out of the zinc bath.

An alternative method is electroplating which is based on electrolytic action. In this process the steel is used as the negative electrode (cathode) and the zinc as the positive electrode (anode). An electric current is then sent from the anode to the cathode through a solution of a zinc salt (electrolyte). This causes the zinc to deposit on the negative pole or steel. The main advantage is that the steel is not heated, and thus does not distort. Limitations are the slow rate of coating and the relative sponginess of the coating obtained.

Two processes are used to deposit a layer of tin on sheet steel. Both are *hot-dip* processes, similar to that used in hot-dip galvanizing. In one process, the sheet usually passes through three tin baths, each purer than the one preceding it. In the other, the sheet is first passed through zinc chloride, then through the molten tin. In both processes, the coated sheet is finally passed through rollers that govern the depth of the deposited layer.

Another heat process for coating steel sheet with zinc is known as *sheradizing.* In this process, the zinc is not molten but is in the form of a fine powder, which is deposited on the sheet. The sheet is then heated until the powder melts to form the zinc coat.

A newer class of processes coats the basis metal of finished parts with a sacrificial layer of zinc, tin, lead, or cadmium, using a mechanical process in which small flakes of coating metal are hammered onto the surfaces of the parts in a tumbling barrel or in a vibratory finishing tub.

Of increasing commercial importance is vacuum metallizing, used to deposit an extremely thin layer of metal on metallic or other surfaces, including architectural glass.

Finally, a number of processes have been developed to coat surfaces by a system of electrical discharges. The basic process involves the passing of an arc from a coating material to a part to be coated. These processes, of which the most widely known is that used to coat razor blades, involve high localized temperatures, and permit coating metallic and nonmetallic parts with any metal, including tungsten. —Felix Giordano

Metal Shaping, any of various processes used to change the dimensions of a metal without chip removal. These processes include casting, forging, stamping, rolling, extruding, spinning, and drawing.

CASTING.—Casting, in which molten metal is poured into a mold and allowed to cool and solidify, is one of the oldest forms of metal shaping. Since unusual shapes with a high degree of complexity can be duplicated by this method, a wide variety of casting processes have been devised.

Sand casting, in which the mold is a cavity in compacted sand, has been used to form metal objects for more than 4,000 years. The first step in sand casting is the creation of a wood or metal pattern in the shape of the part to be produced. The second step is to pack sand around the pattern and then to remove the pattern, leaving the mold. Pouring molten metal into the cavity left by removing the pattern is the third step. After the metal has cooled and solidified, the sand is broken away from the casting. Practically any metal can be cast in sand molds. The principal limitations of sand casting are a rough surface finish, a minimum section thickness of ⅛ inch, and minimum tolerances of ± 1/16 inch.

Shell molding, a variant of sand casting, uses a mixture of sand and resin as the mold medium. The sand-resin mixes are formed over a polished metal pattern and hardened in an oven. Castings produced in shell molds have very smooth surfaces and can reproduce fine detail. At a cost little more than for sand castings, shell molded parts can be produced with tolerances as small as ± 0.004 inch and with sections as thin as 1/16 inch.

Permanent mold castings are produced in reusable molds of refractory material for production runs in substantial volume. The molds are made in several parts to allow removal of the finished casting. The complexity of parts cast in permanent molds is limited, since the molds must be easily removable. Because of the chilling effect of the molds, permanent-mold cast parts have excellent mechanical properties. Hollow castings, called *slush castings,* can be made in permanent molds by pouring off liquid metal after a shell has formed on the mold surface. Permanent-mold cast parts have excellent sand castings in finish and density, can be held to closer tolerances (± 0.025 inch), and have a finer metal grain structure. While both ferrous and nonferrous parts are made by permanent mold casting, aluminum and magnesium account for the highest volume.

Die casting involves the injection of molten metal into closed steel dies clamped together under pressure. The dies are mounted in a machine that incorporates a heated reservior for the metal, hydraulic rams to open and close the dies, and an injection system to feed the molten metal under pressure. In operation, the dies are closed and clamped by the hydraulic cylinder, molten metal is injected, and pressure is retained until the castings cool. The dies are then opened by the hydraulic cylinder, and the castings are ejected by pins in the dies. Machines capable of up to 100 shots (cycles per hour) are available, depending on the size of the part.

Although the surface finish of die-cast parts is excellent and rarely requires finishing operations, application of the process is limited by the problem of removing the part from the die. Complex multipart dies provide some design flexibility, but such dies are expensive, substantially increasing the die cost per part. Although usually applied to small parts, die casting is now being used for parts as large as automobile engine blocks weighing 80 pounds. Machines are available that can cast zinc parts weighing as much as 200 pounds.

Primary benefits to be derived from die casting are the close tolerances that can be held and the high production rates. With ordinary zinc castings, tolerances can be held to within ± 0.001 inch per inch of part length. Closer tolerances can be held if required, but die costs will then increase. Zinc, aluminum, magnesium, copper-base alloys (brass), lead, and tin are the metals most commonly used for die casting.

Centrifugal casting takes place in a rapidly rotating mold, relying on centrifugal force to produce a part with greater accuracy and improved physical properties. Symmetrically shaped parts, such as railroad car wheels, engine cylinder liners, sleeves, gear blanks, and pipe, are ideally suited to centrifugal casting. The mold consists of a metal flask, lined with sand or a refractory material, that resists the centrifugal forces. To cast pipe, the flask and its liner are rotated around its horizontal axis while a measured amount of molten metal is poured into one end. Centrifugal force holds the molten metal against the mold until the metal solidifies. A critical factor in centrifugal casting is selection of the correct spinning speed. Excessive speed produces highly stressed parts that will fracture easily, while slow speeds prevent the metal from adhering to the inside of the mold. Though any metal can be centrifugally cast, the process is most widely used to produce cast-iron pipe and large plumbing fixtures.

Lost wax, or *investment mold, casting* is a precision casting process widely used to produce parts from alloys that are difficult to machine. The process was developed 3,500 years ago, but it was not until the Renaissance that it was rediscovered by Italian metalworkers. Investment molding starts with a wax duplicate of the part to be cast. The wax pattern is placed in a box or flask which is filled with a liquefied refractory plaster. The mold is then baked in an oven to melt out the wax and harden the refractory material. Liquid metal is poured into the cavity left by the wax and allowed to cool. The mold is then broken away to free the castings. Because lost-wax cast parts can be used without machining,

UNITED STATES STEEL CORPORATION

FORD MOTOR COMPANY

ROLLED STEEL (*left*) is tempered as it passes through this machine before being plated with tin. Presses (*above*) shape flat steel sheets for use in the production of automobiles.

the process is widely used to cast the so-called superalloys. These alloys are used for gas-turbine blades, metal-cutting tools, extrusion dies, and pump impellers handling abrasive compound.

FORGING.—Forging, as old a metal-shaping process as casting, consists of heating the metal and hammering it by hand or machine to the desired configuration. Flat dies are used to work heated billets to approximate shape; closed impression dies are used to produce a given shape for a specific part. In either case, forging has the advantage of compacting the material to improve its grain structure and mechanical properties. Steel forgings are widely used for automobile and truck axles, engine crankshafts, connecting rods, and gear blanks.

Drop forging is performed by either a steam or board hammer. Steam hammers are lifted and driven downward by steam pressure; board hammers rely entirely on gravity to deliver the hammer blow. Board hammers are built up to 4-ton capacity, while steam hammers are capable of delivering 25-ton blows. One advantage of the steam hammer is speed, for it can deliver 300 or more blows per minute. Dies for the part being forged are mounted in dovetails cut in the press's hammer and anvil.

Several progressive dies are usually required to forge a part to its final shape. Separate trimming presses are used to remove the *flash* (excess metal extruded around the parting line of the dies as they close) from the forging. For close tolerance, a high-pressure coining process is used to bring the finished forging to exact dimensions.

Forgings are produced in sizes ranging from a few ounces to several tons. Because they can be worked easily, straight carbon steels are the most commonly forged metal. Brass, bronze, aluminum, magnesium, and titanium are also formed by forging to improve their mechanical strength. Alloy steels, too, are forged, but they require additional skill and the design and use of special dies.

Upset forging, press forging, and cold-headed parts are variations of the basic forging operation, and each has its advantages. *Upset forging* uses bars of metal that are gripped between stationary jaws and impacted by the moving die. Common products of the upset-forging process are bolt heads, rivets, and other fasteners. *Press forging*—either hot or cold—starts with a slug of steel that is squeezed to make the metal flow into all recesses in the die. Progressive dies are usually used, mounted in either hydraulic or mechanical presses. *Cold heading* is similar to upset forging, but stock is fed from reels of wire to fully automatic machines. Production rates for smaller cold-headed parts can reach as high as 450 parts per minute.

ROLLING.—By passing the metal between two rolls, which may be either plain or grooved, a wide variety of shapes—such as bars, plates, rods, sheets, slabs, and strips—can be produced. The rolls rotate at the same speed but in opposite directions. As the metal passes between them, it is reduced in thickness and increased in length. (The slight increase in width may be disregarded.) Rolling at atmospheric temperatures is called *cold rolling,* while rolling above the critical temperature of the metal is called *hot rolling.*

COLD PRESSING.—Metal powder parts are made by compacting fine metal powders in a press, then sintering or oven-brazing the compact. The advantage of the process is that unusual metallurgical combinations can be obtained and the density of the part closely controlled. Porous parts can be produced by mixing in materials that dissolve when the compact is sintered. This process is used to produce extremely fine filters and porous bushings that retain lubricants, often used in sealed electric motors.

PRESS FORMING.—Press forming involves a range of operations that include blanking, pressing, stamping, and drawing. Many different types of dies are used for press forming, depending on the metal, the thickness of the part, and the quantity of parts required. For long production runs, such as automobile body panel, matched steel dies are used; but for limited production runs, dies of softer metals, plastics, and even wood are satisfactory. When the softer materials are employed for shallow draws, a rubber pad or blanket replaces the upper die. This is common practice in the aircraft industry where short runs of thin, soft aluminum are required.

Blanking involves deformation of the metal beyond the shear point to produce a clean, sharp break at right angles to the surface. *Pressing, stamping,* and *drawing* operations deform the metal beyond its plastic limit to produce a permanent set. For deep drawing operations, progressive dies are required, and the stock is annealed between draws. Nearly all ferrous and nonferrous alloys can be stamped, but springy materials are avoided.

EXTRUSION.—Extrusion is a process similar to the squeezing of toothpaste from a tube, with heated metal instead of the paste. Most extrusion machines rely primarily on pressure to generate the heat needed although the billet is frequently preheated before insertion in the press. The press consists of a cylinder with a hardened die at one end and a hydraulic ram at the other. The ram applies tremendous pressure on the billet to raise the metal's temperature. Under pressure, the heated metal becomes plastic, flowing through the shaped dies in a continuous length. The extruded metal is cut to length and machined, forged, or otherwise processed to create finished parts.

Formerly, extrusion was limited to lead, copper alloys, aluminum, magnesium, zinc, and other soft metals, but recent advances have made it possible to extrude steels and other hard or high-temperature metals. Although the steels extrude easily, die life is short and special lubricants are required. For some steels, powdered glass is used as a lubricant.

OTHER METHODS.—Although the most common methods of shaping metals have been described, hundreds of other processes are used by industry. Many of these operations involve highly specialized equipment that cannot be adapted to the general metal-shaping field. Among the more exotic procedures that show promise of wider application are explosive forming, high-energy-rate forming, and magnetic forming. *Explosive forming* shapes thick sections or ultrahard alloys on a single die, which is submerged in water and subjected to extreme hydraulic pressure by an explosion. *High-energy-rate forming,* with closed dies, employs a press driven by the sudden release of compressed gas. The energy applied to the metal slug in a few milliseconds induces plastic flow that results in close tolerance parts with improved grain flow. *Magnetic forming* relies on a burst of electromagnetic energy to compress or expand tubular sheet-metal parts on a mandrel or die. This process is limited to ferrous metals or other magnetic alloys.

—Joseph J. Kelleher

Methods Engineering, the branch of engineering that deals with the minimization of the cost of labor involved in performing repetitive jobs in factories or offices by the improvement of work methods. Each operation required to finish a given job or piece of work is analyzed, and unnecessary operations are eliminated. Methods, working conditions, and equipment are standardized, and the operator is trained to work according to a standard pattern. Once the job has been standardized, the time needed for an operator to complete the job is measured. Finally, a pay schedule is computed that provides an incentive for the worker to achieve or exceed the standard measure of performance.

Before any *methods study* is undertaken, the cost of making the study must be determined and balanced against the potential savings. For new jobs, a method is devised to accomplish the work, and then alternative ways of accomplishing the job are sought and evaluated. Motion studies—for which personal observation or motion pictures are employed—are then made if the job is repeated frequently, and the cost can be justified.

Once the work method has been refined as far as possible and standardized, the job is measured by *time study.* Using a stop watch, the engineer observes a typical operator, timing each movement. All the times are added, and allowances are made for stock handling, unavoidable delays, and personal needs to set a standard time for the specific job. These time standards are used to determine compensation or the work volume to be accomplished during a specific time interval.

—Joseph J. Kelleher

Microscope, an optical device that enlarges the image of small objects. The simplest microscope, therefore, is the ordinary magnifying glass. Normally, people see small objects best when they are held about 10 inches from the eye. At 8 or 9 inches the objects appear fuzzy, since the human eye cannot focus at the shorter distance. A microscope brings the object image closer to the viewer's eye while keeping it in focus. *Simple microscopes* consist of a single lens, like a magnifying glass, or several lenses which act as a single positive lens.

A *compound microscope* has two lens systems, each acting as a single positive lens. Compound microscopes have their *objective lens system* mounted at the bottom of a tube and an *eyepiece lens system* at the top. The magnification of a compound microscope is the product of the magnification of the objective lens system and the eyepiece. With an eyepiece magnification of 5 and an objective magnification of 25 the final magnification is 125, usually expressed as 125X.

The *resolving power* of a microscope is the limit of its useful magnification. This defines a microscope's ability to make very small details clearly visible. If a microscope magnifies an object beyond its ability to resolve detail, the image in the eyepiece becomes blurred. Additional magnification will increase the size of the image, but the image also becomes increasingly blurred.

Binocular microscopes have two eyepieces so that the viewer can use both eyes to observe the image. As a result, the operator has depth perception and eye strain is reduced. The image from the single objective lens is split, by a prism system, and divided between the two eyepieces like a pair of binoculars.

Proper illumination is essential for good viewing of highly magnified objects. When the object being viewed is transparent, illumination is provided by an adjustable plane mirror on the base of the microscope which reflects light up through the specimen. A plane mirror is used, with a small light source, to provide parallel rays in the illuminating beam. For intense illumination, a concave mirror or a system of condensing lenses provides a concentrated cone of light. Opaque objects require illumination from above. This is usually accomplished by a light source built into the microscope tube and focused on the specimen through the objective lens system.

Though simple magnifying lenses were used in China and the East before the Christian Era, it was not until about 1600 that a Dutch spectacle maker, Zacharias Janssen, produced a compound microscope that was 6 feet long and contained two lenses. Though the compound microscope had numerous advantages, the baffling problem of optical aberration which distorted the image and created colored fringes restricted development. Anton van Leeuwenhoek, the man who is credited with developing microscopy into a science, used a simple microscope for his studies. The problem of aberration was not fully solved until about 1850 when Charles Chevalier developed an achromatic lens eliminating the color fringe. (See also *Electron Microscope*, page 449.) —Joseph J. Kelleher

Mining, the process or business of extracting geologic materials from the earth's crust. All mining, whether it be by open-pit methods or underground, consists of three operations: breaking the ore, loading the ore, and transporting the ore to the treatment plant.

Whether a mineral deposit is mined by surface or underground methods (or both) depends on a number of factors. Among them are the proximity of the deposit to the surface, the unit value of the minerals to be recovered, the physical characteristics of the material to be mined, and the safety of personnel engaged in mining the deposit.

Surface mining entails the removal of all material overlying the ore body, plus the removal of enough more of this *overburden* to make an opening with sides sloping at an angle flat enough so there will be no sliding of the material from the walls into the pit. Thus, with the safety of the workmen in mind, surface mining requires the ore body to be stripped of its overburden and then be removed.

Overburden does not always have to be broken. When it does, depending on its consistency, it can be broken with a *ripper* or drilled and blasted. A ripper works like an oversized plow, pulled by a powerful tractor. If the overburden requires drilling and blasting, a churn drill, a percussion drill, or a rotary drill, may be used.

The *churn drill* is a heavy blunt drill that is raised and allowed to fall at the end of a wire rope, very much as the dasher moves in an old-fashioned churn. In *percussion drilling,* the sharp end of a steel drill is held against the bottom of the hole, and a sharp blow struck against the other end of the drill rod. The energy is transmitted through the rod to the point, which breaks off a chip of the material being drilled. The drill is raised, turned, and the process repeated. In *rotary drilling,* the drill bit armed with hard material, frequently diamond, is pressed against the bottom of the hole and twisted mechanically. As it turns, the material at the bottom of the hole is ground off.

With all three types, the ground or broken material from the bottom of the hole is removed to avoid wasting energy. It may be washed out with water or blown out with air. The holes drilled will vary from 3 inches to 3 feet in diameter. They may be vertical or at an angle, and anywhere from 3 feet to more than 100 feet long.

When a round of holes has been drilled, they are loaded with explosives and filled with stemming. Holes are drilled in a regular pattern calculated to distribute the explosive throughout the rock mass so the energy liberated when the explosive is detonated will be most effective. The object of drilling and blasting is to break the material into sizes easily handled by the loading and hauling equipment. At the same time, it is not efficient to break it into pieces smaller than necessary.

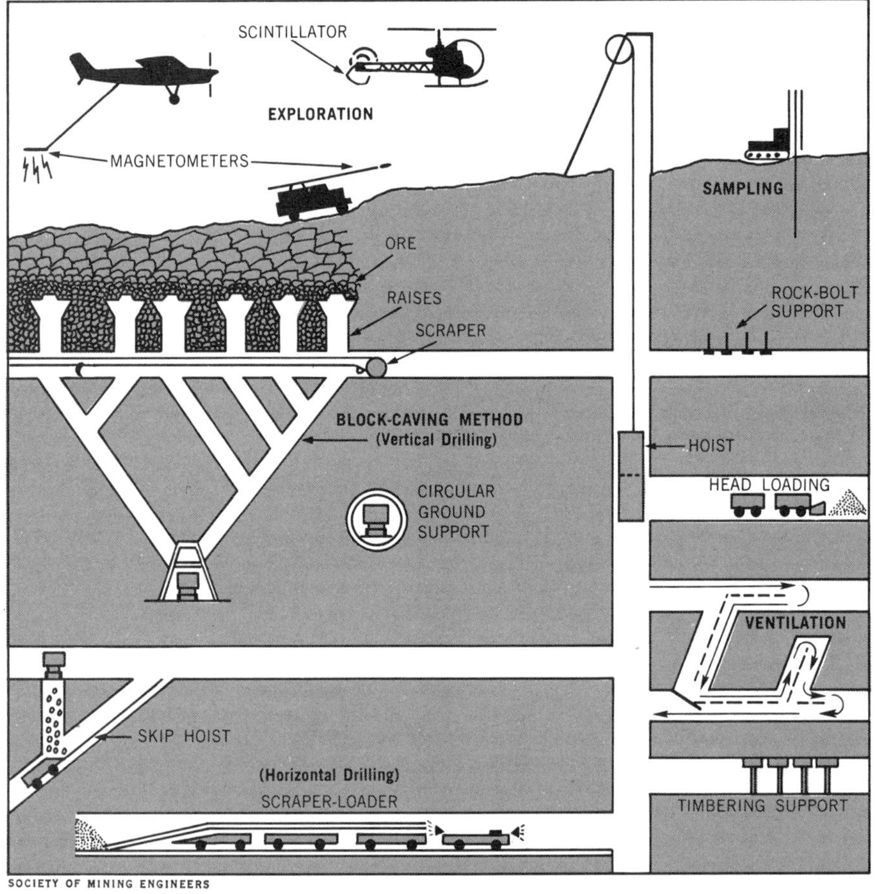

SOCIETY OF MINING ENGINEERS

MINING begins with geophysical exploration. Measuring instruments such as *magnetometers*, recording distortions in the earth's magnetic field that indicate possible ore deposits, and *scintillators*, reacting in the presence of radioactivity, are carried over an area by aircraft. Following the aerial survey, ground crews pinpoint deposits with similar devices and drill or blast holes for samples. If further search locates rich deposits, shafts are lowered and supported by beams and rock bolts, hoists are rigged for carrying men and ore, and a ventilating system is opened. This schematic drawing shows the *block-caving* method of mining, used primarily to obtain copper and molybdenum. Vertical "raises" and horizontal drilling knock the support away from ore-rich strata, and scrapers push the fallen ore through chutes or onto conveyor belts, then into mine cars. It is then transported for processing. A mine of the scale illustrated here may take ten years, from surveying to removing the first ore, to be put into operation.

Explosives used are dynamites of various strengths. Mixtures of ammonium nitrate and fuel oil, commonly called *ANFO,* have displaced the more expensive nitroglycerine-base dynamites to a large extent. Nuclear explosives have not yet been used in stripping operations, but the time may come when they are.

After the overburden has been broken, it is loaded into some means of transport. Loading is accomplished by several types of equipment. Mechanical shovels powered by electricity are most common. Some of these giant shovels with buckets of 200-ton capacity can pick up this load a third of a city block from the operator and deposit it on a spoil pile a block away in the other direction. In one day they use enough electric current to supply a city of 30,000 people. Huge draglines that sit on the bank above the cut can reach farther than the shovel and discharge farther away. However, bucket capacity of the dragline is less than that of a shovel of equal size. Where the character of the overburden permits, bucket-wheel excavators are coming into increased use. These also are gigantic, with a series of 20-cubic-yard buckets mounted on the periphery of a suitable wheel, discharging their loads onto a 6-foot-wide endless belt. The belt in turn discharges onto a stacker or conveyor belt, which may carry the waste material a mile or more before discharging it on the spoil pile.

Transport, as has been mentioned, is sometimes provided by conveyor belts. With the big shovels a more common type of transport is the off-highway truck. Again, these may be tremendous in size. Capacities of 70, 100, and more cubic yards are not uncommon. In many cases, transportation away from the shovel is accomplished in railroad dumpcars of 50- to 100-ton capacity.

Given proper conditions, *stripping* can be accomplished with bowl scrapers. These self-loading transports are used on hauls up to 5 miles when the material to be moved is uniform, relatively soft, and easily dug. Capacities run up to 36 cubic yards. Once the overburden has been removed, mining of the mineralized, valuable portion of the deposit begins. This follows the same general pattern as that of overburden removal, but the equipment is usually not so big. The pit is carried down in a series of benches that may range from 30 to 100 feet in height. The overall slope is maintained at an angle that will preclude slides. Usually, the ore material is more difficult to break than the overburden. Hence, a greater number of small holes are drilled closer together, loaded, and blasted to break a given amount of material. Transportation is by railroad cars, trucks, or conveyor belts. Every mine presents a different problem so the combinations of equipment used to load and haul the ore from its place in the earth to the concentration plant are almost infinite.

Unlike surface mining, *underground mining* does not require the removal of all material overlying the ore body. It does demand its support. To reach the ore body entails the sinking of one or more shafts, the driving of many horizontal openings and secondary vertical openings in or under the ore body, and the support of all the overlying material while the ore is being removed. All this must be carried out with strict attention to safety while the work is going on.

In underground mining, also, the cycle of breaking, loading, and transporting is basic. As in surface mining, the combinations of men, methods, and machinery used are almost infinite. In the simple, unmechanized *coyote-hole mine,* one or two men using hand drill, sledgehammer, and wheelbarrow, mine small amounts of material. In large mechanized mines, as much as 50,000 tons of material per day may be produced by hundreds of men using many complicated machines to break the ore, load it on trucks, rail cars, or conveyors, transport it to a central shaft, and there hoist it to the surface and the ore treatment plant. Mining methods may be classified on the basis of the amount and kind of support provided for overlying material.

Open stoping denotes large openings from which the ore is removed completely, with little or no provision for support. *Room and pillar* mining is characterized by parallel openings driven off a tunnel with pillars of ore left in between. When the limits of the ore body have been reached, the pillars are mined *on the retreat,* allowing the mined-out portion of the mine to cave in behind the work. In the *block-caving* method, blocks of the ore are undermined and allowed to cave. The broken material is then drawn off into cars below the caved areas and hauled away.

Shrinkage stoping is room-and-pillar mining adapted to steeply inclined ore bodies. Mining begins at the bottom of the *room,* which in this case is called a *stope.* The ore is broken overhead from the top of the stope and enough drawn out to leave room in which miners can work. Then, standing on top of the broken ore, the miners take out another cut from the ore over their heads, and again enough is drawn out from the bottom to leave room for the miners to work on the next cut. This is continued until the next level above is reached. At that time all the remaining ore in the stope is drawn off and, depending on the character of the material overlying the ore body, the empty stope is filled or allowed to stand open. When a series of parallel stopes with pillars between has been completed and the empty stopes filled, work begins on recovering the pillars.

In *cut-and-fill stoping,* work begins as it does for shrinkage stoping, but all the ore is drawn out and replaced by filling material. The miners then stand on top of the filling material to take the next cut. When all the ore is drawn off, another layer of fill is introduced, and the process is repeated. —John C. Fox

Ore Treatment, the step in which run-of-mine ore is converted into a concentrate that is a salable product. In general, this means that the valuable constituent in run-of-mine ore delivered to the plant must be separated from the nonvaluable component. Ore treatment usually entails coarse crushing, fine grinding, and separation. Nature seldom, if ever, produces two mineral deposits that are exactly the same.

In coarse crushing, large chunks of material coming from the mine are broken into sizes more convenient to handle in the fine grinding machines. In the fine grinding circuit of the mineral treatment plant, the material is reduced to the grain size of the valuable material. Ideally, fine grinding proceeds only to the point necessary to liberate the valuable mineral from the waste material.

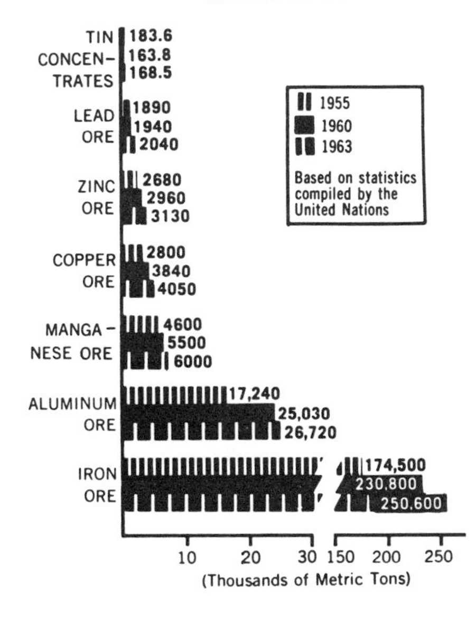

COPPER DEVELOPMENT ASSOCIATION

OPEN-PIT MINING often requires blasting to remove rock from over the ore layer. Here workmen at an Arizona copper mine carefully set an explosive charge at the edge of the pit. When the ore has been exposed, it is scraped away and transported to a refinery.

From this point on there are a number of different methods of separating the two, depending on the chemical or physical properties of each.

Probably the first property used by man to separate the metalliferous portion of the rock mined from the waste portion was the difference in specific gravity. Nature herself uses this method in concentrating gold, tin, platinum, etc., in placer deposits. This is the same process used by the Forty-Niners with their gold pans and sluice boxes. It depends on the fact that a moving stream of water will carry away light (low specific gravity) material and will drop heavy (high specific gravity) material as it loses velocity. In the *jig,* an upward pulse is given to water through a bed of mixed heavy and light material. This pulse must be strong enough to lift and separate the lighter grain, allowing the heavier grains to work their way to the bottom, where they are drawn off periodically.

A multiplicity of separations is obtainable by feeding finely ground ore to a *hindered settling column.* This is a closed vertical tube through which a column of water rises at a given velocity. The speed of the water is controlled so that all but the heavest material in the charge is carried up and over the top into another hindered settling column in which the water is flowing upward, again at a speed that will allow the heavest constituent remaining to settle to the bottom. This cycling continues until only the finest, lightest material is discharged from the last column in the battery. The heavier portions are drawn off at the bottom of the tubes.

Shaking tables also make use of the difference in specific gravity between the various portions of the material under treatment. In shaking tables, a moving blanket of water is allowed to flow down an inclined table over a series of riffles while a horizontal motion is imparted to the whole table by pushing it forward gently, then jerking it back faster. Ground ore is fed at the top of the table and carried down over the riffles by the blanket of water. Separation is achieved through a combination of the differential horizontal motion, the force of gravity, and the carrying power of the water. Heavier material comes off at the end of the table nearest the point at which ore is fed, while lighter material comes off at successive points along the lower edge of the table in different bands, separated according to the specific gravities of the materials that make up the mixture.

In the *heavy-medium method* of separation, the ore is fed into a receptacle containing a heavy liquid, the specific gravity of which is greater than that of the lighter portion of the ore and less than that of the heavier portion of the ore. The light material rises and the heavy material sinks, thus achieving a separation.

Flotation is a variation of the gravity principle. It uses the tendency of an air bubble to stick to the grains of one of the materials making up the mix, thus buoying it up. A stream of air is introduced into the bottom of a tank, called a *cell,* containing a water and ore mixture that is kept in constant agitation. Air bubbles attach themselves to one constituent but not the other; while one is carried to the top, the other falls to the bottom. The material to which the air bubbles are attached rises to the surface, forming a froth that is scraped off the top and collected.

Treatment or conditioning of the mixture with chemicals before addition to the cell, and the addition of other chemicals to the cell, can affect the propensity of air bubbles to attach themselves to grains of mineral. Thus, where the specific gravity of the materials is very close, one can be floated and the other depressed through the addition of proper chemicals to the system. Flotation is used to separate the valuable constituents from the waste rock in ores of copper, lead, zinc, molybdenum, and many other metallic ores. It is also used for a number of the nonmetallics and occasionally for coal.

Some materials will dissolve in a solution in which others will not dissolve. This property is used to effect a separation of valuable and waste portions of an ore. For example, in the cyanide process for gold recovery, when a charge of crushed and ground gold ore is allowed to come in contact with a potassium cyanide solution, the gold forms a complex ion with the cyanide and goes into solution. The rich liquor so formed is then drawn off and passed over powdered zinc. The zinc has a stronger affinity for cyanide than does the gold and displaces it from the gold cyanide solution, allowing the gold to precipitate. Nickel has a strong affinity for ammonia, forming the nickel-ammonia complex. This property is employed to separate nickel from some of its ores by allowing it to come in contact with an ammonium carbonate solution. The nickel ammonia carbonate solution is then heated to drive off carbon dioxide and ammonia and allow a nickel oxide to precipitate. —John C. Fox

Paper, matted or fiber sheets (usually vegetable, although sometimes mineral, animal, or synthetic) formed on a fine screen from a water suspension. Paper derives its name from *papyrus,* a sheet made by pasting together thin sections of an Egyptian reed (*Cyperus papyrus*) of the sedge family, used in ancient times as a writing material.

A subdivision of paper is *paperboard.* The distinction between paperboard and paper is not sharp but, broadly speaking, paperboard is heavier, thicker, and more rigid than paper. In general, all sheets 0.012 or more inches thick are classified as paperboard.

Paper, as we know it, was invented in China in 105 A.D. by a young scholar named Ts'ai Lun. He pounded the inner bark of the mulberry tree into pulp, added water, and dried the flat, matted sheets in the sun for several days.

The first major improvement was made in 1798 when a Frenchman, Nicholas Louis Robert, patented a machine for making paper in a continuous sheet. Both linen and cotton rags were used then, but these were so scarce that papermakers searched continuously for a cheaper, more abundant raw material.

A firm of London stationers, the Fourdrinier brothers, added improvements to Robert's machine. The paper-forming sections of today's papermaking machines still bear the Fourdrinier name. In 1840, a German named Friedrich Keller patented the first practical methods of making paper from wood. He used a machine in which sticks of wood were held against a grindstone. Water was added to the separated fibers to make pulp.

In 1857 an American chemist, Benjamin Tilghman, patented the sulfite process of making wood pulp, using chemical treatment of the wood fibers. This development was followed in 1884 by discovery of the sulfate process by a German chemist named Carl Dahl.

The paper industry in America began with a small mill, built by William Rittenhouse near Philadelphia in 1690, which used rags at the basic material. Discovery of methods of using wood resulted in the establishment of a large number of mills in northeastern and eastern states where vast forests of spruce and fir—then favored woods—were to be found. Today the pulp industry uses almost all species of native trees; consequently, pulp and paper mills are located in nearly all forested sections of the nation.

There are 362 pulp mills and 806 paper mills in 44 states and the District of Columbia. These are the mills that make pulp paper from which consumer goods are made. Mills that make the paper and paperboard into end products, such as book paper, shipping containers, writing paper, envelopes, and grocery bags are termed *converting mills.* There are more than 4,000 such mills in the United States.

MATERIALS.—Most pulp manufactured in the United States is made from wood. Waste paper also is used, by processes that remove ink and other impurities before repulping. Pulp from this source is used chiefly in the manufacture of paperboard. Rags, one of the earliest sources of fiber for pulp, are still used to some extent—in paper money, for example. Some straw is used for making coarse papers, and small quantities of coarse grasses and bagasse, the crushed stalks of sugarcane, are used.

MANUFACTURE.—The several processes for making wood pulp are much alike in principle. The fibers are separated and arranged in new patterns and in combination with other substances to make different products. This is done either chemically or mechanically. The method chosen depends on the type of wood used and the requirements of the end product.

Bark is first removed from the logs, sometimes by hand labor in the woods, but usually by machines or jets of water at the mills.

In plants using the *mechanical pulping* method to make what is called *groundwood pulp,* sticks are reduced to fibers by huge, rough-faced grinding stones. Because lignin is not removed, papers made by this method do not generally keep their brightness and strength as long as papers made from chemical pulp. However, groundwood paper has certain qualities desired for various printing processes. Paper on which newspapers are printed consists largely of groundwood pulp. So are the lower grades of tablet paper and other papers that do not require unusual strength or durability.

In the *chemical processes,* wood fibers are separated from their lignin binder by cooking in any one of several chemical solutions. The most common are *sulfite* (usually an acid calcium bisulfite solution), *soda* (a caustic soda solution), or *sulfate* (a modification of soda with sodium sulfide). Special sulfite and sulfate pulps, called *dissolving pulps,* are used in the manufacture of cellophane, explosives, plastics, and rayon.

Paper made from hardwoods by the soda process is noted for its fine texture and excellent printing qualities. Hardwood fibers are much shorter than softwood fibers. When pulped by this process and mixed with longer fibers, hardwoods make high-quality magazine and book papers that are well adapted for the reproduction of photographs.

Originally, the sulfate process was used mainly for the pulping of softwoods, especially pine. Now the process is used in pulping both softwoods and hardwoods for *kraft* papers, which are used for wrappings, grocery bags, and other durable papers requiring strength. Envelopes for heavy mail, corrugated

KIMBERLEY-CLARK CORPORATION

NEWLY MADE PAPER is wound into a roll at the "dry end" of this machine at the rate of 13 miles per hour. The continuous sheet of paper flowing from this end of the machine was a slushy mixture of pulp and water at the machine's "wet end" only a few moments earlier.

board, and *tag stock* are among the paper products made from sulfate pulp. By bleaching, the uses of this pulp have been broadened in recent years to include tissue, writing, printing, and many other kinds of paper, including bright white grades.

Another way of making wood pulp combines mechanical and chemical methods in what is called the *semichemical process*. It was developed particularly for the pulping of hardwoods, but it has many variations. The pulpwood chips are cooked in a mild chemical solution, and the fibers are then separated mechanically. Semichemical pulps produce stiff resilient products and are used in making corrugated paperboard, egg cartons, and other items.

In each of the chemical methods, pulpwood sticks must be chipped before they are cooked. Rotating knives slice across the logs, cutting off chips from ⅜ to 1⅛ inches long and about ⅛ inch thick. An endless conveyor belt carries them to storage above chemical cooking vats, called *digesters*.

These digesters operate on the same principle as a kitchen pressure cooker. Wood chips in chemical solution are cooked with steam until reduced to a wet, pulpy mass. It is this cooking that dissolves the resins and gluelike lignin and causes the fibers to separate.

No matter which process is used, wood pulp requires additional treatments—washing and sometimes bleaching—before it can be made into paper or other products. *Washing* removes leftover dirt and uncooked or unbroken pieces of wood, and it separates fibers according to size. *Bleaching* gives pulp the desired degree of whiteness and also further purifies it.

All of the processes prescribed thus far take place in the pulp mill. Pulp not used immediately may be shipped or stored. In this event, most of the water is extracted from the screened stock. Then the pulp can be made into sheets, which can be placed in bundles or *laps* until it is shipped to a paper mill and converted into consumer products.

The United States is the world's largest producer of paper and paperboard. Mills produced 43.3 million tons of paper and paperboard in 1965 with 1966 production expected to reach 45.2 million tons. Per capita consumption of these products in the U.S. in 1966 was estimated at 530 pounds. —Charles A. Gillett

Petroleum Refining, the process by which crude oil is converted into usable finished products. It is the manufacturing phase of the oil industry. The methods employed vary widely from one refinery to the next, depending on the crude processed, the nature and location of the market, the type of equipment available, and other factors. However, for simplification, it may be considered that most refining processes fall into one of four basic categories.

The FIRST type of process is *fractionation*, or distillation, This method of physically separating the crude's compound according to their boiling points was the earliest refining process, and it is employed to this day. It was originally accomplished in *shell stills*, networks of drums and pipes in which heated crude, one batch at a time, was separated into gases, raw gasoline, kerosene, light heating oil, heavy fuel oil, and other products. The *continuous-battery system* was evolved from the original process. In this system, the crude was introduced at one end of a series, or battery, of shell stills and was separated into various distillates as it passed through the battery.

Today, crude oil is run through hot coils of pipe in a large furnace, then into a tall steel cylinder known as a *fractionation tower*. As the petroleum vapors rise in the tower, they condense on trays according to the temperature at which each becomes liquid again. The various *fractions* of the crude, or *cuts*, each of which is characterized by a carefully controlled boiling range, can therefore be drawn off from the tower at different levels.

The SECOND basic type of process, this one essentially chemical, consists of transforming, or *converting*, certain of these cuts into products of higher commercial value to meet consumer demands. There are many ways of doing this, but all consist fundamentally of altering the molecular structure of the components. In some cases, the molecules may be *cracked* (broken into smaller molecules) to form lighter, more valuable products. In other cases, small hydrocarbon molecules are combined to form larger molecules in reactions called *polymerization* or *alkylation*, or molecules of hydrogen are combined with hydrocarbon molecules in a reaction called *hydrogenation*.

The most widely used conversion process is *cracking*. The type of cracking first invented (and still used) employs only heat pressure and is called *thermal cracking*. A later development involves the use of a catalyst, a substance that helps other substances to change chemically without itself becoming part of the final product. This is called *catalytic cracking*.

Nearly all fractions produced by the processes mentioned above contain certain objectionable constituents or impurities. The THIRD basic process is, therefore, *treating* or *purification*. Treating comprises the removal of the unwanted components, or their conversion to innocuous or less undersirable compounds. Examples are the catalytic sweetening processes, in which the sulfur compounds that are corrosive and give the product a foul, objectionable odor are combined with hydrogen to form a gaseous material and thus separated from the fraction that is being treated. The sulfur subsequently may be recovered for use or sale.

The FOURTH and last basic category is *blending* of the finished cuts into commercially salable products such as motor gasoline, kerosene, lubricating oils, and bunker fuel oil, according to their specifications.

—Clifton Garvin, Jr.

Pharmaceutical Industry, the industry engaged in the research, development, and production of drugs for the relief and cure of illness. In a sense, the pharmaceutical industry dates back to earliest recorded history. Priest-physicians of ancient Persia were called *magi*, the source of the word "magic." They used herbs to combat disease demons, or *drogues*, the source of the word "drug." Ancient Greek and Roman physicians prepared their own drugs, but in Egypt there were separate drug-gatherers who acquired medicinal herbs, often from far-off lands.

The first known apothecary shop was opened in Arabia in 754 A.D. While most of the remedies used by the ancients were worthless, they did stumble upon some useful ones, including quinine, cocaine, and opium.

Apothecary shops spread across Europe. The Benedictine Monastery on Monte Cassino in Italy had one with its hospital in 1086. In 1597, monks of the monastery of Santa Maria della Scala founded a pharmacy in Rome that continued to sell herbal remedies into modern times.

Paul Ehrlich, an assistant of Robert Koch, led a great medical breakthrough in Germany and coined the word *chemotherapy* for his treatment of illness with drugs, opening new areas for the pharmaceutical industry. In 1910, after 606 trials, Ehrlich developed an arsenic compound that cured syphilis effectively. The compound, first called "606" and later salvarsan or arsphenamine, was the first to cure an infectious disease in man. Important new products were added to the industry with the development of insulin, the lifesaver for diabetics, and vitamins—first from natural then from synthetic sources.

The mass production of pharmaceuticals in the United States began in 1778 when Andrew Cragie, a prominent Boston physician, opened a laboratory in Carlisle, Pennsylvania.

The modern pharmaceutical industry divides its products into two categories: *ethical drugs*, which are sold by prescription, and *proprietary drugs*, which are sold over the counter without prescription. After World War II, the U.S. pharmaceutical industry came into its own. In 1946, ethical drug sales were $424 million. By 1960 they had topped $2 billion, and it is estimated that sales in 1970 will reach nearly $5 billion. The investment in research into new drugs has grown spectacularly, too, from a little more than $25 million after the war to more than $200 million by 1960. By 1970, research spending is expected to reach $450 million.

Products of the U.S. ethical pharmaceutical industry fall into seven basic categories. *Antibiotics* and *sulfonamides* are agents used to treat infectious disease. *Vitamins* are used to treat and prevent malnutrition and *hemanitics* to cure blood deficiencies. *Endocrines*, including corticosteroids and steroid sex hormones, are used to regulate body functions. *Barbiturates* and *tranquilizers* induce sleep and calm emotional disturbances; the

CHAS. PFIZER & CO., INC.

ANTIBIOTIC CREAM will fill these tubes. Research facilities and mass output of modern drug firms aids the control of many painful, dangerous and once untreatable diseases.

increased use of tranquilizers has slowed the use of barbiturates. *Antihistamines* are used mainly for treating allergies and colds. *Biologicals* are immunizing agents, such as the vaccines for preventing poliomyelitis and influenza. Acetylsalicylic acid (aspirin) and other *analgesics,* many of which are sold as proprietary drugs, are also incorporated into ethical preparations.

One of the most spectacular developments in recent years has been the perfecting of synthetic hormone products as oral contraceptives. Research is continuing in this field, and several new products are now in the final testing stages.

As the pharmaceutical industry has produced drugs to control tuberculosis, pneumonia, diphtheria, and many other infectious diseases, research has switched its attention to diseases affecting older persons. Atherosclerosis (so-called hardening of the arteries) has been the target of extensive investigation. Much has been learned about this degenerative process, but so far no real breakthroughs have been made, although several helpful materials have been produced.

Cancer, metabolic diseases, mental illness, and diseases of the kidney, heart, and other vital organs all are being given close attention in the laboratory. —John Price

Photoelectric Devices, light-actuated devices designed to convert light into electric energy. They are used to measure illumination, for counting and sorting, and in the automatic control of machinery.

The *photoelectric cell* is an evacuated electron tube containing two electrodes: an electron-emitting electrode, called the *cathode,* and an electron-collecting wire, called the *anode.* The cathode is often a metal plate covered with a layer of silver that contains an alkali metal such as cesium. In other cells, the light-sensitive coating is placed on most of the inside surface of the glass, leaving an opening, or window, for the admission of light. When light strikes the silver-cesium coating, electrons are emitted from its surface. An external voltage source of about 100 volts is connected between the anode (the positive terminal) and the cathode (the negative terminal). This voltage causes the emitted electrons, called *photoelectrons,* to enter the positively charged anode. The movement of these electrons through the external circuit constitutes the output current of the cell.

In practical use, a photoelectric cell converts a change in illumination into a change in electric current. This current change can then be used to control a motor in the performance of various useful tasks.

Some photoelectric cells are filled with a gas such as argon. Initial photoelectrons then strike gas particles, dislodging additional electrons on impact. This process increases the useful current of the cell as much as a hundredfold. Nevertheless, the current produced in such cells is still quite small. The *photomultiplier tube* increases the current still further. In this tube, the initial photoelectrons are sent down a long, narrow cylinder. The photoelectrons strike the inner wall of the cylinder many times as they move through it. At each such impact, several additional electrons are knocked out of the atoms of the cylinder. This process increases the number of free electrons up to a million times in such tubes. This makes it possible for extremely low light levels to generate relatively large amounts of current.

The photoelectric cell depends for its operation on the *photoelectric effect,* discovered in 1887 by Heinrich Hertz, the first scientist to experiment with radio waves. The photoelectric effect is the ejection of electrons from a material as the result of irradiation with light. The effect has two important characteristics: the number of electrons emitted depends on the light intensity falling on the material; the energy or speed of the emitted electrons depends on the frequency (or color) of the light and not on its intensity. Each metal has a minimum or threshold frequency below which it will not emit photoelectrons. The alkali metals—sodium, potassium, rubidium, and cesium—emit more photoelectrons than do other metals under the same conditions.

The *photovoltaic cell* is also capable of transforming light into electricity. All practical photovoltaic cells consist of a junction of dissimilar semiconductor materials. In the *selenium photovoltaic cell,* a layer of pure selenium metal is deposited onto a metal baseplate. A small amount of cadmium or cadmium oxide is then deposited on the surface of the selenium forming the desired junction. Wire leads are then connected to the two materials of the cell. When light strikes the cell, a current flows from the selenium layer, through the external circuit connected to the cell, into the cadmium layer. The silicon solar cell operates in much the same way. Selenium cells are widely used in photographic exposure meters, automatic exposure devices for cameras, and for the measurement of illumination.

The *photoconductive effect* is also sometimes used in light-sensitive systems. Photoconductivity involves an increase in the electrical conductivity of a material resulting from illumination by visible or infrared light. In some substances, such as lead sulfide, lead telluride, and lead selenide, the incident radiation causes an increase in the number of free electrons in the substance. This makes it easier to send an electric current through the substance. Another type of photoconductive cell is made of two dissimilar semiconductor layers in a manner similar to the selenium photovoltaic cell described above. In fact, many photovoltaic cells can be made to function as photoconductive cells merely by connecting their terminals to a battery. The resulting current then changes in accordance with the amount of light received by the cell.

Photoconductive cells are often used in infrared missile tracking systems. This is possible because the exhaust of a missile gives off most of its energy in the infrared region. Photoelectric devices are also used to control street lights, to produce sound signals from the sound track of movie film, to sound fire alarms, and to do many other important tasks efficiently and effortlessly.

—William C. Vergara

Pottery Industry. See *Ceramics,* page 590.

Production Engineering, the branch of engineering concerned with the preparation of a new product for manufacture, the supervision of the manufacturing plant during production, and the collection of data for use in solving problems in the first two areas.

Preparation for production involves selecting a product worth producing and analyzing critically the details of the design. This includes the simplification of the design wherever possible, the setting of tolerances and size variations of the component parts, and consideration of whether standard hardware can be used. Estimates of labor, material, and tool costs are then made. These estimates are derived from past records gathered during the data-collection phase of the operation. After tools are designed and materials are requisitioned, initial production and performance tests are made of the manufactured product to evaluate the success of the production planning.

Supervision of the manufacturing plant during production involves the formulation of production schedules and follow-up. This requires procurement of material, production schedules for component parts, training of manufacturing personnel, setting of wage incentives, quality control, equipment maintenance, and correction of the initial production design. The last element involves cost reduction and cost control. Cost reduction efforts often require developmental changes in the initial product design, or production plan, in order to reduce manufacturing costs.

The third element, data collection, requires the recording of design standards, manufacturing capacity, factory cost accounting, time and motion studies, labor and equipment studies, and product performance in the field. Accurate data collection is essential if the experience gained in manufacturing existing products is to be transferred to the new products.

—Joseph J. Kelleher

Pumps, devices that move, lift, or compress liquids, gases, or mixtures of both liquids and gases. These machines can be defined as either displacement pumps or accelerating pumps. *Displacement pumps,* either piston or rotary, work by drawing fluid into a casing and displacing it by the movement of mechanical elements. *Accelerating pumps,* either rotary or jet, work by accelerating the fluid with mechanical elements or a stream of fluid.

■DISPLACEMENT PUMPS.—The group rereferred to as displacement machines consists of piston, gear, vane, or screw pumps. These pumps are positive displacement units; that is, for each rotation or reciprocation cycle a fixed amount of fluid is displaced or moved.

Piston pumps consist of a casing or cylinder in which a piston is reciprocated by a power source. During the suction stroke, the piston moves away from the cylinder head, admitting fluid; during the pressure stroke, the fluid in the cylinder is displaced, or forced out. Check valves in the inlet and exhaust ports maintain unidirectional flow through the pump cylinder.

Gear pumps consist of a pair of meshed gears rotating in a housing shaped like a figure eight. Inlet and outlet ports are located at the midpoint of the housing. As the gears are rotated, fluid is trapped between the periphery of the gear teeth and the housing and is moved to the discharge end of the pump. As the gears mesh, fluid between the teeth is displaced and forced out of the discharge port.

Vane pumps have cylindrical rotors, equipped with sliding vanes, eccentrically mounted in a cylindrical housing. Retained in slots in the rotor, the vanes are spring-loaded to press against the housing walls. With the rotor mounted eccentrically, the space between the rotating vanes decreases as the vanes pass from the intake to the exhaust port. Spring-loading the vanes prevents leakage and makes it unnecessary to use check valves.

Screw pumps, as their name implies, incorporate right- and left-hand intermeshed helices on parallel shafts connected by gearing. As the helices are turned, fluid is displaced axially from inlet ports at each end to a discharge port at the center. These units are used primarily to pump oils and other fluids with good lubricating qualities.

■ACCELERATING PUMPS.—Grouped under the classification of fluid accelerating devices are centrifugal, axial-flow, and jet (injector) pumps.

Centrifugal pumps have an impeller, shaped like a ship's propeller, which is mounted in a housing. The fluid flows through the housing in a straight line parallel to the impeller's axis of rotation. These pumps are used to move large volumes of fluid at low pressure.

Jet, or *injector, pumps* employ a small, high-velocity stream of gas or liquid to move a larger volume of fluid. These pumps are primarily applied as feed-water injectors on steam boilers, where a small portion of the stream generated can be employed as the operating fluid. Among the advantages of such pumps are the durability and simplicity that come from having no moving parts.

—Joseph J. Kelleher

Quarrying, the removal of large deposits of rock, usually by the open-pit method, for use in construction, road building, the manufacture of cement and lightweight aggregates, other industrial purposes, or arts and crafts. Methods of quarrying differ depending upon the physical characteristics of the rock and its ultimate use. In quarrying building and monumental stone it is desirable to obtain large unbroken blocks, which later may be cut, dressed, and polished to the desired dimensions. Careful cutting and blasting are necessary to avoid shattering the rock. Quarrying the softer rock is usually done with a *channeling* machine. This machine cuts or saws a large block of rock from the main formation without breaking it. Such careful methods are not needed when the rock is to be crushed, as for road building, for making concrete, or for use as a flux in the steel industry. For such purposes, the rock is usually broken loose by explosives, and hauled to a processing plant where it is crushed, recrushed, conveyed over screens for sizing, and often washed to meet all market requirements.

—Michael Trojan

Quality Control, the process of maintaining production within the control limits that have been established for an operation. Many complex factors in materials, machines, and conditions of operation interact to cause slight variations in the "identical" parts produced. Variations may be slight, but they always exist. Two steel shafts may seem to be 3 inches in diameter, when one is really 2.99 and the other 3.01. Or two may seem to be 2.9998 inches, when one is really 2.99976 and the other 2.99984 inches. Each part will have some tolerance, or allowable variation, that will depend on the application. For example, the tolerance in diameter of the main shaft in a jet aircraft engine will be very small, but that for the diameter of a stewardess' call button in the same plane can be quite large. If a large number of parts are inspected and their size is plotted on a chart, the variations will fall into a pattern called the *normal distribution.* The largest number will be near the center, with a rapid or gradual falling off away from the center. By a combination of sampling inspection and mathematical analysis, it is possible to tell whether the distribution is abnormal because of some defect in the manufacturing process, or the center of distribution is drifting in one direction or the other because of changes.

Quality control is not the same as inspection. Inspection is a method of separating good and bad parts. Quality control is a method of studying the product to detect incipient defects in the process and correct them, preferably before bad parts are produced.

The process of quality control was developed by Walter Shewhart at the Bell Telephone Laboratories in 1924. It is now standard practice in many manufacturing plants, often in combination with other forms of inspection procedures.

—Anderson Ashburn

Radiation Detectors, devices that indicate the intensity of radioactivity by counting the number of subatomic particles that they encounter.

The *Geiger-Müller counter* consists of a positively charged wire enclosed by a negatively charged cylinder. An electron speeding through the cylinder collides with gas particles, knocking other electrons free. The process continues rapidly until an "avalanche" of negatively charged electrons flows into the positively charged wire. This produces an electrical impulse that causes a pointer to deflect or a clicking sound to be emitted. The greater the intensity of the radiation, the greater is the motion of the pointer, or the louder the clicking sound. Geiger-Müller counters find wide application in prospecting for nuclear materials and in nuclear safety programs.

The *scintillation counter* is a more accurate detector of nuclear radiation. When a radiation particle strikes certain materials, such as zinc sulfide, a faint flash of light is given off. This light flash falls on the surface of a material that emits electrons when struck by light. These electrons, in turn, pass through an electron multiplier tube, which produces an electric current large enough to be measured. Thus, a short pulse of electricity from the counter indicates that a charged particle has passed through the scintillator. Scintillation counters are widely used in nuclear physics laboratories.

Recently, radiation detectors called *image intensifiers* have been developed that amplify the light flashes from the scintillating plate without destroying the image. This makes it possible to take photographs of the actual track of the charged particle in the scintillator.

—William C. Vergara

Refrigeration, the removal of heat from an enclosed space and the maintenance there of a controlled temperature lower than that of the surroundings. Refrigeration for the preservation of food and other perishables has been in use for centuries, but mechanical methods have been developed only within the past 75 years.

From America's beginnings, "ice harvesting" was carried on during cold months on frozen lakes, ponds, and rivers. Until the early 1900's, such natural ice was the major source of the cooling employed in "cold storage" as well as in household "iceboxes."

The earliest mechanical refrigeration machines were the "ice makers," whose product supplemented the sometimes unreliable supply of natural ice (following a winter warmer than usual). The first machines used ammonia and brine as refrigerants, in both the compression cycle and the absorption cycle, which differs from the compression cycle only in the way the pressure is increased between the evaporating coils and the condenser. (For description of refrigeration cycle, see *Physics.*)

The absorption refrigeration cycle is a two-pressure, heat-operated cycle that makes use of a vaporizable liquid as the refrigerant and a second liquid as the absorbent. The liquid refrigerant vaporizes in the evaporator, taking in heat at low temperature, and is compressed by a heat-operated device to such a pressure that it condenses at a higher temperature in the condenser.

■ MODERN REFRIGERATION.—By the 1920's, mechanical refrigeration had made great strides. Then, the household refrigerator replaced the old icebox, and the iceman's daily visit became a thing of the past.

Today, mechanical refrigeration is essential to many processes and products. It has contributed to the nation's health through its use in the development and storage of drugs. It has made feasible the storage of sections of living tissue used in surgery, as well as whole blood, preserved in *blood banks.* It is responsible for the entire frozen food industry, established between 1940 and 1950, which makes out-of-season foods available all through the year.

Its applications extend to transport vehicles—railway cars, trucks, and trailers that carry frozen products from packing houses to home **freezers** —as well as to fixed installations. It has contributed immeasurably to the national defense program through provision for rocket fuels as well as many other defense needs.

Today, new areas of refrigeration are in process of development—for example, *cryogenics,* which involves the behavior of matter at temperatures below −250° F. —L. N. Hunter

Rope, a large, strong cord made of twisted or braided fibers or wires. Cordage, some of which is still made by hand, is commonly manufactured on specially designed machines, in three steps: spinning the yarn, forming yarns into strands, and laying the strands into rope.

Fibers that have been combed and drawn (*staggered*) to form a continuous sliver, or loose bundle, are spun or twisted into a *yarn.* The length, strength, and thickness of fibers vary widely, as do the machines. The pins of the combs and the *reach* of the drawing rolls must suit the fiber that is being worked. Short fibers, such as cotton and wool, require close, fine pins. Jute and flax need fine pins but their longer fibers require a longer reach. Manila hemp and sisal, which are coarser and longer, require coarse pins and a still longer reach. The spinning machines, too, vary with the size or weight of the yarn being spun. Yarn is usually *right twist,* although for some special ropes it may be *left twist.*

The strength of yarn is derived from the strength and quantity of the fiber and from the friction between fibers after twisting. Yarn may be used for *tying twine.* One of the largest uses is in agriculture as *binder* or *baler twine.*

Stranding is the next step in the manufacture of rope. Two or more yarns, sometimes several dozen, are gathered together and twisted to form the strand. This is usually left twist, since the strand twist must be opposite to the yarn twist. The strand is principally used in the manfacture of rope. Probably the most familiar strand to the layman is *butcher's twine.*

Rope is made by twisting or *laying* together strands, again in the opposite direction, usually right lay. Diameters range from 3⁄16 inch or smaller, sometimes called *twines,* up to about 3 inches, sometimes called *hawsers.* Much of the tonnage produced goes to the marine, fishing, and pleasure boating trade. Among the other uses are scaffold rope, rigger's falls, tents, and cores for wire rope.

Cables are three ropes twisted together in a fourth step; they may be used in the marine trade or in oil- and water-well drilling.

Other constructions use braided and plaited cordage. These are *over-and-under* methods that offer advantages in certain applications. Braided clotheslines, fishing lines and large plaited hawsers for ships are the most common examples.

■ MATERIALS.—Ropes are made from animal sources, such as hair, sinew, and leather, and hand-crafted by primitive peoples. Rope is also made from mineral sources—for example, asbestos rope and packing, and wire rope from iron, steel, and copper. Wire rope has replaced much fiber cordage, for it resists crushing when overwound on drums on heavy machines. However, wire rope lacks elasticity and the ability to absorb shock loads.

Vegetable fibers, such as Manila hemp (abaca), cotton, jute, and sisal are used for cordage throughout the world. Most of the fibers used are cultivated crops, and the quantity and quality of production vary with the conditions of soil, weather, and disease. The fiber must be separated from leaf or stalk by cleaning away pulp and pith. Jute, for example, is retted (or rotted) in water, and the pulp is washed and beaten out by hand. The fiber is then dried. As a result of increased labor costs in areas where most fibers are grown, the price of vegetable fiber has been rising. Therefore, gummed paper tape, steel or wire strapping, and other materials are replacing many traditional vegetable fibers.

Synthetic fibers, such as nylon, polyesters, acrylics, and polyolefins, are growing in use. Although usually higher in cost per pound than natural fibers, they are more economical, as they do not rot, and are stronger per unit of weight and size.

—H. Davis Daboll

Rubber Manufacture, the treatment of natural and synthetic rubbers to form useful end products. Rubber manufacturing can be traced back to ancient civilization. From a tropical tree (*Hevea brasiliensis*), Amazon natives obtained a liquid, later known as natural rubber *latex,* which they used to make crude waterproof clothes, footwear, and balls. In the modern era, rubber manufacturing is generally traced back to a Scotsman named Charles Macintosh, who discovered that waterproof garments can be produced by inserting a thin layer of rubber between two layers of fabric.

Undoubtedly, the most important discovery in the whole history of rubber manufacturing was made by Charles Goodyear, who in 1839 discovered *vulcanization.* Considering that he experimented for nearly ten years, it is likely, as many historians believe, that he arrived at this discovery through trial and error. However, legend persists that he accidentally dropped a mixture of rubber, white lead, and sulfur on a hot stove. When the mixture cooled, it was found to snap back to its original shape. The characteristic of snapping back to the original shape even in extreme cold, and the fact that it was no longer sticky at high temperatures, were the important qualities imparted to rubber by vulcanization.

The first rubber goods factory in the United States was established in 1832 in Roxbury, Mass. After the discovery of vulcanization, other factories sprang up rapidly. At first these were primarily in New England; it was not until about 1870 that rubber manufacturing began in

FIRESTONE TIRE AND RUBBER COMPANY

GOODYEAR TIRE AND RUBBER COMPANY, INC.

FIRESTONE TIRE AND RUBBER COMPANY

TIRES, made of several parts, are joined together on a revolving drum. In the picture at left, the operator is applying the first ply. A "green" or uncured farm tractor tire, *center*, is hoisted into a curing mold, where it will be shaped into one like the suspended tire. The cured tires, *right*, are being conveyed to the last inspection site, where they will receive a series of final quality checks.

Akron, Ohio, which has since become the rubber capital of the world. Today, rubber manufacturing facilities are spread broadly over the United States, from New England to southern California.

By far the best-known and most important product of the industry is the pneumatic rubber tire. Approximately 40 per cent of the dollar value of the industry is in tire products. In addition to the pneumatic tire, a great variety of industrial and consumer products are manufactured from rubber. Conveyor belts, rubber hose, printing rolls, and chemical tank liners are among the more important industrial products. Rubber footwear, raincoats, foam rubber for automobile and furniture upholstery, and latex rubber thread are among the more important consumer products. Rubber continues to become increasingly more useful in sporting goods, ranging from golf balls to scuba diving suits. A major new use for rubber is developing in rubberized roads. The rubber is blended into the asphalt paving mixture, with the result that surface cracking of the road is eliminated in cold weather and the asphalt surface is prevented from softening in hot weather.

■MIXING.—The first major step in manufacturing a typical rubber product is *mixing the compound*. In this step, the rubber is masticated in roller mills or large internal mixers, the best known of which is called a *Banbury*. While the rubber is being masticated, vulcanizing chemicals, reinforcing fillers, antioxidant chemicals, and processing oils are mixed uniformly with the rubber. It is significant that sulfur, the chemical used by Goodyear, is still the most common vulcanizing chemical, although zinc oxide and organic chemical accelerators have replaced the white lead that he used.

Among reinforcing fillers, the most important industrially are the carbon blacks. Their use in rubber compounds is necessary to achieve the good wear resistance in today's tires. The function of antioxidants is to slow down the aging process; without them, rubber products would get sticky and crack. The processing oils are added to make the rubber softer and easier to form. Reclaimed rubber is also frequently used as a processing aid and as an extender to reduce the cost of the compound.

The second major step in the manufacturing process in the *fabrication* of the rubber into a form suitable for vulcanization. This is done by the processes of calendering, extruding, or molding.

■CALENDERING.—Calendering is the process by which rubber is formed into a flat sheet by passing the mixed rubber compound between heavy steel rollers. A calender has much the same appearance as a mill but generally has three or four rolls rather than the two rolls of a mill. Typically, the rubber is preformed into a sheet of approximately the correct thickness by squeezing it between the first and second roll of the calender; it is then formed into a sheet between the second and third roll. In the calendering operation a fabric can also be passed between the rolls, with the result that the rubber is deposited evenly upon the fabric surface. A 4-roll calender can be operated to coat rubber simultaneously on both sides of the fabric.

■EXTRUDING AND MOLDING.—By extrusion, such products as tread rubber for tires, inner tubes, hose, tubing, and strips for window seals are produced. The *extruder,* or *tuber,* as it is sometimes called, consists of a screw that rotates inside of a heated cylinder to force a previously warmed compound through a die or orifice. The final shape of the rubber product is dependent upon the opening in the die. In the *molding* operation, rubber is forced by compression into a heavy steel mold.

■FABRICATION.—In the fabrication of complex rubber parts such as a pneumatic tire, the various elements of the final product are shaped separately and assembled before vulcanization. For example, in the production of a pneumatic tire, the carcass fabric is first prepared by a calender operation. Then bead wire bundles are prepared by extruding rubber over wire, which is finally coiled into bundles of the right diameter. The tread and sidewall of the tire are formed by extruding a rubber slab of appropriate shape. The tire is assembled by wrapping plies of calendered fabric around a revolving building drum, around the end of which bead wire bundles are then placed. The edges of the calendered fabric are wrapped around the assembled bead wire bundles, and the tread and sidewall slab are then applied over the carcass fabric. The whole process of assembling the tire is accomplished in several minutes by an experienced tire builder. The assembled tire, still in the shape of the drum on which it was built, is finally placed in an automatic forming press, in which it is inflated into its final shape as the press closes and the vulcanization process begins.

The familiar inner tube used in many tires is shaped by an extrusion process. The rubber is extruded from a tubular die, cut to appropriate length, and formed into the shape of the tube by splicing the free ends together. The valve is inserted, and through it the tube is inflated after it is placed in a hot mold for final shaping and vulcanization.

Rubber or canvas shoes are assembled from flat pieces of rubber that have been formed by calendering. The sheet rubber is cut to appropriate size and built by hand on lasts that permit the composite layers of fabric and rubber to be formed to the desired shape and size. After the sole is affixed to the upper with a strip of *foxing* (an adhesive strip that combines the upper with the sole), the shoe is ready for vulcanization.

VULCANIZATION.—The final step in the production of rubber products is the all-important *vulcanization.* This is most commonly accomplished by heating the assembled rubber part in a mold. The heat causes the sulfur to set up cross-links between the rubber chains of molecules. This process of cross-linking is responsible for developing the permanence of shape of the vulcanized product. Not all rubber products, however, are vulcanized in a mold. Rubber shoes, for example, are vulcanized in large pressure vessels containing steam and ammonia. Other rubber products, such as golf balls, are vulcanized in an atmosphere of sulfur chloride. Rubber products may also be vulcanized by atomic radiation without the necessity of using sulfur.

LATEX.—Several important commercial products are made directly from liquid latex rubber by processes quite different from those described above; for example, gloves and balloons are formed by dipping. In the production of foam cushions for upholstery, mattresses, and pillows, the latex is whipped with air and molded into the desired shapes. Another important product of latex is Lastex thread. The latex is formed into a thread and, in a continuous process, is covered with textile fibers. Lastex thread has been widely used in bathing suits and other elastic clothing.

Many exciting new rubber products are being manufactured for use in space and underwater technology. Pressure suits, inflatable space stations, living chambers for ocean engineering work, buoys, and supply hoses require new properties and new techniques for their fabrication. (See also *Rubber.*)

—Marvin C. Brooks

Safes, containers designed to protect valuable articles from theft or fire. There are two types of safes: money safes and fire-resistant safes. The *money safe* is usually square with a circular door, constructed of fire- and drill-resistant metal, and is used for the protection of money, jewels, and valuable documents against theft, holdup, and burglary. For protection of the contents against fire, the money safe is welded into a record safe. To prevent it from being carted away easily, the money safe may be encased in a steel-covered reinforced concrete block.

A *fire-resistive record safe* is a container with the prime purpose of protecting its contents against fire. It is usually rectangular in shape with a rectangular door, having an inside and an outside metal shell with heat-resistive insulation between the shells. To be approved by the Underwriters' Laboratories, Inc., a record safe must pass a rigid fire-exposure test, a drop or impact test, and an explosion hazard test. The Underwriters issue labels of approval for three types of record safes: one that will withstand four hours' exposure to heat reaching 2000° F while the inside cannot exceed 350° F.; one for two hours' exposure at 1850° F; and one for one hour exposure at 1700° F.

—Ken Roberts

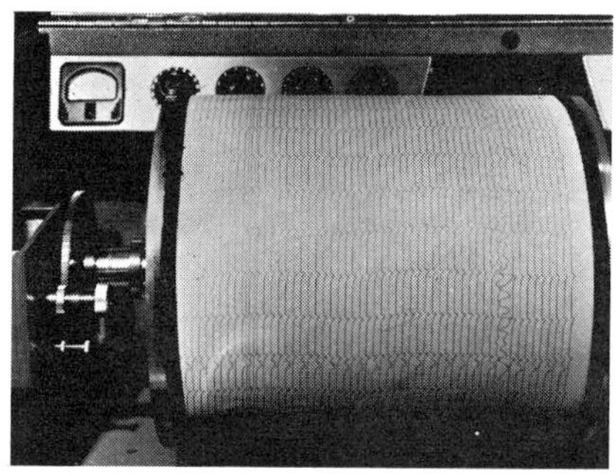

FORDHAM UNIVERSITY

SEISMOGRAPHIC RECORD is traced on a paper wrapped around a rotating cylinder.

Seismograph, a device for detecting and recording, visually or audibly, ground vibrations—especially those caused by earthquakes. As the definition implies, there are two parts to a seismograph, a detector and a recorder.

The *detector* is usually referred to as the *seismometer,* or *pickup.* It is essentially some form of pendulum, that is, a mass (the pendulum *bob*) lightly suspended from a frame attached to the ground. The principle involved is that the mass, because of its inertia, tends to stay at rest while the framework is jolted by ground vibrations, thus setting up a relative motion between the stationary mass and the framework.

In one modern seismometer, the mass is a heavy cylindrical magnet suspended by fine piano wires attached to the frame. Fixed to the base of the frame is a circular coil fitting snugly but freely into the circular magnet. If a ground vibration or wave passes under the seismometer, the coil moves with the ground into or away from the magnet, which remains stationary because of its inertia. Such motion of a coil in a magnetic field produces an electric current in the coil (the principle of the generator), and this current is led off to the recorder, where it is recorded visually or audibly (on tape).

Whether the magnet is the suspended mass and the coil is fixed to the base, or the coil is the suspended mass and the magnet is fixed to the base, makes no difference. Both methods are in equal use today. Such a seismometer may be set up vertically (so as to oscillate vertically), in which case it will respond to vertical ground vibrations; or it may be set up horizontally, in which case it will respond to horizontal vibrations. A complete seismic setup calls for three such seismometers. One is set up vertically, one horizontally on a north-south axis, and one horizontally on an east-west axis. Such an orientation makes it possible to determine the direction from which the vibrations come to the observatory.

The *visible recorder* is usually a drum around which is wrapped a sheet of photographic paper or stylus paper. For *photographic recording,* the current from the pickup is led to a moving-coil galvanometer, where it causes a mirror attached to the moving coil to reflect light from a fixed source back and forth on the drum in keeping with the ground motion. In the *stylus* type (pen and ink or hot-wire stylus), the current is amplified and led to a needle-type galvanometer. The stylus or pen, which is attached to the needle, then traces out the ground motion as the drum rotates. The drum is driven by a synchronous motor, and minute and hour marks are impressed on the record by a clock. Audible recording is done on tape.

—Joseph Lynch

Servomechanism, a combination of elements that control a source of power automatically in order to perform a desired function. All servomechanisms use *feedback,* the same process that helps to control the air temperature in our homes. Suppose the thermostat is set at 70° F. The furnace goes on automatically when the temperature falls below 70° F, and turns off when the temperature rises to about 72° F. The heating plant is merely a source of heat whose effect is to raise the air temperature in the home. This effect—the air temperature—is "fed back" to the thermostat and compared with its temperature setting. Any difference between the two is used to control the source of power, the furnace.

The same principle is widely used to control automatic machinery. Imagine a sheet of steel being pressed to a desired thickness between a pair of rollers. The thickness of the sheet is measured automatically and compared with the desired thickness. Any difference between the two thicknesses controls a motor that increases or decreases the roller spacing to provide the desired thickness.

Servomechanisms are also used to control high-speed aircraft where human reactions would be too slow. They are used in electric-power generating equipment to control voltage and frequency. They control the speed of motors, the temperature of the cooling water in automobile engines, the rate of fuel injection in missiles, and the automatic pointing of guns and radar beams. A servomechanism, in short, is the heart of any self-adjusting process or machine.

—William C. Vergara

Sewing Machine, a device for continuous stitching or sewing. The invention of the sewing machine cannot rightly be credited to one man. In 1790 Thomas Saint, an English cabinetmaker, patented a device for stitching shoes and boots, but his idea was never put to practical use. Others who devised sewing machines during the nineteenth century were Baltasar Krembs in Germany; John Duncan in Scotland; Josef Madersperger in the Austrian Tyrol; John Dodge of Monkton, Vt.; Barthélemy Thimmonier in France; and Henry Lye of Philadelphia, Pa.

About 1832, Walter Hunt of New York built the first sewing machine based on a lock-stitch principle. In this machine, an eye-pointed needle pushed a loop of thread through and beneath the fabric, where a bobbin carried a separate thread through the loop. Hunt, however, he did not apply for a patent. Elias Howe began to work on a sewing machine and was led to the same discoveries as those of Hunt; he patented his machine in 1846. In Boston in 1850, Isaac Merritt Singer, a journeyman mechanic, made the first sewing machine capable of sewing continuously. He was granted patents in 1851 and began to manufacture machines.

Today, more than 4,000 different types of sewing machines are made. Electric models for home use, whether portable or in cabinets, may be of the straight-stitch or swing-needle (zig-zag) varieties. Swing-needle machines can do straight stitching and many variations of zig-zag decorative stitching. One type can also do a chain stitch.

—Rose Marie Burnley

Shipbuilding, the construction of water-borne vessels. Archaeological artifacts recovered in the Mediterranean Sea and in Lake Nemi, Italy, have shown that wooden shipbuilding and boatbuilding were well developed as far back as 50 B.C. It was not until the early nineteenth-century that steel was used in ship construction. Today, wood, steel, fiberglass, and alloys are all used for shipbuilding. Eventually, plastic hulls may be produced for vessels exceeding 60 feet in length although, at present, plastic hulls are still in the process of development.

The art of shipbuilding consists of two phases: design and construction. The factors determining ship design are: the use to which the ship will be put; the speeds required of it; and the depth of the harbor where it will be constructed. These must be considered because they affect a ship's length, beam, depth, hull design, and means of propulsion.

Once such factors are determined, the preliminary designs for the ship may be made. Either a *line-drawing* or a *half-model* of the ship is prepared. The art of drawing ship plans developed slowly. The earliest English manuscript on the subject, known as *Fragments of Ancient English Shipwrightry,* is commonly dated about 1586. Plans were usually drawn on parchment until late in the seventeenth century, when paper gradually came into use. Drafting methods employed in preparing a lines plan showing hull form have changed slightly since 1800. The practice of first making scale drawings of the form of the ship and then "laying down" the full-sized drawings of the shape of the frames and end profiles of the hull was first described in Mediterranean records of the fifteenth and sixteenth centuries.

The use of the half-model as a substitute for a lines plan is supposed to have developed in England as early as 1700. A solid block of wood was shaped to represent one side of the proposed hull. From this, the forms of the cross sections were taken and transferred to a plank or to drawing paper along the hull profile as developed in the block model. The half-model has the advantage of representing the hull-form in three dimensions. Hence, a half-model is sometimes made from the lines plan for study purposes by naval architects. Often, the design is tested in a model tank.

Before construction is started, the line drawing or half-model is reproduced in a full-size drawing. Called *lofting,* this reproduction is done in the *mold-loft,* a large room in which at least the forward and after halves of the hull can be drawn full-size on the floor. Lofting is closely tied to the use of plans or half-models, for only by using such plans can the small-scale design be reproduced to actual size. In large ships, the lofting is very extensive and requires much detailed work—some yards even use photographic methods of lofting. With the completion of *templates,* full-size wooden or paper patterns made in the mold-loft and complete even to drill holes to show the exact position of holes to be drilled or punched in the finished pieces, the actual construction of the ship may be started.

■**INTERNAL MEMBERS.**—The major structural members of ship may be divided into two classes: those that run lengthwise in the ship, and those that are set at right angles to its length. The chief function of the transverse frames is to withstand the pressure of the water. If a ship had adequate transverse frames and could always operate under ideal conditions in which there were no wind or waves, and the load could always be properly distributed, the longitudinal framework would not have to be very strong. Actually, much of the ship's weight may have to be supported by small areas along its length, as when it rides across a wave or when bow and stern are lifted together by two waves. This produces enormous stresses tending to cause the ship to sag, or "break its back."

Therefore, the most important structural member of a ship is its *keel,* the backbone of the ship, which is designed to provide longitudinal strength. It runs from stem to stern at the bottom of the vessel and is usually built to form a continuous structure from one end to the other. Extending upward from the keel to the upper deck for the full length of the vessel is a strong framework, often made of steel, called the *keelson.* Together, keel and keelson form a structure very similar to a bridge and thus provide much of the longitudinal strength of the ship.

■**CONSTRUCTION STEPS.**—The first step in the actual construction of a ship is the preparation of the *berth,* or bed, on which it rests during the building. This must be very solid and strong in order to support the enormous weight placed upon it. First wooden blocks, called *keel-blocks,* are laid. The next step is the erection of bulkhead frames across the middle of the keel, beginning the main transverse structure of the ship. Many other transverse frames are then spaced at other points along the keel. If heavy material or preassembled parts are to be installed, overhead cranes, running on elevated

ABOVE—BETHLEHEM STEEL CORPORATION

TANKER (*above*) with a 27,200,000-gallon capacity takes shape in a shipyard berth five months after the keel-laying. At *left,* a ship's screw dwarfs men working in a drydock.

tracks on both sides of the *ways* (forms upon which the ship rides into the water) transport them and hold them in the exact positions desired until they can be bolted in place. Bolting holds the parts only temporarily until it is convenient to rivet them firmly into position.

When construction of the keel and bottom framing are completed, the floors, bulkheads, and additional frames are installed. Calking, installation of some machinery parts, and tests for watertightness complete the building-berth process. The ship is then ready for launching.

Usually a ship is launched stern-first although, in some parts of the world, the vessel is either set up with the bow to the water and launched bow-first or placed parallel to the shore and launched sidewise. A fourth method of launching is floating: the ship is built in a dry dock, below sea level; water is admitted into the dry dock, gradually floating the ship.

Once floating, the ship is taken to an *outfitting dock*, or *builder's wharf*, and its construction is completed. Deck structures and interior joinerwork are fitted, machinery installed, and, in the case of sailing vessels, masts stepped. With such equipment in place, the ship is given dock and sea trials before delivery to the owner. —Howard I. Chapelle

Shoe Manufacture, a highly mechanized industry, employing over 230,000 workers and producing more than 50,000,000 pairs of shoes each month in the United States. Yet it was long assumed that, because of the complex processes involved, shoemaking would always be done by painstaking handicraftsmen.

In the seventeenth century, when shoemaking as an industry first appeared in North America, it was customary for a cobbler to make regular trips around the countryside, living with each family in turn until he had completed a year's supply of shoes for everyone. As recently as 100 years ago, a good shoemaker, with tools which had changed little since their use by Egyptian artisans in the fourteenth century B.C., could average at the most five or six pairs of shoes a week. By contrast, the giant industry today turns out 70 pairs of shoes a week for each worker. This is accomplished by use of machines wherever practicable and also by the separation of the shoemaking process into small increments, each performed by a skilled craftsman.

The modern shoe industry was born in 1862 with the introduction of the McKay Sewing Machine, the first machine with the capability of stitching the shoe upper to the sole. The increase in mechanization has been rapid, until at the present time there are more than 300 different kinds of machines, some relatively simple, some highly complex. Depending upon the type of operation which they perform, their rate of productivity varies from a few hundred pairs a day to several thousand. There is now not a major operation in shoemaking which is not done better by machine than it once was by hand.

MANUFACTURING PROCESS.—A shoe may undergo as many as 382 different operations in its manufacture and be composed of up to 290 separate items. After the initial step of cutting the upper from a pattern by means of dies, the shoe goes to the Upper Fitting Room, where the lining uppers, box toes, counters, and straps are assembled and permanently attached to one another by stitching, lacing or cementing (adhesive bonding). Other operations are often accomplished here like perforating, eyeletting, lacing, pinking and splitting or rubbing to a uniform thickness. Next, the fitted upper along with other parts of the shoe, including insoles and outsoles, counters, welting, heels and box toes, are assembled in the Stock Fitting Department. Then the assembled shoe undergoes the very important process of lasting.

A *last* is a form, formerly of hard maple wood, but now predominately of plastic, which reproduces the shape of the human foot and determines the exact style and size of the finished shoe. Lasting is a series of operations designed to create a new three dimensional shape from a two dimensional piece or pieces of material. When using leather upper and lining material they are thoroughly moistened, then stretched over the last, attached to the insole, and dried.

The next step accomplishes the attaching of the outsole to the upper. Shoes are traditionally divided into three general types, depending upon the method used for this step. The largest category, in which the sole is attached by cement, accounts for over half of the total shoes produced, including nearly all women's shoes. The second are the stitched shoes, including the Goodyear Welts, the Littleway Lockstitch and McKay constructions, and the Stitchdowns. Third is the vulcanized shoe—including both work and tennis type shoes. A fourth category, vinyl injection molded to fabric or leather uppers, must now be added because of its growing importance.

After attaching of the outsole the shoe is ready for heel attaching (where required) and finishing. By the time it reaches the shipping room it may have passed through 200 different pairs of hands. —Alice Regensburg

Silk Manufacturing, the production of textiles from raw silk. The art of weaving was developed in China more than 4,000 years ago. It then found its way to the Near East and, much later, reached the European continent. By the twelfth century, Italy had become the silk center of the West. In France, silk weaving was begun in 1480 during the reign of Louis XI, although it had been introduced in England somewhat earlier during the reign of Henry VI. Strangely enough, it was not until 1810 that the first silk mill was built in America. The industry did not prosper until a protective tariff was instituted during the Civil War.

When the raw silk is reeled off from cocoons, it is not strong enough to be woven into anything except the sheerest material. Therefore, depending upon the fabric to be woven, two, three, or four silk threads are thrown together. This is done by *throwsters*. Much of the raw silk used for filling, called *tram*, is thrown with a certain twist. The silk yarns are woven on looms very much like those used for cotton. Automatic power looms have long since taken the place of hand-weaving methods in practically all the leading countries.

One of the outstanding features in silk weaving is the Jacquard loom, invented by Joseph Marie Jacquard of Lyons, France, in 1801. Jacquards are composed of a multiple of different weaves forming definite patterns, such as flowers or kindred motifs. The creation of these patterns is by the use of intricate perforated strips of punched cardboards operated in conjunction with the loom (similar to IBM machines). It is on these Jacquard looms that the brocades, brocatelles, damasks, and heavy fabrics for evening wear are woven.

Today, most of the raw silk is consumed in the *haute couture;* that is, in evening dresses where brocades, satins, crepes, chiffons, and georgettes are mostly used, and also in cocktail dresses, in which prints and shantungs are predominant. The demand for silk is also strong for cravat fabrics, as well as for upholstery and curtain materials. Due to high raw silk prices, U.S. consumption has dropped about 40 per cent since 1963. —Hans Vaterlaus

Simple Machines, the most elementary mechanical devices, one or more of which are the basis of all complex machines. The simple machines include the inclined plane or wedge, wheel and axle, pulley, and lever. The inclined plane or wedge becomes a *screw*—which itself is often classed as a simple machine—when it is wrapped around a cylinder, while *gears* are a form of wheel and axle. The pulley is found in cranes and hoists while the lever is found in its commonest form in an ordinary crowbar.

The function of the simple machines is to change either the magnitude or the direction of applied forces. When prehistoric man used a log or branch to move a large stone, he was unknowingly applying the *principle of the lever* to multiply the force available in his muscles.

As shown in the accompanying illustration, the force *F* applied on the log by the prehistoric man is multi-

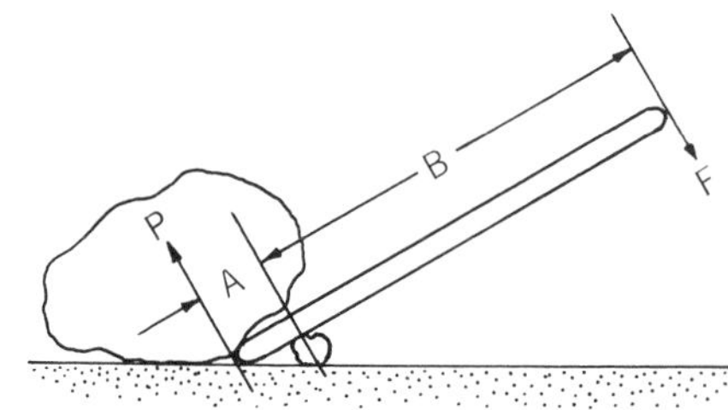

plied into a much larger force P applied to stone. If the distance B is six times longer than the distance A, the force P is six times greater than the force F applied by the man.

This example of a simple machine also illustrates the principle of *conservation of energy*. To lift the stone one inch, the user must exert a force over a distance of B/A inches. Simply expressed, a simple machine can convert a small force applied over a large distance into a large force acting over a small distance.

The *actual mechanical advantage* (AMA) of a simple machine is expressed as the ratio of the force produced to the force applied. In the case of the prehistoric man's lever, the AMA equals P/F.

Not all simple machines multiply the applied force. Some just alter its direction. When a rope is passed over a simple pulley, the input force applied F must be equal to the load it lifts.

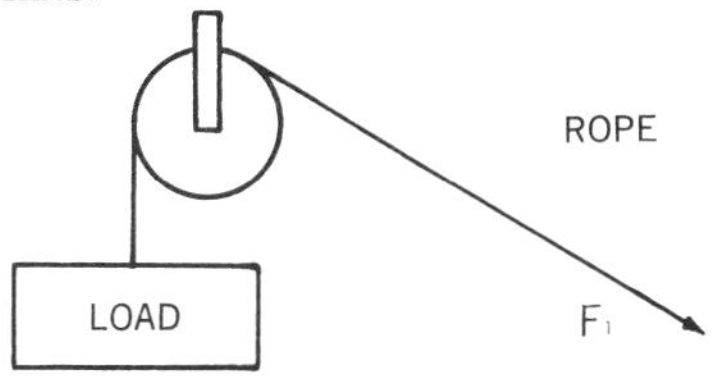

To reduce the force needed to lift the load, an axle or differential pulley is added to the simple pulley. In either case, the force needed to lift the load is reduced but it must be applied over a larger distance than the load moves. To lift the load one foot, the force F_2 must be applied for

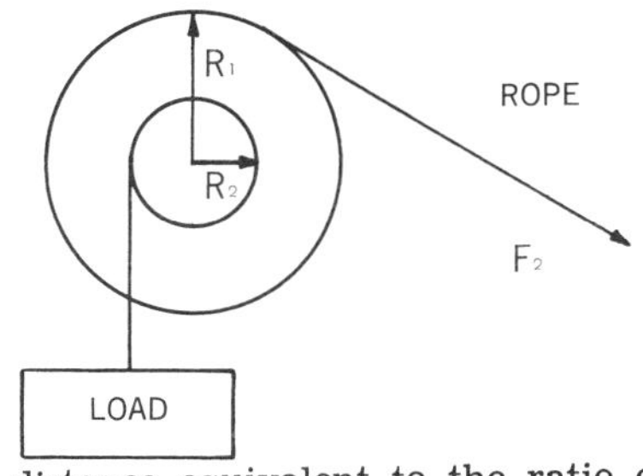

a distance equivalent to the ratio of the radius of the large pulley R_2 to the radius of the axle or small pulley R_1. The AMA of the system equals R_2/R_1.

In actual machines, friction exists and work must be exerted to overcome the friction. The *theoretical* efficiency of an actual machine equals the ratio of the useful work produced to the work input. The actual efficiency equals the *work out/work in*.

Because ideal machines have no friction, their efficiency is 100 per cent. The *ideal mechanical advantage* (IMA) of a simple machine is the ratio of the distance the input force moves to the distance the output force moves. The IMA therefore equals D_{in}/D_{out}.

In an actual machine the IMA is larger than the AMA because of frictional losses and the efficiency of an actual machine is the ratio AMA/IMA. Actual efficiency can be expressed as work out/work in, $F_{out} \times D_{out}/F_{in} \times D_{in}$, or AMA/IMA.

—Joseph J. Kelleher

Soldering, the joining or coating of metals with low-temperature alloys. Mechanical joints made with solder are liquid-tight and gastight, but have little mechanical strength. Depending on the alloy used, solder coating provides improved corrosion resistance, good electrical conductivity, and improved appearances. The most common soldering alloys consist of lead and tin and are available in wire, foil, bar, or ingot form. Soldering takes place when the low-temperature alloy is brought in contact with the heated base metal, where it melts and flows over the surface. The base metal can be heated with a soldering iron, gas torch, or oven, depending on the size of the work and area to be soldered. *Dip soldering,* which involves immersing the parts to be coated or joined in a bath of molten alloy, is used in high-volume manufacturing processes.

To assure adequate adhesion of the solder to the base metal, a *flux* is used to remove oxide coatings and to lower the surface tension. Rosins, zinc chlorides, sal ammoniac, and acids are common fluxes. With the exception of rosin, fluxes should be neutralized or washed away to prevent corrosion of the base metal. *Hollow wire solder* filled with rosin is widely used to join wires in electrical equipment.

—Joseph J. Kelleher

Spinning, the twisting of fibers to make a continuous thread or yarn. Spinning is an art older than recorded history. Spun yarns have been found amid the ruins of ancient civilizations, and hieroglyphs and cave drawings thousands of years old depict men and women spinning yarn by hand. Aborigines in all parts of the world have developed markedly similar methods of spinning.

Spun yarn is composed of a strand of fibers arranged in parallel form and twisted to provide cohesion and strength. Among the most commonly used natural fibers are cotton, wool, silk, flax, and jute. Today, many commercially produced synthetic fibers are spun into yarn on conventional machines.

All spinning was done by hand until about two centuries ago. The fibers were drawn from a supply of stock, loosely wrapped upon a rod which the spinner held under his arm. The fibers were paralleled and formed into a slightly twisted strand by the spinner using the fingers of both hands. A *spindle,* which also served as a yarn carrier, was suspended by the yarn and rotated to provide further twist. As the spun yarn accumulated, it was wound upon the spindle and secured so that it would not unwind as the spindle was again set in motion, and the drafting (drawing out) of the fibers resumed.

The *spinning wheel* or *Saxony wheel,* invented by Johann Jurgens of Brunswick, Germany, in the sixteenth century, was the first machine for spinning yarn from fibers. The *spinning jenny,* employing the principles of the Saxony wheel but equipped with a number of spindles, was invented by James Hargreaves in 1764.

The drafting of fibers with revolving rolls is the invention of James Wyatt, who built a spinning machine in 1730. Lewis Paul, an associate of Wyatt, patented a spinning machine in 1738 and a better model in 1758.

However, Richard Arkwright was the first to develop machine spinning on a commercial basis. Between 1769 and 1775, he patented carding, drawing, roving, and spinning machines. The mills he built and equipped with machines of his own design marked the beginning of what has come to be known as the Industrial Revolution.

The first spinning machine in America was a spinning jenny built by Christopher Tully in 1775 and installed in the Philadelphia mill of Samuel Wetherill.

The *mule frame* was invented by Samuel Crompton in 1779. *Cap spinning* was patented by Charles Danforth in 1828, and John Thorp patented *ring spinning* in 1830. Ring spinning is the predominant system throughout the world today.

A modern spinning frame may have 400 or more spindles. The rapidly revolving rolls take each inch of stock and draft it out into ten, twenty, or even into hundreds of inches of yarn. The spindles rotate at speeds up to 15,000 revolutions per minute to twist the yarn and wind it onto *bobbins* or tubes. (See also *Weaving.*

—Wilmer Westbrook

Spring, mechanical elements that store energy when displaced or deflected. A spring may be made of any elastic material and can take any shape. Although springs are usually made of metal, they can also be made of wood, plastic, glass, or rubber. Springs may be classified according to their shape as coil, torsion bar, leaf, spiral, or disk.

Coil springs are produced by winding a wire around a rod or mandrel to form a helix. Tension springs frequently have the ends of the wire formed into hooks that can be easily connected to machine elements. Compression coil springs are wound loosely, or with space between the coils, to allow for contraction. The ends of compression springs are ground flat, perpendicular to the axis, to assure uniform deflection of the coil during loading.

Torsion bars consist of a rod, shaft, or tube that is loaded by twisting. Energy is stored when one end of the bar is held while the other end is rotated. Torsion bars are used as the springing element in some automobile suspensions.

Leaf springs are cantilever beams that deflect when loaded. Although a single leaf can be used, the usual arrangement is to add several leaves of varying lengths to the main leaf. The deflection characteristics and load-carrying ability of the main leaf can be adjusted by varying the length and number of the auxiliary leaves. One advantage of leaf springs is their ability to resist loads in two directions.

Spiral, or *watch, springs* are wound in an Archimedean spiral in a flat plane. When a watch spring is wound, the force deflects the coils, decreasing the space between successive turns. As the spring unwinds, the energy stored by winding is used to operate the clockwork.

Disk springs, also called *Belleville springs,* consist of a conical washer that deflects when an axial load is applied. These springs occupy very little space and support large loads with little deflection. Several disk springs can be stacked to increase their capacity.

Although springs are used in many ways, their primary function is to store energy. Common examples are clocks and wind-up toys, in which energy is stored by a mechanism that deflects the spring. As the spring returns to its original shape, the energy is released to move the clock hands or drive the toy. —Joseph J. Kelleher

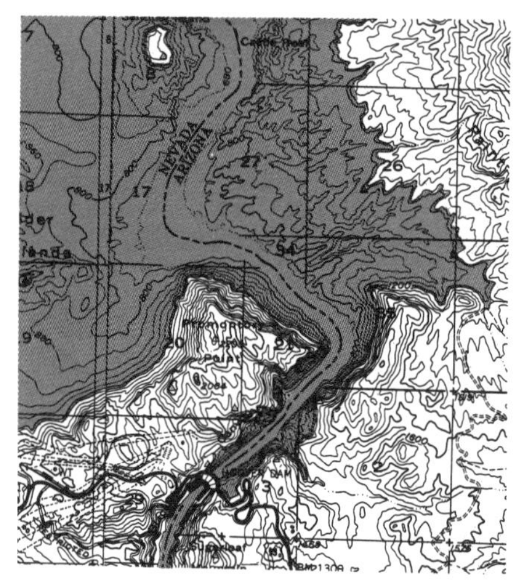

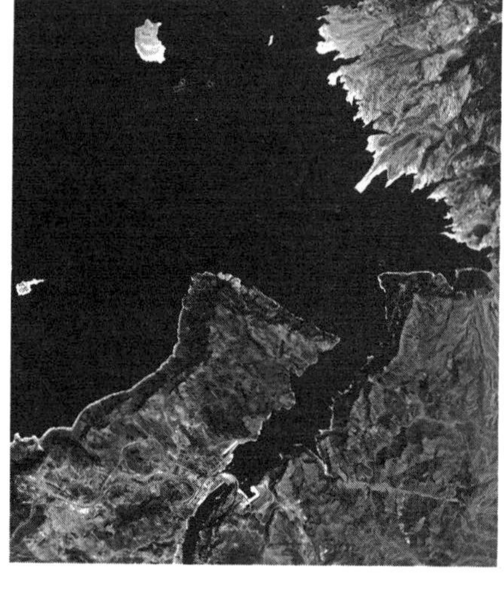

U.S. GEOLOGICAL SURVEY

TOPOGRAPHIC MAP (*left*) of Lake Mead area was plotted from aerial photograph (*right*).

Sugar Processing, the sequence of processes involved in producing sugar from sugarcane and sugar beets. Sugarcane deteriorates rapidly after harvesting. It must, therefore, be promptly processed at a nearby sugar factory or mill. At the factory, the cane is chopped into pieces and passed through huge rollers that crush it, thereby extracting the juice. The fibrous cane residue that remains is called *bagasse.*

Impurities are removed from the juice by heating and adding lime. A syrup develops as water is evaporated from the juice. The syrup is boiled under a vacuum, and a mass of sugar crystals and molasses develops. The sugar crystals, which are separated from the molasses in a centrifugal, are surrounded by a film of molasses and a number of impurities. The mass of crystals is referred to as *raw sugar*—a coarse, sticky, brownish substance.

Molasses and sugarcane bagasse are by-products of sugarcane processing. Molasses is used primarily for livestock feed, and bagasse for fuel at sugar mills and to produce paper.

Raw sugar is shipped to cane sugar refineries where it is converted into a variety of refined sugars. During the refining process, molasses is separated from the raw-sugar crystals. The crystals are dissolved in water to produce a syrup, which is then filtered to remove impurities. Next, the syrup is concentrated by evaporation and boiled under vacuum until refined-sugar crystals develop. These crystals are separated from the syrup by centrifugals and are dried and packaged.

Refined beet sugar is produced in factories that process the sugar beets —not in separate refineries. These factories are located in the sugar beet producing regions. The beets are washed and sliced into thin strips called *cossettes.* The sugar is removed from the cossettes by a diffusion process. The juice is then processed into sugar in a manner similar to the cane-sugar process. Beet pulp and beet molasses are by-products of beet-sugar processing. Both are used as livestock feed.

Approximately 70 per cent of the sugar consumed in the United States is cane sugar; the remainder is beet sugar. Many varieties of refined sugar and styles of packaging are required to meet the demands of consumers. Refined sugar is distributed to food processors in packaged, dry bulk, and liquid forms.

In a number of places throughout the world, crude sugars are still produced for local consumption. These sugars are referred to as noncentrifugal sugars. —Nicholas Kominus

Surveying, the science of locating points on or near the earth's surface. The need for surveying arose in the earliest days of civilized life, and tomb pictures show the ancient Egyptians measuring fields with coils of rope. Similarly, the aid of these so-called *rope-stretchers* and of simple plummet levels was required in laying out the foundations for such great sloping works as the Pyramids.

The Greeks, endeavoring to rationalize earlier rules, created *geometry,* or *land measurement,* and applied their findings to more involved problems—such as surveys for tunnels carried out from two headings or requiring alignment and grade between intermediate shafts. The Romans, in turn, adopted these Greek advances and used the *groma,* or surveyor's cross, a cross-arm sighting device for setting out right angles and in laying out rectangular land plots which, it is said, inspired the American public-land system. For leveling, the *chorobates,* or plummet or water-trough leveling device, was apparently adopted in preference to the Greek *dioptra,* an angle-and-level instrument described by Hero of Alexandria. Surveying, however, was long to remain a simple art, based on similar triangles and right angles. Angle measurements were possible, but there appears to have been little early need or use for such measurements.

The advent of gunpowder and the rise of military engineering in the late Middle Ages and early Renaissance introduced problems of range and elevation. The first printed books on surveying, such as those of Thomas Digges, issued in London in 1556 and 1571, were devoted to military problems. By this time, however, the development of modern methods of computation and of improved angle-measuring and leveling devices was under way. In particular, the *enclosure* movement in Britain led to increased interest in land surveying. Although the graphical method of the "playne table" often sufficed, there was an increased use of the surveyor's *chain,* developed by Edmund Gunter, and of the surveyor's compass.

The French aqueduct and canal projects of the seventeenth century likewise led to improvements in leveling instruments. Following the telescopic-sight plummet level of Henri Picard, the modern dumpy bubble-tube level appears in modern form in Mallet's level of 1702. Angle-measuring instruments, however, were clumsy and costly, and, for land surveying, the compass and chain remained standard until late in the nineteenth-century.

A simplified form of the European *theodolite* was developed in the United States early in the nineteenth century, and this American *transit* remains today the standard American angle-measuring instrument. Equipped with *stadia hairs,* it is also used for smaller topographic surveys.

A major division is made between *plane surveying* and *geodetic surveying.* The latter, extending over large areas, involves the most accurate measurements required in surveying —so accurate that allowance must be made for the curvature of the earth. Carried out largely by government agencies, it provides the basic control for the more detailed topographic maps published by all major nations. The U.S. Coast and Geodetic Survey furnishes basic data to the U.S. Geological Survey, which issues maps—usually at a scale of about one mile to the inch—of most of the country. Aerial photographic surveys are increasingly used in map making.

Surveying provides essential data for the design and layout especially of civil engineering works, and is therefore often undertaken by engineering offices. However, land surveying is usually regarded as a specialty. Land surveyors are licensed by the states, and their activities are limited to this field alone.

—James Kip Finch

Systems Engineering, the branch of engineering dealing with the optimal integration of a variety of machines, instruments, and other devices in order to perform a complex function. Systems engineering is interdisciplinary in nature, requiring a knowledge of all branches of engineering science that might bear upon a specific task.

The design of a satellite surveillance system is a typical application of modern systems-engineering techniques. Such a system might involve search radar sets to detect satellites, tracking radar sets to determine the satellites' precise locations, computers to calculate their orbits, data-processing equipment to reduce raw data to usable form, and a communication network to connect a number of ground stations located hundreds or thousands of miles apart. These and many additional subsystems, such as timing devices, hydraulic equipment, antennas, facilities, buildings, television networks, etc., are selected by systems engineers for their ability to perform a necessary function within the system. Where a needed subsystem does not exist, the systems engineer has it designed and built to his specifications. Many subsystems, in turn, are so complex that they too must be designed as individual systems with major electronic, electrical, pneumatic, and hydraulic components. The radar sets mentioned above would certainly fall into that class.

Systems engineering now plays a vital role in the design of military and commercial aircraft, in space programs, in the design of chemical processing plants, in communication and navigation networks, and in a great variety of complex systems made possible by modern technology.

—William C. Vergara

Transistor, an electronic device, made of a semiconductor material, capable of amplifying or switching electric signals. The transistor was invented in 1948 by William Shockley, Walter H. Brattain, and John Bardeen, scientists of the Bell Telephone Laboratories. Their invention was the greatest contribution to electronics since the invention of the vacuum tube, and for this feat they were awarded the Nobel Prize in physics in 1956.

The *junction transistor* consists of a tiny crystal of semiconductor material, usually of the chemical elements silicon or germanium. The crystal is extremely small in size, about 20-thousandths of an inch square and perhaps 6-thousandths of an inch thick. The crystal is initially of high purity. Small amounts of impurities are then diffused into the crystal at high temperature, a process called *doping*. This produces a sandwich within the crystal consisting of three layers of slightly different semiconductor material. Wires are then attached to each of the three semiconductor layers. These form the output electrodes of the device and are called the *emitter, collector,* and *base*.

Transistors can amplify electric signals because a small electric charge applied in the base electrode controls a much larger charge moving from the emitter to the collector. Unlike the vacuum tube, the transistor requires no heating; furthermore, the transistor operates on a fraction of the power required by the vacuum tube.

Closely related to the transistor is the *junction diode*. It differs from the transistor in that it consists of only two semiconductor layers instead of three. It has the important property of allowing current to flow only in one direction through the crystal. Diodes are widely used in electronic computers, radio receivers, and other electronic equipment.

In 1952 William Shockley developed the theory of a new semiconductor device which has come to be known as the *field effect transistor*. Its main advantage over the junction transistor is the extremely high electrical resistance it presents to the applied signal source. This means that it draws practically no current from the signal source. In that respect, the field effect transistor is similar in operation to the vacuum tube, but the transistor is more efficient.

The *silicon controlled rectifier* is another important transistor-like device. It is made of four parallel semiconductor layers in a single crystal of silicon. Electrodes are attached to three of these layers. The silicon controlled rectifier is used widely in rectification, the transformation of alternating current to direct current in electronic equipment. As a rectifier, it replaces large bulky vacuum tubes and performs the function with great simplicity and efficiency. It is also used as an "on-off" switch, controlled by momentary pulses of current, and as an amplifier which can increase the strength of a signal up to 10,000 times. It is used in battery chargers, in current-limiting circuit breakers, in electronic light flashers, and for speed control of electric motors. The light-sensitive silicon controlled rectifier is used for automatic lamp switching and in many applications in which light must actuate or influence an electronic circuit. A large number of other types of light-sensitive transistors are available for specific applications.

Until recently, the transistor has been unable to function well in extremely-high-frequency circuits. A new discovery, however, called the *metal base transistor,* may make it possible for transistors to be used at frequencies up to 10,000 megacycles per second or higher. Like the junction transistor, the metal base transistor consists of a three-layer sandwich. It differs in that the middle layer is made of a metal instead of silicon. This central metal layer is attached to the base electrode of the transistor, hence the term metal base transistor.

The future appears bright for new and improved types of transistors. Transistors capable of operation at frequencies up to thousands of megacycles are now being perfected. The vision of high-speed, light-actuated devices is now a reality and optical computing and amplifying systems may soon be developed.

One of the most promising new technological developments is known as *microelectronics*. The transistor has given electronics a tiny replacement for the older, larger, and less efficient vacuum tube. Lately, electronics engineers have begun to learn how to reduce the size of other electronic components, such as *resistors* and *capacitors*. This means that entire electronic circuits can be manufactured in a tiny package no larger than an aspirin tablet. One method of fabricating such circuits involves the diffusion of impurities into various portions of a tiny crystal of silicon. In this way, the resistors and capacitors are formed in or upon the silicon crystal concurrently with the formation of transistors. Although microcircuits are still technological infants, they are revolutionizing the electronics industry much more rapidly than did the transistor.

—William C. Vergara

Tufting, a variation of the sewing process used in the production of carpets, rugs, bedspreads, robes, blankets, and wearing apparel.

Carpets are the most important tufted fabric, with 300 million yards produced in 1964—a jump from 21 million yards in 1951, when the first figures were recorded by the U. S. Department of Commerce. During this short time, tufted carpets have replaced most woven carpets, and now more than 80 per cent of all carpeting produced in the United States is tufted.

In the tufting process, a piece of yarn threaded through a needle is pushed through a woven fabric (jute for carpets), caught by a latch and held, and withdrawn through the fabric. On some machines the loops of yarn are cut, producing *cut pile*. Other machines do not cut the loop, and the fabric is *loop pile*. Machines can be interchanged to make both types.

Machine output is high—about 15 times as high as on looms that make woven carpet. One tufting machine 12 or 15 feet wide produces 600 yards of carpeting in one 8-hour day. A machine runs at 550 *courses* (stitches) a minute. Machine gauge (distance between needles) is 1/8, 3/16, or 5/32 inch.

The yarn comes from a creel at the back of a machine that holds an average of 1,600 yarn ends. On machines that run the same style of carpeting for long periods, it is more economical to use yarn on beams, as in weaving. Latex backing is applied to all carpets. It locks the yarn tufts securely into the jute backing, gives carpets dimensional stability, and makes them nonskid.

■**HISTORY.**—Tufting began in 1895, when Catherine Evans (later Whitener), a Georgia farm girl, made a hand-tufted bedspread patterned after an old heirloom bedspread. She made this bedspread by drawing designs on unbleached muslin with a dinner plate and a ruler. Then she sewed stitches along the pattern with cotton yarn, and cut the heavy yarn between each stitch. She made a second hand-tufted bedspread as a wedding gift in 1898 and sold her first bedspread in 1900 for $2.50. Soon she enlisted the help of her neighbors to fill orders for her handiwork. Within ten years, women throughout the area were making bedspreads by the hundreds from her patterns.

Early bedspread operations were conducted in the home. Groups of people bought cotton sheeting by the case and tufting yarn from local mills. As bedspreads found a ready market, sheets marked with patterns were sent to distant rural communities and individual farmhouses. Women—and sometimes men and children—in the families tufted the bedspreads and clipped the stitches.

Many people worked to perfect tufting machines. The first machines were made from converted household sewing machines with single needles. Then came multineedle machines with gooseneck heads open at one end. About 1940, a few tufting machines were built with a fixed support at each end and a needle bar between; they tufted fabric in continuous rows 40 to 50 inches wide. Soon these machines were used to tuft fabrics for robes and small cotton rugs. By 1950, machines were being built to make cotton pile carpeting 9 to 12 feet wide. All of these machines were developed by local people around Dalton, Ga., and Chattanooga, Tenn. Britain has a high concentration of imported tufting machines producing carpets, and is now manufacturing some of its own machines.

Machines with electrically controlled pattern attachments produce many varieties of imaginative patterns from natural and manmade fibers in unlimited bold or soft classic colors. —Richard B. Pressley

Vacuum Tube, a device consisting of several metallic structures in an evacuated glass chamber, capable of amplifying or switching electronic signals. The modern vacuum or electron tube, the triode, was invented in 1906 by Lee De Forest. This was actually a modification of an earlier device, the Fleming valve, or diode, invented by John A. Fleming in 1904.

De Forest's *triode* consists of three essential elements: a metal plate, called the *anode;* a wire filament or cylindrical electrode, called the *cathode;* and a wire-mesh screen, called the *grid.* A glass tube or envelope from which air is evacuated encloses the three elements and gives the structure its name, *vacuum tube.* The triode differed from Fleming's *diode* only in the presence of the grid; this, however, was enough to make the triode a significant advance.

In operation, the cathode is heated red hot and boils off electrons which form in a cloud around the cathode. Some of these electrons are attracted into the plate, which is connected to a positive voltage. In order to reach the plate, however, the electrons must first pass through the openings in the wire-mesh grid. The grid in a vacuum tube acts like a Venetian blind. A blind controls the amount of light that passes through it by varying the openings between the slats. A grid controls the flow of electron current from cathode to anode by varying the amount of charge on the grid. It is a well-known electrical principle that negative charges repel one another. Therefore, if the grid's charge is made sufficiently negative, few electrons can get through to the plate. As the grid's charge becomes less negative, a greater number get through. Since a very small amount of charge on the grid controls a much larger flow of charge (electrons) to the plate, the vacuum tube is able to amplify signals.

Vacuum tubes are often equipped with additional electrodes to alter their characteristics and make them more suitable for high-frequency or other special applications. The *tetrode* contains a second grid, and the *pentode* contains three grids.

Vacuum tubes are used in applications unsuited to transistors, such as high-power transmitters and the extremely-high-frequency circuits of some radar and communication systems. They are also used in applications where transistor costs are still too high, although this economic advantage seems temporary at best.

The most critical element in the Telstar satellite is a device called a *traveling wave tube.* It is a special type of compact and rugged vacuum tube that can amplify electronic signals at extremely high frequencies. For such purposes, the vacuum tube is still unrivaled by the tiny transistor. As transistor technology advances, however, use of the vacuum tube will be increasingly limited to specialized applications for which no transistor is yet available.

—William C. Vergara

Valves, mechanical devices used to control the flow of liquids and gases through pipes or bored passages. According to their application, these devices can be classed as on-off, throttling, or check valves. *On-off valves* allow either uninterrupted flow or provide complete blockage. For such service, *gate* or *plug valves* are commonly used. *Throttling valves* modify the flow by partially blocking the passage, reducing the pressure and volume of fluid passing the valve. *Globe valves* are commonly used for throttling applications. A common application of such throttling valves is the water faucet. *Check valves* limit flow through a pipe to one direction and are widely used in pumps and compressors.

Gate valves and *globe valves* are produced in brass, bronze, cast iron, alloy steel, and stainless steel for pressures up to 5,000 pounds per square inch and temperatures up to 1,200° F. Most sizes are regularly stocked and have been produced to meet material and dimensional standards set by the American Society of Mechanical Engineers (ASME). These valves are generally fitted with flanges or threaded ends for mechanical connection, but special ends are also provided to allow the valves to be welded or sweated to pipe or machinery.

Poppet valves are used where rapid operation and high pressures are encountered. Common applications include internal-combustion engines, intake and exhaust valves, and hydraulic equipment control valves. The primary advantage of the poppet valve is that only a small movement of the valve stem is required to initiate or halt fluid flow.

Reciprocating pumps and compressors use check valves to maintain fluid flow in only one direction. Air or gas compressors employ *reed* or *plate check valves* consisting of thin metal elements operated by the pressure of the gas that is being pumped. During the suction stroke, the flexible metal plate or reed is lifted from the valve seat by the incoming air. During the compression stroke, the plate or reed is clamped to the intake valve seat by the pressure of the air in the cylinder.

Pump check valves are also operated by the pressure differential that exists between the intake and discharge ports of the pump. The design of the valve used is determined by operating pressures and by the type of liquid that is to be pumped. To simplify service, both compressors and pumps frequently use interchangeable intake and discharge valves. —Joseph J. Kelleher

Voting Machine, a mechanical device for automatically recording and totaling the votes of electors. Voting machines are used only in the United States and in Trinidad and Tobago, West Indies. Most of the world still uses paper ballots marked by the voter and deposited in a sealed ballot box for counting after the polls close.

The Meyers Ballot Machine, invented by Jacob Meyers, was the first voting machine used in an official election; it was introduced at Lockport, N.Y., in 1892. These early machines were difficult to operate and not entirely reliable. Development of the Keipers roller interlock in 1912 offered the first practical means of controlling operation of a voting machine to limit electors to one vote for an individual candidate and no more than the legal number of votes for any single office.

Today's voting machine contains thousands of complicated parts, with space for as many as 500 candidates and referendum questions. It provides fast, accurate election results. A major feature is the protection it affords the elector against accidentally invalidating his own ballot.

When a voter enters the machine, it is locked and cannot be operated until an election official releases a latch on the outside of the machine. The voter then moves a large handle that closes the curtains and activates the machine so he can vote in secrecy. On the face of the voting machine, candidates names are listed in party rows or in groups under the office they are seeking. The voter moves a small lever next to the name of his choice, setting the machine to record a vote for that candidate.

After his ballot is cast, the voter pulls back the large lever that in one step records his vote, returns the candidate levers to their original positions, and opens the curtains. At the same time, a paper roll is advanced to hide any write-in vote and to present clean paper for the next voter. Results are obtained by uncovering the counters.

Ninety-nine per cent of the voting machines are made by two companies. Recently, ballot-marking devices and counting equipment have entered the election market. This equipment permits voting on a punched card ballot for tabulation by electronic machines. The special electronic counting systems use paper ballots marked with a fluorescent ink. To date, these devices have been used mostly on an experimental basis. —Howard Burr

Water Treatment, the improvement of the quality of water to make it suitable for use or to control pollution. Water treatment is required for nearly all of the 22 billion gallons of water passing through municipal systems and 160 billion gallons used by U.S. industries each day.

Major forms of treatment include clarification and filtration to remove solid matter; sterilization to prevent disease; softening to remove hardness; demineralization to remove dissolved solids including water "hardeners"; aeration or deaeration to remove undesirable dissolved gases; and the use of chemicals to prevent corrosion and scale formation. Generally, the more water is purified, the greater is its cost to the user.

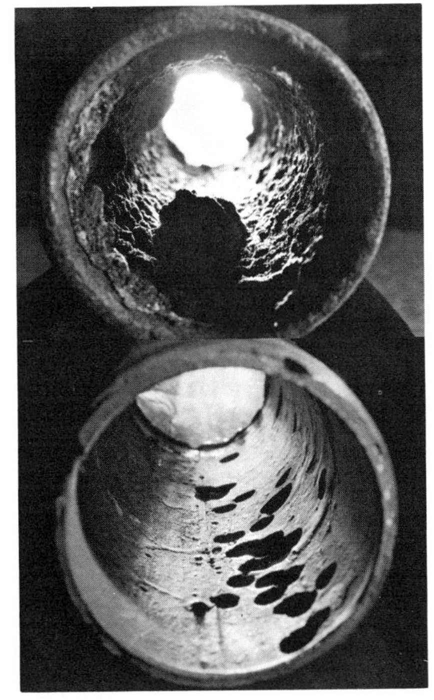

CALGON CORPORATION

WATER treated chemically stops oxidation that clogs (*top*) and pits (*bottom*) pipes.

In *clarification,* water is retained for several hours in large settling basins to allow matter to drop to the bottom, where it collects as sludge. In modern plants, continuously operating equipment removes the sludge deposits. Hundreds of tons of sediment daily may settle out of water supplies withdrawn from turbulent rivers for a typical large community.

Coagulant materials, such as alum, ferrous sulfate, and lime, or organic polymers, are often mixed into the water to speed the settling of small particles. Coagulants attract and entangle particles suspended in the water, causing the formation of tiny clusters, or *floc,* which become heavy enough to drop relatively quickly.

Filtration of water through sand beds or anthracite generally completes clarification by removing fine particles of matter and bacteria that did not settle out. Filtration follows the *softening* treatment where mineral salts must be precipitated out of solution.

Adsorption, by filtering water through granular activated carbon, removes dissolved organic contaminants that do not respond readily to other treatment. Filtration may be combined with adsorption when using beds of granular activated carbon.

In *sterilization,* a biocide—usually chlorine—is added to the water in small quantities to kill disease-causing, or pathogenic, bacteria. Under some circumstances, chlorine dioxide gas is used to sterilize water because of its lesser tendency to accentuate objectionable tastes and odors in drinking water.

Softening of water is the reduction or removal of hardness, caused by the presence of calcium and magnesium salts that can deposit as scale in boilers, cooling water systems, domestic hot-water heaters, and other equipment. A common method of softening water is the addition of lime and soda ash to precipitate the calcium and magnesium salts. Cold lime-soda softening is conducted at normal water temperatures; in hot processes, the temperature is at the boiling point (212° F) or higher. Zeolite softener equipment uses ion-exchange materials to remove calcium and magnesium ions.

Demineralization removes dissolved salts of all types, including hardness, to produce water of distilled quality. Various types of equipment used follow the principle of removing cations (positively charged ions), such as calcium, magnesium, and sodium, in one operation, and anions (negatively charged ions), such as sulfate, chloride, and silica, in another.

Aeration, the mixing of air and water by spraying or forced air, removes dissolved gases such as hydrogen sulfide (which is the cause of a "sulfur" taste) and carbon dioxide (which causes corrosion). It also helps precipitate iron and manganese and increases biochemical oxidation.

Where the corrosive effect of oxygen is undesirable—notably in boiler systems—oxygen, carbon dioxide, ammonia, and other gases are removed by *deaeration.* Water is heated to its saturation temperature (which varies with the operating pressure), the steam and water are mixed, and the gases are vented from the system. Deaeration of cold water is accomplished by vacuum.

Many other types of water treatment are used to solve specific needs. In boilers, phosphates prevent scale by precipitating calcium as a nonadherent sludge. Organic materials, such as tannin and lignin, disperse the sludge in the water so it can be removed by "blowdown." Corrosion control is effected in boilers with caustic soda, and in condensate return systems with volatile alkalies or filming amines that coat metal surfaces with an impervious film. Sodium sulfite and other oxygen-scavenging chemicals are often used to remove any oxygen remaining in the boiler water.

Municipal systems and once-through cooling systems frequently use sodium and sodium-zinc glassy phosphates to prevent scale deposits and reduce corrosion and "red water" caused by the precipitation of iron. Recirculating cooling systems use sodium-zinc glassy phosphates, or zinc ion with chromate, to control corrosion and scale deposits. Algae and other slime growths in cooling and process systems are controlled with biocides, such as chlorine, copper salts, quaternary-ammonium compounds, and potassium permanganate. Biocide materials are often alternated to overcome resistance developed by many organisms within the system.

■POLLUTION CONTROL.—Treatment of water *after* use, before it is discharged, has become a matter of public interest because heavy waste accumulations have caused serious pollution of many bodies of water. In 1964 in the United States, municipal and industrial systems dumped into rivers and lakes organic wastes equivalent to the untreated sewage of about 160 million persons.

Modern municipal sewage treatment plants can remove about 85 to 95 per cent of organic solids, and nearly all of the bacteria. So effective are these processes that, in Baltimore, a large steel company receives its service water supply of up to 150 million gallons per day directly and economically from that city's sewage treatment plant.

Industrial wastes are often more complex, and treatment depends on the specific waste materials being discharged by each plant. These may range from oils to acids to heavy metals to organic wastes. Treatment methods include sedimentation, coagulation, filtration, adsorption, flotation, skimming, and oxidation.

The use and reuse of water within the same plant before it is ultimately discharged as waste is a practice that can be economically attractive in locations where initial treatment of raw water supplies is costly.

—Anthony E. Pizzuto

JACQUARD LOOM, *(left)*, was invented in 1801 by a Frenchman for weaving intricate patterns. The complicated mass of threads that is being fed into the machine from the top determines the pattern. The finished fabric can be seen at the bottom of the machine.

NAVAJO woman weaving a blanket on primitive loom held together with rope.

MODERN HIGH-SPEED looms are just like simple looms but are automatic.

Weaving, the interlacing of two sets of yarns to form a fabric. The concept of weaving has not changed for centuries—some historians say for more than 6,000 years. The operation of a hand loom is based on age-old principles of weaving—the principles still used in modern looms.

First, the many parallel strands of yarn come from a *beam,* which is a large model of a simple sewing spool. These multiple strands of yarn, called *ends,* are the *warp.* It is not uncommon to weave fabric from a beam of more than 10,000 individual warp ends of synthetic fiber. Such a warp is 3,000 or more yards long.

The second part of fabric is the *filling,* the strands of yarn that interlace the warp ends to form the *fabric.* Filling is placed at right angles to the parallel warp ends in continuous rows. Filling yarn is usually placed on a *bobbin;* an average bobbin contains 4,000 yards of yarn.

In the hand loom, the weaver moves the treadle with his feet to move the warp yarn up and down, a process called *shedding.* The long needlelike fingers that support the warp ends are *heddles.* The shafts that support the heddles are *harnesses.* Pattern possibilities increase as the number of harnesses are increased. When one harness is at its lowest position and the second is at its highest, an *open shed* is created. Between each harness movement, the weaver throws a wooden shuttle containing a small bobbin of filling yarn through the open shed, a process called *picking.* After each shedding and each picking, the weaver pulls a handrail holding comblike teeth toward him to push the filling into the woven fabric, a process called *beating up.* The weaver continues this process over and over to weave fabric. Working steadily, the hand weaver of today, like the weaver of centuries ago, produces fewer than 20 picks a minute.

■MECHANIZATION.—With this same slow process, all fabrics were woven from prehistoric time until the textile industry of England pioneered the Industrial Revolution in the eighteenth century. Weaving developments during this great period of invention were most often met with passive resistance or even violence. But they created millions of new jobs, increased standards of living, and improved working conditions.

The first of these inventions was the fly shuttle patented by John Kay in England in 1733, which is still used in most looms today. However, weavers of that day thought it would put them out of work. Consequently a mob broke into Kay's shop, smashed everything, and would have killed him if two friends had not smuggled him out in a sheet.

About 1787, Edmund Cartwright, a minister of the Church of England, patented the first practical power loom. His loom was also the first one that stopped automatically when the yarn broke. Cartwright set up his first factory at Doncaster, using a bull to supply power. He replaced bull power with power from a steam engine shortly afterward.

The American textile industry began in 1790 when Samuel Slater built a loom and started manufacturing cotton cloth in Pawtucket, R.I. Slater, who was from England, built the loom from his memory of the design of looms there.

During the last part of the eighteenth century, Joseph Marie Jacquard, a mechanic of Lyons, France, turned his talents to improving the way of raising harnesses in looms for figure weaving. About 1801, he perfected the harness motion that bears his name today. But his invention also brought violence. Silk weavers of France, fearing that it would deprive them of their livelihood, broke up his device, and he had to flee to save his life. The device was declared public property in 1806, and Jacquard was rewarded with a pension and a royalty on each machine. By 1812, there were 11,000 looms with jacquards in France.

Engineers consider the jacquard one of the most perfect machines ever made. Today the principle and the essential parts are the same as Jacquard originally conceived them.

In 1894, James L. Northrop, an American, invented the rotary battery that changes bobbins of filling automatically. Before this time, weavers had to put up bobbins manually into shuttles on looms that changed shuttles automatically, or had to stop the looms to change bobbins in shuttles. Northrop's invention, like Jacquard's, is used in today's looms and is a marvel of precision—it changes bobbins with looms running at top speed.

Even though weaving today is based on principles thousands of years old, looms are quite efficient. *Fly-shuttle* looms run at 210 to 220 picks a minute. With the newest devices that transfer filling automatically, one weaver, one fixer, and one utility man can operate 100 looms for a total output of 10 yards of fabric 50 inches wide every minute. These looms often run more than a week without stopping. Loom efficiency in well-run mills is 98 per cent, with only 2 per cent second-quality fabric.

There are 2.6 million looms in the world today, including nearly 400,000 in the U. S. The world's total is dropping each year by about 25,000 looms because newer looms are wider, faster, and more productive. The only exception is in Asia, where looms are increasing at about the same rate that they are decreasing in other sections. Even with fewer looms, United States fabric production is gaining.

■TYPES OF LOOMS.—Looms are divided into three classes: cotton, manmade fiber, and woolen.

Three shedding motions are being used in today's looms: (1) cam motions with up to six harnesses, primarily for cotton fabrics with plain or twill designs; (2) dobby motions with 20 and 25 individually controlled harnesses, for cotton and manmade-fiber fabrics with fancy designs; and (3) jacquards for all fibers, with every warp end in the pattern controlled individually, for fabrics varying from brocades to woven portraits as intricate as any painted on canvas.

Loom picking today is much as it has always been. New metals enable picking parts to stand the pressure of high-speed looms. Looms are divided into types: *single-shuttle* looms for only one color or type of filling, and *multishuttle* (box loom) for four colors, or from four to seven colors with each color separate.

Northrop's rotary battery is still used in about 50 per cent of all looms, because the filling yarn and bobbins come directly from the spinning frame to eliminate rewinding, which has to be done with other systems. Three new systems for handling filling at looms have been developed the past 10 years, two of them in the U.S. and the third in the U.S. and many other countries.

The first system, the so-called *loom winder*, winds filling on the loom from a package of yarn weighing up to six pounds. The first production installation was made early in 1956. Today this system is being used in more than 100,000 looms throughout the world.

The second system is an *automatic filling magazine* that holds 96 bobbins. The loom changes these bobbins automatically. The system is being improved to handle bobbins directly from spinning frames. The economic aspects of the device are better than looms with rotary batteries.

The third system is known as *shuttleless looms*. Filling is inserted through the warp shed with flexible steel tapes much like a flexible steel ruler. This principle, *rapier* (swordlike), is not new, but U. S. mills have had such looms only since 1960. Economically, shuttleless looms are much better than fly-shuttle looms, but the range of fabrics that can be woven is limited. Speed is 230 picks per minute on looms 50 inches wide, a gain of about 50 per cent over fly-shuttle looms.

Another weaving system, called a *weaving machine*, is being used for wide looms (90 inches) weaving woolen and worsted fabrics. Filling is fed by a series of small shuttles (up to 16 of them, 3⅛ inches long) from large packages of yarn on the machine. The machine uses four colors of filling. The shuttles are picked through the warp shed one at a time by torque from a highly flexible steel rod. The looms are about 40 per cent more productive than fly-shuttle looms of equal width. (See also *Spinning* and *Tufting.*)

—Richard B. Pressley

Welding, the general term for a number of processes by which two or more objects or parts are joined to form a single, continuous whole. In practice, the term is applied only to those processes in which heat is utilized to produce the joint. The term is further restricted to joints involving only materials identical with or closely related to those of the parts being joined. In this sense, soldering, brazing, and adhesive bonding are not considered welding processes.

Traditionally, welding was restricted to metallic materials, but more recently the term has been extended to cover plastics as well. Not all materials can be welded either to themselves or to one another.

Most welding processes involve melting and resolidification of the materials in the area to be joined. Exceptions are forge welding and, more recently, friction welding. In *forge welding*, the materials are heated to the plastic point and are joined together by the application of external force, as might be exerted by the blows of a hammer, the method normally used by blacksmiths. *Friction welding* is very similar to forge welding, except that the heat is generated by friction of the two parts against each other automatically.

■FUSION WELDING.—All other methods of welding are classified as *fusion welding*. They differ from one another only in the way fusion of the mating surfaces is obtained.

In *gas welding*, heat is generated by the combustion of gases; the flame is applied to the areas to be joined until local melting occurs. The molten metal is restrained from flowing off and, on becoming solid, unites the parts. Often, additional metal, as from welding rods, is used to fill the space between the parts being joined. Gas-welding processes include air-acetylene, oxyacetylene, and oxy-hydrogen, which produce flames of progressively higher temperatures as are needed with metals of progressively higher melting points.

Thermit welding is a process in which aluminum powder is mixed with the oxide of the metal to be welded and is packed around the joint. Heat is applied to some of the mixture, starting a chemical reaction in which the aluminum powder becomes aluminum oxide while the original metal oxide is reduced to pure metal. The reaction produces intense heat that melts the ends being joined. The aluminum oxide floats to the top and is later removed.

Arc welding includes a number of processes in which an electric arc is established between an electrode and the parts to be welded. Electrodes are either carbon or metal, and the processes are known respectively as carbon-arc and metal-arc welding. In *carbon-arc welding*, the consumed carbon is generally volatilized and does not become a part of the weld. Some metal-arc processes involve the use of an electrode, generally tungsten, which does not melt and therefore does not become part of the weld. Other *metal-arc welding* processes use an electrode that is of substantially the same material as the parts being joined, is melted in the process, and becomes part of the weld.

In *induction welding*, the welding heat is obtained by the flow of an induction current in the region of the weld. The resistance of the work to the flow of this induced current causes the metal to heat to the welding point.

Resistance welding also uses an electric current. The parts to be welded are brought into close contact with each other, and a current of high amperage and low voltage is caused to flow through them. Air gaps and surface oxides in the area of the weld act as resistances to the flow of current, resulting in localized heating, with melting of the metals, closing of the air gaps, and expulsion of the oxides. Typical welding techniques based on local electrical resistance are *resistance spot welding*, *resistance seam welding, projection welding, flash welding, upset welding,* and *percussion welding.*

NEWER METHODS.—New welding techniques are electroslag welding, plasma-arc welding, and electron-beam welding.

Electroslag welding is used for depositing large amounts of metal, as might be needed to fill a large space between steel parts. Molten slag is placed between the parts, and electrodes of the metal to be welded are immersed in the slag. A current is then caused to flow through the electrodes. The slag, acting as a resistance, causes the electrodes to melt, the molten metal dropping to the bottom of the opening and gradually solidifying. The slag then rises in the opening, and, as more of the electrode metal is fed into it, continues the process, ultimately filling the gap.

In *plasma-arc welding,* the filler metal in the form of wire or powder is fed into a plasma jet and volatilized, to be condensed and solidified on the ends of the parts to be welded, thus effecting the joint.

In *electron-beam welding,* a flow of electrons is focused on the surfaces to be joined, causing them to melt to a shallow depth. The weld is produced by the solidification of this molten metal upon cooling.

Plastic materials are welded either by the friction process or by localized melting with a hot iron.

—Felix Giordano

Wood Finishing, the process of adding a protective or decorative coating to wood products. Wood finishing is an ancient art, dating back more than 2,000 years when balsams, pitches, shellacs, and oriental oils were used to adorn and protect wood surfaces. Although these purposes remain basically the same, there is quite a difference in the materials and their application today. Wood products that are factory-finished include furniture, wallboard and paneling, some flooring, and a variety of special products.

Paneling and other items with a flat surface can be production-finished on conveyor lines. These wood surfaces may require only two coats (with an intermediate buffing) applied by a *curtain coater.* This machine forms a veil or curtain of finishing material of proper viscosity, through which a panel passes on a conveyor belt to receive a coating.

Hardboard panels may first be given a ground-color coat in a *reverse roller coater,* which forces the pigmented material into the pores. It is then imprinted with a wood grain by an intaglio-type printing plate. A top coat is then applied by a curtain or roller coater, or by mechanical reciprocating spray guns. A dry film, such as polyvinyl with a wood grain or decorative pattern, can be applied to hardboard by high-pressure rotary-press equipment.

Furniture is almost always assembled before finishing, so automatic mechanical methods of finish application are not practical. Skilled operators with spray guns do the work. There are conventional guns that operate under air pressure. Mixing jets of air and fluid causes the material to atomize. By adjusting or changing the tip of the gun, the operator can form different spray patterns.

Hot spray brings the finishing material to the gum in a heated state and, therefore, contains less solvent. Airless spray systems bring the hot fluid to the gun under high fluid pressure, without compressed air. This material is atomized as it leaves the nozzle. *Electrostatic sprays,* used in coating metals and other materials that are electrical conductors, have been used little in coating woods, for a coat of material containing an electrical conductor must first be applied to the wood. Small parts, such as legs, may be *dip finished.* Viscosity must be controlled and the rate of withdrawal accurately timed to achieve a uniform coat.

Appearance plays an important part in furniture finishes. Dyes or pigments are added to materials to produce certain color effects. Woods sometimes are bleached, then stained, thereby enhancing the grain pattern with a uniform color without interference from the original colors in the natural wood, which vary from piece to piece.

Style requirements account for variations in finishing procedures. Artisans, working with dyes, pigments, and deft hands can produce truly artistic finishes to excite the most perceptive furniture buyer or interior designer. Distressing, spattering, antiquing, and padding to accent grain are some of the techniques used to finish furniture.

Nitrocellulose lacquer is still the most popular material for furniture topcoats. It is easy to apply, has good stability, dries and rubs easily, and can be repaired with little difficulty. Amine and epoxidized lacquers, as well as alkyd urea synthetics, are also used for their better heat and mar resistance. However, they have less stability and are more difficult to apply or repair.

In the 1960s, synthetic materials were used in furniture finishing systems in order to improve performance. These so-called super finishes are impervious to most household acids and alkalies, including nail polish remover. They are, however, more difficult to apply and are not easily rubbed or repaired. Thus, their use remains limited.

—Raymond A. Helmers

Woodworking Tools, instruments used to facilitate mechanical operations on wood. Because wood is a universal material, used in some manner by virtually every industry and also in the education and hobby fields, the tools range from simple chisels to giant multipurpose machines costing tens of thousands of dollars. In wood products manufacturing, heavy-duty, high-speed machines are required. In smaller shops, including those involved in custom manufacturing, smaller, less sophisticated tools are needed. School shops, crating departments, pattern and sample shops, and cabinetmakers use machines in the middleweight class. Hobbyists have similar types of tools, but lighter. For construction work and maintenance, tools must be portable.

CROSS-CUT SAWS.—In manufacturing, rough lumber or panels are first cut to specified lengths by powered *cut-off saws,* which are stationary units. The cutting stroke is activated by a button or foot pedal, and there are both *overcutting* and *undercutting* types. *Swing saws* are manually pulled into the wood by the operator. Some machines are built with extra long stroke to cut wide panels. *Multiple panel saws* have two or more saw blades, and pieces are fed by a moving slat bed. These machines can be automated to cut programmed sizes. In smaller shops, the *radial arm saw* is the mainstay for cross-cutting lumber and panels.

In construction and maintenance, the portable electric *circular saw* is important. A combination rip and cross-cut saw blade permits this tool to serve a double purpose.

Blades for production cut-off saws are usually carbide-tipped for long runs between sharpenings. In cutting plywood, panels, or plastic laminate that might splinter or chip easily, a fine-toothed saw blade is required.

RIP SAWS.—Rip sawing (cutting wood parallel to the grain) calls for a blade that gives a slicing, raking cut. In production plants, special machines perform this work. They have power feeds and are adjustable as to width and depth of cut. *Gang rip saws* have two or more blades that cut the board into multiple strips of similar or varying widths.

PLANING AND JOINTING TOOLS.—In volume manufacturing, heavy-duty *facing planers* remove rough surfaces, level up twists and warps in the boards, and equalize their thicknesses. Usually, rough planing precedes ripping. If the stock is to be edge-glued, the boards must be jointed (planed edgewise). *Production jointers* have automatic power feeds and board turnovers. In the custom plant or hobby shop, simple hand-fed jointers are used. The newest method of production jointing uses rip saws, automatically fed with canted powered feed rolls.

EDGE-GLUING AND LAMINATING TOOLS.—Panels for solid wood construction or wood cores are made of strips of wood, glued edge-to-edge. They are produced in *panel-making machines,* which cure the glue by steam heat or high-frequency radio waves. Some types of panels may be made on mechanical rotating clamps, with the adhesive curing at factory temperature. Small shops use hand-operated *cabinet clamps.*

A press is required to make plywood or laminated panels. *Hot presses,* which have one or more openings, can make panels 4 x 8 feet or larger, and are opened and closed by hydraulic pressure. Steam, or a liquid heat transfer medium, under high pressure is forced into the hollow platens, providing the temperature required to cure the adhesive. Similar *cold presses* can be equipped with radio-frequency generators to cure the glue.

Laminating presses need less pressure. Many of these operate on compressed air. Panels are compressed, and stacks are secured between mechanical clamps and removed from the press. The glue then sets overnight at room temperature. Other laminating is done with contact cement. The decorative surface and the core are passed through a *pinch-roller press* that squeezes the elements together. Smaller plants and shops use air-type and pinch-roller presses.

A key machine for fabricating panels is the *double-end tenoner.* It is a continuous-feed machine with adjustable width, equipped with saws, cope, tenon heads, and arbor motors. Virtually any kind of machining of ends or edges of flat stock can be done by this machine. It can tenon, groove, saw, miter, edge-shape, dado, and even edge-sand. Top surfaces can be contoured by cutter heads on a long shaft.

Another multipurpose machine, not limited to flat stock, is the *double-end sawing, boring, and chucking machine,* which operates on cycles. Typically, it grasps the work piece, miter-trims both ends, bores or chucks each end, bores multiple holes in top or sides, and releases the piece. This equipment is widely used in making parts for chairs and upholstery frames.

Single-end tenoners and *miter-saws* are used in smaller shops.

■SHAPING AND ROUTING TOOLS.—Both *single-* and *double-spindle shapers* are used in large and small factories, especially on short runs. A jig or fixture holds the work piece or pieces, one edge of which serves as a pattern guide for the part as it rides against the shaper collar. Automatic shapers clamp the work piece, rotate it to the cutting head, and return it to the operator after the cut is made. Contour profilers perform similar operations. The piece is clamped on a moving table that takes it past a cutterhead, which moves in and out to conform to the pattern. Serpentine drawer fronts are typical parts made on this machine.

Routers can do similar work, but they have a rotating spindle above the work table. This allows the bit to be lowered into the work, and interior as well as exterior patterns thus can be cut in various patterns to specified depth. One make of router has an optical tracer that follows a simple drawing of a part. This in turn guides the travel of the router bit, automatically machining the part. Other routers have been equipped with automatic controls guided by magnetic tape. Numerical positioning control has been used on a router of still another design. Small shops and plastic laminators use portable high-speed routers and trimmers.

■MOLDING TOOLS.—*Molding machines* have one or more cutterheads that produce a continuous pattern lengthwise on a piece of wood. They vary in size, capacity, and speed of operation. In the same family of equipment are *flooring and center-match siding machines.* They take rough lumber and make center- and end-match strip flooring automatically.

■BORING AND MORTISING TOOLS.—*Multiple boring machines* are made in horizontal and vertical models. A boring machine can become a *mortiser* by installing a hollow chisel mortising bit. Smaller shops use *drill presses,* some equipped with multiple spindle attachments. Where portability is required, the *hand electric drill* is used.

■OTHER TOOLS.—There is a wide variety of tools and equipment whose names are synonymous with their functions: band saws, single- and double-end dovetailers, automatic and back-knife turning lathes, single and multiple spindle carvers, embossers, wood benders, veneer clippers and splicers, dowel machines and dowel drivers, corner block machines, chair seat machines, and other tools for home or industry.

■SANDING TOOLS.—*Multiple-drum sanders, wide-belt machines,* and *stroke sanders* are used for production sanding of flat surfaces. A sophisticated arrangement might have a wide-belt machine to cut down and even up the bottom of a panel. Another wide-belt machine would rough-cut down the top, while a third would give it a finer sanding. A final polish would be given by a stroke sander. One type of automatic stroke sander has a pneumatic platen that allows for slight variations in the wood surface, reducing the number of sand-throughs of the veneer. Wide-belt machines are sometimes used in place of knife-type planers.

There are automatic machines to sand moldings, irregular shapes, and turnings. There also are small machines with pneumatic drums or shaped wheels, and sanders whose belt follows a hand block, the shape of which conforms exactly to the molded part to be sanded.

Portable sanders come in models with rotating belts or disks. There are air- and electric-powered units whose pads have either reciprocating or orbital action.

■CUTTING TOOLS.—Home craftsmen and wood sculptors still use *chisels* and *turning tools,* such as *skews, gouges, round-nose chisels,* and *parting tools.* But power tools and machines have replaced these instruments in industry. *Bits* and *cutters* must be engineered for high-speed production. Power-boring bits include *spur bits, augers, countersinks, chisel bits, router bits, hole saws, plug cutters, wing cutters,* and *hollow chisel mortisers.*

The design and engineering of cutters and cutterheads for high-speed production is a demanding science. Electronic balancers and optical comparators are used by toolmakers to assure safe operation and efficient cutting.

GREENLEE BROS. AND CO.

DOUBLE END TENONER, a woodworking plant in itself, trims and machines both ends of a piece of stock in one pass through the machine.

Different wood species and types of cuts require specific cutting angles and optimum number of knife marks per inch for top efficiency. Cutterhead bits and knives may be made of high-speed steel, carbide-tipped or solid carbide, depending on requirements in use. Bits have a shape contoured end to end, so that they will retain the same cutting shape, sharpening after sharpening.

■**FASTENING TOOLS.**—Fastening tools in woodworking have progressed from hand devices to power-operated *staplers, hammers, tackers, nailers,* and *screwdrivers.* These tools drive nails up to two inches long into the hardest woods. One stationary-type machine makes its own fasteners from wire and drives them into the wood, even countersinking and blind fastening. Another drives corrugated fasteners, and there is a portable tool that does likewise. One tacker operates on electric power, although most use compressed air.

Air and electric screwdrivers and impact wrenches are useful woodworking tools. There are automatic screwdrivers and automatic nailing machines that can draw fasteners from a hopper through hollow flexible tube and drive them at the press of a trigger.

Portable power drills and screwdrivers, especially for building, maintenance, and hobby use, are available in cordless units. They are fitted with rechargeable nickel-cadmium batteries.

—Raymond A. Helmers

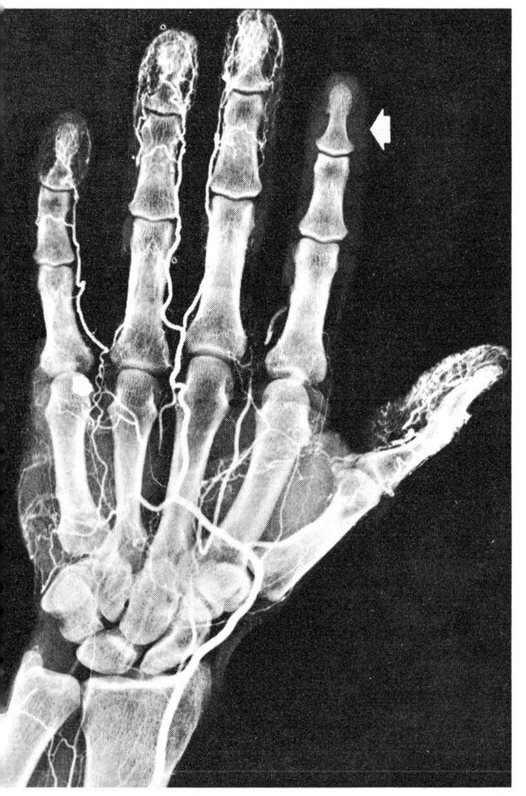

LOOK MAGAZINE

X-RAY shows a circulatory block in the index finger (*arrow*) ; injected radio-opaque material appears only in the other fingers.

X-Rays, high-frequency electromagnetic radiation produced in a vacuum tube when high-speed electrons—accelerated by high voltage—strike a metal target. X-rays are invisible to the human eye. The wavelength of this radiation is about 10^{-7} to 10^{-11} cm, which is about one ten-thousandth the wavelength of visible light. The rays can be detected by photographic plates, Geiger counters, ionization chambers, and chemically coated fluorescent screens.

X-rays were discovered by Wilhelm Roentgen in 1895 while he was experimenting with the *Crookes tube.* Roentgen noticed that a fluorescent screen coated with barium platinocyanide lighted up when electrons from the Crookes tube struck the screen. When Roentgen discovered that the screen glowed, even when the tube was totally encased in a paperboard box, he was baffled, until he realized a new kind of ray—completely different than visible light or ultraviolet light—was being produced. When he announced his discovery, Roentgen labeled them X-rays because X stands for the unknown in scientific research. For his discovery and development of X-rays, Roentgen in 1901 received the first Nobel Prize in physics.

Early X-ray, or Crookes, tube consisted of an evacuated glass tube with a concave cathode and metal anode, or target. A high voltage (30,000 to 40,000 volts) was applied between the anode and the cathode to accelerate electron flow from the cathode to the target. Electrons striking the target were decelerated and their energy absorbed by the atoms of the target. The affected atoms vibrated violently and gave off a high-frequency short-wavelength radiation—the X-ray.

The modern X-ray tube, developed by William D. Coolidge in 1913, uses a hot tungsten filament as a source of electrons, or cathode. The target, or anode, consists of a solid molybdenum or tungsten block capable of withstanding the kinetic energy of the electrons striking it without melting. Large X-ray machines frequently have provision to air- or water-cool the target. The *Coolidge tube* has the advantage of two separate control systems—filament temperature settings control the number of electrons delivered to the anode, while the tube voltage controls the speed of the electrons from cathode to anode.

Though X-rays can cause extensive damage to humans if uncontrolled exposure occurs, they have been a useful medical tool since their discovery. Early shadowgraphs of the human body excited medical interest and were the first practical application for X-rays. Today X-ray machines are found in every hospital and doctor's or dentist's office. The usual procedure is to expose a photographic plate to the X-rays for a permanent picture. However, if continuous observation of moving organs is required, a fluoroscope replaces the negative. High-voltage X-ray machines are frequently used for treatment of cancerous growths that cannot be removed by surgery. *Lead shielding,* which X-rays cannot penetrate, is used to protect personnel from dangerous overexposure during operation. *Film badges* and *dosimeters,* which record the amount of X-ray exposure, are carried to provide an additional safety check for the operating technicians.

—Joseph J. Kelleher

BIBLIOGRAPHY

AINSWORTH, JOHN H. *Paper, the Fifth Wonder,* 2nd ed. Thomas Printing & Publishing Co., 1964.

BERNSTEIN, JEREMY. *The Analytical Engine: Computers—Past, Present and Future.* Random House, Inc., 1964.

CROUSE, WILLIAM H. *Automotive Mechanics,* 5th ed. McGraw-Hill, Inc., 1965.

D'ARCANGELO, AMELIO M. *A Guide to Sound Ship Structure.* Cornell Maritime Press, 1964.

DEGARMO, E. PAUL. *Materials and Processes in Manufacturing.* The Macmillan Co., 1962.

DOLAN, EDWARD F. JR. *The Camera.* Julian Messner, Inc., 1965.

FISHER, DOUGLAS A. The *Epic of Steel.* Harper & Row, Publishers, Inc., 1963.

GREGORY, EDWIN and others. *Steel Working Processes; Principles and Practice of Forging, Rolling, Pressing, Squeezing, Drawing and Allied Methods of Metal Forming.* Transatlantic Arts, Ltd., 1964.

HOWARD, WILLIAM E. AND BARR, JAMES. *Spacecraft and Missiles of the World.* Harcourt, Brace and World, Inc., 1966.

HUNTER, MAXWELL W. *Thrust into Space.* Holt, Rinehart & Winston, Inc., 1965.

JOHNSTON, BETTY J. *Equipment for Modern Living.* The Macmillan Co., 1965.

KELLER, A. G. *A Theatre of Machines.* The Macmillan Co., 1965.

KENT, JAMES A. editor. *Riegel's Industrial Chemistry.* Reinhold Publishing Corp., 1962.

KISSAM, PHILIP. *Surveying Practice.* McGraw-Hill, Inc., 1966.

LAUB, JULIAN M. *Air-conditioning and Heating Practice.* Holt, Rinehart & Winston, Inc., 1963.

LEWIS, ROBERT S. *Elements of Mining.* 3rd ed. John Wiley & Sons, Inc., 1964.

LYTEL, ALLAN H. *ABC's of Computers.* Howard W. Sams & Co., Inc., 1961.

MOORE, HARRY D. AND KIBBEY, DONALD H. *Manufacturing Materials & Processes.* Richard D. Irwin, Publisher, 1965.

ODDO, N. and E. CARINI. *Exploring Simple Machines.* Holt, Rinehart & Winston, Inc., 1965,

PETERSON, HAROLD L. *A History of Firearms.* Charles Scribner's Sons, 1961.

TWORT, ALAN C. *A Textbook of Water Supply.* American Elsevier Publishing Co., 1964.

USHER, ABBOTT P. *History of Mechanical Inventions.* Beacon Press, 1959.

MATERIALS AND STRUCTURES

Introduction.—Structures have long constituted one of man's most evident and outstanding achievements in his efforts to adapt and shape the resources of his environment to the evolving material, social, and economic needs of civilized life. Seven of the outstanding structures of antiquity were known as "The Wonders of the World," and a number of the great constructions of the present century have been awarded similar distinction.

The needs that have challenged the skills and abilities of the master builder to provide such works have been varied. Meanwhile, the resources available with which to meet them, notably the materials, have likewise depended on the environment and age and are reflected in the scope and character of the structure provided.

Structural Requirements.—The varied needs of modern life pose a wide variety of structural problems, from buildings and bridges to dams, canals, and water supplies.

Over the centuries, the provision of shelters for homes and other buildings has evolved from the huts of primitive man to the most typically American structure of the present day, the "skyscraper." The predominantly agricultural way of life and economy of earlier centuries likewise prompted the construction of pioneer irrigation and drainage structures. Later, urban growth similarly demanded water supplies, streets, and other outstanding municipal works. The birth of trade and commerce has further challenged the skill of the structural expert in providing the essential arteries of transportation—roads, tunnels, bridges, canals, harbors, and river improvements. Successive improvements in vehicles, in more recent years, have required the building of the distinguished structures characteristic of the Railroad Age and of the current Highway Era.

Other needs, also, have long stimulated important technical advances. In the later Middle Ages, military interests led to the construction of one of man's most impressive structures: the medieval castle, the fortress-home of the Age of the Nobles. Two of man's outstanding structural achievements were inspired by religious motives: the massive pyramids of Egypt and, in sharp contrast, the daringly slender Gothic cathedral, "a bird-cage in stone" and quite as much a structural as an artistic triumph.

While many of these social and economic needs are thus as old as civilized life itself, the resources available to the structural expert have been vastly increased, notably in the last century and especially since the beginning of the twentieth century. As a result, modern works have set new records for size and scope.

Materials.—Among the physical resources available to the planner and builder, materials have played a vital role in meeting these challenges. Historians usually refer to the Stone Age (about 3500–3000 B.C.), supplemented by an early use of copper and by the Bronze Age. About 1200 B.C., the Iron Age, in turn, followed a limited earlier use of meteoric iron. Actually, metals long remained scarce and costly, and were limited to tools, weapons, and fastenings. It was not until the eighteenth century in Britain that cast iron became available in quantity and at a cost that made possible its use a structural material. The world's first iron bridge was not built until 1779, at Coalbrookdale, England. Wrought iron followed with Henry Cort's puddling process about 1800 and the development of rolling processes to make bars and shapes. A century was to pass, however, before Henry Bessemer's method of producing steel led to the building of the first all-steel bridge in the United States in 1878.

PUBLICITY CONSULTANTS, INC.

THE WORLD'S TALLEST skyscraper, New York's Empire State Building, rests on a steel framework that weighs 60,000 tons.

From the structural standpoint, therefore, man was limited in materials to timber, stone, and brick—essentially the resources of the Stone Age—until well into the nineteenth century. In fact, it was not until the turn of the present century that another material, portland cement, was effectively developed to replace stone in construction.

CEMENT.—The early Greeks had discovered that a mixture of Santorin earth (a volcanic ash from the island of Santorin) with lime produced a "natural" hydraulic cement which, unlike lime itself, would harden in bulk or under water. The Romans, likewise, found that the volcanic ash pozzuolana produced a similar product. They were the first major users of a mixture of sand, broken stone, and cement (concrete) instead of stone or brick masonry.

These materials were of limited, local occurrence, however, and modern hydraulic cement did not become more widely available until the earlier years of the nineteenth century. John Smeaton in Britain had noted that cement made from limestone containing clay possessed hydraulic quantities and, about 1820, Joseph Aspdin developed "portland" cement, so called because it resembled the famous limestone from the Isle of Portland. In the United States, the earlier "natural" cement had been introduced in building the Erie Canal, completed in 1825. Since portland was imported, it remained costly, and stone masonry, using a minimum of mortar, was standard until the birth of the American portland cement industry at the end of the nineteenth century. The New Croton Dam, built for the water supply of New York City in 1895–1907, was the last great stone masonry structure in the United States, the last in which natural cement was used, and the first in which portland was employed.

METALS AND CONCRETE.—Alloy steels, aluminum, and other manufactured materials have also been developed, but their use has been confined largely to building construction. The combination of concrete and steel—reinforced concrete—however, has been widely used in recent years in "thin-shell" specialized engineering structures, such as airport structures.

Labor and Equipment.—These are the chief resources, other than materials, that have affected structural developments. As in the case of metals and cement, construction equipment long remained limited. Man-powered devices prevailed until steam power was applied to construction needs. The steam-powered pile driver, steam-driven pumps, steam drills, the steam shovel, and other modern devices gradually replaced ancient manpower equipment in the nineteenth century. This replacement of man by machine has been particularly characteristic of American practice.

In Europe, skilled labor generally has been plentiful while, relatively speaking, materials have been costly. In the United States, on the other hand, materials have been less costly while there has been a shortage of skilled labor, partly as a result of the availability of cheap land in the last century. Increasing labor costs, therefore, encouraged the development and use of machines. Here again, the change has been particularly rapid in the present century. It is interesting to note that, following the earlier use of steam shovels, there was a notable advance in rock drilling and the machine handling of excavated material when the Chicago Drainage Canal was built in 1892–1900, while the first New York subway, opened in 1904,

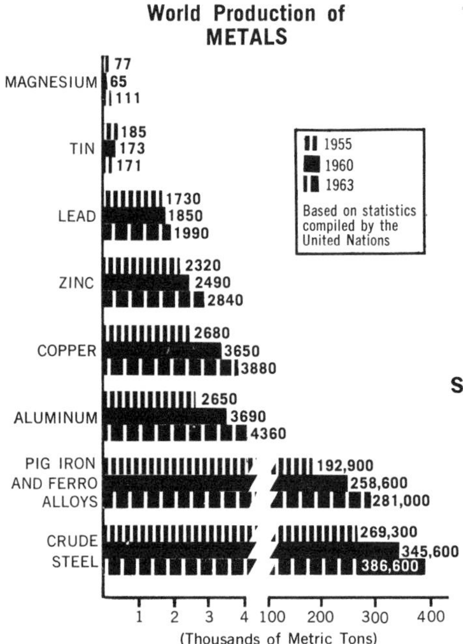

was the last great American pick-and-shovel and horse-and-cart construction project.

One result of mechanized construction is that it has speeded up operations and thus reduced the loss of income on expended capital. For example, the number of men and hoists that could be used on a stone-masonry work was limited. On the Croton Dam, the best record was 17,188 cubic yards placed in one month. This was increased to 84,450 in placing the massive concrete of the Kensico Dam in 1913, while on the Grand Coulee Dam of 1939, with modern machines and belt conveyors, over 400,000 cubic yards were recorded in one month.

The advent of new materials and machine equipment has not only made possible far greater structures than the world has ever known but has also solved a problem that man for centuries was unable to meet. In building the stone arch bridges of earlier days, it was seldom possible with the equipment of the time to unwater, excavate, and build bridge piers more than 8 or 10 feet below water level. While few stone arches failed, foundation failures were frequent, as floods undermined the shallow piers. Today, open cofferdam foundations have been carried down to 80 feet or more below water level, using steel sheet piling and modern pumps and excavating equipment. Similarly, pneumatic and other methods make it possible today to provide adequate foundations under conditions that would have rendered such undertakings impossible a century ago.

These advances, involving major capital expenditures for equipment to avoid or reduce the costs of hand labor, and the emphasis on speed in construction have also brought about a marked change in the organization, planning, and management of construction operations. The building of many structures, such as stone arch bridges, has long required effective planning, direction, and skilled labor, but the earlier years of the rough-and-ready contractor bossing a group of husky laborers has given way to fully planned, scheduled, and organized construction operations under the direction of engineers who specialize in such work. As in industry and manufactures, engineered construction has become a major, exacting, and important field of engineering endeavor, and great works are completed today in far shorter time than was required for many minor undertakings in earlier years. (See *Construction Engineering.*)

Structural Mechanics.—While the major structural forms of the present day—the beam and column, the arch, retaining walls and dams, the truss, and suspension bridges—have long been known and used, their earlier design was based almost solely on an intuitive structural sense and judgment, matured through experience and observation. There was, in fact, little incentive to develop more fully rationalized and exact quantitative, mathematical and scientific techniques of design. Although this has led to greater safety and, especially, made a saving of materials possible, these values were earlier relatively unimportant. Minerals were both plentiful and inexpensive, but they were also far from uniform in their strength qualities. The loads to which earlier works were subjected not only were moderate, but many earlier structures would, by modern standards, be characterized as generously "over-designed." On the other hand, the development of more costly materials, of the truss and other forms, and the greatly increased loads of a Railroad Era inevitably led to a search to replace the earlier purely qualitative techniques by the far more effective and certain methods of the present day. The theoretical studies of engineering science, notably aided by physics, and coupled with practical testing, observation, and analysis by engineers, have been basic in planning and designing our modern record-breaking structures.

The ancient Greeks early noted that a stone beam failed by the breaking of its lower face in tension. In a few notable cases, they reinforced such beams with iron bars imbedded in the lower surface. Galileo, in addition to his purely scientific interests, studied the mechanics of beams and, although his observation was based on erroneous stress assumptions, noted that their strength varied directly with their breadth, but as the square of their depth. Later, the recognition of the elastic behavior of materials led to what became known as "the common theory of flexure," which Charles Augustin de Coulomb, the French military engineer, first expounded about 1780.

The problem of the continuous beam (one built over more than two simple supports) also attracted early attention. However, the first realistic and effective approach to this problem was prompted by the adoption, as the nineteenth century came to a close, of reinforced concrete, in which a notable saving of materials was secured through continuity in design.

DAMS AND WALLS.—Another French engineer, Bernard Forest de Belidor, had advanced theories on retaining and quay walls to which Coulomb added some refinements that later workers, such as William J. M. Rankine in Britain, developed still further. Such walls are subjected to the action of widely varying and varied forces. Only in the last twenty-five years have many of the uncertainties of foundations and earth pressure been resolved and evaluated through the modern engineering science of soil mechanics.

Another retaining structure in which new records of height and size have been made in recent years is the masonry dam of the so-called gravity type. This type was late in development; the first rationalized, truly modern form did not appear until 1866. Even today, such uncertain factors as "up lift" still perplex the dam designer.

ARCHES.—Similarly, although the basic action of the masonry arch has long been understood, the so-called spandrel-filled arch involves so many uncertain forces that design is based largely on arches that have been built by experience, and only an approximate checking through stress analysis is possible. The far later appearance of the modern ribbed arch, however, beginning with the stone Luxembourg span of 1898–1903 but later more fully exploited in reinforced concrete, has led to the development of the modern theory of the elastic arch and to the construction of many notable and daring spans.

TRUSSES.—The truss, so widely used today, also has been known for centuries, but its first recorded use as a bridge form was about 1570. It was later carried forward by Swiss engineers. Nevertheless, the first iron bridges followed the early arch rather than the truss form, substituting cast-iron sections for the stone voussoirs. In the United States, however, the need for many inexpensive bridges and the abundance of timber led first to a remarkable timber truss era. Later, these spans gave way to iron trusses, and various forms were devised to provide for the more effective and economical use of this more costly new material. This advance required a more thorough knowledge and evaluation of the stresses in such framed structures. In 1847, a pioneer American truss builder, Squire Whipple, published the first text on truss analysis.

American engineers in the closing years of the nineteenth century were busy replacing earlier iron bridges with far heavier steel structures to meet the increasing locomotive loads of the Railroad Era. But new forms also were appearing. In 1917, Gustav Lindenthal set a new record with his "continuous" truss of two 775-foot spans for the Sciotoville (Ohio) bridge. The coming of the Highway Era, however, was to bring the great bridge spans of more recent years as well as interesting new truss and arch forms and record-breaking suspension spans.

■**SUSPENSIONS.**—The rise of the suspension bridge again demonstrates the vital role which materials have played in the evolution of structures. Crude suspensions of vines or ropes had been built in many lands, but the adoption and development of this form, which now far exceeds all others in record spans, awaited the advent of more permanent and stronger materials. The pioneer American bridge builder James Finley, in his first suspension of 1801, used links of wrought iron bars to form a "chain" suspension. Louis J. Vicat in France in 1830 adopted iron wire in lieu of bars and later developments have produced steel wire of remarkable tensile strength. The British engineer Thomas Telford reached a record 580-foot span in his "chain" suspension bridge in Menai (Wales) in 1826. Later great spans have been of wire and built in the United States. Washington A. Roebling, son of the American wire pioneer John A. Roebling, set a record in 1883 with the 1,595-foot Brooklyn span at New York. Finally, the cantilever span record of 1,800 feet was surpassed by the Ambassador bridge at Detroit of 1,850 feet in 1929. Within two years this was almost doubled by the completion of the 3,500-foot George Washington bridge in New York, and this, in turn, was followed in 1937 by the 4,200-foot Golden Gate at San Francisco, and the 4,260-foot Verrazano span at New York in 1965. Over 100,000 miles of wire were used in the 26-inch diameter cables of the George Washington span.

Inherently flexible "stiffening trusses" were early used in suspension bridges to avoid excessive local deflection by distributing loads. Such bridges thus became combination structures, and the involved mechanics of what, in effect, was a suspended truss, was not well developed until the earlier years of the present century. The susceptibility of such spans to aerodynamic action also led to some early failures. As spans increased, dead load mounts and stiffening were less necessary but the collapse of the Tacoma (Wash.) bridge in 1940, due to air action, again raised this problem. Today, modern design as well as modern materials make possible a still further span advance in one of the most spectacular structures of modern life.

Structural Engineer.—a specialist in the planning and design of a particular type of structure. As early as 600 B.C., the Greek historian Herodotus coined the title *architekton* for the builder of what would now be called an engineering work, a water-supply tunnel on the island of Samos. This "archtechnician" and the later Roman *architectus* were engaged in both architectural and engineering practice as we know them today. The first specialization arose from the term *ingenium,* or product of genius, which had been used as early as 200 A.D. to describe the engines of war of that day—catapults, battering rams, etc. By the late Middle Ages, the builder of fortresses became known as an *ingeniator,* or engineer. Military engineering became a specialty.

In the Renaissance, architecture began both to emphasize its "fine" art qualities and to draw away from the practical arts, while the rise of new needs, notably canals and other civic works that were unrelated to the artistic approach, tended further to divide the structural profession. It was not until 1802, however, that the last book to treat separately of both professions, architecture and engineering, appeared.

Not only did new engineering needs lead to this divorce, but the architect failed to follow the engineer in his use of new materials, in the development of more fully rationalized techniques of design, and in emphasis on utilitarian and economic values. In recent years architecture has been in transition, turning from a long-prevailing aesthetic approach to design, based on classical examples, to the search for a more practical evaluation of the artistic possibilities inherent in the new structural forms and materials, and to the effective utilization of new services developed by the engineer.

The structural engineer has continued to develop the scientific techniques of design and to further advances in materials, methods, and equipment, in seeking the fullest, most effective and economical use of all available resources in meeting the problems of his day. This involves a full and impartial analysis of the present and possible future needs and requirements of the structure he seeks to provide, and an equally objective evaluation of available resources in a balancing of alternate possibilities and costs. The professional structural engineer, therefore, requires not only special technical training, matured through experience, but also a keen sense of economic factors and values in seeking a synthesis leading to final plans and designs. —James Kip Finch

UNITED STATES STEEL

CATWALKS used by workers stringing steel suspension cables link towers of the Verrazano-Narrows Bridge under construction.

MATERIALS AND STRUCTURES GLOSSARY

Abrasives, materials of extreme hardness that are used to shape metal, wood, and stone by wearing away or grinding. Abrasive grains are used loose, mixed with oil to create a paste or slurry, compressed into wheels or blocks, or glued to a flexible backing such as cloth or paper.

In modern industry abrasives are used to polish, clean, cut, and machine materials such as wood, glass, stone, plastics, and ceramics. The high coefficient of friction of abrasive materials makes them essential in the formulation of nonslip paints and floor tile, brake linings, and clutch disks.

The grinding wheel of bonded abrasive grains has become a basic tool of the metalworking industry. Hand-held pneumatic or electric grinders are used to rough-grind and polish castings, welded joints, and metal stampings. Precision grinding machines are used to finish bearing surfaces, cam profiles, cylinder bores, printing-paper rolls, shafts, and pistons. An essential machine required for every metalworking shop is the tool grinder which is used to sharpen both hand tools and machine tool bits and cutters.

Abrasives may be divided into either natural or manufactured materials. Natural abrasives such as sand, emery, flint, sandstone, and silica have been used since the Neolithic period, which dates from 15,000 to 5000 B.C. The importance of abrasives in prehistoric cultures cannot be overestimated. Tribes or races that learned to cut grooves and drill holes in stones with abrasives had a distinct advantage over their neighbors. The holes and grooves enabled the prehistoric toolmaker to attach handles to his stone weapons securely, thereby producing a superior product. This seemingly simple advantage frequently enabled one tribe or race to conquer and dominate other peoples.

The first use of the grinding wheel was for the milling of grain. The wheel was mounted horizontally and driven by animals, water, or wind. Nearly 1,000 years passed before the grindstone was mounted vertically—on a horizontal axis—and turned by hand to sharpen or polish metal. Drawings in an illuminated manuscript known as the *Utrecht Psalter* (850 A.D.) illustrate the first recorded use of a grinding wheel for metalworking.

Since grinding wheels were produced from blocks of natural stone, grinding by this method was limited by the abrasive quality of the available stone. To take advantage of the wide range of natural abrasives available in granular form, metal and glass-workers in the fifteenth century began using wheels coated with grains. Tallow, wax, or oil was mixed with emery, garnet, or diamond dust and the mixture used to coat wood or metal wheels. Coarser grains of emery were used for metal removal while finer grains or dust, sometimes under pressure, could be used for polishing or light grinding.

The first synthetic abrasive, silicon carbide, was produced by Edward Goodrich Acheson in 1891. Sand (SiO_2), coke (C), sawdust, and salt are mixed and heated by graphite electrodes to 2,400° C. The basic chemical reaction is $SiO_2 + 3C \rightarrow SiC + 2CO$. The function of the sawdust is to create vent holes for the escape of gas and volatile compounds produced by the reaction of the salt with impurities. After firing, the silicon carbide is picked out and crushed to form grains of different sizes. Acheson called the crystals *Carborundum,* a trade name still in use.

The discovery of aluminum oxide in 1897 is credited to C. B. Jacobs. Abrasive aluminum oxide is made by crushing high-quality bauxite, calcinating it, adding coke and iron, and heating the mass to 2,100° C in an arc furnace.

Abrasives, both natural and synthetic, are generally classified as to hardness in terms of the Mohs' scale, devised by the German mineralogist Friedrich Mohs. This classification rates talc as 1 and the diamond as 10. The accompanying table shows the relative hardness of some abrasive materials as ranked on both Mohs' and the Knoop hardness scales. The Knoop scale, developed by the American chemist F. Knoop, is a more precise method for the classification of hardness. Values are determined by pressing a uniform diamond pyramid into the material with a standardized load. The measure of the material's hardness is determined by the size of the indentation.

Relative Hardness Scales

Substance	Mohs' Hardness	Knoop Hardness
Talc	1	
Gypsum	2	30
Calcite	3	135
Fluorite	4	160
Apatite	5	400
Feldspar	6	550
Quartz	7	820
Topaz	8	1350
Aluminum Oxide	9	2000
Silicon Carbide	9 plus or 9.5	2500
Boron Carbide	9 plus or 9.75	2800
Diamond	10	6500

Grinding wheels are produced by mixing the desired abrasive with a binder, then compacting the mix in a press followed by firing in a furnace. After firing, the wheel is dressed and ground to final shape and spin-tested at high speed to check its structural soundness. Materials such as sodium silicate, rubber, shellac, and resinous compounds are used to bond the abrasive grains. Selection of a bonding agent depends on the grinding application. Rigid ceramics are required for precision grinding while rough hand grinding usually calls for resilient bonding materials such as shellac or rubber.

—Joseph J. Kelleher

Acoustical Engineering, the branch of engineering that deals with sound, particularly in theaters, auditoriums, churches, television studios, or other enclosed areas designed primarily for listening to music or speech.

One of the most important acoustical characteristics of a room is its *reverberation time,* the length of time needed for a sound impulse to fall to one-millionth of its initial intensity. A room lined with carpets, drapes, and other soft, sound-absorbing materials may have a reverberation time of a small fraction of a second. Such rooms, free of sound reflections from walls and floors, are sound-muffled or dead. A hard-walled gymnasium, on the other hand, may have a reverberation time of many seconds. In such rooms, sounds are reflected many times from wall to wall, reaching the listener at the same time as new sounds from the speaker and thus producing a confusion of sound.

Optimum reverberation times are somewhat less than one second for speech and between 1 and 3.5 seconds for music. The reverberation time of a room is controlled by lining walls and other reflecting surfaces with the required amount of sound-absorbing material. Recording studios, which must accommodate a wide variety of speech and music, are often equipped with removable wall panels of sound-absorbing and sound-reflecting materials.

Certain curved wall surfaces produce loud and soft sound regions in an auditorium. A sound-reflecting wall far removed from the speaker can produce a loud and annoying echo. Two parallel reflecting walls often produce a sequence of echoes, a phenomenon called *acoustical flutter.* These disturbing effects can be eliminated by careful design and by proper use of sound-absorbing materials.

—William C. Vergara

Adhesives, materials that can bond two other materials together by adhering strongly to the surfaces of both. In addition, the bond itself must be internally strong or else the bonding action will not be effective. Adhesives are usually applied in the form of solutions, emulsions, or soft gels, but they also can be applied as thin layers of solids that become fluid upon heating. (See also *Glue, Cement.*)

—John Price

Alcohols, a class of organic compounds composed of carbon, hydrogen, and oxygen and containing one or more hydroxyl groups (–OH). The suffix *–ol* or the prefix *hydroxy–* in the name of an organic compound indicates that it is an alcohol. Alcohols may be mono–, di–, tri–, or polyhydric, depending on how many hydroxyl groups they contain. In addition, alcohols are classified as *primary, secondary,* or *tertiary,* depending on whether the carbon atom to which the hydroxyl group is attached is linked with one, two, or three other carbon atoms.

The physical and chemical properties of alcohols depend on their molecular weight and the number and type of hydroxyl groups in the molecule. Alcohols are colorless, flammable, and toxic. Those of low molecular weight are liquids while those of high molecular weight are waxy-like solids.

Lower alcohols, such as ethyl alcohol, are generally employed as solvents, antifreezes, and extractants. Higher alcohols, such as cetyl alcohol, are used as antifoaming agents and evaporation retardants in reservoirs. The largest use of alcohols, however, is as intermediates in the production of other chemical compounds.

Fermentation of natural products, chemical syntheses based on hydrocarbons derived from petroleum or natural gas, and chemical treatment of natural fats and oils are the three most important sources of monohydric aliphatic alcohols. The fermentation of sugars and starches to produce alcoholic beverages has been carried out since the beginning of recorded history. However, fermentation is no longer the major source of lower alcohols, although it still is widely used. Synthetic processes are now used industrially to manufacture lower alcohols. These processes include oxidation, hydration, and oxonation of hydrocarbons; reduction of synthesis gas; condensation and reduction of aldehydes derived from alcohols; and reduction of animal fats and vegetable oils. For example, *methanol* (CH_3OH), the simplest alcohol, is produced by the catalytic reduction of synthesis gas (carbon monoxide and hydrogen). A major source of higher monohydric alcohols, which are used to make detergents, is the reduction of animal fats and vegetable oils, either by catalytic hydrogenation or by treatment with reducing metals.

Esters of the lower monohydric alcohols are used extensively as solvents for a wide variety of synthetic products. Esters from higher alcohols and dibasic acids are used as plasticizers for vinyl, cellulosic, and acrylic resins, and for synthetic rubber.

—John Price

Alkalies, substances that have marked basic—in contrast to acidic—properties. Chemically, they include the caustic hydroxides of lithium, sodium, potassium, rubidium, cesium, and (for practical purposes) ammonium salts. Commercially, the alkali industry is centered in the production of sodium carbonate (soda ash), sodium hydroxide (caustic soda), and chlorine—which is included because it is produced along with caustic soda by alkali manufacturers.

Caustic soda and chlorine are produced by the electrolysis of brine (sodium chloride) solutions. Chlorine is produced at the anode (positive electrode) and hydrogen, together with sodium hydroxide (NaOH), at the cathode (negative electrode). Anode and cathode products must be separated. This has led to the development of many ingenious cell designs. (See *Caustic Soda,* Page 1633.)

Sodium carbonate (Na_2CO_3), or *soda ash,* occurs in, and once was extracted from, plant ashes. Most commercial sodium carbonate is produced by the *Solvay process.* In an initial reaction, salt is converted to sodium bicarbonate, which precipitates and is separated. Heating the bicarbonate produces sodium carbonate.

Alkalies are used in the manufacture of glass, soap, various chemicals, cleaners, detergents, pulp and paper, water softeners, and textiles, and in petroleum refining. —John Price

Alloys, metallic substances composed of two or more chemical elements, of which at least one is a metal. Pure metals seldom possess sufficient strength for engineering purposes, such as building structures and machines. By adding the proper kinds and amounts of other chemical elements to pure metals, metallurgists create alloys that have mechanical, physical and/or chemical properties that are more suitable for a given purpose than the unalloyed metals.

ALLOYING METHODS.—Alloys may be made by fusing (melting) the desired components together or by powder metallurgy. The *fusion method* is used to produce by far the largest proportion of alloys. When proper amounts of the constituents are melted together, they dissolve in each other. Upon cooling, the solution crystallizes to form a solid mass. Solidification of a pure metal or a compound occurs at a definite temperature. However, an alloy solidifies over a range of temperatures, and solid crystals of different compositions separate from the liquid at various temperature levels, according to a predictable pattern for a given alloy system. The kinds, sizes, and distribution of the crystals constitute the structure of the alloy. The structure of most alloys produced by fusion consists of crystals representing solid solutions, eutectics, intermetallic compounds, chemical compounds of metals with nonmetals (carbides, nitrides, phosphides, sulfides, oxides, etc.), or combinations of two or more of these. Their proportions depend upon the composition of the liquid solution and the degree of equilibrium attained.

In making alloys by *powder metallurgy,* powders of the required constituents are mixed and then pressed in molds at high pressure to form a *compact* of the desired shape and density. The compact is then heated to a sufficiently high temperature, usually in a furnace having a controlled atmosphere, to coalesce the powders, usually without melting any of the components. This heating operation is called *sintering.* Many alloys that cannot be made in any other practical way can be made by powder metallurgy.

The hardness, toughness, corrosion resistance, and other properties of an alloy depend upon its structure. The structure, and therefore the properties, of an alloy can in most cases be altered, sometimes over wide limits, by heat treatment (controlled heating

Some Effects of Principal Elements Used in Ferrous Alloys

Aluminum—deoxidizes; restricts grain growth.

Chromium—strengthens and hardens; increases corrosion, oxidation and abrasion resistance.

Cobalt—increases hardness.

Copper—improves resistance to atmospheric corrosion.

Manganese—increases shock resistance and hardness; counteracts embrittlement.

Molybdenum—increases hardness, hot strength and resistance to corrosion, abrasion and creep.

Nickel—strengthens and toughens steels.

Phosphorus—strengthens; improves machinability.

Silicon—deoxidizes; increases rust resistance, hardness and strength.

Titanium—stabilizes carbon and improves creep resistance in stainless steels; in carbon and alloy steels, acts as deoxidizer, promotes ferrite formation.

Tungsten—increases hot strength, abrasion resistance and hardness.

Vanadium—increases hardenability and tempering resistance; promotes secondary hardening during tempering and fine grain structure.

Commercially Important Alloys
(Percentage Compositions)

Alloy	Carbon	Chromium	Copper	Iron	Lead	Manganese	Nickel	Tin	Zinc	other
Babbitt metal	—	—	4-5	—	—	—	—	90-92	—	Antimony
Brass	—	—	65-85	—	—	—	—	—	15-35	—
Britannia metal	—	—	4-16	—	—	—	—	80-94	1.5-5	Bismuth, Antimony
Bronze	—	—	80	—	—	—	—	10	0-10	Antimony
Carboloy	—	—	—	—	—	—	—	—	—	Tungsten-Carbide, Cobalt
Duralumin	—	—	4	—	—	trace	—	—	—	Aluminum, Magnesium
Inconel	trace	15	trace	7	—	trace	77	—	—	Silicon, Sulfur
Iron, cast	3.4	—	—	94.3	—	trace	—	—	—	Silicon
Iron, wrought	trace	—	—	98	—	—	—	—	—	Slag
Permalloy	—	trace	—	10-70	—	—	30-90	—	—	Molybdenum
Pewter	—	—	—	—	10-15	—	—	85-90	—	—
Solder	—	—	—	—	67	—	—	33	—	—
Stainless steel	trace	11-26	—	69-89	—	—	0-8	—	—	—
Sterling silver	—	—	7.5	—	—	—	—	—	—	Silver

and cooling cycles at temperatures below the melting point) alone, or by hot and/or cold working combined with heat treatment.

Some alloys can be wrought, that is, shaped by hot or cold working operations such as rolling, drawing, forging, or extrusion. Other alloys cannot be wrought or worked, and must be shaped by casting into molds or by powder-metallurgical methods.

MAJOR ALLOYS.—Commercial alloys are divided into ferrous and nonferrous alloys. As the names indicate, the principal constituent of the ferrous alloys is the metal iron; the nonferrous alloys usually contain iron in very small amounts, if at all.

The principal *ferrous* alloys are the commercial forms known as cast iron and steel. *Cast iron* includes gray iron, malleable iron, nodular iron, and irons resistant to abrasion (chilled irons and white irons), corrosion (alloyed irons), and heat (also called alloyed irons).

Steel is a generic name for a very large number of cast and wrought ferrous alloys, including carbon steels, alloy steels, and stainless steels. The principal alloying elements, alone or in combinations, used for making alloy steels and alloy cast irons are boron, chromium, cobalt, niobium, copper, manganese, molybdenum, nickel, silicon, tungsten, and vanadium.

The most important commercial nonferrous alloys are based on aluminum, copper, lead, magnesium, nickel, tin, titanium, and zinc. Alloys of the precious metals (gold, silver, and the platinum group) form a special class that will not be considered here.

Cast and wrought alloys based on *aluminum* and *magnesium* and wrought forms of *titanium* alloys have a high strength-to-weight ratio. The aluminum and magnesium alloys are the most widely used of the lightweight alloys.

Copper-base alloys include brasses, leaded brasses, bronzes (including silicon bronze and aluminum bronze), copper-nickel alloys, and nickel silver.

Lead-base alloys include solders, arsenical lead, calcium lead, hard lead (lead-antimony alloys), bearing metals (babbitts), type metals (fusible lead alloys that melt at low temperatures), and terne metal.

Nickel is alloyed principally with copper, molybdenum, or chromium, or with all three, to make alloys resistant to many forms of corrosion. Iron, cobalt, and vanadium are also added to some of the alloys.

Tin provides the base for soft solders, pewter, bearing metals, and *white metal* for die castings.

Zinc is alloyed with aluminum and magnesium, or aluminum, magnesium, and copper, to produce alloys used for die casting.

Special-purpose alloys, not classified as ferrous or nonferrous but according to their uses or special properties, include permanent-magnet alloys, magnetically soft materials, alloys with high electrical resistance, electrical-contact materials, low-expansion alloys, hard-facing alloys, and numerous types of cast and wrought heat-resistant alloys.

—Harold McGannon

Aluminum, a metallic element, chemical symbol Al. It is the most abundant metal in the earth's crust (8.13 per cent), and is found in every continent except Antarctica. Aluminum occurs naturally only in a combined form. Its most common compound is aluminum oxide, which is most often found in clay, granite, slate, and marl. It is commercially produced from a gray-white or brownish ore called *bauxite,* which contains from 45 to 60 per cent aluminum oxide, or *alumina.* The ore was named after the town of Les Baux, France, where it was first discovered. Other important sources of ore are in British and Dutch Guiana, Italy, Jamaica, Australia, Hungary, and parts of Africa. Central Arkansas is the prime source in the United States.

Metallic aluminum is also found in *cryolite,* a sodium-aluminum fluoride used primarily as a bath material in the reduction of alumina to aluminum. *Corundum* is a crystallized alumina found in nature in the form of semiprecious and precious stones: amethyst, emerald, ruby, sapphire, topaz, turquoise, and lapis lazuli. Corundum and *alundum,* artificial oxides of aluminum, are abrasive materials and are used for making laboratory crucibles.

A mixture of granulated or powdered aluminum and iron oxide, called *thermit,* is used for welding. When the aluminum is activated by heating, it takes oxygen away from the iron oxide and forms aluminum oxide, leaving melted, pure iron that fills cracks and fuses the ends of the metals to be joined.

To produce aluminum, the bauxite is first ground and dried, then mixed with a solution of sodium hydroxide, which dissolves the alumina to form sodium aluminate. Silica, iron oxide, and other impurities in the bauxite are precipitated out of solution. Hydrated alumina crystals are used to seed the supersaturated sodium aluminate solution, and crystals of alumina form groups that become heavy enough to settle out of solution. After washing to remove traces of impurities, the alumina hydrate crystals are roasted at more than 2000° F to drive off the chemically combined water, leaving a fine white alumina powder. Then, alumina is dissolved in a bath of molten cryolite and aluminum fluoride in a steel container or cell lined with carbon. A carbon electrode is lowered into the mixture, and direct current (DC) power is applied, with the result that the oxygen of the alumina joins the carbon in the anode in the form of carbon dioxide while the metallic aluminum is freed.

QUALITIES.—Aluminum is a soft, silver-colored metal with a melting point of 660° C, and a boiling point of 2500° C. Its most noted quality is lightness combined with high relative strength. It has approximately one-third the density of steel, copper, or zinc, and has high resistance to corrosion because it reacts in the air to form a thin surface coating of oxide that protects the remainder of the metal against further corrosion. Aluminum has high strength at low temperatures, making it particularly suitable for cryogenic uses. It ranks just below silver and copper as a conductor of heat and electricity and is being used increasingly as an electric conductor; on a pound-for-pound basis, aluminum's conductivity is more than double that of copper. It is nontoxic and can be easily rolled, hammered, pressed, drawn, or extruded into thousands of shapes and extremely thin thickness.

INVENTIONAL FABRICATION.—Aluminum was first isolated in pure form by the Danish physicist Hans Christian Oersted in 1825, following unsuccessful attempts at chemically producing the metal by Sir Humphry Davy. It was Davy who called the metal aluminum, the name used in the United States, while it is called *aluminium*

REYNOLDS METALS COMPANY

IN A HYDRAULIC PRESS, aluminum is extruded into a specified shape for use in industry. Aluminum is so easily worked that it may be fashioned into a large variety of shapes.

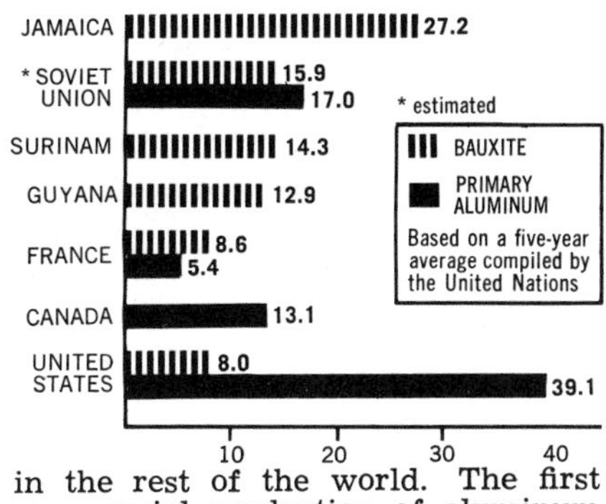

in the rest of the world. The first commercial production of aluminum was in France in 1855, using a chemical method developed by Henri Sainte-Claire Deville. Commercial production was greatly increased by the almost simultaneous discovery in 1886 of an electrolytic method of producing metallic aluminum, as the *Hall process,* by Charles Martin Hall, an American, and Paul L. T. Héroult, a Frenchman.

While most refining produces aluminum that is at least 99.5 per cent pure, a further refinement permits the production of "super purity" aluminum (99.99 per cent) used in making high-octane gasoline, as foil in the electronics industry, and in jewelry. Most aluminum, however, is used in alloy form. Copper, which permits the alloy to be heat treated, also increases strength and hardness, as do magnesium, manganese, and zinc. Corrosion resistance is provided by manganese and magnesium, while silicon lowers an alloy's melting point for improved fluidity for castings. Other alloying elements are beryllium, bismuth, boron, iron, lead, nickel, sodium, tin, titanium, and zirconium.

Extrusion—forcing heated cast billets through a die to produce an infinite variety of shapes—has become one of the most important methods of forming aluminum products. Castings also consume large quantities of aluminum. Forging or hammering and drawing for fine tolerances are other methods of forming aluminum. Plate, sheet, and foil are rolled in a wide range of thickness.

The building and construction industries use the greatest amount of aluminum for products ranging from foil insulation to door and window frames, siding, and anodized exterior facing products. Aluminum powder is used in paint form to protect other metals. The transportation industry is the second largest user of aluminum for truck-trailers, mobile homes, railroad cars, automobiles, and rapid transit systems. Pleasure boats and commercial vessels increasingly employ aluminum for hulls, superstructures, and other components. For many years, aluminum has been used in aircraft and aerospace applications as a lightweight structural material. Lightweight, air-transportable tanks, armored personnel carriers, and trucks use aluminum extensively.

—Robert H. Chamberlin

Aqueduct, a man-made channel or structure for carrying water. Aqueducts may be divided into two major types: the open channel, or grade-line, aqueduct, and the pressure-pipe line. Early irrigation ditches, or nearby streams, often served for public water supplies, but the Romans, the noted aqueduct builders of ancient times, built aqueducts that served not only the city of Rome but the entire Roman world.

Almost invariably these were grade-line works, artificial channels in which the water flowed as in a slow stream, although they were usually of the so-called *cut-and-cover* form, lined and covered with stone masonry. In crossing valleys, these channels were supported on masonry arches such as those carrying the lines to Rome and to Segovia in Spain. In general, these led to public fountains from which the people drew their needs. The Romans also used some lead pipes for delivery for baths and other services, and in a few rare cases built pressure lines down and across valleys, the so-called *inverted siphons,* in lieu of supporting arches.

Suitable metal, however, for such lines was lacking or costly and was later reclaimed for other purposes. Thus the grade-line type long remained standard; no major early pressure lines remain today, although it appears that one notable early, probably Greek, pressure line at Pergamon in Asia Minor may have been built of bronze pipe. Pipes of bored wood logs were later used for the supply system of the city of London as well as for several early American systems. The advent of a true Age of Iron in the eighteenth century, however, made cast-iron pressure pipe possible. The first such line in the United States was built in Philadelphia about 1820.

A number of European cities, notably in Britain, obtain their water through long pipe lines; especially long lines have also been built in Australia. In the United States, earlier works were usually of the cut-and-cover type. However, the 75-mile Catskill line built in 1914 for New York City was almost entirely a grade line and siphon work, while the 240-mile Colorado River aqueduct for a dozen cities in the Los Angeles area involved open canals, cut-and-cover conduit, siphons, pressure tunnels, and waterpower pumping plants to lift the supply over intervening mountan ranges. —James Kip Finch

Artesian Well, a natural source of water in which the water is forced to the surface by hydrostatic pressure. Two geological conditions are essential to create an artesian water supply. First, beneath the well site, there must be layers of water-bearing material entrapped between watertight, or *impervious,* layers of rock or clay. Second, at some location, the layers of entrapped water-bearing material must be elevated above the well site.

Elevation of the water-bearing material produces the hydrostatic pressure required to bring the water to the surface or above it. In some areas, springs are created when faults or cracks in the watertight layers allow the hydrostatic pressure to force water to the surface. If the water-bearing material is elevated far enough above the well site, the hydrostatic pressure is sufficient to shoot a column of water 100 feet or more above the well.

The conditions required for artesian wells are usually found in the valleys of mountain ranges or other hilly country. In the United States, the alternate layers of water-bearing and impervious material are elevated along the foothills of the Rocky Mountains. From there, the layers sweep downward under Kansas, Nebraska, Montana, and the Dakotas, creating a vast pool with sufficient hydrostatic pressure to force water to the surface. —Joseph J. Kelleher

Asbestos, a general name for the fibrous varieties of a number of rock-forming minerals. Products fabricated from various grades of these mineral fibers do not burn, and they resist chemical attack.

Traditional fine *amphibole asbestos* has high acid resistance and withstands temperatures up to 1,000° C. The other chief type, *serpentine asbestos,* is less resistant to heat and acid attack than the amphibole varieties, but it is used more because it is more abundant.

Mining methods are based on removing the host rock and milling. Open-cast mining is the most common, but underground systems also are used. Block-caving methods, used in Canadian mines, consist of undercutting huge blocks several hundred feet down, which are then progressively caved and loaded from the bottom.

After the fibers have been milled, they are graded by length. Those one-half inch or longer are used for textiles; shorter ones go to end products that are less exacting in their standards. Fiber grades are spun into yarns and woven into cloth. Brake linings, heavy packings and gaskets, electrical insulating materials, and protective clothing require the best fibers. Medium and short fibers are used in asbestos shingles, sheet gaskets, pipe, floor tile, less critical packings and gaskets, and as fillers for various other materials.

—John Price

Brass, an alloy of copper and zinc. Brasses are readily extruded, rolled, or forged, have good corrosion resistance, and can be joined by soldering or brazing. Brass is widely used for decorative trim, plumbing fixtures, electrical equipment, pipes and valves, musical instruments, automobile radiators, screws, cartridge cases, and hardware. The ratio of copper to zinc may be varied to control both the color and the malleability of the resultant alloy. An alloy of 80 per cent copper and 20 per cent zinc is very soft and has a golden-red color. A 60/40 copper-zinc ratio results in a yellow metal that is still malleable, while alloys with less than 50 per cent copper are white and become increasingly brittle with a rise in the percentage of zinc.

The more widely used brass alloys are frequently given trade names or names that reflect their application. Some common examples are *Muntz metal* (60 per cent copper, 40 per cent tin), *yellow brass* (66/34), *cartridge brass* (70/30), *low brass* (80/20), *red brass* (85/15), and *commercial bronze* (90/10). The addition of other metals results in a family of metals with names that are frequently misleading. The addition of a small amount of tin to molten brass produces an alloy called *nickel silver,* which contains no silver; while a manganese addition results in *manganese bronze,* which contains no tin. Small amounts of lead are often added to brass to improve its machinability. These *free-cutting brasses* (61 per cent copper, 3 per cent lead, 36 per cent zinc) are readily formed by automatic screw machines for many uses.

—Joseph J. Kelleher

Brick, a shaped, hardened block of clay used as a construction material. Bricks are among the oldest building materials still in common use today. Most are made of clay mixed with some sand, formed into rectangular blocks, and subsequently burned or fired until the clay particles fuse, forming a rigid, cohesive mass. Although brick may be of almost any size or shape, the standard U.S. brick measures $2\frac{1}{4} \times 3\frac{3}{4} \times 8$ inches. Materials other than clay are sometimes formed into similar shape and are also classified as brick.

Common brick, or ordinary building brick, consists of a clay and sand mix burned until the clay particles just begin to fuse.

Facing brick is similar to common brick, except that fewer imperfections are permitted, color variation is limited, and mechanical tolerances are more strict.

To achieve the abrasion resistance and weatherability demanded of a paving material, *paving brick* is fired until the clay particles fuse completely, filling all the voids in the sand-clay mix and rendering the finished product impervious.

Sand-lime brick is made of a mixture containing about 90 per cent sand with a large proportion of very fine grains and 10 per cent lime. Usually hydrated, high-calcium lime is used. The brick is cured in an *autoclave,* a closed vessel containing steam, at 100° to 150° F. The use of sand-lime brick in construction is identical to that of common brick.

Concrete brick is composed of ordinary Portland cement and aggregates formed into blocks, conforming to the standard sizes used by clay-brick manufacturers. Both regular and lightweight aggregates are used. *Refractory bricks* are masonry units manufactured especially for their resistance to heat. The production of refractory brick is similar to that of other brick, but the materials vary: flint clay, aluminum oxide, silica, and various mineral ores are used, depending upon the degree of temperature resistance needed, and upon the nature of the corrosive atmosphere to which they will be exposed.

—William W. Jacobus, Jr.

Bridges, structures providing support for the conveyance of traffic or other services over a valley, body of water, or road with openings for the passage of stream flow, navigation, or other traffic. Crude bridges were built by primitive people the world over, but the first notable bridge builders were the early Romans. In meeting their needs for roads to unite the Italian peninsula and their provinces, the Romans developed both timber and stone arch spans that were to remain the only practicable forms available until the late eighteenth century.

TIMBER BRIDGES.—In many cases, bridges were built first as inexpensive timber structures and later replaced with more costly but more permanent stone arches. The first bridge over the Tiber at Rome, the Pons Sublicius (621 B.C.), made famous in the story of how Horatius held it against the Tarquins, was built of timber. Another early work, Pons Amelius (181 B.C.), was also of timber, but with stone piers. Timber bridges were likewise used in military operations, including Caesar's pile bridge over the Rhine and Trajan's timber arch over the Danube near Hungary's Iron Gate.

In addition to beam and arch forms, framed trusses must also have been developed early to support the roofs of temples and other buildings. Their first recorded use for a bridge, however, was by the Renaissance architect Andrea Palladio in 1570 in a remarkably modern form over the Cismone River. He noted that a friend had seen such a bridge in Germany. Later, Swiss engineers built a number of notable wooden bridges. This form was first extensively used and developed in the early nineteenth century in the United States, where they served for many highway crossings, often as *covered bridges* protected from the weather. Similar trusses and timber trestles were also used for early railroads. Later replaced with iron, these gave way to steel as the nineteenth century came to a close, when increasing loads demanded stronger spans.

STONE ARCHES.—The arch appears early in ancient times in two forms: the *cantilever,* or so-called *false arch,* formed by horizontal overhanging stones, and the true circular arch of wedge-shaped blocks or voussoirs. Both apparently originated in the Near East and were used by the Assyrians. The Etruscans are credited with having introduced the true arch to the Romans.

The Pons Mulvius, now known as the Ponte Molle, which in 109 B.C. replaced a timber bridge on the Flaminian Way, a short distance above Rome, is said to be the earliest Roman arch still standing. It is of the standard Roman form having full-centered circular arches (with a rise equal to half the span) of unequal number, with a large, central span and substantial piers, aesthetically balanced and impressively stable. Its seven spans, each 51 to 79 feet, successfully carried heavy military equipment during World War II.

Many other notable bridges were built not only in Italy but also in the Roman provinces; for example, the high span (130 feet) over the Tagus River in northern Spain about 120 A.D. Here, with high stream banks, this form posed no obstruction to floods, and maximum spans of about 100 feet were reached. Full-centered arches, however, raised two major problems; for low-level crossings, spans were limited, and the arch haunches and massive piers obstructed the passage of floods, causing high-velocity flow through the arch openings. The resulting scour tended to undermine the piers, which, with the limited man-powered equipment of the day, could not be founded at a sufficient depth to avoid this danger.

It was not until the early Renaissance that the low-rise segmental arch was introduced in Italy, notably in the Ponte Vecchio at Florence of 1345 and the Castelvecchio bridge at Verona of 1356. This provided increased arch openings by making possible longer spans for low crossings, and a main span of 160 feet was reached in the latter work.

Conforming to the later aesthetic interests of the Renaissance, which regarded the abrupt angle between such arches and the piers as unpleasing, and believing also that some of the thrust could be more safely turned down to the pier, engineers introduced an elliptical form; an example is the Trinity Bridge at Florence in 1569. Known as the *anse de panier,* or basket-handle, this became a standard form and was widely adopted by the builders of the remarkable French highway system of the eighteenth century. Jean Perronet, the most famous of these French workers, finally revived the segmental form with slender columnar piers in his Pont de la Concorde over the Seine at Paris in 1791.

IRON AND TRUSS BRIDGES.—A remarkable bronze truss of about 130 A.D. once supported the portico of the Pantheon in Rome. But it was destroyed in 1625, and it was not until 1779 that metal again appeared in a major structure. At this time, the advent of a true Iron Age in Britain led to the construction of a 100-foot cast-iron span at Coalbrookdale, and other more notable spans followed in the early nineteenth century. In general, the ancient arch form was used in these bridges, the voussoir stones being replaced by large cast-iron frames allowing larger spans. John Rennie's Southwark Bridge over the Thames at London in 1819 established a new arch record of 240 feet.

Wrought iron in quantity and at a cost making it available for bridges was later produced. By 1850, Robert Stephenson had completed the Britannia Bridge over Menai Strait in four unique spans of 230 to 460 feet, in which the trains traveled inside huge box girders.

Practice in the United States, however, centered largely in the truss forms. As earlier timber bridges were replaced by iron bridges, new truss forms emerged that were better suited to the economical use of the new material. By 1861, the plates and shapes now standard in steel construction had been developed, and the all-wrought-iron truss emerged. Steel later replaced wrought iron. It was first used with wrought iron in the United States in James Eads' famous St. Louis Arch in 1874. The first all-steel bridge was built in 1879.

British practice continued to favor solidly riveted plate constructions built in place, reflecting, it was said, boiler and shipbuilding techniques. It culminated in one of the wonders of the modern world, the Firth of Forth cantilever bridge of 1890, with its two great 1,710-foot spans. In the United States, designers adopted the so-called pin-connected forms of shop-fabricated parts that could be rapidly erected at low cost. It was estimated that by 1889, some 7,000 railroad spans of this type more than 100 feet in length had been built.

■SUSPENSION BRIDGES.—Rope and similar suspensions were built by primitive peoples, but this was not a widely useful form of bridge until wrought iron became available. Bridges were then built using links of iron bars connected with pins. James Finley, an early American builder from Pennsylvania, pioneered in building "chain" suspensions, erecting the first in 1801. Twenty-five years later, the British pioneer Thomas Telford completed his Menai suspension bridge with a span of 580 feet, but the modern wire suspension was soon developed. Louis J. Vicat in France introduced the modern *spinning process,* using iron wires to build up suspension cables, in the Argental span of 1830. A Swiss, Joseph Chaley, reached 810 feet in 1834 at Fribourg, but the span record has since been held by the United States.

The outstanding American builder and wire manufacturer was John A. Roebling, an engineer from Germany, whose first suspensions carried wood troughs for canals over intervening valleys. His famous Niagara span of 822 feet was completed in 1855, and his even more famous Brooklyn Bridge of 1,595 feet at New York was completed after his death by his son, Washington A. Roebling, in 1883.

Not suited to heavy railroad loads, this form found increasing use with the advent of lighter modern motor traffic and, in 1929, when the Ambassador span of 1,850 feet was built across the Detroit River, finally took over the bridge span record. This, in turn, was followed by the great 3,500-foot George Washington span across New York's Hudson River in 1931, the 4,200-foot Golden Gate at San Francisco in 1937, and in 1965 by the 4,260-foot Verrazano-Narrows Bridge at entrance to New York Harbor.

■CONCRETE AND OTHER BRIDGES.—As noted earlier, stone masonry found but rare and limited use in the United States. However, the birth of the American portland cement industry at the turn of the twentieth century made mass-produced concrete available at low cost, and several massive concrete viaducts were built. These included the Walnut Lane of 1908 at Philadelphia, the Rocky River of 1910 at Cleveland, and the great Tunkhannock on the Delaware, Lackawanna & Western Railroad in northern Pennsylvania in 1915 with ten 186-foot arches.

THE GEORGE WASHINGTON BRIDGE stretches 3,500 feet across the Hudson River.

Notable Bridges of the World

Name	Location	Type	Length in Feet	Date Completed
Ambassador	Detroit, Mich.-Sandwich, Ontario over Detroit River	Suspension Highway	span—1850 tot.—7400	1929
Bayonne	Bayonne, N. J.-Staten Island, N. Y. over Kill van Kull	Steel arch Highway	span—1652	1931
Bear Mountain	Peekskill, N. Y. over Hudson River	Suspension Highway	span—1632 tot.—2258	1924
Benjamin Franklin	Camden, N. J.-Philadelphia, Pa. over Delaware River	Suspension Highway	span—1750 tot.—8126	1926
Brooklyn	Manhattan-Brooklyn, N. Y. over East River	Suspension Rail, Highway	span—1595 tot.—6016	1883
Bronx-Whitestone	New York, N. Y. over East River	Suspension Highway	span—2300 tot.—4000	1939
Carquinez	Crockett-Vallejo, Calif. over Carquinez Straits	Cantilever Highway	span—2200 tot.—4482	1927
Chesapeake Bay	Delmarva Peninsula-Maryland mainland over Chesapeake Bay	Suspension Simple truss, Cantilever Highway	span—1600 tot.—7727	1952
Eads	St. Louis, Mo. over Mississippi River	Steel arch Railway	span—520 tot.—6434	1874
Forth	Scotland over Firth of Forth	Cantilever Railway	span—1710	1890
Fribourg	Switzerland over Sarine Valley	Suspension Highway	span—870	1834
George Washington	New York, N. Y.-Fort Lee, N. J. over Hudson River	Suspension Highway	span—3500 tot.—7800	1931
Golden Gate	San Francisco-Marin Co., Calif. over San Francisco Bay	Suspension Highway	span—4200 tot.—9217	1937
Greater New Orleans	New Orleans, La. over Mississippi River	Cantilever Highway	span—1575	1958
Hell Gate	New York, N. Y. over East River	Steel arch Railway	span—1017 tot.—18000	1917
Longview	Longview, Washington over Columbia River	Cantilever Highway	span—1200	1930
Mackinac	Mackinaw City-St. Ignace, Mich. over Mackinac Straits	Suspension Highway	span—3800	1957
Mid-Hudson	Poughkeepsie, N. Y. over Hudson River	Suspension Highway	span—1500 tot.—4072	1930
Plougastel	Brest, France over Elorn River	Concrete arch Rail, Highway	span—612	1929
Salazar	Lisbon, Spain-Seixal, Portugal over Tagus River	Suspension Highway	span—3,323 tot.—10,560	1966
Sando	Sweden over Angermanalven River	Concrete arch Highway	span—866	1943
Stockholm	Stockholm, Sweden over Stockholm Harbor	Concrete arch Rail, Highway	span—866	1943
Storstrom	Sjaellan I.-Falster I., Denmark over Storstrommen	Steel arch Rail, Highway	tot.—10432	1937
Sydney Harbor	Sydney, Australia over Sydney Harbor	Steel arch Rail, Highway	span—1650 tot.—3770	1932
Tappan Zee	Tarrytown, N. Y. over Hudson River	Cantilever Highway	span—1212	1955
Trans-Bay	San Francisco-Oakland, Calif. over San Francisco Bay	Suspension Cantilever Highway	span—2310 span—1400	1937
Triborough	Manhattan-Queens, N. Y. over East River	Suspension Highway	span—1380	1936
Verrazzano-Narrows	Brooklyn-Staten Island, N. Y. over The Narrows	Suspension Highway	span—4260 tot.—7200	1964

Attention abroad turned to reinforced concrete structures, and a far lighter, reinforced, open-spandrel form developed in the Swiss Langwies arch with a 315-foot span in 1915. This form has since been used for highway work in the United States, notably on the Pacific coast. However, the span record remains in Europe with such remarkable tours de force as the Plougastel Bridge at Brest, France, of 1929 with two 612-foot spans, and the Sando Arch at Stockholm with an 866-foot span, completed in 1943.

■**MOVABLE BRIDGES.**—Simple timber draw or beam spans, hinged at the inner end, were used over the moats of medieval fortresses and were early adopted for canal crossings in Holland to provide clearance for navigation. Similar modern steel rolling-lift bridges have been developed to meet needs of low-level crossing and navigation. Usually, the bridges are built in two section, meeting at the center of the span. The two arms rest on curved supports of the *rolling* or *trunnion* form, counterweighted and machine-driven. Other forms have also been developed, notably a *lift* type in which the entire span is counterweighted and lifted vertically between two end towers. The *swing* bridge, rotating on a circular track at the center pier of a two-arm span, is also widely used.

Needless to say, interruptions to traffic over or through such spans is unavoidable, and their use is usually limited to channels that carry only a small volume of shipping. Permanent spans are thus desirable. But in the United States, clearance requirements set by the Federal government may reach 200 feet or more on major navigable waters, requiring costly long, inclined approach spans. As a result, in many cases the alternative of tunnels has been preferred in recent years. —James Kip Finch

Bronze, an alloy of copper and tin. Bronze is rarely used without the addition of elements. The most widespread use of *simple bronze* is for statuary (up to 20 per cent tin) and bells (up to 25 per cent tin). Conventional *binary bronze* contains from 3 to 10 per cent tin, with strength, hardness, and corrosion resistance increasing with the tin content. These alloys are widely used for commercial screening and springs. Minor additions of alloying elements can be used to amplify specific characteristics of bronzes. The alloy's corrosion resistance is dramatically improved by adding up to 3 per cent silicon to produce *silicon bronze.* Traces of phosphorus harden and strengthen bronze, while additions of up to 30 per cent lead improve its ductility and machinability. Many alloys commonly referred to as bronze contain no tin and are not truly bronzes. *Manganese bronze* contains 0.5 to 5 per cent manganese, copper, and other metals —but no tin. Another misnamed alloy, *beryllium bronze* (2 per cent beryllium, but no tin), can be strengthened by heat treatment, which makes it nearly three times stronger than structural steel. —Joseph J. Kelleher

Building Construction, the erection of structures consisting of foundations, frames, walls, and roofs.

■**FOUNDATIONS.**—The foundation of a building is that part of the construction that ties the structure to the site upon which it rests. It must be able to support both the weight of the building itself, called the *dead load,* and the weight of the contents of the building—the furniture, fixtures, and occupants—called the *live load.* In addition, the foundation of a building must be capable of resisting lateral, or sideways, loads that tend to overturn the structure. Both wind and earthquakes impose significant lateral loads upon tall buildings.

Many types of foundations are used, depending upon the height, and hence the weight, of the building and the load-bearing capacity of the soil in the area in which it is built.

The science of *soil mechanics,* which is the study of the behavior of soils under load, is well advanced and engineers can quite accurately predict the pressures (in pounds per square foot) that any particular site can sustain. Thus, when the live and dead loads of a building have been calculated, it is a short step to determine the area of the footings that must be in contact with the soil to support the structure.

In the case of all buildings or poor soil conditions, these calculations may show that the area of the footings needed exceeds the total area of the base of the building. In such cases vertical structural members, called *piles,* which may be of wood, steel, or concrete, are driven into the ground until they reach solid bedrock, or the friction of the soil against the sides of the pile develops the required load-bearing capacity. The building's columns, which transfer vertical loads to the earth, rest atop these piles.

■**FRAMES.**—Until the turn of the century, the exterior and interior walls of a building—usually made of brick or other masonry material—carried the vertical loads to the ground. This type of construction was known as *bearing-wall construction.* Under this system, as taller buildings came into demand, the bearing walls at the base of a building had to be made considerably thicker. Seeking other solutions, engineers developed the skeleton frame with *columns* (vertical members) and *beams* (horizontal members) of steel or concrete. Both of these materials, being much stronger than the common brick they replaced, reduced the volume of material needed to support a tall structure and made possible the multistory commercial and residential buildings that are commonplace today.

■**WALLS.**—With the advent of the skeleton frame, the function of the exterior walls of a building was reduced to that of excluding wind and water from the interior of the building and minimizing the transfer of heat (outward in winter, inward in summer). By the early 1950's, architects were beginning to develop very thin, lightweight exterior walls, often referred to as *curtain walls.*

In this type of construction window glass comprises about half of the exterior wall area. For the remaining area many materials have been used, among them stainless steel, aluminum, porcelain enameled steel, precast concrete, and even architectural bronze. The choice of material for exterior curtain walls is determined by aesthetic appearance, weatherability, and cost.

Interior walls are more properly referred to as *partitions.* The most significant advance in interior partitions has been the development of demountable partitions for office buildings; these can easily be taken down and reused in a new location, thus permitting office arrangements to be changed to meet changing design requirements with little additional expense.

■**ROOFS.**—The roofs of most commercial, industrial, and high-rise residential buildings are flat, unlike the steeply pitched roofs of the typical small dwelling. Thus, they do not shed water as readily as pitched roofs and their construction requires considerable care.

The commonest roof construction for flat roofs is the built-up, multiply roof made up of alternate layers of heavy feltlike paper, and bitumen (tar). The *bitumen,* which is mopped onto the roof deck and onto the subsequent layers of felt, serves both as an adhesive to hold the felts in place and as a waterproof membrane. Often, built-up roofs are topped with a layer of a light-colored mineral aggregate that protects the roof membrane from accidental damage by the traffic of maintenance men. Also, because of its light color, the layer of aggregate increases the reflectance of the roof, thus minimizing the transmission of heat from the sun's rays during summer.

Elastomeric (flexible plastic) materials have recently been developed for roof surfaces. Although these materials possess certain inherent advantages over the conventional built-up, bituminous roof, the cost of the materials has limited their use to rather special applications.

■**BUILDING CODES.**—These are the statutory regulations governing the design and construction of buildings for ensuring the health, welfare, and safety of those who will ultimately occupy the building. The rules embodied in building codes are intended to ensure that the building's structural components, such as beams and columns, are adequate for the loads they must carry; that the building is protected against fire; that emergency exits are large enough for the speedy evacuation of a building in the event of any disaster.

Plans and specifications for a building must be submitted to the appropriate municipal officials, who will examine the design for compliance with the code before granting permission for construction to begin. Authority to enact and enforce building codes stems from the police power of the state government. But, in almost every case, this particular power is delegated to municipal authority. —William W. Jacobus, Jr.

Calcimine, or *kalsomine,* a white or tinted wash made of glue, zinc white, and water, and used especially on plastered surfaces. It is one of the emulsion-type paints made from slated lime and water-washed clays. Like most water-thinned paints, calcimine is sensitive to water and is not washable. Addition of modified oils or resins that are water-soluble but that convert to an insoluble form when the water evaporates, can eliminate this defect. Calcimine was used extensively at one time because of its low cost. However, its use has dwindled over the years with the development of better, more durable paints.
—John Price

Camphor, a crystalline naturally gummy substance used as a medicine and plasticizer. Camphor occurs as colorless to white crystals, granules, or crystalline masses, or as colorless to white translucent tough masses or tablets. It has a characteristic penetrating odor and a pungent, aromatic taste. Camphor has been used for thousands of years as a component of incense and in various domestic uses because of its pleasant odor. Its first important chemical use was discovered a century ago—as a plasticizer for cellulose nitrate to produce *celluloid.*

Natural camphor is obtained by steam distillation from the leaves, twigs, and stems of the camphor tree, *Cinnamomum camphora,* which is found widely in China, Japan, and Taiwan. The trees reach maturity in 45 to 50 years, and camphor is extracted every 5 or 6 years. Because of the limitations on production of camphor by plantations, synthetic methods were developed to manufacture it from pinene, a turpentine derivative. Demand for camphor today has dropped off, however, primarily because of the replacement of cellulose nitrate by other plastics. Also, its use as a pharmaceutical has declined greatly.

Camphor is used widely in liniments and as a mild rubefacient, analgesic, and antipruritic. It is also used in photographic film and as an insect repellent, especially for clothes moths.
—John Price

Calking and Sealing Compounds, a wide range of compositions used in the construction, manufacturing, and transportation industries. They are used to seal joints or voids against water or vapor, air and other gases, dust, sound, heat, and cold. Some compounds are applied after the structure is in place, as in masonry joints; others are applied during manufacture, as in automobile and trailer bodies.

Calks and sealants are usually supplied in knife- and gun-grade compounds. *Knife-grade* materials, the stiffer of the two, lend themselves to application with a putty knife. They are often supplied in extruded shapes ready for use. *Gun-grade* materials are extruded through an opening, using a hand- or pressure-operated gun. They are supplied in bulk form or in cartridges ready for use.
—John Price

Caustic soda, or *sodium hydroxide* (NaOH), one of the high-tonnage industrial chemicals. Consumption in the United States is approximately 5.5 million tons per year, and this amount is expected to increase.

Caustic soda and its companion product, chlorine, are produced by the electrolysis of brine (sodium chloride solutions). Chlorine is evolved at the anode (positive electrode), caustic soda and hydrogen at the cathode (negative electrode). Two variations are widely used in industry—the diaphragm cell and the mercury cell. The *diaphragm cell* keeps anode and cathode products separated by an asbestos diaphragm backed by an iron cathode. To minimize back-diffusion, only part of the sodium chloride is reacted in the anode cell. Caustic soda produced at the cathode is dried and separated from residual sodium chloride. In *mercury cells,* the cation (positively charged ion) forms an alloy or amalgam, which is pumped to another chamber, where it is reacted with water to produce caustic soda and hydrogen. The diaphragm cell is used more extensively, but mercury-cell use is growing, since this method produces a purer product at the same cost.

The chemical industry consumes nearly half of all U.S. caustic production. Pulp and paper manufacturers use substantial amounts of caustic and are expected to become increasingly important customers. It is also used in making rayon, aluminum, textiles, petroleum, soaps and detergents, and cellophane.
—John Price

Cellophane, a transparent, flexible film, usually about one-thousandth of an inch thick, used principally for packaging food and other consumer goods. Invented in 1908 by Jacques Edwin Brandenberger, a Swiss chemist, cellophane was first made commercially in France. E. I. du Pont de Nemours & Company acquired the American rights in 1923 and four years later developed a means of moisture-proofing cellophane, a development that led to its wide use as a protective wrapping material.

The growth in use of cellophane coincided with, and contributed to, the development of the self-service merchandising concept in the United States. Since transparent film combines product visibility and product protection, it has facilitated the prepackaging of many commodities.

Cellophane is made from cellulose, which is derived from wood or cotton in a process that bears some resemblance to papermaking. Among the chemicals added during the process are caustic soda, carbon bisulfide, and sulfuric acid. Moisture-proofing is achieved by a coating of a lacquer-like substance or a synthetic resin. Cellophane manufacture requires a large investment, and a small-capacity plant is not profitable. There are three manufacturers of cellophane in the United States. About 300 other companies "convert" cellophane, that is, print it, make it into bags, or otherwise prepare it for ultimate use.
—E. I. du Pont de Nemours & Co.

Cellulose, the chief carbohydrate found in land plants. It forms the skeletal structure of the cell wall (hence its name) and occurs with polysaccharides and hemicelluloses derived from other sugars, such as xylose, arabinose, and mannose. In the woody part of plants, cellulose is mixed, and sometimes linked, with lignin. Wood normally contains 40 to 50 per cent cellulose, 20 to 30 per cent lignin, and 10 to 30 per cent hemicelluloses.

Commercial cellulose can be produced from cotton linters, but the main source is wood, particularly pine and spruce, from which it is removed by pulping methods designed to remove lignin and other noncellulosic constituents. In addition to mechanical and semimechanical processes that separate wood fibers for use in paper manufacture, there are three chemical pulping methods that are used widely. The most frequently used is the *kraft,* or *sulfate, process,* in which logs are debarked, chipped, and cooked at 160–170° C for 2 to 6 hours in a solution of sodium sulfide and sodium hydroxide. Sodium sulfide, added to the spent cooking liquor, is converted to sodium sulfide when the liquor is burned for recovery of sulfide. In the *soda process,* sodium hyrdoxide alone is used in the cooking liquor. In the *sodium sulfite process,* wood chips are cooked in a calcium bisulfite solution containing a large excess of free sulfur dioxide; lignins are converted to soluble lignosulfonates. Cellulose is used to prepare a wide variety of derivatives in the production of rayon.
—John Price

Cement, any substance that bonds materials together. In the broadest sense, glue, mucilage, epoxy, and paste are all cements, but more specifically the term applies to *hydraulic cement* used in construction. Hydraulic cement is a finely-ground, gray powder, which when mixed with water and sand makes mortar that will dry and hold together stones, bricks, or blocks of masonry. If a larger aggregate, such as gravel or crushed stone, is added to the mixture of water and sand the end product is a rocklike mass called concrete. (See *Concrete.*)

The Romans first discovered that a mixture of volcanic ash and lime, made by burning marble, would bond together rocks and cut stone for building their roads, aqueducts, and other structures. They called the material *caementum,* meaning, literally, pieces of rough uncut rock, in reference to the chips of marble from which the lime came.

Among the many things lost with the decline of the Roman Empire was the formula for making cement. A cement similar to that of the Romans was produced in England in 1756, that would harden under water just as the Romans' did. But the cement used today was not invented until 1824, when an English stonemason, Joseph Aspdin, found that by mixing the lime, silica, and alumina first, then burning and grinding the mixture, he could make a stronger cement. Since the concrete made with his cement looked very much like a rock formation on the Isle of Portland, Aspdin called his discovery *Portland cement.*

Today there are standard formulas for Portland cement, but for many years the various manufacturers used their own specifications and there were nearly a hundred different types.

Lime is the major ingredient of Portland cement, usually comprising about 60 to 64 per cent of the mixture. The lime is derived from limestone, oyster shells, or other natural rocks. *Silica,* obtained from blast-furnace slag, clay, or shale rock makes up 19 to 25 per cent of cement. *Alumina,* which is in the same slag, clay, or shale, comprises 5 to 9 per cent of the mixture.

Cement manufacture begins with the quarrying of the limestone. Huge crushers break the blasted limestone into small pieces about the size of chicken eggs. The crushed rock is then mixed with the silica-alumina-bearing slag, shale, or clay in big rotating cylinders called *ball mills.* Heavy steel or iron balls inside the cylinders batter the mixture into powder. This powder then goes to a tube mill that contains even smaller balls or pebbles of flint. The tube mill produces particles so fine that they can pass through a sieve that water cannot pass through.

Next, this finely ground powder is burned in great kilns as large as 15 feet in diameter and 400 feet long. This burning gives the material its binding quality, producing calcium silicates and calcium aluminates that react chemically with water to form a rocklike mass. The burning kilns also are long rotating steel cylinders, but these are lined with firebrick to protect the steel from temperatures as high as 3,000° F maintained at the lowest end of the kilns. The cement powder, which is sometimes mixed with water to form a soupy *slurry,* takes about two hours to pass through the kilns.

This tremendous heat, besides causing the chemical reaction, also causes the material to emerge from the kilns as *clinker,* larger pieces about the size of children's marbles. The clinker is cooled, a small amount of gypsum is added, and finally, more ball and tube mills regrind it. The end product, Portland cement, is so finely pulverized that 90 per cent will pass through a sieve with 40,000 openings per square inch.

Portland cement is modified in many ways for different purposes. Brick masons add more lime to the cement and sand, so their mortar is smoother and bonds better to the bricks. *Pozzolana,* siliceous material such as blast furnace slag, fly ash, or natural volcanic material, are also added to Portland cement, but mainly as a substitute for some of the more expensive cement when making concrete. There is a naturally occurring argillaceous limestone, which when burnt and pulverized results in natural cement. Natural cement is similar to Portland cement but is slower-setting and its quality is less dependable. A mixture of the two is sometimes used for special purposes.

Although cement traditionally has been packed in bags, most of it today is transported in bulk by huge rail tank cars, trucks, and barges. Cement is measured in barrels (bbl) rather than bags. A barrel of cement is 376 pounds. World production of cement is about 2.7 billion barrels annually. —William W. Jacobus, Jr.

Ceramics, materials or products that are chemically inorganic (except metals or metal alloys) and are usually rendered serviceable by high-temperature processing. The raw materials for most ceramics are oxides and silicates, but many others are also used, including aluminides, beryllides, borides, carbides, nitrides, and silicides. Graphite and ceramic-metal composites are frequently classified as ceramics.

The usual product classification includes abrasives; cements, lime, and gypsum; electronic and technical ware; glass; porcelain enamel; refractories; structural clay products; and whitewares, which include dinnerware and plumbing fixtures. Pottery falls in either the technical or whitewares group, depending on its application.

Ceramic products range from single crystals, such as sapphire, through dense, polycrystalline, glass-free refractory materials, to glass-bonded crystalline aggregates and wholly vitreous (glassy) materials. A number of different manufacturing methods must be employed. Some products, such as volcanic glass (obsidian), depend upon the processes of nature.

Manmade ceramics came with the discovery of fire. Primitive men made simple bricks by adding water to clay to make it plastic and then shaping it by hand or in simple molds. These bricks were dried in the sun and finally hardened by firing.

The origin of such products is lost in the mists of antiquity. However, early man soon learned to refine and improve his raw materials and his forming, drying, and firing procedures. He found that mixtures of several clays often made it easier to attain the sizes and shapes he desired. Certain other materials, such as feldspar, served as fluxes, permitting firing at a lower temperature.

Even today many ceramics are made of three major components: mixtures of clays to provide plasticity; fluxes to assist in forming a glassy bond during firing; and a non-plastic or refractory material, such as quartz, to aid in glass formation and to provide a rigid skeleton.

■PRODUCTS.—Ceramic bodies are usually fabricated by extrusion, soft plastic forming, dry pressing, slip casting, hot pressing, or fusion casting. All but the last two are done at room temperature with raw materials of the proper composition, particle size, and size distribution. Some require the addition of as much as 20 to 30 per cent water. In every case the forming imparts an approximate shape and some degree of strength to the ware. If water has been added, this must be removed by a drying operation, and some shrinkage occurs. Further shrinkage, plus a very great increase in strength, develops when the ware is fired in a furnace or kiln at high temperatures ranging from 1,100° F for enamels to 3,300° F for alumina ceramics, or to even higher temperatures in special cases.

Ceramic products in general exhibit a unique combination of properties. They have great resistance to heat. Many have melting temperatures over 3,000° F. Most of them exhibit extreme mechanical hardness. Cubic boron nitride is as hard as diamond. Ceramic materials have very great intrinsic strength, but their normally brittle behavior is such that even careful design permits only a small fraction of this strength to be realized. Most ceramics are resistant to corrosive chemicals.

—Charles J. Phillips

Charcoal, a porous solid material containing 85 to 98 per cent carbon. It can be produced by the ignition of almost any animal, vegetable, or mineral material containing carbon. *Activated charcoal* is produced by treating charcoal to give it a very large surface area, ranging from 300 to 2,000 square meters per gram of material. There are two types of activated charcoal: *liquid-phase* or *decolorizing* charcoals, which are light fluffy powders; and *gas-phase* or *vapor-adsorbent* charcoals, which are hard, dense granules or pellets.

Decolorizing charcoals are usually made from bones, wood, peat, lignite, soft and hard coals, tars and pitches, asphalt, petroleum residues, and carbon black. Liquid-phase charcoal is used to decolorize sugar solutions, to remove odor and taste from water supplies and to clarify solvents used for dry cleaning. It is also used in the reclaiming of rubber, pharmaceutical manufacture, food and beverage processing, and in purifying oils, waxes, plasticizers, and other chemicals.

Adsorbent charcoals are usually made from coconut shells, coal, peat, and petroleum residues. Gas-phase charcoals are used to recover volatile organic solvents from air, in the purification and separation of gases, and as a catalyst in the oxidation of various organic and inorganic compounds. Charcoal is activated by steam treatment, which tends to remove adsorbed hydrocarbons from the surface.

—John Price

Chemical Engineering, the branch of engineering concerned with the application of the principles of chemistry, together with the principles of economics and human relations, to processes and process equipment.

The manufacturing processes used in the chemical industry are usually resolved into unit chemical processes and unit physical operations. Such steps, essentially, will be the same regardless of the particular branch of the chemical industry involved. In recent years, it has been realized that these processes and operations in themselves are not the true scientific fundamentals of concern to the chemical engineer. Accordingly, there has been increased emphasis on the so-called engineering sciences, including energy transfer, fluid dynamics, kinetics and mass, momentum, and thermodynamics.

Chemical engineers translate the results of laboratory research into large-scale commercial manufacturing plants. A recent survey of chemical engineers in the United States showed that 28 per cent are engaged in research and development work, 15 per cent serve in various administrative functions, 13 per cent in process engineering, 11 per cent in production engineering, 9 per cent in design, 5 per cent in sales and technical service, 5 per cent in consulting work, and the rest in various government and teaching occupations and other work.

Engineers are playing an increasing role in technical management of chemical process industries, many in major executive positions. But the primary consideration of a chemical engineer is the successful commercial development and operation of chemical processes to yield new and better products at lower costs.

The chemical manufacturing industries employ about 40 per cent of today's chemical engineering graduates; petroleum refining takes another 20 per cent, and process equipment and machinery firms about 6 per cent. The rest are employed in such chemical process industries as plastics, rubber, pulp and paper, drugs, nonferrous metals, glass and ceramics, paint and varnish, soap and detergents, synthetic fibers, textile processing, inedible oils and fats, explosives, leather, and food.

—John Price

Civil Engineering, the branch of the engineering profession dealing with the planning, design, and construction of public and private works and the structures, such as bridges, dams, tunnels, etc., incidental thereto, for railroad, highway, canal and harbor, water supply, power, flood control, irrigation, drainage, and similar needs.

For many centuries the three callings known today as architecture, civil engineering, and military engineering, which also included the limited mechanical interests of earlier days, were regarded as one profession. In ancient Egypt, this general "master builder" was known as *chief-of-works* and held an important position in the court. The Greek historian Herodotus was the first to refer to the builder of a water supply (about 530 B.C.) as an *architekton.* The Romans also used the title *architectus.* New interests developed during the Renaissance, however, ultimately led to the modern specialization. Notable among these interests were the advent of gunpowder and cannon, which introduced new techniques, and an emphasis on art and decoration in the design of buildings, which led to architecture being considered a "fine" rather than a practical art. Military workers adopted the title of *ingeniator,* to denote the ingenious builder of engines of war and medieval fortresses, and began to be known as military engineers.

The organization of the Corps des Ponts et Chaussées in 1716 and the famous Ecole in 1747 marked an important step toward the recognition of civil engineering as a separate profession. Architecture and civil engineering, however, remained closely allied throughout the great period of French leadership, and the last book to treat both professions as one appeared as late as 1802. Increased interest in works unrelated to artistic interests together with the rise of more exact mathematical techniques in design ultimately ended this long liaison.

Architecture and engineering had never been very closely allied in Great Britain, and about 1750 John Smeaton, the builder of the famous Eddystone Light, adopted the title *civil engineer* to distinguish his field from that of military workers. The British Institution of Civil Engineers was organized in 1818. However, Smeaton and others were also active in the development of the steam engine; this led to another division, and in 1847 the Institution of Mechanical Engineers was founded. The American Society of Civil Engineers followed in 1852.

Civil engineering, thus, has long been concerned with irrigation, water supply, drainage, highways and streets, canals, railroads, dams, reservoirs and aqueducts, bridges, tunnels, and other structures. While other engineering branches are devoted primarily to the production of consumer goods, civil engineering works are long-lived, permanent, capital investments. They require both an evaluation of benefits and costs and due attention to the forecasting and appraisal of possible future needs. Many also are public works.

The civil engineer produces no mass product. Each work involves individual planning and design to meet most effectively and economically the particular needs and conditions of the site and to take advantage of all available resources. Advances in engineering science have made more accurate design possible, but the wide variation in conditions to be met—from foundations to such adverse natural factors as floods and storms—requires judgment and skill based on practical experience.

—James Kip Finch

Clay, an earthy material that is plastic when wet but becomes hard when fired. It is composed mainly of fine particles of hydrous aluminum silicates and other minerals. There are many different types of clay with a wide range of plasticity—from *fat clay* (very plastic) to *lean clays* (barely plastic). Plasticity is affected by the type of clay mineral, particle size and shape, organic matter, soluble salts, adsorbed ions, and the amount and type of nonclay minerals it may contain.

Kaolinitic clays (these containing the clay mineral kaolinite) include china clay (also called paper clay or kaolin), ball clay, fireclay, and flint clay. Kaolin is white, has very fine particle size, and is nonabrasive and chemically inert. *Kaolins* are used in the manufacture of ceramics, paper, rubber, paint, plastics, insecticides, adhesives, catalysts, and ink. *Ball clays* are composed mainly of kaolinite but are usually much darker than china clay. Ball clay is a fine-grained, very plastic, refractory bond clay, used in whitewares and some sanitary ware.

Fireclay is the term applied to clays that will withstand temperatures of 1,500° C or higher. Fireclays usually are light to dark gray, and contain minor amounts of illite and quartz impurities. Most fireclays are plastic, but some are nonplastic and very hard; these are known as *flint clays.* Fireclays are used extensively by the refractories industry. The foundry uses fireclay to bind sand into molds for casting metals.

Diaspore clay is a hydrated aluminum oxide, very hard and refractory, and is used to make refractory brick. Calcined diaspore clay is used as an abrasive.

Mullite, a high-temperature conversion product of many aluminum minerals, does not spall, resists heating and cooling exceptionally well, and is resistant to slag erosion. It is used in sparkplugs, laboratory crucibles, kiln apparatus, and other special refractories.

Bentonites, which contain montmorillonite, are formed by the alteration of volcanic ash. These clays are used in drilling mud, as a binder for metal-casting forms, to remove coloring matter from oils, and as an adsorbent for a variety of materials.

Attapulgite clays, also called *fuller's earth,* are hydrated magnesium aluminum silicates. They are used as adsorbents, decolorizers, and deodorizers.

—John Price

Coal, a combustible, carbonaceous material, formed beneath the surface of the earth from partly decomposed vegetable matter through a series of complex geologic processes. Although coal is termed a mineral, it differs from ordinary minerals in having an organic origin—that is, it was formed from the remains of living vegetation.

The material representing the first stage of the transformation of vegetable matter into coal is known as *peat.* It forms when plant materials accumulate under water or in a water-saturated environment where their decomposition is retarded. An accumulation of many layers of plant materials formed under such conditions is called a peat deposit.

The processes involved in the transformation of peat into coal are unknown, but it is thought that the conversion was caused mainly by pressures and heat exerted by accumulated overburdens and movements of the earth's crust. Over very long periods of time, this action resulted in both physical and chemical changes in the original plant materials, including a loss of moisture and evolution of carbon dioxide and methane, leaving an ever-increasing proportion of fixed carbon.

■VARIETIES.—Coals are classified according to degree of conversion, which is known as rank, into four general classes: lignite, subbituminous, bituminous, and anthracite.

Lignite, the lowest rank, has a high percentage of moisture and volatile matter and a low percentage of fixed carbon. It has a fibrous or woody structure and low calorific value (moist BTU of less than 8,300 per pound).

Subbituminous coal, which is similar to bituminous in appearance, contains more moisture and usually has a lower heating value. Better grades generally have calorific values ranging between 11,000 and 13,000 BTU's per pound, but the low grades are barely above lignite in heating value. Subbituminous coal also differs from bituminous in being entirely nonagglomerating or noncoking.

Bituminous coal, known as *soft coal,* has a volatile-matter content that ranges from 14 to about 40 per cent, and a calorific value of more than 11,000 BTU's per pound. Because of its wide range of volatile matter and fixed carbon content, bituminous coal is subdivided into three subclasses: *low volatile, medium volatile,* and *high volatile.* Low volatile bituminous approaches anthracite in dry fixed carbon and volatile matter, but the lower grades of high-volatile bituminous are only slightly above the subbituminous types in calorific value. Another distinguishing characteristic of some bituminous coals is their ability to agglomerate, or form coke. Not all bituminous coals have agglomerating properties, however. Varieties of bituminous range from the *banded,* which include the *bright* and *splint coals,* to the *nonbanded,* which are known as *cannel* and *boghead.* Differences in these types are attributed to the materials from which these coals were formed.

Anthracite is the highest-ranked coal. Known as *hard coal,* anthracite is characterized by its low moisture and volatile-matter content and its high fixed carbon. Subclasses range from *meta-anthracite,* which has 98 per cent or more fixed carbon, through *anthracite,* to *semi-anthracite,* which has properties lying between anthracite and low-volatile bituminous coals. Anthracite is dense, has a bright luster, and is uniform in texture. Its calorific value ranges between 13,000 and 14,000 BTU's per pound, somewhat less than the highest-grade bituminous coals. It is noncoking and, when burned, emits almost no smoke.

UNITED ELECTRICAL COAL COMPANIES

STRIP MINING is speeded by this excavator, which can remove 3,500 tons of earth each hour. The excavator removes the top layers of earth; a power shovel then exposes the coal.

■WORLD RESERVES.—Coal is the world's most abundant and widely distributed energy resource. It is found on all continents, in virtually all geographic areas, and in most countries. The major reserves, however, are in Asia, North America, and Europe.

Data on world reserves vary greatly, but the total reserve in all areas, as estimated by the United States Geological Survey in 1960, was 5,115 billion short tons. Of the total, 49 per cent was in Asia, 34 per cent in North America, and 13 per cent in Europe. (The Asian total includes reserves in the European regions of the Soviet Union, for which there is no separate estimate.) The remaining 4 per cent was widely scattered in Africa, Australia, and South and Central America.

The United States, with an estimated 1,660 billion short tons, has the largest reserve. The Soviet Union and China, with 1,323 billion and 1,115 billion short tons respectively, are next in rank, while Germany with 316 billion and the United Kingdom with 188 billion short tons rank fourth and fifth. These five countries account for 90 per cent of the world's coal reserve. Because standards of thickness and depth of the deposits included in the reserve of individual countries vary, the figures above are not completely comparable. They do, however, reflect the approximate relative magnitude of the reserve in these countries.

It is estimated that 830 billion tons of the total United States reserve is recoverable. This estimate is based upon the assumption that one-half of the reserve in the earth will be lost in mining while one-half will be recovered. The total includes 380 billion tons of bituminous coal, 224 billion tons of lignite, 218 billion tons of subbituminous coal, and 8 billion tons of anthracite. Approximately 40 per cent of the total reserve occurs in states east of the Mississippi River, principally in the Appalachian states, Indiana, and Illinois; the remainder occurs principally in the Great Plains and Rocky Mountain states.

■PRODUCTION.—In world production of coal, the United States ranks second only to the Soviet Union. Its output of 547 million tons in 1966 amounted to 17 per cent of the estimated 3,121 million tons of coal produced throughout the world; the Soviet Union produced 648 million tons that year. Virtually all of the United States production, however, was in coals of high rank, while about one-fourth (172 million tons) of the Soviet output was lignite.

The principal coal-producing states are West Virginia, Pennsylvania, Kentucky, Illinois, Ohio, and Virginia, accounting for about seven-eighths of the total United States output. Only 5 per cent of the production comes from west of the Mississippi River, and this coal is mainly subbituminous and lignite.

■CONSUMPTION.—About half the coal consumed in the United States in 1966 was used to generate electric power; about one-fifth was used for producing coke, one of the essential ingredients for manufacturing pig iron in blast furnaces. Most of the remainder was used for fuel by industrial plants and homeowners, although about 10 per cent of the production was exported. Of significance in the distribution pattern of bituminous coal in the past two decades is the change in the quantities used for residential heating and electric-power

World Production of COAL in % (except lignite)

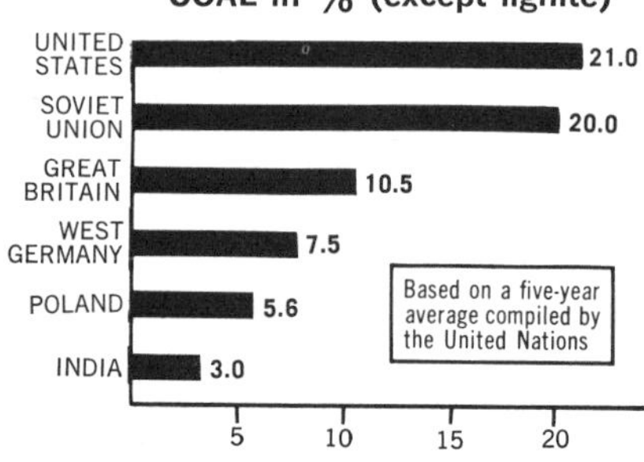

World Production of LIGNITE AND BROWN COAL in %

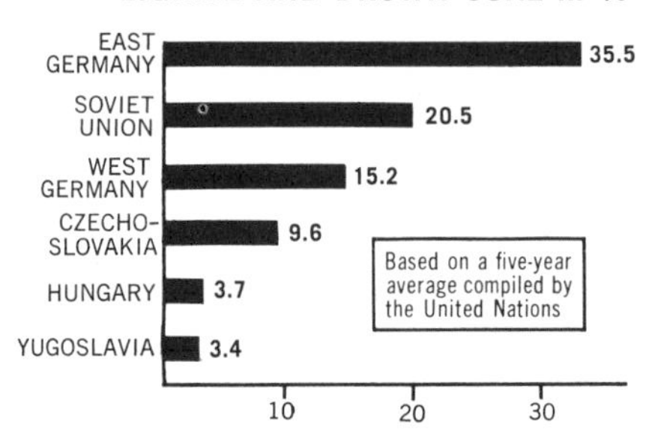

generation. In 1944, 20 per cent of the coal consumed was used for residential heating and 13 per cent for electric power generation, but electric utilities now consume about one-half of the total distributed, while only 6 per cent is used for heating homes.

—Eugene T. Sheridan

Coal Tar, a black, viscous mixture of complex organic compounds, condensed from the volatile matter released when coal is carbonized. The yield and properties of tar vary greatly, but the composition of a particular tar depends principally upon the carbonization temperature and the length of time the coal was carbonized. Tars produced at high temperatures are characterized by large proportions of aromatic, hydrocarbon compounds, which result from the cracking of primary tars, formed at lower temperatures when the volatile matter was first evolved from the coal. Low-temperature tars are composed mainly of primary tars, which are aliphatic hydrocarbon compounds that have not been subjected to cracking. High-temperature tars contain high percentages of naphthalenes and anthracene oil, while low-temperature tars are rich in tar acids.

Virtually all coal tar in the United States is produced by high-temperature carbonization; approximately ten gallons of tar are recovered for each ton of coal carbonized. In some instances, tar is burned for fuel, but usually, it is separated into a number of fractions by several processes involving fractional distillation. In some plants, tar is distilled in batch stills from which the vapors pass into a fractionating column where successive distillate fractions are removed. Modern tar plants, however, use continuous stills that produce a series of distillate fractions simultaneously.

The crude fractions in tar, in order of increasing boiling points, are light oil, middle oil, heavy oil, and anthracene oil.

Light oil, composed of benzene and its homologues, represents from 3 to 5 per cent of the total products recovered. The volume of light oil in tar is too small for economical recovery at most plants, and this fraction usually is recovered with the middle oil. Virtually all the coke-oven benzene, toluene, and xylene—products commonly referred to as coal-tar crudes—are produced from light oil extracted from the coke-oven gas rather than from the light-oil fraction in tar.

Middle oil, composed of tar acids, tar bases, and naphthalene, accounts for about 15 per cent of the total products. Tar acids include phenol, the cresols, and xylenols, while the principal tar bases are pyridine, picoline, and quinoline. These compounds are the basic materials from which a wide variety of chemical materials are synthesized, including synthetic resins, drugs, vitamins, water-repellents, fungicides, weed killers, antioxidants, and rubber-curing agents. Naphthalene, which makes up about two-thirds of the tar bases, is used principally as a source of phthalic anhydride for the production of alkyd resins and many types of plasticizers.

The *heavy* and *anthracene oils,* combined, amount to 15 to 20 per cent of recovered products. The heavy-oil fraction, called *creosote oil,* is used principally in wood preserving. Anthracene oil usually is combined directly with the creosote, but it can be processed for the recovery of anthracene, phenanthrene, and carbazole. Anthracene, the most important of these, can be converted to anthraquinone, used for the synthesis of important classes of dyes and other organic chemicals.

The residue that remains after distillation is called *pitch.* Pitch is used mainly as a binder in the manufacture of carbon electrodes and for producing roofing materials, protective coatings, and fiber pipe. Approximately one-half of the coal-tar pitch produced in the United States, however, is used as fuel in steel plant furnaces.

—Eugene T. Sheridan

Coke, a hard, cellular, carbon residue, formed when coking coals are heated at high temperatures in coke ovens. Coke is irregular in shape, ranges in color from black to metallic gray, and is composed largely of carbon, although most coke contains about 9 or 10 per cent ash.

The process of converting coal to coke, known as carbonization, usually is accomplished in slot ovens and beehive ovens at temperatures above 900° C. *Slot ovens,* which are used almost universally for producing metallurgical coke, recover the volatile constituents of the coal, which subsequently are processed for the recovery of tar, ammonia, light oil, and gas. *Beehive ovens,* so called because of their shape, also produce good metallurgical coke but are used to a lesser extent, as they do not recover the volatile products.

Some coal is also carbonized at high temperatures in retorts, but the coke produced is not suitable for metallurgical use; it is used for domestic heating and for producing gas. Low-temperature carbonizing techniques produce only a char or semicoke, also unsuitable for metallurgical use.

Nearly 1.5 tons of coal are required to produce each ton of coke in a modern coke plant. A total of 92.8 million tons of coal were carbonized in the United States in 1967 to produce 64.6 million tons of coke. Of the total coke production, 1 per cent was produced in beehive ovens and 99 per cent in slot ovens.

Coke is one of the essential raw materials for producing pig iron, and about 90 per cent of all coke produced in the United States is used in pig-iron blast furnaces. Coke is used also in ferroalloy blast furnaces, in foundry cupolas, and for various other industrial purposes, of which the most important are chemical processing, the reduction of ferroalloys in electric furnaces, nonferrous smelting, and mineral wool manufacture. In past years, large quantities of coke were used for residential heating and for the production of producer gas and water gas, but these markets have been taken over by fuel oil and natural gas and only minor quantities are now used for these purposes.

—Eugene T. Sheridan

Concrete, an important construction material made by binding together sand, gravel, or crushed stone with hydraulic cement, which is a mixture of water and Portland cement. (See *Cement.*) In addition to being both fire- and weather-resistant, concrete is a very versatile building material. The initial mixture is plastic and easily workable into almost any shape. When concrete hardens, or *sets,* the sand and gravel, called *fine* and *coarse aggregate,* respectively, are joined together by the cement to form a rocklike mass which has great compressive strength. Well-made concrete, when broken for testing, fractures through the coarse aggregate, demonstrating that the concrete

TWA, EZRA STOLLER

CONCRETE adapts to a variety of uses and forms. For his Kennedy Airport TWA terminal, architect Eero Saarinen created graceful, soaring shapes from reinforced concrete.

is as strong as the rock it contains.

Engineers use concrete in almost every type of construction. The most obvious are highways, bridges, and dams. Less obvious, but just as important, are the massive foundations for skyscrapers, the mostly underwater piers that support the huge steel towers of bridges.

Although strong in compression, concrete is relatively weak under tension. To build concrete structures, such as buildings and bridges, in which the concrete is subject to tensile stresses, engineers embed steel bars in the concrete. Such concrete is called *reinforced concrete*. Reinforcing steel also is used to prevent cracks caused by expansion and contraction due to weather changes. A more recent engineering development is *prestressed concrete*. A prestressed concrete beam is made by putting the concrete member in compression longitudinally, by means of high-strength steel wire or strand. This can be done by casting the concrete around the strands, which are held in tension by jacks (*pre-tensioning*); or by leaving holes running through the beams, inserting the strands after the beam is cast, and then putting the beam into compression, (*post-tensioning*.) Both processes put the beam in a longitudinal compression so that it can better resist transverse tensile stresses.

By comparison, the concrete used for a massive gravity dam (one that relies on its own weight to hold back the water) is plain or *mass* concrete. There is no reinforcing; just enough Portland cement is used to bind it, and the coarse aggregate may be pieces of rock as large as 6 inches.

Aggregate used to make concrete is carefully processed. The particles, from the tiniest grain of sand to the largest piece of gravel or rock, must be clean, hard, and strong. Both fine and coarse aggregate are usually washed to remove any dust, silt, or other impurities that would interfere with the bonding reaction with the cement-water paste. Next, it is screened and carefully separated into various sizes. The quality and type of concrete depends on the proportion and gradations of the various aggregates. The ratio of cement to aggregate is also controlled. Most important is the water-cement ratio. Even when all other components of concrete are correctly proportioned, too much water reduces its strength very markedly. Concrete's strength is derived from the hydration reaction that takes place between the Portland cement and water.

There are many admixtures that are added to concrete at the time of mixing to impart special properties to the concrete, such as air-entraining agents, accelerators, retarders, antifreeze, damp-proofing agents, pozzolana, color pigments, and workability agents. Some of these admixtures accomplish more than one purpose. For example, calcium chloride added to concrete causes it to set faster by increasing the rate of hydration, but it also serves as an antifreeze for cold-weather concreting. *Retarders* do the opposite, slowing the hydration rate and preventing flash-setting, which might occur under certain circumstances. *Air-entraining agents* put in the concrete mix increase the amount of air entrained in the concrete during mixing. Air-entrained concrete is highly resistant to freezing and thawing and the deterioration caused by the deicing chemicals commonly used on highways during the winter. (See Cement.)

Pozzolana, siliceous materials such as diatomaceous earth, tufts, and pumicite, which occur naturally, or fly ash, ground blast-furnace slag, or calcined shale or clay, are being used in increasing amounts in concrete. Pozzolanas themselves are not cementitious, but when very finely pulverized will react with lime in the Portland cement to form compounds that are cementitious. Although pozzolanic concrete generally takes longer to attain its full strength, a large structure such as a dam may take five years to build, so this reduction in early strength is not important. —William W. Jacobus, Jr.

Construction Engineering, the branch of engineering that deals with the building of structures rather than with their planning or design. Paralleling the mechanization of industry, that followed the advent of steam power, was a mechanization of construction that has transformed an ancient, primarily manual art into a highly developed modern engineering specialty. Engineers have long been responsible for the design and specifications for works and, notably in governmental projects, have also planned and directed construction operations. In general, however, the actual building of a project is carried out by a practical builder who is awarded the contract on the basis of the lowest bid.

In earlier years, the direction of laborers in the use of tools and materials constituted the major problem encountered by the contractor. Hand methods were long more or less standardized. Pick-and-shovel excavation, man-powered hoists, and horses and carts were the major construction resources. With the development of modern equipment and machine tools and emphasis on the rapid completion of a project, the selection of the most favorable methods and equipment for a specific construction and their comparative costs, the design of the plant, and the timing and scheduling of the sequence of operations involved, have led to labor-management problems in the field of engineering. Labor has been replaced by capital commitments in equipment. Construction has, in fact, become a highly divided specialty; companies specialize in fields ranging from foundation and underpinning operations, buildings, bridges, dams, and other structural needs, to highway work and subaqueous tunneling. Not only has the scale of modern works also tremendously exceeded those of earlier undertakings but the time required for their completion has been cut to a mere fraction of that prevailing about century ago.

—James Kip Finch

URIS BUILDINGS CORPORATION

FRAMEWORK of steel and mortar rises with the aid of an elevator (*left*) to lift tools.

Copper, an important nonferrous metal, one of the oldest metals known to man; symbol Cu. The discovery of the native form of copper about 8000 B.C. marked the beginning of the Bronze Age, after which this malleable metal was hammered into crude implements and weapons.

Early copper workings on the Sinai Peninsula have been dated as far back as 3800 B.C., and deposits on Cyprus were mined as early as 3000 B.C. These mines were valued possessions of empires that followed, and they were the main source of metal for the Romans. During this time the Egyptians developed the art of metallurgy, and the use of bronze (an alloy of copper and tin) became common.

Copper is widely used because of its chemical, physical, electrical, and mechanical properties. It has low chemical activity and combines in one of three valences: 2+ (cupric) is the most common, but 1+ (cuprous) is also relatively common; the 3+ valence occurs only in a few unstable compounds. With a specific gravity of 8.94, copper is a comparatively heavy metal. Its melting point is 1083° C, its boiling point 2595° C. The thermal and electrical conductivity of copper are both high, and only silver is a better conductor. Copper is one of the strongest of the pure metals. It is moderately hard, extremely tough and wear-resistant, and highly ductile.

Most of the world's copper is obtained from the sulfide ores—chalcocite, covellite, chalcopyrite, bornite, and enargite. Some is obtained from the oxide ores—cuprite, tenorite, malachite, azurite, and brochantite.

Michigan is now the only large-scale producer of copper in the United States. As the demand for copper has grown, the richer ores have been used up. The average U.S. ore now contains about one per cent copper, somewhat lower than that of other countries. However, flotation methods have been developed to upgrade the copper content of low-grade ores. Hence, known reserves will be sufficient to satisfy demand for a long time to come, although the use of lower-grade ores does increase production costs.

Flotation methods produce concentrates containing 20 to 40 per cent copper, and recent improvements promise to increase the yield. The concentrates usually are roasted to reduce their sulfur content (although this step can be bypassed with richer concentrates). Reverberatory furnaces have replaced blast furnaces for smelting concentrates, because they can smelt the finely-divided concentrate without sintering. Furthermore, these furnaces can use a wider range of fuels than can blast furnaces. Smelting yields *copper matte,* a molten solution of copper and iron sulfides, and a slag. The iron oxide is removed in a converter; the iron is further oxidized and formed into a slag by the addition of silica. The sulfur in the matte is oxidized to form sulfur dioxide, which passes off and leaves copper in the converter. When this copper is cast it forms cakes whose surface is roughened and blistered by the escaping gas, hence the name *blister copper.* It is 98 to 99 per cent pure copper. Blister copper is further refined in a furnace and cast into anodes containing 99.0 to 99.3 per cent copper. Electrolytic treatment further refines the metal to 99.98 per cent purity.

Most copper ores contain gold and silver, which are recovered along with other metals as by-products. When the ore is smelted, gold and silver dissolve in the matte. They remain with the copper through converting and furnace refining, and are recovered during electrolytic refining.

World Production of COPPER in %, 1966

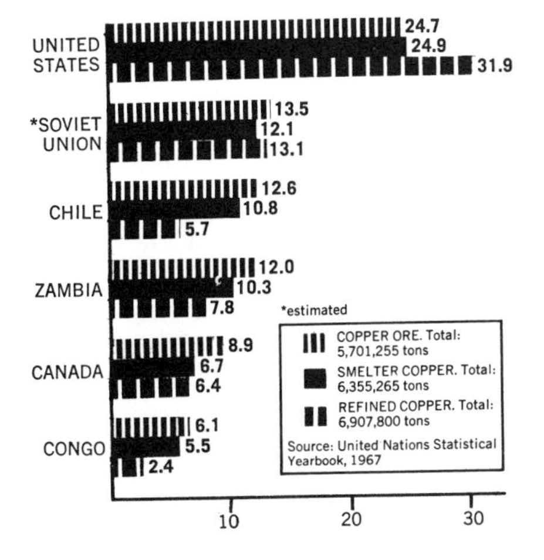

A significant amount of copper is produced by hydrometallurgy. The ore is leached with dilute sulfuric acid or a similar solvent without being concentrated. This process is used mainly for the oxide ores. But leaching does not remove gold or silver—an important part of the metallurgy of copper.

Copper is sold in several commercial grades. The American Society for Testing Materials designates material containing not more than 0.5 per cent of another element or elements as copper. These specifications cover three main commercial varieties: electrolytic, fire-refined, and lake copper. Copper also is specified by the method of processing: tough-pitch, oxygen-free, and deoxidized. Specific kinds of copper include casting copper, phosphorized copper, low- and high-residual phosphorous copper, silver-bearing copper, arsenical copper, selenium-bearing copper, and tellurium copper.

Copper is one of the few metals used more in its pure form than in alloys with other metals. Nevertheless, copper does combine with other metals to form a great number of commercially important alloys, of which brass (copper-zinc) and bronze (copper-tin) in their many varieties are among the oldest and most useful. —John Price

Cork, the bark of the cork oak (*Quercus suber*), which is found in countries along the Mediterranean Sea. Portugal, Spain, France, Italy, Algeria, Morocco, and Tunisia are the main cork-producing countries.

The cork tree has two protective coatings: an outer bark (cork) and an inner bark. The cambium is inside the inner bark, and the cork can therefore be removed without damaging the tree. When the cork oak is about 20 years old—about 9 inches in diameter—cork is removed from the trunk of the tree. The bark grows back again at a faster rate and this *stripping* process can be repeated thereafter at 8 to 10 year intervals. After the cork is boiled, scraped, and dried, it is baled and sold for manufacturing operations.

Written records as far back as 400 B.C. tell of the use of cork. The early Greeks and Romans used it for fishing-net floats, stoppers, and inner soles of shoes. Residents of cork-producing countries for centuries have used cork slabs for roofs and floors.

Cork bark varies in thickness and the yield per tree ranges from 35 to several hundred pounds per stripping. Stripped cork is hauled to central stations where it is boiled in copper tanks for 30 minutes. Heavy weights keep it submerged. Boiling removes tannins, other water-soluble materials, and loose dirt. At the same time, the cork is softened and the rough, hard outside portion, called *hardback,* is loosened so that it can be scraped easily from the good cork. Boiling, scraping, and drying reduce the cork to about two-thirds its original weight, but the volume loss is only about 3 per cent. —John Price

Dams, structural barriers to regulate the flow of or contain a river or stream. Low earth embankments or dikes have been built since ancient times to hold back water for irrigation or drainage. In earlier years in the United States, where timber was plentiful and funds limited, less costly timber and timber crib dams were also widely used. Larger dams of earth or masonry, however, are of recent origin. The first large masonry dams were built for irrigation in Spain in the late sixteenth century, but their modern forms are less than a century old. The greatest dams are products of the twentieth-century. Several other forms—arch, buttress, and movable—are also dams of modern origin.

■EARTH AND ROCK-FILL DAMS.—Several earth dams more than 100 feet high were built before 1900. Usually constructed with a central core-wall of clay or concrete to provide water tightness and keep the lower slope reasonably dry, they were made of selected earth, deposited and compacted in place. However, because of the great disasters due to overtopping, such as the Johnstown flood of 1889 that resulted in the loss of over 2,000 lives, they were considered undesirable for higher structures. In the West, however, the *hydraulic-fill* method was developed and widely adopted; new height records followed. The 250-foot-high Fort Peck Dam on the upper Missouri, intended for flood control of the Mississippi and completed in 1940, involved placing more than 125 million cubic yards of earth and fill.

Rock-fill dams, which have a water-tight diaphragm of steel or concrete, were also developed in the western United States in order to use local materials in remote sites. Composite forms of earth and rock-fill, such as the 400-foot-high Brownlee Dam on the Snake River (Idaho and Oregon), have also been built. A height over 500 feet was reached in the Swift and Trinity dams in Washington and California, respectively. A proposed Russian dam will raise this record to 735 feet.

■GRAVITY DAMS.—The major requirements for a dam are tightness against seepage and stability against water pressure or possible overtopping or destruction by floods. The masonry gravity dam resists water pressure solely because of its weight. The first such dams simply were massive, generally rectangular blocks of masonry. As the mechanics of such structures were more clearly understood, however, the modern triangular, so-called *profile* evolved, increasing in width down from a minimum at the top. The Furens Dam of 1866 in France, 164 feet high, was the first of this rationally designed form. British engineers in India later built a number of similar irrigation dams of record height and length. A new record for height, 297 feet, was reached in the New Croton Dam built for the water supply of New York City in 1907, and the U.S. Bureau of Reclamation, established in 1902, later built a number of outstanding irrigation dams. The Grand Coulee Dam on the Co-

U.S. DEPARTMENT OF THE INTERIOR

HOOVER DAM on the Colorado River provides flood control, water and electric power.

Noted Modern Dams

Name	Location	Height in Feet	Length in Feet	Date Completed
UNITED STATES				
Anderson Ranch	Idaho-Oregon, Boise River	456	1350	1948
Ashokan	New York, Esopus Creek	252	4650	1912
Bull Shoals	Arkansas, White River	278	2349	1948
Center Hill	Tennessee, Caney Fork River	250	2160	1951
Cougar	Oregon, McKenzie River	515	1730	1965
Diablo	Washington, Skagit River	386	1180	1930
El Capitan	California, San Diego River	270	1200	1935
Flaming Gorge	Utah, Green River	502	1285	1964
Fontana	North Carolina, Little Tennessee River	480	2385	1944
Fort Peck	Montana, Missouri River	250	21026	1940
Friant	California, San Joaquin River	319	3488	1944
Garrison	North Dakota, Missouri River	210	12000	1954
Glen Canyon	Arizona, Colorado River	700	1560	1963
Grand Coulee	Washington, Columbia River	550	4173	1942
Hoover (Boulder)	Arizona-Nevada, Colorado River	726	1244	1936
Mansfield	Texas, Colorado River	270	5093	1942
Mud Mountain	Washington, White River	425	700	1948
Oahe	South Dakota, Missouri River	242	9360	1960
Owyhee	Oregon, Owyhee River	417	1010	1932
Palisades	Idaho, Snake River	260	2100	1966
Ross	Washington, Skagit River	545	1275	1949
Saluda	South Carolina, Saluda River	208	7838	1930
San Gabriel #1	California, San Gabriel River	355	1540	1938
Shasta	California, Sacramento River	602	3460	1945
Trinity	California, Trinity River	537	2450	1960
Wolf Creek	Kentucky, Cumberland River	240	5736	1948
FOREIGN				
Aswân	Egypt, Nile River	174	6900	1934
Aswân High	Egypt, Nile River	361	1630	Under construction
Bhakara	India, Sutlej River	680	1700	1960
Bin-el-Ouidane	Morocco, Wadi el Abid	440	950	1954
Burrinjuck	Australia, Murrumbridgee River	264	765	1913
Chambon	France, Romanche River	450	1080	1934
Grande Dixence	Switzerland, Dixence River	932	2296	1962
Karaj	Iran, Karaj River	590	1285	1962
Kurobegawa	Japan, Kurobe River	637	1434	1962
Mauvoisin	Switzerland, Drance de Bagnes	780	1750	1957
Mettur	India, Kaveri River	230	5000	1934
Sakuma	Japan, Tenryu River	512	1960	1956
Tres Marias	Brazil, Sao Francisco River	233	8400	1960
Volta	Ghana, Volta River	370	2100	Under construction
Wu-sheh	Taiwan, Wu-sheh Chi	374	699	1961

lumbia River, 550 feet, also a Federal power and irrigation work, today holds the record for volume, more than 10 million cubic yards of concrete, but the 932-foot-high Grande Dixence Dam, completed in 1962 in Switzerland, holds the height record.

So-called *rubble masonry*, large stones with joints filled with mortar and smaller stones, was used in earlier works, but the production of Portland cement on a large scale led to the wider use of mass-produced concrete. A so-called *cyclopean concrete*, large stones embedded in concrete, was adopted for the Kensico Dam (built 1910–1916) for New York City's water supply.

Large dams must be founded on rock sufficiently hard to support their thrust even under high water pressure. Adequacy of foundations, largely a matter of experience and judgment, thus is critical. Most failures have been due to foundation difficulties, and it may be said of dams that the risk of failure has been the price of progress. The possibility of uplift due to water pressure in the joint between foundation and structure has also received increasing attention in recent years; it may have contributed to some earlier failures.

■ARCH DAMS.—In order to save material, a dam in a narrow valley may be built as an arch with its axis vertical, convex upstream and abutting against solid and reasonably vertical canyon walls. Such an arch dam was apparently built as early as 1843, in the Zola Dam for the water supply of Aix in France. The arched form had been adopted in even the earliest Spanish gravity dams, however, with the idea that this would provide increased stability. Similarly, this plan was followed in early works such as the record-breaking Arrowrock Dam (Idaho), 350 feet high, of 1916. It was soon recognized, however, that in such an *arch-gravity* structure the development of arch action required special precautions to close all vertical joints if a saving in material was to be realized. This problem was met in the Hoover Dam on the Colorado River in 1936, an arch-gravity work that set a new height record of 726 feet. Even this record, however, has been surpassed by the 780-foot-high Mauvoisin Dam in Switzerland of 1957 and the 870-foot-high Vaiont Dam in Italy of 1961.

■SPILLWAY AND OTHER DAMS.—The provision of an adequate overflow or spillway to pass flood flows is a vitally important part of most dams. This is usually provided by a special overflow section of the dam, often built with a curved form, or profile; the flow is discharged downward into a stilling basin or onto a lower apron that breaks its vertical fall to a more horizontal direction. Crest gates or flashboards that can be cleared or removed in flood periods are also used to maintain a higher normal water level.

Some notable dams have also been built of reinforced concrete, using inclined buttresses to support slabs or arches, and various movable types have been devised to meet special needs. —James Kip Finch

Explosives, substances or mixtures of substances that decompose suddenly to release great amounts of energy in the form of heat and expanding gases. They are used for destructive purposes in warfare, but their principal value lies in their peacetime use in mining, construction, road building, quarrying, land clearing, and other useful work. Commercial explosives have become so fundamental to the basic economy that their sales statistics are often considered an accurate indicator of the level of general business activity.

Explosives are useful tools because they concentrate immense force in small, easy-to-handle packages. The key to their power is the fact that they release all of their energy in extremely brief time intervals—usually a few millionths of a second. During its detonation period, for example, a stick of dynamite delivers as much energy as the entire electrical power industry of the United States is generating during that time.

There are two main classes of explosives: low (deflagrating) explosives and high (detonating) explosives. The former burn progressively over a relatively long period of time (though still for very small fractions of a second), whereas the latter act almost instantaneously.

Low explosives, which include black powder and smokeless powder, are now used largely for propellants, where their longer burning time provides a steady thrust to propel a bullet or missile. Other uses include fuzes, fireworks, and some special kinds of blasting.

High explosives include dynamite, TNT (trinitrotoluene), nitrocarbonitrates, and many compounds for special purposes. These are the so-called disruptive materials normally used for blasting today.

Rapid oxidation is the reaction that makes explosives function. A typical explosive compound contains both an oxidant and combustible elements, and it may include additives to prevent deterioration, to lower the freezing point, etc. It does not require air or any other external oxygen supply. Most explosives must be under confinement for proper performance, and many will simply burn if ignited.

In general, explosives used for military purposes differ from those used commercially. Dynamite, for example, is seldom used in military operations, and most military explosives are not satisfactory for civilian work. Some materials, such as TNT, that are primarily military do have limited commercial applications, usually as a constituent of a complex explosive material.

HISTORY.—The early history of explosives is obscure. Some historians say that the Chinese used black powder several hundred years before the Christian era. In the seventh century, a Byzantine fleet is reported to have attacked enemy ships with Greek fire, a concoction of sulfur, rosin, pitch, and saltpeter, much like black powder, which burned with a roaring flame and reportedly had been known to the Greeks more than 1,000 years before. In the thirteenth century, the English chemist Roger Bacon set down the first known instructions for the manufacture of black powder, and by the next century it was being used as a propellant for firearms. In 1627, black powder was used in Hungary for blasting ore, the first known instance of useful work by an explosive.

Gunpowder was vital in colonial America and was one of the first commodities to be manufactured in the colonies. Black powder was probably used for blasting in New England by the early eighteenth century.

E. I. DU PONT DE NEMOURS & CO.

BLASTING AGENT is loaded into vertical holes in the face of a quarry; the explosion rips out 25,000 tons of sandstone, enabling miners to retrieve silica held by the rock.

High explosives originated in 1845 with the discovery of guncotton by Christian F. Schoenbein in Basel, Switzerland. This was followed the next year by the Italian Ascanio Sobrero's invention of nitroglycerin, a major milestone in the history of explosives. Alfred Nobel in 1866 found a way to convert the unpredictable liquid nitroglycerin into a solid, stable material that he called dynamite. Dynamite, which has now been modified into dozens of different types, has been the leading commercial explosive for the past century.

NEW EXPLOSIVES.—Research has been a characteristic of the explosives industry ever since the middle of the nineteenth century. One of the world's first industrial chemical research laboratories was established by Du Pont in Gibbstown, New Jersey, in 1902 to develop new explosives and improve manufacturing processes. Research on explosives led in time to a better understanding of chemistry, particularly organic chemistry, and helped create a base for the modern chemical industry.

Research produced a steady stream of improved and safer dynamites, new techniques for their use, and more efficient blasting accessories—fuzes, blasting machines, detonators, and similar devices. In 1935 a new type of explosive was introduced—*the nitro-carbo-nitrate* family, which is based primarily on ammonium nitrate, a chemical made from atmospheric nitrogen and a coal or oil derivative. Known as blasting agents, these materials are relatively inert and can be set off only by a strong explosive charge. They are not only safe but economical to use when conditions are appropriate. The most recent developments permit the use of fertilizer-grade ammonium nitrate as an explosive at substantial savings in cost. Other new forms of explosives include those in semiliquid or gelatin-like form, which can be loaded into the borehole by pumping, and metalized explosives in which powdered aluminum contributes added energy.

Explosives manufacturing processes vary, depending on the end product. In a typical dynamite plant, operations are conducted in small buildings isolated from each other to minimize hazards. Glycerin is treated in a continuous-process nitrating unit to form *nitroglycerin.* The liquid nitroglycerin is then mixed with ammonium nitrate or other solid materials —usually inert substances, such as wood flour, ground nut shells, and clays—to form a slightly moist, grainy mixture that is loaded into cartridges and shipped to the user.

There are approximately 20 manufacturers of explosives in the United States, operating plants scattered throughout the country. There are at least 20,000 large-scale regular users of explosives in the nation, and most manufacturers sell through a network of distributors. Explosives can be shipped by rail, truck, or other means under regulations of the Interstate Commerce Commission. Annual usage in United States in the 1960's exceeded 1,500,000,000 pounds.

—E. I. du Pont de Nemours & Co.

Felt, a textile composition produced by pressure and friction instead of by customary spinning and weaving techniques. This process causes fibrous materials such as woolens, furs, and artificial staples to form a tough mat.

Fur (or *hat*) *felt* is made from cony, hare, muskrat, nutria, and beaver fur. Small amounts of synthetic fibers, such as protein fibers made from casein, corn, soybeans, or peanuts, sometimes are added. Felting is facilitated by a process called *carroting*—heating the fur while it is still on the pelt with certain oxidizing agents in acid solution. At first this was done with mercury in nitric acid. Fur thus treated turned yellow like a carrot; hence the name. However, mercury treatment is toxic and has been replaced by a method using chlorine compounds. *Wool felt* can be made without a chemical pretreatment. Woolen fibers are generally short to medium length. They are blended and then run through a series of processes: willying, blending, teasing, scribbling, carding, forming, hardening, milling, carbonizing, dyeing, and finishing. Batts of felt thus produced may be superimposed to increase thickness. Manufactured felts are used for upholstery, carpeting, billiard-table covers, padding for clothing, and for heat-insulating and silencing purposes. Impregnated synthetic-fiber felts are used for roofing, housing and shipping sheathing, and for pads under carpets. —John Price

Fibers, Structural, threadlike substances useful as construction materials, including asbestos, rock wool, slag wool, glass wool, and lead wool.

Asbestos is the general name applied to a number of mineral silicates that are incombustible and can be separated into filaments. Largest sources of commercial asbestos are mines in Canada, the United States, and Africa. Asbestos appears in many forms and types, varying from long, soft, silky fibers with a definite orientation of the crystals to a short, harsh, brittle mass fiber with random orientation of the crystals. Asbestos cloth is used for brake linings, clutch facings, gaskets, fireproof curtains, shingles, insulation, and fireproofing.

Mineral wool is the term applied to products known as rock wool, slag wool, and glass wool. Each of these materials is a fluffy, lightweight mass of intermingled vitreous mineral fibers composed of complex silicates. *Rock wool* is made from natural rock or combinations of natural minerals, which are melted and blown into fibers. *Slag wool* is made of iron, copper, or lead blast-furnace slags. *Glass wool* is made from conventional glass batch materials—such as silica sand and soda ash or borax—dolomite, and minor ingredients. Mineral wools are used as thermal insulation, sound-absorbing materials, and as a filter medium.

Lead wool consists of fine strands of metallic lead loosely wound into a rope. It is made by pouring molten lead through a fine sieve. The lead solidifies into fine strands as it falls through the air. Lead wool is used for plugging oil wells to prevent water seepage, and for other caulking purposes. It can be used under water or in gaseous locations where heat (which is not needed for lead-wool caulking) would be dangerous. —John Price

Fibers, Synthetic, manmade materials used in the manufacture of textiles. In the modern world, the use of materials for clothing and other practical purposes has tended steadily away from natural substances toward synthetic replacements. This trend has resulted from the rise of applied technology which has created and made commercially practicable many synthetic materials with functional properties similar to those of their natural counterparts, and by the growth both in the number of people in the world and their standards of living. In response to an expanded demand for consumers' goods, synthetic products, such as coated fabrics of various kinds, and plastic sheetings, such as vinyl, have replaced leather. Plastics in solid form have taken the place of wood and steel in many familiar uses, such as the handles of kitchen knives, knobs and handles, and the fittings of vacuum cleaners.

In no field has the trend toward synthetic materials been more pronounced than in textiles—the woven, knitted, or felted fabrics used since prehistoric times for clothing and home decoration. Synthetic textiles are purely a development of this century. As the century has advanced, their use has increased sharply to the point where they have outdistanced wool and silk and are fast catching up with cotton, in terms of poundage consumed.

■CELLULOSE FIBERS.—The first synthetic fibers were made from cellulose. These are rayon and acetate, known from their basic material as the *cellulose fibers. Rayon,* the first synthetic fiber, was developed experimentally in the latter part of the nineteenth century by Count Hilaire de Chardonnet in France and Charles F. Cross and E. J. Bevin in England. Commercial production of rayon, then known as artificial silk, was undertaken in the United States about 1910. In subsequent years, the manufacture of rayon as a textile fiber spread to virtually every industrialized country in the world. Rayon is made by a complex process in which pure cellulose is extracted from wood pulp, and to a lesser extent from cotton, by subjecting them to treatment with caustic soda and carbon disulfide. The result is a chemical solution which is pumped through a *spinneret,* a device that resembles a tiny shower-bath nozzle. The threadlike streams harden as they emerge, forming filaments that are gathered together into yarn. Among the trade names for rayon fibers are Cupioni, Fortisan, and Zantrel.

Acetate, the second synthetic fiber brought to wide use, is made, like rayon, from cellulose. However, different chemicals are added to the cellulose solution, with the result that acetate has somewhat different properties. Among the trade names for acetate fibers are Celaperm and Chromspun.

■CHEMICAL FIBERS.—The success of rayon and acetate for clothing fabrics and household and industrial uses spurred research into ways of making fibers by chemical means which would have functional properties even more desirable and versatile than those derived from the natural material, cellulose. It was discovered that certain polymers—the technical name for large molecules made by combining smaller molecules—have exceptional fiber-forming properties. In this research, Wallace Carothers is generally credited with the major effort that led to the first of the truly synthetic fibers, *nylon.* This fiber is created by a complex process from various chemical ingredients of which the most notable, for one form, are hexamethylene diamine and adipic acid. Another kind of nylon uses the chemical, caprolactam.

AMERICAN VISCOSE CORPORATION

MANMADE FIBERS come in two forms—continuous filaments and staple fibers shown here. Staple fibers are continuous filaments that are cut into short lengths for spinning into yarn.

Introduced in 1939 by Du Pont, nylon was used during World War II for tow ropes for gliders and other military uses, where its great strength and durability were essential. When the war ended in 1945, as fast as its production could be expanded, nylon found its way into a steadily broadening variety of apparel. Today, nylon is produced in ever-increasing quantities for use in virtually every kind of apparel and decorative fabric.

The success of nylon led to increased interest in the profit possibilities of other synthetic fibers among large chemical companies in the United States, Europe, and Japan. In rapid succession since 1950, new synthetic fibers have been introduced, each with its own advantages for use in textiles. Among them have been the *acrylics,* fibers made largely from a chemical called acrylonitrile. Soft and bulky like wool, with wool's warmth, the acrylic fibers are also highly resistant to chemicals; when made into cloth, they offer good shape and crease retention. These characteristics along with quick drying give acrylic garments a high degree of easy-care convenience for consumers. Among the trade names for acrylic fibers are Acrilan, Creslan, and Orlon.

Another important group of synthetic fibers is the *polyesters,* fibers extruded from a chemical solution made up of dihydric alcohol and terepthalic acid. Blended with cotton or rayon, polyester fabrics have been second only to nylon in yielding apparel fabrics that are durable, wrinkle-resistant, and quick-drying. These qualities and easy maintenance are the basis of the popularity of the so-called *wash and wear* apparel. Among the trade names for polyester fibers are Dacron, Fortrel, Kodel, and Vycron.

Other synthetic fibers that have won acceptance in apparel are *spandex fibers,* such as Lycra, with *snap-back* properties of stretch and recovery that give elasticity to corsets, brassieres, and bathing suits; the *olefin fibers* such as Marvess made from a paraffinlike derivative of petroleum; and *glass fibers,* such as fiberglass, which are extremely fine filaments of glass that have performed well in curtains and other decorative fabrics.

Along with the manmade fibers noted above, other fibers have been found advantageous in certain specialized textile uses. *Saran,* the generic name for vinylidene chloride fibers, is extruded as a stiff yarn with good chemical and flame resistance. Among its established uses are draperies and outdoor furniture fabrics. *Metallic yarns* are widely used for decorative effects in apparel and home furnishings fabrics. They are made by bonding aluminum foil between clear layers of plastic, or by vacuum depositing of aluminum on the surface of a plastic film.

The man-made textile materials have taken their place with cotton and wool as the principal materials for the endless variety of uses that the world's billions of people have for cloth. Indeed, it can be expected that, as the manmade fibers are improved by continuous research and testing, their use will almost completely supersede that of natural fibers. In this connection, it must be remembered that the major natural fibers, cotton and wool, are becoming more and more expensive to produce, while applied technology steadily reduces the cost of the synthetics.

—Jerome Campbell

Flux, a material added during soldering or brazing to improve the quality of the joint. *Soldering fluxes* remove oxide or other obstructing films so the parts will accept solder more readily. Activated rosin flux often forms the core of solder. Other fluxes include zinc chloride, ammonium chloride, and combinations of these. *Brazing fluxes* remove any oxides that may form during heating, brazing, or cooling and protect the joint from oxidation. Fluxes may be chemical formulations or protective and cleansing atmospheres. Borates and fluoroborates; chlorides of sodium, potassium, and lithium; borax and boric acid; and alkalis, together with wetting agents and water, are commonly used. *Protective atmospheres,* used in furnace brazing, include fuel gas, dissociated ammonia, hydrogen, inorganic vapors, or inert gases such as helium or argon. —John Price

Fuller's Earth, a natural earthy material, usually a clay, that decolorizes mineral or vegetable oils. In ancient times, certain earths were used to clean raw wool by adsorption. The process became known as *fulling,* and the cleaning agent was called fuller's earth. The clay materials usually present in fuller's earth include attapulgite, montmorillonite, and kaolinite.

Fuller's earths apparently decolorize by selective adsorption of coloring matter and other impurities. These are held strongly within the clay structure and can be removed only by drastic treatment. Much fuller's earth is used to decolorize petroleum products, cottonseed oil, tallow, soy oils, and other products. For such applications the earth must not only decolorize, it also must not impart any taste or odor and must not retain too much oil that cannot be reclaimed. Some fuller's earths are used for oil-well drilling muds, insecticide carriers, and fillers. The main deposits of fuller's earth are in the United States, Japan, and England.

—John Price

Fur and Leather, pliable materials made from the hides, skins, or pelts of animals. Fur is a form of leather made from the skins of mammals, with part of the hair left on. Both fur and leather are processed so as to inhibit the decaying of the skin. The treatment alters the skin protein. It is necessarily less enduring for fur than for leather, since the hair cannot be destroyed in the process and the skins must be kept flexible and soft.

In *fur dressing,* the skins are first soaked in brine to soften the skin and to inhibit bacteria. The skin is then hammered with many needles to soften it, and the flesh is scraped off by hand with sharp knives. The pelt is then converted by *oil tanning,* called either *pickling* or *biting.* In these processes, the skins are soaked for several hours in solutions of chemicals such as sodium chloride, sulfuric acid, formaldehyde, and alum. This halts or delays putrefaction of the skins. The skins are then thoroughly dried, after which vegetable, animal, and mineral oils and greases are worked into the leather by a wooden piston *tramping machine,* or *kicker.*

The peltry is then rolled in circular vats with sawdust, which absorbs the oil and grease. The fur is combed, beaten, and sometimes sheared. Generally, the furs are dyed by rotating them in vats until the dyes are fixed. The dyes are usually anilines, although some furs, such as mouton, require vegetable dyes. Almost every type of mammal skin has been used for fur, but the favored fur animals come from the rodent family (beaver, muskrats, nutria, etc.), and the weasel family (sable, marten, mink, otter, kolinsky, etc.).

In contrast to fur, *leather manufacturing* removes all of the hair, and the hide or skin is more thoroughly preserved. Hide consists of the epidermis, dermis, and flesh, but only the dermis is used in leather. The hide, or skin, is first soaked in lime and water or various chemical depilatories. The hair and flesh are then scraped off by machine, and the hide is cut up to get pieces of uniform thickness in a process called *rounding.* Next, the hide is *bated* by washing with a weak acid solution (usually dilute sulfuric acid) and then by adding enzymes to the bating water. In bating, the enzymes destroy fat and oil and soften the hide. The hide is then pickled for a day or two in a chemical solution that further preserves the skin. After that, the leather is ready for *tanning.*

Tannin concentrates and extracts are usually used, and some sort of glucose is added to the tanning solution. Hides are often put in a series of vats with increasingly strong solutions so that the hides tan thoroughly and evenly. Tanning may take from several days to a year, depending on the tannin used and the type of leather desired. Tanning causes a reaction of the skin proteins with certain chemicals to produce a decay-resistant hide, although the exact nature of the reaction is not understood. Chrome is the only inorganic tannin of importance. Vegetable tannins, extracted from the bark, wood, or fruit of trees, are numerous. The most common sources of tannin are quebracho (44 per cent of the total used), chestnut wood extract (30 per cent), and mangrove (5.4 per cent).

After the leather has been tanned, it is colored and *fat-liquored.* Dyes and oils are worked into the leather in a vat, and the leather is then dried under increasing temperatures for nearly a day. *Sammying,* which consists of soaking and then milling in sawdust, completes the process. Next, the leather is finished; sometimes it is buffed or polished, sometimes shellacked or varnished.

One major virtue of leather is that it is porous, but it is also a poor heat

conductor. Nearly any skin can be tanned, including the skins of mammals, reptiles, birds, and fish. Cattle hides are the largest source of leather, and the shoe industry still uses most of the product. Luggage, clothing, and accessories are made of leather from the skins of ostrich, alligator, shark, and many other exotic animals.
—John T. Schlebecker

Gas, a combustible fluid supplied and utilized in the gaseous state as a fuel to produce heat, light, or power for domestic, commercial, and industrial applications. Each of the many types may be classified in one of three groups: natural gases, manufactured gases, or liquefied petroleum gases.

■NATURAL GAS.—The U.S. Bureau of Mines defines natural gas as a naturally occurring mixture of hydrocarbon and nonhydrocarbon gases found in porous formations beneath the earth's surface, often in association with crude petroleum. In the United States, this is the fuel used by over 98 per cent of the approximately 39 million customers who purchase gas distributed through street mains by gas utility companies. It was first discovered in the U.S. at Fredonia, New York, in 1821.

Natural gas is found in porous rock formations, not in cavelike spaces under the earth. Gas-bearing rock consists of carbonate rocks (principally limestone) or of sandstone, which act as a kind of rigid sponge holding the gas between the grains. The gas is trapped and held by a covering of impervious or nonporous rock.

The gas contained in these underground fields is obtained by drilling wells through the covering earth and cap rock and allowing the gas to be forced out by its own pressure. In 1967, 9,059 exploratory gas and/or oil wells were drilled in the U.S., and of these 17.6 per cent, or 1,595, were productive. Gas wells now average over 5,000 feet in depth and cost an average of about $100,000 each. Some approach three miles in depth.

Natural gas is found in 31 states, of which six states account for over 90 per cent of U.S. production. By order of rank these are: Texas, Louisiana, Oklahoma, New Mexico, Kansas, and California. Of the net U.S. production of 17,474 billion cubic feet in 1966, Texas alone was responsible for almost 41 per cent. The largest producing gas field in the world is the Hugoton, which extends from the Texas Panhandle, through Oklahoma, and into Kansas.

In 1967 the transportation of gas from the source to the ultimate consumers required 828,270 miles of pipelines, of which 8 per cent were located in the gas fields, 27 per cent were transmission lines, and 65 per cent were local distribution systems. This represents an increase of almost 67 per cent over 1955 statistics. In addition, new materials and techniques now permit the use of pipelines with much larger diameters—up to 42 inches. The first pipeline in the United States, built in 1872, was only two inches in diameter, and stretched five and one-half miles, from Newton to Titusville, Pennsylvania.

Natural gas, as distributed, usually contains 80 to 95 per cent methane (CH_4) and lesser amounts of ethane (C_2H_6) and propane (C_3H_8). Most of the remainder is nitrogen. The heating value ranges from 900 to 1,200 BTU's per cubic foot. With a specific gravity from 0.58 to 0.79, it is therefore lighter than air and dissipates rapidly in the event of leakage. As produced, it is basically odorless and colorless, but for safety reasons odorants are usually added, so that any leakage may be readily detected. The gas is nontoxic (nonpoisonous) to humans and animals, and not harmful to house plants.

Natural gas can be converted into a liquid by lowering its temperature to 250° F below zero and putting it under pressure. In this form it is called *liquefied natural gas,* and the process is referred to as *liquefaction.* It is done to facilitate storage and transportation, since the liquefied gas requires only 1/600 of the storage volume that would be necessary for its containment as a gas.

■MANUFACTURED GAS.—A manmade gas, manufactured gas is produced from coal, coke, oil, or by reforming natural or liquefied petroleum gases. The following are a few of the most widely used types:

Coal gas is made by the distillation of the volatile matter from coal in retorts. It is high in hydrogen and methane and has some carbon monoxide and illuminants. If steam is forced through the glowing coal or coke, it reacts chemically to form *water gas,* which consists of hydrogen and carbon monoxide. Since water gas burns with a blue flame, it is also called *blue gas.* Enrichment with a higher BTU oil gas increases the light-giving qualities, as well as the heat content per cubic foot. *Carbureted water gas* is made by enriching water gas with thermally cracked oil, natural gas, or liquefied petroleum gas. The major constitutents are hydrogen and carbon monoxide. All of the gases made from coal or coke contain sufficient percentages of carbon monoxide to be poisonous if inhaled. As supplied by utility companies they have heat contents of 475 to 550 BTU's per cubic foot.

Oil gas is made by thermal decomposition (cracking) of oils, which may vary from naphtha to heavy residuum carbon oils. The volatile hydrocarbons and hydrogen formed may be controlled to provide mixtures that can supplement or replace 500 BTU manufactured gases or 1,000 BTU natural gases.

Acetylene gas (C_2H_2) is primarily used for metal cutting or welding operations and as an illuminant for lighting. It is formed by the action of water on calcium carbide (CaC_2). Today, however, much of it is made from *synthesis gas,* a mixture obtained by the partial burning of natural gas under controlled conditions. For storage and transportation, acetylene is dissoved in acetone under pressure and put into steel cylinders filled with porous packing.

■LIQUEFIED PETROLEUM GAS.—Referred to as *LP-gas,* liquefied petroleum gas is sometimes better known to the consumer as "bottled gas," "tank gas," "LPG," or simply "propane" or butane." The hydrocarbon constituents are contained in natural gas, natural gas liquids, and crude oil, and consists of combinations of propane, butane, isobutane, propylene, and butylene. These hydrocarbons can be liquefied under moderate pressure at normal temperatures, but are gaseous when used at normal atmospheric conditions. When the gas is stored on the customer's premises, it is kept in liquid form in steel cylinders that may be replaced when they are empty or may be refilled from tank trucks. Some utility companies mix LP-gas with air and distribute it through their street mains.

The two major LP-gases, propane (C_3H_8) and butane (C_4H_{10}), both contain approximately 21,600 BTU's per pound. However, when drawn from the tank as a gas, propane has a heating value of 2,500 BTU's per cubic foot as compared with 3,200 for butane gas. The gases are heavier than air and are nontoxic.

In addition to its suitability as a fuel for home and industry, LP-gas is frequently used for automobile and other internal-combustion engines, and large quantities are consumed in the petrochemical industry as a raw material for making plastics and synthetic rubber. In 1966 approximately 13.6 billion gallons were consumed; this is a 106 per cent increase over the 1956 figures.
—Edgar A. Jahn

Glass, a hard, brittle, usually transparent or translucent material that does not crystallize as it solidifies. In the United States, the term *glass* is restricted to inorganic materials. In other countries, organic materials that possess the characteristic noncrystalline structure are called organic glasses.

The average composition of the earth's crust would make an acceptable glass. When cooled rapidly, magma solidifies to form *natural glass.* Among the rocks classed as glasses are obsidian, pumice, and tachylyte. Another form of natural glass is tektite, which is probably meteoritic in origin.

Glass is a noncrystalline, or amorphous, material. Its atomic structure is one of disorder or ramdomness (Fig. 1). To produce a glass, the ingredients must be present in the cor-

Fig. 1 — Fused Silica Glass

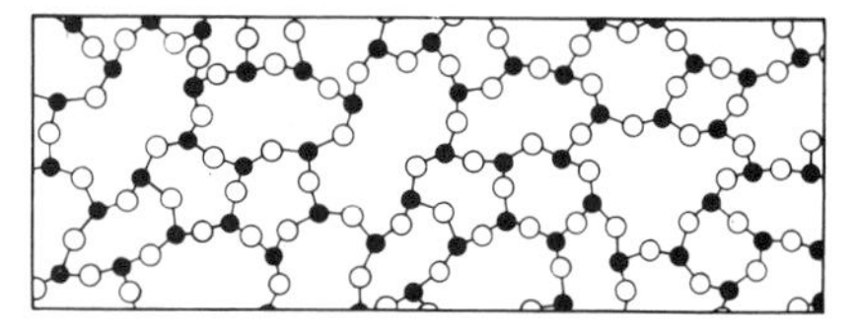

Fig. 2 — Crystalline Silica

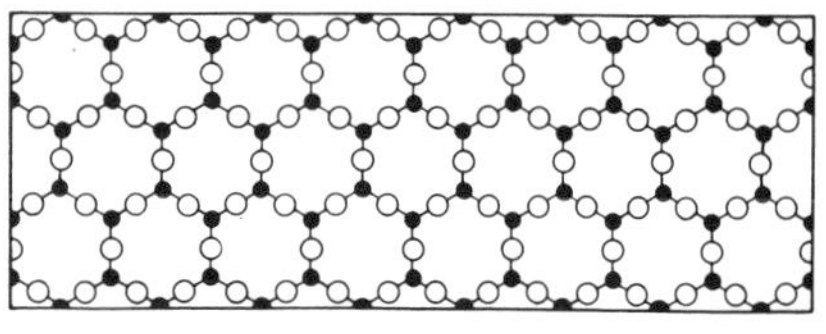

● Silicon ○ Oxygen

Glass Characteristics and Products

Type	Characteristics	Typical Products
Soda-lime	Easily formed to many shapes Low in cost	Windows Bottles Lamp bulb envelopes
Lead-alkali	High clarity High refractive index High electrical resistivity	Art glass Thermometer tubing Optical parts
Borosilicate	Good heat shock resistance High chemical durability	Laboratory ware Cooking ware Pipeline Boiler gauge glasses
Aluminosilicate	Good heat shock resistance High service temperature	Ignition tubes High-speed aircraft windows
96 per cent silica	Excellent heat shock resistance Good ultraviolet transmittance	High-temperature laboratory ware Spacecraft windows Furnace view ports
Fused silica*	Best heat shock resistance Best ultraviolet and acoustical transmittance	High-speed wind-tunnel windows Ultrasonic delay lines.

* Fused silica, fused quartz, and silica glass are essentially the same material.

rect proportions. After these ingredients have melted and combined, the resulting fluid must be cooled rapidly enough so that randomness, which is characteristic of the fluid state, is frozen in. If the mixture of ingredients is not correct of if the cooling rate from the molten condition is too slow, the atoms will arrange themselves in a regular pattern that is characteristic of the crystalline solid state. (Fig. 2). Glass is sometimes called an undercooled fluid. Its structure is random like that of a fluid, but it is frozen and, therefore, has some of the properties of a solid, such as hardness and rigidity.

■CHEMICAL COMPOSITION.—The principal ingredient of most glass is silica sand. This is mixed with soda, lime, borax, alumina, potash, or any of several other materials. Nearly all of the known elements may be used in glass. More than 100,000 different glass compositions have been melted, and more than 1,000 kinds of glass are manufactured commercially. Almost all glass falls into one of six categories, which are based on chemical composition. The accompanying table contains a brief description of the characteristics of each type and some of the products made from each. *Soda-lime glass,* once known as *crown glass,* is the oldest type and still accounts for approximately 90 per cent of the glass produced today. It was first made about 4,000 years ago, but the exact date and place are still matters of speculation. Nor is it known how sand, soda, lime, and heat happened to be brought together by early man to form glass. *Lead-alkali glass,* once known as *flint glass,* was first made in 1675 and used for art glass because of its brilliance. The finest art glass is still made of this type; it is also used for optical parts, thermometers, and electric lamp parts.

While these older glass compositions are still used extensively, it is the four newer types (*borosilicate, aluminosilicate, 96-per-cent silica,* and *fused silica*) that have done most to increase the total number of glass products to nearly 50,000. Because of their improved heat resistance and chemical durability, these newer types perform under severe conditions that the older soda-lime and lead-alkali glasses could not withstand.

When additional strength is needed, glass may be tempered. This is a heat treatment that increases strength from two to four times. Soda-lime, borosilicate, and aluminosilicate glass may be tempered. Typical uses for tempered glass are automobile windows, doors for buildings, and food service ware.

Glass-ceramics are one of the more unusual discoveries of recent years. These materials are mixed, melted, and formed as glasses. Then they are converted, by a process of controlled nucleation and crystal growth, to polycrystalline ceramics. Because they are formed as glass, a greater range of shapes is possible than with other ceramic materials. After conversion, glass-ceramics are approximately four times as strong as glass. They were first used in nose cones for radar-guided ground-to-air missiles, and later for cooking ware.

Chemically strengthened glass is another discovery that opens still more possibilities for glass products. The process most frequently used involves the exchange of ions in the glass surface. After the ion-exchange treatment, the glass is up to ten times as strong as it was before. It can be bent and will withstand impacts better than any other kind of glass. It was first used as a rear-window material for convertible coupes. It is also used for laboratory apparatus such as pipettes, centrifuge tubes, and other equipment.

■NEW USES.—The growth of technologies in other fields continually provides new challenges and new opportunities for glass and glass-derived materials. For instance, space exploration spurred the development of spacecraft windows, low-loss antenna shields, and high-reliability electronic parts. The exploration of hydrospace (the depth of the oceans) makes use of submersible vessels and floats made of both glass and glass-ceramic materials.

Indeed, glass, once considered suitable only for windows and bottles, now appears to be a material with no limitations to its usefulness. Yesterday's limitations are a challenge stimulating today's and tomorrow's inventions. A typical example is photochromic glass. This material darkens automatically on exposure to light and clears when light source is removed. Such a glass shows promise for self-adjusting sunglasses and for control of sunlight in buildings and vehicles. —George W. McLellan

Glue, an adhesive used for bonding separate pieces together. The term at one time referred solely to the high-molecular-weight protein extracts from hides, bones, and fish. However, the use of various protein glues as woodworking and furniture adhesives, and later as plywood adhesives, has tended to establish the term *glue* for all adhesives used by these industries.

Animal, vegetable, casein, blood albumin, liquid, and synthetic resin are the principal classes of woodworking glues. Of these, *casein,* which has a skim-milk base, is used most frequently in small workshops. *Animal glues* and *liquid glues* are used to some extent by cabinetmakers. The other glues require special equipment, such as heating facilities and presses, which tends to limit them to commercial use. Some *thermosetting resins* require temperatures higher than 300° F and pressures as high as 250 pounds per square inch, depending on the wood.

Animal glue is a crude, impure amber-colored form of commercial gelatin. It does not exist as such in the living organism, but is produced by the hydrolysis of animal collagen. It gelatinizes in aqueous solution. Animal glues are prepared by hot aqueous extraction from pretreated collagenous materials, chiefly animal hides and bones. Casein is the main protein of milk. It is made by coagulation by rennet or by acid precipitation. —John Price

Granite, a natural plutonic rock formation with visibly crystalline texture, composed mainly of quartz and alkali feldspar with subordinate plagioclase and dark-colored minerals, such as biotite and horneblende. More generally, the term is applied to plutonic rocks rich in feldspar and quartz. Commercially, the term *granite* is extended to include any phaneritic rock rich in feldspar, with or without quartz and mafic (dark) minerals.

Because of its hardness, durability, and pleasing colors (flesh red, whitish, or gray), granite is an important building and ornamental stone. It occurs in a wide range of textures and structures.

Granite is thought to be formed by three processes—slow crystallization of deeply buried granitic melts; metamorphic recrystallization of volcanic or sedimentary rocks; and metasomatic transformation of various sedimentary or igneous rocks by introduction of certain elements, such as alkalis and silica, or by removal of others, such as iron, magnesium, and calcium. —John Price

Gypsum, the most common of the sulfate minerals, hydrated calcium sulfate ($CaSO_4 \cdot 2H_2O$) which occurs naturally in many parts of the world. The white or slightly yellowish mineral is best known as the raw material, which when heated, loses almost all of its water to produce plaster of Paris. Gypsum is a very soft mineral and can be scratched with the human finger nail. Its hardness on Mohs' scale is 2.

The mineral is found in several forms: transparent crystals called *selenite;* a translucent, fine-grained mass called *alabaster*; a silky, fibrous form called *satin spar,* which looks somewhat like asbestos; a dull colored rock called *rock gypsum*; and an unconsolidated earthy form that is usually quite impure called *gypsite.*

Gypsum is widely used in the building industry to make many types of wallboard. In these applications the mineral is calcined (heated) to an even greater extent than for making plaster of Paris, removing even more of its water. The mineral is also used as a *retarder* in making Portland cement. Ground gypsum can be used as a fertilizer providing lime to the soil, but ground limestone has largely taken over its agricultural applications.

—William W. Jacobus, Jr.

Hydrochloric Acid, an aqueous solution of hydrogen chloride which attacks all common metals; also known as *muriatic acid;* chemical formula HCl. Pure hydrochloric acid is colorless; the acid containing impurities of iron, chlorine, or organic substances is yellow. Pure hydrogen chloride is a colorless, pungent gas at normal temperature and pressure. It fumes strongly in moist air and in concentrated form it is toxic if inhaled.

Basilius Valentinus is generally credited with the first production of hydrogen chloride, in the fifteenth century. The *Leblanc process,* invented by Nicolas Leblanc about 1790, was the first major commercial method of preparation. In this process, hydrogen chloride and salt cake (sodium sulfate) are produced as coproducts of the reaction of sulfuric acid and salt. Although the *Solvay process,* discovered by Ernest Solvay about 1863, has replaced the Leblanc process for making salt cake, the salt-sulfuric acid reaction is still important because of the industrial demand for both salt cake and hydrochloric acid. Hydrogen chloride is also produced by burning hydrogen in chlorine, which yields a high-purity product (99.7 per cent after purification) that is desirable for organic synthesis and in the manufacture of reagent-grade acid. This method is an outlet for chlorine obtained as a by-product in the manufacture of caustic soda. An increasing amount of hydrochloric acid is also produced as a by-product in the chlorination of organic chemicals, although contaminants such as chlorine, air, excess reactants, organic products, and moisture must be removed to yield a high-grade product.

Hydrochloric acid is used widely in industry for metal cleaning, sugar refining, synthetic rubber manufacture, glucose and corn sugar production, and reactivation of petroleum wells. It is also used in the production of metal chlorides, as a neutralizing agent, and in the production of organic chlorides. —John Price

Iron and Steel, the two most important structural materials. Iron is the fourth most common element and the second most abundant metal in the earth's crust, of which it comprises 5 per cent. The core of the earth is believed to consist of iron and nickel, making iron the most plentiful element in the earth as a whole. Its symbol is Fe; atomic number 26, atomic weight 55.84, and melting point 1,535° C.

Iron occurs abundantly near the earth's surface in compounds (ores) that are widely distributed throughout the world. Its chief commercial ores are *hematite, magnetite, limonite,* and *siderite.* Iron rarely appears alone but is found free in meteorites, which usually also contain 7 to 15 per cent nickel. Meteorites were man's first source of iron, which he chipped off in small quantities as early as 4000 B.C. Although copper, bronze, lead, and some other metals were smelted before iron, iron was smelted sporadically over many centuries before the true Iron Age began.

THE IRON AGE.—This term refers more to a stage of cultural development than to a period of time. The Iron Age originated in the Caucasus Mountains in a region known as Chalybia well before 1600 B.C., for by then the Hittites, whose tutors were the Chalybes, were skillful ironworkers. From the Hittites, the Iron Age spread to the Assyrians, Babylonians, Philistines, Egyptians, and other peoples of the Middle East, where it was well established by 1350 B.C. It spread eastward to China and India and was carried westward by the Phoenicians and the Greeks. It reached central Europe about 900 B.C. and Britain about 450 B.C.

Iron ore was smelted in primitive furnaces by means of charcoal kept burning under a forced draft from hand or foot-operated bellows. Oxygen in the ore combined with carbon in the charcoal to form carbon monoxide gas, releasing a relatively pure lump of iron intermixed with gangue from the ore that formed a slag.

This was *wrought iron,* with a low carbon content. With some exceptions, it was the only form produced until the development of the blast furnace in the Rhine provinces in the fourteenth century.

PIG IRON.—In the blast furnaces, taller furnace stacks and a more powerful draft from water-driven bellows enabled the process to produce molten iron, which could be cast in various forms. When the iron poured from the furnace, it was run into a large mold that fed a number of smaller lateral molds. Because this figuration resembled a sow nursing pigs, the iron became known as *pig iron.* Pig iron output in early blast furnaces was 25 tons weekly. Now daily rates up to 5,000 tons are expected in fully automated furnaces.

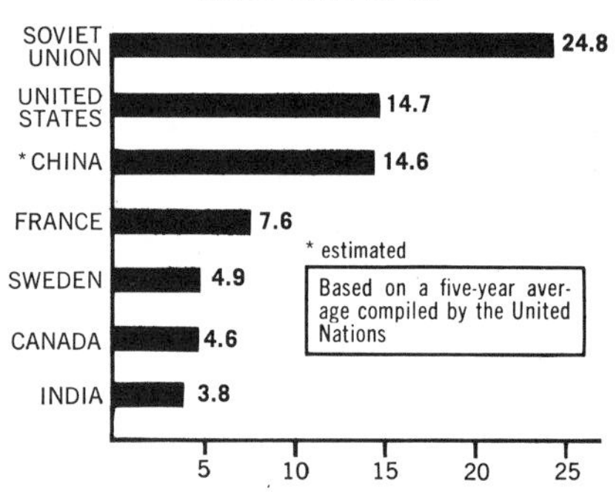

Pig iron contains about 4 per cent carbon and certain other elements, all of which tend to make it brittle. Pig iron can be converted into wrought iron by remelting and purifying it in charcoal hearths or puddling furnaces. Wrought iron is tough, ductile, and malleable. Although it was the chief form of iron for many centuries, it is little used today. *Cast iron,* as used today, is hard and has high wear and corrosion resistance; it lends itself well to casting in intricate shapes as well as in massive forms. Because of these properties, it is extensively used in pipes, engine blocks, cylinders, and all kinds of machinery parts. Iron castings (including malleable iron) consume 5 to 10 per cent of U.S. pig iron production.

STEEL.—Most of the pig iron is refined into *steel. Low carbon steel* contains less than 0.2 per cent carbon; *medium carbon steel* from 0.2 to 0.6 per cent; and *high carbon steel* from 0.6 to 1.8 per cent. All three also contain minimal amounts of other elements.

Alloy steels are those whose properties are enhanced by the presence of one or more alloying elements, chiefly manganese, molybdenum, nickel, chromium, silicon, tungsten, and vanadium. The major classifications are: high-strength low-alloy steels, constructional alloy steels, alloy tool steels, stainless steels, heat-resisting steels, and electrical steels. By varying the chemical composition, heat treatment, and other processing methods, manufacturers can produce carbon and alloy steels in a greater number of grades and for a wider variety of purposes than any other metal.

Because of its versatility, and the abundance of iron ore and other necessary raw materials in many parts of the world, steel has become the basic metal of industrial society. It meets over 95 per cent of civilization's metal requirements. (See also *Iron and Steel Production.*)

—Douglas Alan Fisher

Kaolin, a fine, usually white clay resulting from the extreme weathering of aluminous materials, such as feldspar, that contain kaolinite as a main constituent. Kaolin remains white upon firing, and because of its excellent firing properties and refractoriness, is used extensively in the ceramics industry. It is also used as an adsorbent, as a filler or extender in paper and rubber products and pigments, and as an ingredient of medicines.

Kaolinite-type minerals form under acid conditions at low temperatures and pressures. Kaolinite is a principal component of lateritic soils. Its formation is retarded by the presence of calcium. In addition, the presence of a kaolinitic marine sediment is evidence of a kaolinitic source area, since this mineral does not form in the sea. Such a sediment also indicates a relatively rapid accumulation of material. —John Price

Kapok, the seedpod of a tree of the bombax family (*Ceiba petandra*), that grows in the American tropics, Philippine Islands, Java, Indonesia, Ecuador, West Africa, and Ceylon. The tree produces pods containing seeds covered with silky hair, called *silk cotton*. The fiber has a beautiful, silky luster, is almost white to yellowish brown in color, and is very light and fluffy. Fiber length ranges from three-fifths of an inch to one and one-fifth inches, and the diameter from 30 to 60 microns. Under a microscope, the fiber resembles a smooth, transparent rod.

Because of its buoyancy and moisture resistance, kapok is used extensively to stuff cushions, pillows, and mattresses, and for life jackets. It is also used widely as an acoustical insulating material. —John Price

Kerosene, a flammable hydrocarbon oil usually obtained by the distillation of petroleum. It is removed from the fraction of normal crude oils that boils between 350° and 550° F. It is refined so that it will have a specific gravity of about 0.80 and will remain liquid at temperatures down to about −25° F. Its chemical composition is usually confined to the stable paraffins and naphthalenes (molecules containing 10–14 carbon atoms).

The specifications for kerosene are rigid; otherwise, the material would not perform satisfactorily. The hydrocarbon composition must be paraffinic in order to avoid excess smoke when the liquid is burned. Viscosity must be about 2 centipoises to allow satisfactory feeding by a wick. The content of nonvolatile impurities must be held at a low level to avoid clogging the wick. Sulfur content must be less thn 0.2 per cent, and to reduce the hazard of explosion, a flash point of about 120° to 140° F is specified.

Kerosene, in addition to its use in lamps, is used for small cooking stoves and heaters, especially where more volatile gasoline is not suitable. It is also used as a solvent for insecticide emulsions and for paints. It is widely used as a fuel for jet engines and sometimes as a fuel for reciprocating engines. The specific impulse of kerosene is not high enough to allow it to be used as a modern rocket fuel, and it has been replaced by liquid hydrogen. About 5 per cent of the world's crude oil output is sold as kerosene. —John Price

Lacquer, any of various coatings for surfaces, produced by dissolving a cellulose derivative (usually nitrocellulose) and other modifying materials in a solvent, with a pigment added if desired. Lacquers dry by evaporation of the solvent. Use of highly volatile solvents produces extremely fast drying, and lacquers usually are applied by spray. Drying time can be extended enough to allow brush application, but such lacquers are not common. Nitrocellulose is not soluble in conventional paint thinners, so a mixture of solvents is used, usually containing esters, aromatic hydrocarbons, and petroleum thinners. Other cellulosic derivatives used in lacquers include cellulose acetate, acetate-butyrate, and butyrate—all of which are used where good weather resistance is required—and ethyl cellulose, which gives a flexible film but is too soft for many uses. Lacquers are used as the finish on automobiles and as coatings for furniture and other factory-finished items.

Originally, lacquers were produced from the juice of a tree of the sumac family. However, these materials, called Oriental or Chinese lacquers, are not commonly used today. —John Price

Lampblack, a complex mixture of substances with high carbon content and high molecular weight, prepared from the partial combustion of bulk liquid hydrocarbons. It is a fine powder with great tinting power, producing a dark blue-gray when mixed with zinc white in a 1:100 ratio. As a black pigment, lampblack is used in paint, enamel, lacquer, rubber, leather, concrete, printing inks, and paper. It is usually mixed to a stiff paste with such vehicles as drying or semi-drying oils, lacquer solutions, resin solutions, water, sizes, or unvulcanized rubber.

Lampblack has been used since antiquity. At one time the Chinese prepared it by burning chips of resinous wood and collecting the soot in chimneys. Raw materials for modern processing are creosote oil, with high aromatic and low phenol content, or high-grade petroleum fuels. Oils with high olefin content are preferred because they break down more readily and yield more product. The oil is burned with a deficiency of air to produce a soot that is collected in a settling chamber. —John Price

Lead, one of the oldest metals known to man; chemical symbol Pb. The earliest know specimen of the metal, a statue found at the Dardanelles, dates from 3000 B.C. It is a heavy metal (specific gravity 11.34) with a bright bluish color that tarnishes to dull gray. Lead occurs in nature usually in association with other metals, notably silver and zinc. Of the many minerals that contain lead, *galena* (lead sulfide), *cerrusite* (lead carbonate), and *anglesite* (lead sulfate) are commercially the most important. Galena is the most common of the three.

Commercial ores may contain as little as 3 per cent lead, although about 10 per cent is most common. Before smelting, the ore is concentrated to 40 per cent and zinc-bearing minerals, which interfere with smelting, are removed. Concentration is usually done by flotation. Sulfur is removed from the concentrate by roasting. Both pyrometallurgical and electrolytic methods are used to refine lead blast-furnace bullion.

In both methods, copper is first removed by cooling the melt below the freezing point of copper, and then skimming it. Most lead blast-furnace bullion is low enough in bismuth to allow pyrometallurgical refining in kettles and reverberatory furnaces. Tin, arsenic, and antimony are slagged at high temperatures by selective oxidation. Gold and silver are removed by the *Parkes process,* which utilizes their selective affinity for zinc. Residual zinc is removed with caustic by the *Harris process;* with chlorine, by the *Betterton process;* or by vacuum distillation. If at this point the bullion contains excess bismuth, it is treated with calcium by the *Kroll-Betterton* process. When bismuth content is originally high, copper-drossed bullion is cast into anodes that are electrolytically refined by the *Betts process,* which uses pure lead starting cathodes and an aqueous lead fluosilicate electrolyte. Precious metals and other impurities remain on the anode; high-purity lead is deposited on the cathode.

Lead is used in the manufacture of storage batteries. Tetraethyl lead is used as an antiknock additive in automotive fuels. Lead is used extensively in plumbing and piping and as a vibration isolator and ornamental trim in architectural applications. Because of its good corrosion resistance, lead is used in construction, especially in the chemical industry, and as a sheathing for underground electrical cables. Lead forms an oxide coating, which protects it from acid attack; hence, it is used in the manufacture of sulfuric acid. Alloys formed with copper, tin, arsenic, antimony, cadmium, bismuth, and sodium are all important commercially. Lead blocks radiation effectively; hence, it is used as a shield for X-ray equipment. Basic

World Production of LEAD in %, 1967

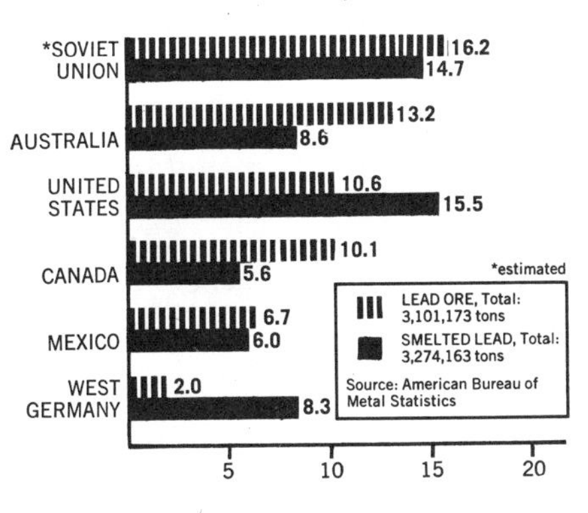

white lead (lead carbonate) is used widely as a paint pigment, and lead-bearing glazes are used in ceramics. Red lead (lead oxide) is used as a protective priming coat for metal surfaces. Lead alloys are used in soldering and welding. The metal is used in ammunition. Because lead compounds are highly toxic, precautions must be taken to prevent lead poisoning. —John Price

Lime, a general term for many different products of calcined limestone, such as quicklime and hydrated lime. *Quicklime* is made of calcium oxide ($CaCO_3$) in vertical-shaft kilns to drive off carbon dioxide. It is often treated with water to make *hydrated lime,* or calcium hydroxide, $Ca(OH)_2$. This process is called *slaking,* but *slaked lime* usually refers to material with water in excess of that needed to form the hydroxide. *Milk of lime* is a suspension and solution of hydrated lime in water. *Hydraulic lime,* made from limestone containing silica and alumina, sets under water.

Lime is used in mortar, stucco, and plaster by the building industry; as a refractory for lining open-hearth furnaces; as a soil additive, for agricultural purposes or to stabilize roadbeds; as a raw material for making glass and other chemical products; and in water purification, sewage treatment, pulp and paper manufacture, refining of sugar and petroleum, and tanning. —John Price

Limestone, a sedimentary rock composed primarily of the carbonates of calcium and magnesium. It is formed chiefly by accumulation of organic remains, such as shells or coral. Fossils contained in limestone have provided much knowledge of invertebrate paleontology and, hence, of the evolution of life and the history of the earth. More specifically, the term limestone applies to carbonate rocks dominated by the mineral calcite ($CaCO_3$), as opposed to dolomite, $CaMg(CO_3)_2$.

Limestones are composed largely of calcium oxide (CaO), and carbon dioxide (CO_2). Magnesium also is a common constituent; if it exceeds 2 per cent, the rock is termed *magnesium limestone.* Small amounts of aluminum and silica often are present. Iron oxide may be present, either as carbonate or in minerals such as clays. The chief minerals in limestone are aragonite and calcite, and dolomite in the dolomitic limestones. Limestone is used extensively in building materials, agriculture, and metallurgy. When calcined, it produces lime. —John Price

Linen, a fabric made of fibers from the flax plant. It is one of the oldest fabrics known to man. Mummies wrapped in linen 4,000 years old have been found in Egypt. Linens are mentioned frequently in the Bible.

Soil, water, and climatic conditions limit the area of flax production. Today Belgium, France, and the Soviet Union are the major producers of flax used for fabrics. Flax seed planted in early spring matures within 100 days, growing to a height of about three feet. Harvesting machines pull the plant up by the roots, leaving the long fibers intact. After drying and seed removal, the fibers have to be dislodged from the woody core of the plant and the pectic gum that holds them together. Called retting, this process involves soaking in chemically treated water.

Linen manufacture combines art, science, and craft. Several technical mill operations are required to crush straw, separate waste, and comb and draw the fibers before they can be spun into yarn. The spinning, wrapping, and weaving methods are not essentially different from those employed for other fibers, though linen requires endless care, precautions, and control of moisture.

Belgium and Ireland are the main producers of the linen fabrics used in the United States. The well-known traditional linen weave is widely used for clothing, table linens, and drapery fabrics. Linen is often dyed in distinctive shades or printed with patterns. In addition, a large variety of weaves are made in linen—intricate novelties, heavy and patterned textures, intricate sheer casements, rich velvets, and smooth damasks. —Dorothy Sparling

Lubricant, a substance capable of reducing friction between surfaces. Lubricants are one of the essential commodities of our civilization. Automobiles, aircraft, ships—in fact, all machinery—would stop running without lubricants. Since prehistoric times man has been aware of the need for lubrication and has used a wide range of substances to reduce friction. The axles of chariots found in Egyptian tombs have traces of lubricant which, upon analysis, proved to be mutton or beef tallow.

In common experience, lubricants are usually associated with oil or some other liquid. However, gases and solids are also used, especially at extremely high temperatures, or where replenishment of liquid lubricants is impractical.

Until recently, all lubricants were of animal or vegetable origin. Mutton and beef tallow, goose grease, fish oil, cottonsed oil, castor oil, and other vegetable oils were the only lubricants available in quantity. Though some petroleum products were in use prior to 1859, the drilling of Col. E. A. Drake's well in Titusville, Pa., in that year signaled the birth of the petroleum industry. Though mineral petroleum products make up the bulk of the lubricant supply, synthetic lubricants and chemical additives are becoming increasingly important.

Petroleum is the source of a wide variety of lubricants ranging from light machine oils to heavy gear oil. Different lubricants are produced by various refining processes which extract the specific fluids from crude oil.

Straight petroleum oils are classified according to their viscosity. General practice calls for the use of the lowest practicable viscosity since the lighter fluid produces less friction and improved cooling. Changes in viscosity with temperature—as temperature rises the viscosity drops—require heavier oils to be used for high temperature operation and lighter oils at low temperature.

Multigrade oils have been developed by the use of additives which enable the oil to provide proper lubrication over wider ranges of temperature and operating conditions. The Society of Automotive Engineers (SAE) has set standards for the different viscosity ranges and assigned ratings. For example, SAE 10W oil has a viscosity of 60 to 90 centistrokes while SAE 30W has a viscosity of 180 to 280 centistrokes. Multigrade oils for auto lubrication are frequently designated as SAE 10W-30, indicating a viscosity range of 60 to 280.

Additives improve the lubricant's general qualities or produce some characteristic required for special applications. Viscosity-index improvers are added to produce the widely used multigrade lubricants. To enable oils to operate at low temperatures, pour-point depressants such as metallic soaps are added. Antioxidants are used to increase the resistance of oil to oxidation by the atmosphere, thus retarding the development of acids and sludges. Antifriction and antiwear additives, such as molybdenum disulfide, assure lubrication during start-up or under extreme operating conditions. Detergents or dispersants prevent the buildup of sludge and carbon deposits by keeping particles of carbon suspended in the lubricating fluid.

Synthetic oils developed during World War II as a substitute for mineral-based lubricants have proved superior in some applications. Their primary advantage is that they have a larger operating range than petroleum lubricants. Another advantage of the synthetics is that water-based lubricants can be formulated for use where fire hazards are present. —Joseph J. Kelleher

Magnesium, a silvery white metal, chemical symbol Mg. It is the lightest structural metal, its specific gravity being 1.74, only two-thirds that of aluminum. Magnesium is the sixth most abundant element, making up about 2.5 per cent of the earth's crust—only iron and aluminum are more abundant among the structural metals. More than 60 minerals contain magnesium, but only a few are commercially important. In the United States, *brucite* (magnesium hydroxide), *dolomite* (calcium magnesium carbonate), *magnesite* (magnesium carbonate), natural brines, sea water, and sea-water bitterns are the important sources of magnesium.

The electrolytic and silicothermic processes are the two most important methods of producing magnesium in the United States. The electrolysis of magnesium chloride to yield chlorine and magnesium is the basis of the *electrolytic process.* Sea water, which contains about 0.13 per cent magnesium, is the main raw material, although magnesite, dolomite, and natural brines also have been used. Sea water is pumped into large set-

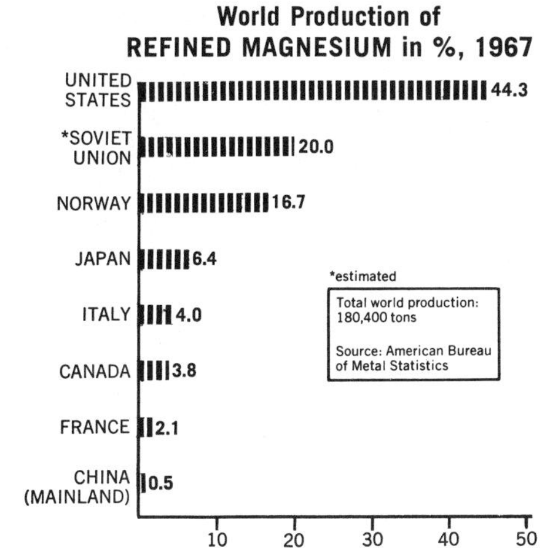

tling tanks where it is mixed with lime obtained by roasting oyster shells. The lime converts the magnesium into insoluble magnesium hydroxide, which is filtered out and treated with hydrochloric acid to produce a magnesium chloride solution. The water is evaporated, and the dry magnesium chloride is transferred to electrolytic cells that break it up into magnesium metal and chlorine.

In the *silicothermic* or *ferrosilicon process,* ferrosilicon (an alloy of iron and silicon) is mixed with calcined dolomite ore and pressed into small briquets. These briquets are charged in a steel retort and heated under vacuum to about 2,200° F. The silicon reduces the magnesium oxide, which is formed when the dolomite is calcined, to a vapor of magnesium metal, which condenses in the cooler end of the retort. Magnesium is removed from the retort in the form of crystals, which are melted and cast into ingots.

Magnesium is very active chemically. It can displace hydrogen from boiling water, and many metals can be prepared by reduction of their salts and oxides with magnesium. The metal reacts with most nonmetals and acids, but it reacts slightly, if at all, with alkalis and many organic chemicals. Magnesium serves as a catalyst for organic condensation, reduction, addition, and dehalogenation reactions. It has long been used to synthesize complex organic compounds by the *Grignard reaction.* Magnesium has excellent working characteristics and can be cast, fabricated, and machine tool worked by virtually every metal-forming method known.

Because of its light weight, magnesium was first used on a large scale for structural applications in the aircraft industry. Since then it has been used to make a wide variety of industrial equipment, household goods, office equipment, and sporting goods. It is an alloying element with aluminum, zinc, and certain other nonferrous elements. Magnesium is used as an oxygen scavenger in the production of nickel and copper alloys and as a powerful reducing agent for the production of titanium, zirconium, beryllium, uranium, and hafnium.

—John Price

Marble, a stone material widely used for monuments, structures, and art objects. Geologists define marble as metamorphic rock composed of recrystallized limestone or dolomite. However, stonemasons regard any limestone that is hard enough to take a polish as marble. Early architects recognized marble's value as a structural stone because of its beauty, durability, and its capability of being carved effectively.

Marble does not have the open pores and small cavities of limestone, and therefore can be polished. Pure *calcite* or *dolomite marble* is white, and was considered to be the most beautiful variety by ancient architects. But impurities, which often occur in marble, give it many beautiful colors. This colored marble—often banded or mottled in gray, black, green, and shades of yellow, red, and pink—is cut into thin slabs and used as decorative facing stone for both the exterior and interior of buildings.

The Greeks and the Romans were famous for their architectural uses of marble and some of the most beautiful marbles are still quarried in Italy. Vermont has the most abundant deposits of marble in the United States, although it also occurs in California, Tennessee, New York, Georgia, and Texas. Marble is quarried in huge blocks and then cut into usable pieces and slabs with diamond-toothed saws and ingenious toothless saws. The latter, called *gang saws,* are merely a series of parallel steel bands, which saw back and forth on a block of marble as a slurry of sand and water flows over them. Although slower than the diamond saw, the gang saw eventually wears its way through the block, slicing it into slabs like a giant egg slicer.

—William W. Jacobus, Jr.

Mechanical Engineering, the branch of engineering that deals with the design of equipment that generates, transmits, or uses power. The mechanical engineer actually fits no precise definition—he may design supersonic aircraft; develop new ways to harvest, pack and ship farm crops; or work as the chief of a drilling crew on an oil-well rig.

The general concept of the profession is that the mechanical engineer works with equipment in motion rather than at rest. His work involves machines rather than stationary structures. The first society of mechanical engineers was formed in Britain in 1847. In 1880 John E. Sweet, professor of practical mechanics at Cornell University, called a meeting of 30 prominent American engineers to form an organization of mechanical engineers. The meeting was held in the offices of the *American Machinist,* a journal for machinists, and the group founded the American Society of Mechanical Engineers (ASME).

Though the term is less than 100 years old, mechanical engineering has been practiced since antiquity. The caveman who used a pole to pry loose a heavy stone discovered the principle of the lever, a basic mechanical device. The Greek mathematician and physicist Archimedes was one of the first mechanical engineers, for he discovered the laws of mechanics and applied them to the theory of simple machines, such as the lever, pulley, and wedge. Leonardo da Vinci was another genius who applied mechanical engineering to the design of toys and military equipment far in advance of his day.

—Joseph J. Kelleher

Nitric Acid, a corrosive liquid inorganic acid, chemical formula HNO_3. The pure material fumes strongly in moist air and is miscible with water in all proportions.

Nitric acid has been known since the Middle Ages, when it was called *aqua fortis.* It was first prepared about 1100 A.D. by heating potassium nitrate with copper sulfate. Henry Cavendish, in 1785, synthesized the acid by treating sodium nitrate with sulfuric acid. The process has been used industrially, with Chile saltpeter as the source of potassium nitrate. Early in this century, various processes were developed to produce nitrous oxide (NO)—which can be converted to the acid by adding water—by subjecting air to an arc discharge. Modern synthesis depends on the catalytic oxidation of ammonia or the formation of nitrous oxide from air in a regenerative furnace.

Nitric acid is a strong monobasic acid. It reacts readily with alkalis, oxides, and basic materials to form salts. Reactions with ammonia to form ammonium nitrate and with sodium carbonate to form sodium nitrate are widely used in industry. It is also used extensively as an oxidizing agent and as a nitrating agent, forming esters and nitro compounds with organic materials. Nitric acid is one of the constituents of *aqua regia* (hydrochloric acid is the other), which is used to dissolve gold and platinum—the ratio is three parts nitric to one part hydrochloric. Nitric acid, when converted into inorganic and organic nitrates and nitro compounds, is used in fertilizers, explosives, dye intermediates, and various synthetic organic compounds.

—John Price

Nonferrous Metals, metals that do not contain iron in any significant amount. The most important metals from the standpoint of industrial consumption in the United States are aluminum, copper, lead, zinc, sodium, nickel, tin, magnesium, antimony, and titanium. Others of lesser importance are uranium, cobalt, silver, mercury, bismuth, molybdenum, beryllium, tungsten, zirconium, gold, tantalum, and niobium.

Most nonferrous metals are extracted from their ores and concentrates by chemical reactions that take place at high temperatures. This method, called *pyrometallurgy,* is one of the oldest known methods for recovering metals from ore minerals. However, electrometallurgy is used for recovering most aluminum, magnesium and some zinc; *hydrometallurgy* is used for some nickel.

Reduction of ores and concentrates results in two or more products: the reduced metal and a residue compound of unreduced *gangue* minerals (silicon dioxide, aluminum oxide, calcium oxide, and ferrous oxide) and the oxidized form of the reducing agent. Before the extractive process is complete, the reduced metal must be separated from this material. High-temperature processes can make this separation in several simple ways, provided the products can be liquefied or selectively vaporized.

Pyrometallurgy of nonferrous metals takes place in three steps: preparation, reduction, and refining.

Preparatory processes convert the raw material (the ore or an ore concentrate) into a chemical form suitable for further processing. One important process, the roasting of sulfides, consists of burning metallic sulfides in air or oxygen to convert them to metallic oxides, sulfates, or both together with sulfur dioxide. Another method, *sintering,* consists of consolidating finely divided particles into relatively large aggregates that are then reduced. Oxides of titanium, zirconium, and other refractory metals are prepared for reduction by *chlorination,* converting the oxides to metal chlorides. *Drying* and *calcining,* to remove free water and to decompose hydrates and carbonates, is another preparatory method.

Reduction consists of changing compounds to the metallic state and separating the free metal from the residue. Pyrometallurgical reduction requires a reducing agent, a substance that will combine with the unwanted element in the metal compound.

The volatile metals (zinc, cadmium, and mercury, and alkali and alkaline-earth metals) are reduced to the gaseous form of the metal. The gaseous metal can be condensed to a liquid or solid in a condenser separated from the reactants and residue, there simplifying purification. Most other nonferrous metals are reduced to a liquid metal in a blast furnace. Tungsten and molybdenum are reduced to solid metal because their melting points are so high. Titanium and zirconium also are reduced to solid metal because no container can hold the liquid metal without contaminating it.

Volatilization, drossing, and slag-refining are the three main purification methods. *Volatilization* is used where either the metal or its impurity can be removed as a gas. *Drossing* consists of bringing the base metal to a temperature at which the impurity becomes insoluble and can be physically removed. *Slag-refining* is the addition of a slag or molten salt that selectively absorbs impurities.

—John Price

Paint, a liquid that contains a binder, a pigment, a solvent, and small amounts of additives. When a thin film of this mixture is applied to a surface, it hardens and adheres tenaciously, so that it protects and decorates that surface. Paint has been used by man for centuries; cave paintings have been dated as old as 20,000 years.

As late as the beginning of the twentieth century, paints were made from naturally occurring ingredients by skilled craftsmen, each of whom developed his own formula. However, modern paints based on technical research, new ingredients, and standardized production methods were developed after World War I.

Paints can be classified according to end use—for example, exterior, interior, house, household appliance, furniture, automotive, or marine; or by composition—for example, enamel, varnish, or lacquer. The end use determines the composition and method of manufacture of the paint. Those used on exterior surfaces are formulated with durable binders and color-retentive pigments to withstand sun, rain, wind, and extreme temperatures. Finishes for interior surfaces must provide decorative effects and also resist abrasion, washing, and the effects of chemicals, atmosphere, —and temperature. In either case, the coat of paint must provide a suitable base for application of successive coats in the future.

A *varnish* is a transparent or clear liquid made by cooking rosin, gums, or other resins with oils. An *enamel* is a very smooth, fine-textured film. A *lacquer* is a paint that dries simply by evaporation of the solvent; it is generally fast-drying. *Shellac* in an alcohol solution is used for floors and other wood surfaces.

The *binders* used most frequently are vegetable oils, rosin, gums, shellac, and also alkyd, phenolic, epoxy, vinyl, acrylic, and styrene-butadiene resins. Vegetable oils such as linseed, soya, and castor oils are used as binders for exterior house paints and also as modifying components of alkyd resins, which are used for many exterior and interior house paints and for automotive and appliance paints. Phenolic, vinyl, and epoxy resins are used to provide chemical resistance and toughness to many appliance finishes, and for protection of chemical plants. Acrylic resins are used to provide extreme durability for automotive and appliance finishes, as well as for exterior house paints. Latex emulsions of styrene-butadiene, vinyl, and acrylic resins are used for many water-based interior and exterior house paints. Many of the resins used in appliance and automotive finishing require heat or catalysts, or both, for curing to a tough, dry film.

Pigments include many naturally occurring minerals, such as iron oxide, talc, marble, diatomaceous earth, and barytes, as well as manufactured forms, such as titanium dioxide, zinc oxide, lead and zinc chromates, lead sulfate, copper phthalocyanine (green and blue), toluidine red, molybdate orange, and ultramarine blue. Those which provide color and opacity are generally classified as *prime pigments;* titanium dioxide white pigment is probably the most important and most widely used. Those which do not add color or opacity are known as *extenders;* they are used to provide or control gloss, adhesion, abrasion resistance, and durability.

Solvents are added to keep the binder in solution as a liquid so that paint can be easily applied. Solvents include ketones, esters, alcohols, and aromatic and aliphatic hydrocarbons. When the paint is applied, the solvent evaporates, and its speed of evaporation controls many properties of the paint. In latex paints, the water content is not a "solvent" but merely acts to keep the latex particles apart until the paint is applied. When the water evaporates, the latex resin particles coalesce, or combine, to form a paint film.

Additives are used in paint to provide various properties: driers to speed drying time; flocculating and thickening agents to control flow and application; inhibitors for stability in the paint can; and fungicides to control mold and mildew growth.

Paint is manufactured in batch lots. The process starts with the grinding or dispersing of the pigment in a mixture of solvent and binder. In this operation, the pigment particles are reduced to the desired size and are coated with the binder solution. The grinding mill provides impact or shear force, or both, to disperse the pigment in a solvent vehicle. Then the remainder of the binder solution, solvents, and additives are added, and the mixture is blended until it is homogeneous.

Paints are applied by brushing, rolling, spraying, dipping, flow-coating, and many adaptations of these, including electrostatic spraying, reverse roller coating, curtain coating, and airless, steam, and hot spraying. The method used for any particular application depends on the kind, shape, and size of the surface to be coated, the application speed desired, the film thickness required, and the viscosity of the finish itself.

—Oliver R. Volk

Petrochemicals, chemicals derived from crude oil or its fractions, or from natural gas. Basically, petrochemicals are compounds of hydrogen and carbon, but a vast number of these chemicals also incorporate oxygen, sulfur, nitrogen, chlorine, or other elements. For many years, coal was the chief source of synthetic organic materials. Today, however, most organic chemicals are made from petroleum and natural gas.

The petrochemical industry is made up of many different chemical and petroleum companies. It is generally considered to have started about 1918 with the production of isopropanol, or rubbing alcohol, from refinery gases. The subsequent development of catalytic processes for making high-quality gasolines and of a high-temperature thermal-cracking process provided large quantities of many important chemical building blocks and spurred rapid growth in the chemical industry. In 1966 about 55 billion pounds of petroleum-derived chemicals were sold in the United States, worth some $12 billion.

Petrochemicals are used in manufacturing an almost limitless number of products. Synthetic rubber, man-made fibers, plastics, paints, detergents, and fertilizers are some of the petrochemicals that have shown the most spectacular growth. Synthetic rubber from petroleum supplies about three-fourths of the rubber needs of the United States. Petrochemical-based synthetic fibers, including nylons, polyesters, polypropylenes, and acrylics, contribute greatly to modern living. Plastics such as polyethylene, polypropylene, polystyrene, polyesters and polyvinyl chlorine have created enormous new markets for squeeze bottles, packaging films, foam insulators, utensils, boats, piping, flooring, synthetic leather, luggage, and so on. Petrochemicals appear in paints as solvents, resins, driers, and emulsifiers. Synthetic detergents for both home and industrial uses owe their great effectiveness to petrochemical ingredients.

Petrochemistry plays a key role in feeding the world's rapidly increasing population. It makes possible large quantities of low-cost nitrogenous fertilizers essential for efficient food production. It is also a source of herbicides, insecticides, and fungicides that, when properly used, control agricultural pests without danger to mankind. Indirectly, it contributes greatly to food production by freeing land that would otherwise be required to grow natural fibers and rubber. In the future, petrochemistry may provide food directly. A new process, awaiting development, shows promise of producing a high-protein, high-vitamin food concentrate from petroleum.

—C. C. Garvin, Jr.

Petroleum, a complex mixture of hydrocarbons of widely varying properties obtained from natural oil wells. Crude oil, the raw material for gasoline, heating oil, lubricants, synthetic fibers, plastics, and a host of other

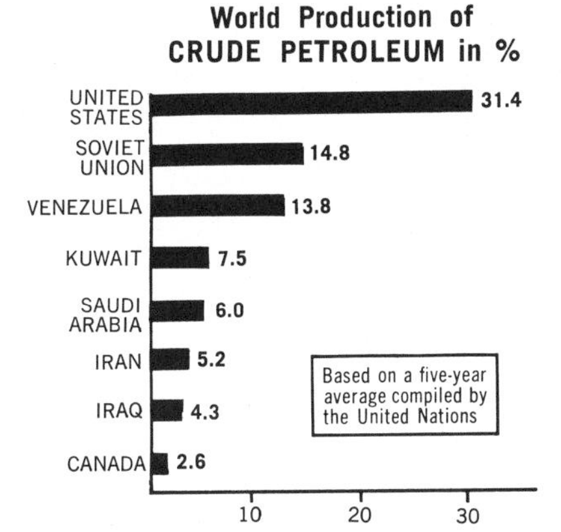

petrochemical products, is a plentiful resource found in North and South America, Europe, Africa, Asia, and Australia, and under the sea on the continental shelves. Although the world has large oil reserves, new reserves are constantly being sought.

Exploration begins with the petroleum geologist, who examines areas that might be expected to contain oil. He looks for porous sedimentary rocks, such as sandstone, limestone, and dolomite, because oil collects in the pore spaces in these rocks. He also looks for places where oil might be trapped. Such places of entrapment might include upward folds or arches in the rock layers, fractures in the earth's crust where porous layers have been cut off against nonporous layers, and buried porous lenses, such as beach sands, which pinch out between nonporous layers. The geologist studies maps and reports showing the rock formations that occur at the earth's surface, aerial photographs that reveal surface features over large areas, and data obtained from wells already drilled in the area that may indicate the nature of the underlying strata.

If the geologist's preliminary studies indicate favorable rock structure, more advanced work involving *geophysics* is then undertaken to learn more about the nature of the rocks below the earth's surface. The geophysicist's chief tools are the *gravimeter,* the *seismograph,* and the *magnetometer.* The gravimeter measures differences in the pull of the earth's gravity from place to place. The seismograph measures the speed and intensity of artificial shock waves reflected from subsurface formations. The magnetometer measures variations in the earth's magnetic field. From studying the data gathered with these tools, the geophysicist can tell much about the nature and structure of the subsurface formations.

■DRILLING.—If this preliminary geological and geophysical work indicates that there may be oil in the rocks below the earth's surface, an exploratory hole is then undertaken, for it is only by drilling that oil can be discovered. Drilling may be accomplished by the *cable-tool* or the more common *rotary* method.

In cable-tool drilling, a heavy bit and stem on the end of a cable are alternately raised and dropped so that the bit pounds and pulverizes the rock. At intervals, the tools are removed and water is flushed into the hole so that the drill cutting may be bailed out. The cable-tool method is used primarily for drilling shallow holes in soft formation.

Most oil wells are drilled by the rotary method, using an augerlike bit attached to the lower end of a "string" of connected lengths of steel drill pipe. The drill pipe and bit are rotated by a turntable at the surface; as it turns, the bit bores through the rock. As the hole deepens, lengths of drill pipe are added at the top of the string. *Drilling mud,* composed of water, clay, and chemical additives, is pumped down through the inside of the drill pipe, forced out through the bit, and returned to the surface carrying rock cuttings from the well. The cuttings are studied under the microscope by the geologist in order to learn more about the rocks penetrated by the drill. In both drilling systems, a tall derrick supports the equipment that must be lowered into the well.

During the past 25 years, techniques have been perfected for deflecting the drill from the vertical into a gradual curve as the hole goes deeper. *Directional drilling* can penetrate rocks inaccessible by vertical drilling, such as oil deposits lying under built-up urban areas. Directional drilling is also used in offshore areas where as many as 30 wells can be drilled from a single platform, each one directed to a different part of the oil field.

■OFFSHORE.—Drilling under the sea involves specialized techniques and equipment. The drilling platform or vessel must be able to withstand storm winds and waves, and all equipment must withstand exposure

PAN AMERICAN PETROLEUM CORPORATION

ROTARY RIG for oil-well drilling drives a screw-shaped bit into the ground when derrick pulleys drop this huge block, which carries the bit at the end of its "string."

to salt water. Some drilling platforms are large, permanent, self-contained units, spacious enough to house personnel and supplies; others are only large enough to support the drilling equipment, with crew and supplies housed on a floating tender. Mobile drilling platforms include some that have legs which can be extended to the bottom for support, others that float freely and must be kept in position by complex navigational manuevers, and still others that may be partly submerged for greater stability. Wells have been drilled as far as 150 miles from shore and in water as deep as 1200 feet.

■RESERVES.—An exploratory well in an area where no oil has been produced is called a *wildcat.* It may cost from $50,000 to $3 million to drill, yet eight out of nine wildcats drilled in the United States are unsuccessful, or dry holes. If oil is discovered in a wildcat, additional wells called *development* wells are then drilled in order to outline the size and characteristics of the oil field.

Oil reserves in a field are estimated from this more detailed drilling. Reserves are described as *proved* or *probable,* depending upon how reliable the estimates are. Proved reserves are the amount of oil and gas that can be recovered economically under present conditions. A few decades ago, as little as 20 per cent of the oil in a field was recoverable, but modern recovery techniques developed through petroleum engineering have increased this to more than 80 per cent for some fields.

■DISTRIBUTION.—A vast network of pipelines carries most of the world's crude oil, natural gas, and petroleum products on its initial steps to market. In the oil fields, gathering lines of perhaps 2-inch diameter lead from wells to storage tanks; larger pipelines lead into trunk lines that may be as much as 30 inches in diameter. The trunk lines move the oil to refineries or to shipping points, and also move refined products to market. Pumping stations keep the oil moving through the pipelines at a speed of 2 or 3 miles an hour. These stations may be as far apart as 150 miles, depending on the terrain and the type of oil to be moved.

Tankers play an important role in the transportation of oil, particularly in hauling oil from the Middle East and South America to Europe and North America. Barges transport much oil on inland and coastal waterways, and railroad tank cars also carry many petroleum products. Tank trucks that supply the neighborhood service stations and bring heating fuels to millions of homes are the final link in the transportation chain.

—Zeb Mayhew

Petroleum Engineering, the application of mathematics and natural sciences to the task of extracting crude oil and natural gas from the earth. The petroleum engineer is concerned with the planning and developing of oil and gas fields, with selecting producing methods, and designing production equipment which is controlled by computers. The petroleum engineer may rely on technological devices to stimulate a well as a means of increasing well productivity. He seeks to prolong the natural flow of wells as long as possible, thereafter installing artificial lift equipment to continue the well's yield.

In drilling, the petroleum engineer is responsible for designing equipment and planning the drilling operation, and preparing cost estimates. Each drilling rig is specially designed for its depth range and area of work. After the rig is assembled and a location for the well is chosen, it is the engineer's job to guide the drilling operation. He selects drilling bits, designs the casing strings that form the *wall* of the bore hole, and specifies the type of *drilling mud* to be pumped down the hole to lubricate the bit, to lift the rock chips out of the hole, and to subdue high pressures.

Reservoirs of oil and gas consist of interconnected pore spaces in underground rock such as sandstone and limestone. The petroleum engineer studies the reservoir to determine its extent and its physical characteristics. By analysis of cores of the reservoir rock and the contained fluid, and by the study of pressure and other measurements, he determines where wells should be drilled and how best to remove the oil and gas from the reservoir.

Oil is driven out of reservoir rock by the pressure of expanding gases associated with the oil, by the pressure of water surrounding the reservoir, or by a combination of the two. It is the job of the petroleum engineer to determine which driving force is present and to plan the most efficient use of such natural forces. He may find it necessary to inject gas and/or water to supplement the natural energy in order to maintain the driving forces or pressure at economically desirable levels.

Often when wells become old and production rates approach uneconomic limits, *secondary-recovery* techniques are employed to obtain still more oil. This consists of injecting even more gas or water into the reservoir and driving the oil to the wells. Other added recovery techniques are being developed in the petroleum industry's laboratories and being tested in the field. They include the use of additives to improve water displacement efficiency, *miscible displacement* using a solvent, and thermal methods. The latter include steam injection and combustion in the reservoir to improve flow properties of heavier oil. The engineer is responsible for carrying out these various additional recovery operations.

—Douglas Ragland

Plastics, synthetic organic materials that can be molded or cast in a variety of intricate shapes. Almost all plastic materials and modern synthetic textile fibers are made from resins. Resins, in turn, are made from chemicals derived from coal, petroleum, natural gas, and other sources.

The first plastic was *cellulose nitrate,* or *celluloid,* invented in 1868 by John Wesley Hyatt, a printer in Albany, New York. A prize had been offered for the invention of a material that could replace ivory in making billiard balls. The suppression of the slave trade from Africa had also taken the profit out of the importation of ivory, which was used not only for billiard balls but also for women's combs, piano keys, and knife handles. Hyatt noted that the end of the Civil War had left surpluses of guncotton, or nitrocellulose, which is a mixture of cellulose fibers, nitric acid, and sulfuric acid. By experimenting, he found he could mix this material with camphor to obtain a whitish solid material that could be pressed into blocks and then machined to the desired shapes. The only trouble was its extreme flammability—the material had to be machined under water. Manufacturers in Leominster, Mass., who specialized in products of ivory, bone, horn,

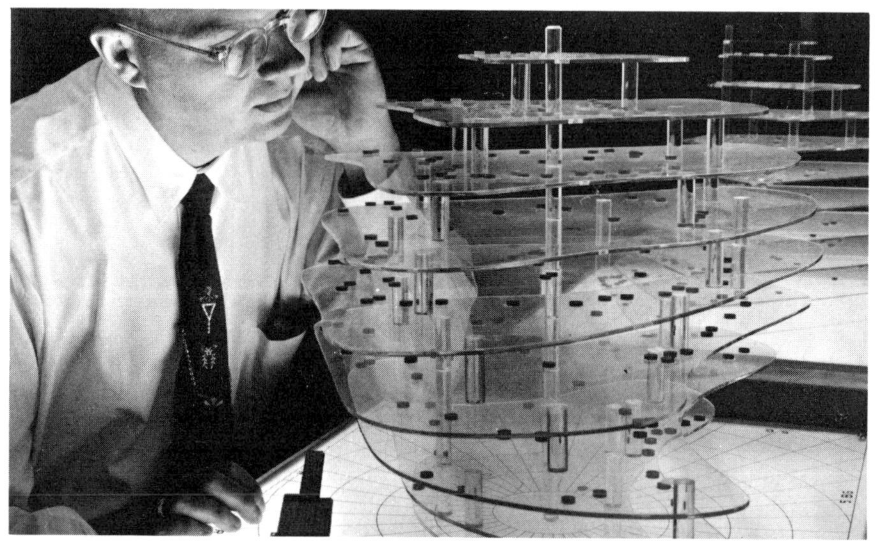

E. I. DU PONT DE NEMOURS & CO.

COLORS IN PLASTICS are analyzed for hue, lightness, and saturation to determine mathematically the proportion of dye to add to the material to obtain the desired color.

Common Plastic Materials

Plastic	Year	Common Uses	Outstanding Properties
Acrylonitrile-Butadiene-Styrene (ABS) (thermoplastic)	1948	Pipe and pipe fittings; football helmets; utensil and tool handles; automotive parts.	High impact and tensile strength; resistant to alkalies, salts, and many aliphatic hydrocarbons.
Acrylics	1936	Aircraft windows; skylights; outdoor signs; camera lenses.	Weather resistant; excellent electrical insulator.
Casein (thermosetting)	1919	Buttons and buckles; game counters; adhesives; toys.	Takes a brilliant surface polish and withstands dry cleaning.
Cellulosics (thermoplastic)			
Cellulose Acetate	1927	Spectacle frames and toys.	Relatively light; among the toughest of plastics; good insulators; available in a wide variety of colors.
Cellulose Acetate Butyrate	1938	Steering wheels and tool handles.	
Cellulose Propionate	1945	Telephone hand sets.	
Ethyl Cellulose	1935	Electrical parts.	
Cellulose Nitrate	1868	Shoe heel covers.	
Epoxy (thermosetting)	1947	Protective coatings for appliances; adhesives.	Good flexibility; excellent resistance to corrosion.
Melamine (thermosetting)	1939	Tableware; laminated surfaces; buttons; hearing-aid cases.	Hard and scratch resistant; unaffected by detergents.
Nylon (thermoplastic)	1938	Tumblers; slide fasteners; brush bristles; fishing line.	High impact, tensile, and flexural strength; can be boiled and steam sterilized.
Phenolics (thermosetting)	1909	Radio and television cabinets; electric insulation; handles; dials; knobs.	Hard, rigid, and strong; excellent electrical insulators.
Polyesters (thermosetting)	1942	Impregnate cloth or mats; reinforced plastics for boats and automobile bodies.	Highly resistant to most solvents, acids, bases, and salts.
Polyethylene (thermoplastic)	1942	Flexible ice cube trays and toys; squeezable bottles; bags for packaging.	Strong and flexible; highly resistant to chemicals.
Polypropylenes (thermoplastic)	1957	Heat sterilizable bottles; packaging film and sheets; valves.	Good heat resistance; excellent resistance to chemicals.
Polystyrene (thermoplastic)	1938	Refrigerator food containers; instrument panels; wall tile; portable radio cases.	Good dielectric qualities; can be produced with a smooth surface or a special texture.
Urea (thermosetting)	1929	Lamp reflectors; appliance housings; electrical devices; stove knobs and handles.	Hard and scratch resistant; good electrical qualities.
Urethanes (thermoplastic and thermosetting varieties)	1954	Foams used for cushioning, sponges, and appliance insulation; solids used in abrasive wheels, bristles, and tire treads.	Foams have good shock and sound absorption and heat insulation; solids have unusual abrasion and tear resistance.
Vinyls (thermoplastic)	1927	Raincoats; inflatable water toys; phonograph records; electric plugs; floor and wall coverings; upholstery.	Strong and abrasion resistant; fade resistant colors; resist penetration of water.

and tortoise shell, adopted Hyatt's new material and, despite many fires and explosions, made their town the first center of the plastics industry. Today, cellulose nitrate clings to only one exclusive market—dice for gamblers, who seem to prefer it for its weight and "feel," and perhaps also for superstitious reasons.

The second plastic was invented in 1909 by Leo Hendrik Baekeland, a Belgian chemist living in New Jersey. He mixed phenol, a coal derivative, with formaldehyde, an embalming fluid, to produce phenol-formaldehyde. This black substance, very difficult to burn and impossible to melt, proved to have good electrical properties. The inventor called it *Bakelite.*

New plastics developed slowly in the next decade. About 1915, *cellulose acetate* was evolved from the flammable cellulose nitrate by substituting acetic acid for the nitric acid. The new material had celluloid's qualities of appearance and formability but without the fire hazard. Later, it was found that cellulose acetate could be extruded as a thread. This fiber found textile uses as *rayon.*

■ **MODERN PLASTICS.**—Other resins were invented, creating new plastic materials for fibers, paints, and adhesives. At first, they were based on cellulose, but then resins were discovered that could be made from petrochemicals. Today, there are approximately 40 basic kinds of resins, each with as many as 10,000 varieties. There now are more different kinds of plastic, with a wider range of predetermined properties, than the combined varieties of metal, wood, glass, rubber, and leather. They run the gamut from plastics that melt like wax to those that can stand thousands of degrees of temperature; from those that bend like tissue paper to those that, reinforced by glass fibers or asbestos, can be formed into plastic boats, automobile bodies, and aircraft parts.

Not only can plastics be made from resins to exact engineering strengths and resistances, but several different plastics can be made out of a single raw material. The first *polyethylene* made in England before 1936 was derived from corncobs; today it is made from natural gas or petroleum. In India, which is short of petrochemicals, sugar is converted into alcohol and polyethylene is made from that. One petrochemical gas, ethylene, can be made into four different kinds of plastics and resins by changing the manufacturing processes.

Thermoset plastics are those that *cure,* or harden, in heat, like waffles or hard-boiled eggs, and cannot be remelted and molded over again. *Thermoplastics* are those that melt in heat and harden in cool temperatures, like wax candles, and can be remelted and remolded. Between the two, in recent years, have come some, like the *urethanes* which may be made to react like either the thermoset plastics or the thermoplastics under certain conditions.

■ **APPLICATIONS.**—All plastics are light in weight. Steel weighs three times as much as aluminum, and aluminum weighs three times as much as plastics; therefore, steel weighs nine times as much as plastics. On the basis of per-cubic-inch production, more plastics are made than aluminum and all other nonferrous metals put together. In 1968, 16 billion pounds of plastics were produced in the United States, and by 1970 a production of more than 19 billion pounds is anticipated.

Plastics may be molded into products and parts; extruded like toothpaste into films and pipe, used for coating papers and fibers; foamed into insulation and heat-and-cold-resistant products such as drinking cups; laminated to other materials or to each other; cast into sheets like glass; formed from sheet into automobile bodies and housings for business machines; and blow-molded into bottles for bleaches, detergents, and milk. The thermoplastics have *plastic memory,* that is, after they are taken as flat sheets and formed into shapes, they return to flat sheet form, or try to, when heated. This quality is useful in making skin-pack for hardware and other items. Plastics can be painted, although most of them are colored through by pigments added during manufacture. They are warm to the touch, being good thermal insulators, and they are easily cleaned. Some are resistant to certain chemicals, such as cleaning fluids, but not to others. All may be made flame-resistant, but some require additives to assure this. Thus, the selection of the proper plastic for a given application is a job for a chemist working with a designer and a process engineer.

Most plastics are *polymers,* that is, they are made of individual units joined together into chains. The individual unit is called a *monomer,* and it is composed of hydrogen and carbon and some other elements that must be removed. The monomer is heated in kettles or in continuous processes, and the extraneous elements and as much hydrogen as possible are removed to produce the polymer. The polymer may be in powder form or may be fed through an extruder and chopper to make pellets that are used for molding and extrusion. All thermoplastics, except the cellulosics, are polymers. Nylons are *polyamides,* acetals are *polyacetals.* The carbon and hydrogen atoms in the molecules are fused to make long, strong chains of molecules, producing plastics and resins.

When two or more different monomers are made into a single superstrong plastic, a *copolymer* results. Three different monomers yield a *terpolymer,* and *quadrapolymers* are made from four different monomers.

PROCESSES.—Plastic materials are processed in a number of ways. The thermoplastics are *injection-molded* by melting the pellets or powders and squirting them into a cold mold where they harden into pieces. The *melt* was formerly pushed into the mold with a ram, but is now extruded with a screw, using *adiabatic,* or frictional, heat to melt it for flow. In the case of rigid vinyl, carbonates, and ABS (Acrylonitrile-Butadiene-Styrene) polymers, this is important because these plastics can degrade in properties if held in melt condition for too long a time. The higher the heat resistance of the polymer, the better the end product will be, but the more difficult it will be to process and the more sophisticated the machinery required to process it into a product or product part.

Extrusion works like a toothpaste tube with a corkscrew in the cylinder. The materials melt in the screw and are pushed out of a die (flat, round, or profile-shaped) into a cold and frequently water-quenched bath, where the material becomes solid and available for further processing. These dies produce film, sheet, and pipe, and are used for coating paper, textiles, and metals.

An extruded sheet may be thermoformed. It is reheated and sucked or blown into a mold to make, for example, a refrigerator door's interior.

Casting is usually done from the liquid monomer. This, or extrusion, is usually how acrylic pieces are made. *Foaming* may be done by casting or by extrusion, depending on the material and the purpose of the product. Thicker sections usually are made by casting with the use of a catalyst or foaming agent. *Calendering* is done on huge and expensive machines and is largely used for soft vinyls for upholstery. *Coating* may be done by either calendering or extrusion, with the plastic spread over the paper or cloth, woven or nonwoven. Coating of metals or other materials may be done by heating the metal piece, dipping it into a bath of plastic powder or liquid, pulling it out, and letting it cool. Nylons, vinyls, and epoxies have been used in this manner, which is called *fluidized bed coating.*

Traditionally, the thermoset plastics have been molded by *compression molding,* or *transfer molding.* The cold powder is put into a high-frequency heating chamber (which softens it) and poured into a hot mold; after a short time, the press is opened and the phenolic, melamine, or urea part is taken out. In 1961, injection molding machines for thermosets became available. Most thermoplastic injection machines can be converted to molding of thermosets. —Hiram McCann; David Nason

Plywood, a cross-banded assembly of layers (*plies*) of wood or of veneer with a lumber core. The plies are joined by an adhesive. Usually, the grain of one ply is approximately at right angles to the grain of the next ply. An odd number of plies generally is used.

Plywood, compared with other wood products, has greater uniformity in strength in the direction of its two major axes and is more apt to preserve its dimensions with changes in its water content. It has increased resistance to end checking and splitting and a greatly reduced tendency to twist and warp.

Thin hardwood veneers often are used with cores of less costly softwood in the manufacture of furniture and wall and ceiling panels. Plywoods made of inexpensive woods are used extensively in concrete forms, sheathing, subflooring, boxes, and crates. Molded, die-pressed, or bag-molded plywoods are used in the manufacture of boat hulls, aircraft parts, chairs, and similar shell-like constructions. —John Price

Potash, a generic term applied to any potassium salt sold for its potassium content. Chief salts used commercially are potassium chloride, potassium sulfate, and a mixture of potassium sulfate and magnesium sulfate. At one time a potassium compound, potassium carbonate, was produced frcm solutions leached from wood ashes evaporated in iron pots; hence, the term *pot ashes.* In 1857, soluble potash minerals were recognized as valuable for fertilizer use, and these minerals since have been the source of potash for fertilizer and chemical use.

Usually the potassium content of minerals is stated in terms of potassium oxide (K_2O) because it was originally thought that potassium was effective as a fertilizer only in this form. More recently, there has been a move to quote potash content in terms of the element potassium rather than the oxide. Although there are many potash-containing minerals, most of the world's known reserves consist of sylvite, carnallite, kainite, langbeinite, niter, and polyhalite. —John Price

Pottery, one of man's oldest uses of ceramic materials. It is quite possible that the art of basketmaking led to the first use of clay as a lining. By some unknown set of circumstances, it was then discovered that moist clay could be shaped by hand, without weeds or grasses as a support, and, when heated, would become permanently hard and durable. The methods used to make pottery today are not fundamentally different from those used thousands of years ago. In fact, at a very early stage, primitive man decorated his pottery with quite elaborate painted and incised designs. The study of pottery fragments, called *potsherds,* permits archaeologists to assess the stages of development of ancient peoples.

Cruder, less pure clays are used to make stoneware, earthenware, art ware, flowerpots, and similar objects. China and porcelain are special types of pottery in which several types of purer clays are employed. The beauty of the pottery made in China during the Chou dynasty (1122–249 B.C.) has, in fact, led to the very use of the term "china." Later, the Chinese developed porcelain by adding feldspar to the clays. Both china and porcelain are hard, translucent, and nonporous, even without a glaze coating. Earthenware, on the other hand, is porous and must be glazed to become impervious.

Regardless of differences in raw materials, all pottery is made by one of three methods: pressing, jiggering, or casting.

Pressing is used only for irregularly shaped objects. The moist clay is shaped by hand pressure or by pressing into plaster molds.

Jiggering, used for making symmetrical objects, involves the use of a *potter's wheel* or its equivalent. Moist clay of the proper weight is placed on the revolving wheel and shaped by hand or by a jiggering tool. Heated dies are often used to assist in spreading the clay to the proper shape, and a water mist may be used for lubrication. The manufacture of plates, cups, and bowls by this method is now quite highly automated.

Casting involves the use of a clay slip—a suspension of clay particles in water with a consistency somewhat like that of paint. This slip is poured into a plaster of paris mold of the proper shape. The mold absorbs much of the water from the slip next to it. When this layer is thick enough to permit handling, the excess slip is poured out, and the mold is opened. This green ware is then dried and fired. Drying and firing must also be used for pressed or jiggered ware. —Charles J. Phillips

Quartz, an important rock-forming mineral that occurs as a subordinate constituent of many igneous, metamorphic, and sedimentary rocks; chemical composition, silicon dioxide (SiO_2). The most important and widespread of all minerals, it is the main constituent of sandstone, quartzite, and unconsolidated sands and gravels. Ordinary quartz is colorless and transparent, with a vitreous luster. *Amethyst* is a purple or bluish-violet; *citrine* is orange-brown, produced by heat treatment of amethyst. *Rose quartz* is a massive type found in pegmatites. Quartz also may be smoky yellow to dark smoky brown, varying to brownish-black and almost opaque.

Quartz may be classified into two broad categories: *coarse crystalline,* in which individual grains are visible to the naked eye, and *fine crystalline,* in which grains are visible only under a microscope. Fine-crystalline quartzes are further divided into two main groups based on particle shape. The *fibrous* variety includes chalcedony, carnelian, and agate, and the *fine granular* variety includes flint and jasper. —John Price

Reservoir, a basin for the storage of water. Reservoirs are essential to the effective and economical control and utilization of water resources for almost every need—from irrigation to flood control, from domestic or industrial water supply and water power to transportation by river or canals. In early *basin irrigation,* low earth embankments were used to retain flood waters for later release to lower areas. Similar, so-called *tanks* of notable size were built quite early in India. Reservoirs of greater size, particularly those formed in river or stream valleys, however, required both the construction of higher dams and provision for the passage of flood flows.

It appears that such constructions were first undertaken in Spain in the sixteenth century. In recent years many huge reservoirs of this type have been built, especially in the United States. These include such outstanding works as the Hoover and Grand Coulee irrigation projects of the U.S. Bureau of Reclamation and those for the water supply of several of our larger cities, such as the numerous basins on the Catskill and Croton watersheds for New York City.

Reservoirs of smaller capacity are also built to help equalize demand or meet emergency needs. The domestic use of water varies throughout the day. When possible, it is usually economical to provide *equalizing reservoirs* large enough to carry peak loads, and thus permit the main supply aqueduct to be designed for average load.

In hydroelectric developments, sufficient *pondage* is desirable to store flow during low-demand hours for later use in carrying peak loads. Reservoirs are likewise essential in river canalization projects, and the provision of a summit water supply is a critical requirement for canals that pass over a divide. *Pumped storage* has been a more recent development in the power field; in the hours of low electrical demand, water is pumped to a high-level reservoir from which it can later be released to provide electric power in time of heavy demand.

Reservoir capacity is usually measured in *acre-feet,* that is, one foot of water over one acre, which is equivalent to 43,560 cubic feet or about 325,000 gallons. Lake Mead, formed by the Hoover, or Boulder Canyon Dam, holds a little over 30 million acre-feet; but, by far, the largest storage is provided by the Kariba Dam in Rhodesia, which contains 130 million acre-feet. —James Kip Finch

Rosin, a solid resinous material obtained from the oleoresinous wood of pine tree stumps. It contains chiefly resin acids and smaller amounts of nonacid compounds. It varies in color from pale yellow to dark red or darker, depending on the source and the method of collecting and processing. It is translucent and brittle at ordinary temperatures, has a slight odor, and tastes like benzine. It is insoluble in water but is soluble in most oils and organic solvents.

Rosin may be used in its natural form, known as *unmodified rosin,* or it may be given chemical treatment (hydrogenation, disproportionation, or polymerization) to increase its stability and improve its physical properties. After such treatment the material is known as *modified rosin.*

Three commercially important methods of obtaining rosin are solvent extraction of pine stumps, turpentining of living trees, and separation from tall oil. Rosin products are used in paper sizing, lubricants, insulation, linoleum, adhesives, soldering fluxes, binders, soaps, and finishes. —John Price

Rubber, any of a wide variety of natural and synthetic hydrocarbon materials that are commercially valuable for their highly elastic, moisture-resistant, electrical insulating, and wear-resistant properties. Slightly less than one-half of the rubber now consumed in the world is in the form of *natural* or *crude rubber* obtained from the sap of tropical plants. About 99 per cent of the world's supply of natural rubber is produced from a milklike sap called *latex,* which is released by "scarring" or tapping the bark of a tree (*Hevea brasiliensis*), of the castor-bean family. Less than one per cent is produced from *Guayule,* a plant native to Mexico.

Prior to the 1900's, most rubber came from the Amazon basin in Brazil, where *brasiliensis* grew. However, in 1876, Henry A. Wickham, later knighted for his enterprise, conceived the idea that rubber trees could also be made to grow on coffee plantations in India, where the climatic conditions were similar. He succeeded in carrying back from Brazil to England nearly 100,000 *Hevea* seeds, which were promptly planted in the Kew Botanical Gardens near London. About 3,000 of the trees survived and were transported to Britain's southeastern Asian possessions of Ceylon, Malaya, Sumatra, and India, where they were transplanted and grew well. In 1967 about 40 per cent of the world's natural rubber production of 2,452,500 long tons was grown in Malaysia, 31 per cent in Indonesia, 9 per cent in Thailand, and 7 per cent in Africa (Liberia and Nigeria).

World Production of NATURAL RUBBER in %, 1967

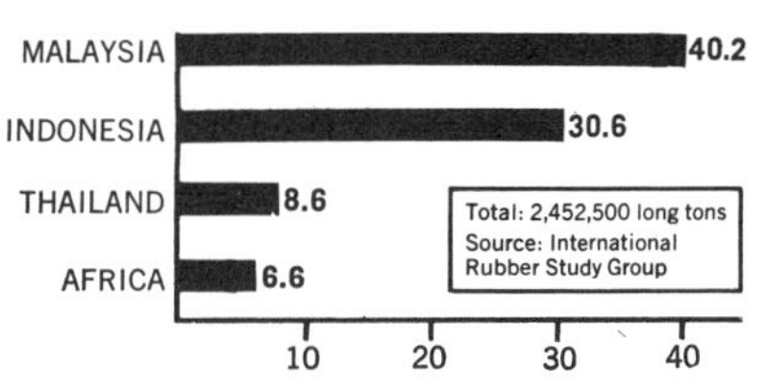

Other areas that produce natural rubber include India, Vietnam, Brazil, Cambodia, Brunei, Territory of Papua in New Guinea, and Burma. Newer plantations have been planted in the Philippines, Guatemala, Costa Rica, and Mexico. It is interesting to note that, almost without exception, all natural rubber plantations are within 15 degrees north or south of the equator.

World* Production of RUBBER

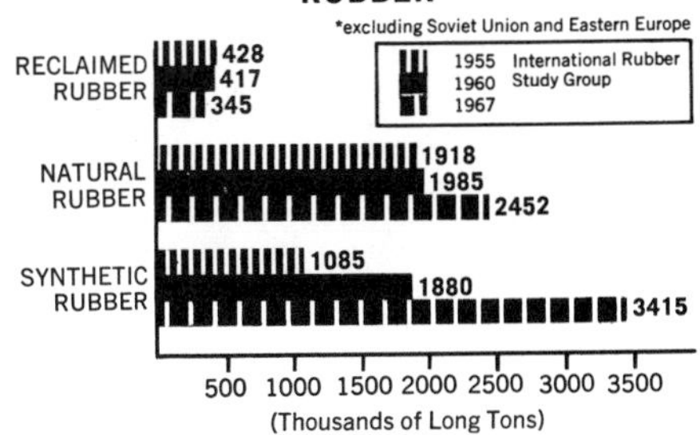

■SYNTHETIC RUBBER.—Since natural rubber is a polymer (a long chain of simple molecules), researchers saw an opportunity to duplicate it by joining simple chemical compounds, called monomers, together in a similar manner. Many different man-made polymers were made with general characteristics similar to those of natural rubber, but each having specific properties superior to it, as in resistance to chemicals, oils, sunlight, or heat, or in lower permeability to gases.

As early as 1879, a French chemist, Guy Bouchardat, used isoprene, a five-carbon molecule (C_5H_8) derived from natural rubber, and succeeded in converting it into a somewhat rubberlike substance. However, only in relatively recent years has science been able to exactly duplicate the polymer structure of natural rubber, using isoprene made synthetically from petroleum.

World* Production of SYNTHETIC RUBBER in %, 1967

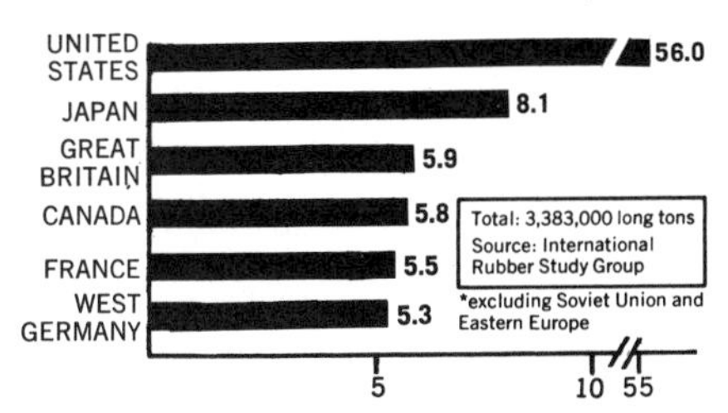

Necessity was proved the mother of invention twice in the history of the development of synthetic rubber. First, the Germans developed a rubber substitute, although a not-too-successful one, called *methyl rubber,* when their rubber supplies were cut off during World War I. Little work was done in the next twenty-five years except on an experimental basis. Again in World War II, when the Allied forces were cut off from their supply of natural rubber by the Japanese, a large-scale rubber industry was created, this time with complete success, by the United States. Today, large rubber and petroleum companies are spending millions of dollars to synthesize new monomers and rubberlike polymers.

A synthetic rubber plant with an annual capacity of 50,000 long tons or more can be built in about one year on a few acres of land. In contrast, *Hevea* trees take seven years to reach maturity and have in recent years yielded, on an overall average, one-third of a long ton (approximately 700 lbs.) of natural rubber per acre per year. Some efficient plantations yield 1,000 pounds per acre annually. Selected clones (trees developed by grafting a bud from a high-yield tree to a seedling) combined with scientific planting, fertilization, and tapping techniques, have yielded as high as 2,000 to 3,000 pounds per acre in a year.

Two-thirds of all the natural and synthetic rubber consumed in the United States goes into the manufacture of tires and other automobile products, which utilize rubber's characteristic properties of elasticity, shock absorption, and outstanding resistance to moisture and wear. Rubber's excellent electrical resistance make it useful for electrical insulation on wires and cables, and it is molded into miscellaneous mechanical goods. Its high resistance to abrasion, when compounded with other filler materials such as carbon black or clays, has won it a market in shoe products, flooring, and athletic goods. It can also be used in its latex form or dissolved in solvents to produce dipped goods, to impregnate paper or textiles, to serve as an adhesive, or to be whipped with air into a foam used in mattresses.

■TYPES OF SYNTHETIC.—A *general-purpose rubber* is one that has large-volume application, such as the manufacture of tires and commonly used molded articles. *Special-purpose rubbers* are polymers designed for specific applications that require distinctive properties, such as resistance to swelling in oils or solvents or the ability to retain strength and flexibility at extreme high or low temperatures. The special-purpose types are produced in smaller quantities than the general-purpose rubbers.

The synthetic-rubber types in large-scale commercial production include the following, in order of their decreasing volume of production:

Styrene-butadiene or S-type (SBR) rubbers constitute about 68 per cent of all the synthetic rubbers produced in the United States. They are copolymers of the petrochemical monomers butadiene (about 75 per cent) and styrene (about 25 per cent), and are used principally for making passenger-car tires and for other general-purpose molding uses.

Polybutadiene Rubber (see Stereo Rubbers).

Neoprene (CR) was the first synthetic rubber made in volume in the United States. It is called a special-purpose rubber today. Its use is confined principally to applications that require resistance to oil and gasoline, or for products that are exposed to sunlight or oxidation, such as hose, electrical insulation, gaskets, etc. Its monomer, chloroprene, is made from acetylene and hydrogen chloride.

Butyl rubber (IIR) is produced by copolymerizing the petroleum-derived monomer isobutylene and a small amount of another unsaturated hydrocarbon, such as isoprene, which permits this rubber to be vulcanized. As this rubber is not unsaturated to the extent natural or SBR are, it resists heat, aging, oxidation, and acids, and is a good electrical insulator. Its outstanding property of impermeability to gases ideally suits it for use in automobile inner tubes, which is its largest present application. It is used also in vibration mounts because of its ability to absorb shock.

A new group of synthetics is known as *"stereo" rubbers.* Their monomer units are lined up in the polymer in regularly repeated arrangements rather than at random as in most synthetic rubbers.

Polybutadiene (BR) was the second stereo rubber to be introduced. It is a highly elastic and abrasion-resistant rubber that duplicates natural rubber in its low heat-buildup characteristics. It is supplanting *Hevea* rubber in heavy-duty truck tires, and it improves the wear of SBR in passenger tires, so it is classified as a general-purpose rubber.

Polyisoprene (IR), the first stereo rubber, was introduced in 1954. The production of polyisoprene marked science's first success in chemically duplicating the natural rubber molecule and, as a result, essentially duplicating all its processing and end-use characteristics. For this reason it is like natural rubber, a general-purpose rubber.

Buna N, or N-type (NBR) rubbers are copolymers of butadiene and acrylonitrile. They are similar to neoprene in being highly resistant to oils, gasoline, and numerous solvents and are, therefore, classified as a special-purpose rubber. They can also be used in blends with polyvinyl-chloride resins to yield rubber stocks that are highly resistant to oxidation and abrasion.

Ethylene-propylene (EPDM) rubbers are the newest of the synthetics that have a potential for becoming general-purpose rubbers. They consist mainly of the petroleum-derived gases, ethylene and propylene, plus a small amount of termonomer, sufficient to permit vulcanization. These rubbers have excellent resistance to ozone, sunlight, and weathering. EPDM rubbers have a very good flexibility at low temperatures and high electrical-insulation properties. In addition to their good potential for tires, they are being used increasingly in electrical insulation, auto parts, footwear, hose, sponge, and coated fabrics.

■CONSUMPTION.—The United States in the late 1960s consumed about 2.5 million long tons of rubber, or about 27 pounds for every man, woman, and child. Approximately 75 per cent of this total was made up of synthetic rubber. Highly industrialized nations of Europe consume only 10 to 15 pounds per capita. Less developed countries, such as India and China, use only a fraction of a pound per person.

Rubber consumption in the United States continues to expand at a rate slightly over 3 per cent per year. Despite the fact that it has grown more than half again as large in the ten-year span from the mid-1950's to the mid-1960's, U.S. consumption continues to take a smaller percentage of the world's total. Until the late 1940's the United States consumed approximately two-thirds of the world total; by the mid-1950's, consumption had dropped to about half of the world total; by the mid-1960's, it had dropped again to less than 40 per cent of the world total. This trend reflects the much more rapid growth in the other industrialized nations.

Rubber was already a billion-dollar-a-year business in the United States when it initiated the large-scale production of synthetic rubber at the start of World War II. Including the production of synthetic rubber and all fabrication operations, the rubber industry has since grown to an $11 billion-a-year industry encompassing almost 1,500 companies. Many of these companies are expanding into diversified, but often related, fields of chemicals, plastics, metals, and textiles, and even into nuclear energy and space. (See also *Rubber Manufacture.*)

—Clayton F. Ruebensaal

Salt, the chemical compound sodium chloride (NaCl). In the United States, salt is produced in the South, chiefly in Louisiana, and in such Northern states as New York, Michigan, and Ohio. A more recent, and increasingly important, source of salt for the United States is the Bahamas.

Refined salt produced in the United States is obtained from underground mines. Salt in these mines is dissolved in water to form brine, from which it is recovered aboveground by evaporation. Salt from the Bahamas is so-called *solar salt,* which is produced by the evaporation of sea water. Northern rock salt usually is 95-97 per cent pure, southern rock salt better than 99 per cent pure, and solar salt 99.6 per cent pure. The recent trend has been toward shipping bulk salt from the mines to regional depots, where it is dried, crushed, screened, and packaged for final use.

Two-thirds of the U.S. salt output is used by the chemical industry. The largest single use is as raw material for the production of chlorine and caustic soda. Salt is used extensively for deicing roads and highways. It is also used in the meatpacking, tanning, food processing, rubber, oil metal, and paper industries, as well as for seasoning foods and as an animal feed supplement. —John Price

Sand, a loose, granular material made up of small particles of minerals or of rock and minerals. There are no generally accepted particle-size limits for sand, although geologists use a maximum of 2 millimeters and a minimum of 1/16 millimeter. Most sand is formed naturally by the disintegration of rocks. Many deposits contain varying amounts of clay and silt; some contain pebbles.

Sand is used in construction as a fine aggregate for concrete, mortar, and plaster. *Black sands,* such as those in Florida, contain *ilmenite* (iron titanium oxide) and *rutile* (titanium dioxide) in sufficient quantities for commercial production. *Green sands* contain *glauconite*, an iron potassium silicate, and have been used as fertilizer because of their potash content. *Silica sands* are composed almost exclusively of quartz (silicon dioxide)—most contain more than 95 per cent, some more than 99 per cent. Silica sand is used in glass making, as molding sand, refractory sand, filter sand, and grinding and polishing sand; it is often called *industrial sand.* Natural molding sand contains enough clay and other bonding material to form molds in which metal is cast. Synthetic molding sand contains added amounts of *fire clay, bentonite,* or other bonding materials.

—John Price

Sandstone, a sedimentary rock consisting usually of quartz sand united by some cementing medium such as silica, iron oxide, or calcium carbonate. Often, many other minerals and even fragments of other rocks are included. Sandstone may take the form of *shale,* which is composed of particles smaller than 1/16 millimeter in diameter, or *conglomerate,* a form containing fragments larger than 2 millimeters in diameter. However, sandstone usually occurs in particle sizes between these two limits.

There are four generally recognized classes of sandstone: *arkose, graywacke, subgraywacke,* and *orthoquartzite.* Distinctions among these classes are made on the basis of mineral content and texture. Sandstone particles vary greatly in composition, including such common minerals as quartz, feldspar, and several clay minerals. Several other minor minerals are found in smaller quantities, including garnet, tourmaline, zircon, rutile, staurolite, magnetite, pyrite, and chromite.

Sandstones with silica cement are used for structural purposes. Some sandstone, such as *novaculite oilstone,* has been quarried to make grindstones, pulpstones, and sharpening stones.

—John Price

Sanitary Engineering, the branch of engineering concerned primarily with public health. A specialist in the more general field of civil engineering, the sanitary engineer is engaged primarily in meeting two basic problems of urban life: water supply and the disposal of street drainage and of domestic and industrial wastes. These have become more pressing and difficult problems with accelerated municipal and industrial growth.

In early centuries, water was obtained from rivers and irrigation ditches. As early as 700 B.C., a supply was brought by tunnel to Jerusalem from a spring outside the city walls. The Greeks followed with several similar works, and the Romans built aqueducts not only for Rome but throughout the Roman world. In many later cities, however, shallow wells were also used locally for supply. Drainage received some early attention, but early "sewers" were, in effect, only street storm-water drains. Cesspools were used for domestic sewage, contaminating local well supplies until recurring cholera epidemics, as in Paris and London in the mid-nineteenth century, forced attention to this problem.

One of the first modern sewage systems was built in Hamburg, Germany, by a British engineer in 1850. Other, notably British, workers followed and by 1900 the relative advantages of "separate" and "combined" systems for drainage and domestic sewage were debated. While local wells were eventually abandoned, contamination through the practice of waste disposal in nearby streams led to typhoid epidemics that were common in American cities until early in the twentieth century. However, after Robert Koch discovered the typhoid bacteria in 1883, filtration to clarify turbid waters was developed and further perfected to provide an almost perfect "bacterial efficiency." The use of chlorine to kill bacteria followed and is widely adopted today. Nevertheless, many cities still face the problem of sewage and waste disposal and will utimately be forced to provide suitable plants to treat their wastes.

In general, sewage plants are designed to accelerate the action of the natural factors that utimately lead to the clarification of polluted waters. This usually involves the provision of conditions leading to the rapid growth of the beneficial bacteria that break down organic matter, as through the *activated sludge* process. This is followed by aeration to restore oxygen content. Industrial wastes which also vary with the industry, constitute a major problem. Manufacturers will undoubtedly be forced to give this source of pollution increasing attention in the future. The sanitary engineers thus face challenging problems, requiring not only the effective design of notable structural and mechanical works but also a thorough understanding of bacterial science and chemistry.

—James Kip Finch

Sericulture, the production of silk by raising silkworms. Silk has been a shining symbol of elegance and beauty ever since Hsi-ling-shi, the 14-year-old bride of Emperor Huang-ti, who ruled over China about 2640 B.C., first disclosed the secret of the silk cocoon. For nearly 3,000 years, the Chinese successfully guarded this secret. After that, silkworm eggs were smuggled into Persia, and that was the end of China's monopoly. Japan also got hold of the eggs and today is by far the largest producer of raw silk, with Red China, India, and South Korea next. In moderate quantities, raw silk is also produced in Italy, in the Near East, and in the Soviet Union.

Silk is a natural or live fiber, the filament which a silkworm spins for its cocoon. The silkworm is actually the caterpillar of the silkmoth *(Bombyx mori)* and its cocoon is the shell it constructs to protect itself during its growth from caterpillar to chrysalis (or pupa) to moth. A single cocoon is made of a continuous filament of silk, which the silkworm extrudes from its body. Along with the silk filament, the silkworm emits a gummy substance called *sericin.*

Sericulture involves the care of the little animal that produces the silk filament from egg through to cocoon. The breeder moths are selected with the utmost care. The eggs undergo many tests to ensure perfect, disease-free worms. They are put in cold-storage under strict governmental supervision until spring, when the mulberry trees begin to leaf. Then they are incubated until they hatch, in about a week, into tiny antlike silkworms. These have to be kept under rigidly clean conditions on trays that must be constantly refilled with the freshest mulberry leaves every two or three hours, for the silkworms eat voraciously day and night for five weeks. During this time they grow to about 70 times their original size. When the silkworm has eaten its fill, it creeps to straw that has been provided for it in individual cells and begins spinning its cocoon.

In the natural course of events, the worm inside the cocoon would develop into a chrysalis and the latter into a moth. The moth would then burst the cocoon and break the single long silk filament into many short ones. Therefore, it is necessary to destroy the worm inside the cocoon to enable the silk to be reeled off in one strand. This is done by stifling the worm with heat.

The next process is the unwinding (or reeling) of the cocoon, today mostly done by machinery. Hot water in basins is still used to melt the sericin, making it possible to unwind the cocoon. Because a single filament is far too fine for reeling, several cocoons are unwound at the same time. Their filaments are drawn together by passing them through a tiny porcelain eye, and the melted sericin now glues the silk filaments into a single thread. Later, the silk is re-reeled and twisted into skeins. The sericin is boiled off, either before or after weaving, to uncover the natural beauty of the silk.

—Hans Vaterlaus

Slate, a dense, fine-grained rock produced by the compression of clays, shale, and other rocks so as to produce a characteristic cleavage. The chief minerals of slates are muscovite, chlorite, and quartz. Lesser constituents include tourmaline, rutile, epidote, sphene, hermatite, and ilmenite.

In the United States, slate is quarried in Maine, New York, Vermont, Pennsylvania, Maryland, and Virginia. The major slate-producing areas overseas are in England, North Wales, Scotland, Ireland, France, Bohemia, and Germany.

Slate is still processed by hand with a chisel and mallet. Big slabs are split along cleavages into separate slates whose thickness depends on the size required and the quality of the rock. Slates are trimmed to final size by hand or by machine-driven rotating knives. Slates are commonly used for blackboards and are used widely for roofing in thickness of about 5 millimeters.

—John Price

Soaps and Detergents, substances used widely for cleaning, washing, and textile processing. They produce their effects by changing the surface tension of a solvent. Soaps actually are a special class of detergent, although the latter term is usually applied only to synthetic materials.

Soaps are the alkali-metal or ammonium salts of straight-chained carboxylic acids, the molecules of which usually contain 10-18 carbon atoms. Metallic soaps are akaline-earth or other metal salts; they are insoluble in water and find use in nonaqueous systems, especially as additives in lubricating oils and rust inhibitors. Soaps are prepared from naturally occurring triglycerides (animal and vegetable fats) by hydrolysis. At one time this was done by the action of water and alkalis, such as sodium hydroxide (NaOH) and potassium hydroxide (KOH), on fats in a soap kettle at high temperature. More common now is the direct hydrolysis of fats by water at high temperature. This allows isolation and rectification of the fatty acids, which are neutralized to form soap.

Rapid development of synthetic detergents has cut heavily into the use of soaps, except in toilet bars. A branched-chain detergent, alkylbenzene sulfonate (ABS), has been one of the most widely used materials in the synthetic group. However, ABS is not sufficiently biodegradable—efficient sewage-treatment plants degrade only 40 to 60 per cent of influent ABS. As a result, rivers and even drinking water supplies are contaminated with foam from residual detergent materials. Linear alkylate sulfonate (LAS) and linear alkylphenols (LAP) have been tried as an alternative. These materials, which have straight carbon chains in place of the branched chains in ABS, improve biodegradability. Under favorable conditions, sewage treatment will degrade more than 90 per cent of influent linear detergents. However, under anaerobic conditions LAS and LAP are not appreciably easier to degrade than ABS. Furthermore, at least one-third of the United States population discharges sewage into essentially anaerobic cesspools and septic tanks. Under unfavorable soil conditions, waste detergents leak into groundwater supplies. Synthetic primary or secondary alcohols, products that degrade faster than LAS or LAP, are therefore being considered as possible substitutes. —John Price

Sulfur, a commercially important, nonmetallic element that makes up less than 0.1 per cent of the earth's crust; chemical symbol S. Although sulfur was discovered before recorded history, it was not until 1777 that Antoine Laurent Lavoisier first recognized it as an element. The largest known free-sulfur deposits are in Texas and Louisiana, where sulfur is found associated with limestone and cap rock formations over salt domes. It is also found near volcanic regions in Japan, Sicily, and Mexico.

The *Frasch process,* developed in 1891 by Herman Frasch, is used to extract sulfur from underground deposits such as those in Texas and Louisiana. A hole is drilled down to the deposit, and three concentric pipes are lowered into the ore bed. Superheated (165° C) water forced down the largest pipe melts the sulfur, which has a melting point of 112.8° C. Hot compressed air is pumped down the smallest pipe, and a frothy mixture of water, air, and molten sulfur—99.5-99.9 per cent pure—is forced to the surface through the intermediate pipe.

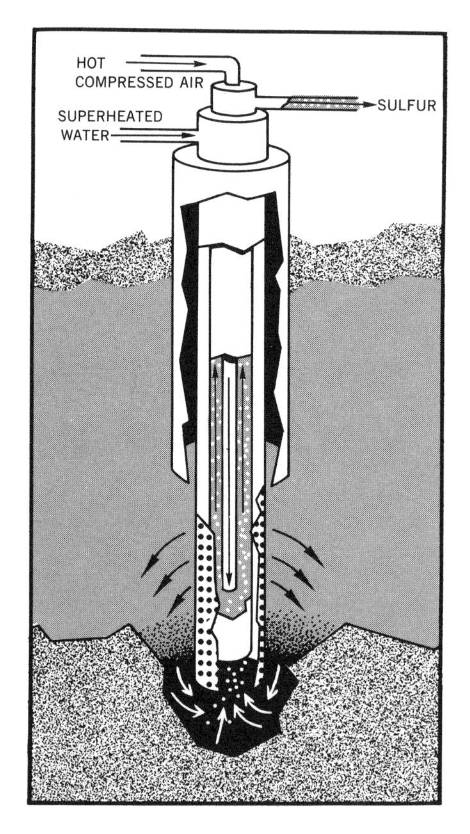

FRASCH PROCESS uses compressed air and superheated water to obtain sulfur for the fertilizer, petroleum, and other industries.

Another major source is so-called *sour gas,* a natural gas containing hydrogen sulfide. Most natural gas is sweet, but the natural gas in France and Canada must be sweetened before it can be marketed. Large quantities of sulfur are obtained as by-products of the sweetening process.

The largest single use of sulfur is in the production of sulfuric acid, which is widely used in the fertilizer industry. Other uses are in paper manufacturing and bleaching; in the manufacture of titanium and other pigments; in pickling steel; in the production of nylon, rayon, and other cellulosic fibers; in petroleum refining; in rubber vulcanizing; and in a wide range of chemical manufacturing processes. —John Price

Sulfuric Acid, a strong, colorless, oily mineral acid, also known as *vitriolic acid* or *oil of vitriol;* chemical formula H_2SO_4.

Sulfuric acid is produced on a large scale by two processes: the *contact process* and the *lead-chamber process.* The contact process, the more important of the two, produces acid of any desired strength, whereas lead-chamber acid is relatively dilute (60-78 per cent), hence less useful.

In the contact process, sulfur is burned or iron pyrites roasted in air to produce sulfur dioxide (SO_2), which is then oxidized to sulfur trioxide (SO_3) in the presence of a catalyst (vanadium pentoxide or platinum). Sulfuric acid is produced by the reaction of sulfur trioxide and water. Addition of more sulfur trioxide produces *oleum* (polysulfuric acid, $H_2S_2O_7$), which can be treated with water to produce acid of any desired concentration. The lead-chamber process involves oxidation of sulfur dioxide by nitric acid in the presence of water. The reaction is carried out in one or more large lead-lined rooms.

The concentrated acid is a strong oxidizing agent, especially at high temperatures. It reacts with metals, carbon, sulfur, and other oxidizable materials. Because of its high boiling point, it can react with salts at high temperatures to liberate volatile acids such as hydrochloric acid (HCL). The concentrated acid is a strong dehydrating agent and reacts vigorously with water, producing much heat. It also extracts hydrogen and oxygen to form water from organic materials, decomposing them and leaving carbon.

The major use of sulfuric acid is in the fertilizer industry. Phosphate fertilizers are made by treating insoluble phosphate rock with sulfuric acid. It is also used in the manufacture of ammonium sulfate, another fertilizer. Petroleum refining consumes large quantities of the acid, as does the manufacture of such chemicals as sulfates, nitric and hydrochloric acids, drugs, dyes, and explosives. It is used to pickle steel, to make storage batteries, paints, plastics, and various textiles, and has wide applications in the metallurgical industry. —John Price

Tin, a metal first used by man more than 4,000 years ago; chemical symbol Sn. *Cassiterite* (tin dioxide) is the only tin-bearing mineral of commercial importance. There are no high-grade tin ores; most of the world's tin comes from low-grade alluvial deposits, which average 0.5 pound of cassiterite per cubic yard (3,000 pounds). Deposits containing up to 4 per cent tin are found in Bolivia and Cornwall, England.

Cassiterite is recovered from deposits by dredging, water jets, and gravel pumps on level ground, hydraulic methods where a head of water is available, and open-pit mining. The fine grains of cassiterite are 2½ times as dense as gravel, so concentration is a simple matter of screening and gravity separations. Concentrates usually contain 70-77 per cent tin, and lower-grade concentrates are upgraded before further treatment. Sulfur, arsenic, lead, antimony, and bismuth are removed by roasting. The addition of salt to form volatile or soluble chlorides helps to remove lead and silver. Excess iron and copper are leached with hydrochloric acid.

Primary smelting is carried out between 1,200° and 1,300° C, during which the amount of reducing agent is limited to give incomplete reduction. This produces metallic tin with low iron content. The resulting rich slag is treated further to remove most of the tin. Crude smelted tin is partly melted to remove iron, copper, and other impurities that form solid compounds appreciably above the smelting point of tin. Refined metal from most smelters is more than 99.8 per cent tin. Secondary tin from metal scrap amounts to about one-third of the total tin consumed in the United States. Most of it comes from tin-bearing alloys, which smelters rework into alloys and chemicals. Much high-purity tin is recovered by the detinning of tinplate scrap with a hot caustic solution and electrolytic treatment.

Half of the tin consumed is used in alloys. Soft solders are alloys of tin and lead, containing 20-70 per cent tin. Lead-free solders (with silver, antimony, or zinc instead of lead) are used for special applications. *Bronzes* are the most ancient of alloys and are still important structural metals. True *copper bronzes* include wrought *phosphor bronzes* (up to 10 per cent tin) and leaded-tin *cast bronzes* (5-10 per cent tin). *Babbitt metal,* used in bearings, is tin containing 4-8 per cent of copper and antimony. Other tin alloys include *pewter, Britannia metal,* and *type metal.*

Pure tin can be applied to all the common metals as a coating by hot-dipping or electrodeposition. Tin protects metal surfaces that oxidize or corrode easily. Tin-coated steel (tinplate for cans) is still an important product for tin, although other metals and plastics are taking over some of

World* Production of TIN in %, 1966

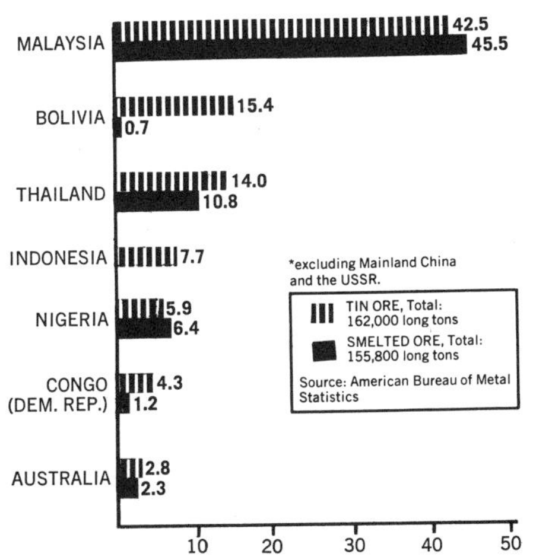

the market. Hot-dip tin coatings are applied to copper wire and sheet and to steel and cast-iron parts.

Tin-cadmium coatings, which resist salt vapors, are widely used in the aircraft industry, and tin-zinc coatings are used widely in the radio, television, and electronics industries. Tin-copper coatings are used for jewelry, handbag frames, wire goods, and hardware because of their lustrous finish. —John Price

Titanium, a silvery-gray paramagnetic metal; chemical symbol Ti. It is stronger than steel but much lighter; its specific gravity is only 4.5, about 56 per cent that of steel. Titanium retains its properties from −320° to 1,000° F. With low electrical and thermal conductivity, it has outstanding resistance to corrosion in oxidizing media, and is impervious to atmospheric or salt-water corrosion. The most important sources of the metal are *rutile* (titanium dioxide) and *ilmenite* (a combination of iron oxide and titanium dioxide).

Titanium dioxide is widely used as a white pigment for exterior paints and as a whitener and filler in paper. Titanium metal, because of its low weight and high strength, has been used extensively in jet engines and airframes. The metal's corrosion resistance is valued in equipment, especially in the chemical and petroleum industry.

Titanium was discovered independently by William Gregor (who called it menaccanite) in 1791 and Martin Heinrich Klaproth in 1795. Jöns Jakob Berzelius first isolated crude titanium in 1825, but is was not until 1906 that M. A. Hunter separated enough metal for study. W. A. Kroll, in 1928, made the first metallic titanium and, in 1937, invented the process that bears his name—dry, high-temperature reduction of titanium halide with magnesium. Titanium production presents several formidable problems. The liquid metal seems to be a universal solvent; it dissolves or is contaminated by every known refractory. The metal must be reduced from its ore with extremely high purity. since contaminants destroy its

World Production of TITANIUM CONCENTRATES in %

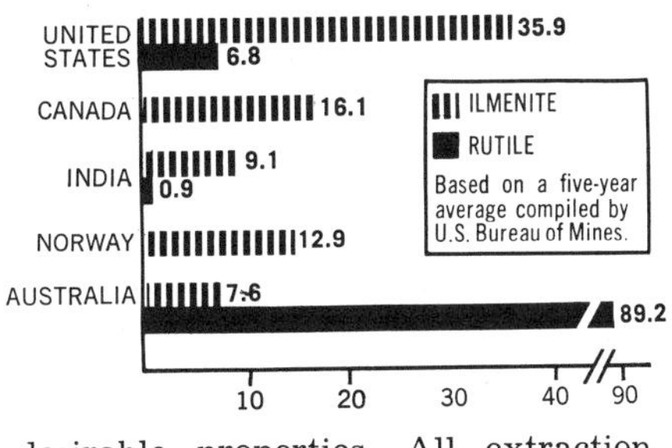

desirable properties. All extraction and ingot melting must be done in vacuum or under a protective helium or argon blanket to prevent absorption of atmospheric nitrogen and oxygen.

The first step in reducing the metal from its ore is to chlorinate an oxide-carbon mixture to obtain titanium tetrachloride, which is treated with magnesium metal in a heat-resistant steel vessel at red heat under an inert-gas blanket. This yields commercially pure spongy titanium metal, which is further purified by remelting. —John Price

Tunnel, any underground passageway. Natural structures, such as long caves created by running water, fit this description, but from an engineering viewpoint, they are not tunnels. Civil engineers designate as a tunnel an underground passageway built without removing the overlying soil and rock. This rules out structures built by digging a trench, building the tunnel tube and then refilling the trench. This purist definition ignores the fact that the end product is a manmade underground passageway, a tunnel. Oddly enough, the first major tunnels driven by modern man were for boats. Since water transportation was most important, large tunnels were driven through hills for canals. The Languedoc Canal in France contained a tunnel built in the late seventeenth century. The first large tunnels in the United States were built for canals, but the railroads quickly superseded water transportation. By 1850, more than 50 tunnels had been built for railroads in the United States. Most of these were relatively short—less than a mile—and it was not until the 1860's, when power rock drills were developed, that longer hard-rock tunnels were driven.

■CONSTRUCTION.—Tunnel construction even today remains both difficult and dangerous. Paradoxically, tunneling is most difficult in rock or earth that is easy to dig through. Driving a tunnel through the soft, silty bottom of a riverbed, material that can be dug with a garden trowel, is far more challenging for engineers than blasting through solid granite. The rock tunnel is apt to be a self-supporting structure, whereas the subaqueous, so-called *soft-ground tunnel* must be supported completely with a steel and concrete lining.

Conventional *hard-rock tunneling* involves drilling holes with air-operated drills into the rock face. In a large tunnel many holes are drilled to prescribed depths in a set pattern. Explosives placed in the holes are detonated electrically, blasting the section of rock into small pieces. Machines remove the broken rock, called *muck,* and drillers return for another round. If the rock is weak and fractured, crews install steel or wood supports as the tunnel progresses. Depending on its purpose, and the quality of the rock, tunnels sometimes are lined with concrete. Softground tunnels, such as the under-river type, are usually shield-driven. The work area, called the *heading,* is often kept under air pressure. A tunneling shield resembles a huge barrel with its ends cut out. Its forward end has a strong cutting edge that forms the tunnel's outer perimeter, as mammoth hydraulic jacks push the shield forward, cookie-cutter fashion, workmen, called *sandhogs,* excavate—usually by hand—within the shield as it moves ahead. Other crews at the rear of the shield install *liner rings,* sectional steel or concrete rings bolted to the ring behind it. A liner ring is installed when the shield is not advancing and its jacks are retracted. Once the ring is in place, the jacks bear on that ring to advance the shield still further.

Water is one of the tunneler's major problems even in hard rock, but particularly in soft ground. Pumping compressed air into the tunnel, so the air pressure in the tunnel is greater than the pressure of the water in the soil being tunneled through, makes it possible to tunnel through very unstable soil. But working under air, sometimes two and three times normal atmospheric pressure, is very dangerous. Like deep-sea divers, sandhogs can work under such pressures for only limited periods. The body must be gradually returned to normal pressure or men suffer the diver's disease, the *bends,* which can be fatal. Also, a tunnel under air pressure can have a *blowout* in which the pressurized air in the tunnel works through the soil and bursts out into the river, causing the tunnel to flood with river water and mud.

Two recent developments have eased somewhat the plight of the tunneler. Giant tunneling machines with rotating cutterheads are taking the place of the sandhogs and blasters. These machines can work in harder and harder rock, as new steels are developed for the cutting teeth. By combining the boring principle with that of the shield, machines are now available that mechanically can bore through soft ground under air pressure. Engineers in the last decade have developed a sophisticated version of one of the oldest tunneling methods. The *cut-and-cover,* or *trench-type,* tunnel is built by digging a trench, building the tunnel in the trench, and then backfilling. The technique is used to build large tunnels in deep water. The tubes for the tunnel are built in dry docks. The mammoth steel and concrete sections have ends capped off. Dredges excavate a trench in the river bottom and the tube sections are floated into position and sunk into place. Divers connect the tubes and their caps are removed forming a continuous tunnel.

THE PORT OF NEW YORK AUTHORITY

THE LINCOLN TUNNEL'S third tube, under construction, reveals its lining of cast-iron plates. The tunnel provides a 1.6-mile roadway deep beneath the Hudson River.

■NOTABLE TUNNELS.—Japan has begun construction of a 22.6-mile undersea tunnel to connect the islands of Honshu and Hokkaido. Besides using tunnels to breech water barriers, the world's engineers have long been busy driving tunnels through mountains for railroads and, more recently, highways. The Alps are literally riddled with tunnels. The longest vehicular tunnel, the 7.25-mile Mont Blanc Tunnel, was opened to traffic in 1965. Some famous rail tunnels in the Alps are the 12.3-mile Simplon Tunnel; the 11.5-mile Apennine Tunnel; and the 9.3-mile St. Gotthard Tunnel.

Auto tunnels, until the 1960s, have been relatively short because of the problem of exhaust gases. But Mont Blanc broke that precedent, relegating Great St. Bernard to second place. Then in 1967, the Swiss opened St. Bernardino, a 4.1 mile bore. A highway tunnel is planned for the Alps that will be 11 miles long.

The longest tunnels that have been built are for water supply. The world's longest is the Delaware River Aqueduct, a 14-foot diameter rock tunnel 85 miles long. Another tunnel, also part of New York City's huge upstate water supply system, is the 44-mile West Delaware Aqueduct, the second longest ever built. California also is famous for its water tunnels. The Colorado River Aqueduct, which brings water hundreds of miles across the state, has over a dozen tunnels, some as long as 18 miles.

—William W. Jacobus, Jr.

Notable Vehicular and Railway Tunnels

Name	Location	Length in Miles	Date Completed
VEHICULAR			
Baltimore Harbor	Baltimore, Md.	1.2	1957
Brooklyn Battery	New York, N. Y.	1.7	1950
Detroit-Canada	Detroit, Mich.-Windsor, Ontario	.9	1930
Holland	New York, N. Y.-Jersey City, N. J.	1.7	1927
Kanmon	Japan	2.2	1944
Liberty Tubes	Pittsburgh, Pa.	1.2	1924
Lincoln	New York, N. Y.-Weehawken, N. J.	1.6	1937
Mersey	Liverpool-Birkenhead, England	2.2	1934
Mont Blanc	Italy-France	7.3	1965
San Bernardino	Switzerland	4.1	1967
Sumner	Boston, Mass.	1.1	1894
Great St. Bernard	Italy-Switzerland	3.6	1964
RAILWAY			
Arlberg	Italy	11.5	1934
Apennine	Austria	6.4	1884
Arthur's Pass	New Zealand	5.3	1912
Grenchenberg	Switzerland	5.3	1915
Hoosac	Hoosac, Mass.	4.8	1873
Loetschberg	Switzerland	9.5	1913
Moffat	Winter Park, Colo.	6.2	1928
Mont Cenis	France-Italy	8.5	1871
Mt. Royal	Montreal, Canada	9.0	1916
Roger's Pass	British Columbia	5.0	1912
Severn	England-Wales	4.5	1886
Shimizu	Japan	6.1	1930
Simplon	Switzerland-Italy	12.3	1905
St. Gotthard	Switzerland	9.3	1881
Tanna	Japan	4.9	1934
Vosges	France	7.0	1937

Varnish, a solution of resinous materials in a solvent, used as a surface coating. It is applied as a liquid, which changes to a hard solid, either by evaporation of the solvent or by some chemical reaction.

Spirit varnish dries by evaporation of the solvent, which usually is alcohol. *Shellac varnish,* made by dissolving shellac in alcohol, is a common spirit varnish. *Oleoresinous varnish* is made by treating a drying oil with a resin, usually with heat, and dissolving the reaction product in a petroleum solvent. These varnishes dry by evaporation of the solvent and by polymerization of the drying oil.

Varnish coatings protect wood from abrasion, staining, and weathering, and reduce the penetration of water and other liquids without obscuring the wood's grain or changing its color significantly. Varnishes protect masonry from damage by moisture penetration and freezing. They are also used as insulation coatings for wires and as vehicles for paint. *Asphalt varnishes,* which contain a bituminous material and a drying oil in a solvent, are used for insulation and as a heat- or corrosion-resistant coating for metals. —John Price

Waxes, unctuous, viscous to solid substances used for protective coatings. Waxes fall into two broad categories: those produced from petroleum, and those derived from animal or vegetable sources. *Petroleum waxes* make up about 90 per cent of all wax used in industry. They may be either crystalline or microcrystalline. *Crystalline wax* is made from distillate lubricating fractions; *microcrystalline wax* is made from residual lubricating fractions of crude oil. Crystalline waxes melt at 120° to 150° F; microcrystalline waxes at 150° to 175° F. Petroleum wax is used to coat paper products and to blend with other waxes in candle manufacturing. It is used in the manufacture of electrical equipment and in many polishes for home and industry. Softer waxes, such as *petroleum jelly,* are used for medicinals.

Animal and *vegetable waxes* are esters of high-molecular-weight monohydroxyl alcohols and carboxylic acids. They are found in the cuticles of plants, in honeycombs and other insect cellular fabrications, as a coating on leaves of many trees and grasses, in the bodies of various land and marine animals, in seed envelopes, on the hair of animals, and associated with certain bacilli. Crude waxes often are mixtures of various organic substances of such complexity that it is impossible to separate completely and identify all components. Basically, however, waxes are known to be simple esters of steroidal or open-chain alcohols composed of an even number of carbon atoms (usually 24-36) esterified with acids of similar carbon content. Waxes are characterized by: *solidification point,* which is not necessarily the same as their melting point; *acid value,* which is the amount of free acid present, as measured by the number of milligrams of potassium hydroxide needed to neutralize free fatty acids present in one gram or substance; *saponification value,* the number of milligrams of potassium hydroxide needed to complete saponification of one gram of material; *iodine value,* which measures the degree of unsaturation—it is the number of grams of iodine absorbed by 100 grams of material, and the *Reichert-Meissel value,* which is a measure of the amount of low-molecular-weight acids present.

Carnauba wax, a coating on the leaves of the Brazilian palm *Corypha cerifera,* is one of the most important vegetable waxes. It is hard, has a relatively high solidification point (remaining solid in hot weather), repels water, and can be polished to a high luster. It is used extensively in floor and automobile polishes. *Japan wax* comes from the fruit coat of sumac berries. It is tough and can be kneaded without crumbling. *Spermaceti wax,* another commercially important wax, comes from the head of the sperm whale. *Lanolin* is purified wool wax or grease; it is used in salves and jellies, and in certain soaps and cosmetics. —John Price

Zinc, a malleable, ductile gray metal; chemical symbol Zn. It was discovered in Europe during the Middle Ages, although it was known much earlier in Asia. Zinc is 25th in order of abundance among the elements, making up 0.004 per cent of the earth's crust.

Pure, freshly polished zinc is bluish-white and lustrous. It tarnishes superficially in moist air, taking on its usual grayish color. Pure zinc is malleable and ductile enough to be rolled or drawn, but small amounts of contaminating metals can render it brittle. Zinc is a good conductor of heat and electricity, although its conductivity is only one-fourth that of silver, the best metal in this respect. Zinc is fairly active chemically. Its electromotive force is higher than that of hydrogen, so it will displace hydrogen from acid solution, liberating it as a gas, while zinc is dissolved to form Zn^{++} ions.

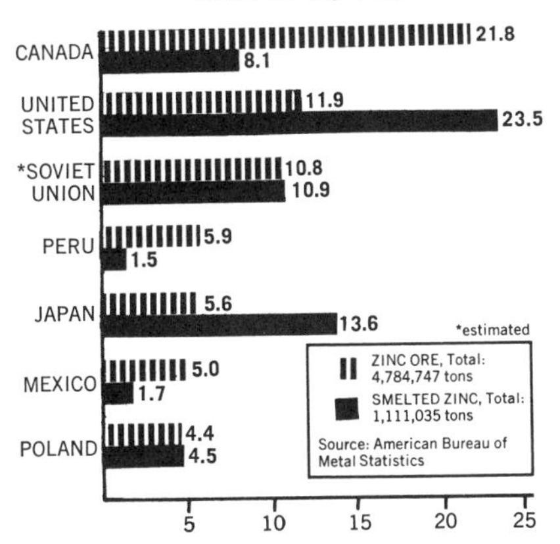

The chief zinc ores are *zinc blende* or *sphalerite,* and *marmatite* (both zinc sulfide). Other ores include *calamine* (a hydrated zinc silicate), *smithsonite* (zinc carbonate), *willemite* (zinc silicate), *zincite* (zinc oxide), and *franklinite* (an oxide of zinc, iron, and manganese). Much zinc is also separated as a by-product during the processing of iron ores.

Zinc is usually refined from its sulfide ores. Concentrates shipped to smelters contain 50 to 60 per cent zinc. Usual impurities are iron, lead, and cadmium. Sulfide ores are roasted before smelting to convert zinc to the oxide. Low-grade ores of the oxide type sometimes are concentrated in a rotary kiln, where the ore is mixed with enough coal to reduce the zinc oxide. Before pyrometallurgical extraction, concentrates are sintered to bring them to the proper density and particle size.

Several different processes have been developed to smelt zinc; the *horizontal retort* (Belgian) process; the *vertical retort* (New Jersey) process; the *electrothermic* (St. Joe) process; the *blast-furnace* process; and the *electrolytic* process. The first four processes depend on the highly endothermic reaction by which carbon reduces zinc at 1,000° C or more, which is above the metal's boiling point. Enough carbon must be present to prevent oxidation of the carbon beyond the monoxide, since carbon dioxide will reoxidize the zinc vapor during the condensing step.

In the electrolytic method, zinc is produced from zinc sulfate solution. Anodes are made of lead alloy, cathodes of aluminum sheets, from which zinc is stripped periodically. This method produces high-purity zinc directly, whereas zinc produced by the carbon-reduction method must be refined to yield a high-grade product. One widely used method entails fractional distillation in reflux refining columns. Two lead columns remove lead, iron, and other high-boiling impurities, and a cadmium column removes cadmium and other low-boiling impurities. The purest grade of zinc, special high-grade, is more than 99.99 per cent pure.

The most important use of zinc is as a protective coating for other metals. Galvanizing is the process of coating iron or steel with zinc by immersion, electrolytic deposition, spraying, or treating with powdered metal near its melting point. Zinc is also used in brass and die-casting alloys and in dry cells. Rolled zinc is used in roofing, gutters, and flashing. —John Price

BIBLIOGRAPHY

BALL, MAX W. *This Fascinating Oil Business.* Bobbs-Merrill Co., Inc., 1966.

BRADY, GEORGE S. *Materials Handbook.* 9th ed. McGraw-Hill, Inc., 1963.

COOK, J. GORDON. *The Miracle of Plastics.* Dial Press, Inc., 1964.

CULLEN, ALLAN H. *Rivers in Harness; the Story of Dams.* Chilton Co., 1962.

DAVIES, J. D. *Structural Concrete.* The Macmillan Co., 1964.

GIES, JOSEPH. *Bridges and Men.* Doubleday & Co., Inc., 1963.

LEONARDS, GERALD A. *Foundation Engineering.* McGraw-Hill, Inc., 1962.

MCGUINNESS, WILLIAM J. and others. *Mechanical & Electrical Equipment for Buildings.* 4th ed. John Wiley & Sons, Inc., 1964.

MCMILLAN, PETER W. *Glass-Ceramics.* Academic Press, Inc., 1964.

MERDINGER, CHARLES. *Civil Engineering Through the Ages.* Society of Military Engineers, 1963.

SALVADORI, MARIO and HELLER, ROBERT. *Structure in Architecture.* Prentice-Hall, Inc., 1965.

SANDSTROM, GOSTA. *Tunnels.* Holt, Rinehart & Winston, Inc., 1963.

STAMM, ALFRED J. *Wood and Cellulose Science.* The Ronald Press Co., 1964.

STOUT, EVELYN E. *Introduction to Textiles.* 2d ed. John Wiley & Sons, Inc., 1965.

WEIGHTS AND MEASURES

		Unit	Abbreviation	Units of Same System	Metric/U.S. Equivalent
Length					
	U.S.	mile	mi	5,280 feet, 320 rods, 1,760 yards	1.609 kilometers
		furlong	fur	0.125 mile, 40 rods, 660 feet	201.2 meters
		rod	rd	5.50 yards, 16.5 feet	5.029 meters
		yard	yd	3 feet, 36 inches	0.914 meter
		foot	ft or ′	12 inches, 0.333 yard	30.480 centimeters
		inch	in or ″	0.083 foot, 0.027 yard	2.540 centimeters
	Metric	myriameter	mym	10,000 meters	6.2 miles
		kilometer	km	1,000 meters	0.62 mile
		hectometer	hm	100 meters	109.36 yards
		decameter	dkm	10 meters	32.81 feet
		meter	m	0.001 kilometer	39.37 inches
		decimeter	dm	0.1 meter	3.94 inches
		centimeter	cm	0 01 meter	0.39 inch
		millimeter	mm	0.001 meter	0.04 inch
Area					
	U.S.	square mile	sq mi or m^2	640 acres, 102,400 square rods	2.590 square kilometers
		acre	a or ac	4,840 square yards, 43,560 square feet	4,047 square meters
		square rod	sq rd or rd^2	30.25 square yards, 0.006 acre	25.293 square meters
		square yard	sq yd or yd^2	1,296 square inches, 9 square feet	0.836 square meter
		square foot	sq ft or ft^2	144 square inches, 0.111 square yard	0.093 square meter
		square inch	sq in or in^2	0.007 square foot	6.451 square centimeters
	Metric	square kilometer	sq km or km^2	1,000,000 square meters	0.3861 square mile
		hectare	ha	10,000 square meters	2.47 acres
		are	a	100 square meters	119.60 square yards
		centare	ca	1 square meter	10.76 square feet
		square centimeter	sq cm or cm^2	0.0001 square meter	0.155 square inch
		square millimeter	sq mm or mm^2	0.01 square centimeter	0.002 square inch
Volume					
	U.S.	cubic yard	cu yd or yd^3	27 cubic feet, 46,656 cubic inches	0.765 cubic meter
		cubic foot	cu ft or ft^3	1,728 cubic inches, 0.0370 cubic yard	0.028 cubic meter
		cubic inch	cu in or in^3	0.00058 cubic foot, 0.000021 cubic yard	16.387 cubic centimeters
	Metric	decastere	dks	10 cubic meters	13.10 cubic yards
		stere	s	1 cubic meter	1.31 cubic yards
		decistere	ds	0.10 cubic meter	3.53 cubic feet
		cubic centimeter	cu cm or cm^3 or cc	0.000001 cubic meter	0.061 cubic inch
		cubic millimeter	cu mm or mm^3	0.001 cubic centimeter	0.0001 cubic inch
Weight					
	Avoirdupois	short ton		20 short hundredweight, 2,000 pounds	0.907 metric ton
		long ton		20 long hundredweight, 2,240 pounds	1.016 metric tons
		short hundredweight	cwt	100 pounds, 0.05 short ton	45.359 kilograms
		long hundredweight	cwt	112 pounds, 0.05 long ton	50.802 kilograms
		pound	lb or lb av or #	16 ounces, 7,000 grains	0.453 kilogram
		ounce	oz or oz av	16 drams, 437.5 grains	28.349 grams
		dram	dr or dr av	27.343 grains, 0.0625 ounce	1.771 grams
		grain	gr	0.036 dram, 0.002285 ounce	0.648 gram
	Troy	pound	lb t	12 ounces, 240 pennyweight, 5,760 grains	0.373 kilogram
		ounce	oz t	20 pennyweight, 48 grains	31.103 grams
		pennyweight	dwt or pwt	24 grains, 0.05 ounce	1.555 grams
		grain	gr	0.042 pennyweight, 0.002083 ounce	0.0648 gram
	Apothecaries'	pound	lb ap	12 ounces, 5,760 grains	0.373 kilogram
		ounce	oz ap or ℥	8 drams, 480 grains	31.103 grams
		dram	dr ap or ʒ	3 scruples, 60 grains	3.887 grams
		scruple	s ap or ℈	20 grains, 0.333 dram	1.295 grams
		grain	gr	0.05 scruple, 0.002083 ounce, 0.0166 dram	0.0648 gram
	Metric	metric ton	MT or t	1,000,000 grams	1.1 short tons
		quintal	q	100,000 grams	220.46 pounds
		kilogram	kg	1,000 grams	2.2046 pounds
		hectogram	hg	100 grams	3.527 ounces
		decagram	dkg	10 grams	0.353 ounce
		gram	g or gm	0.001 kilogram	0.035 ounce
		decigram	dg	0.10 gram	1.543 grains
		centigram	cg	0.01 gram	0.154 grain
		milligram	mg	0.001 gram	0.015 grain
Capacity					
	U.S. Liquid Measure	gallon	gal	4 quarts (231 cubic inches)	3.785 liters
		quart	qt	2 pints (57.75 cubic inches)	0.946 liter
		pint	pt	4 gills (28.875 cubic inches)	0.473 liter
		gill	gi	4 fluidounces (7.218 cubic inches)	118.291 milliliters
		fluidounce	fl oz or f℥	8 fluidrams (1.804 cubic inches)	29.573 milliliters
		fluidram	fl dr or fʒ	60 minims (0.225 cubic inch)	3.696 milliliters
		minim	min or ♏	1/60 fluidram (0.003759 cubic inch)	0.061610 milliliter
	U.S. Dry Measure	bushel	bu	4 pecks (2,150.42 cubic inches)	35.238 liters
		peck	pk	8 quarts (537.605 cubic inches)	8.809 liters
		quart	qt	2 pints (67.200 cubic inches)	1.101 liters
		pint	pt	1/2 quart (33.600 cubic inches)	0.550 liter
	British Imperial	bushel	bu	4 pecks (2,219.36 cubic inches)	0.036 cubic meter
		peck	pk	2 gallons (554.84 cubic inches)	0.0009 cubic meter
		gallon	gal	4 quarts (277.420 cubic inches)	4.545 liters
		quart	qt	2 pints (69.355 cubic inches)	1.136 liters
		pint	pt	4 gills (34.678 cubic inches)	568.26 cubic centimeters
		gill	gi	5 fluidounces (8.669 cubic inches)	142.066 cubic centimeters
		fluidounce	fl oz or f℥	8 fluidrams (1.7339 cubic inches)	28.416 cubic centimeters
		fluidram	fl dr or fʒ	60 minims (0.216734 cubic inch)	3.5516 cubic centimeters
		minim	min or ♏	1/60 fluidram (0.003612 cubic inch)	0.059194 cubic centimeter
	Metric	kiloliter	kl	1,000 liters	1.31 cubic yards
		hectoliter	hl	100 liters	3.53 cubic feet
		decaliter	dkl	10 liters	0.35 cubic feet
		liter	l	0.001 kiloliter	61.02 cubic inches
		deciliter	dl	0.10 liter	6.1 cubic inches
		centiliter	cl	0.01 liter	0.6 cubic inch
		milliliter	ml	0.001 liter	0.06 cubic inch

PART 3

Index

A

D

E

M

N

O

P

Q

R

S

T

U

V

W